Second Edition

Dictionary of Mining, Mineral, and Related Terms

Compiled by
the
American Geological Institute

Published
by the
American Geological Institute
in cooperation with the
Society for Mining, Metallurgy, and Exploration, Inc.

As the Nation's principal conservation agency, the Department of the Interior has responsibility for most of our nationally owned public lands and natural resources. This includes fostering the sound use of our land and water resources; protecting our fish, wildlife, and biological diversity; preserving the environmental and cultural values of our national parks and historical places; and providing for the enjoyment of life through outdoor recreation. The Department assesses our energy and mineral resources and works to assure that their development is in the best interests of all our people by encouraging stewardship and citizen participation in their care. The Department also has a major responsibility for American Indian reservation communities and for people who live in island territories under U.S. administration.

U.S. Department of the Interior
Bruce Babbitt, *Secretary*

U.S. Bureau of Mines
Rhea Lydia Graham, Director

Composed in Palintino using Ventura Publisher 3.0.

Printed offset on Domtar Windsor, an acid-free paper, by Braun-Brumfield, Inc.

Bound in Industrial Coatings Group, Inc. Arrestox.

ISBN 0-922152-36-5

Foreword

The need for a revised mining dictionary is obvious when one considers the technical advances, environmental regulations, and other changes that have occurred since the Bureau's previous mining dictionaries were published. The Bureau had pioneered efforts in this field, beginning in 1918 with Fay's *Glossary of the Mining and Minerals Industry*," and continuing to the 1968 publication, *A Dictionary of Mining, Mineral, and Related Terms.*

To develop a modern mining dictionary, the U.S. Bureau of Mines initiated a collaborative project with the American Geological Institute. The Bureau's staff, the Institute's staff and members, and many minerals experts throughout the Nation contributed their expertise to the work. In the 5-year project's final phase, more than 100 Bureau personnel were involved in the technical review and publication production process. I would like to thank all those who assisted in producing this new mining dictionary, especially the Dictionary Revision Group members and the Bureau's engineers, scientists, and editors.

The CD-ROM version of the Dictionary Revision is one of the last publications of the U.S. Bureau of Mines. Although the agency is closing, the Bureau's minerals information functions will transfer to the U.S. Geological Survey.

I believe this new dictionary will benefit the Nations by continuing one of the Bureau of Mines important missions, that of providing information on minerals, the building blocks of our society, and mining, one of the world's oldest industries.

Rhea Lydia Graham
Director
January 1996

Preface to the print edition

This edition of the *Dictionary of Mining, Mineral, and Related Terms*, published by the American Geological Institute in cooperation with the Society for Mining, Metallurgy, and Exploration, Inc., is the print version of a CD-ROM, released by the U.S. Bureau of Mines in early 1996. Based on the first edition published by the Bureau in 1968 (Paul W. Thrush, editor), the compilation and content review was coordinated by the American Geological Institute (USBM Contract Number J0101017). This publication represents the dedication and efforts of many individuals, including Bureau personnel, members of the mining and minerals community, and the staff of the American Geological Institute.

Preface

Technological developments and environmental laws and regulations that affect mining have proliferated during the past 25 years. Concurrently, the need for a modern mining dictionary has grown -- one that incorporates not only standard mining-related terms but also terms in peripheral areas, such as the environment, pollution, automation, health and safety. The new edition of the Dictionary of Mining and Mineral Related Terms is the culmination of a 5-year effort between the U.S. Bureau of mines and the American Geological Institute (USBM CONTRACT NUMBER J0101017) that will serve the needs of those engaged in minerals-related activities. It is organized to aid the user in appreciating the essential role that minerals and products play in our quality of life.

The Bureau's development of mining dictionaries dates back to Albert Fay's *Glossary of the Mining and Mineral Industry,* which first appeared in December 1918. That glossary contained about 18,000 terms. In 1968, the Bureau published *A Dictionary of Mining, Mineral,and Related Terms,* edited by Paul W. Thrush, with about 55,000 terms. The 1968 dictionary contained many new mining terms and terms from such related areas as metallurgy, ceramics, and glassmaking. That edition was as complete as possible with regard to technical and regional terms, historical terms, foreign terms that attained general usage in the United States, and terminology from the entire English-speaking world. For the past three decades that work has stood as the definitive authority on mineral-related terms.

The 1995 edition reflects a departure from the previous one in scope and in format. This edition, containing some 28,500 terms, is not meant to exhaustive in its coverage. It focusses on mining-related terms and excludes such related categories as ceramics, glass, metallurgy, petroleum, and other specialized disciplines. Geological terms which relate to mining are included, as are minerals which have a commercial value or which are associated with such minerals. Many chemicals and materials that are not usually connected with mining or minerals processing do not appear, nor di the chemical elements unless they are classified as minerals. Abbreviations and acronyms have largely been excluded, because they usually are explained and defined within the context of an individual report. The front material,m however, includes a list of abbreviations used in the definitions. New terms on marine mining, leaching, and automation appear in this edition as do a plethora of pollution and environmental terms, many of which have a legal definition based on law or regulation.

The task of deciding which terms should be deleted from this edition, how to ensure the collection of new terms since 1968, and how to cull terms for inclusion was formidable. Terms from the 1968 edition were categorized by computer, and each category was reviewed by at least one subject specialist. The reviewers judged which terms should be retained or deleted, and they revised definitions as necessary and defined new terms. Final judgment of the inclusion of existing terms or the addition of new ones was left to the collective discretion of a panel of experts called the Dictionary Review Group. This group also examined the Society of Mining Engineer's *Mining Engineering Handbook,* 1993 edition, to ensure that the most modern terms and their definitions would be considered.

Acknowledgments

Specialists in many aspects of mining have volunteered their help in bringing the widely used *Dictionary of Mining, Mineral, and Related Terms* up to date, by reviewing definitions, adding new terms, recommending corrections, and citing references, Hundreds of specialists contributed to the new edition. The mining and minerals community owes special gratitude to the members of the Dictionary Revision Group: Robert L. Bates; V.A. Cammarota; M. Elizabeth Clare; John DeYoung, Jr.; James F. Donahue; Charles D. Hoyt; Tim O'Neil; Eugene Palowitch; John W. Padan; Gloria Ruggiero; Al Schreck; Robert Tuchman; and Dirk Van Zyl.

ABBREVIATIONS

Abbreviations of certain terms, place names, and units of measure are used in the definitions as follows:

Abbrev. - abbreviation
adj. - adjective
Ant. - antonym
CF: - compare
e.g. - for example
esp - especially
et al. - and others
etc. - and so forth
Etymol. - etymology
i.e. - that is
Pl. - plural
Pron. - pronunciation
q.v. - which see
Sing. - singular
sp gr - specific gravity
specif. - specifically
Syn: or syn. - synonym
v. or V. - verb
var. - variant

Place Names

Arg - Argentina.
Aust - Australia.
Belg - Belgium.
Berks - Berkshire, England.
Bol - Bolivia.
Braz - Brazil.
Brist - Bristol coalfield, England.
Can - Dominion of Canada.
Cent. Am. - Central America.
Ches - Cheshire, England.
Clev - Cleveland iron district, England.
Colom - United States of Colombia.
Corn - Cornwall, England.
Cumb - Cumberland coalfield, England.
Derb - Derbyshire coalfield, England.
Dev - Devonshire, England.
E. Ind - East Indies.
Eng - England.
Forest of Dean - Forest of Dean coalfield, England.
Fr - French.
Ger - German.
Gr. Brit - Great Britain.
Glouc - Gloucestershire coalfield, England.
Hid - Hidalgo, Mex.
Hind - Hindustan.
Ire - Ireland.
It - Italian.
Kent - Kent, England.
Lanc - Lancashire coalfield, England.
Leic - Leicestershire, England.
Mex - Meyaco.
Mid - Midland coalfield, England.
Newc - Newcastle coalfield, England.
N.S.W. - New South Wales, Australia.
N.Z. - New Zealand.
Norf - Norfolk, England.
N. of Eng. - North of England.
N. Staff - North Staffordshire coalfield, England.
Northumb - Northumberland coalfield, England.
N. Wales - North Wales.
Pac - Pacific Coast, U.S.A.
Pat - Patagonia, South America.
Port - Portuguese (mostly in Brazil).
Prov - Provincial, U.S., unless otherwise, specif.
Pr - Prussian.
Russ - Russia.
Scot - Scotland.
Shrop - Shropshire, England.
S. Afr - South Africa.
S. Am - South America.
S. Staff - South Staffordshire, England.
S. Wales - South Wales, Great Britain.
Som - Somerset, England.
Sp - Spanish origin but not necess. used in Spain.
Sp Am - Spanish America.
Staff - Staffordshire, England.
Suff - Suffolk, England.
Sw - Swedish.
Trans - Transvaal, Republic of South Africa.
U.K. - United Kingdom of Great Britain and N. Ire.
U.S. - United States of America.
U.S.S.R. - Union of Soviet Socialist Republics.
Venez - Venezuela.
W. Afr - West Africa.
War - Warwickshire, England.
Wis - Wisconsin, U.S.A.
York - Yorkshire, England.

United States

AL - Alabama
AK - Alaska
AZ - Arizona
AR - Arkansas
CA - California
CO - Colorado
CT - Connecticut
DE - Delaware
FL - Florida
GA - Georgia
HI - Hawaii
ID - Idaho
IL - Illinois
IN - Indiana
IA - Iowa
KS - Kansas
KY - Kentucky
LA - Louisiana
ME - Maine
MD - Maryland
MA - Massachusetts
MI - Michigan
MN - Minnesota
MS - Mississippi
MO - Missouri
MT - Montana
NE - Nebraska
NV - Nevada
NH - New Hampshire
NJ - New Jersey
NM - New Mexico
NY - New York
NC - North Carolina
ND - North Dakota
OH - Ohio
OK - Oklahoma
OR - Oregon
PA - Pennsylvania
RI - Rhode Island
SC - South Carolina
SD - South Dakota
TN - Tennessee
TX - Texas
UT - Utah
VT - Vermont
VA - Virginia
WA - Washington
WV - West Virginia
WI - Wisconsin
WY - Wyoming
CZ - Canal Zone
DC - District of Columbia
GU - Guam
PR - Puerto Rico
VI - Virgin Islands

Canada

AB - Alberta
BC - British Columbia
LB - Labrador
MB - Manitoba
NB - New Brunswick
NF - Newfoundland
NT - Northwest Territories
ON - Ontario
PE - Prince Edward Island
PQ - Quebec
SK - Saskatchewan
UT - Yukon Territory

Units of Measure

Abbreviations for Compound Units

Two types of compound unit abbreviations are common.

1. In the metric system, many units comprise a basic unit, such as "m" (meter) modified by a prefix that acts as a "power of ten" multiplier. Thus "cm" is a "centi" "meter", one-hundreth of a meter or 10^{-2} meter.

Several compound metric units are widely used in this volume:

mum - micrometer(s); cm - centimeter(s); Kj - kilojoule(s); km - kilometer(s); Kpa - kilopascal(s); MJ - megajoule(s); Ml - milliliter(s); mm - millimeter(s); Mpa - megapascal(s).

2. Complex units are developed by multiplying ("•") or dividing ("/") units. The metric units for momentum, for example, are kgm/s, which is read Newton meters per second.

Abbreviations for Units of Measure

degrees C - degrees Celsius
degrees F - degrees Fahrenheit
A - ampere(s)
atm - standard atmosphere
AU - astronomical unit
bbl - barrel(s)
Bcf - billion cubic feet (gas flow)
Bcfd - billion cubic feet per day (gas flow)
Bcfy - billion cubic feet per year (gas flow)
Btu - British thermal unit(s)
Bunit - billion units(s)
c - cycle
c - centi (one-hundredth); prefix only
cal - calorie(s)
cd - candela(s)
Ci - curie(s)
cmil - circular mil
cp - candlepower
cpm - count(s) per minute
cps - cont(s) per second
d - day(s)
d - deci (one-tenth); prefix only
D - darcy(s)
Db - decibel(s)
dyn - dyne(s)
Ev - electron volt(s)
F - fermi(s)
F - farad(s)
fc - footcandle(s)
fl oz - fluid ounce(s)
ft - foot/feet
ft/s - foot/feet per second
ft^2 - square foot/feet
ft•lbf - foot pound(s) (force)
g - gram(s)
g, G - Gal (gravity constant)
G - giga (one billion); prefix only
gal - gallon(s)
Gs - gauss
h - hour(s)
ha - hectare(s)
H.E. - high explosive(s)
Gs - gauss
h - hour(s)
ha - hectare(s)
H.E. - high explosive(s)
hp - horsepower
hp•h - horsepower hour
Hz - cycle(s) per second
in Hg - inch(es) of mercury
in - inch(es)
in H_2O - inch(es) of water
J - joule(s)
k - kilo (one thousand); prefix only
K - kelvin
l - liter(s)
L - liter(s) (preferred form)
L - lambert(s)
lb - pound(s)
lbf - pound(s) force
lbt•ft - pound(s) force foot
lt - long ton(s)
m - meter
m - mili (one-thousandth); prefix only
M - mega (one million)
Mcf - thousand cubic feet (gas flow)
Mcfd - thousand cubic feet per day
mho - mho(s)
mi - mile(s)
mil - thousandth of an inch
Mmbl - million barrels
Mmcf - million cubic feet (gas flow)
Mmcfd - million cubic feet per day (gas flow)
Mmcfy - million cubic feet per year (gas flow)
mol - mole(s)
mol wt - mole weight
mol % - mole percent
mpg - mile(s) per gallon
mph - mile(s) per hour
Mx - maxwell(s)
n - nano (one-billionth); prefix only
n - refractive index
N - Newton(s)
nmi - nautical mile(s)
Oe - oersted(s)
ohm•cmil/ft - ohm circular mil per foot
oz - ounce(s)
p - pico (one-trillionth); prefix only
P - poise(s)
Pa - pascal(s)
pct - percent
ppb - part(s) per billion
ppm - part(s) per million
psi - pound(s) (force) per square inch
psia - pound(s) force per square inch, absolute
psig - pound(s) force per square inch, gauge
r - revolution(s)
R - röntgen(s)
rad - radian(s)
rpm - revolutions per minute
s - second(s)
S - siemens(s)
sr - steradian(s)
st - stere(s)
st - short ton(s)
St - stoke(s)
std ft^3 - standard cubic foot/feet
t - metric ton(s)
t/w-h - metric ton(s) per worker hour
t/w-d - metric ton(s) per worker day
T - tesla(s)
tr oz - troy ounce(s)
$unit^{-1}$ - reciprocal unit
$unit^2$ - square unit (or unit squared)
$unit^3$ - cubic unit
W - watt(s)
Wb - weber(s)
wt % - wt pct, weight percent
yd - yard(s)
yr - year(s)

Legislation Related to Mining and Minerals

Clean Air Act of 1970
Clean Air and Water Act Amendments of 1977
Comprehensive Env. Response Comp., & Liability Act (CERCLA)
Deep Seabed Hard Mineral Resources Act of 1980
Endangered Species Act of 1973
Federal Land Policy & Management Act of 1976
Federal Mine Safety and Health Act of 1977
Federal Water Pollution Control Act of 1972
Law of the Sea Treaty
Mineral Leasing Act for Acquired Lands
Mining Law of 1872
National Environmental Policy Act of 1969 (NEPA)
Outer Continental Shelf Lands Act of 1953
Resource Conservation & Recovery Act of 1976 (RCRA)
Submerged Lands Act of 1953
Superfund Amendments & Reauthorization Act of 1986
Surface Mining Control & Reclamation Act of 1977 (SMCRA)
Toxic Substances Control Act of 1976
Uranium Mill Tailings Radiation Control Act, 1978
Water Quality Act of 1987

Organizations and Acronyms

Air Pollution Control Association (APCA)
American Association of Petroleum Geologists (AAPG)
American Chemical Society (ACS)

American Conference of Governmental Industrial Hygienists (ACGIH)
American Electrochemical Society (AES)
American Electromechanical Society (AES)
American Electroplaters' Society (AES)
American Foundrymen's Association (AFA)
American Foundrymen's Society (AFS)
American Gas Association (AGA)
American Geological Institute (AGI)
American Heritage Center (AHC)
American Industrial Hygiene Association (AIHA)
American Institute of Chemical Engineers (AIChE)
American Institute of Mining, Metallurgical, and Petroleum Engineers, Inc. (AIME)
American Iron and Steel Institute (AISI)
American National Standards Institute (ANSI) (U.S. rep. to ISO)
American Petroleum Institute (API)
American Public Health Association (APHA)
American Society for Metals (ASM)
American Society for Testing and Materials (ASTM)
American Society of Civil Engineers (ASCE)
American Society of Mechanical Engineers (ASME)
American Society of Safety Engineers (ASSE)
Association of Iron Ore Exporting Countries (AIOEC)
Bituminous Coal Institute (BCI)
Bituminous Coal Operators' Association (BCOA)
Canada Centre for Mineral and Energy Technology (CANMET)
Canada Institute of Mining and Metallurgy (CIM)
Central Selling Organization (CSO)
Commission of the European Communities (CEC)
Commonwealth Scientific and Industrial Research Organization (CSIRO)
Conseil Intergouvernemental des Pays Exportateurs de Cuivre (CIPEC)
Coordinating Research Council (CRC)
Council for Scientific and Industrial Research (CSIR)
Council on Environmental Quality (CEQ)
Environmental Protection Agency (EPA)
EROS Data Center
European Coal and Steel Community (ECSC)
European Community (EC)
European Free Trade Association (EFTA)
Federal Emergency Management Administration (FEMA)
General Services Administration (GSA)
Geological Society of America (GSA)
GeoRef Information Services
Geoscience Information Society (GIS)
Illinois Institute of Technology Research Institute (IITRI)
Independent Petroleum Association of America (IPAA)
Institut National de l'Environnement Industriel et des Risques (INERIS)
Institute of Electrical and Electronics Engineers (IEEE)
Institute of Makers of Explosives (IME)
Institute of Radio Engineers (IRE)
Institution of Mining and Metallurgy (IMM)
Instrument Society of America (ISA)
Intergovernmental Panel on Climate Change (IPCC)
International Copper Study Group
International Energy Agency (IEA)
International Labor Organization (ILO)
International Lead and Zinc Study Group
International Nickel Study Group
International Organization for Standardization (ISO)
International Tin Council (ITC) (defunct)
International Union of Pure and Applied Chemistry (IUPAC)
International Union of Pure and Applied Physics (IUPAP)
Joint Army-Navy-Air Force (JANAF)
Lawrence Livermore Laboratories (LLL)
Manufacturing Chemists Association (MCA)
Massachusetts Institute of Technology (MIT)
Mine Safety and Health Administration (MSHA)
Minerals Management Service (MMS)
National Academy of Sciences (NAS)
National Coal Board (U.K.) (NCB)
National Independent Coal Operators Association (NICOA)
National Institute for Occupational Safety and Health (NIOSH)
National Institute of Health (NIH)
National Institute of Standards and Technology (NIST)
National Labor Relations Board (NLRB)
National Mine Rescue Association
National Petroleum Association (NPA)
National Petroleum Refiners Association (NPRA)
National Research Council (NRC)
National Science Foundation (NSF)
National Stone Association (NSA)
National Technical Information Service (NTIS)
Northwest Mining Association (NSA)
Nuclear Regulatory Commission (NRC)
Occupational Safety and Health Administration (OSHA)
Occupational Safety and Health Review Commission (OSHRC)
Office of Science and Technology Policy (OSTP)
Office of Surface Mining Reclamation and Enforcement (OSM)
Office of Technology Assessment (OTA)
Organization of Arab Petroleum Exporting Countries (OAPEC)
Organization of Petroleum Exporting Countries (OPEC)
Society for Mining, Metallurgy, and Exploration, Inc. (SME)
Society of Automotive Engineers (SAE)
Society of Exploration Geophysicists (SEG)
Standard Industrial Classification (SIC)
Tennessee Valley Authority (TVA)
The Minerals, Metals, and Materials Society (TMS)
U.S. International Trade Commission Division of Minerals and Metals
United Mine Workers of America (UMWA)
United Nations Conference on Trade and Development (UNCTD)
United States Bureau of Land Management
United States Bureau of Mines (USBM)
United States Geological Survey (USGS)
United Steelworkers of America

a (a) Symbol in structural petrology for the direction of tectonic transport, similar to the direction in which cards might slide over one another. Striations in a slickensided surface are parallel to direction *a*. (AGI, 1987) (b) A crystallographic axis: In the isometric system each axis is designated *a*; in hexagonal, tetragonal, and trigonal systems the nonunique axes are *a*; in the orthorhombic system *a* is always shorter than *b* with *c* either the longest or the shortest axis; in the monoclinic and triclinic systems *a* may be determined by one of several conventions.

aa A Hawaiian term for lava consisting of a rough assemblage of clinkerlike scoriaceous masses. It is contrasted with pahoehoe used to designate the smoother flows. Pron. ah-ah. (Fay, 1920; Hess)

***a* axis** (a) One of the three crystallographic axes used as reference in crystal description. It is oriented horizontally, front to back. (b) One of three reference axes used in describing a rock fabric possessing monoclinic symmetry, such as progressive simple shear. The a axis is the direction of tectonic transport, i.e., the direction of shear. Syn: *a direction*. The letter "a" usually appears in italics. Cf: *b axis; c axis.*:

abandoned mine *abandoned workings.*

abandoned workings Excavations, either caved or sealed, that are deserted and in which further mining is not intended and open workings that are not ventilated and inspected regularly. Syn: *abandoned mine*. (Federal Mine Safety, 1977)

abandonment Abandonment of a mining claim may be by failure to perform work, by conveyance, by absence, and by lapse of time. The abandonment of a mining claim is a question of intent. To constitute an abandonment of a mining claim, there must be a going away and a relinquishment of rights, with the intention never to return and with a voluntary and independent purpose to surrender the location or claim to the next comer. Cf: *forfeiture.*

Abbé jar In mineral processing, a porcelain jar used for laboratory batch grinding tests in ceramic ware. (Pryor, 1958)

Abbé refractometer An instrument to determine the index of refraction of a liquid between two high-index glass prisms. Cf: *refractometer.*

Abbé theory The visibility of an object under the microscope is directly proportional to the wavelength of light, and inversely to the aperture of lens. (Pryor, 1963)

Abbé tube mill A gear-driven tube mill supported on a pair of riding rings and distinguished by an Archimedes spiral, through which the ore is fed and discharged. Grinding is effected by flint pebbles fed into the mill. See also: *ball mill.* (Liddell, 1918)

ABC system A method of seismic surveying by which the effect of irregular weathering thickness may be determined by a simple calculation from reciprocal placement of shotholes and seismometers. The method was originally used to solve refraction problems arising from irregularities in the top of the high-velocity layer. (AGI, 1987)

Abel's reagent Etching agent consisting of 10% chromium trioxide in water. Used in the analysis of carbon steels. (Bennett, 1962)

abernathyite A tetragonal mineral, $K(UO_2)(AsO_4)\cdot 4H_2O$; in small yellow crystals; in the Temple Mountains, UT.

aberration (a) The failure of a lens or mirror to bring the light rays to the same focus. When aberration is due to the form of the lens or mirror, it is called spherical aberration. When due to the different refrangibility of light of different colors, it is called chromatic aberration. When present in magnifiers it often causes inaccurate decisions as to flawlessness or color of gems. (b) Distortion produced by a lens. It is spherical if a flat image appears closer to the viewer in the middle than toward the edges of the field of view. It is chromatic if the visible spectrum is spread to give both a red and a blue image. Cf: *achromatic; aplanatic lens; aplanachromatic lens.* See also: *chromatic aberration.*

abime A large, steep-sided vertical shaft opening at the surface of the ground. (AGI, 1987)

ablation breccia *solution breccia.*

A.B. Meco-Moore A bulky machine that cuts a deep web of coal up to 6 ft (1.8 m) and is used in cyclic mining in medium to thick seams. It runs on the floor of the seam and does not require a prop-free front. It carries two horizontal jibs, one cutting at floor level and the other at a height depending on seam conditions. (Nelson, 1965)

Abney level A surveying instrument for taking levels up steep slopes; also used as a clinometer. (Hammond, 1965)

abnormal place A working place in a coal mine with adverse geological or other conditions and in which the miner is unable to earn a wage, based on the pricelist, equal to or above the minimum wage. A term generally associated with stalls or pillar methods of working. (Nelson, 1965)

abraser A device for assessing the wear resistance of surfaces. The specimen to be tested is rubbed alternately by the flat faces of two weighted abrasive wheels that revolve in opposite directions through frictional contact with the specimen and exert a combined abrasive, compressive, and twisting action twice in each revolution of the specimen holder. (Osborne, 1956)

abrasion (a) The mechanical wearing away of rock surfaces by friction and impact of rock particles transported by wind, ice, waves, running water, or gravity. Syn: *corrasion.* Cf: *attrition.* (AGI, 1987) (b) The wearing away of diamonds, drill-bit matrices, and drill-stem equipment by frictional contact with the rock material penetrated or by contact with the cuttings produced by the action of the drill bit in drilling a borehole. (Long, 1960)

abrasion hardness Hardness expressed in quantitative terms or numbers indicating the degree to which a substance resists being worn away by frictional contact with an abrasive material, such as silica or carborundum grits. Also called abrasion resistance; wear resistance. (Long, 1960)

abrasion index The percentage of a specially prepared 3-in by 2-in (76-mm by 51-mm) sample of coke remaining on a 1/8-in (3.2-mm) mesh British Standards test sieve after the sample of coke has been subjected to a standardized abrasion procedure in a rotating drum. (BS, 1961)

abrasive (a) Any natural or artificial substance suitable for grinding, polishing, cutting, or scouring. Natural abrasives include diamond, emery, garnet, silica sand, diatomite, and pumice; manufactured abrasives include esp., silicon carbide, fused alumina, and boron nitride. (AGI, 1987) (b) Tending to abrade or wear away. (AGI, 1987)

abrasive blasting respirator A respirator designed to protect the wearer from inhalation or impact of, and abrasion by, materials used or generated in abrasive blasting. (ANSI, 1992)

abrasive formation A rock consisting of small, hard, sharp-cornered, angular fragments, or a rock, the cuttings from which, produced by the action of a drill bit, are hard, sharp-cornered, angular grains, which grind away or abrade the metal on bits and drill-stem equipment at a rapid rate. Syn: *abrasive ground.* (Long, 1960)

abrasive ground *abrasive formation.*

abrasive hardness test Test employing a rotating abrasive wheel or plate against which specimens are held. The specimens are abraded for a given number of revolutions, and the weight of material lost is a measure of the abrasive hardness. (Lewis, 1964)

abraum salts *abraumsalze.*

abraumsalze Ger. Mixed sulfates and chlorides of potassium, sodium, and magnesium overlying the rock salt in the Stassfurt salt deposits. Syn: *abraum salts; stripping salt.* (Holmes, 1928)

abriachanite An earthy, amorphous variety of crocidolite asbestos. (Dana, 1914)

absolute (a) In chemistry, free from impurity or admixture. (Hess) (b) In physics, not dependent on any arbitrary standard. (Hess) (c) Frequently used in the trades to indicate a thing as being perfect or exact. Abbrev. abs. (Crispin, 1964)

absolute age The geologic age of a fossil organism, rock, or geologic feature or event given in units of time, usually years. Commonly used as a syn. of isotopic age or radiometric age, but may also refer to ages obtained from tree rings, varves, etc. Term is now in disfavor as it implies a certainty or exactness that may not be possible by present dating methods; i.e., two absolute ages for the same pluton may disagree by hundreds of millions of years. Cf: *relative age.* Syn: *actual age.* (AGI, 1987)

absolute atmosphere An absolute unit of pressure equal to 1 million times the pressure produced on 1 cm^2 by the force of 1 dyn. (Fay, 1920)

absolute bulk strength A measure of available energy per unit volume of explosive. Syn: *bulk strength.* See also: *relative bulk strength.*

absolute chronology Geochronology in which the time-order is based on absolute age, usually measured in years by radiometric dating, rather than on superposition and/or fossil content as in relative chronology. (AGI, 1987)

absolute daily range During the 24 h of the day the difference between the maximum easterly and maximum westerly values of the magnetic declination at any point. (Mason, 1951)

absolute humidity The content of water vapor in air, expressed as the mass of water per unit volume of air. Cf: *relative humidity.* See also: *humidity.* (AGI, 1987)

absolute isohypse A line that has the properties of both constant pressure and constant height above mean sea level. Therefore, it can be any contour line on a constant-pressure chart, or any isobar on a constant-height chart. (Hunt, 1965)

absolute ownership In law, an unqualified title to property and the unquestioned right to immediate and unconditional possession thereof. Applies to mining claims and properties. (Standard, 1964; Hess)

absolute permeability A measure of possible flow of a standard liquid under fixed conditions through a porous medium when there is no reaction between the liquid and the solids. This measure is arbitrarily taken for isothermal viscous flow. It can be duplicated with gases if tests are so conducted that extrapolation to infinite pressure can be made; specific permeability. (Hess)

absolute potential True potential difference between a metal and the solution in which it is immersed. (Pryor, 1963)

absolute pressure (a) Total pressure at a point in a fluid equaling the sum of the gage pressure and the atmospheric pressure. (Webster 3rd, 1966) (b) Pressure measured with respect to zero pressure, in units of force per unit of area. (CTD, 1958)

absolute roof The entire mass of strata overlying a coal seam or a subsurface point of reference. See also: *nether roof.*

absolute scale *Kelvin temperature scale.*

absolute temperature Temperature reckoned from absolute zero. See also: *temperature.* (Handbook of Chem. & Phys., 2)

absolute time Geologic time measured in terms of years by radioactive decay of elements. Cf: *relative time.* See also: *geochronology.*

absolute viscosity *viscosity coefficient.*

absolute weight strength A measure of available energy per gram of explosive. Syn: *weight strength.* See also: *relative weight strength.*

absolute zero The temperature at which a gas would show no pressure if the general law for gases would hold for all temperatures. It is equal to -273.16 °C or -459 °F. Cf: *temperature.* (Handbook of Chem. & Phys., 2)

absorbed water Water held mechanically in a soil mass and having physical properties not substantially different from those of ordinary water at the same temperature and pressure. (ASCE, 1958)

absorbent formation A rock or rock material, which, by virtue of its dryness, porosity, or permeability, has the ability to drink in or suck up a drilling liquid, as a sponge absorbs water. Syn: *absorbent ground.* (Long, 1960)

absorbent ground *absorbent formation.*

absorbents Substances, such as wood meal and wheat flour, that are forms of low explosive when mixed with metallic nitrates and tend to reduce the blasting power of the explosives, making them suitable for coal blasting. (Cooper, 1963)

absorber (a) An apparatus in which gases are brought into intimate contact with an extended surface of an absorbing fluid so that they enter rapidly into solution. (Hess) (b) The resistance and capacitance in series that is placed across a break in an electrical circuit in order to damp any possible oscillatory circuit and would tend to maintain an arc or spark when a current is interrupted. Syn: *spark absorber.* (CTD, 1958) (c) Any material that absorbs or stops ionizing radiation, such as neutrons, gamma rays, alpha particles, and beta particles. (Lyman, 1964)

absorptiometer A device for measuring the solubility of a gas in a liquid. (Bennett, 1962)

absorption (a) The phenomenon observed when a pleochroic mineral is rotated in plane polarized light. In certain positions, the mineral is darker than in others, owing to the absorption of light. (b) In hydrology, a term applied to the entrance of surface water into the lithosphere by all methods. (AGI, 1987) (c) The reduction of light intensity in transmission through an absorbing substance or in reflection from a surface. In crystals, the absorption may vary with the wavelength and with the electric vector of the transmitted light with respect to crystallographic directions. (d) Any mechanism by which energy, e.g., electromagnetic or seismic, is converted into heat. (e) Taking up, assimilation, or incorporation, e.g., of liquids in solids or of gases in liquids. Cf: *adsorption.* Syn: *occlusion.* (f) The entrance of surface water into the lithosphere by any method.

absorption hygrometer A type of hygrometer with which the water vapor content of the atmosphere is measured by means of the absorption of vapor by a hygroscopic chemical. The amount of vapor absorbed may be determined in an absolute manner by weighing the hygroscopic material, or in a nonabsolute manner by measuring a physical property of the substance that varies with the amount of water vapor absorbed. The lithium chloride humidity strip and carbon-film hygrometer element are examples of the latter. (Hunt, 1965)

absorption loss (a) The loss of water occurring during initial filling of a reservoir in wetting rocks and soil. (Hammond, 1965) (b) That part of the transmission loss due to dissipation or the conversion of sound energy into some other form of energy, usually heat. This conversion may take place within the medium itself or upon a reflection at one of its boundaries. (Hy, 1965)

absorption rate (a) The rate, expressed in quantitative terms, at which a liquid, such as a drilling circulation medium, is absorbed by the rocks or rock materials penetrated by the drill bit. (Long, 1960) (b) The amount of water absorbed when a brick is partially immersed for 1 min; usually expressed either in grams or ounces per minute. Also called suction rate; initial rate of absorption. (ACSG, 1961)

absorption spectra Specific wavelengths of electromagnetic radiation have precisely the energy to cause atomic or molecular transitions in substances they are passing through; their removal from the incident radiation produces reductions in intensity of those wavelengths, or absorption spectra, characteristic of the substance under study. Cf: *emission spectra.*

absorption spectrum The array of absorption bands or lines seen when a continuous spectrum is transmitted through a selectively absorbing medium. (AGI, 1987)

absorption tower A tower in which a liquid absorbs a gas.

abundant vitrain A field term denoting, in accordance with an arbitrary scale established for use in describing banded coal, a frequency of occurrence of vitrain bands comprising 30% to 60% of the total coal layer. Cf: *dominant vitrain; moderate vitrain; sparse vitrain.* (AGI, 1987)

abutment A surface or mass provided to withstand thrust, for example, the end supports of an arch or bridge. In coal mining, (1) the weight of the rocks above a narrow roadway is transferred to the solid coal along the sides, which act as abutments of the arch of strata spanning the roadway; and (2) the weight of the rocks over a longwall face is transferred to the front abutment (the solid coal ahead of the face) and the back abutment (the settled packs behind the face). See also: *overarching weight; pressure arch; load transfer.* Syn: *arch structure.* (Nelson, 1965)

abutment load In underground mining, the weight of rock above an excavation that has been transferred to the adjoining walls. (Pryor, 1963)

abutment pillars Pillars intended to support vertical load in excess of the weight of the strata directly above them. Generally, these abutment pillars are large pillars adjacent to smaller pillars, sometimes called yield pillars, which are incapable of carrying the weight of the strata above them. (SME, 1992)

abysmal *abyssal.*

abysmal sea That part of the sea occupying the ocean basins proper. (Fay, 1920)

abyss (a) A very deep, unfathomable place. The term is used to refer to a particularly deep part of the ocean, or to any part below 3,000 fathoms (18,000 ft or 5.5 km). Syn: *pit; pot; pothole; chasm; shaft.* (Hunt, 1965) (b) Syn: *pit; pot; pothole; chasm; shaft.*

abyssal (a) Pertaining to an igneous intrusion that occurs at considerable depth, or to the resulting rock; plutonic. Cf: *hypabyssal.* (AGI, 1987) (b) Pertaining to the ocean environment or depth zone of 500 fathoms (3,000 ft or 915 m) or deeper; also, pertaining to the organisms of that environment. (AGI, 1987) (c) Of, or pertaining to, deep within the Earth, the oceanic deeps below 1,000 fathoms (6,000 ft or 1.83 km), or great depths of seas or lakes where light is absent. See also: *plutonic.* (AGI, 1987) (d) In oceanography, relating to the greatest depths of the ocean; relating to the abyssal realm. Syn: *abysmal.* (CTD, 1958)

abyssal deposit A deposit of the deep sea, accumulating in depths of more than 1,500 fathoms (9,000 ft or 2.7 km) of water; these deposits comprise the organic oozes, various muds, and red clay of the deepest regions. (CTD, 1958)

abyssal injection The process by which magmas, originating at considerable depths, are considered to have been driven up through deep-seated contraction fissures.

abyssal plain An area of the ocean floor with a slope of less than 1 in 1,000 or flat, nearly level areas that occupy the deepest portions of many ocean basins. (Schieferdecker, 1959)

abyssal realm The deep waters of the ocean below 1,000 fathoms or 6,000 ft (1.83 km).

abyssal theory A theory of mineral-deposit formation involving the separation and sinking of ore minerals below a silicate shell during the cooling of the Earth from a liquid stage, followed by their transport to and deposition in the crust as it was fractured (Shand, 1947). Modern thought ascribes more complex origins to mineral deposits. (AGI, 1987)

abyssal zone The marine-life zone of the deep sea embracing the water and bottom below a depth of 6,000 ft (1.83 km). (Stokes, 1955)

abyssobenthic Relating to that part of the abyssal realm that includes the ocean floor; pertaining to or living on the ocean floor at great depths. (CTD, 1958)
abyssolith *batholith.*
abyssopelagic (a) Relating to that part of the abyssal realm that excludes the ocean floor; floating in the depths of the ocean. (CTD, 1958) (b) Pertaining to that portion of the deep waters of the ocean that lie below depths of 6,000 ft (1.83 km). (AGI, 1987)
acanthite A monoclinic mineral, $4[Ag_2S]$; dimorphous with argentite, pseudohexagonal, in slender prisms; sp gr, 7.2 to 7.3; a source of silver.
accelerated weathering test A test to indicate the effect of weather on coal, in which the coal is alternately exposed to freezing, wetting, warming, and light; the alternation may be varied to suit. This test may be applied to other bituminous materials. (Hess)
accelerator (a) A machine that accelerates electrically charged atomic particles, such as electrons, protons, deuterons, and alpha particles, to high velocities. (Lyman, 1964) (b) A substance added to increase the rate of a chemical reaction. (Nelson, 1965)
accelerometer A seismometer with response linearly proportional to the acceleration of earth materials with which it is in contact. (AGI, 1987)
accented contour *index contour.*
acceptor A charge of explosives or blasting agent receiving an impulse from an exploding donor charge. Syn: *receptor.* (Meyer, 1981)
accessory (a) Applied to minerals occurring in small quantities in a rock. The presence or absence of these minor minerals does not affect the classification or the naming of the rock. (Holmes, 1928) (b) Fragments derived from previously solidified volcanic rocks of related origin; i.e., the debris of earlier lavas and pyroclastic rocks from the same cone. See also: *accessory mineral.* (c) Said of pyroclastics that are formed from fragments of the volcanic cone or earlier lavas; it is part of a classification of volcanic ejecta based on mode of origin, and is equivalent to resurgent ejecta. Cf: *auxiliary.* (AGI, 1987)
accessory element *trace element.*
accessory mineral Any mineral the presence of which is not essential to the classification of the rock. Accessory minerals generally occur in minor amounts; in sedimentary rocks they are mostly heavy minerals. Cf: *essential mineral.* Syn: *accessory.* (AGI, 1987)
accessory plate (a) The quartz wedge inserted in the microscope substage above the polarizer in order to estimate birefringence and to determine optical sign of uniaxial minerals. Cf: *quartz wedge.* (Pryor, 1963) (b) The selenite plate that gives the sensitive tint of a specimen between crossed nicols. (Pryor, 1963) (c) The mica plate that retards yellow light. (Pryor, 1963) (d) In polarized-light microscopy, an optical device that may be inserted into the light train to alter light interference after passage through, or reflection by, a crystalline material; e.g., quartz wedge, mica plate, gypsum plate, or Bertrand lens. (e) In polarized-light microscopy, an optical compensator that may be inserted into the light train to alter birefringence after light passage through or reflection by an anisotropic material; e.g., quartz wedge, mica plate, gypsum plate, or Berek compensator. Syn: *gips plate; glimmer plate; compensator.* Cf: *Berek compensator; gypsum plate.*
access road A route constructed to enable plant, supplies, and vehicles to reach a mine, quarry, or opencast pit. In remote and isolated regions, the provision of an access road may be very costly. (Nelson, 1965)
accidental inclusion *xenolith; xenocryst.*
accordion roller conveyor A roller conveyor with a flexible latticed frame that permits variation in length.
accretion vein A vein formed by the repeated filling of a channelway and its reopening by the development of fractures in the zone undergoing mineralization.
accumulation (a) In coal mining, bodies of combustible gases that tend to collect in higher parts of mine workings and at the edge of goaves and wastes. They are found in cavities, at ripping lips, at other sheltered places protected from the ventilating current, and at the higher sides of rise faces. (Mason, 1951) (b) The concentration or gathering of oil or gas in some form of trap. Commercial accumulation is a volume or quantity sufficient for profitable exploitation. (AGI, 1987)
accumulative rock *cumulate.*
accumulator (a) A cylinder containing water or oil under pressure of a weighted piston for hydraulic presses, hoists, winches, etc. It is between the pumps and the presses, keeps a constant pressure on the system, and absorbs shocks. (b) A storage battery. (c) In oceanography, a spring of rubber or steel attached to a trawling warp, to lessen any sudden strain due to the trawl catching. (CTD, 1958)
accumulator conveyor Any conveyor designed to permit accumulation of packages or objects. Usually roller, live roller conveyor, roller slat conveyor, or belt conveyor.
accuracy The degree of conformity with a standard, or the degree of perfection attained in a measurement. Accuracy relates to the quality of a result, and is distinguished from precision, which relates to the quality of the operation by which the result is obtained. (AGI, 1987)
acetamide A trigonal mineral, CH_3CONH_2. Syn: *acetic acid amine; ethanamide.*
acetic acid amine *acetamide.*
acetylene The most brilliant of illuminating gases, C_2H_2. It may be produced synthetically from its elements, by incomplete combustion of coal gas, and commercially from calcium carbide, CaC_2. It also may be produced by reaction with water. Used in manufacturing explosives. Formerly used as an illuminating gas in mines and around drill rigs. When combined with oxygen, acetylene burns to produce an intensely hot flame and hence now is used principally in welding and metal-cutting flame torches. Syn: *ethyne; ethine.* Cf: *gas.* (Standard, 1964; Bennett, 1962; Long, 1960)
acetylene lamp *carbide lamp.*
acetylene tetrabromide Yellowish liquid; $CHBr_2CHBr_2$; sp gr, 2.98 to 3.00; boiling point, 239 to 242 °C with decomposition (at 760 mm); also, boiling point, 151 °C (at 54 mm); melting point, 0.1 °C; and refractive index, 1.638. Used for separating minerals by specific gravity; a solvent for fats, oils, and waxes; a fluid in liquid gases; and a solvent in microscopy. (CCD, 1961)
achavalite Former name for iron selenide, FeSe.
Acheson graphite Artificial graphite made from coke by electric furnace heating. (Bennett, 1962)
Acheson process A process for the production of artificial or synthetic graphite. It consists of sintering pulverized coke in the Acheson furnace at 2,760 to 3,316 °C. (Henderson, 1953)
achirite Former name for dioptase.
achroite A colorless variety of elbaite tourmaline used as a gemstone.
achromatic In microscopy, a compound lens that does not spread white light into its spectral colors. Cf: *aberration.* See also: *aplanachromatic lens.*
acicular (a) A mineral consisting of fine needlelike crystals; e.g., natrolite. (Nelson, 1965) (b) Slender needlelike crystal. (c) Refers to needlelike crystals. Cf: *equant; sagenitic; tabular; rodlike.*
acicular bismuth *aikinite.*
acicular powder In powder metallurgy, needle-shaped particles. (ASM, 1961)
aciculite *aikinite.*
acid (a) A solution of pH less than 7.0 at 25 °C. (b) A substance containing hydrogen that may be replaced by metals with the formation of salts. (CTD, 1958)
acid Bessemer converter One lined with acid refractories.
acid bottom and lining The inner bottom and lining of a melting furnace, consisting of materials like sand, siliceous rock, or silica brick, which give an acid reaction at the operating temperature. Syn: *acid lining.* (ASM, 1961)
acid clay (a) A clay that is used mainly as a decolorant or refining agent, and sometimes as a desulfurizer, coagulant, or catalyst. (b) A clay that yields hydrogen ions in a water suspension; a hydrogen clay.
acid cure In uranium extraction, sulfation of moist ore before leaching. (Pryor, 1958)
acid-dip survey A method of determining the angular inclination of a borehole in which a glass, test-tubelike bottle partly filled with a dilute solution of hydrofluoric acid is inserted in a watertight metal case. When the assemblage is lowered into a borehole and left for 20 to 30 min, the acid etches the bottle at a level plane from which the inclination of the borehole can be measured. Cf: *Kiruna method.* Syn: *acid-dip test; acid test; acid-etch tube.* (Long, 1960)
acid-dip test *acid-dip survey.*
acid drainage Water with a pH of less than 6.0 and in which total acidity exceeds total alkalinity; discharged from an active, inactive, or abandoned surface coal mine and reclamation operation.
acid electric furnace An arc furnace having an acid refractory hearth.
acid embrittlement A form of hydrogen embrittlement that may be induced in some metals by acid treatment. (ASM, 1961)
acid-etch tube A soda-lime glass tube charged with dilute hydrofluoric acid, left in a borehole for 20 to 30 min to measure inclination as indicated by the angle of etch line on the tube. May be fitted in a clinometer. Syn: *acid-etch vial; culture tube; etch tube; sargent tube.* See also: *acid-dip survey.* (Pryor, 1963; Long, 1960)
acid-etch vial *acid-etch tube.*
acid flux Metallurgically acid material (usually some form of silica) used as a flux. (Bennett, 1962)
acid-forming materials Earth materials that contain sulfide minerals or other materials that, if exposed to air, water, or weathering processes, form acids that may create acid drainage.
acidic (a) A descriptive term applied to those igneous rocks that contain more than 60% silica. Acidic is one of four subdivisions of a widely used system for classifying igneous rocks based on their silica content: acidic, intermediate, basic, and ultrabasic. (AGI, 1987) (b) Applied loosely to any igneous rock composed predominantly of light-colored minerals having a relatively low specific gravity. Cf: *felsic.* Syn: *silicic.* (AGI, 1987)

acidization The process of forcing acid into a limestone, dolomite, or sandstone in order to increase permeability and porosity by dissolving and removing a part of the rock constituents. It is also used to remove mud injected during drilling. The general objective of acidization is to increase productivity. Syn: *acidizing.* (AGI, 1987)
acidize To treat a limestone or dolomitic formation with dilute hydrochloric acid to enlarge its void spaces. (Wheeler, 1958)
acidizing *acidization.*
acid leach Metallurgical process for dissolution of metals by means of acid solution. Examples include extraction of copper from oxide- or sulfide-bearing ore and dissolution of uranium from sandstone ores. Acid leaching can occur on heap-leach pads or in situ.
acid lining *acid bottom and lining.*
acid mine drainage (a) Acidic drainage from bituminous coal mines containing a high concentration of acidic sulfates, esp. ferrous sulfate. See also: *acid water.* (b) Drainage with a pH of 2.0 to 4.5 from mines and mine wastes. It results from the oxidation of sulfides exposed during mining, which produces sulfuric acid and sulfate salts. The acid dissolves minerals in the rocks, further degrading the quality of the drainage water. (AGI, 1987)
acid mine water (a) Mine water that contains free sulfuric acid, mainly due to the weathering of iron pyrites. A pit water, which corrodes iron pipes and pumps, usually contains a high proportion of solids per gallon, principally the sulfates of iron, chiefly ferrous and alumina. See also: *acid water.* (Nelson, 1965) (b) Where sulfide minerals break down under chemical influence of oxygen and water, the mine drainage becomes acidic and can corrode ironwork. If it reaches a river system, biological damage may also result. (Pryor, 1963)
acid neutralizers Calcium carbonate, $CaCO_3$, magnesium carbonate, $MgCO_3$, and china clay, which neutralize free acids, thereby preventing explosives from decomposing in storage. They also have a cooling effect and tend to reduce the sensitivity of the explosive. (Cooper, 1963)
acid open-hearth steel Low-phosphorus pig iron treated in an acid (silica or sand)-lined furnace. (Mersereau, 1947)
acid ore *siliceous ore.*
acid process A steelmaking process—Bessemer, open-hearth, or electric—in which the furnace is lined with a siliceous refractory, and for which pig iron low in phosphorus is required, as this element is not removed. See also: *acid steel; basic process.* (CTD, 1958)
acid-recovery operator In the coke products industry, a person who recovers sulfuric acid used in processing coke-gas byproducts by cooking sludge with steam in acid regenerator pots. Syn: *acid regenerator.* (DOT, 1949)
acid refractory material A general term for those types of refractory material that contain a high proportion of silica; e.g., silica refractories (greater than 92% SiO_2) and siliceous refractories (78% to 92% SiO_2). The name derives from the fact that silica behaves chemically as an acid and at high temperatures reacts with bases such as lime or alkalies. (Dodd, 1964)
acid refractory product Refractory product made of clay-silica mixture or pure silica. (Rosenthal, 1949)
acid regenerator *acid-recovery operator.*
acid rock drainage Drainage that occurs as a result of natural oxidation of sulfide minerals contained in rock that is exposed to air and water. It is not confined to mining activities, but can occur wherever sulfide-bearing rock is exposed to air and water. Abbrev. ARD.
acid slag Slag that contains substantial amounts of active silica.
acid sludge Products of refining of tar, shale oil, and petroleum in which sulfuric acid reacts to form a sulfonic acid mixture, green acids, and mahogany acids and salts. Used in the flotation process and in proprietary collector agents for flotation of iron ores. (Pryor, 1958)
acid soil A soil with a pH of less than 7.0. (AGI, 1987)
acid steel Steel melted in a furnace with an acid bottom and lining and under a slag containing an excess of an acid substance, such as silica. See also: *acid process.* (ASM, 1961)
acid strength Related to ability to liberate hydrogen ions to solution, and hence to electrical conductivity of equivalent aqueous solutions of acids. (Pryor, 1963)
acid test (a) *acid-dip survey.* (b) A severe or decisive trial, as of usability or authenticity. (Long, 1960)
acid water Water charged naturally with carbon dioxide. Also applied to natural waters containing sulfur compounds, esp. sulfates. See also: *acid mine water; acid mine drainage.*
aciform Needle-shaped.
aciniform A mineral aggregate shaped like a cluster of grapes. Also, full of small kernels like a grape. Syn: *acinose; acinous.*
acinose (a) Grapelike; applied to the structure of clustered mineral aggregates. Syn: *aciniform; acinous.* (b) Granulated; like grape seeds; applied to the texture of some mineral aggregates.
acinote Former name for actinolite.
acinous *aciniform; acinose.*
aclinal A little-used term said of strata that have no inclination; horizontal. Syn: *aclinic.* (AGI, 1987)
aclinic *aclinal.*
aclinic line The line through those points on the Earth's surface at which the magnetic inclination is zero. The aclinic line is a particular case of an isoclinic line. (Hunt, 1965)
acmite A brown variety of aegirine having pointed terminations. See also: *aegirine; pyroxene.*
acopolado Mex. Ore containing 50 to 60 oz/st (1.56 to 1.88 kg/t) of silver. (Hess)
acoustic Used when the term that it modifies designates something that has the properties, dimensions, or physical characteristics associated with sound waves. (Hy, 1965)
acoustical well logging Any determination of the physical properties or dimensions of a borehole by acoustical means, including measurement of the depth of fluid level in a well. (AGI, 1987)
acoustic attenuation log In theory, a log designed to measure the manner in which the energy of elastic waves is dissipated in passing through rock. Although no practical log of this type has yet evolved, the belief that a log of this parameter would permit the estimation of the permeability of formations would seem to ensure such a development since no log has been developed to record permeability. (Wyllie, 1963)
acoustic dispersion The change of speed of sound with frequency. (Hunt, 1965)
acoustic impedance The acoustic impedance of a given surface area of an acoustic medium perpendicular, at every point, to the direction of propagation of sinusoidal acoustic waves of given frequency, and having equal acoustic pressures and equal volume velocities per unit area at every point of the surface at any instance, is the quotient obtained by dividing (1) the phasor corresponding to the acoustic pressure by (2) the phasor corresponding to the volume velocity. See also: *impedance.* (Hunt, 1965)
acoustic interferometer An instrument for making physical observations upon standing waves. It may be used, e.g., to measure velocity, wavelength, absorption, or impedance. (Hunt, 1965)
acoustic log A continuous record made in a borehole showing the velocity of sound waves over short distances in adjacent rock; velocity is related to porosity and nature of the liquid occupying pores. (AGI, 1987)
acoustic-radiation pressure A unidirectional steady-state pressure exerted upon a surface exposed to an acoustic wave. Such a steady pressure is usually quite small in magnitude and is really observable only in the presence of very intense sound waves. (Hunt, 1965)
acoustic radiometer An instrument for measuring acoustic-radiation pressure by determining the unidirectional steady-state force resulting from reflection or absorption of a sound wave at its boundaries. (Hunt, 1965)
acoustic resistance Product of longitudinal wave velocity and density, being the property that controls the reflective power at a boundary plane. (Schieferdecker, 1959)
acoustics The study of sound, including its production, transmission, reception, and utilization, esp. in fluid media such as air or water. With reference to Earth sciences, it is esp. relevant to oceanography. The term is sometimes used to include compressional waves in solids; e.g., seismic waves. (AGI, 1987)
acoustic scattering The irregular reflection, refraction, or diffraction of sound waves in many directions. (Hy, 1965)
acoustic sounding The indirect evaluation of water depth, using the principle of measuring the length of time necessary for a sound wave to travel to the bottom, reflect, and travel back to the water surface. (Hunt, 1965)
acoustic-strain gage An instrument for measuring strains; e.g., in concrete linings to shafts or roadways. It contains a length of fine wire under tension, the tension being varied by the strain to which the gage is subjected. The measurement made is that of the frequency of vibration of the wire when it is plucked by means of an electromagnetic impulse, and this measurement can be made with great accuracy. The gage is highly stable, and readings can be made over a period of years without any fear of zero drift. See also: *electrical resistance strain gage; mechanical extensometer.* (Nelson, 1965)
acoustic theodolite An instrument designed to provide a continuous vertical profile of ocean currents at a specific location. (Hunt, 1965)
acoustic wave (a) The waves that contain sound energy and by the motion of which sound energy is transmitted in air, in water, or in the ground. The wave may be described in terms of change of pressure, of particle displacement, or of density. (AGI, 1987) (b) Used increasingly to study the physical properties of rocks and composition of gases. Investigations may be made both in situ and in the laboratory. (Nelson, 1965)
acquired lands Defined by the U.S. Department of the Interior as "lands in Federal ownership which were obtained by the Government through purchase, condemnation, or gift, or by exchange for such purchased, condemned, or donated lands, or for timber on such lands.

They are one category of public lands." Public land laws are generally inapplicable to acquired lands. (SME, 1992)

acre (a) A measure of surficial area, usually of land. The statute acre of the United States and England contains 43,560 ft^2 (4,840 yd^2; 4,047 m^2; or 160 square rods). The so-called Scotch acre contains about 6,150 yd^2 (5,142 m^2), and the Irish acre 7,840 yd^2 (6,555 m^2). There are various special or local acres in England (as in Cheshire or among the hop growers), varying from 440 yd^2 (368 m^2) to more than 10,000 yd^2 (8,361 m^2). (Standard, 1964) (b) Can. In Quebec, a linear measure that equals the square root of 43,560, or approx. 208.7 ft (63.6 m). (Fay, 1920) (c) For the calculation of coal reserves, a convenient rule is to allow 1,200 st/ft (coal thickness) per acre (8,821 t/m/ha). For known and dependable areas, 1,500 st/ft per acre (11,027 t/m/ha) may be used. (Nelson, 1965)

acreage rent Royalty or rent paid by the lessee for working and disposing of minerals at the rate of so much per acre.

acre-foot The quantity of water that would cover 1 acre, 1 ft deep (1 ha, 13.6 cm deep). One acre-foot contains 43,560 ft^3 (1,233 m^3).

acre-inch The volume of water, soil, or other material that will cover 1 acre, 1 in deep (1 ha, 1.1 cm deep). (AGI, 1987)

acre-yield The average quantity of oil, gas, or water recovered from 1 acre (0.4 ha) of a reservoir. (AGI, 1987)

actetic acid amine *acetamide.*

actinide A chemical element with atomic number greater than 88; all are radioactive. Syn: *actinide element.*

actinide element (a) One of the group of chemical elements of increasing atomic number, starting with actinium (atomic number 89) and extending through atomic number 103. These elements occupy one single place in the extended periodic table, in the same group into which the rare-earth elements (lanthanides) are classified. See also: *actinide.* (CCD, 1961; Gaynor, 1959) (b) One of the radioactive elements, atomic numbers 89 to 103. (Hurlbut, 1964)

actinolite A monoclinic mineral, $2[Ca_2(Mg,Fe)_5Si_8O_{22}(OH)_2]$ in the hornblende series $Mg/(Mg+Fe^{2+}) = 0.5$ to 0.89 of the amphibole group; forms a series with tremolite; green, bladed, acicular, fibrous (byssolite asbestos), or massive (nephrite jade); prismatic cleavage; in low-grade metamorphic rocks. Syn: *actinote; strahlite.* Cf: *tremolite.*

actinote *actinolite.*

activated alumina Highly porous, granular aluminum oxide that preferentially absorbs liquids from gases and vapors, and moisture from some liquids. (McGraw-Hill, 1994)

activated carbon Carbon, mostly of vegetable origin, and of high adsorptive capacity. Syn: *activated charcoal.* (Bennett, 1962)

activated charcoal *activated carbon.*

activated clay A clay whose adsorbent character or bleaching action has been enhanced by treatment with acid. (CCD, 1961)

activated coal plow With a view to applying the coal plow to seams too hard to be sheared by the normal cutting blade, German mining engineers have developed various types of power-operated cutters. One consists of a series of compressed-air picks mounted above each other; another, of a resonance pattern, houses two high-speed motors eccentrically mounted and rotating in opposite directions. The latter imparts a vibration to the cutting edge equivalent to 2,500 blows per minute with a stroke of 3/16 to 1/4 in (4.8 to 6.4 mm) and a force of approx. 200 st (181 t). (Mason, 1951)

activated plow *Huwood slicer.*

activating agent (a) A substance that when added to a mineral pulp promotes flotation in the presence of a collecting agent. Syn: *activator.* (BS, 1962) (b) Reagent used particularly in differential mineral flotation to help cleanse the mineral surface so that a collector may adhere to it and permit or aid its floatability. Frequently used to allow floating minerals that had been previously depressed. (Mitchell, 1950)

activation (a) In the flotation process of mineral dressing, the process of altering the surface of specific mineral particles in a mineral pulp to promote adherence of certain reagents. (Pryor, 1963) (b) The changing of the passive surface of a metal to a chemically active state. Cf: *passivation.* (ASM, 1961) (c) In the flotation process of ore beneficiation, the process of altering the surface of specific mineral particles in an ore pulp to promote adherence of certain reagents. (Henderson, 1953) (d) The process of making a material radioactive by bombardment with neutrons, protons, or other nuclear particles. See also: *activation analysis.* (Lyman, 1964)

activation analysis A method for identifying and measuring the chemical elements in a sample to be analyzed. The sample is first made radioactive by bombardment with neutrons, charged particles, or other nuclear radiation. The newly radioactive atoms in the sample give off characteristic nuclear radiations that can identify the atoms and indicate their quantity. See also: *activation.* (Lyman, 1964)

activator (a) In flotation, a chemical added to the pulp to increase the floatability of a mineral in a froth or to refloat a depressed (sunk) mineral. Also called activating reagent. (CTD, 1958) (b) A reagent that affects the surface of minerals in such a way that it is easy for the collector atoms to become attached. It has the opposite effect of a depressor. Cf: *depressor.* (Newton, 1959) (c) A substance that is required in trace quantities to impart luminescence to certain crystals. (CCD, 1961) (d) Ions that are photon emitters. (Van Vlack, 1964) (e) Any agent that causes activation. See also: *activating agent.* (Bennett, 1962)

active agent Surface-active substance that immunizes solids against a parting liquid. (Hess)

active earth pressure The minimum value of lateral earth pressure exerted by soil on a structure, occurring when the soil is allowed to yield sufficiently to cause its internal shearing resistance along a potential failure surface to be completely mobilized. See also: *surcharge.* Cf: *passive earth pressure.* (AGI, 1987)

active entry An entry in which coal is being mined from a portion thereof or from connected sections. (USBM, 1960)

active fault One liable to further movement. Cf: *passive fault.* (Carson, 1965)

active layer (a) The surficial deposit that undergoes seasonal changes of volume, swelling when frozen or wet, and shrinking when thawing and drying. (AGI, 1987) (b) A surface layer of ground, above the permafrost, that is frozen in the winter and thawed in the summer. Its thickness ranges from several centimeters to a few meters. (AGI, 1987)

active mining area (a) The area, on and beneath land, used or disturbed in activity related to the extraction, removal, or recovery of coal from its natural deposits. This term excludes coal preparation plants, areas associated with coal preparation plants, and post-mining areas. (SME, 1992) (b) The area in which active mining takes place relative also to extraction of metal ores, industrial minerals, and other minerals of economic value.

active workings All places in a mine that are ventilated and inspected regularly. (Federal Mine Safety, 1977)

activity (a) In nuclear physics, the rate of decay of atoms by radioactivity. It is measured in curies. (Bennett, 1962) (b) The ideal or thermodynamic concentration of a substance, the substitution of which for the true concentration, permits the application of the law of mass action. See also: *ionization constant.* (CTD, 1958)

actual age *absolute age.*

actual breaking strength The breaking load obtained from a tensile test to destruction on a sample of rope. (Hammond, 1965)

actual horsepower The horsepower really developed, as proved by trial. (Standard, 1964)

actual performance curve A performance curve showing the results actually obtained from a coal preparation treatment. (BS, 1962)

actuated roller switch A switch placed in contact with the belt conveyor immediately preceding the conveyor it is desired to control. In the centrifugal sequence control switch, a driving pulley bears against the driving belt; as the latter moves, the pulley rotates and the governor weights attached to the pulley shaft are flung out and so complete an electrical pilot circuit and thus start the subsidiary belt. (Nelson, 1965)

acute bisectrix (a) The line that bisects the acute angle of the optic axes of biaxial minerals. (Fay, 1920) (b) The angle $<90°$ between the optic axes in a biaxial crystal, *bxa.* Cf: *optic angle.*

adamantine (a) Like the diamond in luster. (Webster 3rd, 1966) (b) Diamond hard. A commercial name for chilled steel shot used in the adamantine drill, which is a core-barrel type of rock-cutting drill with a cutting edge fed by these shots. Cf: *vitreous.* (Pryor, 1963)

adamantine luster Diamondlike luster. (Hurlbut, 1964)

adamellite *quartz monzonite.*

adamic earth A term used for common clay, in reference to the material of which Adam, the first man, was made; specif. a kind of red clay. (AGI, 1987)

adamite A rare hydrous zinc arsenate, $Zn_2(AsO_4)(OH)$, occurring granular or in crusts and crystallizing in the orthorhombic system. Weakly radioactive; variable color—yellowish, greenish, or violet, rarely colorless or white; found in the oxidized zone of zinc orebodies. Associated with smithsonite, calcite, malachite, hemimorphite, limonite, and azurite. Small amounts of uranium have been found in some specimens of adamite. (Fay, 1920; Crosby, 1955)

adamsite A greenish-black muscovite found in a schist at Derby, VT; has been called margarodite. (Dana, 1914)

Adam's snuffbox Hollow, roughly rectangular pebble lined with goethite; Lenham beds, Netley Heath, Surrey, U.K. (Arkell, 1953)

ada mud A conditioning material that may be added to drilling mud in order to obtain satisfactory cores and samples of formations. (Williams, 1964)

adapter trough A short section of a shaker conveyor trough that serves as a connecting link between any two sizes of trough. (Jones, 1949)

added diamonds As used by the diamond-bit manufacturing industry, the number or carat weight of new diamonds that must be added to the resettable diamonds salvaged from a worn bit in order to have enough to set a new bit. (Long, 1960)

additive A correction applied to times of seismic reflections measured from an arbitrary time origin. The additive is normally applied for the purpose of translating the time origin to correspond to the datum elevation chosen for computation, and it is algebraic in sign. (AGI, 1987)

addlings A term used in the northern and parts of other coalfields in Great Britain to describe earnings or wages. (Nelson, 1965)

Adeline steelmaking process A process of producing precision castings of steel or steel alloys, which comprises first forming the steel or steel alloy in molten form by the aluminothermic process, by igniting a mixture of iron ore and aluminum; then running the molten metal into a mold prepared by packing a refractory mold composition around a model made of wax or other comparatively low-melting-point substance and heating to melt out the wax and consolidate the mold; and finally centrifuging the mold. (Osborne, 1956)

adelite (a) An orthorhombic mineral, $CaMg(AsO_4)(OH)$; occurs with manganese ores. (b) The mineral group adelite, austinite, conichalcite, duftite, and gabrielsonite.

ader wax *ozocerite.*

adhesion (a) The molecular force holding together two different substances that are in contact, as water in the pore spaces of a rock. Cf: *cohesion.* (b) Shearing resistance between soil and another material under zero externally applied pressure. (ASCE, 1958) (c) In the flotation process, the attachment of a particle to air-water interface or to a bubble.

adhesive slate A very absorbent slate that adheres to the tongue if touched by it. (Standard, 1964)

adiabatic calorimeter A calorimeter that practically remains unaffected by its surroundings and neither gains nor loses heat. (Osborne, 1956)

adiabatic compression Compression in which no heat is added to or subtracted from the air and the internal energy of the air is increased by an amount equivalent to the external work done on the air. The increase in temperature of the air during adiabatic compression tends to increase the pressure on account of the decrease in volume alone; therefore, the pressure during adiabatic compression rises faster than the volume diminishes. (Lewis, 1964)

adiabatic efficiency A compression term obtained by dividing the power theoretically necessary to compress the gas and deliver it without loss of heat, by the power supplied to the fan or compressor driveshaft.

adiabatic expansion Expansion in which no heat is added to or subtracted from the air, which cools during the expansion because of the work done by the air. (Lewis, 1964)

adiabatic temperature The temperature that would be attained if no heat were gained from or lost to the surroundings. (Newton, 1959)

adiabatic temperature change The compression of a fluid without gain or loss to the surroundings when work is performed on the system and produces a rise of temperature. In very deep water such a rise of temperature occurs and must be considered in the vertical temperature distribution. (Hy, 1965)

adinole An argillaceous sediment that has undergone albitization as a result of contact metamorphism along the margins of a sodium-rich mafic intrusion. Cf: *spilosite; spotted slate.* (AGI, 1987)

adipite An aluminosilicate of calcium, magnesium, and potassium having the composition of chabazite. (Dana, 1914)

adipocerite *hatchettite.*

adipocire *hatchettite.*

a direction *a axis.*

adit (a) A horizontal or nearly horizontal passage driven from the surface for the working or dewatering of a mine. If driven through the hill or mountain to the surface on the opposite side, it would be a tunnel. Syn: *drift; adit level.* See also: *tunnel.* (Lewis, 1964) (b) As used in the Colorado statutes, it may apply to a cut either open or undercover, or open in part and undercover in part, dependent on the nature of the ground. (c) A passage driven into a mine from the side of a hill. (Statistical Research Bureau, 1935)

adit end The furthermost end or part of an adit from its beginning or the very place where the miners are working underground toward the mine. (Hess)

adit level Mine workings on a level with an adit. See also: *adit.* (Hess)

adjacent sea A sea adjacent to and connected with the oceans, but semienclosed by land. The North Polar, Mediterranean, and Caribbean Seas are examples. Syn: *marginal sea.* (AGI, 1987)

adjustment of error Method of distributing the revealed irregularities over a series of results. (Pryor, 1963)

adjutage Nozzle or tube from which hydraulic water is discharged. Syn: *ajutage.*

admission *admittance.*

admittance (a) In a crystal structure, substitution of a trace element for a major element of higher valence; e.g., Li^+ for Mg^{2+}. Admitted trace elements generally have a lower concentration relative to the major element in the mineral than in the fluid from which the mineral crystallized. Cf: *capture; camouflage.* Syn: *admission.* (AGI, 1987) (b) The reciprocal of impedance or the ratio of complex current to voltage in a linear circuit. (AGI, 1987)

adobe A fine-grained, usually calcareous, hard-baked clayey deposit mixed with silt, usually forming as sheets in the central or lower parts of desert basins, as in the playas of the southwestern United States and in the arid parts of Mexico and South America. It is probably a wind-blown deposit, although it is often reworked and redeposited by running water. (AGI, 1987)

adobe charge A mud-covered or unconfined explosive charge fired in contact with a rock surface without the use of a borehole. Syn: *bulldoze; mudcapping.* (Atlas, 1987)

adobe flat A generally narrow plain formed by sheetflood deposition of fine sandy clay or adobe brought down by an ephemeral stream, and having a smooth, hard surface (when dry) usually unmarked by stream channels. (AGI, 1987)

adobe shot Ordinarily referred to as a dobe shot. A stick or part of a stick of dynamite is laid on the rock to be broken and covered with mud to add to the force of the explosion. A mudcap shot. (Hess)

adsorption (a) Adherence of gas molecules, or of ions or molecules in solution, to the surface of solids with which they are in contact, as methane to coal or moisture to silica gel. Cf: *absorption.* (AGI, 1987) (b) The assimilation of gas, vapor, or dissolved matter by the surface of a solid or liquid. (c) The attachment of a thin film of liquid or gas, commonly monomolecular in thickness, to a solid substrate.

adsorption analysis Separation by differential adsorption. (Pryor, 1958)

adular *adularia.*

adularescence (a) A milky white to bluish sheen in gemstones. (CMD, 1948) (b) The changeable white to pale bluish luster of an adularia cut cabochon. (Webster 3rd, 1966) (c) A floating, billowy, white or bluish light, seen in certain directions as a gemstone (usually adularia) is turned, caused by diffused reflection of light from parallel intergrowths of another feldspar of slightly different refractive index from the main mass. Syn: *schiller.*

adularia A colorless, moderate- to low-temperature variety of orthoclase feldspar typically with a relatively high barium content. Syn: *adular.*

adularia moonstone Precious moonstone, a gem variety of adularia.

advance (a) The work of excavating as mining goes forward in an entry and in driving rooms; to extract all or part of an area; first mining as distinguished from retreat. (BCI, 1947) (b) Rate at which a drill bit penetrates a rock formation. (Long, 1960) (c) Feet drilled in any specific unit of time. (Long, 1960) (d) The linear distance (in feet or meters) driven during a certain time in tunneling, drifting, or in raising or sinking a shaft. (Fraenkel, 1953)

advance development S. Afr. Development to provide an ore reserve in advance of mining operations. (Beerman, 1963)

advanced gallery In tunnel excavation, a small heading driven in advance of the main tunnel.

advanced materials New materials being developed that exhibit greater strength, higher strength-density ratios, greater hardness, and/or one or more superior thermal, electrical, optical, or chemical properties, when compared with traditional materials (Sorrel, 1987) and with properties needed to perform a specific function and often entirely new functions. (SME, 1992)

advance gate Gate road that is driven simultaneously with the longwall coal face, when the advancing longwall technique is used, but which is maintained some 10 to 20 yd (9 to 18 m) or more in advance of the face. The area immediately ahead of the coal face is therefore preexplored, and steps can be taken to cope with minor disturbances and thus prevent a serious loss of output. (Nelson, 1965)

advance overburden Overburden in excess of the average overburden-to-ore ratio that must be removed in opencut mining. (Mining)

advance per round The length, measured along the longitudinal axis of the working, tunnel, or gallery, of the hollow space broken out by each round of shots. For raises, it is upward advance; for sunk shafts, downward advance. (Fraenkel, 1953)

advance stope A stope in which sections of the face or some pillars are a little in advance of the others. This is achieved either by beginning the stoping of the section that is to be advanced earlier, or by proceeding more quickly. (Stoces, 1954)

advance stripping The removal of overburden required to expose and permit the minable grade of ore to be mined. The removal of overburden is known as stripping.

advance wave The air-pressure wave preceding the flame in a coal-dust explosion. The bringing of the dust into suspension is accomplished by such a wave and the violent eddies resulting therefrom. Syn: *pioneer wave.* (Rice, 1913-18)

advance working Mine working that is being advanced into the solid, and from which no pillar is being removed. See also: *first working.* (Fay, 1920)

advancing Mining from the shaft out toward the boundary. See also: *working out.* (Stoces, 1954)

advancing longwall A longwall mining technique, most commonly found in European coal mines, where the gate roads are advanced while the longwall face is advanced toward the mining limits. The gate roads are maintained throughout the worked-out portion of the longwall panel.

adventurine Spelling variant of aventurine.

adverse To oppose the granting of a patent to a mining claim.

adverse claim A claim made to prevent the patenting of part of the ground within the area in question; e.g., an adverse claim is made by a senior locator to exclude the part of his or her claim that is overlapped by the claim of a junior locator, when the junior locator is applying for patent. (Lewis, 1964)

adverse intent The terms "claim of right," "claim of title," and "claim of ownership," when used in the books to express adverse intent, mean nothing more than the intention of the dissessor to appropriate and use the land as his or her own to the exclusion of all others, irrespective of any semblance or shadow of actual title. (Ricketts, 1943)

advertised out A term used to express the result of the action of a joint owner of a mining claim who by proper notices causes the interest of the co-owner to be forfeited for failure to perform his or her share of the assessment work.

aedelforsite A name given to (1) a mixture of wollastonite, quartz, and feldspar from Edelfors, Sweden; (2) impure wollastonite from Giellebak, Sweden (called also gillebackit); and (3) impure laumontite, under the impression that they were new minerals. Syn: *edelforsite.* (Hess)

aedelite *prehnite.*

AED process An electrostatic process under development, in which fine-size dry coal is passed through an ionized field that selectively charges the coal and the liberated mineral matter. The output of the ionizer is then fed into an electrostatic separator where the coal and impurities are separated.

aegirine A sodium-ferric iron silicate, $NaFe^{3+}Si_2O_6$, occurring commonly in soda-rich igneous rocks; monoclinic; Mohs hardness, 6 to 6.5; sp gr, 3.40 to 3.55. Syn: *acmite; aegirite.* See also: *pyroxene.* (Dana, 1959)

aegirine-augite A monoclinic mineral, $(Ca,Na)(Ca,Mg,Fe)Si_2O_6$, in the range 20% augite to 20% aegirine end members of the pyroxene group. Formerly called acmite-augite, aegirineaugite.

aegirite Former spelling of aegirine. See also: *pyroxene; aegirine.*

aenigmatite (a) A triclinic mineral, $Na_2Fe_5^{2+}TiSi_6O_{20}$; a rare titanium-bearing silicate; black color; found associated with alkalic rocks. (Dana, 1959) (b) The mineral group aenigmatite, rhönite, serendibite, and welshite. Cf: *enigmatite.*

aeolian *eolian.*

Aeonite Trade name for a bitumen allied to wurtzilite. Similar to elaterite. (Tomkeieff, 1954; English, 1938)

aerate (a) To expose to the action of the air; to supply or to charge with air. (Standard, 1964) (b) To charge with carbon dioxide or other gas, as soda water. (Standard, 1964)

aeration (a) The introduction of air into the pulp in a flotation cell in order to form air bubbles. (BS, 1962) (b) In mineral beneficiation, use of copious air bubbled into mineral pulps (1) to provide oxygen in cyanidation, (2) to prevent settlement of solids, and (3) to remove aerophilic minerals in froth flotation by binding them into a mineralized froth that is temporarily stabilized by frothing agents. (Pryor, 1958) (c) The process of relieving the effects of cavitation by admitting air to the section affected. (Seelye, 1951)

aeration zone The zone in which the interstices of the functional permeable rocks are not (except temporarily) filled with water under hydrostatic pressure; the interstices are either not filled with water or are filled with water that is held by capillarity. (Rice, 1960)

aerator An apparatus for charging water with gas under pressure, esp. with carbon dioxide. (Standard, 1964)

aerial Relating to the air or atmosphere. Subaerial is applied to phenomena occurring under the atmosphere as subaqueous is applied to phenomena occurring underwater. (Fay, 1920)

aerial cableway An arrangement of overhead cable supporting a traveling carriage from which is suspended a skip or container that can be lowered and raised at any desired point. (Nelson, 1965)

aerial mapping The taking of aerial photographs for making maps and for geologic interpretation. (AGI, 1987)

aerial photograph Any photograph taken from the air, such as a photograph of a part of the Earth's surface taken by a camera mounted in an aircraft. Syn: *air photograph.* (AGI, 1987)

aerial photomosaic *mosaic.*

aerial railroad A system of cables from which to suspend cars or baskets, as in hoisting ore. See also: *aerial tramway.* (Standard, 1964)

aerial ropeway System of ore transport used in rough or mountainous country. A cable is carried on pylons, and loaded buckets are (1) towed from loading point to discharge, (2) suspended from a carriage running on this cable and then returned empty along a second cable, or (3) the whole cable moves continuously carrying buckets that hang from saddle clips and are loaded and discharged automatically or by hand control. Syn: *overhead ropeway.* See also: *bicable; monocable; aerial tramway; telpher.* (Pryor, 1963)

aerial spud A cable for moving and anchoring a dredge. (Fay, 1920)

aerial survey (a) A survey using aerial photographs as part of the surveying operation. (AGI, 1987) (b) The taking of aerial photographs for surveying purposes. (AGI, 1987)

aerial tramway A system for the transportation of material, such as ore or rock, in buckets suspended from pulleys or grooved wheels that run on a cable, usually stationary. See also: *tramway; aerial railroad; aerial ropeway.* (Fay, 1920; Peele, 1941)

aerobe An organism that lives in the presence of free oxygen. The oxygen is usually used in the cell's metabolism. See also: *aerobic.* (Rogoff, 1962)

aerobic (a) Said of an organism (esp. a bacterium) that can live only in the presence of free oxygen; also, said of its activities. Syn: *aerobe.* (AGI, 1987) (b) Said of conditions that can exist only in the presence of free oxygen. Cf: *anaerobic.* (AGI, 1987)

aeroclay Clay, particularly china clay, that has been dried and air separated to remove any coarse particles. (Dodd, 1964)

aerodynamical efficiency This furnishes a measure of the capacity of a fan to produce useful depression (or positive pressure in the case of a forcing fan) and indicates the extent to which the total pressure produced by the fan is absorbed within the fan itself. (Sinclair, 1958)

aerodynamic diameter The diameter of a unit density sphere having the same terminal settling velocity as the particle in question. (ANSI, 1992)

aerodynamic fan A fan that consists of several streamlined blades mounted in a revolving casing. The cross section and spacing of the blades are designed aerodynamically. This design ensures that the air flows without recirculation between the blades and leaves the rotor in a steady and regularly distributed stream. This appreciably reduces frictional, conversion, and recirculation losses. Fans of a convenient size can handle large volumes of air at the highest pressures likely to be required in mine ventilation.

aerodynamic instability Flutter that may occur in a structure exposed to wind force. This form of instability can be guarded against by suitable design. (Hammond, 1965)

aeroembolism (a) The formation or liberation of gases in the blood vessels of the body, as brought on by a change from a high, or relatively high, atmospheric pressure to a lower one. (Hunt, 1965) (b) The disease or condition caused by the formation or liberation of gases in the body. The disease is characterized principally by neuralgic pains, cramps, and swelling, and sometimes results in death. Syn: *decompression sickness.* (Hunt, 1965)

aerofall mill A short, cylindrical grinding mill with a large diameter, used dry, with coarse lumps of ore, pebbles, or steel balls as crushing bodies. The mill load is flushed with an air stream to remove finish mesh material. (Pryor, 1963)

aerofloc Synthetic water-soluble polymer used as a flocculating agent. (Bennett, 1962)

aerofoil-vane fan An improved centrifugal-type mine fan. The vanes, of aerofoil section, are curved backward from the direction of rotation. This fan is popular in British coal mines, and total efficiencies of about 90% have been obtained. See also: *mine-ventilation fan.* (Nelson, 1965)

aerohydrous (a) Enclosing a liquid in the pores or cavities, as some minerals. (Standard, 1964) (b) Characterized by the presence of both air and water. (Standard, 1964)

aeroides Pale sky-blue aquamarine beryl.

aeromagnetic prospecting A technique of geophysical exploration of an area using an airborne magnetometer to survey that area. Syn: *airborne magnetic prospecting.* (AGI, 1987)

aerometer An instrument for ascertaining the weight or the density of air or other gases. (Webster 3rd, 1966)

aerosite Former name for pyrargyrite.

aerosol (a) A suspension of ultramicroscopic solid or liquid particles in air or gas, as smoke, fog, or mist. (Webster 3rd, 1966) (b) Particles, solid or liquid, suspended in air. (ANSI, 1992) (c) A sol in which the dispersion medium is a gas (usually air) and the dispersed or colloidal phase consists of solid particles or liquid droplets, e.g., mist, haze, most smoke, and some fog. (AGI, 1987)

Aerosol Trade name of strong wetting agent based on sulfonated bi-carboxy-acid esters. (Pryor, 1963)

aerugite A grass-green to brown nickel arsenate, perhaps $Ni_{17}As_6O_{32}$; an analysis gave 48.77% nickel. It is an oxidized vein mineral. (Hess)

aerugo (a) Copper carbonate, due to weathering of the metal; esp., the patina adhering to old bronzes. (Hess) (b) Copper rust; verdigris; esp., green copper rust adhering to old bronzes. (Standard, 1964)

aeschynite An orthorhombic mineral, $(Ce,Ca,Fe,Th)(Ti,Nb)_2(O,OH)_6$; radioactive; occurs in black sands and pegmatites.

aethiops mineral A former name for metacinnabar; isometric HgS.

aetite (a) A nodule consisting of a hard shell of hydrated iron oxide within which yellow iron oxide becomes progressively softer toward the center, which may be hollow. (Fay, 1920) (b) *eaglestone.*

affinity In ion exchange, relative strength of attachment of competing ions for anchorage on a resin. (Pryor, 1963)

A-frame (a) Two poles or legs supported in an upright position by braces or guys and used as a drill mast. (Long, 1960) (b) An open structure tapering from a wide base to a narrow load-bearing top. (Nichols, 1954)

A-frame headgear A steel headgear consisting of two heavy plate A-frames, set astride the shaft mouth. They are braced together and carry the heavy girders that support the winding sheaves platform. It is a completely self-supporting and rigid structure that leaves usable space around the shaft collar and includes a guide-tower structure built over the shaft collar. A number of these headgears have been erected in the Republic of South Africa. (Nelson, 1965)

African emerald (a) A deceiving name for green fluor; also for green tourmaline. (b) An emerald from the Transvaal. It is usually quite yellowish green; often dark and dull. Hardness, 7.5; sp gr, 2.72 to 2.79; refractive index, 1.58 to 1.59; birefringence, 0.007. Syn: *Transvaal emerald.* (c) A term variously used for southern African emeralds (beryl), green tourmaline, and other green gemstones from this region.

afterblast During an explosion of methane and oxygen, carbon dioxide and steam are formed. When the steam condenses to water a partial vacuum is created, which causes an inrush or what is known as an afterblast. (Cooper, 1963)

afterblow Continued blowing of air through Bessemer converter after flame has dropped, for removal of phosphorus in steel production. (Pryor, 1963)

afterbreak In mine subsidence, a movement from the sides, the material sliding inward, and following the main break, assumed to be at right angles to the plane of the seam. The amount of this movement depends on several factors, such as the dip, depth of seam, and nature of overlying materials. (Lewis, 1964)

afterburst (a) A tremor as the ground adjusts itself to the new stress distribution caused by new underground openings. (b) In underground mining, a sudden collapse of rock subsequent to a rock burst.

aftercooler A device for cooling compressed air between the compressor and the mine shaft. By cooling and dehumidifying the air, and thus reducing its volume, the capacity and efficiency of the pipeline are increased. See also: *air-conditioning process; intercooler.* (Nelson, 1965)

afterdamp The mixture of gases that remain in a mine after a mine fire or an explosion of combustible gases. It consists of carbonic acid gas, water vapor (quickly condensed), nitrogen, oxygen, carbon monoxide, and in some cases free hydrogen, but usually consists principally of carbonic acid gas and nitrogen, and is therefore irrespirable. See also: *blackdamp; damp.* (Fay, 1920)

aftergases Gases produced by mine explosions or mine fires. (Fay, 1920)

aftershock An earthquake that follows a larger earthquake or main shock and originates at or near the focus of the larger earthquake. Generally, major earthquakes are followed by many aftershocks, which decrease in frequency and magnitude with time. Such a series of aftershocks may last many days for small earthquakes or many months for large ones. Cf: *foreshock.* (AGI, 1987)

aftersliding In mine subsidence, an inward movement from the side, resulting in a pull or draw beyond the edges of the workings. (Briggs, 1929)

afwillite A monoclinic mineral, $Ca_3Si_2O_4(OH)_6$; it is formed as portland cement is hydrated under special conditions, and where calcium silicate is autoclaved (as in sand-lime brick manufacture).

agalite A fine fibrous variety of talc pseudomorphous after enstatite. Syn: *asbestine.*

agalmatolite A soft, waxy stone—such as pinite, pyrophyllite, or steatite—of a gray, green, yellow, or brown shade; used by the Chinese to simulate jade for carving small images, miniature pagodas, and similar objects. Syn: *figure stone; pagodite; lardite; lard stone.*

agardite A hexagonal mineral, $(RE,Ca)Cu_6(AsO_4)_3(OH)_6 \cdot 3H_2O$; mixite group. Lanthanum, yttrium, or cerium may predominate among the rare earths.

agaric mineral (a) A soft, pulverulent hydrated silicate of magnesium in Tuscany, IT, from which floating bricks can be made. (Fay, 1920) (b) A light, chalky deposit of calcium carbonate formed in caverns or fissures in limestone. Syn: *rock milk.* (Webster 3rd, 1966)

agate (a) A kind of silica consisting mainly of chalcedony in variegated bands or other patterns; commonly occupying vugs in volcanic and other rocks. (AGI, 1987) (b) A translucent cryptocrystalline variety of variegated chalcedony commonly mixed or alternating with opal and characterized by colors arranged in alternating stripes or bands, in irregular clouds, or in mosslike forms; occurs in virtually all colors, generally of low intensity, in vugs in volcanic rocks and cavities in some other rocks. Cf: *onyx.* See also: *banded agate; chalcedony; clouded agate; moss agate.*

agate jasper An impure variety of agate consisting of jasper with veins of chalcedony. Syn: *jaspagate.*

agate opal Opalized agate. (Fay, 1920)

agatized wood A variety of silicified wood which resembles any variety of agate. *silicified wood.*

age (a) The formal geochronologic unit of lowest rank, below epoch, during which the rocks of the corresponding stage were formed. (AGI, 1987) (b) A term used informally to designate a length of geologic time during which the rocks of any stratigraphic unit were formed. (AGI, 1987) (c) A division of time of unspecified duration in the history of the Earth, characterized by a dominant or important type of life form; e.g., the age of mammals. (AGI, 1987) (d) The time during which a particular geologic event or series of events occurred or was marked by special physical conditions; e.g., the Ice Age. (AGI, 1987) (e) The position of anything in the geologic time scale; e.g., the rocks of Miocene age. It is often expressed in years. See also: *geologic age.* (AGI, 1987)

Agecroft device A device placed in the rail track to arrest a forward runaway tram. The front axle of a descending tram traveling at normal speed depresses the catch and allows it to drop back in time for the back axle to pass over. Should the tram be traveling at excessive speed, the tail end of the catch arrests the rear axle. (Mason, 1951)

agent (a) The manager of a mining property. (Zern, 1928) (b) On a civil engineering contract, the responsible representative of the contractor, acting for him or her in all matters. (Hammond, 1965) (c) Before nationalization in Great Britain, the term referred to the chief official of a large coal mine or group of mines under the same ownership. After nationalization, the equivalent term is group manager. (Nelson, 1965) (d) A chemical added to pulp to produce desired changes in climate of the system. (Pryor, 1963)

age ratio The ratio of daughter to parent isotope upon which the age equation is based. For a valid age determination, (1) the isotope system must have remained closed since solidification, metamorphism, or sedimentation, (2) the decay constant must be known, and (3) the sample must be truly representative of the rock from which it is taken. (AGI, 1987)

agglomerate belt flotation A coarse-fraction concentration method used in milling pebble phosphate in which conditioned feed at 70% to 75% solids is placed on a flat conveyor belt traveling at a rate of about 75 ft/min (22.9 m/min). Water sprayed on the surface of the pulp aerates the pulp, causing agglomerates of phosphate particles to float to the side of the belt for removal. The silica fraction travels the length of the belt and is permitted to flow off the opposite end. Baffles are positioned at appropriate points along the belt to stir the material so that trapped phosphate particles are given an opportunity to float. Concentrate from the first belts or rougher operation is cleaned on a second belt for further silica removal. Tailings from the cleaner belt are recycled to the rougher circuit. (Arbiter, 1964)

agglomerate screening A coarse-fraction concentration method used in milling pebble phosphate that is based on flowing reagentized feed over a submerged sloping, stationary screen. Agglomerated phosphate particles float on top of the screen and are recovered at the lower end. Sand particles pass through the screen and are removed as a tailings fraction. Each screen section is approx. 3 ft (0.9 m) wide by 4 ft (1.2 m) long and treats 2 to 3 st/h (1.8 to 2.7 t/h) of feed. (Arbiter, 1964)

agglomerating value A measure of the binding qualities of coal but restricted to describe the results of coke-button tests in which no inert material is heated with the coal sample. Cf: *agglutinating value.* (AGI, 1987)

agglomeration (a) In beneficiation, a concentration process based on the adhesion of pulp particles to water. Loosely bonded associations of particles and bubbles are formed that are heavier than water; flowing-film gravity concentration is used to separate the agglomerates from nonagglomerated particles. Agglomeration also refers to briquetting, nodulizing, sintering, etc. (Gaudin, 1939) (b) *kerosine flotation.*

agglutinate A welded pyroclastic deposit characterized by vitric material binding the pyroclasts, or sintered vitric pyroclasts. Also spelled agglutinite. (AGI, 1987)

agglutinating power *caking index.*

agglutinating value A measure of the binding qualities of a coal and an indication of its caking or coking characteristics. Applicable with reference to the ability of fused coal to combine with an inert material such as sand. Cf: *agglomerating value.* (AGI, 1987)

agglutinating-value test A laboratory test of the coking properties of coal, in which a determination is made of the strength of buttons made by coking a mixture of powdered coal and 15 to 30 times its weight of sand.

agglutination *cementation.*

aggradation (a) The building up of the Earth's surface by deposition; specif., the upbuilding performed by a stream in order to establish or maintain uniformity of grade or slope. See also: *gradation.* Cf: *degradation.* Syn: *upgrading.* (AGI, 1987) (b) A syn. of accretion, as in the development of a beach. The spread or growth of permafrost, under present climatic conditions, due to natural or artificial causes. (AGI, 1987)

aggregate (a) A mass or body of rock particles, mineral grains, or a mixture of both. (AGI, 1987) (b) Any of several hard, inert materials, such as sand, gravel, slag, or crushed stone, mixed with a cement or bituminous material to form concrete, mortar, or plaster, or used alone, as in railroad ballast or graded fill. The term can include rock material used as chemical or metallurgical fluxstone. See also: *chippings; coarse aggregate; fine aggregate; lightweight aggregate.* (AGI, 1987)

aging A change in the properties of a substance with time. See also: *overaging; precipitation hardening.* (Nelson, 1965)

Agitair flotation machine Uses air to separate aerophilic and hydrophilic particles. Low-pressure air bubbles lift aerophilic particles to an overflow, leaving hydrophilic particles behind. (Pryor, 1963)

agitation dredging Consists of pumping the discharge directly into the sea and using the tide to carry the fines to deeper water areas. Agitation dredging is employed only during ebb tide in tidal estuaries having swift tidal flows that will disperse the accumulations of silt. (Carson, 1965)

agitation ratio In older type gravity concentrators, such as tables and vanners, the ratio between the average diameter of a mineral particle and the diameter of a gangue particle that travels at equal speed.

agitator (a) A tank in which very finely crushed ore is agitated with leaching solution. Usually accomplished by means of a current of compressed air passing up a central pipe and causing circulation of the contents of the tank. Sometimes called a mixer. (CTD, 1958) (b) A device used to stir or mix grout or drill mud. Not to be confused with shaker or shale shaker. (Long, 1960) (c) A device used to bring about a continuous vigorous disturbance in a pulp; frequently used to assist bubble formation. (BS, 1962) (d) Pac. *settler.*

aglaite A pseudomorph of spodumene in which the spodumene has been replaced by muscovite either as pinite or as visible plates. Also called pihlite and cymatolite in the belief that the material was a new mineral. (Hess)

agmatite Migmatite with appearance of breccia. Cf: *contact breccia.* (AGI, 1987)

agnesite An early name for bismutite, Cornwall, U.K. (Fay, 1920)

agonic line An isogonic line that connects points of zero magnetic declination. Its position changes according to the secular variation of the Earth's magnetic field. See also: *isogonic line.* (AGI, 1987)

agreement The formal document by which the contractor and the authority mutually agree to comply with the requirements of the drawings, specification, schedule, conditions of tendering, and general conditions of contract and the tender. See also: *tender; contract.* (Nelson, 1965)

agricolite A former name for eulytite.

agricultural geology The application of geology to agricultural needs, e.g., mineral deposits used as fertilizers or the location of ground water. Syn: *agrogeology.* (AGI, 1987)

agricultural lime (a) Either ground quicklime or hydrated lime whose calcium and magnesium content is capable of neutralizing soil acidity. (ASTM, 1994) (b) Lime slaked with a minimum amount of water to form calcium hydroxide. (CCD, 1961)

agrite A brown, mottled calcareous stone. (Schaller, 1917)

agrogeology *agricultural geology.*

aguilarite An orthorhombic mineral, Ag_4SeS.

ahlfeldite A monoclinic mineral, $NiSeO_3 \cdot 2H_2O$; forms a series with cobaltomenite; rose colored; vitreous luster; no cleavage; conchoidal fracture; strongly pleochroic, X rose, Y pale green, Z brown green; from Pacajake, Bolivia. (Am. Mineral., 1945)

A-horizon In a soil profile, the uppermost zone from which soluble salts and colloids have been leached and in which organic matter has accumulated. See also: *B-horizon.*

aikinite (a) An orthorhombic mineral, $PbCuBiS_3$; sp gr, 6.1 to 6.8; an ore of lead, copper, and bismuth. Syn: *acicular bismuth; aciculite; needle ore; acicular bismuth; aciculite.* (b) Wolframite pseudomorphous after scheelite.

aimafibrite *hemafibrite.*

AIME American Institute of Mining and Metallurgical Engineers. (Statistical Research Bureau, 1935)

air (a) The mixture of gases that surrounds the Earth and forms its atmosphere; composed by volume of 21% oxygen and 78% nitrogen; by weight about 23% oxygen and 77% nitrogen. It also contains about 0.03% carbon dioxide, some aqueous vapor, argon, and other gases. (Hartman, 1961) (b) The current of atmospheric air circulating through and ventilating the workings of a mine. (c) Atmospheric air delivered under compression to bottom of drill hole through the drill stem and used in place of water to clear the drill bit of cuttings and to blow them out of the borehole. (Long, 1960) (d) Air piped under compression to work areas and used to operate drilling or mining machinery. See also: *air circulation.* (Long, 1960)

airafibrite *hemafibrite.*

air-avid surface A surface that seems to prefer contact with air to contact with water. A particle (or mineral) of this sort will adhere to an air bubble and float out of a flotation pulp; otherwise, the particle will not float. Also called water-repellent surface; hydrophobic. Cf: *water-avid surface.* (Newton, 1959)

air barrage The division of an opening in a mine by an airtight wall into two sides; one side is used as an air intake, the other side as a return.

air bell In froth flotation, the small air pocket inducted or forced into the pulp at depth; e.g., bell and the two-walled semistable bubble after emergence from pulp into froth have different characteristics and gas-to-liquid, area-to-volume relationships, hence the distinction. These bubbles vary in attractive and retaining power for aerophilic grains and are a critical component of the flotation process. Syn: *air bubble.* (Pryor, 1963)

air belt In a cupola furnace, an annular air space around the furnace, from which air is forced into the furnace. (Henderson, 1953)

airblast (a) A term improperly used by some diamond drillers as a syn. for air circulation. See also: *air circulation.* (Long, 1960) (b) A disturbance in underground workings accompanied by a strong rush of air. The rush of air, at times explosive in force, is caused by the ejection of air from large underground openings, the sudden fall of large masses of rock, the collapse of pillars, slippage along a fault, or a strong current of air pushed outward from the source of an explosion. (Long, 1960)

airblasting A method of blasting in which compressed air at very high pressure is piped to a steel shell in a shot hole and discharged. (BS, 1964)

air block Air trapped in the upper end of an unvented inner tube of a double-tube core barrel, which, when sufficiently compressed, acts like a solid and stops further advance of core into the inner tube. Syn: *air cushion.* (Long, 1960)

airborne electromagnetic prospecting Electromagnetic surveys carried out with airborne instruments. (Dobrin, 1960)

airborne magnetic prospecting *aeromagnetic prospecting.*

airborne magnetometer An instrument used to measure variations in the Earth's magnetic field while being transported by an aircraft. See also: *magnetometer.* (AGI, 1987)

air box (a) A rectangular wooden pipe or tube made in lengths of 9 to 15 ft (2.7 to 4.6 m) for ventilating a heading or a sinking shaft. (Fay, 1920) (b) A box for holding air. (Fay, 1920) (c) The conduit through which air for heating rooms is supplied to a furnace. (Standard, 1964)

air breakers A method of breaking down coal by the use of high-pressure compressed air. (McAdam, 1958)

air brick A hollow or pierced brick built into a wall to allow the passage of air.

air bridge A passage through which a ventilating current is conducted over an entry or air course; an overcast. See also: *air crossing.*

air bubble *air bell.*

air chamber A vessel installed on piston pumps to minimize the pulsating discharge of the liquid pumped. The chamber contains air under pressure and is fitted with an opening on its underside into which some of the liquid from the pump is forced upon the delivery stroke of the piston. The air acts as a cushion to lessen the fluctuation of the liquid flow between the suction and delivery strokes of the piston. (Crispin, 1964)

air change (a) The quantity of infiltration of ventilation air in cubic meters per second divided by the volume of the room gives the number of so-called air changes during a given interval of time. Tables of the recommended number of such air changes for various rooms are used for estimating purposes. (b) The act of instituting a different pattern of air flow in a mine.

air channels In a reverberatory furnace, flues under the hearth and fire bridge through which air is forced to avoid overheating. (Henderson, 1953)

air circulation (a) A large volume of air, under compression, used in lieu of a liquid as a medium to cool the bit and eject drill cuttings from a borehole. Syn: *air flush.* See also: *airblast.* (Long, 1960) (b) The general process of moving air around the openings of a mine. See also: *air.*

air classification (a) In powder metallurgy, the separation of powder into particle-size fractions by means of an airstream of controlled velocity; an application of the principle of elutriation. (ASM, 1961) (b) Sorting of finely ground minerals into equal settling fractions by means of air currents. These are usually controlled through cyclones, which deliver a coarse spigot product and a relatively fine vortical overflow. See also: *infrasizer.* (Pryor, 1958) (c) A method of separating or sizing granular or powdered materials, such as clay, through deposition in air currents of various speeds. This principle is widely used in continuous pulverizing of dry materials, such as frit, feldspar, limestone, and clay. See also: *air classifier; air elutriator.* (Enam. Dict., 1947)

air classifier An appliance for approx. sizing crushed minerals or ores by means of currents of air. See also: *air classification; air elutriator.* (CTD, 1958)

air cleaning A coal-cleaning method that utilizes air to remove the dust and waste from coal. Air cleaning requires that the coal contain less than 5% of surface moisture as a rule. It is effective only in the coarse sizes (plus 10 to 28 mesh) and is best suited to coals having a sharply defined line between coal and refuse material. Predrying to reduce the moisture content of the coal ahead of the air treatment is not uncommon. It is a less expensive and also a less accurate method of cleaning coal than the wet-cleaning method. (Kentucky, 1952)

air compartment An airtight portion of any shaft, winze, raise, or level used for ventilation. (BS, 1963)

air conditioning The simultaneous control, within prescribed limits, of the quality, quantity, temperature, and humidity of the air in a designated space. It is essentially atmospheric environmental control. Control of only one or two of these properties of the atmosphere does not constitute air conditioning. The definition and correct usage require that the purity, motion, and heat content of the air must all be maintained within the prescribed limits. (Hartman, 1982)

air-conditioning process When conditioning is designed to perform only one or a limited number of functions, then it should be so designated. Air-conditioning processes include dust control, ventilation, dehumidification, cooling, heating, and many others. See also: *aftercooler; air receiver; compressed air; duplex compressor; rotary compressor; turbocompressor.* (Hartman, 1982)

air-cooled blast-furnace slag The material resulting from solidification of molten blast-furnace slag under atmospheric conditions. Subsequent cooling may be accelerated by application of water to the solidified surface. (ASTM, 1994)

air course (a) Ventilating passage underground. (Pryor, 1963) (b) A passage through which air is circulated, particularly a long passageway driven parallel to the workings to carry the air current. See also: *airway.* (Fay, 1920)

air coursing The system of colliery ventilation, introduced about 1760, by which the intake air current was made to traverse all the underground roadways and faces before passing into the upcast shaft. (Nelson, 1965)

air creep Stain formed by air entering at edges of mica sheets and penetrating along cleavage planes. (Skow, 1962)

air crossing A bridge where a return airway passes over (overcast) or under (undercast) an intake airway. It is generally constructed with concrete blocks, structural steel, and/or sheet metal, and is made airtight to prevent intermixing of the two air currents. The mining law requires an air crossing to be so constructed as not to be liable to be damaged in the event of an explosion. Syn: *air bridge; bridge; overcrossing; overgate.* See also: *overcast; undercast.* (Nelson, 1965)

air current (a) The flow of air ventilating the workings of a mine. Syn: *airflow; air quantity.* (BS, 1963) (b) A body of air moving continuously in one direction. (Jones, 1949)

air cushion Air trapped in the bottom of a dry borehole by the rapid descent of a tight string of borehole equipment. Syn: *air block.* (Long, 1960)

air cyclone Primarily a vessel for extracting dust from the atmosphere. (Nelson, 1965)

air decking The use of air space or a void within a blast hole between an explosive charge and inert stemming to enhance the shock wave detonation force.

air dome A cylindrical or bell-shaped container closed at the upper end and attached in an upright position above and to the discharge of a piston-type pump. Air trapped inside the closed cylinder acts as a compressible medium, whose expansion and contraction tends to reduce the severity of the pulsations imparted to the liquid discharged by each stroke of a pump piston. Syn: *bonnet; pressure dome.* Cf: *dome.* (Long, 1960)

air door A door erected in a roadway to prevent the passage of air. When doors are erected between an intake and a return airway, they may be known as separation doors. Syn: *door; separation door; trapdoor.* (BS, 1963)

Airdox A system for breaking down coal by which compressed air, generated locally by a portable compressor at 10,000 psi (69.0 MPa), is used in a releasing cylinder, which is placed in a hole drilled in the coal. Thus, slow breaking results, with no flame, in producing a larger percentage of lump coal than is made by using explosives. Its principal advantage is that it may be used with safety in gaseous and dusty mines. See also: *compressed air blasting.* (Lewis, 1964)

air drain A passage for the escape of gases from a mold while the molten metal is being poured. (Standard, 1964)

air-dried Said of minerals naturally dried to equilibrium with the prevailing atmosphere. (Pryor, 1963)

air-dried basis An analysis expressed on the basis of a coal sample with moisture content in approximate equilibrium with the surrounding atmosphere. (BS, 1960)

air drift (a) An opening driven for ventilation purposes, often inclined and driven in stone. (b) A drift connecting a ventilation shaft with the fan.

air drill (a) A small diamond drill driven by either a rotary or a reciprocating-piston air-powered motor; used principally in underground workings. (Long, 1960) (b) As used by miners, a percussive or rotary-type rock drill driven by compressed air. Cf: *air rig.* (Long, 1960)

air-dry (a) Dry to such a degree that no further moisture is given up on exposure to air. Most air-dry substances contain moisture that can be expelled by heating them or placing them in a vacuum. (Webster 3rd, 1966; Fay, 1920) (b) Said of timber the moisture content of which is in approximate equilibrium with local atmospheric conditions. (CTD, 1958)

air duct (a) Tubing that conducts air, usually from an auxiliary fan, to or from a point as required in the mine. (BS, 1963) (b) An air box, canvas pipe, or other air carrier for ventilation. (Hess)

air elutriation Method of dividing a substance into various particle sizes by means of air currents. (Bennett, 1962)

air elutriator An appliance for producing, by means of currents of air, a series of sized products from a finely crushed mineral (e.g., for the paint or abrasive industries). See also: *air classification; air classifier.* (CTD, 1958)

air endway A narrow roadway driven in a coal seam parallel and close to a winning headway chiefly for ventilation; it usually acts as a return and is connected at intervals of 10 yd (9 m) or so to the headway by crosscuts.

air-float table Shaking table in which ore is worked dry, air being blown upward through a porous deck so as to dilate the material. (Pryor, 1958)

airflow *air current.*

airflow-equalizing device A flow-equalizing device that is fitted to tube breathing apparatus. There are two kinds in general use, one consisting of a flexible, corrugated rubber tube and the other a canvas fabric bag. On inspiration, air is drawn partly from the equalizer, which is reduced in volume, and partly from the tube. On expiration, the equalizer restores itself to its original volume and in doing so draws air through the tube. Thus the air is kept flowing very nearly in a continuous stream, and the wearer, without the aid of bellows or rotary blower, experiences very little resistance to breathing. (Mason, 1951)

airflow meter An instrument that measures and provides readout of the flow of air in a pipe or hose in cubic meters per second.

air flush *air circulation.*

air flushing The circulation of air through the drilling apparatus during drilling to cool the bit and to remove the cuttings from the hole. (BS, 1963)

airfoil fan A fan with an airfoil-shaped blade that moves the air in the general direction of the axis about which it rotates. (Strock, 1948)

air furnace Malleable-iron furnace.

air gate (a) Mid. An underground roadway used principally for ventilation. (b) An air regulator.

airhammer A tool in which a hammerhead is activated by means of compressed air. The airhammer is called a jackhammer in coal mining and jackleg-hammer in hardrock mining. The tool is used to drill blastholes to grade or take up bottom or to advance a stope. (Crispin, 1964)

air heater An appliance to warm the air as it enters the downcast shaft or intake drift. In countries where the winter is very cold, nearly all mines are equipped with air heaters of the oil-fired, gas-fired, or electric type.

air hoist (a) Hoisting machinery operated by compressed air. (Fay, 1920) (b) A small portable hoisting machine usually mounted on a column and powered by a compressed air motor. Also called tugger. (Long, 1960)

airhole (a) A small excavation or hole made to improve ventilation by communication with other workings or with the surface. See also: *cundy.* (BS, 1963) (b) A venthole in the upper end of the inner tube of a double-tube core barrel to allow air and/or water entrapped by the advancing core to escape. (Long, 1960) (c) A void, cavity, or flaw in a casting or bit crown. (Long, 1960)

air horsepower The rate at which energy is consumed, in horsepower or kilowatt units, in moving air between two points.

air hp Abbrev. for air horsepower.

airing Operation in which air is blown through molten copper in a wire bar or anode furnace. Sulfur is removed as SO_2 and impurities are slagged off. (Pryor, 1963)

air intake (a) The airway or airways through which fresh air is brought into a mine. (b) A device for supplying a compressor with clean air at the lowest possible temperature.

air jig A machine in which the feed is stratified by means of pulsating currents of air and from which the stratified products are separately removed. (BS, 1962)

air lancing (a) Removing or cutting away loose material by means of compressed air, using an air lance; airblasting. (Henderson, 1953) (b) In founding, a cleaning operation, as cleaning sand from molds and castings, using an air lance; airblasting. (Henderson, 1953) (c) Opening passages for molten materials.

air leakage (a) The short-circuiting of air from intake to return airways (through doors, stoppings, wastes, and old workings) without doing useful work in flowing around the faces. The total air leakage is usually within the range of 35% to 53% of that passing through the surface fan. (Nelson, 1965) (b) The air that escapes from compressed-air lines by leakage from joints, valves, hoses, etc.

air leg (a) A cylinder operated by compressed air, used for keeping a rock drill pressed into the hole being drilled. (Hammond, 1965) (b) A device, incorporating a pneumatic cylinder, providing support and thrust for a jackhammer. (BS, 1964)

air-leg support An appliance to eliminate much of the labor when drilling with handheld machines. It consists of a steel cylinder and air-operated piston, the rod of which extends through the top end of the cylinder and supports the drilling machine. The air leg and machine can be operated by one worker. Syn: *pneumatic drill leg*. (Nelson, 1965)

air level Eng. A level or airway (return airway) of former workings used in subsequent deeper mining operations for ventilation.

air lift An apparatus used for pumping water from wells either temporarily or for a permanent water supply; for moving corrosive liquids such as sufuric acid; for unwatering flooded mines; for elevating mill tailings, sands, and slimes in cyanide plants; and for handling the feed to ball mills. In operation, compressed air enters the eduction pipe and mixes with the water. As the water and air rise, the air expands and is practically at atmospheric pressure at the top of the discharge pipe. The efficiency of the air lift is calculated on the basis of the foot-pounds of work done in lifting the water, divided by the isothermal work required to compress the air. (Lewis, 1964)

air-lift dredge Dredge in which solids suspended in a fluid are lifted. By injecting air into a submerged pipe beneath the water surface, the density of the fluid column inside the pipe can be lessened, forcing the fluid column to rise in the tubular pipe. Syn: *airlift sampler*. (Mero, 1965)

airlift sampler *air-lift dredge.*

air-line lubricator *line oiler.*

airline respirator An atmosphere-supplying respirator in which the respirable-gas supply is not designed to be carried by the wearer (formerly called supplied-air respirator). (ANSI, 1992)

air lock (a) A casing at the top of an upcast shaft to minimize surface air leakage to the fan. It consists of a large double casing enveloping the whole of the upcast-shaft top and extending into the headgear. Some are fitted with power-operated doors and allow high-speed winding with little leakage. A modern light-alloy structure raised through spring-loaded attachments by the top of the cage on ascending has proved efficient. Syn: *shaft casing*. (Nelson, 1965) (b) A system of doors arranged to allow the passage of workers or vehicles without permitting appreciable airflow. (BS, 1963)

airman A worker who constructs brattices. Syn: *brattice worker*. (Hess)

air mat A mat made of porous material, usually canvas, and used to subdivide and distribute air in certain pneumatic-type flotation machines. (Hess)

air mover A portable compressed-air appliance, which may be used as a blower or exhauster. It converts the compressed air into a large induced volume of moving air. The compressed air is fed through a side inlet and is expanded at a high velocity through an annular orifice. It is useful for emergency ventilation in workings where auxiliary fans cannot be installed. Syn: *injector; static air mover*. See also: *auxiliary ventilation*. (Nelson, 1965)

air-operated winch A small, compressed-air drum haulage or hoist used for lifting, dragging, or skidding work in mines. With capacities ranging from 660 to 4,400 lb (300 to 2,000 kg), these winches have powerful piston motors and are capable of continuous operation. They are easy to move from job to job and are used for shaft sinking and moving wagon drills at quarry and opencast operations. (Nelson, 1965)

air photograph *aerial photograph.*

air pit *air shaft.*

airplane-strand wire rope A small 7- or 19-wire galvanized strand made from plow steel or crucible-steel wire. (Hunt, 1965)

air-power-operated mine door Mine doors help to keep the air flow in shafts and mine working areas constant. In cases of explosions, doors "give" to relieve the pressure, then close automatically. The doors are mobile and can be set up in any location. They are opened and closed by a compressed-air cylinder and are designed to be used where haulage equipment operates on a trolley wire. (Best, 1966)

air pressure (a) For rock drills, the air pressure ranges from 70 to 90 psi (480 to 620 kPa), the most economical pressure for such machines being from 90 to 95 psi (620 to 655 kPa), when high drilling speed is attained. (Hammond, 1965) (b) To operate the percussive tool and to flush the hole of cuttings, surface mounted drills use 690 to 1,725 kPa air pressure. (Cumming, 1981)

air-pressure drop The pressure lost or consumed in overcoming friction along an airway.

air propeller A rotating set of blades designed to impart momentum to an air mass.

air pump A pump for exhausting air from a closed space or for compressing air or forcing it through other apparatus. Cf: *vacuum pump*. (Webster 3rd, 1966)

air-purifying respirator A respirator in which ambient air is passed through an air-purifying element that removes the contaminants. Air is passed through the air-purifying element by means of the breathing action or by a blower. (ANSI, 1992)

air quantity The amount of air flowing through a mine or a segment of a mine, in cubic meters per second. Air quantity is the product of the air velocity times the cross-sectional area of the airway. See also: *air current*. Syn: *air volume*.

air ramming A method of forming refractory shapes, furnace hearths, or other furnace parts by means of pneumatic hammers. (Harbison-Walker, 1972)

air receiver A vessel into which compressed air is discharged to be stored until required. See also: *air-conditioning process*. (CTD, 1958)

air-reduction process *roasting and reaction process.*

air regulator An adjustable door installed in permanent air stoppings or in an airway without a stopping to control ventilating current.

air requirements The quantity of air required by law or practical considerations to maintain adequate ventilation of a mine. This quantity will depend on (1) the length of face room in production, (2) the average distance from the shafts to the faces, (3) the gas emission rate, (4) the depth of the workings, and (5) the volumetric efficiency of the mine ventilation. See also: *air volume; ventilation planning*.

air rig A drill machine powered by an airdriven motor. Cf: *air drill*. (Long, 1960)

air rod puller *rod puller.*

air rotary drilling Drilling technique that utilizes compressed air to lift the cuttings up the borehole and to cool the bit. Used when possible for environmental monitoring, because no drilling fluids are introduced into the formation. Feasible only in consolidated or semiconsolidated formations. (Driscoll, 1986)

air-sand process *Fraser's air-sand process.*

air seal A method for the prevention of the escape of warm gases from the entrance or exit of a continuous furnace, or tunnel kiln, by blowing air across the opening. (Dodd, 1964)

air separation In powder metallurgy, the classification of metal powders into particle size ranges by means of a controlled airstream. (Rolfe, 1955)

air separator A machine for the size classification of the fine ceramic powders, such as china clay; the velocity of an air current controls the size of particle classified. (Dodd, 1964)

air set (a) The property of a material to develop high strength when dried; e.g., air-setting mortars. (ARI, 1949) (b) In a material such as a castable refractory, refractory mortar, or plastic refractory, the ability to harden without the application of heat. (AISI, 1949)

air shaft A shaft used wholly or mainly for ventilating mines, for bringing fresh air to places where miners are working, or for exhausting used air. It may be used as an intake (downcast) shaft or a return (upcast) shaft. See also: *downcast; upcast*. Syn: *air pit*.

air shooting In seismic prospecting, a technique of applying a seismic pulse to the ground by detonating explosive charges in the air. (AGI, 1987)

air shot A shot prepared by loading (charging) in such a way that an airspace is purposely left in contact with the explosive for the purpose of lessening its shattering effect. (Fay, 1920)

air shrinkage The volume decrease that a clay undergoes in drying at room temperature.

air-slaked Slaked by exposure to the air; as lime. (Standard, 1964)

air-slaking Exposure of quicklime to the atmosphere to give slow hydration. (Pryor, 1963)

air slit York. A short heading driven more or less at right angles to and between two headings or levels for ventilation. See also: *stenton*.

air slug A mass of air under compression entrapped in the liquid circulated through a borehole drill string or a liquid-piping system. (Long, 1960)

air sollar A compartment or passageway carried beneath the floor of a heading or of an excavation in a coal mine for ventilation. See also: *sollar*.

air-space ratio The ratio of a volume of water that can be drained from a saturated soil under the action of force of gravity to a total volume of voids. (ASCE, 1958)

air-sparged hydrocyclone A separator consisting of two concentric right-vertical tubes, a conventional cyclone header at the top, and a froth pedestal at the bottom. The inner tube is a porous-wall tube. The slurry is fed tangentially through the cyclone header to develop a radial swirl flow. Air is sparged through the jacketed, inner porous tube wall and is sheared into small bubbles by the swirl flow. Hydrophobic coal particles in the slurry attach to the air bubbles and report as overflow product. The hydrophilic refuse particles remain wetted and report as underflow product.

air split The division of the main current of air in a mine into two or more parts. See also: *split.*
air stack A chimney formerly used to ventilate a mine.
air stain Gas trapped beneath mica cleavage surfaces in flattened pockets, tiny bubbles, or groups of closely spaced bubbles. (Skow, 1962)
air starter A starter used on large coal haulers that permits the elimination of all batteries except the 6-V units for the headlights. These starters are operated by compressed air supplied at 100 psi (690 kPa) from a storage tank on the tractor. Trucks can stand idle for 4 or 5 days and there is still enough air in the tanks to start the engines. (Coal Age, 1966)
air-stowing machine The machine used for blowing the stone chippings into the waste area in pneumatic stowing. It consists of a steel paddle wheel revolving in an adjustable casing. Stowing dirt is fed continuously from a hopper to the machine, which in turn blows the material through pipes 5 to 6 in (13 to 15 cm) in diameter into the waste area. See also: *pneumatic stowing.* (Nelson, 1965)
air streak In mica, a series of air inclusions connected (or nearly connected) to form a relatively long, thin streak. Also known as silver streak. (Skow, 1962)
air survey In mining, a check on ventilation, gas, and dust in a mine. (Pryor, 1963)
air-swept ball mill *ball mill.*
air-swept mill A tumbling mill used in dry grinding, from which finished material is removed by means of regulated air currents that can be so controlled as to produce a closed circuit. (Pryor, 1958)
air swivel A device similar to a water swivel but designed to conduct air under compression into a rotating drill stem when air instead of a liquid is used as an agent to flush drill cuttings out of a borehole. Cf: *water swivel.* (Long, 1960)
air table A shaking table used when water is scarce to effect gravity concentration of sands. Air is blown upward through a porous deck, over which a layer of finely crushed ore passes. The heavy and light minerals stratify and gravitate to separate discharge zones. Syn: *pneumatic table.* (Pryor, 1958)
air-track drill A heavy drilling machine for quarry or opencast blasting. It has continuous tracks and is operated by independent air motors. It tows its rotary compressor and drills holes 3 in or 4 in (7.6 cm or 10.2 cm) in diameter at any angle, but it is chiefly used for vertical holes up to 80 ft (24.4 m) in depth. (Nelson, 1965)
air transport A method employed in some mines in which material is transported and stowed pneumatically through pipelines. (Stoces, 1954)
air trunk A large pipe or shaft for conducting air, such as for ventilation or to a furnace. (Fay, 1920)
air tub The cylinder on a blowing engine that pumps a blast of wind or air. (Fay, 1920)
air turbolamp A lamp coupled to the compressed air mains, which may be at any pressure between 275 kPa and 700 kPa. It consumes 0.025 m^3/s of free air. The electrical power is produced by a small turboalternator with a six-pole permanent magnet rotor.
air valve The valve that controls the alternate admission and release of compressed air to each cell of a Baum-type washbox. (BS, 1962)
air velocity The rate of motion of air in a given direction; in mine ventilation it is usually expressed in meters per second. This is usually measured conducting a vane anemometer traverse over a selected cross section, the area of which is also measured.
air vessel A small air chamber fixed to the pipeline on the discharge side of a reciprocating pump that acts as a cushion to minimize the shock produced by the pulsations of the pump. (Nelson, 1965)
airveyor A device for handling dusty materials, built on the principle of a pneumatic cleaner. The system used is a suction system, whereby the material (soda ash, salt cake, cement, or powdered lime) is drawn from the car through a flexible hose into a vacuum tank designed to recover a large percentage of the dust floating in the air. (Hess)
air-void ratio The ratio of the volume of air space to the total volume of voids in a soil mass. (ASCE, 1958)
air volume In mining, the quantity (Q) of air flowing in cubic meters per second. It is obtained by multiplying the average velocity (V) in meters per second by the area (A) of the airway in square meters; i.e., Q = AV. See also: *air requirements; air quantity.*
air washer Air washers make use of water sprays or cooling coils for evaporative and sensible cooling of mine air. Their use has largely been limited to shallow coal mines in the United States, where it is desirable to reduce the dry-bulb temperature of the intake air during the hot summer months to prevent slaking of the roof due to excessive expansion. An air washer is essentially a heat exchanger and is similar to the type of unit employed for heat transfer with refrigeration or evaporative-cooling systems.
air wave The acoustic-energy pulse transmitted through the air as a result of a pressure (sound) source; e.g., explosion, near-surface seismic shot, or supersonic aircraft. (AGI, 1987)
airway Any underground gallery or passage through which a portion of the ventilation passes; i.e., the air is carried. Syn: *air course; wind road.* (BCI, 1947)
Airy hypothesis A concept of the mechanism of isostasy, proposed by George Bedell Airy, that postulates an equilibrium of crustal blocks of the same density but of different thickness; thus the topographically higher mountains would be of the same density as other crustal blocks but would have greater mass and deeper roots. Cf: *Pratt hypothesis.* (AGI, 1987)
Ajax A high-strength, high-density, gelatinous permitted explosive having good water resistance; used for dry and wet conditions both in rock and in the breaking of hard coal. See also: *Polar Ajax.* (Nelson, 1965)
ajkaite A pale-yellow to dark reddish brown, sulfur-bearing fossil resin found in brown coal. Also spelled ajkite. (AGI, 1987)
ajutage Roman term designating size of water delivery pipes and outlet spouts. Syn: *adjutage.* (Sandstrom, 1963)
akaganéite A tetragonal mineral, ferric oxyhydroxide beta-FeO(OH,Cl); rust colored; occurs in soils.
akhtenskite A hexagonal mineral (epsilon-MnO_2), trimorphous with pyrolusite and ramsdellite. *pyrolusite.* (Fleischer, 1995)
Akins' classifier Used for separating fine-size solids from coarser solids in a wet pulp; consists of an interrupted-flight screw conveyor operating in an inclined trough.
akrochordite A monoclinic mineral, $Mn_4Mg(AsO_4)_2{\cdot}4H_2O$; forms reddish-brown rounded aggregates.
aksaite An orthorhombic mineral, $MgB_6O_7(OH)_6{\cdot}2H_2O$.
alabandite An isometric mineral, manganese sulfide, MnS; iron-black; in epithermal vein deposits; an ore of manganese. Formerly called alabandine. Syn: *manganblende; manganese glance.*
alabaster A massive form of gypsum; very fine grained; commonly snow-white and translucent but may be delicately shaded or tinted with light-colored tones. Because of its softness, it can be easily carved and polished. Widely used for ornamental purposes. Chemically it is $CaSO_4{\cdot}2H_2O$. It is a beautifully banded form of stalagmitic calcite occurring in Algeria and in Egypt. Syn: *oriental alabaster; onyx marble.* (CMD, 1948)
aladzha Impure ozokerite containing an admixture of country rocks and found in the region of the Caspian Sea. (Tomkeieff, 1954)
alamosite A monoclinic mineral, $PbSiO_3$. Syn: *lead silicate; lead metasilicate.*
alaskite In the United States, a commonly used term for a granitic rock containing few, if any, dark minerals. The term is used to designate granitoid rocks in which quartz constitutes 20% to 60% of the felsic minerals and in which the ratio of alkali feldspar to total feldspar is greater than 90%; i.e., the equivalent of alkali granite. Alaskite is a commercial source of feldspar near Spruce Pine, NC. Cf: *aplogranite.* (AGI, 1987)
albertite A dark brown to black asphaltic pyrobitumen with conchoidal fracture occurring as veins 1 to 16 ft (0.3 to 4.9 m) wide in the Albert Shale of Albert County, NB, Can. It is partly soluble in turpentine, but practically insoluble in alcohol. It was earlier called albert coal. (AGI, 1987)
albert shale An early name for albertite. (Tomkeieff, 1954)
albite (a) A triclinic mineral, $NaAlSi_3O_8$; feldspar group, with up to 10 mol % CaAl replacing NaSi; a member of the plagioclase and the alkali feldspar series; prismatic cleavage; a common rock-forming mineral in granite, intermediate to felsic igneous rocks, low-temperature metamorphic rocks, and hydrothermal cavities and veins; can be used as a glaze in ceramics. (b) The pure sodium-feldspar end member in the plagioclase series. Syn: *sodium feldspar; white feldspar; white schorl.*
albite-epidote-amphibolite facies The set of metamorphic mineral assemblages (facies) in which basic rocks are represented by hornblende + albite + epidote. Equivalent to Eskola's epidote-amphibolite facies, it is of uncertain status, transitional between the greenschist facies and the amphibolite facies. It is generally believed to be favored by the higher pressures of regional metamorphism. (AGI, 1987)
albite porphyrite *albitite.*
albitite A porphyritic igneous rock, containing phenocrysts of albite in a groundmass chiefly consisting of albite. Muscovite, garnet, apatite, quartz, and opaque oxides are common accessory minerals. Syn: *albitophyre; albite porphyrite.* (AGI, 1987)
albitization Introduction of, or replacement by, albite, usually replacing a more calcic plagioclase. (AGI, 1987)
albitophyre *albitite.*
albond A kaolinitic clay found in Dorsetshire, England. It is used as a low-percentage addition to natural molding sands. (Osborne, 1956)
alchemy The immature chemistry of the Middle Ages, characterized by the pursuit of the transmutation of base metals into gold, and the search for the alkahest and the panacea. (Standard, 1964)
alcohol C_2H_5OH; made from grain. Not to be confused with methyl hydroxide or methanol. Syn: *ethanol.* (Crispin, 1964)

alexandrine sapphire Incorrect name for alexandritelike sapphire; also for so-called synthetic alexandrite.
alexandrite effect The property of some chrysoberyl and other minerals and stones to appear green in sunlight, but red under incandescent illumination.
alexandritelike tourmaline *chameleonite.*
alexeyevite A waxlike resin from Kaluga, Russ, which resembles compact turf. Also spelled alexjejevite.
algae Photosynthetic, almost exclusively aquatic, plants of a large and diverse division (Algae) of the thallophytes, including seaweeds and their fresh-water allies. It ranges in size from simple unicellular forms to giant kelps several meters long, and displays extremely varied life-cycles and physiological processes, with, e.g., different complexes of photosynthetic pigments. Algae range from the Precambrian. An individual plant is called an alga. (AGI, 1987)
algal Of, pertaining to, or composed of algae. (AGI, 1987)
algal coal Coal composed mainly of algal remains, such as Pila, Reinschia, etc. Also called boghead coal. (Tomkeieff, 1954; AGI, 1987)
algal limestone A limestone composed largely of the remains of calcium-carbonate-producing algae, or one in which such algae bind together the fragments of other calcium-carbonate-producing forms. (AGI, 1987)
algal reef An organic reef in which algae are or were the principal organisms producing calcium carbonate, e.g., off the coast of Bermuda. The reefs may be up to 10 m high and more than 15 m across. (AGI, 1987)
algal stromatolite *stromatolite.*
algarite A general term for a bitumen derived from algae. (Tomkeieff, 1954)
alga sapropel Equivalent to peat of the humic coal series. (AGI, 1987)
Algerian onyx *oriental alabaster.*
alginate Designates the characteristic maceral of boghead coal. In reflected light it is very difficult to recognize the cellular structure of the algae. The reflecting power of the maceral is much weaker than that of vitrinite and is also weaker than that of sporinite in coals of low rank. In transmitted light, alginite sometimes shows structure (of colonies of algae). The color is yellow to orange. Alginite is best recognized by luminescent microscopy; it shows marked luminescence of varying color—silvery blue, green, yellowish brown. The humic constituents either are not luminescent or show a different kind of luminescence to the alginite. Syn: *algite.* (IHCP, 1963)
algite The constituent petrological unit, or maceral, of algal material present in considerable quantity in algal or boghead coal. See also: *alginate.* (AGI, 1987)
algodonite (a) Arsenide of copper occurring as a white incrustation in the Algodona silver mine, Chile. (CTD, 1958) (b) An orthorhombic mineral, Cu_6As; pseudohexagonal.
Algoman orogeny Orogeny and accompanying granitic emplacement that affected Precambrian rocks of northern Minnesota and adjacent Ontario about 2.4 billon years ago; it is synonymous with the Kenoran orogeny of the Canadian classification. (AGI, 1987)
alidade (a) In mine surveying, a movable arm used to read horizontal angular distances. (Pryor, 1963) (b) A device having a level bubble combined with a quarter or a half circle graduated in degrees that is used by drillers to determine the inclination of a drill stem and/or borehole at the collar of the borehole. Also called angle level; angle rule; clinometer; clinometer rule. (Long, 1960) (c) An instrument used in planetable surveying, consisting of a telescope or sighting device pivoted to swing through a vertical graduated arc atop a vertical stand attached to a steel rule, one edge of which is parallel with the sight line of the telescope. (Long, 1960) (d) Sometimes incorrectly used as a syn. for transit; theodolite. (Long, 1960) (e) A rule equipped with simple or telescopic sights, used for determining the directions of objects, specif., a part of a surveying instrument consisting of a telescope or other sighting device, with index and reading or recording accessories. (AGI, 1987) (f) A surveying instrument used with a planetable for mapping; e.g., peep-sight alidade and telescopic alidade. (AGI, 1987)
alien filling Filling material brought from the surface or from some place other than the mine. (Stoces, 1954)
align (a) To position a drill so that its drill stem is centered on a point and parallel to a predetermined angle and compass direction. Also called line in; lineup. (Long, 1960) (b) To reposition a drill and bring its drill stem over the center and parallel with a newly collared drill hole. (Long, 1960)
alignment (a) The planned direction of a tunnel or other roadway driven irrespective of coal seam or orebody structure; the planned direction of longwall panels or face lines. (Nelson, 1965) (b) Formation or position in line, or, more properly, in a common vertical plane. (Seelye, 1951) (c) The laying out of the axis of a tunnel by instrumental work. (Stauffer, 1906) (d) *coplaning.* (e) In railway or highway surveying, the ground plan, showing the alignment or direction of the route to be followed, as distinguished from a profile, which shows the vertical element. (Seelye, 1951) (f) The act of laying out or regulating by line; adjusting to a line.
alinement clamp A setscrew-equipped, universal-type clamp from one side of which a slotted angle-iron wand, about 18 in (45.7 cm) long, extends outward from a clamping device at 90°. May be made to fit any size drill rod and is used in pairs, leapfrog fashion, to orient successive rods in a specific compass direction as these are lowered into a borehole being surveyed by the acid-bottle method. By this means, the bearing and inclination of a drill hole may be determined in formation or under conditions where a Maas- or other-type magnetic compass cannot be used. (Long, 1960)
aliphatic Of, relating to, or derived from fat; fatty; acyclic. Applied to a large class of organic compounds characterized by an open-chain structure and consisting of the paraffin, olefin, and acetylene hydrocarbons and their derivatives (as the fatty acids). (Webster 3rd, 1966)
alipite A discredited term referring to massive apple-green, hydrated magnesium-nickel silicate similar to genthite.
alisonite A massive, deep indigo-blue copper-lead sulfide, $Cu_2S{\cdot}PbS$. It contains 53.63% copper and 28.25% lead. Tarnishes quickly. (Hess)
Aliva concrete sprayer A compressed-air machine for spraying concrete on the roof and the sides of mine roadways. Used in coal mines for the fireproofing of roadways, for reducing air leakages, and for spraying tunnels supported by roof bolts. See also: *guniting; gunite.* (Nelson, 1965)
alive Said of coal when it makes a rustling sound as it bursts, cracks, and breaks off while under pressure. The rising of methane from the coal causes a similar sound. Cf: *dead.*
alkali (a) Any strongly basic substance, such as a hydroxide or carbonate of an alkali metal (e.g., sodium, potassium). Plural: alkalies. Adj. alkaline; alkalic. Said of silicate minerals that contain alkali metals but little calcium; e.g., the alkali feldspars. (AGI, 1987) (b) Any substance having marked basic properties; i.e., being capable of furnishing to its solution or other substances the hydroxyl ion, OH^-. (Stokes, 1955)
alkali bentonite A bentonite containing easily exchangeable alkali cations and having original properties that are not permanently destroyed by the action of sulfuric acid, but can be restored by treatment with an alkali salt followed by regulated dialysis. This group includes Wyoming-type bentonite and other similar bentonites. (Davis, 1940)
alkalic (a) Said of an igneous rock that contains more alkali metals than is considered average for the group of rocks to which it belongs. (AGI, 1987) (b) Said of an igneous-rock series that contains more sodium and/or potassium than is required to form feldspar with the available silica. (AGI, 1987) (c) Said of an igneous-rock series containing less than 51% silica when the weight percentages of CaO and of $K_2O + Na_2O$ are equal. (AGI, 1987) (d) Said of an igneous rock belonging to the Atlantic suite. Syn: *alkali; alkaline.* (AGI, 1987)
alkalic feldspar *alkali feldspar.*
alkali earth One of a group of elements (Group II) forming divalent cations; esp. calcium, strontium, and barium, but also includes beryllium, magnesium, and radium.
alkali feldspar (a) Those feldspars composed of mixtures or crystal solutions of potassium feldspar, $KAlSi_3O_8$, and sodium feldspar, $NaAlSi_3O_8$, with little or no calcium feldspar, $CaAl_2Si_2O_8$. (b) The subgroup of the feldspar group including albite, anorthoclase, microcline, orthoclase, and sanidine. Syn: *alkalic feldspar.* Cf: *plagioclase.*
alkali flat A level area or plain in an arid or semiarid region, encrusted with alkali salts that became concentrated by evaporation and poor drainage; a salt flat. See also: *playa.* (AGI, 1987)
alkali garnet A general term for members of the sodalite group that are closely related crystallographically and chemically to the true garnets. (English, 1938)
alkali granite (a) A coarse-grained, plutonic rock carrying free quartz and alkali feldspar. (CMD, 1948) (b) A granitoid rock with accessory sodic amphibole or sodic pyroxene.
alkali-lime series Igneous rocks that contain soda-lime (plagioclase) feldspars. (Hess)
alkali metal A metal in group IA of the periodic system; namely, lithium, sodium, potassium, rubidium, cesium, and francium. They form strong alkaline hydroxides; hence, the name. Syn: *alkaline metal.* (ASM, 1961)
alkalimeter (a) An apparatus for measuring the strength or the amount of alkali in a mixture or solution. (Webster 3rd, 1966) (b) An apparatus for measuring the amount of carbon dioxide (as that liberated from a weighed sample of carbonate-containing material by reaction with acid). (Webster 3rd, 1966)
alkaline Adj. of alkali. *alkalic.*
alkaline-earth bentonite A bentonite containing easily exchangeable alkaline-earth cations and, either before or after acid treatment, capable of being made to assume properties of an alkali bentonite by treatment with an alkali salt followed by regulated dialysis.
alkaline-earth metal A metal in group IIA of the periodic system; namely, beryllium, magnesium, calcium, strontium, barium, and ra-

dium; so called because the oxides or earths of calcium, strontium, and barium were found by the early chemists to be alkaline in reaction. (ASM, 1961)

alkaline metal *alkali metal.*

alkalinity The extent to which a material exhibits the property of yielding hydroxyl ions in a water solution. See also: *pH.* (CTD, 1958)

alkali soil A saline soil having 15% or more exchangeable sodium.

alkali subbentonite A bentonite containing easily replaceable alkali bases but having original properties that are destroyed by acid treatment. (Davis, 1940)

alkane A member of the paraffin series, such as methane, ethane, etc. (Pryor, 1963)

alkene A member of the hydrocarbon group series (CnH_2n); e.g., ethylene, propylene. (Pryor, 1963)

alkinite A discredited term referring to a compound of lead, copper, bismuth, and sulfur.

alkyne One of a group of organic compounds containing a carbon-to-carbon triple bond; e.g., acetylene, allylene. Also spelled alkine. (McGraw-Hill, 1994)

allactite A monoclinic mineral, $Mn_7(AsO_4)_2(OH)_8$; vitreous; resembles the axinites. Also spelled allaktit.

allagite A heavy dull-red or green altered carbonated rhodonite. Syn: *diaphorite.* (Fay, 1920)

allanite A monoclinic mineral, $2[(Ca,Y,Ce)_2(Fe,Al)_3O(OH)(Si_2O_7)(SiO_4)]$; epidote group; massive, pitchy, slightly radioactive, and metamict; a minor accessory in felsic igneous rocks and pegmatites. Formerly called orthite; cerine; bucklandite; treanorite. Syn: *yttro-orthite.*

allargentum A mineral, $Ag_{1-x}Sb_x$ with x=0.09 to 0.16.

allcharite Former name for a variety of goethite.

alleghanyite A monoclinic mineral, $Mn_5(SiO_4)_2(OH)_2$; humite group; dimorphous with ribbeite; in skarns.

allemontite A mixture of stibarsen, SbAs, and arsenic or antimony. Syn: *arsenical antimony.* Cf: *antimonial arsenic.*

Allen cone A conical tank used in mineral flotation to separate sand from slime using a float-controlled spigot on peripheral overflow. See also: *cone classifier*. (Pryor, 1963)

Allen-O'Hara furnace A horizontal, double-hearth furnace for calcining sulfide ores. (Fay, 1920)

allevardite *rectorite.*

alley stone *aluminite.*

alliaceous Said of minerals that have an odor of garlic when rubbed, scratched, or heated; e.g., arsenical minerals.

alligator (a) *safety clamp.* (b) Any of several types of machines for metalworking, rock crushing, etc., in which work is accomplished by two massive jaws, one or both of which move as, e.g., alligator shears (preferably, lever shears) or an alligator crusher (preferably, lever crusher). (Henderson, 1953) (c) A prolonged, steel hingelike device by means of which the abutting ends of a flat drive belt can be fastened or laced together. (Long, 1960)

all-mine pig Iron smelted entirely from raw ore. (Standard, 1964)

allochem A collective term introduced by Folk (1959) for one of several varieties of discrete and organized carbonate aggregates that serve as the coarser framework grains in most mechanically deposited limestones, as distinguished from sparry calcite (usually cement) and carbonate-mud matrix (micrite). Important allochems include silt-, sand-, and gravel-size fragments torn up and reworked from the deposit; ooliths; pellets; lumps; and fossils or fossil fragments (carbonate skeletons, shells, etc.). Syn: *allochemical.* (AGI, 1987)

allochemical *allochem.*

allochroite (a) A calcium-chromium garnet. (Fay, 1920) (b) A reddish brown variety of andradite garnet.

allochromatic (a) Descriptive of crystals that exhibit electrical conductivity under the influence of light. (Hess) (b) A gem stone with a coloring agent extraneous to its chemical composition. Opposite of idiochromatic. (Hess) (c) Color produced by a chromophore that is not essential to mineral composition.

allochromatic mineral Mineral that would be colorless if chemically pure, but which commonly exhibits a range of colors due to the presence of small quantities of one or more coloring elements. Chief among these elements are those having atomic numbers 22 to 29; namely, titanium, vanadium, chromium, manganese, iron, cobalt, nickel, and copper. Corundum, beryl, spinel, and quartz are examples of allochromatic gemstones. See also: *idiochromatic mineral.* (Anderson, 1964)

allochthon (a) A mass of rock that has been moved from its place of origin by tectonic processes, as in a thrust sheet or nappe. Many allochthonous rocks have been moved so far from their original sites that they differ greatly in facies and structure from those on which they now lie. Ant. autochthon. Syn: *allochthonous.* Also spelled allochthone. (AGI, 1987) (b) A mass of redeposited sedimentary materials originating from distant sources. (AGI, 1987)

allochthonous (a) Originated by Gumbel and applied to rocks, the dominant constituents of which have not been formed in place. Cf: *autochthonous.* (Holmes, 1920) (b) Coal formation according to the drift theory. (Nelson, 1965) (c) Formed or produced elsewhere than in its present place; of foreign origin, or introduced. The term is widely applied; e.g., to coal or peat that originated from plant material transported from its place of growth, or to an allochthon on a low-angle thrust fault. The term is similar in meaning to allogenic, which refers to constituents rather than whole formations. Ant. autochthonous. See also: *allochthon.* (AGI, 1987)

allochthonous coal Coal originating from accumulations of plant debris that have been transported from their place of growth and deposited elsewhere. The debris can be differentiated as coming from near or from far, and likewise whether it represents recent (dead or still living) or already fossilized material. Syn: *drift coal.* See also: *drift theory.* (AGI, 1987; Tomkeieff, 1954)

allochthonous peat Drift peat of lacustrine character. It is subdivided into Gyttja type and Dry type. (Tomkeieff, 1954)

alloclasite A monoclinic mineral (Co,Fe)AsS; steel gray; dimorphous with glaucodot. Formerly called alloclase.

alloclastic breccia *volcanic breccia.*

allogenic Generated elsewhere; applied to those constituents that came into existence outside of, and previously to, the rock of which they now constitute a part; e.g., the pebbles of a conglomerate. Cf: *authigenic.* (Holmes, 1928)

allogonite *herderite.*

allomeric Of the same crystalline form but of different chemical composition. Syn: *isomorphous.* Cf: *allomorphous.* (Henderson, 1953)

allomorph (a) Syn: *paramorph (obsolete); pseudomorph.* (AGI, 1987) (b) A polymorph or dimorph. Adj. allomorphous.

allomorphism Changes produced in minerals without gain or loss of components; e.g., the change from kyanite to sillimanite. See also: *paramorphism.*

allomorphite Obsolete term for barite, esp., pseudomorphous after anhydrite.

allomorphous Of the same chemical composition but of different crystalline form. Cf: *allomeric.* (Henderson, 1953)

allopalladium (a) A nearly silver-white palladium, found in hexagonal plates in the Harz Mountains, Germany. (Fay, 1920) (b) Former name for stibiopalladinite. (c) Palladium crystallizing in the hexagonal system (as opposed to isometric palladium).

allophane (a) $Al_2O_3{\cdot}SiO_2{\cdot}nH_2O$. A clay mineral composed of hydrated aluminosilicate gel of variable composition; P_2O_5 may be present in appreciable quantity. Syn: *riemannite.* (McGraw-Hill, 1994) (b) A mineral gel, amorphous hydrous aluminum silicate; soft; has pale tints; in soils developed from volcanic glass and ash. It changes from glassy to earthy upon dehydration.

allophanite An obsolete syn. of allophane. (AGI, 1987)

allophanoids Clays of the allophane, halloysite, and montmorillonite groups. (English, 1938)

allothimorph A constituent of a metamorphic rock which, in the new rock, has not changed its original crystal outlines.

allotriomorphic *xenomorphic.*

allotriomorphic-granular *xenomorphic.*

allotropic Applied by Berzelius to those substances that exist in two or more forms, such as diamond and graphite. See also: *polymorphism.* (AGI, 1987)

allotropism *allotropy.*

allotropy (a) The existence of a substance, esp. an element, in two or more different modifications usually in the same phase, such as different crystalline forms of carbon, iron, phosphorus, and sulfur. (Webster 3rd, 1966) (b) Polymorphism in a chemical element; e.g., isometric and hexagonal carbon (diamond and graphite), monoclinic and orthorhombic sulfur (rosickyite and sulfur). An allotrope is one of the crystal forms. Syn: *allotropism.* Adj. allotropic.

allowable bearing value The maximum pressure that can be permitted on foundation soil giving consideration to all pertinent factors, with adequate safety against rupture of the soil mass or movement of the foundation of such magnitude that the structure is impaired. Also called allowable soil pressure. (ASCE, 1958)

allowable pile-bearing load The maximum load that can be permitted on a pile with adequate safety against movement of such magnitude that the structure is endangered. (ASCE, 1958)

allowable stress If a member is so designed that the maximum stress as calculated for the expected conditions of service is less than some certain value, the member will have a proper margin of security against damage or failure. This certain value is the allowable stress, of the kind, and for the material and condition of service in question. The allowable stress is less than the "damaging stress" because of uncertainty as to the conditions of service, nonuniformity of material, and inaccuracy of stress analysis. The margin between the allowable stress and the damaging stress may be reduced in proportion to the certainty with which the conditions of service are known, the intrinsic reliability of the material, the accuracy with which the stress produced by the loading can be calculated, and the degree to which failure is unattended by danger or loss. Cf: *factor of safety*. Syn: *working stress*. (Roark, 1954)

alloy A substance having metallic properties, and composed of two or more chemical elements, of which at least one is a metal. (ASM, 1961)

alloyage The act or process of alloying; specif., in minting, of alloying the precious metals with baser ones to form a harder alloy. (Standard, 1964)

alloy sludger A laborer who salvages sludge from furnace pots for use in recovery of metals. Also called sludger. (DOT, 1949)

alloy system All the alloys that can be made by mixing two metals from a binary alloy system, three metals from a ternary alloy system, and so on. The limits of temperature and composition within which the constituents in a system are stable are represented by the constitutional diagram. (CTD, 1958)

alluaudite (a) A monoclinic mineral, $(Na,Ca)Fe^{+2}(Mn,Fe^{+2},Fe^{+3},Mg)_2(PO_4)_3$; alluaudite group; forms a series with ferroalluaudite; in pegmatites. (b) The mineral group alluaudite, caryinite, ferroalluaudite, hagendorfite, maghagendorfite, and varulite.

alluvia Seldom-used plural of alluvium. (AGI, 1987)

alluvial (a) Said of a placer formed by the action of running water, as in a stream channel or alluvial fan; also said of the valuable mineral, e.g., gold or diamond, associated with an alluvial placer. (AGI, 1987) (b) Pertaining to or composed of alluvium, or deposited by a stream or running water; e.g., an alluvial clay or an alluvial divide. (AGI, 1987)

alluvial clay A clay that has been deposited by water on land, usually in association with rivers or streams. (ACSB, 1939)

alluvial cone An alluvial fan with steep slopes; it is generally higher and narrower than a fan, and is composed of coarser and thicker material believed to have been deposited by larger streams. The term is sometimes used synonymously with alluvial fan. Cf: *alluvial fan*. Syn: *debris cone; dry delta; wash*. (AGI, 1987)

alluvial deposit *alluvium*.

alluvial fan A low, outspread, gently sloping mass of loose rock material, shaped in plan view like an open fan or a segment of a cone; deposited by a stream (esp. in a semiarid region) at the place where it issues from a narrow mountain valley upon a plain or broad valley, or where a tributary stream is near or at its junction with the main stream, or wherever a constriction in a valley abruptly ceases or the gradient of the stream suddenly decreases; it is steepest near the mouth of the valley where its apex points upstream, and it slopes gently and convexly outward with gradually decreasing gradient. Cf: *alluvial cone; bajada*. Syn: *fan; detrital fan; talus fan; dry delta*. (AGI, 1987)

alluvial flat A small alluvial plain bordering a river, on which alluvium is deposited during floods. Cf: *alluvial plain*. Syn: *river flat*. (AGI, 1987)

alluvial mining The exploitation of alluvial deposits by dredging, hydraulicing, or drift mining. See also: *placer mining*. (Nelson, 1965)

alluvial ore deposit *placer*.

alluvial plain A level or gently sloping tract or a slightly undulating land surface produced by extensive deposition of alluvium, usually adjacent to a river that periodically overflows its banks; it may be situated on a flood plain, a delta, or an alluvial fan. Cf: *alluvial flat*. Syn: *wash plain; river plain; bajada*. (AGI, 1987)

alluvial slope A surface underlain by alluvium, which slopes down and away from the sides of mountains and merges with a plain or a broad valley floor; an alluvial surface that lacks the distinctive form of an alluvial fan or a bajada. See also: *bajada*. (AGI, 1987)

alluvial tin Stream tin, or cassiterite pebbles in the gravel along the courses of valleys and rivers on the bedrock. Generally, the purest tin ore. Syn: *stream tin*. (Fay, 1920)

alluvial values In placer mining, the minerals recoverable from the alluvium. These include cassiterite, gold, diamond, gemstones, zirconia, rutile, monazite, and platinum. (Pryor, 1963)

alluviation (a) The deposition or formation of alluvium or alluvial features (such as cones or fans) at places where stream velocity is decreased or streamflow is checked; the process of aggradation or of building up of sediments by a stream along its course, or of covering or filling a surface with alluvium. (AGI, 1987) (b) A hydraulic effect on solids suspended in a current of water, whereby the coarsest and heaviest particles are the first to settle out, and the finest muds the last, as gradient or velocity of a stream is decreased. (AGI, 1987)

alluvion *alluvium*.

alluvium (a) A general term for clay, silt, sand, gravel, or similar unconsolidated detrital material, deposited during comparatively recent geologic time by a stream or other body of running water, (1) as sediment in the bed of the stream or on its flood plain or delta, (2) as a cone or fan at the base of a mountain slope; esp., such a deposit of fine-grained texture (silt or silty clay) deposited during time of flood. Syn: *alluvial deposit; alluvion*. (AGI, 1987) (b) A driller's term for the broken, earthy rock material directly below the soil layer and above the solid, unbroken bed or ledge rock. Etymol: Latin alluvius, from alluere, to wash against. Plural: alluvia; alluviums. (AGI, 1987)

almagra Sp. A deep-red ocher originally from Andalusia, Spain, similar to Indian red. Used as a pigment and in polishing glass and metals. Also spelled almagre. (Standard, 1964)

almandine (a) An isometric mineral, $8[Fe_3^{2+}Al_2Si_3O_{12}]$; pyralspilite subgroup of the garnet group, with Fe replaced by Mg, Mn, and Ca; in red to brownish-black dodecahedral and trapezohedral crystals, or massive; Mohs hardness, 7-1/2; occurs in medium-grade metamorphic rock and felsic igneous rocks; used as a gemstone and an abrasive. Formerly called almandite; alamandine; almond stone. (b) A violet or mauve variety of ruby spinel; a reddish-purple to purplish-red spinel. (c) A reddish-purple sapphire (almandine sapphire).

almandine ruby A violet-colored magnesium spinel. Syn: *ruby spinel*. (Dana, 1914)

almandite Former spelling of almandine.

almeria ore A Spanish hematite. (Osborne, 1956)

almond furnace A furnace in which the slags of litharge left in refining silver are reduced to lead by being heated with charcoal. (Fay, 1920)

almond stone Former name for almandine.

alnoite A lamprophyre chiefly composed of biotite or phlogopite and melilite as essential minerals, commonly with olivine, calcite, and clinopyroxene. Perovskite, apatite, nepheline, and garnet may be present. Its name (Rosenbusch, 1887) is derived from Alnö, Sweden. Also spelled allnöite; alnöite. (AGI, 1987)

Aloxite Trade name for fused crystalline alumina or artificial corundum used as an abrasive. (English, 1938)

alpha (a) The first letter (α) of the Greek alphabet. Commonly used as a prefix to show that a mineral, the condition of a metal, or other thing or property is one of several closely related species, or one of a series; beta (β), the second letter, and gamma (γ), the third letter are used likewise, e.g., alpha rays, beta rays, and gamma rays and alpha quartz and beta quartz. (Hess) (b) Adj. Of or relating to one of two or more closely related minerals and specifying a particular physical structure (esp. a polymorphous modification); specif. said of a mineral that is stable at a temperature lower than those of its beta and gamma polymorphs (e.g., "alpha cristobalite" or "α-cristobalite," the low-temperature tetragonal phase of cristobalite). Some mineralogists reverse this convention, using α for the high-temperature phase (e.g., "alpha carnegieite," the isometric phase of carnegieite stable above 690 °C). (AGI, 1987) (c) In crystallography the angle between the *b* and *c* axes.

alpha alumina A white, anhydrous, nonhygroscopic powder, Al_2O_3, produced when precipitated Al(OH), is calcined at 1,000 °C. It is the natural product of the Bayer process and other processes used (or proposed) to treat bauxite, clay, or other aluminum-bearing materials. (Newton, 1959)

alpha celsian A silicate of aluminum and barium, $BaAl_2Si_2O_8$. An artificial feldspar, similar to anorthite, but containing barium instead of calcium. Hexagonal prisms. Uniaxial, negative. (English, 1938)

alpha chalcocite *digenite*.

alpha hyblite A porcelain-white, hydrous acidic sulfosilicate of thorium with some uranium, iron, and lead; isotropic. An alteration product of thorite. From Hybla, ON, Can. (English, 1938)

alpha mercuric sulfide *vermilion*.

alpha quartz A quartz polymorph stable below 573 °C; a common constituent of crustal rocks. Syn: *low quartz*.

alpha zinc sulfide Colorless when pure; hexagonal; ZnS; mol wt., 97.43; sp gr, 3.98 to 4.1; Mohs hardness, 3.5 to 4.0; luster, resinous; transformation temperature from beta zinc sulfide to alpha zinc sulfide, 1,020±5 °C: sublimes at 1,180 °C or 1,185 °C; melting point, 1,850 °C (at 150 atm); insoluble in water and in acetic acid; and very soluble in other acids. Occurs as the brownish-black mineral wurtzite, which is unstable compared with its stable dimorph, the mineral sphalerite (beta zinc sulfide), to which it inverts during alteration and from which it is formed by heating sphalerite to the transformation temperature. Can be crystallized from acid solutions above 250 °C. See also: *wurtzite; zinc sulfide*. Cf: *beta zinc sulfide*. (Handbook of Chem. & Phys., 2; Dana, 1944)

alpine (a) Pertaining to, characteristic of, or resembling the European Alps or any lofty mountain or mountain system, esp. one modified by intense glacial erosion. Spelled Alpine when referring specif. to European Alps. (AGI, 1987) (b) Characteristic or descriptive of the mountainous regions lying between timberline and snowline; said of the climate, flora, relief, ecology, etc. Less strictly, pertaining to high elevations and cold climates. (AGI, 1987) (c) A general term for topographical and structural features that resemble in grandeur and complexity those of the European Alps, regardless of the age or location of the mountains and features so described. (AGI, 1987)

Alpine diamond *pyrite*.

alquifou A coarse-grained variety of galena used by potters in preparing a green glaze.

alshedite A variety of titanite containing yttria; found in Sweden. (Standard, 1964)

alstonite A triclinic mineral, $BaCa(CO_3)_2$; pseudo-orthorhombic and trimorphous with barytocalcite and paralstonite. Formerly called bromlite.
altaite An isometric mineral, PbTe; in veins with gold, sulfides, and other tellurides.
alteration Any change in the mineralogic composition of a rock brought about by physical or chemical means, esp. by the action of hydrothermal solutions; also, a secondary, i.e., supergene, change in a rock or mineral. Alteration is sometimes considered as a phase of metamorphism, but is usually distinguished from it because of being milder and more localized than metamorphism is generally thought to be. (AGI, 1987)
altered mineral A mineral that has undergone chemical change by geologic (esp. weathering or hydrothermal) processes.
altered rock A rock that has undergone changes in its chemical and mineralogic composition since its original formation. (AGI, 1987)
altered stone (a) Said of a stone that has undergone chemical and/or mineralogical changes under geologic processes. (b) Any stone of which the appearance, esp. the color, has been changed by any artificial means whatsoever. For example, heat is often used to improve or alter color. Such change may be either external or internal. See also: *treated stone; stained stone.*
alternate pillar and stope *square-set stoping.*
alternate polarity Arrangement in magnetic separator whereby ore travels alternately through normal concentration and entropy fields, thus stirring attracted material and shaking out entrained nonmagnetics. (Pryor, 1958)
alternating-current ampere Current that will produce heat at the same rate as a direct-current ampere, when flowing through a given ohmic resistance. (Kentucky, 1952)
alto A term used in the southwestern United States for a bluff, height, or hill. Etymol: Sp., high ground. (AGI, 1987)
alum (a) Any hydrous, alkali aluminum sulfate mineral, including kalinite, potassium alum, sodium alum, mendozite, tschermigite, and lonecreekite. Syn: *potash alum.* (b) A former name for kalinite and potassium alum. (c) Any salts that are double sulfates of aluminum, chromium, iron, or manganese and one of the alkali metals.
alum cake A product of the action of sulfuric acid on clay, consisting chiefly of silica and aluminum sulfate. (Webster 3rd, 1966)
alum earth An argillaceous rock, commonly a shale, containing marcasite or pyrite which, as it decomposes, forms sulfuric acid that attacks the shale and produces alum. Many such rocks are carbonaceous. See also: *alum shale.* (Hess)
alumina An oxide of aluminum, Al_2O_3; the mineral corundum; an important constituent of clay minerals, $Al_2Si_2O_5(OH)_4$, determining their suitability for firebrick and furnace linings. Synthetic alumina is used as the feed material in aluminum smelters; it is also used in the preparation of paints called lakes, in dyeing, and in calico printing; in granular form it is used for abrasives and grinding or cutting tools of high tensile strength. Most alumina is made via the Bayer process from hydrated aluminum oxides, as found in bauxite, diaspore, and gibbsite. Aluminum oxide can also be made in an electric furnace by fusing bauxite or corundum. Suitably doped alumina is the feed material for boules of synthetic ruby and sapphire made by the Verneuil flame-fusion process. Fused alumina is crushed and used as an abrasive, a refractory, a heating element for electrical heaters, and as a filtering medium.
aluminate A compound having the general formula, $MAlO_2$ or M_3AlO_3, in which M indicates a monovalent metal. Mineral aluminates, such as $MgAl_2O_4$, are termed spinels. (Bennett, 1962)
alumina trihydrate $A_2O_3{\cdot}3H_2O$ or $Al(OH)_3$; monoclinic; white; crystalline powder, balls, or granules; sp gr, 2.42; obtained from bauxite and used as a source of aluminum. (CCD, 1961; Lee, 1961)
aluminite A monoclinic mineral, $Al_2(SO_4)(OH)_4{\cdot}7H_2O$; pseudo-orthorhombic, formerly called websterite. Syn: *alley stone; argil.*
aluminosilicate refractory A general term that includes all refractories of the fireclay, sillimanite, mullite, diaspore, and bauxite types. (Dodd, 1964)
aluminous abrasive An abrasive produced by fusing aluminum oxide. (Mersereau, 1947)
aluminous ore Ore in which the gangue consists principally of alumina. (Osborne, 1956)
aluminum A light, silvery-white, ductile metal with high electrical conductivity and good resistance to corrosion. Obtained from bauxite. Symbol, Al. It is the lightest of the metals in general use commercially and is the basis for light alloys used in the construction of modern aircraft and rockets; aluminum coatings are used for telescope mirrors, decorative paper, packages, and toys. The oxide, alumina, occurs naturally as ruby, sapphire, corundum, and emery. (Handbook of Chem. & Phys., 3)
aluminum detonator *Briska detonator.*
aluminum silicates Varying proportions of Al_2O_3 and SiO_2. Occur naturally in clays. Used in the glass and ceramics industry. (CCD, 1961)
alumite *alunite.*
alumocalcite A variety of opal with alumina and lime as impurities. (Fay, 1920)
alumogel An amorphous aluminum hydroxide that is a constituent of bauxite. Formerly called cliachite, diasporogelite, and sporogelite.
alum rock *alunite.*
alum salt Natural salt from which alum can be made. See also: *halloysite; kaolinite.* (Sanford, 1914)
alum schist *alum shale.*
alum shale An argillaceous, often carbonaceous, rock impregnated with alum, originally containing iron sulfide (pyrite, marcasite) which, when decomposed, formed sulfuric acid that reacted with the aluminous and potassic materials of the rock to produce aluminum sulfates. Syn: *alum earth; alum schist; alum slate.* (AGI, 1987)
alum slate *alum shale.*
alunite (a) A trigonal mineral, $KAl_3(OH)_6(SO_4)_2$; massive or disseminated; in pale tints; formed from sulfuric acid acting on potassium feldspar in volcanic regions (alunization), and around fumaroles. Formerly called alumstone, alum rock, alumite. (b) A mineral group including jarosite.
alunitization Introduction of, or replacement by, alunite. (AGI, 1987)
alunogen A triclinic mineral, $Al_2(H_2O)_{12}(SO_4)_3{\cdot}5(H_2O)$; in fibrous masses or crusts; white, tinged yellow to reddish, with sharp acid taste; in acid environments filling crevices in coals, slates, gossans, and fumaroles. Syn: *feather alum; hair salt.*
alvanite A monoclinic mineral, $(Zn,Ni)Al_4(VO_3)_2(OH)_{12}{\cdot}2H_2O$; forms light-blue-green rosettes in the vanadium deposits of Karatau, Kazakhstan.
alveoli The lungs can be thought of as two elastic bags containing millions of little distensible air sacs. These air sacs or alveoli are all connected to the air passages, which branch and rebranch like the twigs of a tree. (Hunt, 1965)
alvite A zirconium mineral; a source of hafnium, containing 16% HfO_2; tetragonal. Obtained from Alve, Norway. (Kirk, 1947; Webster 2nd, 1960)
amalgam (a) A naturally occurring alloy of silver with mercury; mercurian silver. It is found in the oxidation zone of silver deposits and as scattered grains in cinnabar ores. Syn: *argental mercury.* Cf: *goldamalgam.* (AGI, 1987) (b) A general term for alloys of mercury with one or more of the well-known metals (except iron and platinum); esp. an alloy of mercury with gold, containing 40% to 60% gold, and obtained from the plates in a mill treating gold ore. (AGI, 1987)
amalgamate (a) To unite (a metal) alloy with mercury. (Standard, 1964) (b) To form an amalgam with; as, mercury amalgamates gold. (Standard, 1964)
amalgamated claims Eng. Mining claims adjoining one another that have been grouped into one claim for more economical working.
amalgamating barrel A short, cylindrical vessel or barrel, with solid ends turned to fit bearings, used for amalgamating battery accumulations and other material. See also: *amalgamator.* (Fay, 1920)
amalgamating table A sloping wooden table covered with a copper plate on which the mercury is spread in order to amalgamate with the precious metal particles. (CTD, 1958)
amalgamation (a) The production of an amalgam or alloy of mercury. (b) The process by which mercury is alloyed with some other metal to produce an amalgam. It was used at one time for the extraction of gold and silver from pulverized ores, but has been superseded by the cyanide process. (Barger, 1939)
amalgamation process A process of gold or silver recovery in which the ore, finely divided and suspended in water, is passed over a surface of liquid mercury to form an amalgam that is subjected to fire-refining processes for the recovery of the gold or silver. Syn: *amalgam treatment.* (Henderson, 1953)
amalgamator An apparatus used in mining for bringing pulverized ore into close contact with mercury to extract free metal from it by amalgamation. See also: *amalgamating barrel.* (Standard, 1964)
amalgam barrel A small, cylindrical batching mill used to grind auriferous concentrates gently with mercury. (Pryor, 1963)
amalgam pan A muller mill with a horizontal rotating disk bearing on a fixed plate. Gold-bearing material and mercury flow pulpwise between them. (Pryor, 1963)
amalgam plate A sheet of metal with an adherent film of mercury that seizes gold from flowing pulp. (Pryor, 1963)
amalgam retort The vessel where mercury is distilled from gold or silver amalgam. (Nelson, 1965)
amalgam treatment *amalgamation process.*
amang A term used in Malaysia for the heavy iron and tungsten minerals (and associated minerals) found with placer cassiterite deposits. (AGI, 1987)
amarantite A triclinic mineral, $Fe(SO_4)(OH){\cdot}3H_2O$.

amargosite The trade name for a bentonite from the Amargosa River, CA. Syn: *montmorillonite.* Also called natural soap and soaprock.
amatrice (a) *variscite.* (b) A green gem cut from variscite and its surrounding matrix of gray, reddish, or brownish crystalline quartz or chalcedony.
amause *trass.*
amazonite A bright, apple- or blue-green variety of microcline; may be carved for art objects. Syn: *amazonstone.*
amazonstone *amazonite.*
amber (a) A mineraloid; amorphous hydrocarbons from resins secreted by trees or shrubs upon injury, derived by oxidation and polymerization of nonvolatile terpenoids; in sedimentary rocks and on beaches, e.g., Baltic Sea. See also: *chemawinite.* (b) A hard, brittle fossil resin, yellow to brown, that takes a fine polish; may contain fossil insects and plant matter. Syn: *succinite; bernstein; electrum.* See also: *resin.* (c) A group of fossil resins containing considerable succinic acid and having highly variable C:H:O ratios; e.g., almashite, simetite, delatynite, and ambrosine. See also: *copal.*
Amberine A local trade name for a yellowish-green variety of chalcedony from Death Valley, CA. (English, 1938)
amberite Former spelling of amber.
amber mica *phlogopite.*
amberoid A gem material consisting of small fragments of genuine amber artificially united or reconstructed by heat and pressure; may be characterized by an obvious flow structure or a dull spot left by a drop of ether. Also spelled ambroid. Syn: *pressed amber.*
amber opal A brownish-yellow variety of opal stained by iron oxide.
ambient (a) The environment surrounding a body but undisturbed or unaffected by it. (Hy, 1965) (b) Encompassing on all sides; thus, ambient air is the air surrounding. (Strock, 1948)
ambrosine A yellowish to clove-brown amber found in the phosphate beds near Charleston, SC; it may be a modern resin that has been subjected to the action of salt water. Rich in succinic acids. (Fay, 1920; Tomkeieff, 1954)
amenability Characteristic reactions of minerals to basic methods of mineral processing, studied in preliminary testwork on unknown ores. (Pryor, 1963)
American Permissible explosive used in coal mines. (Bennett, 1962)
American-Belgian furnace A direct-fired Belgian furnace used in the United States, conforming essentially to the Liege design. (Fay, 1920)
American forge *Champlain forge.*
American jade (a) Nephrite in Wyoming. (b) *californite.*
American ruby A red pyrope garnet in Arizona and New Mexico.
American system *churn drill.*
American Table of Distances The quantity-distance table, prepared and approved by the Institute of Makers of Explosives (IME), for storage of explosive materials to determine safe distances from inhabited buildings, public highways, passenger railways, and other stored explosive materials.
amesite An apple-green silicate mineral belonging to the phyllosilicate group and occurring in foliated hexagonal plates. See also: *magnesium kaolinite.* (Kirk, 1947; Webster 3rd, 1966)
amethyst (a) A transparent to translucent, purple to pale-violet variety of quartz common as a semiprecious gemstone. The color results from a hole defect associated with ferric iron substitution for silicon. Syn: *bishop's stone.* (b) A term applied to a deep-purple variety of corundum and to a pale reddish-violet variety of beryl.
amethystine A color designation meaning violet to purplish, used as in amethystine glass and amethystine sapphire.
amethystine quartz A phenocrystalline variety of quartz colored purplish or bluish violet by manganese. See also: *lavendine.* (Standard, 1964)
amethyst point Amethyst crystal, from a geode, commonly possessing only the six (or possibly three) rhombohedral terminal faces; generally with gradational color, the best at the apex commonly grading to colorless at its base.
amianthinite *asbestos.*
amianthus Syn: *asbestos.* Applied esp. to a fine silky variety such as chrysotile. Also spelled amiantus.
amiantoid (a) Having the appearance of asbestos. (Standard, 1964) (b) An olive-green, coarse, fibrous variety of asbestos. (Standard, 1964)
amigo A stick, tied to the end of a rope, on which workers sit when being raised or lowered in shafts. (Hess)
aminoffite A tetragonal mineral, $Ca_2(Be,Al)Si_2O_7(OH){\cdot}H_2O$; in colorless crystals in Sweden.
ammine One of a group of complex compounds formed by coordination of ammonia molecules with metal ions. (McGraw-Hill, 1994)
ammiolite A red or scarlet earthy substance, probably a mixture of copper antimonate and cinnabar; said to occur in some Chilean ore deposits.
ammite A 17th and 18th century term for a sedimentary rock now called oolite. Obsolete syn: ammonite.
ammonal An explosive used mainly for heavy quarry blasts in dry boreholes. It consists of TNT, ammonium nitrate, and powdered aluminum. (Nelson, 1965)
ammonia A colorless, gaseous alkaline compound; NH_3; lighter than air; pungent smell and taste. Byproduct of gas and coke production. Used in making fertilizers and explosives.
ammonia dynamite Dynamite in which part of the nitroglycerin is replaced by ammonium nitrate; used in mining. See also: *extra dynamite.* (Bennett, 1962)
ammonia gelatin An explosive of the gelatin-dynamite class containing ammonium nitrate. (Webster 3rd, 1966)
ammonia gelatin dynamite *gelatin dynamite.*
ammonia niter Ammonium nitrate, NH_4NO_3; nitrammite. Also spelled ammonia nitre. (Spencer, 1952)
ammonia stillman In the coke products industry, one who extracts ammonia from liquor for use in producing ammonium sulfate by circulating substances through stills and auxiliary equipment. Also called pump-and-still operator; byproducts stillman. (DOT, 1949)
ammoniojarosite A trigonal mineral, $(NH_4)Fe_3(SO_4)_2(OH)_6$; alunite group. It occurs in pale yellow lumps of tabular grains on the west side of the Kaibab fault, southern Utah.
ammonium chloride NH_4Cl; isometric; and colorless. When dissolved in water, it is used as an electrolyte for some primary cells. Obtained as a byproduct in gas manufacture. Used as a flux in soldering. Also called sal ammoniac. (Crispin, 1964)
ammonium hydroxide A solution of ammonia in water, NH_4OH. (CTD, 1958)
ammonium nitrate Used in explosives and as a fertilizer. NH_4NO_3; mol wt, 80.04; colorless. (Bennett, 1962)
ammonium nitrate gelignites These explosives are similar to the straight gelatins except that the main constituent is ammonium nitrate instead of sodium nitrate. Ammonium nitrate is a more active explosive ingredient than sodium nitrate; therefore it can be substituted for nitroglycerin in much larger quantities and still give explosives of high weight strength. The nitroglycerin content is usually 25% to 35%, and the ammonium nitrate content ranges from about 30% to 60%. Ammonium nitrate gelignites are characterized by plastic consistency; high densities of 1.5 to 1.6 g/cm^3; medium velocity of detonation of 2,500 m/s; and good fume properties. The ammonium nitrate gelignites are useful all-purpose explosives and are widely used in metal mines, nongassy coal mines, quarries, tunneling, and construction work. Its wide range of strengths enables a suitable grade to be selected for blasting almost every variety of rock from hard to soft. (McAdam, 1958)
amoibite A former named for gersdorffite.
amorphous (a) Said of a mineral or other substance that lacks crystalline structure, or whose internal arrangement is so irregular that there is no characteristic external form. Ant. crystalline. (AGI, 1987) (b) The state of a solid lacking crystal structure, specif. lacking long-range order. (c) A term formerly used to describe a body of rock occurring in a continuous mass, without division into parts. Cf: *massive.*
amorphous coal A somewhat inaccurate term for a coal in which distinct plant material is not discernible. (Tomkeieff, 1954)
amorphous graphite Very fine-grained, generally sooty graphite from metamorphosed coalbeds. The word amorphous is a misnomer because all graphite is crystalline. The term has also been applied to very fine particles of flake graphite that can be sold only for low-value uses (such as foundry facings), and to fine-grained varieties of Ceylon lump graphite. (AGI, 1987)
amorphous metal Metal in which the regular arrangement of atoms characteristic of the crystalline state has been destroyed. (CTD, 1958)
amorphous mineral A mineral with no definite crystalline structure. (Nelson, 1965)
amorphous peat A type of peat in which the original structure of the plants has been destroyed as the result of decomposition of the cellulose matter. It is heavy, compact, and plastic when wet. See also: *fibrous peat.* (Tomkeieff, 1954)
amorphous phosphorus *phosphorus.*
amosite (a) A monoclinic mineral in the cummingtonite-grunerite series. (Sinclair, 1959) (b) A commercial asbestos composed of asbestiform gedrite, grunerite, or anthophyllite of the amphibole group; has typically long fibers.
ampangabeite A former name for samarskite.
ampelite An obsolete term for a black carbonaceous or bituminous shale. (AGI, 1987)
ampelitis An ancient name applied to a variety of bituminous earth used as an insecticide sprinkled over vines. (Tomkeieff, 1954)
ampere The practical unit of electric current. The current produced by 1 V acting through a resistance of 1 Ω. (Webster 3rd, 1966)
ampere-hour The quantity of electricity carried past any point of a circuit in 1 h by a steady current of 1 A; 1 A·h equals 3,600 C. (Webster 3rd, 1966)
ampere volt A watt. (Standard, 1964)

amphibole A mineral group; characterized by double chains of silica tetrahedra having the composition $A_{0-1}B_2Y_5Z_8O_{22}(OH,F,Cl)$, where (A=Ca,Na,K,Pb,B), (B=Ca,Fe,Li,Mg,Mn,Na), (Y=Al,Cr,Fe,Mg,Mn,Ti), and (Z=Al,Be,Si,Ti); in the orthorhombic or monoclinic crystal systems, including actinolite, anthophyllite, arfvedsonite, cummingtonite, hornblende, richterite, glaucophane, grunerite, anthophyllite, riebeckite, tremolite, and others. All display a diagnostic prismatic cleavage in two directions parallel to crystal faces and intersecting at angles of about 54° and 124°. Some members may be asbestiform. See also: *pyroxene.*

amphibole-magnetite rock A granular, more or less banded rock containing grunerite, other ferruginous silicates, and magnetite; produced by metamorphism of ferruginous cherts, such as taconite and jaspillite.

amphibolide A general term, for use in the field, to designate any coarse-grained, holocrystalline igneous rock almost entirely composed of amphibole minerals. Syn: *amphibololite.* (AGI, 1987)

amphibolite A crystalloblastic rock consisting mainly of amphibole and plagioclase with little or no quartz. As the content of quartz increases, the rock grades into hornblende plagioclase gneiss. Cf: *feather amphibolite.* (AGI, 1987)

amphibolite facies The set of metamorphic mineral assemblages (facies) in which basic rocks are represented by hornblende + plagioclase, the plagioclase being oligoclase-andesine or some more calcic variety. Epidote and almandine are common in amphibolites. The facies is typical of regional dynamothermal metamorphism under moderate to high pressures (in excess of 300 MPa) with temperatures in the range 450 to 700 °C. (AGI, 1987)

amphibololite *amphibolide.*

amphigene *leucite.*

amphoteric Having both acidic and basic properties. (CTD, 1958)

amygdale *amygdule.*

amygdaloid An extrusive or intrusive rock containing numerous amygdules. Said of a rock having numerous amygdules. Syn: *amygdaloidal; mandelstone.* See also: *amygdule.* (AGI, 1987)

amygdaloidal (a) Said of rocks containing amygdules and of the structure of such rocks; e.g., certain basaltic lava sheets on Keweenaw Point, Lake Superior, which have amygdules filled with native copper, and are important sources of the metal. Syn: *amygdaloid; amygdule.* (AGI, 1987) (b) Almond-shaped. (Zern, 1928)

amygdaloidal rock A rock containing amygdules, or the structure of a rock resulting from its presence. (Schieferdecker, 1959)

amygdule A gas cavity or vesicle, in an igneous rock, that is filled with such secondary minerals as calcite, quartz, chalcedony, or a zeolite. The term amygdale is preferred in British usage. Syn: *amygdaloidal.* See also: *amygdaloid.* (AGI, 1987)

amyl alcohol $C_5H_{11}OH$; a frothing agent. (Pryor, 1963)

amyl xanthate A powerful collector agent used in the flotation process. (Pryor, 1958)

Anaconda method A bunch-blasting method in which 6 to 15 fuses, cut to respective lengths 2 in (5.1 cm) longer than required, are tied together near one end by two ravelings of fuse spaced about 5 to 6 in (12.7 to 15.2 cm) apart. A special cutter cuts the fuses off evenly between the two ties, leaving the fuses tied together and offering a smooth face of cut ends. Another bunch is made from the fuses of the remaining holes in the round. By using a short notched fuse as a spitter, the flame is directed against the cut end of one bunch of fuses. As soon as this bunch ignites, it is held close to the face of the second bunch, moving slowly to contact all fuses with the flame from the first bunch. Bunches should be held at least 6 in back from the end to avoid burning the hands. By this method, all the holes of a round are fired in only two groups and by one spitter. (Lewis, 1964)

anaerobic (a) Said of an organism (esp. a bacterium) that can live in the absence of free oxygen; also, said of its activities.—n. anaerobe. (AGI, 1987) (b) Said of conditions that exist only in the absence of free oxygen. Cf: *aerobic.* (AGI, 1987)

analcime An isometric mineral, $16[Na(H_2O)(AlSi_2O_6)]$; zeolite group; in white to slightly tinted radiating aggregates or granular masses; a late primary or hydrothermal mineral in mafic igneous rocks, an alkaline lake precipitate, and in silicic tuffs and tuffaceous sandstones. Formerly called analcite.

analcimite An extrusive or hypabyssal igneous rock consisting mainly of analcime and pyroxene (usually titanaugite). Feldspathoids, plagioclase, and/or olivine may be present. Apatite, sphene, and opaque oxides may be present as accessories. (AGI, 1987)

analcimization Replacement of feldspars or feldspathoids by analcime, usually in igneous rocks during late magmatic or postmagmatic stages. Syn: *analcitization.* (AGI, 1987)

analcitization *analcimization.*

analog computer A computer that operates with numbers represented by directly measurable quantities (such as length, voltage, or resistance) in a one-to-one correspondence; a measuring device that operates on continuous variables represented by physical or mathematical analogies between the computer variables and the variables of a given problem to be solved. (AGI, 1987)

analogous (a) Corresponding to or resembling something else in some way, as in form, proportion, etc. (b) Designating that pole (end) of a pyroelectric crystal to which heating gives a positive charge. Cf: *antilogous.*

analytical chemistry Study of the qualitative or quantitative composition of materials. (Pryor, 1963)

analytic group A rock-statigraphic unit formerly classed as a formation but now called a group because subdivisions of the unit are considered to be formations.

analyzer The part of a polariscope that receives the light after polarization and exhibits its properties. In a petrographic microscope, it is the polarizing mechanism (Nicol prism, Polaroid, etc.) that intersects the light after it has passed through the object under study. See also: *polarizer.* (AGI, 1987)

anamigmatization High-temperature, high-pressure remelting of preexisting rock to form migma. Cf: *anatexis.* (AGI, 1987)

anamorphic zone The zone deep in the Earth's crust in which rock flowage takes place. The term, originated in 1898 by Van Hise, is now little used. Cf: *anamorphism.* (AGI, 1987)

anamorphism Intense metamorphism in the anamorphic zone in which rock flowage takes place and simple minerals of low density are changed into more complex ones of greater density by silication, decarbonization, dehydration, and deoxidation. The term was originated by Van Hise in 1904. Cf: *anamorphic zone; katamorphism.* (AGI, 1987)

anastomosing (a) Pertaining to a network of branching and rejoining fault or vein surfaces or surface traces. (AGI, 1987) (b) Said of the channel pattern of a braided stream. (AGI, 1987)

anatase A tetragonal mineral, $4[TiO_2]$; trimorphous with rutile and brookite; brown, greenish-gray, or black; in hydrothermal veins around granite pegmatites, as an alteration of titanium minerals, and as detrital grains. Formerly called octahedrite. See also: *octahedrite; titanium dioxide; xanthitane.*

anatectic *anatexis.*

anatexis Melting of preexisting rock. This term is commonly modified by terms such as intergranular, partial, differential, selective, or complete. Adj. anatectic. Cf: *palingenesis; syntexis; anamigmatization.* (AGI, 1987)

anauxite A clay mineral near kaolinite, but containing excess silica, probably as interlayered sheets. Monoclinic. (AGI, 1987; Dana, 1959)

Anbauhobel A rapid plow for use on longwall faces. It is suitable for seams from 2 to 8 ft (0.6 to 3.9 m) thick, with reasonably good roof and floor. The plow travels along the face at a speed of 75 ft/min (22.9 m/min) with a cutting depth from 1-1/2 to 3 in (3.8 to 7.6 cm); the broken coal is loaded by the plow-shaped body onto an armored conveyor. The machine can be operated independently of the face conveyor. See also: *plow-type machine; Rehisshakenhobel.* (Nelson, 1965)

anchaduar Fillings of old workings in a mine, and said to carry gold of recent deposition. This is a product that deposits in most of the old stopes throughout the mine. In some instances, the whole stope for 20 ft (6.1 m) wide is filled. It is apparently siliceous material with more or less pyrite. (Hess)

anchorage That portion of any beam or structure designed to resist pulling out or slipping of the beam or structure when subjected to stress. (Nelson, 1965)

anchor bolt (a) A bolt with the threaded portion projecting from a structure, generally used to hold the frame of a building secure against wind load or a machine against the forces of vibration. Also called holding-down bolt; foundation bolt. (Hammond, 1965) (b) A bolt or other device used to secure a diamond-drill base to a solid foundation. It may or may not be threaded. (Long, 1960) (c) A lag screw used to anchor the drill base to a platform or sill. (Long, 1960)

anchor charge Means of fastening an explosive charge in a seismic shot hole to allow several charges to be preloaded. At each stage the bottom charges are fired first, the upper charges being held down by anchors. (AGI, 1987)

anchor jack *jack.*

anchor prop *stell prop.*

ancylite A mineral, $SrCe(CO_3)_2(OH){\cdot}H_2O$; in pegmatites.

andalusite An orthorhombic mineral, Al_2SiO_5; trimorphous with kyanite and sillimanite; Mohs hardness, 7-1/2; in aluminous shales and slates subjected to high-temperature, low-stress metamorphism; transparent green varieties used as gems. Syn: *cross-stone.*

andersonite A trigonal mineral, $Na_2Ca(UO_2)(CO_3)_3{\cdot}6H_2O$; bright yellow-green; secondary.

Anderton shearer loader A widely used cutter loader in which the ordinary jib of the longwall coal cutter is replaced by a shear drum which cuts a web from 16 to 22 in (40.6 to 55.9 cm) depending on its width. The machine travels on an armored conveyor and requires a prop-free front for working. It shears the coal in one direction and the front coal is loaded by a plow deflector, and then returns along the face (without cutting) and loads the remainder of the broken coal. The ordinary Anderton is suitable for coal seams more than 3 ft 6 in (1.1 m) thick. (Nelson, 1965)

andesine A triclinic mineral, $(Na,Ca)[(Si,Al)AlSi_2O_8]$, NaSi 50 to 70 mol %, CaAl 50 to 30 mol %; of the plagioclase series of the feldspar group with prismatic cleavage; white to gray; a common rock-forming mineral in andesites, differentiated gabbros, some anorthosites, and as detrital grains.

andesinite A coarse-grained igneous rock almost entirely composed of andesine. It was named by Turner in 1900. Cf: *anorthosite.* Not recommended usage. (AGI, 1987)

andesite A dark-colored, fine-grained extrusive rock that, when porphyritic, contains phenocrysts composed primarily of zoned sodic plagioclase (esp. andesine) and one or more of the mafic minerals (e.g., biotite, hornblende, pyroxene), with a groundmass composed generally of the same minerals as the phenocrysts, although the plagioclase may be more sodic, and quartz is generally present; the extrusive equivalent of diorite. Andesite grades into latite with increasing alkali feldspar content, and into dacite with more alkali feldspar and quartz. It was named by Buch in 1826 from the Andes Mountains, South America. (AGI, 1987)

andorite An orthorhombic mineral, $PbAgSb_3S_6$. Syn: *sundtite.*

andradite An isometric mineral, $8[Ca_3Fe_2Si_3O_{12}]$; never pure; garnet group; in yellow, green, red, brown, or black dodecahedral and trapezohedral crystals, or may be massive; in calcareous metasediments and placers. Varieties include topazolite, demantoid, melanite, aplome, and bredbergite.

andre A direction of coal face roughly halfway between the main (bord) and secondary (end) cleavages; on the cross. (Mason, 1951)

Andreasen pipette An instrument used in the determination of the particle size of clays by the sedimentation method. (Dodd, 1964)

Andrews' elutriator A device for particle size analysis. It consists of (1) a feed vessel or tube, (2) a large hydraulic classifier, (3) an intermediate classifier, and (4) a graduated measuring vessel. (Dodd, 1964)

andrewsite An orthorhombic mineral, $(Cu,Fe)Fe_3(PO_4)_3(OH)_2$; in bluish-green globules with radial structure.

anemoclastic Broken off by wind erosion and rounded by wind action.

anemogram A continuous record of wind speed and direction given by an anemograph. (Hammond, 1965)

anemograph A self-recording anemometer giving a continuous trace of the direction and velocity of surface wind. In the Dines tube anemograph, the wind pressure acts upon the opening of a tube arranged as a vane to face in the direction of the wind. Pressure is transmitted through the tube to a float carrying a pen, the height of which indicates the wind velocity. (Hammond, 1965)

anemolite (a) An upturned form of calcite stalactite; its form is supposed to have been caused by air currents. (English, 1938) (b) A stalactite with one or more changes in its growth axis. Syn: *helictite.*

anemometer An instrument for measuring air velocity. It consists of a small fan from 3 to 6 in (7.6 to 15.2 cm) in diameter that is rotated by the air current. By simple gearing, the number of revolutions of the fan is recorded on dials. It is held in the mine airway for the exact number of minutes (N), the instrument being moved steadily over the entire area. The difference between the initial and the final readings on the dials, divided by N, gives the velocity of the air in feet per minute. Instruments are available for velocities from near zero to 6,000 ft/min (1.83 km/min), also with extension and remote control handles. See also: *vane anemometer; self-timing anemometer*. (Nelson, 1965)

aneroid barograph Consists essentially of an aneroid barometer and a revolving drum. The movement of the evacuated spring can is transmitted and magnified through a system of levers so that it is finally traced by means of a stylo on the graph paper attached to the revolving drum. The drum is rotated by clockwork, and can be of either the 24-h or the 7-day type. The graph paper is usually marked off in hourly intervals, so that a complete record of the atmospheric pressure at any instant may be obtained. These barographs are used extensively in mining and in meteorological offices. (Morris, 1958)

aneroid barometer An instrument for measuring atmospheric pressure, built first by Lucien Vidie in about 1843. Basically, variation in pressure with changes in altitude is determined by the movements of the elastic top of a metallic box from which the air has been partly exhausted. Used generally in measuring altitude. (AGI, 1987)

ANFO (a) An explosive material consisting of ammonium nitrate and fuel oil. (b) A blasting product, with approx. 94.5% industrial-grade ammonium nitrate and 5.5% No. 2 grade diesel fuel oil for a nearly oxygen-balanced mix; available in bulk form for onsite mixing of the AN and fuel or in 50-lb (23-kg) premixed bags as pourable forms. A heavy ANFO product is comprised of up to 45% to 50% ammonium nitrate emulsion mixed with prilled ANFO to increase the bulk density of ANFO; it has improved strength and provides good water resistance in comparison to ANFO. (SME, 1992)

angelellite A triclinic mineral, $Fe_4(AsO_4)_2O_3$; brownish-black, encrusted on andesite in northwestern Argentina.

angle beam A two-limbed beam used for turning angles in shafts, etc. (Zern, 1928)

angle brace A brace used to prevent mine timbers from riding or leaning; a brace across an interior angle. (Fay, 1920)

angle-cut Drill holes converge, so that a core is blasted out. This leaves an open or relieved cavity or free face for the following shots, which are timed to ensue with a fractional delay. (Pryor, 1963)

angledozer (a) A bulldozer whose blade can be turned at an angle to the direction of travel. (Carson, 1961) (b) A power-operated machine fitted with a blade, adjustable in height and angle, used for digging and side casting, and for spreading loose excavated material; used at opencast pits and dumping sites. (Nelson, 1965) (c) A bulldozer with a blade that can be pivoted on a vertical center pin, so as to cast its load to either side. Syn: *angling dozer.* (Nichols, 1976)

angle drilling *inclined drilling; inclined borehole.*

angle hole A borehole that is drilled at an angle not perpendicular to the Earth's surface. Syn: *incline hole.* (Long, 1960)

angle level *alidade.*

angle of attack In mine fan terminology, the angle made by the direction of air approach and the chord of the aerofoil section. (Roberts, 1960)

angle of bite In rolling metals where all the force is transmitted through the rolls, maximum attainable angle between roll radius at the first contact and the roll centers. If the operating angle is less, it is called the contact angle or roll angle. (ASM, 1961)

angle of dip The angle at which strata or mineral deposits are inclined to the horizontal plane. In most localities, earth movements subsequent to the deposition of the strata have caused them to be inclined or tilted. Syn: *dip.* See also: *apparent dip.* (Nelson, 1965; Fay, 1920)

angle of draw (a) In coal mine subsidence, this angle is assumed to bisect the angle between the vertical and the angle of repose of the material and is 20° for flat seams. For dipping seams, the angle of break increases, being 35.8° from the vertical for a 40° dip. The main break occurs over the seam at an angle from the vertical equal to half the dip. (Lewis, 1964) (b) The angle between the limit line and the vertical. Cf: *draw.* (Nelson, 1965)

angle of external friction The angle between the abscissa and the tangent of the curve representing the relationship of shearing resistance to normal stress acting between soil and the surface of another material. (ASCE, 1958)

angle of extinction In polarized-light microscopy, the angle between an extinction direction and a crystallographic direction—e.g., crystal face, cleavage plane—of an anisotropic mineral. An extinction angle of 0° is called "parallel extinction," an angle of 45° "symmetrical extinction," other angles "oblique extinction;" of diagnostic value in mineral identification. Syn: *extinction angle.*

angle of friction The angle between the perpendicular to a surface and the resultant force acting on a body resting on the surface, at which the body begins to slide. (Hammond, 1965)

angle of inclination The angle of slope from the horizontal.

angle of nip In a rock-crushing machine, the maximum angle subtended by its approaching jaws or roll surfaces at which a piece of ore of specified size can be gripped. See also: *nip.* (Pryor, 1963)

angle of obliquity The angle between the direction of the resultant stress or force acting on a given plane and the normal to that plane. (ASCE, 1958)

angle of polarization (a) That angle, the tangent of which is the index of refraction of a reflecting substance. (Fay, 1920) (b) The angle of reflection from a plane surface at which light is polarized. (Hess)

angle of pull The angle between the vertical and an inclined plane bounding the area affected by the subsidence beyond the vertical. Applied to slides of earth. (Fay, 1920)

angle of reflection (a) The angle that a reflected ray of light, on leaving the exterior or interior surface of an object, such as a transparent stone or crystal, makes with the normal to that surface. (b) An erroneous term for the Bragg angle of X-ray diffraction.

angle of repose *angle of rest.*

angle of rest The maximum slope at which a heap of any loose or fragmented solid material will stand without sliding or come to rest when poured or dumped in a pile or on a slope. Syn: *angle of repose.* See also: *natural slope.*

angle of shear The angle between the planes of maximum shear, which is bisected by the axis of greatest compression. (Rice, 1960)

angle of slide The slope, measured in degrees of deviation from the horizontal, on which loose or fragmented solid materials will start to slide; it is a slightly greater angle than the angle of rest.

angle of swing The number of degrees through which the dipper or shovel bucket moves horizontally from the filled position to the dumping position. (Carson, 1961)

anglesite An orthorhombic mineral, $4[PbSO_4]$; sp gr, 6.2 to 6.4; in the supergene parts of lead-ore veins; a minor ore of lead. Formerly called lead vitriol, lead spar.

angle to the right Horizontal angle measured clockwise from the preceding line to the following one. (Seelye, 1951)
angle trough A short curved section of a shaker conveyor trough inserted in a trough line to change the angle of direction. Up to 15° of turn, the angle trough does not employ any means of support other than connection to adjacent troughs. For a greater degree of turn, a fulcrum jack and a swivel device are employed with the trough section. (Jones, 1949)
angling Rope will only coil closely on the drum within the distance between the centers of the pulleys. Spread or diagonal coiling will result outside this distance unless the drum is grooved: this is known as outside angling and with a grooved drum may amount to 1°. After the normal line between the pulley and the drum is passed, the coils attempt to get back to this normal line. This produces friction crushing between the coils and a danger of coils mounting one over the other; this is known as inside angling and should be kept below 2°. The amount of angling for a given distance between the pulleys will depend upon the distance between the headgear pulleys and the drum. Grooving the drum reduces the difficulties associated with angling. Syn: *outside angling.* (Sinclair, 1959)
angling dozer *angledozer.*
angstrom (a) A unit of linear measurement in the centimeter-gram-second system. It equals 10^{-10} m, 10^{-8} cm, 10^{-4} μm, or 10^{-1} nm. Such ultramicroscopic distances as the dimensions of atoms, molecules, unit cells, and short wavelengths are expressed in angstroms. (Webster 2nd, 1960; Webster 3rd, 1966) (b) Either of two units of wavelength: (1) 10^{-10} m, called the absolute angstrom; or (2) the wavelength of the red spectrum line of cadmium divided by 6,438.4696, which is called the international angstrom. (Webster 3rd, 1966)
angular Having sharp angles or borders; specif. said of a sedimentary particle showing very little or no evidence of abrasion. (AGI, 1987)
angular cutter A milling cutter on which the cutting face is at an angle with regard to the axis of the cutter. (Crispin, 1964)
angularity test *slope test.*
angular unconformity An unconformity in which the older underlying strata dip at a different angle (generally steeper) than the younger overlying strata. See also: *disconformity.* (AGI, 1987; AGI, 1987)
anhedral Said of those minerals of igneous rocks that are not bounded by their own crystal faces, but have an imperfect form impressed on them by the adjacent minerals during crystallization. Cf: *euhedral; subhedral.* Syn: *allotriomorphic; xenomorphic.*
anhydride (a) A compound formed from an acid by removal of water. (McGraw-Hill, 1994) (b) An oxide of a nonmetallic element or an organic radical, capable of forming an acid by uniting with the elements of water, or of being formed by the abstraction of the water, or of uniting with basic oxides to form salts.
anhydrite An orthorhombic mineral, $CaSO_4$; massive; primarily in evaporite deposits, hot sulfate volcanic waters, and veins; hydrates to gypsum. Formerly called cube spar.
anhydrock A sedimentary rock composed chiefly of anhydrite. (AGI, 1987)
anhydrous Said of a substance, e.g., magma or a mineral, that is completely or essentially without water. An anhydrous mineral contains no water in chemical combination. (AGI, 1987)
anhydrous ammonia Purified ammonia gas, NH_3, liquefied by cold and pressure.
anidiomorphic *xenomorphic.*
aniline point An approximate measure of the aromatic content of a mixture of hydrocarbons. It is defined as the lowest temperature at which an oil is completely miscible with an equal volume of aniline. (Francis, 1965)
anilite An orthorhombic mineral, Cu_7S_4; alters to a digenitelike crystal solution upon grinding.
animikite A silver ore consisting of a mixture of sulfides, arsenides, and antimonides, with striking intergrowths and in granular masses; contains nickel and lead. Cf: *macfarlanite.*
anion (a) A negatively charged ion, such as a hydroxide, chloride, or sulfate ion; opposite of cation. (Webster 3rd, 1966) (b) An atomic particle with a negative charge; one attracted to the anode. Cf: *cation.*
anion exchange capacity A measure of the ability of a clay to adsorb or exchange anions; usually expressed in milliequivalents or anions per 100 g of dry clay. (ACSG, 1963)
anionic collector A flotation reagent in which the reactive group is acid in character. In these collectors the hydrocarbon group is in the anion. The most common anionic collectors are fatty acids (carboxylic acids). They occur naturally as complex mixtures in which the hydrocarbon chain is saturated or unsaturated. (Fuerstenau, 1962)
anionic current Negative-ion electrical current.
anionic detergent A detergent in which the anion (negative ion) is the active part. (ASM, 1961)
anionic flotation A flotation process employing anionic collectors. Anionic collectors are those in which the negative ion (anion) is the effective part. Opposite of cationic flotation, which employs cationic, or positive, ion collectors.
anisodesmic (a) An obsolete syn. for heterogranular. (b) Said of crystals with unequal dimensions, including those with significant flattening, elongation, or both. Ant. isometric. Cf: *equant; tabular.* (c) *heterogranular.* (d) Said of crystal structures with chemical bond strengths that are directionally unequal; e.g., micas.
anisometric (a) Having unsymmetrical parts; not isometric; applied to crystals with three unequal axes. (Webster 3rd, 1966) (b) Of or relating to a rock of granular texture but having mineral constituents of unequal size. (Webster 3rd, 1966) (c) A textural term applied to granular rocks in which the grains are of different sizes. Obsolete. The term "seriate" expresses the same texture when the crystals vary gradually or in a continuous series. Cf: *isodiametric; isometric.* (Johannsen, 1931-38)
anisotropic Having physical properties that vary in different directions. Specif. in optical crystallography showing double refraction. Characteristic of all crystalline substances, including minerals, except those belonging in the isometric system, which are isotropic. Opposite of isotropic. (Fay, 1920; AGI, 1987)
anisotropic fabric A fabric in which there is preferred orientation of the minerals of which the rock is composed.
anisotropy (a) The property of being anisotropic, or exhibiting properties (such as velocity of light transmission, conductivity of heat or electricity, or compressibility) with different values when measured along axes in different directions. (Webster 3rd, 1966) (b) The condition of having different properties in different directions as in geologic strata that transmit sound waves with different velocities in the vertical and in the horizontal directions. (AGI, 1987) (c) Optically descriptive of crystalline materials having light velocities and indices of refraction dependent upon the crystallographic direction of the electric vector (vibration direction) during transmission or reflection; it includes all nonisometric crystals. Cf: *uniaxial; extinction; isotropy.* (d) In geostatistics, the situation where a variogram exhibits a longer range (i.e., better correlation) in one direction than in another.
ankerite A trigonal mineral, $Ca(Fe,Mg,Mn)(CO_3)_2$; dolomite group; forms series with dolomite and with kutnohorite; associated with iron ores; commonly forms thin veins in some coal seams. Cf: *ferroan dolomite; cleat spar.* See also: *pearl spar.*
annabergite A monoclinic mineral, $2[Ni_3(AsO_4)_2{\cdot}8H_2O]$; vivianite group with cobalt replacing nickel toward erythrite; occurs as light-green soft coatings of fine striated crystals, or earthy; an oxidation product of nickel and cobalt arsenides, the green crusts being a distinctive guide to nickel ores. Formerly called nickel ocher. Cf: *nickel bloom.*
annealed wire rope A wire rope made from wires that have been softened by annealing. (Zern, 1928)
annealing (a) Heating to and holding at a suitable temperature and then cooling at a suitable rate for such purposes as reducing hardness; improving machinability; facilitating cold working; producing a desired microstructure; or obtaining desired mechanical, physical, or other properties. When applied to ferrous alloys, the term "annealing", without qualification, implies full annealing. When applied to nonferrous alloys, annealing implies a heat treatment designed to soften a cold-worked structure by recrystallization or subsequent grain growth or to soften an age-hardened alloy by causing a nearly complete precipitation of the second phase in relatively coarse form. (ASM, 1961) (b) The variation of the cooling rate at different temperatures of porcelain, glass, and other ceramic ware containing large quantities of vitreous material to prevent defects such as dunting, crazing, cracking, crystallization, etc. (c) The process by which glass and certain metals are heated and then slowly cooled to make them more tenacious and less brittle. Important in connection with the manufacture of steel castings, forgings, etc. (Fay, 1920) (d) The process of heating metal shapes to a red heat or above, prior to cleaning.
annealing color The hue taken by steel in annealing. (Standard, 1964)
annealing oven A oven for heating and gradually cooling metals or glass to render them less brittle. Also called annealing furnace. (Standard, 1964; Fay, 1920)
annerodite A black mixture of samarskite with parallel overgrowths of columbite. Also spelled aanerodite.
annite A monoclinic mineral, $KFe_3AlSi_3O_{10}(OH,F)_2$; mica group; trioctahedral.
annivite A variety of tennantite with arsenic partly replaced by bismuth and antimony.
annual labor Same as assessment work, on mining claims.
annual layer (a) A sedimentary layer deposited or presumed to have been deposited during the course of a year; e.g., a glacial varve. (AGI, 1987) (b) A dark band (in a salt stock) of formerly disseminated anhydrite crystals that accumulated upon being freed by solution of the enclosing salt. (AGI, 1987)
annual value The annual value of a property is the estimated annual surplus of revenue over expenditure in process of liquidating the mineral reserves. In the usual case, that of a property owned by a

company, it is the dividend estimated maintainable annually over the whole computed life, the regular distribution of mining profit. (Truscott, 1962)

annular bearing A ring bearing that carries the radial load of a shaft. If a ball bearing, the balls are held in a race and run on a hard band around the shaft. (Petroleum Age, 1923)

annular-drainage pattern A drainage pattern in which streams follow a roughly circular or concentric path along a belt of weak rock, resembling in plan a ringlike pattern. It is best displayed by streams draining a maturely dissected structural dome or basin where erosion has exposed rimming sedimentary strata of greatly varying degrees of hardness, as in the Red Valley, which nearly encircles the domal structure of the Black Hills, SD. (AGI, 1987)

annular kiln A kiln having compartments. (Standard, 1964)

anode (a) The positive pole of an electrolytic cell. (Webster 3rd, 1966) (b) The terminal at which current enters a primary cell or storage battery; it is positive with respect to the device and negative with respect to the external circuit. (McGraw-Hill, 1994) (c) The electropositive pole. (AGI, 1987) (d) The electrode at which electrons leave a device to enter the external circuit; opposite of cathode. See also: *electrode.* (Webster 3rd, 1966) (e) The negative terminal of a primary cell or of a storage battery that is delivering current. (Webster 3rd, 1966)

anode compartment In an electrolytic cell, the enclosure formed by a diaphragm around the anodes. (ASM, 1961)

anode copper Specially shaped copper slabs used as anodes in electrolytic refinement, and resulting from the refinement of blister copper in a reverberatory furnace. (ASM, 1961)

anode effect The effect produced by polarization of an anode in the electrolysis of fused salts. It is characterized by a sudden increase in voltage and a corresponding decrease in amperage due to the anode being virtually separated from the electrolyte by a gas film. (ASM, 1961)

anode furnace A copper- or nickel-refining furnace, in which blister copper or impure nickel is refined.

anode metals Metals used for electroplating. They are as pure as commercially possible, uniform in texture and composition, and have the skin removed by machining. In addition to pure single metals, various alloys are produced in anode form, such as Platers' brass and Spekwite, the latter yielding a white plate harder than nickel. (Brady, 1940)

anode mud A deposit of insoluble residue formed from the dissolution of the anode in commercial electrolysis. Sometimes called anode slime. In copper refining, this slime contains the precious metals that are recovered from it. (ASM, 1961; CTD, 1958)

anode scrap Remnants of anode copper retrieved from electrolytic refining of the metal. (Pryor, 1963)

anode slime Metals or metal compounds left at, or falling from, the anode during electrolytic refining. The plural form is often used.

anodic zone In the electrical self-potential method of geophysical prospecting, if the chemical composition of the soil or subsoil is such as to give electrical polarization, the zone of electropositive potential is the anodic zone. (AGI, 1987)

anomaly (a) A departure from the expected or normal. (AGI, 1987) (b) The difference between an observed value and the corresponding computed value. (AGI, 1987) (c) A geological feature, esp. in the subsurface, distinguished by geological, geophysical, or geochemical means, which is different from the general surroundings and is often of potential economic value; e.g., a magnetic anomaly. (AGI, 1987) (d) Any deviation from conformity or regularity. A distinctive local feature in a geophysical, geological, or geochemical survey over a larger area. An area or a restricted portion of a geophysical survey, such as a magnetic survey or a gravity survey, that differs from the rest of the survey in general. The anomaly might be associated with petroleum, natural gas, or mineral deposits, or provide a key to interpreting the underlying geologic structure. Drilling for economic mineral deposits might be conducted in the area of a geophysical anomaly. In seismic usage, anomaly is generally synonymous with subsurface structure or material properties, but it is also used for spurious or unexplainable seismic events or for local deviations of observed signals which cannot be conclusively attributed to a unique cause. See also: *hydrochemical anomaly.* (AGI, 1987) (e) A gravity anomaly is the difference between the theoretical calculated gravity and the observed terrestrial gravity. In comparing any set of observed data with a computed theoretical curve, the difference of an observed value and the corresponding computed value, or the observed minus the computed value. Excess observed gravity is a positive anomaly, and a deficiency is a negative anomaly. See also: *Bouguer anomaly; free-air anomaly; isostatic anomaly.* (AGI, 1987) (f) A crystallographic anomaly is the lack of agreement between the apparent external symmetry of a crystal and the observed optical properties. (Schieferdecker, 1959) (g) Any departure from the normal magnetic field of the Earth is a magnetic anomaly. It may be a high or a low, subcircular, ridgelike or valleylike, or linear and dikelike. (AGI, 1987)

anorthic Obsolete syn. for a triclinic crystal system. See also: *triclinic.*

anorthite (a) A triclinic mineral, $4[CaAl_2Si_2O_8]$; plagioclase series of the feldspar group, with up to 10 mol % NaSi replacing CaAl; white to gray; in ultramafic intrusive igneous bodies and skarns. Syn: *calcium feldspar; calcium plagioclase; calciclase; lepolite.* (b) A pure calcium end member of the plagioclase series.

anorthoclase A triclinic mineral, $4[(Na,K)AlSi_3O_8]$; feldspar group; occurs in tabular crystals with prismatic cleavage; colorless or white; in felsic volcanic rocks. Cf: *orthoclase.*

anorthosite A plutonic rock composed almost entirely of plagioclase, usually labradorite. It is a monomineralic equivalent of gabbro but lacks monoclinic pyroxene. Cf: *andesinite.*

anorthositization Introduction of, or replacement by, anorthosite. (AGI, 1987)

anoxia Oxygen deficiency in the blood cell or tissues of the body in such degree as to cause psychological and physiological disturbances. Anoxia may result from a scarcity of oxygen in the air being breathed or from an inability of the body tissues to absorb oxygen under conditions of low ambient pressure. Also called hypoxia. Syn: *oxygen deficiency.* (Hunt, 1965)

anthill In blast-hole drilling, the cuttings around the hole collar. (Krumlauf)

anthoinite A triclinic mineral, $WAlO_3(OH)_3$; in tungstic ochers in central Africa and Tasmania.

anthonyite A monoclinic mineral, $Cu(OH,Cl)_2{\cdot}3H_2O$; lavender colored; from the Centennial Mine in Calumet, MI. Cf: *calumetite.*

anthophyllite An orthorhombic mineral, $4[(Mg,Fe)_7Si_8O_{22}(OH)_2]$; amphibole group; commonly lamellar or fibrous, green to clove-brown; in schists from metamorphosed ultramafic rocks; a nonspinning grade of asbestos.

anthra From Greek anthrax, coal; also, a precious stone; combining forms used commonly to denote substances resembling or derived from coal, or fossils found in coal measures. (Standard, 1964)

anthracene Obtained by the distillation of coal tar. Used in the manufacture of dyestuffs. (Crispin, 1964)

anthracene oil A heavy green oil that distills from coal tar above 270 °C and is the principal source of anthracene, phenanthrene, and carbozole. (Webster 3rd, 1966)

anthracite (a) A hard, black lustrous coal containing a high percentage of fixed carbon and a low percentage of volatile matter. Commonly referred to as hard coal, it is mined in the United States, mainly in eastern Pennsylvania, although in small quantities in other States. (BCI, 1947) (b) The rank of coal, within the anthracitic class of Classification D 388, such that on the dry and mineral-matter-free basis, the volatile matter content of the coal is greater than 2% but equal to or less than 8% (or the fixed carbon content is equal to or greater than 92% but less than 98%), and the coal is nonagglomerating. (ASTM, 1994) (c) Coal of the highest metamorphic rank, in which fixed-carbon content is between 92% and 98% (on a dry, mineral-matter-free basis). It is hard and black, and has a semimetallic luster and semiconchoidal fracture. Anthracite ignites with difficulty and burns with a short blue flame, without smoke. Syn: *hard coal; stone coal; kilkenny coal.* See also: *solid smokeless fuel.* (AGI, 1987)

anthracite-coal-base carbon refractory A manufactured refractory comprised substantially of calcined anthracite coal. (ASTM, 1994)

anthracite coal sizes The sizes by which anthracite coal is marketed. The sizes are called broken, egg, stove, chestnut, pea, and buckwheat. Size is graded according to the size of round mesh a piece will pass through.

anthracite duff In Wales, fine screenings used in making pitch-bonded briquets and for mixing with bituminous coal to be burned in cement kilns, on chain grate stokers, and as powdered fuel. (Hess)

anthracite fines The product from an anthracite coal-preparation plant, usually below 1/8 in (3.2 mm). See also: *duff; fines; grain.* (Nelson, 1965)

anthracite silt Minute particles of anthracite too fine to be used in ordinary combustion. (Webster 3rd, 1966)

anthracitic Pertaining to anthracite. (AGI, 1987)

anthracology (a) The science of coal. (Tomkeieff, 1954) (b) Coal petrography, a branch of geology dealing with the physical constitution of coal in much the same way that petrography deals with the mineral composition of rocks. It is concerned with the physical variations in coal that make it possible to classify coal material by type. (AGI, 1987)

anthracometer An instrument for determining the amount of carbon dioxide in a mixture of gases. (Standard, 1964)

anthraconite *bituminous limestone; stinkstone; swinestone.*

anthracosilicosis Massive fibrosis of the lungs marked by shortness of breath from inhalation of carbon and quartz dusts. Also called miner's phthisis. See also: *anthracosis.* (Webster 3rd, 1966)

anthracosis A deposition of coal dust within the lungs from inhalation of sooty air. Syn: *blacklung; collier's lung.* Cf: *anthracosilicosis.* See also: *mining disease.* (Webster 3rd, 1966)
anthrafilt Anthracite used for filtration purposes. (Jones, 1949)
anthrafine Sizes of anthracite smaller than barley. (Jones, 1949)
anthrasilicosis Variant of anthracosilicosis. (Webster 3rd, 1966)
anthraxolite (a) A highly graphitic coal. One specimen contained 97.7% fixed carbon. (AGI, 1987) (b) Anthracitelike asphaltic material occurring in veins in Precambrian slate of the Sudbury district, ON, Can. (AGI, 1987) (c) Probably fragmentary coalified wood.
anthraxylon The vitreous appearing components of coal, which in thin section are shown to have been derived from the woody tissues of plants—such as stems, limbs, branches, twigs, roots, including both wood and cortex—changed and broken up into fragments of greatly varying sizes through biological decomposition and weathering during the peat stage, and later flattened and transformed into coal through the coalification process, but still present as definite units. (AGI, 1987)
anthraxylous coal A bright coal (composed of anthraxylon and attritus in which the translucent cell-wall degradation matter or translucent humic matter predominates) in which the ratio of anthraxylon to attritus is from 3:1 to 1:1. Cf: *attrital coal.* (AGI, 1987)
antibreakage device A cushioning device to reduce the impact of coal in motion against objects with which it may come into contact, with a view to avoiding fracture of the coal. (BS, 1962)
anticlinal (a) Pertaining to an anticline. Cf: *synclinal.* (AGI, 1987) (b) Inclining in opposite directions. Having or relating to a fold in which the sides dip from a common line or crest. Of or pertaining to an anticline. The opposite of synclinal. (Webster 3rd, 1966) (c) The crest of an anticlinal roll may be the apex of a vein. (Fay, 1920) (d) Said of strata assuming an arch-shaped form. (Gordon, 1906)
anticlinal axis (a) The medial line of an upfolded structure, from which the strata dip on either side. (b) If a range of hills or a valley is composed of strata that on the two sides dip in opposite directions, the imaginary line that lies between them and toward which the strata on each side rise is called an anticlinal axis. See also: *axis.* (AGI, 1987)
anticlinal bend An upwardly convex flexure in which one limb dips gently toward the apex and the other limb dips more steeply away from it. Cf: *unicline; monocline.* (AGI, 1987)
anticlinal mountain A mountain whose geologic structure is that of an anticline. Cf: *synclinal mountain.*
anticlinal valley A valley that follows an anticlinal axis. The term was used as early as 1862 by C.H. Hitchcock. Cf: *synclinal valley.* (AGI, 1987)
anticline (a) A fold, generally convex upward, whose core contains the stratigraphically older rocks. Ant. syncline. See also: *antiform.* (AGI, 1987) (b) Applied to strata that dip in opposite directions from a common ridge or axis, like the roof of a house; the structure is termed an anticline or saddleback. (AGI, 1987) (c) In this type of fold (anticline) the sides or limbs of the fold typically slope away from the plane of the axis of either side. Every anticlinal axis pitches in two directions; i.e., toward the two ends of the fold. (AGI, 1987)
anticlinorium A series of anticlines and synclines, so grouped that taken together they have the general outline of an arch; opposite of synclinorium. (Webster 3rd, 1966)
antiferromagnetic Spontaneous magnetic orientation of atoms with equal magnetic moments aligned in opposite directions. (Van Vlack, 1964)
antiferromagnetism A state where *d* electrons are ordered in an antiparallel array, giving materials small positive values for magnetic susceptibility and weak attraction to an external magnetic field. Cf: *ferrimagnetism; ferromagnetism; superexchange.*
antiform A fold whose limbs close upward in strata for which the stratigraphic sequence is not known. Cf: *anticline.* Ant. synform. (AGI, 1987)
antifriction bearing A bearing consisting of an inner and outer ring, separated by balls or rollers held in position by a cage. (Nichols, 1976)
antigorite A monoclinic mineral, $(Mg,Fe)_3Si_2O_5(OH)_4$; kaolinite-serpentine group; polymorphous with clinochrysotile, lizardite, orthochrysotile, parachrysotile; greasy variegated green; used as an ornamental stone. See also: *baltimorite.*
antilogous Designating the pole (end) of a pyroelectric crystal that is negative while the crystal is being heated and positive as it cools. Cf: *analogous.* (Standard, 1964)
antimagmatist *transformist.*
antimonate (a) A salt or ester of antimonic acid; a compound containing the radical SbO_4^{-3}, SbO_3^{-1}, or $Sb_2O_7^{-4}$ (diantimonate) in which antimony has a +5 valence. (AGI, 1987) (b) A salt containing pentavalent antimony and oxygen in the anion. (Webster 3rd, 1966) (c) A mineral characterized by inclusion of antimony and oxygen; e.g., swedenborgite, $NaBe_4SbO_7$.
antimonial arsenic A native compound of arsenic and antimony of which the antimony forms a comparatively small part. Cf: *allemontite.* (Hess)
antimonial copper *chalcostibite.*
antimonial red silver *pyrargyrite.*
antimonial silver (a) Silver ore or alloys containing variable quantities of antimony. (Bennett, 1962) (b) *dyscrasite.*
antimonite (a) A salt or ester of antimonious acid or antimonous acid; a compound containing the radical SbO_3^{-3} or SbO_2^{-1} in which antimony has a +3 valence. (AGI, 1987) (b) *stibnite.*
antimonpearceite A monoclinic mineral, $(Ag,Cu)_{16}(Sb,As)_2S_{11}$. Cf: *arsenpolybasite.*
antimony Metallic antimony is an extremely brittle metal with a flaky, crystalline texture. Symbol, Sb. Sometimes found native, but more frequently as the sulfide, stibnite (Sb_2S_3). Used in semiconductors, batteries, antifriction alloys, type metal, small arms, tracer bullets, cable sheathing, flame-proofing compounds, paints, ceramics, glass, and pottery. Antimony and many of its products are toxic. (Handbook of Chem. & Phys., 3)
antimony blende *kermesite.*
antimony crudum The name given to the molten, high-grade sulfide that drains away from the gangue residue when stibnite (antimony sulfide) is melted by liquation. (Newton, 1959)
antimony glance *stibnite.*
antimony ocher Any of several native antimony oxides; e.g., stibiconite, cervantite.
antimony regulus An impure product of the smelting process; largely antimony sulfide. (Standard, 1964)
antimony star The fernlike marking on the upper surface of the metal antimony when well crystallized. (Fay, 1920)
antimony trioxide *valentite.*
antinode A point, line, or surface in a standing wave system where some characteristic of the wave field has maximum amplitude. Antinodes, like nodes, may be of several types, such as pressure or velocity. (ASM, 1961)
antipathy of minerals The incompatibility of certain rock-forming minerals, according to the theory of fractional crystallization, results from their being too far apart in a crystallization sequence to be associated in such quantities as to make up the entire rock. Thus, a rock made up of quartz and calcic plagioclase is unknown among igneous rocks. (Hess)
antiperthite An intergrowth of a sodic and a potassic feldspar, generally considered to have formed during slow cooling by the unmixing of sodium and potassium ions in an originally homogeneous alkalic feldspar. In an antiperthite, the potassic member (usually orthoclase) forms thin films, lamellae, strings, or irregular veinlets, within the sodic member (usually albite). Cf: *perthite.* (AGI, 1987)
antistatic Descriptive of materials that normally have high insulating qualities, e.g., rubber hoses and belts that have been rendered conductive to reduce risk of sparks or electric shocks in mines, or other places where there is a fire risk. (Pryor, 1963)
antistress mineral A term suggested for minerals such as cordierite, the feldspars, the pyroxenes, forsterite, and andalusite, whose formation in metamorphosed rocks is believed to be favored by conditions that are not controlled by shearing stress, but by thermal action and by hydrostatic pressure that is probably no more than moderate. Cf: *stress mineral.* (AGI, 1987)
antithetic fault A fault that dips in the opposite direction from the direction in which the associated sediments dip. Opposite of synthetic fault. Syn: *antithetic shear.* (AGI, 1987)
antithetic shear *antithetic fault.*
antitropal ventilation Ventilation by a current of air traveling in the opposite direction to that of the flow of mineral out of the mine. See also: *ascensional ventilation; descensional ventilation; homotropal ventilation.* (BS, 1963)
antiturbidity overflow system A system fitted to a drag suction hopper dredge which disperses entrained gases from the overflow in a settling tank and discharges the degassed overflow below the surface. The resulting plume is normally compact and does not appear at the surface. Abbrev., ATOS. (Cruickshank, 1987)
antlerite An orthorhombic mineral, $4[Cu_3SO_4(OH)_4]$; forms emerald to blackish-green striated crystals or parallel aggregates; may be reniform or massive; in oxidized parts of copper veins; an ore of copper in desert regions. Syn: *vernadskite.*
AN-TNT slurry Mixture of ammonium nitrate and trinitrotoluene used as an explosive. (Lewis, 1964)
antofagasite *eriochalcite.*
antozonite A dark-violet to black semiopaque variety of fluorite that emits a strong odor when crushed; commonly causing nausea among miners, perhaps owing to free fluorine; produced by alpha bombardment, as in the inner bands of halos surrounding uraninite and thorite inclusions.
anvil (a) The stationary serrated jaw piece or plate of a safety clamp, adjustable pipe wrench, or jaw-type rock crusher. Also sometimes incorrectly used as a syn. for drive hammer. Also called anvil block; anvil heel; anvil jaw; heel. (Long, 1960) (b) An iron block placed

between a stamp-mill mortar box and the foundation block; generally used in light mortars and concrete foundations. (Fay, 1920) (c) In drop forging, the base of the hammer into which the sow block and lower die part are set. (ASM, 1961) (d) A block of steel upon which metal is forged. (ASM, 1961)

anvil block A massive block of cast iron placed beneath the anvils of steam and other heavy hammers to absorb vibration. It is often embedded in masonry or concrete. (Crispin, 1964)

anvil jaw *anvil.*

anvil stone Eng. Blue building stone, forming a bed of irregular anvil-shaped blocks. (Arkell, 1953)

anvil vise A vise with an anvil on one jaw. (Standard, 1964)

apartalite *zincite.*

apatelite A hydrous ferric sulfate, found in yellow nodules in clay. (Fay, 1920)

apatite (a) Any hexagonal or monoclinic pseudohexagonal mineral with the general formula $A_5(XO_4)_3(F,Cl,OH)$, where A = (Ba,Ca,Ce,K,Na,Pb,Sr,Y) and X = (As,C,P,Si,V). Syn: *calcium phosphate.* (b) A mineral group fluorapatite, chlorapatite, hydroxylapatite, carbonate-fluorapatite (francolite), and carbonate-hydroxylapatite (dahllite).

apex (a) The highest point of a vein relative to the surface, whether it crops out or not. The concept is used in mining law. See also: *apex law.* (AGI, 1987) (b) The tip, summit, or highest point of a landform, as of a mountain; specif. the highest point on an alluvial fan, usually the point where the stream that formed the fan emerged from the mountain or from confining canyon walls. Syn: *culmination.* (AGI, 1987) (c) The highest point of a stratum, as a coalbed. (Standard, 1964) (d) The top of an anticlinal fold of strata. (e) In U.S. mining law, used to designate the highest limit of a vein. (Ballard, 1955) (f) The top of an inclined haulage plane. See also: *brow; landing.* (Nelson, 1965) (g) Point in the center of the face of a concave, noncoring bit. (Long, 1960) (h) In a classifier or hydrocyclone, the underflow aperture through which the coarser and heavier fraction of the solids in a pulp is discharged in accordance with its minimum cross section. (Pryor, 1963)

apex law (a) This law gives the owner of a properly located claim on a vein the right to an indefinite extension on the dip of the vein beyond the vertical planes through the side lines of the claim. In order to secure this right, the owner must lay out the end lines of the claim parallel and of substantial length. A triangular claim would have no apex right and cannot be patented. (Lewis, 1964) (b) Obsolescent mining law allowing the owner of a lode to follow it in depth, regardless of the vertical extension of the legal surface boundaries. (Pryor, 1963) (c) In U.S. mining law, the individual whose claim contains the apex of a vein may follow and exploit the vein indefinitely along its dip, even if it passes downdip under adjoining surface property lines. Syn: *law of extralateral rights.* See also: *apex.* (AGI, 1987)

aphanesite *clinoclase.*

aphanite Any fine-grained igneous rock whose components are not distinguishable with the unaided eye; a rock having aphanitic texture. Cf: *aphanitic.* Syn: *cryptomere; felsite; felsitoid.* (AGI, 1987)

aphanitic (a) Said of the texture of an igneous rock in which the crystalline components are not distinguishable by the unaided eye; also said of a rock or a groundmass exhibiting such texture. Cf: *aphanite; phaneritic.* Syn: *fine-grained.* (AGI, 1987) (b) A crystalline texture with individual crystals too small to be visible to the unaided eye. Syn: *cryptocrystalline.*

aphanophyre A porphyritic igneous rock having a groundmass which the unaided eye cannot distniguish as either crystalline or noncrystalline. (CIPW, 1903)

aphothonite A steel-gray argentiferous variety of tetrahedrite. (Standard, 1964)

aphrite A foliated or scaly white pearly variety of calcite. Syn: *earth foam; foam spar.* (Standard, 1964; Fay, 1920)

aphrodite *stevensite.*

aphyric Said of the texture of a fine-grained or aphanitic igneous rock that lacks phenocrysts. Also, said of a rock exhibiting such texture. (AGI, 1987)

aplanachromatic lens A lens free from both chromatic aberration and spherical aberration. See also: *achromatic.* Cf: *aplanatic lens; aberration; achromatic.*

aplanatic lens A lens free from spherical aberration. Cf: *aberration.* See also: *aplanachromatic lens.*

aplite A light-colored igneous rock characterized by a fine-grained saccharoidal (i.e., aplitic) texture. Aplites may range in composition from granitic to gabbroic, but the term aplite with no modifier is generally understood to mean granitic aplite, consisting essentially of quartz, potassium feldspar, and acid plagioclase. The term, from a Greek word meaning simple, was in use before 1823. Syn: *haplite.* The rock has been used in glass manufacture. (AGI, 1987)

aplitic (a) Pertaining to the fine-grained and saccharoidal texture characteristic of aplites. (AGI, 1987) (b) Said of an igneous rock having such a texture. (AGI, 1987)

aplogranite A light-colored rock of granitic texture consisting essentially of alkali feldspar and quartz, with subordinate biotite; muscovite may be present or absent. Cf: *two-mica granite; alaskite.* (Holmes, 1920)

Apocal A nongelatinous permissible explosive. Used in coal mining. (Bennett, 1962)

Apold-Fleissner process A method of roasting carbonate iron ore in a shaft furnace. The ore sinks continuously down the furnace while a current of hot air or flue gas, with a low carbon dioxide content, is passed through the body of the ore and a current of cold air is passed upward through the lower part of the shaft, this part acting as a cooling chamber for the ore and as a preheating flue for the air, which rapidly oxidizes the ferrous oxide in the upper regions of the furnace. The quantity and temperature of the hot gases and cold air are carefully regulated, so as to keep the carbon dioxide content of the flue gas at a minimum and thereby ensure thorough roasting of the ore at the lowest possible temperature. A furnace roasting 181 to 408 t/d requires about 176,400 to 220,500 kg·cal/t (736 to 923 kg·kJ/t), giving a heat efficiency of 73%. (Osborne, 1956)

apomagmatic Said of a hydrothermal mineral deposit at an intermediate distance from its magmatic source. The term is little used. Cf: *telemagmatic; cryptomagmatic.* (AGI, 1987)

apophyllite A mineral group, $2[KFCa_4(Si_8O_{20})\cdot 8H_2O]$ (fluorapophyllite) with F replaced by (OH) (hydroxyapophyllite) and K replaced by Na (natroapophyllite); occurs in square micaceous crystals as secondary minerals in cavities in igneous rocks. Syn: *fisheye stone.*

apophysis *tongue.*

aporhyolite A rhyolite, the groundmass of which was once glassy but has become devitrified.

Appalachian coalfield The coal-producing area extending from northern Pennsylvania to Alabama, in and adjacent to the Appalachian Mountains.

Appalachian orogeny (a) Late Paleozoic Era diastrophism beginning perhaps in the Late Devonian Period and continuing until the end of the Permian Period. (AGI, 1987) (b) A period of intense mountain-building movements in the late Paleozoic Era, during which the deposits in the Appalachian and Cordilleran geosynclines were folded to form the Appalachian and Palaeocordilleran mountains. Equivalent to the Armorican and Hercynian movements in Europe. Syn: *Appalachian revolution.* (CTD, 1958)

Appalachian revolution *Appalachian orogeny.*

apparent cohesion (a) In soil mechanics, the resistance of particles to being pulled apart, due to the surface tension of the moisture film surrounding each particle. Also called moisture film cohesion. (Hunt, 1965) (b) Cohesion in granular soils due to capillary forces. See also: *cohesion.* (ASCE, 1958)

apparent density (a) The weight (W) of an object or material divided by its exterior volume (V_e) less the volume of its open pores (V_p). Apparent density = $W/(V_e - V_p)$. (ACSG, 1963) (b) Weight per apparent volume. See also: *density.* Cf: *bulk density.* (Van Vlack, 1964)

apparent dip The dip of a rock layer as measured in any exposed section, or direction, not at a right angle to the strike. It is a component of, and hence always less than, the true dip. See also: *angle of dip; true dip; dip.* (Stokes, 1955)

apparent movement of a fault The apparent movement observed in any chance section across a fault is a function of several variables: the attitude of the fault; the attitude of the disrupted strata; the attitude of the surface upon which the fault is observed; and the true movement (net slip) along the fault. (AGI, 1987)

apparent plunge The inclination of a normal projection of lineation in the plane of a vertical cross section. Cf: *plunge.* (AGI, 1987)

apparent porosity The ratio of the volume of open pore space in a specimen to the exterior volume.

apparent resistivity The measured electrical resistivity between two points on the Earth's surface, which corresponds to the sensitivity the ground would have if it were homogeneous.

apparent specific gravity (a) Specific gravity of a rock as measured by water displacement, taking into account the effect of sealed pore spaces as well as constituent minerals. See also: *specific gravity.* (b) The ratio of the weight in air of a given volume of the impermeable portion of a permeable material (e.g., the solid matter including its impermeable pores or voids) at a stated temperature to the weight in air of an equal volume of distilled water at a stated temperature. (ASCE, 1958) (c) This property is determined by the standard method of dividing the weight of a rock by the weight of an equal volume of water. The term apparent specific gravity is used because water cannot penetrate the closed pore spaces inside the rock, and hence the specific gravity measured by water displacement methods includes the effect of internal pore spaces as well as that of the constituent minerals. (Lewis, 1964)

apparent superposition The actual or visible order in which strata lie in any locality. (Standard, 1964)

apparent velocity The velocity with which a seismic-signal wavefront appears to travel along the surface of the Earth. It exceeds the actual velocity if the wave train is not traveling parallel to the surface. (AGI, 1987)

apparent volume True volume plus closed-pore volume. (Van Vlack, 1964)

apparent width The width of a vein or other tabular formation as determined by borehole intercepts. This width will always be greater than the true width if the borehole intersects the vein at any direction other than perpendicular to the surface of the vein. Cf: *true width.* (Long, 1960)

appliances of transportation As applied to a coal mine, these include parts of the locomotive, mobile conveyor, and elevator transportation systems for the removal of coal.

Appolt oven An oven for the manufacture of coke, differing from the Belgian oven in that it is divided into vertical compartments.

appraisal The estimation or fixing of a money value on anything, such as a gemstone. Differs from valuation and evaluation.

approach distance The linear distance, in the direction of feed, between the point of initial cutter contact and the point of full cutter contact. (ASM, 1961)

appropriation In the mining law, the posting of notice at or near the point where the ledge is exposed; next, the recording of the notice; next, the marking of the boundaries. (Ricketts, 1943)

approved permissible flame safety lamp A flame safety lamp that has been approved for use in gaseous coal mines.

approximate original contour The surface configuration achieved by backfilling and grading of the mined area so that the reclaimed area, including any terracing or access roads, closely resembles the general surface configuration of the land prior to mining and blends into and complements the drainage pattern of the surrounding terrain, with all highwalls, spoil piles, and coal refuse piles eliminated.

Apricotine Trade name for yellowish-red, apricot-colored quartz pebbles; may be of gem quality; near Cape May, NJ.

apron (a) A canvas-covered frame set at such an angle in a miner's rocker that the gravel and water in passing over it are carried to the head of the machine. (b) An amalgamated copper plate placed below a stamp battery, over which pulp passes. The free gold contained in the pulp is amalgamated by mercury on the plate. (c) A broad shallow vat used for evaporating. (Webster 3rd, 1966) (d) A receptacle or endless belt for conveying material (such as rock) by means of a cableway and trolley. Syn: *traveling apron.* (Webster 3rd, 1966) (e) The front gate of a scraper body. (Nichols, 1976) (f) *morainal apron.*

apron conveyor (a) A series of overlapping metal plates or aprons running in an endless chain for transferring material from one place to another. Often used to feed raw material from a bin. (ACSG, 1963) (b) A conveyor so contrived as to provide a moving platform on which materials can be carried. Syn: *hinged apron.* (Zern, 1928)

apron feed A method of feeding material forward on an articulated platform. (Nelson, 1965)

apron feeder A feeder in which the material is carried on an apron conveyor and in which the rate of feed is adjusted either by varying the depth of material or the speed of the conveyor, or both. See also: *conveyor-type feeder.* Also called plate-belt feeder; plate feeder. (BS, 1962)

apron plate Sheet of copper or special alloy set in front of a stamp battery and coated with mercury to trap and amalgamate gold. (Pryor, 1963)

apron rope The operating rope for the blade front of a scraper. (Hammond, 1965)

apron wall That part of a panel wall between the windowsill and the support of the panel wall. (ACSG, 1961)

apyrous (a) Not changed by extreme heat, e.g., mica; distinguished from refractory. (Standard, 1964) (b) Noncombustible. (Webster 3rd, 1966)

AQ A letter name specifying the dimensions of bits, core barrels, and drill rods in the A-size and Q-series wireline diamond drilling system having a core diameter of 27 mm and a hole diameter of 48 mm. (Cumming, 1981)

aqua ammonia Ammonia water; esp., a solution of ammonia containing 10% ammonia by weight.

aqua regia A very corrosive, fuming, yellow liquid made by mixing nitric and hydrochloric acids, usually in the proportion of one part by volume of pure nitric acid with three parts by volume of pure hydrochloric acid. Used in dissolving metals such as gold and platinum and in etching. Syn: *nitrohydrochloric acid; nitromuriatic acid.*

aquarium test A test conducted by detonating a standard quantity of explosives under water and measuring both the detonation and gas pressures using transducers; useful for evaluating the relative strengths of various explosives. (Du Pont, 1977)

aqueous (a) Of, or pertaining to, water. (AGI, 1987) (b) Made from, with, or by means of water; e.g., aqueous solutions. (AGI, 1987) (c) Produced by the action of water; e.g., aqueous sediments. (AGI, 1987)

aqueous fusion Melting in the presence of water, as a magma. (AGI, 1987)

aqueous liquor In the ion-exchange (IX) process, the feed to the exchange columns. In solvent extraction, the aqueous feed containing the metal values to be extracted into the organic phase.

aquiclude A body of relatively impermeable rock that is capable of absorbing water slowly but does not transmit it rapidly enough to supply a well or spring. Cf: *confining bed.* See also: *aquitard.* (AGI, 1987)

aquifer (a) A formation, a group of ions, or a part of a formation that is water bearing. (AGI, 1987) (b) A stratum or zone below the surface of the Earth capable of producing water, as from a well. (AGI, 1987) (c) An underground stratum that will yield water in sufficient quantity to be of value as a source of supply. An aquifer is not a stratum that merely contains water, for this would apply to all strata in the groundwater area. An aquifier must yield water. See also: *aquitard.* (Carson, 1961)

aquifer test In situ procedure, such as single-well (bail test or slug test) and multiple-well pumping tests, used to determine hydraulic properties of an aquifer. (Freeze, 1979)

aquifuge (a) Suggested by Bedier, as the opposite of aquifer. (AGI, 1987) (b) A rock that contains no interconnected openings or interstices and therefore neither absorbs nor transmits water. Cf: *confining bed.* (AGI, 1987)

aquitard Low-permeability bed, in a stratigraphic sequence, of sufficient permeability to allow movement of contaminants, and to be relevant to regional ground-water flow, but of insufficient permeability for the economic production of water. See also: *aquifer; aquiclude.* Cf: *confining bed.* (Freeze, 1979)

aragonite (a) An orthorhombic mineral, $4[CaCO_3]$; acicular, pyramidal, tabular, reniform, columnar, or stalactitic habit; formed from hot carbonated water in springs, cavities in basalt, or biogenetically in shells and pearls (mother of pearl). Syn: *aragon spar.* (b) The mineral group aragonite, cerussite, strontianite, and witherite.

aragonite sand Sand-size grains of predominantly aragonite ($CaCO_3$) found in shallow, tropical waters. Aragonite forms by chemical precipitation in sea water due to the presence of SO_4 ions.

Aragon spar Former name for aragonite.

aramayoite A triclinic mineral, $Ag(Sb,Bi)S_2$; iron black with perfect cleavage; at Chocaya, Bolivia.

arbitrage An operation that involves a purchase in one market with the simultaneous sale of an equivalent quantity in another market, (e.g., the London Metal Exchange and the New York Commodity Exchange), and the necessary foreign exchange transaction to protect against any change in the parities between the two currencies involved. (Wolff, 1987)

arborescent Applied to minerals having a treelike form, esp. when fairly massive. If the mineral formation is so thin as to resemble a painting of a tree, it is generally called dendritic. Syn: *dendriform; dendritic.* (Fay, 1920)

arcanite An orthorhombic mineral, K_2SO_4.

arc cutter A device consisting of a bit attached to knuckle-jointed rods used to drill a curved borehole or branched holes from a parent borehole. Syn: *Thompson arc cutter.* Cf: *whipstock.* (Long, 1960)

arc furnace A furnace in which material is heated either directly by an electric arc between an electrode and the work, or indirectly by an arc between two electrodes adjacent to the material. (ASM, 1961)

arch (a) A portion of rock left standing at the intersection of a mine wall and roof, to support the roof. (b) Curved roof of underground opening. See also: *dome.* (c) A curved structural member used to span openings or recesses; also built flat. Structurally, an arch is a piece or assemblage of pieces so arranged over an opening that the supported load is resolved into pressures on the side supports and practically normal to their faces. (ACSG, 1961) (d) A part of a furnace; a crown. (ASTM, 1994) (e) To heat a pot in a pot arch. (ASTM, 1994) (f) One of the five chambers of a brick kiln; also, the fire chamber in certain kinds of furnaces and ovens. (Webster 3rd, 1966) (g) The roof of a reverberatory furnace.

arch blocks Applied to the wooden voussoirs used in framing a timber support for the tunnel roof, when driving a tunnel on the so-called American system. These blocks are made of plank, superimposed in three or more layers, and a breaking joint. (Stauffer, 1906)

Archean Said of the rocks of the Archeozoic. (AGI, 1987)

arched Corn. Said of the roads in a mine, when built with stones or bricks.

Archeozoic The earlier part of Precambrian time, corresponding to Archean rocks. Also spelled: Archaeozoic. (AGI, 1987)

arch forms Forms or patterns on which sprung arch bricks are laid to ensure the proper arch contour.

arch girder A normal H-section steel girder bent to a circular shape. The usual form consists of halves joined together at the crown by bolts and two fishplates. The arch girder is usually splay legged or straight legged in shape, but horseshoe shapes are also in use. See also: *steel support; wood stilt.* (Nelson, 1965)

Archimedes' principle The statement in fluid mechanics that a fluid buoys up a completely immersed solid so that the apparent weight of the solid is reduced by an amount equal to the weight of the fluid that it displaces. (AGI, 1987)

arching (a) Arch. (b) Curved support for roofs of openings in mines; constructed archways in masonry. (c) The development of peripheral cracks around an excavation due to the difference in stress between the skin rock and the rock in the stress ring. See also: *V-arching*. (Spalding, 1949) (d) The folding of schists, gneisses, or sediments into anticlines. (e) The transfer of stress from a yielding part of a soil or rock mass to adjoining less yielding or restrained parts of the mass. (ASCE, 1958) (f) The fretting away of the periphery of a rock tunnel, usually converting it from a rectangular to a circular or elliptical section. The effect in the back is sometimes referred to as the "natural arch." The putting in of a lining built to an arch shape should not be referred to as arching but as "lining" or "putting in the arch.". (Spalding, 1949)

arching action The natural process by which a fractured, pulverulent, or plastic material acquires a certain amount of ability to support itself partially through the resolution of the vertical component of its weight into diagonal thrust. (Woodruff, 1966)

arching to a weakness *V-arching.*

arch rib The main load-bearing member of a ribbed arch. (Hammond, 1965)

arch set Steel assemblies used to support mine workings. (Pryor, 1963)

arch structure *abutment; pressure arch.*

arcose *arkose.*

arc shear machine *universal machine.*

arc shooting A method of refraction seismic prospecting in which the variation of travel time (velocity) with azimuth from a shot point is used to infer geologic structure. The term also applies to a refraction spread placed on a circle or a circular arc with the center at the shot point. (AGI, 1987)

Arctic suite A group of basaltic and associated igneous rocks intermediate in composition between rocks of the Atlantic suite and the Pacific suite. Cf: *Atlantic suite; Pacific suite.* (AGI, 1987)

arcwall machine *slabbing machine.*

arc welding A group of welding processes wherein coalescence is produced by heating with an electric arc or arcs, with or without the application of pressure and with or without the use of filler metal. (Coal Age, 2)

Ardeer double-cartridge test *sensitivity to propagation.*

ardennite (a) A yellow to yellowish-brown vanadiosilicate of aluminum and manganese that crystallizes in the orthorhombic system. (Fay, 1920) (b) An orthorhombic mineral, $Mn_4(Al,Mg)_6(SiO_4)_2(Si_3O_{10})[(As,V)O_4](OH)_6$.

areal geology The branch of geology that pertains to the distribution, position, and form of the areas of the Earth's surface occupied by different types of rock or by different geologic units, and to the making of geologic maps. (Fay, 1920)

areal map A geologic map showing the horizontal area or extent of rock units exposed at the surface. (AGI, 1987)

areal pattern A dispersion pattern resulting from widespread rock alteration. Such patterns may outline the boundaries of a group of deposits and thus limit the area that it is necessary to prospect in detail.

area of airway In mine ventilation, the cross-sectional area of the entry or duct through which the air flows; expressed in square meters.

area of influence of a well The area surrounding a well within which the piezometric surface has been lowered when pumping has produced the maximum steady rate of flow. (ASCE, 1958)

area of settlement The surface area affected by subsidence. (Briggs, 1929)

arenaceous Said of a sediment or sedimentary rock consisting wholly or in part of sand-sized fragments, or having a sandy texture or the appearance of sand; pertaining to sand or arenite. Also said of the texture of such a sediment or rock. The term implies no special composition and should not be used as a syn. of siliceous. Syn: *sandy.* (AGI, 1987)

areng A Bornean term for a yellowish gravelly earth, sometimes containing diamonds.

arenite (a) A general name for sedimentary rocks composed of sand-sized fragments irrespective of composition; e.g., sandstone, graywacke, arkose, and calcarenite. (AGI, 1987) (b) A clean sandstone that is well sorted, contains little or no matrix material, and has a relatively simple mineralogic composition; specif. a pure or nearly pure, chemically cemented sandstone containing less than 10% argillaceous matrix and inferred to represent a slowly deposited sediment well-washed by currents.—Etymol: Latin arena, sand. Adj. arenitic. See also: *lutite.* (AGI, 1987)

Arents tap An arrangement by which molten lead from the crucible of a shaft furnace is drawn through an inverted siphon into an exterior basin from which it can be ladled without disturbing the furnace. Syn: *siphon tap.* (Fay, 1920)

arfvedsonite A monoclinic mineral, $Na_3(Fe^{+2},Mg)_4Fe^{+3}Si_8O_{22}(OH)_2$, of the amphibole group; dark green to black; in silica-poor igneous rocks.

Argall furnace A reverberatory roasting furnace the hearth of which has a reciprocating movement whereby the ore is caused to move forward by the action of rabbles extending across the hearth. (Fay, 1920)

Argall tubular furnace A tubular roasting furnace consisting of four brick-lined steel tubes 30 ft (9.1 m) long nested together inside two steel tires, which revolve upon steel-faced carrying rolls. (Fay, 1920)

argental mercury *amalgam.*

argentate (a) A salt in which silver acts as an acid radical; e.g., ammonium argentate (fulminating silver). (Standard, 1964) (b) Having a silvery appearance. (CTD, 1958)

argentation The act or process of coating or plating with silver. (Standard, 1964)

argentiferous Containing silver.

argentiferous galena *silver lead ore.*

argentiferous lead Lead that contains silver. (CMD, 1948)

argentine (a) A lamellar variety of calcite with a pearly white luster. (b) Silver-coated white metal. (Standard, 1964) (c) A finely divided tin moss or sponge obtained from a solution of tin by precipitation with zinc. (Standard, 1964) (d) Adj. pertaining to, containing, or resembling silver; silvery. (AGI, 1987)

argentite An orthorhombic mineral, Ag_2S; isometric above 180 °C; dimorphous with acanthite; massive or as coating; metallic lead-gray; soft, sectile; sp gr, 7.3; in veins with other silver and sulfide minerals; commonly pseudomorphous after acanthite; an important ore of silver. Syn: *silver glance; vitreous silver; argyrite.*

argentojarosite A trigonal mineral, $AgFe_3(SO_4)_2(OH)_6$; alunite group; yellow to brown-yellow.

argentopyrite An orthorhombic mineral, $AgFe_2S_3$; dimorphous with sternbergite.

argil (a) Potter's clay; white clay. (Standard, 1964) (b) *aluminite.*

argillaceous (a) Pertaining to, largely composed of, or containing clay-size particles or clay minerals, such as an argillaceous ore in which the gangue is mainly clay; esp. said of a sediment (such as marl) or a sedimentary rock (such as shale) containing an appreciable amount of clay. Syn: *clayey.* See also: *argillic.* (AGI, 1987) (b) Pertaining to argillite. (AGI, 1987)

argillaceous hematite A brown to deep-red variety of natural ferric oxide containing an appreciable portion of clay (or sand). Syn: *ironstone clay.*

argillaceous limestone A limestone containing an appreciable amount (but less than 50%) of clay; e.g., cement rock. (AGI, 1987)

argillaceous ore Ore in which the gangue is mainly clay. (Osborne, 1956)

argillaceous rock A sedimentary rock composed of clay-grade particles; i.e., composed of minute mineral fragments and crystals less than 0.002 mm in diameter; containing much colloidal-size material. In addition to finely divided detrital matter, argillaceous rocks consist essentially of illite, montmorillonite, kaolinite, gibbsite, and diaspore. (CMD, 1948)

argillation The development of clay minerals by the weathering of aluminum silicates. Cf: *kaolinization.* (AGI, 1987)

argillic Pertaining to clay or clay minerals; e.g., argillic alteration in which certain minerals of a rock are converted to minerals of the clay group. Cf: *argillaceous.* (AGI, 1987)

argillite A compact rock, derived either from mudstone (claystone or siltstone), or shale, that has undergone a somewhat higher degree of induration than mudstone or shale but is less clearly laminated and without its fissility, and that lacks the cleavage distinctive of slate. (AGI, 1987)

argillization The replacement or alteration of feldspars to form clay minerals, esp. in wall rocks adjacent to mineral veins. Cf: *kaolinization.* (AGI, 1987)

argon A colorless, odorless, monatomic, inert gas. Symbol, Ar. Obtained by the fractionation of liquid air. Used in electric light bulbs and in fluorescent tubes. (Handbook of Chem. & Phys., 3)

argulite A variety of asphaltic sandstone. (Tomkeieff, 1954)

argyrite Former name for argentite; also called argyrose.

argyrodite An orthorhombic mineral, Ag_8GeS_6; pseudocubic; forms a series with canfieldite.

argyropyrite A discredited term for a silver-iron sulfide, probably argentopyrite.

argyrose Former name for argentite.

argyrythrose Former name for pyrargyrite.

arite A nickel mineral between nickeline and breithauptite in composition.

arithmetic-mean particle diameter A measure of the average particle size obtained by summing the products of the size-grade midpoints times the frequency of particles in each class, and dividing by the total frequency. (AGI, 1987)

Arizona ruby A deep-red or ruby-colored variety of pyrope garnet of igneous origin, Southwestern United States.

arizonite (a) A hexagonal mineral, $Fe_2Ti_3O_9$; in irregular metallic steel-gray masses in pegmatite veins near Hackberry, AZ. Formerly called pseudorutile. (b) A mixture of hematite, rutile, ilmenite, and anatase. (AGI, 1987) (c) An ore of micaceous iron, silver iodide, gold, iron sulfides, and antimony in a vein in Yavapai County, AZ. (AGI, 1987) (d) A hypabyssal rock with 80% quartz, 18% alkali feldspar, and accessory mica and apatite in Arizona (not a rock name in the IUGS classification). (AGI, 1987)
Arkansas diamond A diamond from Murfreesboro, AR.
Arkansas stone A variety of novaculite found in the Ouachita Mountains of western Arkansas. Also, a whetstone made of Arkansas stone. See also: *novaculite.* (AGI, 1987)
arkansite A brilliant, iron-black variety of brookite from Magnet Cove, AR. (Fay, 1920)
arkose A feldspar-rich sandstone, typically coarse-grained and pink or reddish, that is composed of angular to subangular grains that may be either poorly or moderately well sorted; usually derived from the rapid disintegration of granite or granitic rocks, and often closely resembles granite; e.g., the Triassic arkoses of the Eastern United States. Quartz is usually the dominant mineral, with feldspar (chiefly microcline) constituting at least 25%. Cement (silica or calcite) is commonly rare, and matrix material (usually less than 15%) includes clay minerals (esp. kaolinite), mica, and iron oxide; fine-grained rock fragments are often present. Arkose is commonly a current-deposited sandstone of continental origin, occurring as a thick, wedge-shaped mass of limited geographic extent (as in a fault trough or a rapidly subsiding basin); it may be strongly cross-bedded and associated with coarse granite-bearing conglomerate, and it may denote an environment of high relief and vigorous erosion of strongly uplifted granitic rocks in which the feldspar was not subjected to prolonged weathering or transport before burial. Arkose may also occur at the base of a sedimentary series as a thin blanketlike residuum derived from and resting on granitic rock. Etymol: French, probably from Greek archaios, ancient, primitive. Syn: *arkosic.* Cf: *graywacke; feldspathic sandstone; subarkose.* Also spelled arcose. (AGI, 1987)
arkose quartzite *arkosite.*
arkosic Having the character of arkose.
arkosic sandstone A sandstone with considerable feldspar, such as one containing minerals derived from coarse-grained quartzo-feldspathic rocks (granite, granodiorite, gneiss) or from highly feldspathic sedimentary rocks; specif. a sandstone containing more than 25% feldspar and less than 20% matrix material of clay, sericite, and chlorite. See also: *arkosite.* (AGI, 1987)
arkosite A quartzite with a notable amount of feldspar. Syn: *arkose quartzite.* See also: *arkosic sandstone.*
arm The inclined member or leg of a set or frame of timber.
armangite A trigonal mineral, $Mn_{26}As_{18}O_{50}(OH)_4(CO_3)$; black; near Långban, Sweden.
arm conveyor A conveyor consisting of an endless belt, or one or more chains, to which are attached projecting arms, or shelves, for handling packages or objects in a vertical or inclined path.
armenite (a) A hexagonal mineral, $BaCa_2Al_6Si_9O_{30}\cdot 2H_2O$; osumilite group. (b) Former name for azurite, Armenian stone.
armor An outer cable covering that may be either metallic or nonmetallic.
armored apron An apron in which each pan is provided with a separate wearing plate.
armored cable A cable that is wrapped with metal, usually steel wires or tapes, primarily for physical protection. See also: *cable.*
armored flexible conveyor A heavy, chain-type flexible conveyor capable of being advanced with the face without dismantling. It is designed either to carry a coal cutter or a cutter loader or to guide and hold a plow against the face. It may be advanced by horizontal hydraulic rams that are fixed at about 20-ft (6-m) intervals on the waste side of the conveyor. It is often employed on prop-free-front faces with hand filling, and it has a capacity of about 200 to 300 st/hr. Syn: *Panzer conveyor.* See also: *conveyor; chain conveyor; face conveyor.* (Nelson, 1965)
armored relict An unstable relict enveloped by a crystal or by a reaction shell which revented its reaction with the other constituents of the rock. See also: *unstable relict.* (Schieferdecker, 1959)
Armstrong air breaker *compressed-air blasting.*
arnimite Orthorhombic $Cu_5(SO_4)_2(OH)_6\cdot 3H_2O$; perhaps the mineral antlerite.
arochlors Chlorinated diphenyl materials that are useful as vehicles for pigments used in glass decoration since they volatilize without leaving a carbon residue. Arochlors provide a grinding and dispersing medium for nonaqueous slurries of pigments and ceramic bodies; also, they can be used in combination with waxes to provide moisture-proof coatings. (Lee, 1961)
aromatic compound A compound derived from the hydrocarbon benzene, C_6H_6, distinguished from that derived from methane, CH_4. (Standard, 1964)
aromatic hydrocarbon A compound of carbon and hydrogen that contains in its molecular structure a closed and saturated ring of carbon atoms; e.g., benzene, naphthalene, and anthracene. (Hackh, 1944)
aromatite A bituminous stone resembling a fragrant gum resin in color and odor. It was a precious stone in ancient Arabia and Egypt. (AGI, 1987)
arquerite A soft, malleable, silver-rich variety of amalgam containing about 87% silver and 13% mercury; from Coquimbo, Chile.
arrastra *arrastre.*
arrastre A circular rock-lined pit in which broken ore is pulverized by stones attached to horizontal poles fastened in a central pillar and dragged around the pit. Also spelled arrastra. (Weed, 1922; CTD, 1958)
arrested decay A stage in coal formation when biochemical action ceases. (Tomkeieff, 1954)
arrester (a) Any mechanical contrivance or device used to stop or slow up motion. (Crispin, 1964) (b) Mechanism for the purification of a gas stream that may contain suspended liquids or solids. (Bennett, 1962)
arrival dealings Dealing in ores, concentrates, and metals in transit from source to market. (Pryor, 1963)
arrojadite A monoclinic mineral, $KNa_4CaMn_4Fe_{10}Al(PO_4)_{12}(OH,F)$; dark green, forms a series with dickinsonite.
arrow A sharp-pointed, thin metal rod about 1 to 2 ft (0.3 to 0.6 m) long with a ring at the other end, used in surveying; a thin metal peg. (Mason, 1951)
arroyo (a) A term applied in the arid and semiarid regions of the Southwestern United States to the small, deep, flat-floored channel or gully of an ephemeral stream, usually with vertical or steeply cut banks of unconsolidated material at least 60 cm high; it is usually dry, but may be transformed into a temporary watercourse or short-lived torrent after heavy rainfall. Cf: *dry wash.* (AGI, 1987) (b) The small intermittent stream or rivulet that occupies such a channel.—Etymol: Spanish, stream, brook; gutter, watercourse of a street. See also: *wadi; nullah.* (AGI, 1987)
arsenargentite Possibly a silver arsenide.
arsenate (a) A salt or ester of an arsenic acid; a compound containing one of the three radicals in which arsenic has a +5 valence: ortho-arsenate, AsO_4; meta-arsenate, AsO_3; pyro-arsenate, As_2O_7. (b) A mineral characterized by pentavalent arsenic and oxygen; e.g., mimetite $Pb_5(AsO_4)_3Cl$. Cf: *vanadate.*
arsenic A metallic, steel-gray, brittle element. Symbol, As. Found native in realgar and orpiment, and combined with heavy metals. Used in bronzing, pyrotechny, insecticides, and poisons, and as a doping agent in transistors. Gallium arsenide is used as a laser material to convert electricity directly into coherent light. Arsenic and its compounds are poisonous. (Handbook of Chem. & Phys., 3)
arsenical antimony *allemontite.*
arsenical nickel *nickeline; niccolite.*
arsenical pyrite *arsenopyrite.*
arsenic bloom *arsenolite; pharmacolite.*
arsenicite *pharmacolite.*
arsenic trioxide A white, odorless, tasteless powder; AsO_3. Used in the manufacture of pigments, glass, and other arsenic compounds, ceramic enamels, and aniline colors; mixed with soda ash for boiler compounds. Syn: *white arsenic; arsenious oxide.* (CCD, 1961)
arsenious oxide *arsenic trioxide.*
arsenite A mineral characterized by trivalent antimony and oxygen; e.g., trigonite, $Pb_3Mn(AsO_3)_2(AsO_2(OH)$.
arsenobismite A yellowish-green mineral, $Bi_2(AsO_4)(OH)_3$.
arsenoclasite An orthorhombic mineral, $Mn_5(AsO_4)_2(OH)_4$; red, from Långban, Sweden. Also spelled arsenoklasite.
arsenolamprite An orthorhombic mineral arsenic; dimorphous with arsenic.
arsenolite An isometric mineral, As_2O_3; dimorphous with claudetite. Syn: *arsenic bloom.*
arsenopyrite (a) A monoclinic mineral, 8[FeAsS]; pseudo-orthorhombic, prismatic, and metallic silver-white to steel gray; the most common arsenic mineral and principal ore of arsenic; occurs in many sulfide ore deposits, particularly those containing lead, silver, and gold. Syn: *mispickel; arsenical pyrite; white pyrite; white mundic.* (b) The mineral group arsenopyrite, glaucodot, gudmundite, osarsite, and ruarsite.
arsenosulvanite An isometric mineral, $Cu_3(As,V)S_4$; forms a series with sulvanite. Syn: *lazarevicite.*
arsenpolybasite A monoclinic mineral, $(Ag,Cu)_{16}(As,Sb)_2S_{11}$. Also spelled arsenopolybasite. Cf: *antimonpearceite.*
arsenuranocircite Syn: *heinrichite* It is not clear which name has priority as applied to a natural mineral. See also: *metaheinrichite.* (Hey, 1961)
arsenuranylite The orthorhombic mineral, $Ca(UO_2)_4(AsO_4)_2(OH)_4\cdot 6H_2O$; typically orange-red.
arterial road A main road with secondary roads joining it. (Hammond, 1965)

artesian (a) Refers to ground water under sufficient hydrostatic head to rise above the aquifer containing it. (AGI, 1987) (b) Pertaining to underground water that is confined by impervious rock or other material under sufficient pressure to raise it above the upper level of the saturated rock or other material in which it occurs, if this rock or material is penetrated by wells or natural fissures. Formerly, the term was applied only to water under sufficient pressure to raise it to the surface of the Earth. (Stokes, 1955)

artesian aquifer An aquifer that contains artesian water. (AGI, 1987)

artesian basin A geologic structural feature or a combination of such features in which water is confined under artesian pressure. (AGI, 1987)

artesian discharge The process of discharge from a well by artesian pressure, and also the quantity of water discharged. The artesian pressure is aided by the buoyancy of the natural gas that enters some wells with the water. (Stokes, 1955)

artesian leakage The slow percolation of water from artesian formations into the confining materials of a less permeable but not of a strictly impermeable character. Such percolation causes a reduction in artesian pressure, depending on the relative impermeability of the materials in the confining formations. (AGI, 1987)

artesian spring A spring, the water from which issues under artesian pressure, generally through some fissure or other opening in the confining bed that overlies the aquifer. (AGI, 1987)

artesian water (a) Ground water that is under sufficient pressure to rise above the level at which it is encountered by a well, but that does not necessarily rise to or above the surface of the ground. (AGI, 1987) (b) Ground water that is confined within a permeable bed and that rises under pressure to approx. the height of the intake. If the outlet (well or spring) is appreciably below the height of the intake, the water will flow out under pressure. If even with or above the height of the intake, the water will rise in the well but it will not flow out. (Bateman, 1951)

artesian well (a) A well in which the water level rises above the top of the aquifer, whether or not the water flows at the land surface. (AGI, 1987) (b) Formerly, only applied to a well drilled to a depth where, owing to the structure of the strata, the water pressure was high enough to raise the water to the surface. (Standard, 1964) (c) Often applied to any deep well, even where pumping is necessary, as in an ordinary driven well. See also: *well.* (Standard, 1964)

arthurite (a) An apple-green monoclinic mineral, $CuFe_2(AsO_4,PO_4,SO_4)_2(O,OH)_2{\cdot}4H_2O$. (b) The mineral group arthurite, earlshannonite, ojuelaite, and whitmoreite.

articulite *itacolumite.*

artificial aging Aging above room temperature. See also: *precipitation heat treatment.* (ASM, 1961)

artificial brine Brine produced from an underground deposit of salt or other soluble rock material in the process of solution mining. Cf: *brine.* (AGI, 1987)

artificial horizon A device for indicating the horizontal, as a bubble, gyroscope, pendulum, or the flat surface of a liquid. It is sometimes simply called a horizon. Syn: *false horizon.* (AGI, 1987)

artificial island An island that is constructed by humans rather than formed by natural means, usually in waters less than 30 m deep. In the mining industry they are commonly used to support the construction of service or ventilation shafts for underground mines extending offshore. (Cruickshank, 1987)

artificial liquid fuel Fuel created by the hydrogenation of coal; the destructive distillation of coal, lignite, or shale at low temperature; and by a recombination of the constituents of water gas in the presence of a suitable catalyst.

artificial refractories Materials manufactured in electric furnaces and used for special purposes; e.g., zirconium carbide, titanium carbide, and silicon carbide. (Newton, 1959)

artinite A snow-white monoclinic mineral, $Mg_2(CO_3)(OH)_2{\cdot}3H_2O$.

asbestiform Said of a mineral that is fibrous, i.e., like asbestos.

asbestine (a) A silicate of magnesium much used in paint. It serves as an aid in holding paint pigment in solution and in binding paint films together. Also marketed under such names as French chalk and talc. Syn: *agalite.* (Crispin, 1964) (b) Of, pertaining to, or having the characteristics of asbestos; incombustible. (Webster 3rd, 1966)

asbestos (a) A commercial term applied to silicate minerals that separate readily into thin, strong fibers that are flexible, heat resistant, and chemically inert, thus making them suitable for uses (as in yarn, cloth, paper, paint, brake linings, tiles, insulation, cement, fillers, and filters) where incombustible, nonconducting, or chemically resistant material is required. Since the early 1970's, there have been serious enviromental concerns about the potential health hazards of asbestos products, which has resulted in strong enviromental regulations. (b) Any asbestiform mineral of the serpentine group (chrysotile, best adapted for spinning and the principal variety in commerce) or amphibole group (esp. actinolite, anthophyllite, gedrite, cummingtonite, grunerite, riebeckite, and tremolite). (c) A term strictly applied to asbestiform actinolite. Syn: *asbestus; amianthus; earth flax; mountain flax; rock wool.*

asbestosis A lung disease caused by breathing asbestos dust. (Harrington, 1937)

asbestos minerals Certain minerals that have a fibrous structure, are heat resistant and chemically inert, possess high electrical insulating qualities, and are of sufficient flexibility to be woven. The two main groups are serpentine and amphiboles. Asbestos proper is actinolite. Chrysotile is fibrous serpentine; amosite is fibrous anthophyllite; crocidolite is fibrous soda-amphibole. Used in fireproof buildings, insulating, paint materials, brake linings, and clutches, and as insulation against heat, electricity, and acid. (Pryor, 1963)

asbestos yarn Yarn consisting of asbestos fiber; asbestos and vegetable fibers; asbestos and vegetable fibers and wire; or asbestos and vegetable fibers with an insert of cotton or other yarn reinforcement. Metallic asbestos yarn is yarn consisting of plain asbestos yarn twisted with brass, copper, or other fine wire. (Hess)

asbestus *asbestos.*

asbolan A hexagonal mineral, $(Co,Ni)_{1-y}(MnO_2)_{2-x}(OH)_{2-2y+2x}{\cdot}nH_2O$; a soft, black, earthy aggregate commonly classed as a variety of "wad," the cobalt content reaching as high as 32% (40% cobalt oxide). Syn: *asbolane; asbolite; black cobalt; cobalt ocher; wad.*

asbolane A form of wad; a soft, earthy manganese dioxide, containing up to about 32% cobalt oxide. Sometimes referred to as earthy cobalt. Syn: *asbolite; cobalt ocher.* See also: *asbolan.* (CMD, 1948)

asbolite *asbolan; asbolane.*

ascensional ventilation A mine ventilation system in which the fresh intake air flows down to the bottom end of the workings and then ascends along the faces to the main return. See also: *descensional ventilation; antitropal ventilation; homotropal ventilation.*

ascension theory A theory of hypogene mineral-deposit formation involving mineralizing solutions rising through fissures from magmatic sources in the Earth's interior. Cf: *descension theory.* (AGI, 1987)

ascharite *szaibelyite.*

aschisite An igneous rock with the same chemical composition as its parent magma, i.e., undifferentiated. Cf: *diaschistic.*

aschistic Said of the rock of a minor intrusion that has a composition equivalent to that of the parent magma, i.e., in which there has been no significant differentiation. Cf: *diaschistic.* (AGI, 1987)

ash (a) The inorganic residue after burning, esp. of coal. Ignition generally alters both the weight and the composition of the inorganic matter. See also: *ash yield; extraneous ash; inherent ash.* (AGI, 1987) (b) Fine pyroclastic material (under 2.0-mm diameter; under 0.063-mm diameter for fine ash). The term usually refers to the unconsolidated material but sometimes is also used for its consolidated counterpart, tuff. Syn: *dust; volcanic ash; volcanic dust; pumicite.* (AGI, 1987) (c) Inorganic residue remaining after ignition of combustible substances, determined by definite prescribed methods. (ASTM, 1994)

ashcroftine A tetragonal mineral, $K_5Na_5(Y,Ca)_{12}Si_{28}O_{70}(OH)_2(CO_3)_8{\cdot}3H_2O$; occurs in small pink needles at Narsarsuk, Greenland.

ash curve A graph that shows a relation between the specific gravity of fractions of a coal sample floated in liquids of step-by-step increased density, and the percentage of ash in each such fraction. Syn: *tromp curve.* (Pryor, 1963)

ash drawers Early name applied to tourmaline because of its polar electrostatic property.

ash error The difference between the percentage ash of a product of a separation and that shown by the washability curve (based on the reconstituted feed) of a product with the same properties (usually percentage of ash). (BS, 1962)

ash fusibility A measure, in terms of temperature, of fusion of coal ash prepared and tested under standard conditions. (BS, 1960)

ash-fusion temperature The temperature at which a special test cone made from particles of ash obtained from the coal will (1) begin to deform, i.e., soften, or (2) completely deform or fuse into a blob. (Nelson, 1965)

ashlar Rectangular pieces of stone of nonuniform size that are set randomly in a wall. (AGI, 1987)

ash-specific gravity curve The curve obtained from the float-and-sink analysis by plotting the ash contents of successive fractions against specific gravity. (BS, 1962)

ashstone An indurated deposit of fine volcanic ash. (AGI, 1987)

ash yield The percentage of material remaining after a fuel is burned; that portion of a laboratory sample remaining after heating under standard conditions to constant weight; i.e., until all the combustible matter has been burned away. See also: *ash; extraneous ash; inherent ash.* (Nelson, 1965)

asparagolite *asparagus stone.*

asparagus stone A yellow-green variety of apatite. Syn: *asparagolite.*

aspect (a) The gross or overall lithologic or biologic characteristics of a stratigraphic unit as expressed at any particular point. (AGI, 1987) (b) The angle made by a target with the line joining it to the observation point is known as the aspect of the target. (Hunt, 1965)

asperolite A variety of chrysocolla, containing more than the usual percentage of water.
asphaltic Pertaining to or containing asphalt; e.g., asphaltic limestone or asphaltic sandstone impregnated with asphalt, or asphaltic sand representing a natural mixture of asphalt with varying proportions of loose sand grains. (AGI, 1987)
asphaltic ore Asphaltlike ore carrying invisible uranium values. (Ballard, 1955)
asphaltic rock Any rock naturally impregnated with asphalt. It is generally sandstone or limestone.
asphaltite Any one of the naturally occurring black solid bitumens that are soluble in carbon disulfide and fuse above 230 °F (110 °C). Examples are uintahite, glance pitch, and grahamite. (AGI, 1987)
asphalt rock A porous rock, such as a sandstone or limestone, that is impregnated naturally with asphalt. Syn: *asphalt stone; rock asphalt.* (AGI, 1987)
asphalt stone *asphalt rock.*
asphyxiate To suffocate; to choke. (Mason, 1951)
aspirating *dedusting.*
aspirator An apparatus for moving or collecting gases, liquids, or granular substances by suction. (Webster 3rd, 1966)
assay (a) To analyze the proportions of metals in an ore; to test an ore or mineral for composition, purity, weight, or other properties of commercial interest. Syn: *crucible assay.* (AGI, 1987) (b) The test or analysis itself; its results. (AGI, 1987)
assay balance A very sensitive balance used in the assaying of gold, silver, etc., for weighing the beads. It usually has magnifying lenses for reading the graduations. *balance.* (Webster 3rd, 1966)
assayer Person who analyzes ores and alloys, esp. bullion, to determine the value and properties of their precious metals. (DOT, 1949)
assay foot In determining the assay value of an orebody, the multiplication of its assay grade by the number of feet along which the sample was taken. Cf: *assay inch; assay value.* (AGI, 1987)
assay grade The percentage of valuable constituents in an ore, determined from assay. Cf: *assay value; value.* (AGI, 1987)
assay inch In determining the assay value of an orebody, the multiplication of its assay grade by the number of inches along which the sample was taken. Cf: *assay foot; assay value.* (AGI, 1987)
assay office A laboratory for examining ores, usually gold and silver ores, in order to determine their economic value. (Standard, 1964)
assay plan Map of a mine showing the assay, stope, width, etc., of samples taken from positions marked. Used to control grade and quality of ore mined and milled. (Pryor, 1963)
assay plan factor In sampling, a term used to describe the rate that the head value bears to the mine sampling. This percentage figure is useful in reducing any extant or subsequent mine-sampling average to that which in actual production it will likely prove to be. In South Africa this is generally known as the "mine call factor." Syn: *correction factor.* (Truscott, 1962)
assay split Agreed average value, as between buyer's and seller's assay, used as pricing basis in sale of mineral. (Pryor, 1963)
assay ton A weight of 29.166+ g, used in assaying to represent proportionately the assay value of an ore. Because it bears the same ratio to 1 mg that a ton of 2,000 lb bears to the troy ounce, the weight in milligrams of precious metal obtained from an assay ton of ore equals the number of ounces to the ton. Abbrev. AT. (AGI, 1987)
assay value (a) The quantity of an ore's valuable constituents, determined by multiplying its assay grade or percentage of valuable constituents by its dimensions. Cf: *assay inch; assay foot.* The figure for precious metals is generally given in troy ounces per ton of ore, or per assay ton. See also: *assay grade; value.* (AGI, 1987) (b) The monetary value of an orebody, calculated by multiplying the quantity of its valuable constituents by the market price. Syn: *average assay value.* (AGI, 1987)
assay walls The outer limits to which an orebody can be profitably mined, the limiting factor being the metal content of the country rock as determined from assays. (Hess)
assembled stone Any stone constructed of two or more parts of gem materials, whether genuine, synthetic, imitation, or a combination thereof; e.g., a doublet or triplet. Syn: *composite stone; imitation.*
assembly rod An external bolt holding a machine together. (Nichols, 1976)
assessment *assessment work.*
assessment drilling Drilling done to fulfill the requirement that a prescribed amount of work be done annually on an unpatented mining claim to retain title. (Long, 1960)
assessment labor Refers to the annual labor required of the locator of a mining claim after discovery (and not to work done before discovery). (Ricketts, 1943)
assessment work The annual work upon an unpatented mining claim on the public domain necessary under U.S. law for the maintenance of the possessory title thereto. This work must be done each year if the claim is to be held without patenting. Syn: *assessment; location work.* (Lewis, 1964)
asset Property with cash sale value. In mining, the dominant asset is the proved ore reserve. (Pryor, 1963)
assigned protection factor (a) The expected workplace level of respiratory protection that would be provided by a properly functioning respirator or a class of respirators to properly fitted and trained users. Abbrev. APF. (ANSI, 1992) (b) The minimum anticipated protection provided by a properly functioning respirator or class of respirators to a given percentage of properly fitted and trained users. (NIOSH, 1987)
assimilation The incorporation and digestion of solid or fluid foreign material, such as wall rock, in magma. The term implies no specific mechanisms or results. Such a magma, or the rock it produces, may be called hybrid or contaminated. See also: *hybridization.* Cf: *differentiation.* Syn: *magmatic assimilation; magmatic dissolution.* (AGI, 1987)
assistant mine foreman A person employed to assist the mine foreman in the performance of his or her duties and to serve in his or her place, in the absence of the mine foreman.
Assmann psychrometer A wet-and-dry-bulb hygrometer in which air is drawn over the thermometer bulbs by an integral fan. (BS, 1963)
association *rock association.*
association placer location A placer location made by an association of persons in one location covering 160 acres (64 ha) is not eight locations covering 20 acres (8 ha) each. It is in law a single location, and as such a single discovery is sufficient to support such a location; the only assessment work required is as for a single claim. (Ricketts, 1943)
assured mineral *reserves.*
astatic Not taking a fixed or definite position or direction; as an instrument in which a negative restoring force has been applied so as to aid any deflecting force, thereby rendering the instrument more sensitive and/or less stable. (AGI, 1987)
astatic gravimeter A gravity meter or gravimeter constructed so that a high sensitivity is achieved at certain positions of the elements of the system; i.e., equilibrium between a negative restoring force and the force of gravity at such positions. See also: *gravimeter.* (AGI, 1987)
astatic pendulum A pendulum having almost no tendency to take a definite position of equilibrium. (AGI, 1987)
astatization The application of a restoring force to a moving element of a physical system in such a manner as to drive the moving element away from its rest position and to aid any deflecting force, so as to increase sensitivity. (AGI, 1987)
asteria Any gemstone that, when cut en cabochon in the correct crystallographic direction, displays a rayed figure (a star) by either reflected or transmitted light; e.g., star sapphire. Syn: *star stone.* See also: *star sapphire.*
asteriated (a) Like a star, with rays diverging from a center. (b) Said of a mineral, crystal, or gemstone that exhibits asterism; e.g., asteriated beryl. Syn: *star.*
asteriated quartz Quartz having whitish or colored radiations within the crystals. See also: *star quartz.* (Standard, 1964)
asteriated topaz Asteriated yellow variety of corundum, wrongly called Oriental topaz. (Schaller, 1917)
asterism (a) Starlike rays of light observed in some minerals when viewed from certain directions, particularly if the mineral is cut en cabochon. Minerals having this feature are called asteriated or star. Asteriated beryl, chrysoberyl, crocidolite, emerald, quartz, ruby, and sapphire are known. (Hess) (b) A starlike effect observed in certain minerals either by transmitted or by reflected light. (AGI, 1987) (c) Elongation of Laue X-ray diffraction spots produced by stationary single crystals as a result of internal crystalline deformation. The size of the Laue spot is determined by the solid angle formed by the normals to any set of diffracting planes; this angle increases with increasing crystal deformation, producing progressively elongated (asteriated) spots. Measurements of asterism are used as indicators of deformation in crystals subjected to slow stress or to shock waves. Cf: *corundum cat's eye.*
asthenolith A body of magma that was formed by melting in response to heat generated by radioactive disintegration. (AGI, 1987)
asthenosphere The layer or shell of the Earth below the lithosphere, which has reduced yield strength, permitting viscous or plastic flow under relatively small stresses; it is a zone in which isostatic adjustments take place, magmas may be generated, and seismic waves are strongly attenuated. It is a part of the upper mantle. Syn: *zone of mobility.* See also: *stereosphere.* (AGI, 1987)
ASTM coal classification A system based on proximate analysis in which coals containing less than 31% volatile matter on the mineral-matter-free basis (Parr formula) are classified only on the basis of fixed carbon; i.e., 100% volatile matter. They are divided into five groups: above 98% fixed carbon; 98% to 92% fixed carbon; 92% to 86% fixed carbon; 86% to 78% fixed carbon; and 78% to 69% fixed carbon. The first three of these groups are called anthracites, and the last two are called bituminous coals. The remaining bituminous coals, the subbituminous coals, and the lignites are then classified into groups as

determined by the calorific value of the coals containing their natural bed moisture; i.e., the coals as mined but free from any moisture on the surface of the lumps. The classification includes three groups of bituminous coals with moist calorific value from above 14,000 Btu/lb (32.5 MJ/kg) to above 13,000 Btu/lb (30.2 MJ/kg); three groups of subbituminous coals with moist calorific value below 13,000 Btu/lb to below 8,300 Btu/lb (19.3 MJ/kg); and two groups of lignitic coals with moist calorific value below 8,300 Btu/lb. The classification also differentiates between consolidated and unconsolidated lignites and between the weathering characteristics of subbituminous and lignitic coals. See also: *coal classification systems.* (Francis, 1965)

astochite A blue to gray-violet variety of amphibole; at Wermland, Sweden. Locally known as blue rhodonite. Syn: *soda richterite.*

astrakanite *blodite.*

astridite An ornamental stone, consisting mainly of chromojadeite. From Manokwari, New Guinea. (English, 1938)

astringent (a) A taste that puckers the mouth; descriptive of certain minerals, such as alum. (Fay, 1920) (b) Causing contraction, shrinking, or puckering. (Webster 3rd, 1966) (c) Said of a clay containing an astringent salt.

astrophyllite (a) A triclinic mineral, $(K,Na)_3(Fe,Mn)_7Ti_2Si_8O_{24}(O,OH)_7$; forms a series with kupletskite. (b) A mineral group.

asymmetrical (a) Without symmetry. (b) Said of mineral crystals having no center, plane, or axis of symmetry.

asymmetrical vein A vein with unlike mineral sequences on either side.

asymmetric class The class of crystal forms without any symmetry. (Fay, 1920)

asymmetric fold A fold in which one limb dips more steeply than the other. If one limb is overturned, the term "overturned fold" or "overfold" is used. Cf: *symmetrical fold.* (AGI, 1987)

asymmetric unit The whole group of prototype atoms that, where repeated by the symmetry operations of a space group, generate a crystal structure. Cf: *unit cell.*

atacamite An orthorhombic mineral, $4[Cu_2Cl(OH)_3]$; trimorphous with paratacamite and botallackite; grass green, in fine crystal aggregates, fibrous or columnar; a supergene mineral in oxidized zones of copper deposits in desert regions; a source of copper. Syn: *remolinite.*

ataxic Said of an unstratified mineral deposit. Cf: *eutaxic.* (AGI, 1987)

atelestite A monoclinic mineral, $Bi_8(AsO_4)_3O_5(OH)_5$; yellow.

at grade *graded.*

Atkinson The resistance of a section of roadway in which there is a pressure of 1 lb/ft^2 (6.9 kPa) throughout the section, when an amount of 1,000 ft^3/s (1 kilocusec or 28.3 m^3/s) of dry air at 60 °F (15.6 °C) and 30 in (762 mm) barometer is passing. See also: *Atkinson's friction coefficient.* (Nelson, 1965)

Atkinson's friction coefficient The measure of the pressure expended per 1,000 ft/min per square foot of surface traversed in order to create motion under the conditions prevailing. It is expressed as pounds per square foot per 1,000 ft/min. See also: *Atkinson.* (Nelson, 1965)

Atlantic suite One of two large groups of igneous rocks, characterized by alkalic and alkali-calcic rocks. Harker (1909) divided all Tertiary and Holocene igneous rocks of the world into two main groups, the Atlantic suite and the Pacific suite, the former being so named because of the predominance of alkalic and alkali-calcic rocks in the nonorogenic areas of crustal instability around the Atlantic Ocean. Because there is such a wide variety of tectonic environments and associated rock types in the areas of Harker's Atlantic and Pacific suites, the terms are now seldom used to indicate kindred rock types; e.g., Atlantic-type rocks are widespread in the mid-Pacific volcanic islands. Cf: *Arctic suite; Mediterranean suite.* (AGI, 1987)

Atlas ore *malachite.*

Atlas spar Syn: *satin spar.* Also called Atlas pearls, Atlas stone.

atmosphere (a) The gaseous envelope surrounding the Earth. The mixture of gases that surrounds the Earth, being held thereto by gravity. It consists by volume of 78% nitrogen, 21% oxygen, 0.9% argon, 0.03% carbon dioxide, and minute quantities of helium, krypton, neon, and xenon. The atmosphere is so compressed by its own weight that half is within 5.5 km of the Earth's surface. (AGI, 1987) (b) A unit of pressure. A normal atmosphere is equal to the pressure exerted by a vertical column of mercury 760 mm in height at 0 °C, and with gravity taken as 980.665 cm/s^2. It equals 14.66 psi (101 kPa). (AGI, 1987) (c) In a furnace, the mixture of gases resulting from combustion. (d) The kind of air prevailing in any place, as within a kiln during firing. (Kinney, 1962)

atmosphere-supplying respirator A class of respirators that supply a respirable atmosphere, independent of the workplace atmosphere. (ANSI, 1992)

atmospheric condenser A condenser using water at atmospheric pressure. (Strock, 1948)

atoll texture (a) A texture sometimes observed in a thin section of a rock, in which a ring of one mineral occurs with another mineral or minerals inside and outside the ring. (b) In mineral deposits, the surrounding of one mineral by a ring of one or more other minerals; commonly results from replacement of pyrite by another mineral, with the outermost pyrite unaffected and constituting the "atoll." Syn: *core texture.* Cf: *tubercle texture.*

atom According to the atomic theory, the smallest particle of an element that can exist either alone or in combination with similar particles of the same element or of a different element. The smallest particle of an element that enters into the composition of a molecule. (Webster 3rd, 1966)

atomic charge Electrical charge density due to gain or loss of one or more electrons. (Pryor, 1963)

atomic distance Distance between two atom centers. (Pryor, 1963)

atomic moisture meter A device to monitor the moisture in coal passing through a preparation plant, by using radiation that is sensitive to hydrogen atoms. The coal is bombarded with neutrons, some of which strike hydrogen atoms and bounce back to a detector tube, thus providing a continuous measure of moisture content. This meter permits the moisture content of coal to be measured instantaneously, continuously, and automatically.

atomic number The number of protons in the nucleus of an atom. See also: *atomic weight.* (Lyman, 1964)

atomic plane (a) Any one of the layers into which atoms form themselves in an orderly pattern during the growth of a crystal. (b) In a crystal, any plane with a regular array of atomic units (atoms, ions, molecules, or radicals); it has potential to diffract X-rays, to parallel a crystal face, or to permit cleavage.

atomic scattering factor Describes the "efficiency" of X-ray scattering of a given atom in a given direction; equal to the amplitude of the wave scattered by an atom divided by the amplitude of the wave scattered by one electron.

atomic susceptibility Change in magnetic moment of 1 g·atom on application of magnetic field of unit strength. (Pryor, 1963)

atomic volume (a) The space occupied by a quantity of an element as compared with its atomic weight. Obtained by dividing the specific gravity of the element by its atomic weight; also called specific volume. (Standard, 1964) (b) The volume occupied by 1 g·atom of an element. (CTD, 1958)

atomic weight The average relative weight of the atoms of an element referred to an arbitrary standard of 16.0000 for the atomic weight of oxygen. The atomic weight scale used by chemists takes 16.0000 as the average atomic weight of oxygen atoms as they occur in nature. The scale used by physicists takes 16.00435 as the atomic weight of the most abundant oxygen isotope. Division by the factor 1.000272 converts an atomic weight on the physicists' scale to the corresponding atomic weight on the chemists' scale. See also: *atomic number.* (AGI, 1987)

atomization (a) In powder metallurgy, the dispersion of a molten metal into particles by a rapidly moving stream of gas or liquid. (ASM, 1961) (b) A patented process for producing a metallic dust, such as zinc dust. (Fay, 1920)

atomized metal powder Metal powder produced by the dispersion of molten metal by a rapidly moving gas, or liquid stream, or by mechanical dispersion. (ASTM, 1994)

atomizer A spray device for producing a very fine mist for the suppression of airborne dust in mines. It is normally operated by compressed air. Syn: *jet mixer; line oiler.*

atopite A yellow or brown variety of romeite containing fluorine. See also: *romeite.*

attached ground water That portion of the subsurface water adhering to the pore walls. It is assumed to be equal in quantity to the pellicular water, and it is measured by specific retention. (AGI, 1987)

attack rate Planned rate of ore extraction from mineral deposit. (Pryor, 1963)

attapulgite A light-green, magnesium-rich clay mineral, named from its occurrence at Attapulgus, GA, where it is quarried as fuller's earth. Crystallizes in the monoclinic system. Syn: *palygorskite.* (AGI, 1987; Kirk, 1947)

attendance signaling system A signaling system that operates between the surface lamp room and the underground office, indicating the workers in attendance at the beginning of the shift. See also: *self-service system.* (Nelson, 1965)

attenuation (a) A reduction in the amplitude or energy of a signal, such as might be produced by passage through a filter. (AGI, 1987) (b) A reduction in the amplitude of seismic waves, as produced by divergence, reflection and scattering, and absorption. (AGI, 1987) (c) That portion of the decrease in seismic or sonar signal strength with distance that is dependent not on geometrical divergence, but on the physical characteristics of the transmitting medium. Cf: *damping.* (AGI, 1987)

Atterberg limits In a sediment, the water-content boundaries between the semiliquid and plastic states (known as the liquid limit) and between the plastic and semisolid states (known as the plastic limit). See also: *consistency limits.* Cf: *plastic limit; plasticity index.* (AGI, 1987)

Atterberg scale A proposed particle-size scale or grade scale for the classification of sediments based on a decimal system beginning with

2 mm. The limits of the subclass are obtained by taking the square root of the product of the larger grade limits. The subdivision thus made follows the logarithmic rule. This is the accepted European standard for classification of particle size. (AGI, 1987)

Atterberg test A method for determining the plasticity of clay in terms of the difference between the water content when the clay is just coherent and when it begins to flow as a liquid. (Dodd, 1964)

attitude The relation of some directional feature in a rock to the horizontal plane. The attitude of planar features (bedding, foliations, joints, etc.) is described by the strike and the dip. The attitude of a linear feature (fold axis, lineation, etc.) is described by the strike of the horizontal projection of the linear feature and its plunge. (AGI, 1987)

attrital coal A bright coal (composed of anthraxylon and of attritus in which the translucent cell-wall degradation matter or translucent humic matter predominates) in which the ratio of anthraxylon to attritus is less than 1:3. See also: *anthraxylous coal; attritus.* (AGI, 1987)

attrition (a) The act of wearing and smoothing of rock surfaces by the passage of water charged with sand and gravel, by the passage of sand drifts, the descent of glaciers, etc. (AGI, 1987) (b) The wear and tear that rock particles in transit undergo through mutual rubbing, grinding, knocking, scraping, and bumping, with resulting comminution in size. Cf: *abrasion.* Syn: *corrasion.* (AGI, 1987)

attrition mill (a) Mill that grinds abrasively, using rubbing action rather than impact shattering to disintegrate material. (Pryor, 1963) (b) A disintegrator depending chiefly on impact to reduce the particle size of the charge. Attrition mills are sometimes used in the clay building materials industry to deal with the tailings from the edge-runner mill. (Dodd, 1964)

attritious wear Wear of abrasive grains in grinding such that sharp edges gradually become rounded. A grinding wheel that has undergone such wear usually has a glazed appearance. (ASM, 1961)

attritus (a) A composite term for dull gray to nearly black coal components of varying maceral content, unsorted and with fine granular texture, that forms the bulk of some coals or is interlayered with bright bands of anthraxylon in others. It is formed of a tightly compacted mixture of altered vegetal materials, esp. those that were relatively resistant to complete degradation. Cf: *attrital coal.* Syn: *durain.* The term was introduced by R. Thiessen in 1919. (AGI, 1987) (b) Thin bands of dull coal interlaminated with the bright, glossy coal bands called anthraxylon. Microscopically it consists of intimately mixed, tightly compacted remains of varied morphological form and origin. Attritus is a collective term, not directly comparable with any one of the microlitho types of the Stopes-Heerlen nomenclature but consists of an intimate association of varying proportions of macerals of the vitrinite, exinite, and inertinite groups. It is present in practically all types of coal. In bright-banded coal it is secondary in importance to anthraxylon, but in splint coal it is the dominant component, and nonbanded attrital coals consist entirely of attritus. (IHCP, 1963) (c) The dull-gray to nearly black, frequently striped portion of material that comprises the bulk of some coals, and the alternating bands of bright anthraxylon in well-banded coals. It was derived from all sorts of comminuted and macerated plant matter, esp. from the plants that were more resistant to complete decomposition. It consists of humic degradation and opaque, charred, resinous, and mineral matter; fats, oils, waxes, cuticles, spores, arid spore exines, and other constituents of the plants forming the coal. (AGI, 1987) (d) Coal components consisting of a mixture of microscopic fragments of vegetable tissues. It is classified into opaque attritus and transparent attritus. Generally, it corresponds to cull coal or durain. (Tomkeieff, 1954)

audiofrequency Any frequency corresponding to a normal audible sound wave (ranges roughly from 15,000 to 20,000 Hz). (Hunt, 1965)

augelite A monoclinic mineral, $Al_2(PO_4)(OH)_3$.

augen In foliate metamorphic rocks such as schists and gneisses, large lenticular mineral grains or mineral aggregates having the shape of an eye in cross section, in contrast to the shapes of other minerals in the rock. See also: *augen structure.* Etymol: Ger., eyes. (AGI, 1987)

augen gneiss A general term for a gneissic rock containing augen. See also: *cataclasite.* (AGI, 1987)

augen schist A metamorphic rock characterized by recrystallized minerals occurring as augen or lenticles parallel to and alternating with schistose streaks. See also: *mylonite gneiss.* (AGI, 1987)

augen structure In some gneissic and schistose metamorphic rocks, a structure consisting of minerals like feldspar, quartz, or garnet that have been squeezed into elliptical or lens-shaped forms resembling eyes (augen), which are commonly enveloped by essentially parallel layers of contrasting constituents such as mica or chlorite. Cf: *augen; flaser structure.* (AGI, 1987)

auger (a) A drill for seismic shotholes or geophone holes modeled after the conventional carpenter's screw auger. Hence, any seismic shothole drilling device in which the cuttings are continuously removed mechanically from the bottom of the bore during the drilling operation without the use of fluids. A rotary drilling device used to drill shotholes or geophone holes in which the cuttings are removed by the device itself without the use of fluids. Cf: *hand auger; hand boring.* (AGI, 1987) (b) Any of various augerlike tools designed for boring holes in wood or for boring into soil and used esp. for such purposes as mining coal, prospecting, drilling for oil or water, and digging postholes. Also, a tool for drilling holes in coal for blasting. (Webster 3rd, 1966) (c) Drilling using an auger. See also: *coal auger; bucket auger; twist drill; horizontal auger.* (Long, 1960)

auger bits Hard steel or tungsten-carbide-tipped cutting teeth used in an auger run on a torque bar or in an auger-drill head run on a continuous-flight auger. (Long, 1960)

auger boring The hole and/or the process of drilling a hole using auger equipment. (Long, 1960)

auger head *auger mining.*

auger hole A hole drilled with power-driven augers. (Williams, 1964)

auger mining A mining method often used by strip-mine operators where the overburden is too thick to be removed economically. Large-diameter, spaced holes are drilled up to 200 ft (61 m) into the coalbed by an auger. Like a bit used for boring holes in wood, this consists of a cutting head with screwlike extensions. As the auger turns, the head breaks the coal and the screw carries it back into the open and dumps it on an elevating conveyor; this, in turn, carries the coal to an overhead bin or loads it directly into a truck. Auger mining is relatively inexpensive, and it is reported to recover 60% to 65% of the coal in the part of the bed where it is used. Syn: *auger head.*

augite A monoclinic mineral, $8[(Ca,Na)(Mg,Fe,Al,Ti)(Si,Al)_2O_6]$; pyroxene group; dark-green to black with prismatic cleavage; a common rock-forming mineral in igneous and metamorphic rocks. Syn: *basaltine; fassaite; pyroxene.* Cf: *pigeonite.*

augite bronzite An obsolete term for a pyroxene between enstatite and augite in composition.

augite diorite A diorite in which augite is a prominent mafic mineral.

augite syenite A syenite in which augite is a prominent mafic mineral.

auralite (a) Altered iolite. (Standard, 1964) (b) Altered cordierite.

aureole (a) A circular or crescentic distribution pattern about the source or origin of a mineral, ore, mineral association, or petrographic feature. It is encountered principally in magnetic and geochemical surveys. Cf: *dispersion pattern.* (AGI, 1987) (b) Discoloration of a mineral, viewed in thin section, in the form of a ring. Most haloes of this sort are caused by radiation damage by alpha particles emitted from uranium- and thorium-bearing mineral inclusions. (AGI, 1987) (c) A zone surrounding an igneous intrusion, in which the country rock shows the effects of contact metamorphism. Syn: *contact zone; metamorphic aureole.* (AGI, 1987) (d) A zone of alteration or other chemical reaction surrounding a mineral in a rock. (e) *halo.*

auri-argentiferous Applied to minerals containing both gold and silver. (Standard, 1964)

aurichalcite An orthorhombic mineral, $4[(Zn,Cu)_5(CO_3)_2(OH)_6]$; forms soft scaly greenish-blue crusts in oxidized zones of copper-zinc ore deposits; a guide to zinc ore.

auricupride An orthorhombic mineral, Cu_3Au. Syn: *gold cupride; cuproauride.*

auriferous Refers to a substance that contains gold, esp. gold-bearing mineral deposits. (AGI, 1987)

auriferous pyrite Iron sulfide, in the form of pyrite, containing gold, probably in solid solution. (CTD, 1958)

aurobismuthinite A doubtful sulfide containing bismuth, gold, and silver; lead-gray in color. It may be a mixture of $(Bi,Au,Ag_2)S$, or possibly of a gold-silver alloy, and bismuthinite, Bi_2S_3. From Nacozari, Sonora, Mex. (English, 1938)

aurosmirid A silver-white crystal solution of gold and osmium in isometric iridium.

aurostibite An isometric mineral, $AuSb_2$; pyrite group.

aurous Of, pertaining to, or containing gold in the univalent state; e.g., aurous chloride (AuCl).

austempering The isothermal transformation of a ferrous alloy at a temperature below that of pearlite formation and above that of martensite formation. Austempering is the isothermal transformation used to form a unique acicular matrix of bainitic ferrite and stable high-carbon austenite in hardenable cast irons. (ASM, 1991)

austenite A solid solution of one or more elements in face-centered cubic iron. Unless otherwise designated (such as nickel austenite), the solute is generally assumed to be carbon. (ASM, 1961)

austenitic stainless steel The so-called 18-8 grades contain from 16% to 26% chromium and 6% to 20% nickel, are not hardenable by heat treatment, and are nonmagnetic in the annealed condition. (Henderson, 1953)

austenitizing Forming austenite by heating a ferrous alloy into the transformation range (partial austenitizing) or above the transformation range (complete austenitizing). (ASM, 1961)

austinite An orthorhombic mineral, $CaZn(AsO_4)(OH)$; adelite group; forms a series with conichalcite.

Australian bentonite Trade name for highly plastic clays from Trida, N.S.W. (New South Wales, 1958)

Australian cinnabar A variety of chrome red.
authigene A mineral or rock constituent that was formed in place; e.g., a mineral of an igneous rock; the cement of a sedimentary rock if deposited directly from solution; or a mineral resulting from metamorphism. Syn: *authigenic mineral.* Ant. allogene. (AGI, 1987)
authigenesis (a) The process by which new minerals form in place within a rock during or after its formation, as by replacement or recrystallization, or by secondary enlargement of quartz overgrowths. (AGI, 1987) (b) Any process involving crystal growth in situ, i.e., subsequent to the origin of its matrix or surroundings but not a product of transformation or recrystallization, customarily reserved for low-temperature sedimentary environments. Ant. allogenesis. Adj. authigenic; authigenous. Adv. authigenous.
authigenetic *authigenic.*
authigenic Formed or generated in place; specif. said of rock constituents and minerals that have not been transported or that crystallized locally at the spot where they are now found, and of minerals that came into existence at the same time as, or subsequently to, the formation of the rock of which they constitute a part. The term, as used, often refers to a mineral (such as quartz or feldspar) formed after deposition of the original sediment. Syn: *authigenetic.* Ant. allogenic. Cf: *autochthonous.* See also: *authigenesis.* (AGI, 1987)
authigenic mineral *authigene.*
authorized fuels In Great Britain, under the regulations made by the Minister (Smoke Control Areas-Authorized Fuels-Regulations, 1956), authorized fuels include coke of all kinds, anthracite, low-volatile steam coals, Phurnacite, Coalite, Rexco, etc., as well as oil, gas, and electricity. (Nelson, 1965)
authorized person An authorized person is either one appointed or permitted by the official designated by State mining laws to be in charge of the operation of the mine or one appointed to perform certain duties incident to generation, transformation, and distribution or use of electricity in the mine. This person shall be familiar with construction and operation of the apparatus and with hazards involved.
autochthon A body of rocks that remains at its site of origin, where it is rooted to its basement. Although not moved from their original site, autochthonous rocks may be mildly to considerably deformed. Cf: *allochthon; stationary block.* Also spelled autochthone. (AGI, 1987)
autochthonous Formed or produced in the place where now found. Applied to a rock the dominant constituents of which have been formed in situ; e.g., rock salt. Cf: *allochthonous; authigenic.* (Holmes, 1920)
autochthonous coal Coal believed to have been formed from accumulations of plant debris at the place where the plants grew. Two modes of origin are distinguished: terrestrial and aquatic. Also called indigenous coal. See also: *in situ origin theory.* (Stutzer, 1940)
autochthonous peat Peat that formed in place by the gradual accumulation of plant remains in water. It is subdivided into low-moor peat and high-moor peat. (Tomkeieff, 1954)
autochthony An accumulation of plant remains in the place of their growth. The term itself can be distinguished between autochthonous elements of growth (euautochthony) and autochthonous elements of sedimentation (hypautochthony). (IHCP, 1963)
autoclastic Having a broken or brecciated structure, formed in the place where it is found as a result of crushing, dynamic metamorphism, or other mechanical processes; e.g., a fault breccia, or a brecciated dolomite produced by diagenetic shrinkage followed by recementation. Cf: *cataclastic.* (AGI, 1987)
autogenous (a) In the dense-media separation process, fluid media partly composed of a mineral species selected from material being treated. (Pryor, 1965) (b) Selectively sized lumps of material used as grinding media. (Pryor, 1965)
autogenous grinding The secondary grinding of coal or ore by tumbling in a revolving cylinder with no balls or bars taking part in the operation. (Nelson, 1965)
autogenous roasting Roasting in which the heat generated by oxidation of the sulfides is sufficient to propagate the reaction. (Newton, 1938)
autohydration The development of new minerals in an igneous rock by the action of its own magmatic water on already existing magmatic minerals. (Schieferdecker, 1959)
autoinjection *autointrusion.*
autointrusion (a) A process wherein the residual liquid of a differentiating magma is injected into rifts formed in the crystallized fraction at a late stage by deformation of unspecified origin. Syn: *autoinjection.* (AGI, 1987) (b) Sedimentary intrusion of rock material from one part of a bed or set of beds in process of deposition into another part. (AGI, 1987)
autolith (a) An inclusion in an igneous rock to which it is genetically related. Cf: *xenolith.* Syn: *cognate inclusion.* (AGI, 1987) (b) In a granitoid rock, an accumulation of iron-magnesium minerals of uncertain origin. It may appear as a round, oval, or elongate segregation or clot. (AGI, 1987)
automatic ash analysis Analysis in which the coal sample passes first to a conditioning unit, which dries and grinds it, then to an X-ray analysis unit. The analysis is based on the difference in the reflection of X-rays by the combustible and noncombustible components of the sample. The reflection is compared photoelectrically with a reference sample. (Nelson, 1965)
automatic belt takeup A device used with certain types of belt conveyors for the taking up or storage of belt during reversible operation. (Jones, 1949)
automatic clip An appliance for attaching and detaching mine trams or cars without manual effort. It is generally attached at inby clipping stations and detached at the shaft bottom. See also: *clip; coupling; haulage clip.* (Nelson, 1965)
automatic clutch A clutch whose engagement is controlled by centrifugal force, vacuum, or other power without attention by the operator. (Nichols, 1976)
automatic coupling A device that automatically couples cars when they bump together. (Zern, 1928)
automatic cyclic winding A system of automatic winding in which the complete installation operates without human aid and winding continues automatically as long as coal is available at the shaft bottom and is cleared at the bank. Syn: *cyclic winding.* See also: *pushbutton winding control; Ward-Leonard control; manual winding control; semiautomatic control.* (Nelson, 1965)
automatic dam *boomer.*
automatic door (a) A mine door operated by pressure of the locomotive wheels on a switch along the rails approaching the doors, which closes the door automatically after the trip has passed. These doors are preferable to regular mine doors. However, they must be carefully maintained to keep them in a safe operating condition. (b) A wooden door arranged to close automatically when released, by installing the door with a slight lean in the direction of closing.
automatic doors Air doors on a haulage road that are automatically operated by a passing vehicle or train of tubs, or other means. (BS, 1963)
automatic feed (a) A hydraulic-control system of valves that when once set and without the manual assistance of a drill runner will reduce or increase feed pressure applied to a drill stem as hardness of rock penetrated changes. (Long, 1960) (b) A pneumatic rock drill equipped with a power-actuated feed mechanism. (Long, 1960)
automatic feed sampler An automatic, timed sampling device used at mill feeds and other plants. (Nelson, 1965)
automatic heat-treating machine *Gilman heat-treating machine.*
automatic pump control The starting and stopping of a pump by a mechanism actuated by the level of water in the suction well or pump, or by the level or pressure of water in a discharge tank. (BS, 1963)
automatic pumping An arrangement to stop and start a mine pump automatically by means of a float switch. (Nelson, 1965)
automatic reclosing relays Relays used to automatically reclose electrically operated circuit breakers. They limit the duration of power failures in many instances where faults clear themselves quickly. Most reclosing relays attempt to close a breaker three times before locking it out. The time interval between reclosures is predetermined. Lockout means that after the third attempt fails to keep the breaker in, the relay will not function until it is reset manually. Such relays can be designed to operate more than three times before locking out, with the number of reclosures depending on the requirements and design of the system. (Coal Age, 1966)
automatic recorder Appliance for recording the working time of machines such as cutter loaders, conveyors, etc. A vibrating type, fitted on the equipment itself, marks on a chart a straight line when the machine is idle and an oscillating one when working. (Nelson, 1965)
automatic sampler An instrument designed to take samples of mine gases or water at predetermined times or intervals.
automatic sampling Automatic removal of samples at timed intervals from a passing stream of ore, pulp, or solution. (Pryor, 1963)
automatic sprinkler A water sprinkling device closed by a metallic alloy that melts at a low temperature. In case of fire the alloy melts, releasing a water spray. These devices are used in wood-lined shafts and timbered bottoms, sometimes by legal requirements. (Zern, 1928)
automatic winding This term includes at least three different systems: (1) fully automatic winding in which no driver, banksman, or onsetter is employed; (2) pushbutton automatic winding, similar to the above except that the operation is started by a pushbutton by the banksman or onsetter; and (3) cyclic winding in which the driver takes off the brakes and throws over the control lever at the beginning of the wind. (Sinclair, 1959)
autometamorphism (a) A process of recrystallization of an igneous rock under conditions of falling temperature, attributed to the action of its own volatiles, e.g., serpentinization of peridotite or spilitization of basalt. (AGI, 1987) (b) The alteration of an igneous rock by its own residual liquors. This process should rather be called deuteric because

it is not considered to be metamorphic. See also: *deuteric.* Cf: *autometasomatism.* (AGI, 1987)
autometasomatism Alteration of a recently crystallized igneous rock by its own last water-rich liquid fraction, trapped within the rock, generally by an impermeable chilled border. Cf: *autopneumatolysis; autometamorphism.* (AGI, 1987)
automolite A dark-green to nearly black variety of gahnite. (Schaller, 1917)
automorphic (a) Said of the holocrystalline texture of an igneous or metamorphic rock, characterized by crystals bounded by their own rational faces. Also said of a rock with such a texture. The term idiomorphic is more common in U.S. usage. Cf: *xenomorphic.* (AGI, 1987) (b) A synonym of euhedral, obsolete in U.S. usage, but generally preferred in European usage. Syn: *automorphic-granular; euhedral.* (AGI, 1987)
automorphic-granular *automorphic.*
autopneumatolysis Autometamorphism involving the crystallization of minerals or the alteration of a rock by gaseous emanations originating in the magma or rock itself. Cf: *autometasomatism.* (AGI, 1987)
autospray A device for controlling dust carried by loaded conveyors. A liquid medium is sprayed on the conveyor load only when moving and not when stationary, or when the belt is running unloaded. The spray control is placed centrally beneath the conveyor belt and a load causes the belt to deflect and rotate the driving pulley, which causes the controller valve to open. A belt stoppage or no load causes the valve to close. (Nelson, 1965)
autostoper A stoper or light compressed-air rock drill, mounted on an air-leg support that not only supports the drill but also exerts pressure on the drill bit. (Nelson, 1965)
autotransformer A special-type of transformer whose use in mines is limited to apparatus for starting induction motors of the squirrel cage type. The winding is a common one for primary and secondary, and the two circuits are electrically in contact with each other. (Mason, 1951)
autotroph Organism capable of growth exclusively at the expense of inorganic nutrients. See also: *chemolithotroph; photolithotroph.*
autrometer (a) An automatic multielement-indexing X-ray spectrograph, capable of the qualitative and quantitative determinations of as many as 24 elements in a single sample. Choice of the elements may be made from magnesium through all the heavier elements. The device measures the intensity of an emitted wavelength band from a standard sample and compares it with the intensity of a like band from an unknown sample. These data are presented in the form of a ratio of one intensity to the other. (Nelson, 1965) (b) An automatic multielement-indexing X-ray spectrograph.
autunite (a) A tetragonal mineral, $2[Ca(UO_2)_2(PO_4)_2 \cdot 10\text{-}12H_2O]$; radioactive; yellow to pale green; fluorescent; forms scaly or foliated aggregates; results from oxidation or hydrothermal alteration of uranium minerals; an ore of uranium. Syn: *calcouranite.* (b) The mineral group autunite, fritzscheite, heinrichite, kahlerite, novacekite, sabugalite, saleeite, sodium autunite, torbernite, trögerite, uranocircite, uranospinite, and zeunerite.
auxiliary (a) Tools or other equipment, such as a pump, drill rods, casing, core barrel, bits, water swivel, safety clamp, etc., required for use with a drill machine to carry on specific drilling operations. (Long, 1960) (b) A helper or standby engine or unit. Cf: *accessory.* (Nichols, 1976)
auxiliary anode A supplementary anode placed in a position to raise the current density on a certain area of the cathode to get better plate distribution. (ASM, 1961)
auxiliary cylinder A cylinder, operated by compressed air, that is used to assist the main engine of a compressed-air shaker conveyor, esp. where the conveyor cannot develop a sufficient amount of forward acceleration because of grades. The auxiliary cylinder is attached to the conveyor by a driving chain and to a prop by a fixing chain. (Jones, 1949)
auxiliary fault A branch fault. A minor fault ending against a major fault. (AGI, 1987)
auxiliary mineral In Johannsen's classification of igneous rocks, any light-colored, relatively rare mineral, or mineral occurring in small quantities, such as apatite, muscovite, corundum, fluorite, and topaz. (AGI, 1987)
auxiliary operations In metallurgy, diverse operations, such as storing in bins, conveying (by conveyors, feeders, elevators, or pumps), sampling, weighing, reagent feeding, and pulp distribution. (Gaudin, 1939)
auxiliary ventilation A method of supplementing the main ventilating current in a mine by using a small fan to draw air from the main current and force it through canvas or metal pipe to some particular place, such as the ends of drifts, crosscuts, raises, entries, or other workings driven in a mine. See also: *air mover; exhaust ventilation; ventilation tubing; forced auxiliary ventilation; overlap auxiliary ventilation; piped air; reversible auxiliary ventilation; two-fan auxiliary ventilation.* (Lewis, 1964)
available alumina The theoretical amount of extractable aluminum oxide, Al_2O_3, present in a bauxite. The amount of alumina in a bauxite that is present in a form that allows it to be extracted by a refining plant.
available energy That part of the total energy that can be usefully employed. In a perfect engine, that part which is converted to work. (Strock, 1948)
available lime (a) Those constituents of a lime that enter into a desired reaction under the conditions of a specific method or process. (ASTM, 1994) (b) Represents the total free lime (CaO) content in a quicklime or hydrate and is the active constituent of a lime. It provides a mean of evaluating the concentration of lime. (Boynton, 1966)
available power The rate at which a given source would deliver energy to a load having an impedance that is the conjugate of the source impedance is designated as the available power of that source. (Hunt, 1965)
available power loss The available power loss of a transducer connecting an energy source and an energy load is the transmission loss measured by the ratio of the source power to the output power transducer. (Hunt, 1965)
available relief (a) The vertical distance between the altitude of the original surface after uplift and the level at which grade is first attained. (AGI, 1987) (b) The relief that is available for erosion.
available silica The amount of silica present in a flux that is not slagged by impurities in the flux itself. (Newton, 1959)
avalanche A large mass of snow, ice, soil, or rock, or mixtures of these materials, falling, sliding, or flowing very rapidly under the force of gravity. Velocities may sometimes exceed 500 km/hr. (AGI, 1987)
avalanche protector Guardplates that prevent loose material from sliding into contact with the wheels or tracks of a digging machine. (Nichols, 1976)
avanturine Alternate spelling of aventurine.
aventurescence (a) A word used to describe the metallic spangled effect seen, in reflected light, in aventurine and aventurine feldspar. A sort of schiller but more scintillating. (b) A display of bright or strongly colored reflections from included crystals in some translucent mineral specimens.
aventurine (a) A glass containing opaque sparkling particles of foreign material, which is usually copper or chromic oxide. With copper particles, it is called gold aventurine, and with chromic oxide particles, it is called chrome aventurine or green aventurine. A glass containing gold-colored inclusions. (Webster 3rd, 1966; AGI, 1987) (b) A translucent quartz that is spangled throughout with scales of mica or of some other mineral. Syn: *aventurine quartz.* (Webster 3rd, 1966) (c) As an adj., having the brilliant spangled appearance of aventurine. Applied esp. to transparent or translucent quartz or feldspar containing shiny inclusions. (Webster 3rd, 1966; AGI, 1987) (d) A variety of albite with reddish reflections from exsolved hematite in certain planes. See also: *goldstone.* Syn: *sunstone; love stone.*
aventurine feldspar Orthoclase, albite, or oligoclase that is more or less transparent, with fiery reflections from enclosed flat mineral particles, which are probably hematite or goethite. Sunstone is aventurine oligoclase. (Hess)
aventurine glass A glass supersaturated with either iron, chromium, or copper oxide (or a combination of the oxides) that is melted and cooled under controlled conditions to cause the excessive oxides to crystallize, forming platelike crystals or spangles. See also: *goldstone.*
aventurine quartz *aventurine.*
average assay value *assay value.*
average clause Eng. A clause that, in granting leases of minerals (coal, ironstone, and clay in particular), provides that lessees may, during every year of the term, make up any deficiency in the quantity of coal, etc., stipulated to be worked, so as to balance the dead or minimum rent.
average loading The average number of tons of a specified material to be carried by a conveyor per hour, based on total operating-shift tonnage. (NEMA, 1961)
avicennite An isometric mineral, Tl_2O_3; black; forms minute crystals.
AW Letter name specifying the dimensions of bits, core barrels, and drill rods in the A-size and W-group wireline diamond drilling system having a core diameter of 30.1 mm and a hole diameter of 48 mm. Syn: *AX.* (Cumming, 1981)
awaruite An isometric mineral, Ni_2Fe to Ni_3Fe. Syn: *native nickel-iron.*
AX Letter name specifying the dimensions of core, core barrels, and casing in the A-size and X-series wireline diamond drilling system having a core diameter of 30.1 mm and a hole diameter of 48 mm. The AX designation for coring bits has been replaced by the AW designation. Syn: *AW.* . (Cumming, 1981)
axes (a) Crystallographic directions through a crystal; used as lines of reference. (Hurlbut, 1964) (b) Reference coordinates *a, b, c,* in crystallography, crystallographic axes. (c) Directions of apparent isotropy in anisotropic crystals, optic axes. (d) Elements of rotational symmetry,

symmetry axes. (e) In ellipsoids representing the Fletcher indicatrix of refractive indices, semiaxes represent optic directions. Singular: axis. Also called optic axes.

axial angle (a) The acute angle between the two optic axes of a biaxial crystal. Its symbol is 2V. (AGI, 1987) (b) The axial angle in air (symbol 2E) is the larger angle between the optic axes after being refracted on leaving the crystal. (AGI, 1987) (c) *optic angle.*

axial compression In experimental work with cylinders, a compression applied parallel with the cylinder axis. It should be used in an appropriate sense only in the interpretation of deformed rocks. (AGI, 1987)

axial element In crystallography, the ratio of a unit distance along a crystallographic axis and the corresponding angle between axes. Syn: *lattice parameter.*

axial figure (a) The interference figure that is obtained in convergent light when an optic axis of the mineral being observed in thin section or as a fragment coincides with the axis of the polarizing microscope. When a thin section of a uniaxial mineral that was cut at right angles to an optic axis is examined between crossed nicols (that is, between two polarizers, the polarization planes of which are at right angles to each other) an equal-armed shadowy cross and a series of spectrally colored, circular bands are seen. If the mineral is biaxial, two shadowy parabolic curves called isogyres and opening away from each other in a series of spectrally colored, oval bands appear. (Hess) (b) In polarized light microscopy, an interference figure in which an optic axis is centered in the field of view.

axial flow In pumping or in ventilation, the use of a propeller or impeller to accelerate the load along the axis of the impeller. (Pryor, 1963)

axial-flow compressor A compressor in which air is compressed in a series of stages as it flows axially through a decreasing tubular area. (Pryor, 1963)

axial-flow fan (a) A type of mine fan in which the mine air enters along the axis parallel to the shaft and continues in this direction to the point of exhaust. The axial-flow fan may have fixed blades (fixed-pitch fan) or adjustable blades (variable-pitch fan). Two, four, or six aerofoil section blades (like an aircraft wing) are usually employed. Also called a screw fan. Cf: *radial-flow fan; mixed-flow fan.* See also: *contra-rotating axial fan; mine-ventilation fan.* (b) The compressed-air auxiliary fan consists essentially of a single-stage axial-flow fan in which the rotor also forms the rotor of a compressed-air turbine. The exhaust from the turbine is added to the ventilating air. The result is a light and very compact machine, capable of the same duties as the smaller sizes of electric auxiliary fans. (Roberts, 1960)

axial line *axis.*

axial plane (a) A more or less planar surface that intersects a fold in such a manner that the limbs of the fold are symmetrically arranged with reference to it. (b) The plane of the optic axes of an optically biaxial crystal. (c) A crystallographic plane that includes two crystallographic axes. (AGI, 1987) (d) Of geologic structures, a plane that intersects the crest of the trough or a fold such that the limbs, or sides, of the fold are more or less symmetrically arrayed with reference to it.

axial-plane cleavage Cleavage that is closely related to the axial planes of folds in the rock, either being rigidly parallel to the axes, or diverging slightly on each flank (fan cleavage). Most axial-plane cleavage is closely related to the minor folds seen in individual outcrops, but some is merely parallel to the regional fold axes. Most axial-plane cleavage is also slaty cleavage. (AGI, 1987)

axial-plane folding Large-scale secondary folding of preexisting folds, in response to stresses that varied considerably from those that caused the original folding. The axial planes of the original folds are folded. (AGI, 1987)

axial-plane foliation Foliation that developed parallel to the axial plane of a fold and perpendicular to the principal deformational pressure. (AGI, 1987)

axial-plane separation The distance between axial surfaces of adjacent antiforms and synforms where the folds occur in the same layer or surface. (AGI, 1987)

axial priming A system for priming blast agents in which a core of priming material extends through most or all of the blasting agent charge length.

axial ratio The lengths of crystallographic axes defined in terms of their ratios with, by convention, *a* set at unity where one axis is unique and *b* set at unity where all three axes are required.

axial stream (a) The main stream of an intermontane valley, which flows along the lowest part of the valley and parallel to its long dimension, in contradistinction to the streams that flow down the mountains on either side. (b) A stream that follows the axis of an anticline or a syncline.

axial trace The intersection of the axial plane of a fold with the surface of the Earth or any other specified surface.

axinite (a) The mineral group ferroaxinite, magnesioaxinite, manganaxinite, and tinzenite. (b) Triclinic borosilicates with the formula $A_3Al_2BSi_4O_{15}(OH)$ where A = (Ca,Fe,Mg,Mn). Syn: *glass schorl.*

axiolite A term proposed by Zirkel for a variety of elongated spherulite in which there is an aggregation of minute acicular crystals arranged at right angles to a central axis rather than from a point.

axis (a) The central or dominating region of a mountain chain, or the line that follows the crest of a range and thus indicates the most conspicuous part of the uplift. (b) The centerline of a tunnel. (Nichols, 1976) (c) Intersection of the axial plane of a fold with a particular bed; axial line. (d) A straight line about which a body or a three-dimensional figure rotates or may be supposed to rotate; a straight line with respect to which a body, figure, or system of points is either radially or bilaterally symmetrical. (Webster 3rd, 1966) (e) In crystallography, one of the imaginary lines in a crystal that are used as coordinate of axes of reference in determining the positions and symbols of the crystal planes. Cf: *crystallographic axes; coordinate system.* (f) Often used synonymously with anticlinal; thus, the Brady's bend axis for Brady's bend anticlinal. See also: *anticlinal axis; synclinal axis.* (g) The trace of the axial surface of a fold on the fold profile plane (obsolete). (h) A line that follows the trend of large landforms, e.g., the crest of a ridge or mountain range, or the bottom or trough of a depression. Plural: axes.

axis of acoustic symmetry For many transducers, the three-dimensional directivity is such that it may be represented by the surface generated by rotating a two-dimensional directivity pattern about the axis corresponding to the reference bearing of the transducer. This axis may then be described as an axis of acoustic symmetry or as the acoustic axis. (Hy, 1965)

axis of symmetry An imaginary line in a crystal, crystal structure, or crystal lattice, about which it may be rotated to an identical configuration. If identity occurs once during a complete rotation of 360°, the axis is a monad, twice a diad, thrice a triad, four times a tetrad, or six times a hexad. Syn: *symmetry axis.*

axonometric projection A method of projection which has the advantage of containing a true plan, and can therefore be set up from drawings already in existence for other purposes. The plan is turned through 45°, vertical lines being drawn from the angles on the plan to show the elevations. See also: *oblique projection; isometric projection.* (Hammond, 1965)

axotomous In crystallography, having cleavage perpendicular to an axis; said of minerals. (Standard, 1964)

axstone A variety of nephrite jade. Also spelled axestone.

azimuth Direction of a horizontal line as measured on an imaginary horizontal circle, the horizontal direction reckoned clockwise from the meridian plane of the observer, expressed as the angular distance between the vertical plane passing through the point of observation and the poles of the Earth and the vertical plane passing through the observer and the object under observation. In the basic control surveys of the United States, azimuths are measured clockwise from south, a practice not followed in all countries. Cf: *bearing.* (AGI, 1987)

azorite A variety of altered zircon. *zircon.* (Crosby, 1955)

Aztec stone (a) A greenish variety of smithsonite. (b) A green variety of turquoise.

azulinhas Small cloudy sapphires occuring with diamonds in Brazil.

azurchalcedony *azurlite.*

azure *lapis lazuli; lazurite.*

azure malachite *azurmalachite.*

azure quartz *sapphire quartz.*

azure spar *azurite; lazulite.*

azure stone A term applied to lapis lazuli (lazurite) and to other blue minerals such as lazulite and azurite.

azurite (a) A monoclinic mineral, $2[Cu_3(OH)_2(CO_3)_2]$; forms vitreous azure crystals; a supergene mineral in oxidized parts of copper deposits associated with malachite; an ore of copper. Syn: *azure spar; chessylite; blue copper; blue copper ore; blue malachite.* (b) A compact semiprecious stone derived from compact azurite and used as a decorator material. (c) A trade name for a sky-blue gem variety of smithsonite.

azurite malachite *azurmalachite.*

azurlite Chalcedony colored blue by chrysocolla; and used as a gemstone. Syn: *azurchalcedony.*

azurmalachite An intimate mixture or intergrowth of azurite and malachite, commonly massive and concentrically banded; used as an ornamental stone. Syn: *azure malachite; azurite malachite.*

B

baaken S. Afr. A boundary mark.
Babcock and Wilcox mill Dry-grinding mill in which steel balls rotate in a horizontal ring, through which the feed is worked downward. (Pryor, 1963)
Babel quartz A variety of quartz, named for the fancied resemblance of the crystal to the successive tiers of the Tower of Babel. Syn: *Babylonian quartz.*
Babylonian quartz *Babel quartz.*
bacalite (a) A variety of amber. (Tomkeieff, 1954) (b) A variety of quartz in Baja California, Mex.
back (a) A system of joints in coal oblique to the bedding, at an angle of about 35° to 75°. Backs are usually perfectly tight and have polished cheeks which suggest a certain amount of movement. Back may be applied to the principal cleat. See also: *backs; cleat.* (b) The roof or upper part of any underground mining cavity. (c) The ore between a level and the surface, or between two levels. See also: *back of ore.* (d) That part of a lode which is nearest the surface relative to any working part of a mine, thus the back of the level or stope is that part of the unstoped lode which is above. See also: *back of lode.* (e) A joint, usually a strike joint, which is perpendicular to the direction of working. (f) As applied to an arch, the outer or upper surface. (g) The pavilion of a gemstone.
back acter Front-end equipment fitted to an excavator, comprising a jib with an arm and bucket. Although designed primarily for vertically sided trenching, it is also useful for bulk excavation below track level. (Nelson, 1965)
back and underhand stoping milling system *combined overhand and underhand stoping.*
back arch A concealed arch carrying the backing or inner part of a wall where the exterior facing material is carried by a lintel. (ACSG, 1961)
back balance (a) A type of self-acting incline in a mine. A balance car is attached to one end of a rope, and a carriage for the mine car is attached to the other end. A loaded car is run on the carriage and is lowered to the foot of the incline, raising the balance car. The balance car in its descent raises the carriage when the carriage is loaded only with an empty car. (Fay, 1920) (b) The means of maintaining tension on a rope transmission or haulage system, consisting of the tension carriage, attached weight, and supporting structure. (Fay, 1920)
backblast *backlash.*
back break Rock broken beyond the limits of the last row of holes marking the outer boundary in a blast.
back brusher (a) A ripper engaged in taking down the roof in roadways some distance back from the face. Syn: *second ripping.* (Nelson, 1965) (b) Back ripper.
back-bye work General work performed behind the working faces, as opposed to work done at the faces. This is commonly referred to as "outby work."
back casing Eng. A temporary shaft lining of bricks laid dry, and supported at intervals upon curbs. When the stonehead has been reached, the permanent masonry lining is built upon it inside of the back casing.
backcast stripping A stripping method using two draglines, one of which strips and casts the overburden while the other recasts a portion of the overburden. (Woodruff, 1966)
back coal Scot. Coal that miners are allowed to carry home.
back coming Scot. Working away the pillars that are left when mining coal inby. Robbing pillars; back working. See also: *back work.*
back-end man A worker who works behind the coal-cutter as it moves along the face. Duties may include cleaning the cuttings from behind the machine and setting props to support the roof or overhang of coal. Cf: *coal-cutter team.* (Nelson, 1965)
back entry The air course parallel to and below an entry or the entry used for secondary purposes in two-entry system of mining. Locally, any entry not having track in it. (BCI, 1947)
backfill (a) Waste sand or rock used to support the roof or walls after removal of ore from a stope. (Pryor, 1963) (b) Sand or dirt placed behind timber, steel, or concrete linings in shafts or tunnels. (Nelson, 1965) (c) The process of sealing and filling, and/or the material used to seal or fill, a borehole when completed, to prevent its acting as a course along which water may seep or flow into rock formations or mine workings. (Long, 1960) (d) Material excavated from a site and reused for filling, for example, the use of stones or coarse gravel for filling draining trenches. See also: *fill.* (Nelson, 1965)
back filling (a) Rough masonry built in behind the facing or between two faces; similar material used in filling over the extrados of an arch; also, brickwork used to fill in space between studs in a frame building, sometimes called brick nogging. (ACSG, 1961) (b) The filling in again of a place from which the rock or ore has been removed. (Ballard, 1955)
back-filling system Filling lower or older workings with the waste from newer workings. See also: *overhand stoping; square-set stoping.* (Hess)
backfire (a) A fire started to burn against and cut off a spreading fire. (Nichols, 1976) (b) An explosion in the intake or exhaust passages of an engine. (Nichols, 1976)
backfolding Folding in which the folds are overturned toward the interior of an orogenic belt. In the Alps, the backward folds are overturned toward the south, whereas most of the folds are overturned toward the north. Syn: *backward folding.* (AGI, 1987)
background (a) The abundance of an element, or any chemical property of a naturally occurring material, in an area in which the concentration is not anomalous. (AGI, 1987) (b) The slight radioactivity shown by a counter, due to normal radioactivity from cosmic rays, impurities in the counter, and trace amounts of radioactivity in the vicinity.
backhand In bituminous coal mining, one who assists either the machineman or machine loader to move and set up a coal cutting or loading machine at the working face. (DOT, 1949)
backhaul A line that pulls a drag scraper bucket backward from the dump point to the digging. (Nichols, 1954)
backhaul cable In a cable excavator, the line that pulls the bucket from the dumping point back to the digging. (Nichols, 1976)
back heading (a) Eng. The companion place to a main winning. (SMRB, 1930) (b) *back entry.*
backhoe The most versatile rig used for trenching. The basic action involves extending its bucket forward with its teeth-armed lip pointing downward and then pulling it back toward the source of power. (Carson, 1961)
back holes In shaft sinking, raising, or drifting, the holes that are shot last.
backing (a) Timbers fixed across the top of a level supported in notches cut in the rock. (b) The action of a roof layer of combustible gases flowing uphill against the direction of the ventilation. (BS, 1963)
backing deals Boards from 1 to 4 in (2.5 to 10 cm) thick and of sufficient length to bridge the space between timber or steel sets or between rings in skeleton tubing. Usually, planks 9 to 12 in (23 to 30 cm) in width are used. Round poles, either whole or split, light steel rails, ribbed sheet metal, and reinforced concrete slabs are sometimes used in place of planks. Backing deals tighten the supports against the ground and also prevent the collapse of material between the timber or steel sets or rings. See also: *lagging.* (Nelson, 1965)
backing off A term used to describe the operation of removing excessive body metal from badly worn bits. (Fraenkel, 1953)
backing sand Reconditioned sand used for supporting the facing sand, and forming the main part of a foundry mold. (Osborne, 1956)
backjoint (a) A joint plane more or less parallel to the strike of the cleavage, and frequently vertical. (Zern, 1928) (b) A rabbet or chase left to receive a permanent slab or other filling. (Webster 3rd, 1966)
backlash (a) The return or counterblast, as the recoil or backward suction of the air current, produced after a mine explosion. Also called backblast; suction blast. (b) The reentry of air into a fan. (c) The violent recoil and whipping movement of the free ends of a rope or wire cable broken under strain. (Long, 1960) (d) Lost motion, play, or movement in moving parts such that the driving element (as a gear) can be reversed for some angle or distance before working contact is again made with the secondary element. (ASM, 1961)
back leads Applied to black sand leads on coastlines which are above high-water mark. (Fay, 1920)
back mine Scot. A passage in a mine crosscut toward the dip of the strata. (Standard, 1964)

back-off shooting The firing of small explosive charges for releasing stuck drilling tools in a borehole. The shock of detonation causes the joint to expand and unscrew slightly. All rods above the joint can then be removed from the hole. (Nelson, 1965)
back of lode The portion of a lode lying between a level driven in a lode and the surface. See also: *back.*
back of ore The ore between two levels which has to be worked from the lower level. See also: *back.*
back-out switch *hoist back-out switch.*
back pressure (a) Resistance transferred from rock into drill stem when bit is being fed at a faster rate than the bit can cut. (Long, 1960) (b) Pressure applied to the underside of the piston in the hydraulic-feed cylinder to partially support the weight of the drill rods and hence reduce pressure on the bit. (Long, 1960) (c) Rock pressures affecting the uppermost portion or roof in an underground mine opening. (Long, 1960)
back prop The name given to the raking strut that transfers the load from the timbering of a deep trench to the ground. These struts are provided under every second or third frame according to the type of ground being excavated. (Hammond, 1965)
back ripper *back brusher.*
back rippings The taking down of a thickness of roof beds in roadways some distance back from the face. The thickness of roof excavated may vary from 1 ft (0.3 m) or so to 6 ft (1.8 m) and more. This work is necessary where there has been a gradual reduction in height, as a result of roof sag, and opening height must be maintained. See also: *second ripping.*
backrush The seaward return of the water following the uprush of waves. For any given tide stage, the point of farthest return seaward of the backrush is known as the limit of backrush or limit of backwash. See also: *backwash.* (AGI, 1987)
backs (a) The height of ore available above a given working level. If the orebody has been proved by shaft sinking to a depth of 300 ft (91 m) from the surface, the orebody is said to have 300 ft (91 m) of backs. See also: *back.* (b) A quarryman's term for one set of joints traversing the rock, the other set being known as cutters. (Nelson, 1965) (c) A system of joints in coal or stratified mineral oblique to the bedding at an angle of 35° to 75°. See also: *slips.* (BS, 1964) (d) Slips; used to denote a slip met with first at floor level. Syn: *hugger.* (TIME, 1929-30)
backs and cutters Jointed rock structures, the backs (joints) of which run in lines parallel to the strike of the strata, the cutters (cross joints) crossing them at about right angles. (Standard, 1964)
backscatter The emergence of radiation from that surface of a material through which it entered. Also used to denote the actual backscattered radiation. (NCB, 1964)
backshift (a) The afternoon or night shift; any shift that does not fill coal or is not the main coal-production shift. (Mason, 1951) (b) N. of Eng. The second or middle shift of the day; varies from 9 to 10:30 a.m. until 4:30 to 6 p.m. in different pits. (Trist, 1963)
back shot A shot used for widening an entry; it is placed at some distance from the head of an entry. (Fay, 1920)
backsight (a) A sight or bearing on a previously established survey point (other than a closing or check point), taken in a backward direction. (AGI, 1987) (b) A reading taken on a level rod held in its unchanged position on a survey point of previously determined elevation when the leveling instrument has been moved to a new position. It is used to determine the height of the instrument prior to making a foresight. Syn: *plus sight.* Abbrev: BS. Ant: *foresight.* (AGI, 1987)
backsight hub A mark or stake placed at some distance behind the position a drill will occupy in a specific compass direction from the borehole marker for an incline hole to enable the driller to set the drill and drill the borehole in the intended direction. Also called back hub; backsight. See also: *picket.* (Long, 1960)
back skin Newc. A leather covering worn by workers in wet workings.
back slip A joint in a coal seam that is inclined away from the observer from floor to roof. It would be a face slip from the opposite direction. Cf: *face slip.* (Nelson, 1965)
back slope (a) S. Wales. A slope with the stalls branching off and working the seam with back slips along the face. (Nelson, 1965) (b) In geology, the less sloping side of a ridge. Contrasted with escarpment, or steeper slope; esp., the slope more nearly parallel with the strata. Also called structural plain. (Standard, 1964) (c) The term is used where the angle of dip of the underlying rocks is somewhat divergent from the angle of the land surface. The slope at the back of a scarp; e.g., the gentler slope of a cuesta or of a fault block. It may be unrelated to the dip of the underlying rocks. Also spelled: backslope. (AGI, 1987) (d) Syn: *dip slope.*
back splinting The working of the top portion of a thick seam that was left as a roof when the bottom portion was worked. The top coal is recovered by working over the goaf or packs of the first working. See also: *back work.* (Nelson, 1965)
backstamp The maker's name and/or trademark stamped on the back of pottery flatware or under the foot of hollowware. (Dodd, 1964)
backstay A drag or trailer fixed at the back of a haulage train (or set) as a safety device when going uphill. See also: *drag.* Cf: *bull.* (Mason, 1951)
backstone Eng. Shaly mudstone used for cooking slabs, quarried near Delph, Yorkshire. Also, a bed in the Staffordshire Coal Measures. (Arkell, 1953)
back stope To mine a stope from working below.
back stopes Overhead stopes; stopes worked by putting in overhead holes and blasting down the ore. (CTD, 1958)
backstroke jigging A process in which strong suction is advocated at all times with the dense-medium process, since none of the bone medium must be allowed to get over into the washed coal. (Mitchell, 1950)
backup gear *reverse-feed gear.*
back vent Scot. An air course alongside the pillar in wide rooms.
backwardation The situation when the cash or spot price of a metal is greater than its forward price. A backwardation occurs when a tight nearby situation exists in a metal. The size of the backwardation is determined by differences between supply and demand factors on the nearby positions compared with the same factors on the forward position. There is no official limit to the backwardation. The backwardation is also referred to as the "back.". (Wolff, 1987)
backward folding *backfolding.*
backwash (a) In uranium leaching, flushing from below of colloidal slime from ion exchange column after adsorption cycle. The cleaning of sand filters. (b) Water or waves thrown back by an obstruction such as a ship, breakwater, cliff, etc. (AGI, 1987) (c) The return flow of water seaward on a beach after the advance of a wave. See also: *backrush.* (AGI, 1987)
back work (a) Any kind of operation in a mine not immediately concerned with production or transport; literally work behind the face; repairs to roads. (Mason, 1951) (b) *back coming; back splinting.*
bacon Eng. Fibrous carbonate of lime, also known as beef and horse-flesh; Isle of Portland. See also: *beef.* (Arkell, 1953)
bacon stone (a) Eng. Calcspar colored with iron oxide, Bristol. (Arkell, 1953) (b) An old name for a variety of steatite (rock gypsum), alluding to its greasy luster. See also: *speckstone.*
baculite A crystallite that appears as a dark rod. (AGI, 1987)
bad air Air vitiated by powder fumes, noxious gases, or respirable dust.
baddeleyite A monoclinic mineral, ZrO_2; may contain some hafnium, titanium, iron, and thorium.
badlands A region nearly devoid of vegetation where erosion has produced, usually in unconsolidated or poorly cemented clays and silts, a dense and intricate drainage pattern with short steep slopes and sharp crests and pinnacles. Specif., the Badlands of the Dakotas.
bad top A coal mining term indicating a weak roof. Bad top sometimes develops following a blast. (Kentucky, 1952)
baeumlerite A colorless chloride of potassium and calcium, $KCl{\cdot}CaCl_2$. Intergrown with halite and tachyhydrite. Orthorhombic. Syn: *chlorocalcite.* From Leintal, Germany. (English, 1938)
baff ends Long wooden edges for adjusting linings in sinking shafts during the operation of fixing the lining. (Zern, 1928)
baffle board A board fitted across a compartment in an ore washer to retain the heavy ore and allow the light material to flow away. (Nelson, 1965)
baffle plate (a) A loading plate attached to the frame of a belt conveyor to prevent spillage at any loading point. (Jones, 1949) (b) A tray or partition placed in a tower, a heat exchanger, or other processing equipment to direct or to change the direction of flow of fluids. (c) A metal plate used to direct the flames and gas of a furnace to different areas so that all portions of it will be heated; a deflector.
baffler A partition in a furnace so placed as to aid the convection of heat; a baffle plate. (Fay, 1920)
baffle tube A pipe of sufficient length to lower the temperature of hot gases before the gases enter a furnace. (CTD, 1958)
baffle wall A refractory wall used to deflect gases or flames from the ware and to provide better heat distribution in the furnace structure.
bag (a) A paper container roughly 2.5 to 5 cm in diameter and 20 to 46 cm long, used for placing an inert material, such as sand, clay, etc., into a borehole for stemming or tamping. Also called a tamping bag. (b) A long tube fastened at the upper end to a pipe leading from a smelter, and gathered and tied at the lower end. The smoke passes through the cloth, which catches the solids. The bag is periodically untied and the dust is shaken out. See also: *baghouse.* (c) A cavity in coal containing gas or water. See also: *bag of gas.* (Tomkeieff, 1954) (d) Flexible pipe or hose. Also called bagging. (Mason, 1951) (e) S. Staff. A quantity of combustible gases suddenly given off by the coal seam. See also: *bag of foulness.* (f) York. A miner's term for a variety of inferior coal. (Tomkeieff, 1954)

bag filter An apparatus for removing dust from dust-laden air, employing cylinders of closely woven material that permit passage of air but retain solid particles. Syn: *filter.* (BS, 1962)

baghouse Chamber in which exit gases from roasting, smelting, melting, or calcining are filtered through membranes (bags) that arrest solids such as fine particulates. See also: *bag.* (Pryor, 1963)

bag of foulness N. of Eng. A cavity in a coal seam filled with combustible gases under a high pressure, which, when cut into, are given off with much force. See also: *bag.* (Fay, 1920)

bag of gas Eng. A gas-filled cavity found in seams of coal. See also: *bag.*

bag powder Originally applied to black powder loaded in bags, but now applied to a number of explosives so packed. The bags are long, cylindrical units about 6 in (15 cm) in diameter and weighing 12-1/2 lb (5.67 kg) apiece. (Carson, 1961)

bag process A method of recovering flue dust and also sublimed lead, whereby furnace gases and fumes are passed through bags suspended in a baghouse. The furnace gases thus are filtered, and the particles in suspension collected. (Fay, 1920)

bahada *bajada.*

baikalite A dark-green variety of diopside containing iron; found near Lake Baikal, Russia.

baikerinite A thick, tarry hydrocarbon that makes up about one-third of baikerite and from which it may be separated by alcohol. See also: *baikerite.*

baikerite (a) A waxlike mineral from the vicinity of Lake Baikal, Russia, apparently about 60% ozocerite. (Fay, 1920) (b) A variety of ozocerite. See also: *baikerinite.*

bail (a) As used by churn drillers, to remove a liquid from a borehole by use of a tubular container attached to a wire line. See also: *bailer.* (b) The handle on a bucket, cage, or skip by means of which it may be lifted or lowered. (Long, 1960) (c) A large clevis. (Long, 1960) (d) To dewater a mine with a skip or bailer. (e) As used by the diamond and rotary-drilling industries, (1) a U-shaped steel rod with the open ends formed into eyes fitting over two lugs projecting from the sides of a water swivel, or (2) a U-shaped steel rod with open ends attached to an open-sided, latch-equipped, circular collar, that fits around a drill rod and under the base of a water swivel. Both types of bails are designed to permit circulation of fluid through the drill rod string while the rods are suspended on the hoist line or while the rods are being raised or lowered a few feet with the hoisting cable. (Long, 1960)

bailer (a) A long cylindrical vessel fitted with a bail at the upper end and a flap or tongue valve at the lower extremity. It is used to remove water, sand, and mud-laden or cuttings-laden fluids from a borehole. When fitted with a plunger to which the bailing line is attached, it sucks the liquid in as it is lifted and is then called a sand pump or an American pump. Syn: *bucket.* (Long, 1960) (b) A metal tank, or skip, with a valve in the bottom, used for dewatering a mine. (c) *sludger; swab.* (d) In bituminous coal mining, a laborer who scoops water from drainage ditches in a mine with a bucket and empties it into a water car, a ditch flowing to a natural outlet or to a pumping station. Also called water bailer. (DOT, 1949) (e) A cylindrical steel container with a valve at the bottom for admission of fluid, attached to a wire line and used in cable-tool drilling for recovering and removing water, cuttings, and mud from the bottom of a well. See also: *bail; bailing.* (AGI, 1987)

bailiff Eng. A name formerly used for manager of a mine.

bailing (a) Removal of the cuttings from a well during cable-tool drilling or of liquid from a well by means of a bailer. (Inst. Petrol., 1961) (b) Dewatering a mine. See also: *bailer.* (c) Removing rock dust and other material loosened in the drilling by means of a bucket or ball. (Mersereau, 1947)

bajada (a) A broad, continuous alluvial slope or gently inclined detrital surface extending from the base of mountain ranges out into and around an inland basin, formed by the lateral coalescence of a series of alluvial fans, and having an undulating character due to the convexities of the component fans; it occurs most commonly in semiarid and desert regions, as in the Southwestern United States. A bajada is a surface of deposition, as contrasted with a pediment (a surface of erosion that resembles a bajada in surface form), and its top often merges with a pediment. Etymol: Sp., descent, slope. Cf: *alluvial slope; alluvial fan.* Syn: *bahada; alluvial plain; piedmont plain.* (AGI, 1987) (b) *ladderway.* (c) Sp. Compound alluvial fans. (AGI, 1987)

baja de metales Peru. Lowering of ores from mine to mill.

bajo Colom. Low-lying alluvial mines that have to be unwatered by artificial means; generally deposits in present riverbeds.

bakerite A monoclinic mineral, $Ca_4B_4(BO_4)(SiO_4)_3(OH)_3{\cdot}H_2O$; gadolinite group; in white compact nodules resembling marble or unglazed porcelain; in the Mohave Desert, CA.

baking (a) A stage in the heating of a clay when the clay particles have lost plasticity and have formed a moderately hard mass composed of particles adhering together, the mass remaining porous. See also: *vitrifying.* (Nelson, 1965) (b) The process of firing shaped clay articles in kilns, in order to give the clay permanent hardness. (CTD, 1958) (c) Heating to a low temperature in order to remove gases. (ASM, 1961) (d) The hardening of rock material by heat from magmatic intrusions or lava flows. Prolonged baking leads to contact-metamorphic effects. (AGI, 1987)

bal A Cornish name for a mine; a cluster of mines.

balance (a) The counterpoise or weight attached by cable to the drum of a winding engine to balance the weight of the cage and hoisting cable and thus assist the engine in lifting the load out of the shaft. (Fay, 1920) (b) An instrument for weighing. (Fay, 1920) (c) *assay balance; balance pit.* (d) A beam device specif. designed and calibrated to determine specific gravity by weighing methods, as in determining the specific gravity of drilling mud. (Long, 1960)

balance bob A counterbalance to take the excess weight of the pitwork, or timber beams, in a shaft; used with the Cornish type of reciprocating pump. (CTD, 1958)

balance brow (a) A self-acting inclined plane down which the cars of coal are lowered and the empties elevated upon a carriage or platform. Also called balance plane; back balance. (b) Eng. An inclined roadway in which a balance is used to assist the haulage. Also called dilly brow. (SMRB, 1930)

balance car (a) In quarrying, a car loaded with iron or stone and connected by means of a steel cable with a channeling machine operating on an inclined track. Its purpose is to counteract the force of gravity and thus enable the channeling machine to operate with equal ease uphill and downhill. (b) A small weighted truck mounted upon a short inclined track, and carrying a sheave around which the rope of an endless haulage system passes as it winds off the drum. (Zern, 1928)

balanced cutter chain A cutter chain that has the same number of bottom and top picks. It usually cuts more freely in hard material and is often used for cutting at higher than floor level. See also: *unbalanced cutter chain.* (Nelson, 1965)

balanced direct-rope haulage A modified form of direct-rope haulage, in which a power-driven reversible pulley (surge pulley) is used instead of a drum. The full trams are hauled up on one end of the rope while the empties go down on the other end. It involves a double track or a bypass midway on the haulage plane. The descent of the empty trams assists in balancing the load being hauled upwards. (Nelson, 1965)

balanced draft Applied to combustion units in which forced and induced drafts are adjusted to give atmospheric pressure in the combustion chamber to avoid the infiltration of unwanted cold air. (Nelson, 1965)

balanced hoisting Arrangement of cages or skips in mine shaft in which the winding drum raises one and at the same time lowers the other, thus reducing power consumption. See also: *balanced winding.* (Pryor, 1963)

balanced ventilation A system of ventilation in which the districts (each with its separate split) are so arranged with regard to length and resistance, that the use of ventilation regulators is unnecessary. Regulators, although sometimes unavoidable, reduce the efficiency and increase the power required to ventilate the mine. (Nelson, 1965)

balanced vibrating conveyor A vibrating conveyor in which the center of gravity of the complete assembly is held constant by having movement of the trough offset by opposite movement of some other element.

balanced winding The conventional method of winding in a mine shaft. As the cage containing the loaded cars ascends, the other cage containing the empties descends, and thus the cages and cars are balanced. Balanced winding also implies the use of a balance rope, and thus, ignoring friction, the only load to be hoisted is the coal or mineral. See also: *winding; balanced hoisting.* (Nelson, 1965)

balance pit Eng. A pit or shaft in which a balance (counterweight) rises and falls. See also: *balance.*

balance plane *balance brow.*

balance rope A steel-wire rope, generally of the same weight per foot as the main winding rope, that is attached to the bottom of the cages, and extends down to form a loop in the shaft bottom or sump. Its function is to balance out the difference in weight of the upgoing or downgoing main ropes during the wind. See also: *winding.* (Nelson, 1965)

balance sheet A record showing the present financial obligations and resources of the company, in terms of cost or book value. (Hoover, 1948)

balance shot In coal mining, a shot for which the drill hole is parallel to the face of the coal that is to be broken by it.

balas *balas ruby.*

balas ruby A pale rose-red or orange variety of spinel in Badakhshan (Balascia) Province of northern Afghanistan. Syn: *balas; ballas; false ruby.*

balata belt A belt with normal multi-ply construction, and in which balata is used to impregnate the plies and provide cover. It cannot be used in high temperatures but possesses a very high resistance to water absorption and is thus well suited for wet conditions. (Nelson, 1965)

Balbach process Electrolytic separation of gold from silver, using the alloy as anode, graphite plate cathodes, and silver nitrate solution as bath. (Bennett, 1962)
bald Without framing; said of a mine timber that has a flat end.
bald-headed anticline An anticline whose crest has been eroded prior to deposition of an unconformably overlying sedimentary unit. Cf: *breached anticline.* (AGI, 1987)
balistite *ballistite.*
balkstone (a) Eng. A provincial name given to an impure stratified limestone. (Fay, 1920) (b) Sandstone used for whetstone. Also called balkerstone. (Arkell, 1953)
ball (a) A rounded mass of spongy iron, prepared in a puddling furnace; a loup. (Fay, 1920) (b) A mass of tempered fire clay, used for forming the crucible in crucible-steel production. See also: *ballstone.* (CTD, 1958)
ball-and-socket reamer A borehole-reaming device consisting of a bit attached to a ball-and-socket or a knuckle-joint member, that in turn is connected to the drill rods and used in borehole-deviation drilling. Also called arc cutter. (Long, 1960)
ball and test A deep well pump valve in which a ball fits into a seat and prevents the backflow of oil or water. Each standing valve and each traveling valve has a ball and seat. (Hess)
ballas (a) A hard, spherical aggregate of many very small diamond crystals, usually cryptocrystalline, arranged radially and more or less concentrically around a central point. Because of their structure, ballas are classed as industrials that are occasionally used in diamond-drill bits and other diamond tools. See also: *shot bort.* (Long, 1960) (b) A dense, globular aggregate of minute diamond crystals, having a confused radial or granular structure lacking through-going cleavage planes, giving it a toughness that makes it useful as an "industrial diamond". Cf: *bort; carbonado.* (c) A term incorrectly applied to a rounded, single crystal of diamond. (d) *balas ruby.*
ballast (a) Broken stone, gravel, water, or other heavy material used to provide weight in a ship or other machine and therefore improve its stability or control its draft. Jettisoned ballast may be found in samples of marine sediments. (AGI, 1987) (b) Gravel, broken stone, expanded slag, or similar material used as a foundation for roads, esp. that laid in the roadbed of a railroad to provide a firm bed for the ties, distribute the load, and hold the track in line, as well as to facilitate drainage. (AGI, 1987)
ballast car A freight car (as for carrying ballast) that may be unloaded from the side or bottom. (Webster 3rd, 1966)
ballast engine A steam engine used in excavating and for digging and raising stones and gravel for ballast. (Webster 3rd, 1966)
ballast shovel A spoon-pointed iron shovel having a thick body. (Standard, 1964)
ball bearing A friction-reducing device consisting of hard steel balls in a circular race; also applied to some pieces of equipment, such as a swivel-type double-tube core barrel, in diamond drilling using ball bearings as load-bearing members on rotating parts. See also: *bearing.* (Long, 1960)
ball burnishing (a) See also: *ball sizing.* (b) Removing burrs and polishing small stampings and small machined parts by tumbling. (ASM, 1961)
ball clay A highly plastic, sometimes refractory clay, commonly characterized by the presence of organic matter, having unfired colors ranging from light buff to various shades of gray, and used as a bonding constituent of ceramic wares; pipe clay. It has high wet and dry strength, long vitrification range, and high firing shrinkage. Ball clay is so named because of the early English practice of rolling the clay into balls weighing 30 to 50 lb (13 to 22 kg) and having diameters of about 10 in (25 cm). (AGI, 1987)
ballers White sand with large spheroidal masses of calciferous sandstone called sand ballers or giants' marbles, some being 3 to 6 ft (approx. 1 to 2 m) in diameter. (Possibly a variant of "bollars," a dialect form of boulders.)
ball grinder A pulverizer or disintegrator consisting of metal balls enclosed in a rotating cylinder. (Fay, 1920)
ball head *ball stamp.*
balling (a) A process that occurs in the cementite constituent of steels on prolonged annealing at 650 to 700 °C. (CTD, 1958) (b) The operation of forming balls in a puddling furnace. Syn: *nodulizing.* (CTD, 1958)
balling formation Rock or formations that, when drilled, produce cuttings and sludge that tend to collect on, and adhere to, borehole walls and drill-stem equipment in sticky or gummy masses. Cf: *gummy; sticky.* (Long, 1960)
balling furnace (a) A kind of reverberatory furnace used in alkali works. (Fay, 1920) (b) A furnace in which piles or fagots of wrought iron are placed to be heated preparatory to rolling. (Fay, 1920)
balling tool A tool used in collecting the iron in a puddling furnace into a mass, preparatory to taking it to the hammer or squeezer; a rabble. (Fay, 1920)
ball ironstone (a) A sedimentary rock containing large argillaceous nodules of ironstone. (AGI, 1987) (b) Nodular iron ore.
ballistic mortar test A laboratory instrument used for measuring the relative weight strength of an explosive material. Also, a test in which a standard weight of explosive is placed within a small borehole fitted with a projectile. The mortar, suspended on a pendulum, recoils upon detonation. The recoil is a measure of the weight strength in percentage (relative to a standard whose value is 100) or pendulum deflection. (Meyer, 1981)
ballistite A smokeless powder consisting essentially of soluble cellulose nitrates and nitroglycerin in approx. equal parts. Syn: *balistite.* (Webster 3rd, 1966)
ball jasper (a) Jasper showing concentric red and yellow bands. (b) Jasper occurring in spherical masses.
ball mill A rotating horizontal cylinder with a diameter almost equal to the length, supported by a frame or shaft, in which nonmetallic materials are ground using various types of grinding media such as quartz pebbles, porcelain balls, etc. Syn: *air-swept ball mill.* See also: *Abbé tube mill; cannonball mill; jar mill.*
ball mill grindability test Crushed particles of a given size range are placed in a ball mill; the reduction in size of particles for a given number of revolutions of the mill is interpreted in terms of a grindability index. (Lewis, 1964)
ball milling A method of grinding and mixing material, with or without liquid, in a rotating cylinder or conical mill partially filled with grinding media such as balls or pebbles. (ASTM, 1994)
ball mill method A grindability method based on the principle that all coals are ground to the same fineness, about that required for pulverized fuels, and then using the relative amounts of energy required for this reduction in size as a measure of grindability. (Mitchell, 1950)
Ball-Norton magnetic separator Dry separator for coarse ore, in which one or two nonmagnetic drums rotate outside a series of fixed magnets alternating in polarity. (Pryor, 1963)
balls (a) Common name for nodules, esp. of ironstone. (Arkell, 1953) (b) In fine grinding, crushing bodies used in a ball mill. Cast or forged iron or steel, or alloy of iron with molybdenum or nickel, are used, mainly spherical; various other shapes are favored locally, e.g., concave.
ball sizing Sizing and finishing a hole by forcing a ball of suitable size, finish, and hardness through the hole or by using a burnishing bar or broach consisting of a series of spherical bands of gradually increasing size coaxially arranged. Also called ball burnishing, and sometimes ball broaching. (ASM, 1961)
ball stamp A rock-crushing stamp whose stem is the piston rod of a steam cylinder. Syn: *ball head.* (Fay, 1920)
ballstone (a) An ancient term for ironstone, North Staffordshire, U.K. (b) A large crystalline mass of limestone containing coral in position of growth, surrounded by shale and impure bedded limestone. See also: *caballa ball.* (c) A nodule or large rounded lump of rock in a stratified unit; specif. an ironstone nodule in a coal measure. Syn: *ball.*
ball vein A stratum in which siderite concretions occur; also, the ore itself. (Hess)
bally seating Underclay with nodular concretions. (Arkell, 1953)
balmaiden Corn. A woman employed in the mines. (Standard, 1964)
balnstone Eng. Stone in the roof of a coal seam; roof stone. (Arkell, 1953)
baltimorite A grayish-green and silky, fibrous, or splintery variety of serpentine; near Baltimore, MD. See also: *antigorite.*
banakite A basaltic rock composed of olivine and clinopyroxene phenocrysts in a groundmass of labradorite with alkali feldspar rims, olivine, clinopyroxene, some leucite, and possibly quartz. Banakite grades into shoshonite with an increase in olivine and clinopyroxene and with less alkali feldspar, and into absarokite with more olivine and clinopyroxene. It was named by Iddings in 1895 from the Bannock (or Robber) Indians. (AGI, 1987)
banalsite An orthorhombic mineral, $BaNa_2Al_4Si_4O_{16}$; feldspar group.
band (a) Shale or other rock interstratified with coal, e.g., dirt band, sulfur band, etc. (b) A thin stratum or lamina of conspicuous lithology or color. A group of such layers is described as being banded. Cf: *parting.* (c) Any well-defined and widespread thin rock deposit that is of value in correlation. (d) Slate or other rock interstratified with coal, commonly called middle band in Arkansas; also, dirt band, sulfur band, or other band, as the case may be. (Fay, 1920) (e) Applied to a stratum or lamina conspicuous because it differs in color from adjacent layers; a group of layers displaying color differences is described as being banded. (AGI, 1987)
band chain A steel or invar tape of a minimum length of 100 ft (30.5 m) used for accurate surveying, graduated in feet. See also: *reglette.* (Hammond, 1965)
band conveyor *belt conveyor.*

banded The property of rocks having thin and nearly parallel bands of different textures, colors, or minerals. Banded coal has alternating bands of different types. (Johannsen, 1931-38; Pryor, 1963)
banded agate Agate in colors disposed in parallel or subparallel bands, more or less wavy or sinuous. Most agate in the trade is dyed, and bands are of differing tones due to varying capacity to absorb the dye. See also: *agate; onyx.*
banded coal (a) The common variety of bituminous and subbituminous coal. It consists of a sequence of irregularly alternating layers or lenses of homogeneous black material having a brilliant vitreous luster; grayish-black, less brilliant, striated material usually of silky luster; and generally thinner bands or lenses of soft, powdery, and fibrous particles of mineral charcoal. The difference in luster of the bands is greater in bituminous than in subbituminous coal. Also called bright-banded coal; common-banded coal. (b) Coal composed of roughly parallel, dull and bright layers. (BS, 1960)
banded differentiate Any igneous rock made up of bands of differing chemical or mineral composition, usually an alternation of two rock types; a layered intrusion. The structure has been attributed to rhythmic crystal settling during convection.
banded ingredient One of the four distinctive and visibly differing portions forming the mass of an ordinary bituminous coal that can be recognized and separated macroscopically by hand, and microscopically in thin sections, and that are not, in themselves, chemical entities; i.e., vitrain, clarain, fusain, and durain. See also: *rock type.* (AGI, 1987)
banded iron formation Iron formation that shows marked banding, generally of iron-rich minerals and chert or fine-grained quartz. Abbrev: BIF. (AGI, 1987)
banded ironstone A term used in South Africa for iron formation consisting essentially of iron oxides and chert occurring in prominent layers or bands of brown or red and black. This usage of the term ironstone is at variance with that applied in the United States and elsewhere. Syn: *ironstone.* (AGI, 1987)
banded obsidian Obsidian with differently colored irregular bands.
banded ore Ore composed of bands as layers that may be composed of the same minerals differing in color, textures, or proportions, or they may be composed of different minerals. Syn: *banded texture.* (AGI, 1987)
banded peat Peat composed of bands of vegetable debris alternating with bands of sapropelic matter. (Tomkeieff, 1954)
banded-quartz hematite *itabirite.*
banded quartz-hematite ore Braz. In the Itabira Region of Minas Gerais, schistose, specular hematite forming alternate bands with sugary quartz. Some of the beds are auriferous and contain gold-palladium alloys with manganese oxides, native copper, and talc. Writers have given the rocks various names, such as iron-glance schist, jacutinga, quartz itabirite, and bandererz. (Hess)
banded structure (a) An outcrop feature developed in igneous and metamorphic rocks as a result of alternation of layers, stripes, flat lenses, or streaks differing conspicuously in mineral composition and/or texture. (AGI, 1987) (b) A term applied to veins having distinct layers or bands. This may be due to successive periods of deposition or replacement of some earlier rock. (Fay, 1920) (c) A structure developed in many igneous and metamorphic rocks owing to layers that differ noticeably in mineral composition or texture. (d) A segregated structure of nearly parallel bands aligned in the direction of working. (ASM, 1961)
banded texture *banded ore.*
banded vein A vein made up of layers of different minerals parallel with the walls. Also called ribbon vein. (Fay, 1920)
band scale An arrangement by which colliers are paid an agreed sum for removing a dirt band, in addition to the usual tonnage rate. The payment varies with the thickness of the band. (Nelson, 1965)
band wander In concentration on shaking table, the movement of a segregated band of mineral so that it no longer discharges from the table deck at the desired point and therefore is not correctly collected. See also: *wander.* (Pryor, 1958)
bandylite A tetragonal mineral, $CuB(OH)_4Cl$; occurs as dark blue crystals in Chile.
bandy metal Shale with thin sandstone bands. (Arkell, 1953)
bank (a) A large pile of mineral material on the ground surface, as in heap leaching. (b) Several like pieces of equipment set close together, as a bank of flotation cells, hydrocyclones, or generators. (c) The surface around the mouth of a shaft. (Zern, 1928) (d) The whole or sometimes only one side or one end of a working place underground. (e) A hill or brow. (f) A road along the coal face formed by the coal on one side and the waste or packs on the other; thus, a double-unit face has a right and left bank. (g) A generally steeply sloping mass of any earthy or rock material rising above the digging level from which the soil or rock is to be extracted from its natural or blasted position in an open-pit mine or quarry. Syn: *bench face.* (h) Terracelike bench from which ore is obtained in an open-pit mine.
Banka drill A portable, manually operated system used in prospecting alluvial deposits to depths of 50 ft (15.2 m) or more. Also known as an Empire drill.
bank claim A mining claim on the bank of a stream.
bank coal Coal contained in, and sometimes salvaged from, the bank. (BCI, 1947)
bank engine Eng. An engine at the mouth of a mine shaft. (Standard, 1964)
banker off Aust. The worker who attends to taking skips off the cage. (Fay, 1920)
banket (a) A general term for a compact, siliceous conglomerate of vein-quartz pebbles about the size of a pigeon's egg, embedded in a quartzitic matrix. The term was originally applied in the Witwatersrand area of South Africa to the mildly metamorphosed gold-bearing conglomerates containing muffin-shaped quartz pebbles and resembling an almond cake made by the Boers. Etymol: Afrikaans, a kind of confectionery. (AGI, 1987) (b) Originally applied by the Dutch settlers to the gold-bearing conglomerates of the Witwatersrand. It is now used more widely for similar conglomerates and conglomeratic quartzites. (CTD, 1958)
bank gravel Gravel found in natural deposits, usually more or less intermixed with sand, silt, or clay. (AGI, 1987)
bank head (a) The upper end of an inclined plane, next to the engine or drum, made nearly level. (Zern, 1928) (b) The mouth and immediate environs of a coal mine. (Webster 3rd, 1966)
bank height The vertical height of a bank as measured between its highest point or crest and its toe at the digging level or bench. Cf: *berm.* Also called: bench height; digging height.
banking (a) The bringing of a cage to a stop at the rail level (the pit top or bank) and the replacement of loaded mine cars by empty ones and the release of the cage for its return journey. (b) Closing down a blast furnace which is still full of burden.
bank measure (a) The quantity of an excavation measured in place in the bank before being disturbed. (Carson, 1961) (b) Volume of soil or rock in its original place in the ground. (Nichols, 1954)
bank mining Surface mining in which the material mined is removed from above the surrounding land surface. (AIME, 2)
bank of cells A row of flotation cells in line. (Pryor, 1965)
bank of ovens A row of ovens for converting coal into coke. (Fay, 1920)
bank protection Devices for minimizing scour. These include brushwood held in place by wooden pegs, embankments, grass and withy planting, groins, mattresses, revetments, and riprap. (Hammond, 1965)
bank pump An auxiliary pump placed on the bank of a stream or a lake and used to pump water to a distant drill. Also called supply pump. (Long, 1960)
bank right The right to divert water for working a bank claim. (Pryor, 1963)
bank slope The angle, measured in degrees of deviation from the horizontal, at which the earthy or rock material will stand in an excavated, terracelike cut in an open-pit mine or quarry. Syn: *bench slope.*
bank slope stability A slope is subject to the influence of gravity and possible pressure of ground water, which tend to cause sliding or caving. It is also subject to surface erosion from running water, wind, and alternate freezing and thawing, or wetting and drying. Weathering causes changes in particle size and composition. Bank slope stability can be attained by benching, by growth of vegetation, and by artificial protections, such as masonry walls, drainage systems to intercept or remove ground water, and fences to catch rolling pieces. See also: *stability.* (Nichols, 1956)
banksman The person in charge of the shaft and cage or skip at the surface of a colliery; the person at the surface who operates the signals from the cage or skip to the winding engineman. See also: *cager.* (Mason, 1951)
bank water In placer mining, applied to streams brought to the pit in ditches, not under pressure. (Hess)
bankwork Eng. A system of working coal in South Yorkshire.
bank yards Yards of soil or rock measured in its original position, before digging. (Nichols, 1954)
banos Mex. Water collected in old mine workings.
banqueria Bol. In alluvial mining, a thick bed of blocks of granite, schists, and quartz.
bantams Small pebbles of a banded garnet-quartz rock; usually associated with diamond in the concentrate obtained when washing the diamond-bearing gravels from the Vaal River in the Republic of South Africa. The occurrence of bantams in a gravel deposit is considered a good indicator of diamond. (Chandler, 1964)
baotite A tetragonal mineral, $Ba_4(Ti,Nb)_8Si_4O_{28}Cl$.
bar (a) A placer deposit, generally submerged, in the slack portion of a stream. Also, an accumulation of gravel along the banks of a stream; bar diggings. (b) A mass of inferior rock in a workable deposit of granite. (AGI, 1987) (c) A fault across a coal seam or orebody. (AGI,

1987) (d) A banded ferruginous rock; specif. jaspilite. (AGI, 1987) (e) A vein or dike crossing a lode. (f) Any band of hard rock crossing a lode. (Arkell, 1953) (g) A unit of pressure equal to 1,000,000 dyn/cm^2, 1,000 mb (100 kPa), or 29.53 in (750 mm) of mercury. (Hunt, 1965) (h) A bank of sand, gravel, or other matter, esp. at the mouth of a river or harbor, often obstructing navigation. (Webster 2nd, 1960) (i) An offshore ridge or mound of sand, gravel, or other unconsolidated material submerged at least at high tide, esp. at the mouth of a river or estuary, or lying a short distance from, and usually parallel to, the beach. (Hunt, 1965) (j) A drilling or tamping rod. (Fay, 1920) (k) A strap or beam used to support the roof between two props or other supports. (Mason, 1951) (l) A length of steel pipe equipped with a flat cap at one end and a jackscrew on the opposite end by means of which the pipe may be wedged securely in a vertical or horizontal position across an underground workplace to serve as a base on which a small diamond or rock drill may be mounted. Syn: *drifter bar; drill bar; drill column.* (Long, 1960) (m) A heavy steel rod with either pointed or flattened ends used as a pry or as a tool by miners to dislodge loose rock in roof or sidewalls of an underground workplace. Syn: *scaling bar.* (Long, 1960) (n) A piece of material thicker than sheet, long in proportion to its width or thickness, and whose width-thickness ratio is much smaller than that of sheet or plate, as low as unity for squares and rounds. (ASM, 1961)

baralyme A compressed pill consisting of a blended mixture of barium octohydrate and calcium hydroxide. It is used as a carbon dioxide absorbent in rebreathing (diving) systems. (Hunt, 1965)

bararite A hexagonal mineral, $(NH_4)_2SiF_6$; dimorphous with cryptohalite; occurs over a burning coal seam.

bar-belt conveyor A conveyor similar to a plate-belt conveyor but in which spaced steel rods arranged transversely are employed in place of the steel plates. (BS, 1962)

barbertonite A hexagonal mineral, $Mg_6Cr_2(CO_3)(OH)_{16}{\cdot}4H_2O$; manasseite group; rose-pink to violet; dimorphous with stichtite.

barbosalite A hydrous ferrous ferric phosphate, $Fe^{2+}Fe^{3+}{}_2(PO_4)_2(OH)_2$; occurs as black grains from Brazil. Syn: *ferro-ferri-lazulite.* (Spencer, 1955)

bar channeler A reciprocating drill mounted on a bar by means of which holes are drilled close together in line by shifting the drill from point to point along the bar. Thereafter, the webs between the holes are removed with a reciprocating chisel-pointed broaching tool that is substituted for the drill. This method of channeling is generally employed in the harder rocks, such as granites. (Hess)

bar coal cutter A coal cutter in which the cutting member was a projecting rotating bar armed with picks throughout its length. The bar cut a kerf in the seam as the machine traveled along the face. The first patent for a bar machine was taken out in 1856. The cutter is now obsolete. (Nelson, 1965)

bar diggings A term applied in the Western United States to diggings for gold or other precious minerals located on a bar or in the shallows of a stream, and worked when the water is low. See also: *bar mining.* (AGI, 1987)

bar drill A small diamond- or other-type rock drill mounted on a bar and used in an underground workplace. Also called bar and used in an underground workbar rig. (Long, 1960)

bare (a) To cut coal by hand; to hole by hand. (Mason, 1951) (b) The uncased portion of borehole. Also called called barefoot; blank; naked; open; open hole. See also: *blank hole.* (Long, 1960) (c) To remove overburden. (Arkell, 1953) (d) Eng. To strip or cut by the side of a fault, boundary, etc.; to make bare.

barefoot Said of an oil well without a liner in the oil-bearing rock. See also: *blank hole.* (Hess)

bare motor A motor without a pulley, belt-tightening base, or slide rails. (NEMA, 1961)

barequear Colom. In placer mining, to extract as much of the pay gravel as possible, without method, leaving the overburden untouched.

barequeo Colom. Extracting the rich ore by crude means.

barequero Colom. A placer miner who uses crude methods of alluvial washing. A spoiler.

barer A worker who removes surface soil or overburdens in a quarry. (Arkell, 1953)

barfe Saturday N. of Eng. The Saturday on which wages are not paid.

bar flight conveyor *drag-chain conveyor; flight conveyor.*

barge loader In the quarry industry, a laborer who controls the movement of a barge in a river as it is loaded with crushed rock. (DOT, 1949)

barges Scot. Sheets of iron, zinc, or wood, used in wet shafts or workings for diverting the water to one side.

bar grizzly A series of spaced bars, rails, pipes, or other members used for rough sizing of bulk material passed across it to allow smaller pieces to drop through the spaces. See also: *grizzly.*

barilla An impure sodium carbonate and sulfate obtained by burning various species of land or marine plants; soda ash. See also: *copper barilla; coro-coro.* (Standard, 1964)

baring (a) The small coal made in undercutting coal seams. (Webster 3rd, 1966) (b) A making bare; an uncovering. See also: *strip.* (Webster 2nd, 1960) (c) The surface soil and useless strata overlying a seam of coal, clay, ironstone, etc., that have to be removed preparatory to working the mineral. *overburden.*

barite (a) An orthorhombic mineral, $4[BaSO_4]$; has nearly pseudocubic cleavage; occurs as interpenetrant masses of crystals with sand and clay (desert roses); sp gr, 4.5; in veins or in residual masses on limestone; the principal source of barium. Syn: *barytes; dreelite; heavy spar; cawk.* (b) The mineral group anglesite, barite, and celestine.

barite rosette *petrified rose.*

barium A silvery-white, metallic element, belonging to the alkaline earth group. Symbol, Ba. Found chiefly in barite or heavy spar and witherite. All barium compounds that are water or acid soluble are poisonous. Used in paint, X-ray diagnostic work, glassmaking, oilwell drilling fluids, and pyrotechny. (Handbook of Chem. & Phys., 3)

barium feldspar *paracelsian.*

Barkhausen effect Observed result of magnetizing a ferromagnetic substance by means of a slow magnetic field increase. Orientation of domains proceeds in abrupt steps. (Pryor, 1963)

barley A stream size of anthracite known also as buckwheat No. 3, sized on a round punched plate. It passes through 1/4-in (6.4-mm) holes. At some mines, it has to pass over 3/32-in (2.4-mm) holes and at others over 1/16-in (1.6-mm) holes. The American Society of Mechanical Engineers has recommended that with a screen with circular holes, barley shall pass through 3/16-in (4.8-mm) holes and pass over 3/32-in holes. See also: *anthracite coal sizes; bird's-eye.* (Fay, 1920)

bar mining The mining of river bars, usually between low and high waters, although the stream is sometimes deflected and the bar worked below water level. See also: *bar diggings.*

barnesite (a) A monoclinic mineral, $(Na,Ca)_2V_6O_{16}{\cdot}3H_2O$. (b) Trademark for a rare-earth oxide used in glass polishing.

barograph A barometer that makes a continuous record of changes in atmospheric pressure. It is usually an aneroid type. See also: *barometer.* (AGI, 1987)

barometer An instrument that is used to measure atmospheric pressure. It may be either a mercury barometer or an aneroid barometer. See also: *barograph.* (AGI, 1987)

barometric leg In filtering system, use of a loop more than 30 ft (9.1 m) high between receiving vessel and vacuum pump, to protect latter against carryover of liquid. (Pryor, 1963)

barometric leveling A type of indirect leveling in which differences of elevation are determined from differences of atmospheric pressure observed with altimeters or barometers. (AGI, 1987)

barometric pressure The barometric pressure of the air at any point is that exerted by the weight of the atmosphere above that point. It therefore varies with the elevation of the point above or the depth below sea level. Barometric pressure is measured by the mercury barometer, and is of the order of 30 in (762 mm) of mercury at sea level. (Spalding, 1949)

baroque (a) Any pearl of very irregular form. (b) A baroque pearl; said of a pearl, or of a tumble-polished gem material, of irregular shape.

barotrauma A generic term for injury caused by pressure. Although squeeze is a colloquialism, it is an excellent descriptive term for all of the phenomena that occur when a rigid closed space within the body or on its surface fails to equalize with external pressure during descent, or is for some reason vented to lower pressure than that acting at the depth. (Hunt, 1965)

barranca A precipice; as used in some parts of Spanish America, a ravine or small canyon. Also spelled barranco.

barrandite A mineral intermediate between strengite and variscite.

barrel (a) As used in the petroleum industry, a volumetric unit of measurement equivalent to 42 U.S. gal (159.0 L). (AGI, 1987) (b) The cylindrical part of a pump from which the movement of the piston causes a liquid or gas to be forcibly ejected. Also, the cylindrical part of a hydraulic jack or of a hydraulic-feed mechanism on a diamond drill. (Long, 1960) (c) The drum of a hoist. (Long, 1960) (d) A cylindrical container or drum having a capacity of 55 gal (208.2 L). (Long, 1960) (e) The water passage in a culvert. (Nichols, 1976) (f) Commonly, although incorrectly, used as a syn. for core barrel. See also: *drum.* (Long, 1960)

barrel copper Pieces of native copper occurring in sizes large enough to be extracted from the gangue, and of sufficient purity to be smelted without mechanical concentration. Syn: *barrel work.* (AGI, 1987)

barrel of oil A volumetric unit of measurement equivalent to 42 U.S. gal (159.0 L). (AGI, 1987)

barrel washer A washer comprising a cylinder rotating slowly about an axis that is slightly inclined to the horizontal, and into which the raw coal, with a current of water or of a suspension, is fed near its upper end. The clean coal is carried by the water or suspension to the lower end of the cylinder over a scroll that conveys the reject to the upper end of the cylinder. (BS, 1962)

barrel work Syn: *barrel copper.* Used in the Lake Superior mining region. (AGI, 1987)
barren (a) In leaching ores, said of a chemical solution from which valuable solute has been removed by precipitation, ion exchange, or solvent extraction before reuse. (Pryor, 1958) (b) Said of rock or vein material containing no minerals of value, or of strata without coal, or containing coal in seams too thin to be workable.
barren ground Strata containing seams of coal that are not of a workable thickness. In metal mining, ground that does not contain ore. See also: *dead bed.*Cf: *dead ground.*
barren hole *blank hole.*
barren measures Coal measures without workable seams. (Standard, 1964)
barren mine A mine may be fully developed and yet, owing to the barrenness of the ore, it would be impossible to work it with profit. (Ricketts, 1943)
barren solution A solution in hydrometallurgical treatment from which all possible valuable constituents have been removed; it is usually recycled back to plant for reuse in process. See also: *cyanide.* (Pryor, 1958)
barricade (a) The process of building a set of barriers to isolate a sufficient quantity of good air to protect mine workers from the asphyxiating gases formed after a fire or explosion. Miners wait behind the barrier until rescued. Used as an alternative to an escape attempt. (b) An artificial mound of earth, usually as high as the eaves of a magazine roof, that is erected to deflect the force of an explosion upward and to protect the enclosed building from flying objects. (c) Timber formwork to contain the material during hydraulic flushing in steep ore workings. (Nelson, 1965)
barrier Blocks of coal left between the workings of different mine owners and within those of a particular mine for safety and the reduction of operational costs. It helps to prevent disasters of inundation by water, of explosions, or fire involving an adjacent mine or another part of a mine and to prevent water running from one mine to another or from one section to another of the same mine. See also: *barrier pillar.* (Mason, 1951)
barrier gate Eng. *tailgate.*
barrier materials Materials such as lead and concrete that are used for protection from X-rays or gamma rays in radiographic installations. (Osborne, 1956)
barrier pillar (a) A solid block or rib of coal, etc., left unworked between two collieries or mines for security against accidents arising from an influx of water. See also: *barrier; pillar; barrier pillar.* (Zern, 1928) (b) Any large pillar entirely or relatively unbroken by roadways or airways that is left around a property to protect it against water and squeezes from adjacent property, or to protect the latter property in a similar manner. (Zern, 1928) (c) Incorrectly used for a similar pillar left to protect a roadway or airway, or group of roadways or airways, or a panel of rooms from a squeeze. (Zern, 1928)
barrier system N. of Eng. An approved method of working a colliery by pillar and stall, where solid ribs or barriers of coal are left in between working places. (Fay, 1920)
bar rig A small diamond or other rock drill designed to be mounted and used on a bar. Also called bar drill. (Long, 1960)
barring The end and side timber bars used for supporting a rectangular shaft. The bars are notched into one another to form a rectangular set of timber. Common sizes are from 9 to 12 in (23 to 30.5 cm) deep and from 3 to 6 in (7.6 to 15.2 cm) thick and may be made from larch, white pine, or red pine. See also: *cribbing; steel rectangular shaft supports.* (Nelson, 1965)
barring down (a) Loosening ore in a bin by means of a bar, so it will flow through the chute. (b) Prying off loose rock after blasting to prevent danger of fall. (Pryor, 1958)
barrings A general term for the setting of bars of timber for supporting underground roadways or shafts. (Nelson, 1965)
barring scrap Prying adhering scrap metal from runners, ladles, or skimmers. (Fay, 1920)
barro A Spanish and South American term for clay, loam, marl, or the overburden of alluvial gold deposits.
barroisite A dark green amphibole intermediate between hornblende and glaucophane. (English, 1938)
barrow (a) A wicker basket in which salt is put to drain. (Webster 2nd, 1960) (b) A box with two handles at one end and a wheel at the other. (Zern, 1928) (c) A vehicle in which ore, coal, etc., is wheeled; a push cart.
barrowman In mining, one who pushes shallow-bodied cars (barrows) or wheelbarrows used for transporting coal or ore along underground haulageways that are too low for ordinary mine cars. Also called buggyman. (DOT, 1949)
barrowway (a) A level through which coal or ore is wheeled. (b) Rails laid between the flat or siding and the coal face. (SMRB, 1930)
Barry mining *Nottingham system.*
barsanovite *eudialyte.*
bar screen *grizzly.*
bar timbering A method of timbering mine roadways by means of horizontal and upright bars. See also: *timber set.* (Nelson, 1965)
bar tin Solid, commercial tin. (Bennett, 1962)
Bartlett table A three-shelf table driven by an eccentric that gives it a vanning motion. Ore and water are fed on the upper shelf, giving two products, heads and tailings. The latter are retreated on the second shelf, and the tailings go to the third or lower shelf for retreatment. (Liddell, 1918)
Barvoys process A sink-float process in which the medium is a suspension of clay from the raw coal and minus 200- or 300-mesh barite in water, with the volume of the clay usually equal to about twice that of the barite. Barite clay and coal suspensions can be regulated to get effective washing gravities from 1.2 to 1.8. Sizes from run-of-mine to one-eighth inch may be cleaned by this process, which has been widely adopted in Europe. Also known as the Sophia-Jacoba process in German publications. (Mitchell, 1950)
barylite An orthorhombic mineral, $BaBe_2Si_2O_7$; forms hard (6 to 7 on the Mohs scale), colorless crystals; at Långban, Sweden; Franklin, NJ; and Park County, CO.
barysilite A white trigonal mineral, $Pb_8Mn(Si_2O_7)_3$; occurs at Långban, Sweden, and Franklin, NJ.
barysphere The interior of the Earth beneath the lithosphere, including both the mantle and the core. However, it is sometimes used to refer only to the core or only to the mantle. Syn: *centrosphere.* (AGI, 1987)
baryta Barium oxide; BaO. (AGI, 1987)
barytes *barite.*
barytocalcite (a) A monoclinic mineral, $BaCa(CO_3)_2$; trimorphous with alstonite and paralstonite. (b) A mixture of calcite and barite.
basal arkose An arkosic sandstone basal to a sedimentary sequence, resting unconformably on a granitic terrane; the arkosic equivalent of a granitic basal conglomerate. (AGI, 1987)
basal cleavage Mineral cleavage parallel to the basal pinacoid that is normal to the *c* crystallographic axis. Also called pinacoidal cleavage.
basal conglomerate A well-sorted, lithologically homogeneous conglomerate that forms the bottom stratigraphic unit of a sedimentary series and that rests on a surface of erosion, thereby marking an unconformity; esp. a coarse-grained beach deposit of an encroaching or transgressive sea. It commonly occurs as a relatively thin, widespread or patchy sheet, interbedded with quartz sandstone. (AGI, 1987)
basal pinacoid A pinacoid of two parallel faces that intersect only the *c* crystallographic axis. Also called a base. Syn: *basal plane.*
basal plane (a) A plane perpendicular to the *c*, or principal, axis in a tetragonal or hexagonal structure. (ASM, 1961) (b) Syn: *basal pinacoid.* (AGI, 1987)
basal reef S. Afr. A gold-bearing reef regarded as the principal carrier of gold in the Orange Free State. It has been associated with the Elsburg series of the central Witwatersrand and occurs below what has become known as the leader reef. (Beerman, 1963)
basalt A general term for dark-colored mafic igneous rocks, commonly extrusive but locally intrusive (e.g., as dikes), composed chiefly of calcic plagioclase and clinopyroxene; the fine-grained equivalent of gabbro. Nepheline, olivine, orthopyroxene, or quartz may be present. Adj. basaltic. Cf: *tholeiite.* (AGI, 1987)
basal thrust plane The sole fault underlying a series of overthrusts.
basaltic Pertaining to, made of, or resembling basalt; as, basaltic lava. See also: *basalt.*
basaltic dome *shield volcano.*
basaltic hornblende A variety of hornblende containing ferric (oxidized) iron in basalts and other basic igneous (volcanic) rocks.
basaltic layer *sima.*
basaltine *augite.*
basaluminite A mineral, $Al_4(SO_4)(OH)_{10}·5H_2O$; in veinlets lining crevices in ironstone. Cf: *felsöbanyaite*
basanite (a) An extrusive rock composed of calcic plagioclase, augite, olivine, and a feldspathoid (nepheline, leucite, or analcime); essentially, a feldspathoidal olivine basalt. Some basanites have been the source of sapphires or rubies. (AGI, 1987) (b) A touchstone consisting of flinty jasper or finely crystalline quartzite. Syn: *touchstone; Lydian stone.* (AGI, 1987) (c) A black variety of jasper. (AGI, 1987)
base (a) As used by drillers, a line of stakes set by an engineer or drill foreman to be used as a guide to line up and point the drill in a specific compass direction. A line in a survey which, being accurately determined in length and position, serves as the origin for computing the distances and relative positions of remote points and objects by triangulation. See also: *base line.* (b) A compound, e.g., lime, ammonia, or caustic alkali, or an alkaloid, capable of reacting with an acid to form a salt either with or without elimination of water. (c) Foundation or supporting structure on which a drill is mounted. (Long, 1960) (d) *basal pinacoid; base course.*

base box A unit of quantity in the tin plate trade consisting of 112 sheets measuring 14 in by 20 in (35.6 cm by 50.8 cm) or the equivalent in area; consequently 31,360 in^2 (20.2 m^2) of tin plate.

base bullion Crude lead containing recoverable silver, with or without gold. See also: *work lead.* (ASM, 1961)

base charge (a) The main explosive charge in the base of a detonator. (Meyer, 1981) (b) The charge loaded into the bottom of vertical holes in quarrying, usually applicable to 3-in (7.62-cm) diameter holes and larger. (Nelson, 1965) (c) The detonating component in a detonator, initiated by the priming charge. (BS, 1964)

base course (a) A layer of specified or selected material of planned thickness constructed on the subgrade or subbase to serve one or more functions, such as distributing load, providing drainage, minimizing frost action, etc. (ASCE, 1958) (b) *base.*

base exchange (a) The clay particle with its cations may be regarded as a kind of salt in which the colloidal clay particle is the anion. Certain cations may replace others, making the clay more flocculent. The cation replacement is known as the base exchange. Syn: *ion exchange; cation exchange.* (AGI, 1987) (b) The physicochemical process by which one species of ions adsorbed on soil particles is replaced by another species. See also: *zeolite process.* (ASCE, 1958)

base failure *slope failure.*

base flow Water entering drainage system from underground sources. (Pryor, 1963)

base fracture In quarrying, used to describe the condition of the base after a blast. It may be a good or bad base fracture. (Streefkerk, 1952)

baselevel (a) The lowest level to which a land surface can be eroded by running water. (Mather, 1964) (b) To reduce by erosion to or toward a baselevel. (Standard, 1964)

baseleveled plain A baseleveled surface is any land surface, however small, that has been brought approx. to a baselevel, either general or local, by the process of gradation. When such a surface has considerable extent, it becomes a baselevel plain. Syn: *peneplain.* (AGI, 1987)

baselevel plain A flat, comparatively featureless area or lowland, the elevation of which cannot be materially reduced by the erosive force of running water. (AGI, 1987)

base line (a) A line taken as the foundation of operations in trigonometrical and geological surveys. See also: *base.* (b) A surveyed line established with more than usual care, that serves as a reference to which surveys are coordinated and correlated. See also: *base.* (AGI, 1987) (c) The initial measurement in triangulation, being an accurately measured distance constituting one side of one of a series of connected triangles, and used, together with measured angles, in computing the lengths of the other sides. (AGI, 1987) (d) One of a pair of coordinate axes (the other being the principal meridian) used in the U.S. Public Land Survey system. It consists of a line extending east and west along the true parallel of latitude passing through the initial point, along which standard township, section, and quarter-section corners are established. (AGI, 1987) (e) An aeromagnetic profile flown at least twice in opposite directions and at the same level, in order to establish a line of reference of magnetic intensities on which to base an aeromagnetic survey. (AGI, 1987) (f) The center line of location of a railway or highway; the reference line for the construction of a bridge or other engineering structure. Sometimes spelled: baseline. (AGI, 1987)

base map (a) A map on which information may be placed for purposes of comparison or geographical correlation. Base map was at one time applied to a class of maps now known as outline maps. It may be applied to topographic maps, also termed "mother maps," that are used in the construction of many types of maps by the addition of particular data. (AGI, 1987) (b) A map of any kind showing essential outlines necessary for adequate geographic reference, on which additional or specialized information is plotted for a particular purpose; esp. a topographic map on which geologic information is recorded. (AGI, 1987)

basement In geology, an underlying complex that behaves as a unit mass and does not deform by folding. (AIME, 1960)

basement complex A series of rocks generally with complex structure beneath the dominantly sedimentary rocks. In many places, these are igneous and metamorphic rocks of either Early or Late Precambrian, but in some places these may be much younger, as Paleozoic, Mesozoic, or even Cenozoic. See also: *complex.* Syn: *fundamental complex.* (AGI, 1987)

basement rock (complex) (a) A name commonly applied to metamorphic or igneous rocks underlying the sedimentary sequence. (AGI, 1987) (b) Metamorphic and igneous Precambrian rocks. (AGI, 1987)

base metal (a) Any of the more common and more chemically active metals, e.g., lead, copper. (AGI, 1987) (b) The principal metal of an alloy, e.g., the copper in brass. Cf: *noble metal.* (AGI, 1987) (c) In plural form, a classification of metals usually considered to be of low value and higher chemical activity when compared with the noble metals (gold, silver, platinum, etc.). This nonspecific term generally refers to the high-volume, low-value metals copper, lead, tin, and zinc. (d) The metal base to which a coating or plating is applied. (e) The chief constituent of a metal alloy; e.g., brass is a copper-base alloy.

base ore Ore in which gold is associated with sulfides, as contrasted to free-milling ores, in which the sulfides have been removed by leaching. (Newton, 1959)

base plug A tapered cylinder, generally of wood, placed in a borehole and into which a deflection drive wedge may be driven in a random or oriented position. Syn: *deflecting plug; deflection plug.* (Long, 1960)

base price There is a minimal market price for each metal below which it cannot fall without putting the average producer out of business; this price has been called the base price. (Hoover, 1948)

base rock (a) As used by some drillers, the solid rock immediately underlying the overburden material. (Long, 1960) (b) As used by drillers in the Midwestern United States, the igneous rock formations underlying the sedimentary rocks. Also called basement; basement rock; pavement. (Long, 1960)

base station An observation point used in geophysical surveys as a reference, to which measurements at additional points can be compared. (AGI, 1987)

basic (a) Said of an igneous rock having a relatively low silica content, sometimes delimited arbitrarily as 44% to 51% or 45% to 52% ; e.g., gabbro, basalt. Basic rocks are relatively rich in iron, magnesium, and/or calcium, and thus include most mafic rocks as well as other rocks. Basic is one of four subdivisions of a widely used system for classifying igneous rocks based on their silica content: acidic, intermediate, basic, and ultrabasic. Cf: *femic.* (AGI, 1987) (b) Said loosely of dark-colored minerals. Cf: *silicic; mafic.* (AGI, 1987) (c) Said of a plagioclase that is calcic.

basic bottom and lining The inner bottom and lining of a melting furnace consisting of materials like crushed burnt dolomite, magnesite, magnesite bricks, or basic slag, that give a basic reaction at the operating temperature. (ASM, 1961)

basic flowsheet A diagram of the various stages in the treatment of the raw coal in a preparation plant, usually either a process flowsheet or an equipment flowsheet. (BS, 1962)

basic front In granitization, an advancing zone enriched in calcium, magnesium, and iron, that is said to represent those elements in the rock being granitized that are in excess of those required to form granite. During granitization, these elements are believed to be displaced and moved through the rock ahead of the granitization front, to form a zone enriched in minerals such as hornblende and pyroxene. Syn: *mafic front; magnesium front.* (AGI, 1987)

basic grade Used to define steel produced by the basic open-hearth process. (Hammond, 1965)

basic lava Lava poor in silica, generally less than 52% total SiO_2; typically dark and heavy, as basalt.

basic lining A lining for furnaces, converters, etc., formed of nonsiliceous material, usually limestone, dolomite, lime, magnesia, or ferrous oxide. (Fay, 1920)

basic lining process An improvement of the Bessemer process in which, by the use of a basic lining in the converter and by the addition of basic materials during the blow, it is possible to eliminate phosphorus from the pig iron and keep it out of the steel. (Fay, 1920)

basic oxygen process A steelmaking process in which oxygen is forced at supersonic speed through a retractable water-cooled lance, accelerating the burning off of unwanted elements in a charge of molten iron and scrap. (Encyclopaedia Britannica, 1964)

basic price (a) As used by the drilling and mining industries, a guaranteed price to be paid for a specific quantity of materials, or type of service. (Long, 1960) (b) As applied to the price of metals, it is that figure at which the price is a minimum. See also: *normal price.* (Fay, 1920)

basic process A steelmaking process, either Bessemer, open-hearth, or electric, in which the furnace is lined with a basic refractory, a slag rich in lime being formed and phosphorus removed. See also: *acid process; dephosphorization.* (CTD, 1958)

basic refractories Refractories that consist essentially of magnesia, lime, chrome ore, or forsterite, or mixtures of two or more of these. (Harbison-Walker, 1972)

basic refractory lining A furnace lining, composed of material low in acidic minerals, such as silica, and high in basic minerals, such as lime, chromite, dolomite, magnesite, and magnesia.

basic slag Slag rich in bases, such as metallic oxides; specif., slag rich in lime, made during the basic Bessemer or basic open-hearth steel processes, and, from the quantity of phosphorus contained in it, valuable as an artificial fertilizer. (Standard, 1964)

basic solvent One that accepts protons from solute. (Pryor, 1963)

basic steel Steel melted in a furnace with a basic bottom and lining and under a slag containing an excess of a basic substance, such as magnesia or lime. (ASM, 1961)

basin (a) A natural depression of strata containing a coalbed or other stratified deposit. Cf: *depression.* (b) The lowest part of a mine or area of coal lands. (Hudson, 1932) (c) A general region with an overall history of subsidence and thick sedimentary section. (Wheeler, 1958)

basin-and-range Said of a topography, landscape, or physiographic province characterized by a series of tilted fault blocks forming longitudinal, asymmetric ridges or mountains and broad, intervening basins; specif. the Basin and Range physiographic province in the Southwestern United States. See also: *basin range.* (AGI, 1987)

basining The bending down or settling of part of the Earth's crust in the form of a basin, as by rock deformation or by solution of underground deposits of salt or gypsum. (AGI, 1987)

basin range A mountain range that owes its elevation and structural form mainly to faulting and tilting of strata and that is flanked by alluvium-filled basins or valleys. Etymol: from the Great Basin, a region in the Southwestern United States characterized by fault-block mountains. See also: *basin-and-range.* (AGI, 1987)

basis price The price agreed between the seller and the buyer of an option at which the option can be exercised. This price is also called the "strike price." The price is normally the current market price of the metal. See also: *option.* (Wolff, 1987)

basket (a) A type of single-tube core barrel made from thin-wall tubing with the lower end notched into points, which are intended to pick up a sample of granular or plastic rock material by bending in on striking the bottom of the borehole or a solid layer. Also used as a fishing tool to recover an article lost or dropped into a borehole. Syn: *basket core lifter; sawtooth barrel.* Cf: *calyx.* (Long, 1960) (b) Wire-mesh strainer in the top of a core barrel to strain out bits of debris, which might clog up the water ports in the core barrelhead. (Long, 1960)

basket centrifuge A device for dewatering in which wet coal is thrown by centrifugal force against a perforated containing surface that permits the outward passage of water and retains the coal. (BS, 1962)

basket core A sample of rock or rock material recovered by using a basket tube or core barrel. See also: *basket.* (Long, 1960)

basket core lifter A type of core lifter consisting of several fingerlike springs brazed or riveted to a smooth-surfaced ring having an inside diameter slightly larger than the core size being cut. Also called basket lifter; finger lifter. See also: *basket.* (Long, 1960)

basonomelane A variety of hematite containing titanium oxide. See also: *ilmenite.*

bass Eng. A black carbonaceous shale, Yorkshire, Lancashire, South and North Staffordshire. (Nelson, 1965)

basset (a) The outcropping edge of a geological stratum. (Webster 3rd, 1966) (b) The shallow or rise side of a working. (c) Coal outcrop. (Pryor, 1963) (d) An obsolete term for the noun outcrop and the verb to crop out. (AGI, 1987)

bassetite A yellow monoclinic mineral, $Fe(UO_2)_2(PO_4)_2 \cdot 8H_2O$; meta-autunite group.

bastard (a) Said of an inferior or impure rock or mineral, or of an ore deposit that contains a high proportion of noncommercial material. (AGI, 1987) (b) Said of any metal or ore that gives misleading assays or values. (AGI, 1987) (c) Said of a vein or other deposit close to and more or less parallel to a main vein or deposit, but thinner, less extensive, or of a lower grade. (AGI, 1987)

bastard asbestos Miners' term for picrolite (antigorite), a mineral associated in places with chrysotile asbestos.

bastard cauk Inferior barite; Derbyshire lead mines, U.K. (Arkell, 1953)

bastard emerald *peridot.*

bastard freestone Any inferior or impure rock; the Inferior Oolite, Bath, U.K.

bastard ganister A silica rock having the superficial appearance of a true ganister but characterized by more interstitial matter, a greater variability of texture, and often an incomplete secondary silicification. (AGI, 1987)

bastard quartz (a) *bull quartz.* (b) A round or spherical boulder of quartz embedded in soft or decomposed rock. (AGI, 1987)

bastard shale *cannel shale.*

bastite An olive-green, blackish-green, or brownish variety of serpentine mineral resulting from the alteration of orthorhombic pyroxene (esp. enstatite), occurring as foliated masses in igneous rocks; characterized by a schiller (metallic or pearly) on the chief cleavage face of the pyroxene. Syn: *schiller spar.*

bastnaesite Alternate spelling of bastnäsite. See also: *bastnasite.*

bastnasite A greasy, wax-yellow to reddish-brown weakly radioactive mineral, $(Ce,La)(CO_3)F$, most commonly found in contact zones, less often in pegmatites; found associated with allanite, cerite, tysonite, fluorite, and tornebohmite; hexagonal; obtained from Ryddarhyttan and Finbo, Sweden; Pikes Peak, CO, and Mountain Pass, CA. Syn: *bastnaesite.* (Crosby, 1955)

bastonite A greenish-brown mica that is closely related to phlogopite. (Standard, 1964)

batch A quantity of material destined for or produced by one operation. (Webster 3rd, 1966)

batch charger A mechanical device for introducing batch to the furnace. Syn: *batch feeder.* (ASTM, 1994)

batch feeder *batch charger.*

batch furnace A furnace in which each charge is placed, heated, and withdrawn on completion of work. (Pryor, 1963)

batch grinding The grinding of a charge of mineral (dry or wet) in a closed ball mill. (Pryor, 1963)

batch house The place where batch materials are received, handled, weighed, and mixed, for delivery to melting units. (ASTM, 1994)

batch mill A grinding mill, usually cylindrical, into which a charge of ore and water is placed and is ground to completion of the required comminution. (Pryor, 1963)

batch sintering Presintering or sintering in such a manner that the products are furnace-treated in individual batches. (Osborne, 1956)

batch smelter Any smelter that operates as a periodic unit, being charged, fired, and discharged according to a predetermined cycle. See also: *smelter.* (ASTM, 1994)

batch test A laboratory test on a small quantity of mineral under close control. (Pryor, 1963)

batch treatment Treatment of a parcel of material in isolation, as distinct from the treatment of a continuous stream of ore. (Pryor, 1958)

bate (a) To enlarge a colliery road by lowering the floor. Cf: *bating.* (Pryor, 1963) (b) Eng. Cleavage in slates, esp. in the Sheerbate stone. (Arkell, 1953) (c) Grain, hem, secondway in other rocks. Also spelled bait. (Arkell, 1953)

batea *pan; gamella.*

bate barrel Leic. After drawing a number of barrels of water out of a sump, the first barrel for which there is not sufficient water to fill it.

bath (a) The molten material in any furnace. (Standard, 1964) (b) A medium such as water, air, sand, or oil for regulating the temperature of something placed in or on it; also, the vessel containing such a medium. (Webster 3rd, 1966)

batholith A large, generally discordant plutonic mass that has more than 40 sq mi ($104\ km^2$) of surface exposure and no known floor. Its formation is believed by most investigators to involve magmatic processes. Also spelled: bathylith. Syn: *abyssolith.* Adj: *batholithic.* (AGI, 1987)

batholithic Pertaining to, originating in, or derived from a batholith.

bathotonic reagent A substance tending to diminish surface tension. See also: *depressant.* (Nelson, 1965)

bathvillite An amorphous, opaque, very brittle woody resin; forms fawn-brown porous lumps in torbanite; at Bathville, Scotland.

bathyal (a) Pertaining to the benthonic environment on the continental slope, ranging in depth from 200 to 2,000 m. (AGI, 1987) (b) Pertaining to the bottom and overlying waters between 100 and 1,000 fathoms (600 to 6,000 ft or 183 to 1,830 m). (AGI, 1987) (c) Of or pertaining to the deeper parts of the ocean; deep sea. (Webster 3rd, 1966)

bathyal zone In oceanography, the slope from the continental shelf at 100 fathoms (183 m) to the abyssal zone at 1,000 fathoms (1,830 m). Also called bathyal district. (Webster 3rd, 1966)

bathyclinograph In oceanography, an instrument for measuring vertical currents in the deep sea. (Webster 3rd, 1966)

bathyconductograph A device to measure the electrical conductivity of seawater at various depths from a moving ship. Abbrev. bc. (Hy, 1965)

bathygram In oceanography, a record obtained from sonic sounding instruments. (Webster 3rd, 1966)

bathylith *batholith.*

bathymeter This instrument measures temperature, pressure, and sound velocity to depths up to 7 miles (11.2 km). The device is completely transistorized and uses frequency modulation for telemetering. (Hunt, 1965)

bathymetric (a) Relating to the measurement of depths of water in oceans, seas, and lakes. Syn: *bathymetrical.* (Webster 3rd, 1966) (b) Relating to the contour of the bottoms of oceans, seas, and lakes. (Webster 3rd, 1966) (c) Relating to the distribution in depth of marine or lacustrine organisms. (Webster 3rd, 1966)

bathymetrical *bathymetric.*

bathymetric chart Chart showing depths of water by means of contour lines or by color shading. (Hy, 1965)

bathymetric contour *isobath.*

bathymetry In oceanography, the measurement of depths of water in oceans, seas, and lakes; also, the information derived from such measurements. (Webster 3rd, 1966)

bathyorographical In oceanography, of or relating to ocean depths and mountain heights. (Webster 3rd, 1966)

bathypelagic In oceanography, of, relating to, or living in the deeper waters of the ocean, esp. those several hundred feet below the surface—distinguished from abyssal and pelagic. (Webster 3rd, 1966)

bathyscaphe In oceanography, a navigable submersible ship that is used for deep-sea exploration, has a spherical watertight cabin attached to its underside, and uses gasoline and shot for ballast. (Webster 3rd, 1966)

bathysmal In oceanography, of or relating to the bottom of the deeper parts of the sea, esp. those parts between 100 fathoms (183 m) and 1,000 (1,830 m) fathoms deep. (Webster 3rd, 1966)

bathysophical In oceanography, of or relating to a knowledge of the depths of the sea or of the things found there. (Webster 3rd, 1966)

bathysphere In oceanography, a spherical diving apparatus, made large enough to contain two people and instruments; capable of resisting tremendous pressure, and therefore of descending to great depths; it is used in oceanography for the investigation of deepwater faunas. (CTD, 1958)

bathysystem A coined word for a permanent sea floor installation. (Hunt, 1965)

bathythermogram In oceanography, a record obtained with a bathythermograph. (Webster 3rd, 1966)

bathythermograph An instrument, which may be lowered into the sea from a vessel at anchor or underway, to record temperature as a function of depth. The temperature-sensing device is a Bourdon tube, the depth finder is a bellows system. Accuracy of temperature is ± 0.1 °F (± 0.056 °C); depth ± 10 ft (3.0 m). Abbrev. bt. (Hy, 1965)

bathythermosphere In oceanography, a bathythermograph. (Webster 3rd, 1966)

bathyvessel In oceanography, a ship (as a submarine or bathysphere) designed for exploration of or navigation in water far below the surface of a sea or lake. (Webster 3rd, 1966)

bating Eng. Lowering a drift or road. See also: *bate.*

batten (a) A strip of wood used for nailing across two other pieces (as to hold them together or to cover a crack). (Webster 3rd, 1966) (b) A piece of square-sawn converted timber, between 2 in and 4 in (5.1 and 10.2 cm) in thickness and from 5 to 8 in (12.7 to 20.3 cm) in width. Used for flooring or as a support for laths. (CTD, 1958) (c) A bar fastened across a door, or anything composed of parallel boards to secure them and to add strength and/or reduce warping. (CTD, 1958)

batter (a) Recessing or sloping a wall back in successive courses; opposite of corbel. (ACSG, 1961) (b) A paste of clay or loam. (Webster 2nd, 1960) (c) The inward slope from bottom to top of the face of a wall. (Nichols, 1976) (d) A pile driven at an angle to widen the area of support and to resist thrust. (Nichols, 1976)

batter boards Horizontal boards placed to mark a line and a grade of a proposed building or slope. (Nichols, 1976)

battered set A set of mine timbers in which the posts are inclined.

batter level An instrument for measuring inclination from the vertical. (Standard, 1964)

battery (a) *blasting machine; exploder.* (Nelson, 1965) (b) A number of similar machines or similar pieces of equipment placed side by side on a single or separate base and by means of common connections as a unit. (Long, 1960) (c) Mine support in which timbers are placed in groups of 3 to 12 or more. The battery may be strengthened by binding with wire. (Spalding, 1949) (d) A wooden platform for miners to stand upon while at work, esp. in steeply dipping coal beds. (e) A series or row of coke ovens. (Mersereau, 1947) (f) A bulkhead or structure of timber for keeping coal in place. (Hess) (g) The plank closing the bottom of a coal chute. (h) A series of stamps, usually five, operated in one box or mortar, for crushing ores; also, the box in which they are operated. (Hess) (i) A stamper mill for pulverizing stone. (Gordon, 1906) (j) Timbering in which the sticks are placed from foot to hanging wall, touching each other, in a solid mass of 3 to 12 or more. The battery may be further strengthened by binding around with wire. (Spalding, 1949) (k) In steeply pitching seams, a wooden structure built across the chute to hold back blasted coal. (Korson, 1938) (l) A number of stamps for crushing and pulverizing ores. (Nelson, 1965) (m) Section of ore dressing (reduction) plant. (Pryor, 1963) (n) A combination of chemically activated accumulators, which, after charging, may be used for a considerable time as a source of direct-current electricity. Also called storage battery. (Long, 1960)

battery charging station *locomotive garage.*

battery of holes A number of charges, in drill holes, fired simultaneously with an electric current. Also called multiple shot. (Fay, 1920)

battery of ovens A row or group of ovens for making coke from coal.

battery ore A type of manganese ore, generally a pure crystalline manganese dioxide (pyrolusite or nsutite), that is suitable for use in dry cells. (AGI, 1987)

battery starter In anthracite and bituminous coal mining, one who charges and sets off explosives in large lumps of coal or where these lumps have accumulated and blocked the flow of coal down chutes from the storage structures (batteries). Also called batteryman; chute tender; starter. (DOT, 1949)

battery starting The use of unconfined explosives to start the flow of coal down a breast or chute in an anthracite mine. (CFR, 4)

baulk A beam. (Mason, 1951)

Baumé gravity Designating or conforming to either of the scales used by the French chemist, Antoine Baumé (1728-1804). One scale, which is used with liquids heavier than water, sinks to 0° (B or Bé, symbols for Baumé) in pure water and to 15° (B or Bé) in a 15% salt solution. The other scale, for liquids lighter than water, sinks to 0° (B or Bé) in a 10% salt solution and to 10° (B or Bé) in pure water. (Webster 2nd, 1960)

baumhauerite A lead- to steel-gray triclinic mineral, $Pb_3As_4S_9$.

Baum jig A washbox in which the pulsating motion is produced by the intermittent admission of compressed air to the surface of the water following a principle introduced by Baum. Also called Baum box; Baum-type washbox. See also: *jig.* Syn: *Baum washer.* (BS, 1962)

baum pot (a) A cavity left in roof strata over coal as a result of the dropping downward of a cast of a fossil tree stump after removal of the coal. (AGI, 1987) (b) Eng. Nodule in the roof of the Halifax hard bed coal. Cf: *potlid.* (Arkell, 1953)

Baum washer *Baum jig.*

bauxite An off-white, grayish, brown, yellow, or reddish brown rock composed of amorphous or microcrystalline aluminum oxides and oxyhydroxides, mainly gibbsite $Al(OH)_3$, bayerite $Al(OH)_3$, boehmite $AlO(OH)$, and diaspore $AlO(OH)$ admixed with free silica, silt, iron hydroxides, and esp. clay minerals; a highly aluminous "laterite." It is massive, pisolitic, earthy; occurs as weathered surface deposits after prolonged leaching of silica from aluminous rocks under tropical to subtropical weathering, also transported deposits. Bauxite is the chief ore of aluminum.

bauxite brick A firebrick composed essentially of hydrated alumina and ferric oxide. Such bricks are used for the lining of furnaces where a neutral material is required. (Osborne, 1956)

bauxite cement A cement made from bauxite and lime in an electric furnace; it hardens rapidly. Sometimes called ciment fondu. (Nelson, 1965)

bauxite pneumoconiosis Found in workers exposed to fumes containing aluminum oxide and minute silica particles arising from smelting bauxite in the manufacture of corundum. Syn: *Shaver's disease.* (NSC, 1996)

bauxitic Containing much bauxite; e.g., a bauxitic clay containing 47% to 65% alumina on a calcined basis, or a bauxitic shale abnormally high in alumina and notably low in silica. (AGI, 1987)

bauxitic clay (a) A clay consisting of a mixture of bauxitic minerals, such as gibbsite and diaspore, with clay minerals, the former constituting not over 50% of the total. The opposite of this would be an argillaceous bauxite. (ACSB, 1939) (b) A natural mixture of bauxite and clay, containing not less than 47% nor more than 65% alumina on a calcined basis. (Harbison-Walker, 1972)

bauxitization Development of bauxite from primary aluminum silicates (such as feldspars) or from secondary clay minerals under aggressive tropical or subtropical weathering conditions of good surface drainage, such as the dissolving (usually above the water table) of silica, iron compounds, and other constituents from alumina-containing material. (AGI, 1987)

bavenite A white hydrous silicate of aluminum, calcium, and beryllium, $Ca_4Be_2Al_2Si_9O_{26}(OH)_2$; orthorhombic; earthy, radiating fibrous; platy prismatic crystals. From Baveno, Italy; Mesa Grande, CA. (English, 1938)

***b* axis** One of the three crystallographic axes used as reference in crystal description. It is oriented horizontally, right to left. The letter b usually appears in italics. Cf: *a axis; c axis.*

bay (a) An open space in a mine for storage of equipment, conducting repairs, or disposal of waste. (b) A recess in the shore or an inlet of a sea or lake between two capes or headlands, not as large as a gulf, but larger than a cove. (Hunt, 1965) (c) A portion of the sea that penetrates into the interior of the land. It is usually wider in the middle than at the entrance. It may be similar to a gulf, but smaller. (Hunt, 1965) (d) A portion of the sea partly surrounded by ice. (Hunt, 1965) (e) The discharge point of a hopper, usually applied to coal or ore hopper rail cars.

bayerite A dimorph of gibbsite, long known as a synthetic product, now found as a naturally occuring mineral, $Al(OH)_3$, from Portole, Istria. The naturally occurring bayerite from Fenyoro, Hungary, was found by X-ray study to be gibbsite. (Am. Mineral., 1945; Am. Mineral., 1945)

Bayer process A process for extracting alumina from bauxite ore before the electrolytic reduction of alumina. (ASM, 1961)

bayldonite A monoclinic mineral, $PbCu_3(AsO_4)_2(OH)_2{\cdot}H_2O$; forms minute grass-green to blackish-green mammillary masses; in Cornwall, U.K.

bayleyite A yellow monoclinic mineral; $Mg_2(UO_2)(CO_3)_3{\cdot}18H_2O$; radioactive.

bay salt A coarse-grained variety of common salt obtained by evaporating seawater in shallow bays or pits by the heat of the sun. (Standard, 1964)

bazzite A hexagonal mineral, $Be_3(Sc,Al)_2Si_6O_{18}$; the scandium analog of beryl; blue; at Baveno, Italy; Val Strem, Switzerland; and in central Kazakhstan.

bc-joint *longitudinal joint.*

b direction *b axis.*

beach concentrate A natural accumulation in beach sand of heavy minerals selectively concentrated (by wave, current, or surf action) from the ordinary beach sands in which they were originally present as accessory minerals; esp. a beach placer. See also: *beach placer.* (AGI, 1987)

beach deposit Concentrations of mineral formed by the grinding action of natural forces (wind, wave, or frost) and the selective transporting action of tides and winds. (Pryor, 1963)

beach drift The movement of material along the shore by the action of the uprush and backwash of waves breaking at an angle with the shore. Syn: *longshore drift.* (Schieferdecker, 1959)

beach mining The exploitation of the economic concentrations of the heavy minerals rutile, zircon, monazite, ilmenite, and sometimes gold, which occur in sand dunes, beaches, coastal plains, and deposits located inland from the shoreline. High-grade concentrate is usually obtained from low-grade material by the use of suction dredges and spiral concentrators. (Nelson, 1965)

beach ore *beach placer.*

beach placer A placer deposit on a present or ancient sea beach. There may be a series of paleo-beach placers owing to changes of shoreline. See also: *black sand; placer; beach concentrate.* Syn: *beach ore.*

beach ridge A low, essentially continuous mound of beach or beach-and-dune material (sand, gravel, shingle) heaped up by the action of waves and currents on the backshore of a beach beyond the present limit of storm waves or the reach of ordinary tides, and occurring singly or as one of a series of approx. parallel deposits. The ridges are roughly parallel to the shoreline and represent successive positions of an advancing shoreline. (AGI, 1987)

bead (a) The globule of precious metal obtained by the cupellation process in assaying. (Webster 3rd, 1966) (b) In blowpipe analysis of minerals, a drop of a fused material, such as a "borax bead," used as a solvent in color testing for various metals. The addition of a metallic compound to the bead will cause the bead to assume the color that is characteristic of the metal. See also: *blowpiping.* (AGI, 1987)

beads In ion exchange, sized resin spheres, usually +20 mesh, so constituted as to capture ions from pregnant solutions under stated loading conditions and to relinquish them under other (eluting) conditions. Two types are anionic and cationic. See also: *resin.* (Pryor, 1963)

bead tests In mineral identification, borax or other flux is fused to a transparent bead by heating in a blowpipe or other flame in a small loop formed from platinum wire. When suitable minerals are flux melted in this bead, characteristic glassy colors are produced in an oxidizing or reducing flame and serve to identify specific chemical elements.

beaker decantation A method of sizing finely ground, insoluble, homogeneous material or classifying ore particles. A weighed quantity is dispersed in liquid and allowed to settle for a timed period, a liquid fraction then being decanted. The treatment is repeated several times, the settled fraction now representing one size group (if homogeneous) or settled group (if minerals of various densities are present). The decanted fluid is similarly treated for progressively lengthened settling periods. (Pryor, 1963)

beam (a) A bar or straight girder used to support a span of roof between two support props or walls. See also: *crossbar.* (Mason, 1951) (b) The walking beam; a bar pivoted in the center, which rocks up and down, actuating the tools in cable-tool drilling or the pumping rods in a well being pumped. (Hess)

beam action In crushing, seizure of rock slab between approaching jaws so as to present crushing stress above unsupported parts of the rock, thus inducing shear failure rather than failure under compression. (Pryor, 1963)

Beaman stadia arc A specially graduated arc attached to the vertical circle of an alidade or transit to simplify the computation of elevation differences for inclined stadia sights (without the use of vertical angles). The arc is so graduated that each division on the arc is equal to 100 (0.5 sin 2A), where A is the vertical angle. Named after William M. Beaman (1867-1937), U.S. topographic engineer, who designed it in 1904 (AGI, 1987)

beam building A process of rock bolting in flat-lying deposits where the bolts are installed in bedded rock to bind the strata together to act as a single beam capable of supporting itself and thus stabilizing the overlying rock. (Lewis, 1964)

beam compass An instrument for describing large arcs. It consists of a beam of wood or metal carrying two beam heads, adjustable for position along the beam, and serving as the marking points of the compass. Syn: *trammel.* (CTD, 1958)

beam engine An early type of vertical steam engine. It operated the Cornish pump. (Nelson, 1965)

bean ore A loose, coarse-grained pisolitic iron ore; limonite occurring in lenticular aggregations. See also: *pea ore.*

bean rock Shingle cemented by tufa, Ventnor, U.K.

beans A cleaned and screened anthracite product 7/8 in by 3/8 in (22.2 mm by 9.5 mm). (Nelson, 1965)

bear (a) To bear in; underholing or undermining; driving in at the top or at the side of a working. (b) Eng. A calcareous or clay ironstone nodule, Derbyshire. (Arkell, 1953) (c) The mass of iron, which, as a result of wear of the refractory brickwork or blocks in the hearth bottom of a blast furnace, slowly replaces much of the refractory material in this location. Syn: *salamander.*

bearer Eng. A band of hard limestone consisting of numerous stromatoporoids, mainly a ramose species, Wenlock Limestone, Dudley. (Arkell, 1953)

bearer bar One of the bars that support the grate bars in a furnace. (Fay, 1920)

bearers Heavy timbers placed in a shaft at intervals of 30 to 100 ft (9.1 to 30.5 m) to support shaft sets. They are usually put beneath the end plates and dividers, and rest in hitches cut in the wall. Also used to support pumping gear. Syn: *biard.*

bearing (a) The part of a beam or girder that actually rests on the supports. (CTD, 1958) (b) Undercutting the coal face by holing. (Nelson, 1965) (c) The horizontal angle between the meridian (true or magnetic) and any specified direction. The angle is measured from either the north or the south point, as may be required to give a reading of less than 90°, and the proper quadrant is designated by the letter N or S, preceding the angle, and the letter E or W, following it; as, N. 80° E. Cf: *azimuth.* (Seelye, 1951) (d) The points of support of a beam, shaft, or axle, i.e., bearing points. (e) The direction of a mine drivage usually given in terms of the horizontal angle turned off a datum direction, such as the true north and south line. (Nelson, 1965) (f) In Texas land surveys, a reference point to identify a land corner or a point on a survey line. (Seelye, 1951) (g) A part in which a shaft or pivot revolves. (Nichols, 1976) (h) *ball bearing.*

bearing bed A bed that contains, or is likely to contain, ore minerals; one that is productive as opposed to dead or barren. (Arkell, 1953)

bearing capacity (a) The load-per-unit area that the soil or solid rock can support without excessive yield. See also: *foundation investigation.* (Nelson, 1965) (b) See also: *ultimate bearing capacity.*

bearing door A door so placed as to direct and regulate the amount of air current necessary for the proper ventilation of a district of a mine. See also: *separation door.* (Nelson, 1965)

bearing-in The depth of an undercut, or holing, from the face of the coal to the end of the undercut. (Fay, 1920)

bearing-in shots Boreholes tending to meet in the body of the rock; intended to unkey the face when charged and fired. (Stauffer, 1906)

bearing plate A plate of the thickness and area required to distribute a given load, such as a plate under a beam flange resting on a wall. If the plate is 2 in (5.1 cm) or more in thickness, it is called a slab. (Crispin, 1964)

bearing pressure The load on a bearing surface divided by the area upon which it rests. (Hammond, 1965)

bearing set In a mine shaft, a specially substantial set of timbers used at intervals to support the linings and ordinary bearers. They are tied into the surrounding rock to give extra strength.

bearing stake A stake set on a line to indicate the horizontal direction an inclined borehole is to be drilled. (Long, 1960)

bearing stratum The earth formation that has been selected as the most suitable to support a given load. (Hammond, 1965)

bearing strength The maximum bearing load at failure divided by the effective bearing area. In a pinned or riveted joint, the effective area is calculated as the product of the diameter of the hole and the thickness of the bearing member. (ASM, 1961)

bearsite A monoclinic mineral, $Be_2(AsO_4)(OH)\cdot 4H_2O$.

bears' muck Eng. Soft, bluish earth. Used by well sinkers in Cambridgeshire and Huntingdonshire. See also: *caballa ball.* (Arkell, 1953)

beater A laborer who shovels or dumps asbestos fibers and sprays them with water to prepare them for the beating process that reduces fibers to pulp for making asbestos paper. (DOT, 1949)

beater mill Mill used for impact crushing of easily broken minerals. An armature carrying swinging hammers, plates, or disks hits the falling stream of rock, dashing particles against one another and against the casing of the mill. *hammermill.* (Pryor, 1963)

beaverite A trigonal mineral, $Pb(Cu,Fe,Al)_3(SO_4)_2(OH)_6$; canary yellow; in minute plates; in Beaver County, UT.

Becke line In the "Becke test" a bright line, visible under a microscope under plane polarized light, that separates substances of different indices of refraction.

beckerite (a) A brown resin, occurring with amber. (English, 1938) (b) A brown variety of retinite having a very high oxygen content (20% to 23%).

Becke test In polarized-light microscopy, a method or test for determining relative indices of refraction between two adjacent mineral grains or between a mineral grain and its host medium; e.g., Canada balsam, Lakeside cement, an epoxy resin, or an immersion oil of known index of refraction. On defocusing by increasing the working distance between the microscope stage and the objective lens, the Becke line moves toward the higher index of refraction. Cf: *van der Kolk method.*

Becorit system An overhead monorail system. See also: *monorail.* (Sinclair, 1963)

becquerelite An orthorhombic mineral, $Ca(UO_2)_6O_4(OH)_6 \cdot 8H_2O$, with barium and potassium substitution for calcium; amber-yellow; a radioactive product of uraninite and ianthinite alteration.

bed (a) The smallest distinctive division of a stratified series, marked by a more or less well-defined surface or plane from its neighbors above and below; a layer or stratum. (Fay, 1920) (b) A deposit, as of ore, parallel to the stratification. (Standard, 1964) (c) A bed (or beds) is the smallest formal lithostratigraphic unit of sedimentary rocks. The designation of a bed or a unit of beds as a formally named lithostratigraphic unit generally should be limited to certain distinctive beds whose recognition is particularly useful. Coalbeds, oil sands, and other beds of economic importance commonly are named, but such units and their names usually are not a part of formal stratigraphic nomenclature *(NACSN, 1983, Art. 26).* (d) That portion of an outcrop or face of a quarry that occurs between two bedding planes. (Fay, 1920) (e) The level surface of rock upon which a curb or crib is laid. (Fay, 1920) (f) All the coal, partings, and seams that lie between a distinct roof and floor. (Hess) (g) Perhaps the most common term in geology, meaning layer or stratum. Quarrymen usually mean by beds not the stone beds in the geologist's sense but the partings between them. (Arkell, 1953) (h) A stockpile, as of ore, concentrates, and fluxes, built up of successive layers so that transverse cutting yields a uniform mixture for furnace feed until the material is all consumed. (i) In mineral processing, a heavy layer of selected oversized mineral or metal shot maintained on screen of jig. (Pryor, 1963) (j) That part of conveyor upon which the load or carrying medium rests or slides while being conveyed. (k) In bulk material conveyors, the mass of material being conveyed. (l) A base for machinery. (Nichols, 1976)

bed claim Aust. A mining claim lying on the bed of a stream.

bedded Applied to rocks resulting from consolidated sediments and accordingly exhibiting planes of separation designated bedding planes. (Fay, 1920)

bedded deposit (a) A term usually applied to mineral deposits that are found parallel with the stratification of sedimentary rocks and usually of contemporaneous origin. The term is used to describe layerlike deposits of replacement origin. See also: *bedded formation.* (Stokes, 1955; Fay, 1920) (b) Syn: *blanket deposit.* (AGI, 1987)

bedded formation A formation that shows successive beds, layers, or strata owing to the manner in which it was formed. See also: *bedded deposit.* (Fay, 1920)

bedded rock One of the two subdivisions of competent rock. To be classed as bedded rock, the rock within each bed, in addition to being elastically perfect, isotropic, and homogeneous, must have a bed thickness that is small compared with the roof span, and the bond between beds must be weak. Most sedimentary rocks and some stratified metamorphic rocks fall in this group.

bedding (a) A quarry term for a structure occurring in granite and other crystalline rocks that tend to split in well-defined planes more or less horizontal or parallel to the land surface. Syn: *sheeting.* (Wheeler, 1958) (b) The storing and mixing of different ores in thin layers in order to blend them more uniformly in reclamation. (AGI, 1987) (c) The layer of heavy and oversized material placed above the screen in jigging. Also called ragging. (Pryor, 1965) (d) Pieces of soft metal placed under or around a handset diamond as a cushion or filler. Also called backing; calking. (Long, 1960) (e) Ground or supports in which pipe is laid. (Nichols, 1976) (f) The arrangement of a sedimentary rock in beds or layers of varying thickness and character; the general physical and structural character or pattern of the beds and their contacts within a rock mass, such as cross-bedding and graded bedding; a collective term denoting the existence of beds. Also, the structure so produced. The term may be applied to the layered arrangement and structure of an igneous or metamorphic rock. See also: *stratification.* Syn: *layering.* Cf: *bedding plane.* (AGI, 1987) (g) The initial filling of a thickener for continuous operation.

bedding cleavage Cleavage that is parallel to the bedding. (Billings, 1954)

bedding down Formation of layer of valueless and inert rock at points in a new flowline where material will settle from the stream of ore being treated, for example, between bottom of thickener and its rakes. (Pryor, 1963)

bedding fault A fault that is parallel to the bedding. (AGI, 1987)

bedding fissility A term generally restricted to primary foliation parallel to the bedding of sedimentary rocks; i.e., it forms while the sediment is being deposited and compacted. It is the result of the parallelism of the platy materials to the bedding plane, partly because they were deposited that way and partly because they were rotated into this position during compaction. (AGI, 1987)

bedding glide Overthrusting in which a bed, such as a coal seam, is disrupted and thrust laterally along the roof or floor parting, giving a duplication of coal. (Nelson, 1965)

bedding joint (a) A thin layer differing in composition with the beds between which it occurs. (Schieferdecker, 1959) (b) A joint parallel to the bedding planes formed by tectonic processes. (Schieferdecker, 1959)

bedding plane (a) In sedimentary or stratified rocks, a surface that separates each layer from those above or below it. It usually records a change in depositional circumstances by grain size, composition, color, or other features. The rock may tend to split or break readily along bedding planes. See also: *plane.* (b) Surface on which rock-forming mineral has been deposited. Syn: *bedding.* (Pryor, 1963) (c) A separation or weakness between two layers of rock caused by changes during the building up of the rock-forming material. See also: *bed joint.* (Nichols, 1976)

bedding thrust A thrust fault that is parallel to the bedding. (Billings, 1954)

bede A miner's pick. (Pryor, 1963)

Bedford limestone One of the finest and best known building stones to be found in the United States. It gets its name from its shipping point, Bedford, IN. (Crispin, 1964)

bed joint (a) A horizontal crack or fissure in massive rock. See also: *bedding plane.* (Webster 3rd, 1966; Fay, 1920) (b) One of a set of cracks or fissures parallel with the bedding of a rock. (Webster 3rd, 1966) (c) A horizontal joint between courses of brick. (ARI, 1949) (d) The horizontal layer of mortar on (or in) which a masonry unit is laid. (ACSG, 1961)

bedrock (a) Solid rock exposed at the surface of the Earth or overlain by unconsolidated material, weathered rock, or soil. (b) In Australia, the stratum upon which the wash dirt rests is usually called bedrock. It usually consists of granite or boulder clay (glacial) and, much more rarely, basalt. When the stratum consists of slates or sandstones (Silurian or Ordovician), it is usually called reef rock. (Eng. Min. J., 1938) (c) A general term for the rock, usually solid, that underlies soil or other unconsolidated, superficial material. A British syn. of the adjectival form is "solid," as in solid geology. *stonehead.* (AGI, 1987) (d) *bottom rock; rock base.*

bedrock test A borehole drilled to determine the character of bedrock and the character and depth of overburden overlying such bedrock. Syn: *testing bedrock.* (Long, 1960)

bed separation The thin cavities formed along bedding planes due to differential lowering of strata over mine workings; e.g., a shale with its greater bending capacity will subside and separate from a higher bed of sandstone. Roof supports are so set as to keep bed separation to a minimum. (Nelson, 1965)

bed vein A vein following the bedding in sedimentary rocks, or a mineralized permeable stratigraphic zone below an impervious bed. Cf: *blanket vein; manto; sheet ground.*

beech coal Charcoal made from beechwood.

beeches Scot. Strips of hardwood fastened to pump rods to save them from wear at the collars. (Fay, 1920)

beef A quarry worker's term, used originally in Purbeck, southern England, for thin, flat-lying veins or layers of fibrous calcite, anhydrite, gypsum, halite, or silica, occurring along bedding planes of shale, giving a resemblance to beef. It appears to be due to rapid crystallization in lenticular cavities. See also: *bacon.* (AGI, 1987)

beehive coke Coke manufactured in beehive, rectangular, or similar forms of ovens in a horizontal bed, where heat for the coking process is secured by combustion within the oven chamber. (ASTM, 1994)

beehive coke oven A coke oven with a brick bottom, side walls, and a domed roof.

beehive kiln An intermittent kiln, circular in plan, with fireboxes arranged around the circumference. Such kilns find use in the firing of blue engineering bricks, pipes, some refractory bricks, etc. Syn: *round kiln.* (Dodd, 1964)

Beethoven exploder A machine for the multishot firing of series-connected detonators in tunneling and quarrying. See also: *exploder.* (Nelson, 1965)

beetle stone A nodule of coprolitic ironstone, so named from the resemblance of the enclosed coprolite to the body and limbs of a beetle. Syn: *septarium.* (Fay, 1920)

beetling stones Flat rocks on which clothes were beetled (stamped or pounded).

before breast A miner's term for that part of the orebody that still lies ahead. See also: *breast.* (AGI, 1987)

beidellite A monoclinic mineral, $(Na,Ca_{0.5})_{0.3}Al_2(Si,Al)_4O_{10}(OH)_2 \cdot nH_2O$; smectite group; aluminum rich and poor in magnesium and iron; a common constituent of soils and certain clay deposits (e.g., metabentonite). Cf: *montmorillonite.*

Beien kep gear An improved type of kep in which the kep shoes are withdrawn without previously raising the cage and thus reducing the decking time. The operation may be automatic, except for cage release, because the arrangement allows the kep shoes to trip without the position of the hand lever being altered. See also: *kep.* (Nelson, 1965)

Beien machine A pneumatic stowing machine that consists of a paddle wheel with six compartments working inside an adjustable airtight casing. This wheel is driven at a speed of 15 to 30 rpm by means of an air turbine through gearing. Two sizes of Beien machines are used having capacities of 30 yd^3 and 60 yd^3 (22.9 m^3 and 45.9 m^3) of stowing material per hour, respectively. The dirt falls from the paddle wheel into the airstream in the pipe underneath the paddle box and passes along 6-in (15.2-cm) diameter pipes to the outlet, where a detachable deflector guides the stream of dirt into the required place in the pack hole. (Mason, 1951)

Beilby layer Flow layer resulting from incipient fusion during polishing of mineral surface, and therefore not characteristic of true crystal structure. (Pryor, 1958)

Belfast truss A bowstring design of girder fabricated entirely from timber components. (Hammond, 1965)

Belgian coke oven A rectangular variation of the beehive coke oven. See also: *Belgian oven.*

Belgian effective temperature A temperature scale used in Belgium for measuring the environmental comfort in mines. (Roberts, 1960)

Belgian oven A rectangular oven with end doors and side flues for the manufacture of coke. See also: *Belgian coke oven.* (Fay, 1920)

Belgian press *double-roll press.*

Belgian process A process (no longer used) to smelt zinc in which roasted zinc ore is mixed with a reducing material, such as coal or coke, placed in cylindrical retorts, and heated in a furnace, such that the escaping zinc vapor is condensed from the open end of the retort. (Fay, 1920)

Belgian silex A very hard, tough, more or less cellular quartzite resembling French buhrstone and the most favored natural mill-lining material for most purposes. It is imported in rectangular blocks that are more or less shaped to fit the curve of a mill. (AIME, 1960)

Belgian zinc furnace A furnace in which zinc is reduced and distilled from calcined ores in tubular retorts. These furnaces may be classified as direct fired and gas fired, but there is no sharp division between these systems, which merge into one another by difficultly definable gradations. Each class of furnace may be subdivided into recuperative and nonrecuperative, but heat recuperation in connection with direct firing is rare.

belgite *willemite.*

belite (a) A constituent of portland cement clinkers. (English, 1938) (b) A calcium orthosilicate found as a constituent of portland-cement clinkers; specif. larnite. Syn: *felite.* (AGI, 1987)

belith Original spelling of belite. (Hey, 1955)

Belknap chloride washer process This coal washer uses a calcium chloride solution of a comparatively low density and depends on mechanically induced upward currents to obtain a separation at the desired specific gravity. It produces a clean, dustless, nonfreezing coal. (Mitchell, 1950)

bell (a) A cone-shaped mass of ironstone or other substance in the roof of a coal seam. Bells are dangerous as they tend to collapse suddenly and without warning. See also: *pot bottom.* (Nelson, 1965) (b) A gong used as a signal at mine shafts. (c) *cone.*

bellan (a) Eng. Dusty lead ore. (Arkell, 1953) (b) A form of lead poisoning to which miners are subject. Also spelled belland; bellund.

belland *bellan.*

bell and hopper *cup and cone.*

bellcrank drive A device used to drive an auxiliary shaker conveyor without changing the direction of the main conveyor. It consists essentially of two driving arms, placed at right angles to each other and supported at their pivot point by a fulcrum jack. When these driving arms are attached to the main and auxiliary conveyors, the reciprocating motion of the main conveyor is transmitted to the auxiliary conveyor, which can then discharge its load onto the main conveyor. (Jones, 1949)

belled Eng. Widened; said of the enlarged portion of a shaft at the landing for running the cars past the shaft, and for caging.

bell holes A conical cavity in a coal mine roof caused by the falling of a large concretion; or, as of a bell mold.

bellies Widenings in a vein. See also: *belly.*

bellingerite A triclinic mineral, $Cu_2(IO_3)_6 \cdot 2H_2O$; forms light green or bluish green crystals; at Chuquicamata, Chile.

bellite (a) An explosive consisting of five parts of ammonium nitrate to one of metadinitrobenzene, usually with some potassium nitrate. (Fay, 1920) (b) A lead chromo-arsenate in delicate velvety, red to orange tufts. (Webster 2nd, 1960)

bell jar *jar collar.*

bell-metal ore Corn. An early name for tin pyrite (stannite), so called because of its bronze color. See also: *stannite.* (Fay, 1920)

bell mold Som. A conical-shaped patch of a mine roof, probably originating with the fossils called sigillaria, or the roots of trees. Syn: *caldron.* See also: *bell; caldron bottom.* (Fay, 1920)

bellows (a) An instrument with an air chamber and flexible sides used for compressing and/or directing a current of air. (Crispin, 1964) (b) An expansible metal device containing a fluid that will volatilize at some desired temperature, expand the device, and open or close an opening or a switch; used in controls and steam traps. (Strock, 1948)

bell pit mining Obsolete method of winning coal or bedded iron from shallow deposits, in which mineral was extracted and dragged to a central shaft. (Pryor, 1963)

Bell process *Bell's dephosphorizing process.*

bell screw An internally threaded bell-shaped iron bar for recovering broken or lost rods in a deep borehole. Syn: *box bell; screw bell; bell tap.* (Fay, 1920)

Bell's dephosphorizing process The removal of phosphorus from molten pig iron in a puddling furnace lined with iron oxide and fitted with a mechanical rabble to agitate the bath. Red-hot iron ore is added. See also: *Krupp washing process.* Syn: *Bell process.* (Fay, 1920)

bell sheave Aust. A sheave in the shape of a truncated cone used in connection with the main-and-tail system of rope haulage at curves, so as to keep the rope close to the ground. (Fay, 1920)

bell socket *bell tap; screw bell.*

bell tap A cylindrical fishing tool having an upward-tapered inside surface provided with hardened threads. When slipped over the upper end of lost, cylindrical, downhole drilling equipment and turned, the threaded inside surface of the bell tap cuts into and grips the outside surface of the lost equipment. Also called bell; box bill; die; die collar; die nipple. See also: *bell screw; screw bell; die.* Syn: *bell socket; outside tap.* (Long, 1960)

bell top Term used to describe a good roof that has a clear ringing sound. (Kentucky, 1952)

bellund *bellan.*

bell work (a) Derb. A system of working an ironstone measure by upward underground excavations around the shafts (raises) in the form of a bell or cone. (Fay, 1920) (b) A method used in working salt deposits. (Standard, 1964)

belly (a) *pocket.* (b) A bulge, or mass of ore in a lode. (Fay, 1920) (c) Widened places in a borehole caused by sloughing of loose material from the borehole sidewalls. See also: *bellies.* (Long, 1960)

belly pipe A flaring-mouthed blast pipe in an iron furnace. (Standard, 1964)

Belomorite Trade name for moonstone from the White Sea. (Spencer, 1946)

belonesite A white, transparent magnesium molybdate, $MgMoO_4$, crystallizing in the tetragonal system. (Fay, 1920)

belonite An elongated or acicular crystallite having rounded or pointed ends. (AGI, 1987)

belovite A hexagonal mineral, $(Sr,Ce,Na,Ca)_5(PO_4)_3(OH)$; apatite group; occurs in pegmatites.

belt (a) *link-plate belt.* (b) Can. Regional surface zone along which mines and prospects occur. (Hoffman, 1958) (c) A continuous strap or band for transmitting power from one wheel to another, or (rarely) to a shaft, by friction. See also: *stitched canvas conveyor belt.* (Standard, 1964) (d) A zone or band of a particular kind of rock strata exposed on the surface. Cf: *zone.* (Fay, 1920) (e) An elongated area of mineralization. (AGI, 1987)

belt capacity The load that a belt conveyor is able to carry, depending upon the area of cross section of load on belt and the speed of the belt. (Nelson, 1965)

belt cleaner A device attached to a belt conveyor to clean or remove dirt or coal dust from the belt surface. Rotary bristle brushes are sometimes used, driven either by gearing from the conveyor or by an independent high-speed motor. Another device consists of a short scraper conveyor with rubber-faced scrapers attached at intervals. The scraper belt is driven via a chain drive from the main conveyor drum. (Nelson, 1965)

belt conveyor (a) A moving endless belt that rides on rollers and on which materials can be carried. The principal parts of a belt conveyor are (1) a belt to carry the load and transmit the pull, (2) a driving unit, (3) a supporting structure and idler rollers between the terminal drums, and (4) accessories, which include devices for maintaining belt tension and loading and unloading the belt, and equipment for cleaning and protecting the belt. See also: *hatch conveyor; underground mine conveyor.* (Kentucky, 1952; Nelson, 1965) (b) *troughed belt conveyor.*

belt conveyor structure The framework for supporting the bottom strand of a belt conveyor. (Nelson, 1965)

belt creep Gentle slip. (Pryor, 1963)

belt dressing A compound used to improve adhesion or flexibility. (Pryor, 1963)

belt feeder Short loop of conveyor belt, or articulated steelplate, used to draw ore at a regulated rate from under a bin or stockpile. (Pryor, 1963)

belt flotation A method sometimes used to recover diamond particles 1 mm or smaller.

belt friction *friction.*

belt horsepower That power developed with all auxiliary equipment (such as pump and fans) attached; it is consequently lower than flywheel horsepower. (Carson, 1961)

belting One of the main parts of a belt conveyor. The belting consists of piles of cotton duck impregnated with rubber, with top and bottom covers of rubber. The carrying capacity of the belt will vary, depending on the running speed and the width of the belt. (Sinclair, 1959)

belt loader A machine whose forward motion cuts soil with a plowshare or disk and pushes it to a conveyor belt that elevates it to a dumping point. Syn: *elevating grader*. (Nichols, 1976)

beltman *conveyor man.*

belt of variables The belt of marine deposition extending from the coast (high watermark) to a depth of about 100 fathoms (183 m); i.e., corresponding roughly with the continental shelf (in the wide sense, to include the shore); passing into the mud belt at the inner mud line. (Challinor, 1964)

belt press A device for dewatering mineral slurries, esp. waste from coal washing, which consists of a moving belt constructed of relatively open cloth. Belt presses are widely used in the minerals industry.

belt protection device A device fitted to a belt conveyor to give an alarm or to cause the conveyor to stop in the event of a defect, such as belt slip, breakage, tearing, misalignment, or overload. (BS, 1965)

belt slip The difference in speed between the driving drum and the belt conveyor.

belt table A table incorporating a belt conveyor so arranged as to provide working space on one or both sides of the belt.

belt tensioning device A device fitted to a belt conveyor that automatically takes up any slack or stretch in the belting. A gravity takeup device is sometimes fitted immediately behind the driving unit, thus eliminating slack that would otherwise occur. The main disadvantage of gravity takeup is that it gives the belt three entire bends. (Nelson, 1965)

belt tripper A device or mechanism that causes the conveyor belt to pass around pulleys for the purpose of discharging material from it.

belt-type conveyor A conveyor consisting of an endless belt used to transport material from one place to another.

Belugou imperfection coefficient In coal testing, a parameter, B, applied to the ash curve: B = (p75 - p25)/2(p50 - 1), where pX is the specific gravity of particles of which the fraction separated is X percent, (p75 – p25) is the statistical intermediate or inquartile range, and p50 is the effective density of separation in a process in which a dense medium, vertical current, or jigging action is used. Equation is used to define shape of a Tromp curve. (Pryor, 1963)

belyankinite A mineral, $Ca_{1-2}(Ti,Zr,Nb)_5O_{12}{\cdot}9H_2O(?)$; amorphous, yellowish-brown.

Belynski's reagent A 1% copper sulfate solution recommended as an etchant for revealing dendritic structures in high-carbon steels. (Osborne, 1956)

bementite (a) A light gray or grayish-brown, common manganese mineral, $Mn^{2+}{}_8Si_6O_{15}(OH)_{10}$. (Kirk, 1947) (b) An erroneous name for danburite.

ben (a) Scot. Inward; toward the workings; the workman's right to enter the pit. (b) The live or productive part of a lode. (Arkell, 1953) (c) A mountain peak; a word occurring chiefly in the names of many of the highest summits of the mountains of Scotland, as Ben Nevis.

bench (a) A terrace on the side of a river or lake having at one time formed its bank. See also: *bench gravel.* (b) In an underground mine, a long horizontal face or ledge of ore in a stope or working place. (CTD, 1958) (c) A layer of coal; either a coal seam separated from nearby seams by an intervening noncoaly bed, or one of several layers within a coal seam that may be mined separately from the others. (AGI, 1987) (d) One of two or more divisions of a coal seam, separated by slate, etc., or simply separated by the process of cutting the coal, one bench or layer being cut before the adjacent one. (e) The horizontal step or floor along which coal, ore, stone, or overburden is worked or quarried. See also: *benching; opencast*. (Nelson, 1965) (f) A stratum of coal forming a portion of the seam; also, a flat place on a hillside indicating the outcrop of a coal seam. (BCI, 1947) (g) In tunnel excavation, where a top heading is driven, the bench is the mass of rock left, extending from about the spring line to the bottom of the tunnel. (Stauffer, 1906) (h) A part of the face of a large excavation that is advanced not as part of the round but as a separate operation. (BS, 1964) (i) A ledge that, in open-pit mine and quarries, forms a single level of operation above which minerals or waste materials are excavated from a contiguous bank or bench face. The mineral or waste is removed in successive layers, each of which is a bench, several of which may be in operation simultaneously in different parts of, and at different elevations in, an open-pit mine or quarry. Cf: *berm.* (j) *siege.*

bench-and-bench Ark. That plan of mining coal in a room that requires the blasting of the two benches of coal alternately, each a little beyond the other. Syn: *bench working*. (Fay, 1920)

bench blasting A mining system used either underground or in surface pits whereby a thick ore or waste zone is removed by blasting a series of successive horizontal layers called benches.

bench claim A placer claim located on a bench above the present level of a stream. (Hess)

bench coal A coal seam cut in benches or layers. (Tomkeieff, 1954)

bench cut (a) In vertical shaft sinking, blasting of drill holes so as to keep one end of a rectangular opening deep (leading), thus facilitating drainage and removal of blasted rock. (Pryor, 1963) (b) Benches in tunnel driving are often drilled from the top with jackhammers. The vertical shotholes are generally spaced 4 ft (1.2 m) apart in both directions, fired by electric delay detonators, one row at a time. When bench shotholes are drilled horizontally with the drifter drills mounted on a bar, the charges are fired in rotation, starting from the upper center. In some cases, a bench may be drilled both vertically and horizontally, particularly where the benches are exceptionally high or when the headroom above the bench is inadequate for handling drill steels long enough to bottom the shotholes to grade. The lifters are drilled by machines mounted on a bar across the bottom of the tunnel, in which case the upper vertical holes will all be fired before the horizontal charges. (Hammond, 1965)

bench diggings River placers not subject to overflows.

benched foundation Foundation excavated on a sloping stratum of rock, which is cut in steps so that it cannot slide when under load. Syn: *stepped foundation*. (Hammond, 1965)

benches A name applied to ledges of all kinds of rock that are shaped like steps or terraces. They may be developed either naturally in the ordinary processes of land degradation, faulting, and the like; or by artificial excavation in mines and quarries.

bench face *bank.*

bench flume A conduit on a bench, cut on sloping ground. (Seelye, 1951)

bench gravel A term applied in Alaska and the Yukon Territory to gravel beds on the side of a valley above the present stream bottom, which represent part of the stream bed when it was at a higher level. See also: *bench; bench placer*. (AGI, 1987)

bench height The vertical distance from the top of a bench to the floor or to the top of the next lower bench.

benching (a) A method of working small quarries or opencast pits in steps or benches, in which rows of blasting holes are drilled parallel to the free face. The benching method has certain dangers since the quarrymen must work on ledges at some height. It is possible to work benches up to 30 ft (9.1 m) high using tripod or wagon drills. See also: *bottom benching; top benching*. (Nelson, 1965) (b) The breaking up of a bottom layer of coal with steel wedges in cases where holing is done above the floor. (Nelson, 1965) (c) Ches. The lower portion of the rock salt bed worked in one operation. (Fay, 1920) (d) *bench.*

benching iron An item of surveying equipment, comprising a triangular steel plate with pointed studs at the corners. These studs are driven into the ground in the desired position. The plate is used either as a temporary bench mark or as a change point in running a line of levels. (Hammond, 1965)

bench mark (a) A relatively permanent metal tablet or other mark firmly embedded in a fixed and enduring natural or artificial object, indicating a precisely determined elevation above or below a standard datum (usually sea level); it bears identifying information and is used as a reference in topographic surveys and tidal observations. It is often an embossed and stamped disk of bronze or aluminum alloy, about 3.75 in (9.5 cm) in diameter, with an attached shank about 3 in (7.6 cm) in length, and may be cemented in natural bedrock, in a massive concrete post set flush with the ground, or in the masonry of a substantial building. Abbrev: BM. (AGI, 1987) (b) A well-defined, permanently fixed point in space, used as a reference from which measurements of any sort (such as of elevation) may be made. (AGI, 1987)

bench of timbers A term used to describe the header when it is complete with legs. Syn: *set*. (Kentucky, 1952)

bench placer A bench gravel that is mined as a placer. Syn: *bench gravel; river-bar placer; terrace placer*. (AGI, 1987)

bench scrap The scrap mica resulting from rifting and trimming hand-cobbed mica. (Skow, 1962)

bench slope *bank slope.*

bench working The system of working one or more seams or beds of mineral by open working or stripping in stages or steps. Syn: *bench-and-bench.* (Zern, 1928)

benchy Forming frequent benches; said of a lode. (Standard, 1964)

Bendelari jig A jig fitted with a flexible rubber diaphragm that is worked by an eccentric motion, thus producing a jigging cycle (pulsion suction). See also: *jig*. (Pryor, 1963)

bending stress The stress produced in the outer fibers of a rope by bending over a sheave or drum. (Zern, 1928)

bend pulley An idler pulley that is used solely for the purpose of changing the direction of travel of the belt other than at the terminals of the conveyor. (NEMA, 1961)

bend radius The inside radius of a bent section. (ASM, 1961)
bends Caisson disease, brought on by too sudden return to normal pressure after working in a pressurized shaft or tunnel. (Pryor, 1963)
bend shaft A shaft that supports a bend wheel or pulley.
bend tangent A tangent point where a bending arc ceases or changes. (ASM, 1961)
bend test A test for determining relative ductility of metal that is to be formed, usually sheet, strip, plate, or wire, and for determining soundness and toughness of metal. The specimen is usually bent over a specified diameter through a specified angle for a specified number of cycles. (ASM, 1961)
bend wheel A wheel used to interrupt and change the normal path of travel of the conveying or driving medium. Most generally used to effect a change in direction of conveyor travel from inclined to horizontal or a similar change.
beneficiate To improve the grade by removing associated impurities; to upgrade.
beneficiation (a) The dressing or processing of coal or ores for the purpose of (1) regulating the size of a desired product, (2) removing unwanted constituents, and (3) improving the quality, purity, or assay grade of a desired product. (b) Concentration or other preparation of coal or ores for smelting by drying, flotation, or magnetic separation. (c) Improvement of the grade of coal or ores by milling, flotation, sintering, gravity concentration, or other processes.
Bengal amethyst An archaic trade name for purple sapphire.
benitoite A hexagonal mineral, $2[BaTiSi_3O_9]$; occurs in double pyramids; colorless, white, or blue, resembling sapphire; in veins in serpentine schist.
benjaminite A monoclinic mineral, $(Ag,Cu)_3(Bi,Pb)_7S_{12}$.
benstonite A trigonal mineral, $(Ba,Sr)_6(Ca,Mn)_6Mg(CO_3)_{13}$; shows flat, simple rhombohedral crystals and cleavage; fluoresces yellow or red; in veins with calcite.
bent (a) In tunnel timbering, two posts and a roof timber. (Nichols, 1976) (b) A transverse structure consisting of legs, bracing, and feet used for the purpose of supporting a gallery or conveyor frame at a fixed elevation.
Benthic Division A primary division of the sea that includes all of the ocean floor. The Benthic Division is subdivided into the Littoral System (the ocean floor lying in water depths ranging from the high watermark to a depth of 200 m or the edge of the continental shelf), and the Deep-Sea System (ocean floor lying in water deeper than 200 m). The systems are further subdivided into the Eulittoral Zone (0 to 50 m), Sublittoral Zone (50 to 200 m), Archibenthic Zone (200 to 1,000 m), and the Abyssal-Benthic Zone (1,000 m and greater). (Hy, 1965)
benthonic (a) Refers to the bottom of a body of standing water. (AGI, 1987) (b) Pertaining to benthos; also, said of that environment. Syn: *demersal.* (AGI, 1987)
benthos All plants and animals living on the ocean bottom. (Hy, 1965)
bentonite A montmorillonite-type clay formed by the alteration of volcanic ash. It varies in composition and is usually highly colloidal and plastic. Swelling bentonite is so named because of its capacity to absorb large amounts of water accompanied by an enormous increase in volume. Occurs in thin deposits in the Cretaceous and Tertiary rocks of the Western United States. It is used for making refractory linings, water softening, decolorizing oils, thickening drilling muds, and preparing fine grouting fluids. As a mud flush, bentonite is used at a concentration of about 3 lb/ft^3 (48.1 kg/m^3) of water. Syn: *Denver mud; volcanic clay.* See also: *clay.* (Nelson, 1965)
bentonitic clay Bentonite.
bent sieve A stationary screen constructed in the form of an arc of a circle and arranged as a chute over which the clean coal from a cyclone washer passes to the orthodox rinsing screen. In the United States, the bent screen is used in magnetite recovery from cyclone washers. (Nelson, 1965)
benzenol *phenol.*
benzol indicator A portable instrument designed specif. for measuring low concentrations of benzol, which are potentially dangerous to the health of personnel. (Best, 1966)
beraunite A monoclinic mineral, $Fe^{2+}Fe^{3+}{}_5(PO_4)_4(OH)_5{\cdot}4H_2O$; a secondary mineral in iron deposits; an alteration product of primary phosphates in pegmatites.
berdan A circular, revolving, inclined iron pan in which concentrates are ground with mercury and water by an iron ball. (Gordon, 1906)
Berek compensator In polarized-light microscopy, an optical device of variable compensation for analysis of birefringence. Cf: *accessory plate.*
bergalite A pitchy black dike rock containing small phenocrysts of haüyne, apatite, perovskite, melilite, and magnetite in a groundmass of the same minerals with nepheline, biotite, and brown interstitial glass; from Kaiserstuhl, Oberbergen, Baden, Germany. Syn: *bergalith.* (Holmes, 1928; Johannsen, 1931-38)
bergalith *bergalite.*
bergbutter Various salts, commonly halotrichite. (Hey, 1955)
bergenite Occurs naturally at Bergen an der Trieb, Saxony, with other uranium minerals; named from locality, the older name being rejected as implying a barian phosphuranylite rather than the barium analogue. Syn. for barium-phosphuranylite. (Hey, 1961)
Berglof process A method of direct reduction of iron ore. The reduction of the ore was carried out in interchangeable containers. The ore was heated to the reduction temperature in one container, and then this container was moved into the reducing zone. (Osborne, 1956)
berkeleyite *lazulite.*
berkelium The element having the atomic number 97, the discovery of which was announced by Thompson, Ghiorso, and Seaborg in 1950. They produced an isotope of 4.5 h half-life, berkelium 243, by helium ion bombardment of americium 241. Symbol, Bk; valences, 3 and 4; and the mass number of the most stable isotope, 249. (Handbook of Chem. & Phys., 2)
Berlin blue In optical mineralogy, an anomalous interference color of the first order; e.g., some epidotes.
Berlin iron A soft iron containing phosphorus, which makes very fine smooth castings and is used for ornaments and jewelry. (Standard, 1964)
berm (a) A horizontal shelf or ledge built into the embankment or sloping wall of an open pit or quarry to break the continuity of an otherwise long slope and to strengthen its stability or to catch and arrest slide material. A berm may be used as a haulage road or serve as a bench above which material is excavated from a bank or bench face. Cf: *bench.* (b) The space left between the upper edge of a cut and the toe of an embankment. (Seelye, 1951) (c) An artificial ridge of earth. (Nichols, 1954) (d) Terraces that originate from the interruption of an erosion cycle with rejuvenation of a stream in the mature stage of its development and renewed dissection, leaving remnants of the earlier valley floor above flood level. (AGI, 1987) (e) A nearly horizontal portion of the beach or backshore formed by the deposit of material by wave action. Some beaches have no berms; others have one or several. See also: *bank height.* (AGI, 1987)
bermanite A monoclinic mineral, $Mn^{2+}Mn^{3+}{}_2(PO_4)_2(OH)_2{\cdot}4H_2O$; in pegmatites.
berm interval Vertical distance from crest of berm to its underlying toe, as in a bank or bench.
bernardinite Originally described by Stillman as fossil resin, but later shown by Stanley-Brown to be a fungus impregnated by resinous material. (Tomkeieff, 1954)
bernstein (a) A fossil resin found in and in association with lignite beds of Eocene age. Commonly occurring in many European localities, and is esp. abundant in areas bordering the Baltic coast. See also: *amber.* (AGI, 1987) (b) German name for amber. (Tomkeieff, 1954)
berthierine A monoclinic mineral, $(Fe^{2+},Fe^{3+},Mg)_{2\text{-}3}(Si,Al)_2O_5(OH)_4$ of the kaolinite-serpentine group; dark steel-gray; in low-temperature vein deposits. Syn: *martourite.*
berthonite A discredited name for bournonite.
bertrandite An orthorhombic mineral, $4[Be_4Si_2O_7(OH)_2]$; colorless to clear pale yellow and shows heart-shaped twins; in pegmatites; a source of beryllium. See also: *beryllium disilicate.*
Bertrand lens A removable lens in the tube of a polarized-light microscope used to converge light to form an interference figure.
Bertrand process A heavy-fluid coal cleaning process that utilizes a calcium chloride solution as separating medium and is applicable only to deslimed feed. It differs from the Lessing process in that the raw coal is introduced into the system countercurrent fashion, from water to separating solution, the purified coal and the waste being withdrawn in a similarly countercurrent fashion. Coal containing less than 1% ash is said to be obtained by this process. (Gaudin, 1939)
beryl A hexagonal mineral, $Be_3Al_2Si_6O_{18}$; green, blue-green, and other pale tints; in granite pegmatites, mica schists, and an accessory mineral in felsic igneous rocks; the chief source of beryllium. Transparent and colored gem varieties include emerald, aquamarine, morganite, heliodor, golden beryl, bixbite, and vorobievite.
beryllides A group of intermetallic compounds of potential interest as special ceramics. Cell dimensions and types of structure have been reported for the beryllides of titanium, vanadium, chromium, zirconium, niobium, molybdenum, hafnium, and tantalum. (Dodd, 1964)
berylliosis An occupational disease caused by the inhalation of fumes liberated during the reduction of beryllium. Beryllium is thought to play the principal role, aggravated by fluorine, and to affect all organs, particularly the larger protective glands, rather than the respiratory apparatus alone. (Hess)
beryllite A hydrous silicate of beryllium, $Be_3SiO_4(OH)_2{\cdot}H_2O$, as an alteration product of epididymite. (Spencer, 1955)
beryllium An element belonging to the alkaline earth metals. Symbol, Be. Beryl and bertrandite are the most important commercial sources of the element and its compounds. Aquamarine and emerald are the precious forms of beryl. Used in nonsparking tools, high-speed aircraft, missiles, spacecraft, communications satellites, and X-ray li-

thography for microminiature integrated circuits, as well as in nuclear reactors and computers. Beryllium and its salts are toxic.

beryllium aluminate $BeAl_2O_4$; mol wt, 126.97; orthorhombic; sp gr, 3.76; source of beryllium. Syn: *chrysoberyl.* (Bennett, 1962)

beryllium carbide Be_2C; decomposes above 2,950 °C. Used as a moderator in nuclear application. Molecular weight, 30.04; yellow; hexagonal; and sp gr, 1.90 at 15 °C. (Lee, 1961; Bennett, 1962)

beryllium disilicate $Be_4Si_2O_7(OH)_2$; mol wt, 238.23; orthorhombic; sp gr, 2.6. Syn: *bertrandite.* (Bennett, 1962)

beryllium nitride Be_3N_2; molecular weight, 55.05; colorless; isometric; and melting point, 2,200 ± 100 °C. (Bennett, 1962)

beryllium oxide A white powder; hexagonal; BeO. Used in the preparation of beryllium compounds and in ceramics and refractories. Melting point, 2,570 °C; and sp gr, 3.02. Bodies high in BeO have extremely high thermal conductivity (in the range of metals) and also possess high mechanical strength. Used in nuclear reactors because of its refractoriness, high thermal conductivity, and ability to act as a moderator for fast neutrons, reducing them to thermal speeds. Beryllia ceramics are used for electronic components and for crucibles for melting uranium and thorium. Syn: *bromellite.* (CCD, 1961; Lee, 1961)

beryllonite A monoclinic mineral, $NaBePO_4$; colorless or yellow; forms transparent, topazlike pseudo-orthorhombic crystals.

beryloid In crystallography, the dihexagonal bipyramid, common in crystals of beryl.

beryloscope *emerald glass.*

beryl preferential stain process A quick, simple method for determining the amount of beryl in a mineral sample. The samples are placed in a hot solution of sodium hydroxide, which etches the beryl grains in the sample; they then are stained an intense blue with another reagent to enable counting under a microscope.

berzelianite An isometric mineral, Cu_2Se; silver-white in fresh break; sp gr, 6.7.; dimorphous with bellidoite.

berzeliite An isometric mineral, $(Ca,Na)_3(Mg,Mn)_2(AsO_4)_3$; commonly massive; bright yellow to orange-yellow. Syn: *pyrrhoarsenite.*

Bessemer Any product of the Bessemer process, such as Bessemer steel, iron, etc.; named from Henry Bessemer, who patented the process in 1855; used also attributively as Bessemer converter, flame, or method. (Standard, 1964)

Bessemer afterblow In the basic Bessemer process of steelmaking, the continuation of the blowing cycle after the oxidation of the silicon, manganese, and carbon content of the charge is complete, and during which the phosphorus and sulfur contents of the charge are reduced. (Henderson, 1953)

Bessemer blow In the Bessemer process of steelmaking, the period of the blowing cycle during which the oxidation of the silicon, manganese, and carbon content of the charge takes place. (Henderson, 1953)

Bessemer converter A pear-shaped steel shell lined with a refractory material containing a number of holes or ports in the bottom or side through which air is blown through the molten pig iron charge. The converter is mounted on trunnions about which it may be tilted to charge or tap. Molten pig iron is charged into the converter, and air is blown through the molten metal to oxidize the impurities, thus making steel. (Henderson, 1953)

Bessemer iron ore *Bessemer ore.*

Bessemer matte In the extraction of copper from sulfide ores, the liquid that remains in the converter at the end of the blow. It is essentially molten nickel sulfide or a solution of copper and nickel sulfides. (Newton, 1959)

Bessemer ore An iron ore containing very little phosphorus (generally less than 0.045%). Named for suitability in the Bessemer process of steelmaking. Syn: *Bessemer iron ore.*

Bessemer pig iron Pig iron with sufficiently low phosphorus (0.100% maximum) to be suitable for use in the Bessemer process. (ASM, 1961)

Bessemer process A method, historically important but no longer in use, in which molten pig iron is charged in a Bessemer converter and air is blown through the molten metal to oxidize the impurities, thus making steel. This process is no longer in use.

beta In crystallography, the angle between the *a* and *c* axes. Cf: *alpha; gamma.*

beta chalcocite *chalcocite.*

betafite An isometric mineral, $(Ca,Na,U)_2(Ti,Nb,Ta)_2O_6(OH)$; pyrochlore group; forms a series with pyrochlore; name assigned to members of the series having uranium greater than 15%; radioactive; in granitic pegmatites. Cf: *uranpyrochlore.*

beta particle An elementary particle emitted from a nucleus during radioactive decay. It has a single electrical charge and a mass equal to 1/1837 that of a proton. A negatively charged beta particle is physically identical to an electron. If the beta particle is positively charged, it is called a positron. (Lyman, 1964)

beta quartz Quartz formed at a temperature between 573 °C and 870 °C. The commonest examples are the bipyramidal quartz crystals found as phenocrysts in quartz porphyries. (Hess)

beta ray A ray of electrons emitted during the spontaneous disintegration of certain atomic nuclei. (ASM, 1961)

beta tin Metallic tin in its common, massive form. (Bennett, 1962)

betatron A doughnut-shaped accelerator in which electrons are accelerated by a changing magnetic field. (Lyman, 1964)

beta zinc sulfide Colorless when pure; isometric; ZnS; mol wt, 97.43; sp gr, 4.102 at 25 °C, and ranges from 3.90 to 4.11; Mohs hardness, 3.5 to 4.0; luster, resinous to adamantine; transformation temperature to alpha zinc sulfide, 1,020±5 °C; sublimes at 1,180°C or 1,185 °C; melting point, 1,850 °C (at 150 atm or 15.2 MPa); insoluble in water; and very soluble in acids. Occurs as the mineral sphalerite, which is nearly colorless, white, yellow, red, green, brown, and black; has perfect dodecahedral cleavage; and is soluble in hydrochloric acid. Sphalerite is the principal ore of zinc, a source of cadmium, and a source of sulfur for manufacturing sulfur dioxide, sulfuric acid, and other sulfur compounds. Also called sphalerite; zinc blende; zincblende; blende; black jack. See also: *sphalerite; zinc sulfide.* Cf: *alpha zinc sulfide.* (CCD, 1961; Handbook of Chem. & Phys., 2; Dana, 1944)

betechtinit *betekhtinite.*

betekhtinite Orthorhombic needles, $Cu_{10}(Fe,Pb)S_6$, in ores from Mansfeld, Germany. Syn: *betechtinit.* (Spencer, 1958)

Bethell's process A process for creosoting timber (such as track sleepers) to extend its useful life. The timber is first dried, then placed in a cylinder and subjected to partial vacuum, and finally impregnated with creosote under pressure. See also: *open-tank method; timber preservation.* (Nelson, 1965)

between-laboratory tolerance The maximum acceptable difference between the means of two determinations carried out by two different laboratories on representative samples taken from the same bulk sample after the last stage of the reduction process. (BS, 1961)

beudantite (a) A trigonal mineral, $PbFe^{3+}{}_3(AsO_4)(SO_4)(OH)_6$; crystallizes in green to black rhombohedra. (b) A mineral group.

bevel cut Any style of cutting with a large table joined to the girdle by one, or possibly two, bevels and a pavilion that may be step cut, brilliant cut, or any style. Used mostly for opaque stones and intaglios. Bevel-cut shapes include round, square, rectangular, oblong, oval, pendeloque, navette, heart, diamond, horseshoe, shield, pentagon, and hexagonal shapes. The style is used predominantly for less valuable gems. Syn: *table cut.*

bevel gear (a) A cone-shaped gear encircling the drive rod in a diamond-drill swivel head, which meshes with a matching gear attached to the drive shaft from the drill motor. By means of these gears, the drill-string equipment can be made to rotate. Syn: *miter gear.* (Long, 1960) (b) Any gear, the teeth of which are inclined to the shaft axis of the gear. (Long, 1960) (c) A gearwheel that transmits power between two shafts that meet at an angle. If at a right angle and the wheel is of the same size, it is called a miter gear. (Crispin, 1964)

beyerite A tetragonal mineral, $(Ca,Pb)Bi_2(CO_3)_2O_2$; forms minute tetragonal crystals and earthy masses; a secondary bismuth mineral.

B.H. bit (a) A noncoring or blasthole bit. (Long, 1960) (b) A CDDA standard-size noncoring bit having a set outside diameter of 1 in (2.54 cm). Normally referred to as a 1-in B.H. bit. (Long, 1960)

B-horizon The layer of a soil profile in which material leached from the overlying A-horizon is accumulated. Syn: *zone of accumulation; zone of illuviation.* See also: *A-horizon.*

biard *bearers.*

biat Eng. A timber stay or beam in a shaft. See also: *bearer.* Also spelled byat.

biaxial The optical character of crystals belonging to the orthorhombic, monoclinic, and triclinic systems, which exhibit double refraction, but have two directions of single refraction and isotropy, i.e., two optic axes. Cf: *uniaxial.*

biaxial stone A mineral that has crystallized in the orthorhombic, monoclinic, or triclinic system and hence has two optic axes. Cf: *uniaxial stone.*

bicable An aerial ropeway using stationary track ropes along which carriers are hauled by an endless haulage rope. See also: *aerial ropeway.* (Nelson, 1965)

bicylindroconical drum A winding drum with a cylindrical middle portion and two conical outer portions; used sometimes where the weight of the winding rope is large compared with the coal or mineral load. The heavily loaded upgoing rope winds on the small diameter, while the downgoing rope winds off the large diameter. The effect is to compensate for the heavy torques due to rope unbalance and acceleration. See also: *cylindroconical drum; winding drum.* (Nelson, 1965)

bieberite A monoclinic mineral, $CoSO_4{\cdot}7H_2O$; melanterite group; flesh-red to rose-red; esp. in crusts and stalactites. Syn: *cobalt vitriol.*

bifurcating feeder One that separates objects moving in a single lane and delivers them to two lanes of movement.

Big Coal D Nongelatinous permissible explosive; used in coal mining. (Bennett, 1962)

bigging N. of Eng. A built-up pillar of stone or other debris in a working place or heading to support the roof; e.g., "bigging the gob" means building a pack in a worked-out place.
big-stone bit Bits set with diamonds as large or larger than eight stones per carat in size. (Long, 1960)
bikitaite A monoclinic and triclinic mineral, $LiAlSi_2O_6 \cdot H_2O$; may be in zeolite group; colorless or white; in granular aggregates with eucryptite in lithia pegmatites.
bilateral transducer A transducer capable of transmission in either direction between its terminations. Syn: *reversible transducer.* (Hy, 1965)
Bilharz table A side-bump table having a surface made of a plane, endless, traveling belt. The Corning, Luhrig, and Stein tables are similar. (Liddell, 1918)
bilinite A monoclinic mineral, $Fe^{2+}Fe^{3+}{}_2(SO_4)_4 \cdot 22H_2O$; hallotrichite group; white to yellowish; in radially fibrous masses.
billy Aust. A name used in the Clermont district of Queensland for a bed of quartzite that caps the coal measures.
billy coal Staff. Miners' term for a thin, unworkable coal seam occurring above or below a workable seam. (Tomkeieff, 1954)
bimagmatic Generic term for porphyritic rocks in which the minerals occur in two generations. (AGI, 1987)
bin A container for storing material.
binary alloy An alloy containing two component elements. (ASM, 1961)
binary cycle A cycle in which two different media are employed, one superimposed on and augmenting the cycle of the other. (Strock, 1948)
binary explosive An explosive based on two ingredients, such as nitromethane and ammonium nitrate, which are shipped and stored separately and mixed at the blast site to form an explosive mixture.
binary granite Two-mica granite.
binary system A chemical system containing two components, e.g., the MgO-SiO_2 system. (AGI, 1987)
bind (a) Shale or mudstone occurring in coal measures. Obsolete. (BS, 1964) (b) To prevent normal operation of drill-string equipment in a borehole, such as by constriction or friction created by swelling or caving ground, settlement or balling of cuttings, an obstruction, or an offset or crooked hole, or as the result of insufficient clearance cut by use of undergage bits or reaming shells. (Long, 1960) (c) To cause to cohere; to give consistency to by means of an agent, such as by drilling mud in a loose, sandy, or fragmented formation. (Long, 1960) (d) A British coal miner's term for any fine-grained, well-laminated rock (such as shale, clay, or mudstone, but not sandstone) associated with coal. See also: *blaes.* (AGI, 1987)
binder (a) A substance used to produce cohesion in loose aggregate, as the crushed stones in a macadam road. (b) A material added to coal or iron ore during the process of briquetting or pelletizing to facilitate adhesion between the particles. (c) Corn. Beds of grit in shale, slate, or clay. (d) Streak of impurity in a coal seam, usually difficult to remove. (e) The material that produces or promotes consolidation in loosely aggregated sediments; e.g., a mineral cement that is precipitated in the pore spaces between grains and that holds them together, or a primary clay matrix that fills the interstices between grains. (AGI, 1987) (f) Soil binder. (AGI, 1987) (g) A term used in Ireland for a bed of sand in shale, slate, or clay. (AGI, 1987) (h) A coal miner's term used in Pembrokeshire, England, for shale. (AGI, 1987)
binderless briquetting The briquetting of coal by the application of pressure without the addition of a binder. (BS, 1962)
bindheimite An isometric mineral, $Pb_2Sb_2O_6(O,OH)$; stibiconite group; yellow to reddish-brown; in the oxidation zone of lead-antimony ore deposits.
binding bolt *anchor bolt.*
bin feeder A worker who rods or bars ore that sticks as it passes through the bin door. (Fay, 1920)
bing To put coal in wagons or in stacks at the surface. (Fay, 1920)
bin gate A device for complete shutoff or control of gravity-impelled flow of materials from a bin, bunker, hopper, or other container. Syn: *bucket gate; bunker gate.* See also: *regulating gate.*
Bingham model One of many rheological models of material behavior. Rheology is the study of change in form and the flow of matter, embracing elasticity, viscosity, and plasticity. A Bingham material is elastic until the yield point is reached; flow occurs beyond the yield point. (SME, 1992)
bing hole Derb. A hole or chute through which ore is thrown.
bing ore The purest lead ore and with the largest crystals of galena. (Hess)
bing place Derb. The place where ore is stored for smelting.
binnite A silver-bearing variety of tennantite.
bioassay A determination of the concentration of a substance in biological fluids and tissue by analysis of urine, feces, blood, bone, tissue, etc. (ANSI, 1992)
biochemical deposit A precipitated deposit resulting directly or indirectly from vital activities of an organism, such as bacterial iron ore or coralline limestone.
biochemical prospecting (a) *biogeochemical prospecting.* (b) Prospecting by means of vegetation. The root systems of trees are actually powerful sampling mechanisms that represent samples of solutions from a large volume of earth. Much of the mineral content from these solutions is found in the leaves. Analysis of leaves may serve as a guide to prospectors. (Lewis, 1964)
bioclastic Said of rocks consisting of fragmental organic remains.
biogenic Said of a rock resulting from the physiological activities of organisms, e.g., a coral reef.
biogeochemical anomaly An area where the vegetation contains an abnormally high concentration of metals. (Hawkes, 1962)
biogeochemical prospecting Geochemical exploration based on the chemical analysis of systematically sampled plants in a region, to detect biological concentrations of elements that might reflect hidden orebodies. The trace-element content of one or more plant organs is most often measured. Syn: *biochemical prospecting.* Cf: *geobotanical prospecting.* (AGI, 1987)
biogeochemistry A branch of geochemistry that deals with the effects of life processes on the distribution and fixation of chemical elements in the biosphere. (AGI, 1987)
bioherm A moundlike or circumscribed mass of rock built up by sedentary organisms such as corals, mollusks, and algae. Cf: *biostrome.* Syn: *reef knoll.*
bioleaching The catalytic action of bacteria, such as *Thiobacillus ferroxidans* and *Thiobacillus thiooxidans* to accelerate chemical oxidation reactions by as much as 10^6 times those of chemical reactions alone; esp. useful in leaching copper and uranium systems. (SME, 1992)
biolite (a) A group name for minerals formed by biologic action. (AGI, 1987) (b) *biolith.* (c) An old term for a concretion formed through the action of living organisms. (AGI, 1987)
biolith A rock of organic origin; a biogenic rock.
bioluminescence The emission of visible light by living organisms. (Hy, 1965)
biomechanical deposit A deposit due to the detrital accumulation of organic material, as in the cases of limestones and coal. (AGI, 1987)
bionomics *ecology.*
biopelite *black shale.*
biophile (a) An element that is required by or found in the bodies of living organisms. The list of such elements includes carbon, hydrogen, oxygen, nitrogen, phosphorus, sulfur, chlorine, iodine, bromine, calcium, magnesium, potassium, sodium, vanadium, iron, manganese, and copper. All may belong also to the chalcophile or lithophile groups. (Hess) (b) Said of those elements that are the most typical in organisms and organic material. (AGI, 1987) (c) Said of those elements that are concentrated in and by living plants and animals. (AGI, 1987)
bioreactor A tank equipped for temperature, pH, and reagent control used to employ bacteria to oxidize or reduce ores of gold and other metals and render them amenable for metal extraction by leaching.
biosphere (a) All the area occupied or favorable for occupation by living organisms. It includes parts of the lithosphere, hydrosphere, and atmosphere. (AGI, 1987) (b) All living organisms of the Earth and its atmosphere. (AGI, 1987)
biostrome A bedded, blanketlike mass of rock composed mainly of the remains of sedentary organisms; an organic layer, such as a bed of shells or corals, or even a coal seam. Cf: *bioherm.* (AGI, 1987)
biotite (a) A monoclinic mineral, $K_2Mg_6(Si_6Al_2O_{20})(OH,F)_2$; mica group, with Fe^{2+} replacing Mg and Fe^{3+} replacing Al; in masses with perfect basal cleavage; dark brown, dark green, black; a common rock-forming mineral in crystalline rocks, either as an original crystal in igneous rocks or as a metamorphic product in gneisses and schists; a detrital constituent of sedimentary rocks. (b) A general term to designate all ferromagnesian micas. Syn: *black mica; dark mica; magnesia mica.* See also: *iron mica.*
biotite gneiss A gneiss in which biotite is the prominent dark mineral.
biotitite A jet black igneous rock consisting essentially of biotite. Near Libby, MT, such a rock has been altered to vermiculite by hot waters. (Johannsen, 1931-38)
Biot number The heat-transfer ratio hr/k, where h is the heat-transfer coefficient, r is the distance from the point or plane under consideration to the surface, and k is the thermal conductivity. The Biot number is a useful criterion in assessing thermal-shock resistance. (Dodd, 1964)
bipolar electrode An electrode that is not mechanically connected to a power supply but is placed in an electrolyte, between the anode and the cathode, such that the part nearer the anode becomes cathodic and the part nearer the cathode becomes anodic. Syn: *intermediate electrode.* (ASM, 1961)
bipyramid A closed crystal form consisting of a positive and negative pyramid. Cf: *pyramid.*

bird A geophysical measuring device such as a magnetometer, plus the housing in which it is towed behind an aircraft. (AGI, 1987)
bird's-eye (a) Mixed screened anthracite passing a 1/2-in (12.7-mm) screen, but retained on a 1/8-in (3.2-mm) screen. May be subdivided into buckwheat, rice, and barley. See also: *anthracite coal sizes; barley; bird's-eye coal.* (CTD, 1958) (b) Eng. Applied to various rocks with small spots, in some places to a concretionary slate, and in Guernsey to a spotted variety of diorite or gabbro. (Arkell, 1953)
bird's-eye coal Sometimes applied to anthracite coal when very small fractures are numerous and freshly broken surfaces display rounded or oval eyelike forms, many of which have convex surfaces. See also: *anthracite coal sizes; bird's-eye.* (AGI, 1987)
bird's-eye limestone A very fine-grained limestone containing spots or tubes of crystalline calcite.
bird's-eye porphyry A name given by prospectors and miners to a fine-grained igneous rock having small phenocrysts, particularly if they are quartz, from a fancied resemblance to birds' eyes. (Hess)
bird's-eye slate A quarryman's term for slate containing abundant deformed or squeezed concretions. (Holmes, 1928)
Bird solid-bowl centrifuge A fine-coal dewatering machine that consists of a tank or truncated conical shell, which is revolved at the desired speed by means of a drive sheave. A screw conveyor rotates inside the cone or bowl at a slightly lower speed in the same direction of rotation. The feed entrance, in the center of the large end of the truncated cone, is high enough to allow formation of a pool of slurry. Adjustable effluent-discharge parts are so located in the large end of the bowl that the level of liquid is maintained at the desired height. The solids are steadily moved forward by the screw conveyor as fast as they are deposited, being carried above the level of the pool for an interval before leaving the bowl. Discharge of both solids and effluent is continuous.
birefracting *birefringent.*
birefractive *birefringent.*
birefringence (a) The numerical difference between the refractive indices of a mineral. This difference results in a display of interference colors when thin sections or small fragments of anisotropic minerals are viewed between crossed polars. Isometric minerals and amorphous materials are isotropic and have the same refractive index in every direction; they have no birefringence and show no interference colors. See also: *colors.* (b) The property of anisotropic crystals to split a beam of light into two polarized rays that traverse the crystal at different velocities as they pass through it and produce characteristic optical effects that are recognizable with the proper instruments or, in some cases (e.g., calcite), by the eye alone. Syn: *double refraction.* Cf: *transmitted light.*
birefringent Said of a crystalline substance that displays birefringence; such materials have more than one index of refraction. Syn: *birefractive; birefracting.*
birne *boule.*
birnessite A monoclinic mineral, $Na_4Mn_{14}O_{27}·9H_2O$; black or dark-brown, named for Birness, Scotland.
Birtley coal picker An electric picker that distinguishes between good coal and slate by their different electrical conductivities. It is said to be more accurate than the human slate picker, who, when fatigued, may fail to remove all the impure material. (Mitchell, 1950)
Birtley contraflow separator A pneumatic table for dry cleaning coal. It consists of perforated deck plates arranged in a series of lateral steps with a longitudinal inclination. A centrifugal fan provides a constant upward blast of air through the deck. The usual layering takes place, the refuse sinking to the deck plates. The capacity of the table ranges from 6 st/h per foot (17.8 t/h/m) of width for sizes 1-1/2 to 2 in (3.8 to 5.1 cm), down to 2 st/h/ft (5.9 t/h/m) for fines below 1/16 in. They are built in any width up to a maximum of about 8 ft (2.4 m). (Nelson, 1965)
bischofite A monoclinic mineral, $MgCl_2·6H_2O$.
biscuit (a) Unglazed ceramic ware that has been fired in a biscuit or bisque oven or film. (b) A small cake of primary metal, such as uranium, made from uranium tetrafluoride and magnesium in a bomb reduction. (ASM, 1961)
bishop's stone *amethyst.*
bisilicate In metallurgy, a slag with a silicate degree of 2. (Newton, 1938)
bismite A straw-yellow monoclinic mineral, Bi_2O_3; earthy to powdery; in oxidized parts of bismuth ores. Syn: *bismuth ocher.*
bismoclite A tetragonal mineral, BiOCl.
bismuth A white crystalline, brittle metal with a pinkish tinge. Symbol, Bi. The most important ores are bismuthinite or bismuth glance (Bi_2S_3) and bismite (Bi_2O_3). Also obtained as a byproduct in refining lead, copper, tin, silver, and gold ores. Forms low-melting alloys that are used in fire detection and extinguishing systems; used as a catalyst for making acrylic fibers and as a carrier for fuel in atomic reactors; extensively used in cosmetics and in medicine. (Handbook of Chem. & Phys., 3)
bismuth blende *eulytite.*
bismuth glance *bismuthinite.*
bismuth gold A pinkish-white native alloy of bismuth and gold, approx. Au_2Bi; contains 65.5% gold. *maldonite.* (Hess)
bismuthine *bismuthinite.*
bismuthinite An orthorhombic mineral, $4[Bi_2S_3]$; metallic; lead-gray to tin-white with an iridescent tarnish; commonly associated with other ore minerals; a source of bismuth. Syn: *bismuth glance; bismuthine.*
bismuth ocher *bismite.*
bismuth selenide Bi_2Se_3; black; orthorhombic; and melting point, 706 °C. Of some interest for thermoelectric applications. Also called bismuth triselenide. (Lee, 1961)
bismuth spar *bismutite.*
bismuth telluride Bi_2Te_2S; hexagonal rhombohedral; gray; and a thermoelectric material. Because it loses its semiconducting properties above 100 °C, it is of value chiefly in cooling devices. Also called bismuth tritelluride. Syn: *tetradymite.* (Lee, 1961)
bismutite A tetragonal mineral, $Bi_2(CO_3)O_2$; earthy or amorphous. Syn: *bismuth spar.*
bismutotantalite An orthorhombic mineral, $Bi(Ta,Nb)O_4$.
bisphenoid In crystallography, a form apparently consisting of two sphenoids placed together symmetrically. Cf: *disphenoid.* (Fay, 1920)
bit Any device that may be attached to, or is, an integral part of a drill string and is used as a cutting tool to bore into or penetrate rock or other materials by utilizing power applied to the bit percussively or by rotation. See also: *detachable bit; drag bit.* (Long, 1960)
bit blank A steel bit in which diamonds or other cutting media may be inset by hand peening or attached by a mechanical process such as casting, sintering, or brazing. Syn: *bit shank; blank; blank bit; body; shank.* (Long, 1960)
bit clearance (a) Technically, the difference between the outside diameter of a set bit and the outside set diameter of the reaming shell. Loosely, the term is used to denote the clearing action of a bit, which is a function of the waterways and the mode in which the diamonds or other cutting media are set in the cutting face of the bit, and also the difference between the outside set diameter of a bit and the outside diameter of the bit shank. (Long, 1960) (b) Incorrectly and loosely used as a syn. for diamond exposure. See also: *diamond exposure.* (Long, 1960)
bit contour The configuration of the crown or cutting face of a bit as seen in cross section. (Long, 1960)
bit core The central, removable, and replaceable portion or pilot of a noncoring or other type of bit. Cf: *core.* (Long, 1960)
bit count *diamond count.*
bit crown *crown.*
bit-crown metal *diamond matrix.*
bit die *bit mold.*
bit disc A bit with two or more rolling discs that do the cutting. Used in rotary drilling through certain formations. (Porter, 1930)
bit drag A bit with serrated teeth used in rotary drilling. (Hess)
bit face That part of the bit crown that comes in contact with the bottom of a borehole. It does not include that part of the bit crown that contacts the walls of the borehole. (Long, 1960)
bit feed *feed rate.*
bit life The average number of feet of borehole a bit may be expected to drill in a specific type of rock under normal operating or specified conditions. (Long, 1960)
bit load The weight or pressure applied to a bit in drilling operations, expressed as the number of pounds or tons of weight applied. Syn: *bit pressure; bit weight; drilling pressure; drilling weight; drill pressure; load.* (Long, 1960)
bit matrix *diamond matrix.*
bit mold A steel, carbon, or ceramic die in which the shape of a bit crown is incised and provided with pips, grooves, or holes in which diamonds are set and held by suction or an adhesive. Filling the die with a matrix alloy by a casting or a powder metal-sintering process affixes the shank to a diamond-inset bit crown having a shape conforming to that incised in the die. Syn: *bit die; crown die; crown mold.* (Long, 1960)
bit performance The achievement of a bit as gaged by the overall cost of using a specific bit per a unit measure of borehole drilled, or by the total number of feet of borehole drilled per bit. (Long, 1960)
bit pressure *bit load.*
bit reaming shell Obsolete name for reaming shell. (Long, 1960)
bit ring (a) *setting ring.* (b) Obsolete name for core bit. (Long, 1960)
bit shank (a) The threaded part of a bit. (Long, 1960) (b) Sometimes incorrectly used as a syn. for bit blank. (Long, 1960)
bitter lake A salt lake whose waters contain in solution a high content of sodium sulfate and lesser amounts of the carbonates and chlorides ordinarily found in salt lakes; a lake whose water has a bitter taste. Examples include Carson Lake, NV, and the Great Bitter Lake in Egypt. (AGI, 1987)

bittern (a) The bitter liquid remaining after seawater has been concentrated by evaporation until most of the sodium chloride has crystallized out. (AGI, 1987) (b) A natural solution, in an evaporite basin, that resembles a saltworks liquor, esp. in its high magnesium content. (AGI, 1987)
bitter salt *epsomite.*
bitter spar (a) A pure, crystalline dolomite that consists of 1 part or equivalent of calcium carbonate and 1 part of magnesium carbonate. Syn: *pearl spar.* (Fay, 1920) (b) *dolomite.*
bit thrust The hydraulic pressure applied to a drill bit when drilling, as shown in pounds per square inch by the pressure gages on the hydraulic-feed cylinders of a diamond drill or the total pressure in pounds as calculated by multiplying the recorded hydraulic pressure by the square-inch area of the piston in the hydraulic-feed cylinder. Syn: *drilling thrust.* (Long, 1960)
bitumen (a) A general name for various solid and semisolid hydrocarbons. In 1912, the term was used by the American Society for Testing and Materials to include all those hydrocarbons that are soluble in carbon disulfide, whether gases, easily mobile liquids, viscous liquids, or solids. (b) A generic term applied to natural flammable substances of variable color, hardness, and volatility, composed principally of a mixture of hydrocarbons substantially free from oxygenated bodies. Bitumens are sometimes associated with mineral matter, the nonmineral constituents being fusible and largely soluble in carbon disulfide, yielding water-insoluble sulfonation products. Petroleums, asphalts, natural mineral waxes, and asphaltites are all considered bitumens. (AGI, 1987)
bitumen cable A cable notable for its resistance to moisture, but not suitable for high temperatures. The wires are tinned to prevent reaction with the sulfur in the bitumen. Outside the bitumen are layers of tape and jute, and one or two layers of steel armoring; outside each layer of steel armoring are layers of serving compound. (Mason, 1951)
bitumenite Cannel coal from Torbane, Scotland. See also: *torbanite.* Also spelled bituminite.
bitumen lapideum An old name for mineral coal. (Tomkeieff, 1954)
bituminiferous Yielding or containing bitumen. (Standard, 1964)
Bituminite High explosive used in mines. (Bennett, 1962)
bituminous (a) Containing bitumen. (b) Pertaining to bituminous coal. (AGI, 1987) (c) Having the odor of bitumen; often applied to minerals. (d) Yielding volatile bituminous matter on heating (for example, bituminous coal). (AGI, 1987) (e) Containing much organic, or at least carbonaceous, matter, mostly in the form of the tarry hydrocarbons, which are usually described as bitumen.
bituminous coal (a) Coal that ranks between subbituminous coal and anthracite and that contains more than 14% volatile matter (on a dry, ash-free basis) and has a calorific value of more than 11,500 Btu/lb (26.7 MJ/kg) (moist, mineral-matter-free) or more than 10,500 Btu/lb (24.4 MJ/kg) if agglomerating (ASTM). It is dark brown to black in color and burns with a smoky flame. Bituminous coal is the most abundant rank of coal; much is Carboniferous in age. Cf: *medium-volatile bituminous coal; low volatile bituminous coal.* Syn: *soft coal.* (AGI, 1987) (b) A coal that is high in carbonaceous matter, having between 15% and 50% volatile matter. Soft coal. (BCI, 1947) (c) A general term descriptive of coal other than anthracite and low-volatile coal on the one hand and lignite on the other. (BS, 1964) (d) A coal with a relatively high proportion of gaseous constituents; dark brown to black in color and burns with a smoky luminous flame. The coke yield ranges from 50% to 90%. The term does not imply that bitumen or mineral pitch is present. See also: *coking coal.* (Nelson, 1965)
bituminous grout A mixture of bituminous material and fine sand that will flow into place without mechanical manipulation when heated. (ASTM, 1994)
bituminous limestone A dark, dense limestone containing abundant organic matter, believed to have accumulated under stagnant conditions and emitting a fetid odor when freshly broken or vigorously rubbed, e.g., the Bone Spring Limestone of Permian age in west Texas. Syn: *stinkstone; anthraconite.* See also: *swinestone.* (AGI, 1987)
bituminous ores Iron ores in which the gangue consists principally of coaly matter; e.g., black band ironstone. (Osborne, 1956)
bituminous rock Natural or rock asphalt, but the term is sometimes used to describe a rock in which the percentage of impregnation is comparatively low. (Nelson, 1965)
bituminous shale A shale containing bituminous material; coaly shale. Cf: *oil shale.*
bit wall That portion of the bit between the crown and the shank of the bit. (Long, 1960)
bit weight (a) Total weight, in carats, of the diamonds set in a diamond bit. (Long, 1960) (b) Weight or load applied to a diamond bit during a drilling operation. See also: *bit load.* (Long, 1960)
bityite A monoclinic mineral, $CaLiAl_2(AlBeSi_2O_{10})(OH)_2$; mica group; perfect basal cleavage.

bivalent (a) Having a valence of 2. (Webster 3rd, 1966) (b) Having two valences; e.g., cobalt has two forms, with valences of 2 and 3, respectively. (Handbook of Chem. & Phys., 2)
bixbite A red variety of beryl; in the Wah Wah and Topaz Mountains of Utah.
bixbyite An isometric mineral, $(Mn,Fe)_2O_3$; forms black cubes. Formerly called partridgeite and sitaparite.
black alkali An old term for an alkali soil whose sodium tends to disperse organic matter and give a black color. Cf: *white alkali.* (AGI, 1987)
black amber A name given to jet that is found with amber. It becomes faintly electric when rubbed. See also: *jet; stantienite.*
black andradite garnet *melanite.*
black ash (a) Any of various dark-colored products obtained in industrial processes. (Webster 3rd, 1966) (b) A black mass containing chiefly soda in the form of sodium carbonate and usually also sodium sulfide with some carbon, and produced esp. for recovery of its soda content by concentrating and burning black liquor in rotary furnaces. (Webster 3rd, 1966)
blackband (a) A dark, earthy variety of the mineral siderite, occurring mixed with clay, sand, and considerable carbonaceous matter, and frequently associated with coal. Syn: *blackband ore.* (AGI, 1987) (b) A thin layer (up to 10 cm in thickness) of blackband interbedded with clays or shales in blackband ironstone. (AGI, 1987) (c) *blackband ironstone.*
blackband ironstone A dark variety of clay ironstone containing sufficient carbonaceous matter (10% to 20%) to make it self-calcining (without the addition of extra fuel). Syn: *blackband.* See also: *clay ironstone.* (AGI, 1987)
blackband ore *blackband.*
black bat A piece of bituminous shale embedded in the rock immediately over the coal measure and liable to fall of its own weight when the coal beneath it has been removed. Cf: *kettle bottom.*
black body As applied to heat radiation, this term signifies that the surface in question emits radiant energy at each wavelength at the maximum rate possible for the temperature of the surface and, at the same time, absorbs all incident radiation. Only when a surface is a black body can its temperature be measured accurately by means of an optical pyrometer. (Dodd, 1964)
black box A separate and self-contained electronic unit or element of an electronic device which can be treated as a single package. (NCB, 1964)
black chalk A bluish-black carbonaceous clay, shale, or slate, used as a pigment or crayon. (AGI, 1987)
black cobalt *asbolan.*
black concentrate The mixture of amalgam gold and magnetite obtained from behind the riffles in a gold sluice. (CTD, 1958)
black copper A name given to the impure metallic copper produced in blast furnaces running on oxide ores or roasted sulfide material; an alloy of copper with one or more other metals generally containing several percent of iron, commonly lead, and many other impurities; also contains 1% to 3% sulfur. See also: *tenorite.*
black copper ore An earthy, black, massive, or scaley form of copper oxide, CuO. See also: *melaconite; tenorite.* (Hess)
black coral An intense black to dark brown coral used in beads, bracelets, art objects, etc.
black cotton In India, soil from 6 to 10 ft (1.8 to 3.0 m) in thickness overlying the coal measures, which, in dry weather, shrinks and produces mud cracks.
blackdamp Generally applied to carbon dioxide. Strictly speaking, a mixture of nitrogen and carbon dioxide. The average blackdamp contains 10% to 15% carbon dioxide and 85% to 90% nitrogen. It is formed by mine fires and the explosion of combustible gases in mines, and hence forms a part of the afterdamp. An atmosphere depleted of oxygen rather than containing an excess of carbon dioxide. Being heavier than air, it is always found in a layer along the floor of a mine. It extinguishes light and suffocates its victims. Hence, it is sometimes known as chokedamp. See also: *afterdamp; damp.* (Fay, 1920; Korson, 1938)
black diamond (a) A variety of crystalline carbon, related to diamond, but showing no crystal form. Highly prized as an abrasive because of its hardness. Occurs only in Brazil. Syn: *carbonado.* (CTD, 1958) (b) A term frequently applied to coal. (c) A black gem diamond. (d) Dense black hematite that takes a polish like metal.
black flux A reducing flux composed of powdered carbon and alkali-metal carbonate. (Webster 3rd, 1966)
black garnet *andradite.*
black gold (a) A slang American term referring to crude oil. (b) Syn: *Maldonite.* (c) Placer gold coated with a black or dark-brown substance (such as a film of manganese oxide) so that the yellow color is not visible until the coating is removed. (AGI, 1987)

black granite A commercial term for crystalline rock that when polished is dark gray to black. It may be a diabase, diorite, or gabbro. (AGI, 1987)
black gunpowder A mixture of potassium nitrate (saltpeter), sulfur, and charcoal in varying proportions. A typical composition is 70% to 75% saltpeter, 10% to 14% sulfur, and 14% to 16% charcoal. It is designated according to grain size: mealed; superfine grain (FFG); fine grain (FG); large or coarse grain (LG); large grain for rifles (RLG); and mammoth. Syn: *black powder.* (CCD, 1961)
black hematite *psilomelane.*
black iron Malleable iron, untinned; distinguished from tinned or white iron. (Standard, 1964)
black iron ore *magnetite.*
blackjack (a) A thin stratum of coal interbedded with layers of slate; a slaty coal with a high ash content. (AGI, 1987) (b) A syn. of sphalerite, esp. a dark variety. See also: *sphalerite.* (AGI, 1987)
black lead (a) An obsolete name for graphite, still used in naming lead pencils, which are really made of graphite. Syn: *plumbago.* (Tomkeieff, 1954) (b) Graphite, in impure crystalline form. (Pryor, 1963) (c) Used for coating patterns and the faces of cast-iron chilling molds. (Crispin, 1964)
blacklead ore An early name for the black variety of cerussite. (Fay, 1920)
black light (a) A prospector's and miner's term for ultraviolet light, used in exploration and evaluation to detect mineral fluorescence. (AGI, 1987) (b) An instrument, usually portable, that produces ultraviolet light for this purpose. See also: *lamp.* (AGI, 1987)
black liquor The alkaline spent liquor from the digesters in the manufacture of sulfate or soda wood pulp.
black liquor recovery furnaces Smelting or recovery furnaces in which evaporated black liquor is burned to a molten chemical smelt.
blacklung *anthracosis.*
black magnetic rouge A polishing material consisting of 99% Fe_3O_4. See also: *black rouge.* (Osborne, 1956)
black manganese (a) *hausmannite.* (b) A term applied to dark-colored manganese minerals; e.g., pyrolusite, hausmannite, and psilomelane.
black metal A black shale associated with coal measures. (AGI, 1987)
black mica *biotite.*
black muck Lanc. A dark-brown powdery substance, consisting of silica, alumina, and iron; found in iron mines.
black mud A mud formed in lagoons, sounds, or bays, in which there is poor circulation or weak tides. The color is black because of iron sulfides and organic matter. (AGI, 1987)
black ocher *wad; bog manganese.*
black opal Precious opal with play of color (commonly red or green) displayed against a dark gray (rarely black) body color; e.g., the fine Australian blue opal with flame-colored flashes.
black ore (a) Eng. Partly decomposed pyrite containing copper. (Fay, 1920) (b) In uranium mining, the term may mean ore containing a high proportion of pitchblende, uraninite, coffinite, or vanoxite. (c) Cumb. A variety of hematite in hard pieces, some kidney shaped, reaching the size of one's hand, in a moderately soft, dark-red, brown, or nearly black mass of smit clay and manganese oxide, the whole having a most confused appearance.
black oxide of manganese *pyrolusite.*
black pigment Lampblack obtained by burning common coal tar. (Fay, 1920)
black powder A deflagrating or low-explosive granular compound of sulfur, charcoal, and an alkali nitrate, usually potassium or sodium nitrate. Syn: *black gunpowder.*
black roast In fluidization roasting (fluosolids process), the conversion of iron sulfide to magnetite. (Pryor, 1963)
black rouge A precipitated black magnetic iron oxide. Used mainly in plate printing inks and in paints, but has small abrasive applications. See also: *black magnetic rouge.* (AIME, 1960)
black sand (a) An alluvial or beach sand consisting predominantly of grains of heavy, dark minerals or rocks (e.g., magnetite, rutile, garnet, or basaltic glass), concentrated chiefly by wave, current, or surf action. It may yield valuable minerals. See also: *beach placer.* (AGI, 1987) (b) An asphaltic sand. (AGI, 1987)
black shale (a) A dark, thinly laminated carbonaceous shale, exceptionally rich in organic matter (5% or more carbon content) and sulfide (esp. iron sulfide, usually pyrite), and often containing unusual concentrations of certain trace elements (U, V, Cu, Ni). It is formed by partial anaerobic decay of buried organic matter in a quiet-water, reducing environment (such as in a stagnant marine basin) characterized by restricted circulation and very slow deposition of clastic material. Fossil organisms are preserved as a graphitic or carbonaceous film or as pyrite replacements. Syn: *biopelite.* (AGI, 1987) (b) Usually a very thin-bedded shale, rich in sulfides (esp. pyrite, which may have replaced fossils) and rich in organic material, deposited under barred basin conditions causing anaerobic accumulation. (AGI, 1987) (c) Generally, a fine-grained, finely laminated carbonaceous shale, sometimes canneloid, often found as a roof to a coal, or in place of a coal, resting on a fire clay. Syn: *black metal.* (Tomkeieff, 1954)
black silver *stephanite.*
black smoker A hydrothermal vent at the crest of an oceanic ridge; e.g., the East Pacific Rise at the mouth of the Gulf of California. Waters blackened by sulfide precipitates jet out at 1 to 5 m/s at temperatures of at least 350 °C. The term refers to uprushing black turbulent suspension. Cf: *white smoker.* (AGI, 1987)
black telluride *nagyagite.*
black tellurium (a) A rare gray metallic mineral, a sulfotelluride of gold and lead with some antimony. (CTD, 1958) (b) *nagyagite.*
black tin Eng. Dressed tin ore ready to be smelted; from Cornwall. See also: *cassiterite.* (Standard, 1964)
black wad An early name for several minerals, including graphite and the softer manganese oxides.
bladder pump A positive displacement pump in which compressed air is forced down an input column to squeeze a water-filled bladder, thereby forcing water up a discharge column to the ground surface. The bladder refills by gravity flow at the end of each lifting cycle, because the bladder unit is below the static water lever. *positive displacement pump.*
blade (a) The shape of a solid, as one in which the ratios of breadth to length and thickness to breadth are each less than 2:3. (b) Having the appearance of blades, e.g., flat crystals strongly elongated in one direction.
bladed Decidedly elongated and flattened; descriptive of some minerals. (Fay, 1920)
bladed structure Consisting of individual minerals flattened like a knife blade. (Hess)
blade mill Trommel washer with lifting blades, which aid in disintegration and scrubbing of passing feed. (Pryor, 1963)
blaes (a) A Scottish term for a gray-blue carbonaceous shale that weathers to a crumbly mass and eventually to a soft clay. See also: *bind.* (AGI, 1987) (b) A Scottish term for a hard, joint-free sandstone. Syn: *blaize.* (AGI, 1987)
Blair process An improved form of the Chenot process for making sponge iron by heating crushed iron oxide and coal in retorts.
Blaisdell excavator An apparatus for automatically discharging a sand tank having a central bottom opening. It consists of a central vertical shaft carrying four arms fitted with round plow disks. Sand is plowed toward a central opening and discharged on a conveyor belt. Syn: *Blaisdell vat excavator.* (Liddell, 1918; Fay, 1920)
Blaisdell loading machinery An apparatus for loading sand tanks. It consists of a rapidly revolving disk with curved radial vanes. The disk is hung on a shaft in the tank center, and the sand dropped on the disk is distributed over the entire tank area. (Liddell, 1918)
Blaisdell vat excavator *Blaisdell excavator.*
blaize *blaes.*
Blake breaker A jaw breaker or particular kind of jaw crusher. (Nelson, 1965)
Blake furnace A furnace, the hearth of which consists of terraces rising from the outer edge to the center. The hearth is circular and revolves when in operation. (Fay, 1920)
blakeite (a) Anhydrous ferric tellurite as reddish-brown microcrystalline (cubic?) crusts from Goldfield, NV. (Spencer, 1946) (b) Titanozirconate of thorium, uranium, calcium, iron, etc., described as zirkelite from Ceylon [now Sri Lanka], but differing in chemical composition and also apparently in crystalline form from the original zirkelite from Brazil. (Spencer, 1952)
Blake jaw crusher The original crusher of jaw type. A crusher with one fixed jaw plate and one pivoted at the top so as to give the greatest movement on the smallest lump. Motion is imparted to the lower end of the crushing jaw by toggle joint operated by eccentric. This machine, or some modification of it, is used for reducing run-of-mine ore or coal to a size small enough to be taken by the next crusher in the series during the first stage of crushing. (Liddell, 1918; Newton, 1959)
Blake Morscher separator *electrostatic separator.*
blank (a) An interval in a borehole from which core was not recovered or was lost, or in which no minerals of value were encountered. (Long, 1960) (b) *bit blank.* (c) In powder metallurgy, a pressed, presintered, or fully sintered compact, usually in the unfinished condition and requiring cutting, machining, or some other operation to produce the final shape. (ASM, 1961) (d) A quartz plate with approx., or exactly, the correct edge dimensions, but not yet finished to final thickness (frequency). Ordinarily applied to pieces of quartz that are in the process of being machine lapped or that are diced out, but not yet lapped. (Am. Mineral., 1947)
blank bit *bit blank.*
blanket (a) A textile material used in ore treatment plants for catching coarse free gold and some associated minerals; e.g., pyrite. The blanket is taken up periodically and washed in a tub to remove the gold concentrate, from which the gold is recovered by amalgamation. Cf:

tabular. (Nelson, 1965) (b) *blanket deposit; blanket vein.* (c) Soil or broken rock left or placed over a blast to confine or direct throw of fragments. (Nichols, 1976) (d) A thin, widespread sedimentary body whose width-thickness ratio is greater than 1,000:1 and may be as great as 50,000:1. Syn: *sheet.* (AGI, 1987)

blanket deposit (a) A horizontal, tabular orebody; manto; bedded vein. (AGI, 1987) (b) A sedimentary deposit of great areal extent and relatively uniform thickness; esp. a blanket sand and associated limestones. See also: *blanket; blanket vein.* (AGI, 1987)

blanket feed A method for charging batch designed to produce an even distribution of batch across the width of the furnace. (ASTM, 1994)

blanketing (a) The material caught upon the blankets used in concentrating gold-bearing sands or slimes; also the process involved. (b) Can. Staking but not recording claims. (Hoffman, 1958)

blanket sand A blanket deposit of sand or sandstone of unusually wide distribution, typically an orthoquartzitic sandstone deposited by a transgressive sea advancing for a considerable distance over a stable shelf area; e.g., the St. Peter Sandstone of the East-Central United States. Syn: *sheet sand; blanket sandstone.* (AGI, 1987)

blanket sandstone *blanket sand.*

blanket shooting Applied to a method of blasting on a face not exceeding 30 ft or 35 ft (9.1 m or 10.7 m) in height. It involves leaving at the quarry face a mass of shattered rock several feet in thickness that serves as a buffer, preventing the rock from being thrown far from its source, and also rendering the shot more effective. Syn: *buffer shooting; shooting against the bank.* (Fay, 1920)

blanket sluice A sluice in which coarse blankets are laid to catch the fine but heavy particles of gold, amalgam, etc., in the slime passing over them. The blankets are removed and washed from time to time to obtain the precious metal. (Fay, 1920)

blanket strake A trough over which gold pulp flows. It is lined with a blanket for catching coarse gold and associated minerals. See also: *strake.* (Nelson, 1965)

blanket vein A horizontal or sheet deposit. See also: *blanket; blanket deposit.* Cf: *bed vein.*

blanket washer In ore dressing, smelting, and refining, one who cleans flannel blankets over which a mixture of finely ground gold ore and cyanide solution from Chilean mills is passed to collect free particles of gold not dissolved by the cyanide. (DOT, 1949)

blank hole (a) A borehole in which no minerals or other substances of value were penetrated. Syn: *barren hole; dry hole.* (Long, 1960) (b) The uncased portion of a borehole. Syn: *bare; barefoot; naked.* (Long, 1960)

blankoff To line a specific portion of a borehole with casing or pipe for the purpose of supporting the sidewalls or to prevent ingress of unwanted liquids or gas. Syn: *case; case off* Cf: *seal off.* (Long, 1960)

blank pipe Unperforated pipe or casing set in a borehole. (Long, 1960)

blank reaming shell A reaming shell in which no reaming diamonds or other cutting media are inset on the outside surface. (Long, 1960)

Blanton cam A device used for locking the cam on the camshaft in a stamp mill. (Fay, 1920)

Blasjo cut This is a cut with a single V where all the holes on one side are parallel and meet the holes from the other side at an angle that may be as low as 30°. (Langefors, 1963)

blast (a) The ignition of a heavy explosive charge. Syn: *shot.* (b) A miner's term for compressed air underground. (Nelson, 1965) (c) Scot. A fall of water in the downcast shaft to produce or quicken ventilation. (d) A suffix signifying a texture formed entirely by metamorphism. (AGI, 1987) (e) The operation of increasing the diamond exposure on a bit face by removing some of the matrix metal through the abrasive action of grains of sand carried in a high-pressure stream of air. Also called sandblast. (Long, 1960) (f) An increase in firing temperature of a kiln immediately before ending the firing operation. (g) The period during which a blast furnace is in blast; i.e., in operation.

blast coupling The degree to which an explosive fills the borehole. Bulk-loaded explosives are completely coupled. Untamped cartridges are decoupled. (Dick, 2)

blast draft The draft produced by a blower, as by blowing in air beneath a fire or drawing out the gases from above it. A forced draft. (Fay, 1920)

blaster (a) A device for detonating an explosive charge. (AGI, 1987) (b) One who sets off blasts in a mine or quarry. A shot firer. Syn: *shooter.* (Hess) (c) *blasting unit.*

blast furnace A shaft furnace in which solid fuel is burned with an airblast to smelt ore in a continuous operation.

blast furnace dust A dust recovered from blast furnace gases, some of which is valuable for its potash content. (Hess)

blast furnace gas A low-grade producer gas, made by the partial combustion of the coke used in the furnace and modified by the partial reduction of iron ore. The gas contains more carbon dioxide and less hydrogen than normal producer gas made from coke and has a lower calorific value. (Francis, 1965)

blast-furnace slag The nonmetallic product, consisting essentially of silicates and aluminosilicates of lime and of other bases, that is developed simultaneously with iron in a blast furnace. Syn: *slag.* (ASTM, 1994)

blast hearth A hearth in connection with which a blast is used, as in reducing lead ore. (Fay, 1920)

blasthole A hole drilled in a material to be blasted, for the purpose of containing an explosive charge.

blasthole charger A portable unit consisting of a prilled explosive reserve tank feeding into an air-activated loading tube. The equipment should be grounded to guard against buildup of static electricity and possible accidental explosive detonation. The blasthole charger permits rapid loading of prilled explosives into blastholes drilled in any direction.

blasthole drill Any rotary, percussive, fusion-piercing, churn, or other type of drilling machine used to produce holes in which an explosive charge is placed. Syn: *shothole drill.* (Long, 1960)

blasthole driller *churn-drill operator.*

blasting agent An explosive material that meets prescribed criteria for insensitivity to initiation. It is a material or mixture consisting of fuel and oxidizer used in blasting, but not otherwise defined as an explosive. The finished mixture used for shipment or transportation cannot be detonated by a No. 8 test detonator cap when unconfined.

blasting barrel A piece of iron pipe, usually about 1/2 in (1.3 cm) in diameter, used to provide a smooth passageway through the stemming for the miner's squib. It is recovered after each blast and used until destroyed. (Fay, 1920)

blasting cap (a) A detonator containing an ignition explosive mixture, a primary initiating charge, and a high-explosive base charge, encapsulated in an aluminum or copper shell. Caps are initiated either electrically or nonelectrically. See also: *waterproof electric blasting cap.* (b) A small sensitive charge placed in the larger explosive charge by which the larger charge is detonated. See also: *electric detonator.* (BCI, 1947)

blasting cartridge A cartridge containing an explosive to be used in blasting. (Fay, 1920)

blasting circuit A shotfiring cord together with connecting wires and electric blasting caps used in preparation for the firing of a blast in mines, quarries, and tunnels.

blasting compounds Explosive substances used in mining and quarrying. (Hess)

blasting cord *shot-firing blasting cord.*

blasting curtain A screen erected to prevent damage to equipment and supports in the vicinity of the blasting point. See also: *curtain; shot-firing curtain.* Syn: *blasting cord.* (Nelson, 1965)

blasting fuse (a) A slow-burning fuse used in blasting operations. (Standard, 1964) (b) A fine core of gunpowder enclosed in the center of jute, yarn, etc., for igniting an explosive charge in a shothole. See also: *safety fuse.* (Nelson, 1965)

blasting galvanometer An instrument that provides a simple means for testing electric blasting circuits, enabling the blaster to locate breaks, short circuits, or faulty connections before an attempt is made to fire the shot. With its use, misfires may be prevented to a great extent. To test a circuit, one wire should be placed on one terminal of the instrument and the other wire on the other terminal. If the needle is not deflected, it indicates that the circuit is broken; if it is an electric blasting cap that is being tested, this should be discarded. (Pit and Quarry, 1960)

blasting gelatin A high explosive, consisting of nitroglycerin and nitrocotton. It is a strong explosive, and a rubberlike, elastic substance, unaffected by water. Taken as a standard of explosive power. Cf: *dynamite.* (Fay, 1920; BS, 1964)

blasting hole well driller *churn-drill operator.*

blasting log A written record of information about a specific blast as may be required by law or regulations. (Atlas, 1987)

blasting machine A portable dynamo that generates enough electric current to detonate electric blasting caps when the machine rack bar or handle is given a quick, downward push. Syn: *battery.* See also: *dynamo exploder; M.E. 6 exploder.* (Long, 1960)

blasting mat A mat of woven steel wire, rope, scrap tires, or other suitable material or construction to cover blastholes for the purpose of preventing flying rock missiles. (Meyer, 1981)

blasting needle A pointed instrument for piercing the wad or tamp of a charge of explosive, to permit introducing a blasting fuse. (Standard, 1964)

blasting off the solid Blasting the working face in a coal mine without providing a second free face by cutting or shearing before blasting. (CFR, 4)

blasting oil *nitroglycerin.*

blasting reflection mechanism *reflection mechanism.*

blasting supplies A term used to include electric blasting caps, ordinary blasting caps, fuse, blasting machines, galvanometers, rheostats, etc., in fact, everything used in blasting, except explosives. Cf: *blowing tools.* (Fay, 1920)

blasting switch A switch used to connect a power source to a blasting circuit. It is sometimes used to short-circuit the leading wires as a safeguard against premature blasts.

blasting timer An instrument that utilizes a powerline as a source of electrical current and that closes the circuits of successive blasting caps with a delay time interval. The timer provides for the circuits of 15 charges and affords positive control of the duration of intervals. (Streefkerk, 1952)

blasting tube A tube of explosives, as nitroglycerin, for blasting. (Standard, 1964)

blasting unit A portable device including a battery or a hand-operated generator designed to supply electric energy for firing explosive charges in mines, quarries, and tunnels. Syn: *blaster; exploder; shot-firing unit.* See also: *single-shot blasting unit; multiple-shot blasting unit.*

blasting vibrations The energy from a blast that manifests itself in earthborne vibrations that are transmitted through the Earth away from the immediate blast site. Syn: *ground vibrations.* (Atlas, 1987)

blast nozzle A fixed- or variable-sized outlet of a blast pipe. (Fay, 1920)

blastogranitic A relict texture in a metamorphic rock in which remnants of the original granitic texture remain. (AGI, 1987)

blastophitic Said of a relict texture in a metamorphic rock in which traces of an original ophitic texture remain. (AGI, 1987)

blastoporphyritic Said of a relict texture in a metamorphic rock in which traces of an original porphyritic texture remain. (AGI, 1987)

blast pattern The array of drilled holes on the surface or underground to be loaded and detonated in sequence; a pattern is indicated by the distance between holes in a row (spacing) and between rows (burden).

blast roasting Roasting conducted in a Dwight-Lloyd machine, in which roasting is accompanied by sintering. The charge is placed in small boxes and ignited; air is then drawn through to burn off sulfur. Syn: *Carmichel-Bradford process.* (CTD, 1958)

blast site The area where explosive material is handled during the loading and detonation of blast holes; in surface blasting, it includes 50 ft (15.2 m) in all directions from perimeter holes; underground, it includes 15 ft (4.6 m) of solid rib. (Atlas, 1987)

bleaching clay A clay or earth that, either in its natural state or after chemical activation, has the capacity for adsorbing or removing coloring matter or grease from liquids (esp. oils). Syn: *bleaching earth.* (AGI, 1987)

bleaching earth *bleaching clay.*

bleb A small, usually rounded inclusion of one mineral in another; e.g., blebs of olivine poikilitically enclosed in pyroxene.

bled ingot In steelmaking, an ingot that has lost its molten center while cooling. (Standard, 1964)

bleeder (a) A connection located at a low place in an air line or gas line, or container, so that by means of a small valve the condensed water or other liquid can be drained or bled off from the line or container without discharging the air or gas. (Long, 1960) (b) A fine-adjustment valve (needle valve) connected to the bottom end of a hydraulic feed cylinder in the swivel head of a diamond drill. By means of the bleeder, the speed at which the hydraulic piston travels can be minutely controlled. (Long, 1960) (c) A pipe on top of an iron blast furnace through which gas escapes.

bleeder entries Widely used for draining methane in coal mines in the United States where the room-and-pillar method is employed. They are panel entries driven on a perimeter of a block of coal being mined and maintained as exhaust airways to remove methane promptly from the working faces to prevent buildup of high concentrations either at the face or in the main intake airways. They are maintained, after mining is completed, in preference to sealing the completed workings. (Hartman, 1961)

bleeder pipe A pipe inserted in a seal to relieve gas pressure from a sealed area.

bleeding (a) The process of giving off oil or gas from pore spaces or fractures; it can be observed in drill cores. (AGI, 1987) (b) The exudation of small amounts of water from coal or a stratum of some other rock. (AGI, 1987)

bleeding surface Any face, such as the walls of a well or borehole or the sides of a fracture, that traverses a reservoir rock or aquifer, permitting the stored liquid or gas to seep (or to bleed) into the opening. (AGI, 1987)

bleed off A coal mining term used when feeders or blowers act as the means by which gas is "bled off" or dissipated to the adjoining strata or to the surface. (Kentucky, 1952)

blende (a) Without specific qualification, it means zincblende or the sulfide of zinc (sphalerite), which has the luster and often the color of common resin and yields a white streak and powder. The darker varieties are called blackjack by English miners. Other minerals having this luster are also called blendes, such as antimony blende, ruby blende, pitchblende, and hornblende. Sphalerite (blende) is often found in brown shining crystals, hence its name among German miners, from the word blenden, meaning to dazzle. (Fay, 1920) (b) A miners' term for sphalerite. (c) Various minerals, chiefly metal sulfides, with bright or resinous but nonmetallic luster, e.g., zinc blende (sphalerite), antimony blende (kermesite), bismuth blende (eulytite), cadmium blende (greenockite), pitchblende (uraninite), hornblende.

blended unconformity An unconformity having no distinct surface of separation or sharp contact, as at an erosion surface that was originally covered by a thick residual soil, which graded downward into the underlying rocks and was partly incorporated in the overlying rocks; e.g., a nonconformity between granite and overlying basal arkosic sediments derived as a product of its disintegration. Syn: *graded unconformity.* (AGI, 1987)

blending Mixing in predetermined and controlled quantities to give a uniform product. (BS, 1962)

blending conveyor A conveyor running beneath a line of ore bins or stockpiles, and so set that each bin or stockpile can deliver onto the conveyor at a controllable rate from individual feeders. Syn: *paddle-type mixing conveyor; screw-type mixing conveyor.* (Pryor, 1963)

blind (a) To drill with the circulation medium (water or drill mud) escaping into the sidewalls of the borehole and not overflowing the collar of the drill hole. (Long, 1960) (b) An underground opening not connected with other workings nearby and at about the same elevation. (Long, 1960) (c) Said of a mineral deposit that does not crop out. The term is more appropriate for a deposit that terminates below the surface than for one that is simply hidden by unconsolidated surficial debris. Syn: *blind vein.* (AGI, 1987)

blind apex The near-surface end of a mineral deposit, e.g., the upper end of a seam or vein that is truncated by an unconformity. Syn: *buried outcrop.* (AGI, 1987)

blind bit *noncoring bit.*

blind borehole process In the underground gasification of coal, a borehole is drilled to a blind end having no outside connection. A tube of smaller diameter is inserted nearly the full length through which air is passed to supply a gasification reaction at the far end of the hole. The hot gases return around the outside of the tube. See also: *underground gasification.* (Nelson, 1965)

blind drift A horizontal passage, in a mine, not yet connected with the other workings. See also: *blind level.*

blinde *blende.*

blind header A concealed brick header in the interior of a wall, not showing on the faces. (ACSG, 1961)

blind hole A borehole in which the circulating medium carrying the cuttings does not return to the surface. (Long, 1960)

blinding (a) In leaching, reduced permeability of ion-exchange resins due to adherent slimes. In sieving, blocking of screen apertures by particles. (Pryor, 1963) (b) A matting of, or stoppage by, fine materials during screening that interferes with or blinds the screen mesh. (c) Compacting soil immediately over a tile drain to reduce its tendency to move into the tile. (Nichols, 1976)

blind joint In apparently massive rock that is being quarried, a plane of potential fracture along which the rock may break during excavation. (AGI, 1987)

blind lead A vein having no outcrop. See also: *blind lode; blind vein; lead.*

blind level (a) One not yet holed through to connect with other passages. Syn: *blind drift.* (Pryor, 1963) (b) A cul-de-sac or dead end. (Pryor, 1963) (c) A level for drainage, having a shaft at either end, and acting as an inverted siphon.

blind lode A lode showing no surface outcrop, and one that cannot be found by any surface indications. See also: *blind lead; blind vein.*

blind road Mid. Any underground roadway not in use, having stoppings placed across it. Syn: *blind way.*

blind roaster A muffle furnace for roasting ore out of contact with the products of combustion. (Standard, 1964)

blind seams Incipient joints.

blind shaft A sublevel shaft, connected to the main (daylight to depth) shaft by a transfer station. A winze. (Pryor, 1963)

blind shearing Scot. A side cutting without undercutting.

blind vein A vein that does not continue to the surface. See also: *blind; blind lode; blind lead.* (Fay, 1920)

blind way *blind road.*

blind zone *shadow zone.*

blip Echo trace on radar or sonar indicator screen. (Hy, 1965)

Bliss sandstone Massive, compact, fine-textured, fossiliferous gray sandstone ranging from almost white to brown. It may be either Cambrian or Ordovician, or both, at any given locality. It represents a period of slow intermittent deposition of sandy material. Found in New Mexico and in Texas. (Hess)

blister (a) In quarrying, an unconfined charge of explosive used to bring down dangerous ground that cannot be made safe by barring and that is too inaccessible to bore. (South Australia, 1961) (b) A protrusion, more or less circular in plan, extending downward into a coal seam. It represents the filling of a streambed pothole worn into the upper surface of the coal-forming material. (AGI, 1987) (c) Copper as a smelter product before it is refined. (Hoffman, 1958) (d) A defect in metal, on or near the surface, resulting from the expansion of gas in a subsurface zone. Very small blisters are called pinheads or pepper blisters. (ASM, 1961)

blister bar A wrought-iron bar impregnated with carbon by heating in charcoal. Used in making crucible steel. (CTD, 1958)

blister copper An impure intermediate product in the refining of copper, produced by blowing copper matte in a converter. (ASM, 1961)

blistered copper ore A reniform variety of chalcopyrite. (Fay, 1920)

blister steel Raw steel that has been cooled very slowly and that has a blistered appearance. The blisters are formed by gas escaping from within the metal. (Camm, 1940)

blister wax *blower wax.*

bloating phenomena The expansion of certain nonmetallic materials by heating until the exterior of the particle or shape becomes sufficiently pyroplastic or melted to entrap gases generated on the interior by the decomposition of gas-producing components.

block (a) A division of a mine, usually bounded by workings but sometimes by survey lines or other arbitrary limits. (b) A short piece of timber placed between the mine roof and the cap of a timber set and directly over the cap support. A wedge driven between the roof and the timber holds the set in place. See also: *blocking and wedging.* (c) A pillar or mass of ore exposed by underground workings. See also: *blocking out.* (Nelson, 1965) (d) Portion of an orebody blocked out by drives, raises, or winzes, so that it is completely surrounded by passages and forms a rectangular panel. If its character, volume, and assay grade are thus established beyond reasonable doubt, it ranks as proved ore in the mine's assets. (Pryor, 1963) (e) The wedging of core or core fragments or the impaction of cuttings inside a bit or core barrel, which prevents further entry of core into the core barrel, thereby producing a condition wherein drilling must be discontinued and the core barrel pulled and emptied to forestall loss of core through grinding or the serious damage of the bit or core barrel. See also: *core block; plug.* (Long, 1960) (f) An obstruction in a borehole. (Long, 1960) (g) *sheave.*

block caving A general term that refers to a mass mining system where the extraction of the ore depends largely on the action of gravity. By removing a thin horizontal layer at the mining level of the ore column, using standard mining methods, the vertical support of the ore column above is removed and the ore then caves by gravity. As broken ore is removed from the mining level of the ore column, the ore above continues to break and cave by gravity. The term "block caving" probably originated in the porphyry copper mines, where the area to be mined was divided into rectangular blocks that were mined in a checkerboard sequence with all the ore in a block being removed before an adjacent block was mined. This sequence of mining is no longer widely used. Today most mines use a panel system, mining the panels sequentially or by establishing a large production area and gradually moving it forward as the first area caved becomes exhausted. The term "block caving" is used for all types of gravity caving methods. There are three major systems of block caving, and they are differentiated by the type of production equipment used. (1) The first system based on the original block cave system is the grizzly or gravity system and is a full gravity system wherein the ore from the drawpoints flows directly to the transfer raises after sizing at the grizzly and then is gravity loaded into ore cars. (2) The second system is the slusher system, which uses slusher scrapers for the main production unit. (3) The last system is the rubber-tired system, which uses load-haul-dump (LHD) units for the main production unit. Block caving has the lowest cost of all mine exploitation systems, with the exception of open pit mining or in situ recovery. See also: *top slicing.* (SME, 1992)

block caving into chutes *chute caving.*

block claim Aust. A square mining claim whose boundaries are marked out by posts.

block diagram A plane figure representing a block of the Earth's crust (depicting geologic and topographic features) in a three-dimensional perspective, showing a surface area on top and including one or more (generally two) vertical cross sections. The top of the block gives a bird's-eye view of the ground surface, and its sides give the underlying geologic structure. (AGI, 1987)

blocked-out ore (a) Ore, the amount, content, and minability of which have been proven by development work or by drilling developed ore. Syn: *developed reserve.* (AGI, 1987) (b) A body of ore exposed, explored, and sampled for valuation purposes on all four sides of the panel formed by driving, winzing, and raising. (Pryor, 1963) (c) *reserves.*

block faulting A type of normal faulting in which the crust is divided into structural or fault blocks of different elevations and orientations. It is the process by which block mountains are formed. (AGI, 1987)

block field A thin accumulation of usually angular blocks, lying on bedrock without a cliff or ledge above as apparent source. Block fields occur on high mountain slopes above the tree line. Syn: *felsenmeer.*

block hole (a) A small hole drilled into a rock or boulder into which an anchor bolt or a small charge or explosive may be placed. (Long, 1960) (b) Used by drillers, miners, and quarry workers for a method of breaking undesirably large blocks of stone or boulders by the discharge of an explosive loaded into shallow holes drilled into the blocks or boulders. (Long, 1960) (c) A relief hole designed to remove part of the burden from a subsequent shot; used in coal mining.

blockholer A person whose duty it is to break up and reduce to safe and convenient size, by blasting or otherwise, any large blocks or pieces of rock that have been blown down by the miners. (Fay, 1920)

block hole shot *pop shot.*

blockholing The breaking of boulders by loading and firing small explosive charges in small-diameter drilled holes.

blocking In a crusher, obstruction of the crushing zone by clayey material or by rock that refuses to break down and pass to discharge. Syn: *packing.*

blocking and wedging A method of holding mine timber sets in place. Blocks of wood are set on the caps directly over the post supports and have a grain of block parallel with the top of the cap; wedges are driven tightly between the blocks and the roof. See also: *block.*

blocking out (a) Exposing an orebody by means of development openings, on at least three sides, in preparation for continuous extraction; the opening of a deep lead deposit. See also: *block.* (Nelson, 1965) (b) As applied to coal reserves, acquiring coal and mining rights in contiguous areas to form a continuous area and in a desirable shape for planned future mining. (c) Aust. Laying or staking out gold-bearing gravel deposits in square blocks in order to facilitate systematic washing. (d) In economic geology, delimitation of an orebody on three sides in order to develop it, i.e., to make estimates of its tonnage and quality. The part so prepared is an ore block. (AGI, 1987)

block kriging Estimating the value of a block from a set of nearby sample values using kriging.

blockmaking Applied to the various processes involved in roofing slate manufacture, which include drilling and wedging, cutting, sawing, etc. (AIME, 1960)

block mica Mica with a minimum thickness of 0.007 in (0.18 mm) and a minimum usable area of 1 in^2 (6.45 cm^2), full trimmed unless otherwise specified. (Skow, 1962)

block mountain A mountain that is formed by block faulting. The term is not applied to mountains that are formed by thrust faulting. Syn: *fault-block mountain.* (AGI, 1987)

block movement A general failure of the hanging wall. In the gold mines of South Africa and the Michigan copper mines, block movements have been experienced. (Nelson, 1965)

block off (a) To fill and seal undesirable openings, fissures, or caving zones in a borehole by cementation or by lining the borehole with pipe or casing. Also called blank off; case off; seal off. (Long, 1960) (b) To secure a mine opening against the flow or escape of gas, air, or liquid by erecting rock, concrete, steel, wood, or cloth barriers. (Long, 1960) (c) To erect barriers to prevent workers from entering unsafe areas in underground workings. (Long, 1960)

block out To delineate the area in which a desirable mineral occurs by systematic core drilling or by underground openings. (Long, 1960)

block riffles These consist of timber blocks, 8 to 12 in (20.3 to 30.5 cm) square, set in transverse rows in a sluicebox; they are arranged so that in contiguous rows the joints are staggered to prevent the development of longitudinal cracks. It is usual to separate adjacent rows by means of a strip of ordinary riffle scantling. (Griffith, 1938)

block structure Used in quarrying to describe granite that has three sets of joints occurring at right angles to each other. (Streefkerk, 1952)

block system (a) A pillar mining system in which a series of entries, panel entries, rooms, and crosscuts are driven to divide the coal into blocks of approx. equal size, which are then extracted on retreat. Development openings are most commonly driven between 15 ft and 20 ft (4.6 m and 6.1 m) wide. Pillars are most commonly 40 to 60 ft (12.2 to 18.3 m) wide and from 60 to 100 ft (18.3 to 30.5 m) long. (Woodruff, 1966) (b) A system of control in which a number of units, for example, powered supports, are operated as a group. (NCB, 1964)

block system of stoping and filling *overhand stoping; Brown panel system.*

block-tin lining Copper vessels are lined or coated with tin by the application of molten tin upon clean copper with the aid of fluxing. (CCD, 1961)

blocky rock Rock ore that breaks into large blocks. (Sandstrom, 1963)

blödite A monoclinic mineral, $Na_2Mg(SO_4)_2{\cdot}4H_2O$. Also spelled bloedite, blodite. Formerly called astrakhanite. Syn: *astrakanite; magnesium blodite.*

Bloman tube breathing apparatus This differs from the smoke helmet in that there is neither helmet nor bellows. Fresh air is passed to the wearer through a corrugated reinforced rubber tube by means of a rotary blower. A mouthpiece having an inhalation valve, an exhalation valve, and a noseclip takes the place of the helmet. It is held in position by straps attached to a head harness. The mouthpiece can be replaced by a full-face mask. This apparatus is fitted with an equalizing device that enables the wearer to continue breathing comfortably, even should the rotary blower stop. (Mason, 1951)

blomstrandine *priorite.*

bloodstone (a) A variety of chalcedony or jasper, dark green in color, interspersed with small red spots. Used as a gem. Also called heliotrope. Cf: *plasma.* (Sanford, 1914) (b) A red variety of quartz.

blooey line A pipe or flexible tube conducting cuttings-laden air or gas from the collar of a borehole to a point far enough removed from the drill rig to keep the air around the drill dust free. (Long, 1960)

bloom (a) A mineral that is frequently found as an efflorescence, cobalt bloom, for example. Syn: *efflorescence.* (Webster 3rd, 1966) (b) To form an efflorescence; as, salts with which alkali soils are impregnated bloom out on the surface of the Earth in dry weather following rain or irrigation. (Webster 2nd, 1960) (c) The fluorescence of petroleum or its products. (Webster 3rd, 1966) (d) A semifinished hot-rolled product, rectangular in cross section, produced on a blooming mill. For iron and steel, the width is not more than twice the thickness, and the cross-sectional area is usually not less than 36 in^2 (232 cm^2). Iron and steel blooms are sometimes made by forging. (ASM, 1961) (e) A surface film resulting from attack by the atmosphere or from the deposition of smoke or other vapors. (ASTM, 1994) (f) A lump or mass of molten glass. (Webster 2nd, 1960)

bloomer *blooming mill.*

bloomery A forge for making wrought iron, usually direct from the ore. Syn: *cinder plate; Merrit plate.* (Fay, 1920)

blooming mill The mill or equipment used in reducing steel ingots to blooms. (ASM, 1961)

blow (a) A sudden escape of gas from coal or associated strata into mine workings. See also: *outburst.* (b) A large outcrop of ore, commonly of low grade. (Nelson, 1965) (c) To lift: said of a floor that lifts owing to pressure from gas or strata. (Mason, 1951) (d) In blasting, a shot that blows part of the unfired explosive out of the hole. See also: *blown-out shot.* (Pryor, 1963) (e) To fire shots. (Mason, 1951)

blow count The number of blows that must be delivered by a specific-weight, freely falling drive hammer dropping a specific distance to force a drive sampler a unit distance into a soil material. (Long, 1960)

blower (a) A fan employed in forcing air either into a mine or into one portion of a mine. A portable blower, also known as a tubing blower or room blower, is used in ventilating small dead-end places like rooms and entries or gangways. (Jones, 1949) (b) The sudden emission of combustible gases from the coal seam or surrounding rock. Blowers vary considerably in violence and magnitude from small emissions that make a hissing noise to severe outbursts. (Nelson, 1965) (c) Eng. A worker who blasts or fires shots in a mine, or who drills the holes and charges them, ready for firing. (Fay, 1920)

blower fan A fan to direct part of an air circuit through a tubing to a particular working face. See also: *mine ventilation auxiliary fan.*

blower system A system in which the pressure-generating source is located at the entrance and raises the pressure of the air above atmospheric. (Hartman, 1982)

blower wax A pale yellow, soft variety of ozocerite that is squeezed out of the veins under the influence of pressure of the surrounding rocks. Syn: *blister wax.* (Tomkeieff, 1954)

blowhole (a) A minute crater formed on the surface of thick lava flows. (Fay, 1920) (b) A hole in a casting or a weld caused by gas entrapped during solidification. See also: *gas evolution.* (ASM, 1961)

blow in To put a blast furnace in operation. Syn: *blowing in.* (Fay, 1920)

blowing (a) Oxidation of molten metal or matte in a converter or other smelting furnace, in order to remove carbon and sulfur and to convert impurities to slag. (b) The bursting of pots from too rapid heating. (ACSG, 1963)

blowing engine An engine for forcing air into blast furnaces under pressure, commonly about 1 psi (6.9 kPa). (Weed, 1922)

blowing in Starting a blast furnace. Syn: *blow in.*

blowing on taphole Blowing air through the hole at casting, to clean the hearth of iron and cinder. (Fay, 1920)

blowing road S. Staff. An intake, or fresh-air road in a mine.

blowing tools A small set of blasting implements. Cf: *blasting supplies.* (Standard, 1964; Fay, 1920)

blowing-up furnace A furnace used for sintering ore and for the volatilization of lead and zinc. (Fay, 1920)

blowing ventilation Mine ventilation in which the air flows from the fan at the portal toward the working face.

blown metal Pig iron purified by blowing air through it.

blown-out shot A shot that dissipates the explosive force by blowing out the stemming instead of breaking down the coal. It may be caused by insufficient stemming, overcharging with explosive, or a burden that is too much for the charge to dislodge. See also: *gun; blow.* Syn: *cannon shot; gunned shot.* (Nelson, 1965)

blowout (a) A large mineralized outcrop beneath which the vein is smaller, e.g., a great mass of quartz that conceals a vein only a few feet wide. (b) A shot or blast that goes off like a gun and does not shatter the rock; a windy shot. (c) A large outcrop beneath which the vein is smaller is called a blowout. (slang). (Fay, 1920) (d) The high-pressure, sometimes violent, and uncontrolled ejection of water, gas, or oil from a borehole. (Long, 1960) (e) Used by prospectors and miners for any surface exposure of strongly altered discolored rock associated, or thought to be associated, with a mineral deposit. (AGI, 1987) (f) Used by miners and prospectors for a large, more or less isolated, usually barren quartz outcrop. Known in Australia as blow. (Hess) (g) To put a blast furnace out of blast, by ceasing to charge fresh materials, and continuing the blast until the contents of the furnace have been smelted. (Fay, 1920) (h) To smelt the iron-bearing materials in the furnace, adding domestic coke so that the stockline is about normal. (Camp, 1985) (i) A general term for a small saucer-, cup-, or trough-shaped hollow or depression formed by wind erosion on a preexisting dune or other sand deposit, esp. in an area of shifing sand or loose soil, or where protective vegetation is disturbed or destroyed; the adjoining accumulation of sand derived from the depression, where recognizable, is commonly included. Some blowouts may be many kilometers in diameter. (AGI, 1987)

blowout shot An improperly placed or overcharged shot of black blasting powder in coal (where used), frequently results in a mine explosion. (von Bernewitz, 1931)

blowover (a) Sand blown by onshore winds across a barrier and deposited on its landward side or as a veneer in the lagoon; e.g., along the Gulf Coast of Texas. Cf: *washover.* (AGI, 1987) (b) The process of forming a blowover. (AGI, 1987)

blowpipe reaction (a) The decomposition of a compound or mineral when heated by the blowpipe, resulting in some characteristic reaction, as a coloring of the flame or a colored crust on a piece of charcoal. (Standard, 1964) (b) A method of analysis in mineralogy. (Fay, 1920)

blowpiping (a) A rapid method for the determination of the approximate composition of minerals and ores. Blowpipe tests are merely qualitative; i.e., they indicate the presence of the different constituents, but not the proportions. A blowpipe consists of a plain brass tube capable of producing a flame of intense heat that may be either oxidizing or reducing. Illuminating gas from a Bunsen burner is the fuel commonly used. The color, nature, and smell of the encrustations suggest the nature of the elements present. See also: *bead; borax bead test.* (Nelson, 1965) (b) The use of a bent tube with a condensation trap and a small hole to direct one's concentrated breath into a small flame from a gas or alcohol lamp to produce intense heat in both oxidizing and reducing flames for the purpose of soldering metals or of performing qualitative analyses on powdered mineral samples. See also: *oxidizing flame; reducing flame.*

blowup (a) Eng. An explosion of combustible gases in a mine. (b) To allow atmospheric air access to certain places in coal mines, so as to generate heat, and ultimately to cause gob fires.

blue annealing Heating hot-rolled ferrous sheet in an open furnace to a temperature within the transformation range, and then cooling it in air in order to soften the metal. The formation of a bluish oxide on the surface is incidental. (ASM, 1961)

blue asbestos A name for crocidolite, the asbestiform variety of riebeckite.

blue band A thin, persistent bed of bluish clay that is found near the base of the No. 6 coal throughout the Illinois-Indiana basin.

blue-black ore Corvusite, extremely high-grade vanadium ore with blue-black color. (Ballard, 1955)

blue brittleness Brittleness exhibited by some steels after being heated to some temperature within the range of 300 to 650 °F (149 to 343 °C), particularly if the steel is worked at the elevated temperature. (ASM, 1961)

blue cap The characteristic blue halo, or tip, of the flame of a safety lamp when combustible gases are present in the air. See also: *top; cap.*

blue chalcedony *sapphirine.*

blue chalcocite *digenite.*

blue copper *azurite; covellite.*

blue copper ore *azurite.*

blue earth *blue ground.*

blue gold (a) A gold-iron alloy containing 25% to 33.3% iron. (Camm, 1940) (b) A bluish collodial solution of gold prepared by reducing a solution of gold chloride with hydrazine hydrate. (Camm, 1940)

blue granite *larvikite.*

blue ground Unoxidized slate-blue or blue-green kimberlite, usually a breccia (as in the diamond pipes of South Africa) that is found below

the surficial oxidized zone of yellow ground. Cf: *hardebank.* Syn: *blue earth.* (AGI, 1987)
blue iron earth *vivianite.*
blue ironstone A bluish iron-bearing mineral; specif: crocidolite and vivianite.
blue lead (a) A term for metallic lead in the lead industry to distinguish it from lead compounds with color designations, such as white lead, orange lead, and red lead. See also: *lead.* (CTD, 1958) (b) A synonym of galena, esp. a compact variety with a bluish-gray color. Syn: *galena; blue lead ore.* (AGI, 1987) (c) A bluish, gold-bearing lead or gravel deposit found in Tertiary river channels of the Sierra Nevada, CA. Pronounced "blue leed.". (AGI, 1987)
blue lead ore An old name for a compact variety of galena with a bluish-gray color. Pron: leed.
blue malachite An erroneous name for azurite.
blue metal A term used in England for a hard bluish-gray shale or mudstone lying at the base of a coalbed and often containing pyrite. (AGI, 1987)
blue mud (a) An ocean-bottom deposit containing up to 75% terrigenous materials of dimensions below 0.03 mm. The depth range occurrence is about 750 to 16,800 ft (229 to 5,120 m). Colors range from reddish to brownish at the surface, but beneath the surface, the colors of the wet muds are gray to blue. (AGI, 1987) (b) A common variety of deep-sea mud having a bluish-gray color due to presence of organic matter and finely divided iron sulfides. Calcium carbonate is present in amounts up to 35%. (AGI, 1987)
blue needles Applied in the grading of quartz crystals to needlelike imperfections, often definitely oriented, which show up with a bluish-white color under the carbon arc. The color is due to the selective scattering of blue light by the minute imperfections. (Am. Mineral., 1947)
blue ocher *vivianite.*
blue powder A mixture of finely divided and partly oxidized metallic zinc formed by the condensation of zinc vapor into droplets; also, any similar zinc byproduct (such as dross, skimmings, or sweepings). (Webster 3rd, 1966)
blue quartz *sapphire quartz.*
blue-rock phosphate The hard, bluish-gray, Ordovician bedded phosphates of central Tennessee.
blue room The first room in a baghouse. (Fay, 1920)
blue schorl (a) Blue tourmaline. (Fay, 1920) (b) The earliest name for anatase (octahedrite).
blue spar *lazulite.*
bluestone (a) A commercial name for a building or paving stone of bluish-gray color; specif. a dense, tough, fine-grained, dark blue-gray or slate-gray feldspathic sandstone that splits easily into thin, smooth slabs and that is extensively quarried near the Hudson River in New York State for use as flagstone. The color is due to the presence of fine black and dark-green minerals, chiefly hornblende and chlorite. The term is applied locally to other rocks, such as dark-blue shale and blue limestone. Cf: *flagstone.* (AGI, 1987) (b) A miners' term for chalcanthite. (c) A term applied locally to rocks such as dark-blue shale, blue limestone, and bluish metabasalt (greenstone). (d) A highly argillaceous sandstone, of even texture and bedding, formed in a lagoon or lake near the mouth of a stream.
blue tops Grade stakes whose tops indicate finish grade level. (Nichols, 1976)
blue vitriol *chalcanthite.*
blue-white diamond A diamond that appears blue or bluish in transmitted white light or against a white background; it reflects white light when viewed edge up at right angles to the table. (Hess)
bluntin Derb. A dark tough vein filling that dulls the drills readily.
Blyth elutriator Laboratory apparatus in which mineral particles suspended in water are syphoned through vertical tubes of increasing cross section, the fraction failing to rise under determined conditions of upward flow reporting as a subsieve fraction. (Pryor, 1963)
BM *bench mark.*
B.M.A.G.A. apparatus Used in the United States for obtaining additional information on the yields of coke, tar, and gas that can be expected in high-temperature practice. This is a vertical cylinder of mild steel holding up to 2 hundredweight (91 kg) of coal and operated at temperatures up to 1,000 °C. (Francis, 1965)
board *bord.*
board-and-pillar *pillar-and-breast.*
board-and-wall *pillar-and-breast.*
board coal Eng. Coal having a fibrous or woody appearance. (Fay, 1920)
board gates York. Headings driven in pairs generally to the rise, out of which banks or stalls are opened and worked. (Hess)
board of trade unit The work done when a rate of working of 1 kW is maintained for 1 h. The British unit of electrical energy; kilowatt-hour. Abbrev: B.O.T. unit. (Nelson, 1965)
board run The amount of undercutting that can be done at one setting of a coal mining machine, usually about 5 ft (1.5 m), without moving forward the board upon which the machine works. (Fay, 1920)
boart *bort.*
boartz *bort.*
boat A gold dredge.
bobbin (a) Aust. A catch placed between the rails of the upline of an incline to stop any runaway trucks. It consists of a bent iron bar, pivoted in such a manner that the downhill end is slightly heavier than the uphill end, which is capable of being depressed by an upcoming truck, but rises above the level of the truck axle as soon as the truck is past. Syn: *monkey; monkey chock.* (Fay, 1920) (b) A spool or reel. (Fay, 1920)
bocca A volcanic crater or vent. (Standard, 1964)
body (a) An orebody, or pocket of mineral deposit. (Zern, 1928) (b) *bit blank.* (c) The fluidity of a drilling mud expressed in the number of seconds in which a given quantity of mud flows through a given aperture, such as the aperture in a Marsh funnel. (Long, 1960) (d) The term used to indicate the viscosity or fluidity of a lubricating oil; e.g., a heavy-body oil is thick and viscous and a light-body oil is thin and fluid. (Long, 1960) (e) The load-carrying part of a truck or scraper. (Nichols, 1976) (f) The fatty, inflammable property that makes a coal combustible; e.g., bituminous coal has morebody than anthracite. (AGI, 1987)
boggildite A fluoride, $Sr_2Na_2Al_2(PO_4)F_9$, from the Greenland cryolite deposit. (Spencer, 1955)
boghead cannel Cannel coal rich in algal remains. See also: *torbanite.* (Tomkeieff, 1954)
boghead cannel shale A coaly shale rich in fatty or waxy algae. (AGI, 1987)
boghead coal (a) A variety of bituminous or subbituminous coal resembling cannel coal in appearance and behavior during combustion. It is characterized by a high percentage of algal remains and volatile matter. Upon distillation it gives exceptionally high yields of tar and oil. See also: *cannel coal; torbanite; kerosine shale.* (b) A nonbanded coal with the translucent attritus consisting predominately of algae, and having less than 5% anthraxylon. (AGI, 1987)
boghedite *torbanite.*
bogie (a) A rail truck or trolley of low height, used for carrying timber or machine parts underground, or for conveying the dirt hoppit from a sinking pit to the dirt heap. It may also be used as a wagon spotter. See also: *timber trolley.* (Nelson, 1965) (b) A weighted truck run foremost or next to the rope in a train or trip. (c) A two-axle driving unit in a truck. Also called tandem drive unit; tandem. (Nichols, 1976) (d) Also spelled bogey; bogy. (e) York. A small truck or trolley upon which a bucket is carried from the shaft to the spoil bank.
bog iron (a) A general term for a soft, spongy, and porous deposit of impure hydrous iron oxides formed in bogs, marshes, swamps, peat mosses, and shallow lakes by precipitation from iron-bearing waters and by the oxidizing action of algae, iron bacteria, or the atmosphere; a bog ore composed principally of limonite that is commonly impregnated with plant debris, clay, and clastic material. It is a poor-quality iron ore, in tubular, pisolitic, nodular, concretionary, or thinly layered bodies, or in irregular aggregates, in level sandy soils, and esp. abundant in the glaciated northern regions of North America and Europe (Scandinavia). Syn: *limnite; morass ore; meadow ore; marsh ore; lake ore; swamp ore.* Cf: *goethite.* (b) A term commonly applied to a loose, porous, earthy form of "limonite" occurring in wet ground. Syn: *bog ore.*
bog iron ore (a) Loose, porous form of limonite occurring in wet ground, often mixed with vegetable matter, $Fe_2O_3{\cdot}nH_2O$. (Pryor, 1963) (b) A deposit of hydrated iron oxides found in swamps and peat mosses. See also: *iron ore; lake ore.* Cf: *limnite; limonite.* Syn: *marsh ore; meadow ore.* (Schieferdecker, 1959)
bog lime *marl.*
bog manganese (a) *earthy manganese; wad.* (Fay, 1920) (b) A bog ore consisting chiefly of hydrous manganese oxides; specif. wad formed in bogs or marshes by the action of minute plants.
bog-mine ore *bog ore.*
bog mine ore *bog ore.*
bog muck A vernacular name for peat. (Tomkeieff, 1954)
bog oak Oak immersed in peat bogs, semifossilized and blackened to resemble ebony by iron from the water combining with the tannin of the oak. (CMD, 1948)
bog ore (a) A spongy variety of hydrated oxide of iron and limonite. Found in layers and lumps on level sandy soils that have been covered with swamp or bog. Includes bog iron ore, bog manganese ore, and bog lime, a calcareous deposit of similar origin. See also: *brown iron ore.* (AGI, 1987) (b) A poorly stratified accumulation of earthy metallic-mineral substances, mainly oxyhydroxides, formed in bogs, marshes, swamps, and other low-lying moist places, by direct chemical precipitation from surface or near-surface percolating waters; specif. "bog iron" and "bog manganese". Cf: *lake ore.* Syn: *bog mine ore; bog iron.*

bog peat Peat consisting mainly of mosses. (Francis, 1965)
Bohemian garnet Yellowish-red gem variety of the garnet pyrope; occurs very commonly in the Mittelgebirge, Czech Republic. See also: *pyrope.*
Bohemian ruby A jeweler's name for rose quartz when cut as a gem. (Fay, 1920)
böhmite An orthorhombic mineral, aluminum oxyhydroxide [γ-AlO(OH)]; grayish, brownish, or reddish, in some bauxites and laterites; an ore of aluminum, dimorphous with diaspore. Also spelled boehmite.
bohr magneton The net magnetic moment arising from electron spins. (Van Vlack, 1964)
boiler burner unit A boiler designed esp. for gas or oil and sold integrally with the burner. (Strock, 1948)
boiler circulating pump A pump, usually of the single-stage, single-entry, overhung type, that must have low suction loss and high-temperature features since it draws water directly from the boiler drums at high saturation pressure and temperature. (Sinclair, 1958)
boiling furnace A water-jacketed reverberatory furnace for decarbonizing iron by a process in which the carbonic oxide escapes with an appearance of boiling. (Standard, 1964)
boiling point (a) The temperature at which a liquid begins to boil or to be converted into vapor by bubbles forming within its mass. It varies with pressure. (Standard, 1964) (b) The temperature at which a cooling gas becomes a liquid. (Hurlbut, 1964)
boilum Hard calcareous or siliceous nodules of irregular shape, found in the shales and underclays of the Coal Measures. See also: *boylom.* (Arkell, 1953)
bolar *bole.*
bole Any of several varieties of compact earthy clay (impure halloysite), usually red, yellow, or brown because of the presence of iron oxide, and consisting essentially of hydrous silicates of aluminum or less often of magnesium. It is a waxy decomposition product of basaltic rocks, having the variable composition of lateritic clays. Syn: *bolar.* (AGI, 1987)
boleite A deep blue pseudoisometric hydrous oxychloride of lead, copper, and silver from Boleo, Lower California. A tetragonal form of percylite. (Fay, 1920)
Boliden gravimeter An electrical stable gravimeter with a moving system suspended on a pair of bowed springs. The moving system carries electrical condenser plates at each end, one to measure the position of the moving system, the other to apply a balancing force to bring the system to a fixed position. Syn: *Lindblad-Malmquist gravimeter.* (AGI, 1987)
Bolognan stone *Bologna stone.*
Bologna spar *Bologna stone.*
Bologna stone A nodular, concretionary, or round variety of barite, composed of radiating fibers; phosphorescent when calcined with charcoal. Syn: *Bolognian stone; Bolognan stone; Bologna spar.*
bolognian stone *Bologna stone.*
bolson (a) A term applied in the desert regions of the Southwestern United States to an extensive flat alluvium-floored basin or depression, into which drainage from the surrounding mountains flows toward a playa or central depression; a basin with internal drainage. Syn: *playa basin.* (AGI, 1987) (b) A temporary lake, usually saline, formed in a bolson. (AGI, 1987) (c) Mex. A pocket of ore. Etymol: Sp., bolsón, large purse.
Bolsover experiment Applied to a method of working by single panels. Single 100-yd (91-m) panels are advanced, leaving 100-yd-wide coal pillars between them. The pillars are then worked on the retreat after the advancing faces have reached a limit line. (Nelson, 1965)
bolt A rod used in roof bolting. See also: *slot-and-wedge bolt; wedge-and-sleeve bolt.* (Nelson, 1965)
bolthole S. Staff. A short narrow opening made to connect the main workings with the airhead or ventilating drift of a coal mine. Also called bolt.
bolt-hole brush A special round brush used to remove porcelain enamel bisque from in and around small openings in the ware. See also: *brush.* (ASTM, 1994)
bolting Separation of particles of different sizes by means of vibrating sieves. (Bennett, 1962)
bolting silk (a) In oceanography, a silk cloth of very fine and regular mesh, used in the construction of tow nets for the smaller members of the surface fauna. (CTD, 1958) (b) Also used to cover a lap for polishing rock and mineral specimens for microscopic examination.
Bolton's reagent An etching reagent for cast iron that contains picric acid, nitric acid, and water. (Osborne, 1956)
boltwoodite A monoclinic mineral, $HK(UO_2)SiO_4 \cdot 1\text{-}1/2H_2O$; analogous to sklodowskite, having potassium in place of magnesium; radioactive; yellow.
bolus alba *kaolin.*
bomb (a) A more or less rounded mass of lava from a few inches to several feet in diameter, generally vesicular, at least inside, thrown from the throat of a volcano during an explosive eruption. (b) An ellipsoidal, discoidal, or irregularly rounded mass of lava ejected at a high temperature during a volcanic eruption. Bombs range upwards in size from the largest lapilli. They are characterized by a well-defined crust and are often cellular or even hollow internally. (Holmes, 1928) (c) A missile containing an explosive, as dynamite. (d) A heavy-walled reaction vessel or autoclave. Used to carry out reactions at high pressure and high temperature. (Hurlbut, 1964)
bomb calorimeter A strong steel vessel used for determining the heat produced during combustion; used, for example, for determining the calorific value of a fuel. (Nelson, 1965)
bonanza (a) A rich body of ore or a rich part of a deposit; a mine is in bonanza when it is operating profitably. Also, discontinuous locally rich ore deposits, esp. epithermal ones. Etymol: Sp., prosperity, success. (AGI, 1987) (b) In miners' phrase, good luck, or a body of rich ore. (c) Part of a precious mineral deposit that is esp. rich. (Bateman, 1951)
bonattite A monoclinic mineral, $CuSO_4 \cdot 3H_2O$; blue; partly dehydrated from chalcanthite.
bond and lease An agreement between a mine owner and tributor that gives the latter the option of buying the mine before the lease expires. (Nelson, 1965)
Bond and Wang theory A theory of crushing and grinding; the energy (h) required for crushing varies inversely as the modulus of elasticity (E) and specific gravity (S), and directly as the square of the compressive strength (C) and as the approximate reduction ration (n). The energy in horsepower hours required to crush a short ton of material is given by the following equation, in which all quantities are in feet per second units: $h = [0.001748C^2 / SE] [(n + 2) (n - 1) / n]$. The theory is due to F. C. Bond and J. T. Wang. (Dodd, 1964)
Bondaroy's yellow An antimony yellow developed by Fourgeroux de Bondaroy in 1766: 12 parts white lead; 3 parts potassium antimonate; 1 part alum; 1 part sal ammoniac. (Dodd, 1964)
bond clay A clay that, because of its plasticity, serves to bond relatively nonplastic materials in the fabrication of ceramic or other molded products (green bond). Also, a clay that, on firing to furnace or vitrification temperature, bonds adjacent ceramic materials that vitrify at a still higher temperature (fired bond). (AGI, 1987)
bonded refractories Refractories in which the constituents are held together by a suitable bonding material, as distinguished from fused refractories. (Henderson, 1953)
bonder (a) A brick that is half as wide again as a standard square (rectangular or arch); such bricks are sometimes used to begin or end a course of bonded brickwork. (Dodd, 1964) (b) In mining, one who welds copper connections in place between the joints of track rails, used for trolley locomotives, to complete the electrical circuit between the sections of rails. Syn: *bondman.* (DOT, 1949)
bondman *bonder.*
Bond's third theory In crushing, the total work useful in breakage that has been applied to a stated weight of homogeneous broken material is invariably proportioned to the square root of the diameter of the product particles. Syn: *work index.* (Pryor, 1963)
bone (a) A hard coallike substance high in noncombustible mineral matter; often found above or below, or in partings between, layers of relatively pure coal. (Hess) (b) In the anthracite-coal trade, a carbonaceous shale containing approx. 40% to 60% of noncombustible materials. Syn: *bone coal; bony coal.* (Hess) (c) A tough, fine-grained, gray, white, or reddish quartz. (Hess) (d) A layer of hard, impure coal which sometimes grades uniformly into the adjacent softer coal and sometimes is sharply separated from it. Bone is usually a mixture of clay shale particles with the coal, the clay particles being well distributed. (Kentucky, 1952)
bone ash The white porous residue containing chiefly tribasic calcium phosphate from bones calcined in air and used esp. in making cupels, pottery, and glass and in cleaning jewelry; also, synthetic tribasic calcium phosphate used similarly. (Webster 3rd, 1966)
bone bed Applied to strata or layers that contain innumerable fragments of fossil bones, scales, teeth, coprolites, and other organic remains. (Fay, 1920)
bone coal (a) Coal with a high ash content, almost rock. See also: *bone.* Syn: *true middlings.* (BCI, 1947) (b) Coal that has a high ash content. It is hard and compact. Syn: *bony coal.* (AGI, 1987) (c) Argillaceous partings in coal, sometimes called slate. (AGI, 1987)
bone phosphate The calcium phosphate of bones and of phosphatic rocks, such as found in North Carolina; so called in commerce. See also: *phosphorite.* (Standard, 1964)
bone phosphate of lime Tricalcium phosphate, $Ca_3(PO_4)_2$. The phosphate content of phosphorite may be expressed as percentage of bone phosphate of lime. Abbrev: BPL. (AGI, 1987)
bone turquoise *odontolite.*
bonnet (a) A covering over a mine cage, which serves as a roof to shield it from objects falling down the shaft, thereby protecting the riders. Syn: *cage cover.* (b) A cap piece for an upright timber. (Zern,

1928) (c) The metal casing of a miner's flame safety lamp, with openings at the top and a hook for carrying the lamp. The bonnet protects the inner gauze from damage and from the impact of high-velocity air. See also: *safety lamp.* (Nelson, 1965) (d) Syn: *air dome.* (Long, 1960) (e) The cap over the end of a pipe. (Strock, 1948)

bony Coal containing slaty material in its composition. (Korson, 1938)

bony coal *bone.*

bonze Undressed or untreated lead ore. (Nelson, 1965)

boobey Som. A box holding 6 to 8 hundredweight of coal in which waste rock is sent to the surface. (Fay, 1920)

book clay Clay deposited in thin, leaflike laminae. Syn: *leaf clay.* (AGI, 1987)

booked mica Lumps of mica in which laminae have not been separated into thin sheets. (Pryor, 1963)

book mica Crystals of crude mica obtained from a mine in various shapes and sizes. Also called book. Syn: *mine-run mica.* (Skow, 1962)

book structure The alternation of ore with gangue, usually quartz, in parallel sheets. Cf: *ribbon.* (AGI, 1987)

boom (a) A spar or beam projecting out over the drill floor from the tripod or derrick, by means of which heavy drill tools and equipment may be moved and safely handled. (Long, 1960) (b) A long, adjustable steel arm on a drill jumbo on which drifter or other types of pneumatic drills are mounted. (Long, 1960) (c) A cantilevered or overhanging member or structure that supports or contains the component parts of a conveyor. It may be fixed, hinged, or pivoted. (d) A pipe fixed across the last supports in a tunnel face to anchor the tail sheave of a scraper loader installation. (Nelson, 1965) (e) In a revolving shovel, a beam hinged to the deck front, supported by cables. (Nichols, 1976) (f) Any beam attached to lifting or excavating equipment. See also: *dragline.* (Nelson, 1965) (g) Any heavy beam that is hinged at one end and carries a weight-lifting device at the other. (Nichols, 1976)

boom cat *stripping-shovel operator.*

boom conveyor Any type of conveyor mounted on a boom.

boom ditch (a) The ditch from the dam used in booming. (b) A slight channel cut down a declivity into which is let a sudden head of water to cut to the bedrock and prospect from the apex of any underlying lode.

boomer (a) In placer mining, an automatic gate in a dam that holds the water until the reservoir is filled, then opens automatically and allows the escape of such a volume of water that the soil and upper gravel of the placer are washed away. When the reservoir is emptied the gate closes and the operation is repeated. On a smaller scale it may be used to furnish water periodically for sluicing. Syn: *automatic dam; flop gate.* (Hess) (b) A sonar transducer, used in the exploration of bottom substrata. (Hy, 1965) (c) Originally, an oilfield worker who migrated from one boom field to another; now, commonly, a member of a drill crew who works one job a short time, quits, and moves on to another locality to seek employment. Also called drifter. (Long, 1960) (d) A combination ratchet and lever device used to tighten a chain or line about a loaded truck or wagon to hold the load in place. (Long, 1960)

boomerang sediment corer This free-instrument-type device can be dropped over the side of a moving ship, where it will sink rapidly to the ocean floor, take a core of sediment, release ballast, and automatically return to the surface for retrieval. (Hunt, 1965)

booming The accumulation and sudden discharge of a quantity of water (in placer mining, where water is scarce). In California, the contrivances for collecting and discharging water are termed "self-shooters," an idea suggested by the sudden and violent manner in which the water makes its escape. In booming, snowmelt or water from small or ephemeral streams is collected behind a dam with a discharge gate. Placer ore is placed below the dam and when the water is released, the ore is washed through sluice boxes, ground sluices, etc., in one large surge. The dam discharge gate is closed to again begin collecting water for the next cycle. Syn: *hushing.*

boom man In bituminous coal mining, one who manipulates the controls of a loading boom (conveyor) to regulate the height of the loading end of a boom, thus controlling the flow of coal from shaking screens or picking tables into railroad cars at the tipple. Also called boom operator; loader headman. (DOT, 1949)

boort *bort.*

boose (a) Eng. Lead ore that separates easily from its matrix and does not have to be buddled, Durham, Yorkshire, and Derbyshire. Hooson defines it as veinstuff and ore mixed. Syn: *booze; bowse.* (Arkell, 1953) (b) Derb. Gangue rock mixed with ore. See also: *bouse.*

booster An explosive of special character used in small quantities to improve the performance of another explosive, the latter forming the major portion of the charge. (Nelson, 1965)

booster conveyor Any type of powered conveyor used to regain elevation lost in gravity roller or wheel conveyor lines. Syn: *humper.*

booster drive An auxiliary drive at an intermediate point along a conveyor.

booster fan A fan installed in an underground opening. A booster fan can be used as the main mine fan but is more commonly used to improve or augment the ventilation in a segment of the mine. Booster fans are illegal in U.S. coal mines but are used in metal mines and coal mines in other countries. (Hartman, 1982)

booster pump (a) A pump used to increase the pressure of fluids, such as to increase the pressure of water delivered to a drill when the source pressure is too low to be used for drilling operations. (Long, 1960) (b) A pump that operates in the discharge line of another pump, either to increase pressure or to restore pressure lost by friction in the line or by lift. (Nichols, 1976)

booster station In long-distance pumping of liquids or mineral slurries, an intermediate pump station. (Pryor, 1963)

boot (a) A projecting portion of a reinforced concrete beam, acting as a corbel to support the facing material, such as brick or stone; the lower end of a bucket elevator. (Hammond, 1965) (b) A leather or tin joint connecting the blast main with the tuyere or nozzle in a bloomery. (c) A suspended enclosure in the nose of a tank protecting a portion of the surface and serving as a gathering opening. (ASTM, 1994) (d) The bottom of a bucket elevator, which receives feed for delivery into an elevating bucket. (Pryor, 1963)

boothite A monoclinic mineral, $CuSO_4 \cdot 7H_2O$; melanterite group; of a lighter blue than chalcanthite, from which it differs in its larger proportion of water.

booting The ejection of balled drill cuttings from the collar in long, tubelike masses. (Long, 1960)

bootleg (a) The part of a drilled blasthole that remains when the force of the explosion does not break the rock completely to the bottom of the hole. See also: *socket.* (Atlas, 1987) (b) *trespass.*

bootlegger One engaged in coal bootlegging. Applies to the worker in bootleg holes as well as the worker who cleans the coal in a small, impermanent breaker, and the trucker who conveys the coal to market. Bootleggers call themselves independent miners. (Korson, 1938)

bootlegging The mining and/or selling of coal produced from coal owned by others and without permission or knowledge of the owner.

booze *boose.*

boracite An orthorhombic mineral, $8[Mg_6B_{14}O_{26}Cl_2]$; isometric above 265 °C in hard, glassy, cubic and octahedral crystals; strongly pyroelectric; in evaporites, a source of boron.

Borascu Borate ore. (Bennett, 1962)

borate A salt or ester of boric acid; a compound containing the radical $BO^{3+}{}_3$. Cf: *nitrate; carbonate.* (AGI, 1987)

borax A monoclinic mineral, $4[Na_2B_4O_5(OH)_4 \cdot 8H_2O]$; soft; deposited by evaporation from alkaline lakes, playas, hot springs, and as surface efflorescence or crystals embedded in lacustrine mud. A source of boron. Syn: *tincal.* See also: *octahedral borax.*

borax bead In blowpipe analysis, a drop of borax that when fused with a small quantity of a metallic oxide will show the characteristic color of the element; e.g., a blue borax bead indicates the presence of cobalt. (Standard, 1964)

borax bead test A chemical test to disclose the presence of certain metals in a sample. A clear glassy bead of borax fused in a wire loop will react chemically with the salts of certain metals and yield colors that help to identify the metal; e.g., manganese compounds produce a violet bead, cobalt produces a deep blue, etc. See also: *blowpiping.* (Nelson, 1965)

bord (a) Newc. A passage or breast, driven up the slope of the coal from the gangway, and hence across the grain of the coal. A bord 4 yd (3.7 m) or more wide is called a wide bord, and one less than 4 yd in width is called a narrow bord. Also spelled board. (b) A side gallery parallel with the main road or drift. (Standard, 1964) (c) A road with solid coal sides. (Mason, 1951) (d) A narrow coal drivage in the pillar-and-stall method of working. (Nelson, 1965) (e) A joint in a coal seam. See also: *cleat.* (Nelson, 1965) (f) Eng. A road driven at right angles to the main cleavage planes of the coal. (SMRB, 1930)

bord-and-pillar A method of working coal seams. First bords are driven, leaving supporting pillars of coal between. Next, cross drives connect the bords, leaving supporting coal as rectangular pillars. Finally, the pillars are mined (extracted, won, robbed) and the roof is allowed to cave in. The bordroom is the space from which bord coal has been removed. Syn: *bord-and-wall; stoop-and-room.* See also: *breast-and-pillar; stret; Warwickshire method.* (Pryor, 1963)

bord-and-pillar method A system of mining in which the distinguishing feature is the winning of less than 50% coal on the first working. It is more an extension of the development work than mining. The second working is similar in principle to top slicing. The remainder of the coal is won by a retreating system, the cover being caved after each unit has been worked. The term bord-and-pillar is not used to any great extent in American mining literature, but has a place in English literature. Various names have been applied to this method, such as checkerboard system, Brown panel system, following up the whole with the broken, Lancashire bord-and-pillar system, modified room-and-pillar working, narrow working, North Staffordshire method, rearer method of working inclined seams, rock-chute mining, room system, room system with caving, Warwickshire method of working contiguous seams, wide or square work, and pillar-and-breast. (Fay, 1920)

bord-and-pillar working N. of Eng. A system of mining in which interlacing roadways are driven at right angles into the seam, leaving small square or rectangular pillars of coal of from 30 to 50 yd (27 to 46 m) side length, which are then wholly or partly extracted by a small group. Syn: *tub-and-stall; bord-and-wall.* See also: *room-and-pillar.* (Trist, 1963)

bord-and-wall *bord-and-pillar; bord-and-pillar working.*

bord cleat Eng. The main cleavage planes or joints in a coalbed. (SMRB, 1930)

bord course Aust. A direction at right angles to the main cleat or facing; i.e., the length of a bord.

bord drivage A coal drivage in the pillar-and-stall method of working. (Nelson, 1965)

border facies The marginal portion of an igneous intrusion, which differs in texture and composition from the main body of the intrusion, possibly because of more rapid cooling or assimilation of material from the country rock. (AGI, 1987)

bord gate (a) A main gate leading and at right angles to a bord face. (TIME, 1929-30) (b) York. A heading driven generally to the rise, out of which stalls are opened and worked.

bordroom (a) A heading driven parallel to the natural joints. (Fay, 1920) (b) The space excavated in driving a bord. Used in connection with the ridding of the fallen stone in old bords when driving roads across them in pillar working; thus, "ridding across the old bordroom.". (Zern, 1928) (c) Eng. The width across an old bord. (Fay, 1920)

bordroom man A repairer who cleans and erects supports in old workings in the bord-and-pillar method of coal mining. (Nelson, 1965)

bords and longwork York. A system of working coal. First, the main levels are started on both sides of the shaft and carried toward the boundary. Second, the bord gates are worked in pairs to the rise and continued as far as the boundary, or to within a short distance of a range of upper levels and other bord gates. Lastly, the whole of the pillars and remaining coal are worked out downhill to within a few yards of the levels, and ultimately, all the coal between the levels is removed. (Fay, 1920)

bordways Eng. The direction of a place or a face being taken at right angles to the main cleavage planes of a seam. (SMRB, 1930)

bordways course The direction at right angles to the main cleavage planes. In some mining districts it is termed "on face.". (Zern, 1928)

bore (a) A tunnel, esp. while being excavated. (b) A circular hole made by boring. (Long, 1960)

borehole (a) A hole with a drill, auger, or other tools for exploring strata in search of minerals, for water supply, for blasting purposes, for proving the position of old workings and faults, and for releasing accumulations of gas or water. (b) A circular hole made by boring; esp. a deep hole of small diameter, such as an oil well or a water well. Also called well bore. See also: *hole.* (AGI, 1987)

borehole bottom charge Explosives loaded in the hole bottom at a weight or density in excess of the main charge in order to fragment difficult to break rock or to break an excessive toe burden.

borehole cable Cable designed for vertical suspension in a borehole or shaft and used for power circuits in the mines. (A borehole cable in mining may also be a cable containing signal, telephone, or control circuits.)

borehole casing A steel pipe lining used in a borehole, particularly when passing through loose, running ground. Flush-jointed casing that is smooth inside and outside may be either screwed or welded. (Nelson, 1965)

borehole deformation gage A device for measuring the change in diameter of a hole.

borehole log A record, made by the driller or geologist, of the rocks penetrated in the borehole. In the laboratory, a more detailed log is prepared giving particulars relating to lithology, paleontology, water analysis, etc. See also: *electric log; well log.* (Nelson, 1965)

borehole logging The determination of the physical, electrical, and radioactive properties of the rocks traversed by a borehole. (BS, 1963)

borehole mining The extraction of minerals in the liquid or gaseous state from the Earth's crust by means of boreholes and suction pumps. Boreholes are used for mining petroleum, and for the extraction of liquid solutions of salt, sulfur, etc. See also: *well.* (Nelson, 1965)

borehole pressure The pressure that the hot gases of detonation exert on the borehole wall. It is primarily a function of the density of the explosive and the heat of explosion. Syn: *gas pressure.*

borehole pump (a) Any pump that can be suspended in a borehole; usually a centrifugal pump suspended in a borehole by its pipe range and driven by a shaft inside the pipe. (BS, 1963) (b) A centrifugal pump, electrically driven, and designed in the form of a vertical narrow chamber. It may be used to provide water, for dewatering purposes, or for borehole mining. See also: *sinking pump.* Also called submersible pump. (Nelson, 1965)

borehole samples The samples of the rocks obtained during boring. The diamond and shot drill yield cores, while percussive drills yield sludge and chippings, which are examined to determine the nature of the rocks passed through. Borehole samples may also be required during site investigations. See also: *exploratory drilling; soil core.* (Nelson, 1965)

borehole sealing The complete filling of a borehole with cement to prevent the entry of water into mine workings. (Nelson, 1965)

borehole spacing The distance between boreholes drilled for exploration or sampling purposes. With bedded minerals, the holes may be positioned at the intersection points of coordinates or at the corners of equilateral triangles with sides from 30 to 200 m apart. The spacing is closer with patchy deposits. With metallic ores following belts across country, the holes are spaced along lines crossing the orebody in order to yield cross sections of the ore at definite intervals. In the case of known and semiproved coalfields, boreholes at 1/2- to 1-km intervals may suffice. (Nelson, 1965)

borehole survey (a) The process of determining the course of, and the target point reached by, a borehole, using one of several different azimuth and dip recording apparatuses small enough to be lowered into a borehole; also, the record of the information thereby obtained. Also called drillhole survey; directional survey. (Long, 1960) (b) The process of determining the mineralogical, structural, or physical characteristics of the formations penetrated by a borehole using geophysical logging apparatus small enough to be lowered into a borehole; also, the record of the information thereby obtained. (Long, 1960) (c) *well log.* See also: *surveying.* (AGI, 1987)

borehole surveying Instrumental tests to determine the amount and direction of deflection of a borehole from vertical and horizontal planes. The instrument is lowered into the hole and tested at intervals of depth. The data obtained may be used to construct a scale model showing the actual course taken by the hole. (Nelson, 1965)

bore meal (a) Eng. Mud or fine cuttings from a borehole. (b) In rock drilling, the sludge from a borehole. (Pryor, 1963)

borer A tool such as a drill used for boring.

bore rod Term used primarily by soil and foundation testing engineers for the equipment customarily called a drill rod by drillers and miners. (Long, 1960)

borides A group of special ceramic materials. Typical properties are great hardness and mechanical strength, high melting point, low electrical resistivity, and high thermal conductivity; impact resistance is low, but thermal-shock resistance is generally good. (Dodd, 1964)

boring (a) The cutting or drilling of a hole for blasting, water infusion, exploration, or water or combustible gases drainage. See also: *percussive boring; rotary boring.* (b) The drilling of deep holes for the exploitation or exploration of oilfields. The term "drilling" is used similarly in connection with metalliferous deposits. (CTD, 1958)

boring bar (a) A rod, made in various lengths, usually with a single chisel cutting edge, for hand drilling in rock. The blows are given by a sledge hammer. (Nelson, 1965) (b) A revolving or stationary bar carrying one or more cutters or drills for boring. (Fay, 1920)

boring log *drill log.*

borings Used by the soil and foundation testing profession as a syn. for boreholes and/or the materials removed from a borehole. Cf: *cuttings; sample.* (Long, 1960)

bornite An isometric mineral, [Cu_5FeS_4]; metallic; brownish bronze tarnishing to iridescent blue and purple; brittle; massive; in hypogene and contact metamorphic deposits and mafic rocks; a valuable source of copper. Syn: *erubescite; variegated copper ore; peacock ore; horseflesh ore; poikilit; purple copper ore; variegated ore.*

boron The element is not found free in nature, but occurs as orthoboric acid in volcanic spring waters and as borates in borax and colemanite. The most important source of boron is the mineral rasorite, also known as kernite. Symbol, B. Amorphous boron is used in pyrotechnic flares and in rockets as an igniter; the most important compound, boric acid or boracic acid, is used as an antiseptic; borax is used as a cleansing flux in welding and as a water softener. The isotope boron-10 is used in nuclear reactors; the nitride has lubricating properties similar to those of graphite; and the hydrides have been studied for use as rocket fuels. (Handbook of Chem. & Phys., 3)

boronatrocalcite *ulexite.*

boron carbide Probably not a true compound, but instead a solution of varying amounts of carbon in a slightly distorted boron lattice; symbol, BC; black; hexagonal rhombohedral crystals; ranking next to diamond in hardness, 9.3 on the Mohs scale; and melting point, 2,350 °C. Used in powder form as an abrasive and in molded form as an abrasion resister. Syn: *tetraboron carbide.* (CCD, 1961; Handbook of Chem. & Phys., 2)

boronitrocalcite A former name for ulexite.

boron nitride White; symbol, BN; hexagonal rhombohedral, crystals or powder; the powder has a Mohs hardness of 2; sublimes at about

3,000 °C; anisotropic; some properties vary according to the method of preparation and the crystal form. Used as a refractory; a high-temperature lubricant, as in glass molds; in furnace insulation; and in molten-metal pump parts. (CCD, 1961; Handbook of Chem. & Phys., 2)

boron phosphate Symbol, BPO; sp gr, 2.81; vaporizes at 1,400 °C; related structurally to high cristobalite. It has been used as a constituent of a ceramic body that fires to a translucent porcelain at 1,000 °C. (Dodd, 1964)

boron phosphide Symbol, BP; melting point, greater than 2,000 °C, but readily oxidizes, which limits its potential use. (Dodd, 1964)

boron silicides *silicon borides.*

borrow material Soil or sediment removed from a site for use in construction, such as sandy sediment dredged and pumped to restore an eroded beach, or clay taken to build a levee or dike. (Army Corps of Eng., 1987)

borrow pit (a) The source of material taken from some location near an embankment where there is insufficient excavated material nearby on the job to form the embankment. Borrow-pit excavation is therefore a special classification, usually bid upon as a special item in contracts. It frequently involves the cost of land or a royalty for material taken from the land where the borrow pit is located; it also often requires the construction of a suitable road to the pit. This type of excavation therefore usually runs higher in cost than ordinary excavation. (Hess) (b) An excavated area where borrow has been obtained. (AGI, 1987)

bort (a) Diamond material unsuitable for gems because of its shape, size, or color and because of flaws or inclusions. It also occurs in finely crystalline aggregates and is usually crushed into finer material. Syn: *boart; bortz; boort; boartz; borts; bowr.* See also: *shot bort.* (b) Inferior, coarsely crystalline diamonds, many of which contain black carbon or other minerals; used for core drilling, cutting, and polishing hard materials. (c) Formerly used to mean the Brazilian carbonado or black diamond. (Hess) (d) Industrial diamond. (ASM, 1961) (e) Very hard, flawed or discolored diamonds used in drilling and glass cutting. (Gordon, 1906) (f) S. Afr. Rounded forms of diamond with rough exterior and radiated or confused crystalline structure, but hardness equal to that of diamond. (Beerman, 1963) (g) Originally the term was used as a name for all crystalline diamonds not usable as gems; later it was used to designate those diamonds not usable as gems or toolstones. Currently the term is applied to low-grade industrial diamonds suitable only for use in a fragmented form. (h) A granular to very finely crystalline aggregate consisting of imperfectly crystallized diamonds or of fragments produced in cutting diamonds. It often occurs as spherical forms, with no distinct cleavage, and having a radial fibrous structure. (AGI, 1987) (i) A diamond of the lowest quality, so flawed, imperfectly crystallized, or off-color that it is suitable only for crushing into abrasive powders for industrial purposes (as for saws and drill bits); an industrial diamond. Originally, any crystalline diamond (and later, any diamond) not usable as a gem. (AGI, 1987) (j) A term formerly used as a syn. of carbonado. Cf: *ballas.* Syn: *magnetic bort.* See also hailstone bort. (AGI, 1987):

bort bit *diamond bit.* Also called bortz bit and boart bit. (Long, 1960)

borts *bort.*

bort-set bit *diamond bit.*

bortz *bort.*

bosh (a) The section of a blast furnace extending upward from the tuyeres to the plane of maximum diameter. (ASM, 1961) (b) A lining of quartz that builds up during the smelting of copper ores and thus decreases the diameter of the furnace at the tuyeres. (ASM, 1961) (c) A trough in which bloomery tools (or in copper smelting, hot ingots) are cooled. (Fay, 1920)

bosh jacket A water jacket used for cooling the walls of a shaft furnace. (Fay, 1920)

bosh tank A water tank that receives newly cast copper shapes for rapid cooling. (Pryor, 1963)

boss (a) Arkansas. A coal mine employee not under the jurisdiction of the miner's union. (Fay, 1920) (b) A proturberant and often dome-shaped mass of igneous rock congealed beneath the surface of the Earth and laid bare by erosion. (Webster 3rd, 1966) (c) An igneous intrusion that is less than 40 mi^2 (104 km^2) in surface exposure and is roughly circular in plan. Cf: *stock.* (AGI, 1987)

bossing Scot. The holing or undercutting of a thick seam, as of limestone, the height of the undercutting being sufficient for a person to work in.

Boss process Modification of the pan-amalgamation process; ore slurry flows continuously through a series of pans and settling tanks. (Bennett, 1962)

bostonite A light-colored hypabyssal rock, characterized by bostonitic texture and composed chiefly of alkali feldspar; a fine-grained trachyte with few or no mafic components. The name is derived from Boston, MA, for no clear reason. Not recommended usage. (AGI, 1987)

bostonitic Said of the texture of bostonite, in which microlites of rough irregular feldspar tend to form clusters of divergent laths within a trachytoid groundmass. (AGI, 1987)

bostrichites An early name for prehnite.

botallackite A monoclinic mineral, $Cu_2Cl(OH)_3$; trimorphous with atacamite and paratacamite.

botanical anomaly A local increase above the normal variation in the chemical composition, distribution, ecological assemblage, or morphology of plants, indicating the possible presence of an ore deposit or anthropomorphic contamination. See also: *geobotanical prospecting.* (AGI, 1987)

botanical prospecting Prospecting in which differences in plant growth or plant family serve as a clue to the presence of metals beneath barren rock or a covering of sand and gravel. (Pearl, 1961)

botryogen A monoclinic mineral, $MgFe^{3+}(SO_4)_2(OH){\cdot}7H_2O$; reniform, botryoidal, or globular; hyacinth-red to orange; in secondary sulfate deposits capping sulfide ore deposits. Syn: *red iron vitriol.*

botryoid A form in the shape of a bunch of grapes. Syn: *clusterite.* Adj: botryoidal.

botryoidal Having the form of a bunch of grapes. Said of minerals, e.g., hematite with a surface of spherical shapes; also said of a crystalline aggregate in which the spherical shapes are composed of radiating crystals. Cf: *colloform; colloid minerals; reniform.*

botryolite A radiated, columnar variety of datolite with a botryoidal surface. (Standard, 1964)

bottom (a) The floor or footwall of an underground mine. (b) The landing of a shaft or slope. (c) *gutter.* (d) To complete a borehole. (e) To construct the bottom of or for; said specif. of underdraining a level. (Standard, 1964) (f) To strike bedrock or clay when sinking a shaft. (Standard, 1964) (g) The landing at the bottom of the shaft or slope. (Fay, 1920) (h) The lowest point of mining operations. (Fay, 1920) (i) To underrun (as a gold deposit that is to be worked by the hydraulic method) with a level for drainage. (Webster 3rd, 1966) (j) Surface in a borehole parallel to the face of a drill bit. (Long, 1960) (k) A mass of impure metal formed below the matte, in matting metal ores. (l) The footwall of a metalliferous deposit. (Nelson, 1965) (m) Barren bedrock. (Nelson, 1965) (n) The rock formation below the alluvium on which the gold or tin wash dirt is met. (Gordon, 1906) (o) In gemstones, the pavilion. (p) Pennsylvania. The stratum, rock, or floor on which a coal seam lies. (Standard, 1964) (q) To break the material and throw it clear from the bottom or toe of the borehole. (r) To place a drill bit in contact with the bottom of a borehole. (s) Surface in a borehole parallel to the face of a drill bit. (t) In metal-melting furnaces, this is usually the hearth or crucible. See also: *bottoming.*

bottom bed Eng. Universally applied to the lowest bed in a quarry. Also used in Southeast England for the basal bed of the Tertiary, whether Thanet sand or Reading beds, resting on an eroded surface of the Chalk.

bottom belt conveyor A belt conveyor that carries the coal or ore on the lower strand; often used where height is limited. (Nelson, 1965)

bottom benching The method by which the bench is removed from below as with a power shovel. See also: *benching.* (AIME, 2)

bottom bounce Technique by which sonar impulses are reflected off the ocean bottom one or more times before reaching the target. Also refers to diving. (Hy, 1965)

bottom break The break or crack that separates a block of stone from a quarry floor. (Hess)

bottom canch In leveling an underground roadway, a part taken out below a bed. See also: *canch.* (Hess)

bottom coal Coal below the undercut; it may or may not be removed.

bottom cut (a) A machine cut made in the bottom or floor of a seam before shot firing. See also: *cutting horizon; middle cut; top cut.* (Nelson, 1965) (b) A drill hole pattern. See also: *drag cut.* (Nelson, 1965) (c) In drilling and blasting a tunnel, the lower of two converging lines of horizontally spaced holes. Upper line is draw cut. When blasted simultaneously, a wedge of rock is removed. (Pryor, 1963)

bottom cutter A dinter; a coal cutter for making floor cuts. (Nelson, 1965)

bottom diameter The diameter of a circle tangent to the seating curve at the bottom of the tooth gap of a roller chain sprocket. Equal to the pitch diameter minus the chain-roller diameter. (Jackson, 1955)

bottom-discharge bucket conveyor A conveyor for carrying bulk materials in a horizontal path consisting of an endless chain to which roller-supported, cam-operated, bottom-discharge conveyor buckets are attached continuously.

bottom-discharge conveyor bucket A vessel generally rectangular or square in plan and having a bottom consisting of an undercut gate.

bottom-dump car *mine car.*

bottom-dump scraper A carrying scraper that dumps or ejects its load over the cutting edge. (Nichols, 1976)

bottom-dump semitrailers Suitable for transporting free-flowing materials over a reasonably level haul route that permits a high travel speed. They can be used where the maximum flotation of a large single tire is required and where dumping in windrows over a wide area is practical. See also: *bottom dump truck.* (Carson, 1961)

bottom dump truck A trailer or semitrailer that dumps bulk material by opening doors in the floor of the body. Also called dump wagon. See also: *bottom-dump semitrailers.* (Nichols, 1976)

bottomed (a) A completed borehole, or the point at which drilling operations in a borehole are discontinued. (Long, 1960) (b) Said of shafts and slopes on being driven to completion when reaching base of coal seam.

bottom-emptying skip A skip equipped with a bottom discharge gate. (Sinclair, 1959)

bottom equipment (a) The tools or equipment attached to the lower end of a drill string and normally used at or near the bottom of a borehole. Also, the nondrilling equipment placed and operated at or near the bottom of a borehole, such as a pump unit or strainer. (Long, 1960) (b) Mine equipment used solely for work at the mine bottom, such as rotary dump and switch motor (if used to spot cars in rotary dump).

bottom filler A worker who fills a barrow with ore, coke, or stone, weighs it, and then places it on the cage or elevator to be hoisted to the top of the furnace. (Fay, 1920)

bottom gate The gate road at the lower end of an inclined coal face. See also: *main gate; tailgate; top gate.* (Nelson, 1965)

bottom heading (a) Method of excavating tunnels, drifts, or other mine openings. The bottom heading, which may be either driven in successive stages or holed through, is subsequently enlarged by excavating the top section. (Fraenkel, 1953) (b) Overhand bench.

bottom hole A point at, or near, the bottom of a borehole. (Long, 1960)

bottom-hole pressure (a) The load, expressed in pounds or tons, applied to a bit or other cutting tool while drilling. (Long, 1960) (b) The pressure, expressed in pounds per square inch, produced at the bottom of a borehole by the weight of the column of circulation or other liquid in a borehole. (Long, 1960) (c) The pressure, expressed in pounds per square inch, exerted by gas or liquids ejected from the rocks at or near the bottom of a drill hole. (Long, 1960) (d) Pressure measured in a well opposite the producing formation. If the well is flowing, the flowing bottom-hole pressure will be obtained; if the well is not producing and has not been producing for a sufficient time, the pressure will be the fully built-up, or static, bottom-hole pressure. Cf: *ground pressure.* (Inst. Petrol., 1961)

bottom-hole temperature The temperature of the fluid at or near the bottom of a borehole; significantly lower than the temperature of the formation if borehole fluids have been circulated recently or are being produced with expansion into the well bore. (Long, 1960)

bottoming The downward pinching-out or termination of an orebody, either structurally or by economic grade. See also: *bottom.* (AGI, 1987)

bottom lift (a) The lowest or deepest lift or level of a mine. (Zern, 1928) (b) The deepest columns of a pump. (Zern, 1928) (c) The deepest lift of a mining pump, or the lowest pump. (Fay, 1920)

bottom loading belt A bottom belt conveyor. (Nelson, 1965)

bottom maker A laborer who relines bottoms of ingot soaking pits with coke dust to retard formation of oxide scale on hot ingots. (DOT, 1949)

bottom pillar A large block of solid coal left unworked around the shaft. See also: *shaft pillar.* (Fay, 1920)

bottom-pour ingot assembly One comprising hot tops, wood blocks, ingot mold, mold stool, lateral outlet bricks, lateral bricks, king brick, fountain bricks, funnel brick, and suitable metal supporting devices.

bottom-pour ladle A ladle poured through a refractory nozzle in the bottom.

bottom rock *bedrock.*

bottoms (a) Used in connection with the Orford process for separating nickel and copper as sulfides. When the mixed sulfides are fused with sodium sulfide, the nickel sulfide separates to the bottom. See also: *tile copper.* (CTD, 1958) (b) The material drawn off from the bottom of a tower or still. Any residue accumulating in the bottom of a process vessel.

bottom sample A sample obtained by collecting a portion of material on the bottom of a container or pipeline. (Bennett, 1962)

bottom sampler One of various types of apparatus capable of piercing the sea bottom and retaining a sample of the deposit when brought to the surface. (CTD, 1958)

bottom sheets The steel plates forming the bottom of an oil still or a steam boiler. (Hess)

bottom subsidence *subsidence.*

bott plug A clay ball used for stopping the taphole in a cupola furnace. (Mersereau, 1947)

bott stick A long stick used for inserting the bott plug into the taphole to stop the flow of metal. (Mersereau, 1947)

boudin (a) One of a series of elongate, sausage-shaped segments occurring in boudinage structure, either separate or joined by pinched connections, and having barrel-shaped cross sections. (AGI, 1987) (b) A term applied loosely, without regard to shape or origin, to any tectonic inclusion. Etymol: French, bag; blood sausage. (AGI, 1987)

boudinage A structure common in strongly deformed sedimentary and metamorphic rocks, in which an original continuous competent layer or bed between less competent layers has been stretched, thinned, and broken at regular intervals into bodies resembling boudins or sausages, elongated parallel to the fold axes. See also: *pull-apart structure.* (AGI, 1987)

Bouguer anomaly A gravity anomaly calculated after corrections for latitude, elevation, and terrain. Pron: boo-gay. See also: *anomaly; Bouguer correction.* (AGI, 1987)

Bouguer correction A correction made to gravity data for the attraction of the rock between the station and the datum elevation (commonly sea level); or, if the station is below the datum elevation, for the rock missing between station and datum. The Bouguer correction is 0.01276 ph mgal/ft, or 0.04185 ph mgal/m, where p is the specific gravity of the intervening rock and h is the difference in elevation between station and datum. See also: *Bouguer anomaly.* (AGI, 1987)

Bouguer gravity Gravity values after latitude, elevation, and Bouguer corrections have been applied. Used in the gravitational method of geophysical prospecting. (Nelson, 1965)

Bouguer reduction The correction made in a gravity survey to take account of the altitude of the station and the rock between the station and sea level. (AGI, 1987)

boulangerite A monoclinic mineral, $Pb_5Sb_4S_{11}$; metallic; bluish-gray; massive.

boulder (a) A detached rock mass larger than a cobble, having a diameter greater than 10 in (25.4 cm) or 8 phi units, or about the size of a volleyball, being somewhat rounded or otherwise distinctively shaped by abrasion in the course of transport; the largest rock fragment recognized by sedimentologists. In Great Britain, the limiting size of 8 in (20.3 cm) has been used. (AGI, 1987) (b) *boulder stone.* (c) A general term for any rock that is too heavy to be lifted readily by hand. Also spelled bowlder. (AGI, 1987)

boulder blasting (a) The breaking down of large stones at quarries by small explosive charges. See also: *secondary blasting.* (Nelson, 1965) (b) Secondary blasting of rocks too big to be moved conveniently in the mine's transport system. (Pryor, 1963)

boulder buster An explosive used to break rock fragments by blockholing or mudcapping methods. (Long, 1960)

boulder clay (a) The stiff, hard, and usually unstratified clay of the drift or glacial period that contains boulders scattered through it. Also called till; hardpan; drift clay; drift. See also: *till; moraine.* (Fay, 1920) (b) Glacial drift that has not been subjected to the sorting action of water and therefore contains mixed particles ranging from boulders to clay sizes. (ASCE, 1958)

boulder flat A level tract covered with boulders. (AGI, 1987)

boulder gravel An unconsolidated deposit consisting mainly of boulders. (AGI, 1987)

boulder motion A surface quarry worked only in detached masses of rock overlying the solid rock; sometimes contracted to motion. (Standard, 1964)

boulder quarry A quarry in which the joints are numerous and irregular, so that the stone has been broken naturally into comparatively small blocks. A local term applied to certain marble quarries in the region of Knoxville, TN, where erosion has formed many large cavities and cracks, between which the rock stands up as pinnacles. The cavities are now filled with clay. (Fay, 1920)

boulder stone An obsolete term for any large rock mass lying on the surface of the ground or embedded in the soil, differing from the country rock of the region, such as an erratic. Syn: *boulder.* (AGI, 1987)

boule A fused mass of synthetic material up to 5 cm long, pear or carrot-shaped, particularly as produced by the Verneuil or Crochralshi processes in the production of synthetic sapphire, ruby, spinel, or rutile. Etymol: Fr. "ball." Syn: *birne.* See also: *Verneuil process.*

boulet A small ovoid; an egg-shaped briquette. (BS, 1962)

bounce (a) A sudden spalling off of the sides of ribs and pillars due to excessive pressure; a bump. (Zern, 1928) (b) The rapid up-and-down reciprocating motion induced in a drill string by rod vibration, drill string wrap-up, excessive volume or pressure of circulation media, or the running of a bit on and over small, loose materials on the bottom of a drill hole. (Long, 1960)

bounce cast Casts of short grooves (up to 5 cm) widest and deepest in middle and fading out at both ends; presumably formed by objects grazing against bottom and rebounding. See also: *impact cast.* Cf: *prod mark.* (Pettijohn, 1964)

bound Corn. An area taken up for tin mining; a tin bound. (Standard, 1964)

boundary (a) A line between areas of the Earth's surface occupied by rocks or formations of different type and age; esp. used in connection with geologic mapping; also, a line between two formations or cartographic units on a geologic map. (b) The limit, border, or termination of a coal or mineral take; a line along which workings must stop in the vicinity of a fault or old waterlogged workings. Also called march. (Nelson, 1965)

boundary fault A major fault with a considerable displacement. A number of collieries and coalfields are limited along one side by such a fault. (Nelson, 1965)

boundary films Films of one constituent of an alloy surrounding the crystals of another constituent. (CTD, 1958)

boundary map A map created for the purpose of delineating a boundary line and the adjacent territory. (AGI, 1987)

boundary pillar A pillar left in mines between adjoining properties.

bound gravel A hard, lenticular, cemented mass of sand and gravel occurring in the region of the water table; it is often mistaken for bedrock. (AGI, 1987)

Bourdon tube Pressure gage, made from elliptical curved tube, which straightens somewhat under pressure, and is made to move a measuring needle over a dial. (Pryor, 1963)

bournonite An orthorhombic mineral, $4[PbCuSbS_3]$; shows wheel-shaped twin crystals; a source of lead, copper, and antimony. Syn: *wheel ore; cogwheel ore; endellinite; endellionite; berthonite.*

bouse N. of Eng. Ore mixed with veinstone; second-class ore that must undergo further preparation before going to the smelter. See also: *boose.*

bouse team N. of Eng. The place where bouse is deposited outside a mine, ready to be dressed or prepared for the smelter.

bout Derb. A method of measuring lead ore. (Nelson, 1965)

bouteillenstein A peculiar green and very pure glass, found as rolled pebbles. Also called bottle stone; pseudochrysolite—the latter from its resemblance to olivine. It is not solely a rock, as it may be prehistoric slag or glass. See also: *moldavite.* (Fay, 1920)

boutgate (a) Scot. A road by which the miners can reach the surface. (b) A passage around a shaft at a landing. (c) A traveling road from one seam to another.

bouton (a) Scot. A mass of roof consisting of stone or shale. (b) Scot. A projecting stone in a shaft or underground road.

bowenite (a) A hard, compact, greenish-white to yellowish-green serpentine once thought to be nephrite jade; translucent; massive, fine-grained; consists of a dense feltlike aggregate of colorless antigorite fibers with patches of magnesite, flakes of talc, and grains of chromite. Syn: *bowenite jade.* (b) N.Z. Serpentine rock (serpentinite). Syn: *tangiwai; tangiwaite; tangawaite.*

bowenite jade *bowenite.*

Bowen's reaction series *reaction series.*

bowk (a) S. Staff. A small wooden box in which iron ore is hauled underground. Syn: *hudge.* (b) Aust. An iron bucket used for raising rock, etc., while sinking. Syn: *hudge.* (c) A noise made by the cracking of the strata owing to the extraction of the coal beneath. (d) Bucket; kibble; hoppit, as used in sinking. (Mason, 1951) (e) A large iron barrel used for men's tools and debris when sinking a shaft. (CTD, 1958) (f) The noise made by the escape of gas under pressure.

bowl (a) The bucket or body of a carrying scraper. (Nichols, 1976) (b) The moldboard or blade of a dozer. (Nichols, 1976) (c) Stationary part of a Symons crusher, which surrounds the cone (the grating member). (Pryor, 1963) (d) *spider.*

bowl classifier A hydraulic classifier similar to a thickener, but differs in that the current carries the fine material into the overflow; used to make separations at very fine particle size.

bowlder *boulder.*

bowlingite *saponite.*

bowl scraper A steel bowl hung within a fabricated steel frame, running on four or two wheels. Its bottom edge digs into the ground, the bowl being filled as it is drawn forward by a tractor; soil is ejected at the dump by a tailgate, moved by wire ropes or hydraulically. Towed scrapers transport soil, in addition to spreading and leveling it. See also: *wheel scraper.* (Hammond, 1965)

bowr *bort.*

bowse *boose.*

box (a) A unit in a sluice for washing gravel; a sluicebox. Syn: *box condenser.* (Hess) (b) A dump body. (Nichols, 1976) (c) To place core samples in a lidded, traylike, partitioned container for safekeeping after they have been removed from the core barrel; also, the container in which core samples are placed after they have been removed from a core barrel. Also called corebox; core tray. (Long, 1960) (d) To drill boreholes at the four corners of a square area at equal distances from a centrally located and already completed borehole. (Long, 1960)

box barrow A large wheelbarrow with upright sides. (Webster 2nd, 1960)

box bell *bell screw.*

box canyon (a) A narrow gorge or canyon containing a stream following a zigzag course, characterized by high, steep rock walls and typically closed upstream with a similar wall, giving the impression as viewed from its bottom of being surrounded or boxed in by almost-vertical walls. (AGI, 1987) (b) A steep-walled canyon heading against a cliff; a dead-end canyon. Syn: *cajon.* (AGI, 1987)

boxcar loader (a) Any of several types of conveyors adapted by portable or hinged mounting for use in loading bulk materials into boxcars. Some types operate at high speeds and throw the materials to the ends of the car. See also: *idler disk.* (b) In anthracite and bituminous coal mining, one who loads coal into railroad boxcars by mechanical shovel or conveyor loader. Syn: *thrower belt; car loader; loader engineer; loader runner.* See also: *portable conveyor.* (DOT, 1949)

box check A ventilation control consisting of a stopping or a curtain with a hole, through which a conveyor can pass. The box check serves as a ventilation regulator in a belt haulage entry. See also: *boxes; conveyor airlock.* (MSHA, 1986)

box condenser *box.*

box cut The initial cut driven in a property, where no open side exists; this results in a highwall on both sides of the cut. (Austin, 1964)

box-cut method A method of opencast mining of coal where the dip of the seam is relatively steep. A boxlike excavation is made to the dip, or at an angle to it, and the coal seam is worked to the right and left. See also: *strike working.* (Nelson, 1965)

box-cut spoil That spoil created from the initial excavation of a pit or pits which is placed upon the surface of adjacent lands. The term does not include spoil from subsequent excavations, which is placed in previously excavated pits.

boxes (a) Pennsylvania. Wooden partitions for conducting the ventilation from place to place. See also: *box check.* (b) More or less hollow cuboidal limonitic concretions. (Arkell, 1953) (c) Eng. Pebbles of hard brown sandstone at the base of the Red and Coralline Crags in East Anglia, containing remains of a fossil; so called by the Suffolk phosphate diggers. (Arkell, 1953)

box filling The use of metal trays, instead of shovels, for hand-filling coal into trams. The collier scooped the lumpy coal into the box and discarded the small material, which had little market value. The use of a box was compulsory at many collieries until several decades ago. See also: *fork-filled; loading pan.* (Nelson, 1965)

box groove A closed groove between two rolls, formed by a collar on one roll and fitting between collars on another. (Fay, 1920)

box heading A heading driven through very loose ground with close timbering. (Nelson, 1965)

boxing A method of securing shafts solely by slabs and wooden pegs. (Zern, 1928)

box loader In the quarry industry, one who loads broken rock into a large box, placed on a small truck running on a narrow gauge track, to be hoisted out of the quarry pit. Syn: *grouter; rock loader.* (DOT, 1949)

box scraper *scraper.*

box sluice An open wooden channel or flume for conveying placer sand. The gold or heavy minerals settle at the bottom. The method is cleaner and requires less water than ground sluicing. (Nelson, 1965)

box tap *bell tap.*

box timbering Use of rectangular close frame for lining shafts or drives. See also: *plank timbering.* (Pryor, 1963)

boxwork A network of intersecting blades or plates of limonite or other iron oxide, deposited in cavities and along fracture planes from which sulfides have been dissolved by processes associated with the oxidation and leaching of sulfide ores, esp. porphyry copper deposits. (AGI, 1987)

boydite Local name for probertite. (English, 1938)

boylom Staff. A bluish iron ore. See also: *boilum.* (Arkell, 1953)

Boylston's reagent A 5% solution of nitric acid in absolute ethyl or methyl alcohol, used for the general etching of normal carbon steels. Syn: *nital.* (Osborne, 1956)

BPL Bone phosphate of lime. (AGI, 1987)

BQ Letter name specifying the dimensions of bits, core barrels, and drill rods in the B-size and Q-group wireline diamond drilling system having a core diameter of 36.5 mm and a hole diameter of 60 mm. (Cumming, 1981)

brace A platform or landing at the top of a shaft. The upper brace is the platform built in the headgear above the shaft collar. (Nelson, 1965)

brace head A cross handle attached at the top of a column of drill rods by means of which the rods and attached bit are turned after each drop in chop and wash operations while sinking a borehole through overburden. Also called brace key. (Long, 1960)

brachy axis The shorter lateral axis in the crystals of the orthorhombic, monoclinic, and triclinic systems. (Webster 3rd, 1966)

brachypinacoid (a) A pinacoid parallel to the vertical axis and the brachydiagonal. (Standard, 1964) (b) The pinacoid 010 intersecting the brachy-axis in orthorhombic and triclinic systems (obsolete). Cf: *pinacoid.*

brachytypous In crystallography, comparatively short. (Standard, 1964)

bracing (a) Diagonal or horizontal members used to prevent swaying of structures, i.e., conveyor-supporting structures. (b) Eng. See also: *lacing.*

brackebuschite A monoclinic mineral, $Pb_2(Mn,Fe)(VO_4)_2{\cdot}H_2O$.

Brackelsberg process A process by which fine ores are moistened with water to which a binding medium is added, and the wet mass, without any heating, is rotated in a drum until it forms into spherical lumps of varying size. The moisture is then dried out by evaporation, and the product remains in the form of hard, very porous balls of ore, which are of great reducibility as compared with sintered ore or briquettes. (Osborne, 1956)

bracket A platform over a shaft entrance. (Standard, 1964)

brackish water Water in which salinity values range from approx. 0.50 to 17.00 parts per thousand. (Hy, 1965)

Bradford breaker A machine that combines coal crushing and screening. It consists of a revolving cylindrical screen 8 to 14 ft (2.4 to 4.3 m) in diameter and 13 to 22 ft (4.0 to 6.7 m) in length. It breaks the coal by gravity impact. On reaching the desired size, the coal is discharged through the plates. It can deal with run-of-mine coal up to 12 in (30.5 cm) at a rate of 500 to 600 st/h (454 to 544 t/h), to give a product size of below 1-1/2 in (3.8 cm); other sizes can be produced, depending on the screen plates used. See also: *screen.* (Nelson, 1965)

bradleyite A monoclinic mineral, $Na_3Mg(PO_4)(CO_3)$; in veryfine grains in saline oil shale in Wyoming.

bradyseism A long-continued, extremely slow vertical instability of the crust, as in the volcanic district west of Naples, Italy, where the Phlegraean bradyseism has involved up-and-down movements between 6 m below sea level and 6 m above over more than 2,000 yr (Casertano). Etymol: Greek "bradys" (slow) + "seismos" (earthquake). (AGI, 1987)

brae Insufficiently charred wood, as in charcoal burning. (Standard, 1964)

Bragg angle The angle, θ, at which X-rays diffract in crystalline materials. It satisfies the relationship $n\lambda = 2d \sin\theta$, where d is the distance between diffraction planes of atomic particles in a crystal structure, λ is the wavelength of the X-rays, and n is the order of diffraction when a crystal is placed in an X-ray beam.

Bragg indices The index numbers assigned to a diffracted X-ray beam. They have the same values as Miller indices but are written without closures. Diffraction order may be factored into Bragg indices, e.g., 111, 222, 333... represent n = 1, 2, 3... for diffraction from atomic planes parallel to 111. Cf: *Miller indices.*

braggite A tetragonal mineral (Pt,Pd,Ni)S: steel gray, minute grains in concentrates from the Bushveld norite of the Transvaal, South Africa. A source of platinum and palladium.

braided stream A stream that divides into an interlacing or tangled network of several small branching and reuniting shallow channels separated from each other by branch islands or channel bars, resembling in plan the strands of a complex braid. Such an anastomosing stream is generally believed to indicate an inability to carry all of its load, such as an overloaded and aggrading stream flowing in a wide channel on a flood plain. (AGI, 1987)

brake (a) A device (as a block or band applied to the rim of a wheel) to arrest the motion of a vehicle, a machine, or other mechanism and usually employing some form of friction. (Webster 3rd, 1966) (b) A device, either hand- or power-operated, for applying resistance to the drum or pulley and thus controlling the movement of mine cars or cages. A common form is a brakeshoe, lined with friction material, which is applied to the surface of a wheel or drum, and thus retards or even stops its movement. See also: *winder brake.* Syn: *haulage brake.* (Nelson, 1965) (c) Eng. A stout, wooden lever to which boring rods are attached. It is worked by one or more people. (d) N. Staff. To lower trams on dips by means of a wheel and rope.

brakedrum A rotating cylinder with a machined inner or outer surface upon which a brake band or brakeshoe presses. (Nichols, 1976)

brake incline (a) An incline in which the full trucks descend by gravity and pull up the empty ones. See also: *gravity haulage.* (CTD, 1958) (b) Gravity plane. (Pryor, 1963)

brakeman (a) Person who attends to a brake or brakes, as on a railroad car. (Standard, 1964) (b) Eng. The person in charge of a winding (hoisting) engine for a mine. "Brakeman" is usually used in the United States; "brakesman" is the British usage. The person in charge of hoisting engines, esp. in the United States, is usually called a hoisting engineer. (Fay, 1920) (c) In mining, a laborer who rides on trains or trips of cars hauled by locomotive or hoisting cable or chain, and assists in their transportation to surface or shaft bottom for hoisting; operates or throws switches; couples and uncouples cars, or attaches and detaches cars to and from the cable; opens and closes ventilation doors in mines; directs movement of the train by signaling motorman. May be designated according to type of hauling machine, such as dinkey operator helper. Also called brake holder, car rider, conductor, dukey rider, gang rider, motorman helper, nipper, patcher, rider, rope conductor, rope rider, set rider, snapper, tailend rider, trailer, train conductor, trainman, transfer car helper, trip rider, tub rider. Syn: *conductor.* (DOT, 1949)

brake sieve A jigger operated by a hand lever. (Fay, 1920)

brake wheel (a) A hand wheel for operating a brake, as on a vehicle. (Webster 2nd, 1960) (b) A wheel or pulley on which a friction brake acts. (Webster 2nd, 1960) (c) A heavy wheel provided with cams for controlling the movement of a triphammer. (Webster 2nd, 1960)

braking distance Tbe distance the haulage unit (i.e., train) will travel after the application of the brakes, depending on the speed, the weight of locomotive and train, and the gradient. (Sinclair, 1959)

brammallite A micaceous mineral differing from illite because it contains soda in excess of potash. Found in crevices in coal measure shales from Llandebie, South Wales. Syn: *sodium illite.* (Spencer, 1943)

brances *brasses.*

branch (a) An underground road or heading driven in coal measures; also, a roadway turned from a level, etc. Syn: *branch hole.* (b) A small vein departing from the main lode. Cf: *main hole.*

branch fault A minor fault that branches from a larger fault. (Stokes, 1955)

branch headings Headings that are turned off the main level at intervals for development purposes. They may proceed to the rise or dip and are adopted in longwall and pillar methods of working. See also: *opening out.* (Nelson, 1965)

branch hole *branch.*

branchite A variety of hydrocarbon found in lignite. According to Hintz it is identical with hartite. (Tomkeieff, 1954)

brandisite Clintonite found as monoclinic hexagonal-shaped prisms in metamorphosed limestone.

brannerite A monoclinic mineral, $(U,Ca,Y,Ce)(Ti,Fe)_2O_6$; radioactive; commonly metamict; in placers of the Stanley Basin, UT.

brashings Brittle shale (the coal miner's "slate") interbedded with thin coalbeds; also, the roof of the Pittsburgh coal in western Maryland. See also: *rashings*. (Hess)

brasque A paste made by mixing powdered charcoal, coal, or coke with clay, molasses, tar, or other suitable substance. Used for lining hearths, crucibles, etc. Syn: *steep.* (Webster 2nd, 1960)

brasqued crucible A crucible lined with charcoal or lampblack, and used for the reduction of oxides of metals to the metallic state. The crucible is prepared by ramming it full of lampblack or charcoal, and then excavating a portion of its contents and polishing the lining with a burnisher. (Fay, 1920)

brass balls Nodular pyrite. (Fay, 1920)

brasses Mineral impurities in coal, of yellow metallic appearance, consisting mainly of iron sulfides. Syn: *brances; brassyn.* (BS, 1960)

brassfounder's disease A disease affecting the general system, characterized by chronic poisoning from inhalation of metallic fumes, with symptoms like those of malarial fever. (Standard, 1964)

brass furnace One of two kinds of furnaces for the making and founding of brass: (1) a reverberatory furnace for producing large quantities of the alloy, or (2) a crucible furnace for producing small quantities. (Fay, 1920)

brass ore (a) An early name for a mixture of sphalerite and chalcopyrite. (Hess) (b) An old name for aurichalcite.

brassyn *brasses.*

brassy top Aust. The top part of the Greta coal seam, in which there are large quantities of sulfide of iron.

brattice (a) Ventilating partition, usually of coated fabric, used to direct air to various faces to remove gas and dust. (b) A board or plank lining, or other partition, in any mine passage to confine the air and force it into the working places. Its object is to keep the intake air from finding its way by a short route into the return airway. Temporary brattices are often made of cloth. Also spelled braddish; brettice; brettis; brattish. See also: *brattice cloth.* (c) An airtight partition in a mine shaft to separate intake from return air. See also: *screen.* (BS, 1963) (d) Used as jumpers for removing gas from a roof cavity. (Nelson, 1965) (e) To provide with a brattice for separation or support; often used with up. Syn: *brattice up.* (Webster 2nd, 1960)

brattice cloth (a) Fire-resistant fabric, usually coated, used to erect a brattice. (b) A heavy canvas, often covered with some waterproofing material, for temporarily forcing the air into the face of a breast or heading; also used in place of doors on gangways; then known as "sheets." Syn: *brettice cloth.* See also: *brattice; brattice sheeting.*

brattice man In mining, a worker who builds doors, stoppings and curtains (ventilation walls or partitions in active workplaces) of burlap, canvas, and wood. Also called airman; braddisher; braddish man; canvasman; doorman; ventilation man. (DOT, 1949)

brattice road A road through the goaf supported by chocks or timber packs.

brattice sheeting A curtain or screen of flexible material used to direct or control the flow of ventilating air. See also: *brattice cloth; sheets.* (BS, 1963)

brattice up *brattice.*

brattice worker *airman.*

brattish *brattice.*

braunite A tetragonal mineral, $Mn^{2+}Mn^{3+}{}_6SiO_{12}$; brittle; may contain appreciable iron.

braunkohle Ger. *brown coal.* (Hess)

bravaisite A former name for a micaceous clay later shown to be a mixture of montmorillonite and illite.

Bravais lattice One of 14 ways points may be arrayed periodically in space such that each point is in an identical point environment. Every crystal structure has associated with it a Bravais lattice. Syn: *space lattice; crystal lattice.* Cf: *direct lattice; space group.*
Bravais law *Bravais rule.*
Bravais-Miller indices *Miller-Bravais indices.*
Bravais rule The most prominent faces of a crystal are those parallel to internal planes having the greatest density of lattice points. Syn: *Bravais law.*
bravoite A nickeloan variety of pyrite.
braze To solder with brass or other hard alloys. (Nichols, 1976)
brazilianite A monoclinic mineral, $NaAl_3(PO_4)_2(OH)_4$; yellow-green; also spelled brasilianite, brasilianita.
Brazilian pebble A colorless transparent quartz, such as is used for optical purposes. (Fay, 1920)
Brazilian test A method for the determination of the tensile strength of rock, concrete, ceramic, or other material by applying a load vertically at the highest point of a test cylinder or disk (the axis of which is horizontal), which is itself supported on a horizontal plane. The method was first used in Brazil for testing concrete rollers on which an old church was being moved to a new site. (Dodd, 1964)
brazilite (a) A mixture of baddeleyite, zircon, and altered zircon. (b) A fibrous variety of baddeleyite. (c) An oil shale.
Brazil twin A type of twin found in quartz in which the two crystalline individuals are of opposite kinds, one being right-handed, the other left-handed, with a face of the trigonal prism of the second order as twinning plane. Since one is not derivable from the other by any rotation, there is no twinning axis. (Hess)
brazing (a) Joining metals by flowing a thin layer (capillary thickness) of nonferrous filler metal into the space between them. Bonding results from the intimate contact produced by the dissolution of a small amount of base metal in the molten filler metal, without fusion of the base metal. Sometimes the filler metal is put in place as a thin solid sheet or as a clad layer, and the composite is heated as in furnace brazing. (ASM, 1961) (b) In joining metals, the term "brazing" is used where the temperature exceeds some arbitrary value, such as 800 °F (427 °C); the term "soldering" is used for temperatures lower than the arbitrary value. (ASM, 1961)
breach (a) An opening made by breaking down a portion of a solid body, as a wall, a dike, or a riverbank; a break; a gap. (b) The face of a level or drift. (c) A large cave hole caused by undermining.
breached anticline An anticline whose crest has been deeply eroded, so that it is flanked by inward-facing erosional scarps. Cf: *bald-headed anticline.* (AGI, 1987)
breaching The breaking through of a bar. (Schieferdecker, 1959)
breadth N. Staff. A set of coal pillars formed by rearer workings. (Nelson, 1965)
break (a) A plane of discontinuity in the coal seam such as a slip, fracture, joint, or cleat. The surfaces are in contact or slightly separated. See also: *break detector.* (Nelson, 1965) (b) A fracture or crack in the roof beds as a result of mining operations. See also: *breakes; induced fracture.* (Nelson, 1965) (c) To separate core from solid rock at the bottom of a borehole by a tensional pull applied to the drill string. (Long, 1960) (d) In mineral processing, optimum mesh of grind (m.o.g.), the practical size range to which ore is reduced before concentration. Not synonymous with liberation mesh. (Pryor, 1963) (e) In drilling, to unscrew, as rods, casing, drill pipe, etc. (Long, 1960)
breakage (a) Voluntary or involuntary division of a solid. (BS, 1962) (b) Small material produced by involuntary breakage during mechanical handling or processing. (BS, 1962)
breakage clause Eng. A clause inserted in some mining leases providing for an abatement of royalty or allowance on weight for certain weight of small coal or breakage sent out in every ton of large coal, for example, 120 lb (54.5 kg) in every collier's ton of 2,640 lb (1,200 kg).
breakage of coal *degradation.*
breakaway chain A chain that holds a tractor and a towed unit together if the regular fastening opens or breaks. Syn: *safety chain.* (Nichols, 1976)
breakback The fractures caused by the shattering of a solid rock ledge back of the drill holes in which the charge is placed. (Fay, 1920)
break detector A scraper capable of detecting breaks in a shothole. See also: *break; stemmer.* (Nelson, 1965)
breakdown Of an emulsion, the reunion of the finely dispersed particles and their separation from the medium in which they form an emulsion. Syn: *hoedown.* (CTD, 1958)
breakdown voltage The voltage at which an insulator or dielectric ruptures; or the voltage at which ionization and conduction begin in a gas or vapor. (Hunt, 1965)
breaker capacity The ability of a switch, in a particular situation, to clear safely the heaviest fault current that can flow; it depends upon the amount of power available in the system, size of cables, transformers, etc. Cf: *breaking capacity.* (Mason, 1951)
breaker props Props, or props and cribs, set to break the roof off at a prearranged line during retreat mining, or when blasting down roof.
breakers The row of drill holes above the mining holes in a tunnel face. (Stauffer, 1906)
breaker zone *surf zone.*
breakes Eng. Fissures in old coal workings. See also: *break.*
break-even point (a) Production level at which total cost equals revenue. (Pryor, 1963) (b) Value or selling price of ore, metal, or mined material that just balances total cost of operations; conversely, maximum unit costs above which there is no profit at given market values.
break-in To start drilling operations with a new bit by rotating the bit slowly under a light load for a short time before full speed and load are applied to the bit. (Long, 1960)
breaking Size reduction of large particles. Also called cracking. (BS, 1962)
breaking capacity The capacity of a switch, circuit breaker, or other similar device to break an electric circuit under certain specified conditions. Cf: *breaker capacity.* (CTD, 1958)
breaking-down rolls A rolling mill unit used for breaking-down operations; a rolling mill used for reducing sectional dimensions, mainly thickness—of ingots, billets, and other rough, semifinished products—as a preliminary step to subsequent rolling operations. (Henderson, 1953)
breaking ground (a) The breaking and loosening of rock as a preparatory step to its loading and removal. See also: *excavation.* (Nelson, 1965) (b) Attrition of an ore deposit by hand, explosive, or mechanical breaking methods to reduce it to pieces of ore suitable for transport and treatment. (Pryor, 1963)
breaking in N. of Eng. *hewing.*
breaking-in shot (a) The first borehole fired in "blasting off the solid" to provide a space into which material from subsequent shots may be thrown. Also called opening shot; buster shot. (Fay, 1920) (b) In blasting a solid face, the first hole or group of holes of a round to be fired simultaneously. See also: *burn cut.* (Pryor, 1963)
breaking lag As applied to an electric blasting cap, the time elapsing between the bridge wire receiving the firing impulse and the breaking of the circuit. (Fraenkel, 1953)
breaking point In rock crushers, a deliberate weak link that yields if excessive strain is developed. May be a scarfed toggle, weak cap bolts on a pitman, a shearpin in drive, or a clutch designed to fail at a given load. (Pryor, 1963)
breaking prop Arkansas. One of a row of props of sufficient strength to cause the rock above the coal to break and so limit the area of top brought down by a brushing shot.
breakings Inferior ores arranged ready for crushing. (Nelson, 1965)
breaking stress *fracture stress.*
break in lode A fault.
break line (a) The line in which the roof of a coal mine is expected to break. (b) The line of complete extraction of coal. (c) A line roughly following the rear edges of the pillars that are being drawn or mined. See also: *rib line.*
break-making Providing a crack indication by striking small slits in the longitudinal direction of a row of drill holes in quarries.
breakout (a) To pull drill rods or casings from a borehole and unscrew them at points where they are joined by threaded couplings to form lengths that can be stacked in the drill tripod or derrick. (Long, 1960) (b) An accidental flow of metal through a hole in a furnace lining.
breakrow A row of timbers erected for the purpose of breaking the roof in pillar mining. (Hess)
breakthrough (a) A passage cut through a pillar to allow the ventilating current to pass from one room to another. Larger than a doghole. Also called room crosscut. Syn: *cut-through; crosscut.* (b) The point at which a drill bit leaves the rock and enters either a natural or a constructed opening. (Long, 1960) (c) An opening made, either accidentally or deliberately, between two underground workings. (Long, 1960) (d) In an ion-exchange column used in leaching, the arrival of traces of uranium in the final column during the loading (adsorption) cycle. (Pryor, 1963) (e) *stenton.*
break thrust A thrust fault that cuts across one limb of a fold. (AGI, 1987)
breakup (a) Eng. An excavation commenced from the bottom of a tunnel heading and carried upward, so as to form two interior working faces. (b) Mid. To cut away and remove the floor of an entry or other opening.
breakup value On exhaustion of an ore deposit or cessation of an exploitation, the value of its onsite buildings, equipment, stockpiles, untouched remnants of ore concentrates, etc.; in foundations of plant; and any other assets still having value apart from their original use. (Pryor, 1963)
breakwater An offshore structure (such as a mole, wall, or jetty) that, by breaking the force of the waves, protects a harbor, anchorage, beach, or shore area. (AGI, 1987)

breast (a) In a coal mine, a chamber driven in the seam from the gangway, for the extraction of coal; the face of a working. (b) In Italy, a stall in a steep seam from 12 to 18 yd (11 to 16.5 m) wide. The stalls are carried one above another from the lowest level to the rise. (c) Leic. To take down or get a buttock (face) of coal end on. (d) The end, in unmined rock, of an underground excavation, sometimes called the face; the vertical end surface of a block. See also: *before breast*. (Nelson, 1965) (e) A place where anthracite coal is mined; in the soft coal regions, it is called a room. (BCI, 1947) (f) The face of a working. (g) That part of the bedplate that is back of the crossheads in engines of the Corliss type. (h) The side of the hearth containing the taphole in a blast furnace; the rammed material in which the taphole is installed in a cupola.

breast-and-pillar Pennsylvania. A system of working anthracite coal by bords 10 yd (9.1 m) in width, with narrow pillars 5 yd (4.6 m) wide between them, holed through at certain intervals. The breasts are worked from the dip to the rise. See also: *bord-and-pillar*. (Fay, 1920)

breast auger An auger supported by a breastplate against a miner's body. Used for drilling holes in soft coal. (Fay, 1920)

breast board (a) Planking placed between the last set of timbers and the face of a gangway or heading which is in quicksand or loose ground. (b) The timber or boards placed horizontally across the face of an excavation, or heading, to prevent the inflow of gravel or other loose or flowing material. (Stauffer, 1906)

breast bore Scot. A borehole put in parallel with the seam, that is made and kept in advance of a working place for the purpose of ascertaining the position of old works, tapping water, letting off gas, etc. (Fay, 1920)

breast coal The face of the middle or main layer of coal in a composite seam. (Nelson, 1965)

breast drill A small, portable hand drill customarily used by handsetters to drill the holes in bit blanks in which diamonds are to be set. The upper end of the drill is provided with a plate against which the breast of the operator is pressed to force the bit into the work. Cf: *brace*. (Long, 1960)

breaster In tunnel blasting, holes that parallel the tunnel alignment drilled between the cut and the perimeter holes.

breast eye Lanc. Opening leading from a working face to the surface. (Hess)

breast hole In driving a tunnel, a hole blasted after the bottom cut. (Pryor, 1963)

breasting (a) N. Staff. A short leading stall, worked at right angles to, and forming the face of, the main level. (b) A wide heading or level. (c) Eng. Taking ore from the face or head of a drift. (d) In drift mining, breaking down the gravel underground, and retreating towards the crosscut from which the drifts were driven. (von Bernewitz, 1931) (e) Cumb. A place driven to open out a longwall face. (SMRB, 1930)

breast machine A machine used for undercutting coal in which the main frame and carriage are held stationary by roof jacks while the cutter frame advances into the kerf during the cutting operation. Since cuts do not exceed 44 in (1.1 m) along the face, it is necessary to relocate the machine several times before the entire face can be cut. (Jones, 1949)

breastplate A slightly curved iron plate fastened to the end of a coal auger to enable a miner to press the auger forward using body pressure.

breast stoping A method of stoping employed on veins where the dip is not sufficient for the broken ore to be removed by gravity. The ore remains close to the working face and must be loaded into cars at that point. See also: *overhand stoping*.

breast timber A leaning brace from the floor of an excavation to a wall support. (Nichols, 1976)

breast wall A wall designed to withstand the force of a natural bank of earth, such as of timber used to support the face of a tunnel. Syn: *jamb wall*. (AGI, 1987)

breathing Alternate expansion and contraction of air in breaks that allows fresh oxygen to be drawn in and oxidation to proceed. (Sinclair, 1958)

breathing apparatus (a) A filter self-rescuer (FSR) is a respiratory protective device that filters ambient air of carbon monoxide, converting it to carbon dioxide. Exhaled air is vented back to ambient. Also called a gas mask. Duration of protection is limited, usually, by water contamination of the chemical bed. Used for escape from underground mines in the event of a fire or explosion. Requirement specified in 30 CFR 75.1714-2(e)(2). Apparatus must be certified as providing at least one hour of respiratory protection. Apparatus available in the United States are the Draeger 910 and the MSA W-65, both certified for one hour. (b) A self-contained self-rescuer (SCSR) is a respiratory protective device that provides breathing gas independent of the ambient atmosphere, containing its own oxygen source. Also called an oxygen self-rescuer. Air is exchanged between the user's lungs and a breathing bag. Oxygen consumed by the user is replaced by the apparatus from its oxygen source, stored either in chemical or compressed form. Carbon dioxide produced by the user is removed by a chemical absorbent in the apparatus breathing circuit. Duration of protection is determined by both quantity of stored oxygen and carbon dioxide absorption capacity. SCSRs differ from rescue breathing apparatus in that the breathing bag is not protected by a rigid cover in order to reduce size and weight. Used for escape from underground mines in the event of a fire or explosion. Requirement specified in 30 CFR 75.1714. Apparatus must be certified as providing at least one hour of respiratory protection. Apparatus available in the United States are the CSE SR-100, the Draeger OXY K plus, the MSA Portal-Pack, the Ocenco EBA 6.5, all certified for one hour. (c) A rescue breathing apparatus (RBA) is a respiratory protective device that provides breathing gas independent of the ambient atmosphere, containing its own oxygen source. Air is exchanged between the user's lungs and a breathing bag. Oxygen consumed by the user is replaced by the apparatus from its oxygen source, stored in compressed form in most apparatus. Carbon dioxide produced by the user is removed by a chemical absorbent in the apparatus breathing circuit. Duration of protection is determined by both quantity of stored oxygen and carbon dioxide absorption capacity. RBAs differ from self-contained self-rescuers in that the breathing bag is protected by a rigid cover in order to prevent accidental puncture or tear. Used for entry into underground mines after a fire or explosion in order to reestablish the ventilation system, rescue trapped miners, and put out fires. Requirement specified in 30 CFR 49.1. Apparatus must be certified as providing at least two hours of respiratory protection. Apparatus available in the United States are the Biomarine BioPak 240 and the Draeger BG-174A, both certified for four hours.

breathing cave (a) A cave in which air is alternately blown out and sucked in at the entrance. (Schieferdecker, 1959) (b) A narrow part in a passage through which air blows. (Schieferdecker, 1959)

breccia A coarse-grained clastic rock, composed of angular broken rock fragments held together by a mineral cement or in a fine-grained matrix; it differs from conglomerate in that the fragments have sharp edges and unworn corners. Breccia may originate as a result of talus accumulation, explosive igneous processes, collapse of rock material, or faulting. Etymol: Italian, broken stones, rubble. Syn: *rubblerock*. Adj: brecciated. Cf: *conglomerate; loose ground*. (AGI, 1987)

brecciated Converted into, characterized by, or resembling a breccia; esp. said of a rock structure marked by an accumulation of angular fragments, or of an ore texture showing mineral fragments without notable rounding. (AGI, 1987)

bredigite An orthorhombic mineral, $Ca_7Mg(SiO_4)_4$; pseudohexagonal. Cf: *larnite*.

breeching Mid. Drawing loaded trams downhill underground. Syn: *britching*.

breese *breeze*.

breeze (a) Coke of small size; the undersize remaining after separating the smallest size of graded coke. Also spelled breese. (BS, 1960) (b) The dust from coke or coal. (Mersereau, 1947) (c) An indefinite term that usually means clinker, but that may refer to coke breeze. (Taylor, 1965) (d) Scot. Fine or slack coal.

breeze concrete A concrete made of 3 parts coke breeze, 1 part sand, and 1 part portland cement. It has poor fire-resisting qualities but it is cheap and nails can be driven into it. (Nelson, 1965)

breeze oven (a) An oven for the manufacture of small coke. (Fay, 1920) (b) A furnace designed to consume breeze or coal dust. (Fay, 1920)

breithauptite Nickel antimonide, NiSb. See also: *niccolite*. (Fay, 1920)

brenston *brimstone*.

Breton pan Large steel mortar in which a heavy steel pestle rolls. Once used in grinding and amalgamation of gold ores. (Pryor, 1963)

brettice cloth A variation of brattice cloth. See also: *brattice cloth*.

brettis way Derb. A road in a coal mine, supported by brattices built on each side after the coal has been worked out. See also: *brattice*. (Fay, 1920)

breunnerite A ferroan variety of magnesite used in the manufacture of magnesia bricks.

brewsterite A monoclinic mineral, $(Sr,Ba,Ca)Al_2Si_6O_{16}{\cdot}5H_2O$; zeolite group.

brewsterlinite Liquid plus vapor CO_2 in cavities in minerals; e.g., quartz, topaz, and chrysoberyl. The meniscus vanishes under the warmth of the hand.

Brewster's law (a) The index of refraction of a crystalline substance is equal to the tangent of its angle of polarization. (Pryor, 1963) (b) In optics, where light is reflected from a smooth, transparent, nonconducting surface, it is plane polarized parallel to the reflecting surface. The angle of incidence for maximum polarization is that angle whose tangent is the index of refraction of the reflecting substance. This angle is also called the "Brewster angle."

brick clay An impure clay, containing iron and other ingredients. In industry the term is applied to any clay, loam, or earth suitable for the manufacture of bricks or coarse pottery. Syn: *brick earth*. (CTD, 1958)

brick coal Eng. Small, dirty coal suitable for brick kilns and similar purposes. (Fay, 1920)

brick earth Earth, clay, or loam suitable for making bricks; specif. a fine-grained brownish deposit consisting of quartz and flint sand mixed with ferruginous clay; found on river terraces as a result of reworking by water of windblown material, such as that overlying the gravels on certain terraces of the Thames River in England. Syn: *brick clay.* (AGI, 1987)

brick fuel In Wales, patent fuel. See also: *briquette.* (Fay, 1920)

bricking The walling or casing of a shaft.

bricking curb A curb set in a circular shaft to support the brick walling. See also: *curb.* (Nelson, 1965)

bricking scaffold A staging or platform suspended in a sinking shaft on which masons stand when building brick walling. Syn: *walling scaffold.* (Nelson, 1965)

brick walling A permanent support for circular shafts. On reaching the rockhead, a firm ledge is prepared to receive the first bricking curb or ring. The curb is fixed correctly with reference to the centerline of the shaft. The bricks are then built upwards from the curb, the space behind being firmly packed to the rock sides with bricks and mortar. Concrete is replacing brickwork as a shaft lining. See also: *lining; permanent shaft support.* (Nelson, 1965)

bridal Staff. A contrivance used in coal mining to prevent cars from overturning upon steep inclined planes having a rise of 1 ft in 3 ft or 4 ft (1 m in 3 m or 4 m). (Hess)

bride cake A black, highly carbonaceous slickensided shale with Carbonicola shells, in the Adwalton stone or Flockton thick coal; also, dirty smudgy coal in the roof of the Stanley Main in the Snydale-don Pedro area. Syn: *bright cake.* (Tomkeieff, 1954)

bridge (a) A rock fragment, cavings, or other obstruction that lodges (either accidentally or intentionally) part way down in a drill hole (such as in an oil well). (AGI, 1987) (b) Debris that plugs a borehole at a point above the bottom. Between the underside of the bridge and the bottom of the drill hole, the borehole is free of debris. (Long, 1960) (c) To deliberately plug a borehole at a point some distance above its bottom. (Long, 1960) (d) To form a bridge in a drill hole. (AGI, 1987) (e) In a cave, a solutional remnant of rock that spans a passage from wall to wall. (AGI, 1987) (f) A device to measure the resistance of a wire or other conductor forming a part of an electric circuit. (g) A piece of timber held above the cap of a set by blocks and used to facilitate the driving of spiling in soft or running ground. (h) Refers to the overburden used for spanning the natural gap between the highwall and the spoil, when such is required to establish a temporary machine surface standing area closer to the disposal area than that provided by the virgin ground. (AGI, 1987) (i) In an electric blasting cap, the wire that is heated by electric current so as to ignite the charge. (Nichols, 1976) (j) Sometimes, the shunt connection between the cap wires. (Nichols, 1976) (k) A plankway or elevator used in ironworking to convey fuel or ore to the mouth of a furnace. (Webster 3rd, 1966) (l) A refractory bar, or member, or fire clay placed across the surface of the batch in a tank furnace near the working end to hold back the scum, or gall. (Mersereau, 1947) (m) The structure formed by the end walls of the adjacent melter and refiner compartments of a tank and the covers spanning the gap between the end walls. (ASTM, 1994) (n) *air crossing.*

bridge break The time that elapses between the application of current and the fusion of the bridge wire when using instantaneous blasting caps. (Streefkerk, 1952)

bridge conveyor A conveyor that is supported at one end by a loading unit and at the other end in such a way as to permit changes in the position of either end without interrupting the operation of the loading unit. (NEMA, 1961)

bridged A borehole plugged by debris lodged at some point above the bottom of a hole. The hole may be bridged deliberately by introducing foreign material into the hole or accidentally by rock fragments sloughing off the sidewalls of the borehole. (Long, 1960)

bridge over Collapse of a well bore around the drill stem. (Williams, 1964)

bridge the hole Deliberate plugging of a borehole at a point some distance above the bottom by introduction of some type of foreign material or a plug. See also: *bridge.* (Long, 1960)

bridge tramway Consists of two steel bridge trusses braced together so as to form between them a runway on which a bucket-carrying trolley runs. (Pit and Quarry, 1960)

bridge wall A low separating wall, usually made of firebrick, in a furnace. (Webster 3rd, 1966)

bridgewire A resistance wire connecting the ends of the leg wire inside an electric detonator and which is embedded in the ignition charge of the detonator.

bridging (a) In crushing practice, the obstruction of the receiving opening by two or more pieces wedged together, each of which could easily pass through. (Nelson, 1965) (b) Formation of arches of keyed or jammed particles across the direction of flow (of rock through apertures or of small particles through filter pores). (Pryor, 1963) (c) Arching of the charge across the shaft in a blast furnace or cupola. (d) Premature solidification of metal across a mold section before the metal below or beyond solidifies. (ASM, 1961) (e) Solidification of slag within the cupola at or just above the tuyeres. (ASM, 1961) (f) Welding or mechanical locking of the charge in a downfeed melting or smelting furnace. (ASM, 1961) (g) Closing of a section of a drill hole by loose blocks of rock or by squeezing of plastic shale, etc.

Bridgman sampler A mechanical device that automatically selects two samples as the ore passes through.

bridle bar *bridle rod.*

bridle cable An anchor cable that is at right angles to the line of pull. (Nichols, 1976)

bridle chain (a) One of the chains used for supporting a cage from the winding rope. (Nelson, 1965) (b) One of the safety chains used to support the cage if the shackle should break or to protect a train of cars on a slope should the shackle or drawbar fail. (Fay, 1920)

bridle hitch A connection between a bridle cable and a cable or sheave block. (Nichols, 1976)

bridle rod A steel tie bar used to join the ends of two point rails to hold them to gage in the proper position. Syn: *bridle bar.* (Webster 3rd, 1966)

brier N. of Eng. A beam or girder fixed across a shaft top.

Briggs clinophone An instrument used in measuring borehole deviation which transmits electrical signals, communicating to the surface the position of a plumb bob fitted with a needle relative to four electrodes arranged N.,S.,E., and W.—the needle and electrodes being immersed in the electrolyte. Signals are matched with a similar arrangement of needle and electrodes at the surface, and the needle then indicates the deviation and the direction of deviation. (Sinclair, 1958)

Briggs equalizer This consists of a head harness, mouthpiece, and noseclip, corrugated breathing tube, Briggs equalizing device, 120 ft (36.58 m) of reinforced air tubes, and a strainer and spike. It has neither bellows nor rotary blower but depends entirely on the action of the equalizer for comfortable respiration. The resistance to breathing is so low that reasonably hard work can be done by the wearer over a period of 2 h or more. The air supply tube is attached to the waist by a strong leather body belt. (Mason, 1951)

bright attritus A field term to denote the degree of luster of attrital coal compared with the brilliant luster of associated vitrain. Cf: *dull attritus.* (AGI, 1987)

bright cake *bride cake.*

bright coal (a) A type of banded coal defined microscopically as consisting of more than 5% of anthraxylon and less than 20% of opaque matter; banded coal in which translucent matter predominates. Bright coal corresponds to the microlithotypes vitrite and clarite and in part to duroclarite and vitrinerite. Cf: *dull coal.* Syn: *brights.* (AGI, 1987) (b) The constituent of banded coal that is of a jet black, pitchy appearance, more compact than dull coal, and breaking with a conchoidal fracture when viewed macroscopically, and that in thin section always shows preserved cell structure of woody plant tissue, either of stem, branch, or root. Same as anthraxylon. (AGI, 1987) (c) A coal composed of anthraxylon and attritus, in which the translucent cell-wall degradation matter or translucent humic matter predominates. (AGI, 1987) (d) A type of banded coal containing from 100% to 81% pure bright ingredients (vitrain, clarain, and fusain), the remainder consisting of clarodurain and durain. (AGI, 1987)

bright head (a) York. A smooth parting or joint in coal; a plane of cleavage. (b) The principal cleat in coal. (Arkell, 1953)

brightness The candlepower of a light source divided by the area of the source, and expressed in candles per square inch or candles per square foot. (Sinclair)

brightness meter Visual-type portable photometer operated by visual comparison of brightness. So named because it can be calibrated to indicate the photometric brightness of the object viewed in the sighting telescope. (Roberts, 1958)

brights (a) Coal that reflects a large part of incident light, either in a definite beam or by scattering. Two kinds of bright coal are distinguished: vitrain, which reflects an incident beam in a definite direction and consequently appears light or dark according to whether the beam is or is not reflected into the eye; and clarain, which scatters the light and shows a silky luster at whatever angle it is viewed. (Tomkeieff, 1954) (b) A commercial term for the larger sizes of bright coal. (BS, 1960)

bright sulfur Crude sulfur free of discoloring impurities and bright yellow in color. (USBM, 1965)

brimstone A common name for sulfur. Syn: *brenston.* (Fay, 1920)

brine (a) Water saturated or strongly impregnated with common salt. (b) Sea water containing a higher concentration of dissolved salt than that of the ordinary ocean. Cf: *artificial brine.* (Hunt, 1965)

brine field A section of land under which quantities of rock salt or natural brine of usable strength have been discovered and a well, or any number of wells, has been bored for raising the brine. (Kaufmann, 1960)

Brinell hardness test A test for determining the hardness of a material by forcing a hard steel or carbide ball of specified diameter into it under a specified load. The result is expressed as the Brinell hardness number, which is the value obtained by dividing the applied load in kilograms by the surface area of the resulting impression in square millimeters. Cf: *Vickers hardness test.* (ASM, 1961)

Brinell hardness tester (a) In heat treating, one who determines the hardness of pieces of metal by the Brinell hardness test. Also called Brinell operator. (DOT, 1949) (b) The machine or instrument used to determine hardness.

brine pit A salt well, or an opening at the mouth of a salt spring, from which water is taken to be evaporated for making salt. (AGI, 1987)

brine well A cased drill hole penetrating a salt formation through which water is introduced and brine pumped to the surface.

bring in Can. Develop a mine from prospect stage. (Hoffman, 1958)

briquette A block of compressed coal dust, used as fuel; also, a slab or block of artificial stone. Syn: *brick fuel; coalette; eggette.* Also called boulet; carbonet. Also spelled briquet. See also: *solid smokeless fuel.* (Standard, 1964)

briquetting A process by which coke breeze, coal dust, iron ore, or any other pulverized mineral is bound together into briquettes, under pressure, with or without a binding agent such as asphalt, and thus made conveniently available for further processing or for commercial markets.

Briska detonator An aluminum tube containing a main charge of tetryl (tetranitromethylaniline). On top of this are initiating charges of lead azide and lead styphnate, which are more sensitive than the tetryl. A safety fuse fitted into an open space at the top is used to set off the detonator. Syn: *aluminum detonator.* (Higham, 1951)

Britannia cell In mineral processing, a pneumatic flotation cell 7 to 9 ft (2.1 to 2.7 m) deep. See also: *southwestern cell.* (Pryor, 1963)

britching Scot. *breeching.*

British equivalent temperature *equivalent temperature.*

British thermal unit Heat needed to raise 1 lb (0.45 kg) of water 1 °F (5/9 °C) (equal to 252 cal or 1,054 J). Symbol, Btu. Cf: *heat unit.* (Pryor, 1963)

brittle material A nonductile material that fails catastrophically under dynamic loading conditions. Ceramics are an example of a class of brittle materials. (Hunt, 1965)

brittle mica Group of micas having brittle laminae. Chief member is chloritoid, a basic silicate of aluminum, iron, and magnesium, $(Fe^{2+},Mg,Mn)_2Al_4Si_2O_{10}(OH)_4$. See also: *mica; margarite.* (Pryor, 1963)

brittleness (a) Of minerals, proneness to fracture under low stress. A quality affecting behavior during comminution of ore, whereby one species fractures more readily than others in the material being crushed. See also: *toughness.* (Pryor, 1963) (b) The quality of a material that leads to crack propagation without appreciable plastic deformation. (ASM, 1961)

brittle silver ore *stephanite.*

broach (a) To restore the diameter of a borehole by reaming. (Long, 1960) (b) To break down the walls between two contiguous drill holes. (Long, 1960) (c) A sharp-pointed chisel, used for rough dressing stone. (Webster 2nd, 1960) (d) The perpendicular grooves machined into the bit mold in which inside and outside gage stones are set. (Long, 1960)

broaching (a) Trimming or straightening a mine working. (b) A method of rock excavation employed where it is important that the adjacent rock formation should not be shattered by explosive. A line of closely spaced holes is drilled along the required line of breakage. The rock between the holes is knocked out with a broach and removed with the aid of wedges. See also: *channeler.* (Hammond, 1965) (c) Removing metal stock from a workpiece with a broach. (ASM, 1961)

broaching bit A tool used to restore the dimensions of a borehole that has been contracted by the swelling of the marl or clay walls; also used to break down the intervening rock between two contiguous drill holes. A reamer. (Fay, 1920)

broad coke oven A special design of oven, used mainly for coking certain grades of coal.

broadgate Eng. A main working.

broad lode Where two or more mining claims longitudinally bisect or divide the apex of a vein, the senior claim takes the entire width of the vein on its dip, if it is in other respects so located as to give the right to pursue the vein downward outside of the sideline. In other words, a broad lode bisected by the division sidelines between two mining claims belongs to the claim having the prior location. The term lode has become extensively used in the classification of ore deposits that are not comprehended by the definition of a vein. Such an occurrence is called a broad lode or zone. See also: *broad vein.* (Ricketts, 1943)

broadside shooting A type of refraction seismic shooting used to determine the structure across the strike. The broadside lines are ordinarily laid out in conjunction with the standard-type profiles that run along the strike. The shot points and detector spreads are laid out along parallel lines, which are generally across the strike. The distance between each line of shots and the receiving line is chosen so that it will always be greater than the double offset distance for the refractor being followed. Generally the distance should be only slightly greater so that the primary refracted event will be received as a second arrival. When this spacing is used, the refracting point associated with the shot will be very close to that associated with the detector, and each delay time will be approx. half the intercept time. A single depth point (based on half the intercept time) is then plotted midway between shot and receiver. All depth points are thus placed along the "control lines" that are located halfway between the shooting line and the receiving line. (Dobrin, 1960)

broadstone A paving slab, so called because it is raised broad and thin from the quarries, not more than 2 to 3 in (5.08 to 7.62 cm) thick.

broad vein Where a broad vein apexes so that the boundary line between two claims splits the apex, the extralateral rights go to the senior locator, who takes the entire width of the vein on the dip; i.e., a broad lode that is bisected by the division side line between two mining claims belongs to the claim having the prior location. See also: *broad lode.* (Lewis, 1964)

brochantite A monoclinic mineral, $Cu_4(SO_4)(OH)_6$; has one good cleavage; occurs in the oxidation zone of copper sulfide deposits; a source of copper. Formerly called blanchardite, kamarezite. Syn: *waringtonite.*

brockite A hexagonal mineral, $(Ca,Th,Ce)(PO_4)\cdot H_2O$; rhabdophane group.

bröggerite A thorian variety of uraninite. Also spelled broggerite.

broggite (a) A variety of asphalt from Peru. (Tomkeieff, 1954) (b) A variety of anthraxolite. (Crosby, 1955)

broil An old Cornish mining term referring to a collection of loose rock fragments usually discolored by oxidation, and indicating the presence of a mineral vein beneath the outcrop or gossan. Also spelled bryle; broyl.

broken charge A charge of explosive in a drill hole divided into two or more parts that are separated by stemming. (Fay, 1920)

broken coal In anthracite only; coal that is small enough to pass through a 3-3/8- to 4-in (8.57- to 10.16-cm) square aperture, but too large to pass through a 2-3/4-in or 2-1/2-in (6.99-cm or 6.35-cm) mesh. Smaller than steamboat, and larger than egg coal. See also: *anthracite coal sizes.*

broken ground (a) A shattered rock formation or a formation criss-crossed with numerous, closely spaced, uncemented joints and cracks. Cf: *loose ground; breccia.* (Long, 1960) (b) Rock or mineral formations fragmented by blasting with explosives, such as the broken material in a shrinkage stope. Syn: *broken rock.* (Long, 1960)

broken rock *broken ground.*

brokens (a) Eng. The removal or extraction of pillars previously formed in bord and pillar working. In Durham and Northumberland, the terms robbery and robbing pillars imply incomplete extraction of the pillars. (SMRB, 1930) (b) Robbery. (c) Robbing pillars.

broken stone (a) A diamond that has been shattered in use, or lost a portion of its size by cleaving. (Long, 1960) (b) *crushed stone.*

broken working The working away or removal of blocks or pillars of coal formed by whole workings. See also: *working the broken.* (Peel, 1921)

bromargyrite An isometric mineral, 4[AgBr]; yellow, in surface oxidation deposits of silver ores in arid climates. Formerly called bromyrite. Cf: *iodargyrite.*

bromellite A beryllium oxide with dihexagonal-pyramidal crystals from Langban, Sweden. *beryllium oxide.*

bromine A member of the halogen group of elements and the only liquid nonmetallic element. A heavy, reddish-brown liquid that volatilizes readily at room temperature to a poisonous, red vapor with a strong disagreeable odor. Symbol, Br. Obtained from natural brines from wells; little bromine is extracted today from seawater. Used in antiknock gasoline, fumigants, flameproofing agents, water purification compounds, dyes, medicinals, and sanitizers. (Handbook of Chem. & Phys., 3)

bromite An alternate spelling of bromyrite. See also: *bromargyrite.*

bromlite A former name for alstonite.

bromocyanide process Recovering values from refractory or special gold ores, in which cyanogen bromide (CNBr), or a chemical mixture forming it, is used for treating the ore. (Bennett, 1962)

bromoform A colorless, heavy liquid; $CHBr_3$; odor and taste similar to those of chloroform; sp gr, 2.8887. Used in mineralogic analysis and in assaying. Cf: *methylene iodide; Clerici solution; Sonstadt solution.* Syn: *methenyl tribromide; tribromomethane.* (CCD, 1961)

bromyrite A former name for bromargyrite.

brongniardite A lead-silver sulfantimonide with 26.2% silver, some of which is apparently diaphorite and some canfieldite. Also spelled brongniartite.

bronze An alloy composed mainly of copper and tin. Various other elements may be added in small amounts for certain specific purposes. A number of copper alloys are referred to as bronzes, although they contain no tin. The American Society for Testing and Materials has classified all copper-based alloys on a basis of composition ranges of the principal alloying elements. (Henderson, 1953)

bronze mica *phlogopite.*

bronzite (a) A mineral consisting of a ferriferous variety of enstatite, often having a luster like that of bronze; $(Mg,Fe^{+2})_2Si_2O_6$; orthorhombic. (Webster 3rd, 1966; Dana, 1959) (b) It is often used as a prefix to the names of rocks containing the mineral. Rocks of the gabbro family are the most common ones having the prefix. (Fay, 1920) (c) A name for an orthopyroxene between enstatite and hypersthene in composition; brown or green; commonly has a bronzelike or pearly metallic luster.

bronzitite A pyroxenite composed almost entirely of bronzite. (AGI, 1987)

brood (a) Impurities as extracted with ore. (Nelson, 1965) (b) Corn. The heavier kinds of waste in tin and copper ores. A mixture of tin and copper ore.

Brookfield viscometer An electrically operated, rotating-cylinder viscometer in which the drag is recorded directly on a dial; it has been used in the testing of vitreous-enamel slips. (Dodd, 1964)

brookite An orthorhombic mineral, $8[TiO_2]$; trimorphous with anatase and rutile; a common accessory mineral in igneous and metamorphic rocks, and placers. *titanium dioxide.*

brooming The crushing and spreading of the head of a timber pile not fitted with a driving band when driven into hard ground. (Hammond, 1965)

brouse Derb. A sort of coarse stopping, made of small boughs of trees, and placed in back of shaft timbers to prevent rock from falling.

brow (a) Lanc. An underground roadway leading to a working place, driven either to the rise or to the dip. (b) A low place in the roof of a mine, giving insufficient headroom. (c) A fault plane. (Arkell, 1953) (d) Top of a mine shaft. Also called pit brow. (Pryor, 1963) (e) The projecting upper part or margin of a steep slope just below the crest; the edge of the top of a hill or mountain, or the place at which a gentle slope becomes abrupt. See also: *apex.*

brow bar Mid. A massive curb or beam of timber fixed in the wall of the shaft across the top of an inset or station. Also called browpiece.

brow bin An ore bin made by cutting away the floor of the station close to the shaft. (Higham, 1951)

Brown agitator *Pachuca tank.*

brown clay (a) York. Hessle boulder clay. (Arkell, 1953) (b) *red clay.*

brown coal (a) A low-rank coal which is brown or brownish-black, but rarely black. It commonly retains the structures of the original wood. It is high in moisture, low in heat value, and checks badly upon drying. (AGI, 1987) (b) A light-brown to seal-brown substance intermediate between peat and bituminous coal; usually regarded as a variety of lignite, other varieties being darker or black. It may be distinguished from peat by three rough criteria: (1) many tissues and fibers can be recognized in peat, but only a few fibers or none in brown coal; (2) water can be squeezed out of fresh peat by manual pressure, but not from brown coal; and (3) peat can be cut, but brown coal cannot. Actually, there is no sharp distinction between peat and coal. Some have attempted to assign it a higher rank by defining lignite as containing at least 20% water, brown coal between 10% and 20% water, and bituminous coal less than 10% water. (Hess) (c) A type of low-rank coal intermediate between bituminous coal and peat, and comparatively high in water content. In English-speaking countries, the terms "brown coal" and "lignite" are synonymous; in Germany and other parts of Europe, brown coal is restricted to megascopically compact structural varieties, and lignite is restricted to individual pieces of wood enclosed in brown coal. It may be subdivided into low-grade brown coal, consisting of visible vegetable remains, and high-grade brown coal, a compact, homogeneous, and tough rock. Syn: *braunkohle.* (Tomkeieff, 1954) (d) Coal of the lowest rank, soft and friable, and having a high inherent moisture content. (BS, 1960) (e) Unconsolidated lignitic coal having less than 8,300 Btu (8.76 MJ), (moist, mineral-matter-free).

brown-coal gel *dopplerite.*

brown gummite *clarkeite.*

brown hematite A misnomer; the mineral bearing this name is limonite, a hydrous mixture of minerals, whereas true hematite is an anhydrous oxide mineral. Syn: *limonite; brown iron ore.* Cf: *red hematite.*

Brown horseshoe furnace An annular turret-type furnace for calcining sulfide ores. (Fay, 1920)

brown iron ore (a) Its approximate formula is $2Fe_2O_3{\cdot}3H_2O$, equivalent to about 59.8% iron. Probably a mixture of hydrous oxides. See also: *bog ore.* (Sanford, 1914) (b) *limonite; brown hematite.*

brown ironstone clay *limonite.*

brown matter Brown matter is found in varying amounts in the attrital matter of all splint and semisplint coals; it is occasionally present in the attritus of bright coals. It consists of cell-wall degradation matter and the contents of cells, which in thin sections are brown and semitranslucent. The term has no exact equivalent in the Stopes-Heerlen nomenclature. Constituents with a reflectance between that of vitrinite and fusinite may correspond in part to brown matter. Some brown matter is identical with semifusinite and massive micrinite. (IHCP, 1963)

brown mica *phlogopite.*

brown millerite *celite.*

brownmillerite A mineral, $Ca_2(Al,Fe)_2O_5$; a constituent of Portland cement. Syn: *celite.*

Brown muffle furnace A mechanically raked, roasting, straight-line-type furnace with a series of longitudinal combustion flues placed under the hearth. (Fay, 1920)

brown ocher *limonite.*

Brown-O'Hara furnace A long, horizontal, double-hearth furnace for the treatment of lead ores. (Fay, 1920)

Brown panel system (a) *pillar-and-breast.* (b) Coal mining by long rooms opened on the upper side of the gangway. The breasts are usually 5 to 12 yd (4.6 to 11.0 m) wide and are separated by pillars (solid walls of coal broken by crossheadings for ventilation) 5 to 12 yd thick. The pillars are robbed by mining from them until the roof comes down and prevents further working. Syn: *block system of stoping and filling.* (Hess)

brown rock Tenn. Dark brown to black phosphorite resulting from the weathering of phosphatic limestone. See also: *phosphorite.* (AGI, 1987)

brown spar Any light-colored crystalline carbonate mineral that is colored brown by the presence of iron; e.g., ankerite, dolomite, magnesite, or siderite. (AGI, 1987)

brownstone A brown or reddish-brown sandstone with grains generally coated with iron oxide; specif. a dark, reddish-brown, ferruginous quartz sandstone of Triassic age.

Brown tank A cylindrical tank or vat, tall in proportion to its diameter, with the bottom ending in a 60° cone. Within the tank is a hollow column extending from the bottom to within about 8 in (20 cm) from the top. The apparatus works on the airlift principle, the aerated pulp in the tube flowing upward and discharging at the top, while more pulp flows in at the bottom to take its place. Syn: *Brown agitator; Pachuca tank.* (Liddell, 1918)

brown tongs A long-handled, plierlike device similar to a certain type of blacksmith tongs used to handle wash or drill rods in place of a safety clamp in shallow borehole drilling. Also called adjustable pipe tongs; extension tongs; lowering tongs. (Long, 1960)

brown umber A brown earthy variety of limonite. See also: *limonite.* (Fay, 1920)

browpiece A heavy, upright timber used for underpinning in opening a station for a level in a mine. See also: *brow bar.* (Webster 3rd, 1966)

browse Ore imperfectly smelted, mixed with cinder and clay. (Fay, 1920)

brow-up Lanc. An inclined roadway driven to the rise. Also called brow; up-brow.

brucite (a) A trigonal mineral, $Mg(OH)_2$; brucite group; perfect basal cleavage; an alteration product of periclase in contact-metamorphosed limestone; a magnesia refractory raw material. (b) The mineral group amakinite, brucite, pyrochroite, and theophrastite.

Bruckner cylinder Pac. A form of revolving roasting furnace. See also: *Bruckner furnace.* (Fay, 1920)

Bruckner furnace Horizontal cylindrical furnace revolving on end trunnions. See also: *Bruckner cylinder.* (Pryor, 1963)

brulee A Canadian term used to describe a windfall of dead trees and brush. Syn: *slash.*

Brunauer, Emmett, and Teller method A procedure for the determination of the total surface area of a powder or of a porous solid by measurement of the volume of gas (usually N_2) adsorbed on the surface of a known weight of the sample. The mathematical basis of the method was developed by S. Brunauer, P. H. Emmett, and E. Teller—hence the usual name, B.E.T. method. (Dodd, 1964)

Brunton compass A compact pocket instrument that consists of an ordinary compass, folding open sights, a mirror, and a rectangular spirit-level clinometer, which can be used in the hand or on a staff or light rod for reading horizontal and vertical angles, for leveling, and for reading the magnetic bearing of a line. It is used in sketching mine workings, and in preliminary topographic and geologic surveys on the surface, e.g., in determining elevations, stratigraphic thickness, and strike and dip. Syn: *pocket transit.* (AGI, 1987)

Brunton sampler A mechanical sampling device that automatically selects 1/625 part of the ore passing through the sampler, by means of an oscillating deflector placed in a falling stream of the ore. (Fay, 1920; Pryor, 1963)

brush (a) To remove rock from the roof or floor of an opening to increase the height of working (coal mines). See also: *brushing.* (b) In a coal mine, a road through the goaf, gob, or worked-out areas packed with waste. (CTD, 1958) (c) To clean up fine coal from the floor. (CTD, 1958) (d) Forest of Dean. A rich brown hematite. (e) Mixed load of large and small coal into a colliery tub. (Pryor, 1963) (f) Mid. To mix gas with air in a mine by buffeting it with a jacket. (g) To rip;

to enlarge. (h) To remove bisque in a definite pattern by means of a brush. See also: *bolt-hole brush.*

brush cleaner A device consisting of bristles set in a suitable backing used for cleaning a conveyor belt. It is usually of the rotary type.

brush discharge In high-intensity electrical fields, discharge from sharp points along a conductor. Electricity concentrates at these points and charges ambient molecules of air, which are then repelled, carrying away charge. The phenomenon is exploited in mineral processing in high-tension separation. (Pryor, 1963)

brush hook A short, stout, heavy hooked blade with a sharpened iron edge, attached to an axe handle; typically used by surveyors for cutting brush. (AGI, 1987)

brushing (a) Scot. That part of the roof or floor of a seam removed to form roadways. (b) Digging up the bottom or taking down the top of an entry or room, where the seam of coal is too thin or shallow for the purpose of admitting cars. See also: *brush.* (c) Cutting or blasting down the roof of a coal seam. (Arkell, 1953) (d) Ripping; normally enlarging a road by taking down the roof, but extended to sides and floor as well. Also called canch. (Mason, 1951) (e) Removal of dry enamel by brushing through a stencil or along an edge to produce a design or edging. (Bryant, 1953)

brushing bed Scot. The stratum brushed or rippled. See also: *brush.*

brushing shot (a) A charge fired in the air of a mine to blow out obnoxious gases or to start an air current. (b) A shot so placed as to remove a portion of the roof to increase the height of a haulageway. See also: *brush.*

brushite A monoclinic mineral, $CaHPO_4 \cdot 2H_2O$; generally massive or in slender crystals. Syn: *metabrushite.*

brush rake A rake blade having a high top and light construction. (Nichols, 1976)

brush treatment A method of treating mine timber in which the timber is painted with a preservative or is merely dipped into a tank of preservative. Preservatives used are creosote, zinc chloride, sodium fluoride, and other chemicals. See also: *timber preservation.* (Lewis, 1964)

bruskin Mid. Lump of coal weighing about 1 lb.

bryle *broil.*

bubble chamber A device that marks the paths of charged particles by photographing the train of bubbles they produce as they move through certain superheated liquids. See also: *cloud chamber; spark chamber.* (Lyman, 1964)

bubble pickup Method of testing small grains of minerals to ascertain their response to flotation collector agents. A bubble of air is pressed down on particles under water, and then raised and examined to find whether it has lifted any grains. This is often done by using a single bubble device for determining fundamental aspects of the mineral-bubble interaction.

bubble pipe Tube inserted in pulp at regulated depth, through which compressed air is gently bubbled. The air pressure indicates the pulp density and provides a means of control. (Pryor, 1965)

bubble pulse A pulsation attributable to the bubble produced by a seismic charge fired in deep water. The bubble pulsates several times with a period proportional to the cube root of the charge, each oscillation producing an identical unwanted seismic effect. (AGI, 1987)

bubbles Air introduced near the bottom of a flotation cell containing pulped ore forms coursing bubbles, which rise through the liquid and emerge as mineralized bubbles forming a semistable froth column. This depends for its continuity partly on the surface-active reagents borne by the mineral in the air-water interphase of each bubble and partly on the aid of frothing reagents. (Pryor, 1958)

bucaramangite A resin resembling amber but insoluble in alcohol and yielding no succinic acid. (Fay, 1920)

Buchner funnel A porcelain filter shaped to support filter paper on a flat perforated disk. (Pryor, 1963)

buck (a) A large quartz reef in which there is little or no gold. See also: *bull quartz.* (b) To push coal down a chute toward a mine car. (Zern, 1928) (c) To break up or pulverize, as to buck ore samples. (Webster 3rd, 1966) (d) To bring or carry, as to buck water. (Webster 3rd, 1966)

bucker helper One who breaks ore.

bucket (a) *bailer; calyx.* (b) Tubular container equipped with auger or other-type cutting edges used to make borings in earthy or soft formation by rotary methods. (Long, 1960) (c) An open-top can, equipped with a bail, used to hoist broken rock or water and to lower supplies and equipment to workers in a mine shaft or other underground opening. (Long, 1960) (d) One of the conveying units on a bucket conveyor that lifts the material from a boot or bin when passing over the lower sprocket and is dumped on passing over the upper sprocket. The bucket is often made of perforated metal so that water entrapped will pass through the perforations and back to the boot. (Zern, 1928) (e) A part of an excavator that digs, lifts, and carries dirt. (Nichols, 1976) (f) The dipper or scoop at the end of the arm of a bucket dredge. (Webster 3rd, 1966)

bucket auger A short helical auger incorporating a steel tube to help hold the cuttings on the auger during withdrawal from the drill hole. See also: *auger.* (Long, 1960)

bucket conveyor A conveyor consisting of a continuous line of buckets attached by pivots to two endless roller chains running on tracks and driven by sprockets. The buckets are so pivoted that they remain in an upright position at all times except when tilted into a dumping position by a cam or other device placed at any required position on the track. See also: *bucket elevator; gravity-discharge conveyor elevator; pivoted-bucket conveyor.* (BS, 1962)

bucket dredge A dredge having two pontoons, between which passes a chain of digging buckets. These buckets excavate material at the bottom of the pond (paddock) in which the dredge floats, and deposit it in concentrating devices on the decks. (Pryor, 1963)

bucket drill Originally developed as an aid in making excavations for cesspools and septic tanks; now used mostly in drilling holes for concrete piers on construction jobs. Also called bucket drilling.

bucket elevator (a) An appliance for elevating material, consisting of steel buckets fastened to an endless belt or chain. It is usually set at steep angles, around 70°. The load is picked up by discharge from a chute or by a dredging action in a boot. Its best application is in a plant where space is restricted and the size of the material is less than 2 in (5.1 cm). Syn: *chain elevator.* (Nelson, 1965) (b) See also: *centrifugal discharge bucket elevator; continuous-bucket elevator; double-leg bucket elevator; elevator; gravity-discharge conveyor elevator; internal-discharge bucket elevator; positive-discharge bucket elevator; pivoted-bucket conveyor; grit collector.*

bucket elevator belt A belt fabricated for bucket elevator use, to which an elevator bucket is attached.

bucket factor *fill factor.*

bucket gate *bin gate.*

bucket-ladder dredge (a) A dredge with a digging mechanism consisting of a ladderlike truss on the periphery of which is attached an endless chain that rides on sprocket wheels and on which buckets are attached. (b) A mechanical dredge that uses a chain of heavy buckets rotating over the dredging arm or ladder to excavate and lift material to the dredging platform. Syn: *bucket-line dredge; ladder-bucket dredge.* See also: *dredger.* (Cruickshank, 1987)

bucket-ladder excavator A mechanical excavator working on the same principle as a bucket-ladder dredge, but adapted for use on land. See also: *trench excavator.* (CTD, 1958)

bucket lift The discharge pipe of a lifting pump in a mine. (Standard, 1964)

bucket line An endless line of digging buckets on a dredger or on a bucket elevator. (Pryor, 1963)

bucket-line dredge *bucket-ladder dredge.*

bucket loader (a) A form of portable, self-feeding, inclined bucket elevator for loading bulk materials into cars, trucks, or other conveyors. See also: *bucket elevator; portable conveyor.* (b) A machine having a digging and gathering rotor and a set of chain-mounted buckets to elevate the material to a dumping point. (Nichols, 1976)

bucket pump (a) An iron or wooden receptacle for hoisting ore, or for raising rock in shaft sinking. (Fay, 1920) (b) A reciprocating lift pump formerly much used in shafts and sinkings. (Nelson, 1965)

bucket rig *rotary bucket drill.*

bucket sheave A pulley attached to a shovel bucket, through which the hoist or drag cable is reeved. (Nichols, 1954)

bucket temperature The surface temperature of the sea as measured by a bucket thermometer or by immersing a surface thermometer in a freshly drawn bucket of water. (Hunt, 1965)

bucket thermometer A water-temperature thermometer provided with an insulated container around the bulb. It is lowered into the sea on a line until it has had time to reach the temperature of the surface water, then withdrawn and read. The insulated water surrounding the bulb preserves the temperature reading and is available as a salinity sample. (Hunt, 1965)

bucket tripper A device that tilts or turns the buckets of a pivoted bucket conveyor, causing them to discharge. It may be fixed or movable.

bucket-wheel dredge An hydraulic cutter dredge that uses a bucket wheel excavator in place of the traditional rotary cutter. The bucket wheel is characterized by its high cutting torque in both directions and by a positive feed of the excavated material into the mouth of the dredge pipe. (Cruickshank, 1987)

bucket-wheel excavator A continuous digging machine originally designed and used in large-scale stripping and mining of brown coal deposits in eastern Germany. Its digging mechanism is essentially a boom on which is mounted a rotating vertical wheel having buckets on its periphery. As the rotating wheel is pressed into the material to be dug, the buckets cut, gather, and discharge the material onto a conveyor belt where it is moved to the mined materials transport system.

bucking hammer A rectangular piece of cast iron 5 to 6 in (12.7 to 15.24 cm) across, usually rounded fore and aft with an eye on the back and with a wooden handle; used for grinding ore on a cast-iron bucking board. (Hess)

bucking iron An iron plate on which ore is ground by hand by means of a bucking hammer. Used extensively for the final reduction of ore samples for assaying. See also: *muller.* (Barger, 1939)

bucklandite (a) A black variety of epidote containing iron and having nearly symmetrical crystals. (b) A former name for allanite.

buckle (a) A bend in a piece of drill-stem equipment induced by excessive feed pressure. (Long, 1960) (b) Deformation of component members of a drill derrick, tripod, or mast, caused by attempting to hoist too heavy a load or by applying excessive strain when pulling on stuck casing, etc. (Long, 1960)

buckling length The length of drill rod that will withstand flexure or bending when subjected to a specific feed pressure or compressional load. (Long, 1960)

buckling load The maximum load that can be imposed on a string of drill rods, casing, or pipe, or on a drill tripod, derrick, or mast without the string buckling; also, a part being bent or buckled. (Long, 1960)

buck quartz *bull quartz.*

buck reef A barren vein. (Hess)

buckshot cinder Cinder from an iron blast furnace, containing grains of iron. (Fay, 1920)

buck up (a) To screw two threaded members, such as drill rods, together tightly. (Long, 1960) (b) To shore up with lagging; to brace. (Long, 1960)

buckwheat Coal size designation, used for anthracite only. Buckwheat is divided into four sizes: No. 1, or buckwheat; No. 2, or rice; No. 3, or barley; No. 4, or barley No. 2 or silt (sometimes also called culm or slush). Buckwheat No. 1 passes through a 1/2-in (12.7-mm) woven wire screen and over a 5/16-in (7.9-mm) woven wire screen, and through a 9/16-in (14.3-mm) round punched plate and over a 3/8-in (9.5-mm) round punched plate. The American Institute of Mechanical Engineers has recommended that buckwheat No. 1 shall pass through 9/16-in holes and over 5/16-in holes, a screen with circular holes being used. See also: *anthracite coal sizes.* (Fay, 1920)

buddle (a) Circular arrangement in which finely divided ore, in water, is delivered from a central point and flows gently to the perimeter. The heaviest and coarsest particles settle, while the lightest overflow. Several variants include concave buddle, with peripheral feed and central discharge; and continuous buddles, as differentiated from those that are periodically stopped and cleaned up. (Pryor, 1963) (b) To separate ore from slime or stamp work by means of a buddle. (Standard, 1964)

Buddy A shortwall coal cutter designed for light duty on longwall power-loaded faces and for subsidiary developments. (Mason, 1951)

buddy system In scuba diving, divers with few exceptions should work in pairs. This is probably the greatest single aid toward scuba safety, esp. under unfavorable conditions. The divers should remain in sight of each other. In poor visibility, they should use a buddy line 6 to 10 ft (1.8 to 3.0 m) long. (Hunt, 1965)

buffer (a) A pile of blasted rock left against or near a face to improve fragmentation and reduce scattering from the next blast. (Nichols, 1976) (b) A substance whose purpose is to maintain a constant hydrogen-ion concentration in water solutions, even when acid or alkalies are added. (ASM, 1961)

buffer shooting *blanket shooting.*

bug (a) A bullet or go-devil. See also: *bullet.* (Long, 1960) (b) Syn. for vug and bug hole (slang). (Long, 1960)

bug dust (a) The fine coal or other material resulting from a boring or cutting of a drill, a mining machine, or even a pick. (b) Fine, dry, dustlike particles of rock ejected from a borehole by a current of pressurized air when compressed air, instead of a liquid, is used as a cuttings removal agent. Syn: *cuttings.* See also: *nickings.* (Long, 1960) (c) Fine coal or rock material resulting from dry boring, drilling, or the use of other cutting machines in underground work places. (Long, 1960)

bug dusting Removing bug dust from an undercut. (BCI, 1947)

buggied Pennsylvania. Said of coal moved underground in a small car. (Hess)

buggy (a) A four-wheeled steel car used for hauling coal to and from chutes. (b) A mine car of small dimensions, sometimes used in thin beds. (Hudson, 1932) (c) Slang for a shuttle car. (BCI, 1947) (d) *bug dust.*

bug hole *vug.*

bug light Slang for a miner's electric cap lamp. (BCI, 1947)

buhrmill (a) A stone disk mill, with an upper horizontal disk rotating above a fixed lower one. Grist is fed centrally and discharged peripherally. Stones are dressed periodically, channels being cut to facilitate passage. Also applied to other rubbing mills; e.g., conical porcelain or steel ones in which a grooved cone rotates in a close fit in a fixed casing. Also spelled: burrmill. (Pryor, 1963) (b) A stone mill, consisting of one stationary stone and one revolving stone, for grinding pigment pastes. (Bennett, 1962)

buhrstone (a) A siliceous rock suitable for use as millstones; e.g., an open-textured, porous but tough, fine-grained sandstone, or a silicified fossiliferous limestone. In some sandstones, the cement is calcareous. Syn: *millstone.* (AGI, 1987) (b) A millstone cut from buhrstone. Also spelled burrstone; burstone. Syn: *burr.* (AGI, 1987)

buhrstone mill A grinding mill with two horizontal circular stones, one revolving upon the other, such as in an old-fashioned grain mill. (Mersereau, 1947)

building stone A general, nongeneric term for any rock suitable for use in construction. Whether igneous, metamorphic, or sedimentary, a building stone is chosen for its properties of durability, attractiveness, and economy. See also: *dimension stone.* (AGI, 1987)

bulk Brist. Run-of-mine coal in large quantities.

bulk density (a) The weight of an object or material divided by its volume, including the volume of its pore spaces; specif. the weight per unit volume of a soil mass that has been oven-dried to a constant weight at 105 °C. Cf: *apparent density.* (AGI, 1987) (b) The ratio of the weight of a collection of discrete particles to the volume that it occupies. (BS, 1962) (c) The weight of a material, on being compacted in a defined way, per unit volume (including voids). (Taylor, 1965) (d) The weight per unit volume of any material, including water; the weight in pounds per cubic foot (kg/m^3). See also: *apparent density; density; loading weight.* (Nelson, 1965)

bulk explosives Explosives not individually packaged in a form usable in the field. Includes ammonium nitrate-fuel oil, slurries, water gels, and other similar blasting agents; often loaded directly into blast holes from a bulk delivery truck. (Federal Mine Safety, 1977)

bulkhead (a) A watertight dam containing some form of door or removable plate. See also: *dam.* (BS, 1963) (b) A tight partition of wood, rock, and mud or concrete in mines for protection against gas, fire, and water. (c) A masonry diaphragm built across a subaqueous tunnel, where compressed air is used as a precaution, and to prevent the flooding of an entire tunnel in case of an accident. It is usually kept some distance in the rear of the working face and is provided with two air locks; one of them is an emergency lock near the roof. (Stauffer, 1906) (d) A stone, steel, wood, or concrete wall-like structure primarily designed to resist earth or water pressure, such as a retaining wall holding back the ground from sliding into a channel, or a partition preventing water from entering a working area in a mine. (AGI, 1987) (e) A timber chock in metal mines. (Nelson, 1965) (f) The end of a flume, whence water is carried in iron pipes to hydraulic workings. (g) A solid crib used to support a very heavy roof. See also: *cog; chock.* (h) A panel of brick of lesser cross-sectional thickness built into a wall for ease of replacement or for entrance to the walled chamber. (AISI, 1949)

bulking (a) The increase in volume of a material due to manipulation. Rock bulks upon being excavated; damp sand bulks if loosely deposited, such as by dumping, because the apparent cohesion prevents movement of the soil particles to form a reduced volume. (ASCE, 1958) (b) The difference in volume of a given mass of sand or other fine material in moist and dry conditions; it is expressed as a percentage of the volume in a dry condition. (Taylor, 1965)

bulking agent Chemically inert materials for increasing the volume of a composition; e.g., clay. Also called a filler. (Bennett, 1962)

bulk mining A method of mining in which large quantities of low-grade ore are mined without attempt to segregate the high-grade portions. Cf: *selective mining.* (Newton, 1959)

bulk mix A mass of explosive material prepared for use in bulk form without packaging. (Atlas, 1987)

bulk modulus The number that expresses a material's resistance to elastic changes in volume; e.g., the number of pounds per square inch necessary to cause a specified change in volume. See also: *modulus of elasticity; modulus of rigidity.* (Leet, 1958)

bulk modulus of elasticity The ratio of a tensile or compressive stress, triaxial and equal in all directions (e.g., hydrostatic pressure), to the relative change it produces in volume. (Roark, 1954)

bulk oil flotation (a) A flotation process in which large amounts of oil are used. (b) In this process the separation of mineral from gangue is accomplished by virtue of the fact that minerals of metallic luster, such as sulfides, or hydrocarbons, such as coal and graphite, are wetted preferentially by oil in the presence of water and consequently pass into the interface between oil and water, while gangue or rock is wetted by water and remains in the medium. See also: *flotation.* (Mitchell, 1950)

bulk oil separation A concentration process based on selective wetting of minerals by oil in the presence of water and in the absence of air. (Kirk, 1947)

bulk pit excavation Primarily excavation of considerable length as well as of substantial volume or bulk that must be hauled from the site of operations. Also called enbankment digging. (Carson, 1961)

bulk sample One of the large samples of a few hundredweight or more taken at regular, though widely spaced, intervals. In the case of coal, a car load may be taken at intervals for size analysis and dirt content. (Nelson, 1965)

bulk sampling The taking of large samples, which may consist of large-diameter drill core, the contents of a trench or mine working, or a car or train load of ore material, for metallurgical testing in mine evaluation. (Peters, 1987)

bulk specific gravity Ratio of the weight in air of a given volume of permeable material (including both permeable and impermeable voids normal to the material) at a stated temperature to the weight in air of an equal volume of distilled water at a stated temperature. Also called specific mass gravity. (ASCE, 1958)

bulk strength The strength per unit volume of an explosive calculated from its weight strength and density. See also: *cartridge strength; absolute bulk strength.*

bulk volume A term used relative to the density and volume of a porous solid, such as a refractory brick. It is defined as the volume of the solid material plus the volume of the sealed and open pores present. (Dodd, 1964)

bulk wide-area excavation In this kind of excavation, there is complete access to the site from many directions, and the excavation banks can be sloped flatly on two or more sides. Usually shallower in depth than bulk pit excavations but larger in area. Cf: *bulk pit excavation.* (Carson, 1961)

bull (a) An iron rod used in ramming clay to line a shothole. See also: *clay iron.* (Stauffer, 1906) (b) Aust. *drag; backstay.* (c) N.S.W. To enlarge the bottom of a drilled hole to increase the explosive charge. (New South Wales, 1958)

Bullard Dunn Process Electrolytic method of descaling iron and steel and coating the surface with a protective layer of tin. (Pryor, 1963)

bull clam A bulldozer fitted with a curved bowl hinged to the top of the front of the blade. (Nichols, 1976)

bulldog (a) A type of drill-rod-foot safety clamp built somewhat like a spider and slips, but differing by having the slips or movable jaws attached to, and actuated by, a foot-operated lever. (Long, 1960) (b) A general term applied to rod and/or casing safety clamps having both fixed and movable serrated jaws that contact and securely grip the rods or casing. (Long, 1960) (c) A fishing tool consisting of a steel body, tapered at the top, on which slide two or more wedge-shaped, serrated-face segments. Lowered into a tubular piece of lost equipment, such as casing, the serrated segments are pushed upward toward the narrow part of the body; when the tool is raised, the segments are forced outward, securely gripping the lost equipment. Also called bulldog spear; casing dog; casing spear. (Long, 1960) (d) To pull or move a drill machine or auxiliary equipment by means of a block and tackle or by power derived from a rope used on the drill cathead or hoist drum. Also called cat; snake. (Long, 1960)

bulldoze (a) To level or excavate an earth surface by means of a heavy, adjustable steel blade attached to the front end of a tractor or a wheeled vehicle. (Long, 1960) (b) To reduce broken rock by the use of explosives to a size handy for raising to the surface. See also: *adobe charge; mudcap; secondary blasting.*

bulldozer (a) In nonmetal mining, a laborer who breaks up large stones with a sledge hammer or pneumatic drill so they will pass through grizzly (grating) in a limestone mine. (DOT, 1949) (b) A horizontal machine, usually mechanical, having two bull gears with eccentric pins, two connecting links to a ram, and dies to perform bending, forming, and punching of narrow plate and bars. (ASM, 1961) (c) A cleaning blade that follows the wheel or ladder of a ditching machine. (Nichols, 1976) (d) A tractor on the front end of which is mounted a vertically curved steel blade held at a fixed distance by arms secured on a pivot or shaft near the horizontal center of the tractor. The blade can be lowered or tilted vertically by cables or hydraulic rams. It is a highly versatile piece of earth excavating and moving equipment esp. useful in land clearing and leveling work, in stripping topsoil, in road and ramp building, and in floor or bench cleanup and gathering operations. Also called dozer.

bulled hole A quarry blasting hole, the bottom of which has been enlarged or chambered to receive a heavy explosive charge. See also: *chambering.* (Nelson, 1965)

bullet (a) A small, lustrous, nearly spherical industrial diamond. (Long, 1960) (b) A conical-nosed, cylindrical weight, attached to a wire rope or line, either notched or seated to engage and attach itself to the upper end of the inner tube of a wire-line core barrel or other retrievable or retractable device placed in a borehole. Syn: *bug; go-devil; overshot.* (Long, 1960) (c) A bullet-shaped weight or small explosive charge dropped to explode a charge of nitroglycerin placed in a borehole. (Long, 1960)

bull gear (a) A toothed driving wheel that is the largest or strongest in the mechanism. (Nichols, 1976) (b) A gear or sprocket that is much larger than the others in the same power train. (Nichols, 1976)

Bullgrader Trade name for an International (formerly Bucyrus-Erie) angling bulldozer. (Nichols, 1976)

bullies Fragments of country rock enclosed in a mineral vein. (Arkell, 1953)

bulling The firing of explosive charges in the cracks of loosened rock. The clay stemming is forced around the charge by a bulling bar. See also: *bulled hole.* (Nelson, 1965)

bulling shovel A triangular, sharp-pointed shovel used in ore dressing. Also called vanning shovel. (Fay, 1920)

bullion (a) A concretion found in some types of coal; composed of carbonate or silica stained by brown humic derivatives; often a well-preserved plant structure forms the nucleus. Also called coal ball. (AGI, 1987) (b) Lanc. Nodule of clay ironstone, pyrite, shale, etc., that generally enclose a fossil. (c) Refined gold or silver, uncoined, in the shape of bars, ingots, or comparable masses. (ASM, 1961)

bullion bar Refined gold or silver in the form of bars of convenient sizes and weights for handling and storage. (Henderson, 1953)

bullion content Bullion (gold or silver) weight in a parcel of mineral or metal changing hands. The major value is that of the carrier (e.g., argentiferous lead), but payment is made both for this and for the precious metal. (Pryor, 1963)

bull ladle Usually the largest ladle in a foundry.

bull mica Large clusters of diversely oriented and partially intergrown crystals of muscovite with a little interstitial albite and quartz. (Skow, 1962)

bullnose bit A noncoring bit having a convex, half-hemispherical-shaped crown or face. Also called wedge bit; wedge reaming bit; wedging bit. See also: *plug bit.* Cf: *taper bit.* (Long, 1960)

bull pup A worthless mining claim.

bull quartz White massive quartz, essentially free of accessory minerals and valueless as ore. Syn: *bastard quartz; buck quartz.* (AGI, 1987)

bull's-eye A nodule of pyrite in roofing slate.

bull's-eye tuyere A tuyere discharging in the center of a hemispherical plate. (Standard, 1964)

bull shaker A shaking chute where large coal from the dump is cleaned by hand. (Zern, 1928)

bull wheel (a) The large winding drum on which the drill cable or bull rope of a churn or cable-tool drill is wound. (Long, 1960) (b) Large sheave at the top of the mineshaft headframe over which the cage- or skip-hoist rope passes. (Long, 1960) (c) An underground sheave wheel; particularly, the wheel around which the tail rope is passed beyond each terminal of a tail-rope haulage system. (d) The pulley that rotates the camshaft of a stamp battery. (Nelson, 1965)

bully A developing heading driven to the dip, usually the full dip of the coal seam; worked by rope haulage. (Nelson, 1965)

bullying *springing.*

bumboat A small boat equipped with a hoist and used for handling dredge lines and anchors. (Nichols, 1976)

bump (a) Any dull, hollow sound produced in a coal seam or associated strata as a result of mining operations. See also: *outburst; crump; rock bump; rock burst.* (Nelson, 1965) (b) Sudden failure of the floor or walls of a mine opening, generally accompanied by a loud report and a sharp shock or jar. (Long, 1960) (c) An earth tremor occasioned by a rock failure, when that failure causes no damage to the workings. (Spalding, 1949) (d) A noise caused by a break in the roof underground. (Mason, 1951) (e) The actual movement due to the roof break. (Mason, 1951) (f) A sudden floor uplift due to a break in the floor. (Mason, 1951) (g) In coal mining, shock due to the movement of coal, floor, or roof strata, with sufficient violence to be heard and to shake the workings. (Pryor, 1963) (h) Rebound caused by a sudden release of tension on the drill stem when a core breaks or snaps free of the bottom of the borehole. (Long, 1960) (i) A sharp, upward blow applied to the drivepipe, casing, or drill stem with a drive hammer. (Long, 1960)

bumper (a) A worker who pushes loaded cars or cans into a station for the hooker and removes the empties. (Hess) (b) A device used to loosen the tools when drilling is carried on without jars. (Porter, 1930) (c) A fender for lessening the jar caused by the collision of cars or other moving equipment. (Jones, 1949) (d) *catch.* (e) A machine used for packing molding sand in a flask by repeated jarring or jolting. (ASM, 1961)

bumper block An impeding device at dumping locations where there is a hazard of vehicle overtravel. (Federal Mine Safety, 1977)

bumper post Barrier of heavy steel construction anchored at a track ending to stop rolling railroad cars and prevent their being thrown off center or derailed. (Best, 1966)

bumping table *shaking table.*

bumping trough An appliance for handling broken rock in flat mine stopes. A sheet-steel trough is hung from chains and arrested at one end of its swing by a bump stop, so that the ore slides forward. (Pryor, 1963)

bumps Sudden, violent expulsion of coal from one or more pillars, accompanied by loud reports and earth tremors. Bumps occur in coal mines where a strong, thick, massive sandstone roof rests directly on the coal with no cushioning layer of shale between. The breaking of this strong roof as the seam is mined causes violent bumps and the crushing and bursting of pillars left for support. There are two distinctive types of bumps: (1) pressure bumps, which appear to be due to the unit loading of a pillar being too great for its bearing strength, and where the coal roof and floor are strong, the pillar is ruptured suddenly and with violence; and (2) shock bumps, which are thought to be due to the breaking of thick, massive, rigid strata somewhere above the coalbed, which causes a great hammerlike blow to be given to the immediate roof, which it transmits as a shock wave to the coal pillar or pillars. (Kentucky, 1952; Lewis, 1964)

bunched seismometers Group of seismometers located at short intervals at the same seismometer station and electrically interconnected. Syn: *multiple seismometers*. (Schieferdecker, 1959)

bunchy (a) An orebody containing small scattered masses of bunches of ore. (Weed, 1922) (b) A mine that is sometimes rich and at other times poor. (Hess)

bund Any artificial embankment used to control the flow of water in a river or on irrigated land. The term is applied extensively in India to large low dams and dikes and also to the small ridges between rice fields. Also, an embanked causeway or thoroughfare along a river or the sea. (AGI, 1987)

bunker A vessel for the storage of materials; the lowermost portion is usually constructed in the form of a hopper. Also called bin. See also: *surge bunker; underground bunker*. (BS, 1962)

bunker coal Applied to coal consumed by ocean steamers, tugs, ferryboats, or other steam watercraft. Also called bunkers. (Fay, 1920)

bunker conveyor A high-capacity conveyor that takes peaks of production from another conveyor and retains and/or discharges the material when production drops. Such a conveyor may be laid under or alongside a trunk belt near its discharge end. The floor of the bunker comprises a slow-moving steel plate conveyor operated by hydraulic or other power. A movable plow plate, situated over the trunk belt, diverts the material sideways into the bunker conveyor. See also: *underground bunker*. (Nelson, 1965)

bunker gate *bin gate*.

bunkering capacity The capacity of anything. It may be expressed as a tonnage or as so many hours of normal production. Bunkering capacity may be provided at the surface and at critical points underground. (Nelson, 1965)

bunney (a) A mass of ore not lying in a regular vein. Also spelled bunny, bonny, bonney. (Nelson, 1965) (b) Corn. An isolated body of ore. (Hess)

bunny *bunney*.

bunsenite A pistachio-green isometric mineral, NiO, of the periclase group.

Bunsen photometer A visual photometer in which a simple mirror system enables both sides of the test plate, consisting of a screen of opaque-white paper on which is a grease spot, to be viewed at the same time. That portion of the screen on which the grease lies is translucent to light, so that there is a difference in brightness between the grease spot and the surrounding ungreased paper. When comparing sources, one on either side of the photometric bench, the point of balance is such that, as seen in the mirror, both sides of the screen show equality of contrast between the grease spot and its white surroundings. (Roberts, 1958)

bunton (a) A steel or timber element in the lining of a rectangular shaft. Buntons may be 6 in by 5 in or 6 in (15.2 cm by 12.7 cm or 15.2 cm) square and extend across the shaft at intervals of 4 ft to 8 ft (1.2 m to 2.4 m). They serve to reinforce the barring and also carry the cage guides. Rolled steel joists are now generally used as buntons. See also: *divider; wallplate*. (Nelson, 1965) (b) A timber placed horizontally across a shaft. It serves to brace the wallplates of the shaft lining and also, by means of planks nailed to them, to form separate compartments for hoisting or ladderways.

bunton racking Timber pieces used in the support of rectangular shafts. See also: *wallplate*. (Nelson, 1965)

buoyant weight The apparent weight of a string of drill tools suspended in a liquid-filled borehole. The apparent weight is the weight of the drill string in air less the weight of the liquid displaced by the drill string when suspended in a liquid-filled borehole. (Long, 1960)

buratite An aurichalcite containing calcium monoxide, probably as a mechanical admixture. (Weed, 1918)

burbankite A hexagonal mineral, $(Na,Ca)_3(Sr,Ba,Ce)_3(CO_3)_5$.

burden (a) The distance from the borehole and the nearest free face or distance between boreholes measured perpendicular to the spacing (usually perpendicular to the free face). (b) All types of rock or earthy materials overlying bedrock. See also: *cover; mantle; overburden*. (Long, 1960) (c) Valueless material overlying ore, esp. that removed by stripping. Frequently called overburden. (Webster 2nd, 1960) (d) The resistance that an explosive charge must overcome in breaking the rock adjacent to a drill hole in mining. (Webster 3rd, 1966) (e) The tonnage or cubic yards of rock, ore, or coal that an explosive charge is expected to break. (Nelson, 1965) (f) The distance between the charge and the free face of the material to be blasted. (Fay, 1920) (g) *line of least resistance*. (h) The charge of a blast furnace exclusive of the fuel; also, the ratio of the ore to the total charge.

burdening the furnace Determining the proper proportions of ore, coke, and limestone for a blast furnace charge. (Mersereau, 1947)

burgee Small coal suitable for furnaces or engines. (Arkell, 1953)

Burgers vector In crystal structures, dislocations locally alter coordination polyhedra with the result that regular polyhedra several atomic diameters distant are offset from their regular positions. The Burgers vector is a measure of that offset, being normal to edge dislocations and parallel to screw dislocations.

buried hill A hill of resistant older rock over which later sediments were deposited. The overlying sedimentary beds have the form of an anticline as the result of original dip, unequal compaction, etc. The term was first applied to the underlying beds of the Healdton Field, OK. (AGI, 1987)

buried outcrop *blind apex*.

buried placer (a) Old placer deposit that has been buried beneath a lava flow or other strata. (b) *deep lead*.

burkeite An orthorhombic mineral, $Na_6(CO_3)(SO_4)_2$; in small flat crystals, twins, and nodules; at Searles Lake, CA. Syn: *teepleite*.

burl An oolith or nodule in fireclay. It may have a high content of alumina or iron oxide. (AGI, 1987)

Burleigh A miner's term for any heavy two-person drill. The Burleigh was the first successful machine rock drill. (Hess)

burley clay A clay containing burls; specif. a diaspore-bearing clay in Missouri, usually averaging 45% to 65% alumina. See also: *diaspore clay*. (AGI, 1987)

burn (a) To permit a bit to become overheated in use. (Long, 1960) (b) To calcine. (Long, 1960) (c) To pulverize with very heavy explosive charges. (Nichols, 1976)

burn cut Type of parallel hole or holes cut for tunnel blasting; centrally located and not containing explosives. Outer loaded holes are designed to break the cut.

burned Said of slate or other impurity that adheres tightly to coal. Similarly, coal is said to be "burned to the roof" when it is hard to separate the roof rock from the coal. (Fay, 1920)

burned bit As a result of high-speed, excessive pressure, and poor water circulation, sufficient heat may be generated at the bottom of a borehole to cause a diamond crown to soften, resulting in displacement of diamonds and a ruined bit. (Nelson, 1965)

burned cut A cut made in the face of a heading for which three or four holes are drilled normal to the face and in a triangle or square, 12 to 18 in (30.48 to 45.72 cm) on a side, with another hole in the center. One, two, or three holes are loaded and shot; the others relieve the pressure and induce breaking. A cavity is formed to which other shots in the face readily break. Used for esp. tough ground. Also called Michigan cut; woodchuck cut. (Hess)

burned lime Calcium oxide (quicklime) formed from limestone, or other forms of calcium carbonate, which has been calcined at high temperature to drive off the carbon dioxide. See also: *burnt lime*. (Shell)

burn in (a) To run a bit with too little coolant until the heat generated by the bit fuses the cuttings, core, bit, and the bottom of the borehole. (Long, 1960) (b) To deliberately run a bit with reduced amount of coolant until the core is jammed inside the bit. (Long, 1960)

burning house A furnace in which sulfide ores are calcined to form gaseous SO_2 and leave the metal oxide, or in the case of noble metals, the metal itself.

burning oil A common name for kerosine.

burning out A loose term, usually used to describe the wearing away of furnace linings without a known reason.

burno man A laborer who gets ore ready for a mechanical shovel or a hand shoveler. (Hess)

burnout A situation encountered in coal seams, usually near the outcrop, where the coal has undergone combustion and burned. Initiation sources could include lightning, forest fires, etc.

burn out To salvage diamonds from a used bit by dissolving the matrix alloy with an acid or by use of an electrolytic process. (Long, 1960)

Burnside boring machine This machine has been specially developed for boring in all types of ground, and incorporates a very important feature, that of controlling the water immediately if it is tapped. In boring, the hole is first prepared for the reception of a special rubber ring, two iron plates, and two wedges. When these are properly adjusted, the rubber washer is compressed and powerfully gripped on the sides of the borehole to effect a sound and reliable joint. If during boring operations water should rush out and the bore rods cannot be withdrawn, the two handwheels are screwed in; this presses india-rubber plugs onto the bore rods and effects a watertight joint. (Mason, 1951)

burnt alum Alum that has been dried at 200°C, and powdered; $AlNH_4(SO_4)_2$ or $AlK(SO_4)_2$. A caustic. Syn: *dried alum; exsiccated alum*.

burnt iron (a) Iron which by long exposure to heat has suffered a change of structure and become brittle. It can be restored by careful forging at welding heat. (Fay, 1920) (b) In the Bessemer and open-hearth processes, iron that has been exposed to oxidation until all of its carbon is gone, and an oxide of iron has been formed in the mass. (Fay, 1920)
burnt lime Calcitic lime, CaO, or dolomitic lime, CaO·MgO. See also: *burned lime.*
burnt metal Metal that has become oxidized by overheating, and so is rendered useless for engineering purposes. (CTD, 1958)
burnt stone (a) An antique carnelian, such as is sometimes found in ancient ruins and has apparently been acted on by fire. (Fay, 1920) (b) A gemstone with color changed by heating; e.g., amethyst, which changes from purple to clear; or tiger-eye, which changes from yellowish-brown to reddish brown. Syn: *heat-treated stone.* Cf: *stained stone.* See also: *heated stone.*
burr (a) A term used in England for a rough or hard stone, such as a compact siliceous sandstone esp. hard to drill. Also spelled bur. (AGI, 1987) (b) A knob, boss, nodule, or other hard mass of siliceous rock in a softer rock; a hard lump of ore in a softer vein. (AGI, 1987) (c) *buhrstone.* (AGI, 1987)
burr rock An aggregate of muscovite books and quartz. (Skow, 1962)
burrstone *buhrstone.*
burst An explosive breaking of coal or rock in a mine due to pressure. In coal mines bursts may or may not be accompanied by a copious discharge of methane, carbon dioxide, or coal dust. Also called outburst; bounce; bump. See also: *rock burst.*
burster (a) A hydraulic mechanism that, when inserted into a large-diameter shothole, breaks down the strata by means of pistons operating transversely. (BS, 1964) (b) Scot. A shot in a coal seam that has not been sheared or undercut. Equivalent to "shot off the solid." Also called bursting shot.
bursting The phenomenon sometimes exhibited by refractories containing chrome ore, when exposed to iron oxide at high temperature, of having the exposed face swell and grow until it breaks away from the brick mass. (ARI, 1949)
bursting charge A small charge of fine powder placed in contact with a charge of coarse powder to ensure the ignition of the latter. (Fay, 1920)
bursting time The time between the application of an electric current and the setting off of an explosive charge. In seismic prospecting, it may be necessary to take into account the maximum difference in time lag between the bursting of the earliest and latest detonators in a series. In a series firing current of over 1 A direct current, the maximum difference with submarine seismic detonators is always less than 1 ms. (Nelson, 1965)
Burt filter A stationary, intermittent filter in which the leaves are suspended vertically in a cylindrical vessel set on a considerable incline. The leaves are therefore ellipses. The slime cake is discharged by introducing air and water into the interior of the leaf. There is also a Burt filter of the continuous-rotating-drum type. (Liddell, 1918)
burton Any of several arrangements of hoisting tackle; usually one with a single and a double block. (Webster 3rd, 1966)
Bushveld Complex A great intrusive igneous body in the Transvaal, South Africa, that has undergone remarkable magmatic differentiation. It is by far the largest layered intrusion known. The Bushveld is the leading source of chromite. (AGI, 1987)
Buss table Shaking table for treatment of ore sands, comprising a deck supported by a Ferraris truss moved by eccentric. (Pryor, 1963)
bustamite A triclinic mineral, $(Mn,Ca)Si_3O_9$.
buster shot *breaking-in shot.*
bus wire Expendable heavy-gage bare copper wire used to connect detonators or series of detonators in parallel in underground blasting.
but Scot. Outward; toward the shaft; outbye.
butane (a) A flammable gaseous hydrocarbon, commonly bottled for use as fuel. (b) A gaseous flammable paraffin hydrocarbon, C_4H_{10}, occurring in either of two isometric forms: n-butane, $CH_3CH_2CH_2CH_3$; or isobutane, $CH_3CH(CH_3)_2$. The butanes occur in petroleum and natural gas. (AGI, 1987)
butane flame methanometer An instrument giving a continuous record of the methane concentration in mine air. It uses a small flame burning butane in a gauze-protected enclosure. Instead of observing the cap, thermocouples are used to show the increased temperature above the flame, and the resulting signal is displayed on a recording milliammeter. The instrument runs for at least a week and is accurate to about 0.05% methane. See also: *methane tester type S.3.* (Nelson, 1965)
Butchart table A shaking table, toggle-actuated, with its deck supported in slipper bearings, and carrying curved riffles. (Pryor, 1963)
butlerite Arg. A monoclinic mineral occurring as oriented intergrowths with parabutlerite. A hydrous sulfate of iron. (Am. Mineral., 1945)
bütschliite A trigonal mineral, $K_2Ca(CO_3)_2$; dimorphous with fairchildite. Also spelled buetschliite.
butt (a) Opposite of face, coal exposed at right angles to the face, and in contrast to the face, generally having a rough surface. Also called end in Scotland. (BCI, 1947) (b) The butt of a slate quarry is where the overlying rock comes in contact with an inclined stratum of slate rock.
butt cleat The minor cleat system, or jointing, in a coal seam, usually at right angles to the face cleat. Syn: *end cleat.* See also: *cleat; butt joint.* (AGI, 1987)
butte (a) A conspicuous hill or small mountain with relatively steep slopes or precipitous cliffs, often capped with a resistant layer of rock, and representing an erosion remnant carved from flat-lying rocks. The summit is smaller in extent than that of a mesa; many buttes in the arid and semiarid regions of the Western United States result from the wastage of mesas. Syn: *mesa-butte.* (AGI, 1987) (b) An isolated hill having steep sides and a craggy, rounded, pointed, or otherwise irregular summit; e.g., a volcanic cone (such as Mount Shasta, CA, formerly known as Shasta Butte) or a volcanic butte. Etymol: French, knoll, hillock, inconspicuous rounded hill; rising ground. Pron: bewt. (AGI, 1987)
butt entry (a) An entry driven at right angles to the butt. (BCI, 1947) (b) The gallery driven at right angles with the butt cleat. An end-on entry. (c) A gallery driven parallel with the main cleat of the coal seam. See also: *entry.* (Nelson, 1965)
butterball A clear-yellow, rounded segregation of very pure carnotite found in the soft sandstone of Temple Rock, San Rafael Swell, UT. (AGI, 1987)
Butters and Mein distributor A turbo distributor that spreads sand evenly around a circular leaching tank in gold cyanidation. (Pryor, 1963)
buttgenbackite A hexagonal mineral, $Cu_{19}Cl_4(NO_3)_2(OH)_{32}{\cdot}2H_2O$.
butt heading *butt entry.*
butt joint (a) A joint between two abutting members lying approx. in the same plane. A welded butt joint may contain a variety of grooves. (ASM, 1961) (b) *butt cleat.*
buttock (a) A corner formed by two coal faces more or less at right angles, such as the end of a working face; the fast side; any short piece of coal approx. at right angles to the face; a rib; the rib side. See also: *rib.* (Mason, 1951) (b) Eng. That portion of a working face of coal, next to be taken down. (Fay, 1920) (c) The rib of coal exposed at one or both ends of a longwall face, to enable a cutter loader to commence its run; the coal removed by a cutter loader. See also: *stable; web.* (Nelson, 1965) (d) Coal that has been undercut and is ready to be broken. (Pryor, 1963)
button (a) A globule of metal remaining in an assaying crucible or cupel after fusion has been completed. (ASM, 1961) (b) That part of a weld that tears out in the destructive testing of spot-seam or projection-welded specimens. (ASM, 1961) (c) Globule of lead formed during fire assay of gold or silver ore. (Pryor, 1963)
button balance A small, very delicate balance used for weighing assay buttons.
button fusion test *button test.*
button rope conveyor *rope and button conveyor.*
button test A test designed to determine relative fusibility of frit or powder. So called because the completed specimens resemble buttons.
butt shot In coal mining, a charge placed so that the face or burden is nearly parallel with the borehole.
butt side The side of the working face of a coalbed in which the joints or cleats are least pronounced, as distinguished from the face side in which the joints are most pronounced. (Hess)
butt weld A weld made between two abutting unscarfed ends or edges without overlapping. Both the pin- and box-thread portions of petroleum drill pipe generally are butt-welded electrically to upset end tubing to form a complete section of drill pipe or rod. (Long, 1960)
butt-welded tube A tube made by drawing mild steel strip through a bell, so that the strip is coiled into a tube, the edges being then pressed together and welded. (CTD, 1958)
butyl rubber Synthetic material, copolymer of butadiene and isobutane. (Pryor, 1963)
Buxton test One of a set of tests carried out in a gallery at the Safety in Mines Research Station at Buxton, England. The tests are made to determine the likelihood or limits at which an explosive will ignite gas or coal dust, before it can be placed on the official permitted list. See also: *permitted explosive.* (Nelson, 1965)
BW Letter name specifying the dimensions of bits, core barrels, and drill rods in the B-size and W-group wireline diamond drilling system having a core diameter of 42 mm and a hole diameter of 60 mm. Syn: *BX.* (Cumming, 1981)
BX Letter name specifying the dimensions of core, core barrels, and casing in the B-size and X-series wireline diamond drilling system having a core diameter of 42 mm and a hole diameter of 60 mm. The BX designation for coring bits has been replaced by the BW designation. Syn: *BW.* (Cumming, 1981)

bye water *bank water.*

bypass (a) A short passage used to get by or around a place it is not advisable to cross; e.g., a mine shaft. Also spelled byepass. (b) To pass to the side of an obstruction in a borehole by deflecting the hole. Syn: *drill by.* See also: *wedge off.* (Long, 1960) (c) An alternative path, in a duct or pipe, for a fluid to flow from one point to another, with the direction determined by the opening or closing of valves or dampers in the main line as well as in the bypass. (Strock, 1948) (d) An arrangement of screens and chutes, or of piping, allowing material to be passed around a given part of a flow line. Much used to avoid feeding fine ore through a relatively coarse crusher, thus reducing load, wear, and chance of blockage. (Pryor, 1963) (e) A small passage to permit equalization of the pressure on the two sides of a large valve so that it may be readily opened or closed.

bypit Scot. A pit nearer the outcrop than the engine pit; an air pit.

byproduct A secondary or additional product; e.g., gallium is commercially recovered from the processing of bauxite to alumina. See also: *coproduct.*

byproduct oven A coke oven consisting of a series of long, narrow chambers arranged in rows, and heated by flues in which are burned a portion of the combustible gases generated by the coking of the coal. All of the volatile products are saved and collected as ammonia, tar, gas, etc. (Fay, 1920)

byproducts of coal The products obtained from coal by destructive distillation and other processes. (Cooper, 1963)

byssolite An olive-green asbestiform variety of tremolite-actinolite. The term is used in the gem trade for a variety of quartz-containing, greenish, fibrous inclusions of actinolite or asbestos.

byströmite (a) A tetragonal mineral, $MgSb_2O_6$. (b) A former name for a monoclinic polymorph of pyrrhotite. Also spelled bystromite.

bytownite A triclinic mineral, $(Ca,Na)[(Al,Si)AlSi_2O_8]$ having 90 to 70 mol % Ca and 10 to 30 mol % Na; of the plagioclase series of the feldspar group; prismatic cleavage; white to gray; forms phenocrysts in some basalts and layered mafic-ultramafic intrusions

by-wash A channel or spillway designed to carry surplus water from a dam, reservoir, or aqueduct in order to prevent overflow. (AGI, 1987)

C

cab A compartment for the driver in a mine locomotive, continuous mining machine, shuttler car, scoop, etc. All coal mine locomotives in excess of 10 st (9 t) weight must have a cab at each end or an adequate center cab. (Nelson, 1965)

caballa ball Eng. Ironstone nodule worked for iron in the Weald. Also called bulls. See also: *ballstone; bears' muck; mare ball.* (Arkell, 1953)

caballing An increase in density occurring when the contents of two pipes carrying water with different temperature and salinity characteristics, but having the same density, are mixed; the resulting fluid mixture is slightly more dense than either of the two original types. (Hy, 1965)

cab guard On a dump truck, a heavy metal shield extending up from the front wall of the body and forward over the cab. (Nichols, 1976)

cable (a) A heavy multiple-strand steel rope used in cable-tool drilling as the line between the tools and the walking beam. Syn: *drilling cable.* (AGI, 1987) (b) A term used loosely to signify a wire line. See also: *wire line.* (AGI, 1987) (c) A fiber cable consists of three hawsers laid up left-handed. See also: *wire rope; cable-laid rope.* (Zern, 1928) (d) A ropelike, usually stranded assembly of electrical conductors or of groups of two or more conductors insulated from each other but laid up together usually by being twisted around a central core, the whole usually heavily insulated by outside wrappings; specif., a submarine cable. (Webster 3rd, 1966) (e) A steel rope for hoisting or for aerial trams. (Fay, 1920) (f) A flexible rope composed of many steel wires or hemp fibers in groups, first twisted to form strands, several of which are again twisted together to form a rope. Also called wire cable; wire line; wire rope; steel cable. See also: *wire-line cable.* (Long, 1960) (g) *armored cable; electric cable.* (h) A single concentration of steel wire intended for prestressing. (Taylor, 1965) (i) A nautical unit of horizontal distance, equal to 600 ft (100 fathoms; 182.9 m) and approx. 0.1 nmi (0.18 km). (Hunt, 1965)

cable belt conveyor A conveyor using steel wire ropes to take the tensile pull, which in a conventional conveyor is taken by the belt. Two-stranded steel ropes, one on either side of the conveyor, are used for this purpose. The belt sits on and is supported across the two ropes by means of rubber shoe forms along the belt edges. These belts can be of long lengths, high capacities, and high lifts. (Nelson, 1965)

cable bolt A device or method for reinforcing ground prior to mining. The basic cable bolt support consists of a high-strength cable installed in a borehole 4.12 to 6.35 cm in diameter and grouted with cement. Syn: *cable tendon; cable dowel.* (Schmuck, 1979)

cable bolting Complex electrical plugs and sockets used throughout a mine distribution system to connect mobile machinery to trailing cables, to connect cables with one another, and to connect cables to power centers, switchhouses, and substations. (SME, 1992)

cable control unit A high-speed tractor winch having one to three drums under separate control. Used to operate bulldozers and towed equipment. (Nichols, 1976)

cable dowel *cable bolt.*

cable drill (a) A heavy drilling rig in which a rope is used for suspending the tools in the borehole. See also: *churn drill.* (Nelson, 1965) (b) A churn or percussion drill rig, consisting of a tower (derrick), wire rope for moving tools vertically, a power unit, and a reciprocating device. It drills holes of up to 10 in (25.4 cm) in diameter vertically to considerable depths. (Pryor, 1963)

cable excavator A long-range, cable-operated machine that works between a head mast and an anchor. (Nichols, 1976)

cable hook A round hook with a wide beveled face. (Nichols, 1976)

cable-laid rope (a) A compound-laid rope consisting of several ropes or several layers of strands laid together into one rope, as, for instance, 6 by 6 by 7. (Hunt, 1965) (b) A rope in which both the fibers forming the strands and the strands themselves are twisted to the left. (Long, 1960) (c) Wire cable made of several ropes twisted together; strands of hawser-laid rope, twisted right-handed together without limitation as to the number of strands or direction of twist. A fiber cable-laid rope is composed of three strands of hawser-laid rope, twisted right-handed. (Zern, 1928)

cable railway An inclined track up and down which wagons travel fixed at equal intervals to an endless steel wire rope, either above or below the wagons. (Hammond, 1965)

cable reel A drum on which conductor cable is wound, including one or more collector rings and associated brushes, by means of which an electric circuit is made between the stationary winding on the locomotive or other mining device and the trailing cable that is wound on the drum. The drum may be driven by an electric motor, by a hydraulic motor, or mechanically from an axle on the machine.

cable-reel locomotive A face or gathering locomotive driven by a power cable connected to trolley wires. The cable winds on a reel attached to the locomotive. (Nelson, 1965)

cable-screw conveyor A one-way or closed-circuit conveyor powered by a flexible, torque-transmitting cable of which helical (screw) threads are an integral part. Loads or load carriers engage the thread and advance a distance equal to one pitch each revolution of the cable screw.

cable selvage belt A conveyor belt in which the carrying section is composed of rubber and fabric with attached intermittent transverse metal supports having both ends supported by cables. The cables transmit the driving force, and the center portion functions as the load-supporting medium.

cable shield A metallic shield consisting of nonmagnetic material applied over the insulation of the individual conductors or conductor assembly. (USBM, 1960)

cable splice kit A short piece of tubing or a specially formed band of metal generally used without solder in joining ends of portable cables for mining equipment.

cable system One of the well-known drilling systems, sometimes designated as the American or rope system. The drilling is performed by a heavy string of tools suspended from a flexible manila or steel cable to which a reciprocating motion is imparted by an oscillating "walking beam" through the suspension rope or cable. See also: *churn drill.*

cable-system drill *churn drill.*

cable tendon *cable bolt.*

cable-tool cuttings The rock fragments and sludge produced in drilling a borehole with a churn drill. (Long, 1960)

cable-tool dresser *tooldresser.*

cable-tool drill *churn drill; percussion drill.*

cable-tool drilling A method of drilling, now largely replaced by rotary drilling, in which the rock at the bottom of the hole is broken up by a steel bit with a blunt, chisel-shaped cutting edge. The bit is at the bottom of a heavy string of steel tools suspended on a cable that is activated by a walking beam, the bit chipping the rock by regularly repeated blows. The method is adapted to drilling water wells and relatively shallow oil wells. (AGI, 1987)

cable tools The bits and other bottom-hole tools and equipment used to drill boreholes by percussive action, using a rope, instead of rods, to connect the drilling bit with the machine on the surface. See also: *churn drill.* (Long, 1960)

cableway A system in which the carriers are supported by a cable and are not detached from the operating span. The travel of the carriers is wholly within the span. See also: *aerial cableway.*

cableway excavator A slackline cableway used for excavating a restricted area. (Hammond, 1965)

cableway transporter A transporter crane on which the track for the carrier is a steel wire rope. (Hammond, 1965)

cabochon (a) An unfaceted cut gemstone of domed or convex form. The top is smoothly polished; the back, or base, is usually flat or slightly convex, may be concave, and is commonly unpolished. The girdle outline may be round, oval, square, or any other shape. (b) The style of cutting such a gem. (c) A polished but uncut gem. See also: *en cabochon.*

cabrerite A hydrous arsenate of nickel, cobalt, and magnesium; possibly magnesian annabergite.

cachalong *cacholong.*

cache Fr. The place where provisions, safety or rescue equipment, ammunition, etc., are cached or hidden by trappers, miners, or prospectors, in unsettled regions. (Fay, 1920)

cacholong An opaque or feebly translucent, bluish-white, pale-yellowish, or reddish variety of common opal containing a little alumina. Syn: *cachalong; pearl opal.*
cacoxenite An orthorhombic mineral, $(Fe,Al)_{25}(PO_4)_{17}O_6(OH)_{12}\cdot 75H_2O$.
cactus grab A digging and unloading attachment hung from a crane or excavator. It consists of a split and hinged bucket fitted with curved jaws or teeth which dig into the loose rock while the bucket is being dropped and contract to lift the load while it is being raised. It is used increasingly for mechanical mucking in shaft sinkings. See also: *hoppit.* (Nelson, 1965)
cadacryst *xenocryst.* Also spelled chadacryst.
cadastral control A system of established monuments whose positions are accurately determined and are used in all correlated cadastral surveys. (Seelye, 1951)
cadastral map A large-scale map showing the boundaries of subdivisions of land, usually with the directions and lengths thereof and the areas of individual tracts, compiled for the purpose of describing and recording ownership. It may also show culture, drainage, and other features relating to use of the land. (AGI, 1987)
cadastral survey Survey relating to land boundaries and subdivisions, made to create or to define the limitations of a title, and to determine a unit suitable for transfer. Includes surveys involving retracements for the identification, and resurveys for the restoration, of property lines. (The term "cadastral" is practically obsolete; may be found in older historical records; current usage is "land survey" or "property survey."). (Seelye, 1951)
cadger A little pocket oilcan for miners. (Fay, 1920)
cadmia (a) An impure zinc oxide that forms on the walls of furnaces in the smelting of ores containing zinc. See also: *furnace cadmium.* (Standard, 1964) (b) The chemical compound CdO. (c) *calamine.*
cadmium A soft, bluish-white metal, similar in many respects to zinc, copper, and lead ores. Almost all cadmium is obtained as a byproduct in the treatment of these ores. Symbol, Cd. Used in electroplating, in solder, for batteries, as a barrier to control atomic fission, and in TV tubes. Cadmium and solutions of its compounds are toxic. (Handbook of Chem. & Phys., 3)
cadmium blend The mineral greenockite, CdS. Also called cadmium ocher.
cadmium columbate $Cd_2Cb_2O_7$ is an antiferroelectric and has low losses at high frequency. Syn: *cadmium niobate.* (Lee, 1961)
cadmium niobate $Cd_2Nb_2O_7$; a ferroelectric compound of potential value as a special electroceramic; the Curie temperature is -103 °C. See also: *cadmium columbate.* (Dodd, 1964)
cadmium ocher The mineral greenockite; used as a pigment. (Standard, 1964)
cadmosellite A hexagonal mineral, CdSe; wurtzite structure; resinous to adamantine; black; perfect cleavage; forms fine xenomorphic disseminations cementing sandstone. Also spelled kadmoselite.
cafemic Said of an igneous rock or magma that contains calcium, iron, and magnesium. Etymol: a mnemonic term derived from calcium + ferric (or ferrous) + magnesium + ic. (AGI, 1987)
cage A vertically moving enclosed platform used in a mine shaft for the conveyance of workers and materials, usually designed to take one or two cars per deck and may be single or multidecked.
cage bar Safety device that holds doors shut or keeps trams in position. (Pryor, 1963)
cage chain *bridle chain.*
cage cover *bonnet.*
cage guide Conductor made of wood, iron or steel, or wire rope; used to guide the cages in the shaft and to prevent them from swinging and colliding with each other while in motion. See also: *guides; fixed guides; rope guide.* (Nelson, 1965)
cage mill Also known as a disintegrator; used for secondary crushing of stone and gravel, and for reduction of slag, fertilizers, etc. (Pit and Quarry, 1960)
cager (a) One who directs station operations and movement of cages used to raise and lower workers, mine cars, and supplies between various levels and surface; works at the top of a shaft or at an intermediate level inside a mine. Also called cageman; cage tender; shaft headman; skip tender. See also: *banksman; hitcher; top cager; onsetter.* (DOT, 1949) (b) A power-operated ram for pushing mine cars into or out of cages at the pit top or pit bottom. (Nelson, 1965) (c) One who supervises weighing and the sequence of sending up components of a furnace charge, keeps tally of the number of charges, and signals to the top filler when it is time to hoist. (Fay, 1920)
cager coupler In bituminous coal mining, one who works with a cager, coupling and uncoupling cars at a shaft station. (DOT, 1949)
cage seat Scaffolding, sometimes fitted with strong springs, to take the shock, and on which the cage rests when reaching the pit bottom or other landing. (Fay, 1920)
cage sheet Short prop or catch on which a cage stands during caging or changing cars. (Zern, 1928)
cage shoe One of the fittings bolted to the side of a cage to engage the rigid guides in a shaft. Usually there are two for each guide, one at the top and one at the bottom of the cage. The shoes are usually about 1 ft (0.3 m) long and shaped to fit closely around about three-quarters of the guide, with sufficient clearance for free movement but not sufficient to allow the shoe to come off the guide. (Nelson, 1965)
cage stop Equipment fitted on the cage floor to hold the car in position while traveling in the shaft. Spring- or rubber-mounted stops are commonly used. See also: *kep.* (Nelson, 1965)
cagutte A baguette; an oblong cut diamond. (Hess)
cahnite A tetragonal mineral, $Ca_2B(AsO_4)(OH)_4$; forms white sphenoidal crystals.
Cainozoic *Cenozoic.*
cairn An artificial mound of rocks, stones, or masonry, usually conical or pyramidal, used in surveying to aid in the identification of a point or boundary. (AGI, 1987)
cairngorm Smoky-yellow or brown varieties of quartz, the coloring matter probably due to some organic compound; named from Cairngorm in the Scottish Grampians; the more attractively colored varieties are used as semiprecious gem stones. Also called smoky quartz, smokestone. (CTD, 1958)
caisson drill In sampling placer deposits, a caisson drill is driven by a combination of rotational impact and the weight of the drilling equipment. (SME, 1992)
caisson sinking A method of sinking a shaft through wet clay, sand, or mud down to firm strata. Cast-iron tubbing, attached ring by ring on the surface, is gradually lowered as the shaft is excavated. There is a special airtight working chamber at the bottom of the lining. A cutting shoe at the lower end of the tubbing helps it to penetrate the soft ground. The caisson method is obsolescent, being replaced by the freezing method, etc. See also: *concrete caisson sinking.* Also called drum shaft; drop shaft. (Nelson, 1965)
cajon (a) *box canyon.* (b) A defile leading up to a mountain pass; also, the pass itself. Etymol: Spanish cajón, large box. The term is used in the Southwestern United States. (AGI, 1987)
cake (a) The solid residue left in a filter press or on a vacuum filter after the solution has been drawn off. (b) Solidified drill sludge. (Long, 1960) (c) That portion of a drilling mud adhering to the walls of a borehole. Syn: *wall cake.* (Long, 1960) (d) *cake of gold; mud cake.* (e) To form in a mass such as when ore sinters together in roasting, or coal cakes together in coking.
cake copper Copper cast in a round, cake-shaped mass. See also: *tough cake.* (Hess)
caked dust Dust particles with sufficient cohesion that a light stroke with a brush or a light airblast, such as from the mouth, will not cause the dust to be dispersed.
cake of gold Gold formed into a compact mass (though not melted) by distillation of mercury from amalgam. Also called sponge gold. Syn: *cake.* (Fay, 1920)
cake thickness The measure of the thickness of the filter cake deposited against a porous medium. Cake thickness and water loss constitute the determining factors of filtration qualities.
caking coal Coal that softens and agglomerates on heating and after volatile matter has been driven off at high temperatures; produces a hard gray cellular mass of coke. All caking coals are not good coking coals. See also: *coking coal.* (Tomkeieff, 1954)
caking index A laboratory method of indicating the degree of caking, coking, or binding together of a coal when a sample is heated in a prescribed manner. Syn: *agglutinating power.* (Nelson, 1965)
calabashing Panning tin gravels in a half-calabash gourd. Used in prospecting and alluvial mining in primitive conditions. (Pryor, 1963)
calaite *turquoise.*
calamanco N. of Eng. Red or mottled Paleozoic marls and shales. Also called calaminker. See also: *symon.* Cf: *whintin.* (Arkell, 1953)
calamine (a) A commercial, mining, and metallurgical term for the oxidized ores of zinc (including silicates and carbonates), as distinguished from the sulfide ores of zinc. Syn: *cadmia.* See also: *electric calamine.* (AGI, 1987) (b) A former name for hemimorphite. (c) In Great Britain, a name used for smithsonite. See also: *smithsonite.* (d) A former name for hydrozincite. (e) A special kind of so-called galvanized iron. Also spelled kalamin. Syn: *galmei.* (Standard, 1964)
calamine stone Eng. A carbonate of zinc; smithsonite. (Fay, 1920)
calamine violet An indicator plant which grows only on zinc-rich soils in the zinc districts of Central and Western Europe. (Hawkes, 1962)
calamite An asparagus-green variety of tremolite. (Standard, 1964)
calaverite A monoclinic mineral, $2[AuTe_2]$; brittle: commonly contains silver; sp gr, 9.35; an important source of gold.
calc Prefix meaning containing calcium carbonate.
calc-alkalic (a) Said of a series of igneous rocks in which the weight percentage of silica is between 56 and 61 when the weight percentages of CaO and of $K_2O + Na_2O$ are equal. (AGI, 1987) (b) Said of an igneous rock containing plagioclase feldspar. (AGI, 1987)

calcarenite A limestone consisting predominantly (more than 50%) of recycled calcite particles of sand size; a consolidated calcareous sand. Cf: *calcareous sandstone.* (AGI, 1987)
calcareous Said of a substance that contains calcium carbonate. When applied to a rock name, it implies that as much as 50% of the rock is calcium carbonate. (AGI, 1987)
calcareous crust An indurated soil horizon cemented with calcium carbonate; caliche. (AGI, 1987)
calcareous dolomite A carbonate rock containing 50% to 90% dolomite. (Leighton & Pendexter, 1962) Cf: *calcitic dolomite.* (AGI, 1987)
calcareous dust Limestone, quicklime, hydrated lime, and cement dusts fall in this class. These dusts are more or less soluble in the body fluids, and are eventually absorbed. (Pit and Quarry, 1960)
calcareous ooze A deep-sea pelagic sediment containing at least 30% calcareous skeletal remains; e.g., pteropod ooze. CF: siliceous ooze. (AGI, 1987)
calcareous ore Ore in which the gangue consists mainly of carbonate of lime. (Osborne, 1956)
calcareous peat *eutrophic peat.*
calcareous rock *carbonate rock.*
calcareous sandstone (a) A sandstone cemented with calcite. (AGI, 1987) (b) A sandstone containing appreciable calcium carbonate, but in which clastic quartz is present in excess of 50%. Cf: *calcarenite.* (AGI, 1987)
calcareous sinter *travertine.*
calcareous spar Coarsely crystalline calcium carbonate. See also: *calcite.*
calcareous tufa *tufa.*
calcarinate Adj. Designates the calcium carbonate cement of a sedimentary rock. (AGI, 1987)
calc-dolomite Rock consisting of both calcite and dolomite crystals. (AGI, 1987)
calce Native calcium oxide, CaO, found on Mount Vesuvius, Italy. It formed from limestone enveloped in lava and altered by the heat of the lava. (Hess)
calcedony *chalcedony.*
calc-flinta A fine-grained calc-silicate rock of flinty appearance formed by thermal metamorphism of a calcareous mudstone, possibly with some accompanying pneumatolytic action. See also: *calc-silicate hornfels.* (AGI, 1987)
calcian *calcic.*
calciborite Calcium borate, CaB_2O_4, monoclinic. White radial aggregates in drill cores from limestone skarn, from the Ural Mountains. Named from the composition. See also: *frolovite.* (Spencer, 1958)
calcic Said of minerals and igneous rocks containing a relatively high proportion of calcium; the proportion required to warrant use of the term depends on circumstances. Said of a series of igneous rocks in which the weight percentage of silica is greater than 61 when the weight percentages of CaO and of K2O + Na2O are equal. Syn: *calcian.* (AGI, 1987)
calciclase *anorthite.*
calcinable Capable of being calcined or reduced to a friable state by the action of fire. (Fay, 1920)
calcination (a) The heating of a substance to its temperature of dissociation; e.g., of limestone to CaO and CO_2 or of gypsum to lose its water of crystallization. (AGI, 1987) (b) Heating ores, concentrates, precipitates, or residues to decompose carbonates, hydrates, or other compounds. Cf: *roasting.* (ASM, 1961; Newton, 1959) (c) Heating metals at high temperatures to convert them into their oxides. (Nelson, 1965)
calcine (a) Ore or concentrate after treatment by calcination or roasting and ready for smelting. (CTD, 1958) (b) By heating, to expel volatile matter as carbon dioxide, water, or sulfur, with or without oxidation; to roast; to burn (said of limestone in making lime). (Fay, 1920)
calcined gypsum Gypsum partially dehydrated by means of heat, having the approximate chemical formula, $CaSO_4{\cdot}H_2O$.
calciner *calcining furnace.*
calcining (a) Roasting of ore in oxidizing atmosphere, usually to expel sulfur or carbon dioxide. If sulfur removal is carried to practical completion, the operation is termed sweet roasting; if CO_2 is virtually removed, dead roasting. (Pryor, 1963) (b) Reducing to powder by heating. (Mersereau, 1947)
calcining furnace A furnace or kiln in which ores or metallurgical products are calcined. Syn: *calciner.* (CTD, 1958)
calcioborite An orthorhombic mineral, CaB_2O_4; in white radial aggregates in drill cores from limestone skarn in the Urals, Russia.
calciocelestite A variety of celestite containing calcium. (Standard, 1964)
calcioferrite A monoclinic mineral, $Ca_4Fe(Fe,Al)_4(PO_4)_6(OH)_4{\cdot}13H_2O$; occurs in scales and nodules.
calciornotite *tyuyamunite.*
calciovolborthite An orthorhombic mineral, $CaCu(VO_4)(OH)$; adelite group; moderately radioactive; in the Colorado Plateau in sandstone associated with carnotite and tyuyamunite, or in the oxidized zone of deposits containing vanadium minerals.
calciphyre *calc-silicate marble.*
calcisiltite A limestone consisting predominantly of detrital calcite particles of silt size; a consolidated calcareous silt. (AGI, 1987)
calcite (a) A trigonal mineral, $4[CaCO_3]$; has prolific crystal habits, rhombohedral cleavage; defines hardness 3 on the Mohs scale; effervesces readily in dilute hydrochloric acid; a common and widely distributed rock-forming, authigenic, biogenic, and vein mineral; raw material for Portland cement, agricultural lime, flux for ore reduction, dimension stone, and concrete aggregate; the major mineral in limestone, marble, chalk, spongy tufa, cave deposits, and carbonatite; a cementing mineral in many clastic sedimentary rocks; a minor mineral in some silicate igneous and metamorphic rocks. Coarsely crystalline varieties are called nailhead spar, dogtooth spar (acute scalenohedra), and Iceland spar (optical-grade crystals). Abbrev.: Cc. Cf: *dolomite.* Syn: *carbonate of calcium; calcspar.* (b) The mineral group calcite, gaspeite, magnesite, otavite, rhodochrosite, siderite, smithsonite, and sphaerocobaltite
calcite limestone A limestone containing not more than 5% of magnesium carbonate.
calcite marble A crystalline variety of limestone containing not more than 5% of magnesium carbonate.
calcitic dolomite A dolomite rock in which calcite is conspicuous, but the mineral dolomite is more abundant; specif. a dolomite rock containing 10% to 50% calcite and 50% to 90% dolomite, or a dolomite rock whose Ca/Mg ratio ranges from 2.0 to 3.5. Cf: *dolomitic limestone.* (AGI, 1987)
calcitite A rock composed of calcite; e.g., limestone.
calcitization (a) The act or process of forming calcite, such as by alteration of aragonite. (AGI, 1987) (b) The alteration of existing rocks to limestone, due to the replacement of mineral particles by calcite; e.g., of dolomite in dolomite rocks or of feldspar and quartz in sandstones. (AGI, 1987)
calcitrant Refractory; said of certain ores. (Fay, 1920)
calcium A metallic element of the alkaline-earth group; never found in nature uncombined, occurs abundantly as limestone ($CaCO_3$), gypsum ($CaSO_4 \cdot 2H_2O$), and fluorite (CaF_2). Symbol, Ca. Used as a reducing agent, deoxidizer, desulfurizer, or decarburizer for alloys; as quicklime (CaO), it is the great cheap base of the chemical industry with countless uses. (Handbook of Chem. & Phys., 3)
calcium autunite Artificially prepared autunite in which calcium can be replaced by Na, K, Ba, Mn, Cu, Ni, Co, and Mg. Syn: *autunite.* (Spencer, 1952)
calcium carbide CaC_2; produced commercially by heating quicklime and carbon together in an electric furnace. Used for the generation of acetylene and for making calcium cyanamide.
calcium carbonate (a) White powder or colorless crystals; $CaCO_3$. One of the most stable, common, and widely dispersed of materials. It occurs in nature as aragonite, calcite, chalk, limestone, lithographic stone, marble, marl, and travertine. Referred to as whiting, it has many uses in ceramics to introduce calcium oxide (CaO). Also used as a separator in glass firing. (CCD, 1961; Lee, 1961; Kinney, 1962) (b) Calcium carbonate (molecular weight, 100.09) crystallizes in two crystal systems: hexagonal rhombohedral or hexagonal as calcite, and orthorhombic as aragonite. Hexagonal calcium carbonate (calcite) is colorless, white, yellowish, or rarely pale gray, red, green, blue, or violet; sp gr, 2.710 (at 18 °C); Mohs hardness, 3; melting point, 1,339 °C (at 1,025 atm); decomposes at 898.6 °C; and soluble in water, in acids, and in ammonium chloride solution. Orthorhombic calcium carbonate (aragonite) is colorless, white, yellow, reddish, bluish, or black; sp gr, 2.93, ranging from 2.85 to 2.94; Mohs hardness, 3.5 to 4.0; transforms to calcite at 520 °C; decomposes at 825 °C; and soluble in water, in acids, and in ammonium chloride solution. (Handbook of Chem. & Phys., 2) (c) Source of quicklime and of calcium metal.
calcium chloride process A method used to consolidate floor dust in mine roadways in which calcium chloride is applied with a wetting agent.
calcium feldspar *anorthite.*
calcium mica *margarite.*
calcium minerals Naturally abundant and widely exploited in industry. Main useful ores are calcite, dolomite, anhydrite, and gypsum. Apatite is mined for phosphorus; fluorite for fluorides; and colemanite and ulexite for boron. (Pryor, 1963)
calcium montmorillonite An artificially prepared clay mineral with calcium in place of magnesium. (Spencer, 1943)
calcium phosphate *apatite.*
calcium plagioclase *anorthite.*
calcouranite *autunite.*
calcrete (a) Conglomerate consisting of surficial sand and gravel cemented into a hard mass by calcium carbonate precipitated from solution and redeposited through the agency of infiltrating waters, or deposited by the escape of carbon dioxide from vadose water. (AGI,

1987) (b) A calcareous duricrust; caliche. Etymol: "cal"careous + con"crete." Cf: *silcrete; ferricrete.* (AGI, 1987)

calc-sapropel A deposit composed of sapropel (dominant) and remains of calcareous algae. (Tomkeieff, 1954)

calc-schist A metamorphosed argillaceous limestone with a schistose structure produced by parallelism of platy minerals. (AGI, 1987)

calc-silicate hornfels A fine-grained metamorphic rock containing a high percentage of calc-silicate minerals. See also: *calc-flinta; hornfels; limurite; skarn; tactite.*

calc-silicate marble A marble in which calcium silicate and/or magnesium silicate minerals are conspicuous. Syn: *calciphyre.* (AGI, 1987)

calc-silicate rock A metamorphic rock consisting mainly of calcium-bearing silicates, such as diopside and wollastonite, and formed by metamorphism of impure limestone or dolomite; associated with skarn-type mineral deposits. Syn: *lime-silicate rock.* (AGI, 1987)

calc-sinter *travertine.*

calcspar Coarsely crystalline calcite. Also spelled: calc-spar. See also: *calcite.* Syn: *calcareous spar.*

calc-tufa *tufa.*

calcurmolite A secondary mineral, $Ca(UO_2)_3(MoO_4)_3(OH)_2{\cdot}11H_2O$.

Caldecott cone A conical tank used to settle and discharge as a continuous underflow the relatively coarse sand from an overflowing stream of mineral pulp. See also: *cone classifier; Callow cone.* (Pryor, 1963)

caldera A large, basin-shaped volcanic depression, more or less circular, the diameter of which is many times greater than that of the included vent or vents, no matter what the steepness of the walls or the form of the floor may be. Cf: *crater.*

calderite An isometric mineral, $(Mn,Ca)(Fe,Al)_2(SiO_4)_3$; the dark reddish-brown manganese-iron end member of the garnet group.

caldron *bell mold; caldron bottom; kettle bottom.*

caldron bottom (a) Mud-filled prostrate trunk of sigillaria in the roof of certain coal seams. The trunk is a separate mass of rock, with a film of coal around it. It is liable to collapse without any warning sound. Also called horseback. Syn: *caldron; kettle bottom.* See also: *pot.* (Nelson, 1965) (b) Eng. A cone-shaped mass with slippery surfaces found in the roof of some seams. It sometimes comprises a ring of coal around a core of material differing slightly from the ordinary roof. Cf: *pot bottom.* Also called pothole. (SMRB, 1930)

caledonite An orthorhombic mineral, $Pb_5Cu_2(CO_3)(SO_4)_3(OH)_6$; green (not to be confused with celadonite).

calf reel The churn-drill winch used for handling casing and for odd jobs. Also called casing reel. (Nichols, 1976)

caliche (a) A term applied broadly in the Southwestern United States (esp. Arizona) to a reddish-brown to buff or white calcareous material of secondary accumulation; commonly found in layers on or near the surface of stony soils of arid and semiarid regions, but also occurring as a subsoil deposit in subhumid climates. It is composed largely of crusts of soluble calcium salts in addition to such materials as gravel, sand, silt, and clay. It is called hardpan, calcareous duricrust, or calcrete in some localities, and kankar in parts of India. Syn: *calcareous crust; tepetate.* —Etymol: American Spanish, from a Spanish word for almost any porous material (such as gravel) cemented by calcium carbonate. (AGI, 1987) (b) Gravel, rock, soil, or alluvium cemented with soluble salts of sodium in the nitrate deposits of the Atacama Desert of northern Chile and Peru; it contains sodium nitrate (14% to 25%), potassium nitrate (2% to 3%), sodium iodate (up to 1%) sodium chloride, sodium sulfate, and sodium borate, mixed with brecciated clayey and sandy material in beds up to 2 m thick. (AGI, 1987) (c) A term used in various geographic areas for a thin layer of clayey soil capping a gold vein (Peru); whitish clay in the selvage of veins (Chile); feldspar, white clay, or a compact transition limestone (Mexico); a mineral vein recently discovered, or a bank composed of clay, sand, and gravel in placer mining (Colombia). The term has been extended by some authors to quartzite and kaolinite. (AGI, 1987)

caliente Mex. Silver ore, generally colored with some iron sulfate, the result of weathering. (Hess)

California poppy A local indicator plant for copper in Arizona, observed over the outcrop of the San Manuel copper deposit. Here the distribution of this species is confined to copper-rich soil, and its population density is closely proportional to the copper content of the soil. (Hawkes, 1962)

California-type drag head A device for sand dredging; the drag has a hinged afterbody that adjusts to the angle of the drag arm, which may vary with the depth of water. (Scheffaur, 1954)

California-type dredge A single-lift dredge with stacker. Buckets, which are closely spaced, deliver to a trommel. The oversize is piled behind the dredge by a conveyor (stacker). Undersize is washed on gold-saving tables on the deck; tailings discharge astern through sluices.

californite (a) A compact, massive, translucent to opaque variety of vesuvianite; typically dark-green, olive-green, or grass-green, commonly mottled with white or gray, closely resembling jade; an ornamental stone. Principal sources are Fresno, Siskiyou, and Tulare Counties, CA. Syn: *American jade.* (b) A white variety of grossular garnet from Fresno County, CA.

caliper (a) An instrument used to measure precisely the thickness or diameter of objects or the distance between two surfaces, etc. (Long, 1960) (b) An instrument used in conjunction with a microlog which, when lowered down a borehole, measures and records the internal diameter throughout its depth. (BS, 1963) (c) An instrument consisting of a graduated beam and at right angles to it a fixed arm and a movable arm which slides along the beam to measure the diameter of logs and trees. (Webster 3rd, 1966)

caliper brake Brake in which two brakeshoes are curved to the brake path and anchored near the centerline of the drum. (Sinclair, 1959)

caliper log A well log that shows the variations with depth in the diameter of an uncased borehole. It is produced by spring-activated arms that measure the varying widths of the hole as the device is drawn upward. Syn: *section-gage log.* (AGI, 1987)

calite (a) A heat-resistant alloy of aluminum, nickel, and iron. (Hess) (b) Iron or steel treated by calorizing. (Hess)

calk (a) To drive tarred oakum into the seams between planks and fill with pitch. (Fay, 1920) (b) Limestone or chalk; also spelled caulk. (Arkell, 1953) (c) A variety of barite. (Hey, 1955) (d) To peen and draw metal toward and around a diamond being hand-wet in a malleable-steel bit blank. Also called peen. Syn: *peeler.* (Long, 1960) (e) To wick. (Long, 1960)

calkinsite An orthorhombic mineral, $(Ce,La)_2(CO_3)_3{\cdot}4H_2O$; pale yellow; a source of rare-earth elements.

callaganite A monoclinic mineral, $Cu_2Mg_2(CO_3)(OH)_6{\cdot}2H_2O$; azure-blue.

callainite An apple- to emerald-green, massive, waxlike phosphate, possibly a mixture of wavellite and turquoise.

callis Lanc. A shaly coal. (Nelson, 1965)

Callon's rule A rule stating that when a pillar has to be left in an inclined seam for the support of a shaft or of a surface structure, a greater width should be left on the rise side of the shaft or structure than on the dip side. (Briggs, 1929)

Callow cone A conical free-settling tank. Pulp is fed centrally; the finer solid fraction overflows peripherally, and the coarser fraction is withdrawn at a controlled rate via the apex at the cone's bottom. See also: *Caldecott cone; cone classifier.* (Pryor, 1963)

Callow flotation cell An early form of pneumatic flotation cell, still in limited use. Air is blown in at the bottom of the tank at low pressure, through a porous septum such as a blanket, and mineralized froth overflows along the sides while the tailings progress to the discharge end. (Pryor, 1963)

Callow screen A continuous belt formed of fine screen wire travels horizontally between two drums. Pulp, fed from above, flows through together with the finer solids, while coarser material is discharged as the screen passes over the end drum. (Pryor, 1963)

calomel A tetragonal mineral, $2[Hg_2Cl_2]$; a secondary alteration of mercury-bearing minerals. Syn: *calomelite; calomelano; horn quicksilver; mercurial horn ore.*

calomelano *calomel.*

calomel electrode Half-cell used to measure electromotive force; potential being that of mercury and mercurous chloride in contact with saturated solution of potassium chloride. Used in pH measurement. (Pryor, 1963)

calomelite *calomel.*

calorescence The phenomenon of glowing when a substance is stimulated by heat rays that lie beyond the red end of the visible spectrum. See also: *thermoluminescence.*

calorie The gram calorie (or small calorie) is the quantity of heat required to raise the temperature of 1 g of water from 15 to 16 °C. The mean calorie is one-hundredth part of the heat required to raise 1 g of water from 0 to 100 °C. Cf: *heat unit.* (CTD, 1958)

calorific intensity The temperature of a fuel attained by its combustion. (Newton, 1959)

calorific power The quantity of heat liberated when a unit weight or a unit volume of a fuel is completely burned. (Newton, 1959)

calorific value *gross calorific value; net calorific value.*

calorimeter Any apparatus for measuring the quantity of heat generated in a body or emitted by it, such as by observing the quantity of a solid liquefied or of a liquid vaporized under given conditions. Used in determining specific heat; latent heat; the heat of chemical combinations; etc. (Standard, 1964)

calorimeter room A place at the surface of a mine where drained combustible gases are monitored or their heat content is ascertained. (BS, 1963)

calorizing A process of rendering the surface of steel or iron resistant to oxidation by spraying the surface with aluminum and heating to a temperature of 800 to 1,000 °C. (CTD, 1958)

calumetite An orthorhombic mineral, $Cu(OH,Cl)_2{\cdot}2H_2O$; in azure-blue spherules and sheaves of scales having good basal cleavage; at the Calumet Mine, Calumet, MI. Named from the locality. Cf: *anthonyite.*

calx The friable residue (as a metal oxide) left when a mineral or metal has been subjected to calcination or roasting; e.g., lime from calcium carbonate. (Webster 3rd, 1966)

calyx (a) A steel tube attached to the upper end of a core barrel and having the same outside diameter as the core barrel. The upper end is open except for two web members running from the inside of the tube to a ring encircling the drill rod. The calyx serves as a guide rod and also as a bucket to catch cuttings that are too heavy to be flushed out of the borehole by the circulation fluid. Syn: *bucket; sludge barrel; sludge bucket.* (Long, 1960) (b) Syn: *shot drill.* (c) A pipe or tube equipped with a sawtooth cutting edge, sometimes used to obtain a core sample of a formation being drilled. Cf: *basket.* (Long, 1960) (d) In well drilling, a long cylindrical vessel that guides an annular toothed bit. Its action is like that of a diamond drill. A toothed cutter takes the place of a diamond crown and is rotated by hollow flushing rods with a strong constant flow of water. A core is cut, preserved in a core barrel, and brought to the surface. The drills are made large enough so that the holes are used as shafts. (Hess) (e) *sediment tube.*

calyx boring (a) The process of drilling with a shot drill. (Long, 1960) (b) The hole or core produced by this process. (Long, 1960)

calyx drill A rotary core drill that uses hardened steel shot for cutting rock, which will drill holes from diamond-drill size up to 6 ft (1.8 m) or more in diameter. Drilling is slow and expensive, and holes cannot be drilled more than 35° off the vertical, as the shot tends to collect on the lower side of the hole. Also called shot drill. See also: *core drill.* (Lewis, 1964)

camber A beam, bar, or girder bent like a bow, with the hump towards the strata. (Mason, 1951)

Cambrian The oldest of the systems into which the Paleozoic stratified rocks are divided; also, the corresponding oldest period of the Paleozoic era. (Fay, 1920)

camel back A miner's term sometimes applied to such structures as bells, pots, kettle bottoms, or other rock masses that tend to fall easily from a mine roof. See also: *pot bottom; tortoise.* (AGI, 1987)

camera lucida Mirror or prism attached to the eyepiece of a microscope, enabling an observer to sketch the object displayed. (Pryor, 1963)

Cammett table A side-jerk concentrating table similar to the Wilfley table. (Hess)

camouflage The substitution for a common element in a crystal lattice by a trace element of the same valence. Cf: *admittance; capture.* (AGI, 1987)

camouflet (a) A cavity formed in a borehole by the detonation of an explosive charge placed in it. Also called chamber. See also: *spring; socket.* (Long, 1960) (b) A quarry blasting hole enlarged by chambering. (Nelson, 1965)

campaign (a) The period during which a furnace is continuously in operation. (Fay, 1920) (b) The working life of a tank or other melting unit between major cold repairs. (ASTM, 1994)

camptonite A lamprophyre, similar in composition to nepheline diorite, being composed essentially of plagioclase (usually labradorite) and brown hornblende (usually barkevikite). (AGI, 1987)

campylite A yellowish to brown variety of mimetite crystallizing in barrel-shaped forms. A source of lead. See also: *mimetite.*

camshaft In stamp milling, a strong horizontal revolving shaft to which a number of cams are attached in such a manner that no two of them strike the tappets at the same instant, thus distributing the weight to be lifted. (Fay, 1920)

cam stick In stamp battery crushing, a square-sectioned wooden stick greased on the underside and leather-lined above; it is inserted between cam and tappet. (Pryor, 1963)

can (a) A term used in the tristate zinc and lead district for a bucket used in hoisting. A "can" ranges in capacity from 1,200 to 1,400 lb (544 to 635 kg). (Jackson, 3) (b) In a nuclear reactor, the container in which fuel rods are sealed to protect the fuel from corrosion and prevent gaseous diffusion products from escaping into the coolant. (Hammond, 1965)

cañada (a) A term used in the Western United States for a ravine, glen, or narrow valley, smaller and less steep-sided than a canyon, such as the V-shaped valley of a dry river bed; a dale or open valley between mountains. (AGI, 1987) (b) A term used in the Western United States for a small stream; a creek. Etymol: Spanish caña, cane, reed. (AGI, 1987)

Canadian asbestos *chrysotile.*

Canadian shield The vast region of Precambrian rocks having an areal extent of 2 million square miles (5.2 million km^2) in eastern Canada. (CTD, 1958)

canal (a) An artificial watercourse cut through a land area for use in navigation, irrigation, etc. (Hunt, 1965) (b) That part of a tank leading from the relatively wide fining area to the machine. (ASTM, 1994) (c) *chute; ditch.*

canal ray *positive ray.*

Canamin clay A clay consisting mainly of colloidal aluminum silicate from British Columbia, Canada. (CCD, 1961)

canary (a) Bird traditionally used for the detection of unsafe carbon monoxide or low oxygen levels in early coal mines. (b) Term used for modern, handheld, electronic air quality monitors, which replaced the use of canary birds. (c) Yellow diamond. (Schaller, 1917)

canary ore A yellow, earthy argentiferous lead ore, generally pyromorphite, bindheimite, or massicot, more or less impure. (Fay, 1920)

canary stone A yellow variety of carnelian.

canch (a) A part of a bed of stone worked by quarrying. (b) Eng. Roof or floor removed to make height and side removed to make width. If above the seam, it is called a top canch; if below the seam, a bottom canch. A canch on a roadway close to the face is called a face canch; a canch on a roadway outbye is called a back canch. Also called brushing; ripping. (SMRB, 1930) (c) The face of the roof ripping in a roadway. It follows that the canch is continually being excavated and advanced. See also: *ripping face support.* Also called ripping lip. (Nelson, 1965)

canche A trench with sloping sides and a very narrow bottom. (Zern, 1928)

cancrinite (a) A hexagonal mineral, $Na_6Ca_2Al_6Si_6O_{24}(CO_3)_2$. (b) The mineral group afghanite, cancrinite, davyne, franzinite, guiseppettite, liottite, microsommite, sacrofanite, vishnevite, and wenkite(?).

candite *ceylonite.*

candle coal *cannel coal; kennel coal.*

candlepower (a) The illuminating power of a standard sperm candle. Used as a measure for other illuminants. (Crispin, 1964) (b) The luminous flux emitted by a source of light per unit solid angle in a given direction. It is expressed in terms of the international candle and new candle. (CTD, 1958)

canel *cannel coal.*

canfieldite (a) This name was first given to an isometric silver sulfogermanate, believed to be a new species, but later proved to be identical with argyrodite. The name was then withdrawn and transferred to (b). (English, 1938) (b) An orthorhombic mineral, Ag_8SnS_6; black; forms a series with argyrodite.

Canfield's reagent An etchant, used for revealing phosphorus segregation in iron and steel, containing 1.5 g cupric chloride, 5 g nickel nitrate, and 6 g ferric chloride, in 12 mL hot water. (Osborne, 1956)

canga (a) Braz. A tough, well-consolidated rock consisting essentially of hard blocks and fragments of the rocks of an iron formation, cemented with limonite. Where these fragments are plentiful and are derived from the hard ore outcrops, canga forms a valuable ore, which may run as high as 68% iron. Generally it is phosphoric, but there are considerable areas in which the phosphorus is below the Bessemer limit. Cf: *itabirite.* (b) A ferruginous laterite developed from any iron-bearing rock, commonly basalt or gabbro; e.g., as used in Sierra Leone, canga is equivalent to lateritic iron ore. (AGI, 1987)

can hoisting system A method of hoisting in shallow lead-zinc mines in areas of the United States. Instead of the conventional engine house, operation is controlled at the top of the shaft. The onsetter below hooks the can on, then signals by a lamp attached to the wrist of the hoister sitting above. The can is hoisted, swinging free. At the surface a tail rope is snapped to the underside, a deflection plate is swung into place, and the can is lowered. It capsizes and discharges its load to the surge bin. The empty can is then again hoisted, freed of its tail rope, and wound down the shaft, where it is replaced by a full can. (Pryor, 1963)

canister (a) A hopper-shaped truck, from which coal is discharged into coke ovens. (Fay, 1920) (b) A container with a filter, sorbent, or catalyst, or combination of these items, which removes specific contaminants from the air passed through the container. Also called cartridge. (ANSI, 1992)

cank (a) York. A completely cemented, compact, and fine-grained sandstone, or any fine-grained rock that is hard to drill. (b) A hard, dark gray massive rock consisting largely of ankerite, found in some Coal Measures marine beds. (BS, 1964)

canker (a) Eng. The ocherous sediment in mine waters, being bicarbonate of iron precipitated by the action of the air. (Fay, 1920) (b) Rust; verdigris, or copper rust. (Webster 2nd, 1960)

cannel *cannel coal.*

cannel coal (a) Term used for sapropelic coal containing spores, in contrast to sapropelic coal containing algae, which is termed boghead coal. Viewed microscopically, cannel coal shows no stratification. It is generally dull and has a more or less pronounced waxy luster. It is very compact and fractures conchoidally. There are transitions between cannel coal and boghead coal, and it is not possible always to distinguish macroscopically between them. Such a distinction can, however, be easily made a with microscope, except in high-rank coals. In American nomenclature, cannel coal must contain less than 5% anthraxylon. Cannel coal occurs in layers or lenses up to several centimeters in thickness. Thin seams consisting entirely of cannel coal are known. It occurs widely but in limited amounts. Syn: *gayet.* Analogous term is parrot coal. See also: *sapropelic coal; spore coal; boghead coal.* (IHCP,

1963) (b) A variety of bituminous or subbituminous coal of uniform and compact fine-grained texture with a general absence of banded structure. It is dark gray to black in color, has a greasy luster, and is noticeably of conchoidal or shell-like fracture. It is noncaking, yields a high percentage of volatile matter, ignites easily, and burns with a luminous smoky flame. Syn: *canel; cannel; candle coal; kennel coal.*

cannel shale (a) A shale in which the mineral and the organic matter are approx. in equal proportions. (Tomkeieff, 1954) (b) A black shale formed by the accumulation of sapropels accompanied by a considerable quantity of inorganic material, chiefly silt and clay. Syn: *bastard shale.* (AGI, 1987)

cannes marble Same as griotte marble; a reddish marble with white spots formed by fossil shells (goniatites). (Hess)

cannonball mill A mill for grinding tough materials by attrition, using cannonballs in a rotating drum or chamber. See also: *ball mill.* (Fay, 1920)

cannon shot *blown-out shot.*

canny Corn. Applied to lodes containing calcium carbonate and fluorspar.

cañon *canyon.*

cansa Hydrated Brazilian hematite ore resulting from the weathering of itabirite. (Osborne, 1956)

cantilever A lever-type beam that is held down at one end, is supported near the middle, and supports a load on the other end. (Nichols, 1976)

cantilever crane A transporter crane with one or both ends overhanging. (Hammond, 1965)

cantilever grizzly Grizzly fixed at one end only, the discharge end being overhung and free to vibrate. This vibration of the bar is caused by the impact of the material. The disadvantage of the ordinary bar grizzly is clogging due to the retarding effect of the cross rods. This has been overcome in the cantilever grizzly by eliminating the tie rods except at the head end, where they are essential. The absence of these rods below the point of support also aids in preventing clogging because it permits the bars to vibrate in a horizontal plane, which keeps the material from wedging. (Pit and Quarry, 1960)

cantonite A covellite that occurs in cubes with cubic cleavage and is probably pseudomorphous after chalcopyrite that had replaced galena; from the Canton Mine, Georgia. (Hess)

canvas Usually applied to brattice cloth, which is a heavy canvas of cotton, hemp, or flax, frequently fireproofed. (Jones, 1949)

canvas door A simple square frame of about 2-in by 2-in (5.1-cm by 5.1-cm) pieces tied with diagonal strips and covered with brattice; used for deflecting air currents at inby points where the pressure is low. (Nelson, 1965)

canvas table Inclined rectangular table covered with canvas. The pulp, to which clear water is added if necessary, is evenly distributed across the upper margin. As it flows down, the concentrates settle in the corrugations of the canvas. After the meshes are filled, the pulp feed is stopped, the remaining quartz is washed off with clear water, and finally the concentrates are removed (by hose or brooms). (Liddell, 1918)

canyon (a) A long, deep, relatively narrow steep-sided valley confined between lofty and precipitous walls in a plateau or mountainous area, often with a stream at the bottom; similar to, but larger than, a gorge. It is characteristic of an arid or semiarid area (such as the Western United States) where stream downcutting greatly exceeds weathering; e.g., the Grand Canyon. (AGI, 1987) (b) Any valley in a region where canyons abound. Etymol: anglicized form of American Spanish cañón. Cf: *cañada.* Syn: *cañon.* (AGI, 1987) (c) A precipitous valley; a gorge. Also spelled canon. (d) Mex. A mine-level drift or gallery.

cap (a) A detonator or blasting cap. (Nelson, 1965) (b) To seal, plug, or cover a borehole. (Long, 1960) (c) The roof or top piece in a three-piece timber set used for tunnel support. (Nichols, 1976) (d) A piece of plank or timber placed on top of a prop, stull, or post. (Long, 1960) (e) The horizontal member of a set of timber used as a roadway support. (Nelson, 1965) (f) Another name for crown. (ASTM, 1994) (g) Barren rock and/or soil covering an ore deposit. See also: *cap rock.* (Long, 1960) (h) Overburden consisting of unconsolidated material overlying or covering bedrock. Also called cover; mantle. Syn: *top.* (Long, 1960) (i) *blue cap.*

capacitance (a) The capacity to store electrical energy; measured in farads, microfarads, or micro-microfarads. (Hunt, 1965) (b) In flotation, a property expressible by the ratio of the time integral of the flow rate of material or electric charge to or from a storage, divided by the related potential change. (Fuerstenau, 1962)

capacitive control An alternative to inductive control is to employ a capacitor in series with the choke and therefore to obtain a leading power factor for the circuit. The current in a capacitive circuit is less affected by changes in voltage than that in an inductive circuit. Therefore, should there be a sudden drop in mains voltage, the capacitively controlled lamp is less likely to be extinguished than the inductively controlled lamp. (Roberts, 1958)

capacitor An electric appliance or an adjustable electric appliance used in circuit with a motor to adjust the power factor. (Pryor, 1963; Kentucky, 1952)

capacitor-discharge blasting machine A blasting machine in which electrical energy, stored in a capacitor, is discharged into a blasting circuit containing electric detonators.

capacity (a) As applied to mines, smelters, and refineries, the maximum quantity of product that can be produced in a period of time on a normally sustainable long-term operating rate—based on the physical equipment of the plant, and given acceptable routine operating procedures involving labor, energy, materials, and maintenance. (b) As applied to diamond and rotary drills, the load that the hoisting and braking mechanism of a diamond or rotary drill is capable of handling on a single line, expressed in feet or meters as the depth to which the drill can operate with different size bits. (Long, 1960) (c) In ore dressing, the capacity of a screen is the measure of the amount of material that can be screened in a given time, and is typically measured in tons per square foot per hour per millimeter of aperture. (Newton, 1959)

capacity factor (a) The ratio between the breaking strength of a winding rope and the load suspended on it (excluding the weight of the rope itself). (Nelson, 1965) (b) A method of assessing the size of a rope. The capacity factor of the rope is the static factor of safety of the rope at the capping; i.e., the breaking strength of the rope divided by the weight of the loaded cage or skip and the suspension gear comprising the chains, or equivalent equipment, and a detaching hook. (Sinclair, 1959)

capacity load The maximum load that can be carried safely. (Crispin, 1964)

capacity of the market As applied to mining, the ability of the market to buy, esp. with regard to the quantity that can be placed in the market, and to the prices that can be obtained. (Stoces, 1954)

cap crimper A mechanical device for crimping the metallic shell of a fuse detonator or igniter cord connector securely to a section of inserted safety fuse. See also: *crimper.* (Atlas, 1987)

Cape blue Crocidolite asbestos found near Prieska, South Africa. See also: *crocidolite.* (Pryor, 1963)

cape diamond A diamond with a yellowish tinge. (Pryor, 1963)

capel (a) A wall of a lode; so called by Cornish miners, primarily where the country rock adjacent to the lode has been more or less altered by the same mineralizing agencies through which the lode was formed. Syn: *carrack; cappel; capping.* See also: *capel lode.* (b) A fitting at the end of the winding rope to enable the bridle chains of the cage to be connected by a pin through the clevis.

capel lode Corn. A lode composed of hard unpromising feldspar containing minute particles of chlorite. See also: *capel.*

Cape ruby Brilliantly red garnet, gem stone. Other varieties are carbuncle and Bohemian garnet. Cf: *pyrope.* (Pryor, 1963)

capillarity (a) The action by which a fluid, such as water, is drawn up (or depressed) in small interstices or tubes as a result of surface tension. Syn: *capillary action.* (AGI, 1987) (b) The state of being capillary. (AGI, 1987) (c) A phenomenon observable when making borehole inclination surveys by the acid-etch method, wherein the upper surface of the acid curves upward, forming a concave surface. When the acid bottle is in a vertical or horizontal position, the concave surface is symmetrical, and the resultant etch plane is horizontal. When the bottle is tilted, the concave surface is asymmetric; the resultant etch plane is not horizontal, and the angle so indicated is always greater than the true inclination of the borehole. A capillarity correction is applied. See also: *etch angle; capillarity correction.* (Long, 1960) (d) The action by which the surface of a liquid, where it is in contact with a solid, is elevated or depressed depending upon the relative attraction of the molecules of the liquid for each other and for those of the solid. Esp. observable in capillary tubes, where it determines the elevation or depression of the liquid above or below the level of the liquid in which the tube is dipped. (Webster 3rd, 1966)

capillarity correction The deduction of a specific angular value from the apparent angle, as indicated by the plane of the etch line in an acid-survey bottle, to correct for capillarity effects and thereby determine the true inclination angle of a borehole. Proper values to be deducted from the apparent angles read on acid bottles differing in size may be determined by referring to charts, graphs, or tables prepared for that purpose. See also: *capillarity; capillarity-correction chart.* (Long, 1960)

capillarity-correction chart A chart, graph, or table from which the amount of capillarity correction may be ascertained and applied to an angle reading taken from an acid-etch line in an acid bottle of specific size to determine the true angle of inclination of a borehole surveyed by the acid-etch method. Also called correction chart; test-correction chart. See also: *capillarity correction.* (Long, 1960)

capillary (a) The action by which the surface of a liquid is elevated at the point at which it is in contact with a solid (such as in a lamp wick). See also: *capillarity.* (Shell) (b) Resembling a hair; fine, minute, slender; esp., having a very small or thin bore usually permitting capillary. (Webster 3rd, 1966) (c) Said of a mineral that forms hairlike or threadlike crystals, e.g., millerite. Syn: *filiform; moss; wire; wiry.* (d) Said of tubes or interstices with such small openings that they can retain fluids by capillarity.

capillary action *capillarity.*

capillary attraction The adhesive force between a liquid and a solid in capillarity. (AGI, 1987)

capillary movement The rise of subsoil water above the water table through the channels connecting the pores in the soil. (Nelson, 1965)

capillary pyrite *millerite.*

capillary water (a) Water held in, or moving through, small interstices or tubes by capillarity. The term is considered obsolete. Syn: *water of capillarity.* (AGI, 1987) (b) Water of the capillary fringe. (AGI, 1987)

capital expenditure The amount of money required for the purchase of the right to mine a deposit, for its preliminary development, for the purchase of adequate equipment and plant to operate it, and for working capital. (Hoover, 1948)

capital scrap Scrap from redundant manufactured goods and equipment, collected and processed by merchants. See also: *process scrap.* (Nelson, 1965)

capitan limestone Massive white limestone found in New Mexico and Texas. (Hess)

cap lamp The term generally applied to the lamp on a miner's safety hat or cap. Used for illumination only. See also: *safety lamp; miner's electric cap lamp.* (BCI, 1947)

caple Corn. A hard rock lining tin lodes. See also: *capel.*

cap light (a) Dry-cell type. A self-contained light that permits free use of the hands and may be suitable for gaseous or explosive atmospheres. The headlamp, with focusing lens and bulb, is strapped to the head or hat, and the dry cell battery unit can be clipped to the belt. To prevent explosion, the bulb-socket ejects the bulb automatically in case of breakage. (Best, 1966) (b) Wet-cell type. With rechargeable, wet-cell cap lights, the battery is worn on the belt, and the light unit, which is attached to the cap or head, contains bulbs filled with krypton gas. The head light contains either two separate bulbs or a single bulb with two filaments in parallel, thus assuring the wearer of a constant source of light in the event that one bulb or one filament burns out. (Best, 1966)

Cappeau furnace A modification of the Ropp furnace for calcining sulfide ore. (Fay, 1920)

capped fuse A length of safety fuse to which a blasting cap has been attached.

capped primer A package or cartridge of cap-sensitive explosive which is specif. designed to transmit detonation to other explosives and which contains a detonator.

capped quartz A variety of quartz containing thin layers of clay. (Fay, 1920)

cappel; capping *capel.*

cappelenite A trigonal mineral, $Ba(Y,Ce)_6Si_3B_6O_{24}F_2$: weakly radioactive; occurs in veins in syenite associated with wohlerite, rosenbuschite, catapleiite, orangite, lavenite, elaeolite, and sodalite.

cap piece (a) A piece of wood usually 24 to 36 in (60.96 to 91.44 cm) long, 6 to 8 in (15.24 to 20.32 cm) wide, and 2 to 6 in (5.08 to 15.24 cm) thick, that is fitted over a straight post or timber to afford more bearing surface for the support. All single posts, or timbers including safety posts, should be covered with a cap piece to provide additional bearing surface. (Kentucky, 1952) (b) Arkansas. Usually a piece of wood split from a log. (Fay, 1920)

capping (a) Syn: *overburden.* Usually used for consolidated material. (AGI, 1987) (b) The overburden or rock deposit overlying a body of mineral or ore. (Nelson, 1965) (c) *gossan.* (d) The preparation of capped fuses. (Nelson, 1965; Lewis, 1964) (e) The process of sealing or covering a borehole and/or the material or device so used. (f) The separation of a block of stone along the bedding plane. (g) The attachment at the end of a winding rope. See also: *continental gland-type capping; interlocking wedge-type capping; white-metal cappel; capel.* (Sinclair, 1959) (h) The fixing of a shackle or a swivel to the end of a hoisting rope. (CTD, 1958) (i) The operation of fastening steel rope to a winding cage. (Pryor, 1963) (j) The name given to a method by which the spouting flow of a liquid or gas from a borehole may be stopped or restricted; also, the mechanism attached to borehole collar piping and so used. (Long, 1960)

capping station A special room or building used solely for the preparation of capped fuses. (Nelson, 1965)

cap rock (a) Barren vein matter, or a pinch in a vein, supposed to overlie ore. Syn: *cap.* (b) A hard layer of rock, usually sandstone, a short distance above a coal seam. (c) A disklike plate over part of or all of the top of most salt domes in the Gulf Coast States and in Germany. It is composed of anhydrite, gypsum, limestone, and sometimes sulfur. (AGI, 1987) (d) A comparatively impervious stratum immediately overlying an oil- or gas-bearing rock. (AGI, 1987) (e) Eng. The cap rock of the alum shale, Estuarine sandstones on the Yorkshire coast. (Arkell, 1953)

capsal *capstan.*

cap sensitivity The sensitivity of an explosive to initiation by a detonator. An explosive material is considered to be cap sensitive if it detonates with a No. 8 strength test detonator. (Atlas, 1987)

cap set A term used in square-set mining methods to designate a set of timber using caps as posts, resulting in a set of timber shorter than the normal set.

cap shot A light shot of explosive placed on the top of a piece of shale that is too large to handle, in order to break it. (BCI, 1947)

cap sill The upper horizontal beam in the timber framing of a bridge, viaduct, etc. (Fay, 1920)

capstan (a) A spoollike drum mounted on a vertical axis used for heave hoisting or pulling. It is operated by steam, electric power, or hand pushes or pulls against bars inserted in sockets provided in the upper flange or head. (Long, 1960) (b) Sometimes used as a syn. for cathead. (Long, 1960)

captive mine Aust. A mine that produces coal or mineral for use by the same company. (Nelson, 1965)

captive tonnage The quantity of mineral product from a mine produced solely for use by the parent company or subsidiary.

capture In a crystal structure, the substitution of a trace element for a major element of lower valence; e.g., Ba^{+2} for K^{+}. Captured trace elements generally have a higher concentration relative to the major element in the mineral than in the fluid from which it crystallized. Cf: *admittance; camouflage.* (AGI, 1987)

car (a) A wheeled vehicle used for the conveyance of coal or ore along the gangways or haulage roads of a mine. Also called mine car; tram-car; tub; wagon; mine wagon. (Zern, 1928; Fay, 1920) (b) A wheeled carrier that receives and supports the load to be conveyed. Generally attached to a chain, belt, cable, linkage, or other propelling medium. See also: *tray.*

caracolite A monoclinic mineral, $Na_3Pb_2(SO_4)_3Cl$; pseudohexagonal; forms crystalline incrustations.

Carapella's reagent An etchant consisting of 5 g of ferric chloride dissolved in 96 mL of ethyl alcohol to which has been added 2 mL of hydrochloric acid; used in etching nonferrous metals and manganese steels. (Osborne, 1956)

carat (a) A unit of weight for diamonds, pearls, and other gems; formerly equal to 3-1/6 troy grains (205 mg). The international metric carat (abbreviated M.C.) of 200 mg was made the standard in the United States in 1913, as it was the standard in Belgium, Denmark, Great Britain, France, Germany, Japan, the Netherlands, and Sweden. A carat grain is 1/4 carat. Syn: *international metric carat.* Not to be confused with "karat." Cf: *point.* (Webster 3rd, 1966) (b) Employed to distinguish the fineness of a gold alloy, and meaning 1/24 part. Pure gold is 24-carat gold. Goldsmiths' standard is 22 carats fine; it contains 22 parts of gold, 1 part of copper, and 1 part of silver. (Fay, 1920)

caratage *carat weight.*

carat count The number of near-equal-size diamonds having a total weight of 1 carat or 200 mg; hence, 40 small diamonds weighing 1 carat would be called 40-count diamonds, or 8 diamonds weighing 1 carat would be called 8-count diamonds. (Long, 1960)

carat-goods Diamonds averaging about 1 carat each in weight. (Long, 1960)

carat loss Amount of diamond material lost or worn away by use in a drill bit, expressed in carats. (Long, 1960)

carat weight Total weight of diamonds set in a drill bit, expressed in carats. Also called caratage. (Long, 1960)

carbankerite Any coal microlithotype containing 20% to 60% by volume of carbonate minerals (calcite, siderite, dolomite, and ankerite). (AGI, 1987)

carbargilite Any coal microlithotype containing 20% to 60% by volume of clay minerals, mica, and in lesser proportions, quartz.

carbide (a) A commercial term for calcium carbide formerly used in miner's lamps. (Fay, 1920) (b) The carbide compound of tungsten. (Long, 1960) (c) The bit-crown matrices and shaped pieces formed by the pressure molding and sintering of a mixture of powdered tungsten carbide and other binder metals, such as cobalt, copper, iron, and nickel. See also: *cemented carbide; sintered carbide.* (Long, 1960) (d) A compound of carbon with one or more metallic elements. (ASM, 1961)

carbide insert Shaped piece of a hard metal compound, sometimes inset with diamonds, formed by the pressure molding and sintering of a mixture of powdered tungsten carbide and other binder metals, such as iron, copper, cobalt, or nickel. Inset into holes, slots, or grooves in bits, reaming shells, or core barrels, the hard metal pieces become cutting points or wear-resistant surfaces. Also called carbide slug. (Long, 1960)

carbide lamp A lamp that is charged with calcium carbide and water and burns the acetylene generated. Syn: *acetylene lamp.* (Hess)

carbide miner A push-button mining machine with a potential range of 1,000 ft (304.8 m) into a seam from the highwall, a maximum production of some 600 st (544 t) per shift, and a recovery of 65% to 75% of the coal within the reach of the machine. This unit is a continuous miner working controlled from outside the seam of coal. The operator can control both the vertical and horizontal direction of the cutting heads as shown on an oscilloscope screen. As the cutting head advances into the coal seam, it drags a series of conveyor sections behind it, which in turn deposit the coal into a truck. (Krumlauf)

carbides Compounds of carbon with iron and other elements in steel; e.g., Fe_3C (cementite), Fe_4W_2C, and Cr_4C_2. (CTD, 1958)

carbide slug *carbide insert.*

carbide tool A cutting tool—made of tungsten carbide, titanium carbide, tantalum carbide, or combinations of them, in a matrix of cobalt or nickel—having sufficient wear resistance and heat resistance to permit high machining speeds. (ASM, 1961)

Carbite Trade name for an explosive. (Hess)

carbo- A combining form meaning carbon, as in carbohydrate.

carbo A Latin name for charcoal, later transferred to fossil coal. (Tomkeieff, 1954)

Carbo Clay-bonded silicon carbide; used as refractory. (Bennett, 1962)

carbocher A variety of hydrocarbon containing about 8% rare earths and found enclosed in a mineral kondrikite. From the Khibine Peninsula, Russia. (Tomkeieff, 1954)

carbodynamite A form of dynamite in which fine charcoal is used as the absorbent. (Webster 2nd, 1960)

carbohumin An amorphous carbonaceous substance, a product of decomposition of plants and impregnating plant remains, which undergo transformation into coal. It is assumed to be present in coal in the form of structureless jelly. Syn: *jelly; fundamental jelly; fundamental substance; gélose; jelly; vegetable jelly.* (Tomkeieff, 1954)

carbolic (a) Of, pertaining to, or derived from carbon and oil; of or pertaining to coal-tar oil. (Standard, 1964) (b) Of or pertaining to carbolic acid. (Hess)

carbolic acid White; crystalline; deliquescent; C_6H_5OH; a burning taste; and an odor resembling that of creosote. Contained in the heavy oil of coal tar, from which it is distilled at between 165 °C and 190 °C. It is a caustic poison. Antidotes are epsom salts, alcohol, and heat. See also: *phenol.* (Standard, 1964)

carbolite A byproduct in iron smelting, consisting of calcium-aluminum silicon carbide; used as a substitute for calcium carbide. (Standard, 1964)

carbon (a) A nonmetallic element, found free in nature in three allotropic forms: amorphous, graphite, and diamond. A fourth form, known as "white" carbon, is now thought to exist. Symbol, C. Graphite is one of the softest known materials, while diamond is the hardest. Occurs as a constituent of coal, petroleum, natural gas, and all organic compounds. The isotope, carbon 14, is radioactive and is used as a tracer in biological and organic chemical research. (Handbook of Chem. & Phys., 3) (b) Rand term for thucolite in banket ore. (Pryor, 1963) (c) A gray-to-black, opaque, tough, hard cryptocrystalline aggregate of diamond crystals occurring in irregular shapes and sizes. It is classed as an industrial diamond and formerly was used extensively as a cutting-medium inset in diamond-drill bits. More recently, only occasionally used in diamond bits and other tools. Also called black diamond; carbonado. See also: *diamond.* (Long, 1960)

carbon-14 A radioactive isotope of carbon having the atomic weight of 14, produced by collisions between neutrons and atmospheric nitrogen. It is useful in determining the age of carbonaceous material younger than 30,000 years old. See also: *carbon; carbon-14 dating.*

carbon-14 dating A method of determining an age in years by measuring the concentration of carbon-14 remaining in an organic material, usually formerly living matter, but also water, bicarbonate, etc. The method is based on the assumption that assimilation of carbon-14 ceased abruptly on the death of an organism and that it thereafter remained a closed system. The method is useful in determining ages in the range of 500 to 30,000 years or 40,000 years, although it may be extended to 70,000 years by using special techniques involving controlled enrichment of the sample in carbon-14. Syn: *radiocarbon dating; carbon dating.* (AGI, 1987)

carbonaceous (a) Coaly, containing carbon or coal, esp. shale or other rock containing small particles of carbon distributed throughout the whole mass. (Fay, 1920) (b) Carbonaceous sediments include original organic tissues and subsequently produced derivatives of which the composition is organic chemically. (AGI, 1987)

carbonado Cryptocrystalline diamond; compact, tough, opaque, dark-gray to black, cleavage absent; generally in rounded masses, also in angular broken fragments. Principal source is Bahia, Brazil, but also found elsewhere in South America and Africa. Syn: *black diamond; carbon diamond.* Cf: *ballas.*

carbonado bit *carbon bit.*

carbon adsorption Recovery of dissolved soluble constituents onto activated carbon due to some form of chemical sorption at the active sites. Carbon adsorption is particularly useful for removing gold and silver from cyanide leach solutions or dissolved organics from process solutions. (Van Zyl, 1988)

carbonate (a) A compound containing the acid radical CO_3 of carbonic acid. Bases react with carbonic acid to form carbonates. Cf: *carbonate.* (CTD, 1958) (b) A mineral compound characterized by a fundamental anionic structure of $(CO_3)^{2-}$. Calcite and aragonite, $CaCO_3$, are examples of carbonates. Cf: *borate; nitrate.* (AGI, 1987) (c) A sediment formed by the organic or inorganic precipitation from aqueous solution of carbonates of calcium, magnesium, or iron; e.g., limestone and dolomite. See also: *carbonate rock.* (AGI, 1987) (d) Ores containing a considerable proportion of metal carbonates. (Fay, 1920) (e) Salts of carbonic acid, H_2CO_3. (Henderson, 1953)

carbonate-fluorapatite *dehrnite.*

carbonate hardness Hardness of water, expressed as $CaCO_3$, that is equivalent to the carbonate and bicarbonate alkalinity. When the total alkalinity, expressed as $CaCO_3$, equals or exceeds the total hardness, all the hardness is carbonate. It can be removed by boiling and hence is sometimes called temporary hardness, although this syn. is becoming obsolete. Syn: *hardness.* (AGI, 1987)

carbonate leach (a) Metallurgical process for dissolution of metal values by means of a sodium carbonate solution. Used on high-lime ores. (Ballard, 1955) (b) Dissolution of uranium with an aqueous solution of sodium carbonate in the presence of sufficient oxygen to render uranium hexavalent. (Pryor, 1958) (c) Tungsten autoclave dissolution.

carbonate mineral A mineral formed by the combination of the radical $(CO_3)^{2-}$ with cations; e.g., calcite, $CaCO_3$.

carbonate of barium *witherite.*

carbonate of calcium *calcite.*

carbonate of strontium *strontianite.*

carbonate rock A rock, such as limestone, dolomite, or carbonatite, that consists chiefly of carbonate minerals; specif. a sedimentary rock composed of more than 50% by weight of carbonate minerals. Syn: *calcareous rock.* (AGI, 1987)

carbonate sand A sand derived predominantly from carbonate material such as corals, mollusc shells, algae, etc. (Cruickshank, 1987)

carbonation (a) A process of chemical weathering involving the transformation of minerals containing calcium, magnesium, potassium, sodium, and iron into carbonates or bicarbonates of these metals by carbon dioxide contained in water (i.e., a weak carbonic-acid solution). Syn: *carbonatization.* (AGI, 1987) (b) Introduction of carbon dioxide into a fluid. (AGI, 1987)

carbonatite A carbonate rock of apparent magmatic origin, generally associated with kimberlites and alkalic rocks. Carbonatites have been variously explained as derived from magmatic melt, solid flow, hydrothermal solution, and gaseous transfer. (AGI, 1987)

carbonatization (a) Introduction of, or replacement by, carbonates. (AGI, 1987) (b) *carbonation.*

carbon bit A diamond bit in which the cutting medium is inset carbon. (Long, 1960)

carbon brick Brick usually made from crushed coke and bonded with pitch or tar.

carbon dating *carbon-14 dating.*

carbon diamond *carbonado.*

carbon dioxide (a) Heavy, colorless; irrespirable gas; CO_2; it extinguishes a flame. It is formed in mine explosions and in mine fires and forms part of the afterdamp. (b) Product of complete combustion of carbon fuels. Transported in liquid form in steel cylinders. Used in gaseous form as a fire extinguisher and in solid form as dry ice. (Crispin, 1964)

carbon dioxide blasting A method of blasting coal that has been undercut, topcut, or sheared. Into one end of a seamless high-grade molybdenum-steel cylinder 2 to 3 in (5.08 to 7.62 cm) in diameter and 36 to 60 in (91.44 to 152.4 cm) long is put a cartridge containing a mixture of potassium perchlorate and charcoal with an electric match. The other end is sealed by a metal disk weaker than the shell and held in place by a cap that has holes at about 45° to the axis of the cylinder. The cylinder is filled with liquid carbon dioxide at a pressure of 1,000 psi (6.9 MPa) and inserted in a borehole with the cap holes pointing outward. The heating mixture is lit and raises the gas pressure so that the disk is sheared; the carbon dioxide escaping through the angular holes tends to hold the cylinder in place, and break and push the coal forward. If the gas pressure is not enough to break the coal, the cylinder, if not properly set, will be blown from the borehole. The cylinder can be used over and over. It is claimed that a greater portion of lump coal is obtained than with ordinary explosives. Some smelters loosen slag in the same way. (Hess)

carbon-hydrogen ratio A method of classifying coals by determining the ratio that exists between the carbon and hydrogen present in them. Thus, if a given coal contains 80% carbon and 5% hydrogen, the C/H ratio would be 80:5, or 16. Bituminous coals have a C/H ratio between 14 and 17, and most anthracites have a ratio between 24 and 29. Abbreviation: C/H ratio. See also: *anthracite.* (Nelson, 1965)

Carboniferous The Mississippian and Pennsylvanian periods combined, ranging from about 345 million years to about 280 million years ago; also, the corresponding systems of rocks. In European usage, the Carboniferous is considered as a single period and is divided into upper and lower parts. The Permian is sometimes included. (AGI, 1987)

carbonification Carbonification is the process by which the vegetable substances of peat were transformed in the partial absence of air and under the influence of temperature and pressure throughout geological time into lignite and subsequently into coal. See also: *coalification.* (IHCP, 1963)

carbon-in-leach process A process step wherein granular activated carbon particles much larger than the ground ore particles are introduced into the ore pulp. Cyanide leaching and precious metals adsorption onto the activated carbon occur simultaneously. The loaded activated carbon is mechanically screened to separate it from the barren ore pulp and processed to remove the precious metals and prepare it for reuse. (SME, 1992)

carbon-in-pulp leaching A precious metals leaching technique in which granular activated carbon particles much larger than the ground ore particles are added to the cyanidation pulp after the precious metals have been solubilized. The activated carbon and pulp are agitated together to enable the solubilized precious metals to become adsorbed onto the activated carbon. The loaded activated carbon is mechanically screened to separate it from the barren ore pulp and processed to remove the precious metals and prepare it for reuse. (SME, 1992)

carbonite (a) A native coke, occurring at the Edgehill Mines, near Richmond, VA; it is more compact than artificial coke and some varieties afford bitumen. (Fay, 1920) (b) Coal altered by an igneous intrusion. Syn: *cokeite.* (Tomkeieff, 1954) (c) Fossil coal. (Tomkeieff, 1954) (d) Very brittle, black variety of bitumen, infusible and insoluble in organic solvents, containing about 85% carbon and 6% hydrogen. (Tomkeieff, 1954) (e) A permissible explosive. (Fay, 1920)

carbonitriding Introducing carbon and nitrogen into a solid ferrous alloy above AC_1 in an atmosphere that contains suitable gases such as hydrocarbons, carbon monoxide, and ammonia. The carbonitrided alloy is usually quench-hardened. (ASM, 1961)

carbonization (a) In the process of coalification, the accumulation of residual carbon by the changes in organic matter and decomposition products. See also: *coalification.* (AGI, 1987) (b) The accumulation of carbon by the slow, underwater decay of organic matter. (AGI, 1987) (c) The conversion into carbon of a carbonceous substance such as coal by driving off the other components, either by heat under laboratory conditions or by natural processes. (AGI, 1987)

carbonizing The reduction of a substance to carbon by subjecting it to intense heat in a closed vessel. (Crispin, 1964)

carbon monoxide Colorless; odorless; very toxic gas; CO; burns to carbon dioxide with a blue flame. Formed as a product of the incomplete combustion of carbon (such as in water gas and producer gas; in the exhaust gases from internal-combustion engines, such as automotive; and in the gases from the detonation of explosives). Used chiefly in the synthesis of carbonyls (such as nickel carbonyl in the refining of nickel), phosgene, and many organic compounds (such as hydrocarbons for fuels, methanol and higher alcohols, aldehydes, and formates). This gas is formed during mine fires and after explosions.

carbon monoxide poisoning In diving, this type of accident usually occurs as a result of contamination of the diver's air supply by exhaust gases from an internal-combustion engine. (Hunt, 1965)

Carbon oil Trade name for kerosine.

carbon steel Steel containing carbon up to about 2% and only residual quantities of other elements except those added for deoxidation, with silicon usually limited to 0.60% and manganese to about 1.65%. Also called plain carbon steel; ordinary steel; straight carbon steel. (ASM, 1961)

carbon trash Carbon remains of plant life found in sedimentary strata and often associated with uranium and red-bed copper mineralization.

carbopyrite Any coal microlithotype containing 5% to 20% by volume of iron disulfide (pyrite and marcasite). (AGI, 1987)

Carborundum Trade name for green, often iridescent, artificial carbon silicide, CSi. Hexagonal-rhombohedral plates. It is produced in an electric furnace and used as an abrasive and as a refractory material. Is useful for sharpening tools. Identical with moissanite. See also: *moissanite.* (Webster 2nd, 1960; English, 1938)

carboxylic acid method In flotation, a method for treatment of various oxygen ores using carboxylic acids as collectors with gangue depressants to float base-metal minerals from associated impurities. The process is suitable for processing apatite (phosphate), carbonates or oxides of lead, copper, or zinc; somewhat less useful with other lead minerals and with hemimorphite; and unsuitable for chrysocolla.

carbozite A black liquid, made from a bituminous ore, used for the protection of steel surfaces during transport and storage. This fluid dries rapidly to a hard gloss, which is resistant to acids, alkalies, moisture, sea air, and temperatures up to 200 °C. (Osborne, 1956)

carburan (a) A hydrocarbon related to, or identical with, thucholite, the ash of which contains uranium, lead, and iron. (Tomkeieff, 1954) (b) A variety of anthraxolite, from pegmatites of Karelia, former U.S.S.R. (Crosby, 1955)

carbureted hydrogen An odorless, flammable gas, CH_4. Known in coal mines as combustible gases or gas. See also: *methane.* (Nelson, 1965)

carburization The process of imparting carbon, such as in making cement steel. (Fay, 1920)

carburizing Hard-surfacing of steel by heating above the critical temperature in an inert atmosphere with a source of carbon (e.g., cyanide salts), thus forming a cementite casing above a tough core (which has already been machined). (Pryor, 1963)

carburizing flame A gas flame that will introduce carbon into some heated metals such as during a gas welding operation. A carburizing flame is a reducing flame, but a reducing flame is not necessarily a carburizing flame. (ASM, 1961)

carcass The tension-carrying portion of a conveyor belt. It may be composed of multiple plies of fabric or cord, and simple layers of cord or steel cable, bonded together with rubber.

car chalker In bituminous coal mining, a laborer who chalks on a car the number of rooms or working places from which coal is obtained in order that a production record of all parts of a mine can be maintained. (DOT, 1949)

card concentrator A table made of two planes having a flexible joint between them dividing the table into two nearly equal triangles, forming a diagonal line along which concentrates separate from the tailings. (Liddell, 1918)

cardinal point (a) One of the four principal "points" of a compass. (AGI, 1987) (b) A change in the speed of the ropes on a winding drum, which occurs at certain definite intervals during the winding cycle. (Sinclair, 1959)

Cardox Trade name for an explosive device used principally in coal mining. See also: *carbon dioxide blasting.*

cardoxide A baked mixture of caustic soda and lime, used in the container or regenerator of self-contained mine-rescue or oxygen-breathing apparatus to absorb the exhaled carbon dioxide. It has an advantage over straight caustic soda in that it does not cake, liquefy, or solidify when used. (Lewis, 1964)

Cardox-plant operator In bituminous coal mining, one who recharges steel shells (tubes) known by the trade name Cardox with metal shearing disks, electrical firing elements, and liquid carbon dioxide to prepare them for blasting coal. (DOT, 1949)

Cardox shell Steel shell used in carbon dioxide blasting.

car dropper *car runner.*

card table A shaking table with a grooved deck instead of nailed-on riffles. Used in gravity concentration of sands. (Pryor, 1963)

card tender In the asbestos products industry, one who tends a carding machine that cleans asbestos, cotton, or other fibers; arranges fibers parallel; and transforms them from a roll or lap into a ropelike untwisted strand of cotton (sliver). Also called allye tender; card feeder; card hand; card operator; winder. (DOT, 1949)

car dump *tipple.*

car dumper (a) A mechanical device for tilting a railroad hopper or gondola car over sidewise and emptying its contents. (Fay, 1920) (b) A person who unloads cars by upending or overturning them.

card weight pipe A term used to designate standard or full weight pipe, which is the Briggs standard thickness of pipe. (Strock, 1948)

car filler *mucker.*

car haul A pusher chain conveyor used for moving small cars, such as mine cars, along a track. A form of tow conveyor.

caries texture In ore microscopy, a replacement pattern in which the younger mineral forms a series of scallop-shaped incursions into the host mineral, which resemble filled dental cavities. (AGI, 1987)

carinate fold In geology, an isoclinal fold. See also: *isocline.* (Standard, 1964)

Carinthian process A metallurgical method for treating lead ore, the characteristics of which are the smallness of the charge; the slow roasting, so that for every part of lead sulfide one part of sulfate and at least two parts of oxide are formed; the low temperature at which all of the operations are carried on; and the aim to extract all the lead in the reverberatory. The hearth is inclined toward the flue, and the lead is collected outside the furnace. Syn: *Corinthian process.* (Fay, 1920)

car loader *loader; chute loader; boxcar loader; loading conveyor.*

Carlsbad twin A twinned crystal in which the twinning axis is the *c* axis, the operation is a rotation of 180°, and the contact surface is parallel to the side pinacoid; common in the alkali feldspars. Also spelled Karlsbad twin. (AGI, 1987)

carman A worker who handles mine or railroad cars at a mine. May be designated according to job, such as brakeman; car cleaner; car pincher; car runner; pusher. Also called car handler. (DOT, 1949)

Carmichel-Bradford process *blast roasting.*

carminite A carmine to tile-red lead-iron-arsenate, perhaps $Pb_3As_2O_8{\cdot}10FeAsO_4$. Found in clusters of fine needles; also in spheroidal forms. (Fay, 1920)
carn *cairn.*
carnallite An orthorhombic mineral, $KMgCl_3{\cdot}6H_2O$; milk-white to reddish; a saline residue.
carnallite plant operator In ore dressing, smelting, and refining, one who makes carnallite flux used in magnesium refining—by weighing carnallite ingredients according to formula and mixing them thoroughly, using a shovel. The mixture is then melted in a furnace crucible and poured into cooling pans. (DOT, 1949)
carnegieite A triclinic and isometric compound, $NaAlSiO_4$; a high-temperature polymorph of nepheline.
carnelian A translucent pale to deep- or orange-red variety of chalcedony containing iron impurities. Cf: *sard.* Also spelled cornelian. Syn: *carneol.*
carneol *carnelian.*
car nipper *car runner.*
carnotite A monoclinic mineral $2[K_2(UO_2)_2(VO_4)_2{\cdot}3H_2O]$; bright yellow to lemon- and greenish-yellow; strongly radioactive; commonly occurs mixed with tyuyamunite; widespread in Colorado, Utah, New Mexico, and Arizona; occurs chiefly in crossbedded sandstones of Triassic or Jurassic age, either disseminated or as relatively pure masses around petrified or carbonized vegetal matter. Secondary in origin, having been formed from the action of meteoritic waters on preexisting uranium minerals; a source of uranium and radium. Syn: *yellow ore.*
carpholite An orthorhombic mineral, $MnAl_2Si_2O_6(OH)_4$; in yellow laths elongated in the *c* direction with prismatic cleavage at 68.5°.
car pincher In anthracite, bituminous, and metal mining, a laborer who moves railroad cars into position directly under loading chutes at a breaker or tipple, inserting a pinch bar under the car wheels and bearing down or pulling up on it to force the car forward. Also called car shifter; car spotter; railroad-car shifter; spotter. (DOT, 1949)
carrack Eng. *capel.*
Carrara marble Any of the marbles quarried near Carrara, Italy. The prevailing colors are white to bluish, or white with blue veins; a fine grade of statuary marble is included.
car retarder (a) An appliance for reducing or controlling the speed of mine cars. (Nelson, 1965) (b) A car retarder consists of a brakeshoe located along the track. On an electrical impulse, it is forced against both sides of the car wheels by compressed air. Control can be manual or automatic. Used to control the speed of railroad cars in industrial yards. (Best, 1966)
carriage (a) A term used with shaker conveyor supports. Carriages may be designated as ball-frame, wheel, or roller carriages, depending on their construction. The carriage may or may not be attached solidly to the conveyor troughs. See also: *slope cage.* (Jones, 1949) (b) *cage.* (c) A sliding or rolling base or supporting frame. (Nichols, 1976)
carriage mounting One or more rock drills mounted on a wheeled frame; used in tunneling. (Pryor, 1963)
Carribel explosive A permitted explosive of medium strength, which can be used in wet boreholes provided its immersion time does not exceed 2 to 3 h. Can be used for coal and ripping shots in conjunction with short-delay detonators. (Nelson, 1965)
carrier (a) A rotating or sliding mounting or case. (Nichols, 1976) (b) Container traveling on an aerial ropeway. (Pryor, 1963)
carrollite An isometric mineral, $Cu(Co,Ni)_2S_4$; linnaeite group. Formerly called sychnodymite.
carrousel conveyor A continuous platform or series of spaced platforms that move in a circular horizontal path.
car runner In anthracite and bituminous coal mining, a laborer who runs cars down inclined haulageways from working places to switches or sidings at the shaft or along main haulageways. A runner may be designated according to material hauled, such as culm runner or rock car runner. Syn: *car dropper; car nipper; dropper; load dropper; runner.* (DOT, 1949)
carry (a) Scot. The thickness of roof rock taken down in working a seam. (b) The thickness of seam that can be conveniently taken down at one working.
carryall A self-loading carrier device with a scraperlike, retractable bottom; usually self-propelled and used esp. for excavating and hauling unconsolidated or crushed rock and earthy materials. See also: *scraper.*
Carryall Trade name for a LeTourneau-Westinghouse scraper.
carryall scraper *carryall.*
carrying belt The belt on which coal or ore is transported to the discharge point. The carrying belt is the upper strand except in the case of a bottom belt conveyor. See also: *carrying run.* (Nelson, 1965)
carrying gate Derb. The main haulage road in a mine.
carrying idler (a) In belt conveyors, one of the belt idlers upon which the load-carrying portion of belting is supported. (b) In live roller conveyors, the roll upon which the load is supported while being conveyed.
carrying roller The conveyor roll upon which the conveyor belt or the object being transported is supported.
carrying run That portion of a conveyor in or on which material is conveyed. See also: *carrying belt.*
car slide The ramped loading platform for a scraper loader. (Nelson, 1965)
car spotter A term used for the small hoist employed to haul a trip of empty cars under the loading end of a gathering conveyor or elevator. Also called tugger. See also: *car pincher.* (Jones, 1949)
car stop A contrivance to arrest the movement of a mine car. (Nelson, 1965)
cartographic Of or pertaining to a map. A cartographic unit in geology is a rock or a group of rocks that is shown on a geologic map by a single color or pattern.
cartography The art of map or chart construction, and the science on which it is based. It includes the whole series of map-making operations, from the actual surveying of the ground to the final printing of the map. (AGI, 1987)
cartology A graphic method of coal-seam correlation, involving the mapping and drawing of both vertical and horizontal sections. (AGI, 1987)
carton A lightweight inner container for explosive materials, usually encased in a substantial shipping container called a case. (Atlas, 1987)
cartridge (a) An individual closed shell, bag, or tube of circular cross section containing explosive material. (Atlas, 1987) (b) A cylindrical, waterproof, paper shell, filled with high explosive and closed at both ends, that is used in blasting. (c) A cylindrical, waterproof, paper shell filled with cement or other material used in plugging or sealing cavities or cavey ground encountered in drilling a borehole. See also: *plug.* (d) Cylinder—about 4 in (10 cm) long and 2-1/2 in (6.4 cm) in diameter—of highly compressed caustic lime made with a groove along the side, used in breaking down coal. (e) A single pellet of explosive, which may be 4 oz or 8 oz (113.4 g or 226.8 g). (Nelson, 1965)
cartridge count The number of cartridges in a standard case, which typically contains about 50 lb (22.7 kg) of explosive material.
cartridge fuse A fuse enclosed in an insulating tube in order to confine the arc when the fuse blows. (Crispin, 1964)
cartridge punch A wooden, plastic, or non-sparking metallic device used to punch an opening in an explosive cartridge to accept a detonator or a section of detonating cord. (Atlas, 1987)
cartridge strength A rating that compares a given volume of explosive with an equivalent volume of straight nitroglycerin dynamite, expressed as a percentage. Syn: *bulk strength.* (Dick, 2)
car-type conveyor A series of cars attached to and propelled by an endless chain or other linkage running on a horizontal or slight incline.
car whacker *mine-car repairman.*
Casagrande liquid limit apparatus An appliance to determine the liquid limit of a soil. It consists of a brass dish, handle, and cam mounted on a hard rubber base. The dish falls through a distance of 1 cm per rotation. A sample of soil 1 cm thick is placed in the dish with a groove 11 mm wide at the top and 2 mm at the bottom. The number of jars required to cause the 2-mm gap to close along 1/2 in (12.7 mm) is recorded. (Nelson, 1965)
cascade coal dryer A thermal process for drying fine coal. An example of this type is the Conreur dryer. Coal entering the top of the drying tower is carried down by a series of rollers, being permeated by an ascending stream of hot air. Fixed baffles direct the air to facilitate mingling. The very finest particles may have to be recovered by dry filters or wet scrubbers. The dryer treats coal with a top size ranging from 1/4 to 2 in (0.64 to 5.08 cm). See also: *fluidized bed dryer; thermal drying.* (Nelson, 1965)
cascade control Externally impressed signal series that connects several controllers or resetting devices in series. (Pryor, 1963)
cascade flotation cell Elementary type of flotation cell in which air is entrained by a plunging cascade of pulp; mineralized bubbles are removed farther downstream. (Pryor, 1963)
cascade upgrading *countercurrent decantation.*
cascading Movement of crop load in a ball mill rotating at such a speed that the balls breaking free at the top of the rising load roll quietly down to the toe of the charge. With increased peripheral speed, motion changes to turbulent cataracting and, still faster, to avalanching when the upper layer of crushing bodies breaks clear and falls freely to the top of the crop load. (Pryor, 1963)
cascadite A sodic minette containing biotite, olivine, and augite phenocrysts in a groundmass composed almost entirely of alkali feldspar. Principally a dike rock. (AGI, 1987)
case (a) A small fissure, admitting water into the mine workings. (b) One of the frames, of four pieces of plank each, placed side by side to form a continuous lining in galleries run in loose earth. (Webster 2nd, 1960) (c) To line a borehole with steel tubing, such as casing or pipe. Syn: *case in.* See also: *blankoff.* (Long, 1960) (d) In a ferrous alloy, the outer portion that has been made harder than the inner portion, or core, by casehardening. (ASM, 1961)

cased A borehole lined with some form of steel tubing, such as casing or pipe. See also: *case off.* (Long, 1960)
cased off *case off.*
casehardening (a) The geological process by which the surface of a porous rock, esp. a sandstone or a tuff, is coated by a cement or a desert varnish; formed by the evaporation of a mineral-bearing solution. (AGI, 1987) (b) Hardening a ferrous alloy so that the outer portion, or case, is made substantially harder than the inner portion, or core. Typical processes used for casehardening are carburizing, cyaniding, carbonitriding, nitriding, induction hardening, and flame hardening. (ASM, 1961)
case in *case.*
case liner A plastic or paper barrier used to prevent the escape of explosive materials from a case. (Atlas, 1987)
case off To line a borehole with some form of steel tubing to prevent entry of broken rock materials, gas, or liquids into the borehole. Also called blank off; case. See also: *blankoff.* (Long, 1960)
cash Som. Soft shale or bind in coal mines.
casing (a) Special steel tubing welded or screwed together and lowered into a borehole to prevent entry of loose rock, gas, or liquid into the borehole, to prevent loss of circulation liquid into porous, cavernous, or crevassed ground, and to support the sides of a borehole. See also: *tubing; flush-joint casing.* (Long, 1960) (b) The large-diameter pipe cemented in the hole, such as surface casing, protective casing, and production casing. (Wheeler, 1958) (c) Process of inserting casing in a borehole. (Long, 1960) (d) A structure of wood, metal, or other material that completely encloses the elevating or conveying machinery elements to support them; to afford safety protection; to protect from the weather; to confine dust, gases, or fumes arising from the material being conveyed; or to form a part of the conveyor in the same manner as a trough. (e) A zone of material altered by vein action and lying between the unaltered country rock and the vein. (f) A term applied to thin slabs of sandstone that split out between closely spaced joints. (g) The steel lining of a circular shaft. See also: *conduit.* (CTD, 1958)
casing catcher A safety device equipped with slips or dogs to catch and grip casing if it is dropped while being lowered into or lifted from a borehole. Also miscalled tubing catcher; tubing hanger. (Long, 1960)
casing clamp A mechanical device designed to facilitate the hoisting or suspension of casing in a borehole. Made by forming a half circle in a heavy steel bar. When bolted together, in pairs, the bars fit around the outside and tightly grip the casing. The size of the clamp is determined by the outside diameter of the casing to be handled. See also: *pipe clamp.* (Long, 1960)
casing dog (a) A lifting device consisting of one or more serrated sliding wedges working inside a cone-shaped collar. Used to grip and hold casing while it is being raised or lowered into a borehole. See also: *bulldog; dog.* (Long, 1960) (b) A fishing tool. (Long, 1960)
casing float A rubber-ball-type check valve, generally placed near the bottom of a long string of casing. Its use reduces the load imposed on the hoisting mechanism in lowering casing into a wet borehole. Also called casing valve; float valve. (Long, 1960)
casing off Process of inserting a line of casing into a borehole. See also: *case; case off.* (Long, 1960)
casing point In borehole drilling, the depth to which the casing is entered. (Pryor, 1963)
casing pressure The pressure built up in the casing when closed at the top of the well. It is usually measured by placing a pressure gage on one of the side outlets on the casing head. (Porter, 1930)
Cassel brown A brown earthy substance found in peat and lignite beds and used as a pigment; originally found near Cassel, Germany. Cologne brown or Cologne earth is a similar substance originally found near Cologne, Germany. (Hess)
cassiterite A tetragonal mineral, 4[SnO_2]; rutile group; adamantine; reddish brown to black; forms prismatic crystals, or massive concentric fibrous structure (wood tin); sp gr, 7.03; occurs in veins associated with granite and granite pegmatite, or placers (stream tin); a source of tin. Syn: *tin stone; tin spar; tin ore; black tin.*
cast (a) Secondary rock or mineral material that fills a cavity formed by the decay or dissolution of some or all of the original hard material. Cf: *mold.* (AGI, 1987) (b) A sedimentary structure representing the infilling of an original mark or depression made on top of a soft bed and preserved as a solid form on the underside of the overlying and more durable stratum; e.g., a flute cast or a load cast. Syn: *counterpart.* (AGI, 1987)
castable A refractory mix containing heat-resistant, hydraulic setting cement. A refractory concrete. (AISI, 1949)
castable refractory (a) A refractory aggregate that will develop structural strength by hydraulic set after having been tempered with water and compacted. (ARI, 1949) (b) A mixture of a heat-resistant aggregate and a heat-resistant hydraulic cement; for use, it is mixed with water and rammed or poured into place. (Harbison-Walker, 1972)
cast-after-cast Corn. The throwing up of ore from one platform to another successively. See also: *shamble.*
castanite A former name for hohmannite, $Fe^{+3}_2(SO_4)_2(OH)_2{\cdot}7H_2O$.
cast bit A drill bit in which the diamond-set crown is formed on a bit blank by pouring molten metal into a prepared mold. Also called cast-set bit; cast-metal bit. (Long, 1960)
cast booster A cast, extruded or pressed, solid high explosive used to detonate less sensitive explosive materials. (Atlas, 1987)
casthouse A building in which pigs or ingots are cast. (Fay, 1920)
castillite An impure variety of bornite, containing zinc, lead, and silver sulfides. (Fay, 1920)
casting (a) An object at or near finished shape obtained by solidification of a substance in a mold. (ASM, 1961) (b) Pouring molten metal into a mold to produce an object of desired shape. (ASM, 1961) (c) A process of shaping glass by pouring hot glass into molds or onto tables or molds. See also: *teemer.* (ASTM, 1994) (d) A process for forming ceramic ware by introducing a body slip into a porous mold that absorbs sufficient water (or other liquid) from the slip to produce a semirigid article. (ASTM, 1994)
casting machine A series of iron molds on an endless-belt conveyor to receive and cast molten pig iron into form as it comes from a furnace. (Mersereau, 1947)
casting over (a) A quarryman's term for an operation consisting of making a cut with a steam shovel, which, instead of loading the material on cars, moves it to one side, forming a long ridge. (b) The operation of reestablishing benches that have been covered or caved, and also cutting up a high bank into one or more smaller banks. (Lewis, 1964)
casting pit The space in a foundry in which molds are placed and castings are made. In the Bessemer and open-hearth steelworks, it is the space utilized for casting the molten steel into cast iron ingot molds. (Fay, 1920)
castings One of several terms (and/or letter symbols) commonly used to designate low-quality drill diamonds. (Long, 1960)
casting shrinkage (a) Liquid shrinkage—the reduction in volume of liquid metal as it cools to the liquidus. (ASM, 1961) (b) Solidification shrinkage—the reduction in volume of metal from the beginning to ending of solidification. (ASM, 1961) (c) Solid shrinkage—the reduction in volume of metal from the solidus to room temperature. (ASM, 1961) (d) Total shrinkage—the sum of the shrinkage in definitions a, b, and c above. (ASM, 1961)
casting strain Strain in a casting caused by casting stresses that develop as the casting cools. (ASM, 1961)
casting stress Stress set in a casting because of geometry and casting shrinkage. (ASM, 1961)
casting wheel A large turntable with molds mounted on the outer edge. Used primarily in the base metal industries for cast ingots, anodes, etc.
casting-wheel operator In ore beneficiation, smelting, and refining, one who operates a large rotating casting wheel to pour molten, nonferrous metal, such as copper or lead, into molds mounted on the edge of the wheel. (DOT, 1949)
cast iron Iron containing carbon in excess of its solubility in the austenite that exists in the alloy at the eutectic temperature. For the various forms—gray cast iron, white cast iron, malleable cast iron, and nodular cast iron—the word "cast" is often left out, resulting in the terms gray iron, white iron, malleable iron, and nodular iron, respectively. (ASM, 1961)
castor *castorite.*
castor amine An oil. Used in ore flotation as a selective collector and in rustproofing metal surfaces. (Bennett, 1962)
castorite (a) A natural, colorless silicate of lithium and aluminum. (Bennett, 1962) (b) A transparent variety of petalite. Syn: *castor.*
cast primer (a) A cast unit of explosive commonly used to initiate detonation in a blasting agent. (Dick, 2) (b) A cast unit of explosive, usually pentolite or composition B; commonly used to initiate detonation in a blasting agent.
cast steel Steel as cast; i.e., not shaped by mechanical working. Originally applied to steel made by the crucible process as distinguished from that made by cementation of wrought iron. (CTD, 1958)
cata- A prefix to indicate that the rock belongs to the deepest zone of metamorphism, which is characterized by very high temperature, hydrostatic pressure, and relatively low shearing stress. Cf: *epi-; meta-; meso-.* Syn: *kata-.* (AGI, 1987)
cataclasis Rock deformation accomplished by fracture and rotation of mineral grains or aggregates without chemical reconstitution. (AGI, 1987)
cataclasite A cataclastic rock that has been formed by shattering (or cataclasis), which has been less extreme than in mylonite. See also: *augen gneiss; crush breccia; mylonite; mylonite gneiss.*
cataclastic (a) Pertaining to the structure produced in a rock by the action of severe mechanical stress during dynamic metamorphism; characteristic features include the bending, breaking, and granulation of the minerals. Also said of the rocks exhibiting such structures. See also: *mortar structure.* (AGI, 1987) (b) Pertaining to clastic rocks, the fragments of which have been produced by the fracture of preexisting

rocks by Earth stresses; e.g., crush breccia. Syn: *kataclastic.* Cf: *autoclastic.* (AGI, 1987)
cataclysm (a) Any geologic event that produces sudden and extensive changes in the Earth's surface; e.g., an exceptionally violent earthquake. Syn: *cataclysmic; cataclysmal.* (AGI, 1987) (b) Any violent, overwhelming flood that spreads over the land; a deluge. (AGI, 1987)
cataclysmal *cataclysm.*
cataclysmic *cataclysm.*
catalysis Acceleration or deceleration of a chemical reaction produced by a substance that is unchanged by the reaction.
catalyst A substance capable of changing the rate of a reaction without itself undergoing any net change. (AGI, 1987)
catalytic methanometer A combustible-gases detector depending upon the combustion or oxidation of methane at heated filaments. Usually the gas is drawn through the apparatus by a rubber suction bulb, and the filaments are heated by a battery in the instrument. A version of this principle is the resistance methanometer. (Nelson, 1965)
catalytic oxidation A process that converts the incompletely burned hydrocarbons present in fuel exhaust into harmless gases. It involves burning up the fuel remnants with the aid of catalysts— chemical agents, such as platinum and palladium, that speed up reactions without being consumed themselves.
cataphoresis Movement of charged particles in a fluid medium in response to an electric field. Metallic hydroxides and other positive sols migrate to the cathode and negatives ones to the anode. See also: *electrophoresis.* (Pryor, 1963)
catapleiite A hexagonal mineral, $Na_2ZrSi_3O_9 \cdot 2H_2O$; yellow to yellow-brown; forms thin, tabular hexagonal prisms.
catarinite An obsolete term for an iron meteorite remarkable for a high proportion of nickel.
catastrophe In geology, a sudden, violent change in the physical conditions of the Earth's surface; a cataclysm. (Standard, 1964)
catch (a) Projection in a mine shaft that arrests a cage, skip, or other reciprocating system in the event of fracture or overwind. (Pryor, 1963) (b) One of the catches or rests placed on shaft timbers, to hold the cage when it is brought to rest at the top, bottom, or any intermediate landing. See also: *chair; dog; wing; rests.* (Fay, 1920) (c) In coal work, a device for holding trams in a cage when hoisting. See also: *jack catch.* (Pryor, 1963) (d) One of the stops fitted on a cage to prevent cars from running off. (Fay, 1920)
catcher *core lifter.*
catch gear An appliance fixed in the headgear to limit the drop of a cage after an overwind. The upward speed and momentum of the loaded cage (after its release from the rope) may be such that its subsequent drop is so severe as to fracture the suspension gear, resulting in the cage falling down the shaft. The amount of drop is limited by the catch gear, which consists of a series of catches suspended from beams supported on hydropneumatic buffers to reduce the impact shock. The cage is released by raising it slightly and retracting the catches. See also: *detaching hook; overwind.* (Nelson, 1965)
catchment area (a) The recharge area and all areas that contribute water to it. (AGI, 1987) (b) An area paved or otherwise waterproofed to provide a water supply for a storage reservoir. See also: *drainage basin.* Syn: *gathering ground.* (AGI, 1987)
catch pit (a) In mineral processing, a sump in a mill to which the floor slopes gently, and into which all spillage gravitates or is hosed either for return by pumping to its place in the flowline or for periodical removal. Also called catch sump. (Pryor, 1963) (b) *sump; tailing pit.*
catch point (a) One of a set of spring-loaded points in an upgrade railway line that close behind a rising train. If any rolling stock breaks away it is then automatically diverted to a siding. (Pryor, 1963) (b) Position of intersection of a road cut or fell with natural ground; usually marked with a stake.
catch prop Prop erected in the face to act as a temporary support until permanent supports are brought forward. Also called watch prop; safety prop. (Nelson, 1965; CTD, 1958)
catch scaffold Eng. A platform in a shaft a few feet beneath a working scaffold; to be used in case of accident.
catchwater drain A surface drain to intercept and collect the flow of water from adjoining land, so as to prevent it from reaching a road or mine sidings. See also: *subsoil drainage.* (Nelson, 1965)
cat claw A miner's term applied locally in Illinois to a bed of marcasite from 2 to 6 in (5.08 to 15.24 cm) thick that sometimes occurs between the "clod" roof of a coal seam and the more stratified shale above. The lower surface of the marcasite bed is characterized by very irregular protuberances extending downward 1 to 3 in (2.54 to 7.62 cm) into the clod. Also called cat. (AGI, 1987)
cat dirt (a) Derb. A hard fireclay. (b) Derb. Coal mixed with pyrite.
catear Sp. To search for new mines; to prospect. (Hess)
catenary suspension The overhead suspension of contact wire for electric traction by vertical links of different lengths connected to a catenary wire above it. The contact wire will thus be maintained at a constant height. (Hammond, 1965)
caterpillar An endless chain of plates that functions as a wheel for heavy vehicles. See also: *crawler track.* (Nelson, 1965)
caterpillar chain A short endless chain on which dogs or teeth are spaced to mesh with and move or be moved by a conveyor chain.
caterpillar chain dog A dog or tooth attached to a "caterpillar chain" to provide the driving contact with the conveyor chain.
caterpillar drive A drive equipped with a "caterpillar chain" that engages and propels the "conveyor chain."
catfaced block In New York and Pennsylvania, a bluestone quarryman's term for a mass of waste situated between two closely spaced open joints.
cathead sheave A sheave set on the topmost part of a pile frame. (Hammond, 1965)
cathode The electrode where electrons enter, or current leaves, an operating system, such as a battery, an electrolytic cell, an X-ray tube, or a vacuum tube. In the first of these, the cathode is positive; in the other three, negative. In a battery or electrolytic cell, it is the electrode where reduction occurs. Opposite of anode. See also: *electrode.* (ASM, 1961)
cathode compartment In an electrolytic cell, the enclosure formed by a diaphragm around the cathode. (ASM, 1961)
cathode copper Electrolytically refined copper that has been deposited on the cathode of an electrolytic bath of acidified copper sulfate solution. Such copper is usually remelted in a furnace before being marketed as electrolytic copper. (Camm, 1940)
cathode efficiency Current efficiency at a cathode. (ASM, 1961)
cathode film The portion of a solution in immediate contact with the cathode during electrolysis. (ASM, 1961)
cathodic corrosion Corrosion of the cathodic member of a galvanic couple resulting from the flow of current.
cathole A local term used in southern Michigan for a shallow boggy depression less than 1 acre (0.4 ha) in extent, esp. one formed by a glacier in a till plain. (AGI, 1987)
catholyte The electrolyte adjacent to the cathode in an electrolytic cell. (ASM, 1961)
cation (a) An ion having a positive charge. (Hurlbut, 1964) (b) Any positive ion; named for its attraction to the cathode or negative terminal of an electrolytic cell. Cf: *anion.*
cation exchange The displacement of a cation bound to a site on the surface of a solid, such as in silica-alumina clay-mineral packets, by a cation in solution. Syn: *base exchange.* See also: *ion exchange.* (AGI, 1987)
cationic collector In flotation, an amine or related organic compound capable of producing positively charged hydrocarbon-bearing ions (hence the name cationic collector) for the purpose of floating miscellaneous minerals, including silicates. (Gaudin, 1957)
cationic detergent A detergent in which the cation is the active part. (ASM, 1961)
cationic reagent In flotation, a surface-active substance that has the active constituent in the positive ion. Used to flocculate and to collect minerals that are not flocculated by the reagents, such as oleic acid or soaps, in which the surface-active ingredient is the negative ion. Reagents used are chiefly the quaternary ammonium compounds; e.g., cetyl trimethyl ammonium bromide. (CCD, 1961)
catlinite A hard red clay found in southwestern Minnesota, formerly used by the Dakota Native Americans for making tobacco pipes. Named after George Catlin (1796-1872), American painter. Syn: *pipestone.*
catogene Pertaining to sedimentary rocks, signifying that they were formed by deposition from above, as of suspended material. (AGI, 1987)
cat run A low passage that requires crawling to traverse it. Syn: *crawlway.* (AGI, 1987)
cat's-eye (a) Any gemstone that, when cut en cabochon, exhibits under a single strong point source of light a narrow, well-defined chatoyant band or streak that moves across the summit of the gemstone, shifts from side to side as it is turned, and resembles a slit pupil of the eye of a cat. Internal reflection of light from parallel inclusions of tiny fibrous crystals or from long parallel cavities or tubes causes the cat's-eye. (AGI, 1987) (b) Alternate term for tiger's-eye, the silicified form of crocidolite asbestos; sometimes polished and used as ornaments. (c) A greenish gem variety of chrysoberyl that exhibits chatoyancy. Syn: *cymophane; oriental cat's-eye.* (d) A variety of minutely fibrous, grayish-green quartz (chalcedony) that exhibits an opalescent play of light. Syn: *occidental cat's-eye.* (e) A yellowish-brown silicified variety of crocidolite. Cf: *tiger's-eye.* The term used alone properly applies only to (c).
catskinner Operator of a crawler tractor. (Nichols, 1976)
Cattermole Process An early flotation process (1903) based on adhesion of sulfide minerals to oil. Mineral oil or fatty acid agglomerated heavy minerals into floccules, which were separated by classification from overflowing gangue. (Pryor, 1958)

catty (a) Any of various units of weight used in China and southeast Asia varying around 1-1/3 lb or 600 g; also, a Chinese unit according to a standard set up in 1929 equal to 1.1023 lb or 500 g. (Webster 3rd, 1966) (b) A gold weight that equals 2.9818 troy pounds (1.1129 kg). (Fay, 1920)

catwalk A pathway, usually of wood or metal, that gives access to parts of large machines. (Nichols, 1976)

cauk *cawk.*

cauldron An inclusive term for all volcanic subsidence structures regardless of shape or size, depth of erosion, or connection with the surface. The term thus includes cauldron subsidences, in the classical sense, and collapse calderas. See also: *caldera.* (AGI, 1987)

cauldron subsidence The sinking of part of the roof of an intrusion within a closed system of peripheral faults into which magma has penetrated, often to form ring dikes. See also: *subsidence.*

cauliflowering The tendency of coal to swell and open out when heated, thus exposing a surface out of all proportion to the size of the original coal. See also: *swelling number.* (Nelson, 1965)

caunche In coal mining, removal of part of the roof or floor to increase the height of a roadway. Also spelled canch. (Pryor, 1963)

caustic Capable of destroying the texture of anything or eating away its substance by chemical action; burning; corrosive. (Webster 3rd, 1966)

caustic ammonia Gaseous or dissolved ammonia. (Standard, 1964)

caustic embrittlement Effect on metal of immersion in caustic alkaline solutions. (Pryor, 1963)

caustic soda Sodium hydroxide, NaOH; deliquescent; a soapy feel; its solution in water is strongly alkaline. The molten caustic dissolves such materials as enamels, sand, or glass, which contain a high percentage of silica.

caustobiolite A general name for a fossil combustible substance. Syn: *caustolith.* (Tomkeieff, 1954)

caustobiolith This term designates a rock with a fairly high content of organic carbon compounds or even pure carbon where the latter is, like the carbon compounds, of organic origin. (IHCP, 1963)

caustolith A rock that has the property of combustibility (Grabau). It is usually of organic origin (e.g., coal and peat), but inorganic deposits (e.g., sulfur, asphalt, and graphite) also occur. See also: *caustobiolite.* (AGI, 1987)

caustophytolith A caustobiolith formed by the direct accumulation of vegetal matter; e.g., peat, lignite, and coal. (AGI, 1987)

cave (a) Fragmented rock materials, derived from the sidewalls of a borehole, that obstruct the hole or hinder drilling progress. (b) To allow a mine roof to fall without retarding supports or waste packs. (c) A falling in of the roof strata, sometimes extending to the surface and causing a depression therein. Also called cave-in. (d) The partial or complete failure of borehole sidewalls or mine workings. (Long, 1960) (e) Collapse of an unstable bank. (Nichols, 1954) (f) A natural cavity, recess, chamber, or series of chambers and galleries beneath the surface of the Earth, within a mountain, a ledge or rocks, etc.; sometimes a similar cavity artificially excavated. See also: *cavity.* (Standard, 1964)

caved stope There are two distinct types of caved stopes. In the first, the ore is broken by caving induced by undercutting a block of ore. In the second, the ore itself is removed by excavating a series of horizontal or inclined slices, while the overlying capping is allowed to cave and fill the space occupied previously by the ore. The first type comprises the caving methods of mining, while the second comprises the top-slicing method.

cave hole A depression at the surface, caused by a fall of the roof in a mine. (Fay, 1920)

cave-in Collapse of the walls or roof of a mine excavation. (Pryor, 1963)

cave-in-heave The partial or complete collapse of the walls of a borehole. (Brantly, 1961)

cave line A linear area inby the last solid ground, in a longwall-type mine, where the roof or back caves behind the retreating excavation. (Federal Mine Safety, 1977)

cave marble *cave onyx.*

cave onyx A compact banded deposit of calcite or aragonite found in caves, capable of taking a high polish and resembling true onyx in appearance. See also: *dripstone; onyx marble; travertine.* Syn: *cave marble.* (AGI, 1987)

caver (a) Eng. A thief who steals ore or coal at a mine. (Standard, 1964) (b) The officer appointed to guard a mine. (Standard, 1964) (c) A person whose hobby is exploring caves. Also called a spelunker. (Schieferdecker, 1959)

cavernous Said of an area or geologic formation, such as limestone, that contains caverns, or caves. Said of the texture of a volcanic rock that is coarsely porous or cellular. (AGI, 1987)

cavil (a) To draw lots at stated periods—by miners to determine the places in which they will work for the following period. (Fay, 1920) (b) A type of heavy sledge with one blunt and one pointed end. Used for rough shaping stone at a quarry. (Crispin, 1964)

caving (a) A stoping method in which ore is broken by induced caving. This may be achieved by (1) block caving, including caving to main levels and caving to chutes or branched raises; or (2) sublevel caving. See also: *stope.* (b) In coal mining, the practice of encouraging the roof over the waste to collapse freely so that it fills the waste area, thereby avoiding the need to pack. In metal mining, caving implies the dropping of the overburden as part of the system of mining. See also: *block caving; sublevel caving; top slicing.* (Nelson, 1965) (c) The failure and sloughing in of sidewalls of boreholes, mine workings, or excavations. (Long, 1960) (d) Fall of rock underground. See also: *cavings.*

caving by raising *chute caving.*

caving ground Rock formations that will not stand in the walls of an underground opening without support, such as that offered by cementation, casing, or timber. (Long, 1960)

caving hole A borehole in which fragments of the material making up the walls of the hole slough so much that the borehole cannot be kept open without the use of casing or cementation. (Long, 1960)

cavings Fragments of borehole wall-rock material that fall into a borehole, sometimes blocking the hole, and which must be washed or drilled out before the borehole can be deepened. See also: *caving.* (Long, 1960)

caving system (a) A method of mining in which the support of a great block of ore is removed, allowed to cave or fall, and in falling to be broken sufficiently to be handled; the overlying strata subside as the ore is withdrawn. There are several varieties of the system. See also: *block caving; fall; top slicing and cover caving; top slicing combined with ore caving.* (b) Longwall coal mining in which excavated space (gob) is left to collapse. See also: *sublevel.* (Pryor, 1963)

caving the back *block caving.*

cavitation The formation and instantaneous collapse of innumerable tiny voids or cavities within a liquid subjected to rapid and intense pressure changes. Cavitation produced by ultrasonic radiation is sometimes used to give violent localized agitation. That caused by severe turbulent flow often leads to cavitation damage. (ASM, 1961)

cavitation noise The noise produced in a liquid by the collapse of bubbles that have been created by cavitation. (Hy, 1965)

cavity (a) A natural underground opening or void, which may be small or large. See also: *cave; vug.* (Long, 1960) (b) The bubble formed by a projectile at water entry. (Hy, 1965) (c) A void in a bit caused by a bubble of gas entrapped in the matrix material during the manufacturing process. (Long, 1960)

cavity-filling deposit A deposition of minerals in a cavity or rock opening. (Bateman, 1951)

cawk (a) Eng. Sulfate of barium heavy spar. See also: *barite.* (Fay, 1920) (b) Scot. Chalk; limestone. Also spelled cauk. (Fay, 1920)

caxas Walls of a vein; chest. (Hess)

***c* axis** (a) In crystallography, a symmetrically unique reference vector, oriented vertically by convention. In the monoclinic system, the second setting orients the *c* axis at the nonorthogonal angle beta to the unique *b* axis, the diad. In the triclinic system, all axes are unique with the *c* axis designated by convention. Cf: *a axis; b axis.* (b) One of three orthogonal reference axes, *a*, *b*, and *c* that are used in structural geology. (c) To help describe the geometry of a fabric that possesses monoclinic symmetry, the *c* axis lies in the unique symmetry plane at right angles to a prominent fabric plane; thus in many tectonites the *c* axis is normal to the schistosity. (d) In a kinematic sense, to describe a deformation plan that possesses monoclinic symmetry, such as a progressive shear. Here the *c* axis lies in the unique symmetry plane and normal to the movement plane. In a progressive simple shear, the *c* axis lies normal to the shear plane. Syn: *c direction.*

***c* direction** *c axis.*

C-D principle The convergence-divergence principle used in the Frenkel mixer. (Dodd, 1964)

Ceag Montlucon gas detector This nonautomatic detector has the appearance of a mine official's electric hand lamp. It indicates on an illuminated scale percentages of methane from 0 to 3 in steps of 0.1. When a test for combustible gases is to be made, the projecting front piece is turned part of a revolution; this extinguishes the main light and lights up the illuminated scale. A sample of air is flushed into the detector by means of a small aspirator (or hand pump), the button switch at the side is operated, and the percentage of combustible gases, if any, is indicated on the illuminated scale. (Cooper, 1963)

cedarite *chemawinite.*

ceiling concentration The concentration of an airborne substance that shall not be exceeded during any part of the working exposure. (ANSI, 1992)

celadonite The mineral monoclinic $K(Mg,Fe)(Fe,Al)Si_4O_{10}(OH)_2$; mica group; soft; green or gray-green; earthy; generally occurs in cavities in basaltic rocks. Formerly called kmaite. Syn: *svitalskite.* Cf: *glauconite.*

celestine An orthorhombic mineral, $4[SrSO_4]$; barite group; disseminated through limestone and sandstone; a source of strontium. Also called celestite.

celestite *celestine.*

celite A constituent of Portland cement clinkers. Also spelled celith. See also: *brown millerite.*

cell (a) A compartment in a flotation machine. (Hess) (b) A single element of an electric battery, either primary or secondary. (Crispin, 1964) (c) Battery unit consisting of two electrodes separately contacting an electrolyte so that there is a potential difference between them. (Bennett, 1962) (d) *galvanic cell; local cell.*

cellar Excavated area under a drill-derrick floor to provide headroom for casing and pipe connections required at the collar of a borehole, or to serve as a covered sump. See also: *cave.* (Long, 1960)

cell feed The material supplied to the cell in the electrolytic production of metals. (ASM, 1961)

cell texture A network along grain boundaries, which may originate by segregation on exsolution. A similar texture may form by the replacement of organic forms, esp. cell walls, by ore minerals. (Schieferdecker, 1959)

cellular Said of the texture of a rock (e.g., a cellular dolomite) characterized by openings or cavities, which may or may not be connected. Although there are no specific size limitations, the term is usually applied to cavities larger than pores and smaller than caverns. The syn. vesicular is preferred when describing igneous rocks. Cf: *porous; cavernous; vesicular.* (AGI, 1987)

cellular cofferdam A cofferdam, with a double wall, consisting of steel sheet piling arranged in intercepting rings about 50 ft (15 m) in diameter. The space between the lines of piling is filled with sand. (Hammond, 1965)

cellulose A polymeric carbohydrate composed of glucose units, formula $(C_6H_{10}O_5)x$, making it the most abundant carbohydrate, and with lignin, an important constituent of plant materials, from which coal is formed. (AGI, 1987)

cellulose nitrate *nitrocellulose.*

celsian A rare monoclinic mineral, $BaAl_2Si_2O_8$; feldspar group; the barium analog of anorthite; dimorphous with paracelsian.

Celsius (a) Designation of the degree on the International Practical Temperature Scale; also used for the name of the scale, as "Celsius Temperature Scale." Formerly (prior to 1948) called "Centigrade." The Celsius temperature scale is related to the International Kelvin Temperature Scale by the equation $T_C = T_K - 273.16$. (ASTM, 1994) (b) Symbol, C. Graduated to a scale of 100; of or pertaining to such a scale. On the centigrade thermometer the freezing point of water is 0° (C) and its boiling point is 100° (C). If any degree on the centigrade scale, either above or below 0 °C, is multiplied by 1.8, the result will be, in either case, the number of degrees above or below 32° F, or the freezing point of Fahrenheit. (Standard, 1964)

cement (a) A manufactured gray powder which when mixed with water makes a plastic mass that will set or harden. It is combined with aggregate to make concrete. Nearly all of today's production is portland cement. See also: *cement rock.* (AGI, 1987) (b) To place cement in a borehole to seal off caves or fissures or to fill cavities or caverns encountered in drilling. (c) Mineral material, usually chemically precipitated, that occurs in the spaces among the individual grains of a consolidated sedimentary rock, thereby binding the grains together as a rigid, coherent mass; it may be derived from the sediment or its entrapped waters, or it may be brought in by solution from outside sources. The most common cements are silica (quartz, opal, chalcedony), carbonates (calcite, dolomite, siderite), and various iron oxides. Others include clay minerals, barite, gypsum, anhydrite, and pyrite. Detrital clay minerals and other fine clastic particles may also serve as cements. (AGI, 1987) (d) A term used in gold-mining regions to describe various consolidated, fragmental aggregates, such as breccia, conglomerate, and the like, that are auriferous. (e) A finely divided metal obtained by precipitation. The word in this sense is generally used in combination, such as, cement copper, cement gold, or cement silver. (Standard, 1964)

cementation (a) The diagenetic process by which coarse clastic sediments become lithified or consolidated into hard, compact rocks, usually through deposition or precipitation of minerals in the spaces among the individual grains of the sediment. It may occur simultaneously with sedimentation or at a later time. Cementation may occur by secondary enlargement. Syn: *agglutination.* (AGI, 1987) (b) Filling cavities or plugging a drill hole with cement or other material to stop loss of water or entrance of unwanted liquids, gas, or fragmented rock materials. Also called dental work. (c) The process by which loose sediments or sands are consolidated into hard rock by injection of chemical solutions, thin cement slurries, or self-hardening plastic. Also called cementing. (Long, 1960) (d) The introduction of one or more elements into the outer portion of a metal object by means of diffusion at high temperature. (ASM, 1961) (e) The precipitation of a more noble metal from solution by the introduction of a less noble metal. (f) Usually, the process of raising the carbon content of steel by heating in a carbonaceous medium. Generally, any process in which the surface of a metal is impregnated by another substance. Also called casehardening; carburization; carbonization. (CTD, 1958)

cementation sinking A method of shaft sinking through water-bearing strata by injecting chemicals or liquid cement into the ground. A number of small-diameter boreholes are put down around the shaft—and about 80 ft (24 m) ahead of the shaft bottom—through which cement is forced by means of pumps. The cement, when set, seals the fissures and thus prevents water inflows during sinking. The method is most successful in strong fissured strata and least successful in loose alluvial deposits. See also: *grouting; precementation process.* (Nelson, 1965)

cementation steel Steel made by a process in which bars of wrought iron are packed into a sealed furnace together with charcoal. The resulting material is blister steel. Syn: *cement steel.* (Camm, 1940)

cementation water Water containing dissolved copper or iron sulfates or other metal compounds. (Stoces, 1954)

cement clinker Portland cement as it comes from the kiln.

cement copper Copper precipitated by iron from copper sulfate in mine water. (Bateman, 1951)

cement deposit Cambrian conglomerate occupying supposedly old beaches or channels. It is gold-bearing in the Black Hills, SD.

cemented carbide Generally, a mixture of powdered tungsten carbide and cobalt, subjected to pressure and heat to produce bit crowns, small plates, cubes, or cylinders of material having a much greater hardness than steel. Mixtures also may contain small amounts of titanium, columbium, or tantalum carbide. Cobalt may be replaced by powdered nickel. Also called sintered carbide. See also: *carbide insert; sintered carbide.* (Long, 1960)

cemented carbide tool A tool made from pulverized carbides and fused into a hard tip for heavy-duty or high-speed cutting of metals. (Crispin, 1964)

cement gold Gold precipitated in fine particles from solution by a more active metal. (Fay, 1920)

cement grout A pumpable thin slurry consisting primarily of a mixture of cement, sand, and water; injected into rock formations through boreholes as a sealant. Also called grout; grouting; cement grouting. (Long, 1960)

cement gun A mechanical device for the application of cement, in the form of gunite, to the walls or roofs of mine openings or building walls. Also called gunite gun. (Long, 1960)

cementite (a) Identical with cohenite, a meteoritic material. (Hey, 1955) (b) An orthorhombic FeC_3 that occurs as a phase in steel and changes composition in the presence of manganese or other carbide-forming metals. *cohenite.*

cementitious Having the property of or acting like cement, such as certain limestones and tuffs when used in the surfacing of roads. (AGI, 1987)

cement-modified soil The addition of small quantities of cement (1% to 2%) to fine-grained soils to reduce the liquid limit, plasticity index, and water-absorption tendency. The effect of the cement is to bring individual soil particles into aggregations, thus artificially adjusting the grading of the soil. See also: *soil stabilization.* (Nelson, 1965)

cement mortar Made from four (or less) parts of sand, one of cement, and adequate water. (Nelson, 1965)

cement plug Hardened cement material filling a portion of a borehole. (Long, 1960)

cement rock (a) Any rock that is capable of furnishing cement when properly treated. (Fay, 1920) (b) Scot. Argillaceous limestone-magnesian. See also: *hydraulic limestone.* (Nelson, 1965)

cement silver Silver precipitated from solution, usually by copper. (Fay, 1920)

cement slurry A pourable or pumpable mixture of water, cement, and fine sand— having the consistency of a thick liquidlike heavy cream. (Long, 1960)

cement stabilization The addition of cement to a soil, which acts as a binding agent and produces a weak form of concrete called soil cement. The quantity of cement to be added depends upon the type of soil. Cement can be used with most types of soil, providing the clay fraction is reasonably small and other specified impurities are not present. A small percentage of lime is usually added. With very poor soils, cement stabilization may be uneconomical or impracticable. See also: *soil stabilization.* (Nelson, 1965)

cement steel *cementation steel.*

cement valve A ball, flapper, or clack-type valve placed at the bottom of a string of casing, through which cement is pumped. When pumping ceases, the valve closes and prevents return of cement into the casing. (Long, 1960)

Cenozoic An era of geologic time, from the beginning of the Tertiary period to the present. (Some authors do not include the Quaternary, considering it a separate era.) The Cenozoic is considered to have begun about 65 million years ago. Also spelled: Cainozoic; Kainozoic. (AGI, 1987)

center A temporary timber framework upon which the masonry of an arch of reinforced masonry lintel is supported until it becomes self-supporting. See also: *centers.* (ACSG, 1961)

center adjustment In surveying, a system that allows accurate final centering of the theodolite above (or below) its station by sliding the whole instrument on its stand (tribrach). Important with short sights where small centering errors could introduce serious inaccuracy. (Pryor, 1963)
center brick A special, hollow, refractory shape used at the base of the guide tubes in the bottom pouring of molten steel. The center brick has a hole in its upper face, and this is connected via the hollow center of the brick to holes in the faces (often six in number). The center brick distributes molten steel from the trumpet assembly to the lines of runner bricks. It is also sometimes known as a crown brick or spider. (Dodd, 1964)
center constant In air velocity determination, the ratio of the mean velocity to the velocity measured at the center. This ratio is found to be dependent upon the Reynolds number. See also: *Reynolds number.* (Roberts, 1960)
center core method A method of tunneling whereby the center is left to the last for excavation. (Sandstrom, 1963)
center country Aust. The rock between the limbs of a saddle reef.
center drilling Drilling a conical hole (pit) in one end of a workpiece. (ASM, 1961)
centering of shaft The fixing of the center spot of a proposed shaft at the site selected and the maintenance of the shaft sinking along this plumb line during its entire depth. See also: *plumbing.* (Nelson, 1965)
center-latch elevator and links *elevator.*
centerline (a) A line marked on the roof of a mine roadway, or a plumbline, for controlling the direction in which the roadway is driven. (b) In U.S. public land surveys, the line connecting opposite quarter-section or sixteenth-section corners. (AGI, 1987)
centerman In anthracite and bituminous coal mining, one who locates the centerline of underground openings in a mine, such as entries, rooms, and haulageways, so that the miners can drive the openings in a straight line without calling the mine surveyor. (DOT, 1949)
center of gravity (a) The center of mass of a cut or a fill. (Nichols, 1976) (b) That point in a body or system of bodies through which the resultant attraction of gravity acts when the body or system is in any position; that point from which the body can be suspended or poised in equilibrium in any position. (AGI, 1987)
center of mass (a) The point that represents the mean position of the matter in a body. (Webster 3rd, 1966) (b) The point in a body through which acts the resultant resisting force due to the body's inertia when it is accelerated. Coincident with the center of gravity. (CTD, 1958) (c) In a cut or a fill, a cross section line that divides its bulk into halves. (Nichols, 1976) (d) Also called center of inertia.
center of shear *torsional center.*
center of symmetry In crystallography, an element of symmetry such that, for each and every lattice point, asymmetric unit, or crystal plane, there is another equidistant in the opposite direction. It is represented by *i* or by 1. Adj. centric or centrosymmetric. Syn: *inversion.*
center of torsion *torsional center.*
center of twist *torsional center.*
center prop Eng. A prop set temporarily under the center of a plank to support it before props are set at the ends of the plank. Syn: *middle prop.* (SMRB, 1930)
centers (a) Framed supports, usually arch shaped, upon which are placed the lagging boards used, in building an arch, for supporting the roof of a tunnel. See also: *center.* (Stauffer, 1906) (b) Linear distance between coal-mine entries or crosscuts.
center shot A shot in the center of the face of a room or entry. Also called center cut. (Fay, 1920)
center spinning A method of casting molten metal, in which the molds are spun and centrifugal force helps to fill them. (Pryor, 1963)
center-trace time One of two approaches used in plotting seismic reflection data on time cross sections. Center-trace times are the times picked on the two traces from the respective detector groups nearest the shot and on opposite sides. The average of the two times for each reflection is plotted at the shot-point position. The points thus plotted for adjacent shot points are connected by straight lines. Cf: *trace-by-trace plotting.* (Dobrin, 1960)
centigrade *Celsius.*
centipoise The one-hundredth part of a poise, an absolute unit of fluid viscosity. Viscosity of drill-mud fluid is sometimes expressed in centipoise or millipascal-second units. See also: *poise.* (Long, 1960)
central breaker A breaker where the coal from a number of mines in a district is prepared. Central breakers, representing the last word in mining technology, make it economical for operators to abandon many local breakers. (Korson, 1938)
centralizer A device that lines up a drill steel or string between the mast and the hole. (Nichols, 1976)
centrifugal brake A safety device on a mine hoist drum that applies a brake if the drum speed exceeds the set limit. (Pryor, 1963)
centrifugal casting Casting molten metals in a rapidly revolving mold.
centrifugal clutch Consists of a driving hub having one or more weighted sections fitted with friction lining on the outer radial surfaces that contact a driven hub having a flange covering that portion of the driving hub containing the radial elements. Upon starting, the radial elements of the driving hub have no appreciable drag, but upon accelerating to the operating speed the force produced by the centrifugal action increases rapidly as the square of the speed and the elements grip the driven element, thereby causing it to speed up to the required speed of the driving hub. (Pit and Quarry, 1960)
centrifugal discharge bucket elevator A type of bucket elevator using centrifugal discharge elevator buckets suitably spaced to permit the free discharge of bulk materials. See also: *bucket elevator; centrifugal discharge elevator bucket.*
centrifugal discharge elevator bucket A bucket designed to scoop material from the boot of an elevator and discharge by reason of the combined effect of centrifugal force and gravity.
centrifugal fan (a) A type of fan often used in mines and ventilation activities in which an impeller, generally consisting of numerous blades, discharges the air radially into an expanding scroll casing while imparting an increase in pressure to the air. (Hartman, 1982) (b) *radial-flow fan.*
centrifugal filter *filter.*
centrifugal pump (a) A form of pump in which water is drawn through the eye of a rotating impeller and discharged from its periphery into a chamber containing a series of passages of gradually increasing cross section. The kinetic energy given to the water by its centrifugal discharge is thus largely converted to pressure energy. Cf: *duplex pump.* (BS, 1963) (b) *turbine pump.*
centrifugal replacement Mineral replacement in which the host mineral is replaced from its center outward. Cf: *centripetal replacement.* (AGI, 1987)
centrifugal separation (a) The separation of different particles by centrifugal action as used in cyclone separators and centrifuges. (b) The use of centrifugal force to increase the apparent density of finely divided particles so as to accelerate their movement with respect to ambient fluid. (Pryor, 1963) (c) Accelerated settlement of finely divided particles from pulp, removal of moisture, or classification into relatively coarse and fine fractions by centrifuging. Performed on a laboratory scale in small batches and commercially in a hydrocyclone or centrifugal classifier. See also: *cyclone; cyclone washer.* (Pryor, 1963)
centrifugal ventilation A mine ventilation system in which the air is led through a shaft in the middle of the field into the mine and out again at the periphery of the mining field. (Stoces, 1954)
centrifuge (a) A centrifugal device for dewatering, usually conical or bowl-shaped, in which the containing surface is imperforated. The greater density of the solid particles causes them to collect preferentially in contact with the inside of the containing surface where they are discharged mechanically; the water usually overflows from a position nearer to the axis. (BS, 1962) (b) A rotating device for separating liquids of different specific gravities or for separating suspended colloidal particles, such as clay particles in an aqueous suspension, according to particle-size fractions, by centrifugal force. Colloidal particles that cannot be deposited from suspension by gravity can be deposited by centrifugal force in a supercentrifuge. See also: *cyclone.* Syn: *hydroextractor.*
centrifuging Dewatering of clean coal or refuse with the aid of centrifugal force. See also: *centrifugal separation.* (BS, 1962)
centripetal drainage Drainage more or less radially inward toward a center. (Stokes, 1955)
centripetal pump A pump with a rotating mechanism that gathers a fluid at or near the circumference of radial tubes and discharges it at the axis. (Standard, 1964)
centripetal replacement Mineral replacement in which the host mineral is replaced from its periphery inward. Cf: *centrifugal replacement.* (AGI, 1987)
centroclinal Said of strata and structures that dip toward a common center. Ant: quaquaversal. Cf: *periclinal.* See also: *centrocline.* (AGI, 1987)
centrocline An equidimensional basin characteristic of cratonic areas, in which the strata dip toward a central low point. The term is little used in the United States. Cf: *pericline.* Ant: quaquaversal. See also: *centroclinal.* (AGI, 1987)
centrosphere *barysphere; core of the Earth.*
centrosymmetrical Having a center of symmetry. Centrosymmetric crystal structures cannot exhibit pyroelectricity or piezoelectricity.
ceramet Substance formed of a mixture of metal and ceramic, to give the requisite conductivity to the latter. (CTD, 1958)
ceramic (a) As a singular or plural noun, any of a class of inorganic, nonmetallic products that are subjected to a high temperature during manufacture or use. (ACSG, 1963) (b) As an adj., of or pertaining to (1) ceramic—that is, inorganic or nonmetallic as opposed to organic or metallic; (2) products manufactured from inorganic nonmetallic substances, which are subjected to a high temperature during manufacture

or use; (3) the manufacture or use of such articles or materials, such as ceramic process or ceramic science. (ACSG, 1963)

ceramic cone *pyrometric cone.*

cerargyrite A former name for chlorargyrite, AgCl. Also spelled kerargyrite.

cerhomilite Borosilicate of calcium, beryllium, iron, thorium, and rare earths. (Hey, 1955)

cerianite An isometric mineral $(Ce,Th)O_2$; forms minute greenish-yellow grains; named for its relationship to thorianite and uraninite.

cerite A trigonal mineral, $(Ce,Ca)_{10}(SiO_4)_6(OH,F)_5$; generally brown, massive.

cerium One of the most abundant of the rare earth metals. Symbol, Ce. The minerals monazite and bastnasite are presently its two most important sources. It is used in the manufacture of pyrophoric alloy. The oxide is a constituent of incandescent gas mantles and is emerging as a catalyst in self-cleaning ovens. The sulfate is used as an oxidizing agent in quantitative analysis. Other cerium compounds are used in the manufacture of glass, as a polishing agent, and in carbon-arc lighting, petroleum refining, and metallurgical and nuclear applications. (Handbook of Chem. & Phys., 3)

cermet A material or body consisting of ceramic particles bonded with a metal. According to the American Society for Testing and Materials, the ceramic phase must be present in 15% or more of the body. A ceramic foam or porous ceramic is not a cermet because the bonding of the ceramic structure is not dependent on or due to the metal. (Hunt, 1965)

cerolite *kerolite.*

cerro A term used in the Southwestern United States for a hill, esp. a craggy or rocky eminence of moderate height. Etymol: Spanish. (AGI, 1987)

certified (a) A certified employee is one who has been granted a State certificate of competency for a given job. (BCI, 1947) (b) Evaluated and listed as permissible by the National Institute for Occupational Safety and Health or the Mine Safety and Health Administration. (ANSI, 1992)

Certified Blaster A blaster certified by a government agency to prepare, execute, and supervise blasting. (Atlas, 1987)

cerulene (a) A trade name for a form of calcium carbonate colored green and blue by malachite or azurite, and used as a gemstone. From Bimbowrie, south Australia. (English, 1938) (b) A term used less correctly for a blue variety of satin spar.

ceruleofibrite A former name for connellite. Also spelled caeruleofibrite.

cerussite An orthorhombic mineral, $4[PbCO_3]$; aragonite group; adamantine; sp gr, 6.55; in oxidized and carbonated parts of lead-ore veins; a source of lead. Syn: *white ore; white lead ore; lead carbonate; lead spar.*

cervantite An orthorhombic mineral, $Sb^{3+}Sb^{5}O_4$; may be confused with stibiconite.

cesium A silvery white, soft and ductile alkaline element, of the rare-earth metals. Symbol, Cs. Occurs in lepidolite and pollucite. Reacts explosively with cold water. Because of its great affinity for oxygen, the metal is used as a "getter" in electron tubes and as a catalyst in the hydrogenation of certain organic compounds; it has recently found application in ion propulsion systems. (Handbook of Chem. & Phys., 3)

cetane number An indication of diesel fuel ignition quality. The cetane number of a fuel is the percentage by volume of cetane in a mixture of cetane and alpha methylnaphthalene, which matches the unknown fuel in ignition quality. American diesel oil usually ranges from 30 to 60 cetane. (Nichols, 1976)

ceylanite Original spelling of ceylonite. See also: *ceylonite.* (Hey, 1955)

Ceylonese peridot The trade name for a yellowish-green variety of tourmaline, approaching olivine in color; used as a semiprecious gemstone. Syn: *peridot of Ceylon.* (CMD, 1948)

ceylonite A dark-green, brown, or black variety of spinel containing iron. Syn: *pleonaste; candite; ceylanite; zeylanite.*

C-frame An angling bulldozer lift and push frame. (Nichols, 1976)

chabazite A trigonal mineral, $1[Ca_2(Al_4Si_8O_{24})\cdot 13H_2O]$; zeolite group; pseudocubic cleavage; occurs in cavities in basalts and hydrothermal veins and as alteration of silicic vitreous tuffs in alkaline saline lake deposits.

chadacryst *xenocryst.* Also spelled cadacryst.

chain (a) A measuring instrument that consists of 100 links joined together by rings and is used in surveying. See also: *Gunter's chain.* (Webster 3rd, 1966) (b) A unit of length prescribed by law for the survey of U.S. public lands and equal to 66 ft (20.12 m) or 4 rods. It is a convenient length for land measurement because 10 square chains equal 1 acre (0.4 ha). (AGI, 1987)

chain block A combination of sheaves over which chains are arranged in the same manner as the rope in a block and tackle. Also called chain hoist. (Long, 1960)

chain brow way An underground inclined plane worked by an endless chain.

chain bucket dredger A dredger with a bucket ladder. (Hammond, 1965)

chain bucket loader A mobile loader that uses a series of small buckets on a roller chain to elevate spoil to the dumping point. Also called bucket loader. (Nichols, 1976)

chain casing *chain guard.*

chain coal cutter A coal cutter that cuts a groove in the coal by an endless chain traveling around a flat plate called a jib. The chain consists of a number of pick boxes. Each box holds a cutter pick fastened into the box by a set screw or similar device. The coal cutter pulls itself along the face by means of a rope at a speed ranging from 7 in/min (17.8 cm/min) to 5 ft/min (1.5 m/min) or more. The chain travels around the jib at a speed ranging from 320 to 650 ft/min (97.6 to 198.2 m/min). The cut in the coal ranges from 3-1/2 to 7-1/2 in (8.9 to 19.1 cm) high and up to 8-1/2 ft (2.59 m) in length. See also: *coal-cutter pick.* (Nelson, 1965)

chain conveyor (a) A conveyor comprising one or two endless linked chains with crossbars or flights at intervals to move the coal or mineral. The loaded side of the conveyor runs in a metal trough, while the empty side returns along guides underneath. The material is transported on the conveyor partly by riding on the chain and flights and partly by being scraped along in the trough. The chain conveyor is widely used in coal mines, and capacities range up to 100 st/h (90.7 t/h) with lengths of about 100 yd (91 m). See also: *armored flexible conveyor.* (Nelson, 1965) (b) *drag-chain conveyor.*

chain driller *chain-machine operator.*

chain-driven belt A conveyor similar in design to those driven by ropes, the essential difference being that the tension is taken by chains, either under or alongside the carrying belt. (Nelson, 1965)

chain elevator *bucket elevator.*

chain feeder *conveyor-type feeder.*

chain-feeder operator *mill feeder.*

chain guard An open guard of sheet metal, expanded metal, or similar construction around a chain drive. (Jackson, 1955)

chain hoist (a) A block and tackle in which chain is used instead of rope. (Crispin, 1964) (b) *chain block.*

chain lacing The arrangement of block positions in a cutter chain so that bits inserted in these blocks will occupy certain positions while cutting. (Jones, 1949)

chain machine Coal-cutting machine that cuts coal with a series of steel bits set in an endless chain moved continuously in one direction either by an electric or a compressed-air motor. These machines may be divided into four classes, known as breast machines, shortwall machines, longwall machines, and overcutting machines. (Kiser, 1929)

chain-machine operator In bituminous coal mining, one who operates a chain-driven machine to undercut coal preparatory to blasting it loose from the working face with explosives. Also called chain driller. (DOT, 1949)

chain pillar A pillar left to protect the gangway and airway, and extending parallel to these passages.

chain pitch For a roller chain, the distance in inches between the centers of adjacent joint members. For a silent chain, the distance in inches between the centers of the holes in a link plate. (Jackson, 1955)

chain road Main underground haulage road through which tubs are hauled by an endless chain. (Pryor, 1963)

chain-selvage belt A belt in which the carrying section may be made up of rubber or fabric, woven metal, or other material and along each edge of which is fastened an endless chain with a suitable attachment. The chains carry the driving tension. The center part functions only as a loading supporting medium.

chain silicate Silicate mineral with silica tetrahedra linked by shared oxygens into infinite one-dimensional chains. Single chains characterize pyroxenes; double chains characterize amphiboles; and wider chains grade toward sheet structures.

chain structure A structure or texture found in a number of chromite occurrences, consisting of a series of connected chromite crystals somewhat resembling a chain. (Schieferdecker, 1959)

chain surveying The simplest method of surveying, which has the advantage that the equipment required is inexpensive and hard wearing. It is the ideal method for small areas and has been employed successfully for large surveys. Nevertheless, it has definite limitations when applied to surveys of enclosed or built-up areas. (Mason, 1951)

chain takeup An idler sprocket, or similar device, mounted on an adjustable bracket to adjust the slack in a chain drive. *takeup.* (Jackson, 1955)

chain tension The actual force existing at any point in a conveyor chain.

chain-type conveyor A conveyor using a driven endless chain or chains, equipped with flights that operate in a trough and move material along the trough.

chainwall (a) A method of mining coal in which the roof is supported by pillars of coal between which the coal is mined away. (Standard, 1964) (b) Scot. A system of working by means of wide rooms and long, narrow pillars, sometimes called "room and rance." (c) Scot. A long, narrow strip of mineral left unworked; e.g., along the low side of a level.

chain width For a roller chain, the distance between the link plates of a roller link. This is not the overall width of the chain. For a silent chain, the width over the working-link plates of the chain, exclusive of pinheads, washers, or other fastening devices. (Jackson, 1955)

chair (a) Movable support for a cage, arranged to hold it at the landing when desired. See also: *catch; dog; rests.* (Fay, 1920) (b) Projection that can be set into a guide so that the skip or cage descending in the mine shaft is brought to rest at the correct level. (Pryor, 1963) (c) A cast-iron support bolted to a timber or concrete railway sleeper used to hold a bullhead rail in position. (Hammond, 1965)

chalcanthite (a) A triclinic mineral $2[CuSO_4 \cdot 5H_2O]$; azure blue; metallic taste; occurs in the oxidized supergene zone above copper sulfides in arid regions; a minor ore of copper. Syn: *blue vitriol; copper sulfate; copper chalcanthite; copper vitriol; bluestone; cyanosite.* (b) The mineral group chalcanthite, jokokuite, pentahydrite, and siderotil.

chalcedonite Fibrous quartz with a negative elongation. See also: *chalcedony.* (Hess)

chalcedony (a) A fine-grained or cryptocrystalline variety of quartz; commonly microscopically fibrous; translucent or semitransparent, with a nearly waxlike luster; has lower density and indices of refraction than ordinary quartz. Chalcedony is the material of much chert, flint, and jasper; commonly an aqueous deposit filling or lining cavities in rocks. In the gem trade, the name refers specif. to the light blue-gray or common variety of chalcedony. Varieties include carnelian, sard, chrysoprase, prase, plasma, bloodstone, onyx, and sardonyx. See also: *agate.* Syn: *calcedony, chalcedonite; white agate.* (b) A general name for crystalline silica that forms concretionary masses with radial-fibrous and concentric structure and that is optically negative (unlike true quartz). (c) A trade name for a natural blue onyx.

chalchuite A blue or green variety of turquoise.

chalco- (prefix) A combining form meaning copper.

chalcoalumite A monoclinic mineral, $CuAl_4(SO_4)(OH)_{12} \cdot 3H_2O$.

chalcocite A monoclinic mineral, $96[Cu_2S]$; pseudohexagonal, metallic gray-black with blue to green tarnish; sp gr, 5.5 to 5.8; a secondary vein mineral; an important source of copper. Syn: *redruthite; copper glance; chalcosine; beta chalcocite; vitreous copper; vitreous copper ore.*

chalcocyanite An orthorhombic mineral, $CuSO_4$; white; it becomes blue upon hydration, thus formerly called hydrocyanite.

chalcodite *stilpnomelane.*

chalcogene Said of ore deposits, such as those of copper, connected with a phase of mountain building and plutonism.

chalcolite *torbernite.*

chalcomenite An orthorhombic mineral, $CuSeO_3 \cdot 2H_2O$; dimorphous with clinochalcomenite.

chalcomiklite *bornite.*

chalcophanite A trigonal mineral, $(Zn,Fe,Mn)Mn_4O_7 \cdot 3H_2O$. Formerly called hydrofranklinite.

chalcophile Said of an element tending to concentrate in sulfide minerals and ores. Such elements have intermediate electrode potentials and are soluble in iron monosulfide. Examples are S, Se, As, Fe, Pb, Zn, Cd, Cu, and Ag. Cf: *lithophile.* (AGI, 1987)

chalcophyllite A trigonal mineral, $Cu_{18}Al_2(AsO_4)_3(SO_4)_3(OH)_{27} \cdot 33H_2O$; green; forms tabular crystals and foliated masses.

chalcopyrite (a) A tetragonal mineral, $CuFeS_2$; brass-yellow with bluish tarnish; massive; softer than pyrite; occurs in late magmatic hydrothermal veins and secondary enrichment zones; the most important source of copper. Syn: *copper pyrite; cupriferous pyrite; yellow copper ore; yellow ore; yellow pyrite; yellow copper.* (b) The mineral group chalcopyrite, eskebornite, gallite, and roquesite.

chalcopyrrhotite A former name for cubanite.

chalcosiderite A triclinic mineral, $CuFe_6(PO_4)_4(OH)_8 \cdot 4H_2O$; turquoise group; occurs in sheaflike crystalline incrustations; forms a series with turquoise.

chalcosine *chalcocite.*

chalcostibite A lead-gray copper-antimony sulfide, $CuSbS_2$. Also called wolfsbergite. Syn: *rosite; antimonial copper.* (Fay, 1920)

chalcotrichite A capillary variety of cuprite in fine, slender interlacing fibrous crystals. Syn: *cuprite; plush copper ore; hair copper.*

chalk A soft, earthy, fine-textured, usually white to light-gray or buff limestone of marine origin. It consists almost wholly (90% to 99%) of calcite, formed mainly by shallow-water accumulation of calcareous remains of floating microorganisms (chiefly foraminifers) and of comminuted remains of calcareous algae, set in a structureless matrix of very finely crystalline calcite. The rock is porous, somewhat friable, and only slightly coherent. (AGI, 1987)

chalk rock Any soft, milky-colored rock resembling white chalk, such as talc, calcareous tufa, diatomaceous shale, volcanic tuff, or white limestone. (AGI, 1987)

chalky chert A commonly dull or earthy, soft to hard, sometimes finely porous chert of essentially uniform composition, having an uneven or rough fracture surface, and resembling chalk. Syn: *dead chert.* Obsolete syn: *cotton chert.* Cf: *granular chert.* (AGI, 1987)

challenge feeder Ore feeder used with stamp batteries to regulate the rate of entry of ore to a mortar box. A horizontal plate is turned by linkages operated when the central stamp falls below a prefixed point; it then draws ore from feeding bin. (Pryor, 1963)

chalmersite A former name for cubanite.

chalybite *siderite.*

chamber (a) A miner's working place, sometimes referred to as a room or breast. (Hudson, 1932) (b) A large irregular or rounded body of ore, occurring alone or as an expansion of a vein. (c) A body of ore with definite boundaries, apparently filling a preexisting cavern. (d) A powder-storage room in a mine. (e) To enlarge the bottom of a drill hole by the use of explosives, so that a sufficient blasting charge may be loaded for the final shot. Syn: *spring.* (f) A space or gallery excavated in a quarry or underground mine to receive a large explosive charge. See also: *heading blast.*

chamber-and-pillar *breast-and-pillar.*

chamber-and-pillar system A modification of sublevel stoping by which a series of sublevels are successively caved. See also: *sublevel stoping.* (Hess)

chamber blast A large-scale blast in which explosives in bulk are placed in excavated subterranean chambers. Also called coyote blast; gopher-hole blast. (Webster 2nd, 1960)

chambered lode A portion of the wall of a lode that is fissured and filled with ore. See also: *chamber.*

chambered vein (a) A vein in which the walls, particularly the hanging wall, are irregular and brecciated, owing to the formation of the vein under low pressure at shallow depth. See also: *chambered lode.* (Schieferdecker, 1959) (b) Stockwork.

chambering (a) The process of enlarging a portion of a blast hole (usually the bottom) by firing a series of small explosive charges. It can also be done by mechanical or thermal methods. (b) The enlarging of the bottom of a quarry blasting hole by the repeated firing of small explosive charges. The enlarged hole or chamber is then loaded with the proper explosive charge, stemmed, and fired to break down the quarry face. See also: *concentrated charge.* (Nelson, 1965) (c) A borehole in which portions of the sidewalls are breaking away and forming cavities or small chambers. (Long, 1960)

chambersite An orthorhombic mineral, $Mn_3B_7O_{13}Cl$; the manganese analogue of boracite; occurs in brines.

chambers without filling *sublevel stoping.*

chameleonite A rare variety of tourmaline, olive green in daylight, changing to brownish-red in most artificial light. Syn: *alexandritelike tourmaline.*

chamfer To bevel or slope an edge or corner. Also spelled chanfer.

chamoisite *chamosite.*

chamosite A monoclinic mineral, $2[(Fe^{2+},Mg,Fe^{3+})_5Al(Si_3Al)O_{10}(OH,O)_8]$; chlorite group; a constituent of oolitic iron ores and sedimentary ironstones. Also spelled chamoisite.

chamotte The refractory portion of a mixture used in the manufacture of firebrick, composed of calcined clay or of reground bricks. (Standard, 1964)

Champlain forge A forge for the direct production of wrought iron, generally used in the United States instead of the Catalan forge, from which it differs in using only finely crushed ore and in working continuously. Syn: *American forge.* (Fay, 1920)

chance (a) In coal mining, the opportunity a shot has to break the coal. (Fay, 1920) (b) The opportunity to put in a shot in a good position. (Fay, 1920)

Chance cone *cone classifier.*

Chance sand-flotation process A dense-media process in which coal is separated from refuse in an artificial dense medium of sand suspended in water. The specific gravity of this medium is such that the merchantable coal floats while the refuse sinks to the bottom, the separation being analogous to that of a float-and-sink separation with a heavy liquidlike zinc chloride. Named after Thomas M. Chance, U.S. mining engineer. (Mitchell, 1950)

changehouse A special building at a mine where workers may wash themselves or change from street to work clothes and vice versa. Also called changing house; dry; dryhouse. Cf: *doghouse.* (Long, 1960)

changing bronze The process of changing tuyeres, plates, monkey, etc., at blast furnaces. (Fay, 1920)

changing house *changehouse.*

changkol Malay. A heavy Chinese hoe with an eye in which the handle fits; used in cutting soft rock and earth and for stirring gravel in sluice boxes, etc. (Hess)

channeler (a) A powerful quarrying machine capable of cutting slots in stone at any angle. It is used for cutting dimension stone off the quarry face without explosives. See also: *broaching*. Syn: *channeling machine*. (Nelson, 1965) (b) A machine that cuts a deep groove in rock, ordinarily to free dimension stone from the mass, or to make a smooth side for a canal or other excavation in rock. Cutting is accomplished by a group of reciprocating chisel-pointed bars, operated by steam or compressed air while the machine carrying them travels back and forth on a track. Ordinarily used only in the softer rocks, such as limestone, soapstone, or slate. Also called track channeler; bar channeler. (Hess)
channeling (a) In ion-exchange, fixed-bed work, development of passages in a resin column through which the liquors flow preferentially so that the resin is unequally loaded. (Pryor, 1963) (b) In cyanide sand leaching, cracks in the sand bed through which cyanide solution runs without proper percolating contact with a mass of particles. (Pryor, 1963) (c) The action of a blast furnace in opening up irregular openings for the blast.
channeling machine *channeler.*
channeling-machine operator (a) In bituminous coal mining, one who operates a coal-cutting machine to cut channels (a few inches wide) in coal, after the overlying ground has been removed, to partly detach coal in blocks so that it may be broken loose more easily by blasting. Also called channeler-machine operator; channel-machine operator. (DOT, 1949) (b) In the quarry industry, one who sets up and operates a track-mounted machine that cuts (drills) vertical channels (a few inches wide) in quarrystone in which wedges are driven to crack off a block from the mass. Also called channeler-machine operator; channeler runner; channel-machine operator; channel-machine runner. (DOT, 1949)
channel man In ore dressing, smelting, and refining, one who installs new channel irons to form a supporting framework for a continuous anode. (DOT, 1949)
channel sample Material from a level groove cut across an ore exposure to obtain a true cross section of it. Syn: *groove sample; strip sample.* (Pryor, 1963)
channel sampling *trench sampling.*
channel sand A sand or a sandstone deposited in a stream bed, or some other channel eroded into the underlying rock; it frequently contains oil, gold, or other valuable minerals.
channel slide rail One of a pair of rails used in a method of temporary rail-track advance at a tunnel face. The rails comprise a pair of specially made channels with ramp ends. They fit over the rail section in use and are pushed forward periodically as a power loader clears the rock ahead. The permanent track is extended as space becomes available. (Nelson, 1965)
channel terrace A contour ridge, built of soil moved from its uphill side, that serves to divert surface water from a field. (Nichols, 1976)
channelway An opening or passage in a rock through which mineral-bearing solutions or gases may move. (AGI, 1987)
chap (a) Scot. A customary and rough mode of judging, by sound, of the thickness of coal between two working places, by knocking with a hammer on the solid coal. (Fay, 1920) (b) To examine the face of the coal, etc., for the sake of safety, by knocking on it lightly. (Fay, 1920) (c) Scot. A blow, rap, knock, or stroke. See also: *chapping; sounding.* (Webster 3rd, 1966)
chapeau de fer A French term for gossan or iron hat.
chapelet (a) A machine for raising water, or for dredging, by buckets of an endless chain passing between two rotating sprocket wheels. (Standard, 1964) (b) A chain pump having buttons or disks at intervals along its chain; paternoster pump. (Standard, 1964) (c) A device for holding the end of heavy work, such as a cannon, in a turning lathe. (Standard, 1964)
Chapman-Jouget plane In a detonating explosive column, the plane that defines the rear boundary of the primary reaction zone. The plane is the point within the reaction at which all thermodynamic properties of temperature, pressure, energy, gas, volume, and density are measured and calculated. Syn: *C-J plane.*
Chapman process A method of gold recovery in which cyanidation dissolves the metal from an ore pulp and the aurocyanide is simultaneously absorbed by activated carbon. This last is then retrieved by froth flotation. (Pryor, 1958)
Chapman shield A pair of vertical plates of sheet iron or steel arranged with a ladle between them, which can be moved longitudinally along the front of a furnace; it is mainly used to protect laborers from furnace heat. (Fay, 1920)
chapping Rough guess of distance separating two approaching drives underground made by knocking with a heavy hammer. See also: *chap.* (Pryor, 1963)
chapra A term used in Bihar, India, for a kind of hoe used in mines for scraping waste debris into pans for carrying or loading cars. (Hess)
char The solid carbonaceous residue that results from incomplete combustion of organic material. It can be burned for heat, or, if pure, processed for production of activated carbon for use as a filtering medium. See also: *coke.* (AGI, 1987)
characteristic ash curve The curve obtained from the results of a float-and-sink analysis showing, for any yield of floats (sinks), the ash content of the highest density (lowest density) fraction passing into these floats (sinks), the yield being plotted on the ordinate and the ash content on the abscissa. (BS, 1962)
characteristic curve In general, a curve that defines one or more of the characteristics or properties of a piece of machinery, such as a fan, pump, motor, etc. See also: *mine-ventilation fan characteristics.*
characteristic impedance (a) Of an explosive, the amount of energy transferred to a given rock is a linear function of the product of density and rate of detonation. (Leet, 1960) (b) For rock, density times velocity of longitudinal waves in the rock. (Leet, 1960)
characteristic radiation High-intensity, single-wavelength X-rays, characteristic of the element emitting the rays, that appear in addition to the continuous white radiation whenever the voltage of an X-ray tube is increased beyond a critical value. (ASM, 1961)
characterizing accessory mineral *varietal mineral.*
charcoal blacking Charcoal used in pulverized form as dry blacking or in suspension with clay as a black wash; either dusted or coated on the surface of molds to improve the surface. (Osborne, 1956)
charcoal iron Sulfur-free pig iron made in a charcoal furnace; it has higher quality, higher density, and closer structure than other iron. (Bennett, 1962)
charcoal tinplate Tinplate with a relatively heavy coating of tin (higher than the coke tinplate grades). (Bennett, 1962)
charge (a) The liquid and solid materials fed into a furnace for its operation or prepared for further processing. (ASM, 1961) (b) The explosive loaded into a borehole for blasting; also, any unit of an explosive, such as a charge of nitroglycerin or a charge of detonating composition in a blasting cap. (Fay, 1920) (c) To put an explosive into a hole, to arrange the fuse or squib, and to tamp it. Cf: *load.*
charged hole Hole to be blasted that contains explosive material and a detonator; for blasting operations under the jurisdiction of the Mine Safety and Health Administration (MSHA), charged holes must be detonated within 72 h of charging unless prior approval has been obtained from MSHA (CFR 30, 1988 - 56.6094).
charge limit For an explosion, the maximum weight of charge that can be fired without causing an ignition in gallery tests. (McAdam, 1958)
chargeman (a) A stallman. (Nelson, 1965) (b) A laborer who moves a mixture of concentrate, slag, and fluxing ingredients through a hopper into charge pipes opening into a reverberatory furnace where smelting takes place, using an air-pressure hose. Also called feeder. (DOT, 1949)
charger (a) A remotely controlled device for moving single wagons at a mine surface over a short distance. The device runs on a narrow-gage track alongside the main rails and uses a pair of roller arms, which extend to engage on either side of a wagon wheel. Propelled by a guided chain engaging a power-drive chain wheel, the charger can position a wagon exactly where required. (Nelson, 1965) (b) In the iron and steel industry, one who loads steel ingots into a furnace for heating, withdraws white-hot ingots from the furnace, and positions them on the bed of a mill for rolling, using a traveling electric charging machine. (DOT, 1949) (c) *lidman.*
charge weigher In ore dressing, smelting, and refining, one who weighs out specified amounts of coke, limestone, and copper-bearing scrap materials to make furnace charges for recovery of copper from plant refuse. (DOT, 1949)
charging (a) The loading of a borehole with explosives. (Fay, 1920) (b) The arranging of a fuse or squib, and the tamping of the hole with stemming material. (Jones, 1949) (c) Feeding raw material into an apparatus, such as a furnace, for treatment or conversion. (Bennett, 1962)
charging box A box in which ore, scrap, pig iron, fluxes, etc., are conveyed to a furnace by means of a charging machine. (Fay, 1920)
charging machine A machine for delivering coal, ore, or metals to a furnace, gas retort, coke oven, or other reactor. (Fay, 1920)
charging peel A long arm or extension attached to a charging machine for conveying and dumping scrap into an open-hearth furnace. (Mersereau, 1947)
charging person A laborer who charges an electric-arc furnace with metals, alloys, and other materials. Also called furnace feeder. See also: *furnace charger.* (DOT, 1949)
charging rack A device used for holding batteries for mining lamps and for connecting them to a power supply while the batteries are being recharged.
charging scale A scale for weighing the various materials used in a blast furnace. (Fay, 1920)
chark (a) To burn to charcoal or coke. (Webster 3rd, 1966) (b) Charcoal; coke; cinder. (Webster 3rd, 1966)
charred peat Peat artificially dried at a temperature that causes partial decomposition. (Bennett, 1962)

chart A base map conveying information about something other than the purely geographic; also, a special-purpose map; esp. one designed for purposes of navigation, such as a hydrographic chart or a bathymetric chart. (AGI, 1987)

chart datum The plane to which soundings on a chart are referred, usually low water. (Hy, 1965)

chaser An edge wheel revolving in a trough for crushing asbestos mineral, without destroying the fiber, and for fine crushing of ore.

chaser mill (a) This type of mill usually consists of a cylindrical steel tank that is lined with wooden blocks laid with the end grain up. The rollers are usually wooden—with a speed of 15 to 30 rpm. (b) Occasionally synonymous with an edge-runner mill. (Dodd, 1964)

chasing Following a vein by its range or direction.

chasing the vein Derb. Following a vein along the surface by means of cast holes or prospect pits.

chasm A yawning hollow or rent, as in the Earth's surface; any wide and deep gap; a cleft; fissure. See also: *abyss.* (Standard, 1964)

chasovrite A variety of clay mineral (glinite) from the Chasovyar deposit in the Ukraine. (Spencer, 1958)

chat The finely crushed gangue remaining after the extraction of lead and zinc minerals in the Tri-State District of Missouri, Kansas, and Oklahoma. The term is derived from chert. See also: *chats.*

chatoyance *chatoyancy.*

chatoyancy An optical phenomenon, possessed by certain minerals in reflected light, in which a movable wavy or silky sheen is concentrated in a narrow band of light that changes its position as a mineral is turned. It results from the reflection of light from minute, parallel fibers, cavities or tubes, or needlelike inclusions within the mineral. The effect may be seen on a cabochon-cut gemstone, either distinct and well defined (such as the narrow, light-colored streak in a fine chrysoberyl cat's-eye) or less distinct (such as in the usual tourmaline or beryl cat's-eye). Syn: *chatoyance.*

chatoyant (a) Having a luster resembling the changing luster of the eye of a cat as seen at night. See also: *cat's-eye.* (Fay, 1920) (b) adj. Said of a mineral or gemstone possessing chatoyancy or having a changeable luster or color marked by a narrow band of light. (c) A chatoyant gem.

chatroller An ore-crushing machine, consisting of a pair of cast-iron rollers, for grinding roasted ore. (Fay, 1920)

chats (a) Northumb. Small pieces of stone with ore. (Fay, 1920) (b) Eng. A low grade of lead ore. Also, middlings that are to be crushed and subjected to further treatment. The mineral and rocks mixed together that must be crushed and cleaned before being sold as mineral. Chats are not the same as tailings, as the latter are not thrown aside to keep for future milling. (Fay, 1920) (c) *chat.* (d) Eng. Bowse when broken up on the knockstone ready for the hotchin tubs; Yorkshire lead mines. (Arkell, 1953) (e) A quarrying term for cherty rock used as an abrasive.

chatter (a) Rapid vibrations caused by overfeeding a bit and/or by drill rods rubbing against the sidewalls of a borehole. (Long, 1960) (b) In grinding, a vibration of the tool, wheel, or workpiece producing a wavy surface on the work. (ASM, 1961) (c) The finish produced by such vibrations during grinding. (ASM, 1961)

chattermark A spiral or flutelike, round-topped ridge, sometimes seen on the outside surface of a drill core. (Long, 1960)

check (a) Applied to slit canvas or brattice cloth placed across a passage to prevent the flow of air while still permitting the passage of personnel and equipment. See also: *check curtain.* (Jones, 1949) (b) A brass disk with a miner's lamp number punched on it that a person exchanges for a lamp at the lamp room every time the person enters or leaves the mine. (Nelson, 1965)

check battery A battery to close the lower part of a chute acting as a check to the flow of coal, and as a stopping to keep air in the breasts.

check board A board usually posted at the entrance to a mine or to a section of a mine on which (1) miners hang their identification checks to show whether they are in or out, or (2) the miners' loading checks are hung.

checkboarding To divide property in a manner so that two parties acquire title to alternating and equal-size square sections of land. (Long, 1960)

check curtain (a) *curtain.* (b) Ventilation control consisting of jute or nylon material fastened to the roof and placed across an entry or a crosscut. It is used to direct the air to the working place, yet allow the passage of equipment and persons. (MSHA, 1986)

check dam A dam that divides a drainageway into two sections with reduced slopes. (Nichols, 1976)

checker arch One of the firebrick supports built of arch brick or keys to support the checker work on the second, third, or fourth pass of hot-blast stoves. (Fay, 1920)

checkerboard drilling *checkerboarded.*

checkerboarded An area in which boreholes have been placed at the intersections of equally spaced parallel lines laid out on a square grid or checkerboard pattern. (Long, 1960)

checkerboard system *bord-and-pillar method.*

checking Temporarily reducing the temperature or the volume of the air blast on a blast furnace. (Fay, 1920)

checkout Scot. The meeting of the roof and floor, the coal seam being thereby cut off; to pinch out.

check screen *oversize control screen.*

checksheet A sheet on which are printed illustrations of various drilling equipment assemblies with the component items shown in their relative operating positions; used as a guide in making up a list of the units necessary to do various routine drilling jobs. (Long, 1960)

check survey A survey made to confirm the positions of established survey stations in a mine. (BS, 1963)

check viewer In bituminous coal mining, one who inspects and checks portions of a mine that have been leased to workers to see that the terms of lease, such as mining within specified limits, safety precautions, and production rate, are duly observed. (DOT, 1949)

checkweigher In mining, one who checks, in the interest of miners, the weighing of coal in mine cars or other containers by the company weighmaster. The person estimates the amount of slate, dirt, rock, and other foreign matter in the coal and sees that only authorized deductions are made. Also called check-docking boss; check measurer; checkweighman; justiceman. (DOT, 1949; Fay, 1920)

checkweighman *checkweigher.*

cheeking The removal of the side or sides of a roadway to increase its width. (TIME, 1929-30)

cheeks (a) The sides or walls of a vein. (Fay, 1920) (b) Extensions of the sides of the eye of a hammer or pick. (Fay, 1920) (c) The refractory sidewalls of the ports of a fuel-fired furnace. (Dodd, 1964)

chelate compound The compound formed by the combination of a chelating agent and a metal ion. (ASM, 1961)

chelating agent A substance that contains two or more electron donor groups and will combine with a metal ion so that one or more rings are formed. (ASM, 1961)

chelation The reaction between a metallic ion and a complexing agent, generally organic, with the formation of a ring structure and the effective removal of the metallic ion from the system. It is significant in chemical weathering. (AGI, 1987)

cheleutite (a) A ferruginous, nickeliferous, and slightly cupriferous smaltite. See also: *smaltite.* (Weed, 1918) (b) A copper-bearing variety of smaltite.

Chelsea color filter An effective dichromatic color filter transmitting light of only two wavelength regions—one in the deep red, the other in the yellow green. Useful for discriminating between emerald and its imitations and for detecting synthetic spinels and pastes colored blue with cobalt. (Anderson, 1964)

chemawinite A pale-yellow to dark-brown variety of retinite (amber) in decayed wood at Cedar Lake, MB, Canada. Syn: *cedarite.* See also: *amber.*

chemechol A method of breaking down coal similar to Hydrox and applied on the same lines as air shooting. (Nelson, 1965)

chemical adsorption Surface adherence, accompanied by the formation of primary bonds. (Van Vlack, 1964)

chemical affinity (a) The force that binds atoms together in molecules. (CTD, 1958) (b) The tendency of one substance to form a chemical compound with another. (Hess)

chemical analysis A method of determining the composition of a material employing chemical techniques by which the various elements are quantitatively separated.

chemical brick *chemical stoneware.*

chemical-clay grout A typical grout of this class used in Great Britain is bentonite-sodium silicate, in which the silicate is used to render irreversible the thixotropic nature of the bentonite suspension. The gel is stronger than pure bentonite and is permanent, in that local vibration cannot cause it to liquefy again. Setting time can be controlled by adjustment of the chemical content. (Nelson, 1965)

chemical combination Change in which permanent alteration of properties occurs, accompanied by intake or release of energy. Reaction is governed by laws of mass conservation, definite and multiple proportions, equivalence, and volumetric reaction. (Pryor, 1963)

chemical composition The weight percent of the elements (generally expressed as certain oxide molecules) in a rock. (AGI, 1987)

chemical constitution of coal The elements or component parts of coal. These are determined by chemical analyses that may be performed in different ways. An ultimate analysis provides exact information as to the percentages of the various elements (such as carbon, oxygen, and hydrogen) present in the coal. Another method is by proximate analysis, which determines the relative percentages of carbon, moisture, volatile matter (such as gas and tar), sulfur, and ash. (Nelson, 1965)

chemical denudation The processes in which the salts or the soluble minerals in the Earth are dissolved by water and carried to the sea. (Bennett, 1962)

chemical deposition The precipitation or plating-out of a metal from a solution of its salts through the introduction of another metal or a reagent into the solution. (ASM, 1961)

chemical engineering Developing, building, and operating plants in which materials are chemically worked up to desired end products. (Pryor, 1963)

chemical equilibrium A state of balance between two opposing chemical reactions. The amount of any substance being built up is exactly counterbalanced by the amount being used up in the other reaction, so that concentrations of all participating substances remain constant. (AGI, 1987)

chemical erosion *corrosion.*

chemical extraction Term taking the place of hydrometallurgy; embraces leaching (acid, alkaline, and pressure), ion exchange, solvation precipitation, and calcination. See also: *leaching.* (Pryor, 1963)

chemical lead Lead of more than 99.9% purity, with traces of copper and silver, as originally obtained from the ore; used for manufacturing storage battery plates and chemical piping. (Bennett, 1962)

chemical limestone A limestone formed by direct chemical precipitation or by consolidation of calcareous ooze. (AGI, 1987)

chemically precipitated metal powder Powder produced by the reduction of a metal from a solution of its salts either by the addition of another metal higher in the electromotive series or by other reducing agents. (ASTM, 1994)

chemical mineralogy The investigation of the chemical composition of minerals and its variation, the processes of mineral formation, and the changes minerals undergo when acted upon chemically. Cf: *physical mineralogy; crystallogeny.*

chemical rock A sedimentary rock composed primarily of material formed directly by precipitation from solution or colloidal suspension (such as by evaporation) or by the deposition of insoluble precipitates (such as by mixing solutions of two soluble salts); e.g., gypsum, rock salt, chert, or tufa. It generally has a crystalline texture. Cf: *detrital rock.* (AGI, 1987)

chemical sediment *chemical rock.*

chemical soil consolidation A process for sinking through loose, heavily watered ground. A gel-forming chemical is injected into the loose material that is eventually consolidated. The time delay in the gel formation can be controlled by chemical means, and the rate of injection at waterlike viscosity is rapid. See also: *bentonite; silicatization process.* (Nelson, 1965)

chemical stoneware A clay pottery product that is widely employed to resist acids and alkalies. It is used for utensils, pipes, stopcocks, pumps, etc.; sp gr, 2.2; hardness, scleroscope 100. Stoneware is made from special clays free from lime and iron, low in sand content, with low temperatures, and having sufficient plasticity to permit turning on a potter's wheel. Syn: *chemical brick.* (CCD, 1961)

chemical water treatment A method of treating hard water by adding selected chemical substances that break down the offending impurities, the residue being passed on in solution in harmless or less harmful form, driven off as a gas, or precipitated for subsequent retention in an incorporated filter. The general reagents are lime or soda or a combination of both with or without the addition of zeolites or colloids. (Nelson, 1965)

chemical weathering The process of weathering by which chemical reactions (hydrolysis, hydration, oxidation, carbonation, ion exchange, and solution) transform rocks and minerals into new chemical combinations that are stable under conditions prevailing at or near the Earth's surface; e.g., the alteration of orthoclase to kaolinite. Cf: *mechanical weathering.* Syn: *decomposition.* (AGI, 1987)

chemihydrometry Determination of flow rate and channels taken by water by the introduction of suitable chemicals upstream and measurement of dilution. (Radiotracers and fluorescin are also used for tracing flow direction.). (Pryor, 1963)

chemiluminescence Luminosity caused by chemical changes in a substance. (Standard, 1964)

chemisorption Irreversible sorption, an adsorbate being held as a product of chemical reaction with an absorbent. Activation energy is relatively high. (Pryor, 1963)

chemist A person versed in chemistry. One whose business is to make chemical examinations or investigations, or one who is engaged in the operations of applied chemistry.

chemolithotroph Autotrophic microorganism that derives energy to do metabolic work from the oxidation of inorganic compounds and assimilate carbon as CO_2, HCO_3^-, or CO_3^{2-}; e.g., *Thiobacillus ferrooxidans,* a bacterium that oxidizes ferrous iron to ferric iron for energy. See also: *autotroph.*

chempure tin Purest commercially available tin; 99.9% tin. (Bennett, 1962)

chenevixite A monoclinic mineral, $Cu_2Fe_2(AsO_4)_2(OH)_4 \cdot H_2O$; earthy to opaline; associated with olivenite in copper deposits.

Chenot process The process of making iron sponge from ore mixed with coal dust and heated in vertical cylindrical retorts. (Fay, 1920)

cheralite A monoclinic mineral, $(Ca,Ce,Th)(P,Si)O_4$; monazite group; an intermediate member of a crystal-solution series between $CePO_4$ (monazite) and $CaTh(PO_4)_2$ (a synthetic compound).

cheremchite A variety of sapropelic coal composed of a mixture of structureless humic sapropel and algal remains. Also spelled tscheremchite. (Tomkeieff, 1954)

cherry picker (a) A fishing tool in the modified form of a horn socket. The lower end or mouth is cut away on one side and resembles a scoop; because of its shape, the device, as it is turned, works around and behind an object that has become partly embedded in the wall of a borehole, thus engaging it where a regular horn socket would fail. (Long, 1960) (b) A small hoist to facilitate car changing near the loader in a tunnel. An empty car is either lifted above the track (to allow a loaded car to pass beneath) or swung to one side free of the track. See also: *double-track portable switch.* The equipment is fairly common, particularly for handling large cars. (Nelson, 1965) (c) In tunneling, a small traveling crane spanning tracks that transfers an empty car to a parallel track so that a loaded one can be drawn from the advancing end. (Pryor, 1963) (d) A small derrick made up of a sheave on an A-frame, a winch and winch line, and a hook. Usually mounted on a truck. (Nichols, 1954)

cherry-red heat A common term used on the color scale, generally given as about 1,382 °F (750 °C).

chert A hard, dense, dull to semivitreous, microcrystalline or cryptocrystalline sedimentary rock, consisting dominantly of interlocking crystals of quartz less than about 30 µm in diameter; it may contain amorphous silica (opal). It sometimes contains impurities such as calcite, iron oxide, and the remains of siliceous and other organisms. It has a tough, splintery to conchoidal fracture, and may be white or variously colored. Chert occurs principally as nodular or concretionary nodules in limestone and dolomites, and less commonly as layered deposits (bedded chert); it may be an original organic or inorganic precipitate or a replacement product. The term "flint" is essentially synonymous, although it has been used for the dark variety of chert. See also: *jasper; silexite.* (AGI, 1987)

chertification A type of silicification in which fine-grained quartz or chalcedony is introduced into limestones, such as in the Tri-State mining district of the Mississippi Valley. (AGI, 1987)

chervetite A monoclinic mineral, $Pb_2V_2O_7$; occurs in small crystals at the Mounana uranium mine, Gabon.

chessy copper *chessylite.*

chessylite A term commonly used for azurite. Syn: *chessy copper.* See also: *azurite.*

chesterite Microcline feldspar found in Chester County, PA.

chestnut coal (a) In anthracite only, coal small enough to pass through a square mesh of 1 to 1-1/8 in (2.54 to 2.86 cm), but too large to pass through a mesh of 5/8 in or 1/2 in (1.59 cm or 1.27 cm). Known as No. 5 coal. (b) Arkansas. Coal that passes through a 2-in (5.1-cm) round hole and over a 1-in (2.5-cm) round hole. See also: *anthracite coal sizes.*

chevee A flat gem having a polished concave depression. Cf: *cuvette.*

chevron crossbedding Crossbedding that dips in different directions in superimposed beds, forming a chevron pattern. Also called herringbone crossbedding; zigzag crossbedding. (Pettijohn, 1964)

chevron drain A rubble-filled trench system in the slope of a railway cutting, laid out in herringbone fashion and leading surface water into buttress drains arranged along the line of steepest slope. (Hammond, 1965)

chevron fold A fold with a sharp angular hinge and planar limbs of equal length. Syn: *zigzag fold.* (AGI, 1987)

chews Scot. Coal loaded with a screening shovel; middling-sized pieces of coal. Syn: *chows.*

chiastolite An opaque variety of andalusite containing black carbonaceous impurities arranged in a regular manner so that a section normal to the longer axis of the crystal shows a black Maltese cross. It has long been used for amulets, charms, and other inexpensive novelty jewelry. Syn: *cross-stone; crucite; macle.*

chiastolite slate A rock formed by contact metamorphism of carbonaceous shale, characterized by prominent cleavage or schistosity and the presence of conspicuous chiastolite crystals in a fine-grained groundmass. (AGI, 1987)

chickenfeed An Alaskan term for fine gravel 1/2 in (1.27 cm) or less in diameter.

Chiddy Assay Cupellation assay, for gold content of barren cyanide solution. The gold (and silver) is precipitated together with metallic lead as sponge on aluminum. This metal is cupeled, and the gold prill is weighed. (Pryor, 1963)

chigura A timber used in making a crib.

childrenite A monoclinic mineral, $FeAl(PO_4)(OH)_2 \cdot H_2O$; forms a series with eosphorite.

Chilean lapis A pale- to light-blue lapis lazuli from Chile. *lazurite; lapis matrix.*

Chilean mill A mill having vertical rollers running in a circular enclosure with a stone or iron base or die. There are two classes: (1) those in which the rollers gyrate around a central axis, rolling upon the die as they go (the true Chile mill), and (2) those in which the enclosure or pan revolves, and the rollers, placed on a fixed axis, are in turn revolved by the pan. It was formerly used as a coarse grinder, but is now used for fine grinding. (Liddell, 1918)
chileite An earthy, secondary lead, zinc, and copper vanadate; occurs near Arqueros, Chile; related to psittacinite. See also: *mottramite.*
chilenite An amorphous mixture(?) containing silver and bismuth.
Chile saltpeter A former name for nitratine.
chill (a) A metal insert imbedded in the surface of a sand mold or core or placed in a mold cavity to increase the cooling rate at that point. (ASM, 1961) (b) White iron occurring on a gray iron casting, such as the chill in a wedge test. (ASM, 1961) (c) To harden by suddenly cooling. (Gordon, 1906) (d) Derb. To test the roof with a tool or bar to determine its safety. (Fay, 1920)
chill casting Pouring molten metal into molds so made that it comes into contact at desired places with metal; cooling is thus accelerated, and special hardness is imparted. (Pryor, 1963)
chill crystal Small crystal formed by the rapid freezing of molten metal when it comes into contact with the surface of a cold metal mold. (CTD, 1958)
chilled casting A casting made by contacting it with something that will rapidly conduct the heat from it, such as a cool iron mold, or by sudden cooling by exposure to air or water. (Fay, 1920)
chilled contact That part of a mass of igneous rock, near its contact with older rocks, that is finer grained than the rest of the mass, because it cooled more rapidly. (AGI, 1987)
chilled dynamite The condition of dynamite when subjected to a low temperature not sufficient to congeal it, but which seriously affects the strength of the dynamite. (Fay, 1920)
chilled shot In hard-rock boring with an adamantine or Calyx drill, chilled iron or steel pellets that are driven by the drill bit and do the actual abrasive cutting. (Pryor, 1963)
chilled-shot drill *shot drill.*
chilled-shot drilling A method of rotary drilling in which chilled steel shot is used as the cutting medium. (BS, 1963)
chimney (a) An ore shoot or pipe. See also: *chute.* (b) A term used for limestone pinnacles bounding zinc ore deposits. (c) A vertical or nearly vertical staple shaft between a lower and an upper coal seam. (d) An orebody that is roughly circular or elliptical in horizontal cross section, but may have great vertical extent. (Nelson, 1965) (e) A restricted section in a lode; rising steeply and unusually rich. See also: *pipe.* (Pryor, 1963) (f) A cylindrical vent for volcanic rock. (Standard, 1964)
chimney effect *stack effect.*
chimney rock (a) A column of rock standing above its surroundings, such as an igneous rock filling a pipe-shaped vent. (b) Gulf States. A local name for any rock soft enough when quarried to be cut or sawn readily and refractory enough for domestic chimneys, which may harden on exposure to the air; e.g., some limestone, siliceous bauxite clay, or soapstone. (Hess)
chimney work Mid. A system of working beds of clay ironstone in patches 10 to 30 yd (9.1 to 27.4 m) square and 18 to 20 ft (5.5 to 6.1 m) thick. The bottom beds are first worked out; then miners work the higher ones by standing upon the fallen debris. Cf: *overhand stoping.*
china clay A commercial term for kaolin obtained from china-clay rock after washing, and suitable for use in the manufacture of chinaware. (AGI, 1987)
china-clay rock (a) Cornwall stone. (b) Granite in its most kaolinized form, in which the feldspar is transformed into kaolinite and the rock is so soft that it is readily broken in the fingers. (Arkell, 1953)
chinaman chute Mine opening over the haulage level through which ore from the stope above is drawn to waiting trucks as planking is removed. Usually, an opening between stulls below the shrinkage stope. (Pryor, 1963)
Chinaman pebble N.Z. A pebble or boulder made from a conglomerate of quartz pebbles cemented by chalcedony. Jaspilite, quartz, and Chinaman pebbles are found in many places. (Hess)
China metal (a) York. Shale baked to a hard, white, coarse, porcellaneous substance. (Arkell, 1953) (b) Porcelain.
china stone (a) Partially kaolinized granite containing quartz, kaolin, and sometimes mica and fluorite. It is harder than china-clay rock and is used as a glaze in the manufacture of china. Syn: *petuntze.* Cf: *Cornish stone.* (AGI, 1987) (b) A fine-grained, compact carboniferous mudstone or limestone found in England and Wales. (AGI, 1987)
chingle (a) Scot. A gravel free from dirt. See also: *shingle.* (b) That portion of a coal seam stowed away in the goaves to help support the mine roof.
chiolite A tetragonal mineral, $Na_5Al_3F_{14}$; massive; granular; occurs with cryolite.
chip (a) Small fragment of a diamond, usually thin and tabular in shape. (Long, 1960) (b) To break small fragments from the surface of a diamond or other material. (Long, 1960) (c) Small, angular, and generally flat pieces of rock or other materials. (Long, 1960) (d) An imperfection due to breakage of a small fragment out of an otherwise regular surface. (ASTM, 1994) (e) A small fragment from a crystal; specif. a diamond chip. (AGI, 1987) (f) A piece of rock to be cut into a thin section for microscopic examination.
chip blasting Shallow blasting of ledge rock. (Nichols, 1976)
chip breaker (a) A notch or groove in the face of a tool parallel to the cutting edge, to break the continuity of the chips. (ASM, 1961) (b) A step formed by an adjustable component clamped to the face of a cutting tool. (ASM, 1961)
chipped (a) When referring to the character of diamond wear, it denotes loss of diamond due to chips and fragments having been broken away from the body of the diamond. (Long, 1960) (b) A surface pitted by loss of material in the form of chips. (Long, 1960)
chipping (a) Loosening of shallow rock by light blasting or air-hammers. (Nichols, 1976) (b) The process of handsetting diamond fragments in a bit. (Nichols, 1976)
chippings Crushed angular stone fragments ranging from 1/8 to 1 in (0.32 to 2.54 cm) in size. See also: *aggregate.* (Nelson, 1965)
chip sample A regular series of ore chips or rock chips taken either in a continuous line across an exposure or at uniformly spaced intervals. (AGI, 1987)
chip sampling (a) The taking of small pieces of ore or coal, with a small pick, along a line or at random, across the width of a face exposure. The samples are usually taken daily and often confined to exploration. Reasonable care is taken to chip a weight of material that corresponds to the length of sample line. See also: *bulk sample.* (Nelson, 1965) (b) A variant of channel sampling, in which, owing to extreme hardness of rock, shape of deposit, or other working difficulty, a true channel sample cannot be taken. Often used in preliminary prospecting. (Pryor, 1963)
chirality Symmetrical handedness. A mirror or center of symmetry changes the chirality of asymmetric units. Cf: *improper.*
chiropterite Bat guano.
chisel (a) A tool of great variety whose cutting principle is that of the wedge. (Crispin, 1964) (b) The steel cutting tool used in percussive boring. It ranges from 6 to 12 in (15.2 to 30.5 cm) in length with variously shaped bits to suit the nature of the ground. The chisel is made to strike a series of blows at the bottom of a borehole. Water or mud is circulated to convert the chippings into sludge and to keep the chisel cool. (Nelson, 1965)
chisel bit (a) *chopping bit.* (b) A percussive-type, rock-cutting bit having a single, chisel-shaped cutting edge extending across the diameter and through the center point of the bit face. Also called chisel-edge bit; chisel-point; Swedish bit. (Long, 1960)
chisel draft The dressed edge of a stone, which serves as a guide in cutting the rest.
chiver *shiver.*
chkalovite An orthorhombic mineral, $Na_2BeSi_2O_6$; occurs on the Kola Peninsula, Russia.
chloanthite An arsenic-deficient variety of nickel-skutterudite. Also spelled cloanthite. Syn: *white nickel ore.*
chloraluminite A trigonal mineral, $AlCl_3{\cdot}6H_2O$; occurs in acid fumaroles on Mt. Vesuvius, Italy.
chlorapatite A monoclinic mineral, $Ca_5(PO_4)_3Cl$; apatite group.
chlorargyrite An isometric mineral, 4[AgCl]; sectile; forms waxy white, yellow, or pearl-gray incrustations, darkening to violet on exposure to light; a supergene mineral occurring in silver veins; an important source of silver. Formerly called cerargyrite. Syn: *horn silver.* Cf: *iodargyrite.*
chlorastrolite A mottled green variety of pumpellyite used as a semiprecious stone; forms grains, small nodules, or a radial, fibrous structure in geodes in mafic igneous rocks; resembles prehnite; occurs in the Lake Superior region (esp. on Isle Royale).
chlorate explosive Explosive with a potassium chlorate base, such as the French cheddite, which contains about 80% potassium chlorate and 5% castor oil, with dinitrotoluene constituting nearly all the remainder. Chlorate explosives are characterized by a hot flame on detonation. (Lewis, 1964)
chlorate powder A substitute for blackpowder in which potassium chlorate is used in place of potassium nitrate.
chloride (a) A miner's or prospector's term for an ore containing silver chloride. (AGI, 1987) (b) A compound of chlorine with another element or radical. A salt or ester of hydrochloric acid. (Crispin, 1964)
chloriding Mining thin veins. (Statistical Research Bureau, 1935)
chloridization An ore treatment using chlorine to produce a metal chloride. (Pryor, 1963)
chloridize To convert into chloride; applied to the roasting of silver ores with salt, preparatory to amalgamation.

chloridizing roasting The roasting of sulfide ores and concentrates, mixed with sodium chloride, to convert the sulfides to chlorides. (CTD, 1958)
chlorination process The process in which auriferous ores are first roasted to oxidize the base metals; then saturated with chlorine gas; and finally treated with water, which removes the soluble chloride of gold, to be subsequently precipitated and melted into bars. (Fay, 1920)
chlorinator A machine for feeding either liquid or gaseous chlorine to a stream of water.
chlorine A common nonmetallic halogen element, found in the combined state only, chiefly with sodium as common salt (NaCl). Symbol, Cl. A greenish-yellow irritating toxic gas with a disagreeable odor; a respiratory irritant. Used for producing safe drinking water, paper products, dyestuffs, textiles, petroleum products, medicines, antiseptics, insecticides, foodstuffs, solvents, paints, plastics, and many other consumer products. (Handbook of Chem. & Phys., 3)
chlorinity (a) The total amount in grams of chlorine, iodine, and bromine contained in 1 kg of seawater, assuming that the bromine and iodine have been replaced by chlorine. (Hy, 1965) (b) The number giving the chlorinity in grams per kilogram of seawater sample is identical with the number, giving the mass in grams of atomic weight silver just necessary to precipitate the halogens in 0.3285233 kg of the seawater sample. (Hy, 1965)
chlorite (a) The mineral group chamosite, clinochlore, cookeite, gonyerite, nimite, orthochamosite, pennantite, and sudoite. (b) Chlorites are associated with and resemble micas (the tabular crystals of chlorites cleave into small, thin flakes or scales that are flexible, but not elastic like those of micas); they may also be considered as clay minerals when very fine grained. Chlorites are widely distributed, esp. in low-grade metamorphic rocks, or as alteration products of ferromagnesian minerals.
chlorite schist A schist in which the main constituent, chlorite, imparts a schistosity by parallel arrangement of its flakes. Quartz, epidote, magnetite, and garnet may be accessories, the last two often as conspicuous porphyroblasts. (AGI, 1987)
chlorite slate A schistose or slaty rock composed largely of chlorite.
chloritic sand A sand colored green by sand-size chlorite grains.
chloritic schist A schist containing chlorite.
chloritization The replacement by, conversion into, or introduction of chlorite.
chloritoid A monoclinic or triclinic mineral, $(Fe,Mg,Mn)_2Al_4Si_2O_{10}(OH)_4$; dull green to gray-black; occurs in masses of brittle folia in metamorphosed argillaceous sedimentary rocks. It is related to the brittle micas.
chlormanganokalite A trigonal mineral, K_4MnCl_6; occurs in yellow rhombohedra.
chlorocalcite *baeumlerite.*
chloromelanite (a) A dark green, nearly black variety of jadeite. (Fay, 1920) (b) A crystal solution of roughly equal amounts of diopside, jadeite, and acmite.
chloropal (a) A former name for nontronite. (b) A greenish variety of common opal from Silesia, Poland.
chlorophaeite A mineral closely related to chlorite in composition and found in the groundmass of tholeiitic basalts where it occupies interstices between feldspar laths, forms pseudomorphs after olivine, or occurs in veinlets and amygdules. The fresh mineral is pale green, but when weathered, it may be dark green, brown, or red. (AGI, 1987)
chlorophane A variety of fluorite that exhibits bright-green phosphorescent light if heated. Also called cobra stone.
chlorophoenicite A monoclinic mineral, $(Mn,Mg)_3Zn_2(AsO_4)(OH,O)_6$; occurs in elongated gray-green crystals at Franklin, NJ.
chlorosis The yellowing of the leaves of plants, sometimes caused by a deficiency of iron necessary in the formation of chlorophyll. Has been useful as a guide to ore since nickel, copper, cobalt, chromium, zinc, and manganese are all antagonistic to iron in plant metabolism. Also may indicate where such toxins have been added to soil by industrial or other human activity.
chlorosity The number expressing chlorinity as grams per liter. Obtained by multiplying the chlorinity of a sample by its density at 20 °C. (Hy, 1965)
chlorospinel A grass-green variety of spinel containing copper.
chlorothionite An orthorhombic mineral, $K_2Cu(SO_4)Cl_2$; occurs in bright-blue crystalline crusts on lava, an alteration product at Mt. Vesuvius, Italy.
chlorotile A green orthorhombic hydrated arsenate of copper. Cf: *mixite.*
chloroxiphite A monoclinic mineral, $Pb_3CuCl_2(OH)_2O_2$; dull-olive or pistachio green.
chlorutahlite *utahlite.*
chock (a) A square pillar for supporting the roof; constructed of prop timber laid up in alternate cross-layers, in log-cabin style, the center being filled with waste. Sometimes called crib. See also: *cog; hydraulic chock.* (b) Type of longwall-mining roof support. (c) One of two blocks of hardwood placed across or between rails to prevent tubs, cars, or wagons from running down an incline.
chock and block Newc. Tightly filled up.
chock block Piece of wood, square or rectangular in cross section, usually made of oak, ash, or other hardwood. Also used to denote a shaped piece of wood provided with a handle and designed for placing between the rails to hold back a tub or set of tubs. (TIME, 1929-30)
chock hole A small depression dug in the earth in which a wheel of a truck-mounted drill rig is set to prevent the drill from moving. (Long, 1960)
chocking The supporting of undercut coal with short wedges or chocks. (CTD, 1958)
chog An English term for chocks, or blocks spiked into the corner of a shaft to form a bearing for the side-walling piece, or the blocks used in headings to separate the cap and poling board. See also: *collaring.* (Stauffer, 1906)
choke (a) In crushing practice, a stoppage of the downward flow in a rock-crushing chamber. See also: *choke point.* (South Australia, 1961) (b) A point in a cave or at the base of a pitch blocked by the influx of clay, sand, gravel, or similar material. (AGI, 1987)
choke crushing A recrushing of fine ore due to the fact that the broken material cannot exit a machine before it is again crushed. Cf: *free crushing.*
chokedamp (a) A mine atmosphere that causes choking or suffocation due to insufficient oxygen. As applied to "air" that causes choking, does not mean any single gas or combination of gases. (Fay, 1920) (b) A name sometimes given in England to carbon dioxide. See also: *blackdamp; damp.* (Fay, 1920)
choke fed In comminution, rolls are choke fed when fed all of the material that they will take. The product of choke-fed rolls is never so uniform as when free feeding is used. Choke feeding is used only on feed of diameter about 1/4 in (0.6 cm) or less. Cf: *free fed.* (Newton, 1959)
choke feed A feeding arrangement in which the potential rate of supplying material at the feed point exceeds the rate at which the conveyor will remove material.
choke feeding As deliberately used in roll crushing of ore, feed at a rate greater than can be discharged at the set of the machine, so that the rolls are sprung apart, the angle of nip is increased, and the product contains oversize. (Pryor, 1963)
choke point Bottleneck of any crusher. (Pryor, 1963)
choker A chain or cable so fastened that it tightens on its load as it is pulled. (Nichols, 1954)
choking Stoppage of flow, due to obstructed discharge, sticky material, packed and compacted fines, or bad control. (Pryor, 1963)
chondrodite A monoclinic mineral, $(Mg,Fe)_5(SiO_4)_2(F,OH)_2$; humite group; commonly occurs in contact-metamorphosed dolomites. Also spelled condrodite.
chonolith An intrusive mass that is so irregular in form and its relationship to the invaded formations is so obscure that it cannot be designated a dike, sill, or laccolith.
chop (a) To break up and drill through boulders, other rock, or lost core encountered in sinking a drivepipe or casing through overburden. It is done by impact produced by lifting and dropping a chopping-bit-tipped string of drill rods. (b) Som. A local term for fault.
chop ahead To break up boulders and other rock material below the bottom of casing or drivepipe by using a chopping bit attached to drill rods. See also: *chop.* (Long, 1960)
chop feeder A feeder in which a power-operated, swinging quadrant gate delivers material at a predetermined rate. The action is similar to a reciprocating plate feeder.
chopping A term used to describe the digging action of a dragline when excavation takes place with the bucket heel above the line of the cutting lip. This term is usually used when referring to an operating method in which the dragline bucket excavates above the line of the fairlead and fills above tub level. (Austin, 1964)
chopping bit A steel, chisel-shaped cutting edged bit designed to be coupled to a string of drill rods and used to fragment, by impact, boulders, hardpan, and lost core in a borehole. Also called chisel bit; chisel-edge bit; chisel-point bit; long-shank chopping bit. Cf: *cross-chopping bit.* (Long, 1960)
chordal effect The effect produced by the chain joint centers being forced to follow arcs instead of chords of a sprocket pitch circle. (Jackson, 1955)
chordal pitch The length of one side of the polygon formed by the lines between the joint centers as a chain is wrapped on a sprocket. It is a chord of the sprocket pitch circle and is equal to the chain pitch. (Jackson, 1955)
chorismite A general term for a group of mixed rocks, which are the result of the injection of the crystallization products of intruding magmas into, and/or the mixture of such material with, the enclosing rocks, sedimentary or metamorphic. There are several varieties. The term is not widely used. Cf: *migmatite.*

chows *chews.*
C/H ratio *carbon-hydrogen ratio.*
chrismatite A butyraceous, greenish-yellow to wax-yellow hydrocarbon from Wettin, Saxony, Germany. It has a specific gravity of less than 1 and is soft at 55 to 60 °C. (Fay, 1920)
Christiansen effect In optical mineralogy, a dispersion phenomenon in which the boundary of a mineral grain (Becke line) immersed in a liquid of the same index of refraction appears blue on one side and red to orange on the other. See also: *dispersion.*
christobalite *cristobalite.*
chromate A salt or ester of chromic acid; a compound containing the radical $(CrO_4)^{2-}$. (AGI, 1987)
chromatic aberration In microscope lenses, the splitting of white light to form two images, one red and the other blue. See also: *aberration.*
chromatic color A hue, as distinguished from white, black, or any tone of gray. Opposite of achromatic color.
chromatite A tetragonal mineral, $CaCrO_4$; forms finely crystalline citron-yellow crusts from clefts in limestones.
chromatograph An instrument for analyzing gases and vapors from liquids with boiling points up to 300 °C. The gas chromatograph often arranges the molecules of a gas in increasing size, and as each group emerges from the column, a detector measures the quantity of each. Since all the molecules of one type emerge after the same time interval, it is possible to identify quickly the constituents present. (Nelson, 1965)
chromatographic analysis Separation of components of mixture into zones, one or more of which can be identified by color, etc.: (1) by adsorption column, adsorbing from solute in a tube packed with cellulose, alumina, lime, etc.; (2) by electrochromatography, passing electricity across a column or paper strip down which solvent mixture is flowing, causing migration to the side of a flow line; (3) by electrophoresis, using electric current to aid migration; and (4) by paper partition, separation into bands as suitable solvent flows past a drop of solution, which contains compounds (qualitative and quantitative analysis). (Pryor, 1963)
chromatography A chemical process of separating closely related compounds by permitting a solution of them to filter through an absorbent so that the different compounds become absorbed in separate colored layers comprising a chromatogram. (Hunt, 1965)
chrome A term commonly used to indicate ore of chromium, consisting esp. of the mineral chromite or chromium-bearing minerals, such as chrome mica or chrome diopside. (AGI, 1987)
chrome antigorite A variety of antigorite containing some chromium.
chrome brick A refractory brick manufactured substantially or entirely of chrome ore. (ASTM, 1994)
chrome chert A variety of chert that has replaced the silicate minerals of a chromite peridotite, the more resistant chromite grains remaining unaltered in the siliceous matrix. (Holmes, 1928)
chrome diopside (a) A variety of diopside. Dark-green specimens are seldom either transparent or cut as gems. (b) A bright emerald-green variety of diopside containing a small amount of Cr_2O_3.
chrome garnet *uvarovite.*
chrome idocrase An emerald-green variety of vesuvianite containing chromium; occurs at Black Lake, Quebec, Canada; and Ekaterinburg, Ural Mountains, Russia. Syn: *chrome vesuvian.*
chrome iron ore *chromite.*
chrome mica *fuchsite.*
chrome ocher A chromiferous clay; specif. a bright-green clay material containing 2% to 10.5% Cr_2O_3.
chrome refractory Refractory consisting essentially of refractory-grade chrome ore bonded chemically or by burning. Chrome refractories are nearly chemically neutral, but may react with strong acids or bases. (Henderson, 1953)
chrome spinel Another name for the mineral picotite, a member of the spinel group. *chromian spinel.* (CMD, 1948)
chrome tourmaline A variety of tourmaline found in the Ural Mountains, Russia, and Maryland.
chrome vesuvian *chrome idocrase.*
chromian spinel A variety of spinel containing chromium, $(Mg,Fe)(Al,Cr)_2O_4$. Formerly called picotite. *chrome spinel.*
chromic iron *chromite.*
chromite An isometric mineral, $8[FeCr_2O_4]$; spinel group; dimorphous with donathite; forms crystal-solution series with magnesiochromite in the chromite series of the spinel group, and with hercynite; rarely occurs as a pure end member. End-member chromite contains 68% Cr_2O_3, but natural minerals do not commonly exceed 50%. Occurs in metallic black octahedral crystals; weakly to moderately ferrimagnetic; an accessory in, or layers in, mafic and ultramafic rocks; also in black sands, the major source of chromium. Syn: *chrome iron ore.*
chromitite (a) A rock composed chiefly of the mineral chromite. (AGI, 1987) (b) A mixture of chromite with magnetite or hematite. (AGI, 1987)
chromium An isometric mineral, Cr; rare; occurs in contact zones between ultramafic rocks and marble.
chromium garnet *uvarovite.*
chromography In mineral identification, a polished section is placed in contact with photographic paper, a current is passed, and ions migrating to the paper are developed so as to produce a color print suitable for microscrutiny. It resembles sulfur printing.
chromowulfenite A red variety of wulfenite, containing some chromium. (Fay, 1920)
chronograph An apparatus for electrically recording explosion phenomena with a continuous time record. (Rice, 1913-18)
chronolith *time-stratigraphic unit.*
chronolithologic unit Time-rock unit. See also: *time-stratigraphic unit.*
chronostratic unit *time-stratigraphic unit.*
chronostratigraphic unit *time-stratigraphic unit.*
chrysoberyl An orthorhombic mineral, $BeAl_2O_4$; vitreous; green, brown, yellow, or red (alexandrite variety appears emerald green in sunlight, but red by incandescent light); occurs in granites, granite pegmatites, schists, and alluvial deposits; a gemstone. Gem varieties: alexandrite, chrysopal, cymophane, and golden beryl. Known as cat's-eye when it has chatoyancy. Syn: *cymophane.* See also: *dichroism; beryllium aluminate; chrysopal.*
chrysocolla A monoclinic mineral, $(Cu,Al)_2H_2Si_2O_5(OH)_4{\cdot}nH_2O$; cryptocrystalline or amorphous; soft; bluish green to emerald green; forms incrustations and thin seams in oxidized parts of copper-mineral veins; a source of copper and an ornamental stone.
chrysocolla quartz A translucent chalcedony colored by chrysocolla.
chrysolite A yellowish-green, sometimes brownish or reddish, iron-magnesium silicate. A common mineral in basalt and diorite. When used as a gem, it is called peridot. The name has at various times been applied to topaz, prehnite, and apatite, but is now used only to mean olivine. See also: *olivine.* (Fay, 1920; Hess)
chrysolite cat's-eye Chrysoberyl cat's-eye.
chrysolithus A pale yellowish-green variety of beryl. (Schaller, 1917)
chrysopal (a) A translucent variety of common opal colored apple green by the presence of nickel. (b) *chrysoberyl.* (c) A gemstone trade name for opalescent chrysolite (olivine). See also: *prasopal.*
chrysoprase (a) An apple-green or pale yellowish-green variety of chalcedony containing nickel and valued as a gem. See also: *green chalcedony.* (b) A misleading name used in the gem trade for a green-dyed chalcedony having a much darker color than natural chrysoprase.
chrysotile A monoclinic mineral (clinochrysotile), or orthorhombic mineral (orthochrysotile, parachrysotile), $[Mg_6(OH)_8Si_4O_{10}]$; serpentine group; forms soft, silky white, yellow, green, or gray flexible fibers as veins in altered ultramafic rocks; the chief asbestos minerals. (Not to be confused with chrysolite.) Syn: *Canadian asbestos; serpentine asbestos.*
chrysotile asbestos A fibrous variety of serpentine.
chuck The part of a diamond or rotary drill that grips and holds the drill rods or kelly and by means of which longitudinal and/or rotational movements are transmitted to the drill rods or kelly. See also: *three-jaw chuck.* (Long, 1960)
chuck block In stamp milling, the wooden block or board that is attached to the bottom of the screen so as to raise the depth of the issue and act as a false lip to the mortar. (Fay, 1920)
chucker-on A device for automatic rerailing of tubs or cars. Also called ramp; rerailer. (Mason, 1951)
chuckie stone One of the pebbles or cobbles of sedimentary rock or of igneous rock occurring as an inclusion in a coalbed. One explanation for their occurrence is that they were attached to roots of floating trees rafted into the swamp during periods of high water. (AGI, 1987)
chuco Caliche deposit in Chile composed mainly of sodium sulfate.
chudobaite A triclinic mineral, $(Mg,Zn)_5H_2(AsO_4)_4{\cdot}10H_2O$.
chukhrovite An isometric mineral, $Ca_3(Ce,Y)Al_2(SO_4)F_{13}{\cdot}10H_2O$; a rare-earth mineral in the Kara-Oba molybdenite deposit, central Kazakhstan, and the Clara Mine, Oberwolfach, Germany.
chungkol Malaysia. Heavy hoe used to stir and loosen a bed when sluicing alluvial tin gravels. (Pryor, 1963)
chunked-up Built up with large lumps of coal to increase the capacity of the car. Also called built-up.
chunker I In bituminous coal mining, a laborer who hand loads large lumps of coal into cars at working places in a mine. (DOT, 1949)
chunker II In bituminous coal mining, a laborer who arranges large lumps of coal uniformly on flatcars as they are loaded at the mine surface. (DOT, 1949)
churchillite *mendipite.*
churchite A monoclinic mineral, $YPO_4{\cdot}2H_2O$. Formerly called weinschenkite.
churn A long iron rod used to hand bore shotholes in soft material, such as coal. (Pryor, 1963)
churn drill (a) Portable drilling equipment, usually mounted on four wheels and driven by steam-, diesel-, electric-, or gasoline-powered engines or motors. The drilling is performed by a heavy string of tools tipped with a blunt-edge chisel bit suspended from a flexible manila or steel cable, to which a reciprocating motion is imparted by its suspension from an oscillating beam or sheave, causing the bit to be raised and

dropped, thus striking successive blows by means of which the rock is chipped and pulverized and the borehole deepened; also, the act or process of drilling a hole with a churn drill. Also called American system drill; blasthole drill; cable drill; cable-system drill; churn-drill rig; rope-system drill; shothole drill; spudder; spud drill; wet drill. See also: *percussion drill*. Syn: *cable-tool drill*. (Long, 1960) (b) A long iron bar with a cutting end of steel, used in quarrying, and worked by raising and letting it fall. When worked by blows of a hammer or sledge, it is called a jumper or jump drill.

churn-drill operator In mining and in the quarry industry, one who drills holes with a churn (cable) drill in rock and overlying ground of open-pit mines or quarries to obtain samples, or to provide holes in which explosives are detonated to break up a solid mass. Syn: *blasthole driller; blasting hole well driller*. (DOT, 1949)

churn-drill rig *churn drill.*

churn shot drill A boring rig that combines both churn and shot drillings. The churn drill is used for rapid penetration in barren ground where no core is required. The shot drill is used for taking cores along important rock formations. (Nelson, 1965)

chute (a) A channel or shaft underground, or an inclined trough aboveground, through which ore falls or is shot by gravity from a higher to a lower level. Also spelled shoot. (b) A crosscut connecting a gangway with a heading. (c) A ditch or inclined timber through which the overflow water or mud from a borehole is conducted from the collar of the hole to the sump. The chute may be fitted with baffles and screens to cause the cuttings to settle before reaching the sump. Syn: *canal; ditch*. (Long, 1960) (d) A body of ore, usually of elongated form, extending downward within a vein (ore shoot). See also: *chimney; shoot*. (e) A trough operated mechanically in loading coal underground. Syn: *rock chute*. (Hudson, 1932) (f) A string of rich ore in a lode (used instead of shoot). (Nelson, 1965) (g) Stockpile withdrawing system, such as a belt conveyor. (Pryor, 1963) (h) A metal trough in a breaker, along which the coal slides by gravity. (Hudson, 1932) (i) A steep, three-sided steel tray for the passage of coal or ore from a conveyor into mine cars. It is designed to minimize degradation and spillage of materials. See also: *loading chute*. (Nelson, 1965) (j) Ore pass connecting a stope with the haulage level. (Pryor, 1963) (k) A high-velocity conduit for conveying to a lower level. Syn: *course of ore*. (Seelye, 1951)

chute boss In coal mining, a foreperson who supervises the loading and drawing of coal into and out of chutes, esp. where coal is mined from inclined beds. (DOT, 1949)

chute caving The method involves both overhand stoping and ore caving. A chamber is started as an overhand stope from the head of a chute and is extended up until the back weakens sufficiently to cave. The orebody is worked from the top down in thick slices, each slice being, however, attacked from the bottom and the working extending from the floor of the slice up to an intermediate point. The cover follows down upon the caved ore. Also called caving by raising; block caving into chutes.

chute checker In metal mining, one who keeps a record of the amount of ore drawn from each raise or chute in an orebody being mined by the caving method (lower part of orebody is mined and developed with a system of chutes so that the remaining ore that sloughs, or caves, from lack of support can be drawn off). Also called tallyman. (DOT, 1949)

chute drawer *chute loader.*

chute loader (a) In metal and nonmetal mining, a laborer who loads ore or rock into mine cars underground by opening and closing chute gates. Also called chute drawer; chute man; chute puller; chute trammer; chute tapper. (DOT, 1949) (b) In the quarry industry, one who loads crushed rock from bins into trucks or railroad cars by opening and closing the chute or bin gates by hand or by means of a lever. Also called car loader. (DOT, 1949)

chute operator In the quarry industry, a laborer who loads barges with crushed rock by operating a hand winch to lower a chute through which crushed rock flows from a bin. (DOT, 1949)

chute system A method of mining by which ore is broken from the surface downward into chutes and removed through passageways below. See also: *glory-hole system*. (Hess)

chute trammer *chute loader.*

cienega A marshy area where the ground is wet because of the presence of seepage or springs, often with standing water and abundant vegetation. The term is commonly applied in arid regions such as the Southwestern United States. Etymol: Spanish ciénaga, marsh, bog, miry place. (AGI, 1987)

ciment fondu A slow-setting, rapid-hardening cement containing 40% lime, 40% alumina, 10% silica, and 10% impurities; used in cementing drill holes. Sometimes called bauxite cement.

cimolite A white, grayish, or reddish hydrosilicate of aluminum; soft and claylike or chalklike in appearance. (Fay, 1920)

cinder (a) A loose volcanic fragment that may range from 4 to 32 mm in diameter. Such fragments are usually glassy or vesicular. (Stokes, 1955) (b) A small (1- to 4-cm), commonly vesicular, fragment of lava projected from an erupting volcano; coarser than volcanic ash but smaller than a volcanic bomb. (c) A juvenile vitric pyroclastic fragment that falls to the ground in an essentially solid condition. (d) Slag, particularly from an iron blast furnace. (e) A scale thrown off in forging metal. Cf: *lapilli.*

cinder block A block closing the front of a blast furnace and containing the cinder notch. (Webster 3rd, 1966)

cinder breakout The slag within a furnace escaping through the brickwork; caused by erosion, corrosion, or softening of brick by heat. (Fay, 1920)

cinder coal (a) Coal that has been cindered by heat from an igneous intrusion. Many coal seams have been affected in this way in Scotland and in Durham, England. See also: *metamorphism*. (Nelson, 1965) (b) Aust. A very inferior natural coke, little better than ash. See also: *natural coke.*

cinder cooler In a blast furnace, a watercooled casting, usually of copper, that is pressed into the cinder notch. (Henderson, 1953)

cinder fall The dam over which the slag from the cinder notch of a furnace flows. (Fay, 1920)

cinder notch The furnace hole, about 1.5 to 2 m above the iron notch and 1 m below the tuyeres, through which slag is flushed two to three times between casts. See also: *cinder tap*. (Fay, 1920)

cinder pig Pig iron made from a charge containing a considerable proportion of slag from puddling or reheating furnaces. (CTD, 1958)

cinder pit Large pit filled with water into which molten cinder is run and granulated at cast or flush. (Fay, 1920)

cinder plate *bloomery.*

cinder runner A trough carrying slag from a skimmer or cinder notch to a pit or ladle. See also: *cinder notch*. (Fay, 1920)

cinder tap The hole through which cinder is tapped from a furnace. Also called Lurmann front. (Fay, 1920)

cinder tub A shallow iron truck with movable sides into which the slag of a furnace flows from the cinder runner. (Fay, 1920)

cinnabar A trigonal mineral, 3[HgS]; trimorphous with hypercinnabar and metacinnabar; forms brilliant red acicular crystals and red to brownish-red or gray masses; soft; sp gr, 8.1; occurs in impregnations and vein fillings near recent volcanic rocks and hot springs, alluvial deposits; the chief source of mercury. Syn: *cinnabarite*. See also: *vermillion; vermilion.*

cinnabarite *cinnabar.*

cinnabar matrix A term applicable to various varieties of minerals containing numerous inclusions of cinnabar but esp. to a Mexican variety of jasper.

cinnamite *cinnamon stone.*

cinnamon stone (a) Grossularite, a lime garnet. See also: *essonite; hessonite; hyacinth*. (Hess; Dana, 1959) (b) *grossular.*

cipolino A European term for a marble rich in silicate minerals and characterized by layers rich in micaceous minerals. (Holmes, 1920)

CIPW classification *norm system.*

CIPW system *norm system.*

circle (a) In the central United States, a nearly circular lead and zinc deposit developed in clayey chert breccias in old sinkholes in Paleozoic limestone or in dolomite (broken ground). (Schieferdecker, 1959) (b) In a grader, the rotary table that supports the blade and regulates its angle. (Nichols, 1976)

circle cutting drill (a) A pneumatic drill carried on rotating arms. Used to cut grindstones and pulpstones from a quarry. (AIME, 1960) (b) *ditcher.*

circle haul In strip mining, a haulage system in which the empty units enter the mine over one lateral and leave, loaded, over the lateral nearest the tipple. This system is utilized where laterals are built into the mine from the main road, whether outside the outcrop or on the high-wall side of the mine workings. This system reduces the haul on the coal surface to a minimum, except where there are only two laterals, one at each end of the workings.

circle reverse The mechanism that changes the angle of a grader blade. (Nichols, 1976)

circle spout Eng. A trough or gutter around the inside of a shaft to catch the water running down the sides; a garland.

circuit breaker These differ from straight overcurrent relays in that they are primarily used for ground protection. They are designed to measure fault current in one or two sections. Whether faults will cause flow in one or two directions is determined by system conditions. The two-directional relay is used on transmission lines where ground-fault currents flow in either direction. These relays provide directional as well as overcurrent protection. Other directional relays provide phase protection.

circuits Circular galleries made at different levels in a mine that enable empty trucks to be pushed out of the cage on one side while full ones are pushed in on the other side, thus ensuring a more rapid journey of the cage. Circuits also aid air circulation. Syn: *roundabouts*. (Stoces, 1954)

circuit tester *blasting galvanometer.*

circular arch A roadway support consisting of an H-section girder of circular form and usually made in three parts. The joints are secured by fishplates and bolts. This type of steel arch is useful for withstanding pressures from roof, sides, and floor. With close lagging between the rings, the finished roadway resembles a tube. See also: *steel arch.* (Nelson, 1965)

circular bin discharger A revolving cone with feeder fingers around the base periphery connected at the apex through a universal joint to a revolving arch breaker arm.

circular cutting drill *ditcher.*

circular grading table *rotary sorting table.*

circular picking table An apparatus used for the same purpose as a picking belt and consisting of a flat horizontal rotating annular plate. See also: *picking belt.* (BS, 1962)

circular shaft A shaft excavated as a cylinder. The circular shaft is equally strong at all points; convenient for concrete lining and tubbing, both of which can be made relatively watertight; and offers the least resistance to airflow. (Nelson, 1965)

circular slip A type of landslide that may occur in embankments or cuttings in clay or homogeneous earth. *slip surface of failure.* (Nelson, 1965)

circular tunnel kiln The same as a straight tunnel kiln, except that it has a movable, circular platform instead of cars.

circulating fluid Fluid pumped into a borehole through the drill stem, the flow of which cools the bit, washes away the cuttings from the bit, and transports the cuttings out of the borehole. See also: *reverse circulation.* Also called circulation fluid; circulation medium; drill fluid; drilling fluid. (Long, 1960)

circulating head A casing-to-drill-rod coupling. When attached to the top of the casing, it is used during the process of pumping cement slurries or circulating water through the casing, forcing the fluid to flow out of the casing into the drill hole between the outside of the casing and the walls of the borehole. Also called stuffing box; tight head. (Long, 1960)

circulating load (a) In mineral processing, use of a closed circuit to check mineral issuing from a specific treatment and to return to the head of the treatment those particles that do not satisfy the maintained conditions for release to the next stage of treatment. (Pryor, 1963) (b) In ore dressing, oversize material returned to a ball mill for further grinding. (Newton, 1959)

circulating medium Medium in circulation in or outside a separating bath, at or about the specific gravity of that in the separating bath. (BS, 1962)

circulating pump (a) A pump (usually centrifugal) used to circulate water through the condenser of a steamplant. (Nelson, 1965) (b) A pump used to circulate water in a coal washer or ore concentration plant. (Nelson, 1965) (c) A pump used to circulate mud or water through a drilling column. Also called slush pump. Syn: *mud pump.* (BS, 1963)

circulating scrap Scrap arising at steelworks and foundries during the manufacture of finished iron and steel or of castings; consists of the sheared-off ends of rolled and other worked products, rejected material, etc. See also: *capital scrap.* (Nelson, 1965)

circulating water The water in the water circuit of a preparation plant. (BS, 1962)

circulation (a) The passing of any liquid or gas from the surface to the end of the drill string and back to the surface in the process of drilling a borehole. (Long, 1960) (b) The movement of air currents through mine openings. (Long, 1960) (c) In rotary drilling, the process of pumping mud-laden or other fluid down the drill pipe, through the drilling bit, and upward to the surface through the annulus between the drill-hole walls and the drill pipe. (AGI, 1987)

circulation fluid The fluid pumped through and to the end of the drill string and back to the surface in the process of drilling a borehole. Cf: *drilling mud.* (Long, 1960)

circulation loss The result of drilling or circulation fluid escaping into one or more formations by way of crevices or porous media. (Brantly, 1961)

circulation medium *circulating medium.*

circulation of air The controlled flow of air to and from the faces to secure adequate ventilation of all workings and traveling roads. See also: *dadding.* (Nelson, 1965)

circulation velocity The speed, generally expressed in lineal feet per second, at which a fluid or gas travels upward in a borehole after passing the face of the bit. (Long, 1960)

circulation volume The amount of liquid or gas circulated through the drill-string equipment in drilling a borehole. The amount of liquid circulated is expressed in gallons (or liters) per minute, and the amount of a gas, as air, is expressed in cubic feet (or cubic meters) per minute. (Long, 1960)

circumferentor A surveyor's compass with diametral projecting arms each carrying a vertical slit sight. (Webster 3rd, 1966)

cistern (a) A settling tank for liquid slag, pulp, etc. (b) An artificial reservoir or tank for holding water. (AGI, 1987)

citation Issued by regulatory representatives alleging a specific condition or practice that violates mining, maritime, construction, environmental, or general industry standards. (NSC, 1992)

citrine Not the true topaz of mineralogists, but a yellow variety of quartz, which closely resembles topaz in color though not in other physical characters; it is of much less value than true topaz. Known under a variety of geographical names such as Bohemian topaz, Indian topaz, Madagascar topaz, Madeira topaz, and Spanish topaz. Brazilian topaz is the true mineral. Also called quartz topaz. See also: *Scotch topaz; false topaz; smoky quartz.* (CMD, 1948)

C-J detonation A detonation characterized by the equivalence of the detonation velocity to the velocity of sound in the burned gas plus the velocity of flow of the burned gases. (Van Dolah, 1963)

C-J plane *Chapman-Jouget plane.*

clack (a) A valve part. The hinged, lidlike part of a check, clack, or pump valve. Also called check; flap. See also: *flapper.* (Long, 1960) (b) A clack or pump valve. (Long, 1960)

clack seat The rim or seat on which the hinged lid or flapper of a clack valve closes. (Long, 1960)

clack valve A valve having a lidlike piece hinged on one side within a chamber that permits the flow of a fluid or gas to proceed in one direction only. Usually, the check valve on the pickup end of a drill-pump suction hose is a clack-type valve. Also called chock valve; flap valve; flapper valve; foot valve. (Long, 1960)

clad metal A composite metal containing two or three layers that have been bonded together. The bonding may have been accomplished by corolling, welding, casting, heavy chemical deposition, or heavy electroplating. (ASM, 1961)

claggy (a) Newc. Adhesive. When coal is tightly joined to the roof, the mine is said to have a claggy top. Also spelled cladgy. (Fay, 1920) (b) Newc. Muddy or clayey dirt. (Pryor, 1963)

claim (a) The portion of mining ground held under the Federal and local laws by one claimant or association, by virtue of one location and record. Lode claims, maximum size 600 ft by 1,500 ft (182.9 m by 457.3 m). Placer claims 600 ft by 1,320 ft (182.9 m by 402.4 m). A claim is sometimes called a location. See also: *title; mining claim.* (b) S. Afr. Land on a mining field to which a miner is legally entitled. A Transvaal claim has an area of 64,025 ft^2 (5,947.9 m^2 or 60,000 Cape square feet). It is about 155 ft (47.3 m) along the strike of the reef, and 413 ft (125.9 m) across the line, or along the dip of the reef. An area of 1.44 claims is equal to a South African morgen. In Cape feet, the claim is 150 ft by 400 ft (46.2 m by 122.0 m). Mining maps are often designed in squares of 1,000 Cape feet by 1,000 Cape feet (304.9 m by 304.9 m), which, therefore, contain about 16 claims measured horizontally. (Beerman, 1963) (c) In Australia, a claim is defined as the portion of Crown land that any person or number of persons shall lawfully have taken possession of and be entitled to occupy for mining purposes. No land comprised in any mining lease can be considered to be a claim. A claim is marked out by fixing in the ground posts at each angle of the claim, and it need not be surveyed. A miner is required to hold a miner's right before legally marking out or working a claim. (Nelson, 1965)

claimant In the Federal mining law, means locator. (Ricketts, 1943)

claim jumping The location of a mining claim on supposedly excess ground within the staked boundaries of an existing location on the theory that the law governing the manner of making the original location has not been complied with. (Ricketts, 1943)

claims held in common The phrase "held in common" means a claim whereof there are more owners of a claim than one; the use of the words "claims held in common," on which work done upon one of such claims shall be sufficient, means that there must be more than one claim so held, to make a case where work upon one of them shall answer the statutory requirements as to all of them. (Ricketts, 1943)

claim system A system used mainly in the United States that grew up in the early days of mining in the Western United States following the gold rush of 1849, as an outgrowth of the desire of a prospector to develop a mineral deposit discovered on the public lands and to have the claim confirmed by law. The mining laws of the United States are based on this system, whereas most other mining countries follow the concession system. Cf: *concession system.* (Hoover, 1948)

clam (a) A clip; a haulage clip; an appliance for attaching mine cars to a rope. See also: *clip.* (Mason, 1951) (b) A clamshell bucket. (Nichols, 1976) (c) To mud-in the door of a kiln. (ACSG, 1963)

clammings Entrance to an oven. (Noke, 1927)

clamshell A twin-jawed bucket without teeth; usually hung from the boom of a crane that can be either crawler or wheel mounted. The bucket is dropped in the open position onto the material to be excavated or handled. It is then closed, encompassing material between the two hinged halves.

clamshell loader A grab-type loader activated by cables. Used in mucking operations. (Lewis, 1964)

clan A compositional category for classifying igneous rocks; e.g., the rhyolite-granite clan. A clan may be defined either by mineralogical or by chemical composition. Clans are subdivided into families. (AGI, 1987)

clap-me-down In inclined shaft timbering, a joint in which the end pieces are checked into the cap and sill for a distance of approx. 1 in (2.5 cm), with a bevel on the inner side. (Higham, 1951)

clapotis The wave pattern established when waves are reflected by a barrier so that the crests and troughs occur alternately in the same places with water particle motion limited to vertical movement, while a quarter wavelength away the particle motion is horizontal (back and forth). This is a standing wave phenomenon. (Hy, 1965)

clarain A coal lithotype characterized macroscopically by semibright, silky luster and sheetlike, irregular fracture. It is distinguished from vitrain by containing fine intercalations of a duller lithotype, durain. Its characteristic microlithotype is clarite. Cf: *clarite; fusain; vitrain.* (AGI, 1987)

clarification (a) The cleaning of dirty or turbid liquids by the removal of suspended and colloidal matter. See also: *recirculation of water.* (Nelson, 1965) (b) The concentration and removal of solids from circulating water to reduce the suspended solids to a minimum. (BS, 1962)

clarifier A centrifuge, settling tank, or other device for separating suspended solid matter from a liquid. (Hess)

clarifying tank A tank for clarifying cyanide or other solutions; frequently provided with a filtering layer of sand, cotton waste, matting, etc. (Fay, 1920)

clarinite (a) The major maceral or micropetrological constituent of clarain. It is a heterogeneous material that is generally translucent in thin section, and in which there may be intercalated lenticels of such other ingredients, such as xylinite, fusinite, resinite, suberinite, periblinite, collinite, and ulminite. (AGI, 1987) (b) Strictly, not a maceral, but may be used for repetitive description. (Tomkeieff, 1954)

clarite A coal microlithotype that contains a combination of vitrinite and exinite totalling at least 95%. The proportions of these two macerals may vary widely, but each must be greater than the proportion of inertinite, and neither must exceed 95%. Distinction may be made between spore clarite, cuticular clarite, and resinous clarite. Clarite is widely distributed and very common, particularly in clarain-type coals and occurs in fairly thick bands. Cf: *clarain.* (AGI, 1987; IHCP, 1963)

clarke The average abundance of an element in the crust of the Earth. Cf: *clarke of concentration.* Syn: *crustal abundance.* (AGI, 1987)

clarkeite A mineral, $(Na,Ca,Pb)_2U_2(O,OH)_7$; strongly radioactive; metamict; massive; dense; forms as an alteration product of uraninite. Syn: *brown gummite.*

clarke of concentration The concentration of an element in a mineral or rock relative to its crustal abundance. The term is applied to specific as well as average occurrences. Cf: *clarke.* (AGI, 1987)

Clark riffler A sample-reducing device that splits a batch sample of ground ore into two equal streams as it falls across an assembly of deflecting chutes. (Pryor, 1963)

clarodurain A rock-type coal consisting of the maceral vitrinite (tellenite or collinite) and large quantities of other macerals, mainly micrinite and exinite. Micrinite and exinite are present in larger quantities than vitrinite. Syn: *clarodurite.* Cf: *duroclarain.* (AGI, 1987)

clarodurite The term clarodurain was introduced by G.H. Cady in 1942, and in the modified form, clarodurite was adopted by the Nomenclature Subcommittee of the International Committee for Coal Petrology in 1956 to designate the microlithotype with maceral composition between that of clarite and durite, but closer to durite than to clarite. It occurs in fairly thick bands; is widely distributed; and, like duroclarite, is a common constituent of most humic coal. Syn: *clarodurain.* (IHCP, 1963)

clarofusain A rock-type coal consisting of the macerals fusinite and vitrinite and may contain all other macerals. Fusinite is present in a larger quantity than in fusoclarain. Cf: *fusoclarain.* (AGI, 1987)

clarovitrain A rock-type coal consisting of the maceral vitrinite (collinite or telinite) with smaller amounts of other macerals. Cf: *vitroclarain.* (AGI, 1987)

clasolite *clastic rock.*

class A division of igneous rocks based on the relative proportions of the salic (siliceous and aluminous minerals, quartz, feldspars, and feldspathoids) and femic (ferromagnesian minerals, pyroxene, amphibole, etc.) standard normative minerals as calculated from chemical analyses. (Holmes, 1928)

Class 1.1 explosive Explosive that has a mass explosion hazard or one that will affect almost the entire load instantaneously; previously designated by the U.S. Department of Transportation as a Class A explosive and including, but not limited to, dynamite, nitroglycerin, lead azide, blasting caps and detonating primers.

Class 1.2 explosive Explosive that has a projection hazard but not a mass explosion hazard; previously designated by the U.S. Department of Transportation as a Class A or B explosive.

Class 1.3 explosive Explosive that has a fire hazard and either a minor blast hazard or a minor projection hazard or both, but not a mass explosion hazard; previously designated by the U.S. Department of Transportation as a Class B explosive and defined as possessing a flammable hazard, such as, but not limited to, propellant explosives, photographic flash powders, and some special fireworks.

Class 1.4 explosive Explosive that presents a minor explosive hazard, and explosive effects are confined to the package; no projection of fragments of appreciable size or range is to be expected. An external fire must not cause virtually an instantaneous explosion of almost the entire contents of the package; previously designated by the U.S. Department of Transportation as a Class C explosive and defined as containing Class A or Class B explosives, or both, as components but in restricted quantities.

Class 1.5 explosive Very insensitive explosive that has a mass explosive hazard but is so insensitive that there is very little probability of initiation or of transition from burning to detonating under normal conditions of transport; large quantities, however, have a higher probability of detonation subsequent to burning; previously designated by the U.S. Department of Transportation as a blasting agent.

Class 5.1 substance A material that yields oxygen and causes or enhances the combustion of other materials; previously designated by the U.S. Department of Transportation as an oxidizer.

classical washout A belt of barren ground or thin coal produced by the erosion of a seam by rivers that flowed during or soon after the deposition of the coal. These erosion channels are now filled with sandy sediment. See also: *rock roll.* (Nelson, 1965)

classification (a) The process of separating particles of various sizes, densities, and shapes by allowing them to settle in a fluid. (Mitchell, 1950) (b) Grading of particles too small to be screened in accordance with their size, shape, and density by control of their settling rate through a fluid medium (water, slurry, or air). (Pryor, 1963) (c) The evaluation and segregation of trimmed sheet mica according to grades and qualities. (Skow, 1962) (d) In powder metallurgy, separation of a powder into fractions according to particle size. (ASM, 1961)

classification of crystals Of 32 crystal classes (based on 32 point groups, the possible combinations of symmetry elements intersecting at a point) assigned to 7 crystal systems, only 11 are found in common minerals. Each system may be described in terms of three noncoplanar vectors (crystallographic axes) that are generally nonorthogonal as well. Although mineral assignment to a crystal system may require only examination of external crystal morphology, assignment of crystal class commonly requires X-ray diffraction analysis. The crystal systems are triclinic, monoclinic, orthorhombic, tetragonal, trigonal, hexagonal, and isometric. See also: *crystallographic axes; crystal systems.*

classification of minerals Each mineral species is a unique, naturally occurring combination of chemical composition and crystal system; e.g., graphite is hexagonal carbon and diamond is isometric carbon, and halite is isometric sodium chloride. (a) Thus, minerals may be classified according to their crystal system. (b) Minerals may be classified chemically according to Dana as (1) native elements and alloys; (2) sulfides, selenides, tellurides, arsenides, and antimonides; (3) sulfosalts, sulfarsenides, sulfantimonides, and sulfobismuthides; (4) halides; (5) oxides; (6) oxygen salts, carbonates, silicates, borates, etc.; (7) salts of organic acids; and (8) hydrocarbon compounds. Silicates are subdivided according to the structural arrangements of their $(SiO_4)^{4-}$ tetrahedral groups and the number of corner oxygen ions shared between them (degree of polymerism). (c) Additionally, minerals may be classified into isostructural groups; e.g., spinel group, garnet group, mica group, pyroxene group, and zeolite group. (Structural classification is not entirely congruent with chemical classification, since some structural groups may contain more than one chemical group; e.g., the apatite group has mainly phosphates, but some arsenates, vanadates, and silicates have the apatite structure.) (d) Rutley classifies minerals according to group in accordance with the periodic table as regards dominant economic constituents. (e) Optically, minerals are classified as opaque (metallic luster) and nonopaque (transmit light in thin section). (f) Economically, minerals are classified as metallics if they are the source of metal from ores and nonmetallics if their products are not metals. See also: *classification of crystals.*

classified sand fill Mechanically separated sand or the sand portion of mill tails used as backfill in underground openings. Usually conveyed hydraulically. Also spelled classified sandfill. See also: *backfill; sand fill; classifier.*

classifier (a) A machine or device for separating the constituents of a material according to relative sizes and densities, thus facilitating concentration and treatment. Classifiers may be hydraulic or surface-current box classifiers (spitzkasten). (Webster 3rd, 1966) (b) The term classifier is used in particular where an upward current of water is used to remove fine particles from coarser material. See also: *centrifugal separation.* (Nelson, 1965) (c) In mineral beneficiation, the classifier is a device that takes the ball-mill discharge and separates it into two portions—the finished product, which is ground as fine as desired, and oversize material. See also: *undersize*. (Newton, 1959)

classifier dredge A dredge in which the gravel goes from the trommel to a classifier and then to jigs. (Lewis, 1964)

classing Sorting ore according to its quality. (Gordon, 1906)

clastic Consisting of fragments of minerals, rocks, or organic structures that have been moved individually from their places of origin. Syn: *detrital; fragmental*. (AGI, 1987)

clastic deformation A process of metamorphism that involves the fracture, rupture, and rolling out of rock and mineral particles. In some instances, the crystal structure may be preserved, but the orientation of the fragments becomes confused. In other instances, the rock may be thoroughly pulverized. (Stokes, 1955)

clastic dike A tabular body of clastic material transecting the bedding of a sedimentary formation, representing extraneous material that has invaded the containing formation along a crack, either from below or from above. See also: *sandstone dike; pebble dike*.

clastic rock A consolidated sedimentary rock composed principally of broken fragments that are derived from preexisting rocks (of any origin) or from the solid products formed during chemical weathering of such rocks, and that have been transported mechanically to their places of deposition; e.g., a sandstone, conglomerate, or shale; or a limestone consisting of particles derived from a preexisting limestone. Syn: *fragmental rock; clasolite*. (AGI, 1987)

clathrate A texture found chiefly in leucite rocks, in which the leucite crystals are surrounded by tangential augite crystals in such a way as to suggest a net or a section of a sponge, the felted mass of augite prismoids representing the threads or walls, and the clear, round leucite crystals, the holes. (Schieferdecker, 1959)

claudetite A monoclinic mineral, As_2O_3; dimorphous with arsenolite.

clauncher (a) Eng. A tool for cleaning blast holes. Also called clanger. (Fay, 1920) (b) Derb. A piece of stone that has a joint in back of it, which becomes loose and falls when a tunnel has been driven past it. (Fay, 1920)

clausthalite An isometric mineral, PbSe; forms a crystal-solution series with galena, which it resembles. Syn: *lead selenide*.

clay An extremely fine-grained natural earthy material composed primarily of hydrous aluminum silicates. It may be a mixture of clay minerals and small amounts of nonclay materials or it may be predominantly one clay mineral. The type is determined by the predominant clay mineral. Clay is plastic when sufficiently pulverized and wetted, rigid when dry, and vitreous when fired to a sufficiently high temperature. See also: *clay mineral; fireclay; bentonite*. (ASTM, 1994)

clay back A back slip in a coal seam containing a clayey deposit. See also: *back slip*. (Nelson, 1965)

clay band A light-colored, argillaceous layer in clay ironstone. Also spelled clayband. (AGI, 1987)

clay barrel *triple-tube core barrel*.

clay bit A mud auger; a mud bit; also, a bit designed for use on a clay barrel. See also: *clay-boring bit*. (Long, 1960)

clay book tile Structural clay tile with tongue and groove edges resembling a book in shape. (Hess)

clay-boring bit A special coring bit used to split inner-tube core barrels. The thickness of the bit face is reduced and the inside shoulder is not inset with diamonds to allow a sharp-edged inner barrel to extend through and project a short distance beyond the face of the bit. Also called clay bit; mud bit. (Long, 1960)

clay course A clay seam or clay gouge found along the sides of some veins.

claycrete Weathered argillaceous material forming a layer immediately overlying bedrock. (AGI, 1987)

clay cutter Cutting ring at the entry to a pipe feeding into a suction cutter dredge. Set of cutting blades in dredge trommel used to break clay brought up by dredge buckets. (Pryor, 1963)

clay dauber One who seals kiln doors before burning and kiln fireboxes after burning and assists other workers in knocking out doors and in unsealing fireboxes after cooling. Also called dauber; plaster man. (DOT, 1949)

clayey *argillaceous*.

clayey soil A soil in which clay is the basic constituent. The clay contributes to strength by cohesion, but detracts from stability by volume change and by plastic flow under load. (Nelson, 1965)

clay gall (a) Mud curl or cylinder formed by drying and cracking of thin layers of coherent mud; commonly rolled or blown into sand and buried; flattened upon wetting forming a lenticular bleb of clay or shale. (Pettijohn, 1964) (b) Eng. Clay gall pellet of clay or mudstone, often ocherous, sometimes hollow, found esp. in false-bedded oolitic limestones such as forest marble. (Arkell, 1953)

clay gouge (a) A clayey deposit in a fault zone. See also: *fault gouge*. (AGI, 1987) (b) A thin seam of clay separating masses of ore, or separating ore from country rock. See also: *gouge*. (AGI, 1987)

clay gun Equipment used to fire a ball of fire clay into the tap hole of a blast furnace. See also: *mud gun*. (Pryor, 1963)

clay hole *clay pocket*.

claying Lining a borehole with clay, to keep explosives dry. (Fay, 1920)

claying bar A rod or tool for lining a newly made coal shot hole with clay to seal up any breaks in the walls of the hole. The hole is filled with clay to about one-third of its length. The claying bar is driven in by hammer to the limit and rotated by a tommy bar in the eyelet at the outer end of the bar. See also: *clay iron; bull; scraper and break detector*. (Nelson, 1965)

clay iron An iron rod used for ramming clay into wet drill holes. See also: *bull; claying bar*. (Fay, 1920)

clay ironstone (a) A compact hard, dark, gray or brown, fine-grained sedimentary rock consisting of a mixture of argillaceous material (up to 30%) and iron carbonate (siderite), occurring in layers of nodules or concretions or as relatively continuous irregular thin beds, and usually associated with carbonaceous strata, esp. overlying a coal seam in the coal measures of the United States or Great Britain; a clayey iron carbonate, or an impure siderite ore occurring admixed with clays. The term has also been applied to an argillaceous rock containing iron oxide (such as hematite or limonite). See also: *blackband ironstone*. (AGI, 1987) (b) A sideritic concretion or nodule occurring in clay ironstone and other argillaceous rocks, often displaying septarian structure. (AGI, 1987) (c) *ironstone; iron clay*.

clay loam (a) A fine-textured soil that breaks into clods or lumps that are hard when dry. When the moist soil is pinched between the thumb and finger, it will form a thin ribbon that will break readily, barely sustaining its own weight. The moist soil is plastic and will form a cast that will bear much handling. When kneaded in the hand, it does not crumble readily but tends to work into a heavy compact mass. (Stokes, 1955) (b) A soil containing 27% to 40% clay, 20% to 45% sand, and the remainder silt. (AGI, 1987)

clay maker One who blends and mixes various clays, as shipped from a mine, into a thin, semiliquid form by operating a blunger (mixing machine). Also called blunger machine operator; clay mixer; clay washer; slip maker; slip mixer; wet mixer. (DOT, 1949)

clay marl A chalky clay, or a marl in which clay largely predominates.

clay mineral (a) A colloidal-size, crystalline, hydroxyl silicate having a crystal structure of the two-layer (7 Å) type (kaolinite), or of the three-layer (14 Å) type (smectite), in which layers of silicon and aluminum ions have tetrahedral coordination with respect to oxygen, while layers of aluminum, ferrous and ferric iron, magnesium, chromium, lithium, manganese, and other cations have octahedral coordination with respect to oxygen and to hydroxyl ions. Exchangeable cations may attach to the silicate layers in an amount determined by the excess negative charge within the composite layers. These cations commonly are calcium and sodium, but may also be potassium, magnesium, hydronium, aluminum, or others. The most common clay minerals belong to the kaolinite, smectite, attapulgite, and illite (hydromica) groups. Mixed-layer clay minerals are either randomly or regularly interstratified intergrowths of two or more clay minerals. See also: *clay*. (b) Any mineral found in the clay fraction (less than 4 μm) of a soil or sediment; e.g., rock flour comminuted by glacial grinding. (c) Any kandite mineral of the kaolinite-serpentine group.

claypan (a) A playa formed by deflation of alluvial topsoils in a desert, in which water collects after a rain. (AGI, 1987) (b) A term used in Australia for a shallow depression containing clayey and silty sediment, and having a hard, sun-baked surface. (AGI, 1987) (c) *hardpan*.

clay parting (a) Clayey material bound between a vein and its wall. Also called casing; parting. (Fay, 1920) (b) Seams of hardened carbonaceous clay between or in beds of coal. (Hess)

claypit (a) A sump in which a drilling mud is mixed and stored. (Long, 1960) (b) A pit or sump in which the return fluid from a borehole is collected and stored for recirculation. (Long, 1960) (c) A pit where clay is dug.

clay pocket A clay-filled cavity in rock; a mass of clay in rock or gravel. Syn: *clay hole*. (Hess)

clay rock *claystone*.

clay sapropel Clay deposit containing sapropel. (Tomkeieff, 1954)

clay shale (a) A consolidated sediment consisting of no more than 10% sand and having a silt to clay ratio of less than 1:2 (Folk, 1954, p. 350); a fissile claystone. (AGI, 1987) (b) A shale that consists chiefly of clayey material and that becomes clay on weathering. (AGI, 1987)

clay size Said of that portion of soil or sediment that is finer than 2 to 5 μm.

clay slate (a) A low-grade, essentially unreconstituted slate, as distinguished from the more micaceous varieties that border on phyllite. (AGI, 1987) (b) A slate derived from an argillaceous rock, such as shale, rather than from volcanic ash; a metamorphosed clay, with cleavage developed by shearing or pressure, as distinguished from mica slate. (AGI, 1987)

clay stains Yellowish-brown or rust-colored films from deposits of clay minerals. (Skow, 1962)

claystone (a) A term applicable to indurated clay in the same sense as sandstone is applicable to indurated or cemented sand. Syn: *clay rock.* See also: *mudstone; siltstone.* (b) One of the concretionary masses of clay frequently found in alluvial deposits, in the form of flat rounded disks either simple or variously united so as to give rise to curious shapes.

clay temperer *wet-pan operator.*

clay vein A body of clay, usually roughly tabular in form like an ore vein, that fills a crevice in a coal seam. It is believed to have originated where the pressure was high enough to force clay from the roof or floor into small fissures and in many instances, to alter and to enlarge them. Also called horseback. (AGI, 1987)

clay wash (a) A deposit of clay transported and deposited by water. (b) The agitation of an oil with fuller's earth or some other clay to improve the color or odor of the oil. (Porter, 1930) (c) A thin emulsion of clay and water, sometimes used to strengthen the face of a mold. (Freeman, 1936) (d) Clay thinned with water and used for coating gaggers and flasks. (Crispin, 1964)

clay washer *clay maker.*

clean (a) Free from combustible gases or other noxious gases. (b) A coal seam free from dirt partings. (c) A diamond or other gem stone free from interior flaws. (Hess) (d) A borehole free of cave or other obstructing material. (Long, 1960) (e) A mineral virtually free of undesirable nonore or waste rock material. (Long, 1960) (f) Free of foreign material. In reference to sand or gravel, it means lack of binder. (Nichols, 1976)

Clean Air Act U.S. law: 42 USC Sections 7401-7428 (1979) and resulting regulations in 40 CFR51, administered by USEPA. Its objective is to reduce atmospheric pollution to acceptable limits. Inter alia, it empowers local authorities to declare smoke control areas in which the emission of any smoke from chimneys will constitute an offense. The act became part of Great Britain's national legislation in July 1956, although its main provisions did not become effective until June 1, 1958. See also: *coal smoke; smoke.* (Nelson, 1965)

clean cutting A rock formation, the cuttings of which do not tend to mud up on the face of a diamond or other bit. (Long, 1960)

clean cuttings (a) Rock cuttings that do not ball or adhere to the walls of a borehole. (Long, 1960) (b) Rock cuttings not contaminated by cave material or drill-mud ingredients. (Long, 1960)

cleaned coal Coal produced by a mechanical cleaning process (wet or dry). (BS, 1962)

cleaner Scot. A scraper for cleaning out a shothole. (Fay, 1920)

cleaner cell Secondary cell for the retreatment of the concentrate from a primary cell. Syn: *recleaner cell.* (BS, 1962)

clean hole A borehole free of cave or other obstructing material. (Long, 1960)

cleaning (a) A general term for the methods and processes of separating dirt from coal or gangue from mineral. See also: *coal-preparation plant; roughing.* (Nelson, 1965) (b) The retreatment of the rough flotation concentrate to improve its quality. (Pryor, 1965)

cleaning plant *coal washer; preparation plant.*

cleanout (a) To remove cave or other obstructing material from a borehole. (Long, 1960) (b) A port or opening provided in the body or base of a machine or other mechanism through which accumulated debris may be removed. (Long, 1960)

cleanout auger *cleanout jet auger.*

cleanout jet auger An auger equipped with water-jet orifices designed to clean out collected material inside a driven pipe or casing before taking soil samples from strata below the bottom of the casing. Also called cleanout auger; M.P.F.M. jet auger. (Long, 1960)

cleanup (a) The operation of collecting all the valuable product of a given period of operation in a stamp mill, or in a hydraulic or placer mine. (b) The valuable material resulting from a cleanup. (c) To load all the coal a miner has broken. (d) The cleanup of sluices in placer mining is a similar process that occurs daily or more often. The gold, tin, or other concentrate is shoveled out for further treatment. (Nelson, 1965) (e) To police and tidy up a drill rig and the surrounding area. (Long, 1960)

cleanup barrel A barrel used to batch grind and then amalgamate gold-bearing concentrates and residues. (Pryor, 1963)

clear (a) Translucent diamond with few visible spots or flaws. (Long, 1960) (b) Water that has not been recirculated in drilling and hence is free of drill cuttings and sludge. Also applied to return water when it contains little or no entrained cuttings or sludge. (Long, 1960) (c) A safe working place. (Long, 1960) (d) Transparent, such as in clear quartz, clear glass.

clearance (a) The space between the top or side of a car and the mine roof or wall. (Fay, 1920) (b) Technically, the annular space between downhole drill-string equipment, such as bits, core barrels, casing, etc., and the walls of the borehole with the downhole equipment centered in the hole. Loosely, the term is commonly and incorrectly used as a syn. for exposure. See also: *exposure; inside clearance.* (Long, 1960) (c) The amount of open space around a drill or piece of mining equipment in an underground workplace. (Long, 1960)

clearance space A space in pumps of the piston and ram types, usually quite small, between the cylinder end and the piston at the end of its stroke. The height to which water can be raised on the suction side is influenced by the volume of this space. (Mason, 1951)

clear clay A clay such as kaolin that is free from organic matter and so does not give rise to bubbles if used in a vitreous enamel; such clays are used in enamels when good gloss and clear colors are required. (Dodd, 1964)

clearer A reservoir (in saltmaking) into which brine is conveyed. (Fay, 1920)

clearing The removal of all standing growth, whether bushes or trees. (Carson, 1961)

clearing and grubbing Removal of tree stumps before excavation starts on a construction site. (Hammond, 1965)

clearing hole A hole drilled to a slightly larger diameter than the bolt passing through it. The clearance for black bolts is normally 1/16 in (1.6 mm). (Hammond, 1965)

clear mica Transparent muscovite without stains and with a smooth surface in reflected light.

clear span The clear unobstructed distance between the inner extremities of the two supports of a beam. This dimension is always less than the effective span. See also: *effective span.* (Hammond, 1965)

cleat (a) Term applied to systems of joints, cleavage planes, or planes of weakness found in coal seams along which the coal fractures. See also: *facing; face cleat; bord; butt cleat.* Also spelled cleet. (b) Main joint in a coal seam along which it breaks most easily. Runs in two directions, along and across the seam. (Pryor, 1963) (c) Joints in coal more or less normal to the bedding planes. (BS, 1964) (d) An attachment fastened to a conveying medium to help propel material along the path of travel.

cleat spar York. Crystalline mineral matter, often ankerite, occurring in the cleat cracks of coal. (Arkell, 1953)

cleavage (a) The breaking of a mineral along its crystallographic planes, thus reflecting crystal structure. Cf: *parting.* (AGI, 1987) (b) The property or tendency of a rock to split along secondary, aligned fractures or other closely spaced planes or textures, produced by deformation or metamorphism. (AGI, 1987) (c) In quarrying, the cleavage of rocks is often called the rift. (Nelson, 1965)

cleavage banding A compositional banding that is parallel to the cleavage rather than to the bedding. It results from the mechanical movement of incompetent material, such as argillaceous rocks, into the cleavage planes in a more competent rock, such as sandstone. Ordinarily, the argillaceous bands are only a few millimeters thick. See also: *segregation banding.*

cleavage plane The plane along which cleavage takes place. (Fay, 1920)

cleavages As used by the diamond-cutting and diamond-bit-setting industries, the more or less flat diamond fragments produced by splitting a crystalline diamond along the octahedral plane. Such fragments are used primarily as a material from which special-shaped, diamond-pointed cutting tools are produced. See also: *mêlée.* (Long, 1960)

cleave To split a crystalline substance, such as a diamond, along a cleavage plane. (Long, 1960)

cleavelandite A white, lamellar, or leaflike variety of albite, having an almost pure Ab content and commonly forming fan-shaped aggregates of tabular crystals that show mosaic development and appear as though bent; formed as a late-stage mineral in pegmatites, replacing other minerals. Also spelled clevelandite.

cleaving Splitting a crystal along a cleavage plane. (Hess)

cleaving way Corn. A direction parallel to the bedding planes of a rock. Cf: *roughway; quartering way.* (Fay, 1920)

cleek (a) Scot. To load cages at the shaft bottom or at midworkings. (b) Scot. *haulage clip.*

cleft An abrupt chasm, cut, breach, or other sharp opening, such as a craggy fissure in a rock, a wave-cut gully in a cliff, a trench on the ocean bottom, a notch in the rim of a volcanic crater, or a narrow recess in a cave floor. Obsolete syn: *clift.* (AGI, 1987)

Clerici solution A molecular mixture of thallium malonate and thallium formate. Used as a heavy solution for the separation of minerals. The solution has a maximum density of 4.25 g/cm^3 at 20 °C. It is prepared by adding formic acid to one of two equal quantities of thallium carbonate, and adding malonic acid to the other until each is neutralized. The two solutions are then mixed, filtered, and evaporated until almandite floats. Cf: *Sonstadt solution; Klein solution; bromoform; methylene iodide.* (Hess)

cleveite A variety of uraninite containing a large percentage of UO_3; also rich in helium. Contains about 10% of the yttrium earths. (Fay, 1920)

clevis (a) In coal mining, a spring hook or snap hook used to attach the hoisting rope to the bucket. Also called clivvy. (Pryor, 1963) (b) A U-shaped iron hook used with an iron pin for connecting ropes to the drawbars of cars or, when used with iron links, for coupling cars together. Also used as a connecting link between chains or lines or to hang a sheave in a drill tripod or derrick. (Jones, 1949; Long, 1960)

cliachite (a) A ferruginous bauxite from Cliache, Dalmatia, Croatia. (English, 1938) (b) Colloidal aluminum hydroxide occurring as one of the constituents of bauxite. Also spelled kliachite. See also: *laterite; sporogelite.* Syn: *alumogel*. (English, 1938)

cliff (a) Wales. Shale that is laminated, splitting easily along the planes of deposition. Also called clift. (b) The strata of rocks above or between coal seams. (Standard, 1964)

clift (a) Obsolete var. of cleft. *cleft.* (AGI, 1987) (b) Dialectal var. of cliff. (AGI, 1987) (c) A term used in southern Wales for various kinds of shale, esp. a strong, usually silty, mudstone. (AGI, 1987)

climate In froth flotation, the prevailing balance of chemical energy reached by the reacting electrical, physical, and chemical forces. (Pryor, 1958)

climb The tendency of an inclined diamond-drill hole to follow an upward-curving, increasingly flat course; also, the tendency of a diamond or other rotary-type bit to drill a hole curved in the updip direction when holes are drilled in alternating hard- and soft-layer rock having bedding planes that cross the borehole at an angle other than 90° to the face of the bit. (Long, 1960)

clink One of the internal cracks formed in steel by differential expansion of surface and interior during heating. The tendency for clinks to occur increases with the hardness and mass of the metal, and with the rate of heating. (CTD, 1958)

clinker (a) Fused or partly fused coal ash, a byproduct of combustion. Cf: *core.* (ACSG, 1963) (b) Coal that has been altered by an igneous intrusion. See also: *natural coke.* Syn: *scoria.* (c) Partially fused intermediate product in the manufacture of portland cement.

clinkstone An older term for a feldspathic rock, usually fissile; it is sonorous when stuck with a hammer. Also spelled klinkstone. See also: *phonolite.* (AGI, 1987)

clino A prefix to the name of a mineral species or group to indicate monoclinic symmetry as opposed to "ortho" indicating orthorhombic symmetry. See the root mineral name.

clinoamphibole (a) A group name for amphiboles crystallizing in the monoclinic system. (b) Any monoclinic mineral of the amphibole group; e.g., hornblende, cummingtonite, grunerite, tremolite, actinolite, riebeckite, glaucophane, and arfvedsonite. Cf: *orthoamphibole.*

clinoaugite A collective name for the monoclinic pyroxenes. See also: *clinopyroxene.* (English, 1938)

clinoaxis The inclined crystallographic axis in the monoclinic system, designated *a* or *b* in the first setting and *a* or *c* in the second. Most mineralogists use the second setting and designate the clinoaxis *a*.

clinochlore A monoclinic mineral, $2[(Mg,Fe)_5Al(Si_3Al)_4O_{10}(OH)_8$; chlorite group; occurs in greenschists.

clinochrysotile Monoclinic and orthorhombic forms of chrysotile, as determined by X-rays. See also: *chrysotile.* (Spencer, 1955)

clinoclase A monoclinic mineral, $Cu_3(AsO_4)(OH)_3$; formerly called clinoclasite. Syn: *aphanesite.*

clinoclasite Former name for clinoclase.

clinodome An open crystal form of four sides parallel to the clinoaxis *a* in the monoclinic system. Cf: *dome; orthodome.*

clinoenstenite A name for the pyroxene series clinoenstatite and clinohypersthene. Cf: *enstenite.*

clinoferrosilite A monoclinic mineral, $Fe_2Si_2O_6$; pyroxene group; contains up to 15% $Mg_2Si_2O_6$ toward clinohypersthene; dimorphic with ferrosilite.

clinograph An instrument for making a borehole survey; i.e., to determine if, and in what direction, a borehole has deviated off the true vertical plane. See also: *crooked hole.* (Nelson, 1965)

clinoguarinite (a) Cesaro's name for a monoclinic form of guarinite. See also: *orthoguarinite.* (English, 1938) (b) A former name for hiortdahlite.

clinohedrite (a) Breithaupt's name for tetrahedrite. (English, 1938) (b) A monoclinic mineral, $CaZnSiO_4{\cdot}H_2O$; forms colorless to white or amethystine clinohedral crystals.

clinohypersthene An intermediate member in the series clinoenstatite-clinoferrosilite in the pyroxene group.

clinometer Any of various instruments used for measuring angles of slope, elevation, or inclination (esp. the dip of a geologic stratum or the slope of an embankment); e.g., a simple hand-held device consisting of a tube with a cross hair, a graduated vertical arc, and an attached spirit level so mounted that the inclination of the line of sight can be read on the circular scale by centering the level bubble at the instant of observation. A clinometer is usually combined with a compass (e.g., the Brunton compass). Syn: *inclinometer; plain clinometer.* Cf: *drift indicator.* (AGI, 1987)

clinophone An exceptionally accurate instrument for borehole surveying, designed particularly for use with the freezing and cementation methods of shaft sinking; capable of giving the slope of a borehole to within 1 min of arc. (Hammond, 1965)

clinoptilolite A monoclinic mineral, $(Na,K,Ca)_2Al_3(Al,Si)_2Si_{13}O_{36}{\cdot}12H_2O$; of the zeolite group.

clinopyroxene A group name for monoclinic pyroxenes. Abbrev. cpx. Syn: *monopyroxene.* Cf: *orthopyroxene.*

clinostrengite A discredited name for phosphosiderite, a dimorph of strengite.

clinoungemachite A monoclinic mineral, sodium potassium iron sulfate; possibly dimorphous with ungemachite.

clinozoisite An epidote having the composition of zoisite, $Ca_2Al_3(SiO_4)_3(OH)$; monoclinic; crystals striated. (Dana, 1959)

clintonite (a) A monoclinic mineral, $Ca(Mg,Al)_3(Al_3Si)O_{10}(OH)_2$; mica group. Syn: *seyberite; xanthophyllite.* (b) A group name for the brittle micas.

Clinton ore A red, fossiliferous sedimentary iron ore; e.g., the Clinton Formation (Middle Silurian) or correlative rocks of the east-central United States, containing lenticular or oolitic grains of hematite. It supplies the ironworks at Birmingham, AL. See also: *fossil ore; flaxseed ore.* (AGI, 1987)

clip Connector between an underground tub, car, truck, or tram, and endless rope haulage. A clip pulley has a broad rim into which studs are set, to grip links of a haulage chain. See also: *haulage clip; automatic clip; coupling; clam.* (Pryor, 1963)

clip method The clip method of making wire rope attachments is widely used. Drop-forged clips of either the U-bolt or the double-saddle type are recommended. When clips of the correct size are properly applied, the method uses about 80% of the rope strength.

clod (a) Eng. Deposits interstratified with coal; Yorkshire and Midland Counties. (Nelson, 1965) (b) A hard earthy clay on the roof of a working place in a coal seam; often a fireclay. (CTD, 1958) (c) A miner's term applied to a soft, weak, or loosely consolidated shale (or to a hard, earthy clay), esp. one found in close association with coal or immediately overlying a coal seam. It is so called because it falls away in lumps when worked. An artificially formed aggregate of soil particles. (AGI, 1987) (d) A clod of dirt, of greater or less diameter, thin at the edges and increasing in thickness to the middle. See also: *kettle bottom.* (e) An artificially formed aggregate of soil particles.

clog (a) Mid. A short piece of timber about 3 in by 6 in by 24 in (7.6 cm by 15.2 cm by 61.0 cm) fixed between the roof and a prop. (Fay, 1920) (b) A flat wedge over a post. See also: *lid.* (Nelson, 1965) (c) To obstruct, hinder, or choke up; e.g., the stoppage of flow through a pipe by an accumulation of foreign matter, or the filling up of the grooves in a file when operating on a soft metal. (Crispin, 1964) (d) Eng. Rock filling a fault. (Arkell, 1953)

close-connected Applied to dredges in which the buckets are each connected to the one in front without any intermediate link. (Fay, 1920)

closed circuit (a) A water circuit designed so that the only water added is that necessary to replace the loss of water on the products. (BS, 1962) (b) A system in which coal passes from comminution to a sorting device that returns oversize for further treatment and releases undersize from the closed circuit.

closed-circuit grinding A size-reduction process in which the ground material is removed either by screening or by a classifier, the oversize being returned to the grinding unit. Typical examples are a dry pan with screens, dry milling in an air-swept ball mill, and wet milling in a ball mill with a classifier. See also: *circulating load.* (Dodd, 1964)

closed-circuit operation Retention and retreatment of ore in part of flow a line until it satisfies criteria for release. Used in comminution to reduce overgrinding by passing intermediate particles repeatedly through grinding systems, classifying the product and returning oversize. Used in concentration (e.g., rougher-scavenger-cleaner flotation) to retain a selected fraction of ore in circuit for retreatment (a middling), until it is either upgraded to rank as concentrate or sufficiently denuded of value to be rejected as tailing. (Pryor, 1958)

closed-circuit television System in which television cameras relay pictures of conditions at important points in a plant, thereby aiding workers to watch inaccessible places and exercise extended control.

closed contour A contour line that forms a closed loop and does not intersect the edge of the map area on which it is drawn; e.g., a depression contour indicating a closed depression, or a normal contour indicating a hilltop. (AGI, 1987)

closed fault A fault in which the two walls are in contact. Cf: *open fault.* (AGI, 1987)

closed frame A mine support frame used esp. in inclined shafts where protection from rock pressure is needed on all sides. This completely closed set is provided at the bottom with a sill. The joint is usually effected by tenons, so that when the pressure is exerted in a downward direction the timbers interlock. (Stoces, 1954)

closed joint A joint found in rocks that causes a plane of weakness known variously as a rift or gain. This largely determines the shape of the blocks that may be extracted from a quarry. Also called incipient joint. Syn: *gain.*

closed-spiral auger A soil-sampling auger made by spirally twisting a flat steel ribbon to form a tubelike, hollow-center, corkscrewlike device. (Long, 1960)

closed top *cup and cone.*

closed traverse (a) A surveying traverse whose accuracy can be checked by the fact that, when it is closed, the angles should add up to 360°, and which ends at its starting point. (Hammond, 1965) (b) A surveying traverse that starts and terminates upon the same station or upon a station of known position. Cf: *open traverse.* (AGI, 1987)

closed-water circuit The separation of solids from a washery slurry so that the water can be returned to the plant and used continuously. (Nelson, 1965)

close goods (a) Pure stones, of desirable shapes. (b) Highest class of South African diamonds, as sorted at Kimberly.

close-jointed Applied to rocks in which the joints are very close together. (Fay, 1920)

close-joints cleavage *slip cleavage.*

close nipple A nipple, the length of which is about twice the length of a standard pipe thread and without any shoulder. See also: *nipple.* (Strock, 1948)

close prospecting Detailed analysis of a proven placer deposit that should determine: (1) the volumetric measurements of both overburden and gravel; (2) the estimation of the gold or other mineral contents; (3) the average value of the area in pence, cents, carats, or other unit per unit of volume; and (4) all possible information regarding the nature of the overburden and gravel—i.e., whether it is clayey, free wash, etc.—as well as of the bedrock. (Griffith, 1938)

close-ranged Screened or classified between close maximum and minimum limits of size or settlement. (Pryor, 1965)

close sheathing Consists of planks placed side by side along a continuous frame. Its use is to prevent local crumbling of less compacted soils. Since crevices can exist between planks, it should not be used with fine silts or liquid soils, which can seep through these cracks. Cf: *skeleton sheathing; tight sheathing.* (Carson, 1961)

close sizing (a) In screening, choice of sieve sizes that are fairly close in mesh size to restrict size range of each fraction of the material separated. (b) Sizing with screens.

close timbering The setting of timber sets and lagging very close together when shaft sinking or tunneling through very loose ground or crushed coal in thick seams. See also: *cribbing; forepoling.* (Nelson, 1965)

closing error When calculating or plotting the distances, angles, or coordinates of a closed traverse or one connecting two accurately located points, the discrepancy between starting and finishing point. This error is adjusted in proportion to the magnitude of the angles and distances involved, if it is below a tolerable limit. See also: *error of closure.* (Pryor, 1963)

closing rope Operating rope for opening and closing a grab. (Hammond, 1965)

closterite Dense, laminated, brownish-red algal coal found in Irkutsk, Russia. It consists of an accumulation of spheroidal algal colonies of different sizes, among which are disseminated great numbers of desmid algae, belonging to the living genus, *Closterium.* (Tomkeieff, 1954)

closure (a) A closed anticlinal structure. (b) The difference in the relative position of the bottom and the collar of a borehole expressed in horizontal distance in a specific compass direction. (Long, 1960) (c) The relative inward movement of the two walls of a stope. (d) A cumulative measure of the various individual errors in survey measurements; the amount by which a series of survey measurements fails to yield a theoretical or previously determined value for a survey quantity. (AGI, 1987) (e) Used in structural geology, esp. in connection with potential oil structures, to designate the vertical distance between the highest point of an anticlinal structure of an anticlinal structure or fold and the lowest contour that closes around the structure. It is an approximate measure of the capacity of a structural trap for oil and/or gas. (Stokes, 1955) (f) A portion of brick to close, when required, the end of a course as distinguished from a half brick. *closure.* (AISI, 1949)

closure meter An instrument for indicating the amount of closure that has taken place. Wall closure in mines is measured by this instrument. Also called sag meter. (Spalding, 1949; Spalding, 1949)

clot A group of ferromagnesian minerals in igneous rock, from a few inches to a foot or more in size, commonly drawn out longitudinally, that may be a segregation or an altered xenolith. (AGI, 1987)

clothing Eng. Brattice constructed of a coarse, specially prepared canvas. (Fay, 1920)

clotting The sintering or semifusion of ores during roasting. (Fay, 1920)

cloud chamber A device that displays the tracks of charged atomic particles. It is a glass-walled chamber filled with a supersaturated vapor. When charged particles pass through the chamber, they leave a cloudlike track much like the condensation trail of a plane. This track permits scientists to see the paths of these particles and study their motion and interaction. See also: *bubble chamber; spark chamber.* (Lyman, 1964)

clouded agate Chalcedony with irregular or indistinct patches of color. See also: *agate.* (Hess)

cloudy chalcedony Chalcedony with dark, cloudy spots in a light-gray transparent base. (Schaller, 1917)

cloudy stain In mica, a cloudlike effect that occurs in various colors. (Skow, 1962)

cloustonite Scot. A mineral related to asphalt, occurring in patches in blue limestone and in blue flags at Inganess, Orkney. It is soluble in benzol and at a red heat gives off a large amount of illuminating gas. (Fay, 1920)

clump (a) A bend in a roadway or passage in a coal seam. (CTD, 1958) (b) A large fall of roof. (CTD, 1958) (c) A tough fireclay. (CTD, 1958)

clustered carbide *interspersed carbide.*

clusterite *botryoid.*

cluster mill A rolling mill in which each of the two working rolls of small diameter is supported by two or more backup rolls. (ASM, 1961)

CM Strata containing coalbeds, particularly those of the Pennsylvanian Period. Used as a proper name for a stratigraphic unit more or less equivalent to the Pennsylvanian Period. Abbrev. of Coal Measures. (AGI, 1987)

CMI centrifuge A fine-coal dewatering machine consisting of two rotating elements, an outside conical screen frame, and an inside solid cone, which carries spiral hindrance flights. By a slight difference in the number of teeth in the gears, the screen element moves slightly faster, in the same direction, than the solid cone. Material enters the machine from the top and falls on the solid cone where centrifugal force throws it against the screen. It slides down the screen until it meets the upper end of the hindrance flights, and, in doing so, the water begins to pass through the screen. The flights spiral downward, and, as the screen moves slowly around them in the direction of the downward pitch, the solids gradually find their way to the bottom of the screen basket and the zone of maximum centrifugal force, tending to remove all of the water. See also: *dewatering.* (Kentucky, 1952)

coagulation (a) The binding of individual particles to form flocs or agglomerates and thus increase their rate of settlement in water or other liquid. See also: *flocculation.* (Nelson, 1965) (b) The coalescence of fine particles to form larger particles.

coagulator A soluble substance, such as lime, which, when added to a suspension of very fine solid particles in water, causes these particles to adhere in clusters that will settle easily. Used to assist in reclaiming water used in flotation. (Hess)

coal A readily combustible rock containing more than 50% by weight and more than 70% by volume of carbonaceous material, including inherent moisture; formed from compaction and induration of variously altered plant remains similar to those in peat. Differences in the kinds of plant materials (type), in degree of metamorphism (rank), and in the range of impurity (grade) are characteristic of coal and are used in classification. Syn: *black diamond.* (AGI, 1987)

coal analysis The determination, by chemical methods, of the proportionate amounts of various constituents of coal. Two kinds of coal analyses are ordinarily made: (1) proximate analysis, which divides the coal into moisture (water), volatile matter, fixed carbon, and ash. Percentage of sulfur and heat value in Btus per pound or kilogram, each obtained by separate determination, are usually reported with the real proximate analysis; and (2) ultimate analysis, which determines the percentages of the chemical elements carbon, hydrogen, oxygen, nitrogen, and sulfur. Other elements that may be present are considered impurities and are reported as ash.

coal ash Noncombustible matter in coal.

coal auger A special type of continuous miner. It consists essentially of a large diameter screw drill that cuts, transports, and loads coal onto vehicles or conveyors. The coal auger is used for (1) winning opencast coal without stripping overburden; (2) pillar-and-stall mining; and (3) extraction of pillars or percentage of pillars that would otherwise be uneconomic to work. See also: *auger; twist drill; mole mining.* (Nelson, 1965)

coal ball Nodules of spheroidal, lenticular, or irregular shape containing petrified plant remains and in some cases animal remains. They vary in size from about 1 to 40 cm or more; occasional specimens weigh more than 1 ton. Infrequently, an entire seam in a restricted area consists largely of coal balls. Coal balls consist mainly of calcareous, dolomitic, sideritic, pyritic, or siliceous material surrounding or impregnating plant and animal remains. They occur in brown coals (mainly sideritic balls) as well as in coals of higher rank generally lying within a coal seam but occasionally in the roof. Calcareous, dolomitic, and pyritic coal balls are commonly found in seams having marine strata in the roof. The distribution in seams is variable. They may occur in a broad zone running through a coalbed or be distributed irregularly in nests. Syn: *torf dolomite.* (IHCP, 1963)

coal bank An exposed seam of coal. (Craigie, 1938)

coal barrier A protective pillar of coal. See also: *barrier pillar.* (Nelson, 1965)

coal basin (a) Depression in older rock formations in which coal-bearing strata have been deposited. See also: *concealed coalfield; exposed coalfield.* (b) A coalfield with a basinal structure; e.g., the Carboniferous Coal Measures of England. (AGI, 1987)

coal blasting There are two methods of breaking coal with explosives, namely, blasting cut coal, which is the method most commonly used, and blasting off the solid, or grunching. (McAdam, 1958)

coalbreaker (a) A building containing the machinery for breaking coal with toothed rolls, sizing it with sieves, and cleaning it for market. (b) A machine for breaking coal.

coal briquette Coal made more suitable for burning by a process that forms it into a regular square- or oval-shaped piece. (Bennett, 1962)

coal briquetting *briquette.*

coal bump Sudden outburst of coal and rock that occurs when stresses in a coal pillar, left for support in underground workings, cause the pillar to rupture without warning, sending coal and rock flying with explosive force.

coal burster An appliance for loosening coal by means of high-pressure water and oil. It consists of a round, stainless steel bar with small telescopic rams acting on a steel liner in a shot hole. The bar is connected to a hand- or power-operated pump placed near the face. The high-pressure liquid from the pump causes the rams and liner to exert a pressure sufficient to loosen or break down the coal. It is a safe method of coal breaking without the use of explosives. It has not, however, made the progress originally anticipated. See also: *water infusion.* Syn: *hydraulic cartridge.* (Nelson, 1965)

coal car A car used in hauling coal in or from a mine. (Craigie, 1938)

coal cart A cart for carrying coal. (Craigie, 1938)

coal chute A trough or spout down which coal slides from a bin or pocket to a locomotive tender, or to vessels, carts, or cars. (Fay, 1920)

coal claim A piece of land having, or thought to have, valuable coal deposits on it and legally claimed by one seeking to own it. (Mathews, 1951)

coal classification The grouping of coals according to certain qualities or properties, such as coal type, rank, carbon-hydrogen ratio, and volatile matter. See also: *high-volatile coals.* (Nelson, 1965)

coal classification systems One system classifies coal by the content of volatile matter: with 10% volatile, anthracite; between 10% and 13% lean coal, semianthracite or dry-steam coal; 14% to 20%, variously designated; 20% to 30%, fat or coking coal. Other systems classify by calorific value, and caking and/or coking property. Post-World War II classifications include (1) volatile matter, (2) caking properties on rapid heating, and (3) coking properties. See also: *ASTM coal classification.* (Pryor, 1963)

coal clay Clay found under a coalbed, usually a fireclay. See also: *underclay.*

coal cleaning The sorting, picking, screening, washing, pneumatic separation, and mixing of coal sizes for the market.

coal cleaning equipment Equipment used to remove impurities — such as slate, sulfur, pyrite, shale, fire clay, gravel, and bone — from coal.

coal cleaning plant A plant where raw or run-of-mine coal is washed, graded, and treated to remove impurities and to reduce ash content. Syn: *washery.* (Pryor, 1963)

coal clearing The loading of broken coal at the face into conveyors or mine cars. The clearing shift is the coal-loading shift or stint. Usually the miner has a measured task or stint (stent). (Nelson, 1965; Pryor, 1963)

coal conglomerate A conglomerate made of fragments of coal. (Tomkeieff, 1954)

coal constituent classification In the United States it is generally agreed that the maceral concept of the nomenclature Stopes-Heerlen System fails to comprehend the effect of the stage of coalification on the nature of coal constituents. W. Spackman's interpretation of the maceral concept incorporates the ideas of variable coalification in suggesting a skeletal framework upon which a systematic classification can be built. The maceral concept, as interpreted by Spackman, implements the classification of the products of coalification. In this scheme, macerals possessing similar chemical and physical properties are assembled into maceral groups that can, in turn, be characterized by a comparatively restricted set of properties. Maceral groups possessing similar characteristics can be classified into maceral suites. Syn: *Spackman system.* (IHCP, 1963)

coal cutter (a) The longwall coal cutter is a power-operated machine that draws itself by rope haulage along the face, usually cutting out a thin strip of coal from the bottom of the seam, in preparation for shot firing and loading or a cutter loader. The bar and disk machines are obsolescent and the chain coal cutter is now almost universal. (Nelson, 1965) (b) *machineman.*

coal-cutter pick One of the cutting points attached to a cutter chain for making a groove in a coal seam. The picks are made from quality carbon steel or a hard alloy steel and tipped with fused tungsten carbide, sintered tungsten carbide, or other hard-wearing material. The advent of the coal-cutter pick tipped with tungsten carbide on a heat-treated, alloy-steel shank has resulted in marked improvements in drilling and a reduction in cutting delays. See also: *chain coal cutter; double-ended pick; duckbill pick; tungsten carbide bit.* (Nelson, 1965)

coal-cutter team The miners in charge of a coal cutter. A cutting team varies from two to five with two to three about average. The leading worker is normally stationed in front of the machine and is in charge of the controls, and an assistant follows behind. See also: *machineman.* Cf: *back-end man.* (Nelson, 1965)

coal-cutting machine A machine powered by compressed air or electricity that drives a cutting chain or other device so as to undercut or overcut a seam, or to remove a layer of shale. Percussive cutters are used to bore holes or to make vertical cuts (nicking, shearing); disk, bar, and chain cutters carry small picks that undercut the seam as the machine travels. (Pryor, 1963)

coal-cutting machine operator *machineman.*

coal digging A place where coal is dug. (Craigie, 1938)

coal drill Usually an electric rotary drill of a light, compact design. Aluminum and its alloys usually are used to reduce weight. Where dust is a hazard, wet drilling is employed. With a 1-hp (745.7-W) electric drill, speeds up to 6 ft/min (1.83 m/min) are possible. Light percussive drills, operated by compressed air, and hand-operated drills are also employed. See also: *electric coal drill.* (Nelson, 1965)

coal driller In coal mining, a worker who uses a hand or power drill to drill holes into the working face of the coal into which explosives are charged and set off to blast down the coal. (DOT, 1949)

coal dryer A plant or vessel in which water or moisture is removed from fine coal. Artificial drying of fine coal is not often employed. Fine coal is removed from wash water by dewatering classifiers or by vacuum filtration. See also: *dryer; thermal drying.* (Nelson, 1965)

coal dust (a) The general name for coal particles of small size. In experimental mine testing, particles that will not pass through a 20-mesh screen—1/32-in-square (0.8-mm-square) openings—are not considered as coal dust. (Rice, 1913-18) (b) In 1964, a series of laboratory tests were made with a spark source on aluminum powder and cornstarch (both dusts presenting a more severe explosion hazard than coal dust). It was found that particles passing a U.S. Standard No. 40 sieve (particles less than 0.016 in or 0.4 mm) contributed to an explosion in the laboratory bomb. The 0.016-in particle diameter was recommended as the definition for dust in surface industry. Thus, two definitions of dust exist. For coal mines, dust consists of particles passing a U.S. Standard No. 20 sieve (particles less than 850 μm), and for surface industries, dust consists of particles passing a No. 40 sieve (particles less than 425 μm). The use of two definitions is not incongruous since the potential igniting sources in a coal mine can be much more severe than those in surface industries. (MSHA, 1986) (c) The dust produced by the breakage and crushing of coal underground and at coal preparation plants. It is usually intermixed with a varying proportion of stone dust. Coal dust in mines presents two main dangers: explosion hazard and pneumoconiosis hazard. The explosibility of a coal dust cloud depends upon its fineness, purity, and volatile content. The dust particles believed to be harmful from the pneumoconiosis aspect are those of 5 μm and under. In mines, the most common explosive dust encountered is bituminous coal dust. The U.S. Bureau of Mines has established that coal dust in the absence of gas can explode and that explosions can occur in any shape of mine opening. See also: *dust-free conditions.* (Nelson, 1965; Hartman, 1961)

coal-dust explosion A mine explosion caused by the ignition of fine coal dust. It is considered that an explosion involving coal dust alone is relatively rare. It demands the simultaneous formation of a flammable dust cloud and the means of ignition within it. The flame and force of a combustible gases explosion are the common basic causes of a coal-dust explosion. The advancing wave of the explosion stirs up the dust on the roadways and thus feeds the flame with the fuel for propagation. See also: *colliery explosion; gas explosion; stone-dust barrier.* (Nelson, 1965)

coal-dust index Percentage of fines and dust passing the 0.0117-in (0.30-mm) mesh or 48-mesh. (Bennett, 1962)

coal elevator A building in which coal is raised and stored preparatory to loading on cars, ships, etc. (Mathews, 1951)

coalesced copper Massive copper made from ground, brittle, cathode copper by briquetting and sintering in a reducing atmosphere at high temperatures with pressure. (ASM, 1961)

coalette *briquette.*

coal face (a) The mining face from which coal is extracted by longwall, room, or narrow-stall system. See also: *face.* (Nelson, 1965) (b) A working place in a colliery where coal is hewn, won, got, or gotten from the exposed face of a seam by face workers. (Pryor, 1963)

coalfield (a) An area of country, the underlying rocks of which contain workable coal seams. The distribution of coalfields was largely determined by folding movements and subsequent denudation. The original coal areas were clearly larger than the present coalfields. See also: *coal basin; field.* (Nelson, 1965) (b) A region in which coal deposits of known or possible economic value occur. (AGI, 1987)

coal flotation *flotation; froth flotation.*

coal formation (a) A stratigraphic coal-bearing unit in coal measures. (b) A stratum in which coal predominates. (Craigie, 1938)

coal fuel ratio The content of fixed carbon divided by the content of volatile matter is called the fuel ratio. According to their fuel ratios, coals have been classified as anthracite, at least 10; semianthracite, 6 to 10; semibituminous, 3 to 6; and bituminous, 3 or less.

coal-hoisting engineer In coal mining, one who operates a hoist for raising coal to the surface where separate shafts or compartments are used for handling coal and people. (DOT, 1949)

coalification Those processes involved in the genetic and metamorphic history of coalbeds. The plant materials that form coal may be present in vitrinized or fusinized form. Materials contributing to coal differ in their response to diagenetic and metamorphic agencies, and the three essential processes of coalification are called incorporation, vitrinization, and fusinization. See also: *incorporation; carbonification.* Syn: *incarbonization.* (AGI, 1987)

coalify To change vegetal matter into coal. (Hess)

coaling (a) The making of charcoal. (Craigie, 1938) (b) The process of supplying or taking coal for use, as in coaling a steamer, etc.

coal interface detector *coal interface sensor.*

coal interface sensor Any device that indicates the boundary between the coal and the surrounding strata either at the roof or the floor. Aids the machine operator or control system controls in positioning the coal-cutting head. (Mowrey, 1991)

Coalite A trade name for a smokeless fuel produced by carbonizing coal at a temperature of about 600 °C. It has a calorific value of about 13,000 Btu/lb (30.2 MJ/kg) and is used for domestic purposes. Also called semicoke. See also: *coking coal.* (Nelson, 1965)

Coalite process *Parker process.*

coal land Land of the public domain that contains coalbeds.

coal lateral A railroad that parallels a coal road. (Mathews, 1951)

coal lead Thin vein of coal in a fault zone. Coal leads may indicate the direction of a displaced seam. See also: *drag.*

coal liquefaction The conversion of coal into liquid hydrocarbons and related compounds by hydrogenation at elevated temperatures and pressures. In essence, this involves putting pulverized bituminous coal into an oily paste, which is treated with hydrogen gas under appropriate conditions of temperature and pressure to form the liquid molecules of carbon and hydrogen that constitute oil. Also called coal hydrogenation. Syn: *hydrogenation of coal.* (CCD, 1961; Kentucky, 1952)

coal measures (a) A succession of sedimentary rocks (or measures) ranging in thickness from a meter or so to a few thousand meters, and consisting of claystones, shales, siltstones, sandstones, conglomerates, and limestones, with interstratified beds of coal. (AGI, 1987) (b) A group of coal seams. (AGI, 1987)

Coal Measures A stratigraphic term used in Europe (esp. in Great Britain) for Upper Carboniferous, or for the sequence of rocks (typically, but not necessarily, coal-bearing) occurring in the upper part of the Carboniferous System. It is broadly synchronous with the Pennsylvanian of North America. (AGI, 1987)

coal-measures unit Coal-measures unit strata disclose a rough repetition or cycle of different kinds of rock in the same regular manner. Broadly, the cycle of strata upward is coal, shale, sandstone, and coal. This sequence is sometimes referred to as a unit. See also: *cyclothem.*

coal mine Any and all parts of the property of a mining plant, on the surface or underground, that contribute, directly or indirectly, under one management to the mining or handling of coal. In addition to the underground roadways, staple shafts, and workings, a coal mine includes all surface land in use, buildings, structures and works, preparation plants, etc. A colliery. See also: *mine.* (Nelson, 1965)

coal mine explosion The burning of gas and/or dust with evidence of violence from rapid expansion of gases. (USBM, 1966)

coal mine ignition The burning of gas and/or dust without evidence of violence from expansion of gases. (USBM, 1966)

coal miner One employed in the mining of coal.

coal mine regulations National, state, and local laws, or enforceable rules that govern coal mining.

coal mining The industry that supplies coal and its various by-products. (Nelson, 1965)

coal mining examinations The examinations held in respective coal mining States which must be passed by every person who wishes to become a mine foreman, assistant mine foreman, mine examiner, or electrician. A candidate for a certificate may submit himself or herself for a written and oral examination before a Mining Qualifications Board. Holders of approved degrees or diplomas usually need less mining experience to qualify for first-class certification.

coal mining explosives The statutory requirements regarding the use of explosives in coal mines are very stringent. In gaseous mines only permissible (or permitted) explosives are allowed. See also: *explosive.* (Nelson, 1965)

coal mining methods The methods of working coal seams have been gradually evolved and progressively improved or modified as knowledge and experience were gained and power machines became available. Over the years, a very large number of methods of mining coal have been developed to suit the seam and local conditions, and they may be split, broadly, into longwall, and pillar methods of working. See also: *stowing method.* (Nelson, 1965)

coal oil Crude oil obtained by the destructive distillation of bituminous coal.

coal patch A small settlement near a coal mine.

coal penetrometer An instrument to assess the strength of a coal seam, its relative workability, and the influence of roof pressure. It consists of a steel rod of sectional area 1/4 in^2 (1.6 cm^2) that is pushed into the coal, normal to the coal face, under the action of a light hydraulic ram. The ram is braced against lightweight props erected at the face. When in position, the penetrometer gives a graph of load against penetration at a particular point. Readings are taken at a number of points laterally and vertically along the face, and these can be correlated with the performance of plow-type machines. Thus, the probable performance of a machine in a seam can be estimated without the need for costly trials. Syn: *penetrometer.* (Nelson, 1965)

coalpit Eng. A place where coal is dug. A coal mine.

coal planer A type of continuous coaling machine developed in Germany esp. for longwall mining. It consists of a heavy steel plow with cutting knives, with power equipment to drag it back and forth across a coal face. A parallel conveyor receives and carries away the coal as the planer digs it from the face.

coal plant A fossil plant found in coalbeds or contributing its substance to the formation of coalbeds. Any plant species, the residue of which has entered into the composition of coal under natural geological conditions. (Fay, 1920)

coal plow (a) A cutter loader with knives to slice the coal off the face. (Nelson, 1965) (b) This device carries steel blades that shear or plane off coal to a limited depth and plow it onto the face conveyor. The plow is hauled backward and forward along the coal face by steel ropes or chains operated by winches in the gate roads, and it planes off a thickness of 11.8 in (30 cm) to a height one-third to one-half the seam thickness each time. The coal is conveyed along the face by a double-chain conveyor with double-ended drive; the conveyor sections are articulated to allow for bends in its tracks and are moved bodily forward at each passage of the plow, either by compressed-air jacks or by means of a torpedo or trailer attached by rope to the plow and an auxiliary drum on the winches. Its uses are limited to softer coal seams, or to suitably prepared coal. Also called kohlenhobel. (Mason, 1951)

coal pocket (a) A structure, bunker, or bin for the storage of coal. (Fay, 1920) (b) An arrangement of bins to load trucks or railcars by gravity.

coal preparation The various physical and mechanical processes in which raw coal is dedusted, graded, and treated by dry methods (rarely) or water methods, using dense-media separation (sink-float), jigs, tables, and flotation. The objective is the removal of free dirt, sulfur, and other undesirable constituents.

coal-preparation plant (a) A facility where raw coal is sized and prepared for loadout. In the United States, plant capacities vary from 500 to 2,500 st/h (454 to 2,268 t/h). See also: *cleaning; dense-medium washer; gravity concentration; screen; washery.* (Nelson, 1965) (b) A facility or collection of facilities that include associated support facilities and consist of, but are not limited to: loading facilities; storage and stockpile facilities; sheds, shops, and other buildings; settling basins and impoundments, coal processing and other waste disposal areas; roads, railroads, and other transport facilities. Exempted from the meaning of coal-preparation plant is an operation that a) loads coal; b) does not separate coal from its impurities; and c) is not located at or near the mine site.

coal-preparation process The process adopted for cleaning and sizing coal for the market. Specialists select the best process for any particular run-of-mine coal. Many conflicting factors must be weighed. The cost of a detailed investigation is well repaid in higher recoveries, in flexibility, and in ease of operation and maintenance. (Nelson, 1965)

coal-preparation shift On mechanized longwall faces, the shift during which coal-cutting, boring, and shot-firing operations are performed. (Mason, 1951)

coal-processing waste Earth materials that are combustible, physically unstable, or acid- or toxic-forming, which are wasted or otherwise separated from product coal. They are slurried or otherwise transported from coal-preparation plants, after physical or chemical processing, cleaning, or concentrating of coal.

coal rank Classification according to degree of metamorphism or progressive alteration, in the natural series from lignite to anthracite; higher rank coal is classified according to fixed carbon on a dry basis; lower rank coal according to Btus on a moist basis. (Bennett, 1962)

coal rash Very impure coal containing much argillaceous material, fusain, etc. (AGI, 1987)

coal room (a) Scot. A working face in stope-and-room workings. (Fay, 1920) (b) The open area between pillars where the coal has been removed.

coal sampling The standard method used by the U.S. Bureau of Mines samplers is as follows: A space of 5 ft (1.52 m) in width should be cleared of dirt and powder from top to bottom of the seam being sampled. Down the center of this cleared space, a zone 1 ft (0.3 m) wide is cut to a depth of at least 1 in (2.54 cm) in order to get perfectly clean coal. A cut is then made up the center of this zone to a depth of 2 in (5.2 cm) and a width of 6 in (15.2 cm); or, if the coal is soft, to a depth of 3 in (7.6 cm) and a width of 4 in (10.2 cm). Approx. 5 to 6 lb (2.3 to 2.7 kg) of coal will be obtained for each foot (0.3 m) of thickness of the seam. This should include all bony coal included in the mining operation and exclude all slate or partings, which are thrown out during the operation. The sample obtained should be collected on a waterproof cloth 6 ft by 7 ft (1.83 m by 2.13 m) and then screened, the lumps being broken in a mortar, and all passed through a 1/2-in (12.7-mm) screen. Any impurities, such as slate or pyrite, are crushed to 1/4 in (6.4 mm) or finer and thoroughly mixed with the coal. The coarser materials should be evenly distributed, the sample being then quartered, remixed, and requartered. When the mixing is complete, the sample should be placed in a can with the capacity of 3 lb (1.35 kg) and the top screwed on and sealed with adhesive tape. The can should be labeled with the name of the collector, the location, the date, and any other information necessary for the analysis. See also: *channel sample; sampling.* (Kentucky, 1952)

coal seam A bed or stratum of coal. (Craigie, 1938)

coal-seam correlation The identification of a coal seam; the linking up or matching of a seam exposed in different parts of a mine or coalfield. A coal seam may be correlated by lithology, by fossils, by chemical composition, or by its spore content. Coal-seam correlation is very important in exploration and in penetrating faults. See also: *correlation.* (Nelson, 1965)

coal-sensing probe An obsolete, nucleonic coal-sensing instrument that can measure the thickness of coal left on the roof or floor of a seam after the passage of a mining machine. The principle used is the measurement of the density of the strata underlying the machine by a gamma-ray backscattering unit. Gamma rays from a radioactive source are scattered in all directions by the atomic particles in the coal and rock. The amount of scattered radiation eventually reaching the Geiger counter is, approx., inversely proportional to the density of the scattering medium; i.e., more radiation will come back from coal than from rock. Thus, as the amount of coal between the source and the underlying rock changes, so the amount reaching the Geiger counter and the counting unit (the ratemeter) will change, and consequently the output of the meter can be calibrated in terms of the thickness of the floor coal. This instrument has been replaced by a natural-gamma coal thickness sensor. See also: *manless face.* (Nelson, 1965)

coal separator A machine that separates coal from associated impurities in run-of-mine material. See also: *coal-preparation plant.* (Nelson, 1965)

coal slime A slurry containing particles of such size range that 50% or more (by weight) will pass a 200-mesh sieve (or finer).

coal sludge A slurry that has been partly dewatered by sedimentation, usually to a dilution that will permit further dewatering by mechanical means.

coal slurry Finely crushed coal mixed with sufficient water to form a fluid. To use coal slurry pumped through a pipeline as fuel, expensive drying and dewatering pretreatment has been necessary. Recent tests indicate that coal slurry can be fired in a cyclone furnace as it is received from a pipeline; i.e., a coal and water mixture. See also: *slurry.* (Nelson, 1965)

coal smoke A suspension of very fine particles in air. A coal that breaks down easily when heated gives off its volatile matter very easily and perhaps more quickly than the available draft can supply the air for combustion, with the result that dark smoke containing much unburnt or partly burnt material is given off—a loss of fuel energy. See also: *smoke.* (Nelson, 1965)

Coal special Explosive; used in mines. (Bennett, 1962)

coal split *split seam.*

coal spragger (a) In bituminous coal mining, one who sets short wooden props in a slanting position (sprags) under the upper or overhead section of a bed of coal to hold that section up while the lower section is being mined, or wedges heavy slanting props (sprags) against the coal to prevent it from flying when broken down by blasting. (DOT, 1949) (b) One who places short pointed wooden sprags between the spokes of a mine car wheel to stop the car.

coal stripper In bituminous coal mining, a general term applied to a worker who is engaged in mining coal in a strip mine, one in which the coal is close enough to the Earth's surface to permit the use of power shovels in stripping back the ground and loading the coal into large cars or trucks. Usually designated according to particular jobs. (DOT, 1949)

coal substance Coal excluding its mineral matter and moisture. (BS, 1960)

coal tar Tar obtained by the destructive distillation of bituminous coal, usually in coke ovens or in retorts, and consisting of numerous constituents (such as benzene, xylenes, naphthalene, pyridine, quinoline, phenol, cresols, light oil, and creosote) that may be obtained by distillation. (Webster 3rd, 1966)

coal-tar oil Oil obtained by the distillation of coal tar. Oils are classified into light and heavy oils. A light oil is one having a specific gravity less than 1.000 and contains the coal-tar napthas. Heavy oils sink in water and contain such compounds as creosote, anthracene, anthracene oil, etc. (Porter, 1930)

coal-tar pitch A dark-brown to black residuum from the distillation of coal tar, ranging from a sticky mass to a brittle solid, depending on the degree of distillation. Most coal-tar pitch melts between 60 °C and 70 °C. (Hess)

coal testing Evaluating coals by methods other than chemical, such as determining the relative values of different coals as fuels by burning them under controlled conditions in furnaces, or determining their gas- and coke-producing properties by testing in a retort. The term coal testing is frequently erroneously used, esp. in coal marketing, for coal analysis.

coal thickness sensor Any measurement instrument that is designed to measure the thickness of the coal remaining on the mine roof or floor after coal is removed by mining.

coal tipple *tipple.*

coal type (a) A variety of coal, such as common banded coal, cannel coal, algal coal, and splint coal. The distinguishing characteristics of each type of coal arise from the differences in the kind of plant material that produced it. (AGI, 1987) (b) A coal, particularly a bituminous coal, contains dissimilar bands or layers that are believed to have been formed mainly from selected portions of the plant material forming the seam. These bands, which have been given the terms vitrain, clarain, durain, and fusain, are the different types of coal in that seam. See also: *vitrain; clarain; durain; fusain.* (Nelson, 1965)

coal washer A place where mined coal is treated by sink-float methods or by froth flotation to remove ash, shale, sulfur, and other unwanted products. The resulting clean coal product is graded to size and regulated for maximum ash content. Also called cleaning plant; preparation plant.

coal wheeler In the iron and steel industry, a laborer who shovels coal into a wheelbarrow and pushes it to a furnace. (DOT, 1949)

coal workings A coal mine with its appurtenances; a colliery. Coal works. (Standard, 1964; Fay, 1920)

coalyard A place where coal is stored. (Craigie, 1938)

coarse aggregate The portion of an aggregate retained on the No. 4 sieve, consisting of particles with diameters greater than 4.76 mm. Cf: *aggregate; fine aggregate.* (AGI, 1987)

coarse gold Gold in large grains, as distinguished from gold dust. Also called coarse quartz gold.

coarse-grained Applied to rocks composed of large grains; used mainly in a relative sense, but an average size greater than 5 mm in diameter has been suggested. Cf: *medium-grained; fine-grained.* (Stokes, 1955)

coarse-grained soil A soil in which gravel and sand predominate. Coarse-grained soils are those least affected by moisture-content changes as most surface rain, etc., becomes gravitational water. (Nelson, 1965)

coarse jig A jig used to handle the larger sizes and heavier grades of ore or metal. (Weed, 1922)

coarse metal An iron-and-copper matte containing sulfur; a product of copper smelting in a reverberatory furnace. (Standard, 1964)

coarse roll A large roll for the preliminary crushing of large pieces of ore, rock, or coal. Used in stage crushing.

coarse sand (a) A geologic term for a sand particle having a diameter in the range of 0.5 to 1 mm (1 to 0 phi units). Also, a loose aggregate of sand consisting of coarse sand particles. See also: *sand.* (AGI, 1987) (b) An engineering term for a sand particle having a diameter in the range of 2 mm (retained on U.S. standard sieve No. 10) to 4.76 mm (passing U.S. standard sieve No. 4). (AGI, 1987)

coaxial cable Electrical cable with inner conducting wire covered by alternating layers of insulating and conducting material. (Pryor, 1963)

cob (a) Corn. To break ore with hammers so as to sort out the valuable portion. (b) Derb. A small solid pillar of coal left as a support for the roof.

coba Uncemented sand or gravel underlying the nitrate (caliche) deposits of Chile. See also: *congela.* (AGI, 1987)

cobalt A tough, lustrous, nickel-white or silvery-gray, metallic element. Symbol, Co. Occurs in the minerals cobaltite, smaltite, and erythrite; often associated with nickel, silver, lead, copper, and iron ores, from which it is most frequently obtained as a byproduct. Its alloys have unusual magnetic strength and are used for high-speed, heavy-duty, high-temperature cutting tools, and for dies, in jet turbines and gas turbine generators. Its salts are used in porcelain, glass, pottery, tiles, and enamels to produce brilliant blue colors. (Handbook of Chem. & Phys., 3)

cobalt bloom Hydrated arsenate, $Co_3(AsO_4)_2 \cdot 2H_2O$. See also: *erythrite.* (Pryor, 1963)

cobalt-bonded Particles of a refractory material, such as powdered tungsten carbide, cemented together with cobalt to form a metallike mixture. (Long, 1960)

cobalt glance *cobaltite.*

cobaltiferous wad An impure hydrated oxide of manganese containing up to 30% cobalt; a source of cobalt in Zaire.

cobaltite (a) An orthorhombic mineral, 4[CoAsS]; pseudocubic; metallic; occurs in high-temperature vein deposits associated with smaltite and in metamorphic rocks; an important source of cobalt. Syn: *cobalt glance; white cobalt; gray cobalt.* (b) The mineral group cobaltite, gersdorffite, hollingworthite, irarsite, platarsite, tolovkite, ullmannite, and willyamite.

cobalt lollingite *safflorite.* Also called cobaltiferous lollingite.

cobalt melanterite *bieberite.*

cobalt-nickel pyrite (a) A name applied by Vernadsky to a steel-gray member of the pyrite group containing 11.7% to 17.5% nickel and 6.6% to 10.6% $(Fe,Ni,Co)S_2$; small, pyritohedral crystals; isometric. Probably a mixture of siegenite and pyrite. From Musen, Westphalia, Germany. (English, 1938) (b) As applied by Henglein, a syn. for hengleinite. See also: *hengleinite.* (Hey, 1955)

cobaltoadamite A pale rose-red to carmine variety of adamite in which cobalt replaces some of the zinc. (English, 1938)

cobaltocalcite (a) Replaces the generally accepted name sphaerocobaltite for rhombohedral $CoCO_3$. Not the cobaltocalcite of F. Millosevich, 1910, a red cobaltiferous variety of calcite. (Spencer, 1952) (b) A former name for sphaerocobaltite.

cobalt ocher *erythrite; asbolan; asbolane.*

cobaltosphaerosiderite A peach-blossom-red rhombohedral variety of siderite with moderate substitution of cobalt for iron. Also spelled cobaltospharosiderite.

cobalt pentlandite An isometric mineral, Co_9S_8; pentlandite group.

cobalt pyrites (a) *linnaeite.* (b) Cobaltiferous pyrite containing up to 14% cobalt; an ore of cobalt in Zambia.

cobalt-rich crust An authigenic deposit of iron-manganese oxides enriched with cobalt. These crusts may contain potentially commercial quantities of manganese (20% to 30%), copper, nickel, and cobalt (less than 3% combined), but are primarily evaluated on the basis of their cobalt content. They are found as encrustations on exposed rocky seabeds on island slopes, seamounts, or submerged plateaus in water depths between 800 m and 2,400 m. The crusts may be up to 40 cm thick, but are more commonly 3 to 5 cm. They often occur in association with platinum and phosphorite.

cobalt skutterudite The pure end member, $CoAs_3$, of the skutterudite series. Syn: *skutterudite.* (Hey, 1964)

cobalt vitriol *bieberite; rose vitriol.*

cobbed ore Eng. Ore broken from veinstone by means of a small hammer.

cobbing (a) The separation, generally with a handheld hammer, of worthless minerals from desired minerals in a mining operation; e.g., quartz from feldspar. Syn: *hand cobbing; piking.* (AGI, 1987) (b) Rubble, such as from furnace bottoms, impregnated with copper. (Standard, 1964)

cobbing board A flat piece of wood used in cobbing. (Fay, 1920)

cobbing hammer A special chisel type of hammer used to separate the mineral in a lump from the gangue in the hand-picking of ores. (Nelson, 1965)

cobble (a) A usually rounded or semirounded rock fragment between 3 to 12 in (76 to 305 mm) in diameter; large than a pebble and smaller than a boulder, rounded or otherwise abraded in the course of aqueous, eolian, or glacial transport. Syn: *cobblestone.* (b) Eng. Small lump coal. See also: *cob coal.*

cobble riffle A sluice with a cobble-paved bottom used in placer mining. (Nelson, 1965)

cobbles A graded size of anthracite below large coal—about 5 in (12 cm). (Nelson, 1965)

cobblestone (a) A naturally rounded, usually waterworn stone suitable for use in paving a street or in other construction. Syn: *cobble; roundstone.* (AGI, 1987) (b) A consolidated sedimentary rock consisting of cobble-size particles. (AGI, 1987)

cobbling Eng. Cleaning the haulage road of coal that has fallen off the trams.

cob coal A large round piece of coal.

Coblentzian Upper Lower Devonian. (AGI, 1987)

cobra stone *chlorophane.*

cocarde ore *cockade ore.*

cocinerite (a) A mixture of chalcocite and silver. (b) A silver-gray copper silver sulfide found at Ramos, Mexico; perhaps a variety of stromeyerite.

cockade ore (a) An open-space vein filling in which the ore and gangue minerals are deposited in successive comblike crusts around rock fragments; e.g., around vein breccia fragments. Syn: *cocarde ore; sphere ore.* See also: *ring ore.* (AGI, 1987) (b) Cockscomb pyrite; a form of marcasite.

cockade structure The form taken by cockade ore.

cocker To set supports in herringbone fashion. (Mason, 1951)

cockering Herringbone supports. A method of support by which a center support of beams or bars running longitudinally along the roof of a road is supported systematically by slanted struts or props with their feet spragged in the side of the road, the whole looking like a herringbone. (Mason, 1951)

cockermeg Temporary support for the coal face. A short crosspiece is held to it by two slanting props, one hitched in the floor, the other in the roof. (Pryor, 1963)

cockers *cockermeg.*

cockersprag *cockermeg.*

cockle (a) Corn. Schorl or black tourmaline. (Fay, 1920) (b) Any mineral occurring in dark, long crystals, esp. schorl. (Webster 2nd, 1960) (c) Eng. A black, thready mineral, seeming to be a fibrous talc; occurs in Cornish tin mines. (Arkell, 1953) (d) Eng. An ironstone nodule. (Arkell, 1953) (e) Cornish name for hard siliceous rocks. (Arkell, 1953)

cocko A piece of slate or bony. (Korson, 1938)

cockscomb pyrite *marcasite.*

cockscomb pyrites A crestlike variety of marcasite. See also: *marcasite.* (Webster 3rd, 1966)

cocoa mat A fabric of wood fibers used to distribute water evenly over a smooth surface. (Nichols, 1976)

codorous ore A highly siliceous hematite containing only a trace of phosphorus, but high in potash. (Osborne, 1956)

coefficient of absolute viscosity *coefficient of viscosity.*

coefficient of acidity A ratio, calculated from the normative molecular proportions of the constituents of a rock or slag; e.g., number of atoms of oxygen in SiO_2 / number of atoms of oxygen in the basic oxides.

coefficient of compressibility The decrease in volume per unit volume produced by a unit change of pressure. (Webster 2nd, 1960)

coefficient of elasticity *modulus of elasticity.*

coefficient of friction (a) A numerical expression of the relationship between pressure and the resistance force of friction. (b) A quantity used to calculate the head loss in a fluid or air. The loss is a function of surface roughness, wetted perimeter, and velocity of the fluid or gas.

coefficient of heat transmission The quantity of heat transmitted from fluid to fluid per unit of time per unit of surface area through a material or arrangement of materials under a unit temperature differential between fluids. Commonly used for building materials. Syn: *heat transmission coefficient.*

coefficient of permeability The rate of flow of water under laminar flow conditions through a unit cross-sectional area of a porous medium under a unit hydraulic gradient and a standard temperature, usually 20 °C. See also: *permeability.* (ASCE, 1958)

coefficient of rigidity *modulus of rigidity.*

coefficient of thermal diffusion A thermal property of matter with the dimensions of area per unit time; it corresponds to the thermal conductivity divided by the product of density and heat capacity. (AGI, 1987)

coefficient of traction Represents the percentage of the total engine power that can be converted into forward motion by means of the friction between tire and track. (Carson, 1961)

coefficient of velocity The rate of transformation of a unit mass during a chemical reaction. (Pryor, 1963)

coefficient of viscosity (a) The shearing force per unit area required to maintain a unit difference in velocity between two parallel layers of fluid a unit distance apart. Syn: *coefficient of absolute viscosity.* (ASCE, 1958) (b) The ratio of the shear stress in a substance to the rate of shear strain. See also: *viscosity.* (AGI, 1987)

coeruleolactite A triclinic mineral, $(Ca,Cu)Al_6(PO_4)_4(OH)_8 \cdot 4H_2O$; turquoise group; occurs in white to pale-blue fibrous crusts.

coesite A monoclinic mineral, SiO_2; polymorphous with cristobalite, quartz, tridymite, and stishovite; insoluble in hydrogen fluoride.

coffee shale Drillers' term in the Appalachian basin for well cuttings of dark-colored shale chips mixed wih light-colored mud. (AGI, 1987)

coffer (a) A rectangular plank frame, used in timbering levels. Also spelled cofer. (b) A floating dock; a caisson. (Standard, 1964)

cofferdam (a) A set of temporary walls designed to keep soil and/or water from entering an excavation. (Nichols, 1976) (b) A method of shaft sinking through saturated sand or mud near the surface. A cofferdam is an enclosure, open to the air, that keeps water out of the shaft area to allow excavation to proceed. The enclosing wall is constructed by driving down strips of steel with interlocking edges or concrete piles, reinforced with steel. In general, cofferdams are used only for short lengths and where piles can be driven into an impervious deposit, so that normal pumping will keep the shaft sufficiently dry for working. See also: *drop shaft; piling*. (Nelson, 1965)

coffering A method of shaft sinking through loose, watery, or running ground. It consists in lining the shaft with a thick wall, made of brick and cement or brick and hydraulic lime with puddled clay in all cavities. Used for keeping back surface water but the method is now somewhat obsolete. (Nelson, 1965)

coffin (a) Corn. An old, open-mine working, in which the ore is cast up from platform to platform. See also: *goffan*. (Standard, 1964) (b) A heavily shielded shipping cask for spent fuel elements. Some coffins weigh as much as 75 st (68 t). (Lyman, 1964)

coffinite A naturally occurring uranium mineral, $U(SiO_4)_{1-x}(OH)_{4x}$; sp gr, 5.1; luster adamantine; color black; commonly fine-grained and mixed with organic matter and other minerals. Found in Colorado, Utah, Wyoming, and Arizona. An important ore of uranium in some mines on the Colorado plateau. (CCD, 1961)

cog (a) Straight timbers set in a large bunch. They should be firmly set and as close together as possible. Sometimes 12 to 20 are set at one location. Under conditions where single straight posts will not suffice to control the top, and yet cribs are not needed, the use of cogs may be advantageous. May also be called a battery. See also: *pigsty; cogging*. (Kentucky, 1952) (b) A crib made of notched timbers built up like a log house. A chock, cob, corncob, or crib. If the timbers are squared instead of notched, the structure is called a nog. It is ordinarily filled with waste, and rocks are put between the timbers. See also: *chock*. (Hess) (c) A rock intrusion. (Fay, 1920) (d) To consolidate ingots or shape them by hammering or rolling. (Hess) (e) An inserted tooth as in a cogwheel. Gears are often improperly referred to as cogwheels. (Crispin, 1964)

cogging (a) The operation of rolling or forging an ingot to reduce it to a bloom or billet. (CTD, 1958) (b) The propping of the roof in longwall stalls. Also spelled coggin. See also: *cog*. (Fay, 1920)

cogging mill A blooming mill, usually consisting of a two-high reversing mill with two rolls, 0.6 to 1.2 m in diameter, between which a hot ingot is reduced to blooms or slabs. (Osborne, 1956)

coggle A rounded, waterworn stone, esp. of the size suitable for paving; a cobble; also called cogglestone. Same as cobblestone.

cognate fissure One fissure of a system of fissures that originated at the same time from the same causes as other fissures in the same system. Cognate may similarly apply to fractures and joints. (Stokes, 1955)

cognate inclusion *autolith.*

cognate xenolith *autolith.*

cogwheel ore A miners' name for bournonite. Same as wheel ore. (Dana, 1959)

cohenite (a) An orthorhombic mineral, $(Fe,Ni,Co)_3C$; an accessory in iron meteorites. (b) An iron carbide phase in steel. See also: *cementite*.

cohesion Property of like mineral grains that enables them to cling together in opposition to forces tending to separate them. Cf: *adhesion*. (Hess)

cohesionless soil (a) A soil that when unconfined has little or no strength when air-dried, and that has little or no cohesion when submerged. (ASCE, 1958) (b) A frictional soil, such as sand, gravel, or clean silt. (Nelson, 1965)

cohesive soil (a) A soil that when unconfined has considerable strength when air-dried, and that has significant cohesion when submerged. (ASCE, 1958) (b) A sticky clay or clayey silt as opposed to sand. (Nelson, 1965)

coil load The total amount of heat, in British thermal units per hour, that must be removed from the air by the cooling coils.

coining (a) A closed-die squeezing operation, usually performed cold, in which all surfaces of the work are confined or restrained, resulting in a well-defined imprint of the die upon the work. (ASM, 1961) (b) A restriking operation used to sharpen or change an existing radius or profile. (ASM, 1961) (c) In powder metallurgy, the final pressing of a sintered compact to obtain a definite surface configuration. (Not to be confused with repressing or sizing.). (ASM, 1961)

coinstone bed Cement stone band. Stone suitable for coinstones, quoinstones, and cornerstones, used in building. (Arkell, 1953)

coke (a) Bituminous coal from which the volatile constituents have been driven off by heat, so that the fixed carbon and the ash are fused together. Commonly artificial, but natural coke is also known; e.g., where a dike has intersected a bituminous coalbed and has converted the bordering coal to natural coke. (Sanford, 1914) (b) A derogatory syn. for carbon; carbonado; black diamond. See also: *char*. (Long, 1960)

coke breeze The fine screenings from crushed coke or from coke as taken from the ovens, of a size varied in local practice but usually passing a 1/2-in (12.7-mm) or 3/4-in (19.0-mm) screen opening. (ASTM, 1994)

coke coal (a) N. of Eng. Carbonized or partially burnt coal found on the sides of dikes. See also: *natural coke*. (Fay, 1920) (b) Coal altered by an igneous intrusion. (Arkell, 1953)

coke drawer In the coke products industry, a laborer who removes coke from beehive ovens by hand. (DOT, 1949)

coke dust Coal dust that has been coked by the heat of an explosion and has assumed different forms under different conditions; usually found either near the origin of the explosion or in a room or wide place where the velocity of the explosion is low and there is insufficient oxygen for complete combustion of the coal dust. The volatile matter of coal dust seems to burn first and, if the coal is a coking coal, coke is formed of one kind or another, depending on the position, temperature, size of the dust, and velocity of the explosion. Also called coked dust. (Rice, 1913-18)

coke iron Iron made in a furnace using coke as fuel. (Webster 3rd, 1966)

cokeite Coal altered by an igneous intrusion. Syn: *carbonite*. See also: *natural coke*. (Tomkeieff, 1954)

coke mill A mill used in the foundry for the grinding of coke for the production of blacking. (Osborne, 1956)

coke oven A chamber of brick or other heat-resistant material in which coal is destructively distilled. Coke ovens are of two principal types: (1) beehive ovens, which were originally built round with a spherical top like an old-fashioned beehive. They had an opening in the top and various small openings for draft at the base. The ovens were developed into banks (rows) of joining cubicles; coke in long columnar pieces is characteristic and is still known as beehive coke. Tar, gas, and other byproducts are lost. (2) Byproduct ovens, which were built in rectangular form with the front and back removable, but so arranged that they may be luted to practical gastightness and all byproducts gaseous at the high temperatures may be pumped out. (Hess)

coke person In the foundry industry, a laborer who unloads, stores, and conveys coke within the foundry. (DOT, 1949)

coke tower A high tower or condenser filled with coke. Used in the manufacture of hydrogen chloride gas to give a large surface for the union of a falling spray of water with the rising hydrochloric acid gas. (Fay, 1920)

coking coal Coal that can be converted into useful coke that must be strong enough to withstand handling. There is no direct relation between the elementary composition of coal and coking quality, but generally coals with 80% to 90% carbon on a dry, ash-free basis are most satisfactory. See also: *caking coal; bituminous coal; Coalite*. (AGI, 1987)

coking stoker A mechanical stoker or device for firing a furnace that allows the coal to coke before feeding it to the grate, thus burning the fuel with little or no smoke. (Fay, 1920)

colander shovel An open wirework shovel used for taking salt crystals from an evaporating brine. (Fay, 1920)

cold bed A platform in a rolling mill on which cold bars are stored. (Fay, 1920)

cold blast Air forced into a furnace (e.g., cupola) without being previously heated. See also: *Gayley process*. (Fay, 1920)

cold-cracking Cracks in cold, or nearly cold, metal, due to excessive internal stress caused by contraction. Formation of cracks may be caused by the mold being too hard or the design of a casting being unsuitable. (Hammond, 1965)

cold-draw To draw (as metal) while cold or without the application of heat. (Webster 3rd, 1966)

cold-drawing The process of reducing the cross-sectional diameter of tubes or wire by drawing through successively smaller dies without previously heating the material, thereby increasing its tensile strength. Steel wire for prestressing is made by this process. (Hammond, 1965)

cold-extractable metal *readily extractable metal.*

cold galvanizing Application of powdered zinc, in suspension in an organic solvent, to iron articles. On evaporation of the solvent an adherent coating of zinc remains. (Pryor, 1963)

cold noser *wildcatter.*

cold-nosing Running an unhoused drill in cold weather. (Long, 1960)

cold-roll To roll (metal) without applying heat. (Webster 3rd, 1966)

cold-rolled Said of metal that has been rolled at a temperature close to atmospheric. The cold rolling of metal sheets results in a smooth surface finish. (CTD, 1958)

cold saw A saw for cutting cold metal. (Mersereau, 1947)

cold shot A portion of the surface of an ingot or casting showing premature solidification caused by a splash of metal during pouring. (ASM, 1961)

cold soldering Soldering in which two pieces are joined without heat (as by means of a copper amalgam). (Webster 3rd, 1966)

cold working Shaping of metals at ordinary temperatures; cold-drawing, rolling, stamping. Within limits, in treatment of iron, copper, aluminum, induces work hardening, thus increasing strength. If carried too far, brittleness results. Metal that is brittle when cold is termed cold-short. (Pryor, 1963)

cold zone The preheating zone of a rotary cement kiln.

colemanite A natural hydrated calcium borate, $Ca_2B_6O_{11}·5H_2O$; white or colorless; white streak; vitreous to dull luster; Mohs hardness, 4 to 4.5; sp gr, 2.26 to 2.48; found in California. One of the raw materials in the United States for boric acid, sodium borate, etc. (CCD, 1961)
Cole reagent Solution of 10 g stannous chloride, 95 mL water, 5 mL HCl, and 10 g pyrogallol. Viscose silk impregnated with this turns red to violet in solution containing gold. (Pryor, 1963)
collain (a) A subvariety of euvitrain. It consists of redeposited ulmin compounds precipitated from solution and observable microscopically. (AGI, 1987) (b) Approved by the Heerlen Congress of 1935 as applicable to vitrain in which plant structure is not visible. Adopted as collite, spelled collit in German but retaining the ain ending in English and French usage. Cf: *ulmain.* (AGI, 1987)
collapse Complete cave-in of walls of a borehole or mine workings. (Long, 1960)
collapse breccia A breccia formed by the collapse of rock overlying an opening, as by foundering of the roof of a cave or of the roof of country rock above an intrusion; e.g., a solution breccia. Syn: *founder breccia.* (AGI, 1987)
collapsing strength The load expressed in pounds or tons, which, if exceeded, results in the collapse of a structure, such as a drill tripod, derrick, or A-frame. (Long, 1960)
collar (a) In a mine shaft, the first wood frame of the shaft; sometimes used in reference to the mouth or portal of the tunnel. (BCI, 1947) (b) Supporting framework at top of shaft from which linings may be hung. (Pryor, 1958) (c) The junction of a mine shaft and the surface. (Nelson, 1965) (d) The beginning point of a shaft or drill hole, the surface. (Ballard, 1955) (e) The mouth of a mine shaft. (f) The bar, or crosspiece, in a framed timber set. (Stauffer, 1906) (g) The term applied to the timbering or concrete around the mouth or top of a shaft. (Lewis, 1964) (h) Scot. A frame to guide pump rods; the fastening of pipes in a shaft. (i) The mouth or opening of a borehole or the process of starting to drill a borehole. (Long, 1960) (j) A pipe coupling or sleeve. (Long, 1960) (k) *friction head.* (l) A sliding ring mounted on a shaft so that it does not revolve with it. Used in clutches and transmissions. (Nichols, 1976)
collar distance The distance from the top of the powder column to the collar of the blasthole, usually filled with stemming. (Dick, 2)
collared A started hole drilled sufficiently deep to confine the drill bit and prevent slippage of the bit from normal position.
collar in The act or process of beginning a borehole. (Long, 1960)
collaring (a) The process of beginning the drilling of a borehole, or the excavation of a mine shaft. (Long, 1960) (b) Eng. Timber framing for supporting pump trees in a shaft. See also: *chog.* (Fay, 1920) (c) The term used to indicate that metal passing through a rolling mill follows one of the rolls so as to encircle it. (CTD, 1958)
collaring a hole The formation of the front end of a drill hole, or the collar, which is the preliminary step in drilling to cause the drill bit to engage in the rock. (Fraenkel, 1953)
collaring bit A fishtail-, spudding-, or other-type bit used exclusively for beginning a borehole. (Long, 1960)
collars In rolling mills, the sections of larger diameter separating the grooves in rolls used for the production of rectangular sections. (CTD, 1958)
collar structure A heavy wooden frame erected at the mouth of a rectangular shaft to provide a solid support for the timber sets. A more permanent structure consists of a concrete wall extending from two to eight sets in depth. On this concrete mass is bolted the bearer timbers that support the top heavy set or collar set. The term also applies to the heavy concrete ring at the mouth of a circular concrete-lined shaft. Syn: *shaft collar.* (Nelson, 1965)
collbranite *ludwigite.*
collecting agent A reagent added to a pulp to bring about adherence between solid particles and air bubbles. (BS, 1962)
collective subsidence That condition in sedimentation in which the particles and flocs are sufficiently close together to retard the coarse fast-settling particles while the slow-settling ones are entrapped and carried down with the mass. (Mitchell, 1950)
collector A heteropolar compound containing a hydrogen-carbon group and an ionized group, chosen for ability to adsorb selectively in a froth flotation process and render adsorbing surfaces relatively hydrophobic. A promoter. (Pryor, 1965)
collier (a) Strictly speaking, a person who mines coal with a pick, though commonly applied to anyone who works in or about a colliery. Also called hewer; stallman. (b) A steam or sailing vessel carrying a cargo of coal. (c) A coal merchant or dealer in coal.
Collier Explosive; used in mines. (Bennett, 1962)
collier's lung *anthracosis.*
colliery (a) An entire coal mining plant, generally used in connection with anthracite mining, but sometimes used to designate the mine, shops, and preparation plant of a bituminous operation. (BCI, 1947) (b) A coal mine. (Pryor, 1963) (c) A ship, or ships, used in the coal trade. (Standard, 1964)
colliery bailiff Derb. The superintendent of the colliery. (Fay, 1920)
colliery consumption That part of the coal output at a colliery that is used for steam generation and other purposes connected with the working of the colliery itself. (Nelson, 1965)
colliery explosion An explosion in the workings or roadways of a colliery as a result of the ignition of combustible gases or coal dust or a mixture of both. See also: *coal-dust explosion; methane; stone-dust barrier.* (Nelson, 1965)
colliery plan Gr. Brit. A map of the mine workings, and sections of the shafts and seams being worked, which the colliery manager must keep at the pithead office in accordance with the Surveyors and Plans Regulations, 1956, of the Act. (Nelson, 1965)
colligative properties These are properties only of solutions and include vapor pressure, freezing point, boiling point, and osmotic pressure changes that occur with changes in the characteristics of the solution. Seawater does not follow the general rules of solutions, but departures are proportional. (Hy, 1965)
collimating mark *fiducial mark.*
collimation (a) Alignment axially of parts of an optical system. Collimation error is due to the line of sight of a survey instrument not coinciding with traversing gear, scales, or leveling devices. The collimation line is the line of sight, passing through the intersection of the crosshairs of the reticule. The collimation method is the height-of-instrument method of leveling whereby fore-and-aft readings are made on a leveling staff by an instrument placed intermediately so that the rise or fall between the fore station and the back station is shown by a change in the staff reading. See also: *rise and fall.* (Pryor, 1963) (b) Conversion of a divergent beam of energy or particles into a parallel beam. (ASM, 1961)
collimation line The line of sight of a surveying instrument that passes through the intersection of the cross hairs in the reticule. (Hammond, 1965)
collinite A maceral of coal within the vitrinite group, consisting of homogeneous jellified and precipitated plant material, lacking cell structure and of middle-range reflectance under normal reflected-light microscopy. See also: *vitrinite.* Cf: *ulminite.* (AGI, 1987)
collinsite A triclinic mineral, $Ca_2(Mg,Fe)(PO_4)_2·2H_2O$; fairfieldite group; forms fibrous nodules.
collision blasting Blasting in which different sections of the rocks are blasted out against each other. (Langefors, 1963)
collision waves Two waves that are propagated in opposite directions through the burned gases, and originating at the point where two explosion waves meet.
collite Another name for euvitrain. See also: *collain.* (Tomkeieff, 1954)
colloform Said of the rounded, finely banded kidneylike mineral texture formed by ultra-fine-grained rhythmic precipitation once thought to denote deposition of colloids. Cf: *botryoidal; reniform.* (AGI, 1987)
colloid A substance composed of extremely small particles, ranging from 0.2 to 0.005 µm, which when mixed with a liquid will not settle, but remain permanently suspended; the colloidal suspension thus formed has properties that are quite different from those of the simple, solid-liquid mixture or a solution.
colloidal clay A clay, such as bentonite, which, when mixed with water, forms a gelatinous-like liquid. (Long, 1960)
colloidal fuel A mixture of finely pulverized coal and fuel oil, which remains homogeneous in storage. It has a high calorific value and is used in oil-fired boilers as a substitute for fuel oil alone. (Nelson, 1965)
colloidal mud A drilling mud in which the gelatinous constituents, such as bentonite, will remain in suspension in water for a long time. (Long, 1960)
colloidal particles Particles so small that their surface activity has an appreciable influence on the properties of its aggregate. (ASCE, 1958)
colloidal sulfur Amorphous sulfur in a finely divided condition. Prepared by the action of dilute sulfuric acid on sodium thiosulfate or by the reaction of hydrogen sulfide and sulfurous acid. Also prepared by mixing equivalent solutions of hydrogen sulfide and sulfur dioxide. Forms a clear yellow solution containing very minute suspended particles of sulfur; the addition of alum immediately precipitates the sulfur. Also called milk of sulfur. (Cooper, 1963)
colloid mill Grinding appliance such as two disks set close and rotating rapidly in opposite directions, so as to shear or emulsify material passed between them. (Pryor, 1958)
colloid minerals Minerals deposited as gradually hardening gelatinous or flocculent masses instead of assuming crystalline form; may apply to some deposits of malachite, hematite, and psilomelane. Cf: *botryoidal; reniform.*
collophane Generic designation for massive, amorphous, cryptocrystalline to fine-grained apatite or phosphate that constitutes the bulk of phosphate rock and fossil bone; not a true mineral species; analogous to the terms limonite and bauxite. Syn: *collophanite.*
collophanite *collophane.*

Collum washer Mineral jig with a quick down stroke and retarded return of its plunger. (Pryor, 1963)

Colmol mining machine A machine in which the coal is hewed from the solid by 10 rotating chipping heads in 2 rows of 5, each with the lower row in advance of the upper. Each head consists of a bit supplemented by widely spaced teeth, each tooth being stepped back to the outside of the head. The circular kerfs made by the heads overlap, and as the machine moves forward, the effect is to break the coal ahead of the teeth into the free spaces, thereby minimizing the production of fines. (Mason, 1951)

Cologne umber An earthy black or brown lignite used as a pigment. Etymol. source near Cologne, Germany. (Tomkeieff, 1954)

colombotantalite A noncommittal term for members of the columbite-tantalite series.

color (a) A trace of metallic gold found in a prospector's pan after a sample of soil or of gravel has been panned out. Prospectors say, e.g., the dirt gave so many colors to the panful. (b) The shade or tint of the soil or rock that indicates ores; e.g., gossan coloration. (c) Color is an important property used in megascopic and microscopic determination of minerals. It depends on the selective absorption or reflection of certain wavelengths of light by the mineral during transmission or reflection. The color of metallic (or metal-bearing) minerals is a fairly constant property, whereas that of nonmetallic minerals is generally less so owing to the pigmentation effect of minor impurities. The color of a massive mineral is commonly different from that of its powder or streak. (d) The Munsell notation has come into wide use for the designation of colors of rocks and soils. In this system, a color is specified by the three variables of hue (dominant spectral color), value (brilliance), and chroma (saturation or purity), and written in the order and form: hue-value-chroma.

coloradoite An isometric mineral, HgTe; sphalerite group.

Colorado lapis lazuli Dark blue lapis lazuli (lazurite) from the Sawatch Range, CO.

Colorado ruby An incorrect name for the fiery-red garnet (pyrope) crystals obtained from Colorado. (CMD, 1948)

Colorado topaz True topaz of a brownish-yellow color obtained in Colorado, but quartz similarly colored is sometimes sold under the same name. (CTD, 1958)

colored slates Cambrian and Ordovician slates quarried in the vicinity of Granville, Washington County, NY. Colors include red, purple, green, and black. The slates are much used in decorative flooring.

color grade The grade or classification into which a gem is placed by examination of its color in comparison to the color of other gems of the same variety.

colorimeter An instrument for measuring and comparing the intensity of color of a compound for quantitative chemical analysis, usually based on the relationship between concentration of a chemical solution and the amount of absorption of certain characteristic colors of light. (AGI, 1987)

colorimetric determination An analytical procedure based on measurement, or comparison with standards, or color naturally present in samples or developed therein by the addition of reagents.

color index In petrology, esp. in the classification of igneous rocks, a number that represents the percent, by volume, of dark-colored (i.e., mafic) minerals in a rock. According to this index, rocks may be divided into leucocratic (color index, 0 to 30), mesocratic (color index, 30 to 60), and melanocratic (color index, 60 to 100). Syn: *color ratio.* (AGI, 1987)

colorless Devoid of any color, as is pure water, a pane of ordinary window glass, or a fine diamond; therefore distinctly different from white, as in milk or white jade. As only transparent objects can be colorless, and no opaque object can be colorless, such terms as white sapphire and white topaz are misnomers. Rock crystal is a colorless variety of quartz; milky quartz is a white variety.

color ratio *color index.*

colors (a) The specks of gold seen after the successful operation of a gold pan, when finely crushed ore has been panned to remove the bulk of light minerals. The residual heavy fraction is then scanned for visual evidence of gold by the prospector. (Pryor, 1963) (b) In optical mineralogy, the colors of doubly refracting substances as seen in doubly polarized light (crossed polars). See also: *birefringence.*

colrake A shovel used to stir lead ores during washing. (Fay, 1920)

columbite (a) The mineral group ferrocolumbite, magnocolumbite, and manganocolumbite. (b) Standing alone it generally refers to ferrocolumbite, an orthorhombic mineral, $FeNb_2O_6$, in granites and pegmatites; an ore of niobium. Syn: *niobite; dianite; greenlandite.* Cf: *magnocolumbite.*

columbium *niobium.*

column (a) A round pillar set vertically or horizontally in a heading to support a machine drill. (b) The rising main or length of pipe conveying water from a mine to the surface. (c) *motive column.* (d) A solid core cut from a borehole. (e) The drill-circulation liquid confined within a borehole. (f) In borehole casing, a row of casing sections screwed together and forming a whole.

columnar (a) Composed of columnlike individuals. (Schieferdecker, 1959) (b) A mineral with a form obscurely resembling prisms, e.g., hornblende. See also: *prismatic.* (c) In columns produced by shrinkage joints, as in columnar basalt.

columnar charge (a) A charge of explosives in a blast hole in the form of a long continuous unbroken column. (b) A continuous charge in a quarry borehole. Cf: *deck charge.* (BS, 1964)

columnar crystals Elongated crystals that grew at right angles to a surface. (CMD, 1948)

columnar jointing Parallel, prismatic columns, polygonal in cross section, in basaltic flows and sometimes in other extrusive and intrusive rocks. It is formed as the result of contraction during cooling. Syn: *columnar structure.* (AGI, 1987)

columnar section A geologic illustration that shows in a graphic manner, and by use of conventional symbols for different rock types, the successive rock units that occur throughout a given area or at a specific locality. It may be accompanied by a very brief description of lithology and by appropriate brief notations indicating the thickness, age, and classification of the rocks. See also: *geologic column.* (Stokes, 1955)

columnar structure (a) A mineral fabric consisting of slender crystals of prismatic cross section, as in some amphiboles. (b) *columnar jointing.* (c) Columns, 9 to 14 cm in diameter and 1 to 1.4 m in length, found in some calcareous shales or argillaceous limestones; oval to polygonal in section. Columns are perpendicular to bedding. Possibly a desiccation structure. (Pettijohn, 1964)

column flotation (a) A pneumatic flotation process with a countercurrent flow of rising bubbles against settling ore within the flotation cell. Typically, the cell height is much greater than the cross section of the cell. The feed slurry is input above the midpoint of the column and water sprays are used at the top of the froth column to remove entrained hydrophilic particles from the froth. (Kelly, 1982) (b) Flotation carried out in a column machine utilizing countercurrent flow of air bubbles from the bottom and solid reagent-conditioned material from the top, such that tailings are withdrawn at the column bottom and the concentrate is collected over the column lip. There is no mechanical agitation. (SME, 1992)

column height The length of each portion of a blast hole filled with explosive materials. (Atlas, 1987)

column leaching Simulation of in-situ leaching through the use of a long narrow column in which ore sample and solution are in contact for measuring the effects of typical variables encountered in actual in-situ leach mining. (SME, 1992)

column load A single continuous charge. (Carson, 1961)

column of mud *mud column.*

column of ore A deposit of ore in a lode having a small lateral, but considerable vertical extent. An older term for ore shoot.

column pipe The large cast-iron (or wooden) pipe through which the water is conveyed from the mine pumps to the surface. Syn: *mounting pipe; rising main.*

colusite (a) A variety of tetrahedrite containing 3.21% tin, from Japan. (Spencer, 1952) (b) An isometric mineral, $Cu_{26}V_2(As,Sn,Sb)_6S_{32}$; in bronze-colored tetrahedra, from Butte, MT.

comagmatic Said of igneous rocks that have a common set of chemical and mineralogic features, and thus are regarded as having been derived from a common parent magma. See also: *consanguinity.* Syn: *consanguineous.* (AGI, 1987)

comagmatic region An area in which the igneous rocks are of the same general geologic age, have certain distinguishing characteristics in common, and are regarded as comagmatic. Syn: *petrographic province.comb* In a fissure that has been filled by successive deposits of minerals on the walls, the place where two sets of layers thus deposited approach most nearly or meet, closing the fissure and exhibiting either a drusy central cavity or an interlocking of crystals. See also: *comb texture.*

comb In a fissure that has been filled by successive deposits of minerals on the walls, the place where two sets of layers thus deposited approach most nearly or meet, closing the fissure and exhibiting either a drusy central cavity or an interlocking of crystals. See also: *comb texture.*

comb dung *comedown.*

combed structure In its simplest form this structure consists of a fissure lined with crystals on each side, having their bases on the walls and their apexes directed toward the center. In some cases the fissure is thus altogether filled up with two sets of crystals meeting in the center.

combeite A trigonal mineral, $Na_2Ca_2Si_3O_9$; in nephelinite at Kivu, Republic of the Congo.

combination drill A drill equipped for cable-tool and/or diamond-drilling operations, or for a cable-tool and/or rotary drilling operations. Syn: *combination rig.* (Long, 1960)

combination electric locomotive A mine locomotive that can operate as a trolley locomotive or as a battery locomotive. While operating on a battery, it can be used, under certain conditions, at the coal face. Also it may be used on the main haulage trolley system where, due to higher voltage, higher speeds are possible. (Nelson, 1965)

combination longwall *longwall.*

combination process Method for extracting alumina from high-silica bauxites, in which the bauxite is first subjected to a Bayer process caustic leach. The resulting red mud, containing sodium aluminum silicate, is sintered with limestone plus soda ash and then leached with water to recover alumina and soda.

combination rig (a) A rig comprising a complete cable-tool outfit and a complete rotary outfit. (Porter, 1930) (b) *combination drill.*

combination sampler A universal-type soil-sampling device in which some of the constructional features of two or more special-use samplers are combined. (Long, 1960)

combination stoping *combined overhand and underhand stoping.*

combined carbon The part of the total carbon in steel or cast iron that is present as other than free carbon. (ASM, 1961)

combined moisture Moisture in coal that cannot be removed by ordinary drying. Cf: *free moisture.* (Cooper, 1963)

combined overhand and underhand stoping This term signifies the workings of a block simultaneously from the bottom to its top and from the top to the bottom. The modifications are distinguished by the support used, as open stopes, stull-supported stopes, or pillar-supported stopes. Also known as combined stopes; combination stoping; overhand stoping and milling system. Syn: *back and underhand stoping milling system; combination stoping.*

combined shrinkage stoping and caving In this method, the orebody is worked from the top down in successive layers of much greater thickness than in top slicing. The mass of ore is weakened by a series of shrinkage stopes, which are extended up between the ribs, pillars, or blocks, which are subsequently caved. The intervening blocks are under cut and caved as in block caving. The caver follows the caved ore. Also called overhand stoping with shrinkage and simultaneous caving.

combined side and longwall stoping *overhand stoping.*

combined stresses Any state of stress that cannot be represented by a single component of stress; i.e., one that is more complicated than simple tension, compression, or shear. (ASM, 1961)

combined top slicing and shrinkage stoping In this method, the orebody is worked from the top down in successive slices. In the working of each slice, the unit is worked as a shrinkage stope. The broken ore serves to give lateral support to the sides of the unit and also serves as a working platform from which the back is reached. After working a unit, the cover is caved. No timber mat is used. Also known as the Kimberley method.

combined twinning A rare type of twinning in quartz in which there appears to be a 180° rotation around *c* with reflection over (1120) or over (0001). The crystal axes are parallel, but the polarity of the *a* axis is not reversed in the twinned parts.

comb texture A texture in which individual crystals have their long axis perpendicular to the walls of a vein. See also: *comb.*

combustibility An assessment of the speed of combustion of a coal under specified conditions. (BS, 1960)

combustible Capable of undergoing combustion or of burning. Used esp. for materials that catch fire and burn when subjected to fire. Cf: *flammable.* (Webster 3rd, 1966)

combustible gases *firedamp.*

combustible gases cap A small cap that forms over the flame of a safety lamp when sufficient combustible gases (methane) are present. (CTD, 1958)

combustible gases drainage The collection of combustible gases from coal measures strata, generally into pipes, with or without the use of suction. See also: *methane drainage.* (BS, 1963)

combustible gases fringe The zone of contact between the goaf gases and the ventilation air current at the face. (Roberts, 1960)

combustible gases layer A sheetlike accumulation of combustible gases under the roof of a mine roadway where the ventilation is too sluggish to dilute and remove the gas. Although the term is new, the hazard existed since the earliest days of coal mining. A combustible gases layer may be specified as one in which the gas is 5% or over and of a length greater than the width of the road in which it occurs. *pocket of gas.* See also: *stratification of methane.* (Nelson, 1965)

combustible shale *tasmanite.*

combustion The action or operation of burning; the continuous combination of a substance with certain elements, such as oxygen or chlorine; e.g., accompanied by the generation of light and heat. See also: *ignition temperature.* (Standard, 1964)

combustion arch A flat or curved refractory roof over a furnace to promote combustion by reflection of heat. (AISI, 1949)

combustion engineer An engineer with practical training and knowledge of all kinds of fuels and their combustion characteristics. In general, the engineer lacks the technical qualifications of the fuel technologist. (Nelson, 1965)

combustion method A method for the quantitative determination of certain elements (such as carbon, hydrogen, and nitrogen) in organic compounds by combustion. (Webster 3rd, 1966)

comedown Softish stone occurring in the roof of a coal seam; it easily falls when coal is removed. Syn: *comb dung.* (Arkell, 1953)

comendite A sodic rhyolite containing alkalic amphibole and/or pyroxene.

come out To withdraw or hoist the drill string or tools from a borehole. (Long, 1960)

come water The constant or regular flow of water in a mine proceeding from old workings or from water-bearing rocks.

comfort air conditioning Air conditioning that controls the atmosphere that human beings breathe. (Hartman, 1982)

coming up to grass Eng. Common terms used by miners for the word basset, or outcrop. Also coming up today.

Comleyan Lower Cambrian. (AGI, 1987)

commercial deposit A deposit of oil, gas, or other minerals in sufficient quantity for production in paying quantities. (Williams, 1964)

commercial explosives Explosives designed, produced, and used for commercial or industrial applications rather than for military purposes. (Meyer, 1981)

commercial granite A general term for a decorative building stone that is hard and crystalline. It may be a granite, gneiss, syenite, monzonite, granodiorite, anorthosite, or larvikite. See also: *black granite.* (AGI, 1987)

commercially disposable coal A statistical term referring to saleable coal, less colliery consumption and coal supplied to employees. (BS, 1960)

commercial marble A crystalline rock composed predominantly of one or more of the following minerals: calcite, dolomite, or serpentine, and capable of being polished.

commercial mine A mine operated to supply purchasers in general as contrasted with a captive mine. (Zern, 1928)

commercial ore Can. Mineralized material currently profitable at prevailing prices. (Hoffman, 1958)

commercial quantity A quantity of oil, gas, or other minerals sufficient for production in paying quantities. (Williams, 1964)

commercial quarry (a) Term that includes quarries for aggregate and quarries for the production of limestone for industrial and agricultural purposes. (Streefkerk, 1952) (b) Not owned or controlled by consumer. Contrasted with a captive quarry.

commercial sampling of coal Procedures intended to produce an accuracy such that if a large number of samples are taken from a single lot of coal, 95 out of 100 test results will be within ± 10% of the average of these samples. (Mitchell, 1950)

comminution (a) The gradual diminution of a substance to a fine powder or dust by crushing, grinding, or rubbing; specif., the reduction of a rock to progressively smaller particles by weathering, erosion, or tectonic movements. (AGI, 1987) (b) The breaking, crushing, or grinding by mechanical means of stone, coal, or ore, for direct use or further processing. Syn: *pulverization; trituration.* (AGI, 1987)

common banded coal *banded coal.*

common feldspar *orthoclase.*

common ion effect Change in concentration of an ion in a saturated solution through addition of another electrolyte that yields an ion in common with the solid substance present in excess. The ion product remains constant, but with the increase of concentration of one ion that of the other diminishes correspondingly. Since the solution is already saturated, precipitation occurs, the effect being a reversal of the process of ionization. (Pryor, 1963)

common lead Lead (Pb) having four isotopes (mass numbers 204, 206, 207, and 208) in the proportions generally obtained by analyzing lead from rocks and lead minerals that are associated with little or no radioactive material; commonly considered to be the lead present at the time of the Earth's formation, as distinguished from lead produced later by radioactive decay. (AGI, 1987)

common mica *muscovite.*

common opal Opal without play of color. Most varieties are of no gemological interest or importance; others because of their color or markings are set in jewelry. Cf: *precious opal.*

common pyrite *pyrite.*

common salt A colorless or white crystalline compound consisting of sodium chloride NaCl, occurring abundantly in nature as a solid mineral (halite), or in solution (constituting about 2.6% of seawater), or as a sedimentary deposit (such as in salt domes and beds or as a crust around the margin of a salt lake). See also: *halite; rock salt; salt.*

commutated current Electric current of constant strength of which the direction of flow is reversed at constant intervals of time. (Schieferdecker, 1959)

compacted yards Measurement of soil or rock after it has been placed and compacted in a fill. (Nichols, 1976)

compaction curve The curve showing the relationship between the density (dry unit weight) and the water content of a soil for a given compactive effort. Syn: *moisture-density curve*. (AGI, 1987)
compaction equipment Machines, such as rollers, to expel air from a soil mass and so achieve a high density. Smooth-wheel rollers are best for gravels, sands, and gravels-and-clay soils with reasonably high moisture contents. Pneumatic-tired rollers are best for clays with reasonably high moisture content, and sheepsfoot rollers are the best for clays with low moisture content. See also: *superficial compaction*. (Nelson, 1965)
compaction test A laboratory compacting procedure to determine the optimum water content at which a soil can be compacted so as to yield the maximum density (dry unit weight). The method involves placing (in a specified manner) a soil sample at a known water content in a mold of given dimensions, subjecting it to a compactive effort of controlled magnitude, and determining the resulting unit weight (ASCE, 1958, term 74). The procedure is repeated for various water contents sufficient to establish a relation between water content and unit weight. The maximum dry density for a given compactive effort will usually produce a sample whose saturated strength is near maximum. Syn: *moisture-density test*. (AGI, 1987)
compact rock A rock so closely grained that no component particles or crystals can be recognized by the eye. (Nelson, 1965)
company account Drilling done by a company on its property using its own equipment operated by personnel working for the company. (Long, 1960)
comparator (a) In photographic mapping, a device for measuring accurately the two rectangular coordinates of the image of a point on a photograph. (Seelye, 1951) (b) An apparatus facilitating comparison of test material with known standard, or with other substances. A comparator microscope has a duplicate optical system, so that the observer sees two fields simultaneously (one with each eye). The Lovibond comparator has colored disks that can be matched against colored liquids to give approximate pH value, etc., using the same principle as with a set of pH color tubes in a more permanent and compact style. (Pryor, 1963)
comparator tintometer Instrument in which color of test solution is compared with that of reference cell or tinted glass slide. Also called colorimeter. (Pryor, 1963)
comparison prism A small, right-angled prism placed in a front of a portion of the slit of a spectroscope or a spectrograph for the purpose of reflecting light from a second source of light into the collimator, so that two spectra may be viewed simultaneously. (CTD, 1958)
compartment (a) A space or division in a shaft formed by cross buntons. The main compartments in a winding shaft are two for cages or skips. See also: *rectangular shaft*. (Nelson, 1965) (b) One section or unit in a coal- or mineral-treatment plant. (Nelson, 1965)
compass (a) An instrument or device for indicating horizontal reference directions relative to the Earth by means of a magnetic needle or group of needles; specif. magnetic compass. Also, a nonmagnetic device that serves the same purpose; e.g., a gyrocompass. (AGI, 1987) (b) A simple instrument for describing circles, transferring measurements, or subdividing distances; usually consisting of two pointed, hinged legs (one of which generally having a pen or pencil point) joined at the top by a pivot. (AGI, 1987) (c) A Maas or other compass device formerly used in borehole-survey work. (Long, 1960)
compass deflection (a) The difference, expressed in degrees, between the direction a magnetic compass needle points and true or astronomical north. This is termed magnetic declination. (Long, 1960) (b) Differences, expressed in degrees, between magnetic north directions and the direction a magnetic compass points, owing to local magnetic interferences. This is termed magnetic deviation. (Long, 1960)
compass direction Direction as indicated by a compass without any allowances for compass error. The direction indicated by a magnetic compass may differ by a considerable amount from the true direction referred to a meridian of the Earth. (Hunt, 1965)
compass points The four principal points of the compass—north, east, south, and west—are called the cardinal points. Midway between the cardinal points are the intercardinal points—northeast, southeast, southwest, and northwest. Midway between each cardinal and intercardinal point is a point with a name formed by combining that of the cardinal and intercardinal point, the former being placed first, as north-northeast, east-northeast, and so forth. Midway between the points already indicated are points bearing the name of the nearest cardinal or intercardinal point followed by the word "by" and the name of the cardinal point in the direction in which it lies, as north by east, northeast by north, and so forth. In all, there are 32 points separated by intervals of 11-1/4°. Each of these intervals is subdivided into quarter points. (Hunt, 1965)
compensating error Random error equally likely to be plus or minus, and if of small dimensions, reasonably likely to be compensated by further errors. In contrast, systematic or biased errors all fall on the same side of correct measurement and may therefore accumulate and produce serious discrepancies. (Pryor, 1963)
compensating rope Balance weight ropes having direct connection with hoisting ropes. (Hammond, 1965)
compensation method A procedure for determining the voltage difference between two points in the ground by balancing against a voltage that is adjusted in phase and amplitude to effect the compensation. See also: *compensator*. (AGI, 1987)
compensator An instrument to determine the voltage difference between two points in the ground by the compensation method. Syn: *accessory plate*. (AGI, 1987)
competence The ability of a current of water or wind to transport detritus, in terms of particle size rather than amount, measured as the diameter of the largest particle transported. It depends on velocity: a small, but swift stream, e.g., may have greater competence than a larger but slower-moving stream. Adj: competent. (AGI, 1987)
competent (a) Strata or rock structure combining sufficient firmness and flexibility to transmit pressure and, by flexure under thrust, to lift a superincumbent load. (Standard, 1964) (b) Streams able to transport debris of a given size. (Standard, 1964) (c) Rock formations in which no artificial support is needed to maintain a cave-free borehole. (Long, 1960) (d) Rock capable of withstanding an applied load under given conditions without falling or collapsing. See also: *incompetent*. (Long, 1960)
competent bed (a) A rock formation that, because of massiveness or inherent strength, is able to lift not only its own weight but also that of the overlying rock. (AGI, 1987) (b) A bed that has a physical characteristic such that it responds to tectonic forces by folding and faulting, rather than by crushing and flowing. A competent bed is relatively strong, an incompetent bed, relatively weak. See also: *incompetent bed*. (BS, 1964)
competent rock (a) Rock that, because of its physical and geological characteristics, is capable of sustaining openings without structural support, except pillars and walls left during mining. (b) Rock formations in which no artificial support is needed to maintain a cave-free borehole. (Long, 1960) (c) Rock capable of withstanding an applied load under given conditions without falling or collapsing. (Long, 1960)
complement *rock fracture*.
complementary dikes Associated dikes (or other minor intrusions) composed of different, but related rocks, regarded respectively as leucocratic and melanocratic differentiation products from a common magma; e.g., aplite and lamprophyre; bostonite and camptonite. (Holmes, 1928)
complementary forms In crystallography, two forms combined geometrically to produce a form having higher symmetry; e.g., two equally developed rhombohedra of quartz resembling a hexagonal bipyramid.
complete combustion Occurs when the products of combustion leaving the furnace or appliance do not contain any gaseous combustible matter. (Nelson, 1965)
complex (a) A large-scale field association or assemblage of different rocks of any age or origin, having structural relations so intricately involved or otherwise complicated that the rocks cannot be readily differentiated in mapping, e.g., a volcanic complex. See also: *igneous complex; basement complex*. (AGI, 1987) (b) A unit that consists of a mixture of rocks of two or more genetic classes, i.e., igneous, sedimentary, or metamorphic, with or without highly complicated structure; example: Franciscan Complex. (AGI, 1987) (c) Said of an ore that carries several metals difficult to extract. (von Bernewitz, 1931) (d) An assemblage of rocks of any age or origin that has been folded together, intricately mixed, involved, or otherwise complicated. (Stokes, 1955)
complex crystals Those having many crystal forms and faces.
complex fold A fold that is cross-folded; i.e., a fold, the axial line of which is folded.
complex ore (a) An ore containing two or more metals, as lead-zinc ore. Many complex ores are difficult or costly to treat, e.g., gold ore with arsenic or antimony minerals, or ore composed almost entirely of several sulfide minerals. (b) An ore containing several metals. (Bateman, 1951) (c) Ores named for two or more valuable metals such as lead-zinc ores, gold-silver ores, etc. (Newton, 1959) (d) This term has no precise meaning. It generally signifies an ore that is difficult or costly to treat because of the presence of unusual minerals, e.g., a gold ore with aresenic and antimony minerals, or an ore containing two or more metals, or ore composed almost wholly of several sulfide minerals. (Nelson, 1965)
complex pegmatite A pegmatite body characterized by pneumatolytic-hydrothermal replacement and rare minerals. (Schieferdecker, 1959)
complicated pneumoconiosis A condition superimposed on simple pneumoconiosis by the effect of tuberculosis lesions. (Nelson, 1965)
component of coal Layers or bands that are petrographic entities, recognizable visually as bands or layers of coal that have distinctive

physical appearance and characteristic microstructural features from coal to coal. (IHCP, 1963)

composite dike A dike formed by two or more intrusions of different compositions into the same fissure.

composite explosives Explosives that contain a mechanical mixture of substances that consume and give off oxygen with one or several simple explosives. They can be regarded as mixed explosives with an addition of one or more simple explosives as sensitizers, which makes for easier initiation of the mixture and gives greater assurance of complete transformation. (Fraenkel, 1953)

composite fold *compound fold.*

composite gneiss (a) A banded rock resulting from intimate penetration of magma (usually granite) into adjacent rocks. See also: *injection gneiss; migmatite.* (b) Gneiss that is constituted of materials of at least two different phases. Cf: *venite; veined gneiss.* (AGI, 1987)

composite intrusion Any igneous intrusion that is composed of two or more injections of different chemical and mineralogical composition. Cf: *multiple intrusion.* (AGI, 1987)

composite map A map on which several levels of a mine are shown on a single sheet. Horizontal projection of data from different elevations. (McKinstry, 1948)

composite materials Structural materials of metal alloys or plastics with built-in strengthening agents that may be in the form of filaments, foils, or flakes of a strong material. (Hunt, 1965)

composite sampling scheme One in which different parts, or stages, of the sample are reached by differing methods. (Pryor, 1963)

composite sill A sill composed of two or more intrusions having different chemical and mineralogical compositions.

composite stone *assembled stone.*

composite stones A comprehensive term that includes doublets, triplets, etc., in which a stone consists of two or more parts either of the same or of different materials cemented or otherwise joined together. (Anderson, 1964)

composite vein A large fracture zone, up to many tens of feet in width, consisting of parallel ore-filled fissures and converging diagonals, the walls and the intervening country rock of which have undergone some replacement.

Composition B A mixture of RDX and TNT that, when cast, has a density of 1.65 g/cm^3 and a velocity of 25,000 ft/s (7.6 km/s). It is useful as a primer for blasting agents. (Meyer, 1981)

composition of forces If two or more forces acting on a body can be replaced by a single force the forces are said to have been compounded. This is known as composition of forces. (Morris, 1958)

composition surface A planar or irregular surface by which parts of a twin crystal are united, not necessarily parallel to a crystal face.

compound compression In compound compression, the work of compression is divided into two or more stages or cylinders. In two-stage compression, air is compressed in the first or low-pressure cylinder to a certain point, then forced into an intercooler where it is cooled to approx. its original temperature, then passes into the second or high-pressure cylinder, in which it is compressed to the final or delivery pressure. The ratio of compression in each cylinder of a two-stage compressor is equal to the square root of the overall ratio of compression, i.e., the square root of the final absolute pressure divided by the absolute atmospheric pressure. In three-stage work, the ratio of compression in each cylinder is the cube root of the overall ratio of compression. Also called stage compression. (Lewis, 1964)

compound cradle An apparatus comprising three tiers of blanket tables, a shaking table, and a mercury riffle for catching gold. (Fay, 1920)

compound dredger A type of dredger combining the suction or suction cutter apparatus with a bucket ladder. (CTD, 1958)

compound fault A series of closely spaced parallel or nearly parallel faults. (CTD, 1958)

compound fold A fold upon which minor folds with similar axis have developed. Syn: *composite fold.* (AGI, 1987)

compound lode *compound vein.*

compound shaft A shaft in which the upper stage is often a vertical shaft, while the lower stage, or stages, may be inclined and driven in the deposit. In this type of shaft, underground winding engines are installed to deal with the lower stages, with transfer points and ore bins at the junction of two stages. (Sinclair, 1959)

compound twins In crystallography, individual crystals of one group united according to different laws. (Standard, 1964)

compound vein (a) A vein or lode consisting of a number of parallel fissures united by cross fissures, usually diagonally. (b) A vein composed of several minerals. Syn: *compound lode.*

compound ventilation (a) An arrangement of a number of major ventilation systems serving various large working areas and served by more than two shafts and their associated fans, but integrated to form one ventilation system. Usually adopted in large combined mines. See also: *radial ventilation.* (BS, 1963) (b) Ventilation by means of a number of splits, which is now normal practice. See also: *ventilation.*

compressed air Air compressed in volume and transmitted through pipes for use as motive power for underground machines. Compressed air is costly to transmit long distances, but has certain advantages, namely, it cools the air at the working face and is relatively safe in gassy mines. See also: *air-conditioning process.* (Nelson, 1965)

compressed-air blasting A method originated in the United States for breaking down coal by compressed air. Air at a pressure of 10,000 to 12,000 psi (69 to 83 MPa) is conveyed in a steel pipe to a tube- or shell-inserted shothole. The air is admitted by opening a shooting valve and is released in the hole by the rupture of a shear pin or disk. The sudden expansion of the air in the confined hole breaks down the coal. Syn: *Armstrong air breaker.* See also: *Airdox.* (Nelson, 1965)

compressed-air-driven lamps These lamps are self-contained units and comprise a strong alloy casing within which are a compressed-air turbine and a small alternating-current generator with stationary windings and revolving field magnets. The air enters the casing at one side, passes through a filter and then through a reducing valve that maintains a constant pressure of 40 psi (276 kPa) on the turbine blades. The air escaping from the turbine is used to scavenge the inside of the lamp and remove any combustible gases that might have entered when the lamp was not in use. It is finally discharged through a series of holes of such a size that the pressure inside the lamp casing is 2 to 3 psi (14 to 21 kPa) above atmospheric. Should this pressure be lost due to the lamp glass being broken, the light is extinguished automatically by a spring-loaded diaphragm, which short-circuits the generator unless held open by the excess pressure. Also called air turbolamp. (Mason, 1951)

compressed-air locomotive A mine locomotive driven by compressed air. It is very safe and is much used in gassy mines in Europe. The air is brought down by pipeline from the surface to a charging station near the pit bottom. See also: *locomotive haulage.* (Nelson, 1965)

compressed-air turbines Turbines used for driving coal cutters, belt conveyors, and similar duties. They are not so efficient in their use of the air as piston engines, but possess the merits of extreme simplicity and robustness, and therefore are preferred for coal face use. (Mason, 1951)

compressed pellets Blasting powders manufactured in cartridge form for use in small diameter shotholes. These pellets are particularly useful for horizontal shotholes. (McAdam, 1958)

compression A system of forces or stresses that tends to decrease the volume or to shorten a substance, or the change of volume produced by such a system of forces. (AGI, 1987)

compressional wave (a) A traveling disturbance in an elastic medium characterized by volume changes (and hence density changes) and by particle motion in the direction of travel of the wave. (AGI, 1987) (b) A longitudinal wave (as a sound wave) propagated by the elastic compression of the medium. Syn: *irrotational wave; pressure wave; P wave.* (Webster 3rd, 1966)

compression ratio The ratio of the volume of space above a piston at the bottom of its stroke to the volume above the piston at the top of its stroke. (Nichols, 1976)

compression subsidence That condition in sedimentation in which the flocs or particles are conceived to be in close contact, further subsidence occurring as a direct effect of compression resulting in the elimination of water from the flocs and interstitial spaces. The settling velocity decreases with time of settling. (Mitchell, 1950)

compression zone The surface area affected by compressive strain. Cf: *neutral zone; tension zone.* (Nelson, 1965)

compressive strength (a) The maximum compressive stress that can be applied to a material, such as a rock, under given conditions, before failure occurs. (AGI, 1987) (b) The load per unit area at which an unconfined prismatic or cylindrical specimen of soil will fail in a simple compression test. Syn: *unconfined compressive strength.* (ASCE, 1958)

compressive stress A stress that tends to push together the material on sides of a real or imaginary plane. Cf: *tensile stress.* (Billings, 1954)

compressor (a) A machine, steam or electrically driven, for compressing air for power purposes. Small air compressors may be compound steam and double-stage air. Large compressors may be triple-expansion steam and three-stage air and are always used with condensers. (Nelson, 1965) (b) Any kind of reciprocating, rotary, or centrifugal pump for raising the pressure of a gas. (CTD, 1958) (c) A machine that compresses air. (Nichols, 1976)

comptonite An opaque variety of thompsonite from the Lake Superior region.

computer (a) An automatic electronic device capable of accepting information, applying prescribed processes to it, and supplying the results of these processes. The term is generally used for any type of computer. See also: *analog computer.* (AGI, 1987) (b) In seismic prospecting, one who with one or two assistants, carries on the routine work of transforming the "wiggly lines" on the reflection records into the form in which they are finally used. Where corrected record sections are prepared, he or she must compute the corrections and must assemble the other information to be fed into the playback so that it will turn out properly corrected records. In addition to handling corrections, the computer must mark the records, read and plot times, and otherwise maintain the flow of data. Such individuals are not used in processing modern digital seismic data. (Dobrin, 1960)

computer-assisted mining The process of controlling single or multiple mining machines in which sensors and computers are used to replace or enhance manual control of all or portions of the formerly manually controlled machine operations.

comstockite A mineral, $(Mg,Cu,Zn)SO_4 \cdot 5H_2O$, containing 5.60% ZnO, 9.40% MgO, 9.00% CuO, and 39.07% H_2O; from the Comstock Lode, NV. Syn: *zinc-magnesia chalcanthite.* (Spencer, 1952)

concave bit A tungsten carbide drill bit for percussive boring. The cutting edge is concave, while in the conventional type the edge is convex. The new bit remains sharper for a longer period before regrinding becomes necessary and gives a higher penetration speed. Also called saddleback tip. See also: *plug bit.* (Nelson, 1965)

concave crown *concave bit.*

concealed coalfield A coalfield that is totally buried beneath newer deposits, usually Permian and Trias strata, which repose unconformably on the coal measures in the basin. A good example of a concealed coalfield is that of Kent, in southeast England. See also: *coal basin; exposed coalfield.* (Nelson, 1965)

concentrate The clean product recovered in froth flotation.

concentrated charge (a) The heavy explosive charge loaded into the enlarged chamber at the bottom of a quarry blasthole. See also: *chambering.* (Nelson, 1965) (b) Means that the height of the charge is small compared with the burden that can be given quantitatively. (Langefors, 1963)

concentrating plant *concentrator.*

concentrating table A device consisting of a riffled deck, usually inclined in two directions to the horizontal, to which a differential reciprocating motion in a substantially horizontal direction is imparted; the material to be separated is fed in a stream of water, the heavy particles collect between the riffles and are there conveyed in the direction of the reciprocating motion while the lighter particles are borne by the current of water over the riffles, to be discharged laterally from the table. (BS, 1962)

concentration (a) The ratio of the dry weight of sediment to the weight of water sediment mixture of which it is part. Sediment concentration is commonly expressed in parts per million (ppm). (b) Separation and accumulation of economic minerals from gangue. See also: *ore dressing; preparation.* (Bateman, 1951)

concentration cell An electrolytic cell, the electromotive force of which is due to difference in concentration of the electrolyte or active metal at the anode and the cathode. (Osborne, 1956)

concentration criterion The ratio between the density in a liquid of two minerals that are to be separated (M_h and M_l being the heavy and light one, respectively); $C = M_h - 1/M_l$, where water (sp gr, 1) is the liquid. This ratio indicates the grain size above which separation by gravity methods should be commercially practicable. Above 2.5 fine sands (down to below 200 mesh) can be tabled. At 1.75 the lower limit is 100 mesh; at 1.5 about 10 mesh, and at 1.25 only gravel sizes can be treated. (Pryor, 1958)

concentration of output Essentially, to secure the maximum output of coal from the minimum length of face with due regard to safety and development. To measure the degree of concentration at a colliery the following data are collected: (1) the total length of coalface; (2) the total length of main haulage roads; and (3) the total output. In general, the greater the dispersion of the workings, the greater the manpower employed and the higher the costs of production. See also: *face concentration; geographical concentration; overall concentration.* (Nelson, 1965)

concentration ratio Weight or tonnage ratio (K) of the weight of feed (F) to the weight of concentration (C) produced: $K = F/C$, for a two-product treatment. (Pryor, 1963)

concentration table A table on which a stream of finely-crushed ore and water flows downward; the heavier metallic minerals lag behind and flow off in a separate compartment. (Weed, 1922)

concentrator (a) A plant where ore is separated into values (concentrates) and rejects (tails). An appliance in such a plant, e.g., flotation cell, jig, electromagnet, shaking table. Also called mill; reduction works; cleaning plant. Syn: *concentrating plant.* Cf: *separator.* (Pryor, 1963) (b) An apparatus in which, by the aid of water, air, and/or gravity, mechanical concentration of ores is performed. A concentration plant. (Fay, 1920) (c) A general term for a worker, who tends concentrating tables, vanners, and other types of equipment used to separate valuable minerals from waste material. (DOT, 1949)

concentric fold *parallel fold; similar fold.*

concentric mine cable *portable concentric mine cable.*

concentric pattern Diamonds set in bit face in concentric circles so that a slight uncut ridge of rock is left between stones set in adjacent circles. Cf: *eccentric pattern.* (Long, 1960)

concentric weathering *spheroidal weathering.*

concession *concession system.*

concession system Under this system the state or the private owner has the right to grant concessions or leases to mine operators at discretion and subject to certain general restrictions. It had its origin in the ancient regalian doctrine that all mineral wealth was the prerogative of the crown or the feudatory lord and applies in almost every mining country in the world, except the United States. Syn: *concession.* Cf: *claim system.* See also: *take.* (Hoover, 1948)

conchilite A bowl-shaped body of limonite or goethite growing in an inverted position on mineralized bedrock and resembling the shell of an oyster or clam coated with a rusty deposit. It is roughly oval or circular in plan, with a smooth or irregular and scalloped outline; it ranges from 2.5 cm to 1 m in diameter and from 2 to 7.5 cm in height. (AGI, 1987)

conchoidal Said of a type of mineral or rock fracture that gives a smoothly curved surface. It is a characteristic habit of quartz and of obsidian. Etymol: like the curve of a conch (seashell). (AGI, 1987)

conchoidal fracture A fracture with smooth, curved surfaces, typically slightly concave, showing concentric undulations resembling the lines of growth of a shell. It is well displayed in quartz, obsidian, and flint, and to a lesser extent in anthracite.

concordant (a) Said of intrusive igneous bodies, the contacts of which are parallel to the bedding or foliation of the country rock. Cf: *discordant.* (Billings, 1954) (b) Structurally conformable; said of strata displaying parallelism of bedding or structure. The term may be used where a hiatus cannot be recognized, but cannot be dismissed. (AGI, 1987) (c) Said of radiometric ages, determined by more than one method, that are in agreement within the analytical precision for the determining methods; or of radiometric ages given by coexisting minerals, determined by the same method, that are in agreement. (AGI, 1987)

concrete An intimate mixture of an aggregate, water, and portland cement, which will harden to a rocklike mass.

concrete caisson sinking A shaft-sinking method sometimes used through soft ground down to bedrock. It is similar to caisson sinking, except that reinforced concrete rings are used and an airtight working chamber is not adopted. (Nelson, 1965)

concrete plug A thick layer of reinforced concrete placed in the bottom of a shaft after it has been sunk to the desired depth and permanently lined. The plug resists floor lifting and provides a clean, smooth sump. (Nelson, 1965)

concrete shaft lining *permanent shaft support; shaft wall.*

concrete vibrator Machine that helps the aggregate to consolidate with minimum interstitial porosity. Gives greater strength as less water is incorporated in the mix, and as consolidation is better than with punning. (Pryor, 1963)

concretion (a) A hard, compact mass or aggregate of mineral matter, normally subspherical, but commonly oblate, disk-shaped, or irregular with odd or fantastic outlines; formed by precipitation from aqueous solution about a nucleus or center, such as a leaf, shell, bone, or fossil, in the pores of a sedimentary or fragmental volcanic rock, and usually of a composition widely different from that of the rock in which it is found and from which it is rather sharply separated. It represents a concentration of some minor constituent of the enclosing rock or of cementing material, such as silica (chert), calcite, dolomite, iron oxide, pyrite, or gypsum, and it ranges in size from a small pelletlike object to a great spheroidal body as much as 3 m in diameter. Most concretions were formed during diagenesis, and many (esp. in limestone and shale) shortly after sediment deposition. Cf: *nodule.* (AGI, 1987) (b) A collective term applied loosely to various primary and secondary mineral segregations of diverse origin, including irregular nodules, spherulites, crystalline aggregates, geodes, septaria, and related bodies. Not recommended usage. (AGI, 1987)

concretionary Characterized by, consisting of, or producing concretions; e.g., a concretionary ironstone composed of iron carbonate with clay and calcite, or a zonal concretionary texture (of an ore) characterized by concentric shells of slightly varying properties due to variation during growth. (AGI, 1987)

concretionary and nodular Minerals, usually monomineralic aggregates, which are found in detached masses, the forms being sometimes spherical, sometimes irregular, e.g., flint. (Nelson, 1965)

concussion Shock or sharp airwaves caused by an explosion or heavy blow. (Nichols, 1976)

concussion table An inclined table, which is agitated by a series of shocks, while operating like a buddle. It may be made self-discharging and continuous by substituting for the table an endless rubber cloth, which is slowly moving against the current of water, as in the Frue vanner. Also called a percussion table. (Fay, 1920)

condensation The process by which a vapor becomes a liquid or solid; the opposite of evaporation. (AGI, 1987)

condenser (a) An apparatus used for condensing vapors obtained during distillation; it consists of a condenser tube, either freely exposed to air or contained in a jacket in which water circulates. (CTD, 1958)

(b) An accumulator of electrical energy. Also called capacitor. (Crispin, 1964)
condenser-discharge blasting machine A blasting machine that uses batteries or magnets to energize one or more condensers (capacitors) whose stored energy is released into a blasting circuit, to initiate detonators. (Dick, 2)
condenser maker In ore dressing, smelting, and refining, one who operates an automatic machine in which fireclay condensers are made. (DOT, 1949)
condenser operator In ore dressing, smelting, and refining, one who recovers magnesium particles from dust-bearing gas, using shock-chilling condensers and other dust-collecting apparatus. Also called dust operator. (DOT, 1949)
condensing lens A lens for producing convergent light.
condie *waste.*
conditioned sinter A name given to sinter with lime additions. (Nelson, 1965)
conditioner An apparatus in which conditioning takes place. (BS, 1962)
conditioners Those substances added to the pulp to selectively treat coal or waste surfaces prior to flotation.
conditioning Stage of froth-flotation process in which the surfaces of the coal or associated impurities present in a pulp are treated with appropriate chemicals to influence their reaction when the pulp is aerated. (Pryor, 1965)
condition the hole To circulate a higher-than-normal volume of drill fluid while slowly rotating and lowering the drill string from a point a few feet above the bottom to the bottom of the borehole to wash away obstructing materials before resuming coring operations. (Long, 1960)
conductance The quantity of heat transmitted per unit time from a unit of surface to an opposite unit of surface of material under a unit temperature differential between the surfaces. (Strock, 1948)
conductor (a) Guides of rope or of rigid construction to guide the cages or skips in the shaft. (Mason, 1951) (b) A relatively short length of pipe driven through the unconsolidated zone of top soil as the first step in collaring a borehole. Also called stand pipe. (Long, 1960) (c) *brakeman.* Cf: *surface string.*
conductor-cable locomotive An electric locomotive having a cable on a reel and connected both with the locomotive motor and the trolley wire in the entry, so that the locomotive may be driven into an unwired room. (Zern, 1928)
conduit (a) An airway. (Zern, 1928) (b) Pipe or casing placed in a borehole. See also: *casing; drivepipe.* (Long, 1960)
conduit hole A flat or nearly horizontal hole drilled for blasting a thin piece in the bottom of a level. (Zern, 1928)
cone (a) A conical hill or mountain, as an alluvial cone or a volcanic cone. (b) A device used on top of blast furnaces to enable charge to be put in without permitting gas to escape. Syn: *bell.* (CTD, 1958) (c) The conical part of a gas flame next to the orifice of the tip. (ASM, 1961) (d) The conical hill or conical mountain built by an active volcano. Explosive volcanoes build their cones from debris, ranging in size from dust to huge blocks, thrown out from the vent and have steep slopes approaching or exceeding the angle of repose. Quieter volcanoes that pour out lava have much gentler slopes. (Hess) (e) A three-sided pyramid made of unfired ceramic materials whose composition is such that when heated at a controlled rate they will deform and fuse at a known temperature. It is placed inside a kiln or furnace with ceramic ware to indicate the temperature of the kiln and the fired condition of the ware. See also: *pyrometric cone.* (f) A solid with a circle for a base and with a convex surface that tapers uniformly to a vertex. (Jones, 1949) (g) Geometric pattern of the rock plug or stickup left in the bottom of a borehole drilled by a concave bit. (Long, 1960) (h) Beveled coupling device on a small diamond drill or percussion rock drill used to attach it to a drill column. (Long, 1960)
cone classifier (a) A cone-shaped hydraulic or free-settling classifier. (b) A conical sheet-steel vessel—usually a 60° cone with its point at the bottom—through which water, clear or weighted, flows upward. Ore, coal, or other mineral matter is fed in at the top. The current carries the smaller particles or those of lowest specific gravity over the rim while the others settle. See also: *Callow cone; Caldecott cone; Allen cone; Menzies cone separator; Jeffrey-Robinson cone.* Syn: *Chance cone; cone system.*
cone crusher A machine for reducing the size of materials by means of a truncated cone revolving on its vertical axis within an outer chamber, the anular space between the outer chamber and cone being tapered. See also: *gyratory breaker.* (BS, 1962)
cone cut A cut in which a number of central holes are drilled toward a focal point and, when fired, break out a conical section of strata. (BS, 1964)
cone-face bit *concave bit.*
cone-in-cone structure (a) A secondary structure occurring in marls, limestones, ironstones, coals, etc. It is a succession of small cones of approx. the same size one within another and sharing a common axis. (Holmes, 1928) (b) Coal exhibiting a peculiar fibrous structure passing into a singular toothed arrangement of the particles is called cone-in-cone coal or crystallized coal. Syn: *crystallized coal.* (Fay, 1920)
Conemaughian Upper Middle Pennsylvanian. (AGI, 1987)
cone of depression The depression, approx. conical in shape, that is produced in a water table or in the piezometric surface by pumping or artesian flow. The shape of the depression is because of the fact that the water must flow through progressively smaller cross sections as it nears the well, and hence the hydraulic gradient must be steeper. See also: *water table.* (AGI, 1987)
cone penetration test A soil penetration test in which a steel cone of standard shape and size is pushed into the soil and the force required to advance the cone at a predetermined, usually slow and constant rate, or for a specified distance, or in some designs the penetration resulting from various loads, is recorded. (AGI, 1987)
cone penetrator A 30° to 60° cone having a basal diameter approx. the same size as an a-size diamond-drill rod used to determine the force required to thrust the cone downward into silty or fine to medium-coarse sands, and hence to obtain information that a foundation or soils engineer may use to calculate some of the load-bearing capabilities of such formations. Syn: *cone penetrometer.* See also: *deflection dial.* (Long, 1960)
cone penetrometer A cone penetrator equipped with a device that will register the pressure required to drive the cone downward into the formation being tested. Syn: *cone penetrator; penetrometer.* (Long, 1960)
cone rock bit A rotary drill, with two hardened knurled cones that cut the rock as they roll. Syn: *roller bit.* (Porter, 1930)
cone settler Conical vessel fed centrally with fine ore pulp. "Undersize" is discharged through a flexible pipe (gooseneck), which permits variation of hydrostatic pressure. This apex discharge is thick and carries the larger sized particles. The peripheral top overflow is thin and carries the finer fraction of the solids. (Pryor, 1963)
cone sheet A curved dike or sheet that is part of a concentric set of such forms that dip inward. (Billings, 1954)
cone system A method of separating impurities from coal in a metallic cone containing a mixture of sand and water with a specific gravity higher than that of coal and lower than the impurities. The coal floats, and the impurities sink. See also: *cone classifier.* (Hudson, 1932)
Conewangoan Upper Upper Devonian. (AGI, 1987)
confined detonation velocity The detonation velocity of an explosive or blasting agent under confinement, such as in a borehole. (Dick, 2)
confined groundwater Artesian water.
confined space An enclosed space that has the following characteristics: its primary function is something other than human occupancy; it has restricted entry and exit; and it may contain potential or known hazards. Examples of confined spaces include, but are not limited to, tanks, silos, vessels, pits, sewers, pipelines, tank cars, boilers, septic tanks, and utility vaults. Tanks and other structures under construction may not be considered confined spaces until completely closed. Restricted entry and exit means physical impediment of the body, e.g., use of the hands or contortion of the body to enter into or exit from the confined space. (ANSI, 1992)
confining bed (a) A watertight bed above or below a stratum containing artesian water. (Fay, 1920) (b) An impervious stratum above and/or below an aquifier. (BS, 1963) (c) A body of impermeable or distinctly less permeable material stratigraphically adjacent to one or more aquifers. Cf: *aquitard; aquifuge; aquiclude.* (AGI, 1987)
confluent Said of a stream, glacier, vein, or other geologic feature that combines or meets with another like feature to form one stream, glacier, vein, etc. (AGI, 1987)
conformability The quality, state, or condition of being conformable, such as the relationship of conformable strata. (AGI, 1987)
conformable Successive beds or strata are conformable when they lie one upon another in unbroken and parallel order and no disturbance or denudation took place at the locality while they were being deposited. If one set of beds rests upon the eroded or the upturned edges of another, showing a change of conditions or a break between the formations of the two sets of rocks, they are unconformable. Cf: *unconformable.* (Fay, 1920)
conformal map projection A map projection on which the shape of any small area of the surface mapped is preserved unchanged. (AGI, 1987)
congela A term used in Chile for coba with a high salt content. See also: *coba.* (AGI, 1987)
congelation temperature (a) The freezing point. (b) The temperature at which an oil becomes a solid or is reduced to a standard pasty state.
conglomerate A coarse-grained clastic sedimentary rock, composed of rounded to subangular fragments larger than 2 mm in diameter (granules, pebbles, cobbles, boulders) set in a fine-grained matrix of sand or silt, and commonly cemented by calcium carbonate, iron oxide, silica, or hardened clay; the consolidated equivalent of gravel. The rock or mineral fragments may be of varied composition and range widely in

size, and are usually rounded and smoothed from transportation by water or from wave action. Cf: *breccia.* Syn: *puddingstone.* (AGI, 1987)

conglomerate mudstone *paraconglomerate.*

conglomerite A conglomerate that has reached the same state of induration as a quartzite.

congo bort Congos used industrially as bort. See also: *bort.* (Long, 1960)

congo diamond *congos.*

congo rounds Spherical- or near-spherical-shaped congos. See also: *congos.* (Long, 1960)

congos (a) Originally and commonly used as a name for a variety of diamonds found in the Republic of the Congo diamond district in Africa and more recently as a descriptive term applied to all diamonds having the appearance and characteristics of those produced in the Republic of the Congo. Congos are white to gray-green and yellow, drusy-surfaced, opaque to somewhat translucent diamonds, having shapes corresponding to the many forms characteristic of the isometric (cubic) crystal system. At one time, congos were considered fit only for use in fragmented form, but a considerable number are now used as tool stone and drill diamonds. Syn: *congo diamond.* See also: *congo rounds; diamond.* (Long, 1960) (b) Sometimes designates drill diamonds ranging from one to eight stones per carat in size. (Long, 1960)

congruent (a) In crystallography, any motif that may generate another by rotation or translation, but with no change in chirality. (b) In phase equilibria, the melting of a crystalline compound to a liquid of the same composition. Cf: *incongruent melting.*

congruent forms In crystallography, two forms that may each be derived from the other by rotation about an axis of symmetry. (Fay, 1920)

congruent melting A geologic or metallurgical process in which a binary compound melts at a certain concentration to a liquid of its own composition. Cf: *incongruent melting.*

conical Cone-shaped. In mineralogy, usually an elongated cone as are most icicles.

conical drum A winding drum, cone-shaped at each end, for balancing the load upon the engine during winding operations. See also: *winding drum.* The heavily loaded upgoing rope winds on the small diameter while the lightly loaded downgoing rope winds off the large diameter of the cone. (Nelson, 1965)

conical head gyratory crusher Gyratory-type crusher used for secondary reduction and identified by the shape of its breaking head. The large included angle of the breaking-head surfaces greatly increases the ratio of discharge to feed area; a large ratio permits crushed materials to separate to prevent power-consuming clogging and packing. The crusher's higher gyrating speed and large discharge area make it eminently suitable for fine crushing at a high capacity. (Pit and Quarry, 1960)

conical mill *Hardinge mill.*

conical refraction The refraction of a ray of light at certain points of double-refracting crystals, so that on emerging from the crystal it widens from an apex into a hollow cone (external conical refraction), or on entering diverges into a cone and issues as a hollow cylinder (internal conical refraction). (Standard, 1964)

conichalcite An orthorhombic mineral, $CaCu(AsO_4)(OH)$; adelite group; forms series with austinite, with calciovolborthite, and with cobaltoaustinite; formerly called higginsite.

coning Method of obtaining true sample from a pile of ore by forming a cone with the material flattening the cone, and removing shovelfuls successively onto four separate heaps of which two are rejected. If there is sufficient material the two opposite quadrants are rejected and the remaining two are combined, reconed, and requartered. As the process is repeated and the pile shrinks, it must be crushed to a smaller size to permit accurate blending of the various sized particles during mixing. Syn: *upconing; quartering.* (Pryor, 1963)

coning and quartering A method of sample reduction. Syn: *quartering.* (Nelson, 1965)

conjugated veins Two sets of related veins that dip in different directions.

conjugate fault system A system of two intersecting sets of parallel faults.

conjugate impedance Two impedances having resistive components that are equal and reactive components that are equal in magnitude but opposite in sign are known as conjugate impedances. (Hunt, 1965)

conjugate joint systems Sets of intersecting joints that are sometimes perpendicular or rectilinear, and often mineralized to form vein systems. Joint patterns such as these are believed to be the result of compressive stresses that were relieved by joint formation rather than the formation of a single fissure. (Lewis, 1964)

Conkling magnetic separator A conveying belt that passes under magnets, below which belts run at right angles to the line of travel of the main belt. The magnetic particles (tramp iron) are lifted up against these crossbelts and are thus removed. (Liddell, 1918)

Conklin process A dense-media coal cleaning process in which the separating medium consists of minus 200-mesh magnetite (sp gr, 5.2) in water in the desired proportions (4.4 parts of water to 1 part of magnetite provides an effective specific gravity of about 1.9). This process has the advantage that the medium requires little agitation to keep it in suspension and is easily removed from the clean coal and refuse. (Mitchell, 1950)

connate (a) Originating at the same time as adjacent material; esp. pertaining to waters and volatile materials (such as carbon dioxide) entrapped in sediments at the time the deposits were laid down. (AGI, 1987) (b) Said of fluids derived from the same magma. (AGI, 1987)

connate water Water entrapped in the interstices of a sedimentary rock at the time of its deposition. Cf: *interstitial water; formation water.* (AGI, 1987)

connecting frame A device similar to a guide frame for shaker conveyors, but with provision for insertion of the puller rod. A connecting frame can be inserted between any two standard trough sections to serve as a substitute for a connecting trough on single-arm electric or air devices. (Jones, 1949)

connecting trough A shaker conveyor trough of standard length to which special lugs or plates have been attached to provide a means of connecting the trough to the driving arms of the conveyor drive unit. All motion of the conveyor is transmitted through the connecting trough. The term drive trough is frequently used for this special type of trough. (Jones, 1949)

connecting trough support The means of supporting connecting troughs where they pass over the drive unit. The support is attached to the drive unit frame and is designed to allow the connecting trough freedom of movement in the direction of the panline. Supports may be of the ball frame, wheel, rolled, or rocker arm types. (Jones, 1949)

connecting wire Wire used to extend the firing line or leg wires in an electric blasting circuit. (Atlas, 1987)

connellite A hexagonal mineral, $Cu_{19}Cl_4(SO_4)(OH)_{32}{\cdot}3H_2O$; having SO_4 replaced by NO_3 toward buttgenbachkite; deep blue; formerly known as footeite.

conode Isothermal construction line between two equilibrated phases. See also: *tie line.* (Van Vlack, 1964)

conoscope A polarizing microscope using convergent light with the Bertrand lens inserted, used to test the interference figures of crystals. Cf: *orthoscope.* (AGI, 1987)

Conrad counterflush coring system A system, the notable feature of which is the provision of a reversed mud flush circulation that permits uninterrupted core recovery in the rotary system of drilling. (Sinclair, 1958)

Conrad machine Mechanized pit digger used in checking of alluvial boring. Five-foot-long (1.52-m-long) sections of tubing 24 in (61 cm) in internal diameter are worked into the ground from their mounting on a tractor, the spoil being at the same time removed by means of a bucket or grab. In suitable ground 50 ft (15.2 m) or more depth has been reached. (Pryor, 1963)

consanguineous (a) Said of a natural group of sediments or sedimentary rocks related to one another by origin; e.g., a consanguineous association (such as flysch, molasse, or paralic sediments) interrelated by common ancestry, environment, and evolution. Syn: *consanguinity.* (AGI, 1987) (b) *comagmatic.*

consanguineous association Natural group of sediments or of rocks of related origin. (AGI, 1987)

consanguinity The genetic relationship that exists between igneous rocks that are presumably derived from the same parent magma. Such rocks are closely associated in space and time and commonly have similar geologic occurrence and chemical and mineralogic characteristics. Adj. consanguineous. See also: *comagmatic.* (AGI, 1987)

consertal A syn. of sutured, preferred in European usage, but obsolescent in American usage. (AGI, 1987)

conservation Conserving, preserving, guarding, or protecting; keeping in a safe or entire state; using in an effective manner or holding for necessary uses, as mineral resources. (Hess)

conservative properties Those properties of the ocean, such as salinity, the concentrations of which are not affected by the presence or activity of living organisms, but which are affected only by diffusion and advection. (Hy, 1965)

conset jig Jig developed for Mesabi iron ores in which vertical movement of water is produced by low-pressure inflation and deflation of rubber tubes just below screens. See also: *jig.* (Pryor, 1963)

consistency (a) The degree of solidity or fluidity of bituminous materials. (b) The relative ease with which a soil can be deformed. (ASCE, 1958) (c) A property of a material determined by the complete flow force relation. (ASTM, 1994) (d) The properties of a slip that control its draining, flowing, and spraying behavior. (ASTM, 1994) (e) Percentage of solids in pulp. (Pryor, 1963) (f) Fluidity. (Pryor, 1963)

consistency limits The liquid limit, plastic limit, and shrinkage limit. These all apply to the water content of a clay, each in a certain state as defined by British Standard 1377. See also: *Atterberg limits.* (Nelson, 1965)

consolidated deposits In geology, any or all of the processes whereby loose, soft, or liquid earth materials become firm and coherent. (Stokes, 1955)

consolidated sediment A sediment that has been converted into rock by compaction, deposition of cement in pore spaces, or by physical and chemical changes in the constituents.

consolidation (a) Any process whereby loosely aggregated, soft, or liquid earth materials become firm and coherent rock; specif. the solidification of a magma to form an igneous rock, or the lithification of loose sediments to form a sedimentary rock. (AGI, 1987) (b) The gradual reduction in volume and increase in density of a soil mass in response to increased load or effective compressive stress; e.g., the squeezing of fluids from pore spaces. See also: *lithification.* (AGI, 1987)

consolidation hole Borehole into which chemical solutions or grout are injected to cement or consolidate fragmental rock material. Cf: *grout hole.* (Long, 1960)

consolidation settlement The gradual settlement of loaded clay. (Nelson, 1965)

consolidation test A test in which an undisturbed sample of clay measuring 6 cm in diameter and 2 cm thick is confined laterally in a metal ring and compressed between two porous plates that are kept saturated with water. A load is applied and the clay consolidates, the excess pore water escaping through the porous stones. After each increment of load is applied, it is allowed to remain on the sample until equilibrium is established, and a consolidation curve showing the deformation with time is obtained for each increment. (Nelson, 1965)

consolidation trickling During closing of bed or particles in the suction half of jigging cycle, interstitial burrowing down of fastest moving small particles before the mass of particles becomes too compact for movement. (Pryor, 1963)

constantan A group of copper-nickel alloys containing 45% to 60% copper with minor amounts of iron and manganese, and characterized by relatively constant electrical resistivity irrespective of temperature; used in resistors and thermocouples. (ASM, 1961)

constant error A systematic error that is the same in both magnitude and sign throughout a given series of observations (the observational conditions remaining unchanged) and that tends to have the same effect upon all the observations of the series or part thereof under consideration; e.g., the index error of a precision instrument. (AGI, 1987)

constant-weight feeder (a) An automatic device that maintains a constant rate of feed of ore from the bin or stockpile to the grinding circuit. It is controlled by tilt due to the weight of ore on a balanced length of the belt conveyor; by electrically vibrated chute; by pusher gear; by timed delivery from automatically loaded hoppers. (Pryor, 1963) (b) A feeder intended to deliver a certain weight per unit of time. (ACSG, 1963)

constituent of attritus Constituents are the petrographic entities of the attritus that are recognizable in thin sections only by the microscope. The following constituents may be distinguished in coals: translucent humic degradation matter; brown or semitranslucent matter; opaque matter (granular, massive); resins and resinous matter; spores and pollen; cuticles and cuticular matter; algae and algae matter. (IHCP, 1963)

constitutional change Transformation of a constituent in an alloy; e.g., austenite into pearlite. (Pryor, 1963)

constitutional water Water molecules completely bound into a hydrated crystal, e.g., in gypsum, $CaSO_4 \cdot 2H_2O$.

constitution diagram A graphical representation of the temperature and composition limits of phase fields in an alloy system as they actually exist under the specific conditions of heating or cooling (synonymous with phase diagram). A constitution diagram may be an equilibrium diagram, an approximation to an equilibrium diagram, or a representation of metastable conditions or phases. Cf: *equilibrium diagram.* (ASM, 1961)

constructed wetland A man-made marsh that is designed to be slow-draining so that specific species will flourish, primarily to replace natural wetlands that have been drained and filled prior to development. In mining, some constructed wetlands are designed to fix metals and other contaminants, primarily by the reduction of metal sulfates to sulfides, or the formation of oxides or carbonates. When the term is also applied to systems constructed without plants, they consist of buried substrates through which the contaminated water is passed under low oxygen or reducing conditions.

constructive possession That possession that the law annexes to the legal title or ownership of property, when there is a right to the immediate actual possession of such property, but no actual possession. (Ricketts, 1943)

consulting engineer A specialist employed in an advisory capacity. Normally, this person does not manage or direct any operation, and is at the service of the board rather than of the company's administrative and executive staff. (Pryor, 1963)

consulting mining engineer A highly qualified mining engineer with a wide background of experience in this particular field. The engineer may be asked by a client or company to examine a property and prepare a report and evaluation or to give advice or expert evidence in cases of alleged subsidence damage. (Nelson, 1965)

consumable electrode-arc melting A method of arc melting in which the electrode itself serves to supply the metal; this method is commonly employed for melting titanium and zirconium. (Newton, 1959)

consume To use up; to expend; to waste; as in the chemical and mechanical loss of mercury in amalgamation.

consumption charge That portion of a utility charge based on energy actually consumed, as distinguished from the demand charge. (Strock, 1948)

contact (a) A plane or irregular surface between two types or ages of rock. (AGI, 1987) (b) The surface of delimitation between a vein and its wall, or country rock.

contact angle The angle across the water phase of an air-water-mineral system, used to measure effect of surface conditioning. (Pryor, 1965)

contact bed In geology, a bed lying next to or in contact with a formation of different character. (Fay, 1920)

contact breccia A breccia around an igneous intrusion, caused by wall-rock fragmentation and consisting of both intrusive material and wall rock; intrusion breccia. Cf: *agmatite.* (AGI, 1987)

contact deposit A mineral deposit between two unlike rocks. The term is usually applied to an orebody at the contact between a sedimentary rock and an igneous rock. See also: *contact vein.*

contact erosion valley A valley that has been eroded along a zone of weakness at the contact between two different kinds of rock, as between two different sedimentary formations, between igneous and sedimentary rocks, along a fault, or along an upturned unconformity.

contact goniometer A protractor for measuring the angles between adjacent crystal faces. See also: *goniometer; reflection goniometer.* (Fay, 1920)

contact line The line of intersection of a contact surface with the surface of an exposure or with the surface of bedrock covered by mantle rock; it may be exposed or concealed.

contact logging In this type log, provision is made for electrodes to be pressed firmly against the borehole wall. By doing this, current flowing from the electrodes to the wall of the borehole no longer has to traverse the mud. The path from the electrodes through the mud filter cake that sheaths permeable beds is also reduced to a mininium. The electrode spacing of contact logging devices is very small by comparison with the spacings used in conventional logging devices. Consequently, contact logging devices see very much more detail in the beds they pass through. (Wyllie, 1963)

contact logging device A device that consists of a spring bow very analogous to a section gage. On one arm of the bow is a rubber pad shaped to fit the curvature of the hole. In this pad, slightly recessed, are three electrodes of about diameter 1/2 in (1.3 cm) and located at 1-in (2.54-cm) intervals. These three electrodes are used to record two resistivity curves. One curve is a three-electrode type with a spacing of 1-1/2 in (3.8 cm), and the second is a two-electrode type with a spacing of 2 in. (Wyllie, 1963)

contact-metamorphic Adj. of contact metamorphism.

contact metamorphic Applied to rocks and/or minerals that have originated through the process of contact metamorphism. (AGI, 1987)

contact metamorphism A process taking place in rocks at or near their contact with a body of igneous rock. Metamorphic changes are effected by the heat and materials emanating from the magma and by some deformation connected with the emplacement of the igneous mass. Cf: *thermal metamorphism; metamorphic aureole.* Approx. syn: pyrometasomatism. Adj: contact-metamorphic. See also: *exomorphism; endomorphism.* (AGI, 1987)

contact-metasomatic deposit A deposit formed by high-temperature magmatic emanations along an igneous contact. (Bateman, 1951)

contact metasomatism A mass change in the composition of rocks in contact with an invading magma, from which fluid constituents are carried out to combine with some of the country-rock constituents to form a new suite of minerals. (AGI, 1987)

contact mineral A mineral formed by contact metamorphism.

contact process A process for making sulfuric acid. Sulfur dioxide gas (obtained by burning pyrite) is purified by electrical precipitation, and is passed over a catalytic agent to form sulfur trioxide that combined with water produces sulfuric acid. (CTD, 1958)

contact reef S. Afr. This term generally denotes the Ventersdorp contact reef, a gold-bearing conglomerate beneath the Ventersdorp lavas and frequently overlying mineralized horizons of the Witwatersrand system. (Beerman, 1963)

contact resistance The resistance observed between a grounded electrode and the ground, or between an electrode and a rock specimen. (AGI, 1987)

contact rocks Rocks produced by contact metasomatism. They include both the border rocks of the intrusion and metamorphosed or recrystallized portions of the intruded rocks, esp. limestone. See also: *skarn; tactite.*

contact shoe Collector shoe, which maintains contact between the conducting wire or rail and the electric vehicle being powered. (Pryor, 1963)

contact twin (a) The simplest type of twin, in which two portions of a crystal appear to have been united along a common plane after one portion has been rotated 180° relative to the other portion. The plane of contact (plane of union or the composition face) may or may not be the twinning plane. Syn: *juxtaposition twin.* (Fay, 1920) (b) A twinned crystal wherein the individual twins meet at a surface. Cf: *interpenetration twin; penetration twin.*

contact vein A contact deposit in vein form. See also: *contact deposit.*

contact zone *aureole.*

containerwinding This winding system makes use of a coal receptacle on small rollers that fits closely in the cage. During the previous wind, it is filled near the shaft, and similarly it is emptied very quickly at the surface and returned to the pit bottom on the next wind. (Sinclair, 1963)

contaminant (a) *impurity.* (b) A harmful, irritating, or nuisance airborne material.

contamination Process whereby the chemical composition of a magma is altered as a result of the assimilation of inclusions or country rock. See also: *hybridization; dilution.* (AGI, 1987)

contango The situation when the price of a metal for forward or future delivery is greater than the cash or spot price of the metal. Contangos occur when the metal is in plentiful supply. The size of the contango does not normally exceed the cost of financing, insuring, and storing the metal over the future delivery period. (Wolff, 1987)

contemporaneous Formed or existing at the same time. Said of lava flows interbedded in a single time-stratigraphic unit, and generally of any feature or facies that develops during the formation of the enclosing rocks. (AGI, 1987)

contemporaneous deformation Deformation that takes place in sediments during or immediately following their deposition. Includes many varieties of soft-sediment deformation, such as small-scale slumps, crumpling and brecciation, but in some areas features of large dimensions. (AGI, 1987)

contiguous Adjoining, touching, or connected throughout, as in a group of mining claims.

contiguous claims Mining claims that have a side or end line in common. (Lewis, 1964)

contiguous limonite Limonite in the gangue around and adjoining a cavity or a group of cavities formerly occupied by iron-bearing sulfide.

continental alluvium Alluvium produced by the erosion of a highland area and deposited by a network of rivers to form an extensive plain. (AGI, 1987)

continental basin A closed structural depression of regional extent in the interior of a continent.

continental deposit A sedimentary deposit laid down on land or in bodies of water not directly connected with the ocean, as opposed to a marine deposit; a glacial, fluvial, lacustrine, or eolian deposit formed in a nonmarine environment. See also: *terrestrial deposit.* Syn: *continental sediment.* (AGI, 1987)

continental gland-type capping A wire-rope capping method in which a rope-clamping device is used instead of a capping. The end of the rope is turned back upon itself over grooved block with a suitable radius, and the short end of the rope is clamped on to the main rope above the block. (Sinclair, 1959)

continental margin (a) The zone separating the emergent continents from the deep sea bottom. It generally consists of the continental shelf, the continental slope, and the continental rise. (AGI, 1987) (b) The submerged prolongation of the land mass of the coastal state, consisting of the seabed and subsoil of the shelf, the slope, and the rise.

continental nucleus *shield.*

continental plate Lithosphere underlying a continent that is part of a tectonic plate. (AGI, 1987)

continental platform The platformlike mass of a continent that stands above the surrounding oceanic basins. Syn. for continental shelf. (AGI, 1987)

continental rise The submarine surface beyond the base of the continental slope, generally having a gradient of less than 1:1,000, occurring at depths from 4,500 to 17,000 ft (1.37 to 5.18 km), and leading down to abyssal plains. (AGI, 1987)

continental sediment *continental deposit.*

continental shelf (a) The gently sloping tread around a continent, extending from the low-water line to the depth of approx. 100 fathoms (183 m), at which depth there is a marked increase of slope toward the great depths. (Schieferdecker, 1959) (b) The gently sloping, shallowly submerged marginal zone of the continents extending from the shore to an abrupt increase in bottom inclination. The greatest average depth is less than 60 ft (18.3 m), and the width ranges from very narrow to more than 200 mi (322 km). (AGI, 1987) (c) An area of a coastal State comprising the seabed and subsoil of the submarine areas that extend beyond its territorial sea throughout the natural prolongation of its land territory to the outer edge of the continental margin, or to a distance of 200 nmi (370 km) from the baselines from which the breadth of the territorial sea is measured where the outer edge of the continental shelf does not extend up to that distance. See also: *outer continental shelf.*

continental shield *shield.*

continental slope (a) The declivity from the offshore border of the continental shelf at depths of approx. 100 fathoms (600 ft or 183 m) to oceanic depths. It is characterized by a marked increase in gradient. (AGI, 1987) (b) Continuously sloping portion of the continental margin with gradient of more than 1:40, beginning at the outer edge of the continental shelf and bounded on the outside by a rather abrupt decrease in slope where the continental rise begins at depths ranging from about 4,500 to 10,000 ft (1.4 to 3.0 km). Formerly considered to extend to the abyssal plains. (AGI, 1987)

continental terrace The sediment and rock mass underlying the coastal plain, the continental shelf, and the continental slope. (AGI, 1987)

continuity The concept that where there is no change of state, seawater is incompressible and the liquid matter is neither created nor destroyed. If there is any vertical contraction in a volume of fluid, therefore, there must be a horizontal expansion, so that the original volume is maintained. This is accomplished by motion resulting in changes of the shape of the original parcel of water. (Hy, 1965)

continuous azimuth method A method of traversing by which the azimuth of the survey lines is obtained from the instrument. (BS, 1963)

continuous-bucket elevator This type of elevator has the buckets so shaped and attached to the chain or belt that the back of each serves as a discharge chute for the one immediately succeeding it. Syn: *bucket elevator.* (Pit and Quarry, 1960)

continuous-bucket excavator An excavator consisting of a series of buckets attached to a continuous chain, guided by two or more ladders. The buckets are drawn against the bank face, taking a cut of constant depth, while simultaneously the machine moves slowly along the ground on a bench above or below the bank; often used in opencast mining in soft deposits. (Nelson, 1965)

continuous casting A casting technique in which an ingot, billet, tube, or other shape is continuously solidified while it is being poured, so that its length is not determined by mold dimensions. (ASM, 1961)

continuous charge A charge of explosive that occupies the entire drill hole, except for the space at the top required for stemming. (Fay, 1920)

continuous coal cutter A coal mining machine of the type that cuts the face of the coal without being withdrawn from the cut. (Fay, 1920)

continuous coring A borehole-drilling technique whereby the cuttings-removal agent is countercirculated through an inside flush-coupled-type drill string to deliver both the cuttings and core produced to a tray or container at the surface. (Long, 1960)

continuous cutters Coal-cutting machines such as the shortwall cutter, longwall cutter, and overcutting machines. They are known as continuous cutters because a continuous cut can be made the full width of the face without stopping these machines, while machines of the intermittent variety must be frequently reset. (Kiser, 1929)

continuous deformation Deformation accomplished by flowage of rocks rather than by rupture.

continuous drier A drier in which the wet material moves through the drying cycle in an uninterrupted flow pattern in contrast to a batch drier. (ACSG, 1963)

continuous driving In this operation, the same personnel do the drilling, blasting, and mucking while working continuously round after round. They can in this way—except for the time for ventilation—be at work during the whole shift. Continuous driving is used when the advance per round is low and the mucking or the drilling and blasting do not need more than a part of the shift. (Langefors, 1963)

continuous extraction Extraction (leaching) of solids by liquid that cycles continuously countercurrent to the material it is depleting of the sought value (e.g., gold in cyanide process), the pregnant liquid at a certain stage being stripped of value and returned as barren solution. (Pryor, 1958)

continuous filter *Oliver filter.*

continuous-flight auger A drill rod with continuous helical fluting, which acts as a screw conveyor to remove cuttings produced by an auger drill head. Also called auger. (Long, 1960)

continuous flow respirator An atmosphere-supplying respirator that provides a continuous flow of respirable gas to the respiratory inlet covering. (ANSI, 1992)

continuous furnace Said of a process in which the charge enters at one end, moves through continuously, and is discharged at the other end. Syn: *furnace.* (CTD, 1958)

continuous haulage A process that is designed to move the mined product (usually coal) from a continuous mining machine to a mine belt conveyor system as a continuous flow. One end of the continuous haulage system (the outby end) always remains positioned so that it discharges onto the mine belt; the other end (inby end) is free to move as the mining machine advances so as to be able to receive the product from the machine's conveyor discharge.

continuous mill A rolling mill consisting of a number of stands of synchronized rolls (in tandem) in which metal undergoes successive reductions as it passes through the various stands. (ASM, 1961)

continuous miner A mining machine designed to remove coal from the face and to load that coal into cars or conveyors without the use of cutting machines, drills, or explosives. See also: *Goodman miner; Marietta miner*. (Jones, 1949)

continuous mining Mining in which the continuous mining machine cuts or rips coal from the face and loads it onto conveyors or into shuttle cars in a continuous operation. Thus, the drilling and shooting operations are eliminated, along with the necessity for working several headings in order to have available a heading in which loading can be in progress at all times. See also: *conventional machine mining; plow-type machine*. (Woodruff, 1966)

continuous profiling A seismic method of shooting in which seismometer stations are placed uniformly along the length of a line and shot from holes also spaced along the line so that each hole records seismic-ray paths identical geometrically with those from immediately adjacent holes, so that events may be carried continuously by equal-time comparisons. Cf: *correlation shooting*. (AGI, 1987)

continuous reaction series A reaction series in which early-formed crystals in a magma react with later liquids without abrupt phase changes; e.g., the plagioclase feldspars form a continuous reaction series. Cf: *discontinuous reaction series*. (AGI, 1987)

continuous recording In geophysics, the process of making uninterrupted records or observations over selected periods of time. (AGI, 1987)

continuous ropeway An aerial ropeway that operates on the same principle as the endless rope haulage. The loaded buckets are hauled by an endless rope in one direction and the empty buckets travel back on the return rope alongside. (Nelson, 1965)

continuous sampling Taking a sample from each unit so that increments are taken at regular intervals whenever the coal or coke is handled at the point of sampling. (BS, 1960)

continuous sintering Presintering, or sintering, in such manner that the objects are advanced through the furnace at a fixed rate by manual or mechanical means. Syn: *stoking*. (ASTM, 1994)

continuous smelter Any smelter that is fed constantly and that discharges frit in a continuous stream. The passage of the material through the smelter is generally effected by gravitational flow. (Enam. Dict., 1947)

continuous spectrum (a) The band of all wavelengths in the electromagnetic spectrum (the rainbow colors, red, orange, yellow, green, blue, and violet), merging one into the other, produced by all incandescent solids. (Anderson, 1964) (b) The spectrum of a wave, the components of which are continuously distributed over a frequency range. (Hunt, 1965)

continuous stream conveyor *en masse conveyor.*

continuous vertical retort A type of gas retort in which coal is continuously charged into the top of the retort, coke is extracted from the bottom, and town gas is drawn off. Continuous vertical retorts are also used in the zinc industry. The charge of briquetted coke and roasted concentrate is continuously added through the top and zinc vapor is drawn off and condensed.

contortion (a) The intricate folding, bending, or twisting-together of laminated sediments on a considerable scale, the laminae being drawn out or compressed in such a manner as to suggest kneading more than simple folding; esp. intraformational contortion. Also, the state of being contorted. (AGI, 1987) (b) A structure produced by contortion. (AGI, 1987)

contour (a) An imaginary line, or a line on a map or chart, that connects points of equal value, e.g., elevation of the land surface above or below some reference value or datum plane, generally sea level. Contours are commonly used to depict topographic or structural surfaces; they can also readily show the laterally variable properties of sediments or any other phenomenon that can be quantified. Cf: *structure contour*. Syn: *contour line*. (AGI, 1987) (b) The outline or configuration of a surface feature seen two-dimensionally, e.g., the contour of a mountain pass or a coastline. (AGI, 1987) (c) A line drawn through points of equal elevation on any surface. It is the intersection of a horizontal plane with the surface. (Rice, 1960) (d) A line or a surface at all points of which a certain quantity, otherwise variable, has the same value (as lines of equal elevation on the ground or isothermal surfaces in a heat-conducting solid). (Webster 3rd, 1966) (e) As a verb, to construct (as a road) in conformity to a contour. To provide (as a map) with contours (contour lines). To draw or to plot a contour. (Webster 3rd, 1966) (f) The profile or cross-sectional outline of a bit face. (Long, 1960)

contour diagram (a) A type of petrofabric diagram prepared by the contouring of a point diagram. Its purpose is to obtain easier visualization of the results of the petrofabric study. (AGI, 1987) (b) An equal-area projection of structural data in which the poles have been contoured according to their density per unit area on the projection. (AGI, 1987)

contour gradient A line marked on the ground surface at a given constant slope. (Hammond, 1965)

contour interval (a) The difference in elevation between two adjacent contour lines. (AGI, 1987) (b) The difference in value between two adjacent contours; e.g., the vertical distance between the elevations represented by two successive contour lines on a topographic map. It is generally a regular unit chosen according to the range of values being contoured. Syn: *interval*. (AGI, 1987)

contour line *contour.*

contour map (a) A map showing by contours (or contour lines) topographic, or structural, or thickness, or facies differences in the area mapped. (AGI, 1987) (b) A map that portrays surface configuration by means of contour lines; esp. a topographic map that shows surface relief by means of contour lines drawn at regular intervals above mean sea level, or a structure-contour map that shows the configuration of a specified rock surface underground and the inferred configuration of that surface where it has been removed by erosion. (AGI, 1987)

contour mining Surface mining that progresses in a narrow zone following the outcrop of a coal seam in mountainous terrain, and the overburden, removed to gain access to the mineral commodity, is immediately placed in the previously mined area, such that reclamation is carried out contemporaneously with extraction. (SME, 1992)

contour plan A plan drawn to a suitable scale showing surface contours or calculated contours of coal seams to be developed. These plans are important during the planning stage of a project. See also: *interpolation of contours*. (Nelson, 1965)

contour race A watercourse following the contour of the country.

contraband In coal mining, a term meaning cigars, cigarettes, pipes, and other contrivances for smoking, matches, and mechanical lighters. It is a violation of safety regulations to take contraband below ground or to have contraband in one's possession below ground. (Nelson, 1965)

contract (a) A bargain or agreement voluntarily made upon good consideration, between two or more persons capable of contracting to do, or forbearing to do, some lawful act. (Hoover, 1948) (b) In mining, applies to an agreement between operator and worker to pay the latter so much per foot for excavating drift or stope. These people are known as contract miners and are usually skilled workers. They work harder than people on wages due to the incentive of higher earnings. (Weed, 1922) (c) Agreement between contractor and employing company to construct, erect, install, and operate specified works under agreed conditions. A cost-plus contract is one in which the contractor undertakes a comprehensive activity, part of which may be subcontracted (or let out). A unit contract is one in which company awards a restricted part of the job to the contractor. See also: *agreement*. (Pryor, 1963)

contraction cavities The bulk of the contraction that accompanies the solidification of metals is concentrated in the feeder heads and risers, from which molten metal flows to compensate for contraction in the casting of ingot proper. (CTD, 1958)

contraction vein A vein formed by the filling of a fissure caused by contraction resulting from the drying or cooling of the surrounding rock.

contract loader In bituminous coal mining, one who is paid a certain rate per ton of car of coal mined, and employs one or more loaders whom the loader pays out of personal earnings. (DOT, 1949)

contract miner (a) In anthracite and bituminous coal mining, one who operates electric or compressed-air machines to drill holes into the working face of coal or rock for blasting, and shovels coal into cars after blasting. A contract miner is usually engaged in production work, i.e., the mining of coal only, and is paid on a tonnage basis. In anthracite regions, the miner is paid the wage rate of a consideration mine when encountering obstructions of rock or slate that prevent earning an amount in excess of a fixed or specified rate per day. Also called contract driller; contract drilling-machine operator; contract contractor. (DOT, 1949) (b) In metal mining, one who drills, blasts, and loads ore or rock into cars in a mine. Is usually engaged in production work, i.e., the mining of ore only, and is paid on a contract basis (so much per ton, cubic yard, or cars of ore produced). (DOT, 1949)

contractor (a) The person who signs a contract to do certain specified work at a certain rate of payment. In mining, the contractor is an experienced miner or hard-heading miner. He or she employs other people and the work may proceed on a three-shift basis. (Nelson, 1965) (b) S. Afr. Mine worker undertaking special tasks on a contractual basis such as shaft sinking, development blasting, etc. (Beerman, 1963)

contract person *contract miner.*

contract work Work that is outside the scope of the mine price list and is performed on the basis of an agreement between a miner and the mine manager. The agreement may be only verbal and renewable weekly or monthly. Payment is made according to performance. In development work, the contract rate is usually per yard advance. There may be bonus payments for good work or for extra performance. See also: *piecework*. (Nelson, 1965)

contragradation Stream aggradation caused by an obstruction. Syn: *dam gradation*.

contra-rotating axial fan A modification of the axial-flow fan. It consists of two impellers with aerofoil shaped blades that rotate in opposite directions. The drive is by means of a single motor through differential gears, or two separate motors, one for each impeller. They are placed in the airstream and act as streamlined hubs. These fans are available for auxiliary ventilation in mines. See also: *axial-flow fan*. (Nelson, 1965)

contributory negligence In mining, means that the law imposes upon every person the duty of using ordinary care for his or her own protection against injury. It is not synonymous with assumption of risk. (Ricketts, 1943)

control (a) The dimensional data used to establish the position, elevations, scale, and orientation of the detail of a map and that are responsible for the interpretations placed on a map. (AGI, 1987) (b) A section or reach of an open channel in which natural or artificial conditions make the water level above it a stable index of discharge. It may be either complete (i.e., water-surface elevation above the control is completely independent of downstream water-level fluctuations) or partial; it may also shift. (AGI, 1987) (c) That waterway cross section that is the bottleneck for a given flow and determines the energy head required to produce the flow. In an open channel, it is the point at which flow is at critical depth; in a closed conduit, it is the point at which hydrostatic pressure and cross-sectional area of flow are definitely fixed, except where the flow is limited at some other point by a hydrostatic pressure equal to the greatest vacuum that can be maintained unbroken at that point. (AGI, 1987) (d) Any of the factors determining the nature of geologic formations at a given place. (Webster 3rd, 1966) (e) In geology, the background and the quantity of data that are responsible for the interpretation placed on a map or a cross section. (AGI, 1987) (f) An attempt to guide a borehole to follow a predetermined course through the use of wedges or by manipulation of the drill string. (Long, 1960)

control assay An assay made by an umpire to determine the basis on which a purchaser is to pay the seller for ore. See also: *umpire*.

control chart Graph showing, horizontally, the operating norm and also the upper and lower control limits within which deviations must be held. Should these values exceed the permitted variance, special steps must be taken to locate and correct the upsetting factor or factors. (Pryor, 1963)

controlled blasting Techniques used to control overbreak and produce a competent final excavation wall. See also: *line drilling; smooth blasting; cushion blasting*. Syn: *presplitting*. (Dick, 2)

controlled caving A mining method utilizing the advantages of longwalls, but at the same time without filling. In this method, the working room in front of the working face is protected by close lines of props and cribs, which are portable and easily taken to pieces. As the face proceeds, the cribs are shifted as well as the props with the face, leaving the mined-out room to cave. This method is also called mining with self-filling. (Stoces, 1954)

controlled cooling Cooling from an elevated temperature in a predetermined manner to avoid hardening, cracking, or internal damage, or to produce a desired microstructure. This cooling usually follows a hot-forming operation. (ASM, 1961)

controlled footage The specified maximum number of feet of borehole a single diamond- or other-type bit may be allowed to drill in a specific-type rock, as predetermined by the drill supervisor. (Long, 1960)

controlled gravity conveyor *controlled velocity roller conveyor*.

controlled mosaic A mosaic in which aerial photographs or images have been adjusted, oriented, and scaled to horizontal ground control to provide an accurate representation with respect to distances and distortions. It is usually assembled from photographs that have been corrected for tilt and for variations in flight altitude. See also: *mosaic*. (AGI, 1987)

controlled release A paradigm of mine waste management (based upon the eventual oxidation of exposed sulfidic rock and mine wastes) that states that a slow release of contamination over time may be superior to complete containment. (Connolly, 1990)

controlled splitting When airways are arranged in parallel and a prescribed quantity of air is made to flow through each branch. Cf: *natural splitting*. (Hartman, 1982)

controlled velocity roller conveyor A roller conveyor having means to control the velocity of the objects being conveyed. Syn: *controlled gravity conveyor*. See also: *roller conveyor*.

controller Any mechanical or electrical device that is part of or added to a machine or device for automatic regulation or control.

controlling rate That at which the key machine in a series arranged for continuous ore processing is set to work. The control function may be for quantity passing per time, ratio of size reduction from feed to discharge, or for a necessary physical or chemical change of state of solid or liquid phase of the process. (Pryor, 1963)

controlling system In flotation, that portion of an automatic feedback control system that compares functions of a controlled variable and a command and adjusts a manipulated variable as a function of the difference. It includes the reference input elements, summing point, forward and final controlling elements, and feedback elements. (Fuerstenau, 1962)

control man Person who maintains depth and composition of cryolite bath in aluminum reduction pots within limits favorable to efficient aluminum production. (DOT, 1949)

control on fracture In quarrying, control on fracture is based on the experimental determination of the type and the grade of explosive, the loading ratio, and the pattern of boreholes. (Streefkerk, 1952)

control point Any station in a horizontal and/or vertical control system that is identified on a photograph and used for correlating the data shown on that photograph. (AGI, 1987)

control samples In any continuous process, samples taken often enough (whether by hand or mechanically) so that the operation process may be guided by the samples and weights of the materials involved. (Newton, 1938)

convection (a) A process of mass movement of portions of any fluid medium (liquid or gas) in a gravitational field as a consequence of different temperatures in the medium and hence different densities. The process thus moves both the medium and the heat, and the term convection is used to signify either or both. (b) In hydrothermal systems, the flow of water around and through heated zones adjacent to plutons in response to thermal gradients and controlled by porosity-permeability, salinity, fluid viscosity, and allied factors. The flow is generally down along the periphery, toward the system at depth, and upward along and through its central portions, possibly completing more than one loop. (AGI, 1987)

convection current (a) A thermally produced fluid flow. (b) A closed circulation of material sometimes developed during convection. Convection currents normally develop in pairs; each pair is called a convection cell. (Leet, 1958)

conventional machine mining A system of mining established for many years in British coal mines. The longwall face is undercut, blasted, and loaded by hand to a face conveyor. The conveyor is then moved forward ready for the next day, the packs are built and the back props withdrawn. Such faces still produce about 60% of the total output and is known as conventional machine mining. It has the disadvantage that there are limits to production because it is cyclic mining, e.g., it involves separate operations as enumerated above. See also: *turnover; continuous mining*. (Nelson, 1965)

conventional mining The cycle of operations that includes cutting the coal, drilling the shot holes, charging and shooting the holes, loading the broken coal, and installing roof support. Also known as cyclic mining. (Woodruff, 1966)

conventional mud A drilling fluid containing essentially clay and water. (Brantly, 1961)

convergence (a) The gradual decrease in the vertical distance or interval between two specified rock units or geologic horizons as a result of the thinning of intervening strata; e.g., the reduction in thickness of sedimentary beds (as measured in a given direction and at right angles to the bedding planes), caused by variable rates of deposition or by unconformable relationship. (AGI, 1987) (b) Loss of height when a coal seam is extracted on a longwall face, as the roof lowers and the floor lifts. Convergence is an important factor in thin-seam mining. (c) Applied to the diminishing interval between geologic horizons. In some instances, this is due to an unconformable relationship and in other instances to variable rates of deposition. (AGI, 1987) (d) The line of demarcation between turbid river water and clear lake water, which denotes a downstream movement of water on the lake bottom and an upstream movement of water at the surface. (AGI, 1987) (e) In refraction phenomena, the decreasing of the distance between orthogonals in the direction of wave travel. This denotes an area of increasing wave height and energy concentration. (AGI, 1987) (f) In paleontology, resemblance that cannot be attributed to a direct relationship or to genetic affinity. (AGI, 1987) (g) In oceanography, an area or zone in which the water sinks slowly downward from the ocean surface. (Schieferdecker, 1959)

convergence map *isochore map*.

convergence recorder An appliance for measuring changes in vertical height, usually at the coalface. It consists of a telescopic strut set between the roof and floor and carries a pen that records the movement on a clockwork-driven chart. See also: *romometer*. (Nelson, 1965)

convergent light In optical microscopy, a condensing lens causes light to converge at a point within a sample to display optical interference patterns or to enhance the Becke line.

conversion burners Fuel-burning devices (usually oil or gas) intended for installation in a wide variety of boilers or furnaces. (Strock, 1948)

conversion factor A number facilitating statement of units of one system in corresponding values in another system. (Pryor, 1963)

converter A furnace in which air is blown through a bath of molten metal or matte, oxidizing the impurities and maintaining the temperature through the heat produced by the oxidation reaction. Also used in converting copper matte. (ASM, 1961; Fay, 1920)

converter foreperson A person who supervises workers engaged in converting copper matte to blister copper and directs activities concerned with charging converter, blowing charge, pouring of slag and copper, casting of blister copper, and removal of castings. (DOT, 1949)

converter plant A plant that incorporates into its structure an insoluble element from the soil, and later, when the plant decays, returns that element to the soil in a soluble form. (AGI, 1987)

converter skimmer In ore dressing, smelting, and refining, one who makes blister copper (high-grade crude copper) by oxidizing iron and sulfur impurities in copper matte, using a converter. (DOT, 1949)

converting The process of removing impurities from molten metal or metallic compounds by blowing air through the liquid. The impurities are changed either to gaseous compounds, which are removed by volatization, or to liquids which are removed as slags. (Kirk, 1947)

Convertol process A German process that cleans the coal and also reduces the moisture content to about 10%. Heavy oil is added to a coal slurry containing 50% to 60% water. On mixing, the coal particles become coated with oil and hence resistant to water, whereas the shale particles remain uncoated and easily wetted. By high-speed centrifuging the coal-oil mixture is retained in the centrifuge while the shale particles pass out with the water. The process is not as efficient as froth flotation. (Nelson, 1965)

conveyor (a) A mechanical contrivance generally electrically driven, which extends from a receiving point to a discharge point and conveys, transports, or transfers material between those points. (b) The apparatus, belt, chain, or shaker, which, in conveyor mining, moves coal from the rooms and entries to a discharge point or to the surface. "Mother conveyors" are the conveyors that receive the coal from several unit conveyors in rooms or entries. See also: *underground mine conveyor; transport; armored flexible conveyor; gate conveyor; shaker conveyor; trunk conveyor.* (BCI, 1947)

conveyor airlock A ventilation stopping or separation door through which a conveyor has to run. It consists of at least two well-built partitions, each with some form of airlock designed to pass the belt and yet to reduce to a minimum the leakage of air and the raising of dust. An airlock chute is sometimes used. See also: *box check.*

conveyor chain A chain used in the conveying medium of conveyors.

conveyor creep The downward slippage of a conveyor on an inclined face. With powered supports, this movement is likely to cause ram damage. Anchor stations are necessary to arrest conveyor creep. See also: *stell prop.* (Nelson, 1965)

conveyor dryer An appliance in which the coal or ore is moved through a chamber containing hot gases on a perforated plate or a heavy mesh, stainless-steel continuous belt. (Nelson, 1965)

conveyor elevator A conveyor that follows a path, part of which is substantially horizontal or on a slope less than the angle of slide of the material and part of which is substantially vertical or on a slope steeper than the angle of slide.

conveyor emergency switch A specif. designed cable operated emergency stop switch for use with conveyors or conveying systems that are used with a pull cord running alongside or above the conveyor so that it may be reached from any point along the conveyor. Persons falling against or on top of the conveyor will pull on the cord and deactivate the conveyor movement. (Best, 1993)

conveyor face A longwall face on which the coal is loaded direct onto a face conveyor. The coal may be loaded by hand or mechanically. The face conveyor delivers its load of coal into tubs or cars or onto a gate conveyor. (Nelson, 1965)

conveyor-feeder operator *mill feeder.*

conveyor loader (a) Conveyor that at its extremity has a digging head that moves with the conveyor and works its way under the coal, which, by the unequal shaking of the conveyor, is carried back to the car. Also called shaking-conveyor loader. (Zern, 1928) (b) One who loads on a conveyor. See also: *loader.* (Zern, 1928)

conveyor man (a) Person who sets up and tends chain, belt, or shaker (reciprocating) conveyors to transport coal or metal ore about a tipple at the surface from working the working face in a mine. Also called loading-boom operator. (DOT, 1949) (b) In the quarry industry, a person who tends an endless conveyor belt used to transport rock from the crusher to storage bins. Syn: *beltman.* (DOT, 1949)

conveyor-operator tripper *tripper man.*

conveyor shaker type A conveyor designed to transport material along a line of troughs by means of a reciprocating or shaking motion. See also: *shaker conveyor.*

conveyor shifter A member of a team responsible for advancing the face conveyor as the coal is worked away. In many modern layouts, the armored conveyor is pushed forward by hydraulic rams. (Nelson, 1965)

conveyor track The path, parallel to the face, occupied by a longwall conveyor. The track is advanced every turnover. Syn: *track.* (Nelson, 1965)

conveyor-tripper operator *tripper man.*

conveyor-type feeder Any conveyor, such as apron, belt, chain, flight, pan, oscillating, screw, or vibrating, adapted for feeder service. See also: *apron feeder.*

cookeite A monoclinic mineral, $LiAl_4(Si_3Al)O_{10}(OH)_8$; chlorite group.

cooler arch An opening of truncated-cone shape in the tuyère breast of furnace. The tuyère cooler is placed in it. (Fay, 1920)

coolers Coolers in which atmospheric air is blown by a fan, through a nest of pipes, into a tower or chamber in which it comes into intimate contact with fine particles of water from atomizing nozzles. By the evaporation of some of this water the air rapidly becomes saturated at the wet-bulb temperature, the remaining water running off at the same temperature. This water is collected and pumped back through the nest of pipes, thereby cooling the air before it enters the spray chamber. The entering air then has a lower dry-bulb temperature than the atmosphere and, since its moisture content is unaltered, the wet-bulb is lower also. (Spalding, 1949)

cooling agent A chemical added to an explosive during manufacture to suppress or inhibit the flame produced in blasting. (BS, 1964)

cooling floor A floor upon which hot ore is placed for the purpose of cooling.

cooling load The total amount of sensible and latent heat to be removed from a space to maintain desired conditions. For mines in operation, it is possible to measure the actual amount of heat generated in underground openings by observing temperature changes in a known weight flow rate of mine air. For projected mines and extensions of operating mines, the amount of heat produced must be calculated, knowing which of the sources of underground heat is operative. (Hartman, 1982)

cooling power The rate at which air will remove heat from a body and may be measured dry or wet. The cooling power of air, as determined by the kata thermometer, is one of the basic environmental standards. (Hartman, 1982)

coontail ore Banded ore consisting mainly of fluorite and sphalerite in alternate light- and dark-colored layers; occurs in the Cave-in-Rock district of southern Illinois.

cooperite A tetragonal mineral, (Pt,Pd,Ni)S; sp gr, 9; in ultramafic rocks, such as the Bushveld, Transvaal, South Africa; an ore of platinum and palladium.

Cooper's lines An anastomosing meshwork of minute curved and branching lines produced in rock by shearing under pressure. (Goldman, 1952)

coordinate Any one of a set of numbers designating linear and/or angular quantities that specify the position of a point on a line, in space, or on a given plane or other surface in relation to a given reference system; e.g., latitude and longitude are coordinates of a point on the Earth's surface. The term is usually used in the plural, esp. to designate the particular kind of reference system (such as spherical coordinates, plane coordinates, and polar coordinates). (AGI, 1987)

coordinate system Crystallographers customarily use a right-handed system with the z axis oriented positive upward, the y axis positive to the right, and the x axis positive toward the viewer. Cf: *axis; crystallographic axes.*

coorongite (a) Elastic, bituminous substances derived from algae. (Schieferdecker, 1959) (b) A boghead coal in the peat stage. See also: *elaterite.* (Stutzer, 1940)

coose Lean; said of ores. (Hess)

copal An inclusive term for a wide variety of hard, brittle, semitransparent, yellowish to red fossil resins from various tropical trees (e.g., Copiafera and Agathis), being nearly insoluble in the ordinary solvents and resembling amber in appearance; e.g., Congo copal and kauri. Copal also occurs as modern resinous exudations. Syn: *gum copal.* See also: *amber.*

copaline *copalite.*

copalite An oxygenated hydrocarbon resembling copal from the blue clay of Highgate, near London, England. Syn: *copaline; fossil copal.*

Copaux-Kawecki fluoride process A process for converting beryl to beryllium oxide by sintering a mixture of beryl, soda ash, sodium silicofluoride, and sodium ferric fluoride, leaching with hot water, and adding caustic soda to precipitate beryllium hydroxide, which is calcined to beryllia. (USBM, 1965)

cope (a) Derb. To contract to mine lead ore by the dish, load, or other measure. (Fay, 1920) (b) An exchange of working places between miners. Also spelled coup. (Zern, 1928) (c) Derb. A duty or royalty paid to the lord or owner of a mine. (Fay, 1920) (d) Eng. A superficial deposit covering or coating the substrata. A cold, stiff, and wet clay. (Arkell, 1953) (e) The upper or topmost section of a flask, mold, or pattern. (ASM, 1961)

copel An alloy containing 55% copper and 45% nickel; used for thermocouples. (Newton, 1959; Newton, 1938)

coper Derb. One who contracts to mine lead ore at a fixed rate; a miner.

copi A name for gypsum, generally in weathered state.

copiapite (a) A triclinic mineral, $Fe^{2+}Fe^{3+}{}_4(SO_4)_6(OH)_2{\cdot}20H_2O$; Syn: *ferrocopiapite; yellow copperas; ihleite; knoxvillite.* (b) The mineral group aluminocopiapite, calciocopiapite, copiapite, cuprocopiapite, ferricopiapite, magnesiocopiapite, and zincocopiapite. Syn: *ihleite.*

coping (a) Cutting and trimming marble or other stone by use of a grinding wheel. (b) The top or cover of a wall usually made sloping to shed water. (c) In quarrying, the process of cutting one slab into two without regard to the finish of the edges. (AIME, 1960) (d) The material or units used to form a cap or finish on top of a wall, pier, or pilaster to protect the masonry below from the penetration of water from above. (ACSG, 1961) (e) Shaping stone or other hard nonmetallic material by use of a grinding wheel. (ACSG, 1963)

coping machine A machine consisting of a gearing and a carborundum wheel for cutting and trimming marble slabs. (Fay, 1920)

coplaning The process of moving the head of a theodolite laterally until its vertical axis lies in the produced vertical plane common to two plumblines. Syn: *alignment.* See also: *jiggling in.* (BS, 1963)

coppel *cupel.*

copper (a) A reddish metallic element that takes on a bright metallic luster and is malleable, ductile, and a good conductor of heat and electricity. Symbol, Cu. Occasionally occurs native, and is found in many minerals such as cuprite, malachite, azurite, chalcopyrite, and bornite. Its alloys, brass and bronze, are very important; U.S. coins are now copper alloys. Its oxides and sulfates are used as an agricultural poison and as an algicide in water purification. (Handbook of Chem. & Phys., 3) (b) An isometric native metal Cu; metallic, red, soft, ductile and malleable; sp gr, 8.9; in oxidized zones of copper deposits, formerly a major source of native copper; the only native metal to occur abundantly in large masses; commonly occurs in dendritic clusters or mossy aggregates, sheets, or in plates filling narrow cracks or fissures. See also: *native copper.*

copperas *melanterite; copiapite; goslarite; coquimbite.*

copperasine A sulfate of iron and copper resulting from the decomposition of copper pyrites. (Standard, 1964)

copperas stone Syn. for pyrite, from which copperas is often made. (Fay, 1920)

copper barilla Bol. Native copper in granular form mixed with sand. See also: *coro-coro; barilla.* (Fay, 1920)

copper bottoms A metallic product of very indefinite composition, made (usually) in reverberatory furnaces by smelting rich cupriferous substances without sufficient sulfur to quite satisfy the copper present. (Fay, 1920)

copper chalcanthite *chalcanthite.*

copper compress operator A laborer who compresses copper scrap into bales for use in charging refining furnaces, by operating a hydraulic ram. (DOT, 1949)

copper direct-firing process A metallurgical process for recovering copper from low-grade complex ores in which a mixture of the ore and a small quantity of salt and coke are heated, and the oxides or sulfides reduce to metal that migrates or segregates in the form of thin films or flakes. These are later recovered by conventional flotation procedures.

copper flower Any one of several indicator plants that serve as guides when prospecting for copper ores.

copper glance *chalcocite.*

copperheads Copper-colored spots—generally in a first coat on iron and not easily covered with a second coat. Copperheads are spots of excessive oxidation with red iron oxide producing the color. (Bryant, 1953)

copper ingots Notched bars of commercial copper used for casting purposes. The notches are provided for convenience in breaking the bars. (Mersereau, 1947)

copperization Impregnation with copper, or with some compound containing copper.

copper mica A miners' name for chalcophyllite. (Weed, 1918)

copper nickel *niccolite; nickeline.*

copper-ore germ A mixture of various copper minerals, such as green malachite, green or blue chrysocolla, blue azurite, and red cuprite. (Schaller, 1917)

copper pitch A jet black to brownish pitchlike material carrying from 12.12% to 84.22% CuO and found in the oxidized zone. It has a conchoidal fracture, and where it occurs in large enough pieces may resemble obsidian or anthracite coal. It apparently may be a mixture of the hydrous oxides of copper and iron, oxide and carbonate of copper, oxide and silicate of copper, or more or less hydrated oxides of copper and manganese. All the varieties may have more or less chalcedony mixed with them. (Hess)

copper precipitate Impure copper that has been precipitated from copper-bearing solutions; it may contain iron and arsenic; cement copper. (Camm, 1940)

copper-precipitation drum operator In ore dressing, smelting, and refining, one who precipitates copper from mine water by tumbling mine water and shredded steel cans in a revolving drum. (DOT, 1949)

copper pyrite *chalcopyrite.*

copper rain Minute globules thrown up from the surface of molten copper, when it contains but little suboxide. (Fay, 1920)

copper segregation process The process involves heating oxidized copper ore with a reducing agent and a halide salt at about 700 °C to produce metallic copper, which may then be recovered by ammonia leaching or by flotation with conventional copper sulfide collectors. (Rampacek, 1959)

copper slate Slate impregnated with copper minerals. (Fay, 1920)

copper smoke The gases from the calcination of copper sulfide ore. The gases contain sulfur dioxide, SO_2. (Fay, 1920; Hess)

copper suboxide *cuprite.*

copper sulfate *chalcanthite; copper sulfate pentahydrate.*

copper sulfate pentahydrate $CuSO_4{\cdot}5H_2O$; blue; triclinic; loses $5H_2O$ at 150 °C; white when dehydrated; slowly effloresces in air. Used in ore flotation and as a source of copper. (CCD, 1961; Handbook of Chem. & Phys., 2; Lee, 1961)

copper sulfide (a) A source of copper. (b) *covellite; cupric sulfide; indigo copper.* Cf: *digenite.*

copper titanate $CuTiO_3$. Sometimes added in quantities up to 2% to $BaTiO_3$ to increase the fired density. (Dodd, 1964)

copper uranite *uranite; torbernite.*

copper vitriol *chalcanthite.*

coppite (a) A niobium-containing mineral used as raw material in the production of ferroniobium. (Osborne, 1956) (b) A variety of tetrahedrite. (Hey, 1955)

coprecipitation The carrying down by a precipitate of substances that are normally soluble under the condition of precipitation. (AGI, 1987)

coproduct One of two commodities that must be produced to make a mine economic; both influence output. A byproduct is produced in association with a main product or with coproducts. See also: *byproduct.*

coprolite Petrified excrement.

coquimbite A trigonal mineral, $Fe_2(SO_4)_3{\cdot}9H_2O$; dimorphous with paracoquimbite. Syn: *white copperas.*

coquina A detrital limestone composed wholly or chiefly of mechanically sorted fossil debris that experienced abrasion and transport before reaching the depositional site and that is weakly to moderately cemented, but not completely indurated; esp. a porous light-colored limestone made up of loosely aggregated shells and shell fragments, such as the relatively recent deposits occurring in Florida and used for roadbeds and construction. (AGI, 1987)

coquinoid limestone A limestone consisting of coarse, unsorted, and often unbroken shelly materials that have accumulated in place without subsequent transportation or agitation, and generally having a fine-grained matrix. It is autochthonous, unlike the allochthonous coquina. (AGI, 1987)

coracite An alteration product of uraninite partly changed to gummite. Syn: *uraninite.* (Standard, 1964)

coral A general name for any of a large group of bottom-dwelling, sessile, marine invertebrate organisms (polyps) that belong to the class Anthozoa (phylum Coelenterata), are common in warm intertropical modern seas and abundant in the fossil record in all periods later than the Cambrian, produce external skeletons of calcium carbonate, and exist as solitary individuals or grow in colonies. (AGI, 1987)

coralgal Said of a firm carbonate rock formed by an intergrowth of frame-building corals and algae (esp. coralline algae). The material so formed is an excellent sediment binder in a coral reef. (AGI, 1987)

coral limestone A limestone consisting of the calcareous skeletons of corals, often containing fragments of other organisms and often cemented by calcium carbonate. (AGI, 1987)

coralline (a) Pertaining to, composed of, or having the form of a coral, as coralline limestone. (b) Any organism that resembles a coral in forming a massive calcareous skeleton or base, such as certain algae or stromatoporoids. (AGI, 1987)

coral mud and sand Marine deposits formed around coral islands and coasts bordered by coral reefs, containing abundant fragments of corals. Near the reefs the particle sizes are relatively coarse and the deposit is described as coral sand; farther out, the particles become gradually smaller until the material is a coral mud. (Holmes, 1928)

coral ore A curved, lamellar variety of liver-colored cinnabar from Idria, Austria. (Standard, 1964)

coral rag A well-cemented, rubbly limestone composed largely of broken and rolled fragments of coral-reef deposits; e.g., the Coral Rag of the Jurassic, used locally in Great Britain as a building stone. (AGI, 1987)

coral reef (a) A coral-algal or coral-dominated organic reef; a mound or ridge of in-place coral colonies and accumulated skeletal fragments, carbonate sand, and limestone resulting from organic secretion of calcium carbonate that lithifies colonies and sands. A coral reef is built up around a potentially wave- and surf-resistant framework, esp. of coral colonies, but often including many algae; the framework may constitute less than half of the reef volume. Coral reefs occur today throughout the tropics, wherever the temperature is suitable (generally above about 18 °C, a winter minimum). (AGI, 1987) (b) A popular term for an organic reef of any type. (AGI, 1987)

coral sand Sand-size particles formed from coral fragments. See also: *coral mud and sand.* (Hess)

coral zone The depth of the sea at which corals thrive. (Fay, 1920)

Cordaites A plant group, which is now extinct, that includes the Coniferales (pines and firs) and the Cycadales (cycads). The Cordaites were tall, slender trees that often attained heights of 100 ft (30.5 m). For a considerable height above the ground, the trunk was devoid of branches. The long, straplike leaves now form matted masses among the Coal Measure fossil plants. (Nelson, 1965)

cord-belt conveyor A rubber belt consisting of spaced cotton duck cords embedded in the rubber and protected at the top by a breaker strip with thick rubber cover. The bottom of the belt contains one or two plies of heavy duct, to give transverse strength. See also: *nylon belt.* (Nelson, 1965)

cordierite An orthorhombic mineral, $Mg_2Al_4Si_5O_{18}$; Mohs hardness, 7 to 7.5; an accessory in peraluminous granite, schist, and gneiss; a gem material called saphir d'eau, water sapphire, dichroite, and iolite. Syn: *polychroite.*

cordierite norite Metamorphosed norite containing cordierite. (Holmes, 1928)

cordillera (a) A comprehensive term for an extensive series or broad assemblage of more or less parallel ranges, systems, and chains of mountains, the component parts having various trends but the mass itself having one general direction; esp. the great mountain region of western North America from the eastern face of the Rocky Mountains to the Pacific Ocean, or the parallel chains of the Andes in South America; a mountain province. (AGI, 1987) (b) An individual mountain chain with closely connected, distinct summits resembling the strands of a rope or the links of a chain; e.g., one of the parallel chains of the Rocky Mountains. (AGI, 1987) (c) A term also used in South America for an individual mountain range. Etymol: Spanish, chain or range of mountains, from Latin chorda, cord. (AGI, 1987)

Cordirie process The refining of lead by conducting steam through it, while molten, to oxidize certain metallic impurities. (Fay, 1920)

cordite An explosive compound consisting of cellulose nitrate and a restrainer, such as vaseline, used chiefly as a propellant. (Standard, 1964)

cord of ore About 7 tons, but measured by wagonloads, and not by weight. The expression "cord" is a term used in some parts of Colorado and applied only to low-grade ore; the smelting ore is reckoned by the ton.

Cordtex A detonating fuse suitable for opencast and quarry mining. It consists of an explosive core of pentaerythritol tetranitrate (PETN) contained within plastic covering. It has an average velocity of detonation of 21,350 ft/s (6,500 m/s). This is practically instantaneous. Cordtex detonating fuse is initiated by electric or a No. 6 plain detonator attached to its side with an adhesive tape. (Nelson, 1965)

Cordtex relay A new device to achieve short-interval delay firing with Cordtex. A relay is an aluminum tube with a delay device, and is inserted in a line of Cordtex where required. The relays are made with two delays, 15 ms and 20 ms, respectively. (Nelson, 1965)

corduroy A ribbed and napped textile material used for recovering coarse gold or other heavy metal or mineral from a stream of sand passing over it. A corduroy blanket is replaced about every 4 hours for washing to remove the gold. (Nelson, 1965)

corduroy spar *graphic granite.*

corduroy texture Bands of coarse-grained quartz and albite or microcline in rock. (Hess)

cordylite A hexagonal mineral, $Ba(Ce,La)_2(CO_3)_3F_2$; rare; in pegmatites in nepheline syenites.

core (a) A cylindrical section of rock, usually 5 to 10 cm in diameter and up to several meters in length, taken as a sample of the interval penetrated by a core bit and brought to the surface for geologic examination and/or laboratory analysis. To obtain a core in drilling. (AGI, 1987) (b) The central part of the Earth below a depth of about 1,800 mi (2,900 km), probably consisting of iron-nickel alloy. (AGI, 1987) (c) A hard, unburned central part of a piece of coal or limestone. Cf: *clinker.* Also, an unburned or an overburned piece of limestone in hydrated lime. (Webster 3rd, 1966) (d) A cone or V-shaped mass of rock that is first blasted out in driving a tunnel. (e) The central part of an anticlinal structure, or of a domal structure, or of mountains having a folded or a completely crumpled structure. (Webster 3rd, 1966) (f) *drill core.* Cf: *clinker; bit core.*

core analysis (a) The characteristics of the minerals contained in a specific section of a core sample as determined petrographically, by metallurgical treatments and/or by chemical or cupelling methods. Also called core assay; core values. (Long, 1960) (b) As used by the petroleum industry, a study of a core sample to determine its water and oil content, porosity, permeability, etc. (Long, 1960)

Coreau detonnant Detonating fuse used in blasting. Syn: *Cordtex.* (Pryor, 1963)

core barrel (a) A hollow cylinder attached to a specially designed bit and which is used to obtain and to preserve a continuous section, or core, of the rocks penetrated in drilling. (AGI, 1987) (b) A tube inside a drill pipe and which is supported by a bit to receive the core, in core boring. (Webster 3rd, 1966)

core bit (a) A hollow, cylindrical boring bit for cutting a core in rock drilling or in boring unconsolidated earth material. It is the cutting end of a core drill. (b) A hollow, cylindrical drill bit for cutting a core of rock in a drill hole; the cutting end of a core drill. Syn: *coring bit.* (AGI, 1987)

core block An obstruction inside a bit, reaming shell, or core barrel consisting of impacted core fragments or drill cuttings, which prevents entry of core into the core barrel. See also: *block.* Syn: *core jam.* (Long, 1960)

core boring As used by soil- and foundation-testing engineers, a syn. for core; cuttings; drill sludge. (Long, 1960)

core box (a) The wooden, metal, or cardboard box divided into narrow parallel sections, used to store the cores at the surface as they are extracted from a core barrel or corer. (AGI, 1987) (b) The box in which the core, or mass of sand producing any hollow part of a casting, is made. (Fay, 1920)

core breaker (a) *core lifter.* (b) A sharp-cornered pluglike device inside an annular-shaped bit, which breaks up any core produced into pieces small enough to be washed out of the borehole as cuttings. (Long, 1960)

core catcher (a) Sievelike tray or device on or in which the core is ejected continuously from the upper end of a drill string, and is caught and held when core is recovered by counterflow or reverse-flow continuous core-drilling techniques. (Long, 1960) (b) *core lifter.* (c) A steel spring fitted at the lower end of a soil sampler to keep the sample from dropping out. (Nelson, 1965) (d) In deep boring, a ring of steel of wedge form cut into vertical stripes that encircles and rides on the core when drilling, but wedges the core in the core barrel when drilling ceases and the rods are lifted. (Nelson, 1965)

cored ammonium nitrate dynamite The dynamites of this class come in cartridges 4 in (10.2 cm) and up in diameter and in weight strengths from 20% to 70%. Their water resistance is considered good (the gelatin core being responsible for this), but their fume characteristics are rated as poor. Besides providing increased water resistance, these explosives tend to exhibit the higher velocities characteristic of gelatin explosives (10,500 ft/s, 15,000 ft/s, and 17,000 ft/s) (3,200 m/s, 4,600 m/s, and 5,200 m/s), rather than the low and medium velocities characteristic of other straight ammonia dynamite. In addition, the gelatin core assures propagation of detonation through the entire explosives column. Gelatin cored ammonia dynamites also are very useful when an operator wishes to practice alternate velocity loading to attain a more effective one-two punch in conjunction with the use of short period or millisecond delay, electric blasting caps. (Pit and Quarry, 1960)

cored hole (a) A borehole put down by a core drill. (Nelson, 1965) (b) A cast hole cored with a dry-sand core instead of delivering as a hole directly from the pattern. In general, the term is applied to any hole in a casting that is not bored or drilled in the shop. (Crispin, 1964)

core dressing A solution used to form clear skin at surface of core. (Pryor, 1963)

core drill (a) A rotary drilling rig that cuts and brings to the surface a core from the drill hole. It is equipped with a core bit and a core barrel. (AGI, 1987) (b) A lightweight, usually mobile drill that uses tubing instead of drill pipe and that can core down from the grass roots. (AGI, 1987) (c) A mechanism designed to rotate and cause an annular-shaped rock-cutting bit to penetrate rock formations, produce cylindrical cores of the formations penetrated, and lift such cores to the surface, where they may be collected and examined. See also: *calyx drill; diamond drill; rotary drill; shot drill.* (Long, 1960) (d) The act or process of producing a cylindrical core of rock, using a core-drilling machine and equipment. (Long, 1960) (e) A lightweight, usually mobile drilling machine equipped with a hollow core bit and a core barrel that by rotation cuts out and recovers a rock core sample. (AGI, 1987) (f) A drill that removes a cylindrical core from the drill hole. (Webster 3rd, 1966)

core drilling (a) The process of obtaining cylindrical rock samples by means of annular-shaped rock-cutting bits rotated by a borehole-drilling machine. (Long, 1960) (b) Drilling with a hollow bit and a core barrel to obtain a rock core.

core-drill sampling The act or process of obtaining cylindrical samples of rock in the form of a core. (Long, 1960)

core dryer A form in foundry work that serves to retain the shape of a core while it is being baked. (Crispin, 1964)

core extractor (a) A special tool that works like a screw or hydraulic jack, used to push core out of a core barrel. Also called core plunger; core pusher. (Long, 1960) (b) A fishing tool designed to recover core dropped from a core barrel and resting on the bottom of a borehole. Also called basket; core basket; core fisher; core grabber; core picker. (Long, 1960)

core grouting Material used in and/or the act or process of injecting small fragments of rock or coarse sand into a core barrel to wedge the core inside the barrel when no core lifter is used, as when using straightwall bits or drilling with a shot drill. (Long, 1960)

core hole A boring by a diamond drill or other machine that is made for the purpose of obtaining core samples. (AGI, 1987)

core intersection The point in a borehole where an ore vein or body is encountered, as shown by the core; also, the width or thickness of the orebody, as shown by the core. Also called core interval. (Long, 1960)

core interval *core intersection.*

core jam *core block.*

core library A structure in which boxed cores from numerous recorded localities are stored and kept available for inspection and study. (Long, 1960)

core lifter A spring clip at the base of the core barrel that grips the core, enabling it to be broken off and brought out of the hole. Also called core clip; core gripper; core spring; ring lifter; split-ring lifter. Syn: *catcher; core breaker; core catcher; spring lifter; spring core lifter; lifter.* (BS, 1963)

core load The explosive core of detonating cord, expressed as the number of grains of explosive per foot or grams per meter. (Atlas, 1987)

core loss The portion of rock cored but not recovered. Cf: *core recovery.* (Long, 1960)

core of the Earth (a) The dense central part of the Earth, below a depth of about 1,800 mi (2,900 km). Syn: *centrosphere.* (Schieferdecker, 1959) (b) The Earth is believed to consist of the following: inner core, solid, 860 mi (1,384 km) radius; outer core, liquid, 1,300 mi (2,092 km) thick; mantle, solid, 1,800 mi (2,897 km) thick; and crust, solid, 622 mi (1,001 km) thick. (Hunt, 1965)

core orientation (a) The act or process of using information obtained from magnetic polarity or other measurements of a piece of core in an attempt to determine the downhole bearing of the structural features of the rock formation as displayed in the core. (Long, 1960) (b) To place a piece of core in the same relative plane as it occupied below the surface. See also: *true dip.*

core plug A cylinder containing chemically treated sand and used for stemming shotholes in coal mines. (Nelson, 1965)

core rack (a) A framework built to support several tiers of core boxes. (Long, 1960) (b) Grooved or partitioned tray, supported on legs or sawhorses, on which core is placed when removed from a core barrel for inspection or temporary storage before being placed in boxes. (Long, 1960)

core recovery (a) The proportion of the drilled rock column recovered as core in core drilling. The amount withdrawn generally is expressed as a percentage of the theoretical total in general terms, as excellent, good, fair, or poor. Cf: *core loss.* (AGI, 1987; Long, 1960) (b) The amount of the drilled rock withdrawn as core in core drilling, generally expressed as a percentage of the total length of the interval cored. (AGI, 1987)

core run Technically, the distance cored per round trip, which is expressed in number of feet or in relative terms, as short or long. Core blocks may occur before the core barrel is filled; the barrel then is short of being full, resulting in a short core run. Loosely, the amount of core recovered per round trip. (Long, 1960)

core sample One or several pieces of whole or split parts of core selected as a sample for analysis or assay. (Long, 1960)

core sand Silica sand to which a binding material has been added to obtain good cohesion and porosity after drying for the purpose of making cores. (Osborne, 1956)

core saw A machine capable of rotating at high speed, equipped with a thin metal disk having diamonds inset in its edge. Used somewhat like a bench saw to cut core longitudinally into sections. Cf: *core splitter; diamond-saw splitter.* (Long, 1960)

core shack A roofed and enclosed structure in which core-filled boxes are stored. Also called core house; core shanty. (Long, 1960)

core sludge The slurry produced during abrasion by the cutting bit, or through fracture and grinding of part of the sample during this process. (Pryor, 1963)

core splitter Tool employing a chisel to split core longitudinally in half, rarely in quarter, sections. One-half usually is assayed, and the other half is retained and stored. Term also may be applied to a diamond saw used for the same purpose. See also: *core saw.* (Long, 1960)

core test A hole drilled with a core drill, usually for the purpose of securing geologic information and sometimes with the purpose of investigating geologic structure. (AGI, 1987)

core texture *atoll texture.*

core-type spiral chute A spiral chute having a center core or column about which it is fabricated, with the core serving as the inside guard.

core values Used in a general sense as a syn. for core analysis; core assay. In a strict sense, the term should not be used to designate the mineral content of the core sample unless the valuable mineral is gold, silver, platinum, etc. (Long, 1960)

core velocity The zone of maximum air velocity in a mine roadway, usually at or near the center of the road. (Nelson, 1965)

core wall In a battery wall, those courses of brick, none of which are directly exposed on either side. (AISI, 1949)

core wash (a) The portion of the core lost through erosive action of the drill circulation fluid. (Long, 1960) (b) The act or process of erosion of core by washing action of the drill circulation fluid. (Long, 1960)

coring A variable composition between the center and surface of a unit of structure (such as a dendrite, grain, or carbide particle) resulting from nonequilibrium growth that occurs over a range of temperature. (ASM, 1961)

coring bit *core bit.*

coring tool A tool that is used when a core is required. In drilling, where speed is the aim, cores are not made. When, however, an important bed or horizon is approached, and detailed geological information is required, the coring bit is inserted and core drilling commenced. Also called corer. (Nelson, 1965)

Corinthian process *Carinthian process.*

cork fossil A variety of amphibole or hornblende, resembling cork; the lightest of all minerals. (Fay, 1920)

corkscrew (a) A device resembling a corkscrew, used as a fishing tool. (Long, 1960) (b) A borehole following a spiraled course. (Long, 1960) (c) A cylindrical surface, such as the outer surface of a piece of spirally grooved core. Also called fluted core. (Long, 1960)

corkscrew core *fluted core.*

corncob *taper bit.*

cornelian A translucent red variety of chalcedony. Also spelled carnelian. (CMD, 1948)

corneous manganese (a) *photicite.* (b) A carbonated variety of rhodonite.

corner (a) A point on a tract of land at which two or more surveyed boundary lines meet; e.g., a township corner. (AGI, 1987) (b) A term that is often incorrectly used to denote the physical station, or monument, erected to mark the corner. (AGI, 1987)

corner-fastened tray conveyor *suspended tray conveyor.*

corner-hung tray conveyor *suspended tray conveyor.*

corner racking Square or triangular strips of pinewood fixed vertically down each corner of a rectangular shaft to secure and stiffen the timber sets. (Nelson, 1965)

corners In Wales, bands of clay ironstone.

cornetite An orthorhombic mineral, $Cu_3(PO_4)(OH)_3$; occurs in peacock-blue minute crystals and encrustations in Katanga, Zaire, and Bwana M'Kubwa, Zambia.

Corning table *Bilharz table.*

Cornish diamond Eng. A quartz crystal from Cornwall. (Webster 3rd, 1966)

Cornish engine *Cornish pump.*

Cornish mining ton The weight equal to 21 hundredweight of 112 lb each, or 2,352 lb (1,066.87 kg). (Webster 2nd, 1960)

Cornish pump A single-acting engine in which the power for pumping operations was transmitted through the action of a cumbersome beam. Syn: *Cornish engine.* (Nelson, 1965)

Cornish rolls A geared pair of horizontal cylinders, one fixed in a frame and the other held by strong springs. The distance apart is adjusted by distance pieces of shims. Used for grinding. (Pryor, 1963)

Cornish stone A variety of china stone composed of feldspar, mica, and quartz and used as a bond in the manufacture of pottery. Syn: *Cornwall stone.* Cf: *china stone.* (AGI, 1987)

cornubianite A hornfels formed by contact metamorphism, and consisting of micas, quartz, and feldspar. Cf: *leptynolite.* Etymol: From the classic name for Cornwall, England. (AGI, 1987)

cornubite A triclinic mineral, $Cu_5(AsO_4)_2(OH)_4$; dehydrated from cornwallite.

cornuite (a) A yellow, gelatinous substance, apparently in albumen with 97% water; found in fissures in diatomite deposit of Luneburger Heide, Hanover, Germany. It may be an organic matter derived from the diatoms or a fungus. (Tomkeieff, 1954) (b) A blue, green, hydrous copper silicate; glassy. The colloidal phase of chrysocolla. (English, 1938)

cornwallite A monoclinic mineral, $Cu_5(AsO_4)_2(OH)_4 \cdot H_2O$; emerald-green; dehydrates to cornubite.

Cornwall stone *Cornish stone.*

coro-coro A dressed product of copper works in South America, consisting of grains of native copper mixed with pyrite, chalcopyrite, mispickel, and earthy minerals. See also: *copper barilla; barilla.*

Coromant cut A new drill hole pattern in which two overlapping holes of diameter about 2-1/4 in (5.7 cm) are drilled in the tunnel center and left uncharged. These holes form a slot roughly 4 in by 2 in (10.2 cm by 5.1 cm) to which the easers can break. All the holes in the round are parallel and in line with the tunnel. Short-delay detonators are used for the easer holes and 1/2-s delays for the rest of the round. A pull of 10 ft (3.0 m) per round has been obtained in strong rock with 10.5-ft (3.2-m) holes. Explosive consumption for the easer holes is about 0.2 lb/ft (0.3 kg/m) of hole. (Nelson, 1965)

corona (a) A microscopic zone of minerals, usually arranged radially around another mineral. The term has been applied to reaction rims, corrosion rims, and originally crystallized minerals. (AGI, 1987) (b) A Spanish term meaning crown. Sometimes used in the Southwestern United States as a syn. for diamond bit. (Long, 1960) (c) Rim of alteration product surrounding an earlier formed crystal, commonly the result of reaction with a cooling magma. Syn: *kelyphitic rim; kelyphite; reaction border; kelyphytic rim.* See also: *reaction rim.*

coronadite A monoclinic mineral, $Pb(Mn^{4+},Mn^{2+})_8O_{16}$; cryptomelane group; pseudotetragonal; at the Coronado vein, Clifton-Morenci district, AZ.

coronite A rock containing mineral grains surrounded by coronas. (AGI, 1987)

corrasion (a) A process of erosion whereby rocks and soil are mechanically removed or worn away by the abrasive action of solid materials moved along by wind, waves, running water, glaciers, or gravity. Syn: *abrasion; attrition.* (AGI, 1987) (b) A term formerly used as a syn. of corrosion, or as including the work of corrosion. (AGI, 1987)

corrected effective temperature The scales of effective temperature take into consideration the temperature, humidity, and speed of the air. The effects of radiant heat can be included in an assessment of effective temperature by using the globe thermometer temperature instead of the dry-bulb temperature in those cases when the reading of the globe thermometer is higher than the dry-bulb temperature. In such cases, the result is described as the corrected effective temperature. (Roberts, 1960)

correcting wedge A deflection wedge used to deflect a crooked borehole back into its intended course. See also: *deflecting wedge.* (Long, 1960)

correction chart A chart, graph, or table giving the true angle of the inclination of a borehole for specific apparent angles as read from the etch line in a specific-size acid bottle. See also: *capillarity-correction chart.* (Long, 1960)

correction factor *assay plan factor.*

correction line *standard parallel.*

correctly placed material (a) Material correctly included in the products of a sizing or density separation. (BS, 1962) (b) In cleaning, the material of specific gravity lower than the separation density that has been included in the low-density product, or material of specific gravity higher than the separation density that has been included in the high-density product. (BS, 1962)

correlate (a) To show correspondence in character and stratigraphic position between such geologic phenomena as formations or fossil faunas of two or more separated areas. Adj. belonging to the same stratigraphic position or level. (AGI, 1987) (b) To establish a definite stratigraphic relationship between strata that are separated by distance or by geologic disturbance; e.g., to find which coalbeds in one coalfield or part thereof correspond with (or are the same as) those of another coalfield. (c) To plot or to arrange two surveys, the surveys of two mines, or the underground and the surface, on the same base line or to a common meridian. (Mason, 1951)

correlation (a) The determination of the equivalence in geologic age and/or stratigraphic position of two formations or other stratigraphic units in separated areas; or, more broadly, the determination of the contemporaneity of events in the geologic histories of two areas. Fossils constitute the chief evidence in problems of such correlation. See also: *lithologic correlation.* (b) The identification of a phase of a seismic record as representing the same phase on another record, thus relating reflections from the same stratigraphic sequence or refractions from the same marker. (AGI, 1987)

correlation shooting A seismic shooting method in which isolated profiles are shot and correlated to obtain relative structural positions of the horizons mapped. Cf: *continuous profiling.* (AGI, 1987)

corrensite A clay mineral having 1:1 regular interstratification of trioctahedral chlorite with either trioctahedral vermiculite or trioctahedral smectite.

corridor system *methane drainage.*

corrode (a) To eat away by degrees as if by gnawing. (Webster 3rd, 1966) (b) To wear away or to diminish by gradually separating or destroying small particles or converting into an easily disintegrated substance; esp., to eat away or to diminish by acid or alkali reaction or by chemical alteration. (Webster 3rd, 1966)

corroded crystal A phenocryst that after crystallization is more or less reabsorbed or attacked by the magma, or a crystal in a vein or a pegmatite that is partly dissolved by later solutions. The process is probably much the same in all three instances. (Hess)

corroding lead Lead of purity exceeding 99.94% , suitable for the production of white lead. (CTD, 1958)

corrosion (a) A process of erosion whereby rocks and soil are removed or worn away by natural chemical processes, esp. by the solvent action of running water, but also by other reactions such as hydrolysis, hydration, carbonation, and oxidation. Syn: *chemical erosion.* (AGI, 1987) (b) A term formerly used interchangeably with corrasion for the erosion of land or rock, including both mechanical and chemical processes. The mechanical part is now properly restricted to corrasion and the chemical to corrosion. Verb: corrode. (AGI, 1987) (c) *magmatic corrosion; abrasion.* See also: *attrition.* (AGI, 1987)

corrosion border One of a series of borders of one or more secondary minerals around an original crystal, representing the modification of a phenocryst due to the corrosive action of its magma. Cf: *reaction rim.* Syn: *corrosion zone; resorption border.* (AGI, 1987)

corrosion potential The steady-state irreversible potential of a metal or alloy in a constant corrosive environment. (Schlain)

corrosion rate The rate that a metal or alloy is removed because of corrosion. This may be expressed in terms of loss in weight or loss of thickness in a given period of time. (Corrosion rates in terms of thickness change refer to the loss of metal from one side only.). (Hunt, 1965)

corrosion surface A pitted, irregular bedding surface found only in certain carbonate sediments, characterized by a black manganiferous stain, and presumed to result from cessation of lime deposition and from submarine solution or resorption of some of the previously deposited materials. Syn: *corrosion zone.* (AGI, 1987)

corrosion zone *corrosion surface; corrosion border.*

corrugated Where on a small scale, beds are much wrinkled, folded, or crumpled, they are said to be corrugated. On a larger scale, they are said to be contorted.

corrugated friction socket A fishing tool. (Long, 1960)

corrugated trough A trough with corrugations formed into the bottom to assist coal travel on steep grades or under wet conditions. (Jones, 1949)

cortex In coal, that part of the axis of a vascular plant that surrounds the central cylinder and is separated from the cylinder by the endodermis, and limited on the outside by the epidermis.

corundolite A rock consisting of corundum and iron oxides. See also: *emery rock.*

corundophilite An iron-bearing variety of clinchlore.

corundum A trigonal mineral, Al_2O_3; hematite group; forms hexagonal prisms with basal and rhombohedral parting; red (ruby), blue (sapphire), green (oriental emerald), reddish-brown, white, or gray; defines 9 on the Mohs hardness scale; in nepheline syenite pegmatites and placer deposits. Emery is granular corundum mixed with magnetite and spinel. Synthetic corundum made from bauxite together with other manufactured abrasives have largely replaced natural materials.

corundum cat's eye Corundum showing a bluish, reddish, or yellowish reflection of light, or lighter shade, than the stone itself. Cf: *asterism; star ruby; star sapphire.*

corvusite A monoclinic mineral, $(Na,K,Ca,Mg)_2(V^{5+},V^{4+})_8O_{26}\cdot 6\text{-}10H_2O$; weakly radioactive; associated with carnotite in Colorado and Utah; a source of vanadium. Syn: *blue-black ore.*

cosalite An orthorhombic mineral, $Pb_2Bi_2S_5$; a source of bismuth.

cosedimentation Contemporaneous deposition. (AGI, 1987)

cosmochemistry The study of the origin, distribution, and abundance of elements in the universe. (AGI, 1987)

costean (a) A trench cut across the conjectured line of outcrop of a seam or orebody to expose the full width. (Nelson, 1965) (b) The channel eroded by a flow of water to expose mineral deposits during prospecting work. (Nelson, 1965) (c) In prospecting, to dig shallow pits or trenches designed to expose bedrock. Etymol: Cornish. (AGI, 1987)

costeaning (a) The removal of soil and subsoil by a rushing of water, to expose rock formations in prospecting for reefs or lodes. (CTD, 1958) (b) Proving an ore deposit or vein by trenching across its outcrop at approx. right angles. (Weed, 1922) (c) Tracing a lode by pits sunk through overburden to underlying rock. (Pryor, 1963)

costean pit Corn. A pit sunk to bedrock in prospecting. (Standard, 1964)

cotectic line A special case of the boundary line, in ternary systems, along which one of the two crystalline phases present reacts with the liquid, upon decreasing the temperature, to form the other crystalline phase. Syn: *reaction curve; reaction line.* (AGI, 1987)

cotectic surface A curved surface in a quaternary system, representing the intersection of two primary phase volumes, one or both of which are solid solution series. It is the bivariant equivalent of the univariant cotectic line in ternary systems. (AGI, 1987)

cotter Eng. To mat together; to entangle. Frequently applied to a hard, crossgrained, tough stone or coal, as cottered coal.

cotterite A variety of quartz having a peculiar metallic pearly luster. (Standard, 1964)

cotton ball *ulexite.*

cotton chert An obsolete syn. of chalky chert.

cotton rock (a) A term used in Missouri for a soft, fine-grained, siliceous, white to slightly gray or buff magnesian limestone having a chalky or porous appearance suggestive of cotton. (AGI, 1987) (b) The white or light-colored decomposed exterior surrounding the dense black interior of a chert nodule. (AGI, 1987)

cotton stone A variety of mesolite. See also: *mesolite; cotton rock.* (Fay, 1920)

Cottrell meter This instrument applies the veiling brightness method of producing threshold conditions. When in use the sighting telescope is directed toward some critical detail of the visual task and the veiling brightness is adjusted until it matches the background. The gradient filter is then turned until the target detail is at threshold visibility. (Roberts, 1958)

Cottrell operator In ore dressing, smelting, and refining, one who recovers magnesium dust particles remaining in magnesium gas after processing, using a battery of Cottrell electrical precipitators. Also called agglomerator operator; dust operator. (DOT, 1949)

Cottrell precipitator An electrostatic device whereby negatively charged dust or fume particles are attracted to a positively charged wire electrode enclosed in a flue, the walls of which act as the other electrode. Widely used for treating sulfuric acid mist, cement mill dust, power-plant fly ash, metallurgical fumes, etc. (CCD, 1961)

cotunnite An orthorhombic mineral, $PbCl_2$; soft; acicular crystals.

coulee (a) A term applied in the Western United States to a small stream, often intermittent. Also, the bed of such a stream when dry. (AGI, 1987) (b) A term applied in the Northwestern United States to a dry or intermittent stream valley, gulch, or wash of considerable extent; esp. a long, steep-walled, trenchlike gorge or valley representing an abandoned overflow channel that temporarily carried meltwater from an ice sheet, e.g., the Grand Coulee (formerly occupied by the Columbia River) in Washington State. (AGI, 1987) (c) A small valley or a low-lying area. Etymol: French coulée, flow or rush of a torrent. Pron: koo-lee. Syn: *coulie.* (AGI, 1987) (d) A tonguelike mass of debris moved by solifluction (Monkhouse, 1965, p. 81). (AGI, 1987) (e) A flow of viscous lava that has a blocky, steep-fronted form. Also spelled: coulée. (AGI, 1987)

coulee lake A lake produced by the damming of a water course by lava. (AGI, 1987)

coulie *coulee.*

coulomb attraction The attraction between ions of opposite electric charges. (AGI, 1987)

coulomb damping (a) The dissipation of energy that occurs when a particle in a vibrating system is resisted by a force whose magnitude is a constant independent of displacement and velocity, and whose direction is opposite to the direction of the velocity of the particle. Also called dry friction damping. (Hunt, 1965) (b) *specific damping capacity.*

coulsonite An isometric mineral, $Fe^{2+}V^{3+}{}_2O_4$; spinel group; a source of vanadium; formerly called vanado-magnetite.

Coulter counter A high-speed device for particle size analysis designed by W.H. Coulter and now made by Coulter Electronics, Inc., Chicago. A suspension of the particles flows through a small aperture having an immersed electrode on either side with particle concentration such that the particles traverse the aperture substantially one at a time. Each particle, as it passes, displaces electrolyte within the aperture, momentarily changing the resistance between the electrodes and producing a voltage pulse of magnitude proportional to practical volume. The resultant series of pulses is electronically amplified, scaled, and counted. (Dodd, 1964)

counter (a) A gangway driven obliquely upwards on a coal seam from the main gangway until it cuts off the faces of the workings, and then continues parallel with the main gangway. The oblique portion is called run. (b) A crossvein. (c) An instrument for the detection of uranium and thorium. (Nelson, 1965) (d) A term used for any device that registers radioactive events, i.e., alpha counter, beta counter, Geiger-Müller counter, scintillation counter. The term is correctly used only for devices that actually register number of events, but is often erroneously applied to count rate meters that register events per unit time. (e) An apparatus for recording the number of strokes made by a pump, an engine, or other machinery.

counterboring Drilling or boring a flatbottomed hole, often concentric with other holes. Syn: *counterflush boring; reversed flush boring.* (ASM, 1961)

counter chute A chute through which the coal from counter-gangway workings is lowered to the gangway below. (Fay, 1920)

counter coal Coal worked from breasts or bords to the rise of a counter gangway. (Fay, 1920)

countercurrent Arrangement in which ore, or pulp, proceeds in one direction and is progressively stripped of part of its contained mineral, while the enriched fraction thus produced moves in the opposite direction, the results being central feed, with discharge of high-grade concentrate at one end of the process and low-grade or barren tailing at the other. (Pryor, 1963)

countercurrent braking Braking accomplished by reversing the motor connections, at the same time inserting appropriate resistance in the rotor circuit to adjust the negative torque to the desired value. With this method, complete control of deceleration is obtained, even to a dead stop. Its greatest disadvantage is that it is expensive in current consumption. It is unsuitable for winders sited at depth, owing to the heat given out. (Spalding, 1949)

countercurrent decantation The clarification of washery water and the concentration of tailings by the use of several thickeners in series. The water flows in the opposite direction from the solids. The final products are slurry that is removed as fluid mud and clear water that is reused in the circuit. May be broader than just thickener. Syn: *cascade upgrading.* (Nelson, 1965)

countercurrent principle A means of maintaining the chemical potential at a uniform level during a reaction. (Newton, 1959; Newton, 1938)

counterflow In a heat exchanger, where the fluid absorbing heat and the fluid losing heat are so directed that lower and higher temperature of the one is adjacent to the lower and higher temperature of the other, respectively. Ordinarily, the one fluid is flowing in the opposite direction from the other, hence the term. (Strock, 1948)

counterflush boring *counterboring.*

counter gangway A gangway driven obliquely across the workings to a higher level, or a gangway driven between two lifts and sending its coal down to the gangway below through a chute. (Fay, 1920)

counterhead Mid. An underground heading driven parallel to another, and used as the return air course.

counterpart *cast.*

countersink In a twist drill, the tapered and relieved cutting portion situated between the pilot drill and the body. (Osborne, 1956)

countervein A cross vein running at approx. right angles to the main orebody.

countess Slate, size 20 in by 10 in (50.8 cm by 25.4 cm); a duchess is 24 in by 12 in (61.0 cm by 30.5 cm), and a princess is 24 in by 14 in (61.0 cm by 35.6 cm). Terms descriptive of slate trimmed for roofing. (Pryor, 1963)

counting assay Approximate method of analysis, where particles of value and gangue are similar in shape and size, and their proportions can be assessed by inspection, probably under a low-powered microscope. (Pryor, 1963)

country bank Arkansas. A small mine supplying coal for local use only.

country rock The rock enclosing or traversed by a mineral deposit. Originally a miners' term, it is somewhat less specific than host rock. Syn: *wall rock.* (AGI, 1987)

County of Durham system A combination of the panel and room-and-pillar method of mining. See also: *room-and-pillar.*

coupled wave A type of surface wave that is continuously generated by another wave that has the same phase velocity. Syn: *C-wave.* (AGI, 1987)

couplet Genetically related paired sedimentary laminae, generally occurring in repeating series, as varves, but applied to laminated nonglacial shales, evaporites, and other sediments as well. (AGI, 1987)

coupling (a) A device for connecting tubs or mine cars to form a set or journey. See also: *automatic clip; clip; shackle.* (Nelson, 1965) (b) A connector for drill rods, casing, or pipe with identical box or pin threads at either end. (Long, 1960)

coup plate In coal mining, steel plate on which tubs are turned from one set of rails to another. (Pryor, 1963)

course (a) To conduct the ventilation backward and forward through the workings, by means of properly arranged stoppings and regulators. (Fay, 1920) (b) A seam of coal. (Fay, 1920) (c) To ventilate a number of faces in series. (BS, 1963) (d) An unproductive vein as opposed to a lode. (e) The horizontal direction of a geologic structure. Syn: *course of ore; strike.* (Webster 3rd, 1966)

coursed ventilation Mine ventilation by the same air current, i.e., without splitting of air. (Nelson, 1965)

course of ore (a) A horizontal shoot. An older term. (Nelson, 1965) (b) *chute; course.*

course of vein The strike of a vein; direction of the horizontal line on which it cuts the country rock.

course stacking The method of shovel operation in which no ground is hauled away. The shovel simply stacks the ground on the opposite side from the working cut, or it may turn entirely around, dumping the spoil on a bank behind. (Lewis, 1964)

coursing The control of ventilation in mines, as by doors, brattices, and stoppings. (Standard, 1964)
coursing bubble One rising freely through the cell during froth flotation. (Pryor, 1965)
courthouse A method used by companies for checking the amount of refuse in coal. The refuse is picked daily from a few cars of run-of-mine coal, and when the amount of refuse is considered unreasonable, it is shown to the miner and the laborers. They may be suspended from work if the amount and size of refuse is too high. (Mitchell, 1950)
courthouse inspector In bituminous coal mining, one who examines mine cars of coal for impurities, such as slate, rock, and dirt, by the courthouse system (selecting cars at random for examination). Rejects, on basis of inspection, any group or lot of cars containing too much impurity. (DOT, 1949)
courtzilite A form of asphaltum allied to gilsonite. (Fay, 1920)
cousin Jack Cornish miner, usually far from home, important to U.S. mining. (Pryor, 1963)
covariance A statistical measure of the correlation between two variables. In geostatistics, covariance is usually treated as the simple inverse of the variogram, computed as the overall sample variance minus the variogram value. These covariance values, rather than variogram values, are actually used in kriging matrix equations for greater computational efficiency.
covelline *covellite.*
covellite A hexagonal mineral, CuS; metallic indigo blue with iridescent tarnish; soft; a supergene mineral in copper deposits; a source of copper. Syn: *blue copper; covelline; indigo copper.* See also: *copper sulfide.*
cover (a) The sedimentary accumulation over the crystalline basement. See also: *cover mass.* (AGI, 1987) (b) The vertical distance between any position in strata and the surface or any other position used as a reference. *surface.* (AGI, 1987) (c) The pattern or number of drill holes (pilot holes) deemed adequate to detect water-bearing fissures or structures in advance of mine workings. Syn: *pilot-hole cover.* (Long, 1960) (d) Total thickness of material overlying mine workings or an orebody. See also: *burden; mantle; cover rock.* Cf: *rock cover.* (Long, 1960)
cover brick Common term for arch brick used to line soaking-pit covers.
cover gap The area in advance of mine workings not adequately probed by pilot holes to detect the presence of water-bearing fissures or structures. (Long, 1960)
cover hole One of a group of boreholes drilled in advance of mine workings to probe for and detect water-bearing fissures or structures. (Long, 1960)
cover line The point at which the overburden meets the coal. (BCI, 1947)
cover load The load due to the weight of the superincumbent rock. (Issacson, 1962)
cover mass The material overlying the plane of an angular unconformity. See also: *cover.*
cover rock *cover.*
cover stress The stress induced by the cover load only and which is uninfluenced by the proximity of any excavations. (Issacson, 1962)
cover work Lumps of copper too large to pass the screen, which accumulate in the bottom of the mortar of the stamp. (Fay, 1920)
cow York. The finest crushed lead ore. Also called coe. (Arkell, 1953)
Cowper-Siemens stove A hot-blast stove of firebrick on the regenerative principle. (Fay, 1920)
cowshut Gray marl. Syn: *cushat marl.* (Arkell, 1953)
cow sucker A cylindrical heavy piece of iron attached to a cable or wire line, making it descend rapidly into a borehole when the cable or line is not attached to a string of drilling tools or equipment. Also called bug; bullet; go-devil. (Long, 1960)
coyote hole A small tunnel driven horizontally into the rock at right angles to the face of the quarry. It has two or more crosscuts driven from it parallel to the face. It is in the ends of these crosscuts that the explosive charge is generally placed, and the remaining space in the tunnel is filled up with rock, sand, timbers, or concrete, to act as stemming or tamping. Same as gopher hole.
coyote-hole blasting *coyote shooting.*
coyote shooting A method of blasting using a number of relatively large concentrated charges of explosives placed in one or more small tunnels driven in a rock formation. Syn: *coyote-hole blasting.* (Atlas, 1987)
C.P. Hemborn dust extractor A dust trap in which the clean air flows inwards around the outside of the drill rods, and the dust and chippings are extracted in the airstream passing through the hollow rods. It includes a drum-type dust container with filter units. The appliance requires special rods and bits. See also: *dust trap.* (Nelson, 1965)
cpx Abbrev. for clinopyroxenes. Cf: *opx.*
crab locomotive A trolley locomotive fitted with a crab or winch for hauling mine cars from workings where a trolley wire is not installed. (Nelson, 1965)
crab operator In bituminous coal mining, one who maintains and operates a crab (electric motor equipped with a drum and haulage cable mounted on a small truck) to pull loaded mine cars from working places to haulageways in the mine. (DOT, 1949)
crab winch An iron machine consisting of two triangular uprights between which are two axles, one above the other. These machines are frequently used in connection with pumping gear where mine shafts are not deep. Also called crab. (Fay, 1920)
crackle breccia An incipient breccia having fragments parted by planes of rupture but showing little or no displacement. It is commonly a chemical deposit. (AGI, 1987)
crackled texture A concentric texture of ore minerals in which minute cracks have developed by shrinkage during crystallization of the original colloid.
cracks of gas Puffs or explosions of gas in blast furnaces. (Fay, 1920)
crack wax A dark-colored variety of ozokerite showing a granular fracture. (Tomkeieff, 1954)
cradle (a) A wooden box, longer than wide provided with a movable slide and hopper and mounted on two rockers, for washing gold-bearing earths. Also known as rocker cradle. (b) The part of a car dumper in which the car rests when it is dumped. (c) The balance platform for the cage at the bottom of some shafts. (Mason, 1951) (d) Device by means of which a small diamond or percussive-type drill may be attached to a drill column or arm. Also called saddle. (Long, 1960) (e) The trough-shaped metal support for a mounted pneumatic drill. (CTD, 1958) (f) To wash, as gold-bearing gravel in a mining-cradle. (Standard, 1964) (g) Mounting for a rock drill. (Pryor, 1963)
cradle dump A tipple that dumps cars with a rocking motion. (Fay, 1920)
Craelius drilling machine A small, fairly light boring machine for shallow exploratory borings underground. It drills in any direction (downwards, upwards, horizontally, or obliquely) to depths of from 200 to 1,000 m, but usually only 50 m. It uses coring or solid bits, with or without flushing and can be driven either by hand, any oil engine, compressed air, or electricity. (Stoces, 1954)
crag (a) A steep precipitous point or eminence of rock, esp. one projecting from the side of a mountain. Syn: *craig.* (AGI, 1987) (b) An obsolete term for a sharp, rough, detached, or projecting fragment of rock. (AGI, 1987)
craig *crag.*
cramp (a) A contrivance for holding parts of a frame in place during construction. It usually consists of a steel bar along which slide two brackets between which the work is fixed, one of the brackets being pegged into a hole in the bar while the other is adjustable for position by means of a screw. (CTD, 1958) (b) A locking bar of incorrodible metal used to bind together adjacent stones in a course, and having bent ends, one of which is fastened into each stone. Also called a cramp iron. (CTD, 1958)
crampon An appliance for holding stones or other heavy objects that are to be hoisted by crane. It consists of a pair of bars hinged together like scissors, the points of which are bent inwards for gripping the load, while the handles are connected by short lengths of chain to a common hoist ring. (CTD, 1958)
crandall (a) A stonecutters' hammer for dressing ashlar. Its head is made up of pointed steel bars of square section wedged in a slot in the end of the iron handle. (Standard, 1964) (b) To dress stone with a crandall. (Standard, 1964)
crandallite (a) A trigonal mineral, $CaAl_3(PO_4)_2(OH)_5 \cdot H_2O$; forms compact to cleavable or fibrous masses; formerly called kalkwavellite. (b) The crandallite mineral group of trigonal phosphates and arsenates: arsenocrandallite, arsenoflorencite-(Ce), arsenogorceixite, arsenogoyazite, crandallite, dussertite, eylettersite, florencite-(Ce), florencite-(La), florencite-(Nd), gorceixite, goyazite, lusungite, philipsbornite, plumbogummite, waylandite, and zairite.
crane boom A long, light boom, usually of lattice construction. (Nichols, 1976)
crane ladle A pot or ladle supported by a chain from a crane; used for pouring molten metals into molds. (Fay, 1920)
crane rope Wire rope consisting of 6 strands of 37 wires around a hemp center. (Hunt, 1965)
crate dam A dam built of crates filled with stone.
crater (a) A typically bowl-shaped or saucer-shaped pit or depression, generally of considerable size and with steep inner slopes, formed on a surface or in the ground by the explosive release of chemical or kinetic energy; e.g., an impact crater or an explosion crater. (AGI, 1987) (b) A basinlike, rimmed structure that is usually at the summit of a volcanic cone. It may be formed by collapse, by an explosive eruption, or by the gradual accumulation of pyroclastic material into a surrounding rim. Cf: *caldera.* (Webster 3rd, 1966) (c) The formation of a large funnel-shaped cavity at the top of a well, resulting from a blowout or occasionally from caving. (Brantly, 1961) (d) In blasting, the funnel of rupture, which in bad rock may have very steep sides and a relatively small volume of broken rock. Syn: *lunar crater.* (Stauffer, 1906)

crater cuts These cuts consist of one or several fully charged holes in which blasting is carried out towards the face of the tunnel, i.e., toward a free surface at right angles to the holes. These represent in principle a completely new type of cut and make use of the crater effect that is obtained in blasting a single hole at a free rock surface. The possibility of a uniform enlargement can be counted on. This means that if the scale is enlarged so that the diameter and depth of hole and length of the charge are all doubled, e.g., a crater of double the depth will be obtained. The number of holes can be increased instead of increasing the diameter of the holes. (Langefors, 1963)
crater theory Crater theory defines an optimum burden or distance to a free face at which a spherical explosive charge is buried and produces the greatest volume of broken and excavatable rock. This distance is unique based on rock type and explosive type. The theory also defines the critical depth or spherical charge buried depth at which surface disturbance is barely detectable, resulting in slight surface mounding and minor cracking.
cratogene *shield.*
cratogenic Of or pertaining to a craton.
craton A part of the Earth's crust that has attained stability, and has been little deformed for a prolonged period. The term is now restricted to the extensive central areas of the continents. (AGI, 1987)
craunch A piece of a vein left uncut as a support. (Arkell, 1953)
crawler One of a pair of an endless chain of plates driven by sprockets and used instead of wheels, by certain power shovels, tractors, bulldozers, drilling machines, etc., as a means of propulsion. Also any machine mounted on such tracks.
crawler track An endless chain of plates used instead of wheels by certain power shovels, continuous miners, etc. (Nelson, 1965)
crawlway A low passageway that only permits the passage of a person by crawling. Syn: *cat run.* (AGI, 1987)
cream A rusty impure meerschaum. (Fay, 1920)
creams Sometimes designates a very high-quality drill diamond. (Long, 1960)
crednerite A monoclinic mineral, $CuMnO_2$.
creedite A monoclinic mineral, $Ca_3Al_2(SO_4)(F,OH)_{10}·2H_2O$.
creek claim A claim that includes the bed of a creek. Under the statute of Oregon, a tract of land 100 yd (91.5 m) square, one side of which abuts on a creek or rather extends to the middle of the stream.
creek placers Placers in, adjacent to, and at the level of small streams.
creek right The privilege of diverting water for the purpose of working a creek claim. Syn: *river right.*
creep (a) The slow and imperceptible movement of finely broken up rock material from higher to lower levels, usually due to alternate freezing and thawing, wetting and drying, or other causes. Also the material that has moved. Cf: *crown-in.* See also: *heaving; lift.* (AGI, 1987) (b) Slow deformation of a material that results from long application of a stress. Part of the creep is a permanent deformation, while part of the deformation is elastic and the specimen can recover. Cf: *thrust.* (AGI, 1987) (c) *drag.* (d) A very slow gradual movement of the drill-hoist drum when the brake is worn or not securely set. See also: *heave.* (Long, 1960)
creeper An endless chain, with projecting bars at intervals that catch the car axles and haul them up an inclined plane. Creepers are used on the surface and around the pit bottom. They are also used on relatively flat roadways to retard or propel the cars as required. (Nelson, 1965)
creeping Eng. The settling or natural subsidence of the surface caused by extensive underground mining.
creep limit The maximum stress that a material can withstand without observable creep. (AGI, 1987)
creep recovery The gradual recovery of elastic strain when stress is released. Syn: *elastic aftereffect.* (AGI, 1987)
creep strength The load per unit area leading to a specified steady creep strain rate at a given temperature. (AGI, 1987)
creeshy Scot. Smooth-faced nodules of shale or bind found occasionally in the roof of some coal seams. Also called greasy blaes. (Nelson, 1965)
creeshy clods Peat which on drying breaks into irregular clods that burn with a clear bright flame like a lump of tallow or grease. (Tomkeieff, 1954)
crenitic Said of mineral veins that have been deposited by springs. Etymol: Greek for spring. Obsolete.
crenulation Small-scale folding (wavelength of a few millimeters) that occurs chiefly in metamorphic rocks. Cf: *plication.*
creolite (a) Red-and-white banded jasper from Shasta or San Bernadino County, CA. (b) A silicified rhyolite from Baja California.
creosote As used in wood preservation, a distillate of coal tar produced by high-temperature carbonization of bituminous coal; it consists principally of liquid and solid aromatic hydrocarbons, and contains appreciable quantities of tar acids and tar bases; it is heavier than water; and has a continuous boiling range of at least 125 °C beginning at about 200 °C. Also called creosote oil; creosote distillate.
crept pillars Eng. Pillars of coal that have passed through the various stages of creep. (Fay, 1920)
crest (a) The highest point on a given stratum in an anticline. See also: *crestal plane; culmination.* (AGI, 1987) (b) *crestline.*
crestal plane The plane formed by joining the crests of all beds in an anticline. See also: *crest.*
crestline In an anticline, the line connecting the highest points on the same bed in an infinite number of cross sections. See also: *crest.* (AGI, 1987)
cresylic Mixture of cresol isomers. Frother and froth stabilizing agent in flotation process. Emulsion stabilizer. (Pryor, 1958)
Cretaceous (a) Applied to the third and final period of the Mesozoic Era. Extensive marine chalk beds were deposited during this period. (b) Of the nature of chalk or relating to chalk. (Fay, 1920) (c) System of strata deposited in the Cretaceous Period. (Fay, 1920)
crevasse (a) A wide breach or crack in the bank of a river or canal; esp. one in a natural levee or an artificial bank of the lower Mississippi River. Etymol: American French. (AGI, 1987) (b) A wide, deep break or fissure in the Earth after an earthquake. (AGI, 1987) (c) A fissure in the surface of a glacier or icefall. (AGI, 1987)
crevice (a) A shallow fissure in the bedrock under a gold placer in which small but highly concentrated deposits of gold may be found. (b) The fissure containing a vein. As employed in the Colorado mining statute relative to a discovery shaft, a crevice is a mineral-bearing vein. An older term.
crevicing Collecting gold that is in the crevices of a rock.
crew loader In bituminous coal mining, one of a crew of loaders who shovels coal, blasting from working face, onto a conveyor that transports it from the underground working place to a point where it is loaded into mine cars. (DOT, 1949)
crib A construction of timbering made by piling logs or beams horizontally one above another, and spiking or chaining them together, each layer being at right angles to those above and below it. See also: *curb.*
cribbing (a) The construction of cribs, or timbers laid at right angles to each other, sometimes filled with earth, as a roof support or as a support for machinery. (BCI, 1947) (b) The close setting of timber supports when shaft sinking through loose ground. The timber is usually square or rectangular and practically no ground is exposed. The method is also used for constructing ore chutes. See also: *barring; close timbering; forepoling.* (Nelson, 1965) (c) A method of timbering used primarily to rectify a mistake of removing too great a percentage of the coal on the advance, and has the effect of replacing part of the coal. Some are made by using timbers in pigpen style; first laying timbers one way then placing other timbers across the first. This is continued until the area between the bottom and the roof is filled and wedged tight. Others are made by laying a layer of timbers first in one direction, then another layer across at right angles to the bottom layer. Space between the timbers in a layer varies according to requirements. The hollow type are generally filled with gob. Syn: *penning.* (Kentucky, 1952)
cribble A sieve.
cribs Segments of oak to encircle the shaft. (Peel, 1921)
crichtonite (a) A trigonal mineral, $(Sr,La,Ce,Y)(Ti,Fe,Mn)_{21}O_{38}$; formerly misidentified as a variety of ilmenite. (b) The mineral group crichtonite, davidite, landauite, loeveringite, and senaite.
crick *loose.*
cricks (a) Som. *clay gall.* (b) Vertical joints affecting only the lower strata in a quarry. (Arkell, 1953) (c) Joints in slate with an inclination opposite to the dip of the rock. (Arkell, 1953)
crimp (a) The flattening made by a crimper near the mouth of a blasting cap for holding the fuse in place. (b) To fix a detonator on blasting fuse by squeezing it with special pliers. (Pryor, 1963)
crimper A tool specially made for fastening a cap to a fuse. See also: *cap crimper.* (Stauffer, 1906)
crimson night stone (a) Purple fluorite from Idaho. (Schaller, 1917) (b) A variety of purple fluorite from Utah.
crinkle A small fold, usually a fraction of an inch in wavelength. (AGI, 1987)
crinkled stone A diamond with a shallowish, wavy, or rough surface.
crinoidal limestone A marine limestone composed largely of fossil crinoid remains, such as plates, disks, stems, or columns.
crispite A former name for the sagenite variety of quartz.
cristobalite A mineral: SiO_2. It is a high-temperature polymorph of quartz and tridymite, and occurs as white octahedrons in the cavities of the fine-grained groundmasses of acidic volcanic rocks. Cristobalite is stable only above 1470 °C; it has a tetragonal structure (alpha-cristobalite) at low temperatures and an isometric structure (beta-cristobalite) at higher temperatures. Cf: *tridymite.* (AGI, 1987)
cristograhamite Grahamite, a mineral asphalt, from the Cristo Mine, Huasteca, Mexico.
critical angle (a) *stalling angle.* (b) The least angle of incidence at which there is total reflection when an optic, acoustic, or electromag-

netic wave passes from one medium to another medium that is less refractive. Cf: *total reflection.* (AGI, 1987) (c) The angle at which a ray of light in passing from a dense medium, such as a gemstone, into a rarer medium, such as air, is refracted at 90° to the normal. Any rays reaching the interface at angles greater than the critical angle are unable to pass into the rarer medium and are totally reflected. (d) The angle of incidence at which refracted light just grazes the surface of contact between two different media. (e) The angle of refraction **r** for which sin **r** = 1/**n**, where **n** is the refractive index of a transparent material. Cf: *law of refraction.*

critical area In prospecting work, an area found to be favorable from geological age and structural considerations. Syn: *favorable locality.* (Nelson, 1965)

critical area of extraction The area of coal required to be worked to cause a surface point to suffer all the subsidence possible from the extraction of a given seam. See also: *subcritical area of extraction.* (Nelson, 1965)

critical current As applied to electric blasting caps, the minimum current that can be employed to fire detonators connected in series so that the chance of a misfire will be less than 1 in 100,000. (Fraenkel, 1953)

critical damping The point at which the damping constant and the undamped frequency of a seismometer or seismograph are equal. After deflection, the moving mass approaches rest position without overswing and the motion is said to be aperiodic. See also: *damping.* (AGI, 1987)

critical density The density of a substance at its critical temperature and under its critical pressure; that density of a saturated, granular material below which, under rapid deformation, it will lose strength and above which it will gain strength. (AGI, 1987)

critical diameter (a) For any explosive, the minimum diameter for propagation of a stable detonation. Critical diameter is affected by confinement, temperature, and pressure on the explosive. (Dick, 2) (b) The minimum explosive diameter which produces the propagation of a detonation wave at a stable velocity. It is affected by conditions of confinement, temperature and pressure on the explosive.

critical distance In refraction seismic work, that distance at which the direct wave in an upper medium is matched in arrival time by that of the refracted wave from the medium below having greater velocity. (AGI, 1987)

critical height The maximum height at which a vertical or sloped bank of soil will stand unsupported under a given set of conditions (ASCE, 1958). (AGI, 1987)

critical minerals (a) Minerals essential to the national defense, the procurement of which in war, while difficult, is less serious than those of strategic minerals because they can be either domestically produced or obtained in more adequate quantities or have a lesser degree of essentiality, and for which some degree of conservation and distribution control is necessary. See also: *strategic minerals; essential mineral.* (Hess) (b) Minerals or mineral associations that are stable only under the conditions of one given metamorphic facies and will change upon change of facies. For example, in Eskola's greenschist facies, sericite and chlorite, albite and epidote are critical mineral associations because these combinations cannot persist out of the field of the greenschist facies, although any one of the individual minerals may be found in more than one facies. (Schieferdecker, 1959)

critical point A point representing a set of conditions (pressure, temperature, composition) at which two phases become physically indistinguishable; in a system of one component, the temperature and pressure at which a liquid and its vapor become identical in all properties. Syn: *decalescence point.* (AGI, 1987)

critical pressure (a) The maximum feed pressure that can be applied to a diamond bit without damaging the bit or core barrel. Cf: *total critical load.* (Long, 1960) (b) The minimum load, in pounds per effective diamond cutting point in a bit face, at which the diamonds cut the rock. Below this load, the diamonds slide on the rock surface without penetrating the rock, and the diamonds polish, become dull, and are rendered unfit for further use in that particular ground unless reset. (Long, 1960) (c) The pressure required to condense a gas at the critical temperature, above which, regardless of pressure, the gas cannot be liquefied. (AGI, 1987)

critical slope The maximum angle with the horizontal at which a sloped bank of soil or given height of soil will stand unsupported. Syn: *angle of repose; angle of rest.* (ASCE, 1958)

critical temperature (a) The temperature of a system at its critical point; for a one-component system; that temperature above which a substance can exist only in the gaseous state, no matter what pressure is exerted. See also: *temperature.* (AGI, 1987) (b) Transformation temperature. (c) The temperature at which a change takes place in the physical form of a substance; e.g., the change of diamond to the amorphous form of carbon begins at a temperature of 1,800 °F (982 °C) in the presence of oxygen. (Long, 1960) (d) Synonymous for critical point if the pressure is constant. (ASM, 1961) (e) The temperature above which the vapor phase cannot be condensed to liquid by an increase in pressure. See also: *temperature.* (ASM, 1961)

critical velocity (a) Reynolds' critical velocity is that at which fluid flow changes from laminar to turbulent, and where friction ceases to be proportional to the first power of the velocity and becomes proportional to a higher power. (Seelye, 1951) (b) Kennedy's critical velocity is that of fluid flow in open channels that will neither deposit nor pick up silt. (Seelye, 1951) (c) Belanger's critical velocity is that condition of fluid flow in open channels for which the velocity head equals one-half the mean depth. (Seelye, 1951)

critical void ratio The void ratio corresponding to the critical density. (ASCE, 1958)

crocidolite An asbestiform variety of riebeckite; forms lavender-blue, indigo-blue, or leek-green silky fibers and massive and earthy forms; suited for spinning and weaving. Also spelled krokitolit. Syn: *blue asbestos; Cape blue.*

crocidolite quartz *tiger's-eye.*

crocoisite *crocoite.*

crocoite A monoclinic mineral, $PbCrO_4$; bright-red, yellowish-red, or orange. Syn: *red lead ore; crocoisite.*

crocus A term used in the Milford, NH, quarries to denote gneiss or any other rock in contact with granite.

crocus martis A name used for impure red ferric oxide pigments and polishing powders, usually produced by heating iron sulfate containing calcium sulfate, lime, or other inert filler. Also sometimes applied more generally to other impure oxides of red or yellow color. (CCD, 1961)

crocus of antimony Brownish-yellow; mainly sodium or potassium thioantimonite; Na_3SbS_3 or K_3SbS_3. Obtained as a slag in refining antimony. (Webster 3rd, 1966)

cronstedtite A monoclinic or trigonal mineral, $Fe^{2+}_2Fe^{3+}(SiFe^{3+})O_5(OH)_4$; kaolinite-serpentine group; in low-temperature hydrothermal veins.

crook A self-acting apparatus for running the hudges (boxes on runners) on inclines in step coalbeds.

crooked hole A borehole that has deviated beyond the allowable limit from the vertical or from its intended course. (AGI, 1987)

crookesite A tetragonal mineral, $Cu_7(Tl,Ag)Se_4$; massive or compact.

crop (a) The outcrop of a lode; or the coal of poor quality at the outcropping of a seam. See also: *outcrop.* (Standard, 1964) (b) Deprecated syn. of outcrop. —v. To appear at the surface of the ground; to outcrop. (AGI, 1987)

crop coal (a) Coal of inferior quality near the surface. Cf: *exposed coalfield.* (Fay, 1920) (b) The coal next to the roof in a seam. (Nelson, 1965)

crop fall A caving in of the surface at the outcrop of the bed; caused by mining operations. Applied also to falls occurring at points not on the outcrop of the bed. Synonymous with day fall. (Fay, 1920)

cropline A line following the outcrop. (Austin, 1964)

crop load The mixture of crushing bodies, ore particles, and water being tumbled in the ball mill.

cropping (a) Coal cutting beyond the normal cutting plane. (Mason, 1951) (b) Portions of a vein or other rock formation exposed at the surface. (Fay, 1920) (c) *outcrop.* (d) The operation of cutting off the end or ends of an ingot to remove the pipe and other defects. (CTD, 1958)

cropping coal The leaving of a small thickness of coal at the bottom of the seam in a working place, usually in back water. The coal so left is termed "cropper coal.". (Zern, 1928)

crop tin The chief portion of tin ore separated from waste in the principal dressing beneficiation operation.

cross *crosscut.*

crossarm (a) The top member of a drill derrick of H-frame from which the sheave wheel is suspended. (Long, 1960) (b) Horizontal bar fitted between two drill columns on which a small diamond or other type rock drill can be mounted. (Long, 1960)

cross assimilation The simultaneous exchange of material from magma to wall rock and vice versa, tending to develop the same phases in both.

crossbar The horizontal roof member of a timber set on mine roadways, or a flat supported by props on the face. See also: *beam.* (Nelson, 1965)

cross-bedded Having minor beds or laminae inclined to the main planes of stratification, e.g., cross-bedded sandstone.

crossbedding (a) The quality or state of being crossbedded. A crossbedded structure. (Webster 3rd, 1966) (b) Lamination, in sedimentary rocks, confined to single beds and inclined to the general stratification. Caused by swift local currents, deltas, or swirling wind gusts, and esp. characteristic of sandstones, both aqueous and eolian. Syn: *cross lamination.* (Fay, 1920) (c) Crossbedding is generally truncated by the overlying stratum. However, at the base of the crossbedded formation, the crossbedding is not truncated, but it approaches the contact with the underlying stratum in a broad tangential curve.

(Forrester, 1946) (d) The arrangement of laminations of strata transverse or oblique to the main planes of stratification of the strata concerned; inclined, often lenticular, beds between the main bedding planes. It is found only in granular sediments. (AGI, 1987) (e) Syn: *inclined bedding*. Should be applied to inclined bedding found only in profiles at right angles to the current direction. (AGI, 1987)

cross-bladed chisel bit cross chopping bit.

cross-chopping bit Bit with cutting edges made by two chisel edges crossing at right angles with the intersection of chisel edges at the center of the bit face. Used to chop (by impact) lost core or other obstructions in a borehole. Also called cross bit; cross-bladed chisel bit; cruciform bit. (Long, 1960)

cross conveyor Any conveyor used for transporting ore or waste from one room or working place through a crosscut to an adjacent room or working place. Used principally where the cross conveyor receives ore or waste from a conveyor and delivers it to another conveyor or a car. (Jones, 1949)

crosscut (a) A small passageway driven at right angles to the main entry to connect it with a parallel entry or air course. (b) A tunnel driven at an angle to the dip of the strata to connect different seams or workings. (Nelson, 1965) (c) A crosscut may be a coal drivage. See also: *pillar-and-stall*. (Nelson, 1965) (d) An underground passage directed across an orebody to test its width and value or from a shaft to reach the orebody. See also: *level crosscut; cross*. (Nelson, 1965) (e) A horizontal opening driven across the course of a vein or in general across the direction of the main workings. A connection from a shaft to a vein. Syn: *cut-through*. (Lewis, 1964) (f) In room-and-pillar mining, the piercing of the pillars at more or less regular intervals for the purpose of haulage and ventilation. Syn: *breakthrough*. (Kentucky, 1952) (g) In general, any drift driven across between any two openings for any mining purpose. (h) A borehole directed so as to cut through a rock strata or ore vein essentially at right angles to the dip and strike of the rock strata, a vein, or a related structure. (Long, 1960) (i) *stenton*.

crossed belt A driving belt that has a twist between the driving and the driven pulleys causing a reversal of direction. (Crispin, 1964)

crossed nicols (a) In optical mineralogy, an anisotropic crystal is interposed between the nicol prisms to observe its optical interference effects. The petrographic microscope is normally used with nicol prisms (or equivalent polarizing devices) in the crossed position. (AGI, 1987) (b) Nicols is often capitalized (crossed Nicols). Two nicol prisms placed one in front of the other, or one below the other, and so oriented that their transmission planes for plane-polarized light are at right angles with the result that light transmitted by one is stopped by the other unless modified by some intervening body. (Webster 3rd, 1966) (c) In polarized-light microscopy, the arrangement where the permitted electric vectors of the two nicol prisms are at right angles. See also: *crossed polars*.

crossed polars A common standard configuration used in polarized-light microscopy with the substage polarizing filter (polarizer) permitting plane polarized light with its electric vector in an east-west direction and the above-stage polarizing filter (analyzer) permitting plane polarized light with its electric vector in a north-south direction. Under these conditions all light is absorbed by the two polars. Introduction of any anisotropic transparent material into the light path repolarizes light between the polars so as to generate interference colors and other effects visible in the microscope ocular. See also: *crossed nicols*.

crossed twinning Repeated or polysynthetic twinning according to two twin laws with twin planes angled to one another and with twin domains so intimate that they appear overlapping in thin section; most notable in the feldspar and feldspathoid minerals, esp. microcline. Syn: *quadrille twinning; gridiron twinning; cross-hatched twinning*.

cross entry (a) An entry or set of entries, turned from main entries, from which room entries are turned. (Federal Mine Safety, 1977) (b) A horizontal gallery driven at an angle or at right angles to a main entry. (Nelson, 1965)

cross face A coal face having a general direction between end and bord line. (TIME, 1929-30)

cross fault A fault that strikes diagonally or perpendicularly to the strike of the faulted strata. (AGI, 1987)

cross fiber Veins of fibrous minerals, esp. asbestos, in which the fibers are at right angles to the walls of the vein. Cf: *slip fiber*. (AGI, 1987)

cross frog A frog adapted for railroad tracks that cross at right angles. (Webster 2nd, 1960)

crossgate (a) A gate road driven at an angle off the main gate in longwall mining, to form new intermediate gates or new faces inside a disturbance. Well-sited crossgates result in reduction of inby conveyors and in roadway maintenance. (Nelson, 1965) (b) Eng. *crossheading*. (c) York. Short headings driven on the strike end at right angles to the main gates or roads. (Fay, 1920)

cross gateway Aust. A road, through the goaf, that branches from the main gateway.

cross-hatched twinning *crossed twinning*.

crosshead (a) A runner or framework that runs on guides, placed a few feet above the sinking bucket to prevent it from swinging too violently. (Fay, 1920) (b) A beam or rod stretching across the top of something; specif., the bar at the end of a piston rod of a steam engine, which slides on the ways or guides fixed to the engine frame and connects the piston rod with the connecting rod. (Fay, 1920)

crosshead guide A guide for making the crosshead of an engine move in a parallel line with the cylinder axis. (Standard, 1964)

crossheading (a) A passage driven for ventilation from the airway to the gangway, or from one breast through the pillar to the adjoining working. Also called cross hole; cross gateway; headway. (Fay, 1920) (b) One driven from one drift or level across to another to improve ventilation. (Pryor, 1963) (c) A heading driven at an angle off the main level to cut off stalls or intermediate headings, and form new ones on the face side of the heading. Also called oblique heading; cutting-off road. (Nelson, 1965) (d) Eng. A road in longwall working to cut off the gateways. Syn crossgate; slope. Also called crossbow; crossend. (SMRB, 1930):

crossite A monoclinic mineral, $Na_2(Mg,Fe)_3(Al,Fe)_2Si_8O_{22}(OH)_2$; amphibole group with $Fe^{3+}/(Fe^{3+}+Al)=0.3\text{-}0.7$. Cf: *glaucophane*.

cross joint (a) A joint in an igneous rock oriented more or less perpendicular to the flow lines. Syn: *tension joint*. (AGI, 1987) (b) A joint in sedimentary rocks that crosses more prominent joints at approximate right angles. (AGI, 1987)

cross-joint fan In igneous rock, a fanlike pattern of cross joints that follow the arching of the flow lineation. (AGI, 1987)

cross lamination (a) The structure commonly present in granular sedimentary rocks that consists of tabular, irregularly lenticular, or wedge-shaped bodies lying essentially parallel to the general stratification and which themselves show a pronounced laminated structure in which the laminae are steeply inclined to the general bedding. Syn: *inclined bedding; crossbedding; false bedding*. (AGI, 1987) (b) An arrangement of laminations, transverse to the planes of stratification of the strata concerned. They generally end abruptly at the top, but in general tend to become more or less parallel to the bedding planes below. (AGI, 1987) (c) Cross-stratification with foresets less than 1 cm thick. (Pettijohn, 1964)

cross-linking agent The final ingredient added to a water gel or slurry, causing it to change from a liquid to a gel. (Dick, 2)

cross measure A heading driven horizontally or nearly so, through or across inclined strata.

cross-measure borehole A borehole drilled at an angle through the rock strata generally for the purpose of combustible gases drainage. (BS, 1963)

cross-measure borehole system *methane drainage*.

cross-measure drift (a) A development drift driven across the strata from the surface to intersect and work coal seams. (Nelson, 1965) (b) A development heading driven from a level in one coal seam to intersect and work upper or lower seams. (Nelson, 1965)

cross measure tunnel A roadway or airway driven across pitching measures on, or nearly on, a level to reach a bed of coal or other objective, or to drain off water. (Zern, 1928)

crossover A track device that permits rail traffic to cross over another track which heads in a different direction on the same level. Signal lights are activated to avoid collision on the crossover.

cross-pit conveyors Conveyor structure crossing the benches of open pit mines to reduce the haul distance across the pit in terrace mining operations. (SME, 1992)

cross poling Short poling boards placed horizontally to cover the gap between runners in excavation trench timbering. See also: *runner*. (Hammond, 1965)

cross section (a) A diagram or drawing that shows features transected by a given plane; specif. a vertical section drawn at right angles to the longer axis of a geologic feature, such as the trend of an orebody. (AGI, 1987) (b) An actual exposure or cut that shows transected geologic features.—Adj: cross-sectional. Also spelled: cross-section. (AGI, 1987) (c) A profile portraying an interpretation of a vertical section of the Earth explored by geophysical and/or geological methods. (d) A horizontal grid system laid out on the ground for determining contours, quantities of earthwork, etc., by means of elevations of the grid points. (Seelye, 1951)

cross-sectional area The area of a surface cut by a plane passing through the body and perpendicular to the long axis of the body if one exists. If not, any such area cut by a plane.

cross-sectional method An ore reserve estimation method in which assay and other data are projected to predetermined planes and the areas of influence of the assay data are determined mainly by judgment. This method is helpful not only for ore reserve computations, but also to mine planning. (Krumlauf)

cross spread (a) A seismic spread that makes a large angle to the line of traverse; it is used to determine the component of dip perpendicular to that line. (AGI, 1987) (b) A seismic spread that is laid out in the pattern of a cross. (AGI, 1987)

cross-spur A vein of quartz that crosses a lode.

cross-stone *andalusite; staurolite; chiastolite; harmotome.*

cross stoping *overhand stoping.*

cross-stratification (a) The minor laminations are oblique to the plane of the main stratum that they help to compose. See also: *crossbedding.* (Standard, 1964; Fay, 1920) (b) The arrangement of layers at one or more angles to the dip of the formation. A cross-stratified unit is one with layers deposited at an angle to the original dip of the formation. Many investigators have used crossbedding and cross lamination as synonymous for cross-stratification, but it is proposed to restrict the terms crossbedding and cross lamination to a quantitative meaning depending on the thickness of the individual layers or cross strata. (Stokes, 1955)

crosstie A timber or metal sill placed transversely under the rails of a railroad, tramway, or mine-car track. (Fay, 1920)

cross validation A technique for testing the validity of a variogram model by kriging each sampled location with all of the other samples in the search neighborhood, and comparing the estimates with the true sample values. Interpretation of results, however, can often be difficult. Unusually large differences between estimated and true values may indicate the presence of "spatial outliers," or points that do not seem to belong with their surroundings.

crowd (a) The process of forcing a bucket into the digging, or the mechanism that does the forcing. Used chiefly in reference to machines that dig by pushing away from themselves. (Nichols, 1954) (b) Used by some drillers as a syn. for overfeed. Cf: *overload.* (Long, 1960) (c) As used by handsetters, the uneven calking of a diamond resulting in its being pinched or forced out of its intended position in a bit. (Long, 1960) (d) To place or set diamonds too closely together in the crown of a bit. (Long, 1960)

crowding In power shovel nomenclature, crowding is the thrusting of the dipper stick forward over the shipper shaft; retracting is the reverse of crowding. (Carson, 1961)

crowding battle In froth flotation, a slanted board used to direct the rising mineralized froth toward the overflow lip of the cell. (Pryor, 1963)

Crowe process The treatment of pregnant cyanide solution to remove air before the gold is precipitated with zinc dust. Also called Merrill-Crowe process.

crowfoot (a) A tool with a sideclaw, for grasping and recovering broken rods in deep boreholes. (b) Irregular or zigzag markings found in Tennessee marble. Also called stylolite. (AGI, 1987)

crown (a) The curved roof of a tunnel. (Nichols, 1976) (b) As used by the drilling and bit-setting industries in the United States, the portion of the bit inset or impregnated with diamonds formed by casting or pressure-molding and sintering processes; hence the steel bit blank to which the crown is attached is not considered part of the crown. Syn: *bit crown.* (Long, 1960) (c) A timber crossbar up to 16 ft (5 m) long, supported by two heavy legs, or uprights, one at each end. Crowns may be set at 3-ft (1-m) intervals; sometimes a roof bolt is put up through the center of the crown. (Nelson, 1965) (d) The topmost part of a drill tripod, derrick, or mast. (Long, 1960) (e) The part of a furnace forming the top or roof. (f) The top or highest part of a mountain or an igneous intrusion; the summit. (AGI, 1987) (g) The practically undisturbed material still in place and adjacent to the highest parts of the scarp along which a landslide moved. (AGI, 1987)

crown block A pulley, set of pulleys, or sheaves at the top of a drill derrick on and over which the hoist and/or other lines run. Also called crown pulley; crown wheel. (Long, 1960)

crown die *bit mold.*

crown-in In mining, a falling of the mine roof or a heave of the mine floor due to the pressure of overlying strata. Cf: *creep.* (AGI, 1987)

crowning The heaving or lifting of the floor beds along a roadway to form a ridge or crown along the centerline. (Nelson, 1965)

crown life *bit life.*

crown mold *bit mold.*

crown pillar An ore pillar at the top of an open stope left for wall support and protection from wall sloughing above.

crown tree A piece of timber set on props to support the mine roof. (Zern, 1928)

crown wheel (a) A wheel driven by a pinion, notably in the drive of a ball mill. Largest wheel of any reduction gear. (Pryor, 1963) (b) Syn: *crown block.* (Long, 1960)

croylstone A variety of finely crystallized barite. (Standard, 1964)

crucible The hearth of a blast cupola, or open hearth furnace; a refractory vessel for melting or calcining metals, ores, etc.

crucible assay *assay; lead button.*

crucible clays Ball clays that are relatively refractory; used in producing crucibles that will withstand high temperatures. (CCD, 1961)

crucible steel Steel made by melting blister bar, wrought iron, charcoal, and ferroalloys in crucibles that hold about 100 lb (45 kg). This was the first process to produce steel in a molten condition, hence the product called cast steel. Mainly used for the manufacture of tool steels, but now largely replaced by the electric-furnace process. (CTD, 1958)

crucible swelling number The number that defines, by reference to a series of standard profiles, the size and shape of the residue produced when a standard weight of coal is heated under standard conditions. (BS, 1961)

cruciform bit (a) *cross-chopping bit.* (b) Percussive rock drill bit having four chisel-shaped cutting edges in the form of a cross on the face of the bit. Also called cross bit. (Long, 1960)

crucite (a) *chiastolite; cross-stone.* (b) Pseudomorph of hematite or limonite after arsenopyrite.

crud A solid-stabilized emulsion that tends to collect at the agueous/organic interface in the settler of a solvent extraction circuit. (Kordosky, 1992)

crude A substance in its natural unprocessed state. Crude ore or crude oil, for example. In a natural state; not cooked or prepared by fire or heat; not altered or prepared for use by any process; not refined. Syn. for raw; crude oil. (Webster 3rd, 1966)

crude anthracene Solid product containing anthracene. Obtained on cooling the coal-tar distillate collected above 270 °C. (Bennett, 1962)

crude asbestos Hand selected cross-vein material of longest fibres in native or unfiberized form. It comes in chunks and must be mechanically processed to develop the usefulness of the fibre. (Arbiter, 1964)

crude mica The crude crystals or books as extracted from the mine. (Skow, 1962)

crude ore The unconcentrated ore as it leaves the mine.

crude-ore bin A bin in which ore is dumped as it comes from the mine.

crude sulfur Elemental sulfur that is 99.0% to 99.9% pure and is free from arsenic, selenium, and tellurium. (USBM, 1965)

crudy asbestos Refers to asbestos that has been only partially milled, so that the fiber has not been fluffed but only separated from the rock. Most of the asbestos is still in the form of bundles of fibers like spicules. (AIME, 1960)

crump Ground movement, perhaps violent, due to failure under stress of ground surrounding underground workings usually in coal, so named because of sound produced. See also: *bump.* (Pryor, 1963)

crush (a) A species of fault in coal. (Fay, 1920) (b) Breakage of supports of underground workings under roof pressure. (Pryor, 1963)

crushability The relative ease of crushing a sample under standard conditions. (BS, 1962)

crush belt A belt of intensely crushed rock.

crush border A microscopic, granular metamorphic structure sometimes characterizing adjacent feldspar particles in granite due to their having been crushed together during or subsequent to crystallization. (AGI, 1987)

crush breccia A breccia formed essentially in situ by cataclasis, esp. along a fault. See also: *cataclasite; crush conglomerate.* Cf: *tectonic breccia.*

crush bursts Rockbursts in which there is actual failure at the face accompanied by movement of the walls. (Higham, 1951)

crush conglomerate (a) A conglomerate produced by the crushing of rock strata in the shearing often accompanying folding. (Standard, 1964) (b) Similar to a fault breccia, except the fragments are more rounded in a crush conglomerate. (AGI, 1987) (c) *tectonic conglomerate; pseudoconglomerate; crush breccia.*

crushed gravel The product resulting from the artificial crushing of gravel with substantially all fragments having at least one face resulting from fracture.

crushed stone (a) The product resulting from the artificial crushing of rocks, boulders, or large cobblestones, substantially all faces of which have resulted from the crushing operation. (ASTM, 1994) (b) Term applied to irregular fragments of rock crushed or ground to smaller sizes after quarrying. Syn: *broken stone.* (USBM, 1965)

crushed vein A mineralized zone or belt of crushed material. The crushing was caused by folding, faulting, or shearing. (Fay, 1920)

crusher A machine for crushing rock or other materials. Among the various types of crushers are the ball-mill, gyratory-crusher, Hadsel mill, hammer mill, jaw crusher, rod mill, rolls, stamp mill, and tube mill. (Fay, 1920; Hess)

crusher feeder In quarry industry, one who feeds broken rock into crusher after it is dumped from trucks or cars, by pushing it down a chute with a shovel or bar, or by pushing it directly into crusher from a platform. Also called crusher loader; crusher laborer; stone breaker; trap person. (DOT, 1949)

crusher man (a) In the mineral and nonmineral industry, including coal, quarry products, mineral and nonmineral ores, a person who operates a machine that crushes rock or other material and regulates the flow of such material into and from the crusher to the next point of processing or use. See also: *crusher; crusher feeder; crushing.* (b) In quarrying, a person who operates crusher through which broken quarry rock is run to break it into crushed stone for construction work. (DOT, 1949)

crusher rock (a) Term used in quarrying to describe the weathered overlying rock that occurs at most quarry operations and which is sold for use as road base. (b) The total unscreened product of a stone crusher. (Shell)

crusher rolls Steel or chilled iron roller with parallel horizontal axis and peripheries at a fixed distance apart so that rocks, coal, or other substances of greater thickness cannot pass between without crushing. Rolls may be toothed or ribbed, but for rock, including ores, the surfaces are usually smooth. (Hess)

crusher-run stone Rock that has been broken in a mechanical crusher and has not been subjected to any subsequent screening process. (Taylor, 1965)

crusher setting The distance between roll faces or plates in a crusher. In the case of jaw and roll crushers, the setting controls the maximum size, and to some extent the grading of the product produced. The best setting is usually that which produces 10% to 15% of oversize pieces, which are fed back for recrushing. Gyratory breakers do not permit any marked variation in the setting or in the size of the product. (Nelson, 1965)

crusher stower A machine that crushes ripping stone in headings and projects it through a pipe into gate side packs. It may also be used for filling old roadways or roof cavities. See also: *pneumatic stowing*. (Nelson, 1965)

crush gate A gate in a development face designed to be abandoned with a view to localizing the crush effect consequent on the winning of the coal immediately above or immediately below the development face. (TIME, 1929-30)

crushing Size reduction into relatively coarse particles by stamps, crushers, or rolls. See also: *comminution*. (BS, 1962)

crushing bort Diamond material with radial or confused crystal structure lacking distinct cleavage forms. Color is faintly milky to grayish or dark and is suitable only for crushing into grit powder or dust. The Bakwanga Mine, Republic of the Congo, is the principal source of this material. Diamond fragments from cutting establishments or recovered from waste are frequently classed as crushing bort. (Chandler, 1964)

crushing bortz *bort.*

crushing cycle The sequence of operations in crushing a material, including, e.g., the screening of the primary product and the recirculation of the screen overflow. (BS, 1962)

crushing machine A machine constructed to pulverize or crush stone and other hard and brittle materials; a stone crusher. (Fay, 1920)

crushing mill *stamp mill; crusher.*

crushing roll A machine consisting of two heavy rolls between which ore, coal, or other mineral is crushed. Sometimes the rolls are toothed or ribbed, but for ore their surface is generally smooth. See also: *roll.* (Fay, 1920)

crushing strength (a) The resistance that a rock offers to vertical pressure placed upon it. It is measured by applying graduated pressure to a cube, 1 in (2.54 cm) square, of the rock tested. A crushing strength of 4,000 lb means that a cubic inch of the rock withstands pressure to 4,000 lb (111 kg/cm^3) before crushing. The crushing strength is greater with shorter prisms and less with longer prisms. (Fay, 1920) (b) The pressure or load at which a material fails in compression; used for comparing the strength of walling and lining materials, such as concrete, masonry, stone, packs, etc. (Nelson, 1965) (c) The maximum load per unit area, applied at a specified rate, that a material will withstand before it fails. Typical ranges of value for some ceramic materials are fireclay and silica refractories, 2,000 to 5,000 psi (13.8 to 34.4 MPa); common building bricks, 2,000 to 6,000 psi (13.8 to 41.4 MPa); engineering bricks, class A, above 10,000 psi (69.0 MPa); sintered alumina, above 50,000 psi (344 MPa). (Dodd, 1964)

crushing test (a) A test of the suitability of stone to be used for roads or building purposes; a cylindrical specimen of the stone, of diameter 1 in (2.54 cm) and 1 in long, is subjected to axial compression in a testing machine. Syn: *unconfined compression test*. (CTD, 1958) (b) A radial compressive test applied to tubing, sintered-metal bearings, or other similar products for determining radial crushing strength (maximum load in compression). (ASM, 1961) (c) An axial compressive test for determining quality of tubing, such as soundness of weld in welded tubing. (ASM, 1961)

crush line A line along which rocks under great compression yield, usually with the production of schistosity.

crush movement Compression, thrust, or lateral movement tending to develop shattered zones in rocks. (Fay, 1920)

crush plane A plane defining zones of shattering that result from lateral thrust. (Fay, 1920)

crush zone A zone of faulting and brecciation in rocks. (Fay, 1920)

crust (a) The outermost layer or shell of the Earth, defined according to various criteria, including seismic velocity, density and composition; that part of the Earth above the Mohorovicic discontinuity, made up of the sial, or the sial and the sima. It represents less than 0.1% of the Earth's total volume. Cf: *tectonosphere*. (AGI, 1987) (b) A laminated, commonly crinkled deposit of algal dust, filamentous or bladed algae, or clots (from slightly arched forms to bulbous cabbagelike heads) of algae, formed on rocks, fossils, or other particulate matter by accretion, aggregation, or flocculation. (AGI, 1987)

crustal abundance *clarke.*

crustal plate A portion of the Earth's crust that moves as a relatively rigid unit with respect to adjacent crustal plates that collectively cover the outermost part of the solid Earth. (AGI, 1987)

crustification (a) The layering of crusts of different minerals deposited successively on the walls of a cavity. (b) Suggested for those deposits of minerals and ores that are in layers or crusts and which, therefore, have been deposited from solution.

crustified banding A structure of vein fillings resulting from a succession, often a rhythmic deposition, of crusts of unlike minerals on the walls of an open space.

crustified vein A vein filled with a succession of crusts of ore and gangue material.

crut A short heading excavated into the face of a coal seam; a heading or drift across the strata, or from one deposit to another. Syn: *tunnel.* (Nelson, 1965)

crutt N. Staff.; Som. A road or heading driven in coal measures, turned from a level, etc.

Cryderman loader A clamshell-type loader activated by hydraulic cylinders operated from a traveling base suspended on the stage. Used in shaft sinking operations. (Lewis, 1964)

cryogenic switching elements In information processing, logical switching information processing elements that utilize the variability of the transition to superconductivity as a function of magnetic field strength. (Hunt, 1965)

cryolite A monoclinic mineral, Na_3AlF_6; waxy colorless to white (disappears in water owing to low refractive index); soft; in veinlike cleavable masses in granite at Ivigtut, Greenland. Syn: *Greenland spar; ice stone.*

cryolithionite An isometric mineral, $Na_3Li_3Al_2F_{12}$; forms large colorless rhombic dodecahedra; at Ivigtut, Greenland, and the Ural Mountains, Russia.

cryology (a) In the United States, the study of refrigeration. (AGI, 1987) (b) In Europe, a syn. for glaciology. *glaciology*. (AGI, 1987) (c) The study of ice and snow. (AGI, 1987) (d) The study of sea ice. (AGI, 1987)

cryoluminescence The low-temperature increase of weak luminescence, or its development in normally nonfluorescent material. (AGI, 1987)

cryopedology The study of the processes of intensive frost action and the occurrence of frozen ground, esp. permafrost, including the civil-engineering methods used to overcome or minimize the difficulties involved. (AGI, 1987)

cryosphere The part of the Earth's surface that is perennially frozen; the zone of the Earth where ice and frozen ground are formed. (AGI, 1987)

cryoturbation Frost action, including frost heaving.

crypthydrous Refers to vegetable accumulations laid down on a wet substratum in contrast to those deposited under water. Cf: *phenhydrous.* (AGI, 1987)

cryptoclastic Said of a rock of compact texture, composed of extremely small, fragmental particles that are barely visible under a microscope.

cryptocrystalline (a) Said of the texture of a rock consisting of crystals that are too small to be recognized and separately distinguished even under the ordinary microscope (although crystallinity may be shown by use of the electron microscope); indistinctly crystalline, as evidenced by a confused aggregate effect under polarized light. Also, said of a rock with such a texture. Syn: *microaphanitic; microcryptocrystalline; microcrystalline; microfelsitic*. Cf: *dubiocrystalline*. (AGI, 1987) (b) Said of a rock or rock texture of a crystalline rock in which the crystals are too small to be recognized megascopically. This usage is not recommended until crystallinity can be established by polarized-light microscopy or X-ray diffraction. Syn: *aphanitic.* (c) Descriptive of a crystalline texture of a carbonate sedimentary rock having discrete crystals with maximum diameters variously set at 1 µm, 4 µm, and 10 µm.

cryptoexplosion structure A nongenetic, descriptive term for a roughly circular structure formed by the sudden, explosive release of energy and exhibiting intense, often localized rock deformation with no obvious relation to volcanic or tectonic activity. Many cryptoexplosion structures are believed to be the result of impact of meteorites of asteroidal dimensions; others may have been produced by volcanic activity. The term largely replaces the earlier term cryptovolcanic structure. (AGI, 1987)

cryptographic (a) Denoting a texture of rocks so fine that the individual constituents cannot be distinguished under a microscope. Usually the result of a cryptocrystalline intergrowth of quartz and feldspar. See also: *cryptocrystalline.* (b) Having a graphic texture of intergrowths too small to be resolved with a light microscope.

cryptohalite An isometric mineral, $(NH_4)_2SiF_6$; dimorphous with bararite.
cryptohydrous The conditions under which coal was formed. Decay under water in swamps. (Tomkeieff, 1954)
cryptomagmatic Said of a hydrothermal mineral deposit without demonstrable relationship to igneous processes. The term is little used. Cf: *apomagmatic; telemagmatic.* (AGI, 1987)
cryptomelane (a) A monoclinic mineral, $K(Mn^{4+}, Mn^{2+})_8O_{16}$; pseudotetragonal. Cf: *psilomelane.* (b) The mineral group coronadite, cryptomelane, hollandite, manjiroite, and priderite.
cryptomere *aphanite.* Also spelled kryptomere.
cryptomerous (a) A very fine crystalline texture. (Stokes, 1955) (b) Of or pertaining to cryptomere. (Johannsen, 1931-38)
cryptoperthite Extremely fine-grained perthite with submicroscopic lamellae (1 to 5 μm) detectable only by X-ray diffraction or electron microscopy. The K-rich host may be sanidine, orthoclase, or microcline; the Na-rich phase may be albite or analbite. Cf: *perthite; microperthite.*
cryptovolcanic structure A circular structure lacking evidence of shock metamorphism or of meteorite impact and therefore presumed to be of igneous origin, but lacking exposed igneous rocks or obvious volcanic features; a rock structure produced by concealed volcanic activity. Preferred term: cryptoexplosion structure. (AGI, 1987)
Cryptozoic (a) Eon of hidden life. Syn. of Precambrian. (AGI, 1987) (b) That part of geologic time represented by rocks in which evidence of life is only slight and of primitive forms. Cf: *Phanerozoic.* (AGI, 1987)
crystal (a) A regular polyhedral form, bounded by planes, which is assumed by a chemical element or compound, under the action of its intermolecular forces, which passing, under suitable conditions, from the state of a liquid or gas to that of a solid. A crystal is characterized first by its definite internal molecular structure and second, by its external form. (Fay, 1920) (b) The regular polyhedral form, bounded by plane surfaces, which is the outward expression of a periodic or regularly repeating internal arrangement of atoms. See also: *crystal face.* (AGI, 1987) (c) A body formed by the solidification under favorable conditions of a chemical element, a compound, or an isomorphous mixture and having a regularly repeating internal arrangement of its atoms; esp. such a body that has natural external plane faces as a result of the internal structure. (Webster 3rd, 1966) (d) Quartz that is transparent or nearly so and that is either colorless or only slightly tinged. Also a piece of this material. Also called rock crystal. (Webster 3rd, 1966) (e) A colorless transparent diamond. (Webster 3rd, 1966) (f) As an adj., consisting of or resembling crystal. Syn. for crystalline; clear; transparent. Relating to or using a crystal. (Webster 3rd, 1966) (g) A regular polyhedron exhibited by a chemical element or compound where its atomic particles assume a periodic array under suitable physical and chemical conditions. The external form is a low-energy response to the symmetry of the internal forces with each face parallel to a high-density plane of atomic particles. (h) Any solid material with a periodic internal structure. Syn: *crystalline.* (i) Glass of superior quality and high density and luster (resulting from inclusion of lead salts in old objects), commonly with ornamental cutting, e.g., flint glass. (j) An adj. referring to material properties, e.g., crystal structure (for internal periodicity), crystal solution (as between end members of a mineral series).
crystal aggregate A number of crystals grown together so that each crystal in the group is large enough to be seen by the unaided eye and each crystal is more or less perfect. In gemmology, it differs from a crystalline aggregate, as a homogenous gem stone can be cut only from an individual crystal of a crystal aggregate. Syn: *crystal group.*
crystal axes Imaginary lines passing through a crystal in important symmetry directions, intersecting in the origin at the center of the crystal. The axes are usually three in number, and they are chosen to act as a frame of reference by means of which the relative positions of the crystal faces can be described. (Anderson, 1964)
crystal axis (a) A reference axis used for the description of the vectorial properties of a crystal. There are generally three noncoplanar axes, chosen parallel to the edges of the unit cell of the crystal structure so as to be parallel to symmetry directions if possible. (AGI, 1987) (b) One of three minimal noncoplanar reference lines used to describe the vectorial properties of crystalline materials. Syn: *crystallographic axis.* (c) A line parallel to the intersections of crystal faces. Syn: *zone axis.* (d) A line about which crystal symmetry appears distributed. Syn: *symmetry axis.* (e) A line about which a part of a crystal appears to have rotated in a fashion not permitted by the symmetry group of the crystal. Syn: *twin axis.* Plural: axes, pron: "aks-eez."
crystal bar Hafnium and zirconium produced by the van Arkel and de Boer process. (Thomas, 1960)
crystal casts Fillings of a cavity left by solution or sublimation of a crystal embedded in a fine-grained sediment. (Pettijohn, 1964)
crystal chemistry The study of the relations among chemical composition, internal structure, and the physical properties of crystalline matter. (AGI, 1987)
crystal class (a) One of the 32 crystallographically possible combinations or groups of symmetry operations that leave one point, or origin, fixed. (AGI, 1987) (b) All minerals having the symmetry of one of the 32 point groups belong to the same crystal class. Cf: *point group.*
crystal defect Any deviation from perfect periodicity in a crystal structure. Some defects depend on temperature, mainly point defects; others depend on the specific history of the crystal. The presence of defects alters the physical properties of crystals. Cf: *Frenkel defect; point defect; line defect; volume defect.*
crystal defects Irregularities in a lattice structure that affect resistance to crushing. Microdefects are due to irregular distribution of ions. Macrodefects are incipient strain areas or discontinuities in an otherwise regular lattice. Mosaic defects are orderly blocks of regular lattice that are packed together to form a larger and imperfect particle. (Pryor, 1963)
crystal diamagnetism The abnormal ratio of magnetization to the magnetizing force responsible for it, as observed in some crystals, such as those of bismuth. (Hess)
crystal face (a) One of the several flat or plane exterior surfaces of a crystal. See also: *crystal.* (Long, 1960) (b) A planar surface developed on a crystal during its growth. Crystal faces tend to parallel planes of high lattice-point density (Bravais' law) with the result that they make rational intercepts with the crystallographic axes and may be assigned rational indices, e.g., (*hkl*). Cf: *Miller indices.*
crystal flotation The floating of lighter-weight crystals in a body of magma. Cf: *crystal settling.* Syn: *flotation of crystals.* (AGI, 1987)
crystal form (a) The form or shape in which crystals occur; the cube, the octahedron, and others. (b) All crystal faces related by the symmetry elements of the point group of the crystal structure belong to the same crystal form. Crystal forms are designated by the indices of the unit face enclosed in braces, e.g., *hkl*. Forms are closed if they singly enclose a volume and open if two or more forms are required. The terms "prism," "pyramid," "cube," "octahedron," and "tetrahexahedron" refer to crystal forms. A crystal form is ideal when all faces are the same size. Syn: *crystalline form.*
crystal fractionation Magmatic differentiation resulting from the floating or settling, under gravity, of mineral crystals as they form. Cf: *fractional crystallization.* Syn: *gravitational differentiation.* (AGI, 1987)
crystal group *crystal aggregate.*
crystal growth The study of conditions for growing crystals experimentally, esp. in the control of chemical and physical properties and in application to the growth history of natural crystals. Also the microchemical and isotopic study of crystals for the physical and chemical constraints on their formation.
crystal habit The forms typically appearing on specimens of a mineral species or group, rarely all the forms permitted by its point group. Crystal habits range from highly diverse, e.g., calcite, to almost never showing crystal faces, e.g., turquoise. In addition to describing mineral habits with form names, e.g., prismatic, pyramidal, or tetrahedral, other names for appearances are used, e.g., fibrous, columnar, platy, or botryoidal. Intergrowths are given by specific description. (Pryor, 1963)
crystal indices (a) Numbers or other representations that indicate the inclination of a crystal face to the crystal axes. (b) Numbers based on the rational intercepts of crystal faces with crystallographic axes. The Miller index is the reciprocal of a face's axial intercepts. Indices of crystal faces are enclosed in parentheses (*hkl*), crystal forms in braces {*hkl*}, crystal directions in brackets [*hkl*], and Bragg indices with no closure *hkl*. For crystals with hexagonal and trigonal symmetry, Miller-Bravais indices (*hkil*) may be used although the added intercept and index number are redundant ($h+k+i=0$). Not all mineralogists follow this usage. Cf: *Miller indices.*
crystal lattice (a) The regular and repeated three-dimensional arrangement of atoms that distinguishes crystalline solids from all other states of matter. Essentially the regularity displayed by a crystal lattice is that of a three-dimensional mesh that divides space into identical parallelepipeds. Imagine a number of identical atoms placed at the intersections of such a mesh; then we have what is known as a simple lattice (synonymous with Bravais lattice). (AGI, 1987) (b) A periodic array of points in three dimensions such that each point is in an identical point environment. Fourteen possible lattices that are used to describe the structural patterns are found in all crystalline materials by assigning an asymmetric unit to each lattice point. Syn: *Bravais lattice; space lattice; direct lattice; translation lattice.* Cf: *reciprocal lattice.*
crystalliferous Producing or bearing crystals. (Webster 3rd, 1966)
crystalliform Having a crystalline form. (Standard, 1964)
crystalline (a) Made of crystal. (Webster 3rd, 1966) (b) Resembling a crystal; clear, transparent, pure. (AGI, 1987) (c) Pertaining to or having the nature of a crystal, or formed by crystallization; specif. having a crystal structure or a regular arrangement of atoms in a space lattice. Ant: amorphous. Said of a mineral particle of any size, having the internal structure of a crystal, but lacking well-developed crystal faces or an external form that reflects the internal structure. (AGI, 1987)

(d) Said of a rock consisting wholly of crystals or fragments of crystals; esp. said of an igneous rock developed through cooling from a molten state and containing no glass, or of a metamorphic rock that has undergone recrystallization as a result of temperature and pressure changes. The term may also be applied to certain sedimentary rocks (such as quartzite, some limestones, evaporites) composed entirely of contiguous crystals. (AGI, 1987) (e) Said of the texture of a crystalline rock characterized by closely fitting or interlocking particles (many having crystal faces and boundaries) that have developed in the rock by simultaneous growth. A crystalline rock. Term is usually used in the plural; e.g., the Precambrian crystalline. This usage is not recommended. (AGI, 1987) (f) Referring to a homogeneous solid material that has long-range periodic order of its atomic constituents. Crystalline materials distinctly diffract X-rays. (g) Referring to a rock composed of crystalline minerals, e.g., granite. (h) Referring to the texture of a rock composed of contiguous mineral crystals with or without crystal faces. (i) Referring to underlying rock with coarse texture as opposed to overlying noncrystalline or finely crystalline rock, e.g., Precambrian crystalline basement.

crystalline aggregate An aggregate of crystalline intergrowths, such as granite, that does not show well-defined crystal forms.

crystalline flake graphite *flake graphite.*

crystalline form (a) The external geometrical shape of a crystal. (CMD, 1948) (b) *crystal form.*

crystalline grains Minute crystals or crystalline particles which compose a granular crystalline aggregate. Distinguished from minute fiberlike crystals which compose fibrous crystalline aggregates.

crystalline granular texture A primary texture due to crystallization from an aqueous medium, as in rock salt (halite), gypsum, and anhydrite.

crystalline limestone (a) A metamorphosed limestone; a marble formed by recrystallization of limestone as a result of metamorphism. (AGI, 1987) (b) A calcarenite with crystalline calcite cement formed in optical continuity with crystalline fossil fragments by diagenesis. (AGI, 1987) (c) A limestone formed of abundant calcite crystals as a result of diagenesis; specif. a limestone in which calcite crystals larger than 20 µm in diameter are the predominant components. Examples include the crinoidal limestones whose fragments have been enlarged by growth of calcite. Cf: *marble.* (AGI, 1987)

crystalline quartz A term used to distinguish all the varieties of quartz which are not cryptocrystalline, such as rock crystal, amethyst, citrine, cairngorm, rose quartz, tiger eye, etc.

crystalline rock (a) An inexact, but convenient term designating an igneous or metamorphic rock, as opposed to a sedimentary rock. (AGI, 1987) (b) A rock consisting wholly of relatively large mineral grains, e.g., a plutonic rock, an igneous rock lacking glassy material, or a metamorphic rock. (AGI, 1987) (c) The term has also been applied to sedimentary rocks, e.g., some limestones, that are composed of coarsely crystalline grains or exhibit a texture formed by partial or complete recrystallization. (AGI, 1987)

crystalline structure *crystal structure.*

crystalline tonstein This type of tonstein contains vermicular, prismatic, or tabular kaolinite crystals and may be either light or dark in color according to the proportion of contained carbonaceous matter. Occasionally granular kaolinite may also be recognized. The crystals lie embedded in either a finely crystalline or cryptocrystalline kaolinite groundmass. (IHCP, 1963)

crystallinity (a) The degree to which a rock (esp. an igneous rock) is crystalline (holocrystalline, hypocrystalline, etc.). (AGI, 1987) (b) The degree to which the crystalline character of an igneous rock is developed (e.g., macrocrystalline, microcrystalline, or cryptocrystalline) or is apparent (e.g., phaneritic or aphanitic). (AGI, 1987)

crystallite (a) A broad term applied to a minute body of unknown mineralogic composition, or crystal form that does not polarize light. Crystallites represent the initial stage of crystallization of a magma or of a glass. Syn: *crystallitic.* Cf: *microlite; crystalloid.* (AGI, 1987) (b) Very small crystals in a mass or matrix. (c) A nucleus from which a crystal may grow. (d) Minute spots of double refraction in a glassy matrix.

crystallitic Of, pertaining to, or formed of, crystallites.

crystallization (a) The process through which crystalline phases separate from a fluid, a viscous, or a dispersed state (gas, liquid solution, or rigid solution). (Holmes, 1920) (b) The process of crystallizing. A form of body resulting from crystallizing. (Webster 3rd, 1966) (c) Formation of crystalline phases during the cooling of a melt or precipitation from a solution.

crystallization differentiation The progressive change in composition of the liquid fraction of a magma as a result of the crystallization of mineral phases that differ in composition from the magma. (AGI, 1987)

crystallization interval (a) The interval of temperature (or less frequently pressure) between the formation of the first crystal and the disappearance of the last drop of liquid from a magma on cooling. It usually excludes the late-stage aqueous fluids. (AGI, 1987) (b) When referring to a given mineral, the range or the ranges of temperatures over which that particular phase is in equilibrium with liquid. In the case of equilibria along reaction lines or reaction surfaces, crystallization intervals, as thus defined, include temperature ranges in which certain solid phases are actually decreasing in amount with decrease in temperature. Syn: *freezing interval.* (AGI, 1987)

crystallization nucleus A small particle of any kind around which crystals begin to form when a substance crystallizes.

crystallization systems (a) The 32 possible crystal groups, distinguished from one another by their symmetry, are classified under 6 systems, each characterized by the relative lengths and inclinations of the assumed crystallographic axes. These are isometric, tetragonal, hexagonal, orthorhombic, monoclinic, and triclinic. (Fay, 1920) (b) *crystal systems.*

crystallize (a) To cause to form crystals or to assume crystalline form; esp. to cause to assume perfect or large crystals. To cause to take a fixed and definite form. To become converted into crystals. To solidify by crystallizing. To deposit crystals. To become fixed and definite in form. (Webster 3rd, 1966) (b) Any process by which matter becomes crystalline from a noncrystalline state. Also spelled "crystalize."

crystallized coal *cone-in-cone structure.*

crystallizing force (a) The potentiality, or the expansive force, by which a mineral tends to develop its own crystal form against the resistance of the surrounding solid mass. This may be a differential force that causes the crystal to grow preferentially and more rapidly in one crystallographic direction than in another. (b) Expulsion of foreign constituent from a growing crystal by diffusion or mechanical displacement. (c) Expansion resulting from a crystal phase that is less dense than its noncrystalline melt phase, e.g., freezing water splitting rock or bursting pipes.

crystalloblast A crystal of a mineral produced entirely by metamorphic processes. See also: *idioblast; xenoblast.* Adj: crystalloblastic. (AGI, 1987)

crystalloblastesis Deformation by metamorphic recrystallization. (Knopf, 1938)

crystalloblastic (a) Pertaining to a crystalloblast. (AGI, 1987) (b) Said of a crystalline texture produced by metamorphic recrystallization under conditions of high viscosity and directed pressure, in contrast to igneous rock textures that are the result of successive crystallization of minerals under conditions of relatively low viscosity and nearly uniform pressure (Becke, 1903). Cf: *homeoblastic; heteroblastic.* (AGI, 1987) (c) A metamorphic texture wherein one or more mineral species grows substantially larger than the rock matrix, e.g., garnet schist. Cf: *porphyroblastic; granoblastic.* (d) A crystalline texture owed to metamorphic recrystallization. A characteristic of this texture is that the essential constituents are simultaneous crystallizations and are not found in sequence, so that each may be found as inclusions in all the others.

crystalloblastic series An arrangement of metamorphic minerals in order of decreasing form energy, so that crystals of any of the listed minerals tend to assume idioblastic outlines at surfaces of contact with simultaneously developed crystals of all minerals occupying lower positions in the series. (AGI, 1987)

crystallochemical element An element essential to the composition and the structure of a mineral. (AGI, 1987)

crystallogenesis The production or formation of crystals. Adj: crystallogenic.

crystallogeny (a) The science and the theory of the production of crystals. (Standard, 1964) (b) That branch of materials science that deals with the formation or growth of crystals. Cf: *chemical mineralogy; experimental mineralogy; phase equilibria.*

crystallogram A record, photographic or electronic, of crystal structure obtained by means of X-ray diffraction. Cf: *Laue diagram; Laue photograph; powder pattern.*

crystallographic axes (a) Three axes intersecting at right angles, the vertical one being the x axis and the two horizontal ones the y and z. The position of a crystal face is defined by the ratio of its intercepts with these axes. (Pryor, 1963) (b) The three noncoplanar reference vectors used to describe crystal properties. Depending upon the crystal system, these axes are not necessarily orthogonal nor of equal length, with angle α between axes b and c, β between a and c, and γ between a and b. Axes of equal length are labeled a. Where one axis is unique (hexad, tetrad, triad, or monad), convention sets it vertical and labels it c. A unique diad may be labelled c (first crystallographic setting) or b (second setting, preferred by mineralogists). By convention positive b is plotted to the right and positive a toward the observer. Some crystal classes with a unique triad may be referred to a set of three nonorthogonal axes of equal length at internal angle alpha, i.e., rhombohedral coordinates a_r. In the case of a unique triad or hexad, a fourth redundant axis a_3 may be added for convenience. Cf: *axis; coordinate system; crystal systems; fold.*

crystallographic axis *crystal axis.* Cf: *intercept.*

crystallographic direction (a) Refers to directions in the various crystal systems that correspond with the growth of the mineral and often with the direction of one of the faces of the original crystal itself. (b) Vectors referred to as crystallographic axes. Because of crystalline periodicity, significant directions within a crystal are determined by the rational intercepts with the crystallographic axes and may be rendered in terms of Miller indices enclosed in brackets, e.g., [*hkl*] or [*hkil*].

crystallographic planes (a) Any set of parallel and equally spaced planes that may be supposed to pass through the centers of atoms in crystals. As every plane must pass through atomic centers and no centers must be situated between planes, the distance between successive planes in a set depends on their direction in relation to the arrangement of atomic centers. (CTD, 1958) (b) Those planes that make rational intercepts with crystallographic axes and that may be noted by their intercept reciprocals, the Miller indices enclosed in parentheses, i.e., (*hkl*) or (*hkil*). Such planes may represent crystal faces, cleavage planes, twin planes, lattice points, or planes of atomic particles in a crystal structure.

crystallographic system (a) Any of the major units of crystal classification, embracing one or more symmetry classes. (CTD, 1958) (b) *crystal system; holohedral.*

crystallographic texture A texture of mineral deposits formed by replacement or exsolution, in which the distribution and form of the inclusions are controlled by the crystallography of the host mineral. (AGI, 1987)

crystallography (a) The study of crystals, including their growth, structure, physical properties, and classification by form. (AGI, 1987) (b) The science of the geometry of crystals and crystalline materials that results from symmetry generated by the periodicity of their atomic particles. See also: *symmetry.*

crystalloid (a) Having some or all of the properties of a crystal. (b) A microscopic crystal that, when examined under a microscope, polarizes light but has no crystal outline or readily determinable optical properties. Cf: *crystallite; microlite.* (AGI, 1987)

crystallology The science of crystals and crystalline materials. It embraces crystallography and crystallogeny.

crystalloluminescence The emission of light by a substance during its crystallization. (AGI, 1987)

crystallothrausmatic A descriptive term applied to igneous rocks with an orbicular texture in which early phenocrysts form the nuclei of the orbicules (Eskola, 1938). Cf: *isothrausmatic; heterothrausmatic; homeothrausmatic.* (AGI, 1987)

crystal material Any substance possessing crystal structure but no definite geometric form visible to the unaided eye. Also known as crystalline material.

crystal mush Partially crystallized magma; an aggregate of solid crystals lubricated by compressed water vapor. (AGI, 1987)

crystal optics (a) The science that treats of the transmission of light in crystals. (Fay, 1920) (b) The study and characterization of the optical properties of crystalline materials. Because each mineral species is a unique combination of chemistry and crystal symmetry, use of optical properties of minerals, both opaque and transparent, for their characterization and identification is a well-developed art.

crystal pattern A space lattice of a crystal structure. See also: *space lattice.* (Hackh, 1944)

crystal recovery The recovery of the original properties in a crystal that has been distorted by stress resulting from continued relief from stress, heating, or decrease in the speed of deformation. (Knopf, 1938)

crystal rectifier A point contact between a metal and a crystal (such as copper and galena), or between two crystals (such as zincite and bornite). It has marked unidirectional conductivity. (CTD, 1958)

crystals (a) Trade term for fourth-grade diamonds; colorless diamonds. (Hess) (b) Atomic structures with long-range order. (Euhedral surfaces are not required.). (Van Vlack, 1964) (c) Geometrical forms of planar faces assumed by minerals and other crystalline materials when grown under appropriate conditions. (d) Australian syn. for drill diamonds.

crystal sandstone (a) A sandstone in which the quartz grains have been enlarged by deposition of silica so that the grains show regenerated crystal facets and sometimes nearly perfect quartz euhedra. Crystal sandstones of this nature sparkle in bright sunlight. (AGI, 1987) (b) A sandstone in which calcite has been deposited in the pores in large patches or units having a single crystallographic orientation, resulting in a poikiloblastic or luster-mottling effect. In some rare sandstones with incomplete cementation, the carbonate occurs as sand-filled scalenohedra of calcite—sand crystals. See also: *sand crystal.* (AGI, 1987)

crystal sedimentation *crystal settling.*

crystal settling In a magma, the sinking of crystals because of their greater density, sometimes aided by magmatic convection. It results in crystal accumulation, which develops layering. Cf: *crystal flotation.* Syn: *crystal sedimentation.* (AGI, 1987)

crystal solution The replacement of one chemical element by others in a crystal structure without changing its symmetry. Syn: *solid solution.*

crystal sorting The separation, by any process, of crystals from a magma, or of one crystal phase from another during crystallization of the magma. (AGI, 1987)

crystal spectrometer An X-ray spectrometer employing a crystal grating. (Webster 3rd, 1966)

crystal structure (a) The periodic or repeated arrangement of atoms in a crystal. (AGI, 1987) (b) The arrangement in most pure metals may be imitated by packing spheres, and the same applies to many of the constituents of alloys. (CTD, 1958) (c) The periodic array of atomic particles as represented by their symmetrical disposition. Syn: *crystalline structure.*

crystal system *crystallographic system.*

crystal systems (a) A classification of crystals based on the intercepts made on the crystallographic axes by certain crystal faces (or bounding planes). Syn: *crystallization systems.* Cf: *indices of a crystal face.* (CMD, 1948) (b) The six main symmetry groups into which all crystals, whether natural or artificial, can be classified. See also: *symmetry.* (Anderson, 1964) (c) The classification of point groups (and their associated crystal classes) and space lattices into seven (or six) symmetry systems; e.g., isometric (three equal orthogonal crystal axes), hexagonal (a unique hexad), trigonal (a unique triad), tetragonal (a unique tetrad), orthorhombic (three orthogonal diads), monoclinic (one diad inclined to an axial plane), and triclinic (no symmetry higher than a center) systems. An alternative classification of six systems assigns hexagonal and rhombohedral divisions to the hexagonal system with a different assignment of point groups having unique triads and hexads. Cf: *tetragonal.*

crystal tuff An indurated deposit of volcanic ash dominantly composed of intratelluric crystals and crystal fragments. Cf: *tuff; crystal-vitric tuff; lithic tuff; vitric tuff.*

crystal-vitric tuff A tuff that consists of fragments of crystals and volcanic glass. Cf: *crystal tuff; vitric tuff.* (AGI, 1987)

crystobalite Crystal modification of quartz, which is formed by heating clay silica bodies at temperatures above 1,100 °C. Heating increases the thermal expansion and decreases the danger of crazing. (Rosenthal, 1949)

C.S. jar collar A thick-wall steel collar, the inside surface of which is tapered to fit two serrated-face taper sleeves. The assembly may be fitted at any point over a casing or pipe and serves as a drive collar in sinking casing or pipe by driving and chopping. Also called self-tightening jar collar; self-tightening jar coupling; Simmons jar block, Simmons jar collar. (Long, 1960)

C.T. Nozzle Trade name; a refractory nozzle for steel pouring designed to give a constant teeming rate (therefore, the name). The nozzle consists of an outer fireclay shell and a refractory insert of different composition. Strictly speaking, the term refers to a particular type of insert developed for the teeming of free cutting steels. (Dodd, 1964)

cubanite An orthorhombic mineral, $CuFe_2S_3$; dimorphous with isocubanite. Formerly called chalmersite.

cubbyhole A niche cut in the rib or wall of an underground mine for the storage of explosives or detonators.

cube (a) Scot. A ventilating furnace in a mine. (b) A relatively rare crystal form of diamond having six equal-area faces at right angles to each other. (Long, 1960) (c) A rectangular prism having squares for its ends and faces. (Jones, 1949) (d) A crystal form of six equivalent (not necessarily square) and mutually perpendicular faces, with indices of {100}. (AGI, 1987) (e) A hexahedron, a crystal form of the isometric system consisting of six mutually orthogonal planar faces. Conditions of growth may yield crystal faces that are not perfectly square, but are at mutual right angles. (f) A diamond in cube form. (g) Dice, e.g., pyrite cubes known locally as devil's dice. (h) Pseudocubic forms, e.g., quartz rhombohedra with faces at near right angles, tetragonal prisms capped by a basal pinacoid, orthorhombic prisms or domes terminated by a pinacoid, or three orthorhombic pinacoids of nearly equal areas.

cube ore Eng. An arsenate of iron, $KFe_4(AsO_4)_3(OH)_4 \cdot 6\text{-}7H_2O$, of an olive-green to yellowish-brown color, and occurring commonly in cubes with the copper ores of Cornwall. Syn: *pharmacosiderite.* (Fay, 1920)

cube powder Gunpowder made in large cubical grains and burning more slowly than the small or irregular grains. (Fay, 1920)

cubic (a) Having the form of a cube, as a cubic crystal; or referring to directions parallel to the faces of a cube, as cubic cleavage. See also: *cubic system.* (b) Crystal cleavage with three planes at mutual right angles. (c) An alternate name for the isometric crystal system. See also: *isometric.* (d) In crystal structures, a cation coordination of six equidistant anions.

cubic cleavage (a) Equally good cleavage in three mutually perpendicular directions. (Fay, 1920) (b) Mineral cleavage parallel to the faces of a cube, e.g., in galena or halite.

cubic foot per minute A standard capacity or performance measurement for compressors. (Nichols, 1976)
cubicite A cubic zeolite; analcime. Also spelled cubizite.
cubic packing The loosest manner of systematic arrangement of uniform solid spheres in a clastic sediment or crystal lattice, characterized by a unit cell that is a cube whose eight corners are the centers of the spheres involved. An aggregate with cubic packing has the maximum porosity (47.64%). (AGI, 1987)
cubic plane Any plane normal to one of the crystallographic axes in the isometric system having Miller indices (100).
cubic system The crystal system that has the highest degree of symmetry; it embraces such forms as the cube and the octahedron. See also: *cubic.* (CMD, 1948)
cubo-octahedron (a) A crystal form that has faces of both the cube and the dodecahedron. (b) The combined isometric form of a cube modified by an octahedron or an octahedron modified by a cube.
cuckoo shots Subsidiary shots in the roof of a longwall working, between the coal face and the waste, or in any waste. (Nelson, 1965)
cuddy brae Scot. An inclined roadway, worked in the same manner as a self-acting incline.
cuesta (a) A hill or ridge with a gentle slope on one side and a steep slope on the other; specif. an asymmetric ridge (as in the Southwestern United States) with one face (dip slope) long and gentle and conforming with the dip of the resistant bed or beds that form it, and the opposite face (scarp slope) steep or even cliff like and formed by the outcrop of the resistant rocks, the formation of the ridge being controlled by the differential erosion of the gently inclined strata. (AGI, 1987) (b) A ridge or belt of low hills between lowlands in a region of gently dipping sedimentary rocks (as on a coastal plain), having a gentle slope conforming with the dip of the rocks and a relatively steep slope descending abruptly from its crest.—Etymol: Spanish, flank or slope of a hill; hill, mount, sloping ground. Cf: *hogback; wold; scarp; escarpment.* (AGI, 1987)
cueva Sp. A cave or grotto.
culasse The part of a brillant-cut stone below the girdle.
culet (a) The small lower terminus of a brilliant-cut gem. It is parallel to the table. (Standard, 1964) (b) The small facet that is polished parallel to the girdle plane across what would otherwise be the sharp point or ridge that terminates the pavilion of a diamond or other gemstone. Its principal function is to reduce the possibility of damage to the gemstone. Also spelled collet.
culm (a) A vernacular term variously applied, according to the locality, to carbonaceous shale, or to fissile varieties of anthracite coal. (Rice, 1960) (b) English Anthracite; a kind of coal, of indifferent quality, burning with a small flame, and emitting a disagreeable odor. (c) Anthracite fines that will pass through a screen with 1/8-in holes. (Nelson, 1965) (d) In anthracite terminology, the waste accumulation of coal, bone, and rock from old dry breakers. (Mitchell, 1950) (e) In bituminous coal preparation, culm corresponds to slurry or slime, depending upon the size distribution of the suspended solids. (Mitchell, 1950) (f) Kolm. (AGI, 1987) (g) The anthracite contained in the series of shales and sandstones of North Devon, England, known as the Culm Measures. (AGI, 1987) (h) Coal dust or fine-grained waste from anthracite mines. Syn: *kulm.* (AGI, 1987)
culm bank The deposit on the surface of culm usually kept separate from deposits of larger pieces of slate and rock. Also called culm dump. (Hudson, 1932)
culmiferous Containing culm as coal. (Standard, 1964)
culmination The highest point of a structural feature, e.g., of a dome or anticlinal crest. The axis of an anticline may have several culminations that are separated by saddles. See also: *crest.* Syn: *apex.* (AGI, 1987)
culture tube *acid-etch tube.*
cumengite (a) Same as volgerite. (English, 1938) (b) A tetragonal mineral, $Pb_{21}Cu_{20}Cl_{42}(OH)_{40}$. Also spelled cumengeite.
cummingtonite A monoclinic mineral, $(Fe,Mg)_7Si_8O_{22}(OH)_2$; amphibole group; has $Mg/(Mg + Fe^{2+}) = 0.30$ to 0.69; prismatic cleavage; may be asbestiform; in amphibolites and dacites; fibrous varieties (amosite, magnesium rich, and montasite, iron rich) are used as asbestos.
cumulate An igneous rock formed by the accumulation of crystals that settle out from a magma by the action of gravity; examples include layered igneous deposits such as the Bushveld complex in South Africa and the Stillwater complex in Montana. Syn: *accumulative rock.* (AGI, 1987)
cumulative float curve The curve obtained from the result of a float and sink analysis by plotting the cumulative yield at each specific gravity against the mean ash of the total floats at that specific gravity. (BS, 1962)
cumulative plot Graphic representation of cumulative curve results of screen analysis, in which the cumulative percentage of weight is plotted against the screen aperture, usually both to logarithmic scale. (Pryor, 1963)
cumulative sink curve The curve obtained from the results of a float and sink analysis by plotting the cumulative yield of sinks at each specific gravity against the mean ash of the total sinks at that specific gravity. (BS, 1962)
cumulophyric Said of the texture of a porphyritic igneous rock in which the phenocrysts, not necessarily of the same mineral, are clustered in irregular groups; also said of a rock exhibiting such texture. Syn: *glomeroporphyritic.* (AGI, 1987)
cundy (a) Scot. The spaces from which coal has been worked out, partly filled with dirt and rubbish between the packs. See also: *openset; goaf.* (Fay, 1920) (b) Aust. The passage under a roadway into which an endless rope passes out of the way at the end of its track. Also called conduct. A variation of conduit. (Fay, 1920) (c) Any small passageway made to improve ventilation or facilitate movement of materials. It is generally made through a pack or along the rib side of a longwall face. See also: *airhole.* (BS, 1963)
cup and cone A machine for charging a shaft furnace, consisting of an iron hopper with a large central opening, which is closed by a cone or bell pulled up into it from below. In the annular space around this cone, the ore, fuel, etc., are placed, then the cone is lowered to drop the materials into the furnace, after which it is again raised to close the hole. Syn: *bell and hopper; closed top.* (Fay, 1920)
cupel (a) A small bone-ash cup used in gold or silver assaying with lead. (b) The hearth of a small furnace used in refining metals.
cupellation (a) The process of assaying for precious metals with a cupel. (b) Oxidation of molten lead containing gold and silver to produce lead oxide, thereby separating the precious metals from the base metal.
cupeller One who refines gold and silver in a type of reverberatory furnace known as a cupel. (DOT, 1949)
cupferron A colorless crystalline salt, $C_6H_5N(NO)ONH_4$, that is a precipitant for copper and iron from solutions and is also used in the analysis of other metals, esp. of the uranium group. Used for separating iron and copper from other metals. Precipitates iron quantitatively from strongly acid solutions.
cupola (a) A cylindrical vertical furnace for melting metal, esp. gray iron, by having the charge come in contact with the hot fuel, usually metallurgical coke. (ASM, 1961) (b) A dome-shaped projection of the igneous rock of a batholith. Many stocks are cupolas on batholiths. Cf: *roof pendant.*
cupola furnace A shaft furnace used in melting pig iron (with or without iron or steel scrap) for iron castings. Metal, coke, and flux (if used) are charged at the top, and air is blown in near the bottom. (CTD, 1958)
cupreous manganese *lampadite.*
cupric sulfide *copper sulfide.*
cupriferous Yielding or containing copper. (Standard, 1964)
cupriferous pyrite *chalcopyrite.*
cuprite An isometric mineral, Cu_2O; red (crimson, scarlet, vermillion, brownish-red); sp gr, 6.1; in oxidized parts of copper veins; an important source of copper. Also called ruby copper; ruby copper ore. Syn: *red copper ore; red copper oxide; red glassy copper ore; red oxide of copper; octahedral copper; cuprous oxide; copper suboxide.* See also: *chalcotrichite; plush copper ore.*
cuproapatite A variety of apatite from Chile containing copper. (Standard, 1964)
cuproauride A former name for auricupride. Syn: *gold cupride.*
cuprocalcite A mixture of cuprite and calcite(?).
cuprocopiapite A triclinic mineral, $CuFe_4(SO_4)_6(OH)_2{\cdot}20H_2O$; copiapite group.
cuprodescloizite A former name for mottramite.
cuprojarosite A variety of melanterite containing copper (4.40% CuO) and magnesium (4.29% MgO). Also spelled kuprojarosit. See also: *jarosite.* (Spencer, 1940)
cuprokirovite A magnesian cuprian variety of melanterite.
cupromagnesite A monoclinic copper magnesium sulfate in bluish-green crusts at Mt. Vesuvius, Italy; of doubtful validity.
cupromontmorillonite Interpretation of the Russian name medmontite. See also: *medmontite.* (Spencer, 1952)
cuproplumbite A former name for bayldonite.
cuprorivaite A tetragonal mineral, $CaCuSi_4O_{10}$; in small blue grains at Mt. Vesuvius, Italy.
cuprosklodowskite A triclinic mineral, $(H_3O)_2Cu(UO_2)_2(SiO_4)_2{\cdot}2H_2O$; strongly radioactive; greenish-yellow or grass-green; a secondary mineral resulting from alteration of pitchblende associated with other uranium minerals.
cuprous oxide *cuprite.*
curb (a) A timber frame, circular or square, wedged in a shaft to make a foundation for walling or tubbing, or to support, with or without other timbering, the walls of the shaft. (b) The heavy frame or sill at the top of a shaft. (c) In tunnel construction, a ring of brickwork or of cast iron, at the base of the shaft, surmounting a circular orifice in the roof of the tunnel. A drum curb is a flat ring of cast iron for supporting

the brickwork having the same diameter externally as the shaft of brickwork. Temporary curbs of oak are also used. (Fay, 1920) (d) An iron border to the incorporating bed of a gunpowder mill. (Webster 3rd, 1966) (e) An iron casing in which to ram-load molds for casting. (Webster 2nd, 1960) (f) The walls of a chamber in which sulfuric acid is manufactured. (Webster 3rd, 1966) (g) A wood, cast-iron, or reinforced concrete ring, made in segments, forming a foundation for a masonry or cast-iron circular shaft lining. The curb is set on a firm ledge of rock notched into the periphery of the shaft. It may be removed at a later stage. Syn: *wedging curb; bricking curb; crib; walling curb.* See also: *foundation curb; water ring.* (Nelson, 1965) (h) A socket of wrought iron or steel for attaching a ring hook or swivel to the end of a rope used for mine hoisting or haulage. (CTD, 1958) (i) A coaming around the mouth of a well or shaft. See also: *binder.* (Hess) (j) A shaft support ring for walling or tubbing. (Mason, 1951)

curb tubbing Eng. A solid wood lining of a shaft. Syn: *curb.*

curie A unit of measurement of radioactivity, defined as the amount of a radionuclide in which the decay rate is 37 billion disintegrations per second, which is approx. equal to the decay rate of 1 g of pure radium. (AGI, 1987)

Curie point The temperature at which there is a transition in a substance from one phase to another of markedly different magnetic properties. Specif., the temperature at which there is a transition between the ferromagnetic and paramagnetic phases. (Webster 3rd, 1966)

Curie's law The susceptibility of a paramagnetic substance is inversely proportional to the absolute temperature. A law of magnetism that has been replaced by the Curie-Weiss law. (Webster 3rd, 1966)

Curie temperature The temperature of magnetic transformation below which a metal or alloy is magnetic and above which it is paramagnetic. (ASM, 1961)

curite An orthorhombic mineral, $Pb_2U_5O_{17}{\cdot}4H_2O$; radioactive; orange-red; an alteration product of uraninite.

curlstone Shrop. Ironstone exhibiting cone-in-cone formation.

curly coal A folded and distorted oil shale.

curly stone Shrop. Shale belonging to the coal formation, which on exposure to the air hardens and assumes a peculiar form, sometimes called cone-upon-cone. Also called curlstone. (Arkell, 1953)

current (a) The part of a fluid body esp. as air or water, that is moving continuously in a definite direction, often with a velocity much swifter than the average, or in which the progress of the fluid is principally concentrated. (AGI, 1987) (b) A horizontal movement or continuous flow of water in a given direction with a more or less uniform velocity, producing a perceptible mass transport, set in motion by winds, waves, gravity, or differences in temperature and density, and of a permanent or seasonal nature, esp. an ocean current. (AGI, 1987) (c) The velocity of flow of a fluid in a stream. (AGI, 1987) (d) The swiftest part of a stream. (Webster 3rd, 1966) (e) A tidal or a nontidal movement, often horizontal, of lake or ocean water. Syn: *drift.* (Webster 3rd, 1966) (f) Condition of flowing. Flow marked by force or strength. Syn: *flow; flux.* (Webster 3rd, 1966)

current bedding Any bedding or bedding structure produced by current action; specif. cross-bedding resulting from water or air currents of variable direction. See also: *false bedding.* Syn: *inclined bedding.* (AGI, 1987)

current density The current per unit area perpendicular to the direction of current flow. (AGI, 1987)

current electrode (a) A piece of metal connected to a cable that, when buried in the earth in a shallow hole or lowered into a well, provides enough contact to permit the passage of substantial electrical current into the surrounding earth. (AGI, 1987) (b) A metal contact with the ground used to facilitate current flow through the ground. (AGI, 1987)

current leakage Portion of the firing current bypassing part of the blasting circuit through unintended paths. (Atlas, 1987)

current leakage tester *earth fault tester.*

current-limiting device An electric or electromechanical device that limits current amplitude; duration of current flow; or total energy of the current delivered to an electric blasting circuit. (Atlas, 1987)

current meter (a) Any one of numerous instruments for measuring the speed alone, or both speed and direction, of flowing water, as in a stream or the ocean; it is usually activated by a wheel equipped with a set of revolving vanes or cups whose rate of turning is proportional to the velocity of the current. (AGI, 1987) (b) An instrument, as a galvanometer, for measuring the strength of an electric current. (Standard, 1964)

current rose A graphical representation of currents, usually by 1° quadrangles, using arrows of different lengths for the cardinal and intercardinal compass points to show resultant drift and frequency of set for a given period of time. (Hy, 1965)

curry pit Leic. A hole sunk from an upper to a lower portion of a thick seam of coal through which the return air passes from the stalls to the airway. (Fay, 1920)

cursing in work Can. False affidavit of assessment work on mining claims. (Hoffman, 1958)

curtain (a) A sheet of brattice cloth hung across an entry in such a way as to prevent the passage of an air current but not to hinder the passage of mules or mine cars. In coal mines, curtains are used to deflect the air from the entries into the working rooms and to hold the air along the faces. They are usually made of a number of overlapping strips of heavy curtain material, that should be of fireproof or fire-resistant material. Syn: *check curtain.* See also: *blasting curtain.* (Kentucky, 1952) (b) Also called cover. (Long, 1960) (c) A thin sheet of dripstone hanging from the ceiling or projecting from the wall of a cave. (Schieferdecker, 1959) (d) A rock formation that connects two neighboring bastions. (AGI, 1987) (e) One of a series of steps cut in a valley side and exaggerated by cultivation. (AGI, 1987)

curtain hole *cover hole.*

curtisite (a) A crystalline hydrocarbon, found in a form of greenish deposits from a hot spring in California. (Tomkeieff, 1954) (b) A former name for idrialite.

curvature of gravity A vector quantity calculated from torsion-balance data indicating the shape of the equipotential surface. It points in the direction of the longer radius of curvature. (AGI, 1987)

curved jib A chain coal cutter jib with the outer end bent upward or downward through 90°. Thus, the machine can make a horizontal and also a vertical cut in one operation. Curved jibs make coal preparation easier, but their use is limited because of the excessive strain and wear on the cutter chain. See also: *turret jib; multicut chain.* (Nelson, 1965)

cusec A unit of waterflow or airflow that equals 1 ft^3/s (0.028 m^3/s). (Nelson, 1965)

cushat marl *cowshut.*

cushion A course of some compressible substance, such as soft wood, inserted between more rigid material. In mine support, it can be placed between the footwall or the hanging wall and the concrete, or internally in the support. (Spalding, 1949)

cushion blasting A method of blasting in which an airspace is left between the explosive charge and the stemming, or in which the shothole is of substantially larger diameter than the cartridge. See also: *controlled blasting.* (BS, 1964)

cushion cut A style of faceting gems in which the finished gem is roughly rectangular in outline but with gently outward curving sides and rounded corners. (Sinkankas, 1959)

cushion firing *water-ampul stemming.*

custom mill A mill that depends on purchased ores mostly or entirely for processing rather than on its own organizational source.

custom ore Ore bought by a mill or smelter, or treated for customers. (Hess)

custom plant A mill, concentrator, or smelter that purchases ore or partly processed mineral for treatment in terms of an appropriate contract, priced on tonnage, complexity of operation, permissible losses, and specification of feed, product, and (perhaps) lost tailings. (Pryor, 1963)

custom smelter (a) A smelter which buys ores or treats them for customers. (Hess) (b) A smelter which depends for its intake mostly on concentrate purchased from independent mines and on scrap metal, rather than its own captive mine sources. (Wolff, 1987)

cut (a) An arrangement of holes used in underground mining and tunnel blasting to provide a free face to which the remainder of the round can break. Also, the opening created by the cut hole. (b) To intersect a vein or working. (c) To shear one side of an entry or crosscut by digging out the coal from floor to roof with a pick. See also: *undercut.* (d) Eng. In Somerset, a staple or drop pit. (e) Eng. The depth to which a drill hole is put in for blasting. (f) A term applied where the cutting machine has cut under the coal. (g) The drill-hole pattern for firing a round of shots in a tunnel or sinking shaft, e.g., the burn cut. (Nelson, 1965) (h) A machine cut in a coal seam; e.g., floor cut. (Nelson, 1965) (i) *stint; sump.* (j) An excavation, generally applied to surface mining; to make an incision in a block of coal; in underground mining, that part of the face of coal that has been undercut. (BCI, 1947) (k) In mining, when used in conjunction with shaft and drift, a surface opening in the ground intersecting a vein. (Ricketts, 1943) (l) Depth to which material is to be excavated (cut) to bring the surface to a predetermined grade; the difference in elevation of a surface point and a point on the proposed subgrade vertically below it. (Seelye, 1951) (m) To excavate coal. (n) To drive to or across a lode. (Gordon, 1906) (o) The groups of holes fired first in a round to provide additional free faces for the succeeding shots. (BS, 1964) (p) To lower an existing grade. (Nichols, 1976)

cut-and-fill stoping A stoping method in which the ore is excavated by successive flat or inclined slices, working upward from the level, as in shrinkage stoping. However, after each slice is blasted down, all broken ore is removed, and the stope is filled with waste up to within a few feet of the back before the next slice is taken out, just enough room being left between the top of the waste pile and the back of the stope to provide working space. The term cut-and-fill stoping implies a definite and characteristic sequence of operations: (1) breaking a slice of ore from the back; (2) removing the broken ore; and (3) introducing filling. See also: *underhand longwall.*

cut-chain brae Scot. An incline on which cut chains are used.

cut holes (a) The first hole or group of holes fired in a drift or tunnel face. Also known as the cut portion of the blasting round. (Lewis, 1964) (b) In tunneling, easers so drilled and fired as to break out a leading wedge-shaped hole and thus enable the later holes in the complete round of shots to act more effectively. See also: *trimmers; drill-hole pattern.* (Pryor, 1963)

cutinite (a) A variety of exinite. The micropetrologic constituent, or maceral, of cuticular material. Cf: *sporinite.* (AGI, 1987) (b) Maceral of the exinite group consisting of plant cuticle. Cf: *resinite.* (AGI, 1987)

cutinite coal This type of coal consists of more than 50% of cuticle, the fragments of which occur embedded in gelito-collinite, fusinito-collinite, and collinite of fusinitic nature. In addition to cuticle, spores, resin bodies, and fragments of finely fusinized and gelified tissue are present. Leaf parenchyme and stem tissue, bordered by cuticle, may also be seen. Hand specimens of this type of coal are grayish-black, matt or semimatt, finely striated, or sometimes even banded. The coal breaks angularly and generally has high ash. Cutinite coal occurs as thin bands in seams of different geological age, and its use is largely determined by the other forms of coal with which it is associated. (IHCP, 1963)

cutoff (a) In firing a round of shots, a misfire due to severance of fuse owing to rock shear as adjacent charge explodes. (Pryor, 1963) (b) A quarryer's term for the direction along which granite must be channeled, because it will not split. Same as hardway. (c) Cf: *cutoff entry.* (d) The number of feet a bit may be used in a particular type of rock (as specified by the drill foreman). (Long, 1960) (e) An impermeable wall, collar, or other structure placed beneath the base or within the abutments of a dam to prevent or reduce losses by seepage along a construction interface or through porous or fractured strata. It may be made of concrete, compacted clay, interlocking sheet piling, or grout injected along a line of holes. (AGI, 1987) (f) A boundary, oriented normal to bedding planes, that marks the areal limit of a specific stratigraphic unit where the unit is not defined by erosion, pinchout, faulting, or other obvious means. Cutoffs are applicable to map, cross-sectional, and three-dimensional views, and are in effect specialized facies boundaries. (AGI, 1987) (g) Minimum percentage of mineral or metal in an ore that can be mined profitably. (Long, 1960) (h) A device for cutting off; as a mechanism for shutting off the admission of a working fluid (as steam) to an engine cylinder. (Webster 3rd, 1966) (i) The point in the stroke of the piston of a steam engine at which the entrance of live steam is stopped by the closure of the inlet valve. (Long, 1960)

cutoff entry An entry driven to intersect another and furnish a more convenient outlet for the coal. Cf: *cutoff.* See also: *entry.* (Fay, 1920)

cutoff grade The lowest grade of mineralized material that qualifies as ore in a given deposit; rock of the lowest assay included in an ore estimate.

cutoff hole Missed hole resulting from the failure of a blasting cap to detonate owing to the breaking of a fuse or conductor or to some other similar cause. (Fraenkel, 1953)

cutoff shot A shot in a delay round in which the charge has been wholly or partially exposed to the atmosphere by reason of the detonation of an earlier shot in the round. (BS, 1964)

cutout (a) Opening made in a mine working in which a drill or other equipment may be placed so as not to interfere with other mining operations. (Long, 1960) (b) The act or process of removing diamonds from a used or dull bit by dissolving the crown metal by corrosive action of an acid or electrolytic dissolution. Also, the diamonds recovered or salvaged by such means. (Long, 1960) (c) A mass of shale, siltstone, or sandstone filling an erosional channel cut into a coal seam. Cf: *low; roll; washout.* Syn: *horseback; want.* (AGI, 1987)

cut point The value of a property (e.g., density or size) at which a separation into two fractions is desired or achieved. (BS, 1962)

cut shot (a) A shot designed to bring down coal that has been sheared or opened on one side. (Fay, 1920) (b) A shot that initially breaks ground to provide a free face for subsequent shots. (BS, 1964)

cut stone (a) Originally, an artificially broken and shaped carbon; now generally, a faceted diamond or other precious or semiprecious gemstone used as an ornament. Syn: *gem.* (Long, 1960) (b) Structural unit for limestone that consists of blocks that are cut to specified dimensions and surface tooled. (AIME, 1960)

cuttable Diamond material suitable for cutting into gems. (Chandler, 1964)

cutter (a) A joint in a rock that is parallel to the dip of the strata. (CTD, 1958) (b) A crack in a crystal that destroys or lessens its value as a lapidary's stone. (c) On a hydraulic dredge, a set of revolving blades at the end of the suction line. (d) Closed or inconspicuous seams along which rock may separate or break easily. (e) Any coal-cutting or rock-cutting machine; or the person operating it. (f) A solution crevice in limestone underlying Tennessee residual phosphate deposits. (AGI, 1987) (g) *underreamer lug.* Cf: *cutting edge.*

cutter bar That part of a chain mining machine that supports the cutting chain and extends under the coal; the bar provides the track for the cutting chain. (Fay, 1920; BCI, 1947)

cutter chain The endless chain carrying picks that travels around the jib of a chain coal cutter at a speed varying from 320 to 650 ft/min (97.6 to 198.2 m/min). See also: *coal-cutter pick.* (Nelson, 1965)

cutter dredge In alluvial mining, one that loosens the alluvium by means of a cutting ring, at the end of a suction pipe through which the products are pumped up for treatment. (Pryor, 1963)

cutterhead pipeline dredge A hydraulic dredge in which the suction action is augmented by a rotating propeller that operates at the point of suction. The cutterhead performs two functions: it cuts into and loosens compacted soils and soft rock such as coral, and it increases dredge capacity by channeling the soils into the end of the suction pipe. The efficiency of a dredge is based on its capacity to handle soils rather than water, and the cutterhead serves to maintain an optimum ratio of about 1 ft^3 (0.028 m^3) of soil handled per 5 ft^3 (0.14 m^3) of water. (Carson, 1961)

cutter loader A longwall machine that cuts and loads the coal onto a conveyor as it travels across the face. See also: *loader.* (Nelson, 1965)

cutter plow A plow-type cutter loader developed for use in hard coal seams. It has four horizontal stepped precutting blades, which make a precut from 8 to 12 in (20.3 to 30.5 cm), to weaken the coal immediately in front of the machine. It can be single or double ended, and is hauled along the face by winches. The coal is loaded onto a panzer conveyor, which is advanced behind the machine by compressed air rams. See also: *plow.* (Nelson, 1965)

cut-through A passage cut through the coal, connecting two parallel entries. See also: *crosscut; breakthrough; jack hole.* (Rice, 1960)

cutting (a) The opening made by shearing or cutting. (b) Low-grade ore or refuse obtained from beneficiating ore. (c) The operation of making openings across a coal seam as by channeling, or beneath a coal seam as by undercutting. (d) Excavating. (Nichols, 1976) (e) Lowering a grade. (Nichols, 1976) (f) Eng. The end or side of a stall next to the solid coal where the coal is cut with a pick in a vertical line to facilitate breaking down; channeling. (g) N. of Eng. The operation of undercutting coal with a mechanical cutter. The machine, which runs on electricity, employs two cutterpersons. (Trist, 1963)

cutting chain The sprocket chain that carries the steel points used for undermining the coal with chain mining machines. (Fay, 1920)

cutting down (a) The trimming of shaft walls to increase their sectional area. (Zern, 1928) (b) Removing roughness or irregularities of a metal surface by abrasive action. (ASM, 1961)

cutting edge (a) The point or edge of a diamond or other material set in a bit that comes in contact with and cuts, chips, or abrades the rock. Also called cutting point. Cf: *cutting stones.* (Long, 1960) (b) That part of a bit in actual contact with rock during drilling operations. (Long, 1960) (c) The leading edge of a lathe tool where a line of contact is made with the work during machining. Cf: *cutter.* (ASM, 1961)

cutting face That part of a bit containing the cutting points, excluding the points inset as reamers. (Long, 1960)

cutting fluid A fluid, usually a liquid, used in metal cutting to improve finish, tool life, or dimensional accuracy. On being flowed over a tool and a workpiece, the fluid reduces friction, heat generated, and tool wear, and prevents galling. It conducts heat away from the point of generation and also serves to wash chips away. (ASM, 1961)

cutting grain The direction along a plane on which a diamond can be most easily abraded. (Long, 1960)

cutting horizon The position in a coal seam in which a horizontal machine cut is made. The normal cutting horizon is along the bottom of the seam. See also: *bottom cut.* (Nelson, 1965)

cutting machine A power-driven machine used to undercut or shear the coal to facilitate its removal from the face. (BCI, 1947)

cutting motor The motor in a cutting machine that provides power for the operation of the cutting chain. (Jones, 1949)

cuttings (a) The particles of rock produced in a borehole by the abrasive or percussive action of a drill bit; excess material caused by the rubbing of core against core or core against steel; erosive effect of the circulating liquid; or cavings from the borehole. Also called drill cuttings; drillings; sludge. Cf: *borings.* (Long, 1960) (b) The fragmental rock samples broken or torn from the rock penetrated during the course of drilling. (AGI, 1987) (c) Eng. *holings.* (d) *bug dust.*

cutting sand Composed of sharp, solid quartz grains and used as abrasive for sawing stone; usually ungraded and about equivalent to a No. 1 sandblasting sand. (AIME, 1960)

cutting speed (a) The linear or peripheral speed of relative motion between a tool and a workpiece in the principal direction of cutting. (ASM, 1961) (b) *feed rate.*

cutting stones Diamonds set in a bit face having points or edges that will be in contact with, and will cut or abrade, the rock when drilling. Cf: *cutting edge; gage stone.* (Long, 1960)

cutting wheel A cutting disk, the edge of which is impregnated with an abrasive, such as diamond dust or aloxite. It is rotated at high speed in a coolant and used to cut rock specimens into suitable thin sections for microscopic inspection in transmitted light with a polarized-light microscope or, after polishing, with a reflected-light (ore) microscope.

cutty clay Plastic clay formerly used in England for making tobacco pipes; "pipe clay.". (AGI, 1987)

cuvette A large-scale basin in which sedimentation has occurred or is taking place, as distinguished from a tectonic basin due to folding of preexisting rocks; e.g., the Anglo-Parisian cuvette of Southeast England and Northeast France, in which Cenozoic rocks accumulated and were later folded into several distinct but smaller basins, such as the London Basin and the Paris Basin. Etymol: French, small tub or vat. Sometimes misspelled curvette. (AGI, 1987)

Cuylen conveyor A single-chain conveyor with an open side to facilitate power loading. (Sinclair, 1959)

Cuyuna The name of an iron range in Minnesota. It is composed of the syllables, "Cuy" and "Una," the former being a contraction of the given name of Cuyler Adams, who was active in the early development of that territory, and the last syllable being the name of his dog "Una."

C-wave *coupled wave.*

cyanamide White; crystalline; NH_2CN. Formed variously by the action of cyanogen chloride on ammonia. (Standard, 1964)

cyanicide Any substance present in a pulp that attacks or destroys the cyanide salt being used to dissolve precious metals. (Pryor, 1958)

cyanidation A process of extracting gold and silver as cyanide slimes from their ores by treatment with dilute solutions of potassium cyanide or sodium cyanide. The slimes are subsequently fused and cast into ingots or bullion. (Henderson, 1953)

cyanidation vat A large tank, with a filter bottom, in which sands are treated with sodium cyanide solution to dissolve out gold. (CTD, 1958)

cyanide Usually refers to cyanide solution in circulation in a mill treating gold or silver ores. The stock or solution is of two main types: barren, from which all possible value has been extracted, and pregs or pregnant, which is charged with gold or silver and awaits their removal. (Pryor, 1965)

cyanide hardening Introducing carbon and nitrogen into the surface of a steel alloy by heating in a bath of molten sodium cyanide and usually followed by quench hardening.

cyanide man In ore dressing, smelting, and refining, a person who tends equipment in which finely ground gold or silver ore is treated with a cyanide solution to separate free gold or silver from the gangue (waste material). (DOT, 1949)

cyanide mill *cyanide process.*

cyanide process A process for the extraction of gold from finely crushed ores, concentrates, and tailings by means of cyanide of potassium or sodium used in dilute solutions. The gold is dissolved by the solution and subsequently deposited upon metallic zinc or other materials. Syn: *cyanide mill.* See also: *MacArthur and Forest cyanide process.* (Fay, 1920)

cyanide pulp The mixture obtained by grinding crude gold and silver ore and dissolving the precious-metal content in a sodium-cyanide solution. (CCD, 1961)

cyanide slime Precious metal in the form of finely divided particles precipitated from a cyanide solution used in its extraction from ore. (ASM, 1961)

cyaniding The process of treating finely ground gold and silver ores with a weak solution of sodium or potassium cyanide, which readily dissolves these metals. The precious metals are then obtained by precipitation from solution with zinc, or by adsorbtion on activated carbon. (CTD, 1958)

cyanite A former spelling of kyanite.

cyanochalcite A phosphoriferous variety of chrysocolla from Nijni Tagilks, Perm, Russia.

cyanochroite A monoclinic mineral, $K_2Cu(SO_4)_2 \cdot 6H_2O$; picomerite group; a clear-blue alteration product from Mt. Vesuvius, Italy.

cyanogen (a) A univalent radical; present in hydrogen cyanide and in other simple and complex cyanides (as ferricyanides). (Webster 3rd, 1966) (b) Colorless, flammable, poisonous gas; $(CN)_2$. It has an odor like that of peach leaves, is variously formed (as by heating mercuric cyanide), and polymerizes readily. (Webster 3rd, 1966)

cyanosite A former name for chalcanthite. Also spelled cyanose.

cyanotrichite An orthorhombic mineral, $Cu_4Al_2(SO_4)(OH)_{12} \cdot 2H_2O$; sky-blue to smalt-blue; minutely crystalline or spheroidal. Formerly called lettsomite. Syn: *velvet copper ore.*

cycle of denudation Cycle of erosion. (AGI, 1987)

cycle of erosion The complete series of changes or stages through which a landmass passes from the inception of erosion on a newly uplifted or exposed surface through its dissection into mountains and valleys to the final stage when it is worn down to the level of the sea or to some other base level. The cycle is usually subdivided into youthful, mature, and old-age stages. One type or many types of erosion may be involved, and the landforms produced and destroyed depend to a large extent on the climate, geographic situation, and geologic structure of the landmass. Syn: *cycle of denudation; geographic cycle; geomorphic cycle.* (Stokes, 1955)

cycle of operations In mining operations, such as tunnel driving, shaft sinking, and coal winning, there are certain tasks that must be repeated in cyclic fashion. In tunnel driving, they are (1) drilling the round; (2) charging and firing; (3) loading; and (4) supporting and track extension. This cycle of operations is time analyzed to achieve maximum efficiency and speed. (Nelson, 1965)

cycle of sedimentation (a) A sequence of related processes and conditions, repeated in the same order, that is recorded in a sedimentary deposit. (AGI, 1987) (b) The deposition of sediments in a basin between the beginnings of two successive marine transgressions, comprising the deposits formed initially on dry land, followed by shallow-water and then deep-water deposits that in turn gradually change to shallow-water and then dry-land type during a marine regression. (AGI, 1987) (c) *cyclothem; sedimentary cycle.*

cycle skipping An instrumental phenomenon occurring in acoustic velocity logs. It consists of intervals where the velocity recorded drops sharply to very low values, and equally sharply, returns to a normal scale figure. Such a log is spiky. (Wyllie, 1963)

cycle time The time required for the dipper of a mechanical shovel to push through the bank and fill, swing to the haul unit, unload, and swing back to the digging position. Cycle time is established under standard conditions of a 90° angle of swing and with an optimum depth of cut. (Carson, 1961)

cyclic Adj. of cycle; recurrent rather than secular. (AGI, 1987)

cyclic mining A mining system in which each shift has a specific task to complete on the conveyor face. If the task on any shift is not completed in time, the following shifts are disorganized. In general, the face is machine cut during the night shift; shot-firing and hand-filling of the coal occupy the day shift; and the afternoon shift is responsible for moving the conveyor and roof supports to the new line of face. See also: *conventional machine mining; conventional mining.* (Nelson, 1965)

cyclic surge In classification, periodic upset of correct separating density of pulp, resulting in wrong release of oversize material from the closed grinding circuit. (Pryor, 1963)

cyclic test In batch tests of small quantities of ore during development of a method of concentration, the retention of selected fractions (usually middlings) for admixture with fresh samples. Used to study effect of recycling minerals or solutions, which they may have contaminated; also, to observe effects of increased concentration of such compounds on the process as a whole. See also: *locked test.* (Pryor, 1963)

cyclic twin Repeated twinning of three or more individual crystals according to the same twin law but with the twin axes or twin planes not parallel, commonly resulting in threefold, fourfold, fivefold, sixfold, or eightfold twins, which, if equally developed, display geometrical symmetry not found in single crystals, e.g., chrysoberyl, rutile. Cf: *repeated twinning; polysynthetic twinning.*

cyclic twinning The repeated twinning of three or more individuals according to the same twin law but with the twinning axes not parallel. Often simulates a higher order of symmetry than that of the untwinned crystal. Cf: *polysynthetic twinning.* (AGI, 1987)

cyclic winding *automatic cyclic winding.*

cyclone (a) The conical-shaped apparatus used in dust collecting operations and fine grinding applications. In principle, the cyclone varies the speed of air, which determines whether a given particle will drop through force of specific gravity or be carried through friction of the air. (Enam. Dict., 1947) (b) A classifying (or concentrating) separator into which pulp is fed, so as to take a circular path. Coarser and heavier fractions of solids report at apex of long cone while finer particles overflow from central vortex. Also called hydrocyclone. See also: *cyclone washer; centrifugal separation; centrifuge.* (Pryor, 1965)

cyclone angle Included angle of conical section of hydrocyclone. (Pryor, 1963)

cyclone classifier A device for classification by centrifugal means of fine particles suspended in water, whereby the coarser grains collect at and are discharged from the apex of the vessel, while the finer particles are eliminated with the bulk of the water at the discharge orifice. (BS, 1962)

cyclone dust collector An apparatus for the separation by centrifugal means of fine particles suspended in air or gas. (BS, 1962)

cyclone furnace A forced circulation heat treatment furnace designed to operate at a maximum temperature of 760 °C, which is either gas fired or electrically heated. The gas circulates at the rate of 0.89 m/s. (Osborne, 1956)

cyclone overflow A finer classified fraction, which leaves via the vortex finder of a hydrocyclone. (Pryor, 1963)

cyclone separator A funnel-shaped device for removing material from an airstream by centrifugal force. (ASM, 1961)

cyclone size Diameter of cylindrical section of hydrocyclone. Also diameter of inlet orifice if round. Dimensions or area in given inches or square inches if opening is rectangular. (Pryor, 1963)

cyclone underflow A coarser sized fraction, which leaves via apex aperture of hydrocyclone. (Pryor, 1963)

cyclone washer Cyclone washing of small coal originates from the Netherlands. Clean separation is effected with the aid of centrifugal force. The heavier shale particles move to the wall of the cyclone and are eventually discharged at the bottom while the lighter coal particles are swept toward the central vortex and are discharged through an outlet at the top. The washer may be used for cleaning coal up to 3/4 in (1.9 cm). The coal is normally deslimed at about 0.5 mm before cleaning. The separating medium is water and ground magnetite, the bulk of which is recovered and returned to the circuit. A 20-in (50.8-cm) cyclone has a feed capacity of about 50 t/h of coal sized between 1/2 in (12.7 mm) and 1/2 mm. See also: *centrifugal separation; cyclone.* (Nelson, 1965)

cyclosteel Steel produced by blowing iron-ore powder into a hot gas. (CTD, 1958)

cyclothem A series of beds deposited during a single sedimentary cycle of the type that prevailed during the Pennsylvanian period. The cyclothem, which ideally consists of 10 members (in western Illinois, the fifth member is a coal layer), indicates an unstable coastal environment in which marine submergence and emergence occurred. A cyclothem ranks as a formation in the scale of stratigraphic nomenclature. Syn: *coal-measures unit.*

cyclotron A particle accelerator in which charged particles receive repeated synchronized accelerations by electrical fields as the particles spiral outward from their source. The particles are kept in the spiral by a powerful magnet. (Lyman, 1964)

cylinder cuts In cylinder cuts the blasting is performed toward an empty hole in such a way that, as the charges in the first, second, and subsequent holes detonate, the broken rock is thrown out of the cut. The opening is successively and uniformly (cylindrically) enlarged in its entire length. (Langefors, 1963)

cylindrical drum *parallel drum.*

cylindrical land Land having zero relief. (ASM, 1961)

cylindrical mill *tube mill.*

cylindrical structure A vertical structure in sandstone, a few centimeters to several decimeters in diameter and several decimeters in length, with a structureless interior, attributed to a rising water column or a spring channel. Syn: *sandstone pipe.*

cylindrite A triclinic mineral, $Pb_4FeSn_4Sb_2S_{16}$; forms cylinders that separate under pressure into distinct shells or folia, or is massive; at Poopo, Bolivia.

cylindroconical drum A combination of a cone and a cylinder. The ascending rope is wound on the smaller diameter of the cone; as the engine reaches full speed after the period of acceleration, the rope is wound on the larger cylindrical part. For deep shafts the rope is wound back on itself for the last part of the hoisting period, thus reducing the width of the drum. See also: *bicylindroconical drum.* (Lewis, 1964)

cymoid loop The splitting of a vein along its dip or strike into two branches, both of which curve away from the general trend and then unite to resume a direction parallel to but not in line with the original trend. See also: *cymoid structure.* (AGI, 1987)

cymoid structure A vein, or a vein-shaped structure, shaped like a reverse curve. See also: *cymoid loop.* (AGI, 1987)

cymophane *cat's-eye; chrysoberyl.*

cymrite A monoclinic mineral, $BaAl_2Si_2(O,OH)_8{\cdot}H_2O$; pseudohexagonal; at the Benalt manganese mine, Wales. Named from Cymru, the Welsh name for Wales.

cyprine A variety of vesuvianite or idocrase, of a blue tint, which is supposedly due to copper. (Fay, 1920)

cypritic steel A steel containing approx. 15% chromium and 9% copper; claimed to be resistant to corrosion in the atmosphere and to tap water, but its corrosion-resistant properties are inferior to the conventional austenitic chromium-nickel steels of the 18-8 type. (Osborne, 1956)

cyrilovite A tetragonal mineral, $NaFe_3(PO_4)_2(OH)_4{\cdot}2H_2O$; in pegmatite at Cyrilov, Moravia, Czech Republic.

cyrtolite A variety of zircon.

Czochralski's reagent An etchant for iron or steel, consisting of a solution of 10% to 20% ammonium persulfate in water. (Osborne, 1956)

Czochralski technique A method of growing single crystals of refractory oxides, and of other compounds, by pulling from the pure melt; the compound must melt congruently. (Dodd, 1964)

D

dachiardite A monoclinic mineral, $(Ca,Na_2,K_2)_5Al_{10}Si_{38}O_{96}\cdot 25H_2O$; zeolite group; in a pegmatite at San Piero, Elba, Italy.
dacite A fine-grained extrusive rock with the same general composition as andesite, but having a less calcic plagioclase and more quartz; according to many, it is the extrusive equivalent of granodiorite. Syn: *quartz andesite.* The name, given by Stache in 1863, is from the ancient Roman province of Dacia (now part of Romania). (AGI, 1987)
dacker Eng. Insufficient ventilation of a mine; dead air.
dactylitic A term applied to a rock texture produced by a symplectic intergrowth, in which one mineral is penetrated by fingerlike projections from another mineral; also, said of a rock exhibiting such texture. See also: *dactylotype intergrowth; symplectic.* (AGI, 1987)
dactylotype A textural term applied by Shand in 1906 to the intergrowth of sodalite with orthoclase in borolanite and its associates. The sodalite is altered to pinitic mica and appears in threadlike or vermicular aggregates closely packed in a matrix of orthoclase. (Holmes, 1928)
dactylotype intergrowth (a) A mineral intergrowth in which thin successive layers resemble a fingerprint pattern, as in some orthoclase-nepheline intergrowths. (Hess) (b) A symplectic intergrowth in which fingerlike projections of one mineral penetrate another. See also: *dactylitic.* (AGI, 1987)
dad N. of Eng. In coal mining, to mix (combustible gases) with atmospheric air to such an extent that the mixture is incapable of exploding. Also called dash. See also: *dashing.*
dadding The circulation, control, and utilization of air produced by the fan to ventilate the mine workings. See also: *circulation of air.* (Nelson, 1965)
Daelen mill An early type of universal rolling mill provided with both vertical and horizontal rolls so that a part could be rolled on all sides in one operation. (Osborne, 1956)
Daeves's reagent An etchant used to distinguish carbides in chromium steels and tungstides in high-speed steels. The solution contains 20 g of potassium ferricyanide and 10 g of potassium hydroxide in 100 mL of water. (Osborne, 1956)
dahllite A resinous, yellowish-white carbonate-apatite mineral or association sometimes occurring as concretionary spherulites. Now called carbonate-hydroxylapatite. (AGI, 1987)
Dahlstrom's Formula Classification through the hydrocyclone. (Pryor, 1963)
dakeite *schröckingerite.*
dalles (a) The rapids in a deep, narrow stream confined between the rock walls of a canyon or gorge; e.g., The Dalles of the Columbia River where it flows over columnar basalt. (AGI, 1987) (b) A steep-sided part of a stream channel, near the dalles proper, marked by clefts, ravines, or gorges; e.g., along the Wisconsin River, WI.—Etymol: French plural of dalle, gutter. Syn: *dells.* (AGI, 1987)
Dalton's law In a mixture of gases, the total pressure is equal to the sum of the pressures that the gases would exert separately. See also: *partial pressure.* (Standard, 1964)
dalyite A triclinic mineral, $K_2ZrSi_6O_{15}$; at Ascension Island in the Atlantic Ocean.
dam (a) A barrier to keep foul air or water, from mine workings. See also: *stopping; bulkhead.* (Fay, 1920) (b) An airtight barrier to isolate underground workings that are on fire. (CTD, 1958) (c) The wall of refractory material, forming the front of the forehearth of a blast furnace, that is built on the inside of a supporting iron plate (dam plate). Iron is tapped through a hole in the dam, and cinder through a notch in the top of the dam. See also: *Lurmann front.* (Fay, 1920)
dam gradation *contragradation.*
damkjernite A hypabyssal rock composed of phenocrysts of biotite and titanaugite in a fine-grained groundmass of pyroxene, biotite, perovskite, and magnetite, with interstitial nepheline, microcline, and calcite. The name, given by Brögger in 1921, is for the locality Damkjern (or Damtjern), Fen complex, Norway. Also spelled: damtjernite. (AGI, 1987)
damp Any mine gas, or mixture of gases, particularly those deficient in oxygen. Damp is probably derived from the German dampf, meaning a fog or vapor. See also: *afterdamp; blackdamp; chokedamp; combustible gases; firedamp; stinkdamp; white damp.* (Nelson, 1965)
damping (a) In seismology, a resistance, contrary to friction, independent of the nature of the contacting surface. Being proportional to the speed of motion, it diminishes with the latter to nothing. (Schieferdecker, 1959) (b) A force opposing vibration, damping acts to decrease the amplitudes of successive free vibrations. Damping may result from internal friction within the system, from air resistance, or from mechanical or magnetic absorbers. Cf: *attenuation.* (AGI, 1987) (c) The loss of amplitude of an oscillation, owing to absorption. See also: *critical damping; damping factor.* (AGI, 1987)
damping constant In damped seismographs, this term is by definition equal to one-half the ratio of the damping resistance (force per unit velocity) to the moving mass. It has the dimensions of a frequency. (AGI, 1987)
damping down In pyrometallurgy, reduction of air supply to a furnace, to lower temperature or reduce working rate. (Pryor, 1963)
damping factor The ratio of the observed damping to that required for critical damping. See also: *damping.* (AGI, 1987)
damping ratio (a) The damping ratio for a system with viscous damping is the ratio of the actual damping coefficient to the critical damping coefficient. (Hunt, 1965) (b) The ratio of two equiphase peak amplitudes within one period of a damped seismograph or seismometer. The ratio is always greater than unity since the greater amplitude is divided by the succeeding amplitude. (AGI, 1987)
dam plate In a blast furnace, the cast-iron plate that supports the dam or dam stone in front. (Fay, 1920)
damp sheet S. Staff. A large sheet placed as a curtain or partition across a gate road to stop and turn an air current.
dampy Mid. Mine air mixed with so much carbonic acid gas as to cause the lights to burn badly or to go out.
damsite testing Boreholes drilled to determine petrological and structural features of the rock or overburden materials at or near the area on which the foundations of a dam will rest. (Long, 1960)
dam stone The wall of firebrick or stone enclosing the front of the hearth in a blast furnace. (Fay, 1920)
dan (a) Mid. A tub or barrel, sometimes with and sometimes without wheels, in which mine water is conveyed along underground roadways to the sump or raised to the surface. (b) A small box or sledge for carrying coal or waste in a mine.
danaite A cobaltoan variety of arsenopyrite.
danalite An isometric mineral, $Fe_4Be_3(SiO_4)_3S$; vitreoresinous; forms series with genthelvite and helvite.
danburite An orthorhombic mineral, $CaB_2Si_2O_8$; resembles topaz in habit, appearance, and properties; in marbles, low-temperature veins, and placers.
Danian Lowermost Paleocene or uppermost Cretaceous. (AGI, 1987)
Daniell cell A primary cell, with a constant electromotive force of about 1.1 V, having as its electrodes: (1) copper in a copper sulfate solution, and (2) zinc in dilute sulfuric acid or zinc sulfate—the two solutions being separated by a porous partition. (Webster 3rd, 1966)
Danish flint pebbles Pebbles for grinding media, of superior hardness, toughness, and uniformity, found on the shores of Greenland. (AIME, 1960)
danks' puddler A revolving mechanical puddler. See also: *puddling.* (Fay, 1920)
dannemorite A monoclinic mineral, $Mn_2(Fe^{2+}, Mg)_5Si_8O_{22}(OH)_2$; amphibole group; has $Fe^{2+}/(Mg + Fe^{2+}) = 0.5$ to 1.0; columnar or fibrous; at Dannemora, Sweden.
d'Ansite An isometric mineral, $Na_{21}Mg(SO_4)_{10}Cl_3$.
dant (a) Soft sooty coal found in face and back slips or cleats; fine slack coal. (CTD, 1958) (b) To reduce, as a metal, to a lower temper. (Standard, 1964)
daourite A pink variety of elbaite.
dap (a) A notch cut in a timber to receive another timber. (Zern, 1928) (b) *legs.*
daphnite A magnesian variety of chamosite.
darapskite A monoclinic mineral, $Na_3(SO_4)(NO_3)\cdot H_2O$.
Darby process A method of carburizing open hearth steel that consists of treating the molten steel with carbon in the form of charcoal, graphite, or coke. (Osborne, 1956)

darg (a) A specified day's work, usually at the coal face. See also: *stint.* (Nelson, 1965) (b) A task, or a fixed quantity of coal, agreed to be produced per shift for a certain price. (CTD, 1958) (c) Scot. To work by the day. (Fay, 1920) (d) A north German name for meadow or moor peat buried under clay. (Tomkeieff, 1954) (e) Peat formed from marine vegetation. (Holmes, 1928)
dark ground Indirect illumination of stage of microscope, causing objects to be brightly displayed by oblique rays against a dark background. (Pryor, 1963)
dark mica *biotite.*
dark mineral Any one of a group of rock-forming minerals that are dark-colored in thin section, e.g., biotite, hornblende, augite. (AGI, 1987)
dark red silver ore *pyrargyrite.*
dark ruby silver *pyrargyrite.*
dark sulfur Crude, dark-colored sulfur containing up to 1% oil or carbonaceous material.
darlingite A variety of lydian stone from Victoria, Australia. (English, 1938)
dashing Eng. Increasing the amount of air in mines to prevent explosions of mine gases. See also: *dad.*
dashkesanite A chlorine-rich variety of hastingsite.
dashpot (a) An appliance for damping out vibration. It consists of a piston attached to the object to be damped and fitting loosely in a cylinder of oil. See also: *hydrabrake retarder.* (Nelson, 1965) (b) A similar device for closing the valves in a Corliss engine, actuated by atmospheric pressure or by a contained spring. (Webster 2nd, 1960)
dasymeter An instrument for testing the density of gases. It consists of a thin glass globe, which is weighed in the gas or gases under observation, and then in an atmosphere of known density. (Osborne, 1956)
dating Age determination of naturally occurrring substances or relicts by any of a variety of methods based on the amount of change, happening at a constant measurable rate, in a component. The changes may be chemical, or induced or spontaneous nuclear, and may take place over a period of time. (AGI, 1987)
datolite A monoclinic mineral, $CaBSiO_4(OH)$; gadolinite group; in cracks and cavities in diabase or basalt; may be used as a minor gem. Also spelled datholite. Syn: *humboldtite; dystome spar.*
datum (a) The top or bottom of a bed of rock, or any other surface, on which structure contours are drawn. (AGI, 1987) (b) Sea-level datum.—Pl: datums. (AGI, 1987) (c) Any numerical or geometric quantity or value that serves as a base or reference for other quantities or values; any fixed or assumed position or element (such as a point, line, or surface) in relation to which others are determined, such as a level surface to which depths or heights are referred in leveling. Pl: datums; the plural data is used for a group of statistical or inclusive references, such as geographic data for a list of latitudes and longitudes. See also: *datum plane; geoid.* (AGI, 1987)
datum level Any level surface, such as mean sea level, used as a reference from which elevations are reckoned; a datum plane. (AGI, 1987)
datum plane (a) A horizontal plane used as a reference from which to reckon heights or depths. (Hunt, 1965) (b) A permanently established horizontal plane, surface, or level to which soundings, ground elevations, water-surface elevations, and tidal data are referred; e.g., mean sea level is a common datum plane used in topographic mapping. Syn: *datum level; reference level; reference plane.* See also: *datum.* (AGI, 1987)
datum water level (a) The level at which water is first struck in a shaft sunk on a reef or gutter. (Zern, 1928) (b) Ground or surface water level used as a reference for all other measurements.
dauberite *zippeite.*
daugh (a) Scot. The floor of a coal seam or where holing is done. (Nelson, 1965) (b) Underclay; soft fireclay. (Arkell, 1953)
daughter element The element formed when a radioactive element undergoes radioactive decay. The latter is called the parent. The daughter may or may not be radioactive. (CCD, 1961)
Dauphiné diamond Rock crystal variety of quartz.
Dauphiné law The law governing a twinning observed in the hexagonal system commonly shown by quartz in which two righthand or two lefthand crystals interpenetrate after one has revolved 180° about the twinning axis. (Hess)
Dauphiné twinning Transformation twinning about the [0001] direction with an irregular contact surface in quartz. Syn: *electrical twinning.*
D'Autriche Method A method of geometrically determining the detonation velocity of an explosive material by using a lead witness plate and detonating cord ignited by the test explosive. (Meyer, 1981)
Dautriche test *velocity of detonation.*
davidite A trigonal mineral, $(La,Ce)(Y,U,Fe)(Ti,Fe)_{20}(O,OH)_{38}$; crichtonite group; radioactive; metamict; in high-temperature hydrothermal veins, pegmatites, and mafic igneous rocks; occurs in all stages of intergrowth and exsolution with ilmenite and hematite, and is an ore of uranium. Syn: *ferutile.*
davidsonite A greenish-yellow variety of beryl.
Davis bit *Davis cutter bit.*
Davis calyx drill A rotary drill similar to the diamond core drill except that the annular groove is cut either by a steel chisel or by a plain hollow rod using chilled shot. When the core is of sufficient length to be withdrawn, some grit is added to the mud flush, which becomes wedged tightly between the core and base of the barrel. When the rods are raised the core is broken off and brought to the surface. (Nelson, 1965)
Davis cutter bit An annular-shaped, sawtoothlike bit used on shot drills to cut core in soft formations in which shot is ineffective as a cutting medium. Syn: *Davis bit.* (Long, 1960)
Davis furnace A long, one-hearth reverberatory furnace, heated by lateral fireplaces for roasting sulfide ore. (Fay, 1920)
Davis magnetic tester An instrument for testing the magnetic content of ores and for checking the efficiency of wet magnetic separators recovering magnetite and ferrosilicon in heavy-media processes. (Nelson, 1965)
davisonite A mixture of crandallite and an apatite. Syn: *dennisonite.*
Davis wheel A railway tire consisting of a soft plate and boss, and a wear- resistant tread of water-toughened manganese steel, cast integrally within. (Osborne, 1956)
davreuxite A monoclinic mineral, $MnAl_6Si_4O_{17}(OH)_2$.
Davy lamp A safety lamp invented by Sir Humphrey Davy in 1815 for the protection of coal miners. Its safety feature consisted of a fine-wire gauze enclosing the flame to keep it from coming in contact with mine gas. See also: *flame safety lamp; safety lamp.* (Fay, 1920)
davyne A hexagonal mineral, $(Na,Ca,K)_8Al_6Si_6O_{24}(Cl,SO_4,CO_3)_2$; cancrinite group; vitreous to pearly. Also spelled davina.
dawsonite An orthorhombic mineral, $NaAl(CO_3)(OH)_2$; white; forms thin incrustations of radiating bladed crystals.
day (a) A term used to signify the surface; e.g., driven to day, meaning to daylight, therefore to the surface. (b) Wales. The surface of the ground over a mine. (c) In mining, generally a period of 8 h for work on the three-shift system, or 24 h if referring to the output or to machinery. (CTD, 1958)
day box *powder chest.*
day coal The topmost stratum of coal; so called from its being nearest to daylight. (Standard, 1964)
daylight (a) When an underground mine working meets the surface it is said to daylight. (Long, 1960) (b) The maximum clear distance between the pressing surfaces of a hydraulic press with the surfaces in their usable open position. Where a bolster is supplied, it shall be considered the pressing surface. (ASM, 1961)
dc (direct chill) casting A continuous method of making ingots or billets for sheet or extrusion by pouring the metal into a short mold. The base of the mold is a platform that is gradually lowered while the metal solidifies, the frozen shell of metal acting as a retainer for the liquid metal below the wall of the mold. The ingot is usually cooled by the impingement of water directly on the mold or on the walls of the solid metal as it is lowered. The length of the ingot is limited by the depth to which the platform can be lowered; therefore, it is often called semicontinuous casting. (ASM, 1961)
deactivation In froth flotation, treatment of one or more species of mineral particles to reduce their tendency to float.
dead (a) Said of a mine, vein, or piece of ground that is unproductive. (b) Said of coal that is under no pressure, does not warp and burst, and makes no sound. Cf: *alive.* (Stoces, 1954) (c) In economic geology, said of an economically valueless area, in contrast to a quick area or ore; barren ground. (AGI, 1987)
dead air (a) Stagnant air. (BS, 1963) (b) The air of a mine when it contains carbonic acid gas (blackdamp), or when ventilation is sluggish. (Fay, 1920)
dead band In flotation, the range through which an input can be varied without initiating response. (Fuerstenau, 1962)
dead bed Unproductive stratum or vein as opposed to bearing or quick bed. Syn: *dead vein.* See also: *barren ground.* (Arkell, 1953)
dead-burned (a) The state of a basic refractory material resulting from a heat treatment that yields a product resistant to atmospheric hydration or recombination with carbon dioxide. (ASTM, 1994) (b) Completely calcined. (AISI, 1949)
dead-burned dolomite A refractory product, $CaO{\cdot}MgO$, produced by calcination of dolomite or dolomitic limestone. (AGI, 1987)
dead-burned magnesia A sintered product consisting mainly of magnesia in the form of dense, weather-stable refractory granules.
dead burnt Calcination of limestone, dolomite, or magnesite to the point where associated clay vitrifies and reduces slaking quality. (Pryor, 1958)
dead chert *chalky chert.*
dead end (a) An entry, gangway, level, or other mine passage extending beyond the mine workings into solid coal or ore; a stub. Syn: *stub*

entry. (b) Underground passageway either blocked or not holed through. (Pryor, 1963) (c) The unworked end of a drift or working. (Hess) (d) An unventilated underground mine passage extending some distance beyond other mine workings into solid rock. (Long, 1960) (e) A term used in coal mining for the termination of all electric wiring (except cables to equipment) outby the last crosscut where ample ventilation will reduce the possibility of an electric arc causing an explosion. (Kentucky, 1952) (f) The end of a drilling line or cable made fast to some stationary part of the drill rig or to a deadman. (Long, 1960)

dead ground (a) Rock in a mine that, although producing no ore, requires removal in order to get to productive ground. (b) In mining subsidence, ground that has settled and no further movement is expected. (Nelson, 1965) (c) Portions of ore deposit too low in value to repay exploitation. Cf: *barren ground.* (Pryor, 1963)

deadhead (a) To return to the commencement of a cut without excavating; usually for the commencement of a new cut after completion of its predecessor. (Austin, 1964) (b) An extra length given to a cast object, as a cannon, to put pressure on the molten metal below so that dross and gases may rise into it; a sullage piece; a sinking head. (Standard, 1964) (c) That part of a casting filling up the ingate; a sprue. (Standard, 1964) (d) Can. Logs forced into the bottom of a waterway during timber drives. (Hoffman, 1958)

deadheading Traveling without load, except from the dumping area to the loading point. (Nichols, 1976)

dead hole (a) One that extends into solid coal beyond the part that can be broken by the maximum safe charge of explosive. (Zern, 1928) (b) A shothole so placed that its width at the point (toe), measured at right angles to the drill hole, is so great that the heel is not strong enough to at least balance the resistance at the point (toe). (Zern, 1928) (c) A shallow hole in an iron casting. (Standard, 1964)

deadline (a) A row of marked empty powder kegs or other danger signal placed by the fireboss to warn miners not to enter workings containing gas. (Fay, 1920) (b) The part of a block-and-tackle cable from the traveling block to the deadline anchor. (Long, 1960)

deadline anchor The fixed point on a drill rig or deadman to which a deadline of a block and tackle is attached. (Long, 1960)

dead load The downward pressure on a structure caused by gravity only, such as the weight of a long string of drill rods suspended from the sheave in a drill derrick. Syn: *static load.* See also: *live load.* (Long, 1960)

dead lode A lode not containing valuable minerals in paying quantity. (Fay, 1920)

deadman (a) A wooden block used to guard the mouth of a mine against runaway cars. (Fay, 1920) (b) A buried log, timber, concrete block, or the like serving as an anchor to which a pulling line can be attached. (Long, 1960)

dead pressing Desensitizing of an explosive, caused by excessive pressure or high density.

dead quartz Quartz carrying no valuable mineral.

dead rent Of a mineral lease, the rent that must be paid whether or not minerals are being extracted. (Pryor, 1963)

dead roast (a) A roasting process for complete elimination of sulfur or other volatiles. Syn: *sweet roast.* (ASM, 1961; Newton, 1959) (b) In fluidization roasting, restriction of entering air to permit oxidation of sulfides, while not allowing process to proceed to any marked degree of sulfate roasting. (Pryor, 1963)

dead roasting Sulfide ores are dead roasted when all the sulfur possible to drive off by roasting has been eliminated. (Weed, 1922)

dead rock The material removed in the opening of a mine that is of no value for milling purposes. Waste rock.

dead soft The state of metal that has been fully annealed. (Light Metal Age, 1958)

dead steel (a) Fully killed steel, which sinks quietly in the ingot mold during solidification. (b) Steel that fails to respond to heat treatment because it has been worked at excessively high temperatures; e.g., 1,300 to 1,350 °C. (Osborne, 1956)

dead time In flotation, the interval of time between initiation of an input and the start of the resulting response. It may be qualified as effective if extended to the start of the buildup time, theoretical if the dead band is negligible, and apparent if it includes the time spent with an appreciable dead band. (Fuerstenau, 1962)

dead true A core barrel or drill rod that does not oscillate or vibrate when rotated at high speed is said to be dead true. (Long, 1960)

dead vein *dead bed.*

dead veins Veins barren of economic minerals.

deadweight (a) The weight of a vehicle or carrier itself as distinguished from carried or live load. (Crispin, 1964) (b) The difference, in tons, between a ship's displacement at load draft and light draft. It comprises cargo, bunkers, stores, fresh water, etc. (CTD, 1958)

dead work (a) Work that is not directly productive—the removal of rock, debris, or other material that is not directly productive of coal—though it may be necessary for exploration and future production. Unfinished work. See also: *stonework.* (BCI, 1947) (b) Unproductive or stone work; the handling of stone or dirt as a preliminary step to winning and working the coal seam. The aim is to keep the dead work per yard of face or ton of coal to the minimum practicable figure. See also: *unproductive development.* (Nelson, 1965) (c) Any kind of miner's work other than actual coal getting and transport. (Mason, 1951) (d) Exploratory or preparatory work, such as cleaning falls of roof, removing rock, etc., during which little or no coal is secured. (Hudson, 1932) (e) The development of a mine when no ore is being raised. (Gordon, 1906) (f) Work done by a contractor not provided for in the yardage or tonnage contract rates. (Mason, 1951) (g) S. Afr. Necessary work to reach and exploit the valuable portions of the mine. Shaft sinking, crosscutting, driving of levels, etc., belong to dead work. (Beerman, 1963)

dead zone That part of the mined strata that has completely settled down after subsidence. (Briggs, 1929)

dealer An operator on the stock exchange who buys and sells on his or her own account and who makes a profit from differences in prices rather than from commissions. (Hoover, 1948)

debacle (a) A breakup on a river, esp. on the great rivers of the former U.S.S.R. and of North America. (AGI, 1987) (b) The rush of water, broken ice, and debris in a stream immediately following a breakup. Syn: *ice run.* (AGI, 1987) (c) Any sudden, violent, destructive flood, deluge, or rush of water that breaks down opposing barriers and sweeps before it debris of all kinds.—Etymol: French débâcle. (AGI, 1987)

Deblanchol rotary furnace A cylindrical refractory-lined shell, provided with a gas flue leading to a recuperator at one end, and a fuel and air port at the other. Air for combustion is preheated in the recuperator, and oil firing is adopted. The furnace may be used for melting gray iron and nonferrous metals. (Osborne, 1956)

debris Any surficial accumulation of loose material detached from rock masses by chemical and mechanical means, as by decay and disintegration. It consists of rock fragments, soil material, and sometimes organic matter. The term is often used synonymously with detritus, although debris has a broader connotation. Etymol: French débris. Pl: debris. Syn: *rock waste.* (AGI, 1987)

debris bag A dirt-filled bag used for pack walls and chocks. See also: *sandbag.* (Nelson, 1965)

debris cone *alluvial cone.*

debris deposits Refuse from hydraulic mining operations, tailings.

decalescence A phenomenon associated with the transformation of alpha iron to gamma iron on the heating (superheating) of iron or steel, revealed by the darkening of the metal surface owing to the sudden decrease in temperature caused by the fast absorption of the latent heat of transformation. Syn: *point of decalescence.* (ASM, 1961)

decalescence point *critical point.*

decantation The settlement of a solid from a liquid, and removal of the clear liquid.

decanter (a) An apparatus for sorting and classifying tailings from gold-washing operations. (b) A vessel used to decant or to receive decanted liquids. (Webster 3rd, 1966)

decarbonation The process of driving off carbon dioxide from a carbonate mineral, e.g., magesite, $MgCO_3$, to form periclase, MgO.

decarburization The loss of carbon from the surface of a ferrous alloy as a result of heating in a medium that reacts with the carbon at the surface. (ASM, 1961)

decay The general disaggregation of rocks; it includes the effects of both the chemical and mechanical agents of weathering with, however, a stress on the chemical effects. (Stokes, 1955)

decay distance The distance between an area of wave generation and a point of passage of the resulting waves outside the area. (Hy, 1965)

dechenite Natural PbV_2O_6; not established as a valid mineral species.

decibar The pressure exerted per square centimeter by a column of sea water 1 m tall is approx. 1 decibar. The depth in meters and the pressure in decibars, therefore, are expressed by nearly the same numerical value. (Hy, 1965)

decibel The unit for measuring sound intensity. (Crispin, 1964)

decision function Rule made to control a specific sampling investigation, which defines the point at which no further observations are to be made, and the nature of the decision that is to be agreed upon. In a series of sampling operations, each successive decision function depends on those that have preceded it. (Pryor, 1963)

deck (a) One of the separate compartments or platforms into which a cage is divided to hold cars. See also: *multideck cage.* (Nelson, 1965) (b) The surface of a concentrating table. (Nelson, 1965) (c) The refractory top of a car used in a tunnel kiln or bogie kiln. (Dodd, 1964)

deck charge (a) A charge that is divided into several separate components along a quarry borehole. Cf: *columnar charge.* (BS, 1964) (b) A charge separated by stemming. (Carson, 1961)

decke *nappe.*

decking (a) The operation of changing the tubs on a cage at top and bottom of a shaft. Also called caging. (Fay, 1920) (b) Separating charges of explosives by inert material and placing a primer in each charge. (Nichols, 1976)

decking level The level at which a cage comes to rest at the pit head and pit bottom for unloading and loading mine cars. (Nelson, 1965)

deck load A charge of dynamite spaced well apart in a borehole and fired by separate primers or by a detonating cord. (Nichols, 1976)

deck loading A method of loading blast holes in which the explosive charges, called decks or deck charges, in the same hole are separated by stemming, air cushion, or a plug. (Atlas, 1987)

deck screens Two or more screens, usually of the vibrating type, placed one above the other for successive processing of the same run of material. (Nichols, 1976)

declaratory statement In practical mining operations, a term applied to the statutory certificate of location, and a certificate or statement of the location, containing a description of the mining claim, verified by the oath of the locator, performing, when recorded, a permanent function. It is the beginning of the locator's paper title, is the first muniment of such title, and is constructive notice to all the world. (Ricketts, 1943)

declared efficiency The efficiency assigned by the maker under certain specified conditions. (Nelson, 1965)

declination (a) The horizontal angle in any given location between true north and magnetic north; it is one of the magnetic elements. Syn: *variation; magnetic variation.* (AGI, 1987) (b) Angular elevation of a star above celestial equator when truly north of observer. (Pryor, 1963) (c) Angular deviation of magnetic compass from true north, observed in conditions where no local deviation affects it. (Pryor, 1963) (d) The angular change in the course of a borehole induced by deflection techniques, usually expressed in degrees. (Long, 1960) (e) Sometimes a syn. for inclination. See also: *inclination.* (Long, 1960)

declination maps Maps on which isogonic lines are shown. (Mason, 1951)

declining conveyor A conveyor transporting down a slope. See also: *retarding conveyor.*

décollement Detachment structure of strata owing to deformation, resulting in independent styles of deformation in the rocks above and below. It is associated with folding and with overthrusting. Etymol: French, unsticking, detachment. Cf: *disharmonic folding.* Syn: *detachment.* (AGI, 1987)

decomposing furnace A furnace used in the conversion of common salt into sulfate of soda, aided by the action of sulfuric acid.

decomposition *chemical weathering.*

decompression The process of reducing high air pressure gradually enough so as not to injure people who have been working in it. (Nichols, 1976)

decompression illness A condition among underwater workers and mine rescue teams that is caused by ascending too quickly from deep dives.

decompression sickness *aeroembolism.*

decorative stone (a) A term sometimes used alternately with ornamental stone. (b) Natural material used as architectural trimmings in columns, fireplaces, and store fronts; may be set in silver- or gold-filled jewelry, as curio stones; e.g., malachite and marble. Cf: *gemstone.*

decoupling A method for decreasing the ground motion generated by an underground explosion. The method involves the firing of the explosive in the center of an underground cavity so that the surrounding medium is not in close proximity to the explosive. (Lyman, 1964)

decrepitate (a) To roast or calcine (as salt) so as to cause crackling or until crackling stops. (Webster 3rd, 1966) (b) A mineral is said to decrepitate when it flies to pieces with a crackling noise on being heated. (Hess)

decrepitation (a) Method of differential disintegration of closely sized mineral, part of which explodes and is separable by finer screening. (Pryor, 1963) (b) The breaking up with a crackling noise of mineral substances upon exposure to heat, as when rock salt is thrown into fire. (c) An obsolete method of tunneling, called fire setting.

decussate texture A microtexture in metamorphosed rocks, in which axes of contiguous crystals lie in diverse, crisscross directions that are not random but rather are part of a definite mechanical expedient for minimizing internal stress. It is most noticeable in rocks composed largely of minerals with a flaky or columnar habit. (AGI, 1987)

dedolomitization (a) A process resulting from metamorphism, wherein part or all of the magnesium in a dolomite or dolomitic limestone is used for the formation of magnesium oxides, hydroxides, and silicates (e.g., brucite, forsterite) and resulting in an enrichment in calcite. (AGI, 1987) (b) Diagenetic or weathering processes wherein dolomite is replaced by calcite.

dedusting A cleaning process in which dust and other fine impurities are removed. Dedusting is accomplished both by pneumatic means and by dry screening. Syn: *aspirating.* (Mitchell, 1950)

deenergize To disconnect any circuit or device from the source of power. (NCB, 1964)

deep (a) Workings below the level of the pit bottom or main levels extending therefrom. (b) Forest of Dean; Lanc. A vein, seam, mine, or bed of coal or ironstone. (c) Term used to designate ocean bottom depressions of great depth, usually deeper than 6,000 m. (Hy, 1965)

deep cell count A method for examining the mineral particle content of drilling water. In this method, a glass cell is filled with the water, a little acid is added, and the sample is placed under a microscope. Dark ground illumination is used, which shows up the suspended particles. The number of these is counted, and this number, multiplied by a factor, gives the number of particles per cubic centimeter. (Higham, 1951)

deep coal Eng. Coal seams lying at a depth of 1,800 ft (549 m) or more below the surface. (Fay, 1920)

deep drawing The process of cold working or drawing a sheet of strip metal, by means of dies, into shapes involving considerable plastic distortion of the metal; e.g., automobile mudguards, electrical fittings, etc. (CTD, 1958)

deep hole In continuous wire-line core drilling, a term applied to boreholes 3,000 ft (915 m) or more in depth.

deep-hole blasting Blasting a quarry or opencast face by using small- or medium-diameter holes drilled from top to bottom of the face. (Nelson, 1965)

deep lead Alluvial deposit of gold or tin stone buried below a considerable thickness of soil or rock. Cf: *lead.*

deep level (a) Trans. The first mining properties developed from the surface were stopped from trespassing beyond their side lines projected downward. The next mine on the dip of the lode became known as the "deep-level" mine or "deep." (b) S. Afr. The distinction of deep level and ultradeep level is a vague one, and has changed with the times. Ultradeep is now a mining level at a vertical depth of 9,000 ft (2.7 km) and over. (Beerman, 1963)

deep mining The exploitation of coal or mineral deposits by underground mining methods. "Deep" is often interpreted as meaning 5,000 ft (1.5 km) or more, where stresses are high enough to cause sloughing of development openings, not to mention walls and faces in stopes. However, tectonic horizontal stresses are greater than gravitational force in many areas. Hence deep conditions can exist at lesser depths. Indeed, severe rock bursting caused the closure of a Canadian mine operating at depths of 500 to 700 ft (152 to 213 m). Also, rocks with low strength will produce deep failure patterns at modest depths, e.g., in Saskatchewan potash mines. (Nelson, 1965)

deep placer A sandy or gravelly bed or bottom of an ancient stream covered by lava.

deep scattering layer Applied to widespread strata in the ocean that scatter or return vertically directed sound as in echo depth sounding. These layers, which are evidently of biological origin, are located at depths ranging from 150 to 200 fathoms (274 to 366 m) during the day, with most of them migrating to or near the surface during the night. Abbrev: dsl. (Hy, 1965)

deep seated *plutonic.*

deep-seated deposit An ore deposit formed at an estimated depth of 12,000 ft (3.66 km) or more, at temperatures ranging from 300 to 575 °C; e.g., the tin deposits of Cornwall, England. The deposits are commonly tubular or veinlike in form, though some are irregular in shape. (Lewis, 1964)

deep-sea terrace The benchlike feature bordering an elevation of the deep-sea floor at depths greater than 300 fathoms (1,800 ft or 549 m). (Schieferdecker, 1959)

deepside The working of 5 to 10 yd (4.6 to 9.1 m) of the coal seam on the dip side of an advance gate. It gives some protection from crush along the rib side and also accommodates dirt from the gate instead of conveying it to the surface. See also: *self-stowing gate.* (Nelson, 1965)

deep sinker Aust. A tall drinking glass; also the drink it contains, so called in fanciful allusion to the shaft of a mine.

deep well A borehole put down through an upper impervious bed into a lower pervious one, from which a supply of water is obtained. See also: *well.* (Nelson, 1965)

deep-well pump (a) Any kind of pump delivering from a well, shaft, or borehole. (BS, 1963) (b) An electrically driven pump located at the low point in the mine to discharge the water accumulation to the surface. (c) Consists of a series of centrifugal pump impellers mounted on a single rotating shaft. The casings are termed bowls and the impellers are of the axial or mixed-flow type. Available in capacities ranging from 25 to 10,000 gal/min (94.6 to 37,854 L/min). It can be used in wells from 25 to 800 ft (7.6 to 244 m) in depth and from 6 to 24 in (15.2 to 61 cm) in diameter. (Carson, 1961)

deep-well turbine A simple type of vertical centrifugal pump having one or more stages or bowls, which are supported from the motor head on the surface by means of screwed or flanged column pipe sections, each usually 10 ft (3 m) long. The line shafting from the motor to the impellers is sectional to correspond with the column section, and may operate in a sectional extra-heavy enclosing tube if oil is used as a lubricant, or may be exposed to the water when the pump is built to be water-lubricated. (Pit and Quarry, 1960)

deep winding (a) Broadly, shaft winding from depths of about 3,000 ft (915 m) and deeper (coal mining). In the case of shafts deeper than about 5,000 ft or 1.52 km (gold and metal mining), two-stage hoisting may be used. See also: *winding.* (Nelson, 1965) (b) Hoisting from depths below 5,000 ft in one lift. (Spalding, 1949) (c) Deep hoisting.

Deerparkian Middle Lower Devonian. (AGI, 1987)

deficient coal Ark. Coal more difficult to mine than the standard, and for which the miners are paid an extra price. (Fay, 1920)

deflagrate To burn; burst into flame; specif., to burn rapidly with a sudden evolution of flame and vapor.

deflagrating mixture An explosively combustible mixture, as one containing niter. (Standard, 1964)

deflagration An explosive reaction such as a rapid combustion that moves through an explosive material at a velocity less than the speed of sound in the material. (Atlas, 1987)

deflation The removal of loose dry particles by the wind, as along a sand-dune coast or in a desert; a form of wind erosion. (AGI, 1987)

deflecting plug (a) *base plug.* (b) Sometimes used by petroleum drillers as a syn. for deflecting wedge. (Long, 1960)

deflecting wedge A class of devices intentionally placed in a borehole to change its course. All such devices are basically long, tapered, concave metal plugs that can be set at a predetermined point and bearing in a borehole to deflect or change its course. Also called correcting wedge, deflecting plug, deflection wedge, Hall-Rowe wedge, spade-end wedge, Thompson wedge. See also: *correcting wedge; wedge.* (Long, 1960)

deflection A change in the intended course of a borehole produced intentionally or unintentionally by various conditions encountered in the drill hole or by the operational characteristics of the drilling equipment used. Syn: *deviation.* (Long, 1960)

deflection angle (a) The angular change in the course of a borehole produced accidentally or intentionally. (Long, 1960) (b) A vertical angle, measured in the vertical plane containing the flight line, by which the datum of any model in a stereotriangulated strip departs from the datum of the preceding model. (AGI, 1987) (c) A horizontal angle measured from the forward prolongation of the preceding line to the following line; the angle between one survey line and the extension of another survey line that meets it. A deflection angle to the right is positive; one to the left is negative. (AGI, 1987)

deflection bit A taper bit, generally a bullnose type, used to drill down past the deflecting wedge when deflecting a borehole. (Long, 1960)

deflection dial The load-indicating gage on a penetrometer, which is a soil-testing device used to determine some of the load-bearing characteristics of silt and sandy soils. See also: *cone penetrator.* (Long, 1960)

deflection plug *base plug.*

deflection point Point of deflection on a refraction time-versus-distance graph separating two segments that correspond to different wave paths. (Schieferdecker, 1959)

deflection wedge A wedge-shaped tool inserted in a borehole to direct the bit along a prescribed course. Also called whipstock (undesirable usage). See also: *deflecting wedge.* (BS, 1963; Long, 1960)

deflectometer An instrument for gaging any deflections of a structure. (Hammond, 1965)

deflector A device across the path of a conveyor placed at the correct angle to deflect objects or discharge bulk material. Also called a plow.

deflector sheet A sheet of brattice or other material erected in a roadway or face to remove a combustible gases layer. It is usually set at an angle of about 45° from the horizontal and inclined in the direction of airflow. See also: *pocket of gas.* (Nelson, 1965)

deflector-wedge ring An annular steel ring attached to the upper end of a deflecting wedge, having a slightly smaller diameter than that of the borehole in which the wedge is inserted, serving as a stabilizing ring to hold and center the wedge in the borehole. Also called rose ring. (Long, 1960)

deflocculant (a) Any organic or inorganic material that is used as an electrolyte to disperse nonmetallic or metallic particles in a liquid, (i.e., basic materials such as calgonate, sodium silicate, soda ash, etc., are used as deflocculants in clay slips). (b) A basic material such as sodium carbonate or sodium silicate, used to deflocculate. Syn: *deflocculating agent.* (ACSG, 1961)

deflocculate (a) To disperse a clay suspension so that it has little tendency to settle and has a low viscosity. (ACSG, 1963) (b) To break up from a flocculated state; to convert into very fine particles. Cf: *peptize.* (AGI, 1987)

deflocculating (a) The thinning of the consistency of a slip by adding a suitable electrolyte. (ASTM, 1994) (b) The process of making clay slips or suspension using electrolytes or deflocculants.

deflocculating agent An agent that prevents fine soil particles or clay particles in suspension from coalescing to form flocs. Syn: *deflocculant; dispersing agent.* (ASCE, 1958)

deflocculation A state of colloidal suspension in which the individual particles are separate from one another, this condition being maintained by the attraction of the particles for the dispersing medium (for example, hydration) or by the assumption of like electrical charges by the particles, thus resulting in their mutual repulsion, or both. It is generally possible to deflocculate a gel to such an extent that it loses its gel strength entirely, thus becoming a Newtonian fluid, in which case it is known as a sol. The relative contribution of hydration and electrostatic repulsion to the deflocculation of a suspension accounts in large measure for the wide variation in viscosities and gel strengths of suspensions partially flocculated by different means; as, e.g., a partial flocculation of drilling fluid by cement on one hand, and by salt water on the other. Some suspensions can be deflocculated repeatedly by mechanical agitation alone, thus giving a reversible gel-sol, sol-gel transformation known as thixotropy. (Brantly, 1961)

deformation (a) A general term for the process of folding, faulting, shearing, compression, or extension of the rocks as a result of various Earth forces. (AGI, 1987) (b) *strain.*

deformation bands Parts of a crystal that have rotated differently during deformation to produce bands of varied orientation within individual grains. (ASM, 1961)

deformed crossbedding Crossbedding with foresets overturned or buckled in the down current direction, usually prior to deposition of the overlying bed. Foreset dip angle may also be altered by subsequent tectonic folding. (Pettijohn, 1964)

deformed crystal A crystal bent or twisted out of its normal shape, so that the angle between its crystal faces may differ widely from those on the regular form. See also: *distorted crystal.*

defrother An agent, e.g., butanol, that destroys or inhibits froth. (Pryor, 1958)

degasification Progressive loss of gases in a substance leading to the formation of a more condensed product. Applied primarily to the formation of solid bitumens from liquid bitumens, but also used in connection with coal formation. (Tomkeieff, 1954)

degasifier A substance that can be added to molten metal to remove soluble gases that might otherwise be occluded or entrapped in the metal during solidification. (ASM, 1961)

degassing (a) Removing gases from liquids or solids. (ASM, 1961) (b) In pyrometallurgy, addition of deoxidants (phosphorus, aluminum, silicon, etc.) to remove hydrogen from molten metals before casting. (Pryor, 1963)

degassing equipment (a) The equipment for extracting gas from an oil-well drilling fluid. The presence of gas reduces the density of the fluid. (Nelson, 1965) (b) The pumps and equipment used in methane drainage. (Nelson, 1965)

degaussing Method of demagnetization in which a substance is passed through a coil that carries alternating current of progressively diminishing strength. (Pryor, 1963)

degradation (a) The general lowering of the surface of the land by erosion, esp. by the removal of material through the action of flowing water. (b) Breakage of coal incidental to mining, handling, transport, or storage. (c) The excessive crushing of coal during cutting, loading, and transportation. All face machines cause degradation, and this has become a problem at collieries where the market calls for the larger sizes. The degradation of a coking coal is of lesser importance. See also: *gradation; fragmentation.* Cf: *aggradation.* Syn: *breakage of coal.* (Nelson, 1965)

degradation screens Screens used for removing the small sizes, caused by breakage in handling, from sized coal just before it is loaded for shipment. Degradation screening is usually necessary where a sized coal is picked, mechanically cleaned, stored, conveyed, or otherwise handled so that breakage occurs after it is sized on the main screens. This applies particularly to domestic coal, which should reach the consumer in as attractive condition as possible. (Mitchell, 1950)

degraded illite Illite that has lost much of its potassium as the result of prolonged leaching. (AGI, 1987)

degreasing Removal of oil and grease films from metal surfaces before electroplating, galvanizing, or enameling. (Pryor, 1963)

degreasing machine An electrically driven machine including high-pressure pump and special cleaning solution for removing grease and oil from underground mine machines as a prevention of mine fires.

degree of compaction The degree of compaction of a soil sample. (Hammond, 1965)

degree of consolidation The ratio, expressed as a percentage, of the amount of consolidation at a given time within a soil mass, to the total amount of consolidation obtainable under a given stress condition. (ASCE, 1958)

degree of liberation In mineral dressing, the degree of liberation of a certain mineral or phase is the percentage of that mineral or phase occurring as free particles in relation to the total of that mineral occurring in the free and locked forms. (Gaudin, 1939)

degree of locking In mineral dressing, the degree of locking of a mineral is the percentage occurring in locked particles in relation to the total occurring in the free and locked forms. (Gaudin, 1939)

degree of packing Of an explosive, the loading weight per unit of nominal volume, which is always known. Its unit is kilogram per cubic decimeter. The degree of packing defined in this way is 6% greater than the density of the explosive in the drill hole. (Langefors, 1963)

degree of saturation (a) The percentage of the volume of water-filled voids to the total volume of voids in a soil. (Nelson, 1965) (b) Ratio of weights of water vapor in air at given conditions and at saturation, with temperature constant. Specific humidities are usually employed. Measured in percent. (Hartman, 1961)
degree of size reduction Ratio of the surface areas or sizes of the broken or crushed material to those of the feed material. (BS, 1962)
degree of sorting (a) The measure for the spread of grain-size distribution. (Schieferdecker, 1959) (b) A measure of the spread or range of variation of the particle-size distribution in a sediment. It is defined statistically as the extent to which the particles are dispersed on either side of the average; the wider the spread, the poorer the sorting. It may be expressed by sigma phi. (AGI, 1987)
degrees Kelvin Absolute temperature on the centigrade scale, or degrees C plus 273.16. (Strock, 1948)
degrees Rankine Absolute temperature on the Fahrenheit scale, or degrees F plus 459.6. (Strock, 1948)
Dehottay process A variation of the freezing method of shaft sinking, in which liquid carbon dioxide is pumped into the ground instead of brine. See also: *Oetling freezing method.* (Nelson, 1965)
dehrnite A hydrous phosphate of calcium, sodium, and potassium; hexagonal; crystalline crusts and minute crystals; grayish- to greenish-white. The mineral from Dehrn, Nassau, Germany, is richer in sodium, conforming nearly to the formula $7CaO{\cdot}Na_2O{\cdot}2P_2O_5{\cdot}H_2O$, whereas the mineral found near Fairfield, UT is described as $14CaO{\cdot}2(Na,K)_2O{\cdot}4P_2O_5 3(H_2O,CO_2)$. (English, 1938)
dehumidification The process of removing moisture from mine air to increase its cooling capacity—an important factor in environmental health and comfort in deep mining. See also: *dry kata cooling power; effective temperature.* (Hartman, 1982)
dehydrate (a) To render free from water. (Webster 3rd, 1966) (b) The process of driving water from a hydrated mineral, e.g., gypsum, $CaSO_4{\cdot}2H_2O$, to anhydrite, $CaSO_4$.
dehydrated Freed from water or lacking water.
dehydrated stone One from which the normal water content has been evaporated, usually by natural processes.
dehydrator A device or material that will remove water from a substance. See also: *dryer.*
de-ionization Removal of ions from solution by chemical means. Syn: *demineralization.* (ASM, 1961)
Deister table Proprietary type of shaking table used in mineral processing. (Pryor, 1963)
delaflossite A trigonal mineral, $CuFeO_2$; in the oxidized zone of copper deposits.
delatorreite Former name for todorokite.
delatynite A variety of amber rich in carbon, low in succinic acid, and lacking sulfur, at Delatyn in the Carpathian Mountains.
delawarite An aventurine feldspar from Delaware County, PA.; a pearly orthoclase. Syn: *lennilite.* (Hess)
delay A distinct pause of predetermined time between detonation or initiation pulses, to permit the firing of explosive charges separately. (Atlas, 1987)
delay action In blasting, firing of a round of shots in planned sequence so that cut or relief holes are blown first. Delay-action electric detonators have largely replaced safety fuses for this purpose, successive shots being separated by milliseconds. (Pryor, 1963)
delay blasting The practice of initiating individual explosive decks, boreholes, or rows of boreholes at predetermined time intervals using delay detonators, as compared to instantaneous blasting where all holes are fired essentially simultaneously. (Atlas, 1987)
delay detonator An electric or nonelectric detonator used to introduce a predetermined time lapse between the application of a firing signal and detonation. (Dick, 2)
delayed filling Filling in which the mined-out rooms are filled later, generally on a large scale and when the neighboring sections are already being mined. (Stoces, 1954)
delayed pillar extraction A pillar method of working in which the coal pillars are not extracted until the whole workings have been driven to the boundary. It is sometimes adopted when a seam a short distance above is worked simultaneously. Delayed pillar working increases the difficulty of ventilation, and the amount of deadwork is increased because of the crushing of coal pillars. (Nelson, 1965)
delayed quench One in which the material is not quenched immediately on coming from the solution heat-treat furnace. This allows precipitation to proceed to a point at which mechanical properties and corrosion resistance are lowered. (Light Metal Age, 1958)
delay electric blasting cap An electric blasting cap with a delay element between the priming and detonating composition to permit firing of explosive charges in sequence with but one application of the electric current. It detonates about 1 to 2 s after the electric current has passed through the bridge. It is made in two kinds, first and second delay, and is used in connection with regular, waterproof, or submarine electric blasting caps for blasting in tunnels, shafts, etc., where it is desirable to have charges fired in succession without the necessity of the blaster returning betweeen shots. (Fay, 1920)
delay element An explosive train component consisting of a primer, a delay column, and a relay transfer charge assembled in a single housing to provide a controlled time delay. (Meyer, 1981)
delay firing The firing of several shots in sequence, at designed intervals of time, usually by means of delay detonators, detonating relays, or sequence switches. (BS, 1964)
delay interval The nominal period between the firing of successive delay detonators in a series of shots. (BS, 1964)
delay period A designation given to a delay detonator to show its relative or absolute delay time in a given series.
delay rental A payment, commonly made annually on a per acre basis, to validate a lease in lieu of drilling. (Wheeler, 1958)
delay series A series or sequence of delay detonators designed to satisfy specific blasting requirements. There are basically two types; millisecond (MS) and long period (LP). (Atlas, 1987)
delay tag A tag, band, or marker on a delay detonator that denotes the delay series, delay period and/or delay time of the detonator. They are often color coded for convenience.
delay time In seismic refraction work, the additional time required to traverse any raypath over the time that would be required to traverse the horizontal component at the highest velocity encountered on the raypath, as it refers to either the source or receiver end of the trajectory. Syn: *intercept time.* (AGI, 1987)
delessite A magnesian variety of chamosite.
delfman Eng. A miner or worker in a stone quarry.
delhayelite An orthorhombic mineral, $(Na,K)_{10}Ca_5Al_6Si_{32}O_{80}(Cl_2,F_2,SO_4)_3{\cdot}18H_2O$; forms in laths in a melilite nephelinite lava at Mt. Shaheru, Kivu Province, Congo.
deliquescent Capable of becoming liquid by the absorption of moisture from the air; e.g., calcium chloride crystals. (Standard, 1964)
delivery column *rising main.*
delivery date The date on which a metal has to be delivered to fulfill the contract terms. Also called prompt date. (Wolff, 1987)
delivery drift A drift or adit connected to a shaft from a point on the surface at a lower level than the shaft top and used as an outlet into which mine pumps discharge, so reducing the height through which the water must be lifted. Syn: *jackhead.* (BS, 1963)
delivery gate Eng. A road into which a face conveyor delivers the coal. (SMRB, 1930)
delivery table (a) A conveyor that transports material from the discharge of a machine. (b) A table onto which a chute discharges.
dellenite (a) An extrusive rock between rhyolite and dacite in composition, and, broadly, the extrusive equivalent of granodiorite. (Webster 2nd, 1960; Fay, 1920) (b) *plagioclase rhyolite.* (c) A rhyodacite from Dellen Lake, Sweden.
dells *dalles.*
delorenzite *tanteuxenite.*
delphs York. The working places in ironstone quarries. (Nelson, 1965)
Delprat method *overhand stoping.*
delrioite A monoclinic mineral, $CaSrV_2O_6(OH)_2{\cdot}3H_2O$; pale yellow-green microcrystalline efflorescence on sandstone occurring in Montrose County, CO.
delta iron The polymorphic form of iron stable between 1,403 °F (762 °C) and the melting point (about 1,532 °F or 833 °C). The space lattice is the same as that of alpha iron and different from that of gamma iron. (CTD, 1958)
deltaite A mixture of crandallite and hydroxylapatite
deluge water system A method of fire control in which water is sprayed or sprinkled in sufficient volume to overwhelm the fire and put it out. (Federal Mine Safety, 1977)
Demag cappel A rope cappel used in Koepe winding, particularly in Germany. The rope is led along the side of the eye and secured by a hinged retaining arm lined with rubber, and then turned around the eye and held in position by pressure exerted by knee-action links. (Nelson, 1965)
Demag drag-belt shuttle conveyor Consists of a single length of belting, half the length of a double unit face, which is shuttled backward and forward along the face by means of low-type winches at each end of the face, interlocked and fitted with limit switches. The coal is plowed off the belt at the loader gate onto the gate conveyor. (Sinclair, 1959)
demagnetize To disperse, by means of a suitable magnetic field, solids in a dense medium that have flocculated magnetically.
demand respirator An atmosphere-supplying respirator that admits respirable gas to the facepiece only when a negative pressure is created inside the facepiece by inhalation. (ANSI, 1992)
demantoid A transparent, green variety of andradite, having a brilliant luster and used as a gem. Also called Uralian emerald. (Dana, 1959)
demersal *benthonic.*

demidovite A phosphoriferous variety of chrysocolla from Tagilsk, Perm, Russia.
demineralization (a) Water softening by use of zeolites or resins to remove cations. (Pryor, 1963) (b) *de-ionization.*
demonstrated resources A term for the sum of measured resources plus indicated resources. (USGS, 1980)
dempy A mine or part of a mine that is prone to outbursts and accumulations of noxious gases. (Nelson, 1965)
demulsification Breakdown into separate phases of a relatively stable emulsion, by such means as flocculation with a surface-active agent or removal of an emulsifying agent. (Pryor, 1963)
demurrage The detention of a vessel, railroad car, or other vehicle beyond an allotted time and for which a fee is usually charged.
dendriform Resembling a tree, descriptive of some minerals. Syn: *arborescent; dendritic.*
dendrite Any mineral forming branching moss-, fern-, or treelike patterns, e.g., some native silver and gold. Syn: *dendrolite.*
dendritic Said of a mineral that has crystallized in a branching pattern. Syn: *arborescent; dendriform.*
dendritic and arborescent A mineral in treelike or mosslike forms; e.g., manganese oxide. (Nelson, 1965)
dendritic drainage The pattern of stream drainage in a region underlain by horizontally bedded rock, in which the valleys extend in many directions without systematic arrangement and have a dendritic (treelike) arrangement.
dendritic markings (a) Superficial dendrites on rock surfaces, joint faces, or other fractures, e.g., manganese oxyhydroxides on rock fracture surfaces. (b) Inclusion of a dendrite in another rock or mineral, e.g., chlorite in silica to form moss agate.
dendrolite *dendrite.*
Denison core barrel *Denison sampler.*
Denison sampler A large-size, swivel-type double-tube core barrel designed for soil-testing work to obtain relatively undisturbed corelike samples of soft rock and/or soil formations. The inner tube is provided with a thin wall liner and a finger- or basket-type core lifter or core-retaining device. Also called Denison core barrel. (Long, 1960)
denningite A tetragonal mineral, $(Mn,Zn)Te_2O_5$; colorless to pale green; forms tetragonal plates and platy masses; at Sonora, Mexico.
dennisonite A former name for davisonite. *davisonite.*
dense (a) Said of a fine-grained, aphanitic igneous rock whose particles average less than 0.05 to 0.1 mm in diameter, or whose texture is so fine that the individual particles cannot be recognized by the unaided eye. (AGI, 1987) (b) Said of a rock whose constituent grains are crowded close together. The rock may be fine or coarse grained. (AGI, 1987) (c) Said of a rock or mineral possessing a relatively high specific gravity. (AGI, 1987)
dense graded aggregate Graded mineral aggregate which contains a sufficient number of very small particles to reduce the void spaces in the compacted aggregates to a minimum. (API, 1953)
dense liquid A homogeneous liquid or solution of specific gravity greater than that of water (e.g., zinc chloride and calcium chloride) that can be used in industry or in the laboratory to divide coal or other minerals into two fractions of different specific gravities.
dense-media separation (a) Heavy-media separation, or sink float. Separation of sinking heavy from light floating mineral particles in fluid of intermediate density. Abbreviation: DMS. See also: *heavy-media separation.* (Pryor, 1965) (b) Separation of relatively light (floats) and heavy (sinks) particles, by immersion in a bath of intermediate density. This is the dense or heavy media, a finely ground slurry of appropriate heavy material in water. Barite, magnetite ferrosilicon, and galena are in principal use.
dense medium A fluid formed by the artificial suspension in water of heavy particles (e.g., magnetite, barite, and shale) that can be used in industry or in the laboratory to divide coal into fractions of different specific gravities. (BS, 1962)
dense-medium jigging This method involves two essential features: (1) the circulation in the jig of a middling of approx. 3/16 in (4.8 mm) or smaller in size, with sp gr, 1.7 to 2.0—which fills the interstices of the jig bed and in effect converts the jig into a float-and-sink machine; and (2) the use of a suction stroke to hold the medium in the bed and prevent its washing over with the coal. (Mitchell, 1950)
dense-medium process A process for the washing of coal, in which the desired separation is effected in a dense medium. (BS, 1962)
dense-medium recovery The collection, for reuse, of medium solids from dilute medium, usually understood to include the removal, in whole or in part, of contaminating fine coal and clay. Syn: *medium-solids recovery.* (BS, 1962)
dense-medium washer A machine for cleaning coal and other materials that uses a dense fluid in which the coal floats and shale sinks. The fluid consists of water intimately mixed with sand (or finely ground magnetite or even shale) and agitated to maintain its consistency. The fluid has an effective specific gravity of 1.3 to 1.9. In general, coal from about 8 in (20.3 cm) down to 1 in (2.54 cm) is washed by dense medium, below 1 in by Baum washer, and below 0.75 mm (where cleaning is necessary) by froth flotation. Magnetite as the dense medium solid is preferred as it can be easily recovered by magnetic separators and also the upper limit of the specific gravity is higher (up to 2.0). See also: *coal-preparation plant; washery.* (Nelson, 1965)
dense noncrystalline tonstein This type of tonstein consists almost entirely of fine-grained kaolin groundmass, showing weak aggregate polarization, containing isolated corroded crystals of kaolinite. Such bands are commonly more than 100 mm thick and light in color. (IHCP, 1963)
densimeter An apparatus used to determine the relative density, or specific gravity, of a dense media.
densiscope An apparatus to obtain the specific gravity of pearls as an indication, but not proof, of genuineness (cultured pearls tend to be denser).
densitometer An instrument for the measurement of the density of an image produced by light, X-rays, gamma rays, etc., on a photographic plate; used in some dust-sampling instruments. (Nelson, 1965)
density (a) The mass of a substance per unit volume. (Webster 3rd, 1966) (b) The quality or state of being dense; closeness of texture or consistency. (Webster 3rd, 1966) (c) The distribution of a quantity (as mass, electricity, or energy) per unit usually of space (as area, length, or volume). (Webster 3rd, 1966) (d) The ratio of the mass of any volume of a substance to the mass of an equal volume of a standard substance; water is used as the standard substance. (Long, 1960) (e) Having the quality of being dense, hard, or compact. (Long, 1960) (f) Weight of a substance in grams per cubic centimeter (at specified temperature when close accuracy is needed). For liquids and solids, it equals specific gravity. Density fluids are heavy liquids used in float-sink tests. Of a particle, the true density is its mass (m) divided by volume (v) excluding pores; its apparent density is its mass divided by volume (m/v) including open but excluding closed pores. Of a mass of particles (powder), the apparent density is mass divided by volume (m/v); the bulk density mass divided by volume (m/v) under stated freely poured conditions; and the tap density mass divided by volume (m/v) after vibrating or tapping under stated conditions. See also: *apparent density; bulk density.* (Pryor, 1963) (g) Mass per unit volume. Cf: *specific gravity.* (h) Although density is defined as mass per unit volume, the term is frequently used in place of unit weight in the field of soil mechanics. See also: *unit weight.* (ASCE, 1958)
density contrast The difference in density of a valuable mineral and the host rock. (Lewis, 1964)
density current A current caused by differences in densities, for example, an excess of evaporation, cooling, or dilution in a restricted basin or an open sea. (Schieferdecker, 1959)
density logger An instrument for direct measurement of formation densities in boreholes. This tool furnishes a log of backscattered gamma radiation, which is a simple function of formation density. (Dobrin, 1960)
density of dust cloud The number of ounces of coal dust per cubic foot (or grams per cubic meter) of space, suspended in the air or gases in a specified zone. (Rice, 1913-18)
density of gases The vapor density of a gas, or its density relative to hydrogen, is the number of times a volume of the gas is heavier than the same volume of hydrogen, the volume of both gases being at the same temperature and pressure. (Cooper, 1963)
density of seams (a) An indication of the spacing of seams in the strata; the seam density is said to be high if the seams are close together, or low if they are widely separated. (BS, 1963) (b) The ratio of the sum of the thickness of a number of adjacent seams to the thickness of an arbitrarily chosen sequence of strata. (BS, 1963)
density ratio In powder metallurgy, the ratio of the determined density of a compact to the absolute density of metal of the same composition, usually expressed as a percentage. (ASM, 1961)
dental excavation A controlled blasting technique used to minimize damage, in which the blasting of small, specially designed rounds over partial faces is used in extremely sensitive situations. (SME, 1992)
dental work The act or process of filling cracks, crevices, or caverns encountered in drilling a borehole with cement or grout; also, the cracks, etc., so filled. (Long, 1960)
denudation The sum of the processes that result in the wearing down of the surface of the Earth, including wear by running water, solution, and wind action.
Denver cell A flotation cell of the subaeration type. Design modifications include receded-disk, conical-disk, and multibladed impellers, low-pressure air attachments, and special froth withdrawal arrangements.
Denver jig Pulsion-suction diaphragm jig for fine material, in which makeup (hydraulic) water is admitted through a rotary valve adjustable as to the portion of jigging cycle over which controlled addition is made. Used in coal preparation for the removal of pyritic sulfur from thickener underflow material prior to its treatment by froth flotation. See also: *jig.* (Pryor, 1963; Mitchell, 1950)

Denver mud *bentonite.*
deoxidation The process of extracting the oxygen content of a dissolved oxide, or of removing dissolved oxygen, with the aid of a reducing agent. (Henderson, 1953)
deoxidize To remove oxygen by chemical reaction, generally with carbon. (Mersereau, 1947)
deoxidized copper Copper from which cuprous oxide has been removed by adding a deoxidizer, such as phosphorus, to the molten bath. (ASM, 1961)
deoxidizer A substance that can be added to molten metal to remove either free or combined oxygen. (ASM, 1961)
deoxidizing (a) The removal of oxygen from molten metals by use of suitable deoxidizers. (ASM, 1961) (b) Sometimes refers to the removal of undesirable elements other than oxygen by the introduction of elements or compounds that readily react with them. (ASM, 1961) (c) In metal finishing, the removal of oxide films from metal surfaces by chemical or electrochemical reaction. (ASM, 1961)
dependent shot A charge of explosives in a borehole that depends for its effect upon the result of one or more previously fired shots.
dephosphorization Elimination of phosphorus from steel, in basic steelmaking processes. Accomplished by forming a slag rich in lime. See also: *acid process; basic process; Bessemer process; open-hearth process.* (CTD, 1958)
dephosphorizing Removal of part or all of residual phosphorus from steel in basic smelting. (Pryor, 1963)
depleted fuel *spent fuel.*
depletion The act of emptying, reducing, or exhausting, as the depletion of natural resources. In mining, specif. said of ore reserves. See also: *economic depletion*. (Fay, 1920)
depletion allowance A proportion of income derived from mining or oil production that is considered to be a return of capital not subject to income tax. (AGI, 1987)
depocenter An area or site of maximum deposition; the thickest part of any specified stratigraphic unit in a depositional basin. (AGI, 1987)
deposit (a) Anything laid down. Formerly applied only to matter left by the agency of water, but now includes mineral matter in any form that is precipitated by chemical or other agent, as the ores in veins. (b) Mineral deposit or ore deposit is used to designate a natural occurrence of a useful mineral, or an ore, in sufficient extent and degree of concentration to invite exploitation. (c) Earth material of any type, either consolidated or unconsolidated, that has accumulated by some natural process or agent. The term originally applied to material left by water, but it has been broadened to include matter accumulated by wind, ice, volcanoes, and other agents. Cf: *sediment.* (AGI, 1987) (d) An informal term for an accumulation of ore or other valuable earth material of any origin. (AGI, 1987) (e) Verb. To lay down or let drop by a natural process; to become precipitated. (AGI, 1987)
deposition (a) The process of natural accumulation of rock material thrown down or collected in strata by water, wind, or volcanic action; also, the material thus deposited. Opposite of denudation. (Standard, 1964) (b) The precipitation of mineral matter from solution, as the deposition of agate, vein quartz, etc. (Fay, 1920)
deposit type A class representing all the recognized mineral deposits that are defined by physical and genetic factors that can be consistently differentiated from those of other classes or deposit types. (Barton, 1995)
depressant In the froth flotation process, a reagent that reacts with a particle surface to render it less prone to stay in the froth, thus causing it to wet down as a tailing product. Depressants act by complexing elements at surface lattices of minerals that might carry a charge attractive to conditioning agents; by destroying collector coating; by surface modification of particles. See also: *bathotonic reagent; surface activity.* (Pryor, 1963)
depressed water level The lowest level of ground water during drainage or pumping. (BS, 1963)
depression (a) Any relatively sunken part of the Earth's surface; esp. a low-lying area surrounded by higher ground and having no natural outlet for surface drainage, as an interior basin or a karstic sinkhole. (AGI, 1987) (b) A structurally low area in the crust, produced by negative movements that sink or downthrust the rocks. Cf: *basin; uplift.* (AGI, 1987)
depression contour A closed contour, inside of which the ground or geologic structure is at a lower elevation than that outside, and distinguished on a map from other contour lines by hachures marked on the downslope or downdip side. (AGI, 1987)
depressor A substance (usually inorganic) that inhibits flotation of the mineral. Cf: *activator.* (Newton, 1959)
depth S. Afr. The word alone generally denotes vertical depth below the surface. In the case of incline shafts and boreholes, it may mean the distance reached from the beginning of the shaft or hole, the borehole depth, or inclined depth. (Beerman, 1963)
depth contour *isobath.*
depth indicator A dial or other appliance on a winding apparatus that indicates to the person in charge the position of the cage in the shaft. The indicator must be in addition to any mark on the rope or drum. See also: *visual indicator.* (Nelson, 1965)
depth marker A small metal tag or wooden block placed in the core box at the bottom of the core recovered from each run, on which is marked the depth at which the core was cut in the borehole. (Long, 1960)
depth of cut The thickness of material removed from the workpiece in a single pass. (ASM, 1961)
depth of focus Depth of an earthquake or explosion below the Earth's surface. (Schieferdecker, 1959)
depth of soil exploration Soil sampling is usually carried down to include all deposits likely to have a bearing on the stability of mine structures. Shear tests are made in each bed below the foundation to a depth of at least 1 - 1/2 times the breadth of the foundations. See also: *site investigation.* (Nelson, 1965)
depth of stratum The vertical distance from the surface of the Earth to a stratum. (AGI, 1987)
depth per bit The length of borehole that can be drilled with a steel bit until it must be resharpened. (Streefkerk, 1952)
depth point In seismic work, a position at which a depth determination of a mapped horizon has been calculated. (AGI, 1987)
deputy (a) An underground official in a mine of coal, stratified ironstone, shale, or fire clay, with statutory responsibility for the safe and proper working of a district of the mine. Also called examiner; fireman (undesirable usage). See also: *fireman.* (BS, 1963) (b) Within limits, the deputy is also in charge of the workers in the district. (Nelson, 1965) (c) Eng. In Northumberland and Durham, the person who sets timbers or props in a coal mine is sometimes called a deputy. (Nelson, 1965) (d) N. of Eng. A junior official responsible for safety precautions and mining operations in a face district. (Trist, 1963) (e) N. of Eng. A person who fixes and withdraws the timber supporting the roof of a mine, attends to the safety of the roof and sides, builds stopping, puts up bratticing, and looks after the safety of the miners. (Fay, 1920) (f) Eng. In the Midland coalfield, an underground official who looks after the general safety of a certain number of stalls (rooms) or of a district, the deputy does not set timber but verifies that it is properly done. (Fay, 1920) (g) A mine boss. (Fay, 1920)
deputy surveyor A person appointed by the Surveyor General of the United States to make proper surveys of lode or placer mining claims, prior to the issuing of a patent. (Fay, 1920)
derail A safety device for derailing mine cars, usually installed on grades to protect miners working below. See also: *drop log.*
derailing drag *backstay.*
derail unit This device locks to rails to derail cars. Wedge construction eliminates spiking. It protects workers in railroads and mines against wild cars, switching cars, or sudden car movement. Some types are equipped with a warning flag. (Best, 1966)
derbylite A monoclinic mineral, $(Fe,Ti)_7SbO_{13}(OH)$; forms minute prismatic crystals or twins.
Derbyshire spar Fluorite, found abundantly in Derbyshire, England. See also: *fluorspar.* Syn: *Derby spar.* (Fay, 1920)
Derby spar A popular name for fluorite in Derbyshire, England. Syn: *Derbyshire spar.*
derbystone An amethyst-colored variety of fluorite.
derivative rock A rock composed of materials derived from the weathering of older rocks; a sedimentary rock, or a rock formed of material that has not been in a state of fusion immediately before its accumulation. (AGI, 1987)
derivative structure Representation of crystal structures in terms of a master structure, e.g., feldspar as derivative of coesite with aluminum replacing tetrahedral silicon and charge balance maintained by intertetrahedral alkali and alkali-earth ions.
derived fossil A fossil that is not native to the rock in which it is found, e.g., a fossil found as a pebble in a conglomerate.
derived fuel A fuel obtained from a raw fuel by some process of preparation for use, for example, coke, charcoal, benzene, and gasoline. (Nelson, 1965)
dermatitis A skin disease caused by the application of dust or liquids. In coal mining, the dusts may be coal or stone dust and the liquids may be mine waters, oil or grease, perspiration and acids or alkalis. The majority of cases occur in deep and hot mines having high wet-bulb temperatures. (Mason, 1951)
derrick (a) The framed wood or steel tower placed over a borehole to support the drilling tools for hoisting and pulling drill rods, casing, or pipe. Sometimes incorrectly called a tower. (Long, 1960) (b) The framework over a borehole, used primarily to allow lengths of drill rod to be added to the drilling column. (BS, 1963) (c) A three- (or more) legged framework for supporting drill rods and tackle in deep boring; a temporary three-legged headframe, or headgear, for a shaft. (Mason, 1951)

derrick crane A crane in which the top of the post is supported by fixed stays in the rear and the jib is pivoted like the boom of a derrick. See also: *derricking jib crane.* (Fay, 1920; CTD, 1958)
derricking jib crane A jib crane in which the inclination of the jib, and hence the radius of action, can be varied by shortening or lengthening the tie ropes between the post and the jib. (CTD, 1958)
derrick rope The rope used for supporting and hoisting the boom on jib cranes and excavators. (Hammond, 1965)
desalting Any process for making potable water from sea water or other saline waters. Distillation is the oldest method. Others involve electrodialysis, freezing, extraction, and ion exchange. Also called desalination.
descensional ventilation A ventilation system in which the downcast air is conducted to the top end of the workings (in inclined workings) and it then flows downhill from level to level. In deep mines, the system helps to keep the faces cool. See also: *ascensional ventilation; homotropal ventilation; antitropal ventilation.*
descension theory A theory of formation of supergene mineral deposits involving the descent from above of mineral-bearing solutions. The theory originated with the Neptunian school of thought of the 18th century, which postulated an aqueous origin for all rocks. Cf: *ascension theory.* (AGI, 1987)
descloizite (a) An orthorhombic mineral, 4[PbZn(VO_4)(OH)] having Zn replaced by Cu toward mottramite; greasy; varicolored; in oxidized zones of ore deposits; a source of vanadium. Syn: *vanadite.* (b) The mineral group arsendescloizite, cechite, descloizite, mottramite, and pyrobelonite.
descriptive gemology The classification, composition, properties, trade grades, sources, and the methods of recovery, fashioning, and use of gem minerals and gem materials and their substitutes. See also: *gemology.*
descriptive mineralogy That branch of mineralogy devoted to the description of the physical and chemical properties of minerals. (Fay, 1920)
deseaming Removal by chipping of surface blemishes from ingots or blooms. (Pryor, 1963)
desert crust (a) A hard layer, containing calcium carbonate, gypsum, or other binding matter, exposed at the surface in a desert region. (AGI, 1987) (b) Desert varnish. (AGI, 1987) (c) Desert pavement. (AGI, 1987)
desert glass *obsidian; moldavite.*
desert lands All lands exclusive of timber lands and mineral lands that will not, without irrigation, produce some agricultural crop. (Ricketts, 1943)
desert pavement A natural residual concentration of wind-polished pebbles, boulders, and other rock fragments, mantling a desert surface where wind action and sheetwash have removed all smaller particles, and usually protecting the underlying finer-grained material from further deflation. The fragments commonly are cemented by mineral matter. Syn: *desert crust.* See also: *lag gravel.* (AGI, 1987)
desert rat In the Western United States, a prospector, esp. one who works and lives in the desert, or who has spent much time in arid regions. The name is derived from a small rodent common throughout much of the Great Basin and Southwestern United States. (Fay, 1920)
desert rose A radially symmetrical group of crystals with a fancied resemblance to a rose, formed in sand, soft sandstone, or clay. These crystals are commonly calcite, less commonly barite, gypsum, or celestine.
desert varnish A thin dark shiny film or coating, composed of iron oxide accompanied by traces of manganese oxide and silica, formed on the surfaces of pebbles, boulders, and other rock fragments in desert regions after long exposure, as well as on ledges and other rock outcrops. It is believed to be caused by exudation of mineralized solutions from within and deposition by evaporation on the surface. See also: *patina.* (AGI, 1987)
desiccant A substance having an affinity for water. Used for drying purposes. (Bennett, 1962)
desiccate To dry; to remove moisture; to preserve by drying. (Webster 3rd, 1966)
desiccation A drying out, as in loss of water from sediments, or evaporation from water bodies in arid regions, producing evaporites.
desiccation crack *mud crack.*
desiccator A short glass jar fitted with an airtight cover and containing some desiccating substance (as calcium chloride), above which is placed the material to be dried or to be protected from moisture. (Webster 3rd, 1966)
design A type of diamond-drill fitting that, when standardized, has specific dimensions and thread characteristics establishing interchangeability of parts made by different manufacturers, and size by specific dimension of the set core-bit inside diameter. Design characteristics supplement the group characteristics that provide for integration of ranges. The design characteristics of drill fittings are established by the second letter in two-letter names and by the third letter in three-letter names. Letters denoting design may establish interchangeability of all parts, as in the M-design core barrel, or only of certain parts, as in the X-design core barrel. Cf: *group; range.* (Long, 1960)
designated size The particle size at which it is desired to separate a feed by a sizing operation. (BS, 1962)
designed borehole deflection The turning of a borehole along a different course at depth. This may be achieved, but not without difficulty. The cutting bit is guided upon its new course by the curved surface of a deflecting wedge that is positioned with the aid of a modified Oehman instrument. In petroleum drilling, much use is made of holes that are deflected at a predetermined depth. The technique is known as whipstocking. (Nelson, 1965)
design horsepower The specified horsepower multiplied by a service factor. It is the value used to select the chain size for a chain drive. (Jackson, 1955)
desilication The removal of silica from a rock or magma by the breakdown of silicates and the resultant freeing of silica, or by reaction between a body of magma and the surrounding wall rock. (AGI, 1987)
desiliconizing A practice of jetting oxygen into pig iron before it is charged into the steel furnace; this oxidizes and removes most of the silicon. (Newton, 1959)
desilverization The process of removing silver (and gold) from lead after softening. See also: *Parkes process; Pattinson process.* (CTD, 1958)
desliming The removal of slimes from coal or a mixture of coal and water, however accomplished. (BS, 1962)
desliming screen A screen used for the removal of slimes from larger particles, usually with the aid of water sprays. (BS, 1962)
deslurrying Fines removal by wet methods. (BS, 1962)
desmine A former name for stilbite.
desmite The amorphous groundmass, which is transparent in thin sections, binding together the constituents of bituminous coal of high grade. Applies to the transparent variety of residuum found in high-grade coals. (Tomkeieff, 1954)
desmosite A banded adinole. (AGI, 1987)
desorption The reverse process of adsorption whereby adsorbed matter is removed from the adsorbent. The term is also used as the reverse process of absorption.
destinezite *diabandite.*
destressed area (a) In strata control, a term used to describe an area where the force is much less than would be expected after considering the depth and type of strata. Cf: *overstressed area.* Syn: *zone of substantial deformation.* (Mason, 1951) (b) A region of low stress behind the walls of a stoped-out region. (Issacson, 1962)
destressing In deep mining, relief of pressure concentrations induced by mining or caused by geological factors. Performed by drilling and blasting to loosen the zones of peak stress. The peak load surrounding the excavation walls is thus transferred deeper into the undisturbed rock, and a protective barrier is formed. (Pryor, 1963)
destructive distillation The distillation of solid substances accompanied by their decomposition. The destructive distillation of coal results in the production of coke, tar products, ammonia, gas, etc. (CTD, 1958)
destructive testing Testing methods, the use of which destroy or impair the part or product insofar as its intended use is concerned, but which give proof or an indication of the strength or quality of similar or duplicate parts or products. Such tests involve the subjection of the test piece to various influences, of destructive magnitude, such as impact, stress, pressure, cyclic movement, etc. See also: *nondestructive testing.* (Henderson, 1953)
desulfurization of steel The removal of a high proportion of sulfur from steel by injection of calcium or magnesium. (Nelson, 1965)
desulfurize To free from sulfur; to remove the sulfur from an ore or mineral by some suitable process, as by roasting.
detachable bit A drilling bit that is threaded or tapered and is removable from the drill steel; not formed as an integral part of the drill steel. The all-steel bit can be resharpened, but the tungsten carbide insert type may be nonresharpenable. Also known as rip bit or knockoff bit. See also: *bit; hot miller.*
detached head pulley *head pulley.*
detaching hook An appliance that releases automatically the winding rope from the cage should an overwind occur. See also: *wedge guide.* (Nelson, 1965)
detachment *decollement.*
detail drawing A large-scale drawing showing all small parts, details, dimensions, etc. (Nichols, 1976)
detailed soil survey The final soil tests at site as guided by the general soil survey. The tests may be performed in situ by mobile laboratory units, or the samples are sent to the nearest soils laboratory. See also: *general soil survey; preliminary soil survey.* (Nelson, 1965)
detaline system A nonelectric system of initiating blasting caps in which the energy is transmitted through the circuit by means of a low-energy detonating cord. (Dick, 2)
detector (a) *magnetic detector.* (b) *seismometer.* (c) The component of a remote-sensing system that converts electromagnetic radiation into

a signal that can be recorded. See also: *pickup*. Syn: *radiation detector*. (AGI, 1987) (d) *sensor*.

determinative gemology The science of differentiating (1) between the various gemstones, (2) between gemstones and their substitutes, and (3) among such substitutes. See also: *gemology*.

determinative mineralogy That branch of mineralogy that comprises the measurement of the nature, composition, and classification of minerals by means of physical tests (e.g., density, hardness), chemical analyses both qualitative and quantitative, spectrochemical analyses including both absorption and emission spectra, electron probe microanalyses, autoradiography, thermal analyses, optical tests in both transmitted and reflected light, electron microscopy, diffraction of X-rays or electrons, and crystallographic analyses.

detinning Treatment by chlorination of tinbearing scrap for recovery of tin as its chloride. (Pryor, 1963)

detonate To cause to explode by the application of sudden force. (Standard, 1964)

detonating cord A flexible cord made of wound hemp or jute threads covered with plastic containing a center core of high explosive (PETN) and used to initiate other explosives.

detonating fuse A fuse consisting of high explosive that fires the charge without the assistance of any other detonator. It consists of a high-explosive core of pentaerythritol tetranitrate (PETN) enclosed in tape and wrapped with textile countering yarns. Usually, this fuse is then reinforced or completely enclosed in a strong waterproof plastic outer cover. The finished external diameter is normally about 0.2 in (5 mm). Primacord is the best known brand. See also: *Cordtex; safety fuse*. (Fay, 1920; McAdam, 1958; Nichols, 1976; Nelson, 1965)

detonating gas A gaseous mixture that explodes violently on ignition (as two volumes of hydrogen with one volume of oxygen, forming water). (Webster 2nd, 1960)

detonating powder Any powder or solid substance that when heated or struck explodes with violence and a loud report. (Webster 2nd, 1960)

detonating primer A name applied for transportation purposes to a device consisting of a detonator and an additional charge of explosives, assembled as a unit.

detonating rate The velocity with which the explosion wave travels through the column of charge. (Streefkerk, 1952)

detonating relays A device for obtaining short-delay blasting in conjunction with the detonating fuse. It consists essentially of two open-ended delay detonators coupled together with flexible neoprene tubing. (McAdam, 1958)

detonating tube A eudiometer for making explosions. (Webster 2nd, 1960)

detonation (a) An explosive decomposition or explosive combustion reaction that moves through the reactant(s) at greater than the speed of sound in the reactant(s) to produce (1) shock waves and (2) significant overpressure, regardless of confinement. (b) An extremely rapid explosion; the firing of an explosive charge by fuse or electric detonator. (Nelson, 1965) (c) The action of converting the chemicals in an explosive charge to gases at a high pressure, by means of a self-propagating shock wave passing through the charge. (BS, 1964)

detonation pressure The pressure produced in the reaction zone of a detonating explosive and is a function of explosive density and detonation velocity.

detonation traps Devices that prevent a detonation initiated in one part of a system from propagating to another. (Van Dolah, 1963)

detonation velocity (a) The velocity at which a detonation progresses through an explosive. (b) *velocity of detonation*.

detonator A device for producing detonation in a high-explosive charge, and initiated by a safety fuse or by electricity. Syn: *percussion cap*. See also: *blasting cap; electric detonator*. (BS, 1964)

detonator case A container for carrying detonators in mines. It is so constructed that, when closed, a detonator or the leads of a detonator cannot come into contact with either the metal of the case or any metal outside the case. (Nelson, 1965)

detrital Pertaining to or formed from detritus; said esp. of rocks, minerals, and sediments. See also: *clastic*. (AGI, 1987)

detrital deposits Placer or detrital deposits are composed of minerals that have been released by weathering and later have been transported, sorted, and collected by natural agencies into valuable deposits. Such minerals are usually of high specific gravity and are resistant to abrasion and weathering. Examples are gold, diamonds, platinum, tin (cassiterite), monazite, magnetite, and ilmenite, these last two being the common constituents of black sand. (Lewis, 1964)

detrital fan *alluvial fan*.

detrital mineral Any mineral grain resulting from mechanical disintegration of parent rock; esp. a heavy mineral found in a sediment or weathered and transported from a vein or lode and found in a placer or alluvial deposit. (AGI, 1987)

detrital rock A rock composed primarily of particles or fragments detached from preexisting rocks either by erosion or by weathering; specif. a sedimentary rock having more than 50% detrital material. Cf: *chemical rock*. (AGI, 1987)

detritus A collective term for loose rock and mineral material that is worn off or removed by mechanical means, as by disintegration or abrasion; esp. fragmental material, such as sand, silt, and clay, derived from older rocks and moved from its place of origin. Cf: *debris*. (AGI, 1987)

deuteric Referring to reactions between primary magmatic minerals and the water-rich solutions that separate from the same body of magma at a late stage in its cooling history. Syn: *epimagmatic*. See also: *autometamorphism*. (AGI, 1987)

deuteromorphic A general term applied to crystals whose shapes have been acquired or modified by mechanical or chemical processes acting on the original forms. (AGI, 1987)

De-Vecchis process A method for the smelting of pyrites that entails the roasting and magnetic concentration of the raw material followed by reduction in a rotary kiln or electric furnace. The product may be briquetted and reduced in the blast furnace, but is better smelted in an electric furnace. (Osborne, 1956)

develop (a) To open a mine and ore; more or less, to search, prospect, explore. (von Bernewitz, 1931) (b) To traverse a mineralized body horizontally by drives and vertically by shafts or winzes to prove its extent. (CTD, 1958) (c) To open up orebodies by shaft sinking, tunneling, or drifting. (Ballard, 1955)

developed ore *developed reserve*.

developed reserve Ore that has been exposed on three sides and for which tonnage and quality estimates have been made; ore essentially ready for mining. Cf: *proved reserve*. Syn: *developed ore; ore in sight; blocked-out ore; assured mineral*. (AGI, 1987)

development (a) The preparation of a mining property or area so that an orebody can be analyzed and its tonnage and quality estimated. Development is an intermediate stage between exploration and mining. (AGI, 1987) (b) To open up a coal seam or orebody as by sinking shafts and driving drifts, as well as installing the requisite equipment. (Nelson, 1965) (c) Work of driving openings to and in a proved orebody to prepare it for mining and transporting the ore. (Lewis, 1964) (d) The amount of ore in a mine developed or exposed on at least three sides. (CTD, 1958) (e) S. Afr. The work done in a mine to open up the paying ground or roof and, in particular, to form drives or haulages around blocks of ore, which are then included under developed ore reserves. (Beerman, 1963) (f) A geologic term, applied to those progressive changes in fossil genera and species that have followed one another during the deposition of the strata of the Earth. (g) In construction of a water well, the removal of fine-grained material adjacent to a drill hole, enabling water to enter the hole more freely. (AGI, 1987) (h) Exploitation of ground water. (AGI, 1987)

development drift (a) A main tunnel driven from the surface, or from a point underground, to gain access to coal or ore for exploitation purposes. (Nelson, 1965) (b) Slant.

development drilling Delineation of the size, mineral content, and disposition of an orebody by drilling boreholes. (Long, 1960)

development drivages The shafts, tunnels, laterals, crosscuts, and staple pits to prove and render accessible the coal or ore to be extracted. See also: *productive development; unproductive development*. (Nelson, 1965)

development engineer In bituminous coal mining, one who operates a hoist to raise and lower workers, rock, and supplies during development work (sinking shafts and driving horizontal underground passages prior to the actual mining of coal from a seam). (DOT, 1949)

development miner *miner*.

development plan A plan showing the proposed development of the mine workings, and kept for operational purposes. (BS, 1963)

development rock S. Afr. The rock broken during development work in payable ground, which contains both valuable and barren rock and is, therefore, included in the tonnage sent to the reduction plant of a mine. (Beerman, 1963)

development sampling Sampling for the establishment of reserves and conducted primarily upon the exposures along the development drivages. See also: *reserve*. (Nelson, 1965)

development work Work undertaken to open up orebodies as distinguished from the work of actual ore extraction. Sometimes development work is distinguished from exploratory work on the one hand and from stope preparation on the other. (AGI, 1987)

Devereaux agitator An upthrust propeller, stirring pulp vigorously in a cylindrical tank, used in leach agitation of minerals. (Pryor, 1958)

deviate To change the course of a borehole. Cf: *walk; wander*. (Long, 1960)

deviating Syn. for deflecting. (Long, 1960)

deviation (a) The departure of a drilled hole from being straight. The hole may be either vertical or inclined, and the departure may be in any direction. Deviation may be intentional, as in directional drilling, or undesirable. Syn: *deflection*. (AGI, 1987) (b) In more general use, the angle of departure of a well bore from the vertical, without reference to direction. (AGI, 1987) (c) The distance, measured in a horizontal plane, between two surveyed points in a borehole or between the collar and any point below the collar in a borehole. Also called dislocation; throw. (Long, 1960)

devilline A monoclinic mineral, $CaCu_4(SO_4)_2(OH)_6{\cdot}3H_2O$; emerald-green to verdigris-green. Formerly called devillite; herrengrundite; lyellite; urvolgyite.

devillite *devilline.*

devil's dice Cubes of fully or partially oxidized and hydrated pseudomorphs of pyrite in alluvial workings.

devil's dough A hard, gray-white siliceous rock. (Arkell, 1953)

devitrification (a) Deferred crystallization, which, in glassy igneous rocks, converts obsidians and pitchstones into dull cryptocrystalline rocks (commonly called felsites) consisting of minute grains of quartz and feldspar. Such devitrified glasses reveal their originally vitreous nature by traces of perlitic and spherulitic textures. (b) The process by which glassy rocks break down into definite minerals, which are commonly minute, chiefly quartz and feldspar. (c) Any change from a glassy state to a crystalline state after solidification. (d) In ceramics, a surface defect manifested by loss of gloss as a result of crystallization.

devitrify To destroy the glasslike character of volcanic glasses by changing from the vitreous state to the crystalline state.

devolatilization Progressive loss of volatiles by the substance undergoing coalification process. (Tomkeieff, 1954)

Devonian The fourth period, in order of decreasing age, of the periods making up the Paleozoic era. It followed the Silurian period and was succeeded by the Mississippian period. Also, the system of strata deposited at that time. Sometimes called the Age of Fishes. (Fay, 1920)

De Vooy's process The sink-float or dense-media process used for coal cleaning. The separating fluid is a clay-barite water pulp. (Pryor, 1963)

dewater To remove water from a mine; an expression used in the industry in place of the more technically correct word, unwater. (Hudson, 1932)

dewatering (a) The removal of water from a drowned shaft or waterlogged workings by pumping or drainage as a safety measure or as a preliminary step to resumption of development in the area. Cf: *unwatering.* (Nelson, 1965) (b) The draining of an aquifer when adjacent wells or mine workings are pumped. (c) The mechanical separation of a mixture of coal and water into two parts, one which is relatively coal-free, the other relatively water-free, with respect to the original mixture. (Mitchell, 1950) (d) The mechanical separation of solid matter from water in which it is dispersed, by such equipment as thickeners, classifiers, hydrocyclones, filters, and centrifuges. Coarser coal sizes may be dewatered by slotted screens or perforated bucket elevators. (e) The process in which solid material, either submerged or containing liquid, is conveyed or elevated in a manner that allows the liquid to drain off while the solid material is in transit.

dewatering classifier A settling tank for clarifying washer circulating water or for concentrating gold slimes before cyaniding. The tank may have a continuously working rake that moves the sludge toward the outlet pipe in the bottom. See also: *dryer.* (Nelson, 1965)

dewatering elevator Similar to the continuous bucket elevator, it is often used in sand and gravel plants where the dredge line discharges to a sump. The dewatering elevator digs the material from the sump, allowing the water to drain out through perforations in the backs of the buckets while being elevated, and discharges to the plant for further processing. (Pit and Quarry, 1960)

dewatering screen A screen used for the separation of water from solids. (BS, 1962)

deweylite A mixture of a disordered clinochrysotile or lizardite with a talclike mineral.

dewindtite An orthorhombic mineral, $Pb_3(UO_2)_6H_2(PO_4)_4O_4{\cdot}12H_2O$; strongly radioactive; canary yellow; associated with torbernite and other secondary uranium minerals.

dewpoint The temperature to which air must be cooled, at constant pressure and constant water vapor content, in order for saturation to occur. Since the pressure of the water vapor content of the air becomes the saturation pressure, the dewpoint may also be defined as the temperature at which the saturation pressure is the same as the existing vapor pressure. Also called saturation point. (AGI, 1987)

dewpoint hygrometer An instrument for determining the dewpoint; a type of hygrometer. (Hunt, 1965)

dextral fault *right-lateral fault.*

dextrin A carbohydrate, $C_6H_{18}O_5$, hydrolyzed from starch by dilute acids. Used in flotation as depressant. (Pryor, 1958)

d'Huart reagent An etching reagent that reveals not only the macrostructure and faults, such as piping, segregation, particularly sulfur and phosphorus, and cracks, but also slip lines in mild steel that has been stressed beyond its elastic limit. Composition is 100 mL of distilled water, 100 mL of concentrated hydrochloric acid, and 40 g of crystallized chromic acid, 16 g of anhydrous nickel chloride. (Osborne, 1956)

diabandite A ferroan variety of clinchlore. Syn: *destinezite.*

diabase In the United States, an intrusive rock whose main components are labradorite and pyroxene and that is characterized by ophitic texture. As originally applied by Brongniart in 1807, the term corresponded to what is now recognized as diorite. The word has come to mean a pre-Tertiary basalt in Germany, a decomposed basalt in England, and a dike-rock with ophitic texture in the United States and Canada (Johannsen, 1939). Cf: *trap.* Syn: *dolerite.* (AGI, 1987)

diabasic Composed of or resembling diabase. (AGI, 1987)

diablastic Pertaining to a texture in metamorphic rock that consists of intricately intergrown and interpenetrating constituents, usually with rodlike shapes. (AGI, 1987)

diachronism The transgression, across time planes or biozones, by a rock unit whose age differs from place to place; the state or condition of being diachronous. (AGI, 1987)

diachronous Said of a rock unit that is of varying age in different areas or that cuts across time planes or biozones; e.g., said of a sedimentary formation related to a narrow depositional environment, such as a marine sand that was formed during an advance or recession of a shoreline and becomes younger in the direction in which the sea was moving. Syn: *time-transgressive.* Cf: *synchronous.* (AGI, 1987)

diad An axial rotation of 180°. Syn: *twofold.* Cf: *axis of symmetry.*

diadochite A hydrated ferric phosphate and sulfate mineral, brown or yellowish in color. (Fay, 1920)

diagenesis Any change occurring within a sediment after its deposition and during and after its lithification, exclusive of weathering. It includes such processes as compaction, cementation, replacement, and crystallization, under normal surficial conditions of pressure and temperature.

diagenetic deposits Deposits consisting dominantly of minerals crystallized out of sea water, such as manganese nodules. (Hunt, 1965)

diagnostic mineral (a) A mineral, such as olivine or quartz, whose presence in an igneous rock indicates whether the rock is undersaturated or oversaturated. There are also diagnostic minerals in sedimentary and metamorphic rocks. Syn: *symptomatic mineral.* (b) A mineral whose presence permits certain deductions pertaining to a geologic history of a rock or sediment.

diagonal fault *oblique fault.*

diagonal joints (a) Joints diagonal to the strike of the cleavage. (Zern, 1928) (b) In igneous rocks, joints that occur at 45° to the flow lines and are caused by shear. (Lewis, 1964)

diagonal-slip fault *oblique-slip fault.*

dial (a) A compass used for surface and underground surveying. It is fitted with sights, spirit levels, and a vernier, and mounted on a tripod. Syn: *mining dial.* (Pryor, 1963) (b) Corn. To make a mine survey. (Pryor, 1963)

dialing (a) The process of running an underground traverse with a mining dial. Also spelled: dialling. (CTD, 1958) (b) Surveying, usually magnetic, using miner's dial. (Pryor, 1958)

diallage A dark green or bronze-colored monoclinic pyroxene, which in addition to the prismatic cleavages has others parallel to the vertical pinacoids. Mohs hardness, 4; sp gr, 3.2 to 3.35. Used also as a prefix to many rocks containing the mineral. See also: *pyroxene.* (AGI, 1987; Webster 3rd, 1966; Fay, 1920)

dialogite A former name for rhodochrosite.

dialysis A method of separating compounds in solution or suspension by their differing rates of diffusion through a semipermeable membrane, some colloidal particles not moving through at all, some moving slowly, and others diffusing quite readily. Cf: *osmosis.* See also: *electrodialysis.* (AGI, 1987)

diamagnetic Having a small, negative magnetic susceptibility. All materials that do not show paramagnetism or magnetic order are diamagnetic. Typical diamagnetic minerals are quartz and feldspar. Cf: *paramagnetic.* (AGI, 1987)

diamagnetism The property of certain substances by virtue of which they are repelled from both poles of a magnet and tend to set with the longer axis across the lines of magnetic force. Cf: *ferrimagnetism; paramagnetism.* (Standard, 1964)

diamantiferous *diamondiferous.*

diametric rectifier circuit A circuit that employs two or more rectifying elements with a conducting period of 180 electrical degrees, plus the commutating angle. (Coal Age, 1960)

diamond (a) An isometric mineral, a form of carbon, C; crystallizes in octahedra, dodecahedra, or cubes, commonly with curved edges and striated faces; rarely twinned; has octahedral cleavage and conchoidal fracture. Fresh cleavages have adamantine luster, but crystal faces are commonly greasy; colorless when pure but pale tints to black (bort) with impurities. The hardest natural substance, it defines 10 on the Mohs hardness scale and 15 on the Povarennykh scale, but ranges from 42 to 46 on a linearized Mohs scale. Its high refractive index ($n = 2.42$) and strong dispersion give fire to faceted gems. Diamond occurs in kimberlite pipes and dikes, also in river and beach placers. See also: *congos.* (b) A crystalline material resembling diamond such as rock crystal (quartz) locally known as "Bristol diamond," "Herkimer diamond," "Lemont diamond," "Lake George diamond," or "Arkansas diamond." See also: *industrial diamonds; manmade diamond.* (c) A pointed wooden or iron arrangement placed between rails, just before a curve or switch, where tram cars are liable to be derailed, to force them to remain on the rails. (Fay, 1920)

diamond ballas An important industrial variety of diamond. The stones are spherical masses of minute diamond crystals arranged more or less radially. They have no well-defined cleavage planes and thus have great resistance to abrasion. While the term, ballas, was first applied to such stones from Brazil, diamonds of similar structure known as Cape and African ballas are found. In color, ballas ranges from white to varying shades of black. While Cape and African ballas are not as hard as the Brazilian, they include some fine and unusual stones. Production is small. Rarely, if ever, used for diamond drilling but very valuable for diamond tools. (Cumming, 1951)

diamond bit A rotary drilling bit studded with bort-type diamonds. Also called boart bit; boart-set bit. Syn: *bort bit; bort-set bit.* (AGI, 1987; Long, 1960)

diamond boring Precision boring with a shaped diamond (but not with other tool materials). (ASM, 1961)

diamond chip A thin, tabular chip of an uncut diamond crystal, weighing less than 0.75 carat. (AGI, 1987)

diamond chisel A cutting chisel having a diamond or V-shaped point. (Fay, 1920)

diamond cleavage The plane along which a diamond crystal can be split easily. The four planes paralleling the faces of an octahedron are those generally referred to as the cleavage planes, or diamond cleavage. All crystalline diamonds are more or less brittle and will be fractured by a sufficiently violent blow, but the irregular surface of a fracture cannot be mistaken for the brilliant flat surface produced by cleaving. The carbon has no cleavage, and in ballas cleavage is absent or very poorly defined. (Long, 1960)

diamond cleaving The act or process of splitting diamonds into smaller pieces, which may be more readily used as tool points, gems, or drill diamonds. (Long, 1960)

diamond concentration The ratio of the area of a single-layer bit face covered by the inset diamonds or, in an impregnated bit, the bulk proportion of the crown occupied by diamonds. (Long, 1960)

diamond content The number of carats of diamonds inset in the crown of a diamond bit. Also called stone content; stone weight. (Long, 1960)

diamond core drill A rotary-type drill machine using equipment and tools designed to recover rock samples in the form of cylindrical cores from rocks penetrated by boreholes. See also: *core drill; diamond drill.* (Long, 1960)

diamond coring The act or process of obtaining a core sample of rock material using a diamond-inset annular bit as the cutting tool. This tubular bit and attached core barrel are rotated at a speed under controlled pressure by means of hollow steel, flush-jointed rods through which water is pumped to cool the bit and remove rock cuttings. With the advance of the bit, a cylindrical core of rock passes up into the core barrel, where it is held by a core lifter or other device. (Long, 1960; Cumming, 1951)

diamond count (a) The number of diamonds set in the crown of a specific diamond bit. Also called bit count; stone count. (Long, 1960) (b) Sometimes incorrectly used to indicate the average size of the diamonds inset in a specific bit. See also: *carat count.* (Long, 1960)

diamond crown The cutting bit in diamond drilling. It consists of a steel shell containing small cavities in its face and edges into which black diamonds are set. In some types of crown the diamonds can be removed and reset for further use. Grooves, called waterways, are usually provided in the face of the crown to allow the passage of the drilling fluid. For surface-set bits in diamond drilling, it is recommended that 2 to 20 stones per carat should be used in soft ground (such as shale); 10 to 80 stones per carat in medium ground (such as sandstone); and 20 to 150 stones per carat in hard ground (such as granite). See also: *burned bit.* (Nelson, 1965)

diamond cubic With respect to atomic arrangements, similar to the diamond in having the two face-centered cubic arrangements of atom centers either of which is displaced with respect to the other by one-fourth of the diagonal of the unit cube. (Henderson, 1953)

diamond cutter (a) An individual skilled in the art of shaping diamonds as gems. (Long, 1960) (b) A tool in which a single diamond, shaped as a cutting point, is inset. (Long, 1960)

diamond cutting One of the three processes by which diamonds are prepared for use as ornaments or in the arts, the others being diamond cleaving and diamond polishing. (Fay, 1920)

diamond drill (a) A drilling machine with a rotating, hollow, diamond-studded bit that cuts a circular channel around a core, which can be recovered to provide a more or less continuous and complete columnar sample of the rock penetrated. (AGI, 1987; Long, 1960) (b) Diamond drilling, a common method of prospecting for mineral deposits. Also called adamantine drill; diamond core drill; rotary drill. See also: *core drill; hydraulic circulating system.* (AGI, 1987; Long, 1960)

diamond-driller helper One who assists in the erection and operation of a core drill that bores into rock, earth, and other minerals to obtain core samples. Also called core-driller helper; core-drill-operator helper; diamond-point-drill-operator helper; drill-runner helper; shot-core-drill-operator helper; test-borer helper; test-hole-driller helper; wash-driller helper. (DOT, 1949)

diamond drilling The act or process of drilling boreholes using bits inset with diamonds as the rock-cutting tool. The bits are rotated by various types and sizes of mechanisms motivated by steam, internal-combustion, hydraulic, compressed-air, or electric engines or motors. A common method of prospecting for mineral deposits. See also: *diamond drill.* (Long, 1960)

diamond-drill sample The core brought to the surface in the core barrel. The cuttings in the uprising drilling fluid will also provide sampling material. Syn: *core recovery.* (Nelson, 1965)

diamond dust (a) Finely fragmented or powdered diamonds used as a cutting, grinding, and polishing abrasive or medium. (b) A diamond powder produced in the cutting of gems.

diamond exposure The proportional mass of a diamond protruding beyond the surface of a matrix metal in which the diamond is inset. Cf: *bit clearance.* Syn: *stone exposure.* (Long, 1960)

diamond grade The worth of a diamond as based on an individual sorter's interpretation of somewhat arbitrary standards of color, presence of flaws, soundness, and shape. (Long, 1960)

diamondiferous Any substance containing diamonds, generally applied to rock or alluvial material containing diamonds, but may also refer to diamond-impregnated substances, such as the crown of a diamond-impregnated drill bit.

diamond impregnated Having diamonds distributed throughout a matrix. (Long, 1960)

diamond life The amount of cutting a diamond will accomplish before being completely worn away by abrasion. In bits, diamond life usually is expressed in the number of feet drilled in a specific rock before the inset diamonds become too dulled to continue cutting or are lost by rollout or completely worn away by abrasion. (Long, 1960)

diamond matrix (a) A metal or metal alloy forming the material in which the diamonds inset in a bit crown are embedded. Also called bit-crown metal; bit-crown matrix; bit matrix; crown metal; matrix. (Long, 1960) (b) The rock material in which diamonds are formed naturally and occur, such as in kimberlite. (Long, 1960)

diamond needle A small-diameter hollow metal tube attached to a flexible rubber tube through which air is pulled by a suction or vacuum pump. The suction created at the tip of the metal tube enables a bit setter to pick up and place a small diamond in a bit mold with greater facility than with tweezers. Called a needle because the metal tube generally is made by using a discarded hypodermic needle. Also called diamond pickup needle; diamond pickup tube; diamond pipe. (Long, 1960)

diamond pipe (a) Term used for an occurrence of kimberlite in volcanic pipes large enough and sufficiently diamondiferous to be minable. The size and shape of these pipes depend on the position of the planes of structural weakness in the country rock through which the molten kimberlite passed. They may be columnar, tabular, or irregular in shape, and where mining is deep enough the diamond pipe is found to decrease in area and assume a dikelike habit. (Chandler, 1964) (b) *diamond needle.*

diamond powder *diamond dust.*

diamond pressure The proportional amount of the total feed pressure applied to a diamond bit theoretically borne by an individual diamond inset in the face of the bit. Also called pressure per diamond; pressure per stone; stone pressure. (Long, 1960)

diamond-pyramid hardness test An indention hardness test employing a 136° diamond-pyramid indenter and variable loads enabling the use of one hardness scale for all ranges of hardness from very soft lead to tungsten carbide. See also: *Vickers hardness test.* (ASM, 1961)

diamond saw A circular metal disk having diamonds or diamond dust inset in its cutting or peripheral edge. Employed to cut rocks and other brittle substances. See also: *diamond wheel.* (Long, 1960)

diamond-saw splitter *core saw.*

diamond scale Instrument on which diamonds are weighed with weight units calibrated in carats; scales vary from a folding 50-carat-capacity type, small enough to fit in a coat pocket when closed, to those large enough to weigh several thousand carats at one time. (Long, 1960)

diamond scrap As used in the diamond-drilling industry; broken diamonds and diamond fragments deemed unfit for reuse in a diamond bit. In other industries using diamond-pointed tools, any piece of diamond salvaged from a tool and deemed unfit for reuse in the same kind of tool. (Long, 1960)

diamond screen A perforated metal or wirecloth sieve used to sort diamonds or fragments of diamonds according to size. (Long, 1960)

diamond-set bit A rock-boring or rock-cutting tool, the cutting points of which are inset diamonds. (Long, 1960)

diamond-set inserts Small, shaped metallic slugs inset with diamonds designed to be brazed or welded into slots or depressions machined in a metal bit or reaming-shell blank. (Long, 1960)

diamond-set ring A powdered metal-alloy band encircling a reaming shell in which diamonds are inset mechanically. (Long, 1960)

diamond spar corundum.

diamonds per carat The number of relatively equal size diamonds having a total weight of 1 carat. Also called stone per carat. (Long, 1960)

diamond tin Large bright crystals of cassiterite. (Fay, 1920)

diamond-tooth saw A circular saw for cutting stone with points of the teeth made of pieces of diamonds. (Mersereau, 1947)

diamond washer An apparatus used for washing diamondiferous gravel.

diamond wheel (a) A grinding wheel in which crushed and sized industrial diamonds are held in a resinoid, metal, or vitrified bond. (ASM, 1961) (b) *diamond saw.*

dianite *columbite.*

diaphaneity (a) The quality or state of being diaphanous. Specif., the ability of a mineral to transmit light. Cf: *transparent; semitransparent; translucent; opaque.* (Webster 3rd, 1966; Fay, 1920) (b) Degrees of transparency of minerals. Cf: *transparent; translucent; opaque.* (c) *transparency.*

diaphanous Allowing light to show or to shine through. (Webster 3rd, 1966)

diaphorite (a) *allagite.* (b) A monoclinic mineral, $Pb_2Ag_3Sb_3S_8$.

diaphragm A porous or permeable membrane separating anode and cathode compartments of an electrolytic cell from each other or from an intermediate compartment. (ASM, 1961)

diaphragm jig In the gravity concentration of minerals, a jig with a flexible diaphragm used to pulse water. The Bendelari, Pan-American, Denver, and Conset are examples. (Pryor, 1958)

diaphragm pump A positive displacement pump used for lifting small quantities of water and discharging them under low heads. It has a plunger arm operating either on an eccentric shaft or a rocker arm thrusting on a rubber diaphragm stretched over a cylinder. As the diaphragm is depressed, the water and air in the cylinder are forced out through the discharge side of the pump. As the diaphragm is lifted, a vacuum is created in the cylinder, and water is forced in. (Carson, 1961)

diaphragm-type washbox A washbox in which the pulsating motion is produced by the reciprocating movement of a diaphragm. (BS, 1962)

diaphthoresis *retrograde metamorphism.*

diaphthorite A crystalline rock in which minerals characteristic of a lower metamorphic grade have developed by retrograde metamorphism at the expense of minerals peculiar to a higher metamorphic grade. (AGI, 1987)

diapir A dome or anticlinal fold in which the overlying rocks have been ruptured by the squeezing-out of plastic core material. Diapirs in sedimentary strata usually contain cores of salt or shale; igneous intrusions may also show diapiric structure.

diapir fold An anticline in which a mobile core, such as salt, has ruptured the more brittle overlying rock. Syn: *piercement dome; piercement fold.*

diaschistic Said of the rock of a minor intrusion that consists of a differentiate, i.e., its composition is not the same as that of the parent magma. Cf: *aschisite; aschistic.* (AGI, 1987)

diaspore An orthorhombic mineral, AlO(OH); white, colorless, or pale tints; in bauxite and emery deposits; a source of aluminum. Formerly spelled disaporite. Syn: *kayserite.*

diaspore clay A high-alumina refractory clay consisting essentially of the mineral diaspore. It has been interpreted as a desilication product of associated flint clay and other kaolinitic materials. Commercial diaspore of first-grade quality contains more than 68% alumina. See also: *burley clay.* (AGI, 1987)

diasporogelite A colloidal form of aluminum hydroxide in bauxite. Syn: *sporogelite; cliachite.* (English, 1938)

diastem A relatively short interruption in sedimentation, involving only a brief interval of time, with little or no erosion before deposition is resumed; a paraconformity of very small time value. (AGI, 1987)

diasterism Asterism seen by transmitted light. See also: *asterism; epiasterism.*

diastrophism The processes of deformation in the Earth's crust that produce its continents and ocean basins, plateaus and mountains, and major folds and faults. Syn: *tectonism.*

diathermanous (a) Transmitting infrared radiation. (Webster 3rd, 1966) (b) Allowing the free passage of the rays of heat as a transparent body allows free passage of light. (Standard, 1964)

diatom A microscopic unicellular plant with an envelope (frustule) or outer skeleton of hydrated silica, close to opal in composition, and usually in two parts. Diatoms inhabit both fresh water and salt water, and in places their frustules form masses of diatomaceous earth or shale hundreds of feet thick.

diatomaceous Composed of or containing diatoms or their siliceous remains. (AGI, 1987)

diatomaceous earth *diatomite.*

diatomite A light-colored soft friable siliceous sedimentary rock, consisting chiefly of opaline frustules of the diatom, a unicellular aquatic plant related to the algae. Some deposits are of lake origin, but the largest are marine. Owing to its high surface area, high absorptive capacity, and relative chemical stability, diatomite has a number of uses, esp. as a filter aid and as an extender in paint, rubber, and plastics. The term is generally reserved for deposits of actual or potential commercial value. Syn: *diatomaceous earth; kieselguhr; guhr; tripoli.* Obsolete syn: infusorial earth; tripoli-powder. See also: *tripolite.* Cf: *diatomite.* (AGI, 1987)

diatom ooze A deep-sea deposit, resembling flour when dry, largely composed of the frustules of diatoms and containing a small but variable proportion of calcareous organisms and mineral particles. (Holmes, 1928)

diatomous Having a single distinct diagonal cleavage; applied to certain crystals. (Standard, 1964)

diatom saprokol A saprokol containing a large amount of diatoms. (Tomkeieff, 1954)

diatreme A breccia-filled volcanic pipe that was formed by a gaseous explosion. (AGI, 1987)

dibutyl carbinol 2-methyl-l-butanol; a frother used in the flotation process. (Pryor, 1958)

dice mineral A Wisconsin term for small cubic galena. (Fay, 1920)

dicey clay Any clay or mudstone with a cuboidal fracture, as in the Kimmeridge clay. (Arkell, 1953)

dichroic colors A term loosely used to refer to either the two colors observable in a dichroic stone or the three colors in a trichroic stone. Syn: *twin colors.* See also: *dichroscope.*

dichroism (a) Pleochroism of a crystal, which is indicated by two different colors or two shades of the same color. In plane-polarized light, dichroic minerals change color upon rotation. Cf: *trichroism; pleochroism.* (b) Color change owing to change in the spectrum of illumination; e.g., alexandrite, which is green in sunlight but red by tungsten incandescent light. See also: *chrysoberyl.* (c) The property of some surfaces to reflect light of one color while transmitting light of another.

dichroite A former name for iron-rich cordierite that may have been the navigation stone of the Vikings; reveals maximum light polarization in the southern sky.

dichromate A salt containing the divalent $(Cr_2O_7)^{2-}$ radical.

dichroscope (a) An instrument designed to detect two of the different colors emerging from pleochroic (that is, dichroic or trichroic) minerals. Contains a rhomb of Iceland spar and a lense system in a short tube, and exhibits the two colors side by side. See also: *dichroic colors.* (b) An instrument to detect two colors transmitted by pleochroic minerals and display them side-by-side.

dickensonite A monoclinic mineral, $(K,Ba)(Na,Ca)_5(Mn,Fe,Mg)_{14}Al(PO_4)_{12}(OH,F)_2$; forms a series with arrojadite.

dickinsonite A green, hydrous phosphate mineral, chiefly of manganese, iron, and sodium. (Fay, 1920)

dickite A monoclinic mineral, $Al_2Si_2O_5(OH)_4$; kaolinite-serpentine group; polymorphous with halloysite, kaolinite, and nacrite, each having a different stacking order of identical layers (polytypy); commonly in hydrothermal veins.

diclinic A crystal having two of the three axes inclined to the third and perpendicular to each other. (Standard, 1964)

didymium (a) The name applied to commercial mixtures of rare-earth elements obtained from monazite sand by extraction followed by the elimination of cerium and thorium from the mixture. The name is used like that of an element in naming mixed oxides and salts. The approximate composition of didymium from monazite, expressed as rare-earth oxides, is 46% lanthana, La_2O_3; 10% praseodymia, Pr_6O_{11}; 32% neodymia, Nd_2O_3, 5% samaria, Sm_2O_3; 0.4% yttrium earth oxides; 1% ceria, CeO_2; 3% gadolinia, Gd_2O_3; and 2% others. The mineral bastnaesite could also be a source of didymium mixtures. (CCD, 1961) (b) The name didymium has also been applied to mixtures of the elements praseodymium and neodymium because such mixtures were once thought to be an element; it was assigned the symbol, Di. (CCD, 1961)

didymolite A former name for a plagioclase mineral.

die (a) *bell tap.* (b) A piece of hard iron, placed in a mortar to receive the blow of a stamp or in a pan to receive the friction of a muller as ore is crushed between the die and the stamp or muller. (Fay, 1920)

die-casting alloys Alloys that are suitable for die casting and that can be relied on for accuracy and resistance to corrosion when cast. Aluminum-, copper-, tin-, zinc-, and lead-base alloys are those generally used. (CTD, 1958)

die collar *bell tap.*

dielectric (a) A material that offers relatively high resistance to the passage of an electric current but through which magnetic or electrostatic lines of force may pass. Most insulating materials, for example, air, porcelain, mica, and glass, are dielectrics; and a perfect vacuum would constitute a perfect dielectric. (NCB, 1964) (b) An insulator. A term applied to the insulating material between the plates of a capacitor. (Hunt, 1965)

dielectric constant The numerical expressions of the resistance to the passage of an electric current between two charged poles. It is the ratio of the attraction of two oppositely charged poles as measured in a vacuum to their attraction in a substance. The dielectric constant, which corresponds to permeability in magnetic materials, is a measure of the polarizability of a material in an electric field. This property determines the effective capacitance of a rock material and consequently its static response to any applied electric field, either direct or alternating. The dielectric constant of a vacuum is unity. (Hess; Dobrin, 1960)

dielectric heating A method of high-frequency heating in which the object to be heated, which must be nonconducting, is placed in a high-frequency alternating field where it is heated by the continually reversed polarization of the molecules. Applied in the foundry for drying sand cores. (Osborne, 1956)

dielectric separation Method of ore treatment based on differences between dielectric constants of minerals suspended in an intermediate nonconducting fluid, when subjected to electric fields. Of limited use in laboratory work. (Pryor, 1958)

dielectric strength The maximum potential gradient that a dielectric material can withstand without rupture. (Lowenheim, 1962)

dienerite An isometric mineral, Ni_3As; in gray-white cubes at Radstadt, Salzburg, Austria.

die nipple *bell tap.*

diesel hammer A pile driving drophammer operated by a type of diesel engine. (Hammond, 1965)

dieseling In a compressor, explosions of mixtures of air and lubricating oil in the compression chambers or other parts of the air system. (Nichols, 1976)

diesel particulate matter (a) Exhaust material, excluding water, that results from the incomplete combustion of fuel and lubricating oil in a diesel engine. The particulates collected on a filter after dilution of the exhaust with ambient air, are carbonaceous solid chain aggregates with adsorbed or condensed organic compounds. (SME, 1992) (b) The fumes (solid condensation particles) and adsorbed gases that are emitted from a diesel engine as a result of the combustion of diesel fuel. Abbrev. DPM. DPM is a complex mixture of chemical compounds, composed of nonvolatile carbon, hundreds of thousands of different adsorbed or condensed hydrocarbons, sulfates, and trace quantities of metallic compounds. DPM is of special concern because it is almost entirely respirable, with 90% of the particles, by mass, having an equivalent aerodynamic diameter of less than 1.0 µm. This means that the particles can penetrate to the deepest regions of the lungs and, if retained, cause or contribute to the development of lung disease. Of equal concern is the ability of DPM to adsorb other chemical substances, such as (1) potentially mutagenic or carcinogenic polynuclear aromatic hydrocarbons (PAHs); (2) gases, such as sulfur dioxide and nitrogen dioxide; and (3) sulfuric and nitric acids. DPM carries these substances into the lungs, where they may be removed and transported by body fluids to other organs, where they may cause damage.

diesel rig Any drill machine powered by a diesel engine. (Long, 1960)

diesel truck In opencast mining, a powerful and robust diesel-engined vehicle carrying from a few to more than 100 cubic yards of earth or rock. Also used in trackless transport in tristate mines. (Pryor, 1963)

die steels Steels of plain-carbon or alloy types; they must be of high quality, which is usually attained by special methods of processing. Essentially, they are steels used in making tools for cutting, machining, shearing, stamping, punching, and chipping. (USBM, 1956)

Dietert tester An apparatus for the direct reading of a Brinell hardness after impression without the aid of magnification or conversion tables. (Osborne, 1956)

dietzite A monoclinic mineral, $Ca_2(IO_3)_2(CrO_4)$; dark golden-yellow; forms prismatic, tabular, fibrous, or columnar crystals; at Atacama, Peru.

difference in gage of drill bits The difference in diameter of the bits when passing from one length (change) of drill steel to the next longer one of a set. (Fraenkel, 1953)

difference of potential The difference in electrical pressure existing between any two points in an electrical system or between any point of such a system and the Earth. Determined by a voltmeter. (Fay, 1920)

differential compaction The uneven settling of homogeneous earth material under the influence of gravity (as where thick sediments in depressions settle more rapidly than thinner sediments on hilltops) or by differing degrees of compactability of sediments (as where clay loses more interstitial water and comes to occupy less volume than sand). (AGI, 1987)

differential curvature A quantity represented by the acceleration due to gravity times the difference in the curvatures in the two principal planes; i.e., $g(1/p_1 - 1/p_2)$ where p_1 and p_2 are the radii of curvature of the two principal planes. (AGI, 1987)

differential erosion Erosion that occurs at irregular or varying rates, caused by the differences in the resistance and hardness of surface materials; softer and weaker rocks are rapidly worn away, whereas harder and more resistant rocks remain to form ridges, hills, or mountains. (AGI, 1987)

differential fault *scissor fault.*

differential grinding Application of comminution in such a way as to accentuate differences in grindability between the various mineral species in the ore. Therefore, in suitable cases, the relatively tough mineral particles remain coarse while the more friable ones are finely ground. (Pryor, 1958)

differential pressure flowmeter An instrument for measuring water and water-ore slurries in ore dressing and coal dressing processes.

differential pumping engine A compound direct-acting pumping engine, generally of the horizontal class. (Fay, 1920)

differential settlement Nonuniform settlement; the uneven lowering of different parts of an engineering structure, often resulting in damage to the structure. See also: *settlement.* (AGI, 1987)

differential thermal analysis (a) A method of analyzing a variety of minerals, esp. clays and other aluminiferous minerals. The method is based upon the fact that the application of heat to many minerals causes certain chemical and physical changes and is reflected in endothermic and exothermic reactions. By comparing the changes in temperature of a mineral heated at a definite rate with that of a thermally inert substance (alumina, for example) heated under the same conditions, a curve or pattern is obtained that is characteristic of the particular mineral under examination. (Henderson, 1953) (b) Thermal analysis carried out by uniformly heating or cooling a sample that undergoes chemical and physical changes, while simultaneously heating or cooling in identical fashion a reference material that undergoes no changes. The temperature difference between the sample and the reference material is measured as a function of the temperature of the reference material. Abbrev: DTA. (AGI, 1987)

differential weathering Weathering that occurs at different rates, as a result of variations in composition and resistance of a rock or differences in intensity of weathering, and usually resulting in an uneven surface where more resistant material protrudes above softer or less resistant parts. Syn: *selective weathering.* (AGI, 1987)

differentiate A rock formed as a result of magmatic differentiation. (AGI, 1987)

differentiated Said of an igneous intrusion in which there is more than one rock type, owing to differentiation. (AGI, 1987)

differentiation *magmatic differentiation.* Cf: *assimilation.*

diffraction The cooperative scattering of any electromagnetic radiation where it encounters an obstacle, esp. the edge of an obstacle, resulting in constructive and destructive interference. Also, a single event resulting from constructive interference. See also: *optical diffraction; X-ray diffraction.* Syn: *wave diffraction.* Cf: *reflection.*

diffraction grating An optical device having equidistant fine lines (on the order of wavelengths of visible light) scribed on glass for transmission, or on metal for reflection diffraction, of monochromatic light.

diffraction pattern (a) Diffracted X-rays recorded on film, giving a means of identification of a powder. (b) A record of diffracted X-rays on film or paper showing angles of diffraction of monochromatic radiation; used for characterization or identification of a crystalline substance. Cf: *Laue photograph.*

diffuser (a) The inner shell and water passages of a centrifugal pump. (Nichols, 1976) (b) *evasé.*

diffuser chamber A chamber in a turbine pump consisting of a number of fixed blades. On leaving the impeller, the water is guided outward by these blades with the minimum of eddying and swirling. See also: *turbine pump.* (Nelson, 1965)

diffusion of gases The property that all gases possess of mixing with each other. (Nelson, 1965)

diffusivity The relative rate of flow per unit area of a particular constituent of a mixture divided by the gradient of composition, temperature, or other property considered to be causing the diffusion.

dig (a) To mine coal; applied to bituminous workings. See also: *gouge.* (b) To excavate; make a passage into or through, or remove by taking away material. (c) Crushed strata. (Nelson, 1965)

dig-down pit A pit that is below the surrounding area on all sides. Also called sunken pit. (Nichols, 1976)

digenite An isometric mineral, Cu_9S_5; blue to black; in veins with chalcocite; a source of copper. Syn: *blue chalcocite; alpha chalcocite.* Cf: *copper sulfide.*

digger (a) One that digs in the ground, as a miner or a tool for digging. (Webster 3rd, 1966) (b) A worker who is paid by the ton for coal produced; a miner in the stricter sense. Originally the digger mined or undermined the coal; now the term is applied to the worker who merely shoots out the coal. (Fay, 1920) (c) A machine for removing coal from the bed of streams, the coal having washed down from collieries of culm banks above. (Zern, 1928)

digger edges The formed serrated edges of the buckets used for digging purposes on a bucket loader.

digger tools The formed tools interspaced with the buckets of a bucket loader to aid in digging action.

digging Mining operations in coal or other minerals.

digging bit According to English drillers, a noncoring bit usually similar to a steel drag or mud bit. (Long, 1960)

digging cycle Complete set of operations a machine performs before repeating them. (Nichols, 1954)

digging height *bank height.*

digging line On a shovel, the cable that forces the bucket into the soil. Called crowd in a dipper shovel, drag in a pull shovel, and dragline and closing line in a clamshell. (Nichols, 1976)

digging resistance The resistance that must be overcome to dig a formation. This resistance is made up largely of hardness, coarseness, friction, adhesion, cohesion, and weight. (Nichols, 1956)

diggings Applicable to all mineral deposits and mining camps, but as used in the United States it is usually applied to placer mining only. See also: *bar diggings.*

digital map A map using data in a software format so that the maps have the characteristic of layered features on an overlay generated by computer-aided drafting and design to plot these features. (SME, 1992)

dihedral Having two sides, as a figure; having two faces, as a crystal. (Fay, 1920)

dihydrite *pseudomalachite.*

dike (a) An earthen embankment, as around a drill sump or tank, or to impound a body of water or mill tailing. (b) A tabular igneous intrusion that cuts across the bedding or foliation of the country rock. Also spelled: dyke. Cf: *sill; sheet.* See also: *dikelet.* (AGI, 1987)

dikelet A small dike. There is no agreement on specific size distinctions. (AGI, 1987)

dike ridge A wall-like ridge created when erosion removes softer material from along the sides of a dike.

dike rock The intrusive rock comprising a dike. (AGI, 1987)

dike set A group of parallel dikes. Cf: *dike swarm.*

dike swarm A group of dikes, which may be in radial, parallel, or en echelon arrangement. Their relationship with the parent plutonic body may not be directly observable. Cf: *dike set.* (AGI, 1987)

dilatancy An increase in the bulk volume of a granular mass during deformation, caused by a change from close-packed structure to open-packed structure, accompanied by an increase in the pore volume. The latter is accompanied by rotation of grains, microfracturing, and grain boundary slippage. (AGI, 1987)

dilatational wave *P wave; compressional wave.*

dilation Deformation by a change in volume but not shape. Also spelled: dilatation. (AGI, 1987)

dilational transformation A phase transformation requiring change in coordination about a cation, e.g., quartz with silica tetrahedra to stishovite with silica octahedra. Cf: *reconstructive transformation; displacive transformation; rotational transformation.*

dilation vein A mineral deposit in a vein space formed by bulging of the walls, contrasted with veins formed by wall-rock replacement. (AGI, 1987)

diligence The attention and care legally required of a person (for example a claim holder) while that person has temporary possession of a property. With regards to mining claims, the courts have said that due diligence requires that "the exploration for minerals should be made within a reasonable time" and that, "The failure to make such exploration within a reasonable time, and to make it with such thoroughness and certainty as to determine the existence of mineral or oil, would be fatal to the agreement (claim)". Legal requirements for "diligence" may include annual improvements to the claim and the filing of reports and notices. (Webster 3rd, 1966)

dillenburgite An impure variety of chrysocolla containing copper carbonate.

dilly rider In bituminous coal mining, a laborer who rides and attends a dilly (light wagon, truck, or water cart) used to haul coal or water underground or at the surface of a mine, loading, unloading, and cleaning it. (DOT, 1949)

diluent (a) That which dilutes or makes more fluid; a fluid that weakens the strength or consistency of another fluid upon mixing. (Fay, 1920) (b) Waste rock in ore. (Hess) (c) In solvent extraction, the inert liquid used to dissolve the extractant. (Newton, 1959)

dilute medium Medium of specific gravity below that in the separating bath and usually occurring as a result of spraying the bath products for the removal of adhering medium solids. (BS, 1962)

dilution The contamination of ore with barren wall rock in stoping. The assay of the ore after mining is frequently 10% lower than when sampled in place. See also: *contamination.* (Nelson, 1965; Long, 1960)

dimensional rated capacity The weight of a specified material per foot of belt length that a belt conveyor will transport. (NEMA, 1961)

dimension stone Any rock suitable for construction purposes, as distinguished from crushed stone or aggregate.

dimetric system Same as tetragonal system.

dimorphism The property of a chemical compound to crystallize in either of two different crystal structures, e.g., $CaCO_3$ as trigonal calcite and as orthorhombic aragonite. Noun: dimorph. Adj: dimorphic. Cf: *trimorphism; polymorphism.*

dimorphite An orthorhombic mineral, As_4S_3; orange-yellow; a volcanic product closely related to orpiment.

dingot An oversized derby (possibly a ton or more) of a metal produced in a bomb reaction, such as uranium from uranium tetrafluoride and magnesium. The term ingot for these metals is reserved for massive nits produced in vacuum melting and casting. See also: *biscuit.* (ASM, 1961)

Ding's magnetic separator In its earlier form, a mineral separator to which the material was fed by a vibrating conveyor and passed through successive zones of magnetic influence. The zones were covered by the rims of rotating disks, which became magnetized, carried the particles having magnetic susceptibility out of the fields, were demagnetized, and dropped the concentrate beyond the edge of the belt. Now made with rollers having an induced magnetism; dried, finely crushed ore passed over the rollers in a thin stream from which particles attracted by the magnet are drawn out. (Hess; Liddell, 1918)

dinite A yellowish hydrocarbon having a low melting temperature; in lignite.

dinkey A small locomotive used to move cars in and about mines and quarries. (Fay, 1920)

dint To cut into the floor of a roadway to obtain more headroom. (Fraenkel, 1953)

dioctahedral Said of layered silicates having two-thirds of the voids in the octahedral layer filled, generally with trivalent cations. Cf: *trioctahedral.*

diopside A monoclinic mineral, $CaMgSi_2O_6$; pyroxene group; white to light green; in metamorphic rocks, esp. contact metamorphosed limestones; where transparent, a semiprecious gemstone. Symbol: Di or di. See also: *malacolite.*

dioptase A trigonal mineral, $CuSiO_2(OH)_2$; emerald green; in the oxidized zones of copper deposits; a source of copper. Also called emerald copper, Zaire emerald.

diorite A group of plutonic rocks intermediate in composition between acidic and basic, characteristically composed of dark-colored amphibole (esp. hornblende), acid plagioclase (oligoclase, andesine), pyroxene, and sometimes a small amount of quartz; also, any rock in that group; the approximate intrusive equivalent of andesite. Diorite grades into monzonite with an increase in the alkali feldspar content. Etymol: Greek diorizein, to distinguish, in reference to the fact that the characteristic mineral, hornblende, is usually identifiable megascopically. Cf: *dolerite; gabbro.* See also: *diabase.* (AGI, 1987)

dioxide ore A term that has been used somewhat in the Western United States for manganese ore. (Hess)

dip (a) The angle at which a bed, stratum, or vein is inclined from the horizontal, measured perpendicular to the strike and in the vertical plane. See also: *pitch; hade; angle of dip; apparent dip.* Cf: *plunge.* (Lewis, 1964) (b) To be inclined or dip at an angle. (c) The angle of a slope, vein, rock stratum, or borehole is measured from the horizontal plane downward. (Long, 1960) (d) The direction of the true or steepest inclination. (Mason, 1951) (e) The lower workings of a mine. (Hudson, 1932) (f) The slope of layers of soil or rock. (Nichols, 1976) (g) A dip entry, dip room, etc. A heading driven to the full rise in steep mines. (Fay, 1920) (h) In terrestrial magnetism, the angle formed by the lines of total magnetic force and the horizontal plane at the Earth's surface; reckoned positive if downward. See also: *apparent dip; full dip.* (Hy, 1965) (i) In mines, the increase in depth of a moored mine case, due to current force against the case and cable. (Hy, 1965)

dip calculation Any of a number of methods of converting observed seismic arrival time values to the dip of a reflector; most commonly the conversion of delta T values to dip values by a conversion factor based upon the geometry of the seismic array and approximate seismic propagational velocity. (AGI, 1987)

dip compass An instrument to measure magnetic intensity by means of a magnetic needle fixed to swing in a vertical plane so that it can readily be deflected downward by magnetic materials. Used to explore for subsurface deposits containing magnetic materials. May also be called dip needle, dipping compass, dipping needle, doodle bug magnetometer. (Long, 1960)

dip-corrected map A map that shows stratified formations in their original position before movement.

dip cut In cutting out blocks of stone, the cut that follows a line at right angles to the strike.

dip entry An entry driven downhill so that water will stand at the face. If it is driven directly down a steep dip it becomes a slope. See also: *entry; slope.*

dip equator *aclinic line.*

dip face A face proceeding toward the dip of the seam. (Briggs, 1929)

dip fault A fault that strikes approx. perpendicular to the strike of the bedding or cleavage. Cf: *oblique fault; strike fault.* (Billings, 1954)

diphead A drift inclined along the dip of a coal seam. (Webster 3rd, 1966)

diphead level (a) A mine level connecting an engine shaft with the rooms or chambers. (Standard, 1964) (b) The main level, drift, or slope.

dip joint A joint that strikes approx. perpendicularly to the strike of the bedding or cleavage. (Billings, 1954)

dip meter (a) An instrument used to record the amount and direction of the dip of strata exposed in the sides of a borehole. (BS, 1963) (b) *dipmeter.*

dipmeter A dipmeter measures both the amount and direction of dip by readings taken in the borehole and can be operated by using either self-potential or resistivity measurements. (Sinclair, 1963)

dip needle An obsolete type of magnetometer used for mapping high-amplitude magnetic anomalies. It consists of a magnetized needle pivoted to rotate freely in a vertical plane, with an adjustable weight on the south side of the magnet. See also: *Hotchkiss superdip.* (AGI, 1987)

dipole (a) Coordinate valence link between two atoms. (Pryor, 1963) (b) Electrical symmetry of a molecule. When a molecule is formed by sharing two electrons between a donor atom and an acceptor, it is more positive at the donor end and more negative at the acceptor end, and has a dipole moment of the order of 10^{-18} electrostatic unit. Dipole moment is also the couple required to maintain the dipole at right angles to an electrical or magnetic field of unit intensity. (Pryor, 1963)

dipole moment Product of the dipole charge and the dipole length. (Van Vlack, 1964)

dippa Corn. A small pit sunk on a lode to catch water; a pit sunk on a bunch ore.

dipper (a) A digging bucket rigidly attached to a stick or arm on an excavating machine; also the machine itself. (b) N. of Eng. A downthrow, or a fault.

dipper dredge A dredge in which the material excavated is lifted by a single bucket on the end of an arm, in the same manner as in the ordinary steam shovel. (Fay, 1920)

dipper dredger A dredger consisting of a single large bucket at the end of a long arm, swung in a vertical plane by gearing. The bucket capacity may be up to about 12 cubic yards. See also: *dredger.* (CTD, 1958)

dipper factor *fill factor.*

dipper stick (a) The straight shaft that connects the digging bucket with the boom on an excavating machine or power shovel. (Nichols, 1954) (b) Standard revolving dipper shovel. (Nichols, 1954)

dipping needle A needle, consisting of a steel magnet, similar to that in a miners' dial, but pivoted at the center so as to be free to rotate vertically. It is used to locate the presence of shallow deposits of magnetic ores. The magnetometer has now replaced the dipping needle for large-scale prospecting work. Syn: *dip compass.* See also: *geophysical exploration.* (Nelson, 1965; Long, 1960)

dipping weight *pickup.*

dip reading An angular measurement taken in an inclined borehole by using one of several types of borehole-surveying devices or techniques. (Long, 1960)

dip separation The distance or separation of formerly adjacent beds on either side of a fault surface, measured along the dip of the fault. Cf: *dip slip.* (AGI, 1987)

dip shift In a fault, the shift or relative displacement of the rock units parallel to the dip of the fault, but outside the fault zone itself. Cf: *dip slip; strike shift.* (AGI, 1987)

dip shooting A system of seismic surveying in which the primary concern is determining the dip and position of reflecting interfaces rather than in tracing such interfaces continuously. (AGI, 1987)

dip slip In a fault, the component of the movement or slip that is parallel to the dip of the fault. Cf: *dip separation; strike slip; oblique slip; total displacement; dip shift.* (AGI, 1987)

dip-slip fault A fault on which the movement is parallel to the dip of the fault. Cf: *strike-slip fault.* (AGI, 1987)

dip slope A landform developed in regions of gently inclined strata, particularly where hard and soft strata are interbedded. A long, gentle sloping surface that parallels the dip of the bedding planes of the strata below ground. See also: *back slope.* Cf: *stripped plain.* (CTD, 1958)

dip split A current of intake air directed into or down a dip.

dip switch (a) A slant or piece of track connecting the back entry or air course of a dipping coal seam with the main entry or gangway. (Fay, 1920) (b) Circuit board component that consists of several switches used to alter circuit performance.

dip test As used in the diamond-drilling industry, an angular measurement of the inclination of a borehole taken with a clinometer. See also: *acid-dip survey.* (Long, 1960)

dip throw The component of the slip measured parallel with the dip of the strata. (Fay, 1920)

dip valley A valley trending in the direction of the general dip of the rock layers of a region.

dip workings (a) The workings on the lower side of the level or gate road in an inclined seam. Dip workings may present water problems and require pumping. Also called deep workings. (Nelson, 1965) (b) Underhand excavations in which miner works downward and lifts spoil to removal point. Not self-draining. (Pryor, 1963)

dipyramid A closed form consisting of orthorhombic, trigonal, tetragonal, or hexagonal positive and negative pyramids. Syn: *bipyramid.* Cf: *bipyramid; pyramid.*

dipyre A variety of scapolite having marialite:meionite between 3:1 and 3:2. Syn: *mizzonite; dipyrite.*

dipyrite *dipyre; pyrrhotite.*

direct-acting controller One in which an increasing measured value in the input signal produces an increasing controller output, and vice versa. (Pryor, 1963)

direct-acting haulage *direct-rope haulage.*

direct-arc furnace One in which an arc is struck between an electrode and the material charged into the furnace. (Pryor, 1963)

direct attack A method of effecting extinction of mine fires using water or the effluent of chemical fire extinguishers. When a mine fire is readily accessible to the firefighting personnel, extinction of it may be achieved by direct application of some substance that will cool down the hot mass below its ignition temperature, or, in the case of oils, will arrest the volatilization process by sealing or emulsifying the oil surface. (Mason, 1951)

direct firing (a) The combustion of coal effected by burning directly on a grate. (Fay, 1920) (b) A method of firing wherein the products of combustion come in contact with the ware. (ACSB, 1948)

direct flushing Flushing in which the water rises along the rod on its outer side; i.e., between the walls of the borehole and the rod, and with such a velocity that the broken rock fragments are carried up by this water current. (Stoces, 1954)

direct haulage The system in which an engine with a single drum and rope draws loaded trucks up an incline. The empties run downhill dragging the rope after them. (CTD, 1958)

direct initiation The placing of the detonator in the last cartridge to be inserted in the shothole with the active end of the detonator pointing inward. This position tends to minimize the risk of gas ignition. See also: *inverse initiation.* (Nelson, 1965)

direction (a) Angle to the right (clockwise) from an arbitrary zero direction. Used chiefly in triangulation. (Seelye, 1951) (b) *trend.*

directional drilling (a) The art of drilling a borehole wherein the course of the hole is planned before drilling. Such holes are usually drilled with rotary equipment and are useful in drilling divergent tests from one location, tests that otherwise might be inaccessible, as controls for fire and wild wells, etc. (AGI, 1987) (b) Drilling in which the course of a borehole is controlled by deflection wedges or other means. The technique of directional drilling is used (1) to deflect a deviated borehole back onto course and (2) to deflect a borehole off course, either to bypass an obstruction in the hole or to take a second core. Syn: *slant drilling.* (BS, 1963)

directional solidification The solidification of molten metal in a casting in such a manner that feed metal is always available for the portion just solidifying. (ASM, 1961)

directional work *directional drilling.*

direction-finding methods Electromagnetic exploration methods in which one determines the direction of the magnetic field associated with the currents. (Schieferdecker, 1959)

direction indicator Any one of a number of geophysical devices used to determine the deviation of a borehole from vertical. (AGI, 1987)

direction of dip *line of dip.*

direction of strata (a) The strike or line of bearing. (b) The direction of the line formed by the intersection of the individual stratum with the horizontal plane. The direction of this line is customarily referred to north. See also: *strike.*

direction of tilt (a) The azimuth of the principal plane of an aerial photograph. (AGI, 1987) (b) The direction of the principal line on a photograph. (AGI, 1987)

directivity index A measure of the directional properties of a transducer. It is the ratio in decibels of the average intensity of response over the whole sphere surrounding the projector or hydrophone to the intensity or response on the acoustic axis. (Hy, 1965)

direct labor A method of carrying out mining works in which the owners, Board, or Authority, carry out the scheme by employing labor and purchasing the necessary equipment. The method is in contrast to work entrusted to outside contractors for performance at a fixed sum. (Nelson, 1965)

direct lattice A symmetrical array of points in direct space; used when comparison is made with the direct lattice. Syn: *Bravais lattice; crystal lattice.*

direct oxidation The reaction of metals with dry gases, leading to the formation of oxides or other compounds on the surface; it does not occur to a pronounced extent except at elevated temperature. (CTD, 1958)

direct plot In a graph of particle distribution (screen analysis), a plot in which the abscissa shows the size and the ordinate shows the percentage of sample of that size. (Pryor, 1958)

direct raw-water cooling system A cooling system in which water, received from a constantly available supply, such as a well or water system, is passed directly over the cooling surfaces of the rectifier and discharged. (Coal Age, 1960)

direct-reading capillary chart A graduated scale printed on transparent paper, which, when used in the prescribed manner, enables one to determine the true angle a borehole is inclined from readings taken directly on the etch plane in an acid bottle. This eliminates the need for a protractor or goniometer and for a capillarity-correction chart. (Long, 1960)

direct-rope haulage (a) A system of incline haulage, comprising one rope and one drum. The engine hauls up the journey of loaded cars, then the empties are connected to the rope and returned to the bottom by gravity. See also: *balanced direct-rope haulage.* Also called direct-acting haulage. (Nelson, 1965) (b) Haulage in which a loaded truck is pulled up the slope by a hoist while an empty one descends, perhaps passing halfway on a loop of single track. Also called brake incline; engine plane. (Pryor, 1963)

direct shipping ore *natural ore.*

Dirigem A copyrighted trade name for green synthetic spinel.

dirt band A thin stratum of shale or other inorganic rock material in a coal seam. Syn: *shale band; dirt bed; dirt parting; stone band.* (AGI, 1987)

dirt bed (a) Eng. A thin stratum of soft, earthy material interbedded with coal seams. Syn: *dirt band.* (Fay, 1920) (b) Old soil in which trees, fragments of timber, and numerous plants are found. (Fay, 1920)

dirt bing Scot. A debris heap; a waste heap.

dirt parting *dirt band.*

dirt scraper A road scraper or a grading shovel, used in leveling or grading ground. (Fay, 1920)

dirt slip *clay vein.*

dirty coal Scot. A coal seam with thick partings of blaes or fireclay; a very ashy coal.

disability glare The glare resulting in reduced visual performance and visibility caused by the action of stray light, which enters the eye and scatters within. It causes a "veiling luminescence" over the retina, which, in turn, has the effect of reducing the perceived contrast of the objects being viewed. Cf: *discomfort glare.* See also: *glare.*

discard (a) The material extracted from the raw coal and finally thrown away. Also called dirt; stone. (BS, 1962) (b) The portion of an ingot cropped off to remove the pipe and other defects. Also called crop. (CTD, 1958)

discharge (a) The production or output from crushing or processing machines, such as ball mills, or thickeners. (b) The outflow from a pump, drill hole, piping system, or other mechanism. (c) The quantity of water, silt, or other mobile substances passing along a conduit per unit of time; rate of flow in cubic feet per second, gallons per day, etc. (d) The rate of flow at a given moment, expressed as volume per unit of time. (AGI, 1987)

discharge chute A chute used to receive and direct material or objects from a conveyor.

discharge head The sum of static and dynamic head. The vertical distance between intake and free delivery of pump is static head. Allowance for friction, power loss, propeller slip, and issuing velocity is made for calculating the overall discharge head. (Pryor, 1963)

discharge station A place where bulk materials are removed from a conveyor.

discomfort glare A sensation of annoyance, or in extreme cases pain, caused by high or nonuniform distribution of brightness in the field of view. Discomfort glare is a measure of discomfort or annoyance only. Cf: *disability glare.* See also: *glare.*

disconformity An unconformity in which the bedding planes above and below are essentially parallel, indicating a considerable interval of erosion (or sometimes of nondeposition), and usually marked by a visible and irregular or uneven erosion surface of appreciable relief. The term formerly included what is now known as paraconformity. Syn: *parallel unconformity; nonangular unconformity.* See also: *angular unconformity.* (AGI, 1987)

discontinuity (a) An abrupt change in the physical properties of adjacent materials in the Earth's interior. (Mather, 1964) (b) Any interruption in the normal physical structure or configuration of a part, such as cracks, laps, seams, inclusions, or porosity. A discontinuity may or may not affect the usefulness of a part. (ASM, 1961)

discontinuity lattice *lattice.*

discontinuous deformation Deformation of rocks accomplished by rupture rather than by flowage.

discontinuous reaction series A reaction series in which early-formed crystals react with later liquid by means of abrupt phase changes; e.g., the minerals olivine, pyroxene, amphibole, and biotite form a discontinuous reaction series. Cf: *continuous reaction series.* (AGI, 1987)

discordance A lack of parallelism between contiguous strata, e.g., angular unconformity. (Standard, 1964; Fay, 1920)

discordant (a) Said of a contact between an igneous intrusion and the country rock that is not parallel to the foliation or bedding planes of the latter. (AGI, 1987) (b) Structurally unconformable; said of strata lacking conformity or parallelism of bedding or structure. Cf: *concordant.* Syn: *unconformable.* (AGI, 1987)

discordant bedding *crossbedding.*

discovery (a) In mining, the term may be defined as knowledge of the presence of the valuable minerals within the lines of the location or in such proximity thereto as to justify a reasonable belief in their existence. But in all cases there must be a discovery of mineral, in both lode and placer claims, as distinguished from mere indications of mineral. In other words, in a lode location there must be such a discovery of mineral as gives reasonable evidence of the fact either that there is a vein or lode of rock in place carrying the valuable mineral; or, if it be claimed as placer ground, that it is valuable for such mining. (Ricketts, 1943) (b) Pac. The first finding of the mineral deposit in place upon a mining claim. A discovery is necessary before the location can be held by a valid title. The opening in which it is made is called discovery shaft, discovery tunnel, etc. The finding of mineral in place as distinguished from float rock constitutes discovery. See also: *mine.* (Fay, 1920)

discovery claim A claim containing the original discovery of exploitable mineral deposits in a given locale, which may lead to claims being made on adjoining areas. (AGI, 1987)

discovery vein The original mineral deposit on which a mining claim is based. Cf: *secondary vein; discovery claim.* (AGI, 1987)

discretization In kriging, the process of approximating the area of a block by a finite array of points.

disequilibrium assemblage An association of minerals not in thermodynamic equilibrium. (AGI, 1987)

dish (a) *pan; gold pan.* (b) The landowner's part of the ore. (Fay, 1920) (c) Gold-bearing gravel or other material found by panning.

disharmonic fold A fold that varies noticeably in profile form in the various layers through which it passes. Ant: harmonic fold. (AGI, 1987)

disharmonic folding Folding in which there is an abrupt change in fold profile when passing from one folded surface or layer to another. It is characteristic of rock layers that have significant contrasts in viscosity. An associated structure is décollement. Ant: harmonic folding. Cf: *decollement.* (AGI, 1987)

disintegrate (a) To break up by the action of chemical and/or mechanical forces. (AGI, 1987) (b) To separate or decompose into fragments; to break up; hence, to destroy the wholeness, unity, or identity. (Ballard, 1955)

disintegration The breaking up and crumbling away of a rock, caused by the action of moisture, heat, frost, air, and the internal chemical reaction of the component parts of rocks when acted upon by these surface influences. See also: *mechanical weathering; chemical weathering.* Cf: *putrefaction.*

disintegrator (a) A mill for comminuting materials to a fine dry powder such as by impact breaker. (Nelson, 1965) (b) A machine for reducing by means of impact the particle size of the coal or pitch binder, or both. Also called beater. Cf: *impact mill; hammermill.* (BS, 1962)

disk *tappet.*

disk-and-cup feeder A reagent dispenser used in the flotation process. Cups, mounted around the periphery of a slowly rotating disk driven by a fractional horsepower motor, dip into a reservoir of reagent and upon rising deliver a closely controlled quantity to the process, usually to conditioners. (Pryor, 1963)

disk coal cutter A coal cutter whose cutting unit consists of a disk or wheel, armed at its periphery with cutters. The first disk machine, with detachable picks, was patented in 1861. The disk coal cutter is obsolescent. (Nelson, 1965)

disk fan An axial-flow fan with a series of blades formed by cutting and bending flat sheets or plates. When rotated, the disk imparts to the air a motion along the axis of the fan shaft. (Strock, 1948)

disk feeder A feeder consisting of a rotating horizontal metal disk under the opening of a bin such that the rate of turning or opening of the gate governs the quantity delivered. Also called rotary table feeder, rotary feed table. See also: *plate feeder.* (ACSG, 1963)

disk filter A continuous dewatering filter in which the membrane (filter cloth) is stretched on segments of a disk. These disks rotate through a tank of slurry. The vacuum inside the disk draws the liquid through the cloth to discharge; the solids forming a cake on the filter cloth are lifted clear of the slurry tank and separately discharged, by application of air pressure behind the filter cloth. (Pryor, 1963)

disk grizzly *grizzly.*

disk mill A laboratory grinding mill with two circular plates almost parallel, of which one is fixed while the other rotates. Ore fed centrally between the plates is ground and discharged peripherally. The disk breaker (obsolescent) had two saucer-shaped disks working in similar fashion. (Pryor, 1963)

dislocation (a) Displacement. (AGI, 1987) (b) The shifting of the relative position of a boulder in a borehole or of the rock on either side of a crack or fissure cutting across a borehole. (Long, 1960) (c) The offset in a borehole. Also called deviation; throw. (Long, 1960) (d) A general term to describe a break in the strata, for example, a fault. A washout is a disturbance but not a dislocation. (Nelson, 1965) (e) The displacement of rocks on opposite sides of fracture. (Pryor, 1963) (f) In metallurgy, the structural defect in metal or crystal produced by distortion. (Pryor, 1963) (g) A linear crystal defect. (Van Vlack, 1964)

disorder A state where different ions are distributed randomly in identical structural positions. See also: *long-range order.* Ant: order. Cf: *crystal defect; volume defect.*

dispatcher (a) An employee who controls or keeps track of the traffic on haulageways and informs workers when to move trains or locomotives. (BCI, 1947) (b) *motor boss.* (c) A person or electronic device that routes haulage trucks to shovels or directs trucks from shovels to one of several destinations: ore pass, crusher, or spoil embankment.

dispatching system A system employing radio, telephones, and/or signals (audible or visual) for orderly and efficient control of the movements of trains of cars in mines.

dispersant *dispersing agent.*

dispersed element An element that is generally too rare and unconcentrated to become an essential constituent of a mineral, and that therefore occurs principally as a substituent of the more abundant elements. (AGI, 1987)

dispersed pattern In geochemical prospecting, a pattern or the distribution of the metal content of soil, rock, water, or vegetation. (AGI, 1987)

disperse medium Homogeneous phase (gas, liquid, or solid) through which particles are dispersed to form a relatively stable sol. Mainly descriptive of colloidal dispersion. See also: *disperse system.* (Pryor, 1963)

disperse system A two-phase system consisting of a dispersion medium and a dispersed phase; a dispersion. (Webster 3rd, 1966)

dispersibility of dust The ease with which dust is raised into suspension. (Sinclair, 1958)

dispersing agent (a) A material that increases the stability of a suspension of particles in a liquid medium by deflocculation of the primary particles. Syn: *deflocculating agent.* (ASM, 1961) (b) Dispersant, deflocculating, or peptizing agent. One that acts to prevent adherence of particles suspended in fluid, and delays sedimentation. (Pryor, 1963) (c) Reagents added to flotation circuits to prevent flocculation, esp. of objectionable colloidal slimes. Sodium silicate is frequently added for this purpose, and there is some indication that it has value in coal froth flotation where a high percentage of clay slimes is present. (Mitchell, 1950)

dispersion (a) The fairly permanent suspension of finely divided but undissolved particles in a fluid. (API, 1953) (b) The creation of a dispersion by deflocculation. (BS, 1962) (c) The separation of polychromatic light (e.g., white light; sunlight) into its component wavelengths. (d) The degree of inequality of refractive index and refraction of light of various colors. Syn: *refractive index.* (e) Change in the angle between optic axes in biaxial crystals due to change in refractive indices with change in wavelength of light. (f) Change in the orientation of optical directions with respect to crystallographic directions in monoclinic or triclinic minerals. See also: *index of refraction; optic axis.* (g) Distortion of the shape of a seismic-wave train because of variation of velocity with frequency. (h) Advance or recession of peaks and troughs from the beginning of the seismic wave as it travels. (i) Breaking down or separation of soil aggregates into single grains.

dispersion halo A region surrounding an ore deposit in which the ore-metal concentration is intermediate between that of the ore and that of the country rock. (AGI, 1987)

dispersion pattern The pattern of distribution of chemical elements, esp. trace elements, in the wall rocks of an orebody or in the surface materials surrounding it. Cf: *aureole; halo.* (AGI, 1987)

dispersoid A body dispersed in a liquid.

disphenoid (a) In crystallography, a solid bounded by eight isosceles triangles. (Standard, 1964) (b) A closed crystal form of four faces, each an isosceles triangle and derived from a bipyramid by suppressing alternate faces. It differs from a tetrahedron, the four faces of which are equilateral triangles, by lower symmetry. See also: *sphenoid.* Syn: *bisphenoid.* Adj. disphenoidal.

displaced seam A coal seam that has been dislocated by a fault. (Nelson, 1965)

displacement (a) The lateral movement of a point, usually at the surface, during subsidence. (Nelson, 1965) (b) The volume displaced by the net area of the piston multiplied by the length of the stroke. (Lewis, 1964) (c) Sometimes used as a syn. for offset deflection; deviation; dislocation; throw. (Long, 1960) (d) A general term for the change in position of any point on one side of a fault plane relative to any corresponding point on the opposite side of the fault plane. (Ballard, 1955) (e) The capacity of an air compressor, usually expressed in cubic feet of air per minute. (Long, 1960) (f) The word displacement should receive no technical meaning, but is reserved for general use; it may be applied to a relative movement of the two sides of the fault, measured in any direction, when that direction is specified; for instance, the displacement of a stratum along a drift in a mine would be the distance between the two sections of the stratum measured along the drift. The word dislocation will also be most useful in a general sense. (Fay, 1920) (g) The volume of liquid delivered by a single stroke of a pump piston. (Long, 1960) (h) Any shift in the position of an image on a photograph that does not alter the perspective characteristics of the photograph. It may be caused by the relief of the objects photographed, the tilt of the photograph, changes of scale, or atmospheric refraction. Cf: *distortion.* (AGI, 1987) (i) A general term for the relative movement of the two sides of a fault, measured in any chosen direction; also, the specific amount of such movement. Syn: *dislocation.* (AGI, 1987)

displacement pump One in which compressed air or steam, applied in pulses, drives out water entering the pump chamber between pulses, a nonreturn valve preventing reverse flow. (Pryor, 1963)

displacement-type float A device for measuring the liquid level in sumps or vessels. It consists of a float, whose vertical height is greater than the level range being measured and whose weight is such that it would sink in the fluid if not supported. It is placed in a float chamber and supported in such a way that as the liquid level rises around the displacer float it creates a buoyant force equal to the weight of the liquid displaced. This force is measured, and since it is proportional to level, the force measurement becomes a level measurement. The device is used on sumps containing high-gravity slurries. (Nelson, 1965)

displacive transformation A change in crystal summetry as a result of changes in bond length or bond angles (as contrasted to reconstructive transformations). The short-range order is unchanged; the long-range order is changed. Cf: *dilational transformation; reconstructive transformation; rotational transformation.* (Van Vlack, 1964)

disposable respirator A respirator for which maintenance is not intended and that is designed to be discarded after excessive resistance, sorbent exhaustion, physical damage, or end-of-service-life renders it unsuitable for use. Examples of this type of respirator are a disposable half-mask respirator or a disposable escape-only self-contained breathing apparatus. (ANSI, 1992)

disrupted seam A coal seam intersected by a fault or where its continuity is excessively broken. (Nelson, 1965)

disruptive Applied to that kind of force exerted by an explosive that tends to shatter the rock into fragments. (Fay, 1920)

disseminated Said of a mineral deposit (esp. of metals) in which the desired minerals occur as scattered particles in the rock, but in sufficient quantity to make the deposit an ore. Some disseminated deposits are very large. Cf: *impregnated.* (AGI, 1987)

disseminated crystals Crystals that are found not attached to the mother rock, sometimes with well-developed faces and doubly terminated.

disseminated deposit A type of mineral deposit in which the minerals occur as small particles or veinlets scattered through the country rock. (Nelson, 1965)

dissociation constant The equilibrium constant for a dissociation reaction, defined as the product of activities of the products of dissociation divided by the activity of the original substance. When used for ionization reactions, it is called an ionization constant; when it refers to a very slightly soluble compound, it is called a solubility product. (AGI, 1987)

dissolution (a) The act or process of dissolving or breaking up, as a separation into component parts. (Webster 3rd, 1966) (b) The taking up of a substance by a liquid with the formation of a homogeneous solution. (CTD, 1958)

dissue Corn. To break the rock from the walls of a rich lode in order to move the ore without taking with it much gangue. (Standard, 1964)

distance blocks Wooden blocks placed in between the main spears and the side pump rods by which the proper distance between them is adjusted. (Fay, 1920)

distance lag In flotation, a delay attributable to the transport of material or the finite rate of propagation of a signal or condition. Syn: *velocity lag.* (Fuerstenau, 1962)

distaxy A mineral overgrowth not in crystallographic continuity with its core or nucleus. Cf: *epitaxy; syntaxy.*

disthene A former name for kyanite.

disthenite A metamorphic rock composed almost entirely of kyanite (disthene) and some quartz, often associated with magnetiferous quartzite and amphibolite. (AGI, 1987)

distillation (a) The process of decomposition whereby the original chitinous material of certain fossils has lost its nitrogen, oxygen, and hydrogen, and is now represented by a film of carbonaceous material. Syn. for carbonization. (AGI, 1987) (b) The process of heating a substance to the temperature at which it is converted to a vapor, then cooling the vapor, and thus restoring it to the liquid state. See also: *destructive distillation; fractional distillation.* (Shell) (c) A process of evaporation and recondensation used for separating liquids into various fractions according to their boiling points or boiling ranges. (CTD, 1958)

distillation furnace A reverberatory heating furnace in which the charge is contained in a closed vessel and does not come in contact with the flame. (Fay, 1920)

distinctive mineral *varietal mineral.*

distorted crystal A crystal whose faces have developed unequally, some being larger than others. Some distorted crystal forms are drawn out or shortened, but the angle between the faces remains the same. See also: *deformed crystal.*

distortion (a) The change in shape and size of a land area on a map due to the flattening of the curved Earth surface to fit a plane. Distortion is inevitable and is controlled in the development of a projection to produce the characteristics of equal area, conformality, or equidistance. (AGI, 1987) (b) Any shift in the position of an image on a photograph that alters the perspective characteristics of the photograph. It may be caused by lens aberration, differential shrinkage of film or paper, or motion of the film or camera. Cf: *displacement.* (AGI, 1987)

distortional wave *shear wave; S wave; transverse wave; secondary wave.*

distributing magazine A place or building, either near the mine entrance or underground, in which explosives are stored for current use. Only one day's supply should be kept at such points. The main supply of explosives is kept in a magazine generally a safe distance from the mine or any mine buildings. (Kentucky, 1952)

distribution curve *partition curve; Tromp distribution curve.*

distribution factor *partition factor.*

distributive fault *step fault.*

distributive province The environment embracing all rocks that contribute to the formation of a contemporaneous sedimentary deposit, including the agents responsible for their distribution. Cf: *provenance.* (Schieferdecker, 1959)

distributor box Box that receives feed from launder, pipe, or pump and splits it into parallel mill circuits. Box attached to deck of shaking table, which receives sands and distributes them along top of deck at feed end. (Pryor, 1963)

district (a) In the States and Territories of the United States west of the Missouri River (prior to 1880), a vaguely bounded and temporary division and organization made by the inhabitants of a mining region. (b) A limited area of underground workings. (Nelson, 1965) (c) A coal mine is generally divided into sections or districts for purposes of ventilation and daily supervision. (Nelson, 1965) (d) An underground section of a coal mine served by its own roads and ventilation ways; a section of a coal mine. (CTD, 1958)

disturbance A term used by some geologists for a minor orogeny, e.g., the Palisades disturbance. Schuchert (1924) used revolution for a major orogeny at the end of an era, and disturbance for an orogeny within an era; this usage is obsolete. (AGI, 1987)

disturbed area An area where vegatation, topsoil, or overburden is removed or upon which topsoil, spoil, coal processing waste, underground development waste, or noncoal waste is placed by surface coal mining operations. Those areas are classified as disturbed until reclamation is complete and the performance bond or other assurance of performance is released.

disused workings Workings that are no longer in operation but that are not classified as abandoned. (BS, 1963)

ditch (a) A drainage course in a mine. (BCI, 1947) (b) An artificial channel to convey water for use in mining. Cf: *flume.* (c) The drainage gutter along gangways and openings in anthracite mines. (d) In rotary drilling, a trough carrying mud to a screen. (Nichols, 1976) (e) The artificial course or trough in which the drill circulation fluid is conducted from the collar of the borehole to the sump. To dump and discard contents of a bailer, without taking a sample, into a ditch leading away from the collar of a borehole. Syn: *canal; chute; ditch.* Cf: *trench.* (Long, 1960)

ditch drain A gutter excavated in the floor of a gangway or airway to carry the water to the sump, or out to the surface.

ditcher (a) A mobile tracked machine fitted with an endless chain of buckets used for shallow vertically sided trenching. (Nelson, 1965) (b) A drill mounted on a frame that rotates about a central axis. It is used to cut circular trenches for the production of large grindstones. Also called circle cutting drill. (Fay, 1920)

ditching The digging or making of a ditch by the use of explosives. See also: *propagated blast.*

ditching dynamite A nitroglycerin type explosive esp. designed to propagate sympathetically from hole to hole in ditch blasting.

ditch water The stale or stagnant water collected in a ditch.

ditch wiring The method of connecting electric blasting caps in such a way that the two free ends can be connected at one end of the line of holes. (Fay, 1920)

dithiocarbamate A flotation collector agent of the general formula $X_2N.CS.SM$, X being hydrogen, aryl, or alkyl radical. (Pryor, 1963)

dithionate process A process for extracting manganese from low-grade oxide ores. The manganese ore is leached with dilute sulfur dioxide gas in the presence of calcium dithionate solution, the manganese being recovered from solution by precipitation with slaked lime and then nodulized or sintered. (Osborne, 1956)

dithiophosphates In mineral processing, flotation collector agents, marketed as Aerofloats by the American Cyanamid Co.

dithizone Diphenylthiocarbazone. Used in geochemical prospecting to detect traces of certain metals. (Pryor, 1963)

diurnal fluctuations Variations occurring within a 24-h period and related to the rotation of the earth. (Hy, 1965)

diurnal inequality (a) The departure easterly or westerly from the mean value of the declination for the day. (Mason, 1951) (b) In tides, the difference in height and/or time of the two high waters or of the two low waters of each day; also, the difference in velocity of either of the two flood currents or of the two ebb currents of each day. (Hy, 1965)

diurnal variation (a) The daily variation in the earth's magnetic field. (AGI, 1987) (b) In tides, having a period or cycle of approx. 1 lunar day (24.84 solar hours). The tides and tidal currents are said to be diurnal when a single flood and single ebb occur each lunar day. (Hy, 1965)

diver Small plummet, so adjusted as to density that by rising or falling it can be used to show whether specific gravity of pulp is above or below a desired control point. If pulp is opaque, diver can initiate magnetic signal, or in a pulp containing magnetic material can carry radioactive marking material. (Pryor, 1963)

diversion valve A valve that permits flow to be directed into any one of two or more pipes. (Nichols, 1976)

diversity factor The ratio of the sum of the individual maximum loads during a period to the simultaneous maximum loads of all the same units during the same period. Always unity or more. (Strock, 1948)

divided cell A cell containing a diaphragm or other means for physically separating the anolyte and catholyte. (Lowenheim, 1962)

divider Cross-steel or timber piece in a circular or rectangular shaft. Such pieces serve to divide the shaft into compartments and may also carry the cage guides, etc. See also: *bunton.* (Nelson, 1965)

diviner Dowser. (AGI, 1987)

diving bell A watertight, bell-shaped steel chamber that can be lowered to or raised from a freshwater or seawater bed by a crane. It is open at the bottom and filled with compressed air, so that persons can prepare foundations and undertake similar construction work underwater. (Hammond, 1965)

divining A method of searching for water or minerals by holding a hazel fork (or other device) in the hands, and the free end is said to bend downward when a discovery is made. In the Middle Ages, the divining rod was closely associated with the mine surveying profession. The water diviner has not succeeded when submitted to impartial scientific tests. (Nelson, 1965)

divining rod Traditionally, a forked wooden stick, cut from a willow or other water-loving plant, used in dowsing. It supposedly dips downward sharply when held over a body of ground water or a mineral deposit, thus revealing the presence of these substances. Syn: *witching stick; wiggle stick; dowsing rod; twig.* Cf: *water witch; waterfinder.* (AGI, 1987)

divisional plane A general term that includes joints, cleavage, faults, bedding planes, and other surfaces of separation. (AGI, 1987)

division method One of three recognized methods for determining the average velocity of airflow in a mine roadway by anemometer. This is the precise method of determining the mean velocity of the air current. Cf: *single-spot method; traversing method.*

dixanthogen A breakdown product of xanthate collectors (flotation agents) with some residual value for that purpose. (Pryor, 1963)

dixenite A trigonal mineral, $CuMn_{14}Fe(AsO_3)_5(SiO_4)_2(AsO_4)(OH)_6$; forms nearly black aggregates of thin folia; at Långban, Sweden.

djalmaite A former name for uranmicrolite. See also: *microlite.*

djurleite A monoclinic mineral, $Cu_{31}S_{16}$; X-ray pattern is similar to, but distinct from, chalcocite.

D.L.T. reagents Condensation products of ethanolamine and higher fatty acids, used as flotation agents (collectors). (Pryor, 1963)

dneprovskite *wood tin.*

dobie man *blaster.*

Dobson prop A hydraulic prop that is basically a self-contained hydraulic jack with an integral pump unit built into the prop. It is designed to yield at 25 st (22.7 t) and has a setting load of 6 st (5.4 t). (Nelson, 1965)

Dobson support system A self-advancing support for use on longwall faces. One unit embodies three props. The front prop, which is attached to the face conveyor, carries two roof bars side by side that give cantilever support over the conveyor track. The two rear props are mounted on a common floor bar and carry a single roof bar that passes between the two front bars. The front prop is attached to the rear structure only by the advancing ram within the box structure of the floor bar. (Nelson, 1965)

docket A pay ticket containing particulars of shifts worked, coal filled, yardage driven, and other work done, including the total wages less deductions. (Nelson, 1965)

doctor (a) To treat a poor-quality carbon with substances such as oil, wax, gutta-percha, solder, gum, or resin, to camouflage its defects, hence changing its appearance to make it look like a better grade stone. Also called dope. (Long, 1960) (b) A makeshift, temporary repair. (Long, 1960) (c) As used in the mining industry, to salt. (Long, 1960)

Dodd buddle A round table resembling in operation a Wilfley table, and also like the Pinder concentrator except that it is convex instead of concave. The table does not revolve but has a peripheral jerking motion imparted to it circumferentially by means of a toggle movement. (Liddell, 1918)

dodecahedral cleavage In isometric minerals, a cleavage parallel to the faces of a rhombic dodecahedron 110, e.g., sphalerite.

dodecahedron (a) An isometric form composed of 12 equal rhombic faces, each parallel to 1 axis and intersecting the other 2 axes at equal distances, specif. named the rhombic dodecahedron. See also: *pyritohedron.* (Fay, 1920) (b) Any solid with 12 symmetrically equivalent faces, e.g. deltoid; pyritohedron. (c) The isometric form 110, the rhombic dodecahedron. (d) Brazilian diamonds with the dodecahedral form, also called Brazilian stone.

dodecant Each 12th of crystal space defined by a trigonal or hexagonal *c* axis and its orthogonal three coplanar *a* axes. Cf: *octant.*

Dodge crusher Similar to the Blake crusher, except the movable jaw is hinged at the bottom. Therefore the discharge opening is fixed, giving a more uniform product than the Blake with its discharge opening varying every stroke. This type of jaw crusher gives the greatest movement on the largest lump. (Liddell, 1918)

Dodge pulverizer A hexagonal barrel revolving on a horizontal axis, containing perforated die plates and screens. Pulverizing is done by steel balls inside the barrel. (Liddell, 1918)

dog (a) An iron bar, spiked at the ends, with which timbers are held together and steadied. (b) A short, heavy iron bar, used as a drag behind a car or trip of cars when ascending a slope to prevent them running back down the slope in case of an accident; a drag. (c) *casing dog.* (d) A trigger that limits the advance of a traversing table. (ACSG, 1963) (e) Any of various devices for holding, gripping, or fastening something. See also: *chair; dog; catch; wing.* (Webster 3rd, 1966) (f) A drag for the wheel of a vehicle. (Webster 3rd, 1966) (g) A device attached to the workpiece by means of which the work is revolved. (ACSG, 1963)

dog-and-chain An iron lever with a chain attached by which props are withdrawn. (Fay, 1920)

dogger (a) A large, irregular nodule, usually of clay ironstone, sometimes containing fossils, found in a sedimentary rock, as in the Jurassic rocks of Yorkshire, England. (AGI, 1987) (b) An English term for any large, lumpy mass of sandstone longer than it is broad, with steep rounded sides. (AGI, 1987)

doghole A small opening from one place in a coal mine to another; smaller than a breakthrough. Syn: *monkey hole.*

doghole mine Name applied to small coal mines that employ fewer than 15 miners. The so-called dogholes are most numerous in Kentucky, but there are many in Virginia and West Virginia.

dogholes *doghole mine.*

dog hook (a) A strong hook or wrench for separating iron boring rods. (Fay, 1920) (b) An iron bar with a bent prong used in handling logs. (Fay, 1920)

doghouse (a) The structure enclosing the drill platform and machine. (Long, 1960) (b) A small shelter in which members of a drill crew change clothing. Cf: *changehouse.* (Long, 1960) (c) *forechamber.* (d) Any enclosure or small chamber in a mine used for storage or resting.

dog iron A short bar of iron with both ends pointed and bent down so as to hold together two pieces of wood into which the points are driven; or one end may be bent down and pointed, while the other is formed into an eye, so that if the point be driven into a log, the other end may be used to haul on. (Zern, 1928)

dogleg (a) An abrupt angular change in course or direction, as of a borehole or in a survey traverse. Also, a deflected borehole, survey course, or anything with an abrupt change in direction resembling the hind leg of a dog. (AGI, 1987) (b) An abrupt bend or kink in a wire rope or cable. (Long, 1960) (c) An abrupt bend in a path, piping system, or road. (Long, 1960)

dogleg severity Same as deflection angle; hole curvature. (Long, 1960)

dogs (a) Eng. In the plural; bits of wood at the bottom of an air door. (Fay, 1920) (b) See also: *dog.*

dog spike A spike generally used to fasten rails to the sleepers when laying track. Their length should be 1/2 in (1.27 cm) less than the depth of the sleeper into which they are being pounded. (Sinclair, 1959)

dog-tooth spar Calcite with sharp scalenohedral termination.

dogtooth spar Calcite in acute scalenohedral crystals facing like dogs' teeth into an open cavity or vein. See also: *scalenohedron.*

doit Eng. Foulness, or damp air. (Fay, 1920)

dole A division of a parcel of ore. Also spelled dol. (Fay, 1920)

dolerite (a) In the United States, a syn. of diabase. (AGI, 1987) (b) In British usage, the preferred term for what is called diabase in the United States. Etymol: Greek doleros, "deceitful," in reference to the fine-grained character of the rock that makes it difficult to identify megascopically. Cf: *diorite; trap.* Syn: *whin.* (AGI, 1987)

doleritic (a) Of or pertaining to dolerite. See also: *ophitic.* (AGI, 1987) (b) A preferred syn. of ophitic in European usage. (AGI, 1987)

dolerophanite A monoclinic mineral, $Cu_2(SO_4)O$; brown; reported at Mt. Vesuvius. Also spelled dolerophane.

dolina *doline.*

doline A syn. of sinkhole. Also spelled: dolina. Etymol: German transliteration from Slovene "dolina," "valley.". (AGI, 1987)

dollie *dolly.*

dolly (a) A trucklike platform, with an attached roller, used in shifting heavy loads. (b) A counterbalance weight sometimes used in a hoisting shaft. (Nelson, 1965) (c) To break up quartz with a piece of wood shod with iron, in order to be able to wash out the gold. (Fay, 1920) (d) A tool used to sharpen drills. (Stauffer, 1906) (e) *car.* (f) A wooden disk for stirring the ore in a dolly tub, in order to concentrate the ore by the tossing and packing process. See also: *dolly tub.* (Standard, 1964)

dolly tub A large wooden tub used for the final washing of valuable minerals separated by water concentration in ore dressing. See also: *tossing; dolly.* (CTD, 1958)

dolly wheels Pairs of wheels used to support rods of a Cornish pump working on a slope. (Pryor, 1963)

doloma Calcined dolomite, that is a mixture of the oxides CaO and MgO. (Dodd, 1964)

dolomite (a) A trigonal mineral, $[CaMg(CO_3)_2]$; forms saddle-shaped rhombohedra having rhombohedral cleavage; white to pale tints; in large beds as dolostone and dolomitic marble, also in veins and in serpentinite; a source of magnesium and dimension stone. Syn: *bitter spar; pearl spar; magnesian spar; rhomb spar.* (b) The mineral group ankerite, dolomite, kutnohorite, minrecordite, and norsethite. (c) A carbonate sedimentary rock consisting of more than 50% to 90% mineral dolomite, depending upon classifier, or having a Ca:Mg ratio in the range 1.5 to 1.7, or having an MgO equivalent of 19.5% to 21.6%, or having a magnesium-carbonate equivalent of 41.0% to 45.4%. Dolomite beds are associated and interbedded with limestone, commonly representing postdepositional replacement of limestone. Syn: *dolostone; dolomite rock.*

dolomite limestone *dolomitic limestone.*

dolomite marble A crystalline variety of limestone, containing in excess of 40% of magnesium carbonate as the dolomite constituent.

dolomite rock *dolomite.*

dolomitic (a) Dolomite-bearing, or containing dolomite; esp. said of a rock that contains 5% to 50% of the mineral dolomite in the form of cement and/or grains or crystals. (AGI, 1987) (b) Containing magnesium; e.g., dolomitic lime containing 30% to 50% magnesium. (AGI, 1987)

dolomitic limestone (a) A limestone that has been incompletely dolomitized. (AGI, 1987) (b) A limestone in which the mineral dolomite is conspicuous, but less abundant than calcite. Syn: *dolomite limestone.* Cf: *magnesian limestone.*

dolomitization The process by which limestone is wholly or partly converted to dolomite rock or dolomitic limestone by the replacement of the original calcium carbonate (calcite) by magnesium carbonate (mineral dolomite), usually through the action of magnesium-bearing water (seawater or percolating meteoric water). It can occur penecontemporaneously or shortly after deposition of the limestone, or during lithification at a later period. Syn: *dolomization.* (AGI, 1987)

dolomization *dolomitization.*

doloresite A monoclinic mineral, $H_8V_6O_{16}$; an alteration product of montroseite in sandstone from the Colorado Plateau; named for the Dolores River, CO.

dolostone A term applied by some petrologists to rock consisting primarily of the mineral dolomite. Syn: *dolomite.*

dolphin A fixed mooring in the open sea formed of a number of piles, or a guide for ships entering a narrow harbor mouth. (Hammond, 1965)

domain (a) A substructure in a ferromagnetic material within which all of the elementary magnets (electron spins) are held aligned in one direction by interatomic forces; if isolated, a domain would be a saturated permanent magnet. (ASM, 1961) (b) A region within a grain of magnetically ordered mineral, within which the spontaneous magnetization has a constant value characteristic of the mineral composition and temperature. Syn: *magnetic domain.* (AGI, 1987)

dome (a) Roof of a furnace that is roughly hemispherical in shape. (b) The steam chamber of a boiler. Cf: *air dome.* (Long, 1960) (c) An uplift or anticlinal structure, either circular or elliptical in outline, in which the rocks dip gently away in all directions. A dome may be small, such as a Gulf Coast salt dome, or many kilometers in diameter. Domes include diapirs, volcanic domes, and cratonic uplifts. Type structure: Nashville Dome, TN. See also: *pericline; arch; salt dome.* Syn: *dome structure; structural dome; quaquaversal fold.* Less-preferred syn: swell. Cf: *basin.* (AGI, 1987) (d) A general term for any smoothly rounded landform or rock mass, such as a rock-capped mountain summit, that roughly resembles the dome of a building; e.g., the rounded granite peaks of Yosemite, CA. The term is also applied to broadly up-arched regions, such as the English Lake District or the Black Hills of South Dakota. (AGI, 1987) (e) A large magmatic or migmatitic intrusion whose surface is convex upward and whose sides slope away at low but gradually increasing angles. Intrusive igneous domes include laccoliths and batholiths; the term is used when the evidence as to the character of the lower parts of the intrusion is insufficient to allow more specific identification. (AGI, 1987) (f) An open crystal form of four parallel faces that intersect the *c* axis and one other; incorrectly called a horizontal prism. Adj. domatic. (g) A symmetrical structural uplift having an approx. circular outline in plan view, and in which the uplifted beds dip outward more or less equally in all directions from the center, which is both the highest point of the structure and locally of the uplifted beds. (h) A mountain having a smoothly rounded summit of rock that resembles the cupola or dome on a building. (AGI, 1987) (i) An open crystal form consisting of two parallel faces that truncate the intersections of two sets of pinacoids and are astride a symmetry plane.

domestic coal (a) Coal for use around colliery in miners' houses or for local sale. (Zern, 1928) (b) Sized coal for use in houses. See also: *house coal.* (Zern, 1928) (c) Coal used in country of origin; not for foreign consumption. (Zern, 1928)

domestic sampling Routine sampling by mine officials for systematic control of mining operations. See also: *development sampling.* (Nelson, 1965)

dome structure *dome.*

dome theory A theory that strata movements caused by underground excavations were limited by a kind of dome that had for its base the area of excavation, and that the movements diminished as they extended upward from the center of the area. See also: *harmless depth theory; normal theory.* (Nelson, 1965)

domeykite An isometric mineral, Cu_3As; forms reniform and botryoidal masses and disseminated grains. See also: *white copper.*

dominant vitrain A field term to denote, in accordance with an arbitrary scale established for use in describing banded coal, a frequency of occurrence of vitrain bands comprising more than 60% of the total coal layer. Cf: *abundant vitrain; moderate vitrain; sparse vitrain.* (AGI, 1987)

donarite An explosive consisting of 70% ammonium nitrate, 25% trinitrotoluol, and 5% nitroglycerin. (Hackh, 1944)

donbassite A possible mineral species in the chlorite group.

donkey engineer In anthracite and bituminous coal mining, a general term for the attendant of a small auxiliary engine, powered by steam or compressed air, used to drive pumps to drain sumps (pits in which excess water is collected) or supply water to boilers, or to operate a hoist for a shallow shaft. Also called donkey runner. (DOT, 1949)

donkey hoist A small auxiliary hoisting drum and engine operated by steam, by compressed air, and sometimes by an electric motor or an internal-combustion engine. (Long, 1960)

donkey pump Any of several kinds of combined pump and steam engine. It may be operated independently of the engine; used to supply water to a boiler, drain sumps, etc.

donkey runner *donkey engineer.*

donor An exploding charge producing an impulse that impinges upon an explosive acceptor charge. (Meyer, 1981)

doodlebug (a) The essential treatment plant of a small dredge set on a pontoon. There is usually a hopper into which the dragline dumps its spoil and which may have a grizzly arrangement, according to the nature of the gravel. A water supply washes the contents of the hopper into a revolving screen, feeding the fines over riffled tables and rejecting the stones and oversize by means of a stacker. This treatment plant or washing unit can be floated in the excavation dug by a dragline and is the ideal unit to install when small-scale operations are to be carried out below water level or where it is not necessary to use dry opencast paddock methods. (Harrison, 1962) (b) Any one of a large number of unscientific devices with which it is claimed water, mineral, and oil deposits can be located. (AGI, 1987) (c) A popular term for any of various kinds of geophysical prospecting equipment.

dook (a) Scot. A mine or roadway driven to the dip, usually the main road. See also: *slope.* (b) Som. An underground inclined plane.

door A hinged or sliding frame or piece of wood, metal, stone or other material, generally rectangular, used for closing or opening an entrance or exit. Doors are placed in air passages of mines to prevent the ventilating current from taking a short cut to the upcast shaft, and to direct the current to the working face. See also: *air door; ventilation doors.* (Standard, 1964)

door boy *trapper.*

doorheads Scot. The roof or top of the workings at a shaft.

door tender One whose duty it is to open and close a mine door before and after the passage of a train of mine cars; a trapper. (Zern, 1928)

door trapper *door tender.*

door-type sampler A soil-sampling tube or barrel equipped with an auger-type cutting shoe and made to be rotated to obtain samples of sand, gravel, and other granular material. The body of the sampler is essentially a tube in which a small opening or window is machined and equipped with a covering, which can be latched shut while the sample is being taken. When the sampler is removed from the ground, the latch is released and the sample removed through the door or window. Syn: *window-type sample.* (Long, 1960)

dope (a) Individual, dry, nonexplosive ingredients that comprise a portion of an explosive formulation. (b) Absorbent material, as sawdust, infusorial earth, mica, etc., used in certain manufacturing processes, as in making dynamite. (Webster 2nd, 1960) (c) Heavy grease or other material used to protect or lubricate drill rods and/or open gears, chain and sprockets, etc. Also called gunk; rod dope; rod grease. (Long, 1960) (d) To apply a lubricant to drill rods, rod couplings, open gears, etc. (Long, 1960) (e) To doctor a drill diamond. See also: *doctor.* (Long, 1960) (f) A rubberlike compound applied to granite surfaces before inscriptions are cut in the granite. (AIME, 1960) (g) A viscous liquid put on pipe threads to make a tight joint. (Nichols, 1976) (h) Slang for mold lubricant. (ASTM, 1994)

Doppler A self-contained electronic system that makes use of Doppler's principle of frequency shift of waves emanating from a moving source. In this system, a pulsed or continuous wave is sent diagonally downward fore and aft, forward and backward, and the frequencies are compared in order to obtain the true ground speed. The heading is obtained from a special magnetic compass and is maintained by a directional gyro used as an integrating device. The distance thus determined has a precision better than one part in a thousand, which is sufficient for most geophysical surveys. (Dobrin, 1960)

dopplerite (a) A black gelatinous matter in peat and soft brown coal consisting of humic acids or their salts; has a detrimental effect on briquettes and coke. Syn: *torf-dopplerit; trof-dopplerit; Weichbraunkohlen-dopplerit; peat gel; brown-coal gel.* (b) A gel in peat composed of ulmins derived from plant carbohydrates by bacterial destruction of proteins. (c) An asphalt found in New Zealand and parts of Siberia.

dopplerite sapropel A variety of sapropel that contains much humic acid. (Tomkeieff, 1954)

dore Gold and silver bullion that remains in a cupelling furnace after the lead has been oxidized and skimmed off. Syn: *dore bullion.*

dore bullion *dore.*

dore metal *dore silver.*

dore silver Crude silver containing a small amount of gold, obtained after removing lead in a cupelling furnace. Syn: *dore bullion; dore metal.*

dornick (a) A piece of rock of approx. fist size. (b) A boulder of iron ore.

Dorr mill A tube mill designed for operation as a closed-circuit wet-grinding unit. See also: *tube mill.* (Dodd, 1964)

Dorr rake classifier A mechanical classifier consisting of an inclined settling tank and a rake-type conveying agitating mechanism. Feed introduced at the low end of the tank flows over a distributing apron toward the high end of the tank. The heavier materials of sand size settle into the rake zone and are raked up the slope and out the tank; slime and finer sands are carried over the rear wall in suspension. (Taggart, 1945)

Dosco miner A heavy, crawler-tracked, 200-hp (149-kW) cutter loader designed for longwall faces in seams over 4-1/2 ft (1.37 m) thick, and takes a buttock 5 ft (1.52 m) wide. The maximum cutting height is 7-1/2 ft (2.29 m). Dimensions: length 17-3/4 ft (5.41 m), width 4-1/2 ft (1.37 m), and height 3-3/4 ft (1.14 m). The cutterhead consists of seven cutter chains mounted side by side and can be moved up and down radially to cut the coal from roof to floor. It delivers the coal onto the face conveyor by a short cross conveyor. Capacity is more than 400 st (363 t) per machine and more than 4 st (3.63 t) output per worker per shift. (Nelson, 1965)

dose (a) A special charge used in a blast furnace, designed to cure furnace troubles. (Fay, 1920) (b) The amount of ionizing radiation energy absorbed per unit mass of irradiated material at a specific location, such as a part of the human body. Measured in reps, rems, and rads. (Lyman, 1964)

dot chart (a) A graphic aid used in the correction of station gravity for terrain effect, or for computing gravity effects of irregular masses. It can also be used in magnetic interpretation. (AGI, 1987) (b) A transparent graph-type chart used in the calculation of the gravity effects of various structures. The dots on the chart represent unit areas. (AGI, 1987)

double In rotary drilling, two pieces of drill rod left fastened together during raising and lowering. Also called couple; couplet. (Nichols, 1976)

double A. One of several terms (or letter symbols) used to designate medium-quality drill diamonds. (Long, 1960)

double-acting pump Scot. A pump that discharges at both forward and backward stroke.

double-acting ram *two-way ram.*

double-action press A press handling two operations each revolution. It carries two rams, one inside the other, so actuated that one motion immediately follows the other. (Crispin, 1964)

double-action pump A pump whose water cylinders are equipped with intake and discharge valves at each end; hence liquid is delivered by the pump on both the forward and the backward strokes of the pump piston. (Long, 1960)

double bank (a) To take up a claim parallel with and adjoining another claim containing an auriferous vein or deposit. (Fay, 1920) (b) Working with double sets or relays of persons. (Fay, 1920)

double block (a) A pair of multiple-sheave blocks reeved with rope or lines; a block and tackle. (Long, 1960) (b) Two pulleys or small sheaves mounted on a single shaft within a frame or shell. (Long, 1960)

double-burned Burned at a high temperature. This does not mean two firings. (AISI, 1949)

double-burned dolomite (a) Dolomite, with additions of oxides of iron, burned at a high temperature. This does not mean two firings. (ARI, 1949) (b) Clinkered dolomite. (AISI, 1949)

double core barrel A core barrel with an inner tube to hold the core. The inner tube does not rotate during drilling, thereby giving a better core recovery. Syn: *double-tube core barrel.* (Nelson, 1965)

double crib Eng. Two crib sets are placed back to back to form a two-compartment crib-lined raise. This technique is employed in weak ground in place of a double compartment separated by only a single dividing member.

double-cut sprocket For double-pitch roller chains, a sprocket having two sets of effective teeth. Tooth spaces for the second set are located midway between those of the first set. (Jackson, 1955)

double-deck gangway A method of silling or working out 10 ft (3 m) or so above the haulage level and forming a double-deck gangway. Chutes are constructed at intervals for ore transfer into mine cars. (Nelson, 1965)

double-double unit conveyor A longwall conveyor layout in which the center or main gate serves two double units, one on each side. The gate belts from each double unit deliver the coal onto cross-gate belts, which in turn deliver to the main gate conveyor and then by trunk conveyor or cars to the pit bottom. (Nelson, 1965)

double drum Hoisting device having two cable spools or drums rotating in opposite directions. (Long, 1960)

double-drum hoist A hoist with two drums that can be driven separately or together by a clutch. See also: *main-and-tail haulage.* (Nelson, 1965)

double-duo mill Has two pairs of rolls, mounted in one stand, one pair of rolls being higher than, and in advance of the other. (Osborne, 1956)

double-ended A term applied to any cutter loader that can cut both ways on a longwall face without turning at each end. This requires cutting units at both ends of the machine and duplication of other essential parts. (Nelson, 1965)

double-ended pick A diamond-shaped coal-cutter pick that is held in a special holder and chain. Both ends of the pick are used and then discarded. The type is used widely in the United States. See also: *coal-cutter pick.* (Nelson, 1965)

double-engine plane Loads are raised or lowered on a slope by a stationary engine and wire rope, as in an inclined shaft. There is a double track, or three rails and turnout; the descending trip assists the engine to raise the ascending trip, thus eliminating dead load, except rope. (Peele, 1941)

double entry (a) A pair of entries in flat or gently dipping coal so laid out that rooms can be driven from both entries; twin entries. See also: *entry.* (Fay, 1920) (b) A system of ventilation by which the air current is brought into the rooms through one entry and out through a parallel entry or air course. (Fay, 1920) (c) *main entry.*

double-entry room-and-pillar mining *room-and-pillar.*

double-entry zone test A test in which coal dust is placed in each of two connected parallel entries. (Rice, 1913-18)

doubleheader Applied to quarry equipment consisting of two independent channeling machines on a single truck, operated by one person.

double headings The driving of two coal headings, parallel and side by side, for development purposes. Usually a pillar 10 to 20 yd (9.14 to 18.29 m) wide is left between them. Formerly it was the practice at many coal mines to drive only one heading from which the stalls were turned off right and left. Two headings simplify ventilation and provide a second egress in an emergency. (Nelson, 1965)

double helical bag conveyor Closely spaced parallel tubes with right- and left-hand rounded helical threads rotating in opposite directions, on which bags or other objects are carried while being conveyed. Syn: *helical bag conveyor.*

double-inlet fan A centrifugal fan in which air enters the impeller on both sides. Also called double-width fan. (BS, 1963)

double jack (a) A two-hand heavy hammer, usually weighing about 10 lb (4.54 kg). Cf: *single jack.* (Long, 1960) (b) A double or twin-screw drill column. (Long, 1960)

double jigback An aerial ropeway in which two parallel track ropes are used, each carrying a carriage. (Nelson, 1965)

double-leg bucket elevator A type of bucket elevator having the carrying and return runs enclosed in separate casings between the head and boot. See also: *bucket elevator.*

double leg en masse conveyor A conveyor or elevator in which the carrying and return runs are operated in separated parallel and adjacent casings.

double load A charge in a borehole separated by a quantity of inert material for the purpose of distributing the effect, or for preventing part of the charge blowing out at a seam or fissure, in which case the inert material is placed so as to include the seam. (Fay, 1920)

double packing A form of strip packing that removes the localized high roof pressure from the vicinity of a roadway into a region in the goaf. It consists of two parallel packs adjacent to, and on each side of, the roadway, with the packs immediately at the roadsides built of such a width as to offer less resistance than wider and stronger packs (called buttress packs) more remote from the roadway. The principle of double packing was developed by D.W. Phillips in Great Britain. See also: *gate side pack; strip packing; yield-pillar system.* Cf: *single packing.* (Nelson, 1965)

double parting A bypass for mine cars. See also: *junction.* (Nelson, 1965)

double-pitch roller chain A roller chain having double the pitch of a standard roller chain, but otherwise having standard pins, bushings, and rollers. (Jackson, 1955)

double-pulley-drive conveyor A conveyor in which power is transmitted to the belt by two pulleys. (NEMA, 1961)

double-refracting spar *Iceland spar.*

double refraction Refraction shown by certain crystals that split the incident ray into two refracted rays, polarized in perpendicular planes. See also: *birefringence.* (Standard, 1964; AGI, 1987)

double-roll breaker (a) A coalbreaker that relies on the impact of special teeth for the bulk of reducing, rather than on the compression between the rolls. An important feature is adjustment, which may be made during operation. The machines are flexible enough to produce top size ranging from 6 to 14 in (15.4 to 35.56 cm). (Mitchell, 1950) (b) *double-roll crusher.*

double-roll crusher A machine for breaking down ore, rock, or coal and to discharge the crushed material below. See also: *spring-roll crusher; roll crusher; single-roll crusher.* (Nelson, 1965)

double-roll press A press in which pressure is applied by the mating of one or more pairs of indented rolls of equal diameter, revolving in opposite directions. Syn: *Belgian press.* (BS, 1962)

double-room system *room-and-pillar.*

double-round nose The cross-sectional view of the cutting face portion of a coring bit when its profile is a full half circle, the radius of which is one-half the wall thickness or kerf of the bit face. Cf: *single-round nose.* (Long, 1960)

double setting A leveling procedure whereby observations are duplicated by resetting the instrument to detect errors of measurement immediately. Also called dual setting. (BS, 1963)

double-shift places At collieries where there is only one recognized coal-winning shift in 24 h, it is a general practice to have double shifts (and sometimes treble) of workers in development headings that require a speedy advance. (Nelson, 1965)

double-spaced neutron log This method employs two neutron logging tools with different spacings between the source and the detector or two detectors in the same tool at different spacings. The spacings usually differ by 6 to 10 in (15.24 to 25.4 cm). The long-spaced log is run slowly and with a large time constant so that its statistical variation is not excessive, for the counting rate is much lower than that of the regular-spaced log. This technique has proved to be a potent technique for discriminating gas sands from oil sands in Venezuela. (Wyllie, 1963)

double spiral cut A cylindrical drill-round cut whose spiral hole pattern gives the widest opening and permits opposite holes to be ignited successfully. This gives the best cleaning of the opening and safety in the advance is increased, since one section of the double spiral can give breakage irrespective of the other. (Langefors, 1963)

double stall An earlier system of working thick seams in South Wales. Two narrow stalls are turned off the heading and after advancing some 8 to 12 yd (7.32 to 10.97 m) (so as to leave a pillar of coal next to the heading) are connected and the coal between them worked as a single face. Double stalls are intermediate between pillar-and-stall and longwall. (Nelson, 1965)

doublet An assembled gem substitute composed of two pieces of material fused or cemented together. If both parts are of the species being imitated, it is a genuine doublet; if one part, it is a semigenuine doublet; if it contains no parts of the species being imitated, it is a false doublet; or if no part is a mineral, it is an imitation doublet. Cf: *triplet.*

double-track portable switch A tub-changing arrangement for a tunnel face. The double-track loop is superimposed on the tunnel track and equipped with ramps, clamps, and spring switches so arranged that the loaded cars take one track outward while the empties take the other track inward. Syn: *portable shunt.* (Nelson, 1965)

double-trolley system A system of electric traction where, instead of the running rails, a second insulated contact wire is used for the return or negative current. (CTD, 1958)

double-tube core barrel A double-tube core barrel having the upper end of the inner tube coupled to the core-barrel head by means of an antifriction device, such as a roller or ball bearing; hence, the inner tube tends to remain stationary when the outer tube, which is rigidly coupled to the core-barrel head, is rotated. (Long, 1960)

double-unit conveyor A longwall conveyor layout from 200 to 280 yd (183 to 256 m) long, developed between two tailgates with a main gate in the center of the face. The main gate conveyor is served by two face conveyors and may act as an intake or a return airway. The tailgates may serve as supply roads. See also: *double-double unit conveyor; main gate.* (Nelson, 1965)

double wedge cut A drill hole pattern consisting of a shallow wedge within an outer wedge, which is often used to obtain deep pull in hard rock. See also: *wedge cut.* (Nelson, 1965)

double wicket A method of working in which rooms are driven from adjacent headings to meet at their extremities. (Zern, 1928)

double working N. of Eng. Two hewers (miners) working together in the same heading. Syn: *hewing double.* (Fay, 1920)

doubly plunging fold A fold that plunges in opposite directions from a central point. In a doubly plunging anticline, the plunge is away from this point; in a doubly plunging syncline, the plunge is toward this point. (Billings, 1954)

doubly refractive Causing double refraction. See also: *birefringent.*

dough Alternate spelling of daugh.

doughnut The cylinder of coal formed by a coal auger. (Nelson, 1965)

Douglas furnace A horizontal, revolving cylindrical furnace having a central flue. (Fay, 1920)

Douglas process *Hunt and Douglas process.*

douk Eng. A soft dark clay found in veins. Probably derived from the Saxon deagan, meaning to knead or mix with water. Syn: *dowk.*

dousing *dowsing.*

dousing rod Commonly used by drillers as a name applied to a wooden wand, rod, forked tree limb, or twig (usually witch hazel) supposedly useful in locating formations bearing water, oil, or mineral. Also called divining rod; doodlebug; dowsing rod. (Long, 1960)

doverite *synchysite.*

Dow cell The Dow electrolytic cell is a steel shell about 16 ft (4.88 m) long, 5 ft (1.52 m) wide, and 6 ft (1.83 m) deep. The electrolyte contains about 60% NaCl, 15% $CaCl_2$, and 25% $MgCl_2$; it is maintained at a temperature of 700 to 750 °C by controlled firing underneath the cell. (Newton, 1959)

dowk *douk.*

downbuckle *tectogene.*

downcast (a) The shaft through which the fresh air is drawn or forced into the mine; the intake. See also: *air shaft; intake.* (Fay, 1920) (b) That side of a fault on which the strata have been displaced downward in relation to the upthrow or upcast side. (CTD, 1958)

downcast shaft The shaft down which the fresh air enters the mine or workings. See also: *upcast shaft.* (Nelson, 1965)

downcomer A pipe to conduct something downward, such as: (1) a pipe for leading the hot gases from the top of a blast furnace downward to the dust collectors and flue system, and (2) a tube larger in diameter than the water tubes in some water-tube boilers for conducting water from each top drum to a bottom drum under the influence of thermal circulation. (Webster 3rd, 1966)

downdip Parallel to or in general direction of the dip of a bed, rock stratum, or vein. (Long, 1960)

downdraft A downward current of air or other gas (as in a mine shaft, kiln, or carburetor). Syn: *downcast.* (Webster 3rd, 1966)

downdrift In a mine drift, the direction of predominant water movement.

downhole (a) A borehole drilled at any angle inclined downward in a direction below the horizon. (b) adj. In a borehole; e.g., downhole equipment.—adv. Deeper; e.g., to perforate downhole. (AGI, 1987)

downline A line of detonating cord or plastic tubing in a blast hole that transmits the detonation from the trunkline or surface delay system down the hole to the primer.

downslope The land surface between the projected outcrop of the lowest coalbed being mined along each highwall and a valley floor.

downspouts Lanc. Pipes fixed down the side of a shaft for conducting water from one level or sump to another.

downstream face The dry side of a dam. (Nichols, 1976)

down-the-hole drill A percussive or hammer drill in which the bit-driven mechanism is located immediately behind the drill bit and is small enough in diameter to permit it to enter and follow the bit down into the hole drilled.

down-the-hole extensometer A device used to measure differential strains in a drill hole.

downthrow (a) The downthrown side of a fault. (AGI, 1987) (b) The amount of downward vertical displacement of a fault. Cf: *upthrow; heave.* (AGI, 1987)

downthrow fault A fault that displaces the strata downward relative to the workings approaching it. It would be an upthrow fault to workings on the opposite side.

downthrow side The lower side of a fault.

Downtonian Uppermost Silurian or lowermost Devonian. (AGI, 1987)

downward continuation (a) Interpretation method in which the values of a component of the magnetic field at lower levels are computed from the values at the surface. (Schieferdecker, 1959) (b) The process of determining, from values measured at one level, the value of a potential (e.g., gravitational) field at a lower level. (AGI, 1987)

downward course In mining, the course of the vein from the surface downward. Also called course downward.

downward enrichment *supergene enrichment.*

downward percolation *sand leaching.*

Dow process A process for the production of magnesium by electrolysis of molten magnesium chloride. (ASM, 1961)

dowsing The practice of locating ground water, mineral deposits, or other objects by means of a divining rod or a pendulum. A dowser may claim also to be able to diagnose diseases, determine the sex of unborn babies, etc. Syn: *dousing; water witching.* Cf: *rhabdomancy.* (AGI, 1987)

dowsing rod *divining rod.*

Dowson gas A mixture of producer gas and water gas obtained by passing steam and air over heated coal or coke in a Dowson producer.

Dowson producer A furnace used for the manufacture of producer gas. (Fay, 1920)

Dowty hydraulic tub retarder A retarder that consists of lengths of steel channel with attached rubbing strips that operate on the face of the wheels above center. The action is controlled by a hydraulic cylinder containing opposed pistons. The hydraulic pressure is supplied from an accumulator in which pressure is maintained by means of a 5-hp (3.7-kW) electric motor-driven pump that is sufficient for 10 retarder unit. (Mason, 1951)

Dowty prop A prop that is in effect a self-contained hydraulic jack consisting of two tubes, the upper one telescoping into the lower. The upper (or inner) tube acts both as a reservoir for the oil and as a container for the pump, yield valve, and other accessories. (Nelson, 1965)

Dowty roofmaster A self-contained, oil-operated steel support for use on a mechanized long-wall face. It has support frames constructed of rigid roof and floor members supported by yielding hydraulic props. Two- and three-prop units are connected alternately to the armored conveyor by means of jacks mounted in the floor members, to carry long and short cantilever roof beams, respectively. See also: *self-advancing supports.* (Nelson, 1965)

dozer Abbrev. for bulldozer; shovel dozer. See also: *bulldozer.* (Nichols, 1954)

dozer shovel A tractor equipped with a front-mounted bucket that can be used for pushing, digging, and truckloading. (Nelson, 1965)

dozzle *core.*

D.P. reagents Flotation reagents made by DuPont are D.P. 243, a 50% aqueous paste of lorolamine (lorol being a mixture of primary straight-chain alcohols) and D.P.Q., lauryl trimethyl ammonium bromide. Others include D.P.Q.B., D.P.C., D.P.N., and D.P.L.A. (Pryor, 1963)

dradge Corn. The inferior portions of ore separated from the best ore by cobbing. (Fay, 1920)

draft A survey line in a traverse. Syn: *leg.* (BS, 1963)

draft engine Corn. An engine used for pumping.

draft gage An instrument used to measure the small pressure differentials below atmospheric; e.g., an inclined manometer to measure the pressure difference between a flue and the atmosphere for combustion control. (ACSG, 1963)

draft hole An opening through which air is supplied to a furnace. (Fay, 1920)

draftsman In petroleum production, one who specializes in drawing subsurface contours in rock formations from the data obtained by a geophysical prospecting party. The draftsman plots maps and diagrams from computations based on recordings of seismograph, gravimeter, magnetometer, and other petroleum prospecting instruments, and from prospecting and surveying field notes. (DOT, 1949)

drag (a) The frictional resistance offered to a current of air or water; resistance created by friction. (Fay, 1920) (b) Fragments of ore torn from a lode by a fault. Such fragments are scattered along the line of the fault and are usually inclosed within crushed or brecciated pieces of the rock traversed by that fault. (c) The flexuring of strata associated with faults. In a normal fault, the coal seam often bends upward on the downthrow side and downward on the upthrow side. Thus, drag is an indication of direction of displacement of the beds. Also called terminal curvature. See also: *coal lead*. (Nelson, 1965) (d) In an inclined stope, the weight of the arch block is resolved into two components, one at right angles to the dip, which tends to close the opening, and one parallel to the dip, which tends to produce movement of the hanging wall with respect to the footwall. This movement is known as drag, or creep. Syn: *creep*. (Higham, 1951) (e) See also: *drag ore*. (AGI, 1987) (f) An appliance to be attached to the rear of a loaded train of cars to prevent the cars from running down the incline or grade in case the cable should break. See also: *backstay*. (Fay, 1920) (g) The uptilted or downtilted curve in rock beds or strata adjacent to a fault. (Long, 1960) (h) The force exerted by a flowing fluid on an object in or adjacent to the flow. (AGI, 1987) (i) The bending of strata on either side of a fault, caused by the friction of the moving blocks along the fault surface; also, the bends or distortions so formed. Cf: *bull*. Syn: *trailer*. (AGI, 1987)

drag angle The angle at which the leading surface of a cutting plane or point meets the surface to be cut. If less than 90°, the angle is said to be negative; if over 90°, it is called a positive rake or drag angle. Cf: *rake*. (Long, 1960)

drag bit (a) A noncoring or full-hole boring bit that scrapes its way through strata that must not be too hard. It may be a two-, three-, or four-bladed pattern with various curves and cutaways. The drilling fluid passes down through the hollow drill stem to the cutting point. See also: *roller bit*. (Nelson, 1965) (b) Various kinds of rigid steel bits provided with fixed (as contrasted to the movable or rolling cutting points of a roller bit) and sometimes replaceable cutting points, which are rotated to drill boreholes in soft to medium-hard rock formations. See also: *bit; fishtail bit; mud bit*. (Long, 1960)

drag bolt A coupling pin. (Standard, 1964)

drag brake On a revolving shovel, the brake that stops and holds the drag (digging) drum. (Nichols, 1954)

drag breccia Fragments of rock in the brecciated zone of a fault. (Long, 1960)

drag bucket A bucket widely used in sampling sea-floor rock deposits in all depths up to and exceeding 30,000 ft (9.1 km). See also: *drag dredging*. (Mero, 1965)

drag cable In a dragline or hoe, the line that pulls the bucket toward the shovel. (Nichols, 1976)

drag-chain conveyor A type of conveyor having one or more endless chains that drag bulk materials in a trough. See also: *chain conveyor; drag conveyor; portable drag conveyor*. Syn: *bar flight conveyor*.

drag classifier Inclined trough that receives ore pulp, and classifies it into settling solids and relatively fine pulp overflow. The settled material is continuously dragged up slope and out by a continuous belt, perhaps provided with transverse scrapers. (Pryor, 1963)

drag conveyor A conveyor in which an endless chain, having wide links carrying projections or wings, is dragged through a trough into which the material to be conveyed is fed; it is used for loose material. See also: *chain conveyor*. (CTD, 1958; Nelson, 1965)

drag cut (a) A cut on which groups of holes are drilled at increasing heights above floor level and at increasing angles from the free face. The shots are fired to break out successive wedges of strata across the width of the face. (BS, 1964) (b) A drill-hole pattern widely used in high-speed drilling. The cut holes are inclined downward to cut a wedge along the floor, the other holes being drilled to break to the cut holes. Also called horizontal cut. (Nelson, 1965) (c) A cut in which the cut holes are angled in the vertical plane toward a parting in order to breakout the ground along the parting. Drag cut rounds are suitable for small drifts 6 to 7 ft (1.83 to 2.13 m) wide or where shallow pulls are sufficient, but the drag cut does not find much application in large-scale drifting practice. See also: *bottom cut*. (McAdam, 1958)

drag dip Local change of attitude as a result of drag near a fault.

drag dredging A method in which the bucket is lowered to the sea floor and dragged over the ocean floor for some distance in order to collect samples. Dredge and trawl hauls normally can only give a rough indication of heavy or light concentrations of the minerals within an area. (Mero, 1965)

drag engineer *slope engineer*.

drag fold A minor fold, usually one of a series, formed in an incompetent bed lying between more competent beds, produced by movement of the competent beds in opposite directions relative to one another. Drag folds may also develop beneath a thrust sheet. They are usually a centimeter to a few meters in size. (AGI, 1987)

drag head The underwater end of a hydraulic dredging system that comes in contact with bottom sediments and through which a dredge pump recovers a slurry of water and sediment. (Padan, 1968)

dragline A type of excavating equipment that casts a rope-hung bucket a considerable distance; collects the dug material by pulling the bucket toward itself on the ground with a second rope; elevates the bucket; and dumps the material on a spoil bank, in a hopper, or on a pile. See also: *boom; excavator*.

dragline boom A crane boom used with a drag bucket. (Carson, 1961)

dragline dredge An excavation system involving a digging bucket, cable, and boom, which permits the recovery of sediments and rocks from trenches, canals, and pits that contain or are covered by water.

dragline engineer *slope engineer*.

dragline excavator A mechanical excavating appliance consisting of a steel scoop bucket that is suspended from a movable jib; after biting into the material to be excavated, it is dragged toward the machine by means of a wire rope. (CTD, 1958)

dragline scraper An apparatus for moving soil, gravel, or other loose material. It ordinarily consists of a scraper attached to an endless cable or belt operated by a drum or sprocket wheel, and can be drawn back and forth by the operator at the drum.

drag loader *dragman*.

dragman One who operates a scraper loading machine, known as a drag, to load ore into cars or chutes. Also called drag loader; drag operator. (DOT, 1949)

dragon S. Staff. A barrel in which water is raised from a shallow shaft.

dragonite A rounded quartz pebble representing a quartz crystal that has lost its brilliancy and angular form; in gravels, once believed to be a fabulous stone obtained from the head of a flying dragon.

dragon's skin Miner's term for part of a fossil tree trunk, such as Lepidodendron or Sigillaria, with a leaf-scar pattern suggesting scales.

drag operator *dragman*.

drag ore Crushed and broken fragments of rock or ore torn from an orebody and contained in and along a fault zone. See also: *trail of a fault*. Syn: *drag*. (AGI, 1987)

drag-out loss Misplacement of relatively fine material due to its adherence to a coarser fraction being settled and dragged out in mechanical classification or heavy-media separation. (Pryor, 1963)

drag rake *negative rake*.

drags Steel bars with a hook at one end and prongs at the other, which are inserted in the drawbar at the rear of the tub ascending an incline so as to prevent it running back. (Mason, 1951)

drag scraper (a) A digging and transporting device consisting of a bottomless bucket working between a mast and an anchor. (Nichols, 1976) (b) A towed bottomless scraper used for land leveling. Called leveling drag scraper to distinguish from cable type. (Nichols, 1976)

dragshovel A shovel equipped with a jack boom, a live boom, a hinged stick, and a rigidly attached bucket, that digs by pulling toward itself. Also called hoe; backhoe; pullshovel. (Nichols, 1976)

dragstaff A pole projecting backward and downward from a vehicle, to prevent it from running backward. See also: *backstay; drag*. (Fay, 1920)

drag-stone mill A mill in which ores are ground by means of a heavy stone dragged around on a circular or annular stone bed. See also: *arrastre*. (Webster 3rd, 1966)

drag tank *dredging sump*.

drain A conduit or open ditch for carrying off surplus ground or surface water. Closed drains are usually buried. (Seelye, 1951)

drainage The manner of gravity flow of water or the process of channelization, for removal at a point remote from a mining operation. See also: *drain tunnel; water hoist*. (Nelson, 1965)

drainage basin (a) The area from which water is carried off by a drainage system; a watershed or a catchment area. (Seelye, 1951) (b) *basin*.

drainage head (a) The furthest or highest spot in a drainage area. (Nichols, 1976) (b) Difference in elevation between two points in an area to be drained.

drainage level *water level*.

drainage trench A channel cut alongside a mine roadway to provide for drainage and enable the proper ballasting of the rail track. The trench may be lined with precast concrete sections to a carefully laid gradient. (Nelson, 1965)

drainage tunnel *drain tunnel*.

drained shear test A shear test on a clay sample after completed consolidation under normal load, carried out in drained conditions. The strengths given by drained tests are higher than those from undrained tests. (Nelson, 1965)

drain hole (a) A borehole drilled into a water-bearing formation or mine workings through which the water can be withdrawn or drained. (Long, 1960) (b) Any hole provided in the base covering or housing on a machine through which oil or liquids can be withdrawn. (Long, 1960)

drainman A laborer who regulates flow of tailings, through flumes or pipes (mixture of waste materials and water resulting from treatment of ore for recovery of valuable minerals) in back filling (filling of working places from which all ore has been mined) in such manner that water will be drawn off and the sand left for filling purposes. (DOT, 1949)

drain tunnel A tunnel constructed for disposing of mine water. Long tunnels have been driven in some mining districts for the purpose of passing under the lower workings of several mines and tapping the water for the entire group. Where topographic features permit, a drain tunnel—more properly called an adit—may also be driven to serve a single mine. The chief advantages of a drain tunnel lie in saving the cost of pumping and eliminating the danger of the mine being flooded through failure of the pumps. Also called drainage tunnel. Cf: *drainage.* See also: *water level.* (Lewis, 1964)

dranyam A new cutter loader devised by Maynard Davies and developed at the Central Engineering Establishment of the National Coal Board of Great Britain. A shearer drum is carried on a vertical shaft in contrast to the horizontal shaft in the Anderson shearer. (Nelson, 1965)

draper washer Vertical-current separator (obsolete) used to separate shale from coal. (Pryor, 1963)

draping Warping in the beds overlying a reef, as a result of differential compaction.

draft (a) S. Staff. The quantity of coal hoisted in a given time. (b) The pressure required to supply air to a furnace and to remove the flue gases from the furnaces. Natural draft is produced by a chimney, while artificial draft is produced by fans and is controlled by the speed of the fans, by variation in the pitch of the fan blades, or by dampers. (Francis, 1965)

dravite A trigonal mineral, $3[NaMg_3Al_6(OH,F)_4(BO_3)_3Si_6O_{18}]$; tourmaline group; forms series with schorl and with elbaite; in triangular or hexagonal prisms; pyroelectric and piezoelectric; in metamorphic and metasomatic lime-rich rocks; slices are used to measure transient blast pressures.

draw (a) The horizontal distance on the surface ahead of an underground coal face over which the rocks are influenced by subsidence. See also: *angle of draw.* (Nelson, 1965) (b) The break in strata from a coal face to the surface; the angle between this break and the vertical. (Mason, 1951) (c) To remove broken ore by gravity from stopes, chambers, or ore bins by aid of chutes or conveyors. (Pryor, 1963) (d) To mine out or rob the pillars in a mine, after the rooms are worked out. (e) To pull bit-bank metal toward a diamond by peening and calking when handsetting a diamond bit. (Long, 1960) (f) *pull.* (Lewis, 1964) (g) The effect of creep upon the pillars of a mine. (Fay, 1920) (h) To raise ore, coal, rock, etc., to the surface; to hoist. (Fay, 1920) (i) To transport by hand; to put; to tram. (Mason, 1951) (j) To allow ore to run from working places and stopes through a chute into trucks. (CTD, 1958) (k) To withdraw timber props from overhanging coal, so that it falls ready for collection. (CTD, 1958)

drawability A measure of the workability of a metal subject to a drawing process. This term is usually expressed to indicate a metal's ability to be deep-drawn. (ASM, 1961)

draw a charge Remove explosives. (Zern, 1928)

draw bead (a) A bead or offset used for controlling metal flow. (ASM, 1961) (b) Riblike projections on draw rings or hold-down surfaces for controlling metal flow. (ASM, 1961)

drawcut (a) In underground blasting, cut holes that are inclined upward. (Lewis, 1964) (b) In rock blasting, bottom cut. (Pryor, 1963) (c) *drag cut.*

drawer (a) Scot. A person who takes ore or rock from the working face to the shaft, or terminus of the horse or haulage road. One who pushes trams or drives a horse underground. (Fay, 1920) (b) Derb. A person who hoists ore or rock by means of a windlass, or otherwise, from a shaft. (Fay, 1920) (c) Putter; trammer; wagoner; a person who moves tubs either manually or with a machine. (Mason, 1951)

draw firing Removal of the load from a furnace for a short time, prior to the completion of burning, to equalize heating of all areas. Also called draw burning. (Bryant, 1953)

drawgear The term includes drawbars, chains, shackles, detaching hooks, etc., used in haulage, winding, and hoisting. (Nelson, 1965)

draw hole An aperture in a battery through which the coal or ore is drawn. (Fay, 1920)

drawing (a) Recovering the timbers, chocks, etc., from the goaves. This work is commonly performed with the use of the dog and chain. (Fay, 1920) (b) Knocking away the sprags from beneath the coal after holing. (Fay, 1920) (c) Raising coal through a shaft or slope. (Fay, 1920) (d) In hydraulic mining, throwing the water beyond the dirt to be removed and causing it to flow toward the giant. Cf: *goosing.* (Fay, 1920) (e) Removing or pulling out the crown bars in a tunnel. (Stauffer, 1906) (f) The movement of tubs. (Pryor, 1963) (g) Forming recessed parts by forcing the plastic flow of metal in dies. (ASM, 1961) (h) Reducing the cross section of wire or tubing by pulling it through a die. (ASM, 1961) (i) A misnomer for tempering. (ASM, 1961) (j) Continuous forming of sheet, tube or fibrous glass from molten glass. (Van Vlack, 1964)

drawing an entry Removing the last of the coal from an entry. (Fay, 1920)

drawing down Reduction of cross section of steel by forging. (Pryor, 1963)

drawing lift The lowest lift of a Cornish pump, or that lift in which the water rises by suction (atmospheric pressure to the point where it is forced upward by the plunger). Also called drawlift.

drawing small When a winding rope, from the effects of wear and tear, has become less in diameter or in thickness from that cause, it is said to be "drawing small.". (Fay, 1920)

drawing timber The removal of timbers and supports from abandoned or worked out mine areas. This work is highly specialized and should be attempted only by the most experienced persons. Generally, timbers are pulled by a timber puller that permits the operator to be under a safe roof while doing this work. In some cases, where so much weight is resting on the timber that it cannot be removed safely, it must be shot out by use of explosives, and the roof allowed to fall. See also: *sylvester.* Syn: *timber drawing.* (Kentucky, 1952)

draw kiln Scot. A limekiln in which the process of calcination is carried on continuously, the raw limestone and fuel being put in at the top and the lime withdrawn at the bottom. (Fay, 1920)

drawman *grizzly worker.*

drawn The condition in which an entry or room is left after all the coal has been removed. See also: *rob.* (Fay, 1920)

drawn clay Clay that is shrunk or decreased in volume by burning. (Fay, 1920)

drawn tube A tube produced by drawing a tube bloom through a die. (Light Metal Age, 1958)

drawpoint (a) A spot where gravity fed ore from a higher level is loaded into hauling units. (Nichols, 1976) (b) Heavy chisel cut across the face of a bit blank a short distance from a diamond to serve as a starting point for calking the metal toward and around a diamond being handset. (Long, 1960)

draw slate A soft slate, shale, or rock approx. 2 in (5.08 cm) to 2 ft (0.61 m) in thickness, above the coal, and which falls with the coal or soon after the coal is removed. (Fay, 1920)

draw works In rotary drilling, that part of the equipment functioning as a hoist to raise or lower drill pipe and in some types, to transmit power to the rotary table. See also: *hoist.* (AGI, 1987)

dredge (a) Large floating machine used in underwater excavation for developing and maintaining water depths in canals, rivers, and harbors; raising the level of lowland areas and improving drainage; constructing dams and dikes; removing overburden from submerged orebodies prior to openpit mining; or recovering subaqueous deposits having commercial value. Cf: *grab sampler.* (b) *dradge.* (c) Very fine mineral matter held in suspension in water. (d) A type of bag net used for investigating the fauna of the sea bottom. (CTD, 1958) (e) In dry process enameling: (1) the application of dry, powdered frit to hot ware by sifting; and (2) the sieve used to apply powdered porcelain enamel frit to the ware. Also called dredging. (ASTM, 1994) (f) Any of various machines equipped with scooping or suction devices used in deepening harbors and waterways and in underwater mining. (Webster 2nd, 1960)

dredge claims The bed of an unnavigable river is open to location and patent as public land, when the opposite banks thereof have not passed into private ownership. Proprietors bordering on such streams, unless restricted by the terms of their grant from the government, hold to the center of the stream, notwithstanding the running of meander lines on the banks thereof, as the true boundary of the land is the thread of the stream. (Ricketts, 1943)

dredgemaster In metal mining, a person who supervises and operates a dredge that is used to mine metal-bearing sands or gravels (gold, tin, or platinum) at the bottom of lakes, rivers, and streams. Also called dredgeman. (DOT, 1949)

dredge pump A heavy-duty-type centrifuged pump with chrome-carbide or manganese steel liners. In silts or rounded sand grains their life is often a matter of months, but where sharp-grained sands or large gravel sizes are being handled, casing and impeller lives may be figured in hours. (Carson, 1965)

dredger (a) A vessel specially equipped for dredging. See also: *bucket-ladder dredge; dipper dredger; grab dredger; sand-pump dredger; suction-cutter dredge.* (CTD, 1958) (b) Person who dredges. (Webster 3rd, 1966) (c) A dredging machine. (Webster 3rd, 1966)

dredger excavator An excavator working on the same principle as the bucket-ladder dredger but designed to work on land. (CTD, 1958)

dredge sump N. of Eng. A small reservoir at the bottom of a shaft, in which the water collects and deposits any sediments or debris. See also: *settling pit.* (Fay, 1920)

dredging Removing solid matter from the bottom of an area covered by water.

dredging conveyor A scraper partially immersed in a vessel containing liquid used for removing any solids that may settle therein. (BS, 1962)

dredging sump A tank, forming part of the water circuit, in which slurry or small coal settles and is removed continuously by means of a scraper chain or scraper buckets. Also called drag tank; sludge sump. (BS, 1962)

dredging tube The large tube of a dredging machine that operates by suction. (Standard, 1964)

dredging well The opening through a dredging vessel in which the bucket ladders work. See also: *bucket-ladder dredge.* (CTD, 1958)

dreelite *barite.* Also spelled dreeite.

dress (a) To resharpen and restore to size the worn teeth on a roller or diamond bit. See also: *face.* (b) To restore a tool to its original shape and sharpness by forging or grinding. (Crispin, 1964) (c) To clean ore by breaking off fragments of the gangue from the valuable mineral. See also: *ore dressing.* (d) To shape dimension stone.

dressing (a) A general term for the processes of milling and concentration of ores. Syn: *ore dressing.* (b) The shaping of dimension stone. (c) Separating rock from lumps of coal by chipping with a hammer or similar means. (d) Can. Developing claims to take them out of wildcat class. (Hoffman, 1958)

dressing a mine A method of fraud carried out by a representative of the seller, by systematically mining out all the low-grade or barren spots in the vein, leaving only the high-grade spots exposed. (Hoover, 1948)

dribble Material that adheres to the conveying medium and, being carried beyond the discharge point, drops off along the return run.

dribbling In underground excavation, fall of small stone and debris from roof, warning that a heavy fall may be imminent. (Pryor, 1963)

dried alum *burnt alum.*

drier A device used for removing water from damp material by evaporation, supplemented usually with forced circulation of air.

drier man In salt production, one who tends operations of rotary driers through which crushed salt is run to drive off contained moisture prior to grinding, examining the salt discharged from the driers to see that evaporation of moisture is complete. (DOT, 1949)

dries Seams in the rock, which are usually invisible in the freshly quarried material, but which may open up in cutting or on exposure to the weather. See also: *dry.*

drif *drift.*

drift (a) An entry, generally on the slope of a hill, usually driven horizontally into a coal seam. Syn: *surface.* See also: *adit.* (BCI, 1947) (b) The deviation of a borehole from its intended direction or target. Cf: *walk.* (Long, 1960) (c) A general term, used esp. in Great Britain, for all surficial, unconsolidated rock debris transported from one place and deposited in another, and distinguished from solid bedrock; e.g., specif. for glacial deposits. Any surface movement of loose incoherent material by the wind; accumulated in a mass or piled up in heaps by the action of wind or water. See also: *fill.* (AGI, 1987) (d) Apparent offset of aerial photographs with respect to the true flight line, caused by the displacement of the aircraft owing to cross winds, and by failure to orient the camera to compensate for the angle between the flight line and the direction of the aircraft's heading. The photograph edges remain parallel to the intended flight line, but the aircraft itself drifts farther and farther from that line. (AGI, 1987) (e) A time variation common to nearly all sensitive gravimeters, due to slow changes occurring in the springs or mountings of the instrumental systems; this variation is corrected by repeated observations at a base station and in other ways. (AGI, 1987) (f) A horizontal opening in or near an orebody and parallel to the course of the vein or the long dimension of the orebody. (Beerman, 1963) (g) A passageway driven in the coal from the surface, usually above drainage, following the inclination of the bed. (Hudson, 1932) (h) Forest of Dean. A hard shale. (i) To make a drift; to drive. (Webster 3rd, 1966) (j) A horizontal gallery in mining and civil engineering driven from one underground working place to another and parallel to the strike of the ore. It is usually of a relatively small cross section. Larger sections are usually called tunnels. (Fraenkel, 1953) (k) A heading driven obliquely through a coal seam. (CTD, 1958) (l) A heading in a coal mine for exploration or ventilation. (CTD, 1958) (m) An inclined haulage road to the surface. (CTD, 1958) (n) In oil well surveying, the angle from a drill hole to the vertical. See also: *inclination.* (Cumming, 1951) (o) A flat piece of steel of tapering width used to remove taper shank drills and other tools from their holders. (ASM, 1961) (p) A tapered rod used to force mismated holes in line for riveting or bolting. Sometimes called a driftpin. (ASM, 1961) (q) A gradual change in a reference that is supposed to remain constant. An instrument such as a gravimeter may show drift as a result of elastic aging, long-term creep, hysteresis, or other factors. (AGI, 1987) (r) A general term applied to all rock material (clay, silt, sand, gravel, boulders) transported by a glacier and deposited directly by or from the ice, or by running water emanating from a glacier. Drift includes unstratified material (till) that forms moraines, and stratified deposits that form outwash plains, eskers, kames, varves, glaciofluvial sediments, etc. The term is generally applied to Pleistocene glacial deposits in areas (as large parts of North America and Europe) that no longer contain glaciers. The term drift was introduced by Murchison in 1839 for material, then called diluvium, that he regarded as having drifted in marine currents and accumulated under the sea in comparatively recent times; this material is now known to be a product of glacial activity. Cf: *glacial drift.* (AGI, 1987) (s) One of the wide, slower movements of surface oceanic circulation under the influence of, and subject to diversion or reversal by, prevailing winds; e.g., the easterly drift of the North Pacific. Syn: *drift current.* The slight motion of ice or vessels resulting from ocean currents and wind stress. The speed of an ocean current or ice floe, usually given in nautical miles per day or in knots. Sometimes used as a short form of littoral drift. *current.* (AGI, 1987) (t) In South Africa, a ford in a river. The term is used in many parts of Africa to indicate a ford or a sudden dip in a road over which water may flow at times. Syn: *drif.* (Afrikaans). (AGI, 1987)

drift and pillar N. Staff. A system of working coal similar to the room and pillar system. (Fay, 1920)

drift angle The angular deviation of a borehole from vertical and/or its intended course. See also: *deflection angle.* (Long, 1960)

drift angle buildup The rate of the increase in the drift angle that is generally expressed as the number of degrees increase for a specific drilled footage; e.g., 2° per 100 ft (30.5 m). (Long, 1960)

driftbolt (a) A bolt for driving out other bolts or pins. (Webster 3rd, 1966) (b) A metal rod, for securing timbers, resembling a spike but with or without point or head. (Webster 3rd, 1966)

drift coal *allochthonous coal.*

drift coalfields Coalfields formed by forests on higher ground being carried away by floods into lakes. (Mason, 1951)

drift copper Native copper transported from its source by a glacier. (AGI, 1987)

drift current *drift.*

drift curve Graph of a series of gravity values read at the same station at different times and plotted in terms of instrument reading versus time. (AGI, 1987)

drift driller In metal mining, one who operates a heavy, mounted, compressed-air, rock-drilling machine in driving drifts (horizontal passages running parallel to the vein opened up to facilitate mining of the ore). (DOT, 1949)

drifted (a) A borehole, the course of which has deviated or departed from the intended direction or did not reach its intended target. (Long, 1960) (b) Inward-bulged casing that has been straightened by the use of a drift. See also: *drift.* (Long, 1960) (c) A horizontal underground passage parallel to or along a vein or related structure. (Long, 1960)

drift epoch *glacial epoch.*

drifter (a) An air-driven, percussive rock drill; also called leyner; liner. (Long, 1960) (b) A drill crewman, miner, or laborer who travels from place to place, only working a short period of time at each place. Cf: *boomer.* (Long, 1960) (c) A person skilled in the use of air-driven, percussive rock drills and other processes utilized in excavating horizontal underground passages or tunnels. (Long, 1960) (d) An excavator of mine drifts. (Webster 3rd, 1966)

drifter bar *bar; drill column.*

drifter drill The heaviest form of hammer drill made in various sizes depending upon the severity of the work to be done. The heaviest type weighs more than 200 lb (91 kg) and is used for holes up to 20 ft (6.1 m) in depth. Must be mounted on a column or bar. (Lewis, 1964)

drift frame *square set.*

drift indicator Various types of mechanical or photographic devices used to determine the compass bearing and inclination of the course of a borehole. Cf: *clinometer.* (Long, 1960)

driftman In bituminous coal mining, one who is engaged in driving a drift, a horizontal passageway underground following the coal vein in a mine. (DOT, 1949)

driftmeter An instrument for determining the inclination of a drill pipe from the vertical and the depth of measurement. (AGI, 1987)

drift mine (a) A placer or gravel deposit worked by underground mining methods. (Webster 3rd, 1966) (b) A mine that opens into a horizontal or practically level seam of coal. This type of mine is generally the easiest to open as the mine opening enters into the coal outcrop. (Kentucky, 1952) (c) One opened by a drift. (Pryor, 1963)

drift mining (a) A term applied to working alluvial deposits by underground methods of mining. The paystreak, varying from 2 to 8 ft (0.6 to 2.4 m), sometimes greater, is reached through an adit or a shallow shaft. Wheelbarrows or small cars may be used for transporting the gravel to a sluice on the surface. If relatively large, the deposit is removed in a system of regular cuts or slices taken across the paystreak, working generally in a retreating fashion from the inner limit of the gravel. Drift mining is more expensive than sluicing or hydraulicking; consequently it is used only in rich ground. See also: *placer mining.*

(Lewis, 1964) (b) The working of relatively shallow coal seams by drifts from the surface. The drifts are generally inclined and may be driven in rock or in a seam. Drift mining may be viewed as intermediate between opencast coal mining and shaft or deep mining. See also: *development drift; surface drift.* (Nelson, 1965)

drift peat A peat deposit associated with or embedded in glacial drift. (Fay, 1920)

drift salt Fluffy, flaky salt particles due to wind and wave action, which produce a mist over the surface of solar salt ponds. The mist contains minute particles of salt, which are driven to the lee shore and deposited as a scale. (Kaufmann, 1960)

drift set A strong timber set in a drift that may form the anchorage for the timber sets of the stope above. (Nelson, 1965)

drift slicing Side slicing as a method of stoping massive deposit. Alternative to top slicing. (Pryor, 1963)

drift stope The excavation of the development drift together with the stope in overhand stoping. Employed in cases where the hanging wall is strong. (Nelson, 1965)

drift stoping *sublevel stoping.*

drift theory That theory of the origin of coal that holds that the plant matter constituting coal was washed from its original place of growth and deposited in another locality where coalification then came about. See also: *allochthonous coal.* (AGI, 1987)

drikold *dry ice.*

drill (a) Any cutting tool or form of apparatus using energy in any one of several forms to produce a circular hole in rock, metal, wood, or other material. See also: *calyx drill; churn drill; core drill; diamond drill; rock drill; rotary drill; shot drill.* (Long, 1960) (b) To make a circular hole with a drill or cutting tool. (Long, 1960)

drillability (a) The relative speed at which a material may be penetrated by a drill bit. High drillability denotes easy penetration at a fast rate. (Long, 1960) (b) The specific value of the drilling properties of a rock expressed in terms of the drilling rate under certain technical conditions. (Fraenkel, 1953)

drill ahead (a) To sink a borehole into solid or unconsolidated rock material, such as overburden or glacial till, to a considerable depth below the bottom of the casing or drivepipe. (Long, 1960) (b) To restart or resume drilling operation. (Long, 1960) (c) To drill boreholes in advance of mine workings to explore for or locate old mine workings or a water-bearing formation. (Long, 1960)

drill bar A drill column that is set horizontally instead of vertically in an underground workplace. See also: *bar; drill column.* (Long, 1960)

drill base Metal or wood framework on which a drilling machine is mounted. (Long, 1960)

drill bit (a) One of a number of different types of detachable cutting tools used to cut circular holes in rock, wood, metal, etc. Also called drill crown in Africa and England. (Long, 1960) (b) Any device at the lower end of a drill stem, used as a cutting or boring tool in drilling a hole; the cutting edge of a drill. Cf: *core bit.* Syn: *bit; rock bit.* (AGI, 1987)

drill boom An adjustable arm projecting from a drill carriage to carry a drill and hold it in position. (BS, 1964)

drill bort *drill diamond.*

drill by Syn: *bypass.* Also called drilled by and drilling by. (Long, 1960)

drill cable In a strict sense, the term should only be used to designate the heavy rope or cable used as the connecting link between the drill stem and the walking beam on a churn drill. However, the term now is commonly used to signify any cable or wire rope used in hoisting drill rods, casing, and other borehole-drilling equipment used with a drill machine, such as a calyx drill, diamond drill, etc. Also called drilling line; drill line. (Long, 1960)

drill capacity The lineal feet of drill rod of a specified size that a hoist on a diamond or rotary drill can lift or that the associated brake is capable of holding on a single line; also sometimes used to designate the size of a drill machine, based on the depth to which it is capable of drilling. See also: *lifting capacity.* (Long, 1960)

drill carriage A movable platform, stage, or frame that incorporates several rock drills and usually travels on the tunnel track; used for heavy drilling work in large tunnels. See also: *drill frame.* (Nelson, 1965)

drill collar A length of extra heavy wall drill rod or pipe connected to a drill string directly above the core barrel or bit, the weight of which is used to impose the major part of the load required to make the bit cut properly. A drill collar is usually of nearly the same outside diameter as the bit or core barrel on which it is used. Not to be confused with guide rod. (Long, 1960)

drill column A length of steel pipe equipped with a flat cap at one end and a jackscrew on the opposite end by means of which the pipe can be wedged securely in a vertical or horizontal position across an underground opening to serve as a base on which to mount a small diamond or rock drill. Syn: *drifter bar; drill bar; drill stem.* See also: *bar.* (Long, 1960)

drill core A solid, cylindrical sample of rock produced by an annular drill bit, generally rotatively driven but sometimes cut by percussive methods. Syn: *core.* (Long, 1960)

drill cradle The metal channel on which a heavy drill is fed forward as drilling proceeds. (BS, 1964)

drill cuttings *well cuttings; cuttings; sludge.*

drill diamond Industrial diamond used in diamond-drill bits and reaming shells for coring, cutting, or reaming rock. Drill diamonds usually contain obvious imperfections and inclusions, although the finer grades approach toolstones in quality. Also called drill bort; drilling bort; drilling diamond; drilling. Cf: *toolstone.* (Long, 1960)

driller (a) A person who has acquired enough knowledge and skill to operate and assume the responsibility of operating a drill machine. Also called drill runner; runner; tool pusher. Syn: *drillman.* See also: *machine driller.* (Long, 1960) (b) The person in charge of the rig and crew during one tour and who handles the drilling controls. (Brantly, 1961) (c) A drilling machine. (Standard, 1964) (d) Can. Property being diamond drilled as compared to one undergoing underground development. (Hoffman, 1958) (e) N. of Eng. Uses an electric or pneumatic twist drill to make shotholes in the coal. Shotholes in the gateway caunches are usually put on by the stoneman. (Trist, 1963)

driller's log A description of the borehole based on the daily logs from the driller.

drill extractor Tool for retrieving broken piece of drill from borehole. (Pryor, 1963)

drill feed The mechanism for advancing the drill bit during boring. (Nelson, 1965)

drill fittings Devices, parts, and pieces of equipment used downhole in drilling a borehole. Also called downhole equipment. (Long, 1960)

drill frame A drill mounting often made at the mine to suit the tunnel requirements. It usually comprises two girders strapped together to form a replica of the tunnel shape but smaller in size. The structure is mounted on wheels and provision is made for clamping the drills to various parts of the frame according to the drill-hole pattern in use. It contains a central opening to allow the passage of the loading machine, cars, or conveyor. (Nelson, 1965)

drill gage The width across the cutting bit or diameter of the drilled hole. With tungsten-carbide bits it is possible to drill long holes without the loss of gage. (Nelson, 1965)

drill hole (a) A hole in rock or coal made with an auger or a drill. (b) Technically, a circular hole drilled by forces applied percussively; loosely and commonly, the name applies to a circular hole drilled in any manner. (Long, 1960) (c) Used by diamond drillers as a syn. for borehole. Cf: *borehole.* (Long, 1960)

drill-hole counting When the results of a survey indicate a possible ore deposit, test holes may be drilled and a special adaptation of a scintillation counter, called a drill-hole counter, may be lowered in a hole in an attempt to locate, outline, and assay an orebody. The drill-hole counter can distinguish between formations by their radiation intensity. (Dobrin, 1960)

drill-hole pattern The number, position, depth, and angle of the shot holes forming the complete round in the face of a tunnel or sinking pit. A good drill-hole pattern will ensure the maximum possible pull and the fragmentation for easy loading without excessive scatter of material. See also: *cut holes.* (Nelson, 1965)

drill-hole record A description of the borehole based on the daily logs from the driller and the samples and the report of the geologist. (Nelson, 1965)

drill-hole returns The circulation fluid and entrained cuttings overflowing the collar when drilling a borehole. (Long, 1960)

drill-hole survey *borehole survey.*

drilling (a) The act or process of making a circular hole with a drill. See also: *drill.* Cf: *boring.* (Long, 1960) (b) The operation of tunneling or stoping, whether with a compressed-air rock drill, a jackhammer, or a drifter. (CTD, 1958) (c) Use of a compressed-air rock drill to prepare rock for blasting. (Pryor, 1963) (d) The operation of making deep holes with a drill for prospecting, exploration, or valuation. (Pryor, 1963)

drilling bit The cutting device at the lower end of cable drilling tools or rotary drill pipe, the function of which is to accomplish the actual boring or cutting. (AGI, 1987)

drilling cable *cable.*

drilling column The column of drill rods to the end of which the bit is attached. (BS, 1963)

drilling conditioning period Time spent in circulating a higher-than-normal volume of fluid through the drilling string while slowly rotating and lowering the string from the last few feet above to the bottom of a borehole to wash away any obstructing material before resuming coring operations. (Long, 1960)

drilling jig (a) A device very accurately made of cast or wrought iron that becomes a guide for the drilling of holes. The work is fastened in the jig, and the drill is guided through holes drilled in the face of the jig itself. The use of a jig makes interchangeable work easily obtainable.

(Crispin, 1964) (b) A portable drilling machine worked by hand. (Fay, 1920)
drilling life *bit life.*
drilling machine A hand-operated, or power-driven machine for boring shot holes or boreholes, in coal, ore, mineral, or rock. See also: *drifter drill; percussive drill; rotary drill; rotary-percussive drill.* (Nelson, 1965)
drilling mud A suspension, generally aqueous, used in rotary drilling and pumped down through the drill pipe to seal off porous zones and to counterbalance the pressure of oil and gas; consists of various substances in a finely divided state among which bentonite and barite are most common. Oil may be used as a base of water. Cf: *circulation fluid; mud-laden fluid.* (AGI, 1987)
drilling pattern The relation of drilled holes to each other and any free faces as part of the blast design.
drilling platform Auxiliary equipment for drilling at heights above head level. The drilling platform is generally assembled and dismantled for each series of drilling operations. (Fraenkel, 1953)
drilling pressure *bit load.*
drilling rate (a) The depth of penetration achieved per unit of time with a given type of rock drill, bit diameter, air pressure, etc. See also: *penetration rate.* (Fraenkel, 1953) (b) The overall rate of advancement of the borehole. (BS, 1963)
drilling rig A general term for the derrick, power supply, draw works, and other surface equipment necessary in rotary or cable-tool drilling. See also: *rig.* (AGI, 1987)
drillings (a) *drill diamond.* (b) Incorrectly used as a syn. for cuttings. (Long, 1960) (c) Sometimes designates drill diamonds ranging from 4 to 23 stones per carat in size. (Long, 1960)
drilling thrust *bit thrust.*
drilling time (a) In rotary drilling, the time required for the bit to penetrate a specified thickness (usually 1 ft or 0.3 m) of rock. The rate is dependent on many factors. (AGI, 1987) (b) The elapsed time, excluding periods when not actually drilling, required to drill a well. (AGI, 1987)
drilling up Preliminary digging out the clay in the taphole of a furnace. This is done usually by hand, air, or electric drill. (Fay, 1920)
drilling weight Also called drilled weight. *bit load.* (Long, 1960)
drill log The record of the events and the type and characteristics of the formations penetrated in drilling a corehole. Syn: *boring log.* Cf: *log.* (Long, 1960)
drillman *driller.*
drill mounting An appliance to provide a feed pressure and a support for the drilling machine usually in tunnels. Four main types of drill mountings are in use, namely, the post, the air leg, the drill frame, and the drill carriage. (Nelson, 1965)
drill output The volume of rock (in tons) corresponding to the footage drilled per hour. (Streefkerk, 1952)
drill pattern The placement of a number of boreholes in accordance to a predetermined geometric arrangement. (Long, 1960)
drill pressure Also called drilling pressure. *bit load.*
drill rate (a) The number of feet of borehole drilled in a specified interval of time; e.g., drilling rate was 80 ft/d (24.4 m/d). (Long, 1960) (b) Price, expressed in dollars, per foot of borehole completed in accordance with terms specified in a drill contract. Syn: *feed rate.* (Long, 1960)
drill rig A drill machine complete with all tools and accessory equipment needed to drill boreholes.
drill-rod bit A noncoring bit designed to be coupled to a reaming shell threaded to couple directly on a drill rod instead of a core barrel. (Long, 1960)
drill runner (a) The tunnel miner who normally handles the rock drills for blasting purposes. (Nelson, 1965) (b) *driller.*
drill sampling (a) A method of sampling a deposit by means of a drill or borehole. The boreholes may be spaced at the corners of squares or triangles at distances according to the nature and extent of the deposit. See also: *exploratory drilling.* (Nelson, 1965) (b) The sampling of gravel deposits or extensive low-grade ore deposits by use of drills. (Hoover, 1948)
drill sharpening machines Machines for sharpening detachable bits and for making shanks. (Lewis, 1964)
drill sludge *cuttings.* Also called drilling sludge.
drill speed May be used by drillers as a syn. for drill bit revolutions per minute; drill rate; feed rate; feed ratio; feed speed; rate of penetration. (Long, 1960)
drill-split longwalling The act of mining tabular ores 2 to 5 m thick using a drill-split tool. (Lombardi, 1994)
drill-split narrow-vein mining The act of mining narrow veins down to 0.6 m using a drill-split tool on a carriage that is independently mobile and remote-controlled. (Lombardi, 1994)
drill-split tool A device that combines a drill and a splitter, which is a unique tool designed to apply radial and axial loads to a rock mass, into a single tool that drills and splits rock in recurring cycles. (Lombardi, 1994)
drill steel (a) A round or hexagonal steel rod for boring in coal, ore, or rock. It consists of shank, shaft, and bit. It forms an important part of jackhammers and drifters. (Nelson, 1965) (b) Hollow steel connecting a percussion drill with the bit. (Nichols, 1976) (c) *stem.*
drill-steel set A series of integral drill-steel sizes consisting of starter and follower bits, necessary for drilling a hole to a certain depth. The length increment is usually determined by the wear of the bit and the feed length of the feeding device. (Fraenkel, 1953)
drill stem (a) In standard drilling, a cylindrical bar of steel or iron screwed onto the cable tool bit to give it weight. (AGI, 1987) (b) In rotary drilling, a string of steel pipe screwed together and extending from the rig floor to the drill collar and bit at the bottom of the hole. The drill pipe transmits the rotating motion from the rotary table to the bit and conducts the drilling mud from the surface to the bottom of the hole. See also: *drill string.* (AGI, 1987)
drill-stem test A procedure for determining the potential productivity of an oil or gas reservoir by measuring reservoir pressures and flow capacities while the drill pipe is still in the hole, the well is still full of drilling mud, and usually the well is uncased. The tool consists of a packer to isolate the section to be tested and a chamber to collect a sample of fluid. If the formation pressure is sufficient, fluid flows into the tester. Abbrev: DST. (AGI, 1987)
drill string The assemblage of drill rods, core barrel and bit or drill rods, drill collars, and bit in a borehole, which is connected to and rotated by the drill machine on the surface at the collar of the borehole. Also called drill stem. See also: *string.*
drill thrust *bit load.*
drip feeder (a) Oil reservoir set to discharge lubricant at steady rate in drops per minute. (Pryor, 1963) (b) Reagent feeder sometimes used in flotation process to meter chemicals into pulp. (Pryor, 1963)
dripping fault A fault down which small quantities of water seep into mine workings. A dripping fault is a hazard, as mining operations may loosen or open it and cause an inrush of water. (Nelson, 1965)
dripstone A general term for any cave deposit of calcite or other mineral formed by dripping water, including stalactites and stalagmites. See also: *cave onyx; dropstone.*
drivage A general term for a roadway, heading, or tunnel in course of construction. It may be horizontal or inclined but not vertical. (Nelson, 1965)
drive (a) To excavate horizontally, or at an inclination, as in a drift, adit, or entry. Distinguished from sinking and raising. (b) A tunnel or level in or parallel to and near a mineralized lode or vein, as distinct from a crosscut, which only gives access normal to the lode. (Pryor, 1963) (c) An underground passage for exploration, development, or working of an orebody. (Nelson, 1965) (d) To advance or sink drive pipe or casing through overburden or broken rock formation by chopping, washing, or hammering with a drive hammer or by a combination of all three procedures. (Long, 1960)
drive collar *pipe drivehead.*
drivehead (a) The driving mechanism for a conveyor. The expressions head-end drive, intermediate drive, and tail-end drive, indicate the position of the drivehead or heads. (Nelson, 1965) (b) A heavy iron cap or angular coupling fitted to top of pipe or casing to receive and protect the casing from the blow delivered by a drive block when casing or pipe is driven through overburden or other material. Also called drive cap; driving cap. Syn: *pipe drivehead; drive collar.* (Long, 1960) (c) The swivel head of a diamond- or rotary-drill machine. (Long, 1960)
drivepipe (a) A thick-walled outside-coupled pipe, fitted at its lower end with a sharp steel shoe. It may be driven through overburden or other material by repeated pile-driverlike blows delivered to the upper end of the pipe by a heavy drive block. (Long, 1960) (b) Casing pipe driven into deep drill hole to hold back water or prevent caving. In shallow drilling of alluvials, bottom pipe of string that may be battered down. Drivehead and drive shoe are also used in this work. (Pryor, 1963) (c) Pipe driven short distance into dumps or unconsolidated ground to obtain samples. See also: *conduit.* (Pryor, 1963)
drivepipe ring (a) A heavy sleevelike device attached to a drill floor to steady and guide the pipe or casing being driven. (Long, 1960) (b) A device for holding the drivepipe while being pulled from well. (Fay, 1920)
drive sample A dry sample of soft rock material, such as clay, soil, sand, etc., obtained by forcing, without rotation, a short, tubular device into the formation being sampled by hydraulic pressure or the piledriver action of a drive hammer. (Long, 1960)
drive sampler A short tubelike device designed to be forced, without rotation, into soft rock or rock material, such as clay, sand, or gravel, by hydraulic pressure or the piledriver action of a drive hammer to procure samples of material in as nearly an undisturbed state as possible. Cf: *piston sampler.* See also: *thick-wall sampler.* (Long, 1960)

drive sampling The act or process of obtaining dry samples of soft rock material by forcing, without rotation, a tubular device into the material being sampled by pressure generated hydraulically, mechanically, or by the piledriver action of a drive hammer. (Long, 1960)
drive shaft (a) Main driving shaft on which the drive and conveyor sprocket wheels or pulleys are mounted. This shaft is connected to the drive unit through a coupling, sprocket wheel, gear, or other form of mechanical power transmission. (b) A shaft used to support the end of a conveyor screw in a trough end and as a driving connection between a conveyor screw and the power transmitting medium.
drive unit The mechanism that imparts the reciprocating motion to a shaker conveyor trough line. The term is frequently shortened to drive, such as shaker drive, uphill drive, etc. (Jones, 1949)
drive wedge A metal wedge, driven into a wooden or soft-metal base plug in a borehole, that acts as a fixed point on which and by means of which a deflection wedge may be set and oriented. (Long, 1960)
driving (a) Extending excavations horizontally or near the horizontal plane. Cf: *sinking; raising.* (Nelson, 1965) (b) The making of a tunnel or level (a drive) in a mineralized lode or vein, as distinct from making one in country rock (crosscutting). (CTD, 1958) (c) Breaking down coal with wedges and hammers. (CTD, 1958) (d) A long narrow underground excavation or heading. (Fay, 1920) (e) Eng. In the Bristol coalfield, a heading driven through rock. (Fay, 1920)
driving cap Steel cap placed above line of casing pipes of drill hole to protect threaded top of pipe while driving them deeper. Driving shoe gives protection to the bottom pipe of line. (Pryor, 1963)
driving head The driving mechanism of a belt conveyor. It consists of an electric motor or compressed-air turbine connected through a train of reduction gearing to the drum or drums. Motion is imparted to the belt by the frictional grip between it and the drums. (Sinclair, 1959)
driving on line The keeping of a heading or breast accurately on a given course by means of a compass or transit. Also, called driving on sights.
drop (a) Granulated material obtained by pouring melted material into water. (Standard, 1964) (b) In the roof of a coal seam, a funnel-shaped downward intrusion of sedimentary rock, usually sandstone. See also: *stone intrusion.* (AGI, 1987) (c) The vertical displacement in a downthrow fault; the amount by which the seam is lower on the other side of the fault. (CTD, 1958) (d) In an air lift, the distance the water level sinks below the static head during pumping. (Lewis, 1964) (e) The small downward descent of the upper section of a drill rod, casing, or pipe into a lower like section when the threads of the box- and pin-threaded parts match, so that upper and lower sections may be screwed together without cross-threading. (Long, 1960) (f) The sudden descent of a bit that occurs when a bit encounters a cavity or cuts through a hard rock and enters a very soft rock. (Long, 1960) (g) To lower drill-string equipment into a borehole. (Long, 1960) (h) To lower the cage to receive or discharge the car when a cage of more than one deck is used. (Fay, 1920) (i) To allow the upper lift of a seam of coal to fall or drop down. (Fay, 1920) (j) To lose equipment in a borehole. (Long, 1960)
drop ball A method of breaking oversize stones left after quarry blasting. The balls weigh from 30 hundredweight (1,360 kg) to 2 st (1.8 t) (many use old cones from gyratory breakers) and are dropped from a crane on to the oversize stone. The drop height varies from about 20 to 33 ft (6.1 to 10.1 m). The method is economical and avoids secondary blasting. (Nelson, 1965)
drop-bottom cage A cage so designed that the middle section of the floor drops a few inches when the cage is lifted from the keps. The mine car is thus kept stationary and secure. (Nelson, 1965)
drop-bottom car A mine car so constructed that all the haulage motor has to do is to pull the loaded trip across the dump. A trigger trips the flaps in the bottom of the car, allowing the coal to drop out, and a second one closes the flaps as the car leaves the dump. See also: *mine car.* (Kentucky, 1952)
drop box Placed at intervals along tailings line to compensate for slope in excess of that required to keep the pulp moving gently through its launders or pipes. (Pryor, 1958)
drop cut The initial cut made in the floor of an open pit or quarry for the purpose of developing a bench at a level below the floor.
drop doors Hinged doors at the bottom of a cupola furnace, which drop down to allow the furnace to be cleaned. (Mersereau, 1947)
drophammer (a) A forging hammer that depends on gravity for its force. (ASM, 1961) (b) A pile driving hammer that is lifted by a cable and obtains striking power by falling freely. (Nichols, 1976)
drop log A timber that in an emergency can be dropped by a remote control across a mine track at the top or bottom of an incline to derail cars.
drop on Portable rail crossing used to transfer wagons from one track to another. (Hammond, 1965)
dropped core Pieces of core not picked up or those pieces that slip out of the core barrel as the barrel is withdrawn from the borehole. (Long, 1960)
dropper (a) A branch vein pointing downwards. See also: *leader.* (b) A spar dropping into the lode. (Zern, 1928) (c) *feeder.* (d) A branch leaving a vein on the footwall side. (Zern, 1928) (e) *car runner; car dropper.*
dropping pillars and top coal Aust. The second working, consisting of drawing the pillars, and in thick seams breaking down the upper portion of the seam that was left temporarily in position. (Fay, 1920)
dropping stones *stalagmite.*
drop pit A shaft in a mine, in which coal is lowered by a brake wheel.
drops Drops of 12 in (30.5 cm) or more in a line of sluices that are formed by allowing the discharge end of one box to rest on the head of the succeeding sluice, instead of telescoping into it. This method ensures a drop of 12 in or more (depending on the depth of the sluice box) at the end of each sluice, which usually is sufficient to disintegrate fairly stiff clay. (Griffith, 1938)
drop shaft A monkey shaft down which earth and other matter are lowered by means of a drop (that is, a kind of pulley with brake attached); the empty bucket is brought up as the full one is lowered. See also: *cofferdam.* (Zern, 1928)
drop-shaft method This sinking system consists in the use of a cutting shoe on the bottom of a shaft lining that is being continually augmented as the shoe descends, the material inside the lining being excavated. (Sinclair, 1958)
drop sheet N. of Eng. A door, made of canvas, by which the ventilating current is regulated and directed through the workings. See also: *curtain.* (Fay, 1920)
drop staple Eng. An interior shaft, connecting an upper and lower seam, through which coal is raised or lowered. (Fay, 1920)
dropstone (a) An oversized clast in laminated sediment that depresses the underlying laminae and may be covered by "draped" laminae. Most dropstones originate through "ice-rafting"; other sources are floating tree roots and kelp holdfasts. (b) A stalagmatitic variety of calcite. Cf: *dripstone.*
drop sulfur Granulated material obtained by pouring the melted material into water. (Standard, 1964)
drop warwicks Steel joists hinged to a substantial cross joist in the roof that are held up by a stirrup during normal running. If a tram runs away down the incline, the stirrup is disengaged by means of a wire operated from the top of the incline; one end of the hinged joist falls into the rail track and arrests the runaway. (Mason, 1951)
drop ways Openings connecting parallel passages that lie at different levels. (AGI, 1987)
drop weights A method of breaking oversize stones after primary blasting at a quarry. See also: *drop ball.* (Nelson, 1965)
dross (a) The scum that forms on the surface of molten metals largely because of oxidation, but sometimes because of the rising of impurities to the surface, and which contains metal and metal oxides. (ASM, 1961) (b) Small coal that is inferior or worthless and often mixed with dirt. (Nelson, 1965) (c) Refuse or impurity formed in melted metal. A zinc-iron alloy forming in a bath of molten zinc while galvanizing iron. (Standard, 1964)
dross bing Pile of refuse from a washer. (Zern, 1928)
drowned Flooded; said of mines underwater.
drowned level (a) A level that is underwater. See also: *blind level.* (Hess; Fay, 1920) (b) Part of a drainage drift that, being below both discharge and entry levels, is constantly full of water. Also called inverted siphon. (BS, 1963)
drowned waste Old workings full of water.
Drucker-Prager criterion A soil and rock failure criterion, which accounts for the general effect of all three principle stresses by using the invariant of the stress tensor. Use is limited to numerical formulations, such as finite element analysis. (Desai, 1984)
druid stone One of the large sandstone blocks formerly scattered on the English chalk downs and used in Stonehenge and other Druid temples and circles. Syn: *sarsen stone.*
drum (a) The large cylinder or cone on which the rope is coiled when hoisting a load up a shaft. (CTD, 1958) (b) A metal cask, for shipment of material, having a liquid capacity of 55 gal (208 L). See also: *barrel.* (Fay, 1920) (c) In a conical mill, the cylindrical central section. (Pryor, 1963) (d) A general term for a roller around which a belt conveyor is lapped. It may be a driving, jib, loop, tension, or holding-down drum. (Nelson, 1965) (e) The spoollike part of a hoisting mechanism on which the cable or wire line is wound. (Long, 1960) (f) A cylindrical or polygonal rim type of wheel around which cable, chain, belt, or other linkage may be wrapped. A drum may be driven or driving. The face may be smooth, grooved, fluted, or flanged.
drum counterweight rope Balance rope direct from drum drive. (Hammond, 1965)
drum curb *curb.*
drum feeder *roll feeder.*

drum filter Cylindrical drum, which rotates slowly through trough-shaped bath, fed continuously with thickened ore pulp.

drum horn Wrought-iron arms or spokes projecting beyond the surface or periphery of a flat-rope drum, between which the ropes coil or lap. See also: *spider*. (Fay, 1920)

drumman *slope engineer.*

drumming The process of sounding the roof of a mine to discover whether rock is loose. (Fay, 1920)

drummy (a) Loose coal or rock that produces a hollow, loose, open, weak, or dangerous sound when tapped with any hard substance to test condition of strata; said esp. of a mine roof. (Fay, 1920; BCI, 1947) (b) The sound elicited when bad (loose) roof is tested by striking with a bar. (Hudson, 1932)

drum pulley A pulley wheel used in place of a drum. See also: *Koepe system*. (Fay, 1920)

drum rings Cast-iron wheels, with projections, to which are bolted the staves or laggings forming the surface for the hoisting cable to wind upon. The outside rings are flanged, to prevent the cable from slipping off the drum. (Fay, 1920)

drum runner *incline man.*

drum separator A slowly rotating cylindrical vessel that separates run-of-mine coal into clean coal, middlings, and refuse. It consists of different and adjustable specific gravities. The low gravity medium in one compartment separates a primary float product (clean coal), the sink material being lifted and sluiced into the second compartment where middlings and true sinks (stone) are separated. (Nelson, 1965)

drum shaft *caisson sinking.*

druse (a) An irregular cavity or opening in a vein or rock, having its interior surface or walls lined (encrusted) with small projecting crystals usually of the same minerals as those of the enclosing rock, and sometimes filled with water; e.g., a small solution cavity, a steam hole in lava, or a lithophysa in volcanic glass. Cf: *geode; miarolitic cavity; vug*. (AGI, 1987) (b) A mineral surface covered with small projecting crystals; specif. the crust or coating of crystals lining a druse in a rock, such as sparry calcite filling pore spaces in a limestone. Etymol: German. Adj: drusy. (AGI, 1987)

drusy (a) Pertaining to a druse, or containing many druses. Cf: *miarolitic*. (AGI, 1987) (b) Pertaining to an insoluble residue or encrustation, esp. of quartz crystals; e.g., a drusy oolith covered with subhedral quartz. (AGI, 1987)

dry (a) Miner's changehouse, usually equipped with baths, lockup cubicles, and means of drying wet clothing. (Pryor, 1963) (b) A borehole in which no water is encountered or a borehole drilled without the use of water or other liquid as a circulation medium. Also called dry hole; duster. (Long, 1960) (c) A borehole that did not encounter a mineral-, oil-, or gas-producing formation. Also called blank hole; dry hole; duster. (Long, 1960)

dry air Air with no water vapor. (Strock, 1948)

dry ash-free basis An analysis expressed on the basis of a coal sample from which the total moisture and the ash have in theory been removed. (BS, 1960)

dry assay Any type of assay procedure that does not involve liquid as a means of separation. Cf: *wet assay*. See also: *volumetric analysis*. (AGI, 1987)

dry block The intentional act or process of running a core bit without circulating a drill fluid until the cuttings at and inside the bit wedge the core solidly inside the bit. Also called dry blocking. (Long, 1960)

dry blower *dry washer.*

dry blowing A process sometimes used where water is scarce. The separation of free gold from the accompanying finely divided material is effected by the use of air currents. See also: *dry cleaning*. (Nelson, 1965)

dry bone *dry-bone ore.*

dry-bone ore (a) An earthy, friable, honeycombed variety of smithsonite in veins or beds in stratified calcareous rocks accompanying sulfides of zinc, iron, and lead. (b) A variety of hemimorphite. Syn: *dry bone*.

dry bulb temperature Temperature indicated by a conventional dry thermometer; a measure of the sensible heat content of air. (Hartman, 1982)

drycleaned coal Coal from which impurities have been removed mechanically without the use of liquid media. (BS, 1960)

dry cleaning The cleaning of coal or ore by air currents as opposed to wet cleaning by water currents. Appliances for the dry cleaning of coal were first introduced about 1850 and since that date a variety of methods have been developed. See also: *Kirkup table*. (Nelson, 1965)

drycleaning table An apparatus in which drycleaning is achieved by the application of air currents and agitation to a layer of feed of controlled depth on the table surface. See also: *Kirkup table*. (BS, 1962)

dry copper Underpoled copper from which oxygen has been insufficiently removed when refining, so that it is undesirably brittle when worked cold or hot. (Pryor, 1963)

dry cyaniding *carbonitriding.*

dry delta *alluvial cone; alluvial fan.*

dry density The weight of a unit volume of a dry sample of soil, after the latter has been heated at a temperature of 103 °C. (Hammond, 1965)

dry density/moisture ratio The relationship between the density of a sample of soil in a dry state and its moisture content for a given degree of compaction. Such relationship can be determined from a curve that will reveal the optimum moisture content. (Hammond, 1965)

dry diggings (a) Placers not subject to overflow. (b) Placer mines or other mining districts where water is not available. (Standard, 1964)

dry distillation *destructive distillation; pyrolysis.*

dry drilling Drilling operations in which the cuttings are lifted away from the bit and transported out of a borehole by a strong current of air or gas instead of a fluid. Cf: *dry running*. (Long, 1960)

dry ductor Compressed-air drill that traps and removes drilling dust instead of sludging it with added water. (Pryor, 1963)

dryer An apparatus for drying coal. Dryers are of various types, such as revolving kilns, flash, and fluidized bed. See also: *coal dryer; dehydrator.*

dry fatigue A condition often appearing in wire rope and often caused by shock loads in winding. These shock loads are produced by picking up the cage from the pit bottom with slack chains or by lifting heavy pithead gates or covers. (Sinclair, 1959)

dry galvanizing A process in which steel is fluxed in hot ammonium chloride and subsequently dried by hot air before being passed through a bath of molten zinc. (Hammond, 1965)

dry grinding Any process of reducing particle size without the liquid medium. See also: *grinding*.

dry hole A drill hole in which no water is used for drilling, as a hole driven upward. See also: *blank hole*. (Standard, 1964)

dryhouse *changehouse.*

dry ice Solid carbon dioxide.

dry ice test A test for detecting glass imitations; if a crystalline substance such as a gem mineral is placed on a piece of dry ice (solid CO_2), a squeaking noise can be heard. This is not true of noncrystalline substances such as glass or plastic.

dry joint Positive separation at the plane of contact between adjacent structural components to allow relative movement arising from differences in temperature or shrinkage. (Hammond, 1965)

dry kata cooling power A measure of the rate of heat loss from the bulb of the kata thermometer. Although the cooling power as obtained by this instrument is not a measure of the capacity of an atmosphere to cool the human body, nevertheless, it is useful for comparing different atmospheres and provides a convenient index of the comfort condition of a working place in a mine. Experience indicates that a face will be reasonably comfortable for working if the dry kata cooling power is above 7 on the kata thermometer and the air velocity above 1 m/s. See also: *dehumidification; effective temperature*. (Nelson, 1965)

dry lake *playa.*

dry milling The comminution of materials in a suitable mill without the presence of a liquid, either by rods, balls, or pebbles, or autogenously, by the material itself; used if the subsequent process is a dry process.

dry mineral matter free basis An analysis expressed on the basis of a coal sample from which the total moisture and the mineral matter have in theory been removed. (BS, 1960)

dry mining (a) In dry mining every effort is made to prevent the ventilating air picking up moisture, and throughout the ventilation circuit there is a wide gap between wet- and dry-bulb temperatures. Dry-bulb temperatures are therefore comparatively high. (Spalding, 1949) (b) Refers to the reasonably dry footing required for the equipment. Dry stripping and placer mining with standard earthmoving heavy equipment depends on good footing for the equipment and short hauls to keep costs in line. Syn: *dry diggings*. (SME, 1992)

dry ore Said of lead or copper ore that contains precious metals (gold and silver) but insufficient lead or copper to be smelted without the addition of richer ore. See also: *natural ore*.

dry placer Gold-bearing alluvial deposit found in arid regions. In some deposits the gold is in the cementing material that binds the gravel together. Because of the lack of water, various machines have been devised for the dry washing of these deposits; such machines commonly include some form of pulverizer and jigs or tables that use compressed air instead of water in their operation. Also called desert placer. (Lewis, 1964)

dry process (a) A method of treating ores by heat as in smelting; used in opposition to wet process, where the ore is brought into solution before extraction of the metal. See also: *wet process*. (Fay, 1920) (b) The process of making Portland cement in which the raw materials are ground and burned dry. (Mersereau, 1947) (c) Process whereby dry powdered enameling materials are applied to a preheated surface. (Van Vlack, 1964) (d) The method of preparation of a ceramic body wherein the constituents are blended dry, following which liquid may be added as required for subsequent processing. (ASTM, 1994)

dry puddling A process of decarbonization on a siliceous hearth in which the conversion is effected rather by the flame than by the reaction of solid or fused materials. As the amount of carbon diminishes, the mass becomes fusible and begins to coagulate (come to nature), after which it is worked together into lumps (puddle balls, loups) and removed from the furnace to be hammered (shingled) or squeezed in the squeezer, which presses out the cinder, etc., and compacts the mass at welding heat, preparatory to rolling. Silicon and phosphorus are also largely removed by puddling, passing into the cinder. See also: *puddling.* (Fay, 1920)

dry rotary drilling *dry drilling.*

dry running To unknowingly or knowingly drill with a bit when the flow of the coolant and cuttings-removal fluid past the bit has been inadvertently or deliberately cut off. Cf: *dry drilling.* (Long, 1960)

dry sample A sample obtained by drilling procedures in which water or other fluid is not circulated through the drill string and sampling device; hence the in situ characteristics of the sample have not been altered by being mixed with water or other fluid. Cf: *drive sample.* (Long, 1960)

dry sampler Various auger and/or tubular devices designed to obtain unwetted samples of soft rock material, such as clay, sand, soil, etc., by drilling procedures wherein water or other fluid is not circulated during the operation. (Long, 1960)

dry sand (a) A stratum of dry sand or sandstone encountered in well drilling. A nonproductive sandstone in oilfields. (Fay, 1920) (b) Green sand dried in an oven to remove moisture and strengthen it (a dried-sand mold is a mold of green sand that is treated as above). (Freeman, 1936)

dry screening The screening of solid materials of different sizes without the aid of water. (BS, 1962)

dry separation (a) The elimination of the small pieces of shale, pyrite, etc., from coal by a blast of air directed upon the screened coal. See also: *wind method.* (b) *dedusting.*

dry sweating A process by which impure blister copper is exposed to long oxidizing heating below fusion point. (Standard, 1964)

dry unit weight The weight of soil solids per unit of total volume of soil mass. Also called unit dry weight. See also: *unit weight.* (ASCE, 1958)

dry wall stone Thin-bedded limestone or sandstone suitable only for mortarless (dry) fencing or retaining walls.

dry wash A wash that carries water only at infrequent intervals and for short periods, as after a heavy rainfall. Cf: *arroyo.* (AGI, 1987)

dry washer (a) A machine for extracting gold from dry gravel. It consists of a frame in which there is a rectangular bellows made of canvas; the upper part of the bellows is made by a plane set at an angle of about 20°, across which are riffles. On the top of the machine is a screen on which gravel is shoveled. The screened gravel falls to a riffled plane from which it feeds to the riffles on the bellows. The screen and upper riffles are shaken by an eccentric worked with a crank, and the same crank actuates the bellows, which blow the dust from the gravel passing over the riffles. The gold is caught behind the riffles. Only gravel in which no moisture can be seen can be worked successfully by a dry washer. (Hess) (b) A person who operates a dry washer. Syn: *dry blower.*

dry well A deep hole, covered and usually lined or filled with rocks, that holds drainage water until it soaks into the ground. (Nichols, 1976)

dual-drive conveyor A conveyor having a belt drive mechanism in which the conveyor belt is in contact with two drive pulleys, each of which is driven by a separate motor. (NEMA, 1961)

dual haulage In strip mining, the use of two types of haulage at the same mine for transporting coal from the face to the preparation plant. Usually, coal is transported from the loading shovel to a transfer station by motorized units, and rail haulage is used to haul the coal from this point to the preparation plant. (Toenges, 1938)

dualin A variety of dynamite consisting of four to five parts nitroglycerin, three parts sawdust, and two parts potassium nitrate. (Webster 2nd, 1960)

dubiocrystalline *cryptocrystalline.*

duchess Slate size of 24 in by 12 in (61.0 cm by 30.5 cm). (Pryor, 1963)

duck A fabric material, usually of woven cotton but of synthetic fibers also, used to construct conveyor belts and filter cloths. (Pryor, 1963)

duckbill The name given to a shaking-type combination loading and conveying device, so named from the shape of its loading end and which generally receives its motion from the shaking conveyor to which it is attached. (BCI, 1947)

duckbill loader *shaker-shovel loader.*

duckbill operator In bituminous coal mining, one who operates a small power shovel that has a round-nosed scoop, called a duckbill, to load coal into cars in a mine. (DOT, 1949)

duckbill pick A duckbill-shaped coal-cutter pick that is forged by the roller type of machine from dies and is the type largely used today. The machine shaping of the pick ensures uniformity. It gives a constant clearance as the point wears down and is particularly suitable for fused-carbide tipping. (Nelson, 1965)

duckfoot A pipe bend at the bottom of a shaft column or rising main fitted with a horizontal base sufficiently strong for the weight of the rising main to rest upon it. Also called duckfoot bend. (BS, 1963)

duckfoot bend *duckfoot.*

duck machine An arrangement of two boxes, one working within the other, for forcing air into mines. (Zern, 1928)

duck's-nest tuyère A tuyère having a cupped outlet. (Standard, 1964)

ducktownite An intimate mixture of pyrite and chalcocite or the matrix of a blackish copper ore containing grains of pyrite, Tennessee.

ducon Abbrev. for dust concentrator, which is a device used to collect dry cuttings ejected from a borehole in which air or gas is used as a circulation medium. (Long, 1960)

duct A pipe or tubing used for auxiliary ventilation in a mine. Generally constructed of coated fabric, metal, or fiberglass.

duct fan An axial-flow fan mounted in, or intended for mounting in, a section of duct. See also: *tube-axial fan; vane-axial fan; mine ventilation auxiliary fan.*

ductile (a) Said of certain metals and other substances that readily deform plastically. (b) Said of a rock that is able to sustain, under a given set of conditions, 5% to 10% deformation before fracturing or faulting. (AGI, 1987) (c) In mineralogy, capable of considerable deformation, esp. stretching, without breaking; said of several native metals and occasionally said of some tellurides and sulfides. (AGI, 1987) (d) Pertaining to a substance that readily deforms plastically. (AGI, 1987) (e) Capable of being permanently drawn out without breaking; such as, a ductile metal. (Webster 3rd, 1966)

ductile cast iron High-carbon ferrous product containing spheroidal graphite. (Bennett, 1962)

ductile crack propagation Slow crack propagation that is accompanied by noticeable plastic deformation and requires energy to be supplied from outside the body. (ASM, 1961)

ductility The ability of a material to deform plastically without fracturing.

ducting Sections of air duct. See also: *ventilation tubing.* (BS, 1963)

dudgeonite A calcian variety of annabergite, $(Ni,Ca)_3(AsO_4)_2{\cdot}8H_2O$, in which about one-third of the nickel is replaced by calcium; Pibble Mine, Scotland.

due The amount of royalty or ore payable to the lord of the manor or owner of the soil. (Fay, 1920)

duff (a) Fine dry coal (usually anthracite) obtained from a coal-preparation plant. The size range is 3/16 to 0 in (4.8 to 0 mm). See also: *gum; slack.* (Nelson, 1965) (b) Smalls, usually with an upper size limit of 3/8 in (9.5 mm). (BS, 1960) (c) Term used among British miners for a fine mixture of coal and rock. (Tomkeieff, 1954) (d) Aust. The fine coal left after separating the lumps; very fine screenings; dust. (e) Coal dust and other unsaleable small coal produced in anthracite mines. See also: *anthracite fines.* (Pryor, 1963)

duffer Aust. An unproductive claim or mine. (Hess)

Duff furnace A furnace used for the manufacture of producer gas. (Fay, 1920)

dufrenite A monoclinic mineral, $Fe^{2+}Fe^{3+}_4(PO_4)_3(OH)_5{\cdot}2H_2O$; forms green botryoidal crusts commonly associated with limonite in gossans. Syn: *kraurite; green iron ore.*

dufrenoysite A monoclinic mineral, $Pb_2As_2S_5$; sp gr, 5.6; in crystalline dolomite in the Binnental, Switzerland.

duftite An orthorhombic mineral, $PbCu(AsO_4)(OH)$; adelite group; sp gr, 6.4; at Tsumeb, Namibia.

duin A gold-washing dish used in Jashpur, India.

dukeway Som. A method of hoisting coal on an incline from the working face to the pit bottom by a rope attached to the winding engine at surface in such a way that while the cage is going up, the empty trams are running down the incline, and as the cage descends the loaded cars are brought up to the shaft.

dull (a) Brist. Slack ventilation; insufficient air in a mine. (b) As applied to the degree of luster of minerals, means those minerals in which there is a total absence of luster, as chalk or kaolin.

dull attritus A field term denoting the degree of luster of attrital coal as it compares to the brilliant luster of vitrain associated in the same locality. Cf: *bright attritus.* (AGI, 1987)

dull-banded coal Coal consisting of vitrain and durain with more or less minor clarain and minor fusain. (AGI, 1987)

dull coal (a) Any coal that absorbs the greater part of incident light instead of reflecting it. Stopes recognizes two kinds of dull coal—durain and fusain. (Tomkeieff, 1954) (b) The constituent of banded coal macroscopically somewhat grayish in color, of a dull appearance, less compact than bright coal, and breaking with a rather irregular fracture. It consists mainly of two kinds of material; thin black bands interlayed by a lighter colored granular-appearing matter. Microscopically, it consists of smaller anthraxylon constituents together with a few other constituents, such as cuticles and barklike constituents embedded in a general matrix, the attritus. (AGI, 1987) (c) A variety of banded coal containing from 20% to 0% of pure, bright ingredients (vitrain, clarain, and fusain), the remainder consisting of clarodurain and durain. (AGI, 1987)

dulong A Malayan term for hardwood pan shaped like a section of the surface of a sphere and used as a miner's pan in prospecting, sample washing, and manual concentration of cassiterite. See also: *pan.* (Pryor, 1963)

dumb barge A barge similar to a hopper barge, frequently used to take dredged material from a dredger to the dumping ground. (Hammond, 1965)

dumb'd Choked or clogged, as a grate or sieve in which the ore is beneficiated. (Fay, 1920)

dumb drift (a) A passage leading from an airway to a point in a shaft some distance above an inset to allow the ventilating current to bypass a station where skips or cages are loaded. (BS, 1963) (b) A roadway driven through the waste in longwall mining to provide packing material. (Nelson, 1965)

dumb furnace A ventilating furnace in which the foul flammable air from remote parts of the mine enters the upcast higher up than the hot gases from the fire. (Webster 2nd, 1960)

dumb screen A chute in which there are no meshes or bars for separating the coal, and down which the run-of-mine coal passes from the tubs direct to the railway wagon. It is used in small mines where the coal is sold as loaded underground or at mines where the coal is conveyed by wagon to a central coal-preparation plant. (Nelson, 1965)

dummy (a) A short piece of core or core-size cylinder of rubber or other material placed in the core lifter in an empty core barrel to guide the first part of a newly cut core into the core lifter. Syn: *guide core.* (Long, 1960) (b) A cathode, usually corrugated to give variable current densities, which is plated at low current densities to remove preferentially impurities from a plating solution. (ASM, 1961) (c) A substitute cathode that is used during adjustment of operating conditions. (ASM, 1961)

dummy elevator A second elevator for boosting tailings to higher stacking levels.

dummy gate N. of Eng. A small gate made on the face between the mother gate and tailgate for the purpose of getting stone to make strip packs for roof support (when goaf roof is supported and not allowed to cave). (Trist, 1963)

dummy locator One whose name is used by a locator to secure for the latter's benefit a greater area of mineral land than is allowed by law to be appropriated by a single person, and any location made in pursuance of such a scheme or device is without legal support and void. (Ricketts, 1943)

dummy maker In bituminous coal mining, a laborer who fills paper cartridges (cylinders) with clay, adobe, or rock dust, used for stemming (tamping clay or other material on top of explosives) drill holes in the working face to be blasted down. (DOT, 1949)

dumontite A monoclinic mineral, $Pb_2(UO_2)_3O_2(PO_4)_2 \cdot 5H_2O$; yellow; secondary; radioactive.

dumortierite An orthorhombic mineral, $Al_7(BO_3)(SiO_4)_3O_3$; fibrous or radiating; Mohs hardness, 7 to 8-1/2; in schists, gneisses, and pegmatite dikes.

dumortierite quartz A massive, blue, violet-blue, or blue-black, opaque variety of quartz colored by intergrown crystals of dumortierite.

Dumoulin process A method whereby copper is deposited on a rotating mandrel and later stripped off as a long strip, which is then drawn into wire without recasting. (Liddell, 1918)

dump (a) A pile or heap of ore, coal, or waste at a mine. (b) The point where a face conveyor discharges its coal into mine cars. (Nelson, 1965) (c) The tipple by which the cars are dumped. Cf: *tipple.* (d) The fall available for disposal of refuse at the mouth of a mine.

dump bailer A bailer used in borehole-cementation work, provided with a valving device that empties the contents of the bailer (cement) at the bottom of a borehole. Also called liquid dump bailer. (Long, 1960)

dumpcart A cart having a body that can be tilted or a bottom opening downward for emptying the contents without handling. (Webster 3rd, 1966)

dumped fill Excavated material transported and dumped in a heap, generally to preestablished lines and grades. Should be kept free of tree stumps, organic matter, trash, and sod if any future use of the filled area is contemplated. (Carson, 1961)

dump equipment One of many conveyances that carry and then dump rock or ore. Generally trucks in surface mining and shuttle cars in underground mining. (SME, 1992)

dumper A wheeled car with an elevated turntable on which is a track. A mine car run on the upper, horizontally revolving track can be dumped sidewise or endwise. Also called hurdy girdy. (Zern, 1928)

dump hook A chain grabhook having a lever attachment for releasing it. (Webster 3rd, 1966)

dump house The building where the loaded mine cars are emptied into the chutes. (Fay, 1920)

dumping bucket A lifting bucket with a tilt or drop bottom. (Standard, 1964)

dumping wagon *dumper.*

dump leaching Term applied to dissolving and recovering minerals from subore-grade materials from a mine dump. The dump is irrigated with water, sometimes acidified, which percolates into and through the dump, and runoff from the bottom of the dump is collected and mineral in solution is recovered by a chemical reaction. Commonly used to recover gold and copper. See also: *heap leaching.*

dumpling A mass of ground left undisturbed until the final stages of excavation, when it is removed. In the intermediate stages it may be used as a support for timbering to the excavations. (Hammond, 1965)

dump motorman In bituminous coal mining, one who operates a mine locomotive (motor) to haul cars of dirt, rock, slate, or other refuse to the dump at the surface of an underground mine. Also called dirt-dump engineer; refuse engineer. (DOT, 1949)

dump room Space available for disposing of waste from a mine.

dump skip A skip with an attachment that dumps the load automatically. (Fay, 1920)

dump wagon A large-capacity side-, bottom-, or end-discharge wagon (or skip) on tired wheels or crawler tracks; usually tractor towed. (Nelson, 1965)

dundasite An orthorhombic mineral, $PbAl_2(CO_3)_2(OH)_4 \cdot H_2O$; secondary; in aggregate tufts of minute, radiating needles.

dunite Peridotite in which the mafic mineral is almost entirely olivine, with accessory chromite almost always present. Named by Hochstetter in 1864 from Dun Mountain, New Zealand. (AGI, 1987)

Dunkard series Continental strata, including thin coal seams, similar to the Pennsylvanian, but of Permian age, occurring in North America. Strata of the same age are marine in Kansas, but include marginal red beds with gypsum, and thick salt deposits were formed later in the Kansas Basin. (CTD, 1958)

duns A term used in SW England for a shale or massive clay associated with coal. (AGI, 1987)

dunstone (a) A term used near Matlock, England, for a hard granular yellowish or cream-colored magnesian limestone. (AGI, 1987) (b) A term used in Wales for a hard fireclay or underclay, and in England for a shale. (AGI, 1987)

Duobel Trademark for high-velocity permissible explosives furnished in seven grades based upon velocity and cartridge count; poor water resistance. Used for mining coal where lump coal is not a factor. (CCD, 1961)

duoflex checker system A checker arrangement for hot-blast stoves. The gas used is only partially cleaned and may contain from 0.5 to 1.5 g of dust per cubic meter of gas. The top zone of the checkers is formed of straight-walled vertical passages, and the middle zone of vertical passages in each of which two opposite walls are continuously curved and the other two are straight, while the bottom zone is formed of vertical passages in each of which all four walls are continuously curved. (Osborne, 1956)

duplex breaker A breaker having more than one crushing chamber.

duplex channeler A type of channeling machine that cuts two channels simultaneously.

duplex compressor Two compressors, side by side, and made in the combination of simple steam and simple air cylinder, simple steam and compound air cylinders, or compound steam and compound air cylinders. See also: *air-conditioning process.* (Lewis, 1964)

duplexing (duplex process) Any two-furnace melting or refining process. (ASM, 1961)

duplex pick A coal-cutter pick that allows a cut to be made in either direction without turning the pick. It is drop forged with a tip of fused tungsten carbide. (Nelson, 1965)

duplex pump A positive displacement pump with two water or liquid cylinders side by side and geared so that the piston strokes in the cylinders alternate. Such a pump may be either single or double action, depending on the number and placement of intake and discharge valves on the cylinder and may be designed so as to deliver a low volume of liquid at high pressures. Cf: *centrifugal pump; triplex pump.* (Long, 1960)

duplex pump Displacement pump for handling pulps. Two cylinders are so geared that one piston falls while other rises. Can lift small tonnages to good heights. (Pryor, 1963)

duplex steel Steel produced by first refining it in a Bessemer converter and afterward completing the process in an open-hearth furnace. (Mersereau, 1947)

duplex Talbot process A combination of the duplex and the Talbot continuous process. Molten steel from the Bessemer converter, already freed of its carbon, silicon, and manganese contents, is charged into the Talbot furnace. As this molten steel is poured through the oxidized slag, the phosphorus is removed almost immediately. Sometimes pig iron is poured in afterwards, which raises the carbon content of the bath and aids in its deoxidation. A portion of the heat can usually be tapped about an hour after this addition. (Osborne, 1956)

duplex wire Two insulated-copper leading wires wrapped together with paraffined cotton covering. (Fay, 1920)

duplicate sampling The placing of alternate samples of coal or ore in different containers that are then analyzed separately. Each container thus holds a representative subsample taken at intervals throughout the sampling intervals or period. (Nelson, 1965)

Dupont process A heavy-liquid minerals separation process in which organic liquids of high specific gravity, known as parting liquids, are used. With sp gr 1.00 to 2.96 and very low viscosities, they serve ideally for the medium in the sink-and-float separation of solid materials. This process is used to clean run-of-mine anthracite, refuse banks, or mixtures of the two. The sizes of anthracite coal that can be cleaned are No. 1 buckwheat, and larger. This includes sizes up through broken. See also: *parting liquid.* (Mitchell, 1950)

durability (a) The capacity of a gem to withstand abrasion, impact, and chemical alteration. (b) The rate of deterioration of foundry sand due to dehydration of its contained clay.

durain The term was introduced by M.C. Stopes in 1919 to designate the macroscopically recognizable dull bands in coals. Bands of durain are characterized by their gray to brownish-black color and rough surface with dull or faintly greasy luster; reflection is diffuse; they are markedly less fissured than bands of vitrain, and generally show granular fracture. In humic coals, durain occurs in bands up to many centimeters in thickness. Widely distributed, but with exceptions not abundant. Cf: *fusain; vitrain.* See also: *attritus; hards.* (IHCP, 1963)

durangite A monoclinic mineral, $NaAl(AsO_4)F$; occurs with cassiterite at the Barranca tin mine, Durango, Mexico.

duricrust A general term for a hard crust on the surface of, or layer in the upper horizons of, a soil in a semiarid climate. It is formed by the accumulation of soluble minerals deposited by mineral-bearing waters that move upward by capillary action and evaporate during the dry season. See also: *ferricrete; silcrete; calcrete; caliche.* Etymol: Latin durus, hard, + crust. Cf: *hardpan.* (AGI, 1987)

durionizing A process of electrodepositing hard chromium on the wearing surfaces of parts as a protection against wear by friction. (Osborne, 1956)

durite (a) Term for the microlithotype consisting principally of the following groups of macerals: inertinite (micrinite, fusinite, semifusinite, and sclerotinite) and exinite, particularly sporinite. Durite contains at least 95% inertinite and exinite. The proportions of these two groups of macerals may vary widely, but each must be greater than the proportions of vitrinite and neither must exceed 95%. Durite E and durite I connote durites rich in exinite and inertinite, respectively. It is found in many coals, in fairly thick bands, principally in durains and the duller type of clarain, generally common. (IHCP, 1963) (b) A coal microlithotype that contains a combination of inertinite and exinite totalling at least 95%, and containing more of each than of vitrinite. Cf: *durain.* (AGI, 1987)

duroclarain A rock-type coal consisting of the maceral vitrinite (telinite or collinite) and large quantities of other macerals, mainly micrinite and exinite. Micrinite is present in lesser quantities than is true with clarodurain. Cf: *clarodurain.* (AGI, 1987)

duroclarite (a) This term was introduced in 1956 by the Nomenclature Subcommittee of the International Committee for Coal Petrology to designate the microlithotype with maceral composition between those of clarite and durite but closer to clarite than durite. Further specification is that the proportion of vitrinite must exceed that of inertinite. It occurs in fairly thick bands, and is widely distributed and, like clarodurite, is a common constituent of most humic coals. The technological properties of duroclarite are intermediate between those of clarite and durite, but because of the predominance of vitrinite over inertinite they resemble those of clarite more closely than those of durite. (IHCP, 1963) (b) Coal microlithotype intermediate between clarite and durite; vitrinite, exinite, and inertinite each exceed 5% and the last is less abundant than vitrinite. (AGI, 1987) (c) A coal microlithotype containing at least 5% each of vitrinite, exinite, and inertinite, with more vitrinite than inertinite and exinite. It is a variety of trimacerite, intermediate in composition between clarite and durite, but closer to clarite. Cf: *clarodurite.* (AGI, 1987)

durovitrain Vitrain with minute inclusions of durain. Cf: *vitrodurain.* (AGI, 1987)

Durville process A casting process that involves rigid attachment of the mold in an inverted position above the crucible. The melt is poured by tilting the entire assembly, causing the metal to flow along a connecting launder and down the side of the mold. (ASM, 1961)

dussertite A trigonal mineral, $BaFe_3(AsO_4)_2(OH)_5$; crandallite group; forms green to yellow-green rosettes or crusts.

dust *ash.*

dust barrier *stone-dust barrier.*

dust catcher (a) A device attached to the collar of a borehole to catch or collect dry, dustlike rock particles produced in dry-drilling a borehole. Cf: *ducon.* (Long, 1960) (b) Any device in which dust is collected or extracted from furnaces, gases, etc.

dust chamber (a) An enclosed flue or chamber filled with deflectors, in which the products of combustion from an ore-roasting furnace are passed, the heavier and more valuable portion settling in the dust chamber and the volatile portions passing out through the chimney or other escape. (Fay, 1920) (b) Room air system, flue or dust extractor, where larger particles can drop out of stream of gas and be periodically removed. Used in conjunction with cyclones, electrostatic precipitators, and bag filters. (Pryor, 1963)

dust cloud Coal or other dust particles carried in suspension in the air by currents and eddies. (Rice, 1913-18)

dust cloud flammability The flammability of a coal-dust cloud is its ability to promote spreading flames away from the source of ignition. (Sinclair, 1958)

dust collection Removal from atmosphere of mill or from transfer points where dust is thrown up. Partially closed ventilating systems are used, which incorporate bag filters, Cottrells, cyclones, washing chambers, and spray towers. (Pryor, 1963)

dust collector (a) A device used on a roof bolting machine while in operation for separating solid particles from air and accumulating them in a form convenient for handling. (b) Device used to collect dust produced in percussion rock drilling, this device has an exhauster operated with compressed air from the available air system. Air laden with dust and cutting is drawn from the boreholes through the bit holes, the hollow drill steel, the adapter, and the suction hose into a filter. The filtered air is excavated with the spent compressed air through an exhaust port, and dust and cuttings settle in a removable storage tank.

dust consolidation The binding of coal dust on roadway surfaces to prevent it becoming airborne by any disturbance. One method is to spread calcium chloride over the dust so that it absorbs water and forms a pasty cake that does not rise into suspension when workers travel on the roadway. See also: *stone dust.* (Nelson, 1965)

dust counter A portable apparatus (as the Koltze tube, an impinger, etc.) used to measure dust concentration in a mine or mill, as a health precaution. (Pryor, 1963)

dust explosion An explosion which consists of a sudden pressure rise caused by the very rapid combustion of airborne dust. Ignition of suspensions of combustible dusts can occur in the following ways: (1) initiation by flame or spark, (2) propagation by a gas explosion or blasting, and (3) spontaneous combustion. Little is known about the last-named mechanism, which is relatively rare in mines. The most frequent causes of major coal mine explosions in the United States today are electric arcs, open flames, and explosives. See also: *coal-dust explosion.* (Hartman, 1982)

dust extraction The removal of solid particles suspended in gas or ambient air. (BS, 1962)

dust extractor An appliance to collect or precipitate suspended dust. Dust extraction is often necessary at coal-preparation plants, loading stations, and also underground. The appliance may be a cyclone, fabric filter, spray tower, scrubber, or an electrostatic separator. See also: *dust precipitator; dust trap.* (Nelson, 1965)

dust firing The burning of coal dust in the laboratory of the furnace. (Fay, 1920)

dust-free conditions In Great Britain, the arbitrary standards laid down by the National Coal Board in 1949 as representing comparative dust-free conditions in coal mines. These are as follows: stone dust, 450 particles/cm^3 (size range, 0.5 to 5.0 μm); anthracite, 650 particles/cm^3 (size range, 1 to 5 μm); and coal, 850 particles/cm^3 (size range, 1 to 5 μm). (Nelson, 1965)

dust gold Pieces of gold under 2 to 3 pennyweights (3.1 to 4.7 g); very fine gold.

dust hopper A hopper placed underneath the scraper, rapping roller, or other belt cleaner, to collect the dust and dirt as it is removed from the belt; any tank or vessel to receive and retain dust. (Nelson, 1965)

dusting loss (a) Shortfall in expected weight of sands or finely ground materials due to wind action or loss when transported in open trucks. (Pryor, 1963) (b) In laboratory sampling, the loss of part of a sample undergoing test, through leakage of particles into the atmosphere. (Pryor, 1965)

dust-laying oil Crude oils, heavy asphalt oils, tars, solutions of petroleum asphalt in gas oils, liquid asphalt, and emulsions of oils and water, used for laying dust on roads. (Fay, 1920)

dustless zone A section of the mine entry from which dust has been removed as completely as possible by scraping or sweeping, aided by a compressed-air blast. (Rice, 1913-18)

dustman One who dumps the dust catcher or loads the dust at blast furnaces. (Fay, 1920)

dustpan dredge A dredge containing a suction head that is pulled over the underwater ground much as a dustpan would be. About 8 in (20 cm) high, the dustpan may be from 20 to 40 ft (3.6 to 7.1 m) long and is supplied with jets along its face to stir up the bottom surface. (Carson, 1965)

dust plan A plan kept with the book in which stone-dust samples are recorded. It shows the sampling zones in each roadway, distinguished by color, letter, number, or mark, and identified with that roadway. (Nelson, 1965)

dustplate A vertical iron plate, supporting the slag runner of an iron blast furnace. (Fay, 1920)

dust precipitator On a large scale, sinter plant gas may be cleaned by precipitators with very high efficiencies. The dust is precipitated in a dry state, suitable for pelletizing and feeding back onto the sinter strand. See also: *thermal precipitator.* (Nelson, 1965)

dustproofing A surface treatment, as with oil or calcium chloride solution, to prevent or reduce the dustiness of coal in handling. (BS, 1962)

dust sampling The taking of air samples to assess their degree of dustiness, either on a mass basis or on particle count in a known volume of air. Numerous instruments have been developed for this purpose. Dust sampling is also necessary to assess the efficiency of stone dusting. See also: *sampling instrument; konimeter; size selector; thermal precipitator; tyndallometer.*

dust-sampling impinger A portable instrument for collecting dust samples so that corrective measures can be taken for dust control and the prevention of respiratory diseases. Dust-laden air is impinged in sampling flasks by manual, compressed air, or electrical suction devices. Dust counts are made from the collected air at laboratories with microscopes and counting cells. (Best, 1966)

dust suppression The prevention or reduction of the dispersion of dust into the air, for example, by water sprays. (BS, 1962)

dust-suppression jib A coal-cutter jib designed to conduct water through ducts, or other arrangement, to the back of the kerf, to suppress dust and reduce the gas-ignition hazard. See also: *whale-type jib.* (Nelson, 1965)

dust-suppression person A person employed in coal mines to apply measures to allay coal dust on mine roadways and along the coal faces. The worker also may be in charge of dust suppression in rock drivages. See also: *rock dust.* (Nelson, 1965)

dust-suppression system With this system, dust can be suppressed before it becomes airborne. A series of nozzles discharge a chemical compound in a fine spray to materially reduce the amount of water or other liquids necessary to saturate fly ash and eliminate dust. The compound also aids in the diffusion of the liquid dust suppressant, allowing it to penetrate deeper into the material. This system can be used at any point in the handling of bulk materials, wherever dust is a hazard. (Best, 1966)

dust trap An appliance for the dry collection of dust during drilling in rock. The rock chippings, dust, and air are sucked from the borehole through a rubber hose to a drum-type container with filters. The drum is discharged and the filters renewed periodically. In some of the newer types, the dust is extracted through the hollow drill rods. See also: *C.P. Hemborn dust extractor; Holman dust extractor.* (Nelson, 1965)

dust wetting agent Chemical compounds that aid in the control of dry dusts such as coal and silica to help prevent explosions and respiratory injury to workers. These compounds are of two types. One type is used in a dry state and controls dust by absorbing moisture from the air. The second type is an agent for increasing the wetting effectiveness of water by breaking the surface tension and permitting the water-compound mixture to thoroughly cover the treated area. (Best, 1966)

Dutch drop A haulage term for flying switch. (Fay, 1920)

Dutch State Mines Process A sink-float process used principally for coal cleaning. The process uses a water suspension of loess (a natural claylike material) in special trough-type separators provided with a clean coal weir and a reject drag conveyor. (Kirk, 1947)

dutch twill A type of wire cloth weave; a weave in which the first shute wire crosses over the first and second warp wires, under the third and fourth warp wires, and the second shute wire crosses under the first warp wire, over the second and third warp wires, under the fourth and fifth, etc. (Henderson, 1953)

duttonite A monoclinic mineral, $VO(OH)_2$; forms minute pale-brown scales as an alteration of montroseite in sandstone on the Colorado Plateau.

duty of giants The amount of gravel that can be moved by a water cannon, or giant, in a 24-h day by a specified flow of water. The duty of giants varies considerably with local conditions, such as the height of the gravel banks, the nature of the gravel bedrock, head of water obtainable, size of jet, etc. (Griffith, 1938)

duty of the miner's inch The number of cubic yards of gravel that can be broken down and sent through the sluice by 1 miner's inch of water for 24 h. It depends upon the height of the bank, the character of the gravel and the bedrock, the grade of the bedrock, the type of sluice, and the pressure of the water. In well-rounded gravel without large stones, the duty of the miner's inch is from 4.5 to 6 yd^3 (3.3 to 4.6 m^3) of gravel for 24 h. Under less favorable conditions, the duty may range from 2.8 to 4.6 yd^3 (2.1 to 3.5 m^3) for 24 h. See also: *miner's inch.* (Lewis, 1964)

duxite An opaque, dark-brown variety of retinite containing about 0.5% sulfur in lignite at Dux, Bohemia, Czech Republic; is similar to muckite, walchowite, and neudorfite.

dwarf Brinell tester A portable ball hardness tester in which the load is applied by means of a vise or lever. It carries a special lens for measuring the impression's diameter and from which the Brinell hardness value can be read directly. (Osborne, 1956)

Dwight-Lloyd machine Sintering machine in which feed moves continuously on articulated grates pulled along by chains in belt-conveyor fashion. Controlled combustion on these grates causes the minerals to sinter. (Pryor, 1963)

Dwight-Lloyd process Blast roasting in which air currents are drawn downward through the ore mass. (Bennett, 1962)

Dwight-Lloyd roaster A multihearthed circular furnace, through which horizontal rabbles revolve and move the feed across each hearth, so that it falls peripherally to the one below and then works inward to central discharge for next hearth below. Rising heat and air provide the roasting conditions. (Pryor, 1963)

dycrasite An orthorhombic mineral, Ag_3Sb; metallic silver-white, commonly tarnished; sp gr, 9.74; in veins with galena, native silver, and silver sulfosalts; an ore of silver.

dyed stones Minerals artificially dyed to improve their color or to imitate a more valuable stone; they usually fade or discolor and may include chalcedony, turquoise, jadeite, opal, serpentine, and alabaster.

dye line print A contact print that has largely replaced the blueprints. (Pryor, 1963)

dyke The British spelling of dike.

dynamic braking A method of retarding an electric winder or haulage, in which a direct current is injected into the alternating-current winder motor stator during the deceleration period; the motor then acts as an alternator, and the negative load of the winding cycle is absorbed as electric power and wasted as heat in the controller. See also: *electric braking.* (Nelson, 1965)

dynamic damping Usually found in seismographs or seismometers where damping of motion is desired that is in proportion to the velocity of the moving mass. (AGI, 1987)

dynamic geology A general term for the branch of geology that deals with the causes and processes of geologic phenomena; physical geology. (AGI, 1987)

dynamic head The head of fluid that would statically produce the pressure of a given moving fluid. Syn: *velocity head.* (Standard, 1964)

dynamic load (a) An alternating or variable load. (Osborne, 1956) (b) *live load.*

dynamic metamorphism The total of the processes and effects of orogenic movements and differential stresses in producing new rocks from old, with marked structural and mineralogical changes due to crushing and shearing at low temperatures and extensive recrystallization at higher temperatures. It may involve large areas of the Earth's crust, i.e., be regional in character. Cf: *dynamothermal metamorphism; regional metamorphism.* Syn: *dynamometamorphism.* (AGI, 1987)

dynamic method *modulus of elasticity.*

dynamic penetration test *penetration test.*

dynamic positioning Maintainence of a drill ship's position through the use of outboard engines on opposite sides of the vessel. The position is maintained by automatic centering in a circle of sonar reflectors placed around the drilling target, either on the bottom or suspended by taut wire buoys. Several drilling ships are now equipped with this facility. (Mining)

dynamic viscosity *viscosity coefficient.*

dynamite (a) An industrial explosive that is detonated by blasting caps. The principal explosive ingredient is nitroglycerin. Diethyleneglycol dinitrate, which is also explosive, is often added as a freezing-point depressant. A dope, such as wood pulp, and an antacid, such as calcium carbonate, are also essential. See also: *blasting gelatin.* Oxidizers, such as ammonium nitrate, and fuels, such as vegetable fiber, are usually added. (b) A general term for detonable explosives in which the principal constituent, nitroglycerin, is contained within an absorbent substance. "Detonable" is a significant part of the definition since there are compositions that contain significant amounts of nitroglycerin but that are not detonable and are not considered to be dynamite.

dynamo exploder A powerful exploder usually operated by a vertical rack, which, on a downward movement, drives an armature. At the end of the stroke of the rack bar an internal short-circuiting device opens and the current generated by the rapidly revolving armature passes into the shot-firing circuit. See also: *blasting machine; exploder.* (Nelson, 1965)

dynamogranite Augen gneiss containing much microline and orthoclase.

dynamometamorphism Dynamic metamorphism.

dynamometer Appliance used in engineering to measure power as output, input, or transitional. (Pryor, 1963)

Dynamon A permissible explosive of the ammonium nitrate group. (Stoces, 1954)
dynamothermal metamorphism A common type of metamorphism involving the effects of directed pressures and shearing stress as well as a wide range of confining pressures and temperatures. It is related both geographically and genetically to large orogenic belts, and hence is regional in character. Cf: *regional metamorphism; dynamic metamorphism.* (AGI, 1987)
Dynobel No. 2 A high-strength, low-density permitted explosive; no water resistance. It is used for coal blasting in a machine-cut seam of medium hardness in dry conditions. (Nelson, 1965)
dyscrasite A natural antimonide of silver, Ag_3Sb; color and streak, silver-white; luster, metallic; usually tarnished; Mohs hardness, 3.5 to 4; sp gr, 9.74; found in Germany, France, and Canada. An ore of silver. Syn: *antimonial silver*. (CCD, 1961)
dyscrystalline *microcrystalline.*
dysluite A brown variety of gahnite containing manganese and iron in Massachusetts and New Jersey.
dysprosia *dysprosium oxide.*
dysprosium oxide A rare-earth oxide; white; Dy_2O_3; isometric; sp gr, 7.81 (at 27 °C); and melting point, 2,340±10 °C. Used as a nuclear-reactor control-rod component and a neutron-density indicator. Syn: *dysprosia.* (Lee, 1961; Handbook of Chem. & Phys., 2)
dystome spar *datolite.*
dystomic Having an imperfect fracture or cleavage. (Fay, 1920)
dzhalindite An isometric mineral, $In(OH)_3$; an alteration of indite; associated with cassiterite ores.
dzhezkazganite Possibly an amorphous lead rhenium sulfide found in the Dzhezkazgab copper ores, Kazakhstan.

E

eaglestone A concretionary nodule of clay ironstone about the size of a walnut that the ancients believed an eagle takes to her nest to facilitate egg-laying. Syn: *aetite.* (Webster 3rd, 1966; Fay, 1920)
ear The inlet or intake of a fan. (Fay, 1920)
earlandite A mineral, $Ca_3(C_6H_5O_7)_2 \cdot 4H_2O$; forms as fine-grained nodules in sediments under the Weddell Sea off the Atlantic coast of Antarctica.
earth amber (a) Amber that is mined rather than from the sea. Also called earth stone. (b) Amber with its surface deteriorated in luster, transparency, and color.
earth auger (a) A hand-boring tool for testing clays, soils, or shallow deposits. See also: *auger.* (Nelson, 1965) (b) A dry-sampling device consisting of a helical-fluted rod encased by a cylindrical tube. The fluted rod is equipped with cutting edges, and the cuttings collect and are retained within the tube. (Long, 1960)
earth balsam A variety of asphalt. (Tomkeieff, 1954)
earth borer An auger for boring into the ground, working in a cylindrical box to retain the cut earth until the tool is withdrawn. (Standard, 1964)
earth current A light electric current apparently traversing the Earth's surface but that in reality exists in a wire grounded at both ends, due to small potential differences between the two points at which the wire is grounded. Syn: *telluric current.* (Standard, 1964)
earth dam A dam constructed of earth material (such as gravel, broken weathered rock, sand, silt, or soil). It has a core of clay or other impervious material and a rock facing of riprap to protect against wave erosion. (AGI, 1987)
earth drill An auger. (Nichols, 1976)
earthed Applied to conductors, connected to the general mass of earth in such a manner as will ensure at all times an immediate discharge of electrical energy without danger. (Nelson, 1965)
earthed system Electrically, one with one neutral point or pole connected to earth. (Pryor, 1963)
earth fault Electrical short circuit from live conductor to earth. (Pryor, 1963)
earth fault lockout system An electrical system whereby a circuit is monitored to prevent application or restoration of supply if an earth fault exists. (BS, 1965)
earth fault meter An instrument for measuring the insulation fault at low voltage without polarization. This instrument is more informative in checking detonators in loaded holes than the insulation meter. (Langefors, 1963)
earth fault protection A system of protection designed to cause the supply to a circuit or system to be interrupted when the leakage current to earth exceeds a predetermined value. Also called earth leakage protection. (BS, 1965)
earth fault tester An apparatus used to prevent or reduce current leakage to the ground when blasting in conducting orebodies, in wet shale or clay, and in underwater blasting, esp. in salt water. The apparatus has no battery and can be used when loading the hole to check if the conducting wires have become damaged during this operation. Syn: *current leakage tester.* (Langefors, 1963)
earth flax An early name for fine silky asbestos. See also: *amianthus.*
earthflow A mass-movement landform and process characterized by downslope translation of soil and weathered rock over a discrete basal shear surface (landslide) within well-defined lateral boundaries. Little or no rotation of the slide mass occurs during displacement. Earthflows grade into mudflows with increasing fluidity. Also spelled: earth flow. (AGI, 1987)
earth foam A soft or earthy variety of calcite. *aphrite.*
earthing a conductor Establishing an electrical connection between a conductor and the earth. An important safeguard in electrical installations. (Nelson, 1965)
earthing system An electrical system in which all the conductors are earthed. (Nelson, 1965)
earth leakage protection A protective system that operates as a result of leakage of current from electrical machines to earth. For electrical apparatus in mines, the usual method of leaking protection is known as the core balance system. This depends for its action on the balance of the currents in three phases. When a fault occurs, the balance is disturbed and the resulting magnetic effect in the transformer core induces a current in the secondary circuit, so energizing the tripping coil and operating the tripping mechanism on the circuit breaker. It may be operated by a leakage current as low as 5% of the full load current of the circuit. (Nelson, 1965)
earth pillar *hoodoo; pillar.*
earthquake (a) A local trembling, shaking, undulating, or sudden shock of the surface of the earth, sometimes accompanied by fissuring or by permanent change of level. Earthquakes are most common in volcanic regions, but often occur elsewhere. Syn: *temblor.* (Fay, 1920) (b) Groups of elastic waves propagating in the earth, setup by a transient disturbance of the elastic equilibrium of a portion of the earth. (AGI, 1987)
Earth's crust *crust.*
Earth shell *shell.*
earth slide Downslope movement of part of an earth embankment sufficient to break up blocks and pulverize the material so that it moves in a somewhat fluid manner. Cf: *earth slump.*
earth slope The angle of superficial slope naturally assumed by rock debris, earthy detritus, etc., when piled up in mounds or ridges. (Standard, 1964)
earth slump Downslope movement of part of an earth embankment in blocklike masses without other apparent deformation. Cf: *earth slide.*
earth stone A term that may be applied to mined amber to distinguish it from sea amber. Cf: *earth amber.*
earth wave An obsolete syn. of seismic wave. (AGI, 1987)
earth wax *ozocerite.*
earthy (a) Consisting of minute particles loosely aggregated; claylike, dull. (b) In mineralogy, roughish to the touch, dull and lusterless. Porous aggregates of a mineral, such as the clays, scatter incident light so completely that they seem to be without luster and are described as dull or earthy. (c) Composed of or resembling earth or soil; e.g., an "earthy limestone" containing argillaceous material and characterized by high porosity, loosely aggregated particles, and close association with chalk. (d) Said of a type of fracture similar to that of a hard clay.
earthy breccia A breccia in which rubble, sand, and silt plus clay each constitute more than 10% of the rock. (AGI, 1987)
earthy fracture A fracture resembling that of a lump of hard clay.
earthy lead ore An earthy variety of cerussite.
earthy manganese *wad; bog manganese.*
easement (a) In surveying, an easement curve is a transitional curve. (Pryor, 1963) (b) An incorporeal right existing distinct from the ownership of the soil, consisting of a liberty, privilege, or use of another's land without profit or compensation; a right-of-way. (Fay, 1920)
easer One of a number of holes surrounding a cut and fired immediately after it. (BS, 1964)
easer holes Holes drilled around the cut to enlarge the cut area so that the trimmers may break out the ground to the required dimensions. The positioning and number of the easer holes will depend upon the pattern of the cut shots. (McAdam, 1958)
Eastman survey instrument A particular make of mechanical and photographic borehole-drift indicators; the single-shot models are small enough to be used in EX diamond-drill holes. See also: *drift indicator.* (Long, 1960)
easy way Scot. Easiest plane of splitting in granite, Aberdeenshire. Cf: *hard way.* (Arkell, 1953)
eat out (a) N. of Eng. To turn a heading or holing to one side in order to mine the coal on the other side of a fault without altering the level course of the heading. (Fay, 1920) (b) Said of a seam when the district or working place reaches a fault, or the boundary of old workings, or any other barren part of a mine. (CTD, 1958)
ebb channel Tidal channel in which the ebb currents are stronger than the flood currents. (Schieferdecker, 1959)
ebb current The movement of the tidal current away from shore or down a tidal stream. (Schieferdecker, 1959)
ebuliscope Instrument for observing liquids' boiling point, esp. for determining a mixture's strength by the temperature at which it boils. (Osborne, 1956)

eccentric bit A modified form of chisel used in drilling, in which one end of the cutting edge is extended further from the center of the bit than the other.

eccentric pattern A mode of arranging diamonds set in the face of a bit in such a manner as to have rows of diamonds forming eccentric circles so that the path cut by each diamond slightly overlaps that of the adjacent stones. Cf: *concentric pattern.* (Long, 1960)

ecdemite A tetragonal mineral, $Pb_6As_2O_7Cl_4$.

echelon cell Wedge-shaped glass cell used in absorption spectrography. (Pryor, 1963)

echelon pattern A delay pattern that causes the true burden, at the time of detonation, to be formed at an oblique angle from the original free face.

echogram A graphic recording of various sonic devices that shows ocean bottom profiles and delineates the bedding planes and dissimilar rock contacts to a depth of 1,500 ft (457 m) into the sediments. (Mero, 1965)

echo sounder In oceanography, a sounding apparatus using echo delay, for determining automatically the depth of sea beneath a ship. (CTD, 1958)

eckermannite A monoclinic mineral, $Na_3(Mg,Fe)_4AlSi_8O_{22}(OH)_2$; amphibole group with $Mg/(Mg + Fe^{2+}) = 0.5$ to 1.0 and $Fe^{3+}/(Fe^{3+} + Al) =$ 0 to 0.5; fibrous; in nepheline syenites; an asbestos mineral.

eclogite A coarse-grained, deep-seated ultramafic rock, consisting essentially of garnet (almandine-pyrope) and pyroxene (omphacite). Rutile, kyanite, and quartz are typically present.

ecology The study of the relationships between organisms and their environment, including the study of communities, patterns of life, natural cycles, relationships of organisms to each other, biogeography, and population changes. Adj: ecologic; ecological. Syn: *bionomics.* (AGI, 1987)

economic coal reserves The reserves in coal seams that are believed to be workable with regard to thickness and depth. In most cases, a maximum depth of about 4,000 ft (1.2 km) is taken, and a minimum thickness of about 2 ft (0.6 m). The minimum economic thickness varies according to quality and workability. See also: *thin seam.* (Nelson, 1965)

economic depletion The reduction in the value of a mineral deposit as the minerals reserves. See also: *depletion.* (Fay, 1920)

economic geology The study and analysis of geologic bodies and materials that can be utilized profitably by humans, including fuels, metals, nonmetallic minerals, and water; the application of geologic knowledge and theory to the search for and the understanding of mineral deposits. (AGI, 1987)

economizer An arrangement to preheat the feedwater before it enters the steam boiler. The water flows through a bank of tubes placed across the flue gases as they leave the boiler. (Nelson, 1965)

ecosystem A dynamic community of biological organisms, including humans, and the physical environment with which they interact. (SME, 1992)

eddy A circular movement of water. Eddies may be formed where currents pass obstructions or between two adjacent currents flowing counter to each other. (Hy, 1965)

eddy-current brake Arrangement by which internal currents are induced in a mass of metal as it moves relative to a magnetic field. (Pryor, 1963)

eddy-current testing A nondestructive testing method in which eddy-current flow is induced in the test object. Changes in the flow caused by variations in the object are reflected into a nearby coil or coils for subsequent analysis by suitable instrumentation and techniques. (ASM, 1961)

edelfall A German term for a shoot of precious-metal ore.

edelforsite *aedelforsite.*

edelite *prehnite.*

edenite A monoclinic mineral, $NaCa_2(Mg,Fe)_5Si_7AlO_{22}(OH)_2$; amphibole group with $Mg/(Mg + Fe^{2+}) = 0.5$ to 1.0; named for the type locality, Eden, NY.

edge dislocation In a crystal, a row of atoms or ions marking the edge of a crystallographic plane extending only part way. Cf: *line defect.*

edge seam mining The working of steeply inclined coal seams, many features of which are comparable to metal mining. See also: *stope.* (Nelson, 1965)

edgewise conglomerate A conglomerate exhibiting edgewise structure; e.g., an intraformational conglomerate containing elongated calcareous pebbles that are transverse to the bedding. See also: *edgewise structure.* (AGI, 1987)

edgewise structure A primary sedimentary structure characterized by an arrangement of flat, tabular, or disk-shaped fragments whose long axes are set at varying steep angles to the bedding. It may be due to running water or to sliding or slumping soon after deposition. See also: *edgewise conglomerate.* (AGI, 1987)

edinite *prase; mother-of-pearl.*

edisonite (a) Titanic acid, rutile, occurring in golden-brown, orthorhombic crystals. (Fay, 1920) (b) An early name for rutile.

eduction pipe The exhaust pipe from the low-pressure cylinder to the condenser. (Fay, 1920)

Edwards roaster Furnace with series of horizontal stepped hearths each equipped with stirring rabbles. Used to sweet-roast or desulfurize pyritic concentrates, notably gold-bearing sulfides. Moist-to-wet feed progresses step by downward step, meeting hot gases produced toward discharge end from burning pyrite. (Pryor, 1963)

effective belt tension That portion of the total tension in a conveyor belt effective in actually moving the loaded belt. It is often referred to as horsepower pull. Effective tension is the difference between tight side tension and slack side tension.

effective breaking force A product of the weight, strength and the degree of packing, calculated per volume of a given drill hole. (Langefors, 1963)

effective diameter (a) The size of an excavation within its stress ring; it includes not only the actual hole in the rock but the destressed loose and semiloose rock that surrounds it. (Spalding, 1949) (b) Particle diameter corresponding to 10% finer on the grain-size curve. Also called effective size. (ASCE, 1958)

effective grounding In mining, effective grounding means that the path to ground from circuits, equipment, or conductor enclosures is permanent and continuous and has carrying capacity ample to conduct safely any currents liable to be imposed upon it. The path to ground associated with high-voltage alternating-current systems will have impedance low enough to limit potential above ground to a maximum of 100 V during the flow of ground fault current and to facilitate operation of the circuit protective devices. On low-voltage systems the sustained voltage above ground, appearing on the frames of power utilizing equipment during existence of a ground fault, will not be greater than 35 V; except when ground circuit check systems requiring higher voltage are used, a maximum of 100 V for a duration of 0.2 s is permissible. When bonded or mechanically connected track is available, such track is considered the grounding medium for direct current equipment only.

effective permeability A measure of the ability of a rock to transmit a given fluid when the rock contains more than one fluid. (Inst. Petrol., 1961)

effective piece weight The weighted average weight of the pieces of sink material as found by separating a given coal product at any required specific gravity. Syn: *piece weight.*

effective pillar area The area of solid coal within the fractured and crushed edges of the pillar. (Nelson, 1965)

effective porosity (a) The ratio of the volume of liquid that a given mass of saturated rock or soil will yield by gravity to the volume of that mass. (AGI, 1987) (b) The property of rock or soil containing intercommunicating interstices, expressed as a percent of bulk volume occupied by such interstices. (AGI, 1987) (c) The ratio of the volume of the voids of a soil mass that can be drained by gravity to the total volume of the mass. (ASCE, 1958)

effective screen aperture The cut point (equal errors or partition size) at which a screening process operates in dividing the material treated into two size fractions. (BS, 1962)

effective screening area Total area of the apertures expressed as a percentage of the useful area of a screen. Syn: *open area.* (BS, 1962)

effective span The distance between the centers of support, or the clear distance between supports plus the effective depth of the beam or slab, the lesser value being taken. (Taylor, 1965)

effective teeth The number of sprocket teeth that engage the chain rollers during one revolution of the sprocket. Applies to sprockets for double-pitch roller chains. (Jackson, 1955)

effective temperature A measure of warmth that is often employed to assess the health and comfort conditions of mine workings, which are a function of dry- and wet-bulb temperatures and air velocity. See also: *dehumidification; dry kata cooling power.* (Nelson, 1965)

effective unit weight The unit weight of a soil that, when multiplied by the height of the overlying column of soil, yields the effective pressure due to the weight of the overburden. See also: *unit weight.* (ASCE, 1958)

efficiency engineer A technical officer who examines processes, methods, and operations in a mine, mill, or smelter, and connecting links, with a view to their improvement of maintenance at an agreed operating standard.

efficiency of a rectifier The ratio of the power output to the total power input. (Coal Age, 1960)

efficiency of screening The weight of underflow (excluding oversize) expressed as a percentage of the total weight of material below the reference size in the feed. (BS, 1962)

efficiency of separation In coal washing this may be expressed as: Efficiency, percent = Actual yield of clean coal × 100 / Theoretical yield at the ash content of the clean coal. The efficiency of separation thus expresses as a percentage, that proportion of the float coal obtained by float-and-sink analysis that will be recovered in practice by a particular washer. The theoretical yield is derived by plotting the cumulative yield of the reconstituted feed coal against the appropriate cumulative ash content and reading off the yield corresponding to the ash content of the clean coal actually obtained. See also: *washery*. (Nelson, 1965)

efficiency of sizing The weight of material correctly placed above or below the reference size, expressed as a percentage of the weight of corresponding material in the feed. (BS, 1962)

efficient airway size For a given air quantity, the efficient airway size is that size for which the combined capital and operating cost is minimal. (Hartman, 1982)

efficient structure A structure in which the load-bearing members are arranged in such a way that the weights and forces are transmitted to the foundations by the cheapest means consistent with safety and permanency. (Nelson, 1965)

efflorescence (a) A whitish fluffy or crystalline powder, produced as a surface encrustation on a rock or soil in an arid region by evaporation of water brought to the surface by capillary action or by loss of water of crystallization on exposure to the air. It may consist of one or several minerals, commonly soluble salts such as gypsum, calcite, natron, and halite. Syn: *bloom*. (AGI, 1987) (b) The process by which an efflorescent salt or crust is formed. (AGI, 1987)

efflorescent In mineralogy, forming an incrustation or deposit of grains or powder that resembles lichens or dried leaves; not uncommonly due to loss of water of crystallization. (Fay, 1920)

effluent A liquid, solid, or gaseous product, frequently waste, discharged or emerging from a process.

effusion (a) The property of gases that allows them to pass through porous bodies; i.e., the flow of gases through larger holes than those to which diffusion is strictly applicable. (Osborne, 1956) (b) The emission of relatively fluid lava onto the Earth's surface; also, the rock so formed. Cf: *extrusion*. (AGI, 1987)

effusive *extrusive*.

efydd A Welsh term for copper.

egg coal (a) A particular size of anthracite coal that passes through 3-1/4-in to 3-in (8.26-cm to 7.62-cm) round holes and over 2-1/16-in (5.24-cm) round holes. See also: *anthracite coal sizes*. (Jones, 1949) (b) A particular size of bituminous coal that passes through 4-in (10.2-cm) round holes and over 1-1/2-in (3.81-cm) round holes (sizes are not uniform but vary with the coalfield). (Jones, 1949)

eggette *briquette*.

eggstone *oolite*.

eglestonite An isometric mineral, $Hg_6Cl_3O(OH)$ or Hg_4Cl_2O; forms brownish-yellow (darkening to black on exposure) dodecahedra with montroydite, calomel, and other mercury minerals at New Idria, CA, and Terlingua, TX.

egueïite A monoclinic(?) material, $CaFe_{14}(PO_4)_{10}(OH)_{14}{\cdot}21H_2O$(?); forms yellowish-brown nodules with fibrous lamellar structure embedded in clay with trona and thernardite in the Egueï region of Chad and Sudan. (Not yet adequately described for a mineral species.)

Egyptian alabaster Banded calcite from near Thebes, Egypt. See also: *onyx marble*.

Egyptian emerald Emeralds from the ancient Egyptian mines of Jabal Sukayt and Jabal Zab-rah east of Aswan, in mica schist and talc schist.

Egyptian jasper (a) A banded jasper found as pebbles scattered over the surface of the Egyptian desert chiefly between Cairo and the Red Sea; used as a broochstone and for other ornamental purposes. Syn: *Egyptian pebble*. (b) Any jasper in which the colors run in zones.

Egyptian pebble *Egyptian jasper*.

Egyptian peridot Term properly applied only to peridot (olivine) from St. John's Island in the Red Sea.

Ehrhardt powder Any of a series of explosive mixtures containing potassium chlorate, together with tannin, powdered nutgalls, or cream of tartar. Used for blasting, shells, etc. (Fay, 1920)

Eichhorn-Liebig furnace A handworked muffle furnace. (Fay, 1920)

eightling A multiple twin consisting of eight individuals. Cf: *cyclic twin; trilling; fourling; fiveling*.

Eimco drill jumbo Drills or drifters mounted on a horizontal drill bar supported by the rocker shovel of an Eimco loader.

eisener hut Ger. Name for iron hat or gossan.

eisenwolframite *ferberite*.

eisinwolframite *ferberite*.

eitelite A trigonal mineral, $Na_2Mg(CO_3)_2$; associated with trona, nahcolite, or in places searlesite, from oil-well cores in northeastern Utah.

ekanite A tetragonal mineral, $ThCa_2Si_8O_{20}$; radioactive; metamict; from the gem pits of Eheliyagoda, Raknapura district, Sri Lanka; as fine crystals at the Mont St. Hilare Quarry, PQ, Canada.

ela A term in Sri Lanka for a drain, as around a gem pit.

elaeolite An alternative spelling of eleolite.

elastic (a) Capable of sustaining stress without permanent deformation; the term is also used to denote conformity to the law of stress-strain proportionality. An elastic stress or elastic strain is a stress or strain within the elastic limit. (Roark, 1954) (b) A physical property of minerals that may be bent without losing cohesion and that return to their original shape when released, e.g., micas. Cf: *plastic; flexible*.

elastic aftereffect *creep recovery*.

elastic axis The elastic axis of a beam is the line, lengthwise of the beam, along which transverse loads must be applied in order to produce bending only, with no torsion of the beam at any section. Strictly speaking, no such line exists except for a few conditions of loading. Usually the elastic axis is assumed to be the line that passes through the elastic center of every section. The term is most often used with reference to an airplane wing of either the shell or multiple-spar type. Cf: *torsional center; flexural center; elastic center*. (Roark, 1954)

elastic bitumen *elaterite*.

elastic boundary The boundary of an underground opening that requires no support. The material around this boundary may be considered to be in the elastic state, and no pressure need be exerted against the boundary to prevent the material from fracturing and falling into the opening. (Woodruff, 1966)

elastic center The elastic center of a given section of a beam is that point in the plane of the section lying midway between the flexural center and center of twist of that section. The three points may be identical and are usually assumed to be so. Cf: *flexural center; torsional center; elastic axis*. (Roark, 1954)

elastic deformation Deformation of a substance, which disappears when the deforming forces are removed. Commonly, that type of deformation in which stress and strain are linearly related, according to Hooke's law. Cf: *plastic deformation*. (AGI, 1987)

elastic design Design of a structure based on working stresses which are about one-half to two-thirds of the elastic limit of the material. For redundant frames, this method of design is replaced by the plastic design. See also: *plastic design*. (Hammond, 1965)

elastic discontinuity A boundary between strata of different elastic moduli and/or density at which seismic waves are reflected and refracted. (AGI, 1987)

elasticity The property or quality of being elastic; said of a body that returns to its original form or condition after a displacing force is removed. See also: *elasticity of bulk; Hooke's law*. (AGI, 1987)

elasticity of bulk (a) The property possessed by all substances by which they tend to recover their original volume after being compressed or extended. (Hess) (b) The elasticity for changes in the volume of a body caused by changes in the pressure acting on it. The bulk modulus is the ratio of the change in pressure to the fractional change in volume. See also: *elasticity*. (CTD, 1958)

elastic limit The greatest stress that can be developed in a material without permanent deformation remaining when the stress is released. (AGI, 1987)

elastic mineral pitch *elaterite*.

elastic modulus *modulus of elasticity*.

elastic rebound Elastic recovery from strain. (AGI, 1987)

elastic surface waves Waves that travel only on a free surface where the solid elastic materials transmitting them are bounded by air or water. (Leet, 1960)

elastic zone In explosion-formed crater nomenclature, the remote zone that undergoes no measurable permanent deformation. (Min. Miner. Eng., 1966)

elaterite A massive, amorphous, dark-brown metamorphic hydrocarbon ranging from soft and elastic to hard and brittle; melts in a candle flame without decrepitation; conchoidal fracture. Also called liverite. See also: *coorongite; wurtzilite*. Syn: *elastic bitumen; elastic mineral pitch; mineral caoutchouc*.

elbaite (a) *red schorl; rubellite; ilvaite*. (b) A trigonal mineral, $3[Na(Li,Al)_3Al_6(OH,F)_4(BO_3)_3Si_6O_{18}]$; tourmaline group; occurs in triangular and hexagonal prisms; varicolored; commonly zoned, pyroelectric and piezoelectric; in granites and granite pegmatites; and used as a gemstone (pink rubellite, blue indicolite, green verdolite, colorless achroite, zoned pink-white-green watermelon tourmaline).

el conveyor A trough-type roller or wheel conveyor consisting of two parallel rows of rolls or wheels set at a 90° included angle, with one row providing a sloped carrying surface and the other acting as a guard. See also: *roller conveyor; troughed roller conveyor; wheel conveyor*.

El Doradoite A trade name in El Dorado County, California, for a blue quartz that may be cut as a gemstone.

electric air drill A type of tripod drill operated by compressed air supplied by a portable motor-driven compressor that accompanies the drill. (Fay, 1920)

electrical conductivity A measure of the ease with which a conduction current can be caused to flow through a material under the influence of an applied electric field. It is the reciprocal of resistivity and is measured in mhos per meter. (AGI, 1987)

electrical double layer Helmholtz layer. Zone that surrounds a particle in aqueous suspension or other electrolyte. Transition zone between the monomolecular zone of shear immediately coupled ionically to the discontinuity lattice at the particle's surface and the normal aqueous phase that exists from 50 to 5,000 Å and beyond. This zone of change contains a superconcentration of ions drawn from the normal population of the liquid phase. See also: *zeta potential*. (Pryor, 1963)

electrical engineer An engineer in charge of all electrical plant and associated labor at a mine or colliery. He or she has an assistant in charge of all the underground electrical equipment, operations, and labor. The electrical engineer is under the authority of the colliery manager. (Nelson, 1965)

electrical method A geophysical prospecting method that depends on the electrical or electrochemical properties of rocks. The resistivity, spontaneous-polarization, induced-polarization, and inductive-electromagnetic methods are the principal electrical methods. (AGI, 1987)

electrical plan A plan, drawn to the same scale as the working plan, that shows the position of all electrical apparatus installed underground except signals and telephones. (Nelson, 1965)

electrical precipitation The removal of suspended particles from gases by the aid of electrical discharges, using alternating or direct current. Alternating current agglomerates the suspended particles into larger aggregates, causing rapid settling, esp. if the gases are quiescent. Direct current is used when large volumes of rapidly moving gas, such as occur in smelter flues, are treated. The suspended particles within a strong electric field of constant polarity become charged and are then attracted to a plate (electrode) of opposite charge. (Fay, 1920)

electrical prospecting Prospecting that makes use of three fundamental properties of rocks. One is the resistivity, or inverse conductivity. This governs the amount of current that passes through the rock when a specified potential difference is applied. Another is the electrochemical activity with respect to electrolytes in the ground. This is the basis of the self-potential method. The third is the dielectric constant. This gives information on the capacity of a rock material to store electric charge, and it must be taken into consideration when high-frequency alternating currents are introduced into the earth, as in inductive prospecting techniques. Electrical methods are more frequently used in searching for metals and minerals than in exploring for petroleum, mainly because most of them have proved effective only for shallow explorations. (Dobrin, 1960)

electrical protection Protection is provided by fuses or other suitable automatic circuit-interrupting devices for preventing damage to circuits, equipment, and personnel by abnormal conditions, such as overcurrent, high or low voltage, and single phasing.

electrical puncturing A rock fracturing technique, applied to secondary fragmentation in quarries, that is characterized by an almost instantaneous action and is accompanied by a mechanical weakening of the dielectric and a lowering of the resistance of the puncture path. If, after puncturing, a high-frequency current continues to pass between the contacts, the action of the conduction current and electric field will rapidly heat the rock, leading to thermal puncture, in which the dielectric is transformed into a good conductor along the puncture path. Further intensive heating will give rise to thermal stresses sufficient to fracture the rock. (Min. Miner. Eng., 1965)

electrical resistance inclinometer An instrument to indicate when a long hole in a coal seam is deviating into the roof or floor. It may be used in underground gasification and pulsed-infusion shotfiring. It uses, among other things, a pellet of mercury to indicate the gradient by its position along a tube. (Nelson, 1965)

electrical resistance strain gage An appliance for measuring strain that may be employed in roof-control research. It makes use of the change in electrical resistance of a thin wire when stretched under the influence of strata strain. Syn: *resistance strain gage*. See also: *acoustic-strain gage; mechanical extensometer*. (Nelson, 1965)

electrical rock fracture A rock-fracturing technique in which electrical energy is used directly in fracturing the rock, either by heating it in a variable electric or electromagnetic field set up in the rock by a high-frequency electric current, or by the direct puncturing of the rock by an electric current. (Min. Miner. Eng., 1965)

electrical slate Slate principally of the mica variety. It should have high mechanical and dielectric strength, be readily machinable, and have low porosity. (USBM, 1965)

electrical twinning A type of quartz twinning in which the two or more intergrown parts are related as by a rotation of 180° about the common Z = *c* axis. The separate individuals of the twin are either all right handed or all left handed. Electrical twinning cannot be detected by optical tests, but can be recognized by etching, X-ray study, pyroelectric tests, or the distribution of the *x* {5161} or *s* {1121} faces. Syn: *Dauphiné twinning; orientational twinning*.

electrical well logging The process of recording the formations traversed by a drill hole, based on the measurements of two basic parameters observable in uncased holes; namely, the spontaneous potential and the resistivity of the formations to the flow of electric currents. The detailed study in situ of the formations penetrated by a drill hole, based on measurements made systematically by lowering an apparatus in the hole responding to the following physical factors or parameters: (1) the resistivities of the rocks; (2) their porosity; (3) their electrical anisotropy; (4) their temperature; and (5) the resistivity of the drilling muds. (AGI, 1987)

electric blasting The firing of one or more charges electrically, whether electric blasting caps, electric squibs, or other electric igniting or exploding devices are used. (Fay, 1920)

electric blasting cap (a) A device for detonating charges of explosives electrically. It consists essentially of a blasting cap, into the charge of which a fine platinum wire is stretched across two protruding copper wires, the whole fastened in place by a crimp or plug. The heating of the platinum wire bridge by the electric current ignites the explosive charge in the cap, which in turn detonates the high explosive. (b) Detonator fired electrically. See also: *electric detonator*. (Pryor, 1963)

electric braking A system in which a braking action is applied to an electric motor by causing it to act as a generator. (Nelson, 1965)

electric cable The conducting wires through which an electric current is conveyed to points in and about a mine, where it is required for lighting or motive power. See also: *armored cable*. (Nelson, 1965)

electric calamine Zinc silicate or calamine; so called on account of its strong pyroelectric properties, and to distinguish it from smithsonite. See also: *calamine*. (Webster 2nd, 1960)

electric cap lamp This lamp consists of a flat portable battery that is strapped around the miner's waist and is connected by an insulated cord to a small electric light and reflector that is fastened on the front of the miner's cap. See also: *safety lamp*. (Lewis, 1964)

electric coal cutter A coal cutter operated by an electric motor; used in coal mines. (CTD, 1958)

electric coal drill An electric motor-driven drill designed for drilling holes in coal for placing blasting charges. Syn: *coal drill*.

electric detonator A detonator requiring electrical energy to activate the explosive train, detonating the base charge. See also: *blasting cap; electric blasting cap*.

electric ear System used to control grinding rate in a ball mill; a microphone listens to the grinding sound and maintains this by varying the rate of new feed to the mill. (Pryor, 1963)

electric-eye method A method of finding large diamonds in which the dry crushed ore is passed in a thin layer on a moving belt through a band of intense polarized light, which, if reflected from a large diamond, actuates a photoelectric cell, which calls attention to the large diamond. (Chandler, 1964)

electric furnace A furnace using electricity to supply heat. (Mersereau, 1947)

electric fuse A metallic cup, usually containing fulminating mercury, in which are fixed two insulated conducting wires held by a plug, the latter holding the ends of the wires near to but not touching each other. At this plug is a small amount of a sensitive priming. When an electric current is sent from the battery through these conductors, the resulting spark fires the priming, then the fulminate and the charge of the explosive proper. (Stauffer, 1906)

electric gathering mine locomotive An electric mine locomotive, the chief function of which is to move empty cars into, and to remove loaded cars from, the working places. Syn: *gathering locomotive; gathering mine locomotive*. See also: *gathering motor*.

electric haulage mine locomotive An electric mine locomotive used for hauling trains of cars, which have been gathered from the working faces of the mine, to the point of delivery of the cars.

electric hoist *electric winder*.

electric-hoist man *hoistman*.

electric ingot process A continuous method of melting and casting metal with progressive solidification. The molten metal is completely protected from the atmosphere. There is minimum segregation, and as no refractory linings are used, there is no contamination. Sound ingots with high yield and no pipe are produced, and as the method possesses extreme flexibility it is possible to make small as well as relatively large ingots. (Osborne, 1956)

electric lamp *cap lamp; hand electric lamp*.

electric log The generic term for a well log that displays electrical measurements of induced current flow (resistivity log, induction log) or natural potentials (spontaneous-potential curve) in the rocks of an uncased borehole. An electric log typically consists of the spontaneous-potential curve and one or more resistivity or induction curves. The Archie equations form the basis for interpretation of electric logs. Abbrev: E-log. Informal syn: resistivity log. See also: *borehole log; well logging*. (AGI, 1987)

electric logging (a) A technique in which electrical measurements are made, and recorded at the surface, while a series of electrodes or coils is caused to traverse a borehole. The resulting curves can be used for geological correlation, and, under favorable circumstances, for the recognition of some rock properties and for indicating the nature and amount of the fluids in the pores of the rock. (Inst. Petrol., 1961) (b) The act or process of taking resistivity, porosity, electrical anisotropy, etc., measurements in a borehole using an electromagnetic teleclinometer or other electrode device. Also called electrical logging. See also: *self-potential log*. (Long, 1960)

electric master fuse *multifuse igniter.*

electric mine locomotive An electric locomotive designed for use underground. See also: *locomotive; mine locomotive; electric haulage mine locomotive; electric permissible mine locomotive; electric gathering mine locomotive.*

electric mule Electric motor. (Korson, 1938)

electric permissible mine locomotive An electric locomotive carrying the official approval plate of the U.S. Mine Safety and Health Administration.

electric powder fuses These fuses were designed so that electrical shotfiring methods could be used for initiating blasting powder. The powder fuse consists of a thick paper tube containing a small charge of blasting powder, with an ordinary low-tension fusehead fixed at one end. On passing electric current through the fusehead, it flashes and sets off the blasting powder in the tube, which can then initiate the main charge of blasting powder in the shot hole. (McAdam, 1958)

electric precipitation A method of collecting particulate particles used chiefly in air polution control that consists of inducing an electric charge on the dust particles and collecting them on an oppositely charged device.

electric prospecting instruments Geophysical prospecting instruments that measure the electrical characteristics of rocks. (Nelson, 1965)

electric resistance The opposition of an electric circuit to the flow of current. (Kentucky, 1952)

electric resistance strain gage This gage consists essentially of a grid of fine wire cemented to a paper membrane, which can be attached to the surface under investigation. The ends of the wire grid are spot welded to a metal strip for the terminal connections. The use of these gages depends upon the fact that certain alloys show a linear relationship between applied strain and electrical resistance, so that if a wire constructed from one of these alloys is fixed to the surface of an object subject to variable strain, the change of resistance in the wire will be a measure of the change of strain in the object. (Issacson, 1962)

electric rotary drill A hand-held rotary drill driven by an electric motor, which may be used in rock or coal. It may be of fan-cooled design with several rod speeds to suit different rocks. The use of aluminum or aluminum alloys is not favored where methane is liable to be present. This drill produces considerably less dust than the percussive drill. (Nelson, 1965)

electric shock Paralysis of the nerve center that controls breathing or a regular heartbeat. Some symptoms of electric shock are sudden loss of consciousness, absence of respiration or respiration that cannot be detected, weak pulse, and burns. (Kentucky, 1952)

electric shovel Most of the larger modern machines are electrically driven and are equipped with the Ward-Leonard system of control, which allows alternating current of fairly high voltage to be carried to the shovel over a very flexible electric cable. This cable is usually carried on a sled back of the shovel or on a reel on the shovel base. The current drives an alternating current motor, which is connected to, and drives, direct current generators, one for each of the operations of the shovel, and an exciter. Each direct current generator and the direct current motor which it drives are in a closed circuit. The field in each circuit is regulated by magnetic contactors or by rotating controls actuated by master controllers at the operator's position.

electric-shovel craneman *shovel craneman.*

electric slope engineer In bituminous coal mining, one who operates a hoist powered by electricity to haul loaded and empty cars along a haulage slope to surface of mine. (DOT, 1949)

electric sponge An electric centrifugal pump consisting of a small vertical centrifugal pump so designed that it will draw water if it is only 2 to 3 in (5.1 to 7.6 cm) deep. It is placed in the water at the bottom of a shaft and lifts the water up to a horizontal centrifugal pump placed about 50 ft (15.2 m) above. (Lewis, 1964)

electric squib A small shell containing an explosive compound that is ignited by the electric current brought in through the lead wires. Used for firing single small holes loaded with black powder. (Lewis, 1964)

electric steel Steel made in an electric furnace. (Mersereau, 1947)

electric traction The haulage of vehicles by electric power, derived from overhead wires, third rail, storage batteries, or diesel-driven generators mounted on the vehicles. (Hammond, 1965)

electric wheel A wheel containing the motor and all the required gearing so that it is an independent drive unit. (Woodruff, 1966)

electric winder A winder or hoist driven by an electric motor. (Nelson, 1965)

electroacoustic transducer A transducer for receiving waves from an electric system and delivering waves to an acoustic system or vice versa. (Hy, 1965)

electrocast process A method of producing refractory materials in the desired form by mixing the raw materials in the requisite proportions, heating to fusion in an electric furnace, and then casting. (Osborne, 1956)

electrochemical Chemical action employing a current of electricity to cause or to sustain the action. (Crispin, 1964)

electrochemical corrosion Corrosion that occurs when current flows between cathodic and anodic areas on metallic surfaces. (ASM, 1961)

electrochemical equivalent The weight of an element, compound, radical, or ion involved in a specified electrochemical reaction during the passage of a unit quantity of electricity, such as a faraday, an ampere-hour, or a coulomb. (Lowenheim, 1962)

electrochemical series *electromotive force series.*

electrode (a) In arc welding, a current-carrying rod that supports the arc between the rod and work, or between two rods as in twin carbon-arc welding. It may or may not furnish filler metal. (ASM, 1961) (b) In resistance welding, a part of a resistance welding machine through which current and, in most cases, pressure are applied directly to the work. The electrode may be in the form of a rotating wheel, rotating roll, bar, cylinder, plate, clamp, chuck, or modification thereof. (ASM, 1961) (c) A conductor (as a metallic substance or carbon) used to establish electrical contact with a nometallic portion of a circuit (as in an electrolytic cell, a storage battery, an electron tube, or an arc lamp). See also: *anode; cathode.* (Webster 3rd, 1966)

electrode burnoff rate The rate at which an electrode is consumed by an arc in units of mass per time per arc power. (Wood, 1965)

electrode configuration Pattern in which the electrodes are set up. (Schieferdecker, 1959)

electrode consumption rate The rate at which an electrode is consumed by an arc in units of mass per time per arc current. (Wood, 1965)

electrode melting rate The rate at which an electrode is consumed by an arc in units of mass per time. (Wood, 1965)

electrodeposition The deposition of a substance upon an electrode by passing electric current through an electrolyte. Electroplating (plating), electroforming, electrorefining, and electrowinning result from electrodeposition. Syn: *electrolytic deposition.* (ASM, 1961)

electrode potential (a) The potential difference at the surface of separation between the electronic and electrolytic conductors that make up the electrode. In the terminology of corrosion it is sometimes called the open-circuit potential. (b) The potential of a half-cell as measured against a standard reference half-cell. (ASM, 1961)

electrode spacing Distance between successive electrodes. (Schieferdecker, 1959)

electrodialysis Dialysis assisted by the application of an electric potential across the semipermeable membrane. Two important uses of electrodialysis are in water desalination and in removing electrolytes from naturally occurring colloids such as proteins. Cf: *electro-osmosis.* (AGI, 1987)

electroendosmosis The movement of fluids through porous diaphragms caused by the application of an electric potential. (Lowenheim, 1962)

electroextraction The application of electrolysis to recover metal from its salts. Syn: *electrowinning.* (Nelson, 1965)

electrofiltration The electromotive force set up between the two sides of the sheet when an electrolyte is forced through a sheet of some pervious solid dielectric. This electromotive force is proportional to the pressure and to the electrical resistivity of the liquid, and inversely proportional to its viscosity. (Lewis, 1964)

electrofiltration potential An electrical potential that is caused by movement of fluids through porous formations. Syn: *streaming potential; electrokinetic potential.* (AGI, 1987)

electrogalvanizing The electroplating of zinc upon iron or steel. (ASM, 1961)

electrokinetic potential *electrofiltration potential; zeta potential.*

electrolysis A method of breaking down a compound in its natural form or in solution by passing an electric current through it, the ions present moving to one electrode or the other where they may be released as new substances. (AGI, 1987)

electrolyte (a) A nonmetallic electric conductor (as a solution, liquid, or fused solid) in which current is carried by the movement of ions instead of electrons with the liberation of matter at the electrodes; a liquid ionic conductor. (Webster 3rd, 1966) (b) A substance (as an acid, base, or salt) that, when dissolved in a suitable solvent (as water) or when fused, becomes an ionic conductor. (Webster 3rd, 1966) (c) For ceramic applications, an electrolyte is a substance capable of dissociating partly or completely into ions in water. For clay dispersions, the basic electrolytes promote deflocculation while the acidic electrolytes produce the opposite effect, flocculation. (Lee, 1961)

electrolytic copper Copper that has been refined by electrolytic deposition, including cathodes, which are the direct product of the refining operation; refinery shapes cast from melted cathodes; and, by extension, fabricators' products made therefrom. Usually when this term is used alone, it refers to electrolytic tough pitch copper without elements other than oxygen being present in significant amounts. (ASM, 1961)

electrolytic deposition The production of a metal from a solution containing its salts by the passage of an electric current through the solution. In electrorefining, the operation is carried out in an electrolytic cell in which the metal is deposited upon the cathode or starting sheet. See also: *electrodeposition.* (Henderson, 1953)
electrolytic dissociation Dissociation in a solvent of molecules of the dissolving substance as cations and anions. Syn: *ionization.* (Pryor, 1963)
electrolytic dissolusion The act or process of dissolving the diamond matrix metal in the crown of a bit utilizing the chemical decompositional effects of a direct electrical current on a metal object submerged in an acidic solution. (Long, 1960)
electrolytic iron A very pure iron produced by an electrolytic process. It has excellent magnetic properties and is often used in magnet cores. (Crispin, 1964)
electrolytic lead Lead refined by the Betts process; it has purity of about 99.995% to 99.998% lead. (CTD, 1958)
electrolytic polishing Producing a smooth bright surface on metal by immersion as an anode in an electrolytic bath. (Webster 3rd, 1966)
electrolytic process (a) A process employing the electric current for separating and depositing metals from solution. (Fay, 1920) (b) As used by the diamond-bit-setting industry, the process in which the chemical decompositional effects of subjecting metal objects immersed in an acidic solution to a flow of direct electric current is utilized to dissolve the metal in the crown of a worn diamond bit to free and salvage the diamonds. (Long, 1960)
electrolytic reduction Removal of oxygen (or decrease of its active valency in the case of a positive element) by electrical means. (Pryor, 1963)
electrolytic refining Suspension of suitably shaped metal ingots as anodes in an electrolytic bath, alternated with sheets of the same metal in a refined state or inert metal, which act as starters or cathodes. Impurities remaining on the anodes are detached as anode slime or are dissolved in the electrolyte from which they must be systematically removed (stripped). (Pryor, 1963)
electrolytics The extraction and refining of metals by the use of electric currents. (Newton, 1959)
electrolytic zinc Zinc exceeding 99.9% purity, produced by electrodeposition. (Pryor, 1963)
electrolyze To decompose a compound, either liquid, molten, or in solution, by an electric current.
electromagnetic brake One in which rubbing surfaces are pressed together when electric current is passed through a solenoid; also, a braking system that uses magnetic attraction generated by an electromagnet as a braking force. (Pryor, 1963)
electromagnetic damping Commonly found in seismometers of the induction type. It may be used in mechanical seismographs by employing a copper plate moving between two permanent magnets. Induction seismometers depend upon voltage generated by motion of coil in the magnetic field. (AGI, 1987)
electromagnetic detector An instrument used in aerial geophysical prospecting for the direct detection of conducting ores. An alternating electromagnetic field is transmitted from an aircraft. This field is received by the conducting body in the Earth and reradiated with some change in phase. The resultant field is picked up by the device, towed behind the aircraft, and compared with the transmitted field. The phase shift is measured automatically and recorded as a profile during flight. See also: *geophysical exploration.* (Nelson, 1965)
electromagnetic geophone The simplest, most widely used type of geophone. It consists of a coil and a magnet, one rigidly fixed with respect to the Earth and the other suspended from a fixed support by a spring. Any relative motion between the coil and magnet produces an electromotive force across the coil's terminals that is proportional to the velocity of the motion. (Dobrin, 1960)
electromagnetic methods Group of electrical exploration methods in which one determines the magnetic field that is associated with the electrical current through the ground. (Schieferdecker, 1959)
electromagnetic prospecting A geophysical method employing the generation of electromagnetic waves at the Earth's surface; when the waves penetrate the Earth and impinge on a conducting formation or orebody, they induce currents in the conductors, which are the source of new waves radiated from the conductors and detected by instruments at the surface. (AGI, 1987)
electromagnetic separation A process of removing magnetic materials from relatively nonmagnetic materials, using electromagnets which travel along a conveyor, over a drum, or into a revolving screen. See also: *electrostatic separator; tramp iron.* (Nelson, 1965)
electromagnetic spectrum The entire range of electrical energy, extending from the extremely long radio rays at one end to the extremely short X-rays at the other. The visible spectrum (visible light) is included.
electromagnetic surveying The act or process of using a geophysical method of systematically measuring electromagnetic waves in a specific area of the Earth's surface or in an area adjacent to boreholes. See also: *electromagnetic prospecting.* (Long, 1960)
electromagnetism The totality of electric and magnetic phenomena, or their study; particularly those phenomena with both electric and magnetic aspects, such as electromagnetic induction. (AGI, 1987)
electrometallurgy A term covering the various electrical processes for the industrial working of metals; e.g., electrodeposition, electrorefining, and operations in electric furnaces. (CTD, 1958)
electromotive force series The elements can be listed according to their standard electrode potentials. The more negative the potential, the greater the tendency of the metals to corrode but not necessarily at higher rates. This series is useful in studies of thermodynamic properties. A hydrogen gas electrode is the standard reference and is placed equal to zero. All potentials are positive or negative with respect to the hydrogen electrode. Also known as the emf series. Syn: *electrochemical series.* (Hunt, 1965)
electron One of the constituent elementary particles of an atom. A charge of negative electricity equal to about 1.602×10^{-19} C and having a mass when at rest of about 9.107×10^{-28} g or 1/1,837 that of a proton. Electrons surround the positively charged nucleus of the atom and determine the chemical properties of the atom. (Webster 3rd, 1966; Lyman, 1964)
electron beam melting A melting process in which heat is supplied by a beam of electrons directed at the metal in high vacuum. (Thomas, 1960)
electron capture A mode of radioactive decay in which an orbital electron is captured by the nucleus. The resulting nuclear transformation is identical with that in β^+ emission. (AGI, 1987)
electronegative Descriptive of element or group that ionizes negatively, or acquires electrons and therefore becomes negatively charged anion. (Pryor, 1963)
electronic CO detector A portable, lightweight instrument for detecting carbon monoxide in mine air. Most of these instruments allow instantaneous reading of the carbon monoxide content but can also be used in an automated recording and monitoring system.
electronic filter An air cleaner in which particulate matter in the airstream is electrically charged, then attracted to surfaces oppositely charged. (Strock, 1948)
electronic high-level indicator An electronic device that signals an operator when a bin is filled to capacity.
electronic liquid density instrument A glass float on the end of a thin rod suspended in a liquid which is supported by two flat springs so that it is constrained to precise vertical motion. The float-rod assembly carries a coil similar to the voice-coil of a dynamic loudspeaker and a differential transformer core. Vertical movement of the float is detected by the electrical response of the differential transformer. The coil moves in a strong, radial, magnetic field, and when the float is buoyed up by the liquid, the reaction force between the coil and the field is used to pull it down. Thus, balance is achieved at a null position by adjusting the coil current while observing the null indicator. (Hunt, 1965)
electronic microscope An instrument similar to the ordinary light microscope, but producing a much magnified image, which is received on a fluorescent screen and is recorded by using a camera. Instead of a beam of light to illuminate the material, a parallel beam of electrons is used. Its magnification is up to about 100,000 times. (Nelson, 1965)
electronics The utilization based on the phenomena of conduction of electricity in a vacuum (thermionic valves), in a gas (thyratrons), and in semiconductors (transistors). (NCB, 1964)
electronic sentry A device for mounting on any direct-current mobile mining machine that receives its power through a portable cable. The device cuts off the power from the machine and its trailing cable in the event of a ground fault, short circuit, or break in the cable, and prevents electrical flow as long as the trouble exists. (Nelson, 1965)
electronic sorting *LaPointe picker.*
electronic tramp iron detector An appliance to prevent large pieces of tramp iron from entering a primary breaker when the ore feed is by conveyor. The appliance is straddled across the conveyor and when the tramp metal (magnetic or nonmagnetic) of dangerous size passes under the detector it automatically stops the conveyor and sounds an alarm, and will not restart motion until the tramp material is removed. (Nelson, 1965)
electronic weighing *weighing-in-motion system.*
electro-osmosis The motion of liquid through a membrane under the influence of an applied electric field. See also: *osmosis; thermo-osmosis.* (AGI, 1987)
electrophoresis (a) Movement of colloid particles toward an oppositely charged electrode through a solution. (Pryor, 1963) (b) The movement toward electrodes of suspended charged particles in a fluid by applying an electromotive force to the electrodes that are in contact with the suspension. See also: *cataphoresis.* (AGI, 1987)
electroplate To plate with an adherent continuous coating by electrodeposition; esp., to plate with a metal. (Webster 3rd, 1966)

electroplating Electrodepositing metal (may be an alloy) in an adherent form upon an object serving as a cathode. (ASM, 1961)
electropneumatic lighting A method of lighting in which the well glass surrounding the bulb is flushed out with compressed air through a special valve before a self-contained generator commences to run; afterwards the exhaust from the turbine is passed through the lamp fitting with a small back pressure of 1-1/2 to 2 psi (10.3 to 13.8 kPa), preventing ingress of methane. The equipment can be used underground where the use of electricity is prohibited and for both roadway and face lighting. (Sinclair, 1958)
electropositive (a) Positively charged; having more protons than electrons. An electropositive ion is a cation. (Pryor, 1963) (b) Term used to describe substances that tend to pass to the cathode in electrolysis. (Mersereau, 1947)
electrorefining The process of anodically dissolving a metal from an impure anode and depositing it in a purer state at the cathode. (Lowenheim, 1962)
electroscope Any of various instruments for detecting the presence of an electric charge on a body, for determining whether the charge is positive or negative, or for indicating and measuring the intensity of radiation by means of the motion imparted to charged bodies (as strips of gold leaf) suspended from a metal conductor within an insulated chamber. (Webster 3rd, 1966)
electrostatic capacity Quantity of electricity needed to raise system one unit of potential. (Pryor, 1963)
electrostatic cleaning process A method of cleaning small sizes of coal, namely, 0.1 to 2 mm, by passing the material over a slowly rotating roller through a high-voltage electrostatic field between the earthed roller and an adjacent wire. Coal loses its charge very slowly and is carried further around by the roller than the impurities, thus effecting separation with reasonable efficiency. (Nelson, 1965)
electrostatic precipitator The most efficient of the dust samplers, the electrostatic precipitator is a medium-volume instrument. Air is drawn through a metal tube serving as a collecting surface (the anode) in which a platinum wire mounted axially acts as the ionizing and precipitating electrode (the cathode). A potential of about 10,000 V direct current is maintained across the tube and wire. The assembly mounting and collecting tube contains a small fan to induce air flow. (Hartman, 1961)
electrostatics Science of electric charges captured by bodies that then acquire special characteristics because of their retention of such charges. Electrostatic bunching is particle cling during the laboratory screening of dry material in which frictional electric charge is set up. (Pryor, 1963)
electrostatic separation A method of separating materials by dropping feed material between two electrodes, positive and negative, rotating in opposite directions. Nonrepelled materials drop in a vertical plane; susceptible materials are deposited in a forward position somewhat removed from the vertical plane. (ASM, 1961)
electrostatic separator A vessel fitted with positively and negatively charged conductors that may be used for extracting dust from flue gas or for separating mineral dust from gangues. (Nelson, 1965)
electrostatic strength As applied to electric blasting caps, a measure of the detonator's ability to withstand electrostatic discharges without exploding. (Fraenkel, 1953)
Electrotape A trade name for a precise electronic surveying device that transmits a radio-frequency signal to a responder unit, which in turn transmits the signal back to the interrogator unit. The time lapse between original transmission and receipt of return signal is measured and displayed in a direct digital readout for eventual reduction to a precise linear distance. It operates on the same principle as the tellurometer. See also: *tellurometer.* (AGI, 1987)
electrowinning An electrochemical process in which a metal dissolved within an electrolyte is plated onto an electrode. Used to recover metals such as cobalt, copper, gold, and nickel from solution in the leaching of ores, concentrates, precipitates, matte, etc.
electrum (a) A part of the series isometric native gold-silver (Au-Ag); deep to pale yellow; argentiferous gold containing more than 20% silver. Also spelled elektrum. Syn: *gold argentide.* (b) An ancient Greek name, now obsolete, for amber. Also spelled elektron.
electrum metal An alloy of gold and silver; contains from 55% to 88% gold. (Pryor, 1963)
element (a) A substance that cannot be decomposed into other substances. (AGI, 1987) (b) A substance all of whose atoms have the same atomic number. The first definition was accepted until the discovery of radioactivity (1896), and is still useful in a qualitative sense. It is no longer strictly correct, because (1) the natural radioactive decay involves the decomposition of one element into others, (2) one element may be converted into another by bombardment with high-speed particles, and (3) an element can be separated into its isotopes. The second definition is accurate, but has the disadvantage that it has little relevance to ordinary chemical reactions or to geologic processes. (AGI, 1987) (c) In crystallography, any point, line, or plane about which crystal structure, crystal faces, or crystal symmetry, including translation, is symmetrically arrayed. Cf: *operation.*
elementary particle Applied to any particle that cannot be further subdivided.
eleolite A dark, translucent massive or coarsely crystalline variety of nepheline; greasy luster; may be used as an ornamental stone. Also spelled: elaeolite; elaolite.
eleolite syenite *nepheline syenite.*
elevating conveyor Any conveyor used to discharge material at a point higher than that at which it was received. Term is specific. applied to certain underground mine conveyors.
elevating grader A grader equipped with a collecting device and elevator, by which the loosened material can be loaded to spoil banks or into vehicles for transport. See also: *belt loader.* (Nelson, 1965)
elevation A general term for a topographic feature of any size that rises above the adjacent land or the surrounding ocean bottom; a place or station that is elevated. The vertical distance from a datum (usually mean sea level) to a point or object on the Earth's surface; esp. the height of a ground point above the level of the sea. The term is used synonymously with altitude in referring to distance above sea level, but in modern surveying practice the term elevation is preferred to indicate heights on the Earth's surface, whereas altitude is used to indicate the heights of points in space above the Earth's surface. (AGI, 1987)
elevation correction In gravity measurements, the corrections applied to observed gravity values because of differences of station elevation to reduce them to an arbitrary reference or datum level, usually sea level. The corrections consist of (1) the free-air correction, to take care of the vertical decrease of gravity with increase of elevation, and (2) the Bouguer correction, to take care of the attraction of the material between the reference datum and that of the individual station. In seismic measurements, the corrections are applied to observed reflection time values due to differences of station elevation in order to reduce the observations to an arbitrary reference datum or fiducial plane. (AGI, 1987)
elevator (a) An apparatus used to facilitate the removal of coal from shuttle cars or low conveyors into mine cars. (BCI, 1947) (b) A type of conveyor for raising coal, stone, ore, or slurry, usually at the coal preparation plant or mill. Normally it comprises a series of steel buckets attached to an endless chain. See also: *bucket elevator.* (Nelson, 1965) (c) A cage hoist. (Nichols, 1976) (d) A device for raising or lowering tubing, casing, or drive pipe, from or into well. (e) An endless belt or chain conveyor with cleats, scoops, or buckets for raising material. (Webster 3rd, 1966) (f) A cage or platform and its hoisting machinery, as in a building or mine, for conveying persons or goods to or from different levels. See also: *hoist.* (Webster 3rd, 1966) (g) A vertical or steeply inclined conveyor. (BS, 1962) (h) A machine that raises material on a belt or a chain of small buckets. (Nichols, 1976) (i) A hinged circle or latch block provided with long links to hang on a hoistlike hook and used to hoist collared pipe, drill pipe and/or casing, and drill rods provided with elevator plugs. Some large elevators are fitted with slips for use on uncollared or flush-outside tubular equipment. (Long, 1960) (j) A term sometimes and incorrectly used as a syn. of lifting bail. See also: *hydraulic dredge; vertical reciprocating conveyor.* (Long, 1960) (k) An apparatus used to facilitate the removal of coal from shuttle cars or low conveyors into mine cars. (BCI, 1947)
elevator bucket A vessel generally rectangular in plan and having a back suitably shaped for attachment to a chain or belt and a bottom or front designed to permit discharge of material as the bucket passes over the head wheel of a bucket elevator.
elevator dredger A dredger fitted with a bucket ladder. (Hammond, 1965)
elevator plug A short steel plug provided with a pin thread by means of which it may be coupled to the upper end of a stand of drill rods. Its diameter is greater than that of the drill rod to which it is attached, and hence it provides a shoulder that can be grasped by an elevator. When each stand of rod is provided with an elevator plug and an elevator is used in lieu of a rod-hoisting plug, the handling of rods is facilitated and a round trip can be made in less time. Syn: *rod plug.* (Long, 1960)
elevator rope A rope used to operate an elevator. (Zern, 1928)
Elie ruby A small-grained variety of pyrope garnet in the trap tuff of Kincraig Point, near Elie, Fife, Scotland.
eliquate (a) To liquate; smelt. (Webster 3rd, 1966) (b) To part by liquation. (Webster 3rd, 1966)
elkerite A name applied to a subgroup of pyrobitumens rich in oxygen and partly soluble in alkali. They resemble an earthy brown coal and probably represent a product of intense weathering of bitumens. (Tomkeieff, 1954)
ellestadite A hexagonal mineral, $Ca_5(SiO_4,PO_4,SO_4)_3(F,OH,Cl)$; apatite group; it is chlorellestadite if (Cl>OH,F), fluorellestadite if (F>OH,Cl), or hydroxylellestadite if (OH>F,Cl). Hydroxylellestadite occurs as veinlets in blue calcite associated with wilkeite, idocrase, and similar contact metamorphic minerals at Crestmore, Riverside County, CA.

elliptical polarization In optics, elliptically polarized light consisting of upward-spiraling vibration vectors, the surface of which is elliptical rather than circular, as in circular polarization. It is caused by the inconstant lengths of vibration vectors of mutually perpendicular plane-polarized waves whose path differences differ in phase by amounts other than $(n+1)/4\lambda$ on emergence from a crystal. (AGI, 1987)

Elmore jig A plunger-type jig of either single or mulitple compartments. Its distinguishing features are (1) an automatic control in the form of a cylinder that measures the specific gravity of the mixture of coal and refuse; (2) the refuse draw is a star gate under the overflow lip in each compartment, which extends the full width of the jig; and (3) the hutch is commonly collected with a screw conveyor and discharged through the refuse elevator. Used both for treatment of nut and slack sizes of bituminous coal. (Mitchell, 1950)

elpasolite An isometric mineral, K_2NaAlF_6; associated with pachnolite in cryolite-bearing pegmatites of the Pikes Peak region, El Paso County, CO.

elpidite An orthorhombic mineral, $Na_2ZrSi_6O_{15}{\cdot}3H_2O$; in fibrous, columnar, prismatic crystals in albitized nepheline-syenite pegmatites at Narsarsuk, Greenland; Kola Peninsula, Russia; Mont St. Hilare, PQ, Canada; and Tarbagatai, Kazakhstan.

Elsner's equation In dissolution of gold by dilute aerated cyanide solution this reads: $4Au + 8NaCN + O_2 + 2H_2O = 4NaAU(CN)_{2+4}NaOH$. Analogous equation is given for silver. Other mechanisms have been suggested by Janin and Bodlaender, the latter requiring two stages of reaction with the intermediate formation of hydrogen peroxide. See also: *MacArthur and Forest cyanide process*. (Pryor, 1963)

Eltran method Electrical exploration method in which an electrical transient is sent into the Earth and the change in shape of this transient is studied. (Schieferdecker, 1959)

eluant Liquid used to displace captured ions from the zeolite or resin on which they are held; also, in ion exchange processes, solution used for elution. In chromatography, the solution used to displace absorbed substances. (Pryor, 1963)

elutriation (a) A method of mechanical analysis of a sediment, in which the finer, lightweight particles are separated from the coarser, heavy particles by means of a slowly rising current of air or water of known and controlled velocity, carrying the lighter particles upward and allowing the heavier ones to sink. (AGI, 1987) (b) Purification, or removal of material from a mixture or in suspension in water, by washing and decanting, leaving the heavier particles behind. Syn: *water separation*. (AGI, 1987)

elutriator An appliance for washing or sizing very fine particles, based on the principle that large grains settle at a faster rate through a liquid than small grains of the same material. The medium is commonly an upward current of water. See also: *hydraulic classifier; Stokes' law*. (Nelson, 1965; Pryor, 1963)

eluvial (a) Said of an incoherent mineral deposit, such as a placer, resulting from the decomposition or disintegration of rock in place. The material may have slumped or washed downslope for a short distance but has not been transported by a stream. (AGI, 1987) (b) Pertaining to eluvium; residual. (AGI, 1987)

eluvium (a) An accumulation of rock debris produced in place by the decomposition or disintegration of rock; a weathering product; a residue. (AGI, 1987) (b) Fine soil or sand moved and deposited by the wind, as in a sand dune. Cf: *alluvium*. (AGI, 1987)

elvan Cornish term for pneumatolized granite rocks containing tourmaline, fluorite, or topaz. (Pryor, 1963)

emaldine *emildine.*

emanation deposit An ore deposit of gaseous magmatic origin.

embankment A linear structure, usually of earth or gravel, constructed so as to extend above the natural ground surface and designed to hold back water from overflowing a level tract of land, to retain water in a reservoir, tailings in a pond, or a stream in its channel, or to carry a roadway or railroad; e.g., a dike, seawall, or fill. (AGI, 1987)

embayment (a) Penetration of microcrystalline groundmass material into phenocrysts, making their normal euhedral boundaries incomplete. An irregular corrosion or modification of the outline of a crystal by the magma from which it previously crystallized or in which it occurs as a foreign inclusion; esp. the deep corrosion into the sides of a phenocryst. The penetration of a crystal by another, generally euhedral, crystal. Such a crystal is called an embayed crystal. (AGI, 1987) (b) A downwarped area containing stratified rocks, either sedimentary or volcanic or both, that extends into a terrain of other rocks, e.g., the Mississippi Embayment of the U.S. Gulf Coast. (AGI, 1987)

embolite (a) Sectile, ductile; occurs as gray, yellowish or greenish-gray hornlike masses, waxy coatings, or crusts, as a secondary mineral in the oxidized zone of silver deposits; commonly associated with native silver, manganese oxides, and secondary lead and copper minerals. (b) The chief source of silver in some Chilean mines occurring as yellow-green incrustations and masses. It occurs in Australia at Broken Hill, New South Wales, and at Silver Reef, Victoria; widespread in the silver mining districts of the United States. Syn: *horn silver.*

embrittlement Reduction in the normal ductility of a metal because of a physical or chemical change. (ASM, 1961)

emerald (a) A brilliant green gem variety of beryl, highly prized as a gemstone. The color, which is caused by chromium or vanadium impurity, ranges from medium-light to medium-dark tones of slightly bluish green to slightly yellowish green. Syn: *smaragd*. (b) Any of various gemstones having a green color; e.g., oriental emerald (sapphire), copper emerald (dioptase), Brazilian emerald (tourmaline), or Uralian emerald (demantoid variety of andradite garnet). (c) Said of a gemmy and richly green-colored mineral; e.g., emerald jade (jadeite), emerald spodumene (variety hiddenite), or emerald malachite (dioptase).

emerald copper *dioptase.*

emerald cut A step cut in which the finished gem is square or rectangular and the rows (steps) of elongated facets on the crown and pavilion are parallel to the girdle with sets on each of the four sides and in some cases at the corners; commonly used on diamonds to emphasize the absence of color and on emeralds and other colored stones to enhance the color. Cf: *step cut.*

emerald filter *emerald glass.*

emerald glass The trade name for a color filter through which genuine emeralds and some other genuine stones appear reddish to violetish while glass imitations and some genuine stones appear greenish. Syn: *beryloscope*. See also: *Walton filter.*

emeraldine A misnomer for chalcedony stained green with chromic oxide. It is a deeper green than nickel-stained chalcedony and, unlike the nickel types, shows a red residual color under the dichromatic filter.

emerald jade Semitransparent to translucent jadeite of emerald color. Syn: *imperial jade.*

emerald loupe *Walton filter.*

emerald malachite *dioptase.*

emerald nickel *zaratite.*

emerald triplet An assembled stone commonly consisting of a crown and pavilion of rock crystal bound together by transparent green cement or a thin piece of green sintered glass; immersed in water and viewed sideways, the top and bottom are colorless with a line of color along the girdle. Green or colorless beryl may be used for the crown and possibly for the pavilion. Glass may be used for the pavilion, and sometimes for the crown as well, but the trade still calls it an emerald triplet. Syn: *Soudé emerald*. See also: *tripletine*. Cf: *triplet.*

emerandine A misnomer for dioptase.

emergence (a) A change in the levels of water and land such that the land is relatively higher and areas formerly under water are exposed; it results either from an uplift of the land or from a fall of the water level. Ant: submergence. (AGI, 1987) (b) The point where an underground stream appears at the surface to become a surface stream. Syn: *resurgence; rise; rising*. (AGI, 1987)

emery An impure mineral of the corundum or aluminum oxide type used extensively as an abrasive before the development of electric-furnace products. See also: *emery rock*. (ASM, 1961)

emery rock A granular rock that is composed essentially of an impure mixture of corundum, magnetite, and spinel, and that may be formed by magmatic segregation or by metamorphism of highly aluminous sediments. Syn: *emery; corundolite*. (AGI, 1987)

emildine A variety of spessartine garnet, with yttrium substituting for manganese up to 2%, in pegmatites at Elk Mountain, NM. *spessartine.*

emilite *emildine.*

emission spectra Monochromatic light from quantized electron transitions in thermally excited ions or atoms. Cf: *absorption spectra.*

emission spectrum A spectrum regarded as characterizing the body that emits the rays rather than one through which they pass. (Standard, 1964)

emission standards The maximum amount of pollutant permitted to be discharged from a single polluting source. (NSC, 1992)

emissivity The ratio of radiant energy emitted by a body to that emitted by a perfect black body. A perfect black body has an emissivity of 1; a perfect reflector an emissivity of 0. (Strock, 1948)

emmonsite A triclinic mineral, $Fe_2(TeO_3)_3{\cdot}2H_2O$; in yellow-green microcrystalline masses, fibrous crusts, patches of minute acicular crystals, or thin scaly coatings in the oxidation zones of gold and silver telluride districts of North and Central America.

Emory picker A chute with narrow opening for the cleaning of coal. The slate, traveling slowly because of friction, falls into the openings and thus is removed from the coal, which, rolling freely down the incline, is carried over the narrow gap. (Zern, 1928)

Empire drill (a) A light, hand-operated churn drill for testing placers from 10 to 125 ft (3.0 to 38.1 m) deep, though it is more commonly used for shallower holes. It consists of a string of 4-in (10.2-cm) casing, to the lower end of which is screwed a toothed cutting shoe. To the upper part, projecting above the ground, is fastened a round steel platform on which workers stand while operating the drilling tools. The casing can be turned by workers or a horse on the end of a long sweep fastened to the platform. The core of material inside the casing is loosened and brought to the surface by a drill pump on the end of a string of rods. Cf: *Banka drill.* (Lewis, 1964) (b) A term often misused as a synonym for churn drill. (Long, 1960)

emplacement A process by which igneous rock intrudes, or an orebody is formed in older rocks. (AGI, 1987)

emplectite An orthorhombic mineral, $CuBiS_2$; metallic, grayish to tin-white, acicular to prismatic crystals with longitudinal striations; sp gr, 6.3 to 6.5; associated with chalcopyrite and other sulfides in silver veins or bismuth-rich parts of contact metamorphic deposits; may be confused with bismuthinite.

emplectum *emplectite.*

empressite An orthorhombic mineral, AgTe; forms fine granular, pale-bronze masses; sp gr, 7.61; associated with galena and native tellurium at the Empress Josephine Mine, Kerber Creek district, CO.

empty An empty car, truck, tub, box, or wagon. (Mason, 1951)

empty-car puller In bituminous coal mining, a laborer who pulls empty cars from cage or detaches them from hoisting cable when hoisting of loaded cars is done on one side of the shaft or haulage slope and lowering is done on the other. (DOT, 1949)

empty rope Any winding or hauling rope from which the load upon it has been removed. (Fay, 1920)

empty track A track for storing empty mine cars. (Fay, 1920)

empty trip Applies to empty coal, ore, and waste cars returning for another load. (Fay, 1920)

Ems method The condensation of dust and fumes from calcining furnaces by use of large flues filled with parallel rows of sheet iron. (Fay, 1920)

emulsification The phenomenon of holding finely divided particles of a liquid in suspension within the body of another liquid. Banka method. (Shell)

emulsifier (a) *mud mixer.* (b) A saponifying or other agent added to water and oil or water and resins, causing them to form an emulsion. (Long, 1960)

emulsion (a) A liquid mixture in which a fatty or resinous substance is suspended in minute particles almost equivalent to molecular dispersion. (Fay, 1920) (b) A suspension of one finely divided liquid phase in another. (ASM, 1961)

emulsion texture An ore texture showing minute blebs or rounded inclusions of one mineral irregularly distributed in another. (AGI, 1987)

enantiomorphism Having crystal forms that, while possessing neither a plane nor a center of symmetry, may occur in two positions that are mirror images of one another. The two positions cannot be converted into each other by any rotation, but are related to each other as are the right and left hand, hence designated right- and left-handed forms. Enantiomorphous crystals cause circular polarization of light, e.g., quartz.

enantiomorphous In crystallography, similar in form but not superposable related to each other as the right hand is to the left, hence, one the mirror image of the other. (AGI, 1987)

enargite An orthorhombic mineral, Cu_3AsS_4; dimorphous with luzonite, metallic gray-black; in vein and replacement copper deposits as small crystals or granular masses; an important ore of copper and arsenic; may contain up to 7% antimony; localities include Butte, MT; Chuquicamata, Chile; Cerro de Pasco, Colquijirca, Peru; Tsumeb, Namibia; and Bor, Serbia.

en cabochon Cut in a style characterized by a smooth-domed, but unfaceted, surface; e.g., a ruby cut en cabochon in order to bring out the star. Etymol: French. See also: *cabochon.* Commonly used for garnets (carbuncles) and for those gems that depend for their beauty largely upon minute oriented inclusions (e.g., crocidolite, star ruby, or sapphire), the plan of the stone being circular or oval.

encroachment (a) To work coal or mineral beyond the boundary that divides one mine area from another; to work coal from a barrier pillar that has been left as a safety measure. Also called trespass. (Nelson, 1965) (b) The advancement of water, replacing withdrawn oil or gas in a reservoir. (AGI, 1987)

encrustation (a) A crust or coating of minerals formed on a rock surface; e.g., calcite on cave objects, soluble salts on a playa, or manganese-rich crusts on the ocean or lake floor. (AGI, 1987) (b) The process by which a crust or coating is formed. Syn: *incrustation.* (AGI, 1987)

end (a) The secondary cleavage more or less at right angles to the bord or face cleat. (b) A direction parallel to the main natural line of cleat or cleavage in coal. Also called end line. (TIME, 1929-30) (c) Solid rock face at the termination of a tunnel. (Pryor, 1963)

end-bump table A mechanically operated, sloping table by which heavy and light materials are separated. The end motion imparted to the table tends to drive all minerals up the slope of the table, but a flow of water carries the light materials down faster than the mechanical motion carries them up. The heavy materials settle to the bottom and finally reach the upper end and are delivered into a proper receptacle. The Gilpin County, Imlay, and Golden Gate concentrators are the chief types. Syn: *Imlay table.* (Liddell, 1918)

end cleat *butt cleat.*

end clinometer A clinometer designed to be fitted only to the bottom end of a drillrod string, as contrasted with a line clinometer that can be coupled into the drillrod string at any point between two rods. (Long, 1960)

end-discharge tippler A framework to discharge the coal or mineral from a mine car or a wagon by elevating the rear end and to deliver the load from its front end onto a screen, chute, or bunker below track level. (Nelson, 1965)

end-dump car *mine car.*

end dumping Process in which earth is pushed over the edge of a deep fill and allowed to roll down the slope. (Carson, 1961)

endellinite *bournonite.*

endellionite *bournonite.*

endellite A monoclinic mineral, $Al_2Si_2O_5(OH)_4{\cdot}2H_2O$; kaolinite-serpentine group; soft, colorless to white; commonly tinted by impurities. (It is called halloysite in European literature.) Formerly called hydrated halloysite, hydrohalloysite, hydrokaolin.

end face A coal face that is at right angles to the main cleats in the seam. (Nelson, 1965)

end-fired furnace A furnace with fuel supplied from the end wall. (ASTM, 1994)

endgate (a) Gate at the front end of a car as it travels toward the dump. This gate has hooks that are engaged at the dump by stirrups that lift it, so that when the dump pitches forward the coal slides under the uplifted endgate and is discharged onto a chute or over a dump pile. (Zern, 1928) (b) A gate leading to and at right angles to an end face. Also called ending. (TIME, 1929-30)

endgate car A mine car constructed with one hinged end that lifts up as the car is tilted down, permitting the coal, ore, and waste to run out. See also: *mine car.* (Kentucky, 1952)

ending (a) A road driven at right angles to the end cleat. (Mason, 1951) (b) Eng. An adit driven in a direction with the grain of the coal. (Fay, 1920)

endings A pillar method of working. See also: *narrow work.* (Nelson, 1965)

endless chain A device for hauling coal in which a chain passes from the engine along one side of the road, around a pulley at the far end, and back again on the other side of the road. Empty cars, attached to one side of the chain by various kinds of clips or hooks, are hauled into the mine; loaded cars attached to the other side of the chain are hauled out of the mine. (Korson, 1938)

endless rope A rope that moves in one direction, one part of which carries loaded cars from a mine at the same time that another part brings the empties into the mine. (Zern, 1928)

endlichite An arsenatian variety of vanadinite, intermediate in composition between vanadinite and mimetite.

endlines The boundary lines of a mining claim that cross the general course of the vein at the surface. If the side lines cross the course of the vein instead of running parallel with it, they then constitute endlines. When a mining claim crosses the course of the lode or vein instead of being along such lode or vein, the endlines are those that measure the width of the claim as it crosses the lode. (Fay, 1920)

endlines not parallel Extralateral rights are allowed on a claim whose endlines converge, but they are not allowed in case the endlines diverge. Converging endlines on a claim would have the disadvantage of giving the owner of such a claim a continually diminishing length of vein on working down the dip. (Lewis, 1964)

end member (a) One of the two or more simple compounds of which an isomorphous (solid-solution) series is composed. For example, the end members of the plagioclase feldspar series are albite, $NaAlSi_3O_8$, and anorthite, $CaAl_2Si_2O_8$. Syn: *minal.* (AGI, 1987) (b) One of the two extremes of a series; e.g., types of sedimentary rock or of fossils. (AGI, 1987)

endogene *endogenetic.*

endogenetic Derived from within; said of a geologic process, or of its resultant feature or rock, that originates within the Earth, e.g., volcanism, volcanoes, extrusive rocks. The term is also applied to chemical precipitates, e.g., evaporites, and to ore deposits that originate within the rocks that contain them. Cf: *exogenetic; hypogene.* Syn: *endogene; endogenic; endogenous.* (AGI, 1987)

endogenetic effects *endomorphism.*

endogenic *endogenetic.*

endogenous *endogenetic.*

endometamorphism *endomorphism.*

endomorph A crystal surrounded by another crystal of a different mineral species. Adj. endomorphic, endomorphous.

endomorphic Pertaining to contact metamorphism that takes place within the cooling intrusive rock; resulting from the reaction of the wall rock upon the peripheral portion of an intrusion.

endomorphic metamorphism *endomorphism.*

endomorphism Changes within an igneous rock produced by the complete or partial assimilation of country-rock fragments or by reaction upon it by the country rock along the contact surfaces. It is a form of contact metamorphism with emphasis on changes produced within the igneous body rather than in the country rock. The term was originated by Fournet in 1867. Cf: *exomorphism.* Partial syn: endogenetic effects. Syn: *endometamorphism; endomorphic metamorphism.* (AGI, 1987)

end-on Working a seam of coal, etc., at right angles to the cleat, or natural planes of cleavage. (Fay, 1920)

end-on working Working of coal seam at right angles to the natural cleats, joints, or slips. (Pryor, 1963)

endoscope In gemology, an instrument that affords a magnified image of the drill hole of a pearl, used to distinguish between genuine and cultured pearl. A tiny beam of light is directed into the walls of the drill hole to reveal whether the structure of the pearl's core is concentric (genuine) or parallel (cultured).

endostratic formation Bedding in clays resulting from alternating, desiccation, and saturation by groundwater. (Hess)

endothermic Accompanied by the absorption of heat. Opposite of exothermic. (CTD, 1958)

endplate In timbering, where both a cap and a sill are used, and posts act as dividers, the posts become the endplates. *sideplate.* (Fay, 1920)

end-port furnace A furnace with ports for fuel and air in the end wall. (ASTM, 1994)

ends York. Headings driven on the end or end-on. (Fay, 1920)

end slicing *top slicing combined with ore caving.*

end span A span that is a slab or a continuous beam at its interior support. (Hammond, 1965)

endurance The ability of a metal or a fabricated structure to recover from or to withstand repeated stress loadings or fluctuations. (Pryor, 1963)

endurance limit That stress below which a material can withstand hundreds of millions of repetitions of stress without fracturing. It is considerably lower than rupture strength. Syn: *fatigue limit.* (AGI, 1987)

endwall (a) The brick, concrete, or stonework construction at the sides of an excavation built to carry a flat or arched roof. Also called sidewall. (Spalding, 1949) (b) The vertical refractory wall, farthest from the furnace chamber, of the downtake of an open-hearth steel furnace. (Dodd, 1964) (c) One of the two vertical walls terminating a battery of coke ovens or a bench of gas retorts; it is generally constructed of refractory bricks and heat-insulating bricks with an exterior facing of building bricks. (Dodd, 1964)

en echelon Said of geologic features that are in an overlapping or staggered arrangement, e.g., faults. Each is relatively short, but collectively they form a linear zone, in which the strike of the individual features is oblique to that of the zone as a whole. Etymol: French en échelon, in steplike arrangement. (AGI, 1987)

enelectrite Minute, colorless, monoclinic, lath-shaped crystals, presumably a hydrocarbon; found in chemawinite (variety of amber), Cedar Lake, MB, Canada. (Tomkeieff, 1954)

energizing coil Primary coil that is used in inductive methods to set up electric currents in the Earth. (Schieferdecker, 1959)

energy (a) The ability of a body to perform work. (Shell) (b) The capacity for producing motion. Energy holds matter together. It can become mass, or it can be derived from mass. It takes such forms as kinetic, potential, heat, chemical, electrical, and atomic energy, and it can be changed from one of these forms to another. (Leet, 1958) (c) Kinetic energy is that due to motion, and potential energy is that due to position. In a stream, for example, the total energy at any section is represented by the sum of its potential and kinetic energies. (Seelye, 1951)

engineering geology Geology as applied to engineering practice, esp. mining and civil engineering. As defined by the Association of Engineering Geologists (1969), it is the application of geologic data, techniques, and principles to the study of naturally occurring rock and soil materials or ground water for the purpose of ensuring that geologic factors affecting the location, planning, design, construction, operation, and maintenance of engineering structures, and the development of ground-water resources, are properly recognized and adequately interpreted, utilized, and presented for use in engineering practice. Syn: *geologic engineering.* (AGI, 1987)

engineering system Any member of any assemblage of members such as a composite column, a coupling, a truss, or other structure.

engine pit Eng.; Scot. A shaft used entirely for pumping purposes.

engine plane (a) A system of rope haulage in which the loads are raised or lowered on the slope by a steam or electric hoist. In the simplest form only one track and one rope are required, and power is used for raising the load. Double engine planes have two separate tracks or three rails and a passing turnout. (Lewis, 1964) (b) A roadway, horizontal or inclined, on which tubs or cars are hauled by rope haulage. (Nelson, 1965) (c) Direct rope haulage. (Pryor, 1963)

engine tenter N. Staff. *brakeman.*

English cupellation A method of refining silver in which a small reverberatory furnace with a movable bed and a fixed roof is used. The bullion is charged gradually, and the silver is refined in the same furnace where the cupellation is carried on.

englishite An orthorhombic mineral, $K_3Na_2Ca_{10}Al_{15}(PO_4)_{21}(OH)_7 \cdot 26H_2O$; occurs with crandallite, wardite, and other phosphate minerals in variscite nodules at Fairfield, UT, and in the Tip Top pegmatite, SD.

English method A method of smelting lead ore in which the characteristics are a large charge of lead ore, a quick roasting, a high temperature throughout, and the aim to extract all the lead in the reverberatory. The hearth inclines toward the middle of one of the sides, the lead collects in the furnace and is tapped at intervals into an outside kettle. (Fay, 1920)

English process In copper smelting, the process of reduction in a reverberatory furnace, after roasting, if necessary.

English zinc furnace A furnace in which zinc is reduced and distilled from calcined ores in crucibles. (Fay, 1920)

engorgement The clogging of a furnace. (Fay, 1920)

enhydrite (a) A mineral or rock having cavities containing water. (b) *enhydros.*

enhydros A hollow nodule or geode of chalcedony containing water, possibly in large amounts. Syn: *enhydrite.*

enhydrous Containing water; having drops of included fluid; as enhydrous chalcedony. (Standard, 1964)

enigmatite *aenigmatite.*

enlarging shots Boreholes driven after the face of the rock has been unkeyed, and two or three free faces have thus been provided. (Stauffer, 1906)

en masse conveyor A conveyor comprising a series of skeleton or solid flights on an endless chain or other linkage that operates in horizontal, inclined, or vertical paths within a closely fitted casing for the carrying run. The bulk material is conveyed and elevated en masse in a substantially continuous stream with a full cross section of the casing. Also called chain conveyor. Syn: *continuous stream conveyor.*

en masse feeder *conveyor-type feeder.*

enriched uranium *uranium.*

enrichment *supergene enrichment.*

Ensign-Bickford hot-wire lighter A fuse lighter similar to a Fourth-of-July sparkler, that burns for 2-1/2 min, sufficient time to light 30 to 50 fuses. The lead splitter is a lead tube of diameter about 1/8 in (3.2 mm) in filled with a slow-burning powder that burns at the rate of 36 s/ft (118 s/m) with a hot splitting flame. (Lewis, 1964)

Ensign-Bickford master fuse lighter A shell, similar to a shotgun cartridge, that contains an ignition compound in the base. As many as seven fuses can be pushed into the shell until the fuses contact the ignition compound. The lighting of one fuse, which burns into the shell, sets off the compound and ignites the other six fuses. (Lewis, 1964)

enstatite An orthorhombic mineral, $2[MgSiO_3]$; pyroxene group; dimorphous with clinoenstatite; a common rock-forming mineral in basalt, gabbro, norite, pyroxenite, and peridotite. Formerly called amblystegite, bronzite, chladnite, ficinite, hypersthene (in part), orthobronzite, orthoenstatite, orthohypersthene (in part), paulite, peckhamite, protobasite, shepardite, and victorite. See also: *pyroxene.* Symbol, En.

enstenite A group name for the orthopyroxenes of the $MgSiO_3$-$FeSiO_3$ isomorphous series. It includes enstatite, hypersthene, and orthoferrosilite. Cf: *clinoenstenite.* (AGI, 1987)

entrainment The process of picking up and carrying along, as the collecting and movement of sediment by currents, or the incorporation of air bubbles into a cement slurry. (AGI, 1987)

entropy (a) A measure of the unavailable energy in a system; i.e., energy that cannot be converted into another form of energy. (AGI, 1987) (b) A measure of the mixing of different kinds of sediment; high entropy is approach to unmixed sediment of one kind. (AGI, 1987) (c) Ratio of amount of heat added to air to the absolute temperature at which it is added. Measured in Btu. (Hartman, 1961) (d) Specific entropy is the ratio of entropy to weight of substance. (Strock, 1948)

entry (a) In coal mining a haulage road, gangway, or airway to the surface. (b) An underground passage used for haulage or ventilation, or as a manway. Back entry, the air course parallel to and below an entry. Distinguished from straight entry, front entry, or main entry. Dip entry, an entry driven downhill so that water will stand at the face directly down a steep dip slope. Gob entry, a wide entry with a heap of refuse or gob along one side. Slab entry, an entry that is widened or slabbed to provide a working place for a second miner. Double entry, a system of opening a mine by two parallel entries; the air current is brought into the rooms through one entry and out through the parallel entry or air course. Cutoff entry, an entry driven to intersect another and furnish a more convenient outlet for the coal. Single entry, a system of opening a mine by driving a single entry only, in place of a pair of entries. The air current returns along the face of the rooms, which must be kept open. Triple entry, a system of opening a mine by driving three parallel entries for the main entries. Twin entry, a pair of entries close together and carrying the air current in and out, so laid out that rooms can be worked from both entries. Also called double entry. (c) A coal heading. To develop a coal mine in the United States, one or more sets of main entries are driven into the take. Each set consists of four to eight coal headings, connected at intervals by crosscuts. From these, and usually at right angles, butt entries, three to six in number, are driven at intervals of up to 1,500 yd (1.37 km). Between the sets of butt entries, face entries, three to four in number, are driven at intervals of up to 500 yd (0.46 km) to form a block or panel. The entries to split the panels may be 12 to 20 ft (3.7 to 6.1 m) wide and at 50- to 100-ft (15.2- to 30.5-m) centers. Each entry is made as productive as possible, and productivity is often higher in the entry work than in pillar extraction. See also: *pillar-and-stall.* (Nelson, 1965)

entry air course A passage for air parallel to an entry.

entry conveyor *underground mine conveyor; entry table.*

entry driver A combination mining machine designed and built to work in entries and other narrow places, and to load coal as it is broken down. An undercutting frame and two vertical shearing frames serve to undercut and shear the sides of the coal, so that the ram equipped with bars and operated by hydraulic jacks can break down the coal. The height at which the ram operates against the coal, when the undercut and shearing are completed, is adjustable. A conveyor in the undercutting frame carries the broken-down coal back to another conveyor mounted on a turntable so that the coal can be loaded into a mine car, or slate can be deposited on the gob side of the entry. The entire machine is mounted in a pan. (Kiser, 1929)

entry driver operator In bituminous coal mining, one who operates a type of coal cutter known as a heading machine that is adapted to the driving of underground haulageways in coal from one part of the mine to another or to the surface. Also called entry driving machine operator. (DOT, 1949)

entryman (a) A miner who works in an entry. (Fay, 1920) (b) One who enters upon public land with intent to secure an allotment under homestead, mining, or other laws. (Webster 3rd, 1966) (c) In anthracite and bituminous coal mining, one who is engaged in driving a haulageway, airway, or passageway from one place to another in the mine or to the surface. Also called heading driver. (DOT, 1949)

entry stumps Pillars of coal left in the mouths of abandoned rooms to support the road, entry, or gangway until the entry pillars are drawn. In Arkansas, these pillars are called entry stumps even when the rooms are first driven, before any pillars are pulled or the rooms abandoned. (Fay, 1920)

entry table A conveyor that transports material to the feeding position of a machine.

envelope (a) The outer or covering part of a fold, esp. of a folded structure that includes some sort of structural break. Cf: *core.* (AGI, 1987) (b) A metamorphic rock surrounding an igneous intrusion. (c) In a mineral, an outer part different in origin from an inner part.

environmental assessment An analysis of environmental conditions which may involve baseline environmental analyses and data gathered with regard to zoological, botanical, geologic, and economic factors. This data may be utilized for environmental impact statements. Abbrev.: EA. (SME, 1992)

environmental audit An evaluation of environmental conditions at a particular facility or site. Major items that could be relevant to an environmental audit for a mining facility may include information on permits, surface and mineral rights, mine ownership and violations, archaeological sites, hydrology issues, air pollution, waste disposal, impoundments, mine fires, underground injections and previously mined areas. (SME, 1992)

Environmental Impact Statement A statement which is prepared by a Federal agency with regard to a permit, and is required under the National Environmental Policy Act (NEPA). The EIS may include but is not limited to information relating to the purposes and needs to which the agency is responding by the preparation of the EIS, alternatives, and the environmental consequences which may arise from the proposed action. (SME, 1992)

eolian (a) Pertaining to the wind; esp. said of such deposits as loess and dune sand, of sedimentary structures such as wind-formed ripple marks, or of erosion and deposition accomplished by the wind. (AGI, 1987) (b) Said of the active phase of a dune cycle, marked by diminished vegetal control and increased dune growth. Etymol: Aeolus, god of the winds. Syn: *aeolian.* (AGI, 1987)

eolian deposit Wind-deposited accumulations, such as loess and dune sand. (Stokes, 1955)

eolianite A consolidated sedimentary rock consisting of clastic material deposited by the wind; e.g., dune sand cemented below groundwater level by calcite. (AGI, 1987)

eolian placer A placer concentrated by wind action.

eon (a) The formal geochronologic unit of highest rank, next above era. The Phanerozoic Eon encompasses the Paleozoic, Mesozoic, and Cenozoic Eras. (AGI, 1987) (b) One billion years. Also spelled aeon. (AGI, 1987)

eosphorite A monoclinic mineral, $MnAl(PO_4)(OH)_2 \cdot H_2O$; forms a series with childrenite; pink to rose red; in granite pegmatites associated with manganese phosphates.

Eötvös balance A sensitive torsion balance for measuring variations in the density of the underlying rocks; it records the horizontal gradient of gravity. (Webster 3rd, 1966)

Eötvös torsion *torsion balance.*

Eötvös torsion balance *torsion balance.*

epaulet A five-sided step-cut gem resembling a shoulder ornament (epaulet) in outline.

epeirogenesis *epeirogeny.*

epeirogeny As defined by Gilbert (1890), a form of diastrophism that has produced the larger features of the continents and oceans, for example, plateaus and basins, in contrast to the more localized process of orogeny, which has produced mountain chains. Epeirogenic movements are primarily vertical, either upward or downward, and have affected large parts of the continents, not only in the cratons but also in stabilized former orogenic belts, where they have produced most of the present mountainous topography. Some epeirogenic and orogenic structures grade into each other in detail, but most of them contrast strongly. Adj. epeirogenic. Syn: *epeirogenesis.* (AGI, 1987)

epi- A prefix signifying on or upon. Cf: *cata-.* (AGI, 1987)

epiasterism Asterism seen by reflected light, as in star ruby or sapphire cut en cabochon to reveal asteria. The effect is created when light is reflected from suitably oriented inclusions within the stone.

epibenthic dredge A bottom sampler consisting of a pair of sheet-metal skis attached to a light framework for a silk or nylon net. Removable rakers in front of the net stir up the bottom as the dredge advances, permitting the net to capture the benthic fauna and flora contained in the sediment. A bottom-walking wheel connected to a small counter indicates the distance over the bottom the device travels during a haul. (Hunt, 1965)

epibenthos In oceanography, animals and plants found living below low tidemark and above the 100-fathom (183-m) line. (CTD, 1958)

epicontinental Situated upon a continental plateau or platform, as an epicontinental sea. (Fay, 1920)

epidiabase A name proposed as a replacement for epidiorite. (AGI, 1987)

epididymite An orthorhombic mineral, $NaBeSi_3O_7(OH)$; dimorphous with eudidymite; forms colorless tabular crystals in nepheline-syenite pegmatites with albite, elpidite, and analcime at Mont St. Hilare, PQ, Canada; Narsarsuk, Greenland; and Langesundfjord, Norway.

epidiorite A metamorphosed gabbro or diabase in which generally fibrous amphibole (uralite) has replaced the original clinopyroxene (commonly augite). It is usually massive but may have some schistosity. See also: *epidiabase.* (AGI, 1987)

epidosite A metamorphic rock consisting of epidote and quartz, and generally containing other secondary minerals such as uralite and chlorite. (AGI, 1987)

epidote (a) A basic silicate of aluminum, calcium, and iron. One form is $CA_2(Fe^{3+},Al)_3(SiO_4)_3(OH)$; monoclinic; Mohs hardness, 6 to 7; sp gr, 3.25 to 3.5; and a common secondary constituent of igneous rocks. (Pryor, 1963) (b) A monoclinic mineral, $2[Ca_2FeAl_2O(OH)(Si_2O_7)(SiO_4)]$; green; forms a series with clinozoisite; a common rock-forming mineral with albite and chlorite in low-grade metamorphic rocks and an accessory in some igneous rocks; may be used as a minor gemstone; formerly called pistacite, arendalite, delphinite, thalalite. (c) The mineral group allanite, allanite-(Y), clinozoisite, epidote, hancockite, mukhinite, piemontite, and zoisite.

epidotization The hydrothermal introduction of epidote into rocks or the alteration of rocks in which plagioclase is albitized, freeing the anorthite molecule for the formation of epidote and zoisite, often accompanied by chloritization. These processes are characteristically associated with metamorphism. (AGI, 1987)

epigene (a) Said of a geologic process, or of its resultant features, occurring at or near the Earth's surface. Cf: *hypogene.* Syn: *epigenic.* (AGI, 1987) (b) Pertaining to a crystal that is not natural to its enclosing material; e.g., a pseudomorph. (AGI, 1987)

epigenesis (a) The change in the mineral character of a rock as a result of external influences operating near the Earth's surface, e.g., mineral replacement during metamorphism. (AGI, 1987) (b) The changes, transformations, or processes, occurring at low temperatures and pressures, that affect sedimentary rocks subsequent to their compaction, exclusive of surficial alteration (weathering) and metamorphism; e.g., postdepositional dolomitization. (AGI, 1987)

epigenetic (a) Said of a mineral deposit formed later than the enclosing rocks. Cf: *syngenetic.* (b) Produced on or near the Earth's surface, e.g., epigenetic valleys, etc. (c) In ore petrology, applied to mineral deposits of later origin than the enclosing rocks or to the formation of secondary minerals by alteration. Syn: *epigenic.*

epigenic *epigene; epigenetic.*

epigenite An orthorhombic(?) mineral, $(Cu,Fe)_5AsS_6$(?); metallic gray; forms prismatic crystals resembling arsenopyrite implanted on barite at the Neugluck Mine near Wittichen, Germany.

epiianthinite *schoepite.*

epimagmatic *deuteric.*

epineritic environment The marine bottoms to a maximum depth of 20 fathoms (36.6 m). (Schieferdecker, 1959)

epiphysis An apophysis or tongue of an intrusion which is detached from its source. Also spelled epiphesis. See also: *tongue.* (AGI, 1987)

epiplankton In oceanography, plankton found in depths of less than 100 fathoms (183 m). (CTD, 1958)

epistilbite A monoclinic mineral, $Ca[Al_2Si_6O_{16}]\cdot 5H_2O$; zeolite group; dimorphous with goosecreekite; forms radiating spherical aggregates of prismatic crystals in cavities in basalt and andesite, or with beryl in pegmatites.

epistolite A triclinic mineral, $Na_2(Nb,Ti)_2Si_2O_9\cdot nH_2O$; forms soft, pearly white rectangular plates in curved folia in the Julianehaab district of Greenland and the Lovozero alkali massif, Kola Peninsula, Russia.

epitaxy (a) Induced orientation of the crystal lattice of an electrodeposit at the plane of contact with the undisturbed underlying metal. (ASM, 1961) (b) Orientation of one crystal with that of the crystalline substrate on which it grew; e.g., halite growing on a cleavage plane of mica because the mesh of the net of halite nearly coincides in shape and size with the pseudohexagonal net of the mica substrate. Adj: epitactic, epitaxic, epitaxial. Cf: *distaxy; topotaxy; syntaxy.*

epithermal Said of a hydrothermal mineral deposit formed within about 1 km of the Earth's surface and in the temperature range of 50 to 200 °C, occurring mainly as veins. Also, said of that depositional environment. Cf: *hypothermal deposit; mesothermal; leptothermal; telethermal; xenothermal.* (AGI, 1987)

epizone According to Grubenmann's classification of metamorphic rocks (1904), the uppermost depth zone of metamorphism, characterized by low to moderate temperatures (less than 300 °C) and hydrostatic pressures with low to high shearing stress. Modern usage stresses pressure-temperature conditions (low metamorphic grade) rather than the likely depth of zone. Cf: *mesozone; katazone.* (AGI, 1987)

epoch (a) The formal geochronologic unit, longer than an age and shorter than a period, during which the rocks of the corresponding series were formed. (AGI, 1987) (b) A term used informally to designate a length (usually short) of geologic time; e.g., glacial epoch. (AGI, 1987)

epsomite An orthorhombic mineral, $MgSO_4\cdot 7H_2O$; bitter tasting; forms efflorescences of prismatic crystals, botryoidal masses, or incrustations on cave and mine walls from oxidizing sulfide minerals; also lacustrine deposits, mineral springs, and fumaroles. Syn: *epsom salt; bitter salt; hair salt.* See also: *kieserite.*

epsom salt *epsomite.*

eq.s. explosive An unsheathed explosive incorporating cooling agents, which is equivalent in safety (relating to the ignition of methane-air mixture) on a charge-weight basis to an explosive having a sheath of cooling agents around it. Abbrev. for equivalent-to-sheathed explosive. (BS, 1964)

equal-errors cut point The density at which equal portions of the feed material are wrongly placed in each of two products of a specific-gravity separation. Syn: *wolf cut point.* (BS, 1962)

equal-errors size The separation size at which equal portions of the feed material are wrongly placed in each of two products of a sizing operation. (BS, 1962)

equal-falling particles Particles possessing equal terminal velocities. They are the oversize material and form the underflow of a classifier. See also: *Stokes' law; terminal velocity.* (Nelson, 1965)

equalization of winding load The balancing of the weight of the winding rope, which varies considerably during a winding cycle. See also: *balance rope; winding; winding drum.* (Nelson, 1965)

equal lay Ropes of which the layers of wires in strands have all been laid to the same length of lay. Also known as parallel lay. See also: *Warrington.* (Hammond, 1965)

equant (a) Said of a crystal having the same or nearly the same diameter in all directions. Cf: *anisodesmic; tabular; prismatic.* Syn: *equidimensional; isometric.* (AGI, 1987) (b) Said of a sedimentary particle whose length is less than 1.5 times its width. (AGI, 1987) (c) Said of a rock in which the majority of grains are equant. (AGI, 1987) (d) Refers to crystals with roughly equal dimensions. Cf: *tabular; lathlike; rodlike; acicular.*

equant element A fabric element all of whose dimensions are approx. equal. Cf: *linear element; planar element.* (AGI, 1987)

equation of motion The Newtonian law of motion states that the product of mass and acceleration equals the vector sum of the forces.

equiaxed crystals Polyhedral crystals formed by spontaneous crystallization in the interior of a mass of metal in a mold. Distinguished from columnar crystals and chill crystals. (CTD, 1958)

equidimensional *equant.*

equiform Said of crystals that have the same (or nearly the same) shape. (AGI, 1987)

equigranular A textural term applied to rocks, the essential minerals of which are all of the same order of size. (Holmes, 1920)

equilibrium (a) A perfect balance of physical forces such that when two or more forces act upon a body, the body remains at rest. (Morris, 1958) (b) The state in which a reversible chemical reaction is proceeding at the same rate in each direction. Metastable equilibrium is a steady unsatisfied state that will undergo further change on addition of the phase necessary to complete its stability. Physical equilibrium can connote stable coexistence of a substance in two or more phases, such as solid, liquid, and/or vapor. (Pryor, 1963) (c) In geology, a balance between form and process, e.g., between the resistance of rocks along a coast and the erosional force of the waves. (AGI, 1987) (d) That state of a chemical system in which the phases do not undergo any change of properties with the passage of time, provided they have the same properties when the same conditions are again reached by a different procedure. (AGI, 1987)

equilibrium diagram *phase diagram.* Cf: *constitution diagram.*

equilibrium eutectic The composition within any system of two or more crystalline phases that melts completely at the lowest temperature; the temperature at which such a composition melts. (ACSG, 1963)

equilibrium moisture content The moisture content of a soil when the water is static. (Nelson, 1965)

equilibrium moisture of coal The moisture content retained at equilibrium in an atmosphere over a saturated solution of potassium sulfate at 30 °C, and 96% to 97% relative humidity. When the sample, before such equilibrium, contains total moisture at or above the equilibrium moisture, the equilibrium moisture may be considered as equivalent to inherent or bed moisture, and any excess may be considered as extraneous moisture.

equipment flowsheet A diagram indicating, preferably by symbols, the units of plant to be used in the various operational steps carried out within a coal-preparation plant. (BS, 1962)

equipotential line (a) A line along which water will rise to the same elevation in piezometric tubes. (ASCE, 1958) (b) A line along which the potential is everywhere constant for the attractive forces concerned.

equipotential-line method A technique used in electrical prospecting requiring artificial currents. It is based on the principle that if two electrodes are inserted in the ground and an external voltage is applied across them, there will be a flow of current through the earth from one electrode to the other. If the medium through which the current flows is homogeneous in its electrical properties, the flow lines will be regular and in a horizontal plane, symmetrical about the line joining the electrodes. Any inhomogeneities in these properties will cause distortions in the lines of current flow. Such distortions indicate the existence of buried material with either higher conductivity than its surroundings, so that it attracts the flow lines toward itself, or with lower conductivity, so that it tends to force the lines into the surrounding medium. (Dobrin, 1960)

equipotential surface A surface on which the potential is everywhere constant for the attractive forces concerned. The gravity vector is everywhere normal to a gravity equipotential surface; the geoid is an equipotential. Syn: *gravity equipotential surface; niveau surface; level surface.* (AGI, 1987)

equivalent Corresponding in geologic age or stratigraphic position; esp. said of strata or formations (in regions far from each other) that are contemporaneous in time of formation or deposition or that contain the same fossil forms. n. A stratum that is contemporaneous or equivalent in time or character. (AGI, 1987)

equivalent circuit An electrical network, the frequency response of which is identical to that of a quartz oscillator plate. (Am. Mineral., 1947)

equivalent diameter (a) The diameter of a hypothetical sphere composed of material having the same specific gravity as that of the actual particle and of such size that it will settle in a given liquid at the same terminal velocity as the actual particle. Also called equivalent size. (ASCE, 1958) (b) Twice the equivalent radius. (AGI, 1987)

equivalent evaporation The quantity of water that would be evaporated by a given apparatus if the water is received by the apparatus at 212 °F (100 °C), and vaporized at that temperature under atmospheric pressure. It is expressed in kilograms per hour. (Strock, 1948)

equivalent freefalling diameter *equivalent particle diameter.*

equivalent grade In textural classification, refers to the arithmetic mean size. (AGI, 1987)

equivalent length The resistance of a mine airway obstruction, duct or pipe elbow, valve, damper, orifice, bend, fitting, or other obstruction to flow, expressed in the number of feet of straight airway, duct, or pipe of the same cross section that would have the same resistance. (Strock, 1948)

equivalent orifice A term that compares the resistance of air of a mine to the resistance of a circular opening in a thin plate through which the same quantity of air flows under the same pressure as in the mine.

equivalent particle diameter A concept used in evaluating the size of fine particles by a sedimentation process; it is defined as the diameter of a sphere that has the same density and the same freefalling velocity in any given fluid as the particle in question. Cf: *particle size.* Syn: *equivalent freefalling diameter*. (Dodd, 1964)

equivalent radius (a) The radius of a spherical particle of density 2.65 (the density of quartz) which would have the same rate of settling as the given particle. (AGI, 1987) (b) A measure of particle size, equal to the computed radius of a hypothetical sphere of specific gravity 2.65 (quartz) having the same settling velocity and same density as those calculated for a given sedimentary particle in the same fluid; one half of the equivalent diameter. (AGI, 1987)

equivalent temperature A composite of mean radiant temperature and air temperature; also defined as the mean temperature of the environment effective in controlling the rate of sensible heat loss from a black body in still air when the surface temperature and size of the black body are comparable to those of the human body. Where the enclosure surface (mean radiant temperature) and air temperatures are equal, this temperature is also the British equivalent temperature; when not equal, the British equivalent temperature is that temperature at which a body with an 80 °F (26.7 °C) surface temperature will lose sensible heat at the same rate as in the given environment. Syn: *British equivalent temperature*. (Strock, 1948)

equivolumnar wave *distortional wave; S wave; transverse wave.*

era The formal geochronologic unit next in order of magnitude below an eon, during which the rocks of the corresponding erathem were formed; e.g., the Paleozoic Era, the Mesozoic Era, and the Cenozoic Era. Each of these includes two or more periods, during each of which a system of rocks was formed. Long-recognized Precambrian Eras are the Archeozoic (older) and Proterozoic (younger). (AGI, 1987)

eremeyevite Former spelling of jeremejevite, also spelled eremeevite.

ericaite An orthorhombic mineral, $(Fe,Mg,Mn)_3B_7O_{13}Cl$; forms a series with boracite; dimorphous with congolite; Mohs hardness of 7 to 7-1/2; forms pseudocubes in halite-anhydrite deposits in the Harz Mountains, Germany.

Erinide A U.S. trademark name for a yellowish-green synthetic spinel.

erinoid A casein plastic used for molding many common objects and possibly for inferior gem imitations; sp gr, about 1.33; refractive index, 1.53 to 1.54.

eriochalcite An orthorhombic mineral, $CuCl_2{\cdot}2H_2O$; forms a sublimate of soft bluish-green woollike aggregates on the sides of fumaroles. Syn: *antofagasite; erythrocalcite.*

erosion The group of physical and chemical processes by which earth or rock material is loosened or dissolved and removed from any part of the Earth's surface. It includes the processes of weathering, solution, corrosion, and transportation. The mechanical wear and transportation are effected by rain, running water, waves, moving ice, or winds, which use rock fragments to pound or to grind other rocks to powder or sand.

erosional unconformity An unconformity that separates older rocks that have been subjected to erosion from younger sediments that cover them; specif. disconformity. (AGI, 1987)

erosion channel *classical washout.*

erosion surface (a) A land surface shaped and subdued by the action of erosion, esp. by running water. The term is generally applied to a level or nearly level surface. (AGI, 1987) (b) An area that has been flattened by subaerial or marine erosion to form an area of relatively low relief at an elevation close to the base level (sea level) existing at the time of its formation. Relics of such surfaces may now be found far above sea level owing to the falling base level, below the present ocean surface. (Hunt, 1965)

erosion thrust A thrust fault along which the hanging wall moved across an erosion surface.

erratic A rock fragment carried by glacial ice or by floating ice, deposited at some distance from the outcrop from which it was derived, and generally though not necessarily resting on bedrock of different lithology. Size ranges from a pebble to a house-size block. (AGI, 1987)

error curve A partition curve drawn to defined conventional scales with the portion showing recoveries over 50% reversed to enclose an error area. Syn: *tromp error curve.* (BS, 1962)

error of closure (a) Of a traverse, the amount by which the computed position of the last point of the traverse fails to coincide with the initial point; i.e., the length of line necessary to close the traverse. Frequently, also, the ratio of the linear error of closure to the perimeter (also known as the error of the survey). (Seelye, 1951) (b) Of angles, the amount by which the sum of the measured angles fails to equal the true sum. (Seelye, 1951) (c) Of azimuths, the amount by which the measurement of the azimuth of the first line of a traverse, made after completing the circuit, fails to equal the initial measurement. (Seelye, 1951) (d) Of a level circuit, the amount by which the last computed elevation fails to equal the initial elevation; or the amount by which the differences of elevation in a circuit fail to add up (algebraically) to zero. (Seelye, 1951) (e) Of a horizon, the amount by which the sum of the angles measured around the horizon differs from 360°. (Seelye, 1951) (f) Of a triangle, the amount by which the sum of the three angles of a triangle differs from the true sum; i.e., 180° plus the spherical excess. (Seelye, 1951)

erubescite *bornite.*

eruptive Said of a rock formed by the solidification of magma; i.e., either an extrusive or an intrusive rock. Most writers restrict the term to its extrusive or volcanic sense. (AGI, 1987)

erythrine *erythrite.*

erythrite A monoclinic mineral, $2[Co_3(AsO_4)_2{\cdot}8H_2O]$; forms a series with annbergite and with hoernesite; occurs in soft pink to crimson crystals, globular or reniform masses, or earthy encrustations as a weathering product of cobalt ores in the oxidized parts of nickel-arsenic-silver-bearing veins; used as ore indicator for cobalt and possibly silver. Syn: *erythrine.* See also: *cobalt bloom; red cobalt; cobalt ocher; peachblossom ore.*

erythrocalcite *eriochalcite.*

escape (a) Eng. A second or additional shaft by which miners may get out of the mine in case of accident to the other shafts. Also an upcast; escape pit; escapeway. (b) A wasteway for discharging the entire flow of a stream. (Seelye, 1951)

escape shaft A shaft driven esp. to permit egress from the mine in case of emergency. (BCI, 1947)

escapeway An opening through which the miners may leave the mine if the ordinary exit is obstructed. (Fay, 1920)

escarpment (a) A long, more or less continuous cliff or relatively steep slope facing in one general direction, breaking the continuity of the land by separating two level or gently sloping surfaces, and produced by erosion or by faulting. The term is often used synonymously with scarp, although escarpment is more often applied to a cliff formed by differential erosion. (AGI, 1987) (b) A steep, abrupt face of rock, often presented by the highest strata in a line of cliffs, and generally marking the outcrop of a resistant layer occurring in a series of gently dipping softer strata; specif. the steep face of a cuesta. Cf: *cuesta.* (AGI, 1987)

Eschka's mixture A mixture of two parts magnesium and one part dried sodium carbonate; used as a reagent for determining sulfur in coal or coke. (Hackh, 1944)

eschynite *aeschynite.*

eskebornite A tetragonal mineral, $CuFeSe_2$; chalcopyrite group; forms a series with chalcopyrite; is metallic brass yellow; variably magnetic; with chalcopyrite, clausthalite, and naumannite in dolomite veins, Eskeborn adit, Tilkerode, Harz Mountains, Germany.

eskolaite A trigonal mineral, Cr_2O_3; hematite group; easily confused with hematite and magnetite; in a chromium-bearing tremolite skarn at the Outokumpu Mine, Finland.

Esperanza classifier A classifier of the free-settling type in which the settled material is removed by dragging it up an inclined plane by means of a continuous belt of flat blades or paddles. It is continuous in its operation. (Liddell, 1918)

esplanade A broad bench or terrace bordering a canyon, esp. in the plateau areas of the southwestern United States.

espley rock A conglomerate or breccia with rapid lateral passage through grit to fine sandstone; cement usually ferruginous with some lime and alumina. Characteristically developed amid variegated clays of Etruria Marl group of Upper Coal Measures in the English Midlands. (Arkell, 1953)

essential mineral A mineral component of a rock that is necessary to the classification and nomenclature of the rock, but that is not necessarily present in large amounts. Cf: *accessory mineral.* Syn: *specific mineral.* (AGI, 1987)

essexite An alkali gabbro primarily composed of plagioclase, hornblende, biotite, and titanaugite, with lesser amounts of alkali feldspar and nepheline. Essexite grades into theralite with a decrease in potassium feldspar and an increase in the feldspathoid minerals. Its name is derived from Essex County, MA. (AGI, 1987)

essonite A yellow-brown or reddish-brown transparent gem variety of grossular garnet containing iron. Syn: *cinnamon stone; hyacinth; jacinth. hessonite.*

estramadurite A massive variety of apatite found in Estramadura, Spain. A phosphate ore. (Hey, 1955; English, 1938)

estuarine deposit A sedimentary deposit laid down in the brackish water of an estuary, characterized by fine-grained sediments (chiefly clay and silt) of marine and fluvial origin mixed with a high proportion of decomposed terrestrial organic matter; it is finer grained and of more uniform composition than a deltaic deposit. (AGI, 1987)

estuary (a) The seaward end or the funnel-shaped tidal mouth of a river valley where fresh water comes into contact with seawater and where tidal effects are evident. (AGI, 1987) (b) A portion of an ocean or an arm of the sea affected by fresh water; e.g., the Baltic Sea. (AGI, 1987) (c) A drowned river mouth formed by the subsidence of land near the coast or by the rise of sea level. (AGI, 1987)

etch angle The angle formed between the true horizon and the actual plane of the etch ring in an acid bottle as measured before capillarity corrections. Also called apparent angle. See also: *capillarity.* Cf: *apparent dip.* (Long, 1960)

etch figure A marking, commonly in the form of minute pits, produced by a solvent on a crystal surface; the form varies with the mineral species and the solvent, but reflects the symmetry of the structure; also called etching figure.

etching A process of engraving in which lines, frosting, or roughening are produced by an acid or mordant. Often used in studying the composition and structure of metals, sandgrains, and crystals.

etch line A line of demarcation between the etched and unetched portions of the inside of an acid bottle, used to determine the inclination of a borehole by an acid-dip survey. (Long, 1960)

etch method A method, using a soda-lime glass tube partially filled with a dilute solution of hydrofluoric acid, of determining the angle at which a borehole is inclined at any specific point of its course below the collar. See also: *acid-dip survey.* (Long, 1960)

etch pattern Regular surface marking developed by solvent action on smooth surface of alloy or crystal, and characteristic for that specific substance. The reagent used is an etchant, usually of an acid in water or alcohol. (Pryor, 1963)

etch ring *etch line.*

etch time The time required for a dilute solution of hydrofluoric acid of a specific strength to etch the inside of an acid bottle enough so that the line of demarcation between the etched and unetched portions of the acid bottle is clearly discernible. Also known as etching time. (Long, 1960)

etch tube *acid-etch tube.*

ethanamide *acetamide.*

ethanol *alcohol.*

ethical gemology The study of the correct and incorrect nomenclature of gems, with emphasis on clarifying names and terms that may mislead or deceive purchasers.

ethine See acetylene.:

ethmolith A crosscutting intrusive body of plutonic rock that narrows downward.

ethylenediamene Used as an electrolyte to transform coal into a tan-gray substance with a relatively high hydrogen-to-carbon ratio.

ethylene glycol A highly explosive liquid $HOCH_2CH_2OH$; somewhat volatile; nonfreezing; explosive base. Used as an antifreeze. (Lewis, 1964)

ethyne *acetylene.*

etindite (a) Leucite nephelinite. (AGI, 1987) (b) A dark-colored extrusive rock intermediate in composition between leucitite and nephelinite with phenocrysts of clinopyroxene in a dense groundmass of leucite, nepheline, and clinopyroxene. The name is not included in the IUGS classification of extrusive igneous rocks.

eubitumen A collective name for fluid, viscid, and solid bitumens that are easily soluble in organic solvents. Petroleum, ozokerite, elaterite, and asphalt are examples. (Tomkeieff, 1954)

eucairite An orthorhombic mineral, CuAgSe; soft; sp gr, 7.6 to 7.8; in most deposits of selenium minerals, associated with unangite, klockmannite, and clausthalite, e.g., Smaland, Sweden; Copiapo, Chile; and the Harz Mountains, Germany. Also spelled eukairite.

eucalyptus oil Frothing agent used in flotation. Essential oil distilled from leaves of eucalyptus trees. (Pryor, 1963)

euchlorine An orthorhombic mineral, $(K,Na)_8Cu_9(SO_4)_{10}(OH)_6$(?); emerald green; forms thin incrustations on lava and fumarole deposits on Mt. Vesuvius, Italy.

euchroite An orthorhombic mineral, $Cu_2(AsO_4)(OH){\cdot}3H_2O$; vitreous; transparent to translucent; emerald or leek green; associated with azurite, malachite, or olivenite in the oxidation zones of copper deposits.

euclase A monoclinic mineral, $4[BeAlSiO_4(OH)]$; vitreous; Mohs hardness, 7-1/2; in auriferous sands in Austria, Russia, Brazil, Peru, and Tasmania Australia; and used as a minor gemstone.

eucolite A variety of eudialyte that is optically negative and rich in calcium. Also spelled eukolite.

eucryptite A trigonal mineral, $LiAlSiO_4$; transparent, vitreous; fluoresces pink in UV light; Mohs hardness = 6-1/2; forms with albite in pegmatites. If heated, beta-eucryptite expands normal to the *c* axis, but contracts parallel to *c*.

eucrystalline *macrocrystalline.*

eudialyte A trigonal mineral, $Na_4(Ca,Ce)_2(Fe,Mn,Y)ZrSi_8O_{22}(OH,Cl)_2$(?); weakly radioactive; in nepheline syenite and granite, commonly associated with arfvedsonite, sodalite, feldspar, aegirine, catapleiite, and astrophyllite. Cf: *eucolite.* Syn: *barsanovite.*

eudidymite A monoclinic mineral, $NaBeSi_3O_7(OH)$; dimorphous with epididymite; vitreous, in transparent to translucent tabular crystals with lamellar twinning having one perfect and one imperfect cleavage; in nepheline-syenite pegmatites with albite, elpidite, and analcime.

eudiometer An instrument for the volumetric measurement and analysis of gases. (Webster 3rd, 1966)

euhedral (a) Said of a mineral grain that is completely bounded by its own rational faces, and whose growth during crystallization or recrystallization was not restrained or interfered with by adjacent grains. (AGI, 1987) (b) Said of the shape of such a crystal. Syn: *idiomorphic.* Cf: *anhedral; subhedral. automorphic.* (AGI, 1987)

Eulerian methods of current measurement A measurement of the rate of flow past a geographically fixed point; current meter methods. (Hy, 1965)

eulite *ferrosilite.*

eulytite An isometric mineral, $Bi_4(SiO_4)_3$; vitreous, transparent to translucent tetrahedral crystals associated with native bismuth near Schneeberg, Saxony, Germany. Syn: *bismuth blende.*

euosmite A brownish-yellow resin with low oxygen content and a characteristic pleasant odor found in brown coal.

europia *europium oxide.*

europium oxide Rare-earth oxide; pale rose; Eu_2O_3; melting point, above 1,300 °C; and sp gr, 7.42. Used as a nuclear-control-rod material and in fluorescent glass. Syn: *europia.* (Lee, 1961; Handbook of Chem. & Phys., 2)

eusapropel A well-matured organic ooze. (Tomkeieff, 1954)

euscope A grain-size comparator. (Osborne, 1956)

eustasy The worldwide sea-level regime and its fluctuations, caused by absolute changes in the quantity of seawater, e.g., by continental icecap fluctuations. (AGI, 1987)

eustatic Pertaining to worldwide changes of sea level that affect all the oceans. Eustatic changes may have various causes, but the changes dominant in the last few million years were caused by additions of water to, or removal of water from, the continental icecaps. (AGI, 1987)

eutaxic Said of a stratified mineral deposit. Cf: *ataxic.* (AGI, 1987)

eutaxitic Said of the banded structure of certain extrusive rocks, which results in a streaked or blotched appearance. Also, said of a rock exhibiting such structure, e.g., a eutaxite. The bands or lenses were originally ejected as individual portions of magmas, were drawn out in a viscous state, and formed a heterogeneous mass in response to welding. (AGI, 1987)

eutectic Said of a system consisting of two or more solid phases and a liquid whose composition can be expressed in terms of positive quantities of the solid phases, all coexisting at an (isobarically) invariant point, which is the minimum melting temperature for the assemblage of solids. Addition or removal of heat causes an increase or decrease, respectively, of the proportion of liquid to solid phases, but does not change the temperature of the system or the composition of any phases. See also: *eutectoid.* (AGI, 1987)

eutectic point The lowest temperature at which a eutectic mixture will melt. Syn: *eutectic temperature.* (AGI, 1987)

eutectic ratio The ratio of solid phases crystallizing from the eutectic liquid at the eutectic temperature. It is such as to yield a gross composition for the crystal mixture that is identical with that of the liquid. It is most frequently stated in terms of weight percent.

eutectic temperature *eutectic point.*

eutectic texture A pattern of intergrowth of two or more minerals, formed as they coprecipitate during crystallization, e.g., the quartz and feldspar of graphic granite. See also: *exsolution texture.* Syn: *eutectoid texture.* (AGI, 1987)

eutectoid The equivalent of eutectic, when applied to a system all of whose participating phases are crystalline. (AGI, 1987)

eutectoid texture *eutectic texture.*

eutomous In mineralogy, having distinct cleavage; cleaving readily. (Fay, 1920)

eutrophic Said of a body of water characterized by a high level of plant nutrients, with correspondingly high primary productivity. (AGI, 1987)

eutrophic peat Peat rich in plant nutrients, nitrogen, potassium, phosphorus, and calcium. Synonymous with calcareous peat. (Tomkeieff, 1954)

euxamite Radioactive radium mineral found in Brazil. (Bennett, 1962)

euxenite An orthorhombic mineral, $(Y,Ca,Ce,U,La,Th)(Nb,Ta,Ti)_2O_6$; forms a series with polycrase; brilliant to vitreous brown to black; in pegmatites and placers commonly with monazite in Canada; Madagascar; Norway; and Pennsylvania. A source of uranium, niobium, and tantalum. Formerly called loranskite.

euxinic (a) Pertaining to an environment of restricted circulation and stagnant or anaerobic conditions, such as a fjord or a nearly isolated or silled basin with toxic bottom waters. Also, pertaining to the material (such as black organic sediments and hydrogen-sulfide muds) deposited in such an environment or basin, and to the process of deposition of such material (as in the Black Sea). (AGI, 1987) (b) Pertaining to a rock facies that includes black shales and graphitic sediments of various kinds. Etymol: Greek euxenos, hospitable. (AGI, 1987)

evaluate To fix a valuation, but not to appraise.

evaluation The fixing of a evaluation, not an appraisal. Used in preference to the word valuation which is often confused with appraisal.

evansite An amorphous mineral, $Al_3(PO_4)(OH)_6 \cdot 6H_2O$(?); may contain small amounts of uranium and thorium; associated with limonite and allophane.

evaporate To convert into vapor. See also: *evaporite.* (AGI, 1987)

evaporation (a) The process, also called vaporization, by which a substance passes from the liquid or solid state to the vapor state. Limited by some to vaporization of a liquid, in contrast to sublimation, the direct vaporization of a solid. Also limited by some (e.g., hydrologists) to vaporization that takes place below the boiling point of the liquid. The opposite of condensation. (AGI, 1987) (b) The process by which a substance is converted from a liquid state into a vapor. Specif., the conversion of a liquid into vapor in order to remove it wholly or partly from a liquid of higher boiling point or from solids dissolved in or mixed with it. Cf: *distillation; sublimation.*

evaporation gage A graduated vessel of glass for determining the rate of evaporation of a liquid placed in it, in a given time and exposure. (Fay, 1920)

evaporative cooling (a) The conversion of sensible heat to latent heat with addition of moisture and practically no change in total heat content of air. (Hartman, 1982) (b) Cooling by the evaporation of water, heat for which is supplied by the air; feasible where the wet-bulb depression is marked, and consequently widely used in dry climates. (Strock, 1948)

evaporite A nonclastic sedimentary rock composed primarily of minerals produced from a saline solution as a result of extensive or total evaporation of the solvent. Examples include gypsum, anhydrite, rock salt, primary dolomite, and various nitrates and borates. The term sometimes includes rocks developed by metamorphism or transport of other evaporites. Syn: *evaporate; saline deposit; saline residue.* (AGI, 1987)

evaporite-solution breccia A solution breccia formed where soluble evaporites (rock salt, anhydrite, gypsum, etc.) have been removed. See also: *solution breccia.* (AGI, 1987)

evasé A passage of gradually increasing area through which the air discharged by a fan must pass. The velocity of the air is gradually reduced and much of the kinetic energy is transformed into equivalent pressure energy. See also: *volute.* (Nelson, 1965)

evening emerald Olivine (peridot or chrysolite varieties) that loses some of its yellow tinge and appears more greenish (like an emerald) in artificial light, used as a gem. Syn: *night emerald.*

evenkite A monoclinic mineral n-tetracosane $C_{24}H_{50}$; forms waxlike crystals in geodes in a vein cutting vesicular tuff, Evenki district, lower Tunguska River, Siberia, Russia.

everlasting lamps N. of Eng. Natural jets of combustible gases or small blowers that continue to burn as long as gas is given off. (Fay, 1920)

evolutionary operation A method of process operation that introduces tightly controlled variations designed to transfer laboratory-proved improvements into changes leading to better commercial production. (Pryor, 1963)

EW Letter name specifying the dimensions of bits, core barrels, and drill rods in the E-size and W-group wireline diamond-drilling system having a core diameter of 21.5 mm and a hole diameter of 37.7 mm. Cf: *EX.* (Cumming, 1981)

EX Letter name specifying the dimensions of core, core barrels, and casing in the E-size and X-series wireline diamond-drilling system having a core diameter of 21.5 mm and a hole diameter of 37.7 mm. The EX designation for coring bits has been replaced by the EW designation. Syn: *EW.* (Cumming, 1981)

excavating cableway Cableway fitted with a bucket suitably designed for excavating. (Hammond, 1965)

excavation (a) The act or process of removing soil and/or rock materials from one location and transporting them to another. It includes digging, blasting, breaking, loading, and hauling, either at the surface or underground. See also: *breaking ground.* (AGI, 1987) (b) A pit, cavity, hole, or other uncovered cutting produced by excavation. (AGI, 1987) (c) The material dug out in making a channel or cavity. (AGI, 1987)

excavation deformation The zone around any excavation within which a structure might be disturbed by rock movements resulting from that excavation. (Nelson, 1965)

excavator The term embraces a large number of power-operated digging and loading machines. Variants are the grab, skimmer, trencher, rotary digger, bucket wheel, and grader. See also: *bulldozer; continuous-bucket excavator; dragline; power shovel; walking dragline.* (Nelson, 1965; Pryor, 1963)

excavator base machine A tracted prime mover to which can be fitted a variety of front-end excavating and lifting appliances. (Nelson, 1965)

excellent fumes Fumes that contain a minimum of toxic and irritating chemicals. (Nichols, 1976)

exception A reservation or exception of the minerals in a tract of land conveyed is a separation of the estate in the minerals from the estate in the surface, and it makes no difference whether the word used is excepted or reserved. (Ricketts, 1943)

excess air In practice, complete combustion cannot be obtained without slightly more air than is theoretically necessary. The amount of this excess air varies with the design and mechanical condition of the appliance, but ranges from 15% upward. (Nelson, 1965)

excessive location A mining claim in excess of the width allowed by law.

excess spoil Spoil in excess of that necessary to backfill and grade affected areas to the approximate original contour. The term may include box-cut spoil where it has been demonstrated for the duration of the mining operation, that the box-cut spoil is not needed to restore the approximate original contour.

excitation The addition of energy to a system, thereby transferring it from its ground state to an excited state. Excitation of a nucleus, an atom, or a molecule can result from absorption of photons or from inelastic collisions with other particles or systems.

excitation time The minimum time for which electric current must flow in the fusehead of a detonator to ensure its ignition. (BS, 1964)

Exclusive Economic Zone An area beyond and adjacent to the territorial sea, subject to the specific legal regime established in Part V, Articles 55-75 of the United Nations Convention on the Law of the Sea, under which the rights and jurisdiction of the coastal State and the rights and freedoms of other States are governed by the relevant provisions of the Convention. Abbrev. EEZ. EEZ resources rights include both living and non-living resources of the subsoil and superjacent water of the zone. The EEZ will not extend beyond 200 nmi (364 km) from the baseline from which the territorial sea is measured. Cf: *continental shelf.* (Holser, 1988)

exclusive prospecting license Grant of right to prospect a designated area for a limited period. Abbrev. EPL. (Pryor, 1963)

exempted claim A mining title on which exemption from otherwise essential activity has been granted. (Pryor, 1963)

exemptions Exemption laws are grants of personal privileges to debtors, which may be waived by contract or surrender or by neglect to claim before sale. (Ricketts, 1943)

exfoliate To peel off in concentric layers, as some rocks weather.

exfoliation (a) The process by which concentric scales, plates, or shells of rock, from less than a centimeter to several meters in thickness, are successively spalled or stripped from the bare surface of a large rock mass. It is caused by physical or chemical forces producing differential stresses within the rock, as by expansion of minerals as a result of near-surface chemical weathering, or by the release of confining pressure of a once deeply buried rock as it is brought nearer to the surface by erosion. It often results in a rounded rock mass or dome-shaped hill. Cf: *spheroidal weathering; spheroidal parting.* Syn: *spalling; scaling; sheeting; sheet jointing.* (AGI, 1987) (b) The property of some silicate minerals, e.g., vermiculite, or rocks, e.g., perlite, to expand permanently when heated to form an irregular or vesicular structure. Cf: *intumescence.*

exfoliation dome A large dome-shaped form, developed in massive homogeneous coarse-grained rocks, esp. granite, by exfoliation; well-known examples occur in Yosemite Valley, CA. (AGI, 1987)

exhalation (a) The streaming-forth of volcanic gases; also, the escape of gases from a magmatic fluid. (AGI, 1987) (b) Any vapor or gas arising from substances or surfaces exposed to the atmosphere. (c) Any gas or vapor formed beneath the surface of the Earth and escaping either through a conduit or fissure, or from molten lava or a hot spring; an emanation. (d) An exhaling or sending forth, as of steam or vapor. Something that is exhaled or given off or that rises in the form of a gas, fumes, or steam, for example, a foul exhalation from the marsh. (Webster 3rd, 1966)

exhaust fan In coal mining, a fan that sucks used air from a mine and thereby causes fresh air to enter by separate entries to repeat the cycle. (BCI, 1947)

exhausting auxiliary fan An auxiliary fan that exhausts air from the face of a tunnel through ducting or piping and discharges it into the return side of the airway off which the tunnel branches. See also: *extraction ventilation; auxiliary ventilation; suction fan.* (Nelson, 1965)

exhaustion In mining, the complete removal of ore reserves.

exhaust ventilation A system of ventilation in which the fan draws air through the workings by suction. Opposite of forced ventilation. See also: *auxiliary ventilation.* (BS, 1963)

exinite (a) M.C. Stopes in 1935 used the term exinite for the constituent represented by the exines of spores in coal. C.A. Seyler in 1932, however, used the term with its present meaning designating the following group of macerals—sporinite, cutinite, alginite, resinite. The macerals grouped under the term exinite are not necessarily exines but appear to have similar technical properties. The term liptinite was introduced by A. Ammosov in 1956. Little information is so far available on the technological behavior of pure exinite. By comparison and extrapolation it has proved possible, however, to deduce that in coals with more than 35% volatile matter exinite is the maceral group richest in volatile matter and in hydrogen (about 80% and about 9%, respectively). In coals with 18% to 25% volatile matter, exinite is more resilient than the vitrinite; in coals with more than 25% volatile matter, it has even greater resilience than micrinite. Exinite, therefore, increases the strength of bands in which it occurs and in broken coal concentrates in particles greater than 1 mm. (IHCP, 1963) (b) The micropetrologic constituent, or maceral, of spore exines and cuticular matter. See also: *sporinite; cutinite.* (AGI, 1987) (c) A coal maceral group including sporinite, cutinite, alginite, resinite, and liptodetrinite, derived from spores, cuticular matter, resins, and waxes. Exinite is relatively rich in hydrogen. It is a common component of attrital coal. Cf: *inertinite; vitrinite.* Syn: *liptinite.* (AGI, 1987)

exinoid A coal constituent similar to material derived from plant exines. (AGI, 1987)

existent corner A claim corner whose position is evidenced by a monument or its accessories as described in the field note record, or whose location can be identified by the aid of acceptable testimony. (Seelye, 1951)

exogene *exogenetic.*

exogenetic (a) Said of processes originating at or near the surface of the Earth, such as weathering and denudation, and of rocks, ore deposits, and landforms that owe their origin to such processes. Cf: *endogenetic.* Syn: *exogene; exogenic; exogenous.* (AGI, 1987) (b) Said of energy sources and objects of extraterrestrial origin, as solar radiation, cosmic rays, meteorites, and cosmic dust. (AGI, 1987)

exogenic *exogenetic.*

exogenous *exogenetic.*

exogenous inclusion *xenolith.*

exometamorphic A descriptive term for those changes produced by contact metamorphism in the wall rock of the intrusion; opposite of endomorphic.

exometamorphism *exomorphism.*

exomorphic metamorphism *exomorphism.*

exomorphism Changes in country rock produced by the intense heat and other properties of magma or lava in contact with them; it is a form of contact metamorphism in the usual sense. The term was originated by Fournet in 1867. Cf: *endomorphism.* Syn: *exometamorphism; exomorphic metamorphism.* (AGI, 1987)

exotic Applied to a boulder, block, or larger rock body unrelated to the rocks with which it is now associated, which has been moved from its place of origin by one of several processes. Exotic masses of tectonic origin are also allochthonous; those of glacial or ice-rafted origin are generally called erratics. (AGI, 1987)

exotic limonite Limonite precipitated in rock that did not formerly contain any iron-bearing sulfide. Cf: *indigenous limonite.* (AGI, 1987)

expanded blast-furnace slag The lightweight cellular material obtained by controlled processing of molten blast-furnace slag with water or with water and other agents, such as steam or compressed air or both. (ASTM, 1994)

expanding electrode test A geophysical test based on the resistivity method to determine underground geological structure. (Nelson, 1965)

expanding reamer A reamer capable of slight adjustment in diameter by means of a coned internal plug acting in a partially split length of the tool. (Nelson, 1965)

expanding waterway A channel or groove incised into and across the face of a bit, the depth and/or width of which gradually increases from the inside to the outside walls of the bit. (Long, 1960)

expansion bit A drill bit that may be adjusted to cut various sizes of holes. The adjustment of some types may be accomplished by mechanical means while the bit is inside the borehole. Also called paddy; paddy bit. (Long, 1960)

expansion bolt A bolt equipped with a split casing that acts as a wedge; used for attaching to brick or concrete. (Crispin, 1964)

expansion cutter A borehole drill bit having cutters that may be expanded to cut a larger size hole than the size of the bit in its unexpanded state; also, a device equipped with cutters that may be expanded inside casing or pipe to sever, or cut slits or holes in, the casing or pipe. Cf: *paddy.* (Long, 1960)

expansion dome Imaginary dome of rock above underground working, matched by a similar inverted dome below the stope. The dome lies inside the zone of stress due to an unsupported ground, but it is partially destressed owing to expansion and peripheral transfer of load. (Pryor, 1963)

expansion fissure In petrology, one of a system of fissures that radiate irregularly through feldspar and other minerals adjacent to olivine crystals that have been replaced by serpentine. The alteration of olivine to serpentine involves considerable increase in volume, and the stresses so produced are relieved by the fissuring of the surrounding minerals. This phenomenon is common in norite and gabbro. (AGI, 1987)

expansion loop Either a bend like the letter U or a coil in a line of pipe to provide for expansion or contraction. (Fay, 1920)

expansion ring A hoop or ring of U-section used to join lengths of pipe so as to permit of expansion. (Fay, 1920)

expected tonnage The calculated tonnage of recoverable ore in the mine. (Lewis, 1964)

experimental face A normal longwall face on which new machines, such as a cutter loader, may be put to work to gain experience and perhaps improved. Such trials may disclose weaknesses, and they would also indicate the best support system, turnover, and other operating factors. See also: *standby face.* Syn: *trial face.* (Nelson, 1965)

experimental mineralogy The study of chemical and isotopic variations in minerals as a function of temperature and pressure. Cf: *chemical mineralogy; crystallogeny; phase equilibria.*

exploder (a) A cap or fulminating cartridge, placed in a charge of gunpowder or other explosive, and exploded by electricity or by a fuse. Also called detonator. Syn: *battery.* (Fay, 1920) (b) Electric shot-firing apparatus specially designed to provide a source of electric energy of sufficient power to fire electric detonators. Each type of exploder is designed to fire a specific number of shots in series, and exploders are rated accordingly; e.g., single-shot exploders, 30-shot exploders, and 100-shot exploders. See also: *Beethoven exploder.* Syn: *blasting unit.* (McAdam, 1958) (c) A chemical employed for the instantaneous explosion of powder. (Zern, 1928)

exploding bridge wire A wire that explodes upon application of current. It takes the place of the primary explosive in an electric detonator.

exploit (a) Excavate in such a manner as to utilize material in a particular vein or layer, and waste or avoid surrounding material. (Nichols, 1976) (b) Turn a natural resource to economic account. For example, to exploit a mineral deposit. (Webster 3rd, 1966)

exploitation (a) The process of winning or producing from the Earth the oil, gas, minerals, or rocks that have been found as the result of exploration. (AGI, 1987) (b) The extraction and utilization of ore.

exploration (a) The search for coal, mineral, or ore by (1) geological surveys; (2) geophysical prospecting (may be ground, aerial, or both); (3) boreholes and trial pits; or (4) surface or underground headings, drifts, or tunnels. Exploration aims at locating the presence of economic deposits and establishing their nature, shape, and grade, and the investigation may be divided into (1) preliminary and (2) final. See also: *preliminary exploration.* Also called prospecting. (Nelson, 1965) (b) A mode of acquiring rights to mining claims.

exploration drilling Drilling boreholes by the rotary, diamond, percussive, or any other method of drilling for geologic information or in search of a mineral deposit. (Long, 1960)

exploratory drift A drift that is driven in an ore deposit for the purpose of exploring the deposit both horizontally and vertically to see whether it will be worth working. (Stoces, 1954)

exploratory drilling The drilling of boreholes from the surface or from underground workings, to seek and locate coal or mineral deposits and to establish geological structure. Exploratory drilling is frequently done from underground workings, the holes being drilled upward, horizontally, or downward as required. For underground drilling, roller bits, diamond crowns, or tungsten-carbide bits may be used and can be coring or noncoring. Rotary boring is the predominant method for exploratory work from the surface, particularly where cores of significant deposits are required. See also: *borehole samples; drill sampling; underground exploration; diamond drilling.* (Nelson, 1965)

exploratory work Mining operations to determine the size of a deposit, and also its character along the strike as well as its dip. This is done by making drives and inclines. These openings follow the deposit both in strike and dip. They are designed in such a way so as to make it possible to use them for mining proper should the exploration turn out favorably. Syn: *exploration.* (Stoces, 1954)

explorer's alidade *gale alidade.*

exploring drift The working drift approaching old workings whose exact position is uncertain, bored as a precaution against an unexpected holing. (Peel, 1921)

explosibility curves Curve lines drawn through coordinating points, indicating ignition or propagation, in which the rectilinear coordinates of the diagram are factors of volatile fixed-carbon ratios, total incombustible, density of dust, size of dust particles, and combustible gases, if any, in the air current. (Rice, 1913-18)

explosibility limit The addition of inert dust to coal dust decreases its explosibility, and when enough has been added an explosion cannot occur. The point at which explosion cannot occur is said to be the explosibility limit of the coal in question. (Rice, 1913-18)

explosible Capable of being exploded. (Webster 3rd, 1966)

explosion dust The dust deposited from the cloud raised by the explosion, which settles after the explosion has died down, only part of which may be traversed by the flame. (Sinclair, 1958)

explosion-hazard investigation The investigation of a mine to determine the possibility of an explosion occurring by reason of the kind, size, purity, and dryness of the coal dust along the mine passages and the presence or absence of combustible gases. It also determines the degree of the hazard of an explosion from natural conditions and of one arising through the neglect or ignorance of the mine personnel. The purpose of such investigations is to make specific recommendations for reducing that hazard to a negligible point by appropriate methods and continued diligence. (Rice, 1913-18)

explosion pressure The pressure developed at the instant of an explosion. (Streefkerk, 1952)

explosion proof (a) The term "explosion-proof casing" or "enclosure" means that it is so constructed and maintained as to prevent the ignition of gas surrounding it by any sparks, flashes, or explosions of gas that may occur within such enclosures. (Fay, 1920) (b) Said of electrical apparatus so designed that an explosion of flammable gas inside the enclosure will not ignite flammable gas outside. Also called flameproof. (CTD, 1958) (c) Fitting, motor, switch, or fixture so made and maintained as to preclude possibility of sparks, arcs, or heat sufficient to initiate explosion in surrounding air or mine dust. (Pryor, 1963)

explosion-proof motor The U.S. Bureau of Mines has applied the term "explosion proof" to motors constructed so as to prevent the ignition of gas surrounding the motor by any sparks, flashes, or explosions of gas or of gas and coal dust that may occur within the motor casing. See also: *permissible motor.* (Fay, 1920)

explosions from molten iron An explosion caused by molten iron coming in contact with water or wet material. (Fay, 1920)

explosion-tested equipment In explosion-tested equipment, housings for electric parts are designed to withstand internal explosions of methane-air mixtures without causing ignition of such mixtures that surround the housings.

explosion tuff A tuff whose pyroclastic fragments are in the place in which they fell, rather than having been washed into place after they landed. (AGI, 1987)

explosion wave The wave or flame that passes through a uniform gaseous mixture with a permanent maximum velocity. The rate of the explosion wave is a definite physical constant for each mixture. The explosion wave travels with the velocity of sound in the burning gas, which itself is moving rapidly forward en masse in the same direction, so that the explosion wave is propagated far more rapidly than sound travels in the unburned gas. (Fay, 1920)

explosive (a) Any chemical compound, mixture, or device that is capable of undergoing a rapid chemical reaction, producing an explosion; a cap sensitive mixture. (b) Any mixture or chemical compound by whose decomposition or combustion gas is generated with such rapidity that it can be used for blasting or in firearms; e.g., gunpowder, dynamite, etc. See also: *explosive factor; Morcol; permitted explosive; sheathed explosive.* (c) In coal mining, there are two main classes permissible and nonpermissible; i.e., those safe for use in coal mines and those that are not. See also: *coal mining explosives.* (CTD, 1958)

explosive antimony A black powder obtained either by the rapid cooling of antimony vapor or by the electrolysis of a solution of antimony chloride in hydrochloric acid, using a platinum cathode and an antimony anode. When scratched, the hard black mass deposited on the cathode will explode. The mass may consist of a solid solution of antimony trichloride in metallic antimony. (Camm, 1940)

explosive cooling agent A substance added to a permitted explosive to cool the explosion flame and thus reduce the risk of igniting mine gases. The agent may be sodium chloride and sodium bicarbonate. (Nelson, 1965)

explosive drilling A technique developed for deep-hole drilling in esp. strong and abrasive rocks. In this method, a series of small underwater explosions are used to break the rock at the bottom of the hole, the fragments from each explosion being washed away by the flushing water. (Min. Miner. Eng., 1965)

explosive dusts Dusts that are combustible when airborne. They include metallic dusts (magnesium, aluminum, zinc, tin, iron), coal (bituminous, lignite), and sulfide ores. (Hartman, 1982)

explosive factor The ratio between the burden of a shothole in tons or cubic yards and the weight of explosive charge in pounds; i.e., tons (cubic yards) per pound of explosive. The factor is dependent on the rock and the fragmentation required, but 5 st/lb (10 t/kg) is about average in quarry blasting. See also: *loading ratio.* (Nelson, 1965)

explosive force A force represented with separate values for the heat liberated by the explosive decomposition and the detonating rate. (Streefkerk, 1952)

explosive limits The limits of percentage composition of mixtures of gases and air (or oxygen) within which an explosion takes place when the mixture is ignited. (Inst. Petrol., 1961)

explosive loading factor *powder factor.*

explosively anchored rockbolt A device to give better support in underground mining operations. It can be anchored more firmly than conventional bolts because the principle of explosive forming enables the anchor to grip the sides of the borehole along its entire length, if necessary. The key to the design is a seamless steel anchoring tube, welded to the threaded end of the bolt. Exploding a small charge inside the tube makes it expand to fit tightly in the borehole. Water, wax, or a similar buffer surrounds the charge to distribute the force of the explosion evenly and prevent it from rupturing the tube. Its use may permit mining of mineral deposits formerly considered uneconomic because of the hazards encountered in loose rock formations.

explosiveness of dust The ability of a dust to produce violent effects; it is measured by the pressure produced after the explosion has traveled a fixed distance under standard conditions. (Sinclair, 1958)

explosive oil *nitroglycerin.*

explosive ratio The weight of explosive per cubic feet (cubic meter) of rock broken. Also called powder factor. See also: *loading ratio.* (Nelson, 1965)

explosives casting In explosives casting, large amounts of low-cost ammonium nitrate mixtures are loaded into medium-sized drill holes in a usual ratio of more than 1 lb of powder per cubic yard (0.59 kg/m^3) of overburden. The explosive charges are detonated through millisecond-delay electric blasting caps. When the shot is fired, a large part of the overburden is blasted into the pit away from the high wall and up on the spoil pile where it attains a favorable angle of repose. (Woodruff, 1966)

explosive sensitiveness The ease with which an explosive will detonate or explode. (Nelson, 1965)

explosive shattering This method consists in soaking the ore thoroughly in water and then heating to 180 °C under a pressure of 150 psi (1.03 MPa). The pressure is then suddenly released, and the absorbed water is converted to steam, which shatters and liberates mineral particles without harmful overgrinding of the ore. (Newton, 1938)

explosive store A surface building at a mine where explosives and detonators may be kept. (Nelson, 1965)

explosive strength A measure of the amount of energy released by an explosive on detonation and its capacity to do useful work. Several methods of expressing explosive strength are used, but in most cases the figures are calculated from the deflection of a freely suspended ballistic mortar in which small explosive charges are fired. (Nelson, 1965)

explosive stripping A method in which, by using an excess of explosives in the strip mine bench, up to about 40% of the overburden can be removed from the coal seam by the energy of the explosive, thereby requiring no excavation. (Encyclopaedia Britannica, 1964)

exponential model A function frequently used when fitting mathematical models to experimental variograms, often in combination with a nugget model.

exposed coalfield (a) A coalfield where the coal measures crop out at the surface all around its margin or boundary. South Wales, England, is a good example of an exposed coalfield. See also: *coal basin; concealed coalfield.* (Nelson, 1965) (b) Deposits of coal that crop out at the surface, as along the rim of a coal basin. Cf: *crop coal.* (AGI, 1987)

exposure (a) An area of a rock formation or geologic structure that is visible (hammerable), either naturally or artificially, i.e., is unobscured by soil, vegetation, water, or the works of humans; also, the condition of being exposed to view at the Earth's surface. Cf: *outcrop.* (AGI, 1987) (b) The nature and degree of openness of a slope or place to wind, sunlight, weather, oceanic influences, etc. The term is sometimes regarded as a syn. of aspect. (AGI, 1987) (c) The proportional mass of a diamond or other cutting medium protruding beyond the surface of the metal in which it is inset in the face of the bit. Sometimes incorrectly called clearance. (Long, 1960) (d) The total quantity of light received per unit area on a sensitized plate or film, usually expressed as the product of the light intensity and the time during which the light-sensitive material is subjected to the action of light. A loosely used term generally understood to mean the length of time during which light is allowed to act on a sensitive surface. The act of exposing a light-sensitive material to a light source. An individual picture of a strip of photographs. (AGI, 1987)

exsiccated alum *burnt alum.*

exsolution The process whereby an initially homogeneous solid solution separates into two (or possibly more) distinct crystalline phases without addition or removal of material, i.e., without change in the bulk composition. It generally, though not necessarily, occurs on cooling. Syn: *unmixing.* (AGI, 1987)

exsolutional Applied to those sedimentary rocks that solidify from solution either by precipitation or by secretion.

exsolution texture A general term for the texture of any mineral aggregate or intergrowth formed by exsolution. It is generally fairly homogeneous, ranging from perthitic to geometrically regular. See also: *eutectic texture.* (AGI, 1987)

extendable conveyor (a) For bulk materials it is usually of troughed design and may be lengthened or shortened while in operation. Commonly used in underground mine conveyor work. (b) For packaged materials, objects, or units, the conveyor may be one of several types including roller, wheel, and belt conveyors. Construction is such that the conveyor may be lengthened or shortened within limits to suit operating needs. See also: *telescoping conveyor.*

extended charges Explosive charges spaced at intervals in a quarry or opencast blast hole. See also: *deck loading*. (Nelson, 1965)
extensible conveyor A conveyor capable of being lengthened or shortened while in operation. (NEMA, 1961)
extensible discharge trough Consists of two or more shaker conveyor troughs, nested, to be installed on the discharge end of the pan line so as to provide for adjustment of the position of the discharge point. After adjustment, they are locked in place. (Jones, 1949)
extension Part of and physically associated with a known mineral deposit, but outside of the identified parts. (Barton, 1995)
extensional fault *tension fault.*
extension fracture A fracture that develops perpendicular to the direction of greatest stress and parallel to the direction of compression; a tension fracture. See also: *extension joint*. (AGI, 1987)
extension joint A joint that forms parallel to the direction of compression; a joint that is an extension fracture. See also: *extension fracture*. (AGI, 1987)
extension ore Ore believed to exist ahead of ore exposed in the face of a drift. See also: *probable ore.*
extensometer (a) Instrument used for measuring small deformations, deflections, or displacements. (Obert, 1960) (b) An instrument for measuring changes caused by stress in a linear dimension of a body. (ASM, 1961)
external sintered tip pick *sintered carbide-tipped pick.*
extinction In polarized-light microscopy with crossed polars and an anisotropic mineral in the light train, when the two electric vectors (permitted light or vibration directions) of a randomly oriented crystal are each parallel to those of the polars, no light is transmitted and the crystal is at extinction. Isotropic crystals and anisotropic crystals viewed parallel to an optic axis remain extinct upon stage rotation, while randomly oriented anisotropic crystals go extinct four times upon stage rotation of 360°. Cf: *anisotropy; extinction angle; extinction direction; inclined extinction; optic axis; parallel extinction; undulatory extinction.*
extinction angle The angle between a crystallographic direction and a position of maximum extinction of an anisotropic crystal as viewed with a polarized-light microscope or polariscope. It can be diagnostic in mineral identification. Syn: *angle of extinction*. Cf: *inclined extinction; extinction.*
extinction direction In polarized-light microscopy, the angular position of extinction with respect to a crystallographic axis, a crystal face, a cleavage plane, or a twin plane. Cf: *extinction.*
extinctive atmosphere An atmosphere created behind mine seals when the supply of oxygen is completely cut off, thereby bringing combustion to an end. (Roberts, 1960)
extractable metal Metal that can be extracted from a sample by a specified chemical treatment, ordinarily only part of the metal that is in the sample. Often abbreviated as "Ex Zn, Ex Cu," etc., indicating the metal of interest. The process is referred to as partial extraction, or selective extraction. Syn: *readily extractable metal.*
extractant In solvent extraction, the active organic reagent that forms an extractable complexion with the metal. (Newton, 1959)
extraction (a) The process of mining and removal of coal or ore from a mine. (Nelson, 1965) (b) Used in relation to all processes of obtaining metals from ores. Broadly, these processes involve breaking down ore both mechanically (crushing) and chemically (decomposition), and separating the metal from the associated gangue. Extractive metallurgy may be conveniently divided into beneficiation, pyrometallurgy, hydrometallurgy, and electrometallurgy. (c) A designation for that part of the metallic content of the ore obtained by a final metallurgical process, e.g., the extraction was 85%. Cf: *recovery*. (d) The process of dissolving and separating out specific constituents of a sample by treatment with solvents specific for those constituents. (e) In chemical engineering, the operation wherein a liquid or solid mixture is brought into contact with an immiscible or partially miscible liquid to achieve a redistribution of solute between the phases.
extraction metallurgy Process of producing metal from ores or their concentrates. (Pryor, 1963)
extraction ratio Ratio of the mined volumes to the total volumes. Also called strip ratio. (Obert, 1960)
extraction ventilation The ventilation of a tunnel face (or mine) by an exhaust fan. See also: *exhausting auxiliary fan*. Syn: *exhaust ventilation*. (Nelson, 1965)
extraction water Superheated water pumped into wells to melt and to extract molten sulfur from salt domes. (Goldman, 1952)
extractive metallurgy The extraction of metals from their ores or from the naturally occurring aggregates of minerals by various mechanical and chemical methods. The major divisions of extractive metallurgy may be classified as mineral dressing, pyrometallurgy, hydrometallurgy, and electrometallurgy. (Kirk, 1947)
extrados The exterior arc of an arch, as in a tunnel. (Sandstrom, 1963)
extradosal The fractured ground outside of the fracture zone. In many mines, extradosal bursts occur more frequently than intradosal; i.e., the extradosal ground ahead of the working face serves as an abutment that supports the superincumbent rock to the surface. Cf: *intradosal*. (Lewis, 1964)
extra dynamite Differs from straight dynamite in that a portion of the nitroglycerin content is replaced with sufficient ammonium nitrate to maintain the grade strength, manufactured in grade strengths of from 20% to 60%. It is lower in velocity and water resistance than straight dynamite, but is less sensitive to shock and friction and less flammable. See also: *ammonia dynamite; low-density explosive*. (Carson, 1961)
extraflexible hoisting rope A rope consisting of 8 strands of 19 wires each with a large hemp center. (Hunt, 1965)
extrahazardous Unusually dangerous; specif. used in insurance in classifying occupational risks, as mining is extrahazardous.
extralateral Situated or extending beyond the sides; specif. noting the right of a mine owner to the extension of a lode or vein from his or her claim beyond the sidelines, but within the vertical planes through the endlines.
extralateral rights Among numerous provisions, the statute (30 USC 26) provides that extralateral rights to veins, lodes, and ledges that come to an apex within the boundary lines and dip downward so as to extend outside the vertical planes through the side lines belong to the owner of such lode location. The extralateral portion of the vein is that part that extends on its downward dip through the vertical planes along the side lines.
extralite An explosive mixture consisting of ammonium nitrate and carbonate, liquid and solid hydrocarbons, and zinc chlorate. (Standard, 1964)
extraneous ash Ash in coal that is derived from inorganic material introduced during formation of the seam, such as sedimentary particles, or filling cracks in the coal. See also: *ash; ash yield; inherent ash*. Syn: *secondary ash; sedimentary ash*. (AGI, 1987)
extraneous electricity Electrical energy, other than actual firing current or the test current from a blasting galvanometer, that is present at a blast site and that could enter an electric blasting circuit; such electricity can include lightning, current from high tension powerline, and static charge carried on a person.
extraordinary ray Light passing through anisotropic crystals is doubly refracted with one or both ray directions not parallel to their wave normals. Such light rays do not follow Snell's law [n=(sin i)/(sin r)] of ordinary refraction and are termed "extraordinary." Also written E-ray or e-ray. Cf: *law of refraction.*
extrapolation Projection of a graphic curve beyond the line of points established from plotting data.
extra-special improved plow A grade of wire rope used for winding, with a tensile strength of between 115 st/in^2 and 125 st/in^2 (16.2 t/cm^2 and 17.6 t/cm^2). (Mason, 1951)
extraterrestrial mining The mining of metals and minerals on locations other than the Earth through space colonization of planets and moons. (SME, 1992)
extraterritorial rights Sometimes affect employment in alien countries by giving immunity from some laws. May affect working conditions. (Pryor, 1963)
extrusion (a) The emission of relatively viscous lava onto the Earth's surface; also, the rock so formed. Cf: *effusion*. (AGI, 1987) (b) The operation of producing rods, tubes, and various solid and hollow sections, by forcing heated metal through a suitable die by means of a ram; applied to numerous nonferrous metals, alloys, and other substances. (CTD, 1958) (c) The act or process of extruding; thrusting or pushing out; also, a form produced by the process; a protrusion. (Webster 3rd, 1966; Webster 2nd, 1960) (d) The emission of magmatic material (generally lavas) at the Earth's surface; also, the structure or form produced by the process, such as a lava flow, a volcanic dome, or certain pyroclastic rocks. (AGI, 1987) (e) Lava or mud forced out, as through a vent or fissure, onto the Earth's surface. (Webster 3rd, 1966) (f) Plastic clay forced through a mouthpiece of a pugmill or a press, forming a rod or a tube, which can be cut to the desired length. (Rosenthal, 1949)
extrusive Said of igneous rock that has been erupted onto the surface of the Earth. Extrusive rocks include lava flows and pyroclastic material such as volcanic ash. An extrusive rock. Cf: *intrusive*. Syn: *effusive; volcanic; eruptive*. (AGI, 1987)
exudation The action by which all or a portion of the low-melting constituent of a compact is forced to the surface during sintering. Sometimes referred to as "bleed out." Syn: *sweating*. (ASTM, 1994)
ex vessel A price quoted ex vessel used in connection with a port name means all costs paid until free of the ship's tackle at the port designated. (Hess)
eye (a) The top or mouth of a shaft. (BS, 1963) (b) The central or intake opening of a radial-flow fan. (BS, 1963) (c) The entrance to a mine working at which daylight can be seen from within. (Pryor, 1963) (d) The hole in a pick or hammer head, that receives the handle. (Fay, 1920) (e) The opening at the top of a beehive coke oven for charging.

(Mersereau, 1947) (f) The opening in the bottom of a pot furnace through which the flame enters. (ASTM, 1994)

eye agate Agate displaying concentric bands, commonly of various colors, about a dark center suggesting an eye. Also called Aleppo stone.

eyebolt A rod or bolt having an eye or loop at one end and threaded at the other end. (Long, 1960)

eyesight A window or other opening in a tuyere through which the operator can see into the melting zone of the blast furnace. (Mersereau, 1947)

ezcurrite A triclinic mineral, $Na_4B_{10}O_{17}·7H_2O$; associated with kernite, borax, and tincalconite at the Tincalayu borax mine, Salta province, Argentina.

fabianite A monoclinic mineral, $CaB_3O_5(OH)$; colorless; fluoresces brownish-yellow under UV light; Mohs hardness, 6; associated with halite, anhydrite, and howlite in a rock-salt drill core at Rehden, Diepholz, Germany.

Fabian system May be described as the father of freefall drilling systems, all others having originated from it, although it is not now used in its original form. See also: *free fall.*

fabric (a) The spatial arrangement and orientation of the components (crystals, particles, cement) of a sedimentary rock. Cf: *packing.* (AGI, 1987) (b) The complete spatial and geometrical configuration of all those components that make up a deformed rock. It covers such terms as texture, structure, and preferred orientation. Syn: *rock fabric.* (AGI, 1987)

fabricator A company which transforms refined metal (and sometimes scrap as well) into semifabricated products, (e.g., wire, cable, tubes, strip, rods) for sale to an end-consumer. Also called semifabricator. (Wolff, 1987)

fabric element A component of a rock fabric that acts as a unit in response to deformative forces.

fabric-type dust collector A collector that utilizes a fabric or cloth to remove dust particles from the air. The basic idea is the same as that employed in vacuum cleaners, but there is usually an automatic or self-cleaning feature for recovering the dust. Fabric-type dust collectors should not be subjected to excessively abrasive or corrosive materials, or high temperatures that might injure the fabric, unless special materials have been employed for that purpose. Bags and tubes employing glass filter fabric are capable of handling gases with temperatures up to 550 °F (288 °C), and also can withstand the action of many corrosive gases. Fabric-type collectors fall into two groups on the basis of design. One uses the fabric in a closed bag or a series of small-diameter bags commonly called tubes, while the other has the fabric on a frame like a screen. (Pit and Quarry, 1960)

face (a) The surface of an unbroken coal bed at the advancing end of the working place. (b) Sedimentary beds are said to face in the direction of the stratigraphic top of the succession (or to be directed toward the younger rocks or to the side that was originally upward), so that an overturned bed facing to the east may have a dip of 45° to the west. Folds are said to face in the direction of the stratigraphically younger rocks along their axial surfaces and normal to their axes; this coincides with the direction toward which the beds face at the hinge (a normal upright fold faces upward, an overturned anticline faces downward, and an asymmetric fold faces its steeper flank). Faults are said to face in the direction of the structurally lower unit. (AGI, 1987) (c) The principal cleavage plane of coal, at right angles to the stratification. (d) The exposed surface of a coal or ore deposit in the working place where mining is proceeding. See also: *coal face; face height; working face.* (e) An edge of rock used as a starting point in figuring drilling and blasting. (Nichols, 1976) (f) The part of a bit in contact with the bottom of a borehole, when drilling is in progress, that cuts the material being drilled; cutting face. (Long, 1960) (g) To dress a bit. (h) The bottom of a drill or borehole. (i) The original upper surface of a layer of rock, esp. if it has been raised to a vertical or a steeply inclined position. (j) The plane surface of a mineral crystal. (k) The surface exposed by excavation. The working face, front, or forehead is the face at the end of the tunnel heading, or at the end of the full-size excavation. (l) A cleat or back. (Fay, 1920) (m) The smooth surface of the coal as contrasted to butt. (n) The main cleavage; bord cleat. (Mason, 1951) (o) The more or less vertical surface of rock exposed by blasting or excavating, or the cutting edge of a drill hole. (Nichols, 1976) (p) The width of a roll crusher. (Nichols, 1976) (q) The outer surface of a pulley in contact with a belt; the outer surface of a gear, roll, or drum usually expressed in terms of inches of width. (r) The end of a drive. (Gordon, 1906) (s) In structure, the original upper surface of a stratum esp. if it has been raised to a vertical or a steeply inclined position. (AGI, 1987)

face-airing The operation of directing the intake air to and along the working face of a mine. The term was used in the early part of the 18th century to describe the coursing of air naturally induced in a coal mine. See also: *circulation of air.* (Nelson, 1965)

face area The working area in from the last open crosscut in an entry or a room, including the pillar being extracted or longwall being mined.

face belt conveyor A light belt conveyor employed at the face. It is the type generally used in conventional machine mining. (Nelson, 1965)

face belt joints Three types of face belt joints are used: (1) hinged-plate type, which is attached to the belt by means of copper rivets and interconnected by means of pins; (2) wire-hook joints, the most popular type—the hooks are inserted by means of a hand-operated machine and are connected by a flexible steel pin; and (3) spliced joint, in which a portion on each side of the belt is cut away so as to provide a splice, and this is secured by cramped-type pins, which are inserted and knocked over by hand. (Mason, 1951)

face boss In bituminous coal mining, a foreman in charge of all operations at the working faces where coal is undercut, drilled, blasted, and loaded. Also called face foreman. (DOT, 1949)

face cleat The major joint or cleavage system in a coal seam. Cf: *butt cleat.* See also: *cleat; face.*

face concentration The ratio of pithead output (tons) to length of face (yard) or tons per yard of face. The management objective is to keep this figure as high as practicable. See also: *concentration of output.* (Nelson, 1965)

face conveyor (a) Any type of conveyor employed at the working face that delivers coal into another conveyor or into a car. (Nelson, 1965) (b) *underground mine conveyor.* (c) A conveyor, generally 10 to 100 ft (3.0 to 30.5 m) in length, used in room and pillar mining to move material from the face to a room or section conveyor. See also: *armored flexible conveyor; gate conveyor.* (NEMA, 1961)

faced crystal Applied in the trade to a natural mass of quartz bounded by one or more of the original crystal faces. (Am. Mineral., 1947)

face-discharge bit A bit designed for drilling in soft formations and for use on a double-tube core barrel, the inner tube of which fits snugly into a recess cut into the inside wall of the bit directly above the inside reaming stones. The bit is provided with a number of holes drilled longitudinally through the wall of the bit, through which the circulation liquid flows and is ejected at the cutting face of the bit. Also called bottom-discharge bit; face-ejection bit. (Long, 1960)

face-ejection bit *face-discharge bit.*

face entry An entry driven at right angles to the face cleat and parallel to the butt cleat. (BCI, 1947)

face equipment Face equipment is mobile or portable mining machinery having electric motors or accessory equipment normally installed or operated inby the last open crosscut in an entry or room.

face half and half Eng. A longwall face crossing the main cleavage planes of the seam at an angle of 45°. (SMRB, 1930)

face hammer Used for rough dressing stones. It has one blunt end and one cutting end. (Crispin, 1964)

face haulage The transportation of mined coal from the working face to an intermediate haulage. It is accomplished by shuttle cars, conveyors, locomotives, and mine cars, or by combinations of such equipment. *primary haulage.* (Woodruff, 1966)

face height The vertical height of a quarry or opencast face from top to toe; i.e., the height of overburden and coal, ore, or stone. A face height is chosen that can be reached by the excavator so that all scaling of loose material can be accomplished by the machine, thus eliminating the necessity for workers to go over the face on ropes to bar off loose ground. Where the height exceeds this figure, a form of benching may be adopted. See also: *face.* (Nelson, 1965)

face left *face right.*

facellite *kaliophilite.* Also spelled phacellite, phacelite.

face loading pan A shaker conveyor pan or trough that has been widened for one-half of its length to provide a greater loading surface when used at the face. (Jones, 1949)

face mechanization On a longwall face, the term implies the use of some type of cutter loader with perhaps self-advancing ports, giving a quicker turnover and higher productivity. In the case of a rock drivage the term would mean the regular use of a shovel loader. See also: *conventional machine mining.* (Nelson, 1965)

face of hole The bottom of a borehole. (Long, 1960)

face-on (a) A location where the face of the breast or entry is parallel to the face cleats of the seam. *face.* (b) Working of a mine in a direction parallel to the natural cleats. Cf: *end-on.* (CTD, 1958)
face on end *end face.*
face right Position of the vertical circle of a theodolite with respect to the telescope, viewed from the eyepiece end. (Pryor, 1963)
face run N. of Eng. The time during which a coal-getting machine is moving along the face. (Trist, 1963)
face sampling The cutting of pieces of ore and rock from exposed faces of ore and waste. The faces may be natural outcrops or faces exposed in surface trenches and pits. Face samples may be taken by cutting grooves or channels of uniform width and depth across the face or sections of the face or by picking off small pieces all over the face, more or less at random.
face shovel equipment An excavator base machine fitted with boom and bucket for excavating and loading material from an exposed face above track level. (Nelson, 1965)
face signal N. of Eng. A wire stretched along the face to control, directly or indirectly, the running of the face conveyor. (Trist, 1963)
face signaling The system for transmitting signals from points on a conveyor face to the operator at the control panel near the main gate. See also: *signaling system.* (Nelson, 1965)
face slip (a) The front slip of a coal seam. Cf: *back slip.* (Fay, 1920) (b) An inclined joint in coal sloping away from the hewing face. (Arkell, 1953)
face stone A diamond inset in the face portion of a bit. Cf: *kerf stone.* (Long, 1960)
facet (a) One of the polished plane surfaces of a gemstone, cut so as to enhance the stone's brilliance and beauty. (b) To cut facets. (c) A nearly planar surface produced on a rock fragment by abrasion, as by wind sandblasting, by the grinding action of a glacier, or by a stream that differentially removes material from the upstream side of a boulder, cobble, or pebble inclined at 50° or less to the direction of the impinging current. (d) Asymmetrically scalloped rock surfaces. Syn: *flute.* (AGI, 1987) (e) Any plane surface produced by erosion or faulting and intersecting a general slope of the land; e.g., a triangular facet. (f) Any part of a landscape defined as a unit for geographic study on the basis of homogeneous topography.
faceted boulder A boulder that has been ground flat on one or more sides by the action of natural agents, such as by glacier ice, streams, or wind. Cf: *faceted pebble.* (AGI, 1987)
faceted pebble A pebble on which facets have been developed by natural agents, such as by wave erosion on a beach or by the grinding action of a glacier. Cf: *faceted boulder.* (AGI, 1987)
faceted spur The end of a ridge that has been truncated or steeply beveled by stream erosion, glaciation, or faulting. See also: *truncated spur.*
face timbering The placing of safety posts at the working face to support the roof of the mine. The safety post is the most important timber in a mine because exposure is greater at this point than at any other since the newly exposed top is always of unknown quality. See also: *timbering.* (Kentucky, 1952)
faceting machine A mechanical device for holding stone during grinding and polishing facets at the exact angles that theoretically produce the most brilliant stone. Rarely used in fashioning diamonds or other valuable stone where preservation of weight is more important than maximum brilliancy.
face transfer point *transfer point.*
face wall A wall built to sustain a face cut into the earth, in distinction to a retaining wall, which supports earth deposited behind it. (Zern, 1928)
facies A term of wide application, referring to such aspects of rock units as rock type, mode of origin, composition, fossil content, or environment of deposition. Cf: *lithofacies.*
facies change A lateral or vertical variation in the lithologic or paleontologic characteristics of contemporaneous sedimentary deposits. It is caused by, or reflects, a change in the depositional environment. (AGI, 1987)
facies contour A line indicating equivalence in lithofacies development; e.g., a particular value of the sand-to-shale ratio.
facies fauna A group of animals characteristic of a given stratigraphic facies or adapted to life in a restricted environment; e.g., the black-shale fauna of the Middle and Upper Devonian of the Appalachian region of the United States. (AGI, 1987)
facies fossil A fossil, usually a single species or a genus, that is restricted to a defined stratigraphic facies or is adapted to life in a restricted environment. It prefers certain ecologic surroundings and may exist in them from place to place with little change for long periods of time. (AGI, 1987)
facies map A map showing the distribution of different types of rock attributes or facies occurring within a designated geologic unit.
facing (a) Aust. The main vertical joints often seen in coal seams; they may be confined to the coal, or continue into the adjoining rocks. See also: *cleat.* (b) A face slip or joint as opposed to a back slip. The plane is inclined towards the observer from floor to roof. (Nelson, 1965) (c) A powdered substance (such as graphite) applied to the face of a mold or mixed with sand that forms it, to give a smooth surface to the casting. (Webster 3rd, 1966) (d) Applied to the original direction of a layer. (AGI, 1987) (e) In machining, generating a surface on a rotating workpiece by the traverse of a tool perpendicular to the axis of rotation. (ASM, 1961) (f) In founding, special sand placed against a pattern to improve the surface quality of the casting. (ASM, 1961) (g) Any material, forming a part of a wall, used as a finishing surface. (ACSG, 1961)
facing of strata The direction of the top of beds is esp. important in steeply dipping or overturned beds. The tops of the beds or their facing is determined by ripple marks, graded bedding, crossbedding, mud cracks, pillow structure, etc. (AGI, 1987)
fact cut A type of cut gem bounded by planar faces as distinguished from cabochon cut or other unfaceted cut. Also called faceted cut.
factor of safety The ratio of the ultimate breaking strength of the material to the force exerted against it. If a rope will break under a load of 6,000 lb (2,724 kg), and it is carrying a load of 2,000 lb (908 kg), its factor of safety is 6,000/2,000 = 3, or f.s 3. See also: *allowable stress.* (Brantly, 1961)
Fagersta cut This cut is drilled with handheld equipment. The empty hole is drilled in two steps, the first as an ordinary hole and the second as an enlargement of this pilot hole. The cut is something between a four-section cut and a double-spiral cut. The Fagersta cut is drilled with light equipment, which makes it suitable for use in mines and in small drifts, where drilling with heavy machines is not profitable. (Langefors, 1963)
faheyite A hexagonal mineral, $(Mn,Mg)Fe_2Be_2(PO_4)_4 \cdot 6H_2O$; forms tufts, rosettes, or botryoidal masses of fibers coating primary minerals, such as muscovite and quartz; in pegmatites, Minas Gerais, Brazil.
fahlband A term originally used by German miners to indicate a band of sulfide impregnation in metamorphic rocks. The sulfides are too abundant to be classed as accessory minerals, but too sparse to form an ore lens. Fahlbands have a characteristic rusty-brown appearance on weathering. (AGI, 1987)
fahlerz Ger. A gray copper ore. Sometimes called fahl ore. Syn: *tetrahedrite; tennantite.* (Fay, 1920; AGI, 1987)
fahlunite Altered cordierite.
Fahrenheit Commonly used thermometer scale in which the freezing point of water is 32° and the boiling point is 212°. To convert from the Fahrenheit scale to the centigrade or Celsius scale, subtract 32 and multiply by 5/9. Symbol, F. See also: *temperature.* (Crispin, 1964; CTD, 1958)
Fahrenwald machines These include (1) a hydraulic classifier and (2) a flotation cell marketed as the Denver Sub A. (Pryor, 1963)
failed hole A drill hole loaded with dynamite that did not explode. (Fay, 1920)
failure by rupture *shear failure.*
fairchildite A hexagonal mineral, $K_2Ca(CO_3)_2$; dimorphous with büschliite; in fused wood ash in partly burned trees.
fairfieldite (a) A triclinic mineral, $Ca_2(Mn,Fe)(PO_4)_2 \cdot 2H_2O$; transparent pearly to subadamantine; in granite pegmatites in Fairfield County, CT, as an alteration of dickinsonite and a pseudomorph after rhodochrosite. (b) The mineral group cassidyite, collinsite, fairfieldite, messelite, and roselite-beta. Syn: *leucomanganite.*
fairlead (a) A device that lines up cable so that it will wind smoothly onto a drum. (Nichols, 1954) (b) Applies to the swivel pully on the drag rope of a dragline excavator. (Hammond, 1965) (c) A block, ring, or strip of plank with holes that serves as a guide for the running rigging or for any ship's rope and keeps it from chafing. (Webster 3rd, 1966)
fairy stone (a) A cruciform-twinned crystal of staurolite, used as a curio stone without fashioning for adornment. The term is also applied as a syn. of staurolite, and esp. to the variety occurring in the form of a twinned crystal. See also: *staurolite.* (AGI, 1987) (b) Any of various odd or fantastically shaped calcareous or ferruginous concretions formed in alluvial clays. (AGI, 1987) (c) A fossil sea urchin. (AGI, 1987) (d) A stone arrowhead. (AGI, 1987)
fake reflection Accidental lineup of disturbances on a seismogram simulating a reflection. (Schieferdecker, 1959)
Falconbridge process Recovery of nickel from a nickel-copper matte. After crushing and roasting to remove sulfur, the copper is acid-leached, filtered off, and electrolyzed. The residual solids are melted, cast as anodes, and refined to produce nickel. (Pryor, 1963)
falding furnace A mechanically raked muffle furnace having three hearths with combustion flues under the lowest hearth. (Fay, 1920)
fall (a) A mass of rock, coal, or ore that has collapsed from the roof or sides of a mine roadway or face. Falls of ground are responsible for the greatest proportion of underground deaths and injuries. (Nelson, 1965) (b) A length of face undergoing holing or breaking down for loading. (c) The rolling of coal from the face into the room, usually as

the result of blasting; sometimes the amount blasted down. Locally, also the caved roof after the coal is extracted. (BCI, 1947) (d) To blast, wedge, or in any other way to break down coal from the face of a working place. (CTD, 1958) (e) A system of working a thick seam of coal by falling or breaking down the upper part after the lower portion has been mined. Cf: *caving system.* (f) A mass of roof or side that has fallen in any subterranean working or gallery, resulting from any cause whatever. (g) The collapse of the roof of a level or tunnel, or of a flat working place or stall; the collapse of the hanging wall of an inclined working place or stope. (CTD, 1958) (h) To crumble or break up from exposure to the weather; clays, shales, etc., fall.

fallers Movable supports for a cage. Also called fangs; keps. (Nelson, 1965)

fallhammer test A test used to determine impact sensitivity of an explosive conducted by allowing standard hammer weights to fall on an amount of confined explosive charge and measuring the fall height required to decompose or detonate the charge. (Meyer, 1981)

falling-head test A soil permeability test in which the borehole is filled up with water and the rate at which the water falls is observed. (Mining)

falling-pin seismometer A limit recorder of the intensity of ground vibrations initiated by a quarry or opencast blast. It consists essentially of a level glass base on which a number of pins 1/4 in (0.64 cm) in diameter and of lengths ranging from 6 to 15 in (15 to 38 cm) are stood upright. The pins stand inside hollow steel rods so that each pin can fall over independently of the others. The longer the pin, the less energy required to topple it. In practice it has been accepted that if the shorter pins (up to 10 in or 25 cm) remain standing, then there is no possibility of structural damage to a building by a quarry blast. See also: *vibrograph.* (Nelson, 1965)

falling slag Blast furnace slag that contains sufficient calcium orthosilicate to render it liable to fall to a powder when cold. Such a slag is precluded from use as a concrete aggregate by the limits of composition specified in British Standards 1047. (Dodd, 1964)

fall line An imaginary line or narrow zone connecting the waterfalls on several adjacent near-parallel rivers, marking the points where these rivers make a sudden descent from an upland to a lowland, as at the edge of a plateau; specif. the fall line marking the boundary between the ancient, resistant crystalline rocks of the Piedmont Plateau and the younger, softer sediments of the Atlantic Coastal Plain in the Eastern United States. It also marks the limit of navigability of the rivers. Syn: *fall zone.* (AGI, 1987)

fall of ground Rock falling from the roof into a mine opening. See also: *fall.* (Weed, 1922)

fall ridder *bordroom man.*

fall table A hinged platform to cover the mouth of a shaft. (Nelson, 1965)

fall velocity *settling velocity.*

fall zone *fall line.*

false amethyst An early name for violet-colored fluorite when cut as a gem. Other colors of the same mineral were called false emerald, ruby, sapphire, or topaz. (Fay, 1920)

false anticline An anticlinelike structure produced by compaction of sediment over a resistant mass, such as a buried hill or reef. (AGI, 1987)

false bedding *crossbedding; current bedding.*

false bottom (a) An apparent bedrock underlying an alluvial deposit that conceals a lower alluvial deposit; e.g., a bed of clay or sand cemented by hydrous iron oxides, on which a gold placer deposit accumulates, and under which there is another alluvial deposit resting on bedrock. (AGI, 1987) (b) A strip of wire screening nailed to a wooden frame that fits into the bottom of a sluice box to trap fine sand containing gold. When the frame is removed, the fine sand, containing gold, is scraped up and placed in pans for washing down. False bottoms are employed for saving both fine and coarse gold. (Griffith, 1938) (c) Aust.; U.S. A bed of drift lying on the top of other alluvial deposits, beneath which there may be a true bottom or a lower bed of wash resting directly upon the bedrock. (Fay, 1920) (d) A floor of iron placed in a puddling machine. (Fay, 1920) (e) A flat, hexagonal or cylindrical piece of iron upon which the ore is crushed in a stamp mill; the die. In Victoria, Australia, it is called stamper bed. (Fay, 1920) (f) An insert put in either member of a die set to increase the strength and improve the life of the die. (ASM, 1961)

false chrysolite *moldavite.*

false cleavage (a) *strain-slip cleavage.* (b) A secondary cleavage superposed on slaty cleavage. (c) Closely spaced surfaces, a millimeter or so apart, along which a rock splits. The surfaces are either minute faults or the short limbs of small folds. (AGI, 1987) (d) A quarrying term for minor cleavage in a rock, such as slip cleavage, to distinguish it from the dominant cleavage. Geologically, the term is misleading and should be avoided. (AGI, 1987)

false diamond Any colorless mineral that if cut and polished makes a brilliant gem; e.g., zircon, corundum, and topaz. Having less dispersion, lower refractive index, and birefringence, they are easily distinguished from diamond.

false equilibrium The growth of a metastable or monotropic phase under conditions apparently indicating true equilibrium, as in the development of andalusite crystals in which sillimanite actually represents the stable phase. (AGI, 1987)

false form *pseudomorph.*

false galena An obsolete term for blende. See also: *sphalerite.*

false gate A gate carried forward in the seam thickness only (which must be over 3 ft or 0.9 m), with cut-throughs as required to the main gate. The false gate has a short conveyor that takes the face conveyor coal and delivers it to the main gate conveyor through a crosscut a short distance behind the face. This layout enables the main gate rippings to be worked on three shifts. (Nelson, 1965)

false gossan A laterally or vertically displaced, rather than indigenous, iron-oxide zone. It may be confused with the iron oxide of a gossan, which is weathered from underlying sulfide deposits. (AGI, 1987)

false horizon *artificial horizon.*

false hyacinth *essonite.*

false lapis (a) *lazulite.* (b) Agate, jasper, or other varieties of chalcedony dyed blue. Syn: *false lapis lazuli; Swiss lapis; German lapis.*

false lapis lazuli *false lapis.*

false-leg arches Temporary arch legs used adjacent to the face conveyor in an advance gate to allow the conveyor to be moved forward and still maintain the gate supports. The conveyor side half-arch is temporarily replaced by props and crossbars (false legs). When the conveyor has passed, the half-arch is bolted back in position. (Nelson, 1965)

false ruby Some garnets, e.g., Arizona ruby, Bohemian ruby, Cape ruby, and some spinels, e.g., balas ruby, ruby spinel, are ruby colored. See also: *balas ruby; Elie ruby.*

false stratification An old term for cross-stratification. (AGI, 1987)

false stull A stull so placed as to offer support or reinforcement for a stull, prop, or other timber.

false topaz (a) A transparent yellow variety of quartz, specif. citrine. See also: *citrine.* Cf: *gold topaz; topaz; Madeira topaz.* (b) A yellow variety of fluorite.

famatinite A tetragonal mineral, Cu_3SbS_4; stannite group; occurs with enargite, tetrahedrite-tennantite, chalcopyrite, and covellite in copper ores worldwide, notably at Mankayan, Luzon, Philippines; Famatina, Argentina; and Butte, MT.

family The basic unit of the clan of igneous rocks. (AGI, 1987)

fan (a) To drill a number of boreholes each in a different horizontal or vertical direction from a single drill setup. (Long, 1960) (b) An accumulation of debris brought down by a stream descending through a steep ravine and debouching in the plain beneath, where the detrital material spreads out in the shape of a fan, forming a section of a very low cone. See also: *alluvial fan.* (AGI, 1987)

fan cleavage Cleavage that, if studied over a large enough area, dips at different angles so that, like the ribs of a fan, it converges either upward or downward. (AGI, 1987)

fan cut A cut in which holes of equal or increasing length are drilled in a pattern on a horizonal plane or in a selected stratum to break out a considerable part of it before the rest of the round is fired; the holes are fired in succession in accordance with the increasing angle they form in relation to the face. (BS, 1964)

fancy A term applied to semiprecious stones prized for qualities other than intrinsic value.

fancy lump coal (a) Soft coal from which all slack and nut coal has been removed. (Fay, 1920) (b) Ark. Semianthracite coal of larger size than grate coal. (Fay, 1920)

fan drift (a) The short tunnel connecting the upcast shaft with the exhaust fan. See also: *ventilation.* (b) The enclosed airtight passage, road, or gallery from the mine to the fan. (Mason, 1951) (c) The passage or duct for the intake of a ventilating fan on a mine. (CTD, 1958) (d) An airway leading from a mine shaft, or airway, to a fan. (BS, 1963)

fan drift doors When there are two fans at a mine it is necessary to install isolation doors for each drift leading to a fan in order to prevent the working fan from drawing air from the outside atmosphere. With modern fan layouts, the fan drift may be 5 m or more square and pass 300 m^3/s of air and sometimes more. Modern fan drift doors can be fixed in any position from fully open to fully closed and can be manipulated by one person from outside the fan drift. Doors of the butterfly type are often used and can be opened manually or by power. See also: *main separation door.*

fan efficiency The ratio obtained by dividing useful power output by power input. This is expressed as a percentage. Fan efficiency is understood to mean that air power is calculated from volume flowing and total pressure, on the assumption that the air does not change in volume. The velocity head, present in the air leaving the evasé, is considered as a loss. The power input is that supplied to the fan shaft and thus includes the loss in the fan bearings but excludes all losses in the drive.

fan exhaust An electric fan used for the removal of enamel dust from the spray booth, or fumes from the pickle room, thus safeguarding the health of the worker. (Enam. Dict., 1947)

fan fireman In bituminous coal mining, a person who tends and fires the boiler generating steam for driving fans used for mine ventilation. (DOT, 1949)

fan fold A fold with a broad hinge region and limbs that converge away from the hinge. (AGI, 1987)

fang bolt Used for attaching ironwork to timber. The nut is a plate with teeth, which bite into the wood. To tighten, the bolt is turned while the nut remains stationary. (Crispin, 1964)

fanglomerate A sedimentary rock consisting of slightly waterworn, heterogeneous fragments of all sizes, deposited in an alluvial fan and later cemented into a firm rock; it is characterized by persistence parallel to the depositional strike and by rapid thinning downdip. (AGI, 1987)

fan laws The general fan laws are the same for either axial-flow or centrifugal fans. These laws are as follows: (1) air quantity varies directly as fan speed; quantity is independent of air density (twice the volume requires twice the speed); (2) pressures induced vary directly as fan speed squared, and directly as density (twice the volume develops four times the pressure); (3) the fan-power input varies directly as the fan speed cubed and directly as the air density (twice the volume requires eight times the power); and (4) the mechanical efficiency of the fan is independent of the fan speed and density.

fanner Scot. A small portable hand fan. (Fay, 1920)

Fanning's equation Frictional pressure drop (Δp_t) of fluid flowing in a pipe; $\Delta p_t = 2f(v^2/g)(l/d)$, where f is a function of the Reynolds number, v = rate of flow, g is acceleration due to gravity, l and d are length and diameter of pipe. (Pryor, 1963)

Fann viscosimeter A specific make of viscosimeter. See also: *viscometer.* (Long, 1960)

fan rating The head, quantity, power, and efficiency to be expected when a fan is operating at peak efficiency. (Hartman, 1982)

fan scarp A fault scarplet or small fault scarp entirely in piedmont alluvium or in an alluvial fan. (AGI, 1987)

fan shaft (a) The ventilating shaft to which a mine fan is connected. (BS, 1963) (b) The spindle on which a fan impeller is mounted. (BS, 1963)

fan shooting A type of seismic survey in which detectors are laid out along an arc so that each detector is in a different direction at roughly the same distance from a single shot point. It was used in the 1920's and 1930's to detect the presence of shallow salt domes intruding low-velocity sediments. Syn: *arc shooting*. (AGI, 1987)

fan static head *fan static pressure.*

fan static pressure (a) The total ventilating pressure required to circulate the air through a mine less the natural ventilation pressure. Also called fan useful pressure. (Nelson, 1965) (b) The difference between fan total pressure and fan velocity pressure. Syn: *fan static head*. (Hartman, 1982)

fan structure The fold structure of an anticlinorium. (AGI, 1987)

fan total head Equal to the fan static head plus the velocity head at the fan discharge corresponding to a given quantity of air flow. See also: *total ventilating pressure*. (Hartman, 1961)

fan total pressure The algebraic difference between the mean total pressure at the fan outlet and the mean total pressure at the fan inlet. (BS, 1963)

fan velocity pressure The velocity pressure corresponding to the average velocity at the fan outlet. (BS, 1963)

Far East Rand S. Afr. The area between Boksburg and Heidelburg, Transvaal, limited in the north and east by the outcrops or sub-outcrops of the Main Reef, but not yet limited in the south. (Beerman, 1963)

farewell rock The highest rock formation of the Millstone Grit of South Wales, occurring immediately beneath the Coal Measures. Since all workable coal seams occur in the Coal Measures, it is useless to search for coal in these rocks, hence the miner's term. (Nelson, 1965)

Farrar process Method of case-hardening iron by use of ammonium chloride, manganese dioxide, and potassium ferrocyanide. (Pryor, 1963)

farringtonite A monoclinic mineral, $Mg_3(PO_4)_2$, with iron and silicon present; colorless, wax-white, or yellow; peripheral to olivine nodules in the Springwater pallasite meteorite discovered near Springwater, SK, Canada.

fascicular schist A schist with elongated ferromagnesian minerals lying in a plane but otherwise unoriented. (Knopf, 1938)

fashioned gemstone Gemstone that has been cut and polished.

Fashoda garnet Dark-red to brownish-red pyrope garnet from Tanzania, the clear blood-red specimens used as gems.

fassaite (a) *augite.* (b) Ferrian aluminian diopside or augite.

fast cord Igniter cord consisting of three central paper strings coated with a black powder composition and held together with cotton countering. These are then enclosed in an extruded layer of plastic incendiary composition and finished with an outer plastic covering. The overall diameter of fast igniter cord is approximateiy 0.10 in (2.5 mm). (McAdam, 1958)

fast country Solid or undisturbed rock. Syn: *fast ground*. (Arkell, 1953)

fast-delay detonation A loosely applied term for any method for the firing of blasts involving the use of the blasting timer or millisecond-delay caps. (Streefkerk, 1952)

fast end (a) The part of the coalbed next to the rock. (Fay, 1920) (b) A gangway with rock on both sides. See also: *loose end*. (Fay, 1920) (c) The limit of a stall in one direction, or where the face line of the adjoining stall is not up to or level with, nor in advance of, it. (Fay, 1920)

fast feed *fast gear.*

fast-feed gear *fast gear.*

fast gear (a) As used by drillers in referring to the feed gears in a gear-feed swivel head, the pair of gears installed in the head that produces the greatest amount of bit advance per revolution of the drill stem. Also called fast feed; high feed. (Long, 1960) (b) As used by drillers in referring to the speed at which the drill motor rotates the drill stem or hoist drum, the transmission gear position giving the fastest rotation per engine revolutions per minute. (Long, 1960)

fast ground *fast country.*

fast junking *junking.*

fast line That portion of the cable or wire line, reeved through a block and tackle, that runs from the stationary block to the hoisting drum on a drill machine. Cf: *deadline*. (Long, 1960)

fast powder Dynamite or other explosive having a high-speed detonation. (Nichols, 1976)

fast side (a) The rock adjoining the coal. (Arkell, 1953) (b) The end of the face where there is a solid face more or less at right angles. (Mason, 1951)

fat clay A cohesive and compressible clay of high plasticity, containing a high proportion of minerals that make it greasy to the feel. It is difficult to work when damp, but strong when dry. Cf: *lean clay*. Syn: *long clay*. (AGI, 1987)

fathogram A continuous profile of the depth obtained by echo soundings. (AGI, 1987)

fathometer An instrument used in measuring the depth of water by the time required for a sound wave to travel from surface to bottom and for its echo to be returned. It may also be used for measuring the rise and fall of the tides in offshore localities. (Hunt, 1965)

fatigue The weakening or failure of a material after many repetitions of a stress that of itself is not strong enough to cause failure. (AGI, 1987)

fatigue limit The maximum stress below which a material can presumably endure an infinite number of stress cycles.

fatigue of metals A deterioration in the crystalline structure and strength of metals due to repeated stresses above a certain critical value. See also: *annealing*. (Nelson, 1965)

fatigue ratio The ratio of the fatigue limit for cycles of reversed flexural stress to the tensile strength.

fat stone Stone with fracture surfaces having a greasy luster. Syn: *nepheline.*

fatty amber A yellowish amber resembling goose fat; not as opaque as cloudy amber. Syn: *flohmig amber.*

fatty luster Having the brilliancy of a freshly oiled reflecting surface; characteristic of slightly transparent minerals, e.g., serpentine and sulfur. Syn: *greasy luster.*

Fauck's boring method An earlier percussive boring method used largely in Europe for exploration, etc. The cutting tool is given a rapid but very short stroke, and the hole is flushed by water passing down through the hollow rods. No beam is used, but the rope to which the boring tools are suspended has an up-and-down motion imparted to it by an eccentric. The arrangement gives up to 250 strokes/min with a stroke length as low as 3-1/4 in (8.3 cm). (Nelson, 1965)

fauld The tymp arch or working arch of a furnace. (Fay, 1920)

faulding boards Catches in a mine shaft to facilitate the stopping of the cage at intermediate coal seams. (CTD, 1958)

fault (a) A fracture or a fracture zone in crustal rocks along which there has been displacement of the two sides relative to one another parallel to the fracture. The displacement may be a few inches or many miles long. (b) A break in the continuity of a body of rock. It is accompanied by a movement on one side of the break or the other so that what were once parts of one continuous rock stratum or vein are now separated. The amount of displacement of the parts may range from a few inches to thousands of feet. Various descriptive names have been given to different kinds of faults, including closed fault; dip fault; dip-slip fault; distributive fault; flaw fault; gravity fault; heave fault; hinge fault; horizontal fault; longitudinal fault; normal fault; oblique fault; oblique slip fault; open fault; overthrust fault; parallel displacement fault; pivotal fault; reverse fault; rotary fault; step fault; strike fault; strike-slip fault; thrust fault; transcurrent fault; translatory fault; underthrust; vertical fault; want. (c) In coal mining, a sudden thinning or disappearance of a coal seam. Also known as a want or pinchout. (Kentucky, 1952)

fault basin A region depressed relative to the surrounding region and separated from it by bordering faults.

fault bench A small fault terrace. (AGI, 1987)

fault block (a) A mass bounded on at least two opposite sides by faults. It may be elevated or depressed relative to the adjoining region, or it may be elevated relative to the region on one side and depressed relative to that on the other side. (AGI, 1987) (b) A body of rock bounded by one or more faults. (AGI, 1987) (c) The displaced mass of rocks on either side of a fault plane. See also: *footwall; hanging wall.* (Nelson, 1965)

fault-block mountain *block mountain.*

fault breccia (a) The assemblage of angular fragments resulting from the crushing, shattering, or shearing of rocks during movement on a fault; a friction breccia. It is distinguished by its cross-cutting relations, by the presence of fault gouge, and by blocks with slickensides. (AGI, 1987) (b) Angular to subangular fragments of crushed rock, up to several meters in size, filling a fault. Syn: *fault stuff.* Cf: *tectonic breccia.*

fault casing A layer of hardened clay lining the fault plane and often showing groovings and striae due to the rock movement along the fault plane. See also: *fault gouge.* (Arkell, 1953)

fault complex An intricate system of interconnecting and intersecting faults of the same or different ages.

faulted mountain *block mountain.*

fault embayment A depressed region in a fault zone or between two faults, invaded by the sea. The Red Sea and Tomales Bay on the San Andreas fault in California are examples.

fault escarpment *fault scarp.*

fault fissure A fissure that is the result of faulting. It may or may not be filled with vein material. (AGI, 1987)

fault-fold A structure that is associated with a combination of folding and nearly vertical faulting, in which crustal material that has been fractured into elongate strips tends to drape over the uplifted areas to resemble anticlines and to crumple into the downthrown areas to resemble synclines. (AGI, 1987)

fault gap A depression between the offset ends of a ridge developed on a resistant rock layer that has been displaced by a transverse fault. (Stokes, 1955)

fault gouge A layer of hardened clay lining a fault plane, commonly showing grooves and striae indicating the direction of most recent movement. Syn: *fault casing.* Cf: *slickensides.* See also: *clay gouge; gouge.*

fault groove One of the undulations on a fault surface, deeper than fault striae but similarly formed. They record larger movements and have greater significance as indicating the direction of movement. (Stokes, 1955)

fault growth Intermittent, small-scale movement along a fault surface that, accumulated, results in considerable displacement. (AGI, 1987)

fault heave The amount of lateral movement of the strata at a fault. The fault throw and heave are essential elements of a fault and form basic values when exploring and driving to recover the disrupted coal seam. See also: *fault shift.* (Nelson, 1965)

faulting The process of fracturing and displacement that produces a fault. (AGI, 1987)

fault inlier An isolated exposure of the overridden rock in a region of thrust faulting. It is surrounded by rocks of the overriding block and is thus separated from other surface exposures of rock like itself. (Stokes, 1955)

fault line The intersection of a fault surface or a fault plane with the surface of the Earth or with any artificial surface of reference. Syn: *fault trace; fault trend.*

fault-line scarp An escarpment that is the result of differential erosion along a fault line rather than the direct result of movement along the fault; e.g., the east face of the Sierra Nevada in California.

fault-line valley A valley that follows the line of a fault. Fault valleys are usually straight for long distances. (Stokes, 1955)

fault mosaic An area divided by intersecting faults into blocks that have settled in varying degrees. (Stokes, 1955)

fault plane A fault surface without notable curvature. See also: *plane.* Cf: *fault surface.*

fault scarp An escarpment that owes its relief to a line of faulting, the escarpment occurring on the upthrown side of the fault. See also: *scarp.*

fault set A group of faults that are parallel, or nearly so, and that are related to a particular deformational episode. Cf: *fault system.* (AGI, 1987)

fault shift The lateral movement of the rocks at a fault. In a normal fault it represents the barren ground on a plan of the area (coal mining). Syn: *fault heave.* See also: *shift.* (Nelson, 1965)

fault striae The scratches on faulted surfaces caused by forced movement of particles or projecting hard points against the fault walls. They may indicate the direction of movement on the fault. Cf: *slickensides.* (Stokes, 1955)

fault strike The direction, with respect to north, of the intersection of the fault surface, or of the shear zone, with a horizontal plane. *strike.* (AGI, 1987)

fault stuff Rock filling a fault. See also: *fault breccia.* (Arkell, 1953)

fault surface The surface of a fracture along which dislocation has taken place. Cf: *fault plane.*

fault system Two or more fault sets that were formed at the same time. Cf: *fault set.*

fault terrace An irregular, terracelike tract between two fault scarps, produced on a hillside by step faulting in which the downthrow is systematically on the same side of two approx. parallel faults. Cf: *fault bench.* (AGI, 1987)

fault trace *fault line; rift.*

fault trap A trap, the closure of which results from the presence of one or more faults. (AGI, 1987)

fault trend *fault line.*

fault trough *rift valley.*

fault-trough coast *fault embayment.*

fault vein A mineral vein deposited in a fault fissure. (Fay, 1920)

fault wedge A wedge-shaped block of rock between two faults. (Stokes, 1955)

faulty structure Irregularities of gemstone crystallization and subsequent breakage or separation, such as cleavage cracks, clouds, or feathers.

fault zone A fault that is expressed as a zone of numerous small fractures or of breccia or fault gouge. A fault zone may be as wide as hundreds of meters. Cf: *step fault.* Syn: *distributive fault.* Less preferred syn., shatter belt. (AGI, 1987)

fauna The entire animal population, living or fossil, of a given area, environment, formation, or time span. Cf: *flora.* Adj. faunal. (AGI, 1987)

faunal Of or pertaining to a natural assemblage of animals.

faustite A triclinic mineral, $(Zn,Cu)Al_6(PO_4)_4(OH)_8{\cdot}4H_2O$; turquoise group; in fine grained apple-green masses in Eureka County, NV.

Faust jig A plunger-type jig, usually built with multiple compartments. It has three distinguishing features: (1) plungers on both sides of the screen plate, are accurately synchronized; (2) the refuse is withdrawn through kettle valves near the overflow lips in the respective compartments; and (3) the hutch is commonly discharged periodically by the operator by means of suitable hand valves operated from the working floor. This jig is used extensively on slack sizes of bituminous coal. (Mitchell, 1950)

Fauvelle A system of drilling that provides for the continuous removal of the detritus from the well by means of a water flush or current of water.

favas Braz. In the diamond fields, brown pebbles consisting of a hydrated phosphate, or of titanium and zirconium oxides; regarded as good indications of the presence of diamonds.

favorable locality The experienced prospector always seeks a favorable locality, which may be in the neighborhood of a mining district or else in a locality that contains favorable rocks and structures and appears as if it might contain the mineral sought. See also: *critical area.* (Nelson, 1965)

fayalite An orthorhombic mineral, $4[Fe_2SiO_4]$; olivine group, with iron replaced by magnesum toward forsterite or by manganese toward tephroite; greenish-yellow to yellowish-amber; a common rock-forming mineral in iron-rich metasediments, quartz syenites and granites, ferrogabbro, and some felsic and alkaline volcanic rocks.

Fayol's theory *harmless depth theory.*

fea Shrop. Workable measures, usually ironstone. Syn: *fey.* (Arkell, 1953)

feasibility studies Studies gathering together the information that is required for a decision whether and how to proceed further. A study of this kind may vary from a preliminary estimate of mill cost to a very complete survey that may include a market analysis, mining plan with ore grades and mining cost, metallurgical testing, process development, plans for the mill, cash flow analysis, etc. (SME, 1985)

feather alum *alunogen; halotrichite.*

feather amphibolite A metamorphic rock in which porphyroblastic crystals of amphibole (usually hornblende) tend to form stellate or sheaflike groups on the planes of foliation or schistosity. Cf: *amphibolite.* Syn: *garbenschiefer.* (AGI, 1987)

feather edge (a) The thin end of a wedge-shaped piece of rock or coal. (b) A sharp edge, such as that produced when a brick is cut lengthwise from corner to corner to produce a triangular cross section. See also: *knife edge.* (ACSG, 1963)

feather ends Firebricks with tapered ends. (Osborne, 1956)

feather gypsum *satin spar.*

feather metal Copper granulated by being poured molten into cold water. (Webster 3rd, 1966)

feather ore (a) *jamesonite.* (b) A capillary, fibrous, or feathery habit of an antimony-sulfide mineral, specif. jamesonite, but including stibnite and boulangerite.

feather quartz Imperfect quartz crystals that meet at an angle of a crystallographic plane so that a cross section somewhat resembles a feather.
feathers of litharge Litharge crystals.
fecal pellet An organic excrement, mainly of invertebrates, occurring esp. in modern marine deposits but also fossilized in some sedimentary rocks; usually of a simple ovoid form less than a millimeter in length, or more rarely rod-shaped with longitudinal or transverse sculpturing, devoid of internal structure, and smaller than a coprolite. Also spelled faecal pellet. (AGI, 1987)
Federov stage *universal stage.*
feed (a) Material treated for removal of its valuable mineral contents. Also, feed to any machine or process along a mill's flow line. Also called mill-head ore. (Pryor, 1963) (b) The process of supplying material to a conveying or processing unit. (Nichols, 1976) (c) The forward motion imparted to the drills or cutters of a rock-drilling or coal-cutting machine. (d) A mechanism that pushes a drill into its work. (Nichols, 1976) (e) The longitudinal movements imparted to a drill stem to cause the bit to cut and penetrate the formation being drilled. (Long, 1960) (f) *drill feed.* (g) The distance that the drill stem on a diamond drill may be advanced before the rods must be rechucked. (Long, 1960) (h) In stonecutting, sand and water employed to assist the saw blade in cutting.
feed control System of valves or other mechanical device controlling the rate at which longitudinal movements are imparted to the diamond- or rock-drill stem and/or the cutting teeth on a coal-cutting machine. (Long, 1960)
feed-control valve A small valve, usually a needle valve, on the outlet of the hydraulic-feed cylinder on the swivel head of a diamond drill; used to control minutely the speed of the hydraulic piston travel and, hence, the rate at which the bit penetrates the rock being drilled. Also called drip valve; needle valve. (Long, 1960)
feed-end blocks In rotary kilns, special fire-clay shapes or rotary-kiln blocks so installed as to reduce the kiln diameter.
feeder (a) Small fissures or cracks through which methane or other gases escape from the coal. As working faces are advanced, fresh feeders are encountered in each fall of coal. (Kentucky, 1952) (b) Any flow of water or gas entering a mine. (BS, 1963) (c) A conveyor or bunker structure for delivering coal or other broken material at a controllable rate. See also: *feeder conveyor; Lofco car feeder.* (Nelson, 1965) (d) A flow of water from the strata or from old workings. (Nelson, 1965) (e) A small ore vein leading to a larger one. See also: *feeder vein.*
feeder and catchers tables A pair of reversible conveyors, entry and exit, which provide for repeat feeding of metal being processed through a rolling mill.
feeder breaker A portable feeding and crushing unit which can move on its own or often be installed in a stationary position. The unit has a feed chain (often referred to as a flight chain) inside a shallow built-on hopper that drags the mined ore into a rotating breaker head (shaft), which has various sizes of breaker heads (known as piks). These piks rotate downward as the chain drags the stone or ore into it, crushing the ore to various sizes depending on the height of the mounted rotating breaker head. Syn: *stamler.*
feeder circuit A feeder circuit is a conductor or group of conductors and associated protective and switching devices installed on the surface, in mine entries, or in gangways, but not extending beyond the limits set for permanent mine wiring.
feeder connection The opening or surrounding blocks in a furnace wall to receive the channel leading to the feeder. (ASTM, 1994)
feeder conveyor (a) A short auxiliary conveyor designed to receive coal from the face conveyor and load it onto the gate conveyor. Also called stage loader. See also: *feeder; gate-end feeder; gate-end loader.* (Nelson, 1965) (b) Any conveyor that transports material to another conveyor. (NEMA, 1961)
feeder trough In a duckbill, the trough that is attached to the conveyor pan line and serves as a base on which the feeder trough rides. (Jones, 1949)
feeder vein A small ore-bearing vein joining a larger one. See also: *feeder.*
feed gear The gearing or assemblage of three to four pairs of matched gears in a gear-feed swivel head of a diamond drill by means of which the drill string coupled to the feed screw is made to advance and penetrate the formation. See also: *gear.* (Long, 1960)
feedhead *swivel head.*
feeding baffle A door or gate that can be opened or closed to regulate the discharge of material from a hopper, bin, or chute. (Nelson, 1965)
feed pressure (a) Total weight or pressure, expressed in pounds or tons, applied to drilling stem to make the drill bit cut and penetrate the formation being drilled. (Long, 1960) (b) Pressure, expressed in pounds per square inch, required to force grout into a rock formation. Cf: *injection pressure.* (Long, 1960)
feed pump The pump that provides a steam boiler with feedwater. (Nelson, 1965)
feed rate Rate at which a drilling bit is advanced into or penetrates the rock formation being drilled, expressed in inches per minute, inch per bit revolution, number of bit revolutions per inch of advance, or feet per hour. Also called cutting rate; forward speed. Syn: *penetration rate; bit feed; cutting speed; drill rate; penetration feed.* (Long, 1960)
feed ratio The number of revolutions a drill stem and bit must turn to advance the drill bit 1 in (2.54 cm) when the stem is attached to and rotated by a screw- or gear-feed-type drill swivel head with a particular pair of the set of gears engaged. (E.g., when a screw-feed swivel head of a diamond drill equipped with three pairs of gears, having a feed ratio of 100, 200, and 400, is operated with the 100-pair engaged, the drill stem must revolve 100 times to advance the bit 1 in, if the 200-pair is engaged, the drill stem rotates 200 times/in (508 times/cm) advanced, and if the 400-pair is engaged the stem must rotate 400 times to advance the bit 1 in). (Long, 1960)
feed shaft A short shaft or countershaft in a diamond-drill gear-feed swivel head rotated by the drill motor through gears or a fractional drive and by means of which the engaged pair of feed gears is driven. Syn: *feed spindle.* (Long, 1960)
feed speed Normally used by drillers to denote feed ratios. (Long, 1960)
feed spindle *feed shaft.* Sometimes incorrectly used as a synonym for drive rod and/or feed screw. (Long, 1960)
feed travel The distance a drilling machine moves the steel shank in traveling from top to bottom of its feeding range. (Nichols, 1976)
feedwater Water that is often purified, heated to nearly boiler temperature, and deaerated before being pumped into a steam boiler by the feed pump. (Nelson, 1965)
fee engineer Person (usually an engineer) who looks after the interest of the owner of mineral rights. Specific duties are to check on the amount of ore mined by the lessor (operator), see that no undue waste is permitted, and see that royalties are paid according to contract.
feigh Refuse or dirt from ore or coal. (CTD, 1958)
feinig A jewel diamond having its grain in regular layers. Cf: *naetig.* (Brady, 1940)
feldspar (a) A monoclinic or triclinic mineral with the general formula XZ_4O_8 where (X= Ba,Ca,K,Na,NH_4) and (Z= Al,B,Si); a group containing two high-temperature series, plagioclase and alkali feldspar; colorless or white and clear to translucent where pure; commonly twinned; 90° or near 90° prismatic cleavage; Mohs hardness, 6. Constituting 60% of the Earth's crust, feldspar occurs in all rock types and decomposes to form much of the clay in soil, including kaolinite. Also spelled felspar, feldspath. Cf: *sanidine.* See also: *moonstone.* (b) The mineral group albite, andesine, anorthite, anorthoclase, banalsite, buddingtonite, bytownite, celsian, hyalophane, labradorite, microcline, oligoclase, orthoclase, paracelsian, plagioclase, reedmergnerite, sanidine, and slawsonite.
feldspar convention *rational analysis.*
feldspar jig A small coal washer to deal with the 1/2- to 0-in (1.3- to 0-cm) range. It works on the same basic principle as the Baum washer, but in view of the small-size material a feldspar (sp gr, 2.6) bed is provided on the perforated grid plates to prevent the bulk of the feed from passing straight through the perforations. Stratification of the raw feed takes place in the usual way. (Nelson, 1965)
feldspar sunstone *sunstone.*
feldspar-type washbox A washbox to clean small coal, in which the pulsating water is made to pass through a layer of graded material, such as feldspar, situated on top of the screen plate. *jig.* (BS, 1962)
feldspathic Said of a rock or other mineral aggregate containing feldspar.
feldspathic emery Emery similar to spinel emery but contains in addition from 30% to 50% plagioclase feldspar. Pure magnetite often is found in streaks within this mass. (AIME, 1960)
feldspathic graywacke A graywacke characterized by abundant unstable materials; specif., a sandstone containing generally less than 75% of quartz and chert and 15% to 75% detrital clay matrix, and having feldspar grains (chiefly sodic plagioclase, indicating a plutonic provenance) in greater abundance than rock fragments (indicating a supracrustal provenance) (Pettijohn, 1954; 1957). Cf: *lithic graywacke.* (AGI, 1987)
feldspathic sandstone A feldspar-rich sandstone; specif., a sandstone intermediate in composition between an arkosic sandstone and a quartz sandstone, containing 10% to 25% feldspar and less than 20% matrix material of clay, sericite, and chlorite. See also: *arkose.* (AGI, 1987)
feldspathide *feldspathoid.*
feldspathization The introduction of feldspar into a rock, or the replacement of other rock-forming minerals by feldspar. Material for the feldspar may come from the country rock or be introduced by magmatic or other solutions.
feldspathoid (a) A group of comparatively rare rock-forming minerals consisting of aluminosilicates of sodium, potassium, or calcium and

having too little silica to form feldspar. Feldspathoids are chemically related to the feldspars, but differ from them in crystal form and physical properties; they take the places of feldspars in igneous rocks that are undersaturated with respect to silica or that contain more alkalies and aluminum than can be accommodated in the feldspars. Feldspathoids may be found in the same rock with feldspars but never with quartz or in the presence of free magmatic silica. See also: *foid; lenad.* (AGI, 1987) (b) A mineral of the feldspathoid group, including leucite, nepheline, sodalite, nosean, hauyne, lazurite, cancrinite, and melilite. Syn: *feldspathide.* (AGI, 1987)

feldstone A rock having a fine granular structure, and composed chiefly of feldspar and quartz. (Gordon, 1906)

felite A constituent of portland cement clinker. Also spelled felith. *belite; larnite.*

fell (a) One of the many names for lead ore. (b) The finer pieces of ore that pass through the riddle in sorting.

feloids A group of minerals comprising the feldspars and feldspathoids. Cf: *feldspathoid.* (English, 1938)

felsenmeer *block field.*

felsic A mnemonic adj. derived from (fe) for feldspar, (l) for lenad or feldspathoid, and (s) for silica, and applied to light-colored rocks containing an abundance of one or all of these constituents. Also applied to the minerals themselves, the chief felsic minerals being quartz, feldspar, feldspathoid, and muscovite. Syn: *acidic; silicic.* Cf: *mafic.*

felsite A general term for any light-colored, fine-grained or aphanitic extrusive or hypabyssal rock, with or without phenocrysts, and composed chiefly of quartz and feldspar; a rock characterized by felsitic texture. Syn: *felstone; aphanite.* Cf: *felsophyre.* (AGI, 1987)

felsitic A textural term ordinarily applied to dense, light-colored igneous rocks composed of crystals that are too small to be readily distinguished with the unaided eye; microcrystalline. It may also be used as a microscopic term for the groundmass of porphyritic rocks that are too fine-grained for the mineral constituents to be determined with the microscope; cryptocrystalline.

felsitoid An informal term applied to any light-colored igneous rock in which the mineral grains are too small to be distinguished by the unaided eye. Cf: *felsite.* Syn: *aphanite.* (AGI, 1987)

felsöbanyaite An orthorhombic(?) mineral, $Al_4(SO_4)(OH)_{11}·5H_2O$; massive; soft; snow white; associated with marcasite, stibnite, and barite at Felsöbánya, Romania. Also spelled felsöbanyite. Cf: *basaluminite.*

felsophyre A general term for any porphyritic felsite. Syn: *aphanophyre.* Cf: *vitrophyre; granophyre.* (AGI, 1987)

felspar A British spelling of feldspar following an error by Kirwan. (Hess)

felstone An obsolete synonym of felsite. (AGI, 1987)

felted *felty.*

felty Said of the texture of dense, holocrystalline igneous rocks consisting of tightly appressed microlites, generally of feldspar, interwoven in irregular, unoriented fashion. Cf: *trachytic.* Syn: *pilotaxitic; felted.*

femic Said of an igneous rock having one or more normative, dark-colored iron-, magnesium-, or calcium-rich minerals as the major components of the norm; also, said of such minerals. Etymol. a mnemonic term derived from ferric + magnesium + ic. Cf: *basic; salic; mafic; felsic.* (AGI, 1987)

femmer Fragile, weak, or slender as in the case of a thin, soft roof bed over a coal seam. (Nelson, 1965)

fence diagram Three or more geologic sections showing the relationship of wells to subsurface formations. The scales diminish with distance from the foreground to give proper perspective. When several sections are used together, they form a fencelike enclosure, hence the name. Similar in some respect to a block diagram, but it has the advantage of transparency, which is not possible in a block diagram. (AGI, 1987)

fender A thin pillar of coal, adjacent to the gob, left for protection while driving a lift through the main pillar.

fenite A quartzo-feldspathic rock that has been altered by alkali metasomatism at the contact of a carbonatite intrusive complex. The process is called fenitization. Fenite is mostly alkalic feldspar, with some aegirine, subordinate alkali-hornblende, and accessory sphene and apatite. (AGI, 1987)

fenster *window.* Etymol. German, window. (AGI, 1987)

ferberite A monoclinic mineral, $FeWO_4$; wolframite series with manganese replacing iron toward hübernite; weakly ferrimagnetic; sp gr, 7.5; in quartz veins; an ore of tungsten. Syn: *eisinwolframite.*

ferganite A very rare, strongly radioactive, possibly orthorhombic, sulfur-yellow translucent mineral, $U_3(VO_4)_2·6H_2O$; found associated with other uranium minerals. Ferganite may be a leached or weathered product of tyuyamunite. A source of vanadium. (Crosby, 1955; Osborne, 1956)

ferghanite *ferganite.*

fergusonite Monoclinic (beta) and orthorhombic minerals $(Ce,Nd,La,Y)NbO_4$, further speciated according to the predominant rare-earth element; dull to vitreous brownish black; in pegmatites associated with euxenite, monazite, gadolinite, and other rare-earth minerals in North Carolina, South Carolina, Virginia, Texas, Norway, Sweden, and Africa.

fermentation The process of decomposition of carbohydrates with the evolution of carbon dioxide or the formation of acid, or both. (Tomkeieff, 1954)

fermorite A hexagonal mineral, $(Ca,Sr)_5(AsO_4,PO_4)_3(OH)$; apatite group; in veinlets in manganese ore at Sitipar, Chindwara district, India.

fernandinite (a) A vanadium ore. (Osborne, 1956) (b) A monoclinic mineral, $Ca_3V_{40}O_{100}·50H_2O$; dull-green, cryptocrystalline to fibrous masses in the oxidized zone of the abandoned Minasragra vanadium deposit near Cerro de Pasco, Peru.

Ferraris screen A screening machine, utilizing inclined supports, developed in southern Europe for screening small sizes of ore and sand. The wooden screen frame is set horizontally and supported on flexible wooden staves inclined at about 65° from the horizontal. The connecting rod also is inclined to the screen frame, so as to be approx. at right angles to the supports. (Mitchell, 1950)

Ferraris truss Supporting batten used originally as a slanting support under a shaking screen. When the screen was pushed forward the radial motion of the truss caused it to rise slightly, giving a throwing motion to the load and aiding the gravity-assisted return as the reciprocating action of the screen vibrator was reversed. The principle is used in shaking tables such as the James. (Pryor, 1963)

ferrate Any of various classes of compounds containing iron and oxygen in the anion. (Webster 3rd, 1966)

Ferrel's law The statement that the centrifugal force produced by the rotation of the Earth (Coriolis force) causes a rotational deflection of currents of water and air to the right in the Northern Hemisphere and to the left in the Southern Hemisphere. (AGI, 1987)

ferreto zone Reddish-brown or reddish zone in permeable near-surface material that is produced under conditions of free subsurface drainage by the deposition of secondary iron oxide.

ferric Of, pertaining to, or containing iron in the trivalent state; e.g., ferric chloride, $FeCl_3$. (Standard, 1964)

ferric furnace A high, iron blast furnace, in the upper part of which crude bituminous coal is converted into coke. (Fay, 1920)

ferricopiapite A triclinic mineral, $Fe_5(SO_4)_6(OH)_2·20H_2O$; copiapite group.

ferricrete (a) A conglomerate consisting of surficial sand and gravel cemented into a hard mass by iron oxide derived from the oxidation of percolating solutions of iron salts. (AGI, 1987) (b) A ferruginous duricrust. Etymol. "ferr"uginous + con"crete." Cf: *calcrete; silcrete.* (AGI, 1987)

ferricrust (a) A general term for an indurated soil horizon cemented with iron oxide, mainly hematite. (AGI, 1987) (b) The hard crust of an iron concretion. (AGI, 1987)

ferride A member of a group of elements that are similar chemically to iron. The group includes chromium, cobalt, manganese, nickel, titanium, vanadium, and other elements.

ferrierite Monoclinic and orthorhombic minerals, $(Na,K)_2Mg(Si,Al)_{18}O_{36}(OH)·9H_2O$, of the zeolite group; transparent to translucent, vitreous to pearly; in spherical aggregates of thin, blade-shaped crystals at Kamloops Lake, BC, Canada and Leavitt Lake, CA.

ferrifayalite *laihunite.*

ferriferous Containing iron. Syn: *ferruginous.*

ferrimagnetism (a) Unbalanced orientation of magnetic moments. Intermediate between ferromagnetism and antiferromagnetism. (Van Vlack, 1964) (b) Strong magnetic susceptibility caused by the overlap of adjacent *d* and *p* orbitals resulting in an unequal number of electrons aligned with spins antiparallel; e.g., magnetite, pyrrhotite, and maghemite. Cf: *antiferromagnetism; diamagnetism; ferromagnetism; paramagnetism; superexchange.*

ferrimolybdite A possibly orthorhombic mineral, $Fe_2^3(Mo^6O_4)_3·8H_2O$; forms yellow earthy powder, incrustation, or silky, fibrous, and radiating crystals from reaction of pyrite with molybdenite under supergene conditions; relatively common in porphyry copper deposits of Southwestern United States in old mine dumps with jarosite and goethite. Syn: *molybdic ocher; iron molybdate; molybdite.*

ferrinatrite A trigonal mineral, $Na_3Fe(SO_4)_33H_2O$; vitreous; soft; in the Atacama Desert, northern Chile. Formerly called ferronatrite.

ferrisicklerite An orthorhombic mineral, $Li(Fe,Mn)PO_4$; forms a series with sicklerite; an alteration product of weathered triphyllite-lithiophilite in granite pegmatites, Keystone and Custer districts, Black Hills, SD; also Varutriisk, Sweden.

ferrisymplesite An amorphous mineral, $Fe_3(AsO_4)_2(OH)_3·5H_2O$; occurs with erythrite and annabergite at Cobalt, ON, Canada.

ferrite (a) Native iron, such as the terrestrial iron from Disko Island, Greenland. (English, 1938) (b) A general term applied to grains, scales, and threads of unidentifiable, more or less transparent or amorphous, red, brown, or yellow iron oxide in the groundmass of a porphyritic rock (Johannsen, 1939). Cf: *opacite.* Syn: *ferrospinel.* (AGI, 1987) (c) A term used by Tieje (1921) for a cemented iron-rich sediment whose particles do not interlock. (AGI, 1987)

ferritization The metasomatic alteration of other minerals into ferrite.

ferritungstite An isometric mineral, $(K,Ca,Na)(W^6,Fe^3)(O,OH)_6 \cdot H_2O$; in minute yellow octahedra and platy aggregates; in Uganda, Rwanda, Zaire, Portugal, France, Washington, and Nevada. Syn: *tungstic ocher.*

ferriturquoise A variety of crystallized turquoise containing 5% Fe_2O_3; from Lynchburg, VA. (Spencer, 1943)

ferroactinolite A monoclinic mineral, $Ca_2(Fe^{2+},Mg)_5Si_8O_{22}(OH)_2$; amphibole group; has $Mg/(Mg+Fe^{2+})$ = 0 to 0.5; forms a series with tremolite and actinolite. Formerly called ferrotremolite.

ferroalloy An alloy of iron and one or more other elements useful for making alloy additions in steel or ironmaking.

ferroalluaudite A monoclinic mineral, $NaCaFe(Fe,Mn)_2(PO_4)_3$; alluaudite group; forms a series with alluaudite; in granite pegmatites.

ferroan dolomite (a) Dolomite having up to 20% of magnesium replaced by iron or manganese. (b) A mineral composition intermediate between those of dolomite and ankerite.

ferroan spinel *ceylonite.*

ferroantigorite A hypothetical composition, $(Mg,Fe)_3Si_2O_5(OH)_4$, used to describe chlorite compositions; intermediate to antigorite and greenalite.

ferroboron A boron iron alloy containing 12% to 14% boron. (USBM, 1965; USBM, 1960)

ferrocarpholite An orthorhombic mineral, $(Fe,Mg)Al_2Si_2O_6(OH)_4$; forms series with carpholite and magnesiocarpholite; dark green; in quartz veins near Tomata, Celebes Island, Indonesia.

ferrochrome An alloy of iron and chromium. (Pryor, 1963)

ferrocopiapite *copiapite.*

ferroeckermannite A monoclinic mineral, $Na_3(Fe^{2+},Mg)_4AlSi_8O_{22}(OH)_2$; amphibole group; has $Mg/(Mg+Fe^{2+})$ = 0 to 0.49; forms a series with eckermannite.

ferroedenite A monoclinic mineral, $NaCa_2(Fe,Mg)_5[Si_7AlO_{22}](OH)_2$; amphibole group; has $Mg/(Mg+Fe)$ = 0 to 0.49; forms a series with edenite.

ferroelectric Spontaneous electrical polarization with all dipoles in the same direction. The polarity can be reversed by an external electrical field. (Van Vlack, 1964)

ferro-ferri-lazulite *barbosalite.*

ferrogabbro A gabbro in which the pyroxene or olivine, or both, are exceptionally high in iron.

ferrohexahydrite A monoclinic mineral, $FeSO_4 \cdot 6H_2O$; hexahydrite group; transparent; forms at Vesuvius, Italy, as fine acicular crystals that are unstable under normal atmospheric conditions.

ferrohortonolite A mineral in the olivine series composed of 70% to 90% fayalite and 30% to 10% forsterite.

ferrohypersthene *ferrosilite.*

ferrolite Black iron slag, said to be satisfactory for fashioning into gemstones.

ferromagnesian Containing iron and magnesium. Applied to certain dark silicate minerals, esp. amphibole, pyroxene, biotite, and olivine.

ferromagnesian mineral Any mineral having a considerable portion of iron and magnesium in its composition. Cf: *ferrous mineral.*

ferromagnetic material (a) Material that in general exhibits the phenomena of hysteresis and saturation and whose permeability is dependent on the magnetizing force. Microscopically, the elementary magnets are aligned parallel in volumes called domains. The unmagnetized condition of a ferromagnetic material results from the overall neutralization of the magnetization of the domains to produce zero external magnetization. (ASM, 1961) (b) The three substances, iron, nickel, and cobalt, are so considerably more magnetic than any other substances that they are grouped by themselves; they are termed ferromagnetic.

ferromagnetism (a) Spontaneous magnetic orientation of all magnetic moments in the same direction. The orientation can be reversed by an external magnetic field. (Van Vlack, 1964) (b) Strong magnetic susceptibility caused by overlap of adjacent *d* orbitals; e.g., iron, nickel, cobalt, and numerous alloys, both ferrous and nonferrous. Cf: *antiferromagnetism; ferrimagnetism.*

ferromanganese A ferroalloy containing about 80% manganese and used in steelmaking. (McGraw-Hill, 1994)

ferromolybdenum A molybdenum-iron alloy produced in the electric furnace or by a thermite process. It is used to introduce molybdenum into iron or steel alloys and as a coating material on welding rods. (USBM, 1965)

ferrosalite *hedenbergite.*

ferroselite An orthorhombic mineral, $FeSe_2$; marcasite group; occurs with chalcopyrite and pyrite as a cement in sandstones in the Tuva, Siberia, Russia.

ferrosilicon Alloy of iron and silicon. (Pryor, 1963)

ferrosilite An orthorhombic mineral, $(Fe,Mg)_2Si_2O_6$; pyroxene group; dimorphous with clinoferrosilite; forms a series with enstatite. Symbol, Fs. Formerly called iron hypersthene; orthoferrosilite. Syn: *ferrohypersthene.*

ferrospinel A synthetic ferrimagnetic substance having spinel structure, containing iron; conducts electricity poorly. See also: *ferrite; hercynite.*

ferrotschermakite A monoclinic mineral, $Ca_2(Fe,Mg)_3Al_2(Si_6Al_2)O_{22}(OH)_2$; amphibole group, with $Mg/(Mg + Fe)$ = 0 to 0.49; forms a series with tschermackite; a fairly common constituent of eclogites and amphibolites.

ferrous (a) The term or prefix used to denote compounds or solutions containing iron in which iron is in the divalent (+2) state. (b) Of, relating to, or containing iron.

ferrous metallurgy The metallurgy of iron and steel. (Newton, 1959)

ferrous metals A classification in the United States of metals commonly occurring in alloys with iron, such as chromium, nickel, manganese, vanadium, molybdenum, cobalt, silicon, tantalum, and columbium (niobium).

ferrous mineral Any mineral having a considerable portion of iron in its composition. Cf: *ferromagnesian mineral.*

ferrous oxide This lower oxide, FeO, tends to be formed under reducing conditions; it will react with SiO_2 to produce a material melting at about 1,200 °C, hence the fluxing action of ferruginous impurities present in some clays if the latter are fired under reducing conditions. Melting point, 1,420 °C; sp gr, 5.7. (Dodd, 1964)

ferrovanadium An alloy of iron and vanadium. (Pryor, 1963)

ferroxdure A sintered oxide consisting mainly of the oxide $BaFe_{12}O_{19}$; used for the production of permanent magnets. (Osborne, 1956)

ferruccite An orthorhombic mineral, $NaBF_4$; forms minute crystals with sassolite as a sublimate around fumaroles at Mt. Vesuvius, Italy.

ferruginate (a) Said of a sedimentary rock cemented with iron-bearing minerals, generally limonite; also, said of the iron-bearing cement. (b) To stain a rock with an iron-bearing compound. (AGI, 1987)

ferruginous (a) Pertaining to or containing iron; e.g., a sandstone that is cemented with iron oxide. Cf: *ferriferous.* (AGI, 1987) (b) Said of a rock having a red or rusty color due to the presence of ferric oxide (the quantity of which may be very small). (AGI, 1987)

ferruginous chert A sedimentary deposit consisting of chalcedony or of fine-grained quartz and variable amounts of hematite, magnetite, or limonite.

ferruginous ores Gangue; principally oxides, silicates, or carbonates of iron. (Newton, 1938)

ferruginous rock Rocks of this group are usually carbonate of iron that has partially or wholly replaced limestone. (Mason, 1951)

ferruginous sandstone A sandstone containing iron oxide as the cementing material, as grains, or both.

Fersman's law Parallel orientation of feldspar prism edges with the edge between two adjacent rhombohedral planes of quartz in graphic granite so that the *c* axis of the quartz forms an angle of 42° 16′ with the *c* axis of the feldspar. (Hess)

fersmite An orthorhombic mineral, $(Ca,Ce,Na)(Nb,Ta,Ti)_2(O,OH,F)_6$; weakly radioactive; in syenites, may be associated with pyrochlore, alkali hornblende, apatite, sphene, magnetite, zircon, xenotime, or allanite; also in some marbles with columbite and monazite.

ferutile *davidite.*

fervanite A monoclinic mineral, $Fe_4(VO_4)_4 \cdot 5H_2O$; golden brown; in uranium-vanadium deposits of the Colorado Plateau.

Fery radiation pyrometer An instrument in which heat radiated from the hot body is focused, by means of a concave mirror, on a small central hole behind which a small thermocouple is placed in front of two small, inclined mirrors. The instrument is sighted onto the hot body and focused by rotating a screw until the lower and upper halves of the image coincide; the electromotive force generated by the thermocouple is indicated on a galvanometer. The instrument, once focused, gives continuous readings and may be connected to a recording indicator. (Osborne, 1956)

fetch The unobstructed distance that the wind can travel to any point when raising waves. (Hammond, 1965)

fetid Having a disagreeable odor caused by the occurrence of certain bituminous substances or of hydrogen sulfide. This odor is apparent when some varieties of limestone and quartz are broken or are rubbed vigorously.

fetid calcite A variety of calcite that emits an offensive odor when dissolved in dilute hydrochloric acid. The odor is due to trace sulfides and other impurities. *swinestone.*

fettle (a) To cover or line the hearth of (a reverberatory furnace) with fettling. (Webster 3rd, 1966) (b) To clean and smooth (as a metal or plastic) after casting or molding. (Webster 3rd, 1966) (c) To remove fins, mold marks, and rough edges from dry, or nearly dry, ware. (ACSG, 1963)

fettling (a) Protecting the bottom of the open-hearth furnace with loose material, such as ore, sand, etc.; also, the material so used.

(Henderson, 1953) (b) The process of repairing a steel-furnace hearth, with dead-burned magnesite or burned dolomite, between tapping and recharging the furnace. (Dodd, 1964)

feverstein *firestone.*

fey *fea.*

fiber The smallest single strand of asbestos or other fibrous material. (Mersereau, 1947)

fibril A single fiber, which cannot be separated into smaller components without losing its fibrous properties or appearance. (Campbell)

fibroblastic The texture of metamorphic rocks resulting from the development during recrystallization of minerals with a fibrous habit. See also: *nematoblastic.*

fibroferrite A monoclinic mineral, $Fe(SO_4)(OH)\cdot 5H_2O$; forms fine fibrous crusts or masses associated with melanterite, copiapite, jarosite, and other secondary sulfates.

fibrogenic dust *pulmonary dust.*

fibrolite (a) A fibrous, felted variety of sillimanite. (b) One of three crystalline forms of aluminum sillicate, Al_2SiO_5, the others being andalusite (low temperature) and kyanite (high temperature). Sillimanite occurs commonly as felted aggregates of exceedingly thin fibrous crystals (hence the name fibrolite) in contact metamorphosed aluminous sediments such as mudstones, shales, etc. Crystals of a pale sapphire blue are used as gems. (CTD, 1958)

fibrolite cat's-eye A pale greenish variety of sillimanite having fibrous inclusions; when cut, produces a chatoyant effect, but not a well-defined cat's-eye.

fibrous (a) Applied to minerals that occur as fibers, such as asbestos. Syn: *asbestiform.* (b) Consisting of fine threadlike strands, e.g., satin spar variety of gypsum.

fibrous aggregate A crystalline aggregate composed of closely packed fibers.

fibrous anthraxylon Thin strands of anthraxylon having the appearance of threads in thin sections.

fibrous calcite Translucent calcite composed of fibrous crystals, which, like fibrous gypsum, with which it is often confused, causes a silky sheen. When cut cabochon, it produces a girasol or chatoyant effect, but not a true cat's-eye. Also like fibrous gypsum, it is called "satin spar" but less correctly.

fibrous gypsum *satin spar.* Cf: *fibrous calcite.*

fibrous peat (a) Peat composed of the fibrous remains of plants. It is fibrous, spongy, moderately tough, and nonplastic. It does not shrink much on drying. Also called woody peat. See also: *pseudofibrous peat; amorphous peat.* (Tomkeieff, 1954) (b) Firm, moderately tough peat in which plant structures are only slightly altered by decay. It shrinks little on drying. (Francis, 1954)

fibrous structure (a) If the crystals in a mineral aggregate are greatly elongated and have a relatively small cross section, the structure or texture is fibrous. The fibers may be parallel, as in crocidolite and sometimes in gypsum and cerussite. When the fibers are very fine, they may impart a silky luster to the aggregate, as in crocidolite and satin-spar gypsum. There is also a feltlike type. Fibrous crystals may radiate from a center, producing asteriated or starlike groups, either coarse or fine, as frequently observed in pyrolusite, wavellite, natrolite, and tourmaline, and sometimes in stibnite and other minerals. Also called fibrous texture. (CMD, 1948) (b) In forgings, a structure revealed as laminations, not necessarily detrimental, on an etched section or as a ropy appearance on a fracture, not to be confused with the silky or ductile fracture of a clean metal. (c) In wrought iron, a structure consisting of slag fibers embedded in ferrite. (d) In rolled-steel plate stock, a uniform, fine-grained structure on a fractured surface, free of laminations or shale-type discontinuities. As contrasted with definition b., it is virtually synomous with silky or ductile fracture. (ASM, 1961)

fibrous texture In mineral deposits, a pattern of finely acicular, rodlike crystals, e.g., in chrysotile and amphibole asbestos. (AGI, 1987)

fibrous wax A fibrous variety of ozocerite, a natural paraffin wax. See also: *ozocerite.*

fichtelite A monoclinic mineral, dimethyl-isopropyl-perhydropenanthrene, $C_{19}H_{34}$; translucent white; in fossil wood or conifers.

fiducial interval A measure of confidence in precision of a set of sample data. For a given numerical value of fiducial interval, the number of samples required from a given deposit to give an accurate measure of its value can be determined. (Lewis, 1964)

fiducial mark In photogrammetry, an index or point used as a basis of reference; one of usually four index marks connected with a camera lens (as on the metal frame that encloses the negative) that form an image on the negative or print such that lines drawn between opposing points intersect at and thereby define the principal point of the photograph. Syn: *collimating mark.* (AGI, 1987)

fiducial time A time marked on a record to correspond to some arbitrary time. Such marks may aid in synchronizing different records or may indicate a reference, such as a datum plane. (AGI, 1987)

field (a) A region or area that possesses or is characterized by a particular mineral resource; e.g., goldfield, coalfield. (AGI, 1987) (b) A broad term for the area, away from the laboratory and esp. outdoors, in which a geologist makes observations and collects data, rock and mineral samples, and fossils. (AGI, 1987) (c) That space in which an effect, e.g., gravity or magnetism, occurs and is measurable. It is characterized by continuity; i.e., there is a value associated with every location within the space. (AGI, 1987) (d) A section of land containing, yielding, or worked for a natural resource; e.g., a coalfield, an oilfield, or a diamond field. A large tract or area, as large as many square miles, containing valuable minerals. See also: *coalfield.* (Webster 3rd, 1966) (e) The immediate locality and surroundings of a mine explosion. (f) A colliery, or firm of colliery proprietors.

field capacity The quantity of water held by soil or rock against the pull of gravity. It is sometimes limited to a certain drainage period, thereby distinguishing it from specific retention, which is not limited by time. Syn: *field-moisture capacity; normal moisture capacity.* (AGI, 1987)

field classification of rocks A classification of rocks made in the field. It is based on features distinguishable in hand specimens by using a hand lens, a knife, an acid bottle, etc. The classification may be refined or modified by subsequent examination with a microscope or other techniques that are generally used in a laboratory. (Stokes, 1955)

field compaction trial Tests carried out under site conditions to determine the best combination of (1) type of compaction equipment; (2) thickness of loose soil layer; (3) number of passes; and (4) moisture content (where variation is possible) in order to achieve the highest possible soil densities. (Nelson, 1965)

field geology Geology as practiced by direct observation in the field; original, primary reconnaissance; field work. (AGI, 1987)

field investigation In reference to experimental-mine tests, the investigation made at a mine when a large sample is taken for testing at the experimental mine; this investigation includes the taking of road dust, rib dust, mine air, and standard coal samples, and the noting of conditions affecting the safety of the mine. (Rice, 1913-18)

fieldite A zinciferous variety of tetrahedrite. Cf: *goldfieldite.*

field-laboratory operator Person who analyzes mine water for acid, copper, and iron content by removing samples of water that flow to and from the precipitation drum, and who performs routine chemical tests. (DOT, 1949)

field-moisture capacity *field capacity.*

field work Work done, observations taken, or other operations, as triangulation, leveling, making geological observations, etc., in the field or upon the ground.

fiery dragon *toadstone.*

fiery heap Eng. A deposit of rubbish and waste or unsalable coal that ignites spontaneously. (Fay, 1920)

fiery mine (a) A mine in which the seam or seams of coal being worked give off a large amount of methane. (Fay, 1920) (b) Mine in which there is danger of explosion due to coal dust or flammable gas. (Pryor, 1963) (c) A gassy mine; a mine where gas ignitions and outbursts have occurred in the past. (Nelson, 1965)

fifth wheel (a) The weight-bearing swivel connection between highway-type tractors and semitrailers. (Nichols, 1976) (b) A wheel used to automatically operate the dump mechanism of mine ore cars.

figure cuts *V-cut.*

figure stone *agalmatolite.*

filar micrometer An instrument, that in its usual form consists of an ocular containing a fine wire that can be moved across the field by means of a thumbscrew for the purpose of measuring size. (Hess)

filiform Having the shape of a thread or filament; e.g., native silver. Syn: *wiry.* Cf: *capillary.*

filiform texture Threadlike crystals of one mineral embedded in another mineral; e.g., rutilated quartz.

filigree (a) Delicate ornamental work, used chiefly in decorating gold and silver. (Crispin, 1964) (b) Naturally occurring native metals (e.g., gold, silver, or copper) in lacelike form.

fill (a) Manmade deposits of natural earth materials (e.g., rock, soil, and gravel) and waste materials (e.g., tailings or spoil from dredging), used (1) to fill an enclosed space, such as an old stope or chamber in a mine, (2) to extend shore land into a lake or bay, or (3) in building dams. See also: *hydraulic fill; backfill.* (AGI, 1987) (b) Soil or loose rock used to raise the surface of low-lying land, such as an embankment to fill a hollow or ravine in railroad construction. Also, the place filled by such an enbankment. (AGI, 1987) (c) The depth to which material is to be placed to bring the surface to a predetermined grade. (AGI, 1987) (d) Any sediment deposited by any agent so as to fill or partly fill a valley, sink, or other depression. (AGI, 1987) (e) Manmade deposits of natural soils and waste material. (ASCE, 1958) (f) Material deposited or washed into a cave passageway. Fill is generally prefixed by a word describing its dominant grain size; e.g., sand fill, silt fill, clay fill, gravel fill, etc. (AGI, 1987) (g) Material used to fill a cavity or a passage. An embankment to fill a hollow or a ravine, or the place filled by such an embankment. Also, the depth of the filling material when it is in place. As a verb, to make an embankment in or to raise the level of a low place with earth, gravel, or rock. (Webster 3rd, 1966) (h) Tailings, waste, etc., used to fill underground space left after extraction of ore. Termed "hydraulic fill" if flushed into place by water. See also: *pack.* (Pryor, 1963) (i) Detrital material partly or completely filling a cave. Syn: *drift.* (AGI, 1987) (j) The unit charge of batch into a tank or pot. (ASTM, 1994) (k) Eng. To load trams in the mine. (Fay, 1920) (l) An earth or broken rock structure or embankment. (Nichols, 1976) (m) Soil that has no value except bulk. (Nichols, 1976)

filled valley A wide-basin valley, in an arid or semiarid region, that contains abundant alluvium in the form of fans, flood plains, and lake deposits. (AGI, 1987)

filler *mineral filler.*

filler and drayer A worker who fills tubs at the coal face and pushes them to the main haulage road. (CTD, 1958)

fill factor The approximate load the dipper is carrying, expressed as a percentage of the rated capacity. The fill factor is commonly called the dipper factor for shovels or the bucket factor for draglines. (Woodruff, 1966)

filling (a) The waste material used to fill up old stopes or chambers. (Weed, 1922) (b) The loading of trams, conveyors, tubs, or trucks with coal, ore, or waste; the place where loading occurs. (CTD, 1958) (c) Allowing a mine to fill with water. (Weed, 1922) (d) Clogging of the abrasive coat by swarf. It may be reduced in many operations by using an opencoat construction or a lubricant. See also: *swarf.* (ACSG, 1963) (e) Loading of mineral into mine trucks; shoveling onto conveyors; gob stowing; packing old stopes with waste. (Pryor, 1963) (f) Material such as waste, sand, ashes, and other refuse used to fill in worked-out areas of excavation. (Stoces, 1954) (g) Backfill.

film (a) A term used in flotation meaning a coating, layer, or thin membrane. (b) A thin layer of a substance, at the most a few molecules thick, generally differing in properties from other layers in contact with it. (CTD, 1958)

film coefficient The heat transferred by convection per unit area per degree temperature difference between the surface and the fluid. Also called unit convection conductance; surface coefficient. (Strock, 1948)

film flotation Early stage in development of modern flotation process for concentration of minerals, notably sulfides. The containing pulp was agitated with oil that then floated up, carrying selected minerals. This mineralized film was then overflowed or skimmed off. See also: *froth flotation; flotation.* (Pryor, 1963)

film mica Knife-trimmed mica split from the better qualities of block mica to any specified range of thicknesses between 0.0012 in and 0.004 in (30.5 µm and 102 µm). (Skow, 1962)

film-sizing tables Tables used in ore dressing for sorting fine material by means of a film of flowing water. These tables may be considered as surface tables, from which the products are removed before they have found a bed, so that the washing is always done on the same surface; also building tables or buddles, on which the products are removed after they have formed a bed. These use the relative transporting power of a film of water flowing on a quiet surface, which may be either rough or smooth, to act upon the particles of a water-sorted product. The smaller grains, of high specific gravity, are moved down the slope slowly or not at all by the slow undercurrent; the larger grains, of lower specific gravity, are moved rapidly down the slope by the quick upper current. (Liddell, 1918)

filter (a) A device for separating suspended solid particles from liquids, or fine dust from air. It incorporates a membrane on which the solids are retained. *bag filter; drum filter; Genter filter; plate-and-frame filter.* See also: *vacuum filtration.* (b) To subject to the action of a filter; to pass a liquid or a gas through a filter for the purpose of purifying, or separating, or both. To act as a filter; to remove from a fluid by means of a filter; to percolate. (Webster 3rd, 1966) (c) An electronic device in seismic instruments that permits selection of frequency characteristics appropriate for the ground motion it is desired to record. (AGI, 1987) (d) A layer, or a combination of layers, of pervious materials designed and installed in such a manner as to provide drainage and yet prevent the movement of soil particles by flowing water. Also called protective filter. (ASCE, 1958) (e) To utilize a filter, as to clarify or purify a liquid or gas.

filter aid (a) A low-density, inert, fibrous, or fine granular material used to increase the rate and improve the quality of filtration. (ASM, 1961) (b) Diatomaceous earth, used either to coat a filter cloth or as a thick filtering layer that can be plowed off with its load of cake from a rotating drum filter. (Pryor, 1963)

filter bed (a) A general name for a pond or tank with a bottom or bed used for filtering purposes. (CTD, 1958) (b) A pond or tank having a false bottom covered with sand, and serving to filter river or pond water. (c) A fill of previous soil that provides a site for a septic field. (Nichols, 1976)

filter cake (a) The compacted solid or semisolid material separated from a liquid and remaining on a filter after pressure filtration. (Inst. Petrol., 1961) (b) The layer of concentrated solids from the drilling mud left behind on the walls of a borehole, or on a filter paper in filtration tests on mud. (Inst. Petrol., 1961)

filter-cake texture The physical properties of a cake as measured by toughness, slickness, and brittleness. (Brantly, 1961)

filter cloth The fabric used as a medium for filtration; e.g., nylon cloth, blanket cloth, finely woven wire mesh, or finely woven glass thread. Syn: *press cloth.* (BS, 1962)

filtered light Light that has passed through a colored-glass filter, absorbing some hues and permitting others to pass through.

filter feed trough A tank containing the pulp to be filtered, generally fitted with an agitator to maintain the solids in the pulp in suspension, and in which the drum or disk of a rotary vacuum filter is partially immersed. (BS, 1962)

filtering stone Any porous stone, such as sandstone, through which water is filtered.

filter loss The amount of fluid delivered through a permeable membrane in a specified time. (Brantly, 1961)

filter press A form of pressure filter, noncontinuous in operation; used in coal preparation for the removal of water from slurries, tailings, and similar products. (BS, 1962)

filter pressing A process of magmatic differentiation wherein a magma, having crystallized to a mush of interlocking crystals in liquid, is compressed by Earth movements and the liquid moves toward regions of lower pressure, thus becoming separated from the crystals. Syn: *filtration differentiation.* (AGI, 1987)

filter pump An aspirator for hastening the process of filtering by creating a partial vacuum. (Standard, 1964)

filter stick Short glass tube with filtering septum; used in laboratory sampling. (Pryor, 1963)

filter-type respirator A protective device that removes dispersoids from the air by physically trapping the particles on the fibrous material of the filter. It offers no protection against gases or vapors, or atmospheres deficient in oxygen. Many workers, however, are subjected to dusts, fumes, and mists in sufficient quantity to impair their health. Common examples are the dusts of cement, coal, flour, limestone, silica, and asbestos encountered in mining, grinding, and crushing operations; the metallic fumes of welding, smelting, and refining processes; and the mists formed by the disintegration of a liquid in such work as spray-coating, atomizing, and chromium-plating. (Best, 1966)

filtrate The liquid product from the filtration process. (BS, 1962)

filtration Removal of suspended and/or colloidal material from a liquid by passing the suspension through a relatively fine porous medium, e.g., a canvas or other fabric diaphragm; the process is activated by suction or pressure, and commonly includes filter aids. The products are clear liquid and a filter cake. (AGI, 1987)

filtration differentiation *filter pressing.*

filtration rate The measure of the amount of filtrate passing through or into a porous medium. Filter loss and cake thickness constitute the determining factors of filtration qualities.

filty Som. A local term for combustible gases. (Fay, 1920)

final controlling element In process control, the element that directly changes the value of the manipulated variable. (Fuerstenau, 1962)

final exploration The detailed investigation of a coal or mineral area on which a preliminary report was favorable. The final exploration of an area may involve a costly boring program, surveys, and sampling. See also: *preliminary exploration.* (Nelson, 1965)

fine (a) Composed of or constituting relatively small particles; e.g., fine sandy loam. Ant. coarse. (AGI, 1987) (b) Sometimes used to designate high-quality drill diamonds. (Long, 1960) (c) A measure of the amount of gold in electrum.

fine aggregate The portion of an aggregate consisting of particles with diameters smaller than approx. 1/4 in (4.8 mm). Cf: *aggregate; coarse aggregate.* (AGI, 1987)

fine gold (a) Almost pure gold. The value of bullion gold depends on its percentage of fineness. See also: *fineness; float gold.* (b) In placer mining, gold in exceedingly small particles. Syn: *greasy gold.* (Hess)

fine-grained (a) Said of a crystalline rock, and of its texture, in which the individual minerals are relatively small; specif. said of an igneous rock whose particles have an average diameter less than 1 mm. Syn: *aphanitic.* (AGI, 1987) (b) Said of a sediment or sedimentary rock, and of its texture, in which the individual constituents are too small to distinguish with the unaided eye; specif. said of a sediment or rock whose particles have an average diameter less than 1/16 mm (62 µm, or silt size and smaller). The term is used in a relative sense, and various size limits have been used. Cf: *coarse-grained; medium-grained.* (AGI, 1987) (c) Said of a soil in which silt and/or clay predominate. (AGI, 1987)

fine-grained rocks Rocks in which the crystals are very fine-grained, or else the whole or part is glass. These are the volcanic rocks. (Mason, 1951)

fine grinder A machine for the final stage of size reduction; i.e., to -100 mesh. *fine grinding; pulverizer.*

fine grinding Fine grinding is usually performed in a mill rotating on a horizontal axis and containing balls, rods, or pebbles (grinding media), which serve to grind the ore in the mill. The different mills used in fine grinding are known as ball mill, pebble mill, Hardinge mill, tube mill, autogenous, etc. (Newton, 1959)

fine industrials *toolstone.*

fine metal The higher grades of copper regulus or matte obtained in the English process of copper smelting. Included are the following four varieties, which are distinguished by appearance and copper content: (1) blue, containing about 60% copper; (2) sparkle, about 74% copper; (3) white, about 77% copper; and (4) pimple, about 79% copper. (Fay, 1920)

fineness (a) The proportion of pure silver or gold in jewelry, bullion, or coins; often expressed in parts per thousand; American silver coin is nine-tenths or 0.900 fine; English gold coin is eleven-twelfths or 0.9166 fine. (Webster 3rd, 1966) (b) A measure of specific surface area or particle-size distribution. (Taylor, 1965) (c) The state of subdivision of a substance. (CTD, 1958)

fineness factor A measure of the average particle size of clay and ceramic material, computed by summing the products of the reciprocal of the size-grade midpoints and the weight percentage of material in each class (expressed as a decimal part of the total frequency). The measure is based on the assumption that the surface areas of two powders are inversely proportional to their average particle sizes. Syn: *surface factor.* (AGI, 1987)

fineness modulus (a) A means of evaluating sand and gravel deposits, consisting of passing samples through standardized sets of sieves, accumulating percentages retained, dividing by 100, and comparing the resultant fineness-modulus number to various specification requirements. (AGI, 1987) (b) An empirical factor obtained by adding the total percentages of a sample of the aggregate retained on each of a specified series of sieves, and dividing the sum by 100. (AIME, 1960) (c) One-hundredth of the sum of the cumulative values for the amount of material retained on the series of Tyler or U.S. sieves including half sizes up to 100 mesh. (Dodd, 1964)

fines (a) Finely crushed or powdered material, e.g., of coal, crushed rock, or ore, as contrasted with the coarser fragments; esp. material smaller than the minimum specified size or grade, such as coal with a maximum particle size less than 3.2 mm, or ores too pulverulent to be smelted in the ordinary way; or material passing through a given screen or sieve. See also: *anthracite fines.* (AGI, 1987) (b) An engineering term for the clay- and silt-sized soil particles (diameters less than 0.074 mm) passing U.S. standard sieve No. 200. (AGI, 1987)

fine sand (a) A geologic term for a sand particle having a diameter in the range of 0.125 to 0.25 mm (125 to 250 µm, or 3 to 2 phi units). Also, a loose aggregate of sand consisting of fine sand particles. See also: *sand.* (AGI, 1987) (b) An engineering term for a sand particle having a diameter in the range of 0.074 mm (retained on U.S. standard sieve No. 200) to 0.42 mm (passing U.S. standard sieve No. 40). (AGI, 1987) (c) A soil term used in the United States for a sand particle having a diameter in the range of 0.10 to 0.25 mm. (AGI, 1987)

fine silt A geologic term for a silt particle having a diameter in the range of 1/128 to 1/64 mm (8 to 16 µm, or 7 to 6 phi units). In Great Britain, the range 1/100 to 1/20 mm has been used. Also, a loose aggregate of silt consisting of fine silt particles. (AGI, 1987)

fine silver Pure silver, 1,000 parts fine or 100% silver.

finger (a) One of the cutting edges on a finger bit. See also: *finger bit.* (Long, 1960) (b) A pair or set of bracketlike projections placed at a strategic point in a drill tripod or derrick, generally at a level with one of the work platforms, to keep a number of lengths of drill rods or casing in place when they are standing in the tripod or derrick. Also one of the flexible prong parts of a basket lifter. (Long, 1960) (c) A minor structure radiating from a major structure. (AGI, 1987)

finger bar Pivoted length of wood used to support a unit in a stamp battery. See also: *cam stick.* (Pryor, 1963)

finger bit A steel rock-cutting bit having fingerlike, fixed or replaceable, steel-cutting points affixed. (Long, 1960)

finger board A board with projecting dowels or pipe fingers located in the upper part of the drill derrick or tripod to support stands of drill rod, drill pipe, or casing. Cf: *finger.* (Long, 1960)

finger chute Steel rails hinged independently over an ore chute, to control rate of flow of rock. (Pryor, 1963)

finger grip A finishing tool designed to recover a broken drill rod or dropped tool from a borehole. (Long, 1960)

finger lifter A basket-type core lifter. (Long, 1960)

finger raise Steeply sloping openings permitting caved ore to flow down raises through grizzlies to chutes on the haulage level.

finished steel Steel that has been processed beyond the stages of billets, blooms, sheet bars, slabs, and wire rods, and is ready for the market. (ASM, 1961)

finish grade The final grade required by specifications. (Nichols, 1976)

finishing jig The jig used to save the smaller particles of ore in a concentrator. (Weed, 1922)

finishing lime A type of refined hydrated lime, which has been milled to be suitable for plastering, particularly a finish coat. (Boynton, 1966)

finishing rolls The last roll, or the one that does the finest crushing in ore dressing, esp. in stage crushing. (Fay, 1920)

fink truss A frequently used symmetrical steel roof truss that is effective over a maximum span of 50 ft (15.2 m). (Hammond, 1965)

finnemanite A hexagonal mineral, $Pb_5(AsO_3)_3Cl$; occurs as prisms forming crusts lining crevices in granular hematite at Långban, Sweden.

fior di persicor A white marble with veins and clouds of purple or red, from Albania. (Fay, 1920)

fiorite Siliceous sinter, named from Mount Santa Fiora, Tuscany, Italy. Syn: *pearl sinter.*

fire (a) To explode or blow up. The expression "the pit has fired" signifies that an explosion of combustible gases has taken place. (Fay, 1920) (b) To blast or explode with ammonium nitrate-fuel oil (ANFO), dynamite, or other explosive. (c) A word shouted by miners as warning just before a shot is fired. (Fay, 1920) (d) Fuel in a state of combustion, as on a hearth, in a grate, or furnace. A manifestation of rapid combustion or combination of materials with oxygen. (e) Flashes of different spectral colors seen in diamonds and other gemstones with high birefringence as a result of dispersion. Cf: *play of color.*

fire agate *goldstone.*

fire assay The assaying of metallic ores, usually gold and silver, by methods requiring a furnace heat; commonly involves the processes of scorification, cupellation, etc. (Standard, 1964)

fireback The back wall of a furnace or fireplace. (Fay, 1920)

fire bank The spoil heap at the surface of a colliery, when burning or heated by spontaneous combustion. (Nelson, 1965)

fireblende *pyrostilpnite.*

fire boss (a) A person designated to examine the mine for gas and other dangers. In certain states, the fire boss is designated as the mine examiner. See also: *gasman.* (Federal Mine Safety, 1977) (b) A State-certified supervisory mine official who examines the mine for combustible gases and other dangers before a shift comes into it and who usually makes a second examination during the shift; in some States, it is used loosely to designate assistant or section foreman. See also: *fireman.* Also called examiner; mine examiner. See also: *gas watchman.* (BCI, 1947)

firebox One of the small refractory-lined chambers, built wholly or partly in the wall of a kiln, for combustion of the fuel. (Dodd, 1964)

firebreak A strip across the area in which no combustible material is employed, or in which, if timber supports are used, sand (not waste rock) is later filled and packed tightly around them. Where timber is not used in stope supports, the firebreaks are simply stretches in the levels or winzes in which timber lagging is replaced by some other substance, such as steel or concrete. (Spalding, 1949)

fire breeding S. Staff. Said of any place underground showing indications of a gob fire. (Fay, 1920)

firebrick (a) Bricks made from a very refractory clay to withstand intense heat. (Mersereau, 1947) (b) An aluminosilicate brick of fireclay composition. (Van Vlack, 1964)

fire bridge The separating low wall between the fireplace and the hearth of a reverberatory furnace. (Fay, 1920)

fire classification The following explains the National Fire Protection Association classifications. Class A fires are defined as those in ordinary solid, combustible materials, such as coal, wood, rubber, textiles, paper, and rubbish. Class B fires are defined as those in flammable liquids, such as fuel or lubricating oils, grease, paint, varnish, and lacquer. Class C fires are defined as those in (live) electric equipment, such as oil-filled transformers, generators, motors, switch panels, circuit breakers, insulated electrical conductors, and other electrical devices. (USBM, 1963)

fireclay (a) A siliceous clay rich in hydrous aluminum silicates, capable of withstanding high temperatures without deforming, disintegrating, or becoming soft and pasty. It is deficient in iron, calcium, and alkalies, and approaches kaolin in composition, the better grades containing at least 35% alumina when fired. (AGI, 1987) (b) A term formerly, but inaccurately, used for underclay. Although many fireclays commonly occur as underclays, not all fireclays carry a roof of coal and not all underclays are refractory. Syn: *firestone; refractory clay.* See also: *sagger; underclay.* (AGI, 1987)

fired Eng. Said of a mine when an explosion of combustible gases has taken place.

firedamp (a) A combustible gas that is formed in mines by decomposition of coal or other carbonaceous matter and that consists chiefly of methane; also the explosive mixture formed by this gas with air. The term "combustible gases" is now used for firedamp. (Webster 3rd, 1966) (b) A stone, brick, or concrete airtight stopping to isolate an underground fire, and to prevent the inflow of fresh air and the outflow of foul air. See also: *seal; methane.* (Nelson, 1965)

fire door (a) A fireproof door in a building or in a mine, as a door to enclose an area in which there is a mine fire. (Fay, 1920) (b) The door or opening through which fuel is supplied to a furnace or stove. (Fay, 1920)

fired stone Certain gems, e.g., zircon, topaz, or corundum, can be heated under controlled conditions to change their color to one that is more attractive. See also: *heated stone.*

fire feeder (a) An apparatus for feeding the fire of a furnace. (Fay, 1920) (b) A stoker.

firefighting plan A plan or chart showing the positions of items of firefighting equipment. Separate plans are used for surface buildings and underground workings. (BS, 1963)

fire grate The grate that holds the fuel in many forms of heaters and furnaces. (Fay, 1920)

fire-heavy Eng. Words marked upon the scale of a mercurial barometer to indicate when considerable combustible gases may be expected to be given off in the mine, and to show that extra vigilance is required to keep the ventilation up to its full strength.

fire kiln An oven or place for heating anything. (Fay, 1920)

fireman (a) In a metal mine, a miner whose duty it is to explode the charges of explosive used in headings and working places. (CTD, 1958) (b) In a fiery mine, the official who checks the underground explosive risk. (Pryor, 1963) (c) In a coal mine, an official responsible for safety conditions underground. See also: *deputy.* (CTD, 1958) (d) Eng. A person whose duty it is to examine with a safety lamp the underground workings, (1) to ascertain if gas is present, (2) to see that doors, brattic-ing, stoppings, etc., are in good order, and (3) generally to see that the ventilation is efficient. See also: *fire boss; gas watchman.*

fire marble *lumachelle.*

fire opal A transparent to translucent yellow, orange, red, or brown variety of opal that gives out fiery reflections in bright light with or without a play of color. Cf: *gold opal.* Syn: *sun opal; pyrophane.*

fire-refined copper Copper that has been refined by the use of a furnace process only, including refinery shapes and, by extension, fabricator's products made therefrom. Usually when this term is used alone, it refers to fire-refined, tough pitch copper without elements other than oxygen being present in significant amounts. See also: *fire refining.* (ASM, 1961)

fire refining Includes a number of processes used for the removal of impurities from impure metals produced by the smelting process. Impurities are removed by introducing air into the molten metal or exposing the metal to air, and by the addition of various fluxes and the removal of impurities as gases, drosses, or liquid slags. Lead, tin, and some types of impure copper are also fire-refined. (Kirk, 1947)

fire rib S. Staff. A solid rib or wall of coal left between workings to confine gob fires. (Fay, 1920)

fire runner In bituminous coal mining, a person who enters a mine immediately after blasting to search for any fires that might have been started by a blast. Also called shotfirer runner; shot runner. (DOT, 1949)

fire sand (a) Refractory oxide or carbide used for furnace linings. (Bennett, 1962) (b) A sand so free from fluxes that it is highly refractory. See also: *foundry sand.* (Freeman, 1936)

fire seal A strip across an area through which neither fire nor noxious gases can penetrate. It involves not only sealing of stopes but levels also. Syn: *sealing.* (Spalding, 1949)

fire setting An ancient method of tunneling through rock. A fire was built against the face of the mineral, which was then quenched with water, thus causing cracking. (Pryor, 1963)

fire stink The smell given off when heating or spontaneous combustion occurs in the waste or elsewhere underground. (Nelson, 1965)

firestone (a) Any fine-grained siliceous stone formerly used for striking fire; specif. flint. Syn: *feverstein.* (AGI, 1987) (b) A nodule of pyrite formerly used for striking fire. (AGI, 1987) (c) A fine-grained siliceous rock that can resist or endure high heat and is used for lining furnaces and kilns, such as certain Cretaceous and Jurassic sandstones in southern England. See also: *fireclay.* (AGI, 1987) (d) In a slag hearth, a plate of iron covering the front of the furnace except for a few inches of space between it and the bedplate. (e) Quartz in which cracks have been artificially produced by heating to create diffraction colors. Syn: *iris quartz.*

fire styth *fire stink.*

fire up A command to start operating a drill either to collar a borehole or to restart work on the first working shift of a day. (Long, 1960)

firing (a) The igniting of an explosive charge. (b) Starting up a furnace or kiln. (c) The process of initiating the action of an explosive charge or the operation of a mechanism that results in a blasting action.

firing a mine Eng. Maliciously setting fire to a coal mine. (Fay, 1920)

firing cable *shot-firing cable.*

firing circuit *shot-firing circuit.*

firing current An electric current of recommended magnitude and duration to sufficiently energize an electric detonator or a circuit of electric detonators. (Atlas, 1987)

firing expansion The increase in size that sometimes occurs when a refractory raw material or product is fired; it is usually expressed as a linear percentage expansion from the dry to the fired state. Firing expansion can be caused by a crystalline conversion (e.g., of quartz into cristobalite, or of kyanite into mullite plus cristobalite), or by bloating.

firing impulse As applied to electric blasting caps, the minimum impulse of current required to fire a detonator. (Fraenkel, 1953)

firing key A special key that fits the exploder used in electric firing of blasting charges; carried by authorized shot firer. (Pryor, 1963)

firing line Scot. An appliance used in former times for clearing a room of combustible gases. A prop was being set up near the face, a ring was fixed in it near the roof, and a cord or wire was passed through the ring. Attaching a lamp to one end of the cord, the miner withdrew to a distance, and by pulling the cord raised the lamp to the height necessary to explode the accumulated combustible gases. (Fay, 1920)

firing machine (a) A designation for an electric blasting machine. (Fay, 1920) (b) An apparatus for feeding a boiler furnace with coal. A mechanical stoker. (Fay, 1920)

firing point Eng. The point or mixture at which combustible gases mixed with atmospheric air explodes. The percentages of gas vary from 6% to 13%, with the maximum explosibility at about 11%. (Fay, 1920)

firmament stone *precious opal.*

first aid Emergency, crude repair of a bit made by a drill runner at the drill site. (Long, 1960)

first arrival The first energy to arrive from a seismic source. First arrivals on reflection records are used for information about a surficial low-velocity or weathering layer; refraction studies are often based on first arrivals. Syn: *initial impulse; first break.* (AGI, 1987)

first break *weight break; first arrival.*

first bye (a) A diamond with a faint greenish tint. (Schaller, 1917) (b) A classification of gem diamond.

first-class lever A bar having a fulcrum (pivot point) between the points where force is applied and where it is exerted. (Nichols, 1976)

first-class ore An ore of sufficiently high grade to be acceptable for shipment to market without preliminary treatment. Cf: *second-class ore.* Syn: *shipping ore.* (AGI, 1987)

first mining In the room-and-pillar method, the part of the coal that is won from the rooms, as distinguished from the second part, which is the extraction of the remaining pillars. (Stoces, 1954)

first-order red plate *selenite plate.*

first red plate *gypsum plate.*

first ripping The ripping work carried out as the roadway is being formed and driven forward. See also: *second ripping.* (Nelson, 1965)

first water Gems, particularly diamonds, of the highest value, irrespective of size. In diamonds, the term applies to stones that are flawless, without color or almost bluish white. A slight amount of color detracts from the value, and the stones are said to be off color.

first way Rift; reed; cleavage way. See also: *easy way.* (Arkell, 1953)

first weight The first indication of roof pressure that takes place after the removal of coal from a seam. (CTD, 1958)

first working The removal of the coal in driving the entries and rooms. See also: *advance working.* Cf: *second working.* (Kentucky, 1952)

firth A long, narrow arm of the sea; also, the opening of a river into the sea. Along the Scottish coast, it is usually the lower part of an estuary (e.g., Firth of Forth), but sometimes it is a fjord (e.g., Firth of Lorne) or a strait (e.g., Pentland Firth). Etymol. Scottish. Syn: *frith.* (AGI, 1987)

fir-tree bit A rotary bit in which a number of cutting edges are arranged behind a pilot bit to enlarge the hole to the required diameter. (BS, 1964)

Fischer-Tropsch process Hydrogenation of carbon monoxide to form hydrocarbons from coal or natural gases. (Pryor, 1963)

fish (a) To join two beams, rails, etc., by long pieces at their sides. (Zern, 1928) (b) The article recovered and/or the act or processes involved in the recovery of lost drilling tools, casing, or other articles from a borehole. Also called fishing. (Long, 1960)

fished joint A rail joint made by means of fishplates. (Hammond, 1965)

Fisher subsieve sizer An apparatus using a gas-permeability method for determination of the average particle diameter of powders. A sample, equal in weight (grams) to the true density of the material, is compacted between two porous plugs in a metal tube, to a known porosity. Air or a suitable gas, under a constant pressure head, is passed through the compressed sample, and rate of flow is measured by a calibrated flowmeter. The average particle diameter of the powder is indicated directly on a self-calculating chart by the liquid height in one arm of the flowmeter tube. No dispersion is required, and the results are unaffected by particle shape. (Osborne, 1956)

fisheye (a) A little-used name for moonstone, also for opal with a girasol effect. (b) A popular trade term for any transparent faceted stone so cut that its center is lacking in brilliancy. (c) A diamond cut too thin to present the maximum effect of brilliancy.

fisheye stone *apophyllite.*

fishing Searching for and attempting to recover, by the use of specially prepared tools, a piece or pieces of drilling equipment (such as sections of pipe, cables, or casting) that have become detached, broken, or lost from the drill string or have been accidentally dropped into the hole. (AGI, 1987)

fishing tap A thread-cutting tool to cut threads inside a casing or other hollow part that is to be fished from a borehole. (Long, 1960)

fishplates Specially shaped steel plates for joining the end of one rail to the next rail in the track. The fishplates are fixed (one on each side) to overlap the rail ends and bolted through the rails. (Nelson, 1965)

fishtail (a) An abrupt and ragged termination of a coalbed that is considered to have resulted from a washout during the peat stage. The more or less leathery peat is believed to have been separated parallel to its bedding, permitting wedges of sand and silt to be forced into the separations in such a manner that, after the coalification has taken place, a cross section shows splayed and ragged coal separated by sandstone wedges. (Raistrick, 1939) (b) The act or process of rotatively drilling a borehole with a fishtail bit. Also called fishtailing. (Long, 1960) (c) In roll forging, the excess trailing end of a forging. It is often used, before being trimmed off, as a tong hold for a subsequent forging operation. (ASM, 1961)

fishtail bit A rotary bit used to drill soft formations. The blade is flattened and divided, the divided ends curving away from the direction of rotation. It resembles a fishtail. Also called drag bit. Cf: *noncoring bit.* (AGI, 1987)

fishtail structure A coal seam structure sometimes observed along the fringe of a washout. It was probably produced by the water forcing open layers of the coaly mass and the injection of fine sand or silt into the splayed partings—the veins of coal branching out like a fishtail. (Nelson, 1965)

fissile (a) Capable of being easily split along closely spaced planes; exhibiting fissility. (AGI, 1987) (b) Said of bedding that consists of laminae less than 2 mm thick. (AGI, 1987)

fissility A general term for the property possessed by some rocks of splitting easily into thin layers along closely spaced, roughly planar, and approx. parallel surfaces, such as bedding planes in shale or cleavage planes in schist; its presence distinguishes shale from mudstone. The term includes such phenomena as "bedding fissility" and "fracture cleavage." Etymol. Latin fissilis, that which can be cleft or split. Adj. fissile. (AGI, 1987)

fission The spontaneous or induced splitting, by particle collision, of a heavy nucleus into a pair (only rarely more) of nearly equal fission fragments plus some neutrons. Fission is accompanied by the release of a large amount of energy. Cf: *fusion.* (AGI, 1987)

fissionable Said of nuclei, such as uranium and plutonium, that are capable of fission. (AGI, 1987)

fissure A fracture or crack in rock along which there is a distinct separation. It is often filled with mineral-bearing material. See also: *fissure vein.* (AGI, 1987)

fissure system A group of fissures of the same age and of more or less parallel strike and dip. (AGI, 1987)

fissure vein A mineral mass, tabular in form as a whole but frequently irregular in detail, occupying a fracture in rock. The vein material is different from the country rock and has generally been produced by the filling of open spaces along the fissure. See also: *true vein.*

fitchered Said of a drill hold sufficiently crooked to make a drill stick.

fitter Broadly, a skilled person who can repair and assemble machines in an engineering shop. (Nelson, 1965)

fitting Hand or bench work involved in the assembly of finished parts by a fitter. (CTD, 1958)

fittings Auxiliary and accessory tools and equipment needed to drill a borehole using either percussive, churn, rotary, diamond, or other types of drills. (Long, 1960)

fiveling (a) A twinned crystal formed by fivefold cyclic twinning. (AGI, 1987) (b) A crystal twin consisting of five individuals. Cf: *trilling; fourling; eightling.*

fix (a) A position determined from terrestrial, electronic, or astronomical data. (AGI, 1987) (b) The act of determining a fix. (AGI, 1987) (c) To fettle or line the hearth of puddling furnace with a fix or fettling, consisting of ores, scrap, and cinder, or other suitable substances. (Fay, 1920)

fixation (a) The act or process by which a fluid or a gas becomes or is rendered firm or stable in consistency, and evaporation or volatilization is prevented. Specif., that process by which a gaseous body becomes fixed or solid on uniting with a solid body, as the fixation of oxygen or the fixation of nitrogen. (b) A state of nonvolatility or the process of entering such a state, as the fixation of a metal or the fixation of nitrogen in a nitrate by bacteria. (Standard, 1964)

fix-bitumens A group name for all authigenic, nonfluid bitumens; divided into stabile protobitumens and metabitumens. (Tomkeieff, 1954)

fixed carbon (a) In the case of coal, coke, and other bituminous materials, the solid residue, other than ash, obtained by destructive distillation; determined by definite prescribed methods. (ASTM, 1994) (b) A calculated figure: 100, less the sum of the percentages of moisture, volatile matter, and ash. (BS, 1960)

fixed-clip monocable An aerial ropeway in which a moving endless rope both supports and transports carriers that are permanently fixed to it. The length of the line may be several miles. Individual loads are limited to about 2 hundredweight (91 kg), and total capacity seldom exceeds about 15 st/h (13.6 t/h). (Nelson, 1965)

fixed-electrode method A geophysical surveying method used in the self-potential system of prospecting, in which one electrode remains stationary while the other is grounded at progressively greater distances from it. The method indicates a mineral body directly beneath the greatest anomaly and has been extensively and successfully used.

fixed-flexible-type carrying idler Consists of a flexible-type carrying idler mounted in a rigid frame, which fixes the position of the points or roll support. (NEMA, 1961)

fixed ground water Ground water in material having interstices so small that the water is held permanently to the walls of the interstices, or moves so slowly that it is not available for withdrawal at useful rates. Outside the zone of saturation, material with infinitely small openings can hold water indefinitely against the pull of gravity, whereas within the zone of saturation there is apparently always movement, even though at very low rates. (AGI, 1987)

fixed guides Wood bars or steel rails fixed vertically to cross buntons in a shaft. The cage shoes travel along the guides and therefore prevent the cage from swinging and doing damage in the shaft. Some skips are fitted with rubber-tired rollers running on 6-in by 4-in (15-cm by 10-cm) steel channel guides. Guide shoes may be fitted to act as alternative guides in case of breakdown of rollers. Fixed guides are used when shaft space is limited; i.e., when the clearances do not permit the use of flexible or rope guides. See also: *steel guides.* (Nelson, 1965)

fixed screen A stationary inclined or curved panel, commonly made of wedgewire, which is used to remove fines and a large proportion of water from a suspension of coal in water. (BS, 1962)

fizelyite A monoclinic mineral, $Pb_{14}Ag_5Sb_{21}S_{48}$(?); soft; forms deeply striated prisms at Kisbánya, Romania, where it is associated with semseyite, pyrite, galena, and sphalerite.

flag (a) Sandstone or sandy limestone rock, usually more or less micaceous, which is fissile along the bedding planes, splitting into slabs. Sometimes misnamed "slate" because it is used for roofing rather than paving. (Arkell, 1953) (b) A thin slab of stone. Syn: *flagstone.* (Fay, 1920) (c) A track signal or target. (Zern, 1928)

flagger In bituminous coal mining, a laborer who attaches a flag to the rear car of a loaded train of cars (if the flag is missing at end of a haulage trip, it denotes that the train has lost one or more cars, and all motormen are warned). Also called flagman. (DOT, 1949)

flagging In geophysical work, the use by surveyors of flags of cloth, paper, or plastic to mark instrument or shot locations.

flaggy (a) Splitting or tending to split into layers of suitable thickness for use as flagstones; specif. descriptive of a sedimentary rock that splits into layers from 1 to 5 cm thick (McKee &Weir, 1953). (AGI, 1987) (b) Said of bedding that consists of layers from 1 to 10 cm thick (Payne, 1942). (AGI, 1987) (c) Pertaining to a flag or flagstone. (AGI, 1987) (d) Said of a soil full of flagstone fragments. (AGI, 1987)

flagstaffite An orthorhombic mineral, $C_{10}H_{22}O_3$, *cis*-terpin hydrate; forms colorless transparent crystals in fossil pine logs near the San Francisco Peaks, north of Flagstaff, AZ.

flagstone (a) A hard sandstone, usually micaceous and fine-grained, that occurs in extensive thin beds with shale partings; it splits uniformly along bedding planes into thin slabs suitable for use in terrace floors, retaining walls, etc. Cf: *bluestone; freestone.* (AGI, 1987) (b) A flat slab of flagstone used for paving; esp. a thin piece split from flagstone. Also, a surface of such stone. (AGI, 1987) (c) A relatively thin flat fragment (of limestone, sandstone, shale, slate, or schist) occurring in the soil, having a length in the range of 6 to 15 in (15 to 38 cm). Syn: *flag; slabstone; grayband.* (AGI, 1987)

flail A hammer hinged to an axle so that it can be used to break or crush material. (Nichols, 1976)

flake copper Very thin scales of native copper. (Weed, 1922)

flake graphite Graphite disseminated in metamorphic rock as thin, visible flakes that are separable from the rock by mechanical means. Syn: *crystalline flake graphite.* (AGI, 1987)

flake mica Finely divided mica recovered from mica and sericite schist and as a byproduct of feldspar and kaolin beneficiation. See also: *scrap mica.* (Skow, 1962)

flake sulfur Pyrite occurring as thin flakes on the natural cleavage surfaces of coal that floats readily on the surface of the wash water in the washing process. Syn: *float sulfur.* (Mitchell, 1950)

flake white A name sometimes given to pure white lead. (Fay, 1920)

flamboyant structure The optical continuity of crystals or grains, as disturbed by a divergent structure caused by slight differences in orientation. (AGI, 1987)
flame A burning mixture of a combustible gas (or vapor) and air. Solid fuels burn with a glow, but with little flame. Flames are normally hot, but under some conditions are relatively cool. Principal types of flame are luminous, nonluminous, long (lazy) flames, and short flames. (Francis, 1965)
flame-coloration tests In mineral identification, qualitative tests made by moistening powdered material with hydrochloric acid, placing a few grains on platinum or nickel-chrome wire, and noting any color imparted to a blue Bunsen flame. Sodium gives a strong yellow flame; calcium, light red; strontium, crimson; barium, green; potassium, lilac; and copper, blue-green. (Pryor, 1963)
flame drill method *jet piercing.*
flame emission spectrometry *flame photometry.*
flame hardening A method for local hardening in which the steel is heated by a mechanically operated oxyacetylene blowpipe, which traverses the object to be hardened at a predetermined rate. (Hammond, 1965)
flame inhibitor A substance, such as hexachloroethane, used for coating limestone dust for use in stone-dust barriers. The inhibitor is dissolved in the waterproofing agent. Tests have indicated its effectiveness in preventing or reducing the propagation of coal-dust explosions. See also: *stone-dust barrier.* (Nelson, 1965)
flame kiln A lime kiln burning wood. (Standard, 1964)
flame opal (a) Opal in which red play of color occurs in more or less irregular streaks. (b) A flash opal with red as the predominant color.
flame photometry Measurement of the intensity of the lines in a flame spectrum by a flame photometer. Syn: *flame emission spectrometry.* (AGI, 1987)
flameproof A term descriptive of electrical machines, switches, and fittings demanded legally for use in fiery mines in Great Britain. Enclosing boxes with accurately fitted wide flanges are used. See also: *explosion proof.* (Pryor, 1963)
flameproof construction A flameproof enclosure for electrical apparatus is one that, under normal working conditions, will withstand the internal explosion of a flammable gas that may exist within it, and which will prevent the transmission of a flame capable of igniting an explosive atmosphere outside the equipment. (Roberts, 1958)
flameproof enclosure An enclosure for electrical apparatus that will withstand, without injury, any explosion of the prescribed flammable gas that may occur within it under practical conditions of operation within the rating of the apparatus (and recognized overloads, if any, associated therewith), and will prevent the transmission of flame such as will ignite the prescribed flammable gas that may be present in the surrounding atmosphere. (BS, 1965)
flame recorder *photographic-paper recorder.*
flame-resistant cable A portable cable that will meet the flame test requirements of the U.S. Mine Safety and Health Administration.
flame safety lamp A lamp, the flame of which is so protected that it will not immediately ignite combustible gases. The original flame safety lamp was developed by Sir Humphrey Davy in 1815 and there are several varieties. The flame is generally surrounded by a cylindrical covering of wire gauze. An explosive or flammable mixture of gas entering the lamp will be ignited by the flame, but the flame of combustion will not pass through the cool gauze and ignite the gas outside the lamp. The illuminating power of these lamps is slightly more than 1 cd, and they will burn for an entire shift with one filling. Each lamp is generally provided with a relighting device, and with a magnetic lock to prevent the lamp from being opened in the mine. The chief disadvantage of this lamp is its low illuminating power. See also: *Davy lamp; safety lamp; electric cap lamp.* (Fay, 1920; Lewis, 1964)
flame spectrum The spectrum of light emitted by a substance by heating it in a flame. (AGI, 1987)
flame spinel Intensely bright orange-red rubicelle (spinel).
flame spread index The numerical designation that indicates the surface flammability of materials as specified by the ASTM E-162 test method "Surface Flammability of Materials Using a Radiant Heat Energy Source." U.S. Federal regulations require a flame spread index of 25 or less for flame-retardant coatings and sealants used in underground mines.
flame test A qualitative analysis of a mineral made by intensely heating a powdered or dissolved sample in a flame and observing the flame's color, which is indicative of the element involved; e.g., green from copper. Cf: *flame photometry.*
flammable Capable of being easily ignited and of burning with extreme rapidity. This adj. is now used technically in preference to inflammable because of the possible ambiguity of the "in" prefix. Certain equipment cannot be used for safety reasons in coal mines in which flammable gases are present. (Webster 3rd, 1966)
flammable fringe In a system where air (or other reactant gas) and a flammable gas are present, that region in which the two gases have mixed to produce a gas capable of propagating flame. (BS, 1963)
flammable mixture of gases A mixture that when once ignited, will allow flame to be self-propagated throughout the mixture, independent of and away from the source of ignition. In coal mines, it is only when methane and air are mixed in certain definite proportions that the mixture is flammable and explosive, and will allow flame to spread in all directions. See also: *limits of flammability.* (Nelson, 1965)
flamper Derb. Clay ironstone in beds or seams.
flamrich screen *resonance screen.*
flange Applied to a vein widening.
flange wheel A truck or trolley wheel having a flange or flanges at the edge to keep it from leaving the rail. (Crispin, 1964)
flank A syn. for limb of a fold.
flank hole (a) A hole bored ahead of a working place, when approaching old workings. (CTD, 1958) (b) A borehole to detect water, gas, or other danger, driven from the side of an underground excavation in a line not parallel with the centerline of the excavation. Also called flank bore; flanking hole. (BS, 1963)
flanking hole A shothole drilled at an acute angle to the coal face for the purpose of trimming it. (BS, 1964)
flanking hole method Holes bored into the face at an angle which may vary from 30° to 60° to the line of face and 6 to 7 ft (1.8 to 2.1 m) long. The distance between shot holes, the angle of the hole, and the charge, depend to a great extent on the hardness of the coal. As the coal grows harder, the burden on each shothole must be reduced by placing the shotholes closer together and reducing the angle of the hole to the face. (McAdam, 1958)
flap (a) *clack.* (b) The hinged, flat disk mounted inside the lower end of a core barrel that closes and holds the sample when the barrel is withdrawn from a boring.
flapper A laborer who flattens copper starting sheets by beating them against a rigid steel or copperplate with a wooden paddle to remove folds, buckles, and creases, which tend to cause short circuits during electrolytic copper refining. Syn: *clack.* (DOT, 1949)
flapper-topped air crossing Eng. An air crossing fitted with a double door or valve giving direct communication between the two air currents when forced open by the blast of an explosion.
flapping (a) Striking through the slag-covered surface of molten copper with a rabble blade just before the bath is poled to hasten oxidation. (ASM, 1961) (b) Striking the surface of molten copper with an iron scraper or rabble to increase the surface exposed to the air. (Mersereau, 1941)
flaps Eng. Rectangular wooden valves about 24 in by 18 in by 1-1/2 in (61 cm by 46 cm by 3.8 cm) thick, hung vertically to the framework of the air chambers of a ventilator. A flap valve.
flap valve Nonreturn valve formed by a hinged flap, which rises as fluid is drawn up through a pipe or chamber and falls back on seating to prevent return flow. (Pryor, 1963)
flare-type bucket A dragline bucket with a bowl of aluminum alloy covered top and bottom with steel wearing plates. Sides and back are of steel plates, and manganese steel is used for the lip and teeth. This bucket has no arch; thus weight is minimized. The sides are flared, permitting heaped loading, and the bucket dumps backward, not forward, thereby giving a somewhat longer dumping range. (Lewis, 1964)
flaser gabbro Coarse-grained blastomylonite formed by dislocation metamorphism of a gabbro. Flakes of mica or chlorite sweep around augen of feldspar and/or quartz with much recrystallization and neomineralization. See also: *mylonite gneiss.* (AGI, 1987)
flaser structure A structure in dynamically metamorphosed rock in which lenses and layers of original or relatively unaltered granular minerals are surrounded by a matrix of highly sheared and crushed material, giving the appearance of a crude flow structure; e.g., flaser gabbro. Cf: *augen structure.* (AGI, 1987)
flash box A box in which a light source, an electromagnet, and a telescope are all mounted in the pendulum apparatus of gravitational recording. (AGI, 1987)
flash coal dryer An appliance in which the moist coal is fed into a column of upward-flowing hot gases and moisture removal is virtually instantaneous. Suspension dryers were widely used in the United States for drying coals from 1/2 in (1.3 cm) downwards in size. See also: *Cascade coal dryer; Raymond flash dryer; suspension dryer; thermal drying.* (Nelson, 1965)
flashes Shallow lakes created by the removal of coal. (Briggs, 1929)
flash flood A local and sudden flood or torrent of relatively great volume and short duration, overflowing a stream channel in a usually dry valley (as in a semiarid area), carrying an immense load of mud and rock fragments, and generally resulting from a rare and brief but heavy rainfall over a relatively small area having steep slopes. It may also be caused by ice jams and by dam failure. See also: *freshet.* (AGI, 1987)

flash opal (a) An opal in which the play of color is limited to a single hue. (b) Opal in which the play of color is pronounced in one direction only.

flash over (a) Sympathetic detonation between explosive charges or between charged blast holes. (Dick, 2) (b) The transmission of detonation from a cartridge to another one in line; also, the tendency of a blast hole to be detonated by the shock wave from an adjacent borehole. Syn: *sympathetic detonation.*

flashpoint The minimum temperature at which sufficient vapor is released by a liquid or solid to form a flammable vapor-air mixture at atmospheric pressure.

flash roast Rapid removal of sulfur, as finely divided sulfide mineral is allowed to fall through a heated oxidizing atmosphere. Flash roasting or melting is widely used in the copper industry. (Pryor, 1963)

flash smelting A smelting process in which dried metal sulfide concentrates are blown with oxygen or oxygen-rich air in a hot hearth-type furnace such that the particles react rapidly with the oxygen to generate a large amount of heat, partially (controlled) oxidizing the concentrates and producing a molten matte phase containing the metal values, which will be further processed, and a molten slag.

flask (a) In foundry work, a molding box that holds the sand into which molten metal is poured. The top half or part is its cope, the bottom half is its drag, and it is furnished with locating lugs. (Pryor, 1963) (b) An iron bottle in which mercury is marketed. It contains 76-1/2 lb (41.3 kg). (Fay, 1920) (c) A tinned vessel in which a miner carries oil for a lamp or beverage for a lunch. (d) A necked vessel for holding liquids; esp., a broad, flattened vessel of metal or sometimes glass. (Webster 3rd, 1966)

flat (a) In mine timbering, horizontal crosspiece or cap used in roof support. (b) Of a mining lode, one less than 15° from horizontal in its dip. (Pryor, 1963) (c) A horizontal orebody, regardless of genetic type. (AGI, 1987) (d) A flat coal seam. (Korson, 1938) (e) A railroad car of the gondola type for shipping coal. (f) A dull diamond bit. See also: *going bord.* (Korson, 1938)

flat arch (a) An arch in that both outer and inner surfaces are horizontal planes. (Harbison-Walker, 1972) (b) In furnace construction, a flat structure spanning an opening and supported by abutments at its extremities; the arch is formed by a number of specially tapered bricks, and the brick assembly is held in place by their keying action. Also called a jack arch. (Harbison-Walker, 1972)

flat belt conveyor A type of belt conveyor in that the carrying run of the conveyor belt is supported by flat belt idlers or by a flat surface.

flat belt idler An idler consisting of one or more rolls supporting the belt in a flat position.

flat cut A manner of placing the boreholes, for the first shot in a tunnel, in which they are started about 2 to 3 ft (0.6 to 0.9 m) above the floor and pointed downward so that the bottom of the hole is about level with the floor.

flat double cabochon *lentil.*

flat drill A rotary end-cutting tool constructed from a flat piece of material provided with suitable cutting lips at the cutting end. (ASM, 1961)

flat ends Thin cleavages from the faces of a diamond crystal.

flat hole A borehole following a near horizontal course. (Long, 1960)

flat idler A belt idler that supports the belt in a flat position. (NEMA, 1956)

flatiron One of a series of short, triangular hogbacks forming a spur or ridge on the flank of a mountain, having a narrow apex and a broad base, resembling (when viewed from the side) a huge flatiron; it usually consists of a plate of steeply inclined resistant rock on the dip slope. (AGI, 1987)

flat joint In igneous rock, a joint dipping at an angle of 45° or less. Rarely applied to joints dipping more than 20°. (AGI, 1987)

flat lode A lode which varies in inclination from the horizontal to about 15°.

flatnose shell A cylindrical tool with a valve at the bottom, for boring through soft clay. (Fay, 1920)

flat rods A series of horizontal or inclined connecting rods, running up upon rollers, or supported at their joints by rocking arms, to convey motion from a steam engine or water wheel to pump rods at a distance. (Fay, 1920)

flat rope A steel rope made up of a number of loosely twisted four-strand ropes placed side by side, the lay of the adjacent strands being in opposite directions to secure uniformity in wear and to prevent twisting during winding. The strands are sewn together with steel wire. (Nelson, 1965)

flats (a) Subterraneous beds or sheets of traprock or whin. (Fay, 1920) (b) Narrow decomposed parts of limestones that are mineralized. (Fay, 1920)

flats and pitches (a) A phrase descriptive of the structure of the lead and zinc deposits in dolomite of the Upper Mississippi Valley region of the United States, esp. in Wisconsin. The flats are nearly horizontal solution openings; the pitches are the inclined, interconnecting joints. (AGI, 1987) (b) A slump structure of both horizontal and steeply inclined cracks in sedimentary strata. Syn: *pitches and flats.* (AGI, 1987)

flat sheet (a) A steel plate laid on the floor at the face of a tunnel or heading before blasting to provide a smooth floor for shoveling the broken rock into tubs. Syn: *turnsheet.* (Nelson, 1965) (b) Blanket deposit.

flattened strand rope A wire rope designed to give a greater wearing surface than ordinary round ropes and yet have about the same strength and flexibility. They have roughly 50% more wearing surface than ordinary round ropes, owing to the Lang lay of wires. See also: *wire rope.* (Lewis, 1964)

flatting (a) York. Horizontal vein of spar or barytes in the lead mines. Also called flatting bed. (Arkell, 1953) (b) A process for truing-up handmade fireclay refractories while they are still only partially dry. Handmaking is now little used except for some special shapes.

flatworking Scot. A working of moderate inclination. See also: *flat; flat lode.* (Fay, 1920)

flaw (a) A general term for any internal or external imperfection of a fashioned diamond or other gemstone. It includes cracks, inclusions, visibly imperfect crystallization, internal twinning, and cleavage. (b) An old term for a steep, transverse strike-slip fault.

flawless Said of a diamond or other gemstone that is free from flaws of any description when observed by a trained eye under efficient illumination with a fully corrected magnifier of not less than 10 power.

flaxseed ore An iron-bearing sedimentary deposit, e.g., the Clinton ore, composed of disk-shaped hematitic oolites that have been somewhat flattened parallel to the bedding plane. Cf: *fossil ore.* (AGI, 1987)

fleches d'amour Acicular, hairlike crystals of rutile, TiO_2, embedded in quartz; a semiprecious gemstone. Also called love arrows, a literal translation. Cf: *rutilated quartz.*

fleck To scale or peel off suddenly; applies to shaley beds in the roof or to coal slab at the face. (Nelson, 1965)

fleckschiefer Ger. A type of spotted slate characterized by minute flecks or spots of indeterminate material. See also: *fruchtschiefer; garbenschiefer.* (AGI, 1987)

fleet The movement of a rope sidewise when winding on a drum. See also: *fleet angle.* (Zern, 1928)

fleet angle (a) The included angle between the rope, in its position of greatest travel across the drum, and a line drawn perpendicular to the drum shaft, passing through the center of the head sheave or head-sheave groove. (b) Of hoisting gear in mine shaft headworks, the angle between the sheave and extreme paying-off position on the winding drum; in good practice the angle is below 3°. (Pryor, 1963) (c) As used by diamond drillers and miners, the angle between the two ends of a hoist drum as a base and the sheave wheel in a drill tripod or derrick or the headframe pulley as the apex. (Long, 1960)

fleet wheel (a) A grooved wheel or sheave that serves as a drum and about which one or more coils of a hauling rope pass. (Zern, 1928) (b) Surge wheel. (Mason, 1951)

fleischerite An isometric mineral, $Pb_3Ge(SO_4)_2(OH)_6 \cdot 3H_2O$; linnaeite group; occurs in fibrous aggregates and crusts associated with cerussite, memetite, and plumbojarosite in the upper oxidation zone of the Tsumeb Mine, Namibia.

Fleissner process A thermal drying, batch-type process, in which the action of high-pressure steam on a lump of lignite produces the following effects. The lump is heated inside and out to an approx. uniform temperature by its envelope of condensing steam. As the temperature rises and the pressure increases part of the colloidal water is expelled from the lump as a liquid. The lump shrinks as water leaves and the cells collapse, and when the pressure is lowered, more water leaves by evaporation caused by the sensible heat stored in the lump. When the pressure is lowered further by vacuum, additional moisture is evaporated, which cools the lump. (Mitchell, 1950)

flexible (a) Capable of being flexed. Capable of being turned, bowed, or twisted without breaking. (Webster 3rd, 1966) (b) Said of a mineral that can bend without breaking and will not return to its original shape; e.g., talc and chlorite. Cf: *elastic; malleable.*

flexible ducts *ventilation ducts.*

flexible guides *winding guide.*

flexible joint Any joint between two pipes that permits one of them to be deflected without disturbing the other pipe. (Fay, 1920)

flexible sandstone A fine-grained, thin-layered variety of itacolumite. (AGI, 1987)

flexible silver ore *sternbergite.*

flexible-type carrying idler Consists of one or more idler rolls arranged to form a catenary trough. This may be accomplished by mounting a single roll on a flexible shaft or by linking a series of rolls with individual rigid shafts. (NEMA, 1961)

flexing The bending of the conveyor belt that takes place as it wraps around the pulleys. The ply nearest the face of the pulley is under the minimum stress and the ply farthest from the face is under the maximum stress. Flexing stresses increase with a decrease in pulley diameters.

Flexlok A patented circle brick, with book ends, used in domestic furnaces, cupola furnaces, and acid tank linings.

flexural center With reference to a beam, the flexural center of any section is that point in the plane of the section through which a transverse load, applied at that section, must act if bending deflection only is to be produced, with no twist of the section. Also called shear center. Cf: *torsional center; elastic center; elastic axis.* (Roark, 1954)

flexural slip The slipping of sedimentary strata along bedding planes during folding. (AGI, 1987)

flexure A general term for a fold, warp, or bend in rock strata. A flexure may be broad and open, or small and closely compressed. Cf: *monocline.* Syn: *hinge.*

flexure correction A correction necessary in pendulum observations of gravity. The vibrating pendulum produces oscillations of the receiver case, of the pillar, and of the surface soil. Rather complex, coupled vibration phenomena arise and the period of the pendulum itself changes. Numerous methods have been suggested to correct for this influence or to eliminate it. Since the correction is of the order of 10×10^{-7} to 40×10^{-7} s on solid rock or cement and may increase to as much as 500×10^{-7} s on marshy ground, it must be determined accurately. (AGI, 1987)

flexure-slip fold A flexure fold in that the mechanism of folding is slip along bedding planes or along surfaces of foliation. There is no change in thickness of individual strata, and the resultant folds are parallel. (AGI, 1987)

flight (a) The metal strap or crossbar attached to the drag chain of a chain-and-flight conveyor. (Jones, 1949) (b) A term sometimes applied to one conveyor in a tandem series.

flight conveyor A type of conveyor comprising one or more endless propelling media, such as chain, to which flights are attached and a trough through which material is pushed by the flights. See also: *grit collector; underground mine conveyor.* Syn: *bar flight conveyor.*

flight line A line drawn on a map or chart to represent the planned or actual track of an aircraft during the taking of aerial photographs. (AGI, 1987)

flight pattern In an aeromagnetic survey or in another airborne geophysical survey, the planned flying route used. (AGI, 1987)

flinders diamond A Tasmanian term for a variety of topaz. (Fay, 1920)

flint (a) A term that has been considered as a mineral name for a massive, very hard, somewhat impure variety of chalcedony (minute crystals of quartz with submicroscopic pores). It is usually black or of various shades of gray, breaking with a conchoidal fracture, and striking fire with steel. Mohs hardness, 7; and sp gr, 2.65. Syn: *firestone.* Cf: *chert.* (b) Pulverized quartz of any type; e.g., "potters' flint" made by pulverizing flint pebbles into powder. (c) A term that is widely used as a syn. of chert or for a homogeneous, dark gray or black variety of chert.

flint clay A smooth, flintlike refractory clay rock composed dominantly of kaolin, which breaks with a pronounced conchoidal fracture and resists slaking in water. It becomes plastic upon prolonged grinding in water, as in an industrial wet-pan unit. (AGI, 1987)

flint mill A device, formerly used to provide light for miners at work, in which flints on a revolving wheel produce a shower of sparks incapable of igniting combustible gases. See also: *steel mill.*

Flintshire furnace A reverberatory furnace with a depression, well, or crucible in the middle of the side of the hearth; used for the roasting and reaction process on lead ores. (Fay, 1920)

Flintshire process Method of smelting galena concentrates in a reverberatory furnace with a crucible well in its hearth. (Pryor, 1963)

flinty crush rock *ultramylonite.*

flinty slate A touchstone consisting of siliceous slate. (AGI, 1987)

flipping turn System of pulleys incorporated in the return-side tracking of a belt conveyor, which turns it through 180°, so that any adherent abrasives do not come in contact with idler pulleys. (Pryor, 1963)

flitching (a) Widening of underground roadway by removing rock from sides. (Pryor, 1963) (b) The working of 2 to 5 yd (1.8 to 4.6 m) or more of the rib side coal in a narrow stall or heading. See also: *skipping.* (Nelson, 1965)

flitter Collier who moves a coal cutter to a new working place; to flit is to shift equipment. (Pryor, 1963)

flitting wagon A low truck or trolley used in pillar methods of working to transport face machines from one heading or bord to another. (Nelson, 1965)

float (a) A general term for loose fragments of ore or rock, esp. on a hillside below an outcropping ledge or vein. Syn: *floater.* Cf: *float mineral; float ore.* (b) Fine gold that floats in panning and other operations and is lost. (von Bernewitz, 1931) (c) A timber platform, faced with boiler iron on both sides, and provided with rings at the corners for lifting. It is used in shaft work to prevent the crushing of the bottom timbers by flying fragments of rock. (Stauffer, 1906) (d) The tendency of the bit in a flat-angle borehole to follow an increasingly flat course as the depth of the borehole increases. (Long, 1960) (e) In mineral concentration, the response of a specific mineral to the flotation process. (Pryor, 1963) (f) The fine dust that does not settle out of the air current in the grinding mills but is filtered out by fine cloth bags. Also, the fine dust collecting on the roof and timbers in a mine. (g) Metal particles so fine that they float on the surface of water in crushing or washing, as float gold. (h) Various forms of ball-and-seat valves commonly inserted in casing and rod strings in such a manner as to keep drilling fluid out of the casing or rod string when lowered into a borehole. Also called float valve. (Long, 1960) (i) To lift a material by the buoyant action of a strong current or flow of a liquid medium; also, that material buoyant enough to float on the surface of a liquid medium. (Long, 1960) (j) The buoyant part of an apparatus for indicating the height of water in a steam boiler or of liquid in a tank.

floatability In mineral concentration, word used in connection with response of a specific mineral to flotation process. (Pryor, 1963)

float-and-sink analysis Use of a series of heavy liquids diminishing (or increasing) in density by accurately controlled stages for the purpose of dividing a sample of crushed coal into gravimetric fractions either equal-settling or equal-floating at each stage. The floats at a given specific gravity are defined as the percentage floating at that density and the sinks have a defined higher density. Each product (minus one density and plus another) is analyzed after weighing and the ash sulfur and Btu content is found. From this testing, a washability curve is drawn that relates density with ash sulfur and Btu content, in the form of cumulative float, sink, and specific gravity curves. The ash curve plots ash against density for successive fractions. The densimetric curve plots specific gravity against cumulative weight. The Mayer curve (M-curve) plots cumulative weight against that of a constituent (e.g., ash). (Pryor, 1963)

float coal Small, irregularly shaped isolated deposits of coal imbedded in sandstone or in siltstone. They appear to have been removed from the original bed by washout during the peat stage and to have been carried a short distance and redeposited. Syn: *raft.* (AGI, 1987)

float copper Fine scales of metallic copper that do not readily settle in water.

float dust Fine particles of coal suspended in the air.

floater A single fragment of float. See also: *float.* (AGI, 1987)

float gold Particles of gold so small and thin that they float on water and may be carried off by it.

floating Said of large particles in a sedimentary rock that are not in contact with each other and are contained in a fine-grained matrix; e.g., quartz grains disseminated in limestone.

floating cable In seismic operations in a watercovered area, a cable connecting geophones suspended by floats. (AGI, 1987)

floating cone A method of designing optimum extraction sequences for an open pit mine. "Cones" of material are built using an ore block as a base and economic net value of the cone is calculated. The process is repeated for each ore block in a deposit, considering cone overlaps. The term "floating" is derived from the "movement" of the cone throughout the model.

floating control system As used in flotation, a system in that the rate of change of the manipulated variable is a continuous function of the actuating signal. (Fuerstenau, 1962)

floating pipeline A pipe supported on pontoons that is used for removing spoil from a suction dredger. (Hammond, 1965)

floating reef An isolated, displaced rock mass in alluvium. Cf: *float.* (AGI, 1987)

floating strainer A buoyant pump suction end that draws its water from near the surface of the free-water level and thus pumping almost clear water. Serves to decant. A floating strainer may be used in dealing with bodies of water other than in properly constructed sumps. (Nelson, 1965)

float mineral Small fragments of any ore carried away from the ore bed by the action of water or by gravity alone, often leading to the discovery of mines; See also: *floater; float ore; float.* (Standard, 1964)

float ore Scattered fragments of vein material broken from outcrops and dispersed in soil; a type of float. See also: *float; float mineral; shode.* (AGI, 1987)

floats Fractions with a defined upper limit of specific gravity and so described; e.g., floats, sp gr, 1.40. (BS, 1962)

floatstone (a) A miners' term for cellular or honeycomb quartz detached from a lode. (b) A lightweight, porous, friable variety of opal that floats on water and occurs in white or grayish, spongy, and concretionary or tuberous masses; also spelled float stone. Syn: *swimming stone.* (c) A carbonate rock containing a few bioclasts or other fragments more than 2 mm in diameter, widely spaced, and embedded in sand- or mud-size carbonate sediment that forms over 90% of a rock.

float sulfur *flake sulfur.*

float valve (a) Syn. for a ball-and-seat type apparatus inserted in a pipe, casing, or drill-rod string being lowered into a borehole. See also: *float.* (Long, 1960) (b) A valve operated by a float. (Long, 1960)

floc (a) A loose, open-structured mass formed in a suspension by the aggregation of minute (colloidal) particles. (ASCE, 1958) (b) A small aggregate of tiny sedimentary grains. (AGI, 1987) (c) A flocculent mass formed by the aggregation of a number of fine suspended particles. Syn: *floccule.* (Webster 3rd, 1966)

flocculate (a) To cause particles to aggregate in large particles. Usually accomplished with polymer, both natural or synthetic. (Webster 3rd, 1966) (b) To thicken a clay suspension by addition of synthetic polymer. (c) The addition of a suitable electrolyte to a clay suspension to cause the clay particles to agglomerate and settle. (d) Something that has flocculated; a flocculent particle or mass; a floc; a floccule. (Webster 3rd, 1966)

flocculating (a) The thickening of the consistency of a slip by adding a suitable electrolyte. (ASTM, 1994) (b) The agglomeration of clay particles in a clay suspension by adding an electrolyte.

flocculating agent A reagent added to a dispersion of solids in a liquid to bring together the fine particles to form flocs. These reagents usually consist of long chain polymers, both natural and synthetic. (BS, 1962)

flocculation The process by which a number of individual, minute suspended particles are tightly held together in clotlike masses, or are loosely aggregated or precipitated into small lumps, clusters, or granules; e.g., the joining of soil colloids into a small group of soil particles, or the deposition or settling out of suspension of clay particles in salt water. See also: *coagulation; floc.* (AGI, 1987)

floccule A small, loosely aggregated mass of material suspended in or precipitated from a liquid. One of the flakes of a flocculent precipitate. Syn: *floc.* (Webster 3rd, 1966)

flocculent Coalescing and adhering in flocks. A cloudlike mass of precipitate in a solution. From the Latin floccus, meaning lock of wool.

flocculent deposit An aggregate or precipitate of small lumps formed by chemical precipitation. (Hy, 1965)

flocculent structure An arrangement composed of flocs of soil particles instead of individual soil particles. See also: *soil structure; honeycomb structure; single-grained structure.* (ASCE, 1958)

Flodin process A direct process for manufacturing steel, by means of which iron with a carbon content from 0.2% upwards can be produced by smelting, in a specially constructed electrical furnace, a mixture of hematite and coal, or charcoal, the process being continuous. The reduced metal accumulates at the bottom of the furnace, from which it is tapped. Both sulfur and phosphorus are reduced to a low figure without additional refining, while the manganese and silicon contents are controlled in the same way as in the ordinary open-hearth process. (Osborne, 1956)

floe rock Rock occurring in or taken from a body of talus; usually refers to ganister. (ARI, 1949)

flohmig amber *fatty amber.*

floocan *flucan.*

flood basalt *plateau basalt.*

floodgate (a) A gate for shutting out, admitting, releasing, or otherwise regulating a body of water, such as excess water in times of flood; specif. the lower gate of a lock. See also: *sluice.* (AGI, 1987) (b) A stream stopped by or allowed to pass by a floodgate. (AGI, 1987)

flood plain The surface or strip of relatively smooth land adjacent to a river channel, constructed by the present river in its existing regimen and covered with water when the river overflows its banks. It is built of alluvium carried by the river during floods and deposited in the sluggish water beyond the influence of the swiftest current. (AGI, 1987)

flookan *flucan.*

floor (a) The rock underlying a stratified or nearly horizontal ore deposit, corresponding to the footwall of more steeply dipping deposits. (b) The bottom of a coal seam or any other mineral deposit. Cf: *roof.* (Arkell, 1953) (c) Plank-covered, or steel-mesh-covered, level work area at the base of a drill tripod or derrick around the collar of a borehole in front of the drill. See also: *platform.* (Long, 1960) (d) Loose plank laid parallel with rock drift at the heading before blasting a round of holes to facilitate the loading of broken rock. (e) A horizontal, flat orebody. (f) The bed or bottom of the ocean. A comparatively level valley bottom; any low-lying ground surface. (AGI, 1987) (g) That part of any underground gallery upon which a person walks or upon which a tramway is laid. (h) A plank platform underground. (i) The upper surface of the stratum underlying a coal seam. (CTD, 1958)

floor break The break or crack that separates a block of stone from the quarry floor. Also called floor cut. (Hess)

floor burst A type of outburst generally occurring in longwall faces and preceded by heavy weighting due to floor lift. Gas that evolved below the seam seems to collect beneath an impervious layer of rock, and a gas blister forms beneath the face, giving the observed floor lift. Later, the floor fractures and the combustible gases escape into the mine atmosphere. See also: *outburst.* (Roberts, 1960)

floor cut (a) A machine cut made in the floor dirt immediately below the coal seam. See also: *undercutting; bottom cut.* (Nelson, 1965) (b) A cut by means of which a block of stone is separated from the quarry floor. See also: *floor break.*

floor lift The upward heave of the floor beds after a coal seam has been extracted. See also: *creep.* (Nelson, 1965)

floor sill A large timber laid flat on the ground or in a level, shallow ditch to which are fastened the drill-platform boards or planking. (Long, 1960)

floor station A survey station secured in the floor of a mine roadway or working face. (BS, 1963)

flop gate An automatic gate used in placer mining when there is a shortage of water. This gate closes a reservoir until it is filled with water, when it automatically opens and allows the water to flow into the sluices. When the reservoir is empty the gate closes, and the operation is repeated. See also: *boomer.*

flora The entire plant population of a given area, environment, formation, or time span. Cf: *fauna.* (AGI, 1987)

florencite A trigonal mineral, $(La,Nd,Ce)Al_3(PO_4)_2(OH)_6$; crandallite group; further speciated according to which rare-earth element predominates; weakly radioactive; pale yellow; in schists, carbonatites, pegmatites, and placer sands.

florspar *fluorite.*

flos ferri An arborescent variety of aragonite occurring in delicate white coralloid masses that commonly encrust hematite, forming picturesque snow-white pendants and branches.

floss White cast iron for converting into steel. (Webster 2nd, 1960)

floss hole (a) A small door provided at the bottom of a flue or chimney for ash removal. (Osborne, 1956) (b) A tap hole. (Fay, 1920)

flot (a) Ore lying between the beds or at certain definite horizons in the strata. (Arkell, 1953) (b) Eng. Veins that branch off laterally, Alston Moor lead mines. (Arkell, 1953)

flotation (a) *froth flotation.* (b) The method of mineral separation in that a froth created by a variety of reagents floats some finely crushed minerals, whereas other minerals sink. Formerly the term flotation with descriptive adjectives was used for all processes of concentration in which levitation in water of particles heavier than water was obtained. Thus, if some particles were retained in an oil layer or at the interface between an oil layer and a water layer, the process was spoken of as bulk-oil flotation; if the particles were retained at a free water surface as a layer one particle deep, the process was skin flotation; and if the particles were retained in a foamy layer several inches thick, the process was froth flotation. Froth flotation is the process that has survived the test of time, and the term flotation is now used universally to describe froth flotation. See also: *bulk oil flotation; film flotation; selective flotation; skin flotation.* Syn: *flotation process.* (AGI, 1987; Gaudin, 1957)

flotation agent A substance or chemical that alters the surface tension of water or that makes it froth easily. See also: *surface activity; depressant.* (Nelson, 1965)

flotation cell Appliance in which froth flotation of ores is performed. It has provision for receiving conditioned pulp, aerating this pulp and for separate discharge of the resulting mineralized froth and impoverished tailings. Types of cell include (1) agitation (impeller, and splashing, now obsolete); (2) pneumatic (in which air blown in agitates pulp), such as Hallimond laboratory cell, Callow, McIntosh, Forrester, Southwestern, and Britannia; (3) vacuum cells (Elmore and Clemens, obsolescent); (4) subaeration with mechanized stirring and pressure-input air (M.S. cell, Agitair); (5) subaeration, self-aerating mechanized cell (Fagergren, Denver, M.S.S.A., Humboldt, Boliden, K.B., etc.). (Pryor, 1963)

flotation conditioning time In ore processing, the period during which pulp is agitated with a given chemical, or combination of chemicals, in the series of conditioning operations that precede separation of various minerals in the ore by froth flotation. (Pryor, 1958)

flotation middlings Intermediate flotation products that may be retreated. (BS, 1962)

flotation of crystals *crystal flotation.*

flotation oil Oil, such as creosote oil, pine oil, or turpentine. Used to wet a particular component of a powdered material and cause it to concentrate in an airy froth. (Bennett, 1962)

flotation plane Plane of a liquid surface in which a body floats. (Hess)

flotation process *flotation; oil flotation.*

flotation reagent Any of several reagents used in the froth flotation process. They include pH regulators, slime dispersants, resurfacing agents, wetting agents, conditioning agents, collectors, and frothers. (Pryor, 1963)

flotation regulator An acid or an alkali used to control the pH of flotation solutions. (Bennett, 1962)

flotation time The time necessary to make the separation into concentrate and tailing depends on such factors as particle size and reagents used, and must be known for determination of the size and number of flotation cells in the plant. (Fuerstenau, 1962)

Flotol A synthetic reagent of the general nature of pine oil; used as frother in flotation process. (Pryor, 1963)

flour agate (a) Any moss agate. (b) A term applied to any moss agate or mocha stone with flowerlike markings. (c) Translucent chalcedony from Oregon containing inclusions of minerals, some red, brown, or yellow and green, arranged in flowerlike forms, commonly both red and green.

flour copper Very fine scaly native copper that floats on water and is very difficult to save in milling. See also: *float copper*. (Weed, 1922)

floured The finely granulated condition of quicksilver, produced to a greater or lesser extent by its agitation during the amalgamation process. The coating of quicksilver with what appears to be a thin film of some sulfide, so that when it is separated into globules these refuse to reunite. Also called flouring. *sickening*. (Fay, 1920)

flour gold *float gold.*

flour gypsum *gypsum.*

flour salt Very fine-grained vacuum pan salt. (Kaufmann, 1960)

floury alumina Fine-grained, highly calcined, powdery textured alumina; also dusty, easily airborne. Traditionally produced by European Bayer plants.

flow (a) The plastic deformation of solids: flowage; solid flow; rock flowage; plastic flow. Cf: *fracture*. (AGI, 1987) (b) That which flows or results from flowing. A mass of matter moving, or that has moved in a stream, such as a lava flow. Syn: *current*. (Fay, 1920) (c) The movement of a fluid, such as air, water, or magma (lava). (AGI, 1987) (d) A tabular-shaped body of lava that consolidated from magma on the surface of the Earth. (AGI, 1987) (e) In ceramics, the flux used to cause color to run and blend in firing. (Fay, 1920) (f) A mass movement of unconsolidated material that exhibits a continuity of motion and a plastic or semifluid behavior resembling that of a viscous fluid; e.g., creep; solifluction; earthflow; mudflow; debris flow; sturzstrom. Water is usually required for most types of flow movement. (AGI, 1987) (g) The mass of material moved by a flow. The smallest formal lithostratigraphic unit of volcanic flow rocks. A flow is a discrete, extrusive, volcanic body distinguishable by texture, composition, order of superposition, paleomagnetism, or other objective criteria. It is part of a member and thus is equivalent in rank to a bed or beds of sedimentary-rock classification. Many flows are informal units. Designation of flows as formal units should be limited to those that are distinctive and widespread. (AGI, 1987)

flowage *flow.*

flowage differentiation (a) The retarding effect produced by walls on the movement under the influence of pressure of a mush of crystals in a magmatic liquid, which may give rise to magmatic differentiation and also to the concentration of ore minerals. (Schieferdecker, 1959) (b) The tendency of suspended crystals to concentrate in the high-velocity zone of a magma that is moving by laminar flow. (AGI, 1987)

flowage fold A minor fold that is the result of the flowage of rocks toward a synclinal axis, toward which the minor folds are overturned.

flowage structure *flow structure.*

flow banding A structure of igneous rocks that is esp. common in silicic lava flows. It results from movement or flow, and is an alternation of mineralogically unlike layers.

flow breccia A breccia that is formed contemporaneously with the movement of a lava flow; the cooling crust becomes fragmented while the flow is still in motion. (AGI, 1987)

flow characteristic The rate at which a metal powder will flow through an orifice in a standard instrument, and/or according to a specified procedure. (Rolfe, 1955)

flow cleavage A syn. of slaty cleavage, so called because of the assumption that recrystallization of the platy minerals is accompanied by rock flowage. Cf: *fracture cleavage*. (AGI, 1987)

flower of iron *flos ferri.*

flowers of sulfur A light-yellow, pulverulent modification of sulfur formed when sulfur vapor is condensed. (Standard, 1964)

flow folding Folding in incompetent beds that offer so little resistance to deformation that they assume any shape impressed upon them by the more rigid rocks surrounding them or by the general stress pattern of the deformed zone. Cf: *ptygmatic folding*. Syn: *incompetent folding*. (AGI, 1987)

flow gneiss A gneiss with a structure produced by flowage in an igneous mass before complete solidification.

flow gradient A drainageway slope determined by the evelation and distance of the inlet and outlet, and by required volume and velocity. (Nichols, 1976)

flowing film concentration In metallurgy, a concentration based on the fact that liquid films in laminar flow possess a velocity that is not the same in all depths of the film. There is no flow at the bottom but maximum at or very near the top resulting from the internal friction of one layer upon another. By this principle, lighter particles are washed off while the heavier particles accumulate and are intermittently removed. This is basis of the stationary table, which has been known for thousands of years. Vanners and round tables have been developed from this basic principle, whereas bumping and shaking tables jointly utilize flowing film and other principles. (Gaudin, 1939)

flow layer An igneous-rock layer, differing mineralogically or structurally from the adjacent layers, which was produced by flowage before the complete solidification of the magma. Cf: *flow line; schlieren.*

flow line A lineation of crystals, mineral streaks, or inclusions in an igneous rock, indicating the direction of flow before consolidation. Cf: *flow layer*. (AGI, 1987)

flowmeter (a) A device installed in a drilling-fluid circulation system that registers the number of gallons (liters) of liquid circulated per minute and also indicates when the flow past the bit ceases. (Long, 1960) (b) A device which registers rate of flow, and perhaps quantity, of gases, liquids, and fluid pulps. Used in mineral dressing to measure rates and quantities of pregnant solutions in cyanide and to control liquid additions to pulps. (Pryor, 1963)

flow rate Weight of dry air flowing per unit time. Measured in kilograms per second. (Hartman, 1982)

flow roll A rounded, pillowlike body or mass of sandstone occurring within or just above finer-grained sediment or commonly within the basal part of a sandstone overlying shale or mudstone. Its shape approaches that of an elongate, flattened ellipsoid (short axis more or less vertical), and is presumed to have formed by deformation, such as by large-scale load casting or mud flowage accompanied by subaqueous slump or foundering of sand channels. (AGI, 1987)

flowsheet A diagram showing the progress of material through a preparation or treatment plant. It shows the crushing, screening, cleaning, or refining processes to which the material is subjected from the run-of-mine state to the clean and sized products. The size range at the various stages may also be shown. (Nelson, 1965)

flow stretching The orientation and possible deformation of crystals with their long axes in the direction of plastic flow in metamorphic rocks.

flow structure (a) A structure of igneous rocks, generally but not necessarily volcanic rocks, in which the stream lines or flow lines of the magma are revealed (1) by alternating bands or layers of differing composition, crystallinity, or texture, or (2) by subparallel arrangement of prismatic or tabular crystals. Syn: *flow texture; fluidal structure; fluxion structure*. (b) A primary sedimentary structure resulting from subaqueous slump or flow. (AGI, 1987)

flow texture *flow structure; fluidal texture.*

flow unit One of the nearly contemporaneous subdivisions of a lava flow (usually basaltic) that consists of two or more parts that were poured one over the other during the course of a single eruption.

F.L.P. In Great Britain, tests of every type of apparatus are made in explosive atmospheres before it is approved and allowed to use the official letters F.L.P. (flameproof). (Mason, 1951)

flucan A narrow band of crushed rock or clayey material found along a fault zone or vein of ore. See also: *breccia; gouge; selvage; pug*. Also spelled fluccan; flookan; flukan; floocan. (Nelson, 1965)

flue (a) S. Wales. A furnace, such as a large coal fire at or near the bottom of an upcast shaft for producing a current of air for ventilating the mine. (Fay, 1920) (b) A tube or passageway in a steam boiler for hot gases or water (depending on whether the boiler is a fire-tube or water-tube boiler). (c) A passage or channel through which the products of combustion of a boiler or other furnace are taken to the chimney. (CTD, 1958) (d) Lanc. Shale. (Arkell, 1953)

flue dust Dust passing into the flues of a smelter or metallurgical furnace and which, unless caught, passes out into the atmosphere. It is composed of particles of unchanged or oxidized ore, volatilized lead that has been converted into oxide, carbonate and sulfate ash, fuel, and volatilized products of arsenic, zinc, bismuth, etc. (Hess)

fluellite An orthorhombic mineral, $Al_2(PO_4)F_2(OH){\cdot}7H_2O$; has one distinct cleavage; occurs with secondary iron phosphate minerals in pegmatites. Formerly called kreuzbergite.

fluid (a) The quality, state, or degree of being fluid: a liquid or gaseous state. Cf: *gas*. (Webster 3rd, 1966) (b) The physical property of a substance that enables it to flow and that is a measure of the rate at which it is deformed by a shearing stress, as contrasted with viscosity: the reciprocal of viscosity. (Webster 3rd, 1966) (c) In mineral transport, the term is not confined to liquids and slurries, but is also used for finely divided solids that flow readily in air currents, fluosolids reactors, or through dry ball mills. (Pryor, 1963)

fluidal structure *flow structure.*

fluidal texture (a) *flow texture*. (b) A metamorphic texture in which narrow stripes or lenticles of a mineral, present as grains approx. 0.01 mm in diameter, are connected with porphyroclasts of the same mineral and extend across regions in which another mineral shows a dominantly mosaic texture. This texture has been given a specific genetic connotation relating to superplasticity (Harte, 1977). (AGI, 1987)

fluid-bed reactor A single or multi-stage reactor, generally used for gas-solid contacting, in which the solid component is a reactant or a catalyst and is in a continuous fluidized state. The gas is injected into the reactor to provide rapid and uniform mixing for reactants to facilitate heat transfer and completion of the reaction.

fluid column The number of feet of drilling fluid standing in a borehole while the drill is operating and/or the number of feet of drilling fluid remaining in a borehole with the drill string withdrawn. (Long, 1960)

fluid conveyor coupling A device for overcoming the starting resistance of a conveyor fed by a constant-speed motor. It is used to allow the motor to reach full speed before starting the conveyor. (Nelson, 1965)

fluid cut *fluid wash.*

fluid energy mill A size-reduction unit depending for its action on collisions between the particles being ground, the energy being supplied by a compressed fluid, (e.g., air or steam) that enters the grinding chamber at high speed. Such mills will give a product of 5 µm or less; they have been used for the fine grinding of frits, kaolin, zircon, titania, and calcined alumina, but the energy consumed per ton of milled product is high. (Dodd, 1964)

fluid inclusion A cavity, with or without negative crystal faces, containing one or two fluid phases, and possibly one or more minute crystals, in a host crystal. If two fluid phases are present, the vapor phase (bubble) may show Brownian motion. See also: *inclusion; negative crystal.*

fluidized bed dryer A coal dryer that depends on a mass of particles being fluidized by passing a stream of hot air through it. As a result of the fluidization, intense turbulence is created in the mass, including a rapid drying action. The dry coal is withdrawn from the opposite side of the chamber. Fine particles in the feed become entrained in the air and are recovered in a cyclone, while the finest particles may need removal by dry filters or wet scrubbers. The dryer has a high capacity and many are in use in the United States. See also: *cascade coal dryer; flash coal dryer; thermal drying.* (Nelson, 1965)

fluidized roasting Oxidation of finely ground ore minerals by means of upward currents of air, blown through a reaction vessel with sufficient force to cause the bed of material to become fluid. Reaction between mineral and air is maintained at a desired exothermic level by control of oxygen entry, by admission of cooling water, or by added fuel. Fluidized beds are used in the mineral industry for a number of concentrates, including copper, lead, zinc, and carbonaceous gold ores.

fluid lubricated The core barrelhead bearings and/or other rotating members in a drill string cooled and lubricated by water or mud-laden fluid circulated as the drilling fluid. (Long, 1960)

fluid pressure (a) The force, expressed in pounds per square inch, exerted by the weight of the column of drilling fluid measured at any given depth in a borehole. Cf: *bottom-hole pressure.* (Long, 1960) (b) The pressure exerted by fluid contained in rock. (AGI, 1987)

fluid ton Thirty-two ft^3 (0.9 m^3) of fluid. A unit to correspond with the short ton of 2,000 lb (908 kg), and of sufficient accuracy for many hydrometallurgical, hydraulic, and other industrial purposes, it being assumed that the water or other liquid under consideration weighs 62.5 lb/ft^3 (1,002.7 kg/m^3). (Fay, 1920)

fluid volume The amount of drilling fluid circulated through the drill string, generally expressed in gallons per minute. (Long, 1960)

fluid wash The wearing away of core and parts of a drill string or bit exposed to the erosive forces of the rapid passage of the circulated drilling fluid. Also called fluid cut. (Long, 1960)

flukan *flucan.*

fluke A rod used for cleaning drill holes before they are charged with explosives. (Fay, 1920)

fluken (a) Corn. *gouge clay.* (b) A crossvein composed of clay. (Arkell, 1953)

flume (a) An artificial inclined channel used for conveying water for industrial purposes, such as irrigation, transportation, mining, logging, or power production; or for diverting the water of a stream from its channel for the purpose of washing or dredging the sand and gravel in the dry bed or to aid in engineering construction. (AGI, 1987) (b) An inclined channel, usually of wood and often supported on a trestle, for conveying water from a distance to be utilized for power, transportation, etc., as in placer mining, logging, etc. Syn: *sluice; race.* (c) To divert by a flume, such as the waters of a stream, in order to lay bare the auriferous sand and gravel forming the bed. Cf: *ditch.*

flumed The transportation of solids by suspension or flotation in flowing water.

flummery A smooth porcellanous rock resembling porridge in color and texture; Carboniferous limestone, Hunts quarry, Porthywaen, and Vale of Clwyd, north Wales. Also called flummery stone.

fluobaryt A compact mixture of fluorite and barytes. (Hey, 1955)

fluoborite A hexagonal mineral, $Mg_3(BO_3)(F,OH)_3$; colorless; occurs as fibrous masses of small prisms in skarns and veinlets in massive franklinite ore at Sterling Hill, NJ; also at Norberg, Sweden; Nocera, Italy; and Broadford, Scotland. Formerly called nocerite.

fluocerite A hexagonal mineral, $(Ce,La)F_3$; further speciated on the basis of the dominant rare-earth element; weakly radioactive; in pegmatites associated with gadolinite, allanite, and bastnäsite in the Front Ranges, CO.

fluometry *fluorimetry.*

fluor (a) The original spelling of fluorite; still used chiefly in Great Britain. (b) A prefix to mineral names indicating that some or all of the hydroxyl radical in its composition has been replaced by fluoride; e.g., micas and amphiboles. See also: *fluorite; fluormica.*

fluoramphibole Artificial amphibole with fluorine replacing the hydroxyl of hydroxyl amphibole. (English, 1938)

fluorapatite A hexagonal mineral, $Ca_5(PO_4)_3F$; apatite group; variably colored; commonly fluorescent or phosphorescent; defines 5 on the Mohs hardness scale; a common accessory mineral in igneous and metamorphic rocks, pegmatites, veins, and carbonate rocks; an ore of phosphate in phosphorites. Syn: *apatite.*

fluoredenite *fluoramphibole.*

fluorene An organic compound formed through the burning of pyritous shale in Bohemia, Czech Republic. Later renamed kratochvilite. (Tomkeieff, 1954)

fluorescence (a) The emission of visible light by a substance exposed to ultraviolet light. It is a useful property in examining well cuttings for oil shows and in prospecting for some minerals. (AGI, 1987) (b) The absorption of radiation at one wavelength, or a range of wavelengths, and its reemission as radiation of longer, visible wavelengths. (c) A type of luminescence in which the emission of light ceases when the external stimulus ceases; also, the light so produced. (AGI, 1987) (d) Quantized electromagnetic radiation as a material drops from a higher to a lower energy state. Fluorescence stops when the excitation energy stops. Cf: *phosphorescence; luminescence.*

fluorescent Having the property to produce fluorescence. (Long, 1960)

fluorescent lamp (a) Commonly and improperly designates an electric lamplike device emitting ultraviolet radiations or black light. Cf: *black light, lamp.* (Long, 1960) (b) A glass globe or tube, the inner surface of which is coated with a fluorescent substance that produces visible light when excited by an electric current. (Long, 1960)

fluorescent penetrant inspection A type of nondestructive testing wherein a penetrating type of oil or other liquid combined with fluorescent material particles is applied over a surface and flowed into cracks, crevices, or other surface defects or irregularities; the excess is then removed, and the article is examined under ultraviolet light. (Henderson, 1953)

fluorhectorite A variety of hectorite in which fluorine partially replaces hydroxyl. Cf: *montmorillonite; smectite; saponite.*

fluoride A mineral with fluorine as its chief anion.

fluorimetry Method of analysis based on the intensity of fluorescence measured when using ultraviolet light. (Pryor, 1963)

fluorine A nonmetallic element, the lightest of the halogens; it is a pale yellow, corrosive gas, that is highly toxic. Symbol, F. It is the most electronegative and reactive of all the elements. Fluorine occurs chiefly in fluorite (CaF_2) and cryolite (Na_2AlF_6), but is widely distributed in other minerals. Used in producing uranium and many high-temperature plastics. (Handbook of Chem. & Phys., 3)

fluorite An isometric mineral, CaF_2; perfect octahedral cleavage; transparent to translucent; defines 4 on the Mohs hardness scale; in veins as a gangue mineral; in carbonate rocks; an accessory in igneous rocks. Syn: *fluorspar; fluor; Derbyshire spar.*

fluormica A group name for the fluorite-rich micas, natural or artificial. *fluor.*

fluorometer A device for measuring the intensity of fluorescence. (Bennett, 1962)

fluoroscope An instrument consisting of a fluorescent screen and a source of ionizing radiation. Used to examine the image formed by opaque objects placed in the beam. (ASM, 1961)

fluoroscopy An inspection procedure in which the radiographic image of the subject is viewed on a fluorescent screen; normally limited to low-density materials or to thin sections of metals because of the low-light output of the fluorescent screen at safe levels of radiation. (ASM, 1961)

fluorphologopite *fluor.*

fluorspar *fluorite.*

fluortremolite *fluor.*

fluosolids system A method of roasting, applied to finely divided material, in which air with sufficient strength is blown through a heated bed of mineral to keep it fluid, while reaction is controlled by continuous adjustment of rate of feed, cooling water, and added fuel (including oxygen in air). Train of appliances includes instrument controls, air compressor, dust-collecting cyclones, and feed pump. (Pryor, 1963)

flush (a) To operate a placer mine, where the continuous supply of water is insufficient, by holding back the water and releasing it periodically in a flood. (Webster 3rd, 1966) (b) To fill underground spaces, as in coal mines, with material carried by water, which after drainage forms a compact mass. (Webster 2nd, 1960) (c) *hydraulic mine filling.*

flushing (a) A drilling method in that water or some other thicker liquid, such as a mixture of water and clay, is driven into the borehole through the rod and bit. The water rises along the rod on its outer side, between the walls of the borehole and the rod, with such a velocity that the broken rock fragments are carried up by this water current (direct flushing). Alternatively, water enters the borehole around the rod and issues upwards through the rod (indirect flushing). (Stoces, 1954) (b) In a colliery, diversion of ventilation to clear foul atmosphere (a dangerous method.). (Pryor, 1963) (c) *hydraulic stowing.* (d) In oil-well production, use of gravitated ground water to force oil or gas to the surface. (Pryor, 1963)

flushing fluid *flush.*

flush-joint casing Lengths (usually 10 ft or 3 m) of steel tubing provided with a box thread at one end and a matching pin thread on the opposite end. Coupled, the lengths form a continuous tube having uniform inside and outside diameters throughout its entire length. Syn: *casing.* (Long, 1960)

flush water Water used to assist the flow of materials in chutes or launders. Cf: *top water.* (BS, 1962)

flute (a) A groove parallel or nearly parallel to the axis of a cylindrical piece, such as the grooves of a split-ring core lifter or the grooves in a core-barrel stabilizer ring. Also applied to grooves or webs following a corkscrewlike course around the outside surface of a cylindrical object, like the spiraled webs on an auger stem or rod. (Long, 1960) (b) A primary sedimentary structure, commonly seen as a flute cast, consisting of a discontinuous scoop-shaped or lobate depression or groove generally 2 to 10 cm in length, usually formed by the scouring action of a turbulent, sediment-laden current of water flowing over a muddy bottom, and having a steep or abrupt upcurrent end where the depth of the mark tends to be the greatest. Its long axis is generally parallel to the current. Cf: *facet.* (AGI, 1987)

fluted core Core, the outside surface of which is spirally grooved or fluted. Syn: *corkscrew core.* (Long, 1960)

fluting A peculiar method of surface decay by which granite or granite gneisses are left with a corrugated or fluted surface. In a large subangular fragment of granite, one side contains a dozen of these little channels, from 1 to 4 in (2.5 to 10.2 cm) deep and from 3 to 10 in (7.6 to 25.4 cm) apart from center to center. These channels run straight down the face of the rock. (AGI, 1987)

fluvial (a) Of or pertaining to a river or rivers. (AGI, 1987) (b) Existing, growing, or living in or about a stream or river. (AGI, 1987) (c) Produced by the action of a stream or river; e.g., sand and gravel deposits. Syn: *fluviatile.* Etymol. Latin fluvius, river. (AGI, 1987)

fluviatile *fluvial.* Geologists tend to use the term for the results of river action (e.g., fluviatile dam, or fluviatile sands) and for river life (e.g., fluviatile fauna). (AGI, 1987)

fluvioglacial *glaciofluvial.*

fluviolacustrine Of or pertaining to sedimentation partly in lake water and partly in streams, or to deposits laid down under alternating or overlapping lacustrine and fluviatile conditions.

fluvioterrestrial Consisting of or pertaining to the land and its streams. (AGI, 1987)

flux (a) *fluxstone.* (b) To cause to become fluid; to treat with a flux, esp. to promote fusion; to become fluid. Syn: *flow.* See also: *current.* (c) Any chemical or rock added to an ore to assist in its reduction by heat; e.g., limestone with iron ore in a blast furnace. (d) In metal refining, a material used to remove undesirable substances; e.g., sand, ash, or dirt, as a molten mixture.

fluxed pellet An iron-ore pellet made by mixing minor amounts of a ground flux (forsterite, calcite, dolomite, or lime) with the magnetite or hematite concentrate to decrease smelting times and coke consumption.

flux fusion A method of growing refractory crystals; similar in many respects to crystallization for aqueous solutions, but the fluxes are salts with relatively high melting temperatures.

flux-gate magnetometer An instrument used for detailed studies of the Earth's magnetic field on a local basis that uses the flux-gate, which consists of two identical saturable cores of high permeability, with identical, but oppositely wound, coils. An alternating current in these coils magnetizes them first with one polarity, then in the opposite sense. If an additional field is present, such as the Earth's field, it will add to the flux in one coil while decreasing that in the other, resulting in different voltage drops across the two coils. This difference is proportional to the unvarying field, which can be measured by noting the average voltage difference between the two halves of the flux gate. In use, a part of the Earth's field is balanced out by an additional winding surrounding both cores and carrying direct current. In airborne use, the recording flux gate is kept aligned with the magnetic field by the use of two additional flux gates. When these are at right angles to the Earth's field, they generate no voltage, but if they depart from this position, they can generate voltages which operate motors returning them to proper alignment. In this fashion, the recording element is held always parallel to the total field. (Hunt, 1965)

fluxing Fusion or melting of a substance as a result of chemical action. (Harbison-Walker, 1972)

fluxing lime Lump or pebble quicklime used for fluxing in steel manufacture. The term may be applied more broadly to include fluxing of nonferrous metals and glass. It is a type of chemical lime. (Boynton, 1966)

fluxing ore An ore containing an appreciable amount of valuable metal, but smelted mainly because it contains fluxing agents that are required in the reduction of richer ores. (Weed, 1922)

fluxing stone Consists of pure limestone or sometimes dolomite and is used in iron blast furnaces and foundries. Usually material below 2 in (5.1 cm) in diameter is eliminated. The most desirable size is between 4 in and 6 in (10 cm and 15 cm). (USBM, 1965)

fluxion banding Banding in rock consisting of flow layers.

fluxion structure *flow structure.*

flux spoon A small ladle for dipping up a sample of molten metal for testing. (Fay, 1920)

fluxstone (a) Limestone, dolomite, or other rock or mineral used in metallurgical processes to lower the fusion temperature of the ore, combine with impurities, and make a fluid slag. Syn: *flux.* (AGI, 1987) (b) Crushed limestone or dolomite used as a flux in the smelting of iron ore. (c) In iron-ore pelletizing, a mixture of ground limestone and dolomite added to the iron ore before the concentrate is rolled into green balls and fired at 1,315 °C in a pelletizing furnace. Also spelled fluxing stone; flux stone.

fly gate (a) An opening in a chute that can be opened or closed at will. In a chute for coal, a fly gate may be inserted so that if rock is deposited in the chute, it may be trapped out by opening the fly gate. (Zern, 1928) (b) An opening or bypass in a chute or hopper that can be opened or closed at will for diverting ore, rock, or coal from one bin or conveyor to another. Sometimes called "to fly."

Flygt pump A submersible pump developed in Sweden.

flying cradle Eng. *cradle.*

flying reef Aust. A broken, discontinuous, irregular vein.

flying veins A pattern of mineral-deposit veins overlapping and intersecting in a branchlike pattern. (AGI, 1987)

flyrock Rocks propelled from the blast area by the force of an explosion. See also: *throw.* (Meyer, 1981)

flysch (a) A marine sedimentary facies characterized by a thick sequence of poorly fossiliferous, thinly bedded, graded deposits composed chiefly of marls, sandy and calcareous shales, and muds, rhythmically interbedded with conglomerates, coarse sandstones, and graywackes. (AGI, 1987) (b) An extensive, preorogenic sedimentary formation representing the totality of the facies deposited in different troughs, by rapid erosion of an adjacent and rising mountain belt at a time directly previous to the main paroxysmal (diastrophic) phase of the orogeny; specif., the Flysch strata of late Cretaceous to Oligocene age along the borders of the Alps, deposited before the main phase (Miocene) of the Alpine orogeny. (AGI, 1987) (c) A term that has been loosely applied to any sediment with most of the lithologic and stratigraphic characteristics of a flysch, such as almost any turbidite. Ethymol. dialectal term of German origin used in Switzerland for a crumbly or fissile material that slides or flows. Cf: *molasse.* (AGI, 1987)

flywheel A heavy wheel used in a rotating system to reduce surges of power input or demand by storing and releasing kinetic energy as it changes its rate of rotation. (Pryor, 1963)

foam drilling A method of dust suppression in which thick foam is forced through the drill by means of compressed air and the foam and dust mixture emerges from the mouth of the hole in the form of a thick sludge. With this method the amount of dust dispersed into the atmosphere is almost negligible and the amount of water used is about 1 gal/h (3.8 L/h). Approx. 30 to 50 ft (9 to 15 m) of drilling can be done with one filling of the unit. (Mason, 1951)

foaming agent A material that tends to stabilize a foam. *frothing agent.* (ASM, 1961)

foaming earth Soft or earthy aphrite. See also: *earth foam.*

foam injection The injection of foam into shotholes and connecting breaks to displace any combustible gases present and to minimize further combustible gases emission into the shotholes, thereby reducing the risk of ignition of the gas during shot firing. (BS, 1964)

foam plug A secondary method of fighting underground fires, devloped in Great Britain in 1956. It consists of filling the fire area with soap bubbles which are moved forward by the air current. The foam is produced by passing the air current through a cotton net, saturated with a dilute solution of detergent, which is stretched across the mine roadway. The air passing through the net forms bubbles 1/2 to 1-1/2 in (1.3 to 3.8 cm) in diameter which honeycomb and form a plug of foam that tends to quench the fire and reduce its temperature to a point where it can be attacked directly and without protective clothing. See also: *high-expansion foam.* (Nelson, 1965)

foam spar *aphrite.*

foamy amber Frothy amber. Almost opaque chalky white amber. Will not take a polish.

f.o.b. *free on board.*

focal sphere An arbitrary reference sphere drawn about the hypocenter or focus of an earthquake, to which body waves recorded at the

Earth's surface are projected for studies of earthquake mechanisms. (AGI, 1987)

focus (a) The initial rupture point of an earthquake, where strain energy is first converted to elastic wave energy; the point within the Earth which is the center of an earthquake. Syn: *hypocenter.* (AGI, 1987) (b) The point at which rays of light converge to form an image after passing through a lens or optical system or after reflection by a mirror. (AGI, 1987)

focused logging device Logging device designed to focus lines of current flow. (Wyllie, 1963)

fodder Eight pigs of cast iron. (Webster 2nd, 1960)

fog quenching Quenching in a fine vapor or mist. (ASM, 1961)

foid A term proposed by Johannsen; derived by contracting the word feldspathoid; used in his classification of igneous rocks to indicate one of the feldspathoidal group of minerals. Cf: *feldspathoid.* (AGI, 1987)

foig A crack or a break in the roof. (CTD, 1958)

fold (a) A curve or bend of a planar structure such as rock strata, bedding planes, foliation, or cleavage. A fold is usually a product of deformation, although its definition is descriptive and not genetic and may include primary structures. Cf: *anticline; syncline; monocline.* (AGI, 1987) (b) In crystallography, refers to the number of repetitions about a crystallographic axis to return to identity; i.e., equal to a complete rotation of 360°. Modern usage refers to a six-fold axis as a hexad, a four-fold axis as a tetrad, three-fold as triad, two-fold as diad, and one-fold as monad.

fold axis *axis.*

fold breccia A local tectonic breccia composed of angular fragments resulting from the sharp folding of thin-bedded, brittle rock layers between which are incompetent ductile beds; e.g., a breccia formed where interbedded chert and shale are sharply folded. Cf: *tectonic breccia.* (AGI, 1987)

fold fault A fault formed in causal connection with folding.

folding The formation of folds in rocks. (AGI, 1987)

fold mountain A mountain resulting chiefly from large-scale folding of the Earth's crust.

fold system A group of folds showing common characteristics and trends and presumably of common origin.

folia Thin, leaflike layers or laminae; specif., those of gneissic or schistose rocks. Singular form is folium. (AGI, 1987)

foliate A general term for any foliated rock. Adj. of foliation.

foliated Adj. of foliation.

foliation A general term for a planar arrangement of textural or structural features in any type of rock; esp., the planar structure that results from flattening of the constituent grains of a metamorphic rock. Adj. foliate; foliated. Cf: *schistosity.* (AGI, 1987)

follower chart A table showing (1) the size of casing or pipe that should be placed in a borehole drilled with a specific-size bit and (2) which sizes of casing or pipe can be nested inside each other. (Long, 1960)

follower rail The follower rail of a mine switch is the rail on the other side of the turnout corresponding to the lead rail. (Kiser, 1929)

following dirt A thin bed of unconsolidated dirt; a parting between the top of coal seam and the roof. See also: *pug.* (CTD, 1958)

followup tag The cardboard tag placed in the cartons, boxes, or cases of blasting supplies; used for identifying the date and place of manufacture. (Fay, 1920)

Follsain process A method for the sintering of the raw materials for the burden of blast furnaces in which continuous sintering is carried out in a rotating tube furnace; at the discharge end a special tuyere is arranged comprising two concentric close-ended tubes parallel to the furnace axis, the outer tube having one nozzle near its closed extremity, the other having a number of nozzles protruding through the outer tube. The inner tube supplies air heated to 650 to 800 °C; the outer one carries cold air, which keeps the inner tube from softening and becoming deformed and itself becomes somewhat heated by the time it emerges from the nozzle. These jets are directed upon the material to be sintered. The fine iron-bearing material is mixed with a proportion of fuel; under the intensive action of the hot air blast, the fuel raises the temperature of the mixture sufficiently for sintering to occur, whereupon the material is discharged from the furnace. (Osborne, 1956)

fool's gold *pyrite.*

foot (a) The bottom of a slope, grade, or declivity. The lower part of any elevated landform; e.g., the foot of a hill, the foot of a mountain, etc. (AGI, 1987) (b) The lower bend of a fold or structural terrace. Cf: *head.* Syn: *lower break.* (AGI, 1987) (c) *footwall.* (d) The foot is 12 in (30.5 cm) in length on the vein, including its entire width, whether 6 in (15.2 cm) or 60 ft (18.3 m), and its whole depth down toward the Earth's center. (Standard, 1964) (e) Corn. An ancient measure containing 2 gal or 60 lb of black tin. (f) That portion of the displaced material of a landslide that lies downslope from the toe of the surface of rupture (Varnes, 1978). (AGI, 1987)

foot-acre *acre-foot.*

footage block *marker block.*

footage per bit The average number of feet of borehole specific types of drill bits can be expected to drill in a certain rock before the bit becomes dulled and is replaced, discarded, resharpened, or reset. (Long, 1960)

foot clamp *safety clamp.*

footeite *connellite.*

foot hole Any of the holes cut in the sides of shafts or winzes to enable miners to ascend and descend. (Zern, 1928)

footing (a) A relatively shallow foundation by which concentrated loads of a structure are distributed directly to the supporting soil or rock through an enlargement of the base of a column or wall. Its ratio of base width to depth of foundation commonly exceeds unity. Cf: *pier.* (AGI, 1987) (b) The characteristics of the material directly beneath the base of a drill tripod, a derrick, or mast uprights. Also, the material placed under such members to produce a firm base on which they may be set. (Long, 1960)

footman In salt production, a laborer who adjusts the height of the gate in the chute leading from the crusher by means of a lever, to regulate flow of crushed rock salt into vibrating screens that separate salt into various sizes prior to shipment or refining. (DOT, 1949)

footmark *marker block.*

foot rod Scot. An iron rod at the foot of pump rods to which the bucket is attached. (Fay, 1920)

footwall (a) The underlying side of a fault, orebody, or mine working; esp. the wall rock beneath an inclined vein or fault. Syn: *heading wall; heading side; lower plate; foot.* Cf: *hanging wall.* See also: *wall; walls.* (AGI, 1987) (b) The wall or rock under a vein. It is called the floor in bedded deposits. (c) Opposite wall from hanging wall. (d) S. Afr. The wall on the lower side of a reef, lode, or fault. (e) Can. The underside of a vein or lens in relation to the dip of an ore deposit. (f) In metal mining, the part of the country rock that lies below the ore deposit.

footwall of a fault The lower wall of an inclined fault plane. (Ballard, 1955)

footwall shaft *underlay shaft.*

foot-yard In Pennsylvania, a miner's measurement of length, such as the distance a working face is advanced. With the heel of one foot on a mark, a short step is taken and the tip of the forward toe marks the foot-yard. The next measurement is taken by placing the first foot against the toe of the second and repeating the first step and so on. The foreperson checks measurements with a rule. (Hess)

Foraky boring method A percussive boring system comprising a closed-in derrick over the crown pulley of which a steel rope is passed from its containing drum. The boring tools are suspended from the end of the rope and are moved in the hole as required by means of the drum. A walking beam, operated by a driving mechanism, gives the boring tools a rapid vibrational motion. (Nelson, 1965)

Foraky freezing process One of the original freezing methods of shaft sinking through heavily watered sands. Although the principle is the same today, the process has been improved in many respects. See also: *freezing method.* (Nelson, 1965)

foram *foraminifer.*

foraminifer Any protozoan belonging to the subclass Sarcodina, order Foraminifera, characterized by the presence of a test (shell) of one to many chambers composed of secreted calcite (rarely silica or aragonite) or of agglutinated particles. Most foraminifers are marine but freshwater forms are known. They are important microfossils in well logging. Range, Cambrian to present. Syn: *foram.* Plural, foraminifera. (AGI, 1987)

foraminifera Plural of foraminifer.

forbesite A mixture of annabergite and arsenolite.

forble Four lengths of drill rod or drill pipe connected to form a section, which is handled and stacked in a drill tripod or derrick as a single unit on borehole round trips. Also spelled fourble. Cf: *treble.* (Long, 1960)

force That which tends to put a stationary body in motion or to change the direction or speed of a moving body. (AGI, 1987)

forced auxiliary ventilation A system in which the duct delivers the intake air to the face. The forcing system may be used with flexible ducting and simplifies arrangements for protecting the duct from blasting. The forcing system has the added advantage that the fan motor always works in intake air, and no special arrangements about fan drives are necessary. See also: *auxiliary ventilation.* (Roberts, 1960)

forced-caving system A stoping system in which the ore is broken down by large blasts into the stopes, which are kept partly full of broken ore. The large blasts break ore directly into the stopes and have the further effect of shattering additional ore, part of which then caves.

forced ventilation A system of ventilation in which the fan forces air through the workings under pressure. (BS, 1963)

force lines Stress fields can be represented by lines, each of which represents a definite force, so that their distance apart is a measure of the intensity of stress. The concept is similar to that of a magnetic field represented by lines of force. (Spalding, 1949)

force majeure (a) The clause in a metal supply contract which allows the seller not to deliver or the buyer not to take delivery of the metal concentrate or scrap under the contract because of events beyond the seller's control. (Wolff, 1987) (b) An unexpected occurrence beyond the control of parties to a contract, such as an earthquake preventing the timely delivery promised by a metal supply contract, that typically relieves a party from contract performance. The language of the contract, as well as applicable State or Federal laws, determine the applicability of this legal mechanism in any given situation. Also called an Act of God. (Padan, 1968)

force of blow The effective diameter of the piston or hammer, its weight, distance of travel, and the air pressure during the forward movement. The energy of the blow in foot-pounds is equal to: $1/2\ mv^2 = w \times v^2 / 64.4$, where m = the mass; w = the weight in pounds; v = the velocity of the hammer in feet per second. (Lewis, 1964)

force of crystallization *crystallizing force.*

force oscillator An instrument to determine the mechanical resonant frequency of a whole crystal or cut crystal plate. A slowly varying frequency is applied to the crystal from a signal generator, and the resonant frequency voltage developed across the crystal is measured with a voltmeter. Also called drive oscillator.

force pump (a) A pump consisting of a plunger or ram, the up-stroke of which causes the suction valve to open and the water to rise in the suction pipe. On the down-stroke of the plunger, the suction valve closes and the contained water is forced through the delivery valve into the rising main or discharge pipe. (Nelson, 1965) (b) A pump that forces water above its valves. (Zern, 1928) (c) A pump in which the water is lifted by the force due to atmospheric pressure acting against a vacuum. (Crispin, 1964)

forcer (a) A small hand pump used in Cornish mining. (Standard, 1964) (b) The solid piston of a force pump. (Standard, 1964)

forcherite An orange-yellow opal colored with orpiment. (Standard, 1964)

forcing fan A fan which blows or forces the intake air into the mine workings, as opposed to an exhaust fan. A mine exhaust fan may become a forcing fan (with reduced efficiency) when the ventilation is reversed in an emergency. (Nelson, 1965)

forcing lift Scot. A set of pumps raising water by a plunger; a ram pump. Also called forcing set. (Fay, 1920)

forcing set A pump for forcing water to a higher level or to the surface. (CTD, 1958)

forebay (a) A reservoir or pond at the head of a penstock or pipeline. (Seelye, 1951) (b) The water immediately upstream of any structure. (Seelye, 1951)

forechamber An auxiliary combination for gas-fired boilers that provides an incandescent surface for lighting gas instantly when turned on after being shut off for any reason. Also called Dutch oven; doghouse. (Fay, 1920)

fore drift The one of a pair of parallel headings that is kept a short distance in advance of the other. (CTD, 1958)

forehearth (a) A projecting bay in the front of a blast furnace hearth under the tymp. In open-front furnaces, it is from the forehearth that cinder is tapped. (Fay, 1920) (b) An independent settling reservoir into which is discarded the molten material from a furnace and which is heated from an independent source. The heavy metal settles to the bottom and the light slag rises to the surface. (Fay, 1920) (c) A section of a furnace, in one of several forms, from which glass is taken for forming. (ASTM, 1994)

foreign coal Coal received at a preparation plant from a source other than that to which the plant is attached. (BS, 1962)

foreigner Dark, ovoid inclusion of country rock in granite. Syn: *furrener.* (Arkell, 1953)

foreign inclusion A fragment of country rock enclosed in an igneous rock. Syn: *xenolith.*

foreland A stable area marginal to an orogenic belt, toward which the rocks of the belt were thrust or overfolded. Generally the foreland is a continental part of the crust, and is the edge of the craton or platform area. (AGI, 1987)

forelimb The steeper of the two limbs of an asymmetrical anticlinal fold. (AGI, 1987)

forelimb thrust A thrust fault cutting strata on the steeply dipping flank of an asymmetrical anticline.

forellenstein *troctolite.*

Forel scale The basic scale for measuring seawater color. See also: *water color.* (Hy, 1965)

foreman The head person; esp., the overseer of a group of workers. See also: *mine foreman; boss.* (Standard, 1964)

forepole A pointed board or steel strap with a sharp edge, that is driven ahead in loose ground for support purposes. See also: *spile; spill.* (Nelson, 1965)

forepoling (a) A method of advancing a mine working or tunnel in loose, caving, or watery ground, such as quicksand, by driving sharp-pointed poles, timbers, sections of steel, or slabs into the ground ahead of, or simultaneously with, the excavating; a method of supporting a very weak roof. It is useful in tunneling and in extracting coal from under shale or clay. See also: *close timbering; cribbing; poling; running ground.* (AGI, 1987) (b) A method of securing loose ground by driving poles, planks, etc., ahead of and on the top and sides of the timbers. See also: *spile.* (Ballard, 1955)

forepoling girder One of two or more heavy straight girders set over and in advance of the last permanent support in a tunnel. They provide protection to the worker until there is space to erect another support. (Nelson, 1965)

foreset (a) To set a prop under the fore or coal-face end of a bar. (TIME, 1929-30) (b) Timber set used at the working face for roof support. Also called force piece. (Pryor, 1963) (c) Temporary forward support; a middle prop under a bar. (Mason, 1951)

foreshaft sinking The first 150 ft (46 m) or so of shaft sinking from the surface, during which time the plant and services for the main shaft sinking are installed. Sometimes, the main sinking contract does not commence until the foreshaft has been completed. (Nelson, 1965)

foreshift (a) In coal mining, first or morning shift. (Pryor, 1963) (b) Eng. The first shift of hewers (miners) who go into the mine from 2 to 3 h before the drivers and loaders. (Fay, 1920)

foreshock (a) One of the initiating shocks preceding the principal earthquake. (Schieferdecker, 1959) (b) An earthquake that precedes a larger earthquake within a fairly short time interval (a few days or weeks), and which originates at or near the focus of the larger earthquake. (AGI, 1987) (c) A small tremor that commonly precedes a larger earthquake or main shock by an interval ranging from seconds to weeks and that originates at or near the focus of the larger earthquake. Cf: *aftershock.* (AGI, 1987)

foresight (a) A sight on a new survey point, taken in a forward direction and made in order to determine its bearing and elevation. Also, a sight on a previously established survey point, taken to close a circuit. (AGI, 1987) (b) A reading taken on a level rod to determine the elevation of the point on which the rod rests. Syn: *minus sight.* Abbrev. FS. Cf: *backsight.* (AGI, 1987)

foresight hub A stake or mark placed by a responsible individual some distance in front of a drill; used by a driller to point and line up a drill to drill a borehole in a specific direction. Also called front hub. See also: *picket.* (Long, 1960)

forest marble *landscape marble.*

forest moss peat Peat formed in forested swamps. (Tomkeieff, 1954)

forest peat Peat consisting mainly of the remains of trees that grew in low wet areas. (Francis, 1954)

forfeiture Penalty incurred in accordance with governing laws and regulations when mining concessions, claims leases, rights, are not adequately, safely, and consistently developed and exploited. Cf: *abandonment.* (Pryor, 1963)

forge (a) An open fireplace or hearth with forced draft, for heating iron, steel, etc.; e.g., a blacksmith's forge. (Standard, 1964) (b) A hearth or furnace for making wrought iron directly from ore; a bloomery. (Standard, 1964) (c) To form by heating in a forge and hammering; to beat into some particular shape, as a mass of metal. (Fay, 1920) (d) A plant where forging is carried out. (CTD, 1958) (e) Eng. That part of an ironworks where balls are squeezed and hammered and then drawn out into puddle bars by grooved rolls. (Fay, 1920)

forge cinder The dross or slag from a forge. (Fay, 1920)

forge iron Pig iron used for the charge of a puddling furnace. (Mersereau, 1947)

forge pigs Pig iron suitable for the manufacture of wrought iron. (CTD, 1958)

forge roll One of the train of rolls by which a slab or bloom of metal is converted into puddled bars. (Fay, 1920)

forge train In iron puddling, the series of two pairs of rolls by means of which the slab or bloom is converted into bars. (Fay, 1920)

forge welding A group of welding processes in which the parts to be joined are heated to a plastic condition in a forge or other furnace and are welded together by applying pressure or blows. (Hammond, 1965)

fork (a) An appliance used in free-fall drilling that serves to hold up the string of tools during connection and disconnection of the rods. (b) A double-pronged clip on a tub or wagon for the haulage rope or chain. (CTD, 1958) (c) A two-pronged lever used to slide the flat belt from a powerdrive over to an idler pulley (loose pulley). (Pryor, 1963) (d) Corn. Bottom of a drainage sump. (Pryor, 1963) (e) Derb. A piece of wood supporting the side of an excavation in soft ground, esp. if it has a Y-shaped end. (Fay, 1920) (f) Scot. A tool used for changing buckets, or for loading lump coal. (g) To pump water out of a mine. A mine is said to be in fork, or a pump to have the water in fork, when all the water is drawn out of the mine. (Webster 2nd, 1960)

fork-filled Aust. Coal filled into skips with a fork, having the prongs about 1-1/4 in (3.2 cm) apart. This separates the bulk of the slack from the round coal. The course product should not contain more than 10% of fine coal. See also: *box filling.*

form (a) In crystallography, all the crystal faces having a like position relative to the elements of symmetry of a point group; e.g., mirror planes and rotation axes. A form is closed if it encloses a volume, such as a cube or a rhombohedron; it is open if additional forms are necessary to enclose a volume, such as a prism or a pinacoid. A single crystal may exhibit faces of two or more crystal forms which supplement one another, such as a prism and a basal pinacoid; or truncate one another's edges or corners, such as a dodecahedron and a trapezohedron. (b) In geomorphology, a syn. of landform. (c) Those aspects of a particle shape that are not expressed by sphericity or roundness. Form can be described as ratios of the long, intermediate, and short axes, which can be combined into various "form indices," and by terms such as platy, bladed, elongate, or compact.

formanite A tetragonal mineral, $YTaO_4$; dimorphous with yttrotantalite; forms a series with fergusonite; in granite pegmatites and placer deposits.

format An informal rock stratigraphic unit bounded by marker horizons believed to be isochronous surfaces that can be traced across facies changes, particularly in the subsurface; useful for correlations between areas where the stratigraphic section is divided into different formations that do not correspond in time value. See also: *marker*. (AGI, 1987)

formation (a) A persistent body of igneous, sedimentary, or metamorphic rock, having easily recognizable boundaries that can be traced in the field without recourse to detailed paleontologic or petrologic analysis, and large enough to be represented on a geologic map as a practical or convenient unit for mapping and description; the basic cartographic unit in geologic mapping. Syn: *rock formation*. (AGI, 1987) (b) A body of sedimentary rock identified by lithic characteristics and stratigraphic position; it is prevailing but not necessarily tabular, and is mappable at the Earth's surface or traceable in the subsurface. The formation is the fundamental unit in lithostratigraphic classification. (AGI, 1987) (c) A general term applied by drillers to a sedimentary rock that can be described by certain drilling or reservoir characteristics; e.g. hard formation, cherty formation, or porous formation. (AGI, 1987) (d) A naturally formed topographic feature, commonly differing conspicuously from adjacent objects or material, or being noteworthy for some other reason; esp. a striking erosional form on the land surface. Syn: *geologic formation*. (AGI, 1987)

formation drilling Boreholes drilled primarily to determine the structural, petrologic, and geologic characteristics of the overburden and rock strata penetrated. Also called formation testing. (Long, 1960)

formation level Level of the ground surface after completion of excavation. (Hammond, 1965)

formation resistivity factor (a) In geophysical borehole logging, the ratio of the conductivity of an electrolyte to the conductivity of a rock saturated with that electrolyte. Symbol, F. (AGI, 1987) (b) The ratio of the resistivity of the saturated rock to the resistivity of the saturating water in a completely water-saturated clean rock. (Inst. Petrol., 1961)

formation striae Color bands in synthetic corundum or spinel which, because they are invariably distinctive and generally curved, differ from the straight color zones in natural minerals. Also called formation striations.

formation water Water present in a water-bearing formation under natural conditions, as opposed to introduced fluids, such as drilling mud. Syn: *native water*. Cf: *connate water*. (AGI, 1987)

form contour A topographic contour determined (1) by stereoscopic study of aerial photographs without ground control or (2) by other means not involving conventional surveying.

form energy The potentiality of a mineral to develop its own crystal form against the resistance of the surrounding solid medium (Eskola, 1939). (AGI, 1987)

formosa marble A high grade of marble; dark gray and white, variously mottled and blotched with yellow and red; from Nassau and Germany. (Fay, 1920)

fornacite A monoclinic mineral, $(Pb,Cu)_3[(Cr,As)O_4]_2(OH)$; forms prismatic crystals on dioptase. Also spelled furnacite.

forsterite An orthorhombic mineral, $4[Mg_2SiO_4]$; olivine group, with magnesium replaced by iron toward fayalite; trimorphous with ringwoodite and wadsleyite; in dunite and metamorphosed dolomitic limestones.

forsterite refractories Semibasic refractories made from olivine and magnesia; consist essentially of forsterite, including about 50% magnesia, 39% silica, 6% ferrous oxide, and 5% of other oxides. (Henderson, 1953)

forstid Derb. Light waste left after washing ore. Also called forstid ore. (Arkell, 1953)

forward The purchase or sale of metal for delivery at a specified future date. Hence "forward price," "forward contract.". (Wolff, 1987)

forward dealing Purchase of stocks, notably metals, for delivery at an agreed future date and price. (Pryor, 1963)

foshagite A triclinic mineral, $Ca_4Si_3O_9(OH)_2$; forms compact fibrous vein filling associated with vesuvianite and blue calcite at Crestmore, Riverside County, CA.

foshallassite A monoclinic mineral, $Ca_3Si_2O_7{\cdot}3H_2O$(?); occurs on the Kola peninsula, Russia, where it is related to foshagite and centrallassite and named from a combination of those names. Also spelled foshallassite.

fossick (a) Aust. To work out the pillars of abandoned claims, or work over waste heaps in hope of finding gold. (Standard, 1964) (b) Eng. In gold mining, to undermine another's digging. (Fay, 1920)

fossicker (a) Person who searches for small amounts of mineral. (CTD, 1958) (b) One who picks over old mine workings. Fossicking is casual and unsystematic mining. (Pryor, 1963) (c) Aust. A sort of mining gleaner who overhauls old workings and refuse heaps for gold that may be contained therein.

fossil (a) Any remains, trace, or imprint of a plant or animal that has been preserved in the Earth's crust since some past geologic or prehistoric time; loosely, any evidence of past life. (AGI, 1987) (b) Said of any object that existed in the geologic past and of which there is still evidence. (AGI, 1987) (c) Used in such expressions as fossil generating plant in reference to the use of fossil fuel rather than nuclear fuel. Not recommended usage. (AGI, 1987)

fossil copal *copalite*.

fossil erosion surface An erosion surface that was buried by younger sediments and was later exposed by their removal. Sometimes used as a syn. of buried erosion surface. (AGI, 1987)

fossil flour *diatomite*.

fossil fuel Coal, petroleum, or natural gas.

fossil ice (a) Ice formed in, and remaining from, the geologically recent past. It is preserved in cold regions, such as the coastal plains of northern Siberia, where remains of Pleistocene ice have been found. (b) Relatively old "ground ice" in a permafrost region. Also, underground ice in a region where present-day temperatures are not low enough to create it. (c) Crystal of selenite. (Arkell, 1953)

fossiliferous Containing fossils.

fossilization All processes involving the burial of a plant or animal in sediment and the eventual preservation of all, part, or a trace of it. (AGI, 1987)

fossilize To turn into a fossil. (Webster 3rd, 1966)

fossilized Preserved by burial in rock or earthy deposits; turned to a fossil.

fossilized wood *silicified wood*.

fossil ore An iron-bearing sedimentary deposit, in which shell fragments have been replaced and cemented together by hematite and carbonate. Cf: *flaxseed ore*. (AGI, 1987)

fossil paper *mountain paper; mountain leather*.

fossil pineapple Opal pseudomorph after glauberite; from New South Wales. (Schaller, 1917)

fossil resin Any of various natural resins found in geologic deposits as exudates of long-buried plant life; e.g., amber, retinite, and copal.

fossil salt *rock salt*.

fossil turquoise *odontolite*.

fossil wax *ozocerite*.

fother Any of the various units of weight for lead; esp. a unit equal to 19.5 hundredweight (885 kg). (Webster 3rd, 1966)

foul (a) A condition of the atmosphere of a mine so contaminated by gases as to be unfit for respiration. Impure. (b) In a coal seam, a place where the seam was washed out during deposition, leaving a barren area. (Pryor, 1963)

foul-air duct A suction line in a tunnel ventilation system. (Nichols, 1976)

foul gas Coke-oven gas or natural gas containing appreciable amounts of hydrogen sulfide and similar contaminants. (CCD, 1961)

fouling The assemblage of marine organisms that attach to and grow upon underwater objects. (Hy, 1965)

fouling position The point on any rail beyond which a wagon or mine car cannot proceed without becoming an obstruction to another wagon or car traveling on the intersecting rail. (Nelson, 1965)

fouls (a) Eng. A condition in which seams of coal disappear for a certain space and are replaced by some foreign matter. See also: *fault*. (Fay, 1920) (b) The cutting out of portions of the coal seam by wash outs or barren ground. (CTD, 1958)

found (a) To form in a mold, such as articles of cast iron, by melting the metal and pouring; cast. (Standard, 1964) (b) The name for the melting operation that the raw materials undergo in a furnace. (CTD, 1958)

foundation (a) The lower, supporting part of an engineering structure, transmitting the weight of the structure and its included loads to the underlying soil or rock material. It is usually below ground level. (AGI, 1987) (b) A term that is sometimes applied to the upper part of the soil or rock mass in contact with, and supporting the loads of, an engineering structure; the subsoil. (AGI, 1987) (c) Mid. The shafts, machinery, building, railways, workshop, etc., of a mine, commonly called a plant. (d) The ground upon which a structure or substructure is supported. (Taylor, 1965) (e) The lower part of a structure that transmits the load to the earth. (ASCE, 1958) (f) The base and/or the underlying support, either natural or artificial, on which a building, dam, or other structure is constructed. (Long, 1960) (g) *base rock.*

foundation bolt A fastener for connecting a structure or machine to a permanent base. See also: *anchor bolt.*

foundation curb A construction in a sinking shaft that will provide support for the concrete lining. It consists of a wedge-shaped excavation around the shaft in solid ground that is filled up completely with wet concrete. Steel shuttering is used and the concrete is filled in behind. Also called foundation canch; foundation crib; walling curb. See also: *curb; permanent shaft support.* (Nelson, 1965)

foundation investigation A branch of soil mechanics involving the drilling and testing of the deposits underlying a proposed foundation. It includes the estimation of bearing capacities, settlements, and the most suitable type of foundation for the prevailing soil conditions. See also: *bearing capacity; depth of soil exploration.* (Nelson, 1965)

foundation testing Testing in which boreholes are drilled for the purpose of obtaining samples by means of which the characteristics of overburden and/or the rock on which the foundation of a structure will rest can be determined. (Long, 1960)

founder breccia *collapse breccia.*

founding The act or process of casting metals. (Fay, 1920)

foundry sand Sand used by founders in making molds for castings. See also: *fire sand.*

four-cutter bit *roller rock bit.*

fourfold An axis of symmetry requiring four repetitions to complete 360° and return to identity. Syn: *tetrad.*

four-high mill A mill that contains four rolls arranged one above the other; i.e., two small-diameter working rolls supported by larger diameter back-up rolls above and below. (Osborne, 1956)

fourling A crystal twin consisting of four individuals; characteristic twinning exhibited by some varieties of authigenic microcline. Cf: *trilling; fiveling; eightling.*

fourmarierite An orthorhombic mineral, $PbU_4O_{13}{\cdot}4H_2O$; strongly radioactive; in pegmatites as an alteration product of uraninite.

four-piece set Squared timber frame used in underground driving to give all around support to weak ground. A cap is supported by two posts on a sill-piece or sill. See also: *timber set; three-piece set.* (Pryor, 1963)

four-way dip In seismic operations, a dip determined by spreads placed in four directions from a shot point. Three are essential, and the fourth serves as a check. (AGI, 1987)

fowlerite A variety of rhodonite containing up to 10% zinc.

foxbench *iron pan.*

foyaite A nepheline syenite with predominant orthoclase and a trachytic texture. Not a name in the I.U.G.S. classification of plutonic rocks. Syn: *midalkalite.* See also: *nepheline syenite.*

fraction A portion of an unconsolidated sediment or of a crushed consolidated rock sample or of a crushed ore or mineral sample that has been separated by some method, and is distinguished in some manner from all the other portions (or fractions) comprising the whole sample being analyzed. Also a fraction may be separated and defined on the basis of its mineral content, its specific gravity or density, its magnetism or lack of magnetism, or its solubility or insolubility in acid.

fractional crystallization (a) Crystallization in which the early-formed crystals are prevented from equilibrating with the liquid from which they grew, resulting in a series of residual liquids of more extreme compositions than would have resulted from equilibrium crystallization. Cf: *crystallization differentiation.* Syn: *fractionation.* (AGI, 1987) (b) Controlled precipitation from a saline solution of salts of different solubilities, as affected by varying temperatures or by the presence of other salts in solution. (AGI, 1987)

fractional distillation A distillation process for the separation of the various components of liquid mixtures. An effective separation can only be achieved by the use of fractionating columns attached to the still. (CTD, 1958)

fractional shoveling A method of sampling sometimes used at points where coal or mineral is loaded or unloaded by shoveling. Every tenth (or other number) shovelful is deposited separately as sampling material. (Nelson, 1965)

fractionating column An apparatus for separating the high-boiling and low-boiling fractions of a substance, whereby the fractions with the lowest boiling point distill over. The efficiency depends on the column length and on the number of bubble plates used.

fractionation (a) The separation of (1) a substance from a mixture, such as the separation of one isotope from another of the same element; (2) one mineral or group of minerals from a mixture; or (3) one size fraction from a mixture. (AGI, 1987) (b) Separation of chemical elements in nature, by processes such as preferential concentration of an element in a mineral during magmatic crystallization, or differential solubility during rock weathering. *fractional crystallization.* (AGI, 1987)

fractography The study of the surfaces of fractures, esp. microscopic study. (AGI, 1987)

fracture (a) A general term for any break in a rock, whether or not it causes displacement, due to mechanical failure by stress. Fracture includes cracks, joints, and faults. (AGI, 1987) (b) The breaking of a mineral other than along planes of cleavage or parting. A mineral can be described in part by its characteristic fracture; e.g., uneven, fibrous, conchoidal, or hackly. Cf: *parting; uneven fracture.* (AGI, 1987) (c) Deformation due to a momentary loss of cohesion or loss of resistance to differential stress and a release of stored elastic energy. Cf: *flow.* Syn: *rupture.*

fracture cleavage A type of cleavage that occurs in deformed but only slightly metamorphosed rocks and that is based on closely spaced, parallel joints and fractures. Cf: *flow cleavage.* (AGI, 1987)

fractured formation *fractured ground.*

fractured ground Rock formation shattered and crisscrossed with fissures and fractures. Cf: *broken ground.* (Long, 1960)

fracture dome The fracture dome is the zone of loose or semiloose rock which exists in the immediate hanging or footwall of a stope. In some mines it may extend into the walls for a considerable distance. In a rock burst it becomes greatly extended. (Spalding, 1949)

fracture porosity Porosity resulting from the presence of openings produced by the breaking or shattering of an otherwise less pervious rock.

fracture stress The differential stress at the moment of fracture. Syn: *breaking stress.*

fracture system A set or group of contemporaneous fractures related by stress. (AGI, 1987)

fragment (a) A rock or mineral particle larger than a grain. (AGI, 1987) (b) A piece of rock that has been detached or broken from a preexisting mass; e.g., a clast produced by volcanic, dynamic, or weathering processes. (AGI, 1987)

fragmental Formed from fragments of preexisting rocks. See also: *clastic.*

fragmental rock *clastic rock.*

fragmental texture (a) A general textural term applied to rocks composed of fine materials or of sandy, conglomeratic, bouldery, and brecciated materials. The texture of clastic rocks. (AGI, 1987) (b) A texture of sedimentary rocks, characterized by broken, abraded, or irregular particles in surface contact, and resulting from the physical transport and deposition of such particles; the texture of a clastic rock. The term is used in distinction to a crystalline texture. (AGI, 1987) (c) The texture of a pyroclastic rock, such as a tuff or volcanic breccia. (AGI, 1987)

fragmentation (a) The breakage of rock during blasting in which explosive energy fractures the solid mass into pieces; the distribution of rock particle sizes after blasting. See also: *degradation.* (b) Index of the degree of breaking up of rock after blasting. (Fraenkel, 1953)

framboidal texture A texture in which pellets form spheroidal aggregates resembling a raspberry.

frame (a) In trench excavations requiring timbering, the struts separating the boards, together with the walings, which they hold, form a frame. (Hammond, 1965) (b) Eng. A table composed of boards slightly inclined, over which runs a small stream of water to wash off waste from slime tin; a buddle. Also called rack. (Fay, 1920)

frame dam Eng. A solid, watertight stopping or dam in a mine to keep back and resist the pressure of a heavy head of water.

framed dam A barrier, generally built of timber framed to form a water face, supported by struts. (Seelye, 1951)

frame set The legs and cap or crossbar arranged so as to support the roof of an underground passage. Also called framing; set. (Fay, 1920)

framesite A variety of black bort from South Africa showing minute brilliant points possibly due to included diamonds. Cf: *bort.* (Tomkeieff, 1954)

framing table An inclined table used in separating ore slimes by running water; a miner's frame. (Standard, 1964)

France screen A traveling-belt screen in which the screen cloth is mounted on a series of separate pallets, thus avoiding bending the screen as it goes over the pulleys. (Liddell, 1918)

francevillite An orthorhombic mineral, $(Ba,Pb)(UO_2)_2(VO_4)_2{\cdot}5H_2O$; forms a series with curienite; occurs as yellow impregnations in sandstone; from Franceville, Gabon.

Franciscan Complex Jurassic to Early Cretaceous rocks, characteristic of the Pacific coastal ranges of California, composed primarily of sandstones, cherts, serpentinites, and glaucophane schists. The Franciscan should not be visualized as a formation or sequence with ordinary physical, spatial, and temporal coherence, but rather as a disorderly assemblage of various characteristic rocks that have undergone unsystematic disturbance; a mélange. The formation includes deep-water sediments and mafic marine volcanic material, locally accompanied by masses of serpentinite.

Francisci furnace A furnace for the treatment of roasted blende and other fine ore. It consists of a series of superimposed muffles formed by arches of magnesia brick and built into the walls of the furnace and communicating with a common condensation chamber. (Fay, 1920)

franckeite A triclinic mineral, $(Pb,Sn)_6FeSn_2Sb_2S_{14}$; forms imperfect, radiated folia. Occurs at Poopo, Llallagua, and several other tin districts of Bolivia; at Coal River, UT; in Canada; and in Inyo and Santa Cruz Counties, CA.

Francois sinking process The cementation sinking method. The process was introduced into Great Britain in 1911. See also: *cementation sinking*. (Nelson, 1965)

frankdicksonite An isometric mineral, BaF_2; fluorite group.

franklinite An isometric mineral, $8[ZnFe_2O_4]$; magnetite series; spinel group; forms metallic black octahedra with rounded edges; weakly ferrimagnetic; a source of zinc at the Franklin and the Sterling Hill deposits, NJ.

Frasch process (a) A process for mining native sulfur, in which superheated water is forced into the deposits for the purpose of melting the sulfur. The molten sulfur is then pumped to the surface. (AGI, 1987) (b) A desulfurizing process that consists of distilling oil over lead oxide, followed by refining with sulfuric acid. See also: *sulfur mining*. (Fay, 1920)

Frasch sulfur Native sulfur mined by the Frasch hot-water process. (USBM, 1965)

Fraser's air-sand process A dense-media process in which a dry, specific-gravity separation of coal from refuse is achieved by utilizing a flowing dense medium intermediate in density between coal and refuse. The dense medium is formed by bubbling air through a mass of dry sand, 30 to 80 mesh in size. The air dilates and fluidizes the sand mass, causing it to behave somewhat as a heavy liquid. The coal floats on the aerated sand mass and the refuse sinks. Syn: *air-sand process*. (Mitchell, 1950)

Fraunhofer lines Dark lines in the solar spectrum resulting from absorption of light by chemical elements in the chromosphere.

frautschy bottle This water-sampling device is messenger actuated. It is designed to allow free flow while in the cocked position on the downward traverse. When the desired sampling point has been reached, the closures are messenger actuated, resulting in isolation of the sample on the return traverse. Frautschy bottles may be attached to the hydrographic wire at intervals and in such a manner that release of a single messenger from the surface will actuate the entire series. In this way samples from several depths may be obtained in a single operation. (Hunt, 1965)

freboldite A hexagonal mineral, CoSe; nickeline group; in trace amounts with clausthalite, trogtalite, and hastite in dolomite veinlets in the Trogtal quarries, Harz Mountains, Germany.

free (a) Native; uncombined with other elements, such as free gold or free silver (native gold or native silver). (Fay, 1920) (b) Chemically uncombined or readily obtainable in uncombined form by heating, as opposed to bound; e.g., free water or free oxygen.

free acidity Free acidity is considered to be the portion of the total acidity that exists in the form of acid, both ionized and un-ionized. (Felegy, 1948)

free air (a) Air under conditions of atmospheric pressure and temperature. The condition of the air at the intake of the compressor, whatever the temperature and barometric pressure may be. (Lewis, 1964) (b) The total area of open space in a grille through which air can pass. (Strock, 1948)

free-air anomaly A gravity anomaly calculated from a theoretical model and elevation above sea level, but without allowance for the attractive effect of topography and isostatic compensation. See also: *anomaly*. (AGI, 1987)

free-air correction (a) *free-air anomaly*. (b) A correction for the elevation of a gravity measurement, required because the measurement was made at a different distance from the center of the Earth than the datum. The first term of the free-air correction is 0.09406 mgal/ft (0.3086 mgal/m). (AGI, 1987)

free-burning coal (a) A bituminous coal having so little fusibility that enough air for rapid combustion can flow between the lumps and high enough in volatiles and fixed carbon to burn readily. (Hess) (b) Coal that does not cake in the fuel bed and which has a high volatile matter. (BS, 1960)

free cementite Iron carbide in cast iron or steel other than that associated with ferrite in pearlite. (CTD, 1958)

free chalk Eng. A variety of soft marly chalk; Sussex. (Arkell, 1953)

free circulation The circulation of a drilling fluid, the flow of which is not restricted by obstructing materials in the borehole or inside the drill string. (Long, 1960)

free crushing Crushing under conditions of speed and feed such that there is plenty of room for the fine ore to drop away from the coarser part and thereby escape further fine crushing. Cf: *choke crushing*. (Fay, 1920)

free cyanide The cyanide not combined in complex ions. (ASM, 1961)

free-drainage level An adit. A level that drains through an adit.

free end *free face*.

free face (a) A longwall face with no props between the conveyor and the coal. See also: *prop-free front*. (Nelson, 1965) (b) A surface in the vicinity of a shothole at which the rock is free to move under the force of the explosion. (BS, 1964) (c) The exposed surface of a mass of rock, or of coal. Also called free end. (Zern, 1928)

free fall (a) An arrangement by which, in deep boring, the bit is allowed to fall freely to the bottom at each drop or downstroke. (b) The process of operating the drill. Often called Russian, Canadian, or Galician free fall.

free falling In ball milling, the peripheral speed at which part of the crop load breaks clear on the ascending side, and falls clear to the toe of the charge. (Pryor, 1963)

free-falling device A sliding piece in percussive boring designed to reduce the vibration and jarring effects when the downward movement of the chisel is suddenly arrested by striking the bottom of the borehole. The lower portion (which is attached to the chisel) is free to slide up and down in a slot provided in the upper part of the joint. When the chisel strikes the bottom of the hole, the slot allows the rods to continue the downward movement without being jarred by the blow of the chisel. See also: *jar*. (Nelson, 1965)

free fed In comminution, rolls are said to be free fed when fed only enough material to keep a ribbon of ore between the rolls. This results in a remarkably uniform product. Cf: *choke fed*. (Newton, 1959)

free ferrite Ferrite in steel or cast iron other than that associated with cementite in pearlite. (CTD, 1958)

free field stress The stresses existing in rock before the excavation of any mine opening. In general, these stresses are known to be influenced primarily by the weight of the overlying material, the relation of the opening of the rock masses around it (depth of overburden, etc.), the physical characteristics to the surrounding rock, and tectonic forces. (Lewis, 1964)

free gold (a) Gold uncombined with other substances. (b) Placer gold.

free haul The distance every cubic yard is entitled to be moved without an additional charge for haul. (Nichols, 1976)

free-milling Applied to ores that contain free gold or silver, and can be reduced by crushing and amalgamation without roasting or other chemical treatment. (Fay, 1920)

free-milling gold Gold with so clean a surface that it readily cyanides after liberation by comminution.

free-milling ore Ore containing gold that can be readily cyanided.

free moisture (a) Moisture in coal that can be removed by ordinary air drying. Cf: *combined moisture*. (Cooper, 1963) (b) The part of the total moisture that is lost by a coal in attaining approximate equilibrium with the atmosphere to which it is exposed. (BS, 1960) (c) Moisture not retained or absorbed by aggregate. (Taylor, 1965) (d) *moisture content*. (e) Moisture removable by air-drying under standard conditions. Also called surface moisture. (BS, 1962) (f) *free water*.

free on board (a) Price of consignment to a customer when delivered, with all prior charges paid, onto a ship or truck. (Pryor, 1963) (b) Free on rail (f.o.r.) describes similar delivery to rail. (Pryor, 1963)

free particles Particles of ore consisting of a single mineral. (Gaudin, 1939)

free-piston drive sampler A drive-sample barrel in which a piston is free to move upward with the top of the sample during the actual dry-sampling operation. (Long, 1960)

free radical (a) A chemical species that is uncharged and has one or more unpaired electrons. (b) An atom or molecule with at least one unpaired electron.

free settling As opposed to hindered settling in classification, free fall of particles through fluid media. (Pryor, 1965)

free share Som. A certain proportion of a royalty on coal, paid to lessor by lessee. (Fay, 1920)

free silica Quartz occurring in granites. (Mason, 1951)

free split In parallel ventilation flow, the branch or airway that does not require artificial resistance (a regulator) to achieve its design air quantity. See also: *open split*. (Hartman, 1982)

freestone Any stone (esp. a thick-bedded, even-textured, fine-grained sandstone) that breaks freely and can be cut and dressed with equal ease in any direction without splitting or tending to split. The ease with which it can be shaped into blocks makes it a good building stone. The term was originally applied to limestone, and is still used for such rock. Cf: *flagstone*. (AGI, 1987)

free streaming A combustible gases roof layer flowing under the action of buoyancy without ventilation. (BS, 1963)

free wall The wall of a vein filling that scales off cleanly from the gouge. (Schieferdecker, 1959)

free water (a) Water that is free to move through a soil mass under the influence of gravity. Also called ground water; phreatic water. (ASCE, 1958) (b) Water in soil in excess of hygroscopic and capillary water; also called gravity water. (Seelye, 1951) (c) The quantity of water removed in drying a solid to its equilibrium water content.

free-water elevation *water table.*

free-water level The surface of a body of water in contact with the atmosphere; i.e., at atmospheric pressure.

free-water surface *water table.*

free way A direction of easy splitting in a rock. (Fay, 1920)

freeze (a) To permit drilling tools, casing, drivepipe, or drill rods to become lodged in a borehole by reason of caving walls, or impaction of sand, mud, or drill cuttings, to the extent that they cannot be pulled out. Also called bind; seize. (Long, 1960) (b) The act or process of drilling a borehole utilizing a drill fluid chilled to -30 °C to -40 °C, as a means of consolidating, by freezing, the borehole wall materials and/or core as the drill bit penetrates a water-saturated formation, such as sand, gravel, etc. (Long, 1960)

freeze-in (a) Used in much the same sense as definition a. of "freeze." (b) Applicable when drill rods become fastened by solidification or freezing of the drilling fluid in a borehole drilled in permafrost. (Long, 1960) (c) To become or be fixed in ice. (Long, 1960)

freeze proofing A surface treatment, as with calcium chloride solution, to prevent or reduce cohesion of coal particles by ice formation during freezing weather. (BS, 1962)

freeze sinking Use of circulating brine in a system of pipes to freeze waterlogged strata so that shafts can be sunk through them, established, and lined. (Pryor, 1963)

freeze-thaw action *frost action.*

freeze-up In ball milling, the theoretical rate of revolution at which the contents of the mill are centrifugally held at the circumference. (Pryor, 1963)

freezing Consolidation of fine-grained waterlogged soil, enabling excavation to proceed, can be effected by freezing. The process, which dates from 1862, is particularly suitable for shaft sinking. (Hammond, 1965)

freezing interval *crystallization interval.*

freezing method A method of shaft sinking through loose waterlogged sands that are not suitable for the cementation sinking method. Rings of lined boreholes are put down outside the proposed shaft and in them a very cold solution, such as brine, is circulated until an ice wall has been formed sufficiently thick to enable sinking to proceed normally. The method consists of the following stages: (1) forming a protective wall of ice, with its base in an impervious deposit; (2) maintaining the ice wall until the sinking and lining of the shaft has been completed, and (3) thawing out the ground without damage to the shaft. The freezing method has been revived, largely due to the successful use of bulk concrete, backed by corrugated sheets in place of tubbing, for lining the shaft through the frozen ground. This is followed by wall grouting. Freezing was introduced originally in 1883 by F. H. Poetsch. See also: *Oetling freezing method; chemical soil consolidation; silicatization process.* (Nelson, 1965)

freibergite An isometric mineral, $(Ag,Cu,Fe)_{12}(Sb,As)_4S_{13}$; tetrahedrite group; forms series with argentotennantite and with tetrahedrite; occurs as tetrahedra; in Idaho, Colorado, Nevada, and Germany.

freieslebenite A monoclinic mineral, $AgPbSbS_3$. Occurs with pyrargyrite, argentite, and galena in late-stage silver ores at Hiendelaencina, Spain; Freiberg, Germany; Oruro, Bolivia; and Rosebery, Tasmania, Australia.

Fremont etching reagent An etchant consisting of 10 g of iodine and 20 g of potassium iodide in 100 mL of water. (Osborne, 1956)

French chalk A soft, white variety of talc, steatite, or soapstone finely ground into powder.

French drain A covered ditch containing a layer of fitted or loose stone or other pervious material. (Nichols, 1976)

French process A process in which zinc is distilled and the vapor burned to produce the oxide; the purity of the oxide is controlled by the purity of the metal. (Newton, 1959)

Frenier sand pump Spiral ribbon of steel enclosed between two steel disks, mounted on a horizontal hollow shaft into which pulp picked up peripherally is discharged during slow rotation. (Pryor, 1963)

Frenkel defect A type of "point defect" in a crystal structure where an atom or ion is displaced from its normal position to an interstitial one. Cf: *crystal defect; point defect; interstitial; Schottky defect.*

frenzied S. Staff. Said of coal crushed by the creep or subsidence of the cover. (Fay, 1920)

frequency factor In crystallography, the number of different families of planes having the same form. (Henderson, 1953)

frequency rate The rate of occurrence of accidents as determined by multiplying the actual number of injuries in any given period by 200,000 and dividing the product by the number of man-hours exposure. Syn: *incidence rate.*

frequency response (a) The percentage response of a seismic amplifier for various frequencies at a given filter setting. (AGI, 1987) (b) Attenuation as a function of frequency produced by passage of a signal through an element, such as a geophone or filter. (AGI, 1987)

fresh Said of a rock or rock surface that has not been subjected to or altered by surface weathering, such as a rock newly exposed by fracturing. Syn: *unweathered.* (AGI, 1987)

fresh air Air free from the presence of deleterious gases. Pure air.

fresh-air base An underground station, located in the intake airway, that is used by rescue teams during underground fires and rescue operations. The base should be as close to the fire as safety will permit, adequately ventilated, and in constant touch with the surface by telephone. (Nelson, 1965)

freshet (a) A great rise in, or a sudden overflowing of, a small stream, usually caused by heavy rains or rapidly melting snow in the highlands at the head of the stream; a rapidly rising flood, usually of minor severity and short duration. See also: *flash flood.* (AGI, 1987) (b) A small clear freshwater stream or current flowing swiftly into the sea; an area of comparatively fresh water at or near the mouth of a stream flowing into the sea. (AGI, 1987) (c) A small stream flowing swiftly into a lake (as in the spring) and often carrying a heavy silt load during its peak flow. (AGI, 1987)

fresh water (a) Water containing less than 1,000 mg/L of dissolved solids; generally, water with more than 500 mg/L is undesirable for drinking and for many industrial uses (Solley et al., 1983). (AGI, 1987) (b) In general usage, the water of streams and lakes unaffected by salt water or salt-bearing rocks. Syn: *sweet water.* Also spelled freshwater; fresh-water. (AGI, 1987)

fresh-water limestone A limestone formed by accumulation or precipitation in a freshwater lake, a stream, or a cave. It is often algal and sometimes nodular. See also: *underclay limestone.* (AGI, 1987)

fresnoite A tetragonal mineral, $Ba_2TiSi_2O_8$; yellow with yellow fluorescence; resembles idocrase.

fretwork A result of honeycomb weathering, consisting of small pits in a rock surface that become fewer as they grow larger and deeper.

freudenbergite A monoclinic mineral, $Na_2(Ti,Fe)_8O_{16}$; forms pseudohexagonal crystals in an apatite-rich alkali pegmatite at Katzenbuckel, Odenwald, Germany.

Freudenberg plates Iron plates suspended in dust chambers for the purpose of settling dust and condensing fumes that escape from the furnace with the gases. (Fay, 1920)

freyalite A rare-earth-rich variety of thorite from Brevik, Norway.

friability Tendency for particles to break down in size (degrade) during storage and handling under the influence of light physical forces. (Pryor, 1963)

friable (a) Said of a rock or mineral that crumbles naturally or is easily broken, pulverized, or reduced to powder, such as a soft or poorly cemented sandstone. (AGI, 1987) (b) Said of a soil consistency in which moist soil material crushes easily under gentle to moderate pressure (between thumb and forefinger) and coheres when pressed together. (AGI, 1987)

friable amber *gedanite.*

friable formation A rock that breaks easily or crumbles naturally; hence a formation from which good core cannot be obtained easily. (Long, 1960)

friction Mechanical resistance to the relative motion of contiguous bodies or of a body and a medium. (AGI, 1987)

frictional electricity Electrostatic charge developed by rubbing amber, tourmaline, topaz, diamond, and some plastic imitations with a cloth.

frictional force The force required to overcome friction when a set of tubs or a run of wagons is hauled along a level track at uniform speed. For ordinary pit tubs the frictional force is about 40 lb/st (20 kg/t) load, and for mine cars or wagons it is about 28 lb/st (14 kg/t) load. This resistance is sometimes called traction. (Morris, 1958)

frictional grip Adhesion between the wheels of a mine locomotive and the rails of the track, its magnitude depending only on locomotive weight and the coefficient of friction between the wheels and track. (Nelson, 1965)

friction breccia A breccia composed of broken or crushed rock fragments resulting from friction; e.g., a fault breccia. (AGI, 1987)

friction factor A factor that measures an airway's drag on the air moving through it. The friction factor for an airway is found by determining the drop in total pressure over a measured length. (Hartman, 1982)

friction head (a) That part of the hydraulic-feed yoke on a diamond drill containing the bearings by means of which the thrust of the hydraulic-feed pistons is transmitted to the drive rod in the drill swivel head. Syn: *cage; collar.* (Long, 1960) (b) The additional pressure that the pump must develop to overcome the frictional resistance offered by the pipe, by bends or turns in the pipeline, by changes in the pipe diameter, by valves, and by couplings. (Carson, 1961) (c) The pressure required to overcome the friction created by the flow of a confined liquid, such as the flow of a drill fluid through drill rods. (Long, 1960)

friction loss *friction head.*

friction socket A tubular-shaped or slightly inside-tapered fishing tool. The inside surface of the tool is nearly covered with circular pinged protuberances, which, when driven over the lost drill tools, wedge the tools in the socket. (Long, 1960)

friction yielding prop *mechanical yielding prop.*

friedelite A monoclinic mineral, $6[(Mn,Fe)_8Si_6O_{15}(OH,Cl)_{10}]$; pseudotrigonal; in manganese skarns, with other manganese silicates; a source of manganese in Sussex County, NJ.

friesite An orthorhombic mineral, $Ag_2Fe_5S_8$; in cobalt-nickel-silver ores in Saxony, Germany, and Joachimsthal (Jáchymov), Czech Republic.

fringe water Water of the capillary fringe. (AGI, 1987)

fringing reef An organic reef that is directly attached to or borders the shore of an island or continent, having a rough, tablelike surface that is exposed at low tide; it may be more than 1 km wide, and its seaward edge slopes sharply down to the sea floor. There may be a shallow channel or lagoon between the reef and the adjacent mainland. Syn: *shore reef.* (AGI, 1987)

Frisbie's feeder A device whereby a bucket of coal is forced up into the eye of a pot furnace from below. (CTD, 1958)

frith *firth.*

fritting The partial melting of grains of quartz and other minerals, so that each grain becomes surrounded by a zone of glass. Fritting results from the contact action of basalt and related lavas on other rocks (Johannsen, 1931). (AGI, 1987)

frog (a) The point of intersection of the inner rails, where a train or tram crosses from one set of rails to another. The frog is in the form of a V. See also: *turnout.* (CTD, 1958) (b) A combination of rails so arranged that the broad tread of the wheel will always have a surface on which to roll and the flange of the wheel will have a channel through which to pass. See also: *reurailer.* (Zern, 1928)

frog size A track haulage term for any distance from the point of the frog to the spread divided by the width of the spread at the place where the measurement was taken. (Kentucky, 1952)

frohbergite An orthorhombic mineral, $FeTe_2$; marcasite group; forms a series with mattagamaite; sp gr, 8.07; associated with gold, altaite, petzite, and other gold and tellurium minerals in polished ore at Montbray, PQ, Canada.

frolovite A triclinic mineral, $CaB_2(OH)_8$; occurs with calciborite in limestone skarn from Novo-Frolov copper mine, Turinsk district, northern Urals, Russia.

Froment process A flotation process in which a sulfide ore is agitated in water with a little oil and sulfuric acid; the sulfide particles become oiled and attach themselves to, and are floated by, gas bubbles. (Liddell, 1918)

frondelite An orthorhombic mineral, $MnFe_4(PO_4)_3(OH)_5$; forms a series with rockbridgeite; an alteration mineral in granite pegmatites in the Black Hills of South Dakota, New Hampshire, and Minas Gerais, Brazil.

front (a) A designation for the mouth or collar of a borehole. See also: *face.* (b) A metamorphic zone of changing mineralization developed outward from an igneous mass. Cf: *basic front.* (AGI, 1987) (c) The working attachment of a shovel, such as a dragline, hoe, or dipper stick. (Nichols, 1976) (d) In connection with concepts of granitization, the limit to which diffusing ions of a given type are carried; e.g., the simatic front is the limit to which diffusing ions carried the calcium, iron, and magnesium that they removed from the rocks in their paths. The granitic front is the limit to which diffusing ions deposited granitic elements. (Leet, 1958)

front abutment pressure The release of energy in the superincumbent strata above the seam, induced by the extraction of the seam. (Sinclair, 1958)

front-end equipment The attachments to a crane that enable it to work as an excavator, a skimmer, a back acter, or a similar machine. See also: *jib.* (Hammond, 1965)

front-end loader (a) A tractor loader with a digging bucket mounted and operated at the front end of the tractor. (Nichols, 1976) (b) A tractor loader that both digs and dumps in front. See also: *tractor shovel.* (Nichols, 1976)

frontland *foreland.*

froodite A monoclinic mineral, $PdBi_2$; forms minute metallic grains in mill concentrates at the Frood Mine, Sudbury, ON, Canada.

frost action (a) The mechanical weathering process caused by repeated cycles of freezing and thawing of water in pores, cracks, and other openings, usually at the surface. (AGI, 1987) (b) The resulting effects of frost action on materials and structures. Syn: *freeze-thaw action; frost splitting.* (AGI, 1987)

frost-active soil A fine-grained soil that undergoes changes in volume and bearing capacity due to frost action (Nelson & Nelson, 1967). (AGI, 1987)

frost boil (a) A local accumulation of excess water and mud liberated from ground ice by accelerated spring thawing, softening the soil and causing a quagmire. (AGI, 1987) (b) A break in a surface pavement due to swelling frost action; as the ice melts, soupy subgrade materials issue from the break. (AGI, 1987)

frost crack A nearly vertical fracture developed by thermal contraction in rock or in frozen ground with appreciable ice content. Frost cracks commonly intersect to form polygonal patterns in plan view. (AGI, 1987)

frost creep Soil creep resulting from frost action.

Frost gravimeter An astatic gravity meter of the balance type, consisting of a mass at the end of a nearly vertical arm, supported by a main spring inclined to the vertical at about a 45° angle. The beam rises and falls with gravity variation, but is restored to its normal position by a sensitive weighing spring tensioned by a micrometer screw. (AGI, 1987)

frost heaving The uneven upward movement, and general distortion, of soils, rocks, vegetation, and structures such as pavements, due to subsurface freezing of water and growth of ice masses (esp. ice lenses); any upheaval of ground caused by freezing. (AGI, 1987)

frosting (a) A lusterless ground-glass or mat surface on rounded mineral grains, esp. of quartz. It may result from innumerable impacts of other grains during wind action or from deposition of many microscopic crystals; e.g., fine silica secondarily deposited on quartz grains. (AGI, 1987) (b) The process that produces such a surface. (AGI, 1987)

frost line (a) The maximum depth of frozen ground in areas where there is no permafrost; it may be expressed for a given winter, as the average of several winters, or as the greatest depth on record. (AGI, 1987) (b) The bottom limit of permafrost. (AGI, 1987) (c) The altitudinal limit below which frost never occurs; applied esp. in tropical regions. (AGI, 1987)

frost mound A general term for a knoll, hummock, or conical mound in a permafrost region, containing a core of ice; represents a generally seasonal and localized upwarp of the land surface, caused by frost heaving and/or hydrostatic pressure of ground water. (AGI, 1987)

frost pin A short, heavy iron pin used by surveyors to make a hole in frozen ground so that a wooden peg may be driven without breaking. (Fay, 1920)

frost splitting The breaking of rock by water freezing in cracks. Syn: *frost action; frost weathering; frost wedging.*

frost weathering *frost splitting.*

frost wedging *frost splitting.*

froth In the flotation process, a collection of bubbles resulting from agitation, the bubbles being the agency for raising (floating) the particles of ore to the surface of the cell. (Hess)

frother A substance used in a flotation process to make air bubbles sufficiently permanent, principally by reducing surface tension. See also: *frothing agent.* (Hess; Newton, 1959)

froth flotation (a) A flotation process in which the minerals floated gather in and on the surface of bubbles of air or gas driven into or generated in the liquid in some convenient manner. See also: *film flotation.* (b) The separating of finely crushed minerals from one another by causing some to float in a froth and others to remain in suspension in the pulp. Oils and various chemicals are used to activate, make floatable, or depress the minerals. (CTD, 1958) (c) A process for cleaning fine coal, copper, lead, zinc, phosphate, kaolin, etc. with the aid of a reagent; the coal or minerals become attached to air bubbles in a liquid medium and float as a froth. (BS, 1962)

frothing agent (a) A reagent used to control the size and stability of the air bubbles in the flotation process. Syn: *foaming agent; frother.* (BS, 1962) (b) A chemical used in the flotation process to aid collector-coated mineral particles to cling to risen air bubbles. The froth thus formed is transient and should persist only long enough to permit its removal from the flotation cell. Terpenes, pine oil, cresyls, amyl alcohol, and alcohol derivatives are among the agents used. (Pryor, 1963)

frothing collector A collector that also produces a stable foam. (Bennett, 1962)

froth promoter A chemical compound used with a frothing agent. Increases greatly the recovery in a flotation process. (Bennett, 1962)

frothy amber *foamy amber.*

Froude's curve In surveying a curve with offset y, $y = x^3 / 6lr$, x being the distance from the tangent point, l the length of transition, and r the radius of circular arc. (Pryor, 1963)

frozen Said of the contact between the wall of a vein and the mineral deposit filling it, in which the vein material adheres closely to the wall; also, said of the vein material and of the wall. (AGI, 1987)

frozen coal Coal that adheres strongly to the rock above or below it. (Fay, 1920)

frozen ground Ground that has a temperature below freezing and generally contains a variable amount of water in the form of ice. Cf: *permafrost.* (AGI, 1987)

fruchtschiefer A type of spotted slate with spots suggestive of grains of wheat. See also: *fleckschiefer; garbenschiefer.*

Frue vanner An ore-beneficiation apparatus consisting essentially of a rubber belt traveling up a slight inclination. The material to be treated is washed by a constant flow of water while the entire belt is shaken from side to side. (Liddell, 1918)

frustule The siliceous cell wall of a diatom, consisting of two valves, one overlapping the other. It is ornate, microscopic, and boxlike. (AGI, 1987)

fuchsite Bright-green chromium muscovite. Syn: *chrome mica.*

fucoid An informal name applied loosely to any indefinite traillike or tunnellike sedimentary structure identified as a trace fossil but not referred to a described genus. It was once considered to be the remains of the marine alga Fucus, and later was regarded as a feeding burrow of a marine animal and assigned to the plantlike genus Fucoides. The term has been broadly applied to crustacean tracks, worm burrows, molluscan trails, marks made by the tide or waves, and rill marks. (AGI, 1987)

fucosite Bitumen derived from the hydration of fucose pentosane and found among clays and sands in California. (Tomkeieff, 1954)

fuel A substance that can be economically burned to produce heat energy for domestic or industrial purposes. Fuels include compounds of carbon and hydrogen and exclude other substances that can be burned. Fuels can be subdivided into recent plant fuels, fossil fuels, such as peat and coal, and products of distillation of plant or fossil fuels. According to their state of aggregation, fuels can be subdivided into solid, liquid, and gaseous fuels. (Tomkeieff, 1954)

fuel feeder A contrivance for supplying a furnace with fuel in graduated quantities. A mechanical stoker. (Fay, 1920)

fuel ratio The ratio of fixed carbon to volatile matter in coal.

fugitive air Applied to air moving through the fan that never reaches the working faces. It leaks through poor stoppings, around doors and so on, back into the returns without moving anywhere near the active sections. Surveys of some mines show that up to 80% of the air moving through the fan never reaches the working faces. (Coal Age, 1966)

fugitive constituent A substance that was present in a magma but was lost during crystallization, so that it does not commonly appear as a rock constituent.

fugitive dust The particulate matter not emitted from a duct or stack that becomes airborne due to the forces of wind or surface coal mining and reclamation operations or both. During surface coal mining and reclamation operations it may include emissions from haul roads; wind erosion of exposed surfaces, storage piles, and spoil piles; reclamation operations; and other activities in which material is either removed, stored, transported, or redistributed.

fulcrum A pivot for a lever. (Nichols, 1976)

fulguration A sudden glistening of molten gold or silver at the close of cupellation. (Standard, 1964)

fulgurite An irregular, glassy, often tubular or rodlike structure or crust produced by the fusion of loose sand (or rarely, compact rock) by lightning, and found esp. on exposed mountain tops or in dune areas of deserts or lake shores. It may measure 40 cm in length and 5 to 6 cm in diameter. Etymol. Latin fulgur, lightning. Syn: *lightning tube.* (AGI, 1987)

full dip *true dip; dip.*

fuller's earth A clay or claylike material with a high adsorptive capacity, consisting largely of the clay minerals montmorillonite and palygorskite. Used originally in England for whitening, degreasing, or fulling (shrinking and thickening by application of moisture) woolen fabrics; fuller's earth now is extensively used as an adsorbent in refining and decolorizing oils and fats; it is a natural bleaching agent. Its color ranges from light brown through yellow and white to light and dark green; it differs from ordinary clay by having a higher percentage of water and little or no plasticity, tending to break down into a muddy sediment in water. Fuller's earth probably forms as a residual deposit by decomposition of rock in place, as by devitrification of volcanic glass. The term is applied without reference to any particular chemical or mineral composition, texture, or origin. Syn: *walker's earth.* (AGI, 1987)

full-face blast The standard type of heading blast consists of a straight in or main drive, at right angles to the rock face, and a back drive at right angles to the main drive and parallel to the face. The main drive is normally driven at quarry floor level to a depth of 0.6 times the height of rock above the back drive. Apart from exceptional circumstances, the maximum depth of the main drive should be 50 ft (15 m), so that with faces over 85 ft (26 m) high the 0.6 ratio should not be used, but the main drive limited to 50 ft. (McAdam, 1958)

full-face firing With modern drilling equipment, it is now possible, in suitable conditions, to drill small-diameter holes from top to bottom of the face, and where this can be done considerable advantages as to cost and efficiency can be obtained as compared with the bench method. In high faces of 50 ft (15 m) and upward it is not always easy to drill vertical holes to give small burdens because of the breakback of the rock at the crest of the quarry face. It is therefore recommended that larger burdens be taken and that the vertical holes be supplemented by breast holes at quarry floor level. These holes are intended to permit concentrated explosive charges to blow out the toe rock, while the explosive in the vertical holes brings down the rock from the face. (McAdam, 1958)

full gage (a) A cylindrical or tubular object, such as a bit or reaming shell, the outside and/or inside diameters of which are the size specified. Also called full size. (Long, 1960) (b) A borehole the inside diameter of which is uniform enough to allow a new-condition bit to follow portions of the hole drilled by other bits cutting the same X-borehole size without reaming. (Long, 1960) (c) As applied to deflection drilling, the branch borehole is the same diameter as the parent hole. (Long, 1960)

full-seam mining A mining system, brought on by the advent of mechanical loading and mechanical coal cleaning, in which the entire section is dislodged together and the coal separated from the rock outside of the mine by the cleaning plant. (Kentucky, 1952)

full subsidence The greatest amount of subsidence that can occur as a result of mine workings. See also: *percentage subsidence.* (Nelson, 1965)

full teeter A condition of teeter in which the maximum degree of fluidization of the suspension is attained but without disruption of the bed. (BS, 1962)

full velocity *settling velocity.*

full-wave rectifier A rectifier that changes single-phase alternating current into pulsating unidirectional current, utilizing both halves of each cycle. (Coal Age, 1960)

fully developed mine In coal mining, a mine in which all development work has reached the boundaries and further extraction will be done on the retreat.

fulminate An explosive compound of mercury, $HgC_2N_2O_2$, that is employed for the caps or exploders, by means of which charges of gunpowder, dynamite, etc., are fired.

fülöppite A monoclinic mineral, $Pb_3Sb_8S_{15}$; soft metallic gray; occurs with zinkenite and sphalerite in Romania and Hungary.

fulvurite An old syn. for brown coal. (AGI, 1987)

fumarole (a) A hole or a vent from which fumes or vapors issue; a spring or a geyser that emits steam or gaseous vapor. Usually found in volcanic areas. (AGI, 1987) (b) The exhalation from a fumarole consists of water vapor, nitrogen, hydrogen, free hydrochloric acid, hydrofluoric acid, and silicon fluoride. Cf: *solfatara; mofette; soffioni.* (Fay, 1920) (c) A hole in a volcanic region, from which gases and vapors issue at high temperature. (Webster 3rd, 1966)

fume (a) The gas and smoke, esp. the noxious or poisonous gases, given off by the explosion or detonation of blasting powder or dynamite. (Fay, 1920) (b) Consists of very fine particles of metals or metallic compounds that have been volatilized at the high temperatures of furnaces, condensed at lower temperatures, and carried by furnace gases into the flues. In general, all the volatile constituents of the ore charge are represented. See also: *metallurgical fume.* (Fay, 1920)

fume quality A measure of the toxic fumes to be expected when a specific explosive is properly detonated. (Dick, 1983)

fundamental complex *basement complex.*

fundamental jelly Structureless colloidal jelly that forms the base of coals and is assumed to have been produced by the decay of plant materials. See also: *carbohumin.* (Tomkeieff, 1954)

fundamental strength The maximum stress that a substance can withstand, regardless of time, under given physical conditions, without creep.

fundamental substance *fundamental jelly; carbohumin.*

fungus subterraneus An old name for elaterite. (Tomkeieff, 1954)

funicular railway A railway that negotiates a steep gradient; the cars are operated by cables and winches. See also: *mountain railway.* (Hammond, 1965)

funnel box A square funnel forming one of a series of gradually increasing size; for separating metal-bearing slimes according to fineness. See also: *spitzkasten.* (Standard, 1964; Fay, 1920)

funnel brick Funnel-shaped fireclay piece used in the bottom-pour ingot assembly to lead metal to the fountain brick. See also: *bottom-pour ingot assembly.*

funnel intrusion *lopolith.*

funnel joint A joint in a joint set that is concentric, with the joints dipping toward a common center. (AGI, 1987)

fun-tso-ka Coolies working tin mines or other projects, in Malaysia. (Hess)

fur Eng. A deposit of chemical salts and other material (sediment) upon the inner sides of pumps, boilers, etc. Also called: furring.

furgen A round rod used for sounding a bloomery fire. Syn: *tempering bar.* (Fay, 1920)

furnace (a) A structure in which, with the aid of heat so produced, (1) the operations of roasting, reduction, fusion, steam-generation, desiccation, etc., are carried on, or, (2) as in some mines, the upcast air current is heated, to facilitate its ascent and thus aid ventilation. (Fay, 1920) (b) Structure in which materials are exposed to high temperatures. Fuels used to maintain this include alcohols, paraffins, gas, coal, hydrogen, electricity, wood, and sulfur. A furnace is called batch type if its contents are treated in successive charges, or continuous when a stream of material passes through, being changed during transit. The main types are (1) the arc, which uses the heat of an electric arc; (2) the blast furnace; (3) the crucible furnace, a laboratory appliance for heating small charges, or, if large, for melting metals held in bigger crucibles; (4) the induction furnace, heated by electrically induced currents; (5) the muffle, in which the material is placed in a sleeve not in direct contact with the heating atmosphere, so that close control of entering and departing gas is possible; (6) the reverberatory, in which head developed on the roof is reflected onto a horizontal bed below; (7) the revolving furnace, a horizontal cylinder; (8) the roasting furnace in which material is oxidized, kilned to drive off carbon dioxide, or heated to remove moisture. See also: *cupola; continuous furnace; converter.* (Pryor, 1963)

furnace bridge A barrier of firebricks or an iron-plate chamber filled with water thrown across the furnace at the extreme end of the fire bars to prevent fuel from being carried into the flues and to quicken the draft by contracting the section of the current of hot gas. (Fay, 1920)

furnace cadmium The zinc-cadmium oxide that accumulates in the chimneys of furnaces smelting zinciferous ores. See also: *cadmia.* (Fay, 1920)

furnace charger (a) A weighing apparatus for feeding into a furnace mouth the proper proportions of ore, fuel, etc. (Standard, 1964) (b) In the iron and steel industry, a laborer who operates a compressed-air-driven arm to push stock steel rails into a heating furnace. See also: *charging person.* (DOT, 1949)

furnace conveyor The conveyor that moves material through a furnace.

furnace holding-the-iron A condition by which the furnace gives much less than the normal amount of iron at casting although the feeding may have been regular. The taphole runs iron slowly, and the amount of cinder is somewhat scanty. Cf: *furnace losing-the-iron.* (Fay, 1920)

furnace linings Refractory materials used to protect the walls of the furnace from reaction with its molten contents (abrasive, melting, or chemical). Three divisions are (1) acid refractories rich in silica (flint, ganister, fireclay), which react with basic oxides; (2) neutral refractories (chromite, graphite), and (3) basic refractories, rich in oxides of calcium and magnesium and low in silica. (Pryor, 1963)

furnace losing-the-iron Escape of iron from the hearth of a blast furnace into the foundation beneath, indicated by decreased quantity of iron at casting and appearance of slag at the tapping hole. Syn: *losing iron.* (Fay, 1920)

furnace magnesite A mortar material prepared from finely ground, dead-burned magnesite; suitable for use as a joint material in laying magnesite brick, and for patching or daubing furnace masonry. (Harbison-Walker, 1972)

furnace sprayer In ore dressing, smelting, and refining, a laborer who sprays the inner surfaces of furnace walls and roof with a slurry of silica, water, and fireclay to protect brick, using a compressed-air gun. Also called slurry man; sprayer. (DOT, 1949)

furnace stack A chimney built over a furnace for increasing the draft. (Fay, 1920)

furnacite Misspelling of fornacite. *fornacite.*

furrener *foreigner.*

furrowed Having deep grooves or striations.

fusain (a) A coal lithotype characterized macroscopically by its silky luster, fibrous structure, friability, and black color. It occurs in strands or patches and is soft and dirty where not mineralized. Its characteristic microlithotype is fusite. Cf: *vitrain; clarain; durain.* Syn: *mineral charcoal; mother of coal.* Obsolete syn. motherham. (AGI, 1987) (b) Coal material having the appearance and structure of charcoal. It is friable, sooty, generally high in ash content, and consists mainly of fusite. (AGI, 1987)

fuse An igniting or explosive device in the form of a cord, consisting of a flexible fabric tube and a core of low explosive (safety fuse) or high explosive (detonating cord).

fuse auger An instrument for regulating the time of burning of a fuse by removing a certain portion of the composition. It has a movable graduated scale that regulates the depth to which the auger should penetrate. (Standard, 1964)

fuse cutter A mechanical device for cutting safety fuse clean and at right angles to its long axis. (Atlas, 1987)

fused alumina Aluminum oxide, Al_2O_3.

fused bath electrolysis Extraction of metals by electrolytic decomposition of their fused salts; extraction of metals from electrolytically decomposable compounds dissolved in substances inert under the conditions of electrolysis. (Bennett, 1962)

fuse detonator A detonator which is initiated by a safety fuse; also referred to as an ordinary blasting cap. (Atlas, 1987)

fused nip A terminal connection, with a fuse, used on portable electrical mining machinery.

fused quartz Silica glass made from clear pieces of vein quartz.

fused refractories Refractories in which the constituents are held together by heating to either the point of fusion or coalescence. (Henderson, 1953)

fused trolley tap A specially designed holder with enclosed fuse for connecting a conductor of a portable cable to the trolley system or other circuit supplying electric power to equipment in mines.

fuse gage An instrument for cutting time fuses to length. (Standard, 1964)

fusehead That part of an electric detonator consisting of twin metal conductors, bridged by fine resistance wire, and surrounded by a bead of igniting compound that burns when the firing current is passed through the bridge wire. (BS, 1964)

fuse lighter Pyrotechnic devices such as a hot-wire fuse lighter or igniter cords for the rapid and certain lighting of safety fuse. *tchesa stick.*

fuse lock A friction lock by which a miner may fire the free end of a blasting fuse by a lanyard. (Standard, 1964)

fusibility scale A temperature scale based on the fusibility of a standard group of minerals. Used prior to the development of modern furnaces.

fusible plug An insert of metal with low melting point placed in boilers, sprinklers, and other devices to melt when the temperature becomes dangerously high, so that the melting will relieve pressure, allow water flow, or otherwise tend to alleviate the dangerous condition. (Strock, 1948)

fusing point The degree of heat at which any substance begins to melt or liquefy. See also: *melting point.* (Hansen, 1937)

fusinite (a) A maceral of coal within the inertinite group with intact or broken cellular structure, a reflectance (except in meta-anthracite) well above that of associated vitrinite, and a particle size generally greater than about 50 µm except when isolated from other macerals. Cf: *fusain.* (AGI, 1987) (b) A constituent showing well-defined cellular structure of wood or sclerenchyma. The cell cavities vary in size and shape—round, oval, or elongated. Bogen structure is common. Occurs as discrete lenses, thin partings or bands, and as small dispersed fragments; is widely distributed; common. The physical and chemical properties of fusinite vary only slightly in coals of different rank, and consequently its technological properties are fairly constant. (IHCP, 1963) (c) The major maceral, or micropetrologic, constituent of fusain. It consists of wood (xylem or lignified tissue) of which very little is left but woody tracheids or thick-walled elements so highly carbonized as to contain only traces of ulmins. A member of the inertite group. (AGI, 1987)

fusinization A process of coalification in which fusain is formed. Cf: *incorporation; vitrinization.* Also spelled: fusainisation. (AGI, 1987)

fusinoid Fusain and similar material in coal. (AGI, 1987)

fusion (a) A union by or as if by a combination of ingredients achieved by heating, and mixing together. (Webster 3rd, 1966) (b) The union of atomic nuclei to form heavier nuclei, resulting in the release of enormous quantities of energy when certain light elements unite. Also called nuclear fusion. Cf: *fission.* (Webster 3rd, 1966)

fusion button test *button test.*

fusion method A method used to remove certain impurities from diamond concentrate with a particle size of 0.5 to 1.0 mm. The material, mixed with 10 times its weight of flake caustic soda, is placed in crucibles and put in a furnace where a temperature of 650 °C is maintained for 45 min. After furnacing, the material is rinsed to remove the caustic soda and boiled in a glass beaker containing a solution of 1 part hydrochloric acid and 4 parts water. After further rinsing, the diamond, free from satellites, is dried on a hotplate. (Chandler, 1964)

fusion of clay The stage on heating a clay when the material is changed from the solid to the liquid state, but complete liquefaction occurs so gradually with most clays that a fusion range and not a fusion point is obtained. See also: *squotting.* (Nelson, 1965)

fusion-piercing drill A machine designed to use the fusion-piercing mode of producing holes in rock. Sometimes incorrectly called a jet drill. Syn: *jet-piercing drill.* (Long, 1960)

fusion point The temperature at which melting takes place. Most refractory materials have no definite melting points, but soften gradually over a range of temperatures. (Harbison-Walker, 1972)

fusion tectonite Igneous rock consolidated from a flowing magma.

fusion test *pyrometric cone equivalent; button test.*

fusite (a) A coal microlithotype that contains at least 95% fusinite. It is a variety of inertite. Cf: *fusain.* (AGI, 1987) (b) In 1955 the Nomenclature Subcommittee of the International Committee for Coal Petrology resolved to use this term for the microlithotype consisting principally of the macerals fusinite, semifusinite, and sclerotinite. Two varieties of fusite are distinguishable—a fragile and powdery fusite and a hard consolidated fusite in which the cavities are filled by various minerals, carbonates, sulfides, kaolin, and other clay minerals. Widely distributed, but in general not abundant. Occurs in fine bands and lenses of varying thickness. The soft variety of fusite concentrates in the very fine particle sizes. Hard fusite distributes itself in various sizes (depending on the thickness of the original bands or lenticles in the seam), but not in the fines. This form of fusite is usually discarded in the middlings and refuse. (IHCP, 1963) (c) A coal microlithotype containing 95% or more fusinite, plus semifusinite, plus sclerotinite. (Schieferdecker, 1959; AGI, 1987)

fusoclarain A type coal rock consisting of the macerals fusinite and vitrinite; it may contain all other macerals. Fusinite is present in a smaller quantity than in clarofusain. Cf: *clarofusain.* (AGI, 1987)

fusodurain (a) Durain in which much of the microconglomeratic elements consist of fusain. (AGI, 1987) (b) Judged obsolete by the Heerlen Congress of 1935. (AGI, 1987)

fusovitrain (a) Preferred. Coal consisting of material transitional between fusain and vitrain with vitrain being predominant (Heerlen Congress of 1935). Cf: *vitrofusain.* (AGI, 1987) (b) Coal consisting of a mixture of vitrain with fusain fragments. (AGI, 1987)

fusuline *fusulinid.*

fusulinid Any foraminifer belonging to the suborder Fusulinina, family Fusulinidae, characterized by a multichambered elongate calcareous microgranular test, commonly resembling the shape of a grain of wheat. Range, Ordovician to Triassic. Syn: *fusuline.* (AGI, 1987)

future ore *possible ore.*

G

Ga Billions of years before the present. (AGI, 1987)
gabbro A group of dark-colored, basic intrusive igneous rocks composed principally of basic plagioclase (commonly labradorite or bytownite) and clinopyroxene (augite), with or without olivine and orthopyroxene; also, any member of that group. It is the approximate intrusive equivalent of basalt. Apatite and magnetite or ilmenite are common accessory minerals. (AGI, 1987)
gabbroid Said of a rock resembling gabbro. (AGI, 1987)
gable-bottom car *mine car.*
gabrielsonite An orthorhombic mineral, $PbFe(AsO_4)(OH)$; adelite group; forms black adamantine lumps at Långban, Sweden.
gad A heavy steel wedge, 6 in or 8 in (15.2 cm or 20.3 cm) long, with a narrow chisel point for cutting samples or breaking out pieces of loose rock. (Hess)
gadder In quarrying, a small car or platform carrying a drilling machine, so as to make a straight line of holes along its course in getting out dimension stone. Also called gadding car. Syn: *gadding machine.* (Standard, 1964)
gadding machine *gadder.*
gadolinite (a) A monoclinic mineral, $(Y,Ce,La,Nd)_2Be_2Si_2O_{10}$; further speciated according to the predominant rare-earth element; weakly radioactive; occurs with fluorite, allanite, and beryl in granites and granite pegmatites; with xenotime in biotite gneisses. (b) The mineral group bakerite, datolite, gadolinite-(Ce), gadolinite-(Y), hingganite-(Ce), hingganite-(Y), hingganite-(Yb), homilite, and minasgeraisite-(Y).
gagatite Jetlike coalified plant material preserving cellular structures. See also: *gagatization.* (Tomkeieff, 1954)
gagatization In coal formation, the impregnation of wood fragments with dissolved organic substances. See also: *gagatite.* (AGI, 1987)
gage (a) Spacing of tracks or wheels. (Nichols, 1976) (b) The nominal size of an aggregate. It is the minimum size of sieve through which at least 95% of an aggregate will pass. Also spelled gauge. (Taylor, 1965)
gage cock A small cock in a boiler at the water line, to determine the water level.
gage door A wooden door fixed in an airway for regulating the supply of ventilation necessary for a certain district or number of workers. Also called a regulator.
gage factor The percentage charge of resistance divided by the percentage strain. For strain gages in common use, this amounts to about 2.2. (Issacson, 1962)
gage loss The diametrical reduction in the size of a bit or reaming shell caused by wear through use. (Long, 1960)
gager In the iron and steel industry, one who determines whether iron or steel bars, sheets, or wire are being rolled to plant specification, so that the rolls may be adjusted to reduce the metal the desired amount for each pass, using calipers to check the thickness (gage) of the various products. (DOT, 1949)
gage size The width of a drill bit along the cutting edge. (Nichols, 1976)
gage stone Any one of several diamonds set in the crown of a diamond bit in a plane parallel with and projecting slightly beyond the inside and/or outside walls of the bit. Syn: *kicker; kicker stone; reaming diamond.* Cf: *cutting stones.* (Long, 1960)
gaging A heap of rubbish placed at the entrance of a disused roadway underground. (CTD, 1958)
gaging station A particular site on a stream, canal, lake, or reservoir where systematic observations of gage height, discharge, or water quality (or any combination of these) are obtained. (AGI, 1987)
gahnite An isometric mineral, $ZnAl_2O_4$; spinel group; forms series with spinel and with hercynite; Mohs hardness, 7-1/2; forms octahedra or masses in schists, contact-metamorphosed limestones, granite pegmatites; relatively common in replacement ores at Franklin, NJ, and Falun, Sweden. Syn: *zinc spinel.*
gahnospinel A blue magnesian variety of gahnite used as a gem, from Sri Lanka.
gain (a) A cutting made in the side of a roadway underground to facilitate the construction of a dam or air stopping. (CTD, 1958) (b) A crosscut in coal mining. (c) *closed joint.* (d) A notch, mortise, or groove (as in a timber or wall) for a girder or joist. (Webster 3rd, 1966) (e) The ratio of the output power, voltage, or current to the input power, voltage, or current. (Hunt, 1965)
gaize (a) A siliceous rock containing some clay in the Ardennes and Meuse Valley, France. (b) A porous fine-grained micaceous glauconitic sandstone containing much soluble silica among the Cretaceous rocks of France and Belgium; a calcareous sediment cemented by chert or flint.
Gal A unit of acceleration, used in gravity measurements: 1 Gal = 1 cm/s^2. The Earth's normal gravity is 980 Gal. The term is not an abbrev.; it was invented to honor the memory of Galileo. See also: *milligal.* (AGI, 1987)
galactite (a) A variety of natrolite found as colorless acicular crystals. (b) An obsolete syn. of novaculite. (c) An unidentified stone (possibly calcium nitrate) whose milky solution gave rise to several medieval legends and superstitions.
galaxite An isometric mineral, $(Mn^{+2},Fe^{+2},Mg)(Al,Fe^{+3})_2O_4$; spinel group; occurs as black grains in Galax, Alleghany County, NC; Ioi Mine, Shiga Prefecture, and Oashi mine, Tochigi Prefecture, Japan.
gale alidade A lightweight compact alidade, with a low pillar and a reflecting prism through which the ocular may be viewed from above. As used by some geologists, it is commonly equipped with the Stebinger drum. See also: *Stebinger drum.* Syn: *explorer's alidade.*
galeite A trigonal mineral, $Na_{15}(SO_4)_5F_4Cl$; occurs embedded in clay associated with gaylussite and northupite in drill cores from Searles Lake, CA. Cf: *schairerite.*
galena An isometric mineral, 4[PbS]; cubic cleavage; forms cubes and octahedra, also coarse- or fine-grained masses; sp gr, 7.6; occurs with sphalerite in hydrothermal veins, also in sedimentary rocks as replacement deposits; an important source of lead and silver. Also spelled galenite. Syn: *lead sulfide; galenite.* See also: *blue lead.*
galenite *galena.*
galenobismutite An orthorhombic mineral, $PbBi_2S_4$; soft; massive, or in needlelike to lathlike crystals in radiating aggregates, commonly intergrown with bismuthinite and other bismuth minerals; sp gr, 7.04 to 7.15; occurs in high-temperature replacement deposits or veins; may be argentiferous (e.g., the variety alaskaite in Colorado) or selenian (e.g., the variety selenbleiwismuthglanz at Falun, Sweden); a source of bismuth.
gallatin The heavy oil of coal tar used in the Bethell process for the preservation of timber. Also called dead oil. (Standard, 1964)
gallery (a) A mine level, drift, tunnel, or passage. (b) A large, more or less horizontal, passage in a cave. (c) A subsurface collector for intercepting ground water. (d) A horizontal or nearly horizontal underground passage, either natural or artificial. (Stokes, 1955) (e) A subsidiary passage in a cave at a higher level than the main passage. (AGI, 1987) (f) A drift or adit. (g) An underground conduit or reservoir. (Seelye, 1951) (h) Underground road. (Mason, 1951) (i) A passageway, as in a dam. (Seelye, 1951)
gallery of efflux Eng. A drain tunnel or adit.
gallery testing Testing conditions designed to resemble those existing underground as closely as possible, and to reproduce what was considered to be the most dangerous condition, namely, a blownout shot discharging into the most easily ignited mixture of combustible gases and air.
galliard A hard, smooth, close-grained, siliceous sandstone; a ganister. Also spelled: calliard. (AGI, 1987)
galliard balls Large ironstone concretions in sandstones, Yorkshire, U.K.
gallite A tetragonal mineral, $CuGaS_2$; chalcopyrite group; forms grains and inclusions in germanite and renierite, and exsolution lamellae in sphalerite and pyrite-sphalerite assemblages at Tsumeb, Namibia, and Shaba Province, Zaire.
gallium arsenide Dark gray; GaAs; isometric; melting point, 1,240 °C. Used in microwave diodes and in high-temperature rectifiers and transistors. (Lee, 1961; Handbook of Chem. & Phys., 2)
gallium oxide Ga_2O_3; melting point, 1,795±15 °C. (Dodd, 1964)
Galloway stage Multidecked platform suspended near bottom of shaft during sinking. It carries part of the equipment in use and can be raised or lowered as required during blasting, mucking, wall concreting, etc. See also: *pentice.* (Pryor, 1963)

gallows timber A timber framework or set for roof support. (CTD, 1958)
galmei *calamine.*
galvanic cell An electrolytic cell that is capable of producing electric energy by electrochemical action. See also: *cell.* (McGraw-Hill, 1994)
galvanic corrosion (a) Corrosion associated with the current of a galvanic cell consisting of two dissimilar conductors in an electrolyte or two similar conductors in dissimilar electrolytes. Where the two dissimilar metals are in contact, the resulting reaction is referred to as couple action. (ASM, 1961) (b) The corrosion above normal corrosion of a metal that is associated with the flow of current to a less active metal in the same solution and in contact with the more active metal. (Hunt, 1965)
galvanic electromagnetic methods Electrical exploration methods in which electric current is introduced in the ground by means of contact electrodes and in which one determines the magnetic field that is associated with the current. (Schieferdecker, 1959)
galvanize To coat with zinc. (Fay, 1920)
galvanizing Immersion of clean steel or iron in bath of molten zinc for purpose of forming a protective coating. Sherardizing is the process of heating iron articles with zinc dust to a temperature at which a strong adherent coating is formed. Electrolytic galvanizing is the electrodeposition of zinc on the iron. See also: *hot-dip coating.* (Pryor, 1963)
galvanoscope An instrument employed for detecting an electric current and showing its direction. It differs from a galvanometer in being only qualitative. (Standard, 1964)
galvanothermometer An instrument for measuring the heat generated by an electric current or for measuring the current by the heat that it generates. (Standard, 1964)
gamagarite A monoclinic mineral, $(Ba_2(Fe,Mn)(VO_4)_2(OH)$; brackebuschite group; forms prismatic crystals and aggregates of needles with diaspore, ephesite, and bixbyite in manganese ore at Gamagara ridge, Postmasburg, Republic of South Africa.
gamella Braz. A wooden bowl, about 2 ft (0.6 m) wide at the mouth, and 5 in or 6 in (12.7 cm or 15.2 cm) deep, used for washing gold out of the auriferous material collected in sluices and in river sand. Syn: *batea.*
gamma (a) In a biaxial crystal, the largest index of refraction. (AGI, 1987) (b) The interaxial angle between the *a* and *b* crystallographic axes. Cf: *alpha; beta.* (AGI, 1987) (c) The cgs unit of magnetic field intensity commonly used in magnetic exploration. It is equal to 10^{-5} Oe (7.957747×10^{-4} A/m). Syn: *nanotesla.* (AGI, 1987) (d) adj. Of or relating to one of three or more closely related minerals and specifying a particular physical structure (esp. a polymorphous modification); specif. said of a mineral that is stable at a temperature higher than those of its alpha and beta polymorphs (e.g., gamma quartz or γ-quartz). (AGI, 1987)
gamma-gamma log A borehole measurement of gamma rays originating in a gamma-ray source in the instrument and scattering back from the rock formation to a detector shielded from the source. The amount of scattering is proportional to electron density and, therefore, proportional to mass concentration so that the measurement, after certain corrections, yields a density log of the formation penetrated. (AGI, 1987)
gammagraphy In the United States a term for inspection by gamma rays. (Osborne, 1956)
gamma iron The face-centered cubic form of pure iron, which is stable from 1,670 to 2,550 °F (910 to 1,400 °C). (ASM, 1961)
gamma radiation Emission by radioactive substances of quanta of energy corresponding to X-rays and visible light but with a much shorter wavelength than light. May be detected by gamma-ray Geiger counters. (AGI, 1987)
gamma-ray probe A gamma-ray counter device built into a watertight case small enough in diameter to be lowered into a borehole. (Long, 1960)
gamma rays High-energy, short-wavelength, electromagnetic radiation emitted by a nucleus. Energies of gamma rays are usually between 0.010 and 10 million eV. X-rays also occur in this energy range but are of nonnuclear origin. Gamma radiation usually accompanies alpha and beta emissions and always accompanies fission. Gamma rays are very penetrating and are best attenuated by dense materials like lead and depleted uranium. (Lyman, 1964)
gamma-ray spectrometer An instrument for measuring the energy distribution, or spectrum, of gamma rays, whether from natural or artificial sources. It is used in airborne remote sensing for potassium, thorium, and uranium. (AGI, 1987)
gamma-ray spectrometer log A log that measures the relative quantities of potassium, thorium, and uranium present in the rocks penetrated by a borehole. (Wyllie, 1963)
gamma-ray well log The radioactivity log curve of the intensity of broad-spectrum, undifferentiated natural gamma radiation emitted from the rocks in a cased or uncased borehole. It is used for correlation, and for distinguishing shales (which are usually richer in naturally radioactive elements) from sandstones, carbonates, and evaporites. Cf: *spectral gamma-ray log.* See also: *radioactivity log.* (AGI, 1987)
gamma-ray well logging A method of logging boreholes by observing the natural radioactivity of the rocks through which the hole passes. It was developed for logging holes that cannot be logged electrically because they are cased. (AGI, 1987)
gamma sulfur *rosickyite.*
gamma zircon A metamict variety of zircon that is nearly amorphous owing to radiation damage; it has lower density than crystalline zircon.
gang (a) A train or set of mine cars or trams. (b) A mine. (c) A set of miners. (d) Gangue.
gang car A car that may be loaded with a block of stone and placed beneath the blades of a gang saw. It is a substitute for the stationary saw bed. (Fay, 1920)
gang drill A set of drills in the same machine operated together. (Standard, 1964)
ganger A work gang foreman. (Webster 3rd, 1966)
gang filler In the stonework industry, one who attaches and detaches crane slings or hooks to and from blocks or slabs of granite, marble, and stone in loading the stone on gang saw cars or trucks and pulling them under the gang saws. (DOT, 1949)
gang miner In bituminous coal mining, one who works in a group that pools its earnings regardless of the type of work performed (drilling, undercutting, blasting, or loading coal). (DOT, 1949)
gangue The valueless minerals in an ore; that part of an ore that is not economically desirable but cannot be avoided in mining. It is separated from the ore minerals during concentration. Cf: *ore mineral.* (AGI, 1987)
gangway (a) A main haulage road underground. Frequently called entry. (Hudson, 1932) (b) A passageway driven in the coal at a slight grade, forming the base from which the other workings of the mine are begun. (Korson, 1938) (c) A passageway or avenue into or out of any enclosed place, as in a mine. (d) An elevated roadway. (CTD, 1958) (e) Pennsylvania. Generally confined to anthracite mines. (Jones, 1949) (f) Newc. A wooden bridge.
gangway cable A cable designed to be installed horizontally (or nearly so) for power circuits in mine gangways and entries.
ganister (a) A hard, fine-grained quartzose sandstone or quartzite, used in the manufacture of silica brick. It is composed of subangular quartz particles, cemented with secondary silica, and possessing a characteristic splintery fracture. Ganister is distinguished from chert by its more granular texture and by the relatively small quantity of chalcedonic or amorphous silica. (AGI, 1987) (b) A mixture of ground quartz and fireclay used as a furnace lining. Also spelled gannister. (AGI, 1987) (c) *quartzite.*
gannen N. of Eng. A road (heading) down which coal is conveyed in cars running upon rails. An inclined gangway in a coal mine. (Fay, 1920; Standard, 1964)
gantlet A narrowing of two single railway tracks almost into the space of one, as on a bridge or in a tunnel, without breaking the continuity of either track by a switch, the two tracks overlapping each other. (Standard, 1964)
gantry (a) A frame erected on a gold dredge for supporting different parts of the machinery. Also spelled: gauntry; gauntree. (b) A bridge or platform carrying a traveling crane or winch and supported by a pair of towers, trestles, or side frames running on parallel tracks. (Webster 3rd, 1966) (c) An overhead structure that supports machines or operating parts. (Nichols, 1976) (d) An upward extension of a shovel revolving frame that holds the boom line sheaves. (Nichols, 1976)
Gantt chart Construction program for major engineering works, set out in graphic form. Down the vertical axis in sequence are set out the items concerned. The abscissa shows the period covered in days, or weeks, and the period allowed for each item marked by a horizontal line. The chart displays the interrelation between the items, and aids in ensuring that no item is so delayed as to impede progress on a later one that depends on it. (Pryor, 1963)
gap In a fault, the horizontal component of separation measured parallel to the strike of the strata, with the faulted bed absent from the measured interval. (AGI, 1987)
gape Maximum aperture at entry to a coarse crushing machine at which the largest piece of rock fed to it can be gripped and acted on by the breaking system. (Pryor, 1963)
gap packing A method of packing for road maintenance that consists of gate side packs 3 to 5 yd (2.7 to 4.6 m) wide, next a gap of at least the width of the road, and finally a large pack 5 to 7 yd (4.6 to 6.4 m) wide. The waste packs are made at least 2 yd (1.8 m) wide and not less in width than twice the thickness of the seam. The gaps provided in strip packing are kept clear of supports and allow the roof to break up and flow toward them. This puts the strata in tension over the roads and reduces fracture and crush. (Mason, 1951)
gap sensitivity The maximum distance for propagation between standard cartridge sizes separated by an air gap. Cf: *sympathetic detonation.* (Atlas, 1987)

gap test The gap is the greatest distance at which, under certain given conditions, a priming cartridge is capable of initiating a receiving cartridge (receptor). The same explosive is usually used both as primer and receptor, although the gap distance in such a case will also be affected by any change in strength that may occur in the explosive. The gap test can be carried out with the cartridges unconfined or confined, for example, in tubes, in air, or in water. The test gives, for example, information about changes in the explosive due to aging, moisture, temperature, etc. (Fraenkel, 1953)

garbenschiefer (a) A type of spotted slate characterized by concretionary spots whose shape resembles that of a caraway seed. Etymol: German. Cf: *fleckschiefer; fruchtschiefer*. (AGI, 1987) (b) *feather amphibolite.*

Gardner crusher A swing and hammer crusher, the hammers being pieces hung from trunnions between two disks keyed to a shaft. When revolved, centrifugal force throws the hammers out against the feed and a heavy anvil inside the crusher housing. (Liddell, 1918)

gargarinite A hexagonal mineral, $NaCaY(F,Cl)_6$; in albitized granites and associated quartz-microcline veins in Kazakhstan.

gargulho A Brazilian term used in the plateau region of Bahia for a comparatively coarse, clay-cemented, ferruginous conglomerate in which diamonds are found. (AGI, 1987)

garland (a) A channel fixed around the lining within a shaft in order to catch the water draining down the shaft walls and conduct it by pipes or water boxes to a lower level. Also called water curb; water garland. See also: *water ring*. (BS, 1963) (b) A frame to heighten and increase content of a truck. (CTD, 1958)

garnet (a) A group of isometric minerals having the general formula $A_3B_2(SiO_4)_{3-2Dx}(OH)_{4x}$ in which A=(Ca,Fe,Mg,Mn) and B=(Al,Cr,Fe,Mn,Si,Ti,V,Zr) with Si partly replaced by (Al,Fe). (b) The silicate minerals almandine, andradite, calderite, goldmanite, grossular, hibshite, katoite, kimzeyite, knorringite, majorite, pyrope, schlorlomite, spessartine, and uvarovite. Syn: *granat.*

garnet doublet (a) A term applied to the most common doublet, that with a very thin top of red garnet, regardless of the color of the doublet. (b) Any doublet of dark red color regardless of whether any portion of it is garnet, more correctly called a garnet-top doublet. (c) A composite stone made with a garnet top on a glass base. Also called garnet-top doublet, garnet-topped doublet.

garnetite A metamorphic rock consisting chiefly of an aggregate of interlocking garnet grains. Cf: *tactite.* (AGI, 1987)

garnetization Introduction of, or replacement by, garnet. This process is commonly associated with contact metamorphism. (AGI, 1987)

garnet jade A light-green variety of grossular garnet closely approaching fine jadeite in appearance, esp. that in Transvaal, South Africa.

garnetoid A group of nonsilicate minerals that are isostructural with garnet, including the oxide yafsoanite, the arsenates berzeliite and manganberzeliite, and the halide cryolithionite. Cf: *hydrogrossular; hydrogarnet.*

garnierite A general term for hydrous nickel silicates. Syn: *genthite; noumeite.*

garrelsite A monoclinic mineral, $Ba_3NaSi_2B_7O_{16}(OH)_4$; in colorless crystals associated with nahcolite and shortite in core from an oil boring at Ouray, CO.

garronite An orthorhombic mineral, $Na_2Ca_5Al_{12}Si_{20}O_{64}{\cdot}27H_2O$; zeolite group; pseudotetragonal; forms radiating aggregates in amygdules in basalts, commonly associated with other zeolites, e.g., chabazite, thomsonite, and levyne; at the Garron Plateau, County Antrim, Ireland, and 22 localities in eastern Iceland.

gas (a) Combustible gases (methane), a mixture of air and combustible gases, or other explosive gaseous mixture encountered in mining. (b) A fluid of low density and of high compressibility. The specific recognition of a gas as distinct from a liquid of the same composition requires the simultaneous presence of both phases at equilibrium. See also: *fluid; vapor*. (AGI, 1987) (c) In mining, a mixture of atmospheric air with combustible gases. (Standard, 1964) (d) The term normally used by miners to designate any impure air, esp. explosive combinations. (BCI, 1947) (e) The term generally applied to denote combustible gases. (BS, 1963) (f) Any aeriform liquid other than atmosphere air, such as gaseous carbon dioxide (blackdamp), carbon monoxide (whitedamp), methane (combustible gases), and the common combustible petroleum-product gases. Cf: *acetylene.* (Long, 1960) (g) Abbrev. for gasoline. See also: *manufactured gas; natural gas.* (Long, 1960) (h) A fluid (as air) that has neither independent shape nor volume but tends to expand indefinitely. A substance at a temperature above its critical temperature and therefore not liquefiable by pressure alone. (Webster 3rd, 1966) (i) As a verb, to affect or to treat with gas. To subject to the action of gas. (Webster 3rd, 1966)

gas alarm Device or signal system that warns underground workers of dangerous concentration of combustible gases. (Pryor, 1963)

gas analysis An analysis of mine air to give information regarding the oxygen content of the air and the presence of explosive or otherwise undesirable gas or gases. It is a valuable aid in following the changes in mine air during fires and after explosions. (Lewis, 1964)

gas bubble (a) Round inclusion in glass, synthetic corundum, and spinel, which reveal their difference from native corundum, spinel, and most other native gemstones, in which inclusions are negative crystals. Syn: *glass enclosure; gas enclosure.* (b) Relatively rare type of fluid inclusions created where a gas such as carbon dioxide is trapped in minerals growing from gas-saturated or boiling liquids, identified by their tubular gas-filled structures.

gas carburizing The introduction of carbon into the surface layers of mild steel by heating in a current of gas high in carbon—usually hydrocarbons or hydrocarbons and carbon monoxide. (CTD, 1958)

gas centrifuge process A method of isotope separation in which heavy atoms are separated from light atoms by centrifugal force. (Lyman, 1964)

gas classification The separation of powder into particle-size fractions by means of a gas stream of controlled velocity. (ASTM, 1994)

gas concrete Coke formed in gas retorts as distinguished from that made in a coke oven. (Webster 3rd, 1966)

gas conductor A pipe for leading combustion gases from the mouth of a blast furnace to a hot-blast stove. (Fay, 1920)

gas detonation system A system for initiating blasting caps in which the energy is transmitted through the circuit by means of a gas detonation inside a hollow plastic tube. See also: *nonelectric blasting; shock tube system.* (Dick, 2)

gas drain (a) Eng. A heading driven in a mine for the special purpose of carrying off methane from any workings. (b) A tunnel or borehole for conducting gas away from old workings. (CTD, 1958)

gas emission The release of gas from the strata into the mine workings. (BS, 1963)

gas-emission rate The quantity of methane discharged from the strata and coal seams into the ventilating air of a coal mine. The rate may be expressed on a time or tonnage basis. Gas emission varies with (1) the rate of advance of the workings; (2) the face operation such as cutting, blasting, loading, etc.; and (3) the barometric reading. (Nelson, 1965)

gas enclosure *gas bubble.*

gaseous (a) Having the form of or being gas; of or relating to gases. (Webster 3rd, 1966) (b) Lacking substance or solidity. (Webster 3rd, 1966)

gaseous diffusion A method of isotope separation based on the fact that atoms or molecules of different masses will diffuse through a porous barrier at different rates. The method is used to enrich uranium with the uranium-235 isotope. (Lyman, 1964)

gaseous dispersion pattern A dispersion pattern that may be detected by analysis of soil, air, or gas dissolved in underground water, or of gas condensed in the rocks and soil. Gaseous dispersion patterns of interest include those of hydrocarbons and some noble gases resulting from nuclear decay of radioactive elements, and gaseous substances such as Hg, H_2, He, SO_2, CO_2, and CS_2. (Hawkes, 1962; Lewis, 1964)

gaseous fuel Includes natural gas and the prepared varieties, such as coal gas, oil gas, and iron blast furnace gas, as well as producer gas, etc. (Newton, 1959)

gaseous place A place that is likely to be dangerous from the presence of flammable gas. (Fay, 1920)

gaseous transfer Separation from a magma of a gaseous phase that moves relative to the magma and releases dissolved substances, usually in the upper levels of the magma, when it enters an area of reduced pressure. (AGI, 1987)

gas evolution The liberation of gas in the form of bubbles during the solidification of metals. It may be due to the fact that the solubility of a gas is less in the solid and liquid metal respectively, as when hydrogen is evolved by aluminum and its alloys, or to the promotion of a gas-forming reaction, as when iron oxide and carbon in molten steel react to form carbon monoxide. See also: *blowhole; unsoundness.* (CTD, 1958)

gas explosion A major or minor explosion of combustible gases in a coal mine, in which coal dust apparently did not play a significant part. See also: *coal-dust explosion.* (Nelson, 1965)

gas firing The combustion of coal effected by burning in such a way as to produce a combustible gas, which is then burned secondarily in the laboratory or the furnace. (Fay, 1920)

gas-flame coal Coal containing 35% to 40% volatiles (dry, ashless basis). (Tomkeieff, 1954)

gas fluxing (a) The addition of gaseous materials as a flux to promote melting. (b) A rapid upward streaming of free juvenile gas through a column of molten magma in the conduit of a volcano. The gas acts as a flux to promote melting of the wall rocks. (AGI, 1987)

gas grooves Hills and valleys in electrolytic deposits caused by streams of hydrogen or other gas rising continuously along the surface of the deposit while it is forming. (Henderson, 1953)

gash fracture A small-scale tension fracture that occurs at an angle to a fault and tends to remain open. (AGI, 1987)

gash vein A nonpersistent vein that is wide above and narrow below, and that terminates within the formation it traverses. The term was originally applied to vein fillings in vertical solution joints in limestone. (AGI, 1987)

gasification Conversion of coal to gaseous fuel without leaving a combustible residue. (BS, 1960)

gas ignition The setting on fire of a small or large accumulation of combustible gases in a coal mine. The ignition may be caused by a safety lamp, electrical machinery, explosives, frictional sparking, etc. Syn: *ignition of combustible gases.* (Nelson, 1965)

gas indicator A pocket device for the rapid determination of the percentage of a specific gas in the atmosphere of mines, boiler rooms, blast furnaces, etc. (Fay, 1920)

gas-logged strata Rock formations, usually in coal mines, that contain a relatively high proportion of methane. In descensional ventilation, the buoyancy pressure of the methane opposes the ventilating pressure, with a tendency for cavities to contain combustible gases of high concentration. The same may apply to waste cavities with no natural exit to the return. (Nelson, 1965)

gasman An underground official who examines the mine for combustible gases and has charge of their removal. See also: *fire boss; gas watchman.* (Hess)

gas pipe (a) Mid. A short wooden pipe about 4 in by 4 in (10.2 cm by 10.2 cm) inside, having its upper end open to the roof, and the lower end opening into the bratticing so that any gas given off in the roof may be carried away as formed. (Fay, 1920) (b) Any pipe for conveying gas. (Fay, 1920)

gas pocket A cavity in the rocks containing gas, generally above an oil pocket. (Mersereau, 1947)

gas pore A gas bubble in a mineral. (Standard, 1964)

gas pressure The forces generated from the expansion of gases formed from the reaction of explosive materials after detonation; gas pressures produce the heaving action during rock blasting. Syn: *borehole pressure.*

gas producer A furnace in which coal is burned for the manufacture of producer gas. There are two types: (1) the step-grate, natural-draft generator, which is but a development of the ordinary firebox; and (2) the shaft furnace, with or without a grate and worked by a natural or forced draft. The latter type is identical in many respects to a blast smelting furnace. (Fay, 1920)

gas ratio The ratio of the volume at atmospheric pressure of the gas developed by an explosive to the volume of the solid from which it was formed. Many commercial explosives have a gas ratio of about 8. Ammonium nitrate plus fuel oil has a ratio of about 20. (Leet, 1960)

gas reverser In the iron and steel industry, one who reverses gas valves by manipulating levers to throw hot combustion gases from one side of the furnace to the other, to keep the furnace heat evenly distributed, and to prevent burning out on one side. (DOT, 1949)

gas rig A borehole drill, either rotary or churn-drill type, driven by a combustion-type engine using a combustible liquid, such as gasoline, or a combustible gas, such as bottle gas, as the source of the motivating energy. (Long, 1960)

gassing (a) Absorption of gas by a metal. (ASM, 1961) (b) Evolution of gas from a metal during melting operations or on solidification. (ASM, 1961) (c) The evolution of gas from an electrode during electrolysis. (ASM, 1961)

gassing of copper A process that denotes the brittleness produced when copper containing oxide is heated in an atmosphere containing hydrogen. The hydrogen diffuses into the metal and combines with oxygen, forming steam that cannot diffuse out. A high steam pressure is built up at the crystal boundaries, and the cohesion is diminished. (CTD, 1958)

gas spectrum (a) The spectrum, consisting of bright lines or bands, obtained by dispersing the light from a glowing gas or vapor. (Webster 2nd, 1960) (b) An absorption spectrum obtained by passing light through a gas or vapor. (Webster 2nd, 1960)

gas streaming A process of magmatic differentiation involving the formation of a gaseous phase, usually during a late stage in consolidation of the magma, that results in partial expulsion, by escaping gas bubbles, of residual liquid from the crystal network. (AGI, 1987)

gassy A mine is said to be gassy when it gives off methane or other gas in quantities that must be diluted with pure air to prevent occurrence of explosive mixtures. (BCI, 1947)

gas tracers Slowly moving air currents can be directly observed by using smokes. These may range from simple dust clouds, through various chemical smokes, to more refined techniques employing gas and radioactive tracers. Various chemicals have been used, including stannic chloride, titanium tetrachloride, and pyrosulfuric acid. These materials give off white fumes when their vapors come into contact with atmospheric moisture. The method in common use is to carry the chemical in sealed glass ampules, which can be broken when an observation is to be made. (Roberts, 1960)

gas trap One of many devices for separating and saving the gas from the flow and lead lines of producing oil wells. The mixture of oil and gas is allowed to flow through a chamber large enough to reduce the velocity of the mixture to the point at which the oil and gas tend to separate. The gas, seeking the top of the chamber, is drawn off free of oil, while the oil is discharged at the bottom. Also called gas tank. (Ash, 1949)

gastrolith A rounded stone or pebble, commonly highly polished, from the stomach of some reptiles, esp. dinosaurs, plesiosaurs, and crocodilians. Gastroliths are thought to have been used in grinding up food, but marine reptiles may have used them to provide body stability while in the water. (AGI, 1987)

gastropod Any mollusk belonging to the class Gastropoda, characterized by a distinct head with eyes and tentacles and, in most, by a single calcareous shell that is closed at the apex, sometimes spiralled, not chambered, and generally asymmetrical; e.g., a snail. Range, Upper Cambrian to present. (AGI, 1987)

gastunite *weeksite.*

gas watchman In bituminous coal mining, person who makes morning examinations for gas before workers enter the mine. See also: *fireman; fire boss; gasman.* (DOT, 1949)

gas water Water through which coal gas has been passed and which has absorbed the impurities of the gas. (Fay, 1920)

gate (a) Eng. Gateway or gate road. A road or way underground for air, water, or general passage; a gangway. See also: *gate end.* (b) Eng. A road packed out in longwall goaf. When ripped in the waste to provide packing material on a conveyor face, it is called a dummy gate. Also called gate road; gateway; main brow; trail road. (SMRB, 1930) (c) The apparatus at the bottom of an ore chute for filling cars. Also called a chute. (Spalding, 1949) (d) Syn: *swivel head.* Also, the swivel ring of the swivel head of a diamond drill. (Long, 1960) (e) The closing piece in a stop valve. (Standard, 1964) (f) A valve controlling the admission of water to a waterwheel or conduit. (Standard, 1964)

gate belt conveyor Conveyor usually from 26 to 30 in (66.0 to 76.2 cm) wide and troughed so as to centralize the load and minimize spillage. A scraper feeder, consisting of an elevating chain conveyor driven by the gate belt, is often used to transfer the coal from the face belt to the gate belt. (Mason, 1951)

gate conveyor A gate road conveyor that carries coal from one source or face only; i.e., from a single-unit or double-unit face. See also: *conveyor; face conveyor; gathering conveyor.* (Nelson, 1965)

gate end The coal face or inby end of a gate. See also: *gate.*

gate-end box A flameproof enclosure primarily for use at or near the coalface and designed to line up with similar boxes to form a control board. A gate-end box may contain bus bars, isolators, switches, contactors, transformers, and protective devices, for the control of motors, lighting, and other equipment. Syn: *gate-end unit.* (BS, 1965)

gate-end feeder A short conveyor that feeds the coal from the face conveyor onto the gate conveyor. See also: *feeder conveyor.* (Nelson, 1965)

gate-end loader A short conveyor designed to receive the coal from the face conveyors and elevate it to such a height as to be convenient for delivery into mine cars. See also: *feeder conveyor.* (Nelson, 1965)

gate-end switch A flameproof motor-starting contactor for use with coal-face machinery. The essential features are a flameproof casing divided into two separate compartments, the smaller of which contains a hand-operated isolating switch and the main bus bars. The isolating switch is interlocked with the cover of the main compartment so that it cannot be removed unless the switch is in the off position; or it may be fitted with contacts enabling the mechanism to be earthed before work is undertaken on it. (Mason, 1951)

gate-end unit *gate-end box.*

gate interlock A system designed to prevent shaft conveyances from being moved or action signals from being transmitted, unless all shaft gates are closed. (BS, 1965)

gate road (a) Eng. A road connecting a stall with a main road. (Standard, 1964) (b) A road through the goaf used for haulage of coal from longwall working. (Pryor, 1963)

gate road bunker An appliance for the storage of coal from the face conveyors during peaks of production or during a stoppage of the outby transport. It may consist of a length of conveyor chain running in high-capacity pans arranged under the delivery end of the gate conveyor. When the trunk conveyor cannot handle the coal from the gate conveyor, the bunker chain is slowly drawn back carrying about 1 st (0.9 t) of coal per yard of chain. The bunker is later discharged by reversing the process. See also: *underground bunker.* (Nelson, 1965)

Gates canvas table A large form of inclined canvas table in which the pulp is first classified, then distributed along the upper edge of the table. The concentrates are caught in the warp of the canvas; after this is full, treatment must be stopped while the concentrates are swept or sluiced off. (Liddell, 1918)

gate shutter A paddlelike implement used to shut off the flow of metal from a mold and divert it to other molds. (Standard, 1964)

gate side pack A pillar consisting of tightly rammed material enclosed in walls of stone, built on each side of the gate road. See also: *double packing*. (Nelson, 1965)
gateway longwall N. of Eng. A continuous coal face served by gateways (in Durham about 12 yd or 11 m apart). A small group works in each gateway down which the coal is removed by tubs. (Trist, 1963)
gather To assemble loaded cars from several production points and deliver them to main haulage for transport to the surface or pit bottom. (BCI, 1947)
gathering arm loader A machine for loading loose rock or coal. It has a tractor-mounted chassis, carrying a chain conveyor the front end of which is built into a wedge-shaped blade. Mounted on this blade are two arms, one on either side of the chain conveyor, which gather the material from the muck pile and feed it onto the loader conveyor. The tail or back end of the conveyor is designed to swivel and elevate hydraulically so that the coal or stone can be loaded into a car or on to another conveyor. See also: *loader*. (Nelson, 1965)
gathering conveyor Any conveyor that is used to gather coal from other conveyors and deliver it either into mine cars or onto another conveyor. The term is frequently used with belt conveyors placed in entries where a number of room conveyors deliver coal onto the belt. See also: *gate conveyor; trunk conveyor*. (Jones, 1949)
gathering ground *catchment area.*
gathering haulage That portion of the haulage system immediately adjacent to the face. In longwall mining, the face belt or tubs and track along the face constitute the gathering haulage system. (Wheeler, 1946)
gathering locomotive *gathering motor; electric gathering mine locomotive.*
gathering mine locomotive *gathering motor; electric gathering mine locomotive.*
gathering motor A lightweight type of electric locomotive used to haul loaded cars from the working places to the main haulage road and to replace them with empties. Syn: *gathering locomotive; gathering mine locomotive.* See also: *electric gathering mine locomotive.*
gathering motorman In bituminous coal mining, a person who operates a mine locomotive to haul loaded mine cars from working places to sidings, for the formation of larger trips (trains) to be handled by a haulage cable or a main-line locomotive. Syn: *relay motorman*. (DOT, 1949)
gathering pumps Portable or semiportable pumps that are required when water is encountered while opening a new mine, for extending headings or entries in an operating mine, for pumprooms or rib sections lying in the dip, for collecting water from local pools, or for sinking a shaft. They discharge water at an intermediate pumping station or into a drainage ditch or tunnel carrying water outside a mine.
gatton Scot. *gauton.*
gaudefroyite A hexagonal mineral, $Ca_4Mn_3(BO_3)_3(CO_3)(O,OH)_3$; forms prisms associated with marokite, braunite, and hasumannite in calcite from Tachgagalt, Morocco.
Gaudin's equation An equation for the particle size distribution that can be expected when a material is crushed in a ball mill or rod mill; it is of the form P = 100(x/D)m, where P is the percentage passing a sieve of aperture x, D is the maximum size of particle, and m is a constant which is a measure of dispersion. The equation holds good only if the ratio of size of feed to size of balls is below a critical value.
gauge *gage.*
gault Firm compact clay; brick clay. Also spelled galt, golt.
gauntree *gantry.*
gauslinite A local name for burkeite. From Searles Lake, Calif. See also: *burkeite*. Syn: *teepleite*. (English, 1938)
gauss Unit of magnetic induction in the electromagnetic and Gaussian systems of units, equal to 1 Mx/cm^2 (10 nWb/cm^2), or 10^{-4} Wb/m^2. Also known as abtesla (abt). (McGraw-Hill, 1994)
gaussian model A function frequently used when fitting mathematical models to experimental variograms, often in combination with a nugget model.
gauton Scot. A watercourse cut in the floor of a mine or working. Syn: *gatton*. See also: *hasson.*
gawl An irregular or uneven line of coal face. (CTD, 1958)
gayet (a) French name for sapropelic coal, such as torbanite or cannel. (Tomkeieff, 1954) (b) *cannel coal.*
Gayley process The process for the removal of moisture from the blast of an iron blast furnace by reducing the temperature of the blast current so that the moisture is deposited as snow. See also: *cold blast*. (Webster 2nd, 1960; Fay, 1920)
Gay-Lussac's law When gases react, they do so in volumes that bear a simple ratio to one another, and to the volumes of their products if these are gaseous, temperature and pressure remaining constant. Also called law of gaseous volumes. (Cooper, 1963)
Gay-Lussac's tower In sulfuric acid making, a tower filled with pieces of coke over which concentrated sulfuric acid trickles down. On meeting gas issuing from the lead chambers, the coke absorbs nitrous anhydride, which otherwise would be lost. Cf: *Glover's tower*. (Standard, 1964)
gaylussite A monoclinic mineral, $Na_2Ca(CO_3)_2{\cdot}5H_2O$; soft; in flattened and elongated crystals in muds from playa lakes in the Mohave Desert, CA, and the Gobi Desert, Mongolia.
geanticline (a) A mobile upwarping of the crust of the Earth, of regional extent. Ant: geosyncline. (AGI, 1987) (b) More specif., an anticlinal structure that develops in geosynclinal sediments, due to lateral compression. Var: geoanticline. (AGI, 1987)
gear (a) The moving parts or appliances collectively that constitute some mechanical whole or set, linked meshing and fitted together, and serving to transmit motion or change its rate or direction. Commonly used in the plural. (Hess) (b) A gear wheel. (Hess) (c) *feed gear.* (d) The accessory tools and equipment required to operate a drill. (Long, 1960) (e) A set of enmeshing-toothed rotating parts or cogwheels designed to transmit motion. (Long, 1960) (f) A toothed wheel, cone, or bar. (Nichols, 1976)
gearksutite A monoclinic mineral, $CaAl(OH)F_4{\cdot}H_2O$; occurs with fluorite and barite in hydrothermally altered sedimentary and volcanic rocks.
gearman In beneficiation, smelting, and refining, one who tends a coarse or primary crusher that breaks large lumps of ore into a smaller size so that it may be run through smaller crushers or shipped to a plant for extraction of the valuable metal or minerals. (DOT, 1949)
gear ratio The relationship between the speeds of the first and last shafts, respectively, of a train of gears. If a certain force drives a machine at a given speed and the output shaft runs at one-tenth of the speed of the input shaft, then the output force will be 10 times the input. If the gear ratio of a motor-driven machine is 10 to 1, then the turning force of the last shaft will be 10 times that of the motor, apart from force used up in friction. (Mason, 1951)
gear set A device that causes one shaft to turn another at reduced speed. (Nichols, 1976)
Gebhardt survey instrument A borehole surveying instrument often used to test the verticality of the freezing holes in shaft sinking. A vernier scale is used to determine the positions of the pendulum points at successive points, and by summating the results an accurate plan of the course of the borehole can be prepared. (Nelson, 1965)
gedanite A brittle wine-yellow variety of amber with very little succinic acid; lacks the toughness and ability to take as high a polish as the succinic acid-rich varieties; rarely used as a gem except for beads. Syn: *mellow amber.*
gediegen *zinn.*
gedrite An orthorhombic mineral, $(Mg,Fe)_5Al_2[Si_6Al_2O_{22}](OH)_2$; amphibole group, with Mg/(Mg + Fe) = 0.1-0.89; forms series with magnesiogedrite and ferrogedrite; common in schists, gneisses, and metasomatic rocks.
geest (a) Alluvial material that is not of recent origin lying on the surface. An example is the sandy region of the North Sea coast in Germany. (AGI, 1987) (b) *saprolite.*
gehlenite A tetragonal mineral, $Ca_2Al(AlSi)O_7$; melilite group; forms a series with akermanite; a common constituent of feldspathoidal rocks formed by reaction of mafic magmas with carbonate rocks. Syn: *velardenite.*
Geiger counter (a) An ionization chamber that records the number of radioactive particles impinging upon it per minute, thus detecting radioactive substances. (Bateman, 1951) (b) An instrument that detects gamma rays given off by radioactive substances. (AGI, 1987) (c) An ionization chamber with its vacuum and its applied potential so adjusted that a gamma ray or other ionizing particle passing through it causes a momentary current to flow. The surge of current can be amplified and counted so as to measure the intensity of radioactivity in the vicinity of the chamber. (AGI, 1987)
Geiger-Muëller counter tube A gas-filled chamber usually consisting of a hollow cylindrical cathode and a fine wire anode along its axis. It is operated with a voltage high enough so that a discharge triggered by a primary ionizing event will spread over the entire anode until stopped by the reduction of the field by space charge.
Geiger-Müller counter An instrument consisting of a Geiger-Müller tube plus a voltage source and the electronic equipment necessary to record the tube's electric pulses. Also called Geiger counter. (AGI, 1987)
Geiger-Müller probe A Geiger-Müller counter encased in a watertight container, which can be lowered into a borehole and used to log the intensity of the gamma rays emitted by the radioactive substances in the rock formations traversed. Syn: *Geiger probe*. (Long, 1960)
Geiger-Müller tube A radiation detector consisting of a gas-filled tube with a cathode envelope and an axial wire anode. It functions by producing momentary current pulses caused by ionizing radiation. It is a part of the Geiger-Müller counter. Syn: *G-M tube*. (AGI, 1987)
Geiger probe *Geiger-Müller probe.*
Geiger test The act or process of using a Geiger-Müller probe or counter to measure the intensity of the gamma rays emitted by the radioactive substance contained in rocks traversed by a borehole. (Long, 1960)

geikielite A trigonal mineral, $MgTiO_3$; ilmenite group; forms a series with ilmenite; in highly metamorphosed magnesian marbles associated with brucite, spinel, or diopside; in serpentinites with chromium-rich chlorites, and in gem gravels of Sri Lanka.

Geissler tube (a) A sealed and partly evacuated glass tube containing electrodes. Used for the study of electric discharges through gases. (Standard, 1964) (b) A gas-filled discharge tube having various shapes and usually containing a narrowly constricted portion in which the luminosity is intensified. (Webster 3rd, 1966)

gel (a) A translucent to transparent, semisolid, apparently homogeneous substance in a colloidal state, generally elastic and jellylike, offering little resistance to liquid diffusion, and containing a dispersion or network of fine particles that have coalesced to some degree. (AGI, 1987) (b) A nonhomogeneous gelatinous precipitate; e.g., a coagel. (AGI, 1987) (c) A liquefied mud, which became firm and then reabsorbed most of the water released earlier. A gel is in a more solid form than a sol, and can sustain limited shear stress. See also: *thixotropy.* (AGI, 1987)

Gelamite Trademark for a semigelatin high explosive of relatively high weight strength of 65%; very good water resistance. Used in underground mining, in quarrying, in construction, and in general blasting. (CCD, 1961)

gelatin *gelatin dynamite.* Also spelled gelatine. (Long, 1960)

gelatin borehole tube A device used in borehole surveying. A tube, containing a compass floating in molten gelatin, is lowered to the point in the borehole at which its verticality is required. It is left in position until the gelatin sets and is then withdrawn. The compass indicates the direction and a small plumb bob shows the angle of dip. (Nelson, 1965)

gelatin dynamite A type of highly water-resistant dynamite, characterized by its gelatinous consistency, containing nitroglycerin and nitrocellulose; straight gelatin dynamite is a series of gelatins that include sodium nitrate, while ammonium gelatin dynamites use ammonium nitrate. Syn: *ammonia gelatin dynamite; gelatin.*

gelatin extras Explosives in which a portion of the nitroglycerin is replaced with ammonium nitrate. The explosive velocity is reduced, but the substantial resistance to water is retained. Less expensive than gelatin dynamites. (Carson, 1961)

gelatins A general term relating to explosives in which a principal constituent, nitroglycerin, is given a gelatinous consistency by mixing it with nitrocotton. (BS, 1964)

gelation The formation of a gel from a sol, as by coagulation or by precipitation with an electrolyte. (AGI, 1987)

gel cement Cement to which a small percentage of bentonite has been added either dry or mixed with water. Such an addition particularly adapts the slurry for use in cementing casing and recovering lost circulation because it reduces loss of slurry to the formation, makes for a more homogeneous mixture, increases the water-cement ratio, reduces loss of water to the formation, and sets in substantially the same volume as occupied when placed. (Brantly, 1961)

gelignite A general term relating to explosives of the gelatin type in which there is a proportion of wood, metal, and oxygen-containing salts. (BS, 1964)

gélose The colloidal product of plant decay that becomes the principal constituent of coal. See also: *carbohumin.* (Tomkeieff, 1954)

gélosite Constituent of torbanite, consisting of birefringent pale yellow microscopic crushed spheres. Cf: *humosite.* See also: *humosite; matrosite; retinosite.* (Tomkeieff, 1954)

gel strength The ability or the measure of the ability of a colloid to form gels. (Brantly, 1961)

gem A cut-and-polished stone that has intrinsic value and possesses the necessary beauty, durability, rarity, and size for use in jewelry as an ornament or for personal adornment; a jewel whose value is not derived from its setting. Syn: *cut stone.* (AGI, 1987)

gem crystal A crystal from which a gem can be cut.

gem gravel A gravel placer containing an appreciable concentration of gem minerals. (AGI, 1987)

gemmary (a) The science of gems. Syn: *gemology.* (b) A collection of gems; gems considered collectively. (c) A house or receptacle for gems and jewels.

gem material Any rough material, either natural or artificial, that can be fashioned into a jewel. Cf: *gemstone.*

gem mineral Any mineral species that yields varieties with sufficient beauty and durability to be classed as gemstones.

Gemolite An illuminator used to observe inclusions and other imperfections in gemstones effectively. Employs either a monocular or a binocular microscope.

gemologist (a) One who appraises gems. Also spelled gemmologist (U.K.). (b) One who has mastered gemology.

gemology (a) The science of gems. Cf: *gemmary.* Also spelled gemmology (U.K.). (b) The study of fashioned minerals, their imitators and substitutes both natural and synthetic, prized for their beauty and durability. It concerns composition, structure, occurrence, origin, fashioning, and identification of gems, some of which are minerals and some of organic origin, e.g., pearls. Cf: *descriptive gemology; determinative gemology.*

gem pearl (a) A term used for those better qualities of fine pearl that possess a rose or other particularly desirable orient. Does not include white pearl. (b) An iridescent pearl, perfectly spherical, with maximum luster of even intensity, free from all visible blemishes, and of a decided and desirable orient, such as pink rose.

gem stick A stick on the end of which a gem is cemented while being cut. (Standard, 1964)

gemstone (a) A mineral or petrified material that, when cut and polished, can be used in jewelry. Syn: *precious stone.* (b) A term that includes pearl, amber, coral, jet, or any stone of any variety of gem material, of sufficient beauty and durability for use as a personal ornament. See also: *decorative stone; gem material; ornamental stone.* Also spelled gem stone.

gem variety The variety of a mineral species that yields gemstones.

general-crusher foreman In beneficiation, smelting, and refining, one who directs and coordinates all operations concerned with reducing ore to designated size. (DOT, 1949)

general soil survey A general investigation of superficial deposits. The sampling procedure may include augers, boreholes, and trial pits, and tests are made to cover soil identification. This type of survey aims at establishing soil profiles and locating areas requiring special investigation. See also: *detailed soil survey; preliminary soil survey.* (Nelson, 1965)

generation All the crystals of the same mineral species that appear to have crystallized at essentially the same time; e.g., if there are olivine phenocrysts in a groundmass containing olivine, there are said to be two generations of olivine. (AGI, 1987)

generator (a) A source of electricity, esp. one that transforms heat or mechanical work directly into electric energy, as opposed to a voltaic battery. (Standard, 1964) (b) A vessel, chamber or machine in which the generation of a gas is effected, as by chemical action. (Standard, 1964)

genesis (a) The origin or formation of a natural gem mineral. (b) In petrology, the origin and evolution of rocks based on field and textural observation allied with laboratory analyses and experimental studies. Cf: *petrogenesis.* (c) In mineralogy, the origin of stable phases in terms of pressure, temperature, and composition of parent materials. (d) In ore deposits, determination of specific peculiar and exceptional conditions under which economic minerals have been concentrated. Cf: *metallogeny.*

genetic classification Any classification based on manner of origin or line of descent. Genetic classifications are set up to deal with fossils, rocks, and minerals. (Stokes, 1955)

Geneva ruby An artificial ruby. (Fay, 1920)

Genter filter A filter utilized in coal-washing plants for the recovery of fine coal particles.

Genter thickener Cylindrical tank with obtuse conical base around which raking gear moves slowly, pushing settled sludge to a central discharge. In the body of the tank hang radially mounted tube frames covered with filter cloths (socks). These are connected with a central valve and timing mechanism, so set that vacuum is applied for 1 to 10 min to remove filtrate, after which the gathered solids are displaced by a brief flushback so that they fall to the raking zone. (Pryor, 1963)

genthelvite An isometric mineral, $Zn_4Be_3(SiO_4)_3S$; forms series with danalite and with helvite; in carbonatites and alkaline pegmatites.

genthite *garnierite.*

gently inclined Said of deposits and coal seams with a dip of from 5° to 25°. (Stoces, 1954)

genus A category in the hierarchy of plant and animal classification intermediate in rank between family and species. Adj: generic. Plural: genera. (AGI, 1987)

geo- Prefix from the Greek "ge," meaning land, of the land, or Earth. (Pryor, 1963)

geobotanical indicators Some plants develop peculiar diagnostic symptoms that can be interpreted directly in terms of probable excesses of a particular element in the soil. Geobotanical indicators are either plant species or characteristic variations in the growth habits of plant species that are restricted in their distribution to rocks or soils of definite physical or chemical properties. They have been used in locating and mapping ground water, saline deposits, hydrocarbons, and rock types, as well as ores. (Hawkes, 1962)

geobotanical prospecting (a) The visual study of plants, their morphology, and their distribution as indicators of such things as soil composition and depth, bedrock lithology, the possibility of orebodies, and climatic and ground-water conditions. Cf: *biogeochemical prospecting.* See also: *botanical anomaly.* (AGI, 1987) (b) Prospecting in which visual observation of plants is used as a guide to finding buried ore. Whereas biogeochemical methods require chemical analysis of plant organs, the geobotanical methods depend on direct observations of plant morphology and the distribution of plant species. (Hawkes, 1962)

geobotany The study of plants as related specif. to their geologic environment. (Hawkes, 1962)

geocerain *geocerite.*

geocerite A white, flaky, waxlike resin of approximate composition $C_{27}H_{53}O_2$ in brown coal. Also spelled geocerain, geocerin. Cf: *geomyricite.*

geochemical anomaly A concentration of one or more elements in rock, soil, sediment, vegetation, or water that is markedly higher or lower than background. The term may also be applied to hydrocarbon concentrations in soils. See also: *significant anomaly.* (AGI, 1987)

geochemical coherence The phenomenon of the intimate occurring together of certain chemical elements in nature because of their similar chemical properties, as, for example, the group of the lanthanides, zirconium-hafnium, niobium-tantalum, etc. (Schieferdecker, 1959)

geochemical cycle The sequence of stages in the migration of elements during geologic changes. Rankama and Sahama distinguish a major or endogenous cycle, proceeding from magma to igneous rocks to sediments to sedimentary rocks to metamorphic rocks and possibly through migmatites back to magma, and a minor or exogenic cycle proceeding from sediments to sedimentary rocks to weathered material and back to sediments again. (AGI, 1987)

geochemical environment Pressure, temperature, and the availability of the most abundant chemical components are the parameters of the geochemical environment that determine which mineral phases are stable. On the basis of these variables, it is possible to classify all the natural geochemical environments of the Earth into two major groups—primary and secondary. The primary environment extends downward from the lower levels of circulating meteoric water to the deepest level of the crust and may extend into the mantle. It is an environment of high temperature and pressure, restricted circulation of fluids, and relatively low free-oxygen content. The secondary environment is the environment of weathering, erosion, and sedimentation at the surface of the Earth. It is characterized by low temperatures, nearly constant low pressure, free movement of solutions, and abundant free oxygen, water, and carbon dioxide.

geochemical exploration The search for economic mineral deposits or petroleum by detection of abnormal concentrations of elements or hydrocarbons in surficial materials or organisms, usually accomplished by instrumental, spot-test, or quickie techniques that may be applied in the field. Syn: *geochemical prospecting.* (AGI, 1987)

geochemical landscape The pattern, in any given area, in which the net effect of all the dynamic forces concerned in the movement of earth materials will be reflected in the overall pattern of distribution of the elements. (Hawkes, 1962)

geochemical mapping The systematic collection and processing of a very large number of samples accompanied by the proper presentation and interpretation of the resulting analytical data, usually with reference to a topographic map or other geographic coordinate system. (Hawkes, 1962)

geochemical prospecting *geochemical exploration.*

geochemical relief A little-used term for the variation in metal values in varied geographic settings. Geochemical contrast.

geochemical survey A survey involving the chemical analysis of systematically collected samples of rock, soil, stream sediments, plants, or water; this expression may be further modified by indicating specif. the material sampled, as, for example, geochemical soil survey. (Hawkes, 1957)

geochemist An individual who studies the chemistry of earth materials. May be qualified by the term "inorganic" for the study of nonbiological materials, "organic" for the study of plants and hydrocarbons, and "isotope" for the study of nuclides of the elements. Generally, the geochemist is concerned with the distribution of elements in exploration application or with the cycles of elements in basic science.

geochemistry The study of the relative and absolute abundances of the elements and their nuclides (isotopes) in the Earth; the distribution and migration of the individual elements or suites of elements in the various parts of the Earth (the atmosphere, hydrosphere, lithosphere, etc.), and in minerals and rocks, and also the study of principles governing this distribution and migration. Geochemistry may be defined very broadly to include all parts of geology that involve chemical changes, or it may be focused more narrowly on the distribution of the elements, as in Mason's definition; the latter is commonly understood if the term is used without qualification. (AGI, 1987)

geochronic *geochronologic.*

geochronite A monoclinic mineral, $Pb_{14}(Sb,As)_6S_{23}$; forms a series with jordanite; soft; sp gr, 6.4 to 6.5; associated with galena, jamesonite, and boulangerite; in lead ores in Bolivia, Germany, Sweden, and the United States. Syn: *kilbrickenite.*

geochronologic Pertaining to geochronology.

geochronologic unit A division of time traditionally distinguished on the basis of the rock record as expressed by a time-stratigraphic unit. Geochronologic units in order of decreasing rank are eon, era, period, epoch, and age. Names of periods and units of lower rank are the same as those of the corresponding time-stratigraphic units; the names of some eras and eons are independently formed. Syn: *geologic time unit.* (AGI, 1987)

geochronology The study of time in relationship to the history of the Earth, esp. by the absolute age determination and relative dating systems developed for this purpose. See also: *absolute time.* Cf: *geochronometry.* Syn: *geologic chronology.* (AGI, 1987)

geochronometry Measurement of geologic time by geochronologic methods, esp. radiometric dating. Cf: *geochronology.* (AGI, 1987)

geochrony An obsolete syn. of geochronology. (AGI, 1987)

geode (a) A hollow globular or subspherical body, from 2.5 to 30 cm or more in diameter, found in certain limestone beds and rarely in shales; it is characterized by a thin and sometimes incomplete outermost primary layer of dense chalcedony, by a cavity that is partly filled by a drusy lining of inward-projecting crystals (usually of quartz or calcite and sometimes of barite, celestite, and various sulfides) deposited from solution on the cavity walls. Unlike a druse, a geode is separable (by weathering) as a discrete nodule or concretion from the rock in which it occurs and its inner crystals are not of the same minerals as those of the enclosing rock. (AGI, 1987) (b) The crystal-lined cavity in a geode. (AGI, 1987) (c) A term applied to a rock cavity and its lining of crystals that is not separable as a discrete nodule from the enclosing rock. Cf: *vug.* (AGI, 1987)

geodesy (a) The science concerned with the determination of the size and shape of the Earth and the precise location of points on its surface. (AGI, 1987) (b) The determination of the gravitational field of the Earth and the study of temporal variations such as Earth tides, polar motion, and rotation of the Earth. Syn: *geodetics.* (AGI, 1987)

geodetic coordinates Quantities defining the horizontal position of a point on an ellipsoid of reference with respect to a specific geodetic datum, usually expressed as latitude and longitude. These may be referred to as geodetic positions or geographic coordinates. The elevation of a point is also a geodetic coordinate and may be referred to as a height above sea level. (AGI, 1987)

geodetics *geodesy.*

geodetic surveying Surveying in which account is taken of the figure and size of the Earth and corrections are made for Earth curvature; the applied science of geodesy. It is used where the areas or distances involved are so great that results of desired accuracy and precision cannot be obtained by plane surveying. (AGI, 1987)

geodimeter Trade name of an electronic optical device that measures ground distances precisely by electronic timing and phase comparison of modulated light waves that travel from a master unit to a reflector and return to a light-sensitive tube where an electric current is set up. It is normally used at night and is effective with first-order accuracy up to distances of 3 to 25 miles (5 to 40 km). Etymol: acronym for geodetic-distance meter. Cf: *tellurometer.* (AGI, 1987)

geodynamic Pertaining to physical processes within the Earth as they affect the features of the crust. (AGI, 1987)

geognosy An 18th-century term for a science accounting for the origin, distribution, and sequence of minerals and rocks in the Earth's crust. The term was superseded by geology as early ideas were abandoned. It has become restricted to absolute knowledge of the Earth, as distinct from the theoretical and speculative reasoning of geology. (AGI, 1987)

geographical concentration The ratio of face length (X) to length of main haulage roads (L) in the same units; i.e., X/L. See also: *concentration of output.* (Nelson, 1965)

geographic cycle *cycle of erosion.*

Geographic Information System A computer system for managing spatial data. Abbrev. GIS. A GIS, e.g., can provide a simultaneous consideration of geology, geophysics, geochemistry, and mineral deposits in a region for the purposes of mineral exploration. (Bonham-Carter, 1987)

geography The study of all aspects of the Earth's surface including its natural and political divisions, the distribution and differentiation of areas, and often, people in relationship to their environment. (AGI, 1987)

geohydrology A term, often used interchangeably with hydrogeology, referring to the hydrologic or flow characteristics of subsurface waters. (AGI, 1987)

geoid The figure of the Earth considered as a sea-level surface extended continuously through the continents. It is a theoretically continuous surface that is perpendicular at every point to the direction of gravity (the plumb line). It is the reference for astronomical observations and for geodetic leveling. See also: *datum.* (AGI, 1987)

geoisotherm *isogeotherm.*

geologic *geological.*

geologic age (a) The age of a fossil organism or of a particular geologic event or feature referred to the geologic time scale and expressed in terms either of time units (absolute age) or of comparison with the immediate surroundings (relative age); an age datable by geologic methods. (AGI, 1987) (b) The term is also used to emphasize the long-past periods of time in geologic history, as distinct from present-day or historic times. See also: *age*. (AGI, 1987)

geological Pertaining to or related to geology. The choice between this term and geologic is optional, and may be made according to the sound of a spoken phrase or sentence. Geological is generally preferred in the names of surveys and societies, and in English and Canadian usage. Syn: *geologic.* (AGI, 1987)

geological horizon An interface that indicates a particular position in a stratigraphic sequence. In practice it is commonly a very thin bed.

geological province An area throughout which geological history has been essentially the same or one that is characterized by particular structural or physiographic features.

geological section *geologic section.*

geological survey (a) A systematic investigation of an area determining the distribution, structure, composition, history, and interrelations of rock units. Its purpose may be either purely scientific or economic with special attention to the distribution, reserves, and potential recovery of mineral resources. Syn: *geologic survey.* (b) An organization engaged in making surveys; e.g., a state survey or the U.S. Geological Survey.

geologic chronology *geochronology.*

geologic column (a) A composite diagram that shows in a single column the subdivisions of part or all of geologic time or the sequence of stratigraphic units of a given locality or region (the oldest at the bottom and the youngest at the top, with dips adjusted to the horizontal) so arranged as to indicate their relations to the subdivisions of geologic time and their relative positions to each other. See also: *columnar section.* (AGI, 1987) (b) The vertical or chronologic arrangement or sequence of rock units portrayed in a geologic column. See also: *geologic section.* (AGI, 1987)

geologic drilling Drilling done primarily to obtain information from which the geology of the formations penetrated can be determined. (Long, 1960)

geologic engineering *engineering geology.*

geologic formation *formation.*

geologic log A written and/or graphic record of the geologic data obtained from drillhole core and/or cuttings. In noncore drilling, the cuttings are separated at depth intervals of about 1 m or 2 m and examined. In core drilling, the core is kept in sequence and examined. The essential description of core includes lithology, alteration, mineralization, and structural discontinuities. Cf: *geotechnical log*. (Peters, 1987)

geologic map A map on which is recorded geologic information, such as the distribution, nature, and age relationships of rock units (surficial deposits may or may not be mapped separately), and the occurrence of structural features (folds, faults, joints), mineral deposits, and fossil localities. It may indicate geologic structure by means of formational outcrop patterns, by conventional symbols giving the direction and amount of dip at certain points, or by structure-contour lines. (AGI, 1987)

geologic mineralizer Substance that promotes mineral concentration and crystallization during solidification of rock-forming material, particularly in pegmatite dikes. Syn: *ore-forming fluid; mineralizer.* (Bennett, 1962)

geologic section Any sequence of rock units found in a given region either at the surface or below it (as in a drilled well or mine shaft); a local geologic column. Syn: *geological section; stratigraphic section.* See also: *geologic column*. (AGI, 1987)

geologic survey *geological survey.*

geologic thermometer *geothermometer.*

geologic time The period of time dealt with by historical geology, or the time extending from the end of the formative period of the Earth as a separate planetary body to the beginning of written or human history; the part of the Earth's history that is represented by and recorded in the succession of rocks. The term implies extremely long duration of remoteness in the past, although no precise limits can be set. (AGI, 1987)

geologic time scale An arbitrary chronologic arrangement or sequence of geologic events, used as a measure of the relative or absolute duration or age of any part of geologic time, and usually presented in the form of a chart showing the names of the various rock-stratigraphic, time-stratigraphic, or geologic-time units, as currently understood; e.g., the geologic time scales published by Harland et al. (1982), Odin (1982), Palmer (1983), and Salvador (1985). Syn: *time scale.* (AGI, 1987)

geologic time unit The time unit corresponding with a time-stratigraphic unit; e.g., period, epoch, or age. See also: *geochronologic unit.*

geologist One who is trained in and works in any of the geological sciences. (AGI, 1987)

geologize To participate in or talk about geology; to practice geology. (AGI, 1987)

geology The study of the planet Earth—the materials of which it is made, the processes that act on these materials, the products formed, and the history of the planet and its life forms since its origin. Geology considers the physical forces that act on the Earth, the chemistry of its constituent materials, and the biology of its past inhabitants as revealed by fossils. Clues on the origin of the planet are sought in a study of the Moon and other extraterrestrial bodies. The knowledge thus obtained is placed in the service of humans—to aid in discovery of minerals and fuels of value in the Earth's crust, to identify geologically stable sites for major structures, and to provide foreknowledge of some of the dangers associated with the mobile forces of a dynamic Earth. See also: *historical geology; physical geology*. (AGI, 1987)

geology system (a) The formal chronostratigraphic unit of rank next lower than "erathem" and above "series". Rocks encompassed by a system represent a time span and an episode of Earth history sufficiently great to serve as a worldwide reference unit. The temporal equivalent of a system is a "period." The system is the fundamental unit of chronostratigraphic classification of Phanerozoic rocks, extended from a type area or region and correlated mainly by its fossil content. System boundaries either have been ratified by the International Union of Geological Sciences or are under review by a working group of the International Commission on Stratigraphy. The Proterozoic systems are related to geologic events with geochronologic boundaries established by the Subcommission on Precambrian Stratigraphy. (b) Some systems initially established in Europe were later divided or grouped elsewhere into units ranked as systems, but these are more appropriately known as "subsystems" or "supersystems".

geomagnetician One who sets up magnetic observatories and stations in order to chart the Earth's magnetic field and applies data obtained to problems in the fields of telephony, telegraphy, radio broadcasting, navigation, mapping, and geophysical prospecting. Also called terrestrial magnetician. (DOT, 1949)

geomagnetic meridian *magnetic meridian.*

geomorphic (a) Pertaining to the form of the Earth or of its surface features; e.g., a geomorphic province. (AGI, 1987) (b) Pertaining to geomorphology; geomorphologic. (AGI, 1987)

geomorphic cycle *cycle of erosion.*

geomorphogeny That part of geomorphology that deals with the origin, development, and changes of the Earth's surface features or landforms. (AGI, 1987)

geomorphology (a) The science that treats the general configuration of the Earth's surface; specif., the study of the classification, description, nature, origin, and development of present landforms and their relationships to underlying structures, and of the history of geologic changes as recorded by these surface features. In the United States, it has come to replace the term "physiography" and is usually considered a branch of geology; in Great Britain, it is usually regarded as a branch of geography. (AGI, 1987) (b) Strictly, any study that deals with the form of the Earth (including geodesy, and structural and dynamic geology). This usage is more common in Europe, where the term has even been applied broadly to the science of the Earth. (AGI, 1987) (c) The features dealt with in, or a treatise on, geomorphology; e.g., the geomorphology of Texas. (AGI, 1987)

geomyricin *geomyricite.*

geomyricite A white, waxy resin of approximate composition $C_{32}H_{62}O_2$ in brown coal. Syn: *geomyricin.* Cf: *geocerite.*

geophone A seismic detector that produces a voltage proportional to the displacement, velocity, or acceleration of ground motion, within a limited frequency range. Syn: *jug; seismometer; pickup.* Cf: *seismograph; transducer.* (AGI, 1987)

geophysical exploration Exploring for minerals or mineral fuels, or determining the nature of Earth materials by measuring a physical property of the rocks and interpreting the results in terms of geologic features or the economic deposits sought. Physical measurements may be taken on the surface, in boreholes, or from airborne or satellite platforms. See also: *geophysical prospecting.*

geophysical log *well log.*

geophysical prospecting Exploring for minerals or mineral fuels, or determining the nature of earth materials by measuring a physical property of the rocks, and interpreting the results in terms of geologic features or the economic deposits sought. Physical measurements may be taken on the surface, in boreholes, or from airborne or satellite platforms. See also: *geophysical exploration.*

geophysical prospecting surveyor Person who locates and marks sites selected for conducting geophysical prospecting activities concerned with locating subsurface earth formations likely to contain petroleum or mineral deposits. (DOT, 1949)

geophysical survey The exploration of an area in which geophysical properties and relationships unique to the area are mapped by one or more geophysical methods. (AGI, 1987)

geophysicist One who studies seismic, gravitational, electrical, thermal, radiometric, and/or magnetic phenomena to investigate geological phenomena, such as structure and composition of the Earth, forces causing movement and warping of surface, origin and activity of glaciers and volcanoes, and the location and cause of earthquakes; charts ocean currents and tides; takes measurements concerning shape and movements of Earth, and acoustic, optical, and electrical phenomena in the atmosphere; and locates petroleum and mineral deposits. (DOT, 1949)

geophysics A branch of physics dealing with the Earth, including its atmosphere and hydrosphere. It includes the use of seismic, gravitational, electrical, thermal, radiometric, and magnetic phenomena to elucidate processes of dynamical geology and physical geography, and makes use of geodesy, geology, seismology, meteorology, oceanography, magnetism, and other Earth sciences in collecting and interpreting Earth data. Geophysical methods have been applied successfully to the identification of underground structures in the Earth and to the search for structures of a particular type, as, for example, those associated with oil-bearing sands. (AGI, 1987)

georgiadesite A monoclinic mineral, $Pb_{16}(AsO_4)_4Cl_{14}(OH)_6$ or $Pb_{16}(AsO_4)_4Cl_{14}O_2(OH)_2$; forms stubby tablets; sp gr, 7.1; associated with laurionite, matlockite, and fiedlerite on altered slags at Laurium, Greece.

geosphere The solid portion of the Earth, including water masses; the lithosphere plus the hydrosphere. Above the geosphere lies the atmosphere, and at the interface between these two regions is found almost all of the biosphere, or zone of life. (Hunt, 1965)

geostatic pressure *rock pressure.*

geostatistics A methodology for the analysis of spatially correlated data. The characteristic feature is the use of variograms or related techniques to quantify and model the spatial correlation structure. Also includes the various techniques such as kriging, which utilize spatial correlation models.

geosyncline A mobile downwarping of the crust of the Earth, either elongate or basinlike, measured in scores of kilometers, in which sedimentary and volcanic rocks accumulate to thicknesses of thousands of meters. A geosyncline may form in part of a tectonic cycle, in which orogeny follows. Recognition of the plate structure of the lithosphere has led to appreciation that nearly all geosynclinal phenomena are related to ocean opening and closing. Cf: *mobile belt.* See also: *synclinorium.* Ant: geanticline. (AGI, 1987)

geotechnical log A written and/or graphic record of the data obtained from drillhole core. In addition to the data given in geologic logs, geotechnical logs require more detail on discontinuities such as fractures, joints, bedding, rock quality designation, and hydrologic conditions. Cf: *geologic log.* (Peters, 1987)

geotechnics The application of scientific methods and engineering principles to the acquisition, interpretation, and use of knowledge of materials of the Earth's crust for the solution of engineering problems; the applied science of making the Earth more habitable. It embraces the fields of soil mechanics and rock mechanics, and many of the engineering aspects of geology, geophysics, hydrology, and related sciences. Syn: *geotechnique.* (AGI, 1987)

geotechnique *geotechnics.*

geotechnology The application of scientific methods and engineering techniques to the exploitation and use of natural resources. (AGI, 1987)

geotectonic *tectonic.*

geotectonics *tectonics.*

geotherm *isogeotherm.*

geothermal Pertaining to the heat of the interior of the Earth. Syn: *geothermic.* (AGI, 1987)

geothermal gradient The rate of increase of temperature in the Earth with depth. The gradient differs from place to place depending on the heat flow in the region and the thermal conductivity of the rocks. The average geothermal gradient in the Earth's crust approximates 25 °C per kilometer of depth. (AGI, 1987)

geothermic *geothermal.*

geothermic gradient *strata temperature.*

geothermometer An indicator of the temperature at which some reaction took place or some geologic process was active. Syn: *geologic thermometer.*

geotomography The adaptation of computer aided tomography (CAT) scan technology to tomographic analysis of geologic features such as fractures and differing rock types. In geotomography, the software analyzes the energy ray paths between a transmitter and receiver that are placed in separate drill holes, at various locations along a single drill hole, or along an underground opening. Various types of energy waves, such as seismic, acoustic, electromagnetic, or X-rays, can be analyzed by the computer software to create an image. (Jessop, 1992)

gerasimovskite An amorphous mineral, $(Mn,Ca)_2(Nb,Ti)_5O_{12}\cdot 9H_2O(?)$; forms a series with manganbelyankinite; soft; in ussingite pegmatites in the Lovozero massif, Kola Peninsula, Russia.

gerhardite A basic copper nitrate containing 52.9% copper. Crystallization, orthorhombic. Cleavage, yields flexible laminae. Tenacity, fragile, and sectile. Mohs hardness, 2; sp gr, 3.426; luster, vitreous, brilliant; color, deep emerald-green; streak, light green; transparent; soluble in dilute acids. From Jerome, AZ. (Weed, 1918)

gerhardtite An orthorhombic mineral, $Cu_2(NO_3)(OH)_3$; dark to emerald green; soft; occurs with atacamite, brochantite, malachite, and azurite in oxidized zones of copper deposits in arid and semiarid regions.

german A straw tube filled with gunpowder and used as a fuse. Not used in coal mines. (CTD, 1958)

germanate-pyromorphite Synthetic $Pb_5(GeO_4)_3Cl$; forms apatite structure.

German cupellation A method of cupellation using a large reverberatory furnace with a fixed bed and a movable roof. The bullion to be cupelled is all charged at once, and the silver is not refined in the same furnace where the cupellation is carried on. (Fay, 1920)

German cut *pyramid cut.*

German gold Archaic name for amber.

germanite An isometric mineral, $Cu_{26}Fe_4Ge_4S_{32}$; associated with tennantite, sphalerite, enargite, pyrite, and bornite; at the Tsumeb Mine in Namibia.

germanium A grayish-white, metallic element occurring in argyrodite, a sulfide of germanium and silver; and in germanite, zinc ores, and coal. Its presence in coal ensures a large reserve of the element in the future. Symbol, Ge. It is a very important semiconductor material. Also used as an alloying agent, a phosphor, a catalyst, and for infrared detectors and optical equipment. (Handbook of Chem. & Phys., 3)

germanium dioxide (soluble) GeO_2; melting point, 1,115 °C. This oxide is a glass former and provides some unique properties; e.g., greater dispersion, lower melting temperature, and higher transmissivity for infrared radiation. Some germanium oxide complexes and solid solutions have ferroelectric properties. Colorless; sp gr, 4.228 (at 25 °C); hexagonal; soluble in alkalies; and slightly soluble in acids and in water. Used as an ingredient of special glass mixtures. (Lee, 1961; CCD, 1961; Handbook of Chem. & Phys., 2)

germanium nitride Ge_3N_4; decomposes at 800 °C. A special electroceramic of high resistivity. (Dodd, 1964)

German lapis *false lapis.*

German reduction process This process consists in (1) roasting copper ore, (2) melting and obtaining a matte with 30% to 40% copper called coarse metal, (3) roasting the coarse metal, (4) melting and obtaining a matte with 60% to 70% copper called fine metal, (5) roasting the fine metal, and (6) melting and obtaining black copper. (Fay, 1920)

German steel A metal made from charcoal iron obtained from bog iron or from sparry carbonate of iron. (Fay, 1920)

gersdorffite An isometric mineral, NiAsS; cobaltite group; massive; sp gr, 5.9 to 6.0; in sulfide veins intergrown with maucherite, nickeline, and chalcopyrite at the Garson and Falconbridge Mines, Sudbury, ON, Can.; with cobaltite and rammelsbergite in the silver-arsenide ores of Cobalt, ON, Can.; a source of nickel.

Gerstenhofer furnace A shaft furnace, filled with terraces or shelves, through which crushed ore is caused to fall for roasting. (Fay, 1920)

gerstleyite A monoclinic mineral, $Na_2(Sb,As)_8S_{13}\cdot 2H_2O$; dark red; soft; in fine granular aggregates, groups of small plates, or spherules in clay, with borates at the Baker Mine, Kramer district, Kern County, CA.

get cleanup Arkansas. To have an opportunity to load out all the coal a miner has loosened. (Fay, 1920)

getter Substance used to combine with the residual oxygen in an electric bulb or tube. Its use is called gettering. (Pryor, 1963)

getting The actual process of digging clay, by hand or by excavator; getting and transporting form the successive stages of winning. (Dodd, 1964)

geversite An isometric mineral, $Pt(Sb,Bi)_2$; pyrite group; forms tiny grains intergrown with native platinum in concentrates at the Dreikop Mine, eastern Transvaal, Republic of South Africa.

geyerite *löllingite.*

geyser basin A valley or other area that contains numerous springs, geysers, and steaming fissures fed by the same ground-water flow. (AGI, 1987)

geyserite Syn: *siliceous sinter.* Used esp. for the compact, loose, concretionary, scaly, or filamentous incrustation of opaline silica deposited by precipitation from the waters of a geyser. (AGI, 1987)

ghost crystal *phantom crystal.*

ghost reflection In the seismic reflection method, a special type of multiple reflection. This is the reflection that takes place when the energy traveling upward from the shot is reflected downward by the base of the weathered zone or by the Earth's surface. The reflected pulse follows the primary downgoing pulse by a time interval determined by the depth of the shot below the weathering (or the free surface) and the velocity of the material above the shot. For normal shooting depths this interval will range from 0.010 to 0.020 s. (Dobrin, 1960)

giant The nozzle of a pipe used to convey water for hydraulic mining and for the purpose of distributing or properly applying and increasing the force of the water. See also: *hydraulic monitor.* (Ricketts, 1943)

giant granite *pegmatite.*

giant powder (a) A blasting powder consisting of nitroglycerin, sodium nitrate, sulfur, rosin, and sometimes kieselguhr. (Webster 3rd, 1966) (b) Nitroglycerin absorbed by an inert filler such as kieselguhr. (Pryor, 1963)
giant tender *nozzleman.*
gib (a) A temporary support at the face to prevent coal from falling before the cut is complete. (BCI, 1947) (b) Scot. A sprag; a prop put in the holing of a seam while being undercut. (Fay, 1920)
gibber An Australian term for a pebble or boulder; esp. one of the wind-polished or wind-sculptured stones that compose a desert pavement or the lag gravels of an arid region. It is pronounced with a hard g. (AGI, 1987)
Gibbs adsorption theorem A solute that lowers the surface tension of its solvent tends to concentrate at the air-liquid interphase, and vice versa. (Pryor, 1963)
gibbsite A monoclinic mineral, $8[Al(OH)_3]$; pisolitic; in micalike crystals, or stalactitic and spheroidal forms; a constituent of bauxite associated with boehmite and/or diaspore; formed by weathering of igneous rocks, esp. nepheline syenite; also in veins; a source of aluminum and synthetic abrasives. In emery deposits formed by thermal or regional metamorphism of bauxites, gibbsite occurs as an alteration crust on corundum. Syn: *hydrargillite.*
Gibbs phase rule *phase rule.*
Gibraltar stone A light-colored onyx at Gibraltar. See also: *Mexican onyx.*
gieseckite (a) An aluminosilicate of magnesium and potassium, sometimes with appreciable FeO. (Hey, 1955) (b) Green fine-grained micaceous alteration of nepheline.
giessenite An orthorhombic mineral, $Pb_{13}(Cu,Ag)(Bi,Sb)_9S_{28}$(?); forms soft, metallic, needles on galena associated with pyrite, rutile, and tennantite; near Giessen in the Binn valley, Valais, Switzerland.
gig (a) A mine cage or skip. (CTD, 1958) (b) Gravity or self-acting haulage. Also called ginney. (Mason, 1951)
Gilbert The unit of magnetomotive force in the electromagnetic system, equal to the magnetomotive force of a closed loop of one turn in which there is a current of $1/(4\pi)$ abamp. (McGraw-Hill, 1994)
gild To wash over or overlay thinly with gold; coat with gold, either in leaf or powder, or by electroplating; as, to gild a chandelier. To overlay with any other substance for the purpose of giving the appearance of gold. (Standard, 1964)
gillespite A tetragonal mineral, $BaFeSi_4O_{10}$; vitreous; red; translucent; associated with sanbornite, celsian, taramellite, and witherite at Dry Delta, AK, and Fresno and Mariposa Counties, CA.
Gilman heat-treating machine Used for tempering and hardening of drill bits at the mine. Syn: *automatic heat-treating machine.* (Lewis, 1964)
gilsonite *uintaite.*
gim peg A device used in faceting gems; a piece of wood containing a series of holes into which a dop stick (gem stick) can be fixed at various angles, thus regulating the angle of the facet being cut. Syn: *jamb stick.*
gin (a) A pump worked by a windlass. (Standard, 1964) (b) A pile-driving machine. (Standard, 1964) (c) A drum framework and pulleys for hoisting mineral from a shallow shaft. (CTD, 1958) (d) Horse gear for hoisting through a mine shaft. (Pryor, 1963) (e) A small, hand-cranked hoist. (Long, 1960) (f) Eng. A drum and framework carrying pulleys, by which the ore and waste are raised from a shallow pit; a whim. Also called horse gin. A contraction of engine. (Fay, 1920) (g) An old form of hoisting engine. (Mason, 1951)
ginging The process of lining a shaft with bricks or masonry; the lining itself. (CTD, 1958)
ginney A journey set or train of tubs, trams, or trucks, or a self-acting incline, in a coal mine. (CTD, 1958)
ginney tender A person working on an endless chain haulage. (CTD, 1958)
ginny carriage Eng. A small railway truck for transporting constructive materials. (Standard, 1964)
ginorite A monoclinic mineral, $Ca_2B_{14}O_{23}\cdot8H_2O$; transparent to translucent; occurs as pellets embedded in a matrix of sassolite and clay within colemanite-veined basalt at Death Valley, CA; also as minute lozenge-shaped plates aggregated into masses with calcite in veins in sandstone at Sasso Pisano, Tuscany, Italy. Cf: *strontioginorite.*
gin pit A shallow mine, the hoisting from which is done by a gin. (Fay, 1920)
gin race The circular path that a gin horse travels. Syn: *gin ring.* (Standard, 1964; Fay, 1920)
gin ring *gin race.*
gin wheel The cylinder of a gin or winch. (Standard, 1964)
giobertite *magnesite.*
gips plate *gypsum plate; accessory plate.*
giraffe (a) A cagelike mine car esp. adapted for inclines, having the frame higher at one end than at the other. (Standard, 1964) (b) A mechanical appliance for receiving and tripping a car of ore, etc., when it arrives at the surface. (Fay, 1920) (c) A multiple-deck skip. (Fay, 1920)
girasol (a) A name applied to many gemstones with a girasol effect, e.g., moonstone; specif. a translucent variety of "fire opal" with reddish reflections in bright light and a faint bluish-white floating light emanating from the center of the stone. (b) adj. Said of any gem variety, e.g., sapphire, chrysoberyl, that exhibits a billowy, gleaming round or elongated area of light that "floats" or moves about as the stone is turned or as the light source is moved. (c) A name for glass spheres used in the manufacture of imitation pearls.
girdle (a) A thin sandstone stratum. (Standard, 1964) (b) Flattened lenticles or nodules of any hard stone in softer beds. Sometimes extended also to beds. (Arkell, 1953) (c) In stratigraphy: (1) a thin stratum, particularly said of sandstone or coal, esp. when exposed in a shaft or borehole or (2) flattened lenticles or nodules of any hard stone in softer beds. (d) In gemology, the line that encompasses a cut gem parallel to the horizon; or that determines the greatest horizontal expansion of the stone. (e) In structural petrology, on an equal-area projection, a belt or concentration of points representing orientations of fabric elements. If this belt coincides approx. with a great circle of the projection, it is referred to as a great-circle girdle. If the belt of concentration coincides approx. with a small circle of the projection, it is called a small-circle or "cleft girdle"
Girond process In this process, fluorspar, soda ash, carbon, lime, and mill scale were thrown on to the bottom of a hot ladle, and thus sintered. On tapping the steel from the open hearth furnace into the ladle, the resulting boil removed part of the phosphorus. (Osborne, 1956)
girth (a) A brace member running horizontally between the legs of a drill tripod or derrick. (Long, 1960) (b) In square-set timbering, a horizontal brace running parallel to the drift. (Long, 1960)
GIS *Geographic Information System.*
Gish-Rooney method An artificial-current conductive direct-current method of measuring ground resistivity which avoids polarization by continually reversing the current with a set of commutators. (AGI, 1987)
gismondine A monoclinic mineral, $Ca_2Al_4Si_4O_{16}\cdot9H_2O$; zeolite group; pseudotetragonal; vitreous, transparent to translucent; in cavities in leucitic tephrites and related lavas; also zeolite zones in basaltic lavas where it is associated with chabazite, thomsonite, and phillipsite. Also spelled gismondite.
Gjer's soaking pit A cavity lined with refractory material used in metal working to enclose large ingots, in order to preserve them at a high temperature, and thus avoid the necessity of reheating. (Fay, 1920)
glacial (a) Of or relating to the presence and activities of ice or glaciers, as glacial erosion. (AGI, 1987) (b) Pertaining to distinctive features and materials produced by or derived from glaciers and ice sheets, as glacial lakes. (AGI, 1987) (c) Pertaining to an ice age or region of glaciation. (AGI, 1987) (d) Suggestive of the extremely slow movement of glaciers. (AGI, 1987) (e) Used loosely as descriptive or suggestive of ice, or of below-freezing temperature. (AGI, 1987)
glacial action All processes due to the agency of glacier ice, such as erosion, transportation, and deposition. The term sometimes includes the action of meltwater streams derived from the ice. See also: *glacial erosion.* (AGI, 1987)
glacial deposits *glacial drift.*
glacial drift Boulders, till, gravel, sand, or clay transported and deposited by a glacier or its meltwater. See also: *glacial overburden; drift.* Syn: *glacial deposits.*
glacial epoch Any part of geologic time, from Precambrian onward, in which the climate was notably cold in both the Northern and Southern Hemispheres, and widespread glaciers moved toward the equator and covered a much larger total area than those of the present day; specif. the latest of the glacial epochs, known as the Pleistocene Epoch. Syn: *glacial period; drift epoch.* (AGI, 1987)
glacial erosion The grinding, scouring, plucking, gouging, grooving, scratching, and polishing effected by the movement of glacier ice armed with rock fragments frozen into it, together with the erosive action of meltwater streams. See also: *glacial action.* (AGI, 1987)
Glacialite Trade name for a white clay from Enid, OK., marketed as a fuller's earth. (English, 1938)
glacial overburden Glacial-drift materials overlying bedrock. See also: *glacial drift*. (Long, 1960)
glacial period *glacial epoch.*
glacial soil Soil composed of boulder clays, moraines, etc., which were formed by the action of ice during the Pleistocene age. (Mining)
glacial till *till.*
glaciated Said of a formerly glacier-covered land surface, esp. one that has been modified by the action of a glacier or an ice sheet, as a glaciated rock knob. (AGI, 1987)
glaciation (a) The formation, movement, and recession of glaciers or ice sheets. (AGI, 1987) (b) The covering of large land areas by glaciers or ice sheets. (AGI, 1987) (c) The geographic distribution of glaciers and ice sheets. (AGI, 1987) (d) A collective term for the geologic processes of glacial activity, including erosion and deposition, and the

resulting effects of such action on the Earth's surface. (AGI, 1987) (e) Any of several minor parts of geologic time during which glaciers were more extensive than at present; a glacial epoch, or a glacial stage. (AGI, 1987)

glacier A large mass of ice formed, at least in part, on land by the compaction and recrystallization of snow, moving slowly by creep downslope or outward in all directions due to the stress of its own weight, and surviving from year to year. Included are small mountain glaciers as well as ice sheets continental in size, and ice shelves that float on the ocean but are fed in part by ice formed on land. (AGI, 1987)

glacier theory The theory, first propounded about 1840 and now universally accepted, that the drift was deposited through the agency of glaciers and ice sheets moving slowly from higher to lower latitudes during the Pleistocene Epoch. (AGI, 1987)

glaciofluvial Pertaining to the meltwater streams flowing from wasting glacier ice and esp. to the deposits and landforms produced by such streams, as kame terraces and outwash plains; relating to the combined action of glaciers and streams. Syn: *fluvioglacial.* (AGI, 1987)

glaciolacustrine Pertaining to, derived from, or deposited in glacial lakes; esp. said of the deposits and landforms composed of suspended material brought by meltwater streams flowing into lakes bordering the glacier, such as deltas, kame deltas, and varved sediments. (AGI, 1987)

glaciology (a) The study of all aspects of snow and ice; the science that treats quantitatively the whole range of processes associated with all forms of solid existing water. Syn: *cryology.* (AGI, 1987) (b) The study of existing glaciers and ice sheets, and of their physical properties. This definition is not internationally accepted. (AGI, 1987)

glaciomarine Of, or relating to, processes or deposits that involve the action of glaciers and the sea, or the action of glaciers in the sea. (Fay, 1920)

gladite An orthorhombic mineral, $PbCuBi_5S_9$; occurs in soft, metallic prismatic crystals; sp gr, 6.96; in lead-zinc ores at Gladhammar, Kalmar, Sweden.

glady (a) A term used in Devon, England, for a variegated black and white clay having a slippery or smooth texture and often associated with stoneware clays. Also spelled: gladii. (AGI, 1987) (b) Said of a limestone-outcrop area having shallow soil. (AGI, 1987)

glance (a) A mineral having a splendant luster, e.g., chalcocite (copper glance). (b) Any of several sulfide minerals that are mostly dark colored, having a metallic luster. Syn: *lead sulfide.*

gland (a) The outer portion of a stuffing box, having a tubular projection embracing the rod, extending into the bore of the box, and bearing against the packing. (Standard, 1964) (b) The fixed engaging part of a positive-driven clutch. (Standard, 1964)

glare (a) A visual sensation that can result in annoyance, discomfort, loss in visual performance, or reduction of visibilty. There are three types of glare: disability, discomfort, and reflected. Glare is a significant factor in determining the design of underground coal mine illumination systems. (b) See also: *disability glare; discomfort glare.*

glass (a) A state of matter intermediate between the close-packed, highly ordered array of a crystal and the poorly packed, highly disordered array of a gas. Most glasses are supercooled liquids, i.e., are metastable, but there is no true break in the change in properties between the metastable and stable states. The distinction between glass and liquid is made solely on the basis of viscosity, and is not necessarily related, except indirectly, to the difference between metastable and stable states. (AGI, 1987) (b) An amorphous product of the rapid cooling of a magma. It may constitute the whole rock (e.g., obsidian) or only part of a groundmass. Cf: *volcanic glass.* (AGI, 1987)

glass-cloth screens A device of clothlike material woven from glass fibers that is attached to a metal frame to form a box- or basin-shaped receptacle, for filtering out impurities from the incoming stream of molten aluminum before the metal reaches the molds. (Light Metal Age, 1958)

glass electrode A glass-membrane electrode used to measure pH or hydrogen-ion activity. (ASM, 1961)

glass enclosure *gas bubble.*

glassies Octahedral diamond crystals (transparent). (Hess)

glass meteorite *moldavite.*

glass opal *hyalite.*

glass porphyry *vitrophyre.*

glass sand A sand that is suitable for glassmaking because of its high silica content (93% to 99%) and its low content of iron oxide, chromium, cobalt, and other colorants. (AGI, 1987)

glass schorl *axinite.*

glass wool *mineral wool.*

glassy feldspar Two varieties of potassium feldspar occur as transparent colorless crystals, sanidine and adularia; also transparent yellow orthoclase and transparent colorless albite. Syn: *sanidine.*

glauberite A monoclinic mineral, $Na_2Ca(SO_4)_2$; slightly salty tasting; in evaporite deposits and fumarolic crusts; a source of sodium sulfate.

Glauber salt *mirabilite.* Also spelled Glauber's salt.

glaucocerinite A hexagonal mineral, $(Zn,Cu)_{10}Al_6(SO_4)_3(OH)_{32}\cdot 18H_2O$; forms soft, blue, fibrous-botryoidal coating on adamite; associated with malachite, smithsonite, and gypsum at Laurium, Greece.

glaucochroite An orthorhombic mineral, $CaMnSiO_4$; occurs in prismatic crystals associated with nasonite, willemite, garnet, axinite, and barite at Franklin, NJ. Cf: *monticellite.*

glaucodot An orthorhombic mineral, (Co,Fe)AsS; pseudocubic, dimorphous with alloclasite; in metallic prismatic crystals, or massive in cobalt ores with cobaltite and pyrite. Also spelled glaukodot.

glaucolite A blue or green variety of scapolite.

glauconite (a) A monoclinic mineral, $4[(K,Na)(Fe^{3+},Al,Mg)_2(Si,Al)_4O_{10}(OH)_2]$; mica group; basal cleavage; dull, light to dark green; soft; a common authigenic mineral in marine sediments, useful for radiometric ages for host rocks. Syn: *greensand.* (b) A general term applied to green hydrous silicates of potassium and iron. Cf: *celadonite.*

glauconitic sandstone A sandstone containing sufficient grains of glauconite to impart a marked greenish color to the rock; greensand. (AGI, 1987)

glaucophane A monoclinic mineral, $2[Na_2(Mg,Fe)_3Al_2Si_8O_{22}(OH)_2]$; amphibole group with Mg/(Mg+Fe) = 0.5 to 1.0; prismatic cleavage; bluish-gray to lavender blue; common in low-temperature, high-pressure schists associated with lawsonite, pumpellyite, or epidote. Cf: *crossite; riebeckite.*

glaucophane schist A type of amphibole schist in which glaucophane rather than hornblende is an abundant mineral. Epidote frequently occurs, and there are quartz and mica varieties. (AGI, 1987)

glaucopyrite A variety of löllingite containing cobalt. See also: *löllingite.* (Standard, 1964)

glazing barrel A rotating barrel in which gunpowder is glazed with graphite. (Standard, 1964)

glebe (a) A tract of land containing mineral deposits or ore. (b) Obsolete term for a clod of earth, an ore, or an earthy mineral. (Arkell, 1953)

Gleeds A glowing coal or small coke such as that used in nailmaking. (CTD, 1958)

gleet Slime, ooze, slimy alluvial deposit. Also spelled glet.

glessite A brown variety of retinite; sp gr, 1.015 to 1.027; found on the shores of the Baltic Sea.

gley soil Soil developed under conditions of poor drainage, resulting in reduction of iron and other elements and in gray colors and mottles. The term is obsolete in the United States. (AGI, 1987)

glide (a) *slip.* (b) A noncrystallographic shearing movement, as one grain over another. (ASM, 1961) (c) *glide reflection.*

glide direction (a) The direction of gliding along glide planes in a mineral. (AGI, 1987) (b) The crystallographic direction of translation along a glide plane in a space group; may be parallel to a crystallographic axis or along a diagonal. (c) The crystallographic direction of slip along a slip plane in a single crystal during plastic deformation. Syn: *slip direction.* Cf: *slip.*

glide plane (a) In single-crystal deformation, a plane on which translation- or twin-gliding takes place without rupture during plastic deformation. Syn: *slip plane.* Cf: *slip.* (b) The common plane of the two axes of a twin crystal. (Hess) (c) Slip plane or parting of mineral specimen. Direction along which slip may occur under suitably directed pressure; due to weakness of bond in crystalline structure along one of the three axes. (Pryor, 1963)

glide reflection *glide.*

gliding (a) A change of form by differential movements along definite planes in crystals without fracture. (Fay, 1920) (b) The formation of twin crystals. (Hess) (c) *slip.*

gliding plane *slip plane.*

glimmer *mica.*

glimmerite *biotitite.*

glimmer plate *mica plate; accessory plate.*

glimmerton An early name for illite. Syn: *illite.*

glinite A group name for clay minerals from clay deposits. See also: *chasovrite.* (AGI, 1987)

glist (a) A gleam; sparkle. (Fay, 1920) (b) Eng. A dark, shining mineral resembling black tourmaline. (Standard, 1964) (c) An early name for mica (Cornish).

glistening As applied to the degree of luster of a mineral, means those minerals affording a general reflection from the surface, but no image, as talc or chalcopyrite. (Fay, 1920)

Global Positioning System A satellite-based navigational system permitting the determination of any point on the Earth with high accuracy. Abbrev: GPS. GPS is used in mapping, mineral exploration, and GIS data collection. The systems depend on the Navigation Satellite Timing and Ranging (NAVSTAR) GPS developed and operated by the U.S. Department of Defense, which consists of a network of 25 satellites in orbit about the Earth as well as ground operations support. After all the satellites are deployed, GPS will provide all-weather, worldwide, two- and three-dimensional (latitude, longitude, and elevation) positioning capabilities over a 24-h period. (Leick, 1989; Bookhout, 1994)

globe thermometer A thermometer in a hollow spherical black globe, the readings from which show a higher value, due to radiation, than those from a conventional thermometer so that the globe device measures the effectual radiation temperature. (Strock, 1948)

globigerina ooze A widespread, deep-sea deposit largely composed of the shells of foraminifera, among which *Globigerina* is esp. abundant. Other calcareous remains are present (about 10%), together with an inorganic residue (about 3% or 4%) having the composition of red clay. (Holmes, 1928)

globular *spherulitic.*

globular powder Particles having approx. spherical shape. (Osborne, 1956)

globulite A tiny, spherical, incipient crystal visible in some volcanic glasses examined in thin section under a polarized-light microscope.

glockerite A cryptocrystalline variety of lepidocrocite; formed by oxidation of iron sulfides as pale yellow to black crusts, stalactites, or earthy masses. Syn: *vitriol ocher.*

glomerocryst Distinct clusters of megascopic crystals as in a glomerophorphyritic rock. See also: *polycrystal.*

glomeroporphyritic *cumulophyric.*

glomeroporphyry A porphyry in which the phenocrysts are gathered in distinct clusters.

gloom A stove for drying gunpowder; drying oven. (Standard, 1964)

glory hole (a) A funnel-shaped excavation, the bottom of which is connected to a raise driven from an underground haulage level. The ore is broken by drilling and blasting in benches around the periphery of the funnel. This process is also called milling, and the excavation is termed a mill hole or simply a mill. (Lewis, 1964) (b) A vertical pit, material from which is fed by gravity to hauling units in a shaft under the pit bottom. (Nichols, 1976) (c) A combination opencast and underground mining system, in which quarried material gravitates or is moved to a short shaft, from the bottom of which it is delivered to an underground transport system. (Pryor, 1963) (d) Can. Large openpit excavation. (Hoffman, 1958) (e) An opening through which to observe the interior of a furnace. (Standard, 1964)

glory-hole system A method of mining using a system of haulageways beneath the block of ore, which has had its top surface exposed by the removal of the overburden. Over the haulageways are chutes that extend up to the surface, and are spaced at intervals of 50 ft (15.2 m) or at any other convenient distance. The excavation of the ore begins at the top of the chute, and broken ore is removed by loading it out from the chutes into cars on the haulage level. The ore block is worked from the top down. The method is similar in principle to underhand stoping. Also called milling system; chute system.

Glover's tower In sulfuric acid works, a tower through which the acid from the Gay-Lussac tower trickles and yields nitrous anhydride to the gases entering the lead chambers, at the same time cooling them. Cf: *Gay-Lussac's tower.* (Standard, 1964)

glow (a) The incandescence of a heated substance, or the light from such a substance; white or red heat. (Standard, 1964) (b) The light from a phosphorescent mineral. (Hess)

glucine $CaBe_4(PO_4)_2(OH)_4 \cdot 1/2H_2O$, massive and encrusting, with moraesite from a locality in the Urals. Named from the alternative of beryllium-glucinum. (Hey, 1964; Fleischer, 1995)

glushinskite A monoclinic mineral, $Mg(C_2O_4) \cdot 2H_2O$; forms on serpentinite as a reaction between oxalic acid from the lichen *Lecanora atra* and magnesium-rich serpentine minerals; in northeast Scotland.

glycerol Clear; colorless or pale yellow; syrupy liquid; $CH_2OHCHOHCH_2OH$. Used in explosives, as a binder for cements and mixes, and as a lubricant and a softener; used in the manufacture of munitions and as an antifreeze liquid. (CCD, 1961; Handbook of Chem. & Phys., 2; Crispin, 1964; Lee, 1961)

glycerol trinitrate *nitroglycerin.*

glyptic Of minerals, a natural resemblance to the carving or engraving on precious stones.

glyptogenesis The sculpture of the Earth's surface by erosion. (AGI, 1987)

glyptography (a) The art of engraving upon gems. (b) The descriptive science of engraved gems.

G.M.B. Eng. Good merchantable brand, as applied to metal on the Metal Exchange. (Fay, 1920)

gmelinite A hexagonal mineral, $(Na_2,Ca)Al_2Si_4O_{12} \cdot 6H_2O$; zeolite group; forms transparent to translucent, pyramidal or tabular crystals, with other zeolites, in the amygdules of basaltic lava.

G-M tube *Geiger-Müller tube.*

gneiss A foliated rock formed by regional metamorphism, in which bands or lenticles of granular minerals alternate with bands or lenticles in which minerals having flaky or elongate prismatic habits predominate. Generally less than 50% of the minerals show preferred parallel orientation. Although a gneiss is commonly feldspar- and quartz-rich, the mineral composition is not an essential factor in its definition. Varieties are distinguished by texture (e.g., augen gneiss), characteristic minerals (e.g., hornblende gneiss), or general composition and/or origins (e.g., granite gneiss). (AGI, 1987)

gneissic structure In a metamorphic rock, commonly gneiss, the coarse, textural lineation or banding of the constituent minerals into alternating silicic and mafic layers. Syn: *gneissosity.* (AGI, 1987)

gneissosity *gneissic structure.*

goaf (a) That part of a mine from which the coal has been worked away and the space more or less filled up with caved rock. See also: *cundy.* (Fay, 1920) (b) The refuse or waste left in the mine. (Fay, 1920)

goaves Old workings.

gob (a) A common term for goaf. (Fay, 1920) (b) To leave coal and other minerals that are not marketable in the mine. (Fay, 1920) (c) To stow or pack any useless underground roadway with rubbish. (Fay, 1920) (d) To store underground, as along one side of a working place, the rock and refuse encountered in mining. (Hudson, 1932) (e) The space left by the extraction of a coal seam into which waste is packed or the immediate roof caves. (CTD, 1958) (f) A pile of loose waste in a mine, or backfill waste packed in stopes to support the roof. (Ballard, 1955) (g) Coal refuse left on the mine floor. (Korson, 1938) (h) The material so packed or stored underground. (Hudson, 1932) (i) To fill with goaf, or gob; to choke; as a furnace is gobbed, or gobs up. See also: *gobbing.* (Webster 2nd, 1960)

gobber (a) Any device used for gobbing waste material. (Jones, 1949) (b) A person employed to pack rubbish or waste into the gob. (CTD, 1958)

Gobber Trade name for a cutting machine provided with a conveyor for gobbing the unusable cuttings formed during the cutting operation. (Jones, 1949)

gobbing The act of stowing waste in a mine. Also called gobbing-up. See also: *gob.* (Jones, 1949; Fay, 1920)

gobbing slate A thick layer of slate between two seams of coal. The lower seam is mined and the upper seam and the slate are shot down; the coal is loaded out and then the slate is gobbed. (Fay, 1920)

gobbing the bone Cleaning up slate. (Korson, 1938)

gob dump *gob pile.*

gob entry A wide entry with a heap of refuse or gob along one side.

gob fire (a) Fire originating spontaneously from the heat of decomposing gob. Also called breeding fire. (b) A fire occurring in a worked-out area, due to ignition of timber or broken coal left in the gob. (CTD, 1958) (c) Fire caused by spontaneous heating of the coal itself, and which may be wholly or partly concealed. See also: *oxidation of coal; spontaneous combustion.* (Mason, 1951)

gob heading (a) A roadway driven through the gob after the filling has settled. (CTD, 1958) (b) Gob road.

gob pile An accumulation of waste material such as rock or bone, either on the surface or underground. Syn: *gob dump.*

gob-pile orator A more or less unflattering term applied to a talkative miner, used much in the same sense as "soapbox orator.". (BCI, 1947)

gob road Eng. A gallery or road extended through goaf or gob.

gob-road system Eng. A form of the longwall system of working coal, in which all the main and branch roadways are made and maintained in the goaves. (Fay, 1920)

gob room Space left for stowing gob.

gob wall A rough wall of flat stones built to prevent the piles of gob from obstructing the passage of air.

go-devil (a) A device used to scrape and descale pipes carrying solids, pulps, sludges, slurries, and other deposit-forming liquors. (Pryor, 1963) (b) A rude sledge upon which one end of a log is borne, the other end trailing on the ground; tieboy; also, a rough, strong wagon used in the woods and about quarries. (Standard, 1964) (c) An iron rod dropped down a well to explode a charge of nitroglycerin. (Mersereau, 1947) (d) *bullet.*

go-devil plane In the United States, a term for gravity haulage. (Nelson, 1965)

Godfrey furnace A furnace with an annular hearth for roasting sulfide ores; used in Wales. (Fay, 1920)

goethite An orthorhombic mineral, alpha-$Fe^{3+}O(OH)$; polymorphous with akaganeite, feroxyhyte, and lepidocrocite; dull to adamantine, varicolored with yellow ocher streak; a common weathering product of iron-bearing minerals; precipitates in bogs and springs; a major constituent of limonite and gossans, and a source of iron and a yellow ochre pigment. A hydrous oxide mineral of iron. Also spelled göthite, gothite. Formerly called allcharite; xanthosiderite. Cf: *bog iron.* Syn: *yellow ocher.*

goffan (a) Corn. A surface working in which the material is thrown from one platform to another. (Hess) (b) Corn. A long narrow surface working. See also: *coffin.*

Gohi iron Copper-bearing iron, very low in impurities and in carbon (0.02% maximum), containing about 0.25% copper. (Bennett, 1962)

going Scot. Working, for example, a going place. A room in the course of being worked.

going bord (a) A roadway to the coal face in bord and pillar working. (CTD, 1958) (b) Eng. The bord or headway used as a main road for conveying the tubs to and from the face to a flat. See also: *flat.* Also called going headway. (SMRB, 1930) (c) N. of Eng. A bord (room) down which coal is trammed, or one along which the coal from several working places is conveyed into the main haulage. (Fay, 1920)

going headway A headway or bord laid with rails, and used for conveying the coal cars to and from the face. (Zern, 1928)

going in The act or process of lowering the drill string, a string of pipe, or casing into a borehole. (Long, 1960)

going off A borehole, the course of which is deviating from that intended. Also called drifting; walking; wandering. (Long, 1960)

going road A working place in a coal mine which is being pushed forward, as distinct from an old or disused place. (CTD, 1958)

Golconda An ancient and famous group of diamond mines on the Kistna River, India, where the Koh-i-noor and other world-famous diamonds were found. (Hess)

gold (a) An isometric mineral, native 4[Au]; commonly alloyed with silver or copper, possibly with bismuth, mercury, or the platinum-group metals; metallic yellow; soft and malleable; sp gr, 19.3 if pure; occurs in hydrothermal veins with quartz and various sulfides; disseminated in submarine massive effusives and in placers or nuggets, fines, and dust. (b) Found in nature as the free metal and in tellurides; very widely distributed. Symbol: Au. Occurs in veins and alluvial deposits; often separated from rocks and other minerals by sluicing and panning operations. Good conductor of heat and electricity. Used in coinage, jewelry, decoration, dental work, plating, and for coating certain space satellites. It is a standard for monetary systems in many countries. Syn: *palladium gold.* (Handbook of Chem. & Phys., 3)

goldamalgam A variety of native gold containing mercury and silver, with gold averaging approx. 40%, commonly associated with platinum in yellowish-white grains that crumble easily. Cf: *amalgam.*

gold amalgam Former spelling of goldamalgam. See also: *amalgam.*

gold argentide *electrum.*

goldbeaters' skin The prepared outside membrane of the large intestine of cattle used for separating the leaves of metal in goldbeating and sometimes as the moisture-sensitive element in hygrometers. (Webster 3rd, 1966)

gold cupride *cuproauride; auricupride.*

gold dust Fine particles, flakes, or pellets of gold, e.g., as obtained in placer mining.

golden beryl A clear, golden-yellow or yellowish-green gem variety of beryl. Cf: *heliodor.*

golden ocher (a) A native ocher. (Standard, 1964) (b) A mixture of light yellow ocher, chrome yellow, and whiting. (Standard, 1964)

golden stone Greenish-yellow peridot (olivine). Not to be confused with "goldstone."

gold fever A mania for seeking gold; applied specif. to the excitement caused by the discovery of gold in California in 1848-49. (Standard, 1964)

goldfieldite An isometric mineral, $Cu_{12}(Te,Sb,As)_4S_{13}$; tetrahedrite group; metallic; at the Mohawk Mine, Goldfield, NV.

goldfoil Gold beaten or rolled out very thin; gold in sheets thicker than goldleaf. (Webster 3rd, 1966)

gold glass A term sometimes applied to goldstone. Also spelled gold fluss.

goldichite A monoclinic mineral, $KFe(SO_4)_2{\cdot}4H_2O$; in soft, pale-green radiating clusters of prismatic laths and fine-grained crystalline crusts from decomposition of pyrite; associated with coquimbite, halotrichite, and roemerite on Calf Mesa, San Rafael Swell, UT.

Goldich's stability series Mineral species differ widely in their resistance to weathering processes. This series summarizes the relative resistance to weathering of the common rock-forming silicates, and indicates that the minerals crystallized at the highest temperatures, under the most anhydrous conditions, are more readily weathered than those that crystallized last from the lower temperature, more aqueous magmas. (Hawkes, 1962)

goldleaf Extremely fine layers of gold formed by beating or rolling between layers of goldbeaters' skin; used for gilding works of art, fabrics, and books. (Bennett, 1962)

goldmanite An isometric mineral, $Ca_3(V,Al,Fe)_2(SiO_4)_3$; garnet group; forms dark-green dodecahedra and minute grains embedded in vanadium-rich clay; at Laguna uranium mining district, NM.

gold matrix Gold in a matrix of milky quartz. Syn: *gold quartz.*

gold milling A general term applied to the treating of ore to recover gold and silver therefrom.

gold mine A mine containing or yielding gold. It may be either in solid rock (quartz mine) or in alluvial deposits (placer mine).

gold opal A fire opal that exhibits only an overall color of golden yellow.

gold pan *pan.*

gold poachers Roving and enterprising freelance miners and prospectors. (Hoover, 1948)

gold quartz Milky quartz containing small inclusions of gold; may be cut and polished for jewelry. Syn: *gold matrix.*

gold-quartz ores Gold-bearing ore from which the sulfides have been removed by the leaching of ground waters so that the ore consists almost entirely of quartz gangue, some iron oxides, and free gold. (Newton, 1959)

gold sapphire A misnomer for lapis lazuli containing flecks of pyrite. Not to be confused with golden sapphire.

goldschmidtine *stephanite.*

goldschmidtite *sylvanite.*

Goldschmidt's phase rule *mineralogical phase rule.*

Goldschmidt's process (a) The thermite process of welding. (Fay, 1920) (b) The removal of tin from scrap tinplate by dry chlorine. (Hess)

goldstone (a) Aventurine spangled close and fine with particles of gold-colored material. Cf: *aventurine.* (Webster 3rd, 1966) (b) A translucent reddish-brown glass containing a multitude of tiny tetrahedra or thin hexagonal platelets of metallic copper that exhibit bright reflections, producing a popular but poor imitation of aventurine. Syn: *aventurine glass; fire agate.* Cf: *sunstone.*

gold telluride (a) Minerals containing tellurium forming tellurides of gold and silver; e.g., sylvanite, calaverite, and petzite. (Statistical Research Bureau, 1935) (b) One of several natural tellurides of gold and silver, e.g., sylvanite, $(Au,Ag)_2Te_4$; calaverite, $AuTe_2$; and petzite, Ag_3AuTe_2.

gold topaz (a) A misnomer for heat-treated citrine. (b) A misnomer for naturally colored citrine. Syn: *false topaz; topaz-quartz.*

goliath crane A portal type of crane having a lifting capacity of 50 st (45 t) or more, with the crab traveling along the horizontal beam. See also: *portal crane.* (Hammond, 1965)

Gommesson method A specialized method of surveying a borehole, utilized when a magnetic compass cannot be used because of local magnetism. The instrument used is essentially a rigid tube, up to 30 ft (9.2 m) long, which is lowered into a borehole. The tube fits the borehole closely and contains a fine wire under tension. The difference between the arc of the tube, when bent at a crook in the borehole, and the chord of the wire is indicated by a stylus marking, which can be measured. The dip is read by etch tubes, and a directional orientation taken at the surface is carried down the hole by precise alignment of the tube and rods as they are lowered into the borehole. (Long, 1960)

gondola A railroad car with no top, flat bottom, fixed sides, and sometimes demountable ends that is used chiefly for hauling heavy bulk materials. Cf: *high side.* (Webster 3rd, 1966)

gondola car Type of open freight truck used in the United States for mineral transport. (Pryor, 1963)

gone off A borehole that has deviated from the intended course. (Long, 1960)

goniometer (a) In crystallography, an instrument for measuring angles between crystal faces. Types are contact, two-circle, and reflection. See also: *contact goniometer.* (b) The part of an X-ray diffractometer that rotates the sample and detector through the Bragg angles of diffraction. (c) A specimen holder with three rotation axes for orientation of single crystals relative to the X-ray beam in Laue and Wiessberg photography. Syn: *three-circle goniometer.* Cf: *two-circle goniometer; spindle stage; universal stage.*

gonnardite An orthorhombic mineral, $Na_2CaAl_4Si_6O_{20}{\cdot}7H_2O$; zeolite group; forms white, finely fibrous, radiating spherules in cavities in basalt, and as an alteration of nepheline.

gonyerite A possibly orthorhombic mineral, $(Mn,Mg)_5Fe(Si_3Fe)O_{10}(OH)_8$; chlorite group; pseudohexagonal with basal cleavage; soft; with barite, berzeliite, and garnet in small hydrothermal veinlets cutting skarn at Långban, Sweden.

good delivery Under Metal Exchange rulings, description of metal delivered at an agreed purity or of a defined quality. (Pryor, 1963)

goodletite The green-amphibole or green-pyroxene matrix rock in which rubies are embedded; Australia and New Zealand.

Goodman duckbill loader The duckbill assembly consists of five major units: (1) a shovel trough to which is attached the shovel head fitting inside the feeder trough, (2) an operating carrier that controls the connection or coupling between the feeder and shovel troughs, (3) a sliding shoe that moves to and fro on the floor of the seam, (4) a swivel trough, and (5) a pendulum jack. The function of the duckbill is to gather the coal and load it into the shaker conveyor pan column. The shovel is forced into the prepared coal by the forward motion of the pan column. As the shovel is propelled forward, the coal is conveyed back along the shovel trough and then to the pan column. (Mason, 1951)

Goodman miner A continuous miner designed for driving coal headings in medium to thick seams. The machine is crawler-mounted and equipped with two triple-arm rotating cutting units and a chain conveyor. The cut coal is discharged on to the chain conveyor which delivers into shuttle or mine cars. Also called Goodman-type 500 miner. See also: *continuous miner.* (Nelson, 1965)

goods A trade term for a lot, parcel, or shipment of diamonds without regard to quality, composition, or quantity. (Chandler, 1964)

good-shooting coal Arkansas. Coal that can be shot "off the solid" with a large proportion of lump coal and little slack. (Fay, 1920)
gooseberry stone A yellowish-green variety of grossular.
goose dung ore An inferior grade of iron sinter containing silver. Also called goose silver ore. (Fay, 1920)
gooseneck (a) In drilling, the bent-tube part of a water swivel to which the water hose is connected. (Long, 1960) (b) A T-shaped connection for supplying water to the top end of wash rods in penetrating overburden. It is fitted with pipe handles by means of which the wash rods may be turned. (Long, 1960)
goosing In hydraulic mining, driving the gravel forward with the stream from the giant. Opposite of drawing.
gopher An irregular prospecting drift following or seeking the ore without regard to maintenance of a regular grade or section. Also called gopher drift. (Fay, 1920)
gopher hole Horizontal opening in wall of quarry, perhaps chambered or tee-headed in preparation for blasting. Also, an irregular pitting hole made when prospecting. Gophering is random prospecting by such pits or by gopher drift. Also called coyote hole. (Pryor, 1963)
gopher hole blasting Terms applied to the method of blasting in which large charges are fired in small adits driven into the face of the quarry at the level of the floor. See also: *tunnel blasting*. (Lewis, 1964)
gophering (a) A method of breaking up a sandy, medium-hard overburden where blastholes tend to cave in. A series of shallow holes are made by a bar, and an explosive charge is fired in each. The debris is removed, and the holes are deepened and further charges fired until the holes are deep enough to take sufficient explosives to break the entire deposit. (Nelson, 1965) (b) The haphazard working of the easiest and richest portions of an ore deposit by miners with little or no capital. (Nelson, 1965)
gopherman In metal mining, one who extracts ore located in pockets or other parts not accessible for machine drilling in an open pit mine. (DOT, 1949)
gor York. Sticky, dirty clay. Also spelled gore. (Arkell, 1953)
gorceixite A monoclinic mineral, $BaAl_3(PO_4)(PO_3OH)(OH)_6$; crandallite group; pseudotrigonal; forms spheroids in fractured novaculite, Garland County, AR; and as rolled pebbles (favas) in the diamantiferous sands of Brazil and Guyana.
gordonite A triclinic mineral, $MgAl_2(PO_4)_2(OH)_2{\cdot}8H_2O$; paravauxite group; in glassy lath-shaped crystals in crusts with variscite; near Fairfield, UT.
Gordon's rule A rule by which the capacity of hydraulic elevators is computed. It is as follows: M = H x N/C, where M = cubic yards of material lifted per hour, H = available head of water in feet, N = water flow in cubic feet per second, C = the efficient working height of the elevator, taken as head H in hundreds of feet multiplied by 15. (Lewis, 1964)
görgeyite A monoclinic mineral, $K_2Ca_5(SO_4)_6{\cdot}H_2O$; forms small, tabular crystals with glauberite and minor halite in salt deposits at Ischl, Austria. Syn: *mikheevite.*
gorgulho A diamond-bearing quartz and clay gravel of Brazil. (Hess)
gorse A barrel or tub for carrying water underground. (CTD, 1958)
goshenite A colorless, white, or bluish beryl from Goshen, MA. (Schaller, 1917)
goslarite An orthorhombic mineral, $ZnSO_4{\cdot}7H_2O$; in acicular or hair-like crystals, or massive; a decomposition product of sphalerite on recent mine workings at Butte, MT, and Bingham Canyon, UT. Syn: *copperas; white copperas; zinc vitriol.*
gossan An iron-bearing weathered product overlying a sulfide deposit. It is formed by the oxidation of sulfides and the leaching-out of the sulfur and most metals, leaving hydrated iron oxides and rarely sulfates. Syn: *capping; iron hat; chapeau de fer.* Also spelled gozzan. Cf: *oxidized zone; false gossan.* (AGI, 1987)
goth Staff. Sudden bursting of coal from the face, owing to tension caused by unequal pressure. The term "airblast" is sometimes used in metal mines, esp. in South Africa. (Zern, 1928; Fay, 1920)
got-on-knobs S. Staff. A system of working thick coal, being a kind of bord-and-pillar plan, the main roadways being first driven to the boundary. (Fay, 1920)
gotten (a) An abandoned or exhausted mine. (CTD, 1958) (b) Coal ready to be filled underground into tubs or trains. (CTD, 1958)
gouge (a) A layer of soft, earthy or clayey, fault-comminuted rock material along the wall of a vein, so named because a miner can "gouge" it out to facilitate the mining of the vein itself. Syn: *selvage; pug.* See also: *gouge clay; hulk.* (b) The clay or clayey material along a fault or shear zone. Also called clay gouge. *fault gouge.* (c) To work a mine without plan or system.
gouge angle The angle at which the surface of a cutting edge in a drill bit is inclined in relation to the surface of the material being cut. See also: *negative rake; positive rake.* (Long, 1960)
gouge clay Clay infillings in a mineral vein. Cf: *gouge.* (Arkell, 1953)
gouge rake *positive rake.*
gouging (a) In placer mining, an operation similar to ground sluicing. Also called booming. (b) The formation of "crescentic gouges.". (AGI, 1987) (c) The local basining of a bedrock surface by the action of glacier ice. (AGI, 1987) (d) The working of a mine without plan or system, by which only the high-grade ore is mined. Syn: *high-grading.* (AGI, 1987)
gouging shot A gripping shot or opening shot used to make the first opening in a straight-room face, or to start a breakthrough. See also: *shot.* (Fay, 1920)
goutwater Forest of Dean. Mine water containing hydrogen sulfide, H_2S.
Gouy layer Modification of the Helmholtz concept of the electrical double layer which surrounds a particle immersed in an electrolyte. In Gouy's view there is only one diffuse layer. The ionic atmosphere near the surface of the particle is highly charged, but this ionization diminishes gradually outward into the ambient liquid. (Pryor, 1963)
gow caisson A device for sinking shafts of small diameter through silt or clay without excessive loss of ground. (Hammond, 1965)
gozzan *gossan.*
GPS *Global Positioning System.*
grab An instrument for extricating broken boring tools from a borehole. (Fay, 1920)
grabbing crane An excavator consisting of a crane carrying a large grab or bucket in the form of a pair of half-scoops, so hinged as to scoop or dig into the earth as they are lifted. (CTD, 1958)
grab bucket An underwater digging device which, in closing, bites into the sediment and contains it inside the closed shell. The bucket and load are then hoisted to the surface where the shell is opened to dump the load. Includes clamshells, orangepeels, and other variations. (Mero, 1965)
grab-camera An ocean floor sampling system incorporating a large sediment grab with a deep-sea camera. (Hunt, 1965)
grab dredger A dredging appliance consisting of a grab or grab bucket suspended from the jib head of a crane, which does the necessary raising and lowering. Also called a grapple dredger. See also: *dredger.* (CTD, 1958)
graben An elongate, relatively depressed crustal unit or block that is bounded by faults on its long sides. It is a structural form that may or may not be geomorphologically expressed as a rift valley. Etymol: Ger., ditch. Cf: *horst.* Syn: *trough; rift structure.* (AGI, 1987)
grab equipment A clamshell bucket fitted with teeth to assist digging. (Nelson, 1965)
grabhooks Hooks used in lifting blocks of stone. They are used in pairs connected with a chain, and are so constructed that the tension of the chain causes them to adhere firmly to the rock.
grab iron A short railing or handle on heavy equipment used to assist operators and other personnel in climbing up or down.
grab sampler An ocean-bottom sampler that commonly operates by enclosing material from the seafloor between two jaws upon contact with the bottom. See also: *Petersen grab.* Cf: *dredge.* Syn: *snapper.* (AGI, 1987)
grab sampling Collection of specimens of ore more or less at random from a heap, scatter pile, or passing load. Used in connection with examination of the characteristic minerals in the deposit rather than for valuation. (Pryor, 1963)
gradall Essentially a hydraulic backhoe equipped with an extensible boom that performs the three separate functions of excavation, backfill, and grading. (Carson, 1961)
gradation (a) The leveling of the land, or the bringing of a land surface or area to a uniform or nearly uniform grade or slope through erosion, transportation, and deposition; specif. the bringing of a streambed to a slope at which the water is just able to transport the material delivered to it. See also: *degradation; aggradation.* (AGI, 1987) (b) The proportion of material of each particle size, or the frequency distribution of various sizes, constituting a particulate material such as a soil, sediment, or sedimentary rock. The limits of each size are chosen arbitrarily. (AGI, 1987)
grade (a) A coal classification based on degree of purity, i.e., quantity of inorganic material or ash left after burning. Cf: *type; rank.* (AGI, 1987) (b) The relative quantity or the percentage of ore-mineral or metal content in an orebody. Syn: *tenor.* (AGI, 1987) (c) A degree of inclination, or a rate of ascent or descent, with respect to the horizontal, of a road, railroad, embankment, conduit, or other engineering structure; it is expressed as a ratio (vertical to horizontal), a fraction (such as m/km or ft/mi), or a percentage (of horizontal distance). Cf: *gradient.* (AGI, 1987) (d) Height above sea level; actual elevation. Also, the elevation of the finished surface of an engineering project (such as of a canal bed, embankment top, or excavation bottom). (AGI, 1987) (e) A particular size (diameter), size range, or size class of particles of a soil, sediment, or rock; a unit of a grade scale, such as clay grade, silt grade, sand grade, or pebble grade. (AGI, 1987) (f) *metamorphic grade.* (g) The classification of an ore according to the desired or worthless material in it or according to value. (Nelson, 1965) (h) The degree of

strength of a high explosive. Those above 40% nitroglycerin are arbitrarily designated as high-grade dynamites, and those below 40% nitroglycerin as low-grade dynamites. (i) In assaying, the percentage of the sought value or of each valuable species in the ore. (Pryor, 1963) (j) A term used to designate the extent to which metamorphism has advanced. Found in such combinations as high- or low-grade metamorphism. Cf: *rank.* (Leet, 1958) (k) A particular occupational classification of employee in a mine. (Nelson, 1965) (l) To sort and classify diamonds, such as drill diamonds, into quality groupings, each group containing diamonds having somewhat similar characteristics deemed to affect their fitness for use in a specific manner; the least fit are considered as constituting the lowest quality of grade. (Long, 1960) (m) *rank.*

graded (a) Said of a surface or feature when neither degradation nor aggradation is occurring, or when both erosion and deposition are so well balanced that the general slope of equilibrium is maintained. Syn: *at grade.* (AGI, 1987) (b) A geologic term pertaining to an unconsolidated sediment or to a cemented detrital rock consisting of particles of essentially uniform size or of particles lying within the limits of a single grade. Syn: *sorted.* (AGI, 1987) (c) An engineering term pertaining to a soil or an unconsolidated sediment consisting of particles of several or many sizes or having a uniform or equable distribution of particles from coarse to fine; e.g., a graded sand containing coarse, medium, and fine particle sizes. The term is rarely used in geology to refer to the sorting of the sediment, although this is common among engineers. Ant: nongraded. (AGI, 1987)

graded bedding A type of bedding in which each layer displays a gradual and progressive change in particle size, usually from coarse at the base of the bed to fine at the top. It may form under conditions in which the velocity of the prevailing current declined in a gradual manner, as by deposition from a single short-lived turbidity current. Cf: *grading.* (AGI, 1987)

graded coal One of the three main size groups by which coal is sold by the National Coal Board in Great Britain. It consists of coal screened between two screens—with an upper and lower limit ranging from a top size of 1-1/2 to 2 in (38 to 51 mm) to a bottom size of 1/8 to 3/4 in (3.2 to 19 mm). See also: *large coal; smalls.* (Nelson, 1965)

graded sand A sand containing some coarse, fine, and medium particle sizes. It is not a uniform sand. (Hammond, 1965)

graded stream A stream in equilibrium, showing a balance between its transporting capacity and the amount of material supplied to it, and thus between degradation and aggradation in the stream channel. (AGI, 1987)

graded unconformity *blended unconformity.*

gradeline (a) The baseline from which elevations are measured. (b) A line defining the intended grade of a roadway that is being driven. Such a line is used to control the gradient of a roadway. (BS, 1963)

grader (a) A self-propelled or towed machine provided with a row of removing or digging teeth and (behind) a blade to spread and level the material. Syn: *towed grader.* (Nelson, 1965) (b) A trommel-type air-swept circular screen used in asbestos milling where the fine rock and fiber dust are eliminated through medium-size perforated plates. (Arbiter, 1964) (c) A machine with a centrally located blade that can be angled to cast to either side, with independent hoist control on each side. (Nichols, 1976)

grade resistance The force, due to gravity, that resists the movement of a vehicle up a slope. (Carson, 1961)

grade scale A systematic, arbitrary division of an essentially continuous range of particle sizes (of a soil, sediment, or rock) into a series of classes or scale units (or grades) for the purposes of standardization of terms and of statistical analysis; it is usually logarithmic. Examples include Udden scale and Tyler Standard scale. See also: *Udden grade scale; Wentworth scale.* (AGI, 1987)

grade stake A stake indicating the amount of cut or fill required to bring the ground to a specified level. (Nichols, 1976)

gradient The inclination of profile gradeline from the horizontal, expressed as a percentage. Syn: *rate of grade.* (Urquhart, 1959)

gradienter An attachment to a surveyor's transit with which an angle of inclination is measured in terms of the tangent of the angle instead of in degrees and minutes. It may be used as a telemeter in observing horizontal distances. (AGI, 1987)

gradient of equal traction The gradient at which the tractive force required to pull an empty tram inby (slightly uphill) is equal to that required to pull a loaded tram outby. This was formerly termed horse haulage gradient. In general, haulage roads are graded about 0.5% in favor of the loaded trams. (Nelson, 1965)

grading (a) The relative proportions of the variously sized particles in a batch, or the process of screening and mixing to produce a batch with particle sizes correctly proportioned. (b) The commercial operation of sorting coke between two screens such that the ratio of the larger to the smaller screen aperture does not exceed 2.5 to 1; the coke which has been so sorted. (BS, 1960) (c) The gradual reduction, upward in a sedimentary bed, of the upper particle-size limit. It implies pulsatory turbulent-fluid deposition. Cf: *graded bedding.* (AGI, 1987)

grading test *screen analysis.*

Graf sea gravimeter A balance-type gravity meter that is heavily damped in order to attenuate shipboard vertical accelerations. It consists of a mass at the end of a horizontal arm, supported by a torsion-spring rotational axis. The mass rises and falls with gravity variation, but is restored to near its null position by a horizontal reading spring, tensioned with a micrometer screw. The difference between actual beam position and null position gives indication of gravity value after the micrometer screw position has been taken into account. See also: *gravimeter.* (AGI, 1987)

Graham pressure surveying apparatus A barometric surveying instrument that is free from the many defects of the aneroid barometer. The apparatus records the change in pressure of a constant volume of air maintained at a constant temperature.

Graham ratio The ratio of the amount of carbon monoxide produced over the oxygen absorbed varies with the temperature of oxidation of coal and also with the time of coal exposure to oxidation, thereby allowing this ratio to be used as an index of the rate of oxidation in a mine. (Roberts, 1960)

grail Gravel or sand; anything in fine particles. (Standard, 1964)

grain (a) A mineral or rock particle having a diameter of less than a few millimeters and generally lacking well-developed crystal faces; e.g., a sand grain. Also, a general term for sedimentary particles of all sizes (from clay to boulders), as used in the expressions grain size, fine-grained, and coarse-grained. (AGI, 1987) (b) A quarrymen's term for a plane of parting in slate that is perpendicular to the flow cleavage; or for a direction of parting in massive rock, e.g., granite, that is less pronounced than the rift and usually at right angles to it. Cf: *rift.* (AGI, 1987) (c) The second direction of easy splitting of a rock, less pronounced than the rift, but more so than the hardway. (d) A unit of hardness of water, expressed in terms of equivalent $CaCO_3$. A hardness of 1 grain per U.S. gal (17.1 mg/L) equals 17.1 ppm by weight as $CaCO_3$. See also: *anthracite fines.* (AGI, 1987)

grain boundary An interface between two crystals putting all adjacent ions in an irregular crystalline environment. Cf: *lineage.*

grainer medium salt Grainer salt screened to give a mixture of coarse- and medium-size flakes, excluding very coarse and very fine flakes. (Kaufmann, 1960)

grainers Diamonds which in weight will correspond to fourths of a carat; a diamond weighing one-half carat is a two-grainer; one weighing three quarters is a three-grainer; a diamond of one carat in weight is a four-grainer. (Hess)

grainer salt Salt produced by the grainer process of surface evaporation from brine. Product has a characteristic flaky shape consisting of hoppers and hopper fragments. (Kaufmann, 1960)

grain gliding Movement between individual mineral grains.

grain gold Gold that has become granular in the process of heating. (Fay, 1920)

grain growth The growth of a crystal, as from solution on the walls of a geode, in open pore space, or in a magma chamber; crystal growth. (AGI, 1987)

grain size (a) A term relating to the size of mineral particles that make up a rock or sediment. See also: *particle size.* (AGI, 1987) (b) For metals, a measure of the size of grains in a polycrystalline material, usually expressed as an average when the individual sizes are fairly uniform. Grain sizes are reported in terms of number of grains per unit area or volume, as average diameter, or as a grain-size number derived from area measurements. (ASM, 1961) (c) The size or size distribution of refractory particles, which is usually determined by sieve analysis. (ARI, 1949)

grain-size classification A scheme of rock classification based upon the average size of certain chosen components; thus, each clan comprises coarse-, medium-, and fine-grained members. (CTD, 1958)

grain tin (a) The granular or nodular form of cassiterite, tin oxide, SnO_2; also known as stream tin. (Henderson, 1953) (b) Metallic tin of high grade obtained by charcoal reduction. (Henderson, 1953)

graith A set of tools, picks, shovels, wedges, hammers, etc., used for work underground. (CTD, 1958)

gram-atom The atomic weight of an element expressed in grams. (Hackh, 1944)

gram-centimeter A unit of work; the work done in raising the weight of 1 g vertically 1 cm; 981 ergs. (Standard, 1964)

gram-molecule Molecular weight of a compound in grams, derived from that of hydrogen which, though 2.016, is expressed as the whole number 2. The gram-molecule, e.g., of H_2SO_4 is $2 + 32 + (4 \times 16) = 98$. Also called mole. (Pryor, 1963)

grampus The tongs with which bloomery loups and billets are handled. (Fay, 1920)

gram weight Pull of gravitation on a mass of 1 g. This varies slightly with the acceleration (g) due to gravity differences in various localities, but is approx. 981 dyn. (Pryor, 1963)

granat *garnet.*

Granby car A type of automatically dumped car for hand- or power-shovel loading. In this type of car, a wheel attached to the side of the car body engages an inclined track at the dumping point. As the side wheel rides up and over the inclined track, the car body is automatically raised and lowered, activating a side door operating mechanism which raises the door, permitting the car to shed its load. See also: *mine car.* (Pit and Quarry, 1960)

grandite The grossular-andradite series of the garnet group.

granite (a) A plutonic rock in which quartz constitutes 10% to 50% of the felsic components and in which the alkali feldspar/total feldspar ratio is generally restricted to the range of 65% to 90%. Rocks in this range of composition are scarce, and sentiment has been growing to expand the definition to include rocks designated as adamellite or quartz monzonite, which are abundant in the United States. (AGI, 1987) (b) Broadly applied, any holocrystalline, quartz-bearing plutonic rock. Syn: *granitic rock.* Etymol: Latin granum, grain. (AGI, 1987) (c) Commercial granite. (AGI, 1987)

granite gneiss (a) A gneiss derived from a sedimentary or igneous rock and having the mineral composition of a granite. (AGI, 1987) (b) A metamorphosed granite. (AGI, 1987)

granite porphyry A hypabyssal rock differing from a quartz porphyry by the presence of sparse phenocrysts of mica, amphibole, or pyroxene in a medium- to fine-grained groundmass. (AGI, 1987)

granite tectonics The study of the structural features, such as foliation, lineation, and faults, in plutonic rock masses, and the reconstruction of the movements that created them. (AGI, 1987)

granite wash A driller's term for material eroded from outcrops of granitic rocks and redeposited to form a rock having approx. the same major mineral constituents as the original rock; e.g., an arkose consisting of granitic detritus. (AGI, 1987)

granitic Pertaining to or composed of granite. Syn: *granitoid.* See also: *granular texture.* (AGI, 1987)

granitic layer *sial.*

granitic rock A term loosely applied to any light-colored, coarse-grained plutonic rock containing quartz as an essential component, along with feldspar and mafic minerals. Syn: *granite.* (AGI, 1987)

granitification *granitization.*

granitization An essentially metamorphic process by which a solid rock is converted into a granitic rock by the entry and exit of material, without passing through a magmatic stage. Some authors include in this term all granitic rocks formed from sediments by any process, regardless of the amount of melting or evidence of movement. The precise mechanism, frequency, and magnitude of the processes are still in dispute. Syn: *granitification.* Cf: *transfusion.* (AGI, 1987)

granitizer *transformist.*

granitoid *granitic.*

granoblastic Pertaining to a homeoblastic type of texture in a nonschistose metamorphic rock upon which recrystallization formed essentially equidimensional crystals with normally well sutured boundaries. Syn: *granular.* Cf: *crystalloblastic.* (AGI, 1987)

granodiorite A group of coarse-grained plutonic rocks intermediate in composition between quartz diorite and quartz monzonite (U.S. usage), containing quartz, plagioclase (oligoclase or andesine), and potassium feldspar, with biotite, hornblende, or, more rarely, pyroxene, as the mafic components; also, any member of that group; the approximate intrusive equivalent of rhyodacite. The ratio of plagioclase to total feldspar is at least 2:1 but less than 9:10. With less alkali feldspar it grades into quartz diorite, and with more alkali feldspar, into granite or quartz monzonite. (AGI, 1987)

granofels A field name for a medium- to coarse-grained granoblastic metamorphic rock with little or no foliation or lineation. (AGI, 1987)

granolith An artificial stone of crushed granite and cement. (Webster 3rd, 1966)

granophyre (a) An irregular microscopic intergrowth of quartz and alkali feldspar. (AGI, 1987) (b) A fine-grained granitic rock having a micrographic texture. (AGI, 1987) (c) A porphyritic rock of granitic composition characterized by a crystalline-granular groundmass. Cf: *felsophyre; vitrophyre.* (AGI, 1987)

granophyric Of or pertaining to granophyre.

grant Eng. A tract of land leased or ceded for mining purposes.

grantsite A monoclinic mineral, $NaCa(V^{5+},V^{4+})_6O_{16}·4H_2O$; in silky, pearly, or subadamantine fibrous aggregates coating fractures or forming thin seams in sandstone or limestone; near Grants, NM, and in Montrose County, CO.

granular Said of the texture of a rock that consists of mineral grains of approx. equal size. The term may be applied to sedimentary rocks, e.g., sandstones, but is esp. used to describe holocrystalline igneous rocks whose major-phase grain size ranges from 2 to 10 mm. The syn. granoblastic is used for metamorphic rocks. (AGI, 1987)

granular chert A compact, homogeneous, hard-to-soft chert, common in insoluble residues, composed of distinguishable and relatively uniform-sized grains, characterized by an uneven or rough fracture surface and by a dull to glimmering luster; it may appear saccharoidal. See also: *granulated chert.* Cf: *chalky chert.* (AGI, 1987)

granularity The quality, state, or property of being granular; specif. one of the component factors of the texture of a crystalline rock, including both grain size and grain-size distribution. (AGI, 1987)

granular texture A rock texture resulting from the aggregation of mineral grains of approx. equal size. The term may be applied to a sedimentary or metamorphic rock, but is esp. used to describe an equigranular, holocrystalline igneous rock whose particles range in diameter from 0.05 to 10 mm. See also: *granitic.* (AGI, 1987)

granular tonstein This type of tonstein consists predominately of kaolinite grains of lighter or darker shades, often surrounded by collinite. These grains show a cryptocrystalline to finely crystalline structure; the cryptocrystalline material is isotropic. Syn: *Graupen tonstein.* (IHCP, 1963)

granulated blast-furnace slag The glassy, granular material formed when molten blast-furnace slag is rapidly chilled, as by immersion in water. (ASTM, 1994)

granulated chert A type of granular chert composed of rough, irregular grains or granules tightly or loosely held in small masses or fragments. See also: *granular chert.* (AGI, 1987)

granulated slag Molten slag broken up into granules and quick quenches. Three general methods of granulation are: (1) running the molten slag into a pit of water; (2) using a jet of high-pressure water to breakup the stream of molten slag as it falls into the pit; and (3) using a mechanical revolving device with relatively small amounts of water. See also: *slag.* (Camp, 1985)

granulated steel Steel made from pig iron by a process in which the first step is the granulation of the iron.

granulating machine A device for reducing metal or slag in a liquid form to fine grain. (Fay, 1920)

granulation The act or process of being formed into grains, granules, or other small particles; specif. the crushing of a rock under such conditions that no visible openings result. Also, the state or condition of being granulated. (AGI, 1987)

granulator (a) A rock breaker which converts large stone into small aggregate. (Hammond, 1965) (b) A machine that produces body raw material in the form of grains with a minimum of fines. (ACSG, 1963)

granule (a) A rock fragment larger than a very coarse sand grain and smaller than a pebble, having a diameter in the range of 2 to 4 mm. The term "very fine pebble" has been used as a syn. (AGI, 1987) (b) A little grain or small particle, such as one of a number of the generally round or oval, nonclastic (precipitated), internally structureless grains of glauconite or other iron silicate in iron formation; a pseudo-oolith. (AGI, 1987)

granule texture A texture of iron formation in which precipitated or nonclastic granules are separated by a fine-grained matrix. (AGI, 1987)

granulite (a) A metamorphic rock consisting of even-sized, interlocking mineral grains less than 10% of which have any obvious preferred orientation. (AGI, 1987) (b) A relatively coarse, granular rock formed at high pressures and temperatures, which may exhibit a crude gneissic structure due to the parallelism of flat lenses of quartz and/or feldspar. The texture is typically granuloblastic. (AGI, 1987)

granulitic Of, pertaining to, or composed of granulite. (AGI, 1987)

granulitization In regional metamorphism, reduction of the components of a solid rock such as a gneiss to grains. The extreme result of the process is the development of mylonite. (AGI, 1987)

graphic (a) Said of the texture of an igneous rock that results from the regular intergrowth of quartz and feldspar crystals. The quartz commonly occupies triangular areas, producing the effect of cuneiform writing on a background of feldspar. Similiar intergrowths of other minerals, e.g., ilmenite-pyroxene, are less common. (AGI, 1987) (b) A mineral intergrowth, e.g., alkali-feldspar and quartz, with angular planar boundaries resembling cuneiform writing.

graphic gold Crystals of naturally occurring sylvanite ore; a mixed gold-silver telluride, occurring in regularity so as to give the appearance of written symbols. Syn: *graphic tellurium.* (Bennett, 1962)

graphic granite (a) An intergrowth of potash feldspar (orthoclase or microcline) and quartz. Syn: *corduroy spar.* (AIME, 1960) (b) A pegmatite characterized by graphic intergrowths of quartz and alkali feldspar. Cf: *intergrowth.* (AGI, 1987)

graphic intergrowths *graphic granite.*

graphic ore Syn: *sylvanite.* Also called graphic tellurium.

graphic section A drawing that shows the sequence of strata. (BS, 1963)

graphic tellurium *sylvanite; graphic gold.*

graphite A hexagonal and trigonal mineral, native carbon 4[C], polymorphous with chaoite, diamond, and lonsdaleite; scaly, soft, lustrous, metallic; greasy feel; as crystals, flakes, scales, laminae, or grains in veins or bedded masses or disseminations in carbonaceous metamorphic rocks; conducts electricity well, is soft and unctuous; immune to most acids; extremely refractory. Syn: *plumbago; black lead.*

graphitic carbon The portion of the carbon in iron or steel that is present as graphite; distinguished from combined carbon. (Webster 3rd, 1966)

graphitic steel Alloy steel made so that part of the carbon is present as graphite. (ASM, 1961)

graphitite A hard, coallike graphite rock interbedded with Precambrian schists. Syn: *graphitoid.*

graphitization Formation of graphite in iron or steel. If graphite is formed during solidification, the phenomenon is called primary graphitization; if formed later by heat treatment, it is called secondary graphitization. (ASM, 1961)

graphitoid (a) Meteoritic graphite. (b) *graphitite.*

grapple A clamshell-type bucket having three or more jaws. (Nichols, 1976)

grapple dredge (a) A dredge using an orangepeel bucket and operating on the clamshell principle. (Carson, 1961) (b) *grab dredger.* (CTD, 1958)

grass crop Scot. The outcrop of a vein. (Fay, 1920)

Grassellis High explosive; used in mines. (Bennett, 1962)

grasshopper conveyor *oscillating conveyor.*

grass roots A miner's term equivalent to the surface. From grass roots down is from the grass roots to the bedrock.

grass-roots deposit The old fabulous deposit, discovered in surface croppings, easy of exploitation, and capable of financing its own development as it went along. (Hoover, 1948)

grass-roots mining Inadequately financed operation, depending on hand-to-mouth existence. Mining from surface down to bedrock. At grass; at surface. Also known as mining on a shoestring. (Pryor, 1963)

grate (a) A screen or sieve for use with stamp mortars for grading ore. (Webster 3rd, 1966) (b) A frame, bed, or a kind of basket of iron bars for holding fuel while burning. (Webster 3rd, 1966)

grate bar (a) A bar forming part of a fire grate. (Standard, 1964) (b) One of the bars forming a coarse screen or grizzly. (Fay, 1920)

grate coal Formerly, coal passing through bars 3-1/4 to 4-1/4 in (8.3 to 10.8 cm) apart and over 2-1/4 in (6.7 cm) round holes. In Arkansas, the bars are 7 in (17.8 cm) apart and the holes are 3 to 3-1/4 in (7.6 to 8.3 cm) in diameter. (Fay, 1920)

grater A laborer who replaces grates on conveyors after roasted lead ore has been dumped into cars, using hooks. Lead ore is loaded on grates and conveyed through a furnace in which the sulfur is driven off by roasting prior to the ore being melted to separate and recover the lead in another furnace. (DOT, 1949)

graticule (a) A network of lines representing geographic parallels and meridians forming a map projection. (b) A template divided into blocks or cells, for graphically integrating a quantity such as gravity. Graticules are used in computing terrain corrections and the gravitational or magnetic attraction of irregular masses. (AGI, 1987) (c) An accessory to an optical instrument such as a microscope to aid in measurement of the object under study; it is a thin glass disk bearing a scale which is superimposed upon the object. (AGI, 1987) (d) The network of lines representing meridians of longitude and parallels of latitude on a map or chart, upon which the map or chart was drawn. Not to be confused with grid. (AGI, 1987)

grating (a) A coarse screen made of parallel or crossed bars to prevent passing of oversized material. (b) A series of parallel and crossed bars used as platform or walkway floors or as coverings for pits and trenches over which traffic can pass; generally removable to permit access to conveying equipment for servicing. (c) A series of parallel and/or crossed bar units fastened to or propelled by a conveying medium, used for carrying large lump-size bulk material or objects. They usually permit passage of air for cooling or for heat to maintain temperature. (d) The act of sorting ores by means of grates.

Graupen tonstein *granular tonstein.*

gravel (a) An unconsolidated, natural accumulation of rounded rock fragments resulting from erosion, consisting predominantly of particles larger than sand (diameter greater than 2 mm or 1/12 in), such as boulders, cobbles, pebbles, granules, or any combination of these. See also: *pebble.* (AGI, 1987) (b) A popular term for a loose accumulation of rock fragments, such as a detrital sediment associated esp. with streams or beaches, composed predominantly of more or less rounded pebbles and small stones, and mixed with sand that may compose 50% to 70% of the total mass. (AGI, 1987) (c) An engineering term for rounded fragments having a diameter in the range of 1.87 in (47.5 mm) (retained on U.S. standard sieve No. 4) to 3 in (76 mm). (AGI, 1987)

gravel bank A natural mound or exposed face of gravel, particularly such a place from which gravel is dug; a gravel pit. (Hess)

gravel deposit *alluvium.*

gravel mine S. Afr. A mine extracting gold from sand or gravel; also called placer mine. See also: *gravel pit.* (Beerman, 1963)

gravel pit A pit from which gravel is obtained. See also: *gravel mine.* (Standard, 1964)

gravel plain placer Placers along the coastal plain of the Seward Peninsula, AK.

gravel powder Very coarse gunpowder. (Standard, 1964)

gravel pump A centrifugal pump with renewable impellers and lining, suitable for pumping a mixture of gravel and water. Rubber is sometimes used as lining to the pump and pipes owing to its high resistance to abrasion. See also: *sand pump.* (Hammond, 1965)

gravel pumping A method of alluvial mining that consists of (1) excavating and breaking up the gravel bank by using giants or monitors, (2) washing the disintegrated material into a sump, excavated in the bedrock, (3) elevating the mixture from the sump to an elevated line of sluices by means of a gravel pump, and (4) sluicing the gravel for the recovery of its mineral content. (Griffith, 1938)

grave wax A natural paraffin. See also: *hatchettite.*

gravimeter (a) An instrument to measure the value of gravity or for measuring variations in the magnitude of the Earth's gravitational field. Measurements of gravity are accomplished generally by one of three methods: dropped ball, pendulum, or spring gravimeter. The latter type of gravimeter, which is based upon the principle of the weighted spring—where the length or measured variations in the length of the spring are a function of the gravitational field at different locations, is the type widely used today. See also: *Graf sea gravimeter; gravitational prospecting.* (Hunt, 1965) (b) An instrument for determining specific gravities, particularly of liquids. See also: *hydrometer.* (Standard, 1964) (c) An instrument that measures variations in the density of underlying rocks. (BS, 1963) (d) An instrument for measuring variations in the gravitational field, generally by registering differences in the weight of a constant mass as the gravimeter is moved from place to place. Syn: *gravity meter.* See also: *astatic gravimeter.* (AGI, 1987)

gravimetric analysis Quantitative chemical analysis in which the different substances of a compound are measured by weight. (AGI, 1987)

gravimetry The measurement of gravity or gravitational acceleration, esp. as used in geophysics and geodesy. (AGI, 1987)

gravitation The mutual attraction between two masses. See also: *law of gravitation.* (AGI, 1987)

gravitational constant The constant, γ, in the law of universal gravitation. Its value is $6.672 \times 10^{-11} m^3/kg \times s^2$. (AGI, 1987)

gravitational differentiation *crystal fractionation.*

gravitational method A geophysical prospecting method that measures irregularities or anomalies in gravity attraction produced by differences in the densities of rock formations, and interprets the results in terms of lithology and structure. (Nelson, 1965)

gravitational prospecting A method of geophysical prospecting, that embraces the mapping of variations in the Earth's gravitational field. See also: *gravimeter.* (Hammond, 1965)

gravity (a) The force by which substances are attracted to each other, or fall to Earth. See also: *law of gravitation.* (von Bernewitz, 1931) (b) The effect on any body in the universe of the inverse-square-law attraction between it and all other bodies and of any centrifugal force that may act on the body because of its motion in an orbit. (AGI, 1987) (c) The force exerted by the Earth and by its rotation on unit mass, or the acceleration imparted to a freely falling body in the absence of frictional forces. (AGI, 1987) (d) A general term for API gravity or Baumé gravity of crude oil. (AGI, 1987)

gravity anomaly The difference between the observed value of gravity at a point and the theoretically calculated value. It is based on a simple gravity model, usually modified in accordance with some generalized hypothesis of variation in subsurface density as related to surface topography. (AGI, 1987)

gravity balance Sensitive weighing system in which a beam rides on a fulcrum, and supports a load of unknown weight at one end which is counterbalanced by known weights at the other end. (Pryor, 1963)

gravity bar A 5 ft (1.5 m) length of heavy half-round rod forming the link between the wedge-orienting coupling and the drill-rod swivel coupling on an assembled Thompson retrievable borehole-deflecting wedge. (Long, 1960)

gravity-bar screen *grizzly.*

gravity classifying The grading of ores into different sorts and the separation of waste from coal by the difference in the specific gravity of the minerals to be separated. (Stoces, 1954)

gravity concentration Separating grains of minerals by a concentration method operating by virtue of the differences in density of various minerals; the greater the difference in density between two minerals, the more easily they can be separated by gravity methods. The laws of free and hindered settling are important in the theory of gravity concentration. (Newton, 1959)

gravity conveyor Continuous belt, system of rollers, or inclined chute down which loaded material gravitates without the application of power. See also: *roller conveyor; wheel conveyor.* (Pryor, 1963)

gravity corer (a) A sampling device capable of working even under moderately adverse sea conditions. The corer weighs about 650 lb (294 kg) in air and consists essentially of a shaft, weights, and coring tube. (Hunt, 1965) (b) An oceanographic corer that penetrates the ocean floor solely by its own weight. It is less efficient than a piston corer.

There are several varieties, including the Phleger corer and the free corer. (AGI, 1987)

gravity-discharge conveyor elevator A type of conveyor using gravity-discharge buckets attached between two endless chains and which operate in suitable troughs and casings in horizontal, inclined, and vertical paths over suitable drive, corner, and takeup terminals. Syn: *bucket elevator. V-bucket conveyor-elevator.*

gravity-discharge conveyor-elevator bucket An elevator bucket designed to contain material on vertical lifts and scrape material along a trough on horizontal runs. Discharge is effected by gravity.

gravity equipotential surface *equipotential surface.*

gravity fault *normal fault.*

gravity feed Applicable when the weight of the drill rods is great enough to impose an adequate pressure on a bit to make it cut properly. (Long, 1960)

gravity gradiometer An instrument for measuring the gradient of gravity. (AGI, 1987)

gravity haulage A system of haulage in which the set of full cars is lowered at the end of a rope, and gravity force pulls up the empty cars, the rope being passed around a sheave at the top of the incline. The speed of the haulage is controlled by a band brake on the sheave. See also: *brake incline; underground haulage.* Syn: *jig haulage; self-act.* (Nelson, 1965)

gravity inclines Openings made in the direction of the dip of the deposit. The gradient of the gravity incline is determined by the dip of the deposit. The ore mined is transported through them, usually to the next lower level drive. (Stoces, 1954)

gravity instruments Devices for measuring the gravitational force or acceleration or its gradient at any point. They are of three principal types: (1) a static type in which a linear or angular displacement is observed or nulled by an opposing force, (2) a dynamic type in which the period of oscillation is a function of gravity and is the quantity directly observed, or (3) a gradient-measuring type, for example, Ëotvos torsion balance. (AGI, 1987)

gravity meter (a) Sensitive device for measuring gravitational variations. (Bennett, 1962) (b) *gravimeter.*

gravity plane A tramline laid at such an angle that full skips running downhill will pull up the empties. (Fay, 1920)

gravity plane rope haulage *self-acting rope haulage.*

gravity prospecting Mapping the force of gravity at different places with a gravimeter (gravity meter) to determine differences in specific gravity of rock masses, and, through this, the distribution of masses of different specific gravity. (Leet, 1958)

gravity railroad A railroad in which the cars descend by their own weight; an inclined railroad. (Standard, 1964)

gravity road Any road on which cars will descend by gravity. (Jones, 1949)

gravity roller conveyor *roller conveyor.*

gravity screen A perforated steel plate, set at an angle, over which large coal or other material slides by gravity to effect a primary classification. (Nelson, 1965)

gravity separation Separation of mineral particles, with the aid of water or air, according to the differences in their specific gravities. (AGI, 1987)

gravity solution A solution used to separate the different mineral constituents of rocks by their specific gravities, as the solution of mercuric iodide in potassium iodide having a maximum specific gravity of 3.19. (Standard, 1964)

gravity stowing A method of stowing in inclined conveyor faces, in which the material is brought into the upper gate (usually the tailgate) and arranged to slide down on trays which are moved forward as each track is filled. (Nelson, 1965)

gravity takeup *belt tensioning device.*

gravity wheel conveyor *wheel conveyor.*

gray antimony *stibnite.*

grayband A variety of sandstone for sidewalks; flagstone. See also: *flagstone.* (Standard, 1964)

gray cast iron A cast iron that gives a gray fracture due to the presence of flake graphite; often called gray iron. See also: *iron.* (ASM, 1961)

gray cobalt *cobaltite; smaltite.*

gray copper *tetrahedrite.*

gray copper ore *tennantite; tetrahedrite.*

gray hematite *specularite.*

gray iron (a) Pig iron or cast iron in which nearly all the carbon not included in pearlite is present as graphitic carbon. See also: *mottled iron; white iron.* (CTD, 1958) (b) Iron that exhibits a gray fracture surface because fracture occurs along the graphite plates (flakes); it is the result of stable solidification. (McGraw-Hill, 1994)

grayite A thorium phosphate containing a little lead, calcium, and minor uranium and rare earths; gives an X-ray pattern like that of rhabdophane, and when heated above 850 °C, a monazite-type pattern. (Am. Mineral., 1945)

Gray-King test Method of assessing the coking property of coal; 20 g is heated in a silica tube to 600 °C and the residual product is compared with a standard series ranging from noncoking (type A) to highly coking (G), all of which have the same volume as the original. Cokes that expand (swell) on coking receive a subscript indicating the degree of swelling. (Pryor, 1963)

gray manganese ore *manganite; pyrolusite.*

graywacke An old rock name that has been variously defined but is now generally applied to a dark gray, firmly indurated, coarse-grained sandstone that consists of poorly sorted, angular to subangular grains of quartz and feldspar, with a variety of dark rock and mineral fragments embedded in a compact clayey matrix having the general composition of slate and containing an abundance of very fine-grained illite, sericite, and chloritic minerals. Graywacke is abundant within the sedimentary section, esp. in the older strata, usually occurring as thick, extensive bodies with sole marks of various kinds and exhibiting massive or obscure stratification in the thicker units but marked graded bedding in the thinner layers. It generally reflects an environment in which erosion, transportation, deposition, and burial were so rapid that complete chemical weathering did not occur, as in an orogenic belt where sediments derived from recently elevated source areas were poured into a geosyncline. Graywackes are typically interbedded with marine shales or slates, and associated with submarine lava flows and bedded cherts; they are generally of marine origin and are believed to have been deposited by submarine turbidity currents. Cf: *arkose; subgraywacke.* Also spelled: greywacke; grauwacke. (AGI, 1987)

graywether One of numerous fragments or blocks of sandstone and conglomerate, covering large tracts in Dorsetshire and Wiltshire, England, supposed to be remnants of decomposed Tertiary strata. Also called druidical stone; sarsen stone; saracen stone. Syn: *sarsen stone.* (Standard, 1964; Fay, 1920)

grease (a) A semisolid form of lubricant, composed of emulsified mineral oil and soda or lime soap. Additives may be incorporated for special purposes, for example, colloidal graphite. (b) This term should be applied only to fatty or oily matter of animal origin, but mixtures of mineral oil with lime and soda soaps constitute well-known lubricating greases. (c) Term used in the flotation process. (d) As used in engineering for lubrication or protection of metal surfaces, grease is an emulsified oil or saturated fatty acid combined with a suitable alkaline base to form a soap. (Pryor, 1963) (e) Thick oil. (Nichols, 1976) (f) A solid or semisolid mixture of oil with soap or other fillers. (Nichols, 1976)

greased-deck concentration A process in which separation is based on selective adhesion of some grains (diamonds) to quasi-solid grease with adhesion of other grains to water. (Gaudin, 1939)

greaser (a) A person who oils or greases the mine cars. (b) An automatic apparatus that greases the axles of skips as they pass.

grease stone A name for steatite or soapstone.

grease table An apparatus for concentrating minerals, such as diamonds, which adhere to grease. It usually is a shaking table coated with grease or wax over which an aqueous pulp is flowed.

greasing truck An electrically driven service vehicle to transport greases and oil for servicing the underground mine machinery. It may include a compressor, air storage tank, and fittings to place lubricant at the proper points in the mining machinery.

greasy Applied to the luster of minerals. Having the luster of oily glass, as elaeolite. (Fay, 1920)

greasy blaes Scot. *creeshy.*

greasy feel Some minerals or rocks are greasy or soapy to the touch, e.g., talc, graphite, steatite, or soapstone.

greasy gold *fine gold.*

greasy luster As if covered with a thin film of oil or grease, e.g., nepheline, some diamond crystals, and some varieties of serpentine.

greasy quartz *milky quartz.*

Great Falls converter A pear-shaped vessel that resembles the Bessemer converter. It has been largely supplanted by the cylindrical (Peirce-Smith) type converter. (Newton, 1959)

Greathead shield A tunneling device invented by J. H. Greathead, first used in London in 1869, and still widely used today. His invention included a circular cutting edge forced through the ground by hydraulic jacks, a cast-iron lining assembled by bolts, and grouting behind the lining with the aid of compressed air and a special mixer. (Hammond, 1965)

great salt Salt in large lumps or crystals. (Kaufmann, 1960)

greave *ditch.*

green acids Mixed sulfonation products from oil refinery cracking processes; used in detergency and as main constituent of a series of flotation agents chiefly concerned with the concentration of iron minerals. Also refers to the initial solution generated from the acceleration of phosphate concentrate. (Pryor, 1963)

greenalite A monoclinic mineral, $(Fe^{2+},Fe^{3+})_{2-3}Si_2O_5(OH)_4$; kaolinite-serpentine group; forms green granules in the taconite of the Mesabi Range, MN, and the Gogebic iron formation of northern Wisconsin.

greenalite rock A dull, dark green rock, uniformly fine-grained with conchoidal fracture, containing grains of greenalite in a matrix of chert, carbonate minerals, and ferruginous amphiboles. (AGI, 1987)

Greenawalt process A system of sintering powdery metalliferous ores. (Osborne, 1956)

Greenburg-Smith impinger A dust-sampling apparatus evolved by the U.S. Bureau of Mines that makes use of the principle of impingement of the dust-laden air at high velocity on a wetted glass surface, together with that of bubbling the air through a liquid medium. See also: *midget impinger*. (Greenburg, 1922)

green chalcedony Usually some cryptocrystalline variety of quartz stained green. Also may be chalcedony of natural green color. See also: *chrysoprase*.

green charge A mixture of ingredients for gunpowder before the intimate mixing in the incorporating mill. (Webster 3rd, 1966)

green copperas The mineral melanterite, hydrous ferrous sulfate, $Fe^{2+}SO_4 \cdot 7H_2O$. Syn: *green vitriol*. (Fay, 1920)

green hole A furnace taphole in which clay is not properly set, and through which the drill may break and let iron out unexpectedly. (Fay, 1920)

green iron ore *dufrenite*.

greenlandite *columbite*.

Greenland spar *cryolite*.

green lead ore *pyromorphite*.

green marble *verde antique*.

green mud A deep-sea terrigenous deposit characterized by the presence of a considerable proportion of glauconite and $CaCO_3$ in variable amounts up to 50%. (Holmes, 1920)

greenockite A hexagonal mineral, 2[CdS]; forms earthy incrustations and coatings on sphalerite and other zinc ores; rarely in amygdules. See also: *xanthochroite*.

greenovite A rose-colored variety of titanite (sphene) containing up to 3% MnO.

green roof A miner's term for a roof that has not broken down or shows no sign of taking weight. (Fay, 1920)

green salt (a) Uranium tetrafluoride. (Lyman, 1964) (b) A wood preservative consisting of copper, arsenic, and chromium compounds. (Bennett, 1962)

greensand (a) An unconsolidated marine sediment consisting largely of dark greenish grains of glauconite, often mingled with clay or sand (quartz may form the dominant constituent), found between the low-water mark and the inner mud line. The term is loosely applied to any glauconitic sediment. (AGI, 1987) (b) A sandstone consisting of greensand that is often little or not at all cemented, having a greenish color when unweathered but an orange or yellow color when weathered, and forming prominent deposits in Cretaceous and Eocene beds (as in the Coastal Plain areas of New Jersey and Delaware); specif. either or both of the Greensands (Lower and Upper) of the Cretaceous System in England, whether containing glauconite or not. Syn: *glauconite; glauconitic sandstone*. (AGI, 1987)

greensand marl A marl containing sand-size grains of glauconite. (AGI, 1987)

greenschist A schistose metamorphic rock whose green color is due to the presence of chlorite, epidote, or actinolite. Cf: *greenstone*. (AGI, 1987)

greenstone (a) A field term applied to any compact dark-green altered or metamorphosed basic igneous rock (e.g., spilite, basalt, gabbro, diabase) that owes its color to the presence of chlorite, actinolite, or epidote. Cf: *greenschist*. (AGI, 1987) (b) *nephrite*. (c) An informal name for a greenish gemstone, such as fuchsite or chiastolite. (AGI, 1987) (d) Compact, igneous rocks that have developed enough chlorite in alteration to give them a green cast. They are mostly diabases and diorites. Greenstone is partially synonymous with trap. It is often used as a prefix to other rock names. The term is used frequently when no accurate determination is possible. (e) Includes rocks that have been metamorphosed or otherwise so altered that they have assumed a distinctive greenish color owing to the presence of one or more of the following minerals: chlorite, epidote, or actinolite. (f) Freshly quarried stone containing quarry water. (Arkell, 1953) (g) Can. Generalized name given to Precambrian lavas. (Hoffman, 1958)

green top Freshly exposed roof that is unknown in quality. (Kentucky, 1952)

green vitriol A ferrous sulfate; copperas. Also called martial vitriol. See also: *melanterite; green copperas*. (Standard, 1964)

Greenwell formula A formula used for calculating the thickness of tubbing: $T = 0.03 + HD/50{,}000$, where T is the required thickness of tubbing in feet, H is the vertical depth in feet, D is the diameter of the shaft in feet, and 0.03 is an allowance for possible flaws or corrosion. (Sinclair, 1958)

greet stone A term used in Yorkshire, England, for a coarse-grained or gritty sandstone. (AGI, 1987)

greigite An isometric mineral, $Fe^{2+}Fe^{3+}{}_2S_4$; linaeite group; in minute grains and crystals in clays in the Kramer-Four Corners area, San Bernadino County, CA.

greisen A pneumatolytically altered granitic rock composed largely of quartz, mica, and topaz. The mica is usually muscovite or lepidolite. Tourmaline, fluorite, rutile, cassiterite, and wolframite are common accessory minerals. See also: *greisenization*. (AGI, 1987)

greisenization A process of hydrothermal alteration in which feldspar and muscovite are converted to an aggregate of quartz, topaz, tourmaline, and lepidolite (i.e., greisen) by the action of water vapor containing fluorine. (AGI, 1987)

grenatite *leucite; staurolite*.

grenz Horizons in coalbeds resulting from temporary halting of the accumulation of vegetal material. They are frequently marked by a bed of clay or sand. (Raistrick, 1939)

greve A ditch or trench.

grid (a) Two sets of uniformly spaced parallel lines, intersecting at right angles, by means of which the surface of an area is divided into squares when a checkerboard placement of boreholes is desired. Elevations may be taken at line intersections. Cf: *checkerboarded*. (b) In surveying, a triangulation scheme that covers its area with a network of acute-angled triangles drawn between mutually visible points. (Pryor, 1963) (c) A grated opening, as in a mining sieve. (Zern, 1928) (d) A network composed of two sets of uniformly spaced parallel lines, usually intersecting at right angles and forming squares, superimposed on a map, chart, or aerial photograph, to permit identification of ground locations by means of a system of coordinates and to facilitate computation of direction and distance. The term is frequently used to designate a plane-rectangular coordinate system superimposed on a map projection, and usually carries the name of the projection; e.g., Lambert grid. Not to be confused with graticule. (AGI, 1987) (e) A systematic array of points or lines; e.g., a rectangular pattern of pits or boreholes used in alluvial sampling. (AGI, 1987)

gridaw The framing at the top of a shaft for the pulley wheels or sheaves for the hoisting rope. (CTD, 1958)

grid azimuth The angle at a given point in the plane of a rectangular coordinate system between the central meridian, or a line parallel to it, and a straight line to the azimuth point. (AGI, 1987)

gridiron twinning *crossed twinning*.

Griffin mill A grinding mill in which a vertically suspended rolling disk rotates, and under the influence of centrifugal force bears on ore passing between it and a stationary bowl, crushing the passing ore on its way to a peripheral discharge. Syn: *pendulum mill*. (Pryor, 1963)

griffithite A ferroan variety of saponite.

Griffith's theory Griffith's theory of failure is based on the assumption that the low order of tensile strength in common materials is due to the presence of small cracks or flaws. Actual stresses may occur around these flaws, which are of the order of magnitude of molecular cohesion values, while the average tensile strength may be quite low. Mohr's theory predicts that failure of materials is due to failure in shear, whereas Griffith's theory postulates that it is due to failure at crack tips. (Lewis, 1964)

grike A joint fracture in limestone, widened by solution.

grindability (a) Grindability of coal, or the ease with which it may be ground fine enough for use as pulverized fuel, is a composite physical property embracing other specific properties, such as hardness, strength, tenacity, and fracture. (Mitchell, 1950) (b) The effect produced on representative pieces of ore by applying standard methods of comminution, assessed comparatively in terms of size reduction and power used. (Pryor, 1965) (c) Relative ease of grinding, analogous to machinability. (ASM, 1961)

grindability index A measure of the grindability of a material under specified grinding conditions, expressed in terms of volume of material removed per unit volume of wheel wear. (ASM, 1961)

grinder-mill operator (a) In ore dressing, smelting, and refining, one who mixes raw materials, such as bauxite, lime, soda ash, and starch, entering the alumina-extraction process to produce a slurry of proper chemical composition, using a ball mill. Also called ball mill operator. (DOT, 1949) (b) In ore dressing, smelting, and refining, one who grinds ore and separates fine particles from coarse particles in a ball mill and classifier arranged in continuous series. (DOT, 1949)

grinders' asthma Disease of the lungs consequent upon inhaling the metallic dust produced in grinding operations. Also called grinders' rot; phthisis. (Standard, 1964)

grinding (a) Size reduction into fine particles; comminution. See also: *dry grinding; wet grinding*. (b) The process of erosion by which rock fragments are worn down, crushed, sharpened, or polished through the frictional effect of continued contact and pressure by larger fragments. (AGI, 1987) (c) Abrasion by rock fragments embedded in a glacier and dragged along the bedrock floor. (AGI, 1987)

grinding aid An additive to the charge in a ball mill or rod mill to accelerate the grinding process; the additive has surface-active or lubricating properties. Grinding aids find particular use in the grinding of portland cement clinker but, in the United Kingdom, their use is precluded by the conditions laid down in British Standard 12. (Dodd, 1964)

grinding cycle The sequence of operations in grinding a material, including, for example, the screening of the primary product and the recirculation of the screen overflow. (BS, 1962)

grinding mill A machine for the wet or dry fine crushing of ore or other material. The three main types are the ball, rod, and tube mills. The mill consists of a rotating cylindrical drum; the ore enters one hollow trunnion and the finished product leaves the other. Modern practice indicates ball mill feeds of 1/2 in, 3/4 in, and 1 in (1.27 cm, 1.91 cm, and 2.54 cm) for hard, medium, and soft ores respectively and the products range from 35 to 200 mesh and finer. See also: *open-circuit mill.* (Nelson, 1965)

grinding pebbles Pebbles, usually of chert or quartz, used for grinding in mills where contamination with iron must be avoided.

griotte marble A French marble of a beautiful red color and often variegated with small dashes of purple and spots or streaks of white, as in the variety locally known as griotte oeil de perdrix from the French Pyrenees. (Fay, 1920)

gripe A strap brake or ribbon brake on a hoisting apparatus. (Standard, 1964)

gripping hole One whose direction is inclined away from the adjacent free face, or may be defined as one whose width is greater at the toe than at the heel. (Zern, 1928)

gripping shot A shot so placed that the point or inner end of the hole is considerably farther from the face of the coal to be broken than is the heel or outer end of the hole. See also: *shot.* (Fay, 1920)

grip the rib When a cut is so made by a mining machine or a shot is so placed by a miner that the cut or shot enters the coal beyond the proper line of the rib, it is said to grip the rib. (Zern, 1928)

grisley *grizzly.*

grit (a) A coarse-grained sandstone, esp. one composed of angular particles; e.g., a breccia composed of particles ranging in diameter from 2 to 4 mm. (AGI, 1987) (b) A sand or sandstone made up of angular grains that may be coarse or fine. The term has been applied to any sedimentary rock that looks or feels gritty on account of the angularity of the grains. (AGI, 1987) (c) A sandstone composed of particles of conspicuously unequal sizes (including small pebbles or gravel). (AGI, 1987) (d) A sandstone with a calcareous cement. The term has been applied incorrectly to any nonquartzose rock resembling a grit; e.g., pea grit or a calcareous grit. (AGI, 1987) (e) A small particle of a stone or rock; esp. a hard, angular granule of sand. Also, an abrasive composed of such granules. (AGI, 1987) (f) The structure or grain of a stone that adapts it for grinding or sharpening; the hold of a grinding substance. Also, the size of abrasive particles, usually expressed as their mesh number. (AGI, 1987) (g) An obsolete term for sand or gravel, and for earth or soil. The term is vague and has been applied widely with many different connotations. Etymol: Old English greot, gravel, sand. (AGI, 1987)

grit collector An adaptation of any of several types of conveyors used for removing heavy solids from settling tanks or basins. See also: *bucket elevator; flight conveyor; reciprocating flight conveyor; screw conveyor.*

grit number *mesh number.*

gritting In quarrying, a process that gives a smoother surface than rubbing. It is accomplished with silicon carbide or aluminum oxide abrasive bricks attached to revolving buffer heads. (AIME, 1960)

grizzly (a) A device comprised of fixed or moving bars, disks, or shaped tumblers or rollers for the coarse screening or scalping of bulk materials. See also: *bar grizzly; grizzly chute; live roll grizzly.* Syn: *grisley; gravity-bar screen.* (b) A series of iron or steel bars spaced so as to size, sort, or separate the bulk material as it falls into the ore chutes. (Ricketts, 1943) (c) A rugged screen for rough sizing at a comparatively large size (e.g., 6 in or 15.2 cm); it can comprise fixed or moving bars, disks, or shaped tumblers or rollers. (BS, 1962)

grizzly chute A chute with a bar grizzly which separates the fine from the coarse material as it passes through the chute. See also: *grizzly.*

grizzly man *grizzly worker.*

grizzly worker In metal mining, a laborer who works underground at a grizzly over a chute or raise heading to a storage bin or haulage level, dumping ore from cars through the grizzly, and breaking oversized lumps with a sledge hammer so that they will pass through the grizzly. Also called draw man; grizzly man. (DOT, 1949)

grog fire clay mortar Raw fireclay mixed with calcined fireclay, or with broken fireclay brick, or both, all ground to suitable fineness. (ASTM, 1994)

groove (a) The long, tapered, half-round slot in the deflection wedge that acts as a guide in directing the bit to follow a new course in deflecting a borehole. Any of the spiral depressions on a cylindrical object, such as the spiral depression on the surface of fluted core or the rifling in a gun barrel. (Long, 1960) (b) Derb. The place where a miner is working. See also: *grove.* (Fay, 1920) (c) A mine, from the German, grube. (Fay, 1920)

grooved drum Drum having a grooved surface to support and guide a rope. (Hammond, 1965)

groover N. of Eng. A miner. (Standard, 1964)

groove sample *channel sample.*

groroilite A nearly black earthy manganese or wad, streaked with dark red markings, occurring in parts of Europe. (Standard, 1964)

gross calorific value (a) The heat produced by combustion of unit quantity of a solid or liquid fuel when burned at constant volume in an oxygen bomb calorimeter under specified conditions, with the resulting water condensed to a liquid. (ASTM, 1994) (b) The amount of heat liberated by the complete combustion of unit weight of coal under specified conditions; the water vapor produced during combustion is assumed to be completely condensed. (BS, 1960) (c) At constant pressure, the number of heat units that would be liberated if unit quantity of coal or coke was burned in oxygen at constant pressure in such a way that the heat release was equal to the sum of the gross calorific value at constant volume and the heat equivalent of the work that would have been done by the atmosphere under isothermal conditions had the pressure remained constant. (BS, 1961) (d) At constant volume, the number of heat units measured as being liberated per unit quantity of coal or coke burned in oxygen saturated with water vapor in a bomb under standard conditions, the residual materials in the bomb being taken (suitable corrections having been made) as gaseous oxygen, carbon dioxide, sulfur dioxide, nitrogen, liquid water in equilibrium with its vapor and saturated with carbon dioxide, and ash. Syn: *gross heat of combustion.* (BS, 1961)

gross cut The total amount of excavation in a road or a road section, without regard to fill requirements. (Nichols, 1976)

gross heat of combustion *gross calorific value.* (AGI, 1987)

gross recoverable value The part of the total metal recovered multiplied by the price. The proportion recovered varies with the ore and the method used. See also: *net unit value.* (Nelson, 1965)

gross ton The long ton of 2,240 avoirdupois pounds. (Webster 3rd, 1966)

grossular An isometric mineral, $8[Ca_3Al_2Si_3O_{12}]$, rarely pure; garnet group; crystallizes in dodecahedra and trapezohedra; varicolored; in metamorphosed calcareous rocks and skarns. Formerly called grossularite. Syn: *cinnamon stone.*

gross unit value The weight of metal per ton (long or short ton), as determined by assay or analysis, multiplied by the market price of the metal. See also: *net unit value.* (Nelson, 1965)

grothite A variety of titanite containing yttrium or cerium.

grouan *growan.*

ground (a) Rock at the side of a lode; country rock. (Gordon, 1906) (b) The mineralized deposit and the rocks in which it occurs, e.g., payground, payable reef; barren ground, rock without value. (CTD, 1958) (c) A ground is a conducting connection, whether intentional or accidental, between an electrical circuit or equipment and either the Earth or some conducting body serving in place of the Earth. Also called earth.

ground boss (a) A mine foreman. (Hess) (b) *mine captain.*

ground control (a) Maintaining rock mass stability by controlling the movement of excavations in the ground, which can be either rock or soil. (SME, 1992) (b) Accurate data on the horizontal and/or vertical positions of identifiable ground points so that they may be recognized on aerial photographs. (AGI, 1987)

grounded circuit Electrical system earthed at key points to ensure a common potential and eliminate danger to personnel. (Pryor, 1963)

grounded power conductor An insulated or bare cable that constitutes one side of a power circuit and normally is connected to ground. It differs from a ground wire in that a grounded power conductor normally carries the load current while the equipment it serves is in operation.

ground fault An electrical contact between part of the blasting circuit and earth. (Atlas, 1987)

ground geophysical anomaly A geophysical anomaly that is mapped instrumentally at the surface of the ground. (Hawkes, 1962)

grounding transformer *zigzag transformer.*

ground log A device for determining the course and speed made good over the ground in shallow water, consisting of a lead or weight attached to a line. The lead is thrown overboard and allowed to rest on the bottom. The course being made good is indicated by the direction the line tends, and the speed by the amount of line paid out in unit time. (Hunt, 1965)

ground magnetometer A magnetometer primarily suitable for making observations of magnetic field intensity on the surface of the Earth. (AGI, 1987)

groundmass (a) The material between the phenocrysts of a porphyritic igneous rock. It is relatively finer grained than the phenocrysts and may be crystalline, glassy, or both. Cf: *mesostasis.* Syn: *matrix.* (AGI, 1987) (b) A term sometimes used for the matrix of a sedimentary rock. (AGI, 1987)

ground movement (a) Subsidence due to the caving or collapse of underground workings. (Pryor, 1963) (b) Displacement of ground along a fault, bedding plane, or joint caused by mining-induced stress.

ground noise Seismic disturbance of the ground not caused by the shot. (Schieferdecker, 1959)

ground plate (a) A bedplate supporting railroad sleepers or ties. (Standard, 1964) (b) In electricity, a metal plate in the ground forming the earth connection of a metallic circuit. (Standard, 1964)

ground pressure (a) The pressure to which a rock formation is subjected by the weight of the superimposed rock and rock material or by diastrophic forces created by movements in the rocks forming the Earth's crust. Such pressures may be great enough to cause rocks having a low compressional strength to deform and be squeezed into and close a borehole or other underground opening not adequately strengthened by an artificial support, such as casing or timber. Also called rock pressure. Cf: *bottom-hole pressure.* (Long, 1960) (b) The weight of a machine divided by the area in square inches of the ground directly supporting it. (Nichols, 1976)

ground prop The puncheon between the lowest frame and a foot block in a timbered excavation, used to support the weight of the timbering. (Hammond, 1965)

ground roll (a) Low-frequency, low-velocity interface waves encountered in seismic prospecting commonly arising from the ground-air interface, in which case they are known as Rayleigh waves. Ground roll can completely mask desired signals, and means to minimize it commonly must be used. (AGI, 1987) (b) Seismic surface wave generated by the shot. See also: *ground waves.* (Schieferdecker, 1959)

ground sluice (a) A channel or trough in the ground through which auriferous earth is sluiced for placer mining. (Webster 3rd, 1966) (b) To wash down a bank of earth with a stream of water.

ground sluicing To strip ground downslope by means of a directed stream of water to excavate placer material and transport it to a riffled trough in which the valuable mineral is recovered. Cf: *hydraulicking.*

ground spears Wooden rods (one on each side of the pump) by which a sinking pump is suspended.

ground vibrations *blasting vibrations.*

ground water (a) That part of the subsurface water that is in the zone of saturation, including underground streams. (AGI, 1987) (b) Loosely, all subsurface water as distinct from surface water. Also spelled: groundwater; ground-water. Syn: *subterranean water; underground water.* (AGI, 1987)

ground-water discharge The return of ground water to the surface.

ground-water hydrology *geohydrology.*

ground-water level (a) *water table.* (b) The elevation of the water table at a particular place or in a particular area, as represented by the level of water in wells or other natural or artificial openings or depressions communicating with the zone of saturation. Syn: *ground-water table.* (AGI, 1987)

ground-water lowering The process of lowering the water table so that an excavation can be carried out in the dry. This is done by means of well points. (Hammond, 1965)

ground-water province An area or region in which geology and climate combine to produce ground-water conditions consistent enough to permit useful generalizations. (AGI, 1987)

ground-water surface *water table.*

ground-water table *water table; ground-water level.*

ground-water tracers The water seeping into shallow workings or shafts may be traced to the surface source by means of tracer dyes or salts. These substances, however, may be leached out of the water by the soil or strata. Some radioactive isotopes are better tracers because of the high sensitivity with which they can be detected. Tritium, an isotope of hydrogen, is unique because it can be used to label the actual water molecule to be traced and is not chemically removed by the strata. (Nelson, 1965)

ground waves Vibrations of soil or rock. See also: *ground roll.* (Nichols, 1976)

ground wire A bare or insulated cable used to connect the metal frame of a piece of equipment to the mine track or other effective grounding medium. (MSHA, 1986)

group (a) The lithostratigraphic unit next in rank above formation, consisting partly or entirely of named formations. A group name combines a geographic name with the term "group," and no lithic designation is included; e.g., San Rafael Group. (AGI, 1987) (b) A stratigraphic sequence that will probably be divided in whole or in part into formations in the future. See also: *analytic group.* (AGI, 1987) (c) A general term for an assemblage or consecutive sequence of related layers of rock, such as of igneous rocks or of sedimentary beds. (AGI, 1987) (d) A more or less informally recognized succession of strata too thick or inclusive to be considered a formation. (AGI, 1987) (e) A number of shots sufficiently close together to be treated in common in respect to preparation for firing. (BS, 1964)

group level (a) A main haulageway drive built in the solid rock underlying the group of seams that it has to serve, or in the floor of a thick deposit. It is preferable to construct the main haulageway as a subdeposit drive, because drives in the deposit suffer from pressure as soon as mining has progressed a certain distance. (Stoces, 1954) (b) Syn: *subdeposit level.*

grout (a) A pumpable slurry of neat cement or a mixture of neat cement and fine sand, commonly forced into boreholes or crevices in a rock to prevent ground water from seeping or flowing into an excavation, to seal crevices in a dam foundation, or to consolidate and cement together rock fragments in a brecciated or fragmented formation. Also called cement grout. (Long, 1960) (b) A cementitious component of high water-cement ratio, permitting it to be poured or injected into spaces within masonry walls. It consists of portland cement, lime, and aggregate, and is often formed by adding water to mortar. (ACSG, 1963) (c) The act or process of injecting a grout into a rock formation through a borehole or crevice. (Long, 1960) (d) Applied to waste material of all sizes obtained in quarrying stone. (Fay, 1920) (e) A coarse kind of plaster or cement usually studded with small stones after application, sometimes used for coating walls of a building. (Webster 3rd, 1966)

grout core Core obtained by drilling into and through formations into which grout has been injected and allowed to set. (Long, 1960)

grout curtain An area into which grout has been injected to form a barrier around an excavation or under a dam through which ground water cannot seep or flow. (Long, 1960)

grouter (a) In the stonework industry, a laborer who maintains the floors, equipment, machinery, and yard in a clean and unobstructed condition, using shovels, brooms, buckets, and wheelbarrows to collect and remove stone scraps, dirt, and debris to dump for disposal or to remove steel shot from under gangsaws and store it in suitable containers to be washed and reused. Also called mucker. (DOT, 1949) (b) *box loader.*

grout hole A borehole drilled for the express purpose of using it as a means by which grout may be injected into the rock surrounding the borehole. Cf: *consolidation hole.* (Long, 1960)

grouting The injection of grout into fissured, jointed, or permeable rocks in order to reduce their permeability or increase their strength. (AGI, 1987)

grout injection An act or process of forcing grout into crevices in rock formations, usually through a borehole, by pressure pumps. (Long, 1960)

grout injector A machine that mixes the dry ingredients for a grout with water and injects it, under pressure, into a grout hole. Cf: *grout machine.* (Long, 1960)

groutite An orthorhombic mineral, MnO(OH); trimorphous with manganite and feitknechtite; in brilliant submetallic to adamantine wedge-shaped crystals in the banded iron formations of the Cuyuna Range, MN.

grout machine A mechanism by which grout may be pressure-injected into a grout hole. Cf: *grout injector.*

grove (a) Eng. A drift or adit driven into a hillside from which coal is worked. See also: *groove.* (Fay, 1920) (b) Corn. Mine; bal. (Hess)

grovesite A former name for pennantite.

growan (a) An old English term for a coarse-grained granite, grit, or sandstone. (AGI, 1987) (b) A grus developed by the disintegration of a granite. Syn: *grouan. grus.* (AGI, 1987)

growler board A notched or fingered plank or light timber used to align ends of pipe being screwed together, as when laying a waterline. (Long, 1960)

grow-on Quarrymen's term to designate the place where the sheet structure dies out, or the place where two sheets appear to grow onto one another.

growth (a) An increase in dimensions of a compact that may occur during sintering (converse of shrinkage). (ASTM, 1994) (b) As applied to cast iron, the tendency to increase in volume when repeatedly heated and cooled. (CTD, 1958) (c) *make of water.*

grubbing The removal of the root system incident to the surface growth. (Carson, 1961)

grube Ger. A mine.

grub saw A saw made from a coarsely notched blade of soft iron and provided with a wooden back; it is used, with sand, for sawing stone by hand. (Standard, 1964)

grubstake In the Western United States, supplies or funds furnished to a mining prospector on promise of a share in his discoveries. So called because the lender stakes or risks provisions so furnished. (Webster 3rd, 1966)

grubstake contract An agreement between two or more persons to locate mines upon the public domain by their joint aid, effort, labor, or expense, and each is to acquire by virtue of the act of location such an interest in the mine as agreed upon in the contract. (Fay, 1920)

grueso Sp. Lump ore. The term is used at the mercury mines in California.

gruff Eng. A pit or shaft.

grunching Blasting coal out of the solid face as opposed to blasting coal that has been undercut by hand or by coal cutter. (Nelson, 1965)
grundy Granulated pig iron used in making granulated steel. (Webster 2nd, 1960)
grunerite A monoclinic mineral, $(Fe,Mg)_7Si_8O_{22}(OH)_2$; amphibole group, with Mg/(Mg+Fe)=0-0.3; forms series with cummingtonite and with magnesiocummingtonite; fibrous or needlelike, commonly in radial aggregates; characteristic of iron formations in the Lake Superior and Labrador Trough regions. Also spelled grünerite.
grunter A hooked rod to aid in supporting a crucible. (Standard, 1964)
grus The fragmental products of in situ granular disintegration of granite and granitic rocks. Syn: *grush; residual arkose; growan.* Etymol: Ger. Grus, grit, fine gravel, debris. Also spelled: gruss. (AGI, 1987)
grush *grus.*
G stone A name which has been used for pyrophyllite. See also: *pyrophyllite.* (Dodd, 1964)
guag Corn. A place from which the ore has been extracted. Syn: *gunis.* (CTD, 1958)
guardian angel A warning device affixed to the roof or back of a mine that automatically displays a visual indication when rock displacement begins. See also: *safety light.* (Federal Mine Safety, 1977)
guard magnet Permanent magnet or electromagnet used in crushing system to arrest or remove tramp iron ahead of the crushing machinery. (Pryor, 1963)
guardplate A plate in front of an iron furnace, covering the taphole through which the slag is drawn out. (Standard, 1964)
guard screen *oversize control screen.*
gublin bat A black, fissible substance, an iron ore, in which a bituminous shivery earth abounds. (Arkell, 1953)
gudgeon (a) The bearing of a shaft, esp. when made of a separate piece. (Standard, 1964) (b) A metallic journal piece set into the end of a wooden shaft. A reinforced bushing or a thrust-absorbing block. (Standard, 1964; Nichols, 1976)
gudmundite A silver-white to steel-gray sulfantimonide of iron, FeSbS. Isomorphous with arsenopyrite; elongated crystals; orthorhombic. From Gudmundstrop, Sweden. (English, 1938)
guest element (a) A trace element substituting a common element in a rock mineral. (b) *trace element.*
gug Som. A self-acting inclined plane underground; sometimes called a dip incline.
gugiaite A mineral, $Ca_2BeSi_2O_7$, tetragonal, in skarn rocks near Gugia (presumably in China). A member of the melilite family near meliphane but containing little sodium or fluorine; an unnecessary name. Named from the locality. See also: *meliphanite.* (Hey, 1964; Fleischer, 1995)
guhr (a) A white (sometimes red or yellow), loose, earthy, water-laid deposit of a mixture of clay or ocher, occurring in the cavities of rocks. (AGI, 1987) (b) *diatomite.*
guide (a) A pulley to lead a driving belt or rope in a new direction or to keep it from leaving its desired direction. (Long, 1960) (b) The tracts that support and determine the path of a skip bucket and skip bucket bail.
guide core *dummy.*
guide fossil Any fossil that has actual, potential, or supposed value in identifying the age of the strata in which it is found or in indicating the conditions under which it lived; a fossil used esp. as an index or guide in the local correlation of strata. Cf: *index fossil.* (AGI, 1987)
guide frame A frame designed to be held rigidly in place by roof jacks or timbers, and with provisions for attaching a shaker conveyor panline to the movable portion of the frame, which can be used to prevent jumping or side movement of the panline. (Jones, 1949)
guide idler An idler roll with its supporting structure so designed that when it is mounted on the conveyor frame it guides the belt in a defined horizontal path, usually by contact with the edge of the belt. (NEMA, 1956)
guide pulley A loose pulley used to guide a driving belt past an obstruction or to divert its direction. (CTD, 1958)
guide rod A heavy drill rod coupled to and having the same diameter as a core barrel on which it is used. It gives additional rigidity to the core barrel and helps to prevent deflection of the borehole. Also called core-barrel rod; oversize rod. Cf: *drill collar.* (Long, 1960)
guide rope *cage guide.*
guide runner A runner driven ahead of other runners to guide them. (Hammond, 1965)
guides (a) Wood, steel, or wire-rope conductors in a mine shaft, which are engaged by shoes on the cage or skip so as to guide its movement. (b) Timber or metal tracks in a hoisting shaft, which are engaged by shoes on the cage or skip so as to steady it in transit. In collieries rope guides are sometimes used. (Pryor, 1963) (c) The holes in a crossbeam through which the stems of the stamps in a stamp mill rise and fall. (d) A pulley to lead a driving belt or rope in a new direction, or to keep it from leaving its desired direction. (Standard, 1964)
guillies Corn. Worked-out cavities in a mine.
guillotine A machine for breaking iron with a falling weight. (Fay, 1920)
guinea bed War. The shelly, conglomeratic basement limestone bed of the Lower Lias. So called because the stones, if dry, ring when struck. Cf: *sun bed.* (Arkell, 1953)
guinea gold Twenty-two carat gold, of which guineas were coined. (Standard, 1964)
gulch A term used esp. in the Western United States for a narrow, deep ravine with steep sides, larger than a gully; esp. a short, precipitous cleft in a hillside, formed and occupied by a torrent, and containing gold (as in California). (AGI, 1987)
Gulf-type (Vacquier) magnetometer A flux gate or saturable reactor type of recording magnetometer. Used primarily in aircraft and there includes means for keeping the measuring element aligned in the direction of maximum intensity (that is, total field). In this case it records variations in the total field regardless of variations in its direction. Sometimes used for establishing the position of the aircraft as well as the magnetometer itself. (AGI, 1987)
gull A structure formed by mass-movement processes, consisting of widened, steeply inclined tension fissures or joints, resulting from lateral displacement of a slide mass and filled with debris derived from above. Primarily a British usage. (AGI, 1987)
gum (a) *gummings.* (b) Small coal broken out by a coal cutter. See also: *duff.* (Pryor, 1963) (c) Small coal, slack, or duff. (CTD, 1958)
gumbo A term used locally in the United States for a clay soil that becomes sticky, impervious, and plastic when wet. (AGI, 1987)
gumbrine Same as fuller's earth, and similar to floridine from Gumbri, near Kutais, Georgia, Transcaucasia, in the former U.S.S.R. (Spencer, 1946)
gum copal *copal.*
gum dynamite Explosive gelatin. (Standard, 1964)
gummer A person who clears the fine coal, gum, or dirt from the undercut made by a coal-cutting machine. (CTD, 1958)
gummings The small coal or dirt produced by the picks of a coal cutter. Syn: *gum.* (Nelson, 1965)
gummite A mixture of yellow or orange secondary uranium oxides formed by alteration of uraninite.
gummy Applicable when rock or formation being drilled produces cuttings and sludge, which tend to fill the waterways of a bit or to adhere massively to the borehole walls or drill-stem equipment. Cf: *balling formation; sticky.* (Long, 1960)
gun A borehole in which the charge of explosive has been fired with no other effect than to blast off a small amount of material at the mouth of the borehole; also called a bootleg or John Odges. See also: *blown-out shot.* (Fay, 1920)
gunboat A self-dumping box on wheels, used for raising (or lowering) coal in slopes; a monitor, a skip. (Fay, 1920)
gun drill A drill, usually with one or more flutes and with coolant passages through the drill body, used for deep-hole drilling. (ASM, 1961)
gunis *guag.*
gunite A mixture of portland cement, sand, and water applied by pneumatic pressure through a specially adapted hose and used as a fireproofing agent and as a sealing agent to prevent weathering of mine timbers and roadways. Etymol: Gunite, a trademark. To apply gunite; to cement by spraying gunite. See also: *Aliva concrete sprayer.* (AGI, 1987)
gunite gun *cement gun.*
guniting Pneumatically applied portland cement mortar, or gunite. The spraying of mine openings with concrete to provide ground support, present a smooth surface to the air current, and prevent weathering with the mortar. See also: *lining; timber preservation; Aliva concrete sprayer.* (Nelson, 1965)
gunk (a) *rod dope.* (b) Any gummy substance that collects inside the working parts and hinders the operation of a machine or other mechanical apparatus. (Long, 1960) (c) A permanent emulsion that forms in liquid-liquid extraction equipment, often containing dust or other solid matter at the core of each globule.
gunned shot Scot. *blown-out shot.*
gunnie (a) Corn. In mining, measure of breadth or width, a single gunnie being 3 ft (0.91 m) wide. (Standard, 1964) (b) Corn. The vacant space left where the lode has been removed; a crevice. Plural: gunnies. Also spelled gunniss; gunnice. Syn: *house.*
gunningite A monoclinic mineral, $(Zn,Mn)SO_4{\cdot}H_2O$; kieserite group; as efflorescences on sphalerite from the Keno Hill and Galena Hill area, central Yukon Territory, Canada.
gunning the pits Agitation of the drilling fluid in a pit by forcing a portion of the fluid under pressure through a constricted tube or gun, jetting it into the main body of fluid. (Brantly, 1961)
gunniss Corn. *gunnie.*

gun-perforator loader In petroleum production, one who loads explosive powder into gun perforators used in shooting holes through tubings, casings, and earth formations of oil or gas wells to aid in well drilling or producing operations, working either in shop or at well site. Also called the loader; perforator loader. See also: *loader.* (DOT, 1949)

gunpowder paper Paper spread with an explosive compound. It is rolled up for use in loading. (Standard, 1964)

gunpowder press A press for compacting meal powder before granulating into gunpowder. (Standard, 1964)

Gunter's chain A surveyor's chain that is 66 ft (20 m) long, consisting of a series of 100 metal links each 7.92 in (20.1 cm) long and fastened together with rings. It served as the legal unit of length for surveys of U.S. public lands, but has been superseded by steel or metal tapes graduated in chains and links. Named after Edmund Gunter (1581-1626), English mathematician and astronomer, who invented the device about 1620. Syn: *chain; pole chain.* (AGI, 1987)

gurhofite A snow-white variety of dolomite, containing a large proportion of calcium. (Standard, 1964)

gurlet (a) A mason's pickax having one cutting edge and a point. (Standard, 1964) (b) A pickax having a sharply pointed peen and a bladed peen for cutting. (CTD, 1958)

gurmy A mine level; working. (Standard, 1964)

guss (a) A rope used for drawing a basket of coal in a thin seam. (CTD, 1958) (b) Brist. A short piece of rope by which a boy draws a tram or sled in a mine. (Fay, 1920)

gusset A V-shaped cut in the face of a heading. (Stauffer, 1906)

Gusto multiplow A number of small plows attached to a rope or chain which cut backward and forward on the face. They operate in conjuction with an armored conveyor. See also: *multiplow.* (Nelson, 1965)

Gusto scraper box An arrangement of scraper boxes with cutting knives attached to the face side. See also: *scraper box plow.* (Nelson, 1965)

gut To rob, or extract, only the rich ore of a mine. (Weed, 1922)

gutter (a) The lowest and usually richest portion of an alluvial placer. The term is used in Australia for the dry bed of a buried Tertiary river containing alluvial gold. Syn: *bottom.* (AGI, 1987) (b) A gob heading. (Nelson, 1965) (c) A drainage trench. (Nelson, 1965) (d) A small airway made through a goaf or gob. (CTD, 1958) (e) In deep lead mining, the lowest portion of a deep lead filled with auriferous wash dirt. (Eng. Min. J., 1938) (f) A channel or gully worn by running water. (Webster 3rd, 1966)

guttering (a) A channel cut along the side of a mine shaft to conduct the water back into a lodge or sump. (b) A process of quarrying stone in which channels, several inches wide, are cut by hand tools, and the stone block detached from the bed by pinch bars. (c) The formation of more or less vertical breaks at or toward the centerline of a roadway, as a consequence of which falls occur along the groove or gutter. (TIME, 1929-30)

Gutzkow's process A modification of the sulfuric acid parting process for bullion containing large amounts of copper. A large excess of acid is used; the silver sulfate is then reduced with charcoal, or, in the original process, ferrous sulfate. (Liddell, 1918)

guy (a) A wire line or rope attached to the top of a drill derrick or pole and extending obliquely to the ground, where it is fastened to a deadman or guy anchor. See also: *guy line.* (Long, 1960) (b) A rope that holds the end of a boom or spar in place. Syn: *guy rope.* See also: *guy line.* (Long, 1960)

guy anchor The object to which the lower end of a guy is attached. Also called deadman. (Long, 1960)

guyed Held upright and steadied by one or more guys. (Long, 1960)

guy line A guy or several guys. See also: *guy.* (Long, 1960)

guyot A flat-topped submarine mountain rising from the floor of the ocean like a volcano but planed off on top and covered by an appreciable depth of water. (Leet, 1958)

guy ring A ring on the head block or top of a drill pole, derrick, or tripod to which guys are attached. (Long, 1960)

guy rope (a) A rope holding a structure in a desired position. (CTD, 1958) (b) *guy.*

gyprock (a) Massive rock gypsum. (b) A driller's term for a rock of any kind in which he has trouble in making a hole. (AGI, 1987) (c) A rock composed chiefly of gypsum. (AGI, 1987)

gypsiferous Containing gypsum. Also spelled gypseous.

gypsification Development of, or conversion into, gypsum; e.g., the hydration of anhydrite. (AGI, 1987)

gypsite *gypsum.*

gypsum A monoclinic mineral, $8[CaSO_4 \cdot 2H_2O]$; colorless to white in crystals, but massive beds may range from red to yellow to brown, gray, or black; the most common natural sulfate; defines 2 on the Mohs hardness scale; commonly associated with rock salt (halite) and anhydrite; forms beds and lenses interstratified with limestone, shale, and clay, esp. in rocks of Permian to Triassic age; also in volcanic fumarolic deposits; an accessory mineral in metalliferous veins. Syn: *gypsite.*

gypsum plate In polarized-light microscopy, an accessory plate of clear gypsum (replaced by quartz of the appropriate thickness in modern instruments) that gives a first-order red (approx. 1 λ out of phase for 560 nm) interference color with crossed polars when inserted in the tube with its permitted electric vectors at 45° to those of the polarizer and analyzer. It is used to determine fast and slow directions (electric vectors) of light polarization in crystals under view on the microscope stage by increasing or decreasing retardation of the light. Also called a sensitive-tint plate. Cf: *accessory plate; quartz wedge.* See also: *selenite plate.* Syn: *gips plate.*

gyrasphere crusher Heavy-duty fixed path cone crusher; a variant from the standard cone crusher. See also: *Symon's cone crusher.* (Pryor, 1963)

gyratory (a) More or less eccentric, as certain rock crushers. (von Bernewitz, 1931) (b) A widely used form of rock breaker in which an inner cone rotates eccentrically in a larger outer hollow cone. (CTD, 1958)

gyratory breaker A primary crusher consisting of a vertical spindle, the foot of which is mounted in an eccentric bearing within a conical shell. The top carries a conical crushing head revolving eccentrically in a conical maw. There are three types of gyratories—those that have the greatest movement on the smallest lump, those that have equal movement for all lumps, and those that have greatest movement on the largest lump. Syn: *gyratory crusher.* See also: *cone crusher.* (Liddell, 1918)

gyratory crusher *gyratory breaker.*

gyrocompass (a) A north-seeking form of gyroscope used as a vehicle's or craft's directional reference. Also known as gyroscopic compass. (McGraw-Hill, 1994) (b) The gyrocompass is used in underground and borehole surveying. Syn: *gyroscopic compass; gyrostatic compass.* (AGI, 1987)

gyroscopic-clinograph method A method for measuring borehole deviation that photographs time, temperature, and inclination from the vertical on 16 mm film and can take 1,000 readings descending then ascending the hole as a check. The gyroscope maintains the casing on a fixed bearing. (Sinclair, 1958)

gyroscopic compass (a) *gyrocompass.* (b) A magnetic compass whose equilibrium is maintained by the use of gyroscopes. (AGI, 1987)

gyrostatic compass *gyrocompass.*

H

Haanel depth rule A rule of thumb for estimating the depth of a magnetic body, which is valid if the body may be regarded as magnetically equivalent to a single pole. The depth of such a pole is equal to the horizontal distance from the point of maximum vertical magnetic intensity to the points where the intensity is one-third of the maximum value. (AGI, 1987)
Haarmann plow *scraper box plow.*
Haase furnace A muffle furnace of the McDougall type, in which the hearths are separated by suitable flues through which the products of combustion from the fireplace pass.
Haase system A system of shaft sinking through loose ground or quicksand by piles in the form of iron tubes connected together by webs. Their downward movement is facilitated by water under pressure that is forced down the tubes to wash away the loose material from underneath their points. (Nelson, 1965)
habit (a) A general term for the outward appearance of a mineral or rock. (AGI, 1987) (b) The characteristic or typical crystal form, combination of forms, or other shape of a mineral, including irregularities.
hachure One of a series of short, straight, evenly spaced, parallel lines used on a topographic map for shading and for indicating surfaces in relief (such as steepness of slopes), drawn perpendicular to the contour lines. Hachures are short, broad (heavy), and close together for a steep slope, and long, narrow (light), and widely spaced for a gentle slope, and they enable minor details to be shown but do not indicate elevations above sea level. Etymol: French. Syn: *hatching; hatchure.* v. To shade with or show by hachures. (AGI, 1987)
hachure map A map that represents the topographic relief by means of hachures.
hack hammer A hammer resembling an adz, used in dressing stone. (Webster 3rd, 1966)
hackiron A miner's pickax or hack. A chisel or similar tool for cutting metal, as wire, into nails. (Fay, 1920)
hackly The property shown by certain minerals or rocks of fracturing or breaking along jagged surfaces, e.g., broken iron.
hackly fracture A mineral's habit of breaking along jagged, irregular surfaces with sharp edges. (Leet, 1958)
hackmanite A sulfur-rich variety of sodalite.
hacksaw structure Irregular, saw-shaped terminations of crystals (such as of augite) or grains due to intrastratal solution.
hade The complement of the dip; the angle that a structural surface makes with the vertical, measured perpendicular to the strike. It is little used. See also: *dip; rise.* Syn: *underlay.*—v. To incline from the vertical. (AGI, 1987)
Hadsel mill Early form of autogenous grinding mill, in which comminution resulted from the fall of ore on ore during the rotation of a large-diameter horizontal cylinder. (Pryor, 1963)
Haenisch and Schroeder process A method for the recovery of sulfur as liquid sulfurous anhydride from furnace gases. (Fay, 1920)
hafnium silicate A compound analogous to zircon, therefore, the suggested name hafnon. It can be synthesized from the oxides at 1,550 °C. Thermal expansion (150 to 1,300 °C), 3.6×10^{-6}. (Dodd, 1964)
hafnium titanate Special refractory compositions have been made by sintering mixtures of HfO_2 and TiO_2 in various proportions. The melting point of these sintered bodies was approx. 2,200 °C; there appeared to be a phase change at about 1,850 °C. Some of the compositions had negative thermal expansions. (Dodd, 1964)
hag (a) To cut as with an ax; to cut down the coal with the pick. (Fay, 1920) (b) Scot. A cut; a notch. (Fay, 1920) (c) N. of Eng. A quagmire or pit in mossy ground; any broken ground in a bog. (Fay, 1920)
häggite A monoclinic mineral, $V_2O_2(OH)_3$; forms black crystals in sandstone, Wyoming.
hag principle The system under which a skilled miner employs an unskilled helper. (CTD, 1958)
haiarn In Wales, a term for iron.
hailstone bort Variety of bort built of concentric shells of clouded diamond and cementlike material. See also: *bort.* (Tomkeieff, 1954; Hess)
hair copper *chalcotrichite.*
hair pyrite *millerite.*
hair salt Common name for efflorescences of hairlike, acicular, or fibrous crystals of epsomite or other hydrous sulfates in caves and old mine workings. See also: *alunogen.*
hairstone Quartz thickly penetrated with hairlike crystals of rutile, actinolite, or some other mineral. Cf: *hedgehog stone.* (Webster 3rd, 1966)
hair zeolite May be natrolite, scolecite, or mesolite. (Fay, 1920)
half-and-half plane Scot. In a direction midway between plane course and end course. See also: *half-course.* (Fay, 1920)
half bearings Bearings such as are used on railway cars where the load is constantly in one direction and is sufficiently heavy to hold the journal against the bearing. (Crispin, 1964)
half blinded Scot. Two ends driven off a plane, one on each side and not opposite each other by half their width. (Fay, 1920)
half-cell An electrode immersed in a suitable electrolyte designed for measurements of electrode potential. (ASM, 1961)
half-course A drift or opening driven at an angle of about 45° to the strike and in the plane of the seam. See also: *half-and-half plane.* (Fay, 1920)
half end York. *horn coal.*
half headers Term applied to material that amounts to a large cap piece. They are used by sawing a header in two and placing one or more timbers under the half header on the same side of the track. Two timbers are generally placed under the half header and the end allowed to extend out over the haulage. The term half header should not be applied to regular cap pieces. (Kentucky, 1952)
half-life (a) The time in which one-half of the atoms in a radioactive substance disintegrate. (Lyman, 1964) (b) The time in which the quantity of a particular radioactive isotope is reduced to one-half of its initial value. Syn: *half-period.*
half-marrow Newc. Youngsters, of whom two do the work of one loader. (Fay, 1920)
half-period *half-life.*
half-round nose *medium-round nose.*
half set In mine timbering, one leg piece and a collar. (Fay, 1920)
half-value distance The horizontal distance between the points of maximum and half-maximum values in a symmetrical anomaly, usually either gravity or magnetic. It is useful in estimating the depth of the geologic feature that causes the anomaly. (AGI, 1987)
half-wave rectifier A rectifier that changes single-phase alternating current into pulsating unidirectional current, utilizing only one-half of each cycle. (Coal Age, 1960)
half width Half the width of a simple anomaly (esp. a gravity or magnetic anomaly) at the point of half its maximum value. For simple models the maximum depth at which the body causing the anomaly can lie can be calculated from the half width. (AGI, 1987)
halide A fluoride, chloride, bromide, or iodide.
Halimond tube Miniature pneumatic flotation cell, operated by hand. Widely used in ore testing, for examination of small samples under closely controllable conditions of flotation. (Pryor, 1963)
halite An isometric mineral, 4[NaCl]; cubic cleavage; soft; salty tasting; forms disseminated grains or crystals in sedimentary rocks, or aggregates of large cubic crystals; granular to massive, the latter as extensive sedimentary beds ranging in thickness from less than 1 mm to more than 50 m, also in convoluted masses called salt domes; a typical constituent of playa lake deposits in arid regions. Syn: *rock salt; common salt; salt.*
halitic Composed partly or wholly of halite; esp. said of a sedimentary rock containing halite as cementing material, such as halitic sandstone. (AGI, 1987)
hälleflinta Sw. A dense, compact, metamorphic rock consisting of microscopic quartz and feldspar crystals, with occasional phenocrysts and sometimes hornblende, chlorite, magnetite, and hematite. It is associated with gneisses, but is of obscure origin. See also: *porcellanite.* (Fay, 1920)
Hallett table A table of the Wilfley type, except that the tops of the riffles are in the same plane as the cleaning planes and the riffles are sloped toward the wash-water side. (Liddell, 1918)
Hallinger shield A tunneling shield of Hungarian design, successfully employed for tunneling at Dortmund and under the Danube. It is

valuable for working in very soft ground. It incorporates a mechanical excavator and does not entail the use of timbering to protect the miners. (Hammond, 1965)

halloysite (a) A monoclinic mineral, $2[Al_4Si_4(OH)_8O_{10}]$; kaolinite-serpentine group; made up of slender tubes as shown by electron microscopy; a gangue mineral in veins. Syn: *metahalloysite.* (b) Used as a group name to include natural "halloysite minerals" with different levels of hydration, as well as those formed artificially. See also: *alum salt.*

Hall process The first commercially successful method for manufacturing aluminum; the purified oxide is dissolved in fused cryolite and then electrolyzed. Syn: *Heroult process.* (Bennett, 1962)

Hall-Rowe wedge A tapered concave metal plug or wedge that can be set in a drill hole at a predetermined depth and bearing to deflect or straighten an off-course borehole. See also: *wedge.* (Long, 1960)

halo (a) A circular or crescentic distribution pattern about the source or origin of a mineral, ore, mineral association, or petrographic feature. It is encountered principally in magnetic and geochemical surveys. Some halos are primary, formed either at the same time as the host rock or at the same time as associated mineral deposits, and some are secondary, formed by surficial alteration of the associated mineral deposit. Syn: *aureole.* Cf: *dispersion pattern. wall-rock halo.* (AGI, 1987) (b) Discoloration of a mineral, viewed in thin section, in the form of a ring. Most halos of this sort are caused by radiation damage by alpha particles emitted from uranium- and thorium-bearing mineral inclusions.

halocline A steep ascendent of salinity. This has an effect on refraction of sound waves, since sound velocity increases with increasing salinity. (Hy, 1965)

halokinesis *salt tectonics.*

halotrichite (a) A monoclinic mineral, $4[Fe^{2+}Al_2(SO_4)_4{\cdot}22H_2O]$; forms a series with pickeringite in which magnesium replaces ferrous iron; soft; a weathering product of pyritic rocks in mines; also in arid regions and around fumaroles. (b) The mineral group apjohnite, bilinite, dietrichite, halotrichite, pickeringite, and redingtonite. Syn: *feather alum; iron alum.*

haloxylin A mixture of yellow prussiate of potash, niter, and charcoal used as an explosive. (Fay, 1920)

hamburgite An orthorhombic mineral, $8[Be_2(BO_3)(OH)]$; in alkali pegmatites and in placers.

hammada An extensive, nearly level, upland desert surface that is either bare bedrock or bedrock thinly veneered by pebbles, smoothly scoured and polished and generally swept clear of sand and dust by wind action; a rock desert of the plateaus, esp. in the Sahara. The term is also used in other regions, as in Western Australia and the Gobi Desert. Etymol: Arabic, hammadah. (AGI, 1987)

hammer (a) Term for drive hammer, a heavy sleeve-shaped weight used for driving drill pipe or casing into overburden or soft rock. (Long, 1960) (b) To pound or drive with pilehammerlike blows delivered by a drive hammer. (Long, 1960)

hammer breaker An impact type of breaker consisting of a number of swinging bars or steel hammers hinged to a horizontal shaft that rotates at high speed. (Nelson, 1965)

hammer drill (a) A light, mobile, and fast-cutting drill in which the bit does not reciprocate but remains against the rock in the bottom of the hole, rebounding slightly at each blow. There are three types of hammer drills; drifter, sinker, and stoper. (Lewis, 1964) (b) A development of the piston drill in which the drill steel is not attached to the piston but remains in the hole, the piston delivering a rapid succession of light hammer blows. The drill steel is frequently hollow so that air or water may be driven through to cool the bit and clean the hole. Rotation of the bit is automatic. Also known as jackhammer. (Barger, 1939) (c) A percussive drill. (BS, 1964) (d) A rock drill powered by compressed air that reciprocates a free piston, causing it to strike the shank of the drill steel. When of light construction, a hand hammer drill, otherwise supported on a tripod or bar. (Pryor, 1963)

hammermill (a) A pulverizing unit consisting of a rotor, fitted with movable hammers, that is revolved rapidly in a vertical plane within a closely fitting steel casing. The hammers hit falling rock, which is fractured on impact, or by collision with other rocks or with the casing. When sufficiently reduced in size, the pulverized rock escapes through grids in the casing. Syn: *beater mill.* Cf: *disintegrator; impact crusher; impact mill.* (b) Coal crusher in which the blow is induced with the aid of centrifugal force. The coal is broken with the impact and usually dragged across grate bars in the bottom of the unit. See also: *ring crusher.* (Mitchell, 1950)

hammerpick A compressed-air-operated hand machine used by miners to break up the harder rocks in a mine. It consists mainly of a pick and a hammer operated by compressed air. The hammer driving the pick is set in a cylinder, where the compressed air enters and presses the hammer, which in turn drives the wedge-shaped edge of the pick into the rock in short sufficient shocks of from 1,500 to 2,000 blows per minute. See also: *poll pick.* (Stoces, 1954)

hammochrysos A stone, the appearance of which suggested sand veined with gold, perhaps mottled jasper. (Webster 2nd, 1960)

hammock structure The intersection of two vein or fracture systems at an acute angle. (AGI, 1987)

hand Measurement of height of mine haulage animals equivalent to 4 in (10.2 cm).

hand auger A screwlike tool much like a large carpenters' bit or a short cylindrical container with cutting lips attached to a rod and operated by hand and used to bore shallow holes and obtain samples of soil and other relatively unconsolidated near-surface materials. Cf: *auger.* (AGI, 1987)

hand boring The drilling of holes by hand for site investigations or for the exploration of shallow mineral deposits. The hand drill is used for depths of about 15 m and where the ground is loose or not too hard. Cf: *auger.* (Nelson, 1965)

hand cable A flexible cable used principally in making electrical connections between a mining machine and a truck carrying a reel of portable cable. Also called head cable; butt cable.

hand cleaning The removal by hand of impurities from coal, or vice versa. (BS, 1962)

hand cobbing *cobbing.*

hand drilling A historical method of drilling blastholes in rock by hammer and a hand-held steel or bit. Single-jack drilling was done by one miner. In double-jack drilling, one miner held the steel for one or two strikers with hammers. (Peele, 1941)

hand electric lamp A hand lamp, with battery and fitments similar to a cap lamp except that it forms a self-contained unit. (Nelson, 1965)

hand filling (a) Scot. Loading coal from face by hand. (Pryor, 1963) (b) Eng. Loading coal from face by hand, but small coals are loaded separately from large lumps. (Pryor, 1963) (c) *sublevel stoping.*

hand frame An iron barrow used in a foundry. (Fay, 1920)

hand gear The mechanism for opening the valves of a steam engine by hand in starting. (Standard, 1964)

hand hammer Any hammer wielded by hand. A blacksmith's (or miner's) hammer used with one hand as distinguished from a heavier hammer or sledge. (Fay, 1920)

hand hammer drill An ordinary rock drill held in the hand and not mounted on a bar or column. The air leg support is now widely used in tunnels and rock drilling generally. (Nelson, 1965)

hand jig Manually operated moving-screen jig used to treat small batches of ore. The jig box is fixed to a rocking beam and moved up and down in a tank of water. (Pryor, 1963)

hand lamp A portable battery-operated lamp incorporating a tungsten filament light source within a glass of the dome or well-glass type and providing maximum illumination in the horizontal plane. (BS, 1965)

hand lead A lead weight attached to a lead line of up to 100 fathoms (183 m), used in hydrographic surveying. (Hammond, 1965)

handling plant Equipment for the mechanical movement of dirt, ore, coal, or other material either horizontally or up an incline, by some form of conveyor, bucket, chain, or rope. (Nelson, 1965)

hand loader A miner who loads coal by shovel rather than by machine. See also: *loader.* (BCI, 1947)

hand mining The working and winning of coal or mineral by hand and not by machines. Broadly, hand coal mining would imply hand holing, shot firing, and hand filling. (Nelson, 1965)

handpicked coal Coal from which all stones and inferior coal have been picked out by hand; large lumps. (Fay, 1920)

handpicking Manual removal of selected fraction of coarse run-of-mine ore, usually performed on picking belts (belt conveyors) after screening away small material, perhaps washing off obscure dirt, and crushing pieces too large for the worker to handle. Hard sorting (Rand) describes picking of banket when up to 30% of waste rock is removed. See also: *sorting.* (Pryor, 1963)

hand sampling (a) In prospecting, valuation, and control, use of manual methods for detaching and reducing to an appropriate size representative samples of ore. (Pryor, 1963) (b) One of the major breakdowns in ore sampling that includes grab sampling, trench or channel sampling, fractional selection, coning and quartering, and pipe sampling. These methods are used in sampling small batches of ore, etc. Cf: *mechanical sampling.* (Newton, 1959)

hand scraper *stope scraper.*

hand selection The selection by hand of pieces of coal with certain specific qualities according to surface appearance. (BS, 1962)

handset A drilling bit in which the diamonds are set into holes drilled into a malleable-steel bit blank and shaped to fit the diamonds. The hand method has been almost completely superseded by mechanical setting methods. Cf: *mechanical set.* (Long, 1960)

hand specimen A piece of rock of a size that is convenient for megascopic study and for preserving in a study collection. (AGI, 1987)

hand spraying A method of dust prevention used in hand-won faces, or in conjunction with wet cutting in thick seams. The sprays are controlled by the colliers who wet the face and the broken coal before loading. Sprays must be connected with the pipeline through the face by means of flexible hoses; one spray for every 20 yd (18.3 m) of face is usually sufficient. (Mason, 1951)

hand tramming Pushing of cars by manpower. It is limited to mines of small output, to prospects, and to work where mechanical haulage would not be justified. See also: *manual haulage.* (Lewis, 1964)

hand whip A counterpoised sweep for raising water from shallow pits.

handyman A person employed to do various kinds of work. (Fay, 1920)

hang (a) To suspend casing or pipe in a borehole in a clamp resting on blocks at the collar of the hole. (Long, 1960) (b) To suspend drill string or other downhole equipment in the drill derrick or tripod either on the hoisting line or on hooks provided in the crown block for that purpose. (Long, 1960)

hanger (a) *hanging wall; hanging bolts.* (b) Scot. The hook of a miner's lamp. (Fay, 1920) (c) Something that hangs, overhangs, or is suspended. (Webster 3rd, 1966) (d) A frame containing a bearing for a shafting. (Standard, 1964)

hangfire An explosive charge that is not properly detonated and burns and may eventually result in a detonation at some nondetermined time. See also: *hung shot.*

hanging bolts Rods of round iron, used in shaft construction to suspend wallplates. In concrete-lined shafts hanging rods give reinforcement, the top set being concreted into the shaft collar and others hooked on below, with periodic consolidation in strong rock strata as the shaft is deepened. Sometimes called hangers. (Pryor, 1963; Fay, 1920)

hanging coal A portion of the coal seam which, by undercutting, has had its natural support removed. (Fay, 1920)

hanging deal Aust. Planks used to suspend a lower curb from the one above it, in cases where backing deals are necessary. (Fay, 1920)

hanging its water Scot. The bucket is said to hang its water when it fails to pump on account of a faulty valve, or air between the bucket and the valve, the column of water above the bucket being sufficient to prevent the opening of the bucket lids.

hanging-on Eng. The pit bottom, level, or inset at which the cages are loaded. (Fay, 1920)

hanging pulley A small fenced pulley hung from the roof or side of a haulage road in which the tail rope of a main-and-tail haulage is suspended. It keeps the rope (which is not used for direct haulage of cars) clear of the roadway and minimizes friction while in motion. The swinging of hanging pulleys and ropes is a hazard to people traveling on the roadway. (Nelson, 1965)

hanging scaffold Scot. A movable platform in a shaft attached to a winding rope.

hanging sets Scot. Timbers from which cribs are suspended in working through soft strata.

hanging wall (a) The overlying side of an orebody, fault, or mine working, esp. the wall rock above an inclined vein or fault. See also: *wall; walls; top wall.* Syn: *hanger.* Cf: *footwall.* (AGI, 1987) (b) Sticking or wedging of part of the charge in a blast furnace. (Fay, 1920)

hanging wall drift In the United States, a horizontal gallery driven in the hanging wall of a vein. (Nelson, 1965)

hanging wall of a fault The upper wall of an inclined fault plane. (Ballard, 1955)

hang-up Underground, blockage of ore pass or chute by rock. (Pryor, 1963)

haplite *aplite.*

hard (a) Containing certain mineral salts in solution, esp. calcium carbonate; said of water having more than 8 to 10 grains/gal (137 to 171 mg/L) of such matter to the gallon. (Standard, 1964) (b) Solid; compact; difficult to break or scratch. See also: *hardness scale.* (AGI, 1987)

hard bottom A condition encountered in some open-cut mines wherein the rock occasionally will not be broken down to grade because of an extra-hard streak of ground or because not enough explosive is used. This is called a "hard bottom"; it interferes with work and puts undue strain on a shovel. Such unbroken ores usually are drilled with a jackhammer and blasted.

hard coal (a) All coal of higher rank than lignite. (BS, 1960) (b) In the United States, the term is restricted to anthracite. See also: *anthracite.* (BS, 1960)

hard-coal plow A plow type of cutter loader for cutting the harder coal seams. It consists of stepped kerving bits that precut the coal, leaving the unstressed coal to be cut by the following bits. The kerving bits may be either rigid or swiveling. See also: *rapid plow.* Syn: *Westfalia plow.* (Nelson, 1965)

hardebank Unaltered kimberlite below the zone of blue ground. Cf: *blue ground.* (AGI, 1987)

hardened steel Steel that has been hardened by quenching from or above the hardening temperature. (Fay, 1920)

hardener An alloy, rich in one or more alloying elements, added to a melt either to permit closer composition control than is possible by addition of pure metals or to introduce refractory elements not readily alloyed with the base metal. (ASM, 1961)

hardening Metallurgical process in which iron or suitable alloy is quenched by abrupt cooling from or through a critical temperature range. See also: *precipitation hardening.* (Pryor, 1963)

hardening media Liquids into which steel is plunged in hardening. They include cold water, various oils, and water containing sodium chloride or hydroxide to increase the cooling power. (CTD, 1958)

hard ground Rock that is difficult to work.

Hardgrove number Empirical index of grindability of ores or minerals, reached as result of comminution of a test sample under stated conditions of control. (Pryor, 1963)

Hardgrove test This test utilizes a special grindability mill of the ring-and-ball type, in which a 50-gm portion of closely sized coal is ground for 60 revolutions. This method is of the constant-work type; i.e., a fixed amount of work is expended on each coal, and a grindability value is determined from the size composition of the ground material. (Mitchell, 1950)

hardhat Slang term for a safety hat. (BCI, 1947)

hardhead (a) A hard knob or knot formed by extreme cementation of sandstone. (b) A large, smooth rounded stone found esp. in coarse gravel. (c) A tunnel in a coal mine driven through rock. Syn: *hard heading.* (d) A hard, brittle, white residue obtained in refining tin by liquation, containing, among other things, tin, iron, arsenic, and copper. Also, a refractory lump or ore only partly smelted. (ASM, 1961)

hard heading A heading driven in rock. In S. Wales and elsewhere, men employed in hard headings have suffered greatly from silicosis. Syn: *hardhead.* See also: *stone drift.* (Nelson, 1965)

Hardinge mill Cylindroconical ball mill, made in three sections—a flattish cone at feed end followed by a cylindrical drum, and finishing with a steep cone leading to the discharge trunnion. The tricone mill has wedge-shaped liners in the drum section that turn this into a gentle conic frustrum widest at the feed end. (Pryor, 1963)

Hardinge thickener A machine for removing the maximum amount of liquid from a mixture of liquid and finally divided solids. The solids settle out on the bottom of the thickener tank as a sludge, and the clear liquid overflows at the top of the tank. It is used for processing chemical, metallurgical, and coal-washing slurries. (Nelson, 1965)

hard kiln A muffle kiln fired at a temperature between that of enamel and gloss kilns. (CTD, 1958)

hard mica Mica that, when slightly flexed or distorted with thumb pressure, generally does not show any tendency to delaminate. Such mica, in thick pieces, will give an almost metallic sound when tapped or dropped on a hard surface. (Skow, 1962)

hard needles (or inclusions) A term applied in the grading of quartz crystals to fairly large needlelike inclusions or imperfections that appear to be hard. (Am. Mineral., 1947)

hardness (a) Quality of water based on the presence of dissolved calcium or magnesium. See also: *hard water.* Syn: *total hardness.* (b) As used by individuals associated with the drilling and bit-setting industry, the relative ability of a mineral to scratch another mineral or to be penetrated by a Knoop indenter. See also: *hardness test.* (Long, 1960) (c) Of a brittle mineral, the resistance to scratching or abrasion by another mineral, e.g., the Mohs scale of relative hardness, which ranges from 1 (talc) to 10 (diamond), or the Povarennykh scale from 1 (talc [001]) to 15 (bort diamond [111]). See also: *hardness scale.* (d) Resistance of a metal to plastic deformation by indentors of various shapes as defined by the Brinell, Knoop, Rockwell, and Vickers scales.

hardness gage *hardness points.*

hardness pencils *hardness points.*

hardness plates A series of small pieces of minerals of differing hardness, polished flat, and set side by side in cement, for testing hardness of another mineral, which is drawn across one piece after another, beginning with the hardest, until it scratches one.

hardness points Small pieces of minerals of differing hardness, with one end pointed and affixed to small handles of wood, metal, or plastic, to be held in hand and used for testing hardness of another mineral, by ascertaining which points will scratch it. Minerals of hardness 6 to 10 are usually used as points for testing gem stones. Syn: *hardness gage; hardness pencils.*

hardness scale (a) The scale by which the hardness of a mineral is determined as compared with a standard. The Mohs scale is as follows: talc, gypsum, calcite, fluorite, apatite, orthoclase, quartz, topaz, corundum, and diamond. Also called Mohs scale. See also: *hard; hardness; hardness test.* (Fay, 1920) (b) Quantitative units by means of which the relative hardness of minerals and metals can be determined, which for convenience is expressed in Mohs, Knoop, or scleroscope units for minerals and Vickers, Brinell, or Rockwell units for metals. (Long, 1960)

hardness table Any listing of substances as to their comparative hardness.

hardness test A determination of the relative hardness of a mineral, such as scratch hardness, as made on a specimen, using appropriate hardness-testing apparatus and techniques. See also: *hardness; hardness scale.* (Long, 1960)

hardness wheel A hand instrument in which hardness points are set as equidistant spokes of a rimless wheel, permitting a more rapid selection of points in testing hardness.

hardpan (a) A layer of gravel encountered in the digging of a gold placer, occurring 1 to 2 m below the ground surface and partly cemented with limonite. (AGI, 1987) (b) A popular term used loosely to designate any relatively hard layer that is difficult to excavate or drill; e.g., a thin resistant layer of limestone interbedded with easily drilled soft shales. (AGI, 1987) (c) *caliche.*

hard radiation Ionizing radiation of short wavelength and high penetration. (NCB, 1964)

hard rock (a) A term used loosely for igneous or metamorphic rock, as distinguished from sedimentary rock. (AGI, 1987) (b) A rock that is relatively resistant to erosion. (AGI, 1987) (c) Rock that requires drilling and blasting for its economical removal. Cf: *soft rock.* (AGI, 1987)

hard-rock drilling Drilling done in dense and solid igneous or highly silicified rocks, which can be penetrated economically only by diamond bits, as opposed to that done in softer rocks easily cut by roller or wing-type rotary bits. (Long, 1960)

hard-rock geology A colloquial term for geology of igneous and metamorphic rocks, as opposed to soft-rock geology. (AGI, 1987)

hard-rock mine A mine in hard rock; esp. one difficult to drill, blast, and square up. (Hess)

hard-rock miner A worker competent to mine in hard rock. Usually used to indicate an expert miner as compared with one fit only to mine in soft rocks. (Hess)

hard rock minerals Solid minerals, as distinguished from oil and gas, esp. those solid minerals found in hard rocks. (Williams, 1964)

hard-rock mining Mining that takes place in igneous and metamorphic rock by means of drilling and blasting to extract the ore. (SME, 1992)

hard-rock phosphate A term used in Florida for pebbles and boulders of a hard massive homogeneous light-gray phosphorite, showing irregular cavities that are usually lined with secondary mammillary incrustations of calcium phosphate. It is essentially equivalent to the term "white-bedded phosphate" that is used in Tennessee. (AGI, 1987)

hard-rock tunnel boring A technique utilizing a machine called a boring machine to bore large horizontal openings in rock or coal.

hards (a) A commercial term for the larger sizes of dull, hard coal, in contrast to brights. (BS, 1960) (b) *durain.* (c) In the United States, this term is used for anthracite. (Tomkeieff, 1954)

hard seat *seat earth.*

hard solder Any solder that melts only at a red heat; used in soldering silver, etc. (Fay, 1920)

hard sorting *handpicking.*

hard spar A name applied to both corundum and andalusite. (Fay, 1920)

hard vector Due to the arrangement of the molecules within some mineral crystals, such as diamond, the substance is found to be harder in certain planes or directions in relation to the axes of the mineral crystals. These hard planes are referred to as hard vectors. Cf: *soft vector. vector.* (Long, 1960)

hard water Water that does not lather readily when used with soap, and that forms a scale in containers in which it has been allowed to evaporate; water with more than 60 mg/L of hardness-forming constituents, expressed as $CaCO_3$ equivalent. See also: *hardness; total hardness of water; temporary hardness of water.* Cf: *soft water.* (AGI, 1987)

hard way (a) A term used in slate quarrying to describe the third direction at right angles to both slaty cleavage and rift, in which there is no tendency to split. It is known as the hard way and designated locally as the sculp. (b) In granite quarrying, the direction at right angles to both rift and run is called the hard way or head grain. See also: *cutoff; tough way.* Sometimes spelled hardway. (AIME, 1960)

hard white ore Georgia bauxite containing less than 1% ferric oxide. (Fay, 1920)

Hardwick conveyor loader head A dust collector for belt conveyors used at the loading station. The delivery pulley of the main gate conveyor is used to drive a scraper chain. The latter is arranged to run at half the belt speed by means of chains and sprockets, and the scraper chain runs at the bottom of a long hopper to the point where the coal is delivered into the trams. The underbelt fines are collected on the scraper chain after having been released from the belt by means of a snub pulley. The whole arrangement is housed in a sheet-steel cover to which rubber flaps are attached. Side spillage and the escape of dust over the side of the trams is prevented by means of rubber flaps. (Mason, 1951)

hardypick drifting machine A heavy electric rotary drilling machine for blasting work in mine tunneling. It consists of a chassis mounted on continuous tracks, turntable, boom, drilling machines, and various controls, and can be operated by two people. (Nelson, 1965)

Harman process A method for producing direct from ore an iron in the form of either sinter or pig that is suitable for charging in steel furnaces. Ore, limestone, and carbon in the form of coal, coke, or oil coke in the proportions of 40:8:5 are dried, crushed to about 1/16 in (1.6 mm), intimately mixed, and fed into the upper end of a sloping rotary kiln. (Osborne, 1956)

harmful dust Generally, any airborne particulate matter that is fibrogenic (harmful to the respiratory system), carcinogenic (capable of causing cancer), radioactive, or toxic. (Hartman, 1982)

harmless depth theory A hypothesis based largely on the dome theory, which states that there is a certain harmless depth below which mining could be carried on without risk of damage to the surface. Subsidence observations at present working depths do not support this theory. (Nelson, 1965)

harmonic A sinusoidal quantity having a frequency that is an integral multiple of the frequency of a periodic quantity to which it is related. (Hy, 1965)

harmonic crystal A crystal designed to oscillate at an integral multiple of its fundamental frequency. (Am. Mineral., 1947)

harmonic folding Folding in which the strata remain parallel or concentric, without structural discordances between them, and in which there are no sudden changes in the form of the folds at depth. Ant: disharmonic folding. (AGI, 1987)

harmotome A monoclinic mineral, $(Ba,K)_{1-2}(Si,Al)_8O_{16}\cdot 6H_2O$; zeolite group; in hydrothermal veins, vugs in igneous rocks, manganese ores, saline lakes. Syn: *cross-stone.*

harpolith (a) A large, sickle-shaped igneous intrusion that was injected into previously deformed strata and was subsequently deformed with the host rock by horizontal stretching or orogenic forces. (AGI, 1987) (b) Essentially a phacolith with a vertical axis. (AGI, 1987)

harrie Scot. To rob; to take all the coal that can conveniently be mined without attempting to systematically remove the whole. A variation of harry; to strip; despoil; to rob. (Fay, 1920)

harriers Trammers, putters, or drawers employed to convey trucks or tubs from the working face. They may help load the trucks. (CTD, 1958)

harrisite (a) Chalcocite pseudomorphous after galena. (Holmes, 1928) (b) A phanerocrystalline rock composed essentially of black, lustrous, cleavable olivine with anorthite and a little augite. (Holmes, 1928)

Harris process Process for removing arsenic, antimony, tin, and zinc from virgin or secondary lead by agitating the molten metal with molten caustic soda and salt. All undesirable metals are oxidized, and the oxides are dissolved in the caustic, with the exception of silver, which is removed in a subsequent desilvering operation. (CCD, 1961)

harrock Hard chalk. (Arkell, 1953)

harrow A pole with teeth in it, which revolves in a puddling trough to puddle auriferous clay. (Gordon, 1906)

hartine *hartite.*

hartite Hydrocarbon occurring in brown coal as transparent masses or small, waxy triclinic crystals. Syn: *hartine; josen.* (Tomkeieff, 1954)

Hartley gravimeter An early form (1932) of stable gravimeter consisting of a weight suspended from a spiral spring, a hinged lever, and a compensating spring for restoring the system to a null position. (AGI, 1987)

hartleyite A variety of oil shale from Hartley, New South Wales. (Tomkeieff, 1954)

hartsalz Hard salt, a mixture of sylvinite and kieserite, with some anhydrite, found in the Stassfürt salt deposits. (Kaufmann, 1960)

hartschiefer A strongly banded and partly schistose rock; associated with other rocks of mylonitic habit, in which the alternating bands have been produced from ultramylonite by recrystallization and metamorphic differentiation. (Holmes, 1928)

Harvey process Method of carburizing the surface layers of low-carbon steel, followed by rapid chilling. (Pryor, 1963)

harzburgite A variety of peridotite that consists essentially of olivine and enstatite or bronzite.

Harz jig A jig in which pulsion is given intermittently with suction; the periods devoted to them are about equal. See also: *hodge jig; jig.* (Liddell, 1918)

Hasenclever turntable This type of turntable is used as an alternative to the shunt-back or the traverser for changing the direction of mine cars or tubs, either on the surface or underground. A pulley driven by a creeper chain bringing along the cars is on the same vertical axis as the turntable and is so disposed that when a car is on the table two of its wheels rest on its central pulley while the other two rest on the outer edge of the turntable. (Sinclair, 1959)

hassing *hasson.*

hasson Scot. A vertical gutter between water rings in a shaft. Syn: *hassing.* See also: *gauton.*

Hastings beds A series of clay and sand deposits in the Lower Cretaceous of southeast England; the Fairlight Clays at the base of these deposits have been used for brickmaking near Hastings and Bexhill. (Dodd, 1964)

hatch conveyor Any of several types of conveyors adapted to loading or unloading bulk materials, packages, or objects to or from ships or barges. See also: *belt conveyor* Cf: *portable conveyor.*

hatchet stake A small anvil on which to bend sheet metal. (Standard, 1964)

hatchettine A yellowish-white, wax-yellow, or greenish-yellow paraffin wax usually found inside ironstone nodules and geodes in coalbeds or limestone. It melts at 55 to 65 °C, is sparingly soluble in boiling alcohol and cold ether, and is decomposed by concentrated sulfuric acid. Syn: *mineral tallow.* Cf: *ozocerite.*

hatchettite A naturally occurring soft paraffin wax; forms veinlike masses in ironstone nodules associated with coal-bearing strata, South Wales; in limestone cavities, France. Syn: *adipocerite; adipocire; mineral adipocire; mineral tallow.* See also: *grave wax.* Cf: *ozocerite.*

hatchettolite A former name for uran-pyrochlore.

hatching (a) Brist. An underground way or self-acting inclined plane, in a thin seam of coal, extending from 60 to 80 yd (55 to 73 m) to the rise. (Fay, 1920) (b) *hachure.*

hatchure *hachure.*

Hatfield process Dielectric separation process. (Pryor, 1963)

hat rollers Hat-shaped metal guides for ropeways around bends. (Pryor, 1963)

hauchecornite The mineral group arsenohauchecornite, bismutohauchecornite, hauchecornite, tellurohauchecornite, and tucekite.

haul (a) The distance from the coal face to pit bottom or surface; the distance quarry or opencast products must be moved to the treatment plant or construction site; the distance from the shaft or opencast pit to spoil dump. (Nelson, 1965) (b) Average haul; the average distance a grading material is moved from cut to fill. (Nichols, 1976) (c) In the construction of an embankment by depositing material from a cutting, the haul is the sum of the products of each load by its haul distance. (CTD, 1958)

haulabout A steel barge with large hatchways and coal transporters used for coaling ships. (Webster 3rd, 1966)

haulage (a) The drawing or conveying, in cars or otherwise, or movement of workers, supplies, ore, and waste both underground and on the surface. (Lewis, 1964) (b) In dividing the transportation system according to the area served there is (1) primary or face haulage, (2) secondary haulage, and (3) main-line haulage. See also: *transport; intermediate haulage; relay haulage; locomotive haulage; underground haulage.* (Kentucky, 1952) (c) Applied generally to track mining as opposed to conveyor mining, although belt conveyor systems are sometimes referred to as belt haulage. (BCI, 1947) (d) The system of hauling coal out of a mine. (Korson, 1938) (e) S. Afr. A drive used for mechanical transport. (Beerman, 1963)

haulage boss In bituminous coal mining, a foreperson who supervises mine haulage operations underground or at the surface. (DOT, 1949)

haulage brake *brake.*

haulage cars Rail haulage cars for surface or mine shaft operations that are used to carry ore and equipment to and from the digging site. (Best, 1966)

haulage chain In the early days chains were used in haulage in and around mines. Wire rope has displaced them. (Korson, 1938)

haulage clip A device to effect a secure attachment of tub to the haulage top, chiefly with endless rope haulage. The usual type attains the grip on the rope by two jaws that may be tightened by either a screw or a lever movement. The connection of the clip to the tub is usually made to the drawbar by a hook or link. The clip must maintain a sure grip on the rope and be capable of easy and quick manipulation. See also: *automatic clip; clip.* (Nelson, 1965)

haulage conveyor Generally 500 to 3,000 ft (152 to 915 m) in length. It is used to transport material between the gathering conveyor and the outside. Haulage conveyors are commonly classified as either intermediate or main haulage conveyors. See also: *underground mine conveyor.* (NEMA, 1961)

haulage curve A bend in a haulage road, which may be horizontal, vertical, or both. On main haulage roads, curves may be 100 ft (30.5 m) in radius or more. With a good rail track and smooth curves, haulage speeds up to 20 mph (32.2 km/h) can be maintained. A useful rule for determining the minimum radius of curvature for tramming is R = 12 to 15 x W; for locomotive haulage R = 20 to 25 x W, where W is the maximum wheel base of the rolling stock or locomotive. (Nelson, 1965)

haulage drum A large cylinder on to which the steel haulage rope is coiled. The rope is attached to the drum by passing it inside and looping it about the drum shaft, and securing the loose end to the rope by rope clamps. The excess length of rope is coiled around the drum to provide for recapping during its useful life. (Nelson, 1965)

haulage hand A worker fully employed on the haulage system in a mine. (Nelson, 1965)

haulage mine locomotive *electric haulage mine locomotive.*

haulage plant A mechanical installation for the tramming of rock, ore, or coal; operated by ropes, compressed air, or electricity. (Weed, 1922)

haulage rope (a) A rope used for haulage purposes. (Zern, 1928) (b) A wire rope composed of six strands of seven wires each. (Lewis, 1964)

haulage stage A mine roadway along which a load is moved by a single form of haulage without coupling or uncoupling of cars and without transfer from one form of haulage to another. (Nelson, 1965)

haulageway The gangway, entry, or tunnel through which loaded or empty mine cars are hauled by animal or mechanical power. (Fay, 1920)

haulaway An excavation method that involves hauling the spoil away from the hole. (Nichols, 1976)

haulback mining Method of surface mining in which the overburden is hauled from over the ore or coal in trucks to a holding area and hauled back after the ore or coal has been removed. (SME, 1992)

haul-cycle time The time it takes the scraper or truck to haul a load to the dumping area and return to position in the loading area.

haul distance The distance measured along the center line or most direct practical route between the center of the mass of excavation and the center of mass of the fill as finally placed. It is the distance material is moved. (Nichols, 1976)

hauling The drawing or conveying of the product of the mine from the working places to the bottom of the hoisting shaft, or slope. (Zern, 1928)

hauling engine An engine employed to move tubs on an underground engine plane. (Peel, 1921)

haul road A road built to carry heavily loaded trucks at a good speed. The grade is limited on this type of road and usually kept to less than 17% of climb in direction of load movement.

haunt Coal sold at the pithead. (Nelson, 1965)

hausmannite (a) A mineral, $Mn^{2+}Mn_2^{3+}O_4$. In tetragonal octahedrons and twins; also granular massive, particles strongly coherent. Luster submetallic. Color, brownish-black. Syn: *black manganese.* (Fay, 1920) (b) A tetragonal mineral, $Mn^{2+}Mn^{3+}_2O_4$; pseudocubic; in submetallic, commonly twinned crystals in high-temperature hydrothermal veins, contact metamorphic deposits associated with felsic igneous rocks; also an alteration product of manganese ores by meteoric waters; associated with psilomelane in the Batesville district, Arkansas.

hauyne An isometric mineral, $[(Na,Ca)_{4-8}Al_6Si_6(O,S)_{24}(SO_4,Cl)_{1-2D2}]$; sodalite group; crystallizes in dodecahedra and octahedra with dodecahedral cleavage; an accessory mineral in alkaline igneous rocks, esp. extrusives, commonly associated with nepheline or leucite. Also spelled haüyne, hauynite.

hauynite A blue feldspathoid, crystallizing in the cubic system, consisting essentially of silicate of aluminum and sodium with sodium sulfate, $(Na,Ca)_{4-8}Al_6Si_6(O,S)_{24}(SO_4,Cl)_{1-2}$. Also spelled haüynite. Also called hauyne. Syn: *sapphirine.* (Dana, 1959)

haüynitite A plutonic or hypabyssal rock composed chiefly of hauyne and pyroxene, usually titanaugite. Small amounts of feldspathoids and sometimes plagioclase and/or olivine are present. Apatite, sphene, and opaque oxides occur as accessories. See also: *haüynophyre.* (AGI, 1987)

haüynophyre An extrusive rock similar in composition to a leucitophyre but containing haüyne in place of some of the leucite. Other possible phases include nepheline, augite, magnetite, apatite, melilite, and mica. A partial syn. of haüynitite; some rocks are called haüynophyre when haüyne is a conspicuous mineral but not necessarily a major constituent. See also: *haüynitite.* (AGI, 1987)

Hauy's law Every crystal of precise chemical structure and purity has a specific and characteristic shape. (Pryor, 1963)

Hauzeur furnace A double furnace for the distillation of zinc wherein waste heat from one set of retorts is utilized for heating the second set. (Fay, 1920)

Hawaiian peridot A gem-quality variety of olivine (forsterite) in Hawaii; forms phenocrysts in basalt; also in derived sands.

hawaiite (a) A gem variety of olivine from the lavas of the Hawaiian Islands. It contains little iron and is pale green. (b) As defined by Iddings in 1913, an olivine basalt with andesine as the normative plagioclase (thus differing from true basalt, in which the normative plagioclase is more calcic). It generally, but not always, lacks normative quartz, and commonly contains normative and modal olivine. Hawaiite is intermediate in composition between alkali olivine basalt and mugearite, and grades into both. See also: *mugearite; trachybasalt.* (AGI, 1987)

hawk's-eye A transparent colorless quartz containing closely packed, parallel fibers of crocidolite that impart to it a blue color. In form and sheen, it resembles tigereye to which it alters geologically. Differs from sapphire quartz, in which fibers are not parallel. Also spelled hawkeye. Cf: *tiger's-eye.*

hawleyite An isometric mineral, CdS; sphalerite group; dimorphous with greenockite; occurs as yellow powder on sphalerite.

hawser A large rope, varying from 5 to 24 in (12.7 to 61.0 cm) in circumference, of 6 to 9 strands and left-handed twist. (Standard, 1964)

hawser laid (a) Of fiber rope, one with three strands of yarn twisted left-handed, these strands being laid up right-handed. (Pryor, 1963) (b) If wire rope, it is called cable laid. (Pryor, 1963)

Hayden process A series method of electrolytic refining. Unrefined copper anodes are suspended in an acid electrolyte; one side of each then acts as an anode and the other as a cathode. (Pryor, 1963)

Hay mist projector This projector can be made at any colliery workshop from a few short pieces of piping and an old oil drum. It may be fixed in the drum or in an open tank of larger capacity placed 12 to 15 yd (11.0 to 13.7 m) back from the face of the hard heading. To the water in the drum, powdered washing soda is added at the ratio of 4 oz (113.4 g) to 5 gal (18.9 L) of water. About 2 min before firing, compressed air is turned on and ejections of water in the form of a coarse mist fill the heading. This continues for a period of 6 min after the firing. In this manner, the heading is filled with a mist of droplets that outnumber the dust particles; the latter are effectively wetted or become attached to the droplets, with the result that the dust rapidly settles out of the air. This mist projector has a high efficiency, particularly where the ventilation current is low. (Mason, 1951)

hazel N. of Eng. In coal mining, a tough mixture of sandstone and shale; also, freestone, flagstone, or chert.

Hazelett process A method for casting liquid metal or steel continuously into rolls for sheet or plate. The steel is poured on to the outer surface of a broad steel cylinder of very large diameter (up to 6 m) that is supported and revolved by a roller turning inside it. The molten steel is carried a short distance to a roller revolving above the ring, which rolls the almost solidified steel into a thin plate or strip. (Osborne, 1956)

head (a) Any road, level, or other passage driven in coal, etc., for the purpose of proving and working the mine. (b) The top portion of a seam in the coal face. (CTD, 1958) (c) The whole falling unit in a stamp battery, or merely the weight at the end of the stem. (CTD, 1958) (d) The top end of the boring rods above the surface. (e) Core-barrel head. (f) In gravity separation of a feed, the heads are the concentrates. Opposite of tail. (Pryor, 1963) (g) Variously used as a syn. for core-barrel head; drill head; swivel head. (Long, 1960) (h) The attitude or direction in a massive crystalline rock along which fracture is most difficult. It is normal to the grain and rift. (i) The difference in air pressure producing ventilation. (CTD, 1958) (j) In mineral processing, the mill head, or grade of ore, accepted by the mill for treatment. Commonly used in the plural. (k) The height of water above any point or plane of reference. Used also in various compounds, such as energy head, entrance head, friction head, static head, pressure head, lost head, etc. (Seelye, 1951) (l) A unit of pressure intensity usually given in inches or feet (millimeters or meters) of a column of the fluid under consideration. Thus, 1 ft (or m) of water head is the pressure from a column of water 1 ft (or m) high. (Strock, 1948) (m) The upper bend of a fold or structural terrace. Cf: *foot*. Syn: *upper break*. (AGI, 1987) (n) An advance main roadway driven in solid coal. (CTD, 1958) (o) Development openings in a coal seam. (Pryor, 1963) (p) Total head (th). Also called head on pump. (q) So. Staff. A shift or day's work by the stint in heading-out, or driving of dead work. (Fay, 1920) (r) To cut or otherwise form a narrow passage or head. (Fay, 1920) (s) A lift. (Fay, 1920) (t) As applied to rock, natural planes of cleavage at right angles to the grain and the rift of the rock. (Stauffer, 1906) (u) Rubble drift on the cliffs of southern England. (Standard, 1964)

headache post A timber set under the walking beam to prevent it from falling on members of the drilling crew when it is disconnected. (Williams, 1964)

headblock (a) A stop at the head of a slope or shaft to stop cars from going down the shaft or slope. (Fay, 1920) (b) A cap piece. (Fay, 1920) (c) A heavy obstruction placed end-on across the rail to prevent the passage of a runaway mine car. (Hudson, 1932) (d) The crosstie that supports the toes of the switch. (Zern, 1928) (e) Commonly used as a syn. for crown block; sheave wheel. (Long, 1960)

headboard (a) A wedge of wood placed against the hanging wall, and against which one end of the stull is jammed. (Zern, 1928) (b) A horizontal board in the roof of a heading, touching the earth above and supported by a headtree at each side. (Hammond, 1965)

headbox A device for distributing a suspension of solids in water to a machine at a constant rate, or for retarding the rate of flow, as to a top-feed filter or for eliminating by overflow some of the finest particles. (BS, 1962)

head coal Scot. Formerly, the stratum of a coal next to the roof. More usually now, the top portion of a coal seam when left unworked, either permanently or afterwards taken down; the top coal on a loaded wagon. (Fay, 1920)

head end (a) Usually the ultimate delivery end of a conveyor. (b) That part of a mining belt conveyor that includes the head section, a power unit, and, if required, the connecting section and a belt takeup. (NEMA, 1961)

header (a) Pieces of plank—longer than a cap—extending over more of the roof and supported by two props, one at each end. (Ricketts, 1943) (b) An entry-boring machine, called a road header, which bores the entire section of the entry in one operation. (Fay, 1920) (c) The person in charge of driving a heading. (CTD, 1958) (d) A rock that heads off or delays progress. (Fay, 1920) (e) A blasthole at or above the head. (Fay, 1920) (f) A masonry unit laid flat with its greatest dimension perpendicular to the face of the wall; generally used to tie two wythes of masonry together. (ACSG, 1961) (g) A large pipe into which one set of boilers is connected by suitable nozzles or tees, or similar large pipes from which a number of smaller ones lead to consuming points. Headers are essentially branch pipes with many outlets, which are usually parallel. (Strock, 1948)

head flat trimmer In bituminous coal mining, a foreman who is in charge of workers picking impurities from coal as it is dumped into railroad cars at the mine surface. (DOT, 1949)

headframe (a) The steel or timber frame at the top of a shaft, which carries the sheave or pulley for the hoisting rope and serves various other purposes. Also called gallows frame; hoist frame; head stocks. (CTD, 1958; Hess) (b) The shaft frame, sheaves, hoisting arrangements, dumping gear, and connected works at the top of a shaft or pit. Also called headgear. (Pryor, 1963) (c) Includes all the raised structure around the shaft that is used for loading and unloading cages. (Mason, 1951) (d) Can. Gallows over the shaft to which cable for hoisting is attached. (Hoffman, 1958)

headgear (a) *headframe*. (b) That portion of the winding machinery attached to the headframe, or the headframe and its auxiliary machinery. (Fay, 1920)

head grain *hard way*.

headhouse (a) The house or building that encloses the headframe. (Fay, 1920) (b) The structure on a hillside to control the lowering of coal to the tipple. (BCI, 1947)

heading (a) A passage leading from the gangway, commonly at right angles in a coal mine. (Korson, 1938) (b) A smaller excavation driven in advance of the full-size section; it may also be driven laterally, as a cross heading or side drift. A heading may be driven at the top or the bottom of the full-size face. (Stauffer, 1906) (c) The vein above a drift. (Fay, 1920) (d) An interior level or airway driven in a mine. (Fay, 1920) (e) A road in the solid strata but also in the seam; a road in solid coal. (Mason, 1951) (f) In a tunnel, a digging face and its work area; the end of a drift or gallery that is being advanced by the mining operation. (Nichols, 1976) (g) A gangway, entry, or airway. (Hudson, 1932) (h) The gravel bank above a sluice in a placer. (Standard, 1964) (i) A continuous passage between two rooms, breasts, or other working places. (Fay, 1920) (j) A collection of close joints. (Fay, 1920) (k) Sometimes applied to the preliminary drift or pioneer bench in tunnel driving. (Fay, 1920)

heading-and-bench A method of tunneling in hard rock. The heading is in the upper part of the section and is driven only a round or two in advance of the lower part or bench. (Fraenkel, 1953)

heading-and-bench mining A stoping method used in thicker ore where it is customary first to take out a slice or heading 7 to 8 ft (2.1 to 2.4 m) high directly under the top of the ore and then to bench or stope down the ore between the bottom of the heading and the bottom of the ore or floor of the level. The heading is kept a short distance in advance of the bench or stope. Syn: *heading-and-stope mining*.

heading and stall *room-and-pillar*.

heading-and-stope mining *heading-and-bench mining*.

heading blast A method of quarry blasting in which the explosive is confined in small tunnel chambers inside the quarry face. The charges are placed at suitable intervals according to the burden to be blasted. In large blasts, several tunnels and cross tunnels may be employed. See also: *chamber*. Syn: *tunnel blast*. (Nelson, 1965)

heading-overhand bench The heading is the lower part of the section and is driven at least a round or two in advance of the upper part, which is taken out by overhand excavating. Syn: *inverted heading and bench*. (Fraenkel, 1953)

headings (a) Coarse gravel or drift overlying placer deposits. (Fay, 1920) (b) The portion of a vein that is above a level. (c) Highly jointed parts of granite. (Arkell, 1953) (d) In ore beneficiation, the purest ore obtained by washing, as opposed to middlings and tailings. Also called concentrates. See also: *heads*.

heading seam *joint*.

heading side *footwall*.

heading wall *footwall*.

headline In dredging, the line that holds the dredge up to its digging front. This line is anchored well ahead of the dredge pond or paddock and attached to a winch on the dredge. Lateral movement is controlled by sidelines similarly led from winches to land anchorages, usually two on each side. See also: *sideline*. (Pryor, 1963; Fay, 1920)

head motion Vibrator of shaking table that imparts reciprocating motion to the deck. (Pryor, 1963)

head-of-hollow fill A fill structure consisting of any materials, other than a coal processing waste or organic material, placed in the uppermost reaches of a hollow where side slopes of the existing hollow measured at the steepest point are greater than 20°, or the average slope of the profile of the hollow from the toe of the fill to the top of the fill is greater than 10°. In fills with less than 250,000 yd^3 (191,000 m^3) of material, associated with steep slope mining, the top surface of the fill will be at the elevation of the coal seam. In all other head-of-hollow fills, the top surface is the fill, which when completed, is at approx. the same elevation as the adjacent ridge line, and no significant area of natural drainage occurs above the fill draining into the fill areas.

head piles The top poling boards in a heading. (Stauffer, 1906)

head pulley (a) The discharge pulley of the conveyor. It may be either an idler pulley or a drive pulley. A head pulley that is mounted on a boom is termed an extended head pulley; a head pulley that is separately mounted is termed a detached head pulley. (NEMA, 1961) (b) The crowned pulley or idler mounted at the extreme front end or delivery point of a belt conveyor. The belt, after passing around this pulley, begins its travel toward the tail end or foot section of the conveyor. (Jones, 1949)

head-pulley-drive conveyor A conveyor in which the belt is driven by the head pulley without a snub pulley. (NEMA, 1961)

head-pulley-snub-drive conveyor A conveyor in which the belt is driven by the head pulley with a snub pulley. (NEMA, 1961)

headrace Sluice aqueduct, leat, or launder that leads water to head of operation or to waterwheel. A forebay. (Pryor, 1963)

headroom Height between the floor and the roof in a mine opening. (Long, 1960)

headrope In any system of rope haulage, the rope that is used to pull the loaded transportation device toward the discharge point. In scraper loader work, the headrope pulls the loaded scoop from the face to the dumping point. (Jones, 1949)

heads (a) In ore dressing, the feed to a concentrating system. Cf: *tails.* (b) Low-grade material overlying an alluvial placer. (AGI, 1987) (c) In New York and Pennsylvania, a local term applied by bluestone quarrymen to the open joints that run north and south. (d) *headings.* (e) Scot. Large top coal on a loaded hutch. (f) Aust. Small faults. (g) Low-grade wash overlying the wash proper. (Nelson, 1965) (h) Can. Material taken from ore in treatment plant and containing the valuable metallic constituents. Opposite of tails. (Hoffman, 1958)

head section A term used in both belt and chain conveyor work to designate that portion of the conveyor used for discharging the coal. (Jones, 1949)

head shaft The shaft mounted at the delivery end of a chain conveyor, on which is mounted a sprocket that drives the drag chain. The shaft, in turn, is driven by means of a drive chain from the speed reducer of the power unit through a sprocket mounted on the shaft end. (Jones, 1949)

head sheave Pulley in headgear of the winding shaft over which the hoisting rope runs. See also: *winding sheave.* (Pryor, 1963)

head side N. Staff. The rise side of a heading driven on the strike.

headsticks *headgear; headframe.*

headstocks *headframe.*

headsword Corn. Water discharged through the adit level.

head tank Any tank or vessel in the water circuit that is used to control the delivery pressure of the water to the washing units. (BS, 1962)

headtree (a) The horizontal timber at each side of a rectangular heading that supports the headboard. See also: *side trees.* (Hammond, 1965) (b) The cap piece of a heading set. (Stauffer, 1906)

head value The assay value of the heads or mill feed. (Nelson, 1965)

headwall (a) Retaining wall at both ends of a culvert or similar structure. (Hammond, 1965) (b) A culvert sidewall; sometimes only the upstream wall. (Nichols, 1976)

headway The second set of excavations in post-and-stall work. See also: *crossheading.* (Fay, 1920)

headwork (a) Arkansas. The cutting and other work done at the face of an entry. (Fay, 1920) (b) The headframe with the headgear. (Webster 2nd, 1960)

heap (a) Scot. To load up a tub (car) or truck above the top of the sides. (Fay, 1920) (b) The soil carried above the sides of a body or bucket. (Nichols, 1976) (c) Newc. The refuse at the pit's mouth.

heap closure *heap decommissioning.*

heap decommissioning Legal closure of a heap leaching operation. Depends on individual State regulations, but includes requirements for physical stability and chemical effluent requirements for metals and pH. A monitoring period is included. Syn: *heap closure.*

heaped capacity In scraper or truck loading, a term used to describe the volume of material the scraper will hold when the material is heaped. Frequently, sideboards are added to increase the heaped capacity. Heaped capacity will exceed struck capacity by approx. one-third, depending upon the heaping condition assumed. Cf: *struck capacity.* (Carson, 1961)

heap leaching A process used for the recovery of copper, uranium, and precious metals from weathered low-grade ore. The crushed material is laid on a slightly sloping, impervious pad and uniformly leached by the percolation of the leach liquor trickling through the beds by gravity to ponds. The metals are recovered by conventional methods from the solution.

heap matte Matte produced by heap roasting. (Fay, 1920)

heap rinsing Method used to remove soluble constituents remaining within a heap leach pile after the metals concentration decreases to levels below economic limits. Simple water rinsing, chemical, or biological techniques or combinations thereof may be employed. (Van Zyl, 1988)

heap roasting Removal of sulfur from pyritic ore by burning in heaps, perhaps with aid of fuel. (Pryor, 1963)

heap sampling Method of reducing a large sample of ore to yield a representative sample. A conical heap is made by shoveling the material accurately on to the apex so that it runs down equally all around. The heap can then be flattened somewhat by rubbing with a spade, and is shoveled into four equal heaps, the same amount being taken from the base of the cone each time the worker goes around. Of the four smaller heaps thus formed, two are discarded and two retained. These may now be crushed to improve the ease of thorough mixing, and are then formed into another cone in the same way as the first. The process is repeated, with periodic size reduction of the retained portions, until the required small sample has been produced. (Pryor, 1963)

heapstead The buildings and surface works around a colliery shaft. (CTD, 1958)

hearth (a) The bottom portion of certain furnaces, such as a blast furnace, air furnace, and reverberatory furnace, in which molten metal is collected or held. (ASM, 1961) (b) The part of a furnace in which heat is developed for the purpose of melting glass. (CTD, 1958) (c) A plate or table upon which cylinder glass is flattened. (Standard, 1964)

hearth and bosh brick Fireclay brick for use in lining the hearth walls and bosh sections of a blast furnace. (ARI, 1949)

hearth furnaces Furnaces in which the charge rests on the hearth or kiln wall and is heated by hot gases passing over it. Even though hearth furnaces such as the multiple-hearth roasting furnace and rotary kilns operate on the same basis as reverberatory furnaces, they bear little resemblance to them. (Newton, 1959)

hearthplate A cast-iron plate serving as a sole for a refiner's furnace. (Standard, 1964)

hearth roasting A roasting process in which the ore or concentrate enters at the top of a multiple hearth roaster and drops from hearth to hearth in succession until it is discharged at the bottom. In the downward progress of the ore, the sulfide particles are roasted as they come in contact with the heated air. (Newton, 1959)

heart joint Scot. A particular form of attachment joint between the bucket rod and the foot rod of a pump.

heat (a) One operation in a furnace not operating continuously. (Fay, 1920; Newton, 1959) (b) The energy a body possesses because of the motion of its molecules. (Jones, 1949) (c) Energy in transit from a higher temperature system to a lower temperature system. The process ends in thermal equilibrium. (AGI, 1987) (d) The material heated, melted, etc., at one time; as, the foundry runs three heats a day. (Standard, 1964) (e) Form of energy generated or transferred by combustion, chemical reaction, mechanical means, or passage of electricity, and measurable by its thermal effects. (Pryor, 1963)

heat balance (a) In furnaces, heat engines, etc., the distribution of the known input of energy (as heat); also, the method of determining, or the graphical or tabular record of, such distribution. (b) In fluidization roasting, the thermodynamic calculation used to control addition or removal of heat in order to maintain the desired temperature in the reacting vessel. (Pryor, 1963) (c) Equilibrium that exists on the average between the radiation received by the Earth and its atmosphere from the Sun and that emitted by the Earth and its atmosphere. That the equilibrium does exist in the mean is demonstrated by the observed long-term constancy of the Earth's surface temperature. On the average, regions of the Earth nearer the equator than about 35° latitude receive more energy from the Sun than they are able to radiate, whereas latitudes higher than 35° receive less. The excess of heat is carried from low latitudes to higher latitudes by atmospheric and oceanic circulations and is reradiated. (AGI, 1987)

heat capacity That quantity of heat required to increase the temperature of a system by 1° at constant pressure and volume. It is usually expressed in calories per degree Celsius. Syn: *thermal capacity.* (AGI, 1987)

heat conductivity *thermal conductivity.*

heated stone A stone that has been artificially heated to the proper temperature with the intention of improving or completely altering its color. The induced color is permanent in varieties, such as hyacinth, burnt amethyst, etc.; less permanent in blue zircon. See also: *fired stone.* Syn: *heat-treated stone; burnt stone.* Cf: *stained stone.*

heat energy Energy in the form of heat. (Standard, 1964)

heat engine A mechanism (as an external-combustion or an internal-combustion engine) for converting heat energy into mechanical energy. (Webster 3rd, 1966)

heater In the coke products industry, one who regulates the temperature of heating flues and combustion of fuel gas used to heat coal in a byproduct coke oven. (DOT, 1949)

heater drain pump Self-regulating pumps capable of dealing with water at fairly high temperatures and pressures. They are used to return heater condensate to the feed line instead of to waste. (Sinclair, 1958)

heat exchanger Any device that transfers heat from one medium to another or to the environment. (Lyman, 1964)

heating back A chamber back of a forge in which the air intended for the blast is heated. (Standard, 1964)

heating medium A fluid used for conveying heat from a heat source to heat-dissipating devices; includes air, water, and steam. (Strock, 1948)

heatings The heat generated before an actual mine fire occurs. Heatings, or incipient fires, are detected in mines by smell and by analysis of air samples. In mines liable to spontaneous combustion, trained officials and workers are employed to detect and deal with heatings and fires and they become expert in these duties. (Sinclair, 1958)

heating surface (a) That surface in a steam boiler or similar apparatus from which heat passes to the liquid to be evaporated or heated; the fire surface. (Standard, 1964) (b) Broadly, the area intended for transferring heat. (Strock, 1948)

heating tendency The ability of a coal to fire spontaneously. This phenomenon can occur whenever the heat generated from oxidation reactions in a coal exceeds the heat dissipated. This characteristic varies for different types of coals and even for coals of the same classification but of different origin. (Smith, 1963)

heat of combustion The heat of reaction resulting from the complete burning of a substance and expressed variously (as in calories per gram or per mole, or esp. for fuels in British thermal units per pound or per cubic foot). (Webster 3rd, 1966)

heat of compression As air passes down shafts and along inclined workings, it is compressed. Heat is always generated when air is compressed, and although the reverse process of decompression and cooling takes place as the air ascends the upcast shaft, the net effect is to raise the air temperature underground. (Mason, 1951)

heat of crystallization Heat evolved when unit weight of a salt crystallizes from a large amount of a saturated solution. (Osborne, 1956)

heat of hydration The quantity of heat liberated or consumed when a substance takes up water. (Osborne, 1956)

heat of ionization The quantity of heat that is absorbed when 1 g equivalent of a substance is broken up completely into positive and negative ions. (Osborne, 1956)

heat of mixture That quantity of heat evolved when two liquids that do not react together are mixed. It is calculated from the temperature change and the specific heat of the mixture, and expressed in gram-calories per gram of mixture. (Osborne, 1956)

heat of reaction The quantity of heat consumed or liberated in a chemical reaction as heat of combustion, heat of neutralization, or heat of formation. For example, the number of calories of heat absorbed when 1 g at wt of carbon reacts with 1 g mol wt of oxygen to form 1 g mol wt of carbon dioxide. (Hackh, 1944; Newton, 1959)

heat of transformation The quantity of heat accompanying a constitutional change in a solid chemical compound or metal, e.g., the change from gamma to alpha iron. The temperature at which one crystalline form of a substance is converted into another solid modification is known as the transition point or transition temperature. (Osborne, 1956)

heat of wetting Heat evolved or absorbed when a liquid and a solid surface are placed in contact. (Osborne, 1956)

heat pump A mechanical refrigerating system used for air cooling in the warm season of the year, which, when the evaporator and condenser effects are reversed, absorbs heat from the outside air or water in the cold season of the year and raises it to a higher potential so that it also can be used for heating. (Strock, 1948)

heat recuperation The recovery of heat from waste gases. (Fay, 1920)

heat sensitivity A test that determines the flammability of explosives brought into contact with flame or heat.

heat transmission coefficient *coefficient of heat transmission.*

heat-treated stone A (gem) stone that has been artificially heated to change its color. Syn: *heated stone; burnt stone.*

heat unit A unit of quantity of heat; the heat required to raise the unit mass of water through 1 degree of temperature. Cf: *calorie; British thermal unit.* (Standard, 1964; Fay, 1920)

heat value The amount of heat obtainable from a fuel and expressed, for example, in British thermal units per pound. (Shell)

heave (a) A rising of the floor of a mine caused by its being too soft to resist the weight on the pillars. See also: *creep.* (b) Upward movement of soil caused by expansion or displacement resulting from such phenomena as moisture absorption, removal of overburden, driving of piles, and frost action. (c) Horizontal displacement of strata or other rocks along a fault, as opposed to the throw or vertical displacement. (Arkell, 1953) (d) The horizontal component of the slip, measured at right angles to the strike of the fault. Used by J.E. Spurr and A. Geikie for offset. Used by Jukes Brown for strike slip. (Fay, 1920) (e) Cf: *upthrow.* (Arkell, 1953) (f) Displacement of mineral vein by faulting. Lifting of floor of underground working through rock pressure. (Pryor, 1963) (g) Fault or throw in a lode. See also: *throw.* (Gordon, 1906)

heavily watered Scot. Said of a colliery when the escape of water from the strata into the shaft or workings is abundant, requiring powerful pumping machinery.

heaving Refers to the rising of the bottom after removal of the coal. See also: *creep.* (Jones, 1949)

heaving shale An incompetent or hydrating shale that runs, falls, swells, or squeezes into a borehole. (AGI, 1987)

heavy crop Gr. Brit. Collectively, the heavy minerals of a sedimentary rock. (AGI, 1987)

heavy gold Gold occurring as large particles. Cf: *nugget.* (AGI, 1987)

heavy ground (a) Closing or squeezing ground. (b) Dangerous hanging wall, which sounds hollow when rapped, indicating the possibility of a rock fall. (Pryor, 1963)

heavy joist Timber over 4 in (10.2 cm) and less than 6 in (15.2 cm) in thickness and 8 in (20.3 cm) or over in width. (Crispin, 1964)

heavy liquid separation Separation of ore particles by allowing them to settle through, or float above, a fluid of intermediate density. (Pryor, 1963)

heavy-media ore *natural ore.*

heavy-media separation A series of patented processes originally developed for the concentration of ore, but finding increased usage in coal cleaning. Suspension of magnetite (sp gr, 5.0) and ferrosilicon (sp gr, 6.7) are usually used for ore concentration; suspensions of magnetite for coal. The basic features of these processes as applied to coal are in the methods used for handling the magnetic medium. Specifications for magnetite should be somewhat as follows: 100% -100 mesh, 65% to 75% -325 mesh, 85% magnetics, and wet-ground in a ball or rod mill. See also: *dense-media separation.* (Mitchell, 1950)

heavy metals In exploration geochemistry, principally zinc, copper, cobalt, and lead, but under special conditions including one or more of the following metals: bismuth, cadmium, gold, indium, iron, manganese, mercury, nickel, palladium, platinum, silver, thallium, and tin.

heavy mineral (a) An accessory detrital mineral of a sedimentary rock, of high specific gravity, e.g., magnetite, ilmenite, zircon, rutile. Cf: *light mineral.* (b) In igneous petrology, a mafic mineral. (c) Resistant minerals that can be concentrated in the panning of alluvium and used as mineralogical and geochemical guides in prospecting. (Peters, 1987)

heavy soil A fine-grained soil, made up largely of clay or silt. (Nichols, 1976)

heavy spar *barite.*

heavy tiff Barite in southeast Missouri. (Fay, 1920)

heavy water Deuterium oxide, D_2O, in which D is the symbol for deuterium (heavy hydrogen or hydrogen 2). Water in which ordinary hydrogen atoms have been replaced by deuterium atoms. Natural water contains 1 heavy-water molecule per 6,500 ordinary water molecules. Deuterium oxide has a low neutron absorption cross section; hence, it is used as a moderator in some nuclear reactors. (Lyman, 1964)

hecatolite *moonstone.*

hectare A metric unit of land area equal to 10,000 m^2, 100 ares, or 2.471 acres. Abbrev. ha. (AGI, 1987)

hectorite A monoclinic mineral, $Na_{0.3}(Mg,Li)_3Si_4O_{10}(F,OH)_2$; smectite group. Cf: *montmorillonite.*

hedenbergite A monoclinic mineral, $4[CaFeSi_2O_6]$; pyroxene group; forms series with diopside and johannsenite; a common rock-forming mineral in iron-rich metamorphic rocks, limestone skarns, and fayalite-bearing plutonic rocks.

hedgehog stone Quartz crystals containing needles of göthite or some other iron oxide. Cf: *hairstone.* (Fay, 1920)

hedging The establishment of an opposite position on a futures market from that held and priced in the physical commodity. Without hedging, the physical position would be at risk to price fluctuations. (Wolff, 1987)

hedyphane A hexagonal mineral, $Pb_3Ca_2(AsO_4)_3Cl$; apatite group; yellowish-white; at Franklin, NJ, and various localities in Sweden. Cf: *mimetite.*

heel (a) A small body of coal left under a larger body as a support. Known also as heel of coal. (BCI, 1947) (b) The mouth or collar of a borehole. (c) The fixed jaw on an adjustable-wrench safety clamp or on a rock crusher. (Long, 1960) (d) A floor brace or socket for wall-bracing timbers. (Nichols, 1976) (e) The trailing edge of an angled blade. (Nichols, 1976) (f) Any material remaining in a vessel after removal of the main portion of the contents.

heeling in Temporary planting of trees and shrubs. (Nichols, 1976)

heel of a shot (a) In blasting, the face of a shot farthest away from the charge. (Stauffer, 1906) (b) The distance from the mouth of the drill hole to the corner of the nearest free face; or that portion of the hole that is filled with the tamping; or that portion of the coal to be broken that is entirely outside the powder. (Zern, 1928)
heep stead The entire surface plant of the mine. (Gordon, 1906)
height of instrument A surveying term used in spirit leveling for the height of the line of sight of a leveling instrument above the adopted datum, in trigonometric leveling for the height of the center of the theodolite above the ground or station mark, in stadia surveying for the height of the center of the telescope of the transit or telescopic alidade above the ground or station mark, and in differential leveling for the height of the line of sight of the telescope at the leveling instrument when the instrument is level. Abbrev. HI. (AGI, 1987)
heinrichite The arsenic analogue of uranocircite, $Ba(UO_2)_2(PO_4)_2{\cdot}12H_2O$ and its dehydration product with $Ba(UO_2)_2(AsO_4)_2{\cdot}8H_2O$ (metaheinrichite); occur near Lakeview, OR, and in the Black Forest, Germany. See also: *arsenuranocircite; metaheinrichite.* (Hey, 1961)
heintzite A hydrous borate of magnesium and potassium, colorless to white. Occurs in small crystals sometimes aggregated. From Stassfurt, Germany. Syn: *hintzeite; kaliborite.* (Fay, 1920)
helenite A wax near ozocerite but elastic like caoutchouc; yellow; sp gr, 0.915. It occurs at Ropa in Galicia. Syn: *mineral caoutchouc.* (Fay, 1920)
helical bag conveyor *double helical bag conveyor.*
helical conveyor A conveyor for handling coal, grain, cement, or similar bulk materials. It comprises a horizontal shaft, with helical paddles or ribbons, which turn on its center line inside a stationary tube filled with the material. See also: *worm conveyor.* (Hammond, 1965)
helical steel support A continuous screw-shaped steel joist lining for staple shafts. The lining is fixed to the ground by strata bolts rigidly fixed every 120° without any yielding device. Developed in Germany and its use is claimed to effect considerable overall savings. (Nelson, 1965)
helicitic Pertaining to a metamorphic-rock texture consisting of bands of inclusions that indicate original bedding or schistosity of the parent rock and cut through later-formed crystals of the metamorphic rock. The relict inclusions commonly occur in porphyroblasts as curved and contorted strings. The term was originally, but is no longer, confined to microscopic texture. Also spelled: helizitic. Cf: *poikiloblastic.* (AGI, 1987)
helictite A distorted twiglike lateral projection of calcium carbonate, found in caves, etc. Cf: *stalactite; stalagmite. anemolite.* (Standard, 1964)
heliodor A clear yellow variety of beryl found near Rossing, Namibia, and prized as a gemstone. Also spelled helidor. Cf: *golden beryl.*
heliolite A whitish to reddish-gray aventurine oligoclase with internal yellowish or reddish firelike reflections. Syn: *sunstone.* (Hess)
heliotrope (a) A red-spotted, deep-green variety of chalcedony (cryptocrystalline quartz) used as a semiprecious stone. Syn: *bloodstone.* Cf: *plasma.* (b) An instrument used in geodetic surveying to aid in making long-distance (as much as 320 km) observations; composed of one or more plane mirrors so mounted and arranged that a beam of sunlight is reflected toward a distant survey station where it is observed with a theodolite.
helium An inert, monatomic, colorless, odorless element, the lightest of the rare gases. Except for hydrogen, helium is the most abundant element found in the universe. The bulk of the world's supply is obtained from wells. Symbol, He. Widely used in cryogenic research; vital in the study of superconductivity. Helium is used for arc welding, as a cooling medium for nuclear reactors, and as a gas for supersonic wind tunnels; extensively used for filling balloons as it is much safer than hydrogen. One of the recent largest uses for helium has been for pressuring liquid fuel rockets. (Handbook of Chem. & Phys., 3)
helizitic *helicitic.*
helks (a) Large detached crags; a confused pile or range of rocks. (Arkell, 1953) (b) Bare tracts of limestone. (Arkell, 1953)
hellyerite A triclinic mineral, $NiCO_3{\cdot}6H_2O$; occurs with zaratite at the Lord Brassey nickel mine, Heazlewood, Tasmania.
Helmholtz coil A pair of similar coaxial coils with their distance apart equal to their radius, which permits an accurate calculation of the magnetic field between the coils. Used in calibration of magnetometers. (AGI, 1987)
helper-up *pig tailer.*
helvite An isometric mineral, $2[Mn_4Be_3(SiO_4)_3S]$; manganese may be replaced by iron toward danalite, or by Zn toward genthelvite; forms tetrahedra; Mohs hardness, 6 to 6-1/2; in veins with quartz, hornblende, and iron oxide; in pegmatites, and some alkaline igneous rocks.
hemafibrite A brownish-red to garnet-red, transparent to translucent, hydrous manganese arsenate, $Mn_3(AsO_4)(OH)_3{\cdot}H_2O$; soon turns black; Mohs hardness, 3; sp gr, 3.50 to 3.65; rare; from Nordmark, Sweden. Syn: *aimafibrite.* (Larsen, 1934)
Hematine A copyrighted, confusing name for an imitation of hematite. Usually in the form of an imitation cuvette. Apparently an artificially processed friable mineral or other substance. Breaks easily. Mohs hardness, about 6.5; sp gr, 4.8; streak, black.
hematite (a) A trigonal mineral, α-Fe_2O_3; red if earthy, reddish to bluish gray if massive, or bright metallic steel-gray in thin tablets or micalike flakes (specular hematite); invariably has red ocher streak. Kidney ore is massive reniform hematite. Commonly associated with quartz, oxyhydroxides such as goethite or limonite, and magnetite, after which it may be pseudomorphic; nonmagnetic when pure, but may appear magnetic due to residual or included magnetite or maghemite; the most widely mined ore of iron; in sedimentary rocks, Precambrian banded iron formations (including their metamorphosed equivalents), oolitic ironstones, contact-metamorphic deposits, commonly by alteration of magnetite; may be of secondary origin, having formed by oxidation and decomposition of iron silicates and carbonates; also occurs as a primary mineral in veins and replacement deposits associated with igneous intrusions, and in fumarolic deposits from volcanic gases. See also: *iron ore; specularite.* Syn: *specular iron ore; oligist; oligist iron.* (b) The mineral group corundum, eskolite, hematite, and karelianite.
hematite schist *itabirite.*
hemetite Any synthetic imitation of hematite.
hemi- A prefix meaning half.
hemicrystalline *hyalocrystalline.*
hemidome The upper or lower two faces of a dome resulting from symmetry lower than that required for a dome in orthorhombic or monoclinic crystal systems.
hemihedral In crystallography, having lower symmetry, resulting in forms with half the number of faces as the holohedral point group. Cf: *holohedral; merohedral.*
hemimorphic In crystallography, having no transverse plane of symmetry and no center of symmetry, and composed of forms belonging to only one end of the axis of symmetry. (Fay, 1920)
hemimorphism In crystallography, refers to minerals in crystal classes with merohedral symmetry such that crystal forms are different at opposite ends of the crystallographic axes, thus permitting polar crystal properties; e.g., hemimorphite, zincite.
hemiopal *semiopal.*
hemipelagic Sharing neritic and pelagic qualities. (AGI, 1987)
hemipelagic-abyssal Refers to sediments of the deep sea that contain terrestrial detritus. (AGI, 1987)
hemiprism A triclinic crystal form of two parallel faces with intercepts on two crystallographic axes.
hemipyramid A monoclinic crystal form of two parallel faces with intercepts on three crystallographic axes. See also: *pyramid.*
hemitropic Crystals that appear as if composed of two halves of a crystal turned partly round and united. Examples of this structure may often be found in feldspar and cassiterite crystals. (Fay, 1920)
hemloite A triclinic mineral, $(As,Sb)_2(Ti,V,Fe)_{12}O_{23}(OH)$; metallic to submetallic black; in drill core samples associated with rutile, pyrite, molybdenite, and arsenopyrite at the Hemlo gold deposit, ON, Can.
hendersonite An orthorhombic mineral, $Ca_3V_{12}O_{32}{\cdot}12H_2O$; in black fibrous crystals in Montrose County, CA, and San Juan County, NM.
Henderson process The treatment of copper sulfide ores by roasting with salt to form chlorides, which are then leached out and precipitated. (Fay, 1920)
hengleinite A steel-gray iron sulfide with about 20% cobalt and nickel, $(CoNiFe)S_2$. Minute pyritohedral crystals; isometric. Probably a mixture of siegenite and pyrite. Formerly called cobaltnickelpyrite. From Musen, Westphalia, Germany. Syn: *cobalt-nickel pyrite.* (English, 1938)
henry Unit of electrical induction. With an electromotive force of 1 V and current of 1 A, one henry (H) = 10^9 electromagnetic unit. Symbol H. (Pryor, 1963)
henryite An isometric mineral, $Cu_4Ag_3Te_4$; pale blue; associated with hessite, petzite, sylvanite, altaite, rickardite, and pyrite at Bisbee, AZ.
hepatic cinnabar A mixture of cinnabar, bituminous material, and clay; liver-brown; may be flammable.
hepatic pyrite (a) *marcasite.* (b) *pyrite.*
hepatin An amorphous limonite, of a liver-brown color, and containing a small percentage of copper. (Fay, 1920)
hepatite A variety of barite, so called from the fetid odor it exhales when heated. (Standard, 1964)
heptaphyllite Dioctahedral clay mineral. Cf: *octaphyllite.* (AGI, 1987)
heptavalent Having a valence of 7. Also called septavalent; septivalent. (Webster 3rd, 1966)
hercularc lining A German method of lining roadways subjected to heavy pressures. It consists of a closed circular arch of specially shaped precast concrete blocks. The blocks, which are wedge-shaped, are made in two sizes for each lining and erected in such a way that alternate blocks offer their wedge action in opposite directions—the larger blocks toward the center of the roadway and the smaller outward. This arrangement gives a double-wedge effect so that part of the

lateral pressure exerted by the strata on the lining is diverted axially along the roadway. (Nelson, 1965)

Hercules powder Weak form of dynamite, based on nitroglycerin and a semiactive carrying dope. (Pryor, 1963)

Hercules stone *lodestone.*.

hercynite An isometric mineral, 8[$Fe^{2+}Al_2O_4$]; spinel group; forms series with magnesian spinel, with gahnite, and with chromite; massive or fine grained; Mohs hardness, 7.5 to 8; in metamorphosed argillaceous sediments with andalusite, sillimanite, or garnet; also in contact metamorphic deposits in limestone or marble; in some eruptive mafic rocks; may occur with corundum to form emery. Syn: *ferrospinel.*

herderite A monoclinic mineral, 4[$CaBe(PO_4)F$], with OH replacing F toward hydroxylherderite; pseudo-orthorhombic prismatic crystals or radiated fibrous aggregates; in late-stage pegmatites.

Herkimer diamond Small, exceptionally clear, commonly doubly terminated quartz crystals in Herkimer County, NY; in cavities in sandstone and in loose earth or clay; also in mid-Atlantic beaches. Syn: *Lake George diamond.*

Heroult process *Hall process.*

herrerite Copper-stained blue and green smithsonite from Albarradon, Mexico.

Herreshoff furnace A mechanical, cylindrical, multiple-deck muffle furnace of the McDougall type for smelting ores. (Fay, 1920)

herringbone roller conveyor A roller conveyor consisting of two parallel series of rolls having one or both series skewed. See also: *roller conveyor.* Syn: *herringbone table.*

herringbone stoping Method used in flattish Rand stope panels 500 to 1,000 ft (152 to 305 m) long for breaking and moving ore. Stope is divided into 20-ft (6.1-m) panels, each worked by its own gang. A light tramming system delivers severed rock to a central scraper system. (Pryor, 1963)

herringbone table *herringbone roller conveyor.*

herringbone texture In mineral deposits, a pattern of alternating rows of parallel crystals, each row in a reverse direction from the adjacent one. It resembles the herringbone textile fabric. (AGI, 1987)

hertz A syn. for "cycles per second." Abbrev. Hz.

hess S. Staff. Clinker from furnace boilers. (Fay, 1920)

hessite A monoclinic mineral, 4[Ag_2Te], with gold replacing silver toward petzite; soft; metallic gray; fine grained, massive, or compact; sp gr, 8.24 to 8.45; in hydrothermal veins associated with quartz, pyrite, and native gold; a source of silver in California, Colorado, Ontario, Mexico, Chile, Romania, and Zimbabwe.

hessonite An orange to yellow-brown gem variety of grossular. Syn: *essonite; cinnamon stone.*

hetaerolite (a) A very rare double oxide of zinc and manganese, $ZnMn_2^{3+}O_4$, occurring in ore deposits as black tetragonal and fibrous crystals; hausmannite family. (CMD, 1948; Hey, 1955) (b) A tetragonal mineral, $ZnMn_2O_4$; forms bipyramids; may be fibrous; with zinc ores at Franklin, NJ, and near Leadville, CO. Also spelled heterolite.

hetero- A prefix signifying various, or of more than one kind or form.

heteroblastic Pertaining to a type of crystalloblastic texture in a metamorphic rock in which the essential mineral constituents are of two or more distinct sizes. Cf: *crystalloblastic; homeoblastic.* (AGI, 1987)

heterogeneous (a) Having more than one constituent or phase, thus exhibiting different properties in different portions. (Pryor, 1963) (b) A term describing metals and alloys with structures composed of more than one constituent. (Rolfe, 1955) (c) Unlike in character, quality, structure, or composition; consisting of dissimilar elements or ingredients of different kinds; not homogeneous. (Standard, 1964)

heterogenite (a) A cobalt mineral, $Co^{3+}O(OH)$, containing up to 4% CuO. Syn: *stainierite.* (Hey, 1955) (b) Name suggested for all cobaltocobaltic hydroxides of varying purity. (English, 1938)

heterogranular (a) Said of the texture of a rock having crystals of significantly different sizes. (b) Said of a rock having such a texture. Syn: *inequigranular.* (c) *anisodesmic.*

heteromorphic Said of igneous rocks having similar chemical composition but different mineralogic composition. (AGI, 1987)

heteromorphism The crystallization of two magmas of nearly identical chemical composition into two different mineral aggregates as a result of different cooling histories. (AGI, 1987)

heteropic Said of sedimentary rocks of different facies, or said of facies characterized by different rock types. The rocks may be formed contemporaneously or in juxtaposition in the same sedimentation area or both, but the lithologies are different; e.g., facies that replace one another laterally in deposits of the same age. Also, said of a map depicting heteropic facies or rocks. Cf: *isopic.* (AGI, 1987)

heterothrausmatic A descriptive term applied to igneous rocks with an orbicular texture in which the nuclei of the orbicules are composed of various kinds of rock or mineral fragments. Cf: *crystallothrausmatic; isothrausmatic; homeothrausmatic.* (AGI, 1987)

heterotomous Having a cleavage unlike that characteristic of the mineral in its ordinary form, as a variety of feldspar. (Standard, 1964)

heugh (a) Scot. A place where coal or other mineral is worked; a pit or shaft. Also spelled heuch. (b) The steep face of a quarry or other excavation. (c) A glen with rugged sides; a crag. (Standard, 1964) (d) An old English term for coal seams or coal workings. (Tomkeieff, 1954)

heulandite A monoclinic mineral, $(Na, Ca)_{2\text{-}3}Al_3(Al, Si)_2Si_{13}O_{36}\cdot 12H_2O$]; zeolite group, with extensive substitution of NaSi for CaAl and K predominant over Na; forms in cavities in basalt and andesite, as skarn druses, as diagenetic product in silicic vitreous tuffs, in bentonitic clays, and as authigenic mineral in limestone or sandstone.

hewer (a) Eng. In the Newcastle coalfield, one who undercuts the coal with a pick. A coal miner. (Fay, 1920) (b) N. of Eng. One who may use a hand pick but usually uses a pneumatic (windy) pick to win coal. Task consists of breaking in or making a nicking, digging out the coal, and filling onto a conveyor belt or into tubs. (Trist, 1963)

hewettite A monoclinic mineral, $CaV_6O_{16}\cdot 9H_2O$; forms deep red microscopic needles in vanadium deposits near Cerro de Pasco, Peru, and Paradox Valley, CO.

hewing (a) Eng. In the Newcastle coalfield, undercutting or mining the coal. Syn: *breaking in.* (Fay, 1920) (b) The dressing of timber by chopping or by blows from an edged tool. (Crispin, 1964)

hewing double Eng. *double working.*

hexad A crystallographic axis of rotation of 60°, a sixfold axis. Cf: *axis of symmetry.*

hexagonal (a) A geometrical form of six sides; e.g., hexagonal prism, hexagonal pyramid. (b) The crystal system characterized by a unique hexad (sixfold axis of rotation.) Cf: *crystal systems.*

hexagonal close-packed crystals Crystals having atoms at the corners of the hexagonal unit cells that are right prisms with rhombic bases, and at the corners of those (isosceles) triangular prisms that are similarly located halves of the hexagonal unit cells. The two sets of atoms are not crystallographically equivalent. (Henderson, 1953)

hexagonal system In crystallography, that system of crystals in which the faces are referred to four axes—a principal or vertical axis and three lateral axes perpendicular to the vertical axis and intersecting at mutual angles of 60°. Cf: *trigonal; trigonal system.* (Fay, 1920)

hexagonite A lavender variety of manganoan tremolite.

hexahydrite The mineral group bianchite, ferrohexahydrite, hexahydrite, moorhouseite, and nickel-hexahydrite.

hexavalent (a) Having a valence of 6. (Webster 3rd, 1966) (b) Having six valences; e.g., manganese with valences of 1, 2, 3, 4, 6, and 7. (Webster 3rd, 1966; Handbook of Chem. & Phys., 2)

Hexhlet sampler A selective mine dust-sampling instrument. It collects the airborne dust sample in two components. The fraction larger than 5 µm in size is separated from the total cloud in a size selector. The instrument collects some grams of respirable dust by filtration of the mine air through a fine pore ceramic thimble. See also: *thermal precipitator.* (Nelson, 1965)

Heyn's reagent An etching reagent containing 10% copper ammonium chloride in water. (Osborne, 1956)

hiatal (a) Said of the texture of an igneous rock in which the sizes of the crystals are not in a continuous series but are broken by hiatuses, or in which there are grains of two or more markedly different sizes, as in porphyritic rocks. Cf: *seriate.* (AGI, 1987) (b) Pertaining to or involving a stratigraphic hiatus. (AGI, 1987)

hiatus (a) A break or interruption in the continuity of the geologic record, such as the absence in a stratigraphic sequence of rocks that would normally be present but either were never deposited or were eroded before deposition of the overlying beds. (AGI, 1987) (b) A lapse in time, such as the time interval not represented by rocks at an unconformity; the time value of an episode of nondeposition or of nondeposition and erosion together. (AGI, 1987)

Hicks' hydrometer An instrument consisting of a series of colored glass beads of different densities contained in a glass tube for testing the specific gravity of electrolytes. (Osborne, 1956)

hiddenite An emerald-green to yellow green gem variety of spodumene. See also: *spodumene.* Syn: *lithia emerald.*

hide salt Coarse sizes of rock salt, usually No. 2 or No. 1. (Kaufmann, 1960)

high (a) The crest or culmination of a structure, such as a dome or an anticline. Cf: *low.* Syn: *structural high.* (AGI, 1987) (b) Name for the coal of a thick seam. (Fay, 1920) (c) A geophysical anomaly with values greater than normal; e.g., a gravity maximum or a geothermal maximum. (AGI, 1987)

high-alumina refractories Alumina-silica refractories containing 45% or more alumina. (Harbison-Walker, 1972)

high-angle fault A fault with a dip greater than 45°. Cf: *low-angle fault.* (Billings, 1954)

high-carbon steel Carbon steel that contains more than 0.5% carbon. (Hammond, 1965)

high-conductivity copper Metal of high purity, having an electrical conductivity not much below that of the international standard, which is a resistance of 0.15328 Ω for a wire 1 m long and weighing 1 g. (CTD, 1958)

high doors Scot. An upper landing in a shaft. (Fay, 1920)
high-expansion foam A method of fighting underground fires developed in the United States, and somewhat similar to the British foam plug. It involves the formation of a high-expansion noncombustible foam. Large volumes of the foam are drawn or blown over and around the fire until it can no longer be sustained due to lack of oxygen. The foam is made from ammonium lauryl sulfate and 1 gal (3.8 L) of solution is used for each 250 to 300 ft^3/min (7.1 to 8.5 m^3/min) of air passing through the net. (Nelson, 1965)
high explosive An explosive that is capable of detonating. There are two main types: (1) primary explosives, which detonate no matter what type of stimulus is given—these usually are very sensitive and (2) secondary explosives, which detonate normally only when the stimulus is a strong shock—under other types of stimulus they may merely deflagrate.
high feed *fast gear.*
high furnace The ordinary blast furnace. (Fay, 1920)
high-grade (a) Said of an ore with a relatively high ore-mineral content. Cf: *low-grade.* (AGI, 1987) (b) To steal rich or specimen ore. (AGI, 1987) (c) An arbitrary designation for dynamite of 40% strength or over. See also: *grade.* (Fay, 1920)
high-grade mill A plant for treating high-grade ores.
high-grading Theft of valuable pieces of ore. See also: *gouging.* (Pryor, 1965)
high-level placer A placer on an alluvial terrace.
highmoor peat Peat occurring on high moors and formed predominantly of moss, such as sphagnum. Its moisture content is derived from rain water rather than from ground water and is acidic. Mineral matter and nitrogen content are low, and cellulose content is high. Syn: *moorland peat; moor peat; moss peat.* (AGI, 1987)
high-phosphorus ores Ores containing from 0.18% to 1.0% phosphorus. (Newton, 1959)
high pillar *shaft pillar.*
high quartz Phase of quartz stable from 867 to 1,470 °C. Also called beta quartz. See also: *quartz.* (Bennett, 1962)
high-raise miner *miner.*
high-rank coals Coals containing less than 4% of moisture in the air-dried coal or more than 84% of carbon (dry ash-free coal). All other coals are considered as low-rank coals. (BCI, 1947)
high-rank metamorphism Metamorphism accomplished under conditions of high temperature and pressure. See also: *metamorphic grade.* (AGI, 1987)
high-ratio resistance controller This controller gives a high ratio of maximum to minimum resistance, 6,000:1. A high resistance is thus available for reverse-current braking, but the design ensures that there is an ample volume of electrolyte between the electrodes when starting at twice full-load torque. It is similar to the swinging-electrode controller. (Sinclair, 1959)
high reef The bedrock or reef rising from the lowest and richest part of an alluvial placer and forming the slopes of the ancient valley.
high-reef wash Deposits of wash dirt upon the high reef.
high-resolution seismic technique A seismic prospecting technique in which a special recording system yields readable reflections from layers less than 10 ft (3 m) thick at depths as little as 100 ft (30 m). (Dobrin, 1960)
high seas The entire world's oceans except for the portion lying shoreward of the outer limit of the territorial seas. (United Nations, 1983)
high side A deep coal-mine car, i.e., one with high sides. Cf: *gondola.* (Fay, 1920)
high-silica ore *natural ore.*
high sintering Synonymous with advanced sintering at high temperatures, usually the final sintering close to the melting point of the material. (Osborne, 1956)
high-temperature bonding mortar A mixture of refractory materials, either raw or calcined, to which other materials not classified as refractory materials have been added for the purpose of increasing the plasticity, giving air-setting properties, and lowering the temperature at which the bond develops. (Henderson, 1953)
high-tensile steel A type of structural steel having a maximum yield point of 23 st/in^2 as compared with 15.25 st/in^2 for mild steel. See also: *yield stress.* (Hammond, 1965)
high-tension line A high-voltage transmission line.
high-tension separation In mineral processing, the use of high-voltage direct current at between 18,000 V and 80,000 V to charge small particles of dry material as they fall through its field (emanating as a spray or a point discharge). These are then sorted into relatively charge-retaining and charge-losing minerals in accordance with their conducting power. Also called electrostatic separation. (Pryor, 1963)
high velocity *velocity.*
high volatile A bituminous coal The rank of coal, within the bituminous class of the Classification D 388, such that on the dry and mineral-matter-free basis, the volatile matter content of the coal is greater than 31% (or the fixed carbon content is less than 69%), and its gross calorific value is equal to or greater than 14,000 Btu/lb (32.54 MJ/kg) of coal on the moist, mineral-matter-free basis, and the coal is commonly agglomerating. (ASTM, 1994)
high volatile B bituminous coal The rank of coal, within the bituminous class of Classification D 388, such that, on the moist mineral-matter-free basis, the gross calorific value of the coal in British thermal units per pound is equal to greater than 13,000 (30.24 MJ/kg) but less than 14,000 (32.54 MJ/kg) and the coal commonly agglomerates. (ASTM, 1994)
high volatile C bituminous coal The rank of coal, within the bituminous class of Classification D 388, such that, on the moist, mineral-matter-free basis, the gross calorific value of the coal in British thermal units per pound is equal to or greater than 11,500 (26.75 MJ/kg) but less than 13,000 (30.24 MJ/kg) and the coal commonly agglomerates, or equal to or greater than 10,500 (24.42 MJ/kg) but less than 11,500 (26.75 MJ/kg) and the coal agglomerates. (ASTM, 1994)
high-volatile coals Coals containing over 32% volatile matter with a coal rank code No. 400 to 900. See also: *coal classification.* (Nelson, 1965)
high voltage (a) A high electrical pressure or electromotive force. (b) In coal mining, voltages above 1,000 V. Cf: *low voltage; medium voltage.* (Federal Mine Safety, 1977)
highwall The unexcavated face of exposed overburden and coal or ore in an opencast mine, or the face or bank on the uphill side of a contour strip mine excavation.
high-water level The plane of high water. (Schieferdecker, 1959)
highwoodite A dark-colored intrusive rock composed of alkali feldspar, labradorite, pyroxene, biotite, iron oxides, apatite, and possibly a small amount of nepheline. It is essentially a monzonite. Its name, given by Johannsen in 1938, is derived from the Highwood Mountains, MT. Not recommended usage. (AGI, 1987)
hilgardite Hydrated chloroborate of calcium, $Ca_2B_5O_9Cl{\cdot}H_2O$, as colorless monoclinic-domatic crystals in the rock salt of Louisiana. See also: *parahilgardite.* (Spencer, 1940)
hill (a) An arch or high place in a mine. (Fay, 1920) (b) Scot. The surface at a mine. (Fay, 1920) (c) N. of Eng.; Mid. An underground inclined plane. (Fay, 1920) (d) A natural elevation of land of local area and well-defined outline. (Webster 3rd, 1966)
hill-and-dale formation Applied to the ridges and hollows along the surface of dumped material (usually overburden) at an opencast mine. The undulations are leveled out when the land is restored. (Nelson, 1965)
hillock A small, low hill; a mound. Adj: hillocky. (AGI, 1987)
hill peat Peat formed in mountainous districts and characterized by the presence of Sphagnum, Andromeda, heath, pine trees, etc. (Tomkeieff, 1954)
hillside (a) Used to describe quarries when located in high slopes. Cf: *terrain slope.* (Streefkerk, 1952) (b) A part of a hill between its crest and the drainage line at the foot of the hill. Syn: *hillslope.* (AGI, 1987)
hillside placers Gravel deposits intermediate between the creek and bench gravels; their bedrock is slightly above the creek bed, and the surface topography shows no indication of benching.
hillside quarry A quarry cut into and along the hillside; may comprise a single face or a series of benches. If the depth of face is not more than about 30 ft (9 m) it can be worked in one cut, but deeper faces are usually worked in two or more benches. See also: *pit quarry.* (Nelson, 1965)
hillslope *hillside.*
Hilt's law A generalization that states that, in a vertical sequence at any given point in the coalfield, the rank of the coal of the successive seams rises with increasing depth. Although this statement is generally true, there are numerous departures from it. (Tomkeieff, 1954; Nelson, 1965)
hindered settling (a) In classification, when the minerals settle in a thick pulp, as opposed to free settling in which the free particles fall through fluid media. (Newton, 1959) (b) Settlement of particles through a crowded zone, usually in a hydraulic column through which their fall is opposed by rising water. (Pryor, 1960)
hindered-settling ratio The ratio of the apparent specific gravities of the mineral against the suspension (not against the liquid) raised to a power between one-half and unity. (Gaudin, 1939)
hindostan A fine-grained sandstone used extensively in the manufacture of very cheap sharpening stones, esp. axstones. Found in Indiana. (Fay, 1920)
hinge The locus of maximum curvature or bending in a folded surface, usually a line. Syn: *flexure.* (AGI, 1987)
hinged apron *apron conveyor.*
hinged apron pan An apron pan that is made with a hinge construction along each edge so that it may be joined to companion pans by a hinge pin or through a rod.
hinged bar Steel bars placed in contact with the roof and at right angles to the longwall face. They are usually supported by yielding steel props. The bar can be extended to support newly exposed roof by adding another bar, which can be locked onto it by a simple wedge or

pin arrangement. The hinged bar is widely used on conveyor faces in continuous mining. (Nelson, 1965)

hinged-hammer crusher *Williams' hinged-hammer crusher.*

hinge fault A fault on which the movement of one side hinges about an axis perpendicular to the fault plane; displacement increases with distance from the hinge. It is a questionable term. Cf: *scissor fault; rotational fault.* (AGI, 1987)

hinterland (a) A subjective term referring to the relatively undisturbed terrain on the back of a folded mountain range; i.e., the side away from which the thrusting and folding appears to have taken place. (b) The land that lies behind a seaport or seaboard and supplies the bulk of its exports and absorbs the bulk of its imports.

hintzeite *heintzite.*

hircite A yellowish-brown, amorphous hydrocarbon found in Burma, which emits a bad smell on burning. (Tomkeieff, 1954)

Hirschback method A method for draining combustible gases from coal seams in which superjacent entries are developed over the coal seam being mined. The entries are located from about 80 to 138 ft (24 to 42 m) above the seams to be mined and are often supplemented with up or down boreholes drilled perpendicular to the walls of the entries. Also known as the superjacent roadway system. (Virginia Polytechnic, 1960)

hisingerite A monoclinic mineral, $Fe_2Si_2O_5(OH)_4{\cdot}2H_2O$; fine-grained to cryptocrystalline or fibrous; at Hibbing, MN; Blaine County, ID; and in Canada, Greenland, Finland, and Sweden.

histogram A vertical-bar graph representing a frequency distribution, in which the height of bars is proportional to frequency of occurrence within each class interval and, due to the subdivision of the x-axis into adjacent class intervals, there are no empty spaces between bars when all classes are represented in a sample so graphed. Histograms are used to depict particle-size distribution in sediments. (AGI, 1987)

historical geology A major branch of geology that is concerned with the evolution of the Earth and its life forms from its origins to the present day. The study of historical geology therefore involves investigations into stratigraphy, paleontology, and geochronology, as well as the consideration of paleoenvironments, glacial periods, and plate-tectonic motions. It is complementary to physical geology. Not to be confused with history of geology. See also: *geology.* (AGI, 1987)

hit Eng. To find, prove, or cut into a coal seam or fault. (Fay, 1920)

hitch (a) Step cut in rock face to hold timber support in underground working. Syn: *stip.* (Pryor, 1963) (b) N. of Eng. A minor geological fault or roll in the coal seam. (Trist, 1963) (c) Scot.; Eng. A minor dislocation of a vein or stratum not exceeding in extent the thickness of the vein or stratum. (Fay, 1920) (d) A hole cut in the side rock, when this is solid enough to hold the cap of a set of timbers permitting the leg to be dispensed with. (Fay, 1920) (e) A fault. Fractures and dislocations of strata common in coal measures, accompanied by more or less displacement. (Raistrick, 1939) (f) A connection between two machines. (Nichols, 1976) (g) To attach trams to hauling ropes by short chains. (Fay, 1920) (h) A sudden stoppage of pumping machinery. (Standard, 1964) (i) To dig or pick holes or places to receive the ends of timbers. (Standard, 1964)

hitch-and-step (a) S. Wales. A system of regulating the distance between the faces of stalls in longwall work. (Fay, 1920) (b) *stepped longwall.*

hitch cutter A miner who cuts places in the coal, ore, or wall in which to rest or place timbers to prevent rock from falling. (Fay, 1920)

hitcher (a) The person who runs trams into or out of the cages, gives the signals, and attends at the shaft when miners are riding in the cage. See also: *cager.* (Fay, 1920) (b) One who works at the bottom of a haulage slope or plane, engaging the clips or grips by means of which mine cars are attached to a hoisting cable or chain used for haulage up a steep incline to the mine surface. Also called hitcher-on. See also: *hitcher on; onsetter.* (DOT, 1949)

hitcher-on The person employed at the bottom of a shaft or slope to put loaded cars on, and take empty cars off the cage. See also: *hitcher.* (Fay, 1920)

hitch timbering Installing bars in hitches either cut or drilled in the rib, thereby eliminating the need for legs. Hitch holes may be provided for each individual bar. (Coal Age, 1966)

Hi-Velocity gelatin Explosive containing low-density gelatin; used for submarine blasting. (Bennett, 1962)

hjelmite A former name for yttromicrolite.

hod (a) Forest of Dean. A cart or sled for conveying coal in the stalls of thin seams. (Fay, 1920) (b) A tray or trough with a pole handle that is borne on the shoulder, for carrying mortar, brick, or similar load. (Webster 3rd, 1966)

hodge jig Variation on Harz jig in which the plunger (piston) has differential motion. See also: *Harz jig.* (Pryor, 1963)

hoedown *breakdown.*

hoegbomite *högbomite.*

hoelite An orthorhombic mineral, $C_{14}H_8O_2$; forms delicate yellow needles associated with chlorides and sulfur deposited by gases at a burning coalbed on Spitzbergen Island (anthraquinone).

hoernesite A monoclinic mineral, $(Mg_3(AsO_4)_2{\cdot}8H_2O)$; vivianite group, with Mg replaced by Co toward erythrite; in white crystals resembling gypsum; also columnar; a secondary mineral formed by alteration of arsenate minerals. Syn: *hörnesite.*

Hoesch process A method of working the open-hearth furnace in the duplex process so as to reduce as much as possible the amount of manganese lost in the slag in the production of manganese steels. (Osborne, 1956)

hoe scraper In mining, scraper-loader used to gather and transport severed rock. Cables pull a box-sided hoe over the loose ore, which is gathered and dragged (slushed) to the delivery point. (Pryor, 1963)

Hoganas process A method for the production of sponge iron that consists of charging fireclay pots with flat briquettes of a concentrate of iron ore interspersed with layers of carbon, prepared by mixing coal with coke breeze; the pots are charged in batches in a long pit furnace where they are heated to about 1,200 °C. (Osborne, 1956)

hogback (a) Any ridge with a sharp summit and steep slopes of nearly equal inclination on both flanks, and resembling in outline the back of a hog; specif. a sharp-crested ridge formed by the outcropping edges of steeply inclined resistant rocks, and produced by differential erosion. The term is usually restricted to ridges carved from beds dipping at angles greater than 20°. Cf: *cuesta.* (AGI, 1987) (b) A term applied in New England to a drumlin (western Massachusetts) and to a horseback or esker (Maine). (AGI, 1987) (c) The name given by geologists to the ridgy structure of certain districts, which consist of alternate ridges and ravines, occasioned either by the sharp undulations of the subjacent rocks, or more frequently by the erosive action of mountain torrents that cut out the ravines and leave the ridges or "hog's-backs" standing between. This structure occurs most abundantly on the lower slopes and flanks of mountain ranges. (AGI, 1987) (d) A sharp anticlinal, decreasing in height at both ends until it runs out. (AGI, 1987) (e) A ridge produced by highly tilted strata. (AGI, 1987) (f) Local term for drumlins in western Massachusetts. (AGI, 1987) (g) A name applied in the Rocky Mountain Region to a sharp-crested ridge formed by a hard bed of rock that digs rather steeply downward. (h) A ridge or lines of high hills with sharp summits and steeply sloping sides. (Long, 1960) (i) Eng. A sharp rise in the floor of a coal seam. (Fay, 1920)

högbomite A trigonal and hexagonal mineral, $(Mg,Fe)_2(Al,Ti)_5O_{10}$; metallic black with imperfect cleavage and conchoidal fracture; with magnetite, ilmenite, corundum, or ferroan spinel in iron ore at Redstand, Norway, and in emery at Whittles, VA. Also spelled hoegbomite.

hoggan Corn. The food carried by the miner to the mine. (Fay, 1920)

hogger Scot. A leather or canvas delivery pipe at the top of a sinking set of pumps.

hogger pipe N. of Eng. The upper terminal pipe with delivery hose from the mining pump. (Fay, 1920)

hogger pump The topmost pump in a shaft. (Fay, 1920)

hoggin A material composed of screenings or siftings of gravel or a mixture of loam, coarse sand, and fine gravel, used in making filter beds, as a binder, etc. (Webster 2nd, 1960)

hogging moment A bending moment that tends to cause hog. See also: *sagging moment.* (Hammond, 1965)

hohmannite Probably amarantite. See also: *metahohmannite.* (Larsen, 1934)

hoist (a) The windlass mechanism incorporated as an integral part of a power-driven drilling machine used to handle, hoist, and lower drill-string equipment, casing, pipe, etc., while drilling, or to snake the drill from place to place. (Long, 1960) (b) The act or process of lifting drill string, casing, or pipe out of a borehole. (Long, 1960) (c) A power-driven windlass for raising ore, rock, or other material from a mine and for lowering or raising people and material. Also called hoister. Syn: *mine hoist.* (Long, 1960; Fay, 1920) (d) The mechanism by which a bucket or blade is lifted, or the process of lifting it. (Nichols, 1976) (e) A drum on which hoisting rope is wound in the engine house, as the cage or skip is raised in the hoisting shaft. (Pryor, 1963) (f) An engine with a drum, used for winding up a load from a shaft. See also: *winding engine.* (CTD, 1958) (g) The amount of ore, coal, etc., hoisted during a shift. (Fay, 1920) (h) *draw works; elevator.*

hoist back-out switch A switch that permits operation of the hoist only in the reverse direction in case of overwind. Syn: *back-out switch.*

hoist boy In bituminous coal mining for Arkansas and Oklahoma, a general term applied to a hoisting engineer who operates a small hoisting engine, or an oiler who lubricates and cleans the engine. (DOT, 1949)

hoist engineer *hoistman.*

hoist engineman *hoistman.*

hoister A machine used in hoisting the product. (Zern, 1928)

hoisting (a) Winding in a mine. (Nelson, 1965) (b) In power-shovel nomenclature, hoisting is a term applied to two operations: (1) the raising or lowering of the boom, and (2) the lifting or dropping of the dipper stick in relation to the boom. (Carson, 1961)
hoisting block (a) The lower block of a block and fall, bearing the hoisting hook. (Standard, 1964) (b) Used incorrectly as a syn. for sheave wheel. (Long, 1960) (c) A traveling block or sheave. (Long, 1960)
hoisting compartment The section of a mine shaft used for hoisting the mineral to the surface. (Stoces, 1954)
hoisting crab A crab, winch, or windlass for hoisting. (Standard, 1964)
hoisting cycle The periods of acceleration, uniform speed, retardation, and rest. The deeper the shaft, the greater is the ratio of the time of full-speed hoisting to the total hoisting cycle. For shallow shafts there may be very little time at full speed, retardation beginning almost at the end of the accelerating period. (Lewis, 1964)
hoisting drum The flanged cylindrical part of a windlass around and on which the hoist rope or cable is wound. Also called spool. (Long, 1960)
hoisting engineer One who operates a hoisting engine, esp. at a mine or quarry. Also called engineman. See also: *hoistman.* (Fay, 1920; DOT, 1949)
hoisting jack A device for applying hand power to lift an object by means of a screw or lever, or by hydraulic power. (Fay, 1920)
hoisting plug A pin-thread heavy-bodied coupling provided with a swivel-mounted eye in the end opposite the pin-thread end. When attached to the hook on the drill-hoist line, the pin-thread end can be screwed into the rods to hoist and otherwise handle drill-string equipment when making borehole round trips. Also called swivel plug. See also: *plug.* Syn: *screwplug.* (Long, 1960)
hoisting power The capacity of the hoisting mechanism on a drill machine. May be expressed in terms of the number of lineal feet of a specific-size drill rod a hoist can lift on a single line or in terms of the total weight it can handle, figured in pounds or tons. (Long, 1960)
hoisting rope A rope composed of a sufficient number of wires and strands to ensure strength and flexibility. Such ropes are used in shafts, elevators, quarries, etc. (Zern, 1928)
hoisting sheave *winding sheave.*
hoistman In mining, a person who operates steam or electric hoisting machinery used to lower cages (elevators) and skips (large, metal, boxlike containers) into a mine and to raise them to the surface from different levels. May be designated according to type of power used, as electric-hoist man or steamhoist man. Also called cageman; cage runner; hoist engineer; hoist engineman; hoisting engineer; hoisting engineman; hoist operator; operating engineer; shaft driver; shaft engineer; shaft-hoist engineer; shaft hoistman. (DOT, 1949)
hoist operator (a) In petroleum production, one who lowers and raises surveying, servicing, or testing instruments in and out of oil or gas wells on electrical conductor cables, using truck-mounted hoisting equipment. Also called winch operator. (DOT, 1949) (b) *hoistman.*
hoist overspeed device A device that can be set to prevent the operation of a mine hoist at speeds greater than predetermined values and usually causes an emergency brake application when the predetermined speed is exceeded.
hoist overwind device A device that can be set to cause an emergency brake application when a cage or skip travels beyond a predetermined point into a danger zone.
hoist signal code Prescribed signals for indicating to the hoist operator the desired direction of travel and whether people or materials are to be hoisted or lowered in mines.
hoist signal system A system whereby signals can be transmitted to the hoist operator (and in some instances by the operator to the cager) for control of mine hoisting operations.
hoist slack brake switch A device for automatically cutting off the power from the hoist motor and causing the brake to be set in case the links in the brake rigging require tightening or the brakes require relining.
hoist trip recorder A device that graphically records information such as the time and number of hoists made as well as the delays of idle periods between hoists. Syn: *trip recorder.*
holdback On an inclined belt conveyor system, a brake that comes automatically into use in the event of power failure, thus preventing the loaded belt from running downward and piling up rock. (Pryor, 1963)
Holdcroft thermoscope bar Temperature indicator that consists of a series of small bars placed horizontally on a refractory stand. On heating, some bars are bent to varying degrees, while others are unaffected. The temperature is indicated by the bar that is just beginning to sag. The bars are numbered 1 to 40, the temperature range being 600 to 1,550 °C. See also: *thermoscopic bar.* (Osborne, 1956)
holdenite An orthorhombic mineral, $(Mn,Mg)_6Zn_3(AsO_4)_2(SiO_4)(OH)_8$; in zinc deposits at Franklin, NJ.
holdfast Temporary anchorage for guy ropes. (Hammond, 1965)
holding (a) *take.* (b) Also Syn: *mining claim.* under General Mining Law of 1872, (Act. of May 10, 1872; 17 Stat. 91) See also: *undercut.*
holding-down bolt *anchor bolt.*
holding rope Support rope for suspension of a grab used for excavating or handling bulk material. (Hammond, 1965)
hold out! Derb. An exclamation by the banksperson down a shaft to the bottomer, when workers are about to descend the shaft, to let them know that they are not to send up a load of coal, but merely the empty rope or chain. (Fay, 1920)
hole (a) In Joplin, Missouri, a local term for a mine shaft. (Fay, 1920) (b) A drill hole, borehole, or well. See also: *borehole.* (Long, 1960) (c) To undercut a seam of coal by hand or machine. (Fay, 1920) (d) To make a communication from one part of a mine to another. (Fay, 1920) (e) To pick out the soft clay beneath a lode or seam of coal preparatory to wedging or blasting the mass out. (Gordon, 1906) (f) A perforation through the laminae. (Skow, 1962)
hole curvature The amount, expressed in degrees, that a borehole has diverged from its intended course in a distance of 100 ft (30 m). Syn: *hole deviation.* (Long, 1960)
hole deviation *hole curvature.*
hole-in To start drilling a borehole. Also called collar; spud; spud-in. (Long, 1960)
hole layout In quarrying, an arrangement consisting of a combination of vertical and horizontal holes. (Streefkerk, 1952)
holeman Pennsylvania. Person who loads holes with explosives; a charger. (Fay, 1920)
holers The workers employed in the operation of holing the coal. (Peel, 1921)
hole system A system of contract work underground by which the pointing of the holes and blasting are done by company personnel and the rest of the work by the miners. (Fay, 1920)
hole through Successful meeting of two approaching tunnel heads, or of winze and raise. (Pryor, 1963)
holing (a) Cutting. (Mason, 1951) (b) The working of a lower part of a bed of coal for bringing down the upper mass. (Fay, 1920) (c) The final act of connecting two workings underground. See also: *holing-through.* (Fay, 1920) (d) The meeting of two roadways driven expressly to intersect each other. Syn: *thirling.* (BS, 1963) (e) Eng. Shale partings in which the first charges were inserted for blasting, Wenlock limestone, Dudley. (Arkell, 1953) (f) *undercutting.* (g) Eng. *kirving.*
holing about Eng. The operation of establishing an air current between the downcast and upcast shafts. (Fay, 1920)
holing nog A kerf wedge. (Nelson, 1965)
holing pick A pick used in holing coal. (Standard, 1964)
holings Eng. Holing dirt or small coal made by kirving with a coal-cutting machine. Also called scuffings. Syn: *cuttings.* (SMRB, 1930)
holing shovel S. Staff. A short-handled, round-bladed shovel. (Fay, 1920)
holing-through Driving a passage through to make connection with another part of the same workings, or with those in an adjacent mine. See also: *holing.* (Fay, 1920)
Holland-Gaddy formula A coal pillar design formula that predicts the strength of coal pillars based on laboratory tests of coal cube strength combined with pillar height and width specifications. The Holland-Gaddy formula is generally considered to be overly conservative for large pillar width-to-height ratios (over 5). (SME, 1992)
hollandite A monoclinic mineral, $BaMn_8O_{16}$; cryptomelane group; pseudotetragonal; massive, botryoidal, stalactitic, or prismatic with deeply striated faces; commonly associated with pyrolusite. Cf: *romanechite.*
hollow dam A dam built of reinforced concrete, mass concrete, or masonry in which the water pressure is resisted by a sloping slab or vault carried by buttresses. (Hammond, 1965)
hollow-plunger pump A pump used for mining and quarrying, as in muddy and gritty water. (Standard, 1964)
hollow quoin Recessed masonry that carries the heel post of a lock gate. (Hammond, 1965)
hollow-rod churn drill A churn drill in which hollow rods replace the steel wire rope. The drilling fluid is pumped down the inside of the rods, and the chipping and fluid return to the surface on the outside. (Nelson, 1965)
hollows Eng. Old abandoned workings. (Fay, 1920)
Holman Airleg A drill support consisting of a cylinder of about 2-in (5.1-cm) bore in which the piston is actuated by compressed air controlled by a twist-grip control valve. This valve is also used to release the air pressure to allow the piston to be lowered. The control valve also regulates the feeding pressure on the drill. The length of feed of the Holman Airleg is 39 in (99 cm), overall length 57 in (145 cm), and weight 25 lb (11.4 kg). A vent hole near the top of the cylinder allows the air to escape when the piston travels past it, so as to warn the operator that the limit of feed has been reached. The leg is then

readjusted for drilling to be resumed. The leg gives good support to the drill at all heights within reasonable limits, and one person can comfortably handle a drill when mounted in this way. (Mason, 1951)

Holman counterbalanced drill rig A drill rig consisting of a rail-track carriage on which is mounted a counterbalanced boom 10 ft (3 m) in length. For two-drill mounting, a crossbar is attached to the boom, and drill cradles can be fixed over or under the bar by swinging it through 180°. For a four-drill mounting, two additional vertical columns are used; these are jacked to the floor and the drill carriage clamped to the track. The crossbars can be clamped to any required position on the vertical columns. (Mason, 1951)

Holman dust extractor A dust trap in which the dust and chippings created during percussive drilling are drawn back through the hollow drill rod and out through the rear of the machine and along a hose to a metal container with filter elements. The appliance requires a special type of drilling machine, rods, and bits. See also: *dust trap.* (Nelson, 1965)

Holmberg system A method of sintering iron ore. Sintering is carried out in a series of pans, 11 ft 6 in (3.5 m) square, with robust chromium vanadium steel grate bars, which have a life of 5 to 6 years. Advantages include low fuel consumption and absence of moving parts in the sintering zones. (Osborne, 1956)

Holme mud sampler A device that takes samples with a scoop rotating on an axle mounted on a heavy frame that rests firmly on the bottom. The device is lowered open with the entire weight taken by a shackle on a balanced arm. Closure is not effected on touching bottom. On hoisting, the pressure of water on a vane attached to the balanced arm tips a lever allowing a pin to slip out, releasing the shackle. The weight is then transferred to a rope rotating around a large drum. This in turn rotates a small pulley and drags the sampling hemisphere through the bottom via a second pulley to which it is attached by a light wire. The maximum volume of the sample is 5 L, and in practice, usually about 3 to 4 L are collected. (Hunt, 1965)

Holme suction grab A sampling device using force provided by a vacuum chamber (containing air at atmospheric pressure) to suck in a sediment sample. On striking the bottom, the chamber is put into communication with the outside. Water pressure forces the sample into the collecting tube. The pressure chamber itself is a strong brass tube closed at the upper end by a lid and held firmly in position by a clamp. A sampling tube is fixed below the chamber and extends upward into it. Between the upper and lower parts of the central tube is a plug held in position by retaining hooks through slots in the wall of the lower tube. When the device strikes the bottom, a mouth tube rises, disengaging the plug, which flies up the tube. The water then forces material up into the collecting tube. (Hunt, 1965)

holmquistite An orthorhombic mineral, $Li_2(Fe,Mg)_3Al_2Si_8O_{22}(OH)_2$; amphibole group, with $Mg/(Mg+Fe^{2+}) = 0.1$ to 0.89; in granite pegmatites.

Holocene An epoch of the Quaternary period, from the end of the Pleistocene, approx. 10,000 years ago, to the present time; also, the corresponding series of rocks and deposits. When the Quaternary is designated as an era, the Holocene is considered to be a period. Syn: *Recent.* (AGI, 1987)

holocrystalline Said of the texture of an igneous rock composed entirely of crystals, i.e., having no glassy part. Also, said of a rock with such a texture. (AGI, 1987)

holohedral The point group with the highest symmetry of its crystal system. Cf: *hemihedral; merohedral; crystallographic system.*

holohyaline Said of an igneous rock that is composed entirely of glass. (AGI, 1987)

hololeims Coalified remains of entire plants. (Tomkeieff, 1954)

holystone (a) A soft sandstone used to scrub a ship's decks. (Webster 3rd, 1966) (b) To scrub with a holystone. (Webster 3rd, 1966) (c) Eng. Limestone full of holes, white limestones of the Great Oolite near Minchinhampton, Gloucestershire, used for megalithic monuments. (Arkell, 1953) (d) Pumice or friable sandstone used to scrub a ship's deck.

home N. of Eng. In the direction of, or toward, the shaft, as in an underground mine. Outby. (Fay, 1920)

homeoblastic Pertaining to a type of crystalloblastic texture in a metamorphic rock in which the essential mineral constituents are approx. of equal size. Cf: *crystalloblastic; heteroblastic.* (AGI, 1987)

homeothrausmatic A genetic term applied to igneous rocks with an orbicular texture in which the nuclei of the orbicules are formed of inclusions of the same generation as the groundmass. Cf: *isothrausmatic; heterothrausmatic; crystallothrausmatic.* (AGI, 1987)

homocline A general term for a series of rock strata having the same dip, e.g., one limb of a fold, a tilted fault block, or an isocline. Cf: *monocline.* Adj: homoclinal. (AGI, 1987)

homoeomorphism A near similarity of crystalline forms between unlike chemical compounds. (Fay, 1920)

homogeneity In geochemical prospecting, the homogeneity of a geochemical anomaly is a measure of the smoothness, or absence of strong local variations, in the distribution of the indicator element. (Hawkes, 1957)

homogeneous Made up of similar parts or elements; of the same composition or structure throughout; uniform. Opposite of heterogeneous. (Standard, 1964)

homogeneous mass A mass that exhibits essentially the same physical properties at every point throughout the mass. (ASCE, 1958)

homopolar crystal A crystal characterized by covalent bonding—the type of atomic bonding resulting from the sharing of electrons by neighboring atoms. (ASM, 1961)

homoseismal line Line on the Earth's surface connecting points where the seismic wave arrives, generated by an earthquake, at the same time. (Schieferdecker, 1959)

homotropal ventilation Ventilation by a current of air traveling in the same direction as the flow of mineral out of a mine. See also: *descensional ventilation; ascensional ventilation; antitropal ventilation.* (BS, 1963)

hondurasite Formerly called selen-tellurium (Se,Te), a variety of the trigonal mineral form of tellurium, Te.

honeycomb Any substance, as cast iron, worm-eaten wood, etc., having cells suggesting a honeycomb; also applied to certain rock structures. (Webster 2nd, 1960; Fay, 1920)

honeycomb structure An arrangement of soil particles having a comparatively loose, stable structure resembling a honeycomb. See also: *soil structure; flocculent structure; single-grained structure.* (ASCE, 1958)

honeycomb weathering A type of chemical weathering in which innumerable pits are produced on a rock exposure. The pitted surface resembles an enlarged honeycomb and is characteristic of finely granular rocks, such as tuffs and sandstones, in an arid region. (AGI, 1987)

honey stone A mellate of aluminum, $Al_2[C_6(COO)_6]{\cdot}16H2O$, of yellowish or reddish color, and a resinous aspect, crystallizing in octahedrons with a square base. See also: *mellite.* (Fay, 1920; Tomkeieff, 1954)

Honigmann process A continental method of shaft sinking through sand that is water bearing. The shaft is formed by boring in stages, increasing in size from the pilot hole of about 4 ft (1.2 m) in diameter to the final excavation size. Once the shaft is bored, mud flush circulation continues while the lining is lowered. The lining consists of two steel cylinders, one within the other, and the annular space is filled with concrete. The cylinders are lowered into the shaft, and 10-ft (3-m) lengths are added and welded in position at the shaft top. (Nelson, 1965)

hoodoo A fantastic column, pinnacle, or pillar of rock produced in a region of sporadic heavy rainfall by differential weathering or erosion of horizontal strata, facilitated by joints and by layers of varying hardness, and occurring in varied and often eccentric or grotesque forms. Syn: *rock pillar.* (AGI, 1987)

hook block The lower sheave or block, on a crane hoist, to which a swivel hook is attached. (Fay, 1920)

Hooke's law A statement of elastic deformation, that the strain is linearly proportional to the applied stress. See also: *elasticity.* (AGI, 1987)

hook forward method A method of lashing on to the rope in which the chain, after lapping two to five times around the rope, depending on the load and the inclination, is brought forward across the laps and threaded through the hook at the front. This method keeps the last lap tight and prevents the laps from spreading. (Sinclair, 1959)

hook-on The worker who adjusts cables or chains about objects to be lifted; places hook of crane block in bucket bails, and hooks winches to objects to be moved, etc. (Fay, 1920)

hook tender In bituminous coal mining, a laborer who attaches the hook at the end of a hoisting cable to the link of the leading or near car of a trip of cars to be hauled up or lowered down an incline in the mine or at the surface. Also called rope cutter. (DOT, 1949)

Hoolamite indicator A carbon monoxide detector consisting essentially of a small glass tube filled with a powdered chemical; when air is drawn through it, if any of the gas is present, the powder will change color, its degree of change depending upon the amount of carbon monoxide present. This device is very sensitive and will detect gas as low as 0.01%. (Kentucky, 1952)

Hooper jig Pneumatic jig, used in regions where water is scarce, or where the ore must be kept dry, to concentrate values from sands. (Pryor, 1963)

Hoope's process An electrolytic process of aluminum refining that utilizes three liquid layers in the reduction cell. An anode of aluminum-copper alloy is used in a fused fluoride bath. The lighter aluminum, about 99.99% purity, collects at the cathode above the fused bath. (Henderson, 1953)

Hopcalite Catalytic granules consisting of finely divided manganese dioxide mixed with copper oxide and a small quantity of silver oxide, and used in gas mask cannisters to remove carbon monoxide by oxidizing it to carbon dioxide. With the development of Hopcalite, the problem of providing adequate protection against carbon monoxide poisoning was solved. (McAdam, 1955)

hopeite An orthorhombic mineral, $Zn_3(PO_4)_2·4H_2O$; dimorphous with parahopeite; in minute grayish-white crystals with zinc ores, esp. at Broken Hill, Zimbabwe.

Hopfner process A process for the recovery of copper in which a solution of cuprous chloride in sodium or calcium chloride is used to dissolve copper sulfides. The solution is then electrolyzed in tanks with diaphragms. The anodes are impure copper; the cathodes, pure copper. Copper is deposited from the cuprous chloride solution, and cupric chloride is regenerated. (Liddell, 1918)

Hopkinson chain machine An electrically driven chain coal cutter designed and manufactured by Mather and Platt in 1901 with provision for slewing the jib. A large number of these machines were built. For many years the Hopkinson was the only chain coal cutter built in the United Kingdom. (Nelson, 1965)

hopper (a) A vessel into which materials are fed, usually constructed in the form of an inverted pyramid or cone terminating in an opening through which the materials are discharged (not primarily intended for storage). (BS, 1962) (b) Surge bin placed at discharge end of intermittent transporting system that handles dry ore or rock; used to smooth out and regulate delivery from that point. A hopper car is one with bottom discharge gear and insloping side walls. (Pryor, 1963) (c) A storage bin or a funnel that is loaded from the top and discharges through a door or chute in the bottom. (Nichols, 1976) (d) A container or bin for broken ore. (CTD, 1958) (e) A place of deposit for coal or ore. (Fay, 1920)

hopper car A car for coal, gravel, etc., shaped like a hopper, with an opening to discharge the contents. (Standard, 1964)

hopper crystal A crystal with edges grown beyond face centers. Cf: *skeletal crystal.*

hopper dredge A hydraulic dredge that operates in cycles, alternately filling at a dredge site and traveling to and from a disposal or offloading site.

hopperings In gold washing, gravel retained in the hopper of a cradle.

hoppers Pockets at the bottom of a breaker through which the processed coal falls as it is loaded into railroad cars; also the cars. (Korson, 1938)

hopper salt Grainer or solar salt produced in characteristic hollow-faced cubes by surface evaporation. (Kaufmann, 1960)

hopper table Early type of pneumatic table used in ore treatment. (Pryor, 1963)

hoppet A vessel for measuring ore. (Standard, 1964)

hoppit A large bucket used in shaft sinking for hoisting men, rock, materials, and tools. Since about 1955, hoppit sizes have increased to about 80 ft^3 (2.3 m^3) and in some cases to 110 ft^3 (2.5 m^3) and surface tipping facilities have been brought to a high degree of efficiency to cope with large-diameter shafts and fast sinking rates. See also: *cactus grab.* (Nelson, 1965)

horadiam (a) The drilling of a number of horizontal boreholes radiating outward from a common center; a single drill site or drill setup. (Long, 1960) (b) *horizontal-ring drilling.*

horizon (a) An interface indicative of a particular position in a stratigraphic sequence. In practice it is commonly a distinctive, very thin bed or marker. See also: *marker bed.* (AGI, 1987) (b) One of several lines or planes used as reference for observation and measurement relative to a given location on the Earth's surface and referred generally to a horizontal direction (Huschke, 1959); esp. apparent horizon. The term is also frequently applied to artificial horizon. (AGI, 1987) (c) One of the layers of the soil profile, distinguished principally by its texture, color, structure, and chemical content, designated as A-horizon; B-horizon; C-horizon. (ASCE, 1958) (d) An identifiable rock stratum regionally known to contain or be associated with rock containing valuable minerals. Cf: *marker; marker bed.* (Long, 1960) (e) *soil horizon.*

horizon mining A system of mine development that is suitable for inclined, and perhaps faulted, coal seams. Main stone headings are driven, at predetermined levels, from the winding shaft to intersect and gain access to the seams to be developed. The stone headings, or horizons, are from 100 to 200 yd (91.44 to 182.88 m) vertically apart, depending on the seams available and their inclination. The life of each horizon ranges from 10 to 30 years. Connections between horizons at inby points are by staple shafts or drivages in the coal. Also called horizontal mining; continental mining. See also: *lateral.* Cf: *in-the-seam mining.* (Nelson, 1965)

horizontal auger A rotary drill, mechanically driven, for drilling horizontal blasting holes in quarries and opencast pits. See also: *auger; vertical auger drill.* (Nelson, 1965)

horizontal balance A magnetic-field balance instrument much less commonly used than the vertical type. It is quite similar to it in construction except that the magnet points approx. vertically instead of horizontally. (Dobrin, 1960)

horizontal borer A machine, making holes from 2 to 6 in (5.08 to 15.24 cm) in diameter, used for drilling overburden at opencut coal mines. Bits are of the auger or winged types. (Lewis, 1964)

horizontal circle The circular horizontal plate of a theodolite, accurately divided so that horizontal angles can be precisely measured. (Hammond, 1965)

horizontal crosscut *horizontal drive.*

horizontal cut *drag cut.*

horizontal-cut underhand *underhand stoping.*

horizontal departure The amount, expressed in feet or degrees, a borehole has digressed horizontally from the intended target. (Long, 1960)

horizontal dip slip *horizontal slip.*

horizontal displacement (a) A term used by Tolman to designate strike slip. (AGI, 1987) (b) The distance two formerly adjacent points moved horizontally. (c) *strike slip.*

horizontal drive An opening with a small inclination (about 2 to 4 mm for 1 m in length) in the direction toward the shaft for draining the water and to facilitate hauling of the full cars to the shaft. Syn: *horizontal crosscut.* (Stoces, 1954)

horizontal fault A fault in the Earth's crust with no vertical displacement. (Webster 3rd, 1966)

horizontal intensity (a) The intensity of the horizontal component of the magnetic field in the plane of the magnetic meridian. (Hy, 1965) (b) The horizontal component of the vector magnetic-field intensity; it is one of the magnetic elements, and is symbolized by H. Cf: *vertical intensity.* (AGI, 1987)

horizontal load-bearing test *load-bearing test.*

horizontal pendulum A pendulum whose mass is constrained to move horizontally. (Schieferdecker, 1959)

horizontal prism *macrodome.*

horizontal-ring drilling *horadiam.*

horizontal screens Shaking screens with the plates supported in an essentially horizontal position that have been developed to obtain the advantages of low head room requirement. (Mitchell, 1950)

horizontal separation In faulting, the distance between the two parts of a disrupted unit (e.g., bed, vein, or dike), measured in any specified horizontal direction. Cf: *vertical separation. strike slip.* (AGI, 1987)

horizontal slip In a fault, the horizontal component of the net slip. Cf: *vertical slip.* Syn: *horizontal dip slip.* (AGI, 1987)

horizontal takeup A mechanism in which the takeup or movable pulley travels in an approx. horizontal plane. (NEMA, 1961)

horizontal throw The heave of a fault. (Hess)

hornblende granite A felsic plutonic rock, generally adamellite or granodiorite, containing an amphibole (often hornblende) as an essential dark-colored constituent; with decreasing quartz it grades through tonalite into normal diorite.

hornblende schist A schistose metamorphic rock consisting principally of hornblende, with little or no quartz. Unlike amphibolite, it does not need to contain plagioclase. (AGI, 1987)

horn coal (a) Eng. Coal worked partly end-on and partly face-on. (Fay, 1920) (b) A variety of cannel coal from South Wales. (Fay, 1920) (c) A coal that emits an odor when burning like that of burnt horn. (Fay, 1920) (d) Term in use in Saxony, Germany, for a variety of pitch coal similar to cannel coal. Syn: *half end.* (Tomkeieff, 1954)

horn coral Solitary coral. (AGI, 1987)

hörnesite *hoernesite.*

hornfels A fine-grained rock composed of a mosaic of equidimensional grains without preferred orientation and typically formed by contact metamorphism. Porphyroblasts or relict phenocrysts may be present in the granoblastic matrix. See also: *calc-silicate hornfels; pelitic hornfels; magnesian hornfels.* (AGI, 1987)

horn lead *phosgenite.*

horn quicksilver *calomel.*

Hornsey process A method for the low-temperature reduction of iron ore by means of a series of rotary kilns. The kilns are each about 5 ft (1.5 m) in diameter and 30 ft (9.1 m) in length. The first is used for preheating, the second for reduction, and the third for cooling the product. Pulverized coal is used, which makes it readily possible to control the combustion and to maintain constant temperature. (Osborne, 1956)

horn silver *chlorargyrite; embolite.*

horn socket A fishing tool specially designed to recover lost collared drill rods or drill pipe. It consists of a smooth-wall, tapered socket, the larger end down, equipped with a spring latch, which grips the drill rod under the collar when it is slid down over the top of the lost drill rod. When the socket is equipped with a flaring (bell-shaped) mouth, it is called a bell-mouth socket. (Long, 1960)

horn tiff In Missouri, calcite stained with carbonaceous material; sometimes dark enough to be mistaken for sphalerite. (Fay, 1920)

horse (a) Any irregularity cutting out a portion of the vein. See also: *rock fault.* (Fay, 1920) (b) To split into branches, as a vein of ore in a mine. (Standard, 1964) (c) Rock occupying a channel cut into a coalbed. See also: *horseback.* (AGI, 1987) (d) A body of sandstone or shale occupying a channel in a coal seam. See also: *horseback.* (AGI, 1987) (e) In structure, a large block of displaced wall rock caught along a fault, particularly a high-angle normal fault. (AGI, 1987) (f) A mass of country rock lying within a vein. See also: *internal waste.*

horseback (a) *cutout; swell.* (b) A bank or ridge of foreign matter in a coal seam. (AGI, 1987) (c) A large roll in a coal seam. (AGI, 1987) (d) A clay vein in a coal seam. Syn: *kettleback.* See also: *horse; symon fault; washout; slip.* (AGI, 1987) (e) A name applied by some writers to floor rolls in coal mines. (AGI, 1987) (f) Applied in some areas to clay veins; i.e., intrusions of clay into coalbeds. See also: *clay vein; sandstone dike.* (AGI, 1987) (g) Eng. A mass of stone with a slippery surface in the roof. In shape, it resembles a horse's back. (SMRB, 1930) (h) Natural channels cut or washed away by water in a coal seam and filled up with shale and sandstone. Sometimes, a bank or ridge of foreign matter in a coal seam. (Fay, 1920) (i) A portion of the roof or floor that bulges or intrudes into the coal. (Fay, 1920) (j) A mass of country rock lying within a vein or bed. (Fay, 1920) (k) A piece of slate, flat underneath, thick in the middle, and running out to a thin edge upon each side. See also: *kettle bottom.* (Fay, 1920) (l) Eng. A tree branch that has been horizontally embedded, carbonized, and compressed into lenticular shape in shale immediately above a coalbed. (Chem. Indust., 1939) (m) A term used in Maine for a low and somewhat sharp ridge of sand or gravel; also, but not generally, a ridge of rock that rises for a short distance with a sharp edge. A hogback. (Fay, 1920)

horseback excavator In bituminous coal mining, one who excavates horseback (banks or ridges of dirt or rock in the coal seam) in a strip mine with a power shovel. (DOT, 1949)

horseflesh ore *bornite.*

horse gear Bar pulled around by draft animal to actuate winding capstan. Syn: *whim gin.* Also called bullock gear. (Pryor, 1963)

horsepower applied *power upon the air.*

horsepower-hour The work performed or the energy consumed by working at the rate of 1 hp for 1 h (2.68 MJ), being equal to 1,980,000 ft·lbf (2.68×10^6 N·m). Abbrev., hp·h. (Webster 3rd, 1966)

horsepower of ventilation The work done in ventilating a mine or part of a mine is measured by the quantity circulated multiplied by the ventilating pressure required, the quantity being measured in cubic feet per minute (cubic meters per minute) and the pressure in pounds per square foot (kilograms per square meter). The horsepower required is, therefore, this product divided by 33,000. (Sinclair, 1958)

horsepower pull The effort necessary to maintain the normal operating speed of a conveyor under a rated capacity load. To this must be added the effort of acceleration, drive losses, etc., to arrive at a final driving effort. Horsepower pull may be referred to in terms such as effective tension, chain pull, turning effort, gear tooth pressure, etc. See also: *effective belt tension.*

horsetail Said of a major vein dividing or fraying into smaller fissures; also, said of an ore comprising a series of such veins. (AGI, 1987)

horsetail ore Ore in fractures that diverge from a major fracture.

horse transport An old method of transportation in mines in which horses were used to pull the mine cars along the roadways. Stables were installed underground in order for the horses to be kept permanently in the mine. Horse transportation has been replaced today by mechanical transport. (Stoces, 1954)

horse whim A device used for raising ore or water from mines, provided with radiating beams to which horses, oxen, or camels may be yoked. (Sandstrom, 1963)

horsfordite A possibly isometric mineral, Cu_3Sb; silver-white; sp gr, 8.8; reported from Lesbos Island, Greece.

horst An elongate, relatively uplifted crustal unit or block that is bounded by faults on its long sides. It is a structural form and may or may not be expressed geomorphologically. Etymol: German: no direct English equivalent. Cf: *graben.* (AGI, 1987)

hortonolite A magnesian variety of fayalite.

hose coupling A joint between a hose and a steel pipe, or between two lengths of hose. (Hammond, 1965)

hoshiite A nickeloan variety of magnesite.

Hoskold formula Two-rate valuation formula, once much used to determine present value (Vp) of mining properties or shares, with redemption of capital invested. Largely replaced by Morkill's formula. (Pryor, 1963)

host A rock or mineral that is older than rocks or minerals introduced into it or formed within or adjacent to it, such as a host rock, or a large crystal with inclusions of smaller crystals of a different mineral species; a palasome. Ant: guest. (AGI, 1987)

host element A common element that is substituted by a trace element in a rock mineral.

host rock A body of rock serving as a host for other rocks or for mineral deposits; e.g., a pluton containing xenoliths, or any rock in which ore deposits occur. It is a somewhat more specific term than country rock. (AGI, 1987)

hot Applied to a mine or part of a mine that generates methane in considerable quantities. (Fay, 1920)

hotbed An area, adjacent to the runout table, where hot rolled metal is placed to cool. Sometimes called the cooling table. (ASM, 1961)

hot blast Air forced into a furnace after having been heated. (Fay, 1920)

hot-blast man A stove tender at blast furnaces. (Fay, 1920)

hot-blast system The plenum system of ventilation. (Webster 2nd, 1960)

hot-carbonate process A process developed by the U.S. Bureau of Mines in which a hot solution of potassium carbonate is used to absorb impurities from gases and is then regenerated for reuse in a continuous cycle with maximum efficiency and minimum wasted heat. Also called hot-potash process; Benfield process.

hot cell A heavily shielded enclosure in which radioactive materials can be handled remotely through the use of manipulators and viewed through shielded windows so that there is no danger to personnel. (Lyman, 1964)

Hotchkiss superdip Much more sensitive than the common dip needle. The instrument consists of a magnetic needle free to rotate about a horizontal axis and a nonmagnetic bar with a counterweight at the end which is attached to the needle at its pivot, the two axes making an angle that can be varied. It measures changes in the total field and can be used to measure variations in the vertical field if its plane is oriented in a direction perpendicular to the magnetic meridian. See also: *dip needle.* (Dobrin, 1960)

hot crushing strength Compressive strength of brick at high temperature.

hot-dip coating The process of dipping metal components in molten tin or zinc to protect them against corrosion. See also: *galvanizing.* (Hammond, 1965)

hot-dip galvanizing Immersion of iron or steel articles in a bath of melted spelter, to produce a zinc coating. (Pryor, 1963)

hot-drawn Elongation of metal wire, tube, or rod by drawing it while heated through a constricting orifice. Opposite of cold-drawn. (Pryor, 1963)

hot forming Working operations such as bending, drawing, forging, piercing, pressing, and heading performed above the recrystallization temperature of the metal. (ASM, 1961)

hot laboratory A laboratory designed for the safe handling of radioactive materials. Usually contains one or more hot cells. (Lyman, 1964)

hot-laid type A bituminous pavement that is mixed and laid at relatively high temperatures, generally above 250 °F (121 °C). The highest type pavement that can be laid, it has greater durability and lower maintenance than any other type. (Pit and Quarry, 1960)

hot material Any material that, at the time of charging, is at a temperature of 70 °C or higher.

hot-metal ladle A ladle for the transfer of molten iron from a blast furnace to a mixer furnace and from there to a steel furnace; alternatively, the ladle may transfer molten pig iron direct from blast furnace to steel furnace. Such ladles are generally lined with fire clay refractories, but for severe conditions high-alumina and basic refractories have been tried with some success. (Dodd, 1964)

hot-metal mixer A large holding furnace for molten pig iron. The capacity of these furnaces, which are of the tilting type, is up to 1,400 st (1,270 t). Hot metal mixers may be active (that is, the pig iron is partially refined while in the furnace) or inactive (that is, the pig iron is merely kept molten until it is required for transfer to a steelmaking furnace). In either case, the bottom and walls of the furnace are made of magnesite refractories and the roof of silica refractories. (Dodd, 1964)

hot miller A tool operated by compressed air, fitted with cutting wheels that mill the hot cutting edges or rock drill bits to the required angle. See also: *detachable bit.* (Hammond, 1965)

hot-quenching Quenching in a medium at an elevated temperature. (ASM, 1961)

hot rolling The passing of hot steel bars through pairs of steel rolls to form rolled-steel sections. The final dimension of the product is approached in stages by adjusting the height of the rolls. (Nelson, 1965)

hot shortness Embrittlement of steel or wrought iron when hot, usually due to excessive sulfur content. (Pryor, 1963)

hot spot (a) A small portion of a furnace shell that is warmer than the rest. It indicates a thin lining. (Fay, 1920) (b) The zone of highest temperature within a glass-melting furnace. (ASTM, 1994)

Hot Springs diamonds Quartz crystals found near Hot Springs, AR.

hot top A refractory-lined steel or iron casting inserted into the tip of a mold and supported at various heights to feed an ingot as it solidifies. (ASM, 1961)

hot-wire anemometer Instrument particularly suited to the measurement of very low air velocities and the fluctuating velocities that occur in turbulent flow. Basically, it consists of a wire or wires, usually platinum, supported in a frame and heated electrically. When exposed to an air current, the heated wire cools, and as a result, its electrical resistance alters. The heated wire forms one arm of a Wheatstone-bridge-type circuit, and measurements of resistance change may be correlated with the velocity of airflow that caused that change. (Roberts, 1960)

hot working Deforming metal plastically at such a temperature and rate that strain hardening does not occur. The low limit of temperature is the recrystallization temperature. (ASM, 1961)

hourglass structure A type of zoning, esp. common in clinopyroxenes and chloritoids, in which a core, distinguished from the outer part by a difference of color or optical properties, has a cross section resembling that of an hourglass. (AGI, 1987)

house (a) Corn. A large mass of rich tin ore. Also called a carbona. (Arkell, 1953) (b) Eng. *gunnie; turnhouse.*

house coal Coal for use around colliery in miners' houses and for local sale. (BCI, 1947)

house of water Corn. A cavity or space filled with water.

hove (a) Scot. Past participle of heave. The floor of a mine working is said to heave or rise. (Fay, 1920) (b) A lode is hove or thrown in a certain direction by a fault. (Gordon, 1906)

hovel A large conical or conoidal brick structure within which a firing kiln is built. (Webster 3rd, 1966)

howell The upper stage in a porcelain furnace. (Standard, 1964)

howlite A monoclinic mineral, $Ca_2B_5SiO_9(OH)_5$; white; earthy or in small nodules; in the Mojave Desert region of California.

H-piece That part of a plunger lift in which the valves or clacks are fixed. (Fay, 1920)

HQ A letter name specifying the dimensions of bits, core barrels, and drill rods in the H-size and Q-group wireline diamond drilling system having a core diameter of 63.5 mm and a hole diameter of 96 mm. (Cumming, 1981)

hsianghualite An isometric mineral, $Ca_3Li_2Be_3(SiO_4)_3F_2$; occurs with taaffeite in metamorphosed limestone in Hunan Province, China.

huanghoite A trigonal mineral, $BaCe(CO_3)_2F$; as yellow platy masses in hydrothermal deposits near the Huang-Ho River, China. Also spelled huangeite-(Ce).

huascolite A variety of galena in which part of the lead is replaced by zinc. (Standard, 1964)

hub A survey point marked with a stake or metal pin and used as a reference point for locating a specific spot in a predetermined direction. (Long, 1960)

hub-and-groove diameter The outside diameter of the hub, or the diameter at the base of a groove cut in the hub to provide clearance for the link plates. (Jackson, 1955)

Hubbard distributor A continuous distributor consisting of a steel open-topped box filled with stone dust. Resting on the surface of the stone dust is a steel plate 1/4 in (0.64 cm) in thickness fitting loosely into the box and perforated with holes 3/16 in (0.48 cm) in diameter. The plate is connected by a series of chains and levers to a lever between the rails on either the loaded or empty side of the roadway. Each tub passing along depresses the lever and causes the steel plate to be lifted. A counterweight restores the track lever to vertical and the plate falls, causing a puff of stone dust to be ejected through each hole into the ventilating current. Thus stone dusting keeps pace with output. (Sinclair, 1958)

hübnerite A monoclinic mineral, $2[MnWO_4]$, with Mn replaced by Fe toward ferberite in the series commonly known as wolframite; one perfect cleavage; resinous; sp gr, 7.12; in granitic rocks, including pegmatites; in high-temperature quartz veins, and in placers; may be alone or associated with cassiterite, or with sulfides of iron, lead, or zinc. Also spelled huebernite. See also: *sanmartinite; huebnerite.*

hudge (a) A small box or tram without wheels, running on timber slides, drawn by a youngster, in thin and steep seams. (Fay, 1920) (b) An iron bucket for hoisting ore or coal. Syn: *bowk.* (Fay, 1920)

huebnerite A brownish-red tungstate of manganese, $Mn^{+2}WO_4$, one of the end-members of a variable series, commonly known as wolfram or wolframite, (Fe, Mn)WO_4; monoclinic. Syn: *hübnerite; wolframite.* (CMD, 1948)

huel Corn. A mine; a variant of wheal. (Fay, 1920)

Huff separator Type of electrostatic separator used in ore treatment. (Pryor, 1963)

hugger (a) N. of Eng. In coal mining, a back or cleat. (Fay, 1920) (b) Northumb. The principal cleat in coal. See also: *backs.* (Arkell, 1953)

hugger belt conveyor Two belt conveyors whose conveying surfaces combine to convey loads up steep inclines or vertically.

hugger drive A drive employing an auxiliary belt that bears against the surface of the conveying belt as it passes around the drive pulley to increase the pressure between the conveyor belt and the drive pulley.

hulk (a) Corn. To take down and remove the softer part of a lode, before removing the harder part. See also: *gouge.* (b) The removal of the soft gouge. (c) The excavation made by this operation.

Hull cell A special electrodeposition cell giving a range of known current densities for test work. (ASM, 1961)

hulsite A monoclinic mineral, $(Fe,Mg)_2(Fe,Sn)BO_5$; forms small black crystals or tabular masses at the contact of granite and metamorphosed limestone.

humacite A group name for bitumens that vary from gelatinous to hard resinous or elastic. Believed to represent an emulsion of highly acidic (humic acids) hydrocarbons with a varying amount of water (as high as 90%). Insoluble in organic solvents. (Tomkeieff, 1954)

humboldtine A hydrous ferrous oxalate, $Fe^{2+}C_2O_4 2H_2O$, occurring in capillary or botryoidal forms and black shale. Syn: *humboltite; oxalite.* (Tomkeieff, 1954)

humboldtite *datolite; humboldtine.*

Humboldt jig Movable-screen type of ore jig. (Pryor, 1963)

humboltite *datolite; humboldtine.*

humic acid Black acidic organic matter extracted from soils, low-rank coals, and other decayed plant substances by alkalis. It is insoluble in acids and organic solvents. (AGI, 1987)

humic coals (a) A group of coals, including the ordinary bituminous varieties, that have been formed from accumulations of vegetable debris that have maintained their morphological organization with little decay. The majority of them are banded and have a tendency to develop jointing or cleat. Chemically, humic coals are characterized by hydrogen varying between 4% to 6%. (Tomkeieff, 1954) (b) Coals in which the attritus may be composed predominately of transparent humic degradation matter. (AGI, 1987) (c) Introduced in 1906 by H. Potonie to describe coals, the original organic matter of which underwent change chiefly by humification; i.e., through the process of peat formation in the presence of oxygen. Most seams of coal consist principally of humic coal and the technological properties vary with their rank, with their petrographic composition, and with the manner of distribution of mineral inclusions. (IHCP, 1963)

humic degradation matter Finely comminuted degradation matter in coal, largely but not altogether derived from the woody tissues of plants, and like anthraxylon, largely derived from lignin. (AGI, 1987)

humid heat Ratio of the increase in total heat per kilogram of dry air to the rise in temperature, with constant pressure and humidity ratio.

humidifying effect The quantity of water evaporated per unit of time (usually 1 h) times the latent heat of vaporization at the evaporating temperature. (Strock, 1948)

humidity The water-vapor content of the atmosphere. The unmodified term often signifies relative humidity. See also: *absolute humidity; specific humidity.* (AGI, 1987)

humidostat An instrument for regulating the humidity in the atmosphere. Syn: *hygrostat.* (Standard, 1964)

humins In coal, amorphous brown to black substances formed by natural decomposition from vegetable substances; insoluble in alkali carbonates, water, and benzol. (Hess)

humite The mineral group alleghanyite, chondrodite, clinohumite, humite, jerrygibbsite, leucophoenicite, manganhumite, norbergite, ribbeite, and sonolite; monoclinic and orthorhombic fluosilicates of magnesium, iron, and/or manganese, with hydroxyl commonly replacing fluorine; similar physical properties, and structures closely related to those of the olivines; in metamorphosed dolomitic limestones, or skarns associated with ore deposits; commonly as chondrodite and clinohumite; at the Tilly Foster iron mine near Brewster and at Franklin, NJ.

humite group A group of isomorphous minerals consisting of olivine, chondrodite, humite, and clinohumite, and closely resembling one another in chemical composition, physical properties, and crystallization. (Webster 3rd, 1966)

Hummer screen Screen used to size moderately small material, vibrated electrically by solenoid action. (Pryor, 1963)

humocoll Peat derived from humic material and in rank corresponding to saprocoll. (Tomkeieff, 1954)

humonigritite A type of nigritite that occurs in sediments. Cf: *polynigritite; keronigritite.* (AGI, 1987)

humopel An organic mud composed of humic material corresponding in rank to sapropel. (Tomkeieff, 1954)

humosite A microscopical constituent of torbanite; translucent; dark brownish-red; isotropic. See also: *gélosite; matrosite; retinosite.* (Tomkeieff, 1954)

humper *booster conveyor.*

Humphrey's spiral A concentrating device that exploits differential densities of coal and its associated impurities by a combination of sluicing and centrifugal action. The material gravitates down through a stationary spiral trough with six turns (five for ore treatment) of mean radius 8 in (20.32 cm) with a fall per turn of 11 in (27.94 cm). Heavy particles stay on the inside, the lightest ones climb to the outside, and the resulting bands are separated at convenient points.

humus Dark-colored, organic, well-decomposed soil material consisting of the residues of plant and animal materials together with synthesized cell substances of soil organisms and various inorganic elements. (Stokes, 1955)

humus coal (a) Coal composed of anthraxylon in varying proportions and of varying thicknesses, associated with transparent attritus. (AGI, 1987) (b) Amorphous brown to black coal formed from vegetable matter and insoluble under continuous boiling in caustic alkalies; also insoluble in water and benzol. (Hess)

hundredweight A weight commonly reckoned in the United States, and for many articles in England, at 100 avoirdupois lb, but commonly in England, and formerly in the United States, at 112 avoirdupois lb. There is also an older hundredweight, called the long hundredweight, of 120 avoirdupois lb. Abbrev., cwt. (Standard, 1964)

Hungarian cat's eye An inferior greenish cat's eye obtained in the Fichtelgebirge in Bavaria. No such stone occurs in Hungary. (CMD, 1948)

Hungarian mill A rotating, grinding mill used in Hungary for removing small portions of gold from quartz by mixing with mercury; one of the many forms of pan amalgamators.

Hungarian opal (a) A white opal, with a fine color play; found in Slovakia. (b) A name used by the importing trade for any white opal regardless of origin.

Hungarian riffles Riffles used in undercurrents that are small angle irons or pieces of wood shod with iron. Syn: *transverse riffles.* (Lewis, 1964)

hung fire Delay in a blasting explosion caused by dampness of the powder or by too slow combustion of the fuse. (Korson, 1938)

hungry (a) Said of a rock, lode, or belt of country that is barren of ore minerals or of geologic indications of ore, or that contains very low-grade ore. Ant: likely. (AGI, 1987) (b) Said of a soil that is poor or not fertile. (AGI, 1987)

hung shot (a) A shot that does not explode immediately upon detonation or ignition. See also: *hangfire.* (Zern, 1928) (b) A delayed shot. (Hudson, 1932)

Hunt and Douglas process Consists of roasting matte carrying copper, lead, gold, and silver at a very low temperature, forming copper sulfate and oxide, but not silver sulfate. This product is leached with dilute sulfuric acid for copper. The resulting solution is treated with calcium chloride, and the copper is precipitated as subchloride by passing SO_2 through the solution. The cuprous chloride is then reduced to cuprous oxide by milk of lime, regenerating calcium chloride, and the cuprous oxide is smelted. (Liddell, 1918)

huntilite A silver arsenide occurring with native silver at Silver Islet, Lake Superior, MI. (Fay, 1920)

hunting (a) Unstable conditions occur with all fans when they are working against too high a resistance, and with forward-bladed radial-flow fans over most of their range, including the point of maximum efficiency. In these conditions, a drop in volume causes only a slight rise in fan pressure and conditions are only slowly restored to normal. This leads to continual and heavy fluctuations in load, a phenomenon known as "hunting." In extreme cases, a fan may hunt to the point where there is no rise in pressure with decreasing volume. It can then lose its load entirely and never recover it. (Roberts, 1960) (b) Abnormal time lag in automatic control system, in which a corrective change is so much exceeded that overmodulation ensues, the result being oscillation above and below the desired norm. Also called cycling; oscillation. (Pryor, 1963)

hunting coal York. Ribs and posts of coal left for second working. (Fay, 1920)

Huntington-Heberlein process A sink-float process employing a galena medium and utilizing froth flotation as the means of medium recovery. (Chem. Eng., 1949)

Huntington mill A cylindrical vertical tub from 3-1/2 to 6 ft (1.07 to 1.83 m) in diameter, with screen-guarded peripheral apertures through which ore pulp can be discharged after passing through the comminuting zone. Grinding is done by four rolling mullers, which hang inside from a yoke and which press outward when rotating, thus bearing an ore caught between them and the inner wall of the tub. Syn: *pendulum mill.* (Pryor, 1963)

hunting tooth Extra tooth designed for driven wheel so that its total number of teeth is not a multiple of those of the driving pinion. (Pryor, 1963)

huntite A trigonal mineral, $CaMg_3(CO_3)_4$; in white chalky masses in caves; as a weathering product in vugs and veins in magnesium-rich rocks; in magnesite deposits in Nevada.

Hunt's process Treatment of precious metal ores containing copper or zinc, using an ammoniacal cyanide solution and recovering ammonia by boiling. (Liddell, 1918)

hurdle A temporary screen or curtain to deflect the air upwards against the roof to disperse gas. (BS, 1963)

hurdle screen (a) Scot. A temporary screen or curtain for clearing gas out of a pit. Used esp. where gas has collected in potholes or caves in the roof. (Fay, 1920) (b) Scot. A screen used in underground fire-fighting that pushes the smoke back toward the fire and allows the firefighting team to advance within striking distance of the fire. (McAdam, 1955)

hurdle sheet A screen of brattice cloth erected across a roadway below a roof cavity or at the ripping lip to divert the air current upwards to dilute and remove an accumulation of combustible gases. See also: *pocket of gas.* (Nelson, 1965)

hurdy-gurdy (a) *hurdy-gurdy wheel.* (b) A dance house in a mining camp. (Standard, 1964)

hurdy-gurdy drill A hand auger used to drill boreholes in soft rock or rock material, such as soil, clay, coal, etc. (Long, 1960)

hurdy-gurdy wheel A water wheel operated by the direct impact of a stream upon its radially placed paddles. Syn: *hurdy-gurdy.* (Fay, 1920)

hureaulite A monoclinic mineral, $Mn_5(PO_4)_2[PO_3(OH)]_2{\cdot}4H_2O$; in pegmatites.

hurlbarrow Scot. A wheelbarrow. (Standard, 1964)

hurricane air stemmer A mechanical device for the rapid stemming of shotholes. It consists of a sand funnel connected by a T-piece to the charge tube, one end of which is provided with a valve and fittings to the compressed air column. The funnel is filled with sand, which is held uppermost, and the charge tube is inserted into the shothole. The sand is injected by the compressed air, and the tube is gradually withdrawn as the hole is being filled. (Nelson, 1965)

hurry (a) Scot. A screen or sieve. (b) To haul, pull, or push cars of coal in a mine. (Fay, 1920) (c) A chute, slide, or pass as for ore in a mine, or for coal discharged from cars into vessels. (Webster 2nd, 1960) (d) Gr. Brit. A wooden staging on a navigable river from which to load vessels with coal. (Standard, 1964)

hurry gum Scot. The fine material that passes through a screen or sieve.

hush Eng. To clear away (soil) from ore with a rush of water. (Standard, 1964)

hushing *hydraulic prospecting; booming.* (AGI, 1987)

hutch (a) Corn. A cistern or box for washing ore. See also: *jig.* Syn: *washing hutch.* (Fay, 1920) (b) Scot. A basket for coal. (Standard, 1964) (c) The bottom compartment of a coal or ore-dressing jig. (Webster 3rd, 1966) (d) The part of a jig below the screen plate, in which the washbox rests and the pulsating movement of the water takes place. (BS, 1962) (e) A car on low wheels in which coal is drawn and hoisted out of a mine pit. (Webster 3rd, 1966) (f) To wash ore in a box or jig. (g) A basket for measurement and transport of coal. (h) A small train or wagon for removing coal or ore from a mine. (i) An old and varying English measure, as (for coal) 2 Winchester bushels (70.5 L). (Webster 2nd, 1960) (j) Scot. Two hundredweight (91 kg or 96 kg) of pyrites. (Standard, 1964) (k) The bottom compartment of an ore-dressing jig and/or the mineral product that collects there. (Webster 3rd, 1966) (l) The part of a washbox situated below the screenplate in which the controlled pulsating movement of the water takes place. (BS, 1962)

hutch cleading The boards that form the sides, bottom, and ends of a mine car, or hutch. (Standard, 1964)

hutcher One who runs hutches. (Webster 2nd, 1960)

hutching N. of Eng. Term used for tramming. (Nelson, 1965)

hutch mender A repairer of tubs or hutches broken in a mine. (CTD, 1958)

hutch mounting Scot. The ironwork on the frame and box of a wooden hutch. (Fay, 1920)

hutch product The fine, heavy materials that pass through the meshes of the screen in a jig. (Newton, 1959)

hutch road (a) A road through a mine. (Standard, 1964) (b) Scot. A hutch tramway. (Fay, 1920)

hutchwork In mineral processing, the concentrates passing down through the ore jig into the hutch. (Pryor, 1963)

huttonite A very rare, strongly radioactive, colorless to pale cream, monoclinic mineral, $ThSiO_4$, found in sands and gravels with scheelite, cassiterite, uranothorite, zircon, ilmenite, and gold. (Crosby, 1955)

Huwood loader This machine comprises a number of horizontal rotating flight bars working near the floor of the seam and driven through gearing by an electric motor. These bars push into the coal in their extended position and are almost completely concealed inside the loader casing in their retracted position. They push prepared coal up a ramp on to a low, bottom-loaded conveyor belt. The machine is hauled along the face by means of two steel ropes wound on separate drums on the loader; one rope passes up the front of the coal and is held by means of an anchor prop; the other rope is threaded under the cut coal by means of a threader pipe attached to the rear of the coal cutter. Lengths of rope equal to the drum capacity are joined by figure-8 links and are detached and unwound from the drum as the loader proceeds along the face. (Mason, 1951)

Huwood slicer A cutter-loader based on the plow principle and designed to cut coal that is too hard for the ordinary plow. Two vertical blades, fitted one at each end of the machine, carry cutting picks that shear the coal from the face by an oscillating motion. The machine is hauled backward and forward along the face by a chain haulage mounted on the tail end of the conveyor. The conveyor and slicer are held up to the coal by pneumatic rams spaced along the goaf side of the conveyor. The sheared coal is forced from the face by the wedge shape of the machine and is loaded on to the armored conveyor by means of specially shaped ramps. The machine has been designed for seams with a minimum thickness of 4 ft (1.22 m) and has a maximum cutting depth of 14 in (35.56 cm). Syn: *activated plow*. (Nelson, 1965)

Huygen's principle A very general principle applying to all forms of wave motion that states that every point on the instantaneous position of an advancing phase front (wave front) may be regarded as a source of secondary spherical wavelets. The position of the phase front a moment later is then determined as the envelope of all of the secondary wavelets (ad infinitum). This principle is useful in understanding effects due to refraction, reflection, diffraction, and scattering, of all types of radiation, including sonic radiation as well as electromagnetic radiation, and applying also to ocean wave propagation. (Hunt, 1965)

HW Letter name specifying the dimensions of bits, core barrels, and drill rods in the H-size and W-group wireline diamond drilling system having a core diameter of 76.2 mm and a hole diameter of 99.2 mm. Syn: *HX*. (Cumming, 1981)

H wave *hydrodynamic wave.*

HX Letter name specifying the dimensions of core, core barrels, and casing in the H-size and X-series wireline diamond drilling system having a core diameter of 76.2 mm and a hole diameter of 99.2 mm. The HX designation for coring bits has been replaced by the HW designation. Syn: *HW*. (Cumming, 1981)

hyacinth A red-orange variety of zircon; also applied to similarly colored garnet, quartz, and other minerals. Syn: *cinnamon stone; essonite; jacinth.*

hyaline (a) Sometimes used as a prefix ("hyalo-") to names of volcanic rocks with a glassy texture, e.g., hyalobasalt. (AGI, 1987) (b) Said of a mineral that is amorphous. (AGI, 1987)

hyalite A variety of globular or botryoidal opal that shows greenish-yellow fluorescence under ultraviolet light and can be mistaken for uranium-bearing minerals such as autunite. Syn: *glass opal; water opal.*

hyalo- A prefix meaning glassy. (AGI, 1987)

hyalocrystalline Said of the texture of a porphyritic igneous rock in which crystals and glassy groundmass are equal or nearly equal in volumetric proportions, the ratio of phenocrysts to groundmass being between 5:3 and 3:5. Syn: *semicrystalline; hemicrystalline.* (AGI, 1987)

hyalophane A monoclinic or triclinic mineral, $4[(K,Ba)Al(Si,Al)_3O_8]$; feldspar group, intermediate in the series orthoclase-celsian; prismatic cleavage; in manganese ore deposits, or veins and pegmatites.

hyalophitic Said of the texture of an igneous rock in which the mesostasis is glassy and makes up a proportion of the rock intermediate in texture between hyalophilitic and hyalocrystalline. Cf: *intersertal.* (AGI, 1987)

Hybinette process A process used for refining of crude nickel anodes. These are placed in reinforced concrete tanks lined with asphalt. The nickel anodes are dissolved electrochemically and the impurities, such as copper and iron, pass into solution. The cathodes are surrounded by bags of closely woven canvas duck, fastened on wooden frames, and pure nickel electrolyte is passed continuously into them to maintain a higher solution level inside the cathode compartment than outside. By this means, the pure solution flows through the pores of the bags, thus preventing the ions of copper, etc., in the solution in the anode compartment from migrating into the cathode compartment, depositing on the cathode, and preventing the refining process from taking place. The electrolyte in the anode compartments is drawn off continuously and is purified in the copper cementation and iron precipitation departments before being returned to the cathode compartments of the nickel deposition tanks. (Osborne, 1956)

hybrid (a) Pertaining to a rock whose chemical composition is the result of assimilation. (AGI, 1987) (b) A rock whose composition is the result of assimilation. See also: *hybridization*. (AGI, 1987)

hybridization The process whereby rocks of different composition from that of the parent magma are formed, by assimilation. See also: *hybrid; assimilation; contamination*. (AGI, 1987)

hydatogenesis The crystallization or precipitation of salts from normal aqueous solutions; the formation of an evaporite. (AGI, 1987)

hydatogenic Derived from or modified by substances in a liquid condition; said of the genesis of ores and other minerals; opposite of pneumatogenic. Cf: *hydatopneumatogenic*. (Standard, 1964)

hydatopneumatogenic Said of a rock or mineral deposit formed by both aqueous and gaseous agents. Cf: *hydatogenic; pneumatogenic.* (AGI, 1987)

hydrabrake retarder A mine car retarder, based on the principle of the dashpot, consists of individual braking units that can be fastened to the rails at spacings according to need over any desired distance. The unit offers no resistance to motion at very low car speeds, but as the speed increases, the braking force exerted upon it increases accordingly, following the usual oil dashpot characteristic. See also: *dashpot*. (Nelson, 1965)

hydrargillite *gibbsite.*

hydrate (a) A compound or complex ion formed by the union of water with some other substance and represented as actually containing water. (Webster 3rd, 1966) (b) A hydroxide, such as calcium hydrate (hydrated lime). (Webster 3rd, 1966)

hydrated lime A dry powder, $Ca(OH)_2$, obtained by hydrating quicklime.

hydration (a) The chemical combination of water with another substance. (AGI, 1987) (b) The process of adding water, or the elements of water (oxygen and hydrogen combined in the hydroxyl radical), to any substance.

hydraulic (a) Conveyed, operated, effected, or moved by means of water or other fluids, such as a hydraulic dredge, using a centrifugal pump to draw sediments from a river channel. (AGI, 1987) (b) Pertaining to a fluid in motion, or to movement or action caused by water. (AGI, 1987) (c) Hardening or setting under water; e.g., hydraulic lime or hydraulic cement. (AGI, 1987)

hydraulic action The mechanical loosening and removal of weakly resistant material solely by the pressure and hydraulic force of flowing water, as by a stream surging into rock cracks or impinging against the bank on the outside of a bend, or by a jet of water impacting a gravel bank. (AGI, 1987)

hydraulic blasting Fracture using a hydraulic cartridge, a ram-operated device used to split coal. (Pryor, 1963)

hydraulic cartridge (a) A device used in mining to split coal, rock, etc., having 8 to 12 small hydraulic rams in the sides of a steel cylinder. (Fay, 1920) (b) *coal burster.*

hydraulic cement A cement that can set and harden under water.

hydraulic cementing A borehole-cementing operation using a downhole cement injector. (Long, 1960)

hydraulic chock A steel face support structure consisting of from one to four hydraulic legs or uprights. The four-leg chock is mounted in a strong fabricated steel frame with a large head and base plate. It is controlled by a central valve system that operates either on the four legs simultaneously or on the front and rear pairs separately. See also: *chock; yielding prop; self-advancing supports.* (Nelson, 1965)

hydraulic chuck A diamond-drill rod chuck having jaws with clamping and unclamping movements actuated hydraulically instead of by hand-turned setscrews. (Long, 1960)

hydraulic circulating system A method used to drill a borehole wherein water or a mud-laden liquid is circulated through the drill string during drilling. See also: *diamond drill*. (Long, 1960)

hydraulic classifier Tank into which ore pulp is fed steadily and subjected to the sorting effect of a stream of hydraulic water that rises at controlled rate. Heavier or coarser equal settling particles gravitate down and away via a bottom discharge, while lighter ones are carried up and out. Syn: *hydrosizer*. See also: *elutriator*. (Pryor, 1963)

hydraulic conveying Use of flowing water or slow settling fluids based on water mixed with suitable heavy minerals to convey rock, coal, etc., in pipes. (Pryor, 1963)

hydraulic conveyor A type of conveyor in which water jets form the conveying medium for bulk materials through pipes or troughs.

hydraulic cylinder As applied to a diamond drill, a syn. for feed cylinder. (Long, 1960)

hydraulic discharge The direct discharge of ground water from the zone of saturation, as via springs, wells, or infiltration ditches or tunnels.

hydraulic dredge (a) A dredge consisting of a hull on which is mounted a suction pipe and support, a pump with motors and controls, and a discharge line. Commonly used in dredging canals and in providing fill for the creation of land in near-shore or low-lying areas and in sand and gravel dredging operations. (Mero, 1965) (b) A floating pump that sucks up a mixture of water and soil, and usually discharges it on land through pipes. (Nichols, 1976)

hydraulic dredger A suction dredger. (CTD, 1958)

hydraulic drill A hand-held or machine-mounted rotary drill for boring shot-firing holes in coal or rock and operated by hydraulic fluid. The drill outfit includes a skid-mounted powerpack comprising a 5-hp (3.73-kW) flameproof electric motor, pump, and tank. The coal drill weighs about 32 lb (14.5 kg). (Nelson, 1965)

hydraulic elevator An arrangement for lifting gravel and sand up to the drainage level. A jet of water is used to create a powerful suction in a hopper, and the water and gravel are carried up a pipeline and then run down the sluice boxes. This appliance was widely used in various goldfields toward the end of the 19th century. (Nelson, 1965)

hydraulic excavation Excavation by means of a high-pressure jet of water, the resulting waterborne excavated material being conducted through flumes to the desired dumping point. (Hammond, 1965)

hydraulic extraction A term that has been given to the processes of excavating and transporting coal or other material by water energy. Also called hydroextraction or hydraulic mining. See also: *hydromechanization.* (Nelson, 1965)

hydraulic feed A method of imparting longitudinal movement to the drill rods on a diamond or other rotary-type drill by a hydraulic mechanism instead of mechanically by gearing. (Long, 1960)

hydraulic fill Waste material transported underground and flushed into place by use of water. Syn: *mine fill*. See also: *fill*. (Pryor, 1963)

hydraulic-fill dam A dam composed of earth, sand, gravel, etc., sluiced into place; generally the fines are washed toward the center for greater imperviousness. (Seelye, 1951)

hydraulic filling Washing waste material, such as mill tailings and ground waste rock, into stopes with water to prevent failure of rock walls and subsidence. Problems involved in its use are stope preparation, choice and mixing of material, its particle size distribution, wear on pipe, and removal of water that transports the material into the mine. Compressed air may be used to force the filling through pipes. (Lewis, 1964)

hydraulic fluid A fluid supplied for use in hydraulic systems. Low viscosity, low rate of change of viscosity with temperature, and low pour point are desirable characteristics. Hydraulic fluids may be of petroleum or nonpetroleum origin. (Hammond, 1965)

hydraulic fluid coupling A hydraulic fluid coupling transmits power from the driving member to the driven member through oil. A rotating impeller attached to the drive shaft throws oil directly against a turbine converter, which always delivers the same torque as the engine or motor produces. Fluid couplings are particularly advantageous in starting heavy loads since the motor or engine is permitted to run at high efficient speeds while the coupling output shaft gradually accelerates the load to running speed. (Pit and Quarry, 1960)

hydraulic flume transport The transport of coal, pulp, or mineral by the energy of flowing water in semicircular or rectangular channels. The gradient should not be less than 3°. Coal movement in flumes commences at a water velocity of about 3 ft/s (0.9 m/s), but in practice a velocity of at least 6 ft/s (1.8 m/s) is arranged. See also: *hydromechanization.* (Nelson, 1965)

hydraulic flushing *hydraulic stowing.*

hydraulic friction The resistance to flow exerted on the perimeter or contact surface between a stream and its containing conduit, due to the roughness characteristic of the confining surface, which induces a loss of energy. Energy losses arising from excessive turbulence, impact at obstructions, curves, eddies, and pronounced channel changes are not ordinarily ascribed to hydraulic friction. (AGI, 1987)

hydraulic giants Used for working large placer deposits. Also called hydraulic monitors and water cannons. (Lewis, 1964)

hydraulic gradeline In a closed conduit, a line joining the elevations to which water could stand in risers. In an open conduit, the hydraulic gradeline is the water surface. (Seelye, 1951)

hydraulic gradient (a) A line joining the points of highest elevation of water in a series of vertical, open pipes rising from a pipeline in which water flows under pressure. (Webster 3rd, 1966) (b) Loss of hydraulic head per unit distance of flow. (ASCE, 1958) (c) The slope of the hydraulic grade-line. The slope of the surface of water flowing in an open conduit. (Seelye, 1951)

hydraulic gravel-pump mining Consists of the use of high-pressure water jets to disintegrate ore-bearing ground, together with gravel pumps to elevate the spoil to a treatment plant. Initial mining operations consist of the establishment of the mine hole or paddock. This is achieved by sinking or cutting downwards with monitors and removing the spoil by pumping, the pump being lowered as the hole deepens. (Mining)

hydraulic hoisting *hydraulic transport.*

hydraulic jack A jack in which the lifting head is carried on a plunger working in a cylinder, to which oil or water is supplied under pressure from a small hand-operated pump. See also: *hydrostatic press.* (CTD, 1958)

hydraulic jack operator *track-moving machine operator.*

hydraulic jump In fluid flow, a change in flow conditions accompanied by a stationary, abrupt turbulent rise in water level in the direction of flow. It is a type of stationary wave. (AGI, 1987)

hydraulicking Excavating alluvial or other mineral deposits by means of high-pressure water jets. See also: *hydraulic mining; hydromechanization; monitor.* Cf: *ground sluicing.*

hydraulic lime Lime that is combined with silica, alumina, and iron oxide and will set and harden underwater.

hydraulic limestone An impure limestone that contains silica and alumina (usually as clay) in varying proportions and that yields, upon calcining, a cement that will harden underwater. See also: *cement rock.* Syn: *waterlime.* (AGI, 1987)

hydraulic load cell A safety device developed by the U.S. Bureau of Mines for sensing pressure changes, thereby warning in advance of bumps. The cells are embedded in the walls and roofs of coal mines.

hydraulic loading The flushing or slicing of coal or other material broken down by water jets along the floor and into flumes. Coal will flow back toward the flume if sufficient water is available and the gradient is not less than 6° to 7° in favor of the flow. Flexible low-pressure hoses (150 to 200 psi or 1.0 to 1.4 MPa) are sometimes used to assist in the flushing operations. See also: *hydromechanization.* (Nelson, 1965)

hydraulic machine A borehole-drilling machine on which the bit-feeding mechanism is hydraulically actuated. (Long, 1960)

hydraulic main A main (pipe) for collecting and condensing the volatile matter given off in carbonization of coal in the coking process. (Mersereau, 1947)

hydraulic mean depth The cross section of water flowing through a channel or pipe divided by the wetted perimeter of the conduit. Syn: *hydraulic radius.* (Hammond, 1965)

hydraulic mine filling Filling a mine with material transported by water. Cf: *silting; slush. flush.* (Fay, 1920)

hydraulic miner In metal mining, one who tends riffles, sluices, and does other work in connection with the hydraulic placer mining of gold. In this type of mining, gold bearing gravel, usually in a bank, is excavated by the erosive action of a high-pressure stream of water being directed at the bank through a nozzle. The gravel is then forced into sluices where the gold particles sink and are caught by riffles (cleats) along the sluice bottom. (DOT, 1949)

hydraulic mining (a) Mining by washing sand and dirt away with water that leaves the desired mineral. See also: *hydraulicking.* (MacCracken, 1964) (b) The process by which a bank of gold-bearing earth and rock is excavated by a jet of water, discharged through the converging nozzle of a pipe under great pressure, the earth or debris being carried away by the same water, through sluices, and discharged on lower levels into the natural streams and watercourses below; where the gravel or other material of the bank is cemented, or where the bank is composed of masses of pipe clay, it is shattered by blasting with powder. Also used for other ores, earth, anthracite culm, etc. Made unlawful and prohibited in certain river systems where it obstructs navigation and injures adjoining landowners. See also: *placer mining.* (Ricketts, 1943; Fay, 1920) (c) In underground hydraulic mining, the extraction of coal by high-velocity water jets, directed at the seam from a monitor or powerful jet, which can withstand high water pressures. The jets are also used to impel the broken coal along the floor to the point of collection. See also: *hydromechanization; jet-assisted cutting.* (Nelson, 1965)

hydraulic monitor A device for directing a high-pressure jet of water in hydraulicking. It is essentially a swivel-mounted, counterweighted nozzle attached to a tripod or other type of stand and so designed that one person can easily control and direct the vertical and lateral movements of the nozzle. See also: *giant; monitor.*

hydraulic motor A multicylinder reciprocating engine, generally of radial type, driven by water under pressure. (CTD, 1958)

hydraulic permeability The ability of a rock or soil to transmit water under pressure. It may vary according to direction. (AGI, 1987)

hydraulic pipe transport The conveyance of coal by means of water flowing in pipes. Coal may be pumped to the surface in shallow mines, but beyond 150 ft (46 m) or so of depth, there are technical difficulties. Solids handling pumps rarely deliver against heads exceeding about 200 ft (61 m). Two such pumps, placed in series, have been used in Trelewis Drift, Wales, to pump out slurry. See also: *hydromechanization; transport.*

hydraulic power The use of pressure oil or soluble oil and water for operating face machines and steel supports. The fluid is supplied by rotary pumps driven by electricity located at points near the face. Hydraulic power has an advantage in that the space required is considerably less than that for conventional drives. See also: *power pack.* (Nelson, 1965)

hydraulic pressure The total thrust, expressed in pounds or tons, that the hydraulic-feed mechanism on a drill can impose on a drill string; also, the pressure of the fluid within the hydraulic cylinders, generally expressed in pounds per square inch. (Long, 1960)

hydraulic profile A vertical section of the potentiometric surface of an aquifer. (AGI, 1987)

hydraulic prop A prop consisting of two telescoping steel cylinders that are extended by hydraulic pressure that may be provided by a hand-operated pump built into the prop. The prop holds about half a gallon of mineral oil and is fitted with a yield valve that relieves the pressure when the load exceeds that for which the prop is set. A hydraulic prop enables quicker setting and uniform initial loading, and it can be withdrawn from a remote, safe position. The hydraulic prop was first used in a British coal mine in 1947. See also: *steel prop.* (Nelson, 1965)

hydraulic prospecting The use of water to clear away superficial deposits and debris to expose outcrops, for the purpose of exploring for mineral deposits. Syn: *hushing.* (AGI, 1987)

hydraulic radius In a stream, the ratio of the area of its cross section to its wetted perimeter. Symbol: R. Syn: *hydraulic mean depth.* (AGI, 1987)

hydraulic ram (a) A pump that forces running water to a higher level by utilizing the kinetic energy of flow, only a small portion of the water being so lifted by the velocity head of a much larger portion when the latter is suddenly checked by the closing of a valve. Also called ram. (Webster 3rd, 1966) (b) A device for lifting water by the water hammer produced by checking the flow periodically. (Seelye, 1951) (c) The plunger of a hydraulic press. (CTD, 1958) (d) A device whereby the

pressure head produced when a moving column of water is brought to rest is caused to deliver some of the water under pressure. (CTD, 1958)
hydraulic rotary drilling Method of drilling that uses rotating bits lubricated by a stream of mud. (Mersereau, 1947)
hydraulics A branch of science that deals with practical applications, such as the transmission of energy or the effects of flow of water or other liquid in motion. (Webster 3rd, 1966)
hydraulic set The set obtained by the addition of water to hydraulic setting materials.
hydraulic sluicing The process of moving materials by water; colloquially, hydraulicking. (Seelye, 1951)
hydraulic stowing The filling of the waste in mines by waterborne material by pipeline. See also: *pneumatic stowing.* Syn: *hydraulic flushing.* (Nelson, 1965)
hydraulic stowing pipe A steel or iron pipe used for transporting the material in hydraulic stowing. Ordinary pipes wear very rapidly owing to the chippings in the water; therefore, they are lined with abrasion-resistant material. This lining gives a very much longer life to the pipe. (Nelson, 1965)
hydraulic stripping The excavation and removal of overburden by hydraulicking. (Nelson, 1965)
hydraulic transport Movement of ore by water flowing through pipelines. Includes hydraulic hoisting. See also: *pipeline transport.* (Pryor, 1963)
hydraulic underreamer An underreamer with cutting lugs that can be expanded or retracted by a hydraulically actuated device. See also: *underreamer.* (Long, 1960)
hydraulic valve A valve for regulating the distribution of water in the cylinders of hydraulic elevators, cranes, etc. (Crispin, 1964)
Hydrik process A commercial process for the production of hydrogen by reaction of caustic soda on aluminum. (Osborne, 1956)
hydrite Denotes a very common microlithotype in Japanese Tertiary coal. It consists of the macerals vitrinite, degradinite, and exinite. The dull bands of many Japanese Tertiary humic coals consist largely of hydrite and generally occurs alternating with vitrite as microfine bands, one or the other predominating.
hydrobarometer An instrument for determining the depth of sea water by its pressure. (Crispin, 1964)
hydrobiotite A monoclinic mineral, 1:1 interstratified biotite and vermiculite; in clay.
hydrocarbon Any organic compound, gaseous, liquid, or solid, consisting solely of carbon and hydrogen. They are divided into groups of which those of special interest to geologists are the paraffin, cycloparaffin, olefin, and aromatic groups. Crude oil is essentially a complex mixture of hydrocarbons. (AGI, 1987)
hydrocarbon anomaly Very weak oil or gas seeps, so weak that the deposition of material at the surface cannot be recognized without chemical analysis. (Hawkes, 1962)
hydrocerussite A basic carbonate of lead, $Pb_3(CO_3)_2(OH)_2$. It occurs as a secondary mineral found associated with leadhillite, matlockite, cerussite, mendipite, and paralaurionite. (Dana, 1944)
hydrochemical anomaly Anomalous patterns of elements contained in ground or surface water. *anomaly.* (Hawkes, 1962)
hydrochemical prospecting *hydrogeochemical prospecting.*
hydrocyanic acid Unstable; volatile; colorless; extremely poisonous; gas or liquid; HCN; soluble in water, in alcohol, and in ether; only slightly dissociated with water; and an odor resembling that of bitter almonds. Formed by decomposing metallic cyanides with hydrochloric acid. Syn: *hydrogen cyanide; prussic acid.* (Standard, 1964; Handbook of Chem. & Phys., 2)
hydrocyclone A cyclone separator in which a spray of water is used. (ASM, 1961)
hydrodynamic computer codes Computer codes or models that compute the properties or predicted behavior of explosives or materials subjected to supersonic (dynamic) forces.
hydrodynamics The aspect of hydromechanics that deals with forces that produce motion. Cf: *hydrostatics.* (AGI, 1987)
hydrodynamic wave An obsolete term for a type of surface wave that is similar to a Rayleigh wave but has an opposite particle motion. Syn: *H wave.* (AGI, 1987)
hydrodynamometer An instrument for determining the velocity of a fluid in motion by its pressure. (Standard, 1964)
hydroextraction *hydraulic extraction.*
hydroextractor *centrifuge.*
hydrofranklinite *chalcophanite.*
hydrogarnet A member of the garnet group having SiO_4 partly replaced by $(OH)_4$. Cf: *hydrogrossular; garnetoid.*
hydrogenation of coal *coal liquefaction.*
hydrogen cyanide *hydrocyanic acid.*
hydrogen embrittlement A condition of low ductility in metals resulting from the absorption of hydrogen. (ASM, 1961)
hydrogen ion The stripped (naked) proton of hydrogen, H+, or the proton combined with one or more molecules of water, as H_3O^+ or $H(H_2O)^+$. The latter is sometimes called oxonium, hydroxonium, or hydronium ion. H+ is usually spoken of as the proton. H-ion concentration is the pH value. (Pryor, 1963)
hydrogenous (a) Said of coals high in moisture, such as brown coals. (AGI, 1987) (b) Said of coals high in volatiles, such as sapropelic coals. (AGI, 1987)
hydrogen sulfide Colorless; flammable; gas; H_2S. It is readily decomposed. Reacts with bases forming sulfides and with some metals to produce metal sulfides and to liberate hydrogen. Poisonous. Syn: *hydrosulfuric acid.* (Handbook of Chem. & Phys., 2)
hydrogeochemical prospecting Prospecting guided by the composition of ground or surface water. Syn: *hydrochemical prospecting.*
hydrograph A graph showing stage, flow, velocity, or other characteristics of water with respect to time. A stream hydrograph commonly shows rate of flow; a ground-water hydrograph, water level or head. (AGI, 1987)
hydrography (a) The science that deals with the physical aspects of all waters on the Earth's surface, esp. the compilation of navigational charts of bodies of water. (AGI, 1987) (b) The body of facts encompassed by hydrography. (AGI, 1987)
hydrogrossular A group name for the series of hydrogarnets encompassing the series hibschite, katoite, and grossular. Water content ranges from about 1.5% in hibschite to 13% in katoite; grossular is anhydrous. Syn: *hydrogarnet.* Cf: *garnetoid; hydrogarnet.*
hydrohematite A mineral, $Fe_2O_3{\cdot}nH_2O$, probably a mixture of the two minerals haematite and goethite, the former being in excess. It is fibrous and red in mass, with an orange tint when powdered. Also called turgite. See also: *turgite.* (CMD, 1948)
hydrologic cycle The constant circulation of water from the sea, through the atmosphere, to the land, and its eventual return to the atmosphere by way of transpiration and evaporation from the sea and the land surfaces. Syn: *water cycle.* (AGI, 1987)
hydrology (a) The science that deals with global water (both liquid and solid), its properties, circulation, and distribution, on and under the Earth's surface and in the atmosphere, from the moment of its precipitation until it is returned to the atmosphere through evapotranspiration or is discharged into the ocean. In recent years, the scope of hydrology has been expanded to include environmental and economic aspects. (AGI, 1987) (b) The sum of the factors studied in hydrology; the hydrology of an area or district. (AGI, 1987)
hydrolysis (a) The formation of an acid and a base from a salt by interaction with water, caused by the ionic dissociation of water. (CTD, 1958) (b) The decomposition of organic compounds by interaction with water; either cold, or on heating alone, or in the presence of acids or alkalies. (CTD, 1958)
hydromagnesite A monoclinic mineral, $Mg_5(CO_3)_4(OH)_2{\cdot}4H_2O$; pseudo-orthorhombic; in low-temperature veins in serpentinite.
hydromechanics The mechanics of fluids, including hydrostatics, hydrodynamics, hydrokinetics, and pneumatics. (Standard, 1964)
hydromechanization A term applied to hydraulic methods of excavating and transporting coal and other products underground. See also: *hydraulic extraction; hydraulic flume transport; hydraulic loading; hydraulic mining; hydraulic pipe transport; hydraulicking.* (Nelson, 1965)
hydrometallurgy The treatment of ores, concentrates, and other metal-bearing materials by wet processes, usually involving the solution of some component, and its subsequent recovery from the solution. Syn: *wet metallurgy.* (Kirk, 1947)
hydrometamorphism Alteration of rock by material that is added, removed, or exchanged by water solutions, without the influence of high temperature and pressure. Syn: *hydrometasomation; hydrometasomatism.* Cf: *pyrometamorphism.* (AGI, 1987)
hydrometasomation *hydrometamorphism.*
hydrometasomatism *hydrometamorphism.*
hydrometer An instrument used for determining the density or specific gravity of fluids, such as drilling mud or oil, by the principle of buoyancy. See also: *gravimeter; Marsh funnel; specific-gravity hydrometer.* (Long, 1960)
hydrometer method The method employed for the determination of the apparent specific gravity of coal and coke. (Kentucky, 1952)
hydrometrograph An instrument for determining and recording the quantity of water discharged from a pipe, orifice, etc., in a given time. (Osborne, 1956)
hydromica (a) *illite.* (b) A general term for brammallite, hydrobiotite, and illite.
hydromorphic anomaly An anomaly where the dynamic agents are aqueous solutions, which brought the elements to the site of deposition.
hydromuscovite *illite.*
hydronium jarosite A trigonal mineral, $(H_3O)Fe_3(SO_4)_2(OH)_6$; alunite group.
hydrophilic (a) Of, relating to, having, or denoting a strong affinity for water. (Webster 3rd, 1966) (b) Applied to such easily dispersed colloidal clay minerals as montmorillonite that swell in water as the result of water attraction and hydration and that are not easily coagu-

lated. (c) Substance attracted to a water phase rather than to air in an airwater interphase. A group tending to bind water is hydrophilic (opposite of hydrophobic). The hydroxyl (OH) groups in hydroxides are typical, and their hydrophilic solutions in water are hydrosols. See also: *lyophilic.* (Pryor, 1963)

hydrophobic (a) Lacking a strong affinity for water. Opposite of hydrophilic. (Webster 3rd, 1966) (b) Applied to water-repelling substances and surfaces and to easily coagulated colloids.

hydrophone A pressure-sensitive detector that responds to sound transmitted through water. It is used in marine seismic surveying, or as a seismometer in a well. Syn: *pressure detector.* (AGI, 1987)

hydrophotometer A sensitive instrument used in water transparency and light absorption measurements at sea. The instrument, which contains its own light source, can measure fine graduations of transparency of an individual water mass. (Hunt, 1965)

hydroscope An instrument for detecting moisture, esp. in the air. (Standard, 1964)

hydroseparator Essentially, a shallow tank, usually cylindrical, which is kept agitated by hydraulic water and/or stirring devices. Pulp fed to the tank is separately discharged as a free-settling fraction containing the coarser and heavier particles and an overflowing fraction containing the finer, lighter material. (Pryor, 1963)

hydrosizer *hydraulic classifier.*

hydrosphere The waters of the Earth, as distinguished from the rocks (lithosphere), living things (biosphere), and the air (atmosphere). Includes the waters of the ocean; rivers, lakes, and other bodies of surface water in liquid form on the continents; snow, ice, and glaciers; and liquid water, ice, and water vapor in both the unsaturated and saturated zones below the land surface. Included by some, but excluded by others, is water in the atmosphere, which includes water vapor, clouds, and all forms of precipitation while still in the atmosphere. (AGI, 1987)

hydrostat A contrivance or apparatus to prevent the explosion of steam boilers. (Webster 2nd, 1960)

hydrostatic Relating to the pressure or equilibrium of fluids.

hydrostatic balance A balance for weighing a substance in water to ascertain its specific gravity. (Webster 3rd, 1966)

hydrostatic head The height of a vertical column of water whose weight, if of unit cross section, is equal to the hydrostatic pressure at a given point; static head as applied to water. See also: *static head.* (AGI, 1987)

hydrostatic joint Used in large water mains, in which sheet lead is forced tightly into the bell of a pipe by means of the hydrostatic pressure of a liquid. (Strock, 1948)

hydrostatic press A large ram, the surface of which is acted on by liquid in contact with a small ram. See also: *hydraulic jack.* (Hammond, 1965)

hydrostatic pressure (a) Stress that is uniform in all directions, e.g., beneath a homogeneous fluid, and causes dilation rather than distortion in isotropic materials. (AGI, 1987) (b) The pressure exerted by the water at any given point in a body of water at rest. The hydrostatic pressure of ground water is generally due to the weight of water at higher levels in the zone of saturation. (AGI, 1987)

hydrostatic roller conveyor A section of roller conveyor having rolls suitably weighted with liquid to control the velocity of the moving objects. See also: *roller conveyor.*

hydrostatics A branch of physics that deals with the characteristics of fluids at rest and esp. with the pressure in a fluid or exerted by a fluid on an immersed body. Cf: *hydrodynamics.* (Webster 3rd, 1966)

hydrostatic stress A state of stress in which the normal stresses acting on any plane are equal and where shearing stresses do not exist in the material. (AGI, 1987)

hydrostatic test On a boiler, the closing of all openings and pumping water into the boiler at a pressure (such as 50%) greater than the normal operating pressure. The purpose is to locate leaks or prove that there are no leaks. (Strock, 1948)

hydrosulfuric acid *hydrogen sulfide.*

hydrotalcite The mineral group desautelsite, hydrotalcite, pryoaurite, reevesite, stichtite, and takovite.

hydrotasimeter An electrically operated apparatus showing at a distance the exact level of water, as in a reservoir; an electric high- and low-water indicator. (Standard, 1964)

hydrotator A coal washer of the classifier type whose agitator or rotator consists of hollow arms radiating from a central distributing manifold or center head. There may be four or more of these radiating arms, each with one or more downwardly inclined nozzles. When water is discharged from these nozzles, the impulse has the effect of rotating the agitator in a manner similar to a lawn sprinkler. This agitator is suspended in a cylindrical tank and water is pumped through it under pressure, thereby creating a controlled upward current uniform over the entire area of the tank. (Mitchell, 1950)

hydrothermal Of or pertaining to hot water, to the action of hot water, or to the products of this action, such as a mineral deposit precipitated from a hot aqueous solution, with or without demonstrable association with igneous processes; also, said of the solution itself. Hydrothermal is generally used for any hot water but has been restricted by some to water of magmatic origin. (AGI, 1987)

hydrothermal alteration Alteration of rocks or minerals by the reaction of hydrothermal water with preexisting solid phases. (AGI, 1987)

hydrothermal deposit A mineral deposit that originated from hot, ascending aqueous solutions derived from a magma. Cf: *hypothermal deposit.*

hydrothermal solution A hot-water solution originating within the Earth and carrying dissolved mineral substances. Syn: *ore-bearing fluid; ore-forming fluid.* (AGI, 1987)

hydrothermal stage That stage in the cooling of a magma during which the residual fluid is strongly enriched in water and other volatiles. The exact limits of the stage are variously defined by different authors, in terms of phase assemblage, temperature, composition, and/or vapor pressure; most definitions consider it as the last stage of igneous activity, coming at a later time, and hence at a lower temperature, than the pegmatitic stage. (AGI, 1987)

hydrothermal synthesis Mineral synthesis in the presence of water at elevated temperatures. (AGI, 1987)

hydrous (a) Containing water; watery; specif., hydrated. (Webster 3rd, 1966) (b) Minerals that contain water chemically combined. (Gordon, 1906)

hydrous salts Salts containing water of crystallization.

Hydrox A permitted device, used in some English coal mines, that resembles Cardox in that a steel cylinder with a thin shearing disk is used. However, the charge is not liquid carbon dioxide but rather a powder composed chiefly of ammonium chloride and sodium nitrate. It is proportioned to give water, nitrogen, and salt as the products of combustion. On being ignited, this powder is gasified and shears the steel disk, with the gas escaping into the hole. (Lewis, 1964)

Hydrox steel tube An alternative to explosives for breaking down coal in safety lamp mines. The gasification of the Hydrox charge generates sufficient pressure within the shothole to break down the coal. The original plastic disc attached to the charge has been replaced by a metal disc separately seated. The gaseous products from the Hydrox charge are mainly carbon dioxide, nitrogen, and water vapor. The tubes can be recharged underground. The method gives a high yield of +2 in (5.1 cm) in size of coal. (Nelson, 1965)

hydroxybenzene *phenol.*

hydroxyl OH; the characteristic radical of bases, consisting of one atom of hydrogen and one atom of oxygen. The valence of this radical or anion is -1. (Crispin, 1964)

hydroxylapatite A hexagonal mineral, $Ca_5(PO_4)_3(OH)$; apatite group; an uncommon apatite in which hydroxyl predominates over fluorine and chlorine. Formerly spelled hydroxyapatite.

hydrozincite A monoclinic mineral, $2[Zn_5(CO_3)_2(OH)_6]$; forms massive, fibrous, earthy, or compact encrustations with blue luminescence; a secondary mineral in weathered zones of zinc deposits commonly associated with smithsonite or sphalerite; a source of zinc. Syn: *zinc bloom.*

hyetometer *rain gage.*

hygrometer Any of several instruments for measuring the humidity of the atmosphere. See also: *psychrometer.* (Webster 3rd, 1966)

hygrometry Measurement of atmospheric humidity.

hygroscopic water content The water content of an air-dried soil. (ASCE, 1958)

hygrostat A device sensitive to humidity changes and arranged to actuate other equipments when a predetermined humidity is attained. Syn: *humidostat.* (Strock, 1948)

hypabyssal Pertaining to an igneous intrusion, or to the rock of that intrusion, whose depth is intermediate between that of abyssal or plutonic and the surface. This distinction is not considered relevant by some petrologists. Cf: *abyssal; plutonic.* Syn: *subvolcanic.* (AGI, 1987)

hypabyssal rock An igneous rock that has risen from the depths as magma but solidified mainly as such minor intrusions as dikes and sills.

hyper- A prefix from the Greek meaning over, above, or abnormally great.

hyperfusible Any substance capable of lowering the melting ranges in end-stage magmatic fluids. (AGI, 1987)

hypermelanic Said of igneous rocks that consist of 90% to 100% mafic minerals. Cf: *melanocratic; ultramafic.* (AGI, 1987)

hyperon A class of short-lived elementary nuclear particles with masses greater than that of the neutron. (Lyman, 1964)

hypersthene An orthorhombic mineral series, $8[(Mg,Fe)_2Si_2O_6]$; pyroxene group, intermediate between enstatite and ferrosilite; prismatic cleavage; a common rock-forming mineral in intermediate to mafic igneous rocks and high-grade metamorphosed iron formations of the Lake Superior type; abundant in host rocks of copper-nickel ores of Sudbury, ON, Canada, and of chromite deposits of the Bushveld complex, South Africa. See also: *pyroxene.*

hypersthenite Originally defined as a syn. of norite, but now commonly used to mean a rock composed entirely of hypersthene; an orthopyroxenite. Not recommended usage. (AGI, 1987)

hypidiomorphic *subhedral.*

hypidiomorphic texture A texture of igneous rocks in which the greater proportion of the crystallized minerals have subhedral forms. (Hess)

hypnum peat Peat composed mostly of disintegrated plants of hypnum, often associated with other mosses and with intermingled rootlets of sedges and other flowing plants. It is formed chiefly in areas where the ground is only slightly acid, neutral, or slightly alkaline; it is brownish or drab, light, spongy, and matted, and often laminated and porous. (USBM, 1956)

hypobatholithic deposit A mineral deposit found in a deeply eroded mass of intrusive rock with few roof pendants remaining.

hypocenter *focus.*

hypocrystalline Said of the texture of an igneous rock that has crystalline components in a glassy groundmass, the ratio of crystals to glass being between 7:1 and 5:3. Syn: *merocrystalline.* (AGI, 1987)

hypogeal *hypogene.*

hypogeic *hypogene.*

hypogene (a) Said of a geologic process, and of its resultant features, occurring within and below the crust of the Earth. Cf: *epigene; endogenetic.* Syn: *hypogenic; hypogeal; hypogeic.* (AGI, 1987) (b) Said of a mineral deposit formed by ascending solutions; also, said of those solutions and of that environment. Cf: *supergene; mesogene.* (AGI, 1987) (c) A rarely used syn. of plutonic. (AGI, 1987)

hypogene ore Ore deposited from ascending hydrothermal solutions of magmatic origin.

hypogene rock A rock that was formed deep within the Earth under the influence of heat and pressure. (Hess)

hypogenic *hypogene.*

hypohyaline Said of the texture of an igneous rock that has crystalline components in a glassy groundmass, with a ratio of crystals to glass between 3:5 and 1:7. Cf: *hypocrystalline.* (AGI, 1987)

hypothermal deposit Said of a hydrothermal mineral deposit formed at high temperatures and pressures. Cf: *epithermal; hydrothermal deposit; leptothermal; mesothermal; telethermal; xenothermal.*

hypothesis A conception or proposition that is tentatively assumed, and then tested for validity by comparison with observed facts and by experimentation; e.g., the planetesimal hypothesis to explain the evolution of the planets. It is less firmly founded than a theory. (AGI, 1987)

hypothetical resources Undiscovered resources that are similar to known mineral bodies that may be reasonably expected to exist in the same producing district or region under analogous geologic conditions. If exploration confirms their existence and reveals enough information about their quality, grade, and quantity, they will be reclassified as identified resources. (USGS, 1980)

hypsometer An instrument for measuring the elevation above sea level by determining the atmospheric pressure through observing the boiling point of water. (Standard, 1964)

hypsometric map Any map showing relief by means of contours, hachures, shading, tinting, or any other convention. (AGI, 1987)

hypsometry The science of determining, by any method, elevations on the Earth's surface with reference to sea level; e.g., barometric hypsometry in which elevations are determined by means of mercurial or aneroid barometers. (AGI, 1987)

hysteresis (a) A lag in the return of an elastically deformed body to its original shape after the load has been removed. (AGI, 1987) (b) The property that a rock is said to exhibit when its magnetization is nonreversible. Syn: *magnetic hysteresis.* (AGI, 1987) (c) A phase lag of dielectric displacement behind electric-field intensity, due to energy dissipation in polarization processes. (AGI, 1987)

hysteresis loop Entire pattern of magnetization showing how a body with magnetic susceptibility can remain polarized after the disappearance of the original magnetizing force.

hysteretic repulsion Separation by alternating current that depends on magnetic properties of coercive force and remanence.

I

ianthinite An orthorhombic mineral, $(UO_2)_6O_2(OH)_8{\cdot}6H_2O$; acicular with perfect basal cleavage; in cavities, as an alteration product of uraninite at Kasolo, Congo, and Wolsendorf, Bavaria.
ice Water in the solid state; specif. the dense substance formed in nature by the freezing of liquid water, by the condensation of water vapor directly into ice crystals, or by the recrystallization or compaction of fallen snow. It is colorless to pale blue or greenish blue, usually white from included gas bubbles. At standard atmospheric pressure, it is formed at and has a melting point of 0 °C; in freezing it expands about one eleventh in volume. Ice commonly occurs in hexagonal crystals, and in large masses is classed as a rock. (AGI, 1987)
iceberg A large, massive piece of floating or stranded glacier ice of any shape, detached (calved) from the front of a glacier into a body of water. An iceberg extends more than 5 m above sea level and has the greater part of its mass (four-fifths to eight-ninths) below sea level. It may reach a length of more than 80 km. (AGI, 1987)
ice concrete A dense frozen mixture of sand, rock fragments, and ice. (AGI, 1987)
ice-laid drift *till.*
Iceland agate *obsidian.*
Iceland crystal *Iceland spar.*
Iceland spar A transparent, pure, optically clear variety of calcite principally in Iceland in vugs and cavities in basalt; formerly used in optical instruments such as nicol prisms, but has been replaced by artificial materials such as polaroid plates. Syn: *Iceland crystal; optical calcite.*
ice period The period of time from freezeup to breakup of ice. (AGI, 1987)
ice plug An ice obstruction formed by a circulation medium freezing inside the drill-rod couplings while the rods are racked up or standing in the drill derrick or tripod in extremely cold weather. Such plugs may loosen when rods are lowered into the borehole and may be ejected from the open end of the rod with enough force to injure drill crewmembers severely. (Long, 1960)
ice run *debacle.*
ice sheet A glacier of considerable thickness and more than 50,000 km^2 in area, forming a continuous cover of ice and snow over a land surface, spreading outward in all directions and not confined by the underlying topography; a continental glacier. Ice sheets are now confined to polar regions (as on Greenland and Antarctica), but during the Pleistocene Epoch they covered large parts of North America and northern Europe. (AGI, 1987)
ice spar A white transparent variety of orthoclase. Syn: *sanidine.* (Fay, 1920)
ice stone *cryolite.*
icositetrahedron *trapezohedron.*
idaite A hexagonal mineral, Cu_3FeS_4(?); soft; metallic copper red to brown; a decomposition product of bornite or chalcopyrite at the Ida Mine, Kahn, Namibia.
iddingsite An outdated term for a reddish-brown mixture of the kaolinite-serpentine group and iron oxides formed by the alteration of olivine.
ideal gas The nonexistent norm of a gas, which perfectly obeys Boyle's law of gases, in which the pressure of the gas times its volume equals a constant.
ideal section A geologic cross section that combines observed evidence on stratigraphy and/or structure with interpretation of what is not present. It may be the summation or average of several successive cross sections. (AGI, 1987)
identified resources Resources whose location, grade, quality, and quantity are known or estimated from specific geologic evidence. Identified resources include economic, marginally economic, and subeconomic components. To reflect varying degrees of geologic certainty, these economic divisions can be subdivided into measured resources, indicated resources, and inferred resources. (USGS, 1980)
identity (a) In X-ray crystallography, the distance along a crystallographic between like points in a lattice. (b) In geometrical crystallography, the completion of a sequence of symmetry operations, e.g., four rotations of 90° each about a tetrad.
idioblast A mineral constituent of a metamorphic rock formed by recrystallization and bounded by its own crystal faces. It is a type of crystalloblast. The term was originated by Becke in 1903. Cf: *xenoblast.* (AGI, 1987)
idioblastic Pertaining to an idioblast of a metamorphic rock. It is analogous to the term idiomorphic in igneous rocks. (AGI, 1987)
idiochromatic Minerals in which a specific coloring agent is an essential constituent, e.g., copper in malachite, iron in olivine, manganese in rhodochrosite. Cf: *allochromatic.*
idiochromatic mineral Mineral in which the color is due to some essential constitutent of the stone, for example, malachite, peridot, and almandine. In contrast to allochromatic minerals, idiochromatic minerals have a limited range of color. See also: *allochromatic mineral.* (Anderson, 1964)
idiogenite Suggested by Posepny for those ore deposits that are contemporaneous in origin with the wall rock. The word means of the same origin.
idiomorphic A syn. of automorphic, originally proposed by Rosenbusch in 1887 to describe individual euhedral crystals. Though the term lacks priority, it is now commonly applied to the igneous-rock texture characterized by such euhedral crystals, esp. in U.S. usage. (AGI, 1987)
idler disk A device used for holding the belt in proper position on certain types of boxcar loaders. See also: *boxcar loader.*
idler gear (a) A gear meshed with two other gears that does not transmit power to its shaft; used to reverse direction of rotation in a transmission. (Nichols, 1976) (b) Same as neutral gear. (Long, 1960)
idle wheel (a) A wheel interposed in a gear train, either to reverse rotation or to obtain the required spacing of centers, without affecting the ratio of the drive. Also called idler. (CTD, 1958) (b) A pulley to guide a driving belt, to increase its tension, or to increase its arc of contact on one of the working pulleys. (Standard, 1964)
idocrase *vesuvianite.*
idrialite An orthorhombic hydrocarbon mineral, $4[C_{22}H_{14}]$; soft; greenish yellow to light brown with bluish fluorescence; mixed with clay, pyrite, and gypsum associated with cinnabar in the Idria region, Yugoslavia. Its combustibility gave rise to the term "inflammable cinnabar." Syn: *inflammable cinnabar.*
igdloite *lueshite.*
Igewsky's reagent A solution consisting of 5% picric acid in absolute alcohol used as an etching reagent for carbon steels. (Osborne, 1956)
igneous Said of a rock or mineral that solidified from molten or partly molten material, i.e., from a magma; also, applied to processes leading to, related to, or resulting from the formation of such rocks. Igneous rocks constitute one of the three main classes into which rocks are divided, the others being metamorphic and sedimentary. Etymol: Latin ignis, fire. See also: *magmatic; plutonic; pyrogenic; hypabyssal; extrusive.* (AGI, 1987)
igneous breccia (a) A breccia that is composed of fragments of igneous rock. (AGI, 1987) (b) Any breccia produced by igneous processes; e.g., volcanic breccia, intrusion breccia. (AGI, 1987)
igneous complex An assemblage of intimately associated and roughly contemporaneous igneous rocks differing in form or in petrographic type; it may consist of plutonic rocks, volcanic rocks, or both. See also: *complex.* (AGI, 1987)
igneous cycle The sequence of events in which volcanic activity is followed by major plutonic intrusions, and then minor intrusions (e.g., dikes). (AGI, 1987)
igneous metamorphism A high-temperature metamorphic process that includes the effects of magma on adjacent rocks as well as those due to injection pegmatitization (Lindgren, 1933). The term is no longer in common use. Cf: *pyrometamorphism.* (AGI, 1987)
igneous rock Rock formed by the solidification of molten material that originated within the earth.
igneous-rock series An assemblage of temporally and spatially related igneous rocks of the same general form of occurrence (plutonic, hypabyssal, or volcanic), characterized by possessing in common certain chemical, mineralogic, and textural features or properties so that the rocks together exhibit a continuous variation from one extremity of the series to the other. Syn: *rock series.* See also: *series.* (AGI, 1987)

ignescent Applied to a mineral that sparks when struck with steel or iron; e.g., pyrite.

ignimbrite The rock formed by the widespread deposition and consolidation of ash flows and neúes ardentes. Cf: *trass; welded tuff.* (AGI, 1987)

ignitability An assessment of the ease with which a coal can be ignited. (BS, 1960)

ignitability (dust cloud) The relative ignitability of a dust cloud may be defined as the degree of ease with which it can be ignited. (Sinclair, 1958)

igniter (a) A blasting fuse or other contrivance used to fire an explosive charge. (CTD, 1958) (b) In mining, a metal cylinder that connects a main fuse with separate fuses that are only limited by the number of blasts to be fired. (Standard, 1964) (c) A device to relight safety lamps internally by friction. One type uses a waxed strip with igniting matches at intervals, while another type has a small burred wheel operating against a piece of cerium or something of a similar nature. Electrical devices are sometimes employed. (Fay, 1920) (d) One that ignites as (1) a charge, usually of black gunpowder, used to facilitate the ignition of a propelling charge and sometimes of a bursting charge; (2) a device for igniting fuel mixture (as in an internal combustion engine, a jet engine, or a rocket engine); (3) a separately energized electrode used for restriking the arc in an ignitron. (Webster 3rd, 1966)

igniter cord (a) A cord that passes an intense flame along its length at a uniform rate to light safety fuses in succession. (BS, 1964) (b) Two types are manufactured: a fast cord having a nominal burning speed of 1 s/ft (3.3 s/m) and a slow cord having a nominal burning speed of 10 s/ft (33 s/m). The burning speeds are reliable and consistent even under adverse conditions, as, e.g., when burning underwater or in a direction opposite to a strong wind. (McAdam, 1958)

ignition (a) Percussion material or detonating powder. (Standard, 1964) (b) The firing of an explosive mixture of gases, vapors, or other substances by means of an electrical or frictional spark. (CTD, 1958) (c) An outburst or fire or an explosion. (Mason, 1951) (d) The act of igniting, or the state of being ignited; specif., in mechanics, the act of exploding the charge of gases in the cylinder of an internal combustion engine. (Standard, 1964)

ignition arch A flat or curved refractory roof over a furnace at the point of fuel entrance that promotes ignition by reflection of heat. (AISI, 1949)

ignition charge A small charge, usually of black powder, used to facilitate the ignition of the main charge. (Webster 3rd, 1966)

ignition delay Time interval between contact of an oxidant and a combustible and ignition. (Van Dolah, 1963)

ignition of combustible gases *gas ignition.*

ignition point (a) Of solids and liquids, the minimum temperature at which combustion can occur, but at which it is not necessarily continuous. (Pryor, 1963) (b) Of combustible gases, the flashpoint. (Pryor, 1963)

ignition temperature (a) The ignition temperature of a substance is the temperature at which that substance starts to burn. The temperature of ignition varies greatly with different substances. All solid fuels must be heated to their ignition temperature before they will burn continuously by the process known as combustion. (Morris, 1958; Nelson, 1965) (b) The temperature required to effect ignition of a combustible-oxidant system at a specified pressure; in general, the minimum temperature is implied. (Van Dolah, 1963)

ignition tube A heavy-walled test tube of hard glass for examining the behavior of heated substances. (Webster 3rd, 1966)

ihleite *copiapite.*

ijolite A series of plutonic rocks containing nepheline and 30% to 60% mafic minerals, generally clinopyroxene, and including sphene, apatite, and melanite; also, any rock of that series. Melteigite and jacupirangite are more mafic members of the series; urtite is a type rich in nepheline. Named by Ramsay in 1891 for Ijola (Iivaara), Finland. (AGI, 1987)

ikunolite A trigonal mineral, $Bi_4(S,Se)_3$; the selenium-rich analog of laitakarite, similar to joseïte but containing no tellurium; at the Ikuno Mine, Hyogo Prefecture, Japan.

ilesite A monoclinic mineral, $(Mn,Zn,Fe)SO_4{\cdot}4H_2O$; rozenite group; a green, friable crystalline aggregate in Hall Valley, Park County, CO.

ilesmannite An amorphous mineral, $Mo_3O_8{\cdot}nH_2O$(?).

Ilgner flywheel A heavy flywheel used in the Ward-Leonard control of winding engines in mine hoists. It is mounted on the shaft of a motor generator, which draws on this source of energy as the hoist starts to move. (Pryor, 1963)

Ilgner system A modification of the Ward-Leonard system of speed control, in which a heavy flywheel is carried on the motor-generator shaft to smooth out peak loads, which would otherwise be taken from the power supply. The system is used on mine winding engines, etc. See also: *Ward-Leonard control.* (Nelson, 1965)

ill air Scot. Noxious gas, as from underground fires or chokedamp; a stagnant state of the atmosphere underground. (Fay, 1920)

illegal mine A mine that has not obtained the necessary permits and licenses from the appropriate State and Federal agencies before commencement of mining.

illinition A thin extraneous crust formed on minerals. (Standard, 1964)

illiquation (a) Melting or infusing. (Webster 2nd, 1960) (b) Mixture of metallic and earthen substances. (Webster 2nd, 1960)

illite A general term for a group of three-layer (14Å), micalike clays $(K,H_3O)(Al,Mg,Fe)_2(Si,Al)_4O_{10}[(OH)_2,H_2O]$; widely distributed in argillaceous sediments and derived soils; intermediate in composition and structure between muscovite and montmorillonite; contains less potassium and more water than muscovite, but more potassium than kaolinite or montmorillonite; potassium is generally replaced by calcium and/or magnesium; named from studies by Grimm of shales and clays in Illinois. See also: *muscovite.* Cf: *pholidoide; phyllite.* Syn: *hydromica; hydromuscovite; glimmerton.*

ilmenite (a) A trigonal mineral, $FeTiO_3$; ilmenite group; forms two series with geikielite and with pyrophanite. Iron-black. Also called menaccanite; titanic iron ore. See also: *zircon group.* (b) The mineral group geikielite, ilmenite, and pyrophanite. See also: *basonomelane.*

ilmenitite A hypabyssal rock composed almost entirely of ilmenite, with accessory pyrite, chalcopyrite, pyrrhotite, hypersthene, and labradorite. Cf: *nelsonite.* (AGI, 1987)

ilmenorutile A tetragonal mineral, $(Ti,Nb,Fe)_3O_6$; forms a series with strüverite in which tantalum substitutes for niobium; black; sp gr, 5.14. Syn: *iserine.*

ilvaite An orthorhombic and monoclinic mineral, $CaFe_2FeSi_2O_7O(OH)$; in prisms with vertically striated faces; compact, massive, or fibrous; in some magnetite orebodies, in zinc and copper ores, in contact deposits in dolomitic limestone, and in sodalite syenite near Julianehaab, Greenland. Syn: *elbaite; lievrite.*

imandrite A rock composed of quartz and albite, formed by interaction between a nepheline-syenite magma and graywacke. (Holmes, 1928)

imbibition (a) Formation of feldspathic minerals by penetration of alkaline solutions of magmatic origin into aluminum-rich metamorphic rocks. (AGI, 1987) (b) The absorption of a fluid, usually water, by a granular rock or any other porous material, under the force of capillary attraction, and in the absence of pressure. (AGI, 1987)

imbricate structure (a) A sedimentary structure characterized by imbrication of pebbles all tilted in the same direction, with their flat sides commonly displaying an upstream dip. Syn: *shingle structure.* (AGI, 1987) (b) A tectonic structure displayed by a series of nearly parallel and overlapping minor thrust faults, high-angle reverse faults, or slides, and characterized by rock slices, sheets, plates, blocks, or wedges that are approx. equidistant and have the same displacement and that are all steeply inclined in the same direction (toward the source of stress). (AGI, 1987)

imitation *synthetic stone; assembled stone.*

Imlay table *end-bump table.*

immediate roof Lowest layer or layers of rock immediately above an underground opening. See also: *roof; nether roof.*

immersion cup An accessory for a gemological microscope containing a liquid of high refractive index and designed to eliminate reflections from highly polished facets and thus to expedite the observation of determinative inclusions, growth lines, etc.

immersion method In optical mineralogy, the determination of the refractive indices of a mineral by matching them with liquids of known refractive index. See also: *index of refraction.*

immersion objective Oil-immersion lens used in microscopy to help concentrate light on the object under examination. (Pryor, 1963)

imminent danger The existence of any condition or practice that could reasonably be expected to cause death or serious physical harm before such condition or practice can be abated.

immiscible Said of two or more phases that, at mutual equilibrium, cannot dissolve completely in one another; e.g., oil and water. (AGI, 1987)

impact Collision between bodies, the velocity of one or both being changed. In direct impact, the velocity of the moving bodies is perpendicular to the bodies at the point of contact. The impact coefficient, known as the coefficient of restriction, is the ratio of the differences of velocities of the two bodies after impact to the same differences before impact. (Hammond, 1965)

impact breakers The impact breaker or double impeller breaker uses the energy contained in falling stone, plus the power imparted by the massive impellers for complete stone reduction. (Pit and Quarry, 1960)

impact cast Cast of marking produced by object striking the mud bottom. The steeply raised end of impact casts are always oriented downcurrent. Syn: *prod mark; prod cast.* See also: *bounce cast.* (Pettijohn, 1964)

impact crusher Machine in which material is broken by sharp blows. See also: *hammermill.* (Pryor, 1963)

impact factor The factor of from 1 to 2 by which the weight of a moving load is multiplied to calculate its full effect on the structural design of a floor or bridge. (Hammond, 1965)

impact glass *impactite.*

impact grinding Shattering of particles by direct fall upon them of crushing bodies or the use of a device that vibrates a metal object such as a shutter box.

impactite (a) A vesicular, glassy to finely crystalline material produced by fusion or partial fusion of target rock by the heat generated from the impact of a large meteorite, and occurring in and around the resulting crater, typically as individual bodies composed of mixtures of melt and rock fragments, often with traces of meteoritic material; a rock from a presumed impact site. Syn: *impact slag; impact glass.* (AGI, 1987) (b) A term used incorrectly for any shock-metamorphosed rock. (AGI, 1987)

impact loss The head lost as a result of the impact of particles of water; included in and scarcely distinguishable from eddy loss. (Seelye, 1951)

impact mill A crushing unit in which a rapidly moving rotor projects the charged material against steel plates; impact mills find use in the size reduction of such materials as coal, feldspar, perlite, etc. Cf: *disintegrator; hammermill.* (Dodd, 1964)

impact screen One in which the loaded screen is so suspended that it can be caused to swing or rock forward until it is abruptly checked on hitting a stop. (Pryor, 1963)

impact sensitivity Sensitivity of an explosive to detonate when impacted. *fallhammer test.*

impact slag Glassy material produced mainly by the melting of local sediment or rock where a meteorite has struck the earth. See also: *impactite.* (AGI, 1987)

impalpable Extremely fine, so that no grains or grit can be felt. (Webster 3rd, 1966)

impedance The complex ratio of voltage to current in an electrical circuit, or the complex ratio of electric-field intensity to magnetic-field intensity in an electromagnetic field. It is the reciprocal of admittance. See also: *acoustic impedance.* (AGI, 1987)

imperfect combustion A term meaning that not all of the fuel is oxidized to its highest degree; e.g., if carbon monoxide is formed instead of carbon dioxide. Cf: *incomplete combustion.* (Newton, 1959)

imperial jade A high-quality, semitransparent variety of jadeite. (Hess)

Imperial screen An oscillating or vibrating screen on which the ore is thrown upward, as well as moved forward on the screen. (Liddell, 1918)

impermeable Said of a rock that does not permit the passage of fluids under the pressure conditions ordinarily found in the subsurface. Ant: permeable. Syn: *impervious.* Cf: *vitreous.*

impervious *impermeable.*

impinger Dust-sampling apparatus into which a measured volume of dusty mine air is drawn through a jet in such a way as to strike a wetted glass plate, to which dust particles adhere. (Pryor, 1963)

implosion A bursting inward; sudden collapse; opposite of explosion. (Standard, 1964)

impound (a) A reservoir for impounding. Used in connection with the storage of tailings from ore-dressing plants and hydraulic mines. (Webster 3rd, 1966) (b) To collect in a reservoir or sump provided near a borehole the water, drill cuttings, etc., ejected therefrom. (Long, 1960)

impounding dam One in which tailings are collected and settled; also, a water-storage dam. (Pryor, 1963)

impregnated (a) Said of a mineral deposit (esp. of metals) in which the minerals are epigenetic and diffused in the host rock. Cf: *interstitial.* (AGI, 1987) (b) Said of timber that has been soaked in various fluids to enable it better to resist the decomposing influences of the atmosphere. (Crispin, 1964) (c) A metallic material in which fragments of diamond or other hard substances (in unflocculated distribution) are intermixed and embedded. See also: *impregnated bit.* (Long, 1960)

impregnated bit A sintered, powder-metal matrix bit with fragmented bort or whole diamonds of selected screen sizes uniformly distributed throughout the entire crown section. As the matrix wears down, new, sharp diamond points are exposed; hence, the bit is used until the crown is consumed entirely. Cf: *multilayer bit.* See also: *impregnated.* (Long, 1960)

impregnated timber Timber that has been treated either to make it flame resistant or to protect it from destruction by fungi and insects. Cover boards used with steel arches are often vacuum-pressure impregnated with a flame-retardant preservative for safety and to comply with the flame-proofing requirements covering escape roadways. (Nelson, 1965)

impression block A bell-shaped or hollow, tubular device filled with wax or other water-resistant plastic materials, which is lowered onto an article resting on the bottom of a borehole. The plastic material molds itself about the lost article, and by inspecting the impression so formed, the driller can determine which fishing tool is best fitted to recover the lost article. (Long, 1960)

improper In crystallography, any element or operation of symmetry involving a mirror or inversion resulting in a change of chirality of an asymmetric unit. Cf: *chirality; proper.*

improved dial A miner's dial in which a telescope replaces the usual sighting vanes. (BS, 1963)

Improved paragon Trade name for nonrotating wire rope of 18 by 4 over 3 by 24 construction. (Hammond, 1965)

impulse turbine (a) A water turbine, such as the Pelton wheel, in which the driving force is provided more by the speed of the water than by a fall in its pressure. See also: *Pelton wheel.* (Hammond, 1965) (b) A turbine in which the steam is expanded in a series of stationary nozzles that it leaves at a very high velocity, perhaps 4,000 ft/s (1.2 km/s), and then gives up its kinetic energy to blades or buckets attached to the revolving disk that furnishes the power. (Mason, 1951)

impurity Any undesirable substance not normally present in air, water, coal, or other materials or present in an excessive amount. Syn: *contaminant.*

inby (a) Eng. Toward the working face, or interior, of the mine; away from the shaft or entrance; from Newcastle coalfield. Also called inover; inbye; inbyeside. (Fay, 1920) (b) In a direction toward the face of the entry from the point indicated as the base or starting point. (Rice, 1960) (c) The direction from a haulage way to a working face. (CTD, 1958) (d) Opposite of outby. (BCI, 1947)

incandescent Made luminous by heat; white or glowing with heat. (Standard, 1964)

incarbonization *coalification.*

incendivity The property of an igniting agent (e.g., spark, flame, or hot solid) whereby the agent can cause ignition. (Atlas, 1987)

inches of pressure The height in inches of a column of water or of mercury as a measure of hydrostatic pressure. (Standard, 1964)

inching starter In one form, electrical gear that allows power to be applied gently to a stationary ball mill so as to avoid a high starting strain. The mill is said to be inched over as it begins slowly to rotate. (Pryor, 1963)

inch of water A unit of pressure equivalent to 0.036136 psi (248.84 Pa). (Hartman, 1961)

inch pennyweights In valuation of gold lodes, the product of the width of the exposure of ore, measured normal to the containing host rock, and the assay value in pennyweights of a true sample of the ore, cut evenly along the measured line. In evaluation of ore tonnages in base-metal mining, the equivalent measurement is the assay-foot or similar convenient combination across the exposed lode. Abbrev., in. dwt. (Pryor, 1963)

inch-pound The work done in raising 1 lb, 1 in (0.179 kg, 1 cm); a unit of work or of energy. Abbrev., in·lb. (Standard, 1964)

incidence rate (a) The rate of occurrence of accidents as determined by multiplying the actual number of injuries in any given period by 200,000 and dividing the product by the number of man-hours exposure. Syn: *frequency rate.* (CFR, 1992) (b) A statistic used by the mining industries to measure safety performance and compare the performance of different groups or employers. The formula for determining the rate is: the number of injuries or illesses times 200,000 h divided by employee hours worked. The 200,000 hours represents 100 employees working 40 h/wk and 50 wk/yr. This number keeps the value that results from the formula small. The number of employee hours comes from company records. They represent hours worked; not vacation, sick leave, or holiday time. (Brauer, 1994)

incidental vein A vein discovered after the original vein on which a claim is based. (AGI, 1987)

inclination (a) The angular dip of a vein, a bed, etc., measured in degrees from the horizontal plane. (Fay, 1920) (b) Angle between the direction of the magnetic field and the horizontal plane. Syn: *declination.* (Schieferdecker, 1959)

inclinator Instrument to determine the inclination of the magnetic field. (Schieferdecker, 1959)

incline (a) A shaft not vertical; usually on the dip of a vein. See also: *slope.* (Fay, 1920) (b) Any inclined plane, whether above or beneath the surface; usually applied to self-acting planes above ground, as in the bituminous coalfields. (Fay, 1920) (c) In mines, an inclined drift driven upwards at an angle from the horizontal. (Fraenkel, 1953) (d) A sloping tunnel along which rails are laid from one level to another; a mechanically worked inclined haulageway in a coal mine. (CTD, 1958) (e) A slanting shaft. (Gordon, 1906) (f) An opening driven up or down the pitch. Syn: *inclined shaft.* (Hudson, 1932)

incline bogie Scot. A wheeled carriage for inclines, constructed with a horizontal platform so that cars can be run on it and be conveyed up and down the incline or slope. (Fay, 1920)

inclined bedded formation Any bedded formation of rock where the dip of the bedding planes is greater than 10°.

inclined bedding A type of bedding appearing commonly in sandy deposits; the strata, essentially intraformational, dip in the direction of the current flow. Syn: *crossbedding; cross lamination; current bedding.* (AGI, 1987)

inclined borehole A borehole drilled at an angle from the vertical not exceeding 45°. The drilling technique is called angle drilling. (Nelson, 1965)

inclined cableway A monocable cableway in which the track cable has a slope of about 1 in 4 over its whole length, sufficiently steep to allow the carrier to run down under its own weight. (Hammond, 1965)

inclined drilling Drilling blastholes at an angle from the vertical. Also called angle drilling. (Woodruff, 1966)

inclined extinction Alignment of optical extinction at some angle other than 0°, 45°, or 90° to a crystal face or cleavage trace. Cf: *extinction; extinction angle.*

inclined fold A fold, the axial plane of which is not vertical. (AGI, 1987)

inclined gage A common type of gage used with the pitot tube. A straight glass tube with connections at each end is mounted in an inclined position on an aluminum frame, and a scale is placed under the tube. In place of water, a colored oil is used, and the scale is graduated to read directly in inches of water. (Lewis, 1964)

inclined magnetic polarization The inclination angle of the geomagnetic field.

inclined plane (a) A natural or artificial slope used for facilitating the ascent, descent, or transfer from one level to another of vehicles or other objects. See also: *incline.* (Standard, 1964) (b) A slope used to change the direction and speed-power ratio of a force. (Nichols, 1976)

inclined polarization Polarization that is inclined to the linear dimensions of a magnetized body, or to the plumb line or the horizon. (AGI, 1987)

inclined railroad operator *inclined railway operator.*

inclined railway operator In metal mining, one who operates the machinery that drives the haulage cable along a power incline railway used for hauling cars, supplies, workers, and materials to and from one level to another on a steep slope. Also called inclined railroad operator; tramway operator. (DOT, 1949)

inclined shaft *incline.*

inclined traverser A traverser that moves mine cars laterally and vertically by traveling on an inclined plane. It is sometimes used at the pit bottom for the transfer of cars from a higher decking level to a lower decking level on the opposite side rail track. The cars are held upright in a frame and can handle loaded or empty cars to and from the two levels. See also: *traverser.* (Nelson, 1965)

inclined water gage A sensitive form of water gage, giving readings of greater accuracy. It is used mainly for ventilation surveys. See also: *water gage.* (Nelson, 1965)

incline engine A stationary haulage engine at the top of an incline. (CTD, 1958)

incline hole *angle hole.*

incline man In mining, a laborer who controls the movement of cars on a self-acting incline (loaded car going down one track pulls empty cars up on other), hooking cable to loaded or empty cars, starting them down the incline, and applying brake to cable drum by a lever to control their speed of descent. Also called dilly boy; drum runner; monitor operator; plane man; wheel runner; jinnier. (DOT, 1949)

incline repairman In mining, a person who oils, greases, repairs, and replaces pulleys or rollers which support the cable on a haulage slope or plane (incline) underground and at the mine surface. Also called rolley man. Syn: *incline trackman.* (DOT, 1949)

incline shaft A shaft sunk at an inclination from the vertical, usually following the dip of a lode. See also: *turned vertical shaft; underlay shaft.* (Nelson, 1965)

incline trackman *incline repairman.*

inclinometer (a) An instrument for measuring the inclination or slope, as of the ground, Syn: *clinometer.* (Standard, 1964) (b) Any of various instruments for measuring the departure of a drill hole from the vertical; a driftmeter. (AGI, 1987) (c) An instrument that measures magnetic inclination. (AGI, 1987)

included angle Either of the two angles formed at the station by the intersection of the two survey lines. (Mason, 1951)

inclusion (a) Any size fragment of another rock enclosed in an igneous rock. Syn: *xenolith.* (Fay, 1920) (b) A particle of nonmetallic material retained in a solid metal or alloy. Such inclusions are generally oxides, sulfides, or silicates of the host metal, but may also be particles of refractory materials picked up from a furnace or ladle lining. (c) A crystal, aggregate, or minute cavity filled with one or two fluid phases and with or without a crystal phase enclosed in a host crystal. See also: *fluid inclusion; three-phase inclusion; negative crystal.*

inclusions (a) A term applied to crystals and anhedra of one mineral involved in another; and to fragments of one rock inclosed in another, as when a volcanic flow picks up portions of its conduit. (b) Particles of foreign matter, solid, liquid, or gaseous, enclosed within a gem stone. The nature of such inclusions provides a powerful clue to the origin of a stone and enables natural stones to be distinguished from their synthetic counterparts. (Anderson, 1964)

incoalation (a) The process of coal formation that begins after peat formation is completed without there being any sharp boundary between the two processes. From the German inkohlung. (AGI, 1987) (b) *coalification.*

incoherent (a) Said of a rock or deposit that is loose or unconsolidated. (AGI, 1987) (b) Said of electromagnetic radiation (EMR) having a broad spectrum of frequencies; e.g., incoherent (sunlight) EMR vs. coherent (laser) EMR. (AGI, 1987)

incombustible (a) Applies to substances that will not burn. (Mersereau, 1947) (b) Any building material that contains no matter subject to rapid oxidation within the temperature limits of a standard fire test of not less than 2-h duration. Materials that continue to burn after this time period are termed combustible. (ACSG, 1961) (c) *noncombustible.*

incompetent (a) Applied to strata, a formation, a rock, or a rock structure not combining sufficient firmness and flexibility to transmit a thrust and to lift a load by bending; consequently, admitting only the deformation of flowage. See also: *competent.* (Standard, 1964) (b) Soft or fragmented rocks in which an opening, such as a borehole or an underground working place, cannot be maintained unless artificially supported by casing, cementing, or timbering. (Long, 1960)

incompetent bed A bed that, in a particular case of folding, has yielded to the lateral pressure by plastic adjustment and flow. This may result in the bedding being thrown into complex structures or in the development of more regular internal structures, particularly drag folds and fracture cleavage. The bed tends to thicken toward the hinges, and to thin in the limbs, of the folds. See also: *competent bed.* (Challinor, 1964)

incompetent folding *flow folding.*

incomplete combustion A term applied to combustion in which all of the fuel is not burned; e.g., leaving unburned carbon in ashes. Cf: *imperfect combustion.* (Newton, 1959)

incongruent melting (a) Melting accompanied by decomposition or by reaction with the liquid, so that one solid phase is converted into another; melting to give a liquid different in composition from the original solid. An example is orthoclase melting incongruently to give leucite and a liquid richer in silica than the original orthoclase. Cf: *congruent melting.* (AGI, 1987) (b) A crystalline compound that dissociates into another compound and a melt of different composition upon heating. Cf: *congruent melting.*

incorporation A process by which material contributing to coal formation responds to diagenetic and metamorphic agencies of coalification and becomes a part of the coal without undergoing any material modification. See also: *coalification.* Cf: *vitrinization; fusinization.* (AGI, 1987)

increment The quantity of coal or coke taken by a single sweep of the sampling instrument. (BS, 1960)

incrop A former outcrop concealed by or buried beneath younger unconformable strata. (AGI, 1987)

incrustation *encrustation.*

independent subsidence The condition in sedimentation in which each floc or particle settles freely; i.e., its movement is not influenced in any way by other flocs or particles in suspension. (Mitchell, 1950)

independent wire rope core This core may be 6 by 7, 7 by 7, 6 by 19, or 7 by 19 (number of strands laid together and ropes twisted together) construction, and the individual wires shall be of an appropriate grade of steel in accordance with the best practice and design, either bright (uncoated), galvanized, or drawn galvanized wire. See also: *wire-strand core.*

inderborite Hydrous calcium and magnesium borate mineral, $CaMg[B_3O_3(OH)_5]_2{\cdot}6H_2O$, as monoclinic crystals from the Inder borate deposits, Kazakhstan. Named from locality. See also: *metahydroboracite.* (Spencer, 1943)

inderite A monoclinic mineral, $MgB_3O_3(OH)_5{\cdot}5H_2O$; forms nodular aggregates or acicular crystals; named from the locality, Inder, Kazakhstan.

index bed *key bed.*

index contour A contour line shown on a map in a distinctive manner for ease of identification, being printed more heavily than other contour lines and generally labeled with a value (such as figure of elevation) along its course. It appears at regular intervals, such as every fifth or sometimes every fourth contour line (depending on the contour interval). Syn: *accented contour.* (AGI, 1987)

index fossil A fossil that, because of its wide geographic distribution and restricted time range, can be used to identify and date the strata in which it occurs. Cf: *guide fossil.*

index horizon A structural surface used as a reference in analyzing the geologic structure of an area. (AGI, 1987)

index mineral A mineral developed under a particular set of temperature and pressure conditions, thus characterizing a particular degree of metamorphism. When dealing with progressive metamorphism, it is a mineral whose first appearance (in passing from low to higher grades of metamorphism) marks the outer limit of the zone in question. Cf: *typomorphic mineral.* (AGI, 1987)

index of liquidity This is found by the formula: water content of test sample-water content at plastic limit/index of plasticity. This gives a value of 100% for clay at the liquid limit, and zero at the plastic limit, and is the reverse of the consistency index. (Hammond, 1965)

index of plasticity The difference between the water content of clay at its liquid and plastic limits, showing the range of water contents over which the clay is plastic. (Hammond, 1965)

index of refraction A number *n* found by dividing the velocity of light in a vacuum *c* by the velocity of light in a transparent substance. Isometric crystals are isotropic and have one index of refraction; all other symmetries are anisotropic. Hexagonal, trigonal, and tetragonal crystals have two principal indices; orthorhombic, monoclinic, and triclinic crystals have three principal indices. Determination of indices of refraction of a mineral is a major means of mineral identification. See also: *dispersion; interference.* Syn: *immersion method; optical mineralogy; refractive index.* Cf: *transmitted light.*

index properties Properties that can be used to identify the soil type. The properties are of two kinds: (1) soil grain properties and (2) soil aggregate properties. (Nelson, 1965)

index property tests Tests to determine index properties that in turn serve to identify the soil type and indicate its consistency. (Nelson, 1965)

indialite A hexagonal mineral, $Mg_2Al_4Si_5O_{18}$; the high-temperature polymorph of cordierite; in sediments fused by a burning coal seam in India. Named for the locality.

Indian-cut A style of diamond cutting, usually of Indian or other Oriental origin, in which the table is usually double the size of the culet; such stones are generally recut for European or American requirements. (Hess)

Indian jade Aventurine quartz containing inclusions of chromian muscovite. Syn: *regal jade.*

Indian title An Indian's right to occupancy of land, and that right recognized by the United States, constitutes Indian title. (Ricketts, 1943)

India steel A fine natural steel from southern India made directly from the ore; wootz.

indicated ore Ore for which tonnage and grade are computed from information similar to that used for measured resources, but the sites for inspection, measurement, and sampling are farther apart or otherwise less adequately spaced. The degree of assurance is high enough to assume continuity between points of observation. (USGS, 1980)

indicated resources Resources from which the quantity and grade and/or quality are computed from information similar to that used for measured resources, but the sites for inspection, sampling, and measurement are farther apart or are otherwise less adequately spaced. The degree of assurance, although lower than that for measured resources, is high enough to assume continuity between points of observation. (USGS, 1980)

indicator (a) A geologic or other feature that suggests the presence of a mineral deposit, such as a geochemical anomaly, a carbonaceous shale indicative of coal, or a pyrite-bearing bed that may lead to gold ore at its intersection with a quartz vein. A plant or animal peculiar to a specific environment, which can therefore be used to identify that environment. See also: *pencil mark.* (AGI, 1987) (b) *marker bed; marker.* (Long, 1960)

indicator plant (a) Some plants develop peculiar diagnostic symptoms that can be interpreted directly in terms of probable excesses of a particular element in the soil. Geobotanical indicators are either plant species or characteristic variations in the growth habits of plant species that are restricted in their distribution to rocks or soils of definite physical or chemical properties. They have been used in locating and mapping ground water, saline deposits, hydrocarbons, and rock types, as well as ores. (Hawkes, 1962) (b) A plant whose occurrence is broadly indicative of the soil of an area; e.g., its salinity or alkalinity, level of zone of saturation, and other soil conditions. (AGI, 1987)

indicator vein A vein that is not metalliferous itself, but may lead to an ore deposit.

indices of a crystal face The numbers that define the position of a crystal face in space with reference to crystallographic axes. Miller and Bravais-Miller indices, those in current use, are the reciprocals, cleared of fractions, of the intercepts the face makes with the crystallographic axes. Syn: *Miller indices.* Cf: *crystal systems; intercept.*

indicolite A blue variety of elbaite tourmaline.

indigenous limonite Sulfide-derived limonite that remains fixed at the site of the parent sulfide, often as boxworks or other encrustation. Cf: *exotic limonite; relief limonite.* (AGI, 1987)

indigo copper *covellite; copper sulfide.*

indirect flushing Flushing in which the water enters the borehole around the rod and issues upwards through the rod. (Stoces, 1954)

indirect initiation *inverse initiation.*

indirect priming Placement of the blasting cap in the first cartridge going into the borehole with the business end pointing toward the collar. Recommended method of priming charges of permissible dynamite.

indite An isometric mineral, $FeIn_2S_4$; linnaeite group; in minute iron-black grains in cassiterite from the Dzhalind deposit, Little Khingan ridge, the former U.S.S.R.

indium (a) A tetragonal mineral. Symbol, In; native; metallic silvery white; sp gr, 7.31; in meteorites and as a trace constituent in the minerals of other metals, principally zinc, lead, tin, tungsten, and iron minerals; principal source is sphalerite concentrates that may contain up to 10,000 ppm. (b) Used in making bearing alloys, germanium transistors, rectifiers, thermistors, and photoconductors. (Handbook of Chem. & Phys., 3)

individually screened trailing cable A trailing cable with a screen of metallic covering over each power conductor. This is the type now adopted in British coal mines. See also: *trailing cable.* (Nelson, 1965)

individual reduction ratio In crushing practice, this term may be expressed as: Size most abundant in feed/mean size of grading band concerned. (South Australia, 1961)

indoor catches Beams that catch the walking beam of the Cornish pump or engine on its down piston stroke if the string of tools being moved should break. The indoor stroke is the lifting stroke of such a pump. (Pryor, 1963)

indoor stroke Eng. That stroke of a Cornish pump that lifts the water at the bottom or drawing lift.

indraft The act of drawing in, or that which is drawn in; an inward flow; such as an indraft of air. (Standard, 1964)

induced breaks The fine breaks or cracks that occur in the nether roof of a coal seam following the holing of the coal or its removal, and having the same general direction as that of the coal face itself. (TIME, 1929-30)

induced bursts Rock bursts caused by stoping operations to distinguish them from development bursts, which are called inherent. (Spalding, 1949)

induced caving In the block caving mining method, the ore zone is undercut until the ore material breaks apart and falls under gravity load into draw points. If the ore zone is reluctant to fall under gravity loading, the fall is sometimes induced with explosives set in boreholes drilled into the orebody. This induced caving is generally accomplished from drifts above the ore zone. (SME, 1992)

induced cleavage *induced fracture.*

induced current *induction.*

induced fracture Fracture formed in a roof bed as a result of mining operations. For example, on longwall faces fractures are formed in a shale roof parallel to and along successive lines of face. They are induced after coal cutting and become intensified at the end of the loading shift. The distance between the fractures coincides, roughly, with the depth of cut. See also: *break.* Syn: *induced cleavage.* (Nelson, 1965)

induced magnetization The magnetic field spontaneously induced in a volume of rock by the uniform action of an applied field. Its direction and magnitude are parallel and proportional, respectively, to the applied field. In the absence of remanent magnetization, induced magnetization is the magnetic moment per unit volume. See also: *remanent magnetization.* (AGI, 1987)

induced polarization The production of a double layer of charge at a mineral interface, or production of changes in double-layer density of charge, brought about by application of an electric or magnetic field (induced electrical or magnetic polarization). Induced electrical polarization is manifested either by a decay of voltage in the Earth following the cessation of an excitation current pulse, or by a frequency dependence of the apparent resistivity of the Earth. Abbrev: IP. (AGI, 1987)

induction The production of magnetization or electrification in a body by the mere proximity of magnetized or electrified bodies, or of an electric current in a conductor by the variation of the magnetic field in its vicinity. Syn: *induced current.* (Crispin, 1964)

induction balance An apparatus for measuring changes of conductivity, detecting the proximity of metallic bodies, etc., by noting extremely minute changes in an electric current. (Standard, 1964)

induction furnace An alternating-current electric furnace in which the primary conductor is coiled and generates, by electromagnetic induction, a secondary current that develops heat within the metal charge. (ASM, 1961)

induction hardening Quench hardening in which the heat is generated by electrical induction. (ASM, 1961)

induction log A continuous record of the conductivity of strata traversed by a borehole as a function of depth.

induction period A term used with reference to instantaneous caps to describe the time between the bridge break and the detonation of the base charge. (Streefkerk, 1952)

induction time The interval between the bursting and lag times of a detonator. (BS, 1964)

inductively coupled plasma spectroscopy An atomic emission spectroscopy analytical technique where liquid solutions are passed through a quartz tube surrounded by a high-frequency induction coil for heating the sample to high temperatures. It is an important method for measuring trace element concentrations.

inductive method An electrical exploration method in which electric current is introduced into the ground by means of electromagnetic induction and in which the magnetic field associated with the current is determined. (AGI, 1987)

indurated Said of a rock or soil hardened or consolidated by pressure, cementation, or heat. (AGI, 1987)

induration (a) The hardening of a rock or rock material by heat, pressure, or the introduction of cementing material; esp. the process by which relatively consolidated rock is made harder or more compact. See also: *lithification.* (AGI, 1987) (b) The hardening of a soil horizon by chemical action to form a hardpan. (AGI, 1987)

industrial calorific value The calorific value obtained when coal is burned under a boiler. (Kentucky, 1952)

industrial degree-day A degree-day unit based on a (usually) 45 °F or 55 °F (7.2 °C or 12.9 °C) mean daily temperature so as to be applicable to industrial buildings maintained at relatively low temperatures. (Strock, 1948)

industrial diamonds (a) Crystalline and/or cryptocrystalline diamonds having color, shape, size, crystal form, imperfections, or other physical characteristics that make them unfit for use as gems. Industrial diamonds usually are grouped as toolstones, drill diamonds, fragmented bort, ballas, and carbons. Also called industrials; industrial stones. See also: *diamond.* (Long, 1960) (b) Impure diamond used in borehole drilling and the grinding industry. Also called black diamond; bort; boart; carbonado. (Pryor, 1963)

industrial minerals Rocks and minerals not produced as sources of the metals, but excluding mineral fuels. See also: *non-metallic minerals.*

inequigranular *heterogranular.*

inequilibrium Uranium is soluble in acid waters and tends to be removed in solution, but radium is much less soluble and its compounds tend to remain behind in the leached outcrop. Therefore, the outcrop may be radioactive due to the presence of the gamma-emitting elements RaC and RaD, even though much of the uranium has been lost in solution. In this case a radiometric assay may indicate a high counter reading, but the uranium content may be low. Uranium minerals deposited less than a million years ago may be in inequilibrium because daughter products have not accumulated in their equilibrium amounts. Hence, counter readings may indicate less uranium than is actually present. See also: *radiometric assay.* (Ballard, 1955)

inert anode An anode that is insoluble in the electrolyte under the conditions obtained in the electrolysis. (Lowenheim, 1962)

inert dust (a) Any dust that contains only a small amount of combustible material. (Rice, 1960) (b) Dust that has no harmful effect. (Hartman, 1961)

inert gas (a) A gas that is normally chemically inactive, esp. in not supporting combustion. (Webster 3rd, 1966) (b) One of the helium group of gases comprising helium, neon, argon, krypton, xenon, and sometimes radon. Also called a noble gas; a rare gas. (Webster 3rd, 1966)

inertia The reluctance of a body to change its state of rest or of uniform velocity in a straight line. Inertia is measured by mass when linear velocities and accelerations are considered and by moment of inertia for angular motions (that is, rotations about an axis). (CTD, 1958)

inertinite A coal maceral group including micrinite, macrinite, sclerotinite, fusinite, semifusinite, and inertodetrinite. They are characterized by a relatively high carbon content and a reflectance higher than that of vitrinite. They are relatively inert during the carbonization process. Syn: *inerts.* (AGI, 1987)

inert primer A cylinder of inert material that enshrouds a detonator, but that does not interfere with the detonation of the explosive charge. (BS, 1964)

inerts Constituents of a coal that decrease its efficiency in use; e.g., mineral matter (ash) and moisture in fuel for combustion or fusain in coal for carbonization. Syn: *inertinite.* (BS, 1962)

inesite A triclinic mineral, $Ca_2Mn_7Si_{10}O_{28}(OH)_2 \cdot 5H_2O$; forms small prismatic crystals, fibers, or radial aggregates; associated with some manganese deposits; although related to zeolites, contains no aluminum.

infective jaundice Mine workers served by drifts and adits are subject to occasional attacks of this disease, which is often fatal. It is caused by a micro-organism, the principal carrier of which is the sewer rat. If the skin is scratched, the germ can enter the bloodstream of the miner. Preventive measures include clearing up of all stores, food, and other waste to deprive the rats of food and of a systematic extermination by a pest control officer. Also known as Weil's disease. See also: *mining disease.* (Sinclair, 1958)

inferred ore Ore for which estimates are based on an assumed continuity beyond measured and/or indicated ore. Inferred resources may or may not be supported by samples or measurements.

inferred reserve base The in-place part of an identified resource from which inferred reserves are estimated. Quantitative estimates are based largely on knowledge of the geologic character of a deposit for which there may be no samples or measurements. The estimates are based on an assumed continuity beyond the reserve base, for which there is geologic evidence. (USGS, 1980)

inferred resources Resources from which estimates are based on an assumed continuity beyond measured and/or indicated resources, for which there is geologic evidence. Inferred resources may or may not be supported by samples or measurements. (USGS, 1980)

infilling (a) Material used for filling in; filling. (Standard, 1964) (b) Material, such as hardcore, used for making up levels; e.g., under floors. (CTD, 1958)

infiltration The flow of a fluid into a solid substance through pores or small openings; specif. the movement of water into soil or porous rock. Cf: *percolation.* (AGI, 1987)

infiltration vein An interstitial mineral deposit formed by the action of percolating waters. Cf: *segregated vein.* (AGI, 1987)

inflammable cinnabar *idrialite.*

inflatable seal A seal made from polyvinyl chloride reinforced with glass fiber. It is inflated by compressed air and can cover or seal roadways up to 12 ft wide and 10 ft in height (about 4 m wide and 3 m in height). It is used to isolate a fire, or heating, and reduce the volume of smoke and gases so that erection of stoppings can proceed in respirable air without workers being hampered by breathing apparatus. (Nelson, 1965)

inflation *tumescence.*

influence line An influence line usually pertains to a particular section of a beam, and is a curve so drawn that its ordinate at any point represents the value of the reaction, vertical shear, bending moment, or deflection produced at the particular section by a unit load applied at the point where the ordinate is measured. An influence line may be used to show the effect of load position on any quantity dependent thereon, such as the stress in a given truss member, the deflection of a truss, the twisting moment in a shaft, etc. (Roark, 1954)

in fork (a) Eng. When pumps are working after the water has receded below some of the holes of the windbore, they are said to be in fork. (b) Of mine pumps, sucking air and water. (Pryor, 1963)

infrared Pertaining to or designating that part of the electromagnetic spectrum ranging in wavelength from 0.7 µm to about 1 mm. Cf: *visible light.* (AGI, 1987)

infrared gas analyzer An instrument used for routine gas analysis for the determination of methane and other gases. The results are accurate to 0.1%. (Sinclair, 1958)

infrared photography This technique is employed in air survey during misty weather, using special film that is more sensitive to infrared rays than to light rays. See also: *photogrammetry.* (Hammond, 1965)

infrasizer An apparatus for sizing air elutriation of very fine particles. See also: *air classification.* (Osborne, 1956)

infrastructure (a) Structure produced at a deep crustal level, in a plutonic environment, under conditions of elevated temperature and pressure, which is characterized by plastic folding, and the emplacement of granite and other migmatitic and magmatic rocks. This environment occurs in the internal parts of most orogenic belts, but the term is used esp. where the infrastructure contrasts with an overlying, less disturbed layer, or superstructure. (AGI, 1987) (b) The basic facilities, equipment, roads, and installations needed for the functioning of a system. (AGI, 1987)

infusible Said of a mineral crystal or fragment that will not melt in the hottest flame produced by a hand-held blowpipe or blowtorch, i.e., around 1,500 °C. The bronzite variety of pyroxene, e.g., has a melt point of approx. 1,400 °C and is said to be practically infusible; quartz, with a melt point of about 1,710 °C, is infusible.

infusion gun *water infusion gun.*

infusion shot firing A technique of shot firing in which an explosive charge is fired in a shothole, which is filled with water under pressure and in which the strata around the shothole have been infused with water. (BS, 1964)

infusorial earth An obsolete syn. of diatomite. Syn: *infusorial silica.* (AGI, 1987)

infusorial silica *infusorial earth.*

ingate The point of entrance from a shaft to a level in a coal mine. See also: *inset.* (Standard, 1964)

ingot (a) A mass of cast metal as it comes from a mold or crucible; specif., a bar of gold or silver for assaying, coining, or export. Cf: *pig.* (Standard, 1964) (b) A mold in which an ingot may be cast. (Standard, 1964) (c) A casting suitable for working or remelting. (ASM, 1961)

ingot header In ore dressing, smelting, and refining, one who pours molten aluminum, copper, or other nonferrous metals into solidifying ingots to compensate for shrinkage that occurs when ingots cool in their molds. Also called billet header; casting header; header; ingot pipe filler; pipe-out man. (DOT, 1949)

ingot iron Iron of comparatively high purity, produced, in the same way as steel, in the open-hearth furnace, but under conditions that keep down the carbon, manganese, and silicon content. See also: *iron*. Cf: *wrought iron*. (CTD, 1958)

ingotism A defect common to almost all metal ingots in which metal crystals (dendrites) tend to grow at right angles to the walls of the mold and form planes of weakness at their junctions; these make the ingot tender and it tends to tear apart when rolled. (Newton, 1959)

ingot mold The mold or container in which molten metal is cast and allowed to solidify in order to form an ingot. (CTD, 1958)

ingot pitch The chemical condition in which metal is fit to be cast into ingots. (Standard, 1964)

ingot saw A saw that is run at a high rate of speed and has a fusing action at its cutting edge; used in cutting hot ingots. (Standard, 1964)

ingot structure The general arrangement of crystals in an ingot, which consists typically of chill, columnar, and equiaxed crystals. According to the relations between the mass and the temperature of the molten metal and mold, respectively, one or two types of crystals may be absent. (CTD, 1958)

ingredient The primary and higher order reactants of the resins and the chemical constituents of the molding compound, such as plasticizer, lubricant, solvent, catalyst, stabilizer, fire retardant, hardener, and coloring material.

ingress (a) A place for entering; a way of entrance. (Webster 2nd, 1960) (b) In underground bituminous mining there are three methods of ingress—by drift, shaft, or slope. Drift mines are opened by driving horizontally from the side of an elevation into the seam; shaft mines by sinking a vertical shaft through the overlying strata into the seam; and slope mines by driving an inclined entry through the overlying strata through the surface into the seam. (BCI, 1947)

inhabited building (As in the American Table of Distances for storage of explosives). A building regularly used in whole or in part as a habitation for human beings, or any church, schoolhouse, railroad station, store, or other structure where people are accustomed to assemble, except any building or structure occupied in connection with the manufacture, transportation, storage, or use of explosive materials. (Cote, 1991)

inhaul cable In a cable excavator, the line that pulls the bucket to dig and bring in soil. Also called digging line. (Nichols, 1976)

inherent ash Widely used to designate the part of the ash content of a coal that is structurally part of the coal itself and cannot be separated from it by any mechanical means, usually amounts to about 1%. Also called dirt; fixed ash; constitutional ash. Opposite of free ash. See also: *ash; ash yield; extraneous ash*. (Mitchell, 1950; Pryor, 1963)

inherent floatability Property considered by some physicists to be possessed by certain naturally occurring minerals, which readily respond without pretreatment to levitation by the froth-flotation process; by other workers considered due to slight surficial contamination during mining and transport. (Pryor, 1963)

inherent mineral matter The portion of the mineral matter of coal organically combined with the coal. It contains elements that have been assimilated by the living plant, such as iron, phosphorus, sulfur, calcium, potassium, and magnesium. (AGI, 1987)

inherent moisture (a) In general, the moisture that is present in the coal in the bed. (Mitchell, 1950) (b) Of coal, that remaining after natural drying in air. (Pryor, 1963) (c) Maximum moisture that a sample of coal will hold at 100% humidity and atmospheric pressure. (Bennett, 1962) (d) *moisture content*.

inherited structure An original structural feature of the country rock that has been faithfully preserved after replacement by ore. (Schieferdecker, 1959)

inhibitor (a) In drilling, a substance that, when added to cement, has the capacity to slow down or lengthen the normal time required for that specific cement to set; also, a substance added to drilling mud to check or slow down organic or chemical deterioration or change in the physical characteristics of the drilling mud. (Long, 1960) (b) A substance that when present in an environment substantially decreases corrosion.

initial depression The total water gage actually produced by a mine fan. See also: *theoretical depression*. (Nelson, 1965)

initial dip *original dip*.

initial face In quarrying, the face formed by the blasting of the slope. (Streefkerk, 1952)

initial impulse *first arrival*.

initiation The process of causing a high explosive to detonate. The initiation of an explosive charge requires an initiating point, which is usually a primer and electric detonator, or a primer and a detonating cord or fuse. (Nelson, 1965)

initiator A detonator or detonating cord used to start detonation or deflagration in an explosive material; can refer to a blasting cap or primer.

injected hole A borehole into which a cement slurry or grout has been forced by high-pressure pumps and allowed to harden. (Long, 1960)

injection (a) The process of emplacement of magma in preexisting rocks; magmatic activity; also, the igneous rock mass so formed. (b) The forcing, under abnormal pressure, of sedimentary material (downward, upward, or laterally) into a preexisting deposit or rock, either along some plane of weakness or into a crack or fissure, producing structures such as sandstone dikes. (AGI, 1987)

injection gneiss A composite rock whose banding is wholly or partly caused by lit-par-lit injection of granitic magma into layered rock. See also: *composite gneiss*. Cf: *migmatite; lit-par-lit*. (AGI, 1987)

injection metamorphism Metamorphism accompanied by intimate injection of sheets and streaks of liquid magma (usually granitic) in zones near deep-seated intrusive contacts. Cf: *plutonic metamorphism; lit-par-lit*. (AGI, 1987)

injection pressure The total amount of pressure required to force a liquid or grout into cracks, cavities, and pores in rocks or other substance. (Long, 1960)

injector (a) Any apparatus used to force, under pressure, material into an opening in another material. (Long, 1960) (b) A device used to force-feed water into a boiler by the direct action of steam. See also: *inspirator*. (Long, 1960) (c) Mechanism used for spraying fuel oil into the combustion-type engine or to spray a fine oil mist into a stream of air or steam. See also: *air mover*. (Long, 1960)

inkstone (a) Native copperas (melanterite), or a stone containing it. Used in inkmaking. (Webster 2nd, 1960) (b) A stone slab used in preparing India ink for use. (Webster 2nd, 1960)

inlier An area or group of rocks surrounded by rocks of younger age; e.g., an eroded anticlinal crest. Cf: *outlier*. (AGI, 1987)

in-line valve A valve that proves the cage is in the correct position relative to the decking level. (Sinclair, 1959)

inmost Being at a point, place, or position farthest from the exterior; deepest within; innermost; such as, the inmost depths of a mine. (Standard, 1964)

inner core The central part of the Earth's core, extending from a depth of about 5,100 km to the center (6,371 km) of the Earth; its radius is about one-third of the whole core. The inner core is probably solid, as evidenced by the observation of S waves that are propagated in it, and because compressional waves travel noticeably faster through it than through the outer core. Density ranges from 10.5 to 15.5 g/cm^3. It is equivalent to the G layer. Cf: *outer core*. Syn: *siderosphere*. (AGI, 1987)

inner mantle The lower part of the mantle. (Schieferdecker, 1959)

innermost isoseismal The isoseismal line surrounding the area experiencing the greatest damage to manmade structures during an earthquake.

inoculation The addition of a material to molten metal to form nuclei for crystallization. (ASM, 1961)

inorganic Pertaining or relating to a compound that contains no carbon. Cf: *organic*. (AGI, 1987)

in-over *inby*.

in-pit crusher A crushing system that can be a fully mobile unit or a permanently fixed unit at the point of mining so that the mined material can be transported out of the pit by a conveyor system. (SME, 1992)

inquartation In bullion assay, dissolution of silver from associated gold by use of nitric acid. (Pryor, 1963)

inrush of water A sudden and often overwhelming flow of water into mine workings. Inrushes of water may be caused by striking unsuspected waterlogged old workings that possibly were shown inaccurately on the mine plans. Faults have also been responsible for serious inflows of water. A fault may retain large volumes of water above or at the same level as workings approaching it. It is usual to drive exploring headings in the direction of the suspected water danger. See also: *mud rush; old working; inundation; stopping; tapping; tapping old workings; water inrush; waterlogged*. (Nelson, 1965)

inselberg A prominent isolated residual knob, hill, or small mountain rising abruptly from an extensive erosion surface in a hot, dry region (as in the deserts of southern Africa or Arabia), generally bare and rocky, although partly buried by the debris derived from its slopes. Etymol: German Inselberg, island mountain. Cf: *monadnock*. (AGI, 1987)

insert Formed pieces of sintered cobalt-tungsten carbide mixture (in which diamonds may be inset), brazed into slots or holes in drilling bits or into grooves on the outside surface of a reaming shell to act as cutting points, reaming surfaces, or wear-resistant pads or surfaces of reaming shells or outside surfaces of other pieces of drilling equipment or fittings. Also called inserts. See also: *insert bit; insert set*. Cf: *slug*. (Long, 1960)

insert bit A bit into which inset cutting points of various preshaped pieces of hard metal (usually a sintered, tungsten carbide-cobalt powder alloy) are brazed or hand-peened into slots or holes cut or drilled into a blank bit. Hard-metal inserts may or may not contain diamonds. Also called slug bit. See also: *insert*. (Long, 1960)

inserted rod-type pick *sintered carbide-tipped pick*.

insert reaming shell A reaming shell, the reaming diamonds of which are inset in shaped, hard, metal plates brazed into grooves cut into the outside surface of the shell. (Long, 1960)

insert set Bits or reaming shells set with inserts. See also: *insert.* (Long, 1960)

inset (a) The entrance to underground roads from the shaft; a landing. See also: *ingate.* (Mason, 1951) (b) *phenocryst.* (c) The opening from the mine shaft to a seam of coal. (CTD, 1958) (d) A surface into which diamonds or other cutting points are embedded or set; also, the act or process of embedding such materials in a surface. (Long, 1960)

inside A term often used to designate the interior of a mine. (Fay, 1920)

inside angling *angling.*

inside clearance The difference between the outside diameter of a core and the inside diameter of the core-barrel parts through which the core passes or enters; also, the annular space between the inner and outer tubes in a double-tube core barrel. See also: *clearance.* (Long, 1960)

inside face That part of the bit crown nearest to and/or parallel with the inside wall of an annular or coring bit. (Long, 1960)

inside gage The inside diameter of a bit as measured between the cutting points, such as between inset diamonds on the inside-wall surface of a core bit. (Long, 1960)

inside-gage stone A diamond set in the inside-wall surface of the crown of a diamond core bit so that it cuts sufficient inside clearance to permit the core to pass through the bit shank and into the core barrel without binding. Syn: *inside kicker; inside reamer; inside stone.* (Long, 1960)

inside-haulage engineer In bituminous coal mining, one who operates a mine locomotive to haul trains of cars along underground haulageways in a mine. (DOT, 1949)

inside kicker *inside-gage stone.*

inside parting A side track or parting some distance from the beginning of a long entry, at which cars are left by a gathering driver. Also called a swing parting. (Fay, 1920)

inside reamer *inside-gage stone.*

inside slope (a) A slope on which coal is raised from a lower to a higher entry, but not to the surface. (Fay, 1920) (b) An inside slope is a passage in the mine driven through the seam by which coal is brought up from a lower level. (Korson, 1938)

inside stone *inside-gage stone.*

inside upset A tubular piece having ends that are thickened for a short distance on the inside. See also: *upset.* (Long, 1960)

inside work (a) The drilling of boreholes in underground workplaces; also applied to work done on the surface with the drill machine and tripod completely housed. (Long, 1960) (b) Any work in the mines. Most commonly used in bituminous coal mining.

in situ (a) In the natural or original position. Applied to a rock, soil, or fossil occurring in the situation in which it was originally formed or deposited. See also: *place.* (b) Said of tests done on a rock or soil in place, as compared with collecting discrete samples for testing in the laboratory. (c) *solution mining.*

in situ autoclave A six-sided in situ vat containing liquid or gas for ore treatment and recovery of mineral values requiring elevated temperatures and pressure, usually for long periods of time. (Lombardi, 1994)

in situ gasification Process that can recover the energy of coal seams without the extensive use of traditional mining operations. The primary product brought from underground is hot fully combusted flue gas, 1,100 to 1,800 °F (approx. 600 to 1,000 °C), whose sensible heat contains most of the heating value of the coal, 5,000 to 13,000 Btu/lb (11.6 to 30.2 MJ/kg). Particularly applicable to coal deposits that are not economically or technically feasible to mine by conventional methods because of seam quality or quantity, depth, dip, strata integrity, overburden thickness, etc. (SME, 1992)

in situ leaching (a) A hydrometallurgical process that treats ore for the recovery of mineral values while the ore is in place. It is a true mining technique in that the ore is not extracted from the ground and no mine waste piles or tailings impoundments are created. (SME, 1992) (b) A leaching technique in which ore is leached in place by solution injected into the deposit through wells. Cf: *solution mining.* (Aplan, 1974)

in situ liner A water, gas, or aqueous, chemically impermeable material placed in artificially constructed underground channels, crevices, or slices. (Lombardi, 1994)

in situ origin theory The theory of the origin of coal that holds that a coal was formed at the place where the plants from which it was derived grew. See also: *autochthonous coal.* (AGI, 1987)

in situ processes Activities conducted on the surface or underground in connection with in-place distillation, retorting, leaching, or other chemical or physical processing of coal or ore. The term includes, but is not limited to, in situ gasification, in situ leaching, slurry mining, solution mining, borehole mining, and fluid recovery mining.

in situ soil tests Tests carried out on the ground, in a borehole, trial pit, or tunnel, as opposed to a laboratory test. An in situ soil test may be a vane test, dynamic penetration test, etc. (Nelson, 1965)

in situ vat A five- or six-sided enclosure constructed in the earth by backfilling cuts around an orebody or orebody zone with material that is impervious to solutions so that aqueous leaching can be conducted for the extraction of mineral values from the isolated ore. (Lombardi, 1994)

insoluble (a) Incapable of being dissolved in a particular liquid. (Shell) (b) Term used of solid that does not dissolve under specified attack. No known substance is completely insoluble, so the term refers to systems characterized by very low solubility. (Pryor, 1963) (c) As used in smelter contracts, the terms "insoluble" and "silica" are often used interchangeably, but they are different things. Silica is determined by a special fusion assay. Insoluble is the residue left after the ore has been digested with acid in the course of assaying for some of the metals. The insoluble is generally silica plus something else, often alumina, since this substance is not always dissolved by acids. (Lewis, 1964)

insoluble anode An anode that does not dissolve during electrolysis. (ASM, 1961)

inspector One employed to make examinations of and to report upon mines and surface plants relative to compliance with mining laws, rules and regulations, safety methods, etc. State inspectors have authority to enforce State laws regulating the working of the mines. Federal inspectors have authority to enforce Federal laws in coal mines. See also: *mine inspector.* (Fay, 1920)

inspirator A kind of injector for forcing water by steam. See also: *injector.* (Webster 2nd, 1960)

instantaneous charge weight The weight of explosive detonated at any one precise time.

instantaneous cuts Cuts characterized by the drilling and ignition being done so that all the holes can cooperate and break smaller top angles. They are called instantaneous cuts because they are preferably ignited by instantaneous detonators to ensure a simultaneous detonation of all the charges in the cut. Some examples are Blasjo cut; WP-cut. (Langefors, 1963)

instantaneous detonator A detonator in which there is no designed delay period between the passage of an electric current through the detonator and its explosion. (BS, 1964)

instantaneous fuse Term used to distinguish rapid burning from slow fuse. Ignition rate is several thousand feet per minute, but slower than that of detonating fuse. (Pryor, 1963)

Institution of Mining and Metallurgy The London Institution of Mining and Metallurgy is the central British organization for regulating the professional affairs of suitably qualified mining engineers engaged in production and treatment of nonferrous metals and rare earths. Related bodies are those of Canada (Can. I.M.M.), Australia (Aust. I.M.M.), and Republic of South Africa (Rep. S. Af. I.M.M.). (Pryor, 1963)

Institution of Mining and Metallurgy screen scale Laboratory screens of usual 8-in-round (20-cm-round) size, in which the diameter of each new wire is equal to the distance between successive parallel wires. Therefore, in a 60-mesh screen (having 60 wires/in (152 wires/cm) measured along either the warp or the woof) the aperture is a square measuring 1/120 in (0.21 mm) on the side. The meshes used are 5, 8, 10, 12, 16, 20, 30, 40, 50, 60, 70, 80, 90, 100, 120, 150, and 200. (Pryor, 1963)

instroke The right to raise or take ore from a leased mine through the shaft or tunnel of an adjoining mine. (Ricketts, 1943)

instrumentalities of mining The true meaning of such expressions as shafts, tunnels, levels, uprises, crosscuts, inclines, and sump when applied to mines signifies instrumentalities whereby and through which such mines are opened, developed, prospected, and worked. (Ricketts, 1943)

instrumentation Control by servomechanisms. Use of signaling devices originating with the process to indicate, vary, or regulate performance.

insufflator An injector for forcing air into a furnace. (Webster 3rd, 1966)

insular shelf The zone surrounding an island extending from the line of permanent immersion to about 100 fathoms (600 ft or 183 m) of depth, where a marked or rather steep descent toward the great depths occurs. (AGI, 1987)

insular slope The declivity from the offshore border of the insular shelf at depths of from 50 to 100 fathoms (300 to 600 ft or 91 to 183 m) to oceanic depths. It is characterized by a marked increase in gradient. (AGI, 1987)

insulate To separate or to shield (a conductor) from conducting bodies by means of nonconductors, so as to prevent transfer of electricity, heat, or sound. (Webster 3rd, 1966)

insulating water bottle In oceanography, an instrument used for the accurate determination of the temperature of the sea at moderate depths. Also called Nansen-Pettersson water bottle. (CTD, 1958)

insulator-tube header One who forms heads on porcelain tube insulators by means of hand capping press, inserting clay tube in machine and pulling lever to form the head. (DOT, 1949)

insurance and freight cost Term showing that these items have been paid by the shipper of concentrates, metal, etc. (Pryor, 1963)

intake (a) The ventilating passage in an underground mine through which fresh air is conducted via an adit, drill hole, or downcast shaft to the workings. (b) The passage by which the ventilating current enters a mine. See also: *downcast.* A term that is more appropriate for a shaft; intake for an adit or entry. (Fay, 1920) (c) Scot. Person who works underground at odd work. (Fay, 1920) (d) N. of Eng. Any roadway underground through which fresh air is conducted to the working face. (Trist, 1963) (e) The passage and/or the current of ventilating air moving toward the interior of a mine. (Long, 1960) (f) The suction pipe or hose for a pump. (Long, 1960) (g) In hydraulics, the point at which the water or other liquid is received into a pipe, channel, or pump. (Long, 1960) (h) The headworks of a conduit; the place of diversion. (Seelye, 1951)

intake area That part of the land surface where water passes downward on its way to the zone of saturation in one or more aquifers. See also: *recharge.*

integral pilot A pilot-type noncoring bit having a pilot section that is an integral, nonreplaceable part of the bit. (Long, 1960)

integrated producer A producer of metal who owns mines, smelters, and refineries, and sometimes also fabricating plants. (Wolff, 1987)

integrated train A long string of cars, permanently coupled together, that shuttles continuously back and forth between one mine and one generating plant, not even stopping to load and unload, since rotary couplers permit each car to be flipped over and dumped as the train moves slowly across a trestle.

integrating meter A meter that records the total quantity of liquid or electricity passing through it. (Hammond, 1965)

integration In petrology, the formation of larger crystals from smaller ones by recrystallization. (Goldman, 1952)

integrator A circuit whose output is substantially proportional to the time integral of the input. (NCB, 1964)

intense anomaly An anomaly whose elemental values rise sharply to one or more well-defined peaks.

intensity As applied to color, the comparative brightness (vividness) or dullness or brownishness of a color; its comparative possession or lack of brilliance; therefore, the variation of a hue on a vivid-to-dull scale. See also: *tone.*

intensity of magnetization The magnetic moment per unit volume. (AGI, 1987)

intensity of pressure The pressure per unit area. (Seelye, 1951)

intensity of radiation The energy per unit time entering a sphere of unit cross-sectional area centered at a given place. The unit of intensity is the erg per square centimeter second or the watt per square centimeter. (NCB, 1964)

intensity scale A standard of relative measurement of earthquake intensity. Four such systems are the Mercalli scale, the modified Mercalli scale, the Richter scale, and the Rossi-Forel scale.

interbedded Occurring between beds, or lying in a bed parallel to other beds of a different material. Syn: *interstratified.* Cf: *intercalated.* (Fay, 1920)

interburden Material of any nature that lies between two or more bedded ore zones or coal seams. Term is primarily used in surface mining. (Federal Mine Safety, 1977)

intercalated Said of layered material that exists or is introduced between layers of a different character; esp. said of relatively thin strata of one kind of material that alternates with thicker strata of some other kind, such as beds of shale intercalated in a body of sandstone. Cf: *interbedded.* (AGI, 1987)

intercept (a) The distance along a crystallographic axis to its intersection with a crystal face. This intercept is a rational number because the axial unit length of each mineral is selected to make it so. The ratio of these intersections of a face with each of the crystallographic axes constitutes a parameter, such as Miller indices, that defines the crystal face. Cf: *indices of a crystal face; crystallographic axis; Miller indices.* (b) The part of the rod seen between the upper and lower stadia hairs of a transit or telescopic alidade; e.g., a stadia interval.

intercepting channel A channel excavated at the top of earth cuts, or at the foot of slopes, or at other critical places to intercept surface flow; a catch drain. (Seelye, 1951)

intercepting drain A drain that intercepts and diverts water before it reaches the area to be protected. Also called curtain drain. (Nichols, 1976)

intercepts (a) That portion included between two points in a borehole, as between the point where the hole first encounters a specific rock or mineral body and where the hole enters a different or underlying rock formation. (Long, 1960) (b) In crystallography, the distances cut off on axes of reference by planes. (Fay, 1920)

intercept time *delay time.*

intercooler (a) A radiator in which air is cooled while moving from low-pressure to high-pressure cylinders of a two-stage compressor. See also: *two-stage compression.* (Nichols, 1976) (b) In multistage compression of air, a cooling arrangement between stages. See also: *aftercooler.* (Pryor, 1963) (c) A cooling device used on a turbocharged diesel engine to reduce the air volume between the turbocharger and the cylinders.

intercooling Extraction of heat from a compressed gas between two stages of compression in order to improve the efficiency of compression. (Strock, 1948)

interfacial angle The internal or dihedral angle between two faces of a crystal. It is also the "angle of dip" between faces.

interfacial energy Tension at interfaces between the various phases of a system; may include solid, liquid, and gas interfaces, varying in their combinations and qualities. (Pryor, 1963)

interfacial tension The contractile force of an interface between two phases. (ASM, 1961)

interference The meeting of two wave systems resulting in increased amplitude (constructive interference) if they are in phase, i.e., crest to crest, and decreased amplitude (destructive interference) if they are out of phase, i.e., crest to trough. In polarized-light microscopy (PLM), phase differences are generated when white light passes through an anisotropic (i.e., doubly refracting) crystal or crystal fragment, these differences being determined primarily by birefringence, light wavelength, and crystal thickness. Waves of light in different parts of the visible spectrum interfere both constructively and destructively when resolved in the microscope analyzer to give an interference color. Since anisotropic minerals have a range of birefringence, interference colors are useful as an aid to their identification. See also: *optic sign; index of refraction.*

interference color One of the spectral colors produced by the strengthening or the weakening of certain wavelengths of a composite beam of light in consequence of interference. This is an important characteristic in determining minerals in thin section or in fragments under the polarizing microscope. (Webster 3rd, 1966)

interference figure (a) An optical figure composed of a series of spectrally colored rings combined with a blank cross (if uniaxial) or a series of spectrally colored curves or rings with two black parabolic curves called isogyres (if biaxial). The figure is observed when a properly oriented thin section or fragment of a mineral is examined in convergent light through the polarizing microscope. The interference figure, which is caused by the birefringence of the mineral and by the orientation of the mineral so that it presents an optic axis in the field of the microscope, is one of the most valuable optical aids in identifying minerals. Also called the direction image. (b) An optical pattern produced by conoscopic illumination of anisotropic crystals which appears on a spherical focal surface located above the objective lens of a polarized-light microscope. Cf: *melatope.*

interference methanometer A combustible-gas detector based on the velocity of light. A beam of light is split into two parts that pass respectively through chambers containing pure air and the test air at velocities characteristic of the gases. When methane is present, the light beams are out of step and this movement becomes a measure of the methane concentration. See also: *refractometer.* (Nelson, 1965)

interfluve The area between two rivers flowing in the same general direction. Syn: *interstream area.*

interformational Between formations, such as an interformational unconformity. Cf: *intraformational.*

intergranular corrosion Corrosion that occurs preferentially at grain boundaries of a metal or alloy.

intergrown (a) Of coal and mineral matter, naturally associated and separable only by crushing or grinding. (BS, 1962) (b) In crystallography, a descriptive term for mineral species that have crystallized simultaneously and therefore become intertwined or interlocked. (Pryor, 1963)

intergrowth The state of interlocking of grains of two different minerals as a result of their simultaneous crystallization. Cf: *graphic granite.* (AGI, 1987)

interior angle Horizontal angle between adjacent sides of a polygon, measured within the polygon. (Seelye, 1951)

interior coalfields U.S. Includes Eastern Interior Field, Illinois, Indiana, and Kentucky; Western Interior Field, Great Plains States from Iowa to Arkansas; Southwestern Field, Texas; and Northern Field, Michigan. (Bateman, 1950)

interior span A continuous beam or slab, both supports of which are continuous with adjacent spans. (Hammond, 1965)

interleaved (a) Intercalated in very thin layers. (b) A term used in remote sensing to describe a technique of storing digital image data (pixel interleaved, line interleaved, or band interleaved.)

interlock The clutch in steel sheet piles. (Hammond, 1965)

interlocking controls A system of electrical controls for a system of conveyors that maintains a controlled relationship between the units of the system. Sometimes applied to sequence-starting controls.

interlocking wedge-type capping This type of wire-rope capping is simple to apply. The sleeves are first threaded on the rope, a white metal bob is then formed on the end of the rope by untwisting the wires and cutting out the hemp core, if present, and white metal is run into a mold around the wires. The bob is allowed to cool. Two tapered interlocking steel wedges are then fitted on to the rope clear of the bob so that wedges can move forward toward the bob and grip the rope as the load is applied to the capping. The rope is cleaned of lubricant where the wedges will grip, and the groove in the wedges must be of such a size that a gap is left between the wedges so that they can grip the rope firmly. The edges of the wedges should come opposite a valley between the strands of the rope. The outer socket is now placed over the wedges and the sleeves are lightly tapped into position to hold the parts together. (Sinclair, 1959)

intermediate (a) A secondary or auxiliary horizontal passage driven between levels in a mine, which may extend from a raise or stope and, depending upon its orientation, may be either an intermediate drift or a crosscut. Syn: *sublevel.* (Forrester, 1946) (b) Said of an igneous rock that is transitional between basic and silicic (or between mafic and felsic), generally having a silica content of 54% to 65%; e.g., syenite and diorite. "Intermediate" is one subdivision of a widely used system for classifying igneous rocks on the basis of their silica content; the other subdivisions are acidic, basic, and ultrabasic. Syn: *mediosilicic.* (AGI, 1987)

intermediate constituent A constituent of alloys that is formed when atoms of two metals combine in certain proportions to form crystals with a different structure from that of either of the metals. The proportions of the two kinds of atoms may be indicated by formula, e.g., CuZn; hence, these constituents are also known as intermetallic compounds. (CTD, 1958)

intermediate cooler In a blast furnace, a water-cooled casting, usually of copper, that is installed inside the cinder cooler. (Henderson, 1953)

intermediate cut *middle cut.*

intermediate-duty fireclay brick A fireclay refractory having a pyrometric cone equivalent not lower than No. 29, or having a refractoriness of not more than 3.0% deformation as measured by the load test at 2,460 °F (1,349 °C) (ASTM requirements). (ARI, 1949)

intermediate electrode *bipolar electrode.*

intermediate gate A gate between the central gate and the end gates, particularly in double-double unit layouts. (Nelson, 1965)

intermediate haulage (a) The transportation of mined coal or ore from the face haulage to that point where it is accessible to the main line. It is accomplished by conveyors, belts, or locomotives and mine cars. (Woodruff, 1966) (b) Mine haulage used to collect loads and deliver empties from and to the sections. These are taken to and from central sidetracks served by the main line motor. Locomotives and track are frequently lighter than those on the main line. See also: *haulage.* (Kentucky, 1952)

intermediate haulage conveyor Generally 500 to 3,000 ft (152 to 915 m) in length. It is used to transport material between the gathering conveyor and the main haulage conveyor. (NEMA, 1961)

intermediate layer *sima.*

intermediate loading station *loading station.*

intermediate microcline *mesomicrocline.*

intermediate packs Packs built between gates with wastes on each side and usually supported by packwalls. (TIME, 1929-30)

intermediate principal plane The plane normal to the direction of the intermediate principal stress. (ASCE, 1958)

intermediate principal stress The principal stress whose value is neither the largest nor the smallest (with regard to sign) of the three. (ASCE, 1958)

intermediate section The part of a mining belt conveyor that consists of the framing and the belt idlers supported by the framing, both of which guide and support the belts between the head end and the tail end. There are two general types of intermediate sections: rigid side framed and wire-rope side framed. (NEMA, 1961)

intermediate shaft A shaft that is driven by one shaft and drives another. (Nichols, 1976)

intermediate transfer point The point along a conveyor, which may already be carrying a load, at which coal is delivered from another panel conveyor. (Nelson, 1965)

intermediate vein zone deposit A deposit thought to have been formed at a depth ranging from 4,000 to 12,000 ft (1.2 to 3.7 km) below the surface and at a temperature between 175 °C and 300 °C. Such a deposit may take the form of a fissure vein, a series of parallel fissures called a sheeted zone, a replacement of the wall rock of fissures, or a large disseminated deposit. Much of the gold, silver, copper, lead, and zinc of the Western United States comes from these deposits. (Lewis, 1964)

intermetallic compound An intermediate phase in an alloy system, having a narrow range of homogeneity and relatively simple stoichiometric proportions, in which the nature of the atomic binding can vary from metallic to ionic. See also: *intermediate constituent.* (ASM, 1961)

intermittent cutter Coal-cutting machine of the pick machine and breast machine type. They are called intermittent cutters because they must be frequently reset, whereas with continuous cutters, a continuous cut can be made the full width of the face without stopping the machine. (Kiser, 1929)

intermittent filters These usually consist of a number of filtering leaves that are simply rectangular frames carrying filter cloth on the outer surface. A number of these leaves are mounted in a suitable tank, and the clear liquid passes through the filter cloth and out through pipes leading from the interior of each filter leaf. The solid material forms a cake on the outside of the leaf. (Newton, 1959)

intermontane Lying between mountains.

internal-discharge bucket elevator A type of bucket elevator having continuous buckets abutting, hinged, or overlapping; designed for loading and discharging along the inner boundary of the closed path of the buckets. Syn: *internal elevator.* See also: *bucket elevator.*

internal drainage Drainage that does not reach the ocean by surface streams, such as drainage toward the lowermost or central part of an interior basin. It is common in arid and semiarid regions, as in western Utah. (Stokes, 1955)

internal elevator *internal-discharge bucket elevator.*

internal ribbon conveyor A trunnion-supported, revolving cylinder, the inner surface of which is fitted with continuous or interrupted ribbon flighting.

internal stress Residual stress existing between different parts of metal products, as a result of the differential effects of heating, cooling, or working operations, or of constitutional changes in the solid metal. (CTD, 1958)

internal waste (a) Barren rock between two or more bands (veins) or reef which are mined simultaneously. (Beerman, 1963) (b) *horse.*

international ampere The electric current that, when passed through a solution of silver nitrate in water, will deposit silver at the rate of 0.001118000 g/s. The unit of current in common use. (CTD, 1958)

International Ellipsoid of 1930 Equation relating variation of gravity with latitude, adopted by an international commission as best expressing the normal gravity field of the Earth to the approximation of an ellipsoid of revolution.

international metric carat *carat.*

interpenetration twin Two or more individual crystals twinned into such a position that they penetrate one another. See also: *penetration twin.* Cf: *contact twin; juxtaposition twin.* (Fay, 1920)

interphase In physical chemistry, the transition layer, zone of change, zone of shear, or zeta layer, through which the characteristic qualities of each contacting phase diffuse outward with diminishing strength toward the adjoining phase. Not an interface, since the division is not sharp. (Pryor, 1963)

interpolation Estimation of a statistical value from its mathematical or graphical position intermediate in a series of determined points.

interpolation of contours The process of drawing contour lines by inferring their plan position and trend from spot levels or from other contours, assuming the intervening ground to have uniform slope. Where the spot levels are sparse, the process requires knowledge of the land or lie of the seams. See also: *contour plan.* (Nelson, 1965)

interrupter A device, usually automatic, for rapidly and frequently breaking and making an electric circuit, as in an induction coil. (Standard, 1964)

intersect (a) To cut across or meet, as a borehole cuts through a stratum of rock or encounters a vein. (Long, 1960) (b) In mining, to cut across or meet a vein or lode with a passageway; also, the point at which a vein or lode cuts across an earlier formation. (Long, 1960)

intersection (a) The point at which a deliberate deflection of the trend of a borehole is made. (Long, 1960) (b) The point at which a drill hole enters a specific orebody, fault, or rock material. (Long, 1960) (c) Meeting of two orebodies or veins, or the point at which a vein or orebody meets a fault, dike, or rock stratum. (Long, 1960) (d) The point at which two underground workings connect. (Long, 1960) (e) A method in surveying by which the horizontal position of an unoccupied point is determined by drawing lines to that point from two or more points of known position. Syn: *resection.* (AGI, 1987)

intersection angle The angle of deflection, as measured at the intersection point, between the straights of a railway or highway curve. (CTD, 1958)

intersection point That point at which two straights or tangents to a railway or road curve would meet if produced. See also: *tangent distance.* (Hammond, 1965)

intersection shoot An ore shoot located at the intersection of one vein or vein system with another. It is a common type of ore deposit. (AGI, 1987)

intersertal Said of the texture of a porphyritic igneous rock in which the groundmass, composed of a glassy or partly crystalline material other than augite, occupies the interstices between unoriented feldspar laths, the groundmass forming a relatively small proportion of the rock. Cf: *hyalophitic; hyalocrystalline.* (AGI, 1987)

interspersed carbide Small-size (1/8 inch or 3 mm and larger), irregular-shaped fragments of tungsten carbide slugs mixed with a suitable matrix metal; applied to cutting faces of bits or other cutting tools as a weldment. Also called clustered carbide. (Long, 1960)

interstice (a) An opening or space, as in a rock or soil. Syn: *void; pore.* Adj. interstitial. (AGI, 1987) (b) Small void in the body of a metal. (ASM, 1961)

interstitial Said of a mineral deposit in which the minerals fill the pores of the host rock. Cf: *Frenkel defect; impregnated.* (AGI, 1987)

interstitial water Subsurface water in the voids of a rock. Syn: *pore water.* Cf: *connate water.* (AGI, 1987)

interstratification The state or condition of occurring between strata of a different character. (AGI, 1987)

interstratified (a) Interbedded; strata deposited between or alternatingly with other strata. (Fay, 1920) (b) Of coal and mineral matter, associated in random horizontal layers, usually with a natural cleavage. (BS, 1962)

interstream area *interfluve.*

intertrappean Lying between beds of trap. (Standard, 1964)

interval (a) The distance between two points or depths in a borehole; core intersection. (b) The vertical distance between strata or units of reference. (AGI, 1987) (c) The contour interval is the vertical distance between two successive contour lines on a topographic, structural, or other contour map. Syn: *contour interval.* (AGI, 1987)

interwoven conveyor belt A construction of conveyor belt similar to the solid woven type of belt and having the plies interwoven to the extent that it is impossible to separate them.

in-the-seam mining The conventional system of mining in which the development headings are driven in the coal seam. Cf: *horizon mining.* See also: *unproductive development.* (Nelson, 1965)

into the house Newc. The upstroke of a pump engine. (Fay, 1920)

into the solid Said of a shot that goes into the coal beyond the point to which the coal can be broken by the blast. (Fay, 1920)

intracrystalline Within or across the crystals or grains of a metal. Syn: *transgranular.* (ASM, 1961)

intrados The interior curve of an arch, as of a tunnel lining. (Sandstrom, 1963)

intradosal The fractured ground within the fracture zone. Cf: *extradosal.* (Lewis, 1964)

intraformational (a) Formed within a geologic formation, more or less contemporaneously with the enclosing sediments. The term is esp. used in regard to syndepositional folding or slumping; e.g., intraformational deformation or intraformational breccia. (AGI, 1987) (b) Existing within a formation, with no necessary connotation of time of origin. Cf: *interformational.* (AGI, 1987)

intraformational breccia A rock formed by brecciation of partly consolidated material, followed by practically contemporaneous sedimentation. It is similar in nature and origin to an intraformational conglomerate, but contains fragments showing greater angularity. (AGI, 1987)

intraformational conglomerate (a) A conglomerate in which the clasts are essentially contemporaneous with the matrix, developed by the breaking up and rounding of fragments of a newly formed or partly consolidated sediment (usually shale or limestone) and their nearly immediate incorporation in new sedimentary deposits; e.g., an edgewise conglomerate. (AGI, 1987) (b) A conglomerate occurring in the midst of a geologic formation, such as one formed during a brief interruption in the orderly deposition of strata. It may contain clasts external to the formation. The term is used in this sense esp. in England. (AGI, 1987)

intraformational contortion Intricate and complicated folding, resulting from the subaqueous slumping or sliding of unconsolidated sediments under the influence of gravity. (AGI, 1987)

intratelluric (a) Said of a phenocryst, of an earlier generation than its groundmass, that formed at depth, prior to extrusion of a magma as lava. (AGI, 1987) (b) Said of the period of crystallization occurring deep within the Earth just prior to the extrusion of a magma as lava. (AGI, 1987) (c) Located, formed, or originating deep within the Earth. (AGI, 1987)

intrinsically safe apparatus Apparatus that is so constructed that, when installed and operated under the conditions specified by the certifying authority, any electrical sparking that may occur in normal working, either in the apparatus or in the circuit associated therewith, is incapable of causing an ignition of the prescribed flammable gas or vapor. (BS, 1965)

intrinsically safe circuit A circuit in which any electrical sparking that may occur in normal working under the conditions specified by the certifying authority, and with the prescribed components, is incapable of causing an ignition of the prescribed flammable gas or vapor. (BS, 1965)

intrinsically safe machine A machine that is safe in itself, without having to be placed inside a flameproof enclosure. It implies that the machine cannot produce any spark that is capable of igniting mixtures of combustible gases and air in mines. (Nelson, 1965)

intrinsic safety In a circuit, safety such that any sparking that may occur in that circuit in normal working, or in reasonable fault conditions, is incapable of causing an explosion of the prescribed inflammable gas. (NCB, 1964)

introductory column In casing a borehole, the highest and first column that is inserted. (Stoces, 1954)

intrusion (a) In geology, a mass of igneous rock that, while molten, was forced into or between other rocks. (Fay, 1920) (b) A mass of sedimentary rock occurring in a coal seam. (BS, 1964)

intrusion breccia *contact breccia.*

intrusion displacement Faulting coincident with the intrusion of an igneous rock.

intrusion grouting A method of placing concrete by intruding the mortar component in position; it is then converted to concrete by intruding the mortar component into its voids. One of the chief advantages of the method is that it permits the placing of concrete underwater. (Carson, 1965)

intrusive Of or pertaining to intrusion—both the processes and the rock so formed. n. An intrusive rock. Cf: *extrusive.* (AGI, 1987)

intrusive vein An igneous intrusion resembling a sheet, apparently formed from a magma rich in volatiles. (AGI, 1987)

intumescence The property of some silicate minerals (e.g., stilbite, vermiculite, scapolite) or rocks (e.g., perlite) to expand permanently when heated to form an irregular or vesicular structure. Cf: *exfoliation.*

inundation An inrush of water, on a large scale, that floods the entire mine or a large section of the workings. See also: *tapping old workings.* (Nelson, 1965)

Invar An alloy of nickel and iron, containing about 36% nickel, and having an extremely low coefficient of thermal expansion. It is used in the construction of surveying instruments, such as pendulums, level rods, first-order leveling instruments, and tapes. (AGI, 1987)

invasion (a) Mex. A mining trespass. (Fay, 1920) (b) *transgression.*

inverse distance squared A method for interpolating spatial sample data and determining values between data points. A value interpolated for any spatial point is determined by applying a weighting factor based on distance between the spatial point and surrounding sample data. Selection of sample points to include in the calculation may be determined by minimum and/or maximum distance, azimuth orientation, and the minimum and/or maximum number of the nearest sample data points. Abbrev. IDS.

inverse initiation The placing of the detonator at the back of the shothole. This is the usual practice when using delay detonators to minimize the danger of cutoff holes. See also: *direct initiation.* Also known as indirect priming. Syn: *indirect initiation.* (Nelson, 1965)

Inverse Square Law Law that governs the distance-dependence of physical effects, such as intensity of light, magnetism, and gravitational force. The effect at a point due to an emitting source varies as the square of the distance between them. (Pryor, 1963)

inversion (a) Construction of a geophysical model from a set of measurements; e.g., using numerous gravity measurements to infer subsurface density distributions, or using slip vectors and spreading rates to define global plate motions. Inversion models are inherently ambiguous. (AGI, 1987) (b) A change of crystalline phase brought about by a change in temperature or pressure; e.g., the inversion between alpha quartz and beta quartz at 573 °C. (c) *center of symmetry.*

inversion point (a) A point representing the temperature at which one polymorphic form of a substance, in equilibrium with its vapor, reversibly changes into another under invariant conditions. (AGI, 1987) (b) The temperature at which one polymorphic form of a substance inverts reversibly to another under univariant conditions and a specific pressure. (AGI, 1987) (c) More loosely, the lowest temperature at which a monotropic phase inverts at an appreciable rate into a stable phase, or at which a given phase dissociates at an appreciable rate, under given conditions. (AGI, 1987) (d) A single point at which different phases are capable of existing together at equilibrium. Syn: *transition point; transition temperature.* (AGI, 1987)

invert The floor or bottom of the internal cross section of a closed conduit, such as an aqueduct, tunnel, or drain. The term originally referred to the inverted arch used to form the bottom of a masonry-lined sewer or tunnel. (AGI, 1987)

inverted *overturned.*

inverted fold *overturned.*

inverted heading and bench *heading-overhand bench.*

inverted pendulum An instrument in which the acceleration of gravity is determined by measuring the swinging period of a mass that is supported on top of a spring. (Schieferdecker, 1959)

inverted plunge The plunge of folds, or sets of folds, whose inclination has been carried past the vertical, so that the plunge is now less than 90° in a direction opposite from the original attitude. It is a rather common feature in excessively folded or refolded terranes. (AGI, 1987)

inverted relief A topographic configuration that is the inverse of the geologic structure, as where mountains occupy the sites of synclines and valleys occupy the sites of anticlines. (AGI, 1987)

inverted siphon (a) A pipeline crossing over a depression or under a highway, railroad, canal, etc. The term is common but inappropriate, as no siphonic action is involved. The term "sag pipe" is suggested as a substitute. (Seelye, 1951) (b) A pipe or tube in the shape of a siphon, but inverted, as for carrying water across the depression of a ravine to a lower level. (Standard, 1964) (c) *drowned level.*

invert level The datum level of the lowest part of an invert. (Hammond, 1965)

invert strut Flat strut that sometimes is used instead of an arch on the bottom of a tunnel cross section.

invisible light Wavelengths in the electromagnetic spectrum too short or too long to be detected by the human eye; e.g., ultraviolet and infrared light. Cf: *visible light.*

inwall (a) The refractory lining of the stack of a blast furnace. See also: *stack.* (Dodd, 1964) (b) The interior walls or lining of a shaft furnace. (Fay, 1920)

inwall brick Fireclay brick for use in lining the inwall section of a blast furnace. (ARI, 1949)

inyoite A monoclinic mineral, $Ca_2B_6O_6(OH)_{10}{\cdot}8H_2O$; forms large, soft, colorless, tabular crystals associated with colemanite and other boron minerals in Inyo County, CA; dehydrates to meyerhoffite.

iodargyrite A hexagonal mineral, AgI; soft, sectile, waxy; normally a secondary mineral; in the oxidized zone of silver deposits from primary ores containing argentite, tennantite or tetrahedrite, and native silver; a source of silver; in arid regions where volcanic rocks are common—Southwestern United States, Mexico, Chile, and Broken Hill, New South Wales, Australia. Formerly called iodyrite or iodite. Cf: *chlorargyrite; bromargyrite.*

iodembolite A name to replace iodobromite whose composition is not definite. Used to designate minerals of the cerargyrite group containing chlorine, bromine, and iodine, Ag(Cl,Br,I). (English, 1938)

iodide metal Hafnium produced by the van Arkel and de Boer process. (Thomas, 1960)

iodide process The process developed by van Arkel and de Boer; used for refining zirconium and hafnium by the decomposition of iodide on a hot wire. Syn: *van Arkel and de Boer process.* Also called crystal bar process. (Thomas, 1960)

iodide process for producing titanium Process involving the reaction of impure titanium metal with iodine to form the volatile tetraiodide, which is then decomposed on a hot wire at temperatures between 2,000 °F and 2,730 °F (1,093 °C and 1,500 °C) to form high-purity titanium and iodine.

iodimetry Volumetric analysis involving either titration with a standardized solution of iodine, or the release by a substance under examination of iodine in soluble form, so that its concentration can be determined by titration, using starch as an indicator. The method is used with substances that can oxidize potassium iodide to release free iodine, or conversely with substances which combine with free iodine. The basis of reaction is $I_2 + 2Na_2S_2O_3 \rightarrow 2NaI + Na_2S_4O_6$. Also called iodometry. (Pryor, 1963)

iodine A nonmetallic, bluish-black, lustrous solid element, volatilizing at ordinary temperatures into a blue-violet gas with an irritating odor. Symbol, I. Occurs sparingly in sea water, in saltpeter and nitrate-bearing earth (known as caliche), in brines, and in brackish waters from oil and salt wells. Its compounds are used in medicines, photography, and organic chemistry. (Handbook of Chem. & Phys., 3)

iodine pentoxide method Accurate determinations of very low concentrations of carbon monoxide are required of samples from pits troubled with spontaneous combustion and those in which diesel locomotives are operating. This method involves the passage of a known volume of the sample first through a train of reagents for purifying and drying and then through a heated tube of iodine pentoxide; if carbon monoxide is present, proportional amounts of iodine and carbon dioxide are formed, either of which may be determined. (Sinclair, 1958)

iodine pentoxide test An iodine pentoxide (Hoolamite) test consists of a glass tube filled with a reagent mixed with fuming sulfuric acid and carried by crushed pumice through which a known volume of the air to be sampled is discharged from a rubber bulb after preliminary drying. The bulb is squeezed 10 times; the carbon monoxide present changes the color of the reagent from grey to green, and the concentration is obtained by reference to a color chart supplied with the instrument. The instrument reads down to about 0.07% carbon monoxide. (Sinclair, 1958)

iodobromite *iodembolite.*

iodyrite *iodargyrite.*

iolite *cordierite.*

ion An atom or a group of atoms combined in a radical or molecule that carries a positive or a negative electric charge as a result of having lost or gained one or more electrons. It may exist in solution, usually in combination with molecules of the solvent, or out of solution; it may be formed during electrolysis and migrate to the electrode of opposite charge, or it may be formed in a gas and be capable of carrying an electric current through the gas. (Webster 3rd, 1966)

ion exchange Reversible exchange of ions contained in a crystal for different ions in solution without destruction of crystal structure or disturbance of electrical neutrality. The process is accomplished by diffusion and occurs typically in crystals possessing one- or two-dimensional channelways where ions are relatively weakly bonded. Also occurs in resins consisting of three-dimensional hydrocarbon networks to which many ionizable groups are attached. Syn: *base exchange.* See also: *cation exchange.* (AGI, 1987)

ion-exchange column A tube packed with particles or beads of resin chosen for their ability to capture specific ions from an aqueous solution as it passes through the column.

ionic activity measurement Use of an electrode reversible to the ion under test to form a half cell. This is connected by a salt bridge to a reference electrode, and the resulting electromotive force is measured. (Pryor, 1963)

ionic bond (a) Electrostatic force holding ions together in a crystal. (Hurlbut, 1964) (b) A chemical bond between atoms, one of which is an electron donor and the other is an electron acceptor. (Gaudin, 1957)

ionic equilibrium The situation when, for a prescribed temperature, pressure, concentration of reactants, and pH, the rate of dissociation of molecules into ions is approx. in balance with that of their recombination. (Pryor, 1963)

ionic migration Movement of a charged particle through an electrolyte toward an electrode of opposite charge sign. The losses in a neutral salt around two electrodes during the passage of electric current are in ratio to velocities of ions migrating from these electrodes. Ionic velocities are stated in centimeters per second for a potential gradient of 1 V/cm. (Pryor, 1963)

ionic mobility Velocity in a dilute solution of an ion where the potential difference across this is 1 V/cm. For hydrogen ions, the velocity is 0.00326 cm/s. (Pryor, 1963)

ionic transport number Fraction of total current carried by one ion during electrolysis (ion migration). (Pryor, 1963)

ionite A term referring to two different materials found in the Ione Valley of northern California: (1) a clay mineral, possibly a variety of kaolinite, found as scales in the Ione sandstone formation, and (2) an impure fossil hydrocarbon found in lignite, which is brownish-yellow, dissolved by chloroform, and yields a brown, tarry oil on destructive distillation.

ionization The process of adding electrons to, or knocking electrons from, atoms or molecules, thereby creating ions. High temperatures, electrical discharges, and nuclear radiation can cause ionization. (Lyman, 1964)

ionization chamber An instrument that detects and measures ionizing radiation by observing the electrical current created when radiation ionizes gas in the chamber, making it a conductor of electricity. (Lyman, 1964)

ionization constant The ratio of the product of the activities of the ions produced from a given substance to the activity of the undissociated molecules of that substance. See also: *activity.* (CTD, 1958)

ionizing radiation Any radiation that directly or indirectly displaces electrons from the outer domains of atoms; e.g., alpha, beta, or gamma radiation. (Lyman, 1964)

ionosphere The outer part of the Earth's atmosphere, beginning at an altitude of about 25 miles (40 km) and extending to the highest parts of the atmosphere: it contains several regions that consist of a series of constantly changing layers characterized by an appreciable electron and ion content. (Webster 2nd, 1960)

ion sieve separation Separation of ions by filtering them through the intermediately sized lattice of a suitable aluminosilicate zeolite, chosen to permit passage only of undersized ions through its rigid structure. (Pryor, 1963)

iozite *wüstite.*

iranite A triclinic mineral, $Pb_{10}Cu(CrO_4)_6(SiO_4)_2(F,OH)_2$; forms a series with hemihedrite in which zinc substitutes for copper; at the Sebarz Mine, northeast of Anarak, Iran.

iridescence The exhibition of interference colors from the surface or interior of a mineral, caused by light interference from thin films or layers of different refractive index. Labradorite and some other feldspars show it. The tarnish on the surface of coal, copper pyrites, etc., is sometimes iridescent. Adj. iridescent.

iridic gold Said to be a native alloy of gold and iridium carrying 62.1% gold, 30.4% iridium, 3.8% platinum, and 2.1% silver. (Hess)

iridioplatinum An alloy usually containing 90% or more of platinum. The remaining percentage is of iridium, which is necessary to produce an alloy sufficiently stiff for use in gem mountings.

iridium (a) An isometric mineral, native Ir; Mohs hardness, 6 to 7; sp gr, 22.2 if pure; occurs native (>80% Ir) and alloyed with osmium (iridosmine) or platinum in mafic to ultramafic rocks and derived alluvial deposits; in rare arsenides and sulfides, such as irarsite $(Ir,Ru,Rh,Pt)AsS$, iridarsenite $(Ir,Ru)As_2$, and iridisite-beta $(Ir,Cu)_3S_8$; placer deposits may include gold. (b) The most corrosion-resistant metal known. Symbol, Ir. It is mainly used as a hardening agent for platinum. (Handbook of Chem. & Phys., 3)

iridosmine A natural alloy of iridium and osmium, (Ir,Os). Analyses show 43% to 77% iridium, 17% to 49% osmium, and a little rhodium, ruthenium, platinum, iron, and copper. Rhombohedral. Syn: *osmite.* (Sanford, 1914; Dana, 1959)

iriginite A canary-yellow mineral, $(UO_2)(Mo_2^{6+}O_7)\cdot 3H_2O$; luster, vitreous; uneven fracture; does not fluoresce. Syn: *priguinite.* (Am. Mineral., 1945)

Irish coal Slate, shale, or rock loaded out from a colliery as coal. (Pryor, 1963)

Irish dividend An assessment on mining stock. (Fay, 1920)

iris quartz A transparent quartz crystal containing minute air-filled or liquid-filled internal cracks that produce iridescence by interference of light. The cracks may occur naturally or be caused artificially by heating and sudden cooling of the specimen. Syn: *firestone; rainbow quartz.*

iron (a) Iron-base materials not falling into the steel classifications. See also: *gray cast iron; ingot iron; malleable cast iron; nodular cast iron; white cast iron; wrought iron.* (ASM, 1961) (b) Colloquially, all derrick and drilling equipment above the heads of the workers on the drill platform. (Long, 1960) (c) Any ferrous metal tool or part that must be fished from a borehole. Also called junk. (Long, 1960) (d) An isometric mineral, native alpha-Fe; metallic steel gray to black; sp gr, 7.3 to 7.9. Occurs as grains in basalt in Disko Island, Greenland; in meteorites, and in placers on South Island, New Zealand; and in Oregon and British Columbia. Nickel is commonly reported in iron in quantities up to several percent; meteoric iron generally contains at least 5%, and up to 25% to 65%, nickel. Because of the instability of iron under oxidizing conditions and the abundance of oxygen in the Earth's crust and atmosphere, practically all terrestrial iron occurs in the divalent (ferrous) or trivalent (ferric) state combined with other metals and nonmetallic elements in silicates, oxides, sulfides, etc. (e) Fourth most abundant element, by weight, making up the crust of the Earth. Symbol, Fe. The most common ore is hematite, Fe_2O_3, from which the metal is obtained by reduction with carbon. Iron is the cheapest and most abundant, useful, and important of all metals. (Handbook of Chem. & Phys., 3)

iron alum *halotrichite.*

iron black Finely divided antimony. (Standard, 1964)

iron clay *clay ironstone.*

iron formation A chemical sedimentary rock, typically thin-bedded and/or finely laminated, containing at least 15% iron of sedimentary origin, and commonly but not necessarily containing layers of chert. Various primary facies (usually not weathered) of iron formation are distinguished on the basis of whether the iron occurs predominantly as oxide, silicate, carbonate, or sulfide. Most iron formation is of Precambrian age. In mining usage, the term refers to a low-grade sedimentary iron ore with the iron mineral(s) segregated in bands or sheets irregularly mingled with chert or fine-grained quartz. Cf: *ironstone; jaspilite.* Essentially synonymous terms are itabirite; banded hematite quartzite; taconite; quartz-banded ore; banded iron formation; calico rock; jasper bar. (AGI, 1987)

iron froth A fine spongy variety of hematite. (Fay, 1920)

iron furnace A furnace in which iron is smelted or worked in any way. (Standard, 1964)

iron glance A variety of hematite; specular iron. See also: *specularite.* (Fay, 1920)

iron hat (a) *gossan.* (b) *safety hat; hardhat.* (c) A weathered ironstone outcrop. (Pryor, 1963)

iron hypersthene *ferrosilite.*

iron man (a) An iron worker; a manufacturer of iron; esp. one engaged in the processing of iron. (Webster 3rd, 1966) (b) A worker who weighs out ground iron ore and adds it to slurry or dry-ground rock as it goes into the cement kiln. (Webster 3rd, 1966) (c) An apparatus on wheels for supporting a glassblower's pontil while blowing large cylinders, as for window glass. (Standard, 1964)

ironmaster Person who conducts or manages the founding or manufacture of iron, esp. on a large scale. (Webster 3rd, 1966)

iron mica A micaceous hematite. *biotite.* (Fay, 1920)

iron molybdate *ferrimolybdite.*

iron monarch An important iron ore (chiefly hematite) deposit in the Middleback Ranges area of South Australia. (Nelson, 1965)

iron olivine *fayalite.*

iron ore Ferruginous rock containing one or more minerals from which metallic iron may be profitably extracted. The chief ores of iron consist mainly of the oxides: hematite, Fe_2O_3; goethite, α-FeO(OH); magnetite, Fe_3O_4; and the carbonate, siderite or chalybite, $FeCO_3$. See also: *bog iron ore; limonite; kidney iron ore; magnetite; siderite; hematite; prereduced iron-ore pellet.* (AGI, 1987)

iron oxide (a) A common ore of iron, sometimes prepared as a fine powder for use by drillers as a drill-mud heavy loader. (Long, 1960) (b) A common compound of iron and oxygen; e.g., rust. (Long, 1960)

iron oxides The basic constituent of the ferromagnetic spinels or ferrites. FeO, Fe_2O_3, and Fe_3O_4 have melting points ranging from that of FeO at 1,420 °C to that of Fe_2O_3 at 1,565 °C. Used extensively for producing colors in glasses, glazes, and enamels. Iron and iron oxides produce brown or reddish colors in ceramic mixtures if they are fired in an oxidizing atmosphere, and greenish or bluish colors if fired in a reducing atmosphere. Iron oxides are fluxing and coloring materials. Larger particles produce brown or black spots, which, particularly in whiteware, are undesirable. Much care is taken to remove iron and iron oxides from the raw materials and from the bodies used for whiteware manufacture. (Lee, 1961; Handbook of Chem. & Phys., 2; Rosenthal, 1949)

iron pan A general terrm for a hardpan in a soil in which iron oxides are the principal cementing agents; several types of iron pans are found in dry and wet areas and in soils of widely varying textures. (AGI, 1987)

iron phosphate *vivianite.*

iron piler A laborer who removes iron from cars, sometimes breaks it, and piles and classifies it according to grade. (Fay, 1920)

iron portland cement Mixture of portland cement and granulated blast furnace slag. (Bennett, 1962)

iron pyrite *pyrite.*

iron-reduction process A process used in the treatment of lead ores. See also: *precipitation process.*

iron refining The reduction of prepared ores of iron to metallic iron, as in the blast furnace. The reduction and purification of semirefined iron, such as pig iron, or a mixture of pig iron, scrap iron, and scrap steel, to form substantially pure iron or steel, as in the electric furnace, open- hearth furnace, puddling furnace, or Bessemer converter. (Henderson, 1953)

iron runner The spout by which iron flows from the taphole of a blast furnace. (Fay, 1920)

iron sampler In the iron and steel industry, a laborer who obtains samples of iron ore as it is brought into the plant, or samples of semifinished or finished metal products, such as iron and steel sheets, rails, rods, or bars, and carries them to the laboratory for routine tests. Also called sampler; test carrier. (DOT, 1949)

iron sand A sand containing particles of iron ore (usually magnetite), as along a coastal area. (AGI, 1987)

iron schefferite *pyroxene.*

iron series *magnetite.*

ironshot (a) Said of a mineral that is streaked, speckled, or marked with iron or an iron ore. (AGI, 1987) (b) Containing small nodules or oolitic bodies of limonite or hematite; e.g., an ironshot rock in which the ooliths are essentially composed of limonite. A limonitic oolith in an ironshot rock. (AGI, 1987)

ironsmith A worker in iron, as a blacksmith. (Standard, 1964)

iron spar *siderite.*

iron stain Strongly colored yellowish, reddish, or brownish deposit of iron oxides. (Skow, 1962)

iron steel A material formed of iron between steel surfaces, or of steel-located iron. (Standard, 1964)

ironstone Any rock containing a substantial proportion of an iron compound, or any iron ore from which the metal may be smelted commercially; specif., an iron-rich sedimentary rock, either deposited directly as a ferruginous sediment or resulting from chemical replacement. The term is customarily applied to a hard, coarsely banded or nonbanded, and noncherty sedimentary rock of post-Precambrian age, in contrast with iron formation. The iron minerals may be oxides (limonite, hematite, magnetite), carbonate (siderite), or silicate (chamosite); most ironstones containing iron oxides or chamosite are oolitic. *clay ironstone; banded ironstone.* Cf: *iron formation.* (AGI, 1987)

ironstone cap A surficial or near-surface sheet or cap of concretionary clay ironstone. (AGI, 1987)

ironstone clay *argillaceous hematite; red hematite.*

iron sulfides *chalcopyrite; pyrite; marcasite; pyrrhotite.*

iron talc *minnesotaite.*

iron vitriol *melanterite.*

ironworker Person engaged in manufacturing iron or ironwork. (Standard, 1964)

ironworks An establishment for the manufacture of iron or of heavy ironwork. (Standard, 1964)

irradiation Exposure to radiation, as in a nuclear reactor. (Lyman, 1964)

irregular polygon Polygon in which neither the sides nor the angles are equal. (Jones, 1949)

irrespirable Not respirable; not fit to be breathed. Said of mine gases. (Standard, 1964; Fay, 1920)

irrespirable atmosphere In a coal mine, atmosphere containing poisonous gases or a lack of sufficient oxygen as a result of combustible gases explosions, coal-dust explosions, combined gas and dust explosions, or mine fires, and which can only be entered by persons wearing breathing apparatus. (McAdam, 1955)

irrotational wave *P wave; compressional wave.*

irruption A nearly obsolete syn. of intrusion. (AGI, 1987)

isanomalic line Line of equal value of an anomaly. (Schieferdecker, 1959)

Isbell table Obsolete type of shaking table. (Pryor, 1963)

iserine *ilmenorutile.*

ishikawaite An orthorhombic mineral, $(U,Fe,Y,Ca)(Nb,Ta)_4(?)$; opaque black; at Ishikawa, Iwaki Province, Japan.

isinglass Transparent sheet mica, commonly muscovite, principally from pegmatite dikes. See also: *mica; muscovite.*

island arc A group of islands having a curving, arclike pattern. Most island arcs lie near the continental masses, but inasmuch as they rise from the deep ocean floors, they are not a part of the continents proper. (Stokes, 1955)

isoanthracite lines (a) A term for lines of equal volatile content (now called isovols) drawn on a map or diagram. (Tomkeieff, 1954) (b) Lines of equal C:H ratio in coal drawn on the map or diagram. (Tomkeieff, 1954)

isobar An imaginary line or a line on a map or chart connecting or marking places on the surface of the Earth where the height of the barometer reduced to sea level is the same either at a given time or for a certain period. (Webster 3rd, 1966)

isobase A term used for a line that connects all areas of equal uplift or depression; it is used esp. in Quaternary geology as a means for expressing crustal movements related to postglacial uplift. (AGI, 1987)

isobath (a) A line on a map or chart that connects points of equal water depth. Syn: *bathymetric contour; depth contour.* (AGI, 1987) (b) An imaginary line on a land surface along which all points are the same vertical distance above the upper or lower surface of an aquifer or above the water table. (AGI, 1987)

isocals Lines of equal calorific value in coal drawn on a map or diagram. (Tomkeieff, 1954)

isocarb On a map or diagram, a line connecting points of equal fixed-carbon content in coal. See also: *isocarbon map.* (AGI, 1987)

isocarbon map A map showing, by contours, the areas having an equal quantity of carbon within an assumed interval of stratigraphic section. See also: *isocarb.* (Ballard, 1955)

isochore (a) A line drawn on a map through points of equal drilled thickness for a specified subsurface unit. Thickness figures are uncorrected for dip in vertical wells, and corrected for hole angle, but not for dip, in deviated wells. Cf: *isopach.* (AGI, 1987) (b) In a phase diagram, a line connecting points of constant volume. (AGI, 1987)

isochore map A map showing drilled thickness of a given stratigraphic unit by means of isochores. Syn: *convergence map.* (AGI, 1987)

isochromatic lines In stress analysis by the photoelastic method, lines of equal difference of principal stress, appearing as colored streaks. See also: *photoelasticity.* (Hammond, 1965)

isochrome map A contour map that depicts the continuity and extent of color stains on geologic formations. (AGI, 1987)

isochronal *isochronous.*

isochrone lines Lines connecting points of equal times. When the relative seismic velocities are known, the isochrones can be translated into depth contours. See also: *refraction shooting.* (Nelson, 1965)

isochronous (a) Equal in duration or uniform in time; e.g., an isochronous interval between two synchronous surfaces, or an isochronous unit of rock representing the complete rock record of an isochronous interval. (AGI, 1987) (b) A term frequently applied in the sense of synchronous, such as an isochronous surface having everywhere the same age or time value within a body of strata. Syn: *isochronal.* Cf: *synchronous.* (AGI, 1987)

isoclinal Adj. of isocline. (AGI, 1987)

isoclinal fold A fold whose limbs are parallel. Syn: *isocline.*

isocline *isoclinal fold.*

isoclinic A line (in a stressed body) at all points on which the corresponding principal stresses have the same direction. (Roark, 1954)

isoclinic line (a) A line drawn through all points on the Earth's surface having the same magnetic inclination. The particular isoclinic line drawn through points of zero inclination is given the special name of aclinic line. (Hunt, 1965) (b) An isomagnetic line connecting points of equal magnetic inclination. (AGI, 1987)

isodesmic Characteristic of a crystal structure with bond strengths roughly equal in all directions.

isodiametric Refers to the hexagonal, trigonal, and tetragonal crystal systems; i.e., having the lateral crystal axes a_1 and a_2 of equal length. Cf: *isometric; anisometric.*

isodimorphous In mineralogy, both isomorphous and dimorphous; said of certain groups of minerals. (Fay, 1920)

isodynamic line Any line joining points of equal magnetic intensity. Applicable to the total intensity or the vertical, horizontal, north-south, or east-west components. So used in terrestrial magnetism literature, esp. in British and Canadian writings. (AGI, 1987)

isoelectric point Zero potential or point of electrical neutrality; the hydrogen-ion exponent at which particles in aqueous suspension are neutral and best able to flocculate. Also called zero point of change. (Pryor, 1965)

isofacial (a) Pertaining to rocks belonging to the same metamorphic facies and having reached equilibrium under the same set of physical conditions. Syn: *isograde.* (AGI, 1987) (b) Pertaining to rocks belonging to the same sedimentary facies; e.g., an isofacial line on a map, along which the thickness of stratum of the same lithologic composition is constant. (AGI, 1987)

isogal In gravity prospecting, a contour line of equal gravity values. (AGI, 1987)

isogam In magnetic prospecting, a contour line of equal magnetic values. (AGI, 1987)

isogam map A chart showing contour lines of equal magnetic field intensity and employed in the magnetic methods of geophysical prospecting. Also called isogal map in gravity surveys. (Nelson, 1965)

isogeotherm A line or surface within the Earth connecting points of equal temperature. Syn: *isogeothermal line; geotherm; geoisotherm.* (AGI, 1987)

isogeothermal line *isogeotherm.*

isogon *isogonic line.*

isogonic line An isomagnetic line connecting points of equal magnetic declination. See also: *agonic line.* Syn: *isogon.* (AGI, 1987)

isograd A line on a map joining points at which metamorphism proceeded at similar values of pressure and temperature as indicated by rocks belonging to the same metamorphic facies. Such a line represents the intersection of an inclined surface with the Earth's surface corresponding to the boundary between two contiguous facies or zones of metamorphic grade, as defined by the appearance of specific index minerals; e.g., garnet isograd, staurolite isograd. (AGI, 1987)

isograde *isofacial.*

isogram A general term for a line on a map or chart connecting points having an equal numerical value of some physical quantity (such as temperature, pressure, or rainfall); an isopleth. (AGI, 1987)

isohume A line constructed on a map, somewhat similar to a contour line, but connecting points of equal moisture content of coal in the bed. (AGI, 1987)

isohyet A line connecting points of equal precipitation. (AGI, 1987)

isolate A culture of an organism isolated by selection procedures. (Rogoff, 1962)

isolated consignment A particular case of a single consignment where the sampling is to be carried out without prior knowledge of a coal's sampling characteristics other than its presumed ash content and size. (Nelson, 1965)

isolator Part of a circuit that can be removed from it in order to break the circuit when there is no current flowing. (Hammond, 1965)

isoline *isopleth.*

isolith (a) An imaginary line connecting points of similar lithology and separating rocks of differing nature, such as of color, texture, or composition. (AGI, 1987) (b) An imaginary line of equal aggregate thickness of a given lithologic facies or particular class of material within a formation, measured perpendicular to the bedding at selected points (which may be on outcrops or in the subsurface). (AGI, 1987)

isolith map A map that depicts isoliths; esp. a facies map showing the net thickness of a single rock type or selected rock component in a given stratigraphic unit. (AGI, 1987)

isomagnetic line A line connecting points of equal value of some magnetic element; e.g., isogonic line; isodynamic line; isoporic line. (AGI, 1987)

isomer (a) One of two or more substances composed of the same molecular formula, but differing in chemical or physical properties owing to the arrangement of the atoms in the molecule. (Hess) (b) In nuclear science, one of two or more nuclides with the same numbers of neutrons and protons in the nucleus, but having different energy. (Lyman, 1964)

isomeric Of, relating to, or exhibiting isomerism. (Webster 3rd, 1966)

isometric (a) A system of crystallization with three axes at right angles and of equal length; nine planes of symmetry; singly refracting. Cf: *anisodesmic.* (Hess) (b) Characterized by equality of measure. (c) The crystal system characterized by three orthogonal axes of equal length. Syn: *cubic; equant.* Cf: *anisometric; isodiametric.*

isometric line *isopleth.*

isometric projection In technical drawing, a three-dimensional view of an object can be drawn to scale with three perpendicular edges at 120° to each other, and with the vertical edges vertical. See also: *oblique projection; axonometric projection.* (Hammond, 1965)

isomorphism The name given to chemical compounds that have analogous composition, similar crystal structures, and closely related crystal forms; e.g., carbonate minerals of the aragonite group—aragonite, witherite, strontianite, and cerrusite—in which the metal ions are different but the several minerals crystallize in the orthorhombic system in closely similar forms. Adj. isomorphous. Noun, isomorph. Cf: *polymorphism.*

isomorphous Originally defined (Mitscherlich, 1819) as having similar crystalline form, but now generally restricted to compounds that form solid solutions by isomorphous substitution; i.e., by the replacement of one ion for another in a crystal structure without alteration in the crystal form. Cf: *isotypic.* Syn: *allomeric.* (AGI, 1987)

isomorphous mixture (a) A solid solution of two or more isomorphous substances. (Fay, 1920) (b) A type of solid solution in which mineral compounds of analogous chemical composition and closely related crystal habit crystallize together in various proportions. Cf: *solid solution.* (Harbison-Walker, 1972)

isomorphous replacement A characteristic of some minerals where substitution for one or more elements by others does not change the crystal structure. An example is the substitution of iron for zinc in sphalerite, wherein the iron content can range up to more than 15% without changing the sphalerite structure. Similarly, iron, manganese, and magnesium ions can replace each other in the calcite structure common to siderite, rhodochrosite, and magnesite.

isomorphous series Descriptive of two minerals with the same crystal structure but different end-member compositions which may show partial or complete crystal miscibility (solid solution) between them. One mineral may belong to more than one isomorphous series; e.g., the garnet grossular forms a series with andradite, with hibschite and katoite, and with uvarovite. There are many isomorphous series among minerals; e.g., plagioclase feldspars, monoclinic pyroxenes, and the spinel and garnet groups. Cf: *solid solution.*

isontic line *isopleth.*

isopach A line drawn on a map through points of equal true thickness of a designated stratigraphic unit or group of stratigraphic units. Cf: *isochore.* Syn: *isopachyte; isopachous line; thickness line; thickness contour.* (AGI, 1987)

isopach map A map indicating, usually by means of contour lines, the varying thickness of a designated stratigraphic unit. Also called isopachous map. (AGI, 1987)

isopachous Of, relating to, or having an isopach; e.g., an isopachous contour. Not recommended usage. (AGI, 1987)

isopachous line *isopach.*

isopachous map A nonrecommended syn. of isopach map. (AGI, 1987)

isopachyte British term for isopach. (AGI, 1987)

isophysical series A series comprising rocks of different chemical composition that were metamorphosed under identical physical conditions.

isopic Said of sedimentary rocks of the same facies, or said of facies characterized by identical or closely similar rock types. The rocks may be formed in different sedimentation areas or at different times or both, but the lithologies are the same; e.g., a facies repeated in vertical succession. Also, said of a map depicting isopic facies or rocks. Cf: *heteropic.* (AGI, 1987)

isopical Relating to synchronous deposits that exhibit the same facies.

isopiestic Constant value of pressure on a surface of the sea. (Hy, 1965)

isopiestic line *equipotential line.*

isopleth (a) In a strict sense, a line or surface on which some mathematical function has a constant value. It is sometimes distinguished from a contour by the fact that an isopleth need not refer to a directly measurable quantity characteristic of each point in the map area; e.g., maximum temperature of a particular point. (AGI, 1987) (b) A general term for a line, on a map or chart, along which all points have a numerically specified constant or equal value of any given variable, element, or quantity (such as abundance or magnitude), with respect to space or time; esp. a contour. Etymol. Greek isos, equal, + plethos, fullness, quantity, multitude. Syn: *isogram; isoline; isontic line; isometric line.* (AGI, 1987)

isopor *isoporic line.*

isoporic line A line drawn through points whose annual change in magnetic declination is equal. Syn: *isopor.* (AGI, 1987)

isopycnic A line connecting points of equal density, particularly of ocean water. A line connecting points of equal atmospheric density may be called an isostere. (Hunt, 1965)

isorads Lines joining points of equal radioactivity, drawn from geiger- or scintillation-counter data to form an isorad map. (AGI, 1987)

isoresistivity plan A plan showing lines of equal resistivity at a certain selected depth. It is prepared from data obtained by the resistivity method of geophysical prospecting. (Nelson, 1965)

isoseism *isoseismal line.*

isoseismal A line on the surface of the Earth joining points of equal seismic disturbance due to any single earthquake. Syn: *isoseismal line.* (AGI, 1987)

isoseismal line A line connecting points on the Earth's surface at which earthquake intensity is the same. It is usually a closed curve around the epicenter. Syn: *isoseism; isoseismal.* (AGI, 1987)

isostannite *kesterite.*

isostasy The condition of equilibrium, comparable to floating, of the units of the lithosphere above the asthenosphere. Crustal loading, as by ice, water, sediments, or volcanic flows, leads to isostatic depression or downwarping; removal of load leads to isostatic uplift or upwarping. Two differing concepts of the mechanism of isostasy are the Airy hypothesis of constant density and the Pratt hypothesis of constant thickness. See also: *isostatic compensation.* Adj. isostatic. (AGI, 1987)

isostatic Subjected to equal pressure from every side; being in hydrostatic equilibrium; relating to or characterized by isostasy. *stress trajectory.* (Webster 3rd, 1966)

isostatic adjustment *isostatic compensation.*

isostatic anomaly (a) The difference between the observed value of gravity at a point after applying to it the isostatic correction and the normal value of gravity at the point. (AGI, 1987) (b) Anomaly on a map of observed gravity anomalies after applying the isostatic correction. Negative isostatic anomalies indicate undercompensation, implying a tendency to rise; positive isostatic anomalies connote overcompensation and a tendency to sink. (AGI, 1987) (c) A gravity anomaly calculated on a hypothesis that the gravitational effect of masses extending above sea level is approx. compensated by a deficiency of density of the material beneath those masses; the effect of deficiency of density in ocean waters is compensated by an excess of density in the material under the oceans. See also: *anomaly.* (AGI, 1987)

isostatic compensation The adjustment of the lithosphere of the Earth to maintain equilibrium among units of varying mass and density; excess mass above is balanced by a deficit of density below, and vice versa. See also: *isostasy.* Syn: *isostatic adjustment; isostatic equilibrium.* (AGI, 1987)

isostatic correction The adjustment made to values of gravity, or to deflections of the vertical, observed at a point, to take account of the assumed mass deficiency under topographic features for which a topographic correction is also made. (AGI, 1987)

isostatic equilibrium The shifting of the rock beneath the Earth's crust in response to the shifting in the weight above the Earth's crust. Syn: *isostatic compensation.* (MacCracken, 1964)

isostructural (a) Refers to minerals that are closely similar in crystallographic, physical, and chemical properties but have little tendency for isomorphous substitution; same as isotypic. (AGI, 1987) (b) Said of minerals that have the same ionic or molecular crystal structure. Isostructure is more rigorous than isomorphism; the latter requires similar crystal forms, the former a one-to-one correspondence of atomic particles. Isostructural minerals may differ markedly in chemical composition and physical properties, e.g., fluorite and uraninite, or may be closely similar, e.g., the calcite group of carbonates.

isotherm A line connecting points of equal temperature. Isotherm maps are often used to portray surface temperature patterns of water bodies. (AGI, 1987)

isothermal (a) A change taking place at a constant temperature. (Strock, 1948) (b) Pertaining to the process of changing the thermodynamic state of a substance, such as its pressure and volume, while maintaining the temperature constant. (AGI, 1987)

isothermal compression (a) Reduction in the volume of a fluid without any change in its temperature. (Standard, 1964) (b) Compression in which there is no change in the temperature of the air; used as a standard against which the conditions of actual compression may be checked. (Lewis, 1964)

isothermal expansion The expansion of air under constant temperature. Since the air does work on expanding, it loses heat; consequently, heat must be added to the air to maintain it at constant temperature. (Lewis, 1964)

isothermal layer A water column through which a constant temperature exists. (Hy, 1965)

isothrausmatic A descriptive term applied to igneous rocks with an orbicular texture in which the nuclei of the orbicules are composed of the same rock as the groundmass. Cf: *crystallothrausmatic; homeothrausmatic; heterothrausmatic.* (AGI, 1987)

isotropic Said of a medium with properties the same in all directions; in crystal optics, said of a crystal with refractive index that does not vary with crystallographic direction. Isometric crystals and amorphous substances, such as glass, are generally isotropic. Noun, isotropy. Cf: *uniaxial; anisotropic.*

isotropic mass A mass having the same property (or properties) in all directions. (ASCE, 1958)

isotropy (a) The condition of having properties that are uniform in all directions. Adj. isotropic. Cf: *uniaxial.* (AGI, 1987) (b) Refers to matter with properties the same in all directions; in optics, a crystal with an index of refraction that does not vary with crystallographic direction.

Isometric crystals and amorphous substances, e.g., glass, are generally isotropic. Cf: *anisotropy.*

isotypic Having analogous composition and closely similar crystal structure, but not capable of intercrystallizing to form solid solutions. Examples are calcite and soda niter; galena and NaBr. Cf: *isomorphous.* (AGI, 1987)

isotypy One-dimensional polymorphism; e.g., alternate stacking of identical layers of micas or clays. Also called isotypism. Cf: *polymorphism.*

isovelocity The phenomenon of sound being the same in all parts of a given water column. (Hy, 1965)

isovol Lines constructed on a map of a coalbed connecting points of equal volatile matter, delineating the distribution of volatile matter of the coal. (AGI, 1987)

Istrian stone A marble found near Trieste, from which Venice is largely built. (Fay, 1920)

itabirite A laminated, metamorphosed oxide-facies iron formation in which (1) the original chert or jasper bands have been recrystallized into megascopically distinguishable grains of quartz and (2) the iron is present as thin layers of hematite, magnetite, or martite (Dorr & Barbosa, 1963). The term was originally applied in Itabira, Brazil, to a high-grade, massive specular-hematite ore (66% iron) associated with a schistose rock composed of granular quartz and scaly hematite. The term is now widely used outside Brazil. Cf: *jacutinga; canga.* Syn: *banded-quartz hematite; hematite schist.* (AGI, 1987)

itacolumite A micaceous sandstone or a schistose quartzite that contains interstitial, loosely interlocking grains of mica, chlorite, and talc; exhibits flexibility when split into thin slabs. Type locality is Itacolumi Mountain in Minas Gerais, Brazil. Syn: *flexible sandstone; articulite.* (AGI, 1987)

Italian asbestos A name often given to tremolite asbestos to distinguish it from Canadian or chrysotile asbestos; extensively quarried in Piedmont and Lombardy, Italy. (CMD, 1948)

ivory The fine-grained, calcareous creamy-white dentine forming most of the tusks of elephants and the teeth or tusks of certain other large animals, such as the walrus; it has long been esteemed for a wide variety of ornamental articles. Elephant ivory is now illegal to use. It is often simulated by bone.

ivory turquoise Odontolite; fossil tooth or bone colored by a blue phosphate of iron. Syn: *bone turquoise.*

Iwan-pattern earth auger A dry sampler equipped with an Iwan auger shoe or cutterhead. *post-hole digger.* (Long, 1960)

Ixiolite A monoclinic mineral $(Ta,Nb,Sn,Fe,Mn)_4O_8$).

J

jacinth *essonite; hyacinth; zircon.*

jack (a) A name for zinc ore; blackjack. *sphalerite.* (b) Cannel coal interstratified with shale. (c) Coaly shale, commonly canneloid. (d) To drill a rock by hand, using a 4-lb (1.8-kg) hammer in one hand and a steel drill in the other ("single jacking"). "Double jacking" is to use a two-handed (sledge) hammer with the steel drill held by another worker. (e) A portable device used for exerting great pressure or for lifting a heavy body through a small distance. Syn: *anchor jack.* (Jones, 1949)

jack boom (a) A boom that supports sheaves between the hoist drum and the main boom in a pull shovel or a dredge. (Nichols, 1976) (b) A boom whose function is to support sheaves that carry lines to a working boom. (Nichols, 1976)

jack catch Safety catch in the rail track to stop tubs running back on inclines. (Mason, 1951)

jack engine Eng. A donkey engine; a small engine employed in sinking a shallow shaft. (Fay, 1920)

jacket An outer casing or cover constructed around a cylinder or pipe, the annular space being filled with a fluid for either cooling, heating, or maintaining the cylinder contents at constant temperature; e.g., the water jackets of an internal combustion engine. (CTD, 1958)

jacket set (a) Set of timbers used in a shaft outside the regular shaft set, as extra protection in heavy ground. (Pryor, 1963) (b) Set like the larger shaft set with dividers omitted, except that the wall and end plates are broken at all joints to facilitate renewing. Used in heavy ground to protect the regular shaft timbers. The jacket set is placed outside the regular timbers, separated from them by short blocks, and is blocked and wedged against the rock. (Lewis, 1964)

jackhammer A percussive type of automatically rotated rock drill that is worked by compressed air. It is light enough to be used without a tripod and to be hand held. (Pryor, 1963)

jackhead *delivery drift.*

jackhead pit (a) A small pit without hoisting appliances, frequently serving as a ventilation shaft. (Standard, 1964) (b) A small shaft sunk within a mine. A winze. (Fay, 1920)

jackhead pump A subordinate pump in the bottom of a shaft, worked by an attachment to the main pump rod. (Fay, 1920)

jackhead set Newc. The set of pumps in the jackhead staple. (Fay, 1920)

jackhead staple Eng. A small mine for the supply of coal for the boilers. (Fay, 1920)

jack hole Eng. In coal mining, a bolthole. See also: *cut-through.* (Standard, 1964; Fay, 1920)

jacking pressure The amount of pressure exerted by a jack to force a cone penetrometer into a soil being tested. (Long, 1960)

jacking up The raising up of masses of machinery and heavy structures by means of jacks. (Crispin, 1964)

jackknife The collapse of a drill tripod or derrick. (Long, 1960)

jackknife rig A truck-mounted diamond or small rotary drill equipped with a hinged derrick. (Long, 1960)

jackknifing A collapsing of square-set timbers by wall pressure or through poor placement. (Fay, 1920)

jackleg (a) Light supporting bar for use with a jackhammer. (Pryor, 1963) (b) An outrigger post. (Nichols, 1976)

jackpipe A hollow iron pipe large enough to slip over the end of the front jack of a cutting machine to make it hold more firmly against the coal. (Fay, 1920)

jack pit N. of Eng. A shallow shaft in a mine communicating with an overcast. See also: *jackshaft; overcast.* Syn: *jacky pit.* (Fay, 1920)

jack post This timber is used where the coal seam is separated by a rock band and one bench is loaded out before the other. If the top bench is worked off first, the jack posts are set between the bottom bench of the coal and the roof. If the bottom bench is cleaned up first, the jack posts are set between the bottom and the top bench. At least two jack posts should be used and as many more as is necessary to keep the top bench of coal or the roof from coming down while the coal is being loaded out. (Kentucky, 1952)

jackroll A windlass worked by hand. (Fay, 1920)

jackscrew (a) A jack in which a screw is used for lifting or exerting pressure; also, the helical-screw part of a jackscrew. Syn: *screwjack.* (Long, 1960) (b) A heavy screw set in the base or frame of a drill machine for the purpose of leveling the drill. (Long, 1960)

jack setter A miner who assists in the operation of an auger-type underground mining machine; duties include seeing that the roof of the mine at or near the machine is in a safe condition. See also: *machine helper.* (Fay, 1920; DOT, 1949)

jackshaft An intermediate driving shaft. See also: *jack pit.* (Standard, 1964)

jacky pit *jack pit.*

jacobsite An isometric mineral, $8[MnFe_2O_4]$; spinel group; occurs with manganese ores.

Jacobs process A method in which bauxite is fused in an electric furnace to form a synthetic corundum.

Jacob staff A straight rod or staff pointed and shod with iron at the bottom for insertion in the ground, having a socket joint at the top, and used instead of a tripod for supporting a compass. (Webster 3rd, 1966)

Jacquet's method Use of electrolytic polishing to complete the finish on metal surfaces. After mechanical polishing they are made the anodes in a suitable electrolyte. (Pryor, 1963)

jacupirangite An ultramafic plutonic rock that is part of the ijolite series; composed chiefly of titanaugite and magnetite, with a smaller amount of nepheline; a nepheline-bearing clinopyroxenite. Its name, given by Derby in 1891, is derived from Jacupiranga, Brazil. (AGI, 1987)

jacutinga A term used in Brazil for disaggregated, powdery itabirite, and for variegated thin-bedded, high-grade hematite iron ores associated with and often forming the matrix of gold ore. Etymol. from its resemblance to the colors of the plumage of Pipile jacutinga, a Brazilian bird. Cf: *itabirite.* (AGI, 1987)

jad (a) Som. A long and deep holing, cutting, or jud, made for the purpose of detaching large blocks of stone from their natural beds. (b) Prov. Eng. To undercut (coal or rock). (Standard, 1964)

jade A microcrystalline gem variety of jadeite or nephrite (actinolite) with a toughness (resistance to breakage) exceeded only by that of carbonado diamond; ranges from nearly white to emerald-green, the latter being the most valuable; finest quality is reported to come from northern Burma (Myanmar) and the Yunnan Province of south China. The emerald-green color is attributed to jadeite and to trace amounts of chromium. There are many imitations of jade, including green-dyed onyx (Mexican jade), aventurine quartz containing fuchsite mica (Indian jade), vesuvianite (California jade), green hydrogrossular from South Africa, green organic or inorganic dyes or substitutes inserted under white jade, glass, dyed quartz, and bowenite or williamsite varieties of serpentine. "Jade cat's-eye" is a contradiction in terms. *jadeite; nephrite; toughness.*

jadeite A monoclinic mineral, $4[NaAlSi_2O_6]$; pyroxene group; apple green to emerald green, white, lavender, tomato red, or brown; tough; associated with albite from high-pressure, low-temperature metamorphosed plagioclase; an ornamental stone called "jade" (along with nephrite).

jadeitite A metamorphic rock consisting principally of jadeite, commonly associated with small amounts of feldspar or feldspathoids. It is probably derived from an alkali-rich igneous rock by high-pressure metamorphism. (AGI, 1987)

jadeolite A deep-green chromiferous syenite cut as a gemstone and resembling jade in appearance. Obsolete. (AGI, 1987)

jag bolt An anchor bolt with a barbed flaring shank, which resists retraction when leaded into stone or set in concrete. Also called hacked bolt; rag bolt. (Webster 3rd, 1966)

jager A bluish-white diamond of modern cut. (Schaller, 1917)

jagging board An inclined board on which ore slimes are washed, as in a buddle. (Standard, 1964)

jagoite A trigonal mineral, $Pb_3FeSi_4O_{12}(Cl,OH)$; in yellow-green micaceous plates associated with another lead-iron silicate (melanotekite) in iron ore at Långban, Sweden.

jailer Som. A small tub or box in which water is carried in a mine.

jalpaite A tetragonal mineral, Ag_3CuS_2; forms intergrowths with argentite in the Black Hawk District, Grant County, NM.

jam The blocking of a core barrel or core bit with core, sometimes deliberately. (Long, 1960)

jamb (a) A vein or bed of earth or stone, that prevents miners from following a vein of ore; a large block. (Fay, 1920) (b) A projecting columnar part (as of a masonry wall) or mass (as of ore). (Webster 3rd, 1966) (c) A vertical structural member forming the side of an opening in a furnace wall. (Harbison-Walker, 1972) (d) A type of brick shape intended for use in the sides of wall openings. (Harbison-Walker, 1972) (e) Sidewall of port of furnace superstructure carrying port crown load. (ASTM, 1994)
jamb cutter In the coke products industry, a laborer who chips carbon and mud from the edges of coke-oven doors with a steel bar prior to the discharge of the coke. (DOT, 1949)
jamb stick *gim peg.*
jamb wall *breast wall.*
James jig Movable sieve box supported on a rubber diaphragm and jigged mechanically up and down. (Pryor, 1963)
jamesonite A monoclinic mineral, $2[Pb_4FeSb_6S_{14}]$; metallic gray to black; in acicular crystals or capillary forms of featherlike appearance, thus the term "feather ore;" in low- to medium-temperature veins with other lead sulfosalts, galena, and carbonates. Syn: *wolfsbergite; feather ore.*
James table Shaking table used in concentration of ground ores by gravity. (Pryor, 1963)
jam out S. Staff. To cut or knock away the coal between holes. (Fay, 1920)
jam riveter A riveting hammer provided with an air-operated telescopic casing to hold the hammer against the work. (Hammond, 1965)
jar (a) An appliance to permit relative movement between the rope and rods in a cable drill. It reduces shocks and the risk of rod or chisel breakages. See also: *free-falling device.* (Nelson, 1965) (b) To loosen or free stuck drill-stem equipment or tools by impacts delivered by quick, sharp, upward-traveling blows delivered by a drive hammer or jars. (Long, 1960)
jar collar A swell coupling, attached to the upper exposed end of a drill rod or casing string, to act as an anvil against which the impact blows of a drive hammer are delivered and transmitted to the rod or casing string; also, sometimes used as a syn. for drive hammer. Also called drive collar; jar head. Syn: *bell jar.* (Long, 1960)
jargon A colorless, yellow, or smoky gem variety of zircon.
jarlite A monoclinic mineral, $NaSr_3Al_3(F,OH)_{16}$; in brownish crystals and spherulites at Ivigtut, Greenland.
jar mill (a) A small batch mill of ceramic material used in ore-testing laboratories in investigation of grinding problems. (Pryor, 1963) (b) Any of the stoneware-lined pebble mills used in milling enamels on a small scale. (Enam. Dict., 1947) (c) *ball mill.*
Jarno taper This taper of 0.6 in/ft (4.2 cm/m) is used by a number of manufacturers for taper pins, sockets, and shanks used on machine tools. (Crispin, 1964)
jarosite A trigonal mineral, $KFe_3(SO_4)_2(OH)_6$; alunite group; amber yellow to dark brown; forms druses of minute crystals, crusts, and coatings; may be fibrous or fine-grained and massive; associated with volcanic rocks; thought to form under solfataric conditions at elevated temperatures and pressures; in some places with alunite; at many localities in the Western United States, Bolivia, Europe, and Russia. See also: *cuprojarosite; kirovite.*
jaspagate *agate jasper.*
jasper A red variety of chalcedony. See also: *chert; chalcedony.* Syn: *jasperite.*
jasper bar *jaspilite.*
jasperine Banded jasper of varying colors.
jasperite *jasper.*
jasperization The conversion or alteration of igneous or sedimentary rocks into banded rocks like jaspilite by metasomatic introduction of iron oxides and cryptocrystalline silica. See also: *jaspilite.* (AGI, 1987)
jasperoid (a) A dense, usually gray, chertlike siliceous rock, in which chalcedony or cryptocrystalline quartz has replaced the carbonate minerals of limestone or dolomite; a silicified limestone. It typically develops as the gangue of metasomatic sulfide deposits of the lead-zinc type, such as those of Missouri, Oklahoma, and Kansas. (AGI, 1987) (b) Resembling jasper. (AGI, 1987)
jasper opal An almost opaque common opal; commonly yellow-brown, reddish brown to red owing to iron oxides; resembles opal but has the luster of common opal. Also called jaspopal.
jaspidean Resembling or containing jasper; jaspery. (AGI, 1987)
jaspilite Interbedded jasper and iron oxides. Cf: *iron formation.* Syn: *jasper bar.* See also: *jasperization.*
jasponyx An opaque onyx, part or all of whose bands consist of jasper.
jaw (a) In a crusher, one of a pair of nearly flat or ribbed faces separated by a wedge-shaped opening. (Nichols, 1976) (b) One or a set of two or more serrate-faced members between which an object may be grasped and held firmly, as in a vise, drill chuck, foot clamp, or pipe wrench. (Long, 1960) (c) In a clutch, one of a pair of toothed rings, the teeth of which face each other. (Nichols, 1976)
jaw breaker *jaw crusher.*
jaw crusher (a) A primary crusher designed to reduce large rocks or ores to sizes capable of being handled by any of the secondary crushers. (Enam. Dict., 1947) (b) A crushing machine consisting of a moving jaw, hinged at one end, which swings toward and away from a stationary jaw in a regular oscillatory cycle. (ACSG, 1963) (c) A machine for reducing the size of materials by impact or crushing between a fixed plate and an oscillating plate, or between two oscillating plates, forming a tapered jaw. Syn: *jaw breaker.* (BS, 1962)
jedding ax A stonecutter's ax with a flat face and a pointed peen. (Webster 3rd, 1966)
jeffersite A kind of vermiculite from West Chester, Chester County, PA. (Fay, 1920)
jeffersonite A variety of monoclinic pyroxene, commonly augite or diopside, containing manganese and zinc; forms large, coarse crystals having rounded edges and uneven faces; at Franklin, NJ.
Jeffrey crusher Crusher used to break softish materials; e.g., limestone and coal. See also: *swing-hammer crusher.* Syn: *whizzer mill.* (Pryor, 1963)
Jeffrey diaphragm jig A plunger-type jig with the plunger beneath the screen. May be either single or multiple compartments. Its distinguishing features are (1) the stroke is produced with a cam operated by a lever and rocker-arm mechanism, (2) the weight of the column of water above the plunger is balanced by means of compressed air, (3) automatic operation is obtained by means of a submerged float that measures the specific gravity of the mass of coal, refuse, and water at the peak of the pulsion stroke, (4) refuse is withdrawn through a star gate extending the full width of the overflow lip, and (5) the slope of the screen plate is readily adjustable by means of heavy screws at the feed end. It is widely used on bituminous coal on sizes ranging up to a maximum of 6 in (15 cm). (Mitchell, 1950)
Jeffrey molveyor An arrangement to keep a continuous miner in full operation at all times. It consists of a series of short conveyors, each mounted on driven wheels and coupled into a train to run alongside the heading or room conveyor. (Nelson, 1965)
Jeffrey-Robinson cone A cone for coal washing; similar to the Callow and Caldecott cones. See also: *cone classifier.* (Hess)
Jeffrey single-roll crusher A simple type of crusher for coal, with a drum to which are bolted toothed segments designed to grip the coal, forcing it down into the crushing opening.
Jeffrey swing-hammer crusher A crusher enclosed in an iron casing in which a revolving shaft carries swinging arms having a free arc movement of 120°. The rotation of the driving shaft causes the arms to swing out and strike the coal, ore, or other material, which, when sufficiently fine, passes through the grated bottom. (Liddell, 1918)
Jeffrey-Traylor vibrating feeder A feed chute vibrated electromagnetically in a direction oblique to its surface. Rate of movement of material depends on amplitude and frequency of vibration.
Jeffrey-Traylor vibrating screen An electric vibrating screen operated by action on an oscillating armature and a stationary coil. (Gaudin, 1939)
jelly *carbohumin; vegetable jelly.*
jenkin (a) A drivage at right angles to the main cleat. (Mason, 1951) (b) Eng. A road driven bordways in a pillar of coal. A jud driven bordways along a pillar of coal with goaf or an old bord on one side is called a "loose end jenkin.". (SMRB, 1930) (c) N. of Eng. A variation of "junking.". (Fay, 1920)
jenkinsite A ferroan variety of antigorite.
Jeppe's tables A series of tables esp. compiled for mining work that includes tables of density, vapor pressure, and absolute humidity. (Roberts, 1960)
Jeppestown shales S. Afr. Part of the Jeppestown Series forming the footwall of the Main Reef on the Central and East Rand. (Beerman, 1963)
jerking table *shaking table.*
jerry man An employee in a mine whose duty it is to clean up falls or refuse, or to make a miner's working place safe. See also: *wasteman.* (Fay, 1920; DOT, 1949)
Jersey fire clay brick A highly siliceous clay brick, semisilica brick. (AISI, 1949)
jeso Beds of decomposed gypsum. (Standard, 1964)
jet (a) A hard black variety of cannel coal or brown coal, compact in texture, having a rough fracture and dull luster that takes a good polish and is thus used in the jewelry trade. Syn: *black amber.* See also: *jet shale.* (b) A sudden and forceful rush or gush of fluid through a narrow or restricted opening; e.g., a stream of water or air used to flush cuttings from a borehole. (AGI, 1987) (c) A black variety of marble. (d) Jet piercing, a thermal method of drilling large-diameter blast holes in hard cherty iron formation (taconite); formerly used mainly on the Mesabi Range in Minnesota, but also on the Marquette Range in Michigan. The rotating drill head is fed a mixture of kerosene and oxygen, ignited to direct a high-temperature flame against the rock, causing the rock to spall into fine particles by thermal expansion. Method has been almost totally replaced by rotary drills using tricone bits. See also: *jet hole.*

jetair flotation machine A multiple-cell machine of the mechanical agitation type.

jet-assisted cutting (a) Cutting rock by the very high force of a water jet against the ore so that the material could be processed for mineral recovery. (SME, 1992) (b) Penetration of high-pressure water into material cracks such that the cracks between the grains are propagated and cutting occurs. See also: *hydraulic mining*. (SME, 1992)

jet coal *cannel coal.*

jet corer Consists of a length of pipe that is lowered from a vessel to obtain samples. High-velocity water is pumped through the pipe, and the jetting action of this water issuing from the lower end of the pipe very effectively cuts a hole in the unconsolidated overburden sediments. Once at bedrock, the pipe is rammed into the rock with sufficient force to obtain a plug several inches in length. (Mero, 1965)

jet drilling Piercing of rock strata by use of high-temperature flame to fuse the rock, together with jets of water to cause decrepitation and to flush the fragments out. (Pryor, 1963)

jet grinding mill Enclosed chamber of relatively small cross section in which gas, at substantially high atmospheric pressures, is circulated at high speed, 400 to 700 ft/s (120 to 210 m/s).

jet hole (a) A borehole drilled by use of a directed, forceful stream of fluid or air. See also: *jet.* (b) A small hole in a nozzle. See also: *jet.* (Long, 1960)

jet hydraulic Stream of water used in alluvial mining. (Pryor, 1963)

jet impact mill *fluid energy mill.*

jetloader A powder loader for loading horizontal drill holes with a diameter of more than 2 in (5.1 cm). It is intended for AN-prills and oil, and is also employed to blow sand into the holes for stemming. (Langefors, 1963)

jet mill Differs from other mills in that the material is not ground against a hard surface. Instead, a gaseous medium is introduced. The gas may convey the feed material at high velocity in opposing streams or it may move the material around the periphery of the grinding and classifying chamber. The high turbulence causes the particles and feed material to collide and grind upon themselves. (Cote, 1991)

jet mixer An apparatus that utilizes the mixing action of a water stream jetted into dry drill-mud ingredients to form a mud-laden fluid. Cf: *atomizer*. (Long, 1960)

jetonized wood A name given to vitrain lamellae in coal. Synonymous with vitrain. (Tomkeieff, 1954; AGI, 1987)

jet piercing The use of high-velocity jet flames to drill holes in hard rocks, as taconite, and to cut channels in granite quarries. It involves combustion of oxygen and a fuel oil fed under pressure through a nozzle to produce a jet flame generating a temperature above 2,600 °C. A stream of water joins the flame, and the combined effect is a thermodynamic spalling and disintegration of the rock into fragments that are blown from the hole or cut. Syn: *thermic drilling.*

jet-piercing drill *fusion-piercing drill.*

jet pump Consists of a centrifugal pump and motor at the ground surface and a jet down in the well below the water level, discharging at high velocity through a contracted section into the lift pipe. The centrifugal pump has two discharge pipes; one leads down to the jet, the other carries the water into the distribution system or into a storage tank. Syn: *jet impact mill.* (Urquhart, 1959)

jet rock (a) A coallike shale containing jet. (Standard, 1964) (b) *jet shale.*

jet shale Bituminous shale containing jet. See also: *jet.* Syn: *jet rock.*

jetstone A term for black tourmaline (schorl) in New South Wales, Australia.

jetters Corn. The horizontal rods or poles connecting the water wheel and the pumps.

jetting The process of sinking a borehole, or of flushing cuttings or loosely consolidated materials from a borehole, by using a directed, forceful stream (jet) of drilling mud, air, or water. (AGI, 1987)

jetting drill A percussive drill for prospecting through superficial deposits. The drill is given a short stroke, 10 to 20 cm, and rotated by hand. Water is pumped down through the hollow steel rods and escapes through openings in the chopping bit. Casing is used, and the drilling rate is from 6 to 12 m per shift. (Nelson, 1965)

jetting pump A water pump that develops very high discharge pressure. (Nichols, 1976)

jetty (a) An engineering structure (such as a breakwater, groin, seawall, or small pier) extending out from the shore into a body of water, designed to direct and confine the current or tide, to protect a harbor, or to prevent shoaling of a navigable passage by littoral materials. Jetties are often built in pairs on either side of a harbor entrance or at the mouth of a river. (AGI, 1987) (b) A British term for a landing wharf or pier used as a berthing place for vessels. (AGI, 1987)

jeweler's shop (a) Corn. Miner's expression for a rich section of ore. (Pryor, 1963) (b) Aust. A very rich patch of gold in either a reef or an alluvial formation.

jews' tin Ancient slabs of tin found in Cornwall, England; so called from the belief that they were made by Jewish merchants and miners from Asia Minor before the present era of mining. (Hess)

jib The lifting arm of a crane or derrick having a pulley at its outer end over which the hoisting rope passes. See also: *front-end equipment.* (Hammond, 1965)

jibbing-in The operation of gradually working the jib of a shortwall coal cutter into the cutting position in the coal seam. Jibbing-in is the first operation before starting the cutting run across the face. Syn: *sumping-in.* (Nelson, 1965)

jib crane A crane having a swinging boom or jib. (Crispin, 1964)

jib end In conveyor systems, the delivery end when a jib is fitted to deliver the load in advance of and remote from the drive. (Nelson, 1965)

jib holeman A person whose work is to make recesses for the cutting disk at the end of coal-cutting machine faces. (CTD, 1958)

jib in The process of starting a cut by swinging the jib of the coal cutter (while the chain is cutting) from the front of the face to the full cutting position. *sump.* (Mason, 1951)

jig (a) A device that separates coal or ore from foreign matter, by means of their difference in specific gravity, in a pulsating water medium. Also spelled gig. See also: *Baum jig; Bendelari jig; conset jig; Denver jig; Harz jig; hutch; jigger; jig washer; Pan-American jig; stripping a jig.* (b) A link or coupling connecting mine wagons.

jig bed The agent used in a jig that consists of the heavy fractions in the coal that behave in some respects like a dense fluid. The pulsation of the water or the motion of the screen keeps the bed open or in suspension during part of the cycle so that heavy minerals entering the jig can settle into the bed. Lighter materials cannot penetrate the jig bed and therefore are forced to remain in the upper part of the jig and eventually discharge over the top. Other agents in use are lead shot, iron punchings, iron shot, pyrite, and magnetite. (Newton, 1959)

jig brow Self-acting inclined track used to lower filled coal tubs and raise empty ones. (Pryor, 1963)

jig bushing Hardened-steel bushing inserted in the face of a jig to serve as a guide for drills. (Crispin, 1964)

jig chain S. Staff. A chain hooked to the back of a skip and running around a post to prevent its too rapid descent on an inclined plane. Cf: *snub.* (Fay, 1920)

jig dips N. Staff. *crossgate.*

jigger (a) *pan conveyor; shaker conveyor; jigging conveyor.* (b) Scot. An apparatus for attaching hutches to a haulage rope. (Fay, 1920) (c) A mechanism that operates with quick up-and-down motion; a jolting device. See also: *jig.* (Crispin, 1964; CTD, 1958) (d) A machine for dressing small ore in which a sieve is dipped or moved about under water. (e) Leic. A coupling hook used between coal cars. (Standard, 1964) (f) Person that concentrates ore by jigging. (Webster 3rd, 1966)

jigger work Eng. Dressed or partly dressed ore obtained from jigging.

jigging (a) The separation of the heavy fractions of an ore from the light fractions by means of a jig. (b) Up-and-down motion of a mass of particles in water by means of pulsion. (Pryor, 1965) (c) *skimping.*

jigging conveyor A series of steel troughs suspended from roof of stope or laid on rollers on its floor and given reciprocating motion mechanically to move mineral. Syn: *jigger; pan conveyor.* (Pryor, 1963)

jigging machine A machine to jig ore. See also: *jig.*

jigging screen A screen or pair of screens to which a combined horizontal and vertical motion is imparted, normally by a crankshaft and connecting rod, the screen decks being horizontal or inclined at a small angle. See also: *shaking screen.* (BS, 1962)

jiggling in A technique for transferring a surface survey down a mine shaft in such a manner as to tie it in to an underground mine survey. See also: *coplaning.*

jig haulage *gravity haulage.*

jig indicator An apparatus resembling a steam engine indicator; used for drawing curved lines illustrating the action of jigs in ore beneficiation. (Webster 2nd, 1960)

jig pin A pin used to prevent the turning of the turn beams. (Standard, 1964)

jig washer A coal or mineral washer for relatively coarse material. The broken ore, supported on a screen, is pulsed vertically in water; the heavy (valuable) portion passes through the screen into a conical receptacle (hutch), and the gangue goes over the side. In coal washing, the heavy (worthless) shale passes downwards, and the lighter coal remains on top. See also: *plunger jig washer; jig.* (Nelson, 1965)

jimboite An orthorhombic mineral, $Mn_3B_2O_6$; the manganese analog of kotoite; with other manganese minerals at the Kaso Mine, Kanuma, Tochigi Prefecture, Japan.

Jim Crow A portable hand-operated appliance for bending or curving rails. It incorporates a strong buttress screw thread. (Nelson, 1965)

jinny road Underground gravity plane. (Pryor, 1963)

jitty Leic. A short heading along which empties, horses, or workers travel. (Fay, 1920)

joaquinite A monoclinic mineral, $Ba_2NaCe_2Fe(Ti,Nb)_2Si_8O_{26}(OH,F)\cdot H_2O$; dimorphous with orthojoaquinite; forms minute honey-colored crystals; with benitoite and neptunite in San Benito County, CA.
jock Scot. An iron rod, usually pronged, attached to the rear end of a train of hutches or cars being drawn up an incline; used to stop their descent in the event of the rope breaking. (Fay, 1920)
jockey (a) Aust. A Y-shaped grip placed in sockets at the end of a skip. The endless rope rests on this when used above the skip. (Fay, 1920) (b) Mid. A self-acting apparatus on the front of a car, for releasing it from the hauling rope at a certain point. (Fay, 1920)
joggle (a) A joint of trusses or sets of timber for receiving pressure at right angles or nearly so. (Zern, 1928) (b) Notches cut in round timbers set above other round pieces in underground timbering. (Pryor, 1963) (c) An offset in a flat plane consisting of two parallel bends in opposite directions by the same angle. (ASM, 1961) (d) A slight step-shaped offset formed into a flat piece of metal (as for providing a flange). (Webster 3rd, 1966)
johachidolite An orthorhombic mineral, $CaAlB_3O_7$; forms transparent grains and lamellar masses; fluoresces an intense blue owing to traces of rare-earth elements; in nepheline dikes, cutting sandstone in the Johachido District, Kenkyohokuco Prefecture, Korea.
johannite A triclinic mineral, $[Cu(UO_2)_2(SO_4)_2(OH)_2\cdot 8H_2O]$; radioactive; forms soft, bitter-tasting crystals that are polysynthetically twinned in two directions; in druses and reniform masses; at Joachimsthal, Czech Republic, and Gilpin County, CO; associated with gypsum.
johannsenite A monoclinic mineral, $4[CaMnSi_2O_6]$; pyroxene group; forms series with diopside and hedenbergite; shows clove-brown to grayish-green columnar, radiating, and spherulitic aggregates of fibers and prisms with black tarnish; in metasomatized limestones with manganese ore. Occurs at Puebla and Hildago, Mexico; Lane County, OR; Franklin, NJ; and Schio, Venetia, Italy.
Johannsen number A number, composed of three or four digits, that defines the position of an igneous rock in Johannsen's classification. The first digit represents the class, the second the order, and the third and fourth the family. Cf: *Johannsen's classification.* (AGI, 1987)
Johannsen's classification A quantitative mineralogic classification of igneous rocks developed by the petrographer Albert Johannsen (1939). Cf: *Johannsen number.* (AGI, 1987)
John Odges *gun.*
Johnson concentrator A cylindrical shell lined with grooved rubber set parallel to its axis, which is inclined, peripheral riffles thus being formed. Used to arrest heavy particles such as metallic gold as auriferous pulp flows gently through while the cylinder revolves slowly and the arrested material rises and drops on to a separate discharge launder. (Pryor, 1963)
Johnston vanner Modified form of Frue vanner. (Pryor, 1963)
johnstrupite *rinkite.*
joint (a) A divisional plane or surface that divides a rock and along which there has been no visible movement parallel to the plane or surface. (Ballard, 1955) (b) A standard length of drill rod, casing, or pipe equipped with threaded ends by which two or more pieces may be coupled together; also, two or more standard lengths of drill rods or pipe coupled together and handled as a single piece in round trips. (Long, 1960) (c) A fracture or parting that cuts through and abruptly interrupts the physical continuity of a rock mass. Not to be confused with bedding or cleavage. (Long, 1960) (d) A line of cleavage in a coal seam. Syn: *heading seam.* See also: *joint plane.*
joint box A cast iron box surrounding an electric cable joint, often filled with insulation after the joint between cables has been made. (Hammond, 1965)
jointing (a) In quarrying, the process of cutting to specified sizes and shapes, with smooth unchipped edges. (AIME, 1960) (b) The condition or presence of joints in a body of rock. (AGI, 1987)
jointing sleeves Insulating thimbles placed over the connected ends of detonator leads coupled in large rounds of shots, and also over the connections between the detonator leads and the shot-firing cable. (BS, 1964)
joint line A visible line on imperfect glassware reproducing the line between separate parts of the mold in which the glass was made. Also called parting line; match mark; miter seam; mold mark; mold seam. Syn: *seam.* (Dodd, 1964)
joint plane A plane along a joint fracture or parting. Not to be confused with bedding and/or cleavage. See also: *joint.* (Long, 1960)
joint rose A rose diagram that shows the azimuth and intensity of jointing in an area.
joints Natural cracks or fractures in rocks. They tend to occur in more or less parallel systems, and when quarry walls are maintained parallel and at right angles to them, they may be utilized as natural partings in the process of block removal. (AIME, 1960)
joint set A group of more or less parallel joints. (Billings, 1954)
joint system Consists of two or more joint sets or any group of joints with a characteristic pattern, such as a radiating pattern and a concentric pattern. (Billings, 1954)
jointy Full of joints; specif. in mining, full of minute cracks or crevices, as rock. (Standard, 1964)
Jolly balance (a) A spring balance used to measure specific gravity of mineral specimens by weighing a specimen both in the air and immersed in a liquid of known density. (McGraw-Hill, 1994) (b) A spiral-spring balance with two specimen pans, one for measuring weight in air and one for weight in water, used to determine specific gravity. Cf: *Westphal balance.*
Jominy test A hardenability test in which a standard test piece, 4 in (10.2 cm) long and 1 in (2.54 cm) in diameter, is heated to a predetermined temperature, rapidly transferred to a jig fixture, and quenched, under standard conditions, by a jet of water impinging at one end. When the specimen is cool, determinations of hardness are made along it from the quenched end. The diagram relating hardness to distance from the quenched end of the specimen is known as a hardenability curve. (Hammond, 1965)
Jones riffle An apparatus used for cutting the size of a sample. It consists of a hopper above a series of open-bottom pockets, usually 1/2 in or 3/4 in (1.27 cm or 1.91 cm) wide, which are so constructed as to discharge alternately, first into a pan to the right, and then into another pan to the left. Each time the sample is passed through the riffle, it is divided into two equal parts; the next pass of one of those parts will give a quarter of the original sample, and so on, until the sample is reduced to the desired weight. (Pearl, 1961)
Jones splitter A device used to reduce the volume of a sample consisting of a belled, rectangular container, the bottom of which is fitted with a series of narrow slots or alternating chutes designed to cast material in equal quantities to opposite sides of the device. Also called sample splitter. Cf: *riffle.* (Long, 1960)
Joosten process Method of soil consolidation used in tunneling through sands and gravels. Solutions of calcium chloride and sodium silicate are forced into the ground, where they mingle and produce a watertight gel. (Pryor, 1963)
Joplin jig A device used for jigging the shaker products of the diamond washer. The products are fed to the jigs, one at a time, and jigged, with frequent stoppages for scraping off the top layer of tailings; more sand is added, and the process is repeated until a product is obtained that consists entirely of concentrates. (Griffith, 1938)
jordanite A monoclinic mineral, $Pb_{14}(As,Sb)_6S_{23}$; pseudohexagonal; forms a series with geochronite in which antimony increases relative to arsenic.
joren A scoop-shaped bamboo basket used in Japan for carrying auriferous gravel.
josëite A trigonal mineral, Bi_4TeS_2; sp gr, 7.9; at San Jose, Minas Gerais, Brazil. A related mineral, josëite-B, contains more tellurium and less sulfur.
josen *hartite.*
josephinite A gray, nickeliferous iron, Fe_2Ni_5 or $FeNi_3$. Massive, granular, and forms the metallic portion of rolled pebbles. From Josephine County, OR. Syn: *native nickel-iron.* (English, 1938)
joule (a) The absolute meter-kilogram-second (mks) unit of work or energy that equals 10^7 erg or approx. 0.7377 ft·lb or 0.2390 g (caloric); the standard in the United States. (Webster 3rd, 1966) (b) The gram-degree centigrade thermal unit; the small calorie. (Standard, 1964)
Joule's law The rate at which heat is produced by a steady current in any part of an electric circuit is jointly proportional to the resistance and to the square of the current. (Webster 3rd, 1966)
jouph holes Derb. Hollows in a vein.
jourado diamond A colorless synthetic spinel used as a simulated diamond.
journal That part of a rotating shaft resting in a bearing. (Long, 1960)
journey (a) Welsh term for train of mine cars moved mechanically. Also called gang; set; rake. (Pryor, 1963) (b) A cycle of work done in glass manufacturing in converting a quantity of material into glass or glass products. (Webster 3rd, 1966)
Joy double-ended shearer A cutter loader for continuous mining on a longwall face. It consists of two cutting heads fixed at each end of a continuous-track-mounted chassis. The heads are pivoted and controlled hydraulically for vertical movement. Each head comprises two bores and a frame or loop cutter that trims the bottom, face, and top. A cross conveyor delivers the coal to the adjacent face conveyor. The machine cuts a web of 5 ft (1.5 m) in seams from 37-1/2 in (95 cm) to 5 ft high. With an overall length of 18 ft (5.5 m), it weighs 15 st (13.6 t). (Nelson, 1965)
Joy extensible conveyor A belt conveyor to serve between a loader or continuous miner and the main transport. It consists of two main units, a head and a tail section, each mounted on crawler tracks and independently driven. In operation, the tail unit (i.e., the receiving end) moves forward with the loading machine, and belting is automatically released from a loop takeup. Fifty feet (15 m) of advance is possible

before additional belting has to be inserted into the conveyor run. Capacity equals 280 st/h (254 t/h) with a 30-in (76-cm) belt. (Nelson, 1965)

Joy extensible steel band An arrangement to provide a link between a continuous miner and the main transport. The equipment is hydraulically driven, and the steel band is coiled on the drivehead. (Nelson, 1965)

Joy loader Loading machine for coal or ore that uses mechanical arms to gather mineral onto an apron pressed into the severed material. A built-in conveyor then lifts it into tubs or onto a conveyor. (Pryor, 1963)

Joy microdyne A wet-type dust collector for use at the return end of tunnels or hard headings. It may be either 6,000 ft^3/min or 12,000 ft^3/min (2.8 m^3/s or 5.7 m^3/s) in capacity. It wets and traps dust as it passes through the appliance, and releases it in the form of a slurry, which is removed by a pump. The microdyne is bolted to the outby end of the exhaust pipe, and the auxiliary fan is bolted to the outby end of the dust collector. (Nelson, 1965)

Joy transloader A rubber-tired, self-propelled loading, transporting, and dumping machine.

Joy walking miner A continuous miner with a walking mechanism instead of crawler tracks. The walking mechanism was adopted to make the machine suitable for thin seams. The lowest crawler-track-mounted machine can operate in a minimum seam height of 4 ft (1.2 m), whereas the walking type can work in a 2.5-ft (0.76-m) seam. Syn: *walking miner.* (Nelson, 1965)

jubilee wagon A small wagon, running on rails, which tips sideways. (Hammond, 1965)

jubs Eng. Top jubs and bottom jubs, soft marly limestone, coarsely oolitic in places, in the Great Oolite at Kingsthorpe, Northamptonshire; also in the same formation at Bedford. (Arkell, 1953)

jud (a) N. of Eng. A block of coal about 4 yd (3.7 m) square, holed and cut ready for breaking down. (Fay, 1920) (b) In whole working, a portion of the coal laid out and ready for extraction; in pillar working (i.e., the drawing or extraction of pillars), the yet unremoved portion of a pillar. (Fay, 1920) (c) Applied to a working place, usually 6 to 8 yd (5.5 to 7.3 m) wide, driven in a pillar of coal. When a jud has been driven the distance required, the timber and rails are removed, and this is termed "drawing a jud.". (Zern, 1928) (d) Som. *jad; lift.*

judge (a) Derb.; Newc. A measuring stick to measure coal work underground. See also: *judge rapper.* (Fay, 1920) (b) Eng. Formerly a youth who proved the holing. (Fay, 1920)

judge rapper The upper end of the vertical arm of a judge. See also: *judge.* (Fay, 1920)

Judson powder A blasting explosive containing sodium nitrate, sulfur, coal, and a little nitroglycerin. (Webster 2nd, 1960)

jug A colloquial equivalent of seismic detector, geophone, etc. See also: *geophone.* (AGI, 1987)

jugglers Timbers set obliquely against pillars of coal to carry a plank partition, making a triangular air passage or manway. (Fay, 1920)

julienite Possibly a tetragonal mineral, $Na_2Co(SCN)_4{\cdot}8H_2O$; a blue thiocyanate that may be an artifact.

jumble Derb. The place where veins intersect. (Fay, 1920)

jumbler Big Jumbler, a bed of limestone in the Lower Lias at Rugby, United Kingdom.

jumbles The thickest oolitic bed in the Carboniferous limestone of the Clee Hills, Shropshire, United Kingdom.

jumbo (a) In mining, a drill carriage or mobile scaffold on which several drills of drifter type are mounted. It is used in tunnels and large headings. (Pryor, 1963) (b) An asbestos-fiberizing machine that is effective for moderately soft ore where crushing or breaking is not required. (Sinclair, 1959) (c) A mobile scaffold to assist drilling in large headings. (BS, 1964) (d) A number of drills mounted on a mobile carriage and used in tunnels. (Nichols, 1976) (e) Drilling platform used in tunneling. (Sandstrom, 1963)

jump (a) *jumping a claim.* (b) A sudden rise in the dip of a coal seam. (Arkell, 1953)

jump correlation Identification of events on noncontiguous seismic records as involving the same interfaces in the Earth. (AGI, 1987)

jumper (a) The borer, steel, or bit for a compressed-air rock drill. (CTD, 1958) (b) A long steel bar, or light aluminum tube with a steel end, used to dress rock faces, pry off loose rock, etc. (Pryor, 1963) (c) Person who jumps a claim; i.e., takes possession of another's mining property. (Pryor, 1963) (d) A steel bar used in manual drilling. (Sandstrom, 1963)

jumper bar A weighted steel bar with a cutting edge; raised and dropped by hand. (CTD, 1958)

jumping a claim (a) Taking possession of a mining claim liable to forfeiture owing to the requirements of the law being unfulfilled. (b) Taking possession of a mine or claim by stealth, fraud, or force. Syn: *jump.* (Fay, 1920) (c) The act of locating a mining claim on supposed excess ground within staked boundaries of an existing claim on the theory that the law governing the manner of making the original location had not been complied with. (Fay, 1920)

jump sheet A flat metal plate used as a turnsheet on which to turn the empty cars. (Lewis, 1964)

junckerite *siderite.* Also spelled junkerite.

junction (a) The point where two or more passageways intersect horizontally or vertically. (AGI, 1987) (b) In ventilation surveys, where three or more airways meet. (Roberts, 1960) (c) The union of two lodes. (Gordon, 1906)

junction box In mining, a stationary piece of enclosed apparatus from which one or more electric circuits for supplying mining equipment are connected through overcurrent protective devices to an incoming feeder circuit.

junk (a) Any foreign metallic material accidentally introduced into a borehole. (Long, 1960) (b) Very poor or low-grade drill diamond. (Long, 1960)

junkerite *siderite.* Also spelled junckerite.

junket (a) Eng. A bucket used for raising rock or ore in a shaft. (Hess) (b) Eng. *kibble.*

junking (a) The process of cutting a passage through a pillar of coal. (CTD, 1958) (b) N. of Eng. An opening cut into, or a narrow slice taken off, a pillar in the room-and-pillar system of working coal. A fast junking is a narrow place driven lengthwise in a pillar of coal, but unholed into the room on either side of the pillar. A loose junking is a similar place driven along the side of the pillar and open to the room along that side. (Fay, 1920)

junk mill A bit designed to grind or cut foreign metallic material or junk in a borehole into pieces small enough to be washed out of the hole or recovered by a basket. Cf: *milling bit; rose bit.* (Long, 1960)

junks (a) Dev. Limestone concretions in slate. (b) Corn. Joints in rocks.

Jupiter process A patented process for making cast steel by melting wrought-steel scrap with about 2.0% ferrosilicon, up to about 0.5% ferromanganese, and about 3.0% aluminum, then casting in molds of a special composition.

Jupiter steel A steel produced by the Jupiter process; it is about as strong and as ductile as forged steel.

Jurassic The second period of the Mesozoic Era (after the Triassic and before the Cretaceous), thought to have covered the span of time between 190 million years and 135 million years ago; also, the corresponding system of rocks. It is named after the Jura Mountains between France and Switzerland, in which rocks of this age were first studied. (AGI, 1987)

jury rig Any temporary or makeshift device, rig, or piece of equipment. (Hunt, 1965)

jutty A small tub or truck used for gathering coal in thin seams. (CTD, 1958)

juvenile (a) Said of an ore-forming fluid or mineralizer that is derived from a magma, via fractional crystallization or other plutonic mechanism, as opposed to fluids of surface, connate, or meteoric origin. (AGI, 1987) (b) A term applied to water and gases that are known to have been derived directly from magma and are thought to have come to the Earth's surface for the first time. (AGI, 1987) (c) In the classification of pyroclastics, the equivalent of essential; derived directly from magma reaching the surface. (AGI, 1987)

juvenile water Water from the interior of the Earth that is new or has never been a part of the general system of ground water circulation. Cf: *magmatic water.*

juxtaposition twin *contact twin.* Cf: *interpenetration twin.*

K

kaemmererite Chromian clinochlore. Alternate spelling of kämmererite.

kaersutite A monoclinic mineral, $NaCa_2(Mg,Fe)_4Ti(Si_6Al_2)O_{22}(OH)_2$; amphibole group with Mg/(Mg+Fe)= 0.5 to 1.0; forms a series with ferrokaersutite; a typical constituent of alkaline volcanic rocks.

kahlerite A tetragonal mineral, $Fe(UO_2)_2AsO_4)_2 \cdot 10\text{-}12H_2O$; autunite group; forms lemon-yellow plates; at Carinthia, Austria.

kalamin *calamine.*

Kaldo steel process A steelmaking process in which oxygen is fed into a large inclined rotating vessel through water-cooled lances, but at velocities somewhat lower than in the L.D. steel process, so that the jet does not completely or continuously penetrate the slag layer. In some respects, it is like a continuously rotating open hearth. It gives better heat utilization than the L.D. steel process but is slower. (Nelson, 1965)

kale Eng. Surface-weathered ironstone or oolite; rottenstone; in Northamptonshire, Rutland, and Lincolnshire. Also spelled keale. (Arkell, 1953)

kaliborite A monoclinic mineral, $KHMg_2B_{12}O_{16}(OH)_{10} \cdot 4H_2O$; formerly called heintzite, hintzeite, paternoite. See also: *heintzite.*

kalicine *kalicinite.*

kalicinite A monoclinic mineral, $KHCO_3$; colorless, white, or yellow. Also spelled kalicine; kalicite.

kalinite A possibly monoclinic mineral, $KAl(SO_4)_2 \cdot 11H_2O$. Syn: *potash alum.* Cf: *alum.*

kaliophilite A hexagonal mineral, $KAlSiO_4$; polymorphous with kalsilite, panunzite, and trikalsilite. Syn: *facellite; phacellite.*

kaliphite A mixture of limonite with oxides of manganese and silicates of zinc and lime. (Osborne, 1956)

kalistrontite A trigonal mineral, $K_2Sr(SO_4)_2$; occurs in prisms and plates with anhydrite and dolomite in a drill core from Alshtan, Bashkir, Russia.

Kalling's solution An etching reagent for developing the microstructure of chromium steels with more than 5% of chromium. It contains 5 g copper chloride, 100 mL hydrochloric acid, 100 mL alcohol, and 100 mL water. (Osborne, 1956)

kalsilite A hexagonal mineral, $2[KAlSiO_4]$; polymorphous with kaliophilite, panunzite, and trikalsilite; forms a partial series toward nepheline; in groundmasses of potassium-rich, silica-poor lavas and as an alteration of blast furnace brick.

kamacite A meteorite mineral consisting of the body-centered cubic alpha-phase of a nickel-iron alloy, with a fairly constant composition of 5% to 7% nickel. It occurs in iron meteorites as bars or girders flanked by lamellae of taenite. See also: *nickel iron.* (AGI, 1987)

kamarezite A grass-green, hydrated, basic copper hydrate and sulfate, $Cu_3(OH)_4SO_2 \cdot 6H_2O$. Syn: *brochantite.* (Standard, 1964)

kämmererite A chromian variety of clinochlore.

kammerling furnace A modification of the Belgian zinc smelting furnace wherein there are two combustion chambers separated by a central longitudinal wall. In principle, the furnace is similar to the Hauzeur, a compound furnace.

kanase Burma (Myanmar). A local custom in the gem mines at Mogok that permits women to work without licenses in streambeds, tailraces, and dumps from mines and washeries and to keep any gems they find. (Hess)

Kanawhan Upper Lower Pennsylvanian. (AGI, 1987)

Kanawha series A group of productive coal measures occurring in the Pennsylvanian of the Appalachian Region and completely developed in Virginia. Sometimes known as the Upper Pottsville series. (CTD, 1958)

kandite *kaolin.*

kankar (a) A term used in India for (1) masses or layers of calcium carbonate, usually occurring in nodules, found in the older alluvium or stiff clay of the Indo-Gangetic plain; or (2) precipitated calcium carbonate in the form of cement in porous sediments or as a coating on pebbles. (AGI, 1987) (b) A limestone containing kankar and used for making lime and building roads. Etymol. Hindi. The term is occasionally applied in the United States to a residual calcareous deposit, such as caliche. Also spelled kunkur. (AGI, 1987)

kaoleen A colloquial term used in south-central Missouri for a chalky, porous, weathered chert with a white to tan or buff color. Etymol. corruption of kaolin, to which the material bears a slight resemblance. (AGI, 1987)

kaolin Former name for kaolinite. The aluminous minerals of the kaolinite-serpentine group. Syn: *bolus alba; kaoline; kandite; kaolinite; white clay.*

kaoline *kaolin.*

kaolinic Of, relating to, or resembling kaolin. (Webster 3rd, 1966)

kaolinite (a) A monoclinic mineral, $2[Al_2Si_2O_5(OH)_4]$; kaolinite-serpentine group; kaolinite structure consists of a sheet of tetrahedrally bonded silica and a sheet of octahedrally bonded alumina with little tolerance for cation exchange or expansive hydration; polymorphous with dickite, halloysite, and nacrite; soft; white; formed by hydrothermal alteration or weathering of aluminosilicates, esp. feldspars and feldspathoids; formerly called kaolin. (b) Kandites in general. (c) Individual kandites not specif. designated. See also: *alum salt; kaolin.*

kaolinite-serpentine The mineral group amesite, antigorite, berthierine, brindleyite, clinochrysotile, cronstedite, dickite, endellite, fraiponite, greenalite, halloysite, kaolinite, kellyite, lizardite, manandonite, nacrite, nepouite, orthochrysotile, parachrysotile, and pecoraite. Cf: *serpentine.*

kaolinization Replacement or alteration of minerals, esp. feldspars and micas, to form kaolin as a result of weathering or hydrothermal alteration. Cf: *argillation; argillization.* (AGI, 1987)

kapel *capel.*

kappa carbide A carbide of iron, $Fe_{23}C_6$, in which all or part of the iron may be replaced by chromium, molybdenum, and/or tungsten, $(Fe,Cr,Mo,W)_{23}C_6$. (Osborne, 1956)

karang Term used in the Malay States for the pay streaks of cassiterite. (Lewis, 1964)

karat One-twenty-fourth part. It is used to designate the fineness of gold; thus, 18-karat gold is 18/24 (or 75%) pure gold and 6/24 (or 25%) other alloying metal or metals. Not to be confused with carat.

karelianite A trigonal mineral, V_2O_3; hematite group; black; in boulders from the Outokumpu orebody in the Karelian schist belt, Finland.

Karlsbad twin *Carlsbad twin.*

karnasurtite A possibly hexagonal mineral, $(Ce,La,Th)(Ti,Nb)(Al,Fe)(Si,P)_2O_7(OH)_4 \cdot 3H_2O(?)$; metamict; at Mt. Karnasurt, Kola Peninsula, Russia.

karpatite A monoclinic mineral (coronene), $C_{24}H_{12}$. Also spelled carpathite. Syn: *pendletonite.*

karpinskite A possibly monoclinic mineral, $(Mg,Ni)_2Si_2O_5(OH)_2(?)$; greenish blue; with talc in serpentinite from the Ural Mountains, Russia. (Not karpinskyite.)

karpinskyite A mixture of leifite and a zinc-bearing smectite. (Not karpinskite.)

karst A type of topography that is formed on limestone, gypsum, and other rocks by dissolution, and that is characterized by sinkholes, caves, and underground drainage. Etymol. German, from the Yugoslavian territory Krš; type locality, a limestone plateau in the Dinaric Alps of northwestern Yugoslavia and northeastern Italy. First published on a topographic map, Ducatus Carnioliae, in 1774. Adj. karstic. Syn: *karst topography.* (AGI, 1987)

karst topography *karst.*

kasoite A potassian variety of celsian.

kasolite A monoclinic mineral, $Pb(UO_2)SiO_4 \cdot H_2O$; radioactive; an oxidation product of uraninite; at Kasolo, Katanga, Zaire. Formerly called droogmansite.

Kast furnace A small, circular shaft furnace with three or four tuyeres, for lead smelting. (Fay, 1920)

kata- *cata-.*

kataclastic *cataclastic.*

kata cooling power A measure of the cooling effect of the ambient air as determined by the kata thermometer. This instrument may be used wet or dry. (BS, 1963)

katamorphism Destructive metamorphism in the katamorphic zone, at or near the Earth's surface, in which complex minerals are broken down and altered through oxidation, hydration, solution, and allied processes to produce simpler and less dense minerals. The term was introduced by Van Hise in 1904. Also spelled catamorphism. Cf: *anamorphism.* See also katazone. (AGI, 1987):

kata thermometer A type of alcohol thermometer used to determine the cooling power of the ambient air and sometimes to measure low air velocities. (BS, 1963)

katazone According to Grubenmann's classification of metamorphic rocks (1904), the lowermost depth zone of metamorphism, which is characterized by high temperatures (500 to 700 °C), mostly strong hydrostatic pressure, and low or no shearing stress. It produces rocks such as high-grade schists and gneisses, granulites, eclogites, and amphibolites. The concept includes effects of high-temperature contact metamorphism and metasomatism. Modern usage stresses temperature-pressure conditions (highest metamorphic grade) rather than the likely depth of zone. Also spelled catazone. Cf: *mesozone; epizone.* See also: *katamorphism.* (AGI, 1987)

katoptrite A monoclinic mineral, $(Mn, Mg)_{13}(Al, Fe)_4Sb_2Si_2O_{28}$; in jet-black crystals in limestone at Nordmark and Långban, Sweden.

katungite A volcanic rock composed chiefly of melilite, with subordinate olivine and magnetite and minor leucite and perovskite; a pyroxene-free melilitite. (AGI, 1987)

kauk A very heavy substance, common in the mines, Derbyshire, U.K. Also spelled cauk.

kayserite *diaspore.*

K-bentonite *potassium bentonite.*

kearsutite Misspelling of kaersutite.

keatite Synthetic tetragonal silica, SiO_2; crystallized hydrothermally at high pressure.

keeleyite *zinkenite.*

keel wedge A long iron wedge for driving over the top of a pick hilt. (Fay, 1920)

keen sand Forest of Dean. Sand forming poor rye soil.

keeper Person in charge of opening and closing the taphole of a blast furnace and who runs iron at cast. (Fay, 1920)

Keewatin A division of the Archeozoic rocks of the Canadian Shield. It is older than the Timiskamian. Also spelled Keewatinian. (AGI, 1987)

keg A cylindrical container made of steel or some other substance, which contains 25 lb (11.4 kg) of blasting powder or gunpowder. Any small cask or barrel having a capacity of 5 to 10 gal (18.9 to 37.9 L). (Fay, 1920)

kehoite Amorphous $(Zn,Ca)Al_2(PO_4)_2{\cdot}5H_2O$; a doubtful mineral species.

keldyshite A triclinic mineral, $Na_{2-2Dx}H_xZrSi_2O_7{\cdot}nH_2O$; forms irregular grains in the Lovozero massif, Kola Peninsula, Russia.

kell A variation of kiln. (Fay, 1920)

Keller automatic roaster A six-deck horizontal furnace for calcining sulfide ores. (Fay, 1920)

Keller furnace A multiple-deck roasting furnace for sulfide ore. It is a modification of the Spence furnace. (Fay, 1920)

kellerite A cuprian variety of pentahydrite.

kellow Black lead or wad, Cumberland, U.K.

kelly The rod attached to the top of the drill column in rotary drilling. It passes through the rotary table and is turned by it, but is free to slide down through it as the borehole deepens. Also called grief stem. (BS, 1963)

kelly bar A hollow bar attached to the top of the drill column in rotary drilling; also called grief joint; kelly joint; kelly stem. (Meyer, 1981)

Kelly filter An intermittent, movable pressure filter. The leaves are vertical and are set parallel with the axis of the tank. Pulp is introduced into the tank (a boilerlike affair) under pressure, and the cake is formed. The head is then unlocked, the leaves are run out of the tank chamber by means of a small track, and the cake is dropped. The carriage and leaves are then run back into the tank, and the cycle begins again. (Liddell, 1918)

kelp The ashes of seaweeds, formerly the source of soda as used in glassmaking and soapmaking; now a source of potash, iodine, and char. (Standard, 1964)

kelvin (a) Board of Trade unit of energy (1 kW/h). (Pryor, 1963) (b) *Kelvin temperature scale.*

Kelvin temperature scale The absolute temperature scale in which the temperature measure is based on the average kinetic energy per molecule of a perfect gas. The zero of the Kelvin scale is -273.16 °C. The temperature scale adopted by the International Bureau of Weights and Measures is that of the constant volume hydrogen gas thermometer. The magnitude of the degree in both these scales is defined as one one-hundredths the difference between the temperature of melting ice and that of boiling water at 760 mm pressure. See also: *temperature.* Syn: *absolute scale.* (Handbook of Chem. & Phys., 2)

kelyphite *corona.*

kelyphitic rim *corona.*

kelyphytic A term applied to the rims or borders consisting of microcrystalline aggregates of pyroxene or amphibole occurring around olivine, where it would otherwise be in contact with plagioclase, or around garnet, where it would otherwise be in contact with olivine or other magnesium-rich minerals.

kelyphytic rim *corona.*

Kema plow A scraper-box type of plow for use on longwall faces. See also: *scraper box plow.* (Nelson, 1965)

kempite An orthorhombic mineral, $Mn_2Cl(OH)_3$; forms minute emerald-green prismatic crystals; at Alum Rock Park, CA.

kennel coal A coal that can be ignited with a match to burn with a bright flame. Syn: *candle coal.* See also: *cannel coal.*

kentrolite An orthorhombic mineral, $Pb_2Mn_2Si_2O_9$; forms a series with melanotekite; massive or in sheaflike prismatic crystals; at Långban and Nordmark, Sweden, and Sassari, Sardinia, Italy.

Kent roller mill A revolving steel ring with three rolls pressing against its inner face. The rolls are supported on springs, and the rings support the roll, so that there is some freedom of motion. The material to be crushed is held against the ring by centrifugal force. (Liddell, 1918)

kentsmithite A local name used in the Paradox Valley, Colorado, for a black, vanadium-bearing sandstone. Syn: *vanoxite.* (AGI, 1987)

kep (a) One of the steel supports on which a cage rests at the pithead during unloading or loading so that the rail track is always at the proper level. During this period, the rope is released from the weight of the cage. The ordinary type of kep gear consists of four steel arms, two for each end of the cage, carried on shafts that are connected to and operated by levers. Normally, the cage must be raised from the kep arms before the latter can be withdrawn to allow the cage to descend the shaft. See also: *Beien kep gear; cage stop.* (Nelson, 1965) (b) One of the retractable rests on which the mine cage is supported during its stop at a shaft landing. Also called catch; chair; keep; landing chair; stop. (Pryor, 1963) (c) One of the bearing-up stops for supporting a cage or load at the beginning or end of hoisting in a shaft. (CTD, 1958)

kepel Corn. Spar or hard stone on each side of the lode.

kerargyrite *cerargyrite.*

keratophyre A name generally applied to all salic extrusive and hypabyssal rocks characterized by the presence of albite or albite-oligoclase and chlorite, epidote, and calcite, generally of secondary origin. Some varieties of keratophyre contain sodic orthoclase, sodic amphiboles, and pyroxenes. Keratophyres commonly are associated with spilitic rocks and interbedded with marine sediments. (AGI, 1987)

kerf (a) The undercut usually made in the coal to facilitate its fall. (BCI, 1947) (b) A horizontal cut in a block of coal, as opposed to a shearing, which is a vertical cut. (BCI, 1947) (c) Undercut in a coal seam from 3 to 7 in (7.6 to 18 cm) thick and entering the face to a depth of up to 4 ft (1.2 m), made by a mechanical cutter. Also called kirve. (Pryor, 1963) (d) The undercut made in a coalbed to assist the action of explosives in blasting. (Hudson, 1932) (e) The annular groove cut into a rock formation by a core bit. See also: *kerve.* (Long, 1960) (f) The space that was occupied by the material removed during cutting. (ASM, 1961) (g) The thickness of the wall of the diamond-insert part of the crown of a core bit. (Long, 1960) (h) Sometimes incorrectly used as a syn. for nose, as applied to a diamond core bit. (Long, 1960)

kerf stone One of the diamonds inset in the kerf of the crown of a diamond bit. Also called face stone. (Long, 1960)

kermesite A triclinic mineral, Sb_2S_2O; pseudomonoclinic; occurs as soft tufts of cherry-red capillary crystals with one perfect cleavage; an alteration product of stibnite. Syn: *antimony blende; pyrostibite; pyrostibnite; red antimony; pyroantimonite.*

kerned stone Corn. Sand blown off the seashore into the country and concreted there.

kernel Atom that has lost the valence electrons of its outermost shell. (Pryor, 1963)

kernel roasting *roasting.*

kernite A monoclinic mineral, $4[Na_2(H_2O)_3B_4O_6(OH)_2]$; soft, colorless to white; in the Mojave Desert and Kern County, CA; a major source of borax and boron compounds in the United States. See also: *Rasorite.*

kerogen (a) A term generally used for organic matter in sedimentary rocks that is insoluble in common organic and inorganic solvents. (b) The solid, organic substance in shales that yields oil when the shales undergo destructive distillation.

kerogen shale *oil shale.*

kerolite A variety of talc with randomly stacked structure; forms a series with nickel-kerolite. Also spelled cerolite.

keronigritite A type of nigritite that is derived from kerogen. Cf: *polynigritite; humonigritite.* (AGI, 1987)

kerosine flotation As sometimes practiced, this method is a combination of bulk oil flotation and froth flotation. By adding large quantities of kerosine to a pulp plus a small amount of frother and agitating vigorously, surfaces of the amenable mineral (coal) are attracted to both the oil and air bubbles, forming heavy flocs. This type of concentrate is more readily dewatered than ordinary froth and therein lies its advantage, plus the fact that coarser particles (6 to 10 mesh) can be handled than in ordinary froth flotation. Also called granulation. Syn: *agglomeration.* (Mitchell, 1950)

kerosine shale Substance originally described as a variety of oil shale but later proved to be similar to torbanite. See also: *boghead coal; torbanite.* (Tomkeieff, 1954)

kerrite A variety of vermiculite.

kersantite A lamprophyre containing biotite and plagioclase (usually oligoclase or andesine), with or without clinopyroxene and olivine. Defined by Delesse in 1851; named for the village of Kersanton, France. (AGI, 1987)

kerve N. of Eng. In coal mining, to cut under. See also: *kerf.* Also spelled kirve. (Zern, 1928)

kesterite A mineral, (Cu,Sn,Zn)S, containing 30.36% copper, 25.25% tin, 11.16% zinc, and 23.40% sulfur. In quartz sulfide ore from Kester, Magadan, Yakutia, northeast Siberia, Russia. Named from locality. Syn: *isostannite; khinganite.* (Spencer, 1958)

kettle A depression in the ground surface formed by the melting of a large, detached block of stagnant ice wholly or partly buried by glacial drift. (AGI, 1987)

kettleback *horseback; slip.*

kettle bottom (a) A smooth, rounded piece of rock, cylindrical in shape, which may drop out of the roof of a mine without warning, sometimes causing serious injuries to miners. The surface usually has a scratched, striated, or slickensided appearance and frequently has a slick, soapy, unctuous feel. The origin of this feature is thought to be the remains of the stump of a tree that has been replaced by sediments so that the original form has been rather well preserved. Sometimes spelled kettlebottom. Also called bell; pot; camelback; tortoise. Syn: *caldron.* See also: *clod; pot bottom.* Cf: *black bat.* (Kentucky, 1952) (b) *horseback; caldron bottom.*

kettle dross Skimmings resulting from the desilverization of lead bullion. It consists principally of lead oxides mixed with metallic lead. (Fay, 1920)

kettleman In ore dressing, smelting, and refining, a person who (1) refines lead in a series of oil-fired kettles, and (2) removes silver and copper from black mud in a gas-fired kettle, preparatory to the separation of gold. (DOT, 1949)

kettle operator In ore dressing, smelting, and refining, a person who melts and fumes antimony in oil-fired kettles to make antimony oxide. (DOT, 1949)

kettnerite A tetragonal mineral, $CaBi(CO_3)OF$; in pegmatite from Krupka, Czech Republic.

kevell Calcspar from Derbyshire lead mines, United Kingdom. Also spelled kebble.

kevil (a) Derb. A veinstone, consisting of a mixture of calcium carbonate and other minerals. (Fay, 1920) (b) N. of Eng. The amount of coal sent out by the various miners during a certain period. (Fay, 1920)

Keweenawan A provincial series of the early Proterozoic in Michigan and Wisconsin. (AGI, 1987)

key The pieces of core causing a block in a core barrel, the removal of which allows the rest of the core in the core barrel to slide out. Syn: *legend.* (Long, 1960)

key bed (a) A well-defined, easily identifiable stratum or body of strata that has sufficiently distinctive characteristics (such as lithology or fossil content) to facilitate correlation in field mapping or subsurface work. (AGI, 1987) (b) A bed, the top or bottom of which is used as a datum in making structure-contour maps. Syn: *key horizon; index bed; marker bed.* (AGI, 1987)

key blocks The first blocks that are removed in opening up a new quarry floor.

keyhole slot A slot enlarged at one end to allow entrance of a chain or bolt that can then be held by the narrow end. (Nichols, 1976)

key horizon *key bed.*

keystone (a) A symmetrically tapered piece at the center or crown of an arch. (b) A filling-in block of cast iron used in some lead smelting furnaces. (Webster 2nd, 1960)

keystoneite Blue chalcedony colored by chrysocolla.

K-feldspar The mineral microcline, orthoclase, or sanidine. Syn: *potassium feldspar.*

K-Fuel process In this process, the feed coal is heated sufficiently to remove all the moisture from the coal and to mildly carbonize it. The carbonized hot coal, in a plastic state, flows to an extruder which agglomerates the product into pellets.

Khari salt A native mineral salt of India, predominantly sodium chloride with large amounts of sodium sulfate, the composition varying greatly with locality where obtained. Synthetic Khari salt has 40% anhydrous sodium sulfate. (Kaufmann, 1960)

khinganite *kesterite.*

khlopinite A titanian variety of samarskite.

kibbal *kibble.*

kibble (a) Steel bucket used during shaft sinking. (Pryor, 1963) (b) To carry in a hoisting bucket, such as ore. Syn: *junket; kibbal.* (Standard, 1964)

kibble rope Eng. A rope or chain for hoisting a kibble or bucket. (Standard, 1964)

kick (a) Assay with metal values. Also, a geophysical indication of ore. As opposed to background values. (Slang only.). (Hoffman, 1958) (b) A small sidewise displacement or offset in a borehole caused by the sidewise deviation of a bit when entering a hard, tilted rock stratum underlying a softer stratum. (Long, 1960) (c) A quick snap of the drill stem caused by the core breaking in a blocked core barrel or sudden release of a momentary bind. Cf: *step.* (Long, 1960)

kick back (a) Arkansas. To break the coal on both sides of the auger hole that contains the powder, usually along a joint in the coal. (Fay, 1920) (b) A track arrangement for reversing the direction of travel of cars moving by gravity. (Fay, 1920)

kicker (a) Ground left, in first cutting a vein, for support of its sides. (b) Also, a gage stone handset in the outside surface or wall of the metal shank of a diamond bit. Syn: *gage stone.* (Long, 1960)

kicker stone *gage stone.*

kicking pieces Short struts to prevent a sill or other member from being pushed out of place. (Stauffer, 1906)

kickoff point The place in a borehole where the first intentional deviation starts. Sometimes abbreviated KOP. (Long, 1960)

Kick's law The amount of energy required to crush a given quantity of material to a specified fraction of its original size is the same no matter what the original size. Cf: *Rittinger's law.* (CCD, 1961)

kidney iron ore A reniform variety of hematite with concentric or radiating structure. Syn: *kidney stone; kidney ore.* See also: *iron ore.*

kidney ore *kidney iron ore.*

kidneys A mineral zone that contracts, expands, and again contracts downwards. (AGI, 1987)

kidney stone (a) *nephrite; kidney iron ore; urolith.* (b) Any reniform nodule.

kidney sulfur *lense.*

kies A general term for the sulfide ores, adopted into English from the original German.

kieselguhr *diatomite.*

kieserite (a) A monoclinic mineral $4[MgSO_4{\cdot}H_2O]$; in evaporite salt deposits from dehydration of epsomite or decomposition of kainite. Syn: *martinsite.* (b) The mineral group dwornikite, gunningite, kieserite, poitevinite, szmikite, and szomolnokite.

kilbrickenite *geochronite.*

kilchoanite A orthorhombic mineral, $Ca_3Si_2O_7$; dimorphous with rankinite, which it replaces at Kilchoan, Ardnamurchan, Scotland.

kilkenny coal *anthracite.*

kill To mix atmospheric air with combustible gases or other gases so as to make them harmless. (Fay, 1920)

killas (a) Corn. The slates or schists that form the country rock of the Cornish tin veins. (b) Corn. Term used in the china-clay mines for the altered schistose or hornfelsic rocks in contact with the granite and often considerably modified by emanations from the latter. (Dodd, 1964)

killed steel Steel treated with a strong deoxidizing agent, such as silicon or aluminum, to reduce the oxygen content to such a level that no reaction occurs between carbon and oxygen during solidification. (ASM, 1961)

killing (a) Allowing the molten steel to remain in the crucible for about 45 min for the escape of the gases. (Mersereau, 1947) (b) In metallurgy, esp. in foundry terminology, a term for deoxidation. (Gaynor, 1959)

kiln (a) A large furnace used for baking, drying, or burning firebrick or refractories, or for calcining ores or other substances. (ASM, 1961) (b) A furnace or oven, which is usually made from refractory brick, used to dry and fire various types of ceramic ware.

kiln burner A worker who is responsible for firing kilns to produce ceramic products. Type of product produced is often attached to the term as, brick kiln burner, tile burner, pipe burner, etc. Also called kiln fireman; kiln operator; baker; kiln tender; kiln firer, etc.

kiln-burner helper Person who assists kiln burner in firing kiln, supplying coal and wood to fireboxes, and cleaning ashes from ashpits. (DOT, 1949)

kiln-car unloader A worker who removes fired ceramic ware from kiln cars.

kiln cleaner Person who prepares kilns for burning. (DOT, 1949)

kiln-dry To dry in a kiln. (Webster 3rd, 1966)

kilneye An opening for removal of lime from a vertical lime kiln.

kilnman A worker who tends a kiln. (Standard, 1964)

kiln placer A worker who places clayware in kiln for drying or firing. Also called kiln setter; sagger filler; kiln loader; kiln stacker.

kilo- One thousand units. The kilodyne is 1,000 dynes; kilogram is 2.2046 lb avoirdupois; kilometer is 1,093.6 yd; kilowatt is 1.341 hp. (Pryor, 1963)

kilocusec A unit of volumetric rate of air flow, expressed in thousands of cubic feet per second. (BS, 1963)

kilocycle One thousand cycles. Abbrev. kc. (Crispin, 1964)

kilometer A length of 1,000 m; equals 3,280.8 ft or 0.621 mile. The chief unit for long distances in the metric system. Abbrev. km. (Standard, 1964)

kiloton A unit for measuring the energy of a nuclear explosion. A 1-kiloton explosion releases energy equal to that in the explosion of 1,000 st (907 t) of TNT. Abbrev. kton. (Lyman, 1964)
kilowatt-hour A unit of work or energy equal to that expended in 1 h at a steady rate of 1 kW or 3.6 x 10^6 J. (Webster 3rd, 1966)
Kimberley method *combined top slicing and shrinkage stoping.*
kimberlite A highly serpentinized porphyritic peridotite, commonly brecciated, which occurs in vertical pipes, dikes, and sills. It is the principal original environment of diamond, but only a small percentage of the known kimberlite occurrences are diamondiferous. See also: *blue ground; yellow ground.* The name is derived from Kimberley, South Africa.
kimzeyite An isometric mineral, $Ca_3(Zr,Ti)_2(Si,Al,Fe)_3O_{12}$; garnet group.
kind Eng. Generally signifies tender, soft, or easy to work; said of certain ores.
Kind-Chaudron process A method of sinking a deep shaft of large diameter in which a pilot bore of smaller diameter is first cut, after which the cut is enlarged to the final diameter, the debris falling into the pilot bore. When water is encountered, a lining with a moss box at the bottom is forced into place. Cf: *Triger process.* (Standard, 1964)
kindly *likely.*
kindly ground Eng. Those rocks in which lodes become productive of mineral of value.
kindred *rock association.*
kinematic viscosity Ratio of absolute viscosity to mass density. Measured in square meters per second. (Hartman, 1982)
kinetic energy The form of mechanical energy a body possesses by virtue of its velocity. The kinetic energy of a body, or the energy of motion, is the work done by it, or against it, in coming to rest. Water flowing through pipes or air flowing through a mine roadway possesses kinetic energy. See also: *velocity head.* (Morris, 1958)
kinetic head The energy of flowing water that is a function of its velocity.
kinetic metamorphism A type of metamorphism that produces deformation of rocks without chemical reconstitution or recrystallization to form new minerals. (AGI, 1987)
kingbolt (a) The bolt with which a cage is attached to the hoisting cable. It supports the suspended cage. (b) A large bolt that holds the upper end of the tripod legs together and from which the sheave-wheel clevis is suspended. (Long, 1960)
king brick Special, hollow, cylindrical fireclay brick, between the bottom fountain brick and the first lateral brick in a bottom-pour ingot assembly. See also: *bottom-pour ingot assembly.*
king pile In a wide excavation where strutting is required, this is a long pile driven at the strut spacing in the center of the trench before excavation is started. (Hammond, 1965)
king post A vertical member of a stamp battery frame that carries the camshaft. (CTD, 1958)
king screen A drum-type screen in which the pulp to be screened is delivered on the outside, the undersize passing through the screen and discharging through the open end. (Liddell, 1918)
kink (a) A sharp angular deflection in a borehole. (Long, 1960) (b) A deflection in a vein or lode that does not interrupt its continuity. (c) Scot. A twist in a rope; a doubling and interlocking of several links in a chain. (Fay, 1920)
kink bands Microscopic to macroscopic deformation lamellae in mineral grains; e.g., quartz, pyroxene, mica, kyanite, resulting from plastic flow under tectonic strain. Also called translation gliding.
kinradite Orbicular jasper containing spherical inclusions of colorless or nearly colorless quartz.
kinzigite A coarse-grained metamorphic rock of pelitic composition occurring in the granulite facies. Essential minerals are garnet and biotite, with which occur varying amounts of quartz, K-feldspar, oligoclase, muscovite, cordierite, and sillimanite. The name is from Kinzig, Schwarzwald, Germany. (AGI, 1987)
kip (a) N. of Eng. A level or gently sloping roadway, at the extremity of an engine plane, upon which the full cars stand ready to be sent up the shaft. The tubs, or cars, usually go to the shaft by gravity. (Fay, 1920) (b) A load of 1,000 lb (454 kg). (ASM, 1961)
kir Natural asphalt at the surface, Russia.
Kirkup table A machine for dry cleaning coal. Raw coal is fed onto a perforated plate inclined at 12.5° to 15° to the horizontal. A pulsating air current is applied to the underside of the inclined plate, which stratifies the material, the coal forming the upper layer with the dirt below, which then passes into separate compartments. The Kirkup table gives a three-product separation; i.e., coal, middling, and reject. See also: *pneumatic jig; pneumatic table; S.J. table.* (Nelson, 1965)
kirovite A magnesian variety of melanterite. Cf: *jarosite.*
Kiruna method A borehole-inclination survey method whereby the electrolytic deposition of copper from a solution is used to make a mark on the inside of a metal container. Cf: *acid-dip survey.* (Long, 1960)
kirve N. of Eng. To undercut. See also: *kerve; kerf.* (Fay, 1920)
kirving (a) Newc. The cutting made at the bottom of the coal by the miner. See also: *holing.* (Fay, 1920) (b) Eng. Undercutting the coal horizontally, usually by hand. Also called laying-in; ligging-in; holing. (SMRB, 1930)
kirvings N. of Eng. Small coal, fragments, or cuttings lying in the undercut made by a cutting machine jib. (Trist, 1963)
kish Free graphite that forms and floats out of molten hypereutectic cast iron as it cools. (ASM, 1961)
kishly York. Hard, dry vein matter, poor in ore; in lead mines. (Arkell, 1953)
kist (a) Eng. The wooden box or chest in which the deputy keeps tools at the flat or inspection station; the station is said to be at the kist. (SMRB, 1930) (b) N. of Eng. The meeting place in a district where workers assemble. (Strictly, the box in which the deputy keeps papers, etc.). (Trist, 1963)
kit A wooden vessel. (Fay, 1920)
kladnoite A monoclinic mineral (phthalimide), $C_6H_4(CO)_2NH$; results from burning waste at the Kladno coal basin, Czech Republic.
klaprothine *lazulite.*
klaprothite A discredited mineral since it is a mixture of wittichenite and emplectite. (Am. Mineral., 1945)
klebelsbergite An orthorhombic mineral, $Sb_4O_4(OH)_2(SO_4)$; forms pale-yellow to orange-yellow platy or acicular crystals on stibnite; at Baia Sprie, Romania.
kleinite A hexagonal mineral, $Hg_2N(Cl,SO_4){\cdot}nH_2O$; yellow to orange; occurs as short prisms darkening on exposure; sp gr, 8.0; in Brewster County, TX.
Klein solution A solution of cadmium borotungstate; sp gr, 3.6; used as a "heavy liquid" for mineral separation. Cf: *Clerici solution; methylene iodide; Sonstadt solution.*
kliachite *cliachite.*
klinkstone *phonolite.*
klintite A reef limestone, particularly the massive core.
klippe An isolated rock unit that is an erosional remnant or outlier of a nappe. The original sense of the term was merely descriptive; i.e., included any isolated rock mass, such as an erosional remnant. Plural, klippen. See also: *nappe.* Etymol. German; rock protruding from a sea or lake floor. (AGI, 1987)
klockmannite A hexagonal mineral, CuSe; slate gray to reddish violet, tarnishing to blue-black; basal cleavage. Occurs at Sierra de Umango, Argentina; Harz Mountains, Germany; and Skrikerum, Sweden.
kloof S. Afr. A mountain pass or cleft; a gorge or narrow valley.
kmaite A former name for celadonite.
Knapp bottom pressure gage An instrument designed for studying harbor surging. A strain gage unit, used in connection with pressure sensitive bellows, comprises the transducer of the pressure head. The four strain gages in the unit are connected to form a bridge circuit that is linked to the recorder by an electrical cable. A direct current voltage is applied to the bridge, and the record is obtained by recording the unbalanced current from a magnetic oscillograph. Any standard strain gage recorder can be used for the recording system. (Hunt, 1965)
knapping machine An instantaneous stonecrushing machine; a stonebreaker. (Standard, 1964)
kneading A filling material composed of clayey rock from which balls are made by a special machine. These balls are driven by an air current through a tube resembling a gun barrel into the excavated area to be filled. By impact and weight, the soft balls are squashed and perfectly fill the excavation. (Stoces, 1954)
knebelite A manganoan variety of fayalite.
knee brace A stiffener between a stanchion or column and a roof truss, to ensure greater rigidity in a building frame under wind load. (Hammond, 1965)
kneeler (a) The triangular connection by which a horizontal motion is changed to a vertical one, as in certain mine pumps. (Standard, 1964) (b) A stone cut to provide a change of direction. (Crispin, 1964) (c) The return of the dripstone at the spring of an arch. (CTD, 1958)
knee pad A protective cushion, usually made of sponge rubber, that can be strapped to the miner's knee.
knee piece (a) A bent piece of piping. (Fay, 1920) (b) An angular piece of timber used in a roof (mine) to strengthen a joint where two timbers meet. (Fay, 1920)
knee timber (a) Timber with natural knees or angles in it. (Webster 3rd, 1966) (b) A piece of timber with an angle or knee in it. (Webster 3rd, 1966)
knell stone N. of Eng. Freestone (sandstone).
knife dog A tool that fits around and grips drill rods or any tubular drilling equipment so they can be pulled or lifted from a borehole where workspace in narrow underground openings is too confined to allow the use of a hoisting plug. (Long, 1960)
knife edge (a) The girdle of a brilliant, cut to a sharp edge and polished. (Hess) (b) A narrow ridge of rock or sand. Syn: *feather edge.*

knife switch A switch that opens or closes a circuit by the contact of one or more blades between two or more flat surfaces or contact blades. (Crispin, 1964)
knipovichite A chromian variety of alumohydrocalcite.
knistersalze Crystals of salt mined at Wieliczka, Poland, that decrepitate violently upon heating, due to excessive enclosed water or gases. (Kaufmann, 1960)
knitted texture A texture typical of the mineral serpentine in rocks where it replaces a clinopyroxene. Cf: *lattice texture.*
knobbing The act of roughdressing stone in the quarry by knocking off the projections and points. (Fay, 1920)
knobbing fire A bloomery for refining cast iron. (Fay, 1920)
knobellite An intergrowth of galena with bismuthinite or stibnite. (Hess)
knock To sound the mine roof for competence by rapping.
knockbark Crushed lead ore. (Arkell, 1953)
knocking (a) S. Wales. Signals made underground by knocking on the coal. (Fay, 1920) (b) Eng. Ore broken with a hammer, esp. the large lumps. (Webster 2nd, 1960)
knockoff bit *detachable bit.*
knockout man A laborer who frees solidified metal castings from inverted molds of a casting wheel by prying the castings from the molds with a long steel bar to drop them into a water pit (bosh) for cooling. (DOT, 1949)
knockstone Eng. Stone bench on which lead ore is buckered or broken small for the hotching tubs; Yorkshire lead mines. Also called binkstone. (Arkell, 1953)
knogging Eng. Small refuse stones used for the inside of walls; Northamptonshire and Worcestershire. (Arkell, 1953)
Knoop hardness Microhardness determined from the resistance of metal to indentation by a pyramidal diamond indenter, having edge angles of 172° 30' and 130°, and making a rhombohedral impression with one long and one short diagonal. Cf: *Mohs scale.* (ASM, 1961)
knot A small concretion; e.g., of galena in sandstone, or a segregation of darker minerals in granite and gneiss. (AGI, 1987)
knots (a) Nodules or concretions of pyrite. (b) Any hard inclusions in a rock. (c) Quarrymen's term for dark gray to black masses, more or less oval or circular in cross section, which are segregations of biotite or hornblende in granite. English quarrymen call them heathen. (d) Diamonds with included small diamond crystals that cause trouble in cutting.
knotted schist *spotted slate.*
knotted slate *spotted slate.*
known mine Land cannot be held to be a "known mine" unless at the time the rights of the purchaser accrued, there was upon the ground an actual and open mine that either had been worked or was capable of being worked.
known to exist A vein or lode is known to exist when it could be discovered by anyone making a reasonable and fair inspection of the premises for the purpose of a location.
known vein A vein or lode is known to exist within the meaning of the mining act when it could be discovered by or is obvious to anyone making a reasonable and fair inspection of the premises for the purpose of making a location of a placer mining claim. This term is not to be taken as synonymous with located vein and refers to a vein or lode whose existence is known as distinguished from one that has been appropriated by location. Hence, a regular location is not necessary before a vein or lode can be a known vein or lode. The time at which a vein or lode must be known to exist to except it from a placer patent is the time at which the application for a patent is made and to contain minerals in such quantity and quality as to justify expenditure for the purpose of extracting them. (Ricketts, 1943)
knox hole A circular drill hole with two opposite vertical grooves that direct the explosive power of the blast. (Fay, 1920)
knox system A system of separating masses of rock by blasting with black blasting powder in reamed drill holes, a considerable air space being left between the charge and the stemming. (Fay, 1920)
knoxvillite *copiapite.*
knuckle The place on an incline where there is a sudden change in grade. The top of a grade or hill on a track over which mine cars are hauled. (Fay, 1920)
knuckle joint A mechanism consisting of some form of two forks coupled together by means of a cube or sphere provided with projecting pins extending through holes provided in the outer end of each branch of the fork. When inserted between the bit and the drill rods, the mechanism can be used to deflect a borehole. A similar mechanism often is used as a connection between two shafts on a machine when the ends of the shafts are placed at an angle to each other. (Long, 1960)
knuckle man In bituminous coal mining, a person who works at the knuckle (top) of a haulage slope, coupling trains of cars and attaching and detaching cars to and from a haulage cable by which they are raised or lowered. Also called knuckle boy. (DOT, 1949)
knurs and fundlers Eng. Two words always used together and applied to lumps of gypsum in marl, Nottinghamshire. (Arkell, 1953)
knurs and knots The stony nodules found lodged in the strata; commonly harder than the rest of the mass of the strata. (Arkell, 1953)
kobeite An amorphous mineral, $(Y,U)(Ti,Nb)_2(O,OH)_6$(?); forms black metamict crystals at Kobe, Japan.
kobellite An orthorhombic mineral, $Pb_{22}Cu_4(Bi,Sb)_{30}S_{69}$; forms a series with tintinaite.
kochenite A fossil resin, like amber. From Kochenthal, Tirol, Austria. (English, 1938)
Koch freezing process A freezing method of shaft sinking, in which a solution of magnesium chloride, $MgCl_2$, is circulated through the freezing tubes instead of brine. The magnesium chloride is cooled to about -30 °C at the refrigerating plant, and this temperature is capable of freezing the ground without the solution itself freezing and choking the tubes. Anhydrous ammonia, carbon dioxide, or sulphur dioxide is used as the refrigerating agent. (Nelson, 1965)
koechlinite An orthorhombic mineral, Bi_2MoO_6; forms minute tabular crystals; at Schneeberg, Saxony, Germany.
koehler lamp A naphtha-burning flame safety lamp for use in gaseous mines. (Fay, 1920)
Koepe hoist Winding system in which both cages in a mine shaft are connected to the same rope, via the drum. (Pryor, 1963)
Koepe sheave The wheel that is used in the Koepe winder instead of a winding drum. It usually consists of a cast-steel hub with steel arms and rim of welded construction. The diameter of the sheave varies from about 16 to 26 ft (4.9 to 7.9 m), depending on the size and type of winding rope and the total load. It is recommended that the ratio of the diameter of the sheave to the diameter of the rope should be 100:1 and that of the sheave to the largest wire in the rope, 2,000:1. A Koepe sheave is fitted with renewable friction linings. See also: *sheave; winding sheave.* (Nelson, 1965)
Koepe system (a) A system of hoisting without using drums, the rope being endless and passing over pulleys instead of around a drum. (Zern, 1928) (b) In this system, the hoisting drum is replaced by a large driving sheave 15 to 20 ft (4.6 to 6.1 m) in diameter. The angle of contact of the rope with the sheave is from 190° to 200° or a little over a half circle. Driving is done through the friction grip of the sheave on the rope; therefore, a tail rope is used around a sheave at the bottom of the shaft to give the necessary pull on the slack side. (Lewis, 1964)
Koepe winder A system where the winding drum is replaced by a large wheel or sheave. Both cages are connected to the same rope, which passes around some 200° of the sheave in a groove of friction material. The Koepe sheave may be mounted on the ground adjacent to the headgear or in a tower over the shaft. The drive to the rope is the frictional resistance between the rope and the sheave. It requires the use of a balance rope. It is often used for hoisting heavy loads from deep shafts and has the advantage that the large inertia of the ordinary winding drum is avoided. The system has been widely used in Europe for many years, and some large projects in Great Britain are being equipped with winders of this type. See also: *multirope friction winder.* (Nelson, 1965)
Koepe winder brake A brake which acts directly on the Koepe sheave and can be applied by the winding engine operator's brake lever and the other safety devices. (Nelson, 1965)
koettigite *köttigite.*
kohlenhobel *coal plow.*
kolbeckite A monoclinic mineral, $ScPO_4{\cdot}2H_2O$; blue to gray. Syn: *sterrettite.*
Kollen garnet Almandine from Kollen, Bohemia, Czech Republic. See also: *Bohemian garnet.*
kolm (a) A variety of cannel coal occurring locally as lenticles in Swedish alum shales and containing 30% of ash rich in rare metals, including uranium and radium. (Tomkeieff, 1954) (b) *culm.* (c) A variety of anthraxolite; a hydrocarbon that resembles oil shale. (Crosby, 1955) (d) A shale impregnated with asphaltite. (Hess)
kolovraite The mineral hydrous vanadate of Ni and Zn; in yellow botryoidal crusts; at Ferghana, Uzbekistan.
komatiite Magnesium-rich ultramafic volcanic rock of high temperature origin. The term was originally applied by Viljoen and Viljoen (1969) to basaltic and ultramafic lavas near the Komati river, Barberton Mountain Land, Transvaal, South Africa. Nickel-copper sulfide mineral deposits may be associated with komatiites.
kong Barren bedrock underlying tin-bearing gravel. (Pryor, 1963)
konimeter Apparatus used to measure dust in mine atmosphere. A measured volume of air is drawn through a jet so as to impinge on a glass surface coated with glycerin jelly. The adherent dust is then examined and the particles are counted under the microscope. Also called Zeiss konimeter. See also: *dust sampling.* (Pryor, 1963)
koninckite A tetragonal mineral, $FePO_4{\cdot}3H_2O$(?); occurs in yellow spheroidal aggregates; at Liége, Belgium.
Kootenai series Part of the Comanchean, of the continental facies, including coal seams, occurring in western Canada. (CTD, 1958)

Korfmann arch saver A machine for withdrawing steel arches by means of a controlled hydraulic system instead of drawing them by hand or winches. (Nelson, 1965)

Korfmann power loader A double-ended cutter loader; i.e., it can cut and load in both directions. It consists of four milling heads and one cutter chain surrounding them; guided by armored conveyor; rope-hauled; cuts at 2.3 ft/s (0.7 m/s) and flits at 10 ft/s (3.0 m/s). The minimum workable seam thickness is 26 in (66 cm) on gradients from 0° to 12°; maximum length of face is 165 yd (151 m); takes 3-1/2 ft (1.07 m) per cut, continuous mining. (Nelson, 1965)

kornelite A monoclinic mineral, $Fe_2(SO_4)_3{\cdot}7H_2O$; forms rose to violet prisms; an alteration of pyrite.

kornerupine An orthorhombic mineral, $Mg_4(Al,Fe)_6(Si,B)_4O_{21}(OH)$; forms columnar crystals with prismatic cleavage; in high-grade metamorphic aluminous rocks.

korzhinskite A mineral, $CaB_2O_4{\cdot}H_2O$; in skarns with other calcium borate minerals.

köttigite A monoclinic mineral, $Zn_3(AsO_4)_2{\cdot}8H_2O$; vivianite group; forms a series with parasymplesite; carmine-red with one perfect cleavage; occurs with smaltite at Schneeberg, Saxony, Germany. Also spelled koettigite.

kotulskite A hexagonal mineral, Pd(Te,Bi).

Kotze konimeter A dust-sampling instrument consisting of an air cylinder and piston actuated by a spring, and so arranged that on release of the piston a known volume of air is impinged at high velocity against a plate coated with vaseline. The dust spot so produced is examined under the microscope, and a count is made of the number of particles. (Greenburg, 1922)

Kourbatoff's reagent An etchant for steel; it consists of seven parts of a solution containing 20% methyl alcohol, 20% ethyl alcohol, 20% iso-amyl alcohol, and 10% butyl alcohol added to three parts of a solution of 4% nitric acid in acetic anhydride. This etchant colors sorbite and troostite, leaving the other constituents unaffected. (Osborne, 1956)

koutekite A hexagonal mineral, Cu_5As_2.

kramerite *probertite.*

krantzite A fossil resin resembling amber and found in small yellowish grains disseminated in brown coal. (Tomkeieff, 1954)

kratochvilite An orthorhombic mineral, $C_{13}H_{10}$.

kraurite *dufrenite.*

krausite A monoclinic mineral, $KFe(SO_4)_2{\cdot}H_2O$; forms soft yellowish-green prismatic crystals and crusts.

kremersite An orthorhombic mineral, $(NH_4,K)_2FeCl_5{\cdot}H_2O$; forms red pseudo-octahedra around fumaroles in Sicily, Italy.

krennerite One of the gold telluride group of minerals, $(Au,Ag)Te_2$; corresponds to the same general formula as sylvanite and calaverite. Silver-white to pale yellow color; sp gr, 8.35. Found in Colorado and Romania. Syn: *white tellurium.* (CCD, 1961)

kriging (a) A weighted, moving-average interpolation method in which the set of weights assigned to samples minimizes the estimation variance, which is computed as a function of the variogram model and locations of the samples relative to each other, and to the point or block being estimated. (b) In the estimation of ore reserves by geostatistical methods, the use of a weighted, moving-average approach both to account for the estimated values of spatially distributed variables, and also to assess the probable error associated with the estimates. (SME, 1992)

kriging standard deviation The standard error of estimation computed for a kriged estimate. By definition, kriging is the weighted linear estimate with the particular set of weights that minimizes the computed estimation variance (standard error squared). The relationship of the kriging standard deviation to the actual error of estimation is very dependent on the variogram model used and the validity of the underlying assumptions; therefore, kriging standard deviations should be interpreted with caution.

Krohnke process The treatment of silver ores preparatory to amalgamation, by humid chloridization with copper dichloride. (Fay, 1920)

kröhnkite A monoclinic mineral, $Na_2Cu(SO_4)_2{\cdot}2H_2O$; azure; in Chilean copper deposits. Also spelled kroehnkite.

Kroll process Process in which purified titanium tetrachloride is reduced to the metallic state with magnesium in an inert atmosphere of helium or argon.

Krupp process (a) *Krupp washing process.* Also called Bell-Krupp process. (Fay, 1920) (b) A cementation process designed for the hardening of surface steel, as for armor plates, where the object is to strengthen the outer portion of the mass from the surface toward the interior. (Standard, 1964)

Krupp-Renn The process for the production of iron and steel from medium-grade ores, such as those containing 44% to 57% iron and having a high silicon content. The process involves a continuous reduction and is carried out in a revolving tube furnace, which is designed for the production of iron. The iron is reduced into a sponge and then converted into low-carbon metallic grains, which are called pellets.

Krupp washing process The removal of silicon and phosphorus from molten pig iron by running it into a Pernot furnace lined with iron oxides. Iron ore may also be added, and the bath is agitated by rotation for 5 to 8 min only. *Krupp process.* See also: *Bell's dephosphorizing process.* (Fay, 1920)

kryptomere *aphanite.* Also spelled cryptomere.

kryzhanovskite An orthorhombic mineral, $MnFe_2(PO_4)_2(OH)_2{\cdot}H_2O$; an oxidation product of phosphoferrite; forms a series with garyansellite.

K-spar *potassium feldspar.*

K.T.A.M. auger tube *K.T.A.M. double-tube auger.*

K.T.A.M. double-tube auger A double-tube soil-sampling device designed to be rotated by hand to obtain soil samples from relatively shallow depths. The inner tube is a swivel type, and its cutting end leads the bottom of the spiral on the outside of the outer tube. (Long, 1960)

kua Specially shaped hoe used for working gravel in the sluice in Japan. (Fay, 1920)

kulm *culm.*

kundaite A variety of grahamite from Estonia. Distinguished by the brown color of its powder and its greater solubility in turpentine and chloroform. (English, 1938; Tomkeieff, 1954)

kunkur *kankar.*

kunzite A gem variety of spodumene; lilac, transparent. Syn: *lithia amethyst.*

kupferschiefer A dark-colored shale once worked for copper in Germany.

kupletskite A triclinic mineral, $(K,Na)_3(Mn,Fe)_7(Ti,Nb)_2Si_8O_{24}(O,OH)_7$; astrophyllite group; forms series with astrophyllite and with cesium kupletskite.

kuprojarosit *cuprojarosite.*

kurnakovite A triclinic mineral, $MbB_3O_3(OH)_5{\cdot}5H_2O$; dimorphous with inderite.

kuroko In Japan, black ore. The kuroko deposits consist of intimately mixed zincblende, galena, and barite, associated (in places) with large masses of pyrite and gypsum. (Hess)

kurtosis In statistics, the technical name describing peakedness of a frequency distribution. It is statistically measured as the coefficient of quartile kurtosis.

kutnohorite A trigonal mineral, $2[Ca(Mn,Mg,Ca,Fe)(CO_3)_2]$; dolomite group; in carbonate veins. Also spelled kutnahorite.

kyack A packsack to be swung on either side of a packsaddle. (Webster 3rd, 1966)

kyanite A triclinic mineral, $4[Al_2SiO_5]$; trimorphous with andalusite and sillimanite; in blades that are distinctly harder across than along; a common rock-forming mineral in schist and gneiss. Also spelled cyanite. Syn: *disthene; sappare.*

kyanophyllite A mixture of paragonite plus muscovite.

kyrosite A variety of marcasite. Syn: *white copper ore.* (Dana, 1914)

L

labeled atom Atom rendered radioactive and thus traceable through a chemical process or a flow line. Also called tagged atom. (Pryor, 1963)
labile Said of rocks and minerals that are mechanically or chemically unstable; easily decomposed. Cf: *unstable*. (AGI, 1987)
labile protobitumens Easily decomposable plant and animal products such as fats and oils or proteins that are found in peat and sapropels. (Tomkeieff, 1954)
lability The property of bituminous emulsions, relating to the ease with which they break when put into use. Labile emulsions are quick breaking. (Nelson, 1965)
labor (a) A shaft, cavity, or other part of a mine from which ore is being or has been extracted; a working, as a labor in a quicksilver mine. See also: *working*. (Standard, 1964) (b) The annual assessment work required on claims calls for $100.00 of labor and improvements. In Australia, claims have continuous labor or are manned throughout the year. (von Bernewitz, 1931) (c) A Spanish term used in early land surveys in Texas for unit of area equal to about 177.14 acres (representing a tract 100 varas square). Pronounced la-bore. (AGI, 1987)
laboratory In the iron and steel industry, the space between the fire and flue bridges of a reverberatory furnace in which the work is performed. Also called "kitchen and hearth.". (Fay, 1920)
La Bour centrifugal pump A self-priming centrifugal pump containing a trap, which always ensures sufficient water for the pump to function, and also a separator to remove the entrained air in the water. (Lewis, 1964)
labradorescence Optical diffraction of monochromatic light, commonly blue, caused by exsolution lamellae in some labradorite samples; appears to come from within the sample. Cf: *labrador moonstone*.
labrador hornblende *orthopyroxene.*
labradorite A triclinic mineral, $(Ca,Na)[(Al,Si)AlSi_2O_8]$; CaAl 70 to 50 mol % and NaSi 30 to 50 mol %; plagioclase series of the feldspar group; a common rock-forming mineral in basalt, gabbro, and anorthosite; also in hornfels and siliceous marble.
Labrador moonstone A variety of labradorite displaying an internal play of colors. *labradorescence.*
Labrador rock *labradorite.*
labuntsovite A monoclinic mineral, $(K,Ba,Na)(Ti,Nb)(Si,Al)_2(O,OH)_7 \cdot H_2O$. Also spelled labuntzovite; labountsovite.
labyrinth (a) A series of canals through which a stream of water is directed for sorting crushed ore according to its specific gravity. (Fay, 1920) (b) A pipe or chamber of many turnings, for condensing metal vapors or fumes. (Standard, 1964)
laccolite *laccolith.*
laccolith A concordant igneous intrusion with a known or assumed flat floor and a postulated, dikelike feeder commonly thought to be beneath its thickest point. It is generally plano-convex in form and roughly circular in plan, less than 5 miles (8 km) in diameter, and from a few feet to several hundred feet in thickness. See also: *phacolith*. Syn: *laccolite*. (AGI, 1987)
lace Eng. Line cut, with the point of a pick, on slickensides.
lacing (a) The timber or other material placed behind and around the main supports. See also: *lagging; lofting*. (Nelson, 1965) (b) Strips or light bars of wrought iron bent over at the ends and wedged between the bars and the roof. (Fay, 1920) (c) Small boards or patches that prevent dirt from entering an excavation through spaces between sheeting or lagging planks. (Nichols, 1976) (d) Bars placed diagonally to space and stiffen members, as in a built-up column. (Crispin, 1964) (e) Eng. Wood placed inside the sets of timber as a tie from prop to prop. Also called stringing piece. See also: *bracing*. (SMRB, 1930) (f) N. Staff. Timbers placed across the tops of bars or caps to secure the roof between the timbers. (Fay, 1920)
lack clay Ire. A thin pan under the moors. (Arkell, 1953)
LaCoste-Romberg gravimeter A long-period vertical seismograph suspended system adapted to the measurement of gravity differences. Sensitivity is achieved by adjusting the system to proximity to an instability configuration. (AGI, 1987)
lacroixite A monoclinic mineral, $NaAl(PO_4)F$; isostructural with durangite; in cavities in granite pegmatites.
lactic acid In flotation, a depressant sometimes used to depress iron minerals. (Pryor, 1963)
lacustrine Pertaining to, formed in, growing in, or inhabiting lakes.
ladder (a) The arm that carries the tumblers and bucket line of a dredge. (Fay, 1920) (b) The continuous line of mud buckets, carried on an oblique endless chain, in a bucket ladder excavator or dredger. (CTD, 1958) (c) The digging boom assembly in a hydraulic dredge or chain-and-buckets ditcher. (Nichols, 1976)
ladder-bucket dredge *bucket-ladder dredge.*
ladder ditcher An excavator that digs ditches by means of a chain of traveling buckets supported by a boom.
ladder drilling Arrangement used in large-scale rock tunneling. Retractable drills with pneumatic power legs are mounted on prefabricated steel ladders in tiers, connected into a holding frame or jumbo. As many as 22 drills can be worked simultaneously by a small labor force. (Pryor, 1963)
ladder lode *ladder vein.*
ladder sollar A platform at the bottom of each ladder in a series. (Standard, 1964)
ladder vein One of a series of mineral deposits in transverse, roughly parallel fractures that have formed along foliation planes perpendicular to the walls of a dike during its cooling or along shrinkage joints in basaltic rocks or dikes. Syn: *ladder lode*. (AGI, 1987)
ladderway (a) *manway*. (b) Mine shaft, raise, or winze between two main levels, equipped with ladders. (Pryor, 1963) (c) The particular shaft or compartment of a shaft containing ladders. Also called ladder road. (Fay, 1920)
lade (a) Scot. A load. (b) A watercourse, ditch, or drain. (c) The mouth of a river.
laded metal Molten glass dipped from a melting pot to a casting table. Also called gathered metal. (Standard, 1964)
lading hole In glassmaking, an orifice through which melted glass is ladled or taken out by a cuvette. (Standard, 1964)
ladle (a) In a smelter or foundry, a steel-holding vessel used in the transfer and transport of molten metal, matte, or slag. (Pryor, 1963) (b) A long-handled, cup-shaped tool for ladling glass out or from one spot to another. Also used for filling open pots. (ASTM, 1994)
ladle addition In foundry work, addition of special metals (e.g., granulated nickel) or compounds (e.g., ferrosilicon) to molten iron in ladle to produce special qualities in castings. (Pryor, 1963)
ladle craneman In the iron and steel industry, a person who places ladles under the tapholes of furnaces and holds them in position while the metal is poured into them. Also called charging floor crane operator; ladle crane operator; steel charger. (DOT, 1949)
ladle filler In metallurgy, a person who transfers molten metal from the furnace into a ladle and skims and fluxes metal preparatory to casting. (DOT, 1949)
ladle furnace A small furnace for calcining or melting substances in a ladle. (Standard, 1964)
ladle liner In ore dressing, smelting, and refining, a person who repairs and relines pouring ladles used to transport molten metals, such as iron, steel, and copper. Also called ladle cleaner; ladle dauber; ladle houseman; ladle mender; ladle patcher; ladle repairman. (DOT, 1949)
ladle lip A concave projection at the upper edge of a ladle to guide the metal in pouring. (Mersereau, 1947)
ladler Worker who pours molten glass from a suspended ladle on casting table for rolling into sheet glass. (DOT, 1949)
Lafond's tables A set of tables and associated information for correcting reversing thermometers and computing dynamic height anomalies; compiled by E.C. Lafond and published by the U.S. Navy Hydrographic Office as H.O. Pub. No. 617. (Hunt, 1965)
lag (a) To place planks, slabs, or small timbers over horizontal, or behind upright, members of mine timber sets to form a ceiling or wall. (Long, 1960) (b) In flotation, any retardation of an output with respect to the causal input. (Fuerstenau, 1962) (c) A flattish piece of wood or other material to wedge the timber or steel supports against the ground and to secure the area between the supports. See also: *lid; wedge*. (Nelson, 1965) (d) To protect a shaft or level from falling rock by lining it with timber (lagging). (CTD, 1958) (e) Time required for circulation liquid to travel downward from the drill pump through the drill string

to the bit or upward from the bottom of the borehole to the collar. (Long, 1960) (f) The lapse of time between the occurrence of an event or condition and its detection on a recording device. (Long, 1960) (g) To provide or cover with lags; as, to lag a boiler with a nonconductor; to lag timbers in a mine. See also: *lags; lagging.* (Standard, 1964; Fay, 1920) (h) A distance class interval used for variogram computation.

lag deposit *lag gravel.*

lagergestein Ger. The enclosed pieces of older rocks in a sedimentary ore deposit. (Schieferdecker, 1959)

lag fault An overthrust, the thrusted rocks of which move differentially so that the upper part of the geologic section is left behind; the replacement of the upper limb of an overturned anticline by a fold fault. Syn: *tectonic gap.* (AGI, 1987)

lagged liner A metal plate with raised areas, to be inserted in the bottom of shaker conveyor troughs and held in place by spot welding. The raised areas assist coal travel on steep grades or under wet conditions. (Jones, 1949)

lagging (a) Lagging wedges and secures the roof and sides behind the main timber or steel supports in a mine and provides early resistance to pressure. If concrete slabs are used, they are made in lengths to fit between the arch webs. The lagging behind steel arches in tunnels may be pyrolith-treated, fire-resisting boards. Also called lacing. (Nelson, 1965) (b) Pieces of timber about 4 ft 6 in by 6 in by 2 in (1.4 m by 15 cm by 5 cm) with one end sharpened or beveled to give the lath an upward trend when being driven into the roof gravels. A number of laths driven into the roof form a protective shield for the miners working in the face. Sometimes called laths. (Eng. Min. J., 1938) (c) In shafts, planks, usually 2 in, placed on the outside of sets. Coeur D'Alene lagging has 2-in by 2-in cleats nailed to the top and bottom of the wall and end plates about 2 in back from the outer edge. The lagging is then cut to fit between the plates and is placed against the cleats and flush with the plates on the outside. (Lewis, 1964) (d) Narrow boards, generally planed, placed horizontally on the arch frames of a center. On this lagging the arch of masonry is built. The term is also applied to poling boards. (Stauffer, 1906) (e) Planks, slabs, or small timbers placed over the caps or behind the posts of the timbering, not to carry the main weight, but to form a ceiling or a wall, preventing fragments or rock from falling through. See also: *lag.* (Fay, 1920) (f) Heavy planks or timbers used to support the roof of a mine, or for floors of working places, and for the accumulation of rock and earth in a stope. (Fay, 1920) (g) Long pieces of timbers closely fitted together and fastened to the drum rings to form a surface for the rope to wind on. (Fay, 1920) (h) The narrow strips supporting an arch of masonry while in construction. (Standard, 1964) (i) The surface or contact area of a drum or flat pulley, esp. a detachable surface or one of special composition. (Nichols, 1976) (j) Boards fastened to the back of a shovel for blast protection. (Nichols, 1976) (k) Covering on boilers, tanks, and pipes used to provide thermal insulation. (Pryor, 1963) (l) Material applied to pulleys to increase traction between the pulley and belt and to decrease wear on both. See also: *backing deals.* (m) Verb. To install lagging. (AGI, 1987)

lagging bar *roof stringer.*

lag gravel (a) A residual accumulation of rock fragments remaining on a surface after the finer material has been blown away by winds. See also: *desert pavement; pebble armor.* (AGI, 1987) (b) Coarse-grained material that is rolled or dragged along the bottom of a stream at a slower rate than the finer material, or that is left behind after currents have winnowed or washed away the finer material. Syn: *lag; lag deposit.* (AGI, 1987)

lagre Fr. In sheet glass manufacture, a sheet of perfectly smooth glass, interposed between the flattening stone and the cylinder that is to be flattened. (Standard, 1964)

lags Eng. Long pieces of timber closely fitted together and fastened to oak curbs or rings forming part of a drum used in sinking through quicksand or soft ground. Cf: *lag.*

lag screw (a) A square-headed, heavy wood screw. It must be tightened down with a wrench because its head is not slotted. (Crispin, 1964) (b) A flat-headed machine screw by which to fasten wood lagging, as on a curve surface. (Standard, 1964)

lag time The total time between the initial application of current and the rupture of the circuit within the detonator. (BS, 1964)

lahar *mudflow.*

laihunite Black monoclinic mineral, $Fe^{2+}Fe_2^{3+}(SiO_4)_2$.

lair Clay; mud; mire. (Arkell, 1953)

laitakarite A trigonal mineral, $Bi_4(Se,S)_3$. Syn: *selenjoseite.*

lake-bed placer In Alaska, a placer in the bed of a present or ancient lake; generally formed by landslides or glacial damming.

Lake copper Copper produced from the Lake Superior ores in which the metal occurs native and is of high purity. (CTD, 1958)

Lake George diamond *Herkimer diamond.*

lake ore (a) *bog iron; bog iron ore.* (b) A disklike or irregular concretionary mass of ferric oxyhydroxide less than 1 m thick; or a layer of porous, yellow bedded limonite formed along the borders of certain lakes. See also: *bog ore.*

Lake Superior agate (a) Any agate from the Lake Superior region. (b) Thompsonite marked or banded like agate from the Lake Superior region.

Lake Superior greenstone *pumpellyite.*

lambda plate *mica plate.*

Lambert's Law *translucency.*

lamella A thin scale, leaf, lamina, or layer; e.g., one of the units of a polysynthetically twinned mineral, such as plagioclase. Plural, lamellae. (AGI, 1987)

lamellar Composed of or arranged in lamellae; disposed in layers like the leaves of a book. Syn: *lamellate.* (AGI, 1987)

lamellar flow Flow of a liquid in which layers glide over one another. Cf: *laminar flow.* (AGI, 1987)

lamella roof A vault or large span built up with short structural members of timber or pressed steel, joined together in a diamond pattern by bolting or other suitable connections. This system, which is a type of stressed-skin construction, was patented in 1925 by a German engineer. (Hammond, 1965)

lamellar pyrite *marcasite.*

lamellar stellate In mineralogy, having or consisting of lamellae arranged in groups resembling stars. (Standard, 1964)

lamellar twinning Multiple parallel twins; e.g., albite twinning in plagioclase. Syn: *polysynthetic twinning.*

lamellate *lamellar.*

lame-skirting Widening a passage by cutting coal from the side of it. Also called skipping; slicing. (Fay, 1920)

lamina The thinnest recognizable layer in a sedimentary rock. Plural, laminae. Cf: *stratum.*

laminar flow Water flow in which the stream lines remain distinct and the flow direction at every point remains unchanged with time. It is characteristic of the movement of ground water. Cf: *turbulent flow; lamellar flow.* Syn: *streamline flow; sheet flow.* (AGI, 1987)

laminar velocity That velocity below which, in a particular conduit, laminar flow will always exist, and above which the flow may be either laminar or turbulent, depending on circumstances. Also known as lower critical velocity. (Seelye, 1951)

laminated In very thin parallel layers.

laminated iron Iron in the form of thin sheets; used as cores of transformers, etc. The losses due to eddy currents with laminated iron cores are lower when compared with solid cores. (Nelson, 1965)

laminated quartz Vein quartz containing slabs, blades, or laminar films of other material. (AGI, 1987)

laminating machine A set of rolls or any apparatus for making thin plates of metal, such as gold, preliminary to beating. (Standard, 1964)

laminating roller The adjustable roller in a rolling mill whereby the thickness of rolled metal sheets is regulated. (Standard, 1964)

lamination (a) The formation of a lamina or laminae. (AGI, 1987) (b) The state of being laminated; specif. the finest stratification of bedding, typically exhibited by shales and fine-grained sandstones. (AGI, 1987) (c) A laminated structure. (AGI, 1987)

laminations Of rocks, bedding in layers less than 1 cm thick; formation with thin layers that vary in grain or composition. (Pryor, 1963)

lamings (a) Partings in coal seams. (Arkell, 1953) (b) N. of Eng. A collier's term for accidents of almost every description to people working in and about the mines. A variation of lame, to cripple or disable. (Fay, 1920)

lamp (a) *safety lamp.* (b) A small handheld electrical device that produces an intense ultraviolet radiation, called "black light." Cf: *black light; fluorescent lamp.* (Long, 1960) (c) An electrical lamplike device producing intense ultraviolet radiations for visually examining drill cores or rock specimens for the presence and/or abundance of fluorescent minerals. (Long, 1960)

lampadite Wad containing 4% to 18% copper oxide and commonly cobalt. Syn: *cupreous manganese.*

lamp cabin (a) A place above ground where the safety and cap lamps are maintained, before being handed to the workers. (b) *lamp room.*

lamp cleaner *lampman.*

lamp cup A means for supporting a flame safety lamp on a tripod to provide a sight for surveying. (BS, 1963)

lamp house *lamp cabin.*

lamping In prospecting, use of a portable ultraviolet lamp to reveal fluorescent minerals. (Pryor, 1963)

lamp keeper *lampman.*

lampman (a) The person in charge of the lamp room at a mine responsible for the maintenance of the safety lamps. (Nelson, 1965) (b) In mining, one who cleans, tests, and repairs lamps used underground by miners. Also called battery charger; lamp cleaner; lamp-house man; lamp keeper; lamp repairer; safety-lamp keeper. (DOT, 1949)

lamp rack A rack upon which electric cap lamp batteries are placed to be charged.

lamp repairer *lampman.*

lamprobolite *oxyhornblende.*
lamproite A group name for dark-colored hypabyssal or extrusive rocks rich in potassium and magnesium; also, any rock in that group, such as madupite, orendite, fitzroyite, verite, cedricite, or wyomingite. (AGI, 1987)
lamp room A room or building at the surface of a mine provided for charging, servicing, and issuing all cap, hand, and flame safety lamps held at the mine. See also: *self-service system.* (BS, 1965)
lamprophyllite A monoclinic mineral, $Na_2(Sr,Ba)_2Ti_3(SiO_4)_4(OH,F)_2$. Syn: *molengraafite.*
lamprophyre A group of dark-colored, porphyritic, hypabyssal igneous rocks characterized by panidiomorphic texture, a high percentage of mafic minerals (esp. biotite, hornblende, and pyroxene), which form the phenocrysts, and a fine-grained groundmass with the same mafic minerals in addition to feldspars and/or feldspathoids; also, any rock in that group. Most lamprophyres are highly altered. They are commonly associated with carbonatites. Cf: *leucophyre.* Adj: lamprophyric. (AGI, 1987)
lamprophyric Said of the holocrystalline-porphyritic texture exhibited by lamprophyres, in which phenocrysts of mafic minerals are contained in a fine-grained crystalline groundmass. (AGI, 1987)
lamproschist Metamorphosed lamprophyre with a schistose structure containing brown biotite and green hornblende. (AGI, 1987)
lamp station (a) Fixed places in the intake airway of a coal mine where the miners' safety lamps are externally examined by a deputy before the workers proceed to their working places. In a safety lamp mine, the lamp station is the only place where flame safety lamps may be opened and relighted. (Nelson, 1965) (b) Locations in gaseous mines where safety lamps are opened, cleaned, and refilled or charged by a qualified attendant. (Hudson, 1932) (c) A place underground, appointed for the examination, by an official, of safety lamps in use. (BS, 1965) (d) A lamp room. (Fay, 1920)
Lanarkian A subdivision of the Coal Measures—based mainly on plant fossils. It represents in part, the millstone grit of South Wales; well developed in South Scotland, where coal seams are present. (Nelson, 1965)
lanarkite A monoclinic mineral, $Pb_2(SO_4)O$; perfect basal cleavage; named for Lanarkshire, Scotland.
Lancashire bord-and-pillar system *bord-and-pillar method.*
Lancashire method A method of working moderately inclined coal seams. The first stage consists of splitting a panel of coal into pillars and as a second stage the pillars are extracted on the retreat by a longwall face. (Nelson, 1965)
lanchut A short sluice used for cleaning tin concentrate. (Hess)
land (a) To set or allow the bottom end of a drivepipe or casing to rest at a preselected horizon in a borehole. (Long, 1960) (b) For reamers, drills, and taps, the solid section between the flutes. (ASM, 1961)
land accretion Reclamation of land from the sea or other low-lying or flooded areas by draining and pumping, dumping of fill, or planting of marine vegetation. (AGI, 1987)
land chain A surveyor's chain of 100 links.
land compass A surveyor's compass.
lander (a) A worker stationed at one of the levels of a mine shaft to unload rock from the bucket or cage and load drilling and blasting supplies to be lowered to the crew. (Webster 3rd, 1966) (b) In the quarry industry, one who supervises and assists in guiding, steadying, and loading, on trucks or railroad cars, blocks of stone hoisted from the quarry floor. Syn: *top hooker.* (DOT, 1949) (c) In metal mining, a laborer who (1) cleans skips by directing a blast of compressed air into them through a hose; (2) records number of loaded skips hoisted to surface; and (3) loads railroad cars with ore from bins by raising and lowering chute doors. (DOT, 1949) (d) In anthracite coal mining, bituminous coal mining, or metal mining, one who works with shaft sinking crew at top of shaft or at a level immediately above shaft bottom, dumping rock into mine cars from a bucket in which it is raised. Also called bucket dumper; landing tender; top lander. (DOT, 1949) (e) Eng. The person who receives the loaded bucket or tub at the mouth of the shaft. Also called banksman. (Fay, 1920)
landerite A pink or rose variety of grossular in dodecahedra. Syn: *rosolite; xalostocite.*
landesite An orthorhombic mineral, $(Mn,Mg)_9Fe_3(PO_4)_8(OH)_3{\cdot}9H_2O$.
landform Any physical, recognizable form or feature of the Earth's surface, having a characteristic shape, and produced by natural causes; it includes major forms such as plain, plateau, and mountain, and minor forms such as hill, valley, slope, esker, and dune. Taken together, landforms make up the surface configuration of the Earth. Syn: *relief feature.* (AGI, 1987)
landform system A group of related natural features, objects, or forces; e.g., a drainage system or a mountain system. See also: *mountain system.*
landing (a) The top or bottom of a slope, shaft, or inclined plane. (Fay, 1920) (b) The mouth of a shaft where the cages are unloaded; any point in the shaft at which the cage can be loaded with people or materials. (Nelson, 1965) (c) The brow or level section at the top of an inclined haulage plane where the loaded tubs are exchanged for empty tubs, or vice versa. See also: *apex.* (Nelson, 1965) (d) A preselected and prepared horizon in a borehole on or at which the bottom end of a drivepipe or casing string is to be set. (Long, 1960) (e) A platform from which to charge a furnace. (Standard, 1964) (f) Level stage in a shaft at which cages are loaded and discharged. (Pryor, 1963)
landing box Scot. The box into which a pump delivers water. (Fay, 1920)
landing shaft S. Wales. A shaft through which coal is raised. (Fay, 1920)
landing tender *lander.*
land pebble *land-pebble phosphate.*
land-pebble phosphate A term used in Florida for a pebble phosphate occurring as pellets, pebbles, and nodules in gravelly beds a few feet below the ground surface. It is extensively mined. Syn: *land pebble; land rock; matrix.* (AGI, 1987)
land plaster Finely ground gypsum used as a fertilizer.
land rock A syn. used in South Carolina for land-pebble phosphate. (AGI, 1987)
Landsat Multispectral data from satellite remote sensing imagery that provides landscape patterns reflecting geologic structures, types of rocks, and vegetation. (SME, 1992)
landscape agate *moss agate.*
landscape marble A close-grained limestone characterized by dark conspicuous dendritic markings that suggest natural scenery (woodlands, forests); e.g., the argillaceous limestone in the Cotham Marble near Bristol, England. Syn: *forest marble.* (AGI, 1987)
landslide A general term covering a wide variety of mass-movement landforms and processes involving the downslope transport, under gravitational influence, of soil and rock material en masse. Usually the displaced material moves over a relatively confined zone or surface of shear. The wide range of sites and structures, and of material properties affecting resistance to shear, result in a great range of landslide morphology, rates, patterns of movement, and scale. Landsliding is usually preceded, accompanied, and followed by perceptible creep along the surface of sliding and/or within the slide mass. Terminology designating landslide types generally refers to the landform as well as the process responsible for it; e.g., rockfall, translational slide, block glide, avalanche, mudflow, liquefaction slide, and slump. Syn: *landsliding; slide; landslip.* (AGI, 1987)
landsliding *landslide.*
landslip A British syn. of landslide. (AGI, 1987)
land subsidence *subsidence.*
lands valuable for minerals As used in the mining law, applies to all lands chiefly valuable for nonmetalliferous deposits, such as alum, asphaltum, borax, guano, diamonds, gypsum, marble, mica, slate, amber petroleum, limestone, and building stone, rather than for agricultural purposes. Such lands are subject to disposition by the United States under the mining laws only. (Ricketts, 1943)
land weight Lanc. The pressure exerted by the subsidence of the cover or overburden.
Lane mill A slow-speed roller mill of the Chilean type. A horizontal spider carrying six rollers revolves slowly in a pan 10 ft (3 m) or more in diameter, making about 8 rpm. (Liddell, 1918)
langbanite A trigonal mineral, $(Mn,Ca)_4(Mn,Fe)_9SbSi_2O_{24}$; in skarns. Also spelled långbanite. (Not langbeinite.)
langbeinite An isometric mineral, $K_2Mg_2(SO_4)_3$; associated with halite and sylvite in marine evaporite deposits; a source of potash. (Not langbanite.)
langite A monoclinic mineral, $Cu_4(SO_4)(OH)_6{\cdot}2H_2O$; in blue-green concretionary crusts; in Cornwall, U.K., and Stredoslovensky, Slovakia.
lang lay rope A rope in which the wires are twisted in the same direction as the strands and the wires are thus exposed to wear for a much greater length than in round rope. The smoother lang lay resists wear to better advantage and is frequently preferred for haulage ropes. Syn: *universal lay rope.* (Lewis, 1964; Sinclair, 1959)
Langmuir's adsorption isotherm The equation for calculating a gas monolayer on a flat surface. (Pryor, 1963)
Langmuir trough Rectangular tank used to measure the surface tension of a monolayer adsorbed at the surface of a liquid. (Pryor, 1963)
Lang's lay rope *winding rope.*
lansfordite A monoclinic mineral, $MgCO_3{\cdot}5H_2O$; forms small stalactites that alter to nesquehonite in mines near Lansford, PA.
lantern Enclosed light (candle or oil) carried by a mine worker. (Hess)
lanthanite An orthorhombic mineral, $(Ce,La,Nd)_2(CO_3)_3{\cdot}8H_2)$; further speciated according to its predominant rare-earth element; in pegmatites and carbonate-rich sediments.
lanyon shield An iron curtain, stiffened by ribs of angle iron, suspended from trolley wheels running on a rail parallel with and in front of a furnace to protect the worker from the furnace heat. (Fay, 1920)
lap (a) One coil of rope on the winding drum of the mine hoist. (Pryor, 1963) (b) Polishing cloth used in preparing polished mineral

specimens by abrasive grinding. (Pryor, 1963) (c) A surface defect, appearing as a seam, caused by folding over hot metal, fins, or sharp corners and then rolling or forging them into the surface, but not welding them. (ASM, 1961) (d) To dimension, smooth, or polish (as a metal surface or body) to a high degree of refinement or accuracy. (Webster 3rd, 1966) (e) An imperfection; a fold in the surface of a glass article caused by incorrect flow during forming. (ASTM, 1994) (f) A tool used for polishing glass. (ASTM, 1994) (g) A rotating disk of soft metal or wood, used to hold polishing powder for cutting or polishing gems or metal.

laper Impure sandy green limestone with shaly partings in the Middle Purbeck beds, Swanage, U.K. Also spelled leaper; leper.

lapidary (a) An artificer who cuts, polishes, or engraves gems or precious stones. (b) Person who is skilled in the nature and kind of gems or precious stones; a connoisseur of lapidary work.

lapilli Pyroclastics that may be either essential, accessory, or accidental in origin, of a size range that has been variously defined within the limits of 2 mm and 64 mm. The fragments may be either solidified or still viscous when they land (though some classifications restrict the term to the former); thus there is no characteristic shape. An individual fragment is called a lapillus. Cf: *cinder.* (AGI, 1987)

lapilliform Having the form of small stones.

lapis lazuli (a) A lazurite-bearing rock; contains lazurite or hauyne (possibly zeolitized), diopside, edenitic amphibole (koksharovite), muscovite, calcite, and pyrite; occurs in various shades of blue; possibly the original sapphire of the ancients; Syn: *azure; lazuli.* (b) Gem-quality lazurite. (Not lazulite.) (c) An ultramarine-colored serpentine in India. See also: *ultramarine.*

lapis matrix Lapis lazuli (lazurite) containing prominent patches of calcite. See also: *Chilean lapis; lazurite.*

LaPointe picker Miniature belt conveyor, on which small ore particles move singly past a Geiger-Müller tube that is set to operate a sorting device. This removes from the passing stream each particle of radioactive ore that reaches the required intensity, therefore sorting out the valuable material. (Pryor, 1963)

lapped (a) Overlapped and fitted together. (Nichols, 1976) (b) The act of polishing or grinding on a lap.

Laramide orogeny A time of deformation, typically recorded in the eastern Rocky Mountains of the United States, whose several phases extended from late Cretaceous until the end of the Paleocene. It is named for the Laramie Formation of Wyoming and Colorado, probably a synorogenic deposit. (AGI, 1987)

Laramide revolution *Laramide orogeny.*

lardalite *laurdalite.*

larderellite A monoclinic mineral, $(NH_4)B_5O_6(OH)_4$.

lardite (a) White hydrated silica, probably a variety of opal; occurring in clay in central Russia. (b) Massive talc. Syn: *steatite; agalmatolite.* (c) A massive variety of muscovite and/or pyrophyllite. See also: *pagodite.*

lard oil An oil produced from animal fats. This oil is an efficient lubricant for use on metal-cutting tools. (Crispin, 1964)

lardstone Massive talc; steatite.

lard stone A kind of soft stone found in China. See also: *steatite; agalmatolite.* (Fay, 1920)

large Eng. The largest lumps of coal sent to the surface, or all coal that is handpicked or does not pass over screens; also the largest coal that passes over screens. (Fay, 1920)

large coal (a) One of the three main size groups by which coal is sold by the National Coal Board of Great Britain. Large coal has no upper size limit and has a lower size limit of 1-1/2 to 2 in (3.8 to 5.1 cm) and embraces large screened coal, cobbles, and treble sizes. See also: *graded coal; smalls.* (Nelson, 1965) (b) Coal above an agreed size without any upper size limit. Also called lump coal. (BS, 1962)

large colliery Gr. Brit. In general, a colliery producing more than 1,500 st/d (1,360 t/d). (Nelson, 1965)

large-diameter boring machine An auger-type coal-cutting machine developed by the U.S. Bureau of Mines for use in anthracite mining. It can drill holes 1 ft (31 cm) in diameter, 300 ft (91 m) long, and larger holes for shorter distances.

large knot A large knot is one whose average diameter exceeds one-third the width of the surface on which it appears; but such a knot may be allowed if it occurs outside the sections of the mine track tie between 6 in and 18 in (15 cm and 46 cm) from each end.

large shake A large shake is one that exceeds one-third the width of the mine track tie. A shake not exceeding this limitation and that does not extend nearer than 1/2 in (1.3 cm) to any surface shall be permissible.

large split A term applying to mine track ties. A large split is a split exceeding 5 in (13 cm) in length. Splits not longer than 5 in are permissible providing satisfactory antisplitting devices have been properly applied.

larnite A monoclinic mineral, 4[beta-Ca_2SiO_4]; gray; in contact metamorphosed limestone. Cf: *bredigite.*

larsenite An orthorhombic mineral, $PbZnSiO_4$; forms colorless to white prisms; in veins at Franklin, NJ.

Larsen's pile A type of pile consisting of hollow cylinders that increase resistance against bending and crumpling. They are esp. useful in shaft sinking in sand and gravel. (Stoces, 1954)

Larsen's spiles Steel sections of various forms, made esp. to resist bending, that are used in place of wooden spiles in forepoling. (Stoces, 1954)

Larson ledge finder A tool used to reach bedrock when the driven pipe has failed.

larvikite An alkalic syenite, grading to monzonite, composed of phenocrysts of two feldspars (esp. oligoclase and alkali feldspar), often intimately intergrown, which comprise up to 90% of the rock, with diopsidic augite and titanaugite as the chief mafic minerals, and accessory apatite (generally abundant), ilmenite, and titaniferous magnetite, and less commonly olivine, bronzite, lepidomelane, and quartz or feldspathoids (less than 10% by volume). Its name, given by Brögger in 1890, is derived from Larvik, Norway. Also spelled laurvikite. Syn: *blue granite.* (AGI, 1987)

laser An active electron device that converts input power into a very narrow, intense beam of coherent visible or infrared light; the input power excites the atoms of an optical resonator to a higher energy level, and the resonator forces the excited atoms to radiate in phase. Derived from "light amplification by stimulated emission of radiation." See also: *maser.* (McGraw-Hill, 1994)

lash To attach a chain to a haulage rope by wrapping or lapping the end of the chain around the rope, the other end being attached to a mine wagon. (Mason, 1951)

lasher (a) A native employed to do lashing. See also: *lashing.* (CTD, 1958) (b) *mucker.*

lasher-on A person employed to lash the chains from the tubs to the endless rope, in underground mechanical haulage. (CTD, 1958)

lashing (a) Any of a number of planks nailed inside of several frames or sets in a shaft to keep them in place. Also called listing. (Webster 2nd, 1960) (b) A binding, generally of light line around the end of a rope. (c) In South Africa, loading broken rock or ore with shovels. (Nelson, 1965) (d) Shoveling rock downstope to ore passes—work performed by a lasher. A "lasher-on" connects tubs or trucks to a rope haulage. Also called mucking. (Pryor, 1963)

lashing chain A short chain to attach tubs to an overrope in endless rope haulage by wrapping it around the rope. The chain may be about 12 ft (3.7 m) long, of low manganese steel, with 3/8-in-diameter (9.5-mm-diameter) standard links. At one end of the chain, a ring 4 in (10.2 mm) in diameter is attached to the drawbar hook of the tub, and to a hook about 3/8 in in diameter to secure it to the rope at the lashing end. On an undulating road, two lashing chains may be necessary—one forward and one rear of the tub. (Nelson, 1965)

lash-up Extemporized engineering rig for a temporary job. (Pryor, 1958)

lasque A thin, flat diamond with a simple facet at the side; used by Indian cutters to cover miniature paintings. Also spelled lask. Also called portrait stone.

last lift N. of Eng. The last rib or jud to come off a pillar. (Fay, 1920)

latch (a) The locking device on a hoist hook, elevator, lifting bail, etc. (Long, 1960) (b) The inner-tube locking and unlocking device in the head of a wire-line core barrel. (Long, 1960)

latches (a) Applied to the split rail and hinged switches. Syn: *switch.* (Fay, 1920) (b) Hinged switch points, or short pieces of rail that form rail crossings. (Fay, 1920)

late magmatic mineral A mineral formed during the late stages of magmatic activity, between the main stage of crystallization and the pegmatitic stage.

latent heat Thermal energy absorbed or emitted in a process (as fusion or vaporization) other than change of temperature. Cf: *sensible heat* (Webster 3rd, 1966)

latent heat of fusion The amount of heat required to change 1 g of a substance at the temperature of its melting point from the solid to the liquid state without changing temperature. (Morris, 1958)

lateral (a) A hard heading that branches off a horizon, in horizon mining, along the strike of the seams. It may be from 14 to 20 ft (4.3 to 6.1 m) wide. At intervals of 1,000 to 1,500 yd (910 to 1,400 m) along the lateral, crosscut roads are driven at right angles to intersect and develop the coal seams. From the crosscuts, conveyor panels are opened out in the seams. In general, the term lateral is also applied to any coal heading driven in a sideways direction. (Nelson, 1965) (b) Belonging to the sides, or to one side. (Fay, 1920) (c) A horizontal mine working. (Fay, 1920) (d) Situated on or at, or pertaining to, a side. (CTD, 1958) (e) A conduit diverting water from a main conduit, for delivery to distributaries. (Seelye, 1951) (f) A secondary ditch. (Seelye, 1951)

lateral cleavage Cleavage parallel to the lateral planes. (Webster 3rd, 1966)

lateral development Any system of development in coal seams or thick orebodies in which headings are driven horizontally across the coal or ore and connected to main haulage drifts, entries, or shafts. There are many variations and modifications depending on the thickness, shape, and inclination of the deposit. See also: *horizon mining.* (Nelson, 1965)

lateral deviation The horizontal distance by which a borehole misses its intended target. (Long, 1960)

lateral draw The angle of draw over a strike face or over workings in a flat seam. (Briggs, 1929)

lateral secretion A theory of ore genesis formulated in the 18th century and passing in and out of use since. It postulates the formation of ore deposits by the leaching of adjacent wall rock. In current usage, convectively driven fluids associated with cooling plutons are thought to have abstracted metals from adjacent host rocks and transported them to new sites of deposition, as in the formation of certain porphyry base-metal deposits. See also: *lithogene; segregated vein.* (AGI, 1987)

lateral support Means whereby walls are braced either vertically or horizontally by columns, pilasters, or crosswalls, or by floor or roof constructions, respectively. (ACSG, 1961)

later arrival A signal that is recorded on a seismogram later than the first arrival of energy. (Schieferdecker, 1959)

laterite Red residual soil developed in humid, tropical, and subtropical regions of good drainage. It is leached of silica and contains concentrations particularly of iron oxides and hydroxides and aluminum hydroxides. It may be an ore of iron, aluminum, manganese, or nickel. Adj. lateritic. Syn: *latosol.*

laterlog The electrical resistivity of coal appears to decrease with ash content. The laterlog measures what is virtually the true resistivity of the coal and may ultimately provide information on seam quality. The laterlog uses a sheet of current that is focused on each formation in succession and so measures the resistivity of that formation only. The mud column or a salty mud has no effect on the measured resistivity. The laterlog may be measured by a seven- or three-electrode arrangement but the former is preferred. (Sinclair, 1963)

latex cement A specialized cementing material consisting of a portland-type cement, latex, a surface-active agent, and water, having a setting time equivalent to a neat portland-cement slurry. Latex cement shrinks less and is tougher, stronger, less permeable, and more durable than portland cement. (Long, 1960)

Latex spray Trade name for a synthetic rubber fluid, which, when sprayed onto underground stoppings, forms a tough nonflammable coating thus preventing air feeding fires or heatings, or air leakages through doors, surface air locks, and air crossings. Also called Latex sealant. (Nelson, 1965)

lath (a) A board or plank sharpened at one end, like sheet piling, used in roofing levels or in protecting the sides of a shaft through a stratum of unstable earth. *spill.* (Fay, 1920) (b) A long, thin mineral crystal. (AGI, 1987)

lath frame A weak lath frame, surrounding a main crib, the space between being for the insertion of piles. (Fay, 1920)

lathlike Refers to crystals with three distinctly different dimensions. Cf: *acicular; tabular; rodlike; equant.*

laths (a) Corn. The boards or lagging put behind a frame of timber. (Fay, 1920) (b) Corn. Pieces of timber about 4 ft 6 in by 6 in by 2 in (1.4 m by 15 cm by 5 cm) with end sharpened or beveled to give the lath an upward trend when being driven into the roof gravels. A number of laths driven into the roof form a protective shield for the miners working in the face. Also called lagging. (Eng. Min. J., 1938)

latite A porphyritic extrusive rock having phenocrysts of plagioclase and potassium feldspar in nearly equal amounts, little or no quartz, and a finely crystalline to glassy groundmass, which may contain obscure potassium feldspar; the extrusive equivalent of monzonite. Latite grades into trachyte with an increase in the alkali feldspar content, and into andesite or basalt, depending on the presence of sodic or calcic plagioclase, as the alkali feldspar content decreases. It is usually considered synonymous with trachyandesite and trachybasalt, depending on the color. The name, given by Ransome in 1898, is derived from Latium, Italy. (AGI, 1987)

latitude correction (a) The north-south correction made to observed magnetic-field intensity in order to remove the Earth's normal field (leaving, as the remainder, the anomalous field). (AGI, 1987) (b) A correction of gravity data with latitude, because of variations in centrifugal force owing to the Earth's rotation and because of differences in the radius owing to polar flattening. The correction for latitude ϕ amounts to $1.308 \sin 2\phi$ mgal/mi = $0.813 \sin 2\phi$ mgal/km. (AGI, 1987)

latosol *laterite.*

latrobite A pink anorthite from Amitok Island, LB, Canada. (Standard, 1964)

latten Metal in thin sheets, esp. (and originally) brass, which in this form is also called latten brass. (Standard, 1964)

lattice An array of points in space such that each point is in an identical point environment. Thus, any straight line drawn between any two points in a lattice and continued will pass at equal intervals through a succession of similar points. Fourteen possible lattices exist. Syn: *Bravais lattice; space lattice.* Cf: *net; row.*

lattice constant *lattice parameter.*

lattice energy Energy required to separate the ions of an ionic crystal to an infinite distance from each other. (Pryor, 1963)

lattice girder An open girder, beam, or column in timber, steel, or aluminum alloy, built up from structural members joined and braced together by intersecting diagonal bars. See also: *space lattice.* (Hammond, 1965)

lattice parameter Lattice parameters are the unit lengths along each crystallographic axis and their interaxial angles. See also: *axial element.*

lattice texture (a) In mineral deposits, a texture produced by exsolution in which elongate crystals are arranged along structural planes. (AGI, 1987) (b) A texture that is typical of the mineral serpentine in a rock where it replaces an amphibole. Cf: *knitted texture.* (AGI, 1987)

lattice water (a) Water that is an integral part of the clay structure. This structural water (OH lattice water) is not to be confused with interlayer water. The lattice water can be removed by heating in the range of about 450 to 600 °C. (ACSG, 1963) (b) Molecular water at specific lattice sites. (Van Vlack, 1964)

laubanite *natrolite.*

laubannite Alternate spelling of laubanite.

Laue diagram *Laue photograph.* Cf: *crystallogram.*

laueite A triclinic mineral, $MnFe_2(PO_4)_2(OH)_2{\cdot}8H_2O$; dimorphous with stewartite.

Laue photograph A collection of X-ray diffraction spots made by a crystal using a Laue camera and white radiation. Syn: *Laue diagram.* Cf: *crystallogram.*

launder (a) A flume, trough, channel, or chute by which water or powdered ore is conveyed in a mining operation. (AGI, 1987) (b) An inclined channel, lined with refractory material, for the conveyance of molten steel from the furnace taphole to a ladle. Also spelled lander. (Dodd, 1964)

launder man In ore dressing, smelting, and refining, a laborer who maintains and repairs the launders (long boxes), used to convey water and mill pulp between the various units of ore-treating equipment in a mill. (DOT, 1949)

launder screen A screen used for the sizing and dewatering of small sizes of anthracite. (Mitchell, 1950)

launder separation process In this process, a stream of fluid carries the material to be separated down a channel provided with draws for separating a heavy-gravity product and means for overflowing a lighter one. If properly constructed and operated, a comparatively solid bed of material will form on the bottom of the launder. Above this bed, a layer of particles will move along by the stream at a comparatively slow speed. Above this, successive layers will move with greater and greater velocity. (Mitchell, 1950)

launder washer A type of coal washer in which the coal is separated from the refuse by stratification due to hindered settling while being carried in aqueous suspension through a trough. Modern launder washers have various mechanisms for continuously removing refuse from the bottom of the trough. Early launder washes were intermittent in operation.

laundry box The box at the surface receiving the water pumped up from below. (Fay, 1920)

Laurasia Hypothetical continent in the Northern Hemisphere that broke up about the end of the Carboniferous Period to form the present northern continents.

laurdalite An alkalic syenite containing more than 10% modal feldspathoids and characterized by porphyritic texture. Also spelled lardalite. The name, given by Brögger in 1890, is for Laurdal, Norway. (AGI, 1987)

Laurentian granite A name that was originally applied to Precambrian granites of the Laurentian Highlands, eastern Canada, and later to the oldest granites near the U.S.-Canadian border northwest of Lake Superior.

laurionite An orthorhombic mineral, $PbCl(OH)$; dimorphous with paralaurionite; in ancient lead-ore slags.

laurite An isometric mineral, RuS_2; pyrite group; occurs with other platinum-group minerals in ultramafic and placer deposits.

lauroleic acid Unsaturated fatty acid, $C_{12}H_{22}O_2$. (Pryor, 1963)

laurvikite *larvikite.*

lausenite A monoclinic mineral, $Fe_2(SO_4)_3{\cdot}6H_2O$.

lautarite A monoclinic mineral, $Ca(IO_3)_2$.

Lauth mill Mill with three rolls, the middle roll being much smaller than the other two. Only the two larger rolls are driven, work being performed between the bottom and middle and middle and top rolls alternately; the roll setting is adjusted between passes. (Osborne, 1956)

lautite An orthorhombic mineral, CuAsS.

lava breccia *volcanic breccia.*

lava dome A dome-shaped mountain of solidified lava in the form of many individual flows, formed by the extrusion of highly fluid lava, e.g., Mauna Loa, HI. See also: *shield volcano.* (AGI, 1987)

lavatory A place where gold is obtained by washing. (Standard, 1964)
lavendine *amethystine quartz.*
lavenite A monoclinic mineral, $(Na,Ca)_2(Mn,Fe)(Zr,Ti)Si_2O_7(O,OH,F)_2$. Also spelled låvenite.
lavialite A metamorphosed basaltic rock with relict phenocrysts of labradorite in an amphibolitic groundmass. The term was originated by Sederholm in 1899, who named it after Lavia, Finland. (AGI, 1987)
lavrovite A chromian variety of diopside. Also spelled lavroffite; lawrowite.
lawn A fine-mesh gauze used as a sieve for clay. (Crispin, 1964)
law of cosines In trigonometry, a law stating that in any triangle the square of one side equals the sum of the squares of the two other sides minus twice the product of these two other sides multiplied by the cosine of the included angle. (Jones, 1949)
law of equal volumes In ore genesis, the statement that during the formation of ore by replacement there is no change in rock volume or form. Syn: *Lindgren's volume law.*
law of extralateral rights *apex law.*
law of gravitation The law, discovered by Sir Isaac Newton, that every particle of matter attracts every other particle of matter, and the force between them is proportional to the product of their masses divided by the square of their distance apart. See also: *gravitation; gravity.* (Standard, 1964)
law of mass action The rate of a chemical reaction is directly proportional to the molecular concentrations of the reacting substances.
law of motion A statement in dynamics that a body at rest remains at rest and a body in motion remains in uniform motion in a straight line unless acted upon by an external force. The acceleration of a body is directly proportional to the applied force and is in the direction of the straight line in which the force acts. For every force there is an equal and opposite force or reaction. (Webster 3rd, 1966)
law of refraction (a) When a wave crosses a boundary, the wave normal changes direction in such a manner that the sine of the angle of incidence between wave normal and boundary normal divided by the velocity in the first medium equals the angle of refraction divided by the velocity in the second medium. (AGI, 1987) (b) Light, upon crossing a boundary between two transparent substances of different optical densities, changes direction according to n_2/n_1=sin i, where n_1 is the refractive index (RI) for the incident light ray making an angle i, and n_2 is the RI for the refracted light ray making an angle r with the boundary ($n_1 < n_2$). Light rays refracted according to this law are called "ordinary" rays. Cf: *critical angle; extraordinary ray; ordinary ray; total reflection.* Syn: *Snell's law.*
law of sines In trigonometry, a law stating that in any triangle (either right or oblique) the sides are proportional to the sines of their opposite angles. (Jones, 1949)
law of superposition A general law upon which all geologic chronology is based: In any sequence of sedimentary strata (or of extrusive igneous rocks) that have not been overturned, the youngest stratum is at the top and the oldest at the base; i.e., each bed is younger than the bed beneath, but older than the bed above it. The law was first clearly stated by Steno (1669). (AGI, 1987)
lawsonite An orthorhombic mineral, $4[CaAl_2Si_2O_7(OH)_2{\cdot}H_2O]$; dimorphous with partheite; pale blue; Mohs hardness, 8; in high-pressure, low-temperature schists.
laxmannite *vauquelinite.*
lay (a) The direction, or length, of twist of the wires and strands in a rope. (Zern, 1928) (b) The length of lay of wire rope is the distance parallel to the axis of the rope in which a strand makes one complete turn about the axis of the rope. The length of lay of the strand, similarly, is the distance in which a wire makes one complete turn about the axis of the strand. (c) To close or withdraw from work; said of collieries. (Standard, 1964) (d) The pitch or angle of the helix of the wires or strands of a rope, usually expressed as the ratio of the diameter of the strand or rope to the length required for one complete twist. (Hunt, 1965) (e) Prov. Eng. A standard of fineness for metals; possible from the Spanish ley. (Hess)
lay-by (a) Siding in otherwise single-track underground tramming road. (Pryor, 1963) (b) A term used for an underground siding at or near a shaft for storing empty mine cars. (c) *bank.*
layer (a) A bed or stratum of rock. (b) One of a series of concentric zones or belts of the Earth, delineated by seismic discontinuities. A classification of the interior of the Earth that designates layers A to G from the surface inward. (AGI, 1987)
layer depth Thickness of the mixed surface layer of water. (Hy, 1965)
layered N. of Eng. Choked up with sediment or mud.
layering *bedding.*
layering number A dimensionless number, the value of which, taken in conjunction with inclination, roughness, and whether the ventilation is ascensional or descensional, determines the mixing and movement of combustible gases roof layers. (BS, 1963)
layering of combustible gases The formation of a layer of combustible gases at the roof of a mine working and above the ventilating air current. (BS, 1963)
layer-loading Procedure for loading coal in railroad cars in horizontal layers. Layer-loading is a simple and inexpensive method for smoothing out the irregularities in coal and consists in shuttling two to six railroad cars, hooked together, past the loading boom two or more at a time. This results in a more uniform product. (Mitchell, 1950)
laying out *setting out.*
lay of rope *winding rope.*
lay of the land *topography.*
layout (a) The design or pattern of the main roadways and workings. The proper layout of mine workings is the responsibility of the manager aided by the planning department. (Nelson, 1965) (b) The map of a mine or part of a mine, usually including future workings arrangement. (BCI, 1947) (c) Diagram showing disposition of machines in a mill's flow line. (Pryor, 1963)
lay rope Ordinary lay rope has the wires twisted in a direction opposite to the twist of the strands in the rope. The pitch of wire is from 2-1/2 to 3 times the diameter of the rope, and the pitch of the strands is from 6-1/2 to 9 times the diameter of the rope, the wires being exposed only in short lengths at intervals. (Lewis, 1964)
lazarevicite *arsenosulvanite.*
lazuli *lapis lazuli.* Also spelled lazule; lasule.
lazulite (a) A monoclinic mineral, $2[MgAl_2(PO_4)_2(OH)_2]$; forms a series with scorzalite; azure blue; in granite pegmatites and quartz veins; may be a blue gemstone. Syn: *azure spar; blue spar; false lapis; berkeleyite; klaprothine.* (Not lazurite.) (b) The mineral group barbosalite, hentschelite, lazulite, and scorzalite.
lazulitic Of, pertaining to, or having the characteristics of lazulite. (Standard, 1964)
lazurite (a) An isometric mineral, $(Na,Ca)_8(Al,Si)_{12}(O,S)_{24}[(SO_4),Cl_2,(OH)_2]$ of the sodalite group; deep blue to greenish blue; a contact metamorphic product in limestone; may be a blue gemstone (lapis lazuli). Also spelled lasurite. (Not lazulite.) See also: *Chilean lapis.* Cf: *ultramarine.* (b) *azure; azurite.*
lazy balk (a) Eng. A timber placed at the top of a hopper, against which the top of the car strikes in dumping, to prevent the car from falling into the hopper. (Fay, 1920) (b) Eng. The balk or girder held in position by a banger. Also called lazy girder. (SMRB, 1930)
lazy bench The bench to one side of the drill tripod or derrick floor where visitors and workers can sit while observing the drilling operation. (Long, 1960)
lazy girder *lazy balk.*
lazy tong conveyor *accordion roller conveyor.*
L.D. steel process Process in which oxygen is blown downwards at high velocity through a watercooled lance onto the surface of the hot metal contained in a basic lined vessel. To offset the intense heat produced, coolant materials are added with the original charge. These may be iron ore, sinter, or roll scale, but usually steel scrap is the main material used. As much as 26% of scrap may be used. After about 20 min, the charge is converted into liquid steel. During the process, tests and analyses are made and materials may be added to bring the metal to the required grade and temperature. See also: *open-hearth process; O.L.P. steel process.* (Nelson, 1965)
leachate A solution obtained by leaching; e.g., water that has percolated through soil containing soluble substances and that contains certain amounts of these substances in solution. Syn: *lixivium.* (AGI, 1987)
leach dump Low-grade ores that are dumped loosely in piles on soil surfaces so that fluids may be sprinkled on the piles to leach recoverable metals. (SME, 1992)
leached zone The part of a lode above the water table, from which some ore has been dissolved by down-filtering meteoric or spring water.
leacher In ore dressing, smelting, and refining, one who dissolves valuable metal out of ore or slime, using chemical solution. (DOT, 1949)
leach hole *sinkhole.*
leaching (a) The separation, selective removal, or dissolving-out of soluble constituents from a rock or orebody by the natural action of percolating water. (AGI, 1987) (b) Dissolution from ore or concentrates after suitable comminution to expose the valuable minerals, by aqueous and chemical attack. If heat and pressure are used to intensify or speed this, the work is called pressure leaching. See also: *chemical extraction; hydrometallurgy.* (Pryor, 1963) (c) The removal in solution of nutritive or harmful constituents (such as mineral salts and organic matter) from an upper to a lower soil horizon by the action of percolating water, either naturally (by rainwater) or artificially (by irrigation). (AGI, 1987) (d) The extraction of soluble metals or salts from an ore by means of slowly percolating solutions; e.g., the separation of gold by treatment with a cyanide solution. Syn: *lixiviation.* (AGI, 1987)

leaching rate test A test designed to assess the value of antifouling compositions by measuring the rate of loss of toxic ingredients from a painted surface during immersion in seawater. (Osborne, 1956)

leach material Material sufficiently mineralized to be economically recoverable by selectively dissolving the wanted mineral in a suitable solvent. See also: *leaching.*

leach pile Mineralized materials stacked so as to permit wanted minerals to be effectively and selectively dissolved by application of a suitable solute.

leach precipitation float A mixed method of chemical reaction plus flotation developed for such copper ores as chrysocolla and the oxidized minerals. The value is dissolved by leaching with acid, and the copper is reprecipitated on finely divided particles of iron, which are then recovered by flotation, yielding an impure concentrate in which metallic copper predominates. Abbrev., L.P.F. (Pryor, 1963)

lead (a) A bluish-white metal of bright luster, very soft, highly malleable, ductile, and a poor conductor of electricity; very resistant to corrosion; a cumulative poison. Symbol, Pb. Rarely occurs in native form; chiefly obtained from galena (PbS). Lead is used in storage batteries, cable covering, plumbing, ammunition, antiknock gasoline, radiation shielding, and to absorb vibration. Other lead compounds are used in paints, fine glass, and lenses. Environmental concern with lead poisoning has resulted in a U.S. national program to reduce the concentration of lead in gasoline. (Handbook of Chem. & Phys., 3) (b) An open watercourse, usually artificial, leading to or from a mill, mine, or reservoir. Syn: *leat.* (c) *ledge; lode.* Cf: *blind lead.* Pron. leed. (d) A placer deposit. Cf: *blue lead; deep lead.* Pron. leed. (e) A defined gutter of auriferous wash. Pron. leed. (Gordon, 1906) (f) A track haulage term for the distance from the point of a frog to the point of the switch. Pron. leed. (Kentucky, 1952) (g) A term sometimes used for the distance between the sheave and the winding drum centers. The greater the lead, other things being equal, the smaller the fleeting angle. Too great a lead results in vibration and whipping of the rope between sheave and drum. Idler or sag rollers are frequently installed where long leads are necessary. Pron. leed. (Nelson, 1965) (h) The distance a bit is held suspended off bottom in a borehole before rotation and downward movement of the drill string is started. Pron. leed. (Long, 1960) (i) Commonly used synonym for ledge or lode. Many mining location notices describe the locator's claim as extending a certain number of feet along and so many feet on each side of the lode, lead, vein, or ledge. Thus Lead, S. Dak., was so named because of the Homestake lead. Blind lead: A lead or vein that does not outcrop or show at the surface. Used esp. at Virginia City, NV. Cf: *lode.* Pron. leed. (Fay, 1920) (j) Properly, placer gravels. Blue lead: A Tertiary river channel at Placerville, CA. So called because of the bluish-gray color of the gravels. Deep lead: Goldbearing gravels deeply covered with debris or lava applied particularly to those of Victoria, Australia. Pron. leed. (Fay, 1920) (k) The longitudinal distance traveled in one revolution by a spiral thread or screw. Pron. leed. (Long, 1960)

lead-acid accumulator A secondary cell battery with an electromotive force of about 2 V. It is suitable for work where a steady voltage is required, and extensively used for motor car lighting, miners' safety lamps, shuttle cars, and battery locomotives. (Morris, 1958)

leadage The distance coal has to be hauled from the mine to its place of shipment. (Standard, 1964)

lead azide A nitrite of lead, $Pb(N_3)_2$, used as an initiating explosive in blasting caps.

lead bath A furnace in which gold or silver ores are smelted with lead. (Standard, 1964)

lead button In the separation of the noble metals from their impurities, lead is fused with the ore. The bullion so formed drops to the bottom of the crucible in the lead button from which the precious metal is extracted by cupellation. Syn: *crucible assay.* (Nelson, 1965)

lead carbonate *cerussite; white lead ore.*

lead edge The surfaces or inset cutting points on a bit that face in the same direction as the rotation of the bit. (Long, 1960)

leader (a) A narrow vein branching upwards at an angle from a much larger vein. See also: *dropper.* (Nelson, 1965) (b) A thin layer of coal, coaly shale, or ironstone that serves as a guide or datum toward workable beds in a mine.

leaders Guides in a pile frame to take the drop hammer of a pile driver. (Hammond, 1965)

lead feldspar Synthetic $PbAl_2Si_2O_8$.

lead fume The fume escaping from lead furnaces and containing both volatilized and mechanically suspended metalliferous compounds. (Fay, 1920)

lead glance *galena.*

leadhillite A monoclinic mineral, $8[Pb_4(SO_4)(CO_3)_2(OH)_2]$; trimorphous with macphersonite and susannite; soft; may fluoresce yellow; in oxidized zones of lead-ore deposits.

leading heading The one of a pair of parallel headings that is kept a short distance in advance of the other. This may be adopted to drain the water and thus secure one dry heading. The term is also applied to a heading that is driven in the solid coal in advance of the general line of face. (Nelson, 1965)

leading place Scot. A working place in advance of the others, such as a heading or a level. (Fay, 1920)

leadings Derb. Small sparry veins in the rock. Syn: *leader.*

leading stone *lodestone.*

leading winning Aust. A heading in advance of the ordinary bords. A leading bank. (Fay, 1920)

leading wire An insulated wire strung separately or as a twisted pair, used for connecting the two free ends of the circuit of the blasting caps to the blasting unit. See also: *leads.*

lead lap (a) A gem cutter's lap, of lead, copper, or iron; also, the entire machine. (Standard, 1964) (b) In mechanics, a lap of lead charged with emery and oil. (Fay, 1920)

lead metacolumbate $Pb(CbO_3)_2$; a ferroelectric material with a Curie temperature of 570 °C. The material can be polarized to obtain piezoelectric properties. Uses include high-temperature transducer applications, sensing devices, and accelerometers. Syn: *lead metaniobate.* (Lee, 1961)

lead metaniobate *lead metacolumbate.*

lead metasilicate *alamosite.*

lead motorman In ore dressing, smelting, and refining, a person who operates a small electric locomotive (motor) to haul pots of molten lead bullion from a blast furnace to refining kettles for the separation of copper, antimony, silver, and other metals contained in the lead bullion. (DOT, 1949)

lead niobate $Pb(NbO_3)_2$; a ferroelectric compound having properties that make it useful in high-temperature transducers and in sensing devices. The Curie temperature is 570 °C. (Dodd, 1964)

lead ocher *massicot; litharge.*

lead of a switch The distance measured on the main line from the point of switch to the point of frog. Also called frog distance. (Kiser, 1929)

lead rail The lead rail of an ordinary mine switch is the turnout rail lying between the rails of the main track. (Kiser, 1929)

leads The wires, forming part of an electric detonator, to which the shot-firing cable is attached. (BS, 1964)

lead selenide *clausthalite.*

lead silicate *alamosite.*

lead spar *cerussite; anglesite.*

lead sulfide *galena; glance.*

lead tantalate $PbTa_2O_6$; a compound believed to have ferroelectric properties and of possible interest as a special electroceramic. The Curie temperature is 260 °C. (Dodd, 1964)

lead tree A crystalline deposit of metallic lead on zinc that has been placed in a solution of acetate of lead. (Standard, 1964)

lead-uranium ratio The ratio of lead-206 to uranium-238 and/or lead-207 to uranium-235, formed by the radioactive decay of uranium within a mineral. The ratios are frequently used as part of the uranium-thorium-lead age method. (AGI, 1987)

lead vitriol *anglesite.*

lead-well man In ore dressing, smelting, and refining, a person who maintains flow of molten lead from the blast furnace to the lead pot for removal to refinery. (DOT, 1949)

lead wires (a) In blasting, the heavy wires that connect the firing current source or switch with the connecting or cap wires. (Nichols, 1976) (b) Two insulated copper wires leading from the battery or igniting apparatus to the primer cartridge in an explosive charge. Also called connecting wires. (Stauffer, 1906)

lead works A place where lead is extracted from the ore. (Fay, 1920)

lead zirconate $PbZrO_3$; a ferroelectric material. It is also used in lead titanate-zirconate (P.Z.T) piezoelectric ceramics. (Dodd, 1964)

leaf A very thin sheet or plate of metal, as gold. (Standard, 1964)

leaf clay *book clay.*

league (a) Any of various linear units of distance, ranging from about 2.42 to 4.6 statute miles (3.89 to 7.4 km); esp. land league (an English land unit equal to 3 statute miles or 4.83 km) and marine league (a marine unit equal to 3 nmi or 5.56 km). (AGI, 1987) (b) Any of various units of land area equal to a square league; esp. an old Spanish unit for the area of a tract 5,000 varas square, equal to 4,428.4 acres (1,792.1 ha) in early Texas land descriptions or equal to 4,439 acres (1,796 ha) in old California surveys. (AGI, 1987)

leak Low-grade mineralized rock into which an orebody degenerates. (Hess)

leakage An unintentional diversion of ventilation air from its designed path. (BS, 1963)

leakage coefficient A numerical expression of a duct's liability to leak. The National Coal Board of Great Britain defines this as the volume of air in cubic feet per minute that would leak from 100 ft (30 m) of a ventilation duct under a uniform pressure of 1 in (2.54 cm) of water gage. (Roberts, 1960)

leakage halo A dispersion pattern formed by the movement of ore-forming fluids in the rock overlying a mineral deposit.

leakage intake An additional intake that is a component part of a system of controlled leakage. (BS, 1963)
leakage intake system A ventilation circuit with two adjacent intake roadways leading to the coalface. The method has been criticized because the air flow may become so sluggish as to cause combustible gases layers. See also: *two intakes*. (Nelson, 1965)
leakage resistance The resistance between the blasting circuit, including lead wires, and the ground. (Atlas, 1987)
leak vibroscope An instrument that detects leaks in water, oil, gas, steam, and air lines by amplifying the sound produced by the escaping fluid. (Osborne, 1956)
lean (a) Of ore, low-grade; submarginal; unpay; of doubtful exploitable value. (Pryor, 1963) (b) A rock in which the minerals sought occur in much less than exploitable amounts. (Long, 1960) (c) *hang*. (d) *low-grade*.
lean clay A clay of low to medium plasticity owing to a relatively high content of silt or sand. Cf: *fat clay*. (AGI, 1987)
lean ore A low-grade ore. See also: *lean; natural ore*. (Nelson, 1965)
leap Eng. A dislocation of strata by faulting.
leapfrog system A system employed with self-advancing supports on a longwall face in which alternate supports are advanced on each web of coal removed. To do this, alternate units have to be moved a distance equal to twice the web thickness—half before snaking and half after snaking. (Nelson, 1965)
lear *lehr*.
learies Eng. Empty places; old workings. (Fay, 1920)
leasable minerals A legal term that for Federally owned lands, or Federally retained mineral interest in lands in the United States, defines a mineral or mineral commodity that is acquired through the Mineral Lands Leasing Act of 1920, as amended; the Geothermal Steam Act of 1970, as amended; or the Acquired Lands Act of 1947, as amended. These Acts are found in Title 30 of the United States Code - Mineral Lands and Mining. The leasable minerals include oil, gas, sodium, potash, phosphate, coal, and all minerals within Acquired Lands. Acquisition is by application for a Government lease and permits to mine or explore after lease issuance. (SME, 1992)
lease (a) A contract between a landowner and another, granting the latter the right to search for and produce oil or mineral substances upon payment of an agreed rental, bonus, and/or royalty. (AGI, 1987) (b) The instrument by which such grant is made. (c) A piece of land leased for mining purposes. See also: *claim; concession system; royalty*.
leaser A Western colloquialism meaning lessee.
Leasing Act Minerals Deposits of coal, phosphate, oil, oil shale, gas, sodium, potassium and sulfur that can be leased from the U.S. Government under the Mineral Leasing Act for acquired lands. (SME, 1992)
leat (a) A mill stream used by small mines for power generation. (Nelson, 1965) (b) A ditch that leads water to mineral workings. (Pryor, 1963)
leatherstone *mountain leather*.
leaving Corn. The mineral left after the good ore has been removed; tailings.
lechatelierite Naturally fused gray siliceous glass; actually a minor rock type varying in composition according to the original sand type, typically 90% to 99.5% silica. Cf: *fulgurite; impactite*. Also spelled lechateliérite.
lechosos opal A variety of precious opal exhibiting a deep green play of color; esp. a Mexican opal exhibiting emerald-green play of color and flashes of carmine, dark violet, dark blue, and purple.
lecontite An orthorhombic mineral, $(NH_4,K)Na(SO_4)\cdot 2H_2O$; occurs in bat guano.
led N. of Eng. A spare tub, or one that is being loaded while another is being emptied. (Fay, 1920)
ledge (a) A narrow shelf or projection of rock, much longer than wide, formed on a rock wall or cliff face. (AGI, 1987) (b) A rocky outcrop; solid rock. (AGI, 1987) (c) An underwater ridge of rocks, esp. near the shore; also, a nearshore reef. (AGI, 1987) (d) A quarry exposure or natural outcrop of a mineral deposit. (AGI, 1987) (e) A bed or several beds in a quarry or natural outcrop, particularly those projecting in a steplike manner. (AGI, 1987) (f) The surface of such a projecting bed. (AGI, 1987) (g) In mining, a projecting outcrop or vein, commonly of quartz, that is supposed to be mineralized; also, any narrow zone of mineralized rock. (AGI, 1987) (h) A mass of rock that constitutes a valuable mineral deposit. (Webster 3rd, 1966) (i) A colloquial syn. of bedrock, used in northern Michigan. (Long, 1960) (j) The only true ledges are deposits of oil-shale, slate, or the like. A ledge is a horizontal layer, therefore a vein or lode is not a ledge. (von Bernewitz, 1931) (k) A rocky formation continuous with and fringing the shore. Syn: *lead*. (Hunt, 1965)
ledge rock True bedrock. Cf: *false bottom*.
Leebar separator A dense medium washer consisting of a static bath. The floats, or clean coal, are removed by means of paddles or chains suspended from bars connected to rotating spokes. The sinks, or shale, are extracted by a scraper device. The bath can be fed directly from the raw coal screens. The separator has been developed for the treatment of large coal. See also: *Nelson Davis separator*. (Nelson, 1965)
Lee configuration (a) An electrical resistivity measuring method using two current electrodes and three equispaced potential electrodes. (AGI, 1987) (b) A configuration employing electrodes, the outer two of which are the current and the inner three of which are the potential electrodes. Syn: *partitioning method*. (AGI, 1987)
leelite A reddish variety of potassium feldspar.
Lee-Norse miner A continuous miner, developed in the United States, for driving headings in medium or thick coal seams. It weighs about 26 st (23.6 t), and makes a cut 8-1/2 ft (2.6 m) wide, gathers the cut coal and loads it into cars or conveyor at a rate up to about 4 st/min (3.6 t/min). It can work in seam heights from 3 ft 8 in (1.1 m) up to about 10 ft (3 m). It consists, mainly, of a boom carrying the cutting head; the gathering head, and at the rear the jib support frame on which the jib can be slewed. The machine is operated by hydraulic motors. (Nelson, 1965)
leer *lehr*.
leering In glassmaking, the process of treating in the annealing oven or leer. (Standard, 1959)
Leet seismograph A portable three-component seismograph designed primarily for registration of vibrations from blasts, traffic, machinery, and general industrial sources. (Leet, 1960)
lefkasbestos A bleached-white variety of asbestiform chrysotile from Mt. Troodos, Cyprus.
left bank The bank of a stream that is to the left of an observer facing downstream.
left lang lay Wire or fiber rope or cable in which the wires or fibers in a strand and the strands themselves are twisted to the left. (Long, 1960)
left-lateral fault A fault on which the displacement is such that the side opposite the observer appears displaced to the left. Cf: *right-lateral fault*. Syn: *sinistral fault*.
left long lay *left lang lay*.
left regular lay Wire or fiber rope or cable in which the individual wires or fibers in the strands are twisted to the right and the strands to the left. Syn: *regular-lay left lay*. (Long, 1960)
left twist *right lay*. Corresponds to a right-hand screw thread. (Hunt, 1965)
leg (a) In mine timbering, a prop or upright member of a set or frame. Also called upright; post; arm. (Pryor, 1963; Nelson, 1965) (b) One of the main upright members of a drill derrick or tripod. (Long, 1960) (c) A term sometimes applied to a centrifugal discharge bucket elevator. Usually a double leg bucket elevator. (d) *draft*. (e) A side post in tunnel timbering. (Nichols, 1976)
legend A brief explanatory list of the symbols, cartographic units, patterns (shading and color hues), and other cartographic conventions appearing on a map, chart, or diagram. On a geologic map, it shows the sequence of rock units, the oldest at the bottom and the youngest at the top. The legend formerly included a textual inscription of, and the title on, the map or chart. Syn: *key*. (AGI, 1987)
leg piece The upright timber that supports the cap piece in a mine. Cf: *legs*. (Fay, 1920)
legrandite A monoclinic mineral, $8[Zn_2(AsO_4)(OH)\cdot H_2O]$.
legs (a) The wires attached to and forming a part of an electrical blasting cap. (Fay, 1920) (b) The uprights of a set of mine timbers. See also: *dap; leg piece*. (Fay, 1920)
legua Sp. Land league used in the original surveys of the Philippines, California, and Texas. It is equal to 2.63 miles or 4.24 km. See also: *league*. (AGI, 1987)
leg wire One of the two wires attached to and forming a part of an electric blasting cap or squib.
Lehigh jig A plunger-type jig with the following distinguishing characteristics: (1) the plunger contains check valves that open on the upstroke to reduce suction; (2) the makeup water is introduced with the feed; (3) the screen plate is at two levels, which have different perforations, to keep the water distribution uniform; (4) the bottom of the discharge end of the jig is hinged. This jig has been used extensively in washing anthracite. (Mitchell, 1950)
lehiite A mixture of crandallite with other minerals.
Lehmann process A process for treating coal by disintegration and separation of the petrographic constituents (fusain, durain, and vitrain). It consists of subjecting the coal to resilient disintegrating or shattering action for a sufficient length of time to break the constituents into granules of various sizes by reason of their respective resistances to shattering impacts and separating the granules into different sizes by screening or equivalent means. (Mitchell, 1950)
lehr An enclosed oven or furnace used for annealing, or other form of heat treatment; particularly used in glass manufacture. It is a kind of tunnel down which glass, hot from the forming process, is sent to cool slowly, so that strain is removed, and cooling takes place without additional strain being introduced. Lehrs may be of the open type (in

which the flame comes in contact with ware), or of the muffle type. Syn: *leer; lear.* (CTD, 1958)

lehr man Person who regulates temperature of a reheating oven (lehr) used to fire-glaze glass articles. Arranges glass articles according to size and shape on lehr conveyor so that maximum quantity will be carried in oven on a long paddle. Also called leer man; lehr operator, glass; lehr tender. (DOT, 1949)

lehrnerite *ludlamite.*

leightonite A triclinic mineral, $K_2Ca_2Cu(SO_4)_4 \cdot 2H_2O$; pseudo-orthorhombic; blue; at Chuquicamata, Chile.

Leitz tyndallometer Measures the intensity of the light scattered at an angle from the incident beam by a dust cloud, and correlates well with the concentration determined by the thermal precipitator or the surface area calculated from such a count. However, it needs to be calibrated for each type of dust cloud, owing to difference in mineralogical content, against the thermal precipitator. (Sinclair, 1958)

Lemberg's solution Logwood digested in an aqueous solution of aluminum chloride; used to distinguish calcite and dolomite. Calcite and aragonite are stained violet after treatment for about 10 min, but dolomite remains unchanged.

lenad A contracted form of the names leucite and nephelite; suggested as an alternative group name for the feldspathoid minerals. Cf: *feldspathoid.*

lengenbachite A triclinic mineral, $Pb_6(Ag,Cu)_2As_4S_{13}$.

lengthening rod A screwed extension rod for prolonging a well-boring auger or bit. (Standard, 1964)

length fast *negative elongation.*

length of lay The distance measured along a straight line parallel to the rope in which the strand forms one complete spiral around the rope or the wires around the strand. See also: *lay.* (Hammond, 1965)

length of shot (a) The depth of the hole in which the powder is placed, or the size of the block of coal to be loosened by a single blast measured parallel with the hole. (b) In open pit mining, the distance from the first drill hole to the last drill hole along the bank.

lengths Eng. In tunnel construction, the successive sections in which a tunnel is executed. Shaft lengths are directly under the working shaft; side lengths are on each side of the shaft length; leading lengths are prolongations of the tunnel from the side lengths; and junction lengths complete the portion of the tunnel extending between two shafts, or between a shaft and an entrance.

length slow *positive elongation.*

lennilite (a) A green variety of orthoclase at Lenni Mills, Delaware County, PA. Syn: *delawarite.* (b) A variety of vermiculite.

lens (a) A geologic deposit bounded by converging surfaces (at least one of which is curved), thick in the middle and thinning out toward the edges, resembling a convex lens. A lens may be double-convex or plano-convex. See also: *lentil; lenticular.* —v. To disappear laterally in all directions; e.g., a unit is said to lens out within a mapped area. (AGI, 1987) (b) In optics, a device that modulates the direction taken by a transient beam of light. (Pryor, 1963)

lense Pyrite, round or oval in plan and lenticular in section, ranging up to 2 to 3 ft (0.6 to 0.9 m) in thickness and several hundred feet in the greatest lateral dimension, that is found in coalbeds. Sometimes called kidney sulfur. (Mitchell, 1950)

lens grinding The process of grinding pieces of flat sheet glass (or pressed blanks) to the correct form of the lens. Cast-iron tools of the correct curvature, supplied with a slurry of abrasive and water, are used. (CTD, 1958)

lensing The thinning-out of a stratum in one or more directions.

lenticle (a) A large or small lens-shaped stratum or body of rock; a lentil. (AGI, 1987) (b) A lens-shaped rock fragment of any size. (AGI, 1987)

lenticular (a) Resembling in shape the cross section of a lens, esp. of a double-convex lens. The term may be applied, e.g., to a body of rock, a sedimentary structure, or a mineral habit. (AGI, 1987) (b) Pertaining to a stratigraphic lens or lentil. Syn: *lentiform.* (AGI, 1987)

lenticular iron ore Impure concretionary hematite.

lenticule A small lentil. (AGI, 1987)

lentiform *lenticular.*

lentil (a) A minor rock-stratigraphic unit of limited geographic extent, being a subdivision of a formation and similar in rank to a member, and thinning out in all directions; a geographically restricted member that terminates on all sides within a formation. Cf: *tongue.* (AGI, 1987) (b) A lens-shaped body of rock, enclosed by strata of different material; a geologic lens. See also: *lenticule; lenticle.* (AGI, 1987)

lentil ore *liroconite.*

leonardite (a) A soft, earthy, medium-brown coallike substance associated with lignitic outcrops in North Dakota. It is a naturally oxidized form of lignite with variations in color and properties depending upon the extent of weathering. Usually, the material occurs at shallow depths, overlying or grading into the harder and more compact lignite. Of little value as a fuel, it has been used in oil-drilling muds, in water treatment, and in certain wood stains. It is frequently referred to as "slack" because of its texture; however, the term leonardite is finding common usage. (b) A weathering product of subbituminous coal or lignite, rich in humic and fulvic acids and soluble in alkaline water. It is a byproduct of mining near-surface coal seams, and is used as a soil conditioner, additive to drilling mud, and binder for taconite iron ore (Fowkes & Frost, 1960). Not to be confused with leonhardite or leonhardtite. (AGI, 1987)

Leon combustible gases tester A combustible gases detector developed in 1902. A form of Wheatstone bridge is used and changes in electrical resistance due to temperature differences are measured. The combustible gases/air sample flows over one set of wires and the gas burns catalytically while the other wires do not come into contact with the sample. (Nelson, 1965)

leonhardite A partially dehydrated variety of laumontite. (Not leonhardtite.)

leonhardtite *starkeyite.* (Not leonhardite.)

leonite A monoclinic mineral, $K_2Mg(SO_4)_2 \cdot 4H_2O$; in marine evaporite deposits. Syn: *magnesium leonite.*

leopardite A variety of quartz porphyry containing small phenocrysts of quartz in a microgranitic groundmass of quartz, orthoclase, albite, and mica. The rock has a characteristically spotted or streaked appearance due to staining by hydroxides of iron and manganese.

leopard rock (a) Can. Pegmatitic rocks associated with the apatite veins of Ontario and Quebec. (b) Syenite gneiss consisting of ellipsoidal lumps measuring several inches across and separated by material that is mainly greenish pyroxene. The rock may be slightly schistose.

Leopard stone Dolomite full of worm castings set in a gray matrix and containing chert nodules, near the base of the Upper Cambrian, Scotland.

Leopoldi furnace A furnace for roasting mercury ores in a batch process, differing from the Bustamente furnace in having a series of brick condensing chambers. (Fay, 1920)

leopoldite *sylvite.*

lepidoblastic Pertaining to a flaky schistosity caused by an abundance of minerals like micas and chlorites with a general parallel arrangement.

lepidocrocite An orthorhombic mineral, 4[gamma-FeO(OH)]; polymorphous with akaganeite, feroxyhyte, and goethite; yellow to orange-red; the weathering product of iron-bearing minerals; forms the pigment brown ocher; such iron oxyhydroxides constitute the rock limonite.

lepidolite (a) A monoclinic, trigonal, or orthorhombic mineral; 1, 2, or $3[K_2Li_4Al_2(Si_8O_{20}(OH)_4]$; mica group; forms a series with muscovite; perfect basal cleavage; pink to purple; in lithium-rich granite pegmatites; a source of lithium. Syn: *lithia mica; lithionite.* (b) A group name for lithium-rich micas.

lepidomelane A black ferrian variety of biotite.

lepidomorphite A variety of phengite, a siliceous variety of muscovite.

lepolite *anorthite.*

leptothermal Said of a hydrothermal mineral deposit formed at temperature and depth conditions intermediate between mesothermal and epithermal; also, said of that environment. Cf: *hypothermal deposit; xenothermal; telethermal.* (AGI, 1987)

leptynolite A fissile or schistose variety of hornfels containing mica, quartz, and feldspar, with or without accessories, such as andalusite and cordierite. The term was originated by Cordier in 1868. Cf: *cornubianite.* (AGI, 1987)

Lerchs-Grossmann optimization A mathematical method based on a block model of an orebody used for determining the most profitable optimum shape for an open pit.

lernilite *vermiculite.*

lesleyite (a) A mixture of a hydromica and corundum. (b) A potassian variety of margarite.

Lessing process A heavy-fluid coal-cleaning process in which a calcium chloride solution having a specific gravity of approx. 1.4 is used for the separation, which takes place in a cylindrical tank 6 to 10 ft (1.8 to 3 m) in diameter with a conical bottom, the total height being nearly 30 ft (9.1 m). The cleaned coal rises to the top where it is removed by a chain scraper and delivered to draining towers. (Gaudin, 1939)

lethal dose A dose of ionizing radiation sufficient to cause death. Median lethal dose (abbreviated MLD or LD_{50}) is the dose required to kill half of the individuals in a group similarly exposed within a specified period of time. The median lethal dose for humans is about 400 rads. (Lyman, 1964)

let into Eng. The recessing of supports into the floor, side, or roof. (SMRB, 1930)

letovicite A triclinic mineral, $(NH_4)_3H(SO_4)_2$; a decomposition product of pyrite in coal.

letter and tracing cutter In the stonework industry, a person who cuts incised or raised letters and simple designs on monumental stones with pneumatic and hand tools. Also called letter cutter; letterer. (DOT, 1949)

letter stone An igneous rock with sheath and core structure giving the appearance of letters on its surface.
lettsomite *cyanotrichite; velvet copper ore.*
leuchtenbergite A pale iron-poor variety of clinochlore.
leucite A tetragonal mineral, $16[KAlSi_2O_6]$; a pseudocubic feldspathoid; forms white to gray trapezohedra in potassium-rich, silica-poor lavas. Syn: *amphigene; grenatite; white garnet; vesuvian.* Cf: *pseudoleucite.*
leucitite A fine-grained or porphyritic extrusive or hypabyssal igneous rock chiefly composed of pyroxene (esp. titanaugite) and leucite, with little or no feldspar and without olivine. (AGI, 1987)
leucitohedron *trapezohedron.*
leucitophyre A porphyritic extrusive rock composed chiefly of leucite, nepheline, and clinopyroxene. Cf: *haüynophyre.* (AGI, 1987)
leucochalcite *olivinite.*
leucocratic Light-colored; applied to igneous rocks that are relatively poor in mafic minerals. The percentage of mafic minerals necessary for a rock to be classified as leucocratic varies among petrologists, but is usually given as less than 30% to 37.5%. Cf: *melanocratic; mesocratic.* Noun, leucocrate. Syn: *light-colored.* (AGI, 1987)
leucomanganite *fairfieldite.*
leucoperthite A loamlike substance, between a resin and wax in character; $C_{50}H_{84}O_3$; very impure and sandy as found in a brown coal at Gesterwitz, near Weissenfels, Germany. It crystallizes in white needles from ether and boiling absolute alcohol, and melts above 100 °C. (AGI, 1987)
leucophane A green to pale-yellow sodium calcium silicate containing beryllium. One of the sources of beryllium. See also: *leucophanite.* (Dana, 1914)
leucophanite A triclinic mineral, $(Na,Ca)_2BeSi_2(O,OH,F)_7$; pseudo-orthorhombic; a source of beryllium. Syn: *leucophane.*
leucophoenicite A monoclinic mineral, $Mn_7(SiO_4)_3(OH)_2$; humite group; pseudo-orthorhombic; purple-pink to raspberry-red; in veins associated with manganese ore deposits.
leucophyllite A variety of muscovite. (Dana, 1914)
leucophyre A term originally applied to altered diabase in which the feldspar has been altered to saussurite, kaolin, and chlorite. This usage is obsolete, but the term is occasionally used for a light-colored hypabyssal rock, being the antithesis of lamprophyre. Not recommended usage. (AGI, 1987)
leucopyrite An oxidized variety of löllingite. See also: *löllingite.*
leucosphenite A monoclinic mineral, $BaNa_4Ti_2B_2Si_{10}O_{30}$; in alkali pegmatites and the Green River Formation of Utah and Wyoming.
leucoxene (a) Fine-grained, opaque white alteration products of ilmenite, mainly finely crystalline rutile. (b) A variety of sphene.
levee An embankment beside a river or an arm of the sea, to prevent overflow. (Standard, 1964)
level (a) A main underground roadway or passage driven along a level course to afford access to stopes or workings and to provide ventilation and a haulageway for the removal of coal or ore. Levels are commonly spaced at regular depth intervals and are either numbered from the surface or designated by their elevation below the top of the shaft. See also: *level interval.* (Nelson, 1965) (b) *mother gate.* (c) An instrument for establishing a horizontal line or plane. (Long, 1960) (d) The act or process of adjusting something with reference to a horizontal line. (Long, 1960) (e) In pitch mining, such as anthracite, there may be a number of levels driven from the same shaft, each being known by its depth from the surface or by the name of the bed or seam in which it is driven. (Jones, 1949) (f) Applied to seams that run like floors in an office building. Under and above the seam lie the rock strata. (Korson, 1938) (g) All openings at each of the different horizons from which the orebody is opened up and mining is started. (Higham, 1951) (h) N.S.W. A drive in a mine. (New South Wales, 1958) (i) In speleology, a series of related passageways in a cave, occurring at the same relative, vertical position. (AGI, 1987) (j) A gutter for the water to run in. (Fay, 1920)
level course (a) A direction along the strike of an inclined coal seam; a coal seam contour line. The productive faces in a coal mine, such as stalls and conveyor faces are, in general, advanced on level course or slightly to the rise. (Nelson, 1965) (b) Scot. In the direction of the strike of the strata, or at right angles to the dip and rise. See also: *strike; true dip.* (Fay, 1920)
level crosscut A horizontal crosscut. See also: *crosscut.* (Nelson, 1965)
level drive A drive that opens up a deposit and makes it accessible along its length and forms the basis for the division of the deposit into levels. (Stoces, 1954)
leveler A buck scraper, drag, or any other form of device for smoothing land. (Seelye, 1951)
level-free (a) War. Old coal or ironstone workings at the outcrop, worked by means of an adit driven into the hillside. (b) A mine that discharges water by gravitation.
leveling The operation of determining the comparative altitude of different points on the Earth's surface, usually by sighting through a leveling instrument at one point to a level rod at another point. Also, the finding of a horizontal line or the establishing of grades (such as for a railway roadbed) by means of a level. Also spelled levelling. (AGI, 1987)
leveling instrument A surveyor's level bearing a telescope. See also: *level.* (Standard, 1964)
leveling practice In leveling, the station is the point at which the staff is held and not the position of the instrument. The operation is one of carrying forward a known level, hence the backsight is a reading taken on the staff at a known elevation and the last sight from each station is called the foresight. All other readings refer to intermediate sights. Leveling sections may be referred to bench marks or to arbitrary levels, but in all cases they must be checked either by closing on the starting point or by starting and finishing on convenient bench marks. (Mason, 1951)
leveling rod A graduated rod used in measuring the vertical distance between a point on the ground and the line of sight of a surveyor's level. (Webster 3rd, 1966)
level interval (a) The vertical distance between the levels turned off the shaft in metal mines for ore intersection and development. The interval varies but may be about 150 ft (46 m). (Nelson, 1965) (b) The horizontal distance between levels turned off main development drifts and varies from 200 to 600 yd (180 to 550 m). Levels are usually designated by numbers, names, or depth from the surface. (Nelson, 1965)
level-luffing crane A crane embodying an automatic device that causes the load to move horizontally with any alteration of the operating radius. (Hammond, 1965)
levelman Person who operates a surveyor's level. (Crispin, 1964)
level of control A measure of mastery over a process of production; in concrete work, it is measured by cube crushing strength and the standard deviation therefrom. See also: *statistical uniformity.* (Hammond, 1965)
level of saturation *water table.*
levels *level.*
level surface *equipotential surface.*
leverman Person who operates brakes, or levers, at the top of an inclined plane. A brakeman. (Fay, 1920)
levigation (a) Separating fine powder from coarser material by forming a suspension of the fine material in a liquid. (ASM, 1961) (b) A means of classifying a material as to particle size by the rate of settling from a suspension. Cf: *trituration.* (ASM, 1961)
levitation In the mineral process of froth flotation, raising of acrophilic particles to the surface of a pulp, by so activating them that they cling to the air-water interface of a rising or coursing air bubble. (Pryor, 1963)
levyne A trigonal mineral, $(Ca,Na_2,K_2)_3Al_6Si_{12}O_{36}\cdot 18H_2O$; zeolite group; in cavities in basalts with other zeolites. Also spelled levynite; levyine; levyite.
lewis hole A series of two or more holes drilled as closely together as possible, then connected by knocking out the thin partition between them, forming thus one wide hole, having its greatest diameter in a plane with the desired rift. Blasts from such holes are wedgelike in their action, and by means of them larger and better-shaped blocks can be taken out than would otherwise be possible. (Fay, 1920)
Leyner-Ingersoll drill *water leyner.*
lherzolite A peridotite containing both clinopyroxene and orthopyroxene in addition to olivine.
liberation Freeing by comminution, or crushing and grinding, of particles of a specific mineral from their interlock with other constituents of the coal or ore. Also called severance. Syn: *unlocking.* (Pryor, 1965)
liberation of intergrown constituents Crushing of intergrown material to free the constituent materials. (BS, 1962)
liberator cells In electrolytic refining of metals, tanks in which the electrolytic solution is reconstituted. (Pryor, 1963)
Liberty-Gel Gelatinous permissible explosive; used in mining. (Bennett, 1962)
libethenite An orthorhombic mineral, $Cu_2(PO_4)(OH)$; olive-green; forms small crystals and druses in copper deposits.
LICADO process A selective agglomeration process under development, in which the liquid carbon dioxide-water interface is used for the differentiation and separation between coal and mineral matter. The resultant clean coal is a low-sulfur and low-ash content product of relatively low moisture content.
licensed material Source material, special nuclear material, or byproduct material received, possessed, used, or transferred under a general or special license issued by the Nuclear Regulatory Commission. (Lyman, 1964)
licensed store A place or building licensed by the local authority for the storage of explosives. See also: *magazine; registered premises.* (BS, 1964)
lid (a) A short, flattish piece of wood or steel plate wedged over a post, timber set, or steel arch. A lid is used to tighten the support against the

ground and also to increase the area supported. See also: *clog; wedge.* (Nelson, 1965) (b) A cap piece used in timbering. Cf: *lag.* (BCI, 1947)

lidman In the coke-products industry, a laborer who lifts lids of charging holes of ovens and chips carbon from edges of holes, using bars with hook-and-chisel ends. Syn: *charger.* (DOT, 1949)

lie! Scot. In mine haulage, a command to stop. (Fay, 1920)

lie (a) Scot. The line, direction, or bearing; as of a vein, lode, or fault. (Fay, 1920) (b) Pass-by; shunt; a storage or bypass arrangement in haulage track. Also spelled lye. (Mason, 1951) (c) To become quiet or inactive; said of a mine that is idle. (Fay, 1920)

liebigite An orthorhombic mineral, $Ca_2(UO_2)(CO_3)_3 \cdot 11H_2O$; olive-green; forms soft, scaly, or granular crystalline aggregates with one cleavage. Syn: *uranothallite.*

lien Right to legal claim on goods or property. (Pryor, 1958)

liesegang banding Banding in color and composition of ores caused by diffusion. (AGI, 1987)

lievrite *ilvaite.*

lifeline A slide wire or cable extending from a work platform in a drill tripod or derrick at an oblique angle downward to an anchor on the ground, which the derrick or tripod worker could grasp and use when sliding to safety in an emergency. (Long, 1960)

life of mine The time in which, through the employment of the available capital, the ore reserves—or such reasonable extension of the ore reserves as conservative geological analysis may justify—will be extracted. (Hoover, 1948)

life of property Theoretically, the mineral or coal reserves divided by the actual or projected average annual production. (Nelson, 1965)

lift (a) The vertical height traveled by a cage in a shaft. (Fay, 1920) (b) The distance between the first level and the surface or between any two levels. (Fay, 1920) (c) Any of the various gangways from which coal is raised at a slope colliery. (Fay, 1920) (d) A certain thickness of coal worked in one operation. (Fay, 1920) (e) To break up, bench, or blast coal from the bottom of the seam upward. (Fay, 1920) (f) The plane approx. parallel with the floor of the quarry, along which the stone is usually split in quarrying. (Fay, 1920) (g) The quantity of ore between one haulage level and the next above or below. (Nelson, 1965) (h) A step or bench in a multiple layer excavation. (Nichols, 1976) (i) The amount a bit is raised off the bottom of a drill hole by excessive pressure created by pump surges or the forcing of too great a volume of circulation fluid through the bit. (Long, 1960) (j) In churn drilling, the vertical movement of the drill tools while drilling. (Long, 1960) (k) In pumping, the difference in the elevation between the surface of the liquid being pumped and the elevation at which the pump stands or the elevation at which the liquid is discharged. (Long, 1960) (l) A certain vertical thickness of coal seams and measures, having considerable inclination, between or in which the workings are being carried on to the rise, all the coal being raised from one shaft bottom. (Fay, 1920) (m) The upheaval of the floor in coal mines. See also: *creep.* (Nelson, 1965) (n) The extraction of a coal pillar in lifts or slices. See also: *jud.* (Nelson, 1965)

lifter (a) In mining, a shothole drilled near the floor when tunneling and fired subsequently to the cut and relief holes. (Pryor, 1963) (b) *core lifter.* (c) In ore grinding, a projection, rib or wave profile on the horizontal liners (body liners) of a ball, tube, or rod mill, designed to aid the crop load in the mill to rise. In a drum-washer or dense-medium separator, a perforated plate, projecting radially inward from the circumference of a horizontal cylindrical vessel, used to stir, lift, or remove material. (Pryor, 1963)

lifter holes Shotholes drilled along the floor of a tunnel for lifting the rock to floor level. They are fired after the cut holes, or by delay detonators in the round. (Nelson, 1965)

lifters *lifter holes.*

lifter spring *core lifter.*

lift gate A lock gate that is raised vertically to open. (Hammond, 1965)

lift hammer *tilt hammer.*

lifting Scot. Drawing hutches or cars out of the working places into the main roads. (Fay, 1920)

lifting block An arrangement of pulleys and rope that enables heavy weights to be lifted with least effort. (Hammond, 1965)

lifting capacity (a) The weight that the hydraulic cylinders in the swivel head of a diamond drill can raise or lift. (Long, 1960) (b) *drill capacity.*

lifting guard Fencing placed around the mouth of a shaft, and lifted out of the way by the ascending cage. (Fay, 1920)

lifting magnet An electromagnet that is hung from a crane and used instead of a hook for lifting iron or steel components. (Hammond, 1965)

lifting set A series of pumps or sets of pumps by which water is lifted from the mine in successive stages. *lift.* (Standard, 1964)

lifting wicket S. of Wales. *lifting guard.*

lift joint A horizontal tension fracture observed in massive rocks, such as granite; thought to originate from the removal of load in quarrying. Cf: *sheeting.*

lift pump A pump for lifting to its own level, as distinguished from a force pump. A suction pump. Also called bucket pump. See also: *suction head.* (Standard, 1964; Fay, 1920)

light alloys The general term for alloys of aluminum and magnesium used for structural purposes. (Hammond, 1965)

light blasting Includes loosening up of shallow or small outcrops of rock and breaking boulders. It may constitute the entire job, be done in connection with dirt excavation, or follow heavy blasting that has failed to cut gradelines or slope lines, or has left chunks too large to load. (Nichols, 1976)

light burden *burden.*

light-colored *leucocratic.*

lightening A peculiar brightening of molten silver, indicating that maximum purity has been attained. Occurs in cupellation. (Standard, 1964; Fay, 1920)

light-extinction method *turbidimeter.*

light figure The visible geometric figure observed when an etched flat surface of quartz is placed over a pinhole-focused light source. (Am. Mineral., 1947)

lighting In metallurgy, annealing. (Standard, 1964)

lightman Person who uses an electric extension light as an aid in detecting blisters and flaws in the inside of green pipe. (DOT, 1949)

light mineral (a) A rock-forming mineral of a detrital sedimentary rock, having a specific gravity lower than a standard (usually 2.85); e.g., quartz, feldspar, calcite, dolomite, muscovite, feldspathoids. Cf: *heavy mineral.* (AGI, 1987) (b) A light-colored mineral. (AGI, 1987)

lightning explosion Eng. An explosion of combustible gases caused by electric current, during a thunderstorm, entering a mine and igniting the gas. (Fay, 1920)

lightning gap A lightning gap is a break about 6 ft (1.8 m) long made at the mine entrance in blasting circuits, used in firing blasts from the outside, to prevent lightning discharges from following the circuits into the mine.

lightning protection A system to enable high electrical discharge from the atmosphere to be conducted safely to earth by one or more conductors. The provision is very important in the case of mine explosive stores and also headgears, tower winders, and chimneys. (Nelson, 1965)

lightning tube *fulgurite.*

light railway A railway built to narrow gage. (Hammond, 1965)

light red silver ore *proustite.*

light ruby ore *proustite.*

light ruby silver *proustite.*

light water Ordinary water, H_2O, as distinguished from heavy water, D_2O, D being the symbol for deuterium (heavy hydrogen or hydrogen 2). (Lyman, 1964)

lightweight aggregate An aggregate with a relatively low specific gravity; e.g., pumice, volcanic cinders, expanded shale, foamed slag, or expanded perlite or vermiculite. Cf: *aggregate.* (AGI, 1987)

lightweight concrete A concrete made with lightweight aggregate. (AGI, 1987)

light-yellow heat A division of the color scale, generally given as about 2,400 °F (1,316 °C).

lignin sulfonic acids Chemicals produced during the sulfite treatment of wood pulps. Of interest in flotation process as a deflocculating agent and protective colloid. (Pryor, 1963)

lignite (a) A brownish-black coal in which the alteration of vegetal material has proceeded further than in peat but not so far as subbituminous coal. (Fay, 1920) (b) Coal of low rank with a high inherent moisture and volatile matter; in this general sense, lignite may be subdivided into black lignite, brown lignite, and brown coal. (BS, 1960)

lignite A The rank of coal, within the lignitic class of Classification D 388, such that, on the moist, mineral-matter-free basis, the gross calorific value of the coal in British thermal units per pound is equal to greater than 6,300 (14.65 MJ/kg but less than 8,300 (19.31 MJ/kg), and the coal is nonagglomerating. (ASTM, 1994)

lignite B The rank of coal, within the lignitic class of Classification D 388, such that, on the moist, mineral-matter-free basis, the gross calorific value of the coal in British thermal units per pound is less than 6,300 (14.65 MJ/kg), and the coal is nonagglomerating. (ASTM, 1994)

lignitic Containing lignite. (Fay, 1920)

lignitic coal *subbituminous coal.*

ligurite An apple-green variety of titanite. (Standard, 1964)

likasite An orthorhombic mineral, $Cu_3(NO_3)(OH)_5 \cdot 2H_2O$; sky blue; at Likasi, Kutanga, Zaire.

likely Said of a rock, lode, or belt of ground that gives indications of containing valuable minerals. Syn: *kindly.* Cf: *hungry.* (AGI, 1987)

lill Eng. Greenish-gray shale; weathering yellow; Wenlock Limestone, Dudley. (Arkell, 1953)

lillianite An orthorhombic mineral, $Pb_3Bi_2S_6$.

Lilly controller A controller used on both steam and electric winding engines that protects against overspeed, overwind, too rapid acceleration, delayed retardation, and against starting in the wrong direction.

It also gives warning of overspeed and indicates by a bell signal when retardation should commence. (Sinclair, 1959)

limb (a) That area of a fold between adjacent fold hinges. It generally has a greater radius of curvature than the hinge region and may be planar. Syn: *flank.* Obsolete syn. shank. (AGI, 1987) (b) The graduated margin of an arc or circle in an instrument for measuring angles, such as the part of a marine sextant carrying the altitude scale. (AGI, 1987) (c) The graduated staff of a leveling rod. See also: *tribrach.* (AGI, 1987)

lime (a) Calcium oxide, CaO; specif. quicklime and hydraulic lime. The term is used loosely for calcium hydroxide (as in hydrated lime) and incorrectly for calcium carbonate (as in agricultural lime). (AGI, 1987) (b) A cubic mineral, CaO. (AGI, 1987) (c) A term commonly misused for calcium in such deplorable expressions as carbonate of lime or lime feldspar. (AGI, 1987) (d) A limestone. The term is sometimes used by drillers for any rock consisting predominantly of calcium carbonate. (AGI, 1987)

lime boil A reaction in an open-hearth furnace caused by the decomposition of limestone and the escape of the carbon dioxide gas. This reaction begins before the ore boil is completed. See also: *ore boil.* (Newton, 1959)

limeburner Person who burns limestone or shells to make lime. (Webster 3rd, 1966)

lime feldspar Misnomer for calcium feldspar. See also: *anorthite.* (AGI, 1987)

lime mica *margarite.*

lime mortar A mortar in which lime is used as a binding agent instead of cement. (Nelson, 1965)

lime pan (a) A playa with a smooth, hard surface of calcium carbonate, commonly tufa. (AGI, 1987) (b) A type of hardpan cemented chiefly with calcium carbonate. Also spelled limepan. (AGI, 1987)

lime pit (a) A limestone quarry. (Webster 3rd, 1966) (b) A pit where lime is made. (Webster 3rd, 1966)

lime rock A term used in the Southeastern United States (esp. Florida and Georgia) for an unconsolidated or partly consolidated form of limestone, usually containing shells or shell fragments, with a varying percentage of silica. It hardens on exposure and is sometimes used as road metal. Also spelled limerock. (AGI, 1987)

lime set An infusible slag, too high in lime, in an iron blast furnace.

lime shells Scot. Calcined limestone.

lime-silicate rock *calc-silicate rock.*

lime slaker Person who mixes lime and water in rotary slaker or open batch tank to make milk of lime (slaked lime). Also called lime mixer; milk-of-lime slaker; slaker. (DOT, 1949)

lime-soda sinter process A process for manufacturing alumina, Al_2O_3. The raw material, such as clay or anorthosite, is sintered with limestone and soda ash to form sodium aluminate and calcium silicate. This sinter is then leached with water, caustic soda solution, or sodium aluminate liquor to dissolve the soluble sodium aluminate. The resulting slurry is then filtered, and the liquor is decomposed as in the Bayer process or is treated with carbon dioxide to precipitate hydrated alumina. When operated in conjunction with the Bayer process to recover alumina and soda from red mud, it is called the combination process.

limestone (a) A sedimentary rock consisting chiefly (more than 50% by weight or by areal percentages under the microscope) of calcium carbonate, primarily in the form of the mineral calcite, and with or without magnesium carbonate; specif. a carbonate sedimentary rock containing more than 95% calcite and less than 5% dolomite. Common minor constituents include silica (chalcedony), feldspar, clays, pyrite, and siderite. Limestones are formed by either organic or inorganic processes, and may be detrital, chemical, oolitic, earthy, crystalline, or recrystallized; many are highly fossiliferous and clearly represent ancient shell banks or coral reefs. Limestones include chalk, calcarenite, coquina, and travertine, and they effervesce freely with any common acid. Abbrev. ls. (AGI, 1987) (b) A general term used commercially (in the manufacture of lime) for a class of rocks containing at least 80% of the carbonates of calcium or magnesium and which, when calcined, gives a product that slakes upon the addition of water. (AGI, 1987)

limestone dust Dust prepared by grinding limestone; used to dilute the coal dust accumulation in a mine so that the dust does not form explosive mixtures with air. (Rice, 1913-18)

limewater Natural water with large amounts of dissolved calcium bicarbonate or calcium sulfate. (AGI, 1987)

limit charge A charge that gives a complete loosening of the rock without throwing it excessively.

limiting creep stress A somewhat loose term used to denote the maximum stress at which a material will not creep by more than a certain amount within the working life of the part. It is also used in some short-time creep tests; e.g., the Hatfield time yield. (Hammond, 1965)

limiting current density The maximum current density that can be used to get a desired electrode reaction without undue interference, such as may come from polarization. (ASM, 1961)

limiting gradient The maximum railway gradient that can be climbed without the help of a second power unit. Syn: *ruling gradient.*

limiting mixture The mixture of coal and rock dusts that will not permit the propagation of an explosion. (Rice, 1913-18)

limit line The line joining the coal face underground and the surface limit of draw; the boundary of a mine. (Nelson, 1965)

limit of draw The point on the surface beyond which no movement occurs. (Nelson, 1965)

limit of proportionality The point on a stress-strain curve at which the strain ceases to be proportional to the stress. Its position varies with the sensitivity of the extensometer used in measuring the strain. (CTD, 1958)

limits of flammability (a) Extreme concentration limits of a combustible in an oxidant through which a flame, once initiated, will continue propagating at a specified temperature and pressure. (Van Dolah, 1963) (b) Usually expressed as the limiting percentages of methane in air, beyond which the mixture is no longer flammable. The lowest percentage of methane in air that yields a flammable mixture is called the lower limit of flammability, and the highest percentage of methane in air to yield a similar mixture is called the higher limit of flammability. These limiting percentages depend on a number of factors, such as the initial temperature and pressure; whether the mixture is at rest or moving; the manner in which the mixture is confined, etc. With methane mixtures at ordinary mine pressures and temperatures, the widest limits of flammability are (1) lower limit of flammability about 5.4% of methane in air; (2) higher limit of flammability about 14.8% of methane in air. See also: *methane.* (Nelson, 1965)

limit switch (a) A device fitted to an electrically driven hoist or winding engine that becomes effective at the end of a wind to prevent the cage overwinding or underwinding. (Nelson, 1965) (b) A control to limit some function. (Strock, 1948)

limnic (a) Said of coal deposits formed inland in freshwater basins, peat bogs, or swamps, as opposed to paralic coal deposits. Cf: *paralic.* (AGI, 1987) (b) Said of peat formed beneath a body of standing water. Its organic material is mainly planktonic. (AGI, 1987)

limnic coal basin A coal basin formed inland from the seacoasts, as opposed to a paralic coal basin. (AGI, 1987)

limnite *bog iron; bog iron ore.*

limonite A rock composed of cryptocrystalline and amorphous hydrated iron oxyhydroxides, generally predominantly goethite with or without adsorbed water, but also akaganeite, feroxyhyte, or lepidocrocite; may be yellow, red, brown, or black; hardness variable; an oxidation product of iron (rust) or iron-bearing minerals and may be pseudomorphous after them; as a precipitate, both inorganic and biogenic, in bogs, lakes, springs, or marine deposits; and as a variety of stalactitic, reniform, botryoidal, or mammillary deposits. It colors many yellow clays and soils and is a minor ore of iron. See also: *iron ore; brown hematite; brown umber.* Cf: *bog iron ore.* Syn: *brown iron ore; brown hematite; brown ocher; meadow ore; yellow ocher.*

limonitic Consisting of or resembling limonite.

limurite A metasomatic rock found at the contact of calcareous rocks with intruded granite and consisting of more than 50% axinite. Other minerals include diopside, actinolite, zoisite, albite, and quartz. See also: *calc-silicate hornfels.* (AGI, 1987)

linarite A monoclinic mineral, $2[PbCu(SO_4)(OH)_2]$; azure blue; in oxidized zones of lead, copper, and silver deposits.

lindackerite A monoclinic mineral, $H_2Cu_5(AsO_4)_4{\cdot}8\text{-}9H_2O$; green; forms rosettes and reniform masses with one perfect cleavage.

Lindblad-Malmquist gravimeter *Boliden gravimeter.*

lindgrenite A monoclinic mineral, $Cu_3(MoO_4)_2(OH)_2$; green; in tabular crystals from Chuquicamata, Chile.

Lindgren's volume law *law of equal volumes.*

lindströmite An orthorhombic mineral, $Pb_3Cu_3Bi_7S_{15}$. Also spelled lindstroemite.

line (a) The limit of a surface; a length without breadth; an outline; a contour. (Fay, 1920) (b) The course in which anything proceeds, or which anyone takes; direction given or assured. (Fay, 1920) (c) A cable, rope, chain, or other flexible device for transmitting pull. (Nichols, 1976) (d) To line pieces up in order to couple them together. (Nichols, 1976) (e) *plumbline.*

lineage An imperfection in a crystal structure characterized by a series of line defects, but not extensive enough to form a grain boundary. Cf: *grain boundary; crystal defect.*

lineal foot A foot in length as distinguished from square foot or cubic foot. (Crispin, 1964)

lineal travel Commonly used as a syn. for peripheral speed, as applied to bit rotation; also, a syn. for rope or cable speed, as applied to hoisting. (Long, 1960)

lineament (a) A significant line of landscape that reveals the hidden architecture of the rock basement. Lineaments are character lines of the Earth's physiognomy. (AGI, 1987) (b) A topographic feature; esp., one that is rectilinear. (Webster 3rd, 1966) (c) A topographic line that is structurally controlled. Lineaments are studied esp. on aerial photographs. Sometimes inappropriately called a linear. (Billings, 1954)

linear Arranged in a line or lines; pertaining to the linelike character of some object or objects.—n. Not recommended as a syn. of lineament. (AGI, 1987)

linear alkylbenzene sulfonate A high-quality, biodegradable detergent, obtained from lignite tar, which is derived from carbonizing coal at low temperatures.

linear element A fabric element having one dimension that is much greater than the other two. Lineations are the common linear elements. Cf: *planar element; equant element.* (AGI, 1987)

linear expansion The increase in one dimension of a soil mass, expressed as a percentage of that dimension at the shrinkage limit to any given water content. (ASCE, 1958)

linear model A function frequently used when fitting mathematical models to experimental variograms, often in combination with a nugget model.

linear parallelism *lineation.*

linear schistosity A schistosity due to the parallel alignment of columnar or acicular crystals in rocks.

linear shrinkage Decrease in one dimension of a soil mass, expressed as a percentage of the original dimension, when the water content is reduced from a given value to the shrinkage limit. (ASCE, 1958)

linear structure *lineation.*

lineation A general, nongeneric term for any linear structure in a rock; e.g., flow lines, slickensides, linear arrangements of components in sediments, or axes of folds. Lineation in metamorphic rocks includes mineral streaking and stretching in the direction of transport, crinkles and minute folds parallel to fold axes, and lines of intersection between bedding and cleavage, or of variously oriented cleavages. Syn: *linear structure; linear parallelism.* (AGI, 1987)

line brattice A partition placed in an opening to divide it into intake and return airways. See also: *brattice.* (Hartman, 1982)

line clinometer A borehole-survey clinometer designed to be inserted between rods at any point in a string of drill rods. Cf: *clinometer; end clinometer; plain clinometer; wedge clinometer.* (Long, 1960)

line defect An imperfection in crystal structure characterized by an atomic layer extending part way into a crystal structure. Cf: *crystal defect; edge dislocation.*

lined gold Gold foil backed with other metal.

line drilling A term used in quarrying to describe the method of drilling and broaching for the primary cut. In this method, deep holes are drilled close together in a straight line by means of a reciprocating drill mounted on a bar. The webs between the holes are removed with a drill or a flat broaching tool; thus a narrow continuous channel cut is made. See also: *broaching; controlled blasting.* (AIME, 1960)

line drop Loss in voltage owing to the resistance of conductors conveying electricity from a power station to the consumer. (Hammond, 1965)

line electrode A series of electrodes put out along a straight line on the surface and electrically interconnected, approx. the condition of continuous electrical contact with the Earth along that line. (Schieferdecker, 1959)

line lubricator *line oiler.*

line map *planimetric map.*

line of bearing (a) The compass direction of the course a borehole follows. (Long, 1960) (b) A syn. for strike as applied to a rock stratum. (Long, 1960) (c) The direction of the strike. *strike.*

line of collimation The line of sight of the telescope of a surveying instrument, defined as the line through the rear nodal point of the objective lens of the telescope and the center of the reticle when they are in perfect alignment. See also: *line of sight.* (AGI, 1987)

line of creep The path that water follows along the impervious surface of contact between the foundation soil and the base of a dam or other structure. Syn: *path of percolation.* (ASCE, 1958)

line of dip (a) The direction in which an inclined borehole is pointed. (Long, 1960) (b) The line of greatest inclination of a stratum from the horizontal plane. (c) The direction of the angle of dip, measured in degrees by compass direction. It generally refers to true dip, but can be said of apparent dip as well. Syn: *direction of dip.* (AGI, 1987)

line of force (a) The straight line in which a force acts. (Standard, 1964) (b) A curve in a field of force drawn so that at every point it has the direction of resultant force; specif., a line of magnetic force. (Standard, 1964)

line of least resistance The shortest distance between the center line of a drill hole and the free rock face. (Fraenkel, 1953)

line of outcrop The intersection of a stratum with the ground surface. (Schieferdecker, 1959)

line of seepage The upper free-water surface of the zone of seepage. Syn: *seepage line; phreatic line.* (ASCE, 1958)

line of sight The sighting or pointing line of a telescope, defined by the optical center of the objective and the intersection of crosshairs. See also: *line of collimation.* (Seelye, 1951)

line of thrust Locus of the points through which the resultant forces in an arch or retaining wall pass. (Hammond, 1965)

line of tunnel The width marked by the exterior lines or sides of a tunnel.

line oiler An apparatus inserted in a line conducting air or steam to an air- or steam-actuated machine that feeds small controllable amounts of lubricating oil into the air or steam. Also called lubricator; potato. Syn: *air-line lubricator; line lubricator; oil pot; oiler; pineapple; pot.* See also: *atomizer.* (Long, 1960)

liner (a) A foot piece for uprights in timber sets. (Nelson, 1965) (b) Timber supports erected to reinforce existing sets that are beginning to collapse due to heavy strata pressure. (Nelson, 1965) (c) A string of casing in a borehole. (Long, 1960) (d) *lining.*

lines Plumblines, not less than two in number, hung from hooks driven in wooden plugs. A line drawn through the center of the two strings or wires, as the case may be, represents the bearing or course to be driven on. (Fay, 1920)

linesman An assistant to a surveyor. (BS, 1963)

lines up The number of lines strung through the traveling block and crown block. (Brantly, 1961)

line timbers Timbers placed along the sides of the track of a working place in rows on some predetermined plan. (Kentucky, 1952)

line up (a) A command signifying that the drill runner wants the hoisting cable attached to the drill stem, threaded through the sheave wheel, or wound on the hoist drum. (Long, 1960) (b) To reposition a drill so that the drill stem is centered over and parallel to a newly collared drill hole. (Long, 1960) (c) Regular linear pattern of peaks or troughs on a seismogram, such as occurs when a reflection comes in. (Schieferdecker, 1959)

lingot An iron ingot mold. (Standard, 1964)

lining (a) A casing of brick, concrete, cast iron, or steel, placed in a tunnel or shaft to provide support. See also: *tunnel lining.* (AGI, 1987) (b) In grinding of rocks and ores, a layer of wear-resistant material, used to protect rock breakers, shells of ball mills, chutes, launders, and other areas subject to abrasion. Special plastics, rubber, and acid-resistant alloys are used in vats, piping, and autoclaves, and for pump impellers exposed to abrasion and chemical attack. See also: *scour protection.* (Pryor, 1963) (c) A cover of clay, concrete, synthetic film, or other material, placed over all or part of the perimeter of a conduit or reservoir, to resist erosion, minimize seepage losses, withstand pressure, and improve flow. (AGI, 1987) (d) Supporting the rock sides of a mine gallery is known as lining, and the material used in so doing is called the lining. (Spalding, 1949) (e) The planks arranged against frame sets. (Zern, 1928) (f) Refractory brickwork of furnace used to protect hearth, bosh, roof, or walls. (Pryor, 1963) (g) The brick, concrete, cast iron, or steel casing placed around a tunnel or shaft as a support. Timber sets are not viewed as a lining. See also: *brick walling; guniting; liner; permanent shaft support.* (Nelson, 1965)

lining mark Eng. A drill hole in the mine roof with a wooden plug driven into it from which to hang a plumbline. (Fay, 1920)

lining sight An instrument consisting essentially of a plate with a longitudinal slot in the middle, and the means of suspending it vertically. It is used in conjunction with a plumbline for directing the courses of underground drifts, headings, etc. (Webster 2nd, 1960)

lining up a mine In surveying, placing the sights for driving entries, drifts, or rooms nearer the working face. (Fay, 1920)

linishing The operation of polishing as carried out on a linisher. This machine is designed for the polishing of flat objects and carries a flat revolving cloth belt whose surface is impregnated with a suitable abrasive material. (Osborne, 1956)

link bar A lightweight steel bar extending faceward from the steel supports behind the conveyor. It supports the area between the conveyor and the coal on longwall faces where cutter loaders are carried on armored flexible conveyors. The joint and locking device may consist of a hinge pin and wedge. The standard bar can carry in cantilever a maximum tip load of about 2 st (1.8 t). In general, linked bars are stronger than corrugated straps and wooden bars, and their use is increasing. (Nelson, 1965)

link conveyor A chain conveyor. (Nelson, 1965)

linked veins An ore-deposit pattern in which adjacent, more or less parallel veins are connected by diagonal veins or veinlets. (AGI, 1987)

Linkenbach table In mineral processing, a stationary round table onto which finely divided pulp (slimes) is fed from the center from a revolving feed box. A light wash of water follows, and lighter particles are washed downslope to a peripheral discharge launder. Behind this a heavier water flush displaces the settled heavier material (concentrates) and flushes it down to a bridging arrangement that delivers it to a separate discharge. (Pryor, 1963)

link-plate belt A grizzly type of belt consisting of two strands of endless chain connected by through rods at each articulation, on which are carried a series of plates or bars mounted in a vertical plane for the purpose of rough screening while conveying. Syn: *belt.*

linnaeite (a) An isometric mineral, $8[Co_3S_4]$; commonly has small amounts of Ni, Fe, Cu; forms a series with polydymite; crystallizes in octahedra with cubic cleavage; metallic; steel gray tarnishing to copper red; in quartz sulfide veins; a source of cobalt. Also spelled linneite. (b) The mineral group bornhardtite, carrollite, daubreelite, fletcherite, greigite, indite, linnaeite, polydymite, siegenite, trüstedite, tyrrellite, and violarite.
linneite Alternate spelling of linnaeite.
linoleic acid An unsaturated fatty acid, $CH_3(CH_2)_4CH = CH(CH_2)_7COOH$, used as a collector in the flotation process. (Pryor, 1963; SME, 1985)
linolenic acid An unsaturated fatty acid, $CH_3CH_2CH = CH\ CH_2\ CH = CH\ CH_2\ CH = CH(CH_2)_7COOH$. (Pryor, 1963; SME, 1985)
linophyric Pertaining to porphyritic rocks in which the phenocrysts are arranged in lines or streaks.
linseed fatty acid Byproduct of manufacture of linseed oil. Flotation agent used as collector, emulsifier, or stabilizer for davidite, a uranium mineral. (Pryor, 1963)
linsey Lanc. A strong, striped shale and a streaky, banded sandstone or siltstone, interbedded in such a manner as to resemble a mixed linen and woolen fabric (linsey-woolsey). Cf: *whintin.* (AGI, 1987)
lintel A horizontal supporting member spanning a wall opening. (Harbison-Walker, 1972)
Linz-Donawitz process A process for making steel from cast iron; it resembles the Bessemer process except for two important differences: (1) oxygen is used rather than air and (2) instead of blowing the gas through tuyeres submerged in the bath (as in the Bessemer converter), the oxygen stream impinges on the surface of the molten iron. (Newton, 1959)
lionite *tellurium.*
lip (a) The digging edge of a dredge bucket. (Fay, 1920) (b) The cutting edge of a fixed-wing bit, such as the cutting edge on a fishtail bit. (Long, 1960)
lip-and-gate builder One who constructs lips and gates that support and regulate flow of molten glass from furnace to glass-rolling machine. (DOT, 1949)
liparite A syn. of rhyolite used by German and Soviet authors. Its name, given by Roth in 1861, is derived from the Lipari Islands, in the Tyrrhenian Sea. Not recommended usage; the much more widely used syn. rhyolite has priority by 1 yr. (AGI, 1987)
lip of shaft Eng. The bottom edge of a shaft circle where it is open to the seam workings. (Fay, 1920)
lip screen (a) A common term applied to stationary screens installed in the loading chutes over which the coal flows as it is loaded into railroad cars for market. (Mitchell, 1950) (b) A small screen or screen bars, placed at the draw hole of a coal pocket to take out the fine coal.
liptinite *exinite.*
liptite Term used to describe a microlithotype consisting mainly of the exinite group of macerals and esp. of sporinite. Contains not less than 95% of exinite (liptinite) with thickness (bands) of exinite greater than 50 µm recorded as liptite (sporite). Liptite (sporite) is a rare constituent of hard coal. Cf: *vitrite.* (IHCP, 1963)
Liqnipel process A process for reducing high-moisture-content lignite from up to 40% moisture to about 10% moisture by pulverizing the raw coal, pelletizing the 14-mesh top size, thermally drying, and then mixing with a binder. The final product exhibits good handling and storage properties, and resists spontaneous combustion.
liquation (a) A method of recovering sulfur by liquefying it under pressure and heat, drawing off the molten sulfur, and allowing it to solidify. (Fay, 1920) (b) The heating of a solid mixture until one of the constituents melts and can be separated from the solid remaining. (Nelson, 1965)
liquid air Air in the liquid state but usually richer in oxygen than gaseous air. A faintly bluish, transparent, mobile, intensely cold liquid. Obtained by compressing purified air and cooling it by its own expansion to a temperature below the boiling points of its principal components, nitrogen (-195.8 °C, at 760 mm) and oxygen (-182.96 °C, at 760 mm). Used chiefly as a refrigerant and as a source of oxygen, nitrogen, and inert gases (as argon). (Handbook of Chem. & Phys., 2)
liquidation grade The amount paid by the smelter or other purchaser per ton of ore mined. (McKinstry, 1948)
liquid bituminous material Material having a penetration at 77 °F (25 °C), under a load of 50 g applied for 1 s, of more than 350. (Urquhart, 1959)
liquid glass *water glass.*
liquid inclusion A partial syn. of fluid inclusion. (AGI, 1987)
liquid-liquid extraction *solvent extraction.*
liquid measure Includes the following: 4 gills (gi) equal 1 pint (pt) or 0.47 L; 2 pt equal 1 quart (qt) or 0.95 L; 4 qt equal 1 gal or 3.785 L; 31-1/2 gal equal 1 bbl or 119.2 L; and 2 bbl equal 1 hogshead (hhd) or 238.5 L. (Crispin, 1964)
liquid-phase sintering Sintering of a compact or loose powder aggregate under conditions where a liquid phase is present during part of the sintering cycle. (ASTM, 1994)
liquid pressure The pressure of a liquid on the surface of its container or on the surface of any body in the liquid is equal to the weight of a column of the liquid whose height equals the depth of the liquid at that certain point. (Kentucky, 1952)
liquids flowsheet A flowsheet to indicate the flow of liquids throughout a series of operations. (BS, 1962)
liroconite A monoclinic mineral, $Cu_2Al(AsO_4)(OH)_4{\cdot}4H_2O$; soft; blue to green; occurs with malachite in the oxidized zone of copper deposits. Syn: *lentil ore.*
liskeardite A possibly orthorhombic mineral, $(Al,Fe)_3(AsO_4)(OH)_6{\cdot}5H_2O$; forms thin encrustations at Liskear, Cornwall, U.K.
list (a) Mid. Dull coal, sometimes dirty. (Tomkeieff, 1954) (b) Derb. and Leic. A hard parting between coal seams. (Arkell, 1953) (c) York. A weak shale or a thin bed of any kind of rock. (Arkell, 1953) (d) Eng. A mine inspector's term for the schedule of particulars of accidents. (Fay, 1920)
listing (a) Fine shale or clay with glossy surfaces or slickensides due to crushing and relative movement. The rock may form a layer over a coal seam as clod; also applied to leather bed. (Nelson, 1965) (b) *lashing.*
list mill In gem cutting, a wheel covered with list or cloth on which gems are polished. Also called list wheel. (Standard, 1964)
list pan A perforated skimmer for skimming molten tin. (Standard, 1964)
listric fault A curved downward-flattening fault, generally concave upward. Listric faults may be characterized by normal or reverse separation. (AGI, 1987)
listy bed Soft freestone between the Chert Vein and House Cap in the Portland beds at Winspit, Purbeck, U.K.
lith- A prefix meaning stone or stonelike.
litharge A tetragonal mineral, PbO; red; dimorphous with massicot, which is yellow. Syn: *lead ocher; lithargite.*
lithargite *litharge.*
lithia amethyst *kunzite; spodumene.*
lithia emerald *hiddenite; spodumene.*
lithia mica *lepidolite.*
lithic (a) A syn. of lithologic, as in lithic unit. (AGI, 1987) (b) Said of a medium-grained sedimentary rock, and of a pyroclastic deposit, containing abundant fragments of previously formed rocks; also said of such fragments. (AGI, 1987) (c) Pertaining to or made of stone; e.g., lithic artifacts or lithic architecture. (AGI, 1987)
lithic arenite (a) A sandstone containing abundant quartz, chert, and quartzite, less than 10% argillaceous matrix, and more than 10% feldspar, and characterized by an abundance of unstable materials in which the fine-grained rock fragments exceed feldspar grains. The rock is roughly equivalent to subgraywacke. (AGI, 1987) (b) *lithic sandstone.*
lithic graywacke A graywacke characterized by abundant unstable materials; specif. a sandstone containing a variable amount (generally less than 75%) of quartz and chert and 15% to 75% detrital-clay matrix, and having rock fragments (primarily of sedimentary or low-rank metamorphic origin) in greater abundance than feldspar grains (chiefly sodic plagioclase, indicating a plutonic provenance). (AGI, 1987)
lithic sandstone A sandstone containing rock fragments in greater abundance than feldspar grains. Syn: *lithic arenite.* (AGI, 1987)
lithic tuff An indurated deposit of volcanic ash in which the fragments are composed of previously formed rocks; e.g., accidental particles of sedimentary rock, accessory pieces of earlier lavas in the same cone, or small bits of new lava (essential ejecta) that first solidify in the vent and are then blown out. Cf: *crystal tuff.* (AGI, 1987)
lithidionite Alternate spelling of litidionite. *litidionite.*
lithifaction *lithification.*
lithification (a) A compositional change in a coal seam from coal to bituminous shale or other rock; the lateral termination of a coal seam due to a gradual increase in impurities. (AGI, 1987) (b) The conversion of a newly deposited, unconsolidated sediment into a coherent, solid rock, involving processes such as cementation, compaction, desiccation, and crystallization. It may occur concurrent with, soon after, or long after deposition. (AGI, 1987) (c) A term sometimes applied to the solidification of a molten lava to form an igneous rock. See also: *consolidation; induration.* Syn: *lithifaction.* (AGI, 1987)
lithionite *lepidolite.*
lithiophilite An orthorhombic mineral, $4[LiMnPO_4]$; forms a series with triphylite; in granite pegmatites; an ore of lithium.
lithiophorite A hexagonal or monoclinic mineral, $(Al,Li)MnO_2(OH)_2$; powdery; black; occurs with other supergene manganese oxide minerals.
lithiophosphate An orthorhombic mineral, Li_3PO_4; in white to colorless masses in pegmatite; Kola Peninsula, Russia.
lithium A soft, silvery-white metallic element of the alkali group, the lightest of all metals. Symbol, Li. Does not occur free in nature; is

found combined in small amounts in nearly all igneous rocks and in the waters of many mineral springs. Lepidolite, spodumene, petalite, and amblygonite are the more important minerals containing it. The metal is corrosive and requires special handling. Used as an alloying agent, in the synthesis of organic compounds, and for nuclear applications. (Handbook of Chem. & Phys., 3)

lithium borosilicate $LiBSiO_4$. Used in low-temperature enamels and as a component of high-temperature, corrosion-resistant coatings. (Lee, 1961)

lithium mica *lepidolite.*

lithium niobate $LiNbO_3$; a ferroelectric compound having the ilmenite structure and of potential interest as an electroceramic. (Dodd, 1964)

lithium tantalate $LiTaO_3$; a ferroelectric compound of potential value as a special electroceramic. The Curie temperature is above 350 °C. (Dodd, 1964)

lithodeme A body of intrusive, pervasively deformed, or highly metamorphosed rock, generally nontabular and lacking primary depositional structures, and characterized by lithic homogeneity. It is mappable at the surface and traceable in the subsurface. For cartographic and hierarchical purposes, it is comparable to a formation. The name of a lithodeme combines a geographic term with a lithic or descriptive term, e.g., Duluth Gabbro. Cf: *suite.* Adj: lithodemic. (AGI, 1987)

lithofacies The rock record of any sedimentary environment, including both physical and organic characters. It is a mappable subdivision of a designated stratigraphic unit, distinguished from adjacent subdivisions on the basis of lithology. See also: *facies.* (AGI, 1987)

lithofracteur Nitroglycerine mixed with siliceous earth, charcoal, sodium, and sometimes barium, nitrate, and sulfur. (Fay, 1920)

lithogene Said of a mineral deposit formed by the process of mobilization of elements from a solid rock and their transportation and redeposition elsewhere. On a local scale, the process may be called a product of lateral secretion; on a larger scale, the deposit may be called a product of regional metamorphism. (AGI, 1987)

lithogenesis The origin and formation of rocks, esp. of sedimentary rocks. Also, the science of the formation of rocks. Cf: *petrogenesis.* Syn: *lithogeny.* Adj: lithogenetic. (AGI, 1987)

lithogenesy The science of the origin of minerals and the causes of their modes of occurrence. (Standard, 1964)

lithogenetic Of or pertaining to the origin or formation of rocks.

lithogeny *lithogenesis.*

lithoglyph A carving or engraving on a stone or gem; also, a stone or gem so engraved. (Standard, 1964)

lithoglyptics The art of gem cutting; the cutting or engraving of precious stones or gems. (Standard, 1964)

lithographiclimestone A compact, dense, homogeneous, exceedingly fine-grained limestone having a pale creamy yellow or grayish color and a conchoidal or subconchoidal fracture. It was formerly much used in lithography for engraving and the reproduction of colored plates. See also: *Solenhofen stone.* Syn: *lithographic stone.* (AGI, 1987)

lithographic stone *lithographic limestone.*

lithographic texture A sedimentary texture of certain calcareous rocks, characterized by uniform particles of clay size and by an extremely smooth appearance resembling that of the stone used in lithography. (AGI, 1987)

lithoid Rocklike or stonelike.

lithoidal Said of the texture of some dense, microcrystalline igneous rocks, or of devitrified glass, in which individual constituents are too small to be distinguished with the unaided eye. (AGI, 1987)

lithologic Adj. of lithology. Syn: *lithic.* (AGI, 1987)

lithologic correlation The matching or linking up of identical rock formations, veins, or coal seams, exposed some distance apart, by lithology. See also: *correlation.* (Nelson, 1965)

lithology (a) The character of a rock described in terms of its structure, color, mineral composition, grain size, and arrangement of its component parts; all those visible features that in the aggregate impart individuality to the rock. Lithology is the basis of correlation in coal mines and commonly is reliable over a distance of a few miles. (Nelson, 1965) (b) The study of rocks based on the megascopic observation of hand specimens. In French usage, the term is synonymous with petrography. (Holmes, 1928)

lithomarge A smooth, indurated variety of common kaolin, consisting at least in part of a mixture of kaolinite and halloysite. (AGI, 1987)

lithophile (a) Said of an element that is concentrated in the silicate rather than in the metal or sulfide phases of meteorites. Such elements concentrate in the Earth's silicate crust in Goldschmidt's tripartite division of elements in the solid Earth. Cf: *chalcophile; siderophile.* (AGI, 1987) (b) Said of an element with a greater free energy of oxidation per gram of oxygen than iron. It occurs as an oxide and more often as an oxysalt, esp. in silicate minerals. Examples are Se, Al, B, La, Ce, Na, K, Rb, Ca, Mn, U. Syn: *oxyphile.* (AGI, 1987)

lithophosphor A mineral, such as barite, that becomes phosphorescent when heated. Syn: *lithophosphore.* (Standard, 1964)

lithophosphore *lithophosphor.*

lithophysae Hollow, bubblelike, or roselike spherulites, usually with a concentric structure, that occur in rhyolite, obsidian, and related rocks. Sing: lithophysa.

lithopone A white pigment composed of a mixture of barium sulfate, zinc sulfide, and zinc oxide. (Hess)

lithospar A naturally occurring mixture of spodumene and feldspar. (USBM, 1960)

lithosphere (a) The solid outer portion of the Earth, as compared with the atmosphere and the hydrosphere. (AGI, 1987) (b) In plate tectonics, an outer layer of great strength relative to the underlying asthenosphere for deformation at geologic rates. It includes the crust and part of the upper mantle and is about 100 km thick. (AGI, 1987)

lithotope An area or surface of uniform sediment, sedimentation, or sedimentary environment, including associated organisms. Cf: *lithofacies.* (AGI, 1987)

lithotype (a) A rock defined on the basis of certain selected physical characters. (AGI, 1987) (b) This term was proposed by C.A. Seyler in 1954 in a communication to the Nomenclature Subcommittee of the International Committee for Coal Petrology to designate the different macroscopically recognizable bands of humic coals. These bands were described by M.C. Stopes in 1919 as the four visible ingredients in banded bituminous coal. The following macroscopic bands are distinguished in humic coals: vitrain, clarain, durain, and fusain. (IHCP, 1963)

lithoxyl Wood opal in which the original woody structure is observable; also petrified (opalized) wood. Also spelled lithoxyle; lithoxylite; lithoxylon. See also: *wood opal.*

litidionite Blue triclinic mineral, $KNaCuSi_4O_{10}$.

litmus A blue dyestuff that turns red when treated by an acid and remains blue when treated by an alkali. (Crispin, 1964)

litmus paper Absorbent paper dipped into a solution of litmus. Used to test solutions to determine whether they are acid or alkaline. (Standard, 1964)

lit-par-lit Having the characteristic of a layered rock, the laminae of which have been penetrated by numerous thin, roughly parallel sheets of igneous material, usually granitic. Etymol: French, bed-by-bed. Cf: *injection gneiss; injection metamorphism.* (AGI, 1987)

Little Demon exploder Trade name for a small exploder that employs a manually operated electromagnetic generator as its source of power. It is approved for firing single shots in gassy mines and is in widespread use. Syn: *one-shot exploder.* (Nelson, 1965)

little giant A jointed iron nozzle used in hydraulic mining. See also: *hydraulic monitor.*

little winds (a) Corn. A sump. (Fay, 1920) (b) An underground shaft, sunk from a horizontal drift. A winze. (Fay, 1920)

littoral Of or pertaining to a shore. A coastal region. (Webster 2nd, 1960)

littoral current A current generated by waves breaking at an angle to the shoreline and that usually moves parallel to and adjacent to the shoreline within the surf zone. See also: *longshore current.* (AGI, 1987)

littoral drift (a) Applied to the movement along the coast of gravel, sand, and other material composing the bars and beaches. (AGI, 1987) (b) Material moved in the littoral zone under the influence of waves and currents. (AGI, 1987)

littoral zone In mine subsidence, the zone that embraces the disturbed strata lying round about and outside the mined strata. (Briggs, 1929)

live boom A shovel boom that can be lifted and lowered without interrupting the digging cycle. (Nichols, 1954)

liveingite A monoclinic mineral, $Pb_9As_{13}S_{28}$. Syn: *rathite-II.*

live load (a) In drilling, a variable load suspended on the hoist line. (Long, 1960) (b) In mechanics, a load that is variable, as distinguished from a load that is constant. Also called dynamic load. (Long, 1960) (c) A load on a structure that may be removed or its position altered. See also: *dead load.* (Nelson, 1965) (d) It may not be a dynamic load, and it does not include wind load or earthquake shocks. (Hammond, 1965)

live lode A lode containing valuable minerals.

liverite Utah. A variety of bitumen, probably elaterite. (Tomkeieff, 1954)

live roll grizzly A device for screening and scalping that consists of a series of spaced rotating, parallel rolls so constructed as to provide openings of a fixed size. See also: *grizzly.*

liver opal *menilite.*

liver ore A miners' term for cinnabar.

liver pyrites A massive form of iron sulfide (marcasite and sometimes also pyrite and pyrrhotite) having a dull liver-brown color.

liver rock A sandstone that breaks or cuts as readily in one direction as in another and that can be worked without being affected by stratification; a dense freestone that lacks natural division planes. (AGI, 1987)

liverstone A variety of barite that gives off a fetid odor when rubbed or heated.

livesite A clay mineral intermediate to kaolinite and halloysite; a disordered kaolinite.

live steam Steam direct from the boiler and under full pressure. Distinguished from exhaust steam, which has been deprived of its available energy. (Webster 3rd, 1966; Fay, 1920)

livingstonite A monoclinic mineral, $HgSb_4S_8$; metallic lead gray.

lixiviant A liquid medium that selectively extracts the desired metal from the ore or material to be leached rapidly and completely, and from which the desired metal can then be recovered in a concentrated form. (SME, 1992)

lixiviation *leaching.*

lixivium *leachate.*

lizardite A trigonal and hexagonal mineral, $Mg_3Si_2O_5(OH)_4$; kaolinite-serpentine group; polymorphous with antigorite, clinochrysotile, orthochrysotile, and parachrysotile; forms a series with nepouite; in platy masses as an alteration product of ultramafic rocks; the most abundant serpentine mineral.

L-joint An approx. horizontal joint plane in igneous rocks. Syn: *primary flat joint.*

llano A term for an extensive tropical plain, with or without vegetation, applied esp. to the generally treeless plains of northern South America and the Southwestern United States. Etymol: Spanish. (AGI, 1987)

load (a) *bit load.* (b) The act or process of placing an explosive in a borehole; also, the explosive so placed. See also: *charge.* (Long, 1960)

load-bearing test Load-bearing tests may be divided into two types, horizontal and vertical. Both types require excavation of a test pit into the region that is being investigated. This test pit must be excavated with a minimum of blasting, with particular attention to the bearing surfaces to avoid disturbance of the foundation rock. Load-bearing tests are being used to an increasing extent as a source of information for the design of heavily loaded surface structures and have supplanted seismic tests where the foundation rock is highly shattered. Syn: *horizontal load-bearing test.*

load-bearing wall A wall carrying any load put upon it together with its own weight and the wind load. (Hammond, 1965)

load binder A lever that pulls two grabhooks together, and holds them by locking over center. (Nichols, 1976)

load cell Consists essentially of a hollow steel cylinder capped top and bottom by a steel plate. Strain gages are cemented on the inner wall of the cylinder in such a way that, as the cylinder is compressed, the strain gages will be distorted, with a resultant change in resistance. This instrument is designed for measuring the load transferred from the hanging wall onto props or other units used for support. (Issacson, 1962)

load controller A device to control the load and prevent spillage on a gathering conveyor receiving coal or mineral from several loading points or subsidiary conveyors. The device is a simplified weightometer and is installed on the main belt a short distance before the intermediate loading point. When the main belt is fully loaded, the scale registers this fact and causes a break in the electrical circuit of the intermediate conveyor, which is stopped. Immediately after that, the flood loading on the main conveyor is finished, and the load controller operates and starts up the subsidiary conveyor. (Nelson, 1965)

load dropper *car runner; car dropper.*

loaded filter A graded filter placed at the foot of an earth dam or other construction, stabilizing the foot of the structure by virtue of its weight and permeability. (Hammond, 1965)

loaded hole A drilled borehole in which explosive materials are loaded without priming the hole or including an initiator.

loaded wheel A grinding wheel that has a glazed or clogged surface from particles of the material being ground. (Crispin, 1964)

loader (a) A mechanical shovel or other machine for loading coal, ore, mineral, or rock. Syn: *mechanical loader.* See also: *loading machine; scraper loader; shaker-shovel loader; shovel loader; cutter loader; gathering arm loader.* (Nelson, 1965) (b) The worker who loads coal, either by hand or by operating a loading machine, at the working face after the coal has been shot down; also keeps the working place in order. Syn: *loader operator.* See also: *hand loader.* (Fay, 1920; BCI, 1947) (c) In anthracite and bituminous coal mining, a laborer who shovels coal or rock, blasted from the working face in a mine, into cars or onto a conveyor belt from which cars are loaded at some point removed from the working place. May be designated as car loader; conveyor loader. Also called coal loader. (d) In metal and nonmetal mining, one who shovels ore or rock, caved or blasted from the working face in a mine, into cars or onto a conveyor belt from which cars are loaded at some point removed from the working place. See also: *gun-perforator loader; tractor shovel.* Syn: *loader engineer; loader runner; loader operator.* (DOT, 1949)

loader boss (a) In bituminous coal mining, a foreperson who supervises a crew of loaders loading coal onto a conveyor or into cars at working places in a mine. Also called loading boss; loading-unit boss. (DOT, 1949) (b) In anthracite coal mining, a foreperson who supervises the loading of railroad cars with prepared coal. (DOT, 1949)

loader engineer *loader; boxcar loader.*

loader gate A gate road equipped with a gate conveyor or a gate-end loader; the gate to which the face conveyors deliver their coal. (Nelson, 1965)

loader-off Eng. A man who regulates the sending out of the full cars from a longwall stall, or gate. (Fay, 1920)

loader operator *loader.*

loader runner *boxcar loader; loader.*

load-extension curve A line plotted from the results of a tensile test on metal, the loads being shown as ordinates and the elongations of the gage length as abscissae, thus relating the extension of the material under test to the applied load. See also: *stress-strain curve.* (Hammond, 1965)

load factor (a) The ratio of the average compressor load during a certain period of time to the maximum rated load of the compressor. (Lewis, 1964) (b) The ratio of the collapse load to the working load on a structure or section. (Taylor, 1965) (c) In electric power engineering, the ratio of average electrical load to peak electrical load, usually calculated over a 1-h period. (Lyman, 1964) (d) Average load carried by an engine, machine, or plant, expressed as a percentage of its maximum capacity. (Nichols, 1976) (e) Ratio of average output during a period to maximum output during the period. Sometimes the ratio of output to maximum capacity. (Strock, 1948)

load fold A plication of an underlying stratum, believed to result from unequal pressure and settling of overlying material. (AGI, 1987)

load indicator A measuring device used to indicate the load or weight suspended on a drill hoisting line or cable. (Long, 1960)

loading In ion exchange, sorption onto resins of metallic ions from a pregnant solution in loading cycle. (Pryor, 1963)

loading boom (a) In coal preparation, an overhanging steel chute for loading coal into rail wagons or lorries; usually capable of vertical movement as loading proceeds to minimize breakage. (Nelson, 1965) (b) An adjustable conveyor used for lowering coal into cars with little breakage. Widely used for handling domestic sizes of bituminous coal. (Mitchell, 1950) (c) A hinged portion of a conveyor that is designed to receive materials at a fixed level and to discharge them at varying levels; usually employed for loading coal into wagons. (BS, 1962)

loading-boom operator *conveyor man.*

loading boss *loader boss.*

loading chute (a) A three-sided tray for loading or for transfer of material from one transport unit to another. See also: *chute.* (Nelson, 1965) (b) A gravity chute used to convey coal from the pocket or the screen to the railroad car. (Mitchell, 1950) (c) Used to direct material onto a conveyor.

loading conveyor Any of several types of conveyors adapted for loading bulk materials, packages, or objects into cars, trucks, or other conveyors. See also: *portable conveyor.*

loading density The number of pounds of explosive per foot length of drill hole. (Nelson, 1965)

loading equipment Mechanical shovels or other machines singly or in combination used to load excavated or stockpiled materials into trucks, mine cars, conveyors, or other transportation or haulage units.

loading head That part of a loading machine that gathers the coal or rock and places it on the machine's elevating conveyor.

loading-head man One who operates the loading device of any type of conveyance equipped with a self-loading head for the mechanical underground loading of coal or other mineral. (Hess)

loading hopper A hopper used to receive and direct material to a conveyor.

loading machine (a) A generic term for any power-operated machine for loading mined coal or any other material into mine cars, conveyors, road vehicles, or bins. (b) A machine for loading materials, such as coal, ore, or rock, into cars or other means of conveyance for transportation to the surface of the mine. See also: *loader.*

loading machine operator A person who operates a mobile loading machine. Syn: *loading-machine runner.*

loading-machine runner *loading machine operator.*

loading pan A box or scoop into which broken rock is shoveled in a sinking shaft while the hoppit is traveling in the shaft. A small hoist is used to lift and discharge the pans into the hoppit on its return to the shaft bottom. See also: *box filling.* (Nelson, 1965)

loading pick Eng. A pick for cleaning coal. (Fay, 1920)

loading point (a) The point where coal or ore is loaded into cars or conveyors, where a conveyor discharges into mine cars, or where a wagon or lorry is loaded. See also: *transfer point.* (Nelson, 1965) (b) N. of Eng. Where coal is transferred from a mother gate or trunk belt conveyor into tubs. (Trist, 1963)

loading pole A nonmetallic pole used to assist the placing and compacting of explosive charges in boreholes. (Atlas, 1987)

loading ramp A surface structure, often incorporating storage bins, used for gravity loading bulk material into transport vehicles. (Nelson, 1965)

loading ratio In quarrying, the number of tons of rock blasted per 1 lb (0.454 kg) of explosive. The harder the rock, the lower the ratio. See also: *explosive factor; explosive ratio.* Also called powder factor. (Streefkerk, 1952)

loadings Eng. Pillars of masonry carrying a winding drum or pulley. (Fay, 1920)

loading shovel A mechanical shovel able to operate as a forklift truck, a crane, or a loader. See also: *shovel loader.* (Hammond, 1965)

loading station A device consisting of one or more plates, or a hopper receiving and placing material on the conveyor belt for transport. When such a loading station is located at the tail end, it is known as a tail-end loading station; when it is located along the intermediate section, it is known as an intermediate loading station. (NEMA, 1961)

loading-unit boss *loader boss.*

loading weight Weight of a powder that is filled into a container under stated conditions. See also: *bulk density.* (Pryor, 1963)

load metamorphism A type of static metamorphism in which pressure due to deep burial has been a controlling influence, along with high temperature. Cf: *thermal metamorphism.* See also: *static metamorphism.* (AGI, 1987)

load-out To load coal or rock that is to be taken out of the mine. (Fay, 1920)

loadstar *lodestone.*

loadstone Alternate spelling of lodestone. See also: *lodestone.*

load transfer The weight of the strata above every excavation is largely transferred to the coal pillars or packs in the vicinity. Attempts are made to control and facilitate this load transfer in the yield-pillar system and in double packing. See also: *abutment; pressure arch.* (Nelson, 1965)

loam beater A rammer used in making a loam mold. (Standard, 1964)

loam cake A disk of dried loam used to cover a loam mold; it has holes through which melted metal is poured and air escapes. (Standard, 1964)

loaming A method of geochemical prospecting in which samples of soil or other surficial material are tested for traces of the metal desired, its presence presumably indicating a near-surface orebody. (AGI, 1987)

Lobbe Hobel An earlier type of rapid plow traveling at 70 ft/min (21.3 m/min) across the face. (Nelson, 1965)

Lobbert lagging A lagging consisting of galvanized steel wire frames fastened to underground haulageway supports by special wire fasteners to provide a continuous lining. (Nelson, 1965)

lobs Steps in a mine. (Hess)

local cell A galvanic cell resulting from inhomogeneities between areas on a metal surface in an electrolyte. The inhomogeneities may be of physical or chemical nature in either the metal or its environment. See also: *cell.* (ASM, 1961)

local current A natural earth current of local origin, such as one arising from the oxidation of a sulfide deposit. A term used in electrical prospecting. (AGI, 1987)

local extension The extension produced in a tensile test after the ultimate tensile stress has been passed and concentrated on part of the gauge length where a neck is formed. (CTD, 1958)

local indicator plant Plants of wide geographic distribution having an affinity for absorbing certain metallic elements from the soil and which can be used in geochemical exploration to indicate presence of local concentrations of metals. Cf: *universal plant indicator.* (Hawkes, 1962)

local metamorphism Metamorphism caused by a local process; e.g., contact metamorphism or metasomatism near an igneous body, hydrothermal metamorphism, or dislocation metamorphism in a fault zone. Cf: *regional metamorphism.* (AGI, 1987)

local shear failure Failure in which the ultimate shearing strength of the soil is mobilized only locally along the potential surface of sliding at the time the structure supported by the soil is impaired by excessive movement. (ASCE, 1958)

local unconformity An unconformity that is strictly limited in geographic extent and that usually represents a relatively short period, such as one developed around the margins of a sedimentary basin or along the axis of a structural trend that rose intermittently while continuous deposition occurred in an adjacent area. It may be similar in appearance to, but lacks the regional importance of, a disconformity. Cf: *regional unconformity.* (AGI, 1987)

local vent A pipe or shaft to convey foul air from a plumbing fixture or a room to the outer air. (Crispin, 1964)

local ventilation Ventilation of the drives and headings in mines by use of the pressure gradient of the main air current. (Stoces, 1954)

locatable minerals A legal term that, for public lands in the United States, defines a mineral or mineral commodity that is acquired through the General Mining Law of 1872, as amended (30 U.S.C. 22-54, 161, 162, 611-615). These are the base and precious metal ores, ferrous metal ores, and certain classes of industrial minerals. Acquisition is by staking a mining claim (location) over the deposit and then acquiring the necessary permits to explore or mine. (SME, 1992)

located Delimited by having the boundaries ascertained and monumented on the ground, identified by having a notice of location posted upon the land, and further proclaimed to the public by having such notice of location recorded in the manner customary under the rules for recording mining claims. (Ricketts, 1943)

location (a) The act of fixing the boundaries of a mining claim, according to law. (b) The claim itself. (c) The act of taking or appropriating a parcel of mineral land. It includes the posting of notices, the record thereof when required, and marking the boundaries so that they can be readily traced. The terms "location" and "mining claim" are synonymous, though a mining claim may consist of several locations. See also: *mining claim.* (Ricketts, 1943) (d) Selecting or defining, on a map or in the field, the alignment of a road, rail track, or site of a shaft or mine; the actual route or site as fixed. See also: *location plan; location survey.* (Nelson, 1965) (e) A spot or place where a borehole is to be drilled; a drill site. (Long, 1960)

location and patent The location of a mining claim and patent for a mining claim are not governed by the same rules. The mining statutes expressly provide for the location of surface ground that must include the lode or claim as discovered; and a patent cannot grant any greater extent of surface ground than the location as made and marked by the surface boundaries. (Ricketts, 1943)

location notice A written notice prominently posted on a claim, giving name of locator and description of its extent and boundaries. (AGI, 1987)

location plan A map, drawn to a suitable scale, showing the proposed mine development, shafts, works, etc., in relation to existing surface features. See also: *location.* (Nelson, 1965)

location survey *location.*

location work Labor required by law to be done on mining claims within 60 days of location, in order to establish ownership. Syn: *assessment work.* (Weed, 1922)

loch An unfilled cavity in a vein. See also: *vug.*

lockage Water consumed in passing from the upper reach of a canal when a vessel passes through a lock. (Hammond, 1965)

locked coil rope Made of specially formed wires assembled in layers of alternate lay about a wire core, which gives a smooth rope, and the entire surface is available for resisting wear. Such ropes are used in the United States for track cables on aerial tramways, and in England as hoisting ropes collieries. See also: *wire rope.* (Lewis, 1964)

locked-cycle test A series of repetitive batch tests in which the middling products generated in one test are added to the subsequent test to simulate the operations of a continuous process in which intermediate-grade materials are recycled. Each test is referred to as a "cycle." When equilibrium is reached in two or more cycles, depending on the sensitivity of the separation, the test is said to be "locked" or "balanced.". (SME, 1985)

locked particles Particles of coal or ore consisting of two or more minerals. (Gaudin, 1939)

locked test In laboratory tests on small quantities of ore, a method in which any selected fraction of the product is added to a fresh batch of the sample, so that the cumulative effect of its retention can be studied under conditions that simulate a continuous process in which middlings are recirculated or partly used water or leach liquor is returned. See also: *cyclic test.* (Pryor, 1963)

locked-wire rope A rope with a smooth cylindrical surface, the outer wires of which are drawn to such shape that each one interlocks with the other and the wires are disposed in concentric layers about a wire core instead of in strands. Particularly adapted for haulage and rope-transmission purposes. (Zern, 1928)

locker A short piece of round timber or iron rod for inserting between the spokes of a tram wheel to retard its movement. Also called lolley; sprag. (Nelson, 1965)

locking bolts Bolts of any type used for locking parts in position. (Crispin, 1964)

locknut The nut securing the feed gears in the feeding mechanism in a gear-feed swivel head on a diamond drill; also, any extra nut used to secure a principal nut. Also called jamnut; jambnut. (Long, 1960)

lock paddle A sluice whereby a lock chamber is emptied or filled. (Hammond, 1965)

lockpin Any pin or plug inserted in a part to prevent play or motion in the part so fastened. (Crispin, 1964)

lock sill A raised portion of the floor of a lock chamber, forming a stop against which the lock gates bear when they are shut. (Hammond, 1965)

lockup clutch A clutch that can be engaged to provide a nonslip mechanical drive through a fluid coupling. (Nichols, 1976)

locomotive An electric engine, operating either from current supplied from trolley and track or from storage batteries carried on the locomotive. The locomotive may be powered by battery, diesel, compressed air, trolley, or some combination such as battery-trolley or trolley-cable reel. Used to move empty and loaded mine cars in and out of the mine. (BCI, 1947)

locomotive arches Arches built of special refractory shapes, supported by water-circulating members.

locomotive brakeman In anthracite and bituminous coal mining, one who works on trains or trips of cars hauled by locomotive or motor, as distinguished from rope haulage. Also called locomotive helper; locomotive patcher; motor brakeman; motor nipper; poleman. (DOT, 1949)

locomotive garage An elongated recess in an intake airway of a mine for servicing locomotives. It contains two or three rail tracks (with pit space under one), good lighting, lifting equipment, oils, benches, and tools. Where battery locomotives are used, the garage will serve as a charging station. Syn: *underground garage; battery charging station.* (Nelson, 1965)

locomotive gradient The statutory maximum gradient for locomotive haulage is 1:15, but ordinarily the practical limit is about 1:25. Roads driven specially for locomotives are normally graded about 1:400 in favor of the load, unless a steeper gradient is required for drainage. (Nelson, 1965)

locomotive haulage The transport of coal, ore, workers, and materials underground by means of locomotive-hauled mine cars. The locomotive may be powered by battery, diesel, compressed air, trolley, or some combination such as battery-trolley or trolley-cable reel. See also: *compressed-air locomotive; haulage; underground haulage.* (Nelson, 1965)

locomotive helper *locomotive brakeman.*

locomotive pan brick Shapes, used to build a burner enclosure, flash wall, and protecting walls for the water legs.

locomotive patcher *locomotive brakeman.*

locomotive resistance The combined resistance caused by the friction of the journal and the wheel tread. It can range from 12 to 20 lb (5.44 to 9.07 kg), but for practical purposes may be taken as 15 lb (6.8 kg), as the locomotive represents only a small portion of the total tractive effort. (Kentucky, 1952)

lode A mineral deposit consisting of a zone of veins, veinlets, disseminations, or planar breccias; a mineral deposit in consolidated rock as opposed to a placer deposit. Syn: *lead.* Cf: *vein; vein system.* See also: *vein or lode claim.* (AGI, 1987)

lode claim (a) That portion of a vein or lode, and of the adjoining surface, that has been acquired by a compliance with the law, both Federal and State. Any dispute as to whether a given parcel of land is a vein or a lode is a question of fact to be determined by those experienced in mining, and it cannot be determined as a matter of law. (Ricketts, 1943) (b) *vein or lode claim.* (c) A mining claim on an area containing a known vein or lode. Cf: *placer claim.* (AGI, 1987)

lode plot A horizontal lode.

lodestone (a) A variety of magnetite showing spontaneous magnetization because of preferential alignment of magnetic domains within a crystal or mass of crystals. If permitted to do so, lodestone will align its polarity with the geomagnetic field, e.g., floating on a piece of wood in water as an early type of compass for navigation. Also spelled loadstone. Syn: *leading stone; Hercules stone; loadstar.* (b) An intensely magnetized rock or ore deposit. See also: *magnetite.*

lodestuff Minerals included in a lode or vein, including economically valueless gangue. See also: *veinstuff.* (Pryor, 1963)

lode tin Tin ore (cassiterite) occurring in veins, as distinguished from stream tin or placer tin. Cf: *stream tin.* (AGI, 1987)

lodge (a) A reservoir of any size used for holding water in a mine. A sump or standage. Also called lodgement. See also: *pound.* (BS, 1963) (b) Eng. A subterranean reservoir for the drainage of the mine made at the shaft bottom, in the interior of the workings, or at different levels in the shaft. (Fay, 1920) (c) Scot. A cabin at the mine shaft for workers. (Fay, 1920) (d) The room or flat at the shaft into which the pushers or trammers empty their loads. (Standard, 1964) (e) S Wales. The local branch of the coal miners' union. (Nelson, 1965) (f) A pump room, near the pit bottom or other main pumping station. (Nelson, 1965)

lodgment level Scot. A room driven from a level a short distance to the dip and used for storage of water. A sump. (Fay, 1920)

loellingite *löllingite.*

loess A widespread, nonstratified, porous, friable, usually highly calcareous, blanket deposit (generally less than 30 m thick), consisting predominantly of silt with subordinate grain sizes ranging from clay to fine sand. It covers areas extending from north-central Europe to eastern China as well as the Mississippi Valley and Pacific Northwest of the United States. Loess is generally buff to light yellow or yellowish brown; often contains shells, bones, and teeth of mammals; and is traversed by networks of small narrow vertical tubes (frequently lined with calcium carbonate concretions) left by successive generations of grass roots, which allow the loess to stand in steep or nearly vertical faces. Loess is now generally believed to be windblown dust of Pleistocene age. Cf: *adobe.* (AGI, 1987)

loeweite Alternate spelling of löweite.

Loewinson-Lessing classification A chemical classification of igneous rocks (into the four main types—acid, intermediate, basic, and ultrabasic) based on silica content. (AGI, 1987)

Lofco car feeder An appliance for controlling mine cars at loading points, for marshalling trains, and for loading cars into cages, tipplers, and other points. It consists of a carrying chain running between the rails. The dummy or live axles of the cars are held firmly by the chain's profile, and an overload slipping action is provided. See also: *feeder.* (Nelson, 1965)

lofthead An overhead cavity caused by a fall of roof. (Nelson, 1965)

lofting (a) S. Wales. An old or disused heading over the top of another one. (Fay, 1920) (b) N. of Eng. *lacing.* (c) Scot. Wood filling up vacant space on top of crowns or gears. (Fay, 1920) (d) Timbers, usually old, laid across the caps of steel frames or sets in a working to support the roof. (Webster 2nd, 1960)

log The record of, or the process of recording, events or the type and characteristics of the rock penetrated in drilling a borehole, as evidenced by the cuttings, core recovered, or information obtained from electric, sonic, or radioactivity devices. Also called logged; logging. See also: *well log.* Cf: *logbook.* (Long, 1960)

Logan slabbing machine This machine consists essentially of two horizontal cutting chains, one working at the base of the coal seam and the other at a distance from the floor; a third cutter chain is mounted vertically on a shearing jib at right angles to the other two and shears off the coal at the back of the cut. The upper jib breaks the coal up into loadable size, and a short conveyor transfers it to the face conveyor. Similar in principle to the A.B. Meco-Moore cutter-loader. (Mason, 1951)

logbook A book in which the official record of events or the type and characteristics of the rock penetrated by the borehole is entered. Also called journal; journal book. Cf: *log.* (Long, 1960)

logging chain A chain composed of links of round bar pieces curved and welded to interlock, with a grabhook at one end and a round hook at the other. (Nichols, 1976)

logging tongs Tongs with end hooks that dig in when the tongs are pulled. (Nichols, 1976)

log washer A slightly slanting trough in which revolves a thick shaft or log, carrying blades obliquely set to the axis. Material is fed in at the lower end, water at the upper. The blades slowly convey the lumps of material upward against the current, while any adhering clay is gradually disintegrated and floated out the lower end. (Liddell, 1918)

Lohmannizing A process by which a protective zinc coating is amalgamated to a base-metal sheet. (Liddell, 1918)

loipon A term proposed by Shrock (1947) for a residual surficial layer produced by intense and prolonged chemical weathering and composed largely of certain original constituents of the source rock. Typical accumulations of loipon are the gossans over orebodies, bauxite deposits in Arkansas, terra rossa deposits of Europe, and duricrust of Australia. Etymol: Greek, residue. Adj: loiponic. (AGI, 1987)

lok batanite Variety of bituminous material derived from a mud volcano. (Tomkeieff, 1954)

löllingite (a) An orthorhombic mineral, 2[$FeAs_2$]; basal cleavage; metallic; sp gr, 7.45; in quartz veins; a source of arsenic. Also spelled loellingite, lollingite. Syn: *leucopyrite; glaucopyrite; geyerite.* (b) The mineral group costibite, löllingite, nisbite, rammelsbergite, safflorite, and seinäjokite.

loma A term used in the Southwestern United States for an elongated, gentle swell or rise of the ground (as on a plain), or a rounded, broad-topped, inconspicuous hill. Etymol: Spanish, hillock, rising ground, slope. (AGI, 1987)

lomonosovite A triclinic mineral, $Na_2Ti_2Si_2O_9{\cdot}Na_3PO_4$; in alkalic pegmatites and sodalite syenites; Kola Peninsula, Russia.

long awn A direction of less than 45° to the main natural line of cleat or cleavage in the coal. Also spelled horn. Cf: *short awn.* (TIME, 1929-30)

long clay A highly plastic clay. Syn: *fat clay.* (AGI, 1987)

long column A column that fails by buckling, as distinct from crushing, when overloaded. Cf: *short column.* (Hammond, 1965)

long hole Underground boreholes and blastholes exceeding 10 ft (3.05 m) in depth or requiring the use of two or more lengths of drill steel or rods coupled together to attain the desired depth. Cf: *long-hole drill.* (Long, 1960)

long-hole blasting Method of blasting, employing diamond drills or extension steel drills with tungsten carbide bits, applied to ore-winning operations where conditions are suitable. The essential requirements from the practical and economic points of view are (1) a large orebody or wide regular vein, (2) a strong country rock, and (3) a good parting between the ore and the rock to avoid undue contamination of the ore. Holes to take cartridges up to 2 in (5.1 cm) in diameter may be drilled. Since the drilling of long holes is relatively expensive, high-strength, high-density gelatinous explosives are usually employed so that the maximum burden can be placed on each hole. For this reason also, the largest diameter of explosive cartridge that can be loaded into the holes should be used to obtain the greatest possible loading density. (McAdam, 1958)

long-hole drill A rotary- or percussive-type drill used to drill underground blastholes to depths exceeding 3 m. Cf: *long hole.* (Long, 1960)

long-hole infusion There are two methods: (1) Long holes are drilled parallel to the coal face. Charges of Hydrobel explosive are spaced along the holes and fired under water pressure. The object is to loosen the coal and allay the dust along the entire face in one operation. See also: *pulsed infusion shot firing.* (2) Two or more holes are drilled about 40 ft (12.2 m) long in the line of advance of the workings. The holes are not charged with explosive and fired but only subjected to water pressure up to 3,000 psi (20.7 MPa) to loosen the coal and allay the dust in the cleats. The holes may be drilled during weekends and of a length about equal to the weekly advance of the face. (Nelson, 1965)

long-hole jetting A hydraulic mining system consisting essentially of drilling a hole down the pitch of the vein, replacing the drilling head with a jet cutting head, and then retracting the drill column with the jets in operation to remove the coal. Coal cut loose by the jets drops to a flume drift and is carried by water to a screening and loading plant. (Coal Age, 1966)

longitudinal fault A fault whose strike is parallel with that of the general structural trend of the region. (AGI, 1987)

longitudinal fissure A fissure that is parallel with the strike of the deposit. (Stoces, 1954)

longitudinal joint A steeply dipping joint plane in a pluton that is oriented parallel to the lines of flow. Syn: *S-joint; bc-joint.* (AGI, 1987)

longitudinal trace A trace on the ground motion record representing the component of motion in a horizontal plane and in the direction of the seismic wave travel direction. Syn: *radial axis.*

longitudinal valley *strike valley.*

longitudinal velocity *modulus of elasticity.*

longitudinal wave An elastic wave in which the displacements are in the direction of wave propagation. Syn: *P wave; primary wave.* (AGI, 1987)

long (lazy) flames Partially aerated flames, particularly when aerated in stages. Nonturbulent flow. (Francis, 1965)

Longmaid-Henderson process Method of recovering copper from burned sulfides, by roasting with rock salt and using the issuing gases, after condensation, to leach the chloridized residue. Copper then dissolved is precipitated from the leach liquor onto scrap iron. (Pryor, 1963)

long piggyback conveyor An appliance to provide a constant flow of coal from a continuous miner to the main haulage system. It consists of a conveyor slung under the tail end of the loader and running on a bogey straddling the heading conveyor so that it can telescope over it. (Nelson, 1965)

long-pillar work Coal winning in three stages in underground mining. First, large pillars are left as the face is advanced by means of drives. Second, parallel drives connect these drives and form large blocks. Finally, the pillars so formed are mined. (Pryor, 1963)

long-range order The property of crystal structures wherein atomic particles show periodicity over large numbers of atomic diameters and each atomic particle has specific relationships with lattice points. Cf: *short-range order; disorder. superlattice; superstructure.*

long run To fill or nearly fill the core barrel with core on a single trip into the borehole. Cf: *short run.* (Long, 1960)

long-running thermal precipitator A dust-sampling instrument designed by the Mining Research Establishment of Great Britain that operates over periods of up to 8 h and collects only respirable dust that is selected aerodynamically during the sampling process. The respirable fraction is selected by drawing the dust through a horizontal duct elutriator that simulates the acting principle of the human nose and respiratory passages in that the larger and faster-falling particles are caught by the processes of settlement and impingement. (Roberts, 1960)

long section A section of land that contains more than 640 acres (256 ha). (Williams, 1964)

long-shank chopping bit A steel chisel-edged chopping bit having a longer and heavier-than-normal shank, designed to give added weight and directional stability when chopping an angle hole through overburden. (Long, 1960)

longshore current A current in the surf zone, moving generally parallel to the shoreline, generated by waves breaking at an angle with the shoreline. Syn: *littoral current.* (Schieferdecker, 1959)

longshore drift *beach drift.*

long tom A trough for washing gold-bearing earth. It is longer than a rocker. (Webster 3rd, 1966; Fay, 1920)

longues tailles Fr. *longwall.*

longwall (a) A long face of coal. (Stoces, 1954) (b) A method of working coal seams believed to have originated in Shropshire, England, toward the end of the 17th century. The seam is removed in one operation by means of a long working face or wall, thus the name. The workings advance (or retreat) in a continuous line, which may be several hundred yards in length. The space from which the coal has been removed (the gob, goaf, or waste) either is allowed to collapse (caving) or is completely or partially filled or stowed with stone and debris. The stowing material is obtained from any dirt in the seam and from the ripping operations on the roadways to gain height. Stowing material is sometimes brought down from the surface and packed by hand or by mechanical means. See also: *longwall advancing; longwall retreating.* Also known as longwork; Shropshire method; combination longwall; and Nottingham or Barry system. Syn: *combination longwall.* (Nelson, 1965; Fay, 1920) (c) Opposite of shortwall. See also: *longwall mining.* (BCI, 1947)

longwall advancing A system of longwall working in which the faces advance from the shafts toward the boundary or other limit lines. In this method, all the roadways are in worked-out areas. See also: *longwall.* (Nelson, 1965)

longwall coal cutter Compact machine, driven by compressed air or electricity, that cuts into the coal face with its jib at right angles to its body. See also: *longwall machine.* (Pryor, 1963)

longwall machine A power-driven machine used for shearing coal on relatively long faces. See also: *longwall coal cutter.*

longwall miner In bituminous coal mining, one who extracts coal from seams by a specialized method known as longwall mining. In this method, all coal is extracted as the work progresses, and the roof caves behind the working face.

longwall mining (a) A full-extraction method for mining large panels of coal. (b) A system of mining in which all the minable coal is recovered in one operation. (Hudson, 1932) (c) A method of mining flat-bedded deposits, in which the working face is advanced over a considerable width at one time. See also: *longwall.* (Pryor, 1963)

longwall peak stoping A method of underland stoping in which rapid advance of the face is maintained, and by working the faces at an angle of 60° to the strike, the peak travels down the dip at twice the rate of the face advance. This method was introduced on the Witwatersrand for stoping below 5,000 ft (1,524 m), where rapid face advance resulting from the closer spacing of holes reduces the incidence of rock bursts. (Higham, 1951)

longwall pillar working The extraction of the coal pillars formed by a pillar method of working, by a longwall face, which can be advancing or retreating. Where the crush is not excessive, this method is more efficient and often safer than extracting each pillar individually. (Nelson, 1965)

longwall retreating A system of longwall working in which the developing headings are driven narrow to the boundary or limit line and then the coal seam is extracted by longwall faces retreating toward the shaft. In this method, all the roadways are in the solid coal seam and the waste areas are left behind. See also: *longwall; mechanized heading development.* (Nelson, 1965)

lonkey Eng. The best Yorkshire sandstone, Rossendale District. Cf: *lunker.* (Arkell, 1953)

loodwin Burma (Myanmar). Ruby mines in which fissures, caves, and hollows in the limestone, filled with detritus from its disintegration, are followed, and their contents, often cemented or buried under recent travertine, are extracted and washed. (Hess)

looking-glass ore Lustrous hematite, Fe_2O_3. Specular iron ore; iron glance. Mohs hardness, 6; sp gr, 5.2. (Pryor, 1963)

lool A vessel to receive ore washings. (Standard, 1964)

loom Eng. A variant of loam, esp. in the Thames Valley; applied to Thanet sand and dredged mud used for cement. (Arkell, 1953)

loop circuit A term used when two positive wires are installed in divergent directions but later come close enough together to be connected. They then form a series or so-called loop circuit. Cf: *parallel blasting circuit.* (Kentucky, 1952)

loop drag An eye at the end of a rod through which tow is passed for cleaning boreholes.

looper In ore dressing, smelting, and refining, a laborer who attaches loops (folded copper sheets or strips) to starting sheets so that sheets can be suspended in electrolytic tanks. Also called looper puncher. (DOT, 1949)

loop haulage A system used when a continuous supply of cars must be supplied to a loading station. In this system, the empty cars are simply coupled to one end of the standing cars, which are pulled forward by a hoist with the rope attached to a clip fixed to the side of the cars. The rope is pulled back by hand for a distance of three or four car-lengths, and the process is repeated frequently as the end of the rope approaches the hoist. (Wheeler, 1946)

looping The running together of ore matter into a mass when the ore is only heated for calcination. Cf: *loup.* (Standard, 1964; Fay, 1920)

looping mill Consists of one or more trains of alternating two-high stands. As the piece being rolled emerges from one stand, it is turned through 180° and entered into the next stand and similarly into the succeeding stands. In some mills, the looping is performed by hand, while in others it is done mechanically by means of repeaters or looping channels. (Osborne, 1956)

loop-type pit bottom A pit bottom layout in which the loaded cars are fed to the cage from one side only and the empties are returned to the same side by means of a loop roadway. The loop arrangement provides more standage room for cars and is more suitable for multideck cages. When the coal reaches the pit bottom from two sides, a double-loop layout may be adopted. (Nelson, 1965)

loopway A double-track loop in a main single-track haulage plane at which mine cars may pass. A loopway located close to the face or loading machine enables a train of empty cars to be stationed in the loop, while the loaded cars are run straight out on the main single track. (Nelson, 1965)

loose Loose, or crick, is a vertical joint affecting only the lower strata in a oolite quarry, Northamptonshire, U.K. Syn: *crick.*

loose end (a) A gangway in longwall working, driven so that one side is solid ground while the other opens upon old workings. Cf: *fast end.* (Fay, 1920) (b) Coal prepared by cutting, or that coal that is certain to be loosened by a shot. (Fay, 1920) (c) The limit of a stall next to the goaf, or where the adjoining stall is in advance. (Fay, 1920)

loose goods Industrial diamonds as purchased from a diamond supplier in bulk. Cf: *loose stone.* (Long, 1960)

loose ground (a) Broken, fragmented, or loosely cemented bedrock material that tends to slough from sidewalls into a borehole. Also called broken ground. Cf: *breccia.* Syn: *loose rock.* (Long, 1960) (b) As used by miners, rock that must be barred down to make an underground workplace safe; also fragmented or weak rock in which underground openings cannot be held open unless artificially supported, as with timber sets and lagging. Cf: *broken ground.* (Long, 1960) (c) A shattered rock formation, or a formation crisscrossed with numerous, closely spaced, uncemented joints and cracks. (Long, 1960)

loose-needle traversing A method of traversing in which the magnetic bearings of survey lines are separately obtained by reference to the magnetic needle. (BS, 1963)

loosening bar An implement for loosening a pattern from a sand mold. (Standard, 1964)

loose rails Aust. Rails that can be lifted and placed across a permanent line when desired to run skips across it. (Fay, 1920)

loose rock *loose ground.*

looses Vertical joints affecting only the lower strata in the oölite quarries, Northamptonshire, U.K. Also called cricks.

loose stone (a) A diamond insecurely bonded in a bit matrix. (Long, 1960) (b) A diamond detached from a bit and lying on the bottom of a drill hole. (Long, 1960) (c) An unset industrial diamond. Cf: *loose goods.* (Long, 1960)

loose yards Measurement of soil or rock after it has been loosened by digging or blasting. (Nichols, 1976)

loosing S. Staff. Lowering a cage, etc., into a shaft or pit. (Fay, 1920)

loparite A possibly orthorhombic mineral, $(Ce,Na,Ca)_2(Ti,Nb)_2O_6$; perovskite group; pseudocubic; Kola Peninsula, Russia.

lopezite A triclinic mineral, $K_2Cr_2O_7$; forms minute orange-red balls in nitratine in Chile.

lopolith A large, concordant, typically layered igneous intrusion, of planoconvex or lenticular shape, that is sunken in its central part owing to sagging of the underlying country rock. Syn: *funnel intrusion.* (AGI, 1987)

loran Any of various long-range radio position-fixing systems by which hyperbolic lines of position are determined by measuring the difference in arrival times of synchronized pulse signals from two or more fixed transmitting radio stations of known geographic position. Loran fixes may be obtained at a range of 1,400 nmi (2,593 km) at night. Cf: *shoran.* Etymol: long-range navigation. (AGI, 1987)

lorandite A monoclinic mineral, $TlAsS_2$; red; forms modified tabular or prismatic crystals with one perfect cleavage.

loranskite A mineral, $(Y,Ce,Ca)ZrTaO_6$(?).

lorettoite A tetragonal artifact, $Pb_7O_6Cl_2$.

lorry (a) York. A movable bridge over a shaft top upon which the bucket is placed after it is brought up for emptying. (Fay, 1920) (b) A car used on mine tramways, or at coke ovens. (Fay, 1920) (c) Gr. Brit. A long wagon having a very low platform and four very small wheels. (Standard, 1964)

Los Angeles abrasion testing machine A machine for measuring abrasion resistance or toughness. It consists of a closed hollow steel cylinder 28 in (71.12 cm) in diameter and 20 in (50.8 cm) long mounted for rotation with its axis horizontal. The sample being tested and a charge of steel spheres are tumbled during rotation by an internal shelf. (AIME, 1960)

lose (a) Eng. To work a seam of coal, etc., up to where it dies out or is faulted out of sight. This is called "losing the coal.". (Fay, 1920) (b) To be unable to work out a pillar on account of thrust, creep, gob fire, etc. (Fay, 1920) (c) A pit shaft is said to be "lost" when it has run in or collapsed beyond recovery. (Fay, 1920)

loseyite A monoclinic mineral, $(Mn,Zn)_7(CO_3)_2(OH)_{10}$; soft; bluish white; at Franklin, NJ.

losing iron *furnace losing-the-iron.*

loss of vend Difference between weight of raw coal and that of salable products, expressed as a percentage. (Pryor, 1963)

loss on ignition As applied to chemical analyses, the loss in weight that results from heating a sample of material to a high temperature, after preliminary drying at a temperature just above the boiling point of water. The loss in weight upon drying is called free moisture; that which occurs above the boiling point of water, loss on ignition. (Harbison-Walker, 1972)

lost circulation The condition during rotary drilling when the drilling mud escapes into porous, fractured, or cavernous rocks penetrated by the borehole and does not return to the surface. (AGI, 1987)

lost closure The amount of closure of the walls of a stope that occurs before supports have been placed and begin to oppose that closure. (Spalding, 1949)

lost core The portion of a core that is not recovered. It may be the soft rock that crumbles and falls from the core barrel or the solid piece or pieces of core that drop to the bottom of a borehole after slipping out of the core barrel while the drill string is being pulled from the drill hole. (Long, 1960)

lost corner A corner whose position cannot be determined, beyond reasonable doubt, either from traces of the monument, or by reliable testimony relating to it; and whose location can be restored only by surveying methods and with reference to interdependent existent corners, by mutual agreement of abutters, or by court decision. (Seelye, 1951)

lost hole A borehole in which the target could not be reached because of caving, squeezing, loose ground, or inability to recover lost tools or junk. (Long, 1960)

lost level Corn. A level or gallery driven with an unnecessarily great departure from the horizontal. (Fay, 1920)

lost river (a) A dried-up river in an arid region. (b) A river in a karst region that drains into an underground channel.

lost thread method *string survey.*

lost water *lost circulation.*

loudspeaker face telephone An intrinsically safe public address system developed for coal face communications. Up to 20 individual units, each containing a telephone handset and speaker, can be coupled together along the face and gate roads by a five-way cable. Instructions, requests, etc., made into any one of the handsets are broadcast simultaneously over all the loudspeakers. See also: *signaling system.* (Nelson, 1965)

loup The pasty mass of iron produced in a bloomery or puddling furnace. Cf: *looping.* (Webster 3rd, 1966)

loupe Any small magnifying glass or lens mounted for use in the hand, held in the eye socket, or attached to spectacles and used to study minerals and rocks. Also spelled lupe, loup, loop.

louver cleaner In ore dressing, smelting, and refining, one who trims carbon anodes and cleans louvers to minimize electrical resistance in magnesium refining cells. (DOT, 1949)

louvers Overlapping and sloping slats arranged to prevent entrance or exit of a portion of an air stream. Louvers are sometimes used as a regulator in place of a sliding or other adjustable door.

love arrows *fleches d'amour.*

love stone *aventurine.*

Love wave (a) A transverse wave propagated along the boundary of two elastic media that both have rigidity; i.e., both media must be capable of propagating transverse waves. (AGI, 1987) (b) A surface seismic wave in which the particles of an elastic medium vibrate transverse to the direction of the wave's travel, with no vertical component. (AGI, 1987) (c) A type of surface wave having a horizontal particle motion that is transverse to the direction of propagation. Its velocity depends only on density and rigidity modulus, and not on bulk modulus. It is named after A.E.H. Love, the English mathematician who discovered it. Syn: *Q wave; Querwellen wave.* (AGI, 1987)

lovozerite The mineral group imandrite, kazakovite, koashvite, lovozerite, petarasite, tisinalite, and zirsinalite.

low (a) A former stream channel in a coalbed, filled with sandstone, clay, and shale. Cf: *cutout.* See also: *washout.* (b) A general term for such features as a structural basin, a syncline, a saddle, or a sag. Cf: *high.* Syn: *structural low.* (AGI, 1987) (c) Indicates that a crystal structure is the low-temperature dimorph for the crystal compound.

low-alumina silica brick Special brick in which the total alumina, titania, and alkali are significantly lower than in regular silica brick.

low-angle fault A fault with a dip of 45° or less. Cf: *high-angle fault.* (AGI, 1987)

low bed A machinery trailer with a low deck. (Nichols, 1976)

low blast A blast delivered to a smelting furnace at low pressure. (Standard, 1964)

low coal Coal occurring in a thin seam or bed. (Fay, 1920)

low-deflagrating explosive Another name for black powder. (Kentucky, 1952)

Lowden drier Mechanized drying floor used for ores or concentrates. Reciprocating rakes move the material gently over steel plates heated from below. (Pryor, 1963)

low-density explosive Explosive designed for use in mining, where it is required to blast with the least amount of shattering and/or to reduce explosive cost. The density of ordinary explosives may be decreased by (1) loose packing, (2) an alteration in the granular state of the ammonium nitrate, and (3) the impregnation of woodmeal or suitable substitutes. By decreasing the density of an explosive the same weight of explosive is used, but owing to its greater bulk the explosive effects are distributed over a greater area, thus producing a less shattering effect.

low-discharge ball mill One with a substantial downslope between the trunnion-high feed end and the peripheral discharge end. This facilitates the brisk movement of ore through the mill. (Pryor, 1963)

low doors Scot. The lowest of two or more landings in a shaft. (Fay, 1920)

low-duty fire clay brick A fire-clay refractory having a pyrometric cone equivalent not lower than 19 and a minimum modulus of rupture of 600 psi (4.1 MPa). (ARI, 1949)

löweite A trigonal mineral, $Na_{12}Mg_7(SO_4)_{13}{\cdot}15H_2O$; white; water soluble. Also spelled loeweite.

lower Pertaining to rocks or strata that are normally below those of later formations of the same subdivision of rocks. The adj. is applied to the name of a chronostratigraphic unit (system, series, stage) to indicate position in the geologic column and corresponds to early as applied to the name of the equivalent geologic-time unit; e.g., rocks of the Lower Jurassic System were formed during the Early Jurassic Period. The initial letter of the term is capitalized to indicate a formal subdivision (e.g., Lower Devonian) and is lowercased to indicate an informal subdivision (e.g., lower Miocene). The informal term may be used where there is no formal subdivision of a system or of a series. Cf: *upper; middle*. (AGI, 1987)

lower break The lower bend of either a terrace or a monocline, also known as the foot or the lower change of dip. See also: *foot*. (Stokes, 1955)

lowering conveyor Any type of vertical conveyor for lowering objects at a controlled speed. See also: *arm conveyor; suspended tray conveyor; vertical reciprocating conveyor*.

lowering skips Used in some river tipples to let the coal down into the barges. Also known as weigh pans. (Mitchell, 1950)

lowering tongs Long-handled, plierlike device similar to a certain type of blacksmith tongs used to handle wash or drill rods in place of a safety clamp in shallow borehole drilling. Syn: *brown tongs; knife dog*.

lower leaf Scot. The lower portion of a seam of coal that is worked in two sections or leaves. (Fay, 1920)

lower limit of flammability The smallest quantity of combustible gas that, when mixed with a given quantity of air (or oxygen), will just support a self-propagating flame. (Francis, 1965)

lower liquid limit In soil mechanics, the moisture content at which soil changes from plastic to liquid. (Pryor, 1963)

lower plastic limit Moisture content of soil at which it changes from a plastic to a semisolid state. (Pryor, 1963)

lower plate *footwall*.

lowest visible red-heat Common division of the color scale—about 887 °F (475 °C).

low explosive An explosive in which the change into the gaseous state is effected by burning and not by detonation as with high explosives. Blasting powder (black powder or gunpowder) is the only low explosive in common use. It requires no detonator but is ignited by means of a safety fuse. Also called propellant. (Nelson, 1965)

low-freezing dynamites Dynamites made by replacing part of the nitroglycerin of straight dynamites with some ingredient to render the dynamite incapable of freezing under ordinary conditions of use. The freezing point is depressed by adding nitro substitution compounds, such as nitrated sugars, nitrotoluene, nitrated polymerized glycerin, or ethylene glycol dinitrate. (Lewis, 1964)

low-freezing explosive *polar explosive*.

low gear (a) *slow gear*. (b) Mining and/or drilling operations carried on at a leisurely pace and at less-than-normal output per worker shift. (Long, 1960) (c) When applied to a screwfeed-type drill, the pair of feed gears in the feed mechanism that advances the bit the least amount for each revolution of drill drive rod and/or coupled drill stem. (d) When applied to speed at which the drill motor rotates the drill stem, the transmission-gear position giving the lowest number of bit revolutions per minute per engine revolutions per minute; corresponds to low gear in an automobile.

low-grade (a) An arbitrary designation of dynamites of less strength than 40%. It has no bearing on the quality of the materials, as they are of as great purity and high quality as the ingredients in a so-called high-grade explosive. (Fay, 1920) (b) Sometimes applied to poor- or low-quality drill diamonds. (Long, 1960) (c) Pertaining to ores that have a relatively low content of metal compared with other richer material from the same general area. Also designates coal high in impurities. Low-grade metamorphism refers to metamorphism at a relatively low temperature and/or pressure. Cf: *high-grade*. (Stokes, 1955) (d) *lean ore*.

low-grade coal Combustible material that has only limited uses owing to undesirable characteristics (for example, ash content or size). (BS, 1962)

low-heat cement A cement in which there is only limited generation of heat during setting, achieved by modifying the chemical composition of normal portland cement. (Hammond, 1965)

low-heat-duty clay A clay that fuses between 1,520 °C and 1,590 °C. (Osborne, 1956)

low-iron magnesite brick A burned magnesite brick, containing 90% or more of magnesia, and 1.5% to 2.5% iron oxide.

low level Scot. The drift or working that is farthest to the dip; also called laigh level.

low-nitrate barren In uranium leach treatment, the bulk of the barren solution after some 75,700 L of high-nitrate solution have been run through the ion-exchange (IX) column. Low in nitrate and uranium and contains some backwash water. (Pryor, 1963)

low powders Explosives containing a small portion of nitroglycerin and a base similar to blasting powder. Intermediate between blasting powder and dynamite in action. *low-grade*. (Fay, 1920)

low-pressure air stower The filling of the waste by means of compressed-air blower. The blower is usually located close to the stowing machine and operated at a pressure below 15 psi (103.4 kPa). For light duties, only one blower is required to operate one stowing machine. For heavier work, or very long lengths of pipeline, two blowers are used in series. The maximum horsepower for two blowers in series does not commonly exceed 200 (149 kW) at large installations. See also: *pneumatic stowing*. (Nelson, 1965)

low-pressure limit The lowest pressure at which flame propagation can be obtained through a combustible-oxidant system at a fixed temperature in a particular chamber. (Van Dolah, 1963)

low quartz Low-temperature quartz; when formed below 573 °C, SiO_2 tetrahedra are less symmetrically arranged than at higher temperatures; inversion is reversible. Syn: *alpha quartz*. (AGI, 1987)

low-rank coals *rank*.

low-rank metamorphism Metamorphism that is accomplished under conditions of low to moderate temperature and pressure. CF: high-rank metamorphism. (AGI, 1987)

low red-heat Temperature of iron at which redness is observable in subdued daylight (525 to 700 °C). Bright red heat is in the range 700 to 1,000 °C and is followed by orange at 1,000 °C, white at 1,300 °C, and blue-white at 1,500+ °C. (Pryor, 1963)

low shaft furnace A short shaft-type blast furnace that can be used to produce pig iron and ferroalloys from low-grade ores, using low-grade fuel. The air blast is often enriched with oxygen. It can also be used for making a variety of other products such as alumina, cement-making slags, and ammonia synthesis gas. (ASM, 1961)

low-temperature carbonization Carbonization carried out at a low temperature (between 500 °C and 700 °C). During the process, the smoke-producing compounds are driven off as tars and oils and collected as valuable byproducts, leaving a coke with about 10% volatile matter. The coke yield is about 1,400 lb/st (700 kg/t) and is used as a domestic fuel. (Nelson, 1965)

low-temperature coke A solid fuel produced by the low-temperature carbonization of coal. (Nelson, 1965)

low-temperature incineration method In this method for the determination of incombustible matter, the mine roadway dust (dust containing carbonates) is incinerated at a temperature of not less than 500 °C and not more than 530 °C until it is constant in weight. This temperature is sufficient to complete the combustion of the organic matter in the dust, but is not high enough to decompose the carbonates. However, this method is unsuitable for dust containing magnesium carbonate because this substance decomposes below 500 °C, and therefore a low result for the incombustible matter would be obtained. The moisture content of the sample may be calculated so that an allowance can be made for the weight loss attributed to moisture. (Cooper, 1963)

low-tension detonator A detonator requiring a minimum current of 1 A for firing and having a resistance of about 1 Ω. (BS, 1964)

low-terrace drift Aust. Gravel and shingle in terraces.

low velocity *velocity*.

low-velocity-layer correction *weathering correction*.

low volatile bituminous coal The rank of coal, within the bituminous class of Classification D 388, such that, on the dry and mineral-matter-free basis, the volatile matter content of the coal is greater than 14% but equal to or less than 22% (or the fixed carbon content is equal to or greater than 78% but less than 86%), and the coal commonly agglomerates. Cf: *bituminous coal*. (ASTM, 1994)

low voltage In coal mining, 660 V or less. Also called low potential. Cf: *high voltage; medium voltage*. (Federal Mine Safety, 1977)

low working voltage Low working voltage in coal mines is one of the many conditions that must be given continual attention. Loss of voltage means a proportional loss in power. Since the quantity of dc power is obtained by multiplying the number of amperes times the number of volts, it follows that for a given amount of power if the volts are lowered the amperes are increased, and the increase in amperes results in an increase in power loss in the mine circuit. Lower operating voltages result in heavier currents in the dc motor circuits, thus heating the motors, cables, and circuit wiring and causing loss in motor speed, inefficient operation, and increased maintenance cost. (Kentucky, 1952)

lozenge A form of cut stone produced by the meeting of the skill and star facets on the benzil of brilliants; or by the meeting of the facets in the horizontal ribs of the crown. (Hess)

lpb To determine the horizontal compass direction that a borehole is trending at a specific depth by means of one of several borehole surveying instruments. (Long, 1960)

LP delays Long period delays used in delay blasting underground and generally available in intervals of seconds.

lublinite A fibrous variety of calcite.

lubrication The act of applying lubricants. There are two main types of lubricants, solid and liquid. Examples of the solid type are graphite, French chalk, and sulfur. Liquid lubricants are by far the more important, and among these, oils and greases are the most common. (Crispin, 1964; Morris, 1958)

Luce and Rozan process A modification of the Pattinson process whereby the molten lead is stirred by the injection of steam; used in desilverizing base bullion. (Fay, 1920)

lucianite An expansive clay of the smectite(?) group at Santa Lucia, near Mexico, D.F.

lucid attrite Variety of attritus that is transparent in thin section. (Tomkeieff, 1954)

lucinite A variety of variscite from Lucin, UT.

Luckiesh-Moss visibility meter This instrument has been used over a wide range of lighting applications. It consists of a pair of similar photographic gradient filters, which increase in density as they are rotated together before the eyes. The filters therefore reduce the apparent brightness of the observed field and at the same time lower the contrast between the object of view and its background. (Roberts, 1958)

luckite A manganoan variety of melanterite at the Lucky Boy Mine, Salt Lake County, UT.

lucky stone *staurolite.*

Lüders line Surface markings that result from strain. Sometimes called Hartmann lines; Piobert lines; stretcher strains. See also: *slip line.* (ASM, 1961)

ludlamite A monoclinic mineral, $(Fe,Mg,Mn)_3(PO_4)_2 \cdot 4H_2O$; green; in small tabular crystals at Truro, Cornwall, U.K. Syn: *lehrnerite.*

ludwigite An orthorhombic mineral, Mg_2FeBO_5; forms a series with vosenite; forms finely fibrous masses. Includes magnesioludwigite, ferroludwigite. Syn: *collbranite.*

lue Prov. Eng. To sift; a miner's term. (Standard, 1964)

lueneburgite Alternate spelling of lüneburgite.

lueshite A monoclinic mineral, $NaNbO_3$; perovskite group; pseudocubic; dimorphous with natroniobite. Syn: *igdloite.*

luffing cableway mast A cableway tower hinged at the base and sustained by adjustable guys so that its inclination can be varied. (Hammond, 1965)

lug (a) A replaceable cutting member on an expansion reamer. (Long, 1960) (b) One of a set of cutting lugs used to cut off a length of casing in a borehole at any desired point below the collar of the hole. Syn: *jaw.*

lug down To slow down an engine by increasing its load beyond its capacity. (Nichols, 1976)

lugeon test Single-hole in situ test of formation permeability performed by measuring the volume of water taken in a section of test hole when the interval is pressurized at 10 bars (150 psi). Used primarily in variably permeable formations under evaluation for grouting. (Houlsby, 1990)

Luhrig vanner Vanning machine with side feed and end shake—a hybrid between the true vanner and the shaking table. (Pryor, 1963)

lumachelle (a) A compact, dark-gray or dark-brown limestone or marble, composed chiefly of fossil mollusk shells, and characterized by a brilliant iridescence or chatoyant reflection from within. Syn: *fire marble.* (AGI, 1987) (b) Any accumulation of shells (esp. oysters) in stratified rocks. Etymol: French, coquina, oyster bed, from Italian lumachella, little snail. (AGI, 1987)

lumber Timber that has been sawed into boards, planks, staves, or other pieces of comparatively small dimensions. In mines, timber is used in the construction of coal chutes, mine cars, mine doors, forms for concrete structures, surface buildings, and for many other purposes. (Jones, 1949)

lumber scale A graduated measuring scale for determining the number of board feet in rough-sawed lumber. (Crispin, 1964)

luminance A measure of surface brightness that is expressed as luminous flux per unit solid angle per unit projected area. (ASM, 1961)

luminance of a surface Luminous intensity per unit of apparent area; i.e., the area as projected on a plane normal to the direction of viewing. It is determined by the incident light flux falling upon the surface, the reflection factor of the surface, and the angle that the surface makes with the direction in which it is viewed. It is independent of distance; i.e., a surface appears equally bright no matter what the distance from which it is seen. (Roberts, 1958)

luminescence (a) The emission of light by a substance that has received energy or electromagnetic radiation of a different wavelength from an external stimulus; also, the light so produced. It occurs at temperatures lower than those required for incandescence. See also: *phosphorescence; fluorescence.* (AGI, 1987) (b) Quantized electromagnetic emissions resulting from electrons in a crystal structure dropping from a higher excited state to a lower one. Cf: *thermoluminescence; triboluminescence.*

luminosity (a) The quality of emitting or of giving out light; shining. (Crispin, 1964) (b) Subjective brightness sensation. (Roberts, 1958)

luminous Radiating or emitting light; bright; clear. (Crispin, 1964)

Lumnite Well-known brand of quicksetting cement for sealing rock cavities, plugging drill holes, etc. Is made from bauxite ore and limestone and is highly resistant to acids and heat. (Cumming, 1951; Hess)

lump coal Bituminous coal in the large lumps remaining after a single screening that is often designated by the size of the mesh over which it passes and by which the minimum size lump is determined. Also, the largest marketable size. (Webster 3rd, 1966; Fay, 1920)

Lump Coal Trademark for permissible dynamites (types C and CC) with very low velocity of detonation. Are used in coal mining where maximum production of large-size coal is desired. (CCD, 1961)

lump ore *natural ore.*

lumpy Describes a gemstone cut with too great a depth in proportion to its width.

lunar crater *crater.*

lunker Scot. A lenticular mass of sandstone or clay ironstone; a big nodule. Cf: *lonkey.* (AGI, 1987)

Lurgi process This process consists of roasting iron ore in a reducing atmosphere, thus forming magnetic oxide of iron that is separated by crushing followed by magnetic separation. The internal structure of the kiln is so designed that the ore falls in a continuous veil through the current of reducing gases. Burners are distributed throughout the periphery of the kiln, so that roasting and reduction can be controlled in the various zones to the required temperature. Blast furnace gas for reduction passes into the center of the lower end of the kiln, while the gas and air for heating pass in from the circumference of the drum, nearer to the center and upper end of the furnace. The ingoing ore is crushed to give a maximum size of 20 mm, while the outgoing concentrate is crushed to 3.8 mm after cooling. (Osborne, 1956)

Lurmann front An arrangement of water-cooled castings through which iron and cinder are tapped from the blast furnace, thus avoiding the use of a forehearth. See also: *dam.* (Fay, 1920)

lurry (a) York. A weighted tram to which an endless rope is attached, fixed at the inby end of the plane, forming part of an appliance for taking up the slack rope. (Fay, 1920) (b) A movable platform on wheels, the top of which is level with the bank or surface. It is run over the mouth of a shaft to receive the bucket when it reaches the top. A variation of lorry. (Fay, 1920)

lusakite A cobaltoan variety of staurolite.

lussatite A fibrous variety of silica, possibly tridymite.

luster (a) The character of the light reflected by minerals; it constitutes one of the means for distinguishing them. There are several kinds of luster: metallic, the luster of metals; adamantine, the luster of diamonds; vitreous, the luster of broken glass; resinous, the luster of yellow resin, as that of eleolite; pearlylike pearl; and silkylike silk. These lusters have different degrees of intensity, being either splendent, shining, glistening, or glimmering. When there is a total absence of luster, the mineral is characterized as being dull. (Fay, 1920) (b) In ceramics, a glaze, varnish, or enamel applied to porcelain in a thin layer, and giving it a smooth, glistening surface. (Standard, 1964)

luster mottling (a) The macroscopic appearance of poikilitic rocks. (AGI, 1987) (b) The shimmering appearance of a broken surface of a sandstone cemented with calcite, produced by the reflection of light from the cleavage faces of conspicuously large and independently oriented calcite crystals, 1 cm or more in diameter, incorporating colonies of detrital sand grains. It may also develop locally in barite, gypsum, or dolomite cements. (AGI, 1987)

lutaceous Said of a sedimentary rock formed from mud (clay- and/or silt-size particles) or having the fine texture of impalpable powder or rock flour; pertaining to a lutite. Also said of the texture of such a rock. Cf: *argillaceous; pelitic.* (AGI, 1987)

lutecite A fibrous variety of chalcedony with fiber elongation perpendicular to the *c* crystallographic axis (opposite to normal chalcedony) and showing other optical anomalies.

luthos lazuli A violet variety of fluorite.
luting Sealing the joint of a retort with fire clay. (Pryor, 1963)
lutite A general name used for consolidated rocks composed of silt and/or clay and of the associated materials which, when mixed with water, form mud; e.g., shale, mudstone, and calcilutite. The term is equivalent to the Greek-derived term "pelite." Etymol: Latin lutum, mud. Also spelled lutyte. See also: *rudite; arenite.* (AGI, 1987)
lutose Covered with clay; miry. (Webster 2nd, 1960)
luxullianite A granite characterized by phenocrysts of potassium feldspar and quartz that enclose clusters of radially arranged acicular tourmaline crystals in a groundmass of quartz, tourmaline, alkali feldspar, brown mica, and cassiterite. Its name is derived from Luxulyan, Cornwall. Also spelled: luxulianite; luxulyanite. Var: luxuliane. (AGI, 1987)
luzonite A tetragonal mineral, Cu_3AsS_4; stannite group; dimorphous with enargite. Formerly considered an arsenian variety of famatinite.
lyddite High explosive based on picric acid, $(NO_2)_3C_6H_2OH$, with 10% nitrobenzene and 3% vaseline. (Pryor, 1963)
Lydian stone A touchstone consisting of a compact, extremely fine-grained, velvet- or gray-black variety of jasper. Etymol: Greek Lydia, ancient country in Asia Minor. Syn: *lydite; touchstone; basanite.* (AGI, 1987)
lydite *Lydian stone.* Also spelled: lyddite.
lye *double parting.*
lying money An allowance to miners on piecework who are rendered idle during a shift owing to circumstances beyond their control, such as a breakdown in power services, or supplies of empty cars. (Nelson, 1965)
lying wall *footwall.*
lynx eye Green iridescent labradorite.
lynx sapphire (a) Dark blue iolite variety of cordierite. (b) Dark blue sapphires in Sri Lanka.
lyophilic (a) Condition of solid-liquid mixture in which surface-active molecules that contain two or more groups have both an affinity for the phase in which one group is dissolved, and a repulsion from this phase for another group or ion. See also: *hydrophilic.* (Pryor, 1963) (b) Having the property of attracting liquids. (Pryor, 1965)
lyophobic Of, relating to, or having a lack of strong affinity between a dispersed phase and the liquid in which it is dispersed; systems such as colloidal metals in water are easily coagulated. Opposite of lyophilic. (Webster 3rd, 1966)
lyosorption Adsorption of liquid to a solid surface. (Pryor, 1963)
lype Scot. An irregularity in the mine roof. A projecting rock in a mine roof that may fall at any time. Usually used in the plural, and sometimes spelled lipe. (Fay, 1920)
Lyster process A flotation process that separates galena and zinc blende by treatment, at a low temperature, with eucalyptus oil or other frothing agent, and with agitation or aeration in a neutral or alkaline, but not acid, solution of the sulfates, chlorides, or nitrates of calcium, magnesium, sodium, potassium, or mixtures of these substances. (Fay, 1920)
lyway This term is commonly used in and around mines in Indiana and Illinois to describe a mine sidetrack or a passing track. (Hess)

M

maacle *macle.*

macadam Crushed stone of regular sizes below 3 in (7.6 cm) for road construction, commonly with tar or asphalt binder. The sizes below 1 in (2.54 cm) are more specif. defined as chippings. See also: *penetration macadam; tarmacadam.* (Nelson, 1965)

MacArthur and Forest cyanide process A process for recovering gold by leaching the pulped gold ore with a solution of 0.2% to 0.8% potassium cyanide, KCN, and then with water. The gold is obtained from this solution by precipitation on zinc or aluminum, or by electrolysis. See also: *Elsner's equation.* (Fay, 1920)

macedonite (a) A tetragonal mineral, $PbTiO_3$; black; in irregular grains in quartz-syenite veins near Prilep, Macedonia. (b) A bluish-black aphanitic igneous rock composed of alkali feldspars and biopyriboles; related to mugearite, a trachyandesite, Macedon district, Victoria, Australia. The name is not part of the IUGS classification of extrusive igneous rocks.

maceral Applied to all petrologic units seen in microscopic sections of coal, as distinct from the visible units seen in the hand specimens. Comparable in rank to mineral as used in petrography. Thus, macerals are organic units composing the coal mass, being the descriptive equivalent of the inorganic units composing rock masses and universally called minerals. Individual macerals have the termination -inite; i.e., vitrinite, as the organic unit making up the lithologic specimen, vitrain. Three groups are recognized: (1) vitrinites; (2) exinites; and (3) inertinites. (AGI, 1987)

macfarlanite A silver ore consisting of a mixture of sulfides, arsenides, etc., and containing cobalt, nickel, and lead. Cf: *animikite.*

macgovernite *mcgovernite.*

machine boss In bituminous coal mining, a foreperson who is in charge of machine workers who undercut the working face of coal prior to blasting, and machine loaders who load the coal into cars after it is blasted down. (DOT, 1949)

machine cut A slot or groove made horizontally or vertically in a coal seam by a coal cutter, as a preliminary step to shot firing. See also: *cutting horizon.* (Nelson, 1965)

machine-cutter helper *machine helper.*

machined (a) A smooth surface finish on metal. (Nichols, 1976) (b) Shaped by cutting or grinding. (Nichols, 1976)

machine design The application of scientific principles to the practical constructive art of engineering, with the object of expressing original ideas in the form of drawings. See also: *theory of machines.* (Nelson, 1965)

machine driller In mining, one who operates any one of several types of heavy, mounted or unmounted, compressed-air drilling machines to drill holes into the working face of ore or rock into which explosives are inserted and set off to blast down the mass. Also called drill engineer; drillman; drill operator; power driller. (DOT, 1949)

machine helper In anthracite and bituminous coal mining, one who assists the machineman in moving and setting up the coal-cutting machine in position for cutting a channel under or along the sides of the coal working face prior to breaking the coal down with explosives. Also called chain-machine helper; coal-cutter helper; cutter helper; cutting-machineman helper; jack setter; machine-cutter helper; machineman helper; miner; assistant; mining-machine-operator helper. (DOT, 1949)

machine holings The small coal or dirt produced by a coal cutter. (Nelson, 1965)

machine loader (a) In anthracite and bituminous coal mining, person who operates an electric loading machine that digs, elevates, and loads coal blasted from the working face into cars by means of a conveyor loaded by a scraping device, a digging conveyor, or a shoveling mechanism. May be designated according to machine used, as Joy loader. Also called loader engineer; loader operator; loader runner; loading-machine man; loading-machine operator; loading-machine runner. (DOT, 1949) (b) Machine loader I is used in anthracite coal mining, bituminous coal mining, and metal mining, to designate one who operates a small electric or air-power shovel in underground working places to load ore or rock into cars after it has been blasted from the working face. Also called loader operator; loading-machine operator; mechanical-shovel operator; shovel operator. (DOT, 1949) (c) Machine loader II is used in anthracite coal mining, bituminous coal mining, and metal mining, to designate one who operates a loading machine consisting of a small hoisting engine and a cable-drawn scraper, or scoop, to load ore or rock into cars after it has been blasted from the working face in underground or open-pit mines. Also called scraper-loader hoistman; scraper-loader operator; scraper operator. (DOT, 1949)

machine loader helper In anthracite coal mining and bituminous coal mining, a laborer who assists a machine loader in loading coal into cars. (DOT, 1949)

machineman (a) A coal-cutter operator. See also: *coal-cutter team.* (Nelson, 1965) (b) A person who sets up and operates an electrically driven or compressed-air-driven coal-cutting machine that is used to cut out a channel along the bottom or side of the working face of coal so that it may be blasted down without shattering the mass. Also called coal cutter; coal-cutting machine operator; machine cutter; cutter operator; cutting machine operator; holer; undercutter. (DOT, 1949) (c) Eng. One who weighs coal, etc., and keeps an account of the number of cars sent to the surface. (Fay, 1920)

machineman helper *machine helper.*

machine miner In bituminous coal mining, a general term applied to workers who are capable of operating one or more coal mining machines used for drilling, loading, and undercutting. Usually designated as machine driller; machine loader I; machine loader II; machineman. Also called machine operator; machine runner; mining machine operator. (DOT, 1949)

machine mines Mines in which coal is cut by machines. (Kiser, 1929)

machine mining (a) Implies the use of power machines and equipment in the excavation and extraction of coal or ore. In coal mines, the term almost invariably signifies the use of coal cutters and conveyors and perhaps some type of power loader working in conjunction with face conveyors. See also: *face mechanization.* (Nelson, 1965) (b) Mechanized mining. (CTD, 1958)

machine nog Eng. A wedge-shaped wood block for supporting machine-cut coal. (SMRB, 1930)

machine operator *machine miner.*

machine rating (a) The amount of power a machine can deliver without overheating. (Crispin, 1964) (b) Also, work capacity of a machine.

machine runner *machine miner.*

machine screw A very commonly used type of screw with clear-cut threads and of a variety of head shapes. It may be used either with or without a nut. (Crispin, 1964)

machine set *mechanical set.*

machine sumper *sumper.*

machine tool Any machine used for cutting metal, such as a boring machine, drill, grinder, planing machine, hobber, shaper, or lathe. (Hammond, 1965)

machine wall The face at which a coal-cutting machine works. (Fay, 1920)

machinist Aust. The person in charge of a coal cutter. (Fay, 1920)

mackintoshite *thorogummite.*

Mac-Lane system This system consists essentially of an inclined rail track with the haulage gear and the loading station at the base. The haulage rope passes round a return sheave in the extending frame at the top of the heap. The tipping gear may consist of a carriage with a portable tippler that conveys a tub of dirt to the top of the heap, where a trigger operates the tippler, thus discharging the tub. For a greater quantity, the carriage incorporates a revolving frame carrying two or three tubs; this gives a broader top to the heap. The main disadvantages of this system are the unsightly conical heaps produced and the tendency to segregation of material, with the large pieces at the base of the heap, which increases the danger of spontaneous combustion. (Sinclair, 1959)

macle (a) A twin crystal; esp. octahedral diamond twins flattened parallel to the twin plane. Also spelled maacle, maccle, mackle. (b) A variety of andalusite with tessellated cross section resulting from segregation of carbonaceous impurities. Syn: *chiastolite.*

Maclean separator A revolving disk-type magnetic separator widely used for the separation of large quantities of ilmenite from tin ore. (Harrison, 1962)
macled (a) Twinned. (b) *tessellated.* (c) Spotted. Also spelled mackled.
macro- A prefix meaning large, long; visibly large, e.g., macrocrystalline. Cf: *micro-.* Syn: *mega-.*
macro-axis In the orthorhombic and triclinic systems, an obsolete term for the longer lateral crystallographic axis, the *b* axis in orthorhombic and mostly the *b* axis in triclinic minerals.
macrocrystalline Said of the texture of a rock consisting of or having crystals that are large enough to be distinctly visible to the unaided eye or with the use of a simple lens; also, said of a rock with such a texture. Syn: *eucrystalline.* See also: *macromeritic; phaneritic.* (AGI, 1987)
macrodiagonal In crystallography, the longer lateral axis in the orthorhombic and triclinic systems. (Standard, 1964)
macrodome (a) In crystallography, a dome parallel to the macrodiagonal. Cf: *dome.* (Standard, 1964) (b) In the orthorhombic system, an obsolete term for a form of four faces parallel to the macro-axis, but intercepting the other two. Syn: *horizontal prism.* (c) In the triclinic system, an obsolete term for a form of two faces parallel to the macro-axis, but intercepting the other two.
macroetch Etching of a metal surface for accentuation of gross structural details and defects for observation by the unaided eye or at magnifications not exceeding 10 diameters. (ASM, 1961)
macrograph A graphic reproduction of the surface of a prepared specimen at a magnification not exceeding 10 diameters. When photographed, the reproduction is known as a photomacrograph. (ASM, 1961)
macromeritic *phaneritic.*
macroporosity Porosity visible without the aid of a microscope, such as pipes and blowholes in ingots. (Newton, 1959)
macroscopic (a) Visible at magnifications of from 1 to 10 diameters. (ASM, 1961) (b) Visible without a microscope or in a hand specimen. See also: *megascopic.* Cf: *microscopic.*
macrostructure (a) The general arrangement of crystals in a solid metal (for example, an ingot) as seen by the naked eye or at low magnification. The term is also applied to the general distribution of impurities in a mass of metal as seen by the naked eye after certain methods of etching. (CTD, 1958) (b) The structure of metals as revealed by examination of the etched surface of a polished specimen at a magnification not exceeding 10 diameters. (ASM, 1961) (c) A structural feature of a rock that is discernible to the unaided eye, or with the help of a simple magnifier. (Fay, 1920) (d) The arrangement of crystals in a metal or in a rock that, with or without etching or other chemical treatment, is discernible to the unaided eye or with the help of a simple magnifier (generally less than 10×).
maculose Applied to the group of contact-metamorphic rocks represented by spotted slates, to denote its spotted or knotted character. The term may be applied either to the rock or to its structures. See also: *spotted slate.* (AGI, 1987)
maculose rock *spotted slate.*
Madagascar aquamarine A strongly dichroic variety of blue beryl obtained, as a gemstone material, from Malagasy Republic. (CTD, 1958)
Madaras system A method of obtaining pig iron, consisting of charging a retort with a mixture of iron ore, coal, and enough water to form a paste; and injecting compressed air at 2 to 4.2 kg/cm^2 pressure and at 815 to 930 °C to burn the coal. In a few minutes the entire mineral charge is heated to 980 to 1,095 °C, which is the optimum range for hydrogen reduction; then hydrogen is injected at a temperature of 815 to 923 °C at 2 to 8 kg/cm^2 pressure so that it penetrates the entire mass and reacts with the iron oxide to produce steam and metallic iron. About 50% of the injected hydrogen reacts with the ore to produce sponge iron. Sulfur is oxidized by the hot air and is eliminated by the hydrogen. (Osborne, 1956)
made ground (a) A recent deposit, as of river silt. (Fay, 1920) (b) Ground formed by filling in natural or artificial pits with hardcore or rubbish. (CTD, 1958)
Madeira topaz Citrine variety of quartz. Cf: *false topaz; Spanish topaz.*
made up Coupled; the assembled component parts of a drill string or pipe system. (Long, 1960)
madogram A plot of mean absolute difference of paired sample measurements as a function of distance and direction. Madograms are not true variograms, and generally should not be used in kriging. If used, the kriged estimates might be "reasonable," but the kriging standard deviations will be meaningless.
madrepore marble A fossiliferous limestone occurring in a variety of colors. It takes a high polish, is used as a marble, and derives its name from its most characteristic fossil, a species of coral.
maenite An intrusive trachytic rock, regarded as a differentiation product of a gabbroic magma. Maenite is a bostonite relatively high in calcium and low in potassium.
mafelsic Said of igneous rocks containing roughly equal amounts of felsic and mafic minerals, color index 40 to 70. (AGI, 1987)
mafic Pertaining to or composed dominantly of the ferromagnesian rock-forming silicates; said of some igneous rocks and their constituent minerals. Cf: *basic; felsic.* See also: *mafite.*
mafic front *basic front.*
mafite Any dark mineral in Johannsen's classification of igneous rocks. See also: *mafic.* (AGI, 1987)
magazine (a) A storage place for explosives. (Jones, 1949) (b) A building specially constructed and located for the storage of explosives. Cf: *powder chest.*
maghemite An isometric mineral, gamma-Fe_2O_3; magnetite series; spinel group; dimorphous with hematite; strongly ferrimagnetic. Syn: *oxymagnite.*
magistral Roasted copper pyrites. (Pryor, 1963)
magma Naturally occurring molten rock, generated within the Earth and capable of intrusion and extrusion, from which igneous rocks are derived through solidification and related processes. It may or may not contain suspended solids (such as crystals and rock fragments) and/or gas phases. Adj: magmatic. (AGI, 1987)
magmatic Of, pertaining to, or derived from magma. See also: *igneous.*
magmatic assimilation *assimilation.*
magmatic blister The swelling up by differential heating of magma; e.g., by local concentrations of radioactive matter. (Schieferdecker, 1959)
magmatic corrosion A process of re-solution in which an early formed phase, such a quartz phenocryst, later becomes corroded or embayed as the result of some change in the conditions affecting the solubility of the phase; sometimes corrosion rims result. Syn: *corrosion.* (AGI, 1987)
magmatic cycle *igneous cycle.*
magmatic deposit *magmatic ore deposit.*
magmatic differentiation The process by which more than one rock type is derived from a parent magma. Cf: *assimilation.* Syn: *differentiation.*
magmatic disseminated ore deposit *disseminated deposit.*
magmatic dissolution *assimilation.*
magmatic injection deposit Straight magmatic mineral (ore) deposit, the formation of which has often been ascribed to injection into the older country rock of liquefied crystal differentiates, of residual liquid segregations, or of immiscible liquid separations and accumulations. An older term. (Schieferdecker, 1959)
magmatic ore deposit An ore deposit formed by magmatic segregation, generally in mafic rocks and layered intrusions, as crystals of metallic oxides or from an immiscible sulfide liquid. Syn: *magmatic deposit.* (AGI, 1987)
magmatic segregation Concentration of crystals of a particular mineral (or minerals) in certain parts of a magma during its cooling and crystallization. Some ore deposits (i.e., magmatic ore deposits) are formed in this way. See also: *differentiation.* (AGI, 1987)
magmatic stoping A process of magmatic emplacement or intrusion that involves detaching and engulfing pieces of the country rock. The engulfed material presumably sinks downward and/or is assimilated. See also: *piecemeal stoping.* (AGI, 1987)
magmatic water Water contained in or expelled from magma. Cf: *juvenile water.* (AGI, 1987)
magmatist One who believes that much granite is a primary igneous rock produced by differentiation from basaltic magma. Cf: *transformist.*
Magnafloat Trademark for iron oxide heavy media systems for the purification of coal, sand, gravel, and other similar materials. (CCD, 1961)
Magnedisc A widely used medium for magnetic recording. It consists of a circular platter of magnetically coated plastic material that rotates on a turntable in the same way as a phonograph record. The magnetic channels, each in contact with a fixed head aligned along a radius from the center, are on concentric circular bands extending inward from the outer edge of the disk. (Dobrin, 1960)
magnesia Magnesium oxide, MgO. A constituent of lime made from dolomitic limestone. (Barger, 1939; AGI, 1987; Mersereau, 1947)
magnesia alum *pickeringite.*
magnesia covering Hydrated magnesium carbonate containing about 15% asbestos, used for heat insulation. Also referred to as 85% magnesia. (Osborne, 1956)
magnesia glass Glass containing usually 3% to 4% of magnesium oxide. Electric lamp bulbs have been mainly made from this type of glass since fully automatic methods of production were adopted. (CTD, 1958)
magnesia mica *biotite.*
magnesian hornfels A fine-grained metamorphic rock derived from a high-magnesium igneous rock; e.g., serpentine. See also: *hornfels.* (AGI, 1987)

magnesian limestone A limestone containing from 5% to 35% $MgCO_3$. Cf: *dolomitic limestone.* (ASTM, 1994)
magnesian marble A crystalline variety of limestone containing not less than 5% nor more than 40% of magnesium carbonate as the dolomite constituent.
magnesian schist A schistose metamorphic rock derived from a rock high in content of magnesium; e.g., serpentine. See also: *schist.* (AGI, 1987)
magnesian spar *dolomite.*
magnesia ramming materials Granular, airsetting mixtures, containing 70% to 80% MgO, used for monolithic furnace linings.
magnesiochromite An isometric mineral, $MgCr_2O_4$; chromite series; spinel group; forms series with spinel and with chromite; crystallizes in black octahedra, but commonly massive. Also spelled magnochromite. Syn: *picrochromite.*
magnesioferrite An isometric mineral, $MgFe_2O_4$; magnetite series; spinel group; strongly ferrimagnetic. Also spelled magnoferrite.
magnesiolaumontite A magnesian variety of laumontite.
magnesite A trigonal mineral, $MgCO_3$; calcite group; rhombohedral cleavage; in veins in serpentinite and peridotite, magnesium-rich schist, and altered dolomitic marbles. Syn: *giobertite; magnesium carbonate.*
magnesite cement Common term for ground magnesite.
magnesite refractory A refractory material, fired or chemically bonded, consisting essentially of dead-burned magnesite; the MgO content usually exceeds 80%. Such refractories are used in the hearths and walls of basic steel furnaces, mixer furnaces, and cement kilns. (Dodd, 1964)
magnesium A light, silvery-white, and fairly tough metal. Symbol, Mg. It does not occur uncombined, is found in large deposits in the form of magnesite, dolomite, and other minerals. Readily ignites upon heating. Used in flashlight photography, flares, and pyrotechnics, including incendiary bombs. Its alloys are essential for airplane and missile construction. (Handbook of Chem. & Phys., 3)
magnesium aluminate $MgAl_2O_4$; melting point, 2,135 °C; sp gr, 3.6; thermal expansion (100 to 1,000 °C), 9.0 X 10^{-6}. This compound is the type mineral of the spinel group. See also: *spinel.* (Dodd, 1964)
magnesium-aluminum garnet *pyrope.*
magnesium bentonite A smectite with exchangeable magnesium.
magnesium blodite *blödite.*
magnesium carbonate *magnesite.*
magnesium chalcanthite *pentahydrite.*
magnesium-chlorophoenicite A monoclinic mineral, $(Mg,Mn)_3Zn_2(AsO_4)(OH,O)_6$; associated with manganese silicates at Franklin, NJ.
magnesium front *basic front.*
magnesium kaolinite *amesite.*
magnesium leonite *leonite.*
magnesium mica *phlogopite.*
magnesium minerals Chiefly magnesite, dolomite, and brucite.
magnesium orthite A magnesian variety of allanite.
magnesium titanate $MgTiO_3$. Used in ceramic dielectric bodies. Although relatively low in dielectric constant, it has a positive temperature coefficient of capacity. Thus, it can be used in conjunction with a negative coefficient material; e.g., titania, to produce a near 0 temperature coefficient product. (Lee, 1961)
magnesium zirconate MgZrO; melting point, 2,150 °C. This compound is sometimes added in small amounts (up to 5%) to other electroceramic bodies to lower their dielectric constant at the Curie point. (Dodd, 1964)
magnetic Of or pertaining to a mineral, object, area, or locale possessing the properties of a magnet. (Long, 1960)
magnetic alloys Alnico is an alloy of aluminum, nickel, and cobalt, with strong magnetic properties similar in all directions. Alcomax is anisotropic, with maximum flux along preferred axis. Hycomax is also anisotropic. Platinax, a cobalt-platinum alloy containing 23.3% cobalt, is isotropic. (Pryor, 1963)
magnetic anomaly Variation of the measured magnetic pattern from a theoretical or empirically smoothed magnetic field. (Hy, 1965)
magnetic azimuth The azimuth measured clockwise from magnetic north through 360°; the angle at the point of observation between the vertical plane through the observed object and the vertical plane in which a freely suspended magnetized needle, influenced by no transient artificial magnetic disturbance, will come to rest. (AGI, 1987)
magnetic bearing The bearing expressed as a horizontal angle between the local magnetic meridian and a line on the Earth; a bearing measured clockwise from magnetic north. It differs from a true bearing by the amount of magnetic declination at the point of observation. (AGI, 1987)
magnetic bort *bort.*
magnetic bottle A magnetic field used to confine a plasma in controlled-fusion experiments. (Lyman, 1964)
magnetic circuit The closed path taken by the magnetic flux in an electric machine or other piece of apparatus. (CTD, 1958)
magnetic clutch One in which connection between drive and driven member is provided by electromagnetic force. (Pryor, 1963)
magnetic correlation The orientation of an underground survey, using the Earth's magnetic field. (BS, 1963)
magnetic crack detection The part to be examined is magnetized either by passing a heavy current through it or by making it the core of a coil through which a heavy current is passed. Small cracks, or nonmagnetic phases such as inclusions, cause the magnetic flux to break the surface, thus forming small magnets. When the part is sprayed with a suspension of iron oxide particles in paraffin, the particles cling to the small magnets and thereby reveal defects. (Hammond, 1965)
magnetic declination In nautical and aeronautical navigation, the continually changing acute angle between the direction of the magnetic and geographic meridians. The term "magnetic variation" is preferred. (AGI, 1987)
magnetic detector An electrical device for indicating the presence of magnetic material in an area.
magnetic dip Vertical angle through which a freely suspended magnetic needle dips from horizontal. (Pryor, 1963)
magnetic domain Aggregation of ferromagnetic atoms into a group, usually a fraction of a micrometer in size, which lies among similar groups with random group orientation. This cancels out any magnetic moment until or unless they all are oriented by an applied magnetic field. (Pryor, 1963)
magnetic elements (a) These are declination, dip, and magnetic intensity in the horizontal plane. (Pryor, 1963) (b) The characteristics of a magnetic field that can be expressed numerically. The seven magnetic elements are declination D, inclination I, total intensity F, horizontal intensity H, vertical intensity Z, north component X, east component Y. Typically, only three elements are needed to give a complete vector specification of the magnetic field. (AGI, 1987)
magnetic feeder Any feeder that uses magnetism to pick up, hold, separate, and deliver objects.
magnetic field (a) Space surrounding a magnet or current-carrying coil, in which appreciable magnetic force exists. Its intensity (H) is the force exerted on a unit pole. (Pryor, 1963) (b) A region in which magnetic forces would be exerted on any magnetized bodies or electric currents present; the region of influence of a magnetized body or an electric current. (AGI, 1987)
magnetic field strength The force exerted on a unit pole is the field strength at that point. (AGI, 1987)
magnetic flocculation Phenomenon that results from residual magnetism of ferromagnetic particles that have bunched together under the influence of their individual polar forces. (Pryor, 1963)
magnetic flowmeter A device used for the flow measurement of abrasive slurries. The calibration is affected by the presence of magnetic constituents in the slurry, and therefore, a pipe coil is also used to compensate the magnetic flowmeter calibration for varying amounts of magnetic material in the slurry. (Nelson, 1965)
magnetic flux (a) Induced strength or flux density in a magnetic field, measured in maxwells: $B = 4\pi/A + H$, where B = the flux density, P = the strength of each magnetic pole, A = the cross-sectional area of a cylinder through which the flux flows, and H = the magnetic intensity in oersteds. See also: *magnetic intensity.* (Pryor, 1963) (b) The surface area times the normal component of magnetic induction B; the number of magnetic field lines crossing the surface of a given area. Expressed in maxwells in the cgs system. (AGI, 1987)
magnetic gradiometer An instrument, designed but not applied, for measuring the gradient of the magnetic intensity. (AGI, 1987)
magnetic hoist A hoisting device that does its lifting by means of an electromagnet. (Crispin, 1964)
magnetic hysteresis *hysteresis.*
magnetic induction (a) Magnetic-flux density, symbolized by B. In a magnetic medium, it is the vector sum of the inducing field H and the magnetization M. B is expressed in teslas in SI and in gauss or gammas in the cgs system. Syn: *magnetic field.* (AGI, 1987) (b) In a magnetic medium, the vector sum of the inducing field H and the corresponding intensity of magnetization I, according to the relationship $B = H + 4\pi I$. (AGI, 1987) (c) A nonrecommended syn. of electromagnetic induction. (AGI, 1987) (d) The process of magnetizing a body by applying a magnetic field. This usage is not recommended. (AGI, 1987)
magnetic intensity A vector quantity pertaining to the condition at any point under magnetic influence (as of a magnet, an electric current, or an electromagnetic wave) measured by the force exerted in a vacuum upon a free unit north pole placed at the point in question. Also called magnetic force. (Webster 3rd, 1966)
magnetic iron ore *magnetite.*
magnetic level coil A device for measuring the liquid level in sumps and other vessels. It consists of a loop of wire that is encased in a fiber glass protective sheath. The loop is inserted in a sump of thickener containing a magnetite or ferrosilicon slurry, and the electrical signal

given off represents the level of the slurry surrounding the loop. (Nelson, 1965)

magnetic meridian The horizontal line that is oriented, at any specified point on the Earth's surface, along the direction of the horizontal component of the Earth's magnetic field at that point; not to be confused with isogonic line. Syn: *geomagnetic meridian.* (Hunt, 1965)

magnetic method A geophysical prospecting method that maps variations in the magnetic field of the Earth that are attributable to changes of structure or magnetic susceptibility in certain near-surface rocks. Sedimentary rocks generally have a very small susceptibility compared with igneous or metamorphic rocks, and most magnetic surveys are designed to map structure on or within the basement, or to detect magnetic minerals directly. Most magnetic prospecting is now carried on with airborne instruments. (Dobrin, 1960)

magnetic mirror A magnetic field used in controlled-fusion experiments to reflect charged particles back into the central region of a magnetic bottle. (Lyman, 1964)

magnetic moment (a) That vector associated with a magnetized mass; the vector product of it and the magnetic field intensity in which the mass is immersed (ignoring the field distortion thereby produced) is a measure of the resulting torque. Also called moment of a magnet. (AGI, 1987) (b) A vector quantity characteristic of a magnetized body or an electric-current system; it is proportional to the magnetic-field intensity produced by this body and also to the force experienced in the magnetic field of another magnetized body or electric current. The magnetic moment per unit volume is the magnetization. (AGI, 1987)

magnetic permeability Ratio of magnetic induction (B) to the inducing field of magnetic intensity (H). With magnetic intensity lines of force per square centimeter in air, and flux density lines in a substance placed in that field, B/H is the magnetic permeability. When this is less than 1.0, the substance is diamagnetic; above 1.0, paramagnetic; and when high, ferromagnetic. (Pryor, 1963)

magnetic plug A drain or inspection plug magnetized for the purpose of attracting and holding iron or steel particles in a lubricant. (Nichols, 1976)

magnetic polarity The orientation of the constituent minerals within the rocks of the Earth's crust conforming to the Earth's magnetic field as it existed at the time the strata were deposited. See also: *core orientation.* (Long, 1960)

magnetic pole (a) Either of two points on the Earth's surface where the lines of magnetic force are vertical; an end of the axis of the Earth's magnetic polarity, not coincident with a geographic pole, and continually changing its position. The north magnetic pole is in northern Canada. (Standard, 1964) (b) Either of two nonstationary regions on the Earth that sometimes move many miles in a day, toward which the isogonic lines converge, and at which the dip is ±90°. (Webster 3rd, 1966) (c) The area on a magnetized part at which the magnetic field leaves or enters the part. It is a point of maximum attraction in a magnet. (ASM, 1961)

magnetic prospecting *magnetic method.*

magnetic pyrite *pyrrhotite.*

magnetic recording Any process by which the output of a detector-amplifier system is recorded on a magnetic recording medium. The advantages of such a system are that the resulting records may be played back and converted into conventional records with phase shifting, mixing, etc., between traces and with filtering variations. Most modern magnetic recording is in digital form. (AGI, 1987)

magnetic roasting process Heating ferrous iron ore in the presence of air in order to oxidize the iron content, present in whatever form, to the magnetic oxide so that in a subsequent operation it can be separated from the gangue by means of a magnetic separator. Also, roasting a hematitic ore with scrap iron to reduce it to magnetite. (Osborne, 1956)

magnetic roll feeder A feeder that uses magnetized, power-operated rolls for separating and delivering objects.

magnetics Branch of science that deals with magnetic phenomena. (Osborne, 1956)

magnetic separation (a) The separation of magnetic materials from nonmagnetic materials, using a magnet. This is an esp. important process in the beneficiation of iron ores in which the magnetic mineral is separated from nonmagnetic material; e.g., magnetite from other minerals, roasted pyrite from sphalerite, etc. (Newton, 1959; Henderson, 1953) (b) The use of permanent magnets or electromagnets to remove relatively strongly ferromagnetic particles from paramagnetic and diamagnetic ores. (Pryor, 1960)

magnetic separator (a) A device used to separate magnetic from less magnetic or nonmagnetic materials. The crushed material is conveyed on a belt past a magnet. (ASM, 1961) (b) For medium solids recovery. A device in which medium solids are caused to adhere, by magnetic means, to a conveying belt or drum, while a current of water removes nonmagnetic particles that contaminate the medium. (BS, 1962)

magnetic storm A worldwide disturbance of the Earth's magnetic field, commonly with amplitude of 50 to 200 gammas. It generally lasts several days, and is thought to be caused by charged particles ejected by solar flares. Magnetic prospecting usually has to be suspended during such periods. (AGI, 1987)

magnetic susceptibility A measure of the degree to which a substance is attracted to a magnet; the ratio of the intensity of magnetization to the magnetic field strength. *susceptibility.* (AGI, 1987)

magnetic unit in prospecting The gamma (Γ), which equals 10^{-5} Oe (7.9577×10^{-4} A/m). (AGI, 1987)

magnetic variation *declination.*

magnetic variometer A geophysical instrument similar to the gravimeter in that absolute values are not measured, but only the differences in vertical magnetic force between field stations and a selected base station. (Nelson, 1965)

magnetism (a) That property of iron, steel, and some other substances, by virtue of which they exert forces of attraction and repulsion according to fixed laws. (Crispin, 1964) (b) The science that is concerned with the conditions and laws of magnetic force. (Crispin, 1964)

magnetite (a) An isometric mineral, $8[FeOFe_2O_3]$; spinel group; forms series with jacobsite and with magnesioferrite; crystallizes in octahedra; metallic; black; strongly ferrimagnetic; an accessory mineral in many igneous rocks; a common detrital mineral; a major mineral in banded iron formations and magmatic iron deposits; an ore of iron. Syn: *black iron ore; lodestone; magnetic iron ore; octahedral iron ore.* See also: *iron ore.* (b) The mineral series magnesioferrite, magnetite, maghemite, franklinite, jacobsite, and trevorite in the spinel group. Syn: *iron series.*

magnetite An igneous rock consisting essentially of magnetite and having an iron content of 65% to 70% or more. Apatite may accompany the magnetite. (Johannsen, 1931-38)

magnetite olivinite A dunite high in content of titaniferous magnetite and containing shreds of biotite. (Holmes, 1928)

magnetite spinellite An eruptive iron ore occurring at Routivara, Sweden, and consisting of magnetite (in part titaniferous), spinel, and smaller amounts of olivine, pyroxene, apatite, and pyrrhotite. The ore contains about 14% titanic oxide.

magnetized A body is said to be magnetized when it possesses or can be made to possess that peculiar property whereby, under certain circumstances, it will naturally attract or repel a similar body in accordance with magnetic laws; e.g., drill rods become magnetized in use and will strongly attract other iron or steel articles. (Long, 1960)

magnetizing force (a) The phenomenon associated with a magnetic flux density at a point. Theoretically, measured by the mechanical force on a unit magnetic pole in an evacuated tunnel along the direction of the magnetic flux; the magnetomotive force per centimeter in this direction. (CTD, 1958) (b) A force field, resulting from the flow of electric currents or from magnetized bodies, that produces magnetic induction. (ASM, 1961)

magnetizing roast A process in which an ore containing pyrite is heated and the magnetic iron oxide so formed is removed by a magnetic separator. In many cases the iron oxide is extracted as gangue. (Nelson, 1965)

magnetometer (a) An instrument for measuring magnetic intensity. In ground magnetic prospecting, an instrument for measuring the vertical magnetic intensity; in airborne magnetic prospecting, an instrument for measuring the total magnetic intensity. Also, an instrument used in magnetic observatories for measuring various components of the magnetic field of the Earth. (AGI, 1987) (b) A sensitive instrument for detecting and measuring changes in the Earth's magnetic field, used in prospecting to detect magnetic anomalies and magnetic gradients in rock formations. See also: *airborne magnetometer.* (Pryor, 1958)

magnetoplumbite (a) A hexagonal mineral, $Pb(Fe,Mn)_{12}O_{19}$; black; ferrimagnetic; at Långban, Sweden. (b) The mineral group hawthorneite, hibonite, magnetoplumbite, and yimengite.

magnetorque A form of transmission based on the principle that a magnetic field of force is produced whose strength is varied so as to transmit a torque of corresponding magnitude. With a system of this kind, control of any particular operation is effected through the agency of the magnetorque clutch, while the motor runs continuously at constant speed. (Mining)

magnetostriction The characteristic of a material that is manifest by strain when it is subjected to a magnetic field; or the inverse. Some iron-nickel alloys expand; pure nickel contracts. (ASM, 1961)

magnetotelluric method (a) An electrical prospecting technique based on an application of telluric currents in which the magnetic fields induced by the alternations in earth currents would be measured simultaneously with the voltage fluctuations between electrodes at the surface. The ratio between the amplitudes of these alternating voltages and the associated magnetic fields would be plotted as a function of frequency. (Dobrin, 1960) (b) An electromagnetic method of surveying, in which natural electric and magnetic fields are measured. Usually the two horizontal electric-field components plus the three magnetic-field components are recorded; orthogonal pairs yield elements of the tensor impedance of the Earth. This impedance is measured at frequencies within the range 10^{-5} Hz to 10 Hz. (AGI, 1987)

magniotriplite A monoclinic mineral, $(Mg,Fe,Mn)_2(PO_4)F$; yellow.

magnochromite A variety of chromite that contains magnesium. Alternate spelling of magnesiochromite. (Standard, 1964)
magnocolumbite An orthorhombic mineral, $(Mg,Fe,Mn)(Nb,Ta)_2O_6$. Cf: *columbite.*
magnoferrite Original form of magnesioferrite. (Dana, 1914)
magnussonite An isometric and tetragonal mineral, $Mn_5As_3O_9(OH,Cl)$; green; forms crusts at Långban, Sweden, and green to brown selvages at Sterling Hill, NJ.
mahogany ore Compact mixture of oxides of iron and copper. (Schaller, 1917)
mailly stone A softer sort of limestone, very dusty. (Arkell, 1953)
main airway (a) The road along which the principal ventilating current passes. (Peel, 1921) (b) Underground ventilation channel directly connected with a shaft or other entry to a mine. (Pryor, 1963)
main-and-tail haulage A single-track haulage system operated by a haulage engine with two drums, each with a separate rope. The engine is usually located at the out end of the system. The main rope is attached to the out end of the set of tubs, and the tail rope passes around a sheave in, and is then attached to, the rear end of the set. To draw the full set out, the main rope is wound in, the tail rope being allowed to run free. To draw the empty set in, the tail rope is wound in, the main rope being allowed to run free. The tail rope is equal to twice the length of the haulage road. The main-and-tail haulage is adopted when the gradient is irregular and the empty set will not run in by gravity. See also: *rope haulage.* (Nelson, 1965)
main arch (a) The refractory blocks forming the part of a horizontal gas retort comprising the division walls and the roof that covers the retorts and the recuperators. (Dodd, 1964) (b) General term for the central part of a furnace roof, particularly used as a syn. for the crown of a glass tank furnace. (Dodd, 1964)
main bord gate York. The heading that is driven slightly to the rise from the shaft. (Fay, 1920)
main bottom Hard rock below alluvial deposits.
main break In mine subsidence, the break that occurs over the seam at an angle from the vertical equal to half the dip. (Lewis, 1964)
main brow Eng. *gate.*
main conveyor *underground mine conveyor.*
main crosscut The crosscut that traverses the entire mining field and penetrates all deposits. There is such a crosscut at each level, and it is the main one for the level in question. It serves the same purpose as the shaft and thus must have correct cross section, and be particularly well constructed, as repairs to its support would hold up the transport of the entire level. (Stoces, 1954)
main drive A main tunnel driven in the rock underlying a lead and about 50 ft (15.2 m) below the wash dirt. It is used as a drainage tunnel for carrying the water from the drainage holes to the shaft sump and also for the transport of cars from the raises. (Eng. Min. J., 1938)
main endings Pairs of narrow coal headings with crosscuts at intervals, driven to form large pillars of coal in panel working. See also: *narrow work.* (Nelson, 1965)
main engine N. of Eng. The surface pumping engine, usually of the Cornish type. (Fay, 1920)
main entry (a) The principal entry or set of entries driven through the mine from which cross entries, room entries, or rooms are turned. (Federal Mine Safety, 1977) (b) A term used in the United States for the principal horizontal gallery giving access to an underground mine and used for haulage, ventilation, etc. Where two entries are driven in parallel, the term "double entry" is used. With three parallel entries, the term "triple entry" is used. (Nelson, 1965) (c) An entry driven at right angles with the face slips of the coal. See also: *entry.* (Fay, 1920) (d) A main haulage road. See also: *main road.* (Fay, 1920)
Maine sampler A drive-type, split-tube soil-sampling device, usually equipped with a flap or clack valve near the cutting shoe. Usually produced in sizes having outside diameters ranging from 4 to 6-5/8 in (10 to 17 cm). (Long, 1960)
Maine-type sampler A soil-sampling device that works essentially on the same principle as a Maine sampler. (Long, 1960)
main facets (a) The bezel and pavilion facets. (b) Any facet extending from the girdle to the table or from the girdle to the cutlet.
main fans Main fans produce the general ventilating current of the mine, and are generally of large capacity and permanently installed. They are assisted by natural ventilation, if present, and, if necessary, by booster fans. They are installed to perform a certain duty, and great attention is paid to their efficiency since this governs the cost of performing the duty. (Roberts, 1960)
main firing The firing of a round of shots by means of current supplied by a transformer fed from a main power supply. (BS, 1964)
main gate The principal or central heading along which the coal is conveyed from two or more conveyor panels. Normally, the main gate is also the intake airway to the face. See also: *double-unit conveyor; bottom gate; mother gate.* (Nelson, 1965)
main haulage (a) That portion of the haulage system that moves the coal from the secondary haulage system to the shaft or mine opening. The method employed is the same for either longwall or room-and-pillar mining. Any one of four methods may be used: (1) mine cars and battery or trolley locomotives, (2) mine cars and a direct rope haulage, (3) a combination of methods 1 and 2, or (4) belt conveyors. (Wheeler, 1958) (b) The system by which coal is transported in trains in or out of a slope or drift entry or from the bottom of a shaft. (BCI, 1947)
main haulage conveyor Generally 500 to 3,000 ft (approx. 150 to 1,000 m) in length. It is used to transport material between the intermediate haulage conveyor and a car-loading point or the outside. (NEMA, 1961)
main haulageway The principal transportation road, drift, tunnel, etc. (BCI, 1947)
main hole The first or primary borehole from which secondary or branch holes are drilled. Also called original hole; parent hole. Cf: *branch.* (Long, 1960)
main intake The trunk or principal intake airway of a mine. The main intake air current is usually split into two or more air currents before reaching the workings. (Nelson, 1965)
main levels The first and leading excavations in mines that are made for the purpose of opening out or winning the material being mined, and that are intended to be the principal roadways of the mine. (Peel, 1921)
main-line locomotive A large, high-powered locomotive that hauls trains of cars over the main haulage system. (BCI, 1947)
main-line motorman In bituminous coal mining, a person who operates a mine locomotive to haul cars over the main haulage tracks underground or at the surface, as distinguished from the branch lines. (DOT, 1949)
main return The principal return airway of a mine. The main return air current represents the total quantity of air; i.e., after the air splits have reunited. (Nelson, 1965)
main road The principal underground road in a district along which mined material is conveyed to the shafts, generally forming the main intake air course of each district. See also: *main entry.* (Fay, 1920)
main rod Corn. *pump rod.*
main roof The rock above the immediate roof. (Obert, 1960)
main rope The rope that pulls a train of loaded cars out along a haulage plane, as opposed to a tail rope that pulls a train of empty cars in, as in main-and-tail haulage. See also: *pull rope.* (Nelson, 1965)
main-rope haulage system This system of haulage is used for hauling loaded trains of tubs or cars up, or lowering them down, a comparatively steep gradient that is not steep enough, in the latter case, for a self-acting incline. In the normal system, a single track only is required. The electrically driven or compressed-air-driven engine has a single drum that runs loose on the forged-steel drum shaft; it is controlled by the brake when lowering the empty train, and is clutched to the shaft by means of a dog clutch when hauling the loaded train up the gradient. (Sinclair, 1959)
main-rope rider In bituminous coal mining, one who has charge of and rides trips (trains) of cars hauled in and out of the mine along the main cable haulageway. (DOT, 1949)
main-rope system A system of underground haulage in which the weight of the empty cars is sufficient to draw the rope in. (Fay, 1920)
main separation door A wooden or steel door erected near the pit bottom to prevent the intake air leaking into the main return airway or upcast shaft; a door to direct the main intake air in toward the workings. It may be fitted with an appliance, or shutter, to ease the opening for traveling purposes. See also: *fan drift doors.* (Nelson, 1965)
mains firing Firing blasts from a mains supply. (McAdam, 1958)
main shaft The line of shafting that receives its power directly from the engine or motor and transmits power to other parts. (Crispin, 1964)
mains lighting A system of underground mine lighting in which the lamps are fed from the main electrical supply. It is used principally at the pit bottom, using filament lamps in either transparent well glass or prismatic bulkhead fittings. (Nelson, 1965)
main-slope engineer In bituminous coal mining, one who operates the hoisting engine for raising or lowering workers, material, coal, and rock along the main haulage slope (incline) of a mine having one or more auxiliary or relay slopes. (DOT, 1949)
maintaining levels In quarrying, consists of the removal by blasting of rock protruding above the level of the quarry floor or bench, to allow the movement of loading and transport equipment. (Streefkerk, 1952)
maintenage Fr. The face of workings in inclined or vertical seams consisting of a series of steps each about 6 ft (1.8 m) high, and forming the working place for one person. (Fay, 1920)
maintenance Proper care, repair, and keeping in good order. (Crispin, 1964)
main tie Tension member connecting the feet of a roof truss, usually at the level of bearing on the wall plate or padstone. (Hammond, 1965)
main transport The conveying or haulage of mined material from the mining area subsidiary transport to the shaft bottom or surface. For the main transport of coal, the trend is toward trunk conveyors or

locomotive haulage. See also: *subsidiary transport; underground haulage.* (Nelson, 1965)

main way A gangway or principal passage. (Fay, 1920)

maitlandite *thorogummite.*

Majac mill A mill for dry-grinding mica by means of fluid energy. It consists of a chamber containing two horizontal directly opposed jets. Mica is fed continuously from a screw conveyor into this chamber. The particle size of the product from this mill can be controlled over a broad range down to the micrometer sizes. (USBM, 1965)

major diameter Formerly called outside diameter. It refers to the largest diameter of a thread on a screw or nut. (Crispin, 1964)

major face Used in widely different meanings in different manufacturing plants; the meaning of the term depends on the special process of cutting used. Thus, the faces of the major rhombohedron r {1011} are spoken of as the major face when BT wafers are cut directly from a faced mother crystal. Also applied to the prism faces that terminate in the faces of the major rhombohedron, etc. (Am. Mineral., 1947)

major mine disaster Defined by the Mine Safety and Health Administration, U.S. Department of Labor, as any accident that results in the death of five or more persons. (Federal Mine Safety, 1977)

major principal plane The plane normal to the direction of the major principal stress. (ASCE, 1958)

major principal stress The largest (with regard to sign) principal stress. (ASCE, 1958)

make A wide portion of a lode. Cf: *pinch.* (Nelson, 1965)

make gas Mid. To yield or produce gas. A seam of coal that gives off combustible gas is said to make gas. (Fay, 1920)

make of water The rate of entry of water into a mine or part of a mine. Also called growth. (BS, 1963)

makeup To assemble; to couple or screw together. Usually applied to the process of assembling the component parts of a drill string or pipe system. (Long, 1960)

makeup bunker *makeup shed.*

makeup medium Medium or medium solids added to the circuit to replace losses during the separating operation. (BS, 1962)

makeup shed A surface building at which explosives, drawn from the magazine, are issued and where the charges are prepared or made up. Syn: *makeup bunker.* (Nelson, 1965)

makeup time The time required to couple together the component parts of a drill or casing string and the lowering of such a string to the working position in a borehole. (Long, 1960)

makeup water (a) Water supplied to a washery to replace that lost from the circuit. (BS, 1962) (b) Water supplied to replenish that lost by leaks, evaporation, etc. (Strock, 1948)

making hole (a) The depth gained in the day's drilling. (Hess) (b) The act of, or portion of work time spent in, actual drilling and advancement of a bore or drill hole.

makings Newc. The small coal hewn out in undercutting or channeling. Also, in some localities, called bug dust. (Fay, 1920)

malachite A monoclinic mineral, $Cu_2CO_3(OH)_2$; dimorphous with georgeite; bright green; occurs with azurite in oxidized zones of copper deposits; a source of copper. Syn: *Atlas ore.*

malacolite (a) A pale-colored, translucent variety of diopside with good (001) parting; also diopside from Sala, Sweden. (b) *diopside.*

malacon A metamict variety of zircon. Also spelled malakon, malacone.

maldonite An isometric mineral, Au_2Bi; pink to silver-white; intermetallic compound. Syn: *black gold.*

male thread *pin thread.*

maletra furnace A hand reverberatory furnace for roasting finely divided ore entirely without the aid of extraneous heat. (Fay, 1920)

malignite A mafic nepheline syenite. Fifty percent of the rock is composed of aegirine-augite; the remainder is nepheline and orthoclase in about equal amounts. Accessories include apatite, biotite, titanite, and opaque oxides.

malinite An aluminous variety of halloysite.

malinowskite A variety of tetrahedrite that contains lead and silver. (Standard, 1964)

mall Eng. A heavy hammer. A drawing or prop mall is a long-handled mall used when withdrawing timber. (SMRB, 1930)

malladrite A trigonal mineral, Na_2SiF_6. (Not mallardite.)

mallan Soft turf containing lumps of ore, Derbyshire, U.K.

mallardite A monoclinic mineral, $MnSO_4{\cdot}7H_2O$; melanterite group. (Not malladrite.)

malleability (a) The characteristic of metals that permits plastic deformation in compression without rupture. See also: *plastic deformation.* (ASM, 1961) (b) Minerals are malleable when slices cut from them may be flattened out under a hammer. Examples are native gold; silver; copper; platinum. (Nelson, 1965)

malleable (a) Said of a mineral, e.g., gold, silver, copper, platinum, that can be plastically deformed under compressive stress, e.g., hammering. (AGI, 1987) (b) A mineral that can be sliced and the slices hammered flat without breaking, e.g., gold, copper. Cf: *sectile; flexible.*

malleable cast iron A cast iron made by a prolonged anneal of white cast iron in which decarburization or graphitization, or both, take place to eliminate some or all of the cementite. The graphite is in the form of temper carbon. If decarburization is the predominant reaction, the product will have a light fracture, (whiteheart malleable); otherwise, the fracture will be dark (blackheart malleable). Pearlitic malleable is a blackheart variety having a pearlitic matrix along with, perhaps, some free ferrite. See also: *iron.* (ASM, 1961)

malleable mineral A mineral that may be flattened or deformed by hammering without breaking, for example, native copper or gold. (Stokes, 1955)

malleable nickel Nickel obtained by remelting and deoxidizing electrolytic nickel and casting it into ingot molds. It can be rolled into sheet and used in equipment for handling food, and for coinage, condensers, and other purposes where resistance to corrosion, particularly by organic acids, is required. (CTD, 1958)

malleablizing Annealing white cast iron in such a way that some or all of the combined carbon is transformed to graphite or, in some instances, part of the carbon is removed completely. (ASM, 1961)

malleate To shape into a plate or leaf by beating or hammering; said of metal. (Standard, 1964)

malmstone (a) A hard, cherty, grayish-white sandstone, specif. the Malmstone from the upper part of the Upper Greensand (Cretaceous) of Surrey and Sussex in England, used as a building and paving material. (AGI, 1987) (b) A marly or chalky rock. (AGI, 1987)

malpais A term used in the Southwestern United States and Mexico for a region of rough and barren lava flows. The connotation of the term varies according to the locality. Etymol: Spanish, mal país, bad land. (AGI, 1987)

maltesite A variety of andalusite, resembling chiastolite, showing in cross section a maltese cross of pure material separated by areas of impure material. (English, 1938)

maltha (a) Various natural tars resulting from the oxidation and drying of petroleum. See also: *mineral tar; pittasphalt.* (AGI, 1987) (b) A black viscid substance intermediate between petroleum and asphalt. Also called malthite. (Webster 3rd, 1966) (c) A variety of ozocerite. (Webster 2nd, 1960)

malthacite A scaly, sometimes massive, white or yellowish clay related to fuller's earth, having a Si-Al ratio of about 4. (AGI, 1987)

mammillary Of, or pertaining to, smoothly rounded masses resembling breasts or portions of spheres; descriptive of the shape of some mineral aggregates, such as malachite or limonite; similar to but a larger size than botryoidal. (Fay, 1920)

mammillary structure *pillow structure.*

mammillated Said of a mineral displaying large spheroidal surfaces, e.g., malachite. (Nelson, 1965)

managerial organization Coordination of functional units and presentation of the results of their achievements in policy-guiding form as facts, conclusions, and recommendations. Efficient integration of the operations managed. (Pryor, 1963)

manandonite An orthorhombic mineral, $Li_2Al_4(Si_2AlB)O_{10}(OH)_8$; kaolinite-serpentine group.

manasseite (a) A hexagonal mineral, $Mg_6Al_2(CO_3)(OH)_{16}{\cdot}4H_2O$; dimorphous with hydrotalcite. (b) The mineral group barbertonite, chlormagaluminite, manasseite, and sjögrenite.

manchado (a) Spotted ore, Spain. (b) A commercial grade of mica which is stained to very slightly spotted, Brazil.

Manchurian jade *soapstone.*

mandelstone *amygdaloid.*

mandrel (a) A miner's pick. (Webster 3rd, 1966) (b) A usually tapered or cylindrical axle spindle, or arbor that is inserted into a hole in a piece of work so as to support the work during machining. (Webster 3rd, 1966) (c) A metal bar that serves as a core around which metal or other material may be cast, molded, forged, bent, or otherwise shaped. (Webster 3rd, 1966) (d) Any of a train of jointed units intended to be pulled through an underground duct as each joint is made to ensure perfect alinement, or through a steel pipe in process of welding to ensure a smooth interior. (Webster 3rd, 1966) (e) The shaft and bearings on which a tool (as a grinding disk or circular saw) is mounted. (Webster 3rd, 1966) (f) A temporary interior support for a thin-walled tube (as a tubular steel pile to be filled later with concrete) being driven into something. (Webster 3rd, 1966)

mandril A miner's pick. (Nelson, 1965)

Manebach law A type of twinning in the monoclinic crystal system. The basal pinacoid is the twinning plane. (Hess)

Manebach twin A monoclinic crystal twinned across the basal pinacoid, according to the Manebach law, e.g., orthoclase.

man engine *man machine.*

manganandalusite A variety of andalusite containing 6.91% Mn_2O_3. It differs from ordinary andalusite in its grass-green color and strong pleochroism. From Vestana, Sweden. (English, 1938)

manganberzeliite An isometric mineral, $(Ca,Na)_3(Mn,Mg)_2(AsO_4)_3$; forms a series with berzeliite and is yellow to orange.

manganblende *alabandite.*

mangandalusite Variety of andalusite containing manganese. Alternate spelling of manganandalusite.

manganepidote *piemontite.*

manganese A gray-white, hard, brittle metallic element. Symbol, Mn. Manganese does not occur uncombined in nature, but its minerals are widely distributed. Pyrolusite (MnO_2) and rhodochrosite ($MnCO_3$) are the most common minerals. The discovery of large quantities of manganese nodules on the ocean floor, containing about 24% manganese, holds promise as a source of manganese. Used to form many important alloys, esp. with steel, aluminum, and antimony; used in dry cells and glass, and in the preparation of oxygen, chlorine, and medicines. (Handbook of Chem. & Phys., 3)

manganese-aluminum garnet *spessartine.*

manganese bronze Alloy of 59% copper, 1% tin, and up to 40% manganese. Practically a brass with high tensile strength and toughness. Used for ship's impellers and such fittings as must withstand corrosion by seawater. (Pryor, 1963)

manganese garnet *spessartite.*

manganese glance *alabandite.*

manganese-hörnesite A monoclinic mineral, $(Mn,Mg)_3(AsO_4)_2 \cdot 8H_2O$; vivianite group.

manganese hydrate *wad.*

manganese minerals Those in commercial production are pyrolusite, psilomelane, braunite, and manganite. Manganese is used chiefly in steel manufacture.

manganese nodules The concretions, primarily of manganese salts, covering extensive areas of the ocean floor. These vary in size from extremely small to some 6 in (15.2 cm) in diameter. They have a layer configuration and may prove a useful source of minerals. (Hy, 1965)

manganese ore A term used for ore containing 35% or more manganese; it may include concentrate, nodules, or synthetic ore.

manganese oxide There are several manganese oxides, the commonest being MnO_2 (pyrolusite). It is used as a coloring oxide (red or purple); mixed with the oxides of cobalt, chromium, and iron, it produces a black. This oxide is also used to color facing bricks, and to promote adherence of ground-coat vitreous enamels to the base metal. See also: *manganite; psilomelane.* (Dodd, 1964)

manganese sicklerite *sicklerite.*

manganese silicate *rhodonite.*

manganese spar *rhodonite; rhodochrosite.*

manganiferous iron ore A term used for iron ores containing 5% to 10% manganese.

manganiferous ore A term used for any ore of importance for its manganese content containing less than 35% manganese but not less than 5%. See also: *natural ore.*

manganite A monoclinic mineral, MnO(OH); trimorphous with feiknechtite and groutite; a hydrothermal vein mineral; an ore of manganese. Syn: *gray manganese ore; sphenomanganite.*

mangan-neptunite A monoclinic mineral, $KNa_2Li(MN,Fe)_2Ti_2Si_8O_{24}$; forms a series with neptunite. See also: *neptunite.*

manganocalcite A variety of calcite that contains manganese carbonate and is closely related to rhodochrosite. Syn: *kutnohorite.* (Standard, 1964)

manganolangbeinite An isometric mineral, $K_2Mn_2(SO_4)_3$; rose-red; forms tetrahedra in cavities in recent lavas on Mt. Vesuvius, Italy.

manganolite (a) *rhodonite.* (b) A general term for rocks composed of manganese minerals, esp. manganese oxides such as wad and psilomelane. (AGI, 1987)

manganophyllite A manganoan variety of biotite.

manganosiderite An intermediate member of the rhodochrosite-siderite series.

manganosite An isometric mineral, MnO; periclase group; emerald-green; forms octahedra that blacken on exposure to air.

manganotantalite An orthorhombic mineral, $MnTa_2O_6$; dimorphous with manganotapiolite; forms series with manganocolumbite and with ferrotantalite.

manganpyrosmalite A hexagonal mineral, $(Mn,Fe)_8Si_6O_{15}(OH,Cl)_{10}$; dimorphous with brokenhillite; forms a series with ferropyrosmalite.

mangrove coast A tropical or subtropical low-energy coast with a shore zone overgrown by mangrove vegetation. Such coasts are common in Indonesia, Papua New Guinea, and other tropical regions. The marine coast of southern Florida is the only significant U.S. example. (AGI, 1987)

manhole (a) A safety hole. (Hudson, 1932) (b) Cubicles cut into the solid strata or built into the gob pile along haulageways in which miners can be safe from passing locomotives and cars. Also called refuge hole. (BCI, 1947) (c) A refuge hole constructed in the side of a gangway, tunnel, or slope. (Fay, 1920) (d) A small and generally very short passage used only for the ingress and egress of the miners. (Fay, 1920) (e) A hole in cylindrical boilers through which a worker can get into the boiler to examine and repair it. (Fay, 1920) (f) A small passage connecting a level with a stope, or with the level next above. (Webster 2nd, 1960)

manhole cover A movable cast-iron plate fitting a cast-iron frame bedded on a rebated concrete slab or kerb over a manhole. Covers over foul drains are formed to prevent escape of foul air. (Hammond, 1965)

manila rope Broadly, rope or cordage formed from twisted fibers obtained from abaca, agave, or hemp plants. (Long, 1960)

manipulated variable In mineral processing, a quantity or condition that is varied as a function of the actuating signal so as to change the value of the controlled variable. (Fuerstenau, 1962)

manipulator (a) A machine for moving and turning over hot billets or blooms of iron or steel in the process of rolling. (Standard, 1964) (b) A mechanical device used for safe handling of radioactive materials. Frequently, it is remotely controlled from behind a protective shield. (Lyman, 1964)

manless coal face A coal face mined by remotely controlled equipment that eliminates the need for workers in dangerous places. (Encyclopaedia Britannica, 1964)

manless coal mining Longwall coal faces equipped for the automatic starting, stopping, and steering of power-loading machines as well as the manipulation of electric trailing cables, including air and water hoses; controlled automatic advancing of face conveyors as well as advancing and setting of roof supports. Each and all operations are correctly phased and accomplished from a remote point. With the advent of electronics and automation techniques, the prospects of manless coal mining are very promising. (Nelson, 1965)

manless face A longwall face on which the coal is cut and brought out to the gate road mechanically, without the aid of miners on the face. The face is unsupported. In general, the coal seam is thin but of high grade, and some type of rapid plow is employed. See also: *coal-sensing probe; ram scraper.* (Nelson, 1965)

manlock An air lock through which workers pass to a working chamber that is under air pressure. (Hammond, 1965)

man machine Corn.; Derb. An obsolete term for a mechanical lift for lowering and raising miners in a shaft by means of a reciprocating vertical rod of heavy timber with platforms at intervals, or of two such rods moving in opposite directions. In the former case, stationary platforms are placed in the shaft, so that the miner in descending, for instance, can step from the moving platform at the end of the downstroke and step back upon the next platform below at the beginning of the next downstroke. When two rods are employed, the miner steps from the platform on one rod to that on the other. Syn: *man engine; movable ladder.* (Fay, 1920)

manmade diamond Diamond produced synthetically. Also called MM diamond. See also: *diamond; synthetic diamond.* (Long, 1960)

Mannheim process Contact method of catalyzing SO_2 to SO_3 in two stages, using first iron oxide and second platinized asbestos as catalysts. (Pryor, 1963)

Manning's formula An empirical formula for the value of the coefficient, C, in the Chezy formula, the factors of which are the hydraulic radius and a coefficient of roughness; a simplification of the Kutter formula. (Seelye, 1951)

man-of-war Staff. A small pillar of coal left in a critical spot; also, a principal support in thick coal workings. (Fay, 1920)

manometer (a) Any instrument that measures gaseous pressure. (Nelson, 1965) (b) Measures pressure or a pressure difference by balancing the applied pressure against the hydrostatic head of a column of liquid of known density. In practice, most manometers measure a pressure difference, so that if an absolute pressure is to be measured, it is essential to have access to an accurate barometer to determine the atmospheric pressure. (Roberts, 1960) (c) An instrument designed to give a continuous record of the pressure of an explosion at the point where the instrument is located. See also: *micromanometer; piezometer; two-liquid differential manometer; U-tube manometer; vernier-reading manometer; water gage.* (Rice, 1913-18; Roberts, 1960)

manometer calibration Many manometers require calibration, and this may be carried out by the (1) static method in which simultaneous readings of the manometer under test and the primary standard are taken when one limb of the manometer and the standard are connected to a variable pressure source, the other limbs being connected to a source of constant pressure, or (2) the dynamic method in which the difference in pressure obtained between a low-speed atmospheric wind-tunnel hole (static pressure) and the atmospheric static pressure is used to carry out the calibration. One limb of the manometer and the low-pressure side of the Chattock-Fry are connected to the tunnel-wall hole (reference variable pressure), while the other limb of the manometer is connected to the atmospheric outlet of the Chattock-Fry. (Roberts, 1960)

manometric efficiency (a) The ratio of the actual head developed to the velocity pressure of air moving at the fan-tip speed, equal to one-half the theoretical head of a radial-tip fan. (Hartman, 1961) (b) An indication of the capability of the fan to produce pressure. It is the ratio of the initial depression to the theoretical depression. Manometric efficiency = 4,380 total water gage / U^2, where U = tip speed in feet per second of fan blades. (Nelson, 1965) (c) The chief value of the manometric efficiency lies in its being a rough check on the mechanical efficiency of the fan. See also: *theoretical depression.* (Lewis, 1964)

manoscope *manometer.*

manoscopy The science of determining the density of gases and vapors. (Standard, 1964)

man-riding car A car or carriage designed for the riding of miners to and from the workings. The car body has a low center of gravity to avoid the risk of overturning and is fitted with track brakes and an overspeed clutch. The train set is arranged to brake from the rear to avoid pileup, and the brakes are applied immediately on overspeed from a preset velocity. (Nelson, 1965)

man-riding conductor A worker appointed by the mine manager to be in charge of the running of a train of man-riding cars. This worker is the responsible person for starting and stopping the vehicles, for giving the proper signals, and for seeing that the safe seating capacity is not exceeded. During the shift this worker is employed on other duties. (Nelson, 1965)

mansfieldite An orthorhombic mineral, $AlAsO_4 \cdot 2H_2O$; variscite group; forms a series with scorodite.

manshift The output or work done by a worker in one shift; a basis for assessing the magnitude of a job to complete. (Nelson, 1965)

Man-Ten steel alloy An alloy containing 0.35% carbon as maximum, from 0.25% to 1.75% manganese, 0.10% to 0.30% silicon, 0.01% to 0.25% copper, 0% to 0.40% molybdenum, and 0% to 0.20% vanadium. Used for bodies and doors of stripping shovel dippers, which have manganese steel for bail, lip, and renewable teeth. (Lewis, 1964)

mantle (a) The soil or other unconsolidated rock material commonly referred to as overburden. *surface.* See also: *burden; cover.* (Long, 1960) (b) A sheath of manganese steel that fits over the iron or steel cone of the breaking (gyrating) head of a gyratory crusher. (Pryor, 1963) (c) That part of a blast furnace that carries the weight of the stack, continuing up from the bosh. (Pryor, 1963) (d) *mantle rock; regolith.* (e) The outer zone in a zoned crystal; an overgrowth. (AGI, 1987) (f) The zone of the Earth below the crust and above the core, which is divided into the upper mantle and the lower mantle, with a transition zone between. (AGI, 1987)

mantle rock *regolith.*

manto A flat-lying, bedded deposit; either a sedimentary bed or a replacement strata-bound orebody. Etymol: Spanish, vein, stratum. Cf: *bed vein.* (AGI, 1987)

manual haulage The practice of pushing tubs, trams, etc., by hand. See also: *hand tramming.* (Sinclair, 1959)

manual takeup A hand-operated mechanism for adjusting the takeup or movable pulley. (NEMA, 1961)

manual-type belt-tensioning device A hand-operated mechanism for adjusting a takeup pulley to vary the tension in a conveyor belt. The most common types are chain-jack, sylvester, rack, and screw. (NEMA, 1956)

manual winding control A system in which the winder is controlled in the conventional manner by the driver, following the usual bell signals from the onsetter and the banksman. The system of control ensures that the speed of the winder follows closely the position of the driver's level, driving or dynamic braking being applied automatically to the motor as needed to preserve the chosen speed. See also: *automatic cyclic winding.* (Nelson, 1965)

manufactured gas A mixture of gaseous hydrocarbons produced from coal or oil. See also: *gas.* (Barger, 1939)

manufactured marble (a) A mixture of marble dust and plastics. (b) A mixture of powdered stone and plastics.

manufactured sand Fine aggregate produced by crushing rock. (AIME, 1960)

manway (a) A compartment, vertical or inclined, for the accommodation of ladders, pipes, and timber chutes. The drivage may be a winze or a raise and its purpose is to give convenient access to a stope. Also called ladderway. (Nelson, 1965) (b) A passageway for the use of miners only; an airshaft; a chute. (Standard, 1964) (c) Eng. *manhole.* (d) A passage in or into a mine used as a footpath for workers. (Korson, 1938) (e) A short heading between two chutes.

map (a) A horizontal projection of surface plants, mine workings, or both, drawn to a definite scale, upon which is shown all the important features of the mine; a plan; a plat. (b) The act of preparing such plans of a mine. (c) A representation to a definite scale on a horizontal plane of the physical features of a portion of the Earth's surface (natural or artificial or both) by means of symbols, which may emphasize, generalize, or omit certain features as conditions may warrant. A map may be derived from ground surveys made by transit, plane table, or camera, or from aerial photographic surveys, or both. (Seelye, 1951)

map projection A method of representing the curved surface of the Earth on a flat map. As the true shape of the Earth is a globe, it is impossible to make a map of large areas of the Earth's surface without some distortion. (Hammond, 1965)

marathon mill A form of tube mill used in the cement industry, in which the pulverizing is done by long pieces of hardened steel shafting. (Liddell, 1918)

marbella A Spanish magnetite with a siliceous gangue. (Osborne, 1956)

marble (a) A metamorphic rock composed essentially of calcite, dolomite, or a combination of the two, with a fine- to coarse-grained crystalline texture. (b) In commerce, any crystalline carbonate rock that will take a polish, including true marble, certain coarse-grained limestones, alabaster, and onyx. (c) *verde antique.* Cf: *crystalline limestone.*

marble handsaw A toothless blade fitted at the back with a block handle, used with sand, for cutting slabs of marble into pieces. (Fay, 1920)

marble saw A toothless blade used with sand in marble cutting. (Standard, 1964)

Marble's reagent An etchant for stainless steels, consisting of 4 g of copper sulfate in 20 mL of hydrochloric acid and 20 mL of water. (Osborne, 1956)

marcasite (a) White iron pyrites, FeS_2, the orthorhombic dimorph of pyrite, having a lower specific gravity, less stability, and a paler color. Often called white iron pyrites, coxcomb pyrites, and spear pyrites. (AGI, 1987) (b) In the gemstone trade, marcasite is either pyrite, polished steel (widely used in ornamental jewelry in the form of small brilliants), or even white metal. (CTD, 1958) (c) An orthorhombic mineral, FeS_2; dimorphous with pyrite; metallic; bronze-yellow to white; an authigenic or supergene mineral from acid solutions. Syn: *white iron pyrites; white pyrite; cockscomb pyrite; spear pyrite; lamellar pyrite.* (d) The mineral group ferroselite, frohbergite, hastite, kullerudite, marcasite, and mattagamite. (e) A gemstone with a metallic luster, esp. pyrite, but including polished steel and white metal. Syn: *radiated pyrite.*

marcus A patented shaker screen with a nonharmonic or quick-return motion. (Zern, 1928)

Marcy mill A ball mill in which a vertical-grate diaphragm is placed near the discharge end. Between this perforated diaphragm and the end of the tube, there are arranged screens for sizing the material, oversize being returned for further grinding while undersize is discharged. (Liddell, 1918)

mare ball York and Lanc. Spherical ferruginous concretions. Cf: *caballa ball.* (Arkell, 1953)

marekanite Obsidian that occurs as rounded to subangular bodies, usually less than 2 in (5.1 cm) in diameter and having indented surfaces. These bodies occur in masses of perlite and are of special interest because of their low water content as compared with the surrounding perlite. The name is from the Marekanka River, Okhotsk, Siberia, Russia. (AGI, 1987)

margarite (a) A beadlike string of globulites, commonly found in glassy igneous rocks. (AGI, 1987) (b) A monoclinic mineral, $4[CaAl_2(Al_2Si_2O_{10})(OH)_2]$; mica group; forms brittle folia; associated with corundum. Syn: *brittle mica; pearl mica.*

margarodite A pearly variety of muscovite.

margarosanite A triclinic mineral, $Pb(Ca,Mn)_2Si_3O_9$.

marginal deposit (a) A magmatic segregation at the bottom and periphery of an intrusive rock; e.g., nickel-copper-sulfide deposits at Sudbury, ON, Canada. (b) Marginal ore deposit.

marginal fissure A fracture, bordering an igneous intrusion, that has become filled with magma. (AGI, 1987)

marginal ore deposit A deposit near the lower limit of commercial workability. (Bateman, 1951)

marginal reserves That part of the reserve base that, at the time of determination, borders on being economically producible. Its essential characteristic is economic uncertainty. Included are resources that would be producible, given postulated changes in economic or technologic factors. (USGS, 1980)

marginal sea An adjacent sea that is widely open to the ocean. See also: *adjacent sea.* (AGI, 1987)

marginal thrust A thrust fault along the margin of an intrusive that dips toward the intrusive. (AGI, 1987)

marginal trench *trench.*

maria glass An early name for both mica and selenite (gypsum).

marialite A tetragonal mineral, $3NaAlSi_3O_8 \cdot NaCl$; scapolite group; forms a series with meionite.

Marietta miner Trade name for a heavy track-mounted continuous miner for operation in thick seams. The front end contains two cutter arms that rotate in opposite directions to sweep the coal it cuts inwards toward the center. The broken coal is taken back through the machine to a chain conveyor. Two cutter chains are arranged at roof and floor level behind the arms to cut down the coal left between the rotating arms. The machine cuts an area 12 ft (3.7 m) wide and 7 ft (2.1 m) high. Power is supplied by two motors, one of 70 hp (52.2 kW) and the other of 25 hp (17.4 kW). It has a continuous capacity from 3 to 3-1/2 st/min

(2.7 to 3.2 t/min). This machine has been subject to several modifications. See also: *continuous miner.* (Nelson, 1965)

marine biology Science that treats of the living organisms of the sea, the chemical and physical characteristics of their environment, and factors affecting their distribution. (Hy, 1965)

marine core drill Used for investigating strata beneath the seabed and taking sample and cores from which dredging conditions may be assessed. (Hammond, 1965)

marine deposit A sedimentary deposit laid down in the sea, usually beyond the seaward edge of the littoral belt. (Stokes, 1955)

marine erosion Erosion by moving seawater, the action of which is largely intensified by detritus carried by it. (Schieferdecker, 1959)

marine geology The marine science that treats of the topographical features of the sea bottom, the phenomena that have developed it, and the types, processes, and distribution of sedimentation. (Hy, 1965)

marine humus Organic matter deposited on the sea bottom. (Tomkeieff, 1954)

marine invasion The spreading of the sea over a land area. (AGI, 1987)

marine mining The mining of marine mineral deposits, classified as unconsolidated deposits such as gravel, mineral sands, or nodules; consolidated deposits such as outcrops, veins, or crusts; and fluid that is seawater or hydrothermal fluids; from, on, or beneath the seabed, whether on the continental shelf or in the deep ocean basins. (SME, 1992)

marine transgression *transgression.*

mariposite A chromian variety of phengite, a siliceous variety of muscovite.

maritime plants Plants that grow naturally under salty conditions on a foreshore and that may materially help to prevent scour and stabilize sand dunes. (Hammond, 1965)

marker (a) An easily recognized stratigraphic feature having characteristics distinctive enough for it to serve as a reference or datum or to be traceable over long distances, esp. in the subsurface, as in well drilling or in a mine working; e.g., a stratigraphic unit readily identified, or any recognizable rock surface such as an unconformity or a corrosion surface. See also: *format.* Syn: *marker bed; marker horizon; marker formation.* Cf: *horizon.* (AGI, 1987) (b) A layer that yields characteristic reflections over a more or less extensive area. (AGI, 1987) (c) A layer that accounts for a characteristic segment of a seismic-refraction time-distance curve and can be followed over reasonably extensive areas. (AGI, 1987) (d) S. Afr. *outcrop.* (e) *marker block.* Cf: *indicator.*

marker bed (a) *marker.* (b) A stratigraphic bed selected for use in preparing structural, paleogeologic, and other maps that emphasize the nature or attitude of a plane or a surface. It is generally selected for lithologic characteristics, but biologic factors and unconformities may control. Syn: *indicator; key bed; marker horizon.* See also: *horizon; marker.* Cf: *indicator.*

marker block A small block on which the footage below the collar of a borehole is marked and inserted between pieces of core at its appropriate place in a core box to indicate the depth in the borehole at which the core was obtained. Also called footage block; footmark; marker. The marker block is placed in the core box on completion of each drilled interval. See also: *marker.* (Long, 1960)

marker formation *marker.*

marker horizon *marker bed; marker.*

market pot In silver refining, the pot at the end of the series of pots used in the Pattinson process, in the direction in which the amount of silver left in the lead is diminishing. It contains the market lead. (Fay, 1920)

markovnikovite Variety of petroleum found in Russia. (Tomkeieff, 1954)

marl (a) An old term loosely applied to a variety of materials, most of which occur as loose, earthy deposits consisting chiefly of an intimate mixture of clay and calcium carbonate, formed under marine or esp. freshwater conditions; specif. an earthy substance containing 35% to 65% clay and 65% to 35% carbonate. Marl is usually gray; it is used esp. as a fertilizer for acid soils deficient in lime. In the Coastal Plain area of Southeastern United States, the term has been used for calcareous clays, silts, and sands, esp. those containing glauconite (greensand marls); and for newly formed deposits of shells mixed with clay. (AGI, 1987) (b) A soft, grayish to white, earthy or powdery, usually impure, calcium carbonate precipitated on the bottoms of present-day freshwater lakes and ponds, largely through the chemical action of aquatic plants, or forming deposits that underlie marshes, swamps, and bogs that occupy the sites of former (glacial) lakes. The calcium carbonate may range from 90% to less than 30% . Syn: *bog lime.* (AGI, 1987) (c) A term occasionally used (as in Scotland) for a compact, impure, argillaceous limestone. Etymol: French marle. (AGI, 1987)

marlite *marlstone.*

marl slate An English term for calcareous shale; it is not a true slate. (AGI, 1987)

marlstone (a) An indurated rock of about the same composition as marl. It has a blocky subconchoidal fracture, and is less fissile than shale. Syn: *marlite.* (AGI, 1987) (b) A term originally applied by Bradley (1931) to slightly magnesian calcareous mudstones or muddy limestones in the Green River Formation of the Uinta Basin, UT, but subsequently applied to associated rocks (including conventional shales, dolomites, and oil shales) whose lithologic characters are not readily determined. Picard (1953) recommended abandonment of the term as used in the Uinta Basin. (AGI, 1987) (c) *marlstone ore.*

marlstone ore A stratified ironstone located in the Midlands (England) and occurring at the top of the Middle Lias series. (Nelson, 1965)

marly Pertaining to, containing, or resembling marl; e.g., marly limestone containing 5% to 15% clay and 85% to 95% carbonate, or marly soil containing at least 15% calcium carbonate and no more than 75% clay (in addition to other constituents). (AGI, 1987)

marmarization *marmorization.*

marmarosh diamond Rock crystal variety of quartz.

marmatite A ferroan variety of sphalerite.

marmolite A thinly foliated variety of serpentine.

marmorization The conversion of limestone into marble by metamorphism. Syn: *marmarization; marmorosis.*

marmorosis *marmorization.*

marm stone Obsolete term for marble. (Arkell, 1953)

marokite An orthorhombic mineral, $CaMn_2O_4$; opaque black with dark red internal reflection; in Morocco.

Marriner process A modification of the cyanide process in which the ore is deadroasted, after which all of it is ground to slime, and the resulting product is treated by agitation. (Liddell, 1918)

marrow (a) Well sinkers' term for a fine-grained floury rock, Oxfordshire, U.K. (b) Up to six workers who pool and share earnings equally, all at the same workplace but not necessarily on the same shift.

Marsaut lamp An earlier type of miners' flame safety lamp fitted with two or three conical gauzes, thus adding to the safety of the lamp when used in high air velocities. The Marsaut lamp is the basis of modern flame safety lamps. (Nelson, 1965)

marsh buggy A special, self-propelled geophysical vehicle designated to operate over marsh or extremely soft ground, usually having wheels with very wide tread or buoyant wheels that will float the vehicle in water. (AGI, 1987)

Marsh funnel An appliance for measuring viscosity. It consists of a copper funnel, about 30 cm long with a 15 cm diameter at the top, that has a 10-mesh screen over half its diameter to remove debris and a 6-mm-diameter exit tube at the bottom through which the rate of flow is timed. It takes 26 s for a quart of clean water to flow through and correspondingly longer for muds of greater viscosity. Gel strength is measured by comparing the rate of flow of freshly agitated mud with that of mud that has been allowed to remain quiescent for 10 min. See also: *hydrometer; specific-gravity hydrometer.* (Nelson, 1965)

marsh gas (a) Methane, CH_4. If the decaying matter at the bottom of a marsh or pond is stirred, bubbles of methane rise to the surface, thus the name marsh gas. (Nelson, 1965) (b) It is nonexplosive until met with air or oxygen. In miners' language synonymous with firedamp. See also: *methane; firedamp.* (BCI, 1947; Fay, 1920)

marshite An isometric mineral. CuI; soft; oil brown; dodecahedral cleavage; at Broken Hill, New South Wales, Australia.

marsh ore *bog iron; bog iron ore.*

marsh pan A salt pan in a marsh. (AGI, 1987)

martenite A synthetic fettling material, used for sintering open-hearth furnace bottoms. The approximate composition (variable) is 5.2% silica, 2.1% alumina, 10.5% ferric oxide, 13.4% lime, 66.5% magnesia (ignition loss, 2.3%). Martenite sinters more rapidly than magnesite, thereby reducing repair time; moreover, it is suitable for hot patching. Martenite is as wear resistant as magnesite and has no deleterious effect on the slag. (Osborne, 1956)

martensite Alpha-iron supersaturated with carbon as a result of quenching austinite (gamma-iron) below 150 °C. (Not martinsite.)

martic A mixture of bituminous matter, such as asphalt, and some foreign material, such as sand. (Mersereau, 1947)

Martin process Used in the manufacture of steel. Also called Siemens-Martin and open-hearth process. (Fay, 1920)

martinsite *kieserite.* (Not martensite.)

martite Hematite pseudomorphous after magnetite octahedra.

martourite *berthierine.*

marundite A coarse-grained pegmatite consisting of corundum and margerite with accessory biotite, plagioclase, apatite, tourmaline, garnet, and kyanite. Allied to plumasite grading into normal pegmatite. (Hess)

mascagnite An orthorhombic mineral, $(NH_4)_2SO_4$; occurs around fumaroles and in guano deposits.

mascot emerald A beryl triplet that simulates emerald.

maser Contracted version of microwave amplification by simulated emission of radiation. A class of amplifier from which the optical laser was developed. See also: *laser.* (Anderson, 1964)

mask (a) A screen, usually made of tracing cloth, to subdue and diffuse the light behind a plumbline or other sighted object. (BS, 1963) (b) *respirator.*

maskeeg *muskeg.*

maskelynite A plagioclase glass in some chondrites and irons formed by preterrestrial impact between meteorites in space.

mason's hammer A square-faced hammer with a peen in line with handle. (Standard, 1964)

mass (a) The quantity of matter in a body, obtained by dividing the weight of the body by the acceleration due to gravity. (b) A large irregular deposit of ore, which cannot be recognized as a vein or bed. See also: *nontabular deposit.* (Nelson, 1965)

mass aqua Borosilicate glass imitative of aquamarine (beryl).

mass copper In the Lake Superior region, a term for native copper occurring in large masses.

mass density Mass of air per unit volume. Measured in kilograms per cubic meter. (Hartman, 1982)

mass detonation A term applied to the unintentional detonation of all or a part of a large quantity of explosive material (bulk truckload, shipload, or caseload) by the explosion of a smaller quantity of explosives or a flame.

mass diagram A plotting of cumulative cuts and fills used for engineering computation of highway jobs. (Nichols, 1976)

mass effect The tendency for hardened steel to decrease in hardness from the surface to the center, as a result of the variation in the rate of cooling throughout the section. Becomes less marked as the rate of cooling required for hardening decreases; i.e., as the content of alloying elements increases. (CTD, 1958)

mass fiber (a) One of the three recognized forms in which asbestos fiber is found in rock deposits. In this form the fibers are usually found intermixed in a matrix, which forms the orebody. Found mostly in the matrix of the deposits of the amphibole varieties of asbestos. The rock forming the orebody is sometimes inclined to be soft, and the fibers intermingled in such a mass deposit consist of patches of apparent slip-fiber forms. Indications of disturbed cross-fiber disposition are often discernible. (Sinclair, 1959) (b) A mass aggregate of interlaced, unoriented, or radiating fibers of chrysotile serpentine asbestos.

mass haul diagram Diagram used in construction work to show the location of digging and filling sites, and the distances over which earth and materials are to be transported. (Pryor, 1963)

massicot An orthorhombic mineral, PbO; dimorphous with litharge; prismatic cleavage; soft; yellow; occurs in oxidized zones of lead ore deposits. Syn: *lead ocher.*

massif A massive topographic and structural feature, esp. in an orogenic belt, commonly formed of rocks more rigid than those of its surroundings. These rocks may be protruding bodies of basement rocks or younger plutonic bodies. Examples are the crystalline massifs of the Helvetic Alps, whose rocks were deformed mainly during the Hercynian orogeny, long before the Alpine orogeny. (AGI, 1987)

massifs long Fr. Pillar in longwall workings. (Fay, 1920)

massive (a) Said of a mineral deposit (esp. of sulfides) characterized by a great concentration of ore in one place, as opposed to a disseminated or vein deposit. (b) Said of any rock that has a homogeneous texture or fabric over a wide area, with an absence of layering, foliation, cleavage, or any similar directional structure. See also: *massive rock.* (c) Said of a mineral that is physically isotropic; i.e., lacking a platy, fibrous, or other structure. Cf: *amorphous.*

massive bedding Very thick homogeneous stratification in sedimentary rocks.

massive eruption Outpouring of lava from a line or system of fissures, so that vast areas have become covered by nearly horizontal sheets of extrusive flows.

massive mineral If the crystalline grains are so small that they cannot be distinguished except under the high magnification of a microscope, the structure is described as compact and the mineral as massive. (CMD, 1948)

massive pluton Any pluton that is not tabular in shape.

massive rock Isotropic, homogeneous rock with a strength that does not vary appreciably from point to point. Typical examples are igneous rocks such as granite and diorite; metamorphic rocks such as marble and quartzite; and some thick-bedded sedimentary rocks. See also: *massive.*

mass movement The unit movement of a portion of the land surface; specif. the gravitative transfer of material down a slope. Syn: *mass-wasting.* (AGI, 1987)

mass number The sum of the neutrons and protons in a nucleus. It is the nearest whole number to the actual atomic weight of the atom. (Lyman, 1964)

mass profile A road profile showing cut and fill in cubic yards or meters. (Nichols, 1976)

mass shooting Simultaneous exploding of charges in all of a large number of holes, as contrasted with sequential firing with delay caps. (Nichols, 1976)

mass spectra Positive ray spectra obtained by means of a mass spectrograph. In such spectra the images due to positive ray particles of different masses are spaced according to the masses of the particles; i.e., according to their atomic weights. (CTD, 1958)

mass spectrometer An instrument for producing and measuring, usually by electrical means, a mass spectrum. It is esp. useful for determining molecular weights and relative abundances of isotopes within a compound. (AGI, 1987)

mass unit weight *wet unit weight.*

mass-wasting *mass movement.*

mast (a) A drill derrick or tripod mounted on a drill unit, which can be raised to its operating position by mechanical means. (Long, 1960) (b) A single pole, used as a drill derrick, supported in an upright or operating position by guys. (Long, 1960) (c) A tower or vertical beam carrying one or more load lines at its top. (Nichols, 1976)

master alloy An alloy, rich in one or more desired addition elements, that can be added to a melt to raise the percentage of a desired constituent. (ASM, 1961)

master hauler S. Wales. The person in charge of haulers in a coal mine and who controls the horse haulage traffic and the allocation of trams. (Nelson, 1965)

master joint A persistent joint plane of greater-than-average extent. (AGI, 1987)

master lode The main productive lode of a district. Cf: *mother lode.*

master pin The only pin in an integrated crawler track that will open the track when driven out. (Nichols, 1976)

mastershifter N. of Eng. Official responsible for the working of a seam during the third (night) shift of the day. (Trist, 1963)

master station A position in the ventilation circuit of a mine specially chosen for the regular and accurate estimation of the total quantity of air circulating. (Spalding, 1949)

mat (a) An accumulation of broken mine timbers, rock, earth, etc., coincident with the caving system of mining. As the ore is extracted, the mat gradually settles and forms the roof of the working levels, slopes, etc. (b) Lusterless or dull surface in a metal, produced by a method of finishing. (Standard, 1964)

match The part of a detonator that is most easily ignited. (Mason, 1951)

materials flowsheet A flowsheet principally concerned with solid materials. See also: *operational capacities.* (BS, 1962)

materials handling The art and science involving movement, packaging, and storage of substances in any form.

materials lock An air lock through which materials are passed into or out of a pneumatic caisson or a shaft being driven under air pressure. See also: *manlock.* (Hammond, 1965)

Mathewson's device An apparatus for separating matte and slag at lead-silver blast furnaces where matte is of secondary importance. (Fay, 1920)

matildite A hexagonal mineral, $AgBiS_2$; forms prismatic crystals; sp gr, 6.9. Syn: *schapbachite; plenargyrite.*

matlockite A tetragonal mineral, 2[PbFCl]; basal cleavage; an alteration of galena or lead-bearing slags.

mat pack Small pack of timber consisting of a number of timbers laid side by side to form a solid mass 2 to 2-1/2 ft square by 4 to 6 in thick (approx. 0.7 m square by 10 to 15 cm thick). Holes are drilled edgeways through the mat, and wires are threaded through to hold it together. Mats are transported underground and built up to form very effective supports. (Spalding, 1949)

matraite A trigonal mineral, ZnS; trimorphous with sphalerite and wurtzite.

matrix (a) The nonvaluable minerals in an ore; the gangue. (b) The rock material in which a fossil, crystal, or mineral is embedded. Syn: *groundmass.* (c) A local term for the phosphate-bearing gravel in the land-pebble deposits of Florida. (AGI, 1987) (d) The metal in which the diamonds inset in the crown of a bit are embedded. (Long, 1960) (e) The material that forms a cushion or binder in the construction of pavement. (f) The finer-grained material between the larger particles of a rock or the material surrounding a fossil or mineral. (BS, 1964) (g) The principal phase or aggregate in which another constituent is embedded. (ASM, 1961) (h) In electroforming, a form used as a cathode. (ASM, 1961)

matrix metal The continuous phase of a polyphase alloy or mechanical mixture; the physically continuous metallic constituent in which separate particles of another constituent are embedded. (ASTM, 1994)

matrosite A microscopical constituent of torbanite; opaque, black mass forming its groundmass. Cf: *gélosite; humosite; retinosite.* (Tomkeieff, 1954)

matte A metallic sulfide mixture made by melting the roasted product in smelting sulfide ores of copper, lead, and nickel. (ASM, 1961)

matte fall Weight of matte expressed as a percentage of the total charge. (Newton, 1959)

matte smelting The smelting of copper-bearing materials, usually in a reverberatory furnace. The valuable product is a liquid, copper-iron sulfide called matte. (Kirk, 1947)

matting The process of smelting sulfide ores into matte. (Weed, 1922)

mattock (a) A miner's pickax. (Fay, 1920) (b) An implement that combines the features of an adz, ax, and pick, and is used for digging, grubbing, and chopping. (Webster 3rd, 1966)

Matura diamond Colorless to faintly smoky gem-quality zircon from the Matara (Matura) district of Sri Lanka; any smokiness is removable by heating.

mature (a) Pertaining to the stage of maturity of the cycle of erosion; esp. said of a topography or region having undergone maximum development and accentuation of form; or of a stream (and its valley) with a fully developed profile of equilibrium; or of a coast that is relatively stable. (AGI, 1987) (b) Said of a clastic sediment that (1) has been differentiated or evolved from its parent rock by processes acting over a long time and with a high intensity and (2) is characterized by stable minerals (such as quartz), deficiency of the more mobile oxides (such as soda), absence of weatherable material (such as clay), and well-sorted but subangular to angular grains. Example: a clay-free mature sandstone on a beach. (AGI, 1987)

maturity (a) A stage in the development of a coast that is characterized by straightening of the shoreline by bridging of bays and cutting back of headlands so as to produce a smooth, regular shoreline consisting of sweeping curves; and, eventually, retrogradation of the shore beyond the bayheads so that it lies against the mainland as a line of eroded cliffs throughout its course. (AGI, 1987) (b) The extent to which a clastic sediment texturally and compositionally approaches the ultimate end product to which it is driven by the formative processes that operate upon it. (AGI, 1987) (c) The stage in the development of a stream at which it has reached its maximum efficiency, having attained a profile of equilibrium and a velocity that is just sufficient to carry the sediment delivered to it by tributaries. (AGI, 1987) (d) The second of the three principal stages of the cycle of erosion in the topographic development of a landscape or region, intermediate between youth and old age (or following adolescence), lasting through the period of greatest diversity of form or maximum topographic differentiation, during which nearly all the gradation resulting from operation of existing agents has been accomplished. (AGI, 1987)

maturity index A measure of the progress of a clastic sediment in the direction of chemical or mineralogic stability; e.g., a high ratio of alumina-soda, of quartz-feldspar, or of quartz + chert-feldspar + rock fragments indicates a highly mature sediment. (AGI, 1987)

maucherite A tetragonal mineral, $Ni_{11}As_8$; forms tabular crystals; occurs in nickel-cobalt-native-silver ore deposits. Syn: *placodine; temiskamite.*

Mawco cutter loader A cutter loader similar to an Anderton shearer except that the drum is replaced by a frame jib 42 in (107 cm) high and 20 in (51 cm) deep. The machine travels on an armored flexible conveyor at a speed of about 4-1/2 ft/min (1.4 m/min). It cuts a 20-in (51-cm) web on the cutting run, and the plow deflector loads the cut coal onto the conveyor. On the reverse run, the deflector loads all the loose coal left on the track, and the conveyor is snaked over behind the machine. The loader is suitable for medium-thickness seams, and the yield of large coal is good. (Nelson, 1965)

maximum and minimum densities In soil tests, the maximum density is found by compacting soil with a Kango hammer; the minimum density is measured by pouring soil into a container of known capacity. This test is useful for determining the relative density of sands, by comparison with field tests. (Hammond, 1965)

maximum angle of inclination The maximum angle at which a conveyor may be inclined and still deliver a predetermined quantity of bulk material within a given time. As the maximum angle is approached, the rate of handling of bulk material is usually decreased.

maximum belt slope The slope beyond which the material on a conveyor tends to roll downhill. The maximum slope on which a conveyor can operate depends on (1) the material carried, (2) the loading or feeding efficiency, (3) the size and type of belt, and (4) the environment. In general, in the case of run-of-mine coal and ore, belt conveyors can operate up to about 18°. If the material conveyed contains large lumps, spillage may result if the belt is too narrow. (Nelson, 1965)

maximum belt tension The total of the starting and operating tensions. In the average conveyor this is considered to be the same as the tight side tension.

maximum carbon dioxide content The recommended maximum allowable concentration of carbon dioxide in mine air is 0.5%. (Hartman, 1961)

maximum charge weight per delay The maximum quantity of explosive charge detonated on one interval (delay) within a blast. The charge detonated within any 8-ms interval over the entire duration of the blast.

maximum demand Upper limit of electric power that may be drawn at any time from the mains without penalty, as agreed by contract. (Pryor, 1963)

maximum density *maximum unit weight.*

maximum dry density The dry density obtained by the compaction of soil at its optimum moisture content. (Hammond, 1965)

maximum microcline Microcline with the most complete ordering possible of aluminum and silicon ions in the tetrahedral sites and the smallest angle for beta (maximum triclinicity). Cf: *mesomicrocline.*

maximum operating belt tension The tension in the carrying run necessary to maintain the normal operating speed of a loaded belt.

maximum per delay The maximum vibration at distant points is that which has been generated by the greatest amount of explosive fired at any one instant. (Leet, 1960)

maximum-pressure arch *pressure arch theory.*

maximum-pressure gage An instrument for registering the maximum pressure occurring during an explosion at the point where the instrument is located. (Rice, 1913-18)

maximum subsidence The maximum amount of subsidence in a subsidence basin. The value of maximum subsidence for a given seam thickness depends on the underground geometry and the thickness and character of the overburden. See also: *subsidence factor.*

maximum unit weight The dry unit weight defined by the peak of a compaction curve. Syn: *maximum density.* See also: *unit weight.* (ASCE, 1958)

Maxton screen A screening machine of the trommel class, rotating on rollers that support the tube. There are radial elevating ribs to prevent wear of screen cloth and to elevate the oversize. Unscreened material is delivered on the inside screen surface, undersize passes through, and oversize is elevated and discharged into a separate launder. (Liddell, 1918)

maxwell The cgs (centimeter-gram-second) unit of magnetic flux. One maxwell = 10^{-8} Wb, or the flux through 1 cm^2 normal to a field of magnetic induction of 1 Gs. (AGI, 1987)

Maxwell's rule A law stating that every part of an electric circuit is acted upon by a force tending to move it in such a direction as to enclose the maximum amount of magnetic flux. (CTD, 1958)

Mayari iron Pig iron made from Cuban ores that contain vanadium and titanium, or pig iron made to duplicate the Cuban iron. (Brady, 1963)

mayenite An isometric mineral, $Ca_{12}Al_{14}O_{33}$; occurs in metamorphosed marly limestone; an important constituent of portland cement clinker.

mboziite A potassian variety of taramite amphibole.

McGinty Three sheaves over which a rope is passed so as to take a course somewhat like that of the letter M. The resulting friction causes the rope to slide with difficulty. It is used for lowering loaded cars from the face to the mouth of a room on a steep roadway. (Zern, 1928)

mcgovernite A trigonal mineral, $(Mn,Mg,Zn)_{22}(AsO_3)(AsO_4)_3(SiO_4)_3(OH)_{20}$; reddish-bronze; forms granular masses with perfect basal cleavage. Also spelled macgovernite.

McKelvey diagram A graphical classification of mineral resources according to economic viability and certainty of existence. Used by the U.S. Bureau of Mines and the U.S. Geological Survey in their definitions of mineral resources and reserves, USGS Circular 831. Named after Vincent E. McKelvey, ninth director of the U.S. Geological Survey. (Barton, 1995)

McLuckie gas detector This nonautomatic detector, or portable air analysis apparatus, can be used underground, or samples of air can be brought out of the mine in small rubber bladders and analyzed at the surface. This apparatus depends for its action on the fact that when methane is burned in air (oxygen) a definite chemical action takes place. When the resulting steam condenses to water there is a reduction in pressure and in the McLuckie detector this reduction of pressure, which is proportionate to the methane present, is indicated by the height of the liquid in one limb of the U-tube which rises up the side of the scale: the scale is graduated from 0% to 3% in steps of 0.1%. (Cooper, 1963)

McNally-Carpenter centrifuge A fine-coal dewatering machine consisting of a conical rotating element with a vertical axis, built up of three rows of perforated stainless steel screen plates or stainless steel wedge or round-wire sections cut and rolled to conform to the surface of the cone. Wet feed is deposited by gravity into the top of the cone-shaped dryer onto a distributing disk that throws the material by centrifugal force onto the stepped screen or basket. The rapid circular movement of the basket forces the water through the screen, while the coal moves toward the bottom. Peripheral velocity and centrifugal force become greater for each particle, breaking down the surface tension of water film on each piece, increasing drying action directly in ratio to the cone circumference and peripheral speed. The most effective drying area is at the bottom of the cone just before the material is discharged from the machine. See also: *dewatering.* (Kentucky, 1952; Mitchell, 1950)

McNally-Norton jig In this jig, raw coal is conveyed to a wash box through sluices. Air pulsations are transmitted through valves to water in a compartment adjacent to the washing bed, causing the water in the wash box to rise. Pulsating water causes the incoming fuel to be loosely suspended in the water and permits heavier refuse to sink to the screen plate while suspended coal spills over into the second compartment. In the second compartment the process is repeated with the remaining refuse sinking to the screen and clean coal discharging to the dewatering screens. (Kentucky, 1952)

McNally-Vissac dryer A convection dryer of the forced-draft type. The heat source is a coal-fired furnace. It consists essentially of a declined reciprocating screen over which the coal travels. Two balanced tandem decks are used. They are suspended from the supporting structure by inclined flexible hangers and actuated in opposition through flexible pitmans from a common eccentric shaft. The removal of moisture is accomplished by passing hot furnace gases, tempered with cold air, downward through the bed of coal as it travels along the screen. An induced-draft fan at the exhaust end provides the motive force for the gases. See also: *thermal drying*. (Mitchell, 1950)

McNamara clamp A drill-rod safety clamp somewhat similar to a Wommer's safety clamp. (Long, 1960)

M-design core barrel Standard-design, double-tube, swivel-type core barrel made in sizes to be used with appropriate standard ranges of diamond-drill fittings. Its distinguishing features are that a 2-1/2° taper core lifter is carried inside a short tubular sleeve (called a lifter case) coupled to the bottom end of the inner tube, and that the lifter case extends downward inside the bit shank to within a very short distance behind the face of the core bit. (Long, 1960)

M-discontinuity *Mohorovicic discontinuity.*

M.E. 6 exploder An exploder approved for firing six shots simultaneously in British coal mines. It contains a 67-1/2-V high-tension dry battery, used to charge a 150-μF condenser, which in turn is discharged through the shotfiring circuit by a firing key. The test circuit and an ohmmeter are incorporated in the exploder, thc ohmmeter pointer moving over a scale to indicate whether or not the external circuit is in order. A pushbutton disconnects the test circuit from the external circuit and makes connection with the firing circuit. See also: *blasting machine*. (Nelson, 1965)

meadow ore *bog iron; bog iron ore; limonite.*

meager feel Moistureless; dry and rough to the touch, such as chalk and magnesite. (Nelson, 1965)

mean An arithmetic average of a series of values; esp. arithmetic mean. Cf: *mode*. (AGI, 1987)

mean birefringence The numeral that represents the average between the greatest strength of double refraction and the least strength of double refraction possessed by a species or variety. The refractive index of sphene, e.g., is 1.885 to 1.990 and 1.915 to 2.050; hence the birefringence ranges from 0.105 to 0.135. The average, or mean, is 0.120. Syn: *refractive index.*

mean calorie One-hundredth of the heat required to raise 1 g of water from 0 °C to 100 °C. (Newton, 1959)

mean depth The cross-sectional area of a stream divided by its width at the surface. (AGI, 1987)

meander (a) One of a series of regular, freely developing sinuous curves, bends, or loops in the course of a stream. It is produced by a mature stream swinging from side to side as it flows across its floodplain or shifts its course laterally toward the convex side of an original curve. Etymol: Greek maiandros, from Maiandros River in western Asia Minor (now known as Menderes River in SW Turkey), proverbial for its windings. (AGI, 1987) (b) To wind or turn in a sinuous or intricate course; to form a meander. (AGI, 1987)

meander belt That part of a floodplain between two lines tangent to the outer bends of all the meanders. It is the zone within which channel migration occurs, as indicated by abandoned channels, accretion topography, and oxbow lakes.

meander line A line run in a survey of a mining claim bordering on a stream or other body of water, not as a boundary of the tract surveyed, but for the purpose of defining the sinuosities of the bank or shore of the water, and as a means of ascertaining the quantity of land within the surveyed area. (Ricketts, 1943)

mean effective pressure In an air compressor, the equivalent average pressure exerted by the piston throughout a stroke. (Lewis, 1964)

mean radiant temperature (mrt) Single temperature of all enclosing surfaces that would result in the same heat emission as the same surface with various different temperatures. (Strock, 1948)

mean refractive index (a) The index of refraction measured for the D line of sodium. (b) For uniaxial crystals: $(2n_{\omega}+n_{\epsilon})/3$. For biaxial crystals: $(n_{\alpha}+n_{\beta}+n_{\gamma})/3$.

mean size The weighted average particle size of any sample, batch, or consignment of particulate material. (BS, 1962)

mean sphere depth The uniform depth to which water would cover the Earth if the solid surface were smoothed off and parallel to the surface of the geoid. Generally accepted as a depth of 2,440 m. (Hy, 1965)

mean stress (a) In fatigue testing, the algebraic mean of the maximum and minimum stress in one cycle. Also called the steady-stress component. (ASM, 1961) (b) In any multiaxial stress system, the algebraic mean of three principal stresses; more correctly called mean normal stress. (ASM, 1961)

measured depth *measured drilling depth.*

measured drilling depth The apparent depth of a borehole as measured along the longitudinal axis of the borehole. The measured drilling depth is always equal to the unoverlapped drilled footage in a borehole. Also called measured depth. Sometimes abbreviated md. (Long, 1960)

measured resources Resources from which the quantity is computed from dimensions revealed in outcrops, trenches, workings, or drill holes; grade and/or quality are computed from the results of detailed sampling. The sites for inspection, sampling, and measurement are spaced so closely and the geologic character is so well defined that size, shape, depth, and mineral content of the resource are well established. See also: *reserves*. (USGS, 1980)

measurement The finding of the number of units of measure in a line, area, space or volume, period of time, etc. (Jones, 1949)

measurement of concentration Gr. Brit. At the National Coal Board collieries, in order to assess the degree of concentration, certain basic data are collected, involving the pithead output, length of main haulage roads, and length of coalface in production. See also: *face concentration; geographical concentration; overall concentration*. (Nelson, 1965)

measures A group or series of sedimentary rocks having some characteristic in common; specif. coal measures. The term apparently refers to the old practice of designating the different seams of a coalfield by its measure or thickness. (AGI, 1987)

measures head A heading or drift made in various strata. (Fay, 1920)

measuring chain A surveyor's chain, containing 100 links of 7.92 in (20.12 cm) each.

measuring chute A bin installed adjacent to the shaft bottom in skip winding. The capacity of the chute is equal to that of the skip used, ranging from 4 to 10 st (3.6 to 9.1 t). Bin-feeding arrangements differ but may be by a steelplate conveyor from a surge bunker that in turn receives the ore or coal from the mine cars or a trunk conveyor. A measuring chute ensures a quick and correct loading of skips without spillage. Immediately the skip is positioned in line, the measuring chute bottom door opens and material is discharged into the skip. See also: *pocket*. Syn: *underground ore bin*. (Nelson, 1965)

measuring day The day when face or other work is measured and recorded for assessing wages. (Nelson, 1965)

measuring element In flotation, that portion of the feedback elements that converts the signal from the primary detecting element to a form compatible with the reference input. (Fuerstenau, 1962)

measuring pocket Storage space near an entry from underground workings to a hoisting shaft; laid out so as to deliver a measured volume into a hoisting skip and to be refilled before the skip returns empty. (Pryor, 1963)

measuring tape A graduated tape, steel or linen, usually in 50-ft or 100-ft (15.2-m or 30.4-m) lengths; used by engineers, builders, surveyors, etc. (Crispin, 1964)

measuring weir A device for measuring the flow of water. It generally consists of a rectangular, trapezoidal, triangular, or other shaped notch in a thin plate in a vertical plane through which the water flows. The weir head is an index of the rate of flow. See also: *notch*. (Seelye, 1951)

mechanical advantage Ratio between the resistance or load raised by a machine, and the applied force. Mechanical advantage divided by velocity ratio gives the efficiency of the machine. (Hammond, 1965)

mechanical air machine A flotation machine that utilizes pulp-body concentration by the agitation froth method and bubble-column action by pneumatic and cascade means. (Taggart, 1927)

mechanical analysis Determination of the particle-size distribution of a soil, sediment, or rock by screening, sieving, or other means of mechanical separation; the quantitative expression of the size-frequency distribution of particles in granular, fragmental, or powdered material. It is usually expressed in percentage by weight (and sometimes by number or count) of particles within specific size limits. See also: *particle-size analysis*. (AGI, 1987)

mechanical classifier One of the machines, such as the Dorr classifier, that are commonly used to classify a ball-mill or rod-mill discharge into finished product and oversize. (Newton, 1959)

mechanical clay A clay formed from the products of the abrasion of rocks.

mechanical cleaning The removal of impurities by mechanical units as compared with hand picking. Broadly, mechanical cleaning may be subdivided into dry cleaning and wet cleaning. (Mitchell, 1950; Nelson, 1965)

mechanical efficiency The ratio of the air-indicated horsepower to the indicated horsepower in a power cylinder, in the case of compression driven by steam or internal-combustion engines, and to the brake horsepower delivered to the shaft in the case of a power-driven machine. (Lewis, 1964)

mechanical equivalent of heat Amount of mechanical energy that can be transformed into a single heat unit; the equivalent of 778 ft·lbf/Btu (1,000 N·m/kJ). (Hammond, 1965)
mechanical extensometer An appliance for measuring strain; often used in roof control investigations. It employs a micrometer dial gage actuated through a lever giving initial magnification of the movement. See also: *acoustic-strain gage; electrical resistance strain gage.* (Nelson, 1965)
mechanical flotation cell A cell in which the solids-water pulp feed is kept agitated, and is circulated by means of an impeller mounted at the bottom of a vertical shaft. The rotating impeller creates vacuum enough to draw air down the standpipe surrounding the impeller shaft, and the impeller disperses the air throughout the pulp in the form of small bubbles. The flotable minerals are carried upward by the bubbles and eventually collect in the froth above the pulp in the machine. Automatic scrapers remove the mineral-laden froth that contains the concentrate, and after the values have been removed, the barren pulp containing the tailing flows out of the cell. (Newton, 1959)
mechanical loader A power machine for loading material. See also: *loader.* (Nelson, 1965)
mechanical mixture A composition of two or more substances, each remaining distinct, and generally capable of separation by mechanical means. (Standard, 1964)
mechanical properties Of metals: the elastic limit, elongation, fatigue range, hardness, maximum stress, reduction in area, shock resistance, and yield point. (Pryor, 1963)
mechanical puddler A wrought-iron (rocking) furnace in which puddling is done by mechanical motion instead of by hand. (Mersereau, 1947)
mechanical puddling *mechanical puddler.*
mechanical rabble A rabble worked by machinery. See also: *rabble.* (Fay, 1920)
mechanical rammer A machine embodying a weight that is lifted and dropped upon the material being rammed. See also: *power rammer.* (Hammond, 1965)
mechanical sampling Mechanical sampling systematically removes a portion of the stream of material for a sample. Mechanical sampling is widely used in cone preparation plants and concentrators where large quantities of materials are to be sampled, while hand sampling is used for smaller amounts. Cf: *hand sampling.* (Newton, 1959)
mechanical sediment Sediment that has been brought to its places of deposition as separate particles by mechanical means. Water, wind, and ice are the agents commonly involved; the resulting rocks are conglomerate, sandstone, siltstone, shale, and certain limestones. (Stokes, 1955)
mechanical set Bits produced by the various means in which diamonds are set in a bit mold into which a cast or powder metal is placed, embedding the diamonds and forming the bit crown, as opposed to handsetting. Also, the act or process of producing diamond bits in such a manner. Also called cast set; machine set; sinter set. Cf: *handset.* (Long, 1960)
mechanical-set bit A diamond bit produced by mechanical methods as opposed to handsetting methods. See also: *mechanical set.* (Long, 1960)
mechanical shovel A loader limited to level or only slightly graded drivages. The machine operates a shovel in front of it and pushes itself forward; when full, the shovel is swung over the machine and delivers into a mine car or tub behind. It will shunt, pull, and push its own cars, delivering them into a shunt or passby when full. See also: *shovel loader.* (Mason, 1951)
mechanical stabilization Mixing two or more poorly graded soils to obtain a well-graded one. (Nelson, 1965)
mechanical weathering The process of weathering by which frost action, salt-crystal growth, absorption of water, and other physical processes break down a rock to fragments, involving no chemical change. Cf: *chemical weathering.* Syn: *disintegration.* (AGI, 1987)
mechanical working Subjecting metal to pressure exerted by rolls, presses, or hammers, to change its form or affect the structure, and therefore the physical properties. (Rolfe, 1955)
mechanical yielding prop A steel prop in which yield is controlled by friction between two sliding surfaces or telescopic tubes. Although crude when compared with the hydraulic prop, the friction yield prop is very robust, is cheap, and requires little maintenance. (Nelson, 1965)
mechanics The branch of physics that treats of the phenomena caused by the action of forces on material bodies. It is subdivided into statics, dynamics, or kinetics; or into the mechanics of rigid bodies and hydromechanics (including hydrostatics and hydrodynamics). (Standard, 1964)
mechanization Essentially, the introduction of power machines to replace manual labor. In coal mining, it may denote the introduction of conventional machine mining to replace hand mining, or continuous mining to replace conventional machine mining. (Nelson, 1965)
mechanization engineer Usually a qualified mining engineer with first-hand experience and knowledge of the various mining machines and the physical conditions most suitable for them. In general, the National Coal Board, Great Britain, appoints a mechanization engineer for each group of collieries. (Nelson, 1965)
mechanization scheme A plan or project to convert a handmining or a conventional machine mining face to mechanized mining; i.e., the use of machines that either load prepared coal or cut and load coal simultaneously (cutter loaders). The scheme may also include the introduction of locomotives, skip winding, etc. (Nelson, 1965)
mechanized Term descriptive of a mine that has a high percentage of machinery for all steps of mining and handling mineral product, from the face to the mine working place, and on to the tipple or treatment plant. (BCI, 1947)
mechanized heading development A pillar method of working suitable for seams 4 ft (1.2 m) and over in thickness. Three or more narrow headings are driven rapidly with machines at about 30-yd (10-m) centers with crosscuts for ventilation. The headings are 10 ft (3 m) or more wide and 5 to 6 ft (approx. 2 m) high or seam thickness. Upon reaching the boundary, the pillars formed by the headings are extracted, again with machines, on the retreat. This method is favored in the United States and the heading work is quite as productive, if not more so, than pillar working. See also: *entry; longwall retreating.* (Nelson, 1965)
mechanized output The coal produced by all coal face machinery that either loads prepared coal or cuts and loads coal simultaneously. Also includes all coal obtained by hand filling on faces where an armored flexible conveyor is used on a prop-free front. (Nelson, 1965)
Meco-Moore cutter loader A heavy 120-hp (89.5-kW) cutter loader. The first Meco-Moore was used in a Lancashire, England, colliery in 1934. See also: *A.B. Meco-Moore.* (Nelson, 1965)
medfordite *moss agate.*
median diameter An expression of the average particle size of a sediment or rock, obtained graphically by locating the diameter associated with the midpoint of the particle-size distribution; the middlemost diameter that is larger than 50% of the diameters in the distribution and smaller than the other 50%. (AGI, 1987)
medical lock An air chamber comprising steel cylinder 18 ft (5.5 m) long and about 6 ft (1.8 m) in diameter, which has airtight doors at one end and is closed at the other. It is used for immediate treatment of sufferers from caisson disease. (Hammond, 1965)
Medina emerald Green glass emerald simulant.
mediosilicic A term proposed by Clarke (1908) to replace intermediate. Cf: *subsilicic; persilicic.* See also: *intermediate.* (AGI, 1987)
Mediterranean suite A major group of igneous rocks, characterized by high potassium content. This suite was so named because of the predominance of potassium-rich lavas around the Mediterranean Sea; specif. those of Vesuvius and Stromboli. Cf: *Atlantic suite; Pacific suite.* (AGI, 1987)
medium Any suspension of medium solids in water. (BS, 1962)
medium band A field term that, in accordance with an arbitrary scale established for use in describing banded coal, denotes a vitrain band ranging in thickness from approx. 1/12 to 1/5 in (2 to 5 mm). (AGI, 1987)
medium draining screen A screen for draining the separating medium from dense-medium bath products. (BS, 1962)
medium-grained (a) Said of an igneous rock, and of its texture, in which the individual crystals have an average diameter in the range of 0.04 to 0.2 in (1 to 5 mm). (AGI, 1987) (b) Said of a sediment or sedimentary rock, and of its texture, in which the individual particles have an average diameter in the range of 1/16 to 2 mm (62 to 2,000 µm, or sand size). Cf: *coarse-grained; fine-grained.* (AGI, 1987)
medium-inclined Said of deposits and coal seams with a dip of 25° to 40°. (Stoces, 1954)
medium pressure When applied to valves and fittings, means suitable for a working pressure of 125 to 175 psi (860 to 1,200 kPa). (Strock, 1948)
medium-recovery screen A composite screen for draining and spraying the product from a dense-medium bath to remove adherent medium solids. (BS, 1962)
medium-round nose A diamond bit the cross-sectional outline of which is partially rounded but not as fully rounded as a double-round nose bit. Syn: *half-round nose; modified-round nose.* (Long, 1960)
medium solids The solid component of a dense-medium suspension. (BS, 1962)
medium-solids preparation Any purification or grinding of the raw dense-medium solids to make them suitable for use. (BS, 1962)
medium-solids recovery *dense-medium recovery.*
medium-solids recovery plant The equipment used to remove adherent medium solids from a product from a dense-medium bath (after drainage of surplus medium), usually by spraying, and to remove contaminating coal and clay from these medium solids. (BS, 1962)
medium-stone bit A bit with diamonds ranging from 8 to 40 per carat in size. (Long, 1960)

medium-thickness seam In general, a coal seam over 2 ft (0.6 m) and up to 4 ft (1.2 m) in thickness. (Nelson, 1965)

medium-volatile bituminous coal The rank of coal, within the bituminous class of Classification D 388, such that, on the dry and mineral-matter-free basis, the volatile matter content of the coal is greater than 22% but equal to or less than 31% (or the fixed carbon content is equal to or greater than 69% but less than 78%), and the coal commonly agglomerates. Cf: *bituminous coal*. (ASTM, 1994)

medium voltage In coal mining, voltage from 661 to 1,000 V. Cf: *high voltage; low voltage.* (Federal Mine Safety, 1977)

medmontite A mixture of chrysocolla and mica. See also: *cupromontmorillonite.*

meehanite High-duty cast iron produced by ladle addition of calcium silicide. (Pryor, 1963)

meerschaum *sepiolite.*

meet (a) Eng. To keep pace with; e.g., to keep sufficient supply of coal at the pit bottom to supply the winding engine. (Fay, 1920) (b) To come together exactly, as in survey lines from opposite directions. (Fay, 1920)

meeting (a) A siding or bypass on underground roads. (Fay, 1920) (b) Newc. The place at middle-depth of a shaft, slope, or plane, where ascending and descending cars pass each other. (Fay, 1920)

meeting post A vertical timber at the outer edge of each of a pair of lock gates, mitered so that the gates fit tightly when closed. (Hammond, 1965)

mega- (a) A prefix meaning large. As a prefix to petrological and other geologic terms, it signifies parts or properties that are recognizable by the unaided eye. Opposite of micro-. (Stokes, 1955) (b) A combining form meaning 1 million times; e.g., megavolt for 1 million volts. (AGI, 1987)

megabar A unit of pressure equal to 1 $Mdyn/cm^2$. (Standard, 1964)

megacycle A unit of 1 million cycles. (CTD, 1958)

megaphenocryst A phenocryst that is visible to the unaided eye. (AGI, 1987)

megascopic Said of an object or phenomenon, or of its characteristics, that can be observed with the unaided eye or with a hand lens. Syn: *macroscopic.* Cf: *microscopic.* (AGI, 1987)

megaseismic region The most disturbed earthquake area. (Schieferdecker, 1959)

megaspore Female spore; part of the reproduction organs of many coal measures plants. See also: *spore.* (Nelson, 1965)

Megator A displacement type of pump operating on the eccentric principle. (Mason, 1951)

megger An electrical measuring instrument comprising a hand-operated generator equipped with a governor, a moving measuring system consisting of a voltage, and a current coil so disposed that the deflection of the moving system is proportional to the ratio of voltage to current. Used to measure insulation resistance and resistance to ground. It has been used to some extent in electrical prospecting. (AGI, 1987)

Meigen's reaction A test to distinguish calcite from aragonite. After boiling for 20 min in a cobalt nitrate solution, aragonite becomes lilac, the color showing in thin section, while calcite and dolomite become pale blue, the color not showing in thin section.

meionite A tetragonal mineral, $3CaAl_2Si_2O_8{\cdot}CaCO_3$; scapolite group; forms a series with marialite.

meizoseismal Of or pertaining to the maximum destructive force of an earthquake. (Standard, 1964)

meizoseismal area The most disturbed area within the innermost isoseismal line. (Schieferdecker, 1959)

meizoseismal curve A curved line connecting the points of the maximum destructive energy of an earthquake shock around its epicenter. (Standard, 1964)

mela- A prefix meaning dark-colored. (AGI, 1987)

melaconite An earthy variety of tenorite. See also: *black copper ore; tenorite.*

melanasphalt An early name for albertite. (Tomkeieff, 1954)

melanchyme A bituminous substance found in masses in the brown coal of Zweifelsruth, Bohemia, Czech Republic. That part of this substance that is soluble in alcohol is termed rochlederite, the residue melanellite. Also spelled melanchym. (Fay, 1920)

melanellite That portion of melanchyme that is insoluble in alcohol; it is black and gelatinous. (Fay, 1920)

mélange (a) Diamonds of mixed sizes. (Hess) (b) An assortment of mixed sizes of diamonds weighing more than 1/4 carat; e.g., larger than those of a mêlée. (c) A body of rock mappable at a scale of 1:24,000 or smaller, characterized by a lack of internal continuity of contacts or strata and by the inclusion of fragments and blocks of all sizes, both exotic and native, embedded in a fragmental matrix of finer-grained material. Neither matrix composition and fabric nor genesis is significant for the definition.

melanic *melanocratic.*

melanite A titanian variety of andradite. Syn: *black andradite garnet; pyreneite.*

melanocratic Applied to dark-colored rocks, esp. igneous rocks, containing between 60% and 100% dark minerals; i.e., rocks, the color index of which is between 60 and 100. Cf: *hypermelanic; leucocratic; mesocratic.* (AGI, 1987)

melanostibite A trigonal mineral, Mn_2SbFeO_6. Originally named melanostibian.

melanotekite An orthorhombic mineral, $Pb_2Fe_2Si_2O_9$; forms a series with kentrolite.

melanovanadite A triclinic mineral, $CaV_4O_{10}{\cdot}5H_2O$; a natural vanadium bronze with perfect prismatic cleavage; at Cerro de Pasco, Peru, and on the Colorado Plateau.

melanterite (a) A monoclinic mineral, $FeSO_4{\cdot}7H_2O$; green; tastes slightly sweet, astringent, and metallic; from the decomposition of pyrite. (b) The mineral group bieberite, boothite, mallardite, melanterite, and zincmelanterite. Syn: *copperas; green vitriol; iron vitriol.*

melatope The narrowest part of an isogyre in an interference figure representing the point of emergence of an optic axis. Cf: *interference figure.*

mêlée (a) A collective term for small round faceted diamonds, such as those mounted in jewelry. The term is sometimes applied to colored stones of the same size and shape as the diamonds. (AGI, 1987) (b) A small diamond cut from a fragment of a larger size. (AGI, 1987) (c) In diamond classification, a term for small round-cut diamonds weighing more than 1/4 carat. Cf: *mélange.* Etymol, French. (AGI, 1987)

melilite (a) A tetragonal mineral in the series akermanite, $Ca_2MgSi_2O_7$-gehlinite, $Ca_2Al(AlSi)O_7$. (b) The mineral group akermanite, gehlenite, hardystonite, and melilite. Jeffreyite, leucophanite, and meliphanite are structurally similar. Also spelled mellilite.

melilitite A generally olivine-free extrusive rock composed of melilite and clinopyroxene (or other mafic mineral) usually comprising more than 90% of the rock, with minor amounts of feldspathoids and sometimes plagioclase. (AGI, 1987)

melinite (a) A high explosive similar to lyddite; said to be chiefly picric acid. (Webster 2nd, 1960) (b) A species of soft, unctuous clay, common in Bavaria, and probably identical with bole. (Standard, 1964)

meliphanite A tetragonal mineral, $(Ca,Na)_2Be(Si,Al)_2(O,OH,F)_7$; structurally similar to the melilites. Also spelled meliphane. Syn: *gugiaite.*

melle Small cut diamonds, usually about one-eighth carat. Generally refers to stones used in jewelry. (Hess)

mellilite *melilite.*

mellite A tetragonal mineral, $Al_2[C_6(COO)_6]{\cdot}16H_2O$; resinous; honey yellow; forms nodules in brown coal. Also spelled melinite, mellilite. (Not melite.) Syn: *honey stone.*

mellorite A silicate of ferric iron, calcium, etc., approaching garnet in composition, but with optical properties similar to those of an orthorhombic pyroxene. Formed by the action of basic slag on silica brick in a steel furnace. (Spencer, 1943; AGI, 1987)

mellow amber *gedanite.*

melonite (a) A trigonal mineral, $NiTe_2$; forms a series with merenskyite; one perfect cleavage; metallic reddish white; soft; sp gr, 7.3. (b) The mineral group berndtite, kitkaite, melonite, merenskyite, and moncheite. Syn: *tellurnickel.* (Not melanite.)

melteigite A dark-colored plutonic rock that is part of the ijolite series and contains nepheline and 60% to 90% mafic minerals, esp. green pyroxene. The name is from Melteig farm, Fen complex, Norway. (AGI, 1987)

melting hole The opening in the floor to a furnace in a melting house. (Mersereau, 1947)

melting house The building in which crucible furnaces for steel making are located. (Mersereau, 1947)

melting point That temperature at which a single, pure solid changes phase to a liquid or to a liquid plus another solid phase, upon the addition of heat at a specific pressure. Unless otherwise specified, melting points are usually stated in terms of 1 kPa. The term can also be used for the isothermal melting of certain mixtures, such as eutectic mixtures. Erroneously used also to refer to the temperature at which some appreciable but unspecified amount of liquid develops in a complex solid mixture that possesses a melting range; e.g., the melting point of granite. Abbrev.: mp or MP. (AGI, 1987)

melting pot A crucible. (Standard, 1964)

melting shop Open-hearth plant. (Newton, 1959)

melting zone The hottest part of a furnace, where melting takes place. (Mersereau, 1947)

member A division of a formation, generally of distinct lithologic character and of only local extent.

membrane filter *molecular filter sampler.*

membrane theory An advanced theory of design for thin shells, based on the premise that a shell cannot resist bending because it deflects. The only stresses that exist, therefore, in any section are shear stress and direct compression or tension. (Hammond, 1965)

menaccanite (a) A variety of ilmenite found as sand at Menaccan, Cornwall, Eng. (b) A black, magnetic sand from Cornwall, England, from which the element, titanium, was first isolated. Also spelled menachanite; manaccanite; menachite. (Hess)
mend Eng. To load, or reload, trams at the gate ends out of smaller trams used only in the working faces of thin seams.
mendipite An orthorhombic mineral, $Pb_3Cl_2O_2$; white; in the Mendip Hills, United Kingdom.
mendozite A monoclinic mineral, $NaAl(SO_4)_2 \cdot 11H_2O$. See also: *soda alum.*
meneghinite An orthorhombic mineral, $Pb_{13}CuSb_7S_{24}$; forms slender prismatic blackish lead-gray crystals.
Menevian European stage: Middle Cambrian (above Solvan, below Maentwrogian). (AGI, 1987)
menilite A concretionary, opaque, dull, grayish variety of opal. Syn: *liver opal.* (Fay, 1920)
meniscus (a) The curved top surface of a liquid column. It is concave upwards when the containing walls are wetted by some liquid (such as water in a vertical glass tube) and convex upwards when wetted with other liquids (such as mercury in a vertical glass tube). (Webster 3rd, 1966) (b) A concavoconvex lens; esp., one of true crescent-shaped cross section. (Webster 3rd, 1966)
men on! Scot. A brief expression to indicate that workers are on the cage to be raised, or lowered, in a shaft. (Fay, 1920)
Menzies cone separator Consists of a 60° cone with a short cylindrical top section. It is provided with a stirring shaft, located in its vertical axis, carrying several sets of horizontal arms with rings of nozzles projecting through the sides of the cone for the admission, at several horizons, of the required water currents. At the base of the cone, a classifier column several feet long is fitted, through which refuse discharges continuously to an inclined refuse conveyor. Water is supplied by a centrifugal pump. See also: *cone classifier.* (Mitchell, 1950)
mephitic air (a) Carbon dioxide. (Webster 3rd, 1966) (b) Air exhausted of oxygen and containing chiefly nitrogen. (Webster 3rd, 1966)
mephitic gas *mephitic air.*
mephitis A noxious exhalation caused by the decomposition of organic remains; applied also to gases emanating from deep sources, such as mines, caves, and volcanic regions. (Standard, 1964)
merchant A metal merchant, as distinct from a producer's agent or broker, often acts as a principal, buying metal or concentrate from producers and others and selling it to others. The merchant will often hold metal on personal account while waiting for a buyer. (Wolff, 1987)
merchant iron Iron in the common bar form, which is convenient for the market. Also called merchant bar. (Standard, 1964)
mercurial horn ore *calomel.*
mercury (a) A liquid mineral, (trigonal below -38.87 °C); metallic silver to tin white; sp gr, 13.6; occurs as minute droplets in cinnabar and in some hot-spring deposits; amalgamates with many metals. (b) Symbol: Hg. Rarely occurs free in nature. Chief ore is cinnabar, HgS. Used in laboratory work for making thermometers, barometers, diffusion pumps, mercury-vapor lamps, advertising signs, and pesticides. Mercury is a virulent poison and is readily absorbed through the respiratory tract, gastrointestinal tract, or unbroken skin. (Handbook of Chem. & Phys., 3)
mercury gatherer A stirring apparatus that causes mercury, which has become floured or mixed with sulfur in amalgamating, to resume the fluid condition, through the agency of mechanical agitation and rubbing. (Fay, 1920)
mercury ore Native mercury; same as cinnabar (sulfide). (Dana, 1955)
mercury switch A glass tube employing mercury to establish electrical contact between circuits when the tube is tilted so that the mercury bridges the gap between contacts, and conversely. (Strock, 1948)
mercury-vapor lamp Consists essentially of a sealed glass tube provided with two electrodes and containing a gas. When an electrical potential difference is applied, a current passes and with a suitable gas, light will be emitted. In the case of mercury vapor, this light is of a bluish color and has proven effective in distinguishing dirt from coal. A special starting electrode close to one of the main electrodes initiates the discharge, and a choke coil in series with the lamp serves to limit the current passing, since the resistance tends to fall with increasing current. (Mason, 1951)
mero- A prefix signifying part or portion. (AGI, 1987)
merocrystalline *hypocrystalline.*
merohedral Any crystal form with fewer faces than the holohedral equivalent for the crystal system. Cf: *hemihedral; holohedral.* See also: *tetartohedral.*
meroleims Coalified remains of parts of plants. (Tomkeieff, 1954)
merosymmetric Any crystal class with less symmetry than the distribution of points in its lattice.
meroxene Biotite with its optic axial plane parallel to its *b* crystallographic axis.
Merrill-Crowe process Removal of gold from pregnant cyanide solution by deoxygenation, followed by precipitation on zinc dust, followed by filtration to recover the resultant auriferous gold slimes. (Pryor, 1963)
Merrill filter A type of plate and frame pressure filter.
merrillite (a) High-purity zinc dust used to precipitate gold and silver in the cyanide process. (b) *whitlockite.*
Merrit plate *bloomery.*
mersey yellow coal *tasmanite.*
merwinite A monoclinic mineral, $Ca_3Mg(SiO_4)_2$; colorless to pale green.
mesa An isolated, nearly level landmass standing distinctly above the surrounding country, bounded by abrupt or steeply sloping erosion scarps, and capped by a layer of resistant, nearly horizontal rock (often lava). Cf: *plateau.* Etymol: Spanish, table. (AGI, 1987)
Mesabi non-Bessemer ore *natural ore.*
mesa-butte *butte.*
Mesa Grande tourmaline Fine-quality tourmaline from a pegmatite near Mesa Grande, San Diego County, CA.
mesh (a) In ventilation, a series of airways that form a closed loop. (Roberts, 1960) (b) The screen number of the finest screen of a specified standard screen scale. (c) The number of apertures per unit area of a screen (sieve).
mesh aperture The dimension or dimensions of the aperture in a screen deck, usually with a qualification as to the shape of aperture; e.g., round-hole, square-mesh, and long-slot. (BS, 1962)
mesh fraction That part of a material passing a specified mesh screen and retained by some stated finer mesh. (Henderson, 1953)
mesh liberation size The particle size at which substantially all of the valuable minerals are detached from the gangue minerals. (Fuerstenau, 1962)
mesh number The designation of size of an abrasive grain, derived from the openings per linear inch in the control sieving screen. Syn: *grit number.* (ACSG, 1963)
mesh of grind Optimum particle size resulting from a specific grinding operation, stated in terms of percentage of material passing (or alternatively being retained on) a given size screen. The mesh of grind is the liberation mesh decided on as correct for commercial treatment of the material. Abbrev., m.o.g. (Pryor, 1963)
mesh structure A structure resembling network or latticework that is found in certain alteration products of minerals. Also called net structure; lattice structure. (Standard, 1964; Fay, 1920)
mesh texture (a) A texture resembling a network, caused by the alteration of certain minerals; e.g., the serpentinization of olivine. Syn: *reticulate texture.* (b) An interlacing network of microveinlets of fibrous serpentine enclosing cores of more weakly birefringent cryptocrystalline serpentine in which relict remnants of olivine may survive. Also called net structure; lattice structure.
meso- A prefix meaning middle. Cf: *cata-; meta-.* (AGI, 1987)
mesocratic Applied to igneous rocks that are intermediate between leucocratic and melanocratic rocks; they contain 30% to 60% dark minerals. Cf: *melanocratic; leucocratic.* (CTD, 1958)
mesocrystalline Said of the texture of a rock intermediate between microcrystalline and macrocrystalline; also, said of a rock with such a texture. (AGI, 1987)
mesogene Said of a mineral deposit or enrichment of mingled hypogene and supergene solutions; also, said of such solutions and environment. Cf: *hypogene; supergene.* (AGI, 1987)
mesokaites Group name for brown coals. (Tomkeieff, 1954)
mesolite A monoclinic mineral, $Na_2Ca_2[Al_2Si_3O_{10}]_3 \cdot 8H_2O$; zeolite group; pseudo-orthorhombic; in cavities in basalt and andesite, geodes, and hydrothermal veins. Syn: *cotton stone; winchellite.*
mesomicrocline Microcline intermediate in structural state between orthoclase and maximum microcline. Syn: *intermediate microcline.* Cf: *maximum microcline.*
mesostasis The last-formed interstitial material between the larger mineral grains in an igneous rock or in a microcrystalline groundmass. Cf: *groundmass.*
mesothermal Said of a hydrothermal mineral deposit formed at considerable depth and in the temperature range of 200 to 300 °C. Also, said of that environment. Cf: *hypothermal deposit; epithermal; leptothermal; telethermal; xenothermal.* (AGI, 1987)
mesothermal deposit A mineral deposit formed at moderate temperature and pressure, in and along fissures or other openings in rocks, by deposition at intermediate depths, from hydrothermal fluids. Mesothermal deposits are believed to have formed mostly between 175 °C and 300 °C at depths of 4,000 to 12,000 ft (1,220 to 3,660 m). Many valuable metalliferous deposits of western North America are of this type. (Stokes, 1955)
Mesozoic An era of geologic time, from the end of the Paleozoic to the beginning of the Cenozoic, or from about 225 million years to about 65 million years ago. (AGI, 1987)
mesozone According to Grubenmann's classification of metamorphic rocks (1904), the intermediate-depth zone of metamorphism, which is characterized by temperatures of 300 to 500 °C and moderate hydro-

static pressure and shearing stress. Modern usage stresses temperature-pressure conditions (medium to high metamorphic grade) rather than the likely depth of zone. Cf: *katazone; epizone.* (AGI, 1987)

messelite A triclinic mineral, $Ca_2(Fe,Mn)(PO_4)_2{\cdot}2H_2O$; fairfieldite group. Syn: *neomesselite; parbigite.* (Not mesolite.)

mestre Port. Mine boss. (Hess)

meta (a) In petrology, indicates a metamorphosed protolith. (b) In mineralogy, indicates a mineral species that is a dehydration product of another mineral species or is a polymorph.

meta- A prefix that, when used with the name of a sedimentary or igneous rock, indicates that the rock has been metamorphosed, e.g., metabasalt. Cf: *cata-; meso-.* (AGI, 1987)

meta-alunogen A monoclinic mineral, $Al_4(SO_4)_6{\cdot}27H_2O$. See also: *alunogen.*

meta-anthracite The rank of coal, within the anthracite class of Classification D-388, such that, on the dry and mineral-matter-free basis, the volatile matter content of the coal is equal to or less than 2% (or the fixed carbon is equal to or greater than 98%), and the coal is nonagglomerating. (ASTM, 1994)

meta-argillite An argillite that has been metamorphosed. (AGI, 1987)

meta-arkose Arkose that has been welded or recrystallized by metamorphism so that it resembles a granite or a granitized sediment. Cf: *recomposed granite.* (AGI, 1987)

meta-autunite (a) A tetragonal mineral, $Ca(UO_2)_2(PO_4)_2{\cdot}$ 2-6H_2O; yellow; an alteration product of uraninite and other uranium-bearing minerals. (b) The mineral group abernathyite, bassetite, chernikovite, meta-ankoleite, meta-autunite, metaheinrichite, metakahlerite, metakirchheimerite, metalodevite, metanovacekite, metatorbernite, meta-uranocircite, meta-uranospinite, metazeunerite, sodium uranospinite, and uramphite.

metabasite A collective term, first used by Finnish geologists, for metamorphosed mafic rock that has lost all traces of its original texture and mineralogy owing to complete recrystallization. (AGI, 1987)

metabentonite (a) Metamorphosed, altered, or somewhat indurated bentonite; characterized by clay minerals (esp. illite) that no longer have the property of absorbing or adsorbing large quantities of water; nonswelling bentonite, or bentonite that swells no more than do ordinary clays. The term has been applied to certain Ordovician clays of the Appalachian region and upper Mississippi River Valley. See also: *potassium bentonite.* (AGI, 1987) (b) A mineral of the montmorillonite group with SiO_2 layers in the montmorillonite structure. (AGI, 1987)

metabitumite Hard black lustrous variety of hydrocarbon found in proximity of igneous intrusions. (Tomkeieff, 1954)

metabolism A term proposed by Barth (1962) for the redistribution of granitizing materials within sediments by mobilization, transfer, and reprecipitation, as opposed to metasomatism involving addition of new materials. (AGI, 1987)

metabolite (a) An old term for altered glassy trachyte. (b) Iron meteorite with the composition of an octahedrite but lacking Widmänstatten figures.

metaborite An isometric mineral, HBO_2; in crystalline aggregates in rock salt in Kazakhstan(?).

metabrushite *brushite.*

metachemical metamorphism A term proposed by Dana to describe metamorphism that involves a chemical change in the affected rocks.

metacinnabar An isometric mineral, HgS; forms black tetrahedra; sp gr, 7.7; a source of mercury. Also called metacinnabarite. Also spelled: metacinnibar

metacinnabarite A mineral of the same composition as a cinnabar, but black in color, and crystallizing in isometric forms (tetrahedral). Used as a source of mercury. (Sanford, 1914)

metaclase Leith's term for a rock possessing secondary cleavage, or cleavage in its modern meaning (1905). Cf: *protoclase.* (AGI, 1987)

metacryst Any large crystal developed in a metamorphic rock by recrystallization, such as garnet or staurolite in mica schists.

metacrystal *porphyroblast.*

metadiabase A contraction of metamorphic diabase, suggested by Dana for certain rocks simulating diabase, but which were possibly produced by the metamorphism of sedimentary rocks. Cf: *metadiorite.*

metadiorite (a) A contraction of metamorphic diorite that was proposed for certain metamorphic rocks that resemble diorite, but which may have been the result of the metamorphism of sedimentary rocks. Cf: *metadiabase.* (Fay, 1920) (b) Metamorphosed gabbro, diabase, or diorite. (AGI, 1987)

metadolomite A metamorphic dolomite, or dolomite marble.

metadurit Durain of a high-rank bituminous coal. See also: *durain.* (Tomkeieff, 1954)

metaglyph A hieroglyph formed during metamorphism. (Pettijohn, 1964)

metahalloysite Dehydrated halloysite. See also: *halloysite.*

metaheinrichite A tetragonal mineral, $Ba(UO_2)_2(AsO_4)_2{\cdot}8H_2O$; meta-autunite group; yellow; a secondary mineral. Syn: *arsenuranocircite; metasandbergite.* Cf: *heinrichite.*

metahewettite A monoclinic mineral, $CaV_6O_{16}{\cdot}3H_2O$; dark-red to yellow-brown; forms tabular crystals impregnating sandstone in southwest Colorado and southeast Utah. Cf: *hewettite.*

metahohmannite A mineral, $Fe_2(SO_4)_2(OH)_2{\cdot}3H_2O$. See also: *hohmannite.*

metahydroboracite A hydrous calcium and magnesium borate, $CaMgB_6O_8(OH)_6{\cdot}XH_20$; like hydroboracite but with more water. Syn: *inderborite.* (Spencer, 1943)

metakahlerite A tetragonal mineral, $Fe(UO_2)_2(AsO_4)_2{\cdot}8H_2O$; meta-autunite group.

metakirchheimerite A possibly tetragonal mineral, $Co(UO_2)_2(AsO_4)_2{\cdot}$ 8 $_2O$; meta-autunite group.

metal In most cases, an opaque, lustrous, elemental substance that is a good conductor of heat and electricity. It is also malleable and ductile, possesses high melting and boiling points, and tends to form positive ions in chemical compounds.

metal bath A bath, such as of mercury or tin, employed for chemical processes requiring high temperatures.

metal drift A drift or heading driven in barren and hard rock. (Nelson, 1965)

metaleucite A name for isometric leucite that is stable above 625 °C. Cf: *pseudoleucite.*

metalimestone A metamorphosed carbonate rock not suitable for use as polished dimension stone (Brooks, 1954). Cf: *metamarble; ortholimestone.*

metalist A metallurgist. (Standard, 1964; Fay, 1920)

metalized slurry blasting The breaking of rocks, etc., using slurried explosive medium containing a powdered metal, such as powdered aluminum.

metallic (a) A term used to describe metal particles, such as gold in ores. (Newton, 1959) (b) (adj.) The adj. indicates that the noun it modifies possesses metallic properties. These properties often include a metallike luster, conduction of electricity, tensile strength, opacity, and malleability, although some metallic materials may possess only a few such characteristics. (c) When used with "mineral" in the context of resources, e.g., metallic mineral, it has a different and special meaning; it refers to the product, not the mineralogy. Thus chalcopyrite, $CuFeS_2$, is metallic (in the sense above), and the copper and iron it contains are metallic minerals in the resource sense. A single mineral, such as chalcopyrite, may also be the source of a nonmetal, sulfur. Adding to the confusion, rutile (TiO_2) is the source of both titanium, which is used as metallic titanium, and titanium oxide, which is used as a nonmetallic mineral pigment. Because many industrial minerals (in a resource sense) tend to be nonmetallic (in either the mineralogical or the resource sense), the terms "industrial minerals" and "nonmetallic minerals" are sometimes carelessly used interchangeably. Cf: *mineral; nonmetallic mineral; industrial minerals.* Syn: *metalliferous.*

metallic element Element that is generally distinguishable from nonmetallic elements by its luster, malleability, and electrical conductivity, and its usual ability to form positive ions. (Henderson, 1953)

metallic iron Metal iron, as distinguished from iron ore.

metallic luster The ordinary luster of metals. When feebly displayed, it is termed submetallic. Gold, iron pyrites, and galena have a metallic luster, e.g., while chromite and cuprite have a submetallic luster. (Nelson, 1965)

metallic minerals Minerals with a high specific gravity and metallic luster, such as titanium, rutile, tungsten, uranium, tin, lead, iron, etc. In general, the metallic minerals are good conductors of heat and electricity. See also: *nonmetallic minerals.* (Nelson, 1965)

metallic ore From a strictly scientific point of view, the terms metallic ore and ore deposits have no clear significance. These are purely conventional expressions, used to describe those metalliferous minerals or bodies of mineral having economic value, from which useful metals can be advantageously extracted. In one sense, rock salt is an ore of sodium, and limestone an ore of calcium, but to term beds of those substances ore deposits would be quite outside of current usage. (Ricketts, 1943)

metalliferous Metal-bearing; specif., pertaining to a mineral deposit from which a metal or metals can be extracted by metallurgical processes. See also: *metallic.* (AGI, 1987)

metalliferous mud Marine deposit of mud formed under anoxic conditions and containing anomalously high quantities of zinc, silver, and copper, and lesser amounts of lead and gold. The term has most often been applied to deposits of muds in the Red Sea which have been formed by submarine precipitation of metallic sulfides from hydrothermal vents. These vents occur along the axis of a spreading center which forms the Red Sea Basin. (Cruickshank, 1987)

metallify To convert into metal. (Fay, 1920)

metallites A word used to embrace all ores or metalliferous material.

metallization The process or processes by which metals are introduced into a rock, resulting in an economically valuable deposit; the mineralization of metals. (AGI, 1987)

metallogenetic epoch The time interval favorable for the genesis or deposition of certain useful metals or minerals. Syn: *minerogenetic epoch.* (AGI, 1987)
metallogenetic province *minerogenetic province.*
metallogenic element An element normally forming sulfides, selenides, tellurides, arsenides, antimonides, and/or sulfosalts, or occurring uncombined as a native element; i.e., an element of primary ore deposits. (Schieferdecker, 1959)
metallogenic province (a) An area characterized by a particular assemblage of mineral deposits, or by one or more characteristic types of mineralization. A metallogenic province may have had more than one episode of mineralization. Syn: *metallographic province.* (b) *minerogenetic province.*
metallogeny The study of the genesis of mineral deposits, with emphasis on its relationship in space and time to regional petrographic and tectonic features of the Earth's crust. The term has been used for both metallic and nonmetallic mineral deposits. Adj: metallogenic. Syn: *ore geology.* Cf: *genesis.* (AGI, 1987)
metallograph An optical instrument designed for both visual observation and photomicrography of prepared surfaces of opaque materials, at magnifications ranging from about 25 to about 1,500 diameters. The instrument consists of a high-intensity illuminating source, a microscope, and a camera bellows. On some instruments, provisions are made for examination of specimen surfaces with polarized light, phase contrast, oblique illumination, dark-field illumination, and customary bright-field illumination. (ASM, 1961)
metallographic province *metallogenic province.*
metallography (a) The science dealing with the constitution and structure of metals and alloys as revealed by the unaided eye or by tools, such as low-power magnification, optical microscope, electron microscope, and diffraction or X-ray techniques. See also: *reflected-light microscope.* (ASM, 1961) (b) The study of the constitution and structure of metals and alloys.
metalloid (a) A nonmetal, such as carbon or nitrogen, that can combine with a metal to form an alloy. (Webster 3rd, 1966) (b) An element—such as boron, silicon, arsenic, or tellurium—intermediate in properties between the typical metals and nonmetals. (Webster 3rd, 1966)
metalloidal luster (a) Reflecting light, somewhat like a metal, but less than metallic luster. (b) Having the luster of a semimetal; e.g., native bismuth or arsenic.
metallometric surveying Geochemical prospecting term used by Russian authors for soil surveys or for the chemical analysis of systematically collected samples of soil and weathered rock. (Hawkes, 1962)
metallometry The geochemical determination of metals. (AGI, 1987)
metallo-organic compound A compound in which a metal combines with organic compounds to form metallo-organic complexes, such as porphyrins and salts of various organic acids. Some metallo-organic compounds are soluble in water, others are not. (Hawkes, 1962)
metallurgical balance sheet Material balance of a process.
metallurgical coke A coke with very high compressive strength at elevated temperatures, used in metallurgical furnaces, not only as a fuel, but also to support the weight of the charge. (ASM, 1961)
metallurgical engineer One who applies engineering principles to the science and technology of metallurgy. Cf: *metallurgist.*
metallurgical fume A mixture of fine particles of elements and metallic and nonmetallic compounds either sublimed or condensed from the vapor state. (Fay, 1920)
metallurgical smoke A term applied to the gases and vapors, and fine dust entrained by them, that issue from the throats of furnaces; consists of three distinct substances: gases (including air), flue dust, and the fume. (Fay, 1920)
metallurgist One who is skilled in, or who practices, metallurgy. Cf: *metallurgical engineer.*
metallurgy (a) The science and art of separating metals and metallic minerals from their ores by mechanical and chemical processes; the preparation of metalliferous materials from raw ore. (b) Study of the physical properties of metals as affected by composition, mechanical working, and heat treatment.
metal mining The industry that supplies the community with the various metals and associated products. Similar to coal mining, it is an extractive industry, and once the raw material, the orebody, is depleted it is not replenishable. See also: *vein miner.* (Nelson, 1965)
metal notch *taphole.*
metal pickling The immersion of metal objects in an acid bath to remove scale, oxide, tarnish, etc., leaving a chemically clean surface for galvanizing or painting. (Nelson, 1965)
metal powder (a) Metallic elements or alloys in finely divided or powder form. (Henderson, 1953) (b) A general term applied by drillers, bit setters, and bit manufacturers to various finely ground metals, which, when mixed, are commonly used to produce sintered-metal diamond bit crowns. Also called powdered metal; powder metal. (Long, 1960)
metal stone (a) Newc. An argillaceous stone, shale, and sandstone. (b) Staff. *ironstone.*
metamarble A marble suitable for use as polished dimension stone; e.g., the Vermont metamarble. Cf: *orthomarble; metalimestone.*
metamic A metal ceramic consisting of high $Cr{\cdot}Al_2O_3$. (Osborne, 1956)
metamict (a) A mineral that has become virtually amorphous owing to the breakdown of the original crystal structure by internal bombardment with alpha particles (helium nuclei) emitted by radioactive atoms within the mineral. Many green zircons, esp. those from Sri Lanka, which are Precambrian in age, and have thus had over 800 million years of this internal bombardment, owe their low refractive index and density to this cause, and may be termed metamict zircons. (Anderson, 1964) (b) Said of a mineral containing radioactive elements in which various degrees of lattice disruption and changes have taken place as a result of radiation damage, while its original external morphology has been retained. Examples occur in zircon, thorite and several other minerals. Not all minerals containing radioactive elements are metamict; e.g., xenotime and apatite are not. (AGI, 1987)
metamictization The process of disruption of the structure of a crystal by radiations from contained radioactive atoms, rendering the material partly or wholly amorphous. (AGI, 1987)
metamict mineral A mineral whose crystal structure has been disrupted by radiation from contained radioactive particles. (AGI, 1987)
metamontmorillonite (a) The product of dehydration of montmorillonite at 400 °C. (Hey, 1961) (b) A dehydrated smectite.
metamorphic Pertaining to the process of metamorphism or to its results. (AGI, 1987)
metamorphic aureole *aureole.* Cf: *contact metamorphism.*
metamorphic deposit An ore deposit that has been subjected to great pressure, high temperature, and alteration by solutions. It may have become warped, twisted, or folded, and the original minerals may have been rearranged and recrystallized.
metamorphic differentiation A collective term for the various processes by which minerals or mineral assemblages are locally segregated from an initially uniform parent rock during metamorphism; e.g., garnet porphyroblasts in fine-grained mica schist. (AGI, 1987)
metamorphic diffusion Migration, by diffusion, of materials from one part of a rock mass to another during metamorphism. Diffusion may involve chemically active fluids from magmatic sources, hot pore fluids, or fluids released from hydrous minerals or carbonates. Ionic diffusion in the solid state may also occur. (AGI, 1987)
metamorphic facies A set of metamorphic mineral assemblages, repeatedly associated in space and time, such that there is a constant and therefore predictable relation between mineral composition and chemical composition. It is generally assumed that the metamorphic facies represent the results of equilibrium crystallization of rocks under a restricted range of externally imposed physical conditions; e.g., temperature, lithostatic pressure, and water pressure. Syn: *mineral facies.* (AGI, 1987)
metamorphic grade The intensity or rank of metamorphism, measured by the amount or degree of difference between the original parent rock and the metamorphic rock. It indicates in a general way the pressure-temperature environment or facies in which the metamorphism took place. For example, conversion of shale to slate or phyllite would be low-grade dynamothermal metamorphism (greenschist facies), whereas its continued alteration to a garnet-sillimanite schist would be high-grade metamorphism (almandine-amphibolite facies). Syn: *metamorphic rank.* See also: *high-rank metamorphism.* (AGI, 1987)
metamorphic overprint *overprint.*
metamorphic rank *metamorphic grade.*
metamorphic rock Any rock derived from preexisting rocks by mineralogical, chemical, and/or structural changes, essentially in the solid state, in response to marked changes in temperature, pressure, shearing stress, and chemical environment, generally at depth in the Earth's crust. (AGI, 1987)
metamorphic water Water driven out of rocks by metamorphism. (Stokes, 1955)
metamorphism The mineralogical, chemical, and structural adjustment of solid rocks to physical and chemical conditions that have generally been imposed at depth below the surface zones of weathering and cementation, and that differ from the conditions under which the rocks in question originated. (AGI, 1987)
metarossite A triclinic mineral, $CaV_2O_6{\cdot}2H_2O$; pearly to dull yellow; a dehydration product of rossite.
metasandbergite *metaheinrichite.*
metasapropel Compact sapropel rock. (Tomkeieff, 1954)
metaschoderite A monoclinic mineral, $Al_2(PO_4)(VO_4){\cdot}6H_2O$.
metaschoepite Formerly called schoepite II. See also: *schoepite; paraschoepite.*
metasediment A sediment or sedimentary rock that shows evidence of having been subjected to metamorphism. (AGI, 1987)

metashale Shale altered by incipient metamorphic reconstitution but not recrystallized and without the development of partings or preferred mineral orientation. (AGI, 1987)
metasomasis *metasomatism.*
metasomatic Pertaining to the process of metasomatism and to its results. The term is esp. used in connection with the origin of ore deposits. (AGI, 1987)
metasomatism The process of practically simultaneous capillary solution and deposition by which a new mineral of partly or wholly different chemical composition may grow in the body of an old mineral or mineral aggregate. The presence of interstitial, chemically active pore liquids or gases contained within a rock body or introduced from external sources is essential for the replacement process, that often, though not necessarily, occurs at constant volume with little disturbance of textural or structural features. Syn: *metasomosis.* Cf: *pyrometasomatism.* (AGI, 1987)
metasomatite A rock produced by metasomatism.
metasome (a) A replacing mineral that grows in size at the expense of another mineral (the host or palasome); a mineral grain formed by metasomatism. (AGI, 1987) (b) The newly formed part of a migmatite or composite rock, introduced during metasomatism. (AGI, 1987)
metasomosis *metasomatism.*
metastrengite Monoclinic $FePO_4{\cdot}2H_2O$, dimorphous with orthorhombic strengite; named to correspond with metavariscite and variscite $AlPO_4{\cdot}2H_2O$. Syn: *phosphosiderite; clinostrengite.* (Spencer, 1952)
metatorbernite A tetragonal mineral, $Cu(UO_2)_2(PO_4)_2{\cdot}8H_2O$; meta-autunite group; strongly radioactive; emerald to apple green; occurs in granite pegmatites. Cf: *torbernite.*
metatyuyamunite An orthorhombic mineral, $Ca(UO_2)_2V_2O_8{\cdot}3H_2O$; yellow; radioactive; a secondary mineral.
meta-uranocircite A monoclinic mineral, $Ba(UO_2)_2(PO_4)_2{\cdot}8H_2O$; meta-autunite group; yellow green; radioactive; a secondary mineral in quartz veins.
meta-uranopilite A mineral, $(UO_2)_6(SO_4)(OH)_{10}{\cdot}5H_2O$; radioactive.
meta-uranospinite A tetragonal mineral, $Ca(UO_2)(AsO_4)_2{\cdot}8H_2O$; meta-autunite group.
metavariscite A monoclinic mineral, $AlPO_4{\cdot}2H_2O$; green; dimorphous with variscite.
metavauxite A monoclinic mineral, $FeAl_2(PO_4)_2(OH)_2{\cdot}8H_2O$; dimorphous with paravauxite; forms acicular crystals or radiating fibrous aggregates.
metavermiculite The product of dehydration of vermiculite at 400 °C. (Hey, 1961)
metavolcanic Said of partly metamorphosed volcanic rock. (Stokes, 1955)
metavoltine A hexagonal mineral, $K_2Na_6Fe_7(SO_4)_{12}O_2{\cdot}18H_2O$.
metaxite (a) A fibrous serpentine mineral; a variety of chrysotile. (AGI, 1987) (b) *micaceous sandstone.*
metazeunerite A tetragonal mineral, $Cu(UO_2)_2(AsO_4)_2{\cdot}8H_2O$; meta-autunite group; grass to emerald green.
meteoric iron (a) Iron of meteoric origin. (AGI, 1987) (b) An iron meteorite. (AGI, 1987)
meteoric stone (a) A stone of meteoric origin; a stony meteorite. (AGI, 1987) (b) A meteorite having the appearance of a stone. (AGI, 1987)
meteoric water Ground water of atmospheric origin.
meteorite A stony or metallic body that has fallen to the Earth's surface from outer space. Adj: meteoritic.
meter (a) An instrument, apparatus, or machine for measuring fluids, gases, electric currents, etc., and recording the results obtained; e.g., a gasmeter, a watermeter, or an air meter. (Standard, 1964) (b) The fundamental unit of length in the metric system equal to 39.37079 in or 3.2808 ft. (Standard, 1964; Fay, 1920)
metering pin A valve plunger that controls the rate of flow of a liquid or a gas. (Nichols, 1976)
methane CH_4; carbureted hydrogen or marsh gas or combustible gases; formed by the decomposition of organic matter. The most common gas found in coal mines. It is a tasteless, colorless, nonpoisonous, and odorless gas; in mines the presence of impurities may give it a peculiar smell. It weighs less than air and may therefore form layers along the roof and occupy roof cavities. Methane will not support life or combustion; with air, however, it forms an explosive mixture, $CH_4{+}2O_2{\rightarrow}CO_2{\cdot}2H_2O$. The gases resulting from a methane explosion are irrespirable. Methane is often referred to as combustible gases because it is the principal gas composing a mixture that, when combined with proper proportions of air, will explode when ignited. Breathing methane causes ill effects only where the air is so heavily laden with it that oxygen is supplanted. See also: *colliery explosion; marsh gas; firedamp; limits of flammability.* (Nelson, 1965; Webster 2nd, 1960; BCI, 1947)
methane drainage (a) Capture of the concentrated methane through boreholes drilled into a coalbed or associated strata. (SME, 1992) (b) Outside the United States three main systems of methane drainage have been developed: (1) the cross-measure borehole method, (2) the superjacent roadway system and (3) the pack cavity system. The cross-measure borehole method which consists of boring holes from 2-1/4 to 3-1/4 in (5.7 to 8.3 cm) in diameter and 150 to 300 ft (45 to 90 m) in length, into the strata above or below the seam, generally close to the working face. This method has the advantage of being suited to a wide variety of conditions and does not require another seam within reasonable distance above or below the seam to be drained, or the use of solid stowing Syn: *cross-measure borehole system.* In the superjacent roadway system, boreholes are drilled from a roadway situated above the seam being worked, the drainage of the methane then taking place from this roadway. In the pack cavity system, corridors are left and supported in the goaf as the face advances, and from these combustible gases is drawn off. Syn: *combustible gases drainage; corridor system.* (Sinclair, 1958; Roberts, 1960) (c) In contrast to the above, methane drainage technology in the United States is conducted from the surface as well as underground. Underground methane drainage is primarily by means of horizontal boreholes drilled into the coalbed to be extracted. Surface methane drainage methods include vertical gob gas vent holes on longwall panels and hydraulically stimulated vertical wells generally drilled several years in advance of mining into virgin coalbeds.
methane monitoring system A system whereby the methane content of the mine air is indicated automatically at all times. When the content reaches a predetermined concentration, the electric power is cut off automatically from each machine in the affected area. The mechanism is so devised that its setting cannot be altered. The system is used, mainly, in conjunction with the operation of continuous miners and power loaders. (Nelson, 1965)
methanephone An instrument for detecting methane in mine air. It contains an electric battery that sustains a small electric glow light. As soon as a certain percentage of methane enters the workings, a tiny explosion occurs in the fuse head, where a fine wire filament is melted and starts a bell ringing continuously. (Fay, 1920)
methane recorder An instrument that gives a continuous record of the methane concentration over a period of time. (Roberts, 1960)
methane removal *water infusion method.*
methane tester A methane detector. See also: *methanometer.* (Nelson, 1965)
methane tester type S.3 A nonautomatic methane detector approved under the regulations for use in coal mines. The instrument is normally calibrated at 1% methane, and this provides an accuracy of ±0.05% over the most important part of the scale; i.e., 0.75% to 1.5%. It weighs 3-1/2 lb (1.6 kg), and the source of power is an Edison cap lamp battery. (Nelson, 1965)
methanometer An instrument for determining the methane content in mine air. See also: *sampling instrument; catalytic methanometer.* (Nelson, 1965)
methenyl tribromide *bromoform.*
method of working The system adopted to work or extract material in a mine. It includes all the operations involved in the cutting, handling, and transport of valuable material and waste rock, support of ground, ventilation of workings, and provision of supplies. The term does not include winding or hoisting, surface handling, and preparation or dressing. See also: *coal mining methods.* (Nelson, 1965)
method study A study to provide the essential data on which mine management can operate in making the most effective use of workpower, machines, and materials. Method study has been applied in the mining industry for many years, although sometimes under different names. See also: *time study; work study.* (Nelson, 1965)
methyl acetone A mixture of methyl acetate and acetone. Used as a solvent. (Crispin, 1964)
methylene iodide A heavy liquid used for mineral separation (sp gr, 3.33); also for refractive index determination (R.I.=1.74). Cf: *Clerici solution; Sonstadt solution; Klein solution; bromoform.*
metore Both capping and gossan. (AGI, 1987)
metra A pocket implement combining the uses of many instruments, such as thermometer, level, plummet, and lens. (Standard, 1964)
metric carat An international unit equal to 200 mg that had been adopted in most European countries and in Japan when it was made the standard in the United States in 1913. Abbrev.: M.C. and cm. See also: *carat.* (Webster 3rd, 1966)
Mexican diamond Rock crystal (quartz).
Mexican onyx Yellowish brown or greenish brown banded calcite. See also: *onyx marble; Gibraltar stone.*
Mexican turquoise Turquoise from the central part of Baja California, Mexico.
Mexican water opal A fire opal from Mexico.
meyerhofferite A triclinic mineral, $Ca_2B_6O_6(OH)_{10}{\cdot}2H_2O$; forms prismatic, commonly tabular, crystals or is fibrous; an alteration product of inyoite from Inyo County, CA.
meymacite An amorphous mineral, $WO_3{\cdot}2H_2O$; resinous; yellow brown.

miargyrite A monoclinic mineral, $AgSbS_2$; soft; metallic; in low-temperature hydrothermal veins; an ore of silver.
miarolite A granite having miarolitic cavities; a textural modification of normal granite. (Johannsen, 1931-38)
miarolithite A chorismite having miarolitic cavities or remnants thereof; a variety of ophthalmite. (AGI, 1987)
miarolitic A term applied to small irregular cavities in igneous rocks, esp. granites, into which small crystals of the rock-forming minerals protrude; characteristic of, pertaining to, or occurring in such cavities. Also, said of a rock containing such cavities. Cf: *drusy.* (AGI, 1987)
miarolitic cavity A cavity of irregular shape in certain plutonic rocks. Crystals of the rock constituents sometimes project into the cavity. Cf: *druse; vug.* (Schieferdecker, 1959)
miascite A mixture of strontianite and calcite. (Hey, 1955)
mica (a) A group of phyllosilicate minerals having the general composition, $X_2Y_{4-6}Z_8O_{20}(OH,F)$ where X=(Ba,Ca,Cs,H_3O,K,Na,NH_4), Y=(Al,Cr,Fe,Li,Mg,Mn,V,Zn), and Z=(Al,Be,Fe,Si); may be monoclinic, pseudohexagonal or pseudo-orthorhombic; soft; perfect basal (micaceous) cleavage yielding tough, elastic flakes and sheets; colorless, white, yellow, green, brown, or black; excellent electrical and thermal insulators (isinglass); common rock-forming minerals in igneous, metamorphic, and sedimentary rocks. See also: *brittle mica.* Syn: *glimmer; isinglass.* (b) The mineral group anandite, annite, biotite, bityite, celadonite, chernykhite, clintonite, ephesite, ferri-annite, glauconite, hendricksite, kinoshitalite, lepidolite, margarite, masutomilite, montdorite, muscovite, nanpingite, norrishite, paragonite, phlogopite, polylithionite, preiswerkite, roscoelite, siderophyllite, sodium phlogopite, taeniolite, tobelite, wonesite, and zinnwaldite.
micaceous (a) Consisting of or containing mica; e.g., a micaceous sediment. (b) Resembling mica; i.e., thinly foliated.
micaceous iron ore Hematite in which the texture is foliated or micaceous; some micaceous varieties are soft and unctuous. (Rice, 1960)
micaceous sandstone A sandstone containing conspicuous layers or flakes of mica, usually muscovite. Syn: *metaxite.* (AGI, 1987)
mica house A shop where hand-cobbed mica is rifted, trimmed, graded, and qualified. Syn: *trimming shed.* (Skow, 1962)
Micanite Trade name for a form of built-up mica used for insulating.
mica peridotite A peridotite consisting principally of altered olivine and biotite.
mica plate An accessory introduced into the optic path of a polarized-light microscope to produce a first-order gray interference color. Syn: *glimmer plate; lambda plate.* See also: *accessory plate.*
mica powder A dynamite in which the dope consists of fine scales of mica. (Fay, 1920)
mica schist A schist whose essential constituents are mica and quartz, and whose schistosity is mainly due to the parallel arrangement of mica flakes.
michenerite An isometric mineral, (Pd,Pt)BiTe; pyrite group; metallic; at Sudbury, ON, Canada and Monchegorsk, Kola Peninsula, Russia.
Michigan cut (a) In the United States, a cut that consists of drilling a hole with a large diameter or a number of holes of smaller diameter at the center of the heading and parallel to the direction of the tunnel. These holes are not charged. The remaining bench holes are then broken out towards these holes. (Fraenkel, 1953) (b) *burned cut.*
Michigan slip A very plastic, tough, fine-grained impure clay, similar to Albany slip clay; used as a bonding and plasticizing agent in grinding wheels, refractories, etc., and as a suspension agent for glassy frit in vitreous enamels. (CCD, 1942)
micrinite (a) A maceral of the inertinite group consisting of granular material without cellular structure; one of the principal components of durain and clarain. (AGI, 1987) (b) Proposed by the Heerlen Congress, 1935, as a substitute for micronite. (Tomkeieff, 1954)
micrinoid A coal constituent similar to material derived from finely macerated vegetation. (AGI, 1987)
micrite (a) A descriptive term originally used for the semiopaque crystalline matrix of limestones, consisting of chemically precipitated carbonate mud with crystals less than 4 µm in diameter, and interpreted as a lithified ooze. The term is now commonly used in a descriptive sense without genetic implication. (AGI, 1987) (b) A limestone consisting dominantly of micrite matrix; e.g., lithographic limestone. (AGI, 1987)
micro- (a) A prefix that divides a basic unit by 1 million or multiplies it by 10^{-6}. Abbrev., µ. (Lyman, 1964) (b) A prefix meaning small. When modifying a rock name, it signifies fine-grained hypabyssal, as in microgranite. Cf: *macro-.* (AGI, 1987)
microampere One-millionth of an ampere; 10^{-6} A; abbrev., µa. (Crispin, 1964)
microaphanitic *cryptocrystalline.*
microballoon One of the tiny hollow spheres of glass or plastic that are added to explosive materials to enhance sensitivity and control density by assuring an adequate content of entrapped air. (Atlas, 1987)
microbar A unit of pressure commonly used in acoustics. One microbar is equal to 1 dyn/cm^2. (Hunt, 1965)
microbreccia (a) A poorly sorted sandstone containing large angular particles of sand set in a fine silty or clayey matrix; e.g., a graywacke. It is somewhat less micaceous than a siltstone. (AGI, 1987) (b) A breccia within fragments of a coarser breccia. (AGI, 1987) (c) A well-indurated, massive rock that has been crushed to very fine grain size through cataclastic flow, commonly in detachment faults. (AGI, 1987)
microchemical Applied to chemical reactions conducted on the stage of a microscope and viewed through the microscope.
microclastic (a) Applied to a clastic or fragmental rock composed of very small particles. (b) Said of coal that is composed mainly of fine particles; e.g., cannel coal. (AGI, 1987)
microcline A triclinic mineral, $KAlSi_3O_8$; feldspar group; pseudomonoclinic; dimorphous with orthoclase; a major rock-forming mineral in granites, pegmatites, and metamorphic rocks; may be a detrital mineral in arkoses and graywackes. Cf: *orthoclase.*
microconglomerate A sedimentary rock composed of relatively coarse sand grains in a very fine silt or clay matrix. (AGI, 1987)
microcosmic salt *stercorite.*
microcryptocrystalline *cryptocrystalline.*
microcrystalline Said of the texture of a rock consisting of or having crystals that are small enough to be visible only under the microscope; also, said of a rock with such a texture. Syn: *cryptocrystalline; micromeritic.* (AGI, 1987)
microelement *trace element.*
microfacies Those characteristic and distinctive aspects of a sedimentary rock that are visible and identifiable only under the microscope (low-power magnification). (AGI, 1987)
microfarad A unit of capacitance; one-millionth of a farad; symbol, µF. (Crispin, 1964)
microfelsitic *cryptocrystalline.*
microfluidal Having a microscopic flow texture. (Standard, 1964)
microgeology (a) The study of the microscopic features of rocks. (b) The study of the relationships of microorganisms to geologic and geochemical processes.
micrograniotoid Having a microscopic granitoid structure. (Standard, 1964)
microgranular (a) Said of the texture of a microcrystalline, xenomorphic igneous rock. Also, said of a rock with such a texture. (AGI, 1987) (b) Minutely granular; specif. said of the texture of a carbonate sedimentary rock wherein the particles are mostly 10 to 60 µm in diameter and are well-sorted, and the finer clay-sized matrix is absent. Also said of a sedimentary rock with such a texture. (AGI, 1987)
micrograph A graphic reproduction of a magnified object as seen through a microscope. When it is a photograph, it is called a photomicrograph. (Stokes, 1955)
micrographic Said of the graphic texture of an igneous rock that is distinguishable only with the aid of a microscope; also, said of a rock having such texture. (AGI, 1987)
microhardness The hardness of microscopic areas or of the individual microconstituents in a metal, as measured by means such as the Tukon, Knoop, or scratch methods. (ASM, 1961)
microhm One microhm equals 10^{-6} Ω, which equals 10^3 electromagnetic units. Symbol, µΩ. (Webster 2nd, 1960)
microite Microite is found in many coals and occurs in large quantities in Gondwana coals and in Permocarboniferous coals of the former U.S.S.R. It is most abundant in coals with little exinite, or coals of high rank in which exinite cannot be recognized, and may occur in very persistent thick bands. It is present in small amounts in Carboniferous coals of the Northern Hemisphere. (IHCP, 1963)
microlaterolog A well log obtained with an arrangement of electrodes similar to a miniature laterolog but disposed in concentric fashion in an insulating pad. The current from a central electrode is focused and flows out in a pattern that resembles the shape of a trumpet. As in the microlog, the electrodes are mounted on a pad that is held against the wall of the hole by springs. The microlaterolog serves a purpose similar to that of a microlog, investigating only a small volume of rock immediately adjacent to the hole. Syn: *trumpet log.* (AGI, 1987)
microlite (a) A microscopic crystal that polarizes light and has some determinable optical properties. Cf: *crystallite; crystalloid.* Syn: *microlith.* (AGI, 1987) (b) A pale-yellow, reddish, brown, or black isometric mineral of the pyrochlore group: $(Ca,Na)_2Ta_2O_6(O,OH,F)$. It is isomorphous with pyrochlore, and it often contains small amounts of other elements (including uranium and titanium). Microlite occurs in granitic pegmatites and in pegmatites related to alkalic igneous rocks, and it constitutes an ore of tantalum. Syn: *djalmaite.* (AGI, 1987)
microlith *microlite.*
microlithotype A typical association of macerals in coals, occurring in bands at least 50 µm wide. Microlithotype names bear the suffix "-ite". See also: *lithotype.* (AGI, 1987)
microlitic Said of the texture of a porphyritic igneous rock in which the groundmass is composed of an aggregate of differently oriented or parallel microlites in a glassy or cryptocrystalline mesostasis. (AGI, 1987)

micromanometer Essentially a U-type gage employing a micrometer to measure the change in inclination of the gage from its zero or datum position. Normally, micromanometers are used in the laboratory for such purposes as the calibration of secondary manometers and, in conjunction with pressure measurement, in low-speed atmospheric wind tunnels. See also: *manometer*. (Roberts, 1960)
micromeritic An obsolete syn. of microcrystalline. See also: *microcrystalline.* (AGI, 1987)
micrometer (a) An instrument for measuring very small dimensions or angles. Used in connection with a microscope or a telescope. There are a great variety of forms, but in nearly all, the measurement is made by turning a very fine screw, which gives motion to a scale, a spider line, a lens, a prism, or a ruled glass plate. (Standard, 1964) (b) A unit of length, equal to one-millionth of a meter. Symbol, µm. (1 µm = 10^{-6}m). Formerly called micron. (c) A micrometer caliper. (Standard, 1964)
micrometer caliper A caliper with a graduated screw attachment for measuring minute distances. (Crispin, 1964)
micrometer-reading manometer *vernier-reading manometer.*
micrometrics The study of very fine particles. (AGI, 1987)
micromillimeter One-millionth of a millimeter; abbrev., mµ. (AGI, 1987)
micron Former term for micrometer.
micronized mica An ultrafine material produced in a disintegrator that has no moving parts but depends on jets of high-pressure superheated steam to reduce the mica to micrometer sizes. Micronized mica is produced in particle size ranges of 10 to 20 µm and 5 to 10 µm. (USBM, 1965)
Micronizer A special type of dry-grinding machine in which micronized mica is produced. It consists of a disintegrator that has no moving parts but depends on jets of high-pressure superheated steam for reducing the mica to micrometer sizes. (USBM, 1965)
micronizer mill Disintegrator, in which feed particles are entrained in a pressure jet (steam or air) and whirled through a cylindrical chamber with sufficient force to break them. (Pryor, 1963)
micropegmatite A less-preferred syn. of granophyre. See also: *granophyre.* (AGI, 1987)
microperthite Exsolution lamellae in alkali feldspar visible only with the aid of a microscope. See also: *cryptoperthite; perthite.*
micropetrological unit *maceral.*
microphotograph *photomicrograph.*
microporosity Porosity visible only with the aid of a microscope. (ASM, 1961)
microscope An instrument used to produce enlarged images; it consists of a lens (or lenses) of the objective and an ocular set into a tube, with or without other accessories, and held by an adjustable arm over an object stage.
microscopic (a) Of, relating to, or conducted with a microscope or microscopy. (Webster 3rd, 1966) (b) So small or fine as to be invisible or not clearly distinguishable without the use of a microscope. Cf: *macroscopic; megascopic.* (Webster 3rd, 1966)
microscopy The art and practice of using a microscope for identification and analysis of objects. See also: *ore microscopy; reflected-light microscope.*
microsecond One-millionth of a second; abbrev.: µsec, µs. (Webster 3rd, 1966)
microsection (a) Any thin section used in microscopic analysis. (AGI, 1987) (b) A polished section. (AGI, 1987)
microseism A collective term for small motions in the Earth that are unrelated to an earthquake and that have a period of 1.0 to 9.0 s. They are caused by a variety of natural and artificial agents, esp. atmospheric events. Syn: *seismic noise.* (AGI, 1987)
microseismic instrument An instrument for observing the behavior of roof strata and supports. The device is inserted in 4-ft (1.2-m) long 1-1/2-in (3.8-cm) diameter holes, drilled at selected points, for listening to subaudible vibrations which are known to precede rock failure. (Nelson, 1965)
microseismic movement *microseism.*
microseismic rate The number of microseisms per unit of time. (Isaacson, 1962)
microseismic region Area in which an earthquake is registered by instruments only. (Schieferdecker, 1959)
microseismometer An apparatus for indicating the direction, duration, and intensity of microseisms. Also called microseismograph. (Standard, 1964)
microspar Calcite matrix in limestones, occurring as uniformly sized and generally loaf-shaped crystals ranging from 5 µm to more than 20 µm in diameter. (AGI, 1987)
microspherulitic Said of the spherulitic texture of an igneous rock that is distinguishable only with the aid of a microscope, owing to the small size of the spherules. Also, said of a rock having such texture. (AGI, 1987)
microstriation Microscopic scratch developed on the polished surface of a rock or mineral as a result of abrasion. (AGI, 1987)
microstructure (a) Structural features of rocks that can be discerned only with the aid of the microscope. (AGI, 1987) (b) The structure of polished and etched metals as revealed by a microscope at a magnification greater than 10 diameters. (ASM, 1961)
microstylolite A type of grain boundary indicating differential solution between two minerals and characterized by fine interpenetrating teeth; often marked by a little opaque material. See also: *stylolite.*
microvitrain The thin vitrainlike bands present in clarain, having a maximum thickness of about 1/10 in (2.5 mm) with a tolerance of 1 mm, and a minimum thickness of 0.05 mm (50 µm). (AGI, 1987)
microvolt One-millionth of a volt; 10^{-6} V; symbol, µV. (Crispin, 1964)
microwatt One-millionth of a watt; symbol, µW. (Webster 3rd, 1966)
midalkalite *foyaite; nepheline syenite.*
Mid-Atlantic Ridge A mountain range that extends parallel to the continental margins in mid-ocean in both the North and South Atlantic Oceans. It rises 6,000 ft above the ocean floor and surfaces as the Azores, Ascension Island, Saint Helena, and Tristan da Cunha islands. (AGI, 1987)
middle Pertaining to a segment of geologic time intermediate between Late and Early, or to rocks intermediate between Upper and Lower. Thus, rocks of the Middle Jurassic Series were formed during the Middle Jurassic Epoch. Cf: *Upper; Lower.*
middle cut A machine cut in the midsection of a coal seam; sometimes adopted in thick seams (over 4 ft or 1.2 m) with a layer of dirt or inferior coal in the middle. A middle cut would be made with a turret coal cutter. See also: *bottom cut; top cut.* Syn: *intermediate cut.* (Nelson, 1965)
middle man A stratum of rock dividing or separating two seams or beds of coal. (Fay, 1920)
middle prop *center prop.*
middles *middlings.*
middletonite A brown, resinous, brittle mineral found between layers of coal at the Middleton collieries, near Leeds, England, and also at Newcastle. (Fay, 1920)
middlings (a) That part of the product of a washery, concentration, or preparation plant that is neither clean mineral product nor reject (tailings). It consists of fragments of coal ore mineral and gangue. The material is often sent back for crushing and/or retreatment. Syn: *middles.* (Nelson, 1965) (b) In two-component ore, particles incompletely liberated by comminution into concentrate or gangue. In complex ores, in addition to incomplete liberation, there may be multiphased particles of middling or intermediate species that react too feebly to treatment to report as concentrate or tailing. (Pryor, 1960)
middlings elevator An elevator that removes material for further treatment or for disposal as an inferior product. (BS, 1962)
middoor Scot. The middle one of three landing places in a shaft. (Fay, 1920)
midfeather In mining, a support to the center of a tunnel. (Standard, 1964)
midge N. of Eng. Lamp (not safety) carried by trammers, etc. (Fay, 1920)
midge stone Moss agate with inclusions resembling a swarm of mosquitoes. Also called gnat stone; mosquito agate.
midget impinger A dust-sampling apparatus almost identical in principle and design with the regular Greenburg-Smith impinger, the main difference being its smaller size and the fact that only a 12-in (30.5-cm) head of water is required for its operation. See also: *Greenburg-Smith impinger.*
mid-ocean ridge A continuous, seismic, median mountain range extending through the North and South Atlantic Oceans, the Indian Ocean, and the South Pacific Ocean. It is a broad, fractured swell with a central rift valley and usually extremely rugged topography; it is 1 to 3 km in elevation, about 1,500 km in width, and over 84,000 km in length. According to the hypothesis of sea-floor spreading, the mid-ocean ridge is the source of new crustal material. See also: *rift valley.* Syn: *mid-ocean rise; oceanic ridge.* (AGI, 1987)
midocean rift *rift valley.*
mid-ocean rise *mid-ocean ridge.*
midworkings (a) Scot. Mine workings above or below in the same mine or colliery. (Fay, 1920) (b) *middoor.*
miemite A variety of dolomite from Miemo, Tuscany, Italy.
miersite An isometric mineral, (Ag,Cu)I; canary yellow.
miesite A brown variety of pyromorphite containing calcium from Mies, Czech Republic.
migma Mobile, or potentially mobile, mixture of solid rock material(s) and magma, the magma having been injected into or melted out of the rock material. Etymol: Greek, mixture. (AGI, 1987)
migmatite A composite rock composed of igneous or igneous-appearing and/or metamorphic materials that are generally distinguishable megascopically. See also: *composite gneiss.* Cf: *chorismite; injection gneiss.* (AGI, 1987)

migmatization Formation of a migmatite. The more mobile, typically light-colored, part of a migmatite may be formed as the result of anatexis, lateral secretion, metasomatism, or injection. (AGI, 1987)

migration (a) The movement of oil, gas, or water through porous and permeable rock. Parallel (longitudinal) migration is movement parallel to the bedding plane. Transverse migration is movement across the bedding plane. (AGI, 1987) (b) The process by which events on a reflection seismogram are mapped in an approximation of their true spatial positions. It requires knowledge of the velocity distribution along the raypath. Also, the seismic correction that is applied. (AGI, 1987; Schieferdecker, 1959) (c) The movement of a topographic feature from one locality to another by the operation of natural forces; specif. the movement of a dune by the continual transfer of sand from its windward to its leeward side. (AGI, 1987) (d) The slow downstream movement of a system of meanders, accompanied by enlargement of the curves and widening of the meander belt. (AGI, 1987) (e) A broad term applied to the movements of plants and animals from one place to another over long periods of time. (AGI, 1987)

migration of oil The movement or seepage of oil through rocks wherever they are sufficiently permeable to allow such passage; of considerable importance in oil geology. (Nelson, 1965)

mikheevite *görgeyite.*

milarite A hexagonal mineral, $K_2Ca_4Al_2Be_4Si_{24}O_{60}{\cdot}H_2O$; osumilite group; forms colorless to greenish prisms. (Not millerite.)

mild and tough Mellowed or ripened by weathering; said of brick clay; opposite of short and rough. (Standard, 1964)

mild earth Eng. Soft, loamy clay suitable for brickmaking, as opposed to stiffer clay below, which is suitable for making tiles and drainpipes. Kimeridge clay, Brill, Buckinghamshire. (Arkell, 1953)

mild steel Steel that contains from 0.12% to 0.25% carbon. Also called low-carbon steel; soft steel. See also: *yield stress.* (Pryor, 1963)

mil-foot A standard of resistance in wire. The resistance of 1 ft (30.5 cm) of wire that is 1 mil (25.4 µm) in diameter. (Crispin, 1964)

milk of sulfur *colloidal sulfur.*

milk-opal A translucent, milk-white to green, yellow, or blue variety of common opal.

milky quartz A milk-white, nearly opaque variety of quartz, commonly with a greasy luster. The milkiness is due to the presence of minute, fluid-filled inclusions. Syn: *greasy quartz.*

mill (a) A mineral treatment plant in which crushing, wet grinding, and further treatment of ore is conducted. Also, separate components, such as ball mill, hammer mill, and rod mill. See also: *ball mill; hammermill; rod mill; grinding mill; pug mill.* (b) A passage connecting a stope or upper level with a level below, intended to be filled with broken ore that can then be drawn out at the bottom as desired for further transportation; an opening in the floor or bottom of a stope, through which the ore or mineral is passed or thrown downward along the footwall to the level. See also: *glory hole.* (Fay, 1920) (c) To fill a winze, or interior incline, with broken ore, to be drawn out at the bottom. (Fay, 1920) (d) A finishing plant where blocks of stone are sawed and trimmed. (e) In quarrying, usually applied to the finishing plant where blocks are sawed into slabs; all other manufacturing processes are classed as shop work. (AIME, 1960) (f) An excavation made in the country rock, by a crosscut from the workings on a vein, to obtain waste for filling. It is left without timber so that the roof may fall in and furnish the required rock. (Fay, 1920) (g) Can. Reducing plant where ore is concentrated and/or metals are recovered. (Hoffman, 1958) (h) A single machine or a complete plant for rolling metals. (i) *cogging mill.* (j) To grind or cut away steel or iron with a toothed or serrated face bit; also, the tool so used. (Long, 1960) (k) Eng. That part of an ironworks where puddle bars are converted into merchant iron; i.e., rolled iron ready for sale in bars, rods, or sheets. See also: *forge.* (Fay, 1920) (l) A preparation facility within which metal ore is cleaned, concentrated, or otherwise processed before it is shipped to the customer, refiner, smelter, or manufacturer. A mill includes all ancillary operations and structures necessary to clean, concentrate, or otherwise process metal ore, such as ore and gangue storage areas and loading facilities. (SME, 1992) (m) By common usage, any establishment for reducing ores by other means than smelting. More strictly, a place or a machine in which ore or rock is crushed. (Fay, 1920)

mill bar A rough bar rolled or drawn directly from a bloom or puddle bar for conversion into merchant iron in the mill.

mill car A car without a roof for carrying hoisting apparatus. (Standard, 1964)

milled A metal object lost in a borehole that has been cut or ground away with a milling bit. (Long, 1960)

Miller-Bravais indices A four-index type of Miller indices, useful but not necessary in order to define planes in crystal lattices in the hexagonal system; the symbols are hkil, in which i = -(h + k). Cf: *Miller indices.* (AGI, 1987)

Miller indices Integers used to designate crystallographic planes. They are found as follows: (1) Determine where the plane intercepts each crystallographic axis in terms of multiples of the axial parameters; (2) take the reciprocals of these numbers and clear of fractions and common factors. These three numbers, designated (hkl), are the Miller indices for that plane. Cf: *Bragg indices; Miller-Bravais indices; crystal face; crystal indices; intercept.* See also: *symbols of crystal faces; indices of a crystal face.*

millerite A brass-yellow to bronze-yellow rhombohedral mineral: NiS. It usually has traces of cobalt, copper, and iron, and is often tarnished. Millerite generally occurs in fine hairlike or capillary crystals of extreme delicacy, chiefly as nodules in clay ironstone. Syn: *capillary pyrite; nickel pyrite; hair pyrite.* (Not milarite.). (AGI, 1987)

mill feeder In ore dressing, smelting, and refining, a laborer who regulates flow of ore, coke, flue scrapings, or other materials from bins, chutes, or belts into crushers, furnaces, or other equipment, or onto conveyor belts leading to equipment. Syn: *conveyor-feeder operator.* (DOT, 1949)

mill furnace A furnace for reheating iron that is to be rerolled, or welded, under a hammer. (Standard, 1964)

millgrit rock Som. Triassic dolomitic conglomerate.

mill head (a) Ore accepted for treatment in a concentrator, after any preliminary rejection such as waste removal. (Pryor, 1965) (b) Assay value, or units of value per ton, in ore accepted for treatment in a concentrating plant or mill. (Pryor, 1963)

mill-head grade The grade of ore as it comes from a mine and goes to a mill. (McKinstry, 1948)

mill-head ore *run-of-mill.*

mill hole An auxiliary shaft connecting a stope or other excavation with the level below. See also: *mill; glory hole.* (Lewis, 1964)

milli- A prefix meaning one thousandth of. It divides a basic unit by 1,000, or multiplies it by 10^{-3}; abbrev., m. Commonly applied to units of measure in the metric and cgs systems; e.g., 1 mg = 0.001 g. (Lyman, 1964)

milliampere One-thousandth of an ampere; abbrev., mA. (Crispin, 1964)

milliangstrom One-thousandth of an angstrom; abbrev., mÅ. (Webster 3rd, 1966)

millibar A unit of atmospheric pressure equal to one-thousandth of a bar or 1,000 dyn/cm^2; abbrev., mb. (Webster 3rd, 1966)

millicurie-hour A measure of gamma-ray exposure expressed as the product of the source in millicuries and the time of exposure in hours. (NCB, 1964)

millidarcy The customary unit of measurement of permeability, equal to one-thousandth of a darcy. See also: *permeability.* Abbrev., md. (AGI, 1987; Webster 3rd, 1966)

millidegree A unit of temperature equal to one-thousandth of a degree; abbrev., mdeg. (Webster 3rd, 1966)

millifarad One-thousandth of a farad; abbrev., mF. (Webster 3rd, 1966)

milligal (a) A unit employed in the gravitational method of geophysical prospecting. It is about one millionth of the average value of the acceleration due to gravity at the Earth's surface; i.e., 1 milligal = 1 cm/s^2. (Nelson, 1965) (b) A unit of acceleration used with gravity measurements; 10^{-3} Gal = 10^{-5}m/s^2. Abbrev: mGal. (AGI, 1987)

milligauss One-thousandth of a gauss; abbrev., mG. (Standard, 1964)

milligram-hour A measure of gamma-ray exposure expressed as the product of the equivalent radium content of the source, in milligrams, and the time of exposure in hours. (NCB, 1964)

millihenry One-thousandth of a henry; a unit of inductance; abbrev., mH. (Webster 3rd, 1966)

millimeter screw micrometer A precision caliper gage that measures the overall dimensions of unmounted fashioned gems more accurately but less conveniently than dial gages.

milling (a) The grinding or crushing of ore. The term may include the operation of removing valueless or harmful constituents and preparation for market. (Nelson, 1965) (b) A combination of open-cut and underground mining, wherein the ore is mined in open cut and handled underground. It is underhand stoping applied to large deposits, wherein the ore is mined near the mouth of winzes or raises, and dropped by gravity to working levels below for transportation to the surface. Usually called glory-hole system. See also: *mill.* (c) The act or process of cutting or grinding away a metal object lost in a borehole with a mill or milling bit. (Long, 1960)

milling bit A bit equipped with hardened serrations or teeth used to grind or cut away metallic materials or junk obstructing a borehole. Also called rose bit. Cf: *junk mill.* (Long, 1960)

milling grade (a) Ore containing sufficient recoverable value to warrant treatment. (Pryor, 1963) (b) S. Afr. An assumed average value of the ore sent to a mill, expressed as a percentage, or in pennyweights per short ton. (Beerman, 1963)

milling ore (a) Any ore that contains sufficient valuable minerals to be treated by any milling process. Syn: *mill ore.* (b) A dry ore that can be amalgamated or treated by leaching and other processes; usually these ores are low grade and free, or nearly so, from base metals.

milling width Width of a lode that is designated for treatment in a mill, as calculated with regard to daily tonnage. Any excess broken during mining (stoping width) should be rejected before milling. (Pryor, 1963)

milling yield S. Afr. The valuable material obtained from milling, expressed as a percentage or in pennyweights per short ton. (Beerman, 1963)

millisecond One-thousandth of a second; abbrev.: msec; ms. (Webster 3rd, 1966)

millisecond delay A type of delay cap with a definite but extremely short interval between the passing of current and the explosion. (Nichols, 1976)

millisecond-delay cap A detonating cap that fires from 0.02 to 0.5 s after the firing current passes through it. (Nichols, 1976)

millisecond-delay detonator *short-delay detonator.*

millivolt One-thousandth of a volt; abbrev., mV. (Crispin, 1964)

millman One who is employed in a mill, such as in an ore-dressing plant. (Fay, 1920)

mill ore *milling ore.*

mill roll One of the rolls through which puddled iron is run prior to being marketed. (Standard, 1964)

mill run (a) A given quantity of ore tested for its quality by actual milling; the yield of such a test. (Craigie, 1938) (b) Average, not esp. selected. (Craigie, 1938) (c) In intermittent treatment of ore, with periodic cleanup, the period of such a run. Bulk test on a sample of ore during development of a treatment process. (Pryor, 1963)

mill sampler A laborer who removes samples of crushed ore, concentrates, or tailings at various stages of processing, and puts them in labeled containers for laboratory analysis. Also called sampler. (DOT, 1949)

mill scale The scale of ferric oxide that peels from iron during rolling. (Standard, 1964)

Mills-Crowe process Method of regeneration of foul cyanide liquor from the gold leaching process. The barren solution is acidified, and gaseous hydrocyanic acid (HCN) is liberated, separated, and reabsorbed in an alkaline solution, such as lime water. (Pryor, 1963)

mill shoe (a) A shoe equipped with a hardened serrated cutting edge used to mill downward over and around a piece of drill-stem equipment lost in a borehole. See also: *mill.* (Long, 1960) (b) A shoe designed either to dress the down-hole tools for access to retrieve them, or to grind up the tool in a drill hole so that drilling can continue.

millstock Term used in the slate industry to include all forms of structural slate used in exterior or interior construction.

millstone A buhrstone; e.g., a coarse-grained sandstone or a fine-grained quartz conglomerate. Also, one of two thick disks of such material used for grinding grain and other materials fed through a center hole in the upper stone. Quarried underground in Virginia; also, produced in New York and (of granite) in North Carolina. See also: *buhrstone.*

millstone grit Any hard, siliceous rock suitable for use as a material for millstones; specif. the Millstone Grit of the British Carboniferous, a coarse conglomeratic sandstone. (AGI, 1987)

mill test The determination of the metallic contents and recoverable metal in any given ore by the milling of a sufficient quantity to afford average milling conditions. See also: *mill run.* (Weed, 1922)

milltons Net tonnage of ore available for milling after eliminating waste and unpayable material. (Beerman, 1963)

mill value S. Afr. The calculated value of ore before crushing. After treatment, there remains a residue of the valuable metal in the sands, slimes, or tailings, which, added to the yield obtained by treatment, should be equal to the mill value. (Beerman, 1963)

mimetene *mimetite.*

mimetesite *mimetite.*

mimetic (a) Said of crystals appearing to have a higher symmetry than their internal structure due to twinning, distortion, or interfacial angles at or close to a more symmetric form. (b) Said of crystals showing highly complex pseudosymmetry. Cf: *pseudosymmetry.* (c) A tectonite with a deformation fabric influenced by preexisting structural anisotropy; also the fabric itself.

mimetite A monoclinic mineral, $Pb_5Cl(AsO_4)_3$; apatite group, with arsenate replaced by phosphate; pseudohexagonal; resinous; sp gr, 7.3; in oxidized zones of lead-ore deposits. Syn: *mimetesite; campylite; mimetene.* Cf: *hedyphane.*

minal *end member.*

minasragrite A monoclinic mineral, $VO(SO_4)\cdot 5H_2O$; a blue efflorescence at Minasragra, near Cerro de Pasco, Peru.

mine (a) An opening or excavation in the ground for the purpose of extracting minerals; a pit or excavation from which ores or other mineral substances are taken by digging; an opening in the ground made for the purpose of taking out minerals, and in case of coal mines, commonly a worked vein; an excavation properly underground for digging out some usual product, such as ore, metal, or coal, including any deposit of any material suitable for excavation and working as a placer mine; collectively, the underground passage and workings and the minerals themselves. (Ricketts, 1943) (b) A work for the excavation of minerals by means of pits, shafts, levels, tunnels, etc., as opposed to a quarry, where the whole excavation is open. In general, the existence of a mine is determined by the mode in which the mineral is obtained, and not by its chemical or geological character. The term also includes only excavations for their minerals or valuable mineral deposits. (Ricketts, 1943) (c) An excavation beneath the surface of the ground from which mineral matter of value is extracted. The word carries the sense of laborers working beneath a cover of ground and thus excludes oil, brine, and sulfur wells. Excavations for the extraction of ore or other economic minerals not requiring work beneath the surface are designated by a modifying word or phrase as: (1) opencut mine—an excavation for removing minerals that is open to the weather; (2) steam shovel mine—an opencut mine in which steam shovels or other power shovels are used for loading cars; (3) strip mine—a stripping; an opencut mine in which the overburden is removed from a coalbed before the coal is taken out; (4) placer mine—a deposit of sand, gravel, or talus from which some valuable mineral is extracted; and (5) hydraulic mine—a placer mine worked by means of a stream of water directed against a bank of sand, gravel, or talus; soft rock similarly worked. A quarry from which rock is extracted becomes a mine when it is carried under cover. Mines are commonly known by the mineral or metal extracted such as bauxite mines, copper mines, silver mines, coal mines, etc. (Hess) (d) The terms mine and coal mine are intended to signify any and all parts of the property of a mining plant, either on the surface or underground, that contribute directly or indirectly to the mining or handling of coal or ore. (Fay, 1920) (e) The term mine, as applied by quarrymen, is applied to underground workings having a roof of undisturbed rock. It is used in contrast with the open pit quarry. (Fay, 1920) (f) To dig a mine; to get ore, metals, coal, or precious stones out of the earth; to dig into, as the ground, for ore or metal; to work in a mine. (Webster 3rd, 1966) (g) An active mining area, including all land and property placed under, or above the surface of such land, used in or resulting from the work of extracting metal ore or minerals from their natural deposits by any means or method, including secondary recovery of metal ore from refuse or other storage piles, wastes, or rock dumps and mill tailings derived from the mining, cleaning, or concentration of metal ores. (CFR, 7)

mine atmosphere The concentration of gases, including oxygen, that are present in a mine. Safe levels are maintained through ventilation. Measured at any point at least 12 in (30.5 cm) away from the back, face, rib, and floor in any mine. (CFR, 1992)

mine bank (a) An area of ore deposits that can be worked by excavations above the water level. (Craigie, 1938) (b) The ground at the top of a mining shaft. (Craigie, 1938)

mine cage Elevator used to transport workers in a lined shaft; available in open, semienclosed or fully enclosed models with a choice of sliding, folding, or rollup doors. Cages are used in either vertical or incline mine shafts. All cages are required to contain necessary safety features. (Best, 1966)

mine captain (a) A superintendent of a mine. (Standard, 1964) (b) The director of work in a mine, with or without superior officials or subordinates. (Fay, 1920) (c) In metal mining, a foreman who supervises the extraction, hauling, and hoisting of ore in a mine. Also called ground boss. (DOT, 1949)

mine car One of the cars that are loaded at production points and hauled to the pit bottom or surface in a train by locomotives or other power. They vary in capacity, and are either of wood or steel construction or combinations of both. Mine cars have been classified into six kinds: (1) the solid or box type, which requires a rotary dump at the unloading terminal; (2) the rocker dump type, which has a V-shaped body rounded at the bottom; (3) the gable-bottom car, which is shaped like a capital W in cross section; (4) the Granby car, a special form of a side-dumping car; (5) bottom-dump cars; and (6) end-dump cars, which are commonly used for hand tramming in small mines. See also: *drop-bottom car; endgate car; gable-bottom car; Granby car; solid car.* (BCI, 1947; Kentucky, 1952; Lewis, 1964)

mine-car repairman In anthracite coal mining; bituminous coal mining; metal mining; nonmetal mining, a person who repairs or replaces damaged parts of mine (pit) cars, such as axles, wheels, bodies, and couplings, by straightening, bolting, riveting, refitting, or making new parts as required. Also called car whacker; pit-car repairer. (DOT, 1949)

mine characteristic The relation between pressure, p, and volume, Q, in the ventilation of a mine. If the resistance, R, of the mine is known, then the mine characteristic can be expressed as $p=RQ^2$. The curve of this equation for a particular mine may thus be plotted on the same axes as the characteristics of a fan. The point of intersection of this curve, termed the mine characteristic, with the pressure characteristic of the fan, indicates the pressure and volume at which the fan would work in ventilating that mine. Knowing the volume and pressure, the power and efficiency are obtained. The suitability of any fan to any mine can be studied, and the effect of possible changes in mine resistance may be predicted. (Roberts, 1960)

mine characteristic curve As a graphical aid to the solution of problems in mine ventilation, the mine head (static and/or total) is often plotted against the quantity. This is called the mine characteristic curve, or simply the mine characteristic. (Hartman, 1982)

mine circulating fan Mines create special problems in proper ventilation by their isolation from fresh air sources, and the presence of dangerous gases and dusts. Large fans are used for the stationary systems, while small portable types provide fresh air in dead-ends and other inaccessible locations. These fans may be driven by electricity or compressed air, and in addition to mine operations, are useful for work in manholes, pipe galleries, silos, tanks, vats, plane fuselages, ship holds, etc. The Mine Safety and Health Administration, Department of Labor can furnish specific recommendations concerning special problems. (Best, 1966)

mine committee Representatives chosen by the union employees to confer with the representatives of the company; corresponds in mining to shop committees in manufacturing. Also called pit committee. (BCI, 1947)

mine conveyor *underground mine conveyor.*

mine cooling load The total amount of heat, sensible and latent, which must be removed by the air in the working places. (Hartman, 1982)

mine development The term employed to designate the operations involved in preparing a mine for ore extraction. These operations include tunneling, sinking, crosscutting, drifting and raising. (Jackson, 3)

mine dial *miner's dial.*

mine door *trapdoor; door.*

mine drainage *drainage; drain tunnel; water hoist.*

mined strata In mine subsidence, the strata lying over the excavated area. (Briggs, 1929)

mine dust (a) Dust from rock drills, blasting, or handling rock. (Fay, 1920) (b) In the quantity inhaled by workers, dust may be classified as dangerous, harmless, and borderline, although the classification is purely arbitrary. Silica is a dangerous dust and aluminum hydroxide is borderline. (Lewis, 1964) (c) Scot. Calcined ironstone screenings. (Standard, 1964) (d) *coal dust.*

mined volume In mine subsidence, the mined area multiplied by the mean thickness of the bed, or of that part of the bed that has been extracted. (Briggs, 1929)

mine examiner *fire boss.*

mine expert *mining engineer.*

mine fan (a) The main fan for a mine; normally situated at the surface. (BS, 1963) (b) A radial- or axial-flow ventilator. See also: *fan; ventilation.* (Nelson, 1965)

mine fan signal system A system that indicates by electric light or electric audible signal, or both, the slowing down or stopping of a mine ventilating fan.

mine feeder circuit A conductor or group of conductors, including feeder and sectionalizing switches or circuit breakers, installed in mine entries or gangways and extending to the limits set for permanent mine wiring, beyond which limits portable cables are used.

mine fill *hydraulic fill.*

mine fire This very dangerous occurrence may arise as the result of spontaneous combustion, the ignition of timbers by gob fires, electric cable defects, or the heating and ignition of conveyor belts due to friction. (Nelson, 1965)

mine fire truck Designed to fight underground fires in mining operations, this low slung railcar is equipped with a water supply and pressure equipment for its fire hoses. When a fire occurs, the car can be sped to the scene along existing rails. The truck is capable of delivering hundreds of gallons of water, depending upon the size and model used. (Best, 1966)

mine foreman (a) The person charged with the responsibility of the general supervision of the underground workings of a mine and the persons employed therein. In certain states, the mine foreman is designated as the mine manager. See also: *foreman.* (b) Generally used to designate that company representative in complete 24-h charge of underground workings and legally held responsible for the safety and welfare of all underground employees. The foreman is generally State certified for competency. (BCI, 1947) (c) A deputy in metal mines. (Nelson, 1965) (d) An official in charge of plant and associated labor on the surface, e.g., screen foreman. (Nelson, 1965)

mine ground A stratum or group of strata containing layers of ironstone.

minehead *pithead.*

mine head In a mine ventilation system, the cumulative energy consumption is called the mine head. A head is in reality a pressure difference, determined in accordance with Bernoulli's principle. (Hartman, 1961)

mine hoist A device for raising or lowering ore, rock, or coal from a mine and for lowering and raising workers and supplies. See also: *hoist.*

mine hoist control This mechanism is designed to prevent accidents in mine cages caused by overspeeding in hoisting and lowering. It also prevents hoisting or lowering beyond the limits for which the controller is set. On electric hoists, it can apply a brake in case of power failure and may regulate brake speed in the event of an emergency stop. (Best, 1966)

mine inspector (a) Person who checks mines to determine the safety condition of working areas, equipment, ventilation, and electricity, and to detect fire and dust hazards. (Webster 3rd, 1966) (b) Generally used as denoting the State mine inspector as contrasted to the Federal mine inspector. See also: *inspector.* (BCI, 1947)

mine iron Pig iron made entirely from ore; distinguished from cinder pig. (Standard, 1964)

mine jeep A special electrically driven car for underground transportation of officials, inspectors, repair, maintenance, and surveying crews and rescue workers.

mine lamp Battery operated lamp; may be attached to a miner's cap to provide illumination in poorly lighted mine areas. Lamps are designed to focus on working areas when attached. A unit consists of a rechargeable battery, bulb, reflector, wires, etc. Mine lamps may be purchased outright or leased. Models include cap lamps, hand lamps, and trip lamps. (Best, 1966)

mine land reclamation The process of transforming mine land into usable conditions; e.g., with regard to residential, recreational, agricultural, commercial or forestry purposes. (SME, 1992)

mine locomotive A low, heavy, haulage engine, designed for underground operation; usually propelled by electricity, gasoline, or compressed air. See also: *electric mine locomotive; trolley locomotive.* (Fay, 1920)

mine mason Mineworker who lines the galleries and other rooms with masonry, works on the repair of mine supports, and builds ventilation doors and dams. (Stoces, 1954)

mine motorman *motorman.*

mine opening *opening.*

mine power center A mine power center is a combined transformer and distribution unit complete within a metal enclosure, usually of explosion-proof design, from which one or more low-voltage power circuits are taken.

mine prop Section of wood, generally part of a small tree trunk, used for holding up rock in the roof of a mine. (Mersereau, 1947)

miner (a) In anthracite and bituminous coal mining, one who performs the complete set of duties involved in driving underground openings to extract coal, slate, and rock, with a hand or machine drill, into which explosives are charged and set off to break up the mass. Also called coal digger; coal getter; coal hewer; digger; faceman. (DOT, 1949) (b) In nonmetal mining, such as limestone, one who drills holes in working face of a limestone mine, and inserts and sets off charges of explosives in holes. May be designated as development miner when working in new areas to drive drifts, shafts, sumps, and entry areas into stopes; high-raise miner when working in vertical areas to gain access into new development areas; or stope miner when working in horizontal openings in limestone strata. Syn: *stope miner.* (DOT, 1949) (c) In mining, one who performs the complete set of duties involved in driving underground openings to extract ore or rock: drills holes in working face of ore or rock, with a hand or machine drill; inserts explosives in drill holes and sets them off to break up the mass; shovels ore or rock into mine cars or onto a conveyor, and pushes mine cars to haulageways, where they are hauled by draft animals, mine locomotive (motor), or haulage cable to the surface, or to the shaft bottom for hoisting; and installs timbering to support the walls and roof, or for chutes or staging. Due to standardization of mining methods and development of mining machinery, these jobs may be performed by several workers. Where conditions are favorable, loading is done by machine. (DOT, 1949) (d) One who mines; such as (1) a person engaged in the business or occupation of getting ore, coal, precious substances, or other natural substances out of the earth; (2) a machine for automatic mining (as of coal); and (3) a worker on the construction of underground tunnels and shafts (as for roads, railways, waterways). (Webster 3rd, 1966) (e) A worker in a coal mine who is paid a certain price for each ton of coal he or she digs or blasts from the solid seam, as distinguished from the laborer who loads the cars, etc. The miner's helpers load the coal; they are also called laborers. (Fay, 1920) (f) Includes all laborers who work in a mine, whether digging coal, timbering, or making places safe. (Fay, 1920) (g) Loosely used to designate all underground employees; technically, and in many cases legally, only those who have served an apprenticeship as helpers or those who are State licensed as miners. (BCI, 1947) (h) A worker who cuts coal in a breast or chamber by contract; the highest-skilled worker of a colliery. (Korson, 1938) (i) One who mines; a digger for metals and other minerals; is not necessarily a mechanic, handcraftsman, or artisan, and the term imports neither learning nor skill. (Ricketts, 1943) (j) Abbrev. for continuous miner. (Nelson, 1965)

mineragraphy *ore microscopy.*

mineral (a) A naturally occurring inorganic element or compound having an orderly internal structure and characteristic chemical composition, crystal form, and physical properties. Cf: *metallic.* (AGI, 1987) (b) In miner's phraseology, ore. See also: *ore.* (Fay, 1920) (c) *mineral species; mineral series; mineral group.* (d) Any natural resource extracted from the earth for human use; e.g., ores, salts, coal, or petroleum. (e) In flotation, valuable mineral constituents of ore as opposed to gangue minerals. (f) Any inorganic plant or animal nutrient. (g) Any member of the mineral kingdom as opposed to the animal and plant kingdoms.

mineral acre The full mineral interest in 1 acre (0.4 ha) of land. (Williams, 1964)

mineral adipocire *hatchettite.*

mineral assessment (a) The process of appraisal of identified and undiscovered mineral resources within some specified region, and the product of that appraisal. (Barton, 1995) (b) The estimation of mineral endowment, meaning the number of deposits or the tonnage of metal that occurs in the region, given some minimum size of accumulation (deposit), minimum concentration (grade), and maximum depth of occurrence. Syn: *predictive metallogeny.* (DeVerle, 1983)

mineral association A group of minerals found together in a rock, esp. in a sedimentary rock. (AGI, 1987)

mineral belt An elongated region of mineralization; an area containing several mineral deposits. (AGI, 1987)

mineral blossom Drusy quartz. (Fay, 1920)

mineral bruto Sp. Raw ore. (Hess)

mineral caoutchouc *elaterite; helenite.*

mineral carbonatado Sp. Carbonate ore. (Hess)

mineral charcoal (a) A pulverulent, lusterless substance, showing distinct vegetal structure, and containing a high percentage of carbon with little hydrogen and oxygen, occurring in thin layers in bituminous coal. Called mother of coal by miners. (Fay, 1920) (b) Another name for fusain. (BS, 1960)

mineral claim A mining claim. (Mathews, 1951)

mineral cleavage Mineral breakage along specific crystallographic planes in all specimens due to fewer or weaker chemical bonds in those directions. Cf: *mineral parting.*

mineral deed A conveyance of an interest in the minerals in, on, or under a described tract of land. The grantee is given operating rights on the land; easements of access to the minerals are normally implied unless expressly negated. (Williams, 1964)

mineral de fusion propia Sp. Self-fluxing ore. (Hess)

mineral deposit (a) A mass of naturally occurring mineral material; e.g., metal ores or nonmetallic minerals, usually of economic value, without regard to mode of origin. Accumulations of coal and petroleum may or may not be included; usage should be defined in context. Syn: *orebody.* (AGI, 1987) (b) A mineral occurrence of sufficient size and grade that it might, under favorable circumstances, be considered to have economic potential. See also: *ore; ore deposit.* (USGS, 1986)

mineral deposit model The systematically arranged information describing the essential attributes (properties) of a class of mineral deposits. The model may be empirical (descriptive), in which instance the various attributes are recognized as essential even though their relationships are unknown; or it may be theoretical (genetic), in which instance the attributes are interrelated through some fundamental concept. See also: *model.* (SME, 1992)

mineral dresser A machine for trimming or dressing mineralogical specimens. (Standard, 1964)

mineral dressing (a) Physical and chemical concentration of raw ore into a product from which a metal can be recovered at a profit. (ASM, 1961) (b) Treatment of natural ores or partly processed products derived from such ores in order to segregate or upgrade some or all of their valuable constituents, and/or remove those not desired by an industrial user. Mineral dressing processes are applied to industrial wastes to retrieve useful byproducts. See also: *mineral processing; ore dressing.* (Pryor, 1960)

mineral economics Study and application of the technical and administrative processes used in management, control, and finance connected with the discovery, development, exploitation, and marketing of minerals. (Pryor, 1963)

mineral endowment The physical aggregate of mineral occurrences in a region above some lower cutoff. (Shanz, 1983)

mineral engineering Term covers a wide field in which many resources of modern science and engineering are used in discovery, development, exploitation, and use of natural mineral deposits. (Pryor, 1963)

mineral entry The filing of a claim for public land to obtain the right to any minerals it may contain. (Craigie, 1938)

mineral facies *metamorphic facies.*

mineral fat *ozocerite.*

mineral fiber (a) Fibrous mineral whose fibers are longer than 5 µm and with an aspect ratio (length over width) equal to or greater than 3:1 as determined by the membrane filter method at 400X to 500X magnification (4-mm objective) phase contrast illumination. (ACGIH, 1993-1994) (b) The smallest elongated crystalline unit that can be separated from a bundle or appears to have grown individually in that shape, and that exhibits a resemblance to organic fibers. (Campbell)

mineral field Scot. A tract of country in which workable minerals are found; a mineral leasehold.

mineral filler A finely pulverized inert mineral or rock that is included in a manufactured product—e.g., paper, rubber, and plastics—to impart certain useful properties, such as hardness, smoothness, or strength. Common mineral fillers include asbestos, kaolin, and talc. (AGI, 1987)

mineral fuel Coal or petroleum. See also: *fossil fuel.* (Pearl, 1961)

mineral group Two or more mineral species having identical or closely related structures; e.g., hematite group or zeolite group. See also: *mineral.*

mineral interests Mineral interests in land means all the minerals beneath the surface. Such interests are a part of the realty, and the estate in them is subject to the ordinary rules of law governing the title to real property. (Ricketts, 1943)

mineral inventory An accounting of the mineral reserves and resources contained in known mineral deposits including inactive mines, operating mines, and undeveloped sites. (Shanz, 1983)

mineralization The process or processes by which a mineral or minerals are introduced into a rock, resulting in a valuable or potentially valuable deposit. It is a general term, incorporating various types; e.g., fissure filling, impregnation, and replacement. (AGI, 1987)

mineralize To convert to a mineral substance; to impregnate with mineral material. The term is applied to the processes of ore formation and also to the process of fossilization. (AGI, 1987)

mineralized bubble In flotation, one of the bubbles that rise from the pulp loaded with particles of desired mineral. (Pryor, 1963)

mineralizer *ore-forming fluid; geologic mineralizer.*

mineralizing agent *ore-forming fluid; geologic mineralizer.*

mineral land Land that is worth more for mining than for agriculture. The fact that the land contains some gold or silver would not constitute it mineral land if the gold and silver did not exist in sufficient quantities to pay to work. Land not mineral in character is subject to entry and patent as a homestead, however limited its value for agricultural purposes. Cf: *stone land.* (Ricketts, 1943)

Mineral Lands and Mining The leasable minerals include oil, gas, sodium, potash, phosphate, coal, and all minerals within acquired lands. Acquisition is by application for a Government lease and permits to mine or explore after lease issuance. (SME, 1992)

mineral lease *mining lease.*

mineral occurrence (a) The presence of useful minerals or rocks in an area under examination. (Shanz, 1983) (b) A concentration of a mineral (usually, but not necessarily, considered in terms of some commodity, such as copper, barite, or gold) that is considered to be valuable or that is of scientific or technical interest. In rare instances (such as titanium in a rutile-bearing black sand), the concentration of the commodity might be less than its average crustal abundance. (USGS, 1986)

mineralogical guide A mineral that is present near an orebody and is related to the processes of ore deposition. Guides help locate ore and may constitute targets for ore search.

mineralogical phase rule Any of several modifications of the fundamental Gibbs phase rule, taking into account the number of degrees of freedom consumed by the fixing of physical-chemical variables in the natural environment. The most famous such rule, that of Goldschmidt, assumes that two variables (taken as pressure and temperature) are fixed externally and that consequently the number of phases (minerals) in a system (rock) will not generally exceed the number of components. The Korzhinskii-Thompson version takes into account the external imposition of chemical potentials of perfectly mobile components, and thereby reduces the maximum expectable number of minerals in a given rock to the number of inert components. Syn: *Goldschmidt's phase rule.* (AGI, 1987)

mineralogist Person who studies the formation, properties, use, occurrence, composition, and classification of minerals; a geologist specialized in mineralogy. Syn: *oryctologist (obsolete).*

mineralography *ore microscopy.*

mineralogy The study of minerals: formation, occurrence, use, properties, composition, and classification. Adj. mineralogic, mineralogical.

mineraloid Minerallike constituent of rocks which is not definite enough in chemical composition or in physical properties to be considered a mineral. Hydrocarbons, volcanic glass, and palagonite are classed as mineraloids. (Hess)

mineral paint *mineral pigment.*

mineral parting Mineral breakage along specific crystallographic planes in some specimens due to twinning, exsolution lamellae, or chemical alteration. See also: *parting.* Cf: *mineral cleavage.*

mineral pigment A mineral material used to give color, opacity, or body to a paint, stucco, plaster, or similar material. See also: *ocher; sienna; umber.*

mineral processing The dry and wet crushing and grinding of ore or other mineral-bearing products for the purpose of raising concentrate grade; removal of waste and unwanted or deleterious substances from an otherwise useful product; separation into distinct species of mixed minerals; chemical attack and dissolution of selected values. Among the methods used are hand sorting (including radioactivation and fluorescence); dense media separation; screening and classification; gravity treatment with jigs, shaking tables, Humphries spirals, Frue vanners, or sluices; magnetic separation at low or high intensity; leach treatment, perhaps using pressure and heat; and (universally) froth flotation. Also called beneficiation; ore dressing; mineral dressing.

mineral province A region in which the source, age, and regional distribution of a complex of minerals in a sediment are related. (Schieferdecker, 1959)

mineral reserves *reserves.*

mineral resin Any of a group of resinous, usually fossilized, mineral hydrocarbon deposits; e.g., bitumen and asphalt. See also: *resin.* (AGI, 1987)

mineral right The ownership of the minerals under a given surface, with the right to enter thereon, mine, and remove them. It may be separated from the surface ownership, but, if not so separated by distinct conveyance, the latter includes it.

mineral rubber *uintaite.*

mineral salt Mined rock salt, halite.

mineral sequence The sequential order of mineral deposition during formation of an ore deposit. Cf: *paragenesis.*

mineral series Mineral species showing continuous variation in their properties with change in composition. A series may be complete; e.g., tennantite-tetrahedrite, or partial; e.g., iron replacing zinc in sphalerite. Syn: *crystal solution; solid solution.* See also: *mineral.*

mineral species (a) Any mineral that can be distinguished from all other minerals by current determinative methods. (Stokes, 1955) (b) A naturally occurring homogeneous substance of inorganic origin, in chemical composition either definite or ranging between certain limits, and possessing characteristic physical properties and usually a crystalline structure. See also: *mineral.*

mineral spring A spring whose water contains enough mineral matter to give it a definite taste, in comparison to ordinary drinking water, esp. if the taste is unpleasant or if the water is regarded as having therapeutic value. This type of spring is often described in terms of its principal characteristic constituent; e.g., salt spring. (AGI, 1987)

minerals separation process A flotation process based on surface-tension phenomena, accelerated by means of addition to the pulp of small quantities of oil and air in minute subdivision. Only about 0.1% oil is added, and the pulp is violently agitated for 1 to 10 min. Innumerable small bubbles of air are thus mechanically introduced, which join the oil-coated particles. These are then removed in a spitzkasten. Exposure to the air after this treatment then aerates any mineral that has not already taken up its oil film, after which a second spitzkasten treatment removes this. An early name for froth flotation. (Liddell, 1918; Pryor, 1963)

mineral stabilizer A fine, water-insoluble, inorganic material, used in admixture with solid or semisolid bituminous materials. (ASTM, 1994)

mineral streaking In metamorphic rocks, lineation of grains of a mineral. Syn: *streaking.* (AGI, 1987)

mineral tallow *hatchettine; hatchettite.* Also called mountain tallow.

mineral tar (a) Tar derived from various bituminous minerals, such as coal, shale, peat, etc. Shale tar. Syn: *mountain tar.* (Standard, 1964) (b) *maltha; pittasphalt.*

mineral turquoise Term to distinguish true turquoise from odonotolite (bone turquoise).

mineral variety Specimens of a mineral species with distinctive physical properties due to: (1) specific history; e.g., Iceland spar, a coarsely crystalline variety of calcite of optical grade, or (2) small chemical variation; e.g., amalgam, a mercurian variety of native silver.

mineral vein *vein.*

mineral water Water that contains an unusually high percentage of some mineral substance that gives the water a distinctive taste and sometimes other properties. Considered to be beneficial in the treatment of various ailments. Also called spa water. (Cooper, 1963)

mineral wax *ozocerite.*

mineral wedging A form of chemical weathering resulting in the formation of new minerals that have greater aggregate volumes than the old ones. These expanding minerals then act as wedges to split adjacent minerals and rocks apart. (AGI, 1987)

mineral wool A substance outwardly resembling wool, presenting a mass of fine interlaced filaments. It is made by subjecting furnace slag or certain rocks, while molten, to a strong blast. Being both insect proof and fireproof, it forms a desirable packing for walls, a covering for steam boilers, etc. Also called mineral cotton, silicate cotton, and slag wool. Syn: *glass wool.* (Standard, 1964; Fay, 1920)

mineral zoning *zoning of ore deposits.*

mine refuse (a) Waste material in the raw coal that has been removed in a cleaning or preparation plant. (b) Notably used to describe colliery rejects; also called tailings. (Pryor, 1963)

mine rescue apparatus The rescuing of miners overcome by a mine fire, or trapped in workings by an explosion, necessitates the use of apparatus that will enable the rescue team to work in irrespirable or poisonous gases. The equipment used is known as mine rescue apparatus. See also: *rescue apparatus.* (Nelson, 1965)

mine rescue car One of a number of railway cars specially equipped with mine rescue apparatus, safety lamps, first-aid supplies, and other materials, maintained by the Mine Safety and Health Administration in various sections of the United States. These cars serve as movable stations for the training of miners in the use of mine rescue apparatus, and in first aid to the injured; as centers for the promotion of mine safety; as emergency stations for assisting at mine fires, explosions, or other disasters. Similar cars are maintained by a number of mining companies. Syn: *mine-safety car.*

mine rescue crew A crew consisting usually of five people who are thoroughly trained in the use of mine rescue apparatus, and are capable of wearing it in rescue or recovery work in a mine following an explosion, or to combat a mine fire. (Fay, 1920)

mine rescue lamp A name given to a particular type of electric safety hand lamp used in rescue operations. It is equipped with a lens for concentrating or diffusing the light beam as occasion may require. (Fay, 1920)

mine resistance (a) The resistance offered by a mine to the passage of an air current. The mine resistance is due to the friction of the air rubbing along the sides, top, and bottom of the air passages. To overcome this friction, the total ventilating pressure must be applied against the airway and this pressure must be equal to the mine resistance. Mine resistance is caused by the dragging of the air against the mine surfaces and other obstructions. The rougher the mine surfaces and the more the obstructions, the greater the resistance to the flow of air. (Kentucky, 1952) (b) Includes any natural ventilation effect present and is calculated from air volume and total pressure. The standard practice in the United Kingdom is to express the resistance of a mine in square feet of equivalent orifice. (Roberts, 1960)

mine road Any mine track used for general haulage. (Fay, 1920)

mine roadway area measurement *tape-triangulation method.*

mine rock A more or less altered rock found in ore channels. Gangue.

minerocoenology The study of mineral associations in the broadest sense, such as the correlation of igneous rocks or magmatic provinces with their ore deposits. (AGI, 1987)

minerogenesis The origin and growth of minerals. (Challinor, 1964)

minerogenetic epoch *metallogenetic epoch.*

minerogenetic province (a) An area in which mineralization has been active at one or more periods. If the mineralization has been chiefly metalliferous, the term metallogenetic is applicable. Syn: *metallogenetic province.* (b) A region characterized by relatively abundant mineralization dominantly of one type. (AGI, 1987)

miners The row of drill holes in a tunnel face, located below the breaking-down holes. (Stauffer, 1906)

miner's bar An iron bar pointed at one end, chisel-edged at the other, used in coal mining. (Standard, 1964)

miner's box A wood or iron box located in or near the working place of a miner in which tools, supplies, etc. are kept. Required by law in some States. (Fay, 1920)

miner's dial An underground surveying instrument for measuring and setting out angles and determining magnetic north. Syn: *mine dial.* (BS, 1963)

miner's dip needle A portable form of dip needle used for indicating the presence of magnetic ores. Also called dipping compass. (CTD, 1958)

miner's electric cap lamp (a) A lamp for mounting on a miner's cap and receiving electric energy through a cord which connects the lamp with a small battery. (b) An electric lamp designed for fixing to a miner's helmet. Its principal parts are: (1) the battery, either lead acid or alkaline; (2) the headpiece, of plastic or aluminum alloy, with switch; and (3) a length of twin-cord cable covered with tough rubber or with neoprene—a fire- and acid-resisting substitute. The lead-acid lamps commonly use either a 4-V, 1.0-A bulb with a light output of about 47.5 lumens or else a 4-V, 0.8-A bulb of 38 lumens output. The headpiece is equipped with an auxiliary, 4 V, 0.46-A bulb. (Nelson, 1965)

miner's hammer A hammer for breaking ore. (Standard, 1964)

miner's hand lamp A self-contained mine lamp with a handle for convenience in carrying.

miner's hard cap Cap made of rigid, strong materials such as vulcanized fiber, glass fiber or plastic, which protects a worker from injury caused by falling objects, large chips, or by striking the head against projecting materials. The cap has a cradle to cushion the shock of blows and a sweat band to absorb perspiration; it is water resistant and nonconductive. A front visor shields the face and eyes from overhead glare, and makes the cap suitable for wear in close, confined spaces where a full brim might interfere. (Best, 1966)

miner's helmet A hat designed for miners to provide head protection and for holding the cap lamp. See also: *miner's hard cap*. (Nelson, 1965)

miner's horn A horn or metal spoon used to collect ore particles in gold washing. (Standard, 1964)

miner's inch (a) The rate of flow of water through an aperture 1 in (2.54 cm) square under a given pressure, generally taken to be that of water standing 6 in (15.2 cm) above the top of the aperture. It is not a universal value but is fixed by statute in several States. A commonly accepted rate is 90 ft^3/h (2.5 m^3/h), or 1-1/2 ft^3/min (0.042 m^3/min). Cf: *sluice head*. (b) A unit used in California, around 1900, for measuring water flow in hydraulicking. It represented the outflow from a 1-in^2 (6.5-cm^2) opening in the side of a box. It varied from 2,000 to 2,600 ft^3/per day (56.6 to 73.6 m^3/d), according to the height of water, etc. (Nelson, 1965) (c) The term is not definite without specification of the head or pressure. It has no fixed meaning and in one locality sometimes is a very different quantity according to miner's measurement in another locality. It has been defined as the amount of water that will pass in 24 hours through an opening 1 in (2.54 cm) square under a head of 6 in. (Ricketts, 1943)

miner's inch day Flow of 1 miner's inch for 24 hours. (Mining)

miner's lamp (a) In nongassy mines, acetylene lamps and various electric lamps; in gassy mines, approved flame safety lamps, electric hand or cap lamps. (Nelson, 1965) (b) Any one of a variety of lamps used by a miner to furnish light; e.g., oil lamps, carbide lamps, flame safety lamps, electric cap lamps, etc. (Fay, 1920)

miner's lung *pneumoconiosis.*

miner's needle A long, slender, tapering, metal rod left in a hole when tamping and afterwards withdrawn, to provide a passage, to the blasting charge, for the squib. (Fay, 1920)

miners' nystagmus An occupational disease that occurs among coal miners, usually those of middle age or elderly, who have worked for a period of 25 to 30 years underground. Its physical symptoms consist of difficulty of seeing in the dark or in poor light, excessive sensitivity to and intolerance of glare, and a rhythmic oscillation of the eyeballs. As a result of these oscillations, there may be apparent movement of the objects looked at and defective visual acuity. Associated with these ocular symptoms are other general disorders, such as headaches and dizziness, particularly after stooping or bending, and the development of psychoneurotic symptoms is common in the later stages of the disease. If the disease is not checked, the nervous disorders may become so severe as to render the miner totally disabled. (Roberts, 1958)

miners' oil An oil, producing little smoke, used in miners' wick-fed open lamps. (Fay, 1920)

miner's pan *pan.*

miner's pick *pick.*

miners' rescue party A team of trained mine rescue workers, from five to eight strong; they operate after explosions, and during and after mine fires. See also: *rescue team*. (Nelson, 1965)

miner's right (a) An annual permit from the Government to occupy and work mineral land. (b) In California, the right of a miner to dig for precious metals on public lands, occupied by another for agricultural purposes.

miners' rules Rules and regulations proclaimed by the miners of any district relating to the location, recording, and the work necessary to hold possession of a mining claim. It was the miners' rules of the early days of the mining industry that were the basis of the present laws. The local mining laws and regulations of 1849 and later are given in vol. 14, 10th Census of the United States, 1880, compiled by Clarence King. (Fay, 1920)

miner's self-rescuer A small form of breathing apparatus for protection against carbon monoxide, worn on a miner's belt. It consists of a canister with a mouthpiece directly attached to it. The wearer breathes through the mouth, the nose being closed by a clip. The canister contains a layer of fused calcium chloride to absorb the water vapor in the air which destroys the efficiency of the other chemical called hopcalite. The self-rescuer affords protection for 30 min, so that miners surviving an explosion may walk out through a mine atmosphere that contains sufficient oxygen but also a fatal percentage of carbon monoxide.

miner's wedge A metallic wedge or plug for splitting off masses of coal. (Standard, 1964)

miner's weight The term used in a coal mining lease as the basis for the price per ton to be paid for mining. It is not a fixed, unvarying quantity of mine-run material, but is such a quantity of material as operators and miners may, from time to time, agree as being necessary or sufficent to produce a ton of prepared coal. (Ricketts, 1943)

mine run (a) The entire unscreened output of a mine. Also called run-of-mine. (Zern, 1928) (b) The product of the mines before being sized and cleaned. (Hudson, 1932) (c) A product of common or average grade. (Webster 3rd, 1966)

mine-run coal Ungraded coal of mixed sizes as it comes from the mine. (Hess)

mine-run mica *book mica.*

mine-safety car *mine rescue car.*

mine signal system Designed esp. for use in mines, these signal lights at individual switches immediately indicate to the motorman whether or not it is safe to proceed. Green and amber lights work automatically with the movement of the switches. May be used with locally controlled switches, or with those operated by a central dispatcher. (Best, 1966)

mine skip Skip used to bring mined material to the surface of a mine shaft; manufactured in various sizes and designs for both vertical and incline shafts, including tipover models and bottom door dump models. (Best, 1966)

mine static head The energy consumed in the ventilation system to overcome all flow head losses. It includes all the decreases in total head (supplied from static head) that occur between the entrance and discharge of the system. (Hartman, 1982)

mine superintendent A mine manager or group manager. (Nelson, 1965)

mine surveyor The official at a mine who periodically surveys the mine workings and prepares plans for the manager. Formerly, the mine surveyor carried out many of the duties now performed by the planning department. *surveyor; mine.* (Nelson, 1965)

mine tin Tin obtained from veins or lodes, as distinguished from stream tin.

mine tons Gross tonnage of ore including waste and unpayable material. (Beerman, 1963)

mine total head The sum of all energy losses in the ventilation system. Numerically, it is the total of the mine static and velocity heads. See also: *total ventilating pressure; ventilating pressure*. (Hartman, 1982)

mine track device One of a variety of track devices to provide maximum safety for haulage trains in mines. Designed to be used in conjunction with switch signals, these devices include electric switch throwers operated by hand contractors on a copper plate, overhead hand controllers, remote control, or trolley contractors. Other safety equipment includes mechanical switches for gaseous or hot mines, derailing switches for trains out of control, and automatic mine-door opening devices. (Best, 1966)

mine tractor A trackless, self-propelled vehicle used to transport equipment and supplies and for general service work. See also: *tractor*.

mine valuation Properly weighing the financial considerations to place a present value on mineral reserves. (Nelson, 1965)

mine velocity head The velocity head at the discharge of the ventilation system. Throughout the system, the velocity head changes with each change in duct area or number and is a function only of the velocity of airflow. It is not a head loss. The velocity head for the system must technically be counted a loss, because the kinetic energy of the air is discharged to the atmosphere and wasted. Therefore, it must be considered a loss to the system in determining overall energy loss. (Hartman, 1982)

mine ventilating fan A motor-driven disk, propeller, or wheel for blowing (or exhausting) air to provide ventilation of a mine. See also: *ventilation*. Syn: *ventilating fan*. (Nelson, 1965)

mine ventilation auxiliary fan A small fan installed underground for ventilating coal faces or hard rock headings that are not adequately ventilated by the air current produced by the mine-ventilation fan. An auxiliary fan is usually from 0.5 to 1.0 m in diameter. It is driven by compressed air or electricity. The auxiliary fan can be used to force or exhaust ventilate the workplace.

mine-ventilation fan A machine possessing rotating air-moving blades to exhaust or push the air volume necessary to ventilate mine workings. See also: *aerofoil-vane fan; axial-flow fan*.

mine-ventilation fan characteristics The behavior of a fan under various conditions cannot be expressed in simple mathematics but may be shown graphically by suitable curves, known as the fan's characteristic curves or characteristics. The curves of interest are generally head versus air quantity, power versus air quantity, and efficiency versus air quantity. (Hartman, 1982)

mine ventilation system An arrangement of connecting airways in a mine together with the pressure sources and control devices that produce and govern airflow. (Hartman, 1982)

mine water Water pumped from mines usually contains impurities, some of which are in suspension, but the majority, which are soluble, cause the water to be hard. The water often contains corrosive agents, such as acids or alkalis. (Cooper, 1963)

mine wireman *wireman.*

mineworks Ancient subterranean passages or mine excavations. (Standard, 1964)

mingled ground Mixed clay and sand or rock. (Arkell, 1953)

minguzzite A monoclinic mineral, $K_3Fe(C_2O_4)_3{\cdot}3H_2O$; green; an oxalate.

miniature current meter A device used to measure the passage of current past a probe on each blade of a propeller type, or each cup of a price-type meter by detecting the change of electrical resistance between that probe and a distant electrode. (Hunt, 1965)

minimum deviation A method for measuring the refractive index of a prism or a liquid in a hollow prism by determining the minimum deflected angle of a light beam.

minimum firing current As applied to electric blasting caps, the limit below which firing will not occur. (Fraenkel, 1953)

minimum ignition energy The minimum ignition energy required for the ignition of a particular flammable mixture at a specified temperature and pressure. (Van Dolah, 1963)

minimum oxygen content The U.S. Mine Safety and Health Administration and other recognized safety and health agencies recommend 19.5% as the minimum oxygen content allowable. (Hartman, 1961)

minimum product firing temperature The lowest product temperature at which an explosive or explosive unit is approved for use. (CFR, 4)

mining The science, technique, and business of mineral discovery and exploitation. Strictly, the word connotes underground work directed to severance and treatment of ore or associated rock. Practically, it includes opencast work, quarrying, alluvial dredging, and combined operations, including surface and underground attack and ore treatment. Cf: *mining geology; mining engineering.* (Pryor, 1963)

mining advancing A method of mining by which the ore or coal is mined as the excavation advances from the shaft or main opening. Cf: *mining retreating.* (Fay, 1920)

mining camp (a) A colony of miners settled temporarily near a mine or a goldfield. (Standard, 1964) (b) A term loosely applied to any mining town. (Fay, 1920)

mining captain Person in charge of mining operations. (Craigie, 1938)

mining case A frame of a shaft, or gallery, composed of four pieces of plank. (Standard, 1964)

mining claim (a) That portion of the public mineral lands that a miner, for mining purposes, takes hold of and possesses in accordance with mining laws. (b) A mining claim is a parcel of land containing valuable minerals in the soil or rock. A location is the act of appropriating such a parcel of land according to law or to certain established rules. See also: *claim; placer claim; location.* (c) In the General Mining Law of 1872, that portion of a vein or lode and of the adjoining surface, or of the surface and subjacent material to which a claimant has acquired the right of possession by virtue of a compliance with such statute and the local laws and rules of the district within which the location may be situated. Independent of acts of the U.S. Congress providing a mode for the acquisition of title to the mineral lands of the United States, the term has always been applied to a portion of such lands to which the right of exclusive possession and enjoyment by a private person or persons, has been asserted by actual occupation, and compliance with the mining laws and regulations. Syn: *holding.* (Ricketts, 1943) (d) Distinction between mining claim and location is that they are not always synonymous and may often mean different things; a mining claim may refer to a parcel of land containing valuable mineral in its soil or rock, while location is the act of appropriating such land according to certain established rules. A mining claim may include as many adjoining locations as the locator may make or purchase, and the ground covered by all, though constituting what is claimed for mining purposes, will constitute a mining claim and will be so designated. (Ricketts, 1943) (e) Title issued by the Government concerned to an individual or group, which grants that individual or group the right to exploit mineral wealth in a specified area by approved methods in accordance with the ruling laws and regulations. (Pryor, 1963) (f) A claim on mineral lands. (AGI, 1987)

mining compass An instrument giving qualitative indications of anomalies in the magnetic field. (Schieferdecker, 1959)

mining dial *dial.*

mining disaster An accident in a mine in which a large number of people are killed. See also: *major mine disaster.* (SME, 1992)

mining disease See also: *anthracosis; infective jaundice; nystagmus; pneumoconiosis; silicosis; simple silicosis.*

mining district A section of country usually designated by name, having described or understood boundaries within which minerals are found and worked under rules and regulations prescribed by the miners therein. There is no limit to its territorial extent and its boundaries may be changed if vested rights are not thereby interfered with. (Ricketts, 1943)

mining ditch A ditch for conducting water used in mining. Cf: *mining sluice.* (Craigie, 1938)

mining engineer (a) A person qualified by education, training, and experience in mining engineering. (Nelson, 1965) (b) If qualified and of standing in the profession, a trained engineer with knowledge of the science, economics, and arts of mineral location, extraction, concentration and sale, and the administrative and financial problems of practical importance in connection with the profitable conduct of mining. Usually a specialist in one or more branches of work. Activities may include prospecting, surveying, sampling and valuation, technical underground management, milling, assaying, ventilation control, layout of workings and plant, geological examination, and company administration. (Pryor, 1963) (c) One versed in, or one who follows, as a calling or profession, the business of mining engineering. Graduates of technical mining schools are given the degree of engineer of mines and authority to sign the letters E.M. after their names.

mining engineering The planning and design of mines, taking into account economic, technical, and geologic factors; also supervision of the extraction, and sometimes the preliminary refinement, of the raw material. Cf: *mining; mining geology.* (AGI, 1987)

mining explosive One of the high explosives used for mining and quarrying. They can be divided into four main classes: gelatins; semigelatins; nitroglycerin powders, and non-nitroglycerin explosives, including water gels, emulsions, and ANFO.

mining geology (a) The study of geologic structures and particularly the modes of formation and occurrence of mineral deposits and their discovery. Syn: *ore geology.* (Nelson, 1965) (b) In coal mining, the study of: rock formations, particularly with reference to the Carboniferous System; the mode of formation of coal seams, their discovery and correlation. See also: *geology.* (Nelson, 1965) (c) The study of the geologic aspects of mineral deposits, with particular regard to problems associated with mining. Cf: *mining; mining engineering.* (AGI, 1987)

mining ground No land can be a mining claim unless based upon a location; otherwise it may be mining ground or a mine. For instance, the bed of a navigable river is not subject to mining location, but if mining is conducted thereon by dredging, it is mining ground; or, where land is covered by an agricultural patent and worked for its mineral deposits, it is mining ground and not a mining claim. Hence, land from which a mineral substance is obtained from the earth by the process of mining may, with propriety, be called mining ground or mining land. See also: *claim; location.* (Ricketts, 1943)

mining hazard Any of the dangers peculiar to the winning and working of coal and minerals. These include collapse of ground, explosion of released gas, inundation by water, spontaneous combustion, inhalation of dust and poisonous gases, etc. (Nelson, 1965)

mining head The mechanism on a continuous mining machine that breaks down the coal.

mining lease A legal contract for the right to work a mine and extract the mineral or other valuable deposits from it under prescribed conditions of time, price, rental, or royalties. Syn: *mineral lease.* (Webster 3rd, 1966)

mining locomotive A small locomotive for use in underground haulage, sometimes consisting of a car bearing a powerful electric motor, built very low and operated through a trolley. May also be operated by electricity from batteries. (Standard, 1964)

mining machine A coal-cutting machine. (Standard, 1964)

mining machine operator *machine miner.*

mining-machine-operator helper *machine helper.*

mining machine truck A truck used for transporting shortwall mining machines. Track-mounted trucks are necessarily limited in use to sections employing track. Crawler-type trucks are capable of transporting the cutting machine without need of track and without benefit of ropes. (Jones, 1949)

mining method Any of the systems employed in the exploitation of coal seams and orebodies. The method adopted depends on a large number of factors, mainly, the quality, shape, size, and depth of the deposit; accessibility and capital available. See also: *coal mining methods; metal mining; stoping methods.* (Nelson, 1965)

mining ore from top down *top slicing and cover caving.*

mining property Property, esp. land, valued for its mining possibilities. (Craigie, 1938)

mining purposes The term "mining purposes" as used in connection with mill-site locations, is very comprehensive, and may include any reasonable use for mining purposes that the quartz lode mining claim may require for its proper working and development. This may be very little, or it may be a great deal. The locator of a quartz lode mining claim is required to do only $100 worth of work each year until obtaining a patent. (Ricketts, 1943)

mining recorder In a mining camp, a person selected to keep a record of all mining claims and properties. (Mathews, 1951)

mining retreating A process of mining by which the ore, or coal, is untouched until after all the gangways, etc., are driven, when the work of extraction begins at the boundary and progresses toward the shaft. Cf: *mining advancing.* (Fay, 1920)

mining right Upon a specific piece of ground, a right to enter upon and occupy the ground for the purpose of working it, either by underground excavations or open workings, to obtain from it the mineral ores which may be deposited therein. (Ricketts, 1943)

mining shield A cover or canopy developed by the U.S. Bureau of Mines for the protection of mine workers and machines at the face of a mechanized coal heading. Hydraulic rams telescope and steer the shield forward as the face advances. It enables continuous miners to operate with greater safety. (Nelson, 1965)

mining sluice An artificial channel or passage for water used in mining. Cf: *mining ditch*. (Craigie, 1938)

mining system Work, as it is commenced on the ground, is such that, if continued, will lead to a discovery and development of the veins or orebodies that are supposed to be in the claim, or, if these are known, that the work will facilitate the extraction of the ores and minerals.

mining theodolite A theodolite having particular features of design that make it suitable as an underground surveying instrument; e.g., incorporating an arrangement for the centering movement to be above the foot screws. (BS, 1963)

mining title A claim, exclusive prospecting license, concession, right, or lease. A grant under laws and mining regulations to a person or group of approved persons of the right to develop and exploit a properly delineated area for its mineral wealth. (Pryor, 1963)

mining-type visibility meter An instrument to facilitate observation of the essential elements of visual tasks in coal mines. It is a brightness meter in which the comparison field is illuminated by a cap lamp headpiece attached outside the instrument. No internal electrical circuit exists other than that which connects the photocell to the microammeter and the meter can therefore be used anywhere in a safety-lamp mine without restrictions. (Roberts, 1958)

mining under The act of digging under coal or in a soft strata in coal seams.

mining width The minimum width necessary for the extraction of ore regardless of the actual width of ore-bearing rock. See also: *stoping width*. (AGI, 1987)

minion The siftings of iron ore after calcination. (Standard, 1964)

minium A tetragonal mineral, Pb_3O_4; red; an alteration product of galena or cerussite. Syn: *red lead*.

minnesotaite A monoclinic mineral, $(Fe,Mg)_3Si_4O_{10}(OH)_2$; talc-pyrophyllite group; in the banded iron formations of Minnesota. Syn: *iron talc*.

minor element (a) Trace element or accessory element. See also: *trace element*. (b) Less commonly, any of the elements present in the range of 0.1% to 1%, between major elements in concentrations greater than 1% and trace elements in concentrations less than 0.1%, or occasionally less than 0.01%.

minus sieve In powder metallurgy, the portion of a powder sample that passes through a standard sieve of specified number. Contrast with plus sieve. (ASM, 1961)

minus sight *foresight*.

minus station Stakes or points on the far side of the zero point from which a job was originally laid out. (Nichols, 1976)

minute of arc A unit of angular measure equal to the 60th part of a degree and containing 60 s of arc. (Webster 3rd, 1966)

minyulite An orthorhombic mineral, $KAl_2(PO_4)_2(OH,F)\cdot 4H_2O$; in phosphate rock in Western Australia and South Australia.

Miocene An epoch of the later Tertiary period, after the Oligocene and before the Pliocene; also, the corresponding worldwide series of rocks. It is considered to be a period when the Tertiary is designated as an era. (AGI, 1987)

mirabilite A monoclinic mineral, $Na_2SO_4\cdot 10H_2O$; one perfect cleavage; tastes cool, then saline and bitter; in evaporite deposits, where it is mined as Glauber salt, and as an efflorescence. Syn: *Glauber salt*. (Not glauberite.)

mire Mud.

mirror plane *plane of symmetry*.

mirror stone *muscovite*.

mischio marble A violet-red breccia from Serravezza, Italy. Also known as African breccia (breche Africaine).

mischmetal A natural mixture of the rare-earth metals cerium, lanthanum, and didymium; e.g., the waste matter from monazite sand after the extraction of thoria may contain large quantities of ceria, lanthana, didymia, yttria, and other substances. This is reduced to the metallic state by converting the oxides to chlorides, and then recovering the metal by electrolysis. The material obtained is an alloy containing about 50% cerium and 45% lanthanum and didymium. (Henderson, 1953)

misenite A monoclinic mineral, $K_2SO_4\cdot 6KHSO_4$(?); in silky white fibers at Cape Miseno, Italy. (Not mizzonite.)

miser A tubular well-boring bit, having a valve at the bottom, and a screw for forcing the earth upward. Also spelled mizer. (Standard, 1964; Fay, 1920)

misfire An explosive charge in a drill hole that has partly or completely failed to explode as planned. Causes include unskilled charging; defective explosive, detonator, or fuse; broken electric circuit or—most dangerous—cutting off of part or all of the charge through lateral rock movement as other holes in the vicinity are fired. Stringent safety precautions cover procedure in minimizing these risks and in dealing with known or suspected misfires. A smoldering fuse may delay explosion, causing a hangfire, so return to workings after a suspected failure is necessary. Another main cause of accident is drilling into or dangerously near a socket—an apparently empty drill hole. See also: *missed hole; hangfire*. (Pryor, 1963)

misfire hole *missed hole*.

mispickel *arsenopyrite*.

misplaced material (a) In mineral processing, material, particularly screen products and tailings, that has been reported in the wrong section. (Pryor, 1963) (b) Material wrongly included in the products of a sizing or density separation; i.e., material that has been included in the lower size or specific gravity product but which itself has a size or specific gravity above that of the cut point, or vice versa. Its weight may be expressed as a percentage of the product or of the feed. Also called tramps (undesirable usage). (BS, 1962) (c) In sizing and screening, undersize contained in the overflow, or oversize contained in the underflow. (BS, 1962) (d) In cleaning, material of specific gravity lower than the separation density that has been included in the high density product, or material of specific gravity higher than the separation density that has been included in the low density product. (BS, 1962)

missed hole A drill hole charged with explosives, in which all or part of the explosive has failed to detonate. Syn: *misfire; misfire hole*. (Fraenkel, 1953)

missed round A round in which all or part of the explosive has failed to detonate. (Fraenkel, 1953)

Mississippian A period of the Paleozoic era (after the Devonian and before the Pennsylvanian), thought to have covered the span of time between 345 and 320 million years ago; also, the corresponding system of rocks. It is named after the Mississippi River valley, in which there are good exposures of rocks of this age. It is the approximate equivalent of the Lower Carboniferous of European usage. (AGI, 1987)

misy An old term for copiapite and related minerals.

Mitchell system for underhand quarrying of panel cores *underhand stoping*.

miter gear *bevel gear*.

mitridaite A monoclinic mineral, $Ca_2Fe_3(PO_4)_3O_2\cdot 3H_2O$; forms red crystals, green stains in granite pegmatites.

M.I.T. sampler A single-tube, drive-type, soil-sampling barrel esp. adapted to sampling deposits of plastic clay where a minimum 5-in (12.7-cm) diameter sample is required. A loop or snare of piano wire is inserted in a groove inside the cutting shoe with the free end of the wire extending through a slot on the side of the sampler to the surface. When pulled, the wire cuts the sample off at the bottom of the cutting shoe. (Long, 1960)

mitscherlichite A tetragonal mineral, $K_2CuCl_4\cdot 2H_2O$; greenish blue.

mix-crystal *solid solution*.

mixed Drill diamonds ranging from 23 to 80 per carat in size. (Long, 1960)

mixed blast process A modification of the basic Bessemer process in which all the nitrogen is removed from the blast, it being made up of a mixture of oxygen and carbon dioxide or oxygen and superheated steam. The oxygen and superheated steam blast is claimed to be the more efficient, the final nitrogen content of the metal being brought to a mean level of 0.0028%. (Osborne, 1956)

mixed cements A product obtained by mixing, or blending, either portland, natural, or pozzolana cement with one another or with other inert substances. (Zern, 1928)

mixed crystal *solid solution*.

mixed dust Dust prepared for testing in a mine by mixing coal dust and inert dust in predetermined proportions. The mixture may also contain water, and different sizes of coal dust may be mixed to produce some desired intermediate size. (Rice, 1913-18)

mixed explosion Occurs when both combustible gases and coal dust are present below their lower limits, but in combination produce sufficient heat of combustion to propagate an explosion. (Sinclair, 1958)

mixed explosive Explosive consisting of an intimate mechanical mixture of substances which consume and generate oxygen but are not in themselves explosive. Inorganic nitrates, chlorates, and perchlorates. Most important is ammonium nitrate. (Fraenkel, 1953)

mixed face In tunneling, digging in dirt and rock in the same heading at the same time. (Nichols, 1976)

mixed-feed kiln Upright lime kiln in which the fuel (coal) is mixed and burned with the limestone charge. (Mersereau, 1947)

mixed-flow fan A mine fan in which the flow is both radial and axial. The Schicht fan is of this type and has the advantage that it can produce a high water gage with a single stage. This fan, however, is not well suited to mines where a large change in equivalent orifice may occur. Cf: *axial-flow fan; radial-flow fan*. (Nelson, 1965)

mixed-flow turbine An inward flow, reaction-type water turbine, in which the runner vanes are so shaped that they are acted on by the water pressure both axially and radially. (Hammond, 1965)

mixed-layer mica (a) *clay mineral.* (b) Layered silicates, generally of the smectite group, in which different layered mineral species are stacked in an ordered or random fashion.

mixed-layer mineral A mineral whose structure consists of alternating layers of clay minerals and/or mica minerals; e.g. chlorite, made up of alternating biotite and brucite sheets. (AGI, 1987)

mixed ore Ore containing both oxidized and unoxidized minerals. See also: *oxidized ore; sulfide ore.* (Nelson, 1965)

mixer (a) An apparatus used to thoroughly mix drilling fluid water with drilling ingredients. May be cement, mud, or lost circulation materials. Also called atomizer; mud mixer. (Long, 1960) (b) *agitator.*

mixer cone A funnel-shaped hopper attached to the body of a mud mixer into and by means of which the dry, powdered, drill-mud and/or cement ingredients are fed into the mud mixer. (Long, 1960)

mix-house person One who mixes sintered lead or zinc ore with such materials as pulverized coal and coke, salt, skimmings, water, and chemical solutions preparatory to smelting. Also called mixer operator. (DOT, 1949)

mixing (a) In powder metallurgy, the thorough intermingling of powders of two or more different materials (not blending). (ASM, 1961) (b) An instrumentation technique used in seismograph recording in which a certain portion of the energy from each amplifier channel is fed to the adjacent channels, giving results somewhat analogous to those obtained from the use of multiple geophones to attenuate noise. (AGI, 1987)

mixing pit A pit in which drill mud is mixed and stored until the mud is cured and needed for use as a drill circulation fluid. (Long, 1960)

mix-in-place A common soil stabilization method in which the soil on the site is first pulverized, then mixed with an admixture or stabilizing agent, compacted and, if necessary, surfaced. All the work is carried out on the site. Cf: *plant mix.* (Nelson, 1965)

mixite (a) A hexagonal mineral, $BiCu_6(AsO_4)_3(OH)_6{\cdot}3H_2O$. Cf: *chlorotile.* (b) The mineral group agardite, goudeyite, mixite, and petersite.

mix-metal Cerium alloy (55% to 65%) containing rare-earth metals and iron. (Bennett, 1962)

mixture A commingling in which the ingredients retain their individual properties or separate chemical nature. See also: *mechanical mixture.* (Standard, 1964)

mizzonite One of the series of minerals forming the scapolite groups, consisting of a mixture of the meionite and marialite molecules. It includes those minerals with 54% to 57% silica, and occurs in clear crystals in the ejected masses on Mount Somma, Vesuvius, Italy. Also called dipyrite. See also: *dipyre.* (CMD, 1948)

MM diamond Synthetic manufactured diamond. (Long, 1960)

moat (a) A ditch or deep trench. (b) To surround with a ditch.

moating A backing of clay, as for the masonry lining of a shaft. (Standard, 1964)

mobile belt An elongated zone of the Earth's crust subjected to relatively great structural deformation. Cf: *geosyncline.* (Schieferdecker, 1959)

mobile conveyor A hand-loaded, hoist-operated hauler used principally in certain central Pennsylvania coal mines. It is essentially a chain-and-flight conveyor of exceptional width, with high sideboards, mounted on wheels for operation on mine track, with motive power supplied by a hoist mounted on the chassis of the machine. The conveyor chain and flights can be moved in the bed of the machine to assist in loading or unloading. The hoist rope can be attached to props at the face and near the dumping point to provide anchorage when the conveyor is moved. (Jones, 1949)

mobile crane A crane driven by gasoline, diesel, or electric motor, traveling on crawler tracks, pneumatic tires or solid rubber tires, and capable of moving in any direction under its own power. (Hammond, 1965)

mobile drill A drill unit mounted on wheels or crawl-type tracks to facilitate moving. (Long, 1960)

mobile equipment Applied to all equipment that is self-propelled or that can be towed on its own wheels, tracks, or skids.See also: *transportable equipment.* (Nelson, 1965)

mobile hoist A platform hoist which is mounted on a pair of pneumatic-tired road wheels, so that it can be towed from one site to another. This type of hoist has been developed for use in house and flat construction. See also: *platform hoist.* (Hammond, 1965)

mobile loader A self-propelling machine capable of lifting material off the bottom and placing it in a mine car, conveyor, or other means of transportation. (Jones, 1949)

Mocha stone Syn: *moss agate.* Also called Mocha pebble. From Mocha, Yemen.

mock lead A Cornish term for zinc blende; also called wild lead. *sphalerite.* (Fay, 1920)

mock ore *sphalerite.*

mock vermilion A basic chromate of lead.

mode (a) The mineral composition of a rock, usually expressed in weight or volume percentages. Adj: modal. Cf: *norm.* (AGI, 1987) (b) The value or group of values that occurs with the greatest frequency in a set of data; the most typical observation. Cf: *mean.* (AGI, 1987)

model (a) A facsimile in three dimensions—a reproduction in miniature of the surface and underground workings of a mine, showing the shafts, tunnels, crosscuts, etc., in all their details. From its very nature, it does not fall within any definition of the word map and it is a misapplication of the term to call it a map, though it may far better serve the purpose in hand. (Ricketts, 1943) (b) A unifying concept that explains or describes a complex phenomenon. See also: *mineral deposit model.* (Barton, 1995)

model analysis The comprehensive testing of scale models of various structures, including harbors and rivers to determine the behavior of the actual structure under consideration. See also: *photoelasticity.* (Hammond, 1965)

modeler (a) Person who shapes plaster of Paris, clay, etc., to form original models used to make molds for producing ceramic ware. (b) Person who builds models used in model analysis.

modeling clay Fine, plastic clay, esp. prepared for artists in modeling by kneading with glycerin, or by other methods. (Fay, 1920)

moderate vitrain A field term that, in accordance with an arbitrary scale established for use in describing banded coal, denotes a frequency of occurrence of vitrain bands comprising from 13% to 30% of the total coal layer. Cf: *sparse vitrain; abundant vitrain; dominant vitrain.* (AGI, 1987)

modified longwall A method used in room-and-pillar mining where the lease requires at least 80% of recovery. Basically it consists of turning the rooms on 70-ft (21.3-m) centers then working the room up 30 ft (9.1 m) wide and butting it off at its completion, then withdrawing the remaining 40 ft (12.2 m) of pillar immediately. (Kentucky, 1952)

modified room-and-pillar working *bord-and-pillar method.*

modified-round nose *medium-round nose.*

modifier In froth flotation, reagent used to control alkalinity and to eliminate harmful effects of colloidal material and soluble salts. See also: *modifying agent.* (Fuerstenau, 1962)

modifying agent In flotation, a chemical that increases the specific attraction between collector agents and particle surfaces, or conversely, that increases the wettability of those surfaces. (Pryor, 1963)

modulating A control adjusting by increments. (Strock, 1948)

modulus of elasticity The ratio of stress to its corresponding strain under given conditions of load, for materials that deform elastically, according to Hooke's law. It is one of the elastic constants. See also: *modulus of rigidity; bulk modulus.* Syn: *elastic modulus; longitudinal velocity.* (AGI, 1987)

modulus of incompressibility The ratio between the pressure in the mass of a soil and the change of volume caused by such pressure. Cf: *Poisson's ratio.* (Hammond, 1965)

modulus of rigidity (a) The rate of change of unit shear stress with respect to unit shear strain, for the condition of pure shear within the proportional limit. For nonisotropic materials such as wood it is necessary to distinguish between the moduli of rigidity in different directions. (Roark, 1954) (b) A modulus of elasticity in shear. Symbol: μ or G. Syn: *shear modulus; rigidity modulus; coefficient of rigidity.* (AGI, 1987)

modulus of rupture (a) Nominal stress at fracture in a bend test or a torsion test. In bending, modulus of rupture is the bending moment at fracture divided by the section modulus. In torsion, modulus of rupture is the torque at fracture divided by the polar-section modulus. (ASM, 1961) (b) The load required to break a piece of material, such as a refractory brick, supported on two spaced and parallel flat bearing edges with the load applied through a third bearing edge placed midspan and on top of the piece. (ARI, 1949)

Moebius process A method of electrolytic refining of silver. Silver plate of 95% to 98% pure forms the anodes, and thin silver plate forms the cathodes. The electrolyte consists of a weak, acidulated solution of silver nitrate. (Fay, 1920)

Moe gage A diamond-weight calculator which estimates to within a few hundredths the weights of brilliant-cut diamonds only, by simple measurements of width and depth of both set or unset diamonds. (Hess)

mofette The exhalation of carbon dioxide in an area of late-stage volcanic activity; also, the small opening from which the gas is emitted. Occurs in Yellowstone National Park. Etymol: French, noxious gas. (AGI, 1987)

moganite Monoclinic silica, SiO_2 (silica-G), with quartz in chert from dry lake beds; also cavity fillings in rhyolitic ignimbrites. Named for Mogán, Canary Islands.

mogensenite Titaniferous magnetite with exsolved ulvöspinel. Cf: *titanomagnetite.*

mohavite A dull white hydrous borate of sodium, $Na_2B_4O_7{\cdot}5H_2O$. Rhombohedral. An alteration film on borax. Locally, octahedral borax (same as tincalconite). From the Mohave Desert, CA. See also: *octahedral borax.* (English, 1938)

Moho Short name for the Mohorovicic discontinuity separating the Earth's crust from the mantle. (Mather, 1964)

mohole The never-completed program to drill through the Earth's crust under the ocean to the Mohorovicic discontinuity in order to provide scientific knowledge of the Earth's mantle. (Hy, 1965)

Mohorovicic discontinuity The boundary surface or sharp seismic-velocity discontinuity that separates the Earth's crust from the subjacent mantle. It marks the level in the Earth at which P-wave velocities change abruptly from 6.7 to 7.2 km/s (in the lower crust) to 7.6 to 8.6 km/s or average 8.1 km/s (at the top of the upper mantle); its depth ranges from about 5 km beneath the ocean floor to about 35 km below the continents, although it may reach 60 km or more under some mountain ranges. The discontinuity probably represents a chemical change from basaltic or simatic materials above to peridotitic or dunitic materials below, rather than a phase change (basalt to eclogite); however, the discontinuity should be defined by seismic velocities alone. It is variously estimated to be between 0.2 and 3 km thick. It is named in honor of its discoverer, Andrija Mohorovicic (1857-1936), Croatian seismologist. Abbrev. Moho. Syn: *M-discontinuity.* (AGI, 1987)

Mohr balance *Westphal balance.*

Mohr circle A graphical representation of the stresses acting on the various planes at a given point. (ASCE, 1958)

Mohr-Coulomb criterion The most popular of numerous rock failure criteria. It assumes that there is a functional relationship between the normal and shear stresses acting on a potential failure surface. (SME, 1992)

Mohr envelope The envelope of a series of Mohr circles representing stress conditions at failure for a given material. According to Mohr's rupture hypotheses, a rupture envelope is the locus of points, the coordinates of which represent the combination of normal and shearing stresses that will cause a given material to fail. Syn: *rupture envelope; rupture line.* (ASCE, 1958)

Mohr's salt A ferrous-ammonium sulfate, $FeSO_4(NH_4)_2SO_4{\cdot}6H_2O$; a light green crystalline salt.

Mohr's theory Mohr's theory of failure utilizes the well-known stress circle and the envelope of a family of circles as criteria for failure of materials subject to biaxial or triaxial stress. Thus Mohr's theory predicts that failure of materials is due to failure in shear, whereas Griffith's theory postulates that it is due to failure at crack tips. (Lewis, 1964)

mohsite A plumboan variety of crichtonite.

Mohs scale Arbitrary quantitative units by means of which the scratch hardness of a mineral is determined. The nonlinear units of hardness are expressed in numbers ranging from 1 through 10, each of which is represented by a mineral that can be made to scratch any other mineral having a lower-ranking number; hence the minerals are ranked from the softest, as follows: talc (1) ranging upward in hardness through gypsum (2), calcite (3), fluorite (4), apatite (5), orthoclase (6), quartz (7), topaz (8), corundum (9), to the hardest, diamond. Cf: *hardness; Knoop hardness.* (Long, 1960)

moil point A solid bar of casehardened steel, pointed at one end, with a shank and upset collar at the other. The moil point, hammered into rock or concrete, produces a small hole that gradually deepens and widens until the sides of the point are in full contact with the rock. The effect is then that of wedging, similar to plug-and-feathering. (Carson, 1961)

moissanite A hexagonal mineral, alpha-SiC (Carborundum); in meteorites. See also: *Carborundum.*

Moissan process A process for the reduction of chromic oxide with carbon in an electric furnace, the hearth of which is lined with a calcium chromite prepared by heating together lime and chromic oxide. (Fay, 1920)

moisture (a) Agreed percentage of water content to be allowed in mineral products that are sold. (b) Essentially water, quantitatively determined by definite prescribed methods which may vary according to the nature of the material.

moisture allowance A deduction from the inital weight of washed coal to allow for the expected loss of water by drainage. (BS, 1960)

moisture content The percentage moisture content equals the weight of moisture divided by the weight of dry soil multiplied by 100. The moisture content of a coal or mineral sample consists of two portions, namely, the free or surface moisture which can be removed by exposure to air, and the inherent moisture which is entrapped in the fuel, and is removed by heating at 200 °F (93.3 °C). Syn: *inherent moisture; water content.* (Nelson, 1965)

moisture-density curve *compaction curve.*

moisture-density test *compaction test.*

moisture equivalent The ratio of the weight of water—which soil, after saturation, will retain against a centrifugal force 1,000 times the force of gravity—to the weight of the soil when dry. The ratio is stated as a percentage. (AGI, 1987)

moisture-holding capacity The quantity of moisture (not removable by mechanical means) contained by a coal in equilibrium with an atmosphere saturated with water vapor. This is employed in some systems of classification as a criterion of rank. (BS, 1960)

moisture man Person who determines the moisture content of ores or concentrates by removing a sample from pile or conveyor, using a metal paddle, and weighing the sample before and after drying. (DOT, 1949)

moisture meter An instrument for determining the percentage of moisture in a substance such as timber or soil, usually by measuring its electrical resistivity. See also: *atomic moisture meter.* (AGI, 1987)

moisture sample A sample taken for the determination of moisture content. (Newton, 1959)

mojavite *tincalconite.*

molasse (a) A partly marine, partly continental sedimentary facies consisting of a thick sequence of fossiliferous conglomerates, sandstones, shales, and marls, characterized by primary sedimentary structures and sometimes by coal and carbonate deposits. It is more clastic and less rhythmic than the preceding flysch facies. (AGI, 1987) (b) An extensive sedimentary formation representing the totality of the molasse facies resulting from the wearing down of elevated mountain ranges during and immediately after the main paroxysmal (diastrophic) phase of an orogeny, and deposited considerably in front of the preceding flysch. Adj: molassic. Cf: *flysch.* (AGI, 1987)

molasses/AN explosive An explosive mixture consisting of about 80 lb (36.5 kg) of ammonium nitrate mixed with 10 pints (4.7 L) of molasses and 5 pints (2.37 L) of water, for quarry and opencast blasting. The molasses and water may be used instead of fuel oil and give a denser mixture with improved fragmentation. (Nelson, 1965)

mold A body of molding sand or other heat-resisting material containing a cavity which when filled with molten metal yields a casting of the desired shape. See also: *die.* Cf: *cast.* (Freeman, 1936)

moldavite (a) A tektite from the Moldau valley, Czech Republic. Syn: *glass meteorite.* (b) A variety of ozocerite from Moldavia. See also: *bouteillenstein.*

moldboard (a) A board on which to ram a pattern. (Standard, 1964) (b) A curved surface of a plow, dozer, or grader blade, or other dirt mover, which gives dirt moving over it a rotary, spiral, or twisting movement. (Nichols, 1954)

molded cameo A cameo produced by casting in a mold material such as ceramics, metals, glass, plastics, or sealing wax. Cf: *pressed cameo.*

moldering An obsolete term for decomposition of organic matter under conditions of insufficient oxygen. (AGI, 1987)

molders' rule A ruler, with measurements sufficiently elongated to compensate for the heat expansion and contraction of a metal. Such rules are used in metal casting to correct dimensions to normal temperatures. (Bennett, 1962)

molding (a) The practice of pouring molten metal into suitable molds. (Hansen, 1937) (b) The pressing of powder to form a compact. (ASTM, 1994)

molding compound A mixture of resins, ingredients, and fillers before processing into the finished product. (Jessop, 1992)

molding frame A template to shape a loam mold. (Standard, 1964)

molding hole An excavation in a foundry floor for large castings. (Standard, 1964)

moldings Derb. Weathered ore at the surface of an outcrop. (Arkell, 1953)

mold plug A truncated-cone-shaped refractory piece which sits in the bottom of an ingot mold. (AISI, 1949)

mole (a) A massive, solid-fill nearshore structure of earth, masonry, or large stone that may serve as either a breakwater or a pier. (Hy, 1965) (b) An egg-shaped device pulled behind the tooth of a subsoil plow to open drainage passages. Also called mole ball. (Nichols, 1976) (c) Weight in grams of a compound in terms of its molecular weight. (Pryor, 1963)

molecular crystal Loosely bound aggregate of stable molecules; e.g., dry ice, solid iodine, sulfur, paraffin, and most of the other crystalline organic solids. (Newton, 1959)

molecular filter sampler Using a porous membrane filter of very small openings, this sampler achieves nearly 100% efficiency at moderate rates of flow. Membranes of three different porosities are used to permit the sampling of dusts of varying concentrations and particle sizes. Flow rate is controlled by the use of calibrated orifices. Syn: *membrane filter.* (Hartman, 1982)

molecular sieve A microporous structure of either crystalline aluminosilicates, such as zeolites, or crystalline aluminophosphates, created by dehydration so that the empty cavities in the structure where water molecules were previously present will accept any material that can penetrate the cavity. The sieving action is a function of the pore size of the structure.

molecule (a) The smallest part of a substance that can exist separately and still retain its composition and characteristic properties. (b) The smallest combination of atoms that will form a given chemical compound.

mole mining A method of working coal seams about 30 in (76.2 cm) thick, using a small continuous miner type of machine, which is remotely controlled from the roadway and without any associated supports. The machine is used to cut and extract sections of coal about 6 ft (1.8 m) wide for a distance of 100 yd (91 m) or so from pillars alongside the roadway. Small ribs of coal, 3 to 6 ft (approx. 1 to 2 m) wide, are left between the sections extracted by the machine. The accurate steering of the machine is a critical feature of this system of mining. See also: *coal auger*. (Nelson, 1965)

molengraafite *lamprophyllite.*

moler A deposit of diatomite of marine origin occurring on the island of Mors, Denmark. It has been worked since 1912 for use as a heat insulating material, as a constituent of special cements, and for other purposes. (Dodd, 1964)

molten slag A waste product of smelting; usually a mixture of silicates. (Kirk, 1947)

moluranite An amorphous mineral, $H_4U(UO_2)_3(MoO_4)_7{\cdot}18H_2O$.

molybdenite A hexagonal and trigonal mineral, MoS_2; polymorphous with jordisite; foliated; soft; metallic lead gray; an accessory in granites and deep veins; an ore of molybdenum.

molybdenum A silvery-white, very hard, metallic element. Symbol, Mo. Does not occur native, but is obtained principally from molybdenite. Wulfenite ($PbMoO_4$), and powellite ($Ca(MoW)O_4$) are also minor commercial ores. Valuable as an alloying agent with steel and nickel. Used for electrodes in electrically heated glass furnaces, in nuclear energy applications, and for missile and aircraft parts. (Handbook of Chem. & Phys., 3)

molybdenum aluminide Mo_3Al; melting point, 2,150 °C. Although its oxidation resistance is poor compared to other aluminides and silicides, it is good compared to molybdenum metal. It is a refractory crucible material for melting certain metals. (Lee, 1961)

molybdenum anhydride *molybdenum trioxide.*

molybdenum borides Five compounds have been reported: Mo_2B, melting point, 2,120 °C, and sp gr, 9.3; Mo_3B_2, dissociates at 2,250 °C; MoB, exists in two crystalline forms: alpha MoB, melting point, 2,350 °C, sp gr, 8.8; and beta MoB, melting point, 2,180 °C, sp gr, 8.4; Mo_2B_5, dissociates at approx. 1,600 °C, sp gr, 7.5; and MoB_2, melting point, 2,100 °C, sp gr, 7.8, and thermal expansion, 7.7×10^{-6}. (Dodd, 1964)

molybdenum carbides MoC, melting point 2,692 °C, sp gr, 8.5; Mo_2C, melting point 2,687 °C, sp gr, 8.9. (Dodd, 1964)

molybdenum disilicide $MoSi_2$; gray; metallic; tetragonal; and melting point, 2,000 °C. It has good oxidation resistance at elevated temperatures; maintains fairly good strength; and has refractory applications. Molecular weight, 152.11; sp gr, 6.31 (at 20.5 °C); insoluble in acids and in aqua regia; and soluble in hydrofluoric acid plus nitric acid. Syn: *molybdenum silicide*. (Lee, 1961; Handbook of Chem. & Phys., 2)

molybdenum mineral Most molybdenum is a byproduct of mining porphyry copper deposits.

molybdenum silicide *molybdenum disilicide.*

molybdenum trioxide MoO_3; white at ordinary temperatures; yellow at elevated temperatures; molecular weight, 143.94; orthorhombic; sp gr, 4.69 (at 21 °C); melting point, 795 °C; boiling point, 1,264 °C or sublimes at 1,155 °C, at 1 atmophere (101 kPa); sparingly soluble in water; very soluble in excess alkali with the formation of molybdates; and soluble in concentrated mixtures of nitric and hydrochloric acids and in mixtures of nitric and sulfuric acids. (CCD, 1961; Handbook of Chem. & Phys., 2)

molybdic ocher *ferrimolybdite.*

molybdic oxide *molybdenum trioxide.*

molybdite *ferrimolybdite; molybdenum trioxide.*

molybdomenite A monoclinic mineral, $PbSeO_3$.

molybdophyllite A trigonal mineral, $Pb_2Mg_2Si_2O_7(OH)_2$.

moment distribution A method of calculating bending moments, in redundant frames and continuous beams, using successive approximations. (Hammond, 1965)

moment of force The turning effect on a body about a point called a pivot or fulcrum. In practice, the turning effect is commonly called leverage. Syn: *turning effect*. (Morris, 1958)

moment of inertia Resistance by a body to angular acceleration about a specific axis of rotation. (Pryor, 1963)

moment of resistance The couple produced by the internal forces in a beam subjected to bending under the maximum permissible stress. (Hammond, 1965)

momentum Mass times velocity. (Pryor, 1963)

momentum grade Grade so situated that the kinetic energy of a train (due to its speed at the foot of the grade) will enable the locomotive to haul the train to the top without a reduction of speed below 10 or 12 mph (16.1 to 19.3 km/h). (Urquhart, 1959)

Momertz-Lintz system A unique winding arrangement where two winding engines are arranged alongside the shaft, the shaft collar forming a common foundation. The ropes are practically vertical and there is less rope oscillation. (Sinclair, 1959)

M.O. Mine safety indicator A system showing, by means of lights, the position of all switchgear underground, with a master control that would show up if a section of a mine had been switched off at the substation, leaving switches going into the mine in the on position. A recorder gives a complete operating picture of a power-loading face. (Nelson, 1965)

monad An axial rotation of 360°, one-fold. See also: *axis of symmetry.*

monadnock A residual hill rising above a peneplain, representing an isolated remnant of a former erosion surface. It is named after Mt. Monadnock, NH. Cf: *inselberg.*

monazite (a) A monoclinic mineral, $(Ce,La,Y,Th)PO_4$; waxy yellow to brown; an accessory in granites; pegmatites, and placers; a source of thorium and rare earths; further speciated according to the dominant rare-earth element. See also: *zircon group.* (b) The mineral group brabandite, cheralite, huttonite, monazite, and rooseveltite.

monazite sand *monazite.*

moncheite A trigonal mineral, $(Te,Bi)_2$; melonite group; steel-gray; in chalcopyrite on the Kola Peninsula, Russia.

monchiquite A lamprophyre containing phenocrysts of olivine, clinopyroxene, and typically biotite or amphibole (barkevikite), in a groundmass of glass or analcime, often highly altered. Nepheline or leucite may be present. Its name is derived from Serra de Monchique, Portugal. (AGI, 1987)

Mond process A process for extracting and purifying nickel. The main features consist of forming nickel carbonyl by reaction of the finely divided reduced metal with carbon monoxide, and decomposing the nickel carbonyl, with deposition of nickel on small nickel pellets. (ASM, 1961)

monetite A triclinic mineral, $CaHPO_4$; yellowish-white; in phosphorites.

mongrel Eng. Irregular gray limestone, Lower Lias, Lyme Regis. (Arkell, 1953)

monheimite A variety of smithsonite containing iron carbonate. (Fay, 1920)

monitor (a) A high-pressure nozzle, mounted in a swivel on a skid frame, used to direct high-pressure water (hydraulic water) on unconsolidated gravels and sands in alluvial mining to break down, wash, and transport them. Syn: *hydraulic monitor*. (Pryor, 1963) (b) A self-dumping car, holding from 5 to 8 tons of coal. It is filled by emptying the mine car into it at the foot of the slope or top of the incline. See also: *gunboat*. (BCI, 1947) (c) A gas sampler for continuous sampling of the atmosphere in a mine, usually sounding an alarm when the gas threshold is exceeded. (Hartman, 1961) (d) An instrument that measures ambient conditions in an area. (Lyman, 1964) (e) In hydraulicking, a high-pressure nozzle mounted in a swivel on a skid frame. (Nichols, 1954)

monitor operator *incline man.*

monkey (a) The word monkey prefixed to a technical term means small, thus: monkey chute, a small chute; monkey drift, a small drift, usually driven in for prospecting purposes. (Fay, 1920) (b) An appliance for mechanically gripping or letting go the rope in rope haulage. (Fay, 1920) (c) A block placed between the rails on an incline to prevent cars from running back. Syn: *monkey chock; bobbin*. (Standard, 1964)

monkey board A single, unrailed, heavy plank, mounted above the drill platform in the derrick or tripod and serving as a walkway or work platform. (Long, 1960)

monkey chock Aust. *monkey; bobbin.*

monkey drift A small drift driven in for prospecting purposes, or a cross-cut driven to an airway above the gangway. (Zern, 1928)

monkey face A term applied to a chain bracket to be inserted between adjacent shaker conveyor troughs to permit suspension of the trough line from the roof. (Jones, 1949)

monkey gangway (a) Pennsylvania. An air course driven parallel with a gangway and heading at a higher level. Used where a seam has considerable pitch or dip. (Fay, 1920) (b) A small gangway parallel to a main gangway. (Hudson, 1932)

monkey hair Caotchouc, or rubber, derived from a milk juice of plants and which fills latex cells and tubes. It is resistant to decomposition and found in many brown coals as a wooly mass. (Stutzer, 1940)

monkey heading A narrow and low passage driven in the coal where miners take refuge while coal is being blasted. Chiefly in mines where seams pitch sharply. (Korson, 1938)

monkey hole *doghole.*

monkey roll Any of the smaller rolls in an anthracite breaker.

monkey shaft A small shaft raise extending from a lower to a higher level. (Fay, 1920)

monkey winch A device for exerting a strong pull; may be used to withdraw steel arches from disused roadways. It consists of a framework containing a hand-operated drum, around which a steel rope 50 ft (15.24 m) in length is wound. To hold the drum in position a ratchet device is used with both gears. A simple reversing mechanism that disengages the ratchet is also attached. The winch is firmly anchored when in use. See also: *winch; sylvester*. (Nelson, 1965)

monkey wrench An adjustable wrench named for its inventor, Thomas Monkey. (Crispin, 1964)

Monnier process The treatment of copper sulfide ores by roasting with sodium sulfate, and subsequent lixiviation and precipitation. (Fay, 1920)

monobasic Containing one hydrogen atom replaceable by a metal with the formation of a salt. (CTD, 1958)

monocable A form of aerial ropeway in which the same rope is used both to support and haul along the overturning skips in which the debris is carried. These rest upon the rope and obtain a sufficient frictional grip on it to be carried up moderate gradients and over pulleys by means of an inverted Vee-shaped saddle lined with wood, rubber, or composition. The rope is driven by a surge wheel in a similar manner to an endless-rope haulage. See also: *aerial ropeway; fixed-clip monocable*. (Sinclair, 1959)

monochromatic Having or consisting only of one color or frequency.

monochromatic light Electromagnetic radiation of a single wavelength or frequency.

monoclinal Adj. of monocline.

monocline A local steepening in an otherwise uniform gentle dip. Cf: *anticlinal bend; homocline; flexure*. Adj: monoclinal. Obsolete syn: unicline. (AGI, 1987)

monoclinic block A quarryman's term for blocks with two parallel sets of sides at right angles and one parallel set not at right angles.

monoclinic system All point groups characterized by lattices with two crystallographic axes at right angles and one axis inclined. See also: *crystal systems*.

monogene (a) *monogenetic*. (b) Said of an igneous rock (such as dunite) composed essentially of a single mineral. Syn: *monomineralic*. (AGI, 1987)

monogenetic (a) Resulting from one process of formation or derived from one source, or originating or developing at one place and time; e.g., said of a volcano built up by a single eruption. (AGI, 1987) (b) Consisting of one element or type of material, or having a homogeneous composition; e.g., said of a gravel composed of a single type of rock. Cf: *polygenetic*. Syn: *monogene*. (AGI, 1987)

monograin A free-flowing high explosive widely used for charging bulled holes and large-diameter (well drill) holes. (Nelson, 1965)

monolith (a) A piece of unfractured bedrock, generally more than a few meters across; e.g., an unweathered joint block moved by a glacier. (AGI, 1987) (b) A large upstanding mass of rock, such as a volcanic spine. (AGI, 1987) (c) One of many large blocks of stone or concrete forming the component parts of an engineering structure, such as a dam. (AGI, 1987) (d) A vertical soil section, taken to illustrate the soil profile. (AGI, 1987)

monolithic refractory Furnace lining made in one piece or formed by casting, ramming, or tamping into position. (Osborne, 1956)

monomaceral Coal microlithotype consisting of a single maceral; i.e., fusite or vitrite. (AGI, 1987)

monomineralic *monogene*.

Monongahelan Upper Pennsylvanian geologic time. (AGI, 1987)

Monongahela series The Upper Productive Coal Measures of the Pennsylvanian, of which they constitute the highest member. (CTD, 1958)

Mono pump This pump consists essentially of a rubber stator in the form of a double internal helix and a single helical rotor which rolls in the stator with a slightly eccentric motion. The rotor maintains a constant seal across the stator and this seal travels continuously through the pump, giving a positive uniform displacement. The Mono pump is manufactured to meet mining conditions. The rotor is made of special abrasion-resisting or noncorroding steel. The length of the stator and rotor provides for a twist of slightly more than 360° to provide for a complete seal. For greater heads the length of stator and rotor are increased so as to provide two or more complete seals in series and the head is then developed in stages. Syn: *progressing cavity pump*. (Mason, 1951)

monopyroxene *clinopyroxene*.

monorail A relatively new underground transport system in which the carriages, or buckets, are suspended from, and run along, a single continuous overhead rail or taut wire rope. The monorail is used in coal mines to transport supplies to the workings. It may be installed alongside the gate conveyor and worked by endless or main rope. See also: *overhead monorail; overhead-rope monorail; Becorit system*. (Nelson, 1965)

monorail crane A traveling crane suspended from a single rail. (Crispin, 1964)

monoschematic Said of a body of rock or a mineral deposit, the fabric of which is identical throughout.

monotower crane A tower crane that rotates through a full circle and is erected on a fixed base. (Hammond, 1965)

monotron An indentation hardness testing machine by which measurements are obtained by the pressure required to indent a specimen with a diamond 5/8 mm in diameter, the depth of indentation remaining constant. (Henderson, 1953)

monotron hardness test A method of determining the indentation hardness of metals by measuring the load required to force a spherical penetrator into the metal to a specified depth. (ASM, 1961)

monovalent (a) Having a valence of 1. (Webster 3rd, 1966) (b) Having one valence; e.g., calcium, which has only one valence of 2. (Webster 2nd, 1960)

montana (a) Sp. Mountain. (b) Mex. Ores scattered through country rock and not found in deposits of any appreciable size.

Montana agate Moss agate from Montana.

Montana ruby Red garnet, pyrope or almandine.

Montana sapphire Electric or steel-blue sapphire from Montana.

montanite A possibly monoclinic mineral, $Be_2TeO_6{\cdot}2H_2O$; yellow to white; forms earthy encrustations.

montan wax A solid bitumen that may be extracted by solvents from certain lignites or brown coals. It is white to brown and melts at 77 to 93 °C. (AGI, 1987)

montasite A variety of amosite; an asbestiform variety of amphibole.

Mont Blanc ruby Rose quartz from Mont Blanc, southeastern France.

montbrayite A triclinic mineral, $(Au,Sb)_2Te_3$; forms tin-white crystals at Montbray, Quebec. Named for the locality.

montebrasite A triclinic mineral, $LiAl(PO_4)(OH_1F)$; amblygonite group; perfect cleavage; occurs in granite pegmatites where it may be a source of lithium.

Montgomery jig A plunger-type jig with the plunger beneath the screen. The distinguishing feature of this jig is the use of two sets of valves beneath the screen plate. Used in washing bituminous coal, both closely sized and slack sizes. (Mitchell, 1950)

Montian European stage: Paleocene (above Danian, below Thanetian). (AGI, 1987)

monticellite An orthorhombic mineral, $MgCaSiO_4$; olivine group; in contact-metamorphosed limestones, rarely in ultramafic rocks. Cf: *glaucochroite*.

monticule A small hill, knob, or mound; esp., a minor volcanic cone.

montiform Mountainlike; having the shape of a mountain.

montmorillonite A monoclinic mineral, $(Na,Ca)_{0.33}(Al,Mg)_2Si_4O_{10}(OH)_2{\cdot}nH_2O$; smectite group; expansive and cation exchangeable; perfect basal cleavage; in bentonite clays formed by alteration of volcanic ash or mafic igneous rocks. The chief constituent of bentonite and fuller's earth. Cf: *beidellite; hectorite*. Syn: *amargosite*.

montre An opening in a kiln wall to permit inspection of the contents. (Standard, 1964)

montroseite An orthorhombic mineral, $(V,Fe)O(OH)$; black; a primary mineral in sandstone-type uranium-vanadium ores associated with roscoelite, pitchblende, coffinite, and secondary uranium and vanadium minerals.

montroydite An orthorhombic mnineral, HgO; sectile; perfect cleavage; sp gr, 11.2; a rare secondary mineral; in the oxidized zone of some mercury deposits.

monument (a) The structure erected to mark the position of a corner. Permanence is implied. In a legal sense, a monument is any physical evidence of a boundary of real property. (Seelye, 1951) (b) Survey point; e.g., a pile of stones indicating the boundary of a mining claim. (AGI, 1987)

monzonite A granular plutonic rock containing approx. equal amounts of orthoclase and plagioclase, and thus intermediate between syenite and diorite. Quartz is minor or absent. Either hornblende or diopside, or both, are present and biotite is a common constituent. Accessories are apatite, zircon, sphene, and opaque oxides. The intrusive equivalent of latite. Syn: *syenodiorite*.

mooihoekite A tetragonal mineral, $Cu_9Fe_9S_{16}$; in magmatic sulfide deposits; a significant copper ore mineral at Noril'sk, Russia.

moonstone (a) A semitransparent to translucent variety of alkali feldspar (adularia) or cryptoperthite that exhibits a bluish to milky-white pearly or opaline luster; an opalescent variety of orthoclase; a gemstone if flawless. Cf: *sunstone*. Syn: *hecatolite*. See also: *feldspar*. (b) A name incorrectly applied to peristerite or to opalescent varieties of plagioclase (esp. albite). (c) A name incorrectly applied (without proper prefix) to milky or girasol varieties of chalcedony, scapolite, corundum, and other minerals.

moonstone glass A type of opal glass resembling the mineral moonstone. (ASTM, 1994)

moor (a) A more or less elevated tract of barren land, having, as a rule, a rather broad, flat, and poorly drained surface, commonly diversified by peat bogs and patches of heath. (b) A common term for peat unfit for use, as opposed to turf, which is dug for fuel. (Tomkeieff, 1954)

moorband *moorband pan*.

moorband pan Eng. A hard ferruginous crust that forms at the bottom of boggy places above a stiff and impervious subsoil.

Moore and Neill sampler A sediment coring device containing a protected glass tube through which water flows freely during descent and which is forced by impact into the sediment. On hauling, a simple valve mechanism closes the top of the tube and the sample may be brought to the surface. The body of the sampler is a brass cylinder into which fits a thinner metal tube holding the glass sampling tube. When the glass tube is in position, it is closed by a rubber bung that comes hard against the upper surface of the main body of the sampler in the center of which is a hole. (Hunt, 1965)

Moore filter press A movable, intermittent vacuum filter consisting of a series, or basket, of leaves fastened together in such a way that it may be dropped in a pulp tank and kept submerged until a cake is formed; it is then transferred by crane to an adjoining wash-solution tank and washed; the basket is then lifted out of the tank and the cake dropped. (Liddell, 1918)

Moore free corer A sediment sampler designed to drop free from a ship to the sea floor, obtain a core, and return to the surface, leaving its expendable weight and casing embedded in the bottom. The free corer consists of two basic assemblies: (1) a recoverable core barrel, check valve, buoyant chamber assembly filled with gasoline; and (2) an expendable weight and casing assembly. When these two assemblies are combined, the core barrel fits loosely inside the casing. The device is dropped over the side of the ship and allowed to fall free to the bottom. A simple release-delay timer made of magnesium releases the core barrel and its buoyant float rises from the weight and casing assembly. (Hunt, 1965)

moorland peat *highmoor peat.*

moorlog (a) Remains of a submerged forest, composed of a tangled mass of brushwood and tree trunks and forming a layer from 3 to 8 ft (0.9 to 2.4 m) thick. (Tomkeieff, 1954) (b) Hard, brown peat dredged by fishing boats from the bed of the North Sea. (Arkell, 1953)

moor peat *highmoor peat.*

moose pasture Can. Derisive term applied to mining country that is largely muskeg. (Hoffman, 1958)

mor In contrast to "mull", a type of forest soil in which the humus layer forms a dense carpet over the soil. Synonymous with raw humus. (Tomkeieff, 1954; AGI, 1987)

morainal apron *apron.*

moraine A mound, ridge, or other distinct accumulation of unsorted, unstratified glacial drift, predominantly till, deposited chiefly by direct action of glacier ice, in a variety of topographic landforms that are independent of control by the surface on which the drift lies. See also: *boulder clay; till.* (AGI, 1987)

Moran and Proctor sampler A simple, split-tube, drive-type soil-sampling barrel. The sampler is equipped with a thin-walled unsplit brass liner, which can be capped and sealed to act as a watertight shipping container for the sample. (Long, 1960)

morass ore *bog iron; bog iron ore.*

Morcol A semigelatinous permitted explosive possessing both high power and good water-resisting properties. It has a density about midway between Dynobel No. 2 and Ajax. (Nelson, 1965)

mordant A substance used in dyeing to fix the coloring matter, as a metallic compound that combines with a organic dye to form an insoluble colored compound, or lake, in the fiber of a fabric. (Webster 3rd, 1966)

mordenite A white, yellowish, or pinkish member of the zeolite group of minerals with the formula $(Ca,Na_2K_2)Al_2Si_{10}O_{24}{\cdot}7H_2O$. (Larsen, 1934)

morenosite An orthorhombic mineral, $NiSO_4{\cdot}7H_2O$; may contain appreciable magnesium; apple-green; in secondary incrustations from the oxidation of nickel-bearing sulfides. Syn: *nickel vitriol.*

morganite A rose-colored, alkali-bearing, gem-quality variety of beryl identical with vorobievite. Syn: *vorobievite.*

morgen The South African measure of land, equal to 640.25 square rods, 92,196 ft^2 (8,565 m^2), 1.44 claims, or 2.1165 English acres (0.857 ha). There are 284 morgens to a square mile and 735.5 morgens/km^2. (Beerman, 1963)

Morisette expansion reamer A reaming device equipped with three tapered lugs or cutters designed so that the drilling pressure necessary to penetrate rock with a noncoring pilot bit forces the diamond-faced cutters of the reamer to expand outward, thereby enlarging the pilot hole sufficiently to allow the casing to follow the reamer as drilling progresses. The casing is rotated with a pipe wrench while the noncoring and expansion bit is turned by the drill, and the casing is allowed to follow down the reamed-out pilot hole about 1-1/2 to 2 in (4 to 5 cm) behind the upper end of the reamer lugs. (Long, 1960)

Morkill's formula Valuation formula used to ascertain present value (Vp) of a mining share. (Pryor, 1963)

morlop A mottled variety of jasper found in New South Wales, Australia; much sought by miners, it commonly occurs with diamonds.

morphological crystallography The study of the external shapes of crystals. (Hurlbut, 1964)

morphologic unit (a) A rock stratigraphic unit identified by its topographic features; e.g., a Pleistocene glacial deposit. (b) A surface, either depositional or erosional, recognized by its topographic character.

morphology (a) The observation of the form of lands. (Standard, 1964) (b) The study of the form and structure of organisms. (AGI, 1987) (c) The form or shape of a crystal or mineral aggregate. Adj. morphological.

mortar (a) The receptacle beneath the stamps in a stamp mill, in which the dies are placed, and into which the rock is fed to be crushed. (Fay, 1920) (b) A vessel in which rock is crushed by hand with a pestle for sampling or assaying. (Fay, 1920)

mortar bed Lime-cemented, valley-flat deposit of clay, silt, sand, and gravel, found in Nebraska and Kansas; a type of caliche. (AGI, 1987)

mortar box The large, deep, cast-iron box into which the stamps fall and the ore is fed in a gold or silver stamp mill; also called stamper box. (Fay, 1920)

mortar structure A structure in crystalline rocks characterized by an aggregate of small grains of quartz and feldspar occupying the interstices between, or forming borders on the edges of, much larger, rounded grains of the same minerals. Long considered a product of cataclasis, the structure may actually be the result of plastic deformation and dynamic recrystallization (Harte, 1977). Syn: *porphyroclastic structure.* See also: *cataclastic.* (AGI, 1987)

mortice *mortise.*

mortification Destruction of active qualities, as in mercury amalgamation. (Standard, 1964)

mortise A rectangular hole cut in one member of a framework to receive a corresponding projection on the mating member. Syn: *mortice.* (CTD, 1958)

morts terrains (a) Fr. Barren or dead ground. (b) The water-bearing strata overlying the Coal Measures.

mosaic (a) An assembly of aerial or space photographs or images whose edges have been feathered and matched to form a continuous photographic representation of a part of the Earth's surface; e.g., a composite photograph formed by joining together parts of several overlapping vertical photographs of adjoining areas of the Earth's surface. See also: *controlled mosaic.* Syn: *aerial photomosaic.* (AGI, 1987) (b) A textural subtype in which individual mineral grains are approx. equal (Harte, 1977). (AGI, 1987)

mosaic silver An amalgam of mercury, tin, and bismuth, used for imitating silverwork. (Standard, 1964)

mosaic structure Slight irregularity of orientation of small, angular, and granular regions of varying sizes in a crystal.

mosaic texture (a) A granoblastic texture in a dynamically metamorphosed rock in which the individual grains meet with straight or only slightly curved, but not interlocking or sutured, boundaries. (AGI, 1987) (b) A texture in a crystalline sedimentary rock characterized by more or less regular grain-boundary contacts; e.g., a texture in a dolomite in which the mineral dolomite forms rhombs of uniform size so that in section contiguous crystals appear to dovetail, or a texture in a orthoquartzite in which secondary quartz is deposited in optical continuity on detrital grains. (AGI, 1987)

moschellandsbergite An isometric mineral, Ag_2Hg_3; a silver-white amalgam; sp gr, 13.5. Named for the locality in Bavaria, Germany.

mosesite An isometric mineral, $Hg_2N(Cl,SO_4,MoO_4,CO_3){\cdot}H_2O$; yellow; secondary. See also: *kleinite.*

moss (a) adj. A fine dendritic growth having the texture of moss; e.g., moss gold. (b) A term used for fractures or fissures in gem stones which produce the appearance of moss, such as in many emeralds.

moss agate A general term for any translucent chalcedony containing inclusions of any color arranged in dendritic patterns resembling trees, ferns, leaves, moss, and similar vegetation; specif. an agate containing brown, black, or green mosslike markings due to visible inclusions of oxides of manganese and iron. See also: *agate; Mocha stone; tree agate.* Syn: *landscape agate; medfordite.*

Mossfield loader A scraper-box type of coal loader developed at the Mossfield colliery (Great Britain) in 1953. It consists of a hinged front, scooplike plate which elevates and deflects the broken coal onto an armored conveyor. On inclined faces, the loader is hauled by a double-drum Pikrose haulage. (Nelson, 1965)

moss form A material in dendritic forms.

mossing During low water in the Salmon River, CA, the algae and other plants growing in the stream are gathered, dried, and burned. The ashes are washed, and some gold is obtained. This process is called mossing. (Hess)

mossite A tantalum-bearing variety of ferrocolumbite or tapiolite named for the locality in Norway.

moss peat Peat derived from water-loving mosses, chiefly sphagnum. See also: *highmoor peat.* (Tomkeieff, 1954)

mossy zinc Granulated zinc obtained when the molten metal is poured into cold water.

mother (a) Gouge clay in a mineral vein. (Arkell, 1953) (b) Shale adhering to quarried limestone. (Arkell, 1953) (c) *mother crystal.*

mother conveyor A term frequently used in connection with conveyors used in gathering service. The mother conveyor receives coal from other conveyors or gathering machines, such as shuttle cars, and delivers it to some central loading point. See also: *underground mine conveyor.* (Jones, 1949)

mother crystal A name given to a mass of raw quartz, either faced or rough, as found in nature. Syn: *mother.* (Am. Mineral., 1947)

mother gate (a) N. of Eng. The main roadway to a coal face up which miners travel; air, power, and supplies pass; and down which coal from the face travels on a conveyor belt. (Trist, 1963) (b) Eng. The main road of a district off which crossheadings are set away in longwall working. See also: *level; main gate.* (SMRB, 1930)

Mother Hubbard bit A heavy drag-type or fishtaillike bit having a long grooved shank, the diameter of which is only slightly less than the width of the cutting edges; it is designed for drilling boreholes in formations that mud-up excessively. (Long, 1960)

mother liquor (a) The magmatic rest solution from which a mineral deposit has received its metal content. (Schieferdecker, 1959) (b) In crystallization, the liquid that remains after the substances readily and regularly crystallizing have been removed. (AGI, 1987)

mother lode (a) A main mineralized unit that may not be economically valuable in itself but to which workable veins are related; e.g., the Mother Lode of California. Cf: *master lode.* (AGI, 1987) (b) An ore deposit from which a placer is derived; the mother rock of a placer. (AGI, 1987)

mother of coal *fusain; mineral charcoal.*

mother-of-emerald *prase.*

mother-of-pearl Iridescent portion of mollusk shells, made of the mineral aragonite. See also: *nacre.* (Hurlbut, 1964)

mother rock A general term for the rock in which a secondary or transported ore deposit originated; mother lode. (AGI, 1987)

motion driver In bituminous coal mining, one who operates the engine that moves an endless cable by which cars are raised or lowered along an inclined haulageway. Syn: *motioner.* (DOT, 1949)

motioner *motion driver.*

motion study A technical investigation of the essential movements of a worker when performing a specific task, and assessing the results with the objective of reducing labor and increasing work performance. The study may also include the layout tasks, availability of tools or materials, and the design of new methods. See also: *time study.* (Nelson, 1965)

motive column The height of a column of air, of the same density as the air in the downcast shaft, which exerts a pressure equal to the ventilating pressure. It is the ventilating pressure expressed in meters of air column. Also called head. Syn: *column; ventilating column.* (Nelson, 1965)

motive power An agency (such as water, steam, wind, or electricity) used to impart motion to machinery. (Webster 3rd, 1966)

motive zone In mine subsidence, that portion of the mined strata which, being still in the process of sinking, goes far to furnish the motive power producing the phenomena. (Briggs, 1929)

motometer A speed counter, as for a steam engine; also a speedometer. (Crispin, 1964)

motor body The boxlike portion at the lower end of a coal-cutting machine. (Fay, 1920)

motor boss In mining, a person who directs locomotive (motor) haulage operations underground and at the surface of a mine. Also called car dispatcher; car distributor; dispatcher; passing boss; traffic man; train dispatcher. Syn: *turn keeper.* (DOT, 1949)

motor brakeperson *locomotive brakeman.*

motor-change man In anthracite and bituminous coal mining, a person who in addition to charging and repairing batteries, removes spent storage batteries from electric mine locomotives and replaces them with freshly charged ones. (DOT, 1949)

motor driver In bituminous coal mining, one who operates a small electric haulage locomotive to haul mine cars underground and at the surface of a mine. (DOT, 1949)

motor hammer drill Usually has a built-in gasoline engine as prime mover, flushing being provided by the exhaust gases or by compressed air produced in the machine. Total weight varies between 50 lb and 120 lb (22.7 kg and 54.4 kg). Motor hammer drills are used for odd-job operations, on forest roads, in prospecting, etc., where it is not worthwhile to lay air supply lines on account of transport difficulties or insufficient volume of work. (Fraenkel, 1953)

motorized grader *grader.*

motorman The person who operates a haulage locomotive. (Jones, 1949)

motor nipper *locomotive brakeman.*

mottle The spotted, blotched, or variegated appearance of any mottled surface, such as of wood or marble; esp., in metallurgy, the appearance of pig iron of a quality between white and gray. (Standard, 1964)

mottled iron Pig iron in which the majority of the carbon is combined with iron in the form of cementite, Fe_3C, but in which there is also a small amount of graphite. The fractured pig has a white crystalline fracture with clusters of dark spots, indicating the presence of graphite. See also: *gray iron.* (CTD, 1958)

mottled limestone Limestone with narrow branching fucoidallike, cylindrical masses of dolomite, often with a central tube or hole; a variegated limestone. It may be organic or inorganic in origin.

mottled silica brick A silica brick having harmless areas of dark cream to reddish brown.

mottled slate Slate in which blotches of red or purplish colors appear on a generally green surface; ascribed chiefly to different forms of iron oxide.

mottled structure Discontinuous lumps, tubes, pods, and pockets of a sediment, randomly enclosed in a matrix of contrasting textures, and usually formed by the filling of animal borings and burrows). (AGI, 1987)

mottramite An orthorhombic mineral, $PbCu(VO_4)(OH)$; descloizite group; zinc replaces copper toward descloizite; conchoidal fracture; a supergene mineral associated with base-metal deposits. Formerly called cuprodescloizite, psittacinite.

mountain brown ore Limonite or brown iron ore. A local name applied in Virginia to the low-grade siliceous variety, which commonly occurs in hard lumps and is found on the mountain slopes at or near the contact of the Cambrian shale and sandstone with the Cambro-Ordovician limestone. Cf: *valley brown ore.* (Sanford, 1914)

mountain building *orogeny.*

mountain chain A complex, connected series of several more or less parallel mountain ranges and mountain systems grouped together without regard to similarity of form, structure, and origin, but having a general longitudinal arrangement or well-defined trend; e.g., the Mediterranean mountain chain of southern Europe. See also: *cordillera.* Cf: *mountain system.* (AGI, 1987)

mountain cork (a) A white or gray variety of asbestos consisting of thick interlaced fibers and resembling cork in texture and lightness (it floats on water). Syn: *rock cork; mountain leather.* (b) A fibrous clay mineral such as sepiolite or palygorskite.

mountain flax A fine silky asbestos.

mountainite A monoclinic mineral, $(Ca,Na_2,K_2)_2Si_4O_{10}{\cdot}3H_2O$; zeolitelike; white; fibrous. See also: *rhodesite.*

mountain leather (a) A tough variety of asbestos occurring in thin flexible sheets made of interlaced fibers. Syn: *rock leather; mountain cork; mountain paper.* (b) A fibrous clay mineral such as sepiolite or palygorskite. Syn: *leatherstone.*

mountain meal *diatomite.*

mountain milk A very soft, spongy variety of calcite. (Standard, 1964)

mountain paper A paperlike variety of asbestos occurring in thin sheets; specif. mountain leather or mountain cork. (AGI, 1987)

mountain pediment A plain of combined erosion and transportation at the foot of a desert mountain range, similar in form to an alluvial plain but beveling solid rock.

mountain railway A railway having such steep gradients that trains are hauled up them by ropes or by a rack locomotive. See also: *funicular railway.* (Hammond, 1965)

mountain range A single, large mass consisting of a succession of mountains or narrowly spaced mountain ridges, with or without peaks, closely related in position, direction, formation, and age; a component part of a mountain system or of a mountain chain. (AGI, 1987)

mountain soap An unctuous variety of halloysite containing some iron oxide and about 24% water. See also: *saponite.* Syn: *rock soap.* (Fay, 1920)

mountain system A group of mountain ranges exhibiting certain unifying features, such as similarity in form, structure, and alignment, and presumably originating from the same general causes; esp. a series of ranges belonging to an orogenic belt. Cf: *mountain chain.* See also: *landform system.* (AGI, 1987)

mountain tar *mineral tar.*

mountain wood A variety of asbestos that is compact, fibrous, and gray to brown in color, resembling wood. Syn: *rockwood.* (Fay, 1920)

mounting In power shovel nomenclature, the mounting consists of a frame on which the entire shovel is supported and on which it moves. (Carson, 1961)

mounting pipe *column pipe.*

mourite A monoclinic mineral, $UMo_5O_{12}(OH)_{10}$; forms spherulitic masses of radiating purple fibers in incompletely oxidized uranium ores.

mousetrap A cylindrical fishing tool, fitted with an inward-opening flap valve at the bottom end, used to recover small metal fragments from the bottom of a borehole. (Long, 1960)

mouth (a) The surface outlet of an underground conduit, as of a volcano. (AGI, 1987) (b) The collar of a borehole. (Long, 1960) (c) An opening resembling or likened to a mouth, such as one affording entrance or exit. (Webster 2nd, 1960) (d) The entrance to a mine. (BCI, 1947) (e) The top of a mine shaft or the point of entrance to a slant, drift, or adit. (Nelson, 1965) (f) The end of a shaft, adit, drift, entry, tunnel, etc., emerging at the surface. (Fay, 1920) (g) The opening in a metallurgical furnace through which it is charged; also, the taphole. (Fay, 1920) (h) The place of discharge of a stream, as where it enters a larger stream, a lake, or the sea. (AGI, 1987) (i) The entrance or opening of a geomorphic feature, such as a cave, valley, or canyon. (AGI, 1987)

mouthing *shaft inset.*

mouth of pit Aust. The top of a shaft. (Fay, 1920)

movable conveyor Any of several types of conveyors designed to be moved in a defined path. See also: *portable conveyor; shuttle conveyor.*

movable jaw The jaw or slip of a safety or foot clamp; it can be raised or lowered into or out of the body or frame of the clamp either to engage or to disengage the drill rods being run into or pulled out of a borehole. Syn: *movable slip.* (Long, 1960)

movable ladder *man machine.*

movable sieve-type washbox A washbox in which the screen plate supporting the bed of material under treatment is moved up and down in water. (BS, 1962)

movable slip *movable jaw.*

move-up Extension; move to a forward position. Syn: *turnover.* (Mason, 1951)

moving annual total In a study of process costs (in large or in detail) a series of costs-per-unit observed and recorded at regular intervals (usually in monthly financial summaries cross referenced to analyzed detail cost). Twelve months are covered and each month the new month's figures are added and those for the corresponding month of the previous year are removed. Therefore, like periods are always compared and seasonal fluctuations are smoothed out. Abbrev., MAT. (Pryor, 1963)

moving grizzly Grizzly in which alternate bars rise and then fall gently, any required lateral (conveying) movement being built into the mechanism, therefore reducing loss of headroom in conveying and screening. Other types are traveling bar, ring-roll, and chain grizzlies. (Pryor, 1963)

moya S. Am. Volcanic mud, sometimes carbonaceous.

moyle An iron with a sharp steel point, for driving into clefts when levering off rock. (Zern, 1928)

M.P.F.M. jet auger An auger equipped with cutting blades; designed so that fluid, under pressure, passing through inclined holes just above the blades, washes away the material loosened by the blades, thereby cleaning the inside of the casing without disturbing the material below the bottom of the casing that is to be sampled. (Long, 1960)

MRE The Mining Research Establishment of the National Coal Board, Great Britain. Its prime purpose is to carry out tests and investigations aimed at increasing the efficiency of coal production while maintaining a high level of safety. (Nelson, 1965)

MSA distributor A high-air-pressure directional machine which operates in the manner of a whitewashing machine and displaces coal dust from the roof and sides. This machine carries 2 st (1.8 t) of stone dust, and traveling at 11 ft/min (3.4 m/min) distributes over 26 lb/min (11.8 kg/min) of dust. (Sinclair, 1958)

MSA methanometer This methane indicator is one in which the sample is made to flow continuously over the filaments while the determination is being made. In this case, two matched filaments form the adjacent arms of the bridge. One of these is specially activated in order to burn the methane while the other filament is inactive and operates at a somewhat lower temperature. This arrangement is designed to compensate for changes in barometric pressure, temperature, humidity, and the presence of carbon dioxide. Facilities for adjustment are provided to compensate for zero drift and change in battery voltage. The meter is provided with two shunts so that two ranges are provided on the scale: 0% to 5% by 0.1% divisions and 0% to 2% by 0.02% divisions. (Roberts, 1960)

MS connector Nonelectric, short-interval (millisecond, ms) delay device for use in delaying a blast; initiated by detonating cord.

mscp Apprev. for mean spherical candlepower. (Mason, 1951)

M-series core barrel *M-design core barrel.*

muck (a) Unconsolidated soil, sand, clay, or loam encountered in surface mining; generally, earth which can be moved without blasting. (Pryor, 1963) (b) Useless material produced in mining. Syn: *mullock.* (Nelson, 1965) (c) A layer of earth, sand or sediment lying immediately above the sand, or gravel containing, or supposed to contain, gold in placer mining districts; it may itself contain some traces of gold. (Fay, 1920) (d) Finely blasted rock, particularly from undergound. (Nichols, 1976) (e) To excavate or remove muck from. (Webster 2nd, 1960)

muck bar Gray, forge pig iron melted in a puddling furnace, then balled, squeezed, and rolled. (Mersereau, 1941)

muck boss In bituminous coal mining, a person who is in charge of a crew of loaders shoveling rock into cars during the driving of new underground passageways from one part of the mine to another. (DOT, 1949)

mucker In mining and quarrying, a laborer who (1) shovels ore or rock into mine cars or onto a conveyor from which mine cars are loaded and at some point are removed from the working face or surfaces of natural stone deposits; or (2) works in a stope shoveling ore into chutes from which it is loaded into cars on haulage level below. Also called car filler; rock passer; shoveler. (DOT, 1949)

mucking The operation of loading broken rock by hand or machine, usually in shafts or tunnels. (Nelson, 1965)

muck iron Crude puddled iron ready for squeezing or rolling. (Fay, 1920)

muckle Soft clay overlying or underlying coal.

muckle hammer A scaling or spalling hammer. (Standard, 1964)

muck saw A saw using an uncharged blade, usually steel, which runs in a bath or stream of carborundum abrasive. Also known as a mud saw.

muck shifting Operations concerned with stripping overburden, valuable gravels, or sands in exploitation of opencast mineral deposits. (Pryor, 1963)

muck soil A soil that contains at least 50% organic matter that is well decomposed. (AGI, 1987)

mucky hole In a furnace, a taphole from which the iron is so pasty that it does not run freely. (Fay, 1920)

mud (a) A sticky or slippery mixture of water and silt- or clay-sized earth material, with a consistency ranging from semifluid to soft and plastic; a wet, soft soil or earthy mass; mire, sludge. (AGI, 1987) (b) An unconsolidated sediment consisting of clay and/or silt, together with material of other dimensions (such as sand), mixed with water. (AGI, 1987) (c) A suspension made by mixing the drill circulation fluid (water) with the fine cuttings produced by the bit when drilling a borehole. (Long, 1960)

mud auger A diamond-point bit with the wings of the point twisted in a shallow, augerlike spiral. Also called clay bit; diamond-point bit. See also: *mud bit.* (Long, 1960)

mud balance An instrument used to measure the density of drill mud. (BS, 1963)

mud barrel (a) A double-tube core barrel with a greater-than-normal clearance between the inner and outer tubes, for use with mud-laden circulation liquids. (Long, 1960) (b) A bailing device to bring to the surface the cuttings formed by the action of the bit at the bottom of a borehole in free-fall or churn drilling. (Long, 1960) (c) A small bailer. (Long, 1960)

mud belt The belt of marine deposits composed largely of detrital clay, and lying between the coarser terrigenous sediments to the landward and the deep oceanic organic oozes on the seaward side. At present, the inner boundary of the inner mud belt is the edge of the continental shelf. (AGI, 1987)

mud bit A pointed-edge, chisellike tool used for boring drill holes through clay or claylike overburden materials. Also called clay bit; diamond-point bit. See also: *mud auger.* (Long, 1960)

mud blasting In this method, sticks of explosive are stuck on the side of a boulder with a covering of mud, and when detonated, very little of the energy of the explosive is used in breaking the boulder. (Higham, 1951)

mud bucket The bucket attached to a dredger. (Standard, 1964)

mud cake The material filling the cracks, crevices, pores, etc., of the rock or adhering to the walls of a borehole. The cake may be derived from the drill cuttings, circulating drill mud, or both; it is formed when the water in the drilling mud filters into porous formations, leaving the mud ingredients as a caked layer adhering to the walls of the borehole. Syn: *cake; mud wall cake.* (Long, 1960)

mudcap A charge of dynamite, or other high explosive, fired in contact with the surface of a rock after being covered with quantity of wet mud, wet earth, or sand, no borehole being used. The slight confinement given the dynamite by the mud or other material permits part of the energy of the dynamite being transmitted to the rock in the form of a blow. A mudcap may be placed on top or to one side, or even under a rock, if supported, with equal effect. Also called adobe; dobie; sand-blast.

mudcap method *secondary blasting.*

mudcapping Method for blasting rock without drilling, in which an explosive is placed on top of the rock and covered by a cap of mud or earth. Syn: *adobe charge.* (Fraenkel, 1953)

mud cast *mud crack.*

mud column The length in feet (meters), as measured from the bottom of a borehole of a drill-mud liquid standing in a borehole either while being circulated during drilling operations or when the drill string is not in the hole. Syn: *column of mud.* (Long, 1960)

mud crack (a) The filling of desiccation cracks in mud, customarily in sandstone; generally preserved as raised ridges (casts) arranged in polygonal patterns on the underside of a sandstone bed. (Pettijohn, 1964) (b) An irregular fracture in a crudely polygonal pattern, formed by the shrinkage of clay, silt, or mud, generally in the course of drying. Syn: *sun crack; mud cast; shrinkage crack; desiccation crack.* (AGI, 1987)

mudding Filling voids with clay in limestone from which sulfur has been extracted. (Bennett, 1962)

mudding off In petroleum production, commonly thought of as reduced productivity caused by the penetrating, sealing, or plastering effect of a drilling fluid. Actually there is little penetration into the capillaries of an ordinary producing formation, and a slight amount of differential back pressure will remove even thick filter cakes. (Brantly, 1961)

mudflow A general term for a mass-movement landform and a process characterized by a flowing mass of predominantly fine-grained earth material possessing a high degree of fluidity during movement. The water content of mudflows may range up to 60%. With increasing fluidity, mudflows grade into loaded and clear streams; with a decrease in fluidity, they grade into earthflows. Syn: *lahar.* (AGI, 1987)

mud gun An apparatus for pushing a clay stopper into the taphole of a blast furnace. A steam cylinder operates a plunger inside a steel tube into which clay is fed from a hopper tube as the plunger is worked back and forth, and is thus forced into the taphole at the end of a cast. See also: *clay gun.* (Fay, 1920)

mud hog (a) A pump used to circulate mud-laden drill fluid during borehole drilling operations. See also: *mud pump; sludge pump.* (Long, 1960) (b) Pressure tunnel worker. (c) A machine for the disintegration of dry or moist plastic clay. It consists of a rotating swing hammer operating close to a series of anvils linked together to form a steeply inclined slat conveyor. (Dodd, 1964)

mud-laden Said of a liquid (usually water) mixed with finely ground earthy or clayey materials. (Long, 1960)

mud-laden fluid The water or oil in which mudlike solids are suspended; used to support the open bore and cool and clear the cuttings from a drill bit. The fluid is circulated while rotary- and/or diamond-drilling a borehole. See also: *drilling mud.* (Long, 1960)

mudline Line of demarcation between fairly clear supernatant water and settling solids in a thickener or other sedimentation vessel. (Pryor, 1963)

mud log A continuous analysis of the drilling mud and well cuttings during rotary drilling for entrained oil or gas. Visual observation, ultraviolet fluoroscopy, partition gas chromatograph, and hydrogen-flame ionization analyzer may be used. A drilling-time log is kept concurrently. (AGI, 1987)

mud logging (a) A method of determining the presence or absence of oil, gas, and salt water in the various formations penetrated by a drill bit. The drilling fluid and the cuttings are continuously tested on their return to the surface, and the results of these tests are correlated with the depth of origin. (Brantly, 1961) (b) A mud log is a recording vs. depth of the parameters being monitored. Basic parameters monitored are: bit weight, rotary speed, rotary torque, mud weight, gas content (trip and background) cutting analysis, pit volume, flow rate, pump pressure (strokes), hole depth, and chlorides. This is a common service used to obtain data from the mud system and drilling parameters.

mud mixer A machine, pump, hopper, or other apparatus used to mix dry ingredients with water or other liquids to prepare a drill mud. Also called atomizer; jet mixer; mixer. Syn: *emulsifier.* (Long, 1960)

mud pot (a) A type of hot spring containing boiling mud, usually sulfurous and often multicolored, as in a paint pot. Mud pots are commonly associated with geysers and other hot springs in volcanic areas, esp. Yellowstone National Park, WY. (AGI, 1987) (b) A geyser that throws up mud. Also called mud geyser. (Standard, 1964; Fay, 1920)

mud pump (a) The circulating pump that supplies fluid to a rotary drill. Also called slush pump. See also: *mud hog.* (Nichols, 1976) (b) *circulating pump.*

mud ring (a) The section of a boiler where scale, alkalies, and sediment collect. (b) The ring or frame forming the bottom of a water leg in a steam boiler. (Webster 3rd, 1966)

mud rush The more or less sudden inflow of peat, moss, sand, gravel, silt, or any other waterlogged material into shallow mine workings. The manager has a duty to take steps to prevent such inrushes as laid down in the Precautions against Inrushes Regulations, 1956. Also called mud run. See also: *inrush of water; running ground.* (Nelson, 1965)

mud scow A flatboat or barge for the transportation of mud, generally used in connection with dredges.

mud snapper An 11-in-long (28-cm-long), 3-lb (1.4-kg) clamshell-type snapper attached to the bottom of a sounding lead by means of a hole drilled in the lead. The jaws are cast bronze and are actuated by a spring. The jaws are held open by engaging two trigger pins within the jaws. The mud snapper and sounding lead may be operated in shallow water by hand lowering or by lowering from a bathythermograph or oceanographic winch. (Hunt, 1965)

mud socket A device attached to drill rods and used to remove sand from a borehole. Cf: *mule shoe.* (Long, 1960)

mudstone (a) An indurated mud having the texture and composition of shale, but lacking its fissility; a blocky or massive, fine-grained sedimentary rock in which the proportions of clay and silt are approx. equal; a nonfissile mud shale. See also: *claystone; siltstone.* (AGI, 1987) (b) A general term that includes clay, silt, claystone, siltstone, shale, and argillite. (AGI, 1987)

mudstone ratio A uranium prospector's term, esp. on the Colorado Plateau, for the ratio of the total thickness of red mudstone to that of green mudstone within an assumed stratigraphic interval. Its value is based upon the premise that uranium-bearing solutions will bleach red mudstone containing ferric iron to green mudstone containing ferrous iron in the course of depositing uranium minerals. (AGI, 1987)

mudstone trap A place where uranium mineralization has been trapped at a mudstone-sandstone interface. (Ballard, 1955)

mud sump In drilling operations, a mud pit. (Long, 1960)

mud up (a) The act or process of filling, choking, or clogging the waterways of a bit with consolidated drill cuttings. Also called sludging; sludging up. (Long, 1960) (b) The act or process of filling the pores or cracks in the rock surrounding a borehole; also, to cause mud to adhere to the walls of a borehole. (Long, 1960)

mud viscosity The property of a mud-laden fluid to resist flow due to internal friction and the combined effects of adhesion and cohesion; e.g., a Marsh funnel (used to measure the viscosity of mud) will discharge 1 quart (0.95 L) of water in about 36 s, whereas an equal volume of an average drilling mud is discharged in 40 to 55 s or more from the same funnel. (Long, 1960)

mud volcano An accumulation, usually conical, of mud and rock ejected by volcanic gases; also, a similar accumulation formed by escaping petroliferous gases. (AGI, 1987)

mud wall cake (a) *mud cake.* (Long, 1960) (b) The formation of mud in the drilling fluid by adhering to the wall of the hole. When the drilling mud particles comes in contact with porous, permeable formation, solid particles immediately enter the openings. This sealing property is dependent upon the amount and physical state of the colloidal material in the mud.

muffle (a) A semi-cylindrical or long, arched oven (usually small and made of fireclay), heated from outside, in which substances may be exposed at high temperature to an oxidizing atmospheric current, and kept at the same time from contact with the gases from the fuel. Cupellation and scorification assays are performed in muffles; on a larger scale, copper ores were formerly roasted in muffle furnaces. (Fay, 1920) (b) An enclosure in a furnace to protect the ware from the flame and products of combustion. (ASTM, 1994)

muffle furnace (a) A furnace with an externally heated chamber, the walls of which radiantly heat the contents of the chamber. (McGraw-Hill, 1994) (b) A furnace in which heat is applied to the outside of a refractory chamber containing the charge. The charge may be held in a muffle, crucible retort, or other enclosure that is enveloped by the hot flame gases, and the heat must reach the charge by flowing through the walls of the container. (CTD, 1958; Newton, 1959)

muffle kiln (a) An arched fireclay-lined furnace in which seggars are placed. (CTD, 1958) (b) A kiln in which combustion of the fuel takes place within refractory muffles, which, in turn, conduct heat into the ware chamber. (ACSG, 1963)

muffler A muffler that concentrates on suppressing sound waves vibrating 200 to 2,000 times a second—the loudest and most objectionable ones created by rock drills. Because the muffler bypasses the lower frequencies, it does not interfere with the column of air that makes a pneumatic drill function.

mugearite Orthoclase-bearing oligoclase basalt, with major olivine, accessory apatite, and opaque oxides. Pyroxene may or may not be present. See also: *trachyandesite.* (AGI, 1987)

mule (a) A small car, or truck, used to push cars up a slope or inclined plane. (Fay, 1920) (b) *pusher.* (c) An extra worker who helps push the loaded cars out in case of an upgrade, etc.; from Joplin, Mo. (Fay, 1920)

mule's foot Kansas. An extension bit used in boring coal. (Hess)

mule shoe A short length of tubing coupled to the bottom of a drill string to wash and clean out sand or mud from a borehole, the washing action being aided by cutting off the bottom end of the tubing at an angle of 45° to its longitudinal axis. Also called mud socket. (Long, 1960)

mule skinner A mule driver. (Fay, 1920)

muleway Heavily timbered passage between levels in a mine for the transfer of unattached mules from one level to the other. (Korson, 1938)

muller (a) Stone, iron shoe, or heavy steel rubbing disk, used to bear down upon rock in comminution. (Pryor, 1963) (b) A heavy grinding wheel that is the crushing and mixing member in a dry or wet pan. (ARI, 1949) (c) *bucking iron.*

mulling Mixing sand and clay particles by a rolling, grinding, rubbing, or stirring action. (ASM, 1961)

mullion structure A wavelike pattern of parallel grooves and ridges, measuring as much as several feet from crest to crest, and formed on a folded surface or along a fault surface. Cf: *slickensides.* (AGI, 1987)
mullite An orthorhombic mineral consisting of an aluminum silicate that is resistant to corrosion and heat; used as a refractory. Also known as porcelainite. (McGraw-Hill, 1994)
mullock *muck.*
multibucket excavator A machine similar to a dredger used for excavating cuttings for roads, railways, or canals. One large machine of this type can dig 100 yd/h (91.4 m/h) on a slope 25 ft (7.62 m) high. (Hammond, 1965)
Multicut chain Trade name for a coal-cutter chain designed for use with curved jibs. It is short pitch and of high flexibility. See also: *curved jib.* (Nelson, 1965)
multideck cage A cage containing two or more compartments or platforms to hold the mine cars supplies or workers. Every effort is made to keep the number of decks as low as possible for a given output in order to cut down the decking time and equipment at shaft top and bottom. See also: *deck.* (Nelson, 1965)
multideck screen A screen with two or more superimposed screening surfaces mounted rigidly within a common frame. (BS, 1962)
multideck sinking platform A sinking platform consisting of several decks to enable various shaft-sinking operations to be performed simultaneously. The bottom deck, in a three-deck platform, is usually suspended from four winch ropes which also act as guides for the kibbles, and the middle and top decks are supported above the bottom deck by rigid supports. The top deck is used for the manipulation of the concrete buckets and for fixing the shuttering. The center deck is used by the workers when placing the concrete, while the bottom deck carries telephone, blasting, lighting, and signaling cable drums. The lower side of the bottom deck may carry the equipment for manipulating the cactus grab. (Nelson, 1965)
multideck table Shaking table with two or more superposed decks, independently fed and discharged but worked by one vibrating mechanism. (Pryor, 1963)
multifuse igniter A device employed to reduce the number of fuses to be lit by the miner before retiring to safety. By means of a multifuse igniter, it is possible to remotely fire stopes or headings primed with plain detonators and safety fuse. (Nelson, 1965)
multihearth furnace Roasting furnace with several hearths vertically superposed. Material is raked downward by horizontally rotating rabbles, so as to work alternately to the periphery and center of successive hearths, encountering roasting heat as it gravitates downward. (Pryor, 1963)
multijib cutter A cutter loader with a number of horizontal jibs similar to a coal cutter. Loose coal is diverted onto the conveyor by gummer and plow plate. The machine is usually used in seams up to about 3 ft (0.9 m) in thickness with a clean roof parting. The depth of cut varies up to 4 ft (1.2 m). Coal degradation is considerable. (Nelson, 1965)
multilayer bit A bit set with diamonds arranged in successive layers beneath the surface of the crown. Cf: *impregnated bit; surface-set bit.* (Long, 1960)
multilock lode Lode that occupies a shear zone. Such a zone has no definite walls, the ore gradually shading off into country rock. It is probable that the gold of some rich alluvial fields came from shear zones. (Nelson, 1965)
multilouvre dryer A dryer whose moving element consists of two strands of roller chain with specially designed flights, suspended in such a way as to provide means for keeping the bed in a constantly flowing mass. The material flows in a shallow bed over the ascending flights and at the same time is gradually moved across the dryer from the feed point to the discharge point. The gases are pulled from the furnace and through the flowing bed of coal. The entire area of the dryer is covered by suction from the exhaust fan. See also: *thermal drying.* (Mitchell, 1950)
multiphase *polyphase.*
multiple-arch dam A lightweight dam constructed of repeated arches with axes sloping at about 45° to the horizontal, the arches being carried on parallel buttress walls. (Hammond, 1965)
multiple-bench quarrying The method of quarrying a rock ledge in a series of successive benches or steps. (Fay, 1920)
multiple detectors Two or more seismic detectors whose combined output energy is fed into a single amplifier-recorder circuit. This technique is used to cancel undesirable near-surface waves. Syn: *multiple geophones; multiple recording groups.* (AGI, 1987)
multiple dike A dike made up of two or more intrusions of the same kind of igneous rock. (Billings, 1954)
multiple-entry system A system of access or development openings, generally in bituminous coal mines, involving more than one pair of parallel entries, one for haulage and fresh air intake and the other for return air. Multiple-entry systems permit circulation of large volumes of air.
multiple-expansion engine An engine driven by steam or compressed air expanding in two or more stages. (Hammond, 1965)
multiple fault A structure consisting of several parallel faults in close proximity with no distortion. See also: *step fault.* (Nelson, 1965)
multiple firing Firing electrically with delay blasting caps in a number of holes at one time. (Mitchell, 1950)
multiple fuse-igniter cartridge Consists of a cardboard cartridge about 2 in long with a 3/4-in (1.9-cm) outside diameter. The closed end is coated with black powder, and the ends of eight fuses are inserted in the cartridge, in contact with the powder. A master fuse is then inserted, which, when lit, burns to the powder and ignites it. The powder flares brightly and lights the eight fuses. (South Australia, 1961)
multiple geophones *multiple detectors.*
multiple intersections (a) The intercepts that cross a vein, orebody, or other geologic feature, accomplished by drilling several auxiliary boreholes from a single, main, or parent borehole with the aid of wedges and similar deflecting devices. (Long, 1960) (b) Intercepting a steeply dipping vein at various depths by changing the angle of the drill head.
multiple intrusion Any type of igneous intrusion that has been produced by several injections separated by periods of crystallization. Chemical composition of the various injections is approx. the same. Cf: *composite intrusion.* (AGI, 1987)
multiple lines A single line reeved around two or more sheaves so as to increase pull at the expense of speed. (Nichols, 1976)
multiple openings Any series of underground openings separated by rib pillars or connected at frequent intervals to form a system of rooms and pillars. (Obert, 1960)
multiple-ply plate Steel plate made up of thicknesses of other plates of steel or steel and wrought iron welded together. (Mersereau, 1947)
multiple recording groups *multiple detectors.*
multiple ribbon belt conveyor A belt conveyor having a conveying surface of two or more spaced strands of narrow flat belts.
multiple-row blasting The drilling, charging, and firing of several rows of vertical holes along a quarry or opencast face. The holes may be spaced in the square pattern with delay detonators in the rows as well as row by row. The spacing of the holes will vary according to their depth, diameter, and the type of rock. See also: *small-diameter blastholes.* (Nelson, 1965)
multiple-seam mining Mining two or more seams of coal, frequently close together, that can be mined profitably where mining one alone would not be profitable. (Coal Age, 1966)
multiple seismometers *bunched seismometers.*
multiple series A method of wiring a large group of blasting charges by connecting small groups in series and connecting these series in parallel. Syn: *parallel series.*
multiple shooting The firing of an entire face at one time. The holes are connected in a single series and all the holes shoot at the same instant. (Kentucky, 1952)
multiple shot *battery of holes.*
multiple-shot blasting unit A multiple-shot blasting unit is designed for firing simultaneous explosive charges in mines, quarries, and tunnels. Syn: *blasting unit.*
multiple shotholes Two or more shotholes that are shot almost simultaneously. They are so spaced as to minimize near-surface interferences that mask desired signals if only one shothole is used. (AGI, 1987)
multiple-shot instrument A borehole-survey instrument capable of taking and recording a series of inclination and bearing readings on a single trip into a borehole. Cf: *single-shot instrument.* (Long, 1960)
multiple-shot survey A borehole survey using a multiple-shot instrument. (Long, 1960)
multiple sill A sill made up of two or more intrusions composed of the same kind of igneous rock. (Billings, 1954)
multiple speed floating control system As used in flotation, a form of floating control system in which the manipulated variable may change at two or more rates each corresponding to a definite range of values of the actuating signal. (Fuerstenau, 1962)
multiple splitting The parting or separation of a thick seam into more than two layers of coal. See also: *simple split seam.* (Nelson, 1965)
multiple-strand chain A roller chain made up of two or more strands assembled as a single structure on pins extending through the entire assembly. (Jackson, 1955)
multiple-strand conveyor (a) Any conveyor which employs two or more spaced strands of chain, belts, or cords as the load-supporting medium. (b) Any conveyor in which two or more strands are used as the propelling medium connecting pans, pallets, etc.
multiple-strand rope A wire rope designed to obviate spinning due to untwisting. It is formed by a series of layers of strands built around a center fiber core. Each layer of strands is given a lay opposite to that on which it is built, each layer thus tending to impart its own twist that is cancelled by the next layer. Therefore, this rope can be used for sinking or where a free load is to be lifted. The stretch with a multiple

strand rope is not so great as with round strand and flattened strand ropes. See also: *multistrand rope; wire rope.* (Nelson, 1965)
multiple twin *repeated twinning.*
multiple-vent basalt *shield basalt.*
multiple wedge *plug-and-feather method.*
multiplow A layout consisting of six or more plows, 220 lb (100 kg) each, 20 yd (18.3 m) apart on one rope or chain, feeding onto an armored conveyor; the load on the conveyor is well distributed. A driving unit is arranged at both ends of the face and operated alternately to impart a to-and-fro movement to the plows. The minimum workable seam thickness is 20 in (50.8 cm) at gradients from 0° to 20°; maximum length of face is about 190 yd (174 m). See also: *Gusto multiplow.* (Nelson, 1965)
multiplying constant The constant, used in stadia work, by which the staff intercept is multiplied to determine the distance between the staff and the theodolite. The value is generally taken as 100. See also: *tachymeter.* (Hammond, 1965)
multirope friction winder A winding system based on the principles of the Koepe winder. The drive to the winding ropes is the frictional resistance between the ropes and the driving sheaves. Multirope friction winders are usually tower mounted, with either cages or skips, and provided with a counterweight. The sheaves are from about 6 to 12 ft (1.8 to 3.7 m) in diameter with a direct coupled or geared drive. Four ropes are favored and these operate in parallel and share the total suspended load. The system was introduced partly because of the difficulty of winding heavy loads from deep shafts with a single large-diameter winding rope. Modern winding ropes have become large and heavy, being 2-1/4 in (5.72 cm) in diameter locked coil, weighing 16.5 st (15 t) for a 1,000-yd (915-m) shaft; therefore, the introduction of the friction winder, with its counterweight, and using four smaller ropes side by side in place of one. Such ropes need be only 1-1/4 in (3.2 cm) in diameter to give equivalent breaking strain. See also: *Koepe winder.* (Nelson, 1965)
multishot firing The firing of a number of shots simultaneously. See also: *M.E. 6 exploder.* (Nelson, 1965)
multishot gyroscopic instrument A borehole surveying instrument that can take a number of readings during its descent and ascent in the borehole. It comprises gyroscopic and photographic recording units; direction and inclination indicators; a timing clock, and other accessories. A movie film enables numerous records to be taken throughout the depth of the borehole. See also: *gelatin borehole tube.* (Nelson, 1965)
multispectral scanner A remote sensing device that is capable of recording data in the ultraviolet, visible, and infrared portions of the spectrum. Syn: *shuttle multispectral infrared radiometer.* See also: *thematic mapper.* (ASPRS, 1984)
multistage fan A fan having two or more impellers working in series. (BS, 1963)
multistrand rope A flexible, nonspinning rope; composed of concentric layers of strands of relatively fine wires, alternate layers of strands being wound in opposite directions over a hemp core. See also: *multiple-strand rope.* (Sinclair, 1959)
multiwheel roller A heavy roller with pneumatic tires used to consolidate embankments. (Hammond, 1965)
mummification The process of preservation of plant tissues under the influence of arrested decay. (Tomkeieff, 1954)
mun Corn. Any fusible metal.
mundic A drillers' term for pyrite. See also: *pyrite.* (AGI, 1987)
Munroe effect The concentration of explosive effect (i.e. jetting) which occurs at a cavity at the end of an explosive charge; this effect is the basis of the design and performance of shaped charges.
muon Contraction of mu-meson. An elementary particle with 207 times the mass of an electron. It may have a single positive or negative charge. (Lyman, 1964)
Murakami's reagent An etching reagent developed for use in the investigation of the structure of iron-carbon-chromium alloys. It consists of a solution of 10 g potassium ferricyanide, 10 g potassium hydroxide, and 100 mL water. (Osborne, 1956)
murasakite A schistose rock composed of piedmontite and quartz. See also: *ollenite.* (AGI, 1987)
murdochite An isometric mineral, $PbCu_6O_8(Cl,Br)_2$; rare; secondary; forms minute black octahedra.
Murex process A flotation process that is not strictly of the same class as others, but still makes use of the principle of selective oiling of sulfide particles. The crushed ore is fed into an agitator and mixed with 4% to 5% of its weight of a paste made of one part of oil or thin tar with three or four parts of magnetic oxide of iron. This oxide must be ground to an impalpable powder. These ingredients, with enough water to make a pulp, are agitated for 5 to 20 min. The paste preferentially adheres to the sulfides because of the oil. The ore is then fed over magnets, and oxide of iron, with the mineral adhering to it, is pulled out. The oil and magnetite are then recovered. (Liddell, 1918)
Muschamp coal miner A cutter loader which is essentially a conversion unit for Anderton shearer machines and designed to produce a reasonable percentage of large coal. The top and bottom of the seam are cut by rotating drums of small diameter while shearing the back of the cut with a narrow-kerf jib and chain. (Nelson, 1965)
muscovado A term applied in Minnesota to rusty-colored outcropping rocks, such as gabbros and quartzites, that resemble brown sugar. Etymol: Spanish, brown sugar. (AGI, 1987)
muscovite A monoclinic mineral, $KAl_2(Si_3Al)O_{10}(OH,F)_2$; mica group; pseudohexagonal; perfect basal cleavage; forms large, transparent, strong, electrically and thermally insulating, stable sheets; a common rock-forming mineral in silicic plutonic rocks, mica schists, gneisses, and commercially in pegmatites; also a hydrothermal and weathering product of feldspar and in detrital sediments. Also spelled: moscovite. Syn: *isinglass; white mica; potash mica; common mica; Muscovy glass; mirror stone.* Sericite is fine-grained muscovite, commonly in connection with hydrothermal alteration, but sericite also includes paragonite and illite. Illite is a common syn. for fine-grained muscovite in clay mineralogy.
Muscovy glass *muscovite.*
mushroom jib A standard form of coal cutter jib with a sprocket at the end remote from the machine. The sprocket carries a vertical turret or bar and is driven by the cutting chain. The bar makes a vertical cut at the back of the normal horizontal cut. See also: *turret jib.* (Nelson, 1965)
muskeg A bog, usually a sphagnum bog, frequently with grassy tussocks, growing in wet, poorly drained boreal regions, often areas of permafrost. Tamarack and black spruce are commonly associated with muskeg areas. Syn: *maskeeg.* (AGI, 1987)
mussel bed A band containing or chiefly composed of mussellike shells, very valuable in the correlation of Coal Measures strata. (Mason, 1951)
Musso process A mixture of iron ore and fuel is reduced in an externally heated rotary retort. The gases are exhausted and constitute the fuel when the process has been started. The gases, after purification, are passed through combustion rings surrounding the retort and are burned according to the method of catalytic combustion. After reduction, the charge is cooled when it is poured through a layer of fluxing material; it is then transferred to a steelmaking furnace. (Osborne, 1956)
mustard gold A spongy type of free gold found in the gossan above gold-silver-telluride deposits. (AGI, 1987)
mutabilite A soft corklike bitumen of porous or resinous consistency. Partly soluble in organic solvents. (Tomkeieff, 1954)
muthmannite A mineral, (Ag,Au)Te; soft; heavy; gray-white; in tabular crystals with perfect cleavage; an ore mineral.
mutu A Malayan term denoting the degrees of fineness of gold.
mutual boundary texture A rock texture showing smooth, regular, curved contacts between minerals. (Schieferdecker, 1959)
mvb Abbrev. for medium-volatile bituminous. See also: *medium-volatile bituminous coal.*
mylonite A compact, chertlike rock without cleavage, but with a streaky or banded structure, produced by the extreme granulation and shearing of rocks that have been pulverized and rolled during overthrusting or intense dynamic metamorphism. Mylonite may also be described as a microbreccia with flow texture. See also: *cataclasite; protomylonite; mylonite gneiss; ultramylonite.* (AGI, 1987)
mylonite gneiss A mylonitic rock that has been partly recrystallized. See also: *augen schist; cataclasite; flaser gabbro; mylonite; phyllonite.* (AGI, 1987)
mylonitization Deformation of a rock by extreme microbrecciation, due to mechanical forces applied in a definite direction, without noteworthy chemical reconstitution of granulated minerals. Characteristically the mylonites thus produced have a flinty, banded, or streaked appearance, and undestroyed augen and lenses of the parent rock in a granulated matrix. Also spelled mylonization. (AGI, 1987)
mylonization *mylonitization.*
myrickite (a) A white or gray chalcedony, opal, or massive quartz unevenly colored by, or intergrown with, pink or reddish inclusions of cinnabar, the color of which tends to become brown. The opal variety is know as opalite. (b) Cinnabar intergrown with common white opal or translucent chalcedony.
myrmekite An intergrowth of plagioclase feldspar (commonly oligoclase) and vermicular quartz, generally replacing potassium feldspar; formed during the later stages of consolidation in an igneous rock or during a subsequent period of plutonic activity. The quartz occurs as blobs. *vermicular quartz.*
myrmekite antiperthite Myrmekitelike intergrowth of predominant plagioclase and vermicular orthoclase. The wormlike forms of orthoclase are, as a rule, broader than those of quartz in the typical myrmekite. (Schieferdecker, 1959)
myrmekite perthite Myrmekitelike intergrowth of microcline and vermicular plagioclase. (Schieferdecker, 1959)

N A Diamond Core Drill Manufacturer's Association letter name for a range of diamond drill fittings intended to be used together with the appropriate casing having an inside diameter of 81 mm or somewhat less. (Cumming, 1981)

nablock A rounded mass, as of flint in chalk, or of ironstone in coal. (Standard, 1964)

nacre The hard, iridescent internal layer of various mollusk shells, having unusual luster and consisting chiefly of calcium carbonate in the form of aragonite deposited as thin tablets normal to the surface of the shell and interleaved with thin sheets of organic matrix. Syn: *pearly; mother-of-pearl.* (AGI, 1987)

nacreous Adj. Applied to the luster of certain minerals, usually on crystal faces parallel to a good cleavage, the luster resembling that of pearls. (CMD, 1948)

nacrite A monoclinic mineral, $Al_2Si_2O_5(OH)_4$; kaolinite-serpentine group; differs from other kandites, such as triclinic kaolinite and monoclinic dickite, in the stacking order of its layers (polytypy); fine-grained, massive; commonly associated with hydrothermal alteration.

nadir (a) The point on the celestial sphere that is directly beneath the observer and directly opposite the zenith. (AGI, 1987) (b) The point on the ground vertically beneath the perspective center of an aerial-camera lens. (AGI, 1987)

nadorite An orthorhombic mineral, $PbSbO_2Cl$; an uncommon secondary mineral; in resinous to adamantine crystals in some base-metal deposits.

naetig Industrial diamond having a grain in all directions instead of in regular layers. Cf: *feinig.* (Brady, 1940)

nagyagite A possibly orthorhombic mineral, $Pb_5Au(Te,Sb)_4S_{5-8}$; occurs with other tellurides in lamellar, soft, heavy, lead-gray crystals; one perfect cleavage but may form anhedral masses. Syn: *black tellurium.*

nahcolite A monoclinic mineral, $NaHCO_3$; forms small, white, highly soluble, prismatic crystals; in nonmarine evaporite deposits; esp. abundant in parts of the Green River lake beds of Colorado.

naif adj. Said of a gemstone having a true or natural luster when uncut; e.g., of the natural, unpolished faces of a diamond crystal. Also spelled naife.

nail (a) *shooting needle.* (b) A slender piece of metal, one end of which is pointed, the other end having a head, either flattened or rounded. It is a common means of fastening together several pieces of wood or other material by striking the head with a hammer. The term penny as applied to nails refers to the number of pounds per 1,000 nails; e.g., six-penny nail means 6 lb (2.7 kg) per 1,000; three-penny means 3 lb (1.4 kg) per 1,000, etc. (Crispin, 1964)

nailhead spar A composite variety of calcite having the form suggested by the name. (Fay, 1920)

naked *bare; blank hole.*

name of lode The name by which a lode is designated in the notice of location, and subsequent addition thereto is immaterial. The same vein or lode may have different names in different mining locations. (Ricketts, 1943)

nanotesla *gamma.*

Nansen bottle An oceanographic water-sampling bottle, made of a metal alloy that is little reactive with seawater, equipped with a rotary valve at each end so that when it is rotated at depth the valves close and lock shut, entrapping a water sample and setting the reversing thermometers. This bottle is named for its designer, Fridtjof Nansen. (Hy, 1965)

nantokite An isometric mineral, CuCl; granular, massive; cubic cleavage; adamantine luster; an uncommon secondary mineral in copper deposits.

naphtha (a) An archaic term for liquid petroleum. (AGI, 1987) (b) Designates those hydrocarbons of the lowest boiling point (under 250 °C) that are liquid at standard conditions, but easily vaporize and become flammable. They are used as cleaners and solvents. (AGI, 1987)

naphtha gas Illuminating gas charged with the decomposed vapor of naphtha. (Standard, 1964)

naphthode Concretion of bituminous limestone rich in carbonaceous matter. (Tomkeieff, 1954)

Napierian logarithm A natural logarithm. (Webster 3rd, 1966)

napoleonite A variety of hornblende. (Standard, 1964)

nappe (a) A sheetlike, allochthonous rock unit, which has moved on a predominantly horizontal surface. The mechanism may be thrust faulting, recumbent folding, or both. The term was first used for the large allochthonous sheets of the western Alps, and it has been adopted into English. The German equivalent, decke, is also sometimes used in English. Etymol: French, cover sheet, tablecloth. Syn: *decke.* See also: *klippe.* (AGI, 1987) (b) Belg. *aquifer.*

napthoid Liquid petroleumlike product found in cavities of igneous rocks and assumed to be a product of thermal distillation of bituminous substances contained in the country rocks. (Tomkeieff, 1954)

narrow (a) A roadway driven in the solid coal with rib sides. All roadways when opening out a pillar method of working are narrow. See also: *working the whole.* (Nelson, 1965) (b) N. of Eng. A gallery, or roadway, driven at right angles to a drift, and not quite so large in area. (Fay, 1920)

narrow gage A railway gage narrower than the standard gage of 4 ft 8-1/2 in (1.44 m). (Hammond, 1965)

narrow place Aust. Working place that is less than 6 yd (5.5 m) wide; these are paid for by the yard in length. (Fay, 1920)

narrow stall A stall driven in solid coal, usually from 6 to 9 ft (1.8 to 2.7 m) wide; it has rib sides in coal. See also: *rib-side; solid road.* (Nelson, 1965)

narrow work (a) The driving of narrow stalls to form coal pillars as the first stage in the pillar-and-stall method of working. (Nelson, 1965) (b) A system of mining in which narrow coal roadways, called endings, are driven along the strike of the seam, from 12 to 15 yd (11.0 to 13.7 m) apart, to a limit line. The long narrow coal pillars between the endings are extracted on the retreat. It has been adopted in parts of Yorkshire and Lancashire, England. See also: *endings; main endings.* (Nelson, 1965) (c) All work for which a price per yard of length driven is paid, and which, therefore, must be measured. Any dead work. (Fay, 1920) (d) Penn. Headings, chutes, crosscuts, gangways, etc., or the workings previous to the removal of the pillars. (Fay, 1920) (e) A working place in coal only a few yards in width. (Fay, 1920)

narrow working *bord-and-pillar method; narrow work.*

nascent Just formed by a chemical reaction, and therefore very reactive. Nascent gases are probably in an atomic state. (CTD, 1958)

Nasmyth hammer A steam hammer, having the head attached to the piston rod, and operated by the direct force of the steam. (Fay, 1920)

Na-spar *soda spar.*

National coarse thread The screw thread of common use, formerly known as the United States Standard thread. (Crispin, 1964)

National Electrical Code A set of rules to guide electricians when installing electrical conductors, devices, and machinery. (Crispin, 1964)

national grid Gr. Brit. A system of rectangular coordinates used by the Ordnance Survey and based upon the Transverse Mercator Projection (which is also known as the Gauss Conformal Projection). (BS, 1963)

national grid coordinates Gr. Brit. Coordinates, referred to the National Grid of the Ordnance Survey, which are specified in meters and consist of two components, an Easting and a Northing. (BS, 1963)

Nationalization Act The Coal Industry Nationalization Act, 1946, which brought all coal mines in the United Kingdom under public ownership. It was passed through Parliament in July 1946 and put into operation on January 1, 1947. (Nelson, 1965)

National Physical Laboratory British government organization that, among other things, tests and certifies calibration of scientific glassware, weights, and measures. Abbrev. NPL. (Pryor, 1963)

native (a) Occurring in nature, either pure or uncombined with other substances. Usually applied to metals, such as native mercury, native copper. Also used to describe any mineral occurring in nature in distinction from the corresponding substance formed artificially. (b) Adj. Applied to earth materials occurring in elemental form; e.g., nugget gold, metallic copper. Syn: *native element.*

native copper (a) Metallic copper, sometimes containing a little silver and bismuth, that occurs as a metasomatic deposit filling cracks and forming the cement of sandstone and conglomerate. Such deposits have been found in the Keweenaw Peninsula, Lake Superior, MI, USA;

Chile; Queensland, Australia; and Zimbabwe. Native copper is also found in the upper workings of copper mines. (b) A mineral in the form of particles and nuggets of very pure metallic copper associated (but not alloyed) with small amounts of silver and arsenic minerals. It is found in small amounts in many copper ores but occurs in commerical quantities in only one place in the world, the Upper Peninsula in Michigan. See also: *copper.* (Newton, 1959)

native element Element that occurs in nature uncombined, such as nugget gold, metallic copper, etc. See also: *native.* (Fay, 1920)

native mercury *quicksilver.*

native nickel-iron *awaruite; josephinite.*

native water *formation water.*

natrite A monoclinic mineral, $NaCO_3$; occurs in very soluble, white, granular masses; a rare mineral in the alkalic complexes of the Kola peninsula, Russia.

natrium *sodium.*

natrochalcite A monoclinic mineral, $NaCu_2(SO_4)_2(OH)\cdot H_2O$; bright emerald-green; forms steep pyramidal crystals; a secondary copper mineral at Chuquicamata, Chile.

natrojarosite A trigonal mineral, $NaFe^{3+}{}_3(SO_4)_2(OH)_6$; alunite group; in earthy masses and minute scales; a common alteration product in pyrite-bearing deposits. Formerly called utahite.

natrolite An orthorhombic mineral, $Na_{16}[Al_{16}Si_{24}O_{80}]\cdot 16H_2O$; zeolite group, which may contain appreciable calcium; dimorphous with tetranatrolite; forms acicular to slender prismatic crystals in cavities and veins, esp. in mafic and alkaline igneous rocks. Syn: *laubanite.*

natromontebrasite A triclinic mineral, $(Na,Li)Al(PO_4)(OH,F)$; amblygonite group; in granitic pegmatites associated with other lithium minerals. Formerly called fremontite.

natron A monoclinic mineral, $Na_2CO_3\cdot 10H_2O$; very soluble; forms earthy or granular crusts and efflorescences; in nonmarine evaporites and on lavas.

natrona *sodium sesquicarbonate.*

natron granite Granite abormally high in soda, presumably from the presence of a soda-rich orthoclase or of anorthoclase. It is also called soda granite.

nattle Eng. To make a slight rattling or tapping noise. Said of a mine when movement or settling is taking place. (Webster 3rd, 1966; Fay, 1920)

natural air crossing An air crossing in which two airways are separated by rock in its natural state. (BS, 1963)

natural asphalt Asphalt before crushing or refining, as mined or quarried in the case of natural rock asphalt, or surface-excavated in the case of lake deposits. (Hammond, 1965)

natural carbon Carbon found in a shape that has not been artificially modified. Also called natural stone. (Long, 1960)

natural cement A hydraulic cement produced by pulverizing and then heating naturally occurring rock (cement rock) containing appropriate proportions of limestone, clay, magnesia, and iron. Ignition temperatures are usually lower than for portland cement. Final pulverizing is necessary as with portland cement.

natural coke Coke made by natural processes, usually by the intrusion of an igneous dike. See also: *cinder coal; clinker; coke coal; cokeite.* (Hess)

natural convection *convection.*

natural diamond This abrasive is the densest form of crystallized carbon, the hardest substance known. It occurs most commonly as well-developed crystals in volcanic pipes or in alluvial deposits. Bort (boart or borts) sometimes refers to all diamonds not suitable for gems, or it may refer to off-color, flawed, or impure diamonds not fit for use for gems or most other industrial applications, but suitable for the preparation of diamond grain and powder for use in lapping or the manufacture of most diamond grinding wheels. This type of bort is also called crushing bort or fragmented bort. (ACSG, 1963)

natural Earth current Electric current in the Earth not due to human activity. (Schieferdecker, 1959)

natural face A name given to the X direction as pencilled on Z sections of unfaced quartz and whose position is determined by X-ray measurements or etching. The name is also given to the artificial prism face (parallel to 1120) thus located, and produced by sawing the section in the YZ plane. Also applied to the natural growth faces on faced raw quartz crystals. (Am. Mineral., 1947)

natural floatability *inherent floatability.*

natural frequency The frequency of free oscillation of a system. For a multiple-degree-of-freedom system, the natural frequencies are the frequencies of the normal modes of vibration. (Hy, 1965)

natural frequency of a foundation The frequency of free vibration of a complete soil-foundation oscillating system. This frequency must differ distinctly from that of any machinery carried by the foundation if resonance is to be avoided. (Hammond, 1965)

natural frequency vibrating conveyor A vibrating conveyor in which the rate of free vibration of the trough on its resilient supports is approx. the same as the rate of vibration induced by the driving mechanism.

natural gamma radiation detector This is a type of sensor that measures bursts of high-energy electromagnetic waves that are emitted spontaneously by some naturally occurring radioactive elements, such as potassium-40, thorium-232, and uranium-238, that are commonly present in shales and clays. (Mowrey, 1991)

natural gamma-ray logging A process whereby gamma rays naturally emitted by formations traversed by a borehole are measured. A tool containing a radiation detector is lowered into the borehole and gamma ray measurements are transmitted to the Earth's surface. The signals are utilized to produce a record of gamma rays detected in correlation with the depth of the detector in the borehole. The record thus obtained in the form of a curve indicating relative number per unit of time of natural gamma rays at different depths, is a conventional natural gamma-ray log, sometimes simply called a gamma-ray log. (Williams, 1964)

natural gas A mixture of the low-molecular-weight paraffin series hydrocarbons methane, ethane, propane, and butane, with small amounts of higher hydrocarbons; also frequently containing small or large proportions of nitrogen, carbon dioxide, and hydrogen sulfide, and occasionally small proportions of helium. Methane is almost always the major constituent. Natural gas accompanying petroleum always contains appreciable quantities of ethane, propane, butane, as well as some pentane and hexane vapors, and is known as wet gas. Dry gas contains little of these higher hydrocarbons. The exact composition of natural gas varies with locality. The heating value of natural gas is usually over 1,000 Btu/ft^3 unless nitrogen or carbon dioxide are important components of the gas. See also: *gas; sour gas.* (CCD, 1961)

natural glass A vitreous, amorphous, inorganic substance that has solidified from magma too quickly to crystallize. Granitic or acid natural glass includes pumice and obsidian; an example of a basaltic natural glass is tachylite. (AGI, 1987)

natural logarithm A logarithm with *e* as a base. (Webster 3rd, 1966)

naturally bonded molding sand A term used by foundrymen to refer to a sand that, as mined, contains sufficient bonding material for molding purposes. (Osborne, 1956)

natural ore (a) In iron mining, the term given to naturally occurring high-grade iron ore; consists of: (1) soft ore, such as porous hematite and limonite (goethite) with minor magnetite and manganese oxides; and (2) hard ores, such as compact, fine-grained, steel-gray hematite, specular hematite, magnetite, or martite. Syn: *direct shipping ore.* (Marsden, 1933/1967) (b) Iron ore that contains moisture, in contrast to "dry ore" that has been dried but not calcined. (USBM, 1965) (c) A term used in the U.S. Lake Superior mining district to indicate iron ore formed by natural processes, as distinguished from iron ore products (such as pellets) produced artificially from hard, low-grade ores of the taconite type. Published prices for natural ores from this district usually specify iron content as "natural," meaning that the analysis is based on gross weight including moisture. Syn: *lump ore; silicious ore; high-silica ore; wash ore; heavy-media ore; manganiferous ore; natural ore concentrate; lean ore; paint-rock ore; Mesabi non-Bessemer ore.*

natural ore concentrate *natural ore.*

natural paper Brownish paperlike deposit formed from the filaments of Conferva. (Tomkeieff, 1954)

natural resin *resin.*

natural sand Sand derived from a rock, in which the grains separate along their natural boundaries. This includes unconsolidated sand or a soft sandstone where little pressure is required to separate the individual grains. (Osborne, 1956)

natural scale Applied to a drawing made to equal vertical and horizontal scales. (Hammond, 1965)

natural slope The maximum angle at which loose material in a bank or spoil heap will stand without slipping. See also: *angle of rest; angle of slide.* (Nelson, 1965)

natural splitting In mine ventilation, a practice that allows the airflow to divide among the branches of its own accord and without regulation, in inverse relation to the resistance of each airway. Cf: *controlled splitting.* (Hartman, 1982)

natural stress relief The failure of the skin rock of an excavation by crushing, shear, or plastic flow. It can occur on free surface rock with explosive force. See also: *arching.*

natural ventilating pressure (a) In a mine, air returning from the workings to the surface via the upcast shaft can be of a higher temperature than the air in the downcast shaft because of heat added to the ventilation current from the strata exposed in the mine. Thus, even in a mine with the fan stopped, the upcast air density is less than the downcast air density. This lack of balance in the two vertical air columns produces a pressure difference across the shaft bottom doors known as natural ventilating pressure. (Roberts, 1960) (b) The ventilating pressure that produces natural ventilation. (BS, 1963)

natural ventilation The ventilation produced in a mine as a result of a difference in density of the air in the upcast and downcast shafts, brought about by natural causes. Natural ventilation is feeble, seasonal, and inconstant. (Nelson, 1965)

nauckite A variety of resin. (Tomkeieff, 1954)

naumannite An orthorhombic mineral, Ag_2Se; pseudocubic; black; lustrous; forms heavy, sectile granular masses; also in thin plates; epithermal; a major source of silver in some deposits.

nautical chart A representation on a horizontal plane, and according to a definite system of projection, of a portion of the navigable waters of the Earth, including the shorelines, the topography of the bottom, and aids and dangers to navigation; it may be derived from hydrographic, topographic, or aerial surveys, or a combination thereof. (Seelye, 1951)

nautical measure One nautical mile or knot equals 6,080.20 ft (1,853.248 m); 3 nmi equal 1 league; and 60 nmi equal 1 degree of longitude (at the equator). (Crispin, 1964)

nautical mile Any of various units of distance, used for sea and air navigation, based on the length of a minute of arc of a great circle of the Earth and differing because the Earth is not a perfect sphere: (1) a British unit that equals 6,080 ft or 1,853.2 m; also called Admiralty mile; (2) a U.S. unit, no longer in official use, that equals 6080.20 ft or 1,853.248 m; and (3) an international unit that equals 6,076.1033 ft or 1,852 m; used officially in the United States since July 1954. (Webster 3rd, 1966)

navajoite A monoclinic mineral, $(V,Fe)_{10}O_{24}{\cdot}12H_2O$; rare; weakly radioactive; soft; a fibrous mineral with a silky luster; associated with corvusite, hewettite, tyuyamunite, rauvite, steigerite, and limonite.

Navier's hypothesis An assumption in the design calculation of beams. It states that the stress at any point due to bending is assumed as being proportional to its distance from the neutral axis. (Hammond, 1965)

NCB The National Coal Board of the United Kingdom. (Nelson, 1965)

NCB boring tower A boring tower developed by the National Coal Board of the United Kingdom to make test drillings for coal from positions off the coast. When drilling is in progress, the tower is resting on the seabed. The base is divided into four airtight sections, which are filled with water when the tower is in position for drilling. The water is pumped out to give buoyancy when the tower is refloated for towing to a new drilling site. The tower is designed to withstand gales of 80 mph (128 km/h) and waves of 30 ft (9.1 m) from crest to trough, and it can be used in any depth of water up to 120 ft (36.6 m). The overall height of the tower is 189 ft (57.6 m), and its total weight is about 570 st (517 t). It has reached over 3,000 ft (915 m) drilling depth with core recovery. The first borehole was put down in the Firth of Forth, Scotland. (Nelson, 1965)

NCB recorder This butane combustible gases recorder has a small flame of burning butane gas, which is controlled to give constant heat output with varying ambient temperature and humidity, and with varying butane gas pressure. The heat output is measured by means of a group of thermocouples in a chimney above the flame. The presence of methane in the atmosphere, which has access to the flame via suitable gauzes, increases the voltage generated by the thermocouples. These changes are recorded on a rotating chart calibrated in percentage methane. (Roberts, 1960)

neap tide In oceanography, a high tide occurring at the moon's first or third quarter, when the sun's tidal influence is working against the moon's, so that the height of the tide is below the maximum in the approximate ratio of 3:8. (CTD, 1958)

nearest neighbor interpolation method A method of assigning a sample value to a point in space. The value assigned is equal to the value of the spatially nearest sample data point. This method is sometimes used as a computer equivalent of the polygonal method of interpolation.

near-gravity material A washability term popularly defined as the percentage of material in the raw coal within ±0.1 of the separating specific gravity. (Nelson, 1965)

near-mesh Near-sized; grains close in cross section to a specified screening mesh, which tend to blind apertures and slow down sizing. Syn: *near-mesh material.* (Pryor, 1965)

near-mesh material Material approximating in size to the mesh aperture. (BS, 1962)

neat Cement slurries containing no aggregate, such as sand or gravel. Syn: *neat cement.* (Long, 1960)

neat cement A slurry composed of any cement and water. Syn: *neat cement.* (Brantly, 1961)

neat lines The excavation lines of a tunnel within which the rock removed is paid for at the agreed contract rate. See also: *overbreak.* (Nelson, 1965)

nebulite A type of mixed rock whose fabric is characterized by indistinct, streaky inhomogeneities or schlieren and in which no sharp distinction can be made between the component parts of the fabric. Adj: nebulitic. (AGI, 1987)

neck (a) A lava-filled conduit of an extinct volcano exposed by erosion; also called chimney rock or plug. See also: *plug.* (b) A pipe of igneous rock crossing bedding planes. (Mason, 1951) (c) The narrow entrance to a room next to the entry, or a place where the room has been narrowed on account of poor roof. (Fay, 1920) (d) A narrow stretch of land, such as an isthmus or a cape. (Webster 3rd, 1966) (e) A narrow body of water between two larger bodies; a strait. (Webster 3rd, 1966)

needle (a) A piece of copper or brass about 0.5 in (1.3 cm) in diameter and 3 ft or 4 ft (0.9 m to 1.2 m) long, pointed at one end, and turned into a handle at the other, tapering from the handle to the point. It is thrust into a charge of blasting powder in a borehole, and while in this position the borehole is tamped solid, preferably with moist clay. The needle is then withdrawn carefully, leaving a straight passageway through the tamping for the miner's squib to shoot or fire the charge. (Fay, 1920) (b) A timber set on end to close an opening for the control of water; it may be either vertical or inclined; a form of stop plank. (Seelye, 1951) (c) A small metal rod for making the touchhole in the powder used for blasting. (d) A hitch cut in the side rock to receive the end of a timber. (e) A needle-shaped or acicular mineral crystal.

needle bearing An antifriction bearing using very small-diameter rollers between wide faces. (Nichols, 1976)

needled Pocketed, as when face bars are set with the face end of the bar pocketed into the coal adjacent to the roof. (TIME, 1929-30)

needle instrument Any surveying instrument controlled by a magnetic needle. See also: *compass.* (Hammond, 1965)

needle ore (a) Iron ore of very high metallic luster, found in small quantities, which may be separated into long, slender filaments resembling needles. (AGI, 1987) (b) *aikinite.*

needles Elongated crystals, tapering at each end to a fine point, as those typical of martensite. (Rolfe, 1955)

needle traverse In a survey with a dial (compass), use of a magnetic needle to read the bearing of lines. Opposite is fast needle traverse or work, and refers to the use of a dial as in traversing with a theodolite, where proximity of iron might deflect the needle. Systems can be combined, using needle readings where iron is absent. Also called swinging needle traverse; loose needle traverse. (Pryor, 1963)

negative crystal (a) A birefringent crystal in which the refractive index of the extraordinary ray is less than that of the ordinary ray. (b) A cavity within a crystal bounded by the crystal faces of that crystal. See also: *inclusion; three-phase inclusion.*

negative element A large structural feature or portion of the Earth's crust, characterized through a long period of geologic time by frequent and conspicuous downward movement (subsidence, submergence), extensive erosion, or an uplift that is considerably less rapid or less frequent than those of adjacent positive elements. (AGI, 1987)

negative elongation Lathlike, rodlike, or acicular crystals in which the slow polarized light ray lies across the long direction of the crystal. Cf: *positive elongation.* Syn: *length fast.*

negative moment *hogging moment.*

negative rake (a) The orientation of a cutting tool in such a manner that the angle formed by the leading face of the tool and the surface behind the cutting edge is greater than 90°. Syn: *drag rake.* See also: *gouge angle.* (Long, 1960) (b) Describes a tooth face in rotation whose cutting edge lags the surface of the tooth face. Cf: *rake.* (ASM, 1961)

negligence In a legal sense, a failure upon the part of a mine operator to observe for the protection of the interests of the miner that degree of care, precaution, and vigilance that the circumstances justly demand, whereby the miner suffers injury. (Ricketts, 1943)

neighborite An orthorhombic mineral, $NaMgF_3$; insoluble; forms rounded grains or pseudo-octahedral crystals; associated with dolomite and quartz in oil shale within the Green River Formation, UT.

nekton A biological division made up of all the swimming animals found in the pelagic division. (Hy, 1965)

Nelson Davis separator A cylindrical dense-medium washer developed in the United States. It uses a magnetite water suspension as medium. The bath resembles a drum in shape, its longitudinal axis being horizontal; within the stationary outer casing there is a rotor divided into compartments. Raw coal is fed near the top of the separator, and separation takes place as the rotor revolves. The machine produces clean coal and shale; the magnetite is recovered. It can handle coal up to 10 in (254 mm) in size, the lower limit being about 1/4 in (6.4 mm). Magnetite consumption runs at about 1/2 lb/st (0.25 kg/t) of feed. See also: *Leebar separator.* (Nelson, 1965)

nelsonite A rock composed essentially of ilmenite and apatite, with or without rutile. The ratio of ilmenite to apatite varies widely. Cf: *ilmenitite.*

nematoblastic Pertaining to the texture of a recrystallized rock in which the shape of the grains is threadlike. See also: *fibroblastic.* (Pettijohn, 1957)

neokerogen Organic debris deposited among marine sediments and modified by bacterial action in such a way as to form the source material of petroleum, or, under certain conditions, to form the kerogen of oil shales. (Tomkeieff, 1954)

neolite A silky, fibrous, stellated, green, hydrous magnesium-aluminum silicate. (Standard, 1964)

Neolithic In archaeology, the last division of the Stone Age, characterized by the development of agriculture and the domestication of farm animals. Correlation of relative cultural levels with actual age (and, therefore, with the time-stratigraphic units of geology) varies from region to region. Adj: pertaining to the Neolithic. (AGI, 1987)

neomesselite *messelite.*

neomineralization Chemical interchange within a rock whereby its mineral constituents are converted into new mineral species; a type of recrystallization. (AGI, 1987)

neomorphic Said of the mineral grains of a rock that have been regenerated by zones of secondary growth in crystalline continuity. The new material may have been deposited from solutions or from molten fluids.

neoprene plug closure The function of the neoprene plug is to provide a completely waterproof seal at the open end of the detonator. Moisture penetration could cause desensitization of the explosive charge in the detonator, and in the case of copper-tubed detonators, moisture could produce a potentially dangerous chemical reaction between the lead azide and the copper. (McAdam, 1958)

nepheline A hexagonal mineral, $(Na,K)AlSiO_4$; feldspathoid group; greasy luster; forms glassy crystals, colorless grains, coarse crystals with prismatic cleavage, or masses without cleavage; occurs in alkalic igneous rocks; an essential constituent of some sodium-rich rocks, e.g., nepheline syenite. Formerly called nephelite; eleolite.

nepheline syenite A plutonic rock composed essentially of alkali feldspar and nepheline. It may contain an alkali ferromagnesian mineral, such as an amphibole (riebeckite, arfvedsonite, barkevikite) or a pyroxene (acmite or acmite-augite); the intrusive equivalent of phonolite. Sodalite, cancrinite, hauyne, and nosean, in addition to apatite, sphene, and opaque oxides, are common accessories. Rare minerals are also frequent accessories. Syn: *foyaite; eleolite syenite; midalkalite.* (AGI, 1987)

nephelinite A fine-grained or porphyritic extrusive or hypabyssal rock, of basaltic character, but primarily composed of nepheline and clinopyroxene, esp. titanaugite, and lacking olivine and feldspar. (AGI, 1987)

nephelinitoid A nepheline-rich groundmass in an igneous rock; the glassy groundmass in nepheline rocks. (AGI, 1987)

nephelinization The process of introduction of or replacement by nepheline. (AGI, 1987)

nephelometry The measurement of the cloudiness of a medium; esp. the determination of the concentration or particle sizes of a suspension by measuring, at more than one angle, the scattering of light transmitted or reflected by the medium. Cf: *turbidimetry.* (AGI, 1987)

nephrite An exceptionally tough, compact, fine-grained, greenish or bluish variety of amphibole (specif. tremolite or actinolite) constituting the less rare or valuable kind of jade. Syn: *kidney stone; greenstone.*

neptunian dike A dike filled by sediment, generally sand, in contrast to a plutonic dike filled by volcanic materials. See also: *sand dike.*

neptunian theory *neptunism.*

neptunism The theory, advocated by A. G. Werner in the 18th century, that the rocks of the Earth's crust all consist of material deposited sequentially from, or crystallized out of, water. Etymol: Neptune, Roman god of waters. Cf: *plutonism.* Syn: *neptunian theory.* (AGI, 1987)

neptunite A monoclinic mineral, $KNa_2Li(Fe,Mn)_2Ti_2Si_8O_{24}$; forms red to black prismatic crystals; occurs in late stages of reduced, silica-deficient environments; e.g., alkaline igneous rocks and veins in serpentinite. See also: *mangan-neptunite.*

neritic Pertaining to the shallow seas; for accumulations of shells, but sometimes for the whole environment of deposition on the continental shelf. (Challinor, 1964)

neritic zone That part of the sea floor extending from the low tide line to a depth of 200 m.

Nernst film In ion exchange, the diffusion-layer supposed to surround a bead of resin. This static film is reduced, or diffusion through it is accelerated, if agitation of the ambient liquor is increased, if temperature is raised, or if concentration of ions in solution is made greater. (Pryor, 1963)

nero-antico A black marble found in Roman ruins, probably from the Taenarian peninsula, Greece.

nesquehonite An orthorhombic mineral, $Mg(HCO_3)(OH)\cdot 2H_2O$; forms low-temperature efflorescences, particularly as an alteration product of lansfordite. Named for a coal mine at Nesquehoning, PA.

ness A British term used esp. in Scotland for a promontory, headland, or cape, or any point or projection of the land into the sea; commonly used as a suffix to a place name, e.g., Fifeness. Also called nose. (AGI, 1987)

nest (a) A concentration of some relatively conspicuous element of a geologic feature, such as a nest of inclusions in an igneous rock or a small, pocketlike mass of ore or mineral within another formation. (AGI, 1987) (b) A fitting of the next-smaller-size casing inside the casing already set in a borehole, or of one tube inside another.

nested variogram model A model that is the sum of two or more component models, such as nugget, spherical, etc. Adding a nugget component to one of the other models is the most common nested model, but more complex combinations are occasionally used.

net (a) Scot. Strapping used for lowering or raising horses in shafts. (Fay, 1920) (b) A plane of points each with identical point surroundings. Cf: *space lattice; lattice.*

net calorific value The heat produced by combustion of unit quantity of a solid or liquid fuel when burned, at a constant pressure of 1 atm (0.1 MPa), under conditions such that all the water in the products remains in the form of vapor. Net heat of combustion at constant pressure is expressed as Q_p (net). (ASTM, 1994)

net-corrected fill Net fill after making allowance for shrinkage during compaction. (Nichols, 1976)

net cut The amount of excavated material to be removed from a road section, after completing fills in that section. (Nichols, 1976)

net drilling time The rotating time actually spent in deepening a borehole. (Long, 1960)

net fill The fill required, less the cut required, at a particular station or part of a road. (Nichols, 1976)

nether The lower part of, as in nether roof; opposite of the term "upper.". (TIME, 1929-30)

nether roof (a) The strata directly over a coal seam. The props set at the face hold only the nether roof. E.g., if the props carry a load of 20 st (18.1 mt) each and are set 4 ft (1.2 m) apart each way, the supported weight is 1.5 st/ft^2 (14.6 t/m^2). See also: *underweight; absolute roof; overarching weight; immediate roof.* (Nelson, 1965) (b) In mine subsidence, the immediate roof of limited depth, such as timber might be expected to support. (Briggs, 1929)

nether strata The roof and strata immediately above the coal. (Mason, 1951)

net slip On a fault, the distance between two formerly adjacent points on either side of the fault, measured on the fault surface or parallel to it. It defines both the direction and relative amount of displacement. (AGI, 1987)

net texture *network structure.*

Nettleton method An indirect means of density determination in which a closely spaced gravity traverse is run over some topographic feature, such as a small hill or valley. When the profile of observed values is plotted, the gravitational effect of the feature itself is calculated at each observation point along the profile and removed from the observed value for that point. The calculation is repeated a number of times, different densities being assumed for each computation. The density value at which the hill is least conspicuous on the gravity profile is considered to be most nearly correct. (Dobrin, 1960)

net unit value The difference between the gross unit recoverable value and the cost of mining, treating, and marketing ore; in other words, the net operating profit. See also: *gross recoverable value; gross unit value.* (Nelson, 1965)

network (a) Esp. in surveying and gravity prospecting, a pattern or configuration of stations, often so arranged as to provide a check on the consistency of the measured values. (AGI, 1987) (b) In ventilation surveys, the multiple development openings, haulage ways, and working faces that constitute the ventilation system of a mine. (Hartman, 1982)

network deposit *stockwork.*

network structure A structure in which one constituent occurs primarily at the grain boundaries, thus partially or completely enveloping the grains of the other constituent. Syn: *net texture.*

Neuenburg saw A plow consisting of a 2-in (5.1-cm) steel plate 6 ft by 20 in (1.8 m by 50.8 cm) of seven pieces hinged together to follow floor rolls; picks on the face edge cut in both directions. The minimum workable seam is 14 in (35.6 cm) on gradients of 35° to 70°. Maximum face length is 80 yd (73.2 m). The machine is used in the Ruhr. (Nelson, 1965)

neuk The tailgate corner of a face behind the face conveyor tension end. (Trist, 1963)

Neumann lamellae Straight, narrow bands parallel to the crystallographic planes in the crystals of metals that have been subjected to deformation by sudden impact. They are actually narrow twin band, and are most frequently observed in iron. (CTD, 1958)

neuropteris A large tree-fern of the coal forest, with trunks about 2 ft (0.6 m) thick, containing several cylinders of wood inside the stem instead of one column of wood as in modern trees. (Nelson, 1965)

neutral atmosphere Atmosphere in which there is neither an excess nor a deficiency of oxygen.

neutral axis The line of zero fiber stress in any given section of a member subject to bending; it is the line formed by the intersection of the neutral surface and the section. (Roark, 1954)

neutral equilibrium A body is said to be in neutral equilibrium if on being slightly displaced it remains in its new position; e.g., a ball placed on a horizontal surface or a cone supported on its side on a horizontal surface. (Morris, 1958)

neutralize To add either an acid or alkali to a solution until it is neither acid nor alkaline. (Gordon, 1906)

neutral lining Furnace lining of neutral refractories. (Osborne, 1956)
neutral point (a) A neutral point in a wye-connected alternating-current power system means the connection point of transformer or generator windings from which the voltage to ground is nominally zero, and which is the point generally used for system grounding. (b) In titration, the point at which hydrogen ions and hydroxyls are approx. balanced, each at about 1 times 10^{-7} molar. Since color-change-indicating dyes do not all react at this point, selection for a given titration must be made with regard to the required point of change. (Pryor, 1963)
neutral pressure The hydrostatic pressure of the water in the pore space of a soil. See also: *pore-water pressure; pore pressure; neutral stress.* (Hammond, 1965)
neutral refractory (a) A refractory that is neither strongly basic nor strongly acid, such as chrome, mullite, or carbon. (ARI, 1949) (b) A refractory that is resistant to chemical attack by both acid and basic slags, refractories, or fluxes at high temperatures. (ASTM, 1994)
neutral salt A salt in which all the hydrogen of the hydroxyl groups of an acid is replaced by a metal. (Standard, 1964)
neutral salt effect Reduction of ionization of a weak acid or base by addition of ionizing salt that contains one of the ions already present; form of common ion effect. (Pryor, 1963)
neutral stress The stress transmitted by the fluid that fills the voids between particles of a soil or rock mass; e.g., that part of the total normal stress in a saturated soil caused by the presence of interstitial water. Syn: *pore pressure; pore-water pressure; neutral pressure.* (AGI, 1987)
neutral surface The longitudinal surface of zero fiber stress in a member subject to bending; it contains the neutral axis of every section. (Roark, 1954)
neutral zone A strain-free area. Cf: *compression zone; tension zone.* (Nelson, 1965)
neutron An uncharged elementary particle with a mass that nearly equals that of the proton. An isolated neutron is unstable and decays with a half-life of about 13 min into an electron, a proton, and a neutrino. Neutrons sustain the fission chain reaction in a nuclear reactor. (Lyman, 1964)
neutron density The number of neutrons per cubic centimeter. (Lyman, 1964)
neutron-gamma log A radioactivity log employing both gamma and neutron-log curves. The neutron log should respond best to porous fluid-filled rock and the gamma best to shale markers. (AGI, 1987)
neutron log Strip recording of the secondary radioactivity arising from the bombardment of the rocks around a borehole by neutrons from a source being caused to move through the borehole. Used, generally in conjunction with other types of logs, for the identification of the fluid-bearing zones of rocks. See also: *radioactivity log; neutron logging.* (Inst. Petrol., 1961; AGI, 1987)
neutron logging A radioactivity logging method used in boreholes in which a neutron source provides neutrons that enter rock formations and induce additional gamma radiation, which is measured by use of an ionization chamber. The gamma radiation so induced is related to the hydrogen content of the rock. (AGI, 1987)
Nevadan orogeny Late Jurassic-Early Cretaceous diastrophism in Western North America.
nevyanskite A former name for iridosmine. See also: *iridosmine.*
Newark Supergroup Continental strata of Lower Jurassic or Upper Triassic age in the Eastern United States, consisting essentially of red sandstone, shale, arkose, and conglomerate, some 14,000 to 18,000 ft (4.3 to 5.5 km) thick. The series includes black shales with fish remains, thin coal seams in Virginia and North Carolina, and basaltic flows and sills. (CTD, 1958)
Newaygo screen A slanting screen in which the material to be screened passes down. The screen is kept in vibration by the impact of a large number of small hammers. (Liddell, 1918)
newjersite Variety of resin. (Tomkeieff, 1954)
Newlyn datum The mean sea level now used as the British Ordnance Datum for leveling. It was determined as the result of several years' observations at Newlyn, Cornwall, England, and differs at various places by more than 0.3 m from levels based on the Liverpool datum, which it supersedes. (Hammond, 1965)
Newmann hearth A modified Scotch hearth in which poking or rabbling is done mechanically. (CTD, 1958)
new miner training In mining, mandatory training given the miners having no previous mining experience; includes instruction in the statutory rights of miners and their representatives, use of self-rescue devices and respiratory devices where appropriate, hazard recognition, emergency procedures, electrical hazards, first aid, walk around training, and other health and safety aspects of the tasks to which the person will be assigned. Cf: *refresher training; task training.* (Federal Mine Safety, 1977)
new sand Newly mixed, but not unused, molders' sand. (Standard, 1964)
new scrap The material generated in the manufacturing process of articles for ultimate consumption; it includes defective castings, clippings, turnings, borings, drosses, slags, etc., that are returned directly to the manufacturing process or sold directly for reprocessing.
Newtonian fluid Term marking the distinction, made in mineral processing that involves agitation, between a truly viscous (Newtonian) liquid and one in which shear or apparent viscosity (pseudoviscosity) varies with the dimensions of the containing system and the speed of agitation. The latter type of fluid is said to be non-Newtonian. (Pryor, 1963)
Newton's law of gravitation *law of gravitation.*
New York rod A leveling rod marked with narrow lines, ruler fashion. (Nichols, 1976)
N-frame brace A diagonal brace in a square set. (Fay, 1920)
ngavite A chondritic stony meteorite composed of bronzite and olivine in a friable, breccialike mass of chondrules. (AGI, 1987)
niccolite A former name for nickeline. Also spelled nicolite. Syn: *arsenical nickel; copper nickel.*
Nicholls' technique A technique used in the determination of elastic constants of rock in situ. Longitudinal and shear waves are generated in rock by small explosive charges in shallow drill holes. Accelerometers and strain gages are employed to measure arrival times for both waves. From wave velocities and measured density, Poisson's ratio, modulus of elasticity, modulus of rigidity, Lame's constant, and bulk modulus can be calculated. (Lewis, 1964)
nickel (a) An isometric mineral, elemental Ni; hard; metallic; silver-white; a native metal, esp. in meteorites; also alloyed with iron in meteorites. (b) A silvery white, hard, malleable, ductile, somewhat ferromagnetic element. Symbol: Ni. It takes on a high polish and is a fair conductor of heat and electricity. Used for making stainless steel and other corrosion-resistant metals and is chiefly valuable for the alloys it forms. Also used extensively in coinage, in desalination plants for converting sea water into fresh water, and in making nickel steel for armor plate and burglar-proof vaults. (Handbook of Chem. & Phys., 3)
nickel antimony glance Sulfantimonide of nickel, crystallizing in the cubic system. Also called ullmannite. (CMD, 1948)
nickel bloom A green hydrated and oxidized patina on rock outcrops indicating the existence of primary nickel minerals; specif. annabergite (a nickel arsenate). The term is also applied to zaratite (a nickel carbonate) and to morenosite (a nickel sulfate). See also: *annabergite.*
nickel carbonyl A volatile compound of nickel, $Ni(CO)_4$, formed by passing carbon monoxide over the heated metal. The compound is decomposed into nickel and carbon monoxide by further heating. It is used on a large scale in industry for the production of nickel from its ores by the Mond process. (CTD, 1958)
nickel glance *gersdorffite.*
nickel green *annabergite.*
nickeliferous Containing nickel.
nickeline (a) A hexagonal mineral, 2[NiAs]; commonly contains antimony, cobalt, iron, and sulfur; one of the chief ores of nickel. Formerly called: niccolite; nickelite; arsenical nickel; copper nickel; kupfernickel. (b) The mineral group breithauptite, freboldite, imgreite, langisite, nickeline, sederholmite, sobolevskite, stempflite, and sudburyite.
nickel iron (a) A mineral, NiFe, containing about 76% nickel and found in meteorites. Isometric. (Dana, 1959) (b) The native alloy of nickel with iron in meteorites. See also: *kamacite; taenite.*
nickelite A former name for nickeline. See also: *nickeline.*
nickel ocher An early name for annabergite. (Fay, 1920)
nickelous oxide (a) NiO; green, becoming yellow. Found in nature as the mineral bunsenite. Soluble in acids and in ammonium hydroxide; insoluble in water; sp gr, 6.6 to 6.8. NiO absorbs oxygen at 400 °C forming Ni_2O_3 which is reduced to NiO at 600 °C. Used in nickel salts and in porcelain painting. (CCD, 1961) (b) Isometric; green to black; molecular weight, 74.71; melting point, 1,990 °C; sp gr, 6.67. Used for painting on china. (Bennett, 1962; Handbook of Chem. & Phys., 2)
nickel oxide Comprises the two nickel oxides, nickelous oxide (NiO) and nickelic oxide (Ni_2O_3), which are used extensively as colorants in glasses, glazes, and enamels. The use of nickel oxide in enamels is generally in the ground coat, in which it is used with cobalt and manganese. It is also used in cover coat enamels to give what is known as a daylight shade for reflector units. Nickelic oxide imparts a color to glass which is dependent upon the character of the alkali present. Nickelous oxide is used in glazes to produce blues, greens, browns, and yellows. Nickel oxide is also one of the principal components of certain type of ferrites, e.g, the nickel-zinc ferrite. See also: *nickelous oxide.* (Lee, 1961)
nickel plating The deposition of a coating of metallic nickel by electrolysis. (Nelson, 1965)
nickel pyrite *millerite.*
nickel-skutterudite An isometric mineral, $NiAs_{2-3}$; isostructural with skutterudite; tin white; in intermediate-temperature hydrothermal veins, particularly in association with Co, Ni, Fe arsenides, sulfarsenides, and native silver. Syn: *white nickel.*

nickel vitriol *morenosite.*
nicking (a) The cutting of a vertical groove in a seam to liberate coal after it has been holed or undercut. (Nelson, 1965) (b) Used in wire-rope terminology to describe the internal crosscutting of wires within a rope. (Sinclair, 1959) (c) The chipping of coal along the rib of an entry or room, which is usually the first indication of a squeeze. (Fay, 1920) (d) A vertical cutting or shearing one side of a face of coal. Also called cut; cutting. (Fay, 1920)
nickings Newc. The small coal produced in making a nicking. See also: *bug dust; makings.* (Fay, 1920)
nicol (a) Nicol prism. (b) Any apparatus that produces polarized light, e.g., Nicol prism or Polaroid. See also: *polar; polarizer.*
nicolite A former name for nickeline. See also: *nickeline.*
Nicol prism A special prism for producing polarized light, made from two pieces of Iceland spar (calcite) cemented together with Canada balsam. Light entering the prism is split into two polarized rays; of these, the ordinary ray is totally reflected at the balsam layer while the extraordinary ray is able to pass through the prism. In a petrological microscope two Nicol prisms are incorporated. See also: *polarizing prism.* (Anderson, 1964)
Niggli's classification (a) A classification of igneous rocks on the basis of their chemical composition, similar in some respects to the norm system. It was proposed in 1920 by the Swiss mineralogist Paul Niggli. (AGI, 1987) (b) A classification of ore deposits, the major groups being plutonic, or intrusive, and volcanic, or extrusive. It was proposed in 1929. (AGI, 1987)
night emerald *evening emerald.*
night pair Corn. Miners who work underground during the night. The night shift. (Fay, 1920)
night shift The coal miners' shift from about 12:00 p.m. to 8:00 a.m. It may be a coal-winning shift, but in general it is a preparation or maintenance shift. (Nelson, 1965)
nigritite A product of the coalification of fix bitumens rich in carbon; insoluble or only slightly soluble in organic solvents. It is subdivided into polynigritite, humonigritite, exinonigritite, and keronigritite. (Tomkeieff, 1954)
nil Nothing; zero. Often used in reporting gold and silver assays. (Webster 3rd, 1966; Fay, 1920)
nine-inch straight A standard 9-in by 4-1/2-in by 2-1/2-in (22.9-cm by 11.4-cm by 6.4-cm) straight brick.
nine-point sample Final sample taken for test when a small quantity of finely ground mineral is required for assay. A suitable quantity of dry material is thoroughly mixed on glazed cloth or paper, if necessary, being rolled lightly with a round bottle to break down any floccules. It is then flattened to a disk and eight equal segments are marked out diametrically with a spatula. Approx. equal quantities are taken from each segment and from the center, making the nine points of withdrawal. (Pryor, 1963)
ningyoite An orthorhombic mineral, $(U,Ca,Ce)_2(PO_4)_2 \cdot 1\text{-}2H_2O$; rhabdophane group; occurs in an unoxidized zone of the Ningyo-toge Mine, Tottori prefecture, Japan.
niobite *columbite.*
niobium A shiny, white, soft, and ductile metallic element. Symbol, Nb (niobium) or Cb (columbium). The name niobium was adopted by the International Union of Pure and Applied Chemistry. Many chemical societies and government organizations refer to it as niobium, but most metallurgists, metals societies, and commercial producers still refer to the metal as columbium. Found in niobite (or columbite), niobite-tantalite, pyrochlore, and euxenite. Used as an alloying agent in carbon and alloy steels, in nonferrous metals, and in superconductive magnets. Syn: *columbium.* (Handbook of Chem. & Phys., 3)
niobium boride One of several compounds that have been reported, including the following: NbB_2; melting point, 3,050 °C; sp gr, 7.0; thermal expansion, 5.9×10^{-6} parallel to *a* and 8.4×10^{-6} parallel to *c*; NbB, melting point, 2,300 °C; sp gr, 7.6; Nb_3B_4 melts incongruently at 2,700 °C; sp gr, 7.3. (Dodd, 1964)
niobium nitride One of three nitrides that have been reported: NbN, Nb_2N, and Nb_4N_3. During reaction between niobium and N_2 at 800 to 1,500 °C the product generally consists of more than one compound. Most of the phases are stable at least to 1,500 °C. (Dodd, 1964)
nip (a) Where the roof and the floor of a coal seam come close together pinching the coal between them. Cf: *want.* (BCI, 1947) (b) The contact ends of a cable for quick attachment to a power cable. (BCI, 1947) (c) The device at the end of the trailing cable of a mining machine used for connecting the trailing cable to the trolley wire and ground. (Jones, 1949) (d) To move a machine along a track by sliding the nip along the trolley wire. (Hess) (e) The seizing of material between the jaws or rolls of a crusher. (Nichols, 1976) (f) *angle of nip.* (g) To cut grooves at the end of a bar, to make it fit more evenly. (h) An undercutting notch in rock, particularly limestone, along a seacoast between high- and low-tide levels produced by erosion or possible solution. (AGI, 1987)
niperyth *penthrite.*
nipple A tubular pipe fitting usually threaded on both ends and under 12 in (30.5 cm) in length. Longer pipe is regarded as cut pipe. See also: *close nipple.* (Strock, 1948)
Ni-resist A cast iron consisting of graphite in a matrix of austenite. It contains 3.0% carbon, 14.0% nickel, 6.0% copper, 2.0% chromium, and 1.5% silicon; it has a high resistance to growth, oxidation, and corrosion. (CTD, 1958)
nital *Boylston's reagent.*
niter An orthorhombic mineral, $4[KNO_3]$; water soluble; has a cooling salty taste; a product of nitrification in most arable soils in hot, dry regions, and in the loose earth forming the floors of some natural caves. Cf: *nitratine.* Syn: *saltpeter.* Also spelled nitre.
niter cake Crude sodium sulfate, a byproduct in the manufacture of nitric acid from sodium nitrate. (Fay, 1920)
nitrate (a) A salt of nitric acid; e.g., silver nitrate or barium nitrate. (Standard, 1964) (b) A mineral compound characterized by a fundamental anionic structure of NO_3^-. Soda niter, $NaNO_3$, and niter, KNO_3, are nitrates. Cf: *carbonate; borate.* (AGI, 1987) (c) Salts formed by the action of nitric acid on metallic oxides, hydroxides, and carbonates. Readily soluble in water and decompose when heated. The nitrates of polyhydric alcohols and the alkyl radicals explode violently. (CTD, 1958)
nitratine A trigonal mineral, $NaNO_3$; rhombohedral cleavage; water soluble with a cooling taste; occurs only in very arid regions. Formerly called soda niter.
nitrification The formation of nitrates by the oxidation of ammonium salts to nitrites (usually by bacteria) followed by oxidation of nitrites to nitrates. It is one of the processes of soil formation. (AGI, 1987)
nitro An abbrev. for nitroglycerin or dynamite.
nitrocalcite A monoclinic mineral, $Ca(NO_3)_2 \cdot 4H_2O$; water soluble; soft; occurs as an efflorescence, e.g., on walls and in limestone caves. Syn: *wall saltpeter.*
nitrocellulose Nitric acid esters of cellulose formed by the action of a mixture of nitric and sulfuric acids on cellulose. The cellulose can be nitrated to a varying extent, ranging from two to six nitrate groups in the molecule. Nitrocellulose having a low nitrogen content, up to the tetranitrate, is not explosive. They dissolve in ether-alcohol mixtures and in so-called lacquer solvents, such as butyl acetate. A nitrocellulose having a high nitrogen content is guncotton, an explosive. The principal nitrocellulose plastic is celluloid. Syn: *cellulose nitrate.* (CTD, 1958)
nitrocotton A chemical combination of ordinary cotton fiber with nitric acid. It is explosive, highly inflammable, and in certain degrees of nitration, soluble in nitroglycerin.
nitrogelatin *gelatin dynamite.*
nitrogen Colorless, tasteless, odorless, relatively inert element. Symbol, N. Nitrogen makes up 78% of the air, by volume. From this inexhaustible source it can be obtained by liquefaction and fractional distillation. Used in the production of ammonia and nitric acid, as a blanketing medium in the electronics industry, as a refrigerant, in annealing stainless steel, in drugs, and for forcing crude oil from oil wells. (Handbook of Chem. & Phys., 3)
nitrogen fixation (a) Extracting nitrogen from the air in commercial quantities for use in agriculture or industry. (b) In a soil, the conversion of atmospheric nitrogen to a combined form by the metabolic processes of some algae, bacteria, and actinomycetes. (AGI, 1987)
nitroglycerin $CH_2NO_3CHNO_3CH_2NO_3$; pale yellow; flammable; explosive; thick liquid; soluble in alcohol; soluble in ether in all proportions; slightly soluble in water; melting point, 13.1 °C; and explosion point, 256 °C. Used as an explosive, in the production of dynamite and other explosives, as an explosive plasticizer in solid rocket propellants, and as a possible liquid rocket propellant. Molecular weight, 227.09; triclinic or orthorhombic when solid; sp gr, 1.5918 (at 25 °C, referred to water at 4 °C); soluble in methanol and in carbon disulfide; very soluble in chloroform; and slightly soluble in petroleum ether. This highly explosive liquid is made by mixing sulfuric acid and nitric acid in a steel tank and then adding glycerin. Its great shattering effect has made it esp. suitable for shooting oil wells. Because of its sensitiveness to shock, liquid nitroglygerin is dangerous to transport and unsuitable for use in mining and quarrying operations. Syn: *glycerol trinitrate; trinitrate glycerol; trinitrin; explosive oil.* (CCD, 1961; Handbook of Chem. & Phys., 2; Lewis, 1964)
nitroglycerin explosive An explosive containing, principally, nitroglycerin, nitrocotton, and inorganic nitrates, with a suitable combustible absorbent giving a balanced composition. (Nelson, 1965)
nitroglycerin powder Explosive usually characterized by a low nitroglycerin content, up to 10% , and a high ammonium nitrate content of 80% to 85%, with carbonaceous material forming the remainder of the composition. This composition produces a powdery consistency and, consequently, nitroglycerin powders have relatively poor water-resistance properties, so that they should be used only in dry conditions. Their storage properties are fairly good, but this is largely dependent on the protection given after manufacture, for example, in the methods

of cartridging and packing. The main application of these explosives is in quarrying and mining where the ground to be blasted is relatively soft. (McAdam, 1958)

nitrohydrochloric acid *aqua regia.*

nitrolite An excellent and cheap explosive in powder form, consisting of the constituents ammonium nitrate + trotyl + nitroglycerin + silicon. (Fraenkel, 1953)

nitromagnesite A monoclinic mineral, $Mg(NO_3)_2{\cdot}6H_2O$; water soluble; white; an efflorescence in limestone caverns.

nitromuriatic acid *aqua regia.*

nitrostarch explosive Nitrostarch explosives have been used to a limited extent for over 50 years. When these explosives were first introduced, nitrostarch was the principal explosive ingredient in their composition. Of recent years, because of the trend toward the low-sensitivity, noncap sensitive nitrocarbonitrates and ammonium nitrate-oil mixtures, certain grades of explosives are being produced with low amounts of sensitizers. Some of these explosives today contain a very large percentage of ammonium nitrate, and nitrostarch is used only in small quantities to act as a sensitizer. (Pit and Quarry, 1960)

nitrosubstitution The act or process of introducing by substitution the nitryl radical, NO_2, in place of one or more replaceable hydrogen atoms, such as in an organic compound. Nitrosubstitution compounds are used in the manufacture of some explosives. (Standard, 1964)

nitrosulfuric acid An exceedingly corrosive mixture of one part by weight of nitric acid and two parts by weight of sulfuric acid. Used in the manufacture of nitroglycerin. (Standard, 1964)

nitrous oxide A gas with the chemical formula, N_2O; molecular weight, 46; sp gr, 1.6. This gas is produced by the blasting of certain nitroglycerine explosives, esp. if there is incomplete detonation. It is also produced in the exhaust of diesel locomotives. It is used as an anesthetic in dentistry and is commonly known as laughing gas. (Morris, 1958)

niveau surface *equipotential surface.*

noble (a) A term used to express great value or purity, as in a noble metal (e.g., platinum); or inertness, as in a noble gas (e.g., helium). See also: *noble gas; noble metal.* (b) An adj. usually modifying "metal" or "gas" and referring to those elements which do not normally combine with oxygen or other non-metallic elements under near-surface conditions; thus they commonly occur as native elements (or alloys). The metals usually included are gold, silver, platinum, palladium, ruthenium, iridium, rhodium, and osmium; the gases are helium, neon, argon, krypton, xenon, and radon. Other elements found in their native states, such as sulfur, copper, or mercury, are not noble because they spontaneously (even though slowly) oxidize.

noble gas A rare inert gas: helium, neon, argon, krypton, xenon, and radon. Syn: *inert gas.* See also: *noble.*

nobleite A monoclinic mineral, $CaB_6O_9(OH)_2{\cdot}3H_2O$; tabular or mammillary; transparent; formed by the weathering of colemanite and priceite in Death Valley, CA.

noble metal A metal with marked resistance to chemical reaction, particularly to oxidation and to solution by inorganic acids. The list includes mercury and the precious and platinum-group metals. Cf: *base metal.* See also: *noble.* (Fay, 1920)

no-cut rounds In blasting underground, drilling all holes straight into the face. (Lewis, 1964)

nodular (a) Composed of nodules; e.g., nodular bedding consisting of scattered to loosely packed nodules in matrix of like or unlike character. (AGI, 1987) (b) Having the shape of a nodule, or occurring in the form of nodules; e.g., nodular ore such as a colloform mineral aggregate with a bulbed surface. (AGI, 1987) (c) Orbicular. (AGI, 1987)

nodular cast iron A cast iron that has been treated while molten with a master alloy containing an element, such as magnesium or cerium, to produce primary graphite in the spherulitic form. See also: *iron.* (ASM, 1961)

nodular structure *orbicular structure.*

nodule (a) A small, irregularly rounded knot, mass, or lump of a mineral or mineral aggregate, normally having a warty or knobby surface and no internal structure, and usually exhibiting a contrasting composition from the enclosing sediment or rock; e.g., a nodule of pyrite in a coalbed, a chert nodule in limestone, or a phosphatic nodule in marine strata. Nodules can be separated as discrete masses from the host material. (AGI, 1987) (b) One of the widely scattered concretionary lumps of manganese, cobalt, iron, and nickel found on the floors of the world's oceans; esp. a manganese nodule. Etymol: Latin nodulus, small knot. Cf: *concretion.* (AGI, 1987) (c) A rounded material accretion built of successive layers, of easily handled size. (Pryor, 1963) (d) A small, rounded, irregularly shaped mass, as those of graphite in malleable cast iron. (Rolfe, 1955)

nodulizing (a) The forming of rounded shapes by the application to fine coal of a gyratory, rotary, or oscillatory motion, without the use of pressure. (BS, 1962) (b) *balling.*

noise (a) Any undesired sound. (NCB, 1964) (b) By extension, any unwanted disturbance such as undesired electric waves in any transmission channel or device. (NCB, 1964) (c) In gravity and magnetic prospecting, disturbances in observed data due to more or less random inhomogeneities in surface and near-surface material. (AGI, 1987) (d) In seismic prospecting, all recorded energy not derived from the explosion of the shot. Sometimes loosely used for all recorded energy except events of interest. (AGI, 1987)

noise level In observed or recorded data, the fluctuations not attributable to signals. (AGI, 1987)

noise reduction rating A measure of a hearing protector's effectiveness in reducing noise such as results from mining operations where drilling, cutting, blasting, etc. create high noise levels. Abbrev. NRR. (MSHA, 2)

nolanite A hexagonal mineral, $(V,Fe,Ti)_{10}O_{14}(OH)_2$; rare; forms tabular prisms and plates; associated with uranium ores at Beaverlodge, SK, Canda, and with native gold and various tellurides at Kalgoorlie, Western Australia.

nominal area Of a screen, the total area of the screen deck exposed to the flow of the material feed. (BS, 1962)

nominal bandwidth In a filter, the difference between the nominal upper and lower cutoff frequencies. This difference may be expressed in cycles per second, as a percentage of the pass-band center frequency, or as the difference between the upper and lower cutoffs in octaves. (Hy, 1965)

nominal capacity A notional figure expressed in tons per hour used in the title of a flowsheet and in general descriptions of the plant, applying to the plant as a whole and to the specific project under consideration. It may be taken as representing the approximate tonnage expected to be supplied to the plant during the hour of greatest load. (BS, 1962)

nominal price An estimate of the price for a future month date which is used to designate a closing price when no trading has taken place in that date. Also used for current price indications in similar circumstances in physical trading. (Wolff, 1987)

nominal screen aperture (a) A nominal mesh aperture used to designate the result of a screening operation. (BS, 1962) (b) A notional size at which it is intended to divide a feed by a screening operation. Syn: *nominal screen size.* (BS, 1962)

nominal screen size *nominal screen aperture.*

nominal size The limit or limits of particle size used to describe a product of a sizing operation. (BS, 1962)

nomogram A type of line chart that graphically represents an equation of three variables, each of which is represented by a graduated straight line. It is used to avoid lengthy calculations; a straight line connecting values on two of the lines automatically intersects the third line at the required value. Syn: *nomograph.* (AGI, 1987)

nomograph *nomogram.*

nonangular unconformity *disconformity.*

nonasphaltic pyrobitumen Any of several species of pyrobitumens, including dark-colored, comparatively hard and nonvolatile solids, composed of hydrocarbons containing oxygenated bodies, infusible and largely insoluble in carbon disulfide. This includes peat, coal, and nonasphaltic pyrobitumen shales. (Tomkeieff, 1954)

nonbanded coal Coal that does not display a striated or banded appearance on the vertical face. It contains essentially no vitrain and consists of clarain or durain, or of material intermediate between the two. (AGI, 1987)

non-Bessemer ore Ore containing up to about 0.18% phosphorus. (Newton, 1959)

noncaking coal Coal that does not form cake; namely hard, splint, cherry, and durain coal. (Tomkeieff, 1954)

noncaving method Any of several stoping methods, including open stopes, sublevel, shrinkage, cut-and-fill, and square set. (Lewis, 1964)

noncoal mine A mine in which the material being mined is incombustible or contains at least 65% by weight of incombustible material, and in which the underground atmosphere in any open workings contains less than 0.25% by volume of flammable gas. (CFR, 2)

noncoking coal A bituminous coal that burns freely without softening or any appearance of incipient fusion. The percentage of volatile matter may be the same as for coking coal, but the residue is not a true coke. (Fay, 1920)

noncombustible Any material that will neither ignite nor actively support combustion in air at 1,200 °F (649 °C) when exposed to fire. See also: *incombustible.* (ACSG, 1963)

nonconformable Pertaining to the stratigraphic relations shown by a nonconformity. (AGI, 1987)

nonconformity (a) An unconformity developed between sedimentary rocks and older rocks (plutonic igneous or massive metamorphic rocks) that had been exposed to erosion before the sediments covered them. The restriction of the term to this usage was proposed by Dunbar & Rodgers (1957). Although the term is well known in the classroom, it is not commonly used in practice (Dennis, 1967). (AGI, 1987) (b) A term that formerly was widely, but is now less commonly, used as a syn. of angular unconformity, or as a generic term that includes angular unconformity. Term proposed by Pirsson (1915). (AGI, 1987)

nonconsumable-electrode arc melting A method of arc melting in which a carbon or tungsten electrode is used and the sponge metal to be melted is fed into the arc at the proper rate. (Newton, 1959)

noncore bit *plug bit.*

noncore drilling Drilling a borehole without taking core. (Long, 1960)

noncoring bit A general type of bit made in many shapes that does not produce a core and with which all the rock cut in a borehole is ejected as sludge. Used mostly for blasthole drilling and in the unmineralized zones in a borehole where a core sample is not wanted. Also called blasthole bit; plug bit. Cf: *fishtail bit; roller bit.* Syn: *blind bit.* (Long, 1960)

nondestructive testing Methods of examination, usually for soundness, which do not involve destroying or damaging the material or part being tested. It includes radiological examination, magnetic inspection, etc. Also called nondestructive inspection. (Rolfe, 1955)

nondiamond core drill A rotary or percussive-type drill equipped with core-cutting tools or bits, the cutting points of which are not inset with diamonds. (Long, 1960)

nonel *nonelectric blasting.*

nonelectric blasting The firing of one or more charges using safety fuse, igniter cord, detonating cord, shock or gas tubing, or similar nonelectric materials to initiate a blasting cap. Syn: *nonel.* See also: *gas detonation system; shock tube system.* (Federal Mine Safety, 1977)

nonelectric delay blasting cap A detonator with an integral delay element and capable of being initiated by miniaturized or regular detonating cord. (CFR, 1992)

nonferrous Of, or relating to, a metal or compound that does not contain appreciable quantities of iron; ores that are not processed primarily for their iron content.

nonferrous alloy Specif., an alloy containing no iron. Generically, any alloy that has as its base any element other than iron. Common commercial nonferrous alloys are based upon aluminum, copper, lead, magnesium, nickel, tin, and zinc. (Henderson, 1953)

nonferrous metallurgy That branch of metallurgy that deals with the broad field of metals other than iron, or alloys other than of iron base. Cf: *ferrous metallurgy.* (Henderson, 1953)

nonferrous metals (a) Classification of metals that are not commonly associated with alloys of iron, including base metals, precious metals, and light metals. (b) In singular form, an alloy that has as its base metal a metal other than iron, e.g., copper, lead, aluminum, etc.

nonflowing well A well that yields water at the ground surface only by means of a pump or other lifting device. See also: *artesian well.* (AGI, 1987)

nonfreezing explosive Explosive to which 15% to 20% of nitroethylene glycol has been added. This acts as a freezing-point depressant and prevents freezing at ordinary temperatures. Polar or Arctic explosives are nitroglycerin explosives of this type. (Higham, 1951)

nongraded sediment (a) In geology, detrital sediment, loose or cemented, containing notable amounts of more than one grade; e.g., loam or boulder clay. Syn: *poorly sorted.* (Stokes, 1955) (b) In engineering, sediment in which the constituent particles are all of nearly the same size. (Stokes, 1955)

nonhardening salt Salt containing substantial quantities of impurities such as calcium and/or magnesium chloride, which are highly deliquescent and prevent caking. (Kaufmann, 1960)

nonluminous flame Hydrogen, carbon monoxide, or aerated coal gas flames. (Francis, 1965)

nonmagnetic rod A drill rod made of brass, aluminum, or other metal unaffected by magnetism. (Long, 1960)

nonmagnetic steel Steel alloyed with 12% or more of manganese, chromium, or nickel. Such an alloy cannot be removed from a passing stream of ore by an ordinary guard magnet. Magnetic permeability is below 1.05. (Pryor, 1963)

nonmetal A chemical element (as boron, carbon, phosphorus, nitrogen, oxygen, sulfur, chlorine, or argon) that is not classed as a metal because it does not exhibit most of the typical metallic properties. An element that, in general, is characterized chemically by the ability to form anions, acidic oxides and acids, and stable compounds with hydrogen. (Webster 3rd, 1966)

nonmetallic (a) Of or pertaining to a nonmetal. (AGI, 1987) (b) Said in general of mineral lusters other than metallic luster. Cf: *submetallic luster.* (AGI, 1987) (c) An industrial mineral; usually used in the plural. (AGI, 1987)

nonmetallic armor A tough outer covering or cable sheath of rubber, rubber compound, or thermoplastic, designed to protect cable conductors and insulation from abrasion or other damage from external sources. (USBM, 1960)

nonmetallic mineral (a) In resource usage, "nonmetallic mineral" refers to the nonmetallic character of the product, not the mineralogy. Thus graphite is a nonmetallic mineral and molybdenite is "metallic" even though the minerals graphite and molybdenite are so similar in appearance that they may be confused in a casual inspection. Examples include: asbestos, barite, cement, feldspar, gem stones, helium, kyanite, perlite, salt, soda ash, and vermiculite, and even extending to mineral fuels. Cf: *metallic.* (b) In mineral usage, the luster of a nonopaque mineral, which transmits light at least through the thinnest edges and in thin section.

nonmetallic minerals Minerals are conveniently divided into metallic and nonmetallic groups, and then arranged in subdivisions according to the elements which form their main constituents. The nonmetallic minerals (carbon, diamond, coals, bitumen, asphalt, boron, sulfur, rock salt, etc.) lack the properties of the metallic minerals such as a bright metallic luster, hardness, density, and good conduction of heat and electricity. (Nelson, 1965)

non-metallic minerals Rocks, minerals, and select naturally occurring and synthetic materials of economic value, exclusive of fuel and metallic ore minerals. The select materials include peat, mineral-derived materials such as lime and cement, and synthetic versions of gemstones, abrasives, graphite, and calcite. Generally, non-metallic minerals undergo no chemical or mineralogical alteration for and in their end-uses; are low-price, high-volume commodities such as construction materials; are higher-priced and large-volume commodities that are raw materials in the chemical and agricultural industry; and very high price but very low volume materials such as gemstones. A characteristic of non-metallic minerals is that, in most cases, they maintain their form and physical properties through processing to final end use. See also: *industrial minerals.* (AGI, 1987; Indust. Miner., 1983)

non-Newtonian flow Flow in which the relationship of the shear stress to the rate of shear strain is nonlinear; i.e., flow of a subsurface in which viscosity is not constant. (AGI, 1987)

nonnitroglycerin explosive Explosive containing TNT instead of nitroglycerin to sensitize ammonium nitrate; a little aluminum powder may also be added to increase power and sensitiveness. Straight TNT-ammonium nitrate explosives usually contain 15% to 18% TNT and 82% to 85% ammonium nitrate. TNT-ammonium nitrate explosives have densities between 1.0 g/cm^3 and 1.2 g/cm^3 and velocities of detonation between 3,200 m/s and 4,000 m/s. They are reasonably free from noxious fumes. This type of explosive is susceptible to moisture and should be used only in dry conditions unless packed in sealed containers. The main use of non-nitroglycerin explosives is in primary blasting in quarries and opencast mining, although they are used in some underground work, particularly in ironstone mining. The combination of high strength and relatively high velocity of detonation also makes them applicable for secondary blasting by plaster shooting. (McAdam, 1958)

nonpareil brick An insulating brick.

nonpermissible explosive An explosive that is not approved in law for use in gassy mines.

nonpolarizable electrode Electrode in which the phenomenon of polarization cannot occur. (Schieferdecker, 1959)

nonproductive formation (a) A rock unit that, because of its stratigraphic position, is presumed to contain no valuable mineral deposits. (Long, 1960) (b) A rock unit in which no minerals of interest are found. (Long, 1960)

nonrotating rope A wire rope composed of 18 strands of 7 wires each; the inner 6 strands are left lay and outer 12 strands are right lay. It is esp. fabricated for use where loads are handled in free suspension, as in lifting of loads with a single line. (Lewis, 1964)

nonsegregating chute A chute, usually used to charge stoker hoppers, so designed as to deliver coal in a mixed state rather than having the large lumps tend to be deposited separate from the fine.

nonselective mining The object of nonselective mining is to secure a low cost, generally by using a cheap stoping method combined with large-scale operations. This method can be used in deposits where the individual stringers, bands, or lenses of high-grade ore are so numerous and so irregular in occurrence and separated by such thin lenses of waste that a selective method cannot be employed. Nonselective methods of stoping include caving, top slicing, some forms of open stoping, and shrinkage stoping under most conditions. (Jackson, 1936)

nonsequence A diastem or other relatively unimportant sedimentary or stratigraphic interruption. Cf: *unconformity; paraconformity.*

nonsignificant anomaly An anomaly that is superficially similar to a significant anomaly but is unrelated to ore. Formerly called false anomaly. (Hawkes, 1962)

nonsparking tool Tool, made from beryllium-copper or aluminum-bronze, that produces no sparks, or low energy sparks, when used to strike other objects. (NSC, 1992)

nonspin cable A wire or fiber cable so constructed as to reduce twisting to a minimum. (Long, 1960)

nonspin differential A differential that will turn both axles, even if one offers no resistance. (Nichols, 1976)

nonspinning rope A rope wire consisting of 18 strands of 7 wires each, in 2 layers; the inner layer consists of 6 strands lang lay rope and left lay around a small hemp core, and the outer of 12 strands regular lay, right-hand lay. Will carry a load on a single end without untwisting. (Hunt, 1965)

nonstranded rope A rope in which the wires are not laid up in strands but in concentric sheaths, and in opposite directions in the different sheaths, which gives the rope nonspinning properties. The outer sheaths are composed of specially shaped interlocking wires, and there is no hemp core in the rope. (Sinclair, 1959)

nonstructural A phase transformation not involving structural rearrangement, e.g., Curie point in magnetism.

nontabular deposit A mineral deposit of irregular shape. See also: *mass.* (Nelson, 1965)

nontectonite Any rock whose fabric shows no influence of movement of adjacent grains; e.g., a rock formed by mechanical settling. Some rocks are transitional between a tectonite and a nontectonite. (AGI, 1987)

nontronite A monoclinic mineral, $Na_{0.33}Fe^{3+}{}_2(Si,Al)_4O_{10}(OH)_2 \cdot nH_2O$; smectite group; expansive, a swelling clay; earthy; occurs in vesicles and veins in weathered basalt and as an alteration product of volcanic glass. Syn: *pinguite.* Formerly called chloropal; gramenite; morencite.

nonuniform flow A flow the velocity of which is undergoing a positive or negative change. If the flow is constant it is referred to as uniform flow. (Seelye, 1951)

nonvitreous A relative term as applied to ceramic products based on the water absorbing characteristics; i.e., brick, tile, etc., which absorb water in excess of that given by the specifications would be described as nonvitreous. See also: *impervious; vitreous; semivitreous.*

nonvitrified *nonvitreous.*

nonweathering coal Coal having a weathering index, as defined by U.S. Bureau of Mines standards, of less than 5%. (AGI, 1987)

nonwetted (a) A term used in the flotation process and applied to certain metallic minerals that are not wetted with water but are easily wetted with oil. (Fay, 1920) (b) As used by diamond-bit setters, a diamond inset in a metal or alloy that has not adhered to or wetted the surface of the diamond. (Long, 1960)

Norbide Trade name for boron carbide, an artificial abrasive; chemical formula, BC. It is markedly harder than silicon carbide and second only to diamond. (AIME, 1960)

nordite An orthorhombic mineral, $(La,Ce)(Sr,Ca)Na_2(Na,Mn)(Zn,Mg)Si_6O_{17}$; forms pale brown crystals on the Kola Peninsula, Russia.

norite A coarse-grained plutonic rock containing basic plagioclase (labradorite) as the chief constituent and differing from gabbro by the presence of orthopyroxene (hypersthene) as the dominant mafic mineral. Cf: *hypersthenite.* (AGI, 1987)

norm (a) The theoretical mineral composition of a rock expressed in terms of normative mineral molecules that have been determined by specific chemical analyses for the purpose of classification and comparison; the theoretical mineral composition that might be expected had all chemical components crystallized under equilibrium conditions according to certain rules. Adj: normative. See also: *normative mineral; norm system.* Cf: *mode.* (AGI, 1987) (b) Optimum operating condition of one or more controlled characteristics in a process such as continuous ore treatment. (Pryor, 1963)

normal (a) Of or pertaining to a solution having a concentration of 1 g-equivalent weight of solute per liter of solution; commonly used term in analytical chemistry. See also: *normal solution.* (CCD, 1961; Webster 3rd, 1966) (b) Used to designate aliphatic hydrocarbons, their derivatives, or alkyl radicals, the molecules of which contain a single unbranched chain of carbon atoms.

normal air A mixture of dry air and water vapor, varying from 0.1% to 3% by volume (usually over 1% in mines). (Hartman, 1982)

normal arc A term specif. intended to differentiate between the arcs that are commonly observed and the low-pressure skittering arcs.

normal calorie The quantity of heat required to raise 1 g of water from 14.5 °C to 15.5 °C. (Newton, 1959)

normal corrosion When used in connection with galvanic corrosion, it may refer to corrosion of the anodic metal when there is no contact with the dissimilar metal. (Schlain)

normal depth The depth of water in an open conduit that corresponds to uniform velocity for the given flow. It is a hypothetical depth under conditions of steady nonuniform flow, the depth for which the surface and bed are parallel; also termed the neutral depth. (Seelye, 1951)

normal dip *regional dip.*

normal displacement *dip slip.*

normal fault A fault in which the hanging wall appears to have moved downward relative to the footwall. The angle of the fault is usually 45° to 90°. There is dip separation, but there may or may not be dip slip. Cf: *reverse fault; thrust fault.* Syn: *gravity fault.* (AGI, 1987)

normal field In magnetic prospecting, the smoothed value of a magnetic field component as derived from a large-scale survey, worldwide or of continental scope. The normal field of the Earth varies slowly with time, and maps of it are as of a certain date. (AGI, 1987)

normal fold *symmetrical fold.*

normal haul A haul whose cost is included in the cost of excavation, so that no separate charge is made for it. (Nichols, 1976)

normalized steel Steel that has been given a normalizing heat treatment intended to bring all of a lot of samples under consideration into the same condition.

normalizing conveyor A conveyor that moves material through a normalizing furnace under heat.

normally consolidated Said of a soil deposit that has never been subjected to an effective pressure greater than the existing overburden pressure and one that is also completely consolidated by the existing overburden. (ASCE, 1958)

normal moisture capacity *field capacity.*

normal pressure Usually equal to the weight of a column of mercury 760 mm in height. Approx. 14.7 psi (101.4 kPa). Syn: *standard pressure.* (Webster 3rd, 1966; Fay, 1920)

normal price As applied to metal prices, it is the average over a long term—sometimes a period greater than the life of a mine. See also: *basic price.*

normal scale *effective temperature.*

normal shift In a fault, the horizontal component of the shift, measured perpendicular to the strike of the fault. See also: *offset; shift.* (AGI, 1987)

normal solution A solution made by dissolving 1 g-equivalent weight of a substance in sufficient distilled water to make 1 L of solution. See also: *normal.* (CTD, 1958)

normal stress The stress component at right angles to a given plane. See also: *stress.* (ASCE, 1958)

normal stress component That component of the stress in a rock mass that acts perpendicular to the lode plane or any other reference plane. (Spalding, 1949)

normal temperature (a) Normal temperature and pressure are taken as 0 °C (273° absolute) and 30 in (760 mm) of mercury pressure. Also called standard temperature. (Cooper, 1963) (b) As applied to laboratory observations of the physical characteristics of bituminous materials, it is 77 °F (25 °C). (Urquhart, 1959)

normal theory A theory claiming that the removal of a coal seam caused the overlying strata to fracture at right angles to the inclination of the beds. Subsidence observations do not support this theory. See also: *dome theory.* (Nelson, 1965)

normal travel-time curve In fan shooting, a time-distance curve obtained along a profile in some nearby area that does not contain geologic structures of the type being sought. (AGI, 1987)

normative The adj. of norm.

normative mineral A mineral whose presence in a rock is theoretically possible on the basis of certain chemical analyses. A normative mineral may or may not be actually present in the rock. See also: *norm.* Syn: *standard mineral.* (AGI, 1987)

norm system A system of classification and nomenclature of igneous rocks based on the norm of each rock. It is used in detailed petrologic studies rather than in ordinary geologic or mining work. Syn: *CIPW system.* See also: *norm.*

Norsk-Staal process A process for the direct production of iron sponge. A mixture of carbon monoxide and hydrogen is used as the reducing agent. The equipment consists of three vertical ovens, for preheating, reducing, and cooling the charge, and apparatus for regenerating the spent gases. The ore is contained in a series of muffle trays, each tray holding about 3 tons of ore. These trays pass down through the preheating oven, where the ore is heated to 1,000 °C, and are then transferred to the reducing oven, where they are raised through and against the downward gas current, then transferred to the cooling shaft, and slowly lowered down it. The transfer of a tray from one oven to another is made without contact with air. The ore is preheated in the first oven by part of the gases from the reduction oven, and the sponge iron is cooled in the third oven to 50 °C by cold gas from a gas holder. (Osborne, 1956)

norstrandite A triclinic mineral, $Al(OH)_3$; cryptocrystalline; a constituent of strongly weathered soils, of laterites and bauxites.

north end York. The rise side of the coal in North Yorkshire.

north-seeking pole The end of a magnet that points approx. north. The other end is the south-seeking pole. (Morris, 1958)

Norwalt separator Trade name for a dense-medium washer for treating coal lump size down to about 1/8 in (3.2 mm). It comprises a shallow circular tank with a flat base and a conical inner shell containing the driving mechanism. The clean coal floats and passes over a weir while the shale sinks to the bottom and is conveyed to an outlet chute. Its capacity ranges from about 50 st/h (45.4 t/h) to over 500 st/h (454 t/h), depending on the size of the vessel. (Nelson, 1965)

Norwegian cut A variation of the ordinary cut that may be said to represent a combination of the latter and the fan cut. The first drill holes are formed with a sharper angle toward the working face, which facilitates breaking. This type of cut has been employed successfully in headings of small section, the cut hole being blasted first, followed by the bench holes. In order to obtain the maximum possible advance the cut may also be deepened after blasting, during the first pause in working, e.g, the whole section then being broken out simultaneously. (Fraenkel, 1953)

nose (a) Scot. A point; a projecting angle of coal or other mineral. Also called ness. (Fay, 1920) (b) The lead face of the crown of a diamond bit. (Long, 1960) (c) A short plunging anticline without closure. (d) To dip or run in the form of a geological nose. (Webster 3rd, 1966)

nosean An isometric mineral, $Na_8Al_6Si_6O_{24}(SO_4){\cdot}H_2O$; sodalite group; occurs in silica- and lime-deficient igneous rocks. Formerly called noselite, nosin, nosite.

nose in Eng. A stratum is said to nose in when it dips beneath the ground into a hillside in a V-form or nose form.

noselite *nosean.*

nose out (a) Eng. A nose-shaped stratum cropping out. (Fay, 1920) (b) To diminish by losing stratum after stratum and getting into the lower part of the measure; said of a coal seam. (Standard, 1964)

nose pipe The inside nozzle of a tuyere. (Standard, 1964)

nosin *nosean.*

nosite *nosean.*

notch (a) An angular recess cut in the ends of a crossbar of a timber set to fit over a corresponding wedge in the upright posts. With the advent of steel arches, the craft of notching is becoming extinct. See also: *Welsh notch.* (Nelson, 1965) (b) Eng. *let into.* (c) A small weir made for use in measuring flow in laboratory models of hydraulic structures. See also: *measuring weir.* (Hammond, 1965) (d) A hollow formed by the undermining of a cliff, as a result of wave erosion and/or solution. (Schieferdecker, 1959)

notch effect Locally increased stress at that point in a structural load-bearing member where the section changes at a sharp angle. (Hammond, 1965)

notcher A machine tool in a steel-fabrication shop used to strip the flanges from the ends of rolled steel joists. (Hammond, 1965)

notching (a) A method of excavating in a series of steps. (Standard, 1964) (b) Cutting out various shapes from the edge of a strip, blank, or part. (ASM, 1961)

notch sensitivity ratio Alternative term for factor of stress concentration in fatigue or fatigue strength reduction factor. (Roark, 1954)

not previously known to exist These words refer to the time of the location and commencement of the tunnel and not to the respective times of the discoveries of the various veins in the tunnel. (Ricketts, 1943)

Nottingham system A longwall method of working coal seams in which the trams run on a rail track along the face and are hand loaded at the sides. It follows that the system can be adopted only in relatively thick seams where the trams can travel along the face without any roof ripping. The method is now replaced by face conveyors. Syn: *Barry mining.* (Nelson, 1965)

noumeite *garnierite.*

novackite A monoclinic mineral, $(Cu,Ag)_{21}As_{10}$; pseudotetragonal; forms steel-gray granular aggregates; at Cerny Dul, Czech Republic.

novaculite (a) A dense, hard even-textured, light-colored cryptocrystalline siliceous sedimentary rock, similar to chert but characterized by dominance of microcrystalline quartz over chalcedony. It was formerly believed to be the result of primary deposition of silica, but in the type occurrence (Lower Paleozoic of the Ouachita Mountains, Arkansas and Oklahoma) it appears to be a thermally metamorphosed bedded chert, distinguished by characteristic polygonal triple-point texture. The origin of novaculite has also been ascribed to crystallization of opaline skeletal material during diagenesis. The rock is used as a whetstone. See also: *Arkansas stone.* Syn: *razor stone; Turkey stone; ouachita stone.* (AGI, 1987) (b) A term used in southern Illinois for an extensive bedded chert. (AGI, 1987) (c) A general name formerly used in England for certain argillaceous stones that served as whetstones. (AGI, 1987)

nowel (a) The inner part of a large mold, corresponding to the core in small work. (Standard, 1964) (b) The bottom or drag of a molding flask, as distinguished from the cope. (Standard, 1964)

noxious gas A gas that is injurious to health. (BS, 1963)

nozzle brick A tabular refractory shape used in a ladle, with a hole through which steel is teemed at the bottom of a ladle, the upper end of the shape serving as a seat for the stopper. (ARI, 1949)

nozzleman In metal mining, one who operates a hydraulic giant or monitor (nozzle) used to direct a high-pressure stream of water against a bank of gold-bearing gravel to erode and force the gravel into sluiceboxes, where the gold separates out and is caught by riffles (cleats). Syn: *giant tender.* (DOT, 1949)

NPN process A modification of the basic Bessemer process. The main feature is the shortening of the blow by increasing the pressure of the blast as much as possible. Normally, the melt is cooled by the addition of scrap or iron ore, but it is claimed that a fairly high temperature can be maintained without an undue increase of the nitrogen content, so that ladle skulls can be avoided. (Osborne, 1956)

NQ A letter name specifying the dimensions of bits, core barrels, and drill rods in the N-size and Q-group wireline diamond drilling system having a core diameter of 47.6 mm and a hole diameter of 75.7 mm. (Cumming, 1981)

N rod bit A Canadian standard noncoring bit having a set diameter of 2.940 in (74.7 mm). More commonly called a 2-15/16 N drill-rod bit. (Long, 1960)

N-truss A bridge or roof truss that has parallel upper and lower chords and an arrangement of web members consisting of tension diagonals and compression verticals, with the vertical struts separating the panels. Also known as a Pratt truss. See also: *Warren girder.* (Hammond, 1965)

nubber (a) Mid. A block of wood about 12 in (30.5 cm) square, for throwing mine cars off the road in case the couplings or ropes break. (Fay, 1920) (b) *stopblocks.*

nuclear-assisted mining The use of a nuclear explosive for fracturing and fragmenting large volumes of ore underground into rubble chimneys, in preparation for block-cave-type mining or in-situ leaching. (SME, 1992)

nuclear log *radioactivity log.*

nuclear magnetism log Primarily a hydrogen log, useful for the following purposes: (1) provides valuable correlating curve to replace the S.P. in holes containing oil or invert muds; (2) provides a means of qualitatively distinguishing zones containing hydrocarbons from zones containing only water; (3) provides a means of measuring quantitatively what proportion of the total fluid-filled porosity in a formation is sufficiently free from the influence of chemical binding forces to be considered mobile and thus potentially recoverable; and (4) provides a means of estimating the permeabilities of formations. (Wyllie, 1963)

nuclear magnetometer *nuclear resonance magnetometer.*

nuclear powerplant Any device, machine, or assembly thereof that converts nuclear energy into some form of useful power, such as mechanical or electric power. In a nuclear electric powerplant, heat produced by a reactor is used to make steam, and the steam drives a turbine generator in the conventional way. (Lyman, 1964)

nuclear reaction A reaction involving the nucleus of the atom, such as fission, neutron capture, radioactive decay, or fusion; and distinct from a chemical reaction, which is limited to changes in the electron structure surrounding the nucleus. (Lyman, 1964)

nuclear resonance magnetometer An instrument that measures the Earth's magnetic field, depending on the magnetic moment of the atom. Hydrogen atoms are generally used, and these can be in a compound such as water. Each hydrogen atom can be viewed as a tiny electromagnet whose strength and direction are determined by the revolution of the electron of the atom about its nucleus. In a magnetic field, atoms of hydrogen have a tendency to align themselves in opposition to the field. If the direction of the field is suddenly changed, there will be a moment pulling the atoms toward the new direction. But each atom is a midget gyroscope, and instead of shifting directly to the new field direction, it will precess about this direction. The frequency of this precession will be a function only of the strength of the magnetic field. Syn: *nuclear magnetometer.* (Hunt, 1965)

nucleation The beginning of crystal growth at one or more points. (AGI, 1987)

nucleometer A Geiger counter employing 20 Geiger tubes to increase the sampling area and overcome the inefficiency of a 1-tube counter. (Ballard, 1955)

nucleon A constituent of the atomic nucleus; i.e., a proton or a neutron. (Lyman, 1964)

nucleus The central point about which matter accumulates to form a larger mass, esp. of a growing crystal or pearl. Plural: nuclei.

nuclide Any species of atom that exists for a measurable length of time. A nuclide can be distinguished by its atomic weight, atomic number, and energy state. The term is used synonymously with isotope. A radionuclide is the same as a radioactive nuclide, a radioactive isotope, or a radioisotope. (Lyman, 1964)

Nuflex Trade name for a nonrotating rope of 17×7 or 34×7 strand construction. (Hammond, 1965)

nugget A large lump of placer gold or other metal. Cf: *heavy gold.* (AGI, 1987)

nugget effect Anomalously high precious metal assays resulting from the analysis of samples that may not adequately represent the composition of the bulk material tested due to nonuniform distribution of high-grade nuggets in the material to be sampled. (SME, 1992)

nugget model A constant variance model most often used in combination with one or more other functions when fitting mathematical models to experimental variograms.

nuisance dust Dust with a long history of little adverse effet on the lungs; does not produce significant organic disease or toxic effect when exposures are kept at reasonable levels.

Nujol In flotation, any of a group of nonionizing hydrocarbon oils that act as collector agents by smearing action, giving aerophilic quality to the surface they selectively coat. (Pryor, 1963)

Nullagine Series Local name in Western Australia for the formation consisting essentially of pre-Cambrian rocks made up mainly of jasperoid quartzites and dolomite. Crocidolite asbestos occurs in this formation as cross-fiber seams in lodes in stratified ferruginous quartzites and shales with occasional bands of dolomite. (Sinclair, 1959)

nullah (a) A term used in the desert regions of India and Pakistan for a sandy river bed or channel, or a small ravine or gully, that is normally dry except after a heavy rain. (AGI, 1987) (b) The small, intermittent, generally torrential stream that flows through a nullah. Etymol: Hindi nala. See also: *wadi; arroyo.* Also spelled: nulla; nallah; nalla. (AGI, 1987)

Numidian marble A general name for marbles of cream, yellow, pink, and red color, found in northern Africa. The quarries were worked by the ancient Romans.

Nummulite limestone A thick bed of limestone, of Eocene age, composed mainly of the remains of the foraminifer *Nummulites.* The formation stretches from the Alps through Iran to China. It is the stone of which the Great Pyramid is built.

Nusse and Grafer PIV/6 drilling machine A rotary machine used for drilling the holes in combustible gases drainage. It is a two-speed, 150- to 250-rpm machine, drill-rod rotation being operated by a 6-hp (4.47-kW) motor. Traversing is done by a 2-hp (1.5-kW) motor, a pinion of which engages a toothed rack that runs the length of the drill frame. A forward drilling thrust of 4 st (3.6 t) is possible. The machine measures 10 ft (3 m) overall and weighs about 1,000 lb (454 kg). With an improved high-speed gearbox, 100 to 130 ft (30 to 40 m) of coal measures strata can be drilled in a shift. (Nelson, 1965)

nut coal (a) An abbrev. for chestnut coal. Also called nuts. (Fay, 1920) (b) Prepared bituminous coal that passes through 2- to 3-in (5.1- to 7.6-cm) round holes and over 3/4-, 1-, or 1-in (1.9-, 2.54-, or 3.2-cm) holes, depending on the screening practice. Anthracite, through 1-5/8-in (4.1-cm) and over 3/16-in (4.8-mm) round holes. (Jones, 1949)

nutcracker *boulder buster.*

NW Letter name specifying the dimensions of bits, core barrels, and drill rods in the N-size and W-group wireline diamond drilling system having a core diameter of 54.7 mm and a hole diameter of 75.7 mm. The NW designation has replaced the NX designation. (Cumming, 1981)

NX The NX designation for coring bits has been replaced by the NW designation. See also: *NW.* (Cumming, 1981)

nylon A generic term for a group of synthetic fiber-forming polyamides. The polymer is melted, extruded, stretched, and finally processed to turn it into a textile yarn having a very high strength, great powers of energy absorption, and high resistance to abrasion and rotting. Its major uses in mining are as a reinforcement for conveyor belting and ventilation ducting. (Nelson, 1965)

nylon belt A rubber belt containing nylon fiber reinforcing. It is stronger than cotton-duck belts of equivalent size and possesses better troughability and fastener holding strength. Nylon belt has the advantage of a long flex life, and the thinner carcass means easier bending. (Nelson, 1965)

nystagmus An eye disease suffered by some miners, in which there is a spasmodic oscillatory movement of the eyeballs; in severe cases, the victim finds difficulty in walking straight. Bad lighting is generally believed to be the main cause, and is possibly aggravated by the workers lying on their sides in thin seams. See also: *mining disease.* (Nelson, 1965)

O

oakum Loosely twisted fiber usually of hemp or jute impregnated with tar or with a tar derivative (such as creosote or asphalt); used in caulking seams (such as the wood hulls and decks of ships) and in packing joints (in pipes, caissons, etc.). (Webster 3rd, 1966)

Oamaru stone A white, granular limestone found in large quantities in Oamaru, New Zealand, and valued as a building stone.

obduction The overriding or overthrusting of oceanic crust onto the leading edges of continental lithospheric plates; plate accretion. See also: *subduction*. (AGI, 1987)

object glass *objective.*

objective The lens (or lenses) that gives an image of an object in the focal plane of a microscope or telescope eyepiece. Syn: *objective lens; object glass.*

objective glass *objective.*

objective lens *objective.*

oblique block A quarry term applied to a block of stone bounded by 3 pairs of parallel faces—4 of the 12 interfacial angles being right angles, 4 obtuse, and 4 acute.

oblique fault A fault that strikes oblique to, rather than parallel or perpendicular to, the strike of the constituent rocks or dominant structure. Cf: *oblique-slip fault; strike fault; dip fault.* Syn: *diagonal fault.* (AGI, 1987)

oblique illumination method *van der Kolk method.*

oblique joint (a) A joint whose strike is oblique to the strike of the strata or metamorphic rocks in which it occurs. (b) A joint that forms an acute angle with dip joints and strike joints. (Lewis, 1964)

oblique offset The distance of a point from a main survey line measured at an angle to the latter that is not a right angle. See also: *offset.* (Hammond, 1965)

oblique projection A pictorial view of an object showing its elevation, plan, or section to scale with parallel lines projected from the corners, at 45° or any other angle, indicating the other sides. See also: *axonometric projection; isometric projection.* (Hammond, 1965)

oblique slip In a fault, movement or slip that is intermediate in orientation between the dip slip and the strike slip. Cf: *strike slip.* (AGI, 1987)

oblique-slip fault A fault in which the net slip lies between the direction of dip and the direction of strike. Syn: *diagonal-slip fault.* Cf: *oblique fault.*

obra The narrow prismatic part of a blast furnace immediately above the crucible. (Fay, 1920)

obsequent fault-line scarp A scarp along a fault line, where the topographically low area is on the upthrown block. Cf: *resequent fault-line scarp.*

observer (a) In seismic prospecting, the person in charge of the recording crew, including the shooters and linemen. The observer must maintain the electronic equipment and decide on the best shooting and detector arrangement as well as the best instrumental settings for getting records of optimum quality. The observer operates the recording equipment in the field, often with the help of an assistant. In conventional recording, or in tape recording when photographic monitors are run, the observer or an assistant develops the record in the recording truck immediately after it is shot. Also called operator. (Dobrin, 1960) (b) In gravity and magnetic prospecting, a person who secures the instrument readings, e.g., on a torsion balance or magnetometer. (AGI, 1987)

obsidian A black or dark-colored volcanic glass, usually of rhyolite composition, characterized by conchoidal fracture. It is sometimes banded or has microlites. Usage of the term goes back as far as Pliny, who described the rock from Ethiopia. Obsidian has been used for making arrowheads, other sharp implements, jewelry, and art objects. Syn: *Iceland agate.* (AGI, 1987)

obsidianite brick Lightweight, siliceous fireclay, acid-resisting brick, burned to a glasslike mass.

obtuse bisectrix (a) That axis that bisects the obtuse angle of the optic axes of biaxial minerals. (Fay, 1920) (b) The angle >90° between the optic axes in a biaxial crystal, bxo. Cf: *optic angle.*

occidental cat's-eye *cat's-eye; tiger's-eye.*

occlude To take in and retain (a substance) in the interior rather than on an external surface; to sorb. Used esp. of metals sorbing gases; e.g., palladium occludes large volumes of hydrogen. (Webster 3rd, 1966)

occluded Contained in pores (said of gas occluded in coal). (Mason, 1951)

occluded gas Any of several gases that enter a mine atmosphere from pores, as feeders and blowers, and also from blasting operations. These gases pollute the mine air chiefly by the absorption of oxygen by the coal, and in addition by chemical combination of oxygen with carbonaceous matter, for example, from decaying timbers, rusting of iron rails, burning of lights, and breathing of humans and animals. These gases include oxygen, nitrogen, carbon dioxide, and methane. (Kentucky, 1952)

occlusion (a) Taking up or incorporation of liquids in solids or of gases in liquids. (AGI, 1987) (b) *absorption.*

occupant An occupant of a tract of land, as the word ordinarily is used, is one who has the use and possession thereof, whether he resides upon it or not. (Ricketts, 1943)

occupation As used in the mining law, it is equivalent to possession, and the right to locate is included in the right to occupy, and incident to a location is the right of possession; but mere occupancy of the public lands and making improvement thereon gives no vested right therein as against a location made in pursuance of law. (Ricketts, 1943)

occurrence *mineral occurrence.*

ocean current (a) The name current is usually restricted to the faster movements of the ocean, while those movements that amount to only a few miles a day are termed drifts. (AGI, 1987) (b) A nontidal current constituting a part of the great oceanic circulation. Examples are gulf stream, kuroshio, and equatorial currents. (AGI, 1987)

oceanic ridge *Mid-Atlantic Ridge; mid-ocean ridge.*

oceanic trench *trench.*

oceanographic dredge Apparatus used aboard ships to bring up quantity samples of the ocean bottom deposits and sediments. (Hunt, 1965)

oceanography The broad field of science that includes all fields of study that pertain to the sea. This includes the studies of boundaries of the ocean, its bottom topography, the physics and chemistry of seawater, the characteristics of its motion, and marine biology. (Hy, 1965)

ocher A name given to various native earthy materials used as pigments. They consist essentially of hydrated ferric oxide admixed with clay and sand in varying amounts and in impalpable subdivision. When carrying much manganese ochers grade into umbers. They are either yellow, brown, or red. The best reds are sometimes obtained by calcining the yellow varieties. They are called burnt ochers. Others are obtained by calcining copperas or as a residue from roasting pyrite. In general, the native yellows and browns are varieties of limonite and the native reds are varieties of hematite. One variety of red ocher is known as scarlet ocher. Their value as pigments depends not only on the depth of color but also on the amount of oil required as a vehicle. Syn: *terra sienna.* Cf: *umber.* (CCD, 1961)

octagon A polygon having eight sides. (Jones, 1949)

octahedral borax A rhombohedral form of hydrous sodium borate, $Na_2B_4O_7{\cdot}5H_2O$, simulating regular octahedrons. From the Lagoong of Tuscany, Italy. See also: *tincalconite.* Syn: *borax.* (English, 1938)

octahedral cleavage In the isometric system, cleavage parallel to the faces of an octahedron. (Fay, 1920)

octahedral copper *cuprite.*

octahedral iron ore *magnetite.*

octahedrite (a) A class of meteorites. (Hey, 1955) (b) The most common iron meteorite contains 6% to 18% nickel in the metal phase; on etching, shows Widmanstätten structure owing to the presence of intimate intergrowths (of plates of kamacite with narrow selvages of taenite) oriented parallel to the octahedral planes. (c) A former name for anatase. See also: *titanium dioxide.*

octahedron (a) A closed crystallographic form with isometric symmetry and eight faces, each an equilateral triangle; sometimes called a regular octahedron to distinguish it from the more general usage defined below. Commonly observed in isometric minerals, such as fluorite, pyrite, magnetite, and diamond. (b) Less precisely, a closed crystallographic form composed of (or bounded by) eight triangular surfaces (a bipyramid), such as in some samples of anatase. Plural: octahedra. Adj.: octahedral.

octant Each eighth of crystal space divided by three noncoplanar axial planes. Cf: *dodecant.*

octant search Used to limit the number of sample data points used for estimating intermediate spatial values. The search neighborhood is divided into eight equal-angle sectors. Constraints on selection of data values to include in the estimation include: minimum and maximum of samples or the number of consecutive empty sectors. If either criteria is below minimum, an interpolated value is not calculated. Applies to any interpolation method where a limited number of sample data points are used to estimate intermediate values.

octaphyllite (a) A trioctahedral clay mineral. (AGI, 1987) (b) A group of mica minerals that contains eight cations per ten oxygen and two hydroxyl ions. (c) Any mineral of the octaphyllite group, such as biotite; a trioctahedral clay mineral. Cf: *heptaphyllite.*

octopus A bin or tank to facilitate the concrete lining of circular shafts. The concrete is mixed on the surface, taken down the shaft in buckets, and discharged into the octopus. The concrete is then led away through flexible rubber pipes to different points around the shaft. (Nelson, 1965)

ocular Eyepiece of a microscope. (Pryor, 1963)

O'Donaghue formula A formula used for calculating the thickness of tubbing: $t = hdF/2C + A$, where: t is the required thickness of tubbing in inches; h is the pressure of water in pounds per square inch; d is the diameter of the shaft in inches; C is the crushing strength of cast iron in pounds per square inch, which may be taken as 95,000; F is the factor of safety adopted between 5 and 10; and A is the allowance for possible flaws and corrosion and may vary from 1/4 to 1 in (6.4 to 25.4 mm), averaging 1/2 in (12.7 mm). (Sinclair, 1958)

O'Donahue's theory A mine subsidence theory based on an extension of the theory of the normal. In it, subsidence is regarded as taking place in two stages. There is, first, a breaking of the rocks in which the lines of fracture tend to run at right angles to the stratification. This is followed by an aftersliding, or inward movement from the sides, resulting in a pull or draw beyond the edges of the workings. (Briggs, 1929)

odontolite A fossil bone or tooth colored deep blue by iron phosphate (vivianite), and rarely green by copper compounds, and resembling turquoise, such as that from the tusks of mammoths found in Siberia. It is cut and polished for jewelry. Syn: *bone turquoise; fossil turquoise.* (AGI, 1987)

oersted (a) The practical, cgs electromagnetic unit of magnetic intensity. A unit magnetic pole, placed in a vacuum in which the magnetic intensity is 1 Oe (79.577 A/m), is acted upon by a force of 1 dyn in the direction of the intensity vector. (AGI, 1987) (b) Commonly used as the cgs unit of magnetic-field intensity. Except in magnetized media, a magnetic field with an intensity (H) of 1 Oe has an induction (B) of 1 Gs (0.1 mT). (AGI, 1987)

Oetling freezing method A method of shaft sinking by freezing wet ground in sections as the sinking proceeds. The permanent lining is also inserted as the shaft is sunk. The freezing equipment is a cylinder equal in diameter to the shaft and 44 in. (1.12 m) in height, with the lower end closed by a plate. The cylinder is in sections, each of which can be removed. Each section is provided with freezing coils. After freezing the ground, two sections are removed, the ground is thawed locally and removed, and a segment of the permanent lining is inserted. The process is repeated. See also: *Dehottay process; freezing method.* (Nelson, 1965)

offcenter waterway A waterway port in a noncoring diamond bit, not located in the center of the bit face. (Long, 1960)

off gate N. of Eng. One of the goaf roadways in longwall workings, which are set about 120 yd (110 m) apart. (Fay, 1920)

off-highway truck A truck of such size, weight, or dimensions that it cannot be used on public highways.

official plat of survey The expression in a patent according to the official plat of survey of the land returned to the general land office by the surveyor general refers to the description of the land as well as to the quantity conveyed. (Ricketts, 1943)

off line (a) A condition existing when the drive rod of the drill swivel head is not centered and parallel with the borehole being drilled. (Long, 1960) (b) A borehole that has deviated from its intended course. (Long, 1960) (c) A condition existing wherein any linear excavation (shaft, drift, borehole, etc.) deviates from a previously determined or intended survey line or course.

off-peak load Electricity drawn at a period when the power station that supplies it is not fully loaded. (Pryor, 1963)

offretite A hexagonal mineral, $(K_2,Ca)_5Al_{10}Si_{26}O_{72}{\cdot}30H_2O$; zeolite group; commonly intergrown with erionite and levyne as a vein or cavity filling in mafic lavas.

offset (a) A short drift or crosscut driven from a main gangway or level. (Fay, 1920) (b) The horizontal distance between the outcrops of a dislocated bed. (Fay, 1920) (c) Of a fault, the horizontal component of displacement, measured perpendicular to the disrupted horizon. See also: *normal shift.* (AGI, 1987) (d) A side (horizontal) measurement of distance perpendicular to a line, usually a transit line. (Seelye, 1951) (e) To collar and drill a borehole at some distance from the designated site to avoid a difficult setup. (Long, 1960) (f) To drill a borehole near one previously drilled, which may have been lost, for purposes of correlation or to determine the lateral extent of mineralization. (Long, 1960) (g) An abrupt change in the trend of a drill hole, usually caused by a small shelflike projection of rock alongside one wall of the drill hole. (Long, 1960) (h) A well drilled near the boundary of a lease opposite a completed well on an adjacent lease. Syn: *offset hole; offset well.* (Long, 1960) (i) To offset a well by drilling the next adjoining location in accordance with a spacing pattern. (Wheeler, 1958) (j) A spur or minor branch from a principal range of hills or mountains. (k) The distance along the strain coordinate between the initial portion of a stress-strain curve and a parallel line that intersects the stress-strain curve at a value of stress which is used as a measure of the yield strength. It is used for materials that have no obvious yield point. A value of 0.2% is commonly used. (ASM, 1961)

offset deposit (a) A mineral deposit, esp. of sulfides, formed partly by magmatic segregation and partly by hydrothermal solution, near the source rock. (AGI, 1987) (b) At Sudbury, ON, Canada, the term refers to dikelike bodies radiating from the Sudbury Complex, thought to have been filled from above by xenolithic rock fragments and massive pyrrhotite-chalcopyrite-pentlandite. (AGI, 1987)

offset digging In a ladder ditcher, digging with the boom not centered in the machine. (Nichols, 1976)

offset hole *offset.*

offset line In surveying, a line established parallel to the main survey line, and usually not far from it; e.g., a line on a sidewalk, 2 ft (0.6 m) from the established street line, or a line parallel to the centerline of a bridge and 50 ft (15.2 m) from it. (Seelye, 1951)

offset ridge A ridge that is discontinuous because of faulting. (AGI, 1987)

offset staff In surveying, a rod, usually 10 links (0.2012 m) long, used in measuring short offsets. (Webster 2nd, 1960)

offset well *offset.*

off-sider A driller or drill-crew-worker working on the opposite shift. Also called drill helper. (Long, 1960)

offtake A length of boring rods unscrewed and detached at the top of a borehole. Also called rod stand; setout. (BS, 1963)

offtake lad *shackler.*

offtake rod One of the auxiliary rods at the top of a winding shaft for guiding and steadying the cages during decking or loading and unloading operations. (Nelson, 1965)

off-the-road hauling Hauling that takes place off the public highways, generally on a mining or excavation site. The hauling units used are generally higher and wider than those used in over-the-road hauling since highway restrictions do not limit size, weight, etc. Cf: *over-the-road hauling.* (Carson, 1961)

off the solid In this type of blasting, coal is blasted from the solid with no precutting or shearing. (McAdam, 1958)

ogie The space before the fire in a kiln. Also called killogie. (Standard, 1959)

ohm The practical mks unit of electric resistance that equals the resistance of a circuit in which a potential difference of 1 V produces a current of 1 A; the resistance in which 1 W of power is dissipated when 1 A flows through it. The standard in the United States. Symbols, Ω and ω. (Webster 3rd, 1966; Zimmerman, 1949)

ohmmeter A type of galvanometer that directly indicates the number of ohms of the resistance being measured. (Crispin, 1964)

Ohm's law The formula expressing Ohm's law is $I = E/R$, in which I is the electric current in amperes; E is the electromotive force in volts; and R is the resistance in ohms. (Handbook of Chem. & Phys., 2)

-oid A suffix meaning "in the form of.". (AGI, 1987)

oikocryst In poikilitic fabric, the enclosing crystal. (AGI, 1987)

oil agglomeration A coal beneficiation process in which an oil is used to preferentially wet the coal particles, which have an affinity to agglomerate into masses that are then selectively removed by screening, e.g. See also: *selective agglomeration.*

oil base The residuum from the distillation of petroleum. When paraffin is obtained from petroleum, the original oil is said to have a paraffin base; when the residue is entirely asphaltic, the original petroleum is said to have an asphaltic base. Some petroleums have both an asphaltic and a paraffin base. (API, 1953)

oil-bearing shale Shale impregnated with petroleum. Not to be confused with oil shale. (Tomkeieff, 1954)

oiled A term used in flotation when a particle is given a water repellent surface. When such a coating has been formed, the particle is said to be oiled or treated and ready to be floated. (Newton, 1959)

oiler (a) In flotation, oil that provides a film around a mineral particle. (b) One of several types of mechanical devices that deliver oil to machines and into air or steam lines in controllable amounts. Also called atomizer; line oiler; lubricator; oil pot; pineapple; pot. (Long, 1960)

oilfield winch An extremely powerful low-speed winch on a crawler tractor. (Nichols, 1976)

oil flotation A process in which oil is used in ore concentration by flotation. Syn: *flotation process.* (Fay, 1920)

oil of vitriol Concentrated sulfuric acid.

oil pot *line oiler; oiler.*

oil pump A hydraulic pump supplying oil under pressure to the hydraulic-feed cylinders and pistons of a hydraulic-type swivel head on a diamond drill. (Long, 1960)

oil shale A kerogen-bearing, finely laminated brown or black sedimentary rock that will yield liquid or gaseous hydrocarbons on distillation. Cf: *bituminous shale.* Syn: *kerogen shale; petroleum-oil cannel coal; petrolo-shale.* (AGI, 1987)

oilstone A fine-grained stone used for sharpening edged tools or other similar metal surfaces. (Fay, 1920)

oil-temper To harden steel by quenching in oil after heating. (Webster 3rd, 1966)

okenite A triclinic mineral, $Ca_{10}Si_{18}O_{46}{\cdot}18H_2O$; white; fibrous; commonly associated with zeolites in basalts.

old age (a) The stage in the development of a stream at which erosion is decreasing in vigor and efficiency, and aggradation becomes dominant as the gradient is reduced. It is characterized by a broad open valley with a flood plain that may be 15 times the width of the meander belt; numerous oxbows, bayous, and swamps; a sluggish current; and slow erosion, effected chiefly by mass-wasting at valley sides. (AGI, 1987) (b) The final stage of the cycle of erosion of a landscape or region, in which the surface has been reduced almost to base level and the landforms are marked by simplicity of form and subdued relief. It is characterized by a few large meandering streams flowing sluggishly across broad flood plains, separated by faintly swelling hills, and having dendritic distributaries; and by peneplanation. (AGI, 1987) (c) A hypothetical stage in the development of a coast, characterized by a wide wave-cut platform, a faintly sloping sea cliff pushed far inland, and a coastal region approaching peneplanation. The stage is probably a theoretical abstraction, since it is doubtful whether stability of sea level is maintained long enough for the land to be so reduced. (AGI, 1987)

oldhamite An isometric mineral, (Ca,Mn)S; pale brown; occurs in some meteorites and slags; rapidly oxidizes in contact with air.

Oldham stone duster A self-contained transportable stone duster. A high-velocity current of air from a fan or blower is mechanically fed from a hopper above, both the fan and the feeding mechanism being driven from the tub axle as it is drawn along by rope haulage, horse, or manually. It delivers about 3/4 lb (0.34 kg) of dust per yard (0.9 m) of travel. (Sinclair, 1958)

Oldham-Wheat lamp A cap lamp designed for full self-service. This lamp, weighing 6-5/8 lb (3.0 kg), has a 4-V lead-acid battery in a hard rubber case with covers of stainless steel or nickel-plated hard brass. The switch is magnetically operated and is situated in a sealed plastic moulding. A 4-W bulb burning 11 h or a 2-W bulb burning 14 h is used. The lamp is of one-piece construction and no dismantling is needed to charge the accumulator. (Sinclair, 1958)

old hole *main hole.*

oldland (a) Any ancient land; specif. an extensive area (such as the Canadian shield) of ancient crystalline rocks reduced to low relief by long-continued erosion and from which the materials of later sedimentary deposits were derived. (AGI, 1987) (b) A region of older land, behind a coastal plain, that supplied the material of which the coastal-plain strata were formed. (AGI, 1987) (c) A term proposed by Maxson and Anderson (1935) for the land surface of the old-age stage of the cycle of erosion, characterized by subdued relief. (AGI, 1987)

Old Red Sandstone A thick sequence of nonmarine, predominantly red sedimentary rocks, chiefly sandstones, conglomerates, and shales, representing the Devonian System in parts of Great Britain and elsewhere in northwestern Europe. (AGI, 1987)

old scrap Scrap derived from consumer goods that have outlived their usefulness in the economy; it includes discarded white goods, automobiles, electrical equipment, machinery, etc.

old silver Silver made to appear old by the application of graphite and grease.

old waste Scot. Old or abandoned workings. (Fay, 1920)

old working Mine working that has been abandoned, allowed to collapse, and perhaps sealed off. Unless proper safeguards are taken, old workings can be a source of danger to workings in production particularly if they are waterlogged and their plan position is uncertain. See also: *inrush of water.* (Nelson, 1965)

oleander-leaf texture Leaf-shaped masses of stromeyerite (or other minerals) in a matrix of chalcocite (or other mineral). (AGI, 1987)

oleic acid $CH_3(CH_2)_7CH{:}CH(CH_2)_7COOH$; a mono-unsaturated fatty acid used in ore flotation; insoluble in water; and soluble in alcohol, ether, and in most other organic solvents. (CCD, 1961)

oligist *hematite.*

oligist iron *hematite.* Also spelled oligiste iron.

Oligocene An epoch of the early Tertiary Period, after the Eocene and before the Miocene; also, the corresponding worldwide series of rocks. It is considered to be a period when the Tertiary is designated as an era. (AGI, 1987)

oligoclase A triclinic mineral, $(Na,Ca)[(Si,Al)AlSi_2O_8]$; plagioclase series of the feldspar group; has NaSi (albite) 10 to 30 mol % and CaAl (anorthite) 90 to 70 mol %; pseudomonoclinic with prismatic cleavage and characteristic polysynthetic twinning commonly visible on cleavage traces; white; may be chatoyant; a common rock-forming mineral in igneous and metamorphic rocks of intermediate to high silica content; forms the entire mass in some anorthosites; less commonly a vein mineral.

oligonite A former name for manganoan siderite, $(Fe,Mn)CO_3$.

oligosiderite A meteorite containing a small amount of metallic iron. (AGI, 1987)

oligotrophic peat Peat poorly supplied with nutrients. (Tomkeieff, 1954)

olivenite An orthorhombic mineral, $Cu_2AsO_4(OH)$; adamantine to vitreous; a secondary mineral in copper deposits.

Oliver filter A continuous-type filter made in the form of a cylindrical drum with filter cloth stretched over the convex surface of the drum. The drum rotates slowly about a horizontal axis, and the lower part is immersed in a tank containing the pulp to be filtered. Arrangement of pressure and suction pipes on the interior of the drum permits the application of suction to the filtering surface. As the filter passes through the tank, it picks up a layer of solid material and emerges carrying a layer of filter cake. Syn: *continuous filter.* (Newton, 1959)

olivine A mineral group including fayalite, Fe_2SiO_4; forsterite, Mg_2SiO_4; liebenbergite, $(Ni,Mg)_2SiO_4$; and tephroite, Mn_2SiO_4; orthorhombic; olive green, grayish green, brown, or black; members intermediate in the forsterite-fayalite crystal solution series are common rock-forming minerals in gabbros, basalts, peridotites, and dunites; alters hydrothermally to serpentine. Fayalite occurs in some granites and syenites, forsterite in thermally metamorphosed dolomites, and tephroite in iron manganese ore deposits and their associated skarns. Syn: *peridot; chrysolite; olivinoid.*

olivine rock *dunite.*

olivinite (a) In the International Union of Geological Sciences classification, Syn: *dunite.* (b) An olivine-rich ore-bearing igneous rock that also contains other pyroxenes and/or amphiboles. Syn: *leucochalcite.* (AGI, 1987)

olivinoid An olivinelike substance found in meteorites. See also: *olivine.* (Standard, 1964)

olivinophyre Porphyry containing olivine phenocrysts. (Fay, 1920)

ollenite A type of hornblende schist characterized by abundant epidote, sphene, and rutile. Garnet is one of the accessories. (AGI, 1987)

O.L.P. steel process A steelmaking method similar to the L.D. except that powdered lime is blown with the oxygen stream (therefore, the letters O.L.P.—oxygen- lime-powder). See also: *L.D. steel process; oxygen-Bessemer.* (Nelson, 1965)

ombrogenous peat Peat, the nature of which is determined by the amount of rainfall. (Tomkeieff, 1954)

omnibus In glassmaking, a sheet-iron cover to protect, from drafts, the glass articles in a leer. (Standard, 1964)

omnidirectional hydrophone A hydrophone whose response is essentially independent of angle of arrival of the incident sound wave. (Hy, 1965)

omphacite A mineral of the pyroxene group intermediate among aegirine, jadeite, and augite; i.e., high in aluminum and sodium, and of high-pressure origin; monoclinic; pale to grass green; occurs in eclogites.

o.m.s. (a) Output (usually in hundredweights) per manshift. It is a method of expressing the productivity of mines, miners, and management. (Nelson, 1965) (b) N. of England. Output (usually tons) per manshift. Interpretation depends on the basis for calculating manshifts, e.g., face o.m.s. is based on manshifts at the face; seam o.m.s. on piecework and bargain work manshifts in the seam; overall (pit) o.m.s. on all manshifts underground, including datal labor. (Trist, 1963)

on air (a) The state of a pump which is operating although having no liquid in its working parts. (BS, 1963) (b) Scot. Said of a pump when air is drawn at each stroke.

oncosimeter An instrument for determining the specific gravity of a molten metal by the immersion of a ball made of another metal and of known weight. (Standard, 1964)

onegite A variety of goethite. (Hey, 1955)

one on two A slope in which the elevation rises 1 ft (m) in 2 horizontal ft (m). (Nichols, 1976)

one-part line A single strand of rope or cable. (Nichols, 1976)
one-piece set A term applied to a single stick of timber, called a post, stull, or prop. Post and prop are applied to vertical timbers, and stull is applied to inclined timbers, or those placed horizontally. See also: *set.* (Lewis, 1964)
one shot (a) A borehole-survey instrument that records a single inclination and/or bearing reading on each round trip into a borehole. (Long, 1960) (b) Single shot.
one-shot exploder *Little Demon exploder.*
one-spot strip mining Consists of three operations: the top material is cast out of the way; pay material is dug and trucked away; and the top is pushed or cast back in. (Nichols, 1976)
one-way ram A hydraulic cylinder in which fluid is supplied to one end so that the piston can be moved only one way by power. Syn: *single-acting ram.* (Nichols, 1976)
one-way ventilation *peripheral ventilation.*
onion-skin weathering *spheroidal weathering.*
onlap (a) An overlap characterized by the regular and progressive pinching out, toward the margins or shores of a depositional basin, of the sedimentary units within a conformable sequence of rocks, in which the boundary of each unit is transgressed by the next overlying unit and each unit in turn terminates farther from the point of reference. Also, the successive extension in the lateral extent of strata (as seen in an upward sequence) due to their being deposited in an advancing sea or on the margin of a subsiding landmass. Ant: offlap. Syn: *transgressive overlap.* Cf: *overlap.* (AGI, 1987) (b) The progressive submergence of land by an advancing sea. Cf: *transgression.* (AGI, 1987)
on line (a) A linear underground excavation advancing in compliance with a predetermined surveyed direction or line. (b) A borehole the course of which is not deviating from the intended direction. (Long, 1960) (c) Said of a diamond drill when its drive rod is centered on and parallel to a borehole. (Long, 1960)
on long awn A face between end and crosscut. (Sinclair, 1960)
onofrite A former name for selenian metacinnabar Hg(S,Se).
on plane Scot. In a direction at right angles to, or facing, the plane or main joints. Syn: *plane course.* (Fay, 1920)
onsetter (a) The person in charge of loading and unloading of cages or skips at the pit bottom, and also the signaling to the pithead. In modern mines, the onsetter is stationed in a cabin and all controls are within reach for the loading and unloading of the cages, shaft signaling, and other car control equipment at the shaft bottom. See also: *cager; hitcher.* (Nelson, 1965) (b) The person in charge of winding operations underground, who is stationed at the shaft side and gives all signals to the winding engineman. (Mason, 1951)
onsetting machine Eng. A mechanical apparatus for loading cages with full tubs and discharging the empties, or vice versa, at one operation. (Fay, 1920)
on short awn A face in a direction between bord and crosscut. (Sinclair, 1960)
on sights (a) Following sights placed by a surveyor. (Fay, 1920) (b) On line.
Ontarian (a) Stage in New York State: Middle Silurian (middle and lower parts of Clinton Group). (AGI, 1987) (b) An obsolete name for the Middle and Upper Ordovician in New York State. (AGI, 1987)
on-the-solid (a) Applied to a blasthole extending into coal farther than the coal can be broken by the blast. (Fay, 1920) (b) That part of a blasthole that cannot be broken by the blast. (Fay, 1920) (c) A practice of blasting coal with heavy charges of explosives, in lieu of undercutting or channeling. (Fay, 1920)
on the track Diamonds inset in the crown of a bit in concentric circles so that the diamonds in any one circle follow the same groove cut into the rock. (Long, 1960)
onyx (a) A chalcedonic variety of quartz with color, chiefly white, yellow, red, or black, in straight parallel bands; used esp. in making cameos. See also: *banded agate; agate; chalcedony; sardonyx; jasponyx.* Cf: *onyx agate.* (b) A name applied incorrectly to dyed, unbanded, solid-color chalcedony; esp. black onyx. (c) Adj. Parallel banded; e.g., onyx marble and onyx obsidian. (d) Jet black translucent layers of calcite from cave deposits, often called Mexican onyx or onyx marble. See also: *travertine.*
onyx agate Banded agate with straight parallel layers of differing tones of gray; not a syn. for onyx. Cf: *onyx.*
onyx marble Translucent, generally layered, cryptocrystalline calcite with colors in pastel shades, particularly yellow, brown, and green. See also: *cave onyx; Egyptian alabaster; oriental alabaster; Mexican onyx; travertine.* (ASTM, 1994)
oolite (a) A sedimentary rock, usually a limestone, made up chiefly of ooliths cemented together. The rock was originally termed "oolith." Syn: *roestone; eggstone.* (AGI, 1987) (b) A term often used for oolith, or one of the ovoid particles of an oolite. Etymol. Greek oon, egg. Cf: *pisolite.* (AGI, 1987)
oolith One of the small round or ovate accretionary bodies in a sedimentary rock, resembling the roe of fish, and having diameters of 0.25 to 2 mm (commonly 0.5 to 1 mm). It is usually formed of calcium carbonate, but may be of dolomite, silica, or other minerals, in successive concentric layers, commonly around a nucleus such as a shell fragment, an algal pellet, or a quartz-sand grain, in shallow, wave-agitated water; it often shows an internal radiating fibrous structure indicating outward growth or enlargement at the site of deposition. Cf: *pisolith.* Syn: *oolite; ovulite.* (AGI, 1987)
oolitic Pertaining to an oolite, or to a rock or mineral made up of ooliths; e.g., an oolitic ironstone, in which iron oxide or iron carbonate has replaced the calcium carbonate of an oolitic limestone. (AGI, 1987)
oolitic limestone An even-textured limestone composed almost wholly of relatively uniform calcareous ooliths, with virtually no interstitial material. It is locally an important oil reservoir (such as the Smackover Formation in Arkansas), and is also quarried for building stone. (AGI, 1987)
oolitic texture The texture of a sedimentary rock consisting largely of ooliths showing tangential contacts with one another. (AGI, 1987)
ooze (a) A soft, slimy, sticky mud. (Fay, 1920) (b) To emit or give out slowly. (Webster 3rd, 1966) (c) A fine-grained pelagic deposit that contains more than 30% of material of organic origin. (AGI, 1987) (d) An unconsolidated deposit composed almost entirely of the shells and undissolved remains of foraminifera, diatoms, and other marine life; diatom ooze and foraminiferal ooze. (Hy, 1965)
opacite A general term applied to swarms of opaque, microscopic grains in rocks, esp. as rims that develop mainly on biotite and hornblende phenocrysts in volcanic rocks, apparently as a result of post-eruption oxidation and dehydration. Opacite is generally supposed to consist chiefly of magnetite dust. Cf: *ferrite.* (AGI, 1987)
opacity The quality of being opaque. See also: *opaque.*
opal An amorphous or microcrystalline mineral, $SiO_2 \cdot nH_2O$; may be tridymite or cristobalite; has a varying proportion of water (as much as 20% but commonly 3% to 9%); occurs in nearly all colors; transparent to nearly opaque; typically shows a marked iridescent "play of color"; differs from quartz in being isotropic; has a lower refractive index than quartz and is softer, more soluble, and less dense; generally occurs massive and may be pseudomorphous after other minerals; deposited at low temperatures from silica-bearing water; occurs in cracks and cavities of igneous rocks, in flintlike nodules in limestones, in mineral veins, in deposits of thermal springs, in siliceous skeletons of various marine organisms (such as diatoms and sponges), in serpentinized rocks, in weathering products, and in most chalcedony and flint. The transparent colored varieties exhibiting opalescence are valued as gemstones. Syn: *opalite.* See also: *opaline.*
opalescence A milky or somewhat pearly appearance or luster of a mineral, such as that shown by opal and moonstone. Cf: *play of color.*
opaline (a) Any of several minerals related to or resembling opal; e.g., a pale-blue to bluish-white opalescent or girasol corundum, or a brecciated impure opal pseudomorphous after serpentine. (AGI, 1987) (b) An earthy form of gypsum. (AGI, 1987) (c) A rock with a groundmass or matrix consisting of opal. adj. Resembling opal, esp. in appearance; e.g., opaline feldspar (labradorite) or opaline silica (tabasheer). (AGI, 1987)
opalite *opal.*
opalized wood Silicified wood. See also: *wood opal.* (AGI, 1987)
opaque Said of a material that is impervious to visible light or has metallic luster. Cf: *transopaque; translucent; transparency; transparent; opacity*
opaque-attrite Attritus that is opaque in thin sections. Syn: *opaque-durit.* (Tomkeieff, 1954)
opaque attritus Refers to coal material of which the most prominent and important constituent is opaque matter and also referred to as opaque matrix, black fundamental matter or matrix and residuum. (IHCP, 1963)
opaque-durit *opaque-attrite.*
open area *effective screening area.*
opencast A working in which excavation is performed from the surface. Commonly called open pit. See also: *bench; quarry.* Cf: *strip mine.* Syn: *opencut; opencut mine.* (Webster 3rd, 1966)
opencast method A mining method consisting of removing the overlying strata or overburden, extracting the coal, and then replacing the overburden. When the overlying material consists of earth or clay it can be removed directly by scrapers or excavators, but where rock is encountered it is necessary to resort to blasting to prepare the material into suitable form for handling by the excavators. See also: *strip mining.* (McAdam, 1958)
opencast mine *opencast.*
opencast working *opencast; strike working.*
open circuit In mineral dressing, a flow line in which the solid particles pass from one appliance to the next without being screened, classified, or otherwise checked for quality; no fraction is returned for retreatment. (Pryor, 1960)
open-circuit mill A grinding mill without classifiers. (Nelson, 1965)

open-crib timbering Shaft timbering with cribs alone, placed at intervals. (Fay, 1920)

opencut (a) To increase the size of a shaft when it intersects a drift so as to form a puddle wall behind the sets of timber. (b) Open pit; surface working in which the working area is kept open to the sky. See also: *opencast.* Syn: *strip mine.* (Fraenkel, 1953; AGI, 1987)

opencut mine An excavation for removing minerals which is open to the weather. See also: *opencast.* (Hess)

opencut mining *surface mining; openpit mining.*

open-drive sampler A drive-type soil-sampling device that is essentially a headpiece, threaded to fit a drill rod, to which is attached a removable length of thin-wall brass or steel tubing. An example is the Shelby tube. (Long, 1960)

open end method A method of mining pillars in which no stump is left; the pockets driven are open on the goaf side and the roof is supported on timbers. (Lewis, 1964)

opener hole The first hole or holes fired in a round blasted off the solid to create an additional free face in a coal mine. (CFR, 4)

open fault A fault in which the two walls are separated. Cf: *closed fault.* (AGI, 1987)

open fire Fire occurring in a roadway or at the coal face in a mine. Such fires may or may not be easily accessible. They may be in the roof of a roadway or seam, or in the kerf of a machine-cut face. However, they are quite distinct in their initiation from gob fires. An open fire may be ignited by a blown-out shot, electrical failure, or from sparks produced by friction. See also: *spontaneous combustion.* (Nelson, 1965)

open fold A fold in which the limbs diverge at a large angle. (AGI, 1987)

open front The arrangement of a blast furnace with a forehearth. (Fay, 1920)

open-graded aggregate Mineral aggregate containing very few small particles so that the void spaces are relatively large. (Shell)

open hearth The form of regenerative furnace of the reverberatory type used in making steel by the Martin, Siemens, and Siemens-Martin processes. See also: *furnace; open-hearth furnace.*

open-hearth furnace A reverberatory melting furnace with a shallow hearth and a low roof. The flame passes over the charge on the hearth, causing the charge to be heated both by direct flame and by radiation from the roof and sidewalls of the furnace. In the ferrous industry, the furnace is regenerative. See also: *open hearth.* (ASM, 1961)

open-hearth process A process for manufacturing steel, either acid or basic, according to the lining of the reverberatory furnace, in which selected pig iron and malleable scrap iron are melted, with the addition of pure iron ore. The latter, together with the air, contributes to the oxidation of the silicon and carbon in the melted mass. The final deoxidation is sometimes produced by the addition of a small quantity of aluminum or ferromanganese, which at the same time desulfurizes and recarburizes the metal to the required extent. See also: *L.D. steel process; Siemens-Martin process.* (Fay, 1920)

open-hearth steel *open-hearth process.*

open hole (a) Coal or other mine workings at the surface or outcrop. Also called opencast; opencut; open pit. (b) A borehole that is drilled without cores. (Nelson, 1965) (c) Uncased portion of a borehole. (Pryor, 1963) (d) A borehole free of any obstructing object or material. (Long, 1960)

opening (a) A widening out of a crevice, in consequence of a softening or decomposition of the adjacent rock, so as to leave a vacant space of considerable width. (Fay, 1920) (b) A short heading driven between two or more parallel headings or levels for ventilation. (Fay, 1920) (c) Surface entrance to mine workings. Syn: *mine opening.* (Hudson, 1932)

opening out The formation of a longwall face by driving headings and cross headings and connecting the faces to form a continuous line of coal face. It may be viewed as the final stage in development, leading to full coal production. In pillar-and-stall mining, opening out would imply the setting off of the main headings and subsidiary drivages for the formation of coal pillars. See also: *branch headings; mechanized heading development.* (Nelson, 1965)

openings (a) The parts of coal mines between the pillars, or the pillars and ribs. (Fay, 1920) (b) A series of parallel chambers or openings, separated by pillars or walls, in slate mining. The width of an opening varies from 35 to 50 ft (11 to 15 m) depending on roof conditions. (Nelson, 1965)

opening shot In blasting into solid rock, the wedging shot, gouging shot, or burn cut. Leading shot fired to open up the rock face by creating a cavity and therefore easing the work done by later shots in a round. (Pryor, 1963)

opening stock Quantity on hand at start of accounting period—ore, concentrates, stores, etc. (Pryor, 1963)

open lagging Lagging placed a few inches apart.

open light A naked light. Not a safety light. (Fay, 1920)

open off (a) To turn stalls off stalls, or to drive branch roadways from crossheadings. (Nelson, 1965) (b) Eng. To begin the longwall system from the shaft pillar, or the far end of the royalty, or from any headings previously driven out for the purpose of commencing such system. (Fay, 1920) (c) To start any new working, as a heading, entry, gangway, room, etc., from another working, as a slope, gangway, etc. (Fay, 1920)

openpit mine A mine working or excavation open to the surface. *strip mine.*

openpit mining (a) A form of operation designed to extract minerals that lie near the surface. Waste, or overburden, is first removed, and the mineral is broken and loaded, as in a stone quarry. Important chiefly in the mining of ores of iron and copper. (Barger, 1939) (b) The mining of metalliferous ores by surface-mining methods is commonly designated as openpit mining as distinguished from the strip mining of coal and the quarrying of other nonmetallic materials such as limestone, building stone, etc. See also: *strip mining.* (Woodruff, 1966)

openpit quarry A quarry in which the opening is the full size of the excavation. One open to daylight. (Fay, 1920)

open pot Fireclay pot for melting glass—open at the top. (Mersereau, 1947)

open rock Any stratum sufficiently open or porous to contain a significant amount of water or to convey it along its bed. (AGI, 1987)

opens Large, open cracks or crevices and small and large caverns. (Long, 1960)

open-sand casting Casting made in a mold simply excavated in sand, without a flask. (Fay, 1920)

openset Scot. An unfilled space between pack walls. See also: *cundy.* (Fay, 1920)

open shop A shop, or mine, where the union price is paid, but where the workers are not all union members. Cf: *union shop.*

open split A split in which no regulator is installed. Syn: *free split.* (Higham, 1951)

open stope (a) An unfilled cavity. (Nelson, 1965) (b) Underground working place either unsupported, or supported by timbers or pillars of rock. (Pryor, 1963)

open-stope method (a) Stoping in which no regular artificial method of support is employed, although occasional props or cribs may be used to hold local patches of insecure ground. The walls and roof are self-supporting, and open stopes can be used only where the ore and wall rocks are firm. (Jackson, 1936) (b) *overhand stoping.*

open-tank method A method of treating mine timber to prevent decay in which the timber is immersed in a tank of hot preservative and then in a tank of cold preservative. The preservatives used are creosote, zinc chloride, sodium fluoride, and other chemicals. See also: *timber preservation; Bethell's process.* (Lewis, 1964)

open timbering The usual method of setting timber or steel supports in mines—they are spaced from 2 to 5 ft (0.6 to 1.5 m) apart, with laggings and struts to secure the ground between each set. The method is used in ground that does not crumble or flow. See also: *close timbering.* (Nelson, 1965)

open-top carrier The main use of this type of bucket elevator has been in handling the product of the larger crushers. Steel buckets of large capacity, which may be as long as 7 ft (2.1 m), are attached rigidly to a heavy flat bar chain, each strand made of two bars with a pitch of 2 and with self-oiling flanged rollers at each intersection. The elevator rises at an angle of about 60°, and the rollers run on ways made of light T-rails. The buckets have overlapping edges, so that there is no spill between them. (Pit and Quarry, 1960)

open-top tubbing A length of tubbing having no wedging crib on the top of it. (Fay, 1920)

open traverse A surveying traverse that starts from a station of known or adopted position but does not terminate upon such a station and therefore does not completely enclose a polygon. Cf: *closed traverse.* (AGI, 1987)

open working Surface working, e.g., a quarry or opencast mine. Among the minerals often exploited by open workings are coal, brown coal, gems; the ores of copper, gold, iron, lead, and tin; and all kinds of stone. Also called open work. (Nelson, 1965)

operating carrier The mechanism used with the automatic duckbill through which the extension and retraction of the shovel trough are controlled. (Jones, 1949)

operating cost The sum of the costs of mining, beneficiation, and administration gives the operating cost of a mine. (Nelson, 1965)

operating engineer *hoistman.*

operating point A ventilation system is composed of a fan and a set of connected ducts. In a mine ventilation system, mine openings comprise the ducts. At a given air density and with the fan operating at constant speed, there is only one head and quantity of airflow that can result. This is an equilibrium condition and is known as the operating point of the system. (Hartman, 1982)

operation In crystallography, the rotation, reflection, or inversion of an attribute of a crystal structure to complete its symmetry. Cf: *element.*

operational capacities Figures given on flowsheets to indicate quantities passing various points in plant per unit time, taking account of fluctuations in the rate of supply and composition (as to size and content of impurity), as follows: (1) design capacity, the rate of feed, defined by limits expressing the extent and duration of load variations, at which specific items of plant subject to a performance guarantee must operate continuously and give the guaranteed results on a particular quality of feed; (2) peak design capacity, a rate of feed in excess of the design capacity, which specific items of plant will accept for short periods without fulfilling the performance guarantees given in respect of them; and (3) mechanical maximum capacity, the highest rate of feed at which specific items of equipment, not subject to performance guarantees, will function on the type and quality of feed for which they are supplied. (BS, 1962)

operative temperature Operative temperature is that temperature of an imaginary environment in which, with equal wall (enclosing areas) and ambient air temperatures and some standard rate of air motion, the human body would lose the same amount of heat by radiation and convection as it would in some actual environment at unequal wall and air temperatures and for some other rate of air motion. (Strock, 1948)

ophicalcite A recrystallized metamorphic rock composed of calcite and serpentine, commonly formed by dedolomitization of a siliceous dolostone. Some ophicalcites are highly veined and brecciated and are associated with serpentinite. (AGI, 1987)

ophiolite A group of mafic and ultramafic igneous rocks ranging from spilite and basalt to gabbro and peridotite, including rocks rich in serpentine, chlorite, epidote, and albite derived from them by later metamorphism, whose origin is associated with an early phase of the development of a geosyncline. The term was originated by Steinman in 1905. (AGI, 1987)

ophite A general term for diabases that have retained their ophitic structure although the pyroxene is altered to uralite. The term was originated by Palasson in 1819. (AGI, 1987)

ophitic Applied to a texture characteristic of diabases or dolerite in which euhedral or subhedral crystals of plagioclase are embedded in a mesotasis of pyroxene crystals, usually augite. Also said of a rock with such a texture. Cf: *poikilitic.* Syn: *doleritic.* (AGI, 1987)

optical anomaly Optical properties apparently at variance with optical rules, such as: anisotropy in isotropic minerals, such as birefringent diamond; biaxiality in uniaxial minerals, such as quartz; and erratic variation in birefringence near optical absorption bands—e.g., some epidote minerals.

optical calcite Crystalline calcite so clear that it has value for optical purposes; e.g., polarizers. Syn: *Iceland spar.*

optical centering device An optical device that enables a theodolite to be accurately positioned over or under a survey station. Also called optical plummet (undesirable usage). (BS, 1963)

optical character The designation as to whether optically positive or optically negative; said of minerals. (Fay, 1920)

optical constant In optical mineralogy, any of the following: indices of refraction, birefringence, optic sign, axial angles, extinction angles, and dispersion of a nonopaque mineral. In ore microscopy (mineragraphy), any of the reflectances and anisotropy of opaque minerals. See also: *reflected-light microscope.*

optical crystallography The study of the behavior of light in crystals. (Hurlbut, 1964)

optical diffraction Constructive interference of monochromatic light; e.g., labradorescence in plagioclase, fire in opal. See also: *diffraction.*

optical flat Glass or other surface rendered truly planar. (Pryor, 1963)

optical glass Carefully made glass of great uniformity and usually special composition to give desired transmission, refraction, and dispersion of light. (CCD, 1961)

optical mineralogy The determination of optical properties of minerals for the purpose of characterization and identification. See also: *index of refraction.*

optical property Any of several effects of a substance upon light. Refractive index, double refraction, (and its strength, birefringence), dispersion, pleochroism, and color are gemmologically the most important optical properties.

optical pyrometer A type of pyrometer that measures high temperature by comparing the intensity of light of a particular wavelength from the hot material with that of a filament of known temperature. It is used to determine the temperature of incandescent lavas. See also: *pyrometer.* (AGI, 1987)

optical sign When a translucent crystal is viewed under microscope, light travels through the mineral at a speed which corresponds with its refractive index, as this is affected by the crystal planes. A uniaxial crystal has a negative optical sign when the velocity of its extraordinary ray exceeds that of the ordinary ray and vice versa. Calcite is negative; quartz positive. For biaxial crystals, the three principal directions of vibration are mutually at right angles. (Pryor, 1963)

optical square A hand-held instrument enabling right angles to be set out accurately on a site. (Hammond, 1965)

optical twinning A type of twinning in quartz in which the parts of the twin are alternately left- and right-handed. So named because it can be recognized by optical tests in distinction to Dauphine (electrical) twinning. Optical twinning as ordinarily applied includes all twin laws in quartz with the exception of the Dauphine. Also called Brazil twinning; chiral twinning. (Am. Mineral., 1947)

optic angle The angle between the two optic axes of a biaxial crystal; its symbol is 2V (less than 90°), $2V_\alpha$, or $2V_\gamma$, depending on whether the optic direction X or Z is in the acute bisectrix. Syn: *axial angle; optic-axial angle.* Cf: *acute bisectrix; obtuse bisectrix.*

optic-axial angle *optic angle.*

optic axis A direction of single refraction in a doubly refracting mineral. Hexagonal and tetragonal minerals have one such axis, and are termed uniaxial; rhombic, monoclinic, and triclinic minerals have two optic axes and are thus biaxial. See also: *uniaxial; dispersion.* Cf: *extinction.* (Anderson, 1964)

optics The sub-field of physics that covers the behavior of light.

optic sign (a) Indicates the type of double refraction in a mineral. In uniaxial minerals, the material is said to be positive when the extraordinary ray has a higher refractive index than the ordinary ray and negative when the ordinary ray has the greater index. In biaxial minerals, which have three basic optical directions, the refractive index of the intermediate or beta ray is the criterion; if its refractive index is nearer that of the low or alpha ray, it is said to be a positive mineral or stone; if it is nearer the high or gamma ray, it is said to be a negative mineral or stone. (b) More technically, in uniaxial minerals, the material is positive when the extreme refractive index (n_ε) is greater than the apparently isotropic one (n_ω) and negative when the extreme refractive index is less. In biaxial minerals, which have extreme refractive indices both above and below the apparently isotropic one (n_β), the material is positive when the lower refractive index (n_α) is closer to the apparently isotropic one and negative when the higher one (n_γ) is closer. Syn: *optical sign; optical character.* See also: *interference.*

optimization Coordination of various processing factors, controls, and specifications to provide best overall conditions for technical and/or economic operation. (Pryor, 1965)

optimum depth of cut That depth of cut required to completely fill the dipper in one pass without undue crowding. (Carson, 1961)

optimum moisture content The water content at which a soil can be compacted to the maximum dry unit weight by a given compactive effort. Also called optimum water content. (ASCE, 1958)

option (a) A privilege secured by the payment of a certain consideration for the purchase, or lease, of mining or other property, within a specified time, or upon the fulfillment of certain conditions set forth in the contract. (b) S. Afr. The word option may refer to shares under option to the holder of option certificates. In regard to mining activities, options are granted to acquire the mineral rights and/or surface rights over some farm at a price fixed in the agreement. This price may be a sum of money or a participation in a mining company still to be formed. The option itself can be acquired for a lump sum or for a payment of so much per morgen a year. The option contract is generally connected with the permission for the option holder to prospect for minerals and briefly referred to as option and prospecting contract. (Beerman, 1963)

optional-flow storage In coal preparation, optional-flow setups are those where coal usually goes to the plant but can be diverted into storage, either in bins or hoppers or on the ground. (Coal Age, 1966)

opx Abbrev. for orthopyroxene. Cf: *cpx.*

oral agreement to locate An agreement to locate need not be in writing. If a party, in pursuance of an oral agreement to locate at the expense of another, locates the claim in his or her own name, he or she holds the legal title to the ground in trust for the benefit of the party for whom the location was made. Such a party could, upon making the necessary proofs, compel the locator of the mining claim to convey the title to him or her, although the agreement to do so was not in writing. Such an agreement is not within the statute of frauds. (Ricketts, 1943)

orange heat A division of the color scale, generally given as about 900 °C.

orangepeel A variant of the clamshell bucket with four or five leaves instead of the clamshell's two. Each leaf ends in a reinforced point. Its digging ability is less than that of the clamshell, and its principal use is for underwater excavation and digging. (Carson, 1961)

orangepeel sampler An apparatus consisting of four movable jaws that converge to a point when closed; used to obtain samples of underwater sediment. (AGI, 1987)

orbicular Adj. Describes rounded to spherical, commonly banded, textures within minerals or rocks; e.g., orbicular diorite.

orbicular structure A structure developed in certain phanerocrystalline rocks (e.g., granite and diorite) due to the occurrence of numerous orbicules. Syn: *spheroidal structure; nodular structure.*

orcelite A hexagonal mineral, $Ni_{5-x}As_2$; rose-bronze; at the Tiebaghe massif, New Caledonia.

ordered solid solution A condition when atoms in a solid solution arrange themselves in regular or preferential positions in the lattice, rather than at random. (Newton, 1959)

order of crystallization The apparent chronological sequence in which crystallization of the various minerals of an assemblage takes place, as evidenced mainly by textural features. See also: *paragenesis.* (AGI, 1987)

order of persistence *stability series.*

order of reaction A classification of chemical reactions based on the index of the power to which concentration terms are raised in the expression for the instantaneous velocity of the reaction; i.e., on the apparent number of molecules which interact. (CTD, 1958)

ordinary kriging A variety of kriging which assumes that local means are not necessarily closely related to the population mean, and which therefore uses only the samples in the local neighborhood for the estimate. Ordinary kriging is the most commonly used method for environmental situations. See also: *kriging.*

ordinary ray (a) In a uniaxial crystal, that ray that travels with constant velocity in any direction within it. (Anderson, 1964) (b) In mineral optics, a light ray that, because of its crystallographic orientation, follows Snell's law, n=sini/sinr, where n is the refractive index, i is the angle of incidence, and r is the angle of refraction. In anisotropic crystals, not all light rays follow Snell's law and are, hence, "extraordinary rays." Cf: *law of refraction.*

ordinate Y-axis; the vertical scale of a graph.

ordnance bench mark Survey station the level of which has been officially fixed with reference to the ordnance datum, the arbitrary mean sea level at Newlyn in Cornwall, England. (Pryor, 1963)

ordnance survey Originally, a military mapping activity; now a precise survey maintained by government which maps land and building features of Great Britain in close detail. (Pryor, 1963)

ordonezite A tetragonal mineral, $ZnSb_2O_6$; brown.

ore (a) The naturally occurring material from which a mineral or minerals of economic value can be extracted profitably or to satisfy social or political objectives. The term is generally but not always used to refer to metalliferous material, and is often modified by the names of the valuable constituent; e.g., iron ore. See also: *mineral; mineral deposit; ore mineral.* Syn: *orebody.* (b) The term ores is sometimes applied collectively to opaque accessory minerals, such as ilmenite and magnetite, in igneous rocks. (AGI, 1987)

ore band Zone of rock rich in ore.

ore-bearing fluid *hydrothermal solution.*

ore bed (a) Metal-rich layer in a sequence of sedimentary rocks. (AGI, 1987) (b) Economic aggregation of minerals occurring between or in rocks of sedimentary origin.

ore bin (a) A receptacle for ore awaiting treatment or shipment. (Fay, 1920) (b) Robustly constructed steel, wooden, or concrete structure which receives intermittent supplies of mined ore and can transfer them continuously by rate-controlled withdrawal systems (bottom gates and ore feeders) to the treatment plant. Thus a buffer stock is held which allows a mine to hoist ore intermittently without bringing milling operations to a standstill. It characteristically receives a weighed-in input of finely broken ore from the final dry-crushing section (usually between 1-in and 3/8-in (2.54-cm and 9.5-mm) maximum particle size). The surge bin is a much smaller one, able to receive a dumped load of run-of-mine ore and to transfer it at a regular rate to the crushing system between arrivals of further skip loads. (Pryor, 1963)

ore blending Method whereby a mine, or a group of mines, served by a common mill, sends ores of slightly varied character for treatment and separate bins or stockpiles are provided. From these, regulated percentages of ores are drawn and blended to provide a steady and predictable feed to the mineral dressing plant. (Pryor, 1963)

ore block A section of an orebody, usually rectangular, that is used for estimates of overall tonnage and quality. See also: *blocking out.* (AGI, 1987)

ore blocked out *developed reserve.*

ore boat A boat constructed esp. for transporting iron ore on the Great Lakes. (Mersereau, 1947)

orebody A continuous, well-defined mass of material of sufficient ore content to make extraction economically feasible. See also: *ore; mineral deposit.* (AGI, 1987)

ore boil A reaction that occurs in an open-hearth furnace in which the carbon monoxide released by the oxidation of carbon causes a violent agitation of the metal as it escapes. (Newton, 1959)

ore bridge A large electric gantry-type crane which, by means of a clamshell bucket, stocks ore or carries it from a stockpile into bins or a larry car on a trestle. (Fay, 1920)

ore-bridge bucket A clamshell grab bucket of 5 to 7 tons capacity. (Fay, 1920)

ore car A mine car for carrying ore or waste rock. (Weed, 1922)

ore chute An inclined passage, from 3 to 4 ft (approx. 1 m) square, for the transfer of ore to a lower level, car, conveyor, etc. It may be constructed through waste fills. See also: *orepass.* (Nelson, 1965)

ore cluster A genetically related group of orebodies that may have a common root or source rock but that may differ structurally or otherwise. (AGI, 1987)

ore control Any tectonic, lithologic, or geochemical features considered to have influenced the formation and localization of ore. (AGI, 1987)

ore crusher (a) A machine for breaking up masses of ore, usually prior to passing through stamps or rolls. See also: *crusher.* (b) *crusher man.*

ore delfe (a) Ore lying underground. (b) Right or claim to ore from ownership of land in which it is found.

ore deposit (a) A body of ore. See also: *mineral deposit.* (USGS, 1986) (b) A mineral deposit that has been tested and is known to be of sufficient size, grade, and accessibility to be producible to yield a profit. (In controlled economies and integrated industries, the "profit" decision may be based on considerations that extend far beyond the mine itself, in some instances relating to the overall health of a national economy.)

ore developed *positive ore.*

ore developing Ore exposed on two sides. First class, blocks with one side hidden; second class, blocks with two sides hidden; third class, blocks with three sides hidden. See also: *probable ore.* (Fay, 1920)

ore dike An injected wall-like intrusion of magmatic ore, forced in a liquid state across the bedding or other layered structure of the invaded formation. (Schieferdecker, 1959)

ore district A combination of several ore deposits into one common whole or system.

ore dressing The cleaning of ore by the removal of certain valueless portions, such as by jigging, cobbing, vanning, etc. See also: *concentration; beneficiation; preparation.* Syn: *mineral dressing; ore preparation.* (Fay, 1920)

ore expectant The whole or any part of the ore below the lowest level or beyond the range of vision. The prospective value of a mine beyond or below the last visible ore, based on the fullest possible data from the mine being examined, and from the characteristics of the mining district. See also: *possible ore; prospective ore.* (Fay, 1920)

ore face An orebody that is exposed on one side, or shows only one face, and of which the values can be determined only in a prospective manner, as deducted from the general condition of the mine or prospect. (Fay, 1920)

ore-forming fluid A gas or fluid that dissolves, receives by fractionation, transports, and precipitates ore minerals. A mineralizer is typically aqueous, with various hyperfusible gases (CO_2, CH_4, H_2S, HF), simple ions (H^+, HS, Cl^-, K, Na, Ca), complex ions (esp. chloride complexes), and dissolved base and precious metals. Syn: *geologic mineralizer; mineralizer; hydrothermal solution.*

ore genesis The origin of ores.

ore geology The branch of applied geology dealing with the genetic study of ore deposits. Syn: *metallogeny; mining geology.* See also: *economic geology.*

oregonite Probably Ni_2FeAs_2; hexagonal. From Josephine Creek, Josephine County, OR. Named from the locality. Also spelled oregonit. (Hey, 1961)

ore grader In metal mining, a person who directs and regulates the storage of iron ores of various grades in bins at shipping docks so that the grade of ore contained in each bin will contain the approximate percentage of iron guaranteed to the buyer (iron and steel mills). (DOT, 1949)

ore guide Any natural feature—such as alteration products, geochemical variations, local structures, or plant growth—known to be indicative of an orebody or mineral occurrence. (AGI, 1987)

ore hearth A small, low fireplace surrounded by three walls, with a tuyere at the back. Three important types are called: ore hearth, waterback ore hearth, and Moffet ore hearth; used in smelting. (Fay, 1920)

oreing down A blocking operation in which ore is added to an open-hearth bath to oxidize the bath and to further reduce the carbon. (Henderson, 1953)

ore in sight (a) A term frequently used to indicate two separate factors in an estimate, namely: (1) ore blocked out; i.e., ore exposed on at least three sides within reasonable distance of each other; and (2) ore that may be reasonably assumed to exist, though not actually blocked out; these two factors should in all cases be kept distinct, because (1) is governed by fixed rules, while (2) is dependent upon individual judgment and local experience. The expression ore in sight as commonly used in the past, appears to possess so indefinite a meaning as to discredit its use completely. The terms positive ore, probable ore, and possible ore are suggested. (b) *developed reserve.*

ore intersection The point at which a borehole, crosscut, or other underground opening encounters an ore vein or deposit; also, the thickness of the ore-bearing deposit so traversed. (Long, 1960)

ore magma A term proposed by Spurr (1923) for a magma that may crystallize into an ore; the sulfide, oxide, or other metallic facies of a solidified magma. (AGI, 1987)

ore microscope *reflected-light microscope.*

ore microscopy The study of opaque ore minerals in polished section with a reflected-light microscope. See also: *reflected-light microscopy; microscopy.* (AGI, 1987)

ore mineral The part of an ore, usually metallic, which is economically desirable, as contrasted with the gangue. See also: *ore.* Cf: *gangue.* (AGI, 1987)

ore partly blocked Said of an orebody that is only partly developed, and the values of which can be only approx. determined. See also: *probable ore.*

orepass A vertical or inclined passage for the downward transfer of ore; equipped with gates or other appliances for controlling the flow. An orepass is driven in ore or country rock and connects a level with the hoisting shaft or with a lower level. See also: *ore chute.* (Nelson, 1965)

ore personal property Ore, or other mineral product, becomes personal property when detached from the soil in which it is imbedded. (Ricketts, 1943)

ore pipe A long and relatively thin deposit commonly formed at the intersection of two planes. See also: *pipe.* (Stokes, 1955)

ore plot A place where dressed ore is kept. (Fay, 1920)

ore pocket (a) Excavation near a hoisting shaft into which ore from stopes is moved, preliminary to hoisting. (Pryor, 1963) (b) Used in a phrase such as a rich pocket of ore, to describe an unusual concentration in the lode. (Pryor, 1963)

ore preparation *ore dressing.*

ore province A well-defined area containing ore deposits of a particular kind; e.g., the porphyry copper deposits of the Southwestern United States. Related to, but not exactly synonymous with, metallogenic province, which need not contain economic ore deposits.

ore reserve (a) The term is usually restricted to ore of which the grade and tonnage have been established with reasonable assurance by drilling and other means. (Nelson, 1965) (b) The total tonnage and average value of proved ore, plus the total tonnage and value (assumed) of the probable ore. (Hoover, 1948) (c) A mine's substantial asset, without which none of the surface works are economically viable. A body of ore that has been proved to contain a sufficient tonnage of amenable valuable mineral to justify the mining enterprise. The British Institution of Mining and Metallurgy, which regulates the professional standards of its membership, considers that the term ore reserves should be restricted to ore of which the quantity and grade have been established with reasonable assurance by a responsible, professionally qualified person. Additional ore insufficiently developed or tested for inclusion in ore reserves should be clearly described in simple terms best suited to the circumstances; modes of mineral occurrence vary too widely to permit standardization of categories. (Pryor, 1963) (d) S. Afr. Orebodies made available for mining through drives connected by winzes (a connection driven down) and raises (a connection driven up), thus forming blocks that are accessible from four sides. Some companies record partially developed ore reserves in which this making of blocks has not been completed. Newcomers in gold mining occasionally speak of ore reserves when they mean the orebodies contained in a mining area and in copper mining this method of expression has been accepted by large concerns. (Beerman, 1963) (e) *reserve.*

ore sampling The process in which a portion (sample of ore) is selected in such a way, that its composition will represent the average composition of the entire bulk of ore. Such a selected portion is a sample, and the art of properly selecting such a sample is called sampling. (Newton, 1959)

ore separator A cradle, frame, jigging machine, washer, or other device or machine used in separating the metal from broken ore, or ore from worthless rock. (Standard, 1964)

ore shoot (a) An elongate pipelike, ribbonlike, or chimneylike mass of ore within a deposit (usually a vein), representing the more valuable part of the deposit. Syn: *shoot.* (AGI, 1987) (b) Concentration of primary ore along certain parts of a rock opening. (Bateman, 1951) (c) A large and visually rich aggregation of mineral in a vein. It is a more or less vertical zone or chimney of rich vein matter extending from wall to wall, and has a definite width laterally. Sometimes called pay streak, although the latter applies more specif. to placers. (d) An area of payable lode surrounded by low values. See also: *shoot.* (Spalding, 1949)

ore sill A tabular sheet of magmatic ore, injected in a liquid state along the bedding planes of a sedimentary or other layered formation.

ore stamp A machine for reducing ores by stamping. (Standard, 1964)

ore-storage drier man One who removes moisture from ore or other material preparatory to roasting or electrolytic processing, using a gas or hot-air drier. Also called drier operator. (DOT, 1949)

ore strand Individual mass of quartz with a halo of alteration and ore minerals, or close assemblage of seams of such quartz and accompanying altered ground. (AGI, 1987)

ore trend A term used on the Colorado Plateau to indicate the extension of an orebody along its major axis; the average trend of ore in a particular area, or the regional trend of mineralization over a large area. The local trend of individual orebodies may vary from the regional trend of so-called mineral belts. (Ballard, 1955)

ore vein A tabular or sheetlike mass of ore minerals occupying a fissure or a set of fissures and later in formation than the enclosing rock. (Schieferdecker, 1959)

ore washer A machine for washing clay and earth out of earthy brown hematite ores. The log washer is a common type. (Fay, 1920)

ore zone A horizon in which ore minerals are known to occur. (Long, 1960)

Orford process A process for separating the copper and nickel in the matte obtained by Bessemerizing. The matte, which consists of copper-nickel sulfides, is fused with sodium sulfide, and a separation into two layers, the top rich in nickel and the bottom rich in copper, is obtained. Also known as top-and-bottom process. (CTD, 1958)

organ A series of closely spaced props placed at the borders of the chamber at the coal face. Such an arrangement protects the future, adjoining chamber from caving. (Stoces, 1954)

organic Being, containing, or relating to carbon compounds, esp. in which hydrogen is attached to carbon whether derived from living organisms or not. Usually distinguished from inorganic or mineral. Cf: *inorganic.* (Webster 3rd, 1966)

organic ash Ash in coal derived from the incombustible material contained in plants. (Tomkeieff, 1954)

organic colloid Any of the depressants used in the flotation process. They include glue, gelatin, albumen, dried blood, casein (proteins), tannin, licorice, quebracho extract, and saponin (complex polyhydroxy carboxylic acids and glucosides). (Pryor, 1963)

organic deposit A rock or other deposit formed by organisms or their remains.

organic efficiency The ratio (normally expressed as a percentage) between the actual yield of a desired product and the theoretically possible yield (based on the reconstituted feed), both actual and theoretical products having the same percentage of ash. (BS, 1962)

organic soil A general term applied to a soil that consists primarily of organic matter such as peat soil and muck soil. (AGI, 1987)

organic sulfur The difference between the total sulfur in coal and the sum of the pyritic sulfur and sulfate sulfur. (BS, 1961)

organic test A test in which organic matter in soil is destroyed by oxidizing agents and the loss measured. This test is used in preparation of soil for a sedimentation test, and gives an indication of the amount of organic matter present. See also: *sedimentation test.* (Hammond, 1965)

organogenic Derived from or composed of organic materials; e.g., a crinoidal limestone.

organolite Rock formed from organic substances, esp. those of vegetable origin, such as coal, oil, resins, and bitumens. (Tomkeieff, 1954)

orichalcum An ancient copper alloy resembling gold in color. (Hess)

orient (a) To place a deflection wedge in a borehole in such an attitude that the concave surface is pointed in a predetermined direction. (Long, 1960) (b) To place a piece of core in the same relative plane as it occupied below the surface. See also: *core orientation.* (Long, 1960) (c) To turn a map or planetable sheet in a horizontal plane until the meridian of the map is parallel to the meridian on the ground. (Seelye, 1951) (d) In a transit, to turn the instrument so that the direction of the 0° line of its horizontal circle is parallel to the direction it had in the preceding, or in the initial, setup. (Seelye, 1951) (e) To place a diamond in a bit mold in such an attitude that when it is embedded in the crown matrix one of its hard vector planes will come in contact with the rock to be abraded or cut by the diamond. (Long, 1960) (f) The characteristic sheen and irridescence displayed by pearl. (Anderson, 1964) (g) To align an optical or crystallographic direction of a mineral with a rotation axis of a microscope stage.

oriental (a) Frequently used in the same sense as precious when applied to minerals; from an old idea that gems came principally from the Orient; e.g., oriental amethyst, oriental chrysolite, oriental emerald, and oriental topaz, all of which are varieties of sapphire. Syn: *precious.* (Fay, 1920) (b) Specially bright, clear, pure, and precious; said of gems. (Standard, 1964)

oriental alabaster Calcium carbonate in the form of onyx marble. Gibraltar stone. Syn: *Algerian onyx; onyx marble.* (Hess; CTD, 1958)

oriental cat's-eye *cat's-eye; tiger's-eye.*

oriental powder An explosive consisting of tan bark, sawdust, or other vegetable fiber, or resins, such as gamboge, impregnated with a nitrate or chlorate and mixed with gunpowder. (Standard, 1964)

orientation (a) Arrangement in space of the axes of a crystal with respect to a chosen reference or coordinate system. See also: *preferred orientation.* (ASM, 1961) (b) In surveying, the rotation of a map (or instrument) until the line of direction between any two of its points is parallel to the corresponding direction in nature. (c) In structural petrology, refers to the arrangement in space of particles (grains or atoms) of which a rock is composed. (AGI, 1987) (d) The act or process of setting a diamond in the crown of a bit in such an attitude that one of its hard vector planes will contact the rock and be the surface that cuts or abrades it. (e) As used in borehole surveying and directional drilling practice, orientation refers to the method and procedure used in placing an instrument or tool, such as a deflection wedge, in a drilled hole so that its directional position, bearing, or azimuth is known. (Long, 1960) (f) The position of important sets of planes in a crystal in relation to any fixed system of planes. (CTD, 1958) (g) The spatial relationship between crystallographic axes and principle optic directions in anisotropic minerals. (h) The characteristic sheen or iridescence displayed by pearl.

orientational twinning *electrical twinning.*

orientation survey In geochemical prospecting, a geochemical survey normally consisting of a series of preliminary experiments aimed at determining the existence and characteristics of anomalies associated with mineralization. This information is then used in selecting adequate prospecting techniques and in determining the factors and criteria that have a bearing on interpretation of the geochemical data. (Hawkes, 1962)

oriented Said of a specimen or thin section that is so marked as to show its original position in space. (AGI, 1987)

oriented bit A surface-set diamond bit with individual stones set so as to bring the hard vector direction or planes of the crystal into opposition with the rock surface to be abraded or cut. See also: *orient.* (Long, 1960)

oriented core A core specimen that can be positioned on the surface as it was in the borehole prior to extraction. Such a core is useful where the general dip of the strata is required from one borehole. A magnetic method may be used to disclose the polarity the core specimen possessed while in situ. See also: *true dip; borehole surveying; oriented sample.* Cf: *core orientation.* (Nelson, 1965)

oriented core barrel An instrument used in borehole surveying, which marks the core to show its position in space. (Hammond, 1965)

oriented diamond A diamond inset in the crown of a bit in such an attitude that one of its hard vector planes will be the surface that cuts or abrades the rock. See also: *orient.* (Long, 1960)

oriented sample *oriented core.*

oriented specimen (a) In structural petrology, a hand specimen so marked that its exact arrangement in space is known. (Billings, 1954) (b) In paleontology, a fossil whose position is known in regards to such features as anterior and dorsal sides, dorsal and ventral sides, the axis of coiling, the plane of coiling, etc. (AGI, 1987)

oriented survey A borehole survey made by lining up a reference mark on the clinometer case with that on the drill rods, which in turn are oriented as they are lowered into the borehole.

orienting coupling A rotatable coupling on a Thompson retrievable wedge-setting assembly that may be set and locked in a predetermined position in reference to the gravity-control member. This places the deflection wedge so as to direct the branch borehole in the desired course. (Long, 1960)

orientite An orthorhombic mineral, $Ca_2Mn^{2+}Mn^{3+}{}_2Si_3O_{10}(OH)_4$; forms minute, brown to black, radiating, prismatic crystals.

orifice (a) In ventilation, a hole in a very thin plate. (Mason, 1951) (b) A hole or opening, usually in a plate, wall, or partition, through which water flows, generally for the purpose of control or measurement. (Seelye, 1951) (c) The end of a small tube, such as the orifice of a Pitot tube, piezometer, etc. (Seelye, 1951) (d) An opening through which glass flows. In a feeder, an opening in bottom of spout formed by orifice ring. (ASTM, 1994) (e) Opening. Formerly applied to discs placed in pipelines or radiator valves to reduce the fluid flow to a desired amount. (Strock, 1948)

orifice meter A form of gas or liquid flowmeter consisting of a diaphragm in which there is an orifice placed transversely across a pipe; the difference in pressure on the two sides of the diaphragm is a measure of flow velocity. (Lowenheim, 1962)

orifice of passage Said of a fan with an orifice comparable to the equivalent orifice of a mine; i.e., the area in a thin plate that requires the same pressure to force a given volume of air through as is required to force the same volume through the fan. Orifice of passage O = 0.389 Q/w.g.f., where Q = volume of air passing in thousands cubic feet per minute; w.g.f.= loss of pressure in the fan in inches of water gage. (Nelson, 1965)

origin The source or ground of the existence of anything, either as cause or as occasion; that from which a thing is derived or by which it is caused; esp., that which initiates or lays the foundation; e.g., the origin of ore deposits. (Standard, 1964)

original dip The attitude of sedimentary beds immediately after deposition. Syn: *initial dip; primary dip.* (AGI, 1987)

original hole *main hole.*

original lead The common lead in a uranium mineral. (AGI, 1987)

original mineral *primary mineral.*

ormolu (a) Gold ground for use in gilding; also metal gilded with ground gold. (Webster 3rd, 1966) (b) A brass made to imitate gold and used in mounts for furniture and for other decorative purposes. Also called mosaic gold. (Webster 3rd, 1966)

ornamental stone *gemstone.*

ornansite A stony meteorite composed of bronzite and olivine in a friable mass of chondri. (Hess)

orocline An orogenic belt with an imposed curvature or sharp bend, interpreted by Carey (1958) as a result of horizontal bending of the crust, or deformation in plan. (AGI, 1987)

orogen A belt of deformed rocks, in many places accompanied by metamorphic and plutonic rocks; e.g., the Appalachian orogen or the Alpine orogen. (AGI, 1987)

orogenesis *orogeny.*

orogenic Adj. of orogeny.

orogeny The process by which structures within fold-belt mountainous areas were formed, including thrusting, folding, and faulting in the outer and higher layers, and plastic folding, metamorphism, and plutonism in the inner and deeper layers. Syn: *orogenesis; mountain building; tectogenesis.* Adj: orogenic; orogenetic. (AGI, 1987)

orometer An aneroid barometer having a second scale that gives the approximate elevation above sea level of the place where the observation is made. (Webster 3rd, 1966)

oronite An enamel paint for protecting metal surfaces from the action of hot vapors. (Fay, 1920)

O'Rourke car switcher A crossover switch that consists essentially of a single-acting cylinder hoist on a crossrail fastened to the roof at right angles to the track. While a car is being loaded, the switcher picks up the empty car next to the locomotive and holds it to one side. As soon as a car is loaded the locomotive pulls the train back past the switcher, and the empty car is placed at the front of the train and pushed under the slide. (Lewis, 1964)

orphaned mine land Abandoned and unreclaimed mines for which no owner or responsible party can be found. The reclamation and environmental conditions of such lands is then defaulted to the State or Federal Government. (SME, 1992)

orpiment (a) A yellow arsenic trisulfide, As_2S_3, containing 61% arsenic; monoclinic. Syn: *yellow arsenic.* (Dana, 1959) (b) A monoclinic mineral, $4[As_2S_3]$; soft; pearly lemon yellow with one perfect cleavage; in powdery foliated masses and coatings, botryoidal or fibrous; a low-temperature alteration of other arsenides; associated with realgar and gold in hot springs. Syn: *yellow ratsbane.*

Orsat gas-analysis instrument An instrument for analyzing flue gases. Although outside its normal field of application, it may be used for analyzing mine air. (Nelson, 1965)

orthite A former name for allanite, esp. when found in slender prismatic or acicular crystals.

ortho- (a) A combining form meaning straight; at right angles; proper. (AGI, 1987) (b) In petrology, a prefix that, when used with the name of a metamorphic rock, indicates that it was derived from an igneous rock, e.g., orthogneiss, orthoamphibolite; it may also indicate the primary origin of a crystalline, sedimentary rock, e.g., orthoquartzite as distinguished from metaquartzite. (AGI, 1987) (c) A prefix to the name of a mineral species or group to indicate orthorhombic symmetry as opposed to "clino" indicating monoclinic symmetry.

orthoamphibole The orthorhombic subgroup of amphiboles including anthophyllite, gedrite, and holmquistite. Cf: *clinoamphibole.*

orthochlorite (a) A group name for distinctly crystalline forms of chlorite (such as clinochlore and penninite). (b) A group name for chlorites conforming to the general formula: $(R^{2+},R^{3+})_6(Si,Al)_4O_{10}(OH)_8$.

orthoclase A monoclinic mineral, $KAlSi_3O_8$; feldspar group; prismatic cleavage; partly ordered, monoclinic potassium feldspar dimorphous with microcline, being stable at higher temperatures; also a general term applied to any potassium feldspar that is or appears to be monoclinic; e.g., sanidine, submicroscopically twinned microcline, adularia, and twinned analbite. It is a common rock-forming mineral and occurs esp. in granites, granite pegmatites, felsic igneous rocks, and crystalline schists, and is commonly perthitic. Syn: *common feldspar; orthose; pegmatolite.* Cf: *microcline; plagioclase; anorthoclase.*

orthoclase gabbro A descriptive name for rocks now known as monzonite, in which the plagioclase is at least as calcic as labradorite. (Holmes, 1928)

orthoclasite An orthoclase-bearing porphyritic intrusive rock, such as granite or syenite. The term is sometimes restricted to rocks containing more than 90% orthoclase. Not recommended usage. (AGI, 1987)

orthoclastic Cleaving in directions at right angles to each other. (Webster 3rd, 1966)

orthodolomite (a) A primary dolomite, or one formed by sedimentation. (AGI, 1987) (b) A term used by Tieje (1921) for a dolomite rock so well-cemented that the particles are interlocking. (AGI, 1987)

orthodome A monoclinic crystal form whose faces parallel the orthoaxis and cut the other axes. Cf: *dome; clinodome.* (AGI, 1987)

orthoferrosilite An orthorhombic mineral, $(Fe,Mg)_2Si_2O6$; pyroxene group; now simply called ferrosilite. Cf: *ferrosilite; clinoferrosilite.*

orthogneiss Applied to gneissose rocks that have been derived from rocks of igneous origin. Cf: *paragneiss.* (CTD, 1958)

orthogonal At right angles. (Hammond, 1965)

orthoguarinite Cesaro's name for an orthorhombic form of guarinite, through superposition of hemitropic lamellae of the monoclinic mineral, clinoguarinite. See also: *clinoguarinite.* (English, 1938)
orthohydrous maceral Maceral having a normal hydrogen content, such as vitrine. (Tomkeieff, 1954)
orthokalsilite An artificial orthorhombic high-temperature polymorph of $KAlSiO_4$.
ortholimestone A term proposed by Brooks (1954) for sedimentary limestone. Cf: *metalimestone; orthomarble.* (AGI, 1987)
orthomagmatic *orthomagmatic stage.*
orthomagmatic stage Applied to the main stage of crystallization of silicates from a typical magma; the stage during which perhaps 90% of the magma crystallizes. Cf: *pegmatitic stage.* Syn: *orthomagmatic; orthotectic stage.*
orthomarble A crystalline limestone that will take a polish; e.g., the Holston orthomarble of Tennessee. Cf: *metamarble; metalimestone; ortholimestone.*
orthomic feldspar Triclinic feldspar, which by repeated twinning (orthomimicry) simulates a higher degree of symmetry with rectangular cleavages; e.g., orthoclase, anorthoclase, and cryptoclase. (English, 1938)
orthophotography The product of a procedure that corrects the distortions in aerial photography due to the instability of the camera platform, the terrain relief, and the angle of the light rays entering the camera lens. The ortho instrumentation attached to a stereo plotting instrument rectifies the image in a transfer process so as to reposition it in its correct planar position. (SME, 1992)
orthophyric Said of the texture of the groundmass in certain holocrystalline, porphyritic, igneous rocks in which the feldspar crystals have quadratic or short, stumpy, rectangular cross sections, rather than the lath-shaped outline observed in trachytic texture. Also, said of a groundmass with this texture, or of a rock having an orthophyric groundmass. (AGI, 1987)
orthopinacoid In the monoclinic system, the form consisting of two planes parallel to the vertical and orthodiagonal axes. See also: *pinacoid.* (Standard, 1964)
orthoprism A monoclinic prism, the orthodiagonal intercept of which is greater than 1. (Standard, 1964)
orthopyroxene The subgroup name for pyroxenes crystallizing in the orthorhombic system, commonly containing no calcium and little or no aluminum; e.g., enstatite, hypersthene, and ferrosilite. Cf: *clinopyroxene.* Syn: *labrador hornblende.*
orthoquartzite A clastic sedimentary rock that is made up almost exclusively of quartz sand (with or without chert), and relatively free of or lacks a fine-grained matrix, derived by secondary silicification; a quartzite of sedimentary origin, or a pure quartz sandstone. The term generally signifies a sandstone with more than 90% to 95% quartz and detrital chert grains that are well-sorted, well-rounded, and cemented primarily with secondary silica (sometimes with carbonate) in optical and crystallographic continuity with the grains. The rock is characterized by stable but scarce heavy minerals (zircon, tourmaline, and magnetite), by lack of fossils, and by prominence of cross-beds and ripple marks. It commonly occurs as thin but extensive blanket deposits associated with widespread unconformities (e.g., an epicontinental deposit developed by an encroaching sea), and it represents intense chemical weathering of original minerals other than quartz, considerable transport and washing action before final accumulation (the sand may experience more than one cycle of sedimentation), and stable conditions of deposition (such as the peneplanation stage of diastrophism); e.g., St. Peter Sandstone (Middle Ordovician) of midwestern United States. See also: *quartzite.* (AGI, 1987)
orthorhombic (a) Any mineral crystallizing with orthorhombic symmetry. (b) *orthorhombic system.*
orthorhombic system In crystallography, that system of crystals whose forms are referred to three unequal mutually perpendicular axes. Syn: *prismatic system; rhombic system; orthorhombic.* (Fay, 1920)
orthoschist A schist derived from an igneous rock. Cf: *paraschist; schist.* (AGI, 1987)
orthoscope A polarizing microscope in which light is transmitted by the crystal parallel to the microscope axis, in contrast to the conoscope, in which a converging lens and Bertrand lens are used. Cf: *conoscope.* (AGI, 1987)
orthose A name for the whole feldspar family before it was divided into separate species. Syn: *orthoclase.* (Fay, 1920)
orthotectic *magmatic.*
orthotectic stage *orthomagmatic stage.*
orthotropic The description applied to the elastic properties of material, such as timber, which has considerable variations of strength in two or more directions at right angles to one another. See also: *isotropic.* (Hammond, 1965)
Orton cone (a) Pyrometric cone made in one of two sizes: 2-1/2 in (6.4 cm) high for industrial kiln control, and 1-1/8 in (3.2 cm) high for pyrometric cone equivalent testing. See also: *pyrometric cone.* (b) Used in the United States for heat recording, Orton cones are similar to Seger cones, but the same numbers do not indicate the same temperatures; e.g., Orton cone 14 corresponds to Seger cone 13. (Rosenthal, 1949)
oryctognosy The description and systematic arrangement of minerals; mineralogy. See also: *mineralogy.* (Fay, 1920)
oryctologist *mineralogist.*
oryctology *mineralogy.*
Osann's classification A purely chemical system of classification of igneous rocks. (AGI, 1987)
osarizawaite A trigonal mineral, $PbCuAl_2(SO_4)_2(OH)_6$; alunite group; the aluminum analog of beaverite; a yellow, powdery secondary crust; at the Osarizawa Mine, Akita prefecture, Japan.
oscillating beam *walking beam.*
oscillating conveyor A type of vibrating conveyor having a relatively low frequency and large amplitude of motion. See also: *vibrating conveyor.* Syn: *grasshopper conveyor.*
oscillating feeder *conveyor-type feeder.*
oscillating grease table An assembly of 4 to 8 metal trays, usually 30 in (76.2 cm) wide and 8 to 15 in (20.3 to 38.1 cm) long, arranged in series in the direction of flow. The trays are detachably mounted in the assembly by steps, so that the overflow from one tray overlaps the next tray by 1 in (2.54 cm) and is 2 to 4 in (5.1 to 10.2 cm) above it. The trays are inclined downward in the direction of the flow at an angle adjustable from 14° to 18°. The entire assembly is mechanically oscillated transversely to the direction of the flow at about 200 strokes/min with an adjustable stroke of about 1/2 in (1.27 cm). The storage bin and feed roller are independently mounted and discharge a layer 1 grain thick. (Chandler, 1964)
oscillation Independent movement through a limited range, usually on a hinge. (Nichols, 1976)
oscillator plate A thin slab of quartz which, by mechanical vibration, controls the frequency of a radio transmitter. (Hurlbut, 1964)
oscillator quartz Flawless quartz, which can be used in the manufacture of oscillator plates.
oscillatory twinning Repeated twinning in which a crystal is made up of thin lamellae alternately in reversed position; polysynthetic twinning; found in some feldspars. Syn: *polysynthetic twinning.* (Fay, 1920)
oscillatory zoning Repetitious concentric compositional variation in minerals resulting from cyclical changes in the chemical environment during crystal growth; e.g., garnet and plagioclase.
oscillogram A record of phenomena observed on an oscillograph. (ASM, 1961)
oscillograph An instrument that renders visible, or automatically traces, a curve representing the time variations of various phenomena. The recorded trace is an oscillogram. (AGI, 1987)
oscilloscope An instrument for showing visual representations of electrical outputs from measuring devices. (Hunt, 1965)
osmite *iridosmine.*
osmium The native element, Os; occurs in magmatic deposits in mafic and ultramafic rocks and placers derived from them.
osmosis The passage of a solvent through a membrane from a dilute solution into a more concentrated one, the membrane being permeable to molecules of solvent but not to molecules of solute. (AGI, 1987)
osteolite A massive, earthy mineral (apatite) consisting of an impure, altered phosphate.
ostracod A minute crustacean with a bean-shaped bivalve shell completely enclosing the body. (AGI, 1987)
otavite A trigonal mineral, $CdCO_3$; calcite group; associated with oxidized base-metal ores.
other rock in place As used in the Mining Law of 1872, means any rocky substance containing mineral matter. (Ricketts, 1943)
other valuable deposits Includes nonmetalliferous as well as metalliferous deposits. (Ricketts, 1943)
otisca process A process that uses an inert heavy liquid with a specific gravity between that of coal and free mineral matter to separate coarse or fine-size coal in a static bath or cyclone separator.
Otisca-T process A selective agglomeration process under development, in which ultra-fine grinding of the feed coal to 15 µm releases almost all the associated impurities prior to agglomeration with a low-molecular-weight hydrocarbon. The agglomerant is then recovered and recycled.
Otto cycle In a four-stroke internal combustion engine two complete revolutions of the crankshaft correspond with the working cycle-inlet stroke (suction downstroke of piston in cylinder); compression upstroke; explosion at peak of compression followed by expansion of hot exploded gases on driving downstroke; rising exhaust stroke to complete the cycle. (Pryor, 1963)
Ouachita stone *novaculite.*
oued *wadi.*
outburst The name applied to the violent evolution of combustible gases (usually together with large quantities of coal dust) from a working face. The occurrence is violent and may overwhelm the work-

ings and fill the entire district with gaseous mixtures. Roadways advancing into virgin and stressed areas of coal are particularly prone to outbursts in certain seams and faults often intersect in the area. See also: *floor burst; blow; bump.* (Roberts, 1960)

outby Nearer to the shaft, and therefore away from the face, toward the pit bottom or surface; toward the mine entrance. The opposite of inby. Also called outbyeside. Syn: *out-over.* (BCI, 1947; Fay, 1920)

outcrop (a) The part of a rock formation that appears at the surface of the ground. (Webster 3rd, 1966) (b) A term used in connection with a vein or lode as an essential part of the definition of apex. It does not necessarily imply the visible presentation of the mineral on the surface of the earth, but includes those deposits that are so near to the surface as to be found easily by digging. (Fay, 1920) (c) The part of a geologic formation or structure that appears at the surface of the Earth; also, bedrock that is covered only by surficial deposits such as alluvium. Cf: *exposure.* Syn: *crop; cropping; outcropping.* (AGI, 1987) (d) To appear exposed and visible at the Earth's surface; to crop out. (AGI, 1987)

outcrop map A special type of geologic map that represents only actual outcrops. Areas without exposures are left blank. (Stokes, 1955)

outcropping *outcrop.*

outcrop water Rain and surface water that seeps downward through outcropping porous and fissured rock, fault planes, old shafts, or surface drifts. (AGI, 1987)

outdoor stroke That stroke of a Cornish pumping engine by which the water is forced upward by the weight of the descending pump rods, etc.

outer continental shelf All submerged lands lying seaward and outside of the area of lands beneath navigable waters as defined in Section 2 of the Submerged Lands Act (Public Law 31, 83rd Congress, 1st Session), and of which the subsoil and seabed appertain to the United States and are subject to its jurisdiction and control. Abbrev. OCS. See also: *continental shelf.* (Acuff, 1981)

outer core The outer or upper zone of the Earth's core, extending from a depth of 2,900 km to 5,100 km, and including the transition zone; it is equivalent to the E layer and the F layer. It is inferred to be liquid because it does not transmit shear waves. Its density ranges from 9 to 11 g/cm^3. The outer core is the source of the principal geomagnetic field. Cf: *inner core.* (AGI, 1987)

outer gage Syn. for outside diameter. (Long, 1960)

outer stone A diamond set on the outside wall of a bit crown. Also called reamer; reamer stone. Syn: *outside stone; kicker stone.* (Long, 1960)

outfall (a) Eng. A seam cropping out at a lower level. (b) The mouth of a stream or the outlet of a lake; esp. the narrow end of a watercourse or the lower part of any body of water where it drops away into a larger body. (AGI, 1987) (c) The vent or end of a drain, pipe, sewer, ditch, or other conduit that carries waste water, sewage, storm runoff, or other effluent into a stream, lake, or ocean. (AGI, 1987)

outlay (a) The act of laying out or expending. (Webster 3rd, 1966) (b) Something that is laid out; expenditure. (Webster 3rd, 1966) (c) The cost of equipping a mine and placing it on a producing basis. (Fay, 1920)

outlet An opening from a mine to the surface. Syn: *upcast.* (Fay, 1920)

outlier (a) An isolated mass or detached group of rocks surrounded by older rocks; e.g., an isolated hill or butte. Cf: *inlier.* (b) Ore or favorable geologic indications distant from the main ore zone of a district.

out of gage (a) Bits and reaming shells having set inside or outside diameters greater or lesser than those specified as standard. (Long, 1960) (b) A borehole the inside diameter of which is undersize or oversize. (Long, 1960)

out-over *outby.*

output (a) The quantity of coal or mineral raised from a mine and expressed as being so many tons per shift, per week, or per year. (Nelson, 1965) (b) The power or product from a plant or prime mover in the specific form and for the specific purpose required. See also: *concentration of output; productivity.* (Nelson, 1965) (c) Amount delivered; e.g., volume of a liquid discharged by a pump; volume of air discharged by a compressor; horsepower delivered by a motor. (Long, 1960) (d) Current or signal delivered by any circuit or device. (NCB, 1964) (e) The terminal or other point at which a current or a signal may be delivered. (NCB, 1964)

output device Machine that prints information computed from its memory or store. (Pryor, 1963)

output shaft A shaft that transmits power from a transmission or clutch. (Nichols, 1976)

outrigger An outward extension of a frame that is supported by a jack or block. Used to increase stability. (Nichols, 1954)

outside angling *angling.*

outside clearance One-half the total difference between the outside diameter of any piece of downhole equipment and the inside diameter of the borehole. (Long, 1960)

outside face The peripheral portion or that part of a bit crown, roller bit cutter, or any cutting edge of a bit in contact with the walls of a borehole while drilling. (Long, 1960)

outside foreman In bituminous coal mining, a person who supervises all operations at the surface of a mine. (DOT, 1949)

outside stone *outer stone.*

outside tap *bell tap.*

outside upset The act or process of thickening a length of tubing at its ends by increasing its outside diameter without changing the inside diameter; a length of tubing or drill rod so processed. See also: *upset.* (Long, 1960)

outside wall That part of a bit crown, bit shank, reaming shell, core barrel, drill rod, casing, or other piece of downhole equipment that when in use, comes in contact with the wall of a borehole. (Long, 1960)

outside work Drilling operations conducted on the surface, as opposed to drilling done in underground or enclosed workplaces. (Long, 1960)

outslope The face of the spoil or embankment sloping downward from the highest elevation to the toe.

outstation A location which provides local monitoring and control, and provides a communications interface between a sensor and the trunk connected to a central station computer. Also called field data station. (SME, 1992)

outtake The passage by which the ventilating current is taken out of the mine; the upcast. The return air course. An outlet. (Fay, 1920)

oval socket A fishing tool used to recover broken drill rods from a borehole. (Long, 1960)

oven A chamber in which substances are artificially heated for the purposes of baking, roasting, annealing, etc.; specif.: (1) a kiln, such as a coke oven; and (2) a leer, which is used in glassmaking. (Standard, 1964)

overaging Aging at a higher temperature, or for a longer time, or both, than required for critical dispersion, thus causing particle agglomeration of the precipitating phase and, as a result, loss of strength and hardness. See also: *aging.* (Henderson, 1953)

overall concentration The ratio of pithead output in tons (P) to length of main haulage roads in yards (L) or tons per yard of main haulage roads; i.e., P/L. See also: *concentration of output.* (Nelson, 1965)

overall drilling time The sum of the times required for actual rock drilling, setting up and withdrawal, moving drills from hole to hole and machine delays. The overall drilling time is a better basis for estimating drilling efficiency than penetration speed alone. (Nelson, 1965)

overall efficiency (a) Of an air compressor, the product of the compression efficiency and the mechanical efficiency. (Fay, 1920) (b) Ratio of power output of an engine to the power input; the measure of the difference between indicated and brake horsepower. (Brantly, 1961)

overall fan efficiency The ratio of the horsepower in the air to the horsepower absorbed by the driving motor of the fan. (BS, 1963)

overall reduction ratio With reference to a crusher, mean size of feed/mean size of product. See also: *reduction ratio.* (South Australia, 1961)

overall ventilation efficiency The ratio between the air horsepower and the indicated horsepower of a driving unit. The percentage is expressed by air horsepower x 100/indicated hp of driving unit. Measurements are taken of the air pressure and volume in the fan drift, and the power absorbed by the driving unit. See also: *volumetric efficiency; thermometric fan test.* (Nelson, 1965)

over-and-under conveyor Two endless chains or other linkage between which carriers are mounted and controlled, so that the carriers remain in an upright and horizontal position throughout the complete cycle of the conveyor.

overarching weight The pressure of the rocks over active mine workings. It is the roof weight that acts on the packs and the solid coal in the working area. See also: *abutment; nether roof; underweight.* (Nelson, 1965)

overbreak Excessive breakage of rock beyond the desired excavation limit. See also: *neat lines.* Cf: *underbreak.*

overbreaking *overhand stoping.*

overburden (a) Designates material of any nature, consolidated or unconsolidated, that overlies a deposit of useful materials, ores, or coal—esp. those deposits that are mined from the surface by open cuts. (Stokes, 1955) (b) Loose soil, sand, gravel, etc. that lies above the bedrock. Also called burden, capping, cover, drift, mantle, surface. See also: *baring; burden; top.* (Stokes, 1955)

overburden bit A special diamond-set bit, similar to a set casing shoe, used to drill casing through overburden composed of sand, gravel, boulders, etc. (Long, 1960)

overburden drilling (a) A technique developed in Sweden that involves the sinking, by percussive-rotary drilling, of a drill casing through the overburden to where it seats in the underlying rock. A rotary percussion drill hole is then continued to the desired depth in the rock. While the casing is being sunk through the overburden it is coupled to the drill rod and rotates and reciprocates with it. The rock bit on the end of the drill rod projects about an inch beyond the end of the ring bit with which the casing is fitted and acts as a pilot bit for the casing bit. (Woodruff, 1966) (b) A drilling method whereby drilling is carried out through subsoil and boulders or underwater to and through bedrock. (Eng. Min. J., 1964)

overcast (a) An enclosed airway that permits an air current to pass over another one without interruption. Syn: *overcrossing; overgate.* Cf: *undercast.* See also: *air crossing.* (Kentucky, 1952) (b) To place the overburden removed from coal in surface mines in an area from which the coal has been mined. (c) Pushed forward, so as to overlie other rocks, such as in thrust faults. See also: *jack pit.* (Standard, 1964)

overcasting A procedure used in certain mining activities including strip mining and in some heavy construction work such as channel excavation. Overcasting may be performed in a simple operation consisting of digging out the material, lifting it from one position, moving it over, and dumping it in the spoil position where it remains, for practical purposes, indefinitely. The mechanics of the operation are called "simple overcasting.". (Woodruff, 1966)

overcharging Adding material in excess of the capacity of the equipment used for processing.

overconsolidated soil deposit A soil deposit that has been subjected to an effective pressure greater than the present overburden pressure. (ASCE, 1958)

overcrossing *air crossing; overcast.*

overcurrent relay Relay used to trip circuit breakers when an abnormal current of two to three times the normal flow is detected in a circuit. Relays are adaptable to transmission lines, buses, feeder circuits, transformers, and motors. (Coal Age, 1966)

overcut (a) A machine cut made along the top or near the top of a coal seam; sometimes used in thick seams or a seam with sticky coal. By releasing the coal along the roof, its mining becomes easier. See also: *overcut; turret coal cutter.* (Nelson, 1965) (b) The process of producing a larger size hole than the outside diameter of a bit and/or reaming shell used, due to the eccentric rotational movements of the bit, core barrel, or drill stem. (Long, 1960)

overcutting machine Coal-cutting machine that is an adaptation of a shortwall machine, designed to make the cut, or kerf, at desired place in the coal seam some distance above the floor. The main difference between an overcutting machine and an ordinary shortwall machine is that the cutter bar in the overcutting machine is mounted at the top of the machine instead of at the bottom. See also: *turret coal cutter.* (Kiser, 1929)

overdense medium Medium of specific gravity greater than that in the separating bath; usually produced in the medium recovery system and used to maintain the desired specific gravity in the bath. (BS, 1962)

overdrilling The act or process of drilling a run or length of borehole greater than the core-capacity length of the core barrel, resulting in loss of the core. (Long, 1960)

overdrive The act of inducing a velocity higher than the steady state velocity in a column of explosive material upon detonation by the use of a powerful primer or booster; it is a temporary phenomenon and the explosive quickly assumes its steady state velocity.

overfired A term related to the condition of a ceramic product which has been heated to a temperature in excess of that required to produce proper vitrification.

overfiring Heating ceramic materials or ware above the temperature required to produce the necessary degree of vitrification. Usually results in bloating, deformation, or blistering of the ware.

overflow stand A standpipe in which water rises and overflows at the hydraulic gradeline. (Seelye, 1951)

overgate *air crossing; overcast.*

overgrinding Comminution of ore to a smaller particle size than is required for effective liberation of values before concentrating treatment. Opposite of undergrinding. (Pryor, 1963)

overhand cut-and-fill In this method, two level drives are first connected, the lower and upper one by a raise, from the bottom of which mining is begun. The work proceeds upwards, filling the mined-out room, but in the filling, chutes are left through which the broken ore falls. In inclined seams the chutes, also inclined, have to be timbered. The lower-level drive is protected either by timbering or vaulting, or by a fairly strong pillar of vein fillings. Stoping in the different cuts always proceeds upwards, but as a whole it proceeds between the two level drives in a horizontal direction. Overhand cut-and-fill, esp. in mining irregular orebodies of greater size, is also called back stoping. (Stoces, 1954)

overhand stope (a) Stope in which the ore above the point of entry to the stope is attacked, so that severed ore tends to gravitate toward discharge chutes and the stope is self-draining. (Pryor, 1963) (b) An overhand stope is made by working upward from a level into the ore above. (McKinstry, 1948)

overhand stoping (a) In this method, which is widely used in highly inclined deposits, the ore is blasted from a series of ascending stepped benches. Both horizontal and vertical holes may be employed. Horizontal breast holes are usually more efficient and safer than vertical upper holes, although the latter are still used in narrow stopes in steeply inclined orebodies. (McAdam, 1958) (b) The working of a block of ore from a lower level to a level above. In a restricted way overhand stoping can be applied to open or waste-filled stopes that are excavated in a series of horizontal slices either sequentially or simultaneously from the bottom of a block to its top. Stull timbering or the use of pillars characterize the method. Filling is used in many instances. Modifications are known as backfilling method; back stoping; block system; block system of stoping and filling; breast stoping; combined side and longwall stoping; crosscut method of working; cross stoping; Delprat method; drywall method; filling system; filling-up method; flatback stoping; longwall stoping; open cut system; open stope and filling; open-stope method; open-stope, timbering with pigsties, and filling; overhand stoping on waste; resuing; rock filling; room-and-pillar with waste filling; sawtooth back stoping; side stoping; slicing-and-filling system; stoping and filling; stoping in horizontal layers; transverse with filling. Syn: *combined side and longwall stoping; Delprat method; overbreaking.* Cf: *back-filling system; chimney work; underhand stoping.*

overhand stoping and milling system *combined overhand and underhand stoping.*

overhand stoping on waste *overhand stoping.*

overhand stoping with shrinkage and delayed filling *shrinkage stoping.*

overhand vertical slice *square-set stoping.*

overhang (a) Cliff overhang. (AGI, 1987) (b) A part of the mass of a salt dome that projects out from the top of the dome much like the cap of a mushroom. (AGI, 1987)

overhaul (a) Describes a condition when a journey travels towards a haulage engine at a faster rate than the rope, which then becomes slack and liable to foul the drum. Also called overrun. (Nelson, 1965) (b) The transportation of excavated material beyond certain specified limits. (Seelye, 1951) (c) In many highway contracts, a movement of dirt far enough so that payment, in addition to excavation pay, is made for its haulage. (Nichols, 1976) (d) Applied to inspection, cleaning, and repairing of machines or plant. (Nelson, 1965)

overhead cableway A type of equipment for the removal of soil or rock. It consists of a strong overhead cable, usually attached to towers at either end, on which a car or traveler may run back and forth. From this car a pan or bucket may be lowered to the surface, subsequently raised and locked to the car, and transported to any position on the cable where it is desired to dump its contents.

overhead conveyor *trolley conveyor.*

overhead monorail This system is popular for use in mines since it can be suspended from the roadway supports as the face advances and can carry supplies over equipment installed in the roadway; transport is by means of endless, main-and-tail, or main-rope winches. They are generally slow-moving and can carry light loads into and around many places inaccessible to other forms of transport. See also: *monorail.* (Sinclair, 1963)

overhead-rope monorail In this system, the loads are carried by bogies running on a taut wire rope instead of steel joists or flat-bottomed rails. See also: *monorail.* (Sinclair, 1963)

overhead ropeway *aerial ropeway.*

overhead shovel A tractor loader that digs at one end, swings the bucket overhead, and dumps at the other end. (Nichols, 1976)

overhead traveling crane A crane that traverses the whole width of a workshop along the rails on which it runs. (Hammond, 1965)

overhead trolley conveyor *trolley conveyor.*

overlap (a) A general term referring to the extension of marine, lacustrine, or terrestrial strata beyond underlying rocks whose edges are thereby concealed or overlapped, and to the unconformity that commonly accompanies such a relation; esp. the relationship among conformable strata such that each successively younger stratum extends beyond the boundaries of the stratum lying immediately beneath. Cf: *onlap.* (AGI, 1987) (b) The area common to two successive aerial or space photographs or images along the same flight strip, expressed as a percentage of the photo area. (AGI, 1987) (c) The portion of a borehole that must be redrilled after caving of the hole, cementing a section of the hole, or bypassing unrecoverable material. (Long, 1960) (d) A reversed fault or thrust. (BS, 1964) (e) The lineal portion of a branch hole that nearly parallels the parent hole. (Long, 1960)

overlap auxiliary ventilation To combine the forcing and exhausting systems, it is not necessary to provide two ducts, one forcing and one exhausting, throughout the length of the heading. An arrangement that serves the same purpose is the overlap system. In this system a main exhausting duct is used within a convenient distance of the face, often about 100 ft (30.5 m). Some of the intake air in the heading, before reaching the end of this duct, enters a short length of tubing and is blown onto the face. The advantages of both systems are thus obtained. Precautions must be taken against recirculation of air by the forcing unit, to prevent concentration of dust, and in collieries, combustible gases, at the face. The two ducts must overlap by a minimum distance which, in practice, is usually taken as 30 ft (9.1 m). See also: *auxiliary ventilation.* Syn: *two-fan auxiliary ventilation.* (Roberts, 1960)

overlap fault (a) *thrust fault.* (b) A fault structure in which the displaced strata are doubled back upon themselves. (AGI, 1987)

overlay (a) Scot. The material above the rock in a quarry. See also: *overburden.* (Fay, 1920) (b) Graphic data on a transparent or translucent sheet to be superimposed on another sheet (such as a map or photograph) to show details not appearing, or requiring special emphasis, on the original. Also, the medium or sheet containing an overlay. (AGI, 1987)

overlay tracing A tracing on which the workings in a seam are shown. A series of such tracings allows the workings in several seams to be seen in their correct horizontal relationship. Also called layover tracing (undesirable usage). (BS, 1963)

overload (a) In general, a load or weight in excess of the designed capacity. The term may be applied to mechanical and electrical engineering plants, to loads on buildings and structures, and to excess loads on haulage ropes and engines. (Nelson, 1965) (b) To apply an excessive pressure, by stretching beyond the yield point, to a drill string and bit. Cf: *crowd.* (Long, 1960)

overloader A loading machine of the power-shovel type for quarry and opencast operations. It may be either pneumatic-tired or continuous-tracked. It need not turn from the face to the truck if the latter can be spotted parallel to the face. The bucket is filled, the machine retracted, and the bucket swung over to the discharge point; used chiefly in sand and gravel pits. (Nelson, 1965)

overmining S. Afr. Mining a grade of ore above the average grade of the ore reserves. This practice has the effect of leaving the lower grade ore in the reserves. The opposite is undermining. (Beerman, 1963)

overpoled copper In refining blister copper by reducing its oxides through stirring a molten bath of metal with a green timber pole, continuation of this process until the desirable characteristic fracture of tough-pitch refined metal is lost. Some reoxidation then becomes necessary. See also: *tough pitch.* (Pryor, 1963)

overprint The superposition of a new set of structural features on an older set. Syn: *superprint; metamorphic overprint.* (AGI, 1987)

override A royalty or percentage of the gross income from production deducted from the working interest. (Wheeler, 1958)

overriding royalty The term applied to a royalty reserved in a sublease or assignment over and above that reserved in the original lease. (Ricketts, 1943)

overrope A winding or hoisting rope. (Fay, 1920)

overrope haulage Usually applied to endless rope haulage in which the rope is carried on top of the mine cars, which may be either clipped or lashed to the rope. See also: *underrope haulage.* (Nelson, 1965)

overrun *overhaul.*

overrun brake A special brake fitted to a towed vehicle that operates as soon as the towing vehicle slows down. (Hammond, 1965)

overrunning clutch A coupling that transmits rotation in only one direction, and disconnects when the torque is reversed. (Nichols, 1976)

oversaturated rock A rock that contains silica in excess of that necessary to form saturated minerals from the bases present. Cf: *saturated rock.*

overshot A fishing tool for recovering lost drill pipe or casing. (Inst. Petrol., 1961)

overside Discharging over the side; e.g., by a dredge. (Standard, 1964)

oversize (a) In reference to a mixture of material screened or classified into two products of definite size limits, the larger is the oversize and the smaller the undersize. See also: *classifier.* (Nelson, 1965) (b) In quarry or opencast blasting, that size of rock or ore which is too large to handle without secondary blasting. (Nelson, 1965)

oversize control screen A screen used to prevent the entry into a machine of large particles that might interfere with its operation. Syn: *guard screen; check screen.* (BS, 1962)

oversize core (a) Core cut by a thin-wall bit, as opposed to a standard-diameter core. (Long, 1960) (b) A core the diameter of which is greater than a standard size. (Long, 1960)

oversize coupling (a) *swelled coupling.* (b) Sometimes used in Canada as a synonym for reaming shell. (Long, 1960)

oversize hole A borehole the diameter of which is excessive because of the whipping action or eccentric rotation of the drill string and bit. (Long, 1960)

oversize rod *drill collar; guide rod.*

overspringing *springing.*

overstressed area In strata control, describes an area where the force is concentrated on pillars. This type of area is said to be overstressed or superstressed. This superstressing is limited by the strength of the seam or pillar. Cf: *destressed area.* (Mason, 1951)

Overstrom table Similar to a Wilfley table but of diamond shape (rhomboid), thus eliminating the waste corners. (Liddell, 1918)

over-the-road hauling Hauling over public highways, usually by a dump truck. Various restrictions, such as weight, width of vehicle, safety features, guard against spillage, etc. must be considered in the type equipment used. Cf: *off-the-road hauling.* (Carson, 1961)

overthrust A low-angle thrust fault of large scale, with displacement generally measured in kilometers. Cf: *underthrust.* Syn: *overthrust fault.* (AGI, 1987)

overthrust block *overthrust nappe.*

overthrust fault *overthrust.*

overthrust nappe The body of rock that forms the hanging wall of a large-scale overthrust; a thrust nappe. Syn: *overthrust block; overthrust sheet; overthrust slice.* (AGI, 1987)

overthrust plane *thrust plane.*

overthrust sheet *overthrust nappe.*

overthrust slice *overthrust nappe.*

overtime The period beyond the normal shift time when a worker, on request by the management, performs emergency tasks that are necessary for safety or efficient operation of the oncoming shift. (Nelson, 1965)

overtopping Flow of water over the top of a dam or embankment. (Nichols, 1976)

overtravel *overwind.*

overtub system An endless-rope system in which the rope runs over the tubs or cars in the center of the rails. This system is generally adopted on undulating roads, where the tension in a heavily loaded rope would cause the rope to lift in swilleys and derail tubs. It is also generally adopted in highly inclined roads, as the lashing chain, often adopted with this method of haulage, obtains a good positive grip on the rope and is easier to detach than a clip. The rope is kept from rubbing on roof supports by holding-down pulleys: six or eight small pulleys are mounted in circular cheeks, allowing chains or clips to be accommodated in the spaces between the pulleys; or large diameter pulleys may be used, of the hat or mushroom shape, often starred to provide recesses for chains and clips. Similar large pulleys direct the rope around curves. Cf: *undertub system.* (Sinclair, 1959)

overturned Said of a fold, or the limb of a fold, that has tilted beyond the perpendicular. The sequence of strata thus appears reversed. Syn: *inverted; inverted fold; reversed.* (AGI, 1987)

overturning skip A type of skip commonly used at metal mines, but not as often at coal mines, because of increased breakage. This skip consists of a rectangular receptacle for the material and a suspending frame of bail to an upper crosspiece of which is attached a suspension gear connecting the rope to the skip. Three guide shoes are generally provided at each side of the bail to keep it vertical. The skip body turns about a horizontal shaft at the lower end of the bail. Two rollers on the upper part are mounted on a shaft and cause the skip to tilt at an angle of 35° at the tipping point in the headgear, where rollers run onto the curved guides. To prevent shocks in the case of an overwind the skips are fitted with overwind guides which glide along rollers fitted to the headgear above the tipping point. (Sinclair, 1959)

overventilation Too much air in the mine workings. (Fay, 1920)

overvoltage The difference between the actual electrode potential, when appreciable electrolysis begins, and the reversible electrode potential. (ASM, 1961)

overvoltage relay Relay that serves primarily the same purpose as an overcurrent relay except that it is connected in the line by a potential transformer which measures the voltage across the line. When an overvoltage exists the relay operates and opens the circuit breaker. (Coal Age, 1966)

overwind (a) To hoist a cage into or over the top of a headframe. Syn: *overtravel.* (Fay, 1920) (b) In hoisting through a mine shaft, failure to bring a cage or skip smoothly to rest at the proper unloading point at the surface. If severe, it can lead to a serious accident unless the special preventive devices function effectively. Overwind can also cause a cage to be lowered into the sump at the bottom of the shaft, also with serious consequences. (Pryor, 1963)

overwinder One of the best known overwinder prevention devices consists of two vertical-screwed spindles, each carrying two traveling nuts and chain driven from the drum shaft so as to rotate in opposite directions. The nuts are prevented from rotating by projections engaging with a fixed plate and therefore travel up and down according to the movement of the cages. The upper nut takes care of overwinding and the lower nut of overspeeding. (Mason, 1951)

overwinding (a) A term applied to a continued pull on the hoisting rope of a cage, after the cage has reached the top of the shaft. The result of this carelessness, or accident, is a broken hoisting rope and all the danger that implies. (Stauffer, 1906) (b) A rope or cable wound and attached so that it stretches from the top of a drum to the load. (Nichols, 1976)

overwind switch A switch that may be used on winders, or haulages, to cause the power to cut off from the driving motor, or engine, and the brakes to be applied. Such a switch may be: (1) situated in the headgear and operated by the conveyance, (2) mounted on the automatic contrivance, or (3) operated by the depth or distance indicator. (BS, 1965)

ovulite *oolith.*

Owen process A flotation process involving the violent agitation of the pulp in cold water to which a small percentage of eucalyptus oil, about 62.5 g, is added. (Fay, 1920)

Owen's borehole surveying instrument A clockwork photographic apparatus that records clinometer and compass readings on sensitized paper. It is used during borehole surveying. (Hammond, 1965)

Owen's jet dust counter An instrument similar to the konimeter but differing in that the air to be sampled undergoes humidification prior to being blown through the jet. The velocity of impingement is about 200 to 300 m/s and the jet is rectangular instead of circular. The prior humidification of the air causes condensation of moisture upon the dust particles by super saturation due to the pressure drop at the jet, and so assists in the deposition and retention of the particles on the slide. The Bausch and Lomb dust counter is the American counterpart of this instrument. (Osborne, 1956)

oxacalcite *whewellite.*

oxalite *humboldtine.*

oxammite An orthorhombic mineral, $(NH_4)_2C_2O_4 \cdot H_2O$; transparent; yellowish-white; forms lamellar and pulverent masses in guano.

oxialyphite A variety of aliphite hydrocarbon containing oxygen; light-yellow; soft. (Tomkeieff, 1954)

oxidate Sediment composed of the oxides and hydroxides of iron and manganese, crystallized from aqueous solution. It is one of Goldschmidt's groupings of sediments or analogues of differentiation stages in rock analysis. (AGI, 1987)

oxidation (a) The firing of a kiln in such a manner that combustion is complete and in consequence the burning gases are amply supplied with oxygen, which causes metals in clay and glazes to give their oxide colors. (ACSG, 1961) (b) Combination with oxygen; increase in content of a molecular compound; increase in valency of the electropositive part of compound, or decrease in valency of the electronegative part. (Pryor, 1963) (c) A reaction in which there is an increase in valence resulting from a loss of electrons. Cf: *reduction.* (ASM, 1961) (d) In fuel practice, the combination of oxygen with a substance, with or without the production of food. (Francis, 1965)

oxidation of coal The absorption of oxygen from the air by coal, particularly in the crushed state; this engenders heat which can result in fire. Ventilation, while dispersing the heat generated, supports oxidation that increases rapidly with a rise in temperature. Fresh air should not gain access to the coal. See also: *gob fire.* (Nelson, 1965)

oxide A compound of oxygen with another element. (CTD, 1958)

oxide discoloration Discoloration of a metal surface caused by oxidation during thermal treatment. (Light Metal Age, 1958)

oxide mineral A mineral formed by the union of an element with oxygen; e.g., corundum, hematite, magnetite, and cassiterite. (Leet, 1958)

oxide of iron An iron ore with oxygen as its main impurity; also iron rust. (Mersereau, 1947)

oxidized deposit A deposit that has resulted from surficial oxidation. (Bateman, 1951)

oxidized ore Metalliferous minerals altered by weathering and the action of surface waters, and converted, partly or wholly, into oxides, carbonates, or sulfates. These compounds are characteristic of metalliferous deposits at the surface and often to a considerable depth. See also: *mixed ore.* (Nelson, 1965)

oxidized zone The portion of an orebody near the surface that: (1) has been leached by percolating water carrying oxygen, carbon dioxide, or other gases; or (2) in which sulfide minerals have been partially dissolved and redeposited at depth, the residual portion changing to oxides, carbonates, and sulfates. Cf: *gossan; sulfide zone.* Syn: *zone of oxidation.* See also: *supergene enrichment.* (AGI, 1987)

oxidizer A material that readily yields oxygen or other oxidizing substances needed for an explosive reaction to take place; solid oxidizers common in industrial explosives are ammonium nitrate and sodium nitrate.

oxidizing flame In blowpiping, the outer, least visible, and less intense part of the flame, from which oxygen may be added to the compound being tested. See also: *blowpiping.* Cf: *reducing flame.*

oxidizing fusion An oxidation process used for fire refining bismuth, gold, and silver; the crude metals are melted down with oxidizing fluxes, so that the impurities are oxidized during the melting period and become part of the slag. (Newton, 1959)

oxidizing smelting *pyritic smelting.*

oxonite An explosive prepared by dissolving picric acid in nitric acid. (Fay, 1920)

oxyacetylene A mixture of oxygen, O_2, and acetylene gas, C_2H_2, in such proportions as to produce the hottest flame known for practical use. Oxyacetylene welding and cutting is used in almost every metalworking industry. (Crispin, 1964)

oxyacetylene cutter An appliance for cutting metals by means of a flame obtained from acetylene and compressed oxygen, which are stored in separate steel cylinders. Oxyhydrogen and oxycoal gas flames are also used. (Nelson, 1965)

oxychloride cement A plastic cement formed by mixing finely ground caustic magnesite with a solution of magnesium chloride. (AGI, 1987)

oxygen A nonmetallic element, normally colorless, odorless, tasteless, nonflammable diatomic gas. Symbol, O. Occurs uncombined in the air to the extent of about 21% by volume and is combined in water, in most rocks and minerals, and in a great variety of organic compounds. Oxygen is very reactive and capable of combining with most elements. Essential for respiration in all plants and animals and for practically all combustion. Oxygen enrichment of steel blast furnaces accounts for the greatest use of the gas. Used in manufacturing ammonia, methanol, and ethylene oxide. (Handbook of Chem. & Phys., 3)

oxygen balance The amount of oxygen in an explosive mixture, expressed in weight percent, liberated as a result of complete conversion of explosive material to CO_2, H_2O, SO_2, Al_2O_3, and other non-toxic gases; referred to as positive oxygen balance; negative oxygen balance is a deficient amount of oxygen leading to incomplete oxidation of explosive materials resulting in the possible formation of toxic gases, such as CO and NO.

oxygen-Bessemer A steelmaking process in which the air blown through the bottom tuyeres is enriched with oxygen. If oxygen alone is used, tuyere wear is excessive. Oxygen plus steam or oxygen plus carbon dioxide can be used. Also called oxy-Thomas. See also: *O.L.P. steel process.* (Nelson, 1965)

oxygen consumption A person working hard requires about 10 ft^3/min (283 L/min) of air to supply adequate oxygen. (Hammond, 1965)

oxygen deficiency *anoxia.*

oxygen-deficient atmosphere A concentration of oxygen in the atmosphere equal to or less than 19.5% by volume. (OSHA, 1910.146)

oxygen-enriched atmosphere An atmosphere containing more than 23.5% oxygen by volume. (OSHA, 1910.146)

oxygen-flash smelting process Employed as an autogeneous matte smelting process for smelting copper-nickel concentrate. (Newton, 1959)

oxygen-free copper Electrolytic copper free from cuprous oxide; produced without the use of residual metallic or metalloidal deoxidizers. (ASM, 1961)

oxygen impingement process A process used in steel making in which pure oxygen is blown down onto the bath in a converterlike vessel. (Osborne, 1956)

oxygen index Volumetric ratio of oxygen to the total gases in a mixture. (Van Dolah, 1963)

oxygen lance A device made up of a welding oxygen bottle and a length of rubber hose attached to a valve which is fitted to a steel pipe, so that when the tip of the lance is ignited it can be used to melt the solidified metal out of the iron tap hole in a blast furnace.

oxygen process A process for making steel in which oxygen is blown upon or through molten pig iron, whereby most of the carbon and impurities are removed by oxidation. (Harbison-Walker, 1972)

oxygen steel The use of oxygen instead of air to convert molten pig iron into steel. The oxygen is used in different ways in different furnaces, but the fastest ones utilize the direct oxidation effects of a relatively pure (99.5%) oxygen. See also: *L.D. steel process.* (Nelson, 1965)

oxyhornblende A hornblende with (OH+F+Cl) less than 1.0. Also called basaltic hornblende. Syn: *lamprobolite.*

oxyhydrogen Of, relating to, or utilizing a mixture of oxygen and hydrogen. (Webster 3rd, 1966)

oxyhydrogen blowpipe A blowpipe in which hydrogen is burned in oxygen. Streams of the two gases in the proportion to form water are forced under pressure from separate reservoirs, and issue together from a jet, igniting just as they issue. The temperature produced which has been estimated at 5,000 °F (2,760 °C), is sufficient to fuse very refractory substances. Also called compound blowpipe. (Standard, 1964; Fay, 1920)

oxymagnite A former name for maghemite, isometric Fe_2O_3; also called ferromagnetic ferric oxide.

oxyphile *lithophile.*

ozarkite A white, massive variety of thomsonite, from Arkansas. (Fay, 1920)

ozocerite A mineral paraffin wax, of dark yellow, brown, or black color with a melting point of 55 to 110 °C and sp gr, 0.85 to 0.95. Is soluble in gasoline, benzene, and turpentine and is found near the Caspian Sea region and in Utah as narrow seams in sandstone. Also called mineral wax; fossil wax; native paraffin; earth wax. Also spelled ozokerite. Syn: *ader wax; earth wax; mineral fat.* See also: *fibrous wax.* Cf: *hatchettine; hatchettite.* (CTD, 1958)

ozone An allotropic, triatomic form of oxygen, O_3; a faintly blue, irritating gas with a characteristic pungent odor, but at -112 °C it condenses to a blue magnetic liquid. It occurs in minute quantities in the air near the Earth's surface and in larger quantities in the stratosphere as a product of the action of ultraviolet light of short wavelengths on ordinary oxygen. Ozone is generated usually in dilute form by a silent electric discharge in oxygen or air. It decomposes to oxygen (as when heated) and it is a stronger oxidizing agent than oxygen. Used chiefly in disinfection and in deodorization (such as in water purification and in air conditioning), in oxidation and bleaching (such as in the treatment of industrial wastes), and in ozonolysis (such as in the manufacture of azelaic acid from oleic acid). (Webster 3rd, 1966)

ozonizer Electrical apparatus that converts atmospheric oxygen to ozone; used in sterilizing water for drinking purposes and for purifying air. (Pryor, 1963)

P

pachnolite A monoclinic mineral, $NaCaAlF_6 \cdot H_2O$; white; distinct cleavage; dimorphous with thomsenolite; an alteration of cryolite.
Pachuca tank A cylindrical tank with a conical bottom. It contains a pipe that is coaxial with the leaching tank and open at both ends; compressed air is introduced at the lower end of this pipe, which behaves as an air lift. The density of the pulp within the pipe is less than that of the pulp surrounding it because of the column of air bubbles contained in the pipe, and the pressure of the denser pulp causes the pulp in the central pipe to rise and overflow, thus circulating the entire charge. Syn: *Brown agitator; Brown tank.* See also: *Patterson agitator.* (Newton, 1959)
Pacific suite One of two large groups of igneous rocks, characterized by calcic and calc-alkalic rocks. Harker (1909) divided all Tertiary and Holocene igneous rocks of the world into two main groups, the Atlantic suite and the Pacific suite. Because there is such a wide variation in tectonic environments and associated rock types in the areas of Harker's Atlantic and Pacific suites, the terms are now seldom used to indicate kindred rock types. Cf: *Mediterranean suite.* (AGI, 1987)
pacite An iron arsenosulfide near arsenopyrite in composition.
pack (a) A pillar, constructed from loose stones and dirt, built in the waste area or roadside to support the roof. See also: *double packing; solid stowing; strip packing.* (Nelson, 1965) (b) A pack built on a longwall face between the gate-side packs is called an intermediate pack. (SMRB, 1930) (c) Waste rock or timber support used for a roof over underground workings or used to fill excavations. Also called fill. (Pryor, 1963) (d) To cause the speedy subsidence of ore in the process of washing by beating a keeve or tub with a hammer.
pack builder (a) Person who builds packs or pack walls. See also: *pack.* (Fay, 1920) (b) In anthracite and bituminous coal mining, a worker who: (1) fills worked-out rooms, from which coal has been mined, with rock, slate, or other waste to prevent caving of walls and roofs; (2) builds rough walls and columns of loose stone, heavy boards, timber, or coal along haulageways and passageways and in rooms where coal is being mined, to prevent caving. Also called packer. See also: *pillar man; timber packer; waller.* (DOT, 1949)
pack cavity system *methane drainage.*
pack drawer In anthracite and bituminous coal mining, a laborer who draws (tears down) stone or timber packs (pillars constructed by pack builders in the working place to support the roof during extraction of coal) to permit the roof to cave behind as the mining of the coal recedes toward the entrance of the working area. (DOT, 1949)
packer (a) A short expansible-retractable device deliberately set in a cased or uncased well bore to prevent upward or downward fluid movement; generally for temporary use. (AGI, 1987) (b) A miner employed in stowing or packing the waste area. Also called gobber. (Nelson, 1965)
packfong Chinese. A silver-white alloy of copper, zinc, and nickel; German silver.
pack hardening Case carburizing, using a solid carburized medium, followed by a hardening treatment. (CTD, 1958)
pack hole The space adjacent to a gate end at the face and between the face end of a gate-side pack and the coal face into which packs will be inserted when the gate is ripped or dinted. (TIME, 1929-30)
packing (a) Occurs in crushing plants when the material in the chamber is so compacted as to be nearly without voids. It occurs when free downward movement is inhibited. (South Australia, 1961) (b) The filling of a waste area with stones and dirt. See also: *solid stowing.* (Nelson, 1965) (c) The method of giving support to a roof by the insertion of waste material placed or built into space from which coal or ore has been extracted. (TIME, 1929-30) (d) The spacing or density pattern of the mineral grains in a rock. Cf: *fabric.* (AGI, 1987) (e) *blocking.* (f) With gyratories, packing copy refers to an accumulation of sticky fines on the diaphragm. (South Australia, 1961)
packing density The bulk density of a granular material, when packed under specified conditions. It is commonly determined, particularly for foundry sands.
packing factor Ratio of true volume to bulk volume. (Van Vlack, 1964)
packing gland An explosion-proof entrance for conductors through the wall of an explosion-proof enclosure, to provide compressed packing completely surrounding the wire or cable, for not less than 1/2 in (1.27 cm) measured along the length of the cable.
packsand A very fine-grained sandstone that is so loosely consolidated by a little calcareous cement as to be readily cut by a spade. (AGI, 1987)
pack wall A dry-stone wall built along the side of a roadway, or in the waste area, of a coal or metal mine. The wall helps to support the roof and also to retain the packing material and prevent it spreading into the roadway. (Nelson, 1965)
pad (a) Ground-contact part of a crawler-type track. (Nichols, 1976) (b) *wallplate.* (c) The refractory brickwork below the molten iron at the base of a blast furnace. (Dodd, 1964)
paddle (a) Numbered wooden marker which shovelers put in the cans of ore that they load. (Hess) (b) A straight iron tool for stirring ore in a furnace. (Standard, 1964) (c) A bat or pallet, as used in tempering clay. (Standard, 1964) (d) A scoop for stirring and mixing, as used in glassmaking. (Standard, 1964)
paddle conveyor *paddle-type mixing conveyor.*
paddle loader A belt loader equipped with chain-driven paddles that move loose material to the belt. (Nichols, 1976)
paddle mixer A form of worm conveyor having two noncontinuous spirals that form paddles; the shafts are contrarotating and the spirals opposite. See also: *paddle-type mixing conveyor.* (BS, 1962)
paddle-type mixing conveyor A type of conveyor consisting of one or more parallel paddle conveyor screws. See also: *blending conveyor; paddle conveyor.*
paddle washer A type of conveyor consisting of one or two inclined parallel paddle conveyor screws in a conveyor trough having a receiving tank and an overflow weir at the lower end and a discharge opening at the upper end.
paddle-wheel fan A centrifugal fan with radial blades. (Strock, 1948)
paddy A borehole drill bit having cutters that expand on pressure. Also called expansion bit; paddy bit. (Long, 1960)
paddy bit *paddy.*
paddy lamp A portable battery-operated lamp attached to the front or rear of a personnel train. (BS, 1965)
padlock sheave (a) The bucket sheave on a dipper or hoe shovel. (Nichols, 1976) (b) A sheave set connecting inner and outer boom lines. (Nichols, 1976)
page (a) A small wooden wedge used in securing the timbering for excavations. (Hammond, 1965) (b) In brickmaking, a track carrying the pallets bearing newly molded bricks. (Standard, 1964)
pagodite Ordinary massive pinite in its amorphous compact texture and other physical characters, but containing more silica. The Chinese carve the soft stone into miniature pagodas and images. See also: *agalmatolite; lardite; pinite.* (CTD, 1958)
Pahrump A provincial series of the Precambrian in California.
paint (a) A term used in the western United States for an earthy, pulverulent variety of cinnabar. (b) A film of molybdenite in fractures and veinlets.
paint gold A very thin coating of gold on minerals.
painting The painting of the mine roof with a coal-tar paint that seals the bottom strata of the roof to prevent air from entering the crevices of the roof. (Kentucky, 1952)
paint mill A machine for grinding mineral paints. (Fay, 1920)
paint rock A soft, incompetent, fine-grained mass of quartz, pyrolusite, and kaolin with subangular fragments of chert, hematite, and goethite. (Woodruff, 1966)
paint-rock ore *natural ore.*
pair A party of co-workers; a gang. Also spelled pare. (Webster 2nd, 1960; Fay, 1920)
pair production The transformation of a high-energy gamma ray into a pair of particles (an electron and a positron) during its passage through matter. (Lyman, 1964)
palagonite Devitrified basaltic glass.
palasome The host rock or mineral in a replacement deposit.
pale brick Brick that is underfired. (Fay, 1920)
paleo- (a) A combining form denoting great age or remoteness in regard to time (Paleozoic), or involving ancient conditions (paleoclimate). Sometimes given as pale- (palevent). Also spelled: palaeo; pal-

aio-. (AGI, 1987) (b) A prefix indicating pre-Tertiary origin, and generally altered character, of a rock to the name of which it is added, such as paleopicrite; by some the prefix has been applied to pre-Carboniferous rocks or features, such as the PaleoAtlantic Ocean. (AGI, 1987)
paleobotany The study of plants of past geological ages through the investigation of fossils. Cf: *paleontology; palynology.*
paleoclimatology The branch of science that treats of climatological conditions during the history of the Earth.
paleocurrent A current, generally of water, that influenced sedimentation or other processes or conditions in the geologic past.
paleoecology The science of the relationship between ancient organisms and their environments. (AGI, 1987)
paleogeography (a) The study and description of the physical geography of the geologic past, such as the historical reconstruction of the pattern of the Earth's surface or of a given area at a particular time in the geologic past, or the study of the successive changes of surface relief during geologic time. (b) The study of the relative positions of land masses as part of tectonic reconstructions of Earth history.
paleogeologic map A map that shows the areal geology of an ancient surface at some time in the geologic past; esp. such a map of the surface immediately below an unconformity, showing the geology as it existed at the time the surface of unconformity was completed but before the overlapping strata were deposited. Paleogeologic maps were introduced by Levorsen (1933). (AGI, 1987)
paleolithologic map A paleogeologic map that shows lithologic variations at some buried horizon or within some restricted zone at a particular time in the geologic past. (AGI, 1987)
paleomagnetism Faint magnetic polarization of rocks that may have been preserved since the accumulation of sediment or the solidification of magma whose magnetic particles were oriented with respect to the Earth's magnetic field as it existed at that time and place. (AGI, 1987)
paleontological facies (a) The paleontological aspect of a particular sedimentary lithology; e.g., nummulitic facies, crinoid facies, etc. (Schieferdecker, 1959) (b) Sedimentary facies differentiated on the basis of fossils. (AGI, 1987)
paleontologist Person who studies the fossilized remains of animals and/or plants. (AGI, 1987)
paleontology A science that deals with the life of past geological periods, based on the study of fossil remains of plants and animals, and gives information esp. about the phylogeny and relationships of modern animals and plants and about the chronology of the history of the Earth. Cf: *paleobotany; paleoclimatology; paleogeography.* (Webster 3rd, 1966)
paleozoology That branch of paleontology dealing with the study of fossil animals, both invertebrate and vertebrate. (AGI, 1987)
palimpsest Said of a structure or texture of metamorphic rocks in which remnants of some pre-existing structure or texture are preserved.
palingenesis Formation of a new magma by the melting of pre-existing magmatic rock in situ. Considered incorrectly by some workers as a syn. of anatexis. Adj: palingenic. (AGI, 1987)
palladinite A poorly defined ocherous coating on palladian gold, probably PdO.
palladium A soft, ductile, steel-white metallic element of the platinum group metals. Symbol, Pd. Found along with platinum and other metals of the platinum group in placer deposits; also found associated with nickel-copper deposits. Used as a catalyst, in dentistry, watchmaking, surgical instruments, and electrical contacts. (Handbook of Chem. & Phys., 3)
palladium amalgam A former name for potarite. See also: *potarite.*
palladium gold Same as porpezite, or gold, containing palladium up to 10%. See also porpezite; gold. (Fay, 1920):
pallas iron *pallasite.*
pallasite (a) Any ultramafic rock, whether of meteoric or terrestrial origin, that contains approx. 60% iron if meteoric, or more iron oxides than silica if terrestrial. (AGI, 1987) (b) A stony-iron meteorite composed essentially of large single glassy crystals of olivine embedded in a network of nickel-iron. Pallasites are believed to have been formed at the interface of the stony mantle and metal core of a layered planetoid. Syn: *pallas iron.* (AGI, 1987)
palleting A light platform in the bottom of powder magazines to preserve the powder from dampness. (Fay, 1920)
pallet molding A method of forming bricks in sanded molds, from which they are dumped on a board called a pallet. (Standard, 1964)
pallet-type conveyor A series of flat or shaped wheelless carriers propelled by and attached to one or more endless chains or other linkage.
Palo-Travis analyser A sedimentation apparatus for determining particle size, based upon the settling of powder through a long sedimentation tube filled with liquid. The instrument consists of the sedimentation tube, a smaller reservoir at the top joined to the tube through a large bore stopcock, and a calibrated capillary mounted concentrically at the bottom of the tube. (Osborne, 1956)
paludal Pertaining to swamps or marshes, and to organic, clay, or other material deposited in a swamp environment. Cf: *palustrine.*
paludification Process of formation of a peat bog. This requires a steady growth of new peat-forming plants in phase with a steady general sinking of the depression in which this occurs. (Pryor, 1963)
palustrine Pertaining to material deposited in a swamp or marsh environment. Cf: *paludal.*
palygorskite (a) A monoclinic and orthorhombic mineral, $(OH)_2(Mg,Al)_4(Si,Al)_8O_{20}{\cdot}8H_2O$; fibrous; in desert soils. (b) A general name for lightweight fibrous clay minerals showing significant substitution of aluminum for magnesium; characterized by distinctive rod-like shapes under an electron microscope. Syn: *attapulgite.*
palynology (a) A branch of science concerned with the study of pollen of seed plants and spores of other embryophytic plants, whether living or fossil, including their dispersal and applications in stratigraphy and paleoecology. (AGI, 1987) (b) The study of the fossilized spores and pollen grains of the plants, esp. those whose remains contributed to the formation of coal seams. Cf: *paleobotany; paleontology.* (Nelson, 1965)
pan (a) A shallow steel or porcelain dish in which drillers or samplers wash drill sludge to a gravity concentrate and separate the particles of heavy minerals from the lighter-density rock powder to ascertain if the rocks traversed by the borehole contain minerals of value. Syn: *tin dish.* (Long, 1960) (b) Hardpan. (c) Fireclay or underclay of coal seams. (d) A trough or section of a pan conveyor or shaker conveyor. (Nelson, 1965) (e) The framework of a belt or chain conveyor. See also: *tray.* (Mason, 1951) (f) A circular steel dish from 10 to 16 in (25 to 40 cm) in diameter at the top, from 2 to 2-1/2 in (5.1 to 6.4 cm) deep, and with sides sloping at 35° to 40° to the horizontal, used for testing and working placer deposits. Syn: *batea; miner's pan; prospecting pan; gold pan.* Cf: *dish.* (g) A carrying scraper. (Nichols, 1976) (h) *panning.*
panabase *tetrahedrite.*
panabasite A former name for tetrahedrite. See also: *tetrahedrite.*
pan amalgamation See pan-amalgamation process.:
pan-amalgamation process Method of recovering silver and gold from their ores, in which a cast iron pan or barrel is used for contacting a slurry of the crushed ore with salt, copper sulfate, and mercury; the released silver and gold form an amalgam with the mercury. Syn: *pan amalgamation.* (Bennett, 1962)
Pan-American jig Mineral jig developed for treatment of alluvial sands. (Pryor, 1963)
pancake (a) *ribbon.* (b) Any of concrete discs that are stacked to form concrete columns for stope support. They are cast at the surface and are usually 30 in (76.2 cm) diameter by 4 in (10.2 cm) thick with reinforcement from wire rope. (Higham, 1951)
panclastite An explosive composed of liquid nitrogen tetroxide mixed with carbon disulfide or other liquid combustible, in the proportion of three volumes of the former to two of the combustible. (Fay, 1920)
pan conveyor (a) A conveyor comprising one or more endless chains or other linkage to which usually overlapping or interlocking pans are attached to form a series of shallow, open-topped containers. Some pan conveyors have been known also as apron conveyors. (b) Jigging conveyor; a trough down which coal slides after mining and loading in dipping seams, with motion being aided by a shaking action. See also: *jigger.* (Pryor, 1963) (c) A trough conveyor or gravity conveyor. (Nelson, 1965)
pandermite (a) *priceite.* (b) A name for firm, compact, porcelainlike masses of colemanite.
pan-edge (a) A runner mill for grinding or mixing granular material. (b) Steel supporting plates on which furnace bottom refractories are placed.
panel (a) A large rectangular block or pillar of coal. (b) A method of working whereby the workings of a mine are divided into sections, each surrounded by solid strata and coal with only necessary roads through the coal barrier. Also spelled pannel. (Mason, 1951) (c) The working of coal seams in separate panels or districts; e.g., single unit panel. See also: *panel working.* (Nelson, 1965) (d) Rectangle of lode ore that is defined by means of levels and winzes and then considered to be proved as regards volume for valuation purposes. (Pryor, 1963) (e) A group of breasts or rooms separated from the other workings by large pillars. (Fay, 1920) (f) A small portion of coal left uncut. (Webster 3rd, 1966)
panel barrier The pillar of coal left between the adjacent panels. These pillars are often worked on the retreat after the coal in the panels has been extracted. In the panel system of bord-and-pillar mining, the panel barrier may be 22 yd (20 m) (minimum) wide and about 300 yd (274 m) apart. In longwall panel mining, the barriers may be made of sufficient width for extraction by a conveyor face on the retreat. Also called panel pillar. See also: *Bolsover experiment.* (Nelson, 1965)
panel slicing (a) In stoping, the process of mining out a panel either from above, below, or one side as described by a qualifying term. (Pryor, 1963) (b) *top slicing and cover caving.*

panel working (a) Working laid out in districts or panels, which are then extracted as single units. The panel system of working may be adopted with pillar-and-stall and longwall methods. See also: *pillar methods of working*. (Nelson, 1965) (b) A system of working coal seams in which the colliery is divided up into large squares or panels isolated or surrounded by solid ribs of coal of which a separate set of breasts and pillars is worked, and the ventilation is kept distinct; i.e., every panel has its own circulation, the air of one not passing into the adjoining one, but being carried direct to the main return airway. (Zern, 1928)

pan feeder *conveyor-type feeder.*

pan-feeder operator *mill feeder.*

panhead A head to a rivet or screw having the shape of a truncated cone. (Hammond, 1965)

panidiomorphic A textural term for rocks in which all or almost all of the mineral constituents are idiomorphic or euhedral.

panman (a) A worker who places in position and tends the operation of underground trough conveyors for the transportation of coal or other minerals. These conveyors are built in sections, and the principal task of the panman is to move the sections from one position to another. (Hess) (b) One engaged in dismantling or building conveyors. Also called panner. (Mason, 1951)

panning A technique of prospecting for heavy metals, such as gold, by washing placer or crushed vein material in a pan. The lighter fractions are washed away, leaving the heavy metals behind in the pan. See also: *pan*. (AGI, 1987)

pantellerite A peralkaline rhyolite or quartz trachyte with normative quartz exceeding 10%. It is more mafic than comendite. (AGI, 1987)

pantograph (a) A type of drawing instrument consisting of rods linked together in the form of a parallelogram, used for copying a drawing to any required scale. (Hammond, 1965) (b) The hinged diamond-shaped structure mounted on the roof of an electric locomotive to collect electric power from an overhead wire. (Hammond, 1965)

pan-type car Doorless car of two-way, side-dump design; built in capacities from 4 to 10 yd^3 (3.1 to 7.6 m^3). The car body is reversible and may be dumped to either side. Dumping is accomplished by means of an external hoist at the dumping point. (Pit and Quarry, 1960)

Panzer conveyor *armored flexible conveyor.*

Panzer-Forderer snaking conveyor A very strong, armored conveyor that is moved forward behind a coal plow by means of a traveling wedge pulled along by the plow or by means of jacks or compressed-air-operated rams attached at intervals to the conveyor structure. (Sinclair, 1959)

papa (a) A bluish white, massive New Zealand clay like pipe clay; used for whitening fireplaces. When hard, it is called papa rock. Etymol: Polynesian. (b) Sp. A nugget of gold or silver. (c) A nodule of mineral.

papagoite A monoclinic mineral, $CaCuAlSi_2O_6(OH)_3$; forms blue crystals; secondary; at Ajo, Pima County, AZ.

paper clay A fine-grained, white, kaolinic clay with high retention and suspending properties, high reflectance, and a very low content of free silica. It is used for coating or filling paper. (AGI, 1987)

paper coal (a) Coal in which cuticular matter may be prominent. (AGI, 1987) (b) A variety of brown coal deposited in thin layers like sheets of paper. (Fay, 1920)

paper shale A shale that easily separates on weathering into thin layers or laminae suggesting sheets of paper; it is often highly carbonaceous. (AGI, 1987)

paper spar A crystallized variety of calcite found in thin lamellae or paperlike plates. (Standard, 1964)

par- *para-.*

para- (a) A prefix applied to the names of metamorpic rocks that have been derived from sediments; e.g., paragneiss. (Stokes, 1955) (b) Prefix meaning beside or nearby. (c) Indicating a polymorph. (d) Indicating a schist or gneiss derived from a sedimentary protolith. (e) A matrix-rich clastic sedimentary rock. (f) In chemistry, a prefix indicating: (1) an isomeric or polymeric modification; such as paracyanogen, paraldehyde, etc.; (2) a modification or a similar compound that is not necessarily isomeric or polymeric; such as, paramorphine; (3) a benzene diderivative in which the substituted atoms or radicals are directly opposite each other on the benzene ring—i.e., occupying the positions 1 and 4—such as paraxylene; or (4) an inactive isomer produced by a combination of its dextro- and levo- modifications—such as, paratartaric acid. A Greek prefix meaning beside. Abbrev., p-. Syn: *par-*. (Standard, 1964; Webster 3rd, 1966) (g) A Greek prefix meaning beside. In the name of a metamorphic rock, such as paragneiss, it means derived from an original sediment. (Webster 3rd, 1966)

parabola The shape taken by the curve of a bending moment diagram for a uniformly distributed load on a beam simply supported. (Hammond, 1965)

parabutlerite An orthorhombic mineral, $Fe^{3+}(SO_4)(OH){\cdot}2H_2O$; dimorphous with butlerite: orange; an alteration product of copiapite.

paracelsian A monoclinic mineral, $BaAl_2Si_2O_8$; feldspar group; pale yellow; dimorphous with celsian; at Candoglia, Italy. Syn: *barium feldspar.*

parachrosis Discoloration in minerals from exposure to weather. (Standard, 1964)

paraconformity An unconformity at which strata are parallel and the contact is a simple bedding plane. (AGI, 1987)

paraconglomerate A term proposed by Pettijohn (1957) for a conglomerate that is not a product of normal aqueous flow, but is deposited by such agents as subaqueous turbidity slides and glacier ice; it contains more matrix than gravel-sized fragments (pebbles may form less than 10% of the rock). Examples include tillites, pebbly mudstones, and relatively structureless clay or shale bodies in which pebbles or cobbles are randomly distributed. Syn: *conglomerate mudstone.* (AGI, 1987)

paracoquimbite A trigonal mineral, $Fe^{3+}_2(SO_4)_3{\cdot}9H_2O$; dimorphous with coquimbite; pale violet; astringent tasting; secondary; in oxidized iron sulfide deposits.

paradamite A triclinic mineral, $Zn_2(AsO_4)(OH)$; rare; dimorphous with adamite and isomorphous with tarbuttite; transparent; vitreous; pale yellow; forms sheaflike aggregates and striated equant crystals; at the Ojuela Mine, Durango, Mexico.

paraffin shale *oil shale.*

paragenesis A characteristic association or occurrence of minerals or mineral assemblages in ore deposits, connoting contemporaneous formation. Cf: *mineral sequence.* (AGI, 1987)

paragenetic (a) Pertaining to paragenesis. (AGI, 1987) (b) Pertaining to the genetic relations of sediments in laterally continuous and equivalent facies. (AGI, 1987)

paragneiss In petrology, a gneiss formed by the metamorphism of a sedimentary rock. Cf: *orthogneiss.*

Paragon Trade name for a nonrotating rope of 12 by 6 over 3 by 24 strand construction. (Hammond, 1965)

paragonite A monoclinic mineral, $NaAl_2(AlSi_3)O_{10}(OH)_2$; mica group; pseudohexagonal with basal cleavage; forms fine-grained, massive, scaly aggregates; occurs in metamorphic rocks and in soils; not common as it is incompatible with potassium feldspar (albite plus muscovite is more stable); rarely identified because of its similarity to muscovite. Syn: *soda mica.*

paragonite schist A variety of schist in which paragonite supplants muscovite or biotite.

paraguanajuatite A trigonal mineral, $Bi_2(Se,S)_3$; paramorphous after orthorhombic guanajuatite.

parahilgardite A mineral trimorphous with monoclinic hilgardite and triclinic hilgardite. See also: *hilgardite.*

parahopeite A colorless hydrous zinc phosphate, $Zn_3(PO_4)_2{\cdot}4H_2O$, triclinic. Minute tabular or prismatic crystals; fan-shaped aggregates. From Broken Hill, Northern Rhodesia; Salmo, BC. (English, 1938)

parajamesonite An orthorhombic mineral, $Pb_4FeSb_6S_{14}$; dimorphous with jamesonite; metallic; black; distinguished by its X-ray pattern.

paralaurionite A monoclinic mineral, $PbCl(OH)$; dimorphous with laurionite; soft; forms white pseudo-orthorhombic prismatic crystals; a secondary mineral in lead deposits.

paralic Said of deposits laid down on the landward side of a coast, in shallow fresh water subject to marine invasions. Thus, marine and nonmarine sediments are interbedded; as exemplified in the lower part of the Coal Measures, the nonmarine (paralic) predominate, with relatively thin marine bands. Cf: *limnic.*

paralic coal basin A coal basin that originated near the sea—as opposed to a limnic coal basin. (AGI, 1987)

parallax (a) In survey work, incorrect reading of a graduation on an instrument if the observer's eye is not truly normal to the graduated plate. (Pryor, 1963) (b) The change in bearing or apparent position of an object produced by a change in the observer's position. (NCB, 1964) (c) The apparent displacement, or change in position, of the crosshairs of a focusing telescope with reference to the image of an object, as the eye is moved from side to side, when the focus of the eyepiece or objective is imperfect. (Seelye, 1951)

parallel blasting circuit An electric blasting circuit in which the leg wires of each detonator are connected across the firing line directly or in parallel through bus wires. Syn: *series-in-parallel circuit.* Cf: *loop circuit.* (Atlas, 1987)

parallel circuit firing A method of connecting together a number of detonators that are to be fired electrically in one blast. The electric detonators are connected to two common points. Each detonator offers a path for the electric current, independent of all the other detonators in the circuit, and therefore calls for a higher amperage than a series circuit in which there is but one path. (Nelson, 1965)

parallel cut Group of parallel holes, not all charged with explosive, that creates the initial cavity to which the loaded holes break in blasting a development round. (Pryor, 1963)

parallel displacement fault A fault along which all straight lines on opposite sides of the fault and outside the dislocated zone that were parallel before the displacement are parallel afterward. (AGI, 1987)

parallel drum A cylindrical form of drum on which the haulage or winding rope is coiled. The drum roll may be plain or grooved. For deep winds, multilayering of rope is often used to reduce the drum size required. Also, for deep winding (3,000 ft or 915 m or more), a balance rope is almost essential with a parallel drum. Syn: *cylindrical drum.* See also: *winding drum.* (Nelson, 1965)
parallel duplex mine cable *portable parallel duplex mine cable.*
parallel entry Usually an intake airway parallel to the haulageway. (USBM, 1960)
parallel extinction In mineral optics, refers to crystal edges or cleavage traces parallel to the optic directions of the mineral. Cf: *extinction.*
parallel firing The firing of detonators in a round of shots by dividing the total supply current between the individual detonators. Cf: *series firing.* (BS, 1964)
parallel flow Flow in the same direction of two or more streams within a stream system.
parallel fold A fold in which beds maintain the same thicknesses throughout. Cf: *similar fold; supratenuous fold.* Syn: *concentric fold.*
parallel growth Two or more crystals with corresponding faces parallel. (Fay, 1920)
parallel lines Lines that lie in the same plane and are equally distant from each other at all points. The term is ordinarily applied to straight lines. (Jones, 1949)
parallelogram Quadrilateral that has opposite sides parallel and opposite angles equal. (Jones, 1949)
parallel ripple mark A ripple mark with a relatively straight crest and an asymmetric profile; specif. a current ripple mark. (AGI, 1987)
parallel series Two or more series of electric blasting caps arranged in parallel. See also: *multiple series.* (Nichols, 1976)
parallel series circuit A method of connecting together a number of detonators to be fired electrically in one blast. The circuit consists of a number of series circuits connected in parallel. Syn: *series-in-parallel circuit.* (Nelson, 1965)
parallel unconformity *disconformity.*
parallel wire method An electrical prospecting method using equipotential lines or curves in prospecting for orebodies. In the parallel wire method, two bare copper wires about 3,000 ft (915 m) long, placed about 2,000 ft (610 m) or more apart, are used as electrodes. Current is supplied from the generator, and the electrodes are connected to the ground at 100 ft (30 m) intervals by iron grounding pins. Equipotential lines are located by two electrodes or wooden rods, to one end of which are fastened metal spikes about 6 to 7 in (15 to 18 cm) long. The electrodes are connected by some 150 ft (46 m) of wire that runs down the rods to the spikes. If a head telephone is placed in the circuit, the absence of sound in the telephone indicates that the two electrodes are at the same potential. By this method, the equipotential lines can be traced. (Lewis, 1964)
paramagnetic Having a small positive magnetic susceptibility. A paramagnetic mineral such as olivine, pyroxene, or biotite contains magnetic ions that tend to align along an applied magnetic field but do not have a spontaneous magnetic order. Cf: *diamagnetic.* (AGI, 1987)
paramagnetism (a) The magnetism of a paramagnetic substance. The property by which the north pole of a magnet that is magnetized by induction is repelled to 180° by the north pole of the inducing magnet. (Standard, 1964) (b) The property possessed by a substance of producing a higher concentration of magnetic lines of force within itself than in the surrounding magnetic field when it is placed in such a field. (Miall, 1940) (c) A property of many substances, related to ferromagnetism, by virtue of which, when placed in a nonuniform magnetic field, they tend to move toward the strongest part. Permanent magnetism is practically absent and the susceptibility, which is much less than that of iron, is constant at any given temperature, but in most substances it is nearly inversely proportional to the absolute temperature. Cf: *ferrimagnetism; diamagnetism.* (Holmes, 1920)
paramelaconite A tetragonal mineral, $Cu^{+}_2Cu^{2+}_2O_3$; purplish black; at Bisbee, AZ.
parameter (a) A constant or variable in a mathematical expression that distinguishes various specific situations. (b) In crystallography, one of the three non-coplanar vectors which describe a lattice. Syn: *lattice parameter.*
paramontroseite An orthorhombic mineral, VO_2; forms by loss of hydrogen and iron from montroseite in an initial stage of oxidation of uranium-vanadium deposits.
paramorph A pseudomorph with the same composition as the original crystal, caused by a phase transformation; e.g., calcite with aragonite morphology. Cf: *pseudomorph.*
paramorphism (a) The alteration of one mineral into another without change of composition, such as augite into hornblende in uralitization. (Fay, 1920) (b) With metamorphism, it describes such thorough changes in a rock that its old components are destroyed and new ones are built up. See also: *allomorphism.* (Fay, 1920)
paramoudra Large flint nodule. See also: *potstone.*
pararammelsbergite An orthorhombic mineral, $NiAs_2$; loellingite group; trimorphous with rammelsbergite and krutovite; metallic tin-white; commonly massive.
pararealgar A monoclinic mineral, AsS; trimorphous with realgar and alpha-AsS; powdery; bright yellow to orange-brown; easily mistaken for orpiment.
paraschist A schist derived from a sedimentary rock. See also: *orthoschist; schist.*
paraschoepite An orthorhombic mineral, $UO_3{\cdot}(2\text{-}x)H_2O$; bright yellow; a dehydration product of schoepite. Formerly called schoepite III. See also: *schoepite.*
parasymplesite A monoclinic mineral, $Fe^{2+}_3(AsO_4)_2{\cdot}8H_2O$; vivianite group; dimorphous with symplesite and isomorphous with köttigite; bluish green.
paratacamite A trigonal mineral, $Cu_2(OH)_3Cl$; forms twinned rhombohedra; massive or powdery; green to green-black; a secondary mineral in copper deposits.
paratellurite A tetragonal mineral, TeO_2; rutile group; soft; waxy; gray-white; dimorphous with tellurite; at Cananea, Mexico.
paratomous Having planes of cleavage inclined to the axis; also, abounding with facets of cleavage. (Standard, 1964)
parautochthonous granite A mobilized portion of an autochthonous granite that has moved higher in the crust or, more usually, into tectonic domains of lower pressure. It shows variable marginal relations, in some places migmatitic in others characterized by a thermal aureole. (Schieferdecker, 1959)
paravauxite A triclinic mineral, $FeAl_2(PO_4)_2(OH)_2{\cdot}8H_2O$; colorless; forms small prismatic crystals; at Llallagua, Bolivia.
parbigite *messelite.*
pargasite A monoclinic mineral, $NaCa_2(Mg,Fe)_4Al(Si_6Al_2)O_{22}(OH)_2$; amphibole group; prismatic cleavage; occurs in dolomitic marbles and in skarns.
Parian marble One of the most famous of ancient statuary marbles; from the island of Paros, Greece.
parisite A trigonal mineral, $6[(Ce,La,Nd)_2CaCO_3)_3F_2]$; vitreous to resinous; forms acute hexagonal bipyramids; in veins, such as the emerald deposits of Columbia; also in alkalic pegmatites. Named for J.J. Paris.
parkerite A monoclinic mineral, $Ni_3(Bi,Pb)_2S_2$; metallic; bronze; has three cleavages; in a magmatic sulfide deposit, Insizwa, South Africa.
Parkerizing Treatment of steel in hot aqueous solution of free phosphoric acid and manganese dihydrogen phosphate, other salts sometimes being used as accelerators. A fine-grained insoluble film of ferric phosphate is formed in a few minutes, which is corrosion resistant. (Pryor, 1963)
Parker process A method for producing low-temperature coke in which each retort is a monobloc iron casting 9 ft (2.7 m) high, containing 12 tubes, which taper from 4-1/2 in (11.4 cm) at the top to 5-1/4 in (13.3 cm) at the bottom. A battery contains 36 retorts in 2 rows of 18. Retorts and combustion chambers are arranged alternately, so that each retort is located in a radiation chamber formed by the walls of adjacent combustion chambers. The retorts are heated only by radiation from these walls, so that there is no overheating and the inside temperature of the retorts can be maintained accurately at 1,112 °F (600 °C). A cooling chamber is fitted below each pair of retorts, of a size sufficient to hold the coke from both. The pairs of retorts are charged and discharged every 4 h. Syn: *Coalite process.* (Francis, 1954)
Parkes process A process used to recover precious metals from lead. It is based on the principle that if 1% to 2% of zinc is stirred into molten lead, a compound of zinc with gold and silver separates out and can be skimmed off. (ASM, 1961)
parmalee wrench A wrench that has a smooth segmented sleeve that when tightly clamped around the tube of a core barrel, will not mar or distort the thin tube when the core barrel is taken apart. (Long, 1960)
parral agitator An agitator using a number of small airlifts disposed about a circular, flat-bottomed tank in such a way as to impart a circular swirling motion to the pulp. (Liddell, 1918)
Parr formula The simplest method for determining the amount of mineral matter present in a coal is to determine the ash and sulfur contents and to make corrections for the changes taking place in these during combustion. The Parr formula for doing this is: total inorganic matter=moisture+1.08 ash+0.55 sulfur, where moisture, ash, and sulfur represent the percentages of these substances found by analysis of the coal. (Francis, 1954)
Parrish arm Long arm made of a flexible board for the suspension of a shaker screen. (Zern, 1928)
Parrish shaker A screening shaker with flexible wooden hangers and flexible drive arms; used for sizing anthracite. (Mitchell, 1950)
Parr's classification of coal A classification system based on the proximate analysis and calorific value of ash-free, dry coal. The heating value of raw coal is obtained, and from these data a table is drawn up, at one end of which are the celluloses and woods of about 7,000 Btu/lb (16.3 MJ/kg). These data are then plotted against the percentage volatile matter in unit coal. (Hess)

parsonsite A triclinic mineral, $Pb_2(UO_2)(PO_4)_2{\cdot}2H_2O$; forms pale citron-yellow crusts, powders, and tiny laths; nonfluorescent; radioactive; a secondary mineral in uraniferous pegmatites and other uranium deposits.

part In founding, a section of a mold or flask specif. distinguished (in a three-part flask) as top part, middle part, and bottom part. (Standard, 1964)

part 90 miner A miner employed at an underground coal mine or at a surface work area of an underground coal mine who has exercised the option under the old section 203b program (36 FR 20601, October 27, 1971), or under 90.3 (Part 90 option; notice of eligibility; exercise of option) of this part to work in an area of a mine where the average concentration of respirable dust in the mine atmosphere during each shift to which that miner is exposed is continuously maintained at or below 1.0 mg/m^3 of air, and who has not waived these rights. (CFR, 1992)

part-face blast Either of two stages of blasts when the height of the rock face is too great to blast in one operation. (McAdam, 1958)

partially fixed An end support to a beam or a column that cannot develop the full fixing moment. (Hammond, 1965)

partial melting (a) Melting of part of a rock; because a rock is composed of different minerals, each with its own melting behavior, melting does not take place at one temperature (as for ice at 0 °C) but takes place over a range of temperatures; melting starts at the solidus temperature and continues, nonlinearly, as the temperature increases to the liquidus temperature when the rock is totally molten. (Fowler, 1990) (b) A situation in which only certain minerals in a rock are melted, due to their lower melting temperature.

partial pressure (a) That part of the total pressure of a mixture of gases contributed by one of the constituents. (Strock, 1948) (b) *Dalton's law.*

partial pyritic smelting Blast furnace smelting of copper ores in which some of the heat is provided by oxidation of iron sulfide and some by combustion of coke. See also: *pyritic smelting.* (CTD, 1958)

partial roasting Roasting carried out to eliminate some but not all of the sulfur in an ore. (CTD, 1958)

partial subsidence Any amount of subsidence that is less than full subsidence; such as with solid or strip packing. (Nelson, 1965)

particle A general term, used without restriction as to shape, composition, or internal structure, for a separable or distinct unit in a rock; e.g., a sediment particle, such as a fragment or a grain, usually consisting of a mineral. (AGI, 1987)

particle diameter The length of a straight line through the center of a sedimentary particle considered as a sphere; a common expression of particle size. See also: *particle size.* (AGI, 1987)

particle mean size *particle size.*

particle size The general dimensions (such as average diameter or volume) of the particles in a sediment or rock, or of the grains of a particular mineral that make up a sediment or rock, based on the premise that the particles are spheres or that the measurements made can be expressed as diameters of equivalent spheres. It is commonly measured by sieving, by calculating settling velocities, or by determining areas of microscopic images. See also: *particle diameter.* Syn: *grain size.* (AGI, 1987)

particle-size analysis Determination of the statistical proportions or distribution of particles of defined size fractions of a soil, sediment, or rock; specif. mechanical analysis. Syn: *size analysis.* See also: *wet analysis.* (AGI, 1987)

particle-size distribution The percentage, usually by weight and sometimes by number or count, of particles in each size fraction into which a powdered sample of a soil, sediment, or rock has been classified—such as the percentage of sand retained on each sieve in a given size range. It is the result of a particle-size analysis. Syn: *size distribution; size-frequency distribution.* (AGI, 1987)

particle-size reduction The process of crushing or grinding material to reduce the particle size. (BS, 1960)

particle sorting Separation of solid particles, in a fluid (air, water, etc.), because of different densities or masses. (Bennett, 1962)

particle velocity A measure of the intensity of ground vibration generated from a blasting event, specif. the time rate of change of the amplitude of ground displacement, given in inches (or millimeters) per second.

particulate Refers to particles collected by filtration from ambient air. (SME, 1992)

parting (a) A lamina or very thin sedimentary layer separating thicker strata of a different type; e.g., a thin layer of shale or slate in a coal bed, or a shale break in sandstone. Strata tend to separate readily at partings. Cf: *band.* (b) A small joint in coal or rock, or a layer of rock in a coal seam. See also: *back.* (c) The physical property of some specimens of mineral species to break along specific crystallographic planes because of twinning or chemical alteration along them; e.g., rhombohedral parting in corundum. Cf: *cleavage; fracture.* Syn: *mineral parting.* (d) Cutting simultaneously along two parallel lines or along two lines which balance each other in the matter of side thrust. (e) The final process after cupellation to remove the silver from bullion bead. (f) A side track or turnout in a haulage road on which empty or loaded cars are collected for distribution to points for loading or for haulage to the surface or to the shaft or slope bottom for hoisting.

parting and connection man In bituminous coal mining, a laborer who directs the movement of mine cars from a parting (a side track). Also called connection man; parting boy. See also: *parting.* (DOT, 1949)

parting boy *parting and connection man.*

parting cleaner In bituminous coal mining, one who only picks out seam partings (layers of rock) in the coal working face prior to blasting, using a long-handled pick. (DOT, 1949)

parting density Density maintained in the bath in dense media separation. (Pryor, 1965)

parting flask A flask used to separate gold and silver, such as by quartation, in assaying procedures. Syn: *parting glass.*

parting glass *parting flask.*

parting liquid Any of several liquids—such as tetrabromethane, ethylene dibromide, pentachlorethane, and trichlorethylene—that are used in the DuPont mineral separation process. See also: *DuPont process.* (Mitchell, 1950)

parting powder A powder made from chalk, bone meal, or similar nonsiliceous material, suitably waterproofed, which is applied to a pattern to ensure a clean strip from the molding sand. (Osborne, 1956)

parting slate A term applied to a thin layer of slate between two seams of coal. (Fay, 1920)

partition curve A curve indicating, for each specific gravity (or size) fraction, the percentage that is contained in one of the products of the separation; e.g., the reject. Syn: *distribution curve.* (BS, 1962)

partition density The density corresponding to 50% recovery as read from a partition curve. Syn: *tromp cut point.* (BS, 1962)

partition factor The percentage of a specific gravity (or size) fraction recovered in one of the products of the separation; e.g., the reject. Syn: *distribution factor.* (BS, 1962)

partitioning method A resistivity method in which a special electrode configuration is used, consisting of five electrodes, instead of the usual number of four, to provide a check on the observations. (Schieferdecker, 1959)

partition size The separation size corresponding to 50% recovery as read from a size partition curve. (BS, 1962)

partly filled stope *square-set stoping.*

parts of line Separate strands of the same rope or cable used to connect two sets of sheaves. (Nichols, 1976)

part-swing shovel A power shovel in which the upper works can rotate through only part of a circle. (Nichols, 1976)

party chief In seismic prospecting, the person who supervises the personnel of the crew and generally is in charge of interpretation of the data. (Dobrin, 1960)

party foreman In seismic prospecting, the person who supervises the work of a field party. Subordinate to a nonresident party chief who is responsible for the interpretation of the data. (AGI, 1987)

party manager (a) In seismic prospecting, this person's function is to handle the operational phases of the work, particularly those involving logistics and access in difficult or remote areas, giving the party chief more time for interpretation of the data. (Dobrin, 1960) (b) In gravity and magnetic prospecting, the person in charge of the operations of a field party. (AGI, 1987)

Pasadenian orogeny Mid-Pleistocene diastrophism. (AGI, 1987)

Pascal's law The component of the pressure in a fluid in equilbrium that is due to forces externally applied is uniform throughout the fluid. (Webster 3rd, 1966)

pascoite A monoclinic mineral, $Ca_3V_{10}O_{28}{\cdot}17H_2O$; forms yellow-orange to dark red-orange crusts and tiny laths; a secondary vanadium mineral in uranium-vanadium deposits of the Colorado Plateau, and at Minasragra, Peru.

pass (a) An inclined opening in a mine, a raise or a winze, through which coal or ore is delivered from a higher to a lower level. At the lower end, the pass is normally provided with a chute or hydraulic gate through which the material is discharged into cars or trams. See also: *chute.* (Nelson, 1965) (b) A raise or winze for workers to travel in from one level to another. (Zern, 1928) (c) The running of a sample through a sample divider. (d) In surface mining, a complete excavator cycle in removing overburden. (BCI, 1947)

passage (a) A cavern opening or underground tunnel having greater length than height or width; large enough for human entrance and larger by comparison than a lead. (AGI, 1987) (b) *pass.*

passby (a) The double-track part of any single-track system of transport. (Mason, 1951) (b) A siding in which cars pass one another underground; a turnout. (Zern, 1928)

passing boss *motor boss.*

passing point (a) On haulage roads, the point at which the loaded trams going outby pass the empty trams going inby. (Nelson, 1965)

(b) In shafts, the point at which the loaded ascending cage or skip passes the empty descending cage or skip. (Nelson, 1965)

passing track A sidetrack with switches at both ends. (Kentucky, 1952)

passivation The changing of the chemically active surface of a metal to a much less reactive state. Cf: *activation*. (ASM, 1961)

passivator A type of inhibitor that changes the electrode potential of a metal, causing it to become more cathodic or electropositive.

passive coefficient of earth pressure The maximum ratio of the major principal stress to the minor principal stress. This is applicable where the soil has been compressed sufficiently to develop an upper limiting value of the major principal stress. (ASCE, 1958)

passive earth pressure The maximum value of lateral earth pressure exerted by soil on a structure, occurring when the soil is compressed laterally, causing its internal shearing resistance along a potential failure surface to be completely mobilized; the maximum resistance of a vertical earth face to deformation by a horizontal force. Cf: *active earth pressure*. (AGI, 1987)

passive fault Fault not liable to further movement. Cf: *active fault*. (Carson, 1965)

passive metal A metal on which an oxide film that prevents further attack on the metal is readily formed. When a metal other than a noble metal has a high resistance to corrosion, it is because of passivity; e.g., chromium, nickel aluminum, tin, and various alloys. See also: *passivity*. (CTD, 1958)

passive state of plastic equilibrium Plastic equilibrium obtained by a compression of a mass. (ASCE, 1958)

passive transducer A transducer whose output waves are independent of any sources of power controlled by the actuating waves. (Hy, 1965)

passivity (a) A metal that is normally active according to its position in the electromotive-force series is said to be passive whenever its electrochemical behavior is that of a less active metal. (b) A metal is passive when it is relatively resistant to corrosion in an environment in which a large decrease in free energy is associated with the corrosion reaction. (c) A condition in which a piece of metal, because of an impervious covering of oxide or other compound, has a potential much more positive than the metal in the active state. (ASM, 1961)

pass pipe An iron pipe connecting the water at the back of one set of tubbing with that of another, or a pipe only in communication with one tub and open to the interior of a shaft. (Fay, 1920)

paste (a) The claylike matrix of a dirty sandstone; e.g., the microcrystalline matrix of a graywacke, consisting of quartz, feldspar, clay minerals, chlorite, sericite, and biotite. (AGI, 1987) (b) The mineral substance in which other minerals are embedded; groundmass (as of a porphyry). (Webster 2nd, 1960) (c) An imitation gemstone made from a certain type of lead glass; loosely applied to all glass imitation gemstones. (Anderson, 1964) (d) A white clay body. (e) In magnetic particle suspension, finely divided ferromagnetic particles in paste form used in the wet method. (f) A slurry of sulfur and water, usually containing 30% to 50% of finely divided elemental sulfur. (g) Material of which a porcelain body is formed. Hard paste (pite dure), composed of china stone and china clay, is true porcelain. Soft paste (pite tendre), composed of glass or frit with white, is artificial porcelain. (h) Comparatively concentrated dispersion (greater than 10% by volume) of fine-solid or semisolid particles in a liquid; often shows elastic or plastic behavior. (Bennett, 1962)

paste fill (a) A class of backfills that has low water content; high densities (≥75% by lot); and consistency, transport, and deposition properties different from those of traditional low-concentration slurries or other types of high-concentration backfill. (Aref, 1992) (b) Paste fill (high pulp density) that does not settle out of suspension at zero flow density and does not produce free water when placed in a stope. Uncemented paste fill can generally be mobilized reasonably easily by pumping if left standing in a pipe for many hours.

paste pumping The transport and placement of high-concentration, low-slump material by positive displacement through pipelines by pumps, similar to those used for concrete pumping. See also: *paste fill*.

pasting The operation of mudcapping or plaster shooting whereby rock is blasted without drilling. An explosive is placed on top of the rock and covered by a cap of mud or similar material.

patch (a) A mine village, usually built and owned by a coal company. (Korson, 1938) (b) A small placer property.

patchy Distributed in an irregular manner, as when ore occurs in bunches or sporadically. (Fay, 1920)

pat coal Scot. The bottom, or lowest, coal sunk through in a shaft. (Fay, 1920)

patent A document that conveys title to the ground, and no further assessment work need be done; however, taxes must be paid. The procedure of obtaining a patent is divided into five steps: (1) a mineral surveyor is paid to make a patent survey, to adjust boundaries and correct errors, in which case an amended location should be made; (2) at least $500 worth of improvement must have been made per claim; (3) the presence of valuable mineral must be proven beyond reasonable doubt; (4) the matter is taken up with the local land office, and the proper notices must be published in the papers for a specified time; and (5) the purchase price of the land is paid and the patent is received. (Lewis, 1964)

patent ax A type of surfacing machine employed to remove irregularities from the surface of blocks of stone.

patented claim A claim to which a patent has been secured from the U.S. Government, in compliance with the laws relating to such claims. See also: *patent*.

patented rope Galvanized steel rope. (Pryor, 1963)

patent survey An accurate survey of a mineral claim by a U.S. deputy surveyor as required by law to secure a patent (title) to the claim. (Fay, 1920)

Patera process A metallurgical process consisting of a chloridizing roasting, leaching with water to remove base metals (some silver is dissolved and must be recovered), leaching with sodium hyposulfite for silver, and the precipitation of silver by sodium sulfide. The process was first carried out by von Patera at Joachimstal (Jachymov), Czech Republic. (Liddell, 1918; Fay, 1920)

paternoster pump A chain pump; named from fancied resemblance of the disks and endless chain to a rosary. (Standard, 1964)

pathfinder In geochemical exploration, a relatively mobile element or gas that occurs in close association with an element or commodity being sought, but can be more easily found because it forms a broader halo or can be detected more readily by analytical methods. A pathfinder serves to lead investigators to a deposit of a desired substance. Often called indicator element, but this latter term is restricted by some authors to elements that are important components of the ores being sought. (AGI, 1987)

path of percolation *line of creep*.

patina Strictly, the green film formed on copper and bronze after long exposure to the atmosphere. By extension, the term is applied to a film of any sort formed on wood, marble, chert, or other material after weathering or long exposure. See also: *desert varnish*. (Stokes, 1955)

patinated chert Chert nodules with weathered or case-hardened surface layers. (AGI, 1987)

patio (a) Mex. Cloth used by miners. (b) Sp. Place where minerals are concentrated. The patio floor is one on which silver and/or gold ore is amalgamated. See also: *patio process; arrastre*. (Pryor, 1963)

patio process The patio process, dating back to the 16th century, was a crude chemical method for the recovery of silver by amalgamation in low heaps with the aid of salt and copper sulfate (magistral). See also: *patio*. (Liddell, 1918)

patronite A monoclinic mineral, VS_4(?); synthetic VS_4 is soft, gray-black, fine-grained; impure material constitutes an important ore mineral in the vanadium deposit at Minasragra, Peru.

pattern (a) As applied to diamond bits, the design formed by spacing and distributing the diamonds in conformance with a predetermined geometric arrangement on the crown of a bit. See also: *concentric pattern; eccentric pattern*. (Long, 1960) (b) The system followed in spacing boreholes. See also: *checkerboarded; pattern shooting*. (Long, 1960)

pattern molder One who makes sand molds for castings; a molder. (Standard, 1964)

pattern shooting In seismic prospecting, the use of a number of energy sources arranged in a definite geometric pattern. (AGI, 1987)

Patterson agitator An agitator of the Pachuca-tank type in which air is replaced by a solution or water, under pressure from a centrifugal pump. See also: *Pachuca tank*. (Liddell, 1918)

Pattinson process A process for separating silver from lead in which the molten lead is slowly cooled, so that crystals poorer in silver solifidy out and are removed, leaving the melt richer in silver. (ASM, 1961)

Pattinson's pots A series of pots for separating silver and lead by making use of the fact that the melting point of their alloys rises as the percentage of silver increases. (Standard, 1964)

Paulin altimeter This instrument measures barometric pressure and is quite accurate for a portable instrument. It can be used for finding the difference in pressure between points at various elevations without checking the setting of the pointer, or it can be checked against a mercury barometer and then used as a portable barometer. It is useful in making a survey of the drop in ventilation pressure throughout a mine.

paulingite An isometric mineral, $(K_2,Ca,Na_2,Ba)_5[Al_{10}Si_{32}O_{84}]\cdot 34\text{-}44H_2O$; zeolite group; forms rhombic dodecahedra; at the Columbia River Rock Island Dam, Wenatchee, WA.

Pauling's rules Generalizations about coordination polyhedra and the ways they fit together in stable ionic crystal structures: (1) A coordination polyhedron of anions forms about each cation. (2) Electronic neutrality is maintained over short atomic distances. (3) Coordination polyhedra tend not to share edges or faces. (4) Highly charged cations minimize sharing of polyhedral elements.

pavement (a) A layer immediately underlying coal or any other workable material. (Arkell, 1953) (b) The floor of a mine. (c) Any construction superimposed on a subgrade to reduce loading stresses and to protect it against the abrasive effects of traffic and weather. (Nelson, 1965) (d) *base rock.* (e) A bare rock surface that suggests a paved road in smoothness, hardness, horizontally, surface extent, or close packing of its units; e.g., boulder pavement, glacial pavement, desert pavement, limestone pavement, erosion pavement. (AGI, 1987)
pavilion Any of the undersides and corners of a brilliant-cut gem; they lie between the girdle and the collet. (Hess)
paving breaker An air hammer that does not rotate its steel. (Nichols, 1976)
paving sand A type of commercial sand with applications divided into three general classes: concrete pavements, asphaltic pavements, and grouting.
pavonite A monoclinic mineral, $(Ag,Cu)(Bi,Pb)_3S_5$; at the Porvenir Mine, Bolivia. The synthetic phase, $AgBi_3S_5$, has the same X-ray pattern.
pawl A tooth or set of teeth designed to lock against a ratchet. (Nichols, 1976)
paxite A monoclinic mineral, $CuAs_2$; pseudo-orthorhombic; forms intergrowths with novakite, koutekite, and arsenic; in Bohemia, Czech Republic.
pay (a) That portion of a formation, deposit, prospect, or mine in which valuable mineral, oil, or gas is found in commercial quantity. (b) Profitable ore. See also: *pay dirt.*
pay dirt (a) Gravel. Of alluvial deposits, sand rich enough to be excavated and treated to recover its valuable contents. See also: *pay streak.* (Pryor, 1963) (b) S. Afr. The same as payable ore, but in an alluvial deposit. Also called pay rock. See also: *pay ore.* (Beerman, 1963) (c) Earth, rock, etc., that yields a profit to a miner. (Webster 3rd, 1966)
pay formation A layer or deposit of soil or rock whose value is sufficient to justify excavation. (Nichols, 1976)
pay gravel (a) Gravel containing sufficient heavy mineral to make it profitable to work. (Nelson, 1965) (b) In placer mining, a rich strip or lead of auriferous gravel. See also: *pay ore.*
pay limit S. Afr. Grade below which the mining of ore is considered to become unpayable. There has been much discussion about mining below the pay limit for technical reasons, as a result of taxation, or to conserve natural resources. (Beerman, 1963)
pay load (a) In any winding or haulage system, the pay load is the weight of coal, ore, or mineral handled as distinct from dirt, stone, or gangue. (Nelson, 1965) (b) The mineral raised up the shaft from an underground mine. (Sinclair, 1959)
pay material The mineral to be recovered. (Austin, 1964)
pay ore Rock that, at current cost of discovery, development, and exploitation, can be mined, concentrated and/or smelted profitably at the ruling market value of products. Ore below this value or cut (the threshhold value) is unpay. Syn: *pay rock.* Cf: *pay streak.* See also: *pay dirt; pay gravel.* (Pryor, 1963)
pay out To slacken or to let out rope.
pay rock *pay ore.*
pay shoot A portion of a deposit composed of pay ore; generally a dipping band within a more continuous vein. See also: *pay streak.*
pay streak (a) The area of economic concentration of gold in a placer deposit. (Bateman, 1951) (b) The part of a vein or area of a placer deposit that carries the profitable or pay ore. See also: *pay dirt; pay shoot.* Cf: *pay ore.*
peachblossom ore *erythrite.*
pea coal In anthracite only, coal small enough to pass through a mesh 3/4 to 1/2 in (1.9 to 1.3 cm) square, but too large to pass through a 3/8-in (9.5-mm) mesh. When buckwheat coal is made, the size marketed as pea is sometimes larger than the above; known also as No. 6 coal. See also: *anthracite coal sizes.* (Fay, 1920)
peacock coal Iridescent coal, the iridescence of which is due to a thin film of some substance deposited on the surface of the coal along minute cracks. (Arkell, 1953)
peacock ore Informal name for an iridescent copper mineral having a lustrous, tarnished surface exhibiting variegated colors, such as chalcopyrite and esp. bornite. Also called peacock copper. See also: *bornite.*
pea gravel Clean gravel, the particles of which are similar in size to that of peas. (AGI, 1987)
pea grit The term pea grit has been used for a coarse pisolitic limestone. Such usage should be discontinued; it is erroneous. The term grit should be reserved for a coarse-grained sandstone composed of angular particles. (Rice, 1960)
pea iron ore A variety of pisolitic limonite or "bean ore" occurring in small, rounded grains or masses. Syn: *pea ore.*
peak load Maximum permitted power draft from an electric supply main. (Pryor, 1963)
peak loading The maximum number of tons of a specified material to be carried by a conveyor per minute in a specified period of time. (NEMA, 1961)
peak particle velocity The maximum rate of change of ground displacement with time.
peak stope Flat stope advanced (overhand if deposit is inclined) in slanted steps, each flat forming a separate working place. (Pryor, 1963)
pea ore (a) Eng. Rounded grains of hydrated peroxide of iron, or silicate of iron, commonly found in cavities of Jurassic limestone. See also: *bean ore.* (b) A variety of pisolitic limonite or bean ore occurring in small, rounded grains or masses. Syn: *pea iron ore.* (AGI, 1987)
pearceite A monoclinic mineral, $Ag_{16}As_2S_{11}$, having copper as an apparent necessary minor component; forms pseudorhombohedral tabular crystals or may be massive; metallic black; brittle; in low-to moderate-temperature silver and base-metal ores.
pearl A dense spherical calcareous concretion, usually white or light-colored, consisting of occasional layers of conchiolin and predominant nacrous layers of aragonite (or rarely calcite); deposited concentrically about a foreign particle within or beneath the mantle of various marine and freshwater mollusks; occurs either free from or attached to the shell.
pearl ash Potassium carbonate, K_2CO_3; esp., an impure product obtained by partial purification of potash from wood ashes. (Webster 3rd, 1966)
pearlite The lamellar mixture of ferrite and cementite in the microstructure of slowly cooled iron-carbon base alloys occurring normally as a principal constituent of both steel and cast iron. (Webster 3rd, 1966)
pearlite iron (a) In general, pearlite iron is gray cast iron consisting of graphite in a matrix of pearlite; i.e., without free ferrite. (CTD, 1958) (b) In particular, pearlite iron is a German proprietary name denoting an iron of low silicon content, which is caused to solidify gray by the use of heated molds. (CTD, 1958)
pearl mica *margarite.*
pearl opal *cacholong.*
pearl sinter A variety of opal. Syn: *fiorite.* See also: *siliceous sinter.* (Fay, 1920)
pearl spar Dolomite occurring in rhombohedrons having a pearly luster. Syn: *ankerite; bitter spar.* See also: *dolomite.* (Fay, 1920)
pearlstone *perlite.*
pearly Applied to minerals having a luster like a pearl; e.g., talc, brucite, and stilbite. See also: *nacre.* Cf: *vitreous.*
peastone *pisolite.*
peat There are two types of peat, low moor (Flachmoor) and high moor (Hochmoor) peat. Low moor peat is the most common starting material in coal genesis. It therefore constitutes a caustobiolith of low diagenetic degree. Peat is formed in marshes and swamps from the dead, and partly decomposed remains of the marsh vegetation. Stagnant ground water is necessary for peat formation to protect the residual plant material from decay. Peat has a yellowish brown to brownish black color, is generally of the fibrous consistency, and can be either plastic or friable; in its natural state it can be cut; further, it has a very high moisture content (above 75%, generally above 90%). It can be distinguished from brown coal by the fact that the greater part of its moisture content can be squeezed out by pressure (e.g., in the hand). Peat also contains more plant material in a reasonably good state of preservation than brown coal. Individual plant elements, such as roots, stems, leaves, and seeds, can commonly be seen in it with the unaided eye. Failing that, treatment of peat with dilute alkali will make visible many of these plant tissues. Further, peat is richer in cellulose than brown coal (reaction with Fehling's solution). Unlike brown coal, peat still contains cellulose, protected by lignin or cutin, which gives a reaction with chlorzinc iodide. Correspondingly, peat shows under the microscope tissues that have not undergone either lignification, suberinization, or cutinization, this is not the case in brown coal. The reflectance of peat is low (about 0.3%). Microscopic examination is best undertaken with transmitted light. (IHCP, 1963)
peat bed An accumulation of peat. (Fay, 1920)
peat blasting A method enabling a road to be built across peat deposits. Hard filling is first dumped over the route to a height equal to the ascertained depth of the peat, into which blasting charges are inserted. By the action of blasting, the peat is displaced outwards and the hard fillings sink into place and can then be consolidated. (Hammond, 1965)
peat bog A bog containing peat; an accumulation of peat. (Webster 3rd, 1966)
peat gel *dopplerite.*
peat hag A pit or quag formed by digging out peat. (Standard, 1964)
peat machine A machine for grinding and briquetting peat. (Webster 3rd, 1966)
peatman A digger or seller of peat. (Webster 3rd, 1966)
peat moss Any moss from which peat has formed or may form. (Webster 3rd, 1966)
peat press A machine for making bricks of peat fuel. (Standard, 1964)

peat spade A spade with an L-shaped blade for cutting out peat in blocks. (Webster 3rd, 1966)
peat-to-anthracite theory The theory that there were progressive stages in the conversion of vegetable matter into the various grades of coal of the Carboniferous system. Thus, peat forms at an early stage in coal formation and lignite at an intermediate stage, and by further compression and alteration, bituminous and anthracite coals were formed. See also: *Hilt's law*. (Nelson, 1965)
pebble (a) A general term for a small, roundish, esp. waterworn stone; specif. a rock fragment larger than a granule and smaller than a cobble, having a diameter in the range of 4 to 64 mm (-2 to -6 phi units, or a size between that of a small pea and that of a tennis ball), being somewhat rounded or otherwise modified by abrasion in the course of transport. In Great Britain, the range of 10 to 50 mm has been used. The term has been used to include fragments of cobble size; it is frequently used in the plural as a syn. of "gravel". Syn: *pebblestone.* (b) Transparent and colorless quartz crystal, such as Brazilian pebble. (Webster 3rd, 1966) (c) Grinding media for ball or semi-autogeneous mills. As a rule, these are either hard-flint pebbles or hard-burned, white porcelain balls.
pebble armor A concentration of pebbles coating a desert area. The pebbles are usually the residual product of wind erosion and are closely fitted together so as to cover the surface in the manner of a mosaic. Also called desert pavement. See also: *lag gravel.* (Stokes, 1955)
pebble dike (a) A clastic dike composed largely of pebbles. (AGI, 1987) (b) A tabular body containing sedimentary fragments in an igneous matrix, as from the Tintic district in Utah; e.g., one whose fragments were broken from underlying rocks by gaseous or aqueous fluids of magmatic origin and injected upward into country rock, becoming rounded owing to the milling and/or corrosive action of the hydrothermal fluids. (AGI, 1987)
pebble gravel An unconsolidated deposit consisting mainly of pebbles. (AGI, 1987)
pebble jack Sphalerite in small crystals or pebblelike grains not attached to rock, but found in clay in wall rock cavities.
pebble mill Horizontally mounted cylindrical mill, charged with flints or selected lumps of ore or rock. Usually long and high discharge. See also: *ball mill.* (Pryor, 1963)
pebble phosphate A secondary phosphorite of either residual or transported origin, consisting of pellets, pebbles, and nodules of phosphatic material mixed with sand and clay, such as occurs in Florida; e.g., land-pebble phosphate and river-pebble phosphate. See also: *phosphorite.* (AGI, 1987)
pebble powder A gunpowder or black powder pressed and cut into large cubical grains so as to make it slow burning. (Webster 3rd, 1966)
pebblestone *pebble.*
pecopteris A fernlike tree of the coal forest, with small ovate pinnules that are attached to the pinnate axis by their whole breadth. (Nelson, 1965)
Pecos ore (a) A gossan containing lead and silver. (Fay, 1920) (b) Tasmania. A yellowish, earthy mixture of oxides of iron, lead, and antimony containing silver; mostly massicot. (Fay, 1920)
pectolite A triclinic mineral, $NaCa_2Si_3O_8(OH)$; isomorphous with serandite; forms compact masses or fibers; commonly associated with zeolites in cavities and veins in mafic rocks.
pedalfer Soil enriched in alumina and iron in regions of high temperature and humid climate that are marked by forest cover. Cf: *pedocal.*
pedestal A relatively slender neck or column of rock capped by a wider mass of rock and produced by undercutting as a result of wind abrasion (as in the Southwestern United States.) or by differential weathering. See also: *pedestal rock.* Syn: *rock pedestal.* (AGI, 1987)
pedestal boulder (a) A class of blocks perched on pedestals of limestone. (AGI, 1987) (b) Isolated masses or rock above and resting on a smaller base or pedestal.
pedestal rock An isolated, residual or erosional mass of rock supported by or balanced on a pedestal. The term is also applied to the entire feature. See also: *pedestal boulder.* (AGI, 1987)
pedestrian-controlled dumper A small dumper controlled by a person walking alongside it. Syn: *power barrow.* (Hammond, 1965)
pediment A broad, gently sloping rock-floored erosion surface or plain of low relief, typically developed by running water in an arid or semiarid region at the base of an abrupt and receding mountain front or plateau escarpment; underlain by bedrock that may be bare, but is more often partly mantled with a thin discontinuous veneer of alluvium derived from the upland masses and in transit across the surface. (AGI, 1987)
pedimentation The process of pediment formation.
pediment pass A narrow, flat, rock-floored tongue extending upslope from the main pediment and penetrating a mountain sufficiently to meet another pediment slope extending into the mountain from the other side.
pedion A crystal form consisting of a single crystal face.
pediplane Broad, rock-cut, thinly alluviated surface formed by the coalescence of adjacent pediments and desert domes.
pedis possessio The actual possession of a piece of mineral land to the extent needed to give the locator room to work and to prevent probable breaches of the peace, but not necessarily to the extent of a mining claim.
pedocal Soil enriched in calcium carbonate, accumulating in regions of low temperature, low rainfall, and prairie vegetation. Cf: *pedalfer.*
pedogenesis The formation of soil from parent material.
pedogeochemical prospecting Synonymous with geochemical soil survey. (Hawkes, 1957)
pedologic horizon *soil horizon.*
pedology The science that treats soils, their origin, character, and utilization. (AGI, 1987)
pedometer A pocket-size instrument that registers the number of steps taken by the person carrying it. (AGI, 1987)
pedosphere The part of the Earth in which soil-forming processes occur.
peeler (a) One of a set of blades that picks up and channels water moved outward by the impeller of a centrifugal pump. (Nichols, 1976) (b) An iron implement with a flattened end and ring handle, which is used by a baller in placing blooms, ingots, etc., in a reheating furnace. (Standard, 1964) (c) *calk.*
Peerless explosive High explosive; used in mines. (Bennett, 1962)
peg (a) A surveyor's mark. (b) To mark out a miner's claim at the four corners by pegs bearing the claimant's name. Sometimes used as "peg out.". (Webster 3rd, 1966; Fay, 1920)
peg adjustment The adjustment of a spirit-leveling instrument of the dumpy-level type in which the line of collimation is made parallel with the axis of the spirit level by means of two stable marks (pegs) the length of one instrument sight apart. (AGI, 1987)
pegleg An abrupt change or sharp bend in the course of a borehole. Also called dogleg. (Long, 1960)
pegmatite An exceptionally coarse-grained igneous rock, with interlocking crystals, usually found as irregular dikes, lenses, or veins, esp. at the margins of batholiths. Most grains are 1 cm or more in diameter. Although pegmatites having gross compositions similar to other rock types are known, their composition is generally that of granite; the composition may be simple or complex and may include rare minerals rich in such elements as lithium, boron, fluorine, niobium, tantalum, uranium, and rare earths. Pegmatites represent the last and most hydrous portion of a magma to crystallize and hence contain high concentrations of minerals present only in trace amounts in granitic rocks. Adj: pegmatitic. Syn: *giant granite.* Cf: *symplectite.* (AGI, 1987)
pegmatitic (a) Said of the texture of an exceptionally coarsely crystalline igneous rock. (AGI, 1987) (b) Occurring in, pertaining to, or composed of pegmatite. Syn: *pegmatoid.* (AGI, 1987)
pegmatitic stage (a) A final stage in the normal sequence of crystallization of a magma at which the residual fluid is sufficiently enriched in volatile materials to permit the formation of coarse-grained rocks (pegmatite) more or less equivalent in composition to the parent rock. Cf: *orthomagmatic stage.* (AGI, 1987) (b) The late stages of magma crystallization in S-type, 2-mica granites.
pegmatitization The process of formation of, introduction of, or replacement by pegmatite.
pegmatoid *pegmatitic.*
pegmatolite *orthoclase.*
peg point A pointed bar in a slide clamp. Used to brace a machine during work. (Nichols, 1954)
peg structure A structure characterized by tiny peg-shaped cavities, some with intricate profiles, penetrating the interior of crystals; typical of melilite. (AGI, 1987)
Pehrson-Prentice process A method of producing steel direct from ore. (Osborne, 1956)
Peirce-Smith converter A cylindrical-type converter having a basic (magnesite) lining; used for treating copper. (Newton, 1959)
Peirce-Smith process A basic converting process for copper matte in a magnesite-lined converter. The iron of the matte is fluxed by silica added before the process begins. (Liddell, 1918)
Peissenberg ram *ram scraper.*
PEL *permissible exposure limit.*
pelagic deposit or sediment Deposit found in deep water far from shore and may be predominantly either organic or inorganic in origin. Such deposits are light colored, reddish or brown, fine grained, and generally contain some skeletal remains of plankton organisms. Those that contain less than about 30% of organic remains are called red clay; those that contain more than about 30% of organic remains are known as oozes. (Hunt, 1965)
pelagochthonous A term applied to coal deposits formed from submerged forests and driftwood. (Tomkeieff, 1954)
Pelatan-Clerici process A continuous process of dissolving silver or gold in cyanide solution and simultaneously precipitating the precious metals with mercury in the same vessel, with an electrical current assisting precipitation. (Liddell, 1918)

peldon An English term for a very hard, smooth compact sandstone with conchoidal fracture that occurs in coal measures.
Pele's hair A natural spun glass formed by blowing-out during quiet fountaining of fluid lava, cascading lava falls, or turbulent flows, sometimes in association with Pele's tear pyroclast. A single strand, with a diameter of less than 1/2 mm, may be as long as 2 m. Etymol: Pele, Hawaiian goddess of fire. (AGI, 1987)
Pele's tears Small, solidified drops of volcanic glass behind which trail pendants of Pele's hair. They may be tear shaped, spherical, or nearly cylindrical. Etymol: Pele, Hawaiian goddess of fire. (AGI, 1987)
pelite (a) A sediment or sedimentary rock composed of the finest detritus (clay- or mud-size particles); e.g., a mudstone, or a calcareous sediment composed of clay and minute particles of quartz. The term is equivalent to the Latin-derived term lutite. (AGI, 1987) (b) A fine-grained sedimentary rock composed of more or less hydrated aluminum silicates with which are mingled small particles of various other minerals; an aluminous sediment. Etymol: Greek pelos, clay mud. See also: *psammite; psephite.* Also spelled pelyte. (AGI, 1987)
pelitic (a) Pertaining to or characteristic of pelite; esp. said of a sedimentary rock composed of clay, such as a pelitic tuff representing a consolidated volcanic ash consisting of clay-size particles. (AGI, 1987) (b) Said of a metamorphic rock derived from a pelite; e.g., a pelitic hornfels or a pelitic schist, derived by metamorphism of an argillaceous or a fine-grained aluminous sediment. (AGI, 1987)
pelitic gneiss A gneiss derived from the metamorphism of argillaceous sediments.
pelitic hornfels A fine-grained, nonfissile metamorphic rock derived from an argillaceous sediment. See also: *hornfels.* (AGI, 1987)
pelitic schist A schistose metamorphic rock derived from an argillaceous sediment. See also: *schist.* (AGI, 1987)
pelletizing A method in which finely divided material is rolled in a drum or on an inclined disk, so that the particles cling together and roll up into small, spherical pellets. The addition of a binder may be required to produce a pellet of acceptable mechanical strength. (Newton, 1959)
Pelletol A waterproof, free-running blasting agent. Pelletol is a high explosive, but is not considered cap sensitive and normally cannot be initiated with a cap, except under perfect confinement in small-diameter boreholes. (Du Pont, 1966)
pellet powder Black powder pressed into cylindrical pellets 2 in (5.1 cm) in length and varying from 1-1/4 to 2 in (3.2 to 5.1 cm) in diameter. Each pellet has a 3/8-in (9.5-mm) hole through its center to permit fuse insertion. (Carson, 1961)
pellet texture A concretionary texture characterized by minute pellets of colloidal or replacement origin and closely resembling oolites. (Schieferdecker, 1959)
pell-mell structure Coarse deposits of waterworn materials in which there is an absence of bedding. (AGI, 1987)
pelter A worker employed in a coal mine to take down pelt (shaly stone) from the roof of a narrow seam, to make enough height for a coal cutting machine. (CTD, 1958)
Pelton wheel An impulse water turbine with buckets bolted to its periphery, which are struck by a high velocity jet of water. This turbine is most efficient under a head of from 900 to 1,000 ft (274 to 305 m) or more. See also: *impulse turbine.* (Hammond, 1965)
pelyte *pelite.*
pena A large stone or boulder. Etymol: Spanish, "rock."
penalty (a) In connection with a contract for purchase of mineral concentrates by a custom smelter, a deduction from an agreed price for failure to reach an agreed assay value or to eliminate specified contaminants; charged at so much per unit of mineral or metal concerned. (Pryor, 1963) (b) In a construction contract, a penalty clause is one that imposes a penalty for failure to complete work to agreed time, specification, etc. (Pryor, 1963)
Penang tin Pig tin of about 99.95% purity, obtained from the Penang Mines in Malaysia. (Bennett, 1962)
Penberthy anoloader A simple powder loader with a high air velocity that is used in Canada in underground work for charging holes with a depth of up to 14 ft (4 m).
pencil-core bit The very-thick-wall, medium-round nose bit that cuts a pencil-size core. The bit is essentially a noncoring bit, and in most instances no attempt is made to recover the very-small-diameter core as a sample. Syn: *pencil-coring crown.* (Long, 1960)
pencil-coring crown *pencil-core bit.*
pencil ganister A variety of ganister characterized by fine carbonaceous streaks or markings; so called from the likeness of these to pencil lines. The carbonaceous traces are often recognizable as roots and rootlets of plants. (AGI, 1987)
penciling Reduction in the fire face area of the brick, in which slag erosion at the joints is pronounced.
pencil mark Aust. A thin bed of dark slate, about the thickness of the lead of a carpenter's pencil, that is parallel to the indicator. See also: *indicator.* (Fay, 1920)
pencil ore Hard, fibrous masses of hematite that can be split up into thin rods. (CMD, 1948)
pencil stone A compact pyrophyllite used for making slate pencils. (Webster 3rd, 1966)
pencil structure A very pronounced lineation, such as that produced by intersecting bedding and cleavage planes in slate.
pendant *roof pendant.*
pendletonite *karpatite.*
pendulum In mechanized mining, the arm that extends between the fulcrum jack and the swivel or angle trough or turn. (Jones, 1949)
pendulum buffer In Vermont, large wooden blocks covered with felt pads that are propelled back and forth by means of a crank and pitman. Used for polishing monumental stone.
pendulum mill *Griffin mill; Huntington mill.*
penecontemporaneous structure Small folds and faults that form in sediments shortly after they are deposited, in igneous rocks as they solidify, and in metamorphic rocks as they recrystallize.
peneplain A nearly horizontal surface of slight relief and very gentle slopes, formed by the subaerial degradation of the land almost to baselevel; the penultimate state of the old age of the land produced by such degradation. By extension, such a surface uplifted to form a plateau and subjected to renewed degradation and dissection. See also: *baseleveled plain.*
peneplanation The subaerial degradation of a region approx. to base level, forming a peneplain.
penetrating pulley A pulley around which a wire cable runs in cutting marble. Its thickness is less than the diameter of the wire and consequently, it can follow the wire as the latter cuts into the stone.
penetration feed *feed rate.*
penetration log The penetration speed of a drill related to the size of the hole and bit, mud pressure, speed of rotation, weight on bit, etc. From the results, which are plotted as penetration curves, the thickness of coal and dirt bands in the borehole can be determined with reasonable accuracy. (Nelson, 1965)
penetration macadam Screened gravel or crushed stone aggregate, bound by bituminous grouting, the binder being introduced after compaction of the aggregate. (Nelson, 1965)
penetration per blow The distance a drive-type soil sampler, casing, drivepipe, pile, or penetrometer is driven into the formation being tested by each blow delivered by a specific-size drivehammer allowed to fall a specific distance. (Long, 1960)
penetration rate The actual rate of penetration of drilling tools. See also: *feed rate; drilling rate.* (BS, 1963)
penetration resistance (a) The number of blows of a hammer of specified weight falling a given distance required to produce a given penetration into soil of a pile, casing, or sampling tube. Also called standard penetration resistance; proctor penetration resistance. (ASCE, 1958) (b) The unit load required to maintain constant rate of penetration into soil by a probe or instrument. (ASCE, 1958) (c) The unit load required to produce a specified penetration into soil at a specified rate by a probe or instrument. For a proctor needle, the specified penetration is 2-1/2 in (6.35 cm) and the rate is 1/2 in/s (1.27 cm/s). (ASCE, 1958)
penetration resistance curve The curve showing the relationship between the penetration resistance and the water content. Also called proctor penetration curve. (ASCE, 1958)
penetration speed The speed at which a drill can cut through rock or other material. See also: *overall drilling time.* (Nelson, 1965)
penetration test A test to determine the relative densities of noncohesive soils, sands, or silts; e.g., the standard penetration test that determines the number of blows required by a standard weight, when dropped from a standard height (30 in or 76.2 cm per blow), to drive a standard sampling spoon a standard penetration (12 in or 30.5 cm); or the dynamic penetration test, which determines the relative densities of successive layers by recording the penetration per blow or a specified number of blows. See also: *cone penetration test.* (AGI, 1987)
penetration twin A twin crystal in which two parts interpenetrate with each other and share a common volume. Syn: *interpenetration twin.* Cf: *contact twin.*
penetrometer An instrument to assess the strength of a coal seam, its relative workability, and the influence of roof pressure. See also: *sounding; coal penetrometer.* (Nelson, 1965)
pennant flag Unproductive grit and sandstone between the Lower and Upper Coal Measures, South Wales and Bristol, England, coalfield. Largely quarried for paving and building. Also called pennan grit; pennant stone. (Arkell, 1953)
pennantite A monoclinic mineral, $Mn_5Al(Si_3Al)O_{10}(OH)_8$; chlorite group; excellent cleavage with flexible laminae; commonly associated with manganese deposits.
pennine A pseudotrigonal variety of clinochlore. See also: *penninite.*
Pennine system Eng. The original and typical series of Carboniferous rocks, comprising the Upper Old Red Sandstone, the Mountain limestone, the Millstone grit, and the Coal Measures. (Standard, 1964)

penning *cribbing.*
penning gate Regulating device used to govern the draft of water from a dam; may incorporate arrangements for holding back sediment or floating detritus. (Pryor, 1963)
penninite A green crystallized chlorite from the Penninic Alps. Composition essentially the same as that of clinochlore, $(Mg,Fe^{2+})_5Al(Si_3Al)O_{10}(OH)_8$. See also: *pennine.* (Fay, 1920)
Pennsylvanian A period of the Paleozoic Era (after the Mississippian and before the Permian), thought to have covered the span of time between 320 million years and 280 million years ago; also, the corresponding system of rocks. It is named after the state of Pennsylvania in which rocks of this age are widespread and yield much coal. It is the approximate equivalent of the Upper Carboniferous of European usage. (AGI, 1987)
pennyweight One-twentieth troy ounce (1.56 g). Used in the United States and in England for the valuation of gold, silver, and jewels. Abbrev.: dwt; pwt. (Standard, 1964)
penroseite An isometric mineral, $(Ni,Co,Cu)Se_2$; pyrite group; cubic cleavage; forms radiating columnar masses; occurs in Bolivian mines near Colquechaca. Formerly called blockite.
penstock (a) A sluice or gate for restraining, deviating, or otherwise regulating the flow of water, sewage, etc.; a floodgate. (Webster 3rd, 1966) (b) The barrel of a wooden pump. (c) A closed conduit for supplying water under pressure to a water wheel or turbine. (Seelye, 1951)
pentagon A polygon having five sides. (Jones, 1949)
pentahydrite A triclinic mineral, $MgSO_4{\cdot}5H_2O$; chalcanthite group; highly soluble. Formerly called allenite.
pentahydroborite A triclinic mineral, $CaB_2O(OH)_6{\cdot}2H_2O$; colorless; forms small anhedra at a skarn in the Ural Mountains, Russia.
pentasol xanthate Collector agent use in flotation, in which the hydrocarbon group is crude and unfractionated amyl alcohol. Symbol, Z-6. (Pryor, 1963)
pentavalent (a) Having a valence of five. (Webster 3rd, 1966) (b) Having five valences. (Webster 3rd, 1966; Handbook of Chem. & Phys., 2)
Pentelic marble One of the most famous of ancient statuary marbles; from Mount Pentelicus, Greece.
penthrit *penthrite.*
penthrite Pentaerythritol tetranite. Used as an explosive. Syn: *penthrit; niperyth.* (Bennett, 1962)
pentice (a) A rock pillar left, or a heavy timber bulkhead placed, in the bottom of a two-or-more-compartment-deep shaft through which to sink it further. A small, auxilliary steam or air hoist, dumping apparatus, and pocket or bin are installed above the pentice; through an opening in it, sinking by short lifts is carried on while the shaft is in use above the pentice. Practiced in the Michigan copper country. (Hess) (b) A cover, protection, or roof over a sinking shaft. The cover contains a trapdoor through which the rope and bowk pass. See also: *Galloway stage.* (Nelson, 1965) (c) In shaft sinking, a solid rock pillar left in the bottom of the shaft for overhead protection of miners while the shaft is being extended by sinking.
pentlandite (a) An isometric mineral, $(Fe,Ni)_9S_8$; octahedral parting; metallic; pale bronze-yellow; nonmagnetic; generally associated with pyrrhotite, less commonly associated with chalcopyrite in magmatic sulfide deposits; the principal sulfide ore of nickel. (b) The mineral group argentopentlandite, cobalt pentlandite, geffroyite, manganese-shadlunite, pentlandite, and shadlunite.
pentolite A mixture of PETN and TNT used primarily for boosters and cast primers; military grade pentolite is usually 50% of each ingredient by weight; commercial pentolite often has a lower PETN content.
pentrough The trough in which the penstock of a water wheel is placed. (Fay, 1920)
peon (a) The movable vertical post of an arrastre. (b) A prop, post, or stall. (c) Mex. Helper; a common laborer.
pepper-and-salt texture Said of disseminated ores, esp. with dark grains in a light matrix. (AGI, 1987)
peptization (a) Liquefaction of a gel; deflocculation and dispersion of solids in a pulp; conversion of a substance to its colloidal state by subdivision. (Pryor, 1963) (b) A dispersion due to the addition of electrolytes or other chemical substances. (Brantly, 1961)
peptize To bring into colloidal solution; to convert into a solution. Syn: *deflocculate.*
peralkaline Said of igneous rocks in which the molecular proportion of alumina is less than that of soda and potash combined.
peraluminous Said of igneous rocks in which the molecular proportion of alumina exceeds that of soda, potash, and lime combined.
percentage extraction The proportion of a coal seam that is removed from a mine. The remainder may represent coal in pillars or coal that is too thin or inferior to mine or is lost in mining. Shallow coal mines working under townships, reservoirs, etc., may extract only about 50% of the entire seam, the remainder being left as pillars to protect the surface. Under favorable conditions, longwall conveyor mining may extract from 80% to 95% of the entire seam. With pillar methods of working, the extraction ranges from 50% to 90%, depending on local conditions. (Nelson, 1965)
percentage ore N.S.W. In most cases, understood to be the percentage of the metallic element present in the ore. (New South Wales, 1958)
percentage subsidence The measured amount of subsidence expressed as a percentage of the thickness of coal extracted. See also: *full subsidence.* (Nelson, 1965)
percentage support The percentage of the total wall area of a mine that will actually be covered by supports. (Spalding, 1949)
perch (a) Any of various units of measure for stonework, (including 24-3/4 ft^3(0.70 m^3) representing a pile 1 rod (5.0 m) long by 1 ft (0.3 m) by 1-1/2 ft (0.46 m); or 16-1/2 ft^3(0.47 m^3); or 25 ft^3 (0.71 m^3). (Webster 3rd, 1966) (b) A measure of length equal to 5-1/2 yd or 16-1/2 ft (5.0 m); a rod, or pole; also, a square rod (25.3 m^2). (Webster 2nd, 1960)
perched ground water Unconfined ground water separated from an underlying main body of ground water by an unsaturated zone. (AGI, 1987)
perched water *perched ground water.*
perched water table The water table of a body of perched ground water. *vertical sand drain.* See also: *water table.* (AGI, 1987)
percolation (a) In the leaching treatment of minerals, a process whereby a solvent flows gently upward or downward through a bed of ore-bearing material sufficiently coarse textured to permit this flow. See also: *sand leaching.* (b) Slow laminar movement of water through small openings within a porous material. Also used as a syn. of infiltration. Flow in large openings such as caves is not included. Cf: *infiltration.* (AGI, 1987)
percolation leaching The selective removal of the metal values from a mineral by causing a suitable solvent or leaching agent to seep into and through a mass or pile of material containing the desired mineral.
percolation rate The rate, expressed as either velocity or volume, at which water percolates through a porous medium. (AGI, 1987)
percussion bit A rock-drilling tool with chisellike cutting edges, which when driven by impacts against a rock surface, drills a hole by a chipping action. (Long, 1960)
percussion cap *detonator; primer.*
percussion drill (a) Drill in which the drilling bit falls with force onto rock. Also, a pneumatic drill in which a piston delivers hammer blows rapidly on the drill shank. Syn: *churn drill; cable-tool drill; percussion machine.* (Pryor, 1963) (b) Sometimes limited to large blasthole drills of the percussion type. (Nichols, 1976)
percussion drilling Act of using a percussion drill. See also: *percussion drill.*
percussion figure A pattern of radiating lines produced on a section of a crystal by a blow from a sharp point.
percussion machine *percussion drill.*
percussion powder Powder so composed as to ignite by a slight percussion; fulminating powder. (Fay, 1920)
percussion sieve An apparatus in which material is sorted according to size. It consists essentially of superimposed, oppositely inclined sieves, both mechanically agitated by vertical lever and having water sluices.
percussion system Applicable to drill machines and/or the methods used to drill boreholes by the chipping action of impacts delivered to a chisel-edged bit. See also: *churn drill; percussion drill.* (Long, 1960)
percussion table Early form of shaking table. See also: *concussion table; shaking table.* (Pryor, 1963)
percussive boring A system of boring using solid or hollow rods or ropes; may be used for exploratory drilling and for blasting purposes. See also: *boring.* (Nelson, 1965)
percussive drill A pneumatic drill that is used widely in mining for exploration and for blasting purposes. See also: *rock drill.* (Nelson, 1965)
percussive drilling (a) A method of drilling whereby repeated blows are applied by the bit, which is repositioned by intermittent rotation. (BS, 1964) (b) A form of drilling in which the rock is penetrated by the repeated impact of a reciprocating drill tool. (Fraenkel, 1953)
percussive machine Any of several types of machine, including heading machines, air picks, and the numerous types of percussive drills. (Mason, 1951)
perfect-discharge elevator In the so-called perfect-discharge elevator, there is an extra set of traction or sprocket wheels on the discharge side, so set that they bend the chains back under the head wheels. As a consequence, the discharging chute may be placed directly under the buckets. This elevator will also handle material that packs, and both types of gravity-discharge elevators may be run much slower than the centrifugal type. (Pit and Quarry, 1960)
perfect frame A structural frame that is stable under loads imposed upon it from any direction, but which would become unstable if one of its members were removed or one of its fixed ends became hinged. (Hammond, 1965)

performance curve Any curve used to show the relationship between properties of coal and results of a specific treatment. (BS, 1962)

perhydrous maceral Maceral having a high hydrogen content, such as exinite and resinite. (Tomkeieff, 1954)

peri- A prefix meaning around or beyond. (AGI, 1987)

periblain A kind of provitrain in which the cellular structure is derived from cortical material. Cf: *suberain; xylain.* (AGI, 1987)

periblinite (a) The micropetrological constituent, or marceral, of periblain. It consists of cortical tissue almost jellified in bulk, but still showing indications of cell structure under a microscope. (AGI, 1987) (b) A distinction of telinite based on botanical origin (cortical tissue). Cf: *suberinite.* (AGI, 1987)

periclase (a) An isometric mineral, MgO; cubic cleavage; colorless to yellow or brown; may be strongly colored by inclusions; occurs in high-temperature metamorphic rocks derived from dolomite. Syn: *periclasite.* (b) The mineral group bunsenite, manganosite, monteponite, periclase, and wüstite.

periclasite *periclase.*

periclinal Said of strata and structures that dip radially outward from, or inward toward, a center, to form a dome or a basin. Cf: *quaquaversal; centroclinal.* (AGI, 1987)

pericline (a) A general term for a fold in which the dip of the beds has a central orientation; beds dipping away from a center form a dome, and beds dipping toward a center form a basin. The term is generally British in usage. See also: *centrocline; dome; quaquaversal.* (AGI, 1987) (b) A variety of albite elongated in the direction of the *b*-axis and often twinned with this as the twinning axis. It occurs in veins as large milky-white opaque crystals. Pericline is probably an albitized oligoclase. (AGI, 1987)

pericline twin A twin crystal, in the monoclinic system, whose twinning axis is the orthoaxis of the crystal. (Fay, 1920)

peridot (a) A transparent to translucent green gem variety of forsterite in the olivine group. Also spelled peridote. Syn: *bastard emerald.* (b) A yellowish-green or greenish-yellow variety of tourmaline, approaching olivine in color. It is used as a semiprecious stone. See also: *olivine.*

peridote *peridot.*

peridotite A general term for a coarse-grained plutonic rock composed chiefly of olivine with or without other mafic minerals such as pyroxenes, amphiboles, or micas, and containing little or no feldspar. Accessory minerals of the spinel group are commonly present. Peridotite is commonly altered to serpentinite. (AGI, 1987)

peridot of Ceylon *Ceylonese peridot; peridot.*

perimeter blasting A method of blasting in tunnels, drifts, and raises, designed to minimize overbreak and leave clean-cut solid walls. Holes in the outside row are loaded with very light, continuous explosive charges and are fired simultaneously, so that they shear from one hole to the other. (Nelson, 1965)

perimeter of airway In mine ventilation, the linear distance in feet of the airway perimeter rubbing surface at right angles to the direction of the airstream.

perimorph A crystal of one species enclosing one of another species. (Webster 3rd, 1966)

period (a) The geochronologic unit lower in rank than era and higher than epoch, during which the rocks of the corresponding system were formed. It is the fundamental unit of the worldwide geologic time scale. (AGI, 1987) (b) A term used informally to designate a length of geologic time; e.g., glacial period. (AGI, 1987) (c) The interval of time required for the completion of a cyclic motion or recurring event, such as the time between two consecutive like phases of the tide or a current. (AGI, 1987) (d) The duration of one complete cycle of a periodic function; the reciprocal of the frequency of such a function. The independent variable is limited to time. (ASM, 1961) (e) The elements between an alkali metal and the rare gas of next highest atomic number, inclusive, occupying one (a short period) horizontal row or two (a long period) horizontal rows in the periodic table. (CTD, 1958) (f) The time required for the power level of a reactor to change by the factor 2.718, which is known as *e*. (Lyman, 1964)

periodic law The physical and chemical properties of the elements depend on the structure of their atoms and are for the most part periodic functions of the atomic number. See also: *periodic table.* (Webster 3rd, 1966)

periodic reverse Pertains to periodic change in the direction of flow of the current in electrolysis. It applies to the process and also to the machine that controls the time for both directions. Symbol, PR. (ASM, 1961)

periodic table An arrangement of elements based on the periodic law and proposed in various forms that are usually either short with only short periods (as in Mendeleev's original table) or long with long as well as short periods (as in most modern tables). See also: *periodic law.* (Webster 3rd, 1966)

peripheral fault A fault along the perimeter of a geologically elevated or depressed region. (AGI, 1987)

peripheral speed The distance a given point on the perimeter of a rotating circular object travels, expressed in feet or meters per second; sometimes incorrectly called lineal travel by some drillers. Syn: *surface speed.* (Long, 1960)

peripheral-turbine pump This pump—sometimes called a regenerative pump—is classified with centrifugal pumps, but is designed to develop several times the head obtained from a centrifugal pump having the same-diameter impeller and the same speed. The maximum head developed does not have the same relation to the impeller diameter and speed of the centrifugal pump; it involves size and spacing of the impeller vanes, fluid channels, and other factors. (Pit and Quarry, 1960)

peripheral ventilation A mine ventilation system in which the upcast shaft for taking air out of a mine is situated at the limits of the mining field or away from the downcast shaft. Also called transverse or one-way ventilation. Syn: *transverse ventilation.* (Stoces, 1954)

peritectic Said of an isothermal reversible reaction in a crystallizing melt or magma in which a liquid phase reacts with a solid phase to produce another solid phase on cooling. (ASM, 1961)

perlite (a) A siliceous volcanic glass having numerous concentric spherical cracks that give rise to an onion-skin structure. Most perlite has a higher water content than obsidian. When perlite is heated to the softening point, it expands, or pops to form a light fluffy material similar to pumice. It is used as lightweight aggregate in concrete, as insulation for liquid fuels, and in potting soils. (b) A pearly volcanic glass.

perlitic (a) Said of the texture of a glassy igneous rock that has cracked owing to contraction during cooling, the cracks forming small spheruloids. It is generally confined to natural glass, but occasionally found in quartz and other noncleavable minerals and as a relict structure in devitrified rocks. (AGI, 1987) (b) Pertaining to or characteristic of perlite. (AGI, 1987)

permafrost A permanently frozen layer of soil or subsoil, or even bedrock, which occurs to variable depths below the Earth's surface in arctic or subarctic regions. It underlies about one-fifth of the world's land area.

permafrost drilling (a) Boreholes drilled in subsoil and rocks in which the contained water is permanently frozen. (Long, 1960) (b) Holes drilled into perenially frozen ground that may be superficial unconsolidated material, bedrock, and ice. When no ice is present, it is called dry permafrost.

permalloy An iron-nickel alloy with high magnetic permeability. (Nelson, 1965)

permanent adjustment The adjustment of a surveying instrument that is made infrequently and not at each setup. See also: *temporary adjustment.* (Hammond, 1965)

permanent expansion Increase in bulk volume as a result of decrease in specific gravity.

permanent hardness Hardness of water that cannot be removed by boiling. Opposite of temporary hardness.

permanent hard water Hard water that cannot be softened by boiling; water containing magnesium sulfate or calcium sulfate. (Bennett, 1962)

permanent magnetism Magnetic property of a substance maintained without external excitation. (Pryor, 1963)

permanent monument A monument of a lasting character for marking a mining claim. It may be a mountain, hill, ridge, hogback, butte, canyon, gulch, river, stream, waterfall, cascade, lake, inlet, bay, arm of the sea, stake, post, monument of stone or boulders, shafts, drifts, tunnels, open cuts, or well-known adjoining patented claims. (Fay, 1920)

permanent pump A permanent main pump is one on which a mine depends for the final disposal of its drainage. As it is usually not moved during the life of the mine, its location, installation, and design require careful consideration. A permanent main pump may discharge on the surface, into an underground sump, or into some other part of a mine.

permanent set The amount of permanent deformation of a material that has been stressed beyond its elastic limit. (AGI, 1987)

permanent shaft support After a shaft has been sunk to a certain depth, the final or permanent lining is inserted. This may consist of: brick walling; concrete blocks shaped to the curvature of the shaft; concrete lining put in liquid form behind shuttering; brick coffering; and cast-iron tubbing. The permanent lining is generally built up in sections, during which operation the temporary lining (such as skeleton tubbing) is removed. Concrete is now widely used as a permanent shaft support. See also: *brick walling; lining; steel rectangular shaft supports.* (Nelson, 1965)

permanent way The completed assembly of rails, sleepers, fixings, and ballast forming the finished track for a railway. (Hammond, 1965)

permanganate A salt of permanganic acid of the type, MnO_4; dark purple; good oxidizing agent; often used as a disinfectant. (Enam. Dict., 1947)

permeability (a) The permeability (or perviousness) of rock is its capacity for transmitting a fluid. Degree of permeability depends upon the size and shape of the pores, the size and shape of their interconnections, and the extent of the latter. It is measured by the rate at which a fluid of standard viscosity can move a given distance through a given interval of time. The unit of permeability is the darcy. See also: *millidarcy; coefficient of permeability.* (AGI, 1987) (b) In geophysics, the ratio of the magnetic induction to the magnetic intensity in the same region. In paramagnetic matter, the permeability is nearly independent of the magnetic intensity; in a vacuum, it is strictly so. But in ferromagnetic matter, the relationship is definite only under fully specified conditions. (AGI, 1987) (c) *coefficient of permeability.* (d) In magnetism, a general term used to express various relationships between magnetic induction and magnetizing forces. These relationships are either absolute permeability, which is the quotient of a change in magnetic induction divided by the corresponding change in magnetizing force, or specific (relative) permeability, the ratio of the absolute permeability to the permeability of free space. (ASM, 1961) (e) In founding, the characteristics of molding materials which permit gases to pass through them. Permeability number is deteremined by a standard test. (ASM, 1961) (f) In powder metallurgy, a property measured as the rate of passage under specified conditions of a liquid or gas through a compact. (ASM, 1961)

permeable Pertaining to a rock or soil having a texture that permits passage of liquids or gases under the pressure ordinarily found in earth materials. Syn: *pervious.* (Stokes, 1955)

permeameter An instrument for measuring permeability. (AGI, 1987)

permineralization A process of fossilization wherein the original hard parts of an animal have additional mineral material deposited in their pore spaces. (AGI, 1987)

permissible (a) Means completely assembled and conforming in every respect with the design formally approved by the U.S. Mine Safety and Health Administration for use in gassy and dusty mines. (b) A machine or explosive is said to be permissible when it has been approved by the U.S. Mine Safety and Health Administration for use underground under prescribed conditions. All flameproof machinery is not permissible, but all permissible machinery is flameproof. (c) A low-flame explosive used in gassy and dusty coal mines. (Nichols, 1976)

permissible blasting device Any device, other than explosives, for breaking down coal that is approved by the U.S. Mine Safety and Health Administration.

permissible blasting unit An electrical device for firing blasts, approved by the U.S. Mine Safety and Health Administration.

permissible dustiness *dust-free conditions.*

permissible explosive Explosive that has been tested for safety in handling and approved for use in mines by the U.S. Mine Safety and Health Administration.

permissible exposure limit An exposure limit published and enforced by the U.S. Occupational Safety and Health Administration (OSHA) as a legal standard. (NSC, 1996)

permissible hydraulic fluid Any of several commercially available, fire-resistant fluids that are water-in-oil emulsions and can be substituted for flammable hydraulic fluids by users of large machinery, whether the equipment is operated underground or on the surface.

permissible lamp Any electric or flame safety lamp that is similar in all respects to a lamp tested and approved by the U.S. Mine Safety and Health Administration. (Hess)

permissible machine Any drill, mining machine, loading machine, conveyor, or locomotive that is similar in all respects to machines tested and approved by the U.S. Mine Safety and Health Administration for use in gassy mines.

permissible mine equipment Equipment that is formally approved by the U.S. Mine Safety and Health Administration after having passed the inspections, the explosion tests, and other requirements specified by the Administration. (All equipment so approved must carry the official approval plate required as identification for permissible equipment.)

permissible mine locomotive *electric permissible mine locomotive.*

permissible motor A motor the same in all respects as a sample motor that has passed certain tests made by the U.S. Mine Safety and Health Administration and installed and used in accordance with the conditions prescribed by the Administration. See also: *explosion-proof motor.* (Fay, 1920)

permissible velocity The highest velocity at which water may be carried safely in a canal or other conduit; the highest velocity throughout a substantial length of a conduit that will not scour. (Seelye, 1951)

permit man A member of a geophysical field party whose duty is to obtain permission from landowners for the party to work on their lands, or from public officials for the party to work along highways. (AGI, 1987)

permitted *permitted explosive; permitted light.*

permitted explosive (a) Explosive that has passed the Buxton tests and has been placed on the British list of authorized explosives, implying that they are reasonably safe to manufacture, handle, transport, and use in safety-lamp mines. Upon detonation, a permitted explosive: (1) gives off the minimum possible quantity of noxious gases, and (2) produces a flame of the lowest possible temperature and shortest possible duration, to lessen the risk of combustible gases ignition. The explosive contains cooling agents, such as sodium chloride and sodium bicarbonate. The first British list of permitted explosives was published in 1899. (Nelson, 1965) (b) A permitted explosive is one that has been approved for use in coal mines where there is any possible risk of igniting combustible gases or coal dust. In Great Britain, an explosive is approved by the Minister of Power and placed on the Permitted List after it has passed the official gallery tests prescribed for the particular class of explosives to which it belongs. These tests are carried out at the Safety in Mines Research Establishment's Testing Station at Buxton. (McAdam, 1958) (c) Permitted explosives are divided into four groups: P.1., normal permitted explosives; P.2., sheathed explosives; P.3., eq.s. explosive; P.4., permitted explosives that have passed additional and more stringent tests. (BS, 1964) (d) The term "permissible explosive" is used in the United States. See also: *permissible explosive.* (Fay, 1920)

permitted light Locked safety lamp or any other means of lighting, the use of which below ground in British coal mines is authorized by Regulations under the Act. See also: *safety-lamp mine.* (Nelson, 1965)

permitting process A process in which an applicant files forms to a regulatory agency with required narratives, maps, mine plans, etc., to ensure in advance of mining that the proposed operation will be in compliance with the applicable environmental standards. (SME, 1992)

Permocarboniferous Strata not differentiated between the Permian and Carboniferous systems, particularly in regions where there is no conspicuous stratigraphic break and fossils are transitional. (AGI, 1987)

Permotriassic Strata not differentiated between the Permian and Triassic systems, particularly in regions where the boundary occurs within a nonmarine, red beds succession. (AGI, 1987)

permutite process *base exchange.* Also spelled permutit process.

perofskite *perovskite.*

Perosa process A process by which beryllium is extracted from beryl.

perovskite (a) An orthorhombic mineral, $CaTiO_3$; may have Ca replaced by rare earths and Ti replaced by niobium and tantalum; pseudocubic; massive or in cubic crystals; yellow, brown, or grayish black; occurs in silica-deficient metamorphic and igneous environments such as skarns; also occurs in mafic and alkaline igneous rocks. Also spelled perofskite. (b) The mineral group latrappite, loparite, leushite, and perovskite.

perpend (a) A header extending through a wall so that one end appears on each side of it; a perpendstone border, bondstone, throughstone; through binder. Also called parping; perpender; perpent. (Fay, 1920) (b) A vertical joint, such as in a brick wall. (Standard, 1964)

perpendicular separation The separation of a fault as measured at right angles to the fault plane. (AGI, 1987)

perpendicular slip The component of the slip of a fault that is measured perpendicular to the trace of the fault on any intersecting surface. (AGI, 1987)

perpendicular throw The distance between the two parts of a disrupted bed, dike, vein, or of any recognizable surface, measured perpendicular to the bedding plane or surface in question. It is measured in a vertical plane at right angles to the strike of the disrupted surface. See also: *throw.*

persilicic A term proposed by Clarke (1908) to replace acidic. Syn: *silicic.* Cf: *subsilicic; mediosilicic.* (AGI, 1987)

persistent Continuous; orebodies are often persistent in depth and metal contents. (von Bernewitz, 1931)

personnel proximity survey A survey of radiation conditions at positions occupied by personnel working near apparatus emitting radioactivity. (NCB, 1964)

persorption Deep sorption of gas by liquid. (Pryor, 1963)

persuader A common term for crowbar, lever, or some such article used as a manual aid in moving heavy objects. (Crispin, 1964)

perthite A variety of alkali feldspar consisting of parallel or subparallel intergrowths in which the potassium-rich phase (commonly microcline) appears to be the host from which the sodium-rich phase (commonly albite) exsolved; such exsolved areas may be visible to the naked eye, typically forming strings, lamellae, blebs, films, or irregular veinlets; where texture is invisible to the eye but can be resolved with a microscope, it is microperthite. Cf: *antiperthite; cryptoperthite; microperthite.*

perthorite A deep-seated igneous rock consisting of alkali feldspar with less than 3% dark minerals. Feldspar, both orthoclase and albite, may be perthitically intergrown as cryptoperthite or as anorthoclase. (Hess)

pervious *permeable.*
petalite A monoclinic mineral, $LiAlSi_4O_{10}$; perfect cleavage; vitreous; resembles spodumene; a source of lithium salts; in granite pegmatites.
petaloid Resembling a flower petal in form, appearance, or texture. Applied to the structure seen in minerals that split into pieces with a smooth polished concave-convex surface that fit into one another somewhat like the petals of an unopened flower bud. (Webster 3rd, 1966; Fay, 1920)
petcock A small drain valve. (Nichols, 1976)
peter To fail gradually in size, quantity, or quality; e.g., the mine has petered out. Also called peter out. (Fay, 1920)
petering out The gradual thinning of a vein until it disappears. (Statistical Research Bureau, 1935)
Petersen grab In the Petersen (or van Veen) type of grab, two semicircular buckets of varying sizes are hinged along a central axis. The buckets are held apart for lowering from a ship to the bottom by some form of catch. On striking the bottom, this is released so that on hoisting, the buckets move around on their axis, take a bite out of the sediment, and come together to form a closed container. With this configuration, the rate at which the grab hits the bottom affects the bite, and when the ship is drifting, a poor sample may be obtained if the grab does not hit the bottom vertically. See also: *grab sampler.* (Hunt, 1965)
PETN Abbrev. for pentaerythritol tetranitrate. See also: *penthrite.* (Bennett, 1962)
petralite An explosive compounded of ammonium carbonate, nitrated wood or charcoal, and saltpeter. (Standard, 1964)
petrifaction A process of fossilization whereby organic matter is converted into a stony substance by the infiltration of water containing dissolved inorganic matter (e.g., calcium carbonate, silica), which replaces the original organic materials, sometimes retaining the structure. Syn: *petrification.* (AGI, 1987)
petrification *petrifaction.*
petrified moss *tufa.*
petrified rose An aggregate or cluster of tabular crystals of barite, forming chiefly in sandstone, enclosing sandy grains within the crystals; sand cemented by barite with the crystal form of the latter. Syn: *barite rosette.* See also: *rosette.*
petrified wood *silicified wood.*
petro- Combining form meaning stone or rock.
petrochemical Any of several materials and compounds present in, or derived from, natural gas or crude petroleum by physical refining or by chemical reaction. (Bennett, 1962)
petrochemistry (a) The study of the chemical composition of rocks. (AGI, 1987) (b) The study of the chemistry of petroleum and its products. (AGI, 1987)
petrofabrics The study of spatial relations, esp. on a microscale, of the structural-textural units that comprise a rock, including a study of the movements that produced these elements. The units may be rock fragments, mineral grains, or cleavages.
petrogenesis A branch of petrology that deals with the origin of rocks. Syn: *lithogenesis; petrogeny.* Cf: *genesis.*
petrogenic element An element that is characteristically concentrated in ordinary rock types as opposed to ore deposits. Cf: *metallogenic element.*
petrogeny *petrogenesis.*
petrographer Person who is versed in or engaged in petrography, or the study of rocks. (Fay, 1920)
petrographic Pertaining to the study of rocks. (Stokes, 1955)
petrographic microscope A microscope specially fitted with optical, esp. polarizers, and mechanical accessories for identifying and studying the properties of minerals in granular form or in thin section.
petrographic province A natural region within which some or all of the igneous rocks present certain well-marked peculiarities in their mineralogical and chemical composition, structure, texture, etc., that set them apart from rocks of other petrographic provinces. Consanguineous, comagmatic. Syn: *comagmatic region.* (Schieferdecker, 1959)
petrography A general term for the science dealing with the description and systematical classification of rocks, based on observations in the field, on hand specimens, and on thin sections. Petrography is thus wider in its scope than lithology, but more restricted than petrology, which implies interpretation as well as description. (Holmes, 1928)
petroleum coke Cokelike material found in cavities of igneous rocks intrusive into carbonaceous sediments. (Tomkeieff, 1954)
petroleum ether A mixture of hydrocarbons boiling from 40 to 60 °C; a mixture of low-boiling liquid alkanes. (Handbook of Chem. & Phys., 1)
petroleum-oil cannel coal *oil shale.*
petrology A general term for the study, by all available methods, of the natural history of rocks, including their origins, present conditions, alterations, and decay. Petrology comprises petrography on the one hand, and petrogenesis on the other, and properly considered, its subject matter includes ore deposits and mineral deposits in general, as well as rocks in the more limited sense in which that term is generally understood. (Holmes, 1928)
petrolo-shale *oil shale.*
petrophysics Study of the physical properties of rock. (AGI, 1987)
petrotectonics *structural petrology.*
petrous Said of a material that resembles stone in its hardness; e.g., petrous phosphates. Little used. (AGI, 1987)
petuntze *china stone.*
petzite An isometric mineral, Ag_3AuTe_2; metallic; steel gray to iron black; massive; sp gr, 8.7 to 9.02; in gold-bearing telluride veins; may be a significant source of gold and silver.
pH The negative logarithm (base 10) of the hydrogen-ion activity. It denotes the degree of acidity or of basicity of a solution. At 25 °C, 7 is the neutral value. Acidity increases with decreasing values below 7, and basicity increases with increasing values above 7. (ASM, 1961)
phacellite *kaliophilite.* Also spelled phacelite.
phacolith A concordant intrusive in the crest of an anticline and trough of a syncline; in cross section, it has the shape of a doubly convex lens. Adj: phacolithic. See also: *laccolith.*
phanerite An igneous rock having the grains of its essential minerals large enough to be seen macroscopically. (AGI, 1987)
phaneritic Said of the texture of an igneous rock in which the individual components are distinguishable with the unaided eye, i.e., megascopically crystalline. Also, said of a rock having such texture. Cf: *aphanitic.* Syn: *macromeritic; phanerocrystalline; phenocrystalline.* (AGI, 1987)
phanerocrystalline *phaneritic.*
Phanerozoic That part of geologic time represented by rocks in which the evidence of life is abundant, i.e. Cambrian and later time. Cf: *Cryptozoic.* (AGI, 1987)
phantom crystal A crystal or mineral aggregate within which an earlier stage of crystallization or growth is outlined by dust, tiny inclusions, or bubbles; e.g., a trigonal scalenohedron of calcite coated with hematite and overgrown with a clear calcite rhombohedron in crystallographic continuity. Syn: *ghost crystal.*
phantom horizon (a) In seismic reflection prospecting, a line drawn on seismic sections so that it is parallel to nearby dip segments thought to indicate structural attitude. It is used where actual events are not continuous enough to be used alone. (AGI, 1987) (b) Horizon on a reflection profile that is obtained by averaging the dips of the reflections within a band, thus indicating the trend of the dip, but not necessarily coinciding with an actual boundary plane. (Schieferdecker, 1959)
pharmacolite A monoclinic mineral, $CaHAsO_4{\cdot}2H_2O$; white to gray; forms silky fibers; occurs in the oxidized parts of arsenical deposits. Syn: *arsenic bloom.*
pharmacosiderite An isometric mineral, $KFe_4(AsO_4)_3(OH)_4{\cdot}6\text{-}7H_2O$; crystallizes in cubes or tetrahedra with cubic cleavage; rarely massive; occurs widely as an oxidation product of arsenical ores. Syn: *cube ore.*
phase (a) The sum of all those portions of a material system that are identical in chemical composition and physical state. (CTD, 1958) (b) A homogeneous, physically distinct portion of matter in a heterogeneous system. (AGI, 1987) (c) An interval in the development of a given process; esp. a chapter in the history of the igneous activity of a region, such as the volcanic phase and major and minor intrusive phases. (AGI, 1987) (d) A lithologic facies, esp. on a small scale, such as a minor variety within a dominant or normal facies, or a facies of short duration or local occurrence; e.g., a marine phase or a fluviatile phase. (AGI, 1987)
phase angle An angle expressing phase or phase difference. (Webster 3rd, 1966)
phase-balance relay Relay that protects an electrical system from faults occurring in any phase of a three-phase system. Quite often a fault current will not be large enough to trip the overcurrent relay, but will operate the phase-balance mechanism. (Coal Age, 1966)
phase control The process of varying the point within the cycle at which anode conduction is permitted to begin. (Coal Age, 1960)
phase converter A machine for converting an alternating current into an alternating current of a different number of phases and the same frequency. (Webster 3rd, 1966)
phase diagram A graph designed to show the boundaries of the fields of stability of the various phases of a system. The coordinates are usually two or more of the intensive variables temperature, pressure, and composition, but are not restricted to these. Syn: *equilibrium diagram.* (AGI, 1987)
phase disengagement In solvent extraction or liquid-liquid extraction procedures, allowing the mixture of aqueous liquor and organic solution phases to separate for individual recovery and further treatment.
phase disengagement rate In solvent extraction technology, the rate of disengagement of phases (aqueous and organic carrier).
phase displacement The angle by which the amount of difference of phase between two alternating-current magnitudes is expressed. (Standard, 1964)

phase equilibria The study and determination of stable phases present under various conditions of pressure, temperature, and composition according to the Gibbs phase rule; used in the study of mineral genesis. Cf: *crystallogeny.*

phase inversion In the Convertol process, replacement of the film of water covering a coal particle by a film of oil. (BS, 1962)

phasemeter A device for measuring the difference in phase of two alternating currents or electromotive forces. (Webster 3rd, 1966)

phase rule The statement that for any system in equilibrium, the number of degrees of freedom is two greater than the difference between the number of components and the number of phases. It may be symbolically stated as $F = (C-P) + 2$. See also: *mineralogical phase rule.* (AGI, 1987)

phase shifter A device employed to alter the phase of a wave. (NCB, 1964)

phase system Any portion of the universe that can be isolated completely and arbitrarily from the rest for consideration of the changes that may occur within it under varied conditions. In a closed system, energy may cross the system boundary, but matter may not. In an open system, both energy and matter may enter or leave as required. An equilibrium system is closed with all phases in their lowest energy states. The variance (degrees of freedom) of an equilibrium system is its number of components minus its number of phases plus two. A steady-state system is open with all phases in their lowest energy states while matter streams through it. Systems may be described by the number of their components; e.g., unary for one component, binary for two, ternary for three, etc. They are commonly defined in terms of their components; e.g., the system CaO-MgO-SiO_2-H_2O is a quaternary system.

phase transformation The inversion of one crystalline assemblage of components from one symmetry to another; e.g., calcite to aragonite.

phenacite *phenakite.*

phenakite A trigonal mineral, Be_2SiO_4; colorless to yellow, red, or brown; a minor gemstone sparsely found in granite pegmatites. It is sometimes confused with quartz. Not to be confused with fenaksite. Syn: *phenacite.* (AGI, 1987)

phengite (a) A variety of muscovite having high silica. (b) A transparent or translucent stone (probably crystalline gypsum) used by the ancients for windows.

phenhydrous (a) Applied to certain conditions under which coal was formed, namely those of open waters into which the plant debris was swept from the adjoining land. (Tomkeieff, 1954) (b) Refers to vegetable matter deposited under water in contrast to that laid down on a wet substratum. Cf: *crypthydrous.* (AGI, 1987)

phenocryst A term for large crystals or mineral grains floating in the matrix or groundmass of a porphyry. Syn: *inset.*

phenocrystalline *phaneritic.*

phenol A soluble, crystalline acidic compound; C_6H_5OH; has a characteristic odor. It is present in coal tar and in wood tar. It is a powerful caustic poison and in a dilute solution, a useful disinfectant. Used chiefly in making resins and plastics, dyes, and pharmaceuticals (such as aspirin). Syn: *benzenol; hydroxybenzene; carbolic acid.* (Handbook of Chem. & Phys., 2)

phi grade scale A logarithmic transformation of the Wentworth grade scale in which the negative logarithm to the base 2 of the particle diameter (in millimeters) is substituted for the diameter value (Krumbein, 1934); it has integers for the class limits, increasing from -5 for 32 mm to +10 for 1/1,024 mm. The scale was developed specif. as a statistical device to permit the direct application of conventional statistical practices to sedimentary data. See also: *Wentworth grade scale.* (AGI, 1987)

Philadelphia rod A leveling rod in which the hundredths of feet, or eighths of inches, are marked by alternate bars of color the width of the measurement. (Nichols, 1976)

phillipite A compact, blue, hydrated copper and iron sulfate, $Fe_2Cu(SO_4)_4{\cdot}12H_2O$, produced by decomposition of chalcopyrite. (Standard, 1964)

phillipsite A monoclinic mineral, $(K,Na,Ca)_{1-2}(Si,Al)_8O_{16}{\cdot}6H_2O$; zeolite group; commonly occurs in complex twinned crystals; in basalt amydules, in pelagic red clays, in palagonite tuffs, in alkaline saline lakes from silicic vitric volcanic ash, in alkaline soils, and around hot springs in Roman baths.

Phleger corer Designed to obtain cores up to about 4 ft (1.2 m) in length, the Phleger corer is utilized where only the upper layers of the sea bottom are to be analyzed. (Hunt, 1965)

phloem In coal, the outer conducting part of the central cylinder or vascular tissues. It consists primarily of sieve tubes and companion cells, phloem fibers or bark fibers, stone cells, and parenchymatous cells. (Hess)

phlogopite A monoclinic mineral, $K_2Mg_6(Si_6Al_2O_{20})(F,OH)_4$; a magnesium-rich end-member of the biotite crystal solution series; mica group; pseudohexagonal with perfect basal cleavage; occurs in crystalline limestones as a product of dedolomitization, in potassium-rich ultramafic rocks, as an alteration mineral in sulfur-rich hydrothermal assemblages, and in kimberlites. Syn: *magnesium mica; amber mica; brown mica.*

pH modifier Proper functioning of a cationic or anionic flotation reagent is dependent on the close control of pH. Modifying agents used are soda ash, sodium hydroxide, sodium silicate, sodium phosphates, lime, sulfuric acid, and hydrofluoric acid. (Fuerstenau, 1962)

pholerite A claylike mineral closely related to or identical with kaolinite. (Fay, 1920)

pholidoide The group of aluminous glauconites grading into normal (ferruginous) glauconite and occurring in sedimentary rocks. Includes skolite and bravaisite. Distinct from pholidolite of Nordenskiold. Cf: *illite.* (English, 1938)

phonolite The extrusive equivalent of nepheline syenite. The principal mineral is soda orthoclase or sanidine. Other major minerals are nepheline and aegirine diopside, usually with other feldspathoidal minerals such as sodalite or haüyne. Accessories include apatite and sphene. Phonolite is an important ore progenitor, as at Cripple Creek, CO. Syn: *clinkstone.*

phosgenite A tetragonal mineral, $4[Pb_2(CO_3)Cl_2]$; forms stubby crystals; may be massive; adamantine; sp gr, 6.13; a secondary mineral in lead deposits and from action of seawater on lead slags and artifacts; commonly associated with cerussite and anglesite. Syn: *horn lead.*

phosphalite Phosphorite that occurs as beds of small concretions resting on clay surfaces or scattered in sands and limestone. (AGI, 1987)

phosphate (a) n. Any mineral containing essential tetrahedral phosphate, $(PO_4)^{3-}$, structural entities; e.g., apatite, amblygonite, or monazite. (b) A mineral commodity supplying phosphorus, usually for agricultural or chemical purposes. The source materials for phosphate are marine phosphorite and, less commonly, guano and apatite-rich igneous rocks. (c) Adj., phosphatic. Pertaining to or containing phosphates or phosphoric acid; said esp. of a sedimentary rock containing phosphatic minerals, such as phosphatic limestone produced by secondary enrichment of phosphatic material, or a phosphatic shale representing mixtures of primary and secondary phosphate and clay minerals. Cf: *vanadate.*

phosphate lands In mining law, a leased area for phosphate lands may not exceed 2,560 acres (1,034 ha). A certain expenditure for mine development and operations is required. A royalty of not less than 2% of the gross value of the output must be paid, and an annual rental, similar to that for coal lands, is imposed. (Lewis, 1964)

phosphate of lime *apatite.*

phosphate rock Any rock that contains one or more phosphatic minerals of sufficient purity and quantity to permit its commercial use as a source of phosphatic compounds or elemental phosphorus. About 90% of the world's production is sedimentary phosphate rock, or phosphorite; the remainder is igneous rock rich in apatite. Syn: *rock phosphate.* (AGI, 1987)

phosphatic nodule Black to brown, rounded mass, variable in size from a few millimeters to 30 or more centimeters. Usually consists of coprolites, corals, shells, and bones, more or less enveloped in crusts of collophane. Found in many horizons of marine origin. Also covering the ocean floors at many locations around the world. (AGI, 1987)

phosphide A compound that is a combination of phosphorus with a metal; e.g., schreibersite, $(Fe,Ni)_3P$.

phosphochalcite *pseudomalachite.*

phosphophyllite A monoclinic mineral, $Zn_2(Fe,Mn)(PO_4)_2{\cdot}4H_2O$; forms tabular crystals with perfect cleavage; vitreous; colorless to pale blue-green; a secondary mineral from pegmatites; possibly in some oxidized base-metal deposits.

phosphor Any material that has been prepared artificially and has the property of luminescence, regardless of whether it exhibits phosphorescence. (CCD, 1961; Lee, 1961)

phosphorate (a) To combine or to impregnate with phosphorus; as phosphorated oil. Syn: *phosphorize.* (Standard, 1964) (b) To make phosphorescent. (Standard, 1964)

phosphor bronze An elastic, hard and tough alloy, composed of 80% to 95% copper, 5% to 15% tin, with phosphorus up to 2.5%. (Nelson, 1965)

phosphorescence (a) Luminescence in which the stimulated substance continues to emit light after the external stimulus has ceased; also, the light so produced. The duration of the emission is temperature-dependent, and has a characteristic rate of decay. Cf: *fluorescence; luminescence.* (Webster 3rd, 1966) (b) A misnomer for the property of emitting light without sensible heat; luminescence. Although light is produced by a biochemical reaction involving phosphorus, bioluminescence is the preferred term. (Hy, 1965)

phosphoric acid A clear, colorless, sparkling liquid or a transparent orthorhombic crystal; H_3PO_4 (orthophosphoric acid), depending on the concentration and the temperature. At ordinary atmospheric temperature (20 °C), the 50% and 75% acids are mobile liquids, the 85% acid is syrupy, and the 100% acid is in crystals; specific gravity, 1.834 (at 18 °C); melting point, 42.35 °C; boiling point, 260 °C; soluble in water and in alcohol; and very corrosive to ferrous metals and alloys. (CCD, 1961; Handbook of Chem. & Phys., 2)

phosphorite A sedimentary rock with a high enough content of phosphate minerals to be of economic interest. Most commonly it is a bedded primary or reworked secondary marine rock composed of microcrystalline carbonate fluorapatite in the form of laminae; pellets; oolites; nodules; skeletal, shell, and bone fragments; and guano. Aluminum and iron phosphate minerals (wavellite, millisite) are usually of secondary formation. See also: *brown rock; bone phosphate; pebble phosphate.* (AGI, 1987)

phosphorize *phosphorate.*

phosphorized copper A general term applied to copper deoxidized with phosphorus. The most commonly used deoxidized copper. (ASM, 1961)

phosphorochalcite *pseudomalachite.*

phosphorogen A substance that promotes phosphorescence in a mineral or other compound. (Hess)

phosphorus A nonmetallic element of the nitrogen group. Symbol, P. Never found free in nature, but is widely distributed in combination with minerals. An important source is phosphate rock, which contains the mineral apatite. Ignites spontaneously, and is very poisonous; must be kept under water. Used in safety matches, pyrotechnics, pesticides, incendiary shells, smoke bombs, tracer bullets, and fertilizers. Syn: *amorphous phosphorus.* (Handbook of Chem. & Phys., 3)

phosphorus copper Copper that contains about 15% phosphorus. Used chiefly as a deoxidizer for molten metals. (Henderson, 1953)

phosphorus steel Steel in which phosphorus is the principal hardening element. (Fay, 1920)

phosphosiderite A monoclinic mineral, $Fe^{3+}PO_4 \cdot 2H_2O$; iron may be replaced by aluminum; dimorphous with strengite; isomorphous with metavariscite; forms tabular crystals or reniform crusts; vitreous; occurs in a wide variety of settings where iron and phosphate are in proximity. Formerly called metastrengite, clinostrengite.

phosphuranylite An orthorhombic mineral, $Ca(UO_2)_3(PO_4)_2(OH)_2 \cdot 6H_2O$; radioactive; deep yellow; earthy or as crusts or tiny scales; associated with autunite and other secondary uranium minerals, esp. in pegmatites.

photicite Described as altered rhodonite; carbonated rhodonite. (Dana, 1914)

photoelasticity A property of certain transparent substances that enables the presence of strain to be detected by examination in polarized light. If models of complicated engineering structures are made of such a substance, the stress distribution in the structure may be resolved. See also: *model analysis; isochromatic lines; stress analysis.* (Hammond, 1965)

photoelectric cell Broadly, any device in which the incidence of light causes a change in the electrical state. (Nelson, 1965)

photofluorography The photography of images produced on a fluorescent screen by X-rays. Varieties include photoradiography, photoroentgenography, miniature radiography. (ASM, 1961)

photogeology The identification, recording, and study of geologic features and structures by means of photography; specif. the geologic interpretation of aerial and space photographs and images and the presentation of the information so obtained. It includes the interpretation of second-generation photographs obtained by photographing images recorded on television-type tubes (the images recording wavelengths outside the visible spectrum). (AGI, 1987)

photogeomorphology Study of earth forms as revealed by aerial photographs.

photogrammetry The art and science of obtaining reliable measurements from photographic images. Methods utilize horizontal, vertical, and oblique views, with or without the aid of the stereoscopic principle and with or without computer-based image processing and analysis. See also: *infrared photography.* (AGI, 1987)

photographic borehole survey A method of checking verticality and/or orientation of a long borehole. A compact camera inserted at a known depth takes a photograph of a magnetic needle and/or a clinometer. Instruments have been developed by Oehman, Owen, and Wright. (Pryor, 1963)

photographic interpretation *photointerpretation.*

photographic-paper recorder A small device for registering photographically the passage of flame. This must not be confused with the photographing of the flame on the manometer record. (Rice, 1913-18)

photointerpretation The extraction of information from aerial photographs and images for a particular purpose, such as mapping the geologic features of an area. Syn: *photographic interpretation.* (AGI, 1987)

photolithotroph Autotrophic microorganism that derives energy to do metabolic work by converting radiant energy into chemical energy and assimilates carbon as CO_2, HCO_3^-, or CO_3^{2-} (photosynthesis). See also: *autotroph.*

photomacrograph *macrograph.*

photomagnetic borehole surveying A method of borehole surveying, consisting essentially of a timing clock, batteries and light bulb, a floating light-transparent compass, an inclination unit, and a photographic film for recording both the position of the compass and the crosshairs of the inclinometer. The instrument is enclosed in a nonmagnetic casing. See also: *multishot gyroscopic instrument.* Syn: *single shot.* (Nelson, 1965)

photometric method A dust-sampling method in which samples of dust are collected on filter paper and then placed in a photometer. The instrument shows the intensity of a beam of light after it has passed through the paper, and the fall in intensity is a direct measure of the dust concentration. With dark dust, such as in coal mines, a rough indication of the dustiness may be obtained by comparison of the depth of tone with a graded series of samples that have been calibrated against some other instrument. There are two methods of collecting samples for photometric estimation: (1) by passing the air through a filter paper, as for gravimetric estimation; or (2) by impingement, as in the konimeter. (Spalding, 1949)

photomicrograph A photographic enlargement of a microscopic image such as a petrologic thin section; a type of micrograph. Less-preferred syn: microphotograph. (Webster 3rd, 1966)

photomultiplier A sensitive detector of light in which the initial electron current, derived from photoelectric emission, is amplified by successive stages of secondary electron emission. (NCB, 1964)

photon A discrete quantity of electromagnetic energy. Photons have momentum but no mass or electrical charge. (Lyman, 1964)

photosensitive Term applied to minerals (e.g., chlorargyrite, utenbogaardite) that are visibly injured by light.

photostat printing A method of reproducing a drawing on opaque paper by printing from a photographic negative, which enables the original drawing to be enlarged or reduced. (Hammond, 1965)

phototheodolite A ground-surveying instrument used in terrestrial photogrammetry, combining the functions of a theodolite and a camera mounted on the same tripod. (AGI, 1987)

phototropism The reversible change in color of a substance produced by the formation of an isomeric modification when exposed to radiant energy (such as light). (Webster 3rd, 1966)

phragmites peat Peat composed of reed grass and other grasses. (Tomkeieff, 1954)

phreatic Pertaining to ground water. (AGI, 1987)

phreatic explosion A volcanic eruption or explosion of steam, mud, or other material that is not incandescent; it is caused by the heating and consequent expansion of ground water due to an underlying igneous heat source. (AGI, 1987)

phreatic gas Any of the vapors and gases of atmospheric or oceanic origin which, coming into contact with ascending magma, may provide the motive force for volcanic eruptions. (CTD, 1958)

phreatic line *line of seepage.*

phreatic surface *water table.*

phreatic water A term that originally was applied only to water that occurs in the upper part of the zone of saturation under water-table conditions (syn. of unconfined ground water, or well water), but has come to be applied to all water in the zone of saturation, thus making it an exact syn. of ground water. (AGI, 1987)

phreatic zone *zone of saturation.*

pH regulator Substance used in flotation processes to control the hydrogen-ion concentration. See also: *pH.* (Hess)

phthanite Siliceous shale. The term is used esp. by European geologists. Also spelled phtanite. (AGI, 1987)

phthisis Miner's occupational disease, a form of lung consumption associated with or aggravated by work in dusty surroundings, such as badly ventilated underground workings. See also: *pneumoconiosis.* (Pryor, 1963)

phyllic alteration Hydrothermal alteration typically resulting from removal of sodium, calcium, and magnesium from calc-alkalic rocks, with pervasive replacement of silicates, muting the original rock texture. It is a common style of alteration in porphyry base-metal systems around a central zone of potassic alteration. See also: *propylitization.* (AGI, 1987)

phyllite (a) A metamorphic rock, intermediate in grade between slate and mica schist. Minute crystals of sericite and chlorite impart a silky sheen to the surfaces of cleavage (or schistosity). Phyllites commonly exhibit corrugated cleavage surfaces. Cf: *illite; phyllonite.* (AGI, 1987) (b) A general term for minerals with a layered crystal structure. (AGI, 1987) (c) A general term used by some French authors for the scaly minerals, such as micas, chlorites, clays, and vermiculites.

phyllite-mylonite *phyllonite.*

phyllitic cleavage Rock cleavage in which flakes are produced that are barely visible to the unaided eye. It is coarser than slaty cleavage and finer than schistose cleavage. (Leet, 1958)

phyllonite A rock that macroscopically resembles phyllite but that is formed by mechanical degradation (mylonization) of initially coarser rocks (e.g., graywacke, granite, or gneiss). Silky films of recrystallized mica or chlorite, smeared out along schistosity surfaces, and formation by dislocation metamorphism are characteristic. Syn: *phyllite-mylonite.* See also: *mylonite gneiss.* Cf: *phyllite.* (AGI, 1987)

phyllonitization The processes of mylonitization and recrystallization to produce a phyllonite. (AGI, 1987)

phylloretin Crystalline hydrocarbon similar to fichtelite and extracted along with fichtelite from fossil pine wood. (Tomkeieff, 1954)

phyllovitrinite Vitrain in which the plant remains are discernible under a microscope. (Tomkeieff, 1954)

phyre A suffix used in naming rocks that are porphyritic, such as vitrophyre, orthophyre, or granophyre.

physical depletion The exhaustion of a mine or a petroleum reservoir by extracting the minerals. (Williams, 1964)

physical geology A broad division of geology that concerns itself with the processes and forces involved in the inorganic evolution of the Earth and its morphology, and with its constituent minerals, rocks, magmas, and core materials. Cf: *historical geology.* See also: *geology.* (AGI, 1987)

physical mineralogy That branch of mineralogy which treats of the physical properties of minerals. Cf: *chemical mineralogy.* (Fay, 1920)

physical oceanography That marine science which treats of the Earth's water mass as a fluid and studies its physical properties of motion, density, temperature, etc. (Hy, 1965)

physical shock A state of collapse that interferes with the normal heart action, respiration, and circulation. This condition is probably due to derangement or lack of proper balance within the sympathetic nervous system and may be caused by any number of things, such as serious injury, loss of blood, severe burns, fright, and many others. It is important to look for shock when rendering first aid since it may cause death even when the injury is less serious. (Kentucky, 1952)

physics The science, or group of sciences, that treats of the phenomena associated with matter in general, esp. in its relations to energy, and of the laws governing these phenomena, excluding the special laws peculiar to living matter (biology) or to special kinds of matter (chemistry). Physics treats of the constitution and properties of matter, mechanics, acoustics, heat, optics, electricity, and magnetism. More generally, it includes all the physical sciences. (Standard, 1964)

physiographic province A region of which all parts are similar in geologic structure and which has consequently had a unified geomorphic history; a region whose pattern of relief features or landforms differs significantly from that of adjacent regions. (AGI, 1987)

phyteral The term was introduced by G. H. Cady in 1942, to designate plant forms or fossils in coal as distinguished from the material of which the fossils may be composed. Phyterals are identified in general botanical terms that are usually morphological, such as spore coat, sporangium, cuticle, resin, wax, wood substance, bark, etc. The initial composition of the phyterals differed; these or other differences produced during diagenesis may or may not be perpetuated by and during carbonification (coalification). Phyterals are recognized with increasing difficulty in high rank coals. In contrast to macerals which represent a purely petrographical concept, the concept phyteral demands strict correlation with certain organs of the initial plant material. (IHCP, 1963)

phytogenous rock Rock formed from plant remains. (Tomkeieff, 1954)

phytolith A rock formed by plant activity or composed chiefly of plant remains. The term was applied by Grabau to a large group including coal, peat, lignites, some types of reef limestones, and oolites. (AGI, 1987)

phytoplankton The plant life division of plankton, including diatoms and algae. Unattached plants that are at the mercy of the currents. (Hy, 1965)

piano wire screen A screen formed by piano wires stretched tightly, lengthwise, on a frame 2 to 3 ft (0.61 to 0.91 m) wide and 4 to 8 ft (1.2 to 2.4 m) high. The screen is set up at an angle of about 45° and crushed material is fed to it from above. The mesh size varies from about 4 to 16. Because there are no cross wires, and because the taut wires can vibrate, there is less tendency for blinding, but some elongated particles inevitably pass the screen. (Dodd, 1964)

picacho A peak or sharply pointed hill or mountain; commonly a volcanic rock. The term is used in desert regions of the Southwestern United States.

pick (a) Steel cutting point used on a coal-cutter chain. See also: *coal-cutter pick.* (Nelson, 1965) (b) A miner's steel or iron digging tool with sharp points at each end. It weighs from 3 to 6 lb (1.4 to 1.7 kg) and has a wood handle, fitted to the center or head, from 2 to 3 ft (0.6 to 0.9 m) in length. Syn: *miner's pick.* (Nelson, 1965) (c) To dress the sides of a shaft or other excavation. (d) To remove shale, dirt, etc., from coal. (e) To select good ore out of a heap. (f) In seismic prospecting, any selected event on a seismic record. (AGI, 1987)

pick-a-back conveyor A short conveyor which takes the coal from, and advances with, a face power loader or continuous miner. It delivers the coal onto a gate conveyor over which it runs on a bogey. See also: *long piggyback conveyor*. (Nelson, 1965)

pick-and-shovel miner *pick miner.*

Pickard core barrel A double-tube core barrel in X-group sizes. The distinguishing feature of the Pickard barrel is that when blocked, the inner barrel slides upward into the head, closing the water ports and stopping the flow of the circulating liquid; no additional drilling can be done without irreparably damaging the bit until the barrel is pulled and the blocked inner tube is cleared. (Long, 1960)

pick boy In bituminous coal mining, a person who carries sharpened picks or bits for coal-cutting machines to the machine operator in underground working places. Also called pick carrier. (DOT, 1949)

pick breaker A breaker developed as the mechanical equivalent of the miner's pick. In the modern type, the picks are mounted on alternating arms, the primary and secondary picks being at different spacings so that breaking is performed in two stages. The breaker and plate belt are usually supplied as a standard unit driven from a common motor. (Nelson, 1965)

pick carrier *pick boy.*

picker (a) An employee who picks or discards slate and other foreign matter from coal in an anthracite breaker or at a picking table, or one who removes high-grade ore, iron, or scrap wood from ore as it passes on a conveyor belt to crushers. (BCI, 1947; DOT, 1949) (b) A mechanical arrangement for removing slate from coal. (c) A miner's needle, used for picking out the tamping of a charge that has failed to explode. Syn: *piercer.*

pickeringite A monoclinic mineral, $MgAl_2(SO_4)_4 \cdot 22H_2O$; hallotrichite group; forms acicular crystals and tufts; astringent taste; a product of surficial acid sulfate attack on aluminous rocks in mines and arid regions. Syn: *magnesia alum.*

picket (a) A sighting hub. See also: *backsight hub; foresight hub.* (Long, 1960) (b) A short ranging rod about 6 ft (1.8 m) long. An iron rod, pointed at one end, and usually painted alternately red and white at 1-foot (30.5-cm) intervals; used by surveyors as a line of sight. See also: *range pole.* (CTD, 1958; Fay, 1920)

picking (a) Operation performed between mine and mill in which waste rock, wood, detritus, steel (tramp iron), or any specially separated mineral is removed from the run-of-mine ore material by hand sorting. Usually done during transit of material on belt conveyors, preferably after very large lumps and smalls have been screened off and the ore to be picked has been sufficiently washed to display a true surface. Also done on a picking table, a rotating circular disc around which hand sorters stand or sit to remove part of the ore fed radially from a central point. Picking can also be mechanized. (Pryor, 1963) (b) The falling of particles from a mine roof about to collapse. (c) Extracting over a prolonged period an undue proportion of the richest ore from a mine, thus lowering the average grade of the remaining ore reserves; "picking the eyes out" of a mine. (d) Rough sorting of ore. (Webster 2nd, 1960)

picking belt A continuous conveyor (e.g., in the form of a rubber belt or of a steel apron, steelplate, or link construction) on which raw coal or ore is spread so that selected ingredients may be removed manually. Syn: *picking table; picking chute; picking conveyor*. (BS, 1962)

picking chute A chute along which workers are stationed to pick slate from coal. See also: *picking belt.*

picking conveyor *picking belt.*

picking out eyes Mining in which only the high-grade spots are taken out. (Hoover, 1948)

picking table A flat, or slightly inclined, platform on which the coal or ore is run to be picked free from slate or gangue. See also: *picking belt; picking conveyor.*

pick lacing The pattern to which the picks are set in a cutter chain. In this respect, it may be a balanced or an unbalanced cutter chain. Pick lacing is important as it has a bearing on the stability of the machine, on dust formation, and even on dangerous sparking. (Nelson, 1965)

pickle (a) An acid dip used to remove oxides or other compounds from the surface of a metal by chemical action. (Lowenheim, 1962) (b) To use such an acid dip.

pickling The process of removing scale or oxide from metal objects by immersion in an acid bath to obtain a chemically clean surface prior to galvanizing or painting. (Hammond, 1965)

pick machine Coal-cutting machine which acts percussively, and cuts with a large chisel fixed at the end of a piston reciprocated by compressed air in much the same way as a rock drill is operated. (Kiser, 1929)

pick mine A mine in which coal is cut with picks. (Kiser, 1929)

pick miner In anthracite and bituminous coal mining, a person who: (1) uses hand tools to extract coal in underground workings; (2) cuts out a channel under the bottom of the working face of coal (undercutting) with a pick, working several feet back into the seam; (3) breaks down a coalface with a pick; (4) bores holes with an augerlike drill for blasting, and inserts and sets off explosives in holes to break down coal; (5) shovels coal into cars and pushes them to a haulageway. Also called hand cutter; hand miner; hand pick miner; pick-and-shovel miner. (DOT, 1949)

pick money An earlier practice whereby miners paid a blacksmith for sharpening their picks. (Nelson, 1965)
pickrose hoist A small haulage engine used for pulling light loads over short distances; used at junctions, loading points, and haulage transfer points. See also: *winch; spotting hoist.* (Nelson, 1965)
pick tongs Tongs for handling hot metal. (Webster 3rd, 1966)
pickup (a) Syn. for lift, as applied to hoisting drill rods from a borehole. (Long, 1960) (b) An angular crosscut, through which coal is hauled from one entry to another. See also: *shoo-fly.* (Hess) (c) *geophone; detector.* (d) Transfer of metal from tools to a part, or from a part to tools, during a forming operation. (ASM, 1961) (e) In Alaska, a gold nugget picked up during mining operations prior to sluicing.
pickup test A laboratory procedure used in investigating the floatability of minerals. A few grains, sized between 60 and 120 mesh, are placed, after suitable surface cleansing, under water in an observation cell which is controlled for pH, reagent concentration, temperature, and conditioning time. An air bubble is pressed down on the particles and then raised; the degree and tenacity with which they cling to it are observed. (Pryor, 1963)
pickwork Cutting coal with a pick, as in driving headings. (Fay, 1920)
picky poke bar A steel bar, usually of 7/8-in (2.22-cm) stock and about 4 ft (1.2 m) long, with each end sharpened, bent out at an angle of 45°, the bends being 3 to 6 in (7.6 to 15.2 cm) from each end. (Hess)
picotite A former name for chromian spinel, $(Mg,Fe)(Al,Cr)_2O_4$.
picral An etching reagent consisting of a 2% to 5% solution of picric acid in ethyl alcohol. It may be used for plain carbon and low-alloy steels. (Osborne, 1956)
picric acid A yellow crystalline compound, $C_6H_3N_3O_7$, obtained variously, such as by the action of nitric acid on phenol. It is used in dyeing and is an ingredient in certain explosives. Also called carbazotic acid; chrysolepic acid; trinitrophenic acid. (Standard, 1964)
picrite basalt Olivine-rich basalt, as formed by the settling of olivine in thick flows and sills. Commonly contains 50% or more olivine. (AGI, 1987)
picrochromite Magnesium chromite, $MgCr_2O_4$; melting point, 2,250 °C; sp gr, 4.41. This spinel can be synthesized by heating a mixture of the two oxides at 1,600 °C; it is formed (usually with other spinels in solid solution) in fired chrome-magnesite refractories. Picrochromite is highly refractory but when heated at 2,000 °C, the Cr_2O_3 slowly volatilizes. See also: *magnesiochromite.* (Dodd, 1964)
picrolite An asbestiform antigorite serpentine.
picromerite (a) A monoclinic mineral, $K_2Mg(SO_4)_2{\cdot}6H_2O$; forms highly soluble masses or crusts around fumaroles; also a rare, advanced desiccation constituent of marine evaporites. Formerly called schoenite. (b) A mineral group including boussingaultite, cyanochroite, mohrite, nickel-boussingaultite, and picromerite.
picture A screen to shelter workers from falling water. (Zern, 1928)
Pidgeon process A process for the production of magnesium by the reduction of magnesium oxide with ferrosilicon. (ASM, 1961)
pie A local term for an intermediate pack without supporting walls. (TIME, 1929-30)
piecemeal stoping A process by which magma eats into its roof by engulfing relatively small isolated blocks, which presumably sink to depth where they are assimilated. See also: *magmatic stoping.* (AGI, 1987)
piece weight *effective piece weight.*
piecework The performance of underground work on the basis of an agreement between a miner and the mine manager. Payment may be made by the yard of advance of a heading or tunnel or per ton or cubic yard of coal or ore removed. In ripping work, payment may be made by the yard advance of excavation to a specified width and height; strip packing may be built at a certain sum per yard advance or cubic yard of filling. See also: *contract work; yardage.* (Nelson, 1965)
piedmont Adj. Lying or formed at the base of a mountain or mountain range; e.g., a piedmont terrace or a piedmont pediment.—n. An area, plain, slope, glacier, or other feature at the base of a mountain; e.g., a foothill or a bajada. In the United States, the Piedmont is a plateau extending from New Jersey to Alabama and lying east of the Appalachian Mountains. Etymol: from Piemonte, a region of NW Italy at the foot of the Alps. (AGI, 1987)
piedmont alluvial plain *bajada.*
piedmontite *piemontite.*
piedmont plain *bajada.*
piedmont scarp A small fault scarp at the foot of a mountain range and essentially parallel to the range. (AGI, 1987)
piel An iron wedge for piercing stone. (Standard, 1964)
piemontite A monoclinic mineral, $Ca_2(Al,Mn,Fe)_3(OH)O(Si_2O_7)(SiO_4)$; epidote group; less common than epidote; occurs in a variety of environments: low-grade regional metamorphic rocks, manganese deposits, and some intermediate to silicic volcanic rocks, perhaps due to metasomatism. Syn: *manganepidote; piedmontite.* Cf: *withamite.*
pier A rectangular or sometimes circular form of column, constructed usually of concrete, hard brickwork, or masonry, and designed to support heavy concentrated loads from arches or a bridge superstructure. (Hammond, 1965)
pier cap The upper or bearing part of a bridge pier; usually made of concrete or hard stone; designed to distribute concentrated loads evenly over the area of the pier. (Hammond, 1965)
piercement Salt plug that rises and penetrates rock formations to shallow depths. (Wheeler, 1958)
piercement dome *diapir.*
piercement fold *diapir.*
piercer A blasting needle. See also: *picker.* (Fay, 1920)
piercing A prospecting method used in soft soil free from stones, in which small drivepipes are used to secure samples of underlying material or to determine the thickness of the soil. (Lewis, 1964)
pier dam Dam or jetty to influence current. Cf: *weir.* Syn: *wing dam.*
pietra della raja It. A fine-grained Permian sandstone suitable for sawing and finishing. (Hess)
piezocrystallization Crystallization of a magma under pressure, such as pressure associated with orogeny. (AGI, 1987)
piezoelectric axis One of the directions in a crystal in which either tension or compression will cause the crystal to develop piezoelectric charges. (Gaynor, 1959)
piezoelectric detector A type of detector that depends upon the piezoelectric effect by which an electric charge is produced on the faces of a properly cut crystal of certain materials, particularly quartz and Rochelle salt, when the crystal is strained. The detector is constructed from a pile of such crystals with intervening metal foil to collect the charge. An inertia mass is mounted on the top of the crystal stack that is included in an electronic circuit. (AGI, 1987)
piezoelectricity The property exhibited by some asymmetrical crystalline materials which when subjected to strain in suitable directions develop electric polarization proportional to the strain. Inverse piezoelectricity is the effect in which mechanical strain is produced in certain asymmetrical crystalline materials when subjected to an external electric field; the strain is proportional to the electric field. Quartz is an industrially important example. (Hunt, 1965)
piezometer An instrument for measuring pressure head; usually consisting of a small pipe tapped into the side of a closed or open conduit and flush with the inside; connected with a pressure gage, mercury, water column, or other device for indicating head. See also: *manometer.* (Seelye, 1951)
piezometric surface *potentiometric surface.*
pig (a) A crude casting of metal convenient for storage, transportation, or melting; esp. one of standard size and shape for marketing run directly from the smelting furnace. Cf: *ingot.* (Webster 3rd, 1966) (b) A mold or channel in a pig bed. (Webster 3rd, 1966) (c) A heavily shielded container (usually lead) used to ship or to store radioactive materials. (d) An air manifold having a number of pipes which distribute compressed air coming through a single large line. (Nichols, 1976)
pig and ore process Modification of the open-hearth process of steel manufacture with pig iron and iron ore as the charge. (Bennett, 1962)
pig and scrap process Modification of the open-hearth process of steel manufacture with pig iron and steel as the charge. (Bennett, 1962)
pig bed A series of molds for iron pigs, made in a bed of sand. Connected to each other and to the taphole of the blast furnace by channels, along which the molten metal runs. (CTD, 1958)
pig caster Person who pours molten metal into hand ladles, and from ladles into molds to form ingots. (DOT, 1949)
pigeonhole (a) A room driven directly into a coal seam from the edge of a strip pit. (b) Any small poorly equipped coal mine. (c) A hole in the shaft house floor through which the bucket or skip is raised or lowered. (Hess) (d) An opening left at the meeting of two sections of arch work, permitting the workers to close the arch and to come out. The pigeonhole itself is closed from below. (Stauffer, 1906)
pigeonhole checker An arrangement of checkerbrick such that each course of brick is laid in spaced parallel rows with the brick end to end; each alternate course above and below has its parallel rows at right angles to the intervening course. (ARI, 1949)
pigeonite A monoclinic mineral, $(Mg,Fe,Ca)(Mg,Fe)Si_2O_6$; pyroxene group; crystallographically distinct from augite; occurs only in quickly chilled lavas. Cf: *augite.*
pig foot (a) An iron clamp shaped like a pig's foot used to attach the jack to the feed chain of a continuous electric coal cutter. (Fay, 1920) (b) A pipe jack with a pig foot at one end. (Fay, 1920)
Piggot corer A device for sampling bottom sediments. A core barrel is driven into unconsolidated material by an explosive charge. (AGI, 1987)
piggyback conveyor *long piggyback conveyor.*
pig handler A laborer who removes metal pigs from molds manually and stamps heat numbers on pigs with hammer and punch. (DOT, 1949)
pig lead Commercial lead in large oblong masses or pigs.

pigment mineral A mineral having economic value as a coloring agent. The most important are the red and yellow ochers and brown sienna, which consists of iron oxides with some impurities, and the brown umbers in which manganese oxide is also present. When the iron-oxide content is high the term oxide is used in preference to ocher. (AGI, 1987; Nelson, 1965)

pig metal Metal, such as brass or copper, in its first rough casting. (Standard, 1964)

pigotite A salt of alumina and organic acid, $4Al_2O_3 \cdot C_{12}H_{10}O_8 \cdot 27H_2O$; formed on the surface of granite under the influence of wet vegetation. (Tomkeieff, 1954)

pigsticker A person delegated to the duty of punching or knocking pig iron out of chills or molds at a blast-furnace or pig-casting machine. (Fay, 1920)

pigsty Timber support used in stopes to hold up the roof, consisting of a square frame of chocked round timbers and filled with waste rock. See also: *cog.* (Pryor, 1963)

pigsty timbering Hollow pillars built up of logs laid crosswise for supporting heavy weights. See also: *cribbing.* (Zern, 1928)

pig tailer A laborer who helps a pusher to push loaded mine cars over long distances and up inclines where mechanical or mule haulage is not used. Also called helper-up. (DOT, 1949)

pike A term used in England for any summit or top of a mountain or hill, esp. one that is peaked or pointed. Also, a mountain or hill having a peaked summit. (AGI, 1987)

Pike process A method for the direct production of steel by passing reducing gases over iron oxide ore, carburizing the reduced ore, and alloying it in an electric furnace. Thus, a reducing gas, heated to 900 °C is passed over iron oxide ore to produce metallic iron and spent gas. The carburized, partially reduced metal is melted, reduced, and alloyed in the electric furnace. (Osborne, 1956)

piking *cobbing.*

pilarite An aluminous variety of chrysocolla. (Standard, 1964)

pile (a) A timber, steel, or reinforced concrete plate or post that is driven into the ground to carry a vertical load (bearing pile) or a horizontal load from earth or water pressure (sheet pile). (Nelson, 1965) (b) A spiked or sharped-edged plank, beam, or even pipe or girder that is forced forward or downward (sinking) into running ground with a view to support. (Mason, 1951) (c) A stack of ore or stones. (Gordon, 1906) (d) A prop of timber. (Gordon, 1906) (e) Long thick laths, etc., answering in shafts in loose or quick ground, the same purpose as spills in levels, piles being driven vertically.

pile dam A dam made by driving piles and filling the interstices with stones. The surfaces are usually protected with planking.

pile drawer *pile extractor.*

pile driver (a) A machine for driving down piles; usually consisting of a high frame with appliances for raising and dropping a pile hammer or for supporting and guiding a steam or air hammer. Also called pile engine. (Webster 3rd, 1966) (b) An operator of a pile driver. (Webster 3rd, 1966)

pile extractor A sheet piling extractor that works on the same principle as the piledriving hammer, except that the force of the blow is upward rather than down. (Carson, 1961)

pile group A number of piles driven or cast in situ, will sustain a much heavier load than a single pile can carry, esp. when connected by a pile cap. (Hammond, 1965)

pilehammer This may be a drophammer, a steam hammer, or a diesel hammer of which the last two are completely automatic. Steam hammers are also able to operate on compressed air. See also: *jetting.* (Hammond, 1965)

pile head The top of a precast concrete pile, protected during driving by packing under a pile helmet and sometimes by a timber dolly. The top of a timber pile is protected by a driving band. (Hammond, 1965)

pile helmet A cast-steel cap covering and protecting the head of a concrete pile during driving. See also: *drophammer.* (Hammond, 1965)

pile sinking A method of sinking a circular or rectangular shaft through 20 to 30 ft (6.1 to 9.1 m) of sand or mud at the surface. It cannot be used for greater depths as each ring of piles reduces the inside dimensions of the shaft. See also: *pile; piling.* (Nelson, 1965)

piling A structure or group of piles. See also: *cofferdam.* (AGI, 1987)

pill A loosely rolled cylinder of burlap and 1/4-in-mesh (0.6-cm-mesh) hardware cloth pushed down into a borehole ahead of a string of drill rods to the point where a large crevice or small cavity has been encountered. At this point the cylinder tends to unroll partially, forming a mat that acts as a barrier against which other hole-plugging agents may collect and help seal off the opening. (Long, 1960)

pillar (a) A column of coal or ore left to support the overlying strata or hanging wall in a mine, generally resulting in a "room and pillar" array. Pillars are normally left permanently to support the surface or to keep old workings water tight. Coal pillars, such as those in pillar-and-stall mining, are extracted at a later period. Syn: *stump.* See also: *barrier pillar; shaft pillar.* (b) A block of ore entirely surrounded by stoping, left intentionally for purposes for ground control or on account of low value. (Spalding, 1949) (c) A column of rock remaining after solution of the surrounding rock. See also: *hoodoo.* (AGI, 1987)

pillar-and-breast A system of coal mining in which the working places are rectangular rooms usually five or ten times as long as they are broad, opened on the upper side of the gangway. The breasts usually from 5 to 12 yd (4.6 to 11.0 m) wide, vary with the character of the roof. The rooms or breasts are separated by pillars of solid coal (broken by small cross headings driven for ventilation) from 5 to 10 yd (4.6 to 9.1 m) or 12 yd (11 m) wide. The pillar is really a solid wall of coal separating the working places. When the object is to obtain all the coal that can be recovered as quickly as possible, the pillars are left thin; but where this plan is likely to induce a crush or squeeze that may seriously injure the mine, larger pillars are left and after the mine has been worked out, the pillars are "robbed" by mining from them until the roof comes down and prevents further working. In the steeply inclined seams of the anthracite regions the pillar-and-breast system is employed by working the bed in "lifts". Also called pillar-and-stall; post-and-stall; bord-and-pillar. Syn: *board-and-pillar; board-and-wall.* (Fay, 1920)

pillar-and-chamber A pillar method of working often adopted in extracting a proportion of thick deposits of salt or gypsum. The method may be adopted where the value of the mineral in the pillars is less than the cost of setting artificial supports. (Nelson, 1965)

pillar-and-room A system of mining whereby solid blocks of coal are left on either side of miner's working places to support the roof until first-mining has been completed, when the pillar coal is then recovered. See also: *room-and-pillar.* (Hudson, 1932)

pillar-and-stall (a) A system of working coal and other minerals where the first stage of excavation is accomplished with the roof sustained by coal or ore. See also: *pillar-and-breast; post-and-stall.* (b) One of the earliest methods of working coal seams in Great Britain. It is employed in thick seams and where valuable surface buildings require protection from damage by subsidence. A number of narrow roadways are driven in the coal seam to a predetermined boundary. There are two sets of roadways, driven at right angles to each other, and thus the seam is divided into a large number of square or rectangular pillars. These pillars are extracted at a later period. The driving of the narrow roadways is termed working the whole while pillar working is known as working the broken. The width of the roadways and their distance apart are governed by the thickness and nature of the coal seam and the type of roof and floor. The main headings are driven forward and connected at intervals by crosscuts or stentons for ventilation and as a second exit. The bords are driven off the main headings at fixed distances apart, and are connected at intervals by walls. The width of the main headings, crosscuts, and bords varies from 3 to 5 yd (2.7 to 4.6 m). The bords are driven from 15 to 60 yd (13.7 to 54.9 m) apart. The walls are about 2 to 3 yd (1.8 to 2.7 m) wide and driven at similar or greater intervals according to the size of pillars to be formed. Modern pillar-and-stall mining is highly mechanized. See also: *crosscut; mechanized heading development; stenton.* Also called bord-and-pillar. (Nelson, 1965)

pillar boss In bituminous coal mining, a person who supervises the work of robbers in removing pillars of coal that were left to support the roof of working places during mining operations. Syn: *rib boss.* (DOT, 1949)

pillar burst Failure of remnants, promontories, as well as pillars, by crushing. (Spalding, 1949)

pillar caving Removal of ore from a series of stopes or rooms, leaving pillars between. Eventually the pillars are forced or allowed to cave under the weight of the roof.

pillar coal Coal secured in pillar robbing. (Fay, 1920)

pillar drive A wide irregular drift or entry, in firm dry ground, in which the roof is supported by pillars of natural earth or by artificial pillars of stone, without using timber. (Fay, 1920)

pillar extraction The recovery or working away of the pillars of coal that were left during the first operation of working in the pillar-and-stall method. Also called pillar mining. See also: *jenkin; jud; pillar robbing; pillar extraction.* (Nelson, 1965)

pillaring back The operation of extracting coal pillars, on the retreating system, in a pillar method of working. (Nelson, 1965)

pillar line (a) The line along which pillars are being mined. (b) Air currents which have definitely coursed through an inaccessible abandoned panel or area or which have ventilated a pillar line or a pillared area, regardless of the methane content or absence of methane in such air.

pillar man A person who builds stone packs in mine workings. See also: *pack builder.* (Fay, 1920)

pillar methods of working Methods of working coal seams, which have been given different names in different coalfields, such as stoop-and-room in Scotland; bord-and-pillar in Durham, England; and single and double stalls in South Wales. There are many modifications of pillar mining, but in general, there are two stages: (1) the driving of narrow roadways and thus forming a number of coal pillars, and (2) the extraction of the pillars—often on the retreating system. Pillar methods of mining are widely used in the United States, while the longwall method is favored in Great Britain. Pillar methods also are used for working stratified deposits of ironstone, rock salt, slate, and other layered minerals. (Nelson, 1965)

pillar mining system Any of several systems, including the room-and-pillar system, the block system, and the bord-and-pillar system. (Woodruff, 1966)

pillar recovery Mining of pillars during retreat mining to increase the overall recovery of the reserve.

pillar road (a) Roadway formed in coal pillars. (Nelson, 1965) (b) Working road or incline in pillars having a range of longwall faces on either side. (Fay, 1920)

pillar robber *robber.*

pillar robbing (a) The systematic removal of the coal pillars or ore between rooms or chambers to regulate the subsidence of the roof. Also called pillar drawing. (Fay, 1920) (b) The removal of ore pillars in sublevel stoping or slicing. See also: *sublevel stoping.* (Fay, 1920) (c) Formerly, in pillar-and-stall mining, the coal pillars left were too small, and miners were satisfied to gain some coal by robbing the pillars, usually from middle portions, the remaining coal being too dangerous to extract. (Nelson, 1965)

pillar robbing and hand filling *sublevel stoping.*

pillar split An opening or crosscut driven through a pillar in the course of extraction.

pillar strength The formula for pillar strength can be expressed as follows: $S = C (L/T)^{1/2}$ where the coefficient, C, is directly dependent upon friction, L is the least pillar width, and T is the thickness. (Coal Age, 1966)

pillar working Working coal in much the same manner as with the pillar-and-stall system.

pillow A rock texture characterized by piles of lobate, pillow-shaped masses; individual pillows range up to several meters across; typical of basalt that has erupted under an appreciable depth of water.

pillow block A metal-cased rubber block that allows limited motion to a support or thrust member. (Nichols, 1976)

pillow lava A general term for lava that exhibits pillow structure, mostly basalts and andesites that erupted and flowed under water. The ocean floor sodium-rich basalts known as spilites are commonly pillowed.

pillow structure A structure, observed in certain extrusive igneous rocks, that is characterized by discontinuous pillow-shaped masses ranging in size from a few centimeters to a meter or more in greatest dimension (commonly between 30 cm and 60 cm). The pillows are close-fitting, the concavities of one matching the convexities of another. The spaces between the pillows are few and are filled either with material of the same composition as the pillows, with clastic sediments, or with scoriaceous material. Grain sizes within the pillows tend to decrease toward the exterior. Pillow structures are considered to be the result of subaqueous extrusion, as evidenced by their association with sedimentary deposits, usually of deep-sea origin. See also: *pillow lava; mammillary structure.* (AGI, 1987)

pilot (a) A cylindrical steel bar extending through and about 8 in (20 cm) beyond the face of a reaming bit. It acts as a guide that follows the original unreamed part of the borehole and hence forces the reaming bit to follow and be concentric with the smaller-diameter, unreamed portion of the original borehole. Syn: *reaming pilot.* See also: *plain pilot.* (Long, 1960) (b) A cylindrical diamond-set plug, of somewhat smaller diameter than the bit proper, set in the center and projecting beyond the main face of a noncoring bit. See also: *pilot bit; stinger.* (Long, 1960)

pilotaxitic Said of the texture of the groundmass of a holocrystalline igneous rock in which lath-shaped microlites (typically plagioclase) are arranged in a glass-free mesostasis and are generally interwoven in irregular unoriented fashion. Cf: *trachytic.* Syn: *felty.* (AGI, 1987)

pilot bit A noncoring bit with a cylindrical diamond-set plug of somewhat smaller diameter than the bit proper set in the center and projecting beyond the main face of the bit. See also: *plug bit; drag bit.* (Long, 1960)

pilot bob The weight attached to a shaft plumbline for the purpose of lowering the line down the shaft. (BS, 1963)

pilot burner A small burner kept lighted to rekindle the principal burner when desired (as in a flash boiler). The light so maintained is called a pilot light or pilot flame. (Webster 3rd, 1966; Fay, 1920)

pilot drill A small drill used to start a hole in order to insure a larger drill running true to center. (Crispin, 1964)

pilot hole (a) A small hole drilled ahead of a full-sized, or larger borehole. (Long, 1960) (b) A borehole drilled in advance of mine workings to locate water-bearing fissures or formations. (Long, 1960) (c) A small tunnel driven ahead of, and subsequently enlarged to the diameter required in the following full-size tunnel. (Long, 1960)

pilot-hole cover *cover.*

pilot lamp A small electric bulb that lights when power is turned on in a circuit. (Hammond, 1965)

pilot method The method of excavating a tunnel by driving a small tunnel ahead, and then enlarging its dimensions.

pilot plant A small-scale processing plant in which representative tonnages of ore can be tested under conditions which foreshadow (or imitate) those of the full-scale operation proposed for a given ore. (Pryor, 1965)

pilot reamer An assemblage of a pilot, a pilot reaming bit, and a reaming barrel. See also: *pilot; pilot reaming bit.* (Long, 1960)

pilot reaming bit A box-threaded, diamond-set, annular-shaped bit designed to be coupled to a pilot and used to ream a borehole to a specific casing size. See also: *pilot.* (Long, 1960)

pilot sampling The taking of preliminary samples of a mineral deposit to study its mode of occurrence and its detailed structure. Syn: *reconnaissance sampling.* (Nelson, 1965)

pilot sequence Sequence control by means of a pilot cable is effected by means of a low-voltage supply from one contactor panel to the next, or by means of a line voltage pilot cable. Each contactor has an auxiliary contact that controls the supply to the next contactor. In the low-voltage system, the secondary of each potential transformer is earthed at the preceding panel through an auxiliary switch which closes with the contactor. Until these secondary potential transformer circuits are completed, by closing the auxiliary contact, the next conveyor cannot start. See also: *sequence starting.* (Sinclair, 1959)

pilot shaft *pilot tunnel.*

pilot tunnel A small tunnel or shaft excavated in the center, and in advance of the main drivage, to gain information about the ground and create a free face, thus simplifying the blasting operations. (Nelson, 1965)

pilot valve (a) A small balanced valve, operated by a governor or by hand, which controls a supply of oil under pressure to the piston of a servometer or relay connected to a large control valve, which it is desired to operate. Also called relay valve. (CTD, 1958) (b) In a compressor, an automatic valve that regulates air pressure. (Nichols, 1976)

pilot wedge A half-cylinder member, about 5 in (12.7 cm) long, coupled to the lower end of a Hall-Rowe deflection wedge, by means of which the deflection wedge may be oriented in a specific manner in reference to a matching half-cylinder surface on the upper end of the wedge (drive wedge). This is driven into the wooden plug placed about 8 ft (2.4 m) below the point in a borehole where a deflection is to be made. Also called wedge pilot. (Long, 1960)

Pilz furnace A circular or octagonal shaft furnace, maintaining or increasing its diameter toward the top, and having several tuyeres; used in smelting lead ores. (Fay, 1920)

pimelite A massive or earthy, apple-green, nickel-bearing phyllosilicate; probably willemseite or kerolite having disordered stacking; $(Ni,Mg)_6Si_8O_{20}(OH)_4$.

pimple metal Crude copper matte of about 78% copper, obtained from the smelting of sulfide copper ores. (Bennett, 1962)

pin *wedge rock.*

pinacoid An open crystal form consisting of two parallel faces. (AGI, 1987)

pinch (a) A marked thinning or squeezing of a rock layer; e.g., a coming-together of the walls of a vein, or of the roof and floor of a coal seam, so that the ore or coal is more or less completely displaced. See also: *nip.* Cf: *make; want.* (Standard, 1964) (b) A thin place or a narrow part of, an orebody; the part of a mineral zone that almost disappears before it widens out in another place to form an extensive orebody. (AGI, 1987) (c) The binding action caused when drillhole walls close in before casing is emplaced, resulting from failure of soft or plastic formations. (Long, 1960)

pinchbar A kind of crowbar with a short projection and a heel or fulcrum at the end; a pinch. Used to pry forward heavy objects. (Standard, 1964)

pinched Where a vein narrows, as if the walls had been squeezed in. Where the walls meet, the vein is said to be pinched out. See also: *pinching out; pinch.*

pinching out Where a lode or stratum narrows down and disappears. See also: *pinch.* (BCI, 1947)

pinch out To taper or narrow progressively to extinction; to thin out. (AGI, 1987)

pinder concentrator A revolving table on which are tapering spiral copper cleats on a linoleum cover. The tailings are washed over the riffles and off the edge, while the concentrates are delivered at the end of the riffles. (Liddell, 1918)

pineapple (a) A cast roller, designed to keep the haulage rope centered between rail tracks. Spiral grooves on the sides return a straying rope to the central grooves. Works in one direction only. (Pryor, 1963) (b) *line oiler.*

pine tar Very viscous; dark brown to black; liquid or semisolid; strong characteristic odor; sharp taste; translucent in thin layers; hardens with aging; sp gr, 1.03 to 1.07; boiling point, ranges from 240 to 400 °C; soluble in alcohol, ether, chloroform, acetone, glacial acetic acid, fixed and volatile oils, and sodium hydroxide; and insoluble in water. Chief constituents are complex phenols; also present are turpentine, rosin, toluene, xylene, and other hydrocarbons. Used in flotation. (CCD, 1961)

ping An acoustic pulse signal projected by an echo-ranging transducer. (Hy, 1965)

pinguite A former name for nontronite. See also: *nontronite.*

pinion Smaller of a pair of toothed wheels, e.g., the pinion geared to the driven crown wheel of a ball mill. (Pryor, 1963)

pinion gear A drive gear that is smaller than the gear it turns. (Nichols, 1976)

pinite A compact, fine-grained, generally impure mica near muscovite in composition; dull-gray, green, or brown; derived from the alteration of other minerals, esp. cordierite, nepheline, scapolite, spodumene, and feldspar.

pinnacle (a) Any high tower or spire-shaped pillar of rock, alone or cresting a summit. A tall, slender, pointed mass; esp., a lofty peak. (AGI, 1987) (b) A sharp pyramid or cone-shaped rock under water or showing above it. (AGI, 1987) (c) In alluvial mining, a spine or pillar in limestone bedrock of an irregular and serrated type, in which it is difficult for dredge buckets to work. (Pryor, 1963)

pinned coupling A drill-rod coupling that has been permanently attached to the body of the rod by a metal dowel (or pin) driven into a small hole drilled at the point in the rod where the coupling is screwed into the body of the rod. (Long, 1960)

pinnel (a) Boulder clay, from Cumberland, Northumberland, and Lancashire, England, and North Wales. (Arkell, 1953) (b) Coarse gravel or sandstone conglomerate. (Arkell, 1953)

pin puller A laborer who removes studs from aluminum reduction pots by operating a motor-driven hydraulic jack. (DOT, 1949)

pintadoite (a) A vanadium ore. (Osborne, 1956) (b) A mineral, $Ca_2V_2O_7 \cdot 9H_2O$; green; forms water-soluble efflorescences; associated with uranium-vanadium deposits of the Colorado Plateau.

pin thread A thread on the outside surface of a cylindrical or tubular member. Syn: *male thread.* (Long, 1960)

pin timbering A roof support method following two basic principles: (1) that of drilling holes vertically or at an angle into the roof and anchoring roof bolts into a strong firm structure above the lower weak layers, thereby suspending the weak roof on bolts from the strong roof above; and (2) the binding of several layers of weak strata into a beam strong enough to support its weight across the working place. The advantage of pin timbering is that support can be provided at the face without posts being in the way of equipment and more freedom is provided for shuttle cars and other equipment in tramming. See also: *timbering; roof bolting.* (Kentucky, 1952)

pintle A vertical pin fastened at the bottom that serves as a center of rotation. (Nichols, 1976)

pintle hook A towing device consisting of a fixed lower jaw, a hinged and lockable upper jaw, and a socket between them to hold a tow ring. (Nichols, 1976)

pin-to-box The currently accepted term for a coupling, one end of which is threaded on the outside (pin) and the opposite end threaded on the inside (box). Formerly designated as a male-to-female coupling. (Long, 1960)

pin-to-pin The currently accepted term for a coupling, both ends of which are threaded on the outside. Formerly designated as a male-to-male coupling. (Long, 1960)

pin-type slat conveyor Two or more endless chains to which crossbars are attached at spaced intervals, each having affixed to it a series of pointed rods extending in a vertical plane on which work is carried. Used principally in spraying or washing operations where the least amount of area of the product is contacted.

pion An elementary particle; the contraction of pi-meson. The mass of a charged pion is about 273 times that of an electron. An electrically neutral pion has a mass 264 times that of an electron. (Lyman, 1964)

pioneer bench The first bench in a quarry which is blasted out. It is usually at the top of the rock to be quarried.

pioneer road A primitive, temporary road built along the route of a job, to provide means for moving equipment and workers. (Nichols, 1976)

pioneer wave U.K. The advance vibration set up by a coal dust explosion. See also: *advance wave; shock wave.* (Nelson, 1965)

piotine *saponite.*

pipe (a) A cylindrical, more or less vertical orebody. Syn: *chimney; ore pipe; shoot; stock.* (AGI, 1987) (b) A vertical conduit through the Earth's crust; e.g., a kimberlite pipe of South Africa, through which magmatic materials have passed. It is usually filled with volcanic breccia and fragments of older rock. As a zone of high permeability, it is commonly mineralized. (AGI, 1987) (c) A tubular cavity from several centimeters to a few meters in depth, formed esp. in calcareous rocks, and often filled with sand and gravel; e.g., a vertical joint or sinkhole in chalk, enlarged by solution of the carbonate material and filled with clastic material. See also: *piping.* (AGI, 1987) (d) The name given to the fossil trunks of trees found in coalbeds. (Fay, 1920)

pipe bit A bit designed for attachment to standard coupled pipe for use in securing the pipe in bedrock. Can be set with diamonds or other abrasive materials. (Long, 1960)

pipe clamp (a) A device similar to a casing clamp, used in the same manner on pipe as a casing clamp is used on casing. See also: *casing clamp.* (Long, 1960) (b) A pipe wrench constructed like a parmalee wrench. (Long, 1960)

pipe clay (a) Originally a clay suitable for making tobacco pipes, but the term is now used to include any white-burning plastic clay. (Nelson, 1965) (b) A mass of fine clay, generally of lenticular form, found embedded in or below a placer gravel bank.

pipe coil A device which measures only the density of the magnetic components of a slurry. This electromagnetic sensing unit is mounted on a section of rubber or stainless-steel pipe which is installed as a section of the slurry-carrying pipeline. All components are exterior to the pipe, and there is no obstruction to flow. The pipe coil is used widely in magnetic taconite and heavy-media plants. By combining this device with other instruments, it is possible to continuously measure the ore-to-media ratio. (Nelson, 1965)

pipe coupling An internally threaded, short, sleevelike member of ordinary steel used to join lengths of pipe. Sometimes incorrectly called pipe collar; pipe sleeve. (Long, 1960)

pipe cutter A tool for cutting wrought iron or steel pipes. The curved end which partly encircles the pipe carries one or more cutting disks. (Crispin, 1964)

piped air Air conducted to workings or a tunnel face through air pipes. See also: *auxiliary ventilation.* (Nelson, 1965)

pipe drivehead (a) A drivehead that is coupled to a pipe. See also: *drivehead.* Syn: *drive collar.* (Long, 1960) (b) Extra thick walled pipe or casing coupling against which the blow of a drive block is delivered when driving or sinking drivepipe or casing. (Long, 1960) (c) An oversize rod or casing coupling on which the blows of a drive block are delivered when casing is being driven or an attempt is being made to jar loose stuck casing or a drill-rod string. (Long, 1960) (d) Incorrectly used as a synonym for drive shoe; drive hammer. (Long, 1960)

pipe elevator A device similar to a casing elevator, used to raise and lower outside-coupled pipe in a borehole. (Long, 1960)

pipe factor (a) Correction made when drilling running ground, alluvial gravels, and sands. The volume actually extracted over a measured depth is compared with that which would be obtained over the true drill pipe area and distance, any discrepancy being due to inrush of sands or forcing out of sand by the pumping action during drilling. (Pryor, 1963) (b) The assumed cross-sectional area of a length of borehole when estimating the in situ volume of a core sample. Also called pipe constant. (Nelson, 1965)

pipe fitting A general term referring to any of the ells, tees, various branch connectors, etc., used in connecting pipes. (Crispin, 1964)

pipe friction The drag created on the outside of a pipe being driven into overburden material, which presses and rubs against the outside surface of the pipe and its couplings. See also: *skin friction.* (Long, 1960)

pipe grab A clutch for catching and raising a well pipe. (Standard, 1964)

pipe jack An iron pipe with a clamp or pigfoot on one end and a curved point on the other. It is wedged between the floor and roof of a mine room to hold the feed chain of a continuous electric coal mining machine. (Fay, 1920)

pipeline transport Long distance pipeline used for hydraulic transport of coal, gilsonite, copper concentrates and similar materials. See also: *hydraulic transport.* (Lewis, 1964)

pipeman (a) A person engaged in laying or repairing pipelines. Also called pipefitter. (BS, 1963) (b) Mine worker who repairs, lengthens, and maintains the pipelines for air and water in mines. (Stoces, 1954) (c) A worker in charge of a pipe, esp. in hydraulic mining. (Webster 2nd, 1960)

pipe prover An apparatus for testing the tightness of a pipeline or system, usually by hydraulic pressure. (Standard, 1964)

piper Sometimes applied to a blower of gas in coal mines. (Nelson, 1965)

pipe sampler A device for sampling a pile of ore, consisting simply of a small iron pipe that is driven into the pile and which, when withdrawn, brings a core of ore with it.

pipe sampling Sampling by means of a drivepipe in accumulations of crushed residues or of material where the larger pieces are not usually greater than 2 in (5.1 cm). The advancing end of the pipe is generally sharpened to provide a cutting edge, and sometimes contracted in diameter so that material once entered will not readily fall out when the pipe is lifted. Also called gun sampling. See also: *drivepipe.* (Truscott, 1962)

pipestone *catlinite.*

pipette analysis The size analysis of fine-grained sediment made by removing samples from a suspension with a pipette. (AGI, 1987)

pipette method A method for the determination of particle size. See also: *Andreasen pipette.* (Dodd, 1964)

piping (a) In hydraulic mining, discharging water from nozzles at auriferous gravel. (b) The act or process of driving standpipe, drivepipe, or casing into and through overburden. (Long, 1960) (c) Erosion by percolating water in a layer of subsoil, resulting in caving and in the formation of narrow conduits, tunnels, or pipes through which soluble or granular soil material is removed; esp. the movement of material, from the permeable foundation of a dam or levee, by the flow or seepage of water along underground passages. See also: *water creep.* (AGI, 1987) (d) The flow of water under or around a structure built on permeable foundations that will remove material from beneath the structure. (Nelson, 1965) (e) The tubular depression caused by contraction during cooling, on the top of iron and steel ingots. See also: *pipe.*

piracy Stream piracy.

pirssonite An orthorhombic mineral, $Na_2Ca(CO_3)_2 \cdot 2H_2O$; forms colorless to white short prisms or tablets; in nonmarine evaporites, particularly the Green River oil shales in Wyoming, and Borax Lake, CA.

pisanite A blue to green cuproan melanterite $(Fe,Cu)SO_4 \cdot 7H_2O$.

pisolite (a) A sedimentary rock, usually a limestone, made up chiefly of pisoliths cemented together; a coarse-grained oolite. Syn: *peastone.* (AGI, 1987) (b) A term often used for a pisolith, or one of the spherical particles of a pisolite.—Etymol: Greek pisos, pea. Cf: *oolite.* (AGI, 1987) (c) An individual unit in a mass of accretionary lapilli. (AGI, 1987)

pisolith One of the small, round or ellipsoidal accretionary bodies in a sedimentary rock, resembling a pea in size and shape, and constituting one of the grains that make up a pisolite. It is often formed of calcium carbonate, and some are thought to have been produced by a biochemical algal-encrustation process. A pisolith is larger and less regular in form than an oolith, although it has the same concentric and radial internal structure. The term is sometimes used to refer to the rock made up of pisoliths. Cf: *oolith.* (AGI, 1987)

pisolitic tuff An indurated pyroclastic deposit composed chiefly of accretionary lapilli or pisolites. (AGI, 1987)

pistacite A pistachio-green ferric-iron-rich variety of epidote. Also spelled pistazite.

pistol pipe In metalworking, the tuyere of a hot-blast furnace. (Fay, 1920)

piston The working part of a pump, hydraulic cylinder, or engine that moves back and forth in the cylinder; it is generally equipped with one or several rings or cups to control the passage of fluid. It ejects the fluid from the cylinder, as in a pump, or receives force from the fluid, which causes a reciprocating motion, as in an engine. (Long, 1960)

piston corer An oceanographic corer containing a piston inside the cylinder which reduces friction by creating suction. There are several varieties, including the Ewing corer, the Mackereth sampler, and the Kullenberg corer. Cf: *gravity corer.* (AGI, 1987)

piston drive-sampler *piston sampler.*

piston sampler A drive sampler equipped with either a free or a retractable-type piston that retreats up into the barrel of the sampler in contact with the top of the soil sample as the sampler is pressed into the formation being sampled. Cf: *drive sampler.* (Long, 1960)

piston speed Total feet or meters of travel of a piston in 1 min. (Nichols, 1976)

piston-type sampler *piston sampler.*

piston-type washbox *plunger-type washbox.*

pit (a) Depression produced in a metal surface by nonuniform electrodeposition or by corrsion. (Lowenheim, 1962) (b) Excavation to hold quantities of water and drilling fluids. (Wheeler, 1958) (c) So. Wales. Long, open-air fire for converting coal into coke for blast-furnace purposes. (Fay, 1920) (d) A mine, quarry, or excavation worked by the open-cut method to obtain material of value. (e) The shaft of a mine; a shaft mine; a trial pit. (Nelson, 1965) (f) The underground portion of a colliery, including all workings. Used in many combinations, as pit car, pit clothes, etc. (g) In hydraulic mining, the excavation in which piping is carried forward. (h) Commonly, a coal mine, but not usually called so by workers, except in reference to surface mining where the workings may be known as a strip pit. (BCI, 1947) (i) *abyss.*

pit ash Ash in coal derived from the dirt bands, adjoining shales or cleat minerals. (Tomkeieff, 1954)

pit bank (a) Eng. The raised ground or platforms upon which the coal is sorted and screened at the surface. See also: *heapstead; pit brow; pithead; pit hill.* (b) Scot. The surface of the ground at the mouth of a pit, or shaft.

pit bar One of the wooden props bracing the sides of a pit. (Standard, 1964)

pit boss A mine foreman who is in direct charge of workers in a specific portion of a pit or mine. Also called shift boss. See also: *pit foreman.* (DOT, 1949)

pit bottom The bottom of a shaft and all the equipment and roadways around it. See also: *loop-type pit bottom.* (Nelson, 1965)

pit brow The pithead, and in particular, the mouth of the shaft. The edge or brow of a pit. See also: *brow.* (Nelson, 1965; Standard, 1964)

pit cage The structure used in mine shafts for transport purposes. See also: *cage.* (Nelson, 1965)

pit-car loader A short, electrically powered, lightweight elevating conveyor designed for use in working places, to facilitate the loading of large cars or to aid in shoveling long distances. The loader shovels into the hopper end and the conveyor carries the coal to the car. (Jones, 1949)

pit-car-loader operator In bituminous coal mining, a person who operates a machine to load coal in mine (pit) cars. (DOT, 1949)

pit-car repairer *mine-car repairman.*

pitch (a) The angle between the horizontal and any linear feature, such as an ore shoot or lineation, measured in the plane containing the linear feature. See also: *rake.* (AGI, 1987) (b) The angle between the horizontal and an axial line passing through the highest or lowest points of a given stratum in an anticline or syncline. (c) Loosely, the grade, rise, or incline of a seam or bed. (d) A vein-form deposit that follows dipping joint planes. This usage is confined largely to the Upper Mississippi Valley lead-zinc deposits. (e) The slope of a roof, in inches (or centimeters), of vertical rise per horizontal foot (or meter). (f) The distance between tooth centers, as in a gear wheel, or the number of teeth per unit of diameter. See also: *pitch line.* (g) The grade of an incline or the rise of a coal seam. (BCI, 1947) (h) The solid or semisolid residue from the partial evaporation of tar. Strictly, pitch is a bitumen with extraneous matter, such as free carbon, residual coke, etc. (Nelson, 1965) (i) The angular inclination of an ore shoot with respect to the surface, measured in the direction of the strike. (Nelson, 1965) (j) Of a lode, angle of deviation from the vertical taken by a section of ore having some special characteristic, such as enhanced value. (Pryor, 1963) (k) The angle that a directional feature, for example, slickensides, in a plane makes with a horizontal line within the plane. (BS, 1964) (l) In dredging, the distance between the center of any pin and that of the pin in the next adjacent bucket. (Fay, 1920) (m) *dip.* (n) The slope of a surface or tooth relative to its direction of movement. (Nichols, 1976) (o) In a roller or silent chain, the space between pins, measured center to center. (Nichols, 1976) (p) The amount of advance of a single-thread screw in one turn, expressed in lineal distance along or parallel to the axis, or in turns per unit of length. (Standard, 1964) (q) The distance between corresponding points on adjacent projections produced on work by a cutting tool. (ASM, 1961)

pitch arm One of the rods, usually adjustable, which determine the digging angle of a blade or bucket. (Nichols, 1954)

pitchblende The massive variety of uraninite, $UO_{(2+x)}$; radioactive; black to dark brown; the most important ore of uranium; occurs widely in hydrothermal veins and the disseminated uranium-vanadium deposits of the Colorado Plateau type. Syn: *pitch ore.* Cf: *uraninite.*

pitch circle The circle passing through the chain joint centers when the chain is wrapped on the sprocket. (Jackson, 1955)

pitch diameter The diameter of a circle that passes through the points of average contact between the teeth of two gears running in mesh, or between the teeth of a sprocket and the roller of its companion chain, or between a male and a female thread that are engaged. (Brantly, 1961)

pitcher One who picks over dumps for pieces of ore. (Webster 3rd, 1966)

pitches and flats *flats and pitches.*

pitching bar A kind of pick used, esp. by miners, in beginning a hole. (Webster 2nd, 1960)

pitching chisel A chisel used for making an edge on the face of a stone. Also called pitching tool. (Webster 3rd, 1966)

pitching seam A highly inclined seam. In coal mining, called edge coal. (Nelson, 1965)

pitch length The length of an ore shoot in its greatest dimension.

pitch line (a) The line on which the pitch of gear teeth is measured; an ideal line, in a toothed gear or rack, bearing such a relation to a corresponding line in another gear with which the former works that the two lines will have a common velocity, as in rolling contact. See also: *pitch.* (Webster 3rd, 1966) (b) The line along which the pitch of a rack is marked out, corresponding to the pitch circle of a spur wheel. (CTD, 1958)

pitch off A quarry worker's term for trimming an edge of a block of stone with a hammer and set. (Fay, 1920)

pitch ore *pitchblende; pitchy copper ore.*

pitchstone A dark, resinous volcanic glass.

pitchwork In coal mining, work done on shares. (Standard, 1964)

pitch working Mine working in a steeply inclined seam.

pitchy adj. Resembling the appearance or properties of pitch.

pitchy copper ore (a) A dark, pitchlike oxide of copper. Syn: *pitch ore.* (b) A mixture of chrysocolla and limonite.

pitchy iron ore (a) An old syn. for pitticite. Syn: *pitticite.* (Fay, 1920) (b) *triplite.*

pit efficiency In order to allow for the friction of the skips on the guides and between the air and the skips in the shaft and for other small losses, it is usual to divide the total static torque at any point of the wind by 0.9 for a new shaft with rope guides, or 0.85 for an old shaft with rigid guides. This factor is generally referred to as pit efficiency. (Sinclair, 1959)

pit eye Bottom of a pit shaft from which the sky is visible. (Pryor, 1963)

pit-eye pillar A barrier of coal left around a shaft to protect it from caving. (Fay, 1920)

pit foreman In bituminous coal mining, a foreman who is in immediate charge of all mining operations in a strip mine. See also: *pit boss.* (DOT, 1949)

pit frame (a) The framework carrying the pit pulley. See also: *headframe.* (Fay, 1920) (b) The framework in a coal mine shaft. (Standard, 1964)

pit guide An iron column that guides the cage in a mine shaft. (Standard, 1964)

pit hand In the iron and steel industry, a general term applied to workers who perform varied duties around the processing furnaces. (DOT, 1949)

pithead (a) Landing stage at the top of a shaft. (Pryor, 1963) (b) The top of a mine shaft including the buildings, roads, tracks, plant, and machines around it. See also: *pit brow.* (Nelson, 1965)

pithead output The total tonnage of raw coal produced at a colliery, as distinct from saleable output. It is the tonnage of coal as weighed before it enters the coal-preparation plant. See also: *run-of-mine.* (Nelson, 1965)

pit hill Eng. *pit bank.*

pit lamp An open lamp worn on a miner's cap, as distinguished from a safety lamp.

pit limit Either the vertical or lateral extent to which the mining of a mineral deposit by open pitting may be economically carried. The cost of removing overburden or waste material versus the minable value of the ore so exposed is usually the factor controlling the limits of a pit.

pitman (a) The worker who regularly examines the condition of mine infrastructure. (Nelson, 1965) (b) A connecting rod, such as in the Blake type of jaw crusher; the vertical member linking the eccentric shaft with the toggles between the frame and the lower end of the movable jaw.

pitman arm An arm having a limited movement around a pivot. (Nichols, 1976)

pitmen Workers employed in shaft sinking or shaft inspection and repair. (CTD, 1958)

pit mining Surface mining in which the material mined is removed from below the surrounding land surface. (AIME, 2)

pitotmeter An instrument that consists essentially of two pitot tubes one of which is turned upstream and the other downstream and that is used to record autographically the velocity of a flowing liquid or gas. (Webster 3rd, 1966)

Pitot-static tube When the Pitot tube and static tube are combined, they form the Pitot-static tube, and as such they can be used as an anemometer. The tubes are usually arranged concentrically. When they are connected to the opposite sides of a manometer, the dynamic or velocity pressure will be measured directly. (Roberts, 1960)

Pitot tube Consists of two concentric tubes bent in an L shape. In operation, the instrument is pointed in the direction of air flow: the inner tube, open at the end directed upstream, measures total head, and the outer tube, perforated with small openings transverse to the air flow, records static head. Each tube is connected to a leg of a manometer, when reading velocity head. (Hartman, 1961)

pit pony A pony used for packing or haulage in a mine. (Webster 3rd, 1966)

pit prop (a) A piece of timber used as a temporary support for a mine roof. (Zern, 1928) (b) Length of timber used as a roof support in longwall mining. Modern variants include expandable steel props which can be hydraulically or mechanically lengthened; used in stratified deposits. (Pryor, 1963)

pit quarry An openpit quarry sunk below ground level. Access is gained by stairs, ladders, or mechanical hoists, and material is conveyed from the quarry by inclined tracks, trucks, derricks, or cableway hoists. These pits may reach depths of several hundred feet. A drainage scheme will in most cases be necessary, as the pit will form a natural sump for both surface and subsoil water. This type of quarry is often used for gravel or soft rock that can be extracted by some form of digging. See also: *hillside quarry.* (AIME, 1960; Nelson, 1965)

pit room (a) The number of working places, or the length of a longwall face, available in a mine for coal production. (Nelson, 1965) (b) The extent of the opening in a mine; pit space. (Fay, 1920)

pit rope Eng. Winding rope; a hoisting rope. (Fay, 1920)

pit sampling (a) Use of small untimbered pits to gain access to shallow alluvial deposits or ore dumps for purpose of testing or valuation. (Pryor, 1963) (b) Sampling shallow deposits by means of trial pits, usually about 2 to 3 ft (0.6 to 0.9 m) in diameter. In reasonably dry ground, depths of 50 ft (15.2 m) or more may be reached. Pit sampling is often used to assist site investigations as it provides the maximum of information regarding the nature of deposits and bedrock. See also: *trial pit.* (Nelson, 1965)

pit sand (a) Sand usually composed of grains that are relatively angular; it often contains clay and organic matter. When washed and screened it is a good sand for general purposes. (Zern, 1928) (b) Sand from a pit, as distinct from river or sea sand. (Arkell, 1953)

pit shale The name given to the shale from a drift opened in the side of the ravine at a level 62 ft (18.9 m) below that of the Pittsburgh coal seam. (Rice, 1913-18)

pit slope The angle at which the wall of an open pit or cut stands as measured along an imaginary plane extended along the crests of the berms or from the slope crest to its toe.

pittasphalt An old name give to viscid bitumen. Syn: *maltha.* See also: *mineral tar.* (Tomkeieff, 1954)

pitticite The mineral amorphous, hydrous, ferric arsenate sulfate. It is brown to yellow and red; earthy; occurs as crusts and botryoidal layers; a common oxidation product of arsenical ores. Also spelled pittizite. Syn: *pitchy iron ore.*

pitting (a) The act of digging or sinking a pit. (Fay, 1920) (b) Testing an alluvial deposit by the systematic sinking of small shafts, the material recovered being subsequently tested. The practice is confined to shallow depths; i.e., down to about 50 ft (15.2 m) in fairly dry soft ground. (Nelson, 1965)

Pittsburgh bed The Pittsburgh coal which outcrops prominently in the vicinity of Pittsburgh and extends under a large area of western Pennsylvania, northern West Virginia, northwestern Maryland, and eastern Ohio. It belongs in the Carboniferous system, Pennsylvanian series, at the base of the Monongahela formation. (Rice, 1913-18)

pit water Water from the underground workings of a mine. (BS, 1962)

pit wood The various kinds of timber used at a mine, mainly as supports. (Nelson, 1965)

pitwork Cornish pumps and other engineering appliances in and near a mine shaft. (Pryor, 1963)

pivot A nonrotating axle or hinge pin. (Nichols, 1976)

pivoted-bucket carrier The highest type of combined elevator and conveyor. It consists of two long-pitch roller chains joined by crossbars on which are hung the buckets in such a way that they can be completely turned over. (Pit and Quarry, 1960)

pivoted-bucket conveyor A type of conveyor using pivoted buckets attached between two endless chains that operate in suitable guides or casing in horizontal, vertical, inclined or a combination of these paths over drive-corner and takeup terminals. The buckets remain in the carrying position until they are tipped or inverted to discharge. Syn: *bucket elevator.*

pivot shaft A tractor dead axle, or any fixed shaft that acts as a hinge pin. (Nichols, 1976)

pivot tube A hollow hinge pin. (Nichols, 1976)

place (a) *in situ.* (b) The part of a mine in which a miner works by contract; known as "place" or "working place." (c) A point at which the cutting of coal is being carried on.

placer A deposit of sand or gravel that contains particles of gold, ilmenite, gemstones, or other heavy minerals of value. The common types are stream gravels and beach sands. See also: *alluvial deposit; beach placer.*

placer claim (a) A mining claim located upon gravel or ground whose mineral contents are extracted by the use of water, by sluicing, hydraulicking, etc. The unit claim is 1,320 ft^2 (122.6 m^2) and contains 10 acres (4.1 ha). See also: *mining claim.* (b) Ground with defined boundaries that contains mineral in the earth, sand, or gravel; ground that includes valuable deposits not fixed in the rock. See also: *claim; lode claim.* (c) The maximum size of a placer claim is 20 acres (8.1 ha). Association claims of two or more persons may be located up to an area of 160 acres (64.8 ha) for eight persons. Placer claims must have a discovery. They should be staked, a location notice posted, and recorded in the same manner as for lode claims, stating the mineral for which the location in made. (Lewis, 1964)

placer digging (a) The action of mining by placer methods. (Craigie, 1938) (b) A place at which placer mining is or may be carried on. (Craigie, 1938)

placer gold Gold occurring in more or less coarse grains or flakes and obtainable by washing the sand, gravel, etc., in which it is found. Also called alluvial gold. See also: *stream gold.* Syn: *wash gold.* (Craigie, 1938)

placer ground Ground where placer mining can be done; i.e., where valuable minerals can be obtained by digging up the earth and washing it for the valued mineral. (Craigie, 1938)

placer location A location of a tract of land for mineral-bearing or other valuable deposits upon or within it that are not found within lodes or veins in rock in place; a claim of a tract of land for the sake of the loose deposits on or near its surface. (Ricketts, 1943)

placer mine (a) A deposit of sand, gravel, or talus from which some valuable mineral is extracted. (Hess) (b) *placer mining.*

placer mining (a) The extraction of heavy mineral from a placer deposit by concentration in running water. It includes ground sluicing, panning, shoveling gravel into a sluice, scraping by power scraper and excavation by dragline, dredge or other mechanized equipment. (Nelson, 1965) (b) Extracting the gold or other mineral from placers, wherever situated—in dry channels and in channels temporarily filled with water. The mineral may be found in deep channels, in navigable streams, or in estuaries or creeks and rivers where the sea ebbs and flows. (Ricketts, 1943) (c) That form of mining in which the surficial detritus is washed for gold or other valuable minerals. When water under pressure is employed to break down the gravel, the term hydraulic mining is generally employed. There are deposits of detrital material containing gold which lie too deep to be profitably extracted by surface mining, and which must be worked by drifting beneath the overlying barren material. The term "drift mining" is applied to the operations necessary to extract such auriferous material. See also: *dredge.* Syn: *placer mine.* (d) The extraction and concentration of heavy metals or minerals from placer deposits by various methods, generally using running water. Cf: *alluvial mining; hydraulic mining; drift mining.* (AGI, 1987)

placodine *maucherite.*

plaffeiite A fossil resin found in Switzerland. (Tomkeieff, 1954)

plagihedral *plagiohedral.*

plagioclase (a) Any of a group of feldspars containing a mixture of sodium and calcium feldspars, distinguished by their extinction angles; crystal; triclinic; Mohs hardness, 6; and sp gr, 2.6 to 2.7. (Bennett, 1962) (b) A series of triclinic feldspars of general formula: $(Na,Ca)Al(Si,Al)Si_2O_8$; at high temperatures it forms a complete crystal solution series from albite, $NaAlSi_3O_8$, to anorthite, An, $CaAl_2Si_2O_8$; the series is arbitrarily subdivided and named according to increasing mole fraction of the An component: albite (An 0% to 10%), oligoclase (An 10% to 30%), andesine (An 30% to 50%), labradorite (An 50% to 70%), bytownite (An 70% to 90%), and anorthite (An 90% to 100%). The Al:Si ratio ranges with increasing An content from 1:3 to 1:1. Plagioclase feldspars are common rock-forming minerals, have characteristic polysynthetic twinning, and commonly display zoning. The term was originally applied to all feldspars having an oblique angle between the two main cleavages. Cf: *alkali feldspar; orthoclase.* Syn: *sodium-calcium feldspar.*

plagioclase rhyolite A porphyritic extrusive rock with phenocrysts of plagioclase and quartz in a groundmass of orthoclase and quartz. Also called plagioliparite. Syn: *dellenite.* (Webster 2nd, 1960)

plagioclastic Having the cleavage of plagioclase; breaking obliquely. (Standard, 1964)

plagiohedral Having an oblique spiral arrangement of faces; specif., being a group of the isometric system characterized by 13 axes of symmetry but no center or planes. Also spelled plagihedral. (Webster 3rd, 1966)

plagionite A monoclinic mineral, $Pb_5Sb_8S_{17}$; metallic black to lead-gray; forms stubby tablets; an uncommon associate of other lead sulfosalts in hydrothermal veins.

plain (a) An extent of level, or nearly level, land; a region not noticeably diversified with mountains, hills, or valleys. (Fay, 1920) (b) A flat, gently sloping or nearly level region of the sea floor. (Hunt, 1965) (c) Archaic. Relatively free of gaseous inclusions. (ASTM, 1994)

plain clinometer A clinometer having only its upper end threaded to fit drill rods. Also called end clinometer. See also: *clinometer.* Cf: *line clinometer; wedge clinometer.* (Long, 1960)

plain concrete Concrete with no reinforcement. (Hammond, 1965)

plain detonator A detonator for use with a safety fuse. It consists of an aluminum tube closed at one end and partly filled with a sensitive initiating explosive. The tube is only partially filled because a plain detonator is always used in conjunction with a safety fuse, and the empty space enables the fuse to be inserted into the tube until it comes into contact with the detonating composition. The safety fuse is then secured in position by indenting the detonator tube, this process being known as crimping. The combination of safety fuse and plain detonator is called a capped fuse. (BS, 1964; McAdam, 1958)

plain pilot A pilot in the surface of which no cutting points, such as diamonds or slugs, are inset. See also: *pilot.* (Long, 1960)

plaiting A texture seen in some schists that results from the intersection of relict bedding planes with well developed cleavage planes. (AGI, 1987)

plan (a) A map showing features—such as mine workings, geological structures, and outside improvements—on a horizontal plane. See also: *colliery plan.* (Nelson, 1965) (b) A scheme or project for mine development. See also: *planning.* (Nelson, 1965) (c) The system on which a colliery is worked, such as longwall, room-and-pillar, etc. (Zern, 1928)

planar Lying or arranged as a plane or in planes, usually implying more or less parallelism, as in bedding or cleavage. It is a two-dimensional arrangement, in contrast to the one-dimensional linear arrangement. (AGI, 1987)

planar cross-bedding (a) Cross-bedding in which the lower bounding surfaces are planar surfaces of erosion. It results from beveling and subsequent deposition. (AGI, 1987) (b) Cross-bedding characterized by planar foreset beds. (AGI, 1987)

planar element A fabric element having two dimensions that are much greater than the third; e.g., bedding, cleavage, and schistosity. Cf: *linear element.* (AGI, 1987)

planar flow structure *platy flow structure.*

planar gliding Uniform slippage along plane surfaces. (AGI, 1987)

planar structure *platy flow structure.*

planation The widening of valleys through lateral corrasion by streams after they reach grade and begin to meander and form floodplains. Also, by the extension, the reduction of divides and the merging of valley plains to form a peneplain; peneplanation.

plane (a) Any roadway, generally inclined but not necessarily so, along which ore or workers are conveyed by mechanical means from one bed to another or to a lower elevation in the same bed. See also: *bedding plane; fault plane; slope.* (Nelson, 1965) (b) A road on the natural floor of a seam. (Mason, 1951) (c) A two-dimensional form that is without curvature; ideally, a perfectly flat or smooth surface. In geology the term is applied to such features as a bedding plane or a planation surface. Adj: planar. See also: *surface.* (AGI, 1987) (d) In crystallography, a plane of symmetry dividing a crystal structure into two mirror images. See also: *symbols of crystal faces.* (e) A level surface bounded by straight lines, such as the faces of crystals. (Gordon, 1906)

plane course Scot. In the direction facing the joint planes. Syn: *on plane.*

plane engineer *slope engineer.*

plane fault A fault with a surface that is planar rather than curved.

plane figure A plane surface bounded either by straight lines or curved lines or by a combination of straight and curved lines. (Jones, 1949)

plane group The 17 possible combinations of symmetry elements which may coexist in 2 dimensions. Cf: *space group.*

plane man *incline man.*

plane of saturation *water table.*

plane of stretching A low-angle gravity (normal) fault resulting from stretching of the solidified top of an igneous intrusion.

plane of symmetry Any plane which divides a crystal, crystal structure, or crystal symmetry such that each side is a mirror reflection of the other. Represented as *m* or 2 and graphically as a solid or heavy line. Syn: *mirror plane.*

plane or rectangular coordinate Either of two perpendicular distances of a point from a pair of rectangular coordinate axes. (Seelye, 1951)

plane-polarized light Light with its electric vector confined to a plane.

planer (a) First developed as a fixed-blade device for continuous longwall mining of narrow seams of friable coal, this machine is pulled along the coal face, planing a narrow cut. Vibrating-blade planers were designed later in an attempt to apply the technique to harder coal; they have also been experimented with in the phosphate mines in western Montana and northern Idaho. (b) A machine provided with a cutting tool having lateral and vertical adjustment that is widely used in stone trimming. Both sides and tops of blocks may be planed to desired dimensions. Some planers may be adjusted to cut curved forms.

planerite A triclinic mineral, $Al_6(PO_4)_2(PO_3OH)_2(OH)_8{\cdot}4H_2O$; turquoise group.

plane schistosity A type of schistosity characterized by the arrangement of tabular and prismatic grains in parallel planes.

plane shear One of four types of slope failure. Plane shear failure results when a natural plane of weakness, such as a fault, a shear zone, or bedding plane exists within a slope and has a direction such as to provide a preferential path for failure. Large intact portions of the slope rock may slide along this plane surface. (Woodruff, 1966)

plane strain A state of strain in which all displacements that arise from deformation are parallel to one plane, and the longitudinal strain is zero in one principal direction. (AGI, 1987)

plane stress A state of stress in which one of the principal stresses is zero.

plane surveying Ordinary field and topographic surveying in which Earth curvature is disregarded and all measurements are made or reduced parallel to a plane representing the surface of the Earth. The accuracy and precision of results obtained by plane surveying may decrease as the area surveyed increases in size. Cf: *geodetic surveying.* (AGI, 1987)

plane table (a) An instrument for plotting the lines of a survey directly from the observations; consisting essentially of a drawing board mounted on a tripod and fitted with a ruler that is pointed at the object observed, usually with the aid of a sighting device, such as a telescope. (Webster 3rd, 1966) (b) An inclined ore-dressing table. (Standard, 1964)

planetary geared drum A drum containing planetary gearing that is used to control the motion of the rope drums on certain types of mining machines. In planetary gearing, which is used when a large ratio of speed reduction with only a few operating gears is required, some or all of the gear wheels in the train of mechanism have a motion about an axis and a revolution about the same axis. (Jones, 1949)

planetary lap A type of machine lap employing a number of geared workholders that rotate with an epicyclic motion between two stationary lapping plates. The crystals being lapped, when contained in pentagonal holes in the workholder, have an imposed rotatory motion. Also known as the Hunt-Hoffman lap or Bendix lap. (Am. Mineral., 1947)

planetary mill Mill used for making very large reductions on slabs by one pass through the mill. The mill consists of two large plain rolls, each surrounded by many small work rolls. (Osborne, 1956)

planetary set gear A gearset consisting of an inner (sun) gear, an outer ring with internal teeth, and two or more small (planet) gears meshed with both the sun and the ring. (Nichols, 1976)

plane tender *slope engineer.*

planet gearing Gearing in which one gear wheel revolves around another. (Mason, 1951)

planimeter An instrument for measuring the area of any plane figure by passing a tracer around its boundary line. (Webster 3rd, 1966)

planimetric analysis Analysis of patterns in a fabric diagram based on distribution of points and areal comparisons. (AGI, 1987)

planimetric map A map that presents only the relative horizontal positions of natural or cultural features, by lines and symbols. It is distinguished from a topographic map by the omission of relief in measurable form. Syn: *line map.* (AGI, 1987)

planimetry (a) The measurement of plane surfaces; e.g., the determination of horizontal distances, angles, and areas on a map. (AGI, 1987) (b) The plan details of a map; the natural and cultural features of a region (excluding relief) as shown on a map. (AGI, 1987)

planisher A device for flattening thin sections cut for microscopic examination. (Standard, 1964)

plank timbering The lining of a shaft with rectangular plank frames. See also: *box timbering.*

plankton The whole community of rifting small plants and animals in layers of the water. This term is frequently used to describe all life forms, regardless of size, which have no means of significant self-locomotion. This community can be divided into the phytoplankton (plants) and the zooplankton (animals). (Hy, 1965)

plankton bloom The rapid growth and multiplication of plankton, usually plant forms, producing an obvious change in the physical appearance of the sea surface, such as coloration or slicks. Also called sea bloom; florescence. (Hy, 1965)

planktonic Relating to the chiefly simple types of floating and surface-dwelling forms of organisms of the ocean waters. (Schieferdecker, 1959)

plank tubbing The lining of a shaft with planks, spiked on the inside of curbs. See also: *tubbing.*

planning The predesign of the detailed layout, main roadways, and workings of a mine or group of mines. The scheme usually involves the introduction of mechanical equipment for the working and transport of the coal or mineral. The selection of mining methods and machines properly adapted to the local conditions is part of planning. (Nelson, 1965)

planning engineer A mining engineer responsible for mine planning. The engineer is attached to the planning department of a large mine or a group of smaller mines and is qualified by training, experience, and technical qualifications to envisage new development work and coordinate the ideas of other experts such as a mechanization engineer, ventilation engineer, mining geologist, etc. (Nelson, 1965)

planometric projection Pictorial view of an object showing it in plan with oblique lines showing the front, side, and thickness. See also: *projection.* (Hammond, 1965)

planosol A great soil group in the 1938 classification system; an intrazonal, hydromorphic group of soils having a leached surface layer above a definite clay pan or hardpan. These soils develop on nearly flat upland surfaces under grass or trees in a humid to subhumid climate. (AGI, 1987)

plant (a) The shaft or slope, tunnels, engine houses, railways, machinery, workshops, etc., of a colliery or other mine. (b) To place gold or any valuable ore in the ground, in a mine, or the like to give a false impression of the richness of the property. To salt, as to plant gold with a shotgun. See also: *salting a mine.* (c) In mining, the mechanical installations, machines, and their housings. Earthworks are sometimes loosely included. (Pryor, 1963) (d) Used to include the machinery, derricks, railway, cars, etc., employed in tunnel work. (Stauffer, 1906)

plant mix The process of soil stabilization in which the soil is carried to a stationary mixer, returned to the site after mixing and then spread. Cf: *mix-in-place.* (Hammond, 1965)

plant-mixed concrete Concrete that is mixed at a central mixing plant and delivered to a site in special equipment designed to prevent its segregation. (Hammond, 1965)

plant-mix method A method of preparing aggregates for bituminous surfaces in which aggregates and bitumen are combined in a plant situated at the road or at a relatively long distance from the road. Also known as the premixed method. (Pit and Quarry, 1960)

plant scrap Scrap metal produced in the plant itself; e.g., sprues and gates in a foundry or defective ingots and hot tops in a steel mill. Also called home scrap. (Newton, 1959)

plasma (a) Gas comprising equal amounts of positively and negatively charged particles; a fourth state of matter (solid, liquid, gas, plasma) capable of conducting magnetic force. (Pryor, 1963) (b) A bright-, leek-, to emerald-green subtranslucent variety of cryptocrystalline (chalcedonic) quartz. The green color is attributed to chlorite. Cf: *bloodstone; heliotrope.* (c) That part of a soil which can be or has been moved, reorganized, and/or concentrated by soil-forming processes.

plasma jet (a) A jet formed by passing a high-speed current of nitrogen or a mixture of nitrogen and hydrogen over a tungsten electrode placed in a specially designed narrow orifice in a cutting torch. An arc is struck between this electrode and the earthed nozzle of the torch, which is cooled by a water jacket. When a plasma jet is used to cut rock, two separate zones of action can be expected. (Min. Miner. Eng., 1965) (b) Ionized gas produced by passing an inert gas through a high-intensity arc, causing temperatures up to tens of thousands degrees centigrade. (Harbison-Walker, 1972)

plastering *mudcapping.*

plaster mill A machine consisting of a roller or set of rollers for grinding lime or gypsum to powder. (Fay, 1920)

plaster pit Derb. A gypsum mine.

plaster shooting (a) A surface blasting method used when no rock drill is available or is not necessary. It consists of placing a charge of gelignite, primed with safety fuse and detonator, in close contact with the rock or boulder and covering it completely with stiff damp clay. The charges vary from 8 to 16 oz/yd^3 (297 to 593 g/m^3) of rock. See also: *popping; snakeholing.* (Nelson, 1965) (b) A form of secondary blasting in which the explosive is detonated in contact with the rock without the use of a shothole. See also: *secondary blasting; mudcapping.* (BS, 1964)

plaster stone *gypsum.*

plastic Said of a body in which strain produces continuous, permanent deformation without rupture. Cf: *elastic.* (AGI, 1987)

plastic and semiplastic explosive Any of several explosives used for commercial purposes. The consistency is such that the explosive can be shaped by moderate pressure to fill a drill hole. The difference between plastic and semiplastic form is primarily dependent on the difference in equipment which has been found necessary in manufacturing cartridges of the explosive. The viscosity of the plastic type makes it possible to produce cartridges by a process of extrusion through tubes. (Fraenkel, 1953)

plastic clay Any clay, but chiefly kaolinite, which, when mixed with water, is easily shaped and retains this shape until fired.

plastic deformation (a) Permanent deformation of the shape or volume of a substance, without rupture. It is mainly accompanied by crystal gliding and/or recrystallization. Syn: *plastic flow; thixotropy.* Cf: *plastic strain.* (AGI, 1987) (b) Deformation by one or both of two grain-scale mechanisms: slip, and twinning. This is a metallurgical definition, increasingly used by geologists. Sometimes called crystal plasticity. (AGI, 1987) (c) Rheological term for deformation characterized by a yield stress, which must be exceeded before flow begins. (AGI, 1987) (d) An elastic deformation of brittle minerals—such as olivine under mantle conditions, or quartz, during metamorphism; deformation occurs along well-defined crystallographic planes in specific directions, which may be preserved as thin deformation lamellae or as deformation twinning. It may be annealed out by recrystallization. Cf: *elastic deformation.* (e) Irreversible deformation of metallic minerals, such as gold or copper. See also: *malleability.*

plastic design The design of steel or reinforced-concrete structural frames which is based on the assumption that plastic hinges form at points of maximum bending moment. See also: *elastic design; plastic modulus.* (Hammond, 1965)

plastic explosive *plastic and semiplastic explosive.*

plastic firebrick A common term for both high duty and super-duty fire clay plastic refractories.

plastic flow *plastic deformation.*

plastic fracture The breakage of a solid material under load when being permanently deformed. (Hammond, 1965)

plastic igniter cord A corklike device for lighting a safety fuse. When the cord is ignited an intense flame passes along its length at a uniform rate and ignites the blackpowder core of an ordinary safety fuse. Two types are made: the fast has a nominal burning speed of 1 s/ft (3.3 s/m); the other is about 10 times as slow. (Nelson, 1965)

plasticity The property of a material that enables it to undergo permanent deformation without appreciable volume change, elastic rebound, or rupture. See also: *plastic deformation; plastic flow; plastic limit; plastic soil; plastic state; plasticity index.* (ASCE, 1958)

plasticity index The water-content range of a material at which it is plastic, defined numerically as the liquid limit minus the plastic limit. Cf: *Atterberg limits; plastic limit.*

plasticizer A material, usually organic, capable of imparting plastic properties to nonplastics or improving the plasticity of ceramic mixtures. Syn: *wetting agent.* (ACSG, 1963)

plastic limit (a) The water-content boundary beyond which a soil can be rolled into a thread approx. 3 mm in diameter without crumbling, i.e., beyond which it is plastic. (b) The water content of a soil or clay material corresponding to an arbitrarily defined boundary between a plastic and a semisolid state. Cf: *Atterberg limits; plasticity index.*

plastic modulus A factor used in the plastic design of steel structures. It is a constant for each particular shape of section. See also: *plastic design.* (Hammond, 1965)

plastic soil (a) A soil that can be rolled into 1/8-in (1.6-mm) diameter strings without crumbling. (Nichols, 1976) (b) A soft, rubbery soil. (Nichols, 1976) (c) A soil that exhibits plasticity. (ASCE, 1958)

plastic solid A solid that undergoes change of shape continuously and indefinitely after the stress applied to it passes its elastic limit.

plastic state The range of consistency within which a soil exhibits plastic properties. Also called plastic range. (ASCE, 1958)

plastic strain In rocks, which are composed of many crystals commonly belonging to several mineral species, the term applies to any permanent deformation throughout which the rock maintains essential cohesion and strength regardless of the extent to which local microfracturing and displacement of individual grains may have entered into the process. Cf: *plastic deformation.*

plastic tamping rod A tamping rod or stemmer, of a rigid nature, made from plastic possessing suitable dielectric properties. A plastic conducive to the building up of heavy charges of static electricity is unsuitable. (Nelson, 1965)

plastic tooling Dies, jigs, and fixtures for metal forming, boring, assembly, and checking; made at a saving of time and labor, of laminated and cast components, and cemented into highly stable industrial tools, chiefly with epoxy and some with polyester resins. Epoxies are strong adhesive resins, particularly useful because of their low shrinkage factor. Polyesters have a cost advantage and are easy to handle. (Crispin, 1964)

plastic yield The term commonly applied to plastic deformation. (Hammond, 1965)

plastic zone In explosion-formed-crater nomenclature, this zone differs from the rupture zone by having less fracturing and only small permanent deformations. There is no distinct boundary between the rupture and plastic zones. (Min. Miner. Eng., 1966)

Plast-Sponge High-quality iron powder made by reduction of iron oxide; used in powder metallurgy. (Bennett, 1962)

plat (a) The map of a survey in horizontal projection, such as of a mine, townsite, etc. (b) A diagram drawn to scale showing land boundaries and subdivisions, together with all data essential to the description of the several units. A plat differs from a map in that it does not show additional cultural, drainage, and relief features. (Seelye, 1951) (c) A platform, floor, or surface in or about a mine used esp. for loading and unloading ore, etc. (Webster 3rd, 1966)

plate (a) A flat iron or steel sheet laid around a mine-shaft collar, at the shaft bottom, or at any level station, to enable mine cars and other equipment to be easily turned and moved about. Also, a cast-iron plate with a circular ridge on which mine rail cars are turned at the junction of roads. (b) A horizontal timber laid on a floor or sloping wall to receive a framework of timbers. (c) A torsionally rigid thin segment of the Earth's lithosphere, which may be assumed to move horizontally and adjoins other lithosopheric plates along zones of seismic activity. See also: *plate tectonics.*

plate amalgamation Use of copper or copper-alloy plates coated with enough mercury to form a soft adherent film, in order to trap gold from crushed ore pulp as it flows over the plates. The resulting amalgam, containing up to 40% metallic gold, is periodically scraped off and more mercury is added to the film. (Pryor, 1963)

plate-and-frame filter A filter press consisting of plates with a gridiron surface alternating with hollow frames, all of which are held by means of lugs, on the press framework. The corners of both frames and plates are cored to make continuous passages for pulp and solution; the filter cloth is placed over the plates. The pulp passageway connects with the large, square opening in the frame; the solution and passageways connect with the gridiron surface of the plate. The Dehne and the Merrill are well-known types. (Liddell, 1918)

plate apron feeder An automatic arrangement by which coal or ore is fed forward on steel plates forming segments linked together in an endless chain. See also: *plate feeder.* (Nelson, 1965)

plateau Broadly, any comparatively flat area of great extent and elevation; specif. an extensive land region considerably elevated (more than 150 to 300 m in altitude) above the adjacent country or above sea level; it is commonly limited on at least one side by an abrupt descent, has a flat or nearly smooth surface but is often dissected by deep valleys and surmounted by high hills or mountains, and has a large part of its total surface at or near the summit level. A plateau is usually higher and has more noticeable relief than a plain (it often represents an elevated plain), and it is usually higher and more extensive than a mesa; it may be tectonic, residual, or volcanic in origin. See also: *tableland.* Cf: *mesa.* (AGI, 1987)

plateau basalt A term applied to those basaltic lavas that occur as vast composite accumulations of horizontal or subhorizontal flows, which, erupted in rapid succession over great areas, have at times flooded sectors of the Earth's surface on a regional scale. They are generally believed to be the product of fissure eruptions. Cf: *shield basalt.* Syn: *flood basalt.* (AGI, 1987)

plateau gravel A sheet, spread, or patch of surficial gravel, often compacted, occupying a flat area on a hilltop, plateau, or other high region at a height above that normally occupied by a stream-terrace gravel. It may represent a formerly extensive deposit that has been raised by earth movements and largely removed by erosion. (AGI, 1987)

plate bearing test A method by which the load bearing capacity of a soil may be estimated. See also: *ultimate bearing pressure.* (Hammond, 1965)

plate cleaner A device for cleaning raw coal which uses the difference in the coefficient of resilience or friction between clean coal and an inclined plate, commonly of steel, and that between refuse and the plate to allow the clean coal to jump over a gap while the refuse falls through. (BS, 1962)

plate conveyor A conveyor in which the carrying medium is a series of steel plates, each in the form of a short trough, joined together with a slight overlap to form an articulated band. The plates are attached either to one center chain or to two side chains. The chains connect rollers running on an angle-iron framework and transmit the drive from the driveheads that can be installed at intermediate points as well as at the head or tail ends. A plate conveyor can negotiate bends down to about 20 ft (6.1 m) radius; available in widths 400, 540, and 640 mm with running speeds from 3 to 4 ft/s (0.9 to 1.2 m/s) with a carrying capacity from 100 to 400 st/h (90.7 to 362.8 t/h). Syn: *steel plate conveyor.* (Nelson, 1965)

plate coordinate In photographic mapping, either of two rectangular coordinates measured on a photograph with reference to the principal point as origin. (Seelye, 1951)

plated crystal A crystal with a conductive surface film of gold, silver, aluminum, or other metal produced by cathode sputtering, evaporation, or chemical methods. The films, to which lead wires may be soldered, take the place of the conventional clamped metal electrodes. (Am. Mineral., 1947)

plate feeder The mechanical plate feeder is a device for feeding material at a fixed and uniform rate. It is generally applied at the tail end of a conveyor or elevator which feeds a plant, but may be applied to feeding any other single unit. It relieves the pressure and drag, with the consequent unnecessary wear on the belt, which is ordinarily experienced if feeding from a hopper directly to a belt. It not only cuts maintenance costs by eliminating uneven wear, but increased output can be obtained by steady feeding. This type of feeder also handles wet aggregate. See also: *disk feeder; reciprocating feeder.* (Pit and Quarry, 1960; ACSG, 1963)

plate former Used for lining shafts, winzes, and rises; usually constructed of comparatively thin steel sheeting, stiffened around the edges with angles. Plates should be of such size that they can be conveniently handled in the skips or buckets used for sinking. (Spalding, 1949)

plate girder A built-up riveted or welded steel girder, having a deep vertical web plate, with a pair of angles riveted along each edge to act as compression and tension flanges. For heavier loads, flange plates are riveted or welded to the angles. (Hammond, 1965)

plate roll A smooth roll for making sheet iron or plate iron, as distinguished from iron having grooves for rolling rails, beams, etc. (Standard, 1964)

plate tectonics A theory of global tectonics in which the lithosphere is divided into a number of plates whose pattern of horizontal movement is that of torsionally rigid bodies that interact with one another at their boundaries, causing seismic and tectonic activity along these boundaries. (AGI, 1987)

plate tongs Tongs for grasping and handling iron or steel plates. (Standard, 1964)

platform (a) The place on top of a breaker where the freshly mined coal is weighed by a weigh boss just before it is dumped into the machinery. (Korson, 1938) (b) A wooden floor on the side of a gangway at the bottom of an inclined seam, to which the coal runs by gravity, and from which it is shoveled into mine cars. (c) A plank or mesh steel-covered level area at the base of a drill tripod or derrick, used as a working space in front of a drill machine around the collar of the borehole. Sometimes the platform is large enough to act as a foundation and anchor for the drill machine. (Long, 1960) (d) A scaffold.

(Fay, 1920) (e) A wood mat used in sets to support machinery on soft ground. Also called pontoon. (Nichols, 1976) (f) An operator's station on a large machine, particularly on rollers. (Nichols, 1976) (g) In the breaker, a flat or slightly inclined floor covered with iron plates onto which coal is run from the main screen bars and cleaned by platform workers. (Korson, 1938) (h) Also a similar floored area in the tripod or derrick on which a laborer stands while working in a tripod or derrick. See also: *floor.* (Long, 1960)

platform gantry A gantry constructed for carrying a portal crane or a similar structure. (Hammond, 1965)

platform hoist A power-driven hoist, having a lifting capacity ranging from 200 lb (90.7 kg) to about 2-1/2 st (2.27 t), which can be raised on a loading platform up to 200 ft (61 m) high. (Hammond, 1965)

platina (a) Twisted silver wire. (Standard, 1964) (b) Crude native platinum. (AGI, 1987)

platinic gold Said to be a native alloy containing 84.6% gold, 2.9% silver, 0.2% iron, 0.9% copper, and the remainder 11.4% platinum. (Hess)

platiniridium An isometric mineral, (Ir,Pt), with Ir 50% to 80% (atomic) of Ir + Pt; forms silver-white grains having sp gr, 22.6 to 22.8; Mohs hardness, 6 to 7.

platinize To coat or combine with platinum, esp. by electroplating. (Standard, 1964)

platinum (a) An isometric mineral, native platinum 4[Pt] with variable Pd, Ir, Fe, Ni; malleable; ductile; metallic; sp gr, 21.45; corrosion resistant; occurs in ultramafic rocks, quartz veins, and in placers. (b) A malleable and ductile silvery-white metal, when pure. Symbol: Pt. Occurs native, accompanied by small quantities of iridium, osmium, palladium, ruthenium, and rhodium. Used in jewelry, wire, vessels for laboratory use, and in many valuable instruments including thermocouple elements. (Handbook of Chem. & Phys., 3)

platinum-group metal (PGM). Any of the minerals native platinum, osmium, iridium, palladium, rhodium, ruthenium, and their alloys, such as osmiridium (Ir,Os), ruthenosmiridium (Ir,Os,Ru), rutheniridosmine (Os,Ir,Ru), and platiniridium (Ir,Pt). Other alloys of PGM are exemplified by stanopalliadinite, $(Pd,Cu)_3Sn_2$(?); and potarite, PdHg. Other sources of PGM are sperrylite, $PtAs_2$; cooperite, (Pt,Pd,Ni)S; stibiopalladinite, Pd_5Sb_2; braggite, (Pt,Pd,Ni)S; vysotskite, (Pd,Ni)S; ruthenarsenite, (Ru,Ni)As; cuproiridsite; $CuIr_2S_4$; cuprorhodsit, $CuRh_2S_4$; malanite, $Cu(Pt,Ir)_2S_4$; and dayingite, $CuCoPtS_4$. Varietal terms include plyxene and ferroplatinum for iron alloys and cuproplatinum for copper alloys.

platinum sponge Metallic platinum in a gray, porous, spongy form; obtained by reducing ammonium chloroplatinate, which occludes large volumes of oxygen, hydrogen, and other gases. (Webster 3rd, 1966)

platting Brick laid flatwise on top of a kiln to keep in the heat. (Fay, 1920)

plattman In bituminous coal mining, a colloquialism of English origin for a pusher who pushes loaded mine cars onto a cage from a platt (an enlarged underground opening at the shaft where cars are gathered prior to hoisting). (DOT, 1949)

plattnerite (a) A tetragonal mineral, PbO_2; rutile group; dimorphous with scrutinyite; iron black; occurs in lead mines. (b) Erroneous spelling of planerite.

platy flow structure An igneous rock structure of tabular sheets suggesting stratification. It is formed by contraction during cooling; the structure is parallel to the surface of cooling and is commonly accentuated by weathering. Syn: *platy structure; planar flow structure.* (AGI, 1987)

platynite A trigonal mineral, $PbBi_2(Se,S)_3$; metallic; iron-black; forms thin plates like graphite; at Falun, Sweden. Also spelled platinite.

platy structure *platy flow structure.*

playa (a) A term used in southwestern United States for a dry, vegetation-free, flat area at the lowest part of an undrained desert basin, underlain by stratified clay, silt, or sand, and commonly by soluble salts. The term is also applied to the basin containing an expanse of playa, which may be marked by ephemeral lakes. See also: *salina; alkali flat; salt flat; salar.* Syn: *dry lake.* (AGI, 1987) (b) *playa lake.* (c) A small, generally sandy land area at the mouth of a stream or along the shore of a bay. Etymol: Spanish, beach, shore, coast. (AGI, 1987)

playa basin *bolson.*

playa lake A shallow, intermittent lake in an arid or semiarid region, covering or occupying a playa in the wet season but drying up in summer; an ephemeral lake that upon evaporation leaves or forms a playa. Syn: *playa.* (AGI, 1987)

play of color A pseudochromatic optical effect resulting in flashes of colored light from certain minerals, such as fire opal and labradorite, as they are turned in white light. Periodic spacings of phases with slightly differing refractive indices act as optical diffraction gratings in these minerals. Cf: *fire; opalescence; pseudochromatism.* Syn: *schiller.*

plenargyrite *matildite.*

plenum (a) A system of ventilation in which air is forced into an inclosed space, such as a room or a caisson, so that the outward pressure of air in the space is slightly greater than the inward pressure from the outside, and thus leakage is outward instead of inward. (b) A mode of ventilating a mine or a heading by forcing fresh air into it. (c) Use of compressed air to hold soil from slumping into an excavation. (Nichols, 1976)

pleochroic *pleochroism.*

pleochroic halo (a) A minute zone of color or darkening surrounding and produced by a radioactive mineral crystal or inclusion. (AGI, 1987) (b) Any of the concentrically colored aureoles in minerals—e.g., micas, fluorite, and cordierite—centered by minute grains of minerals containing radioactive elements, such as zircon and monazite. This discoloration results from crystal structural radiation damage from alpha decay.

pleochroism (a) The property of exhibiting different colors in different directions by transmitted polarized light. (AGI, 1987) (b) More precisely, the property of absorbing differently, light that vibrates in different directions in passing through a crystal. If the crystal is uniaxial the change of color is called dichroism; if the crystal is biaxial, the change of color is called pleochroism. (AGI, 1987) (c) The property of birefringent crystals (minerals) to absorb various wavelengths of light differentially depending on the vibration direction of the light within the crystal. Thus a mineral displaying pleochroism shows various colors or tints when it is traversed by plane polarized light and the orientation of the crystal is varied with respect to the plane of polarization. It is a common and diagnostic property of many minerals, and is easily observed under the petrographic microscope or a dichroscope. (AGI, 1987) (d) The capacity of strongly anisotropic minerals to change absorption colors with changing electric vector in plane-polarized light; e.g., as seen with a polarized-light microscope. Uniaxial minerals may be dichroic and biaxial ones trichroic. Qualitative pleochroism is change of intensity in the same color; quantitative pleochroism shows change of color with change of orientation. Adj: pleochroic. Cf: *dichroism; trichroism.* Syn: *polychroism.*

pleomorphism *polymorphism.* Adj. pleomorphous

pleonaste *ceylonite.*

plessite A fine grained intergrowth of kamacite and taenite.

pliable armored cable A flexible cable having collective armor comprising stranded groups of fine, galvanized, steel wires. (BS, 1965)

pliable support A support composed of elastic materials that either yields to the roof pressure, or permits the subsidence of the roof without the support being completely destroyed and losing its significance. (Stoces, 1954)

plication Intense, small-scale folding. Adj: plicated. Cf: *crenulation.* (AGI, 1987)

ploat Eng. To dress down or remove loose stone from the roof or sides. (SMRB, 1930)

plombierite A mineral, $Ca_5H_2Si_6O_{18}{\cdot}6H_2O$(?).

plot mark A mark made in a bit mold, bit die, or blank bit where a pip or hole is drilled to receive or to encompass a diamond. (Long, 1960)

plotting instrument A large drawing machine by means of which stereoscopic pairs of vertical photographs can be viewed in conjunction with their ground control points and mechanically translated into accurate maps. (Hammond, 1965)

plotting scale A scale used for setting off the lengths of lines in surveying.

plow (a) In coal mining, a cutter loader with knives or blades, which is pulled along the longwall face by a powerful chain. The broken coal is loaded onto an armored flexible conveyor which, with the aid of hydraulic rams, holds the plow up to the coal face and causes the knives to bite into the coal as they are pulled along. The plow is a continuous mining machine. See also: *plow-type machine.* (Nelson, 1965) (b) Applied to V-shaped belt scrapers that are attached to the belt conveyor frame and which press against the return belt. They are intended to remove coal or other material that might stick to the return belt and be crushed as the belt passes over the driving rolls or the return pulley. (Jones, 1949)

plow cut *V-cut.*

plow deflector (a) A steel plate attached to the end of a cutter loader for deflecting cut coal onto the face conveyor. (Nelson, 1965) (b) A device for removing or diverting the dust and dirt off a belt conveyor and thus prevent it being carried back along the return belt. (Nelson, 1965)

plow steel A high-tensile steel used in the manufacture of hoisting ropes.

plow-type machine Plows may be divided into two classes: (1) machines that peel the coal to a depth of from 1 to 12 in (2.54 to 30.5 cm) by knives of various designs and the cut coal is then loaded onto a heavy type scraper chain conveyor; and (2) machines that peel a thin slice up to 2 in (5.1 cm) in thickness, by knives attached to each end of a steel box, and the coal is dragged along the face inside the box. From the aspect of speed of travel, plows may be divided into: (1) slow-moving types of 10 to 20 ft/min (3.0 to 6.1 m/min), which remove a thicker slice; and (2) fast-moving types at about 80 ft/min (24.4 m/min), which take a relatively thin slice. See also: *Anbauhobel; continuous mining; Lobbe Hobel.* (Nelson, 1965)

plucking (a) Describes the sudden jerking or plucking on heavy endless-rope haulage when the rope again takes the load, following rope coils. Instead of slipping smoothly sideways, the rope tends to stick until the pressure of oncoming coils overcomes the friction; these slip suddenly, producing a momentary slackening followed by a sudden jerk or pluck as the rope again takes the load. This may loosen chains or clips and cause derailments and runaway sets. (Sinclair, 1959) (b) The disruption of blocks of rock by a glacier or stream. (Standard, 1964)

plug (a) A watertight seal in a shaft formed by removing the lining and inserting a concrete dam, or by placing a plug of clay over ordinary debris used to fill the shaft up to the location of the plug. (BS, 1963) (b) *hoisting plug.* (c) A steel cylinder placed inside the annular opening in a coring bit to convert it for use as a noncoring bit. The face of the plug may or may not be provided with serrations, inset diamonds, or other types of cutting edges. (Long, 1960) (d) *block.* (e) *cartridge.* (f) A cylindrical piece of wood or an expandable metal apparatus placed in a borehole to act as a base into which the drive wedge of a borehole deflection device is driven. (g) Small wooden pin driven into a hole in the rock roof of a tunnel. The axis of the tunnel is marked on such plugs by tacks, or by small iron hooks from which a plummet lamp may be suspended for sighting upon. (Stauffer, 1906) (h) To plug a well by cementing a block inside casing or capping the well with a metal plate. (Wheeler, 1958) (i) Any block installed within casing to prevent movement of fluids. (Wheeler, 1958) (j) A steel wedge used in quarrying dimension stone. *plug-and-feather method.* (k) A vertical, pipelike body of magma that represents the conduit to a former volcanic vent. Cf: *neck.* (AGI, 1987)

plug-and-feather hole A hole drilled for the purpose of splitting a block of stone. These holes are usually in rows. The plug is a slightly wedge-shaped piece of iron driven between two L-shaped irons, or feathers, inserted in the hole. (Stauffer, 1906)

plug-and-feathermethod A method used in quarrying to reduce large masses of stone to smaller size. By using a hammer drill, a row of shallow holes is made along the line where a break is desired. The feathers consist of two iron strips flat on one side for contact with the wedge, and curved on the other to fit the wall of the drill hole. They are placed in the hole and the plug (a steel wedge) is placed between them. They are sledged lightly in succession until a fracture appears. Wherever possible, such fractures are made parallel with the rift of the stone. Syn: *multiple wedge.* See also: *plug.* (AIME, 1960)

plug bit (a) A diamond bit that grinds out the full width of a hole. (Nichols, 1976) (b) A noncoring diamond-set bit that can be in the form of a bullnose bit, pilot bit, or concave bit. Also called bullnose bit; concave bit; noncore bit; pilot bit. (BS, 1963)

plug box Eng. A wooden water pipe used in coffering.

plug drill A stonecutter's percussion drill. (Webster 3rd, 1966)

plugged (a) A borehole that has been filled or capped with a long plug, or in which a plug has been inserted. (Long, 1960) (b) Cracks or openings in the rocks in the walls of a borehole that have been filled or sealed with cement or other substances. (Long, 1960) (c) A borehole that has been drilled with a plug or noncoring bit. (Long, 1960) (d) A blocked core barrel or bit. (Long, 1960) (e) A coring bit in which a plug has been inserted. See also: *plug.* (Long, 1960)

plugged bit (a) *noncoring bit.* (b) A core bit, the annular opening of which is tightly closed or blocked by a piece or the impacted fragments of a core. (Long, 1960)

plugged crib A curb supporting the walling in a shaft and is itself supported on plugs or bolts driven into the ground around the shaft. The crib may be removed when the walling from below is carried up to it. See also: *strata bolt.* (Nelson, 1965)

plugging (a) The stopping of the flow of water into a shaft by plugs of clay. (Zern, 1928) (b) The material used, the act, or the process of inserting a plug in a borehole to fill it or the cracks and openings in the borehole sidewalls. (Long, 1960) (c) The act or process of drilling a borehole with a noncoring bit. (Long, 1960) (d) The practice of filling holes and cavities in castings with porous silicate mixture (cast iron filler) before the application of cover coats. The filler must be firmly forced into the casting holes, since any entrapped air beneath the filler will expand during firing and force the material out causing blowholes. (Enam. Dict., 1947)

plughole (a) A passageway that is left open, while working on an explosion-proof stopping, for the purpose of maintaining the ventilation of the fire area at or as near the normal quantity as possible, to prevent any increase in the combustible gases content in the air. After the stopping is completed, this hole is plugged up with sandbags in order to completely seal off the mine area. The plughole is generally a tapered passageway of about 3.5 ft (1.1 m) square at the inby side of the stopping and 2.5 ft (0.76 m) square at the outby side. (McAdam, 1955) (b) *block hole.*

plughole stopping A stopping in which the floor and the sidewalls of the passage are built of sandbags, and the roof may be the roof of the roadway or covering boards used between the webs of steel arches, or preferably, corrugated steel sheeting used as lagging behind steel arches. The plughole or passage is generally tapered from the inby end from 3 to 3.5 ft (0.9 to 1.1 m) square to 2.5 ft (0.76 m) square so that, in the event of an explosion, the plug of sandbags in the passage is subjected to a wedging action assisting to retain the plug in place. The plughole may be placed in the most convenient position and although this is often at the top, it is sometimes placed to the side and reasonably near the floor. (Sinclair, 1958)

plugman *pumping engineer.*

plug shot Scot. A small charge exploded in a hole to break up a stone of moderate size.

plug valve A valve or cock opened or closed by the turning of a plug, usually conical in shape. Not to be confused with needle valve or globe valve. (Long, 1960)

plum (a) A large random-shaped stone dropped into a large-scale mass of concrete to economize on the volume of the concrete. (Hammond, 1965) (b) An old form of plumb. (Fay, 1920)

plumb (a) *vertical.* (b) *plumb bob; plumbline.* (c) To carry a survey into a mine through a shaft by means of heavily weighted fine wires hung vertically in the shaft. The line of sight passing through the wires at the surface is thus transferred to the mine workings. An important piece of work: in mine shafts, and in transferring courses or bearings from one level to another. (Fay, 1920)

plumbago (a) A special quality of powdered graphite used to coat molds, and in a mixture with clay, to make crucibles. (b) See also: *black lead; graphite.* (c) Impure graphite or graphitic rock. (d) Minerals resembling graphite; e.g., molybdenite.

plumbago crucible Highly refractory crucible composed of a mixture of about equal parts of refractory clay and graphite. (Osborne, 1956)

plumb bob (a) A small weight or bob, hanging at the end of a cord, which under the action of gravity is oriented in a vertical direction. Also called a plummet. (CTD, 1958) (b) A pointed weight hung from a string. Used for vertical alignment. (Nichols, 1976)

plumber's dope A soft sealing compound for pipe threads. (Nichols, 1976)

plumbic Of, pertaining to, or containing lead, esp. in its higher valence. Cf: *plumbous.* (Standard, 1964)

plumbiferous Containing lead. (Webster 3rd, 1966)

plumbing Transferring a point at one level to a point vertically below or above it by means of a weight (plumb bob or plummet) suspended at the end of a string or wire (plumbline). See also: *centering of shaft; string survey.* (Nelson, 1965)

plumbline A device used to produce a vertical line between a survey instrument and the reference point over (or sometimes under, in underground work) which it is set. Special plumblines are used in a vertical shaft to transfer a fixed or an azimuth angle from the surface to underground workings for the purpose of orientation. Also known as plumb bob; plummet. See also: *Weisbach triangle.* (Pryor, 1963)

plumbocalcite A variety of calcite containing a small amount of lead carbonate.

plumboferrite A trigonal mineral, $PbFe_4O_7$; black; at Jakobsberg, Sweden.

plumbogummite A trigonal mineral, $PbAl_3(PO_4)_2(OH)_5{\cdot}H_2O$; crandallite group; forms yellow to brown encrustations; in Cumberland, United Kingdom.

plumbojarosite A trigonal mineral, $PbFe_6(SO_4)_4(OH)_{12}$; alunite group; forms minute brown tabular crystals with rhombohedral cleavage.

plumbomicrolite An isometric mineral, $(Pb,Ca,U)_2Ta_2O_6(OH)$; pyrochlore group. It occurs in greenish-yellow and orange masses and octahedra from Kivu, Zaire.

plumbous Of, pertaining to, or containing lead, esp. in its lower valence. Cf: *plumbic.* (Standard, 1964)

plumb pneumatic jig Mineral concentrator in which air is pulsed upward through a porous deck by means of a rotary valve. (Pryor, 1963)

plumb post One of the vertical posts at the side of a tunnel resting on sills and carrying the wallplates; collectively, they support the tunnel roof by means of centering. (Stauffer, 1906)

plumites A feathery variety of jamesonite. See also: *plumose antimony.*

plummet *plumbline.*

plumose Having a feathery appearance. (Fay, 1920)

plumose antimony A feather-ore variety of jamesonite or boulangerite; also called feather ore. Also spelled plumites, plumosite.

plumose mica A feathery variety of muscovite.

plumosite A feathery variety of jamesonite or boulangerite.

plump Corn. A corruption of the word pump.

plum-pudding stone *puddingstone.*

plunge (a) The vertical angle between a horizontal plane and the line of maximum elongation of an orebody. (b) The inclination of a fold axis or other linear structure, measured in the vertical plane. Cf: *appar-*

ent plunge; dip. (AGI, 1987) (c) To set the horizontal cross wire of a theodolite in the direction of a grade when establishing a grade between two points of known level. (AGI, 1987) (d) To reverse the direction of the telescope of a theodolite by rotating it 180° about its horizontal axis. Syn: *transit*. (AGI, 1987)

plunger (a) In blasting, a rod designed for thrusting into a drill hole and ascertaining the position of a cartridge. (Standard, 1964) (b) The piston of a force pump. (Fay, 1920) (c) A piston and its attached rod. (Long, 1960)

plunger bucket A pump piston without a valve. Also called plunger lift. (Webster 3rd, 1966)

plunger case The pump barrel, or cylinder, in which a solid piston or plunger works. Also called pole case. (Fay, 1920)

plunger jig washer A washer in which water is forced upward and then downward through a screen by the action of a plunger in an adjoining compartment. Although these machines are still in use, the term "jig washer" is now applied to the fixed-screen, air-pulse jig, which is directly descended from the first Baum washer used in 1892. See also: *jig washer; pneumatic jig*. (Nelson, 1965)

plunger lift Scot. A pump and attached column of pipes, that raises water by means of a ram or piston. (Fay, 1920)

plunger press A press in which the pressure is applied by a plunger, with a reciprocating motion, to charges of feed contained in molds in a vertical or horizontal table. (BS, 1962)

plunger pump (a) Reciprocating pump used for moving water or pulp, in which a solid piston displaces the fluid. (Pryor, 1963) (b) A displacement-type pump may be of various types, such as: (1) the triplex pump, a vertical or horizontal, single-acting plunger type for small heads with three single-acting cylinders in the pump frame driven by a motor mounted on the outside of the frame and connected to the crankshaft of the pump through gearing; (2) the quadruplex or quintuplex pump, a pump having four or five cylinders; and (3) the duplex pump, a crank-and-flywheel type for high heads, with double-acting plungers. (Lewis, 1964)

plunger-type washbox A washbox in which pulsating motion is produced by the reciprocating movement of a plunger or piston. Syn: *piston-type washbox*. (BS, 1962)

plus distance Fractional part of 100 ft or m used in designating the location of a point on a survey line—such as, 4+47.2, meaning 47.2 ft or m beyond Station 4; or 447.2 ft or m from the initial point, measured along a specified line. (Seelye, 1951)

plush copper ore *chalcotrichite; cuprite.*

plus mesh The portion of a powder sample retained on a screen of stated size. (Osborne, 1956)

plus sight *backsight.*

pluton A body of medium- to coarse-grained igneous rock that formed beneath the surface by crystallization of a magma.

plutonic (a) Pertaining to igneous rocks formed at great depths. Cf: *hypabyssal*. (AGI, 1987) (b) Pertaining to rocks formed by any process at great depth. Syn: *abyssal; deep seated; hypogene*. (AGI, 1987)

plutonic metamorphism Deep-seated regional metamorphism at high temperatures and pressures, often accompanied by strong deformation; batholithic intrusion with accompanying metasomatism, infiltration, and injection (or, alternatively, differential fusion or anatexis) is characteristic. Cf: *injection metamorphism*. (AGI, 1987)

plutonic ore deposit Collectively, the major group of ore deposits of magmatic origin that have been formed under abyssal conditions. (Schieferdecker, 1959)

plutonic rock Igneous rock formed deep within the Earth under the influence of high heat and pressure, hypogene rocks; distinguished from eruptive rock formed at the surface. (Hess)

plutonic series A series of different igneous rocks that evolved from the same original magma through various differentiation stages.

plutonism (a) The obsolete belief that all of the rocks of the Earth solidified from an original molten mass. Cf: *neptunism*. (b) A general term for the phenomena associated with the formation of plutons. (AGI, 1987)

pluviometer *rain gage.*

ply (a) U.K. A thin band of shale lying immediately over a coal seam. (b) U.K. A rib or successive ribs; e.g., of clayband with very thin partings. (c) Limy ply; a limestone bed; Edinburgh, U.K.

pneumatic Set in motion or operated by compressed air. (Nelson, 1965)

pneumatic blowpipe A long, 3/4-in-diameter (1.9-cm-diameter) metal pipe, connected to an air supply; used to blow out dust and chippings from vertical blast holes at quarries. The blowpipe is generally used for holes exceeding about 12 ft (3.66 m) deep. A stream of water is sometimes used instead of an air jet. (Nelson, 1965)

pneumatic caisson Closed casing in which air pressure is maintained equal to the pressures of the water and soils on the outside. The deeper the caisson, the higher the pressure that must be maintained. (Carson, 1961)

pneumatic cartridge loader A cartridge loader widely used for underwater blasting, for blasting without removing the overburden, and for long-hole blasting. It is also being used increasingly for tunneling and other sorts of rock blasting. (Langefors, 1963)

pneumatic cleaning Mineral cleaning by machines that utilize air currents as the primary separating medium. The air machines can generally be divided into three types: (1) pneumatic jigs, in which the air current is pulsated; (2) pneumatic tables, in which the refuse is diverted from the direction of flow of the clean mineral by a system of riffles fixed to the deck; and (3) pneumatic launders, in which the products are flowing in the same direction, and the clean mineral is skimmed off the top of the bed and/or the refuse is extracted from the bottom in successive stages. (Mitchell, 1950)

pneumatic concentrator Gravity jig, shaking table, or other device in which suitably ground minerals are separated by gravity during their exposure to a continuous or pulsating current of air. (Pryor, 1963)

pneumatic conveying Use of compressed air to move fairly fine aggregates laterally and/or vertically. (Pryor, 1963)

pneumatic conveyor (a) A pipe or tube through which granular material is transported by airblast. It is used for pulverized coal, crushed rock (pneumatic stowing), cement, etc. The term could also be applied to a conveyor operated by compressed air. (Nelson, 1965) (b) An arrangement of tubes or ducts through which bulk material or objects are conveyed in a pressure and/or vacuum system.

pneumatic drill Compressed-air drill worked by reciprocating piston, hammer action, or turbo drive. (Pryor, 1963)

pneumatic drill leg *air-leg support.*

pneumatic filling A filling method in which compressed air is utilized to blow filling material into a mined-out stope. (Stoces, 1954)

pneumatic flotation cell Machine in which the air used to generate a mineralized froth is blown into the cell, either through a porous septum at or near the bottom, or by pipes that bring low-pressure air to that region. (Pryor, 1963)

pneumatic friction clutch This clutch transmits power through friction shoes carried on the tube of cord and rubber construction. The pneumatic clutch is self-adjusting for wear owing to the natural resilience of the rubber tube. Disengagement is complete and automatic when the air under pressure is released. The clutch is controlled by finger pressure on a valve. The valve can be installed at the place most convenient for the operator. (Pit and Quarry, 1960)

pneumatic hammer A hammer that uses compressed air for producing the impacting blow.

pneumatic hoist A device for hoisting; operated by compressed air. (Standard, 1964)

pneumatic injection A method for fighting underground coal fires. This air-blowing technique involves the injection of an incombustible mineral, like rock wool or dry sand, through 6-in (15.2-cm) boreholes drilled from the surface to intersect underground passageways in the mines.

pneumatic jig (a) Air jig used in desert countries for concentrating ore. (Pryor, 1963) (b) A jigging machine in which an airblast performs the work of separation of minerals. (Standard, 1964) (c) *Kirkup table; plunger jig washer.*

pneumatic lighting (a) Underground lighting produced by a compressed-air turbomotor driving a small dynamo. (Pryor, 1963) (b) The use of compressed air to generate electric light. See also: *air turbolamp*. (Nelson, 1965)

pneumatic lubricator *line oiler.*

pneumatic method In flotation, a method in which gas is introduced under slight pressure near the bottom of the flotation vessel, the device used for introduction being either a submerged pipe or a porous cloth, frit, or rubber surface forming the wall of a wind box. (Gaudin, 1957)

pneumatic mortar Mortar applied to a surface with a cement gun in the same manner as gunite. Such mortar has a cube crushing strength of 3,000 psi (20.7 MPa) at 7 days and of 6,000 psi (41.4 MPa) at 28 days, with a water-cement ratio of 0.45. (Hammond, 1965)

pneumatic pick A compressed-air-operated hand tool used to excavate coal, ore, and rock, with a punching action. Without the pick steel, its length is about 18 in (46 cm) and weight about 24 lb (10.8 kg). It delivers about 2,500 blows/min. The latest type is the water-controlled pick, so designed that the air valve is operated by water pressure. The water assists in suppressing the dust made during cutting. (Nelson, 1965)

pneumatic ram A ram fed by a compressed-air pipeline. The piston is about 8 in (20 cm) in diameter, giving an area of 50 in^2 (323 cm^2) and exerts a pushing force of up to 4,000 lb (1,800 kg). (Nelson, 1965)

pneumatic riveter A compressed-air tool used for driving rivets. See also: *rivet*. (Hammond, 1965)

pneumatic rod puller An air-driven rod puller. See also: *rod puller*. (Long, 1960)

pneumatics The branch of physics that deals with the mechanical properties of gases, such as their pressure, elasticity, density, and also of pneumatic mechanisms; sometimes it includes acoustics. (Standard, 1964)

pneumatic shaft sinking (a) Shaft sinking with the aid of a drop shaft fitted with an air-tight deck to form a working chamber. See also: *manlock.* (Hammond, 1965) (b) The caisson-sinking process now largely obsolescent in mining practice. (Nelson, 1965)

pneumatic stowing A system of filling mined cavities in which crushed rock is carried along a pipeline by compressed air and discharged at high velocity into the space to be packed, the intense projection ensuring a very high density of packed material. For stowing shallow workings—up to 200 yd (183 m) in depth—the stowing plant may be installed on the surface. The air pressure is about 60 psi (414 kPa). For deeper workings, the plant may be installed underground, and the crushed rock taken down from the surface. The stowing pipes are about 5 to 6 in (approx. 13 to 15 cm) in diameter. The system is often employed if important surface structures require protection. The material used is from old dirt heaps, screen dirt, and washery rejects. The material is crushed to -2-1/2 in (-6.35 cm) and preferably without the -1/2-in (-1.27-cm) material. See also: *air-stowing machine; crusher stower; hydraulic stowing; low-pressure air stower.* (Nelson, 1965)

pneumatic table An appliance for the dry cleaning of ore or coal. It consists of a perforated deck, with vertical ribs or riffles, which is reciprocated; the motion keeps the bed of raw coal sufficiently mobile for the blast of air from below to effect a process of stratification (or layering). The coal rises to the surface, with dirt at the base and a central layer of middlings. See also: *Birtley contraflow separator; Kirkup table; Vee table; air table.* (Nelson, 1965)

pneumatic tamper Essentially a long-stroke piston with a mushroom-shaped foot about 4 in (10 cm) in diameter. It operates on compressed air, which is used to lift the piston and footpiece; their combined weight, in falling, supplies the impact. (Carson, 1961)

pneumatic tool Tool operated by air pressure. (Crispin, 1964)

pneumatic transport System composed of: a compressor, which provides airflow; a feeder, which meters the flow of material into a pipeline; and the pipeline— for transporting coarse, dry, noncohesive material. (SME, 1992)

pneumatic water barrel A special type of water barrel for removing water from a shaft sinking. By means of a hose connection to an air pump at the surface, a partial vacuum is created inside the barrel and the water lifts the valve and fills the barrel. The hose is then detached and the barrel is hoisted to the surface and discharged. Also called vacuum tank. (Nelson, 1965)

pneumatogenic Said of a rock or mineral deposit formed by a gaseous agent. Cf: *hydatogenic; hydatopneumatogenic; pneumatolytic.* (AGI, 1987)

pneumatolysis Alteration of a rock or crystallization of minerals by gaseous emanations derived from solidifying magma. Adj: pneumatolytic. (AGI, 1987)

pneumatolytic A term used in different connotations by various authors and perhaps best abandoned. It has been used to describe: (1) the surface effects of gases near volcanoes; (2) contact-metamorphic effects surrounding deep-seated intrusives; (3) that stage in igneous differentiation between pegmatitic and hydrothermal, which is supposed to be characterized by gas-crystal equilibria; and (4) very loosely, any deposit containing minerals or elements commonly formed in pneumatolysis, such as tourmaline, topaz, fluorite, lithium, and tin, and hence presumed to have formed from a gas phase. Cf: *pneumatogenic.*

pneumatolytic metamorphism Contact metamorphism in which the composition of a rock has been altered by introduced gaseous magmatic material.

pneumatolytic stage That stage in the cooling of a magma during which the solid and gaseous phases are in equilibrium. (AGI, 1987)

pneumo- A combining form taken from the Greek meaning lung, and used in connection with the terminology of geologic processes and effects involving gases and vapors. (Stokes, 1955)

pneumoconiosis A disease of the lungs caused by habitual inhalation of irritant mineral or metallic particles. It occurs in any workplaces where dust is prevalent, such as mines, quarries, foundries, and potteries. Also called miner's asthma; miner's consumption; miner's lung. Also spelled pneumonoconiosis; pneumonokoniosis. Cf: *anthracosis; silicosis.* See also: *phthisis.* (Webster 3rd, 1966)

pneumokoniosis *pneumoconiosis.*

pocket (a) A localized enrichment; a crevice in bedrock containing gold; a rich patch of gold in a reef. (b) A rich deposit of mineral, but not a vein. (c) A bin, of a capacity equal to the skip, used at the shaft bottom of an underground mine for quick and accurate skip loading. See also: *shaft pocket; measuring chute.* (Nelson, 1965) (d) A receptacle, from which coal, ore, or waste is loaded into wagons or cars. (Fay, 1920) (e) A ganister quarryman's local term for masses of rock, 30 to 50 ft (9.1 to 15.2 m) in width, that are worked out and loaded, leaving buttresses of untouched rock between them to support the upper masses. (Fay, 1920) (f) A hole or depression in the wearing course of a roadway. (Fay, 1920) (g) A local accumulation of gas. (Hudson, 1932) (h) A bulge, sop, or belly in a lode or bed. See also: *belly.* (Arkell, 1953) (i) A cavity, whether filled with air, water, mineral, or gravel. (Arkell, 1953) (j) In pegmatites, the central openings lined with crystals, including those of gem species. (Sinkankas, 1959)

pocket-and-fender method In pillar extraction, a method in which lifts are mined in the same way as in the open-end method, except that a fender of coal or a series of small coal stumps is left adjacent to the gob as the lift is advanced. After the lift is completed, the fender or stumps of coal are blasted, and sometimes part of this coal is recovered. (Woodruff, 1966)

pocket-and-stump method A method of mining pillars in which a narrow pillar of coal, called the stump, is left along the goaf (worked-out space) to support the roof while driving the pocket. This coal acts as a protection for the miners. When the pocket has been completed, the stump is worked back, then another pocket is driven, and so on. (Lewis, 1964)

pocket compass A magnetic needle enclosed in a nonmagnetic case, the needle being free to swing over a graduated face or dial. The compass is useful for experimental purposes or for direction-finding in desolate parts of the countryside, or during darkness and foggy weather. (Morris, 1958)

pocket conveyor A continuous series of pockets, formed of a flexible material festooned between crossrods, carried by two endless chains or other linkage that operate in horizontal, vertical and inclined paths.

pocket hunter California. A miner or prospector who searches for small gold deposits which occur on the surface in the gold-bearing areas of the State. (Fay, 1920)

pocket of gas A small accumulation of methane in a roof cavity, where it is beyond the reach of the ventilating air current. See also: *deflector sheet; combustible gases layer; hurdle sheet.* (Nelson, 1965)

pocket transit *Brunton compass.*

pod A rudely cylindrical orebody that decreases at the ends like a cigar or a potato. See also: *lens.* (Stokes, 1955)

Podsol *Podzol.*

Podzol A great soil group in the 1938 classification system; a group of zonal soils having an organic mat and a very thin organic-mineral layer overlying a gray, leached A2 horizon and a dark brown, illuvial B horizon enriched in iron oxide, alumina, and organic matter. It develops under coniferous or mixed forests or under heath, in a cool to temperate moist climate. Also spelled Podsol. Spelled "podzol" when used as the soil type belonging to the Podzol group. Etymol: Russian podsol, ash soil. (AGI, 1987)

podzolization The process by which a soil becomes more acid owing to depletion of bases, and develops surface layers that are leached of clay and develop illuvial B horizons; the development of a podzol. Also spelled: podsolization. (AGI, 1987)

poecilitic The original spelling of poikilitic. Now obsolete in American usage, it is still the most accepted European spelling. (AGI, 1987)

Poetsch process (a) The original freezing process of shaft sinking developed by F. H. Poetsch in 1883. See also: *freezing method.* (Nelson, 1965) (b) A process in which brine at subzero temperature is circulated through boreholes to freeze running water through which a shaft or tunnel is to be driven during development of a waterlogged mine. (Pryor, 1963)

poicilitic *poikilitic.*

poidometer An automatic weighing device for use on belt conveyors. The device feeds the material from a hopper in a uniform stream onto a short independent belt conveyor and from there onto the main belt or bin. The weight of material on the measuring belt actuates a scale beam that raises or lowers a gate controlling the rate of flow from the feed hopper to a certain predetermined load per foot of measuring belt. A meter records the travel of measuring belt, and this figure multiplied by the weight per foot of belt, as fixed by the scale beam adjustment, gives the weight of material handled in any given period. See also: *weightometer.* (Nelson, 1965)

poikilit *bornite.* Also called poikgopyrite.

poikilitic A rock texture in which numerous grains of various minerals in random orientation are completely enclosed within a large, optically continuous crystal of different composition. Also spelled poicilitic. Cf: *ophitic.*

poikilitic texture *poikilitic.*

poikiloblastic (a) Said of a metamorphic texture in which small grains of one constituent lie within larger metacrysts. Modern usage favors this meaning. Syn: *sieve texture.* (AGI, 1987) (b) Said of a metamorphic texture due to the development, during recrystallization, of a new mineral around numerous relicts of the original minerals, thus simulating the poikilitic texture of igneous rocks. Cf: *helicitic.* (AGI, 1987)

point (a) A predetermined direction for driving a roadway underground. The point is fixed by roof plugs in the roadway. See also: *alignment; spad.* (Nelson, 1965) (b) One one-hundredth (0.01) part of a carat. When less than one carat, the weight of a diamond is usually expressed in points; e.g., 20 points equals 1/5 carat. (Chandler, 1964)

(c) A pipe through which steam or hot water is brought into contact with frozen gravel to thaw it for mining or dredging. (d) *well point.* (e) In quarrying, a type of wedge that tapers to a narrow, thin edge. (f) The end or bottom of a borehole, as distinguished from the mouth or collar. (Fay, 1920) (g) A tool used in trimming and smoothing rough stone surfaces. (Webster 3rd, 1966) (h) Either of a pair of tapered rails at a turnout that can be adjusted to direct a set of mine cars from a straight rail track to another track branching off at an angle. See also: *catch point; turnout.* (Nelson, 1965)

point agate *point chalcedony.*

point chalcedony White or gray cryptocrystalline quartz flecked with tiny spots of iron oxide, giving the whole surface a uniform soft red color. Syn: *point agate.*

point defect A deviation from ideal crystal structure about a point location; e.g., interstitial, atom missing (Schottky), or combined (Frenkel). Cf: *crystal defect; Frenkel defect; Schottky defect.*

point driver In metal mining, a person who drives steam or water points (specially made pipes with a chisel bit at one end) into the frozen ground of a placer deposit in advance of dredging operations, to thaw the ground so that it can be worked by the dredge for recovery of gold. Syn: *thawing.* (DOT, 1949)

pointed box A box, in the form of an inverted pyramid or wedge, in which minerals, after crushing and sizing, are separated in a current of water. See also: *spitzkasten.*

point group One of 32 geometrically possible arrays of symmetry elements intersecting at a point. These symmetry elements are axes of rotation, both proper and improper ($1 = \bar{i}$, $2 = m$). All minerals having the symmetry of one point group belong to the same crystal class. Cf: *symmetry; crystal class; space group.*

point kriging Estimating the value of a point from a set of nearby sample values using kriging. The kriged estimate for a point will usually be quite similar to the kriged estimate for a relatively small block centered on the point, but the computed kriging standard deviation will be higher. When a kriged point happens to coincide with a sampled location, the kriged estimate will equal the sample value. See also: *kriging.*

point of attack *portal.*

point of compound curvature The point of tangency common to two curves of different radii, the curves lying on the same side of the common tangent. Abbrev., P.C.C. (Seelye, 1951)

point of curvature The point where the alignment changes from a straight line or tangent to a circular curve; i.e., the point where the curve leaves the first tangent. Abbrev., P.C. (Seelye, 1951)

point of decalescence *decalescence.*

point of frog The intersection gagelines of the main track and a turnout. (Kiser, 1929)

point of intersection (a) The point where intersecting lines cross one another. (Jones, 1949) (b) The point where the two tangents to a circular curve intersect. Abbrev., P.I. Also called vertex. (Seelye, 1951)

point of recalescence *recalescence.*

point of switch That point in the track where a car passes from the main line onto the rails of a turnout. (Kiser, 1929)

point of tangency The point where the alignment changes from a circular curve to a straight line or tangent; i.e., the point where the curve joins the second tangent. Abbrev., P.T. (Seelye, 1951)

point plotting In seismology, a procedure in reflection interpretation in which depth points are computed and plotted for each seismogram trace separately. (Schieferdecker, 1959)

point source A single point from which light emanates; e.g., the sun or a lamp filament, or their reflections. In the case of multiple reflections, each is a point source.

poise (a) The unit of absolute viscosity, equal to one dyne-second per square centimeter. Named from the physicist Poiseuille. (AGI, 1987) (b) The second unit of fluid viscosity, often expressed in centimeters or grams. See also: *absolute viscosity; Poiseuille's law.* (Pryor, 1963)

Poiseuille's law A statement in physics that the velocity of flow of a liquid through a capillary tube varies directly as the pressure and the fourth power of the diameter of the tube and inversely as the length of the tube and the coefficient of viscosity. See also: *poise.* (AGI, 1987)

poisoning (a) In ion-exchange terminology, loading of resin sites with unwanted ions, thereby eliminating them as locations for loading. (Pryor, 1963) (b) Fouling of an organic solvent used in stripping pregnant leach liquor. (Pryor, 1963)

Poisson's ratio The ratio of the lateral unit strain to the longitudinal unit strain in a body that has been stressed longitudinally within its elastic limit. It is one of the elastic constants. Symbol: σ. Cf: *modulus of incompressibility.* (AGI, 1987)

poker man A laborer who removes blue powder and ash residue from retorts after molten zinc has been tapped. Also called scraper. (DOT, 1949)

polar (a) Lacking a center of symmetry, with the result that crystals are acentric in their crystal forms and physical properties; i.e., electrostatic or magnetic properties are equal and opposite at the opposite ends of these crystals; e.g., tourmalines. Ant. nonpolar. (b) An optical device, such as nicol prism or polarizing filter, for the production of plane-polarized light. See also: *nicol; Nicol prism.*

Polar Ajax A high-strength, high-density, nitroglycerin gelatin explosive, supplied in both unsheathed and sheathed forms. See also: *Ajax.* (McAdam, 1958)

polar curve A graph showing the distribution of light in a flame safety lamp obtained by plotting the values obtained at intervals of 10° around a full circle. (Mason, 1951)

polar explosive Explosive containing an antifreeze ingredient and distinguished by the prefix polar. Polar and nonpolar explosives of equal grade possess similar characteristics. Explosives that contain nitroglycerin tend to freeze when stored at low temperatures for lengthy periods. Syn: *low-freezing explosive.* (Nelson, 1965)

polariscope An optical device consisting of two polarizers with a space between for a crystal or rock under study. Syn: *stauroscope.*

polarity In crystallography, the property of having differing types of termination at the two ends of a prismatic crystal. May be reflected in pyroelectric properties, conduction of electric current, etc.

polarizability The property of an ion or atom to deform so as to create a dipole from the displacement of its electron cloud.

polarization (a) The difference between the equilibrium value of the potential of an electrode and the value attained when an appreciable current flows through a system. (Schlain) (b) In electrolysis, the condition in the vicinity of an electrode, such that the potential necessary to get a desired reaction is increased beyond the reversible electrode potential. (ASM, 1961) (c) The production of dipoles or higher-order multipoles in a medium. (AGI, 1987) (d) The polarity or potential near an electrode. (AGI, 1987) (e) In seismology, the direction of particle motion of shear (S) waves in a plane perpendicular to the direction of propagation. (AGI, 1987) (f) A process of filtration or reflection by which ordinary light is converted to plane-polarized light in which the electric vector of a light ray is confined to a single plane.

polarized light Light with its electric vector restricted to a plane or to an elliptically or circularly helical path as a result of filtration, reflection, or interaction with a crystal structure.

polarizer In a polarized-light microscope, the polarizing filter or Nicol prism (polar) located below the sample stage. See also: *nicol; analyzer.*

polarizing prism A prism of an anisotropic crystal, commonly calcite, cut and cemented together so as to permit passage of one of the doubly refracted light rays while reflecting the other out of the train of a microscope. Syn: *Nicol prism.*

polar moment of inertia The second moment of area about an axis perpendicular to its plane is known as the polar moment of inertia of a plane section. See also: *moment of inertia.* (Hammond, 1965)

Polaroid A sheet of cellulose impregnated with optically aligned crystals of quinine iodosulfate, which permit passage of light with its electric vector in one plane while absorbing all other impinging light. It is a cheap substitute for Nicol prisms in modern polarized-light microscopes.

Polar Viking A typical nitroglycerin powder explosive, which is now supplied only in the sheathed form. (McAdam, 1958)

polder Dutch. Low fertile land, as in The Netherlands and Belgium, reclaimed from the sea by systems of dikes and embankments.

pole (a) Either of the two regions of a permanent magnet or an electromagnet where most of the lines of induction enter or leave. A point toward which a freely suspended ferromagnetic rod aligns itself. (b) The negative or positive electrical pole in a circuit.

pole chain A surveyor's chain. See also: *Gunter's chain.* (Standard, 1964)

pole figure A stereographic projection representing the statistical average distribution of poles of a specific crystalline plane in a polycrystalline metal, with reference to an external system of axes. In an isotropic metal; i.e., in one having a completely random distribution of orientations, the pole density is stereographically uniform; preferred orientation is shown by an increased density of poles in certain areas. (ASM, 1961)

poleman *locomotive brakeman.*

pole piece A specially shaped piece of magnetic material forming an extension to a magnet; e.g., the salient poles of a generator or motor. (CTD, 1958)

pole strength In measurement of magnetic strength, the number of unit poles in the measured field. One unit pole is the strength in a vacuum required to exert 1 dyn in a 1-cm gap between poles. (Pryor, 1963)

polianite A steel-gray dioxide of manganese, MnO_2, crystallizing in the tetragonal system. It is distinguished from pyrolusite by its hardness and anhydrous character. (CMD, 1948; Dana, 1959)

poling (a) The act or process of temporarily protecting the face of a level, drift, cut, etc., by driving poles or planks along the sides of the yet unbroken ground. Used esp. for holding up soft ground. See also: *forepoling.* Also called spiling. (b) A step in the fire refining of copper to reduce the oxygen content to tolerable limits by covering the bath with coal or coke and thrusting greenwood poles below the surface. There is a vigorous release of reducing gases that combine with the oxygen contained in the metal. If the final oxygen content is too high, the metal is underpoled; if too low, overpoled; and if just right, tough pitch. (ASM, 1961; CTD, 1958)

poling back Carrying out excavation behind timbering already in place. (Hammond, 1965)

poling board (a) A forepoling board, driven horizontally ahead to support the roof when tunneling through running ground. See also: *forepoling.* (Pryor, 1963) (b) In trenching, either of a pair of side boards wedged apart. (Pryor, 1963)

polirschiefer Tripoli slate. Also called polishing slate. (Dana, 1914)

polish An attribute of surface texture of a rock, characterized by high luster and strong reflected light, produced by agents, such as desert or glacial polish, or by artificial grinding and smoothing; e.g., marble or granite. (AGI, 1987)

polished section A slice of rock or mineral that has been highly polished for examination by reflected-light or electron microbeam techniques, a procedure mostly applied to opaque minerals. See also: *thin section; reflected-light microscope.*

polished surface *slickensides.*

polishing Removing the last traces of suspended matter from solutions by passing them through a filter coated with diatomaceous earth or similar material.

polishing cask A barrel in which grained gunpowder is tumbled with graphite to glaze it. (Standard, 1964)

polishing mill A lap of metal, leather, or wood used by lapidaries in polishing gems. (Fay, 1920)

polled stone Som. Stone hewn into shape and faced ready for building. Building stone with one side rough faced, as opposed to hammer-and-punch dressed. (Arkell, 1953)

pollen peat Peat rich in pollen grains. (Tomkeieff, 1954)

poll pick A pick with a head for breaking away hard partings in coal seams or knocking down rock already seamed by blasting. See also: *hammerpick.* (Fay, 1920)

pollucite An isometric mineral, $(Cs,Na)_2Al_2Si_4O_{12}·H_2O$; zeolite group; forms a series with analcime; colorless; occurs in granite pegmatites; a source of cesium and a minor gemstone. Syn: *pollux.*

pollux *pollucite.*

poly- A prefix signifying many. Used in many mineral names, such as polybasite, polycrase, polyhalite, and polyaugite. (CCD, 1961; Spencer, 1955)

polyargyrite A mixture of argentite and tetrahedrite.

polybasite A monoclinic mineral $(Ag,Cu)_{16}Sb_2S_{11}$; forms a series with pearcite; pseudohexagonal; soft; metallic; gray to black; sp gr, 6.0 to 6.2; in low-temperature veins; a source of silver.

polychroilite Altered cordierite. (Dana, 1914)

polychroism *pleochroism.*

polychroite *cordierite.*

polycrase An orthorhombic mineral, $(Y,Ca,Ce,U,Th)(Ti,Nb,Ta)_2O_6$; black; in granite pegmatites. Formerly spelled polykras.

polycrystal A mineral specimen composed of an assemblage of individual crystals of various crystallographic orientations. See also: *glomerocryst; syntaxy.*

polycrystalline An aggregate of crystals of the same species.

polydymite An isometric mineral, $NiNi_2S_4$; linnaeite group; easily confused with violarite.

polygenetic (a) Resulting from more than one process of formation, derived from more than one source, or originating or developing at various places and times; e.g., said of a mountain range resulting from several orogenic episodes. (AGI, 1987) (b) Consisting of more than one type of material, or having a heterogeneous composition; e.g., said of a conglomerate composed of materials from several different sources. Cf: *monogenetic.* (AGI, 1987)

polygon A plane figure bounded by straight lines. (Jones, 1949)

polygonal A two-dimensional form having more than four regular straight sides.

polygonal method An ore-reserve computation method in which an assumption is made that the area of influence of each drill hole extends halfway to the neighboring drill holes. Therefore, thickness and grade must vary uniformly in opposite directions and in such cases errors tend to be compensating. Where the thickness and grade vary in the same direction, the errors will accumulate and cause erroneous results. (Krumlauf)

polyhalite A triclinic mineral, $K_2Ca_2Mg(SO_4)_4·2H_2O$; bitter tasting; varicolored; occurs in salt deposits in Texas, New Mexico, and Germany.

polykras *polycrase.*

polymer (a) A compound formed by the union of two or more molecules of the same simple substance. (Standard, 1964) (b) In the plural use, compounds identical in composition but which vary in molecular weight, such as ethylene (ethene), Ch_2:Ch_2; propylene (propene), CH_3CH:CH_2; and butylene (butene), CH_3CH_2CH:CH_2. (Standard, 1964)

polymerization Union of two or more molecules of given structure to form a new compound with the same elemental proportions but with different properties and a higher molecular weight. (Pryor, 1963)

polymerize To chemically combine small molecules into larger molecules; to undergo polymerization.

polymetallicsulfide A sulfide deposit rich in copper, zinc, lead, silver, or gold, which forms as a result of hydrothermal activity in the vicinity of mid-ocean spreading centers or tectonically active basins. The first discovery of these deposits was from the French submersible Cyana, in 1979, during a joint international biological investigation of thermal springs on the deep seabed. The term derives from the French "sulfides polymetalliques."

polymignite An orthorhombic mineral, $(Ca,Fe,Y,Th)(Nb,Ti,Ta,Zr)O_4$; radioactive; in syenites and granite pegmatites. Also spelled polymignyte.

polymorphism The characteristic of a chemical compound to crystallize in more than one crystal class; e.g.: (1) kyanite, andalusite, and sillimanite; (2) quartz, tridymite, cristobalite, coesite, and stishovite. Allotropy refers specif. to chemical elements crystallizing in more than one class; e.g., graphite, diamond, chaoite, and lonsdaleite. Polymorphism limited to two or three crystal classes is dimorphism or trimorphism, respectively. Individual species are polymorphs (dimorphs, trimorphs). Polytypism refers to variable stacking of identical layer structures in different crystal classes. Adj: polymorphic (dimorphic, trimorphic). Adv: polymorphous (dimorphous, trimorphous). Syn: *pleomorphism.* Cf: *dimorphism; trimorphism; isomorphism; isotypy; polysyngony.* See also: *allotropic.*

polynigritite Variety of nigritite found in a finely dispersed state in argillaceous rocks. Cf: *keronigritite; humonigritite.* (Tomkeieff, 1954)

polynite A montmorillonoid clay mineral in soils. (Spencer, 1958)

polyphase In electricity, having or producing two or more phases, such as a polyphase current. Syn: *multiphase.* (Webster 3rd, 1966)

polysomatic Having a texture consisting of numerous small grains; said of minerals. (Standard, 1964)

polysomatism Minerals having a texture of many small grains.

polysyngony A condition where two or more minerals have the same composition, but different crystal classes owing to changed bond angles; e.g., alpha and beta quartz. Cf: *polytypy; polytropy; polymorphism.*

polysynthetic twinning (a) Two systems of lamellar twinning at an angle with one another. (b) Successive twinning of three or more individuals, according to the same twin law, with parallel composition planes; commonly revealed by visibly striated cleavage planes; e.g., albite twinning in plagioclase feldspar. Cf: *cyclic twinning; twin laminae.*

polythionic acid Any of several acids in a series related to sulfurous and thiosulfuric acid. (Pryor, 1963)

polytropy A condition in which there is no change in the geometrical symmetry of the crystal structure of two related minerals, but a change to permit a variant in the resultant mineral; e.g., orthoclase microcline. Cf: *polytypy; polysyngony.* (Hess)

polytypism (a) A condition in micas and similar clay minerals in which they show growth spirals which are due to lamellae of different orientations. (AGI, 1987) (b) One-dimensional polymorphism resulting from alternate stacking of identical layers; e.g., kaolinite, nacrite, and dickite. Syn: *polytypy.*

polytypy A condition in which the space lattice of two related minerals is completely altered to a new type. This is illustrated by the quartz-tridymite relationship. See also: *polytypism.* Cf: *polysyngony; polytropy.* (Hess)

polyvinyl butyral A resin, with a plasticizer. Provides the interlayer in standard laminated glass made from either polished plate glass or window glass. (Lee, 1961)

polyxene A variety of native platinum alloyed with iron.

Poncelet wheel A kind of undershot waterwheel suitable for falls of less than 6 ft (1.8 m), having the buckets curved so that the water presses on them without impact.

Ponsard furnace A furnace in which the escaping combustion gases, passing through tubular flues, heat the incoming air continuously through the flue walls. (Fay, 1920)

pontil An iron rod used in glassmaking to carry and manipulate hot bottles, etc.; has a projection at the end, varying in shape according to the character of the ware carried. Also called snap; pontee; ponto; ponty; puntee; puntil; punty. (Standard, 1964)

pontoon (a) A float supporting part of a structure, such as a bridge. (Nichols, 1976) (b) A wood platform used to support machinery on soft ground. (Nichols, 1976)

pony set A small timber set or frame incorporated in the main sets of a haulage level to accommodate an ore chute or other equipment from above or below. (Nelson, 1965)

pool (a) To undercut or undermine material, such as coal, esp. in excavating. (Webster 3rd, 1966) (b) A continuous area of porous sedimentary rock that yields petroleum or gas on drilling. (Webster 3rd, 1966)

pool washing screen A screen that is divided into alternate transverse screen cloth panels and metal plate pool sections. Water is directed to the pools, setting up a swirling motion that agitates fines into suspension. See also: *vibratory screen.* (Nelson, 1965)

poor fumes Toxic or irritating chemicals produced by an explosion. (Nichols, 1976)

poorly sorted *nongraded sediment.*

pop (a) A short, secondary drill hole blasted to reduce larger pieces of rock or to trim a working face. Also called pophole; pop shot. See also: *pop shot.* (Pryor, 1963) (b) Explosion in sealed area of a mine. Manometers may record a sudden pressure rise due to such an explosion. (Sinclair, 1958)

pop a boulder To place and explode a stick of dynamite on a boulder so as to break it for easy removal from a mine. (Fay, 1920)

pophole A secondary drill hole. See also: *pop.* (Fay, 1920)

pophole blasting Breaking down large pieces of asbestos by means of short blastholes judiciously placed. (Sinclair, 1959)

pop-off valve A pressure-relief valve. (Long, 1960)

poppet (a) A pulley frame or the headgear over a shaft. A headframe. (b) A valve that lifts bodily from its seat instead of being hinged. See also: *poppet valve; puppet valve.*

poppet head (a) The top of a derrick where the pulley is situated. (Gordon, 1906) (b) *headgear.*

poppet valve A valve shaped like a mushroom, resting on a circular seat, and opened by raising the stem. See also: *poppet; puppet valve.* (Nichols, 1976)

popping The drilling, charging, and firing of a hole in the center of a boulder at quarry and open-cast mines. The hole is charged at the rate of 2 to 3 oz (57 to 85 g) of explosive per yd^3 (74.2 to 111.3 g/m^3) of rock. The charge is pushed to the bottom of the hole and then filled with sand or soil. Also called pop shooting. See also: *snakeholing.* (Nelson, 1965)

poppy stone Red orbicular jasper from California; popular for cutting en cabochon.

pop-shooting A method of drilling a hole just beyond the center of a boulder to be broken so that the charge is centrally situated. Stemming is used. Pop-shooting is economical in explosives, but drilling is required. It is somewhat difficult to control the throw of broken material, but there is little noise to cause annoyance to nearby property owners. See also: *secondary blasting.* (Fraenkel, 1953)

pop shot (a) In mining, a shot fired for trimming purposes. (BS, 1964) (b) In quarrying, a method of secondary blasting. (BS, 1964) (c) A shot by which a boulder in a mine is broken up by placing a stick of dynamite on top of the boulder and exploding it. (Ricketts, 1943) (d) In blasting, an explosion of the charge that simply blows out the tamping. Syn: *block hole shot.* See also: *pop.*

pop valve A pressure-relief valve. (Long, 1960)

porcelain clay A clay suitable for use in the manufacture of porcelain; specif. kaolin. (AGI, 1987)

porcelain earth *kaolinite.*

porcelain jasper A hard, naturally baked, impure clay or porcellanite that, because of its red color, was long considered a variety of jasper. See also: *porcellanite.* (AGI, 1987)

porcelain oven A firing kiln used in baking porcelain. (Fay, 1920)

porcelaneous Resembling unglazed porcelain; e.g., said of a rock consisting of chert and carbonate impurities or of clay and opaline silica. Also spelled: porcellaneous; porcelanous. (AGI, 1987)

porcellanite A dense siliceous rock having the texture, dull luster, hardness, conchoidal fracture, and general appearance of unglazed porcelain; it is less hard, dense, and vitreous than chert. The term has been used for: an impure chert, in part argillaceous; an indurated or baked clay or shale often found in the roof or floor of a burned-out coal seam; and a fine-grained, acidic tuff compacted by secondary silica. Etymol: Italian porcellana, porcelain. Also spelled: porcelanite; porcelainite. See also: *siliceous shale; porcelain jasper; hälleflinta.* (AGI, 1987)

pore A space in rock or soil not occupied by solid mineral matter. Syn: *interstice; void.*

pore pressure *neutral stress.*

pore space The open spaces or voids in a rock taken collectively. See also: *porosity; permeability.*

pore-space filling The deposition of minerals in the voids of rocks or between the grains of loose sediment. (Nelson, 1965)

pore water (a) In soil technology, free water present in a soil. Normally under hydrostatic pressure. The shear strength of adjacent soil depends on this pore pressure, which reduces frictional resistance and soil stability. (b) Subsurface water in the voids of a rock. Syn: *interstitial water.*

pore-water pressure *neutral stress.*

porosimeter An instrument used to determine the porosity of a rock sample by comparing the bulk volume of the sample with the aggregate volume of the pore spaces between the grains. Porosimeters are of various designs, some using liquids and some using gases, at known pressures, to find the volume of openings. (AGI, 1987)

porosity (a) The ratio, P, expressed as a percentage of the volume, Vp, of the pore space in a rock to the volume, Vr, of the rock, the latter volume including rock material plus the pore space; $P = 100\ Vp/Vr$. (Holmes, 1928) (b) The amount of void space in a reservoir usually expressed as percent voids per bulk volume. Absolute porosity refers to the total amount of pore space in a reservoir, regardless of whether or not that space is accessible to fluid penetration. Effective porosity refers to the amount of connected pore spaces; i.e., the space available to fluid penetration. Syn: *total porosity.* (Brantly, 1961)

porosity coefficient Evolved by Professor H. Briggs in 1931 to express the conductance of a waste to air leakage, per foot length of the roadway per foot width of the leakage zone. (Roberts, 1960)

porous Containing voids, pores, cells, interstices, and other openings, which may or may not interconnect. See also: *porosity.* (Long, 1960)

porous ground Any assemblage of rock material that, as a result of fracturing, faulting, mode of deposition, etc., contains a high percentage of voids, pores, and other openings. (Long, 1960)

porous-pot electrode Nonpolarizable electrode consisting of a metal bar immersed in a saturated electrolytic solution which is contained in a porous pot. (Schieferdecker, 1959)

porpezite A native alloy of argentiferous gold with palladium, the palladium content varying up to 10%. From Porpez, Brazil. Syn: *palladium gold.* (Fay, 1920)

porphyrite An obsolete term synonymous with porphyry. The term was originally used to distinguish porphyries that contain plagioclase phenocrysts from those that contain alkali feldspar phenocrysts. (AGI, 1987)

porphyritic (a) Said of the texture of an igneous rock in which larger crystals (phenocrysts) are set in a finer-grained groundmass, which may be crystalline or glassy or both. Also, said of a rock with such texture, or of the mineral forming the phenocrysts. (AGI, 1987) (b) Pertaining to or resembling porphyry. (AGI, 1987)

porphyritic obsidian Volcanic glass having microcrystalline phenocrysts.

porphyritic texture *porphyritic.*

porphyroblast A pseudoporphyritic crystal in a rock produced by metamorphic recrystallization. Adj: porphyroblastic. Syn: *metacrystal; pseudophenocryst.* (AGI, 1987)

porphyroblastic (a) Pertaining to the texture of a recrystallized metamorphic rock having large idioblasts of minerals possessing high form energy (e.g., garnet, andalusite) in a finer-grained crystalloblastic matrix. (AGI, 1987) (b) *pseudoporphyritic.* Cf: *crystalloblastic.*

porphyroclast A rock fragment contained in mylonite.

porphyroclastic structure *mortar structure.*

porphyrogranulitic Said of the texture of a diabase porphyry having phenocrysts of plagioclase and augite in a ground mass of plagioclase laths and augite.

porphyroid Said of or pertaining to a blastoporphyritic or sometimes porphyroblastic metamorphic rock of igneous origin, or a feldspathic metasedimentary rock having the appearance of a porphyry. It occurs in the lower grades of regional metamorphism. (AGI, 1987)

porphyry An igneous rock of any composition that contains conspicuous phenocrysts in a fine-grained groundmass; a porphyritic igneous rock. The term (from a Greek word for a purple dye) was first applied to a purple-red rock quarried in Egypt and characterized by phenocrysts of alkali feldspar. The rock name descriptive of the groundmass composition usually precedes the term; e.g., diorite porphyry. Obsolete syn: porphyrite. (AGI, 1987)

porphyry copper deposit A large body of rock, typically porphyry, that contains disseminated chalcopyrite and other sulfide minerals. Such deposits are mined in bulk on a large scale, generally in open pits, for copper and byproduct molybdenum. Most deposits are 3 to 8 km across, and of low grade (less than 1% Cu). They are always associated with intermediate to felsic hypabyssal porphyritic intrusive rocks. Distribution of sulfide minerals changes outward from dissemination to veinlets and veins. Supergene enrichment has been very important at most deposits, as without it the grade would be too low to permit mining. (AGI, 1987)

porphyry deposit (a) A deposit in which minerals of copper, molybdenum, gold, or less commonly tungsten and tin, are disseminated or occur in a stockwork of small veinlets within a large mass of hydrothermally altered igneous rock. The host rock is commonly an intrusive porphyry, but other rocks intruded by a porphyry can also be hosts for ore minerals. (b) A deposit, usually of copper, molybdenum, or tin, in igneous rock of any composition that contains larger crystals in a fine-grained groundmass. (SME, 1992)

port (a) In drilling, a cylindrical opening through the bit shank from which the circulating fluid is discharged at the bit face into the water ways. (Long, 1960) (b) Any opening in a furnace through which fuel or flame enters or exhaust gases escape. (ASTM, 1994)

portable aggregate plant A plant mounted so that it can be moved over the highways on its own mounting and that performs all the operations of a stationary plant, including crushing, scalping, secondary crushing, screening, washing, and sand separation. Some of these complete plants are mounted on one chassis; others have the more common operations on one chassis with the supplementary equipment on separate portable mountings. (Pit and Quarry, 1960)
portable bucket loader Any of several types of self-propelled multibucket loaders that are considered suitable for miscellaneous light excavating work. These loaders dig their own path, and to do this, have various means of gathering the material to a point where it will be picked up by the buckets as they pass over the lower tumbler. While these loaders are usually used for reclaiming from stockpiles, they can, under favorable conditions, excavate from deposits. These machines always are mounted on crawler treads. (Pit and Quarry, 1960)
portable concentric mine cable A double conductor cable with one conductor located at the center and with the other conductor strands located concentric to the center conductor with rubber or synthetic insulation between conductors and over the outer conductor. Syn: *concentric mine cable.*
portable conveyor (a) A conveyor designed to be moved as a unit. It is commonly wheel mounted and may or may not be sectional. (NEMA, 1961) (b) Any type of transportable conveyor, usually having supports that provide mobility. See also: *boxcar loader; bucket loader; hatch conveyor; loading conveyor; movable conveyor; portable drag conveyor; roller conveyor; trimmer conveyor; unloading conveyor; wheel conveyor.*
portable crane A hoisting device carried by a frame mounted on wheels. (Crispin, 1964)
portable crusher A crusher with temporary support foundations, so that it can be moved in sections, or it can be moved along roadways with minimum dismantling. (SME, 1992)
portable drag conveyor A portable conveyor upon which endless drag chains are used as the conveying medium. Also a term sometimes applied to a portable flight conveyor. See also: *drag-chain conveyor; portable conveyor.*
portable drill Any size drill outfit that is wheel-, skid-, or track-mounted so that it can be moved readily as a unit. (Long, 1960)
portable electric lamp Self-contained lamp (such as a battery-operated lamp) that may be worn on the person or carried about freely.
portable flame-resistant cable A portable cable that will meet the flame tests of the U.S. Mine Safety and Health Administration.
portable loader A loading machine mounted on wheels or crawler tracks. See also: *shovel loader.* (Nelson, 1965)
portable mine blower A motor-driven blower (fan) to provide secondary ventilation into spaces inadequately ventilated by the main ventilating system; the air is directed to such spaces through a duct.
portable mine cable An extra-flexible cable used for connecting mobile or stationary equipment in a mine to a source of electric energy where permanent wiring is prohibited or impractical. (A portable cable for mining service is not always extra flexible and is used also with portable as well as with mobile and stationary equipment.)
portable parallel duplex mine cable A double or triple conductor cable with conductors laid side by side without twisting, with rubber or synthetic insulation between conductors and around the whole. The third conductor, when present, is a safety ground wire.
portable pneumatic core sampler A device developed by the U.S. Navy Ordnance Laboratory for sampling coral and sand bottoms. It consists of a four-legged pyramidal frame about 8 ft (2.4 m) high, a pneumatic hammer with air supply and exhaust hosing, 400 lb (180 kg) of lead weight, an anvil, and a 4-ft-long (1.2-m-long) aluminum barrel with a driving head for cutting through coral. (Hunt, 1965)
portable shunt A tub-changing arrangement for a tunnel face. See also: *double-track portable switch.* (Nelson, 1965)
portable substation *transportable substation.*
portable trailing cable A flexible cable or cord used for connecting mobile, portable, or stationary equipment in mines to a trolley system or other external source of electric energy where permanent mine wiring is prohibited or is impractical. (USBM, 1960)
portage Can. Trail between waterways. (Hoffman, 1958)
portal (a) The rock face at which tunnel driving is started. Syn: *point of attack.* (Fraenkel, 1953) (b) The surface entrance to a drift, tunnel, adit, or entry. (c) The log, concrete, timber, or masonry arch or retaining wall erected at the opening of a drift, tunnel, or adit.
portal crane A type of jib crane carried on a four-legged portal frame, which runs along rails. See also: *goliath crane; platform gantry.* (Hammond, 1965)
portal-to-portal A term encountered in disputes over what constitutes compensable "working time" under Federal laws. Portal literally means "entrance" and, in underground coal mining, portal refers to the mine mouth or entry at the surface. Hence, portal-to-portal as a descriptive term means strictly elapsed time from entry through the portal to exit on return. (BCI, 1947)
port crown Port roof of a tank.
porte et gardin plow *scraper plow.*
porter A long iron bar attached to a forging, or a piece in process of forging, by which to swing and turn it. (Standard, 1964)
porthole The opening or passageway connecting the inside of a bit or core barrel to the outside and through which the circulating medium is discharged. (Long, 1960)
Portland beds *Portland limestone.*
portland cement A calcium-aluminum silicate produced by fusing or clinkering limestone and clay in a kiln so as to drive off carbon dioxide and produce an oxide glass. The clinker is ground very fine and, when mixed with water, will recrystallize and set. It is combined with aggregate to form concrete. The name is from a resemblance to the Portland limestone of England.
portland cement mortar A mixture of portland cement, sand, and water. See also: *cement mortar.* (Nelson, 1965)
portlandite A hexagonal mineral, $Ca(OH)_2$; occurs in skarns; an important constituent of portland cement.
Portland limestone A series of limestone strata, belonging to the Oolite group (Upper Jurassic) on the Isle of Portland, Dorsetshire, England. Most of the building stone used in London is from these quarries.
Portland stone (a) A yellowish-white, oolitic limestone from the Isle of Portland (a peninsula in southern England), widely used for building purposes. (AGI, 1987) (b) A purplish-brown sandstone (brownstone) from Portland, CT. (AGI, 1987)
portrait stone A flat diamond, sometimes with several rows of facets around its edge; used for covering very small portraits. (Standard, 1964)
posepnyte An oxygenated hydrocarbon from the Great Western mercury mine, Lake County, CA. It occurs in plates and nodules, sometimes brittle, occasionally hard; the color is light green to reddish-brown; and sp gr, 0.85 to 0.985. (Fay, 1920)
position block Mining claim that is in a position to contain a lode if it continues in the direction in which it has been proved in other claims, but which itself has not been proved.
positive (a) Electrically, a point at a relatively high potential with respect to another point. A positive ion is one in which a particle, molecular or atomic, has ceased to be neutral owing to loss of one or more electrons. (Pryor, 1963) (b) Positive ore is ore that has been proved to exist by being blocked out in panels sampled at close intervals on all four sides so as to establish its quality and quantity beyond reasonable doubt.
positive confining bed The upper confining bed of an aquifer whose head is above the upper surface of the zone of saturation; i.e., above the water table. Little used. (AGI, 1987)
positive crystal An optically positive crystal. (AGI, 1987)
positive derail A device installed in or on a mine track to derail runaway cars or trips. This device is held open by a spring, necessitating that a worker hold it closed while a trip passes over it. (Hess)
positive-discharge bucket elevator A spaced bucket-type elevator in which the buckets are maintained over the discharge chute for a sufficient time to permit free gravity discharge of bulk materials. See also: *bucket elevator.*
positive displacement pump A pump which discharges the same amount of water for a given power, regardless of the head against which it operates. See also: *bladder pump.* (Driscoll, 1986)
positive drive A driving connection in two or more wheels or shafts that will turn them at approx. the same relative speeds under any conditions. (Nichols, 1976)
positive element A large structural feature or area that has had a long history of progressive uplift; also, in a relative sense, one that has been stable or has subsided much less than neighboring negative elements. (AGI, 1987)
positive elongation Tabular, lathlike, or needle crystals with the electric vector of their slow ray parallel to the long direction of the crystal. Syn: *length slow.* Cf: *negative elongation.*
positive ore (a) Ore exposed on four sides in blocks of a size variously prescribed. Syn: *ore developed.* See also: *developed reserve; proved ore.* (Fay, 1920) (b) Ore which is exposed and properly sampled on four sides, in blocks of reasonable size, having in view the nature of the deposit as regards uniformity of value per ton and of the third dimension, or thickness. (Fay, 1920)
positive rake The orientation of a cutting tool in a manner, so that the angle formed by the leading face of the tool and the surface behind its cutting edge does not exceed 90°; e.g., teeth in a ripsaw. Syn: *gouge rake.* See also: *gouge angle* Cf: *rake.* (Long, 1960)
positive ray Stream of positively charged atoms or molecules that take part in the electrical discharge in a rarefied gas. Positive rays have been studied by allowing them to pass through a perforated cathode onto a photographic plate, being deflected by magnetic and electrostatic fields (Thomson's parabola method) and by means of Aston's mass spectrograph. Syn: *canal ray.* (CTD, 1958)
positive temperature coefficient *temperature coefficient.*

positron Positive electron of the same mass as a negative electron; has only transitory existence. (Pryor, 1963)
possessio pedis The actual possession of a mining claim by the first arrival.
possessory title Title vested in the locator of a mining claim by compliance with the State and Federal mining laws.
possible crystal face Any crystal face permitted by the symmetry of crystal structure but not appearing on a particular mineral specimen.
possible ore An obsolete term for inferred reserves. See also: *reserve.* Syn: *future ore.*
post (a) To bring the survey, maps, and records of a mine up to date. (Fay, 1920) (b) A charge of ore for a smelting furnace. (Webster 3rd, 1966) (c) Any of the distance pieces to keep apart the frames or sets in a shaft; a studdle. (Webster 3rd, 1966) (d) A mine timber, or any upright timber, but more commonly used to refer to the uprights which support the roof crosspieces. Commonly used in metal mines instead of leg which is the coal miners' term, esp. the in the Western United States. Syn: *upright.* (e) A support fastened between the roof and the floor of a coal seam; used with certain types of mining machines or augers. (f) A pillar of coal or ore. (g) An item of kiln furniture. Posts, also known as props or uprights, support the horizontal bats on which ware is sent on a tunnel kiln car. (Dodd, 1964) (h) A discrete portion of bond between abrasive grains in a grinding wheel or other abrasive article. When an abrasive grain held by a post has become worn, the post should break to release the worn grain so that a fresh abrasive grain will become exposed. (Dodd, 1964) (i) A mass of slate traversed by so many joints as to be useless for building purposes. (j) Any of the four vertical timbers of a square set. (Lewis, 1964)
post-and-stall A mode of working coal, in which a certain amount of coal is left as pillars and the remainder is taken away, forming rooms or other openings. The method is also called bord-and-pillar; pillar-and-breast; etc. (Fay, 1920)
post brake A type of brake sometimes fitted on a steam winder or haulage. It consists of two upright posts mounted on either side of the drum and operating on brake paths bolted to the drum cheeks. See also: *winder brake.* (Nelson, 1965)
post drill An auger (or drill) supported by a post. (Fay, 1920)
post hole A shallow borehole. (Long, 1960)
post-hole auger A hand-rotated drilling tool that enables bores to be sunk down to about 20 ft (6 m) in unsupported holes and deeper in cased holes. See also: *shell-and-auger boring.* (Hammond, 1965)
post-hole digger Large auger, rotated mechanically or by hand, used for digging in unconsolidated ground and retrieving a sample. (Pryor, 1963)
posthumous In tectonics, said of a recurrence of forces and movement along lines or over areas affected by similar forces in a previous period; overprint.
posting York. Extracting the post or pillars; pillar robbing.
post jack A jack for pulling posts. See also: *post puller.* (Standard, 1964; Fay, 1920)
postmineral movement Movement usually along a fault that occurs after a mineral has been deposited.
postorogenic intrusion An igneous intrusion that took place after an orogenic event or cycle.
post puller An electric vehicle having a powered drum for handling wire rope used to pull mine props after coal has been removed; used for the recovery of the timber.
post puncher A coal-mining machine of the puncher type supported by a post. (Fay, 1920)
post stone An English term for any fine-grained sandstone or limestone. Also spelled: poststone. (AGI, 1987)
pot (a) A vessel for holding molten metal. (ASM, 1961) (b) An electrolytic reduction cell used to make such metals as aluminum from fused electrolyte. (ASM, 1961) (c) Mud-filled stump of Sigillaria in an upright position in the roof of certain coal seams. The stump became hollow by decay of the central pithy part, the hollow being filled by mud. This stump is now a separate mass of shale and is liable for collaspse without warning. See also: *caldron bottom; pot bottom.* (Nelson, 1965) (d) A colloquial syn. of seismic detector. (AGI, 1987) (e) *abyss; line oiler.*
potarite A tetragonal mineral, PdHg; silver white. Syn: *palladium amalgam.*
potash Potassium carbonate,K_2CO_3; formerly extracted from wood ashes; used as a component of glasses, glazes, and enamels to enhance colorants. Also called pearl ash. Syn: *potassium carbonate.*
potash alum *kalinite; alum; potassium alum.*
potash feldspar *potassium feldspar.*
potash fixation The retention of potassium in clays either by chemical combination in clay minerals or by adsorption. (AGI, 1987)
potash mica *muscovite.*
potash spar *potassium feldspar; spar.* Cf: *soda spar.*
potash syenite A syenite with a large excess of potassium feldspar (microcline, orthoclase) or feldspathoid over sodium feldspar (albite).
potassic Of, pertaining to, or containing potassium; relating to or containing potash. (Standard, 1964)
potassium A highly reactive metallic element of the alkali group; it is soft, light, and silvery. Symbol, K. Occurs abundantly in nature; obtained from the following minerals: sylvite, carnallite, langbeinite, and polyhalite. The greatest demand is for use in fertilizers. (Handbook of Chem. & Phys., 3)
potassium alum An isometric mineral, $KAl(SO_4)_2 \cdot 12H_2O$. Syn: *potash alum.*
potassium aluminosilicate *feldspar.*
potassium apatite A synthetic phosphate with K replacing Ca.
potassium bentonite A clay of the illite group, formed by the alteration of volcanic ash; a metabentonite. Syn: *K-bentonite.* See also: *metabentonite.*
potassium carbonate *potash.*
potassium chloride *sylvite.*
potassium feldspar The minerals microcline, orthoclase, and sanidine. Syn: *potash feldspar; potash spar; K-feldspar; K-spar.*
potassium titanate This compound, which approximates in composition to $K_2Ti_6O_{13}$ and melts at 1,370 °C, can be made into fibers for use as a heat-insulating material. (Dodd, 1964)
potato stone *geode.*
pot bottom A large boulder or concretion having the rounded appearance of the bottom of an iron pot and easily detached from the roof of a coal seam. See also: *camel back; kettle bottom; tortoise.* Cf: *caldron bottom; bell.* Syn: *pot; potstone.*
potch Inferior opal that does not exhibit play of color; found in association with precious opal (Australia).
potential (a) The words potential and voltage are synonymous and mean electrical pressure. The potential of a circuit, machine, or any piece of electrical apparatus means the voltage normally existing between the conductors of such a circuit or the terminals of such a machine or apparatus. In U.S. Bureau of Mines practice: (1) any potential less than 301 V shall be deemed a low potential; (2) any potential greater than 301 V but less than 651 V shall be deemed a medium potential; and (3) any potential in excess of 651 V shall be deemed a high potential. (b) Any of several different scalar quantities, each of which involves energy as a function of position or of condition; e.g., the fluid potential of groundwater. (AGI, 1987)
potential crater zone This is the region in which, if a sufficient quantity of explosive is used, the rock will be shattered and projected outward to form a crater. (Leet, 1960)
potential-determining ion Any ion which leaves the surface of a solid immersed in aqueous liquid before equilibrium (saturation point) has been reached, while an electrical double layer is building up and zeta-potential develops. (Pryor, 1965)
potential difference The difference in electric potential between two points; represents the work involved or the energy released in the transfer of a unit amount of electricity between them. (AGI, 1987)
potential energy The form of mechanical energy a body possesses by virtue of its position. If a body is being dropped from a higher to a lower position, the body is losing potential energy, but if a body is being raised, then it gains potential energy. (Morris, 1958)
potential gradient An ascending or descending value of voltage related to a linear measurement, such as a distance along the Earth surface or ground. (USBM, 1960)
potential ore Inferred reserves. *reserves.* (AGI, 1987)
potentiometric surface An imaginary surface representing the total head of ground water; defined by the level to which water will rise in a well. The water table is a particular potentiometric surface. Syn: *piezometric surface; pressure surface.* (AGI, 1987)
pothole (a) A kettlelike or circular depression, generally deeper than wide, worn into the solid rock in a stream bed at falls and strong rapids by sand, gravel, and stones being spun around by the force of the current. Also called a kettle hole; swallow hole. (Fay, 1920) (b) A kettlelike to irregular steep-walled subcircular interruption of bedding in the Merensky Reef of the Bushveld Complex, South Africa. It is filled with younger material. (c) A term used in Death Valley, California, for a circular opening, about a meter in diameter, filled with brine and lined with halite crystals. (AGI, 1987) (d) An underground system of pitches and slopes. Applied in some cases to single pitches reaching the surface. (e) A rounded, steep-sided depression resulting from downward surface solution. (AGI, 1987) (f) The occurrence, in the nether roof of a coal seam, of an irregularly shaped mass generally broader at its base than elsewhere and with smooth sides (slickensides). (TIME, 1929-30) (g) A circular or funnel-shaped depression in the surface caused by subsidence. (Hudson, 1932) (h) A rounded cavity in the roof of a mine caused by a fall of rock, coal, ore, etc. (i) A vertical pitch open to the surface. (j) *abyss.*
pot kiln A small limekiln. (Webster 3rd, 1966)
pot lead (a) An obsolete term for graphite or black lead. (AGI, 1987; Fay, 1920) (b) Graphite used on the bottoms of racing boats.

potlid Eng. Flattened oval dogger of flaggy sandstone; so called because sometimes the upper or under layers, when split off, resemble potlids. Cf: *baum pot.* (Arkell, 1953)
pot ore Foliated galena. (Arkell, 1953)
pot setting In glassmaking, the placing of a pot in a furnace for the purpose of melting metal. (Standard, 1964)
potstone (a) Impure steatite or massive talc; used in prehistoric times to make cooking vessels. Also spelled pot stone. (b) *paramoudra.*
potter A skilled craftsperson who fabricates ceramic ware using various forming techniques.
pottern ore A term used in early metallurgical practice for an ore that becomes vitrified by heat, like the glazing of earthenware. (Standard, 1964)
potter's asthma *potter's consumption.*
potter's bronchitis *potter's consumption.*
potter's clay A plastic clay free from iron and devoid of fissility; suitable for modeling or making of pottery or adapted for use on a potter's wheel. It is white after burning. (AGI, 1987)
potter's consumption An acute bronchitis often occurring among pottery employees, eventually affecting the lungs. (Standard, 1964)
pottery spar A 200-mesh feldspar produced for use by the manufacturers of chinaware, sanitary ware, ceramic tile, frits, enamel, glazes, electrical insulators, and vitrified grinding wheels. (AIME, 1960)
potting The placing of pots, containing either potassium nitrate or sodium nitrate and sulfuric acid, in the kilns used in the manufacture of sulfuric acid from sulfurous acid obtained from the combustion of sulfur in air. (Fay, 1920)
potty ore Som. Brown iron ore, Brendon Hills. Apparently a color term, since the two varieties of ore are black and potty. (Arkell, 1953)
Poulter method A seismic technique that dispenses with the need for drilling shotholes. In this air-shooting method, dynamite is exploded in arrays of simultaneous blasts with charges several feet above the ground. The principal difficulty involves the hazard of working with aboveground explosives and the damage to property or to the peace of mind of nearby inhabitants. (Dobrin, 1960)
pounceon Wales. Underclay. Apparently a survival of the obsolete form of puncheon (punchin)—a supporting timber in a coal mine or in a building floor timber. Also spelled pounson. (Arkell, 1953)
pound (a) A large, natural fissure or cavity in strata. (b) An underground reservoir of water. See also: *lodge.*
pound-calorie (a) A hybrid term between the English and metric units and defined as the amount of heat required to raise 1 lb (0.454 kg) of water 1 °C. (Newton, 1959) (b) An engineering heat unit, often called centigrade heat unit (chu). Defined as above. Approx. equal to 1.8 Btu (1.9 kJ). (CTD, 1958)
pounder An ore-mill stamp. (Standard, 1964)
pound-foot Unit of bending moment being the moment due to a force of 1 lb (0.454 kg) applied at a distance of 1 ft (30.48 cm). (Hammond, 1965)
pour In founding: (1) the amount of material, as melted metal, poured at a time; and (2) the act, process, or operation of pouring melted metal; such as, make a pour at noon. (Standard, 1964)
poured fitting A connecting device that is fastened to the end of a cable (wire rope) by inserting the cable end in a funnel-shaped socket, separating the wires and filling the socket with molten zinc. (Nichols, 1976)
pouring basin A basin on top of a mold to receive the molten metal before it enters the sprue or downgate. (ASM, 1961)
pouring gate A channel in a mold through which to pour molten metal. (Fay, 1920)
pouring pit refractory In the steel industry, refractory used for the transfer of steel from furnace to ingot. Refractories include ladle brick, nozzles, sleeves, stopper heads, mold plugs, hot tops, and mortars used for the brickwork involved. (AISI, 1949)
pouty In glassmaking, a long iron rod for either drawing out glass or twisting it to a fine thread.
powder A general term for explosives including dynamite, but excluding caps. (Nichols, 1976)
powder barrel A barrel made for the conveyance of gunpowder, usually holding 100 lb (45.4 kg). Cf: *powder keg.* (Standard, 1964; Fay, 1920)
powder box A wooden box in a miner's breast or chamber, in which were kept black powder, cartridge paper, cartridge stick, squibs, lampwick, chalk, and tools. Syn: *tool box.* (Korson, 1938)
powder carrier *powder monkey.*
powder chest A substantial, nonconductive portable container equipped with a lid and used temporarily at blasting sites for storage of explosives other than blasting agents. Unused explosives are returned to the magazine at the end of the shift. Syn: *day box.* Cf: *magazine.* (CFR, 3)
powdered coal Coal that has been crushed fine; may be transported by air to fire a boiler or industrial heating furnace. (BCI, 1947)
powdered ore Aust. Ore disseminated with veinstuff.
powder explosive An explosive containing still smaller quantities of liquid products, compared with plastic and semiplastic explosives, so that the spaces between the solid particles are not filled out entirely. As the result of this, the density of the mass is 20% to 40% lower than that of plastic and semiplastic explosives. (Fraenkel, 1953)
powder factor The quantity of explosive used per unit of rock blasted, measured in lb/yd^3 (kg/m^3) per ton (metric ton) of rock.
powder house A magazine for the temporary storage of explosives. See also: *magazine.*
powder keg A small metal keg for black blasting powder, usually having a capacity sufficient for 25 lb (13.5 kg) of powder. Cf: *powder barrel.* (Fay, 1920)
powderman (a) A person in charge of explosives in an operation of any nature requiring their use. (b) In bituminous coal mining and metal mining, one who handles proper storage of explosives in a powder house at a mine and issues powder, dynamite, caps, detonators, and fuses to miners as needed. At smaller mines, may deliver explosives to miners at working places. Also called powder monkey. See also: *blaster.* (DOT, 1949)
powderman helper *powder monkey.*
powder metal As used in the diamond-drilling industry, the finely divided particles of iron, copper, nickel, zinc, tungsten carbide, etc., which, when mixed with a suitable binding material and subjected to processing by heat and pressure, may be used as a matrix material to form a bit crown. (Long, 1960)
powder-metal bit Any diamond bit, mechanically set, in which finely divided metal powders are used as a matrix to hold the diamonds in place. Also called powder-pressed bit; powder-set bit; sinter bit; sintered-metal bit. (Long, 1960)
powder metallurgy The art of producing and utilizing metal powders for the production of massive materials and shaped objects. (ASM, 1961)
powder-metallurgy technique A metallurgical technique in which metal powder is pressed into a desired shape.
powder-metal process The process of mechanically setting diamonds in a bit in a matrix of finely divided metal powders. The metal powder is first cold pressed to compact it in a bit mold or die and then heated to allow the bonding alloy to melt and bind the powder to the diamonds and bit blank. Hot pressing or coining follows heating of the powder in some modifications of the process. (Long, 1960)
powder mine An excavation filled with powder for the purpose of blasting rocks. (Fay, 1920)
powder monkey (a) A person employed at the powder house of a coal mine whose duty is to deliver powder to the miners. (b) In some metal mines, the person who distributes powder, dynamite, and fuses to the miners at the working faces. This is a nautical term, but is frequently used in the mining industry. (c) In the quarry industry, one who carries powder or other explosives to the blaster and assists by placing prepared explosive in a hole, connecting a lead wire to a blasting machine, and performing other duties as directed. Also called blaster helper; powder carrier; powderman helper. See also: *powderman.* (DOT, 1949)
powder pattern The array of monochromatic X-ray diffractions produced by a mineral powder. Cf: *crystallogram.*
powder porosity Ratio of the volume of voids between particles, plus the volume of pores, to the volume occupied by the powder, including voids and pores. (Pryor, 1963)
powellite A tetragonal mineral, $CaMoO_4$; forms a series with scheelite as tungsten replaces molybdenum; a minor source of molybdenum in Idaho, Michigan, Nevada, and California, and Siberia, Russia.
powellizing process A wood treatment consisting of impregnating the wood with a saccharin solution. It hardens the wood, and renders it fireproof to some extent. (Liddell, 1918)
power (a) Any form of energy available for doing any kind of work; e.g., steam power and water power. Specif., mechanical energy, as distinguished from work done by hand. (Standard, 1964) (b) Used loosely to indicate the electric current in a wire. (Fay, 1920) (c) Rate of doing work. The foot-pound-second (fps) unit of power is the horsepower (hp), which is a rate of working equal to 550 ft·lbf/s (745.7 W). The electrical power unit, the watt, equals 10^7 cm-gram-second (cgs) units; i.e., 10^7 erg/s or 1 J/s. (CTD, 1958)
power arm The part of a lever between the fulcrum and the point where force is applied. (Nichols, 1954)
power barrow *pedestrian-controlled dumper.*
power control unit One or more winches mounted on a tractor and used to manipulate parts of bulldozers, scrapers, or other machines. (Nichols, 1954)
power control winch A high-speed tractor-mounted winch with one to three drums; used chiefly for operating bulldozers, scrapers, and rooters. (Nichols, 1976)
power dragscraper A machine consisting of: (1) a bottomless scraper bucket; (2) a two-drum hoist; (3) two long cables that attach to the front and rear of the scraper; (4) a movable tail block; (5) a short, guyed mast

located behind a ground hopper or other delivery point; and (6) two sheave blocks mounted on the mast to guide the operating cables to the hoist. The tail block is shifted manually from time to time, swinging the scraper in a wide arc until all the material within the operating radius has been taken out. (Pit and Quarry, 1960)

power earth auger A mechanically operated auger for exploring and testing deposits that are not very hard. The drilling rig may be mounted on a lorry or on continuous tracks when greater depths may be reached. (Nelson, 1965)

powered supports In fully mechanized coal mining, a system of pit props connected to a flexible armored conveyor by means of hydraulic rams. (Pryor, 1963)

power factor (a) The ratio of the mean actual power in an alternating-current circuit measured in watts to the apparent power measured in volt-amperes; equal to the cosine of the phase difference between electromotive force and current. (Webster 3rd, 1966) (b) The ratio of the total watts input to the total root-mean-square volt-ampere input to a rectifier or rectifier unit. (Coal Age, 1960) (c) A clause frequently found in electric power contracts, which sets forth that if a customer permits the average power factor of the load used to fall below a specified value, a penalty charge will be made. Power factor is often defined as the ratio of actual power to apparent power and is usually expressed as a percentage. (Kentucky, 1952)

power-factor meter Meter that indicates the relation of the phase between the line current and the line voltage, which actually is the same as the power factor of the load. (Coal Age, 1966)

power grizzly Power-operated machine used mainly for removing dirt and fines from material to be crushed. There are three main types—the live-roll grizzly, the vibrating-bar grizzly, and the bar grizzly feeder. See also: *static grizzly.* (Nelson, 1965)

power-operated supports *self-advancing supports.*

power pack (a) In general, an electrically operated hydraulic pump, placed at the gate end, to supply power to face equipment; e.g., self-advancing supports. The system forms a closed circuit with the oil returning to a reservoir containing about 212 gal (800 L) of oil. The pump can supply 2-1/2 gal (9.5 L) of oil per unit at 2,000 psi (13.8 MPa), which allows a setting load of about 9 st (8 t) per prop. See also: *hydraulic power.* (Nelson, 1965) (b) A unit that converts AC or DC current to AC or DC voltages suitable for the operation of electronic equipment. (NCB, 1964)

power rammer A manually operated compacting machine, weighing about 200 lb (91 kg), raised by an intrinsic internal combustion mechanism and allowed to fall by gravity. See also: *mechanical rammer.* (Hammond, 1965)

power sequence A sequence control system that is suitable for a group of conveyors in tandem. The trunk conveyor contactor is first closed; after a delay of from 3 to 15 s, sufficient for its motor to come up to speed, power is switched on to the contactor of the second conveyor; finally, after a similar delay, power is switched on to the third conveyor or conveyors. All power comes through the number 1 conveyor contactor so that, if this conveyor is stopped, all other conveyors in tandem stop as well. See also: *sequence starting.* (Sinclair, 1959)

power shovel An excavating and loading machine consisting of a digging bucket at the end of an arm suspended from a boom, which extends cranelike from the part of the machine that houses the powerplant. When digging, the bucket moves forward and upward so that the machine does not usually excavate below the level at which it stands. See also: *shovel loader.*

power-shovel mining Power shovels are used for mining coal, iron ore, phosphate deposits, and copper ore. The shovels may be used either for mining or for stripping and removing the overburden, or for both types of work, although at some coal mines the shovels used for stripping are considerably larger than those used for other mining. (Lewis, 1964)

power station An assemblage of machines and equipment, including the necessary housing, where electrical energy is produced from some other form of energy. Steam boilers are fed with coal or oil and the heat generated is used to produce high-pressure steam. The steam then passes to turbines that drive the generators and thus produce electricity. (Nelson, 1965)

power takeoff A place in a transmission or engine to which a shaft can be attached so as to drive an outside mechanism. (Nichols, 1976)

power tongs A mechanically powered wrench used to make up or break out a drill rod, casing, or pipe string. (Long, 1960)

power train All moving parts connecting an engine with the point where work is accomplished. (Nichols, 1976)

power unit (a) Generally applied to any device used to drive or operate machinery around a mine. Specif., it is used for the motor-speed reducer combination used to drive belt and chain conveyors. (Jones, 1949) (b) That part of a mining belt conveyor that consists of a power unit base, an electric motor, an electric controller, a speed reducer with a flexible coupling between motor and speed reducer, a power transmission device to power the drive pulley or pulleys, suitable covers for all moving parts and, if the power unit is of the detachable type, a device for attaching it to the conveyor. (NEMA, 1961)

power upon the air In coal mine ventilation, the horsepower applied is often known as the power upon the air. This may be the power exerted by a motive column due to the natural causes, to a furnace, or it may be the power of a mechanical motor. The power upon the air is always measured in foot pounds per minute. (Kentucky, 1952)

pozzolana (a) A leucitic tuff quarried near Pozzuoli, Italy, and used in the manufacture of hydraulic cement. The term is now applied more generally to a number of natural and manufactured materials, such as ash and slag, that impart specific properties to cement. Pozzolanic cements have superior strength when cured and are resistant to saline and acidic solutions. Also spelled: pozzolan; pozzuolana; pozzuolane. See also: *gaize.* (b) A material that is capable of reacting with lime in the presence of water at ordinary temperature to produce a cementitious compound. Natural pozzolanas are siliceous material of volcanic origin. They include trass and Santorin earth. Blast furnace slag is used to produce artificial pozzolanas. (CCD, 1961)

pozzolana cement A cement produced by grinding together portland cement clinker and a pozzolana, or by mixing together a hydrated lime and a pozzolana. Syn: *Roman cement.* (CCD, 1961)

practical shot In coal mining, a shot for which a hole has been drilled in a direction selected with reasonable care and filled with powder and tamped with the same degree of care. (Fay, 1920)

prairie soil Soil transitional between a pedalfer and a pedocal. (Leet, 1958)

prase (a) A translucent dull green or yellow-green variety of chalcedony. (b) Crystalline quartz containing abundant hairlike crystals of actinolite. Syn: *edinite; mother-of-emerald.*

praseodymium oxide A rare earth that, together with zirconia and silica, produces a distinctive and stable yellow color for pottery decoration. (Dodd, 1964)

praseolite (a) A green alteration product of iolite. (Fay, 1920) (b) A greenish foliated alteration product of cordierite.

prasopal A green chromium variety of common opal from Australia, Hungary, and Brazil. Also spelled prase opal.

Pratt hypothesis A suggested type of hydrostatic support for the Earth's solid outer crust in which crustal density is supposed to be greater under mountains than under oceans. Cf: *isostasy; Airy hypothesis.*

Pratt truss *N-truss.*

Prayon process The most common phosphoric acid dihydrate process for phosphoric acid production using sulfuric acid with naturally occurring phosphate rock. (Becker, 1983)

preaeration Aeration of water or ore pulp before treatment, notably by froth flotation where deoxygenated water is used (e.g., from under a frozen lake). Also used to stabilize ore pulp containing unstable sulfides before cyanidation. (Pryor, 1963)

prebaked anode Anode produced by binding together crushed petroleum coke and coal-tar pitch in a mold under pressure; subsequently baked to 1,000 to 1,200 °C; used in a metallurgical electrical furnace and replaced as a unit when consumed; in the production of aluminum metal, the anode is attached to a metal rod.

preblast Pertaining to the time period prior to the initiation of a blast.

preblast survey Documentation of the existing condition of a structure prior to exposure to blasting vibrations.

Precambrian shield Rocks older than the Cambrian age. Name refers to the great shield-shaped areas of ancient mineral-bearing rocks. These ancient rocks occur in many parts of the world. (Cumming, 1951)

precementation process Grouting the strata to control ground water prior to the start of construction or excavation, such as shaft sinking. Precementation has been used in South Africa to depths of 4,000 ft (1,200 m) and considerable savings have resulted. See also: *cementation sinking.* (Nelson, 1965)

precious Descriptive of the finest variety of a gem or mineral. Syn: *oriental; precious stone.*

precious garnet Brilliantly purple almandine.

precious metal Any of several relatively scarce and valuable metals, such as gold, silver, and the platinum-group metals. (ASM, 1961)

precious olivine *peridot.*

precious opal A gem variety of opal that exhibits a brilliant play of delicate colors by diffraction of light from close-packed 150- to 300-mm spheres of cristobalite-tridymite. The color of the bulk material may be black or white. Cf: *common opal.*

precious serpentine A pale or dark oil-green, massive, translucent serpentine. (Dana, 1914)

precious stone *gemstone.*

precious topaz (a) Genuine topaz as distinguished from topaz-colored quartz (jewelers' topaz). (b) An incorrect term for yellow to brown sapphire.

precious tourmaline Dark-colored gem variety of tourmaline.

precipitant Any agent, as a reagent, that when added or applied to a solution causes a precipitate of one or more of its constituents. (Standard, 1964)

precipitate (a) The operation, act, or process of adding a chemical or chemicals to an aqueous solution to react with a dissolved material in the solution and remove the resulting new solid matter by settling. (b) The solids resulting from the precipitation process.

precipitated sulfur Sulfur precipitated from calcium polysulfide solutions by hydrochloric acid and washed to remove all calcium chloride.

precipitation (a) The process of separating mineral constituents from a solution; e.g., by evaporation (such as halite or anhydrite) or by cooling of magma (to form an igneous rock). (b) Exsolution. (c) Water that falls to the surface from the atmosphere as rain, snow, hail, or sleet. It is measured as a liquid-water equivalent regardless of the form in which it fell. (AGI, 1987)

precipitation barrier Metal-rich water, as it moves away from the source of the metal, ordinarily comes into an environment where changing conditions of some kind cause precipitation of part or all of the metal from the water. Precipitation barriers account for the more than normal decay of hydrochemical anomalies than can be accounted for by simple dilution. They characteristically occur in spring and seepage areas where groundwaters coming to the surface encounter an environment of increased availability of oxygen, sunlight, and organic activity. (Hawkes, 1962)

precipitation hardening Hardening caused by the precipitation of a constituent from a supersaturated solid solution. See also: *aging; hardening.* (ASM, 1961)

precipitation heat treatment Artificial aging in which a constituent precipitates from a supersaturated solid solution. See also: *artificial aging; progressive aging.* (ASM, 1961)

precipitation process (a) The manipulation of physical and/or chemical properties of a solution to cause one of the constituents of that solution to become insoluble. (b) The treatment of lead ores by direct fusion with metallic iron or slag, or ore rich in iron; performed generally in a shaft furnace, rarely in a reverberatory. It is often combined with the roasting and reduction process. See also: *iron-reduction process.* (Fay, 1920)

precipitator In beneficiation, smelting, and refining, a person who (1) tends zinc boxes in which gold or silver that has been dissolved in a cyanide solution is precipitated; and (2) precipitates gold from cyanide solution, except that the cyanide solution is agitated with zinc dust in a mixing cone and precipitate, then turned into a filter press where the precipitate is recovered prior to the drying and refining to secure the gold. (DOT, 1949)

precision The degree of agreement or uniformity of repeated measurements of a quantity; the degree of refinement in the performance of an operation or in the statement of a result. It is exemplified by the number of decimal places to which a computation is carried and a result stated. Precision relates to the quality of the operation by which a result is obtained, as distinguished from accuracy, but it is of no significance unless accuracy is also obtained. (AGI, 1987)

precision depth recorder A device for recording a sonic depth trace. Abbrev., PDR. (Hy, 1965)

precision idler bearing A bearing having ground races and in which the bore and outside diameter tolerances are held to ten-thousandths of an inch and the width tolerance to thousands of an inch. (NEMA, 1961)

precooler load The amount of sensible heat, removed from the air in precooling. (Hartman, 1982)

precutting Method used in machine mining where a coal cutter makes a cut along the face in front of a cutter loader. It may be adopted in hard coal seams or where an improvement in the +2-in (+5.1-cm) coal product is required. (Nelson, 1965)

precutting blade A special blade attached to a plow that operates a little in advance of the main blades of the machine. It may be used in hard coals or to prevent the climbing of the machine, which would leave a layer of coal on the floor. (Nelson, 1965)

predicted 4-hour sweat rate An index devised by the Medical Research Council of Great Britain which is based on the rate of loss of sweat from the body. It is designed to measure the physiological effect of work in near limiting conditions in hot working places, and is based on the assumption that heat stress is a function of the rate of sweating. (Roberts, 1960)

predictive metallogeny *mineral assessment.*

preemption act An act providing for a patent to agricultural lands. The act does not include mineral deposits, as they are expressly reserved.

preference A familiar term under the public land laws meaning exclusive. (Ricketts, 1943)

preferential flotation (a) A name applied to a special type of differential flotation in which a mixture of two flotative, sulfide minerals is given a slight roast so that one may be oxidized, and therefore not float, and the other remain unchanged. (b) Preferential flotation may also be achieved by control of pH or by addition of an activating agent or depressant to the flotation mixer, conditioner, or cell. (AGI, 1987)

preferential wetting Applied to froth flotation when separating fine coal from coal washery slurries. The slurry or mixture is treated with a reagent that has an affinity for the material to be recovered and that will lend itself to subsequent stages in the separation process. (Nelson, 1965)

preferred orientation Feature of a rock in which the grains are more or less systematically oriented by shape. A schist in which the mica plates are parallel to one another shows a preferred orientation; so does a hornblende schist in which the long axes of the hornblende crystals are parallel. See also: *orientation; unoriented.*

preformed rope Wire rope in which the strands are bent to their correct lay before being laid up, so that the rope is unlikely to spin or kink. (Pryor, 1963)

pregnant solution A value-bearing solution in a hydrometallurgical operation. (Pryor, 1965)

pregnant solvent In solvent extraction, the metal-bearing solvent produced in the solvent extraction circuit.

pregs The liquor resulting from leaching of ore to dissolve a valuable constituent. Term connotes such a solution when it has reached a loading sufficient to justify its removal from contact with the ore, for separate treatment to reclaim the contained values (by precipitation, ion exchange, or stripping). After this treatment, the now barren solution is returned to work, or if foul, is run to waste or regenerated. Also called pregnant solution; royals. (Pryor, 1963)

preheat To heat beforehand; as: (1) to heat (an engine) to an operating temperature before operation; and (2) to heat (metal) prior to a thermal or mechanical treatment. (Webster 3rd, 1966)

preheat zone That portion of a continuous furnace through which the ware passes before entering the firing zone. (ASTM, 1994)

prehnite An orthorhombic mineral, $Ca_2Al_2Si_3O_{10}(OH)_2$, in which Fe replaces Al; forms botryoidal or mammillary and radiating aggregates. Occurs in: hydrothermal veins, cavities, and amygdules in basalt; veins in felsic plutonic rocks; and low-grade metamorphic rocks. Commonly associated with zeolites. Syn: *aedelite.*

prehnitoid (a) A variety of mizzonite resembling prehnite. (Dana, 1914) (b) Impure prehnite. (Dana, 1914)

preliminary exploration An investigation carried out along certain broad features of a coal or mineral area, with the object of deciding whether the proposition is such as to warrant a detailed or final exploration, which is often costly. (Nelson, 1965)

preliminary prospecting Prospecting undertaken after scout prospecting has disclosed the existence of values. Preliminary prospecting helps to determine approx. the extent of the payable ground. (Griffith, 1938)

preliminary soil survey A quick investigation of surface or near-surface conditions; no special equipment is employed. Tests are carried out on site for approximate classification of soil and are limited to visual or other simple tests. See also: *detailed soil survey; general soil survey.* (Nelson, 1965)

premature blast The detonation of an explosive charge earlier than warranted. Premature explosion may be due to carelessness, accidental percussion, a faulty fuse, or degenerated explosives. See also: *safety fuse; hangfire.* (Nelson, 1965; Pryor, 1963)

premature block An obstruction or block in a core barrel or bit that prevents the entry of core into the barrel before the bit can be advanced far enough to cut a length of core to fill the barrel. (Long, 1960)

premature firing The detonation of an explosive charge or the initiation of a blasting cap before the planned time. (Meyer, 1981)

premature set (a) The hardening of cement in a shorter time than normal or estimated. (Long, 1960) (b) This may be caused by the addition of catalysts to cement to increase setting time or by downhole temperatures and pressure that cause cement to set prematurely.

premium tin Tin of such high purity as to rate a special bonus in the metal market. (Pryor, 1963)

premix Aggregate that has been coated with bituminous binder before spreading. See also: *penetration macadam; tarmacadam.* (Nelson, 1965)

preparation (a) The treatment of ore or coal to reject waste. See also: *concentration; ore dressing; preparation plant.* (b) The process of preparing run-of-mine coal to meet market specifications by washing and sizing. (Jones, 1949)

preparation boss In bituminous coal mining, a foreman who is in charge of the operations of washing and sizing coal for market at the washery plant. (DOT, 1949)

preparation plant Strictly speaking, a preparation plant may be any facility where coal is prepared for market, but by common usage it has come to mean a rather elaborate collection of facilities where coal is separated from its impurities, washed and sized, and loaded for shipment. Syn: *cleaning plant.* See also: *coal-preparation plant; coal washer.* (BCI, 1947)

preparatory work Mining operations to facilitate mining proper after having explored a deposit and having made it accessible both in strike and dip. This work is executed almost entirely within the deposit and includes making: (1) inclines and transfer stations with manways; (2) sublevel drives between the levels; and (3) various crosscuts, chutes, minor shafts, raises, winzes, and other works. (Stoces, 1954)

prepare (a) To shear or undermine coal so that it can be readily blasted loose. (Fay, 1920) (b) Arkansas. To make a cartridge for a blast. (Fay, 1920) (c) Arkansas. To charge a blasthole. (Fay, 1920)

preplaning The lead or stagger that exists between planing blades in the same vertical plane of a plow. (Nelson, 1965)

prereduced iron-ore pellet A semimetallized pellet developed by the U.S. Bureau of Mines from taconite concentrates. The process involves rolling the concentrates into pellets, then drying, calcining, and roasting the pellets in a reducing (oxygen-deficient) atmosphere. During the heat-hardening stage, the pellets are partly converted to metal. Use of these pellets causes a considerable increase in pig-iron production. See also: *iron ore.*

preselective An arrangement by which a gear level can be moved, but the resulting speed shift will not take place until the clutch or the throttle is manipulated. (Nichols, 1976)

present Eng. Stone of suitable thickness for shaping into a tile stone without frosting; occurs in Stonesfield slate series and Chipping Norton limestone of the Cotswolds. (Arkell, 1953)

present value (a) The present value of a mine may be considered to be a sum of money that may be allowed for the purchase, development, and equipment of a mine, with the expectation of receiving for this capital expenditure, during the estimated life of the mine, the return of this capital plus a substantial profit commensurate with the risk involved in the venture. (Hoover, 1948) (b) The present value of a property is the amount that, if invested now in its purchase, would find its repayment with commensurate profit in the estimated series of annual dividends. Actuarially, it is the discounted sum of each and all those dividends, after allowance for any estimated further capital expenditure on necessary works and equipment. (Truscott, 1962)

preservative For mine timbers that are exposed to severe conditions of damp, ventilation, and stress, any of several chemicals used to impregnate them to resist dry or wet rot. These include copper sulfate, creosote, salt, sodium fluoride and silicofluoride, and zinc as chloride or sulfate. (Pryor, 1963)

presplitting (a) A smooth blasting method in which cracks for the final contour are created by blasting prior to the drilling of the rest of the holes for the blast pattern. Once the crack is made, it screens off the surroundings to some extent from ground vibrations in the main round. (Langefors, 1963) (b) *controlled blasting.*

press cloth *filter cloth.*

pressed amber Synthetic amber produced by consolidating amber fragments under pressure with an oil binder. Syn: *reconstructed amber; amberoid.*

pressed cameo Similar to molded cameo, but pressed. Cf: *molded cameo.*

pressed copal Synthetic copal produced by consolidating copal fragments under pressure with an oil binder.

presser In ceramics, a worker who molds the handle, ears, and decorative reliefs to be applied to a pottery vessel before firing. (Fay, 1920)

pressing machine (a) A machine that forms ceramic shapes by forcing plastic or semiplastic raw materials into a die or mold. (b) A machine in which the whole forming operation is carried out by pressing the plastic glass by a plunger forced into a die or mold. The machine may be operated by hand or it may be fully automatic. (CTD, 1958)

pressure (a) The force exerted across a real or imaginary surface divided by the area of that surface; the force per unit area exerted on a surface by the medium in contact with it. (AGI, 1987) (b) A commonly used short form for geostatic pressure. (AGI, 1987) (c) As used in mine ventilation terminology, it is sometimes defined as the available energy content of the air and as the pressure difference between two points in a ventilation current as the energy lost due to friction between two points. (Roberts, 1960) (d) Force exerted by air per unit area, either gage or absolute. Atmospheric pressure is measured by a barometer. Measured in pounds per square inch, kilopascals, or inches of mercury. (Hartman, 1961)

pressure anemometer (a) An instrument for measuring the velocity of ventilating air currents in mines. (b) An anemometer indicating wind velocity by means of the velocity head exerted. (Standard, 1964)

pressure arch The driving of a narrow roadway results in the development of a pressure arch over the excavation. The strata within the arch bend slightly and cease to support the overlying beds, and the load is transferred to the solid or rock along the sides. The wider the roadway, the greater the span of the arch and its height at the crown. A similar but larger pressure arch is formed across a longwall face, with one leg resting on the solid coal and the other on the solid pack behind the coal face. See also: *abutment; arch structure; load transfer.* (Nelson, 1965)

pressure arch theory The pressure arch theory in roof action suggests that, when a narrow heading is advanced, the layers of rock in the immediate roof deflect slightly and relieve themselves of the load of the overlying strata by transferring it to the sides of the opening by means of a pressure arch. The arch width just short of that which the higher strata cannot span and transfer the load to the sides of the opening is called the maximum-pressure arch. The depth mainly influences the minimum width of the pressure arch, although the type of overburden also plays a part. The following formula has been developed for approximating the minimum width of the maximum-pressure arch (W = minimum width of arch, in feet; D = depth of coal from surface, in feet): W = 3[(D / 20) + 20]. The equation does not apply for overburden less than 400 ft (122 m) or more than 2,000 ft (610 m) thick. (Coal Age, 1966)

pressure balancing When an area of a mine has been sealed off from the remainder of the workings by barriers or stoppings inserted at suitable points, it is important to prevent the circulation of air within the sealed area. This means that external air pressures must be equalized on all the seals. The object of equalizing the atmospheric pressures on the seals is attained by inserting or removing doors or brattice cloths at appropriate places. It is possible to make all the seals contiguous with a common airway by this means, so that, if they are not widely separated, they will be subjected to the same external atmospheric pressure. (Roberts, 1960)

pressure block Pressure formed over the workings by masses of rock being severed from the surrounding formations, creating pressure on the pillars, walls, or other supports. Pressure blocks of large size may result from natural geological phenomena, such as faults, or may occur as a result of mining operations. (Lewis, 1964)

pressure blower A machine or blower having either pistons, cams, or fans for furnishing an airblast above atmospheric pressure. (Standard, 1964)

pressure bump An occurrence when a coal pillar suddenly fails on becoming overloaded by the weight of the rocks above it. Generally, the coal is forced with some violence into the roadways and other open spaces. See also: *rock bump; rock burst; shock bump; bumps.* (Nelson, 1965)

pressure chamber (a) An enclosed space arranged on the access side of a stopping, which seals off an area and is furnished with means of raising or lowering the air pressure within it. (BS, 1963) (b) If the mine area to be sealed off is extensive, and the seals are widely scattered, the fact that they are subject to different pressures may be unavoidable. In this event, pressure chambers may be required on the outby side of seals. Pressure chambers are also of value when the seals cannot be made tight, because of broken or fissured ground. The principle consists of building an outer chamber by erecting a second stopping on the outby side of the seal. The air pressure in the intervening space is then controlled to prevent movement of air across the seal. (Roberts, 1960) (c) A method of driving tunnels and sinking shafts through running sand by holding back the loose material by compressed air. The technique is no longer applied to any great extent in mining. See also: *caisson sinking.* (Nelson, 1965)

pressure creosoting The most effective method of preserving timber by impregnation with creosote under pressure in tanks. (Hammond, 1965)

pressure detector *hydrophone.*

pressure die casting The usual die casting process in which the molten metal is forced into highly finished molds under heavy pressure by plungers, compressed air, or combined methods. (Hammond, 1965)

pressure dome (a) *air dome.* (b) The bonnet on a steam boiler. (Long, 1960)

pressure drilling A process of rotary drilling in which the drilling fluid is kept under pressure in an enclosed system. (Brantly, 1961)

pressure drop The decrease in pressure at which a liquid or gas is made to move between the intake and discharge of a pipeline or drill stem. (Long, 1960)

pressure equalizer In drilling, a diaphragm connected to the fluid column by a series of ports incorporated in the design of some core barrels and preventing the entry of drilling fluids into the core-barrel-head bearings. (Long, 1960)

pressure fan (a) A fan supplying air under pressure. (Webster 3rd, 1966) (b) A fan that forces fresh air into a mine as distinguished from one that exhausts air from the mine. (Fay, 1920)

pressure figure The indistinct six-rayed star produced on a mica plate after pressing with a dull point. Cf: *percussion figure.*

pressure filter (a) A machine for removing solids from tailings; the effluent can be reused in the washery or plant. The tailings are pumped into the filter under pressure; filtration takes place and solids are deposited in the chambers. Gradually the resistance increases until a pressure of 100 psi (690 kPa) is necessary to force more tailings into the press. At this stage, the chambers are almost full of solids. The feed is cut off and the press opened to allow the cakes to fall onto the conveyor beneath the chambers. The output of the pressure filter is low. (Nelson, 1965) (b) A filter in which pressure is applied to increase the rate of filtration. (BS, 1962) (c) A filter in which the liquid to be filtered is forced through filtering material by a pressure greater than its own weight in the filter.

pressure forging Forging done by a steady pressure, as in a hydraulic press. (Standard, 1964)

pressure gage An instrument used to measure the force per unit area exerted by a confined fluid or gas. (Long, 1960)

pressure grouting (a) The act or process of injecting, at high pressures, a thin cement slurry or grout through a pipeline or borehole to seal the pores or voids in rock or to cement fragmented rocks together. (Long, 1960) (b) Forcing a slurry of cement and sand into subgrade or an embankment either by use of compressed air or by hydraulic pressure. (Urquhart, 1959)

pressure head The height of a column of liquid supported, or capable of being supported, by the pressure at a point in the liquid. See also: *static head; total head.* (AGI, 1987)

pressure leaching In chemical extraction of valuable ore constituents, use of an autoclave to speed processing by means of increased temperatures and pressures. (Pryor, 1963)

pressure per diamond The feed pressure or load applied per diamond in a bit. The total load supported by the bit divided by the number of stones set in the bit face expresses the pressure per stone in numerical values. Also called diamond pressure; stone pressure. (Long, 1960)

pressure plate In a clutch, a plate driven by the flywheel or rotating housing, which can be slid toward the flywheel to engage the lined disk or disks between them. (Nichols, 1976)

pressure process Treatment of mine timber to prevent decay by forcing a preservative, such as creosote, zinc chloride, sodium fluoride, and other chemicals, into the cells of the wood. See also: *timber preservation.* (Lewis, 1964)

pressure-quantity survey *ventilation survey.*

pressure ring A ring about a large excavated area, evidenced by distortion of the openings near the main excavation. Shear cracks appear and minor slabbing of the rock takes place. (Lewis, 1964)

pressure shadow The name sometimes applied to a fringe or halo differing from the groundmass that often accompanies a porphyroblast in a schistose rock. (Hess)

pressure stripping The predominant means of extracting precious metal values from loaded activated carbon in the cyanidation process. Loaded carbon is placed in a pressure column along with a solution containing about 1% NaOH. This solution is circulated through the column at 150 °C until the precious metal values have been removed from the carbon absorbant. The loaded strip solution serves as the electrolyte for a subsequent electrowinning step for recovery of the precious metals. (Van Zyl, 1988)

pressure surface *potentiometric surface.*

pressure survey An investigation to determine the pressure distribution or pressure losses along consecutive lengths or sections of a ventilation circuit. See also: *ventilation planning; ventilation survey.* (Nelson, 1965)

pressure testing An indirect method of testing porosity and permeability of formations at elevations of proposed structures.

pressure water loader A cartridge loader in which compressed water, rather than compressed air, is used for loading underwater. (Langefors, 1963)

pressure wave A pressure produced by expanding gases moving at high velocity, the side component of which, equivalent to static pressure, may be recorded by a manometer at the side of the entry or mine passage. Syn: *P wave; compressional wave.* (Rice, 1913-18)

pressure wire Wire leading from any of various points of an electric system to a central station, where a voltmeter indicates the potential of the system at the point. (Webster 2nd, 1960)

pressurized (a) Said of any structure, area, or zone fitted with an arrangement that maintains nearly normal atmospheric pressure. (Nelson, 1965) (b) Said of any structure or area in which the pressure within is held higher than the outside pressure.

prestressing The application of load to a structure so as to deform it in such a manner that the structure will withstand its working load more effectively or with less deflection. (Hammond, 1965)

pretensioning The Hoyer method of prestressing concrete beams, precast in a workshop with the tensioned wires embedded in them and firmly anchored. (Hammond, 1965)

preventive maintenance A system that enables breakdowns to be anticipated and arrangements made to perform the necessary overhauls and replacements in good time.

previtrain The dense woody lenses in lignite that are equivalent to the vitrain in coal of higher rank. (AGI, 1987)

priceite A triclinic mineral, $Ca_4B_{10}O_{19}·7H_2O$(?); earthy to porcelainous white; conchoidal fracture. Syn: *pandermite.*

pricking bar (a) A bar used in opening the taphole of a furnace. (Fay, 1920) (b) A rod used for removing obstructions from tuyeres and blowpipes. (Fay, 1920)

priderite A tetragonal mineral, $(K,Ba)(Ti,Fe)_8O_{16}$; cryptomelane group; red; easily mistaken for rutile in leucite rocks; occurs at Kimberley, Western Australia.

priguinite *iriginite.*

prill In assaying, the bullion bead resulting from cupellation of an auriferous or argentiferous lead button. (Pryor, 1958)

prillion Tin extracted from slag. Also spelled prillon. (Standard, 1964; Fay, 1920)

Primacord A fuse composed of an explosive core within a textile or plastic covering. It detonates every explosive that is in direct contact with it. (Streefkerk, 1952)

Primacord-Bickford fuse A detonating fuse having an explosive of pentaerythritetetranitrate (PETN). Used in large-scale blasting work, esp. in quarries. (Lewis, 1964)

primary (a) Characteristic of or existing in a rock at the time of its formation; pertains to minerals, textures, etc.; original. Ant: secondary. (b) Said of a mineral deposit unaffected by supergene enrichment. (AGI, 1987) (c) Said of a metal obtained from ore rather than from scrap. Syn: *virgin.* (AGI, 1987)

primary anomaly An anomaly formed by primary dispersion.

primary basalt A presumed original magma, from which all other rock types are obtained by various processes of fractional crystallization.

primary blast A blast used to fragment and displace material from its original position to facilitate subsequent handling and crushing. (Atlas, 1987)

primary blasting The blasting of solid rock, ore, or coal; blasting in situ. See also: *secondary blasting.* (Nelson, 1965)

primary breaker A machine that takes over the work of size reduction from blasting operations; may be a gyratory or jaw breaker. Its capacity must be greater than the overall crushing plant capacity. In mines, primary ore crushing to about 7 in (18 cm) may be performed underground. See also: *reduction ratio; tertiary crushing.* (Nelson, 1965)

primary breaking A stage in bituminous coal crushing that occurs at the entrance to a plant and consists of raw feed flowing into the primary breaker for reduction to a maximum top size of 4 in, 5 in, 6 in, or 8 in (10.2 cm, 12.7 cm, 15.2 cm, or 20.3 cm) either for washing or for other preparation purposes. (Mitchell, 1950)

primary cell (a) A group or bank of flotation cells in which the raw feed is given a preliminary treatment, either or both of the products being subsequently retreated. (BS, 1962) (b) A cell that generates or makes its own electrical energy from the chemical action of its constituents; e.g., the voltaic cell, Deaniell cell, LeClanche cell, and dry cell. (Morris, 1958)

primary clay A clay found in the place where it was formed. Cf: *residual clay; secondary clay.* (AGI, 1987)

primary coil (a) The coil through which the inducing current passes in an induction coil or transformer. (Webster 3rd, 1966) (b) A coil, forming part of an electrical machine or piece of apparatus, in which a current flows, setting up the magnetic flux necessary for the operation of the machine or apparatus. (CTD, 1958)

primary creep Elastic deformation that is time-dependent and results from a constant differential stress acting over a long period of time. Cf: *secondary creep.* (AGI, 1987)

primary crusher (a) The first crusher in a series for processing shale or other rocks. See also: *secondary crusher.* (ACSG, 1963) (b) In comminution of ore, a heavy-duty dry crushing machine capable of accepting run-of-mine coarse ore and reducing it in size to somewhere between 4 in and 6 in (10 cm and 15 cm). Heavy-duty connotes the ability both to handle large tonnages daily and to withstand very rough treatment. (Pryor, 1963)

primary crushing In ore dressing, the first stage in which crushers take run-of-mine ore and reduce it to a size small enough to be taken by the next crusher in the series. Ordinarily, the Blake jaw crusher or a gyratory crusher is used. (Newton, 1959)

primary crystal The first type of crystal that separates from a melt on cooling. (ASM, 1961)

primary current distribution The current distribution in an electrolytic cell that is free of polarization. (ASM, 1961)

primary dip *original dip.*

primary dispersion Geochemical dispersion of elements by processes operating within the Earth. Cf: *secondary dispersion.* (Hawkes, 1957)

primary drilling The process of drilling holes in a solid rock ledge in preparation for a blast by means of which the rock is thrown down. (Fay, 1920)

primary environment *geochemical environment.*

primary excavation Digging in undisturbed soil, as distinguished from rehandling stockpiles. (Nichols, 1976)

primary flat joint *L-joint.*

primary flow structure Structure of either linear or platy nature developed in igneous rocks prior to or during consolidation. (Stokes, 1955)

primary fluid inclusion A fluid inclusion containing fluid trapped during original crystallization of its host mineral. (AGI, 1987)
primary foliation The variety of platy flow structure that forms during crystallization of a magma and is due to the parallelism of platy minerals. (Stokes, 1955)
primary gneiss A rock that exhibits planar or linear structures characteristic of metamorphic rocks, but lacks observable granulation or recrystallization, and is therefore considered to be of igneous origin. (AGI, 1987)
primary gneissic banding Banding exhibited by certain igneous rocks of heterogeneous composition; produced by the admixture of two magmas only partly miscible or, in other cases, by magma intimately admixed with country rock into which it has been injected along bedding or foliation planes. (CTD, 1958)
primary haulage A short haul in which there is no secondary or mainline haulage; e.g., a mine is started into a hillside, using mine cars, track, and hand loaders. An empty car is placed for the loader, and the loaded car is taken to the dump manually or by machine, repeating the process for each loader. See also: *face haulage*. (Kentucky, 1952)
primary metal (a) Metal produced: by direct smelting of ore; from a mine product, such as that extracted from mined ore; from reprocessing mine tailings; or from reprocessing smelter or refinery slags or residues. (Camm, 1940) (b) Metal extracted from ores, natural brines, or ocean water. Also called virgin metal. (Newton, 1959) (c) Ingot cast from reduced and perhaps refined metal as distinct from ingot containing recovered scrap metals. (Pryor, 1963) (d) Metal recovered as a principal or byproduct material from the processing of ores; includes metal recovered from ore processing wastes such as tailings, and downstream processing wastes such as slags and residues from the smelting and refining of the metal. Excludes metal recovered from scrap or its processing wastes (secondary metal).
primary mill A mill for rolling ingots or the rolled products of ingots to blooms, billets, or slabs. This type of mill is often called a blooming mill and sometimes a cogging mill. (ASM, 1961)
primary mineral A mineral formed at the same time as the rock enclosing it, by igneous, hydrothermal, or pneumatolytic processes, and that retains its original composition and form. Cf: *secondary mineral.* Syn: *original mineral.* (AGI, 1987)
primary mineral deposit Syngenetic ore deposit.
primary ore Ore that has remained practically unchanged from the time of original formation. (Stokes, 1955)
primary ore mineral An ore mineral that was deposited during the original period or periods of metallization. The term has also been used to designate the earliest of a sequence of ore minerals, as contrasted with later minerals of the same sequence, which some writers have called secondary. To avoid confusion, Ransome proposed the terms hypogene and supergene. Hypogene, as the word implies, indicates formation by ascending solutions. All hypogene minerals are necessarily primary, but not all primary ore minerals are hypogene; e.g., sedimentary hematite is of primary deposition even though it formed as a low-temperature precipitate.
primary phase The only crystalline phase capable of existing in equilibrium with a given liquid; it is the first to appear on cooling from a liquid state and the last to disappear on heating to the melting point. (AGI, 1987)
primary reject elevator A refuse elevator that extracts the first or heavier reject; usually situated at the feed end of a washbox. (BS, 1962)
primary relict A relict mineral that was a constituent of the original rock, whether igneous or sedimentary. (Schieferdecker, 1959)
primary screen A screen used to divide coal (usually raw coal) into sizes more suitable for the subsequent cleaning of some or all of them. (BS, 1962)
primary settling The surface subsidence that manifests itself a few months after mineral extraction and that usually constitutes 60% to 90% of the total subsidence. It varies according to the depth and thickness of the seam, the nature of the overburden, the mining method, and the thoroughness of the filling in the mined-out areas. The primary period is followed by the secondary period, in which the surface subsides gradually for a period of many years or even decades. (Stoces, 1954)
primary shaft The shaft from the surface in which the first stage of hoisting is carried out. (Spalding, 1949)
primary solid solution A constituent of alloys that is formed when atoms of an element B are incorporated in the crystals of a metal A. In most cases, solution involves the substitution of B atoms for some A atoms in the crystal structure of A, but there are cases in which the B atoms are situated in the interstices between the A atoms. (CTD, 1958)
primary source An operation that produces or creates new dust. (Hartman, 1982)
primary structure (a) A structure in an igneous rock that originated contemporaneously with the formation or emplacement of the rock, but before its final consolidation; e.g., layering developed during solidification of a magma. (AGI, 1987) (b) A primary sedimentary structure, such as bedding or ripple marks. (AGI, 1987) (c) The structure preexisting the deformation and reequilibration associated with the emplacement at shallow depth of a metamorphic body of deep origin during an orogeny. Cf: *secondary structure.* (AGI, 1987)
primary washbox The first of a series of washboxes, which receives the feed and from which one product at least is given further treatment. (BS, 1962)
primary washer The first of a series of washers, receiving raw feed, from which at least one product is retreated. (BS, 1962)
primary water supply The principal or original source from which drilling water is obtained, as opposed to recirculated water. (Long, 1960)
primary wave *longitudinal wave.*
primary zone Portion of a lode below that changed by leaching and secondary enrichment, and characteristic of the type of ore most likely to persist into the deeper levels of a mine. (Pryor, 1963)
prime mover (a) A machine that converts fuel or other natural energy into mechanical power. (Nelson, 1965) (b) A tractor or other vehicle used to pull other machines. (Nichols, 1976) (c) Any machine capable of producing power to do work. (Shell)
primer (a) A contrivance, such as a cap, tube, or wafer, containing percussion powder or other compound for igniting an explosive charge; ignited by friction, percussion, or electricity. Syn: *percussion cap.* (Webster 3rd, 1966) (b) The cartridge or that portion of a charge that carries a detonator or is coupled to Cordtex fuse and that detonates or sets off the remainder of the charge. The primer cartridge is placed at one end of the charge with the detonator pointing toward the charge. See also: *direct initiation.* (Nelson, 1965) (c) In blasting, the cartridge in which the cap is placed. (Streefkerk, 1952) (d) Usually the combination of a dynamite cartridge and a detonating cap. (Nichols, 1976)
primer cartridge The explosive cartridge into which a detonator has been inserted. (BS, 1964)
primer charge A boosting charge placed in contact with a detonator to ensure detonation of the main charge. (BS, 1964)
prime virgin mercury A term used for mercury produced by mines.
prime western spelter *prime western zinc.*
prime western zinc Low grade of virgin zinc containing about 98% zinc, 1.60% lead, 0.08% iron, with no limitations on cadmium or aluminum. (Bennett, 1962)
priming (a) The act of placing a detonator in an explosive charge. (Pryor, 1963) (b) The act of adding water to displace air, thereby promoting suction, as in a suction line of a pump. Water used to promote initial suction in a centrifugal or reciprocating pump.
priming cartridge *primer.*
priming coat A coating of binder applied to a surface of natural compacted or stabilized soil before surface dressing. (Nelson, 1965)
priming tube A tube containing fulminating powder for firing a charge. A detonator. (Standard, 1964; Fay, 1920)
priming valve (a) A safety valve on the working cylinder of a steam engine to discharge the priming. (Standard, 1964) (b) A valve connected with the discharge pipe of a force pump through which the pump may be primed. (Fay, 1920)
primitive *primitive circle.*
primitive circle In crystallography, the great circle on a stereographic projection that represents the equator of the spherical projection. The poles of all vertical crystal planes plot on the primitive. Syn: *primitive.* (Fay, 1920)
primitive form A crystal form from which other forms may be derived; e.g., a hexoctahedron has six faces replacing each octahedral face.
primitive unit cell *unit cell.*
princess Roofing slate sized 24 in by 14 in (61 cm by 36 cm). (Pryor, 1963)
principal Primary, or leading function. A principal axis is the longest one in a crystal. The principal valence is that at which an element forms the greatest number of stable compounds. (Pryor, 1963)
principal axis (a) In the tetragonal and hexagonal systems, the vertical crystallographic axis; hence what is the same thing in uniaxial crystals, the optic axis. In the orthorhombic and triclinic crystals, the axis of the principal zone; the axis with the shortest period, often the axis of the principal zone. In monoclinic crystals, the axis *c*, usually the axis of the principal zone excluding the symmetry axis; the symmetry axis *b*. (Fay, 1920; AGI, 1987) (b) That crystallographic axis with unique symmetry in a crystal system, designated *c*, except in the monoclinic system where the second setting is used by mineralogists making *b* the unique axis. (c) In a transducer used for sound emission or reception, a reference direction for angular coordinates used in describing the directional characteristics of the transducer. It is usually an axis of structural symmetry or the direction of maximum response, but if these do not coincide, the reference direction must be described explicitly. (Hy, 1965) (d) In experimental structural geology, a principal axis of stress or a principal axis of strain.
principal axis of strain One of the three mutually perpendicular axes of the strain ellipsoid.

principal axis of stress One of the three mutually perpendicular axes that are perpendicular to the principal planes of stress. See also: *principal axis.*

principal meridian A central meridian on which a rectangular grid is based; specif. one of a pair of coordinate axes (along with the base line) used in the U.S. Public Land Survey system to subdivide public lands in a given region. It consists of a line extending north and south along the astronomic meridian passing through the initial point and along which standard township, section, and quarter-section corners are established. The principal meridian is the line from which the survey of the township boundaries is initiated along the parallels. (AGI, 1987)

principal moment of inertia The moment of inertia of an area about either principal axis. (Roark, 1954)

principal point The geometric center of an aerial photograph, or the point where the optical axis of the lens meets the film plane in an aerial camera. Symbol: p. See also: *fiducial mark.* (AGI, 1987)

principal section (a) In crystallography, the plane passing through the optical axis of a crystal. (Standard, 1964) (b) The optical indicatrix of a biaxial mineral is a triaxial ellipsoid with semiaxes proportional to the refractive indices alpha, beta, and gamma. A principal section is an ellipse containing any two of these semiaxes. The indicatrix of a uniaxial mineral is an ellipsoid of revolution; its principal sections contain the axis of revolution which is proportional to the refractive index epsilon.

principal stress The stress normal to one of three mutually perpendicular planes that intersect at a point in a body on which the shearing stress is zero. (ASM, 1961)

principle of superposition To determine the stress in a member due to a system of applied forces, the system can be split up into several component forces and their moments and reactions added in order to calculate the total stress. (Hammond, 1965)

principle of uniformity *uniformitarianism.*

Prins process A dense-media process in which large-size coal is separated from the refuse in a flowing bed of small coal in a reciprocating launder. Refuse sinks to the bottom. The small coal is screened from the coarse refuse and returned to the head of the launder by a drag conveyor. The floating large coal passes over skimmers in the trough to the discharge chute. (Mitchell, 1950)

Prins washer A combination trough washer and jig in which the feed enters the unit through the central launder, where stratification takes place. The stratified material overflowing the stationary trough is divided at the first opening in the shaking jig, allowing the upper stratum of the material to flow onto the top deck of the jig while the lower stratum enters the jig reclean chamber. (Mitchell, 1950)

priorite An orthorhombic mineral, $(Y,Ca,Fe,Th)(Ti,Nb)_2(O,OH)_6$; now formally named aeschynite-(Y); forms series with aeschynite-(Ce) and with tantalaeschynite-(Y); black; forms with other rare-earth minerals in granite pegmatites and placers. Syn: *blomstrandine.*

prism (a) The volume of a length of embankment or excavation. (Seelye, 1951) (b) The liquid mobile volume of a stream. (Seelye, 1951) (c) An open crystal form with faces and their intersecting edges parallel to the principal crystallographic axis. Prisms have three (trigonal), four (tetragonal), six (ditrigonal or hexagonal), eight (ditetragonal), or twelve (dihexagonal) faces. The nine-sided prisms of tourmaline are a combination of trigonal and hexagonal prisms. (d) A long, narrow, wedge-shaped sedimentary body with width:thickness between 5:1 and 50:1; e.g., a bajada adjacent to an escarpment. It is typical of orogenic sediments formed during periods of intense crustal deformation; e.g., the arkoses found in fault troughs. Cf: *tabular.* Syn: *wedge.*

prismatic (a) Descriptive of a clast with length to width ratio between 1.5:1 and 3:1. Cf: *tabular.* (b) Pertaining to a sedimentary prism. (c) Pertaining to a crystallographic prism. (d) Descriptive of a crystal with one dimension markedly longer than the other two. (e) Descriptive of two directions of cleavage. (f) Descriptive of a metamorphic texture in which a large proportion of grains are prismatic and have approx. parallel orientation, giving a lineated appearance in hand specimens and thin sections. Cf: *equant.* See also: *columnar.*

prismatic compass A small magnetic compass held in the hand when in use and equipped with peep sights and a glass prism so arranged that the magnetic bearing or azimuth of a line can be read (through the prism) from a circular graduated scale at the same time that the line is sighted over. (AGI, 1987)

prismatic plane Any crystallographic plane that is parallel to the principal axis of a crystal.

prismatic quartz Collectors' name for cordierite.

prismatic system *orthorhombic system.*

prismatic telescope A telescope having an eyepiece fitted with a prism that reflects at 90°. (Hammond, 1965)

prismoid Any solid, bounded by planes, whose end faces are parallel. Usually understood to include also figures whose bounding surfaces are warped surfaces. (Seelye, 1951)

prismoidal Adj. of prismoid; used in sedimentary petrology (not prismatic, which is a crystallographic term).

prismoidal formula A formula used in the calculation of earthwork quantities. It states that the volume of any prismoid is equal to one-sixth its length multiplied by the sum of the two end-areas plus four times the mid-area. (CTD, 1958)

probability A statistical measure (where zero is impossibility and one is certainty) of the likelihood of occurrence of an event. (AGI, 1987)

probable ore (a) Indicated reserves. *reserves.* (b) A mineral deposit adjacent to developed ore but not yet proven by development. Cf: *extension ore.* (AGI, 1987)

probable performance curve A performance curve showing the expected results of a coal-preparation treatment. (BS, 1962)

probable reserves Areas of coal or mineral believed to lie beyond the developed reserves but not yet proven by development. See also: *economic coal reserves.* Syn: *theoretical tonnage.*

probe (a) A small tube containing a sensing element of electronic equipment, which can be lowered into a borehole to obtain measurements and data. (Long, 1960) (b) To conduct a search for mineral-bearing ground by drilling or boring. (Long, 1960) (c) To lower drill rods, etc., to locate obstructions and/or to determine the attitude of a piece of junk in a borehole. (Long, 1960)

probertite A monoclinic mineral, $NaCaB_5O_7(OH)_4{\cdot}3H_2O$; colorless; forms radiated columnar crystals; in Kern Country, CA. Syn: *kramerite; boydite.*

probing Thrusting a pointed steel rod down through sand or soft clays to contact a seam or orebody. The point of the rod is examined for traces of coal or mineral. See also: *auger.* (Nelson, 1965)

proceedings The term proceedings is broader than the term action, yet in the mining law it is used in the sense of action and refers to the commencement of an action. It is used to enable a party to institute such proceedings under the different forms of actions allowed by the State and Federal courts. (Ricketts, 1943)

process A series of operations conducted to achieve a result. (Webster 3rd, 1966)

process company Company formed for the purpose of exploiting a patented process. (Hoover, 1948)

process flowsheet A basic flowsheet indicating the main operational steps within a plant, the movement of the various materials between the steps, and the final products obtained, and often also the quantities of material with which the plant must be capable of dealing at various points. (BS, 1962)

processing (a) The methods employed to clean, process, and prepare coal and metallic ores into the final marketable product. (Nelson, 1965) (b) The various artificial methods adopted for strengthening a soil, such as compaction, treatment with bitumen, lime, cement, etc. See also: *stabilizer; soil stabilization.* (Nelson, 1965)

processioner An official land surveyor. (Standard, 1964)

process lag In mineral processing, the delay or retardation in the response of the controlled variable at a point of measurement to a change in value of the manipulated variable. (Fuerstenau, 1962)

process metallurgy Branch of metallurgy that deals with the recovery or extraction of metals from their ores. See also: *extractive metallurgy.* (Henderson, 1953)

process scrap The scrap arising during the manufacture of finished articles from iron and steel, and usually returned to steelworks after sorting and processing by scrap merchants. See also: *capital scrap; circulating scrap.* (Nelson, 1965)

Proctor penetration needle A quick and convenient method for testing the resistance of a fine-grained soil to penetration at a standard rate of 1/2 in/s (1.27 cm/s). Needles from 1 to 0.05 in^2 (6.5 to 0.3 cm^2) area are used, and a spring balance indicates the pressure required for the needle to penetrate the soil. See also: *penetration resistance; penetrometer; soil.* (Nelson, 1965)

prod cast *impact cast.*

prod mark (a) An indicator of slip direction on a slickensided fault surface, consisting of a groove made by a clast. (AGI, 1987) (b) A short tool mark oriented parallel to the current of a stream and produced by an object that plowed into and was then raised above the bottom; its longitudinal profile is asymmetric. The mark deepens gradually downcurrent where it ends abruptly (unlike a flute). Cf: *bounce cast.* Syn: *impact cast.* (AGI, 1987)

produce (a) The marketable ores or minerals produced by mining and dressing. (b) Corn. The amount of fine copper in one hundred parts of ore.

producer (a) Person who extracts ore or coal from mines, rock from quarries, metals from ore by metallurgical processes, etc. See also: *production.* (b) A producing well. (AGI, 1987) (c) A furnace or apparatus that produces combustible gas to be used for fuel; usually of the updraft type, which forces or draws air or a mixture of air and steam through a layer of incandescent fuel (such as coke) with the resulting gas consisting chiefly of carbon monoxide, hydrogen, and nitrogen. (Webster 3rd, 1966) (d) An organism (e.g., most plants) that can form new organic matter from inorganic matter such as carbon dioxide, water, and soluble salts. (AGI, 1987)

product Percent of metal in ore. (Gordon, 1906)

production That which is produced or made; any tangible result of industrial or other labor. The yield or output of a mine, metallurgical plant, or quarry. See also: *producer*. (Standard, 1964)

production checker In metal mining, a person who keeps a record of the number of containers (cars, buckets, or skips) raised to the surface, and the amount of ore contained in each, estimating or weighing the contents. (DOT, 1949)

production gang A team of workers employed at the face on production, covering all face operations, maintenance, and supplies. (Nelson, 1965)

productive Yielding payable ore.

productive development The headings and levels excavated in a coal seam, preparatory to opening out working faces. These drivages are planned to prove and render accessible the maximum area of coal for the minimum yardage of development work. The modern trend is to make in-the-seam development as productive as possible with the aid of machines. See also: *unproductive development*. (Nelson, 1965)

productive land Land that has produced farm crops within the previous 5 years. (Woodruff, 1966)

productivity (a) A term closely allied with, and that may be expressed as, the output per manshift of a face or colliery or metal mine. Productivity will vary with the degree of mechanization and multishift working; it is also a function of the horsepower, of a suitable nature, at the disposal of each miner. See also: *output*. (Nelson, 1965) (b) The efficiency with which economic resources (workers, materials, and machines) are employed to produce goods and services. (Crispin, 1964)

profile (a) The outline produced where the plane of a vertical section intersects the surface of ground; e.g., the longitudinal profile of a stream, or the profile of a coast or hill. Syn: *topographic profile*. (AGI, 1987) (b) A graph or drawing that shows the variation of one property such as elevation or gravity, usually as ordinate, with respect to another property, such as distance. (AGI, 1987) (c) Cross section of a region of cylindrical folds drawn perpendicular to the fold axes. (AGI, 1987) (d) In seismic prospecting, the data recorded from one shot point by a number of groups of detectors. (AGI, 1987) (e) *soil profile*. (f) A vertical section of a water table or other potentiometric surface, or of a body of surface water. (AGI, 1987) (g) A drawing used in civil engineering to show a vertical section of the ground along a surveyed line or graded work. (Webster 3rd, 1966)

profile flying The technique of flying at a constant height above the ground during airborne mineral exploration. Generally, the aircraft maintains a height of 300 ft or 500 ft (91 m or 152 m) above the ground. This often involves a series of skillfully controlled climbs and dives over rolling ground. See also: *vertical aerial photograph*. (Nelson, 1965)

profile shooting A type of seismic refraction in which the shots and detectors are laid out on long straight lines. Successive shots are taken at uniform or almost uniform intervals along each line, and successive detector spreads are shifted about the same distance as the corresponding shot points so as to keep the range of shot-detector distances approx. the same for all shots. Generally, shots are received from opposite directions on each detector spread. The distance range is chosen so that the first, or where desired the second, arrivals will be refracted from a particular formation such as the basement or a high-speed limestone marker. The proper distance is usually determined from time-distance plots based on experimental shooting at the onset of the program. (Dobrin, 1960)

profilograph An instrument for plotting the perimeter profile of an airway on a reduced scale, and primarily used when taking air measurements underground.

profit When one speaks of the interest on a mining investment, the rate mentioned ordinarily consists of the normal rate plus a substantial additional rate that represents the profit that should accrue in proportion to the hazardous nature of the mining business. In this sense, the rate of interest in most forms of mining should be high; to be satisfied with less than 10% annually would show a lack of acumen. (Hoover, 1948)

profit in sight Probable gross profit from a mine's ore reserves, as distinct from the ground still to be blocked out.

proforma invoice Invoice that does not charge for goods marked, but shows cost details. (Pryor, 1963)

prograde metamorphism Metamorphic changes in response to a higher pressure or temperature than that to which the rock last adjusted itself. Cf: *retrograde metamorphism*. (AGI, 1987)

progress chart A chart or graph forming a continuous record, which is kept up to date, of the amount of work done on a major project. It may take the form of a bar graph, divided into sections representing different jobs to be done, estimated and actual completion dates, etc. The chart covers the entire project from the initial site preparation or drivage to completion. (Nelson, 1965)

progressing cavity pump *Mono pump*.

progressive aging In heat treatment, aging by increasing the temperature in steps or continuously during the aging cycle. See also: *precipitation heat treatment*. (ASM, 1961)

progressive failure Rock or material failure in which the ultimate shearing resistance is progressively mobilized along the failure surface. (ASCE, 1958)

project data Basic information needed by engineers concerned with design, site development, machine and housing assembly, plant erection, contract supervision, and coordination when planning, erecting, and bringing into operation a new mine and its attendant services, including the ore treatment plant. (Pryor, 1963)

projected pipe A pipe laid on the surface before building a fill that buries it. (Nichols, 1976)

projection (a) In underground mining, a plan showing the proposed direction and location of entries, rooms, shafts, fans, and watercourses. Such projections commonly cover the entire property to be worked. (b) A systematic, diagrammatic representation on a plane (flat) surface of three-dimensional space relations, produced by passing lines from various points to their intersection with a plane; esp. a map projection. (AGI, 1987) (c) Any orderly method by which a projection is made; the process or operation of transferring a point from one surface to a corresponding position on another surface by graphical or analytical means, so that each point of one corresponds to one and only one point of the other. (AGI, 1987) (d) *exposure*. (e) *outcrop*. (f) In mapping, a geometric (or mathematical) system of constructing the true meridians and parallels, or the plane rectangular coordinates on a map. (Seelye, 1951) (g) A geometrically or mathematically derived portrayal of the surface of the geoid on a plane surface. The requirement for a particular projection is that it show the features of the surface of the Earth with a minimum of distortion of distances, directions, shapes, and areas. (Hy, 1965) (h) The act or result of constructing a figure upon a plane or other two-dimensional surface that corresponds point for point with a sphere, a spheroid, or some other three-dimensional form. (Fay, 1920)

projection balance Shows movement of a pointer by means of an illuminated scale. (Pryor, 1963)

project plans A series of plans of a proposed new mine or reconstruction, which are drawn up for the purpose of obtaining approval of a project. (BS, 1963)

prolapsed bedding A series of flat folds with near-horizontal axial planes contained entirely within a bed with undisturbed boundaries.

prolong Secondary condenser used in the zinc industry.

prong Eng. The forked end of a bucket-pump rod; used for attachment to the traveling valve and seat. (Fay, 1920)

Prony's dynamometer A dynamometer for obtaining data for computing power delivered by turbines and other water wheels, or from the flywheel of an engine, or transmitted by shafting. (Fay, 1920)

proof stress (a) The stress that will cause a specified small permanent set in a material. (ASM, 1961) (b) A specified stress to be applied to a member or structure to indicate its ability to withstand service loads. (ASM, 1961)

propagate To transmit or spread from place to place; as coal dust propagates a mine explosion.

propagated blast A blast consisting of a number of unprimed charges of explosives and only one hole primed, generally for the purpose of ditching, where each charge is detonated by the explosion of the adjacent one, the shock being transmitted through the wet soil. In this method, one detonator fired in the middle of a line of holes is capable of bringing about the explosion of several hundred such charges. (Fay, 1920)

propagation (a) The transfer of a signal through a medium; e.g., sound in air, seismic waves in fluids and solids, electromagnetic waves in a vacuum. (b) In general, propagation is said to occur when the flame of an explosion travels over considerable areas of a mine in such manner as might result in loss of life of workers in the mine. (Rice, 1913-18)

propagation velocity The speed of a wave in the material concerned, such as the propagation velocity of a detonation wave front traveling through an explosive or the propagation velocity of a seismic wave from a blast traveling through the ground.

prop-crib timbering Shaft timbering with cribs kept the proper distance apart by means of props. (Fay, 1920)

prop cutter In mining, a person who operates a power saw to cut to designated- and standard-length timbers and props used to support the walls and roofs of underground passageways and workplaces. Also called: prop sawyer; timber cutter. (DOT, 1949)

prop drawer (a) A sylvester or other appliance for withdrawing props from the waste area in coal mining. See also: *monkey winch*. (Nelson, 1965) (b) A worker who withdraws props and allows the roof to collapse. Props are withdrawn when caving of the roof is adopted. See also: *timber robber*. (Nelson, 1965)

propeller fan (a) Axial-flow ventilating fan used to blow fresh air into mine workings or to extract foul air. (Pryor, 1963) (b) A fan having an impeller, other than of the centrifugal-type, rotating in an orifice, the air flow into and out of the impeller not being confined by any casing. (BS, 1963)

propeller pump This type of pump, often called axial-flow, develops most of its head by the propelling or lifting action of the vanes upon the liquid. These pumps are built in horizontal or vertical casings and are primarily used in handling sludge, dewatering pits, sewage pumping, and similar duties requiring large capacities and heads under 100 ft (30 m). (Pit and Quarry, 1960)

propeller shaft Usually a main drive shaft fitted with universal joints. (Nichols, 1976)

propel shaft In a revolving shovel, a shaft that transmits engine power to the walking mechanism. (Nichols, 1976)

proper In crystallography, any symmetry rotation which does not change the chirality (handedness) of an asymmetric unit; e.g., not involving reflection or inversion. Cf: *improper.*

proper proportion In a transparent gemstone, the proportion of the mass above and below the girdle, as well as the angles of the facets in relation to the girdle, that produces the greatest brilliancy from the particular species. These proportions vary with the refractive index of the gem species.

properties of sections These include the cross-sectional area of a structural member, its moment of inertia, section modulus, and other geometrical properties essential for accurate design calculations. (Hammond, 1965)

property One of the physical and chemical characteristics of a material.

property man In bituminous coal mining, formerly one who kept record of location and has charge of distribution of coal cutting machines, drills, loaders, and other mechanical equipment in and about a mine. Now, one who oversees surface lands and structures.

prop-free In longwall mining of a coal seam, a face with no posts between the coal and the conveyor used to remove it. (Pryor, 1963)

prop-free front (a) In coal mining, longwall working in which support to the roof is given by roof beams cantilevered from behind the working face. This leaves unobstructed room for digging and conveying equipment in a mechanized working. (Pryor, 1963) (b) Such a face is necessary where armored flexible conveyors are used to carry a coal cutter or power loader. See also: *link bar; self-advancing supports.* (Nelson, 1965)

prophylene-glycol dinitrate explosive An explosive containing the liquid ingredients named, in contradistinction to dynamite, which contains nitroglycerin. In commerce, the term dynamite is loosely used to include any mixture containing a liquid explosive. (Fay, 1920)

proportion A statement of equality between two ratios. When one ratio is equal to another ratio, they are said to be in proportion. (Jones, 1949)

proportional control action As used in mineral processing, action in which there is a continuous linear relation between the output and the input. (Fuerstenau, 1962)

proportional counter A gas-filled, radiation-detection tube in which the pulse produced is proportional to the number of ions formed in the gas by the primary ionizing particle. (ASM, 1961)

proportional limit The greatest stress that a material is capable of withstanding without deviation from proportionality of stress to strain (Hooke's law). In the case of rocks, this term and "elastic limit" are restricted to short-time tests; rocks may slowly and permanently deform in periods of long duration, even at stresses below the short-time proportional limit.

proportional plus integral control action As used in mineral processing, action in which the output is proportional to a linear combination of the input and to the time integral of the input. (Fuerstenau, 1962)

proportional plus integral plus derivative control action As used in mineral processing, action in which the output is proportional to a linear combination of the input, the time integral of input, and the time rate of change of input. (Fuerstenau, 1962)

proportioning Measuring by weight or by volume the constituents, before mixing of concrete, mortar, or plaster. (Hammond, 1965)

proppant Hydraulic fracturing and propping agent employed in the gas and oil industry to enable production from deep petroleum reservoirs.

propping The setting of timber props in mine workings. (Nelson, 1965)

prop retriever *prop drawer.*

prop sawyer *prop cutter.*

prop setter In anthracite and bituminous coal mining, a worker who installs props (posts) to support the roofs of underground working places, placing and wedging them at the most effective points. (DOT, 1949)

prop wall Props that are fastened together in a group, like a fence, and placed against the walls to prevent the roof from caving into the stope. (Stoces, 1954)

propylite A hydrothermally altered andesite resembling a greenstone and containing calcite, chlorite, epidote, serpentine, quartz, pyrite, and iron oxides. The term was first used by Richthofen in 1868. Propylite is common in mining districts of the Western United States, generally in the outermost subzone of hydrothermal alteration. See also: *propylitization.* (AGI, 1987)

propylitic alteration *propylitization.*

propylitization The result of low-pressure-temperature alteration around many orebodies. The propylitic assemblage consists of epidote, chlorite, Mg-Fe-Ca carbonates, and sometimes albite-orthoclase, all involved in partial replacement of wall-rock minerals. Syn: *propylitic alteration.* See also: *phyllic alteration.* (AGI, 1987)

prosopite A monoclinic mineral, $CaAl_2(F,OH)_8$; forms tabular crystals or granular masses.

prospect (a) An area that is a potential site of mineral deposits, based on preliminary exploration. (AGI, 1987) (b) Sometimes, an area that has been explored in a preliminary way but has not given evidence of economic value. (AGI, 1987) (c) An area to be searched by some investigative technique, such as geophysical prospecting. (AGI, 1987) (d) A geologic or geophysical anomaly, esp. one recommended for additional exploration. A prospect is distinct from a mine in that it is nonproducing. See also: *prospecting.* (AGI, 1987) (e) A mineral property, the value of which has not been proved by exploration. (Lewis, 1964) (f) To search for minerals or oil by looking for surface indications, by drilling boreholes, or both. (Long, 1960) (g) A plot of ground believed to be mineralized enough to be of economic importance. (Long, 1960) (h) Territory under examination for its mineral wealth. Prospecting is the search for deposits and is performed by aerial survey, magnetometry, surface examination, pitting, trenching, use of a prospector's pan, geochemical testing of soil, drilling (shallow or deep), seismic probe, and resistivity survey. (Pryor, 1963) (i) The gold or other mineral obtained by working a sample of ore. (j) A formation that may be capable of development into a mine, but which is untested. See also: *favorable locality.* (Nelson, 1965) (k) A sample of gold obtained in panning. (Nelson, 1965) (l) A specimen or sample of mineral obtained from a small amount of paydirt or ore. (Craigie, 1938) (m) To work (a mine, ledge, etc.) experimentally in order to ascertain its richness in precious minerals. (Craigie, 1938)

prospect drilling The exploratory drilling of boreholes in the search for minerals and petroleum. (Long, 1960)

prospect drill panner In metal mining, a person who, with a cable drill rig, drills down through gravel to bedrock along a present or an old creek bed that usually has been prospected by a hand-dug hole. The panner saves the drillings and pans them to discover the possible presence of paydirt (gold-bearing gravel), and weighs gold particles recovered. In the event of the discovery of gold in quantities sufficient for profitable removal, the panner moves drill and continues operations to determine the boundaries of the gold-bearing strata. (DOT, 1949)

prospect entry *prospect tunnel.*

prospecting (a) The search for outcrops or surface exposure of mineral deposits. (b) Searching for new deposits; also, preliminary explorations to test the value of lodes or placers already known to exist. (c) The surface discovery of coal or mineral only proves its superficial existence and further work is necessary to establish its quality and extent. The term exploration is sometimes applied to this extension of the discovery work. See also: *exploration.* (Nelson, 1965)

prospecting and mining Generic terms which include the whole mode of obtaining metals and minerals. (Ricketts, 1943)

prospecting claim Aust. A claim larger than the average; allotted to the miner who is the first in a district to discover the presence of gold. (Standard, 1964)

prospecting dish A simple appliance used in the search for gold and other heavy minerals. By means of water washing, the lighter, worthless material is separated from the valuable, heavier minerals, which are made visible by concentration and retention in the dish. Standard dishes with sloping sides are made in sizes ranging from top diameter 10 to 18 in (25 to 46 cm) and from 2 to 4 in (5 to 10 cm) deep, with riffles or grooves to retain the heavy minerals. Syn: *pan.* (Nelson, 1965)

prospecting license Authorization granted by a government to an individual in some countries, permitting the person to prospect for minerals and to register (stake) a claim. (Pryor, 1963)

prospecting pan *pan.*

prospective ore Ore that cannot be included as proved or probable, nor definitely known or stated in terms of tonnage. See also: *possible ore; ore expectant.*

prospector A person engaged in exploring for valuable minerals or in testing supposed discoveries of the same.

prospect shaft A shaft sunk in connection with prospecting operations. (Craigie, 1938)

prospect tunnel A tunnel or entry driven through barren measures, or a fault, to ascertain the character of strata beyond. Syn: *prospect entry.* (Zern, 1928)

prospectus A preliminary printed statement describing a business or other enterprise, and distributed to prospective buyers, investors, or participants, giving detailed information concerning the company's business and financial standing. Common in mining.

protecting magnet Electromagnet or permanent magnet installed ahead of crushing machinery to remove tramp iron that otherwise might enter and damage the appliances. (Pryor, 1963)

protection screen deck A screen plate with large apertures mounted over the screening deck in order to reduce the load and wear thereon. (BS, 1962)

protective alkali In the cyanide process, the use of dissolved lime to maintain a slightly alkaline pulp, therefore ensuring that the cyanide salt retains its potency and does not acidify to hydrocyanic acid, which cannot dissolve gold or silver. See also: *cyanide process.* (Pryor, 1963)

protective alkalinity Lime added to auriferous pulp to ensure alkalinity. Important in the cyanidation process for precious metals. (Pryor, 1958)

protest An objection to the patent proceeding; when made, it calls for a hearing on the matter in the local land office. (Lewis, 1964)

protoamphibole A name for a series of artificial orthorhombic fluoramphiboles having only half the *a*-dimension of anthophyllite. The presence of lithium and absence of calcium appear to be essential to their formation. Named because of a structural relation to protoenstatite. (Hey, 1961)

protoclase Leith's term for a rock possessing what he considered to be primary cleavage; e.g., bedding planes in sedimentary rock, formed concurrently with the rock. Cf: *metaclase.* (AGI, 1987)

protoclastic (a) Said of igneous rocks in which the earlier formed crystals have been broken or deformed because of differential flow of the magma before complete solidification. (AGI, 1987) (b) Said of an igneous rock containing deformed xenocrysts. (AGI, 1987) (c) Said of the texture characteristic of an early stage of cataclasis, with a very small amount of finite strain. (AGI, 1987)

protodolomite (a) Dolomite with calcium and magnesium disordered within layers rather than ordered by layer. (b) An imperfectly crystallized synthetic material of composition near $CaMg(CO_3)_2$.

protogene *protogine.*

protogenous Said of original rocks as opposed to derived rocks, and including saline deposits, coal, igneous rocks, and ore deposits. The term is no longer used. (Holmes, 1928)

protogine A granitic rock, occurring in the Alps, that has gneissic structure, contains sericite, chlorite, epidote, and garnet, and shows evidence of a composite origin or of crystallization (or partial recrystallization) under stress after consolidation. Also spelled protogene. The term, dating from 1806, is obsolete. (AGI, 1987)

protomylonite (a) A mylonitic rock produced from contact-metamorphosed rock, with granulation and flowage being due to overthrusts following the contact surfaces between intrusion and country rock. (AGI, 1987) (b) A coherent crush breccia whose characteristically lenticular, megascopic particles faintly retain primary structures. It is a lower grade in the development of mylonite and ultramylonite. Cf: *ultramylonite.* (AGI, 1987)

proton An elementary particle with a single positive electrical charge and a mass approx. 1,847 times that of an electron. The atomic number of an atom equals the number of protons in its nucleus. (Lyman, 1964)

protoquartzite A well-sorted, quartz-enriched sandstone that lacks the well-rounded grains of an orthoquartzite; specif. a lithic sandstone intermediate in composition between subgraywacke and orthoquartzite. (AGI, 1987)

protore In older writings, any primary mineralized material too low in tenor to constitute ore but from which ore may be formed through secondary enrichment. As commonly employed today, the rock below the sulfide zone of supergene enrichment; the primary material that cannot be produced at a profit under existing conditions but that may become profitable with technological advances or price increases. See also: *oxidized zone; sulfide zone.*

protractor An instrument used in drawing and plotting, designed for laying out or measuring angles on a flat or curved surface, and consisting of a plate marked with units of circular measure. See also: *goniometer.* (AGI, 1987)

proustite A trigonal mineral, Ag_3AsS_3; dimorphous with xanthoconite; rhombohedral cleavage; soft; ruby red; occurs in low-temperature or secondary-enrichment veins; a minor source of silver. Syn: *light ruby silver; light red silver ore.*

prove (a) To determine, by boring from the surface or driving a passageway underground, the location and character of a coalbed or the nature of rock strata. (Hudson, 1932) (b) To establish, by drilling, trenching, underground openings, or other means, that a given deposit of a valuable substance exists, and that its grade and dimensions equal or exceed some specified amounts. See also: *proved reserve.* (AGI, 1987)

proved ore *proved reserve.*

proved reserve An ore deposit that has been reliably established as to its volume, tonnage, and quality by approved sampling, valuing, and testing methods supervised by a suitably qualified person. The proved reserve is the overridingly important asset of a mine, and by its nature is a wasting one from the start of exploitation unless it is increased by further development. Syn: *proved ore.* (Pryor, 1963)

provenance A place of origin; specif. the area from which the constituent materials of a sedimentary rock or facies are derived. Also, the rocks of which this area is composed. Cf: *distributive province.* Syn: *source area.* (AGI, 1987)

prove up (a) To show that the requirements for receiving a patent for government land have been satisfied. (Webster 3rd, 1966) (b) Can. To establish economic value of a property. (Hoffman, 1958)

proving hole (a) A borehole drilled for prospecting purposes. (b) Advance bore or heading into a mineral deposit, made either to check the quality of the ore being approached or to relocate a deposit that has been distorted or dislocated by faulting. (Pryor, 1963)

proving ring A steel ring that has been accurately turned, heat treated, and polished. It is precisely calibrated in a testing machine by measuring its deflection for different loads and can be used for measuring applied loads on a structure. (Hammond, 1965)

proving the area The establishment of the quantity and grade of coal or ore available for working by means of geological surveys, exploratory drilling, or exploring headings. (Nelson, 1965)

proximate analysis (a) The determination of the compounds contained in a mixture as distinguished from ultimate analysis, which is the determination of the elements contained in a compound. Used in the analysis of coal. (Standard, 1964; Fay, 1920) (b) The determination, by prescribed methods, of moisture, volatile matter, fixed carbon (by difference), and ash. The term proximate analysis does not include determinations of chemical elements or determinations other than those named. See also: *chemical constitution of coal.* (ASTM, 1994)

proximity log A Schlumberger log based on the principle of shallow investigation; as its name implies it is markedly affected by material that lies in its immediate proximity. It depends for its operation on the forcing of a more or less horizontal beam of current into the formation. Its vertical resolution is about 6 in (15 cm) and it is almost impervious to the presence of a mud cake on the formation wall. (Wyllie, 1963)

prudent-man (person) test The basic legal standard for discovery under the mining law that states: Where minerals have been found and the evidence is of such a character that a person of ordinary prudence would be justified in the further expenditure of his labor and means, with a reasonable prospect of success in developing a valuable mine, the requirements of the statute have been met. (SME, 1992)

P.R.U. hand pump and densitometer A dust sampling instrument comprising a D.V.P. Mark 11 pump with a swept volume of 90 cm^3. A filter paper is inserted into a bridge behind the inlet nozzle of the pump such that a circle of 1-cm diameter of the filter paper is exposed to the dust. The dust, while passing through the filter paper, produces a stain. The optical density of the stain is determined photoelectrically in a densitometer by the light that falls upon a galvanometer. The dust particle concentration is evaluated by a calibration factor. Its main disadvantage is that it underestimates the number of fine particles. (Nelson, 1965)

Prussian blue *vivianite.*

prussic acid *hydrocyanic acid.*

pry Eng. Cornish miners' term for soft white clay. Also spelled pryan.

pryany lode A lode in which the ore is mixed with gossan or flucan. (Arkell, 1953)

prypole A pole that forms the prop of a hoisting gin and stands facing the windlass. (Webster 3rd, 1966)

psammite (a) A sandstone. The term is equivalent to the Latin-derived term arenite. (AGI, 1987) (b) A term formerly used in Europe for a fine-grained, fissile, clayey sandstone. (AGI, 1987) (c) The metamorphic derivative of arenite. Etymol: Greek psammos, sand. See also: *psephite; pelite.* Adj. psammitic. (AGI, 1987)

psatyrite *hartite.*

psephite (a) A sediment or sedimentary rock composed of large fragments set in a matrix varying in kind and amount; e.g., talus, breccia, shingle, gravel, and esp. conglomerate. The term is equivalent to the Latin-derived term rudite. (AGI, 1987) (b) The metamorphic derivative of rudite. Etymol: Greek psephos, pebble. See also: *psammite; pelite.* Adj: psephitic. (AGI, 1987)

pseudo- A prefix meaning false or spurious. (AGI, 1987)

pseudoanticline An upward buckling of the superficial layers of the ground due either to changes in volume brought about by pedogenic processes or to some other nontectonic cause. (Challinor, 1964)

pseudoboleite A tetragonal mineral, $Pb_5Cu_4Cl_{10}(OH)_8{\cdot}2H_2O$; indigo blue; occurs only in parallel growth on boleite, at Boleo, Baja California, Mex. Also spelled pseudoboléite.

pseudobreccia A partially dolomitized limestone, characterized by: a mottled appearance, that gives the rock a texture mimicking that of a breccia; or by a weathered surface that appears fragmental. It is produced diagenetically by selective grain growth in which localized, patchy, and irregularly shaped recrystallized masses of coarse calcite are embedded in a lighter colored and less altered matrix of calcareous mud. (AGI, 1987)

pseudobrookite An orthorhombic mineral, $Fe_2(Ti,Fe)O_5$; resembles brookite; occurs in cavities in andesites. *brookite.*

pseudochromatism Colors and color plays produced by physical optics as opposed to chromophores; e.g., diffraction, dispersion, and scattering. Syn: *structural color.* Cf: *play of color.*

pseudoconglomerate A rock that resembles, or may easily be mistaken for, a normal sedimentary conglomerate. Examples include a crush conglomerate consisting of cemented fragments that have been rolled and rounded nearly in place by orogenic forces; a pebble dike; a sandstone packed with rounded concretions; and an aggregate of rounded boulders produced in place by spheroidal weathering and surrounded by clayey material. See also: *crush conglomerate.* (AGI, 1987)

pseudocrocidolite Quartz pseudomorphous after crocidolite. Syn: *tiger's-eye; hawk's-eye.* (English, 1938)

pseudocrystalline Composed of detrital crystalline grains little worn and solidly compacted by siliceous or other mineral matrix, so as to resemble a true crystalline rock.

pseudoeutectic texture Intergrowth of sulfide minerals that simulate eutectic texture in metals. See also: *graphic granite.*

pseudofibrous peat Peat that in spite of its fibrous condition, is soft, noncoherent, plastic, and on drying, shows great shrinkage. See also: *fibrous peat; amorphous peat.* (Tomkeieff, 1954)

pseudogalena *sphalerite.*

pseudohexagonal Descriptive of minerals with hexagonal habit without hexagonal symmetry; e.g., hexagonal plates of monoclinic mica.

pseudojade A name that may be applied to any mineral resembling jade in appearance; e.g., bowenite, massive serpentine. (English, 1938)

pseudoleucite Large isometric crystals consisting of mixtures of nepheline and orthoclase, or of analcime formed as breakdown products of leucite; occurs in syenites from Arkansas, Montana, and Brazil. Cf: *leucite; metaleucite.*

pseudomalachite A monoclinic mineral, $Cu_5(PO_4)_2(OH)_4$; trimorphous with ludjibaite and reichenbachite; dark green. Syn: *dihydrite; phosphochalcite; phosphorochalcite; tagilite.*

pseudomorph A mineral sample with the external crystal form of one mineral and the internal chemistry of another; e.g., cubes of geothite after pyrite resulting from oxidation of the ferrous sulfide to ferric oxyhydroxide. Cf: *paramorph.* Syn: *false form; allomorph.*

pseudomorphous quartz Quartz displaying the form and habit of any of several mineral species, which it has assumed through replacement. The most common quartz pseudomorphs are those of calcite, barite, fluorite, and siderite. Silicified wood is quartz pseudomorphous after wood.

pseudomorphous tonstein A type of tonstein characterized by numerous pseudomorphs of kaolinite-feldspar or kaolinite-mica within a kaolinite groundmass. (IHCP, 1963)

pseudophenocryst *porphyroblast.*

pseudophite A compact massive mixture of chlorite minerals resembling serpentine.

pseudoporphyritic (a) Said of the texture of an igneous rock in which larger crystals have developed in a macrocrystalline groundmass, but were formed, at least in part, after the rock solidified (e.g., large potassium-feldspar crystals in a granite). (AGI, 1987) (b) *porphyroblastic.*

pseudosecondary inclusion A fluid inclusion formed by healing of a fracture occurring during growth of the host crystal. (AGI, 1987)

pseudosuccinite Variety of amber differing from Baltic amber in its reaction to solvents. (Tomkeieff, 1954)

pseudosymmetrical Said of crystal structures in which the atoms are only slightly displaced from positions that would be in accord with a higher symmetry. Thus, a monoclinic, pseudotetragonal mineral contains atoms only slightly displaced from positions of tetragonal symmetry. (Hess)

pseudosymmetry (a) Close angular approximation of a mineral with lower symmetry to one of higher symmetry; e.g., pseudohexagonal micas with monoclinic symmetry. (b) Compound twins simulating an external symmetry not found in their atomic structure; e.g., orthorhombic aragonite in pseudohexagonal prisms. Syn: *mimetic.* (c) Abnormal crystal growth along one direction; e.g., elongate native gold cubes with apparent tetragonal symmetry.

pseudotachylyte (a) A dense rock produced in the compression and shear associated with intense fault movements, involving extreme mylonitization and/or partial melting. Similar rocks, such as some of the Sudbury breccias, contain shock-metamorphic effects and may be injection breccias emplaced in fractures formed during meteoric impact. Cf: *ultramylonite.* (AGI, 1987) (b) A dark gray or black rock that externally resembles tachylyte and that typically occurs in irregularly branching veins. The material carries fragmental clasts of adjacent rock units, and shows evidence of having been at high temperature. Miarolitic and spherulitic crystallization has sometimes taken place in the extremely dense devitrified base. Some pseudotachylyte has behaved like an intrusive and has no structures obviously related to local crushing. (AGI, 1987)

pseudotopaz Quartz simulating topaz. From Striegau, Silesia, Poland. (English, 1938)

pseudoviscosity Viscous resistance offered by a slurry, sludge, mud, or suspension of minerals in water as a pulp, due to the specific surface involved, with possibly an element of thixotropy under stated conditions of pH value, agitation, flow, temperature, and solid-to-liquid ratio. The pseudoviscous effect is distinct from viscosity due to molecular shear. (Pryor, 1963)

pseudovolcano A large crater or circular hollow believed not to be associated with volcanic activity; e.g., a crater that is possibly meteoritic in origin but may be the result of phreatic explosion or cauldron subsidence. Adj: pseudovolcanic. (AGI, 1987)

pseudowollastonite Synthetic triclinic $CaSiO_3$ polymorphous with wollastonite-1T, wollastonite-2M, and wollastonite-7T.

psilomelane (a) A general term for massive oxides of manganese not otherwise identified; commonly botryoidal or colloform; a source of manganese in the United States (Arkansas, Virginia, Georgia); also in India, South Africa, and Russia. Cf: *cryptomelane; wad.* (b) *romanechite; manganese oxide.* Cf: *pyrolusite.*

psilomelanite *psilomelane.*

psychrometer An instrument for measuring the vapor pressure and the relative humidity of the air or the quantity of moisture in the air. It consists of a dry-bulb thermometer and a wet-bulb thermometer, the latter having its bulb covered with a layer of muslin kept moist with water. The rate of evaporation from the moist muslin depends upon the quantity of moisture in the air. The more rapid the evaporation, the greater the cooling, and hence the greater the difference in the temperature readings of the two thermometers. Also called: hygrometer. (Standard, 1964)

psychrometry (a) Study of atmospheric humidity and its effect on workers. The psychrometer, or hygrometer, measures the difference between dry-bulb and wet-bulb thermometer readings. (Pryor, 1963) (b) The determination of the psychrometric properties of air at a given state point. (Hartman, 1982) (c) Measurement of the humidity of air. (Nelson, 1965)

pteropod ooze A fine-grained pelagic deposit with more than 30% calcium carbonate of organic origin, of which pteropods are an important constituent. (AGI, 1987)

ptilolite *mordenite.*

ptygmatic *ptygmatic folding.*

ptygmatic folding Primary folding in migmatites (injection gneisses, etc.), caused by the high-temperature and high-pressure processes to which the migmatites owe their origin and composite character. Cf: *flow folding.*

public domain Land owned, controlled, or heretofore disposed of by the U.S. Government. It includes the land that was ceded to the Government by the original 13 States, together with certain subsequent additions acquired by cession, treaty, and purchase. At its greatest extent, the public domain occupied more than 1,820 million acres (737 million ha). See also: *public land.* (AGI, 1987)

public land Land owned by a government, esp. a national government; specif. the part of the U.S. public domain to which title is still vested in the Federal Government and that is subject to appropriation, sale, or disposal under the general laws. (AGI, 1987)

public land and public use There is a clear distinction between public lands and lands that have been severed from the public domain and reserved from sale or other disposition under general laws. Such reservation severs the land from the mass of the public domain and appropriates it to a public use. (Ricketts, 1943)

public limited liability company An association of individuals, at least seven in number, who together subscribe the necessary means or capital—i.e., money, property, or other credit—to engage in a joint undertaking. (Truscott, 1962)

public mineral land Land belonging to the United States containing a deposit of mineral in some form, metalliferous or nonmetalliferous, in quantity and quality sufficient to justify expenditures in the effort to extract it and subject to occupation and purchase under the mining laws. (Ricketts, 1943)

pucherite An orthorhombic mineral, $BiVO_4$; trimorphous with clinobisvanite and dreyerite; reddish brown; a source of vanadium.

pucking cutter A worker employed in a coal mine to cut the floor in cases of creep or upheaval toward the roof. (CTD, 1958)

puddingstone (a) A siliceous rock cut into blocks for furnace linings. (b) *conglomerate.*

puddle (a) Earthy material—such as a mixture of clay, sand, and gravel—placed with water to form a compact mass to reduce percolation. (Seelye, 1951) (b) To place such material. (Seelye, 1951) (c) To compact loose soil by soaking it and allowing it to dry. (Nichols, 1976) (d) The molten portion of a weld. (Webster 3rd, 1966) (e) To work (metal) while molten. (Webster 3rd, 1966) (f) To subject (iron) to the process of puddling. (Webster 3rd, 1966)

puddled steel Steel made in a puddling furnace, a type of reverberatory furnace in which the flame plays down upon the metal. (Camm, 1940)

puddler (a) Worker who converts cast iron into wrought iron by puddling. See also: *puddling*. (Webster 3rd, 1966) (b) A rabble used in puddling. (Webster 3rd, 1966) (c) A puddling furnace. (Webster 3rd, 1966) (d) A system of small pipes admitting compressed air to a tank of water and zinc chloride, to effect a thorough solution for use as a timber preservative. (Webster 2nd, 1960) (e) A machine for breaking up alluvial wash, consisting of a shallow tank in which the arms rotate slowly. The coarse stones are forked out and the pulp passed down sluice boxes along which the gold settles. See also: *tormentor*. (Nelson, 1965)

puddle roll Any of the roughing rolls through which puddle balls are passed to be converted into bars. Collectively called a puddle train. (Standard, 1964; Fay, 1920)

puddling The agitation of a bath of molten pig iron by hand or by mechanical means, in an oxidizing atmosphere, in order to oxidize most of the carbon, silicon, and manganese, and thus produce wrought iron. See also: *danks' puddler*. (CTD, 1958)

puddling furnace A reverberatory furnace for puddling pig iron. (Standard, 1964)

puddling machine A machine used for mixing auriferous clays with water to the proper consistency for the separation of the ore. (Fay, 1920)

puddling process Production of wrought iron from molten pig iron, in an oxidizing atmosphere in a reverberatory furnace of special design. (Pryor, 1963)

puff blowing Blowing chips out of a hole by means of exhaust air from the drill. (Nichols, 1976)

puffed bar In powder metallurgy, a cored bar expanded by internal gas pressure. (Rolfe, 1955)

puffer boy A person employed to operate an engine used for hauling loaded mine cars through haulageways. Also the operator of any small stationary hoisting engine. See also: *puffer man*. (Fay, 1920)

puffer man In bituminous coal mining, a worker who operates a small hoisting engine used for hauling loaded mine cars through haulageways in a mine, or operates a small stationary engine used for hoisting coal or rock in a shallow shaft, esp. for prospecting or development work. Also called: puffer; puffer boy; puffer tender. (DOT, 1949)

puffer tender *puffer man.*

puffstone Eng. Travertine; hard enough to use for building; so called from its cavernous structure. (Arkell, 1953)

pug (a) A parting of clay that sometimes occurs between the walls of a vein and the country rock; gouge. (b) The coal left on the floor by a coal cutter. See also: *following dirt*. (Nelson, 1965) (c) Clay or other material used in packing cracks to prevent leakage; also, to use this material. (d) Crushed strata or clay. See also: *flucan*. (Nelson, 1965)

pug lifter One who removes coal left adhering to the floor by a coal-cutting machine. (CTD, 1958)

pug lifting The breaking and clearing of the coal left adhering to the floor by a longwall coal cutter. (Nelson, 1965)

pug mill (a) A machine for mixing water and clay, consisting of a long horizontal barrel containing a long longitudinal shaft fitted with knives; the knives slice through the clay, mixing it with water, which is added by sprays from the top. The knives are canted to give some screw action, forcing the clay along the barrel and out one end. (AISI, 1949) (b) *paddle-type mixing conveyor.*

pug-mill operator (a) One who prepares ground, sifted, and filtered clay for molding by mixing it with water in a rotary-type mixer called a pug mill. This machine is frequently operated in conjunction with an auger mill and a cutting machine, the same worker tending the operation of all three machines simultaneously. Also called clay pugger; mixing-mill operator. (DOT, 1949) (b) One who mixes ground preheated magnesia and carbon with hot asphalt in a pug mill to form a viscous mixture suitable for processing into pellets. Also called: mixer tender; pug miller; pug-mill tender. (DOT, 1949)

pug tub *settler.*

pull (a) The unit advance during the firing of each complete round of shotholes in a tunnel. (b) To loosen the rock around the bottom of a hole by blasting. Usually used with a negative to describe a blast that did not shatter rock to the desired depth. (Nichols, 1976) (c) The amount of core obtained each time a core barrel is removed from a borehole. (Long, 1960) (d) To draw or remove coal pillars, or pillars of ore. (e) To hoist drill-stem equipment from a borehole. (Long, 1960) (f) Strata movements over large excavated areas will extend to the surface and the disturbed surface area is almost always larger than the area of the underground excavation. The extent of this pull or draw depends on the depth of the workings, the nature of the strata, the thickness of the seam being mined, and the degree of packing support. See also: *draw*. (Lewis, 1964)

pull-apart structure Features produced in beds that have been disrupted and separated during soft-sediment deformation. See also: *boudinage*.

pull drift A small crosscut through barren ground to connect two orebodies. (Hess)

puller-out An operator who charges, pulls out, and otherwise manipulates crucibles. (Mersereau, 1947)

puller rod The rod used between the crank arm or drive arm of the drive unit and the panline of a shaker conveyor. Also called: connecting rod. (Jones, 1949)

pulley (a) A cylinder, with a shaft for mounting it so that it may rotate; used to change the direction or plane of belt travel. If the shaft is designed to be mounted so that it will not rotate, a pulley includes the bearings that provide for rotation of the cylinder on the shaft. (NEMA, 1961) (b) A sheave or wheel with a grooved rim, over which a winding rope passes at the top of a headframe. (Fay, 1920) (c) A wheel that carries a cable or belt on part of its surface. (Nichols, 1976)

pulley man *rollerman.*

pulley oiler In bituminous coal mining, a laborer who oils and greases the pulleys on which run the cables that are used to raise and lower cars along haulage roads underground and at the surface of mines. (DOT, 1949)

pulley repairman *rollerman.*

pull hole In sublevel stoping, term applied to a raise along the haulage level put up to the first sublevel. The raise is enlarged at the bottom into a grizzly chamber immediately over the haulage level and at the top is widened into a funnel-shaped opening. As ore is broken, it drops directly into a pull hole. (Lewis, 1964)

pulling pillars The common expression used for mining the coal in the pillars of a mine; robbing pillars. See also: *pulling stumps*. (Fay, 1920)

pulling stumps The process of taking out the pillars of a coal mine. See also: *pulling pillars*. (Fay, 1920)

pull-over mill A two-high mill in which a piece is rolled in one direction only, and after traveling between the rolls has to be passed back over the top roll for rerolling. (Osborne, 1956)

pull pin A device for throwing mechanical parts in or out of gear, or for readily shifting in or away from a fixed relative position. (Crispin, 1964)

pull rope The rope that pulls a journey of loaded cars on a haulage plane; the rope that pulls the loaded scoop or bucket in a scraper loader layout. See also: *tail rope*. Syn: *main rope*. (Nelson, 1965)

pull shovel A shovel with a hinge- and stick-mounted bucket that digs while being pulled inward. (Nichols, 1976)

pullway The path from the face to the loading point taken by the scraper of a scraper loading unit. (Jones, 1949)

pull wheel A large driving wheel or sprocket. (Nichols, 1954)

pulmonary dust Dust harmful to the respiratory system, including: silica (quartz, chert); silicates (asbestos, talc, mica, sillimanite); metal fumes (nearly all); beryllium ore; tin ore; iron ores (some); carborundum; coal (anthracite, bituminous). Syn: *fibrogenic dust*. (Hartman, 1982)

pulp (a) A mixture of ground ore and water capable of flowing through suitably graded channels as a fluid. Its dilution or consistency is specified either as solid-liquid ratio (by weight) or as a percentage of solids (by weight). (Pryor, 1960) (b) Pac. Pulverized ore or coal mixed with water; also applied to dry, crushed ore. See also: *vacuum filter*.

pulp assay Pac. The assay of samples taken from the pulp after or during crushing.

pulp balance Balance that weighs ore or coal pulp in a container of known volume; graduated to show pulp density directly. (Pryor, 1963)

pulp climate In mineral processing, the general physical and chemical conditions of a pulp, in which the pH, added chemicals, solid-liquid ratio, temperature, particle size range, and ionization of a flotation pulp are held within controlled limits while a considerable number of associated factors of less direct importance to the surface chemistry of the process are, at best, only indirectly monitored. (Pryor, 1963)

pulp density (a) In mineral processing, the amount of solids in a pulp, typically ranging from 10% to 25%, by weight. It has a marked effect on the recovery and grade of concentrate. (Taggart, 1927) (b) The weight of a unit volume of pulp; e.g., if 1 cm^3 of pulp weighs 2.4 g, then the pulp density is 2.4 g/cm^3. (Newton, 1959)

pulp dilution The ratio of water to solids by weight. It is expressed as a ratio; e.g., a pulp dilution of 3 to 1 means that a pulp contains 3 t of water for each ton of solids. (Newton, 1959)

pulpit The special platform upon which the operator of a Bessemer converter stands. (Mersereau, 1947)

pulpit man Person who operates the complex controls of a rolling mill, in which iron and steel ingots or billets are rolled into shapes such as bars, T's, rails, and sheets, by throwing the correct electric switches when signaled or by personal observation. Also called: manipulator operator; mill control operator. (DOT, 1949)

pulpstone A very large grindstone employed in pulp mills for crushing or grinding wood into fiber. (Fay, 1920)

pulsator (a) A motor-driven air compressor that supplies compressed air to an electric channeler. It receives the exhaust from the channeling machine cylinder and thus utilizes the pressure of the exhaust. (b) In mineral processing, a Harz-type jig. See also: *Harz jig*. (Pryor, 1963)

pulsator jig A gravity concentrator utilizing vertical pulsations in a hydraulic medium to separate particles by specific gravity differences.

pulsed infusion A variation of water infusion that has been effective in reducing both explosives consumption and airborne dust concentrations during mining. Water is introduced under pressure into long holes containing explosive charges and forced into the coal seam by detonation of the charges. See also: *infusion shot firing*. (Hartman, 1961)

pulsed infusion shot firing A coal blasting technique that consists of firing an explosive charge in a borehole filled with water under pressure. The water is introduced through an infusion tube that also seals the hole. When the charge is fired, it produces in the water a high-pressure impulse that is transmitted into the numerous water-filled cleavage planes and slips and thus breaks the coal. The energy from the explosive is used more efficiently than when blasting in the conventional manner, and better coal preparation is obtained. See also: *water infusion; long-hole infusion*. (McAdam, 1958)

pulsion stroke In mineral concentration by jigging, the stroke of the plunger device that controls the hydraulic lift of water through the bed of particles. (Pryor, 1963)

pulsometer (a) A steam pump in which an automatic ball valve (the only moving part) admits steam alternately to a pair of chambers, forcing out water that had been sucked in by condensation of the steam after the previous stroke. It can tolerate very dirty water and has been widely used for shaft sinking and miscellaneous pumping duties. (Nelson, 1965) (b) A displacement pump with valves for raising water by steam, partly by atmospheric pressure, and partly by the direct action of the steam on the water, without intervention of a piston. Also called: vacuum pump. (Webster 3rd, 1966)

pulsometer pump Pump with two chambers that are alternately filled and discharged. An automatic ball valve admits steam, which forces out the charge from the filled chamber while the other is filling as its steam condenses. (Pryor, 1963)

pulverization (a) In soil stabilization work, the separation of particles from each other rather than the breaking up of individual particles. Separation of the particles is the first step towards good dispersion of stabilization additives and moisture. (Nelson, 1965) (b) The reduction of metal to fine powder by mechanical means. Syn: *comminution; soil stabilization*.

pulverize To reduce (as by crushing or grinding) to very small particles (as in fine powder or dust). (Webster 3rd, 1966)

pulverized fuel Finely ground coal or other combustible material, that can be burned as it issues from a suitable nozzle, through which it is blown by compressed air. (Pryor, 1963)

pulverizer *fine grinder*.

pulverulent That which may easily be reduced to powder. Said of certain ores. (Weed, 1922; Fay, 1920)

pumice A light-colored, vesicular, glassy rock commonly having the composition of rhyolite. It is often sufficiently buoyant to float on water and is economically useful as a lightweight aggregate and as an abrasive. The adjectival form, pumiceous, is usually applied to pyroclastic ejecta. Cf: *scoria; pumicite*. (AGI, 1987)

pumiceous Adj. form of pumice.

pumicite A very finely divided volcanic ash or volcanic dust ranging in color from white to gray and buff. It is the unconsolidated equivalent of tuff. See also: *ash*. Cf: *pumice*.

pump A mechanical device for transferring either liquids or gases from one place to another, or for compressing or attenuating gases. (AGI, 1987)

pump bob The balance weight used to bring up the plunger in a Cornish pumping engine. (Standard, 1964)

pump chamber An underground pumping station. (Fay, 1920)

pumpellyite (a) A monoclinic mineral, $Ca_2(Mg,Fe,Mn)(Al,Mn,Fe)_2(SiO_4)(Si_2O_7)(OH)_2 \cdot H_2O$; pumpellyite group; individual species named according to the preponderance of Fe, Mg, or Mn; occurs in minute bluish-green fibers or plates in Michigan, California, Haiti, and New Zealand. (b) The mineral group jugoldite-(Fe), okhotskite, pumpellyite-(Fe), pumpellyite-(Mg), pumpellyite-(Mn), and shuiskite. Syn: *Lake Superior greenstone*.

Pumpelly's rule The generalization, made by Pumpelly in 1894, that the axes and axial surfaces of minor folds of an area are congruent with those of the major fold structures of the same phase of deformation. (AGI, 1987)

pumper In bituminous coal mining, a person who works a hand pump to force water, accumulated underground in low places, into a drainage ditch flowing to a natural outlet or pumping station. See also: *ram operator*. (DOT, 1949)

pump fist Eng. The lower end of a plunger case of a pump.

pumping (a) The act of moving a liquid or gas by means of a pump. (b) The operation of filling a sludge pump by an up-and-down motion of the rods or rope. Also called pumping the sludger. (c) In scraper operation, raising and lowering the bowl rapidly to force a larger load into it. (Nichols, 1954) (d) Alternately raising and lowering a digging edge to increase the volume of dirt being transported. (Nichols, 1976) (e) The motion of mercury in a barometer arising from the movement of a ship or from fluctuations of air pressure in a varying wind. (CTD, 1958)

pumping engineer In mining and in the quarry industry, a person who operates one or a battery of pumps to force excess water from a lower level to the surface or to a drainage tunnel. Also called: pitwright; plugman. (DOT, 1949)

pumping head In an airlift, the distance from the surface to the level of the water during pumping; it equals static head plus drop. (Lewis, 1964)

pumping shaft The shaft containing the pumping machinery of a mine. (Standard, 1964)

pump kettle A convex perforated diaphragm fixed at the bottom of a pump tube to prevent the entrance of foreign matter; a strainer. (Fay, 1920)

pump lift (a) The vertical distance that a pump can suck up water. Theoretically, this should be about 34 ft (10.4 m) at sea level; practically, the limit is about 26 ft (7.9 m). (Long, 1960) (b) The vertical distance a pump can force water to flow. (Long, 1960)

pump load The back pressure and/or resistance to flow of fluids that a pump must overcome to force a fluid to flow through a pipeline, drill string, etc. (Long, 1960)

pump pressure The force per unit area or pressure against which a pump acts to force a fluid to flow through a pipeline, drill string, etc.; also, the pressure imposed on the fluid ejected from a pump. (Long, 1960)

pump rod The rod or system of rods (usually heavy beams) connecting a steam engine at the surface or at a higher level with the pump piston below. See also: *balance bob*. Syn: *main rod*. (Fay, 1920)

pump-rod plates Scot. Spear plates; strips or plates of iron bolted to wooden pump rods at the joints for the purpose of making the connection. (Fay, 1920)

pump slip Leakage past the valves and the plunger in a reciprocating pump, which should not be greater than 2% or 3% for a pump in good condition. (Lewis, 1964)

pump slope A slope in which pumps are operated. (Fay, 1920)

pump station (a) In mining, a chamber near the shaft at depth, where a pump is installed. (Pryor, 1963) (b) An enlargement made in the shaft, slope, or entry to receive the pump. Also called pumproom. (c) The site at which one or more pumps are installed along a pipeline for the purpose of forcing a fluid through the line. (Long, 1960)

pump stock Lanc. *pump tree*.

pump sump A tank into which fluids gravitate and from which they are recirculated by means of a pump. (BS, 1962)

pump surge The pulsating effect transmitted to a pipeline or drill string at the completion of each compression stroke of a reciprocating-piston pump. (Long, 1960)

pump tree Eng. A cast-iron (wrought iron was formerly used) pipe, generally 9 ft (2.7 m) in length, of which the water column or set is formed. Syn: *pump stock*.

punch (a) A tool (ram) for knocking out timbers in coal workings. (Standard, 1964) (b) *leg; punch prop*.

punched screen Thin plates through which holes have been punched. These may be round, rectangular, or slotted. (Pryor, 1963)

puncher An early-model pick machine used to undermine or shear coal by heavy blows of sharp steel points attached to a piston driven by compressed air. (Fay, 1920)

punching shear If a heavily loaded column punches a hole through the base on which it rests, the base has failed in punching shear. This is prevented either by thickening the base or by enlarging the foot of the column so as to ensure that the allowable shear stress is not exceeded. (Hammond, 1965)

punch mining (a) Mining in which the rooms are opened off the strip mine highwall. (USBM, 1963) (b) An underground method of extracting coal from finger-shaped areas of reserves not amenable to other mining methods. Openings are driven by continuous mining machines back and forth across the fingers from outcrop to outcrop leaving a pillar of coal between each cut.

punch prop A short timber prop for supporting coal in holing or undercutting; a sprag. (Standard, 1964)

puppet valve A valve that, in opening, is lifted bodily from its seat by its spindle instead of being hinged at one side. See also: *poppet; poppet valve*. (Fay, 1920)

puppy An underground set of pumps. (Fay, 1920)

pure bending In mine subsidence, bending without fracture. (Briggs, 1929)

pure coal *vitrain.*
pure culture A collection of microbial cells of the same species in a container that is devoid of any other form of life. (Rogoff, 1962)
pure oxide Any of a group of refractories including alumina, magnesia, thoria, zirconia, beryllia, and ceria. (Osborne, 1956)
pure oxide ceramic Ceramic product made from any of the pure oxides of nonmetallic materials; i.e., Al_2O_3, MgO, SiO_2, etc.
pure shear A strain in which a rock body is elongated in one direction and shortened at right angles to this in such an amount that the volume remains unchanged. (AGI, 1987)
pure steel The product of a basic open-hearth furnace refined to a point where the impurities are reduced to the lowest practicable minimum, after which copper and molybdenum are added in correct proportions. (Mersereau, 1947)
purlins Timbers spanning from truss to truss, and supporting the rafters of a roof. (Crispin, 1964)
puron High-purity iron. (Osborne, 1956)
purple blende An old syn. for kermesite. (Fay, 1920)
purple copper ore A miners' term for bornite. Syn: *bornite.*
purple ore Sintered pyritic ore.
purpurite An orthorhombic mineral, $MnPO_4$; forms a series with heterosite; deep red or reddish purple; forms small, irregular masses as an alteration product of lithiophilite and triphylite; at Pala, CA; Hill City, SD; Newry, ME; and the Erongo Mountains, Namibia.
pushbutton coal mining A fully automatic and remotely controlled system of coal cutting, loading, and face conveying, including self-advancing roof support systems. See also: *manless coal face.* (Nelson, 1965)
pushbutton winding control A system in which the operation of the winder is similar to automatic cyclic winding, but the starting is instigated by the onsetter and banksman. When everything is ready for winding, the onsetter and banksman press their respective start pushbuttons and the winder starts, accelerates, and banks automatically without the intervention of the winding engineman. With this form of control, loading and discharging of the skips is fully automatic. See also: *automatic cyclic winding; manual winding control.* (Nelson, 1965)
pusher (a) A laborer who pushes loaded mine cars on tracks from underground working places to haulage roads where they are hooked up to a locomotive and hauled to the surface, shaft, or slope bottom for hoisting. A pusher may, at bituminous mines, shift empty and loaded cars in and about the tipple, where coal is prepared for market. Also called: car puller; car shifter; headsman; putter; trailer; trammer. Syn: *mule; wheeler.* (DOT, 1949) (b) One who encourages or hastens the miners. Also called jigger boss. (Ricketts, 1943) (c) A tractor that pushes a scraper to help it pick up a load. (Nichols, 1976)
pusher tractor A bulldozer exerting pressure on the rear of a scraper loader while the loader is digging and loading unconsolidated ground being excavated and moved during opencast mining.
push hole A hole through which glass is introduced to a flattening furnace. (Standard, 1964)
push-pull support system A method of advancing power-operated supports on a longwall face. Double-acting hydraulic jacks are used in conjunction with supports that slide forward on the floor and provide their own abutments for both their forward movement and that of the conveyor. (Nelson, 1965)
push-pull wave A wave that advances by alternate compression and rarefaction of a medium, causing a particle in its path to move forward and backward along the direction of the wave's advance. In connection with waves in the Earth, also known as primary wave, compressional wave, longitudinal wave, or P-wave. (Leet, 1958)
push wave *P wave.*
put To haul by hand. (Mason, 1951)
putrefaction A process of decomposition of organic substances that occurs in the presence of water and with the complete exclusion of air. It is a kind of slow distillation whereby chiefly methane (CH_4) and smaller quantities of other gaseous products, such as hydrogen (H_2), ammonia (NH_3), and hydrogen sulfide (H_2S), are formed. Cf: *disintegration.* (Stutzer, 1940)
PVC belt There are two main types of belts: (1) solid woven carcass impregnated and covered with polyvinyl chloride; and (2) normal multiple construction, which has polyvinyl chloride interlayers and covers. PVC belts are now used widely in coal mines, being not only fire resistant but equal, if not better, in quality than normal rubber belting. See also: *conveyor.* (Nelson, 1965)
P wave A seismic wave that propagates by alternating compressions and rarefactions in an elastic medium; the motion is in the direction of propagation. It is the type that carries sound. Syn: *compressional wave; dilatational wave; irrotational wave; longitudinal wave; pressure wave; push wave.* (AGI, 1987)
pycnite A variety of topaz occurring in massive columnar aggregates. Also spelled pychite.
pycnocline A steep vertical gradient of density. (Hy, 1965)
pycnometer (a) A device for weighing and thus determining the specific gravity of small quantities of oil or other liquids. Also spelled pyknometer. (Hess) (b) A small bottle for determining the specific gravity of grains or small fragments.
pyrabol *pyribole.*
pyralmandite A garnet composition between **pyr**ope and **almandine**. See also: *pyrope.*
pyralspite The **py**rope, **al**mandine, **sp**essartine subgroup of the garnet group. See also: *pyrope.*
pyramid An open crystal form consisting of nonparallel faces that intersect the *c* crystallographic axis and consist of three (trigonal), four (tetragonal), six (ditrigonal, hexagonal), or eight (ditetragonal) faces meeting at a point. Cf: *bipyramid; hemipyramid; dome.*
pyramidal Descriptive of a crystal habit dominated by pyramids or bipyramids.
pyramidal garnet Same as idocrase; a variety of vesuvianite. (Fay, 1920)
pyramid cut (a) In tunnel driving or shaft sinking, a pattern of shotholes drilled so that the middle holes converge and outline a pyramid-shaped volume of rock. These holes are fired first, and thus create a free face or relieving cut. (Pryor, 1963) (b) This cut has received its name from the shape of the initial opening. The three or four holes are so directed that they meet at a point farthest in. The pyramid cut is mainly employed in raises and for shaft sinking but is not recommended for horizontal tunnels where a machine setup for a definite direction of the four holes cannot easily be obtained. Syn: *German cut.* (Fraenkel, 1953) (c) This type of cut usually consists of four holes drilled to meet at a common apex in the center of the face. This arrangement permits a high concentration of explosive to be used, and the pyramid cut is therefore particularly suitable for breaking hard ground. In very hard ground the number of holes forming the cut may be increased to six. The main disadvantage of this type of cut is the difficulty in drilling the holes at the correct angles so that they will meet at the back of the cut. As in the case of the wedge cut, therefore, a hole director should be used. Also called diamond cut. (McAdam, 1958) (d) In underground blasting, a type of cut employed in which the three cut holes in the center may be drilled to form a pyramid. Also applied to four holes meeting in a point. The simultaneous firing of these holes is somewhat equivalent to using a very heavy charge of explosive and makes a powerful blast. (Lewis, 1964) (e) A cut in which four central holes are drilled towards a focal point, and when fired break out a tetrahedral section of strata. (BS, 1964)
pyramid-set A bit crown, the face of which is covered with a series of stubby pyramids, each apex of which is set with a diamond. (Long, 1960)
pyramid structure In crystallography, that of a crystal in which three or more inclined faces cut the three crystal axes. (Pryor, 1963)
pyrargyrite A trigonal mineral, Ag_3SbS_3; dimorphous with pyrostilpnite; rhombohedral cleavage; soft; deep red; in late-primary or secondary-enrichment veins, and an important source of silver. Syn: *antimonial red silver; dark red silver ore; dark ruby silver.*
pyrene A tetracyclic hydrocarbon obtained from the coal-tar fraction boiling above 360 °C; $C_{16}H_{10}$; soluble in carbon disulfide, toluene, and ligroin. (CTD, 1958; Handbook of Chem. & Phys., 2)
pyreneite A black variety of andradite garnet. Syn: *melanite.*
pyrheliometer An actinometer that measures the intensity of direct solar radiation. (AGI, 1987)
pyribole The **pyr**oxene group plus amph**ibole** group. See also: *pyroxene.*
pyricaustate A general name for a fossil combustible substance. (Tomkeieff, 1954)
pyrite (a) An isometric mineral, FeS_2; dimorphous with marcasite; forms a series with cattierite; crystallizes in cubes and pyritohedra; sparks readily if struck by steel; metallic; pale bronze to brass yellow; hardness varies from 6.0 to 6.5; occurs in veins, as magmatic segregation, as accessory in igneous rocks, and in metamorphic rocks, in sedimentary rocks including coal seams; a source of sulfur; may have included gold. Syn: *Alpine diamond; iron pyrite; fool's gold; mundic; common pyrite.* (b) The mineral group aurostibite, bravoite, cattierite, erlichmanite, fukuchilite, geversite, hauerite, insizwaite, krutaite, laurite, malanite, maslovite, michenerite, penroseite, pyrite, sperrylite, testibiopalladite, trogtalite, vaesite, and villamaninite.
pyrites (a) Various metallic-looking sulfide minerals including iron pyrites (pyrite); copper pyrites (chalcopyrite); tin pyrites (stannite); white iron, cockscomb, or spear pyrites (marcasite); arsenical pyrites (arsenopyrite); cobalt pyrites (linnaeite); magnetic pyrites (pyrrhotite); and capillary pyrites (millerite). Without qualification it popularly refers to pyrite. (b) Stones that may be used for striking fire.
pyrites of copper Common name for chalcopyrite. (Weed, 1918)
pyritic Of, pertaining to, resembling, or having the properties of pyrites. (Standard, 1964)

pyritic smelting Smelting of sulfide copper ores, in which heat is supplied mainly by oxidation of iron sulfide. Syn: *oxidizing smelting.* (CTD, 1958)
pyritic sulfur The part of the sulfur in coal that is in the form of pyrites or marcasite. (BS, 1961)
pyritiferous Containing or producing pyrite. (Webster 3rd, 1966)
pyritization Introduction of or replacement by pyrite; e.g., the replacement of original fossil material by pyrite. A common hydrothermal introduction of pyrite specks in rock adjacent to veins.
pyritohedron An isometric closed crystal form of 12 faces, each an irregular pentagon. It is named after pyrite, which characteristically has this crystal form. See also: *rhombic dodecahedron* Cf: *dodecahedron.*
pyroantimonite *kermesite.*
pyroaurite A trigonal mineral, $Mg_6Fe_2(CO_3)(OH)_{16}{\cdot}4H_2O$; hydrotalcite group; dimorphous with sjögrenite; occurs in goldlike submetallic scales, or brown crystals having pearly to greasy luster. A silvery white variety is called igelstromite.
pyrobelonite An orthorhombic mineral, $PbMn(VO_4)(OH)$; descloizite group; forms minute fire-red acicular crystals at Långban, Sweden; a source of vanadium.
pyrobitumen Any of the dark-colored, fairly hard, nonvolatile, carbon-rich material substances composed of hydrocarbon complexes, which may or may not contain oxygenated substances and are often associated with mineral matter. The nonmineral constituents are infusible, insoluble in water, and relatively insoluble in carbon disulfide. (AGI, 1987)
pyrobituminous Pertaining to substances that yield bitumens upon heating. (AGI, 1987)
pyrochlore (a) An isometric mineral, $(Ca,Na)_2Nb_2O_6(OH,F)$; forms a series with microlite; in pegmatites in Maine, California, Colorado, Africa, and Europe; a source of niobium. Syn: *pyrrhite.* (b) The mineral group including the betafite subgroup betafite, plumbobetafite, and yttrobetafite; the microlite subgroup bariomicrolite, bismutomicrolite, microlite, plumbomicrolite, and uranmicrolite; and the pyrochlore subgroup bariopyrochlore, ceriopyrochlore, kalipyrochlore, plumbopyrochlore, uranpyrochlore, and yttropyrochlore.
pyrochroite A trigonal mineral, $Mn(OH)_2$; brucite group; soft; pearly white darkening on exposure; has perfect basal cleavage.
pyroclast An individual particle ejected during a volcanic eruption. It is usually classified according to size. (AGI, 1987)
pyroclastic Produced by explosive or aerial ejection of ash, fragments, and glassy material from a volcanic vent. Applied to the rocks and rock layers as well as to the textures so formed. (Stokes, 1955)
pyroclastic deposit A deposit made up mainly of rock material that has been expelled aerially, normally explosively, from a volcanic vent, such as agglomerate, tuff, and ash. The fragments range in size from bombs and blocks to dust or ash. Such deposits are usually designated according to the lavas to which they correspond in composition. (Stokes, 1955)
pyrogenesis A broad term encompassing the intrusion and extrusion of magma and its derivative. Adj. pyrogenic. (AGI, 1987)
pyrogenetic A term introduced to designate minerals, such as olivine and chromite, developed at high temperature in melts containing only a small proportion of volatile (hyperfusible or fugitive) constituents. See also: *pyrogenic.* (Schieferdecker, 1959)
pyrogenic Said of a process or of a deposit involving the intrusion and/or extrusion of magma. See also: *pyrogenetic; igneous.* (AGI, 1987)
pyrogenic ore mineral An ore mineral that crystallized as a primary magmatic mineral of igneous rocks. (Schieferdecker, 1959)
pyrogenic rock A rock resulting from the cooling of a molten magma; an igneous rock.
pyrognostics The characteristics (such as the degree of fusibility or the flame coloration) of a mineral observed by the use of the blowpipe. (Webster 3rd, 1966)
pyrolite An explosive resembling gunpowder in composition. (Webster 2nd, 1960)
pyrolusite A tetragonal mineral, MnO_2; rutile group; trimorphous with akhtenskite and ramsdellite; soft; metallic; steel gray; massive or reniform; a source of manganese. Cf: *psilomelane.* Syn: *polianite; gray manganese ore.*
pyrolysis Chemical decomposition by the action of heat.
pyrolytic graphite Graphite formed by pyrolysis of a carbonaceous gas. (Van Vlack, 1964)
pyrometallurgy Metallurgy involved in winning and refining metals in which heat is used, as in roasting and smelting. Practically all iron and steel, nickel and tin, most copper, and a large proportion of zinc, gold, and silver, as well as many of the minor metals, are won from their ores and concentrates by pyrometallurgical methods. It is the most important and oldest class of the extractive processes. (ASM, 1961)
pyrometamorphism Metamorphism produced by heat; it is a local, intense type of thermal metamorphism, resulting from unusually high temperatures at the contact of a rock with magma, such as in xenoliths. Cf: *igneous metamorphism; hydrometamorphism.* See also: *thermal metamorphism.* (AGI, 1987)
pyrometasomatic Formed by metasomatic changes in rocks, principally in limestone, at or near intrusive contacts, under the influence of magmatic emanations and high to moderate temperature and pressure. (AGI, 1987)
pyrometasomatism Contact metamorphism. Cf: *metasomatism.*
pyrometer An instrument that measures high temperature, e.g., of molten lavas, by electrical or optical means. See also: *optical pyrometer.* (AGI, 1987)
pyrometric cone A small, slender three-sided pyramid made of ceramic or refractory material for use in determining the time-temperature effect of heating and in obtaining the pyrometric cone equivalent (PCE) of refractory material. Pyrometric cones are made in series, the temperature interval between successive cones usually being 20 °C. The best known series are Seger cones (Germany), Orton cones (United States), and Staffordshire cones (United Kingdom). See also: *cone; orton cone.* (ARI, 1949; Dodd, 1964)
pyrometric cone equivalent The number of that standard pyrometric cone whose tip would touch the supporting plaque simultaneously with a cone of the refractory material being investigated when tested in accordance with ASTM Test Method C-24. Abbrev. PCE. (ASTM, 1994)
pyromorphite A hexagonal mineral, $Pb_5(PO_4)_3Cl$; apatite group, with iron replacing lead and arsenic replacing phosphorous; sp gr, 6 to 7; in oxidized zones of lead-ore deposits. Syn: *green lead ore.*
pyrope (a) An isometric mineral, $8[Mg_3Al_2(SiO_4)_3]$; garnet group, with Fe and Mn replacing Mg and Cr replacing Al. See also: *pyralspite; pyralmandite.* Crystallizes in dodecahedra and trapezohedra; deep red to black; in high-pressure ultramafic and metamorphic rocks; also in placers; a gemstone and an abrasive. See also: *Cape ruby; Bohemian garnet.* Syn: *rock ruby; magnesium-aluminum garnet.* (b) Formerly, a name for any bright red gem, such as ruby.
pyrophane An opal, e.g., hydrophane, artificially impregnated with melted wax. See also: *fire opal.*
pyrophanite A trigonal mineral, $MnTiO_3$; ilmenite group; forms a series with ilmenite; blood red.
pyrophoric sphalerite A variety of sphalerite that gives off sparks or glows when abraded. Some pieces are so sensitive that the effect is obtained by scratching them with a fingernail. (Hess)
pyrophyllite A monoclinic and triclinic mineral, $Al_2Si_4O_{10}(OH)_2$; foliated; soft; in schists and hydrothermal veins in North Carolina, California, Newfoundland, and Japan. Syn: *pencil stone.* See also: *G stone.*
pyrophysalite A coarse opaque variety of topaz from Finbo, Sweden. Also spelled physalite.
pyropissite An earthy nonphosphatic pyrobitumen composed primarily of water, humic acid, wax (a source of "montan wax"), and silica, associated with brown coal called pyropissitic brown coal.
pyroradiation pyrometer A self-contained instrument with the millivoltmeter mounted in the pyrometer tube; the radiant energy is concentrated by means of an objective lens (quartz or fluorite) rather than by a reflecting mirror. (Newton, 1938)
pyroretin A brittle, brownish-black resin that occurs in brown coal near Aussig, Bohemia; sp gr, 1.05 to 1.18. (Fay, 1920)
pyrosmaltite Any member of the hexagonal mineral series, ferropyrosmaltite-manganpyrosmaltite, $(Fe,Mn)_8Si_6O_{15}(OH,Cl)_{10}$.
pyrostibite *kermesite.* Also spelled pyrostibnite.
pyrostibnite *kermesite.* Also spelled pyrostibite. (CMD, 1948)
pyrostilpnite A monoclinic mineral, Ag_3SbS_3; dimorphous with pyrargyrite; red. Syn: *fireblende.*
pyrosulfuric acid (a) A heavy, oily, strongly corrosive liquid $H_2S_2O_7$ that consists of a solution of sulfur trioxide in anhydrous sulfuric acid. It fumes in moist air and reacts violently with water with the evolution of heat. (Webster 3rd, 1966) (b) A solution of sulfur trioxide in sulfuric acid; $H_2S_2O_7$. Colorless to dark brown depending on purity; hygroscopic. (CCD, 1961)
pyroxene (a) A group of chiefly magnesium-iron minerals including diopside, hedenbergite, augite, pigeonite, and many other rock-forming minerals. Although members of the group fall into different systems (orthorhombic, monoclinic, and triclinic), they are closely related in form, composition, and structure. See also: *acmite; aegirite; augite; diallage; enstatite; hypersthene.* (Fay, 1920; AGI, 1987) (b) The mineral group aegirine (Ae), aegirine-augite, clinoenstatite, clinoferrosilite, diopside (Di), donpeacorite, enstatite (En), essenite (Es), ferrosilite (Fs), hedenbergite (Hd), jadeite (Jd), jervisite (Je), johannsenite (Jo), kanoite (Ka), kosmochlor (Ko), natalyite, omphacite, petedunnite (Pe), pigeonite, and spodumene (Sp). Some former names relegated to synonyms include acmite = aegirine, bronzite = enstatite, clinohypersthene = clinoenstatite or clinoferrosilite, diallage = altered diopside or other pyroxene with good (100) parting, eulite = ferrosilite, fassaite = ferrian aluminian diopside or augite, ferroaugite = augite, ferrosalite = hedenbergite, hiddenite = spodumene, hypersthene = enstatite or ferrosilite,

kunzite = spodumene, salite = diopside, titanaugite = titanian augite, uralite = pseudomorphous amphibole after pyroxene, and ureyite = kosmochlor. Pyroxenes (px) are either monoclinic (clinopyroxenes, cpx) or orthorhombic (orthopyroxenes, opx). General formula: AB_2ZO_6: A = Ca, Fe^{2+}, Li, Mg, Mn^{2+}, Na, Zn; B = Al, Cr^{3+}, Fe^{2+}, Fe^{3+}, Mg, Mn^{2+}, Sc, Ti, V^{3+}; Z = Al, Si. Their structures are built from single chains of silica tetrahedra each sharing two oxygens, with a silica:oxygen ratio of 1:3, electrostatic neutrality being maintained by cross-linking cations. Crystals are prismatic with prismatic cleavage at 87° and 93°. Colors are mostly greens, but range from white to black. Etymol: Greek pyros (fire) + xenos (stranger) from a mistaken belief that they were only accidently caught up in lavas. See also: *aegirine.* Cf: *amphibole; pyribole.*

pyroxene perthite Lamellar intergrowths of pyroxene of different kinds, as with the feldspars. Also pyroxene microperthite, pyroxene cryptoperthite. (English, 1938)

pyroxenite A coarse-grained, holocrystalline igneous rock consisting of 90% pyroxenes. It may contain biotite, hornblende, or olivine as accessories. (CTD, 1958)

pyroxenoid Single-chain silicates with individual silica tetrahedra twisted relative to the pyroxene chains, resulting in triclinic symmetry; e.g., the wollastonites, rhodonite, and pectolite.

pyroxmangite A triclinic mineral, $MnSiO_3$; forms a series with pyroxferroite where iron replaces manganese; forms brown cleavable masses near Iva, SC; Homedale, ID; Sweden; and Scotland.

pyrrhite *pyrochlore.*

pyrrhoarsenite *berzeliite.*

pyrrhotine *pyrrhotite.*

pyrrhotite A monoclinic and hexagonal mineral, FeS; invariably deficient in iron; variably ferrimagnetic; metallic; bronze yellow with iridescent tarnish; in mafic igneous rocks, contact metamorphic deposits, high-temperature veins, and granite pegmatites. Where associated with pentlandite and nickel replaces iron, it is a source of nickel. Also spelled pyrrhotine. Syn: *magnetic pyrite; dipyrite.*

quad The composition quadrilateral for the Ca-Mg-Fe pyroxenes (enstatite-ferrosilite-diopside-hedenbergite).

quadrantal bearing A horizontal angle or bearing less than 90°, measured to north, south, east, or west from a survey line. (Hammond, 1965)

quadrant cutter A machine that will make a shear cut as well as a horizontal cut. The central column is wedged tightly between roof and floor and operates similarly to a radial percussive coal cutter. (Nelson, 1965)

quadrant search Similar to octant search, but using four sectors instead of eight sectors. Applies to any interpolation method where a limited number of sample data points are used to estimate intermediate values.

quadrilateral A four-sided plane figure of any shape, having an area equal to the product of the diagonals multiplied by half the sine of the angle between them. (Hammond, 1965)

quadrille twinning *crossed twinning.*

quadrivalent (a) Having a valence of 4. (Webster 3rd, 1966) (b) Having four valences; e.g., chlorine, which has a valence of 1, 3, 5, and 7. (Handbook of Chem. & Phys., 2)

quadruple block A pair of blocks, each having four sheaves, reeved with rope or cable and used to increase the lifting capacity of a drill-hoisting mechanism; a four-sheave block and tackle. (Long, 1960)

quagmire A soft marsh or bog that gives under pressure. Cf: *quaking bog.* (AGI, 1987)

quake sheet A well-defined bed resembling a slump sheet but produced by seismic shock from an earthquake and resulting in load casting without horizontal slip. (AGI, 1987)

quaking bog A peat bog that is either floating or is growing over water-saturated ground, so that it shakes or trembles when walked on. Quagmire is sometimes used as a synonym. (AGI, 1987)

qualitative analysis In chemistry, the process of determining which elements are present. (Standard, 1964)

quality (a) Refers to the nature, and not the amount, of material. In the case of a coal seam, its quality is closely linked with its rank and its chemical composition. In the case of metals, average unit values are determined by systematic sampling and therefore represent a known quantity. See also: *ventilation standards.* (Nelson, 1965) (b) Native values of a gem irrespective of color and cut. (Hess) (c) The ratio by weight of vapor to liquid plus vapor in a mixture, as in steam. (Strock, 1948)

quality control (a) Systematic setting, check, and operation designed to maintain steady working conditions in continuous process such as mineral concentration; to forestall trouble; to check condition of ore, pulp, or products at important transfer points. (Pryor, 1963) (b) Graphic method of exposing abnormalities in sets of figures produced by measurement of repetitive operations or as variances from operating norms. (Pryor, 1963) (c) The maintaining of air within desired limits of purity. (Hartman, 1961)

quantitation Once a dust sample has been collected, it must be evaluated. Of principal concern is quantitation—determining how much dust or how many particles. Certain methods of quantitation are favored for the various sampling methods. The number basis is preferable for evaluating a pulmonary hazard, while the weight basis is preferred for toxic, radioactive, or explosive hazards. Number quantitation is usually employed for impinger, konimeter, molecular filter, and thermal precipitator samples. Weight quantitation is used for filter paper and electrostatic samples. (Hartman, 1961)

quantitative In testing ore, how much of each metal is present. (von Bernewitz, 1931)

quantitative analysis In chemistry, the process of determining the quantity of each element present. Also called elementary analysis. Both volumetric and gravimetric methods are included. (Standard, 1964)

quantitative survey *ventilation survey.*

quantity Deals with the amount, and not the nature, of a substance. In the case of a coal seam, quantity refers to its workable thickness and acreage. In the case of ore, the quantity determines its commercial importance. Unit ore values without the quantity factor have only a qualitative significance. (Nelson, 1965)

quantity control The control of air movement, its direction, and its magnitude. (Hartman, 1982)

quantity-distance table A table listing minimum recommended distances from explosive material stores of various weights to a specific location. (Meyer, 1981)

quaquaversal Dipping outward in all directions from a central point, as a dome in stratified rocks. Cf: *centroclinal; periclinal.*

quaquaversal fold *dome.*

quarey lode *quarry lode.*

quarfeloids A portmanteau word from **quar**tz, **fel**dspar, and feldspath**oids**. Cf: *feloids.*

quarl A large brick or tile; esp., a curved firebrick used to support melting pots for zinc and retort covers. (Webster 3rd, 1966)

quarman *quarryman.*

quarpit An obsolete term for a quarry.

quarrel (a) A stone quarry. (Standard, 1964) (b) Materials from a quarry. (Standard, 1964)

quarrier A worker in a stone quarry. (Standard, 1964)

quarry (a) An open or surface mineral working, usually for the extraction of building stone, as slate, limestone, etc. It is distinguished from a mine because a quarry usually is open at the top and front, and, in ordinary use of the term, by the character of the material extracted. See also: *opencast.* (b) Day work pit. Also called opencast; quarpit. (Pryor, 1963) (c) An underground excavation formed in the roof, or fault, for the purpose of obtaining material for pack walls.

quarry body A dump body with sloped sides. (Nichols, 1976)

quarry drainage Arranging the quarry layout so that pools of water do not collect in the working area. One-half percent grade away from the face will generally keep the floor free of mud and water. (Nelson, 1965)

quarry drill A blasthole drill. (Nichols, 1976)

quarry face The freshly split face of ashlar, squared off for the joints only, as it comes from the quarry, and used esp. for massive work. Distinguished from rock face. (Webster 3rd, 1966)

quarry-faced masonry Masonry in which the face of the stone is left unfinished just as it comes from the quarry. (Crispin, 1964)

quarry floor The lowest level on which stone is loaded. (Streefkerk, 1952)

quarrying (a) The surface exploitation of stone or mineral deposits from the Earth's crust. (Nelson, 1965) (b) Removal of rock that has value because of its physical characteristics. (Nichols, 1954) (c) One of the effects of glaciation whereby blocks of stone, bounded mainly by joint planes, are lifted from the bedrock and carried away by ice. Also called plucking. (Stokes, 1955)

quarrying machine Any machine used to drill holes or cut tunnels in native rock; a gang drill, or tunneling machine, but most commonly a small form of locomotive, bearing a rock-drilling mechanism, and operating on a track laid temporarily along or opposite the ledge to be cut. (Standard, 1964)

quarry lode A vein in a heading that is jointed and blocky, like granite in a quarry, or a heading in granite.

quarry machine *quarrying machine.*

quarryman (a) A person employed at the face of a quarry, stripping, drilling, excavating, and loading rock or economic product. (Nelson, 1965) (b) One who operates a jackhammer to drill holes in quarry stone, and drives wedges into the holes to break or split off slabs or blocks of stone. Also called hammerman; plug-and-feather driller; rockman; rock splitter. (DOT, 1949) (c) In crushed rock quarries, a laborer who performs any one or combination of such duties as: loading rock into boxes to be hoisted out of quarry pit; assisting in moving power shovel from one loading position to another; dumping rock from cars into crusher or storage bins; feeding rock into a crusher; tending belt conveyors that transport crushed rock from crusher to storage bins; loading crushed rock from storage bins into trucks or railroad cars. (DOT, 1949) (d) In building stone quarries, a laborer who performs any one or combination of such duties as: cleaning dirt and mud from surface and sides of stone deposits; chipping irregularities from surface of granite blocks; breaking large pieces of stone into smaller sizes suitable for building purposes with a sledge hammer; attaching hoisting cable hooks or slings to blocks of stone to be hoisted from quarry;

guiding and steadying blocks of stone as they are loaded at the quarry surface on trucks or railraod cars by a derrick. (DOT, 1949)

quarry powder Ammonium nitrate dynamites intended to replace the more costly gelatin dynamites used in quarrying, where blasts of several tons of explosives are used. Cartridges up to 8 in (20 cm) in diameter by 21 in (53 cm) in length, can be enclosed in metal cans to protect against water damage. (Lewis, 1964)

quarry-rid Overburden. Cf: *ridding*. (Arkell, 1953)

quarry sap (a) The moisture contained in newly quarried stone. (Arkell, 1953) (b) *quarry water.*

quarry waste Material discarded after crushing, as being too fine, irregular, or flaky for constructional work. (Nelson, 1965)

quarry water (a) Water that fills the pore spaces of a rock in a quarry. See also: *ground water.* (Fay, 1920) (b) Subsurface water retained in freshly quarried rock. Syn: *quarry sap*. (AGI, 1987)

quartation The separation of gold from silver by dissolving out the latter with nitric acid. It requires not less than three-fourths of silver in the alloy, whence the name, which is also applied to the alloying of gold with silver, if necessary, to prepare it for this method of parting. *parting*. (Fay, 1920)

quarter (a) The act or process of dividing sludge, core, and other pulverized or granular samples into four equal parts. See also: *quartering*. (Long, 1960) (b) Syn. for quadrant as applied to a drill-bit crown. (Long, 1960)

quartering (a) The reduction in quantity of a large sample of material by dividing a heap into four approx. equal parts by diameters at right angles, removing two diagonally opposite quarters and mixing the two remaining quarters intimately together so as to obtain a truly representative half of the original mass. The process is repeated until a sample is obtained of the requisite size. Syn: *coning; coning and quartering*. (Taylor, 1965) (b) To split a piece of core longitudinally into four equal parts. (Long, 1960)

quartering in Lanc. A plan of building or putting together tubbing plates from the top downward, the rings and segments being bolted together as the work of excavation proceeds.

quartering way (a) A quarry term to designate a direction in which a rock cleaves with moderate facility; grain. See also: *roughway*. Cf: *hard way.* (b) The direction of the natural joints in a quarry rock. Cf: *rift*. (c) Grain, second way, bate, hem, sheeting plane.

quarter line Western United States. The survey line by which a section of government land is divided into quarter sections.

quarterly survey An underground survey required by law to be undertaken at least once every three months for the purpose of bringing the working plans and other plans up to date. (BS, 1963)

quarter octagonal A square shaft with corners cut back. (Nichols, 1976)

quarter-point veins Small veins having an intermediate bearing between strike and cross veins.

quarter post A post marking a corner of a quarter section of the U.S. Public Land Survey system. It is located midway between section corners. (AGI, 1987)

quarter section A fourth of a normal section of the U.S. Public Land Survey system, representing a piece of land normally 1/2 mile (0.8 km) square and containing 160 acres (64 ha) nearly as possible. It is usually identified as the northeast, northwest, southeast, or southwest quarter of a particular section. (AGI, 1987)

quartz (a) A trigonal mineral, SiO_2; polymorphous with tridymite, cristobalite, coesite, stishovite, and keatite. Amethyst is a variety of the well-known amethystine color. Aventurine is a quartz spangled with scales of mica, hemitite, or other minerals. False topaz or citrine is a yellow quartz. Rock crystal is a watery clear variety. Rose quartz is a pink variety. Rutilated quartz contains needles of rutile. Smoky quartz is a brownish variety, sometimes called cairngorm. Tigereye is crocidolite (an asbestoslike mineral) replaced by quartz and iron oxide and having a chatoyant effect. The name of the mineral is prefixed to the names of many rocks that contain it, as quartz porphyry, quartz diorite. See also: *alpha quartz; beta quartz; high quartz; low quartz*. (Sanford, 1914; Fay, 1920) (b) Pac. Any hard, gold or silver ore, as distinguished from gravel or earth. Hence, quartz mining, as distinguished from hydraulic mining, etc. (Fay, 1920) (c) A general term for a variety of cryptocrystalline varieties of SiO_2; e.g., agate, chalcedony.

quartz andesite *dacite.*

quartz battery A stamp, or series of stamps, for crushing quartz ore. (Mathews, 1951)

quartz boil An outcrop of a quartz vein.

quartz claim In the United States, a mining claim containing ore in veins or lodes, as contrasted with placer claims carrying mineral, usually gold, in alluvium.

quartz conglomerate A rock made of pebbles of quartz with sand. (Osborne, 1956)

quartz diorite A group of plutonic rocks having the composition of diorite but with an appreciable amount of quartz, i.e., between 5% and 20% of the light-colored constituents; also, any rock in that group; the approximate intrusive equivalent of dacite. (AGI, 1987)

quartz felsite *quartz porphyry.*

quartz glass Glass made by fusing quartz.

quartz gold Gold that is not rounded and waterworn, but irregular and frequently twisted in form, usually very bright, and always of fine quality. (Craigie, 1938)

quartzic *quartziferous.*

quartziferous Quartz-bearing as applied to a rock not defined by the presence of quartz, but containing minor amounts of it; e.g., limestone. Syn: *quartzic.* See also: *quartzose.*

quartz index (a) A derived quantity (qz) in the Niggli system of rock classification, which may be either positive or negative, and is as indicator of a rock's degree of silica saturation. (AGI, 1987) (b) A term used to indicate the mineralogic maturity of a sandstone by measuring the percentage of detrital quartz. It is expressed as the ratio of quartz and chert to the combined percentage of sodic and potassic feldspar, rock fragments, and clay matrix. The index is used as a basis for evaluating the degree of weathering of the source rock and the degree to which the sediment has been transported. Values for sandstones range between 3 and 19. (AGI, 1987)

quartzite (a) A granoblastic metamorphic rock consisting mainly of quartz and formed by recrystallization of sandstone or chert by either regional or thermal metamorphism; metaquartzite. Cf: *orthoquartzite.* (AGI, 1987) (b) A very hard but unmetamorphosed sandstone, consisting chiefly of quartz grains that are so completely cemented with secondary silica that the rock breaks across or through the grains rather than around them; an orthoquartzite. (AGI, 1987) (c) Stone composed of silica grains so firmly cemented by silica that fracture occurs through the grains rather than around them. (USBM, 1965) (d) As used in a general sense by drillers, a very hard, dense sandstone. (Long, 1960) (e) A granulose metamorphic rock consisting essentially of quartz. (Holmes, 1920) (f) Sandstone cemented by silica that has grown in optical continuity around each fragment. Syn: *ganister.* (Holmes, 1920)

quartzitic Of, pertaining to, or consisting of quartzite.

quartz keratophyre Altered sodic diabase (trachyte) with accessory quartz.

quartz latite The extrusive or hypabyssal equivalent of a quartz monzonite. The principal minerals are quartz, sanidine, biotite, sodic plagioclase, and hornblende, commonly as phenocrysts in a groundmass of potash feldspar and quartz (or tridymite, cristobalite), or glass in flows. Accessory minerals are magnetite, apatite, and zircon.

quartz lead A lode or vein of ore with quartz gangue.

quartz liquefier In metallurgy, an apparatus for extracting gold from its ore. By the action of an alkali and high-pressure steam, gold-bearing quartz is converted into a soluble silicate from which gold may be separated by washing. (Standard, 1964)

quartz mill A machine or establishment for pulverizing quartz ore, in order that the gold or silver it contains may be separated by chemical means; a stamp mill. (Standard, 1964; Fay, 1920)

quartz mine (a) A mine in which the deposits of ore are found in veins or fissures in the rocks forming the Earth's crust. Usually applied to lode gold mines, but not to placers. (b) A miner's term for a mine in which the valuable constituent, e.g. gold, is found in siliceous veins rather than in placers. It is so named because quartz is the chief accessory mineral. (AGI, 1987)

quartz monzonite A medium- to coarse-grained plutonic rock containing major plagioclase, orthoclase, and quartz, with minor biotite, hornblende, and accessory apatite, zircon, and opaque oxides. Syn: *adamellite.*

quartzoid A crystal having the form of two six-sided pyramids base to base. (Standard, 1964)

quartz ore A rock containing a large quantity of quartz. (Gordon, 1906)

quartzose (a) Of, pertaining to, or consisting of quartz. (b) Containing quartz as a principal constituent; esp. applied to sediments and sedimentary rocks (e.g., sands and sandstones) consisting chiefly of quartz. Cf: *quartziferous.* Syn: *quartzous; quartzy.*

quartzous *quartzose.*

quartz porphyry A field term for a medium-grained porphyritic igneous rock of felsic but unspecified composition occurring normally as minor stock or dike intrusions, and carrying prominent phenocrysts of quartz. It is a common altered companion to porphyry copper deposits. Syn: *quartz felsite.*

quartz reef A lode or vein of quartz. See also: *reef.*

quartz sinter Siliceous sinter. (Fay, 1920)

quartz syenite A potash or soda syenite with quartz as an accessory, hence on the borderline between syenite and granite.

quartz trachyte A fine-grained igneous rock consisting mostly of alkali feldspar, with normative quartz between 5% and 20%; the volcanic equivalent of quartz syenite. It normally shows trachytic texture. (AGI, 1987)

quartz wedge (a) An optical accessory with varying retardation used in polarized-light microscopy to determine birefringence and optic sign. Cf: *Berek compensator.* (b) In polarized-light microscopy, an accessory plate that gives variable compensation for birefringence. Cf: *accessory plate; gypsum plate.*
quartzy *quartzose.*
Quaternary The second period of the Cenozoic era, following the Tertiary; also, the corresponding system of rocks. It began 2 to 3 million years ago and extends to the present. It consists of two grossly unequal epochs; the Pleistocene, up to about 10,000 years ago, and the Holocene since that time. The Quaternary was originally designated an era rather than a period, with the epochs considered to be periods, and it is still sometimes used as such in the geologic literature. The Quaternary may also be incorporated into the Neogene, when the Neogene is designated as a period of the Tertiary era. (AGI, 1987)
quaternary alloy An alloy containing four principal elements. (Rolfe, 1955)
queane *quene.*
quebracho Aqueous extract of a bark of quebracho tree; contains up to 65% tannin. Used in froth-flotation as depressant for oxidized minerals. (Pryor, 1963)
queen Slate measuring 36 in by 24 in (91.44 cm by 60.96 cm). (Pryor, 1963)
queer A fissure, joint, or small cavity in a rock or quartz vein. Also spelled quere, queere, and qweear (U.K.).
queery Corn. When the lode or rock on which the miner is driving partakes of the character of quarry stone, namely, in detached lumps by natural divisions, it is called queery ground, and is frequently worked with crowbars and levers instead of being blasted or gadded. A "queer of ground" is a detached rock. Also called quarry lode. See also: *queer.*
quench (a) To cool suddenly (as heated steel) by immersion, esp. in water or oil. (Webster 3rd, 1966) (b) To produce a crust or a succession of crusts on molten metal, each crust being removed as it is formed. (Standard, 1964)
quenching Generally means rapidly cooling metals and alloys, or any substance to below the critical range by immersing it in oil or water to harden it. Also applied to cooling in salt and molten-metal baths or by means of an air blast, and to the rapid cooling of other alloys after solution treatment. (CTD, 1958)
quenching oils Oils used in heat treating. Fish oils are often used. Minerals, fish, vegetable, and animal oils are often compounded and sold under trade names. (Crispin, 1964)
quenching tub A tub of water in which to cool, harden, or temper iron or steel. (Standard, 1964)
quene Crevice in lode or vein. Also spelled queane. (Hess)
quenselite A monoclinic mineral, $PbMnO_2(OH)$; occurs in pitch-black crystals with perfect cleavage; at Långban, Sweden.
Querwellen wave *Love wave.*
questal bentonite A colloidal bond which, when added to molding sands in amounts up to 3%, increases porosity and strength (green and dry), and reduces the amount of water needed. (Osborne, 1956)
quick (a) Said of a sediment that, when mixed with water, becomes extremely soft and incoherent and is capable of flowing easily under load or by force of gravity; e.g. quick clay or quicksand. (AGI, 1987) (b) Said of blasting powder that burns or goes off very rapidly. (c) *quicksilver.* (d) Said of an economically valuable or productive mineral deposit, in contrast to a dead ground or area. An ore is said to be quickening as its mineral content increases. Syn: *alive.* Ant: dead. (AGI, 1987)
quickening Descriptive of an ore as its mineral content increases with distance.
quicklime sizes The different sizes depending upon the type of limestone, kind of kiln used, or treatment subsequent to calcining. The sizes commonly recognized are as follows: (1) large lump—8 in (203 mm) and smaller; (2) pebble or crushed—2-1/2 in (64 mm) and smaller; (3) ground, screened, or granular—1/4 in (6.4 mm) and smaller; and (4) pulverized—substantially all passing a No. 20, 850 μm, sieve. (ASTM, 1994)
quickness The property of an explosive by virtue of which it exerts a sharp blow or shattering effect on the material with which it is in contact. The quickest explosive of the dynamite class is the 60% straight dynamite. Quick explosives are the ones particularly desired for mudcapping. For maximum effect for this purpose, they should be of high density and sensitiveness. See also: *quick.* (Fay, 1920)
quicksand A mass or bed of fine sand, that consists of smooth rounded grains with little mutual adherence and that is usually saturated with water flowing upward through the voids, forming a semiliquid, highly mobile mass that yields easily to pressure. See also: *running ground.* Syn: *running sand.*
quicksilver A common name for mercury. Syn: *native mercury.* (Fay, 1920)
quicksilver cradle A wooden box placed in a sloping position, and fixed upon rockers, in which gold-bearing gravel is washed, the gold being caught by mercury in the lower part of the cradle. (Fay, 1920)
quicksilver rock An altered rock consisting mainly of dark opal and chalcedony, commonly associated with ore in California mercury deposits in serpentine.
quick test A shear test of a cohesive soil without allowing the sample to drain. See also: *drained shear test.* (Nelson, 1965)
Quigley gun An air gun which mixes dry, granular, refractory materials with water.
quill shaft A light drive shaft inside a heavier one, and turning independently of it. (Nichols, 1976)
quincite Light carmine-red particles found in a limestone near Quincy, France; color apparently organic; a doubtful mineral. (Dana, 1914)
quinquevalent (a) Having a valence of 5. (Webster 3rd, 1966) (b) Having five valences. Tungsten has five valences which are 2, 3, 4, 5, and 6. (Handbook of Chem. & Phys., 2)
quitclaim (a) A release of a claim; a deed of release; specif., a legal instrument by which some right, title, interest, or claim by one person in or to an estate held by himself or another is released to another, and which is sometimes used as a simple but effective conveyance for making a grant of lands whether by way of release or as an original conveyance. (Webster 3rd, 1966) (b) In the United States, a document in which a mining company sells its surface rights but retains its mineral rights. (Nelson, 1965)
quoin (a) The keystone or a voussoir of an arch. (b) A wedge to support or steady a stone. (c) A large square ashlar or stone at the angle of a wall to limit the rubble and make the corner true and strong; an exterior masonry corner. (d) One of the four facets on the crown, pavilion, or base of a gem.
Q wave *Love wave.*

R

rabatage A system of working steep seams of any thickness. (Nelson, 1965)
rabbit-eye York. Limestone in the Coralline Oolite. Cf: *toad's-eye.* (Arkell, 1953)
rabbittite A monoclinic mineral, $Ca_3Mg_3(UO_2)_2(CO_3)_6(OH)_4{\cdot}18H_2O$; radioactive; forms yellow efflorescence on mine walls.
rabble (a) An iron scraper serving as a rake in removing scoriae from the surface of melted metal. (Fay, 1920) (b) A charcoal burner's shovel. (Webster 3rd, 1966) (c) A mechanical rake for skimming the bath in a melting or refining furnace or for stirring the ore in a roasting furnace by hand or mechanically. (Webster 3rd, 1966)
rabbler (a) *rabble.* (b) One who uses a rabble, as in puddling iron. (Standard, 1964) (c) A scraper. (Standard, 1964)
rabbling Stirring molten metal, ore, or other charge, using a hoelike tool or other device. (ASM, 1961)
rabbling tool A rabble of simple construction for use by hand. Also called rabble rake. See also: *rabble.* (Standard, 1964)
race A small thread of spar or ore.
raceway The term is applied to conduits, moldings, and other hollow material, often concealed, through which wires are fished from one outlet to another. (Crispin, 1964)
rack (a) An inclined trough for washing or separating ore. (Nelson, 1965) (b) A toothed or notched drill-base-slide and meshing-gear pinion used to facilitate the moving of a drill to clear the borehole when hoisting or lowering the drill string; generally limited to larger, skid-mounted machines. (Long, 1960) (c) A framework of wood or metal for the orderly storage of core, pipe rods, etc., in a horizontal position. (Long, 1960) (d) A tilting table on which concentrates are separated from the passing flow of finely ground pulp, the system being arranged to be periodically self-flushing. (Pryor, 1963) (e) A screen composed of parallel bars to catch floating debris. (Seelye, 1951) (f) In electroplating, a frame used for suspending and conducting current to one or more cathodes during electrodeposition. (Lowenheim, 1962)
rack-a-rock Mining explosive based on a mixture of potassium chlorate and nitrobenzene. (Pryor, 1963)
rack back To move a drilling machine away from the borehole collar by sliding it on its base, using the rack-and-gear pinion to facilitate moving the machine. See also: *rack.* (Long, 1960)
racked timbering Timbering braced diagonally as stiffening against deformation. (Hammond, 1965)
rack frame Inclined table used to treat slimes. (Pryor, 1958)
rack gear A toothed bar. (Nichols, 1976)
racking (a) Old term for concentration in sluice boxes. (b) The process of separating ores by washing on an inclined plane. (c) *ragging.*
racking table A table on which to wash ore slimes. See also: *rack.* (Standard, 1964)
rack railroad A cog railway; cog tramway. (Fay, 1920)
rack up (a) To move the drilling machine forward into alignment with the borehole, using the rack-and-gear pinion to facilitate moving the machine. (Long, 1960) (b) To stack and arrange the drill rods in an orderly fashion in the tripod, mast, or derrick, or horizontally on a rack provided on the ground. (Long, 1960) (c) To place core on a rack. (Long, 1960)
radial Said of lines or other linear phenomena converging at a single center or departing from one.
radial arm The movable cantilever supporting the drilling saddle in a radial drilling machine. (Crispin, 1964)
radial axis *longitudinal trace.*
radial dikes A descriptive term for dikes that radiate outward from a center, commonly a volcanic neck or stock.
radial drainage pattern A drainage pattern in which consequent streams radiate or diverge outward, like the spokes of a wheel, from a high central area; it is best developed on the slopes of a young, unbreached domal structure or of a volcanic cone. (AGI, 1987)
radial drill (a) A heavy drilling machine in which the drilling head is capable of radial adjustment along a rigid horizontal arm carried by a stand. See also: *radial percussive coal cutter.* (Nelson, 1965) (b) A small diamond drill having a drilling head that can be adjusted radially along a rigid horizontal arm radiating from a vertical column; usually driven by air and used to drill radial blastholes underground. See also: *radial drilling.* (Long, 1960)
radial drilling The drilling of a number of holes in a single plane and radiating from a common point. Cf: *horadiam.* (Long, 1960)
radial-flow fan A mine fan in which the air enters along the axis parallel to the shaft and is turned through a right angle by the blades and discharged radially. There are three main types with (1) backwardly inclined blades; (2) radial blades; and (3) forward curved blades. In (2) and (3) the blades are made of sheet steel, while in (1) the present tendency is to replace curved sheet-steel blades by blades of aerofoil cross section. The aerofoil bladed radial-flow fan has an efficiency of about 90%. Cf: *axial-flow fan; mixed-flow fan.* (Nelson, 1965)
radial machineman In bituminous coal mining, a person who operates a radial-type coal cutter. The machine remains stationary at the center of the working place and undercuts or shears the coal in an arc rather than making a straight cut by moving across the working face. Also called arcwall machineman. (DOT, 1949)
radial percussive coal cutter A heavy coal cutter for use in headings and rooms in pillar methods of working. The machine weighs about 12 hundredweight (545 kg) and is usually mounted on a light carriage to suit the mine track. A percussive drill, with extension rods, makes a horizontal cut about 5-1/2 ft (1.68 m) deep and 15 ft (4.57 m) wide at any height in the heading. The central column is tightened between roof and floor about 4-1/2 ft (1.37 m) from the face. The machine can also be used for drilling shotfiring holes. (Nelson, 1965)
radial pressure The radial pressure of wire rope is a function of the rope tension, rope diameter, and tread diameter. The radial pressure can be determined by the following equation: $P = 2T/Dd$, where P equals radial pressure in pounds per square inch; T equals rope tension in pounds; D equals tread diameter of sheave or drum in inches; and d equals rope diameter in inches.
radial slicing A method of caving by which all the ground around a central raise might be worked in a series of slices arranged like the spokes of a wheel. (Lewis, 1964)
radial strain The change in length per unit length in a direction radially outward from the charge. (Duvall, 1957)
radial stress (a) Stress normal to the tangent to the boundary of any opening. (Obert, 1960) (b) In the Earth, the stress normal to a spherical surface.
radial velocity In a fan, the quantity of air delivered in cubic meters per second divided by the outlet area of the fan at the periphery.
radial ventilation A ventilation system in which a number of downcast shafts arranged around the periphery of the working area are served by a common upcast shaft within the area, or vice versa. Sometimes known as compound ventilation. (BS, 1963)
radian A unit of plane angular measurement equal to the angle at the center of a circle subtended by an arc equal in length to the radius. One radian equals about 57.29°. (Webster 3rd, 1966)
radiant energy Energy that radiates or travels outward in all directions from a source. (MacCracken, 1964)
radiated Said of an aggregate of acicular crystals that radiate from a central point. Cf: *spherulitic.* (AGI, 1987)
radiated pyrite *marcasite.*
radiating A mineral with crystals or fibers arranged around a center point, for example, stibnite. Also called divergent. (Nelson, 1965)
radiation absorbed dose The basic unit of absorbed dose of ionizing radiation. One radiation absorbed dose (abbrev., rad) equals the absorption of 100 ergs of radiation energy per gram of matter. (Lyman, 1964)
radiation damage (a) A general term for the alteration of properties of a material arising from exposure to X-rays, gamma rays, neutrons, heavy-particle radiation, or fission fragments in nuclear fuel material. (ASM, 1961) (b) Damage done to a crystal lattice (or glass) by passage of fission particles or alpha particles from the nuclear decay of a radioactive element residing in or near the lattice. See also: *metamict mineral.* (AGI, 1987)
radiation detector A device used either on the surface or in drill holes to detect and/or indicate the occurrence or the nearby presence of radioactive minerals. Also called electronic logger; gamma-ray detec-

tor; Geiger counter; Geiger-Müller counter; Geiger-Müller probe; scintillation counter; scintillometer. (Long, 1960)

radiation log *radioactivity log.*

radiation pyrometer (a) A device for ascertaining the temperature of a distant source of heat, such as a furnace. A concentrated group of thermocouples, called a thermopile, is used. Radiant heat from the furnace or object is focused by a lens onto the thermopile. Radiation pyrometers may be used for measuring temperatures to 7,000 °F (3,870 °C). The device is used in automatic control systems in mineral dressing and other processes. (Nelson, 1965) (b) Pyrometer that determines temperature by measuring the intensity of radiation from the hot body.

radiation survey A study of factors in any process or device involving radiation that could cause danger to any persons working near the process or device. (NCB, 1964)

radiation-type gage An instrument for measuring the density or percentage of solids in slurries flowing through pipes. It normally uses a gamma-ray source, usually cesium-137 or cobalt-60, mounted in a lead-shielded holder on one side of the pipe. A radiation detector is mounted on the opposite side. Since the absorption of the gamma radiation, as it passes through the slurry, varies as a function of the density of the slurry, the change in radiation received by the radiation detector is representative of the specific gravity or percentage of solids in the slurry. See also: *differential pressure flowmeter.* (Nelson, 1965)

radioactive (a) Generally, the property possessed by certain elements, such as uranium, of spontaneously emitting alpha, beta, and/or gamma rays by the disintegration of the nuclei of their atoms. (Long, 1960) (b) Of, relating to, caused by, or exhibiting radioactivity. Abbrev., RA. (Webster 3rd, 1966)

radioactive decay (a) The change of one element to another by the emission of charged particles from the nuclei of its atoms. (AGI, 1987) (b) The spontaneous disintegration of the atoms of certain nuclides into new nuclides, which may be stable or undergo further decay until a stable nuclide is finally created. Radioactive decay involves the emission of alpha particles, beta particles, and other energetic particles, and usually is accompanied by emission of gamma rays and by atomic de-excitation phenomena. It always results in the generation of heat. Syn: *radioactive disintegration.* (AGI, 1987)

radioactive disintegration *radioactive decay.*

radioactive dusts Dusts that are injurious because of radiation. They include ores of uranium, radium, and thorium. (Hartman, 1982)

radioactive element Applied to certain unstable atoms, the nuclei of which spontaneously disintegrate, emitting particles and rays, eventually reverting through a series of such emissions into an atom having a stable nucleus and a different atomic number. Radium, e.g., becomes lead-207. (MacCracken, 1964)

radioactive mineral One of six radioactive elements that occur naturally: potassium, rubidium, thorium, uranium, and associated radium, samarium, and lutecium. Thorium commonly occurs in monazite, a sparsely scattered accessory mineral of certain granites, gneisses, and pegmatites. It is concentrated, however, by weathering processes in sands and gravels as commercial placer deposits along rivers and beaches. The most important primary uranium ore minerals are davidite and uraninite, esp. pitchblende, the massive variety. These minerals are of rather underspread occurrence in certain granites and pegmatites and occur as secondary minerals in metallic vein deposits. The secondary uranium minerals, however, are more underspread and more numerous than the primary uranium ore minerals. Secondary uranium minerals are found in weathered and oxidized zones of primary deposits and, also, in irregular flat-lying sandstones, such as those in the Colorado Plateau, where the uranium mineralization was precipitated from solutions. Carnotite, the potassium uranium vanadate of conspicuous yellow color, is perhaps the most important of the secondary uranium ore minerals. Others are tyuyamunite, which is closely related to carnotite, and the torbernites and autunites which are uranium minerals.

radioactive series A succession of nuclides, each of which transforms by radioactive disintegration into the next until a stable nuclide results. The first member is called the parent, the intermediate members are called daughters, and the final stable member is called the end-product. Four radioactive series are the uranium series, the thorium series, the actinium series, and the neptunium series. (Glasstone, 1958)

radioactive tracer element A radioactive isotope of an element used to study a process by observing the intensity of radioactivity.

radioactive waste Equipment and materials from nuclear operations that are radioactive and for which there is no further use. Wastes are generally referred to as high-level (having radioactivity concentrations of hundreds to thousands of curies per gallon or per cubic foot); low-level (in the range of 1 microcurie per gallon or per cubic foot); and intermediate (between these extremes). (Lyman, 1964)

radioactivity The spontaneous decay or disintegration of an unstable atomic nucleus, accompanied by the emission of radiation. See also: *radioactive decay.* (Lyman, 1964)

radioactivity log (a) A log of a borehole obtained through the use of gamma, neutron, or other radioactivity logging methods. (b) The generic name for well logs whose curves derive from reactions of atomic nuclei involving the behavior of gamma rays and/or neutrons. Except for the natural gamma-ray log and the spectral gamma-ray log, they record the response of rocks very near the well bore to bombardment by gamma rays or neutrons from a source in the logging sonde. Most can be obtained in cased, empty, or fluid-filled well bores. Varieties include: density log; neutron log; neutron-activation log; epithermal-neutron log; pulsed-neutron-capture log. Syn: *radiation log; nuclear log.* See also: *gamma-ray well log; spectral gamma-ray log; neutron log.* (AGI, 1987)

radioactivity prospecting Exploration for radioactive minerals utilizing various instruments, generally a Geiger counter or scintillation counter, by measuring the natural radioactivity of earth materials. (Dobrin, 1960)

radioaltimeter Equipment carried in survey aircraft to ensure constant height above ground (not sea) level of 300 ft or 500 ft (91.4 m or 152.4 m)—a critical factor in certain airborne geophysical prospecting and aerial mapping surveys. See also: *profile flying.* (Nelson, 1965)

radiocarbon Radioactive carbon, esp. carbon-14, but also carbon-10 and carbon-11. (AGI, 1987)

radiocarbon dating *carbon-14 dating.*

radiochemistry The chemical study of artificial and naturally occurring radioactive materials and their behavior. It includes their use in tracer studies and other chemical problems. (AGI, 1987)

radioelement A form or sample of an element containing one or more radioactive isotopes.

radiogenic Produced by radioactive transformation. Thus, uranium minerals contain radiogenic lead and radiogenic helium. The heat produced within the earth by the disintegration of radioactive nuclides is radiogenic heat.

radiograph (a) A photographic shadow image resulting from uneven absorption of radiation in the object being subjected to penetrating radiation. (ASM, 1961) (b) A picture produced upon a sensitive surface (such as a photographic film), by a form of radiation other than light; specif., an X-ray or a gamma-ray photograph. *roentgenogram.* (Webster 3rd, 1966)

radiography (a) A nondestructive method of internal examination in which metal or other objects are exposed to a beam of X-ray or gamma radiation. Differences in thickness, density, or absorption caused by internal discontinuities are apparent in the shadow image either on a fluorescent screen or on a photographic film placed behind the objects. (ASM, 1961) (b) The use of penetrating ionizing radiation to examine solid material. When the source of radiation is internal, such as an implanted radioactive tracer, the technique is known as autoradiography. (Lyman, 1964)

radiohalo *pleochroic halo.*

radioisotope (a) An unstable isotope of an element that decays or disintegrates spontaneously, emitting radiation. (Lyman, 1964) (b) Radioisotope is loosely used as a syn. for radionuclide. See also: *radium; radon.* Syn: *unstable isotope.*

Radiolaria (a) Subclass of the Sarcodina consisting of marine protozoans that possess complex internal siliceous skeletons. (b) Silica rock formers. (Mason, 1951)

radiolarian ooze Deposits of siliceous ooze made up largely of radiolarian skeletons and are formed at depths between 13,000 ft and 25,000 ft (4.0 km and 7.6 km). (AGI, 1987)

radio link Radio signal unit used to control or communicate between scattered sections of mine, or to link isolated camp with other places. (Pryor, 1963)

radiolite A spherulite composed of radially arrayed acicular crystals.

radiolite survey instrument A one-shot bore-hole-surveying instrument having the horizontal (compass) and vertical indicator markings painted with a radioactive substance, such as that on the luminous dial of a watch. The positions of these markings are recorded on small, circular, photographic film. (Long, 1960)

radiolitic (a) Said of the texture of an igneous rock characterized by radial, fanlike groupings of acicular crystals, resembling sectors of spherulites. (AGI, 1987) (b) Said of limestones in which the components radiate from central points, with the cement comprising less than 50% of the total rock. (AGI, 1987)

radiometallography The application of X-rays to the study of the internal structure of various materials, esp. metals. (Fay, 1920)

radiometer Essentially a heat-flow meter used to measure long-wave radiation as well as solar radiation. It can be used both for daytime and nighttime measurements and to measure the net heat transfer through a surface. (Hunt, 1965)

radiometric assay A test to determine contained quantity of uranium. The actual uranium present may be more or less than the assay shows. See also: *equilibrium; inequilibrium.* (Ballard, 1955)

radiometric ore sorter A device for separating gangue from uranium-bearing ore, after primary crushing. (Nelson, 1965)

radiometric prospecting Use of portable Geiger-Muller apparatus for field detection of emission count in search for radioactive minerals. (Pryor, 1963)

radiophone An FM apparatus, using the mine haulageway trolley wire for power and antenna, that permits the dispatcher to talk back and forth with his motor crews as they are moving throughout the mine. This saves stopping and starting trips to make telephone calls. (Kentucky, 1952)

radiophyllite *zeophyllite.*

radiore method An electromagnetic method used in mineral exploration in which a high-frequency current is used, ranging from 30,000 to 50,000 Hz, but, if necessary, a frequency as low as 50 to 3,000 Hz, can be made available. The detecting or direction-finding coil, mounted on a tripod, has the form of a pair of spectacles and is equipped with an amplifier and head telephone. When the exciting coil is energized, a current is caused to flow in the conductor and a secondary electromagnetic field is set up around the conductor. The detecting coil is affected by both the primary field from the exciting coil and the secondary field. (Lewis, 1964)

radium A radioactive metallic element; one of the alkaline-earth metals. Symbol, Ra. It occurs in pitchblende ore, in carnotite sands, and in all uranium minerals. See also: *radioisotope.* (Handbook of Chem. & Phys., 3)

radium G A name for lead-206, the stable end-product of the radioactive disintegration of uranium-238 in the uranium disintegration series. Natural lead contains 23.6% of lead-206. Symbol, RaG. Syn: *uranium-lead.* (Handbook of Chem. & Phys., 2)

radiumite A mixture of black pitchblende, yellow uranotile, and orange gummite. (Schaller, 1917)

radius Horizontal distance from the center of rotation of a crane to its hoisting hook. (Nichols, 1976)

radius of curve A term used in laying mine track; the calculated radius of an arc that will connect two pieces of track (at a desired angle of direction from each other) with a smooth curve section.

radius of gyration The value used when calculating the slenderness ratio of pillars and struts. If A is the cross-sectional area in inches of the pillar or strut and I is its moment of inertia, the radius of gyration is $\sqrt{(I/A)}$, generally known as K. (Hammond, 1965)

radius of rupture In crater tests, the average distance from the center of the explosive charge to the periphery of the crater at the surface. (Duvall, 1957)

radius ratio (a) The ratio of the radius of the smaller ion to that of the larger ion. It may not exceed 1. (Hurlbut, 1964) (b) The ratio of the radius of the smaller ion to that of the larger; commonly cation to anion. Radius ratios are used to predict coordination numbers of anions about cations in ionic crystal structures. Cf: *Pauling's rules.*

radon (a) A heavy, radioactive, gaseous element; inert; the heaviest known gas. Symbol, Rn. Formed by the disintegration of uranium. Used similarly to radium in medicine. Radon build-up is a health consideration in uranium mines. (Handbook of Chem. & Phys., 3) (b) Heaviest known gas. Colorless as a gas; yellow to orange-red, phosphorescent, opaque crystals; sp gr of liquid, 4.4 (at -62 °C); and of solid, 4.0; soluble in water; and slightly soluble in alcohol and in organic liquids. All 18 known isotopes from radon-204 to radon-224 are radioactive. Radon-222 emanates from thorium; half-life, 54.5 s; and an alpha particle emitter; and radon-219 or actinon emanates from actinium; half-life, 3.92 s; and an alpha particle and a gamma ray emitter. One part of radon exists in 1 sextillion parts of air. (Handbook of Chem. & Phys., 2)

radon daughter A radioactive element produced in the disintegration of radon. Syn: *radon progeny.* (MSHA, 1986)

radon progeny The short-lived decay products of radon, an inert gas that is one of the natural decay products of uranium. The short-lived radon progeny (i.e., polonium-210, lead-214, bismuth-214, and polonium-214) are solids and exist in air as free ions or as ions attached to dust particles. The U.S. Mine Safety and Health Administration has established radiation protection standards that limit a miner's radon progeny exposure to a concentration of 1.0 WL and an annual cumulative exposure to 4 WLM. Each WLM is determined as a 173-h cumulative, time weighted exposure. Syn: *radon daughter.* (NIOSH, 1987)

raffinate The aqueous solution remaining after the metal has been extracted by the solvent; the tailing of the solvent extraction system.

raft *float coal.*

rafter timbering A method of mine timbering in which the timbers appear like roof rafters.

rafting (a) The transporting of sediment, rocks, silt, and other matter of land origin out to sea by ice, logs, etc., with subsequent deposition of the rafted matter when the carrying agent disintegrates. (Hunt, 1965) (b) Matting or agglomerating of powdered coal. (Bennett, 1962)

rag (a) In British usage, any of various hard, coarse, rubbly, or shelly rocks that weather with a rough irregular surface; e.g. a flaggy sandstone or limestone used as a building stone. The term appears in certain British stratigraphic names, as the Kentish Rag (a Cretaceous sandy limestone in East Kent). Syn: *ragstone.* (AGI, 1987) (b) Any of various hard rocks, as a quartzose mica schist used for whetstones or a hard limestone used in building. (Webster 3rd, 1966) (c) A large roofing slate left rough on one side. (Webster 3rd, 1966) (d) To break (ore) into lumps for sorting; to cut or dress roughly (as a grindstone). (Webster 3rd, 1966)

ragged rolls Rolls with rough surfaces to facilitate the gripping of the steel in the first stages of rolling, as distinguished from the smooth-finishing rolls. (Mersereau, 1947)

ragging (a) The rough washing or concentration of ore or slimes on a rag frame. (Nelson, 1965) (b) In roll crushers, grooves cut in surface to improve grip on feed, and increase angle of nip. Also, in ore concentration in jigs, oversized bedding placed on jig screens. (Pryor, 1963) (c) *bedding.*

raggy stone Thin-bedded or flaggy sandstone. (TIME, 1929-30)

ragstone *rag.*

rail The chain or inner surface of a crawler. (Nichols, 1976)

rail gage The distance or width between the inner edges of the heads of the rails; (1) in coal mining, the rail gage for tub and car tracks ranges from 2 to 3 ft (0.6 to 0.9 m), and 2-1/3 ft (0.7 m) is considered a satisfactory compromise; (2) the standard gage for railway tracks is 4 ft, 8-1/2 in (1.44 m) and, (3) in metal mining, the rail gage ranges from 1-1/2 to 2-1/2 ft (0.46 to 0.76 m). See also: *track gage.* (Nelson, 1965)

rail haulage system A materials transportation system consisting of gondola cars, and the steel rails on which the cars are moved about with a suitably powered traction unit as a locomotive.

rail riffles These may be either longitudinal or transverse and consist of rails of various sizes, placed in sets usually upside down, either longitudinally in the sluice box, or transversally across the box. They wear well, are rigid, and give some security against theft of gold from the sluice boxes. (Griffith, 1938)

rails Specially shaped steel bars which, when laid parallel on crossties and fastened, form a track for vehicles with flanged wheels.

rail track ballast Material placed around and between track ties and tamped under sides and ends of the ties to bring the track to proper grade by filling the space between the bottom of the ties and the graded roadbed.

rainbow chalcedony Eng. A variety of chalcedony of thin concentric layers, which, when cut across, exhibit an iridescence resembling the colors of the rainbow. (Fay, 1920)

rainbow quartz *iris quartz.*

rain chamber A chamber in which fumes, such as those from molten metal, may be condensed by a water shower. (Standard, 1964)

rain gage A device used to measure precipitation (melted snow, sleet, or hail as well as rain). It consists of a receiving funnel, a collecting vessel, and a measuring cylinder. Syn: *pluviometer; hyetometer; snow gage.* (AGI, 1987)

rainwash (a) The washing-away of loose surface material by rainwater after it has reached the ground but before it has been concentrated into definite streams; specif. sheet erosion. Also, the movement downslope (under the action of gravity) of material loosened by rainwater. It occurs esp. in semiarid or scantily vegetated regions. (AGI, 1987) (b) The material that originates by the process of rainwash; material transported and accumulated, or washed away, by rainwater. (AGI, 1987) (c) The rainwater involved in the process of rainwash. (AGI, 1987)

raise (a) A vertical or inclined opening in a mine driven upward from a level to connect with the level above, or to explore the ground for a limited distance above one level. After two levels are connected, the connection may be a winze or a raise, depending upon which level is taken as the point of reference. Syn: *raised shaft* See also: *upraise; raising; rise.* (Lewis, 1964) (b) A mine opening, like a shaft, driven upward from the back of a level to a level above, or to the surface. (Cumming, 1951) (c) To take up the floor or bottom rock in a room, gangway, or entry to increase the height for haulage. (Fay, 1920)

raise borer These machines are used to produce a circular excavation either between two existing levels in an underground mine or between the surface and an existing level in a mine. In raise boring, a pilothole is drilled down to the lower level, the drillbit is removed and replaced by a reamer head having a diameter with the same dimension as the desired excavation and this head then is rotated and pulled back up towards the machine. (SME, 1992)

raise climber Equipment used in an opening (raise) that is mined upward. (MSHA, 1992)

raised shaft *raise.*

raising Excavating a shaft or steep tunnel upward. See also: *raise; rise.* (Nelson, 1965)

rait Mid. To split off the walls or sides of underground workings. Called rosh in Leicestershire. Syn: *rate.* (Fay, 1920)

rake (a) As used by diamond drillers and bit manufacturers, rake is the angle, measured in degrees, formed by the leading face of a cutting tool and the surface behind the cutting edge. See also: *negative rake; positive rake.* (Long, 1960) (b) The inclination of anything from the vertical; said of mineral veins, faults, etc. See also: *pitch.* (c) A timber placed at an angle. See also: *rakers.* (Sandstrom, 1963) (d) Shale containing ironstone nodules. (BS, 1964)

rake blade A dozer blade or attachment made of spaced tines. (Nichols, 1976)

rake classifier A type of mechanical classifier utilizing reciprocal rakes on an inclined plane to separate coarse from fine material contained in a water pulp, overflowing the fine material and discharging the coarse material by means of an inclined raking system.

rakers Slanting props placed at the end of a drift set to keep the timbers steady when blasts go off. See also: *rake.* (Fay, 1920)

rake thickener Equipment for thickening in which the concentrated suspension settles in a container of circular section and is delivered mechanically to one or more discharge points by a series of arms revolving slowly around a central shaft. (BS, 1962)

rake vein (a) A steeply inclined crosscutting irregular mineralized fracture or fissure. (BS, 1964) (b) Rake vein and gas vein are synonymous; it is said that they are lodes filling distinct fissures. Their course in irregular; their dip, as a rule, vertical. (Ricketts, 1943) (c) A vein or lode cutting through the strata.

raking strut A strut set at an angle to the vertical to support timbering during excavation. (Hammond, 1965)

Raky boring method A method of boring somewhat similar to Fauck's. Hollow steel rods 2 in (5.08 cm) in diameter are used with a mud flush. A walking beam, fitted with steel springs, imparts from 80 to 120 short blows/min to the chisel. (Nelson, 1965)

Raleigh's law In 1909 Lord Raleigh established the general law of fluid flow: $R = wV^2$ (f)(wVD/m), where R = resistance of flow, w = density of fluid, V = velocity of flow, D = diameter of pipe, m = viscosity of fluid, f = signifies function. For any particular value of (wVD)/m, using any combination of quantities, there will be a definite corresponding value of R/wV^2. (Lewis, 1964)

ralstonite An isometric mineral, $Na_xMg_xAl_{2-x}(F,OH)_6 \cdot H_2O$; structurally related to the pyrochlore group; occurs as octahedra.

Ralston's classification of coal A classification based on the percentage of carbon, hydrogen, and oxygen in the ash-, moisture-, sulfur-, and nitrogen-free coal. These figures are plotted on trilinear coordinates giving well-defined zones of bituminous coals, lignites, peats, etc. (Miall, 1940)

ram (a) To stem; tamp. (Mason, 1951) (b) Black ram, bog iron ore; gold ram, gold ore. (Arkell, 1953) (c) The plunger of a pump. (Zern, 1928) (d) A mechanical pusher for forcing (discharging) coke from a byproduct coke oven. (Mersereau, 1947) (e) An appliance for exerting a pressure on face equipment, such as steel supports, conveyors, or plows. See also: *hydraulic ram; pneumatic ram.* (Nelson, 1965)

ramdohrite A monoclinic mineral, $Ag_3Pb_6Sb_{11}S_{24}$; forms dark-gray twinned prismatic to lance-shaped crystals; at Potosi, Bolivia. Cf: *andorite.*

rammel Loose stone, or waste rock; loose sandy or stony barren soil; mixed shale and sandstone. Also spelled: rammell.

rammelsbergite An orthorhombic mineral, $NiAs_2$; löllingite group; trimorphous with pararammelsbergite and krutovite; metallic; tin white; sp gr, 7.1; in vein deposits. Syn: *white nickel.*

rammer A rod for charging and stemming shotholes. See also: *stemmer.* (Nelson, 1965)

ramming (a) *stemming; tamping.* (b) *scaling.*

ramming and patching refractories Those which can be rammed to form a monolithic furnace lining or special shapes.

ram operator In bituminous coal mining, a laborer who tends operation of, adjusts, and repairs pumping devices (rams) used in low places in shallow mines to force a portion of the mine water to the surface by utilization of the flow of the entire amount. Nearly obsolete. *pumper.* (DOT, 1949)

ramp (a) A fault that is a gravity (normal) fault near the surface but curves through the vertical to dip in the opposite direction at depth, where the displacement is that characteristic of a thrust. (b) A portion of a thrust fault that cuts across formational contacts in a short distance. (AGI, 1987) (c) An incline connecting two levels in an open pit or underground mine.

ram pump (a) A pump consisting essentially of a plunger or ram which forces the contained water into the discharge pipe. See also: *force pump.* (Nelson, 1965) (b) A single-acting reciprocating pump that has a ram instead of a piston. The ram has a constant diameter and does not fit closely in the cylinder, pumping only by displacement. (Hammond, 1965)

ramp valley (a) A valley produced by the ramping or upthrusting of two masses, one on either side of an intervening strip. (b) A valley bounded by thrust faults.

ram scraper A plow-type machine hauled by an endless chain at a speed of 6 ft/s (1.83 m/s). The ram scraper is pressed toward the face by the tension of the return strand, the return sheave being located inby for this purpose. Syn: *Peissenberg ram.* See also: *manless face.* (Nelson, 1965)

ramsdellite An orthorhombic mineral, MnO_2; trimorphous with akhtenskite and pyrolusite.

rance (a) A prop set against the coal face that is undermined. (Fay, 1920) (b) A dull red marble with blue and white markings. From Belgium, and sold in the United States as Belgian marble. (Webster 2nd, 1960; Fay, 1920)

rance marble (a) A white, hard, shining grit, striped red. (Arkell, 1953) (b) A kind of variegated marble from Hainault, Belgium. (Arkell, 1953)

rand S. Afr. A ridge, range of hills, or highland on either side of a river valley.

Rand An abbrev. of Witwatersrand, the gold fields in the Republic of South Africa.

Randolph process A modification of the series process of copper refining in which the electrodes lie horizontally, the top surface of each one acting as anode, the lower as cathode. Theoretically, it has the advantage of extremely low metal losses and great purity of copper. See also: *Hayden process; Smith process.* (Liddell, 1918)

random The direction of a rake vein.

random error Any error that is wholly due to chance and does not recur; an accidental error. Ant: systematic error. (AGI, 1987)

random line (a) A trial line, directed as nearly as may be toward a fixed terminal point that is invisible from the initial point. (Seelye, 1951) (b) A random traverse; i.e., a traverse run from an initial to a terminal point to determine the direction of the latter from the former. (Seelye, 1951)

random orientation *random set.*

random pattern The setting of diamonds in a bit crown without regard to a geometric pattern—without regular and even spacing. See also: *random set.* (Long, 1960)

random sample (a) A sample take without plan or pattern. (Nelson, 1965) (b) A subset of a statistical population in which each item has an equal and independent chance of being chosen; e.g., a sample chosen to determine (within definied limits) the average characteristics of an orebody. (AGI, 1987)

random set The setting of diamonds in a bit crown without regard to the attitude of their vector properties. Syn: *random orientation.* See also: *random pattern.* (Long, 1960)

random stone A term applied by quarry personnel to quarried blocks of any dimensions.

Randupson process A system of molding in which the molds are made of a mixture of silica sand and cement with water added. (Osborne, 1956)

rang (a) In the CIPW classification of igneous rocks, that division below order. (b) A Ceylonese term for gold; from rangwelle meaning golden sand.

range (a) An area in which a mineral-bearing formation crops out; e.g., the iron range and copper range of the Lake Superior region; a mineral belt. (AGI, 1987) (b) An established or well-defined line or course whose position is known and along which soundings are taken in a hydrographic survey. (AGI, 1987) (c) Any series of contiguous townships (of the U.S. Public Land Survey system) aligned north and south and numbered consecutively east and west from a principal meridian. (AGI, 1987) (d) The distribution of a genus, species, or other taxonomic group of organisms through geologic time. (AGI, 1987) (e) An orderly arrangement or family of diamond-drill fittings, such as casing, core barrels, drill rods, etc., with diameters appropriately related to each other and intended to be used together. Ranges commonly are designated by letter names, using letters such as E, A, B, and N individually or as the first letter in two- and three-letter names. Cf: *group.* (Long, 1960) (f) For a spherical model, the distance at which the model reaches its maximum value, or sill. For the exponential and gaussian models, which approach the sill asymptomatically, it means the "practical" or "effective" range, where the function reaches approximately 95% of the maximum. The nugget model effectively has a sill with a range of zero: the linear model uses "sill/range" merely to define the slope.

range line One of the imaginary boundary lines running north and south at six-mile intervals and marking the relative east and west locations of townships in the U.S. public-land survey; a meridional township boundary line. Cf: *township line.* (AGI, 1987)

range of stress The range between the upper and lower limit of a cycle of stress, such as is applied in a fatigue test. The midpoint of the range is the mean stress. (CTD, 1958)

range pole (a) A 6- to 12-ft (2- to 3-m) wooden or metal pole painted in contrasting colors at 1-ft (0.3-m) intervals. It is used in surveying to mark lines of sight, stations, etc. (b) *picket.* (c) A metal rod, pointed at one end, and usually painted alternately red and white at 1-ft intervals; used by surveyors as a line of sight. Syn: *ranging rod.*

ranging rod *range pole.*

rank (a) Describes the stage of carbonification attained by a given coal. (IHCP, 1963) (b) The place occupied by a coal in a classification. Specifications of the American Society for Testing and Materials cover the classification of coals according to their degree of metamorphism, or progressive alteration, from lignite to anthracite. For a complete description of this classification, consult ASTM Designation: D 388. (ASTM, 1994) (c) When applied to coal, denotes its age in geological formation, not necessarily denoting quality. (BCI, 1947) (d) The position of a coal relative to other coals in the coalification series from brown coal (low rank) to anthracite (high rank), indicating its maturity in terms of its general chemical and physical properties. (BS, 1964) (e) Those differences in the pure coal material due to geological processes designated as metamorphic, whereby the coal material changes from peat through lignite and bituminous coal to anthracite or even to graphite. (AGI, 1987) (f) All coallike fossil fuels form a continuous and progressive series, ranging from lignite, through the various bituminous coals, to anthracite. It is the position of a particular coal in this series that determines its rank. Therefore, lignite is a low-rank coal while anthracite is a high-rank coal. See also: *coalification; grade; Hilt's Law; coal rank.* (Nelson, 1965) (g) A term primarily devised to indicate the position of a fuel in the series peat-anthracite, probably best measured by the percentage of carbon (ashless, dry basis). Thus rank depends on the degree of metamorphism of coal, and increase of rank is, in general, marked by the decrease of volatiles and moisture. (Tomkeieff, 1954) (h) The term rank may also be applied to other series, such as the sapropelic coal series or the bitumen series. Cf: *type.* (Tomkeieff, 1954)

Rankine scale An absolute-temperature scale in which a measurement interval equals a Fahrenheit degree and in which zero degrees is equal to -459.67 °F (-273.16 °C). Named for William J.M. Rankine, a Scottish physicist. See also: *temperature.* (Webster 3rd, 1966)

Rankine's formula An empirical formula giving the collapsing load for a given column. (CTD, 1958)

Rankine's theory The state of stress theory as developed by Rankine in 1860 for application to earth pressures. He formulated that the pressure on a vertical retaining wall restraining earth with a horizontal surface is $(1 - \sin\phi) \div (1 + \sin\phi)$ multiplied by the soil density for each foot depth of soil retained, where ϕ is the angle of internal friction of the soil. The value $(1 - \sin\phi) \div (1 + \sin\phi)$ is the coefficient of active earth pressure. (Hammond, 1965)

rankinite A monoclinic mineral, $Ca_3Si_2O_7$; dimorphous with kilchoanite; rare in contact metamorphic rock, but common in blast furnace slag.

rank variety Variety in coals brought about as a result of progressive metamorphism. More or less arbitrarily, although carefully, selected chemical criteria are used to differentiate coals of different rank. Physical criteria are also used but are more difficult of application. (AGI, 1987)

ransomite A monoclinic mineral, $CuFe_2(SO_4)_4{\cdot}6H_2O$; forms slender sky-blue prisms; at Jerome, AZ.

rap (a) To warn workers in an adjoining working place of a blast, when the working places are separated by only a small pillar, by knocking on the pillar with a tool or bar. (b) To test the roof by tapping it with a stick or bar. (c) To signal by knocking on a steam, water, or air pipe. (Fay, 1920)

rapakivi A hornblende-biotite granite containing large rounded crystals of orthoclase mantled with oligoclase. The name has come to be used most frequently as a textural term where it implies plagioclase rims around orthoclase in plutonic rocks.

rapid blow drilling A drilling method utilizeing a great number of short blows in quick succession rather than a few heavy blows from a relatively considerable altitude. In this method, the bit is fixed either to a rod or to a rope, so that it pounds the bottom in quick succession with short blows and at the same time rotates. (Stoces, 1954)

rapid excavation An improved cycle and system of operation to achieve rapid advance and continuous operation in low-drillability rock. (SME, 1992)

rapid plow A fast moving plow with picks attached. The rapid plow is a continuous longwall cutter loader capable of working unattended on the face. For this reason, it is one of the safest machines in operation. See also: *hard-coal plow.* (Nelson, 1965)

rap-in Som. To wedge down blocks of stone in underground quarries.

rappage Excess in size of a casting because the mold is larger than the pattern when the latter is unduly rapped, as with the hand, for drawing. (Standard, 1964)

rapping In foundry work, loosening of pattern before its withdrawal from molding sand in flask. (Pryor, 1963)

rapping roller An eccentric roller or a roller fitted with devices such as bars welded longitudinally along its outer casing, so arranged as to rap the belt and knock off fine coal or dust adhering to the return belt, or to centralize the coal on the carrying belt. (Nelson, 1965)

rare earths Oxides of a series of 15 metallic elements, from lanthanum (atomic number 57) to lutetium (71), and of two other elements; yttrium (39), and scandium (21). These elements are not esp. rare in the Earth's crust, but concentrations are. The rare earth metals resemble one another very closely in chemical and physical properties, thus making it most difficult to separate them. The rare earths are constituents of certain minerals, esp. monazite, bastnaesite, and xenotime. Abbrev: REE. (AGI, 1987)

rarefaction (a) The process or act of making rare or less dense; increase of volume, the mass remaining the same: now usually of gases; also, the state of being rarified; as, the rarefaction of the atmosphere on a high mountain. (Standard, 1964) (b) Diminution of air pressure below normal, as in alternate half-cycles in the transmission of a sound wave past a point. (CTD, 1958)

rashings A very friable carbonaceous clay, with numerous slickensides and sometimes streaks of coal. Rashings may underlie, overlie, or be interstratified with the coal; a very weak material and breaks up around the face supports. See also: *brashings.* (Nelson, 1965)

Rasorite *kernite.* A trade name for kernite.

rasp An instrument used in a borehole for fishing operations, for reducing the length of the box, or for coupling lost tools. Also called mill; rose bit. (Long, 1960)

raspberry spar (a) *rhodochrosite.* (b) Pink tourmaline.

raspite A monoclinic mineral, $PbWO_4$; dimorphous with stolzite; forms small tabular brownish-yellow crystals having intense adamantine luster; at Broken Hill, N.S.W., Australia.

ratchet A set of teeth that are vertical on one side and sloped on the other; holds a pawl moving in one direction, but allows it to move in another. (Nichols, 1976)

ratchet-and-pawl mechanism A cogwheel (ratchet) with which a single pivoted catch (pawl) engages, thereby preventing any backward turning. (Hammond, 1965)

ratchet drill A hand drill in which the drill holder is revolved intermittently by a lever through a ratchet wheel and pawl. A drill used for boring slate. (Webster 3rd, 1966; Fay, 1920)

ratchet man The worker who operates the duckbill loader when mining with duckbill conveyors. (Lewis, 1964)

rate *rait.*

rate action (nonstandard) As used in flotation, the component of proportional plus rate action or of proportional plus reset plus rate action for which there is a continuous linear relation between the rate of change of the controlled variable and the position of a final control element. (Fuerstenau, 1962)

rated capacity The load that a new wire rope or wire rope sling may handle under given operating conditions and at an assumed design factor.

rate determining step In any series or sequence of chemical reactions used to leach a product from its ore, the slowest in the chain. (Pryor, 1963)

rated horsepower (a) Theoretical horsepower of an engine based on dimensions and speed. (Nichols, 1976) (b) Power of an engine according to a particular standard. (Nichols, 1976)

rated load The kilowatt power output that can be delivered continuously at the rated output voltage. It may also be designated as the 100% load or full-load rating of the unit. (Coal Age, 1960)

rated output current The current derived from the rated load and the rated output voltage. (Coal Age, 1960)

rated output voltage The current specified as the basis of rating. (Coal Age, 1960)

rate of advance (a) The distance the bit penetrates a rock formation in a unit of time, such as inches (centimeters) per minute or feet (meters) per hour. (Long, 1960) (b) In coal mining, the number of feet (meters) between the coal face at the beginning and at the end of a workshift.

rate of grade *gradient.*

rates of reduction In crushing practice, these rates are (1) based on crusher dimensions, wherein the largest cube that will enter is compared with the largest that can be discharged, or (receiving opening) ÷ (discharge opening); the receiving opening is measured from the top of the movable member to the top of the stationary member; and (2) based on actual products; (a) overall reduction ratio = (mean size of feed) ÷ (mean size of product); (b) individual reduction ratio = (size most abundant in feed) ÷ (mean size of grading band concerned). (South Australia, 1961)

rathite *liveingite.*

rathite-II *liveingite.*

rathole (a) A hole drilled at an angle to the main hole by means of a deflection wedge. (Long, 1960) (b) A small-diameter pilot-type hole drilled a short distance ahead of a larger diameter hole to stabilize a smaller diameter bit and core barrel when used to core a limited portion of the borehole. After core drilling is completed, the rathole is reamed out and the larger size hole advanced, usually by some noncoring method. (Long, 1960) (c) A small sump or settling pond in which the larger sized cuttings from a drill hole are collected between the top of the drill hole and the main settling pond. (Long, 1960) (d) A slanting hole, perhaps 25 ft (7.62 m) deep, used for adjusting or lubricating the swivel on a grief stem. The start of a hole for rotary drilling. (Hess)

rathole bit A bit designed and used to drill the first portion of a deflected hole alongside and beyond the deflection wedge; also, a bit used to drill a rathole. Cf: *wedge bit.* See also: *rathole.* (Long, 1960)

rat-holing The act or process of drilling a deflected or pilot hole. See also: *rathole.* (Long, 1960)

rating The maximum capacity of a drill hoist or a prime mover, such as an engine, motor, or pump; generally the maximum safe capacity. (Long, 1960)

rating flume A flume used for purposes of control. (Hammond, 1965)

ratiometer An instrument used to measure the ratio of two differences in potential. (AGI, 1987)

rational analysis (a) The mineralogical composition of a material as deduced from chemical analysis. (b) The resolution into chemical types of a mass of rock or coal. (Francis, 1965)

ratio of absorption (a) The percentage by weight that the absorbed water bears to the dry weight of the stone. (b) The ratio (A), expressed as a percentage, of the volume (Vp) of the pore space in a rock to the weight (W) of the rock when dry, $A = 100\ Vp/W$. (Holmes, 1928)

ratio of concentration (Weight of ore fed) ÷ (Weight of concentrate produced). (Pryor, 1963)

ratio of enrichment The ratio of the percentage of valuable material in the concentrate to the percentage of the valuable material in the original material.

ratio of reduction (a) The relationship between the maximum size of the stone which will enter a crusher, and the size of its product. (Nichols, 1976) (b) The ratio of enrichment with respect to the sought mineral: (Assay value in feed) : (Assay value in concentrate). See also: *reduction ratio.* (Pryor, 1963)

ratio of size reduction Ratio of the upper particle size in the crushed material to the upper particle size of the feed material. (BS, 1962)

rattle boxes Limonite geodes from Chester County, PA. (Schaller, 1917)

rattler (a) York. Cannel coal. (b) Scot. Inferior gas coal; sandy shale. (c) U.K. A variety of gas coal that fetched high prices and was reputed to ignite with a match. It is hard, compact, uniform, bright, brittle, fine-grained, slightly sonorous when struck, and resembling jet but not so brilliant. (Arkell, 1953) (d) A revolving steel drum with a charge of metal spheres used for testing the abrasive resistance of brick. (e) A device for shaking out the cores from small castings, such as a tumbling barrel.

rattler test A method for evaluating the resistance of paving bricks to impact and abrasion. A sample of 10 bricks is subjected to the action of 10 cast-iron balls 3.75 in (9.53 cm) in diameter and 245 to 260 balls 1.875 in (4.76 cm) in diameter in a drum 28 in (71.12 cm) in diameter, 20 in (50.8 cm) long, rotating at 30 rpm for 1 h. The severity of abrasion and impact is reported as a percentage loss in weight. (Dodd, 1964)

rattlesnake ore A gray, black, and yellow mottled ore of carnotite and vanoxite, its spotted appearance resembling that of a rattlesnake. (AGI, 1987)

rauvite A mineral, $Ca(UO_2)_2V_{10}O_{28}{\cdot}16H_2O$; radioactive; purple- black; sandstone impregnation on the Colorado Plateau.

ravelly ground Rock that breaks into small pieces when drilled and tends to cave or slough into the hole when the drill string is pulled, or binds the drill string by becoming wedged or locked between the drill rod and the borehole wall. (Long, 1960)

ravine A small, narrow, deep depression, smaller than a gorge or a canyon but larger than a gully, usually carved by running water; esp. the narrow excavated channel of a mountain stream. Etymol: French, mountain torrent. (AGI, 1987)

raw (a) In ceramics, fresh from a plastic process; unbaked. (Standard, 1964) (b) Not prepared for use by heat. (Webster 3rd, 1966)

raw coal Coal which has received no preparation other than possibly screening. Syn: *raw ore.* (BS, 1962)

raw coal screen A screen used for dividing run-of-mine coal into two or more sizes for further treatment or disposal; usually employed to remove the largest pieces for crushing and readdition to the run-of-mine coal. (BS, 1962)

raw dolomite (a) Dolomite that has not been calcined. (ARI, 1949) (b) Crushed dolomite used for dressing of basic open hearth bottoms and banks. (AISI, 1949)

raw feed coal Coal supplied to a plant or machine, in which it undergoes some form of preparation. (BS, 1962)

raw fuel A fuel used in the form in which it is mined or obtained, for example, coal, lignite, peat, wood, mineral oil, natural gas. (Nelson, 1965)

rawhide hammer A hand hammer having a rawhide head that serves to prevent bruising metal parts against which it is used. (Crispin, 1964)

raw material The ingredients before being processed that enter into a finished product. (Crispin, 1964)

raw mica A term commonly used for unmanufactured mica. (Skow, 1962)

raw ore (a) Ore that is not roasted or calcined. (b) *raw coal.*

raydist A radio system for medium-range precision surveying in which the phases of two continuous-wave signals are compared. It is based on the heterodyne principle and uses low or medium frequencies. It requires a minimum number of frequencies and these frequencies usually need bear no fixed relationship with each other. (Hunt, 1965)

Rayleigh wave (a) A type of seismic surface wave having a retrograde, elliptical motion at the free surface. It is named after Lord Rayleigh, the English physicist who predicted its existence. Syn: *R wave.* (AGI, 1987) (b) A surface wave associated with the free boundary of a solid. The wave is of maximum intensity at the surface and diminishes quite rapidly as one proceeds into the solid. Therefore, it has a tendency to hug the surface of the solid. Such waves have been used quite effectively in detecting surface cracks and flaws in castings. (Hunt, 1965)

Raymond flash dryer A suspension-type dryer that employs the principle of flash drying of fine coal. The coal is transported vertically through a drying column in a stream of hot gases. The source of heat for this system is usually an automatic stoker. In this system, the hot gases are drawn into the drying column by the action of the fan connected to the cyclone collector vent. The coal to be dried is continuously introduced to the hot gas stream. Virtually instantaneous drying occurs. The dried coal and the moisture-laden gases are drawn into the cyclone collector. The dry coal drops to the bottom of the collector and the moisture-laden gases are discharged by the fan to the atmosphere. See also: *thermal drying.* (Kentucky, 1952; Mitchell, 1950)

Raymond mill Grinding mill in which spring-loaded rollers bear against a horizontal rotating bowl—developed for coal pulverization. (Pryor, 1963)

rays (a) Negatively charged particles which leave the cathode in an evacuated tube at between 10,000 mi/s and 90,000 mi/s (approx. 16,000 km/s and 144,000 km/s), depending on voltage. Positive rays are gas ions (e.g., N^+, O^+). X-rays are electromagnetic waves which travel at the velocity of light and are not deflected by magnetic fields. Length between 0.1 and 100 Å (visible light lies between 4,000 and 8,000 Å). Short X-rays are soft and long ones are hard. Rays emitted by radioactive substances are of three types, alpha, beta, and gamma. Alpha rays consist of He^{++} and move at some 10,000 mile/s (approx 16,000 km/s). Beta rays are electrons with speeds between 50,000 mi/s and 180,000 mi/s (approx. 80,000 km/s and 288,000 km/s); gamma rays are not charged. They move at the speed of light, but are shorter than X-rays (0.01 to 1 Å). (Pryor, 1963) (b) In wave propagation a ray is the trajectory that a signal travels from the source to another point (location).

raywork A kind of rubble work; in the United States, any rubble work of thin and small stones.

razorback A sharp, narrow ridge, resembling the back of a razorback hog. (AGI, 1987)

razor saw A narrow saw used in excavating limestone. (Webster 3rd, 1966)

razor stone *novaculite.*

reach (a) An arm of the sea extending up into the land; e.g. an estuary or bay. (AGI, 1987) (b) A continuous and unbroken expanse or surface of water or land. (AGI, 1987) (c) An unstated but specific distance; an interval. (AGI, 1987) (d) The length of a channel, uniform with respect to discharge, depth, area, and slope. (AGI, 1987) (e) The length of a channel for which a single gage affords a satisfactory measure of the stage and discharge. (AGI, 1987) (f) The length of a stream between two specified gaging stations. (AGI, 1987) (g) A relatively long, straight section of water along a lake shore; also, a narrow arm of a lake, reaching into the land. (AGI, 1987) (h) A straight, continuous, or extended part of a stream, viewed without interruption (as between two bends) or chosen between two specified points; a straight section of a restricted waterway, much longer than a narrows. (AGI, 1987)

reactance The part of the impedance of an alternating-current circuit that is due either to capacitance or inductance or to both and that is expressed in ohms. (Webster 3rd, 1966)

reaction border *reaction rim; corona.*

reaction curve *cotectic line.*

reaction line *cotectic line.*

reaction pair Any two minerals, one of which crystallizes at the expense of the other by reaction with a melt; esp. two adjacent minerals in a reaction series.

reaction principle A relationship between liquid and crystals during crystallization, esp. during fractionation, whereby crystals and liquid change composition in response to changing temperature and pressure. Cf: *reaction series.*

reaction rim A rind of one mineral surrounding another and presumably crystallized by reaction of the core mineral with surrounding fluids. Cf: *corrosion border.* Syn: *corona; kelyphytic rim; reaction border.*

reaction series The sequence of minerals produced by reaction between liquid and crystals during crystallization of a complex magma.

Bowen's reaction series has a continuous side (calcic to alkalic plagioclase) and a discontinuous side (olivine-pyroxene-amphibole-biotite).

reaction-zone width In explosives, the distance that detonation advances before the products of combustion expand by an appreciable percentage. (Leet, 1960)

reactive Readily susceptible to chemical change. (Osborne, 1956)

reactive reagent Substance, solution, or gas susceptible to chemical change, or used in influencing such change. (Pryor, 1963)

reactive silica The silica, SiO_2, present within various clay minerals occurring in bauxite. During the Bayer process digestion of bauxite, this silica reacts with comparable amounts of alumina to form insoluble sodium aluminum silicate, which is lost as refinery plant residue.

reactivity (a) A measure of ease of ignition and response to the controls varying the rate of burning. It is used particularly in connection with fuels for transport gas producers and low volatile fuels used for open fires. (Nelson, 1965) (b) An assessment of the speed of reaction of a coal with oxygen under specified conditions. (BS, 1960)

readily extractable metal As used in geochemical prospecting, refers to the content of a metal that can be extracted from weathered rock, overburden, or stream sediment, by weak chemical reagents. Syn: *cold-extractable metal*. (Hawkes, 1962)

Reading jig A plunger-type jig of relatively simple design with only a single plunger being manually controlled. The hutch compartment is round for good water distribution. (Mitchell, 1950)

realgar (a) Arsenic monosulfide, AsS, contains 70.1% elemental arsenic. (Sanford, 1914) (b) A monoclinic mineral, AsS; dimorphous with pararealgar; red to orange; soft; in ore veins, hot springs, and as a volcanic sublimate. Syn: *red arsenic; sandarac; red orpiment.*

real property Includes mining claims, dumps, water rights, and ditches. (Ricketts, 1943)

ream To enlarge the hole by redrilling with a special bit. (Wheeler, 1958)

ream back The act or process of enlarging a squeezed or cave-obstructed borehole to its original size by reaming as the drill string is pulled. Syn: *reverse reaming*. (Long, 1960)

reamer A rotary-drilling tool with a special bit used for enlarging, smoothing, or straightening a drill hole, or making the hole circular when the drill has failed to do so. See also: *reaming bit; reaming shell; reamer*. Also called gage stone. (AGI, 1987)

reamer bit *reaming bit.*

reamer shell (a) A cutter just above a diamond bit, used to assure a full-size hole. (Nichols, 1976) (b) See also: *reaming shell.*

reamer stone *gage stone.*

reaming The act or process of enlarging a borehole. (Long, 1960)

reaming bit A bit used to enlarge a borehole. Also called broaching bit; pilot reaming bit. Syn: *reamer; reamer bit.* (Long, 1960)

reaming diamond *gage stone.*

reaming pilot (a) *pilot.* (b) A smooth bar used to guide a reaming bit or casing bit in the hole. (BS, 1963)

reaming pilot adapter An adapter or coupling in a reaming pilot assembly that attaches the flush-joint casing to the casing reaming shell and the reaming pilot horn by pin and box threads, respectively. (Long, 1960)

reaming pilot horn An adapter or coupling in a reaming pilot assembly attached to the reaming pilot adapter. It passes through the reaming shell and casing bit to which is attached the pilot bit. (Long, 1960)

reaming ring *reaming shell.*

reaming shell (a) A short tubular piece designed to couple a bit to a core barrel. The outside surface of the reaming shell is provided with inset diamonds or other cutting media set to a diameter to cut a specific clearance for the core barrel. See also: *reamer; reamer shell*. Syn: *reaming ring*. (Long, 1960) (b) Sets of two or more shells that are alternated every 50 ft (15.24 m) to keep loss in gauge to the hole uniform. The shell is changed when wear reaches 0.012 in (0.03 cm) below the original set diameter.

rebar *reinforcing bar.*

recalescence A phenomenon, associated with the transformation of gamma iron to alpha iron on the cooling (supercooling) of iron or steel, revealed by the brightening (reglowing) of the metal surface owing to the sudden increase in temperature caused by fast liberation of the latent heat of transformation. Syn: *point of recalescence*. (ASM, 1961)

recarburize (a) To increase the carbon content of molten cast iron or steel by adding carbonaceous material, high-carbon pig iron, or a high-carbon alloy. (ASM, 1961) (b) To carburize a metal part to return surface carbon lost in processing. (ASM, 1961)

recarburizing Introducing spiegeleisen into the converter after the blow to add the desired element. (Mersereau, 1947)

recast To form anew by running, as molten metal, into a mold; cast again; as, to recast a cracked bell. (Standard, 1964)

receiving hopper A hopper used to receive and direct material to a conveyor.

Recent The later of the two geologic epochs comprised in the Quaternary period, in the classification generally used; Holocene. Also, the deposits formed during that epoch. Recent includes the geologic time and deposits from the close of the Pleistocene (glacial) epoch until and including the present.

receptor *acceptor.*

recession Going back, as the gradual retreat of a waterfall or an erosional escarpment, the melting back of a glacier, or the withdrawal of a body of water so that the shoreline moves successively farther away from the higher land. (Stokes, 1955)

recharge (a) The processes by which water is absorbed and added to the zone of saturation, either directly into an aquifer or indirectly by way of another formation; also, the quantity of water so added. (b) Putting water brought from elsewhere into a body of ground water to augment ground-water supply. See also: *intake area.*

reciprocal lattice An array of points, each point at a distance that is the reciprocal of the *d* spacing between planes in the direction normal to each set of parallel planes as measured from the origin. Each reciprocal-lattice point may be associated with Bragg diffraction in a crystal. Cf: *crystal lattice; direct lattice; X-ray diffraction.*

reciprocal strain ellipsoid In elastic theory, an ellipsoid of certain shape and orientation that under homogeneous strain is transformed into a sphere. Cf: *strain ellipsoid*. (AGI, 1987)

reciprocating Having a straight back-and-forth or up-and-down motion. (Nichols, 1976)

reciprocating drill A piston drill often referred to as a hammer drill. (Hammond, 1965)

reciprocating engine Any steam or internal-combustion engine, which has a piston moving under pressure within a cylinder. (Hammond, 1965)

reciprocating feeder (a) A feeder in which the material is carried on a plate subjected to a reciprocating motion and so constructed that when the plate moves in the reverse direction the material remains stationary. The rate of feed is normally varied by adjusting the stroke of the reciprocating plate. See also: *plate feeder*. (BS, 1962) (b) A device used to empty a bin or hopper from the bottom by horizontal reciprocating action of its parts, usually after primary crushing. (ACSG, 1963)

reciprocating flight conveyor A reciprocating beam or beams with hinged flights arranged to advance bulk material along a conveyor trough. See also: *grit collector.*

reciprocating pump A pump consisting of a piston or plunger which lifts water to a higher level by a series of to-and-fro movements. See also: *pump*. (Nelson, 1965)

reciprocating screen dryer Usually an inclined reciprocating screen on which the coal travels and through which the hot gases pass. The screen may eliminate moisture in coal up to 2-1/2 in (6.4 cm) in size. It may also serve as a fine coal dryer to treat coals down to 1/8 in (0.32 cm). (Nelson, 1965)

recirculating water Circulating water that has been in the circuit for more than one cycle, often recovered from a collecting device such as a thickener from which the clarified water is circulated back into the process stream.

recirculation The continuous circulation of all or some part of the same air in part of a mine ventilation system. (BS, 1963)

recirculation of air A term describing a condition in which the ventilating air current is returned to the face repeatedly along a circuitous path. It may happen in the case of auxiliary fans or booster fans. If the intake end of the air pipes of a blowing fan is not placed well to the intake side of the main air current, the foul air from the heading may be recirculated to the face again and again. With an exhausting auxiliary fan, the end of the pipes is kept well to the return side of the main air current. See also: *two-fan auxiliary ventilation*. (Nelson, 1965)

recirculation of water The water used in a condenser or in a washery or other wet process is often repumped into the system by means of a circulating pump. The practice is economical in water and in reagent consumption and also reduces pollution of local streams. Water that is recirculated is clarified to reasonable purity. See also: *washery water*. (Nelson, 1965)

reclaiming (a) Digging from stockpiles. (Nichols, 1976) (b) Reprocessing previously rejected material. (Nichols, 1976)

reclaiming conveyor (a) Any of several types of conveyors used to reclaim bulk materials from storage. (b) The conveyor which receives material from the reclaimer in a blending system.

reclamation (a) The recovery of coal or ore from a mine, or part of a mine, that has been abandoned because of fire, water, or other cause. (b) Restoration of mined land to original contour, use, or condition.

recleaner cell *cleaner cell.*

reclosing circuit breaker A circuit breaker that recloses automatically as soon as the demand for current becomes equal to or less than that for which the circuit breaker is set. (Zern, 1928)

recommended exposure limit An 8-h or 10-h time-weighted average or ceiling of exposure to coal dust concentration; recommended by NIOSH and based on an evaluation of the health effects data. (NIOSH, 1987)

recomposed granite (a) An arkose consisting of consolidated feldspathic residue (produced by surface weathering of an underlying granitic rock) that has been so little reworked and so little decomposed that upon cementation the rock looks very much like the granite itself. It has a faint bedding, an unusual range of particle sizes (unlike the even-grained or porphyritic texture of true granite), and a greater percentage of quartz than is normal for granite. Syn: *reconstructed granite.* (b) A conglomerate that has been recrystallized by metamorphism into a rock that simulates granite, as in the Lake Superior region. Cf: *meta-arkose.*
recomposed rock A rock produced in place by the cementation of the fragmental products of surface weathering; e.g. a recomposed granite. The term has been applied to a rock of intermediate character straddling an unconformable surface between the breccia of the lower formation and the conglomeratic base of the upper formation. (AGI, 1987)
reconnaissance (a) A general, exploratory examination or survey of the main features (or certain specific features) of a region, usually conducted as a preliminary to a more detailed survey; e.g. an engineering survey in preparing for triangulation of a region. It may be performed in the field or office, depending on the extent of information available. (AGI, 1987) (b) A rapid geologic survey made to gain a broad, general knowledge of the geologic features of a region. (AGI, 1987)
reconnaissance map A map incorporating the information obtained in a reconnaissance survey and data obtained from other sources.
reconnaissance sampling *pilot sampling.*
reconnoiter To make a reconnaissance of; esp. to make a preliminary survey of an area for military or geologic purposes. (AGI, 1987)
reconstructed amber *pressed amber.*
reconstructed granite *recomposed granite.*
reconstructed turquoise An imitation turquoise made of finely powdered ivory which is deposited in a solution of copper. (Fay, 1920)
reconstruction The modernization of underground roadways, transport, ventilation systems, and the layout of mine workings. It may include modernization of shafts and winding and also the improvement of surface handling and cleaning or washing equipment. See also: *mechanization.* (Nelson, 1965)
reconstructive transformation An isochemical change in a crystal structure in which chemical bonds are broken and reformed, e.g., tridymite-quartz or diamond-graphite. Cf: *dilational transformation; displacive transformation; rotational transformation; phase transformation.*
record borehole *record hole.*
record hole The first borehole drilled in an area that is cored so that a detail record of the formations penetrated can be obtained. Also called test hole. See also: *stratigraphic hole.* Syn: *record borehole.* (Long, 1960)
recording gage A gage which automatically records the level of water in a stream or tank, or velocity and pressure in a pipe. It is operated by a float or by a submerged air tank fitted with a rubber diaphragm. (Hammond, 1965)
record table Heavy-duty shaking table used to treat relatively coarse sands. Shaking is by double-link eccentric motion, with longer and slower throw than with Wilfley type of table. (Pryor, 1963)
recover (a) To restore a mine or a part of a mine that has been damaged by explosion, fire, water, or other cause to a working condition. (Fay, 1920) (b) *recovery.*
recoverable grade The true mill-head grade of an ore-stream in percent, ounces, or parts per million of a metal or mineral, less extractive metallurgical losses; the proportion of an ore material actually recovered.
recovered sulfur Elemental sulfur produced from hydrogen sulfide obtained from sour natural gas, petroleum refinery gas, water gas, and other fuel gases.
recovery (a) The percentage of valuable constituent derived from an ore, or of coal from a coal seam; a measure of mining or extraction efficiency. (AGI, 1987) (b) The ratio of the footage of core acquired from core drilling a specific length of borehole, expressed in percent. (Long, 1960) (c) The carat weight of diamonds salvaged from a worn bit. (Long, 1960)
recovery plant (a) A plant designed for separating diamond particles from concentrate by various processes, usually including grease belts, jigs, electrostatic separators, and flotation. Also known as picking station. (b) The processing facility where minerals are recovered.
recreational mining Mining as an avocation rather than as a business. (Barton, 1995)
recrystallization The formation, essentially in the solid state, of new crystalline mineral grains in a rock. The new grains are generally larger than the original grains, and may have the same or a different mineralogical composition. (AGI, 1987)
recrystallized silicon carbide A refractory made of about 98% to 99% SiC.
rectangular drainage pattern A drainage pattern in which the main streams and their tributaries display many right-angle bends and exhibit sections of approx. the same length; it is indicative of streams following prominent fault or joint systems that break the rocks into rectangular blocks. (AGI, 1987)
rectangular shaft A shaft excavated to an oblong shape. The majority of shafts sunk in the Republic of South Africa before 1948 were rectangular and timber lined. The shape lends itself to equipping concurrently with sinking; it provides a convenient in-line hoisting arrangement and can easily be divided into separate compartments. However, in the 1950's and 1960's developments were towards the concrete lined circular shaft. See also: *compartment.* (Nelson, 1965)
rectification (a) The process by which electric energy is transferred from an alternating-current circuit to a direct-current circuit. (Coal Age, 1960) (b) The purification of a liquid by redistillation. (CTD, 1958) (c) In electronics and signal processing, the transformation of a signal from an alternating positive and negative signal into an all-positive signal by taking its absolute value.
rectifier Equipment used in mines to convert alternating current to direct current.
rectifying device An elementary device consisting of one anode and its cathode that has the characteristic of conducting current effectively in only one direction. (Coal Age, 1960)
rectorite A clay mineral with regularly interstratified mica and smectite layers. Syn: *allevardite.*
recumbent fold An overturned fold, the axial surface of which is horizontal or nearly so.
recuperator (a) A continuous heat exchanger in which heat is conducted from the products of combustion to incoming air through flue walls. (ASTM, 1994) (b) A system of thin-walled refractory ducts used for the purpose of transferring heat from a heated gas to colder air or gas. (Harbison-Walker, 1972) (c) Preheating equipment for recovering sensible heat from hot spent gases from a furnace and using it for heating incoming charge or fuel gases; essentially, a low-pressure heat exchanger. Syn: *regenerative heating.* (Henderson, 1953)
recurrence horizon A layer of peat marking a sharp change in the character of the peat and resulting from a profound change in climate. (Tomkeieff, 1954)
red antimony *kermesite.*
red arsenic *realgar.*
red beds Sedimentary strata composed largely of sandstone, siltstone, and shale, with locally thin units of conglomerate, limestone, or marl, that are predominantly red in color due to the presence of ferric oxide (hematite) usually coating individual grains; e.g. the Permian and Triassic sedimentary rocks of western United States, and the Old Red Sandstone facies of the European Devonian. (AGI, 1987)
red cake The vanadium concentrate in a milling operation. (Ballard, 1955)
red chalk Hematite mixed with clay.
red clay A brown to red deep-sea deposit, which usually contains manganese nodules or a film of manganese. It is the finest divided clay suspension that is derived from the land and transported by ocean currents, accumulating far from land and at the greatest depths. It has a high proportion of volcanic material due to lesser dilution of this material owing to slowness of accumulation of the clay portion. The color is believed to be caused by oxidation. Syn: *brown clay.* (AGI, 1987)
red cobalt *erythrite.*
red copper ore *cuprite.*
red copper oxide *cuprite.*
redd (a) Eng. To clear away fallen stone or debris. Syn: *rid.* (SMRB, 1930) (b) Northumb. Overburden. Cf: *ridding.* (Arkell, 1953) (c) Scot. To scour through, take down, or rip. (Fay, 1920)
redd bing A pile of waste made of material brought direct from the mine, not waste from washery. (Zern, 1928)
reddingite An orthorhombic mineral, $Mn_3(PO_4)_2{\cdot}3H_2O$; forms a series with phosphoferrite; pink; at Redding, CT.
reddle Red ocher mixed with clay. Also spelled ruddle, raddle.
reddleman A dealer in reddle or red chalk.
red dog Material of a reddish color resulting from the combustion of coal shale and other mine waste in dumps on the surface.
reddsman Scot. One who works at night cleaning up and repairing roadways, etc.
red earth The characteristic soil of most tropical regions. It is leached, red, deep, and clayey.
red glassy copper ore *cuprite.*
red-hard A term applied to some varieties of tool steels that will retain their hardness even when operating at a red heat. (Newton, 1959)
red heart A harmless reddish core, sometimes found in fire clay refractories.
red hematite A compact columnar variety of hematite with a brownish-red to iron-black color; so called to contrast it with limonite and turgite. Syn: *ironstone clay.* Cf: *brown hematite.* (Fay, 1920)
redingtonite A monoclinic mineral, $(Fe,Mg,Ni)(Cr,Al)_2(SO_4)_4{\cdot}22H_2O$; halotrichite group; forms reddish-purple fibrous masses.
Red I plate *selenite plate.*

red iron froth A variety of hematite. (Fay, 1920)
red iron ore *hematite.*
red iron vitriol *botryogen.*
redistilled metal Metal from which the impurities, usually zinc and mercury, have been eliminated by selective distillation.
red lead *minium.*
red lead ore *crocoite.*
redledgeite A tetragonal mineral, $BaTi_6Cr_2O_{16}{\cdot}H_2O$; cryptomelane group. Formerly called chromrutile.
red lime mud A red mud to which lime, caustic soda, or quebracho, has been added. The pH is usually 12.0 to 13.0.
red manganese Any reddish manganese mineral, i.e., rhodonite and rhodochrosite. Also called red manganese ore.
red mercury Alleged to be a compound of mercury and antimony, and described in the press to be an ingredient of explosives or nuclear weapons manufacture or possibly a descriptive term employed to mask illicit trading activity involving controlled substances. (SME, 1992)
red metal (a) A copper matte containing about 48% copper. (Fay, 1920) (b) Any one of several alloys used in the manufacture of silverware. (Fay, 1920)
red mud (a) A reddish-brown terrigenous deep-sea mud that accumulates on the sea floor in the neighborhood of deserts and off the mouths of great rivers; contains calcium carbonate up to 25%. (Hunt, 1965) (b) A clay-water-base drilling fluid containing sufficient amounts of caustic soda and tannates to give a pronounced red appearance. The pH is usually 10.0 to 13.0. (Brantly, 1961)
red ore Hematite ore.
red orpiment *realgar.*
red oxide of copper *cuprite.*
red oxide of zinc *zincite.*
redox potential Oxidation-reduction potential. (AGI, 1987)
redrill To reopen a borehole by redrilling after it has been cemented, caved, or lost because of junk in the hole. Also called drill out, drilled out. Cf: *overlap.* (Long, 1960)
red roast In fluidization roasting, conversion of iron from sulfides to red oxide. (Pryor, 1963)
redruthite Corn. Copper glance. *chalcocite.* (Fay, 1920)
reds High explosive; used in mines. (Bennett, 1962)
red schorl *elbaite; rubellite; rutile.*
redsear In ironworking, to break or crack when red-hot, as iron under the hammer. (Standard, 1964)
red silver A red silver sulfide; esp. pyrargyrite and proustite. Syn: *red silver ore.*
red silver ore *red silver.* Also called pyrargyrite; proustite. (Pryor, 1963)
reduce (a) To lower the oxidation state by adding electrons to a chemical species. (b) In general, to treat metallurgically for the production of metal. (Fay, 1920)
reduced iron Free iron in a fine state of division obtained by reducing ferric oxide by heating it in a current of hydrogen. Also called iron by hydrogen, iron powder, and spongy iron. (Standard, 1964; Fay, 1920)
reduced level Height above specified datum level of a surveyed point. (Pryor, 1963)
reduced natural frequency The natural frequency of vibration of a foundation at an average ground pressure of unity is the reduced natural frequency divided by the square root of the ground pressure. This relationship has been established by Tschebotarioff. (Hammond, 1965)
reducing agent (a) A material that adds hydrogen to an element or compound. (McGraw-Hill, 1994) (b) A material that adds an electron to an element or compound; i.e., decreases the positiveness of its valence. (McGraw-Hill, 1994)
reducing atmosphere (a) An atmosphere having a deficiency of oxygen. (b) An atmosphere of hydrogen or other substance that readily provides electrons. (c) Space from which air has been displaced by hydrogen, carbon monoxide, or other reducing gas. (CTD, 1958)
reducing flame The blue part or inner cone of the flame produced by a blowpipe; characterized by an excess of hydrocarbon over oxygen so as to reduce mineral samples heated in it. See also: *blowpiping.* Cf: *oxidizing flame.*
reducing furnace A furnace in which ores are reduced from oxides, or metal is separated from other substances by a nonoxidizing heat or flame; usually a shaft furnace. Syn: *reduction furnace.* (Fay, 1920)
reducing roast The reduction of metallic oxides, sulfides, or halides by heating in contact with carbon or other reducing agents. (Newton, 1959; Newton, 1938)
reduction (a) Process of reducing a metal compound to the metal and separating it from the slag; sometimes applied to the smelting process. (b) A reaction taking place at the cathode in electrolysis through transfer of electrons to the species being reduced. (c) A decrease in positive valence, or an increase in negative valence by the gaining of electrons. A metallic oxide loses oxygen through the action of reducing gas, reducing its valence. Cf: *oxidation.*
reduction cell A pot or tank in which either a water solution of a salt or a fused salt is reduced electrolytically to form free metals or other substances. (ASM, 1961)
reduction factor The factor relating the allowable stress on a long column with that on a short column in order to prevent buckling. (Hammond, 1965)
reduction furnace *reducing furnace.*
reduction of area (a) The difference between the cross-sectional area of a tension specimen at the section of rupture before loading and after rupture, expressed as a percentage of the original area. (Roark, 1954) (b) Percentage decrease in cross-sectional area of bar or wire after rolling or drawing. (Hammond, 1965)
reduction of levels The calculation of reduced levels from the staff readings recorded in a field book. (Hammond, 1965)
reduction ratio In crushing, the ratio of the size of the largest feed particle to the smallest distance between the roll faces. As used frequently in the field, it is the ratio of the smallest aperture passing all of the feed to that passing all of the product. Another basis of expression is the ratio of the average size of feed to the average size of product. See also: *overall reduction ratio; primary breaker.* (Taggart, 1927)
reduction roasting Lowering of oxygen content of ore by heating in reducing atmosphere. (Pryor, 1963)
reduction smelting A pyrometallurgical process that produces an impure liquid metal and a liquid slag by heating a mixture of ore, flux, and reducing agent (usually coke). (Newton, 1959; Newton, 1938)
reduction to center The offset of a side auxiliary telescope requires a correction to observed horizontal angles, and the offset of a top auxiliary telescope requires a correction to observed vertical angles. The process of computing the correct angle from the observed angle is called reduction to center. (Urquhart, 1959)
reduzate A sediment formed in a strongly reducing environment; e.g., coal, sedimentary sulfides, or sedimentary sulfur. (AGI, 1987)
red vitriol *bieberite; rose vitriol.*
Redwood number Viscosity, defined as rate of flow of oil from a Redwood viscometer. (Pryor, 1963)
red zinc ore *zincite.*
red zinc oxide *zincite.*
reed (a) Scot. Rift, or direction of easiest splitting. (b) Weakness in a sedimentary rock parallel with the bedding. See also: *cleat.* (c) A reed filled with powder to act as a fuse. See also: *spire.*
reedmergnerite A triclinic mineral, $NaBSi_3O_8$; feldspar group; occurs in small colorless prisms having wedge-shaped terminations; from oil wells in Duchesne County, UT.
reef (a) A ridgelike or moundlike structure, layered or massive, built by sedentary calcareous organisms, esp. corals, and consisting mostly of their remains; it is wave-resistant and stands above the surrounding sediment. Also, such a structure built in the geologic past and now enclosed in rock, commonly of differing lithology. (AGI, 1987) (b) A narrow ridge or chain of rocks either at the water surface or too shallow to permit safe passage of a vessel. Cf: *bank.* (c) A provincial term for a metalliferous mineral deposit, esp. gold bearing quartz. See also: *reefing.* (AGI, 1987)
reef cap A deposit of fossil-reef material overlying or covering an island or mountain. (AGI, 1987)
reef drive Aust. A cutting through the bedrock in alluvial mining for the purpose of seeking other underground, gold-bearing gravel channels.
reefing Working auriferous reefs or veins. *reef.*
reef knoll *bioherm.*
reef limestone A limestone consisting of the remains of active reef-building organisms, such as corals, sponges, and bryozoans, and of sediment-binding organic constituents, such as calcareous algae. (AGI, 1987)
reef wash Aust. Gold-bearing drift.
reel A device used for hoisting that has largely been replaced by round ropes. A flat rope is used for the reel, which is wound on an overlapping spiral like a clock spring. The reel is like a conical drum that increases in diameter by the thickness of the rope at each turn. Reels are more suitable for hoisting from a single level than from different levels. (Lewis, 1964)
reel boy In bituminous coal mining, one who works on an electric locomotive—power being transmitted through an electric cable wound around a reel on the locomotive—tending the cable to see that it is wound up and fed from the reel so that it will not pull or break from the point where electric current is supplied. Also called nipper. (DOT, 1949)
reel locomotive A trolley locomotive with a wire rope reel for drawing cars out of rooms. The rope end is pulled by a runner into the face of the room, attached to a car, and reeled out by the locomotive. (Zern, 1928)
reenforcing bar *reinforcing bar.*
Ree's torsion anemometer Consists of a thin square aluminum vane centrally suspended from a horizontal wire mounted in a vertical

frame. The velocity of the air current is obtained from the measurement of the torque that has to be applied to the wire to bring the vane back to its vertical position. The instrument is mounted on a tripod, and the arrangement is such that the torsion can be applied, at a point 2 ft (0.6 m) away from the vane, by means of a shaft and bevel gearing. The instrument has been used to measure low air velocities in mines down to about 10 ft/min (3 m/min) and up to 180 ft/min (55 m/min). Syn: *torsion anemometer*. (Roberts, 1960)

reeve The orderly arrangement of a rope or cable on a system of pulleys or sheaves to assemble block-and-tackle equipment for handling heavy loads. Also called reeved. (Long, 1960)

reeving Threading or placement of a working line. (Nichols, 1954)

reference axes In structural petrology, three mutually perpendicular axes to which structural measurements are referred. *a* is the direction of tectonic transport, *c* is perpendicular to the plane along which differential movement takes place, and *b* lies in this plane but is perpendicular to *a*.

reference electrode Hydrogen electrode used to determine electrode potentials of half-cells. (Pryor, 1963)

reference level *datum plane.*

reference mark A selected distant point from which the bearings to other points can be measured at a survey station. (Hammond, 1965)

reference plane *datum plane.*

reference seismometer In seismic prospecting, a detector placed to record successive shots under similar conditions, to permit overall time comparisons. Used in connection with the shooting of wells for velocity. (AGI, 1987)

reference size Separation size, designated size, or control size used to define analyses of the products of a sizing operation. (BS, 1962)

reference standard Taken or laid down as a standard for measuring, reckoning, or constructing. (Webster 3rd, 1966)

reference station A station for which tidal constants have previously been determined and that is used as a standard for the comparison of simultaneous observations at a second station; also, a station for which independent daily predictions are given in the tide or current tables from which corresponding predictions are obtained for other stations by means of differences or factors. (AGI, 1987)

referencing The process of measuring the horizontal (or slope) distances and directions from a survey station to nearby landmarks, reference marks, and other permanent objects that can be used in the recovery or relocation of the station. (AGI, 1987)

refikite An orthorhombic mineral, $C_{20}H_{32}O_2$; soft; white; in modern resins and lignite at Montorio, Abruzzes, Italy. Also spelled reficite.

refine (a) To free from impurities; to free from dross or alloy; to purify, as metals; to cleanse. (Webster 3rd, 1966) (b) To treat cast iron in the refinery furnace so as to remove the silicon. (Webster 3rd, 1966)

refined iron Wrought iron made by puddling pig iron. (CTD, 1958)

refinery (a) A facility in which relatively crude smelter products such as blister copper are refined and emerge as acceptably pure products. (b) An electrolytic or chemical facility producing pure metals.

refining The purification of crude metallic products. (Fay, 1920)

refining heat A medium orange heat, about 655 °C, which imparts fineness of grain and toughness to steel that is raised to it and afterwards quenched. (Webster 2nd, 1960)

refining temperature A temperature, usually just higher than the transformation range, employed in the heat treatment of steel to refine the structure, particularly the grain size. (ASM, 1961)

reflected-light microscope A compound microscope in which plane-polarized light impinges upon a polished specimen, commonly opaque, the light being reflected back to the objective through a second polarizer, where mineral color and polarization colors are observed in the ocular. Syn: *ore microscope.* See also: *optical constant; polished section; microscopy.*

reflected-light microscopy *ore microscopy.*

reflected wave A (gaseous) pressure wave resulting from a direct wave striking an obstacle or an opposing surface and being reflected backwards. (Rice, 1913-18)

reflection (a) The return of a wave incident upon a surface to its original medium. Cf: *refraction; diffraction; total reflection.* (b) The bounding back of light or other rays as they strike a solid surface. Light incident on a polished planar surface reflects at an angle equal to the incident angle, the proportion of reflected light increasing with increasing refractive index; e.g., for normal incidence, 17% reflects from diamond (n=2.4), and 5% reflects from quartz (n=1.5). (c) In seismic prospecting, the returned energy (in wave form) from a shot that has been reflected from a velocity discontinuity back to a detector; the indication on a record of reflected energy. (d) Misnomer for X-ray diffraction peaks. Also spelled reflexion.

reflection goniometer An instrument that measures angles between crystal faces by reflecting a beam of light from successive faces as the crystal is rotated. See also: *contact goniometer*. Cf: *two-circle goniometer.*

reflection mechanism A rule stating that rock breaks from the surface inward toward the explosive rather than from the explosive charge outward. Syn: *blasting reflection mechanism.* (Lewis, 1964)

reflection method *seismic reflection method.*

reflection shooting A type of seismic survey based on measurement of the travel times of waves that originate from an artificially produced disturbance and are reflected back at near-vertical incidence from subsurface boundaries separating media of different elastic-wave velocities. Cf: *refraction shooting.* (AGI, 1987)

reflection wave A wave that is propagated backward through the burned gas as the result of an explosion wave being completely or partly arrested against the closed extremity, or in a constricted portion of its path, as in a tube, gallery, etc. (Fay, 1920)

reflectivity The ratio of radiant energy reflected by a body to that falling upon it. (Strock, 1948)

refraction The deflection of a ray of light or of an energy wave (such as a seismic wave) due to its passage from one medium to another of differing density, which changes its velocity. Cf: *reflection; diffraction.* (AGI, 1987)

refraction method A seismic method of geophysical prospecting. See also: *refraction shooting.* (Nelson, 1965)

refraction shooting (a) The detonation of heavy charges of explosive in comparatively shallow holes or pits. The effects may be measured over a wide area. The firing creates the shock waves in the seismic method of prospecting. (Nelson, 1965) (b) A type of seismic survey based on the measurement of the travel times of seismic waves that have traveled nearly parallel to the surface of high-velocity layers, in order to map such layers. See also: *isochrone lines.* Cf: *reflection shooting.* (AGI, 1987)

refractive index *index of refraction; dispersion.*

refractometer (a) A combustible gases detector. See also: *interference methanometer.* (Nelson, 1965) (b) An instrument for measuring indices of refraction of transparent substances, both liquid and solid. Cf: *Abbé refractometer.*

refractoriness The capacity of a material to resist high temperature. In the refractories industry, the pyrometric cone equivalent (PCE) is a comparative value used to determine the refractoriness of a material. (Henderson, 1953)

refractory (a) Said of an ore from which it is difficult or expensive to recover its valuable constituents. (AGI, 1987) (b) Exceptionally resistant to heat. (AGI, 1987) (c) A nonmetallic material suitable for use in high-temperature applications. (AGI, 1987)

refractory bonding mortars High-temperature bonding mortars containing various materials and exhibiting various properties, but primarily intended for providing structural bond between refractory units in high-temperature industrial furnace construction. (Henderson, 1953)

refractory brick (a) A brick made from refractory material, such as fire clay, bauxite, diaspore, etc., used to withstand high temperatures. Refractory bricks are made in various sizes and shapes. (b) A brick used as a lining for the interior of fireboxes in furnaces and boilers. Refractory brick is constructed so that it can withstand very high temperatures, but it is not a very good insulator. (API, 1953)

refractory clay *fireclay.*

refractory lining A lining that has high refractory qualities and is therefore suitable for furnace linings and boiler foundations. It is made from a good-quality refractory ore, clay, fireclay, or gannister. (Nelson, 1965)

refractory material A material able to withstand high temperatures and, therefore, used in such applications as lining furnaces.

refractory ore Ore difficult to treat for recovery of the valuable substances. (AGI, 1987)

refractory stone Consists of sandstone, quartzite, mica schist, soapstone, or other rock that will withstand a moderately high temperature without fusing, cracking, or disintegrating. It may be used in solid blocks or crushed and mixed with a binder to form bricks. (USBM, 1965)

refractory ware Usually hollow ware, such as, saggers, pyrometer tubes, crucibles, etc.; also refractory brick and shapes.

refresher training In mining, training given to all miners at least once a year consisting of 8 hours of instruction reviewing the essentials of new miner training. Cf: *new miner training; task training.* (Federal Mine Safety, 1977)

refrigerant A substance that will absorb heat while vaporizing and whose boiling point and other properties make it useful as a medium for refrigeration. (Strock, 1948)

refrigeration (a) In special application to mining, cooling of air before release in lowest levels of deep, hot mine; also, expansion of compressed air for the same purpose. (b) The process of absorption of heat from one location and its transfer to and rejection at another place; arbitrarily expressed in units of (short) tons and is equal to the coil cooling load divided by 12,000; 1 st (0.9 t) of ice in melting in 24 h liberates heat at the rate of 200 Btu/min (211 kJ/min), or 12,000 Btu/h (1.27 MJ/h). (Hartman, 1961)

refrigeration plant (a) A surface plant to form the protective barrier of frozen ground in the freezing method of shaft sinking. The cooling agent used is ammonia which, in its gaseous state, is compressed to about 120 psi (827 kPa) when it passes to the top of the condensers, emerging at the bottom as liquid ammonia under pressure. It then passes through a regulator valve into the coolers where it immediately evaporates. The latent heat of evaporation is extracted from the brine circuit—the brine being passed through the coolers by the brine pumps. The ammonia gas passes back for re-use. The brine emerges from the coolers at a temperature of -4 °F (-21.7 °C) and is pumped down the boreholes to freeze the water around the shaft sinking. (b) A surface plant to cool liquids. These liquids or ice are sent underground to cool the air current in heat exchangers. By this method, the air in deep mines is cooled considerably and the working environment is improved. See also: *deep mining*. (Nelson, 1965)

refuge chamber An airtight, fire-resistant room in a mine used as a refuge in emergencies by miners unable to reach the surface. (MSHA, 1992)

refuge hole A place formed in the side of an underground haulageway in which a worker can take refuge during the passing of a train, or when shots are fired. Also called refuge stalls. See also: *manhole.* (Fay, 1920)

refusal A condition arrived at when driving pipe, casing, piling, etc., when it cannot be driven to a greater depth or made to penetrate the ground a distance of more than 1 ft (30.5 cm) per 100 blows delivered by a drive hammer. (Long, 1960)

refuse (a) Waste material in the raw coal that has been removed in a cleaning or preparation plant. (b) Notably used to describe colliery rejects; also called tailings. (Pryor, 1963)

refuse conveyor An adaptation of a drag chain conveyor.

refuse discharge pipes Pipes used on some washboxes instead of a refuse worm. (BS, 1962)

refuse elevator *reject elevator.*

refuse extraction chamber That part of the washbox into which the refuse extractor discharges. (BS, 1962)

refuse extractor A device used in a washbox to remove the reject from the washing compartments, operated manually or automatically. (BS, 1962)

refuse rotor A reject gate in the form of a rotary (or star) valve. (BS, 1962)

refuse worm A screw conveyor fitted at the bottom of some washboxes to collect the fine reject which has passed through the apertures in the screen plate. (BS, 1962)

regalian doctrine The old doctrine that all mineral wealth was the prerogative of the crown or the feudatory lord. The concession system, in which the state or the private owner has the right to grant concessions or leases to mine operators at discretion and subject to certain general restrictions, had its origin in this doctrine. Almost all mining countries of the world, except the United States, follow this system. (Hoover, 1948)

regal jade *Indian jade.*

regenerated anhydrite Anhydrite produced by dehydration of gypsum that itself was generated by the hydration of anhydrite.

regenerated dense medium Medium obtained from the medium recovery system and purified (wholly or partly) from contaminating fine coal and clay. (BS, 1962)

regeneration (a) In mineral leaching, reconstitution of barren leach solution after it has completed its chemical attack on mineral and its values have been removed. The regeneration of ion exchange resins and activated carbons by the removal of elements or compounds from extraction sites on the resins by special eluants. (Pryor, 1963) (b) A reversing heat exchanger for preheating combustion air (and gaseous fuels) from waste heat of the exhaust gases. (Van Vlack, 1964)

regenerative chambers Separate compartments connected with a furnace; they are arranged for preheating the gas and the air used for fuel. (Mersereau, 1947)

regenerative furnace A furnace in which hot gases, usually waste combustion gases, pass through a set of chambers containing firebrick structures, to which the sensible heat is given up. The direction of hot-gas flow is diverted periodically to another set of chambers and cold incoming combustion gas or air is preheated in the hot chambers.

regenerative heating *recuperator.*

regenerative principle Used in open-hearth furnaces to increase the furnace temperature by preheating the fuel gas and air previous to their combustion in the furnace. (Newton, 1959)

regenerator checkers Brick used in furnace regenerators to recover heat from hot outgoing gases, and later to release this heat to cold air or gas entering the furnace; so called because of the checkerboard pattern in which the bricks are arranged. (Harbison-Walker, 1972)

regime In hydraulics, the condition of a river with respect to the rate of its flow as measured by the volume of water passing different cross sections in a given time. (Webster 3rd, 1966)

regional (a) Extending over large areas in contradistinction to local or restricted areas. (Fay, 1920) (b) In gravity prospecting, contributions to the observed anomalies due to density irregularities at much greater depths than those of the possible structures, the location of which was the purpose of the survey. The term is also employed in an analogous sense in magnetic prospecting. (AGI, 1987)

regional anomaly (a) The more localized departures in the Earth's field from the values that would be predicted if the field were to originate with a single magnet oriented along the magnetic axis. These have maximums as great as 10,000 gamma, which is about a third the total intensity at the equator, and extend over areas as large as a million square miles. The locations of such features do not change with time as do anomalies associated with secular variation. (Dobrin, 1960) (b) The departure of a measured quantity from an expected or theoretical value on a scale larger than the most rapid spatial variations of the measured quantity; typically variations over tens to hundreds of kilometers.

regional dip The nearly uniform inclination of strata over a wide area, generally at a low angle, as in the Atlantic and Gulf coastal plains and parts of the Midcontinent region. Cf: *homocline.* Syn: *normal dip*. (AGI, 1987)

regional metamorphism A general term for metamorphism affecting an extensive region, as opposed to local or contact metamorphism. Cf: *dynamothermal metamorphism; local metamorphism.*

regional unconformity A surface of discontinuity in sedimentary rocks that extends throughout an extensive region. It may record a significant interruption in deposition, tectonics, or erosion of older strata.

registered premises Premises registered with the local authority for the storage of not more than 60 lb (27.2 kg) of explosive. See also: *licensed store; magazine*. (BS, 1964)

reglette A 12-in (0.3-m) scale divided into tenths and hundredths of a foot, used for accurate measurement in conjunction with a steel band that is graduated only in feet. See also: *band chain*. (Hammond, 1965)

regolith The layer or mantle of loose incoherent rock material, of whatever origin, that nearly everywhere underlies the surface of the land and rests on bedrock. It comprises rock waste of all sorts: volcanic ash, glacial drift, alluvium, windblown deposits, organic accumulations, and soils. Syn: *mantle rock.*

regular lay Wire rope or cable in which the individual wire or fibers forming a strand are twisted in a direction opposite to the twist of the strands. Also called ordinary lay. Syn: *standard lay*. (Long, 1960)

regular-lay left lay *left regular lay.*

regular-lay right lay *right regular lay.*

regular polygon A polygon having equal sides, and the angles between these sides are equal. (Jones, 1949)

regular sampling The sampling of the same coal or coke received regularly at a given point. There are two forms of regular sampling, namely, continuous sampling and intermittent sampling. (BS, 1960)

regular ventilating circuit All places in a mine through which there is a positive flow of air without the aid of a blower fan or of ventilation tubing.

regulated feed In contrast with choke feed, feed that is throttled back to a value below the full capacity of the crusher. (South Australia, 1961)

regulated split In mine ventilation, a split where it is necessary to control the volumes in certain low-resistance splits to cause air to flow into the splits of high resistance.

regulating gate A gate used to vary size of opening so as to control the flow of material through the opening. See also: *bin gate.*

regulator (a) A ventilating device, such as an opening in a wall or door; usually placed at the return of a split of air to govern the amount of air entering that portion of a mine. (Kentucky, 1952) (b) A device for creating shock loss to restrict passage of air through an airway. Regulators are usually set in doors as adjustable, sliding partitions that can be varied to the desired opening. In their simplest form, for temporary service in an untraveled part of a mine, regulators consist of doors propped partially open. Where possible, regulators are located on the exhaust side of a split (in a return airway) to minimize interference with traffic. See also: *ventilation*. Syn: *ventilation regulator*. (Hartman, 1982)

regulator door *scale door.*

regulus Impure metal produced during smelting of ores or concentrates. (Pryor, 1963)

reheater An apparatus for reheating a substance, as ingot steel, that has cooled or partly cooled during some process. (Standard, 1964)

reheater load The amount of sensible heat in w (British thermal units per hour), restored to the air in reheating. (Hartman, 1982)

reheating furnace The furnace in which metal ingots, billets, blooms, etc., are heated to bring them to the temperature required for hot-working. (CTD, 1958)

Rehisshakenhobel A plow developed from the Anbauhobel machine and designed for cutting thin coal seams. The plow drives, instead of being on the face side of the conveyor, are on the waste side and the plow chains run in two tubes along the waste side of the conveyor chutes. See also: *Anbauhobel.* (Nelson, 1965)

reinerite An orthorhombic mineral, $Zn_3(AsO_3)_2$; blue to green; at Tsumeb, Namibia. (Not renierite.)

reinforcing bar (a) The basic material used to form grouted roof belts. (b) Iron or steel bars of various cross-sectional shapes used to strengthen concrete. (c) *rebar; reenforcing bar.*

reinforcing steel Steel bars of various shapes used in concrete construction to give added strength. (Crispin, 1964)

reinite A pseudomorph, $FeWO_4$; after scheelite(?).

reiteration In surveying, angular measurement made first with vertical circle of theodolite to right of sighting telescope, then repeated after transiting this through 180°. Also called face right, face left observation. (Pryor, 1963)

reject (a) The material extracted from the feed during cleaning for retreatment or discard. (BS, 1962) (b) The stone or dirt discarded from a coal preparation plant, washery, or other process; has no value. See also: *middlings; refuse; residue; tailings.* (Nelson, 1965)

reject elevator An elevator for removing and draining the reject from a washing appliance. Syn: *refuse elevator.* (BS, 1962)

reject gate The mechanism of the refuse extractor that may be manually or automatically operated to control the rate of removal of reject from the washbox. (BS, 1962)

rejuvenation The renewal of any geologic process, such as the revival of a stream's erosive activity or the reactivation of a fissure.

relative age The geologic age of a fossil organism, rock, geologic feature, or event, defined relative to other organisms, rocks, features, or events rather than in terms of years. Cf: *absolute age.* (AGI, 1987)

relative biological effectiveness The relative effectiveness of a given kind of ionizing radiation in producing a biological response as compared with 250,000 electron-volt gamma rays. Abbrev., rbe. (Lyman, 1964)

relative bulk strength A measure of the energy available per unit volume of explosive as compared to an equal volume of ANFO at a density of 0.81 g/cm^3; it is calculated by dividing the bulk strength of an explosive by the bulk strength of ANFO and multiplying by 100. See also: *absolute bulk strength.*

relative compaction (a) For soil compaction, two types of tests are necessary: (1) determining the dry density of the soil after a standard amount of compaction has been applied, and (2) measuring the density of the soil in the field. The state of compaction is expressed as the relative compaction, and is the percentage ratio of the field density to the maximum density as determined by standard compaction. The percentages of relative compaction are high, since the initial relative compaction is about 80%. (Nelson, 1965) (b) The dry density of a soil in situ divided by the maximum dry density of the soil as established by the Proctor compaction test or any other standard test. See also: *loose ground.* (Hammond, 1965)

relative consistency The ratio of the liquid limit minus the natural water content to the plasticity index. (ASCE, 1958)

relative density (a) The relative density or specific gravity of a substance denotes the number of times the substance is heavier or lighter than water (for the same volume). Relative density and specific gravity mean the same thing. (Morris, 1958) (b) The ratio of the difference between the void ratio of a cohesionless soil in the loosest state and any given void ratio to the difference between its void ratios in the loosest and in the densest states. (ASCE, 1958)

relative humidity The ratio, expressed as a percentage, of the amount of water vapor in a given volume of air to the amount that would be present if the air were saturated at the same temperature. Cf: *absolute humidity; specific humidity.* (AGI, 1987)

relative movement In fault descriptions, the displacement of one block relative to the other, rather than to some fixed point or plane of reference. (AGI, 1987)

relative roughness The dimensionless ratio E/d (where E is the average height of the surface irregularities and d is the diameter of the pipe) is termed the relative roughness. The physical interpretation of this functional equation is that the friction factors of pipes are the same if their flow patterns in every detail are geometrically and dynamically similar. The term E indicates the height of the irregularity above the boundary surface only; hence it is apparent that, dependent upon the thickness of the boundary layer adjacent to the surface, the projection can either lie submerged within the boundary layer or else project outside it. (Roberts, 1960)

relative time Geologic time determined by the placing of events in a chronologic order of occurrence; esp., time as determined by organic evolution or superposition. Cf: *absolute time.* (AGI, 1987)

relative variogram A variogram in which the ordinary variogram value for each lag has been divided by the square of the mean of the sample values used in computing the lag. This is sometimes useful when a "proportional effect" is present; i.e., when areas of higher than average concentration also have higher than average variance. When relative variogram models are used in kriging, the resulting kriging standard deviations represent decimal fractions of the estimated values.

relative weight strength This is a measure of the energy available per weight of explosive as compared to an equal weight of ANFO. It is calculated by dividing the absolute weight strength (AWS) of the explosive by the AWS of ANFO and multiplying by 100. See also: *absolute weight strength.*

relaxation (a) In experimental structural geology, the release of applied stress with time, due to any of various creep processes. (AGI, 1987) (b) In an elastic medium, the decrease of elastic restoring force under applied stress, resulting in permanent deformation. (AGI, 1987) (c) Relief of stress by creep. Some types of tests are designed to provide diminution of stress by relaxation at constant strain, as frequently occurs in service. (ASM, 1961) (d) The decrease of load support and of internal stress because of plastic strain at constant deformation. (AGI, 1987)

relay A device, operated by an electric current, and causing by its operation abrupt changes in an electrical circuit (making or breaking the circuit, changing of the circuit connections, or variation in the circuit characteristics). (NCB, 1964)

relay haulage Single-track, high-speed mine haulage from one relay station to another. Each operator has an exclusive track section between relay stations and can run at full speed since no other haulage equipment is operating on the section. Side track at each relay station permits the operator to pick up or drop off loads or empties, then make the return run. Also called intermediate haulage. See also: *haulage.* (Kentucky, 1952)

relay motorman *gathering motorman.*

release analysis A procedure employed to determine the best results possible in cleaning a coal by froth flotation. (BS, 1962)

released mineral A mineral formed during the crystallization of a magma as a consequence of an earlier phase failing to react with the liquid. Thus the failure of earlier formed olivine to react with the liquid portion of a magma to form pyroxene may result in the enrichment of the liquid in silica, which finally crystallizes as quartz, the released mineral. (AGI, 1987)

release fracture A fracture developed as a consequence of the relief of stress in one particular direction. The term is generally applied to a fracture formed when the maximum principal stress decreases sufficiently that it becomes the minimum principal stress; the fracture is an extension fracture oriented perpendicular to the then-minimum principal-stress direction. (AGI, 1987)

release mesh (a) In liberation of specific mineral from its ore by comminution, the optimum grind. (Pryor, 1963) (b) Specified mesh-of-grind for best conditions for treatment to recover a specific mineral from the ore. (Pryor, 1965)

reliability of method In geochemical prospecting, refers to the probability of obtaining and recognizing indications of an orebody or mineralized district by the method being used. Reliability depends not only on whether a readily detectable target exists and how effective the exploration method is in locating it, but also on the extent to which the anomaly is specif. related to ore and the extent to which it is possible that non-significant anomalies may confuse the interpretation. (Hawkes, 1962)

relic A landform that has survived decay or disintegration, such as an erosion remnant; or one that has been left behind after the disappearance of the greater part of its substance such as a remnant island. The term is sometimes used adjectivally as a synonym of relict, but this usage is not recommended. (AGI, 1987)

relict Pertaining to a mineral, structure, or feature of an earlier rock that has persisted in a later rock in spite of processes tending to destroy it. Also, such a mineral, structure, or other feature. (AGI, 1987)

reliction The slow and gradual withdrawal of the water in the sea, a lake, or a stream, leaving the former bottom as permanently exposed and uncovered dry land; it does not include seasonal fluctuations in water levels. Legally, the added land belongs to the owner of the adjacent land against which it abuts. Also, the land left uncovered by reliction. (AGI, 1987)

relict texture In mineral deposits, an original texture that remains after partial or total replacement. (AGI, 1987)

relief (a) A term used loosely for the physical shape, configuration, or general unevenness of a part of the Earth's surface, considered with reference to variations of height and slope or to irregularities of the land surface; the elevations or differences in elevation, considered collectively, of a land surface. Cf: *topography.* (AGI, 1987) (b) The vertical difference in elevation between the hilltops or mountain summits and the lowlands or valleys of a given region. A region showing a great variation in elevation has high relief, and one showing little variation has low relief. (AGI, 1987) (c) The range of values over an anomaly or within an area; e.g., the gravity relief for the magnitude of a gravity anomaly. (AGI, 1987) (d) An apparently rough surface of a crystal section under a microscope. High relief indicates a great difference in index of refraction between the crystal and its mounting medium. The relief is positive if the refractive index of the mineral is greater than that of the medium, and negative in the reverse case. Syn: *shagreen.* (AGI, 1987) (e) The result of the removal of tool material behind or adjacent to the cutting edge to provide clearance and prevent rubbing (heel drag). (ASM, 1961)

relief feature *landform.*

relief holes (a) Boreholes that are loaded and fired for the purpose of relieving or removing part of the burden of the charge to be fired in the main blast. (b) Holes drilled closely along a line, that are not loaded, and that serve to weaken the rock so that it will break along that line. Syn: *trim holes.* (Nichols, 1976) (c) Singular. A port or passageway through which the core, as it advances into the inner tube of a double-tube core barrel, forces water out of the inner tube to the outside of the barrel through the innertube head. (Long, 1960) (d) Singular. A borehole drilled ahead of underground openings to tap and drain a water-bearing formation. Also called cover hole; pilot hole. (Long, 1960)

relief limonite Indigenous limonite that is porous and cavernous in texture, commonly botryoidal after chalcocite. Cf: *indigenous limonite.* (AGI, 1987)

relief map A map representing topographic relief of an area by contour lines, hachures, hill shading, coloring, or similar graphic means.

relief valve A valve that will allow air or fluid to escape if its pressure becomes higher than the valve setting. Also called pressure relief valve. (Nichols, 1976)

relief well A borehole that is drilled at the toe of an earth dam as a relief for any high water pressure caused by the weight of the dam. (Hammond, 1965)

relieving cut In a round of shots planned for sequential firing when shaft sinking or tunneling, holes fired after cut holes and before lifters and slippers. (Pryor, 1963)

relieving platform A loading deck for lorries on the land side of a retaining wall contructed, for example, as a jetty of steel sheet piling, the relieving platform being supported as a rule partly by the wall and partly by bearing piles. See also: *surcharge.* (Hammond, 1965)

relieving shot A shot fired to dislodge or expose a misfire. (BS, 1964)

relighter flame safety lamp A locked spirit-burning lamp fitted with an internal relighting device. (BS, 1963)

relighting station A place in a mine at which safety lamps can be relighted under controlled conditions. (BS, 1965)

reluctance Magnetic quality analogous to resistance in the flow of electric current. (Pryor, 1963)

remanence Residual magnetism in a ferromagnetic substance (its hysteresis) after removal of an external magnetizing force. (Pryor, 1963)

remanent magnetization (a) Part of the magnetization of a body that does not disappear when the external magnetic field disappears. (Schieferdecker, 1959) (b) That component of a rock's magnetization that has a fixed direction relative to the rock and is independent of moderate, applied magnetic fields such as the Earth's magnetic field. Cf: *induced magnetization.* See also: *hysteresis.* (AGI, 1987)

remnant When a block of ground is stoped in such a way that at some time its remainder is surrounded on all sides by stoped ground, that remainder is termed a remnant. (Spalding, 1949)

remolded soil Soil that has had its natural structure modified by manipulation. (ASCE, 1958)

remolding Disturbance of the interval structure of clay or silt; when remolded, such material will lack shearing strength and gain compressibility. In consequence, driven piles are not recommended in certain clays. See also: *thixotropic fluid.* (Hammond, 1965)

remolding index The ratio of the modulus of deformation of a soil in the undisturbed state to the modulus of deformation of the soil in the remolded state. (ASCE, 1958)

remolding sensitivity The ratio of the unconfined compressive strength of an undisturbed specimen of soil to the unconfined compressive strength of a specimen of the same soil after remolding at unaltered water content. Also called sensitivity ratio. (ASCE, 1958)

remolinite *atacamite.*

remote control (a) The control of plant operation by personnel or computers housed under conditions that can be remote, safe, and convenient. This is a feature of both electrical and electronic automatic control. In the control room, various plants can be started up by pushbutton and the governing conditions can be set. Instrumentation records all relevant data; it also gives warning of unsafe conditions and shuts down the plant if no correction is made. Changeover switches can introduce different operations and sequences in the working of the plant. Indicating lights can show what plant is working and the progress made. Fault indication can show the reason in the event of a shutdown. Also called centralized control. (Nelson, 1965) (b) A term applied to a switch, circuit breaker, starter, or similar apparatus, to denote that its operation can be controlled manually, from a distance, by electrical or other means. (BS, 1965)

remote control support system A self-advancing support system in which the chocks and/or props are advanced and reset on a longwall face from a point in the gate road leading to the face. Hydraulic pressure and valves are commonly employed and the system is largely in the experimental stage. See also: *pushbutton coal mining.* (Nelson, 1965)

remote sensing A branch of geophysics that acquires and interprets airborne or satellite images of the surface using infrared and visible wavelengths of light.

renardite An orthorhombic mineral, $Pb(UO_2)_4(PO_4)_2(OH)_4 \cdot 7H_2O$ that may be the mineral dewindtite; at Kasolo, Zaire.

rending The breaking of coal into lumps with a minimum of smalls. The relative slowness of low explosives makes them suitable for rending coal since they lack the greater shattering power of high explosives. (Mason, 1951)

rendrock A dynamite used in blasting and consisting of nitroglycerin, potassium nitrate, wood pulp, and paraffin or pitch. (Webster 3rd, 1966)

renierite A tetragonal mineral, $(Cu,Zn)_{11}(Ge,As)_2Fe_4S_{16}$; pseudocubic; a possible source of germanium. (Not reinerite.)

reniform Kidney-shaped. Said of a crystal structure in which radiating crystals terminate in rounded masses; also said of mineral deposits having a surface of rounded, kidneylike shapes. Cf: *colloform; colloid minerals; botryoidal.* (AGI, 1987)

Renn-Walz process A method of reclaiming iron and other metals from the waste materials produced in the smelting of zinc and lead ores. The process differs from the Krupp-Renn method in that it is a volatilization process for recovering molten metals in oxide form. The metal vapors are oxidized by excess air and carried off in the flue gases from which they are subsequently filtered. (Osborne, 1956)

rensselaerite A compact fibrous variety of talc pseudomorphous after pyroxene; harder than talc; polishes well; made into ornamental objects; in northern New York and Canada.

rent and royalty (a) The amount paid by a coal mining operator to the owner of the coal for each ton of coal mined and usually expressed in cents per ton. (b) In mining leases, words used interchangeably to convey the same meaning. (Ricketts, 1943)

reopening sealed area There are four methods used in reopening sealed-off areas in a mine: (1) the direct method in which the stoppings are breached and air is circulated around the district without previous inspection by a rescue team; (2) the prior-inspection method in which prior inspection of the whole district by a rescue team is followed by circulation of air around the district; (3) the stage method in which the ventilation is restored and the enclosed gases are removed in successive stages; and (4) the partial-reopening method which is adopted when it is required to recover part of a district but leave the remainder sealed off. (Sinclair, 1958)

repairman A worker whose duty it is to repair tracks, doors, brattices, or to reset timbers, etc., under the direction of a foreman. Also called repairer. (Zern, 1928)

repeated twinning Crystal twinning involving more than two individuals. Syn: *multiple twin.* Cf: *polysynthetic twinning; twin laminae; cyclic twinning.*

replaceable hydrogen Hydrogen atoms in acid molecule that can be replaced by those of metal. (Pryor, 1963)

replaceable insert Diamond inset plates and other geometric forms fastened to and/or supported by the bit blank by brazing or mechanical locking so that in drilling they may be replaced when diamond wear exceeds a specified amount. (Long, 1960)

replaceable pilot A central interchangeable pluglike portion of a noncoring bit protruding or leading the outside portion of such bits. See also: *pilot.* (Long, 1960)

replacement (a) Change in composition of a mineral or mineral aggregate, presumably accomplished by diffusion of new material in and old material out without breakdown of the solid state. (AGI, 1987) (b) A process of fossilization involving substitution of inorganic matter for the original organic constituents of an organism. (AGI, 1987)

replacement bit *reset bit.*

replacement deposit A mineral deposit that has been formed by deposition from mineral solutions taking the place of some earlier, different substance.

replacing switch A device consisting of a united pair of iron plates hinged to shoes fitting over the rails to replace, on the track, derailed railway rolling stock. Also used for mine cars. (Fay, 1920)

replica A filling of mineral material deposited by percolating ground waters in external molds, thus reproducing the original exterior of the fossil shell or other object, with its exact size and shape. (AGI, 1987)

replicate sampling Taking each sample in a number of subsamples by putting increments in turn into different containers, in order to estimate the sampling accuracy. The same total weight of sample is collected whether or not replicate sampling is employed. (BS, 1960)

repose angle The angle between the horizontal and the surface slope of any pile of material formed by free fall of the material.

representation work Assessment work on a mining claim.

representative fraction The scale of a map, expressed in the form of a numerical fraction that relates linear distances on the map to the corresponding actual distances on the ground, measured in the same units (centimeters, inches, feet); e.g., 1/24,000 indicates that one unit on the map represents 24,000 equivalent units on the ground. Abbrev: RF. (AGI, 1987)

representative sample In testing or valuation of a mineral deposit, samples large enough and average in composition as to be considered representative of a specified volume of the surrounding orebody. Blended large samples from different exposures are not necessarily representative, since the mineral structure may have varied so as to introduce special problems from area to area in treatment. (Pryor, 1963)

rerailer A small lightweight device, used in pairs that straddle and are locked to each of the rails to retrack railroad cars and locomotives. Of Y-shaped design, they permit both wheels to be retracked from either or both sides of the rail at the same time. As the car is pulled across the device, the derailed wheels are channeled back onto the tracks. Also called retracker. See also: *frog*. (Best, 1966)

rescue To move live workers or dead bodies from a mine after a mine disaster. Sometimes called recover. The latter applies esp. to putting the mine in shape for operation again.

rescue apparatus A name applied to certain types of apparatus worn by workers, permitting them to work in noxious or irrespirable atmospheres such as obtained during mine fires, following mine explosions, as a result of accidents in ammonia plants, from smelter fumes, etc. Oxygen compressed in a cylinder, a regenerating substance to purify the breathed air, and a closed system constitute the general principle of the apparatus. See also: *mine rescue apparatus.* (Fay, 1920)

rescue-car *mine rescue car.*

rescue station Mine chamber equipped with rescue gear, including oxygen apparatus, and manned by trained rescue workers. (Pryor, 1963)

rescue team A team of workers, from five to eight strong, trained in the use of breathing apparatus and in rescue operations after colliery explosions or mine fires. The team trains every week or so at a rescue station. (Nelson, 1965)

research Word often misused. Two broad meanings are reexamination of previously accepted data in the light of current expansion of basic knowledge; and search in reality, specific to an entirely novel concept and calling for development of new approaches. Wrongly defined when descriptive of original rehash. (Pryor, 1963)

resection (a) A method in surveying by which the horizontal position of an occupied point is determined by drawing lines from the point to two or more points of known position. Syn: *intersection.* (AGI, 1987) (b) A method of determining a plane-table position by orienting along a previously drawn foresight line and drawing one or more rays through the foresight from previously located stations. (AGI, 1987)

resequent fault-line scarp A fault-line scarp in which the structurally downthrown block is also topographically lower than the upthrown block. Cf: *obsequent fault-line scarp*. (AGI, 1987)

reserve (a) The quantity of mineral that is calculated to lie within given boundaries. It is described as total (or gross), workable, or probable working, depending on the application of certain arbitrary limits in respect of deposit thickness, depth, quality, geological conditions, and contemporary economic factors. Proved, probable, and possible reserves are other terms used in general mining practice. (BS, 1963) (b) Sampled ore, developed, blocked out, or exposed on not less than three sides. See also: *development sampling; ore reserve; probable ore; workable.* (CTD, 1958) (c) The amount of payable ore, developed and ready for extraction, or blocked out ahead of immediate requirements. (Weed, 1922)

reserve base That part of an identified resource that meets specified minimum physical and chemical criteria related to current mining and production practices, including those for grade, quality, thickness, and depth. The reserve base is the in-place demonstrated (measured plus indicated) resource from which reserves are estimated. It may encompass those parts of the resources that have a reasonable potential for becoming economically available within planning horizons beyond those that assume proven technology and current economics. The reserve base includes those resources that are currently economic (reserves), marginally economic (marginal reserves), and some of those that are currently subeconomic (subeconomic resources). The term geologic reserve has been applied by others generally to the reserve-base category, but it also may include the inferred-reserve-base category; geologic reserve is not part of this classification system. (USGS, 1980)

reserved coal Coal reserved from lease, as coal under buildings.

reserved lands Defined by the U.S. Department of the Interior as "federal lands which are dedicated or set aside for a specific public purpose or program and which are, therefore, generally not subject to disposition under the operation of all the public land laws.". (SME, 1992)

reserved mineral Economic minerals that are not the property of the landowner but belong to the State. The State confers the right to prospect for and to mine these minerals on any one who applies for this right on the form prescribed and at the competent mining office. Such minerals as coal and iron ores are included in this group. Cf: *unreserved mineral.* (Stoces, 1954)

reserves (a) An estimate within specified accuracy limits of the valuable metal or mineral content of known deposits that may be produced under current economic conditions and with present technology. (Shanz, 1983) (b) That part of the reserve base that could be economically extracted or produced at the time of determination. The term reserves need not signify that extraction facilities are in place and operative. Reserves include only recoverable materials; thus, terms such as extractable reserves and recoverable reserves are redundant and are not a part of this classification system. Syn: *mineral reserves.* See also: *measured resources.* (USGS, 1980)

reset action (nonstandard) In flotation, the component of control action in which the final control element is moved at a speed proportional to the extent of proportional position action. This term applies only to a multiple action including proportional position action. (Fuerstenau, 1962)

reset bit A bit made by reusing the sound diamonds salvaged from a used drill bit and setting them in the crown attached to a new bit blank. Some new diamonds usually are added to those salvaged, since generally not all of the salvaged or recovered stones are reusable. Syn: *replacement bit.* (Long, 1960)

resettable A salvaged diamond or used diamonds in good condition; hence, diamonds that can be used again by being reset in another tool or bit. Also called usable diamond; usables; usable stone. (Long, 1960)

resetting (a) The act or process of producing a reset bit. See also: *reset bit.* (Long, 1960) (b) To rerun a casing string into a borehole by placing its bottom end at a lower point in the hole. (Long, 1960)

residual (a) Characteristic of, pertaining to, or consisting of residuum; remaining essentially in place after all but the least soluble constituents have been removed. (b) Standing, as a remnant of a formerly greater mass of rock or area of land, above a surrounding area that has been generally planed; said of some rocks, hills, mesas, and groups of such features.

residual arkose *grus.*

residual boulder A boulder of local origin produced by weathering and standing in locally derived soil or grus.

residual clay Clay material formed in place by the weathering of rock, derived either from the chemical decay of feldspar and other rock minerals or from the removal of nonclay-mineral constituents by solution from a clay-bearing rock (such as an argillaceous limestone); a soil or a product of the soil-forming processes. Cf: *primary clay; secondary clay.* (AGI, 1987)

residual element An element present in an alloy or other material in small quantities after some type of treatment, but not added intentionally. (ASM, 1961)

residual errors The differences between measured values and the most probable value. (Seelye, 1951)

residual field (a) *residual magnetic field.* (ASM, 1961) (b) The field remaining after subtraction of an average or background field (e.g., gravity, magnetic).

residual gravity In gravity prospecting, the portion of a gravity effect remaining after removal of some type of regional variation; usually the relatively small or local anomaly components of the total or observed gravity field. (AGI, 1987)

residual liquid A late-stage magmatic fluid. Syn: *rest magma.*

residual magnetic field (a) The magnetic field that remains in a part after the magnetizing force is removed. (ASM, 1961) (b) *residual field.*

residual magnetism (a) In magnetic prospecting, the portion of a magnetic effect remaining after removal of some type of regional effect; usually the relatively small or local anomaly components of the total or observed magnetic field. (AGI, 1987) (b) The magnetism remaining in a substance after the magnetizing force has been removed. See also: *remanence.* (CTD, 1958)

residual minerals The rock-forming minerals that are either stable in the surface environment or unstable but react so slowly that they are not appreciably broken down. (Hawkes, 1957)

residual oil (a) The amount of liquid petroleum remaining in a formation at the end of a specified production process. (AGI, 1987) (b) Liquid or semi-liquid products obtained as residues from the distillation of petroleum. They contain the asphaltic hydrocarbons. Residual oils are also known as asphaltum oil, liquid asphalt, black oil, petroleum tailings, and residuum. (CCD, 1961)

residual ore deposit An accumulation of valuable minerals, formed by the natural removal of undesired constituents of rocks or conversion of useless to useful components.

residual placer *residual ore deposit.*

residuals The elements ordinarily present in steel in small quantities without definite intent on the part of the steel maker. (Osborne, 1956)

residual stress The stress that exists in an elastic solid body in the absence of, or in addition to, stresses caused by an external load. Such residual stress may be due to: (1) deformation, caused by cold-working, as in drawing or stamping; (2) change in the specific volume due to thermal expansion, a phase change, or magnetostriction; (3) by the joining together of structural parts by force, such as welding. (Hammond, 1965)

residue (a) The waste or final product from a hydrometallurgical plant. (Fay, 1920) (b) S. Afr. The amount of valuable matter remaining in ore after treatment, in percent or pennyweights per ton. (Beerman, 1963) (c) As applied to proximate analysis of coke, a calculated figure obtained by subtracting the sum of the percentages of moisture in the analysis sample, volatile matter, and ash from 100. (BS, 1961) (d) That which remains after a part has been separated or otherwise treated. (e) *rock fracture.*

residuite (a) The constituent petrological unit or maceral occurring as characteristic unresolvable granular and translucent groundmass in clarain. (AGI, 1987) (b) residuum.:

residuum (a) Weathered material, including the soil, down to fresh, unweathered rock. (Legrand, 1960) (b) Material resulting from the decomposition of rocks in place and consisting of the nearly insoluble material left after all the more readily soluble constituents of the rocks have been removed. (c) The structureless groundmass of microscopically unresolvable constituents, consisting of particles of one to two microns or less, usually opaque, and of a dark color. It is the same as the lower range of fine micrinite. See also: *residue.* (AGI, 1987)

resilience The ability of a material to store the energy of elastic strain. This ability is measured in terms of energy per unit volume. (AGI, 1987)

resilient couplings The resilient type of coupling has many designs but essentially has torsional response to application or variation of the transmitted load. For the all-metal types, the resilient element may be in the form of laminated spring packs or a cylindrical grid member, connecting the driver and driven hubs. Resilience damps shock loads and also provides means of keeping gear teeth in contact, compensating for small errors in gear cutting. Other types use rubber or rubberlike material that may be in the form of a spider, segmental blocks, a number of balls, or a molded disk with metal inserts, providing the connection between the driver and driven hubs. (Pit and Quarry, 1960)

resiliometer A device for testing resilience. (Standard, 1964)

resin (a) One of various hard, brittle, transparent or translucent solids formed esp. from plant secretions and obtained as exudates of recent or fossil origin, such as conifers and certain tropical trees, by condensation of fluids on loss of volatile oils. Resins are yellowish to brown with resinous luster; fusible and flammable; soluble in ether and other organic solvents, but not in water; and represent a complex mixture of terpenes, resin alcohols, and resin acids and their esters. Cf: *amber; fossil resin.* See also: *mineral resin.* Syn: *natural resin.* (b) A synthetic addition or condensation polymerization substance or natural substance of high molecular weight, which under heat, pressure, or chemical treatment becomes moldable. See also: *bead; beads.* (Jessop, 1992; 5971, 1962)

resin-anchored bolts A passive roof-bolting technique in which a rebar-type bolt is anchored in resin. A two-part resin cartridge is placed at the back of a hole and is mixed as the bolt is inserted and rotated. The bolt is forced tight against the roof until the resin sets.

resin-in-pulp An ion exchange process applied in acid-leach slurry from which abrasive particles of sand have been removed. Abbrev., R.I.P. (Pryor, 1963)

resin-in-pulp (RIP) process The method in which pulp is classified to remove the sands, and the resin adsorbs the metal directly from the slime pulp without the necessity of thickening or filtering. It is esp. adapted for ores that do not settle readily, and where thickening and filtration are difficult. (Newton, 1959)

resinite A maceral of coal within the exinite group, consisting of resinous compounds, often in elliptical or spindle-shaped bodies representing cell-filling matter or resin rodlets. Cf: *cutinite; sporinite.* (AGI, 1987)

resinite coal This coal consists of more than 50% of small resin bodies embedded in gelito-collinite, fusinito-collinite, or in collinite of fusinitic nature. The resin bodies differ in shape and may be angular, spheroidal, or lenticular. Varying in size, they may be visible to the unaided eye in a hand specimen of coal or only distinguishable under the microscope. Resinite coal may also contain small quantities of microspores, fine fragments of fusinized tissue, and, not infrequently, broad streaks of vitrinite. Hand specimens of resinite coal are matt or semimatt and in coals of low rank are brown or brownish-black. On fractures perpendicular to the bedding, the resin bodies appear rounded, black, and lustrous; in the bedding planes themselves they frequently appear as matt rodlets. Resinite coals frequently are high in ash. (IHCP, 1963)

resin jack *sphalerite.* Also spelled rosin jack.

resinoid A coal constituent similar to material derived from resin. (AGI, 1987)

resinous (a) Resembling resin, as opal, and some yellow varieties of sphalerite. (Fay, 1920) (b) The luster on fractured surfaces of minerals, e.g., opal, sulfur, amber, and sphalerite, and rocks, e.g., pitchstone. Cf: *vitreous.*

resinous coal Coal in which the attritus may contain a large proportion of resinous matter. Coals of this type are found more often among the younger coals. (AGI, 1987)

resin rodlets A fossil resinous secretion that may be isolated from coal. It was presumably deposited in a resin duct by a secretory epithelium. (AGI, 1987)

resin tin *cassiterite.* Also spelled rosin tin.

resistance (a) When an air current flows through a mine it meets with frictional resistance from the roof, sides, and floor. The amount of this resistance depends upon the extent and nature of the rubbing surface, the area of the airways, and the velocity of the air. See also: *Atkinson.* (Nelson, 1965) (b) In flotation, a property opposing movement of material or flow of energy, and involving loss of potential (voltage, temperature, pressure, and level). (Fuerstenau, 1962) (c) The property of an electrical circuit that opposes the flow of a current and is measured in ohms. Syn: *thermal resistance.* (Mason, 1951)

resistance methanometer A version of the catalytic methanometer with the addition of improved detector elements. Platinum may be used as the filament that both heats the detecting element and acts simultaneously as a resistance type thermometer. Gas is drawn through the instrument by a rubber suction bulb, and the filaments are heated from a dry battery of the mercury type contained in the apparatus. Readings of methane concentration can be taken on the built-in electrical meter. (Nelson, 1965)

resistance of detonator As applied to electric blasting caps, the total resistance of the leg wires and the bridge wire. (Fraenkel, 1953)

resistance strain gage *electrical resistance strain gage.*

resistance to blasting Specific value of the resistance of the rock to the explosive force, determined by trial blasting. It is a function of maximum burden, hole depth, quantity of explosive (degree of packing), and throw. (Fraenkel, 1953)

resistivity (a) Resistance, R, of a block of specified material in terms of units of length 1 and cross section a. Unit volume is 1 cm^3 of the material concerned, and the resistivity measurement is made during electrical prospecting. Specific resistance = (Ra) ÷ 1. (Pryor, 1963) (b) The electrical resistance offered to the passage of a current. Usually expressed in ohm meters, which is the electrical resistance of a column of fluid 1 m long and 1 m^2 in cross section. (Brantly, 1961) (c) The opposite of conductivity of an electrical current passing through fluid-bearing rock formations. (Wheeler, 1958) (d) The electrical resistance between opposite faces of a 1-cm cube of a given substance. The unit of resistivity is ohm/centimeter. (Hy, 1965) (e) The reciprocal of conductivity. Syn: *thermal resistivity.* (Strock, 1948)

resistivity method Any electrical exploration method in which current is introduced into the ground by two contact electrodes and potential differences are measured between two or more other electrodes. (AGI, 1987)

resistivity profile (a) A geophysical survey using the resistivity method. An assembly of electrodes spaced at a constant distance is moved along profiles, resulting in lateral variations in resistivity being shown. In favorable terrain, the test shows the existence of faults that have thrown strata of different resistivity against each other; similar relationships result in the detection of an anticline, a syncline, or an underground channel. (Nelson, 1965) (b) A survey by the resistivity method in which an array of electrodes is moved along profiles to determine lateral variations in resistivity. (AGI, 1987)

resistor A device to provide resistance in an electric circuit, usually to limit the current, dissipate energy, or provide heat. (Kentucky, 1952)

resoiling The replacement of the original topsoil at an opencast site on completion of operations to allow the growing of crops. See also: *surface reinstatement.* (Nelson, 1965)

resolution (a) A measure of the ability of individual components, and of remote-sensing systems, to distinguish detail or to define closely spaced targets. (AGI, 1987) (b) The minimum size of a feature that can be detected. See also: *resolving power.* (AGI, 1987) (c) The separation of a vector into its components. (AGI, 1987) (d) The sharpness with which the images of two closely adjacent spectrum lines, etc., may be distinguished. (AGI, 1987) (e) In gravity or magnetic prospecting, the indication in some measured quantity, such as the vertical component of gravity, of the presence of two or more close but separate disturbing bodies. (AGI, 1987) (f) In seismic prospecting, the ability to indicate separately two closely adjacent interfaces. (AGI, 1987) (g) The ability of an optical or radiation system to separate closely related forms or entities; also, the degree to which they can be discriminated. (ASM, 1961)

resolution limit In gravity and magnetic prospecting, the separation of two disturbing bodies at which some obvious indication in a measured quantity of the presence of two separate bodies ceases to be visible. (AGI, 1987)

resolved-time method A seismic reflection technique that involves the plotting of reflections in time and the representation of horizontal distances along the section in equivalent time units (obtained by dividing the true horizontal distance by the sub-weathering velocity as determined from first-arrival times). Once this transformation of the coordinate system is made, migration is accomplished by swinging arcs of reflection times from successive shot points and drawing lines which are tangent to the respective arcs for the same events from adjacent shot points. For the final mapping of migrated horizons in depth, the times are recorded directly beneath the shot points. These times are converted to depths by using the best available velocity information. (Dobrin, 1960)

resolving power In optical viewing, the minimum distance possible between two separately distinguishable objects. (Pryor, 1963)

resonance (a) A term denoting a variety of phenomena characterized by the abnormally large response of a system having a natural vibration period to a stimulus of the same, or nearly the same, frequency. (AGI, 1987) (b) A buildup of amplitude in a physical system when the frequency of an applied oscillatory force is close to the natural frequency of the system. (AGI, 1987)

resonance screen A high-speed vibrating screen in which the applied force has a frequency equal to the natural frequency of the suspended mass. In its basic form, the vibrating frame of the resonance screen is a mass oscillating between two compression springs, that alternately store and return this energy. (Nelson, 1965)

resonant frequency drilling Drilling that utilizes longitudinal vibration corresponding to the resonant frequency of the drill string in order to "fluidize" the sediments being sampled, thereby achieving efficient penetration. (Padan, 1968)

resorption border A border of secondary minerals, produced by partial resorption and recrystallization, surrounding an original crystal constituent of a rock. Syn: *corrosion border.*

resource A concentration of naturally occurring solid, liquid, or gaseous material in or on the Earth's crust in such form and amount that economic extraction of a commodity from the concentration is currently or potentially feasible. (USGS, 1980)

resource characterization The determination of the shape, size, quality, quantity, and variability of the geologic entity and the limits of variable geologic features, so as to provide the information for synthesis of commonly subtle features into an accurate, predictive description of the resource environment. (SME, 1992)

respirable-size particulate Particulates in a size range that permits them to penetrate deep into the lungs upon inhalation. (NSC, 1996)

respirator (a) A device (such as a gas mask) for protecting the respiratory tract (against irritating and poisonous gases, fumes, smoke, dusts) with or without equipment supplying oxygen or air. (Webster 3rd, 1966) (b) A device for maintaining artificial respiration. (Webster 3rd, 1966) (c) The mining-type respirator is a fitting that covers the nose and mouth to prevent the wearer inhaling excessive quantities of dust. Tunnel miners and workers at sinter plants and blast furnaces are issued respirators for use where danger is known to exist. See also: *filter-type respirator; mask.* (Nelson, 1965) (d) A device worn over the mouth or nose for protecting the respiratory tract from noxious gases or dust.

respirator protection factor (a) A measure of the degree of protection provided by a respirator to the wearer. (FR 166, 1989) (b) The ratio of the ambient concentration of an airborne substance to the concentration of the substance inside the respirator at the breathing zone of the wearer, a measure of the degree of protection provided by a respirator to the wearer. (ANSI, 1992)

respiratory cycle One complete breath—an inspiration followed by an expiration, including any pause that may occur between the movements. (Hunt, 1965)

resplendent Referring to a degree of luster that reflects with brilliancy and gives well defined images; e.g., hematite and cassiterite.

rest magma *residual liquid.*

restoration (a) Restoring the disturbed land to the conditions which existed at the site before any disturbance occurred. (SME, 1992) (b) The process of gaining or recovering land, bringing it into a condition for cultivation or other use. (SME, 1992) (c) Response to any disturbances to the Earth and its environment caused by mining activity. (SME, 1992) (d) Returning the disturbed site "to a form and productivity in conformity with a prior use plan.". (SME, 1992) (e) The Surface Mining Control and Reclamation Act of 1977 (SMCRA) states that, among other provisions, reclamation must "restore the land affected to a condition capable of supporting the uses which it was capable of supporting prior to any mining, or higher or better uses.". (SME, 1992)

restore circulation The action taken to fill or seal the cracks or openings through which drill fluid is escaping from the borehole into the rocks forming the walls of the borehole and by which the drill fluid is made to return to and overflow the collar of the borehole. (Long, 1960)

restrained cable plug and socket (a) A flameproof restrained plug and socket incorporates an interlock to ensure that the power connections are dead when they are separated or until they make contact; the design is such that the enclosure is flameproof at all times when there is contact between the pins and tubes. (BS, 1965) (b) A plug and socket designed to be held together by an operating bolt, or screwed union ring, or other equivalent device, the use of which enables the plug to be readily inserted or withdrawn. (BS, 1965)

restrained plug and socket These are used when the cable is removed from a machine or apparatus frequently. The most common type is the 100-amp British Standard plug and socket, and it is employed to connect the trailing cable to a coal cutter or face conveyor. The gland of the plug is arranged to grip the sheath of the cable and to make connection with the screen and earth core. Power and pilot conductors are connected to the appropriate contact tubes, which make connection with corresponding pins in the socket portion. (Mason, 1951)

restricted earth fault protection As used in mining, a system of earth fault protection in which the fault current is limited, without requiring the use of sensitive earth fault protection. (BS, 1965)

restricted resources That part of any resource category that is restricted from extraction by laws or regulations, but otherwise meets all the requirements of reserves. (USGS, 1980)

rests The arrangement at the top and bottom of a shaft, or intermediate levels, for supporting the shaft cage while changing the tubs or cars. See also: *chair; catch; wing.*

resue (a) To mine or strip sufficient barren rock to expose a narrow but rich vein, which is then extracted in a clean condition. (Nelson, 1965) (b) To open up a stope, not in the vein but in the wall rock. See also: *resuing.* (c) In lode mining, separate removal of undercut barren rock immediately below a lode or vein too narrow for human entry. Following this, the lode is mined and separately removed. Used when the lode is less than 30 in (76 cm) wide. (Pryor, 1963)

resuing (a) A method of stoping wherein the wall rock on one side of the vein is removed before the ore is broken. Employed on narrow veins, less than 30 in (76 cm), and yields cleaner ore than when wall and ore are broken together. (b) A method of stoping in which the ore is broken down first and then the waste or vice versa; usually the one which breaks easier is blasted first. The broken waste is left in the stope as filling, and the ore is broken down on flooring laid on the fill to prevent admixture of ore and waste. Resuing is applicable where the ore is not frozen to the walls and works best if there is considerable difference between the hardness of the ore and of the wall rocks.

resurgence *emergence.*

retaining mesh In sieving or screening, that mesh at which division is made between oversize (arrested on screen) and undersize (passing through meshes). (Pryor, 1963)

retaining ring (a) In drilling, a shoulder inside a reaming shell that prevents entry of the core lifter into the core barrel. (Long, 1960) (b) A term sometimes incorrectly applied to a core lifter. (Long, 1960)

retaining screen The screen that has retained the particles. (Pit and Quarry, 1960)

retaining structure A temporary or permanent structure used for holding dredged material on a limited basis, not to be confused with a confined disposal facility.

retardation In crystal optics, the amount by which the slow wave falls behind the fast wave during passage through an anisotropic crystal plate. Retardation depends on plate thickness and the difference in refractive indices of its two principal directions. (AGI, 1987)

retarding conveyor (a) A chain-type conveyor used on steeply inclined faces, where the problem is not so much to move the coal but rather to restrain its movement downhill. It consists of link chains carrying discs 6 to 8 in (15 to 20 cm) in diameter at every yard (0.9 m). The endless chain runs in an open semicircular trough, and the coal is lowered to the discharge end. The chain returns uphill, in an enclosed tube, to the driving unit at the top end. Its capacity is about 100 tons per hour. (Nelson, 1965) (b) Any type of conveyor used to retard the rate of movement of bulk materials, packages, or objects, where the slope is such that the conveyed material tends to propel the conveying medium. See also: *declining conveyor.*

retentivity The capacity of a material to retain a portion of the magnetic field set up in it after the magnetizing force has been removed. (ASM, 1961)

retgersite A tetragonal mineral, $NiSO_4{\cdot}6H_2O$; dimorphous with nickelhexahydrite; blue green; associated with morenosite, the septehydrate.

Retger's salt Thallium silver nitrate that melts to a yellow liquid at 75 °C having a density of 4.6 g/cm^3; can be diluted and used as a heavy liquid for mineral separation.

reticular *reticulate.*

reticulate (a) Said of a vein or lode with netlike texture; e.g., stockwork. Cf: *stockwork.* (AGI, 1987) (b) Said of a rock texture in which crystals are partially altered to a secondary mineral, forming a network that encloses remnants of the original mineral. Cf: *mesh texture.* See also: *reticulated.* Syn: *reticular.* (AGI, 1987)

reticulated A mineral structure of fibers or columns that cross to resemble a net; e.g., rutile. See also: *reticulate.*

reticulated veins Veins that cross each other, forming a network. See also: *stockwork.*

reticulate texture *mesh texture.*

reticule A set of intersecting very fine lines, wires, etc., in the optical focus of an optical instrument. It is also referred to as graticule. See also: *collimation line*. (Hammond, 1965)

reticulite An extremely attenuate pyroclastic rock consisting of glass threads which join a series of points forming a polyhedral space lattice. It is formed from pumice by the collapse of the walls of adjacent vesicles and the retraction of the liquid into threads which form the perimeters of the former polygonal faces. The threads are usually of triangular cross section, indicating chilling, before rounding could take place. Such rock has generally been known by Dana's name, thread-lace scoria. See also: *thread-lace scoria.* (AGI, 1987)

retiform Netted; reticulate; said of the boundaries of some vein quartz (rare). (Hess)

retigen Bitumen contained in meteorites. The name indicates that this substance on distillation gives rise to resin, in contrast to kerogen which on distillation gives rise to oil. (Tomkeieff, 1954)

retinalite A massive, honey-yellow or greenish variety of serpentine with a waxy or resinous luster.

retinasphalt A light brown resinous substance found in brown coal in Devonshire, England. (Tomkeieff, 1954)

retinite A variety of fossil resin found as rodlets secreted in canals or ducts of coal-forming plants.

retinosite A microscopical constituent of torbanite consisting of translucent orange-red discs. Cf: *gélosite; humosite; matrosite.* (Tomkeieff, 1954)

retonation wave A wave passing back through burned or burning explosion gases toward the origin, at the rate of a sound wave through gases of like temperature, from a point in the explosion wave, usually of high pressure, to an area of lower pressure. (Rice, 1913-18)

retort (a) A vessel used for the distillation of volatile materials, as in the separation of some metals and the destructive distillation of coal. (ASM, 1961) (b) A long semicylinder, now usually of fireclay or silica, for the manufacture of coal gas. (Webster 3rd, 1966) (c) *amalgam retort.*

retorting (a) Removing the mercury from an amalgam by volatizing it in an iron retort, conducting it away, and condensing it. (Fay, 1920) (b) In the sulfur industry, synonymous with sublimation. (Fay, 1920)

retort pressman A person who operates a hydraulic press in which fireclay retorts, used in smelting zinc ores, are made. (DOT, 1949)

retract The mechanism by which a dipper shovel bucket is pulled back out of the digging. (Nichols, 1976)

retractable wedge A type of deflecting wedge that can be retrieved after the deflected drill hole has been completed. Syn: *retrievable wedge.* (Long, 1960)

retracting *crowding.*

retreat To work rooms or pillars to finish coal or ore extraction in an area that has been penetrated to its limits by advance work; workings are generally in the opposite direction of advance work and allow the area to be abandoned as finished. (BCI, 1947)

retreating longwall (a) First driving haulage road and airways to the boundary of a tract of coal and then mining it in a single face without pillars back toward the shaft. (Fay, 1920) (b) *longwall retreating.*

retreating system (a) A method of working a mine that is designed to allow a stope to cave soon after it is worked out, thus relieving the weight on the supports in adjacent stopes. (Lewis, 1964) (b) A method of extracting coal or ore by driving a narrow heading to the boundary, then opening out a face and working the deposit backwards towards the shaft, drift, or main entry. See also: *longwall retreating.* (Nelson, 1965) (c) A stoping system in which supporting pillars of ore are left while deposit is worked outward from shafts toward the boundary, the pillars being removed (robbed) as the work retreats toward the shaft; the unsupported workings are abandoned and left to cave in. (Pryor, 1963) (d) A system of robbing pillars in which the line of pillars being robbed retreats or moves from the boundary toward the shaft or mouth of the mine. See also: *longwall retreating.*

retrievable inner barrel The inner barrel assembly of a wire-line core barrel, designed for removing core from a borehole without pulling the rods. (Long, 1960)

retrievable wedge *retractable wedge.*

retrieving ring A catch ring on a retractable wedge that engages a lifting device on the deflection barrel or bit, enabling the drill runner to remove a deflecting wedge from a borehole after deflection has been effected. (Long, 1960)

retrograde metamorphism The mineralogical adjustment of relatively high-grade metamorphic rocks to temperatures lower than those of their initial metamorphism, characteristically inducing hydration and hydrous minerals. Syn: *retrogressive metamorphism; diaphthoresis.* Cf: *prograde metamorphism.*

retrogressive metamorphism *retrograde metamorphism.*

return (a) Any airway in which vapid air flows from the workings to the upcast shaft or fan. See also: *intake.* (Nelson, 1965) (b) Any airway which carries the ventilating air from the face outby and out of the mine. (BS, 1963) (c) Any surface turned back from the face of a principal surface. (ACSG, 1961) (d) The rate of profit in a process of production per unit of cost. (Webster 3rd, 1966)

return air (a) Air traveling in a return. (BS, 1963) (b) Air that has circulated the workings and is flowing towards the main mine fan; vitiated or foul air. (Nelson, 1965) (c) Air returning to a heater or conditioner from the heated or conditioned space. (Strock, 1948)

return aircourse Portion of ventilation system of mine through which contaminated air is withdrawn and evacuated to surface. (Pryor, 1963)

return circulation That portion of a circulated drill fluid flowing from the face of a bit toward the collar of a borehole. Cf: *return water.*

returning charge Charge made per unit of ore or concentrate treated by smelter in custom smelting. In addition to a basic charge that allows for process costs and agreed percentage loss in recovery, extra charges may be specified, or remitted as premiums, in adjustment of variations from the normal makeup of the parcel treated. (Pryor, 1963)

returning fluid The water, mud, or other circulated medium reaching the borehole collar after having been circulated past the drill bit. (Long, 1960)

return-line corrosion tester A tester developed by the U.S. Bureau of Mines for detecting and controlling corrosion in steam-condensate-return lines of large heating plants. This tester determines types and rates of corrosion and can distinguish among various possible causes. It is assembled from ordinary black iron pipe nipples and couplings, the linings are easily machined, and the corroded linings can be analyzed quickly in any laboratory.

return man In anthracite coal mining, one who resets timbers, shovels up falls of slate, rock, or dirt, and keeps in general repair the airways by which mine air returns to the surface. (DOT, 1949)

returns (a) The drill fluid and entrained sludge that overflows the collar of a borehole. (Long, 1960) (b) In seismic reflection prospecting, the signals reflected back to the surface from layer boundaries in the subsurface. (c) Also used in geophysical prospecting to register passage of waves caused by detonation of dynamite. (Hess)

return water Drill fluid that reaches the surface and overflows the borehole collar after it has been circulated downward through the rods and past the drill bit. (Long, 1960)

retzian An orthorhombic mineral, $(Mn,Mg)_2(Ce,La,Nd)(AsO_4)(OH)_4$; speciated on the basis of predominance of cerium, lanthanum, or neodymium; in dolomite cavities in Sweden.

reussin An impure Glauber's salt (mirabilite), found native. (Standard, 1964)

reussinite A resinlike, reddish-brown oxygenated hydrocarbon, soluble in boiling alcohol and in ether. Found in certain coal deposits.

revdanskite An impure, hydrous nickel silicate from Revda (Revdinsk), Ural Mountains, Russia. Also spelled revdinite, revdinskite, rewdinskit, rewdanskite; rewdjanskit, and refdanskite. (English, 1938; Hey, 1955)

revegetation The process of restoring or replacing the botanical species upon an area disturbed by mineral operations. Revegetation is a customary requirement for reclamation of a mineral operation. (SME, 1992)

reverberate (a) To deflect flame or heat, as in a reverberatory furnace. (Fay, 1920) (b) To reduce by reverberated heat; to fuse. (Fay, 1920)

reverberation (a) The persistence of sound in an enclosed space as a result of multiple reflections after the sound source has stopped. (Hunt, 1965) (b) The sound that persists in an enclosed space, as a result of repeated reflection or scattering, after the source of the sound has stopped. (Hunt, 1965)

reverberatory furnace A furnace, with a shallow hearth, usually nonregenerative, having a roof that deflects the flame and radiates heat toward the hearth or the surface of the charge. Firing may be with coal, pulverized coal, oil, or gas. Two of the most important types are the open-hearth steel furnaces and the large reverberatories employed in copper smelting. (ASM, 1961; CTD, 1958; Newton, 1959)

reversal A local change of approx. 180° in the direction of the regional dip.

reversal of ventilation In the case of a centrifugal fan, the reversal arrangement may consist of an emergency drift connecting the fan with the downcast shaft. The drift is normally sealed off by airtight doors. In the case of an axial-flow fan, it is only necessary to reverse the rotation of the fan. This arrangement entails a reduction in volume and pressure in the reversed airflow. (Nelson, 1965)

reverse bearing In surveying, a sight along the reverse direction of a line; the reciprocal of a given bearing. See also: *backsight.*

reverse bend To bend a line over a drum or a sheave, and then in the opposite direction over another sheave. (Nichols, 1976)

reverse book fashion The manner in which drill core is laid in a core box, starting at the upper-right-hand corner of the box and laying core from right to left in each groove. Cf: *snake fashion.* (Long, 1960)

reverse circulation The circulation of bit-coolant and cuttings-removal liquids, drilling fluid, mud, air, or gas down the borehole outside the drill rods and upward inside the drill rods. Also called countercurrent; counterflush. See also: *circulating fluid.* (Long, 1960)

reverse-circulation core barrel A core barrel designed so that core tends to float within the barrel when the fluid is circulated down the outside of the rods and returned to the surface inside the rods. (Long, 1960)

reverse classification In jigging, stratification of particles by size with largest uppermost; in streaming, rolling effect of transporting current that arranges particles with smallest nearest feed end. (Pryor, 1963)

reverse-current braking A method used in the braking of alternating-current winders. This method absorbs power equal to the energy destroyed and dissipates it in the liquid controller as heat. Two phases of the stator supply are interchanged by bringing back the driver's lever to the off position and then to that for the opposite direction of drum rotation. The amount of braking depends upon the position of the lever, since the lower the resistance in the controller, and therefore in the rotor circuit, the greater the rotor current and the braking torque produced. When the direction of rotation of the stator magnetic field is reversed, the voltage between the stator and the rotor is doubled and the insulation of both must be adequate to prevent breakdown. (Sinclair, 1959)

reversed *overturned.*

reversed bratticing A method of narrow heading ventilation in coal mines by means of a brattice partition. The air is led to the face along the wide section of the heading and the contaminated air returns from the face along the narrow section. In this way, workers in the heading are placed in relatively clean air. (Nelson, 1965)

reversed fault *reverse fault.*

reversed flush boring *counterboring.*

reversed loader A front-end loader mounted on a wheel tractor having the driving wheels in front and steering at the rear. (Nichols, 1976)

reverse fault A fault on which the hanging wall appears to have moved upward relative to the footwall. The dip of the fault is usually greater than 45°. There is dip separation but there may or may not be dip slip. Cf: *normal fault.* Partial syn: thrust fault. Syn: *reversed fault.* (AGI, 1987)

reverse feed To move bit and drill stem backwards away from the borehole bottom while the drill stem is rotated. (Long, 1960)

reverse-feed gear System of gears in drill swivel head that can be engaged to move the bit and drill stem backwards away from the bottom of the borehole while the drill stem is rotated in a clockwise direction. Syn: *backup gear; reverse gear.* (Long, 1960)

reverse gear *reverse-feed gear.*

reverse initiation *inverse initiation.*

reverse laid rope A wire rope with alternate strands right and left lay. (Hunt, 1965)

reverse reaming *ream back.*

reversible auxiliary ventilation In this system, a single duct is provided and is normally operated by a blowing fan. After blasting, airflow is reversed and the fumes and dust are exhausted. Ventilation is again reversed to blowing, when the work at the face is resumed. The usual arrangement is to use two fans, one for forcing, one for exhausting, at the mouth of the heading. This arrangement is particularly suited to underground use as it allows clean air to be drawn from, and contaminated air to be discharged to, separate points in the main airways. See also: *auxiliary ventilation; two-fan auxiliary ventilation.*

reversible endless-rope system A haulage system in which a single rope is used passing around a surge wheel. A single track may be used or, if more than one train is hauled, a single track with passbys, or a three-rail system with passbys, that eliminates facing points, may be used. The system may be operated at higher speeds than normal endless systems since the trains are attached and detached from a rope at rest; it has been used for the haulage of workers at speeds up to 12 mph. Extra rope must be spliced onto the rope, and the return wheel moved forward, when the system is extended. (Sinclair, 1959)

reversible pick *double-ended pick.*

reversible transducer *bilateral transducer.*

reversing clutch A forward-and-reversing transmission that is shifted by a pair of friction clutches. (Nichols, 1976)

reversing doors The system of doors or shutters on or near a surface radial-flow fan for reversing the direction of the air passing through a mine. (BS, 1963)

reversing machine A molding machine having a flask or flasks that may be turned over for ramming the sand. (Standard, 1964)

reversing mill A type of rolling mill in which the stock being mechanically worked by rolling passes backwards and forwards between the same pair of rolls, which are reversed between each pass. See also: *continuous mill; three-high mill.* (CTD, 1958)

reversing shaft A shaft whose direction of rotation can be reversed by the use of clutches or brakes. (Nichols, 1976)

reversing thermometer A mercury-in-glass thermometer used to measure temperatures of the sea at depth. The temperature is recorded when the thermometer is inverted; and the recording is maintained until it is once again upright. A protected thermometer and an unprotected thermometer are usually used as a pair, attached to a Nansen bottle. (AGI, 1987)

revetment (a) A facing, sheathing, or retaining wall of masonry or other materials for protecting a mass or bank of earth, etc., as in fortifications and riverbanks. (Standard, 1964) (b) A wall sloped back sharply from its base. (Nichols, 1976)

revolution An obsolete term for a time of profound orogeny and other crustal movements, on a continentwide or even worldwide scale, the assumption being that such revolutions produced abrupt changes in geography, climate, and environment. *orogeny.*

revolving screen A screen consisting of a cylindrical (sometimes conical) screening surface mounted on a revolving frame for sizing coarse material; it is still common in gravel-washing, coal-washing, and stone-treating plants, but is not widely used in ore dressing. See also: *trommel; trommel screen.* (Newton, 1959)

revolving shovel A digging machine that has the machinery deck and attachment on a vertical pivot, so that it can swing independently of its base. (Nichols, 1976)

revolving washing screens The rotary washing screen is cylindrical in shape and made of three sections—a scrubber, a sand jacket, and a gravel-screening section—mounted on a steel frame. (Pit and Quarry, 1960)

rewash To re-treat a product in the same or in another washer. (BS, 1962)

rewash box A washbox to which the product (or a portion thereof) of a previous washing operation is fed for additional treatment. (BS, 1962)

rewdanskite *revdanskite.*

reworked Said of components derived from an older sedimentary formation and incorporated in a younger one.

Reynolds number A numerical quantity used as an index to characterize the type of flow in a hydraulic structure in which resistance to motion depends on the viscosity of the liquid in conjunction with the resisting force of inertia. It is the ratio of inertia forces to viscous forces, and is equal to the product of a characteristic velocity of the system (e.g., the mean, surface, or maximum velocity) and a characteristic linear dimension, such as diameter or depth, divided by the kinematic viscosity of the liquid; all expressed in consistent units in order that the combinations will be dimensionless. The number is chiefly applicable to closed systems of flow, such as pipes or conduits where there is free water surface, or to bodies fully immersed in the fluid so the free surface need not be considered. (AGI, 1987)

Rf value In paper-strip chromatography, ratio of distance moved by component in solution under test to that of transporting solvent. (Pryor, 1963)

rhabdite *schreibersite.*

rhabdomancy A form of dowsing using a rod or twig. Cf: *dowsing.* (AGI, 1987)

rhabdophane (a) A hexagonal mineral, $(Ce,La,Nd)PO_4 \cdot H_2O$; speciated on the basis of predominance of cerium, lanthanum, or neodymium; Also spelled rhabdophanite. (b) The mineral group brockite, grayite, ningyoite, rhabdophane-(Ce), rhabdophane-(La), rhabdophane-(Nd), and tristramite.

rheid (a) A substance below its melting point that deforms by viscous flow during the time of applied stress at an order of magnitude at least three times that of the elastic deformation under similar conditions. (b) A body of rock showing flow structure.

rheid folding Folding accompanied by slippage along shear planes at an angle to the bedding or older foliation.

rheidity The capacity of material to flow within the earth. (AGI, 1987)

Rhenania furnace A combination of the Hasenclever and O'Hara furnaces, with four hearths, and with a combination flue under the lowest hearth and one over the upper hearth. It has mechanical rabbles. (Fay, 1920)

Rhenish furnace A zinc distillation furnace that is a modified type of the Silesian furnace. (Fay, 1920)

rhenium A rare, silvery-white metal. Symbol, Re. Occurs in very small quantities in platinum ores and in columbite, gadolinite, and molybdenite. Used for filaments for mass spectrographs and ion gages; for thermocouples and photoflash lamps. (Handbook of Chem. & Phys., 3)

Rheolaveur washer A washer wherein raw coal and water is fed into the head of an inclined trough equipped with openings in the bottom for the discharge of rejects. There are three types of Rheolaveurs used in coal washing: (1) the sealed discharge type for coarse sizes, from which the reject falls against an upward current of water and is removed by an automatic gate that controls the feed to a drowned elevator; (2) a system of two, three, or four superimposed troughs for washing fine coal below about 1/2 in (1/3 cm). The troughs are equipped with several bottom discharge devices. The separation of the heavy shale from coal and middlings takes place progressively until finally the pure shale is discharged from the lowest trough, and (3) a system for washing slurry consisting usually of two troughs one above the other and equipped with a number of Rheo boxes of the open discharge type but designed to minimize the loss of coal with the fine shale. (Nelson, 1965)

rheology Study of the flowage of materials, particularly plastic flow of solids and flow of non-Newtonian liquids. (AGI, 1987)

rheomorphism The process by which a rock becomes mobile and deforms viscously as a result of at least partial fusion, commonly accomplished, if not promoted, by addition of new material by diffusion.

rheostat (a) An instrument for testing blasting machines by inserting definite resistance equal to a known number of electric blasting caps of a standard-length wire, using one electric blasting cap as an indicator. (b) An instrument by which a variable or an adjustable resistance may be introduced into a circuit to regulate the strength of a current, as in the field coils of a motor or a generator. (Standard, 1964)

rheostat rope A small rope consisting of 8 strands of 7 wires each. (Hunt, 1965)

rhinestone (a) Quartz and other material cut to imitate diamond. (b) Glass backed with a thin leaf of metallic foil to simulate a diamond. (c) Originally a syn. of quartz crystal. (d) Cut colored glass.

rhodesite An orthorhombic mineral, $(Ca,Na_2,K_2)_8Si_{16}O_{40}{\cdot}11H_2O$; fibrous; resembles zeolites; at Bultfontein Mine, Kimberley, South Africa. See also: *mountainite.*

rhodite *rhodium gold.*

rhodium (a) An element of the platinum group, Symbol: Rh. (b) An isometric mineral, RhPt.

rhodium gold Native gold alloyed with rhodium.

rhodochrome Chromian clinochlore, formerly called kämmererite.

rhodochrosite A trigonal mineral, $MnCO_3$; calcite group, with Mn replaced by Fe toward siderite, Ca toward calcite, Mg, Zn, Co, and Cd; rhombohedral cleavage; in hydrothermal veins, residual manganese deposits, and pegmatites; a minor source of manganese. Syn: *manganese spar.* See also: *dialogite; raspberry spar.*

rhodolite A pale pink, rose, or purple to violet variety of pyrope garnet having good transparency; may be of gem quality.

rhodonite A triclinic mineral, $(Mn,Fe,Mg,Ca)SiO_3$; a pyroxenoid; in metasomatic manganese ore deposits; an ornamental stone, esp. in Russia. Syn: *manganese silicate; manganese spar.* See also: *red manganese.*

rhodotilite *inesite.*

rholites A word employed by Wadsworth to designate smelting materials or fluxes. (Fay, 1920)

rhomb *rhombus; rhombohedron.*

rhombarsenite *claudetite.*

rhombenglimmer *biotite.*

rhombic *rhombus; orthorhombic.*

rhombic dodecahedron The isometric form {*hh*0} having twelve faces in the shape of a rhombus; e.g., garnet. Cf: *pyritohedron.*

rhombic mica *phlogopite.*

rhombic quartz An old name for feldspar. (Fay, 1920)

rhombic system (a) In crystallography, same as the orthorhombic system. Syn: *orthorhombic system.* (Fay, 1920) (b) A former name for the orthorhombic system.

rhomboclase An orthorhombic mineral, $HFe(SO_4)_2{\cdot}4H_2O$; forms colorless to gray rhombic plates; in Slovakia.

rhombohedral division In assigning point groups to six crystal systems, those members of the hexagonal system that may be assigned rhombohedral crystallographic axes a_r belong to the rhombohedral division of the hexagonal system. They have a unique triad, but not all point groups with a unique triad may be assigned rhombohedral axes; hence, not all trigonal point groups are rhombohedral. Cf: *trigonal; trigonal system.*

rhombohedral iron ore *siderite.*

rhombohedral system (a) Same as the hexagonal system, except that the forms are referred to three axes parallel to the faces of the fundamental rhombohedron instead of to the usual four axes. (Fay, 1920) (b) The trigonal division of the hexagonal system, the forms being referred to the same three axes as above. Neither usage has been generally accepted. (Fay, 1920)

rhombohedron A parallelepiped with each face a rhombus. Dolomite crystallizes as rhombohedra, and members of the calcite group cleave as rhombohedra. Adj. rhombohedral.

rhomboid A parallelogram that does not have any right angles, and one pair of opposite sides differ in length from the other pair of opposite sides. (Jones, 1949)

rhomb spar *dolomite.*

rhombus A parallelogram that does not have any right angles, but the sides are all equal in length. (Jones, 1949)

rhönite A triclinic mineral, $Ca_2(Fe,Mg,Ti)_6(Si,Al)_6O_{20}$; aenigmatite group; in silica-undersaturated mafic to intermediate rocks commonly as an alteration product of amphiboles; in Germany and the Czech Republic.

rhums Scot. Bituminous shale.

rhyacolite *sanidine.*

rhyodacite The extrusive equivalent of granodiorite. The principal minerals, sodic plagioclase, sanidine, quartz, and biotite or hornblende, commonly occur as phenocrysts in a finely crystalline groundmass of alkali feldspar and quartz. Accessory minerals are apatite and magnetite, and occasionally augite. (AGI, 1987)

rhyolite A group of extrusive igneous rocks, typically porphyritic and commonly exhibiting flow texture, with phenocrysts of quartz and alkali feldspar in a glassy to cryptocrystalline groundmass; also, any rock in that group; the extrusive equivalent of granite. Rhyolite grades into rhyodacite with decreasing alkali feldspar content and into trachyte with a decrease in quartz. The term was coined in 1860 by Baron von Richthofen (grandfather of the World War I aviator). Etymol: Greek rhyo-, from rhyax, stream of lava. See also: *liparite.* Cf: *quartz porphyry.* (AGI, 1987)

rhyolite glass Obsidian.

rhyolite-porphyry A rhyolite in which some grains or crystals are visibly larger than others. (Sinkankas, 1959)

rhythmic crystallization A phenomenon, observed in igneous rocks, in which different minerals crystallize in concentric layers, giving rise to orbicular structure. (AGI, 1987)

rhythmic driving In this type driving, the drilling, loading, and blasting are carried out in one shift and the mucking and transportation in the following one. This enables every worker to specialize in his or her tasks and machines, which in a highly mechanized job is a necessary condition for making the best use of expensive equipment. It also reduces or eliminates the loss of time for ventilation; in rhythmic driving it is carried out between two shifts. (Langefors, 1963)

rhythmic sedimentation A regular interbanding of two or more types of sediment or sedimentary rocks due to a regular change in the conditions of sedimentation, such as alternation of wet and dry periods. See also: *varved clay.* (CTD, 1958)

rhythmite The couplet of distinct types of sedimentary rock, or the graded sequence of sediments, that form a unit bed or lamina in rhythmically bedded deposits. It implies no limit as to thickness of bed, lamina, or complexity, but the term should exclude groups of beds such as cyclothems and carries no time or seasonal connotation. Cf: *cyclothem; varve.* (AGI, 1987)

rib (a) The side of a pillar or the wall of an entry. (BCI, 1947) (b) The solid coal on the side of a gallery or longwall face; a pillar or barrier of coal left for support. (c) The solid ore of a vein; an elongated pillar left to support the hanging wall in working out a vein. (d) A stringer of ore in a lode. (e) The termination of a coal face. Where solid coal is left, the term fast rib, end, or side, is used; and where the coal face ends at the gob, the term used is loose rib, end, or side. (TIME, 1929-30) (f) *buttock.* (g) A hard zone, bed, or horizon within a formation; a silicified zone in a sedimentary stratum. (Long, 1960) (h) A ridge projecting above grade in the floor of a blasted area. (Nichols, 1976) (i) A ridge, paralleling the long axis of a drill string member, that acts as a wear-resistant surface. (Long, 1960)

ribbed roll A crusher in which the material passes between a moving set of rolls with ribs on their surfaces parallel to the axis of the rolls. See also: *roll.* (ACSG, 1963)

ribbing Enlarging a heading or drift.

ribbon (a) One of a set of parallel bands or streaks in a mineral or rock, e.g., ribbon jasper; when the lines of contrast are on a larger scale, the term banding is used. See also: *stripe.* (AGI, 1987) (b) Said of a vein having alternating streaks of ore with gangue or country rock, or simply of varicolored ore minerals. Cf: *banded; book structure.* (AGI, 1987) (c) A color band in slate that represents original bedding and crosses the superimposed slaty cleavage. Ribbon is generally undesirable and decreases the value of the slate. Syn: *pancake.* (AGI, 1987)

ribbon brake A friction brake having a metal strap that encircles a wheel or drum and may be drawn tightly against it. A band brake. (Standard, 1964; Fay, 1920)

ribbon diagram Geologic cross section drawn in perspective and joining control points along a sinuous line. (AGI, 1987)

rib boss *pillar boss.*

rib dust Dust found on the side walls of a mine. The dust from the roof is generally included with this sample. (Rice, 1913-18)

rib hole Final holes fired in blasting around sides of shaft or tunnel. Also called trimmer. (Pryor, 1963)

rib line A continuous line along which pillars are mined. See also: *break line.* (Lewis, 1964)

rib lining In rod or ball mill, replaceable ribs that project longitudinally from shell liners so as to act as lifters for crop load as mill rotates. (Pryor, 1963)

rib mesh Expanded metal stiffened at intervals with bent steel plates. (Hammond, 1965)
rib pillar A pillar whose length is large compared with its width.
ribs The lines or ridges of cut gems that distinguish the several parts of the work, both of brilliants and roses. (Hess)
rib-side The side of a heading or roadway driven in the solid coal. See also: *narrow stall.* (Nelson, 1965)
rib-side gate A gate road in longwall mining with a rib of solid coal along one side. (Nelson, 1965)
rib-side pack A pack formed by working 5 to 10 yd (4.6 to 9.1 m) of coal along a rib-side of a road and packing the waste. See also: *roadside pack; skipping.* (Nelson, 1965)
rice coal (a) Anthracite coal of a small size; No. 2 Buckwheat coal. See also: *anthracite coal sizes.* (Webster 3rd, 1966) (b) A steam size of anthracite. (Jones, 1949)
Richards' pulsator classifier A classifier operating in such a manner that the pulp grains fall through a sorting column against an upward pulsating current of water. It has no screen. (Liddell, 1918)
Richards' pulsator jig An outcome of the pulsator classifier, in which a pulsating column of water is used in the jig. See also: *pulsator jig.* (Liddell, 1918)
Richards' shallow-pocket hindered-settling classifier A series of pockets through which successively weaker streams of water are directed upward. The material that can settle does so, and is drawn off through spigots. (Liddell, 1918)
richetite A triclinic mineral, $PbU_4O_{13}{\cdot}4H_2O$; strongly radioactive; black; occurs embedded among fine needles of uranophane.
richmondite (a) A discredited mineral term since a number of specimens have proved to be mixtures containing, in order of abundance, argentian tetrahedrite, galena, sphalerite, chalcopyrite, pyrite, and perhaps stromeyerite. (Am. Mineral., 1945) (b) A mixture of sulfide minerals containing silver, lead, zinc, and copper.
rickardite An orthorhombic mineral, Cu_7Te_5; pseudotetragonal; deep purple; at Vulcan and Bonanza, CO; Warren, AZ; and Salvador, Brazil.
ricket An airway along the side of an adit or shaft. Also called ricketing.
rid *redd.*
ridding N. of Eng. Separating ironstone from coal shale. See also: *redd.* (Fay, 1920)
riddle (a) A barrel-shaped, revolving perforated drum in which blank coins are washed and dried after passing through a bath of sulfuric acid. (Standard, 1964) (b) A coarse sieve. The large pieces of ore and rock picked out by hand are called knockings. The riddlings remain on the riddle; the fell goes through. (Webster 3rd, 1966; Fay, 1920) (c) A sieve used to separate foundry sand or other granular materials into various particle-size grades or free such a material of undesirable foreign matter. (ASM, 1961)
ride over Arkansas. A squeeze that extends into the workings beyond the pillar. It is said to ride over a pillar. (Fay, 1920)
rider (a) A thin coal seam above a workable seam, or a seam that has no name. (Nelson, 1965) (b) The rock lying between two lodes or beds; a mass of country rock enclosed in a lode; a horse. (c) An ore deposit overlying the principal vein. (Standard, 1964) (d) A steel or iron crossbeam which slides between the guides in sinking a shaft. It is carried up and down by, but is not attached to, the hoppit, which it guides and steadies.
ridge (a) A long, narrow elevation of the Earth's surface, generally sharp crested with steep sides, either independently or as part of a larger mountain or hill. (b) A long elevation of the deep-sea floor having steeper sides and less regular topography than a rise. (AGI, 1987)
ridge fillet A runner or principal channel for molten metal. (Standard, 1964)
ridge terrace A ridge built along a contour line of a slope to pond rainwater above it. (Nichols, 1976)
Ridgeway filter A horizontal, revolving, continuous vacuum filter. The surface is an annular ring consisting of separate trays with vacuum and compressed air attachments. The filtering surface is on the underside, the trays being dipped into the tank of pulp to form the cake and then lifted out of it. (Liddell, 1918)
riding Said of mine timbering when the sets are thrust out of line, or lean.
Ridley-Scholes bath Dense-media system used to float coal away from shale, the latter falling to the bottom of a wedge-shaped pool of separating fluid and being withdrawn by a rising belt. (Pryor, 1963)
rid-up runners To clean up after a cast, as when the scrap, slag, and iron is removed from runners, troughs, and skimmers, and they are freshly clayed, loamed, or sanded. (Fay, 1920)
riebeckite A monoclinic mineral, $Na_2Ca(MgFe^{+2})_5Si_8O_{22}(OH)_2$; amphibole group with $Mg/(Mg+Fe^{2+}) = 0$ to 0.49 and $Fe^{3+}/(Fe^{3+}+Al) = 0.7$ to 1.0; forms a series with magnesioriebeckite; fibrous; in soda-rich rhyolites, granites, and pegmatites; crocidolite variety is blue asbestos, tigereye is crocidolite replaced by quartz. Cf: *glaucophane.*

Riecke's principle The statement in thermodynamics that solution of a mineral tends to occur most readily at points where external pressure is greatest, and that crystallization occurs most readily at points where external pressure is least. It is applied to recrystallization in metamorphic rocks with attendant change in mineral shapes, such that mass is transferred from contact points to pressure shadows resulting in reduced rock porosity. It is named after the German physicist E. Riecke (1845-1915) although it was actually discovered and described by Sorby in 1863. (AGI, 1987)
riemannite *allophane.*
riffle (a) A natural shallows extending across a stream bed over which the water flows swiftly and the water surface is broken into waves; a shallow rapids of comparatively little fall. (AGI, 1987) (b) The lining of the bottom of a sluice, made of blocks or slats of wood or stones, arranged in such a manner that chinks or slots are left between them, into which heavy mineral grains fall and are held for recovery. (c) The raised portions of the deck of a concentrating table, that serve to trap the heaviest particles. Syn: *ripple.* (BS, 1962) (d) A device used to reduce the volume or weight of a sample consisting of a thin metal plate on which is mounted a series of metal strips to guide or deflect a small portion of the sample material into a separate container. Cf: *sample splitter.* See also: *ripple board.* (Long, 1960) (e) Sample reducing device such as Clark riffler or Jones riffle, which splits a batch sample of ground ore into two equal streams as it falls across an assembly of deflecting chutes. See also: *sludge sampler.* Cf: *Jones splitter.* (Pryor, 1963)
riffle bars Slats of wood nailed across the bottom of a cradle or other gold-washing machine for the purpose of detaining the gold.
riffle box A device designed to reduce a sample of coal or ore to half its original size. The box contains about 12 chutes discharging alternately to opposite sides. The width varies according to the largest particle size. The volume reduction is rapid for dry material of suitable fineness. (Nelson, 1965)
riffler *sample splitter.*
rifle (a) As used by drillers, a borehole that is following or has followed a spiral or corkscrew course; also said of a drill core that has spiral grooves appearing on its outside surface. (Long, 1960) (b) A drill hole, in rock, that has become three-cornered while drilling. (c) Applied to the three-cornered section of a hole drilled by hand. Though the bit is supposed to be turned one-eighth after each blow, to insure a circular hole, the majority of hand-drilled holes are three-cornered. (Stauffer, 1906)
rifle bar A cylinder with curved splines. (Nichols, 1976)
rifle nut A splined nut that slides back and forth on a rifle bar. (Nichols, 1976)
rifling (a) Working coal which was left behind over the waste. (Nelson, 1965) (b) The spiral grooving in the walls of a drill hole and/or on the surface of a drill core. (Long, 1960) (c) A borehole following a spiraled course. (Long, 1960)
rift (a) A regional-scale strike-slip fault, e.g., the San Andreas rift in California, with offset measured in hundreds of kilometers. (AGI, 1987) (b) A trough or valley formed by faulting. (c) In quarrying, a direction of parting in a massive rock, such as granite, at approx. right angles to the grain. Cf: *grain; hard way.* (AGI, 1987) (d) A narrow cleft, fissure, or other opening in rock (such as in limestone), made by cracking or splitting. (AGI, 1987) (e) A planar property whereby granitic rocks split relatively easily in a direction other than the sheeting (parallel to the surface of the Earth.). (AGI, 1987) (f) A term used in slate quarrying to describe a second direction of splitting less pronounced than slaty cleavage and usually at right angles to it. (g) In sedimentary rocks, the horizontal plane of stratification, or the bed of the rock. (Stauffer, 1906) (h) An obscure foliation, either vertical (or nearly so) or horizontal, along which a rock splits more readily than in any other direction. Syn: *fault trace.* (i) A crack, such as in the mid-ocean ridges. (MacCracken, 1964)
rifter-trimmer One who separates blocks of mica into sheets and trims sheets preparatory to processing. Also called full trimmer. (DOT, 1949)
rifting The process of splitting hand-cobbed mica into sheets of usable thicknesses. (Skow, 1962)
rift structure A long, narrow structural trough that is bounded by normal faults; a graben of regional extent. Cf: *rift valley.*
rift valley (a) A valley that has developed along a rift structure. Syn: *fault trough.* (AGI, 1987) (b) The deep central cleft in the crest of the mid-oceanic ridge. Syn: *midocean rift.* See also: *Mid-Atlantic Ridge.* (AGI, 1987)
rift zone (a) A system of parallel crustal fractures; a rift structure. (AGI, 1987) (b) In Hawaii, a zone of volcanic features associated with underlying dike complexes. Syn: *volcanic rift zone.* (AGI, 1987)
rig (a) A drill machine complete with auxiliary and accessory equipment needed to drill boreholes. (Long, 1960) (b) To assemble and set up a tripod, derrick, and/or drill machine and put it in order for use. Syn: *rig up.* See also: *setup.* (Long, 1960) (c) A general term denoting

any machine. More specif., the front or attachment of a revolving shovel. See also: *drilling rig*. (Nichols, 1976)

rigged Drill machine and equipment in place at a drill site and ready to start drilling. (Long, 1960)

rigger One who, with special equipment and tackle, moves and transports heavy machinery, etc. (Crispin, 1964)

rigging (a) Process of setting up a drill and its auxiliary equipment preparatory to drilling. Syn: *setting up*. (Long, 1960) (b) The cables or ropes anchoring a drill derrick, mast, or tripod. See also: *guy*. (Long, 1960) (c) Sometimes used as a term for derrick, mast, or tripod complete with anchor, stay ropes, and cables. (Long, 1960) (d) The equipment or gear such as hoists, tackle, winches, chains, or rope used by riggers in their work. (e) The engineering design, layout, and fabrication of pattern equipment for producing castings; including a study of the casting solidification program, feeding and gating, risering, skimmers, and fitting flasks. (ASM, 1961)

rigging bar A long, extension-type jack bar or drill column for use underground, on which a drilling machine can be mounted. (Long, 1960)

right-angled block In quarrying, a block of stone bounded by three pairs of parallel faces, all adjacent faces meeting at right angles.

right bank The bank of a stream that is on the right when one looks downstream.

right-hand cutting tool A cutter, all of whose flutes twist away in a clockwise direction when viewed from either end. (ASM, 1961)

right-hand feed screw A diamond-drill feed screw that rotates in a clockwise direction. (Long, 1960)

right-hand lay Rope or strand construction in which wires or strand are laid in a helix having a right-hand pitch. See also: *lay*. (Hammond, 1965)

right lang lay Wire or fiber rope or cable in which the individual wires or fibers forming a strand and the strands themselves are both twisted to the right. Also called right long lay. (Long, 1960)

right-lateral fault A fault on which the displacement is such that the side opposite the observer appears displaced to the right. Cf: *left-lateral fault*. Syn: *dextral fault*.

right lay Wire or fiber rope or cable in which the strands formed from a group of individual wires or fibers are twisted to the right. (Long, 1960)

right line A straight line; the shortest distance between two points. (Crispin, 1964)

right long lay *right lang lay*.

right-of-way A grant by Act of Congress, to convey water over or across the public domain, for mining purposes.

right regular lay Wire or fiber rope or cable in which the wires or fibers in the strand are twisted to the left and the strands to the right. Syn: *regular-lay right lay*. (Long, 1960)

right running (a) N. of Eng. Applied to a vein carrying ore in beds often unproductive. (b) N. of Eng. Rake veins extending approx. east and west.

rigid arch A continuous arch which is fully fixed throughout. (Hammond, 1965)

rigid coupling A rod-to-feed-screw sub or rod-to-drive-rod sub by means of which the drill rods are coupled directly to the feed screw or drive rod of the diamond-drill swivel head, and the chuck is discarded or eliminated. Also called screw-to-rod adapter. (Long, 1960)

rigid double tube *rigid-type double-tube core barrel*.

rigid ducts *ventilation ducts*.

rigid foam Formed by mixing isocyanate and a polyether polyol containing a halogenated hydrocarbon agent. Mixing releases heat, causing the foam to expand as much as 30 times the original volume of the liquid. The foam, which becomes cellular and rigid within minutes, is heat resistant and essentially impervious to air and water, and has substantial binding strength. Its characteristics suggest possible uses in mining for insulation, stoppings to control ventilation, and seals to control water and to consolidate broken ground. (Encyclopaedia Britannica, 1964)

rigid frame A framed structure having columns and beams rigidly connected; there are no hinged joints in this type of structure. (Hammond, 1965)

rigid hammer crusher A machine in which size reduction is effected by elements rigidly fixed to a rotating horizontal shaft mounted in a surrounding casing. (BS, 1962)

rigidity The property of a material to resist applied stress that would tend to distort it. A fluid has zero rigidity. (AGI, 1987)

rigidity modulus *modulus of rigidity*. Cf: *modulus of elasticity*. (AGI, 1987)

rigid pavement A road, taxitrack, or hardstanding constructed of concrete slabs. (Hammond, 1965)

rigid solution Solubility of solution of elements in a natural glass as compared with a solid solution that implies crystallinity.

rigid-type double-tube core barrel A double-tube core barrel in which both the outer and inner tubes are rigidly connected to a single headpiece. Syn: *rigid double tube*. (Long, 1960)

rigid urethane foam *rigid foam*.

rig-up *setting up*.

rig up *rig*.

rig-up time The time required to set up and make a drill rig ready for use at the site where a borehole is to be drilled. Also called setup time; rigging time; mobilizing a rig. (Long, 1960)

rill To mine ore in such as way that it runs down a slope to a chute or loading level. Ore is said to be rilled to a chute when it is rolled down a slope left in mining. (Hess)

rill stope Overhand stope so shaped that miners can stand on the ore they have severed, and work horizontally along the side walls of unbroken ore that confine the excavation. The stope is carried as an inverted stepped pyramid, its apex ending in a winze that leads to the tramming level, down to which ore gravitates or is moved. (Pryor, 1963)

rim (a) The border, edge, or face of a cliff, as at the Grand Canyon of Arizona. (AGI, 1987) (b) The outermost portion of a zoned crystal, e.g., a reaction rim. (AGI, 1987)

rim flying A reconnaissance method in which a plane follows an outcrop along steep canyon walls, keeping where possible within 50 ft (15 m) of the face of the cliff. This type of prospecting has been successful in discovering new deposits in the Colorado Plateau region. (Dobrin, 1960)

rimmed steel A low-carbon steel containing sufficient iron oxide to give a continuous evolution of carbon monoxide while the ingot is solidifying, resulting in a case or rim of metal virtually free of voids. Sheet and strip products made from the ingot have very good surface quality. (ASM, 1961)

rimrock (a) The outcrop of a horizontal layer of resistant rock, such as sandstone, at the edge of a plateau, butte, or mesa; the cliff or ledge so formed. (b) The bedrock rising to form the boundary of a placer deposit.

rimrocking Prospecting for carnotite on the Colorado Plateau, where the favorable beds, more or less flat-lying, crop out in cliffs or rims.

rim texture A texture in ores where the metasome forms a narrow rim around grains of the host mineral.

rim walking Prospecting a canyon rim with a Geiger counter. (Ballard, 1955)

rincon (a) A term used in the Southwestern United States for a recess or hollow in a cliff or a reentrant in the borders of a mesa or plateau. Also called a cove. (b) A term used in the Southwestern United States for a small, secluded valley. (c) A bend in a stream.—Etymol: Spanish rincón, inside corner, nook.

ring (a) A complete circle of tubbing plates around a circular shaft. Syn: *tunnel lining*. (b) Troughs placed in shafts to catch the falling water, and so arranged as to convey it to a certain point. (c) See also: *wedging crib*. (d) S. Staff. A circular piece of wrought iron, about 8 in (20 cm) deep, placed on the top of a skip of coal to increase its capacity.

ring arch One composed of a series of straight, unbonded rows, one brick wide.

ring coal (a) An old name for bituminous coal. (Tomkeieff, 1954) (b) Bituminous coal as opposed to stone coal or anthracite. (Arkell, 1953)

ring complex An association of ring dikes and cone sheets. (AGI, 1987)

ring crusher (a) A type of hammermill with a high-speed horizontal shaft upon which a series of steel rings are swung. (ACSG, 1963) (b) Impact mill, beater mill, or hammermill, in which the beaters are loosely swinging rings. (Pryor, 1963) (c) See also: *hammermill*. (Mitchell, 1950)

ring-cut Holes in a ring around one central hole used to carry a cavity forward, usually six. (Pryor, 1963)

ring dike A subcircular to circular dike with steep dip. Ring dikes may be many kilometers long, and hundreds or thousands of meters thick. Their radius is generally from 1 to 20 km. Although some ring dikes may form a nearly complete circle, more commonly they encompass 1/4 to 3/4 of a circle or ellipse. They are commonly associated with alkalic igneous complexes and carbonatites, so are probably related to deep shock effects or to cauldron subsidence. Ring dikes are commonly associated with cone sheets to form a ring complex. Syn: *ring-fracture intrusion*.

ringed out A diamond bit in the face of which has been gouged a circular groove deeper than, and at least as wide as, the diameter of one row of the inset diamonds. (Long, 1960)

ringer A crowbar. (Fay, 1920)

ring fault A steep ring-shaped fault, complete or incomplete. It is associated with cauldron subsidence.

ring-fracture intrusion *ring dike*.

ring holes The group of boreholes radially drilled from a common-center setup. See also: *horadiam; radial drilling*. (Long, 1960)

ring-induction method An inductive method in which the primary coil and the measuring coil are concentric. (Schieferdecker, 1959)
ringing The audible or ultrasonic tone produced in a mechanical part by shock, and having the natural frequency or frequencies of the part. The quality, amplitude, or decay rate of the tone may sometimes be used to indicate quality or soundness. (ASM, 1961)
ring main Closed loop of piping, including provision for entry of material, circulation boost and controlled withdrawal points; used for circulating solids such as pulverized fuel, or fluids such as lime slurry, continuously without settlement or chokeup. (Pryor, 1963)
ring ore Fragments of gangue covered with deposits of other minerals. See also: *cockade ore.*
ring pit A circular pit in which a large wheel is revolved for tempering clay. (Fay, 1920)
ring-roll crusher A type of crusher in which high-speed rolls act on the inside circumference of a vertical cylinder to powder raw material like clay. (Enam. Dict., 1947)
ring-roll grizzly A sturdily built grizzly for handling large pieces of ore. This type transports its material across a series of grooved rollers moved mechanically, or alternatively by the sliding ore. Undersize falls through the grooves. (Pryor, 1963)
ring-roll press A press consisting of rolls of unequal diameter, revolving one within the other and in the same direction. (BS, 1962)
Ringrose methane recorder A recorder that gives a continuous record in the range of 0% to 3%. (Roberts, 1960)
Ringrose pocket methanometer A small instrument that is capable of estimating methane in the range of 0% to 2% . (Roberts, 1960)
ring-shaped occurrences In some areas altered rock has been found as a halo over an orebody and thus serves as a geologic target for guiding prospecting operations. The ratio between the size of the ring and the orebody must not be too large for practical purposes. Such target rings are not always obvious and will only be recognized after much painstaking work and study. Also called bulls-eye alteration patterns. (Lewis, 1964)
ring stone (a) A voussoir showing on the face of the wall. (Webster 3rd, 1966) (b) Eng. Large oolitic grains in very hard crystalline matrix, above the slates at Collyweston, U.K. Cf: *sun bed.* (Arkell, 1953)
ring stress The zone of stress, higher than that pre-existing in the rock, which surrounds all development excavations is called the ring stress. (Spalding, 1949)
ring-stress bursts In stoping, the ring stresses around a level, rise, or winze are so increased by the influence of an approaching stope face that at some point on the periphery the rock fails. The stress ring is broken, and the rock of sides, back, and bottom released thereby expands suddenly and violently into the excavation, causing a rock burst. This rock burst is identical in type with those occurring in development. It is usually extremely local in effect, though a heavy earth tremor is caused. (Spalding, 1949)
ring tension Tension that develops in the wall of a circular tank containing liquid or solid material. (Hammond, 1965)
ring-type reaming shell In drilling, a reaming shell, the inset reaming diamonds of which are set into a cast- or powder-metal band encircling the outside surface of the shell. (Long, 1960)
ring-type wedge A drill-hole deflecting wedge having a short metal sleeve attached to the uppermost end. The outside diameter of the sleeve is the same as that of the lower, full-circle part of the wedge. (Long, 1960)
ring wall The inner firebrick wall of a blast furnace. (Standard, 1964)
rinkite A monoclinic mineral, $(Na,Ca,Ce)_3Ti(Si_2O_7)_2OF_3$; weakly radioactive; forms prismatic crystals in veins containing silicates of cerium metals, yttrium, and niobium; near Barkevik, Norway, in sodalite syenite in the Julianehaab district, Greenland, and in large crystals on the Kola Peninsula, Russia. Also spelled rinkolite. Syn: *johnstrupite.*
Rinman scale A Swedish standard scale for the estimation of slag inclusions in iron and steel. This scale consists of a series of micrographs, designed to show different typical fields of view, and arranged in groups according to the form and distribution of the inclusions and numbered according to their quantity. (Osborne, 1956)
rinneite A trigonal mineral, $K_3NaFeCl_6$; colorless to varicolored; becomes brown on exposure to air and has an astringent taste.
rinsing In the ion-exchange (IX) cycle, applied to pregnant leach liquors, the displacement wash used after the absorption cycle, which moves pregnant liquor still in the column onto the next absorption column in the series. Term also applied to water rinse used after elution cycle, and before acid rinse. (Pryor, 1963)
rinsing water (a) Water used to remove fine particles from larger sizes, usually located over vibrating screens. (BS, 1962) (b) Syn: *spray water.*
Rio Tinto process Heap leaching of cupriferous sulfides after their slow oxidation to sulfates on prolonged atmospheric weathering. (Pryor, 1963)
rip (a) To bring down rock in a roadway to increase headroom. See also: *dint.* (Fraenkel, 1953) (b) *brush.*
riparian (a) Pertaining to or situated on the bank of a body of water, esp. of a watercourse such as a river; e.g., riparian land situated along or abutting upon a stream bank, or a riparian owner who lives or has property on a riverbank. (AGI, 1987) (b) Pertaining to shrubs and trees with root systems that seek deep ground water, as mesquite and greasewood. (AGI, 1987)
riparian rights The rights of a person owning land containing or bordering on a watercourse or other body of water in or to its banks, bed, or waters. Under common law, a person owning land bordering a nonnavigable stream owns the bed of the stream and may make reasonable use of its waters.
rip current A strong surface current of short duration flowing seaward from the shore. It usually appears as a visible band of agitated water and is the return movement of water piled up on the shore by incoming waves and wind. Syn: *rip tide.* Cf: *undertow.* (Schieferdecker, 1959)
rippability A measure of the ease or difficulty with which a rock or earth material can be broken by tractor-drawn rippers or rigid steel tines into pieces that can be economically moved by other equipment, usually scrapers.
ripper (a) Coal miner who breaks down the roof of a gate road to increase headroom or breaks down the roof at the ripping lip, or where the roof has sagged on a roadway due to subsidence. The miner is often paid on yardage of ripping performed. Also known as brusher; stoneman; repairer. See also: *trenchman; stoneman.* (Pryor, 1963) (b) A tool for removing slates, or edging them. (Standard, 1964) (c) An accessory that is either mounted or towed at the rear of a tractor and generally used in place of blasting as a means of loosening compacted soils and soft rocks for scraper loading. The ripper has long, angled teeth that are forced into the ground surface, ripping the earth loose to a depth of 2 ft (0.6 m) or more. (Carson, 1961) (d) *rooter.*
ripping (a) A machine for cutting stone into slabs by passing it on a bed under a gang of saws. (Standard, 1964) (b) The act of breaking, with a tractor-drawn ripper or long-angled steel tooth, compacted soils or rock into pieces small enough to be economically excavated or moved by other equipment such as a scraper or bulldozer. (c) The breaking down of the roof in mine roadways to increase the headroom for haulage, traffic, and ventilation. See also: *brushing; second ripping.* (Nelson, 1965)
ripping bed A machine for cutting stone into slabs by passing it on a bed under a gang of saws. (Standard, 1964)
ripping blasting Where coal seams are worked by the longwall method it is necessary to maintain roadways leading to the face. These roadways should be of sufficient height to permit the easy passage of workers and materials, and this invariably means that some of the stone above the coal must be removed. This operation is known as ripping, and, unless the roof strata are very soft, blasting will be required. The main considerations in ripping blasting are to keep the sides of the roadway square, and to obtain good fragmentation of the stone so that it can be removed easily. (McAdam, 1958)
ripping face support A timber, or timber and steel structure, to provide support at the ripping lip. There are various types: one consists of bent corrugated steel bars behind which wooden planks are wedged; another consists of adjustable stretchers that are fitted across the roadway. See also: *roadhead.* (Nelson, 1965)
ripping lip (a) The edge of the rippings at the face of a roadway. When enlarging a roadway, the ripping lip is the end of the enlarged section and where work is proceeding. See also: *forepoling girder.* (Nelson, 1965) (b) The edge of the nether roof at a gate end at the point up to which the ripping has been taken. (TIME, 1929-30)
ripping scaffold A staging or platform erected over the moving conveyor at a ripping lip of a gate road, on which the miners can stand and work. This implies that the coalface and conveyor loading point are some distance ahead. (Nelson, 1965)
rip plates A means of repairing damaged belting. It consists of two short plates, with teeth on one side to grip the belting, which are fastened on both sides of the belting across the rip or worn place. Short bolts and nuts serve to compress and hold the plates tightly against the belting. (Jones, 1949)
ripple A groove or bar across sluices for washing gold. See also: *riffle.*
ripple board An inclined trough having grooves or strips across its bottom to catch fine gold. See also: *riffle.*
ripple index The ratio of wavelength to amplitude of a ripple mark. It usually ranges from 6 to 22 for ripples produced by water currents or waves and from 20 to 50 for ripples produced by wind. Syn: *ripple-mark index.* (AGI, 1987)
ripple mark (a) An undulatory surface or surface sculpture consisting of alternating subparallel small-scale ridges and hollows formed at the interface between a fluid and incoherent sedimentary material (esp. loose sand). It is produced on land by wind action and subaqueously by currents or by the agitation of water in wave action, and generally trends at right angles or obliquely to the direction of flow of the moving fluid. It is no longer regarded as evidence solely of shallow water. (AGI, 1987) (b) One of the small and fairly regular ridges, of various

shapes and cross sections, produced on a ripple-marked surface; esp. a ripple preserved in consolidated rock and useful in determining the environment of deposition. The term was formerly restricted to symmetrical ripple mark, but now includes asymmetrical ripple mark. The singular form may be used to denote general ripple structure (as well as a specific ripple), and the plural form to describe a particular example.

ripple-mark index *ripple index.*

ripple voltage The alternating component of a substantially unidirectional voltage. (Coal Age, 1960)

riprap (a) A layer of large, durable fragments of broken rock, specially selected and graded, thrown together irregularly (as offshore or on a soft bottom) or fitted together (as on the upstream face of a dam). Its purpose is to prevent erosion by waves or currents and thereby preserve the shape of a surface, slope, or underlying structure. It is used for irrigation channels, river-improvement works, spillways at dams, and sea walls for shore protection. (AGI, 1987) (b) The stone used for riprap. (AGI, 1987)

rip tide *rip current.*

rise (a) A vertical or inclined shaft from a lower to an upper level in a mine. See also: *upraise; raise.* (Eng. Min. J., 1938) (b) To dig upward, as from one level to the next one above; opposite of sink. (Standard, 1964) (c) Upward inclination of a coal stratum. (Standard, 1964) (d) An ascending gallery at the end of a level. See also: *hade.* (Gordon, 1906)

rise and fall A system of reduction of levels by working out the rise or fall of staff readings from each level point to the one following it. See also: *collimation.* (Hammond, 1965)

rise face A face advancing toward the rise of a coal seam. (Briggs, 1929)

rise heading A heading driven to the rise in a long-way workings. See also: *heading.*

riser (a) A shaft excavated from below upward. See also: *raise; rise.* (Fay, 1920) (b) *column pipe; rising main.*

rising (a) An excavation carried from below upward; a rise or riser. (Standard, 1964) (b) Eng. The horizontal division of the stratum, from which the blocks of stone are lifted; e.g., in the Portland quarries. (Arkell, 1953)

rising column *rising main.*

rising current The direction in which a drill circulation fluid is flowing after it has passed the bit and continues toward the collar of a borehole. (Long, 1960)

rising-head test A soil permeability test in which the level of water in a borehole is reduced and then the rate at which the water recovers is observed. (Mining)

rising main (a) The length of steel piping that conveys the water from a pump to the surface or to a higher pump in the shaft. The term rising main is obsolete; delivery column preferred. Syn: *delivery column.* (Nelson, 1965) (b) *column pipe.*

rising shaft Excavating a shaft upwards from mine workings. See also: *rise; staple shaft.* (Nelson, 1965)

rittingerite *xanthoconite.*

Rittinger's law The energy required for reduction in particle size of a solid is directly proportional to the increase in surface area. Cf: *Kick's law.* (CCD, 1961)

Rittinger table A side-bump table with plane surface, actuated by a cam, spring, and bumping post. (Liddell, 1918)

rivelaine A pick with one or two points, formed of flat iron, used to undercut coal by scraping instead of striking. (Fay, 1920)

river bar A ridge or mound of boulders, gravel, sand, and mud found along or in a stream channel at places where decrease in velocity causes deposition of sediment.

river-bar placer (a) Gravel flats and terraces laid down by rivers when flowing at higher levels than at present. The deposits are sometimes gold or tin-bearing. See also: *bench placer.* (Nelson, 1965) (b) A term used in Alaska for placers on gravel flats in or adjacent to the beds of large streams.

river claim A claim that includes the bed of a river.

river drift The gravel deposits accumulated by a river in its torrential stages.

river flat *alluvial flat.*

river mining Mining or excavating beds of existing rivers after deflecting their course, or by dredging without changing the flow of water.

river pebble Applied in Florida to a certain class of phosphatic pebbles, or concretions, found in rivers as distinguished from land pebble phosphate.

river plain *alluvial plain.*

river quartz Rounded, waterworn masses of quartz found in stream gravels.

river right *creek right.*

river run gravel Natural gravel as found in deposits that have been subjected to the action of running water. See also: *alluvial deposit.* (Nelson, 1965)

river sand Sand generally composed of rounded particles, and may or may not contain clay or other impurities. It is obtained from the banks and beds of rivers. (Zern, 1928)

rives in Eng. To crack open or produce fissures.

rivet A round bar of mild steel having a conical, cup- or pan-shaped head, which is driven while red hot into a hole through two plates of steel that have to be joined together. Aluminum, copper, and other materials are also used for rivets. See also: *pneumatic riveter.* (Hammond, 1965)

riveter A worker who forms the head of a rivet, generally with a pneumatic rivet hammer. (Hammond, 1965)

rivet forge A portable forge, used by boilermakers and ironworkers, for heating rivets near the work for which they are required. (Crispin, 1964)

rivet heater A laborer responsible for heating rivets in a portable forge and throwing them with tongs to the rivet catcher. (Hammond, 1965)

rivet snap A punch having a recess in its head shaped to the form of the rivet. See also: *pneumatic riveter.* (Hammond, 1965)

rivet test A test on the steel used for rivets, in which a bar is bent through 180°; if any cracks are formed, the steel is rejected. (Hammond, 1965)

rivet tester A trained worker who can detect sound or loose rivets by testing them with a hammer. (Hammond, 1965)

riving seams Open fissures between beds of rock in a quarry.

R.K. process A method for converting pig iron into a product with a low carbon content, which is suitable as a substitute for steel scrap for remelting in steel furnaces. (Osborne, 1956)

road (a) A roadway in a mine, e.g., gate road, traveling road, dummy road. (Nelson, 1965) (b) Any mine passage or tunnel. (Mason, 1951) (c) Rail track. (Mason, 1951)

roadbed (a) The material fundamental part of a road; primarily, the foundation of gravel, road metal, etc., constituting the bed, but by extension, esp. in railway use, the superstructure also. (Standard, 1964) (b) The foundation carrying the sleepers, rails, chairs, points, and crossings, etc., of a railway track. (CTD, 1958)

road cleaner *track cleaner.*

road dust Dust found on the floor of a mine entry. (Rice, 1913-18)

roadhead The face of a roadway, usually in longwall conveyor mining. The records indicate that the roadhead is the most dangerous place in a coal mine based on accidents from falls of ground. See also: *ripping face support.* (Nelson, 1965)

road-making plant Various types of specialized plant used solely for road construction, including such machines as planers, scarifiers, rollers, pavers, finishers, gritters, and mixers. (Hammond, 1965)

roadman (a) A person employed on the laying and maintenance of rail tracks underground. Also known as a trackman. (Nelson, 1965) (b) A person whose duty it is to keep the roads of a mine in order. (Fay, 1920) (c) In bituminous coal mining, a general term for miners working along haulageways or airways (roads). Usually designated according to job, as repairman; wasteman. (DOT, 1949)

road metal Crushed stone for surfacing macadamized roads, and for the base course of asphalt and concrete roadways; also used without asphaltic binder as the traffic-bearing surface, generally on secondary roads. (AGI, 1987)

road-mix method A method of preparing aggregates for bituminous surfaces in which the aggregates and bitumen are combined on the surface of the road, using the penetration or mixed-in-place method. (Pit and Quarry, 1960)

road roller Power-driven roller of any weight from one-half to 12 tons. (Hammond, 1965)

roadside pack A pack built alongside a roadway. See also: *rib-side pack.* (Nelson, 1965)

roadster Low-priced model of a scraper or a truck. (Nichols, 1976)

roadway An underground drivage. It may be a heading, gate, stall, crosscut, level, or tunnel and driven in coal, ore, rock or in the waste area. It may form part of longwall or bord-and-pillar workings or an exploration heading. A roadway is not steeply inclined. See also: *roadway support.* (Nelson, 1965)

roadway cable An electric cable designed for use in mine roadways. It may be either rubber insulated, sheathed, and armored or paper insulated. (Nelson, 1965)

roadway consolidation To bind the floor dust together with water and calcium chloride flakes, or other chemical, to form a firm plastic carpet. See also: *dust consolidation.* (Nelson, 1965)

roadway support A timber, steel, concrete, or other erection in a roadway to (1) ensure safety by preventing falls of ground, and (2) maintain the maximum possible roadway size by resisting the tendency of the roadway to contract and distort. See also: *steel arch; roadway.* (Nelson, 1965)

roast To heat to a point somewhat short of fusing, with access of air, as to expel volatile matter or effect oxidation. (Fay, 1920)

roaster A reverberatory furnace or a muffle used in roasting ore. (Standard, 1964)

roaster slag Slag resulting from the calcination of white metal in the process of copper smelting. (Standard, 1964)

roasting (a) Heating an ore to effect some chemical change that will facilitate smelting. (ASM, 1961) (b) The operation of heating sulfide ores in air to convert to oxide or sulfate. (CTD, 1958) (c) Calcination, usually with oxidation. Good, dead, or sweet roasting is complete roasting; i.e., it is carried on until sulfurous and arsenious fumes cease to be given off. Kernel roasting is a process of treating poor sulfide copper ores, by roasting in lumps, whereby copper and nickel are concentrated in the interior of the lumps. (Fay, 1920) (d) The heating of solids, frequently to promote a reaction with a gaseous constituent in the furnace atmosphere. See also: *magnetizing roast.* (ARI, 1949)

roasting and reaction process The treatment of metal ore in a reverberatory, by first partly roasting at a low temperature and then partly fusing the charge at a higher temperature, which causes a reaction between the lead oxide formed by roasting and the remaining sulfide, producing sulfurous acid and metallic lead. Syn: *air-reduction process.* (Fay, 1920)

roasting and reduction process The treatment of lead ores by roasting to form lead oxide, and subsequent reducing fusion in a shaft furnace. (Fay, 1920)

roasting cylinder A furnace with a rotating cylinder for roasting, amalgamating, or smelting ore. See also: *Bruckner furnace.* (Standard, 1964; Fay, 1920)

roasting furnace A furnace in which finely ground ores and concentrates are roasted to eliminate sulfur or other elements or compounds; heat is provided by the burning sulfur. The essential feature is free access of air to the charge, by having a shallow bed that is continually rabbled. Many types have been devised; multiple hearth is the most widely used. (CTD, 1958)

roast sintering *blast roasting.*

roast stall A form of roasting furnace, built in compartments or stalls open in front, with flues running up the wall at the back for the purpose of creating a draft.

rob To extract mine pillars previously left for support, often preparatory to closing a mine. Syn: *second mining.*

robber (a) In anthracite and bituminous coal mining, one who breaks down and rips out with a pick, pillars of coal left to support the roof in rooms when the usual mining was being done. Also called pillar robber. (DOT, 1949) (b) An extra cathode or cathode extension that reduces the current density on what would otherwise be a high current-density area on the work. (ASM, 1961)

robbing (a) Removing timber from a mined-out stope to use it again elsewhere. (Stoces, 1954) (b) Extraction of the pillars of ore left to support workings during original stoping. (Pryor, 1963) (c) Scot. Reducing the size of pillars; taking as much as possible off pillars, leaving only what is deemed sufficient to support the roof.

robbing an entry *drawing an entry.*

robbing pillars The mining of coal pillars left to support the roof during development mining, often resulting in cave-ins. See also: *working the broken.*

Robiette process A heat treatment process carried out in a substantially closed furnace, in which a fluid fuel is burnt to partial combustion with a gas containing 70% or more of oxygen to produce a nonoxidizing atmosphere. The treatment is effected continuously in the furnace through which the heating gas and metal are passed in opposed directions. The fuel and gas are partially burnt at the exit end of the furnace, and passed to the cooler entry end of the furnace, and burnt to substantially complete combustion so as to preheat metal entering the furnace. (Osborne, 1956)

Robins-Messiter system A stacking conveyor system in which material arrives on a conveyor belt and is fed to one or two wing conveyors. This part of the system moves so as to form a long ridge; reclaimed by raking gear that works across the ridge, moving slowly forward and shifting material loosened and blended by the rake action by means of a spiral that pushes it to a reclaiming conveyor at the side of the ridge. Used to stockpile ore, concentrates, and coal. (Pryor, 1963)

Robinson and Rodger system A method of obtaining sound steel by fluid compression of the ingot in the mold. The molds are divided in the center, a removable packing piece being placed between the halves of the mold. The packing piece is removed when the metal has set, and the mold is placed horizontally in the press, pressure being applied to the ingot at both ends. (Osborne, 1956)

robinsonite A triclinic mineral, $Pb_4Sb_6O_{13}$.

robot loader A pneumatic loader for inserting cartridges into drill holes.

Robson and Crowder process An early oil flotation process. The oil was added to several times its weight of ore and mixed in a slowly revolving drum or tube. The process at one time had quite a large application. The process used but little water (25% to 30%) and no acid. (Fay, 1920)

Roburite Smokeless and flameless safety explosive consisting of ammonium nitrate and dinitrobenzene or dinitrochloro benzene; used in mines. (Bennett, 1962)

roca (a) Sp. Rock or stone, whether in the ordinary or geological sense. (b) Sp. Rock standing out from the general surface. (c) Sp. A vein or bed of hard rock and stone.

roche Fr. Rock, boulder.

rock (a) An aggregate of one or more minerals, e.g., granite, shale, marble; or a body of undifferentiated mineral matter, e.g., obsidian, or of solid organic material, e.g., coal. (AGI, 1987) (b) Any prominent peak, cliff, or promontory, usually bare, when considered as a mass, e.g., the Rock of Gibraltar. (AGI, 1987) (c) A rocky mass lying at or near the surface of a body of water, or along a jagged coastline, esp. where dangerous to shipping. (AGI, 1987) (d) A slang term for a gem or diamond. (AGI, 1987) (e) Strictly, any naturally formed aggregate or mass of mineral matter, whether or not coherent, constituting an essential and appreciable part of the Earth's crust. Ordinarily, any consolidated or coherent and relatively hard, naturally formed mass of mineral matter; stone. In instances, a single mineral forms a rock, as calcite, serpentine, kaolin, and a few others but the vast majority of rocks consist of two or more minerals. (f) A local term used in New York and Pennsylvania for the more massive beds of bluestone that are not jointed and are, therefore, well-suited for structural purposes. (g) In the geological sense, any natural deposit or portion of the Earth's crust whatever be its hardness or softness, but used by miners to denote sandstone. (TIME, 1929-30) (h) In geology, the material that forms the essential part of the Earth's solid crust, and includes loose incoherent masses, such as a bed of sand, gravel, clay, or volcanic ash, as well as the very firm, hard and solid masses of granite, sandstone, limestone, etc. Most rocks are aggregates of one or more minerals, but some are composed entirely of glassy matter, or of mixtures of glass and minerals. (Hunt, 1965) (i) In the Lake Superior region, crude copper ore as it comes from the mines. The concentrate obtained is called mineral, and contains about 65% metallic copper.

rock asphalt *asphalt rock.*

rock association (a) A group of igneous rocks within a petrographic province that are related chemically and petrographically, generally in a systematic manner such that chemical data for the rocks plot as smooth curves on variation diagrams. See also: *tribe.* Syn: *rock kindred; kindred; association.* (AGI, 1987) (b) The association of mineral deposits with certain rock types. If mineral-producing localities are considered individually, valuable generalizations often can be made, and lithotectonic-plate tectonic classifications of ore deposits and exploration strategies derived from them.

rock base *bedrock.*

rock bit (a) Any one of many different types of roller or drill bits used on rotary-type drills for drilling large-size holes in soft-to-medium-hard rocks; also sometimes applied to drag-type bits. See also: *drag bit; roller bit.* (Long, 1960) (b) In mining, a detachable-type chisel or cruciform bit used on percussive drills to drill small-diameter holes in rock. (Long, 1960) (c) *drill bit.*

rock body A dump body with oak planking set inside a double steel floor. (Nichols, 1976)

rock bolt A bar, usually constructed of steel, that is inserted into pre-drilled holes in rock and secured for the purpose of ground control. Rock bolts are classified according to the means by which they are secured or anchored in rock. In current usage there are mainly four types: expansion, wedge, grouted, and explosive. See also: *roof bolt.*

rock bolting (a) *roof bolting.* (b) The process of rock bolting consists of (1) anchoring the bolt in a hole; (2) applying tension to the bolt to place the rock under compression parallel to the bolt; and (3) placing the bolts in such a pattern that they will properly support the rock structure. Rock may be supported by bolts in five ways: (1) suspension; (2) beam building; (3) reinforcement of arched opening requiring support; (4) reinforcement of an opening otherwise self-supporting; and (5) reinforcement of walls against shear and compressive action. See also: *roof bolting.* (Lewis, 1964)

rockbridgeite An orthorhombic mineral, $(Fe,Mn)Fe_4(PO_4)_3(OH)_5$; forms a series with frondelite; dark green; named for Rockbridge County, VA.

rock bump The sudden release of the weight of the rocks over a coal seam or of enormous lateral stresses due to structural or tectonic folds and thrusts and sometimes both. A rock bump may take the form of a pressure bump or a shock bump. (Nelson, 1965)

rock burst A sudden and often violent breaking of a mass of rock from the walls of a tunnel, mine, or deep quarry, caused by failure of highly stressed rock and the rapid or instantaneous release of accumulated strain energy. It may result in closure of a mine opening, or projection of broken rock into it, accompanied by ground tremors, rockfalls, and air concussions. See also: *burst; pressure bump.* (AGI, 1987)

rock butter A soft yellowish mixture of alum with aluminum and iron oxides; a decomposition product of aluminous rocks. *stone butter.* (Standard, 1964)

rock car runner *car runner.*
rock channeler A machine used in quarrying for cutting an artificial seam in a mass of stone. It is made in several forms, the principal types being the bar channeler (in which the cutters are mounted on a carriage that works along a heavy bar or bars) and the track channeler. (Standard, 1964)
rock chute *chute; rock hole.* Also called slate chute.
rock-chute mining *bord-and-pillar method.*
rock cleavage The property or tendency of a rock to split along closely spaced planar structures, produced by deformation or metamorphism. See also: *schistosity.* (AGI, 1987)
rock cone bit *roller bit.*
rock contractor In anthracite coal mining, one who contracts to mine rock, as distinguished from coal, at a certain price per ton or footage of advance. (DOT, 1949)
rock cork A light-colored variety of asbestos. Syn: *mountain cork; rock leather.* (Standard, 1964; Fay, 1920)
rock cover Thickness of consolidated rock above the roof of an opening (equivalent to cover, minus depth of weathering or of other soil). Cf: *cover.*
rock crusher A machine for reducing rock or ore to smaller sizes. Three principal types are the jaw crusher, the gyratory, and the hammer crusher.
rock crystal (a) Transparent quartz. (ASTM, 1994) (b) Highly polished brown glassware, hand-cut or engraved. (ASTM, 1994)
rock cut A way, esp. for a railroad, cut through a rock or rocky formation. (Mathews, 1951)
rock cuttings *cuttings; sludge.*
rock cycle A sequence of events involving the formation, alteration, destruction, and reformation of rocks as a result of such processes as magmatism, erosion, transportation, deposition, lithification, and metamorphism. A possible sequence involves the crystallization of magma to form igneous rocks that are then broken down to sediment as a result of weathering, the sediments later being lithified to form sedimentary rocks, which in turn are altered to metamorphic rocks. (AGI, 1987)
rock dredge (a) A dredge that excavates rock for the purpose of deepening harbors and waterways; also, a device for sampling underwater outcrops and boulders. (b) A general term for a seabed sampling device consisting of a heavy bucket frame that is pulled across the seabed on a cable controlled by a winch operator on deck.
rock drift A horizontal mine passage cut in rock, esp. along a vein on a principal level of a mine. *crosscut; stope.*
rock drill (a) A machine for boring relatively short holes in rock for blasting purposes. It may be a sinker, jackhammer, drifter, or stoper. See also: *percussive drill; rotary drill.* Syn: *Waugh drill.* (Nelson, 1965) (b) A roller bit. (Long, 1960) (c) A conical bit for drilling hard rock. (AGI, 1987)
rock-drill bit *rock bit.*
rock driller (a) In bituminous coal mining, one who works in rock or slate as distinguished from coal. Also called rock shooter; slate driller. (DOT, 1949) (b) *rock splitter.*
rock drivage A hard heading or stone drift. (Nelson, 1965)
rock dust (a) The general name for any kind of inert dust used in rendering coal dust inert or in filling rock-dust barriers. Equivalent to the British stone dust. (Rice, 1913-18) (b) The dust produced in mines by blasting, drilling, shoveling, and handling rock. Rock dust in suspension varies in particle size and composition. The most dangerous dusts are silica, sericite, and asbestos; but all fine dusts are health hazards when inhaled. The smaller sizes, 10 microns and less, are more dangerous than the larger sizes. Wet drills, sprays, water infusion, and ample ventilation are employed to reduce the dust menace. See also: *dust consolidation; dust-free conditions; stone dust; stone-dust barrier.* (Nelson, 1965)
rock-dust barrier (a) A device that releases a large quantity of inert dust in the air in the path of an explosion, extinguishing the flame. (Rice, 1913-18) (b) A series of troughs or shelves laden with rock dust and so arranged that the air waves from an explosion will trip them and fill the air with rock dust and thus quench the flame of exploding coal dust.
rock duster (a) A machine that distributes rock dust over the interior surfaces of a coal mine by means of air from a blower or pipeline or by means of a mechanical contrivance, to prevent coal dust explosions. Also called rock-dust machine. (b) *rock-dust man.*
rock dusting (a) The dusting of underground areas with powdered limestone to dilute the coal dust in the mine atmosphere and on the mine surfaces, thereby reducing explosion hazards. (b) A very widespread control measure used in coal mines to combat explosive dusts. By machine, inert (combustible) dust is sprayed, dry or wet, on the roof, floor, and ribs in all working places and haulageways, to reduce the explosibility of settled coal dust. The Mine Safety and Health Administration requires rock dusting to within 40 ft (12 m) of the face. The incombustible content of settled dust samples after rock dusting must constitute 65% or more by weight, with an increase of 1% for each 0.1% methane present. A dust as nearly inert, physiologically, as possible, should be employed in rock dusting; limestone (calcium carbonate) is most widely used. (Hartman, 1982)
rock-dusting machine A machine consisting essentially of a flexible hose fed by a powerful blower. It is used in forcing rock dust, usually powdered limestone, onto the floor, walls or ribs, and rooms and entries of a mine, thereby making the coal dust nonexplosive.
rock-dust man In bituminous coal mining, a laborer who sprinkles rock dust by hand or with a machine throughout mine workings as a precaution against explosions. Syn: *rock duster.* Also called rock-dust sprinkler. (DOT, 1949)
rock-dust testing kit This kit is designed to prevent coal-dust explosions. It helps to determine the explosion hazard prior to rock dusting, the fineness of the rock dust as it comes from the pulverizer, and the percentage of combustible matter present in rock and coal dust mixtures after rock dusting. (Best, 1966)
rock-dust zone A section of a mine entry, the ribs, roof, and floor of which have been coated with rock dust. (Rice, 1913-18)
rocker (a) A small digging bucket mounted on two rocker arms in which auriferous alluvial sands are agitated by oscillation, in water, to collect gold. A shortened term for rocker shovel; rocker arm shovel. (b) Used for testing placer deposits and for working pockets and small placer deposits. The gold-bearing gravel is placed on the screen; gold and fine sand are washed through the screen, and remaining stones are cleaned out. A chute directs the material to the upper end of the bottom, which may be covered with small transverse riffles or canvas. Waste material passes over a tailpiece at the end of the rocker. Rockers range in length from 6 to 12 ft (2 to 4 m), and in bottom width from 14 to 20 in (35 to 50 cm), with holes in the screens from 1/4 to 1/2 in (0.6 to 1.2 cm) in diameter. The slope of the rock should be adjusted to the nature of the gravel and is commonly 1 in 12, ranging from 1 in 8 to 1 in 20. Two workers with a rocker can handle from 3 to 5 yd^3 (2.3 to 3.8 m^3) of gravel in place in 10 h if the ground is easily rocked. (Lewis, 1964) (c) A portable sluicebox used by prospectors and fossickers in treating alluvial mineral deposits. Also called rocking cradle. (Pryor, 1963)
rocker arm (a) A lever resting on a curved base so that the position of its fulcrum moves as its angle changes. (Nichols, 1976) (b) A bell crank with the fulcrum at the bottom. (Nichols, 1976)
rocker arm shovel *rocker shovel.*
rocker bottom *rocker.*
rocker dump car Among the smaller capacity cars, the most popular and most widely used are the gravity dump types, such as rocker dump and scoop cars, designed so that the weight of the load tips the body when a locking latch is released by hand. The body of this type is balanced to right itself after the load is discharged. Rocker dump cars range in capacity from 1 yd^3 (0.76 m^3) handloaded types, to units of 10 yd^3 (7.6 m^3) for power shovel loading. (Pit and Quarry, 1960)
rocker shovel A digging and loading machine consisting of a bucket attached to a pair of semicircular runners that when rolled, lifts and dumps the bucket load into a car or other materials transport unit behind the machine.
rocker sieve A miner's cradle or rocker, a cradlelike device for washing out mud from the contents of a dredge. (Mathews, 1951)
rock excavation In situ removal of all firm, unaltered, and unweathered surface geological materials.
rock fabric *fabric.*
rock factor *resistance to blasting.*
rock failure Fracture or failure of a rock that has been stressed beyond its ultimate strength. (AGI, 1987)
rockfall (a) The relatively free falling or precipitous movement of a newly detached segment of bedrock (usually massive, homogeneous, or jointed) of any size from a cliff or other very steep slope; it is the fastest form of mass movement and is most frequent in mountain areas and during spring when there is repeated freezing and thawing of water in cracks in the rock. Movement may be straight down, or in a series of leaps and bounds down the slope; it is not guided by an underlying slip surface. (AGI, 1987) (b) The mass of rock moving in or moved by a rockfall; a mass of fallen rocks. Also spelled rock fall. (AGI, 1987)
rock fault Eng. A replacement of a coal seam over a greater or lesser area by some other rock, usually sandstone. See also: *horse; washout.* (Fay, 1920)
rock-fill dam An earth dam built of any broken rock or similar material that may be available. (Hammond, 1965)
rock filling (a) Waste rock, used to fill up worked-out stopes to support the roof. (Weed, 1922) (b) *overhand stoping.*
rock flour (a) Powdered rock material formed by the grinding-up of rocks beneath a glacier, either deposited as part of the till or washed or blown away and deposited elsewhere as stratified drift or loess. Also called glacier meal. Syn: *rock meal.* See also: *silt.* (Fay, 1920) (b) Fault gouge.

rock flow (a) The movement of solid rock when it is in a plastic state. (Leet, 1958) (b) The term given to a slope failure when there is a general breakdown of the rock mass. When such a rock mass is subjected to shear stresses sufficient to break down the cement or to cause crushing of the angularities and points of the rock blocks, the blocks will move as individuals and the mass will flow down the slope, or will slump into a more stable slope position. (Woodruff, 1966)
rock formation *formation.*
rock-forming mineral A mineral that is common and abundant in the Earth's crust; one making up large masses of rock. From 20 to 30 minerals are usually considered as being the most important. (Stokes, 1955)
rock foundation A foundation that is carried down to the solid rock. The rock is cut and dressed level, loose and decayed portions are removed, and holes filled with concrete. The crushing strength of the rock can be ascertained by tests and the bearing pressure should not exceed one-eighth of the value. (Nelson, 1965)
rock fracture When rock is broken by crushing or impact, the resulting fragments can be divided into two components: (1) the complement, comprising a wide size distribution in accordance with a probability law, and (2) the residue of large incompletely broken pieces. The relative proportions of complement and residue depend upon the mode of fracture. If the rock is completely crushed, only complement is formed, but if the rock is fractured by the impact of a point or wedge, there may be more residue than complement. Syn: *complement.* (Roberts, 1960)
rock glacier An ice-cored mass of angular rock waste, usually heading in a cirque or other steep-walled amphitheater and in many cases grading into a true glacier.
rock glass Obsidian or other volcanic glass.
rock gypsum A massive, coarsely crystalline to finely granular, sedimentary rock of the mineral gypsum with bedding commonly disturbed by expansion during hydration of parent anhydrite. Syn: *gyprock.*
rock hardness The resistance of the rock to the intrusion of a foreign body. (Stoces, 1954)
rockhead (a) The upper surface of bedrock. (b) The boundary between superficial deposits (or drift) and the underlying solid rock. (BS, 1964) (c) *bedrock.*
rock hole A short staple shaft driven from a lower to a higher coal seam and used for the gravity transfer of coal to the haulage road in the lower seam. See also: *roofing hole.* Syn: *rock chute.* (Nelson, 1965; Pryor, 1963)
rock hound An amateur mineralogist or collector.
rocking (a) The process of separating ores by washing on an incline trough. See also: *rocker.* (Fay, 1920) (b) Pushing a resistant object repeatedly, and backing or rolling back between pushes to allow it to reach or cross its original position. (Nichols, 1976)
rocking beam *walking beam.*
rocking cradle Short sluice, hand-oscillated; used in gold prospecting and fossicking. See also: *rocker.* (Pryor, 1958)
rocking lever A beam to give the reciprocal motion in hand boring. (Nelson, 1965)
rock kicker Usually found at sand and gravel processing facilities. A mechanical device consisting of a roller (with rows of metal protrusions along its length) placed at a 45° angle just ahead of a feed hopper (or conveyor transfer point). It is usually run by a small electric motor and as roots, large stones, and clay meet the roller, they are dumped off the belt to a small pit that is periodically cleaned out.
rock kindred *rock association.*
rock leather *mountain leather.*
rock loader (a) Any device or machine used specif. for loading slate or rock inside a mine. However, it is most frequently used with scraper loaders equipped for handling rock. Syn: *rock-loader operator.* (Jones, 1949) (b) *box loader.*
rock-loader operator *rock loader.*
rockman In bituminous coal mining, a foreman who is in charge of the drilling of holes in rock or slate and the charging and tamping of explosives in the holes drilled by miners prior to blasting. See also: *slateman.* (DOT, 1949)
rockman helper *slate-shooter helper.*
rock meal *rock flour.*
rock mechanics (a) Mathematical analysis of the forces acting along joints, faults, and bedding planes of natural rock in situ, esp. in the evaluation of wall strengths, and hence slopes and slope ratios, in open-pit mines. (b) The theoretical and applied science of the physical behavior of rocks, representing a branch of mechanics concerned with the response of rock to the force fields of its physical environment (NAS-NRC, 1966). (AGI, 1987)
rock melt A liquid solution of rock-forming mineral ions at sufficiently high temperatures to be considered molten.
rock milk Soft pulverulent forms of calcite found in caves, or as an efflorescence. *agaric mineral.* (Fay, 1920)
rock miner In anthracite and bituminous coal mining, a miner who works in rock as distinguished from coal. (DOT, 1949)
rock navvy A cranelike loading machine used at opencast pits and quarries. See also: *power shovel.* (Nelson, 1965)
rock pedestal *pedestal.*
rock phosphate *phosphate rock.*
rock pillar *hoodoo.*
rock pressure (a) The pressure exerted by surrounding solids on the support system of underground openings, including that caused by the weight of the overlying material, residual unrelieved stresses, and pressures associated with swelling clays. (AGI, 1987) (b) The compressive stress within the solid body of underground geologic material. (AGI, 1987) (c) *ground pressure; geostatic pressure.* (d) In petroleum geology, the pressure under which fluids, such as water, oil, and gas, are confined in rocks. No particular cause or origin of the pressure is implied. Geophysicists and isostasis, however, have used, and are using, the term rock pressure in the primitive and more correct sense of the pressure exerted on underlying rock by superincumbent strata. To avoid confusion, it is desirable to substitute for the term rock pressure, as now used in oil, gas, and underground water technology, the more appropriate term reservoir pressure. (Stokes, 1955)
rock pressure burst A sudden and violent failure of rock masses under stresses exceeding the elastic strength of the rock. The classification and nomenclature of these occurrences are not clear and are based largely on effects and not on the basic causation factor. See also: *pressure bump; rock bump; rock burst.* (Nelson, 1965)
rock quartz Ordinary crystalline quartz. Also called rock crystal.
rock rake A heavy duty rake blade. (Nichols, 1976)
rock roll Inverted ridges of rock, usually sandstone, extending from the overlying strata into a coal seam, caused by localized streams active during the formation of the coal. See also: *classical washout.* (Nelson, 1965)
rock ruby A fine red variety of pyrope garnet. See also: *pyrope.*
rock salt Coarsely crystalline halite, NaCl, resulting from evaporation of saline water; in massive, fibrous, or granular aggregates; occurs as a nearly pure sedimentary rock, as extensive beds, or in domes or plugs. See also: *common salt; salt.*
rock sediment The combined cuttings and residue from drilling and sedimentary rocks and formations, commonly known as sand pumpings. (Williams, 1964)
rock series *igneous-rock series.*
rockshaft A shaft made for sending down rock for filling the stopes, etc., generally kept nearly full, the rock being trammed away as needed. (Standard, 1964)
rock sharp A mineral expert. (Mathews, 1951)
rock shovel A machine for loading broken rock. See also: *shovel loader.* (Nelson, 1965)
rock silk A silky variety of asbestos. (Fay, 1920)
rockslide (a) A slide involving a downward and usually sudden and rapid movement of newly detached segments of bedrock sliding or slipping over an inclined surface of weakness, as a surface of bedding, jointing, or faulting, or other preexisting structural feature. The moving mass is greatly deformed and usually breaks up into many small independent units. Rockslides frequently occur in high mountain ranges, as the Alps or Canadian Rockies. (AGI, 1987) (b) The mass of rock moving in or moved by a rockslide. Also spelled rock slide. Syn: *rock slip.* (AGI, 1987)
rock slip *rockslide.*
rock slope A slope driven through rock strata.
rock soap *mountain soap.*
rock spar Material filling fracture cleavages in coal, consisting of nonclay mineral matter, probably deposited from solution, and sand, usually calcite or gypsum. (AGI, 1987)
rock splitter In the stonework industry, one who splits large blocks of building granite, marble, and sandstone into slabs or smaller blocks, by drilling holes into the stone and then driving wedges into them until the stone breaks along the line of drilled holes. Also called rock breaker; rock driller. See also: *rock driller; quarryman.* (DOT, 1949)
rock stress (a) *rock pressure.* (b) The problem of determining the stresses that exist in the Earth's crust has long been of interest to engineers and geologists. Many mining problems are directly concerned with stresses that may cause mine openings to collapse. Two phases of occurrence of rock stresses are important: (1) the stresses existing in the rock before the excavation of the mine openings; i.e., the free field stress, and (2) the indirect stresses caused by the mine openings. See also: *free field stress.* (Lewis, 1964)
rock temperature (a) The formational temperature at depth. The rate of increase of temperature with depth (the geothermal gradient) is highly variable over the earth, but averages 25 °C/km. (AGI, 1987) (b) The temperature of the rock in a mine. (Lewis, 1964)
rock tunnel A tunnel, drift, or crosscut driven through rock, usually connecting one coalbed with another; also through barren rock in metal mines.

rock turquoise A matrix of turquoise with small grains of turquoise embedded in it. (Fay, 1920)

rock type (a) One of the three major groups of rocks: igneous, sedimentary, metamorphic. (AGI, 1987) (b) A particular kind of rock having a specific set of characteristics. It may be a general classification, e.g., a basalt, or a specific classification, e.g., a basalt from a particular area and having a unique description. (AGI, 1987) (c) The megascopically recognizable ingredients of coal rock; i.e., vitrain, clarain, durain, and fusain. See also: *banded ingredient.*

rock waste *debris.*

rock weight S. Afr. One (short) ton (0.9 t) of rock in place equals about 12 ft^3 (0.34 m^3). Horizontally, therefore, the weight of an ore reserve covering a claim over a stoping width of 3 ft (0.91 m) is 64,000 $ft^2 \times 3$ ft $\div$ 12 ft^3/st = 16,000 st (14,500 t) at 100% payability. In case the vein dips downward, the resulting amount must be divided by the cosine of the angle of dip. (Beerman, 1963)

Rockwell hardness test A method of determining the relative hardness of metals and case-hardened materials. The depth of penetration of a steel ball (for softer metals) or of a conical diamond point (for harder metals) is measured. See also: *scleroscope hardness test.* (Nelson, 1965)

Rockwell machine Trade name for an apparatus that measures the hardness of metals and alloys, in which a diamond-pointed cone is pressed under a specific load into the metal. The relative resistance to penetration (Rockwell hardness) is indicated by a number (Rockwell number) on a dial. The operation is called a Rockwell hardness test. Syn: *Rockwell tester.* (Long, 1960)

Rockwell tester *Rockwell machine.*

rockwood *mountain wood.*

rock wool *asbestos.*

rod bit A noncoring bit designed to fit a reaming shell that is threaded to couple directly to a drill rod, thus eliminating the core barrel in blasthole drilling. Also called blasthole bit. (Long, 1960)

rod clearance *clearance.*

rod coupling Name for a double-pin-thread coupling used to connect two drill rods together. (Long, 1960)

rod damp *safety clamp.*

rodding (a) Cleaning and descaling of piping by means of scrapers attached to series of jointed rods. (Pryor, 1963) (b) Eng. The operation of fixing or repairing wooden cage guides in shafts. (c) In metamorphic rocks, a linear structure in which the stronger parts, such as vein quartz or quartz pebbles, have been shaped into parallel rods. Whether the structure is formed parallel to the direction of transport or parallel to the fold axes has been debated. (AGI, 1987)

rod dope Grease or other material used to protect or lubricate drill rods. Also called rod grease. Syn: *gunk.* (Long, 1960)

rod drag The rubbing of the rods or drill string on the sidewalls of the borehole. Syn: *rod friction.* (Long, 1960)

rod drop The distance of slump or slag in a long string of rods when released from the drill chuck. (Long, 1960)

rod elevator *elevator; elevator plug.*

rod friction (a) The drag created in the flow of the drilling liquid by contact and constrictional effects created by the inside surface of the drill rods and couplings. Cf: *skin friction; wall friction.* (Long, 1960) (b) *rod drag.*

rodingite A massive dense buff to pink rock typically rich in grossular garnet and calcic pyroxene, and enveloped in serpentinite. Epidote, vesuvianite, and other calcium-rich minerals are commonly present. It is formed by metasomatic alteration of a protolith that, in many cases, was a dike rock, as shown by preservation of structures. The name was applied by Bell in 1911. (AGI, 1987)

Rodio-Dehottay process A method of shaft sinking by the freezing method. It is based on the direct cooling effect of expanding highly compressed carbon dioxide in the freezing pipes. See also: *Koch freezing process.* (Nelson, 1965)

rodlike Refers to elongate crystals. Cf: *acicular; equant; tabular; lathlike.*

rodman (a) A person who uses or carries a surveyor's leveling rod. Also called rodsman. (Standard, 1964) (b) *staff man.*

rod mill (a) A mill for rolling rod. (ASM, 1961) (b) A mill for fine grinding, somewhat similar to a ball mill, but employing long steel rods instead of balls to effect the grinding. (ASM, 1961)

rod millman One who grinds clinker, phosphate rock, or ore in a revolving cylinder partially filled with round steel rods; also, he or she tests a product for fineness by observing how much material is left on sieve of determined mesh, and regulates amount of material entering the mill accordingly. (DOT, 1949)

rodney Eng. A rude platform near the shaft's mouth for a night fire. (Fay, 1920)

rod plug *elevator plug.*

rod proof A test specimen taken from the melt on an iron rod. (ASTM, 1994)

rod pull (a) The number of borehole round trips made in a unit of time. (Long, 1960) (b) The number of lengths of drill rod (two or more standard 10-ft lengths coupled together and handled and stacked as unit lengths) needed to reach the bottom of the borehole. (Long, 1960)

rod puller Various mechanisms, essentially a double-acting air-actuated piston equipped with a rod-gripping device, commonly used to pull drill rods from a borehole in underground workings where a small drill without a hoist is used, or from drill rods stuck in a drill hole. Syn: *air rod puller.* (Long, 1960)

rod reaming shell A reaming shell designed to be coupled directly to a drill rod. See also: *rod bit.* (Long, 1960)

rod reducing bushing A pin-to-box sub used to connect one size rod in a string to a larger or smaller size. (Long, 1960)

rod reducing coupling A pin-to-pin adapter used to connect a rod of one size to one of a larger or smaller size. (Long, 1960)

rods (a) Eng. Vertical or inclined timbers for actuating pumps. (b) Long bars of Swedish iron of the toughest quality, for boring through rocks, etc. (c) *cage guide.*

rod sag The bending of a long drill string due solely to its own weight. Also called rod slack. Syn: *sag.* (Long, 1960)

rod shaft The mine shaft containing the pump rods.

rod shell *rod reaming shell.*

rod slack *rod sag.*

rod slap The impact of drill rods with the sides of a borehole, occurring when the rods are rotating. Syn: *whip.* (Long, 1960)

rod snap A sudden acceleration in rotational speed of the rods followed immediately by a sudden return to the former speed. (Long, 1960)

rod spear A long, tapered, four-sided fishing tool. Used to remove a lost drill rod or other tubular piece of drill equipment from a borehole. (Long, 1960)

rod stand The length of drill rod handled and stacked in the tripod or derrick as a unit piece during round trips. Also called offtake. See also: *double; treble.* (Long, 1960)

rod stock Round steel rod. (Nichols, 1976)

rod string The drill rods coupled to form the connecting link between the core barrel and bit in the borehole and the drill machine at the collar of the borehole. (Long, 1960)

rod stuffing box An annular packing gland fitting between the drill rod and the casing at the borehole collar. It allows the rod to rotate freely but prevents the escape of gas or liquid under pressure. Esp. utilized when drilling with counterflow; when drilling in an area where a high hydrostatic pressure or flow of water may be encountered, as in drilling a cover or pilot hole; or when drilling up holes from an underground drill site. (Long, 1960)

roentgen equivalent man A unit of ionizing radiation, equal to the amount that produces the same damage to humans as 1 R of high-voltage X-rays. Abbrev. rem. (McGraw-Hill, 1994)

roentgenite *röntgenite.*

roentgenogram A photograph made with X-rays. Syn: *roentgenograph; radiograph.* (Webster 3rd, 1966)

roentgenograph *roentgenogram.*

Roesing lead pump An automatic apparatus for discharging lead from the kettle; used in the Parkes process. (Fay, 1920)

Roesing wires Wires suspended in a dust chamber to assist in settling and condensing dust and fumes from furnace gases. (Fay, 1920)

Roesler process A process for separating copper, and in part silver, from gold by fusing with sulfur or with antimony sulfide, obtaining copper or silver sulfide. (Fay, 1920)

roestone A fine-grained oolite resembling the roe of a fish. *oolite.* (Fay, 1920)

Rohbach solution An aqueous solution of mercuric barium iodide with a density of 3.5 g/cm^3; used for separating minerals by density.

Roheisenzunder process A method that makes use of an airstream at a pressure of 4 atmospheres for atomizing molten pig iron into minute particles. The molten metal falling into an air stream formed by an annular slit in a steel cyclone is atomized, the particles falling into a water bath and subsequently dried. (Osborne, 1956)

Rohrbach solution An aqueous solution of mercuric barium iodide; clear, yellow liquid; very refractive; sp. gr., 3.5. Used in separating minerals by their specific gravity and in microchemical detection of alkaloids. (CCD, 1961)

roke Prov. Eng. A vein of ore. A variation of rake. *rake.*

roll (a) An elongate protrusion of shale, siltstone, or sandstone (locally limestone) from the roof into a coal seam, causing a thinning of the seam and sometimes replacing it almost entirely. Cf: *cutout.* A roll is commonly overlain by a thin coal stringer. (AGI, 1987) (b) An elongate upheaval of the floor material into a coal seam, causing thinning of the seam. Syn: *horseback.* (AGI, 1987) (c) Various minor deformations or dislocations of a coal seam, such as washouts, small monoclinal folds, or faults with little displacement. (AGI, 1987) (d) In veins and other types of ore deposit, a thickening or an arcuate change in dip in the orebody. (e) *roll orebody.* (f) A rotating cylinder used to support or guide a portion of conveyor belt. (g) One of two cylinders or grooved rollers between which material is drawn, for reducing its thickness, as the finished rolls of a rolling mill. (BS, 1964) (h) The appearance of other types of mineral deposits in places where the bed or vein thickens or thins. (Mason, 1951) (i) A roughly cylindrical distribution of uranium mineralization occurring usually in the Salt Wash Sandstone. There is some question whether the feature is structural or sedimentary. (j) An inequality in the roof or floor of a mine. (NEMA, 1956) (k) S. Wales. The drum of a winding engine. (l) Cast-iron or steel cylinder, used to break coal and other materials into various sizes. Applies to the type of crushing machinery in which the ore or coal is broken between cylindrical rolls, either plain or fitted with steel teeth, revolving toward each other, drawing the material in between the crushing peripheries, which rotate in a vertical plane. (Fay, 1920; Liddell, 1918) (m) In powder metallurgy, a machine used to apply pressure progressively to form a compact. See also: *crushing roll; ribbed roll; smooth roll; want; washout.* (Rolfe, 1955)

roll compacting The progressive compacting of metal powders by the use of a rolling mill. (ASTM, 1994)

roll crusher A type of secondary or reduction crusher consisting of a heavy frame on which two rolls are mounted. These are driven so that they rotate toward one another. Rock fed in from above is nipped between the moving rolls, crushed, and discharged at the bottom. See also: *double-roll crusher.* (Newton, 1959)

rolled metal Refers to metal, such as silver or stainless steel, which has been clad with a precious metal and rolled to reduce the thickness of the coat. (USBM, 1965)

rolled plate A thin plate of gold spread upon a layer of base metal by soldering the metals in the bar and then rolling the whole out into plate, forming a thinner plate of gold than that of the ware known as gold-filled. Also called rolled gold. (Fay, 1920)

rolled-steel joist An I-beam made from a single piece of steel passed through a hot rolling mill. (Hammond, 1965)

roller (a) A broad pulley or wheel fixed to the floor, roof, or sides of roadways to prevent a haulage rope running against the ground that would cause excessive friction and wear of rails and sleepers. (Nelson, 1965) (b) A component part of a roller chain in which it may serve only to reduce frictional loss occurring as the chain negotiates sprockets. Rollers may also serve as the rolling support for the chain and the load being conveyed. (c) A heavy vehicle used for compacting soil, earth fill, and top layers of spoil dumps to increase the density and bearing capacity of the material. (Nelson, 1965)

roller bearings Hard steel cylinders in bearings that have very low frictional resistance. (Hammond, 1965)

roller bit (a) A rotary boring bit consisting of two to four cone-shaped, toothed rollers that are turned by the rotation of the drill rods. Such bits are used in hard rock in oil well boring and in other deep holes down to 5,000 m and more. See also: *tricone bit; drag bit.* (Nelson, 1965) (b) A type of rock-cutting bit used on diamond and rotary drills. The bit consists of a shank with toothed, circular, or cone-shaped cutter parts affixed to the head of the bit in such a manner that the cutters roll as the bit is rotated. Generally used for drilling 10-cm-size or larger holes in soft to medium-hard rocks, such as shale and limestone. Usually noncoring and not diamond set. Also called cone bit. Cf: *noncoring bit.* See also: *rock bit; roller rock bit.* Syn: *roller cone bit; roller-cutter bit; rolling cutter bit; toothed roller bit.* (Long, 1960)

roller chain (a) Generally, any sprocket-driven chain made up of links connected by hinge pins and sleeves. (Nichols, 1976) (b) Specif., a chain whose hinge sleeves are protected by an outer sleeve or roller that is free to turn. (Nichols, 1976)

roller cone bit *roller bit.*

roller-cone core bit A type of roller bit with cutter cones arranged to cut an annular ring leaving an uncut section in the center as core. (Long, 1960)

roller conveyor A series of rolls supported in a frame over which objects are advanced manually, by gravity or by power. See also: *controlled velocity roller conveyor; el conveyor; gravity conveyor; herringbone roller conveyor; hydrostatic roller conveyor; portable conveyor.* Syn: *gravity roller conveyor.*

roller-cutter bit *roller bit.*

roller gate Hollow cylindrical crest gate controlling a dam spillway. See also: *sector gate; sliding gate.* (Hammond, 1965)

roller grip A device for clutching a traction cable between grooved sheaves or rollers.

rollerman In mining, a laborer who inspects idler rollers or pulleys over which a cable passes along inclined haulageways, oiling or greasing rollers, resetting displaced ones, and repairing or replacing damaged ones. Also called pulley man; pulley repairer; pulley repairman; roller repairman; sheaveman; wheelman. (DOT, 1949)

roller repairman *rollerman.*

roller rock bit A rotary bit fitted with two or more hardened steel or tungsten-carbide-tipped rollers of cylindrical or conical form. Variously known as two-cone bit, three-cone bit, four-cutter bit, etc. See also: *roller bit.* (BS, 1963)

roller screen *revolving screen.*

rolley man *incline repairman.*

roll feeder (a) A smooth, fluted, or cleated roll or drum that rotates to deliver packages, objects, or bulk materials. (b) A circular drum, plain or ribbed, rotating on a horizontal shaft and situated at the mouth of a bunker or hopper to control the rate of discharge of material therefrom. (BS, 1962)

roll-front orebody A roll orebody of the Wyoming type, which is bounded on the concave side by oxidized altered rock typically containing hematite or limonite, and on the convex side by relatively reduced altered rock typically containing pyrite and organic matter. See also: *roll orebody.* (AGI, 1987)

rolling and quartering A sampling method in which the sample is formed into the requisite flat heap by placing it upon a rubber or other smooth sheet and, by lifting the corners of this sheet in proper rotation, rolling the material to and fro. The resultant heap is then quartered and alternate quarters are taken. This method is used with smaller bulk and smaller sizes of material. (Truscott, 1962)

rolling cradle A rod slide equipped with rollers that contact the rods and over which the rods roll on being pulled or lowered into an angle borehole. (Long, 1960)

rolling cutter bit *roller bit.*

rolling plant A rolling mill or establishment for rolling metal into forms. (Standard, 1964)

rolling resistance (a) The sum of the external forces opposing motion over level terrain. (Carson, 1961) (b) The tractive resistance caused by friction between the rails and wheels, which forms the major resistance on level tracks. See also: *tractive force.* (Nelson, 1965)

rolling-up curtain weir A type of frame weir, the frame of which remains upright, being rolled up from the bottom. (Hammond, 1965)

roll jaw crusher A crusher of the same general type as the Blake or Dodge, but the moving jaw has a rolling instead of an oscillating motion. (Liddell, 1918)

rollman In beneficiation, one who tends rolls that are used to crush ore, which has already been broken into small pieces in a crusher, to a fine size preparatory to the extraction of the valuable minerals. (DOT, 1949)

roll operator One who operates conical rolls that separate stone from clay, preventing machine from jamming by regulating flow of clay into it. (DOT, 1949)

roll orebody A uranium and/or vanadium orebody in a sandstone lens or layer, which cuts across bedding in sharply curving forms, commonly C-shaped or S-shaped in cross section. Two types can be distinguished: the Colorado Plateau type, named in 1956, and the Wyoming type, named in 1962. Roll orebodies of the Colorado Plateau type are of highly variable geometry, with their longest dimension in plan view parallel to the axes of buried sandstone lenses representing former stream channels, and surrounded by a wide halo of reduced (altered) rock. Orebodies of the Wyoming type are crescent-shaped in cross section and typically form in relatively thick, tabular, or elongate sandstone bodies, with the tips of the crescent thinning and becoming tangent to mudstone layers above and below. See also: *roll-front orebody.* (AGI, 1987)

roll scale *mill scale.*

roll screen A screen consisting of a number of horizontal rotating shafts, fitted with elements arranged to provide screening apertures. (BS, 1962)

roll sulfur (a) A commercial name for sulfur that has been purified and cast into rolls or sticks. (Standard, 1964) (b) *brimstone.*

Roman cement *pozzolana cement.*

romanechite A monoclinic mineral, $(Ba,H_2O)Mn_5O_{10}$; rare as single crystals; commonly intergrown with other manganese oxides. Formerly called psilomelane, a term now reserved for mixtures. Cf: *hollandite.*

romanite A yellow, black, or green amber from Romania. Also spelled rumanite. (English, 1938)

Roman ocher A native ocher of a deep orange-yellow color. (Standard, 1964)

Roman pearl A hollow sphere of opalescent glass with its interior coated with essence d'orient and then filled with wax.

romeite An isometric mineral, $(Ca,Fe,Mn,Na)_2(Sb,Ti)_2O_6(O,OH,F)$; stibiconite group; forms clusters of minute yellow octahedra. Syn: *weslienite.* See also: *atopite.*

romometer An instrument for measuring changes in vertical height and lateral movements of the roof relative to the floor at the coal face. Syn: *roof movement meter.* See also: *convergence recorder.* (Nelson, 1965)

rondle The crust or scale that forms upon the surface of molten metal in cooling. (Fay, 1920)

röntgen The unit of exposure dose of X-ray or gamma-ray radiation. One röntgen is an exposure dose of X-ray or gamma-ray radiation such that the associated corpuscular emission per 0.001293 g of air produces, in air, ions carrying 1 electrostatic unit of quantity of electricity of either sign. Designated by the symbol R. Also spelled roentgen. (NCB, 1964)

röntgenite Minute wax-yellow to brown, trigonal pyramidal crystals, intergrown with synchysite, parisite, and bastnasite, from Narsarsuk, Greenland. From X-ray and optical data, the composition is deduced as $Ca_2(Ce,La)_3(CO_3)_5F_3$. Also spelled roentgenite. (Spencer, 1955)

roof (a) The rock immediately above a coal seam. It is commonly a shale and is often carbonaceous in character and softer than similar rocks higher up in the roof strata. The roof shale may contain streaks and wisps of coaly material, which tends to weaken the deposit. Roof in coal mining corresponds to hanging wall in metal mining. See also: *roof stone*. (Nelson, 1965) (b) In mine timbering there are two classifications of roof, the immediate roof and the main roof. The immediate roof lies directly over the coal and may be a single layer or several layers of rock material of the same, or different consistencies, and from a few inches to several feet in thickness. This roof requires timbering to support it as the coal is removed. The main roof is the roof above the immediate top, and may vary from a few feet to several hundred, or even thousands of feet in thickness. This roof is generally controlled by leaving pillars of solid coal that will support its weight. (Kentucky, 1952) (c) The country rock bordering the upper surface of an igneous intrusion. Cf: *floor*. (AGI, 1987)

roof bolt (a) A long steel bolt inserted into walls or roof of underground excavations to strengthen the pinning of rock strata. It is inserted in a drilled hole and anchored by means of a mechanical expansion shell that grips the surrounding rock at about 4 ft (1 m) spacing and pins steel beams to the roof. (b) Syn: *rock bolt*. (c) Current roof bolting consists of steel rods, 5/8 to 1 in (2 to 2.5 cm) or more in diameter and 3 to 8 ft (1 to 2.5 m) in length, anchored by a mechanical expansion shell, resin grout, or a combination of both. Grouted bolts may be fully or partially grouted. A steel plate, sometimes in combination with wooden headers or steel straps, fits tightly between the bolthead and mine roof or rib.

roof bolter In bituminous coal mining, one who reinforces roofs of mine haulageways, side drifts, and working places with metal or timber to prevent rock and slate falls. Also called raise driller; stoperman; stoperperson; timberman. (DOT, 1949)

roof bolting A system of roof support in mines. Boreholes usually from 3 to 12 ft (1 to 4 m) long are drilled upward in the roof, and bolts of 5/8 to 1 in (2 to 2.5 cm) or more in diameter are inserted into the holes and anchored at the top by a split cone, mechanical anchor, or resin grout. The bolts are put up in a definite pattern. The idea is to clamp together the several roof beds to form a composite beam with a strength considerably greater than the sum of the individual beds acting separately. See also: *slot-and-wedge bolt; pin timbering; rock bolting*. Cf: *strata bolt*.

roof control The scientific study of the behavior of rock undermined by mining operations and the most effective measures of controlling movements and failure. The subject is comprehensive, including the systematic measurement of the movement of strata and the forces and stresses involved. An attempt is made to correlate data with rock types and the type of excavation. See also: *rock mechanics*.

roof cut A machine cut made in the roof immediately above the seam. A roof cut is sometimes made in a soft band of dirt over the coal, which gives increased height in thin seams. The cut is made with a turret coal cutter. (Nelson, 1965)

roof cutting It is a common occurrence to hear miners talk of gas cutting the roof and causing it to weaken; however, this condition is seldom encountered. There are some seams where gas does cut the roof, generally where top coal is left in gassy seams. The most common cause of roof cutting is its exposure to air. Gunite or painting of the top helps a condition of this kind. (Kentucky, 1952)

roof drill Various hydraulically operated mechanized machines designed to install roof bolts. Two workers can install up to 200 bolts per shift. Units are available in both standard and special design to satisfy requirements in different mines. (Best, 1966)

roof-framy A roof that is tenacious and when allowed to fall breaks down in large blocks or frames of stone. (Peel, 1921)

roofing The wedging of a loaded wagon or horse against the top of an underground passage. (Fay, 1920)

roofing hole In West Wales, a small, steeply inclined stone drivage from a lower to an upper coal seam or for exploration in disturbed ground. Syn: *roof-up*. (Nelson, 1965)

roof jack A screw- or pump-type extension post used as a temporary roof support. (BCI, 1947)

roof layer (a) Uniformly thick layer of rock supported or clamped at the edges by pillars. (Obert, 1960) (b) A layer of combustible gas under the roof of mine workings. (BS, 1963)

roof movement meter *romometer*.

roof pendant A downward projection of country rock into an igneous intrusion. Cf: *cupola*. Syn: *pendant*. (AGI, 1987)

roof pressure The pressure that the overlying rocks exert on the support of mine workings. See also: *ground pressure*. (Stoces, 1954)

roof rock Rock forming the ceiling of a cave passage, underground chamber, mine opening, etc. (AGI, 1987)

roof shale The layer or seam of shale occurring immediately above the Pittsburgh coal seam. Because of its friable nature, this shale or slate is taken down in most mining operations. (Rice, 1913-18)

roof station A survey station fixed in the roof of a mine roadway or working face. (BS, 1963)

roof stone Scot. The stone immediately above a coal seam. See also: *roof*. (Fay, 1920)

roof stringer Used in a weak or scaly top in narrow rooms or entries that have short life. It is done by placing lagging bars running parallel with the working place above the header. It has limited uses because of necessary additional height and because its weight rests on the center of the header. See also: *stringer*. (Kentucky, 1952)

roof testing In the simple testing of the roof, it is struck by a hammer or a heavy object. A loose roof will give off a dull or hollow sound compared with a solid top, which has a clear ring. Good roof that has a clear ringing sound is called "bell top." Also known as sounding, sounding the roof, sounding the top, and jowling. Syn: *top testing*. (Kentucky, 1952)

roof-testing tool Usually a wooden pole with a metallic ball at the upper end.

roof-to-floor convergence The deformation of the coal or ore pillars is estimated by monitoring the closure of the entry. This roof-to-floor convergence is generally measured with a tube extensometer, to an accuracy of 0.001 in (25.4 µm), or a tape measure, to an accuracy of 0.01 ft (3.048 mm). Measurements are repeatedly taken as the mining geometry changes. (SME, 1992)

roof-up *roofing hole*.

roof work A term applied to a vein worked overhead.

room (a) A place abutting an entry or airway where coal or ore has been mined and extending from the entry or airway to a face. (b) A wide working place in a flat mine corresponding to a stope in a steep vein. A chamber. Cf: *stope*. (c) A heading or short stall.

room-and-pillar Said of a system of mining in which typically flat-lying beds of coal or ore are mined in rooms separated by pillars of undisturbed rock left for roof support. See also: *bord-and-pillar working; County of Durham system*. Syn: *heading and stall*. (AGI, 1987)

room-and-pillar mining In coal and metal mining, supporting the roof by pillars left at regular intervals. (Lewis, 1964)

room-and-pillar with waste filling *overhand stoping*.

room boss In bituminous coal mining, a miner who inspects the working face in working places (rooms) to determine whether mining operations are being carried on properly and safety regulations are being observed. Also called wall boss. (DOT, 1949)

room conveyor (a) Any conveyor that carries coal from the face of a room toward the mouth. Normally, a room conveyor will deposit its coal into a car or another conveyor at the mouth of the room, but occasionally it will dump into a cross conveyor at some point between the face and the mouth. (Jones, 1949) (b) *underground mine conveyor*.

room entry Any entry or set of entries from which rooms are turned. A panel entry.

room neck A short passageway from the mine entry to the room in which a miner works.

room system with caving *bord-and-pillar*.

rooseveltite A monoclinic mineral, $BiAsO_4$; monazite group; at Santiaguillo, Bolivia.

root clay *underclay*.

root deposit A lode or vein from which alluvial cassiterite or gold may have been derived.

rooter (a) A towed scarifier; sometimes used to break up a hard surface and prior to the use of bulldozers in removing overburden at quarries and opencast pits. A heavy-duty ripper. (Nelson, 1965) (b) A towed machine equipped with teeth, used primarily for loosening hard soil and soft rock. (Nichols, 1954)

root hook A very heavy hook designed to catch and tear out big roots when it is dragged along the ground. (Nichols, 1976)

root-mean-square value The root-mean-square value of an alternating current or voltage. It is the square root of the mean value of the squares of the instantaneous values taken over a complete cycle. When an alternating current or voltage is specified, it is almost invariably the root-mean-square value that is used. Also used of quantities that alternate over longer periods, for example, a month or year. Also known as effective value. Abbrev., R.M.S. (CTD, 1958)

rope and button conveyor A conveyor consisting of a rope with disks or buttons attached at intervals, the upper flight running in a trough. The coal or other material is dropped into the trough, and the conveyor either is actuated by the weight of the coal in the trough when the trough is inclined forming a retarding conveyor, or moves the coal along the trough where the gradient is insufficient or adverse. In the one case a brake is provided; in the other, the sprockets are actuated by a motor. (Zern, 1928)

rope core An important component of stranded ropes is the core, which may be either of fiber or of wire. In winding ropes it is generally made of manilla, sisal, or hemp. The function of the core is to support the strands and prevent them from bearing hard against one another. An even more important function is as a store for lubricant for the interior of the rope, and during manufacture it is saturated with lubricant. (Sinclair, 1959)

rope cutter *hook tender.*

rope diameter The diameter of a steel wire rope is the maximum obtainable measurement across the outer edges of the strands. The size of fiber ropes is usually specified by their circumference. Modern steel wire winding ropes are large and heavy and may be 2-1/4 in (5.7 cm) in diameter for a moderately deep shaft. (Nelson, 1965)

rope driver In bituminous coal mining, a foreperson who looks after the haulage cable and the equipment of trains of cars by which coal is hauled from the mine. The rope driver superintends the attaching of cars to cable by clipper and directs movement of the cable by signaling a slope engineer through a buzzer system. (DOT, 1949)

rope driving The transmission of power by means of rope gearing, as distinguished from belt drive. (Crispin, 1964)

rope drum Any drum, powered or otherwise, on which rope is wound; e.g., mining machine rope drums, room hoist rope drums, etc. (Jones, 1949)

rope fastening The most suitable fastening between a wire rope and its socket is a white metal capping. Haulage ropes are generally doubled back on themselves around a steel thimble and secured with bulldog clips. (Hammond, 1965)

rope guide Steel rope suspended in a vertical shaft to prevent excessive swinging of the cages or skips. Eight rope guides are generally used for the shaft, four for each cage, and two additional rubbing ropes are installed to prevent possible collision between the cages or skips. The ropes are suspended from girders fixed on the safety hook catchplate platform and kept taut in the shaft by means of weights in the shaft bottom sump. The clearances between the cages, and also between the cage corners and the shaft wall, should be about 12 in (30.5 cm). See also: *fixed guides.* (Nelson, 1965)

rope haulage (a) Means of moving loaded and empty mine cars by use of wire rope; generally used on steep inclines where use of electric mine locomotives is inefficient. (BCI, 1947) (b) Any transportation system employing a steel wire rope to haul the mine cars or trams. See also: *direct-rope haulage; main-and-tail haulage; tail-rope haulage.* (Nelson, 1965)

rope haulage systems Systems of rope haulage may be classified as (1) self-acting or gravity planes; (2) engine planes; (3) tail-rope haulage; (4) endless-rope haulage; and (5) aerial tramways, which are frequently considered by themselves, since they are not applied to transporting material underground. (Lewis, 1964)

rope lay That length of rope in which one strand makes one complete revolution about the core.

rope plucking The sudden jerking or twitching of a haulage rope due to the rope laps slipping to a smaller diameter on the drum. A severe plucking of a rope may be felt faintly more than 800 yd (725 m) distance from the engine. See also: *overhaul.* (Nelson, 1965)

rope rider An employee whose duty it is to see that cars are coupled properly, and to inspect ropes, chains, links, and all coupling equipment. A trip rider. See also: *brakeman.* (Fay, 1920)

rope roof bolt A steel wire rope, with wedge heads fixed to its ends, used instead of the normal steel rod in roof bolting. Also known as cable bolting. The rope has a diameter of about 7/8 in (2.2 cm) and a length from 15 to 20 ft (4.5 to 6 m). (Nelson, 1965)

rope socket A drop forged-steel device, with a tapered hole, which can be fastened to the end of a wire cable or rope and to which a load may be attached. It may be either the open- or closed-end type. (Long, 1960)

ropeway (a) A line or double line of suspended ropes, usually wire, along which articles of moderate weight may be transported on slings, either by gravity or power; much used in mining districts for transportation to watercourses or to railway lines. An aerial tramway. (Standard, 1964) (b) *aerial ropeway.*

Ropp furnace A long reverberatory furnace with a series of plows or rakes that are drawn over the hearth by a continuous cable, moving the ore steadily from the feed to the discharge end. (Fay, 1920)

roquesite A tetragonal mineral, $CuInS_2$; chalcopyrite group; at Charrier, Allier, France.

rosasite (a) A monoclinic mineral, $4[(Cu,Zn)_2(CO_3)(OH)_2]$; forms green to blue spherules in oxidized zones of zinc-copper-lead deposits. (b) The mineral group glaukosphaerite, kolwezite, mcguinnessite, nullaginite, rosasite, and zincrosasite.

Röschen method A combustible-gas drainage method utilizing controlled drainage from the coal seams as they are being mined. This method, which is also known as the pack cavity method, was devised to extract gas from the mined-out areas of advancing longwall mining systems by leaving corridors or cavities at regular intervals in the pack. (Virginia Polytechnic, 1960)

roscherite A monoclinic and triclinic mineral, $Ca(Fe,Mn)_2Be_3(PO_4)_3(OH)_3{\cdot}2H_2O$.

roscoelite A monoclinic mineral, $K(V,Al,Mg)_2(AlSi_3)O_{10}(OH)_2$; mica group; soft; a source of vanadium.

rose beryl The morganite variety of beryl.

rose bit A hardened steel or alloy noncore bit with a serrated face to cut or mill out bits, casing, or other metal objects lost in the hole. Also used to mill off the rose-bit dropper on a Hall-Rowe wedge. Also called mill; milling bit. Cf: *junk mill.* (Long, 1960)

rose copper *rosette copper.*

rose diagram A circular diagram for plotting strikes (with or without dips) of planar features, such as joints, faults, and dikes; so named because clusters of preferred orientations resemble the petals of a rose. (AGI, 1987)

roselite (a) A monoclinic mineral, $Ca_2(Co,Mg)(AsO_4)_2{\cdot}2H_2O$; roselite group; forms a series with wendwilsonite; dimorphous with roselite-beta; perfect cleavage. (b) The mineral group brandtite, roselite, wendwilsonite, and zincrosasite.

rose of cracks The system with radial cracks issuing from the center of the hole as a result of the tangential stresses. (Langefors, 1963)

rose opal A variety of opaque common opal having a fine red color. (CMD, 1948)

rose quartz Crystalline quartz with a rose pink color, due probably to titanium in minute quantity. The color is destroyed by exposure to strong sunlight. Used as a gem or an ornamental stone. See also: *Bohemian ruby.* (Fay, 1920)

rose steel A steel that shows a peculiar fracture and texture in the interior, different from that near the surface. (Standard, 1964)

rose topaz The yellow-brown variety of topaz changed to rose pink by heating. These crystals often contain inclusions of liquid carbon dioxide. (CMD, 1948)

rosette A radially symmetric, sand-filled crystalline aggregate or cluster with a fancied resemblance to a rose; formed in sedimentary rocks by barite, marcasite, or pyrite. See also: *petrified rose.*

rosette copper Disks of copper (red from the presence of suboxide) formed by cooling the surface of molten copper through sprinkling with water. Also called rose copper. (Fay, 1920)

rosette texture A flowerlike or scalloped pattern of a mineral aggregate. (AGI, 1987)

rose vitriol Cobalt sulfate. See also: *bieberite.* Also called cobalt vitriol; red vitriol. (Standard, 1964)

rosickyite A monoclinic mineral, S (gamma sulfur); dimorphous with sulfur. Syn: *gamma sulfur.*

rosin (a) The hard, amber-colored residue left after distilling off volatile oil from pine pitch. (API, 1953) (b) To melt a resin and apply a coat to the right-handed threads of heated rod couplings; the coating sets when cooled, which permits the rods to be used in the same manner as left-hand-threaded rods in fishing operations. Also called rosining. (Long, 1960)

rosin blende A yellow variety of zinc blende, ZnS. When dark in color it is called blackjack.

rosined joints Drill-rod or casing couplings to which hot rosin was applied and that were joined before the rosin cooled. (Long, 1960)

rosing The act or process of milling a metal object in a borehole with a rose bit. (Long, 1960)

rosin jack A yellow variety of sphalerite. Also called resin jack, rosin blende, rosen zinc. Cf: *blackjack.*

Rosin-Rammler equation An equation relating to fine grinding: for most powders that have been prepared by grinding, the relationship between R, the residue remaining on any particular sieve, and the grain size in micrometers (x) is exponential: $R = 100e^{-bx^n}$, where e is the base of the natural logarithm and b and n are constants. (Dodd, 1964)

rosin tin A reddish or yellowish variety of cassiterite. Also spelled resin tin.

rosin zinc Sphalerite of a rosiny appearance. (Hess)

rosite (a) *anorthite.* (b) An impure muscovite as alteration product.

rosiwal analysis In petrography, a quantitative method of estimating the volume percentages of the minerals in a rock. Thin sections of a rock are examined with a microscope fitted with a micrometer that is used to measure the linear intercepts of each mineral along a particular set of lines. This method is based on the assumption that the area of a mineral on an exposed surface is proportional to its volume in the rock mass. (AGI, 1987)

rosolite A rose-pink variety of grossulatire garnet. Also called landerite and xalostocite. From Xalostoc, Morelos, Mexico. (English, 1938)

Ross and Welter furnace A multiple-deck roasting furnace of the annular type; used in Germany. (Fay, 1920)

Ross feeder Mechanism for control of rate of feed of coarse ore in the primary and secondary crushing system. Several heavy loops of chain lie above and bear on ore that rests in the delivery chute at just above its natural angle of repose. When the shaft from which the loops are suspended is rotated by its small motor, ore slides under control. (Pryor, 1963)

Rossie furnace An American variety of hearth for the treatment of galena, differing from the Scotch hearth in using wood as fuel, working continuously, and having hollow walls, to heat the blast. (Fay, 1920)

Rossi-Forel intensity scale A scale for rating earthquake effects. Devised in 1878 by de Rossi (Italy) and Forel (Switzerland). No longer in general use, having been supplanted by Wood and Neumann's Modified Mercalli intensity scale of 1931. (AGI, 1987)

rosthornite A brown to garnet-red resinous material forming lenticular masses in the coal of Carusthia. (AGI, 1987)

rotameter A tapered float rises or falls in a transparent tube in accordance with the velocity of the rising liquid. Variations include spinning floats and magnetic or radioactive ones for use with opaque fluids. Rate-of-flow indicator. (Pryor, 1963)

rotap Laboratory screen shaker widely used in screen sizing analysis. Up to seven 8-in round screens are nested on the appliance and given a shaking, rotary, and tapping motion. (Pryor, 1963)

rotary *rotary table; rotary-drill rig.*

rotary bit As used in a broad sense by drillers, a roller bit. (Long, 1960)

rotary boring A system of boring, using usually hollow rods, with or without the production of rock cores. Rock penetration is achieved by the rotation of the cutting tool. The method is used extensively in exploration, particularly when cores are required. It is the usual method in oil well boring with holes from 6 to 18 in (15 to 45 cm) in diameter. See also: *boring; diamond drilling; rotary drill.* (Nelson, 1965)

rotary breaker A breaking machine for coal, rock, or minerals. It consists of a trommel screen with a heavy cast steel shell fitted internally with lifts that progressively raise and convey the coal and stone forward and break it. As the material is broken the undersize passes through the apertures, so that excessive degradation does not occur. See also: *trommel; Bradford breaker.* (Nelson, 1965)

rotary bucket drill A rotary-type drill on which a rotary bucket is fastened to the kelly bar. The bucket is equipped with a hinged bottom, which has straight-edged cutting blades or teeth. When rotated by the kelly bar, the bucket loads from the bottom; when filled, it is withdrawn from the hole and dumped by unlatching the bottom. Holes 30 to 250 cm in diameter can be drilled with this machine in soft, boulder-free ground. Also called bucket rig; dry-hole digger; rathole rig. See also: *bucket drill.* (Long, 1960)

rotary compressor A compressor designed for a delivery pressure of 100 psi (690 kPa) and ranging in capacity from 60 to 300 ft^3/min (2.1 to 8.5 m^3/min). See also: *air-conditioning process.* (Lewis, 1964)

rotary drier A drier in the shape of an inclined rotating tube used to dry loose material as it rolls through. (ACSG, 1963)

rotary drill Broadly, various types of drill machines that rotate a rigid, tubular string of rods to which is attached a bit for cutting rock to produce boreholes. The bit may be a roller cone bit, a toothed or fishtail drag bit, an auger bit, or a diamond bit. See also: *core drill; rock drill.* Cf: *diamond drill; shot drill.*

rotary-drill cuttings The chips and pulverized rock produced by the abrasive and chipping action of a drag, roller bit, or diamond bit when used on a diamond- or rotary-drill machine to drill a borehole. Cf: *cuttings.* (Long, 1960)

rotary drilling The hydraulic process of drilling that consists of rotating a column of drill pipe, to the bottom of which is attached a drilling bit, and during the operation, circulating down through the pipe a current of mud-laden fluid, under pressure, by means of special slush pumps. The drilling mud and cuttings from the bit are forced upward and outside the drill pipe to the surface. Cf: *cable-tool drilling.* (AGI, 1987)

rotary drill motor The space available in the casing of a pneumatic rotary rock or coal drill is necessarily limited and precludes the use of a reciprocating engine. The power unit used instead is similar in design to the vane compressor. The rotor runs at a very high speed, between 3,500 rpm and 4,000 rpm, and this is reduced by gearing to give a drill spindle speed of about 650 rpm.

rotary-drill rig A rotary drill complete with accessory tools and equipment necessary to drill boreholes. (Long, 1960)

rotary dump An apparatus for overturning one or more mine cars simultaneously to discharge coal. They may rotate either 180° or 360°. (BCI, 1947)

rotary dump car A standard small car in which the car body, of about 2 yd^3 (1.5 m^3) capacity, is mounted on a turntable in the car frame. The car body may be swung by hand to dump over either side or either end. (Pit and Quarry, 1960)

rotary dumper A steel structure that revolves a mine car and discharges the contents, usually sideways, into a bunker or onto a screen. (Nelson, 1965)

rotary excavator Earth-moving machine with vertical wheel that carries digging buckets peripherally. These loosen soil and deliver to short conveyor loader, the assembly being mounted on crawler track. Capacity up to 5,000 st/h (4,500 t/h). Also called bucket wheel excavator. (Pryor, 1963)

rotary fault *rotational fault.*

rotary feed table (a) A feeder comprising a horizontal rotating circular plate mounted under the mouth of a hopper and arranged with an adjustable plow to control the rate of flow of material over the edge of the plate. (BS, 1962) (b) *disk feeder.*

rotary furnace Horizontally mounted cylinder rotating between trunnions through which gas or oil flame is introduced. (Pryor, 1963)

rotary-percussive drill A drilling machine that operates as a purely rotary machine to which is added a percussive action. The specially designed rotary-percussive drilling bit not only gives a greater penetration rate, but is also able to operate longer without deterioration of the cutting edges. A disadvantage is the great size of the air-operated machine, which is usually mounted on a carriage. (Nelson, 1965)

rotary percussive drilling A method of drilling in which repeated blows are applied to the bit, which is continually rotated under power. (BS, 1964)

rotary pump A positive-displacement pump in which the liquid-propelling parts are cams, gears, impeller wheels, etc., rotating within a case, as distinguished from those pumps that move liquids by means of the to-and-fro motion of a piston within a cylinder. Cf: *centrifugal pump.* (Long, 1960)

rotary screen (a) A screen for sizing aggregate and similar material; it is a pierced rectangular plate bent into a cylinder. (Hammond, 1965) (b) *trommel.*

rotary shot drill (a) Any rotary drill used to drill blastholes. (Long, 1960) (b) *seismograph drill.*

rotary smelter Any of the cylindrical smelters that depend on slow rotation about a horizontal axis for agitation of the molten mass. See also: *smelter.* (ASTM, 1994)

rotary sorting table A circular plate conveyor to effect a preliminary grading of coals and removal of stone by hand. A screened-out fraction of the run-of-mine coal is delivered to the table by chute from a conveyor. As the stream of coal revolves on the table, the various grades of coal and the dirt are raked into positions where they are diverted by plows into chutes. The operators are positioned on the inner and outer edges of the table and the coal is not handled but only raked. Syn: *circular grading table.* (Nelson, 1965)

rotary table (a) The geared rotating table that propels the kelly and the drill stem when drilling a borehole with an oilfield-type rotary rig. Also called rotary; table; turntable. (Long, 1960) (b) The mechanism used in some forms of rotary drilling to rotate the drilling column. (BS, 1963)

rotary vane feeder A rotor of cylindrical outline with radial, spaced plates or vanes rotating on a horizontal axis, for controlling the flow of bulk materials.

rotary vibrating tippler A tippler designed to overcome the tendency for coal or dirt to stick to the bottom of the tubs or mine cars. When the tippler is in the inverted position, the car rests upon a vibrating frame that gives it a high-speed vertical jolting motion, which frees any material tending to stick inside the car. (Nelson, 1965)

rotating casing screw conveyor A screw conveyor in which the tubular casing rotates at a different speed or in an opposite direction to the conveyor screw. See also: *screw conveyor.*

rotational fault A fault on which rotational movement is exhibited; a partial syn. of hinge fault. Cf: *hinge fault; scissor fault.* Syn: *rotary fault.* (AGI, 1987)

rotational flow Turbulent flow involving all parts of a moving liquid.

rotational movement Apparent fault-block displacement in which the blocks have rotated relative to one another, so that alignment of formerly parallel features is disturbed. Cf: *translational movement.* See also: *rotational fault.* (AGI, 1987)

rotational shear One of four types of slope failure. Failure by rotational shear produces a movement of an almost undisturbed segment along a circular or spoon-shaped surface and occurs in comprehensive, uniform material. This material would not be affected by geological planes of weakness. Failure of this type can occur from causes that either increase the shear stresses or decrease the shear strength of the material. (Woodruff, 1966)

rotational slide A slide of homogeneous earth or clay in which the slip surface of failure closely follows the arc of a circle. (Nelson, 1965)

rotational transformation An isochemical phase change involving only a shift in angle of chemical bonds, e.g., beta-quartz to alpha-quartz. Cf: *phase transformation; displacive transformation; reconstructive transformation; dilational transformation.*

rotational wave *S wave.*

rotation firing Crushing a small piece of rock with a first explosion, and timing other holes to throw their burdens toward the space made by that and other preceding explosions. Also called row shooting. (Nichols, 1976)

rotation recorder An instrument for measuring any slight rotation of a bridge support under load. See also: *spread recorder.* (Hammond, 1965)

rotch *rotche.*

rotche S. Staff. A soft and moderately friable sandstone. Also called roach; rotch; roche.
rothoffite A brown variety of andradite garnet.
Rot I plate *gypsum plate; selenite plate.*
rotobelt filter Drum-type vacuum filter in which the membrane is a belt, which leaves the drum at discharge point and is returned via pulleys. This arrangement facilitates washing of filter cake from both sides, also discharge. (Pryor, 1963)
roto finish A tumbling method using special chips and chemical compounds. (Osborne, 1956)
rotor Any unit that does its work in a machine by spinning and does not drive other parts mechanically. (Nichols, 1976)
rotor steel process A steelmaking process using the principle of rotation as in the Kaldo process. It has two lances; one above the bath surface uses low-pressure oxygen to burn carbon monoxide from the bath, while the other blows oxygen onto the bath at high pressure to obtain similar fast oxidation as in the L.D. steel process. (Nelson, 1965)
rotten reef S. Afr. Decomposed, soft country rock found in connection with auriferous conglomerates.
rottenstone A soft, light, earthy substance consisting of fine-grained silica resulting from the decomposition of siliceous shale-on limestone. It is used for polishing. Cf: *diatomite.*
rough (a) Highly fractured, broken, or cavey ground. (Long, 1960) (b) An uncut gemstone. Pertaining to an uncut or unpolished gemstone; e.g., a rough diamond. (AGI, 1987)
roughbacks Term used in the dimension stone industry for ends of blocks that are used as byproducts.
rough diamond A diamond in its natural state. (Long, 1960)
rougher cell Flotation cells in which the bulk of the gangue is removed from the ore or coal.
rough ground Highly fractured, fragmented, or cavey rock formations. (Long, 1960)
roughing Upgrading of run-of-mill feed either to produce a low-grade preliminary concentrate or to reject valueless tailings at an early stage. Performed by gravity on roughing tables, or in flotation in rougher circuit. See also: *cleaning.* (Pryor, 1963)
roughing hole A hole to receive slag from a blast furnace, or molten iron when it is undesirable to let it run into pigs. (Standard, 1964)
roughing mill (a) A metal disk charged with an abrasive, used for the first work in grinding gems. (Standard, 1964) (b) A set of roughing rolls. (Standard, 1964)
roughing rolls The rolls of a train that first receive the pile, ingot, bloom, or billet, and partly form it into the final shape. Also called breaking-down or roughing-down rolls. (Fay, 1920)
roughing tool The ordinary tool used by machinists for removing the outer skin and generally for heavy cuts on cast iron, wrought iron, and steel. (Crispin, 1964)
rough stone (a) *rough diamond.* (b) A gemstone that has not yet been polished or cut.
roughway Corn. A quarry term to designate a direction along which there is no natural cleavage in a rock. See also: *cleaving way; quartering way.*
round (a) A planned pattern of drill holes or series of shots intended to be fired either simultaneously or with delay periods between shots; also, the muck pile obtained when the round is blasted. (b) A round generally consists of cut holes, easers, and trimmers. See also: *drill-hole pattern; shot firing in rounds.* (Nelson, 1965) (c) The holes drilled for blast, the advance from a blast, or the ore, or rock from a single blast. (Ballard, 1955) (d) A blast including a succession of delay shots. (Nichols, 1976) (e) In the operation of a blast furnace, one complete charge of ore, coke, and limestone. (Henderson, 1953)
roundabouts *circuits.*
round-face bit A bullnose bit; also, any bit the cutting face of which is rounded, such as a single- or double-round nose bit. (Long, 1960)
round-headed buttress dam A mass concrete dam constructed of parallel buttresses thickened at the upstream end until they conjoin. See also: *multiple-arch dam.* (Hammond, 1965)
round hook A hook that has a smooth inner surface and will slide along a chain. (Nichols, 1976)
rounding tool A forming or swaging tool having a semicylindrical groove; a blacksmith's swage or collar tool. (Standard, 1964)
round kiln *beehive kiln.*
roundstone (a) A rounded rock fragment of any size larger than a sand grain; a group name for pebbles, cobbles, or boulders — any or all of these. (Stokes, 1955) (b) A diamond crystal with an arched facet. (Hess) (c) A term proposed by Fernald (1929) for any naturally rounded rock fragment of any size larger than a sand grain (diameter greater than 2 mm), such as a boulder, cobble, pebble, or granule. See also: *cobblestone.* (AGI, 1987)
round strand rope A rope composed of a number of strands, generally six in number, twisted together or laid to form the rope around a core of hemp, sisal, or manilla, or, in a wire-cored rope, around a central strand composed of individual wires. (Sinclair, 1959)
round trip The process of pulling the drill string from a borehole, performing an operation on the string (such as changing a bit, emptying the core barrel, etc.), and then rerunning the drill string into the borehole. See also: *trip.* (Long, 1960)
row (a) Corn. Coarse, undressed tin ore; refuse from stamping mills. (Arkell, 1953) (b) N. Staff. A seam or bed of coal. (c) A line of points in a lattice. See also: *lattice.*
roweite An orthorhombic mineral, $Ca_2Mn_2B_4O_7(OH)_6$; forms a series with fedorovskite; forms light brown laths; at Franklin, NJ.
rowlandite An amorphous mineral, $Y_4FeSi_4O_{14}F_2(?)$; massive; dark-green where fresh; becomes brick red on alteration.
row shooting In a large blast, setting off the row of holes nearest the quarry face first, and other rows behind it in succession. (Nichols, 1976)
Roxite A nonpermitted gelatinous explosive; medium strength, high density, and good water resistance. Used in tunneling in nongassy mines in rocks of medium hardness. (Nelson, 1965)
royal barren Solution overflowed from first ion-exchange column in a series that is receiving and stripping pregnant uranium liquor; contains some uranyl. (Pryor, 1963)
royalty (a) A lease by which the owner or lessor grants to the lessee the privilege of mining and operating the land in consideration of the payment of a certain stipulated royalty on the mineral produced. (Ricketts, 1943) (b) *overriding royalty.* (c) Ownership of mineral rights under restricted terms. (Wheeler, 1958) (d) Eng. The mineral estate or area of a colliery, or a portion of such property. A field of mining operations. (e) The landowner's share of the value of minerals produced on a property. It is commonly a fractional share of the current market value (oil and gas) or a fixed amount per ton (mining). See also: *take.* (AGI, 1987)
royer In foundry work, centrifugal belt thrower. Short length of conveyor belting travels at 2,000 to 4,000 ft/min (600 to 1,200 m/min) and conditions molding sands by discharging them vigorously so as to mix and partially dry them. Similar arrangement also used in forming storage piles and loading small material to ships. (Pryor, 1963)
rozenite (a) A monoclinic mineral, $FeSO_4 \cdot 4H_2O$. (b) The mineral group including aplowite, boyleite, ilesite, rozenite, and starkeyite.
rubasse (a) Fr. A crystalline variety of quartz containing, distributed through it, spangles of hematite that reflect a ruby red. Also spelled rubace. Also known as Ancona ruby; Mont Blanc ruby. (Standard, 1964; Fay, 1920) (b) Quartz, stained red in cracks to imitate ruby. (Hess)
rubber (a) Guide; binder; conductor. (Mason, 1951) (b) Derb. Fine scythestone; micaceous sandstone. (Arkell, 1953) (c) A gold-quartz amalgamator in which slime is rubbed against amalgamated copper surfaces. (d) A bucking iron or bucking hammer. A broadhead hammer. (Fay, 1920) (e) A building brick made from a sandy clay and lightly fired so that it can be readily rubbed to shape for use in gaged work. The crushing strength of such a brick is about 1,000 psi (6.9 MPa). (Dodd, 1964)
rubber-bushed couplings Two flanged hubs, one equipped with rubber-bushed holes, the other equipped with pins that mesh with the rubber bushings. To prevent excessive wear, the rubber bushings are bushed with nonferrous bushings, which provide satisfactory contact and wearing surfaces. (Pit and Quarry, 1960)
rubber conveyor belt A conveyor belt consisting of a central stress-bearing carcass for transmitting power enclosed in rubber or PVC covers to protect the carcass from abrasion and atmospheric changes. The carcass usually consists of plies of cotton duck fabric, but other constructions used are cotton cords, steel cables, and woven fabrics of synthetic fibers, such as rayon, nylon, Orlon, Dacron, fiberglass, and asbestos. The covers are furnished in various thicknesses and qualities of rubber or PVC compounds.
rubber-lined pipes Pipes prepared for handling corrosive liquids in such processes as acid leaching. Also, pumps and impellers handling ore pulps. See also: *hydraulic stowing pipe.* (Pryor, 1963)
rubberstone A sharp-gritted Ohio or Indiana sandstone used for sharpening shoe knives; also known as shoe stone.
rubber-tired haulage The underground use of tractors and dump truck haulage, of battery or diesel type, and battery-driven shuttle cars. See also: *trackless tunneling.* (Nelson, 1965)
rubbing bed In quarrying, a bed consisting of a circular disk of cast iron of varying diameters, rotating on a vertical axis. Marble slabs or blocks held on the surface of this rotating disk, to which sand and water are applied, are worn down to desired dimensions and smoothness. (AIME, 1960)
rubbing stone (a) Stone used for rubbing, smoothing, or sharpening; in particular for facing building stones by removing toolmarks. (Arkell, 1953) (b) A block of fine-grained abrasive, such as corundum, for the stoning of vitreous enamel. (Dodd, 1964)
rubbing surface The total area of a given length of airway; i.e., the area of top, bottom, and sides added together, or the perimeter multiplied by the length.

rubble (a) A loose mass of angular rock fragments, commonly overlying outcropping rock; the unconsolidated equivalent of a breccia. Cf: *talus.* (AGI, 1987) (b) Loose, irregular pieces of artificially broken stone as it comes from a quarry. Syn: *rubblestone.* (AGI, 1987)
rubble concrete Concrete in which large blocks of stone, roughly squared, are placed and arranged roughly in courses, so that they break joint both horizontally and vertically. The stones are placed with not less than 6 in (15 cm) of space between them so that the concrete may be properly rammed. Care is taken that all voids are filled with concrete. (Nelson, 1965)
rubbleman In the quarry industry, a foreman who directs and supervises the work of drilling and splitting stone. (DOT, 1949)
rubble masonry Uncut stone, used for rough work, foundations, backing, etc. (Crispin, 1964)
rubblerock *breccia.*
rubblestone *rubble.*
rubbly reef Aust. A vein much broken up.
rubellite A pink gem variety of elbaite. Syn: *red schorl.* See also: *elbaite.*
ruberite *cuprite.*
rubicelle A yellow or orange-red variety of spinel; an aluminate of magnesium. Cf: *ruby spinel.* (CMD, 1948)
rubidium A soft, silvery-white, metallic element of the alkali metal group, closely resembling potassium. Symbol, Rb. Widely distributed in small quantities in nature. Obtained commercially from lepidolite and from potassium minerals. Forms amalgams with mercury and alloys with gold, cesium, sodium, and potassium. Has been considered for use in space vehicles, and for use in thermoelectric generators. (Handbook of Chem. & Phys., 3)
rubinblende Miners' term for the red silver sulfides pyrargyrite, proustite, and miargyrite. Syn: *ruby blende.*
rubinglimmer *lepidocrocite.*
Ruble hydraulic elevator A device by which coarse material in placer gravel is separated from the fines and elevated onto a dump.
ruby The red gem variety of corundum, the color due to the presence of chromium in the structure. Cf: *sapphire.*
ruby alumina An abrasive similar to white alumina except ruby red in color because of the presence of chromic oxide. (ACSG, 1963)
ruby arsenic An early name for realgar. See also: *ruby sulfur.* (Fay, 1920)
ruby blende (a) A reddish variety of sphalerite. Syn: *ruby zinc.* (b) *rubinblende.*
ruby cat's eye A girasol ruby with a chatoyant effect. Cf: *sapphire cat's eye.*
ruby copper *cuprite.*
ruby copper ore *cuprite.*
ruby mica An old syn. for goethite. (Fay, 1920)
ruby sand A red-colored beach sand containing garnets, as at Nome, AK. (AGI, 1987)
ruby silver Dark ruby silver is pyrargyrite, and light ruby silver is proustite. (Dana, 1959)
ruby spinel A red gem variety of spinel. Cf: *almandine ruby; balas ruby; rubicelle.*
ruby sulfur *realgar.* Also called ruby arsenic; ruby of arsenic; ruby of sulfur. (Standard, 1964; Fay, 1920)
ruby tin A red variety of cassiterite.
ruby zinc Miners' term for transparent red sphalerite or zincite. Syn: *ruby blende.*
rudaceous Said of a sedimentary rock composed of a significant amount of fragments coarser than sand grains; pertaining to a rudite. The term implies no special size, shape, or roundness of fragments throughout the gravel range, and is broader than pebbly, cobbly, and bouldery. Also said of the texture of such a rock. (AGI, 1987)
rudite A general name used for consolidated sedimentary rocks composed of rounded or angular fragments coarser than sand (granules, pebbles, cobbles, boulders, or gravel or rubble); e.g., conglomerate, breccia, and calcirudite. The term is equivalent to the Greek-derived term, psephite. Etymol: Latin rudus, crushed stone, debris, rubble. See also: *lutite; arenite.* (AGI, 1987)
Ruggles-Coles dryer Rotary drier or kiln, in which material is worked through a horizontal cylinder counter to drying or heating gas blown through by fan. (Pryor, 1963)
ruin Eng. A term occasionally employed in familiar description for certain minerals whose sections or cut faces exhibit the appearance of ruined buildings, as ruin agate, ruin marble, etc. (Fay, 1920)
ruin agate A brown variety of agate displaying on polished surfaces markings that resemble the outlines of ruined buildings.
ruin aragonite Brecciated Mexican onyx (aragonite). (Schaller, 1917)
ruinform Minerals having the form or appearance of ruins.
rule of approximation Applicable to placer mining locations and entries upon surveyed lands to be applied on the basis of 10-acre (4-ha) legal subdivisions. (Ricketts, 1943)
ruling grade The grade that determines tonnage that can be handled by a single locomotive over a particular engine district. (Urquhart, 1959)
ruling gradient *limiting gradient.*
rumanite *romanite.*
Rumford's photometer A photometer consisting of a rod standing vertically in front of a white screen on which are cast shadows of it by the two light sources whose intensities are to be compared. When the shadows are of equal darkness the ratio of the intensities of the sources equals the square of the ratio of their distances from the screen. (CTD, 1958)
run (a) A branching or fingerlike extension of the feeder of an igneous intrusion. Typically spread laterally along several stratigraphic levels. (AGI, 1987) (b) A flat irregular ribbonlike orebody following the stratification of the host rock. (AGI, 1987) (c) The direction in which a vein of ore lies. (d) A caving in of a mine working. (e) Soft ground is said to run when it becomes mud and will not hold together or stand. (f) The escape of any flowing material into a tunnel area; it may be sand, gravel, or mud. (Stauffer, 1906) (g) *grain.* (h) The length of feed or the advance made by a bit in drilling before it becomes necessary to rechuck the rods or empty the core barrel. Cf: *pull.* (i) A round trip in drilling. (Long, 1960) (j) Continuous production, the operation of equipment between major repairs. (k) A test made of a process or material.
runaround (a) A passage driven in the shaft pillar to enable workers to pass safely from one side of the shaft to the other side. See also: *bypass.* (Fay, 1920) (b) A conveyor in the form of a circuit as distinguished from a shape in which the carrying and return runs travel substantially the same path.
runaway The uncontrolled downward rush of trams when the haulage rope breaks or becomes detached while the set is being hauled up an incline. (Nelson, 1965)
runaway switch Catch point or other automatic diverting switch gear; acts when a mine car runs downgrade or out of control by diverting it to a siding. (Pryor, 1963)
runback (a) To drill slowly downward toward the bottom of the hole when the drill string has been inadvertently or deliberately lifted off-bottom during a rechucking operation. (Long, 1960) (b) To retract feed mechanism to its starting position when rechucking. (Long, 1960)
runback water Scot. Water from a set of pumps that is run back and pumped up again to keep the pump from going "on air" while the other pumps are at work.
run dry To drill without circulating a drilling fluid or mud. Cf: *dry block.* (Long, 1960)
runic texture Suggested by Johannsen as an alternative to graphic texture, since the intergrown quartz and feldspar resemble runic characters. (CTD, 1958)
run in (a) To lower the assembled drill rods, core barrel, and bit, or other types of pipe, casing, or drill string into a borehole. (Long, 1960) (b) To drill the first few inches slowly at the beginning of a core run or when collaring a borehole. (Long, 1960) (c) The initial period of operation of any mechanism during which the component parts seat themselves. (Jackson, 1955)
run-in table *entry table.*
run levels To survey an area or strip to determine elevations. (Nichols, 1976)
runner (a) Bearer or carrier girder, beam, or bar. (Mason, 1951) (b) A steel-shod poling board, driven into unbroken but loose ground as excavation proceeds. See also: *cross poling.* (Webster 2nd, 1960) (c) *driller.* (d) A fault slip. (Fay, 1920)
runner box A distribution box that divides the molten metal into several streams before it enters the mold cavity. (ASM, 1961)
runners Vertical timber sheet piles driven to protect an excavation from collapse. See also: *guide runner; cross poling.* (Hammond, 1965)
runner stick A slightly tapering, round stick, used as a pattern for the opening through which molten metal is to be poured into the mold. (Standard, 1964)
running (a) The act or process of operating a drill, drilling with a bit, or lowering casing, drivepipe, or drill string into a borehole. (Long, 1960) (b) Earth and rock that will not stand, esp. when wetted, and falls, flows, or sloughs into a borehole or a workplace in a mine. (Long, 1960)
running block *traveling block.*
running dry The act of drilling without circulating a drilling fluid. (Long, 1960)
running ground (a) Insecure or easily caved wall of excavation. (Pryor, 1963) (b) Ground that is incoherent, for example, soils, sand, peat, moss, or waterlogged material. It may be semiplastic or plastic, such as wet clays. All such deposits deform readily under pressure, and relief is obtained by squeezing into openings, such as mine workings. The miner uses the term running ground to indicate the difficulty of support and sometimes of danger. See also: *forepoling; quicksand; mud rush.* (Nelson, 1965)

running kiln A lime kiln that is fed from above and delivers continually below. (Standard, 1964)
running lift Light mine pump used in sinking, which can be raised or lowered in shaft as required. (Pryor, 1963)
running measures Eng. Sand and gravel containing much water.
running off In founding, the opening of the taphole of a blast furnace and allowing of the molten metal to flow out to the molds. (Standard, 1964)
running rope A flexible rope of 6 strands, 12 wires each, and 7 hemp cores. (Hunt, 1965)
running sand (a) An unconsolidated sand. See also: *run*. (Long, 1960) (b) Quicksand. (Fay, 1920)
running sheave A sheave used as a single-pulley traveling block. (Long, 1960)
runoff (a) That part of precipitation appearing in surface streams. (AGI, 1987) (b) The collapse of a coal pillar in a steeply pitching seam, caused either naturally or by a small shot in connection with pillar robbing. The pillar is said to have run off. (Fay, 1920)
runoff coefficient The percentage of precipitation that appears as runoff. The value of the coefficient is determined on the basis of climatic conditions and physiographic characteristics of the drainage area and is expressed as a constant between zero and one (Chow, 1964). Symbol: C. (AGI, 1987)
runoff pit Catchment to which spillage can gravitate should it be necessary to dump the contents of mill machines such as classifiers, thickeners, and slurry pumps. Provided with a reclaiming pump so that the contents can be returned to the appropriate part of the flow line. (Pryor, 1963)
run of lode Corn. The direction or course of a lode.
run-of-mill Ore finally accepted by a mill for treatment, after waste and dense-media rejection. Original mined ore (run-of-mine) is ore as severed and hoisted. Syn: *mill-head ore*. (Pryor, 1963)
run-of-mine (a) The raw mined material as it is delivered by the mine cars, skips, or conveyors and prior to treatment of any sort. See also: *pithead output*. (Nelson, 1965) (b) Average grade of ore produced from a mine. (New South Wales, 1958) (c) Said of ore in its natural, unprocessed state; pertaining to ore just as it is mined. (AGI, 1987)
runout (a) The unintentional escape of molten metal from a mold, crucible, or furnace. (ASM, 1961) (b) In mineral processing, the dumping of pulp before the contained solids pack down and choke a stalled mechanism in the event of a breakdown or a power failure. (Pryor, 1963)
runout fire A forge in which cast iron is refined. (Fay, 1920)
run-to-waste Drill cuttings that are not collected or saved as a sludge sample and are allowed to collect in the sump; also, the return drill-circulation fluid not returned to a sump for recirculation. (Long, 1960)
Ruoss jig A jig used chiefly on tin dredges to treat the undersize from the main revolving screen. It differs from the Harz jig in that there is no longitudinal division and the screening compartment extends over the whole surface of the jig; the plungers are located in the hutch below the screen are set in a vertical plane and reciprocate horizontally. (Harrison, 1962)
rupture Deformation characterized by loss of cohesion. Frequently flow grades into rupture, with a progressive loss of cohesion, until complete separation occurs. See also: *fracture*. (Challinor, 1964)
rupture envelope *Mohr envelope*.
rupture factor The term is used with reference to brittle materials; i.e., materials in which failure occurs through tensile rupture rather than through excessive deformation. For a member of given form, size, and material, loaded and supported in a given manner, the rupture factor is the ratio of the fictitious maximum tensile stress at failure, as calculated by the appropriate formula for elastic stress, to the ultimate tensile strength of the material as determined by a conventional tension test. (Roark, 1954)
rupture line *Mohr envelope*.
rupture strength The differential stress that a material sustains at the instant of breaking, or rupture. The term is normally applied when deformation occurs at atmospheric confining pressure and room temperature. (AGI, 1987)
rupture zone The region immediately adjacent to the boundary of an explosion crater characterized by excessive in-place crushing and fracturing where the stresses produced by the explosion exceeded the ultimate strength of the medium. Cf: *plastic zone*. (AGI, 1987)
rush (a) A stampede of prospectors and miners into a new discovery area. (b) A place where gold is found in abundance. (Standard, 1964)
rush gold *rusty gold*.
russellite Corn. A tetragonal mineral, Bi_2WO_6; a yellow alteration product with native bismuth and wolframite.
Russell process A metallurgical process similar to the Patera process, except that cuprous sodium hyposulfite is used in addition to the sodium hyposulfite. (Liddell, 1918)
Russia iron A high-grade, smooth, glossy sheet iron, not liable to rust; once made by a process that was long a secret with Russian manufacturers. The sheets were subjected to severe hammering in piles with powdered charcoal between them. (Standard, 1964)
Russian crystal A colorless variety of gypsum.
rust (a) A corrosion product consisting of hydrated oxides of iron. Applied only to ferrous alloys. (ASM, 1961) (b) A mixture of iron filings, ammonium chloride, and sometimes sulfur, moistened and placed between iron surfaces, where it hardens by oxidation, and forms a solid joint called a rust joint. (Standard, 1964) (c) An English term for a black shale discolored by ocher. (AGI, 1987)
rustler In anthracite and bituminous coal mining, a general term applied to any worker who looks after the haulage system, performing the necessary work by which mine cars are raised and lowered to and from the mine surface. May be designated according to job, as clipper; rollerman. (DOT, 1949)
rustless process A process for the manufacture of stainless steels in an electric furnace that uses a chrome ore as a source of chromium with or without the addition of silicoferrochromium, conjointly with stainless steel scrap. The hearth of the furnace is lined with chromite bricks. (Osborne, 1956)
rusty gold Cal. Free gold, that does not readily amalgamate, the particles being covered with a siliceous film, thin coating of oxide of iron or manganese, etc. Syn: *rush gold*. (Fay, 1920)
rute In mining, threadlike veins of ore. (Standard, 1964)
ruthenium A hexagonal mineral, Ru.
rutherfordine An orthorhombic mineral, $UO_2(CO_3)$; radioactive; earthy yellow; secondary.
rutilated quartz Sagenitic quartz characterized by enclosed needles of rutile. Cf: *fleches d'amour; sagenite*. Syn: *Venus hairstone*. See also: *sagenitic quartz*.
rutile (a) A tetragonal mineral, TiO_2, in which titanium replaces iron; trimorphous with anatase and brookite; prismatic; in amphibolites, ecologites, granite pegmatites, veins, and placers; a source of titanium; also a gemstone. Syn: *red schorl; titanic schorl*. See also: *titanium dioxide; zircon group*. (b) The mineral group argutite, cassiterite, paratellurite, plattnerite, pyrolusite, rutile, and stishovite.
R wave *Rayleigh wave*.
Rziha's theory A mine subsidence theory that is a variant or extension of the vertical theory. In this theory, allowance is made for movements beyond the undermined area, but the dip of the beds is considered to be of little or no influence. Rziha maintained that if rock is undercut, it will stay undisturbed if cohesion exceeds gravity and will fall if gravity exceeds cohesion. (Briggs, 1929)

S

Sabalite Trade name for banded variscite; may be banded vashegyite and natrolite; used as a gem stone. Syn: *Trainite.*
sabkha (a) A supratidal environment of sedimentation, formed under arid to semiarid conditions on restricted coastal plains just above normal high-tide level. It is gradational between the land surface and the intertidal environment. Sabkhas are characterized by evaporite-salt, tidal-flood, and eolian deposits, and are found on many modern coastlines, e.g., Persian Gulf, Gulf of California. See also: *tidal flat.* (AGI, 1987) (b) Any flat area, coastal or interior, where, through deflation and evaporation, salts crystallize near or at the surface. (AGI, 1987) (c) In the rock record, a sabkha facies may be indicated by evaporites, absence of fossils, thin flat-pebble conglomerates, stromatolitic laminae, desiccation features such as mud cracks, and diagenetic modifications, for example, disrupted bedding, dissolution and replacement phenomena, and dolomitization. The sabkha environment may have been significant in the formation of certain petroleum and sulfide-mineral deposits. Etymol: Arabic. Also spelled: sabkhah. Syn: *sebkha.* (AGI, 1987)
sabugalite A monoclinic mineral, $HAl(UO_2)_4(PO_4)_4 \cdot 16H_2O$; autunite group; pseudotetragonal; forms crusts of minute yellow platy crystals in pegmatites at Sabugal, Portugal.
sacrificial anodes The anodes used in cathodic protection against corrosion. (Hammond, 1965)
sacrificial protection Reducing the extent of corrosion of a metal in an electrolyte by coupling it to another metal that is electrochemically more active in the environment. (ASM, 1961)
saddle (a) A ridge connecting two higher elevations; a low point in the crestline of a ridge. A minor upfold along the axis of a syncline; a minor downfold along the axis of an anticline. (b) A gold-bearing quartz vein of anticlinal form, occurring esp. in Australian saddle reefs. See also: *saddle reef.* (c) A peculiar formation of sand slate found in shale or sand rock may be surrounded by soapstone. The under or exposed side of a saddle looks like natural rock, but its upper side is smooth, having no particular bond with the sand rock with which it is embedded, and it is liable to fall out of its place, a fall, however, producing no other derangement of the surrounding parts of the room from which it falls. (Ricketts, 1943) (d) A hump-shaped piece of roof rock with a smooth back, insecurely attached to adjacent strata. Also called saddleback. (Hudson, 1932)
saddleback (a) Eng.; Scot. A roll or undulation in the roof or pavement of a seam. See also: *saddle.* (b) A hill or ridge having a concave outline along its crest. (AGI, 1987)
saddle back reef A lode or reef bent archwise. Anticline. (Pryor, 1963)
saddle block In a dipper shovel, the boom swivel block through which the stick slides when crowded or retracted. (Nichols, 1976)
saddle reef (a) A mineral deposit associated with the crest of an anticlinal fold and following the bedding planes, usually found in vertical succession, esp. the gold-bearing quartz veins of Australia. Syn: *saddle vein.* (AGI, 1987) (b) Aust. A bedded vein that has the form of an anticline; an inverted saddle has the form of a syncline. See also: *saddle.* (c) An opening at the crest of a sharp fold in sedimentary rocks, occupied by ore. (Bateman, 1951)
saddle vein *saddle reef.*
safety As applied to mining, means freedom from danger, injury, or damage. (Kentucky, 1952)
safety belt (a) A worker's belt attachable to some fixed object to safeguard against falls. (b) A protective belt or harness with remote anchorage, worn by a worker, for example, a quarryman, working on a face at height. Since the belt allows a drop of about 6 ft (2 m) a shock absorber is provided. (Nelson, 1965) (c) A belt worn by a derrickman or tripodman to prevent injury due to accidental falls from the top of a derrick. (Long, 1960) (d) A belt to which tools are attached to prevent risk of their falling into machines, thickeners, etc. (Pryor, 1963)
safety car Any mine car or hoisting cage provided with safety stops, catches, or other precautionary devices. (Fay, 1920)
safety catch A safety appliance that transfers the weight of the cage onto the guides if the winding rope breaks. (Nelson, 1965)
safety chain A chain used across openings to mantrips and personnel carriers to prevent miners from falling out of a moving vehicle. Syn: *breakaway chain.*
safety check A check valve to slow the excessive travel speed of a piston in a hydraulic cylinder. (Long, 1960)
safety clamp Any of several types of rod clamps used at the collar of a borehole to hold the drill rods while they are being pulled or lowered. Also called alligator; automatic spider; floor clamp; foot clamp. (Long, 1960)
safety department A department that deals with all aspects of mine safety and safety training. See also: *safety engineer.*
safety detaching hook A device that releases automatically the hoisting rope from a cage in the event of an overwind. See also: *detaching hook.* (Nelson, 1965)
safety door A spare or extra door fixed ready for use in a roadway in the event of damage to the existing ventilation door. The safety door is also positioned so that it can be employed in any emergency, for example, explosions or fires. The door may be of steel construction. (Nelson, 1965)
safety engineer An employee whose job is to inspect all possible danger spots in the mine and plant; to cooperate with safety committees in various parts of the organization; to keep informed upon safety literature and to carry on a perpetual educational campaign among workers; to cooperate with agencies such as the U.S. Mine Safety and Health Administration, the U.S. Bureau of Mines, the National Safety Council, and State bureaus and inspectors; to head all rescue work, first-aid instruction courses, and safety-first meetings; and to draw up and enforce a written code of minimum safety requirements for all work at the mine and plant.
safety explosive Explosive that requires a powerful initial impulse and therefore may be handled safely under ordinary conditions. (Bennett, 1962)
safety factor (a) The ratio of breakage resistance to load. (Nichols, 1976) (b) *factor of safety.* (c) Ratio of breaking stress to working stress. (Obert, 1960)
safety fuse A train of powder enclosed in cotton, jute yarn, and waterproofing compounds; used for firing a cap containing the detonating compound, which in turn sets off the explosive charge. The fuse burns at the rate of 2 ft/min (0.6 m/min). Used mainly for small-scale blasting in quarries and metal mines. See also: *blasting fuse; capped fuse; premature blast.* (Nelson, 1965)
safety gate An automatically operated gate placed at the top of a mine shaft, or at landings, to guard the entrance, to prevent anyone from falling into the shaft. (Fay, 1920)
safety glass Most commonly a sandwich of plastic between two sheets of glass; i.e., laminated safety glass. Also called tempered safety glass; wire safety glass. See also: *Triplex glass.* (Van Vlack, 1964)
safety hat (a) A cap or hat with a hard crown worn by miners, will resist blows against it. (BCI, 1947) (b) A hat or cap made of rigid material, designed for the protection of the heads of workers. If worn in a mine equipped with electricity, the material should be electrically nonconducting. Also called safety cap. (c) *tin hat.*
safety hook (a) A hoisting hook with a spring-loaded latch that prevents a load from accidentally slipping off the hook. (Long, 1960) (b) A self-acting detachable hook on a mine cage, which acts in the event of an overwind. *safety detaching hook.* (Pryor, 1963) (c) A safety catch in a mine hoist. (Standard, 1964)
safety inspector *mine inspector.*
safety joint A coarse-threaded joint in the head of a double-tube core barrel. If the core barrel becomes lodged in the borehole, the safety joint, inner tube, and core can be removed by backing off at the safety joint, thereby facilitating the subsequent fishing job. (Long, 1960)
safety lamp In gassy mines, a lamp of an approved type, which is relatively safe to use in atmospheres that may contain flammable gas. Latterly, the term tends to be restricted to naptha-fueled safety lamps, which are issued to mine inspection personnel and used for combustible gas tests. See also: *bonnet; cap lamp; Davy lamp; flame safety lamp; electric cap lamp.* (Nelson, 1965)
safety-lamp keeper *lampman.*

safety-lamp mine In Great Britain, a coal mine in no part of which below ground is the use of lamps or lights other than permitted lights lawful. (Nelson, 1965)
safety latch A latch provided on a hook or elevator to prevent it from becoming detached prematurely. Cf: *safety hook.* (Long, 1960)
safety light A spring-tensioned aluminum pole (of two sections) with a holder to hold a flashlight in a horizontal position with an adjustable screw aligned with the push button switch of the flashlight. If there is conversion of the mine roof (back), the adjustment screw will depress, putting on the light to warn a miner in the area. Sometimes accompanied by a conversion gauge (measures in thousandths). See also: *guardian angel.*
safety lock An offset swivel coupling that supports the weight of the rods when whipstocking. (Long, 1960)
safety platform (a) A platform built in a derrick as a safe working place for workers who must be up the derrick to handle elevators, casing, drill rods, etc. (Long, 1960) (b) A platform with a hinged-door opening, over a shaft while being sunk, esp. where blasted materials are hoisted with a muck bucket. After the bucket is hoisted, the hinged door is closed to prevent any material from falling back onto workers in the shaft.
safety post A timber placed near the face of working places to afford protection for the workers at the face. It must be set like a line timber and with equal care. (Kentucky, 1952)
safety powder A term used for short-flame explosives before the introduction of permissible explosives. (Fay, 1920)
safety rope winch A cable-winding device anchored at the upper grade of an inclined face and having its cable attached to the head of the coal cutter or cutter loader to assist overcoming frictional resistance of the cutter or loader while in operation against the grade.
safety stop (a) An appliance to stop or control cars near the shaft at the pit bottom or at the top of incline haulages. (Nelson, 1965) (b) On a hoisting apparatus, a check by which a cage or lift may be prevented from falling. (Standard, 1964) (c) An automatic device on a hoisting engine designed to prevent overwinding. (Fay, 1920)
safety switch A switch that provides shunt protection in blasting circuits between a blast site and the switch used to connect a power source to the blasting circuit. (CFR, 1991)
safety tools (a) Nonsparking tools made of beryllium-copper alloy for use in explosive atmospheres. (Hess) (b) A tool such as a catching hook, a grappling tong, a fish head, or a bell screw, for recovering broken boring tools, picking up material, etc., at the bottom of boreholes. (Fay, 1920)
safety valve A pressure-relief valve. (Long, 1960)
safflorite (a) Essentially, a diarsenide of cobalt but usually with a considerable amount of iron, and rarely, a small amount of nickel. Syn: *cobalt lollingite.* See also: *smaltite.* (CMD, 1948) (b) An orthorhombic mineral, $4[CoAs_2]$; löllingite group; dimorphous with clinosafflorite; metallic; tin-white tarnishing to dark gray; sp gr, 7.4; in hydrothermal veins.
safreiro *mucker.*
sag (a) A depression in a coal seam, mine floor, or roof. (AGI, 1987) (b) A depression produced by downwarping of beds on the downthrown side of a fault such that they dip toward the fault. (AGI, 1987) (c) The difference between the sagging path a conveyor belt actually takes due to the imposed load of material and its own weight, and the theoretical plane tangent to the top of the supporting idler rolls. (d) See also: *rod sag.*
sag belt tension The minimum tension in any portion of the carrying run of belt necessary to prevent excessive sag of the belt between belt idlers.
sag bolt Bolts installed at intersections to measure roof sag. A sag bolt is a 12-ft (3.5-m) unit put in without a bearing plate. It is securely anchored in the 12-ft horizon with the aid of a heavy nut, and extends about 2 in (5 cm) from the hole. Three 1/2-in (1.3-cm) strips of colored pressure-sensitive tape are wrapped around the extending section of the bolt, beginning with green at the roof line, then yellow and red. The color bands provide a simple, economical means of detecting roof sag at a glance. (Coal Age, 1966)
sag correction A tape correction applied to the apparent length of a level base line to counteract the sag in measuring tape. (Hammond, 1965)
sagenite (a) An acicular variety of rutile in reticulated twin groups. (b) *rutilated quartz; Venus hairstone.* Etymol: Latin "sagena" large fishing net. Adj: sagenitic.
sagenitic Occurring as needles or plates intersecting in a gridlike or grill-like manner. Cf: *acicular.* (AGI, 1987)
sagenitic quartz Transparent quartz containing acicular rutile, tourmaline, goethite, actinolite, or other mineral. See also: *Thetis hairstone; Venus hairstone; rutilated quartz.*
sagger A coarse fireclay, often forming the floor of a coal seam, so called because it is used for making saggers or protective boxes in which delicate ceramic pieces are placed while being fired. Etymol: corruption of "safeguard." Also spelled: seggar. Syn: *sagre.* (AGI, 1987)
sagging moment A bending moment that produces concave bending at midspan of a simply supported beam, generally termed a positive bending moment. It is the opposite of a negative or hogging moment, which would occur at the supports. (Hammond, 1965)
sag meter *closure meter.*
sag pipe A term proposed as a substitute for inverted siphon. (Seelye, 1951)
sagponds Ponds occupying depressions along active faults, owing to uneven settling of the ground.
sagre *sagger.*
sag structure A general term for load casts and related sedimentary structures.
sag tower A pair of floating lightweight sheaves, which give support, at a suitable point, to the ropes leading away from the winding drums. The sheaves are located at a point about one-third of the length of rope between the drum and winding pulleys, measured from the drum. A sag tower suppresses rope oscillation and dampens out the rhythmic swing of the rope. (Nelson, 1965)
sahlinite A monoclinic mineral, $Pb_{14}(AsO_4)_2O_9Cl_4$; forms aggregates of thin yellow scales; at Långban, Sweden.
sahlite Alternate spelling of salite.
Saint Stephen's stone A white or grayish chalcedony with tiny red spots so close together that the appearance at a distance is a uniform rose-red. (Hess)
sal *sial.*
salable coal Total coal mine output less tonnage rejected or consumed during preparation for market. (Pryor, 1963)
salable minerals A legal term that for Federally owned lands, defines mineral commodities that are sold by sales contract from the Federal Government. These are generally construction materials and aggregates such as sand, gravel, cinders, roadbed, and ballast material. The applicable statute is the Mineral Materials Sale Act of 1947, as amended (30 USC 602-604, 611-615). (SME, 1992)
salable output The total tonnage of clean coal produced at a mine as distinct from pithead output. It is the tonnage of coal as weighed after being cleaned and classified in the preparation plant. (Nelson, 1965)
salamander A solid mass of iron, frequently weighing many tons, that is deposited and substantially replaces the firebrick hearth in the bottom of a blast furnace after long periods of operation. Syn: *bear.* (Henderson, 1953)
salamanders' hair *asbestos.*
sal ammoniac An isometric mineral, NH_4Cl; a volatile white crystalline salt forming encrustations around volcanic vents. Syn: *salmiak.*
Salamon-Munro formula A coal pillar design formula that predicts the strength of coal pillars based on a survey of failed and standing coal pillars in South Africa. The coal pillar strength is predicted from coal pillar height and width specifications combined with empirical constants developed from fitting the design equation to 125 case studies. (SME, 1992)
salamstone A pale-red or blue variety of gem corundum in small transparent hexagonal prisms in Sri Lanka.
salar A term used in Southwestern United States and in the Chilean nitrate fields for a salt flat or for a salt-encrusted depression that may represent the basin of a salt lake. Etymol: Spanish, to salt. Pl: salares; salars. See also: *playa.* (AGI, 1987)
salband (a) A term current among miners for the parts of a vein or dike next to the country rock. (b) The selvage of an igneous mass or of a mineral vein. Etymol: German Salband or Sahlband. (AGI, 1987)
saleeite A monoclinic mineral, $Mg(UO_2)_2(PO_4)_2 \cdot 2H_2O$; autunite group; forms pseudotetragonal square yellow plates at Katanga, Zaire.
salesite An orthorhombic mineral, $Cu(IO_3)(OH)$; forms blue-green crystals; in Chile.
salic Said of certain light-colored silicon- or aluminum-rich minerals present in the norm of igneous rocks; e.g., quartz, feldspar, felspathoid. Also, applied to rocks having one or more of these minerals as major components of the norm. Etymol: a mnemonic term derived from "s"ilicon + "al"uminum + "ic." Cf: *femic; mafic; felsic.* (AGI, 1987)
saliferous Salt-bearing; esp. said of strata producing, containing, or impregnated with salt. See also: *saline.* (AGI, 1987)
salimeter A hydrometer specially graduated to indicate directly the percentage of a salt (as common salt) in a brine or other salt solution. (Webster 3rd, 1966)
salina (a) A place where crystalline salt deposits are formed or found, such as a salt flat or pan, a salada, or a salt lick; esp. a salt-encrusted playa or a wet playa. See also: *playa.* (AGI, 1987) (b) A body of saline water, such as a salt pond, lake, well, or spring, or a playa lake having a high concentration of salts. (AGI, 1987) (c) Saltworks. (AGI, 1987) (d) Salt marsh. Etymol: Spanish; saltpit; salt mine; saltworks. Anglicized equivalent: saline. (AGI, 1987)
saline (a) A natural deposit of halite or of any other soluble salt; e.g., an evaporite. See also: *salines.* (AGI, 1987) (b) An anglicized form of

salina. In this usage, a saline may refer to various features such as a playa, a salt flat, a saltpan, a salt marsh, a salt lake, a salt pond, a salt well, or a saltworks. (AGI, 1987) (c) Salt spring. (AGI, 1987) (d) Salty; containing dissolved sodium chloride, e.g., seawater. (AGI, 1987) (e) Having a salinity appreciably greater than that of seawater, e.g., a brine. (AGI, 1987) (f) Containing dissolved salts at concentrations great enough to allow the precipitation of sodium chloride; hypersaline. (AGI, 1987) (g) Said of a taste resembling that of common salt, esp. in describing the properties of a mineral. (AGI, 1987)

saline deposit *evaporite.*

saline deposits *salines.*

saline residue *evaporite.*

salines (a) A general term for the naturally occurring soluble salts, such as common salt, sodium carbonate, sodium nitrate, potassium salts, and borax. Syn: *saline deposits.* (AGI, 1987) (b) A general term for salt mines, salt springs, salt beds, salt rock, and salt lands. (AGI, 1987)

saliniferous Said of a stratum that yields salt. (AGI, 1987)

salinity The total amount of solid material in grams contained in 1 kg of water when all the carbonate has been converted to oxide, the bromine and iodine have been replaced by chlorine, and all organic matter has been completely oxidized. Expressed as grams per kilogram of water or parts per thousand. (Hy, 1965)

salinity bridge An instrument for determining salinity of water (a salinometer) by measuring electrical conductivity of the water sample with a wheatstone bridge. (Hunt, 1965)

salinometer An instrument that measures conductivity of a water sample. This conductivity when compared with that of a sample of known salinity can be converted to an expression of salinity for the unknown. (Hy, 1965)

salite A variety of diopside at Sala, Sweden. Also spelled sahlite.

salitral A term used in Patagonia for a swampy place where salts (esp. potassium nitrate) become encrusted in the dry season. Etymol: Spanish, saltpeter bed. (AGI, 1987)

salmiak *sal ammoniac.*

salmoite *tarbuttite.*

salmonsite (a) A buff hydrous phosphate of manganese and iron, $Fe_2O_3 \cdot 9MnO \cdot 4P_2O_5 \cdot 14H_2O$; orthorhombic; cleavable fibrous masses. An alteration product of hureaulite. From Pala, San Diego County, CA. (English, 1938) (b) A mixture of hureaulite plus jahnsite.

salnatron Crude soda ash. (Standard, 1964)

salt (a) A general term for naturally occurring sodium chloride, NaCl. See also: *halite; common salt; rock salt.* (AGI, 1987) (b) To introduce extra amounts of a valuable mineral into a sample to be assayed or into the working places of a mine, with fraudulent intent. (c) The generic term salt is applied to any one of a class of similar compounds formed when the acid hydrogen of an acid is partly or wholly replaced by a metal or a metallic radical. (Kaufmann, 1960)

salt-and-pepper sand Sand consisting of a mixture of light- and dark-colored grains.

salt anticline A diapiric or piercement structure, like a salt dome except that the salt core is linear rather than equidimensional, e.g., the salt anticlines in the Paradox basin of the central Colorado Plateau. Syn: *salt wall.* (AGI, 1987)

saltation A mode of sediment transport in which the particles are moved progressively forward in a series of short intermittent leaps, hops, or bounces from a bottom surface; e.g., sand particles skipping downwind by impact and rebound along a desert surface, or bounding downstream under the influence of eddy currents that are not turbulent enough to retain the particles in suspension and thereby return them to the streambed at some distance downstream. It is intermediate in character between suspension and the rolling or sliding of traction. Etymol: Latin saltare, to jump, leap. (AGI, 1987)

salt block (a) An installation of vacuum pans or grainers for producing salt by evaporation. (Kaufmann, 1960) (b) Evaporated salt or fine rock salt mechanically compressed into dense blocks, usually 50 lb (20 to 25 kg) in weight, for stock feeding. (Kaufmann, 1960)

salt boot The lower portion of a vacuum pan into which finished salt settles; also, the pit or tank into which the barometric leg of a vacuum pan drops salt. (Kaufmann, 1960)

salt bottom A flat piece of relatively low-lying ground encrusted with salt. (AGI, 1987)

saltbox A small reservoir or tank, usually cylindrical, with a false bottom for drainage and a cleanout door, placed under an evaporator for removal of salt. (Kaufmann, 1960)

salt bridge Usually an inverted glass U-shaped tube filled with a sodium chloride solution, the two legs of which dip into the connecting two vessels of electrolyte, forming an electrochemical cell. (Kaufmann, 1960)

salt cake Commercial term for sodium sulfate, Na_2SO_4. (AGI, 1987)

salt cote *saltpit.*

salt-crust process A method of binding mine roadway dust by spraying the area with salt and water. The salt is subjected to wetting and drying cycles. The deposited dust is bound at first by surface tension and then in the recrystallization of the dissolved salt. (Nelson, 1965)

salt dome A diapir or piercement structure with a central, nearly equidimensional salt plug, generally 1 to 2 km or more in diameter, which has risen through the enclosing sediments from a mother salt bed 5 km to more than 10 km beneath the top of the plug. Many salt plugs have a cap rock of less soluble evaporite minerals, esp. anhydrite. The enclosing sediments are commonly turned up and complexly faulted next to a salt plug, and the more permeable beds serve as reservoirs for oil and gas. Certain salt domes are sources of salt and sulfur. Salt domes are characteristic features of the Gulf Coastal Plain in North America and the North German Plain in Europe, but occur in many other regions. Cf: *salt anticline.* See also: *dome; salt tectonics.* (AGI, 1987)

salt effect The solubility of a precipitate in a solution of an electrolyte that has no ion in common with the precipitate is higher than it is in pure water. This is not the salting-out effect. Cf: *salting out.* (Kaufmann, 1960)

saltern (a) A saltworks where salt is produced by boiling or evaporation of salt or brine. (AGI, 1987) (b) *salt garden.*

salt flat The level, salt-encrusted bottom of a lake or pond that is temporarily or permanently dried up; e.g., the Bonneville Salt Flats west of Salt Lake City, UT. See also: *playa; alkali flat.* (AGI, 1987)

salt furnace A simple form of furnace for heating evaporating pans in a salt plant. (Kaufmann, 1960)

salt garden A large, shallow basin or pond where seawater is evaporated by solar heat. Syn: *saltern.* (AGI, 1987)

salt horse A quarryman's term for aplite. See also: *salt vein.*

salting The fraudulent adulteration of a sample, for example, adding a small amount of gold to a sample to make it appear that the gold content of the rock is much higher than it actually is. Salting may be accidental, caused by the fortuitous segregation of rich mineral during sampling. Sampling methods are conducted to reduce chance segregation to a minimum. (Nelson, 1965)

salting a mine Sprinkling gold or rich ore upon or digging it into the ground, or placing it in samples for assay. Drill holes for uranium may be salted before logging with radiation counters.

salting evaporator An evaporator that produces crystals or other solids, in distinction to one that only concentrates liquids. (Kaufmann, 1960)

salting out (a) The addition of sodium chloride or some other electrolyte to a solution of a monelectrolyte to reduce the solubility of the latter. The rarer converse effect is termed salting in. Cf: *salt effect.* (Kaufmann, 1960) (b) Addition of salt to hasten or to improve the separation of soap from glycerol and weak lye during manufacture. Also, to hasten or to improve the separation of sulfated oil from the residual solution after sulfating. (Kaufmann, 1960)

salt mine A mine in which rock-salt deposits are worked. See also: *salina.* (Standard, 1964)

salt of tartar Potassium carbonate, K_2CO_3.

salt of tin A mordant made by dissolving tin in hydrochloric acid. Syn: *tin salt; stannous chloride.* (Standard, 1964)

saltpan (a) A shallow lake of brackish water. (b) An undrained natural depression in which water gathers and leaves a deposit of salt on evaporation. Cf: *playa. saltpit.* (Webster 3rd, 1966)

saltpeter (a) Potassium nitrate. One of the principal ingredients of black blasting powder. *niter.* Also spelled saltpetre. (Fay, 1920) (b) Any earthy nitrate cave deposit.

saltpeter cave A cave in which the earth fill contains appreciable quantities of nitrates; once mined commercially. Syn: *saltpeter earth.*

saltpeter earth *saltpeter cave.*

saltpit (a) A pit where salt is obtained. Syn: *saltpan.* (b) Reservoir along a salt lake or a seacoast for making solar salt. (Kaufmann, 1960)

salt plug The salt core of a salt dome. (AGI, 1987)

salt prairie *soda prairie.*

salts Reaction products of acids (HX) with bases (M.OH). M.OH + HX = MX + H_2O is the simplest formula. (Pryor, 1963)

saltspar Coarsely crystallized and cleavable halite. (Spencer, 1946)

salt stock A general term for a diapiric salt body of whatever shape. (AGI, 1987)

salt table The flat upper surface of a salt stock, along which groundwater solution leads to the formation of cap rock by freeing anhydrite (Goldman, 1952). (AGI, 1987)

salt tectonics A general term for the study of the structure and mechanism of emplacement of salt domes and other salt-controlled structures. Syn: *halokinesis.* (AGI, 1987)

salt vein A term applied by quarrymen to a coarse granite vein from 2 in (5 cm) to 2 ft (61 cm) or more thick, intersecting granite or any other crystalline rock. See also: *salt horse.*

salt wall *salt anticline.*

salt well A drilled or dug well from which brine is obtained. See also: *brine pit.* (AGI, 1987)

saltworks Any installation where salt is produced commercially, as by extraction from seawater, from wells, or from the brine of salt springs. Syn: *salina; saltern.* (AGI, 1987)

salvage (a) A layer or parting of clay or pug occurring on the wall of a vein. (CTD, 1958) (b) To chemically or electrolytically remove diamonds from used diamond bits. (Long, 1960) (c) To recover lost bits or drill pipe from a borehole. (Long, 1960) (d) To reclaim residual assets left at minesite after the mine has closed.

salvage count Number of resettable diamonds salvaged by cutting out of worn or used diamond bit. (Long, 1960)

salvage value (a) The net worth of diamonds recovered from a used or worn diamond bit or other diamond-inset tool. (Long, 1960) (b) The net worth of on-site equipment, and tangible assets after a mine has been closed.

Salzgitter ore An important iron ore deposit in Germany. It consists of conglomerated oolitic limonite and contains about 30% iron. (Osborne, 1956)

samarium A bright, silvery, lustrous, metallic element of the rare-earth group. Symbol, Sm. It is found along with other members of the rare-earth elements in many minerals, including monazite and bastnasite, which are commercial sources. Used for carbon-arc lighting for the motion picture industry, for permanent magnets, and in optical masers and lasers. (Handbook of Chem. & Phys., 3)

samarium oxide Sm_2O_3 has a melting point of 2,350 °C. This material has a high thermal neutron cross section, making it usable as a nuclear control rod material. It is also used as a phosphor activator. (Lee, 1961)

samarskite A monoclinic mineral, $(Y,Ce,U,Fe)_3(Nb,Ta,Ti)_5O_{16}$; commonly metamict; in granite pegmatites. Syn: *ampangabeite; uranotantalite.*

samite Artificially produced silicon carbide or carborundum.

sample (a) Representative fraction of body of material; removed by approved methods; guarded against accidental or fraudulent adulteration; and tested or analyzed to determine the nature, composition, percentage of specified constituents, etc., and possibly their reactivity. Bulk samples are large (several tons), so taken as to represent the ore for the purpose of developing a suitable treatment. Channel samples, cores, chips, grab, pannings, stope samples, etc., are small ones—made primarily to establish the value of the ore reserve. (Pryor, 1963) (b) A section of core or a specific quantity of drill cuttings that represents the whole from which it was removed. Cf: *borings.* (Long, 1960)

sample cutters In mine valuation and process control, devices that cut a representative fraction from a pile of ore or from a passing stream. (Pryor, 1963)

sample extruder A mechanical device for removing a soil sample from a sampling tube; usually consists of a piston driven by a jackscrew or a hydraulic mechanism. (Long, 1960)

sample grinder One who grinds samples of ore to required fineness (depending on character of ore) to prepare them for analysis by assayer. Also called sample crusher. (DOT, 1949)

sampleite An orthorhombic mineral, $NaCaCu_5(PO_4)_4Cl{\cdot}5H_2O$; forms crusts of minute blue crystals.

sample log (a) Strip of graph paper showing units of depth on which the geologist, using cores and samples, describes the rock formations penetrated by drilling. (Wheeler, 1958) (b) Syn: *drill log.* (c) A log depicting the sequence of lithologic characteristics of the rocks penetrated in drilling a well, compiled by a geologist from microscopic examination of well cuttings and cores. The information is referred to depth of origin and is plotted on a strip log form. See also: *well log.* (AGI, 1987)

sample preparation In coal and coke sampling, the process whereby an analysis sample is obtained from a sample by particle size reduction, mixing and sample dividing in successive stages. A stage of sample preparation refers to the sequence of operation leading up to a sample division. (BS, 1960)

sampler (a) A mechanical device for selecting a certain fractional part of ore to be used as an assay sample; as, for example, split shovel, riffle sampler, Brunton's mechanical sampler, and Vezin's sampler. (Fay, 1920) (b) An instrument designed to take samples of the flame or other explosion gases at predetermined intervals during an explosion. (Rice, 1913-18) (c) A specific device for recovering samples of overburden. See also: *sampler barrel.* (Long, 1960) (d) One whose duty it is to select and prepare samples of materials and products for an assay or analysis. (Fay, 1920)

sampler barrel As used in soil-testing work, one of several tubelike devices used to cut and recover a core sample of soil or soft rock. It can either be a plain tube designed to be driven or pressed into the formation being sampled, or be equipped with cutter heads and helical flutes for taking the sample by rotary methods. Syn: *sampler tube.* (Long, 1960)

sample reduction Reducing a soil, coal, or other sample to manageable size while still obtaining a representative sample. The methods may be divided into manual (for example, quartering) and mechanical; riffle box. (Nelson, 1965)

sampler head An adapter or sub for attaching a sampler to a drill-rod string. (Long, 1960)

sampler liner A thin-wall tube fitted inside the barrel of a sampler. The liner serves as a retainer for the sample and when sealed at either end is used as a container in which the sample can be transported safely. Syn: *sampler tube.* (Long, 1960)

sampler tube *sampler barrel; sampler liner.*

sample splitter A device for separating dry incoherent material (such as sediment) into truly representative samples of workable size for laboratory study. Syn: *riffler.* (AGI, 1987)

sampling (a) The gathering of specimens of ore or wall rock for appraisal of an orebody. Since the average of many samples may be used, representative sampling is crucial. The term is usually modified to indicate the mode or locality; e.g., hand sampling, mine sampling, and channel sampling. (AGI, 1987) (b) Cutting a representative part of an ore (or coal) deposit, which should truly represent its average value. Most usually a trenchlike cut 4 in (10.2 cm) wide and 2 in (5.1 cm) deep is cut into the clean face of ore (or coal) and across its course. Honest sampling requires good judgment and practical experience. (Weed, 1922) (c) Selecting a certain fractional part of ore or coal from cars, stock piles, etc., for analysis. (Fay, 1920) (d) Separation of a representative fraction of ore, pulp, or any product for testing or checking purposes. (Pryor, 1960)

sampling bag Collection devices that use a pump to draw the contaminated air into a contaminant bag. The entire sampling bag containing the contaminated air is sent to the laboratory for analysis. (MSHA, 1990)

sampling instrument A device to determine the methane or dust concentration in mine air to assess safety and health. Instruments are designed to sample instantaneously, or over short periods, or to operate continuously. For methane, warning is required whenever the percentage approaches a danger figure. Dust dangers are not momentary peak concentrations but the bulk quantity of dust breathed over a period. See also: *dust sampling; methanometer.* The term may also be applied to soil-, coal-, or mineral-sampling devices of an instrumental nature. (Nelson, 1965)

sampling pipe A small pipe built into and through a stopping or seal to enable samples to be taken of the air within the sealed area. The analysis of such samples will give an indication of the state of the fire or heating. In the case of a waste heating a sampling pipe may be pushed into the waste, on the return side, to give an indication of the conditions. (Nelson, 1965)

sampling spoon A cylinder with a spoonlike cutting edge for taking soil samples. (Long, 1960)

sampling tip The head of a soil auger or soil-sampling barrel. (Long, 1960)

sampling train The order of sequence in which personal health sampling equipment parts are assembled together to complete the cycle of assimilated breathed atmosphere that a person is exposed to during the working time period. (MSHA, 2)

sampling works A plant and its equipment for sampling and determining the value of ores that are bought, sold, or treated metallurgically.

samsonite A monoclinic mineral, $Ag_4MnSb_2S_6$; steel black but red in transmitted light; at the Samson Mine, Harz, Germany.

Samson loader A loader in which the gathering head has rotating arms that pull the stone or coal onto ramps and push it to a scraper chain conveyor, which conveys it to and delivers it at the end of the jib. The jib can be swiveled horizontally and raised or lowered to suit the tub, car, or conveyor to which it is delivering. The whole machine is self-hauling (automobile) on power-driven tractor crawlers with mechanical steering. It is not applicable in steep inclinations. (Mason, 1951)

Samson stripper A longwall cutter loader of the plow type, with two cutting blades, one at each end, operated by a hydraulic cylinder, which can give a powerful thrust (about 42 st or 38 t) to the blades and cause them to bite into the coal. While this thrust is being exerted, the machine is anchored by means of a vertical jack that engages the roof and floor. The jack is built on sliding bars, and the machine is moved by sliding the jack along the bars to the next position for anchoring the machine. It travels alongside the conveyor but is not connected to it. A loading ramp guides the cut coal onto the conveyor. The machine is employed in seams from 4 to 5 ft (1.2 to 1.5 m) thick with a strong roof and floor. (Nelson, 1965)

sand (a) A rock fragment or detrital particle smaller than a granule and larger than a coarse silt grain, having a diameter in range of 1/16 to 2 mm (62 to 2,000 µm, or 0.0025 to 0.08 in, or 4 to 1 phi units, or a size between that at the lower limit of visibility of an individual particle with the unaided eye and that of the head of a small wooden match), being somewhat rounded by abrasion in the course of transport. In Great Britain, the range of 0.1 to 1 mm has been used. See also: *coarse sand; fine sand.* (AGI, 1987) (b) A loose aggregate, unlithified mineral or rock particles of sand size; an unconsolidated or moderately consolidated sedimentary deposit consisting essentially of medium-grained clastics. The material is most commonly composed of quartz, and when the term sand is used without qualification, a siliceous composition is implied; but the particles may be of any mineral composition or mixture of rock or mineral fragments, such as coral sand consisting of limestone fragments. Also, a mass of such material, esp. on a beach, desert, or in a streambed. (AGI, 1987) (c) Sandstone. (AGI, 1987) (d) Separate grains or particles of detrital rock material, easily distinguishable by the unaided eye, but not large enough to be called pebbles; also, a loose mass of such grains, forming an incoherent arenaceous sediment. Building sand, any hard, granular rock material finer than gravel and coarser than dust. The term indicates material comminuted by natural means. (Fay, 1920) (e) Detrital material of size range from 2 to 1/16 mm in diameter. Very coarse, 1 to 2 mm; coarse, 1/2 to 1 mm; medium, 1/4 to 1/2 mm; fine, 1/4 to 1/8 mm; very fine, 1/8 to 1/16 mm. (AGI, 1987) (f) Granular material, composed mainly of quartz, that will settle readily in water. In the mechanical analysis of soil, sand—according to international classification—has a size between 0.02 mm and 2.0 mm. It has no cohesion when dry or saturated but has apparent cohesion when damp. (Nelson, 1965) (g) The residue after amalgamation on plates. (Nelson, 1965) (h) In gold-ore treatment, the coarser and heavier portions of the crushed ore in a mill or battery. (Nelson, 1965) (i) A driller's term applied loosely to any visibly granular sediment, or to any fluid-productive porous sedimentary unit or objective zone of a well. (AGI, 1987) (j) A tract or region of sand, such as a sandy beach along the seashore, or a desert land. (AGI, 1987) (k) A sandbank or a sandbar. The term is usually used in the plural; e.g., sea sands. (AGI, 1987) (l) A term used in the United States for a rock or mineral particle in the soil, having a diameter in the range of 0.05 to 2 mm; prior to 1947, the range 1 to 2 mm was called fine gravel. The diameter range recognized by the International Society of Soil Science is 0.02 to 2 mm. A textural class of soil material containing 85% or more of sand, with the percentage of silt plus 1.5 times the percentage of clay not exceeding 15; specif. such material containing 25% or more of very coarse sand, coarse sand, and medium sand, and less than 50% of fine sand or very fine sand. The term has also been used for a soil containing 90% or more of sand. (AGI, 1987)

sandarac *realgar.*

sandbag In the roof of a coal seam, a deposit of glacial debris formed by scour and fill subsequent to coal formation. See also: *debris bag.* (AGI, 1987)

sandbag stoppings In many mines a rapid and efficient means of erecting stoppings and walls for the control of ventilation near the face. The walls of doors and air crossings in the workings are often entirely of interlocked sandbags. This method minimizes the use of brattice and gives more permanent results in the workings. (Mason, 1951)

sandbar A bar or low ridge of sand that borders the shore and is built to, or nearly to, the water surface by currents in a river or by wave action along the shore of a lake or sea. Syn: *sand reef.* (AGI, 1987)

sand bearings The supports of a core in the sand of a mold. (Standard, 1964)

sand bed (a) The bed into which molten metal from a blast furnace is run. (Standard, 1964) (b) A floor of a foundry in which large iron castings are made. (Standard, 1964)

sandblasting A method of cleaning metal and stone surfaces with sand sprayed over them through a nozzle at high velocity. Sandblasting is also used to form pits on the intrinsically smooth surfaces of materials, such as glass, requiring a particular finish. (Hammond, 1965)

sand bottle A sand-pouring cylinder used for determining the dry density of soil. (Nelson, 1965)

sandburrs Concretions of sandstone. See also: *burr.* (Arkell, 1953)

sand calcite (a) Calcite crystals containing abundant sand grains. (b) Siliceous calcite.

sand crusher A machine comprising a stationary cylinder into which the sand is charged. Inside this cylinder is fitted a main shaft onto which are bolted retaining discs whose main function is to prevent the loose ball in between each disc from traveling laterally. As the discs rotate on the shaft, the lumpy sand is moved into contact with the balls, which crush the sand to grain size. (Osborne, 1956)

sand crystal A large euhedral or subhedral crystal (as of barite, gypsum, and esp. calcite) loaded with detrital-sand inclusions (up to 60%), developed by growth in an incompletely cemented sandstone during cementation. See also: *crystal sandstone.* (AGI, 1987)

sand diamonds Name in the trade for diamonds occurring in the gravels and old marine deposits on the Gold Coast of Africa. (Griffith, 1938)

sand dike A sedimentary dike consisting of sand that has been squeezed or injected upward into a fissure. See also: *neptunian dike.* (AGI, 1987)

sanded in Drill-string equipment, casing, or drivepipe so firmly fastened in a borehole by reason of caving walls or impaction of sand, mud, or drill cuttings that the article cannot be pulled from the borehole. (Long, 1960)

Sander's process A flotation process that uses, instead of an acid bath in deep pans, a dilute solution of aluminum sulfate in shallow pans. (Liddell, 1918)

sand fill Hydraulic filling, stowing. Use of sand or plant tailings, conveyed underground by water to support cavities left by extraction of ore. (Pryor, 1963)

sand filter A filter for purifying domestic water, consisting of specially graded layers of aggregate and sand, through which the water flows slowly downwards. A similar type of filter is used for treating sewage effluent, but has coarser sand. (Hammond, 1965)

sand flag Fine-grained sandstone that can be readily split into flagstones. (AGI, 1987)

sand floor A sand bed in the floor near the blast furnace into which the molten pig is run to be cast into convenient sizes for handling. (Mersereau, 1947)

sand flotation The use of well-graded sand, mixed with water, as the medium for washing coal, as in the Chance washer. See also: *dense-medium washer.* (Nelson, 1965)

sandhog *sand pump.*

sand holder A cavity in a pump barrel to catch sand and keep it out of the way of the plunger or buckets. (Standard, 1964)

sanding-machine worker One who charges mixing mill and removes mixture of sand and refractory clay to be used in packing around ware in bisque firing. (DOT, 1949)

sand jack A device consisting essentially of a sandbox and a series of plungers for gradually lowering into position a heavy weight, supported by the plungers, by running out the sand below. (Webster 3rd, 1966)

sand leaching Sand leaching or percolation may be practiced wherever the ore is coarse enough to permit free passage of the solvent through the voids. The ore is loaded into large vats or tanks that are then filled with leaching solution. After the solution has been in contact with the ore for a certain time, it is withdrawn, and fresh solution is added. Liquid may be added from the top (downward percolation) or from the bottom (upward percolation). See also: *percolation.* (Newton, 1959)

sandman A laborer who switches flow of sand (waste minerals and water resulting from treatment of ore for removal of valuable minerals) in pipe or flumes from one stope (underground openings from which ore has been mined) to another, so that they will be properly filled with sand (after water has drained off) to support the walls and roof and prevent caving of the ground surrounding the worked out area. (DOT, 1949)

sand mold A body of sand and binder surrounding a cavity for the reception of molten metal in the production of castings. The cavity must have dimensional accuracy, and the sand surrounding it must be of sufficient stability to allow the metal to solidify in the exact shape of the impression. The production of the sand mold involves making a pattern of the part to be cast, and packing molding sand round the pattern, which when withdrawn leaves the cavity into which the metal is poured, cores being inserted to leave cavities where desired in the casting. (Osborne, 1956)

sand muller A machine for mixing sand and binders by a kneading and squeezing action for use in sand molds. The mixture is usually sand, clay, and water, but synthetic chemical binders may be used. (Osborne, 1956)

sand pipe A pipe formed in sedimentary rocks, filled with sand. See also: *pipe.* (AGI, 1987)

sand pump (a) A pump, usually a centrifugal type, capable of handling sand- and gravel-laden liquids without clogging or wearing unduly. Syn: *sandhog; suction bailer.* See also: *gravel pump; sludge pump; swab.* (Long, 1960) (b) A pump for lifting tailings at ore-dressing plants. (Fay, 1920) (c) A cylinder with a valve at the bottom, lowered into a drill hole from time to time to take out the accumulated slime resulting from the action of the drill on the rock. Also called shell pump; sludger. (Fay, 1920) (d) A piston-type bailer. Also called American pump. (Long, 1960)

sand-pump dredger A long pipe reaching down from a vessel into the sand, the latter being raised under the suction of a centrifugal pump and discharged into the vessel itself or an attendant barge. Also called a suction dredger. See also: *dredger.* (CTD, 1958)

sand-pump sampler A sand sampler made and used in the same manner as an American pump or sand pump, a. (Long, 1960)

sand reef *sandbar.*

sand-replacement method The normal method of measuring soil density. In its simplest form, the measurement requires only a container full of dry sand of known density, a balance and apparatus for determination of soil moisture content. (Nelson, 1965)

sandrock (a) A field term for a sandstone that is not firmly cemented. (AGI, 1987) (b) A term used in southern England for a sandstone that crumbles between the fingers. (AGI, 1987) (c) *sandstone.*

sand roll A metal roll cast in a mold of sand; distinguished from a chilled roll, which is cast in an iron mold or chill. (Standard, 1964)

sands (a) The coarser and heavier portions of the crushed ore in a mill. (Fay, 1920) (b) Tailings from the stamp mills of certain copper mines. (Fay, 1920) (c) Particles of crushed ore of such a size that they settle readily in water and may be leached by allowing the solution to percolate. (CTD, 1958)

sands-and-slimes process Any cyanidation process for gold ores that involves separation of two portions in a classifier, and separate treatment of sands by percolation and slimes by agitation. (Nelson, 1965)

sand seam A quarry term for a more or less minute vein or dike of muscovite (white mica) with some quartz, in cases also with feldspar.

sandstone (a) A medium-grained clastic sedimentary rock composed of fragments of sand size set in a fine-grained matrix (silt or clay) and more or less firmly united by a cementing material (commonly silica, iron oxide, or calcium carbonate); the consolidated equivalent of sand. The sand particles usually consist of quartz, and the term sandstone, when used without qualification, indicates a rock containing about 85% to 90% quartz. The rock varies in color, may be deposited by water or wind, and contains numerous primary features (sedimentary structures and fossils). Sandstones may be classified according to composition of particles, mineralogic or textural maturity, fluidity index, diastrophism, primary structures, and type of cement. (AGI, 1987) (b) A field term for any clastic rock containing individual particles that are visible to the unaided eye or slightly larger. Syn: *sandrock.* (AGI, 1987)

sandstone dike (a) A clastic dike composed of sandstone or lithified sand; a lithified sand dike. (AGI, 1987) (b) Stone intrusion. See also: *horseback.* (AGI, 1987)

sandstone grit In geology, a coarse, angular-grained sandstone.

sandstone opal A variety of opal occurring in boulders between layers of sandstone and soft clay in the form of pipes in thickness from 1 mm to 3 cm.

sandstone pipe *cylindrical structure.*

sand trap (a) A device for separating sand and other heavy or coarse particles from a cuttings-laden; drill-circulation fluid overflowing the collar of a borehole. Cf: *shaker.* (Long, 1960) (b) A device, often a simple enlargement, in a conduit for arresting the sand, silt, etc., carried by the water, and generally including means of ejecting them from the conduit. (Seelye, 1951)

sand wall A temporary independent wall separated from a slag-pocket wall by a thickness of sand for the purpose of easy slag removal and the protection of the permanent wall. (AISI, 1949)

sand washer An apparatus for separating sand from earthy substances. (Fay, 1920)

sandy *arenaceous.*

sandy alumina Coarse-grained, porous, granular textured alumina that has not been calcined to the alpha alumina stage (artificial corundum). Resembles fine sand, hence free-flowing, with nondusting qualities. The relatively large surface area of sandy alumina permits its use as an absorbent in dry scrubber units at primary aluminum smelters.

sanguinaria (a) Sp. Dark green bloodstone variegated by red spots. (b) Sp. Hematite.

sanidine A monoclinic mineral, $(K,Na)AlSi_3O_8$; feldspar group; forms a series with albite; prismatic cleavage; colorless; forms phenocrysts in felsic volcanic rocks. See also: *ice spar.* Syn: *rhyacolite.* Cf: *feldspar; glassy feldspar.*

sanmartinite A monoclinic mineral $(Zn,Fe)WO_4$; resembles wolframite; at San Martin, San Luis Province, Argentina. See also: *hübnerite.*

santafeite An orthorhombic mineral, $(Na,Ca,Sr)_3(Mn,Fe)_2Mn_2(VO_4)_4(OH,O)_5 \cdot 2H_2O$; forms black needles on limestone in New Mexico.

Santorin earth A pozzolana from the Greek island of Santorin. A quoted composition is 64% SiO_2, 13% Al_2O_3, 5.5% Fe_2O_3, 1% TiO_2, 3.5% CaO, 2% MgO, 6.5% alkalies, and 4% loss on ignition. (Dodd, 1964)

sap The part of the rock in a quarry that is next to the surface or to joints and crevices and has been somewhat stained and softened by weathering.

saphir d'eau French water sapphire. An intense-blue variety of the mineral cordierite, occurring in waterworn masses in the river gravels of Ceylon; used as a gem stone. (CMD, 1948)

saponification (a) The hydrolysis of esters into acids and alcohols by the action of alkalies or acids—by boiling with water or by the action of superheated steam. It is the reverse process to esterification. (CTD, 1958) (b) Conversion into soap; the process in which fatty substances form soap, by combination with an alkali. A term used in the flotation process. (Fay, 1920)

saponifier Any compound, as a caustic alkali, used in soapmaking to convert the fatty acids into soap. A term used in the flotation process. (Standard, 1964)

saponin Complex polyhydroxy carboxylic flotation reagent used as depressant. It destroys bubble adhesion to collector-coated minerals. (Pryor, 1963)

saponite A monoclinic mineral, $(Ca/2,Na)_{0.3}(Mg,Fe)_3(Si,Al)_4O_{10}(OH)_2 \cdot 4H_2O$; smectite group; soft; massive; plastic; unctuous; in veins and cavities in serpentinite and basalt. Syn: *bowlingite; mountain soap; piotine; soapstone.* Etymol: Greek "sapon" soap.

sappare (a) *sapphire.* (b) *kyanite.* Also spelled sapper.

sapphire (a) A blue gem variety of corundum. Syn: *sappare.* (b) Any gem-quality corundum other than ruby (fancy sapphire). (c) Synthetic alumina single-crystal boules made by the Verneuil process for use as bearings and thread guides.

sapphire cat's eye A girasol sapphire with a chatoyant effect. Cf: *ruby cat's eye.*

sapphire glass Sapphire-blue glass.

sapphire quartz (a) An opaque blue variety of quartz colored by nonparallel fibers of silicified crocidolite. Syn: *azure quartz; blue quartz.* See also: *siderite.* (b) Light to pale sapphire-blue quartz in the Western United States.

sapphirine (a) A rare aluminosilicate of magnesium occurring as disseminated blue chalcedony. Syn: *hauynite; blue chalcedony.* (CMD, 1948) (b) A monoclinic or triclinic mineral, $(Mg,Al)_8(Al,Si)_6O_{20}$; light- to dark-blue to green; in Greenland, Madagascar, and Quebec.

saprocol Indurated sapropel. (Tomkeieff, 1954)

saprolite A soft, earthy, typically clay-rich, thoroughly decomposed rock, formed in place by chemical weathering of igneous, sedimentary, and metamorphic rocks. It often forms a layer or cover as much as 100 m thick, esp. in humid and tropical or subtropical climates; the color is commonly some shade of red or brown, but it may be white or gray. Saprolite is characterized by preservation of structures that were present in the unweathered rock. Syn: *saprolith; sathrolith.* Cf: *geest; laterite.* (AGI, 1987)

saprolith *saprolite.*

sapropel (a) An aquatic ooze or sludge rich in organic (carbonaceous or bituminous) matter. (AGI, 1987) (b) A fluid organic slime originating in swamps as a product of putrification. In its chemical composition, it contains more hydrocarbon than peat. When dry, it is a lusterless, dull, dark, and extremely tough mass that is hard to break up. (Stutzer, 1940)

sapropel-clay A sedimentary deposit in which the amount of clay exceeds that of sapropel. (AGI, 1987)

sapropelic coal (a) Designation of coal of which the original plant material was more or less transformed by putrefaction. Complete seams of sapropelic coals are rare, but layers or bands of varying thickness within seams are more frequent. This type coal is not abundant and proves troublesome in cleaning processes with jig and dense medium washers because of its lower density relative to humic coals of the same rank and the same ash content. (IHCP, 1963) (b) A group of coals, including the cannel and torbanite types, which are largely composed of the indurated jellylike slime derived from macerated organic debris, and known as sapropel, and also of remains of spores and algae. They are typically massive, unbanded coals, which break with a conchoidal fracture and do not show, as a rule, jointing. (Tomkeieff, 1954) (c) *cannel coal.*

saprophyte An organism living on dead or decaying organic material. (Rogoff, 1962)

sapropsammite Sapropel rich in sand. (Tomkeieff, 1954)

sard A translucent brown variety of chalcedony. Syn: *sardius; sardine.* See also: *carnelian; sardonyx.*

sardachate A variety of agate with reddish bands of carnelian; carnelian agate. (Standard, 1964)

sardine *sard.*

sardius *sard.*

sardonyx (a) A variety of chalcedonic quartz. See also: *onyx; sard.* (Fay, 1920) (b) A banded gem variety of chalcedony with straight parallel reddish bands of sard alternating with white or colored bands of another mineral.

sargent tube *acid-etch tube.*

sarmientite A monoclinic mineral, $Fe_2(AsO_4)(SO_4)(OH) \cdot 5H_2O$; forms minute lemon-yellow crystals in Argentina.

sarsen stone *graywether; druid stone.*

sartorite A monoclinic mineral. $PbAs_2S_4$; metallic; dark gray; conchoidal fracture; in sugary dolomite of the Binnental at Lengenbach, Valais, Switzerland.

saryarkite A tetragonal mineral, $(Ca,Y,Th)_2Al_4(SiO_4,PO_4)_4(OH) \cdot 9H_2O$ or $Ca(Y,Th)Al_5(SiO_4)_2(PO_4)_2(OH)_7 \cdot 6H_2O$.

sassoline *sassolite.*

sassolite A triclinic mineral, H_3BO_3; forms small pearly scales as an incrustation, or as tabular crystals around fumaroles or vents of sulfurous emanations. Syn: *sassoline.*

satelite Fibrous serpentine with a slight chatoyant effect, being pseudomorphous after asbestiform tremolite that has been silicified, in Tulare County, CA. Syn: *serpentine cat's-eye.*

satellite imagery The counterpart of an object produced from a satellite by the reflection or re-representation of an object produced by optical, electro optical, optical mechanical, or electronic means. (ASPRS, 1984)

satellites Certain minerals usually associated with diamond, such as ilmenite, garnet, zircon, rutile, corundum, spinel, olivine, and gorceixite. (Chandler, 1964)
sathrolith *saprolite.*
satin spar (a) A white, translucent, fine fibrous variety of gypsum, characterized by chatoyancy or a silky luster. (b) A term used incorrectly for a fine fibrous or silky variety of calcite or aragonite. Syn: *atlas spar; feather gypsum; satin stone.*
satin stone *satin spar.*
satpaevite A possibly orthorhombic mineral, $Al_{12}V_8O_{37}{\cdot}30H_2O$; forms weakly pleochroic yellow flakes in argillaceous anthraxolitic vanadiferous deposits of Kurumsak and Balasanskandyk, Karatau, Kazakhstan.
saturated (a) A rock or soil is saturated with respect to water if all its interstices are filled with water. (AGI, 1987) (b) In petrology, applied to minerals capable of crystallizing from rock magmas in the presence of an excess of silica. Such minerals are said to be saturated with regard to silica and include the feldspars, pyroxenes, amphiboles, micas, tourmaline, fayalite, spessartite, almandine, and accessory minerals, such as sphene, zircon, topaz, apatite, magnetite, and ilmenite. Also applied to igneous rocks composed wholly of saturated minerals. Cf: *undersaturated.* (AGI, 1987) (c) In fatty acids and other organic compounds, a structure in which each carbon valence is combined either with a distinct atom or by polylinkages. (Pryor, 1963) (d) A term describing a membrane that is filled as completely as practicable with bituminous material.
saturated air Air that contains the maximum possible amount of water vapor at that temperature. The amount of water vapor that will saturate a given volume of air increases with the temperature. Therefore, if saturated air is cooled, the excess water vapor condenses in the form of mist. See also: *absolute humidity.* (Nelson, 1965)
saturated mineral A mineral that can form in the presence of free silica, i.e., one that contains the maximum amount of combined silica. (AGI, 1987)
saturated rock (a) A rock having quartz in its norm. (AGI, 1987) (b) An igneous rock composed chiefly of saturated minerals. Cf: *oversaturated rock; undersaturated; unsaturated.* (AGI, 1987)
saturated surface *water table.*
saturated unit weight The wet unit weight of a soil mass when saturated. See also: *unit weight.* (ASCE, 1958)
saturated zone *zone of saturation.*
saturation curve *zero air voids curve.*
saturation pressure That pressure for a given temperature at which the vapor and the liquid can exist in stable equilibrium. (Strock, 1948)
sauconite A monoclinic mineral, $Na_{0.3}Zn_3(Si,Al)_4O_{10}(OH)_2{\cdot}4H_2O$; smectite group.
saukovite A former name for cadmian metacinnabar.
sausserite A tough, compact, and white, greenish, or grayish mineral aggregate consisting of a mixture of albite (or oligoclase) and zoisite or epidote, together with variable amounts of calcite, sericite, prehnite, and calcium-aluminum silicates other than those of the epidote group. It is an alteration product of plagioclase; once thought to be a mineral species.
saussurite A tough, compact, white, greenish, or grayish mineral aggregate, resulting from the alteration of feldspars, and consisting of albite, prehnite, zoisite, epidote, and other calcium-aluminum silicates and calcite. (Fay, 1920; CTD, 1958)
saussuritization The replacement, esp. of plagioclase in basalts and gabbros, by a fine-grained aggregate of zoisite, epidote, albite, calcite, sericite, and zeolites. It is a metamorphic or deuteric process and is frequently accompanied by chloritization of the ferromagnesian minerals. (AGI, 1987)
Savelsberg process *blast roasting.*
saw gang A frame provided with a number of parallel iron bars that are employed to saw stone. See also: *stone saw.* (Fay, 1920)
sawsetter In stonework industry, one who maintains stone cutting saws in operating condition, replacing broken and bent saw blades from gang saws. Also called sawmaker. (DOT, 1949)
sawtooth back stoping *overhand stoping.*
sawtooth barrel *basket.*
sawtooth blasting The blasting of oblique, horizontal holes along a face and so cutting a series of slabs that, in plan, resemble saw teeth. (Nelson, 1965)
saw-toothed *serrate.*
sawtooth floor channeling A method of channeling inclined beds of marble by removing right-angle blocks in succession from the various beds, thus giving the floor a zigzag or sawtooth appearance.
sawtooth stoping In the United States, a form of overhand stoping in which the general line of advance is up the dip. The benches are advanced in a line parallel with the drift. The method permits a large number of machines to be used but requires the miners to work under a comparatively dangerous back. (Nelson, 1965)
Sawyear-Kjellgren process A process for converting beryl to beryllium oxide, based on quenching the melted beryl in cold water. The resultant frit reacts with concentrated sulfuric acid, and is steamed and agitated. The liquid, containing soluble beryllium and aluminum sulfates, is filtered and pumped to a tank where ammonium hydroxide is added. The resulting filtrate is further treated with a chelating agent to prevent impurities from precipitating upon subsequent addition of caustic soda. Hydrolysis follows, and the precipitate, beryllium hydroxide, is filtered off. This precipitate is ignited in an electric furnace to form beryllium oxide. (USBM, 1965)
sawyer (a) In stonework industry, a general term applied to workers engaged in cutting stone with power driven saws. (DOT, 1949) (b) A timber cutter.
sax A slate-cutter's knifelike chopping tool for trimming roof slates, having a pointed pick at the back to make nail holes. Also called slate ax. (Standard, 1964)
Saxonian chrysolite A pale wine-yellow topaz. (Fay, 1920)
scab (a) To trim rough blocks of stone with a pick or broad chisel; used by quarrymen. See also: *scabble.* (Standard, 1964) (b) A person who works at a mine contrary to union orders or during a strike. (Zern, 1928) (c) A defect consisting of a flat volume of metal jointed to a casting through a small area. It is usually set in a depression, a flat side being separated from the metal of the casting proper by a thin layer of sand. (ASM, 1961) (d) A fault in the base metal for vitreous enameling; the scab is a partially detached piece of metal (which may subsequently have been rolled into the metal surface) and is liable to cause faults in the applied enamel coating. (Dodd, 1964)
scabble To dress (as stone) in any way short of fine tooling or rubbing. Cf: *scab.* (Webster 3rd, 1966; Fay, 1920)
scabbler In quarry industry, one who roughs stone slabs in blocks with a scabbling pick to produce a uniform rectangular shape and to reduce shipping weight. (DOT, 1949)
scabbling The rough trimming of blocks of dimension stone.
scabbling hammer A hammer with two pointed ends for picking the stone after the spalling hammer. (Fay, 1920)
scabblings Stone chips produced in dressing stone or ore. (Arkell, 1953)
scabby In founding, blistered or marred with scabs; said of a casting. (Standard, 1964)
scacchite (a) A deliquescent cubic mineral, $MnCl_2$. (Larsen, 1934) (b) A name applied to various minerals, including monticellite, a doubtful selenide of lead, and a brick-red powdery fluoride containing rare earths.
scad A name occasionally applied to a nugget, as of gold.
Scaife process A modified Ugine-Sejournet process for hot extrusion of steel and other metals. The basic difference between the original Sejournet extrusion process and the Scaife modification is one of direction. In the Sejournet, the billet is forced forward through the die with the mandrel projecting through the die to maintain internal shape. In the modified process, the billet is forced into the closed die and the ram pressure squeezes it back over the mandrel. Both are based on the use of molten glass as a lubricant. (Osborne, 1956)
scalar A quantity fully described by a number, such as a speed that has no associated direction. See also: *vector.* (Hy, 1965)
scale (a) The ratio between linear distance on a map, chart, globe, model, or photograph and the corresponding distance on the surface being mapped. It may be expressed in the form either of a direct or verbal statement using different units (e.g., 1/24,000 or 1:24,000, indicating that one unit on the map represents 24,000 identical units on the ground) or a graphic measure (such as a bar or line marked off in feet, miles, or kilometers). (AGI, 1987) (b) Loose, thin fragments of rock, threatening to break or fall from the roof or wall of a mine. To remove such fragments. (c) Crude paraffin wax, obtained by filtering the cooled heavy distillation from petroleum or shale. (Standard, 1964) (d) A fault, in glass or vitreous enamelware, in the form of an embedded particle of metal oxide or carbon. (Dodd, 1964) (e) Newc. A small portion of air abstracted from the main current. Also called scale of air, and sometimes spelled skail. (f) To regulate the air current in a roadway. (BS, 1963) (g) Used among English miners for carbonaceous shale interbedded with thin layers of coal. (Tomkeieff, 1954) (h) The flakes and rubble that fall in after the ore has been removed. (Gordon, 1906)
scale cleaner In bituminous coal mining, one who scales off loose pieces of slate from the roof and walls of haulageways, using a pick or a bar. Also called slate handler. (DOT, 1949)
scale copper Copper in very thin flakes. (Weed, 1922)
scaled distance A factor relating similar blast effects from various size charges of the same explosive at various distances; it is computed by dividing the distance of concern by an exponential root of the explosive quantity detonated per delay.
scale door A door that has an air regulator. Syn: *regulator door.* (BS, 1963)
scalenohedron A closed crystal form, each face of which is a scalene triangle; trigonal scalenohedra have 12 such faces; tetragonal have 8. Adjective: scalenohedral. *dogtooth spar.*

scale of hardness *hardness scale.*

scaler (a) An electronic instrument for counting radiation-induced pulses from Geiger counters and other radiation detectors. (Lyman, 1964) (b) A laborer who knocks the roasted lead ore off grates with a bar as it is dumped from conveyors into cars below, prior to melting, to separate and recover the lead. Lead ore is loaded on grates attached to a conveyor and carried through a furnace in which the sulfur is driven off by roasting. (DOT, 1949)

scales For weighing materials in transit, these include the track scale (carload lots) in which the load is checked by manual operation of weights; platform scales; automatic dump hoppers; conveying weighers that continuously record or register the weight of a portion of passing conveyor belt. (Pryor, 1963)

scale-up In plant design, calculations of required capacities, machine sizes, etc., from data obtained in batch and pilot testing. (Pryor, 1963)

scaling (a) The plucking down of loose stones or coal adhering to the solid face after a shot or a round of shots has been fired. (Nelson, 1965) (b) Removal of loose rocks from the roof or walls. Also called ramming. Cf: *exfoliation.* (Fraenkel, 1953; Stoces, 1954)

scaling bar A long metal bar flattened on one end and used to remove loose material from a mine roof or rib. See also: *bar.*

scaling furnace A furnace or oven in which plates of iron are heated for the purpose of scaling them, as in the preparation of plates for tinning. (Fay, 1920)

scaling of the face In mining, consists of the removal of loose overhanging rock. (Streefkerk, 1952)

scall (a) Eng. Loose ground; foliated ground is frequently called scally ground by miners. Also spelled scal. Probably a variation of scale. (b) Rock easily broken up because of its scaly structure.

scallop (a) Eng. To cut or break off the sides of a heading without holing or using powder. (b) Eng. To get or hew coal off the face. (SMRB, 1930) (c) One of a mosaic of small shallow intersecting hollows formed on the surface of soluble rock by turbulent dissolution. They are steeper on the upstream side, and smaller sizes are formed by faster-flowing water. Syn: *flute; solution ripple.* (AGI, 1987)

scalp (a) The process of removing oversize lumps on a continuous basis from a stream of bulk material. (b) Removing large pieces of mine waste from run-of-mine coal, usually when passing over a screen, on way to the preparation plant.

scalped anticline *breached anticline.*

scalper Heavy screen shielding fine screen for separating differently sized particles. (Bennett, 1962)

scalping (a) The removal, by screen or grizzly, of undesirable fine material from broken ore, stone, or gravel. (Nelson, 1965) (b) A milling term for the removal of a mineral during closed-circuit grinding of the ore. (Pryor, 1960)

scalping screen A coarse primary screen or grizzly; usually a vibrating grizzly. (Nichols, 1976)

scaly Said of the texture of a mineral, esp. a mica, in which small plates break or flake from the surface like scales. (AGI, 1987)

scamy Eng. Applied to freestone in thin layers, mixed with mica.

scamy post N. of Eng. Soft, short, jointy freestone, thinly laminated and much mixed with mica.

scandium A silvery-white metallic element that develops a slightly yellowish or pinkish cast upon exposure to air. Symbol, Sc. It occurs as a principal component in the rare mineral thortveitite; or is extracted as a by-product from uranium mill tailings. It is of interest for use in space missiles. (Handbook of Chem. & Phys., 3)

scanning electroprobe X-ray microanalyzer An instrument for metallurgical research. It can create visual images of the minutest particles in metals and alloys by use of X-ray beams. Magnification is up to 3,000 times, and particles smaller than a millionth of an inch can be examined in detail. (Nelson, 1965)

scantite A gage by which slates are assorted in sizes.

scantling (a) The dimensions of a stone in length, breadth, and thickness. (Standard, 1964) (b) Stones more than 6 ft (2 m) long. (CTD, 1958) (c) A piece of timber of thickness from 2 to 4 in (5 to 10 cm) and of width from 2 to 4-1/2 in (5 to 11 cm). (CTD, 1958) (d) Small timber, such as 2 in by 3 in (5.1 cm by 7.6 cm), 2 in by 4 in (5.1 cm by 10.2 cm), etc., used for studding. (Crispin, 1964)

scapolite (a) A mineral group, $Na_4Al_3Si_9O_{24}Cl$ (marialite), to $Ca_4Al_6Si_6O_{24}(CO_3,SO_4)$ (meionite); a tetragonal isomorphous series; occurs in calcium-rich metamorphic rocks or igneous rocks as the product of alteration of plagioclase feldspar. (b) Any mineral of the scapolite group, intermediate in composition between marialite and meionite (Ma:Me from 2:1 to 1:3), containing 46% to 54% silica, and resembling feldspar when massive, but having a higher density and a fibrous appearance. Intermediate compositions were formerly called dipyre and mizzonite for Ma- and Me-rich samples, respectively. Syn: *wernerite.*

scapolitization Introduction of, or replacement by, scapolite. Plagioclase is commonly so replaced. The replacement may involve introduction of chlorine. (AGI, 1987)

scar In founding, an imperfect spot in a casting. (Standard, 1964)

scarbroite A hexagonal mineral, $Al_5(OH)_{13}(CO_3)\cdot 5H_2O$; compact; fine-grained; in vertical fissures in sandstone at South Bay, Scarborough; in sandstone at Yorkshire, England. Also spelled scarbroeite. Syn: *tucanite.*

scarcement A projecting ledge of rock, left in a shaft as footing for a ladder or to support pitwork. (Fay, 1920)

scares (a) Lenticular pockets of clean coal in sandstone, usually found in the region of a washout. Also called coal scares. (Tomkeieff, 1954) (b) N. of Eng. Thin laminae of pyrite in coal.

scarf (a) A lapped joint made by beveling, notching, or otherwise cutting away the sides of two timbers at the ends, and bolting or strapping them together so as to form one continuous piece, usually without increased thickness. Also called scarf joint. (Standard, 1964) (b) A piece of metal shaped or beveled for a scarf weld. (Standard, 1964)

scarfer In the iron and steel industry, one who tends rolls through which skelp (steel strips for making pipe or tube) or steel sheet is run to bevel edges prior to its being formed into tube. (DOT, 1949)

scarfing (a) Splicing timbers, so cut that when joined the resulting piece is not thicker at the joint than elsewhere. (Fay, 1920) (b) Tapering the ends of two pieces to be joined to avoid an enlarged joint. (Crispin, 1964) (c) Cutting surface areas of metal objects, ordinarily by using a gas torch. The operation permits surface defects to be cut from ingots, billets, or the edges of plate that is to be beveled for butt welding. (ASM, 1961)

scarf joint (a) A butt joint in which the plane of the joint is inclined with respect to the main axis of the members. (ASM, 1961) (b) *scarf.*

scarifier A machine with downward projecting teeth for breaking hard soil at quarries and opencast pits. It may be self-propelled or attached to another vehicle. See also: *rooter.* (Nelson, 1965)

scarify To roughen up, as a road, for repairs. (Crispin, 1964)

scarp (a) A line of cliffs produced by faulting or by erosion. The term is an abbreviated form of escarpment, and the two terms commonly have the same meaning, although scarp is more often applied to cliffs formed by faulting. See also: *fault scarp.* (AGI, 1987) (b) A relatively straight, clifflike face or slope of considerable linear extent, breaking the general continuity of the land by separating surfaces lying at different levels, as along the margin of a plateau or mesa. A scarp may be of any height. The term should not be used for a slope of highly irregular outline. Cf: *cuesta.* (AGI, 1987)

scarring The formation of scars or scaurs in roasting pyrite for sulfuric acid manufacture. (Fay, 1920)

scatter Deviation of portions of a radiation beam by scattering centers in the medium through which the beam passes. In ultrasonics, it occurs by reflection, refraction, or diffraction at any acoustical discontinuity comparable in size with, or larger than, the wave length used; in radiography, the scatter occurs by the Compton, photoelectric, and pair-production processes. (ASM, 1961)

scatter pile In underground mining, ore left adjacent to a longwall face to stop flying ore from being lost when blasting. A secondary use is to confine ventilation. (Pryor, 1963)

scavenge To pick up surplus fluid and return it to a circulating system. (Nichols, 1976)

scavenger (a) Any chemical that is added to a system or a mixture to consume, or to convert to an inactive form, small quantities of impurities or undesired materials. (CCD, 1961) (b) In flotation, a rougher cell in which the tailings, before being rejected as waste, are subjected to a scavenging flotation treatment. Concentrating tables are also used as scavenger machines. (Hess) (c) In metallurgical operations, an active metal added to combine with oxygen and/or nitrogen in the molten metal and so cause removal of impurities into slag. (CCD, 1961) (d) Oxygen, iodine, or more complex materials that, when added to a mixture, combine with free radicals in the mixture and permit the measurement of these radicals. (CCD, 1961)

scavenger cells Secondary cells for the retreatment of tailings. (BS, 1962)

scavenger mining The removal of coal so close to the surface as to undermine the topsoil, resulting in devastation above ground. Usually engaged in by an independent operator working an old mine on a lease from a major corporation. (Korson, 1938)

scavenging In mineral processing, final stage in flotation of mineralized froth before discard of tailing. The cells are so worked as to remove for retreatment as much low-grade rising mineral as possible under the given working conditions. (Pryor, 1963)

scawtite A monoclinic mineral, $Ca_7Si_6(CO_3)O_{18}\cdot 2H_2O$; forms bundles of thin, colorless, tabular crystals at Scawt Hill, County Antrim, Ireland.

scepter quartz Quartz crystals resembling a scepter in shape.

schafarzikite A tetragonal mineral, $FeSb_2O_4$; red to reddish-brown.

schairerite A trigonal mineral, $Na_{21}(SO_4)_7F_6Cl$; colorless; at Searles Lake, CA. Cf: *galeite.*

schallerite A trigonal mineral, $(Mn,Fe)_{16}Si_{12}As_3O_{36}(OH)_{17}$; dimorphous with nelenite; reddish-brown.

schapbachite An isometric mineral, $AgBiS_2$; a high-temperature polymorph of matildite and member of the halite group; not recognized as a valid mineral species; occurs as acicular crystals, granular, or massive. See also: *matildite.*

schaum earth Same as aphrite.

scheelite A tetragonal mineral, $CaWO_4$, with molybdenum replacing tungsten toward powellite $CaMoO_4$; prismatic cleavage; sp gr, 5.9 to 6.1; varicolored, fluoresces bright blue; in limestone and pneumatolitic veins near granite contacts, granite pegmatites; a source of tungsten.

scheererite A whitish, gray; yellow, green, or pale reddish; brittle; tasteless; inodorous hydrocarbon; melts at 44 °C; soluble in alcohol and ether; may be distilled without decomposition, boiling at 92 °C. Syn: *xylocryptite.*

schefferite A mineral, $(Na,Ca)(Fe,Mn)Si_2O_6$; brown to black; a variety of manganoan aegirine.

scheibeite (a) Fossil resin found in brown coal. (Tomkeieff, 1954) (b) A former name for phoenicochroite.

schematic Showing principles of construction or operation, without accurate mechanical representation. (Nichols, 1976)

scheteligite (a) A very rare, weakly radioactive, orthorhombic(?), black mineral, $(Ca,Y,Sb,Mn)_2(Ti,Ta,Nb,W)_2O_6(O,OH)$, found at Torvelona, Norway, in pegmatite with plagioclase, tourmaline, bismuth, euxenite, thortveitite, monazite, alvite, beryl, garnet, and magnetite. Small amounts of uranium may be present. (Crosby, 1955) (b) A name for an incompletely described possible member of the betafite subgroup of the pyrochlore group.

Schicht mixed-flow fan In this fan, the blades are mounted on the curved portion of a dish-shaped rotor and are designed to impart dynamic energy but no pressure or static energy to the air, the dynamic energy being converted to pressure in the diffuser. The fan is suitable for water gauges up to 20 in (50.8 cm) and there is an absence of noise. It is useful where the resistance of the mine is known and not liable to alter materially. (Sinclair, 1958)

schieferspar A variety of calcite occurring in very thin plates or scales. (Fay, 1920)

schiller A phenomenon related to sheen; an almost metallic iridescent shimmer or play of color seen just below the surface in certain varieties of pyroxene, feldspar, etc. Etymol: German. See also: *adularescence; schillerization.*

schiller-fels Enstatite or bronzite peridotite with poikilitic pyroxenes. Orthorhombic pyroxenes possess the poikilitic texture to a peculiar degree, and esp. when more or less altered to bastite, the term schiller is esp. applied to them. (Fay, 1920)

schillerization The development of poikilitic texture by the formation of inclusions and cavities along particular crystal planes, largely by solution somewhat as are etch figures. (Fay, 1920)

schiller obsidian Obsidian with schiller effect.

schiller spar An altered enstatite or bronzite, having approx. the composition of serpentine. Syn: *bastite.* Also spelled schillerspar.

schist A strongly foliated crystalline rock, formed by dynamic metamorphism, that can be readily split into thin flakes or slabs due to the well developed parallelism of more than 50% of the minerals present, particularly those of lamellar or elongate prismatic habit, e.g., mica and hornblende. The mineral composition is not an essential factor in its definition unless specif. included in the rock name, e.g., quartz-muscovite schist. Varieties may also be based on general composition, e.g., calc-silicate schist, amphibole schist; or on texture, e.g., spotted schist. See also: *magnesian schist; pelitic schist.* Cf: *paraschist.* (AGI, 1987)

schistose Said of a rock displaying schistosity. (AGI, 1987)

schistosity The foliation in schist or other coarse-grained, crystalline rock due to the parallel, planar arrangement of mineral grains of the platy, prismatic, or ellipsoidal types, usually mica. It is considered by some to be a type of cleavage. Adj: schistose. (AGI, 1987)

schizolite A light-red variety of manganoan pectolite. *pectolite.*

schlieren (a) Tabular bodies, generally a few inches to tens of feet long, that occur in plutonic rocks. They have the same general mineralogy as the rocks, but because of differences in mineral ratios, they are darker or lighter; the boundaries with the plutonic rock tend to be transitional. Some schlieren are modified inclusions, others may be segregations of minerals. Etymol: German. Sing: schliere. Also spelled schliere. Adj: schlieric. Cf: *flow layer.* (b) Regions of different density in fluid, esp. as shown by special apparatus. (Hunt, 1965) (c) A method or apparatus for visualizing or photographing regions of varying density in a field of flow. (Hunt, 1965)

Schmidt apparatus Apparatus used to determine the position of rest of a freely swinging shaft plumbline. (BS, 1963)

Schmidt-type magnetic field balance This has been the most commonly used magnetic instrument for prospecting on land. It consists of a magnet pivoted near but not at its center of mass, so that the magnetic field of the earth creates a torque around the pivot that is opposed by the torque of the gravitational pull upon the center. The angle at which equilibrium is reached depends on the strength of the field. Readings are taken through an eyepiece by comparing a scale reflected from a mirror on the magnetic element with a fixed scale. The balance may be either the horizontal or vertical type. (Dobrin, 1960)

schmiederite A monoclinic mineral, $Cu_2Pb_2(SeO_3)(SeO_4)(OH)_4$; light blue; at La Rioja, Argentina. Also spelled schmeiderite.

Schneider furnace A distillation furnace for the reduction of zinc ores containing lead, with a recovery of the latter metal as well as the zinc. (Fay, 1920)

schoderite A monoclinic mineral, $Al_2(PO_4)(VO_4)·8H_2O$; yellow-orange.

schoenfliesite The mineral group burtite, natanite, schoenfliesite, vismirnovite, and wickmanite.

schoepite An orthorhombic mineral, $UO_3·2H_2O$; strongly radioactive; perfect cleavage; yellow; an alteration product of uraninite or ianthinite. Formerly called schoepite I. Syn: *epiianthinite.* See also: *metaschoepite; paraschoepite.*

Scholl's method A method for determining the uranium in any of its ores in which the uranium is extracted with dilute nitric acid. This extract is then diluted, filtered, and treated with ferric chloride and sodium carbonate causing the vanadium iron and aluminum to precipitate. The uranium is then precipitated from the filtrate by boiling with caustic soda and purified by solution in nitric acid. Following precipitating with ammonia, the ammonium uranate is ignited to the oxide, and weighed. When this weight is multiplied by the factor 0.847, it gives the weight of uranium.

scholzite An orthorhombic mineral, $CaZn_2(PO_4)_2·2H_2O$; dimorphous with parascholzite; colorless to white.

Schone's apparatus An elutriator consisting of a tall glass vessel tapering toward the bottom, where water enters at a constant rate. Schone's formula is: $V = 104.7\ (S1)^{1.57}D^{1.57}$ where V is the velocity of water (millimeter/second) required to carry away particles of diameter D and specific gravity, S. (Dodd, 1964)

schorlite (a) A black variety of tourmaline. (Dana, 1959) (b) A discontinued term for schorl.

schorlomite An isometric mineral, $Ca_3Ti_2(Fe_2Si)O_{12}$, with Ti decreasing toward andradite in the garnet group.

schorl rock A term used in Cornwall, England, for a granular rock composed essentially of aggregates of needlelike crystals of black tourmaline (schorl) associated with quartz, and resulting from the complete tourmalinization of granite. (AGI, 1987)

Schottky defect (a) An atomic vacancy. (Van Vlack, 1964) (b) A point defect in a crystal structure where an atom is missing from a correct site. Syn: *vacancy.* Cf: *point defect; Frenkel defect.* (c) A defect in anionic crystal in which a single ion is removed from its interior lattice site and relocated in a lattice site at the surface of the crystal. (d) A defect in anionic crystal consisting of the smallest number of positive-ion vacancies and negative-ion vacancies that leave the crystal electrically neutral.

Schramhobel cutter plow *cutter plow.*

schreibersite A tetragonal mineral, $(Fe,Ni)_3P$; highly magnetic; contains small amounts of cobalt and traces of copper and tarnishes to brass yellow or brown. Schreibersite occurs in tables or plates as oriented inclusions in iron meteorites. Syn: *rhabdite.* (AGI, 1987)

schröckingerite A triclinic mineral, $NaCa_3(UO_2)(CO_3)_3(SO_4)F·10H_2O$; radioactive; soft; greenish-yellow color and fluorescence; an alteration product of uraninite; a source of uranium. Also spelled schroeckingerite. Syn: *dakeite.*

schrotterite An opaline variety of allophane rich in aluminum. Material at the type locality is a mixture of glassy hyalophane and earthy variscite.

schuchardtite A name for interlayered Ni-rich vermiculites and chlorites.

schuetteite A hexagonal mineral, $Hg_3(SO_4)O_2$; yellow.

schuilingite An orthorhombic mineral, $PbCu(Nd,Gd,Sm,Y)(CO_3)_3(OH)·1.5H_2O$; light blue.

schultenite A monoclinic mineral, $PbHAsO_4$; forms colorless thin crystalline plates; resembles selenite; at Tsumeb, Namibia.

Schulze elutriator The original type of water elutriator; it has since been improved. (Dodd, 1964)

Schulze-Hardy rule This states that the ion that causes a soluble to coagulate is opposite in sign to the electric charge of the colloidal particle; further, the coagulating power increases with the valency of the ion. (Pryor, 1963)

Schulz's theory A mine subsidence theory that distinguishes between the manner of fracture of shale and sandstone, holding that the former rock breaks along vertical lines irrespective of the angle of dip, and that the latter has a vertical fracture over a rise face and a fracture at right angles to the bed over a dip face. The theory predicts vertical lines of break in either rock for a level seam, and is, indeed, a compromise between the vertical theory and that of the normal. (Briggs, 1929)

Schumann plot Integral plot in graphic representation of sizing analysis. (Pryor, 1963)

schungite *shungite.*

schwatzite A mercurian variety of tetrahedrite. Also spelled schwazite.

scientific alexandrite A synthetic corundum colored by vanadium oxide to resemble true alexandrite in some of its optical characters.

scientific emerald A synthetic beryl glass colored by chromic oxide to resemble true emerald in color.

scintillascope *scintillation counter.*

scintillation A very small light flash excited in certain natural or synthetic crystals by radioactive rays or particles; the basic phenomenon of the scintillation counter in which the photoelectric effect of the scintillation flashes is amplified and measured to give a measure of intensity of radioactivity. (AGI, 1987)

scintillation counter A sensitive instrument for the location of radioactive ore, such as uranium, radium, and thorium. It uses a transparent crystal that gives off a minute flash of light when struck by a gamma ray, and a photomultiplier tube that produces an electrical impulse when the light from the crystal strikes it. The scintillation counter has advantages over the Geiger counter as it is more sensitive, more compact, and can distinguish between types of radiation. The instrument responds to gamma rays emitted from the minerals mentioned and charts their intensity. It is used in aerial geophysical prospecting and the resulting maps are used as a guide for a more detailed ground investigation. See also: *coal-sensing probe.* (Nelson, 1965)

scintillation probe An electronic logging device consisting of a scintillation-type gamma-ray detecting unit built into a container small enough to be lowered into a borehole. (Long, 1960)

scintillometer An instrument for measuring radioactivity, based on emission of light by certain crystals under impact of gamma rays. (AGI, 1987)

scissor fault A fault of dislocation, in which two beds are thrown so as to cross each other. Cf: *hinge fault; rotational fault.* (Zern, 1928)

sclaffery Scot. Liable to break off in thin fragments, as the roof of a mine working.

sclerometer An instrument for determining the degree of hardness of a mineral by ascertaining the pressure on a moving diamond point necessary to effect a scratch. (Standard, 1964)

scleroscope hardness test A test to determine the hardness of metals by measuring the rebound from them of a standard diamond-tipped hammer dropped from a given height. See also: *Shore hardness test; Shore scleroscope.* (Nelson, 1965)

sclerotinite A maceral of coal within the inertinite group, consisting of the sclerotia of fungi or of fungal spores characterized by a round or oval form and varying size. (AGI, 1987)

scobs The dross of metals. (Standard, 1964)

scolecite A monoclinic mineral, $CaAl_2Si_3O_{10}{\cdot}3H_2O$; zeolite group; pseudotetragonal; fibrous to acicular; can show wormlike motion if heated; in cavities in basalt and hydrothermal veins.

sconce A protecting cover or screen; protection; shelter. A metal cover and holder combined for holding a miner's candle, esp. for hanging on wooden timbers. (Webster 3rd, 1966; Fay, 1920)

scone High-grade tin requiring little or no dressing. (Nelson, 1965)

scones Firebricks of a certain standard size. (Osborne, 1956)

scoop (a) Diesel- or battery-powered equipment with a scoop attachment for cleaning up loose material, for loading mine cars or trucks, and hauling supplies. (b) A large-sized shovel with a scoopshaped blade. (Zern, 1928) (c) Coal miner's shovel; also sometimes used to refer to scraper. (BCI, 1947) (d) *scraper bucket.* (e) A device that gathers ore at feed end of ball mill and delivers it into the feed trunnion. (Pryor, 1963)

scoopman In bituminous coal mining, a laborer who places the cable-drawn scoop of a scraper loader in position for it to scrape up coal (blasted from the working face) as it is dragged by the hoisting engine to a point where the coal is dumped into mine cars. (DOT, 1949)

scooptram Similar to a front end loader; a low-profile loader articulating in the center with a large bucket in front (usually five tons or more) that transports ore in an underground mine. The operator sits at sideway controls facing the loader and drives it in either direction as required. Usually used for loading ore cars, shuttle cars, or hauling directly to an ore pocket.

scopulite A rodlike or stemlike crystallite that terminates in brushes or plumes. (AGI, 1987)

score (a) A bill run up by a collier in bad times for the necessaries of life. (Fay, 1920) (b) *task.* (c) To mark with scratches or furrows; e.g., rocks in certain localities were scored by glacial drift. (Standard, 1964)

scoria (a) A bomb-size pyroclast that is irregular in form and generally very vesicular. In less restricted usage, a vesicular cindery crust on the surface of andesitic or basaltic lava, the vesicular nature of which is due to the escape of volcanic gases before solidification; it is usually heavier, darker, and more crystalline than pumice. The adj. form, scoriaceous, is usually applied to pyroclastic ejecta. Cinder is sometimes used synonymously. Cf: *pumice.* (AGI, 1987) (b) A local term for melted or partly melted rock surrounding burned-out coal beds in the Western United States. See also: *clinker.* (AGI, 1987)

scoriaceous Said of the texture of a coarsely vesicular pyroclastic rock (e.g., scoria), usually of andesitic or basaltic composition, and coarser than a pumiceous rock. The walls of the vesicles may be either smooth or jagged. Also, said of a rock exhibiting such texture. Syn: *scoriform; scorious.* Cf: *vesicular.* (AGI, 1987)

scorification The separation of gold or silver by heating it to a high temperature with a large amount of granulated lead and a little borax, in a scorifier. The gold or silver dissolves in the molten lead, which sinks to the bottom of the vessel, while the impurities form a slag with the lead oxide that is produced. (Nelson, 1965)

scorifier A bone ash or fireclay crucible somewhat larger than a cupel; used in assaying and in the metallurgical treatment of precious metals. (Nelson, 1965)

scoriform *scoriaceous.*

scorious *scoriaceous.*

scorodite An orthorhombic mineral, $FeAsO_4{\cdot}2H_2O$; variscite group; colorless or pale-green to brown; in gossans, oxidized zones of metal veins, and around hot springs; a minor source of arsenic.

Scortecci process A process for direct reduction of iron pyrites that depends on the dissociation of pyrites in the absence of air, and in the presence of carbon, with the formation of iron and carbon disulfide. (Osborne, 1956)

scorzalite A monoclinic mineral, $(Fe,Mg)Al_2(PO_4)_2(OH)_2$; lazulite group; forms a series with lazulite; light blue.

scotch (a) A wooden stop-block or iron catch placed across or between the rails of underground roadways, to keep the cars from running loose, or to hold them when standing upon an inclined plane. (b) Leic. The lower lift of coal that is wedged up in driving a heading. (c) To dress, as stone, with a pick or picking tool. (Standard, 1964)

scotch block (a) A wedge or block temporarily fitted to a running rail in order to hold (scotch) the wheel of a railway vehicle. (Hammond, 1965) (b) One form of gas port in an open-hearth steel furnace; the distinguishing feature is that it is monolithic, being made by ramming suitably graded refractory material around a metal template. (Dodd, 1964)

scotching A method of dressing stones either with a pick or with pick-shaped chisels. (Fay, 1920)

Scotch pebble A rounded fragment of agate, carnelian, cairngorm, or other variety of quartz; found in gravels of parts of Scotland and used as a semiprecious stone.

Scotch pig A very pure grade of pig iron. (Standard, 1964)

Scotch topaz Applied in the gem stone trade to yellow transparent quartzes resembling Brazilian topaz in color; used for ornamental purposes. See also: *citrine.* Cf: *topaz.* (CMD, 1948)

scour The erosion of the bed or bank of a river or of a seacoast by the action of flowing water and waves. See also: *scour protection.* (Hammond, 1965)

scour and fill A process of alternate excavation and refilling of a channel, as by a stream or the tides; esp. such a process occurring in time of flood when the discharge and velocity of an aggrading stream are suddenly increased, causing the digging of new channels that become filled with sediment when the flood subsides. Syn: *washout.* (AGI, 1987)

scouring cinder A basic slag that attacks the lining of a shaft furnace. (Fay, 1920)

scour protection The protection of soil or other submerged material against scour, by the use of steel sheet pilings, revetment, riprap, or brushwood, or by combining any such methods as most suited to the site. See also: *lining.* (Hammond, 1965)

scout (a) One who gathers information about the drilling rig of a rival company for the benefit of its employer. Also called snooper. (Long, 1960) (b) An engineer who makes a preliminary examination of promising oil and mining claims and prospects. (Long, 1960) (c) One who goes into a potential area, esp. for oil or gas, to lease or option the land. (Long, 1960)

scout boring Trial bores made to test formations of area being prospected. (Pryor, 1963)

scouter In stoneworking, a quarryman whose function is to split off large portions of rock by means of a jump drill and wedges. (Standard, 1964)

scout hole A borehole penetrating only the uppermost part of an orebody with the intention of delineating its surface configuration. Also, a shallow hole drilled to scout for an indication of ore or to explore an area in a preliminary manner. (Long, 1960)

scout prospecting Prospecting undertaken in new country in which the first step is to scout prospect rivers, streams, and creeks by washing gravel obtained from their beds. (Griffith, 1938)

scovan Corn. A tin-bearing lode.

scovillite A former name for rhabdophane.

scow A device used to a limited extent to load solid blocks of coal. The scow proper is a flat steel plate that is moved underneath the undercut and blocks the coal by means of a hoist and a tail rope. The coal is then wedged down on the scow, and the solid block is hauled by means of the hoist and a headrope to a delivery point where it is transferred to cars. (Jones, 1949)

scowl bowl *scowle.*

scowle Forest of Dean. Ancient ironstone quarry and mine workings. Syn: *scowl bowl.*

scrablag Hardpan; Isle of Man, U.K.

scram (a) To search for and extract ore in a mine that is apparently worked out. (Weed, 1922) (b) An Alabama term for a small soft-coal mine complete in itself.

scram drift *scram drive.*

scram drive (a) Underground drive above the tramming level, along which ore is moved by scrapers (slushers) to a discharge chute. (Pryor, 1963) (b) *scram drift.*

scrap (a) Defective product not suitable for sale. (ASM, 1961) (b) Discarded metallic material, from whatever source, which may be reclaimed through melting and refining. (ASM, 1961) (c) Som. Stone only fit for rough walling. (Arkell, 1953) (d) *diamond scrap.*

scrap baler In the iron and steel industry, one who presses, in a baling press, steel scrap into compact blocks, for remelting in the open-hearth furnace. Also called scrap builder; scrap pressman. (DOT, 1949)

scrap bar The uneven ends of the muck bars. (Mersereau, 1947)

scrap-carbon process Indian scrap-carbon process using 100% steel scrap in which petroleum coke replaces carbon and acid slag replaces silicon. The hearth is protected from erosion by spreading an easily fusible silica sand over the banks before charging, and manganese ore is used instead of iron ore for oxidizing the carbon. (Osborne, 1956)

scraper (a) A rod for cleaning out a shothole prior to charging with explosives. See also: *stemmer; spoon.* (Nelson, 1965) (b) A mechanical contrivance used at collieries to scrape the culm or slack along a trough to the place of deposit. (c) A machine used in mines for loading cars and transporting ore or waste for short distances. There are two basic types of scraper: (1) the hoe or open type, which is particularly suitable for moving coarse, lumpy ore; and (2) the box or closed type, which is particularly suited for handling fine material, esp. on a loading slide. Syn: *box scraper.* (Lewis, 1964) (d) A digging, hauling, and grading machine having a cutting edge, a carrying bowl, a movable front wall (apron), and a dumping or ejecting mechanism. Also called carrying scraper or pan. (Nichols, 1976) (e) An apparatus used to take up coal from the floor of a mine, after it has been shot, and deposit it either in cars or in a conveyor. *spoon.* (BCI, 1947) (f) A rubber-tired device used to move earth in surface mining. (BCI, 1947) (g) *carryall.* (h) *poker man.* (i) The name applied to a bowl scraper or a multibucket excavator. Also known as a scraper excavator. (Hammond, 1965)

scraper and break detector In Great Britain, every shot firer is provided with a scraper for cleaning out shotholes, and in safety lamp mines a break detector must also be provided. The two tools are combined in the scraper and break detector. It can clean out a shothole and detect breaks in the walls of 1/8 in (3 mm) or more in width. The firing of a shot in a hole traversed by a crack exceeding 1/8 in in width is forbidden. (Nelson, 1965)

scraper box plow (a) A layout of rope-drawn scraper boxes with knives on the face side. They are drawn to and fro, and pushed against the face by guides controlled by rams. A haulage of 250 hp (186 kW) must be installed in a semipermanent engineroom; it has a rope diameter of 1-1/4 in (3 cm) and a speed of 3 ft/s (1 m/s). No conveyor is required as the coal is scraped by boxes to the loading point. The maximum workable seam thickness is 20 in (51 cm) on gradients of 0° to 30°; maximum length of face, 220 to 275 yd (200 to 250 m); and advance per shift, 6-1/2 to 8 ft (2.0 to 2.4 m). (Nelson, 1965) (b) *Haarmann plow; Kema plow; Gusto scraper box.*

scraper bucket (a) One of the excavating bowls or buckets that form part of a scraper. (Hammond, 1965) (b) In coal mines, the scraper bucket is a bottomless, three-sided box, with a hinged back. The hinge operates in a forward direction so that on the return journey on the coal face the back opens allowing the box to remain empty. On the loading journey, the coal closes the hinge and the material is drawn or scraped forward to the point of discharge. Syn: *scoop.* See also: *scraper loader.* (Nelson, 1965)

scraper chain conveyor *chain conveyor.*

scraper conveyor (a) A mechanical device for conveying coal, rock, ashes, culm, etc., in a metal trough by means of scrapers attached to a rope or chain. (b) A conveyor consisting of chain-drawn scrapers or flights running in a trough through which they push the material to be transported. Also called drag-link conveyor; flight conveyor; chain conveyor. (BS, 1962)

scraper hoist A power-driven hoist that operates a scraper to move material (rock or coal) to a loading point.

scraper loader (a) A machine used for loading coal or rock by pulling an open-bottomed scoop back and forth between the face and the loading point by means of ropes, sheaves, and a multiple drum hoist. The filled scoop is pulled on the bottom to an apron or ramp where the load is discharged onto a car or conveyor. (Jones, 1949) (b) A combined scraper and transporting machine. Originally towed by a tractor, but now diesel-electric with a direct current motor in each wheel. (Pryor, 1963) (c) A double-drum winch with two steel ropes. The loading capacity of a scraper loader ranges from 30 to 80 tons per hour depending on conditions. The loader is used for transporting and loading coal on longwall faces, for removing and loading stone in tunnels, and for stowing dirt on longwall faces. Syn: *slusher.* See also: *loader.* (Nelson, 1965)

scraper plow One scraper box with picks, rope-drawn and unguided along the face. A 30-hp (22.4-kW) haulage advances with the face, which is made convex to eliminate the need for guides. Speed, 3 to 4 ft/s (approx. 1 m/s). Suitable seam conditions; thickness, 12 to 24 in (30 to 60 cm) at gradients of 0° to 35° (preferably 15° to 25°); and maximum length of face, 65 yd (60 m). Syn: *porte et gardin plow.* (Nelson, 1965)

scraper ripper Strip-mine equipment that breaks, loads, and hauls coal. Features include ripping teeth on the lip for breaking the coal and a flight conveyor for carrying the broken coal away from the lip. As the ripper teeth bite into and loosen the coal, the conveyor sweeps the loose coal upward and prevents buildup ahead of the lip. (Coal Age, 1966)

scrap forgings Forgings formed from wrought-iron scrap. (Standard, 1964)

scrap hoist operator In the iron and steel industry, one who operates a skip hoist to carry scrap material to the furnaces. (DOT, 1949)

scrap mica Mica that because of size, color, or quality is below specifications for sheet mica. Includes flake mica and the mica, except sheet, obtained from pegmatite mining as a sole product or as a byproduct, from the preparation of sheet mica, and from waste in fabricating sheet mica. (Skow, 1962)

scrap picker A person employed on the slag dump to pick out pieces of iron carried to the dump in slag ladles. (Fay, 1920)

scrapping bottom coal Lifting coal that has been left by an undercutting machine. (Zern, 1928)

scrap sorter In metallurgy, a laborer who sorts scrap metal and removes foreign matter preparatory to use in recasting. (DOT, 1949)

scratch A calcareous, earthy, or strong substance that separates from seawater in boiling it for salt. *striation.* (Fay, 1920)

scratch pan A pan in saltworks to receive the scratch. (Fay, 1920)

screaming joint A joint from which air leaks and makes a screaming noise. Compressed air escaping at a screaming joint or hole in an air hose can be the cause of a mine fire. (Sinclair, 1958)

scree A sieve, screen, or strainer; a coal screen. See also: *talus.*

scree bars Scot. Bars of which a scree is constructed. See also: *scree.* (Fay, 1920)

screen (a) A large sieve for grading or sizing coal, ore, rock, or aggregate. It consists of a suitably mounted surface of woven wire or of punched plate; it may be flat or cylindrical, horizontal or inclined, stationary, shaking, or vibratory, and either wet or dry operation. See also: *Bradford breaker; shaking screen; stationary bar screen; vibratory screen.* (Nelson, 1965) (b) A cloth brattice or curtain hung across a road in a mine to direct the ventilation. See also: *brattice.* (Fay, 1920) (c) A perforated sheet placed in the gating system of a mold to separate dirt from molten metal. (ASM, 1961)

screen analysis (a) The size distribution of noncohering particles as determined by screening through a series of standard screens. (Harbison-Walker, 1972) (b) *sieve analysis.* (c) The percentage of a sample retained on each size of a series of standard laboratory screens. (Nelson, 1965) (d) Determination of the particle-size distribution of a soil, sediment, or ore by measuring the percentage of the particles that will pass through standard screens of various sizes. Syn: *grading test.* (AGI, 1987)

screen box A container in which diamond screens are inserted and in which the material (diamond particles) that passes through a sieve or screen collects and is retained. (Long, 1960)

screen chute A discharge chute equipped with a screen section, either stationary or vibrating, to remove the finer portions of the material being handled from the major line of flow.

screen cloth A woven medium suitable for use in a screen deck. (BS, 1962)

screen deck A surface provided with apertures of specified size for carrying out the operation of screening. (BS, 1962)

screened coal Coal that has passed over any kind of a screen and therefore consists of the marketable sizes.

screened lump lime Lump lime after forking or screening to remove the finer portion. The portion removed is usually that which will pass a 1/2-in sieve.

screened trailing cable A flexible cable provided with a protective screen or screens of tinned copper wire, or other conducting material, applied (1) to enclose each power core separately (individual screening), or (2) to enclose all the cores of the cable (collective screening). (BS, 1965)

screening (a) The separation of solid materials of different sizes by causing one component to remain on a surface provided with apertures through which the other component passes. (BS, 1962) (b) Use of one or more screens (sieves) to separate particles into defined sizes. Also called sizing. See also: *sieving.* (Pryor, 1960)

screening machine An apparatus having a shaking, oscillatory, or rotary motion, used for screening or sifting coal, stamped ores, and the like.

screenings (a) Fine coal that will pass through the smallest mesh screen normally loaded for commercial sale for industrial use. (BCI, 1947) (b) The residue from a screening operation. (Nelson, 1965)

screenings crushing A stage in bituminous coal crushing in which units for crushing screenings reduce secondary product to final small, commercial sizes, such as 1- to 3/8-in (2.5- to 1-cm) stoker coal or screenings. (Mitchell, 1950)

screen iron ore Accumulation of surface debris on the lower slopes of iron-bearing hills. The scree material may contain sufficient iron ore to make its mining an economic proposition as at Middleback Ranges of South Australia. (Nelson, 1965)

screen loading chute A type of chute with a bar screen or grizzly bottom that permits fines to fall onto the conveyor belt first, providing a cushion for the large material that passes over the screen.

screen overflow That portion of the feed material discharged from the screen deck without having passed through the apertures. (BS, 1962)

screen pipe A perforated pipe lined with fine-mesh screen, set in portions of a borehole where the walls must be supported and the ingress of water or oil cannot be restricted. Also called well point. (Long, 1960)

screen plate (a) A metal plate with specific-sized openings used to control the fineness of grinding in dry pans and hammer mills. (ACSG, 1963) (b) A plate provided with apertures of specified size for use as a screen deck. (BS, 1962)

screen room That part of a breaker where boys picked slate and bony. (Korson, 1938)

screens Wire meshes with specific-sized openings for grading particles of various sizes. See also: *vibrating screen.* (ACSG, 1963)

screen size A standard for determining the size of particles. The particles are passed through screens with openings of specified size. The size of the particles is determined by the size of the opening through which the particles will not pass. (Long, 1960)

screen sizing Separating various-sized grain into portions, by a screen or sieve. Also called screening; sizing.

screen underflow That portion of the feed material that has passed through the apertures in a screen deck. (BS, 1962)

scree plate Scot. An iron plate at the foot of a screen on which screened coal is discharged.

screw (a) The feed screw in the swivel head of a gear-feed diamond drill. (Long, 1960) (b) Syn. for an auger stem having helical webs. (Long, 1960) (c) A combined symmetry operation involving rotation about an axis and a translation parallel to it. Cf: *space group.* (d) A crystal defect involving a dislocation about which layers of adatoms spiral during crystal growth. Cf: *Burgers vector.*

screw bell (a) A device to withdraw broken rods from a borehole, when the fracture occurs below a joint. The screw bell is lowered to cut a thread on the end of the broken rod and thus secure a grip sufficient to withdraw it safely. (Nelson, 1965) (b) A fishing tool shaped like a bell, bell tap, or bell screw. (Long, 1960) (c) *bell screw; bell socket; bell tap.*

screw conveyor (a) A conveyor screw revolving in a suitably shaped stationary trough, or casing fitted with hangers, trough ends, and other auxiliary accessories. (b) A conveyor in which a spiral blade presses material forward as it rotates in a suitable housing. Syn: *tubular screw conveyor; spiral conveyor.* See also: *grit collector; vertical screw conveyor.* (Pryor, 1963)

screw-down mechanism That mechanism on a mill for lowering and raising the rolls to accommodate the distance between them to the requirements of the article being rolled. (Mersereau, 1947)

screw elevators Vertical screw elevators are used for handling pulverulent materials. A typical installation for delivering bulk cement into a plant consists of a screw feeder, which takes the cement from the bulk-cement car and feeds it to the screw elevator. Both these units are airtight and all joints are fitted with rubber gaskets that prevent the loss of cement and render the operation dustproof. (Pit and Quarry, 1960)

screw fan *axial-flow fan.*

screwfeed A system of gears, ratchets, and friction devices, or some combination of these parts, in the swivel head of a diamond drill, which controls the rate at which a bit is made to penetrate the rock formation being drilled. When controlled by a feed gear, the bit maintains the same penetration rate per revolution regardless of drill-stem revolutions per minute. Also called gear feed, mechanical feed. (Long, 1960)

screw feeder An auger-type screw to transfer material from one piece of equipment to another. (ACSG, 1963)

screwjack *jackscrew.*

screw mixer *screw-type mixing conveyor.*

screw pile A wide helical blade fixed on a shaft and screwed into the ground by means of a winch or capstan. (Hammond, 1965)

screwplug *hoisting plug.*

screw shackle A long cylindrical nut, threaded internally with a right-hand thread at one end and a left-hand thread at the other, used to connect and tighten together the ends of two rods forming a brace or tie. (Hammond, 1965)

screw-type mixing conveyor A type of screw conveyor consisting of one or more conveyor screws, ribbon flight or cut flight, conveyor screws with or without auxiliary paddles. See also: *blending conveyor; screw conveyor.*

scribe An instrument used by surveyors for marking posts, trees, etc. (Fay, 1920)

scrin (a) Derb. Ironstone in irregular-shaped nodules. (b) Derb. A small subordinate vein. Also spelled skrin.

scroll A helical projection on a drill rod or stem to remove cuttings from a hole. (BS, 1964)

scronge S. Wales. Overlying strata loosened or broken by workings underneath. Probably a variation of scrunge, to squeeze.

scrowl (a) Corn. A thin, sometimes calcareous or siliceous rock attached to the wall of a lode. (b) Corn. Loose ore at the point where a lode is disturbed by a cross vein.

scrubber (a) Device in which coarse and sticky ore, clay, etc., is washed free of adherent material, or mildly disintegrated. The main forms are the wash-screen, wash trommel, log washer, and hydraulic jet or monitor. Scrubbers or scrubbing towers are also used to separate soluble gases with extracting liquids, or to remove dust from air by washing. (Pryor, 1963) (b) Device for separating environmentally noxious chemical substances from waste gas streams.

scrubstone Eng. A provincial term for a variety of calciferous sandstone.

scruff (a) A mixture of tin oxide and iron-tin alloy formed as dross on a tin-coating bath. (ASM, 1961) (b) *scum.*

SCSR *self-contained self-rescuer.*

scuffing grind Tumbling of sands with sufficient violence to remove loosely adherent surface coatings without otherwise breaking down the particles to any great extent. (Pryor, 1963)

sculls Incrustations of slag, dross, and metal on the contacting surfaces of vessels that treat or hold molten metals. (Pryor, 1963)

sculp To break slate into slabs suitable for splitting. See also: *hard way.* (Webster 3rd, 1966; AIME, 1960)

sculping Fracturing the slate along the grain, e.g., across the cleavage. (Zern, 1928)

scum (a) Impure or extraneous matter that rises or collects at the surface of liquids, as vegetation on stagnant water, or dross on a bath of molten metal. Sometimes incorrectly used for the word froth in flotation. (Fay, 1920) (b) A surface deposit sometimes formed on clay building bricks. The deposit may be of soluble salts present in the clay and carried to the surface of the bricks by water as it escapes during drying; it is then known as dryer scum. The deposit may also be formed during kiln firing, either from soluble salts in the clay or by reaction between the sulfur gases in the kiln atmosphere and minerals in the clay bricks; it is then known as kiln scum. Cf.: *efflorescence.* (Dodd, 1964) (c) Undissolved batch constituents floating as a layer above the molten glass in a pot or tank furnace. (Dodd, 1964) (d) Areas of poor gloss on a vitreous enamel; the fault may be due to action of furnace gases, to a nonuniform firing temperature, or to a film clay arising from faulty enamel suspension. (Dodd, 1964) (e) The "clouds" appearing around decalcomania formed by varnish residue. (ACSG, 1963) (f) A surface defect appearing as dull patches on otherwise bright surfaces of glazes, glass, or porcelain enamel. (ACSG, 1963)

scun Dev. A small vein.

scythestone A whetstone suitable for sharpening scythes.

sea (a) An ocean, or alternatively a large body of (usually) salt water less than an ocean. (AGI, 1987) (b) Waves caused by wind at the place and time of observation. (AGI, 1987) (c) State of the ocean or lake surface in regard to waves. (AGI, 1987)

seabeach placers Alaska. Placers adjacent to the seashore to which the waves have access.

sea bloom *plankton bloom.*

sea coal (a) Old name for bituminous coal; so named either because it was exported by sea from collieries in coastal districts, or because it was at first applied to coal washed ashore from deposits below sea level. (Arkell, 1953) (b) Coal dug from the earth; so called formerly to distinguish it from charcoal, because it was brought to London by sea. Known formerly as pit coal or earth coal. (Standard, 1964) (c) U.S. Rare. Soft coal as distinguished from anthracite. (Standard, 1964) (d) Archaic. Mineral coal. (Webster 3rd, 1966) (e) Pulverized bituminous coal used as a foundry facing. (Webster 3rd, 1966)

sea current The currents that constitute part of the general oceanic circulation. Syn. for ocean current. (Schieferdecker, 1959)

sea-floor trench *trench.*

sea-foam (a) An early syn. for meerschaum. (b) *sepiolite.*

seal (a) To secure a borehole or excavation against cave-ins and flowing or escaping gas or liquids by the use of cement or other sealants. (Long, 1960) (b) To secure a mine opening against flowing or escaping gas, air, or liquids by injecting grout, by coating rock surfaces with gunite, or by erecting rock, concrete, wood, or cloth barriers. (Long, 1960) (c) A short length of roadway that has been tightly filled with concrete, brickwork, sand, or other material to close off an area against fire, gas, or water. In the case of a fire, the seal cuts off the air supply and also prevents noxious fumes given off from reaching other parts of a mine. Also called stopping. See also: *firedamp*.

sealant *painting*.

Seale construction Wire strand construction having one size of wires for the outer layer with the same number of smaller wires in the underlayer. Both layers have the same length and direction of lay. (Hammond, 1965)

sealed area In mining, portion of underground workings sealed off, usually because of fire (in which case no air is allowed to enter) or because no mineable coal remains in the area. See also: *unsealing*. (Pryor, 1963)

sealed-off area A part of a mine that has been sealed off from the rest of the mine. The object of sealing off a fire area is to: (1) contain the trouble, and to prevent an explosion that may occur inby from extending to other parts of the mine; (2) build up an extinctive atmosphere inside the sealed-off area; and (3) prevent the access of air to the inby side of the seal. (Nelson, 1965)

Seale rope A wire rope that has six or eight strands each, having a large center wire covered by nine small wires that are covered in turn by nine large wires. (Lewis, 1964)

Seale's lay A wire rope with the inner and outer layers consisting of the same number of wires, the outer being larger and lying in the grooves or valleys between the inner wires. Both layers are stranded or laid in one operation. Extra support is given to the outer wires by this method and the wires are in line contact throughout and there is no internal crosscutting of wires. (Sinclair, 1959)

sea level correction The deduction made from a measured length of a base line to establish its true length at sea level. See also: *tape corrections*. (Hammond, 1965)

sealing (a) Shutting off all air from a mine or portion of a mine, a practice used in an emergency to check fire by eliminating oxygen. Also, as a routine shutting-off method for worked-out areas in some mines. (BCI, 1947) (b) Sealing is used to overcome mine fires when other methods have failed. It involves the erection of temporary or permanent seals for the purpose of cutting off the oxygen supply to the area on fire. Sealing causes the fire to extinguish itself by consuming the oxygen in the sealed off area. (Kentucky, 1952) (c) Cutting off the air supply to effect extinction of underground fires by erecting sandbag stoppings at convenient places. The combustion process uses up the available oxygen within the sealed area, the process is arrested and the hot ground cools down gradually as the heat is conducted away by the surrounding cooler strata. See also: *fire seal*. (Mason, 1951) (d) Closing pores in anodic coatings to render them less absorbent. (ASM, 1961) (e) Plugging leaks in a casting by introducing thermosetting plastics into porous areas and subsequently setting the plastice with heat. (ASM, 1961)

sealing-wax wood Pieces of wood full of resin found in brown coal. When ignited they burn, melting and giving off soot and an aromatic odor like sealing wax. (AGI, 1987)

seal off The use of a cement or other sealant in a borehole. Seal off is not synonymous with blankoff and case off, where securing the walls of a borehole is accomplished by setting pipe or casing. Cf: *blankoff; seal*. (Long, 1960)

seam (a) A stratum or bed of coal or other mineral; generally applied to large deposits of coal. (Fay, 1920; BCI, 1947) (b) A particular bed or vein in a series of beds; it is usually said of coal but may also pertain to ore minerals. (AGI, 1987) (c) A thin layer or stratum of rock separating two distinctive layers of different composition or greater magnitude. (AGI, 1987) (d) A joint, cleft, or fissure. Syn: *crevice*. (e) A plane in a coalbed at which the different layers of coal are easily separated. (f) A very narrow vein. (g) *joint line*.

seamanite An orthorhombic mineral, $Mn_3(PO_4)B(OH)_6$; pale- to wine-yellow; in the Chicagoan Mine, near Iron River, MI.

seam contour A line drawn on a plan joining points on the floor or roof of a seam that have the same height above a prescribed datum. (BS, 1963)

seamount An elevation of the sea floor, 1000 m or higher, either flat-topped (called a guyot) or peaked (called a seapeak). Seamounts may be either discrete, arranged in a linear or random grouping, or connected at their bases and aligned along a ridge or rise. (AGI, 1987)

seam-out A shot that merely blows out a soft stratum in the coal or escapes through a seam without loosening the main mass of coal. In Arkansas, called squeal-out. (Fay, 1920)

seamy Full of seams, so as to be difficult to blast.

search coil (a) Sensitive device, using the mine-detector principle, for locating ferromagnetic material that is to be removed before ore treatment. It typically monitors a stream of ore passing along a conveyor belt, which it stops when iron is detected. (Pryor, 1963) (b) Coil that is used in electromagnetic methods for measuring the magnetic field that is associated with the electric current. (Schieferdecker, 1959)

search neighborhood Any area searched during interpolation between sample data points. Applies to any interpolation method where a limited number of sample data points are used to estimate intermediate values.

searlesite A monoclinic mineral, $NaBSi_2O_5(OH)_2$; forms minute spherulites composed of radiating fibers; at Searles Lake, CA.

Searles Lake brine A source of trona, $Na_3(CO_3)(HCO_3){\cdot}2H_2O$. Occurs in Searles Lake, San Bernardino County, CA. (CCD, 1961; Handbook of Chem. & Phys., 2)

sea sand Sand containing alkaline salts that attract and retain moisture and cause efflorescence in brick masonry. (Zern, 1928)

sea slick An area of sea surface, variable in size and markedly different in appearance, with color and/or oiliness; usually caused by plankton blooms. (Hy, 1965)

seasoned Applied to quarrystone after the moisture has dried out.

seasoning A mode of treatment of iron castings that are allowed to remain in storage, or to stand out in the open, for a more or less extended period, e.g., 6 months, to effect a reduction in the residual stresses and consequently in the degree of distortion during subsequent machining. A very similar result can often be obtained by a comparatively short period, e.g., 30 min, of tumbling. Since stress relieving by heat treatment is a more certain process, and seasoning involves much delay and the use of considerable space for storage, stress relieving is more usually employed. (Osborne, 1956)

seasoning timber The drying of the sap and moisture in the woody fibers and thus reducing the timber by shrinkage. It becomes more durable and weighs less. Timber may be air-dried, i.e., dried naturally in air, or kiln-dried, which means dried in kilns under the action of artificial heat. The former is more general. See also: *timber preservation*. (Nelson, 1965)

sea state Numerical or written description of ocean surface roughness. (Hy, 1965)

seat (a) The underclay or fireclay on which a coal seam rests. Also called seating. (Arkell, 1953) (b) The foundation or framework on which a structure rests; e.g., engine seat, cage seat. (Fay, 1920)

seat clay *underclay*.

seat earth (a) A British term for a bed of rock underlying a coal seam, representing an old soil that supported the vegetation from which the coal was formed; specif. underclay. A highly siliceous seat earth is known locally as ganister. Also spelled: seatearth. Syn: *seat rock; seat stone; hard seat*. (AGI, 1987) (b) A bed representing oil soil, usually containing abundant rootlets, underlying a coal seam. (BS, 1964) (c) The soil on which the coal forests fluorished. (Nelson, 1965) (d) Stratum underlying the valuable seam. Floor of a coal seam. See also: *underlying; underclay*. (Pryor, 1963)

seated (a) Placed in position. (Long, 1960) (b) Closed by pressing the closure part of a valve against its seat. (Long, 1960)

seating The surface of the point of support for a heavy load. (Hammond, 1965)

seat of settlement The deposit of soil under a loaded foundation within which the major settlement occurs. See also: *excavation deformation; settlement*. (Nelson, 1965)

seat rock *seat earth*.

seat stone *seat earth*.

sebkainite Crude potassium chloride obtained by solar evaporation of brine from a lake south of Gabes, Tunisia. (Hess)

sebkha *sabkha*.

secant modulus of elasticity Materials such as concrete or prestressing wire have a variable Young's modulus (E) so that the particular value of E adopted must be either the slope of the tangent to the stress-strain curve or that of the secant. The latter is the line that joins the origin of the curve to, for instance, the 0.1% proof stress, expressed on the curve. For a material within its elastic range, the secant will coincide with the tangent. See also: *modulus of elasticity*. (Hammond, 1965)

secondarily enriched deposit Deposits that result from supergene enrichment.

Secondary A term applied in the early 19th century as a syn. of Floetz. It was later applied to the extensive series of stratified rocks separating the older Primary and the younger Tertiary rocks, and ranging from the Silurian to the Cretaceous; still later, it was restricted to the whole of the Mesozoic Era. The term was abandoned in the late 19th century in favor of Mesozoic. (AGI, 1987)

secondary Said of metal obtained from scrap rather than from ore.

secondary air In a combustion chamber, air that meets with primary air to consume the fuel completely and complete combustion. (Newton, 1959)

secondary anomaly *geochemical anomaly.*
secondary ash Ash in coal derived from mineral matter precipitated in cleat clavities, etc. See also: *extraneous ash.* (Tomkeieff, 1954)
secondary beam A beam supported off, and transferring loads to, main beams that are themselves carried directly by the walls or columns. (Hammond, 1965)
secondary blasting Irrespective of the method of primary blasting employed, it may be necessary to reblast a proportion of the rock on the quarry floor so as to reduce it to a size suitable for handling by the excavators and crushers available. Two methods of secondary blasting of rock are available. The first, called the plaster or mudcap method, is to fire a charge of explosive placed on the rock and covered with clay, the shock of the detonating explosive breaking the block. The second technique, known as pop-shooting, is to drill a hole into the block and fire a small charge in this hole, which is usually stemmed with quarry fines. Also called blistering; bulldozing. See also: *plaster shooting; popping; snakeholing; pop shot; boulder blasting.* (McAdam, 1958)
secondary cell A group of flotation cells in which a product from the primary cells is retreated. (BS, 1962)
secondary clay A clay that has been transported from its place of formation and redeposited elsewhere. Cf: *residual clay; primary clay. sedimentary clay.* (AGI, 1987)
secondary consolidation Consolidation of sedimentary material, at essentially constant pressure, resulting from internal processes such as recrystallization. (AGI, 1987)
secondary creep Deformation of a material under a constant differential stress, with the strain-time relationship as a constant. Cf: *primary creep.* Syn: *steady-state creep.* (AGI, 1987)
secondary crusher Crushing and pulverizing machines next in line after the primary crushing to further reduce the particle size of shale or other rock. See also: *primary crusher.* This group of machines includes the finer types of jaw crusher and gyratory crusher, and also crushing rolls, hammer mills, and edge runner mills. (ACSG, 1963; Dodd, 1964)
secondary crushing In ore dressing, the second stage of grinding in which the discharge from the primary crusher is broken down to a size suitable for feed to fine grinding machines. (Newton, 1959)
secondary deposit (a) Made when the sediments already deposited are eroded and redeposited. (Schieferdecker, 1959) (b) A mineral deposit formed when a primary mineral deposit is subjected to chemical and/or mechanical alteration. Secondary deposits are divided into three groups: sedimentary rocks, secondarily enriched ore deposits, and residual or detrital ore deposits. (Hoover, 1948)
secondary dispersion Geochemical dispersion of elements by processes originating at the surface of the Earth; opposite of primary dispersion. Secondary patterns are those formed at the Earth's surface by weathering, erosion, or surface transportation. Secondary patterns have been classified more in detail as halos, fans, and trains, depending on the characteristic shape of the pattern and its geometric relationship to the ore deposit or other source. (Hawkes, 1957; Lewis, 1964)
secondary drilling The process of drilling the so-called "popholes" for the purpose of breaking the larger masses of rock thrown down by the primary blast.
secondary dust source If an operation agitates or disperses dust, it is a secondary source. See also: *primary source.* (Hartman, 1982)
secondary enlargement Deposition, around a clastic mineral grain, of material of the same composition as that grain and in optical and crystallographic continuity with it, often resulting in crystal faces characteristic of the original mineral; e.g., the addition of a quartz overgrowth around a silica grain in sandstone. (AGI, 1987)
secondary enrichment *supergene enrichment.*
secondary environment *geochemical environment.*
secondary fan Any fan installed underground to ventilate tunnels or workings where the air current is sluggish. See also: *mine ventilation auxiliary fan; booster fan.* (Nelson, 1965)
secondary geochemical cycle This cycle is comprised of the processes of weathering, soil formation, erosion, transportation, and sedimentation. (Hawkes, 1962)
secondary grinding Further comminution of material already reduced to sand sizes in rod or ball mills. (Pryor, 1963)
secondary hardness Further increase in hardness produced on tempering high-speed steel after quenching due to precipitation of carbides. (CTD, 1958)
secondary haulage That portion of the haulage system that collects the coal from the various gathering-haulage delivery points and delivers it to the main haulage system. (Wheeler, 1946)
secondary lead Lead derived from salvage of wornout end-product items, such as battery plates, cable covering, pipe and sheet, which are collected, remelted, and refined in secondary smelters to produce refined lead or various lead-base alloys. (USBM, 1965)
secondary metal Metal recovered from scrap by remelting and refining. (ASM, 1961)
secondary metal scrap Metal recovered from scrap by remelting and refining. (McGraw-Hill, 1994)
secondary mineral A mineral formed later than the rock enclosing it, usually at the expense of an earlier-formed primary mineral, as a result of weathering, metamorphism, or exsolution. (AGI, 1987)
secondary mineral deposit A mineral deposit formed when a primary mineral deposit is subjected to alterations through chemical and/or mechanical weathering. Secondary deposits are divided into three groups: sedimentary rocks, secondarily enriched mineral deposits, and residual or detrital mineral deposits. (Lewis, 1964)
secondary porosity Porosity developed after the formation of a deposit and resulting from subsequent fracturing, replacement, solution, or weathering. (AGI, 1987)
secondary products *middlings.*
secondary reject elevator A refuse elevator that extracts the second or lighter reject; usually situated at the discharge end of the washbox. (BS, 1962)
secondary rocks Rocks composed of particles derived from the erosion or weathering of preexisting rocks, such as residual, chemical, or organic rocks formed of detrital, precipitated, or organically accumulated materials; specif., clastic sedimentary rocks. (AGI, 1987)
secondary settling (a) The period following the primary settling in surface subsidence in which the surface subsides gradually. This period may continue for many years or even decades. Cf: *primary settling.* (Stoces, 1954) (b) Residual subsidence.
secondary shaft The shaft that extends a mine downwards from the bottom of the primary shaft but not in line with the primary shaft. (Spalding, 1949)
secondary shooting In quarrying, the reduction in size or dimension of blasted rock by additional or secondary blasting. (Streefkerk, 1952)
secondary splits The main air splits occur at the shaft bottom. In most cases, these splits are again separated at some point inby and these are called secondary splits. See also: *ventilation; splitting.* (Nelson, 1965)
secondary structure A structure that originated after the deposition or emplacement of the rock in which it is found, such as a fault, fold, or joint produced by tectonic movement; esp. an epigenetic sedimentary structure, such as a concretion or nodule produced by chemical action, or a sedimentary dike formed by infilling. Cf: *primary structure.* (AGI, 1987)
secondary sulfide zone *sulfide zone.*
secondary twinning Twinning produced subsequent to the original formation of a crystal.
secondary vein A vein discovered subsequent to the one on which a mining claim was based; an incidental vein. Cf: *discovery vein.* (AGI, 1987)
secondary water Water entering the mine from other workings, as opposed to water inherent in the area worked by the mine. (BS, 1963)
secondary wave *S wave; transverse wave; distortional wave.*
second-class lever A lever whose force is exerted between the fulcrum and the point where it is applied. (Nichols, 1976)
second-class ore An ore that needs preliminary treatment before it is of a sufficiently high grade to be acceptable for shipment or market. Cf: *first-class ore.* Syn: *milling ore.* (AGI, 1987)
second-foot A unit of 1 ft^3/s (0.0283 m^3/s). Usually abbreviated to cusec. (Seelye, 1951; Hammond, 1965)
second-foot-day The volume of water represented by a flow of 1 ft^3/s (0.0283 m^3/s) for 24 hours. It is 86,400 ft^3 (2,445 m^3), or nearly 2 acre-feet (actually 1.9835); a convenient unit in storage computations. (Seelye, 1951)
second mining (a) The recovery of pillar coal after first-mining in chambers has been completed. (Hudson, 1932) (b) The recovery of pillars after development of block pillars by the multiple entry system has completed a panel.
second moment of area The correct term for the moment of inertia (I) of the plane area of a section. (Hammond, 1965)
second or back explosion Aust. Supposed to be due to the ignition of gases developed from highly heated coal dust, and gases sucked out of the faces of coal by the partial vacuum resulting from the primary explosion, or liberated by the fall of roof. Cf: *retonation wave.* (Fay, 1920)
second outlet An emergency exit from a mine to the surface. Also called second opening; escapeway. (Hudson, 1932)
second ripping The first back ripping on a roadway. See also: *back brusher.* (Nelson, 1965)
seconds N.S.W. The second-class ore that requires dressing.
seconds A.P.I. A unit of viscosity in drilling mud as measured with a Marsh funnel according to American Petroleum Institute procedure. (Brantly, 1961)
second weight In mine subsidence, the powerful thrust or pressure, generally 20 to 40 ft (6 to 12 m) from the face, that causes distance from the roof to the floor to diminish rapidly and for packwalls to become compressed or pushed down into a soft bottom. Timber legs or metal supports (if any) in the gates are generally broken or twisted. (Briggs, 1929)

second worker In anthracite coal mining, one who is required to serve a specified number of years before being termed a first-class miner. (DOT, 1949)

second working (a) The operation of getting or working out the coal pillars formed by the first working. (Fay, 1920) (b) In coal mining, unless the pillars of coal are left permanently to support the surface, they are removed. This phase of mining is called the second working or pillar working. When the pillars are removed, nearly all of the coal has been recovered. Cf: *first working.* Also called pillar working. See also: *working the broken.* (Kentucky, 1952)

seconite A finely ground plastic clay that in proportions not lower than 6% gives satisfactory green strength when used as a bond with molding sand. (Osborne, 1956)

secretion (a) The act or process by which animals and plants transform mineral material from solution into skeletal forms. (AGI, 1987) (b) A secondary structure formed of material deposited from solution within a cavity in a rock, esp. a deposit formed on or parallel to the walls of the cavity; e.g., a mineral vein, an amygdule, or a geode. The space may be completely or only partly filled. Cf: *concretion.* (AGI, 1987)

sectile (a) Capable of being cut with a knife without breaking off in pieces. (AGI, 1987) (b) Said of a mineral that can be cut with a knife; e.g., argentite. (c) A physical property of minerals permitting shaving of curls with a knife; e.g., gypsum. Cf: *malleable.*

sectility A mineral is said to be sectile when it may be cut with a knife, but is not malleable, for example, graphite. (Nelson, 1965)

section (a) A portion of the working area of a mine. (BCI, 1947) (b) Representation of features such as mine workings or geological features on a vertical (or inclined) plane. A longitudinal section is parallel to the strike of a vein or geologic plane. A cross section is perpendicular to the strike. (McKinstry, 1948) (c) Detailed measurement, taken vertically, of a coal vein or of strata embracing several veins. (Hudson, 1932) (d) A drawing or diagram of the strata sunk through in a shaft or inclined plane, or proved by boring. (e) The local series of beds constituting a group or formation. (Standard, 1964) (f) A piece of land that is 1 square mile (2.59 km^2) or 640 acres (259 ha) in area forming one of the 36 subdivisions of a township in a U.S. public-land survey. (Webster 3rd, 1966)

sectionalizing circuit breaker *circuit breaker.*

sectional mining belt conveyor A belt conveyor so arranged that it can be lengthened or shortened by the addition or the removal of interchangeable increments or parts. (NEMA, 1961)

sectional tank A water tank built up of standardized pressed steel units having external flanges that are bolted together in an assembly for varying sizes of tank. (Hammond, 1965)

sectional-type conveyor A conveyor that is lengthened or shortened by adding or removing intermediate sections. (NEMA, 1956)

section boss A more or less loosely used term applied to the assistant mine foreman in charge of an area, although used in law in some states in lieu of assistant foreman and certified as such. (BCI, 1947)

section factor *section modulus.*

section foreman In anthracite and bituminous coal mining, a foreman who has complete charge of a section of a mine. Syn: *section man.* (DOT, 1949)

section-gage log *caliper log.*

section man *section foreman.*

section modulus The term pertains to the cross section of a beam. The section modulus with respect to either principal central axis is the moment of inertia with respect to that axis divided by the distance from that axis to the most remote point of the section. The section modulus largely determines the flexural strength of a beam of given material. Also called section factor. (Roark, 1954)

section of rectifier unit A part of a rectifier unit with its auxiliaries that may be operated independently. (Coal Age, 1960)

sector gate A roller gate in which the roller is in the form of a sector of a circle instead of being cylindrical. (Hammond, 1965)

secular Said of a process or event lasting or persisting for an indefinitely long period of time, e.g., secular variation; progressive or cumulative rather than cyclic. (AGI, 1987)

secular variation A relatively large, slow change in part of the Earth's magnetic field caused by the internal state of the planet and having a form roughly to be expected from a simple, but not quite uniformly polarized sphere. (AGI, 1987)

secundine dike A dike that has been intruded into hot country rock. Pegmatites and aplites commonly occur in this mode. (AGI, 1987)

sedentary Formed in place, without transportation, by the disintegration of the underlying rock or by the accumulation of organic material; said of some soils, etc. (Fay, 1920)

sediment (a) Solid fragmental material that originates from weathering of rocks and is transported or deposited by air, water, or ice, or that accumulates by other natural agents, such as chemical precipitation from solution or secretion by organisms, and that forms in layers on the Earth's surface at ordinary temperatures in a loose, unconsolidated form; e.g., sand, gravel, silt, mud, alluvium. (AGI, 1987) (b) Strictly, solid material that has settled down from a state of suspension in a liquid. In the singular, the term is usually applied to material held in suspension in water or recently deposited from suspension. In the plural, the term is applied to all kinds of deposits, and refers to essentially unconsolidated materials. Cf: *deposit.* (AGI, 1987)

sedimentary (a) adj. Pertaining to or containing sediment; e.g., sedimentary deposit or a sedimentary complex. (AGI, 1987) (b) Formed by the deposition of sediment (e.g., a sedimentary clay), or pertaining to the process of sedimentation (e.g., sedimentary volcanism).—n. A sedimentary rock or deposit. (AGI, 1987)

sedimentary ash (a) Mineral matter introduced into the coal substance during its accumulation. See also: *extraneous ash.* (Nelson, 1965) (b) Ash in coal derived from the mud mixed up with plant debris during the formation of coal. (Tomkeieff, 1954)

sedimentary clay A clay that has been geologically transported from the site of its formation and redeposited elsewhere. The English ball clays, for example, are secondary kaolins. Cf: *primary clay.* Syn: *secondary clay.* (Dodd, 1964)

sedimentary cycle *cycle of sedimentation.*

sedimentary ore A sedimentary rock of ore grade; an ore deposit formed by sedimentary processes, e.g., saline residues, phosphatic deposits, or iron ore of the Clinton ore type. (AGI, 1987)

sedimentary petrography The description and classification of sedimentary rocks. Syn: *sedimentography.* (AGI, 1987)

sedimentary petrology The study of the composition, characteristics, and origin of sediments and sedimentary rocks. (AGI, 1987)

sedimentary rock A rock resulting from the consolidation of loose sediment that has accumulated in layers; e.g., a clastic rock (such as conglomerate or tillite) consisting of mechanically formed fragments of older rock transported from its source and deposited in water or from air or ice; or a chemical rock (such as rock salt or gypsum) formed by precipitation from solution; or an organic rock (such as certain limestones) consisting of the remains or secretions of plants and animals. The term is restricted by some authors to include only those rocks consisting of mechanically derived sediment; others extend it to embrace all rocks other than purely igneous and completely metamorphic rocks, thereby including pyroclastic rocks composed of fragments blown from volcanoes and deposited on land or in water. Syn: *stratified rock; derivative rock.* (AGI, 1987)

sedimentary rocks Rocks formed by the accumulation of sediment in water (aqueous deposits) or from air (eolian deposits). The sediment may consist of rock fragments or particles of various sizes (conglomerate sandstone, shale); of the remains or products of animals or plants (certain limestones and coal); of the product of chemical action or of evaporation (salt, gypsum, etc.); or of mixtures of these materials. Some sedimentary deposits (tuffs) are composed of fragments blown from volcanoes and deposited on land or in water. A characteristic feature of sedimentary deposits is a layered structure known as bedding or stratification. Each layer is a bed or stratum. Sedimentary beds as deposited lie flat or nearly flat. *stratified rocks.* (Fay, 1920)

sedimentary tuff (a) A tuff containing a subordinate amount of nonvolcanic detrital material. (AGI, 1987) (b) A deposit of reworked tuff and other detrital material. (AGI, 1987)

sedimentation (a) The act or process of settling particles by mechanical means from a state of suspension in air or water. (AGI, 1987) (b) Method of classification by exploitation of free-falling rates of minute (subsieve) particles. (Pryor, 1965)

sedimentation balance Apparatus used to measure settling rate of small particles dispersed in liquid. One scale-pan is immersed in the mixture, and the balance is adjusted by increasing the counterweight at suitable time intervals. Alternatively, a float is suspended, and the compensating external weight is reduced as the density of the suspension surrounding the float is reduced by settlement of its solids. (Pryor, 1963)

sedimentation test A test used when selecting materials for stabilized road construction and concrete. Soil, after pretreatment, is shaken up in water and allowed to settle out. The change in specific gravity of the suspended matter with time is measured, and the equivalent diameter is calculated from Stokes' law. See also: *organic test; Stokes' law.* (Hammond, 1965)

sedimentation trend The direction in which sediments were laid down. Uranium mineralization often follows such trends, owing to increased porosity, carbon precipitants, and other factors. (Ballard, 1955)

sedimentation unit A layer or deposit formed under essentially constant physical conditions, distinguished from other units by differences in grain size and/or fabric, indicating changes in velocity and/or direction of flow.

sediment dispersion The dilution and settling of sediment in a cloud as it advects from a point source.

sedimentography *sedimentary petrography.*

sedimentology The scientific study of sedimentary rocks and of the processes by which they were formed; the description, classification, origin, and interpretation of sediments. (AGI, 1987)

sediment tube A long open tube fixed above the core barrel in the shot-drill method of exploratory boring. The enlarged space above the sediment tube reduces the upward velocity of the flushing water and the coarse chippings are deposited in the tube where they are retained until drawn up to the surface. Also called calyx; sludge barrel. (Nelson, 1965)

sediment vein A sedimentary dike formed by the filling of a fissure from above with sedimentary material. (AGI, 1987)

Seebeck effect The phenomenon involved in the operation of a thermocouple. Named for Thomas Seebeck, the German scientist, who first observed the phenomenon in 1822. See also: *thermocouple.* (Lyman, 1964)

seed charge A small charge of material added to a supersaturated solution to initiate precipitation. (ASM, 1961)

seed gypsum Gypsum beds of loose small crystals. (New South Wales, 1958)

seeding In chemical treatment, addition of tiny crystals of material to a supersaturated solution to induce nuclear precipitation. (Pryor, 1963)

seepage line *line of seepage.*

Seger cone A small cone, made in the laboratory of a mixture of clay and salt, that softens at a definite, known temperature. It is used in the manufacture of refractories. It has also been used in volcanology to determine the approximate temperature of a molten lava. Syn: *pyrometric cone.* (AGI, 1987)

segger *sagger.*

segregate (a) Pac. To separate the undivided joint ownership of a mining claim into smaller individually segregated claims. (b) In geology, to separate from the general mass, and collect together or become concentrated at a particular place or in a certain region, such as in the process of crystallization or solidification. See also: *segregated vein.* (Webster 2nd, 1960)

segregated vein A fissure whose mineral filling is derived from the country rock by the action of percolating water. Cf: *infiltration vein.* See also: *lateral secretion.* (AGI, 1987)

segregation (a) A secondary feature formed as a result of chemical rearrangement of minor constituents within a sediment after its deposition; e.g., a nodule of iron sulfide, a concretion of calcium carbonate, or a geode. (AGI, 1987) (b) Partial reseparation of a previously mixed batch of material into its constituents, as a result of differences in particle size or density. Segregation can occur in storage bins, on conveyors, and in feeders during dry or semidry processing. (Dodd, 1964)

segregation banding A compositional banding in gneisses that is not primary in origin, but rather is the result of segregation of material from an originally more nearly homogeneous rock. (AGI, 1987)

segregation survey The survey of a mining claim located on lands classified as agricultural.

sehta *cobaltite.*

seidozerite A monoclinic mineral, $Na_4MnTi(Zr_{1.5}Ti_{0.5})O_2(F,OH)(Si_2O_7)$; forms brown-red needles embedded in microcline in a nepheline syenite pegmatite; near Lake Seidozero, Lovozero massif, Kola Peninsula, Russia.

seif dune A long, sharp-crested dune extending in the direction of the wind that constructed it. (Mather, 1964)

S.E.I. photometer In this instrument, the internal comparison lamp is set to a standard brightness as indicated by a photoelectric cell and not by reference to a voltmeter or ammeter. (Roberts, 1958)

seismic (a) Pertaining to, characteristic of, or produced by earthquakes or earth vibration; as, seismic disturbances; seismic records. (Standard, 1964) (b) Pertaining to sound waves generated by earthquakes or artificially by explosives to map subsurface structure. (Wheeler, 1958)

seismic activity *seismicity.*

seismic analysis A quick, easy, and inexpensive method of determining the consolidation of overburden. The process is based on the principle that sound or shock waves travel through different subsurface materials at varying speeds and along different paths. By this method the operator can determine whether overburden can be ripped or whether it will need to be drilled and blasted. (Coal Age, 1966)

seismic area The region affected by a particular earthquake. (AGI, 1987)

seismic belt (a) One of the broad, more or less well-defined, elongate zones in which most earthquakes originate. (Stokes, 1955) (b) An elongate earthquake zone, esp. a zone of subduction or sea-floor spreading. (AGI, 1987)

seismic detector *seismometer.*

seismic diffraction *diffraction.*

seismic drill *seismograph drill.*

seismic event Applied to any definite signal change or amplitude difference on a seismic record. It may be a reflection, a refraction, a diffraction, or a random signal. (AGI, 1987)

seismic explosives Special forms of blasting gelatin, or gelatin and ammonia gelatin dynamites, used in geophysical prospecting by the seismic method; developed to shoot consistently at their characteristic rate of detonation under unusually heavy water pressure. (Lewis, 1964)

seismic focus The place of origin within the Earth of an earthquake; usually some more or less restricted area of a fault surface. If the focus is to be some particular point, it is the central point of the area over which fault movement occurred and caused the earthquake. (Challinor, 1964)

seismicity (a) Measure of frequency and magnitudes of earthquakes in a given area; e.g., the average number of earthquakes per year and per 100 km^2. (Schieferdecker, 1959) (b) The phenomenon of earth movements. (Hy, 1965)

seismic method A geophysical prospecting method based on the fact that the speeds of transmission of shock waves through the Earth vary with the elastic constants and the densities of the rocks through which the waves pass. A seismic wave is initiated by firing an explosive charge (or by equivalent artificial sources) at a known point (the shot point); records are made of the travel times taken for selected seismic waves to arrive at sensitive recorders (geophones). There are two main subdivisions of seismic operations: the reflection method and the refraction method. The seismic method has been applied to a lesser extent to elucidate mining problems, partly due to its high cost. It has been used to investigate the base of drift deposits, and drift-filled channels have been successfully outlined. (Nelson, 1965)

seismic noise *microseism.*

seismic prospecting A method of geophysical prospecting in which vibrations are set up by firing small explosive charges in the ground or by other artificial sources. Precise measurements of the resulting waves are taken, from which the nature and extent of underlying strata are revealed. (Hammond, 1965)

seismic reflection method In this geophysical prospecting technique, the structure of subsurface formations is mapped by making use of the times required for a seismic wave (or pulse), generated in the Earth by a near-surface explosion of dynamite or by other artificial sources, to return to the surface after reflection from the formations themselves. The reflections are recorded by detecting instruments responsive to ground motion, which are laid along the ground near the site of generation of the seismic pulse. Variations in the reflection times from place to place on the surface usually indicate structural features in the rock below. Syn: *reflection method.* (Dobrin, 1960)

seismic refraction method In refraction shooting, the detecting instruments are laid down at a distance from the shothole that is large compared with the depth of the horizon to be mapped. The seismic waves travel large horizontal distances along distinct interfaces in the Earth, and the time required for travel gives information on the velocity and depth of certain subsurface formations. (Dobrin, 1960)

seismic shooting (a) The initiation of seismic waves in the rocks by the firing of an explosive charge at a known point. The disturbance must be capable of accurate timing and must be such that, after traveling considerable distances through varying strata, it produces a sharply defined effect on the seismograph. These requirements may be supplied by the shock produced by detonating a charge of high explosive. The intensity of the shock and its effective range can be controlled by varying the quantity of explosive charge. See also: *reflection shooting; refraction shooting.* (Nelson, 1965) (b) A method of geophysical prospecting in which elastic waves are produced in the Earth by the firing of explosives or by other means. See also: *reflection shooting; refraction shooting.* (AGI, 1987)

seismic shothole A hole drilled for a seismic shot. It is usually a slim hole, although it has also been termed core hole. See also: *slim hole; structure test hole.* (Williams, 1964)

seismic spread *seismometer spread.*

seismic survey An exploration technique utilizing the variation in the rate of propagation of shock waves in layered media. It is used primarily to delineate subsurface geologic structures of possible economic importance. (Long, 1960)

seismic waves The Earth motion produced by a natural (earthquake) or synthetic disturbance on the surface or underground; utilized in the seismic method of geophysical exploration and for investigating the Earth's interior. Three types of waves are produced: (1) longitudinal or P waves; (2) traverse or S waves; and (3) surface or Raleigh and L waves. The speed of propagation is characteristic for each type of rock, depending largely on its compactness. In sandy clay, the speed of the P wave is about 4,000 ft/s (1.22 km/s); in sandstone, 10,000 ft/s (3.05 km/s); and in igneous rock up to 22,000 ft/s (6.71 km/s). (Nelson, 1965)

Seismitron An instrument designed to check ground stability. It amplifies 2.5 million times, and can detect a rock movement as small as

0.000001 in (2.54 μm). Receiving phones are placed in holes in the area being tested. Either earphones or automatic recording apparatus may be used for listening. A rate of 3 or more microseisms per second indicates probable collapse, and any rate over 25 or 30 per minute is considered dangerous. This instrument is also finding use above ground in checking highway cut slopes. (Nichols, 1976)

seismogram The record of Earth motion made by a seismograph. (AGI, 1987)

seismogram synthesis (a) This process produces an artificial reflection record from a continuous-velocity log or an electric log. With this system the log is converted from a depth scale to a time scale and is run through a scanning device that transforms the fluctuations on the log into electrical impulses that vary with time so as to simulate reflections. These impulses are passed through appropriate filters and are then recorded on an oscillograph in the same way as signals from a geophone. (Dobrin, 1960) (b) The theoretically calculated ground motion that would be recorded for a given Earth structure and seismic source.

seismograph (a) An instrument that detects, magnifies, and records motions of the Earth, esp. those caused by earthquakes or explosions. The resulting record is a seismogram. Cf: *seismometer; geophone.* (AGI, 1987) (b) The instrument used to record the reception of the waves in the sound seismic method. It works on the general principle that its frame is shaken by the arrival of the waves, while a pendulum of high inertia, mounted in it, remains stationary. The relative movement of the frame and the pendulum is magnified by optical means in the seismograph and by electrical amplifiers in the geophone. The instrument can also detect and record earthquakes. See also: *geophone; vibrograph.* (Nelson, 1965)

seismograph drill A rotary drill, pump, and hinged mast mounted as an integral drilling unit on a truck body and used primarily to drill vertical shallow holes in which explosives are placed and detonated to produce shock waves from the rock strata, which then are measured by seismic recording instruments. Also called jackknife rig; rotary shot drill; shothole drill. Syn: *seismic drill.* (Long, 1960)

seismograph rod A collared, tapered, V-thread-coupling drill rod used on seismograph drills. (Long, 1960)

seismology (a) The science of earthquakes and attendant phenomena. (Schieferdecker, 1959) (b) A geophysical science that is concerned with the study of earthquakes and measurement of the elastic properties of the Earth. (AGI, 1987) (c) The study of earthquakes, and of the structure of the Earth, by both natural and artificially generated seismic waves. (AGI, 1987)

seismometer An instrument that detects Earth motions. Syn: *seismic detector.* Cf: *geophone; hydrophone; seismograph.* (AGI, 1987)

seismometer spacing Distance between successive seismometer positions. (Schieferdecker, 1959)

seismometer spread A set of seismometers, placed along a straight line, that record the same shot. (Schieferdecker, 1959)

seismoscope An instrument that merely indicates the occurrence of an earthquake. It is considered by some, however, to be the equivalent of a seismometer. (AGI, 1987)

seize (a) To bind wire rope with soft wire, to prevent it from raveling when cut. (Nichols, 1976) (b) *bind; freeze.* (c) To cohere or stick to an inadequately lubricated moving part, such as a bearing, piston, or sliding part, through excessive friction, pressure, or temperature. (Long, 1960) (d) To protect rope ends by binding with yarn, marline, or fine wire. (Long, 1960)

selected fill Dumped fills made up of selected materials. These fills are used when it is desired to utilize a particular property of a soil or rock and this property can be secured solely by selective excavation. (Carson, 1961)

selective agglomeration In coal beneficiation, the separation of coal from associated impurities, usually aided by additions of oily reagents that selectively attach to the coal surfaces. Generally restricted to material of 500 μm top size. See also: *oil agglomeration.*

selective crushing Crushing in such a manner as to cause one ingredient of the feed to be crushed preferentially to others. (BS, 1962)

selective digging Separating two or more types of soil while digging them. (Nichols, 1976)

selective filling Hand filling, during which the miner rejects stone or dirt and loads only clean coal. Similar methods are adopted in metal mining. (Nelson, 1965)

selective flotation (a) A process for the preferential recovery of a particular ingredient of the coal, e.g., a petrological constituent, by froth flotation. (BS, 1962) (b) Generally refers to the surface or froth selecting of the valuable minerals rather than the gangue. Sometimes used to mean differential flotation. See also: *flotation; preferential flotation.* (Fay, 1920)

selective grinding Grinding in such a manner as to cause one ingredient of the feed to be ground preferentially to others. (BS, 1962)

selective mining (a) A method of mining whereby ore of high value is mined in such a manner as to make the low-grade ore left in the mine incapable of future profitable extraction. In other words, the best ore is selected in order to make good mill returns, leaving the low-grade ore in the mine. Frequently called robbing a mine. Cf: *bulk mining.* (b) The object of selective mining is to obtain a relatively high-grade mine product; this usually entails the use of a much more expensive stoping system and high exploration and development costs in searching for and developing the separate bunches, stringers, lenses, and bands of ore. In general, selective methods are applicable where the valuable sections of the deposit are rather large, comparatively few in number, and separated by relatively large volumes of waste. Selective methods of stoping are square-set stoping, open stoping in low-dipping beds, and cut-and-fill stoping. (c) In coal mining, selective methods may be dictated by market demands and prices. It may be desirable to work the different quality coal seams in such proportions as to obtain a uniform and salable blend over a period of years. In metal mining, the stopes may be restricted in both length and width and thus produce a much higher grade of ore. It is not always practicable to resort to selective mining because the mineralization may be so distributed as to necessitate taking the whole orebody in mining operations. (Nelson, 1965)

selective reflection The reflection by a substance, such as an opaque gem, of light rays of only certain wavelengths, the others being absorbed. This cause of color in gems is a sort of selective absorption.

selective weathering *differential weathering.*

selective wetting In mineral processing, development of selective attraction to the water phase of a pulp, as a prelude to flotation of an air-attracted fraction of the contained minerals. (Pryor, 1963)

selectivity index Criterion of trend in a continuous operation such as mineral processing. Abbrev., S.I. (Pryor, 1963)

selector In copper smelting, a kind of converter with horizontal tuyeres, to produce bottoms and a purified copper in one operation. (Webster 3rd, 1966)

select round Sometimes used to designate the best quality of industrials normally used as drill diamonds. (Long, 1960)

selenite (a) Finely crystallized gypsum. (Cooper, 1963) (b) A clear, colorless variety of gypsum, occurring (esp. in clays) in distinct, transparent monoclinic crystals or in large crystalline masses that cleave easily into broad folia. Syn: *spectacle stone.*

selenite plate In mineralogy, a plate of selenite that gives a purplish-red interference color of the first order with crossed polars. Syn: *gypsum plate; unit retardation plate; sensitive-tint plate; first-order red plate; Red I plate; Rot I plate.*

selenium A nonmetallic element and member of the sulfur family. Symbol, Se. It is widely distributed in small quantities, usually as selenides of heavy metals. Obtained from electrolytic copper refining. Used in photocells, exposure meters, and solar cells, and extensively in rectifiers. (Handbook of Chem. & Phys., 3)

selenjoseite *laitakarite.*

selenolite (a) Wadsworth's name for rocks composed of gypsum or anhydrite. (Fay, 1920) (b) A mineral, $Pb_2(SeO_4)(SO_4)$, reported as white needles with cerussite and molybdomenite at Cacheuta, Argentina. (Dana, 1944) (c) Former name for olsacherite. (d) Discredited name for downeyite, SeO_2.

self-act *gravity haulage.*

self-acting door A ventilation door consisting of two halves, so constructed that they are forced apart centrally by the trams as they come in contact with the converging beams that operate them. The door halves move on small pulleys that run on inclined rails so that after the passage of the trams the door closes by gravity. (Nelson, 1965)

self-acting incline (a) In transport by mine car, a brake incline. See also: *brake incline.* (Pryor, 1963) (b) *gravity haulage.*

self-acting plane An inclined plane upon which the weight or force of gravity acting on the full cars is sufficient to overcome the resistance of the empties; in other words, the full car, running down, pulls the other car (empty) up.

self-acting rope haulage A system of rope haulage used for transporting material on the surface and to transfer loaded cars from one elevation to a lower one in mines. Slope must be sufficiently steep so the loaded cars will pull the empty cars up the grade. Syn: *gravity plane rope haulage.* (Lewis, 1964)

self-advancing supports An assembly of hydraulically operated steel hydraulic supports, on a longwall face, that are moved forward as an integral unit by means of a hydraulic ram coupled to the heavy steel face conveyor. Syn: *power-operated supports; walking props.* See also: *hydraulic chock; steel prop.* (Nelson, 1965)

self-aligning carrying idler A belt idler that controls and limits the side runout of the carrying belt within practical limits by means of a swivel mechanism. (NEMA, 1961)

self-aligning return idler A belt idler that controls and limits the side runout of the return belt within practical limits by means of a swivel mechanism. (NEMA, 1961)

self-annealing A term applied to metals, such as lead, tin, and zinc, that recrystallize at air temperature and in which little strain hardening is produced by cold working. (CTD, 1958)

self-centering chuck A drill chuck that, when closed, automatically positions the drill rod in the center of the drive rod of a diamond-drill swivel head. (Long, 1960)

self-cleaning tail pulley A conveyor belt structure tail pulley which is designed with vanes along the length of the tail pulley and often is hollow in the center to allow any spillage along the return belt to end up at the tail pulley, which allows it to fall to the sides to eliminate the debris.

self-cleansing gradient The gradient at which flow in a pipe of a particular diameter will carry away any solids in it. This gradient must not be too steep nor too gradual, and is usually established under local laws affecting drains and sewers. (Hammond, 1965)

self-contained breathing apparatus A self-sufficient breathing unit that permits freedom of movement, unencumbered by air hoses. It offers the wearer respiratory protection in atmospheres that are either oxygen-deficient or too highly toxic to permit the use of gask masks or respirators. The oxygen or air is supplied in compressed form or by chemical generation, and the wearer's exhalations are either purified for re-use or released to the surrounding atmosphere. The equipment is devised to afford protection for special lengths of time, in accordance with the standards set by the Mine Safety and Health Administration. The 2-h apparatus is used for mine rescue and recovery operations; shorter period apparatus is available for industrial uses and auxiliary equipment. (Best, 1966)

self-contained portable electric lamps Electric lamps that are operated by an electric battery; designed to be carried about by the user of the lamp. (Fay, 1920)

self-contained self-rescuer A respiratory device used by miners for the purpose of escape during mine fires and explosions; it provides the wearer a closed-circuit supply of oxygen for a minimum of 10 min and up to 1 h. Syn: *SCSR*. (CFR, 6)

self-diffusion The spontaneous movement of an atom to a new site in a crystal of its own species, such as a copper atom within a crystal of copper. (ASM, 1961)

self-dumping cages Cages in which the cars are generally fitted with end doors; the cage deck is pivoted, and a roller engages with a tipping guide at the surface. As the cage is lifted, toward the end of the wind the deck tilts, the end door is lifted, and the coal is discharged. (Sinclair, 1959)

self-dumping car A mine car that can be side-tipped while in motion on a rail track. A ramp structure is fitted alongside the track opposite the spot where tipping is required. The car is fitted with a spherically contoured wheel that engages the ramp and gradually tilts the car while in motion. A chain attachment to the underframe opens the side of the car when tilted for tipping. The ramp can be retracted when not required. (Nelson, 1965)

self-energizing brake A brake that is applied partly by friction between its lining and the drum. (Nichols, 1976)

self-feeder An automatic appliance for feeding ore to stamps or crushers without the employment of hand labor. (Fay, 1920)

self-feeding portable conveyor Any type of power-propelled conveyor designed to advance into a pile of bulk material, thereby automatically feeding itself.

self-fluxing ores Ores that contain both acid and basic gangue minerals in the proper ratio to form a suitable slag. (Newton, 1959)

self-inductance The property of a circuit whereby self-induction occurs. It is measured by the rate of change of linkages in a circuit that accompanies a rate of change of current in that circuit of one unit per second. (CTD, 1958)

self-issue system A system of storage in lamp-room-operation charging and issue for alkaline-type car lamps, that allows a user access only to the storage racks for the purpose of lamp collection or return. Charging is controlled by a lamp-room attendant. (BS, 1965)

self-loading dumper A dumper provided with a bucket, hinged by arms to the chassis, that scoops up the material and discharges it backwards into the hopper. Hydraulic rams control the lift arms, bucket movement, and dumping operation. (Nelson, 1965)

self-opening reamer An underreamer having cutters that expand when they come in contact with, and are pressed against, surface. Cf: *expansion bit; underreamer*. (Long, 1960)

self-potential *spontaneous.*

self-potential curve *spontaneous potential curve.*

self-potential log Strip recording of natural potentials of complex origin, arising in the immediate neighborhood of liquid-filled boreholes. See also: *electric logging*. (Inst. Petrol., 1961)

self-potential method An electrical exploration method in which one determines the spontaneous electrical potentials (spontaneous polarization) that are caused by electrochemical reactions associated with clay or metallic mineral deposits. Syn: *spontaneous-potential method.* (AGI, 1987)

self-potential prospecting A method of electrical prospecting based on the measurement of natural earth potentials caused by the self-potential effects from orebodies, commonly metallic sulfides. (AGI, 1987)

self-powered scraper A scraper built into a single unit with a tractor. (Nichols, 1976)

self-priming centrifugal pump A pump of the centrifugal type that combines in a single hydraulic stage and with a single hydraulic impeller and casing the dual ability to pump, under vacuum, either liquids or gases. These pumps are advantageously used for sump, bilge, mine water gathering, tankcar unloading, vacuum evaporator applications, chemical processing, and other uses where the liquid is below the pump centerline, or under high vacuum. The suction lift is usually guaranteed at 20 ft (6.1 m) for cold water at sea level. (Pit and Quarry, 1960)

self-reading staff A leveling staff, marked with graduations so that an observer looking through the telescope of a level can read the elevation at which his or her line of sight intersects the staff. (Hammond, 1965)

self-rescuer A small filtering device carried by a miner underground, either on a belt or in a pocket, to provide the miner with immediate protection against carbon monoxide and smoke in case of a mine fire or explosion. The device is used for escape purposes only because it does not sustain life in atmospheres containing deficient oxygen. The length of time a self-rescuer can be used is governed mainly by the humidity in the mine air; e.g., in moist air it will last for a minimum period of 30 min, and in moderately dry atmospheres, for a period of 1 h or more. See also: *Siebe-Gorman self-rescuer*. (McAdam, 1955)

self-service system A system of storage and issue for lead-acid-battery-operated lamps, whereby the user has direct access to the charging racks for the purpose of connecting or disconnecting a lamp from the charging circuit. See also: *attendance signaling system; lamp room.* (BS, 1965)

self-shooter *booming; flop gate.*

self stones Fragments of rocks still possessing the original shape and angles, Derbyshire, U.K.

self-stowing gate Applied to an advance gate that carries forward a waste or skip, from 6 to 10 yd (5.5 to 9.1 m) wide, to take all the broken rock produced by the gate rippings. A short face conveyor is usually used to move the coal and dirt as required. The width of the waste is just sufficient for stowing the dirt produced. See also: *deepside*. (Nelson, 1965)

self-timing anemometer An anemometer that has a timing device incorporated in it. Twenty seconds after being started, the device automatically engages the pointer with the rotating vanes and after an interval of 1 min disengages it. See also: *anemometer*. (Nelson, 1965)

seligmannite An orthorhombic mineral, $CuPbAsS_3$, with As replaced by Sb toward bournonite; forms small, lead-gray, complex crystals.

sellaite A tetragonal mineral, MgF_2; colorless; structurally related to rutile.

selvage (a) The altered, clayey material found along a fault zone; fault gouge. Syn: *selvedge*. See also: *gouge; flucan*. (AGI, 1987) (b) A marginal zone of a rock mass, having some distinctive feature of fabric or composition; specif. the chilled border of an igneous mass (as of a dike or lava flow), usually characterized by a finer grain or sometimes a glassy texture, such as the glassy inner margins on the pillows in pillow lava. Syn: *selvedge; salband*. (AGI, 1987)

selvedge *selvage.*

Selvulize system A cold vulcanizing method for use underground. It is simpler and quicker than hot vulcanizing and there is no fire or explosion risk involved. (Sinclair, 1959)

semianthracite The rank of coal, within the anthracitic class of Classification D 388, such that, on the dry and mineral-matter-free basis, the volatile matter content of the coal is greater than 8% but equal to or less than 14% (or the fixed carbon content is equal to or greater than 86% but less than 92%), and the coal is nonagglomerating. (ASTM, 1994)

semiarid Said of a type of climate in which there is slightly more precipitation (25 to 50 cm) than in an arid climate, and in which sparse grasses are the characteristic vegetation. Syn: *subarid*. (AGI, 1987)

semiautomatic control A system to control the speed of a winder consisting of a cam-operated rheostat in parallel with a manually operated winder controller, the instantaneous cam position being directly related to the position of a cage in the shaft. If the driver left the manual control in the full-speed position, the cam control and associated closed-loop control would take charge and automatically decelerate the winder to creep speed as the cage approaches the surface. Should the driver still defer the operation of the lever, the winder would stop on the operation of an overwind limit switch. By the introduction of a small switch to initiate the wind and another to terminate the wind, fully automatic operation is possible. See also: *automatic cyclic winding*. (Nelson, 1965)

semicircumferentor A surveyor's instrument used for setting out land or buildings to any angle and in preliminary survey work generally and made up of a horizontal graduated semicircle that surrounds a compass and is attached to a base with fixed vertical sights at each end and of a movable arm with vertical sights at each end that pivots on the center of the base. See also: *surveyor's compass.* (Webster 3rd, 1966)

semicontinuous mill One that incorporates some stands in tandem, either for roughing or finishing, an example being a semicontinuous wire rod mill with a continuous roughing train and a looping finishing train. (Osborne, 1956)

semicoring bit A noncoring bit that produces a small-diameter core. (Long, 1960)

semicrystalline *hyalocrystalline.*

semidry mining (a) Underground work in which humidity of ventilating air is kept low, though moisture is used in drilling to allay dust. (Pryor, 1963) (b) In semidry mining, every effort is made to prevent the ventilating air from picking up moisture in the downcast shafts and in the main ways leading to the workings. In the workings themselves, moisture is added freely in order to reduce dust, and the air rapidly becomes saturated. (Spalding, 1949)

semiduplex process The process consists essentially of pouring molten metal from a primary open-hearth furnace on a heated solid charge of heavy and light alloy scrap (20% to 40% of total). The charge is melted and finished under reducing conditions. There is no boil. (Osborne, 1956)

semifusain A coal constituent transitional between vitrain and fusain. It displays gradual disappearance of cell structure, hardness, and yellowish color when observed in thin sections. Same as vitrifusain. (Stutzer, 1940)

semifusinite A constituent intermediate between vitrinite and fusinite showing a well-defined structure of wood and sclerenchyma. The cell cavities, either round, oval, or elongated in cross section, vary in size but are generally smaller and sometimes less well defined than those of fusinite. Occurs as lenses and bands of variable thickness, and as small fragments; associated with fusinite, or included in vitrite, clarite, duroclarite, clarodurite, and durite. It often lies as a transition material between vitrinite and fusinite, and the properties lie between those of fusinite and vitrinite; behaves as a semi-inert diluent in carbonization. (IHCP, 1963)

semigelatin Dynamite containing both ammonium nitrate as the chief explosive ingredient and a certain percent of blasting gelatin to make it plastic enough to remain in holes directed upward. It is more resistant to water than ammonia dynamite, but less resistant than gelatin dynamite. (Lewis, 1964)

semihorizon mining Coal mining that consists of driving cross-measure drifts and developing of the seams. The roads in the seam are equipped with conveyors for transporting coal, and locomotives may be used in the cross-measure drifts when these are horizontal or nearly so. Semihorizon mining may include the longwall-retreating method. (Nelson, 1965)

semikilled steel Steel that is incompletely deoxidized and contains sufficient dissolved oxygen to react with the carbon to form carbon monoxide to offset solidification shrinkage. (ASM, 1961)

semiloose In an excavation, both rock that is only partially detached from the solid and that rings as solid when struck, and rock being still attached to the solid, but parting from it by incipient shear cracks. (Spalding, 1949)

semimetallic pellets *prereduced iron-ore pellet.*

semimuffle furnace A furnace with a partial muffle, in which the products of combustion come in contact with the ware. (ASTM, 1994)

semiopal A loosely used term for common opal, hydrophane, and any partly dehydrated or impure opal, as distinguished from precious opal or fire opal. Syn: *hemiopal.*

semiplastic explosives In these types, the quantities of liquid products are insufficient to render the mixture compressible. When packing cartridges, however, the same high density is obtained as with plastic explosives. The proportions of the various constituents are actually so arranged that the spaces between the grains are filled out. The proportions in question are determined entirely according to the constituents selected. (Fraenkel, 1953)

semiportable electric equipment Electric equipment that is moved infrequently; e.g., room hoists, room conveyors, and gathering pumps.

semiportable electric lamps Electric lamps that are connected to a fixed source of power by a flexible cord whose length limits the movable range of the lamp.

semiprecious Of less commercial value than those called precious; applied esp. to such stones as amethyst, garnet, jade, and tourmaline. (Webster 3rd, 1966)

semiround nose A bit-crown design, in which the radius of the arc forming the rounded portion of the bit face is equal to or greater than the thickness of the bit wall. (Long, 1960)

semisolid bituminous material Material having a penetration at 77 °F (25 °C), under a load of 100 g applied for 5 s, of more than 10; and a penetration at 77 °F, under a load of 50 g applied for 1 s, of not more than 350. (Urquhart, 1959)

semisplint coal (a) Coal intermediate between durain coal and clarain coal (duroclarain). (Tomkeieff, 1954) (b) A coal in which the proportions of anthraxylon and attritus are more or less equal, but the attritus of which is essentially composed of brown and granular opaque matter in varying proportions. Translucent humic matter, spores, pollens, and finely divided fusain are always present in small proportions. Also known as block coal. Cf: *bright coal; splint coal.* (Litton, 1993) (c) A banded coal containing 20% to 30% of opaque attritus and more than 5% anthraxylon. (AGI, 1987)

semitrailer A towed vehicle whose front rests on the towing unit. (Nichols, 1976)

semitranslucent A degree of diaphaneity between translucent and opaque. Passes light through edges of cabochons but very little through thicker parts.

semitransparent Used to describe mineral when objects may be seen through it but without distinct outlines.

semivitreous That degree of vitrification evidenced by a moderate or intermediate water absorption. Also called semivitrified. See also: *impervious; nonvitreous; vitreous.* (ASTM, 1994)

semiwater gas A mixture of carbon monoxide, carbon dioxide, hydrogen, and nitrogen obtained by passing a mixture of air and steam continuously through incandescent coke. Its calorific value is low, about 125 Btu/ft^3 (4.66 MJ/m^3). (Osborne, 1956)

semseyite A monoclinic mineral, $Pb_9Sb_8S_{21}$; metallic; soft; sp gr, 5.8.

senaite A trigonal mineral, $Pb(Ti,Fe,Mn)_{21}O_{38}$; crichtonite group; black; forms rounded crystals and grains in diamond-bearing sands.

senarmontite An isometric mineral, Sb_2O_3; dimorphous with valentinite; resinous; Mohs hardness, 2 to 5; sp gr, 5.2 to 5.3.

sengierite A monoclinic mineral, $Cu_2(UO_2)_2V_2O_8{\cdot}6H_2O$; yellow-green.

sensible cooling effect The difference between the total cooling effect and the dehumidifying effect. (Strock, 1948)

sensible heat (a) Thermal energy, the transfer of which to or from a substance results in a change of temperature. Cf: *latent heat.* (Webster 3rd, 1966) (b) The heat added to a body when its temperature is changed. (c) The sensible heat of a body is the heat given off when it cools to ordinary temperature. (Newton, 1959)

sensitive earth fault protection A system of earth fault protection in which the fault current is limited by design to a low value which generally requires amplification in order to operate an earth fault relay. In the case of three-phase alternating-current systems, the limitation of the leakage current may be effected by either (1) inserting a current-limiting device between the neutral point of the system and earth (single-point earthing), or (2) connecting, in each circuit to be protected, all phases, in star, through current-limiting devices, each star point being connected to earth through an earth leakage protective device (multipoint earthing). (BS, 1965)

sensitive explosive *explosive sensitiveness.*

sensitiveness The property in a high explosive that permits it to be exploded by a shock. The more insensitive an explosive is, the stronger detonator it requires to develop the full strength.

sensitive-tint plate *gypsum plate; selenite plate.*

sensitivity (a) In explosives, a measure of the ease with which a substance can be caused to explode and its capacity to maintain explosion through the length of a borehole. (Nichols, 1956) (b) The least change in an observed quantity that can be perceived on the indicator of a given instrument. (AGI, 1987) (c) The displacement of the indicator of a recording unit of an instrument per unit of change of a measurable quantity. (AGI, 1987) (d) The effect of remolding on the shear strength and consolidation characteristics of a clay or cohesive soil. A sensitive clay is one whose shear strength is decreased to a fraction of its former value on remolding at constant moisture content. See also: *sensitivity ratio.* (AGI, 1987)

sensitivity ratio A measurement of the sensitivity of a clay to the action of remolding. (Hammond, 1965)

sensitivity to propagation Sensitivity to propagation of an explosive can be ascertained by a method called the Ardeer double-cartridge, or ADC, test. The ADC test consists of firing an explosive cartridge with a standard detonator and determining the maximum length of the gap across which the detonation wave will travel and detonate a second, or receptor, cartridge. Both the primer and the receptor cartridges should be of the same composition, diameter, and weight. (McAdam, 1958)

sensor The component of an instrument that converts an input signal into a quantity measured by another part of the instrument. Also called sensing element. (Hunt, 1965)

separate system A drainage system in which sewerage and surface water are carried in separate sewers. See also: *surface-water drain.* (Hammond, 1965)

separate tandem electric mine locomotive *electric mine locomotive.*

separate ventilation An early term for auxiliary ventilation. (Nelson, 1965)

separating bath (a) A vessel containing dense medium in which the feed material is separated on a commercial scale into different fractions according to specific gravity. (BS, 1962) (b) The liquid in a separating bath. (BS, 1962)

separating medium Dense medium of the density required to achieve a given separation. (BS, 1962)
separation The distance between any two parts of an index plane (e.g., bed or vein) disrupted by a fault. See also: *horizontal separation; vertical separation; stratigraphic separation.* (AGI, 1987)
separation coal Eng. Coal that has been prepared by screening or washing.
separation density The effective density at which a separation has taken place, calculated from a specific-gravity analysis of the products; commonly expressed as either partition density or equal errors cut point (density). (BS, 1962)
separation distances Minimum recommended distance between explosive materials and other materials or specific locations.
separation door (a) A door to separate the air in an intake airway from that in a return airway and prevent leakage. It is normally constructed with tongued-and-grooved boards secured by battens, and it is built into brick or concrete walls to form an airtight closure of the airway. Separation doors are usually arranged in twos or threes several yards apart, to reduce leakage when workers or cars are passing along the roadway. See also: *bearing door; steel separation door; ventilation doors.* (Nelson, 1965) (b) *air door.*
separation size A general term indicating the effective size at which separation has taken place, calculated from a size analysis of the product; commonly expressed as either partition size or equal errors size. (BS, 1962)
separation valve Eng. A massive cast-iron plate suspended from the roof of a return airway, through which all the return air of a separate district flows, allowing the air to always flow past or underneath it; but in the event of an explosion of gas, the force of the blast closes it against its frame or seating, and prevents a communication with other districts.
separator (a) A machine for separating, with the aid of water, suspensions, or air, materials of different specific gravity. Strictly, a separator parts two or more ingredients, both valuable, while a concentrator saves but one and rejects the rest; but the terms are often used interchangeably. (b) Any machine for separating materials, such as the magnetic separator for separating magnetic materials from gangue. Cf: *concentrator.* (c) A screen, esp. a revolving screen, for separating things like stones or coal into sizes. (Standard, 1964)
sepiolite A monoclinic mineral, $Mg_4Si_6O_{15}(OH)_2{\cdot}6H_2O$; soft; sp gr, 2 but fibrous dry masses float on water; occurs in veins in calcite and in alluvial deposits formed from weathering of serpentine masses, chiefly in Asia Minor, as meerschaum; may be used in making pipes, ornamental carvings. Syn: *meerschaum; sea-foam.*
septaria Plural of septarium.
septarian Said of the irregular polygonal pattern of internal cracks developed in septaria, closely resembling the desiccation structure of mud cracks; also said of the epigenetic mineral deposits that may occur as fillings of these cracks. (AGI, 1987)
septarian concretion *septarium.*
septarian nodule *septarium.*
septarium (a) A large, roughly spheroidal concretion, 8 to 90 cm in diameter, usually of an impure argillaceous carbonate, such as clay ironstone. It is characterized internally by irregular polyhedral blocks formed by a series of radiating cracks that widen toward the center and that intersect a series of cracks concentric with the margins; these cracks are invariably filled or partly filled by crystalline minerals (most commonly calcite) that cement the blocks together. Its origin involves the formation of an aluminous gel, case hardening of the exterior, shrinkage cracking due to dehydration of the colloidal mass in the interior, and vein filling. The veins sometimes weather in relief, thus producing a septate pattern. Syn: *septarian nodule; septarian concretion; beetle stone; turtle stone.* (AGI, 1987) (b) A crystal-lined crack or fissure in a septarium. Pl: septaria. (AGI, 1987)
septechlorite An alternate name for serpentine minerals reflecting their 7Å basal spacing and chloritelike formulae.
septum Membrane separating two phases, for example, pulp and filtrate. (Pryor, 1963)
sequence control A method of control whereby, once action has been initiated, a number of electrical circuits will automatically function in a prescribed order. (BS, 1965)
sequence interlock An interlock provided between a number of manually controlled electrical circuits, which are required to function in a prescribed order, and which prevents a circuit from being operated unless the preceding circuit has completed its part in the sequence. (BS, 1965)
sequence starting An arrangement whereby the starting of one belt conveyor starts all of its feeder conveyors in a predetermined manner. The purpose of sequence starting is to prevent spilling at transfer points and to reduce the power demand in starting the system. See also: *power sequence; pilot sequence.* (NEMA, 1961)
sequester A sequestering agent forms soluble complex ions with a simple ion, thereby inhibiting the activity of that ion. (Lowenheim, 1962)
serandite A triclinic mineral, $Na(Mn,Ca)_2Si_3O_8(OH)$, with Mn replaced by Ca toward pectolite; pink.
serendibite A triclinic mineral, $Ca_2(Mg,Al)_6(Si,Al,B)_6O_{20}$; aenigmatite group; blue.
serial samples Samples collected according to some predetermined plan, such as along the intersections of gridlines, or at stated distances or times. The method is used to ensure random sampling. (AGI, 1987)
seriate Said of the texture of an igneous rock, typically porphyritic, in which the sizes of the grains vary gradually or in a continuous series. Cf: *hiatal.* (AGI, 1987)
sericite A white, fine-grained potassium mica occurring in small scales as an alteration product of various aluminosilicate minerals, having a silky luster, and found in various metamorphic rocks (esp. in schists and phyllites) or in the wall rocks, fault gouge, and vein fillings of many ore deposits. It is commonly muscovite or very close to muscovite in composition, but may also include paragonite and illite. See also: *muscovite.*
sericitization A hydrothermal, deuteric, or metamorphic process involving the introduction of, alteration to, or replacement by sericitic muscovite. (AGI, 1987)
series (a) Any number of rocks, minerals, or fossils having characteristics, such as growth patterns, succession, composition, or occurrence, that make it possible to arrange them in a natural sequence. (AGI, 1987) (b) A conventional stratigraphic unit that is a division of a system. A series commonly constitutes a major unit of chronostratigraphic correlation within a province, between provinces, or between continents. (AGI, 1987) (c) May be applied to intrusive rocks in the same time-stratigraphic sense. Formal series names are binomial, usually consisting of a geographic name (generally but not necessarily with the adjectival ending -an or -ian) and the word Series, the initial letter of both terms being capitalized. See also: *igneous-rock series.* (AGI, 1987) (d) An arrangement of electric blasting caps in which the firing current passes through each of them in a single circuit. (Nichols, 1976)
series circuit firing A method of connecting together a number of detonators that are to be fired electrically in one blast. Each detonator is connected to the adjacent detonator to form a continuous circuit having two free ends that are then connected to the firing cable. In British coal mines, all rounds of shots must be connected electrically in series. This results in large rounds having a high electrical resistance, requiring high voltage at the exploder that, in turn, increases the chance of misfires due to current leakage. See also: *parallel circuit firing.* (Nelson, 1965)
series firing The firing of detonators in a round of shots by passing the total supply current through each of the detonators. Cf: *parallel firing.* (BS, 1964)
series-in-parallel circuit *parallel blasting circuit; parallel series circuit.*
series parallel firing The firing of detonators in a round of shots by dividing the total supply current into branches, each containing a certain number of detonators wired in series. (BS, 1964)
series shots A number of loaded holes connected and fired one after the other. In contradistinction to simultaneous firing, where the charges are connected electrically, and are all exploded at one time. (Stauffer, 1906)
series ventilation A system of ventilating a number of faces consecutively by the same air current. (BS, 1963)
serpentine (a) In petrology, a metamorphic rock serpentinite composed chiefly or wholly of the mineral serpentine. (b) A group of common rock-forming minerals having the formula $(Mg,Fe,Ni)_3Si_2O_5(OH)_4$; mostly monoclinic, but also orthorhombic; greasy or silky luster; slightly soapy feel; tough conchoidal fracture; commonly compact but may be granular or fibrous (asbestiform); green; invariably secondary, derived by alteration of magnesium-rich silicate minerals (esp. olivines); in both igneous and metamorphic rocks; translucent varieties commonly substitute for jade for ornamental and decorative purposes; fibrous varieties are used for asbestos. (c) The mineral group antigorite, clinochrysotile, orthochrysotile, and lizardite. (d) In former usage, serpentine and antigorite were mineral species and chrysotile was a variety. See also: *kaolinite-serpentine.*
serpentine asbestos *chrysotile.*
serpentine cat's-eye *satelite.*
serpentine jade A variety of the mineral serpentine, resembling bowenite, occurring in China; used as an ornamental stone. (CMD, 1948)
serpentine marble *verde antique.*
serpentine rock *serpentinite.*
serpentinite A rock consisting almost wholly of serpentine-group minerals, e.g., antigorite and chrysotile or lizardite, derived from the alteration of ferromagnesian silicate minerals, such as olivine and pyroxene. Accessory chlorite, talc, and magnetite may be present. Syn: *serpentine rock.* (AGI, 1987)
serpentinization The process of hydrothermal alteration by which magnesium-rich silicate minerals (e.g., olivine, pyroxenes, and/or amphiboles in dunites, peridotites, and/or other ultrabasic rocks) are converted into or replaced by serpentine minerals. (AGI, 1987)

serpierite A monoclinic mineral, $Ca(Cu,Zn)_4(SO_4)_2(OH)_6 \cdot 3H_2O$; dimorphous with orthoserpierite; sky-blue.
serra *sierra.*
serrate (a) Said of topographic features that are notched or toothed, or have a saw-edged profile; e.g., a serrate divide. Syn: *saw-toothed.* (AGI, 1987) (b) Said of saw-toothed contacts between minerals, usually resulting from replacement; e.g., the serrate texture of megacrysts in contact with plagioclase in igneous rocks. (AGI, 1987)
serumite A bonding clay for foundry sands. (Osborne, 1956)
service factor A factor by which the specified horsepower is multiplied to compensate for drive conditions. (Jackson, 1955)
service shaft A shaft employed solely for the hoisting of workers and materials to and from underground. (Nelson, 1965)
serving (a) Fiber cord wrapping around the surface of a wire rope. (Hammond, 1965) (b) Corn. A supply of tin ready for smelting.
servomechanism An automatic feedback control system for mechanical motion; it applies only to those systems in which the controlled quantity or output is mechanical position or one of its derivatives (velocity, acceleration, and so on). Also known as servo system. (McGraw-Hill, 1994)
set (a) A timber frame used for supporting the sides of an excavation, shaft, or tunnel. Syn: *sett.* (Webster 3rd, 1966) (b) *one-piece set; timber set; bench of timbers.* (c) The distance a pile penetrates with one blow from a driving hammer. (Nichols, 1976) (d) A group of essentially parallel planar features, esp. joints, dikes, faults, veins, etc. (AGI, 1987) (e) A train of mine cars; a trip. (Fay, 1920) (f) The failure of a rock subjected to intense pressure below the point of rupture to recover its original form when the pressure is relieved. (Fay, 1920) (g) The discharge opening of a crushing machine to regulate the size of the largest escaping particle. (Pryor, 1960) (h) To fix a prop or sprag in place. (i) To place a diamond in the crown of a bit. (Long, 1960) (j) To place casting in a borehole. (Long, 1960)
set bit A bit insert with diamonds or other cutting media. (Long, 1960)
set casing The cementing of casing in the hole. The cement is introduced between the casing and the wall of the hole and then allowed to harden, thus sealing off intermediate formations and preventing fluids from then entering the hole. It is customary to set casing in the completion of a producing well. (Williams, 1964)
set casing shoe A casing shoe set with diamonds. Often used for a one-shot attempt to drill casing down through overburden to bedrock. Also called casing-shoe bit. (Long, 1960)
set copper An intermediate copper product containing about 3.5% cuprous oxide, obtained at the end of the oxidizing portion of the fire-refining cycle. (ASM, 1961)
set i.d. *set inside diameter.*
set inside diameter The minimum inside diameter of a set core bit. Usually written set i.d. in drilling industry literature. Also called bore; center bore; inside gage. Abbrev. set i.d. (Long, 1960)
set o.d. *set outside diameter.*
set of timber The timbers composing any framing, whether used in a shaft, slope, level, or gangway. Thus, the four pieces forming a single course in the curbing of a shaft, or the three or four pieces forming the legs and collar, and sometimes the sill, of an entry framing are together called a set, or timber set. (Zern, 1928)
set outside diameter The maximum outside diameter of a set bit. Usually written set o.d. in drilling industry literature. Also called outside gage. Abbrev. set o.d. (Long, 1960)
set reaming shell A reaming shell, a portion of the outside surface of which has embedded diamonds, diamond-inset inserts, or other cutting media, having a set diameter slightly greater than the standard set size of the bit to which the shell is coupled. (Long, 1960)
sett (a) A quarryman's term for a square-faced steel tool held in position and struck with a sledge to cause a fracture in a rock mass. (b) Corn. A lease; the boundaries and terms of the mining ground taken by the adventurers. See also: *set.* (c) A timber frame used in underground support. Also spelled set. (Pryor, 1963) (d) A small rectangular dressed stone of granite, quartzite, or whinstone, used as road paving in localized areas subject to esp. heavy traffic. (Hammond, 1965)
setting (a) Those runners, sheeting, or poling boards which are held in place by one pair of timber frames supporting the sides of an excavation. See also: *timbering; top frame.* (Hammond, 1965) (b) The timber frames used at intervals in shaft sinking and close-poled behind. (Stauffer, 1906) (c) *heading.* (d) The act of contracting with miners for work to be done. (Standard, 1964)
setting out Marking out on the ground by means of pegs and lines the proposed positions and dimensions of earthworks, masonry, etc. Also called staking out. Syn: *laying out.* (Pryor, 1963)
setting pattern The geometric arrangement of the inset diamonds in a bit crown. (Long, 1960)
setting plug A cylindrical object, having a diameter equal to the inside set diameter of a specific-size bit, used to measure the inside set diameter of a core bit. (Long, 1960)
setting ring A ringlike sleeve, the inside diameter of which is the same as a specific set outside diameter of a diamond bit or reaming shell; it is used to check the set diameter of a bit or reaming shell. Also called bit gage; bit ring; gage ring; gaging ring; ring gage; setting gage. (Long, 1960)
setting rod A special diamond-drill rod used to set a deflecting wedge in a borehole. (Long, 1960)
setting up (a) *rigging.* (b) In mining, to gather the necessary tools and complete all work preparatory to drilling. Syn: *setup.* (Long, 1960) (c) Hardening of air-setting or hydraulic-setting mortars. (d) The act or process of setting a diamond bit. (Long, 1960)
settled ground Ground that has ceased to subside over the waste area of a mine and has reached a state of full subsidence. (Nelson, 1965)
settlement The lowering of the overlying strata in a mine, owing to extraction of the mined material. See also: *subsidence; differential settlement; seat of settlement.* (AGI, 1987)
settlement date Agreed terms on which payment for a consignment of mineral is made. (Pryor, 1963)
settlement price The last unfulfilled offer to sell at cash price at the close of the second morning on the London Metal Exchange, prevails as the accepted cash price for the metal for the succeeding twenty-four hours. (Wolff, 1987)
settler A separator; a tub, pan, vat, or tank in which a separation can be effected by settling. A tub or vat in which pulp from the amalgamating pan or battery pulp is allowed to settle; the pulp is stirred in water to remove the lighter portions. Syn: *pug tub.* See also: *agitator; settling tank.* (Fay, 1920)
settling box A box or container in which drill cuttings or sludges are accumulated and coarse materials permitted to settle. (Long, 1960)
settling cone A conical tank used to settle coarse solids from the circulating water. (BS, 1962)
settlingite A hard, brittle, pale-yellow to deep-red hydrocarbon (H:C about 1.53) in resinous drops on the walls of a lead mine at Settling Stones, Northumberland, U.K. See also: *settling stones resin.*
settling pit An excavation through which mine water is conducted in order to reduce its velocity, thus allowing sediment to settle and to be cleaned out from time to time. Also called dredge sump; settling sump (undesirable usage). (BS, 1963)
settling pond A pond, natural or artificial, for recovering the solids from washery effluent. (BS, 1962)
settling sand Drillers' term for friable sandstone that caves into wells and settles around the bit. (AGI, 1987)
settling stones resin A resinoid, hard, brittle substance possessing a pale yellow to deep red color, a specific gravity of 1.16 to 1.54, and burning in a candle flame. It was found in an old lead mine in Northumberland, England. See also: *settlingite.*
settling tank (a) A reservoir or tank into which the return water from a borehole collects and the entrained drill cuttings settle. (Long, 1960) (b) A tank in which pulp is held while solids settle from suspension. Also called thickener; settler. (Pryor, 1963)
settling vat A vat in which particles of ore are allowed to settle. (Fay, 1920)
settling velocity The rate at which suspended solids subside and are deposited. Syn: *full velocity.* (AGI, 1987)
setup (a) In surveying, location of theodolite above a station point. (Pryor, 1963) (b) In drilling, location of machine. (Pryor, 1963) (c) *setting up; rigging; rig; rig-up.*
set weight The quantity of diamonds set in a bit, expressed in carats. (Long, 1960)
severance (a) Separation of a mineral or royalty interest from other interests in the land by grant or reservation. A mineral or royalty deed or a grant of the land reserving a mineral or royalty interest, by the landowner before leasing, accomplishes a severance, as does his execution of an oil and gas lease. (Williams, 1964) (b) *liberation.*
severance tax A State tax imposed on the severing of natural resources from the land based on the value or quantity of production. These types of taxes are usually calculated either as a flat rate per unit of production (sometimes called a "unit" or "specific" severance tax), or as a percentage of the value of the resource produced (sometimes called an "ad valorem" or "percentage" severance tax). The tax base for an ad valorem severance tax is generally either the gross or net value of resources produced or sold. (SME, 1992)
severed lands These are lands where the surface estate is held by one party and the mineral estate is held be another. The owner of the mineral estate normally retains the right to enter and occupy the surface estate for the purposes of mineral development and extraction. Also called Split Estate Lands. (SME, 1992)
seyberite *clintonite.*
seybertite A former name for clintonite.
Seyler's classification A classification of coals based primarily upon the carbon and hydrogen content calculated to a pure-coal basis, according to the Parr formula. See also: *carbon-hydrogen ratio.* (Nelson, 1965)

shackle (a) A connecting link or device for fastening parts together, usually in such a manner as to permit some motion. (Crispin, 1964) (b) A connecting device for lines and drawbars consisting of a U-shaped section pierced for cross bolt or a pin. (Nichols, 1976) (c) A short wrought-iron or manganese-steel chain for connecting mine cars to form a journey or train, for transport by rope haulage or locomotive to and from the workings. Syn: *coupling.* See also: *automatic clip*. (Nelson, 1965)

shackler A person employed to attach and detach the shackles between mine cars either at a junction near the face or at the pit bottom. Cars are attached at the junction to form trains for the locomotive or rope haulage. Another shackler detaches them at the pit bottom for loading into the cage. Syn: *offtake lad*. (Nelson, 1965)

shadd (a) Corn. Smooth, round stones on the surface, containing tin ore, and indicating a vein. (b) *shoad.*

shade (a) A color that has been darkened by the addition of black. (Hansen, 1937) (b) A term descriptive of that difference between colors resulting from a difference in luminosity only, the other color constants being essentially equal. A darker shade of a color is one that has a lower luminosity. (Hess)

shading A method of showing relief on a map by simulating the appearance of sunlight and shadows, assuming an oblique light from the northwest so that slopes facing south and east are shaded (the steeper slopes being darker), thereby giving a three-dimensional impression similar to that of a relief model. The method is widely used on topographic maps in association with contour lines. (AGI, 1987)

shadow zone (a) Region in which refraction effects cause exclusion of echo-ranging sound signals. (Hy, 1965) (b) An area in which there is little penetration of acoustic waves. (AGI, 1987) (c) A region 100° to 140° from the epicenter of an earthquake where, due to refraction from the low-velocity zone inside the core boundary, there are no direct arrivals of seismic waves. Syn: *blind zone*. (AGI, 1987)

shaft (a) An excavation of limited area compared with its depth; made for finding or mining ore or coal, raising water, ore, rock, or coal, hoisting and lowering workers and material, or ventilating underground workings. The term is often specif. applied to an approx. vertical shaft, as distinguished from an incline or inclined shaft. A shaft is provided with a hoisting engine at the top for handling workers, rock, and supplies; or it may be used only in connection with pumping or ventilating operations.. Cf: *incline.* (Fay, 1920; Lewis, 1964) (b) A brick or stone stack or chimney. (Standard, 1964) (c) The upper zone of a blast furnace. (Mersereau, 1947) (d) *abyss.*

shaft allowance The difference between the excavation diameter and the finished diameter in the clear; the extra space allowed to accommodate the permanent shaft lining. (Nelson, 1965)

shaft bottom *loop-type pit bottom; pit bottom; single-approach pit bottom.*

shaft cable (a) A specially armored cable of great mechanical strength running down the shaft of a mine. (CTD, 1958) (b) *borehole cable.*

shaft capacity The output of ore or coal that can be expected to be raised regularly and in normal circumstances, per day or week. (Nelson, 1965)

shaft casing The structure enclosing the top of a shaft designed to prevent short circuiting of air into or out of the shaft. See also: *air lock.* (BS, 1963)

shaft cave A cave formed primarily of a shaft or shafts. (AGI, 1987)

shaft collar *collar structure.*

shaft deformation bar A useful contrivance for measuring the deformation in the cross section of a shaft. It consists of a length of 1-1/2-in (3.8-cm) pipe fitted at one end with a micrometer and at the other end with a hard steel cone. The micrometer should have a range of 3 to 4 in (7.6 to 10 cm) and should fit into a bushing in the pipe in some manner. It may thus be removed from the bar for safe keeping or during transport. (Issacson, 1962)

shaft drilling The drilling of small shafts up to about 5 ft (1.5 m) in diameter with a shot drill. In Virginia, shafts up to 6 ft (1.8 m) in diameter have been sunk by core drilling. A ring 4 in (10 cm) wide is formed by roller bits similar to oilfield rotary drilling. Six tricone cutters are used, and a large core is formed. A central hole is drilled in the core, and a small explosive charge is fired to break it for removal. In some cases, it is removed bodily by a core catcher. (Nelson, 1965)

shaft feeder cable A cable mounted in a shaft to transmit electrical power to the shaft bottom and/or to an intermediate level. (BS, 1965)

shaft foot Scot. The bottom of a shaft.

shaft guides *cage guide; fixed guides; rope guide.*

shaft-hoist engineer *hoistman.*

shaft horsepower (a) Actual horsepower produced by the engine after deducting the drag of accessories. Also called flywheel horsepower; belt horsepower. (Nichols, 1976) (b) The shaft horsepower of a winding engine is the average load of coal or ore in pounds (kilograms) raised per wind multiplied by the average number of winds per minute (which may be a fraction) multiplied by the depth of the shaft in feet and the product divided by 33,000. (Sinclair, 1959)

shaft house A building at the mouth of a shaft, where ore or rock is received from a mine. (Weed, 1922)

shaft inset The point where a horizontal tunnel intersects a shaft. Syn: *mounting.* See also: *inset*. (Nelson, 1965)

shaft kip Eng. *kip.*

shaft lighting The lighting of shafts at landing stations is often found to be far from ideal. Work in the cages, during loading and unloading, and examination of shaft gear, are facilitated by the provision of fittings to provide a directional flux distribution in such a way that light is thrown forwards from the pit bottom or inset roadway into the shaft area. There must also be adequate illumination on a vertical plane at the shaft inset. (Roberts, 1958)

shaft lining The timber, steel, brick, or concrete structure fixed around a shaft to support the walls. In modern shafts, a concrete lining is generally favored as a permanent shaft support. (Nelson, 1965)

shaft mine (a) A mine in which the coal seam is reached by a vertical shaft which may vary in depth from less than 100 ft (30 m) to several thousand feet. (Kentucky, 1952) (b) A mine in which the main entry or access is by means of a shaft. Cf: *drift mining*. (Nelson, 1965)

shaft mixer *mixer.*

shaft pillar (a) A large area of a coal seam that is left unworked around the shaft bottom to protect the shaft and the surface buildings from damage by subsidence. All roadways in the shaft pillar are narrow, and coal faces are not opened out until the limit line of the shaft pillar is reached. The area of the shaft pillar is considerably greater than the surface area requiring protection. Syn: *high pillar.* See also: *bottom pillar; pillar*. (Nelson, 1965) (b) A solid block of ore left around the shaft where it crosses the lode, for protection against earth movement. (Pryor, 1963)

shaft plumbing (a) The operation of transferring one or more points at the surface of a vertical shaft to plumb line positions at the bottom of the shaft; a method to ensure that a shaft is sunk in the true vertical line. See also: *centering of shaft*. (Nelson, 1965) (b) Survey operation in which the orientation of two plumb bobs is measured both at the surface and at depth in order to transfer the bearing underground. See also: *Weisbach triangle*. (Pryor, 1963)

shaft pocket (a) Ore storage, excavated at depth, which receives trammed ore pending removal by skip. (Pryor, 1963) (b) Loading pockets of one or more compartments for different classes of ore and for waste built at the shaft stations. They are cut into the walls on one or both sides of a vertical shaft or in the hanging wall of an inclined shaft. See also: *pocket*. (Lewis, 1964) (c) *measuring chute.*

shaft raising *raising.*

shaft section A drawing or log giving details of the structure and the nature of strata intersected by a shaft. (BS, 1963)

shaft set (a) Supporting frame of timber, masonry, or steel that supports the sides of a shaft and the gear. Composed of two wallplates, two end plates, and dividers that form shaft compartments. (Pryor, 1963) (b) A system of mine timbering similar to square sets. The shaft sets are placed from the surface downward, each new set supported from the set above until it is blocked in place. New wallplates are suspended from those of the set above by hanging bolts. Blocking, wedging, and lagging complete the work of timbering. At stations the shaft posts are made much longer than usual to give ample head room for unloading timber and other supplies. (Lewis, 1964)

shaft siding The station or landing-place arranged for the full and empty tubs at the bottom of the winding shaft. (Peel, 1921)

shaft signal Code of electric ringing, or for shallow depths, knocking, among the onsetter or hitcher at the shaft bottom, the banksman at the top, and the engineman who operates the winder. Signals inform the latter as to type of load, etc. A telephone is also installed. (Pryor, 1963)

shaft signal indicator A device, usually mounted in the winding engine house, which gives visual indication of the signals received from the banksman and the onsetter to regulate the movement of conveyances in a shaft, and that retains the indication until cancelled. (BS, 1965)

shaft signal recorder A device that records, on paper or otherwise, the signals given by the banksman and the onsetter and the movements of the winder drum. (BS, 1965)

shaft sinking (a) Excavating a shaft downwards, usually from the surface, to the workable coal or ore. High sinking rates are possible by (1) mechanical mucking, (2) increased winding capacity, (3) improved concrete supply and placing, (4) improved surface layout, and (5) improved methods of blasting. (Nelson, 1965) (b) Excavating a shaft with a shaft-sinking drill. Cf: *raise.*

shaft-sinking drill A large-diameter drill with multiple rotary cones or cutting bits used for shaft sinking. An adaptation from oil well drills. (Eng. Min. J., 1966)

shaft-sinking power supply A supply of compressed air at a working pressure of about 100 psi (690 kPa). The quantity required for a modern high-speed sinking may be 2,000 to 2,500 ft^3/min (56.6 to 70.7 m^3/min). At a new mine where two shafts are being sunk, the power installation may comprise eight slow-speed water-cooled compressors with a total output of almost 5,000 ft^3/min (141.5 m^3/min) at 100 psi. (Nelson, 1965)

shaft-sinking ventilation The ventilation of a sinking shaft is by means of auxiliary fans. Axial flow fans powered by flameproof motors are commonly used. (Nelson, 1965)

shaft space An opening created with the object of relieving pressure on the shaft. (Higham, 1951)

shaft spragger In anthracite and bituminous coal mining, a laborer who controls the movement of mine cars as they are run to the top or to the bottom of the shaft by poking sprags (short metal or wooden rods) between the spokes of the wheels. (DOT, 1949)

shaft station An enlargement of a level near a shaft from which ore, coal, or rock may be hoisted and supplies unloaded.

shaft survey A survey to determine the alignment of a shaft.

shaft tackle *poppet head; headframe; poppet.*

shaft tunnel N. Staff. Headings driven across the measures from shafts to intersect inclined seams.

shaft wall (a) The brick or concrete lining in a shaft to support the surrounding ground. The construction of the permanent shaft wall is normally in concrete with reinforcement at insets and in bad ground. The wall thickness is between 12 in and 36 in (30 cm and 91 cm) depending on the shaft size and water pressure within the strata. Also applied to the rock masses surrounding the shaft. See also: *permanent shaft support*. (Nelson, 1965) (b) The side of a shaft.

shag boss In the stonework industry, a foreperson who supervises the removal of waste stone, loading and unloading of finished and semi-finished stone, and the moving and piling of stone slabs and blocks at a stoneworking mill. (DOT, 1949)

shagreen *relief.*

shake (a) In a coal mine, a vertical crack in the seam and roof. (CTD, 1958) (b) Fissures in rock. (Arkell, 1953) (c) Minute calcite veins traversing limestone or other rocks containing carbonates. These veinlets, unlike vents, have no harmful effect on the building stone. (Arkell, 1953) (d) A cavern, usually in limestone. (e) A close-joint structure in rock, due to natural causes, such as pressure, weathering, etc. Used in the plural.

shaker A mechanically vibrated screen through which a returning drill fluid is passed to screen out larger chips, fragments, and drill cuttings before the drill fluid flows into the sump. Syn: *shale screen; shale shaker*. (Long, 1960)

shaker chutes Metal troughs, operated mechanically, for the loading of coal into mine cars underground. (Hudson, 1932)

shaker conveyor (a) A conveyor consisting of a length of metal troughs, with suitable supports, to which a reciprocating motion is imparted by drives. In the case of a downhill conveyor, a simple to-and-fro motion is sufficient to cause the coal to slide. With a level or a slight uphill gradient, a differential motion is necessary; i.e., a quick backward and slower forward strokes. The quick backward stroke causes the trough to slide under the coal, while the slower forward stroke moves the coal along to a new position. Syn: *jigger*. See also: *conveyor; conveyor shaker type; vibrating conveyor*. (Nelson, 1965) (b) A type of oscillating conveyor.

shaker-conveyor engine A reciprocating engine operated by compressed air which is used to impart the reciprocating motion to a shaker conveyor panline. (Jones, 1949)

shaker screen A screening medium mounted in a rectangular frame, supported in a horizontal or slightly inclined position, and reciprocated longitudinally by a crank or eccentric and connecting rod. The unique feature that differentiates the shaker from all other screens is that the load is made to travel over the screening medium by the shaking motion of the screen. (Mitchell, 1950)

shaker-shovel loader A machine for loading coal, ore, or rock usually in headings or tunnels. It consists of a wide flat shovel that is forced into the loose material along the floor by the forward motion of the conveyor. The shaking motion of the conveyor brings the material backwards, and it is loaded into cars or a conveyor. It works at its maximum efficiency to the rise or in flat tunnels. Also called duckbill loader. See also: *loader*. (Nelson, 1965)

shake wave A wave that advances by causing particles in its path to move from side to side or up and down at right angles to the direction of the wave's advance, a shake motion. (Leet, 1958)

shaking (a) *springing; shaking a hole.* (b) Corn. Washing ore; ore dressing.

shaking a hole The enlargement of a blasthole, by exploding a stick of dynamite, so it will contain a larger amount of explosives for a big blast. Also called a shake blast. See also: *springing*.

shaking conveyor An apparatus that slides under the broken coal and by reciprocating motion moves the coal along to a discharge point. (BCI, 1947)

shaking-conveyor loader The broad tapering shovellike end of a shaking conveyor that is thrust suddenly under the coal and slowly withdrawn so as to carry the coal that has been lifted toward the dumping point. (Zern, 1928)

shaking down The stirring of an open-hearth bath with a rod to assist in oxidizing and removal of carbon. (Henderson, 1953)

shaking screen (a) A screen for sizing coal or other material. It consists of a screening surface of punched plate or wire mesh mounted in a rectangular frame, supported in a horizontal or slightly inclined position and reciprocated longitudinally by a crank or eccentric and connecting rod. The slightly inclined shaking screen is favored. Also called jiggling screen. See also: *trommel; jigging screen; wet screening*. (Nelson, 1965) (b) A suspended screen moved with a back-and-forth or rotary motion with a throw of several inches or more. (Nichols, 1976)

shaking table (a) In ore dressing, flattish tables oscillated horizontally during separation of minerals fed onto them. (Pryor, 1960) (b) In concentration of finely crushed ores by gravity, a rectangular deck with longitudinal riffles. It is shaken rapidly in a compounded to-and-fro motion by a vibrator, in such a way as to move the sands along, while they are exposed to the sweeping action of a stream of water flowing across the deck, which is tilted about its long axis. In dry or pneumatic tabling the feed is dry, and air is blown upward through a porous deck. (Pryor, 1963) (c) A slightly inclined table to which a lateral shaking motion is given by means of a small crank or an eccentric. One form is covered with copper plates coated with mercury for amalgamating gold or silver; other forms are provided with riffles and used in separating alluvial gold. Syn: *jerking table; bumping table.*

shale (a) A fine-grained detrital sedimentary rock, formed by the consolidation (esp. by compression) of clay, silt, or mud. It is characterized by finely laminated structure, which imparts a fissility approx. parallel to the bedding, along which the rock breaks readily into thin layers, and by an appreciable content of clay minerals and detrital quartz; a thinly laminated or fissile claystone, siltstone, or mudstone. It is generally soft but sufficiently indurated so that it will not fall apart on wetting; it is less firm than argillite and slate, commonly has a splintery fracture and a smooth feel, and is easily scratched. Its color may be red, brown, black, or gray. Etymol: Teutonic, probably Old English scealu, shell, husk, akin to German schale, shell. (AGI, 1987) (b) One of the impurities associated with coal seams; this term should not be used as a general term for washery rejects. (BS, 1962)

shale-and-clay feeder One who keeps conveyor belt that feeds dry mill constantly loaded with shale and clay. Also called clay-and-shale feeder; clay feeder; conveyor loader; shale-and-clay-conveyor man; shale feeder. (DOT, 1949)

shale band *dirt band.*

shale break Thin layer or parting of shale between harder strata, primarily a driller's term. Cf: *shell.* (AGI, 1987)

shale dust The dust obtained by drying and grinding shale. (Rice, 1913-18)

shalene A term proposed in place of gasoline for that product distilled from oil shale. (von Bernewitz, 1931)

shale-off shale Scot. Shale yielding oil on distillation. This term was formerly used as signifying argillaceous rock.

shale oil A crude oil obtained from bituminous shales, esp. in Scotland, by submitting them to destructive distillation in special retorts. (Fay, 1920)

shale pit A dumping place for coarse material screened out of rotary drill mud. (Nichols, 1976)

shale screen *shaker.*

shale shaker (a) A cylindrical sieve or vibrating table that removes the drill cuttings from the circulating mud stream. (Wheeler, 1958) (b) *shaker.*

shaley blaes Scot. *bituminous shale.*

shallow ground Aust. Land having gold near its surface. (Standard, 1964)

shallow well A shaft sunk to pump surface water only. See also: *well.* (Hammond, 1965)

shaly Pertaining to, composed of, or having the character of shale; esp. readily split along closely spaced bedding planes, such as shaly structure or shaly parting. Also, said of a fine-grained, thinly laminated sandstone having the characteristic fissility of shale owing to the presence of thin layers of shale; or said of a siltstone possessing bedding-plane fissility. (AGI, 1987)

shamble One of a set of shelves or benches, from one to the other of which ore is thrown successively in raising it to the level above, or to the surface. See also: *shammel.*

shammel (a) A stage for shoveling ore upon, or for raising water. See also: *shamble.* (b) To work a mine by throwing the material excavated onto a stage or bench in the "cast after cast" method, which was the usual way before the art of regular mining by means of shafts had been introduced.

Shand's classification A classification of igneous rocks based on crystallinity, degree of saturation with silica, degree of saturation with alumina, and color index. This system was developed in 1927 by S.J. Shand. (AGI, 1987)

shangie Scot. A ring of straw or hemp put round a jumper in boring to prevent the water in the borehole from splashing out.

shank (a) The steel-threaded portion of a diamond bit to which the crown is attached. Also called bit blank; blank; blank bit. (Long, 1960) (b) The body portion of any bit above its cutting edge. (Long, 1960) (c) That part of the drill steel which is inserted in the chuck of the drill. (Fraenkel, 1953) (d) A bar or standard which connects a rooter tooth with the frame. (Nichols, 1954) (e) A ladle for molten metal, with long handles for use by two or more workers. (Webster 3rd, 1966)

Shanklin sand Eng. A marine deposit of siliceous sands and sandstone of various shades of green and yellow gray. Also called Lower Greensand.

shaped charge An explosive contained in a case so shaped as to concentrate the power of the explosion in one small area. Shaped charges are used in armor-penetrating weapons, such as the bazooka, for tapping open-hearth furnaces, for cutting deep-well linings, and for breaking boulders. (Nichols, 1956)

shaped stone An artificially blunted or shaped carbon or diamond cut to form a point conforming to a specific profile. (Long, 1960)

shape factor Property of a particle that determines the relation between its mass and surface area, and hence its response to frictional restraint. (Pryor, 1965)

shape-firebrick molder One who makes shaped firebricks in steel molds. (DOT, 1949)

shape-silica-brick molder One who makes shaped silica bricks in steel molds. (DOT, 1949)

shard (a) A vitric fragment in pyroclastics; some have a characteristically curved surface of fracture. Shards generally consist of bubble-wall fragments produced by disintegration of pumice during or after an eruption. (AGI, 1987) (b) Syn: *sherd.* (AGI, 1987)

sharp fire Combustion with excess air and short flame. (ASTM, 1994)

sharp gravel Angular flint gravel. (Arkell, 1953)

sharpite An orthorhombic mineral, $Ca(UO_2)_6(CO_3)_5(OH)_4{\cdot}6H_2O$; very radioactive; yellow-green.

sharp sand Sand composed of angular quartz grains, used in making mortar.

sharp stone (a) Drill diamonds or carbon having sharp edges and corners that have not been artificially blunted or rounded through use. (Long, 1960) (b) New-condition, unused carbon or drill diamonds. (Long, 1960) (c) A sedimentary rock made up of angular particles more than 2 mm in its greatest dimension. (AGI, 1987)

shatter belt A less-preferred syn. of "fault zone."

shatter cut *burn cut.*

shattered zone Applied to a belt of country in which the rock is cracked in all directions, resulting in a network of small veins.

shatter index The percentage of a specially prepared sample of coke remaining on a sieve of stated aperture after the sample has been subjected to a standardized dropping procedure. (BS, 1961)

shattery S. Staff. Burnt clay in the vicinity of burnt coal.

shattuckite An orthorhombic mineral, $Cu_5(SiO_3)_4(OH)_2$; blue.

Shaver's disease *bauxite pneumoconiosis.* (NSC, 1996)

shcherbakovite An orthorhombic mineral, $(K,Na,Ba)_3(Ti,Nb)_2Si_4O_{14}$; dark brown.

shear (a) A deformation resulting from stresses that cause or tend to cause contiguous parts of a body to slide relatively to each other in a direction parallel to their plane of contact. It is the mode of failure of a body or mass whereby the portion of the mass on one side of a plane or surface slides past the portion on the opposite side. In geological literature the term refers almost invariably to strain rather than to stress. It is also used to refer to surfaces and zones of failure by shear, and to surfaces along which differential movement has taken place. (AGI, 1987) (b) *shearing.* (c) To make vertical cuts in a coal seam that has been undercut. (Standard, 1964)

shear bursts In deep mining fields, shear bursts are the most common type. By the occurrence of a single shear crack parallel to the face in one of the walls, the wall rock behind the shear plane is able to expand freely into the stope, heavily compressing those supports that until then have not taken stress, throwing still more stress on those that have, and causing the wall rock between the nearest supports and the face to disrupt and fill the place with debris. Shear bursts frequently occur at the working face of a pillar, remnant, or promontory. In such cases, they should not be mistaken for true pillar bursts. (Spalding, 1949)

shear-cake A counterweighted refractory slab used as a gate or door to a small furnace or oven. (ASTM, 1994)

shear cleavage Refers to cleavage where there is displacement of preexisting surfaces across the cleavage plane by movement parallel to it. Syn: *slip cleavage.* (AGI, 1987)

shear cut A vertical cut made by a special type of coal cutter or arc-shearing machine. Syn: *vertical cut.* See also: *turret jib.* (Nelson, 1965)

shearer (a) In bituminous coal mining, one who operates a type of coal-cutting machine that shears (cuts) out a channel down the sides of the working face of coal (as distinguished from undercutting) prior to blasting the coal down. Also called shearing-machine operator. (DOT, 1949) (b) A person who operates the shearing machine on a longwall face.

shearer loader Machine that shears coal or other easily broken mineral from the longwall face of a seam and delivers the broken material continuously to a conveying system. (Pryor, 1963)

shear failure Failure in which movement caused by shearing stresses in a soil mass is of sufficient magnitude to destroy or seriously endanger a structure. Syn: *failure by rupture.* (ASCE, 1958)

shear fold A fold model of which the mechanism is shearing or slipping along closely spaced planes parallel to the fold's axial surface. The resultant structure is a similar fold. Syn: *slip fold.* (AGI, 1987)

shear fracture A fracture that results from stresses that tend to shear one part of a rock past the adjacent part. See also: *shear joint.* Cf: *tension fracture.* (AGI, 1987)

shearing (a) The vertical side cutting that, together with holing or horizontal undercutting, constitutes the attack upon a face of coal. (b) Making a vertical cut or groove in a coal face, breast, or block, as opposed to a kerf, which is a horizontal cut. Called in Arkansas a cut or cutting. See also: *shear.* (BCI, 1947) (c) Vertical cuts applied in coal headings only to provide an additional free face, since in heading work it is usual to employ deeper cuts than on longwall faces, and the shots in headings are much tighter. (McAdam, 1958) (d) The deformation of rocks by cumulative small lateral movements along innumerable parallel planes, generally resulting from pressure, and producing schistosity, cleavage, minute application, and other metamorphic structures.

shearing force A straining action wherein tangentially applied forces tend to produce a skewing type of deformation. Shear forces are usually accompanied by normal forces produced by tension, thrust, or bending. (McGraw-Hill, 1960)

shearing jib A jib of a coal cutter or cutter loader that makes a vertical or shear cut in the coal, ore, or rock. (Nelson, 1965)

shearing machine An electrically driven machine used to cut coal during longwall mining. Usually used in a double-drum configuration.

shearing-machine operator *shearer.*

shear joint Joint that formed as a shear fracture. See also: *shear fracture.*

shear lag On account of shear strain, the longitudinal tensile or compressive bending stress in wide beam flanges diminishes with the distance from the web or webs; this stress diminution is called shear lag. (Roark, 1954)

shear legs (a) A high wooden frame placed over an engine or pumping shaft fitted with small pulleys and rope for lifting heavy weights. (Zern, 1928) (b) A tripod on which miners sometimes stand in drilling. (Standard, 1964)

shear modulus Also known as the modulus of rigidity, it equals the shear stress divided by the shear strain. (Hammond, 1965)

shear of ore Ore shoot or orebody. (von Bernewitz, 1931)

shearpin A small soft-metal pin, connecting or pinning together two parts of a tool that will break by shearing if an excessive load is placed on the pinned components. The shearing of the pin prevents damage to the overloaded components; thus it is a safety device. (Long, 1960)

shear plane A fracture that produces a positive draw. It differs from a weight break in that the character of the fracture is less ragged and heaving takes place over the unwrought coal instead of the waste. (Briggs, 1929)

shear rivet Soft copper rivets used in the Hall-Rowe wedge to connect the drive and pilot wedges; they can be sheared off to leave the drive wedge as a permanent reference in the borehole at the point at which the hole is to be deflected. (Long, 1960)

shear slide A slide produced by shear failure, usually along a plane of weakness, as bedding or cleavage. (AGI, 1987)

shear steel A steel produced by heating blister steel (sheared to short lengths) to a high heat, welding by hammering or rolling, or both, and finally finishing under the hammer at the same or slightly greater heat. (Webster 3rd, 1966)

shear strain Angular displacement of a structural member due to a force acting across it, measured in radius. See also: *shear modulus.* (Hammond, 1965)

shear strength (a) The stress or load at which a material fails in shear. (Nelson, 1965) (b) A measure of the shear value of a fluid. Also a measure of the gelling properties of a fluid. (Brantly, 1961) (c) The internal resistance of a body to shear stress, typically including a frictional part and a part independent of friction called cohesion. (AGI, 1987)

shear stress (a) The shear force operating in a material, measured per unit of cross-sectional area. See also: *stress; shear modulus; punching shear.* (Hammond, 1965) (b) The stress component tangential to a given plane. Also called shearing stress; tangential stress. (ASCE, 1958)

shear structure Any rock structure caused by shearing, e.g., crushing, crumpling, or cleavage. (AGI, 1987)

shear wave *S wave.*

shear zone (a) A tabular zone of rock that has been crushed and brecciated by many parallel fractures due to shear strain. Such an area is often mineralized by ore-forming solutions. Syn: *sheeted zone; sheeted-zone deposit*. (AGI, 1987) (b) Hogback. (Zern, 1928)

sheath (a) To enclose or encase with a covering. (Carson, 1961) (b) A chemical compound or mixture incorporated in a sheathed explosive unit that forms a flame-inhibiting cloud on detonation of the explosive. (CFR, 4)

sheathed explosive A permitted explosive surrounded by a sheath containing a noncombustible powder. The powder acts as a cooling agent and reduces the temperature of the resultant gases of the explosion, and therefore reduces the risk of these hot gases causing a combustible gas ignition. See also: *eq.s. explosive; explosive*. (Nelson, 1965)

sheathed explosive unit A device consisting of an approved or permissible explosive covered by a sheath encased in a sealed covering and designed to be fired outside the confines of a borehole. (CFR, 4)

sheathing driver Essentially a paving breaker, an impact hammer driven by air, designed and adapted for driving wood sheathing. (Carson, 1961)

sheave Grooved pulley wheel much used in underground rope haulage. See also: *Koepe sheave; tail sheave; winding sheave*. (Pryor, 1963)

sheave block A pulley and a case provided with a means to anchor it. (Nichols, 1976)

sheave wheel *sheave.*

shed (a) Eng. A thin, smooth parting in rocks, having both sides polished. (b) Eng. A very thin layer of coal. (c) A divide of land; e.g., a watershed. (AGI, 1987)

shedline The summit line of elevated ground; the line of a watershed.

sheen (a) A subdued and commonly iridescent or metallic glitter that approaches but is just short of optical reflection and that modifies the surface luster of a mineral; e.g., the optical effect still visible in the body of a gem (such as tiger's-eye) after its silky surface appearance has been removed by polishing. (b) A luster that emanates from just beneath the surface of a mineral; e.g., opalescence. Cf: *luster.*

sheep's-foot A tamping roller with feet expanded at their outer tips. (Nichols, 1976)

sheep silver A Scottish term for mica.

sheer legs *shear legs.*

sheet (a) A general term for a tabular igneous intrusion, e.g., dike and sill, esp. if concordant or only slightly discordant. Cf: *intrusive vein.* (AGI, 1987) (b) A term used in the Upper Mississippi lead-mining region of the United States for galena occurring in thin, continuous masses. (AGI, 1987) (c) *blanket.*

sheet deposit A mineral deposit that is generally stratiform, more or less horizontal, and areally extensive relative to its thickness. (AGI, 1987)

sheet drying conveyor A disk type of live roller conveyor equipped with air outlets from a blower to remove dampness from processed sheet metal while being conveyed.

sheeted ground (a) Several closely spaced parallel faults along which the wall rocks are broken into thin sheets. (b) *shear zone.*

sheeted vein A group of closely spaced, distinct parallel fractures filled with mineral matter and separated by layers of barren rock.

sheeted zone *shear zone.*

sheeted-zone deposit *shear zone.*

sheeters Light steel poling boards driven down to protect trench sides from collapse. (Hammond, 1965)

sheet flow *laminar flow.*

sheet ground A term used in the Joplin district, Missouri, and applied to horizontal, low-grade, disseminated zinc-lead deposits, covering an extensive area. See also: *sheet deposit.* Cf: *bed vein.*

sheeting (a) The development, in rock formations, of small, closely spaced, parallel fractures. (b) In a restricted sense, the gently dipping joints that are essentially parallel to the ground surface; they are more closely spaced near the surface and become progressively farther apart with depth. Esp. well-developed in granitic rocks. See also: *bedding.* Cf: *exfoliation.* (AGI, 1987)

sheeting caps A row of caps placed on blocks about 14 in (36 cm) high placed on top of the drift sets when constructing the permanent floor in the stope. Round poles are then laid lengthwise of the stope on the sheeting caps and are covered with lagging. (Lewis, 1964)

sheeting driver An air hammer attachment that fits on plank ends so that they can be driven without splintering. (Nichols, 1976)

sheeting jacks Push-type turnbuckles, used to set ditch bracing. (Nichols, 1976)

sheeting pile *sheet pile.*

sheet iron *sheet.*

sheet-iron pitch The inclination of a coal seam at which loose coal will not move on the natural bottom, but at which it will slide or can be easily pushed along on iron slides placed on the bottom in the chambers or rooms. (Fay, 1920)

sheet jointing *exfoliation.*

sheet metal *sheet.*

sheet-metal gage A gage used for measuring the thickness of sheet metal. (Crispin, 1964)

sheet mica Mica that is relatively flat and sufficiently free from structural defects to enable it to be punched or stamped into specified shapes for use by the electronic and electrical industries. Sheet mica is classified further as block, film, and splittings. (Skow, 1962)

sheet pile A pile with a generally flat cross section, which may be meshed or interlocked with adjacent similar members to form a diaphragm, wall, or bulkhead, and designed to resist lateral earth pressure or to reduce ground-water seepage. Syn: *sheeting pile.* (AGI, 1987)

sheet piles Closely spaced piles of timber, reinforced concrete, prestressed concrete, or steel driven vertically into the ground to support earth pressure, to keep water out of an excavation, and often to form an integral part of a permanent structure. (Hammond, 1965)

sheet-pile wall A wall formed of sheet piles which may be of cantilever design, or anchored back at one or two levels. A retaining wall. (Hammond, 1965)

sheet piling A diaphragm made up of meshing or interlocking members of wood, steel, concrete, etc., driven individually to form an obstruction to percolation, to prevent movement of material, for cofferdams, for stabilization of foundations, etc. (Seelye, 1951)

sheet quarry Often used in granite quarrying to designate a quarry having strong horizontal joints and few vertical ones.

sheets Eng. Coarse cloth curtains or screens for directing the ventilation underground. Syn: *brattice cloth; brattice sheeting.*

sheet sand A sandstone of great areal extent, presumably deposited by a transgressing sea advancing over a wide front and for a considerable distance. See also: *blanket sand.*

sheet structure *sheeting.*

Sheffield process A basic open-hearth process using charges so low in sulfur and phosphorus that they could be used in the acid process; the pig iron charged is hematite iron. The charge contains all the elements required to give the required analysis, plus the usual margin of carbon. The charge contains about 0.5% silicon and a maximum amount of manganese, to ensure correct conditions in the bath. (Osborne, 1956)

Shelby tube A thin-walled soil-sampling tube, 12 to 30 in (30.5 to 76.2 cm) long, attached to a special rod adapter or sub by means of machine screws. The device is designed to take soil samples by pressing or pushing the tube down into the formation sampled. Syn: *Shelby-tube sampler; thin-wall drive sampler.* Cf: *thick-wall sampler; thin-wall sampler.* (Long, 1960)

Shelby-tube sampler *Shelby tube.* Cf: *thick-wall sampler.*

shelf (a) Corn. The solid rock or bedrock, esp. under alluvial tin deposits. (b) Corn. A rock, ledge or rock, reef, or sandbank in the sea. (c) Corn. A projecting layer or ledge of rock on land. (d) Corn. The submerged border of a continent or of an island extending from the shoreline to the depth at which the sea floor begins to descend steeply toward the bottom of the ocean basin. See also: *continental shelf.* (Webster 3rd, 1966) (e) Corn. A ledge of bedrock upon which drift rests. (Statistical Research Bureau, 1935)

shelf angle A mild steel angle section, riveted or welded to the web of a comparatively deep I-beam supporting the formwork for hollow tiles or forming the seating for precast concrete floor or roof units. (Hammond, 1965)

shelf quarry An open pit quarry where the ledge of stone forms a hill and the floor of the quarry worked on a hillside may be little, if any, lower than the surrounding country. In such openings, both transportation and drainage are favorable. (AIME, 1960)

shelf retaining wall A retaining wall of reinforced concrete having a relieving platform built onto its upper part. (Hammond, 1965)

shelf sea The sea overlying the continental shelf.

shell (a) A thin, hard layer of rock encountered in drilling a well. Cf: *shale break.* Syn: *shelly formation.* (AGI, 1987) (b) The crust of the Earth. Also, any of the continuous and distinctive concentric zones or layers composing the interior of the Earth (beneath the crust). The term was formerly used for what is now called the mantle. Syn: *Earth shell.* (AGI, 1987) (c) A sedimentary deposit consisting primarily of animal shells. (AGI, 1987) (d) A steel tube from which air or other gas at high pressure is discharged with explosive force in a shothole; as used with Cardox, Hydrox, and air blasting. (BS, 1964) (e) Incorrectly used by some drillers as a syn. for reaming shell; also incorrectly used as a syn. for the inner or outer tube of a core barrel. (Long, 1960) (f) A metal or paper case that holds a charge of powder. (Fay, 1920) (g) A group of electrons in an atom, all of which have the same principal quantum number. (CTD, 1958) (h) Any thin-wall tubular device. (Long, 1960) (i) A torpedo used in oil wells. (Fay, 1920) (j) A hollow structure or vessel. (ASM, 1961) (k) An article formed by deep drawing. (ASM, 1961) (l) The metal sleeve remaining when a billet is extruded with a dummy flock of somewhat smaller diameter. (ASM, 1961) (m) In shell molding, a hard layer of sand and thermosetting plastic or resin formed over a pattern and used as the mold wall. (ASM, 1961) (n) A tabular casting used in making seamless drawn tube. (ASM, 1961) (o) A pierced forging. (ASM, 1961) (p) In a grinding mill, external cylinder and ends. (Pryor, 1963) (q) The falling away of a 1- to 2-in (2.5- to 5.1-cm) internal layer of refractory from the roof of an all-basic open-hearth steel furnace; the probable cause is the combined effect of flux migration, temperature gradient, and stress. This form of wear is also known as slabbing. (Dodd, 1964) (r) The shell of a hollow clay building block refers to the outer walls of the block. (Dodd, 1964) (s) A curved form of plate constuction applicable to roofs. (Hammond, 1965)

shell-and-auger boring Method of making exploratory shallow bores in soft ground using an auger for clay and a shell, or sandpump, for sands. Syn: *post-hole auger.* (Pryor, 1963)

shell cameo A cameo carved from shell with raised figure cut from white layers and the background cut away to the darker layers.

shell clearance The difference between the outside diameter of a bit or core barrel and the outside set or gage diameter of a reaming shell. (Long, 1960)

shelling *shell.*

shell lime Lump lime which, when flaked, has a characteristic shell-like appearance. It is used mainly in Scotland. (Osborne, 1956)

shell marl (a) A sandy, clayey, or limy deposit, loose or weakly consolidated, containing abundant molluscan shells; a common term in the coastal plain of the Southeastern United States. (AGI, 1987) (b) A light-colored calcareous deposit formed on the bottoms of small freshwater lakes, composed largely of uncemented mollusk shells and precipitated calcium carbonate, along with the hard parts of minute organisms. (AGI, 1987)

shell pump A simple form of sand pump or sludger consisting of a hollow cylinder with a ball or clack valve at the bottom, which is used with a flush of water to remove detritus. See also: *sand pump; sludger.* (Webster 3rd, 1966)

shelly formation A thin and generally hard stratum encountered in drilling. See also: *shell.* (AGI, 1987)

shelter hole In coal mining, a niche in the rib along a haulage road into which one may step to avoid passing trains. Also known as a manhole.

Shelton loader An adapted coal-cutting machine in which the picks of the cutter chain are replaced by loading flights. The machine hauls itself along the face, the jib leading, by means of an anchored rope. The flights push the prepared coal up a ramp onto the face conveyor, and on their return path to the back of the prepared coal; they fold back and then open up for coal loading as they emerge at the end of the jib. It requires well-prepared coal for successful operation, and will load only in one direction; consequently it has to be flitted back along the coal face. The jib can be swung into line with the body for flitting. (Mason, 1951)

shepherd Aust. A miner who preserves legal rights to a claim without working on it. (Standard, 1964)

shepherding Preserving the rights in a mining claim while doing the minimum possible amount of developing. (Nelson, 1965)

sherardize To coat an article of iron or steel with zinc by covering with zinc dust in a tightly closed drum and heating for several hours at 300 to 420 °C so that a zinc-iron alloy is formed at the surface through the action of zinc vapor. (Webster 3rd, 1966)

sherardizing A galvanizing process in which the metal to be coated is heated, with or without tumbling, in contact with zinc dust. (Liddell, 1918)

sherd (a) A fragment or broken piece of pottery. (AGI, 1987) (b) *shard.*

sheridanite A variety of clinochlore of the chlorite group.

sherry topaz (a) A valuable and important variety of topaz the color of sherry wine. (b) An incorrect name for citrine of the same color.

Sherwen shaker An electromagnetic vibrator used in shaking table mechanisms, concrete consolidation, ore feeders, and screens. (Pryor, 1963)

sherwoodite A tetragonal mineral, $Ca_9Al_2V_{28}O_{80}\cdot56H_2O$; blue-black.

shield (a) A framework or diaphragm of steel, iron, or wood, used in tunneling and mining in unconsolidated materials. It is moved forward at the end of the tunnel or adit in process of excavation, and is used to support the ground ahead of the lining and to aid in its construction. (AGI, 1987) (b) A large area of exposed basement rocks in a craton, commonly with a very gently convex surface, surrounded by sediment-covered platforms; e.g., Canadian Shield, Baltic Shield. The rocks of virtually all shield areas are Precambrian. Syn: *continental shield; cratogene; continental nucleus.* (AGI, 1987) (c) In longwall mining, the hydraulically powered roof supports that protect the face workers and advance as the panel is extracted.

shield basalt A basaltic accumulation of smaller size than the plateau or flood basalts, arising from the confluence of lava flows from a large number of small, closely spaced volcanoes. Cf: *plateau basalt.* Syn: *multiple-vent basalt.*

shield volcano A volcano in the shape of a flattened dome, broad and low, built by flows of very fluid basaltic lava or by rhyolitic ash flows. Syn: *lava dome; basaltic dome.* (AGI, 1987)

shift (a) A small fault or slip. See also: *subsidence.* (Arkell, 1953) (b) A fault or dislocation. (c) The maximum relative displacement of points on opposite sides of a fault and far enough from it to be outside the dislocated zone. Also called net shift. See also: *dip shift; normal shift; strike shift; vertical shift.* (d) The number of hours or the part of any day worked. Also called tour. (Long, 1960) (e) The gang of workers employed for the period, such as the day shift or the night shift. (Fay, 1920)

shifter (a) In bituminous coal mining, a general term for workers who assist brattice men, repairmen, timbermen, and other workers not engaged in the actual mining of coal. (DOT, 1949) (b) *track shifter.*

shift gear A gear on a gear-feed swivel head of a diamond drill by means of which the feed-shifter rod may be moved to engage the shifter-rod pin into the selected feed gear. (Long, 1960)

shifting clothes Street clothes into which the miner changes on emerging from a mine. (Korson, 1938)

shift lever A short rod or shaft attached to the shift-gear shaft by means of which the ratio of the driving to the driven gears may be changed in a gear-feed swivel head of a diamond drill or other transmission-gear mechanism. (Long, 1960)

shift work Work performed at a mine and paid for by day wage as opposed to payment by results, namely by tonnage, yardage, or price list. (Nelson, 1965)

shinarump *silicified wood.*

shinbone protectors A form of leggings that protect the shinbone. They are designed in both metal and tough plastic and are secured by leather straps around the legs. (Nelson, 1965)

shindle stone Stone from which shindles or roofing slates are made. Local variant of shingle. (Arkell, 1953)

shingle (a) Coarse, loose, well-rounded waterworn detritus or alluvial material of various sizes; esp. beach gravel, composed of smooth and spheroidal or flattened pebbles, cobbles, and sometimes small boulders, generally measuring 20 to 200 mm in diameter; it occurs typically on the higher parts of a beach. The term is more widely used in Great Britain than in the United States. See also: *chingle.* (AGI, 1987) (b) A place strewn with shingle; e.g., a shingle beach. Etymol: probably Scandinavian, akin to singel, coarse gravel that sings or crunches when walked on. (AGI, 1987)

shingler A machine for squeezing puddled iron. (Webster 3rd, 1966)

shingle structure The arrangement of closely spaced veins overlapping in the manner of shingles on a roof. Syn: *imbricate structure.* (AGI, 1987)

shingling In wrought iron manufacture, the operation by which sinter and other impurities are removed from a bloom and the metal solidly welded together by hammering and squeezing. Also called nobbing. (Henderson, 1953)

shining As applied to the degree of luster of minerals, means those that produce an image by reflection, but not one well-defined, as celestite.

shin plaster *suspension.*

ship auger An auger having a simple spiral (helical) body and a single cutting edge, with or without a screw on the end without a spur at the outer end of the cutting edge, used to obtain soil samples in sticky material. (Long, 1960)

shiplap liners Longitudinal lining plates for ball mill of wedgelike shape. The thin edge of each wedge underlies the thick edge of the preceding plate in the direction of revolution. (Pryor, 1963)

ship observations Meteorological and oceanographic data taken for a specific location, observed from a ship underway or at anchor. (Hy, 1965)

shipper shaft In a dipper shovel, the hinge on which the stick pivots when the bucket is hoisted. (Nichols, 1954)

shipping measure For measuring entire internal capacity of a vessel: One register ton equals 100 ft^3 (1.8 m^3); for measuring cargo; one U.S. shipping ton equals 40 ft^3 (1.1 m^3) or 32.143 U.S. bushels. (Crispin, 1964)

shipping ore (a) Any ore of greater value when broken than the cost of freight and treatment. (b) *first-class ore.*

shirt The inner lining of a blast furnace. (Standard, 1964)

shiver An old English term for soft and crumbly shale, or slate clay approaching shale. Also spelled chiver. Etymol: Middle English scifre. Adj: shivery. (AGI, 1987)

shiver spar A variety of calcite with slaty structure; specif. argentine. Syn: *slate spar.* (AGI, 1987)

shivery post Eng. *scamy.*

shoad (a) Waterworn fragments of vein minerals found on the surface, such as beds of streams, away from the outcrop. (Arkell, 1953) (b) Float ore, which has broken from outcrop and gravitated to a distance. (Pryor, 1963) (c) To shoad, to trace a lode by following up shoad. (Pryor, 1963) (d) Corn. Stream tin, or any surface rubble or talus containing fragments of tin, copper, or lead ore, and signifying proximity of a lode. (Arkell, 1953) (e) *shode.*

shoaling effect Alteration of a wave proceeding from deep water into shallow water. The wavelength decreases and the wave height increases. (Hy, 1965)

shock bump (a) A rock bump caused by the sudden collapse of a thick sandstone or other strong deposit. See also: *rock burst; pressure bump.* (Nelson, 1965) (b) A thump experienced in a mine when the breakage of overlying rocks produces the effect of a hammer blow. (Hammond, 1965) (c) *bumps.*

shock loading In winding, shock loads produced by picking up a cage from the pit bottom with slack chains or by lifting heavy pithead gates or covers. This often causes dry fatigue in the winding ropes. (Sinclair, 1959)

shock losses Head losses resulting from changes in direction of flow or area of duct. They also occur at the inlet or discharge of a system, at splits or junctions of two or more currents of air, and at obstructions in airways. (Hartman, 1982)

shock pressure loss There is a constant ratio between shock pressure loss and the velocity pressure corresponding to the mean velocity of flow. Shock pressure losses can be calculated from the following formula: P_s=XP_v, where P_s equals the total shock pressure loss in inches of water, P_v equals the velocity pressure, in inches of water corresponding to the mean velocity of flow (equals approx. $[V/4{,}000]^2$ at standard air density of 0.075 lb/ft^3 or 1.2 kg/m^3), and X equals an empirical factor or shock factor found by experiment. P_s and P_v are expressed in the same units and are equally affected by density. X is therefore independent of both the density and the units used and is the number of velocity pressures equivalent to the shock pressure loss. (Roberts, 1960)

shock-proof As applied to the current-carrying parts of an electric system, excepting trolley wires, is taken to mean that contact with such parts is prevented by the use of grounded metallic coverings or sheaths.

shock tube system A system for initiating blasting caps in which the energy is transmitted to the cap by means of a shock wave inside a hollow plastic tube. See also: *gas detonation system; nonelectric blasting.* (Dick, 2)

shock wave (a) The wave of air and dust that, in some cases, travels ahead of the flame of a coal dust explosion. It may occur when an ignition takes place near the closed end of a mine roadway, and the reaction products behind the flame cannot escape freely. (Nelson, 1965) (b) The wave sent out through the air by the discharge of the shot initiating an explosion. The wave travels with the velocity of sound and produces to the human ear a noise like the boom of a cannon. (Rice, 1913-18) (c) A compressional wave formed whenever the speed of a body relative to a medium exceeds that at which the medium can transmit sound, having an amplitude that exceeds the elastic limit of the medium in which it travels, and characterized by a disturbed region of small but finite thickness within which very abrupt changes occur in the pressure, temperature, density, and velocity of the medium; e.g., the wave sent out through the air by the discharge of the shot initiating an explosion. In rock, it travels at supersonic velocities and is capable of vaporizing, melting, mineralogically transforming, or strongly deforming rock materials. (AGI, 1987)

shock-wave compression Nonisentropic adiabatic compression in a wave traveling at greater than local sound velocity. (Van Dolah, 1963)

shode (a) Corn. A loose fragment of veinstone. Ore washed or detached from the vein naturally. See also: *float ore.* (b) Eng. To search for ore by tracing the shode. See also: *shoad.*

shoe (a) A trough to convey ore to a crusher. (b) A metal block used in a variety of bending operations to form or support the part being processed. (ASM, 1961) (c) A coupling of rolled, cast, or forged steel to protect the lower end of the casting or drivepipe in overburden, or the bottom end of a sampler when pressed into a formation being sampled. (Long, 1960) (d) A wearing piece in various types of machines used to break rock, such as a column of drill pipes; bottom of crushing stamp; muller of amalgamating pan. (Pryor, 1963) (e) The lower replaceable part of a gravity stamp which falls on the mineral ore or rock. (Nelson, 1965)

shoepite-III *paraschoepite.*

shoestring claim A mining claim in the form of a long narrow strip.

shoestring location A location of a long and narrow strip of mineral land. (Ricketts, 1943)

shonkinite A dark-colored syenite composed chiefly of augite and alkali feldspar, and possibly containing olivine, hornblende, biotite, and nepheline. Its name, given by Weed and Pirsson in 1895, is derived from Shonkin, the Indian name for the Highwood Mountains of Montana. (AGI, 1987)

shoo-fly (a) Miner's work train. (Korson, 1938) (b) Any crosscut between a haulageway and airway through which cars are run. See also: *pickup; slant.*

shoot (a) *ore shoot; pipe; blast; chute.* Also spelled chute. (b) To break coal loose from a seam by the use of explosives; loosely used, also as applied to other coal-breaking devices. (BCI, 1947) (c) To break down by airblasting. (BS, 1964) (d) A body of ore, usually of elongated form, extending downward or upward in a vein. Also called ore shoot. (Long, 1960) (e) The payable section of a lode; an enriched portion of a continuous orebody. (Nelson, 1965) (f) Any considerable and somewhat regular mass of ore in a vein, frequently a rich ore streak in a vein; a chimney; also, a vein branching at a small angle from and reentering the main vein. (Standard, 1964) (g) The valuable minerals are commonly concentrated in certain portions of a vein that have one dimension much longer than the others. This shoot or chimney of ore is usually highly inclined to the horizontal. (Lewis, 1964) (h) To explode a charge in blasting operations. (Hudson, 1932) (i) *blast.* A shot is a single operation of blasting. (j) In seismic exploration, the firing of the explosive by an electrical impulse; also, the process of carrying out a seismic survey, to shoot an area or prospect. (AGI, 1987)

shooter *blaster.*

shooting The use of explosives in rock breaking.

shooting against the bank *blanket shooting.*

shooting boat In marine seismic exploration, a boat equipped to carry explosives, and from which the placing and firing of shots are performed. (AGI, 1987)

shooting by seismograph The making of a seismographic survey. The term shooting derives from the setting off of explosions in the ground. The shock waves from these explosions are recorded by a seismograph, and from these records a contour map can be made. (Williams, 1964)

shooting needle A blasting needle; a metallic rod used in the stemming of a drill hole for the purpose of leaving a cavity through which a charge may be fired. (Fay, 1920)

shooting off-the-solid Mining coal by heavy blasting without undermining or shearing it. In England, called shooting fast. (Fay, 1920)

shooting on-the-free The use of a smaller charge of powder to blow down the face of the coal after it has been undercut, as distinguished from "shooting off-the-solid". (Fay, 1920)

shooting rights The right to enter upon land and make a geophysical survey. (Williams, 1964)

shooting truck In seismic operations, a truck equipped to carry explosives, materials, and equipment for preparing, loading, tamping, and firing explosive charges. (AGI, 1987)

shooting valve The control valve provided for the purpose of admitting compressed air to an airblasting shell and of venting residual air, in the shell and hose, to the atmosphere. (BS, 1964)

shop rivet A rivet driven in a workshop, as distinct from a field or site rivet. (Hammond, 1965)

shoran A precise electronic measuring system for indicating distance from an airborne or shipborne station to each of two fixed ground stations simultaneously by recording (by means of cathode-ray screens) the time required for round-trip travel of radar signals or high-frequency radio waves and thereby determining the position of the mobile station. Its range is effectively limited to line-of-sight distances (or about 40 nmi or 74 km). Shoran is used in control of aerial photography, airborne geophysical prospecting, offshore hydrographic surveys, and geodetic surveying for measuring long distances. Cf: *loran.* Etymol: short-range navigation. (AGI, 1987)

shore drift The coarse material covering the bottom where the agitation of the water at the bottom is effective constitutes shore drift. See also: *littoral drift.* (AGI, 1987)

Shore hardness test A scale of hardness of rocks as determined by the Shore scleroscope test. The scale avoids the limitation of Mohs scale of hardness and gives better assessment of rock hardness. See also: *tungsten carbide bit.*

shore reef *fringing reef.*

Shore scleroscope An instrument comprising a small diamond-shaped hammer that falls freely down a graduated tube of glass from a constant height. The hardness of the surface under test is measured by the height of the rebound. In one type of this instrument, the rebound of the hammer actuates the pointer of a scale so that the height of rebound is recorded. See also: *scleroscope hardness test.* (Hammond, 1965)

shore terrace (a) A terrace produced by the action of waves and currents along the shore of a lake or sea; e.g., a wave-built terrace. (AGI, 1987) (b) Marine terrace. (AGI, 1987)

shore up To stay, prop up, or support by braces.

shoring Timbers braced against a wall as a temporary support. Also, the timbering used to prevent a sliding of earth adjoining an excavation. (Crispin, 1964)

short (a) Said of roof shale that tends to break up or crush under pressure into small fragments and that will not hold in any span over a few inches. Also called tender. (Raistrick, 1939) (b) Brittle; friable; breaking or crumbling readily; inclined to flake off; said of coal. (c) Used to denote a roof that has very little structural strength. (TIME, 1929-30)

short awn A direction of more than 45° to the main natural line of cleat or cleavage in the coal. Also spelled horn. Cf: *long awn.* (TIME, 1929-30)

short coal *short.*

short column A column so short in relation to its cross section that, if overloaded, it will fail by crushing rather than by buckling. Cf: *long column.* (Hammond, 1965)

short-delay blasting (a) A method of blasting where the explosive charges are detonated with a very short delay interval between them. It enables shots to assist one another as in simultaneous firing, and also each shot or group of shots establishes a free or semifree face for the following group of shots. (Nelson, 1965) (b) Method of blasting by which charges are caused to explode in a given sequence with time intervals of 0.001 to 0.1 s. (Fraenkel, 1953)

short-delay detonator (a) A detonator in which the interval of time delay is incremental in milliseconds. Syn: *millisecond-delay detonator.* (BS, 1964) (b) The original 1-s delay detonators are no longer used, and the choice now lies between the one-half-second type and those known as short-delay detonators. These give better fragmentation of rock and consequently better loading rates. (Nelson, 1965)

short-delay electric detonator An electric detonator with a designated delay period of 25 to 1,000 ms. (CFR, 4)

short-flame explosive *permissible explosive.*

short hole Taphole in a furnace that is not properly stopped and that is likely to release the molten charge prematurely. (Henderson, 1953)

short-hole work Diamond drilling where the length of borehole generally does not exceed 100 ft (30 m). (Long, 1960)

short period The time required to drill a few holes for trolley hangers or a few short block holes, or one or two holes for bringing down a piece of loose roof.

short-period delay An electric blasting cap that explodes one-fiftieth to one-half second after passage of an electric current. Syn: *millisecond delay.* (Nichols, 1976)

short-range order (a) Identical first-neighbor coordination of atoms. Typical of glassy structures. (Van Vlack, 1964) (b) A structural state in which cations have coordination polyhedra as predicted by radius ratios but lack the long-range order of crystallinity; e.g., silicate glasses. Cf: *long-range order.* (c) The type of order in which the probability of a given type of atom having neighbors of a given type (for example, that an A atom is surrounded by B atoms) is greater than would be expected on a purely random basis. There is thus a tendency to form small ordered domains, but these do not link together at long distances.

short run To be forced by adverse conditions or core blockage to pull the drill string before the core barrel being used is filled to capacity with core. Cf: *long run.* (Long, 1960)

shorts (a) As applied to asbestos, consist of the very shortest of classified grades; the fibers may vary from microscopical thin filaments to crude bundles of fibers of appreciable thickness. Included may be particles of nonfibrous serpentine ranging from a palpable powder to granules of visible size. (Sinclair, 1959) (b) The product that is retained on a specified screen in the screening of a crushed or ground material. (ASM, 1961) (c) In gold cyanidation, the oversize after the gold-rich zinc from the precipitation boxes has been rubbed through. (Pryor, 1963) (d) The shortage in production under a royalty lease. (CTD, 1958)

short section A section of land according to the U.S. Governmental Survey that contains less than 640 acres. (Williams, 1964)

short shot Colloquialism for weathering (or low-velocity layer) shot in seismic prospecting. (AGI, 1987)

short-term exposure limit A 15-min time-weighted average exposure that should not be exceeded at any time during a work day even if the 8-h time-weighted average is within the threshold limit value. Abbrev: STEL. (ACGIH, 1987-88)

shortwall (a) The reverse of longwall, frequently used to mean the face of a room. (Zern, 1928) (b) A method of mining in which comparatively small areas are worked separately, as opposed to longwall; e.g., room and pillar. (BCI, 1947) (c) A length of coal face intermediate between a stall and a normal longwall face. A shortwall face may be any length between about 5 yd and 30 yd (4.6 m and 27 m) and is generally employed in pillar methods of working. Rooms and stables may also be classified as shortwall faces. See also: *shortwall development.* (Nelson, 1965)

shortwall coal cutter (a) A machine for undercutting coal that has the cutter bar fixed in relation to the main body of the machine. It sumps and cuts across a face in a more or less continuous motion, except when it becomes necessary to stop to change the position of the ropes used to move the machine through the action of rope drums, or when difficulties in cutting are experienced. (Jones, 1949) (b) A coal cutter that has a long, rigid chain jib in line with the body of the machine; it cuts across a heading from right to left, being drawn across by means of a steel wire rope. This machine cannot be readily flitted from one heading to another unless a power-propelled flitting truck is available; otherwise each heading requires its own shortwall cutter. A shortwall cutter will make a 6-ft (1.8-m) cut across a 15-ft (4.6-m) heading in 20 min, including sumping in and out of the cut. (Mason, 1951)

shortwall development A system of coal working sometimes employed in seams 4 ft (1.2 m) or under in thickness, with the aid of machines. Short faces, each 15 to 30 yd (13.7 to 27 m) wide, are driven at 50- to 70-yd (46- to 61-m) centers, with crosscuts to assist coal transport and ventilation. The rippings are used to form roadside packs. The shortwalls are driven to the boundary, and the coal pillars are worked by longwall retreating. See also: *shortwall.* (Nelson, 1965)

shoshonite A trachyandesite composed of olivine and augite phenocrysts in a groundmass of labradorite with alkali feldspar rims, olivine, augite, a small amount of leucite, and some dark-colored glass. Shoshonite grades into absarokite with an increase in olivine and into banakite with more sanidine. Its name, given by Iddings in 1895, is derived from the Shoshone River, WY. (AGI, 1987)

shot (a) Coal that has been broken by blasting or other devices. (BCI, 1947) (b) A single explosive charge fired in coal, stone, or ore. (Nelson, 1965) (c) A detonation (or its equivalent) as used in seismic shooting. See also: *gouging shot.* (Nelson, 1965) (d) Small spherical particles of brittle hard steel used as the cutting agent in drilling a borehole with a shot drill. Also called adamantine shot; buckshot; chilled shot. See also: *gripping shot; blast; shot drill.* (Long, 1960)

shot blasting A method similar to sandblasting for cleansing the surface of metals, using broken shot or steel grit instead of sand. (Hammond, 1965)

shot boring The act or process of producing a borehole with a shot drill. See also: *shot drill.* (Long, 1960)

shot-boring drill *shot drill.*

shot bort (a) Incorrectly used to designate a small spherical-shaped drill diamond. *drill diamond.* (Long, 1960) (b) *ballas.* (c) Variety of bort with little impurity, in milky-white to steel-gray spherical stones with radiating structure and great toughness. (Tomkeieff, 1954) (d) Spheres of translucent diamond with more cohesion than ordinary bort. See also: *bort.* (Hess)

shot break In seismic prospecting, a record of the instant of generation of seismic waves, as by an explosion. Syn: *time break; shot instant.* (AGI, 1987)

shot copper Small, rounded particles of native copper, molded by the shape of vesicles in basaltic host rock, and resembling shot in size and shape. (AGI, 1987)

shotcrete Gunite that commonly includes coarse aggregate (up to 2 cm). (AGI, 1987)

shot datum Seismic calculations are usually reduced to a convenient reference surface or plane. These calculations simulate a condition where the charge is shot on the reference surface and the energy is also recorded on this same reference surface. At this reference surface, the time-depth charts have their origin. (AGI, 1987)

shot depth The distance from the surface to the explosive charge. In the case of small charges, the shot depth is measured to the center of the charge or to the bottom of the hole. In the case of large charges, the distances to the top and to the bottom of the column of explosives are frequently given, and may be reduced to effective shot depth to give the equivalent of a concentrated charge. (AGI, 1987)

shot drill A core drill generally employed in rotary-drilling boreholes of less than 3 in (7.62 cm) to more than 6 ft (1.8 m) in diameter in hard rock or concrete, using chilled-steel shot as a cutting medium. Also called adamantine drill; calyx drill; chilled-shot drill. See also: *core drill; shaft drilling.* Syn: *shot-boring drill.* (Long, 1960)

shot-drilled shaft Shafts of up to 5 ft (1.5 m) in diameter drilled through rock to a maximum depth of 1,200 ft (366 m) by means of a shot drill. The latter makes use of shot for cutting a circular groove in the rock being penetrated, from which solid cores are extracted. (Hammond, 1965)

shot elevation (a) Elevation of the dynamite charge in the shothole. (AGI, 1987) (b) The elevation of the explosive charge in the shot hole. Not to be confused with shothole elevation. (AGI, 1987)

shot fast Coal mined by blasting; shot off the solid.

shot feed A device to introduce chilled-steel shot, at a uniform rate and in the proper quantities, into the circulating fluid flowing downward through the rods or pipe connected to the core barrel and bit of a shot drill. (Long, 1960)

shot firer (a) A worker whose special duty is to fire shots or blasts, esp. in coal mines. A shot lighter. Also called a shooter. (b) In a coal mine, a qualified miner who tests for gas before firing explosive shots. (Pryor, 1963) (c) *blaster.*

Shot Firer A multiple-shot permissible blasting unit introduced in 1948, known as Capacitor Type Permissible Shot Firer. Weighing approx. 1 lb (0.454 kg) and about the size of an ordinary flashlight, the nonmetallic unit is equipped with a belt hook which permits its being carried under supervision of the shot firer. If desired, shots can be fired without removing the unit from the belt. It is capacitor operated, eliminating dependence on speed of operation for energy output. Capacitors supply high voltage, for a few milliseconds, more than ample to fire from 1 to 10 electric blasting caps. (Kentucky, 1952)

shot firing (a) The action of detonating or igniting a charge of explosive, usually in a drilled hole. (BS, 1964) (b) The firing of an explosive charge in a drilled hole to break the material to a suitable size for loading. (Nelson, 1965)

shot-firing blasting cord A two-conductor cable used for completing the circuit between the electric blasting cap (or caps) and the blasting unit or other source of electric energy. Syn: *blasting cord.*

shot-firing cable A pair of insulated copper conductors which lead from the exploder to the detonator wires. It may be either twin-core (both conductors contained in the one cable) or single-core (each conductor contained in a separate cable). Twin-core cables having cores of four strands, each 0.018 in (0.046 cm) in diameter (4/0.018) with a resistance of approx. 5 Ω per 100 yd (91 m), are commonly used. Actual choice of cable must depend upon conditions of use and the relevant regulations. Syn: *twin-core shot-firing cable.* (Nelson, 1965)

shot-firing cable tests The methods of testing twin-core and single-core cables are identical. Two tests are applied, one for insulation and one for continuity, and where large and important charges are being fired, as in tunnel, wellhole, and quarry blasts, tests are made before every blast. For the cable insulation test, an approved circuit tester or ohmeter is connected to one end of the cable, the two conductors at the other end being separated. No current should flow, and the resistance should be infinite. For the continuity test, the two far ends of the cable should be joined. The tester should show that the current is complete, or if an ohmeter is used, this should show the correct resistance of the shot-firing cable. (Nelson, 1965)

shot-firing circuit Extends from the exploder along the shot-firing cable, detonator wires, and finally the detonator. The shot-firing circuit is the path taken by the electric current from the exploder when a shot is detonated. (Nelson, 1965)

shot-firing curtain A steel chain mat suspended from the roof about 9 to 12 ft (2.74 to 3.66 m) from the face of an advancing tunnel to limit damage to equipment and danger from flying debris when shot firing at the face. It consists of a steel frame with chains suspended about 6 in (15 cm) apart. See also: *blasting curtain.* (Nelson, 1965)

shot firing in rounds The firing of a number of shots in a tunnel, or shaft sinking at one operation with instantaneous or delay detonators. (Nelson, 1965)

shot-firing unit *blasting unit.*

shothole (a) A hole drilled for the purpose of shot firing. (BS, 1964) (b) A hole drilled in coal, ore, or rock, usually from 3 to 9 ft (0.9 to 2.7 m) in length (underground), for breaking down the material by means of explosives. (Nelson, 1965) (c) The borehole in which an explosive is placed for blasting. See also: *blasthole.* (d) *shot point.* (e) A borehole drilled with a shot drill. (Long, 1960) (f) In seismic prospecting, a borehole in which an explosive is placed for generating seismic waves. (AGI, 1987)

shothole bridge (a) When an obstruction in the shothole makes it difficult or impossible to get the charge deeper, the hole is said to be bridged. A narrow diameter in the hole due to a resistant bed often makes it difficult to get the charge down the hole. A mechanical device that purposely bridges the hole at a shallow depth in order that the hole may be filled. (AGI, 1987) (b) An obstruction in a shothole that prevents an explosive charge from going deeper. It may be accidental or intentional. (AGI, 1987)

shothole casing Lightweight pipe, usually about 4 in (10 cm) in diameter. A typical joint of casing is 10 ft (3 m) long and has threaded connections on both ends. The primary use of casing is to prevent the shothole from caving and bridging. The lightweight casing may be considered as an expendable item. (AGI, 1987)

shothole drill (a) Generally, a rotary percussion or auger type drill for making shotholes for blasting. (b) Drills for shotholes are of two general types: (1) the rotary drill and (2) the churn drill. Rotary drill methods can be subdivided into (1) mechanical feed and (2) hydraulic feed. Both types provide a means for rotating the pipe, and both make provisions for circulating fluid down through the pipe, thus washing the cuttings away from the bit and conveying them up to the surface in the annular space between the wall of the hole and the string of drill pipe. The churn drill is similar to the larger cable-tool type. It is seldom used except in areas where underground cavities hamper the return flow of the circulating fluid used in the rotary methods. Portable drills, water jets, and airblast equipment and augers are also used in certain areas. Syn: *blasthole drill.* (AGI, 1987)

shothole elevation The elevation of the ground at the top of the shothole. (AGI, 1987)

shothole fatigue Phenomenon causing observed travel times to a fixed receiver point to increase with successive shots in the same hole. (AGI, 1987)

shothole log The drillers' record of the depths, thicknesses, and lithologic characteristics of the formations encountered in the seismic shothole. (AGI, 1987)

shothole plug A plug, usually of wood, used by seismic field parties to plug a hole upon completion of shooting. This prevents caving, protects the public from injury, and protects the exploration company from damage claims that might result from open holes. (AGI, 1987)

shot instant The elevation of the explosive charge in the shot hole. Not to be confused with shothole elevation. Syn: *shot break; time break; shot moment.* (AGI, 1987)

shot metal Metal in the form of small, spherical, or nearly spherical, pellets. It is usually made by causing molten metal to fall, dropwise, from a suitable height into a quenching medium. Also called shot. (Henderson, 1953)

shot moment *time break; shot instant.*

shot-moment line An electric line wrapped around a dynamite charge and connected to a telephone or radio circuit. The explosion breaks the circuit to record the shot instant or moment. Use is largely obsolete. (AGI, 1987)

shot off the solid A method of breaking coal from the solid seam by the use of explosives, when the seam has not previously been cut or sheared to prepare the coal for blasting. Also called shot fast. (BCI, 1947)

shot point The point at which a charge of dynamite is exploded for the generation of seismic energy. In field practice, the shot point includes the hole and its immediately surrounding area. See also: *shothole.* (AGI, 1987)

shot rock Blasted rock. (Nichols, 1976)

shot runner *fire runner.*

shot samples Samples taken for assay from molten metal by pouring a portion into water to granulate it. (Webster 3rd, 1966)

shot-sawed surface Term used to describe the surface finish of building limestone that is deeply scored by using steel-shot abrasive with gangsaws. (AIME, 1960)

shot soil Soil in which small pellets of iron oxide occur or are forming. (AGI, 1987)

shot tamper *tamper.*

shotter Bedded pebbles and sand; glacial outwash gravels. (Arkell, 1953)

shotting The production of shot by pouring molten metal in finely divided streams. Solidified spherical particles are formed during descent and are cooled in a tank of water. (ASM, 1961)

shotty gold Small granular pieces of gold resembling shot.

shoulder (a) A line formed by the intersection of the face or leading surface of a bit crown and the straight-wall side surface of the crown. (Long, 1960) (b) A ledge formed by an abrupt change in the course of a borehole. (Long, 1960) (c) A ledge or projection on drill rods, couplings, pipe, or bits formed at points where an increase or decrease in diameter occurs. (Long, 1960) (d) The side of a horizontal pipe, at the level of the center line. (Nichols, 1976) (e) A short, rounded spur projecting laterally from the side of a mountain or hill. (AGI, 1987) (f) The sloping part of a mountain or hill below the summit. (AGI, 1987)

shoulder cutting S. Staff. Cutting the sides of the upper lift of a working place in a thick coal colliery next to the rib, preparatory to breaking the coal.

shoulder stone The diamonds set in a bit at or along the line formed by the intersection of the face or leading surface of a bit crown and the straight-walled side surfaces of the bit crown or shank. Cf: *kerf stone.* (Long, 1960)

shovel (a) Any bucket-equipped machine used for digging and loading earthy or fragmented rock materials. (b) There are two types of shovels, the square-point and the round-point. These are available with either long or short handles. The round-point shovel is used for general digging since its forward edge, curved to a point, most readily penetrates moist clays and sands. The square-point shovel is used for shoveling against hard surfaces or for trimming. (Carson, 1961) (c) *power shovel.*

shovel craneman In bituminous coal mining, a maintenance mechanic who inspects, oils, greases, adjusts, and repairs machinery of a power shovel used to dig and load coal (after blasting) into cars in a strip mine. May be designated according to type of power, as electric-shovel craneman; steam-shovel craneman. Also called stripping-shovel craneman. (DOT, 1949)

shovel dozer A tractor equipped with a front-mounted bucket that can be used for pushing, digging, and truckloading. (Nichols, 1976)

shovel front In power-shovel nomenclature, the shovel front is composed of a main boom and a secondary boom known as the dipper stick, at the outer end of which is the dipper or shovel bucket. (Carson, 1961)

shovel loader A loading machine mounted on driven wheels by which it is forced into the loose rock at the tunnel face. A bucket hinged to the chassis scoops up the material, which is elevated over and discharged behind the machine. There are two types: (1) the bucket is discharged directly into a mine car behind the machine, and (2) a short conveyor, built into the loader, receives the dirt from the bucket and conveys it back into a car or conveyor. See also: *mechanical shovel; loader; loading shovel.* (Nelson, 1965)

shovel trough In a duckbill, the shovel part of the loading mechanism that is advanced into a coal pile or retracted according to the adjustment of the operating carrier. (Jones, 1949)
show (a) When the flame of a safety lamp becomes elongated or unsteady, owing to the presence of combustible gases in the air, it is said to show. (Zern, 1928) (b) The detectable presence of mineral, oil, or gas in a borehole, as determined by examination of the core or cuttings. (Long, 1960) (c) Visual particles of gold found in panning a gravel deposit. (AGI, 1987)
shower roasting Rapid oxidation-roast of finely ground sulfide ores, which are caused to fall through rising heated air. (Pryor, 1963)
showing (a) The first appearance of float, indicating the approach to an outcropping vein or seam. (Zern, 1928) (b) Can. Surface occurrence of mineral. (Hoffman, 1958)
shrinkage (a) The decrease in volume of a soil or fill material through the reduction of voids by mechanical compaction, superimposed loads, or natural consolidation. (Nelson, 1965) (b) The settling or reduction in volume of earthen fills, cement slurries, or concrete on setting. (Long, 1960) (c) In bitmaking by the powder-metal processes, the difference between the dimensions of the finished bit crown and those of the bit mold. (Long, 1960) (d) The decrease in volume of clayey soil or sediment owing to reduction of void volume, principally by drying. (AGI, 1987)
shrinkage cavity A void left in cast metals as a result of solidification shrinkage. (ASM, 1961)
shrinkage crack A crack produced in fine-grained sediment or rock by the loss of contained water during drying or dehydration; e.g., a desiccation crack or a mud crack. (AGI, 1987)
shrinkage index The numerical difference between the plastic and shrinkage limits. (ASCE, 1958)
shrinkage stope One in which only part of the severed ore is removed during stoping, the balance being temporarily available as support of workings. Used in steeply dipping lodes with strong walls. (Pryor, 1963)
shrinkage stoping A vertical, overhand mining method whereby most of the broken ore remains in the stope to form a working floor for the miners. Another reason for leaving the broken ore in the stope is to provide additional wall support until the stope is completed and ready for drawdown. Stopes are mined upward in horizontal slices. Normally, about 35% of the ore derived from the stope cuts (the swell) can be drawn off ("shrunk") as mining progresses. As a consequence, no revenues can be obtained from the ore remaining in the stope until it is finally extracted and processed for its mineral values. The method is labor intensive and cannot be readily mechanized. It is usually applied to orebodies on narrow veins or orebodies where other methods cannot be used or might be impractical or uneconomical. The method can be easily applied to ore zones as narrow as 4 ft (1.2 m), but can also be successfully used in ore widths up to 100 ft (30 m). Syn: *shrinkage with waste fill.* (SME, 1992)
shrinkage with waste fill *shrinkage stoping.*
Shropshire method *longwall.*
shroud laid rope A rope of four strands laid around a core. (Zern, 1928)
shungite A hard black amorphous material containing >98% carbon, interbedded among Precambrian schists; probably the metamorphic equivalent of bitumen, but possibly merely impure graphite. Also spelled schungite.
shunt A connection between two wires of a blasting cap that prevents building up of opposed electric potential in them. (Nichols, 1976)
shunt back A track arrangement for bringing a wagon or mine car to another track without the need for a curve, turntable, or traverser. Also called back shunt; switchback. (Nelson, 1965)
shutdown time One of the rate provisions in drilling contracts, specifying the compensation to the independent drilling contractor when drilling operations have been suspended at the request of the operator. (Williams, 1964)
shute *chute.*
shutoff valve A device by means of which the flow of gas or fluid can be made to cease—usually not with the intention of metering or regulating the flow. (Long, 1960)
shuts Scot. Movable or hinged supports for the cage at a shaft landing. Also called keps; keeps; chairs; dogs; seats.
shuttle A back-and-forth motion of a machine that continues to face in one direction. (Nichols, 1954)
shuttle car A vehicle on rubber tires or continuous treads to transfer raw materials, such as coal and ore, from loading machines in trackless areas of a mine to the main transportation system. See also: *rubber-tired haulage; trackless tunneling.* (Nelson, 1965)
shuttle-car operator In bituminous coal mining, one who drives an electrically powered truck (shuttle car) in a coal mine to transport coal from the excavation point to the conveyor belt. (DOT, 1949)
shuttle conveyor (a) A conveyor that is moved forward or backward in normal operation to vary the loading or discharge points, or both. It may be designed to move only in a straight path, or in either a straight or a curved path. See also: *movable conveyor.* (NEMA, 1956) (b) Any conveyor, such as belt, chain, pan, apron, screw, etc., in a self-contained structure movable in a defined path parallel to the flow of the material.
shuttle multispectral infrared radiometer *multispectral scanner.*
sial (a) A layer of rocks, underlying all continents, that ranges from granitic at the top to gabbroic at the base. The thickness is variously placed at 30 to 35 km. The name derives from the principal ingredients, silica and alumina. Specific gravity, about 2.7. (AGI, 1987) (b) A petrologic name for the upper layer of the Earth's crust, composed of rocks that are rich in silica and alumina; it may be the source of granitic magma. It is characteristic of the upper continental crust. Etymol: an acronym for silica + alumina. Adj: sialic. Cf: *sialma.* Syn: *sal; granitic layer.* (AGI, 1987)
siallite (a) A group name for the kaolin clay minerals and allophane. (AGI, 1987) (b) A rock composed of siallite minerals. (AGI, 1987)
sialma A mnemonic term derived from (si) for silica, (al) for alumina, and (ma) for magnesium, applied as a compositional term to a layer within the Earth that occupies a position intermediate between sial and sima. (AGI, 1987)
Siam ruby A name sometimes erroneously applied to the dark ruby spinel occurring with the rubies of Thailand.
Siberian aquamarine A blue-green beryl from Siberia, Russia.
Siberian ruby Rubellite; a pink variety of elbaite found in Siberia.
siberite A violet-red or purple variety of elbaite from Siberia. See also: *Siberian ruby.*
Sicilian amber Simetite, a variety of amber.
sickening (a) A scum that forms on the surface of mercury that retards amalgamating, caused by grease, sulfides, arsenides, etc. (Gordon, 1906) (b) The flouring of mercury. See also: *floured.* (Fay, 1920)
sicklerite An orthorhombic mineral, $Li(Mn,Fe)PO_4$ forms a series with ferrisicklerite; dark brown. Syn: *manganese sicklerite.*
sick mercury Mercury that has become contaminated so that it has neither a clean, bright surface nor a spheroidal shape when in globules. Effect produced by sulfur, oil, talc, graphite, sulfides of antimony, arsenic or bismuth, calcium earths. In this state, it cannot be used to amalgamate gold. (Pryor, 1963)
side (a) The more or less vertical face or wall of coal or goaf forming one side of an underground working place. (b) The wall of a vein. (c) Part of a rock mass bordering on a fault plane. (Schieferdecker, 1959)
side adit A side passage sometimes made when the main adit is choked with waste rock.
side arch pups Firebricks of a certain standard size. (Osborne, 1956)
side basse A transverse direction of the line of dip in strata.
sideboard (a) A board used in timbering the sides of a heading. See also: *side trees.* (Hammond, 1965) (b) A board applied to the sides of conveyors, usually of the chain type, to increase the height when coal is being loaded at the face by hand onto the conveyor. Also applied to a baffle plate used with belt conveyors. (Jones, 1949)
side-boom dredge Similar to the hopper dredge except that the discharge, instead of going into hoppers or directly back into the sea, is carried in a discharge pipe hung from a boom, a distance of from 200 to 500 ft (60 to 150 m) directly to port or starboard of the vessel, and there discharged into the atmosphere, dropping vertically from a height of about 50 ft (15 m) onto the surface of the sea. The drag heads of the dredge provide a channel, and the excavated soil is spread over a wide shoal area on either side, without the necessity of hauling it to the sea. (Carson, 1965)
sidecasting Piling spoil alongside the excavation from which it is taken. (Nichols, 1954)
side-discharge shovel A shovel loader, driven by compressed air or by electricity, for loading loose coal or rock. A bucket of capacity 21 ft^3 (0.59 m^3), hinged to the chassis, digs, lifts, and discharges the material sideways onto a scraper or belt conveyor; suitable for stable holes, pillar methods of working, and general repair work. (Nelson, 1965)
side-entrance manhole A deep manhole in which the access shaft is built to one side of the inspection chamber. (Hammond, 1965)
side-fired furnace A furnace with fuel supplied from the side. (ASTM, 1994)
sidehill cut A long excavation in a slope that has a bank on one side, and is near original grade on the other. (Nichols, 1976)
side-hitching An act in which a mule is hooked by its harness to the side instead of the front of a loaded car to give it enough momentum to slide onto a cage. (Korson, 1938)
side-laning S. Staff. The widening of an abandoned gate road, and making it part of the new side of work. (Fay, 1920)
sideline (a) The line connecting the dredger with anchorage on either side and winch on board used to steady the hull in required digging position. See also: *headline.* (Pryor, 1963) (b) A surface line of the claim along the vein. It bounds the side of the claim.
sideline agreement In a mineral claim where the apex law applies (United States), neighboring mine owners may come to a sideline

agreement to adjust or limit the law as it affects their respective properties. (Nelson, 1965)

sidelong reef An overhanging wall of rock in alluvial formation extending parallel with the course of the gutter; generally only on one side of it.

side piles The side poling boards used in driving a heading. (Stauffer, 1906)

sideplate In timbering, where both a cap and a sill are used, and the posts act as spreaders, the cap and the sill are spoken of as the sideplates. See also: *endplate; wallplate.* (Fay, 1920)

side-port furnace A furnace with ports on the sides. (ASTM, 1994)

siderite (a) A trigonal mineral, $FeCo_3$; calcite group, with Fe replaced by Mg toward magnesite and Mn toward rhodochrosite; rhombohedral crystals and cleavage; light to dark brown; in bedded deposits (blackband ores, clay ironstone); occurs in hydrothermal veins, cavities in mafic igneous rocks, pegmatites, and limestone; a source of iron. Syn: *chalybite; spathic iron; sparry iron; rhombohedral iron ore; iron spar; junckerite; junkerite; siderose; white iron ore.* See also: *iron ore.* (b) An obsolete syn. of sapphire quartz. See also: *sapphire quartz.* (AGI, 1987) (c) An obsolete term formerly applied to various minerals, such as hornblende, pharmacosiderite, and lazulite. (AGI, 1987) (d) A general name for iron meteorites, composed almost wholly of iron alloyed with nickel. Obsolete syn: aerosiderite. (AGI, 1987)

siderodot A variety of siderite containing calcium.

sideroferrite A variety of native iron occurring as grains in petrified wood.

siderogel Amorphous FeO(OH) in some bog iron ores.

sideromelane A basaltic glass from the palagonite tuffs of Sicily, Italy. (Fay, 1920)

sideronatrite An orthorhombic mineral, $Na_2Fe(SO_4)_2(OH)\cdot3H_2O$; soft; orange to straw-yellow.

siderophile (a) Said of an element concentrated in the metallic rather than in the silicate and sulfide phases of meteorites, and probably concentrated in the Earth's core relative to the mantle and crust (in Goldschmidt's scheme of element partition in the solid Earth). (AGI, 1987) (b) Said of an element with a weak affinity for oxygen and sulfur, and readily soluble in molten iron. Examples are Fe, Ni, Co, P, Pt, Au. Cf: *lithophile.* (AGI, 1987)

siderophyllite A monoclinic mineral, $KFe_2Al(Al_2Si_2)O_{10}(F,OH)_2$; mica group.

siderosa Sp. Spathic iron ore or siderite.

sideroscope An instrument for detecting small quantities of iron by the magnetic needle. (Webster 2nd, 1960)

siderose *siderite.*

siderosis Pneumoconiosis occurring in iron workers from inhalation of particles of iron. (Webster 3rd, 1966)

siderosphere *inner core.*

sidertil A triclinic mineral, $FeSO_4\cdot5H_2O$; chalcanthite group; light green.

sides A local New York and Pennsylvania term applied by bluestone quarrymen to open joints that extend east and west.

side shearing In salt mining, a vertical cut at each end of the room that permits the explosive to expand with the least resistance, thus promoting efficiency and power economy. (Kaufmann, 1960)

side shelves The shelves fastened along the sides of the entry throughout the explosion zone on which dust is placed in explosion testing in an experimental mine or gallery. (Rice, 1913-18)

side shot A reading or measurement from a survey station to locate a point that is off the traverse or that is not intended to be used as a base for the extension of the survey. It is usually made to determine the position of some object that is to be shown on a map. (AGI, 1987)

side slicing *top slicing combined with ore caving.*

side spit The emission of sparks through the sides of a burning fuse. (Fay, 1920)

side stoping *overhand stoping.*

side thrust (a) The lateral force against the borehole walls resulting from the buckling or sag in the drill rods at one or more points above the bit. (Long, 1960) (b) The lateral force developed when the area covered by the bit is not uniformly hard. (Long, 1960)

sidetracked (a) A term applied when tools or downhole drilling equipment is not recovered from a borehole because of the drilling-by or bypassing techniques used. (Long, 1960) (b) A term applied when a borehole has been deflected, so as to bypass an obstruction. (Long, 1960)

sidetracked hole Drill purposely directed away from a normal, straight course in order to bypass an obstruction or to straighten the hole; or to redirect the deeper portion by redrilling to an alternate bottom-hole location. See also: *directional drilling; deviation.* (AGI, 1987)

sidetracking The deliberate act or process of deflecting and redrilling the lower part of a borehole away from a previous course. (AGI, 1987)

side trees (a) Posts ranging from 3 to 6 in (7.6 to 15.2 cm) in thickness which support both the head trees and sideboards in a heading. (Hammond, 1965) (b) The two posts of a heading set. See also: *headtree.* (Stauffer, 1906)

sidewall core A core or rock sample extracted from the wall of a drill hole, either by shooting a retractable hollow projectile, or by mechanically removing a sample. (AGI, 1987)

sidewall coring tool An eccentric sampling device that gouges a small sample, sometimes in the form of a core, from the sidewall of a borehole. Syn: *sidewall sampler.* (Long, 1960)

sidewalls (a) Walls, usually masonry, at each end of a culvert. (Nichols, 1976) (b) *endwall.*

sidewall sampler *sidewall coring tool.*

sidewall sampling The process of securing samples of formations from the sides of the borehole anywhere in the hole that has not been cased. (AGI, 1987)

siding over A short road driven in a pillar in a headwise direction. (Zern, 1928)

Siebe-Gorman self-rescuer A self-rescuer consisting of a hermetically sealed, quick-release canister with inhalation and exhalation valves fitted to the top, a head strap, a rubber mouthpiece, a chin rest, and a nose clip. It is carried on a miner's belt and weighs only 22 oz (0.62 kg). The air enters at the perforated diaphragm in the bottom of the canister, and passes through layers of filters before it reaches the mouthpiece. The complete respirator is held in position by a head strap. See also: *self-rescuer.* (McAdam, 1955)

siegburgite A fossil resin from the brown coal near Bonn, Germany; it ranges in color from golden yellow to brownish red, and is partly soluble in alcohol and ether. (Fay, 1920)

siege The floor of a pot furnace, often called bench.

siegenite An isometric mineral, $(Ni,Co)_3S_4$; linnaeite group.

Siemens and Halske process A metallurgical process for the recovery of copper. Copper sulfides are dissolved by solutions of ferric sulfate containing free sulfuric acid, and the solution is then electrolyzed in a tank having a diaphragm. Copper is deposited, and ferric sulfate is regenerated. (Liddell, 1918)

Siemens direct process A process for making wrought iron directly from iron ore, without the previous production of pig iron. (Standard, 1964)

Siemens furnace A reverberatory furnace, heated by gas, with the aid of regenerators. (Fay, 1920)

siemensite A highly refractory material, produced by the fusion of chromite, bauxite, and magnesite, in an open electric arc furnace. (Osborne, 1956)

Siemens-Martin process The production of steel in a reverberatory furnace by oxidation of the impurities by oxides added (either the rust on scrap, mill scale, or pure ores). It may be conducted on either acid or basic lining. See also: *open-hearth process.* (Liddell, 1918; Fay, 1920)

Siemens producer A furnace used for the manufacture of producer gas. (Fay, 1920)

Siemens-Silesian furnace A Silesian zinc-distillation furnace employing the Siemens system of heat recuperation. (Fay, 1920)

sienna (a) A brownish orange-yellow clay colored by iron and manganese oxides; used as a pigment. (b) *mineral paint.* Cf: *umber.* (AGI, 1987)

sierra (a) A high range of hills or mountains, esp. one having jagged or irregular peaks that when projected against the sky resemble the teeth of a saw; e.g., the Sierra Nevada in California. The term is often used in the plural, and is common in the Southwestern United States and in Latin America. Syn: *serra.* (AGI, 1987) (b) A mountainous region in a sierra. Etymol: Spanish, saw, from Latin serra, saw. (AGI, 1987)

Sierra Leone A diamond from the Sierra Leone District in Africa. (Long, 1960)

sieve (a) A laboratory vessel, the bottom of which is a woven-wire screen, used to separate soil or sedimentary material according to the size of its particles; it is usually made of brass, with the wire-mesh cloth having regularly spaced square holes of uniform diameter. Cf: *screen.* (AGI, 1987) (b) The screen or grating fixed in a stamp box. (c) Vessel, the bottom of which is porous, with apertures of defined size and shape, allowing contents to be retained as oversize or sieved through as undersize. The chief sieve systems used in laboratory work are rings 8 in (20.3 cm) in diameter with woven wire cloths so specified. (Pryor, 1963) (d) This term is generally reserved for testing equipment; the corresponding industrial equipment is generally called a screen. There are several standard series of test sieves; those most frequently met with in the ceramic industry are British standard sieves (conforming with B.S. 410), United States standard sieves (conforming with National Institute of Standards and Technology LC-584 or ASTM-E11), French standard sieves (AFNOR NF 11-501), and German standard sieves (DIN 4188). In Tyler sieves, the ratio between the mesh sizes of successive sieves in the series is $\sqrt{2}$; thus, the areas of the openings of each sieve are double those of the next finer sieve. (Dodd, 1964)

sieve analysis Determination of the particle-size distribution in a soil, sediment, or rock by measuring the percentage of the particles that will

pass through standard sieves of various sizes. Syn: *sieve classification.* (AGI, 1987)
sieve bend Stationary screen with close-spaced wedge wire bars across wet pulp feed, set around arc of circle. (Pryor, 1963)
sieve classification The separation of powder into particle size ranges by the use of a series of graded sieves. Syn: *sieve analysis.* (ASTM, 1994)
sieve fraction In powder metallurgy, that portion of a powder sample that passes through a standard sieve of specified number and is retained by some finer sieve of specified number. (ASM, 1961)
sieve mesh (a) Standard opening in sieve or screen, defined by four boundary wires (warp and woof). The laboratory mesh is square and is defined by the shortest distance between two parallel wires as regards aperture (quoted in micrometers or millimeters), and by the number of parallel wires per linear inch as regards mesh. Sixty mesh equals 60 wires/in (152 wires/cm). (Pryor, 1963) (b) The length of the side of a hole in a sieve. See also: *mesh.* (Fay, 1920)
sieve scale Term applied to the list of screen apertures, taken in order from the coarsest to the finest. (Pit and Quarry, 1960)
sieve shakers Mechanized devices on which a nest of laboratory sieves can be shaken or electrically vibrated during the size analysis of sands. (Pryor, 1963)
sieve sizes Sieves are standardized in British Standard 410 and sieve size diamond powders in British Standard 1987. (Osborne, 1956)
sieve texture *poikiloblastic.*
sieving (a) Grading in accordance with particle size and shape by means of sieves or screens. (Pryor, 1963) (b) The operation of shaking loose materials in a sieve so that the finer particles pass through the mesh bottom. By using a number of sieves with different meshes, the particles can be graded according to size. Syn: *sifting.* (CTD, 1958)
sifting *sieving.*
sight (a) A bob or weighted string hung from an established point in the roof of a room or entry, to give direction to the miners driving the entry or room. (Fay, 1920) (b) A bearing or angle taken with a compass or transit when making a survey. (Fay, 1920) (c) Any established point of a survey. (Fay, 1920)
sight distance The distance from which an object at eye level remains visible to an observer. (Hammond, 1965)
sighting hub A stake or mark used by a driller as a means of setting and orienting a drill so that the borehole can be drilled to follow a predetermined directional course. (Long, 1960)
sight line Established compass or transit course for alignment of working places, usually marked on the roof. (BCI, 1947)
sights Bobs or weighted strings hung from two or more established points to give direction to miners driving a chamber or gangway. (Hudson, 1932)
sigloite A triclinic mineral, $FeAl_2(PO_4)_2(OH)_3 \cdot 7H_2O$; paravauxite group.
sigma heat *total heat.*
sigma recording methanometer *butane flame methanometer.*
signal code *hoist signal code.*
signaling system The arrangement in use for transmitting signals to stop or start conveyors, rope haulages, locomotives, winders, etc. See also: *face signaling; loudspeaker face telephone.* (Nelson, 1965)
signal system *mine fan signal system; hoist signal system.*
significant anomaly An anomaly that is related to ore and that can be used as a guide in exploration. See also: *geochemical anomaly.* (Hawkes, 1962)
silcrete (a) A term suggested by Lamplugh (1902) for a conglomerate consisting of surficial sand and gravel cemented into a hard mass by silica. Examples occur in post-Cretaceous strata of the United States. (AGI, 1987) (b) A siliceous duricrust. Etymol: "sil"iceous + con"crete." Cf: *calcrete; ferricrete.* (AGI, 1987)
silent chain A roller-type chain in which the sprockets are engaged by projections on the link side bars. (Nichols, 1976)
silex (a) The French term for flint. (AGI, 1987) (b) Silica; esp. quartz, such as a pure or finely ground form for use as a filler. (AGI, 1987) (c) An old term formerly applied to a hard, dense rock, such as basalt or compact limestone. Etymol: Latin, hard stone, flint, quartz. The term was used by Pliny for quartz. (AGI, 1987)
silexite (a) An igneous rock composed essentially of primary quartz (60% to 100%), e.g., a quartz dike, segregation mass, or inclusion inside or outside its parent rock. (AGI, 1987) (b) The French term for chert; specif. chert occurring in calcareous beds. See also: *chert.* (AGI, 1987)
silica The chemically resistant dioxide of silicon, SiO_2; occurs naturally as five crystalline polymorphs: trigonal and hexagonal quartz, orthorhombic and hexagonal tridymite, tetragonal and isometric cristobalite, monoclinic coesite, and tetragonal stishovite. Also occurs as cryptocrystalline chalcedony, hydrated opal, the glass lechatelierite, skeletal material in diatoms and other living organisms, and fossil skeletal material in diatomite and other siliceous accumulations. Also occurs with other chemical elements in silicate minerals.
silica brick (a) Refractory bricks used to line roofs of furnaces, where there is no contact with basic molten material. (Pryor, 1963) (b) Silica cemented with a binding agent, for example, slurried lime. (Pryor, 1963)
silica-firebrick molder One who forms silica brick for use in lining furnaces and ovens of various kinds. (DOT, 1949)
silica refractories Refractories made from quartzite, bonded by lime, and consisting essentially of silica, usually with about 2% of lime, and small quantities of iron oxide, alumina, and alkalies. (Henderson, 1953)
silica rock An industrial term for certain sandstones and quartzites that contain at least 95% silica (quartz). It is used as a raw material of glass and other products. Cf: *silica sand.* (AGI, 1987)
silica sand An industrial term for a sand or an easily disaggregated sandstone that has a very high percentage of silica (quartz). It is a source of silicon and a raw material of glass and other industrial products. Cf: *silica rock.* (AGI, 1987)
silicate (a) A compound whose crystal structure contains SiO_4 tetrahedra, either isolaed or joined through one or more of the oxygen atoms to form groups, chains, sheets, or three-dimensional structures with metallic elements. Silicates were once classified according to hypothetical oxyacids of silicon (see metasilicate and orthosilicate) but are now calssified according to crystal structure (see nesosilicate, sorosilicate, cyclosilicate, inosilicate, phyllosilicate, tectosilicate). (AGI, 1987) (b) A term used in the Joplin district, Missouri, for zinc carbonate.
silicate brick Usually refers to Forsterite brick. Strictly, most bricks are silicate bricks of one kind or other.
silicate degree In the metallurgical nomenclature of slags, the ratio of the weight of oxygen in the acid to the weight of oxygen in the base. (Newton, 1959)
silicate minerals Minerals with crystal structure containing SiO_4 tetrahedron arranged as (1) isolated units, (2) single or double chains, (3) sheets, or (4) three-dimensional networks. (Leet, 1958)
silication The process of converting into or replacing by silicates, esp. in the formation of skarn minerals in carbonate rocks. Cf: *silicification.* Adj: silicated. (AGI, 1987)
silicatization process A special method of sealing off water, for example, reducing its inflow into shafts, by the injection of calcium silicate under pressure. It is sometimes used to reduce the leakage of water through defective lengths of tubing in a shaft. The calcium silicate is highly impervious on solidification, behind the leaking tubbing. (Nelson, 1965)
silicatosis A disease of the lungs thought to be caused by silicates.
siliceous (a) Of, relating to, or derived from silica; containing or resembling silica or a silicate; silicic. Also spelled silicious. (Webster 3rd, 1966) (b) Said of a rock containing free silica or, in the case of volcanic glass, silica in the norm.
siliceous dust Dust arising from the crushing or other dry working of sand, sandstone, trap, granite, and other igneous rocks is included in this class. Siliceous dusts are not soluble in body fluids, and when introduced into the respiratory tract in the form of particles of certain sizes and in sufficiently high concentration, they produce nodular growths that often result in a form of pneumonoconiosis that has been known as silicosis or "stone cutters" consumption. (Pit and Quarry, 1960)
siliceous earth A general term including both diatomaceous earth (diatomite) and radiolarian earth (radiolarite).
siliceous fire clay A fire clay composed mainly of fine white clay mixed with clean, sharp sand, found in pockets. (Nelson, 1965)
siliceous materials Materials that consist mainly of SiO_2 and must be low in metallic oxides and alkalies. (Newton, 1959)
siliceous oozes These are pelagic deposits that contain a large percentage of siliceous skeletal materials produced by planktonic plants and animals. The siliceous oozes are subdivided into two types on the basis of the predominance of the forms represented, namely (1) diatom ooze, containing large amounts of diatom frustules, therefore, produced by plankton plants, and (2) radiolarian ooze, containing large porportions of radiolarian skeletons formed by these plankton animals. (Hunt, 1965)
siliceous ore Another name for gold-quartz ores. (Newton, 1959)
siliceous rocks Generally, rocks high in silica.
siliceous shale A hard, fine-grained rock of shaly structure generally believed to be shale altered by silicification. Syn: *phthanite.* See also: *porcellanite.*
siliceous sinter The lightweight porous opaline variety of silica, white or nearly white, deposited as an incrustation by precipitation from the waters of geysers and hot springs. The term has been applied loosely to any deposit made by a geyser or hot spring. Syn: *sinter; pearl sinter; geyserite; fiorite.* (AGI, 1987)
silicic (a) In petrology, containing silica in dominant amount. (b) In chemistry, containing silicon as the acid-forming element. (c) Said of a silica-rich igneous rock or magma. Although there is no firm agreement among petrologists, the amount of silica is usually said to constitute at least 65% or two-thirds of the rock. In addition to the combined silica in feldspars, silicic rocks generally contain free silica in the form

of quartz. Granite and rhyolite are typical silicic rocks. The synonymous terms "acid" and "acidic" are used almost as frequently as silicic. Syn: *acidic; intermediate; persilicic.* Cf: *basic; ultrabasic.* (AGI, 1987)

silicification (a) The introduction of, or replacement by, silica, generally resulting in the formation of fine-grained quartz, chalcedony, or opal, which may fill pores and replace existing minerals. Cf: *silication.* Adj: silicified. Syn: *silification.* (AGI, 1987) (b) A process of fossilization whereby the original organic components of an organism are replaced by silica, as quartz, chalcedony, or opal. (AGI, 1987)

silicified Adj. of silicification.

silicified wood A material formed by replacement of wood by silica in such manner that the original form and structure of the wood is preserved. The silica is generally in the form of opal or chalcedony. Syn: *petrified wood; woodstone; agatized wood; fossilized wood; opalized wood; shinarump.*

silicinate Said of the silica cement of a sedimentary rock. (AGI, 1987)

silicious ore *natural ore.*

silicon A nonmetallic element that is the second most abundant on Earth, being exceeded only by oxygen. Symbol, Si. Silicon is not found free in nature, but occurs as the oxide and silicate. Sand, quartz, rock crystal, amethyst, agate, flint, jasper, and opal are some of the forms in which the oxide appears. Hornblende, orthoclase, kaolin, and biotite are a few of the numerous silicate minerals. Used in the electronics and space-age industries; used to make concrete, brick, and glass. Miners often develop a serious lung disease, silicosis, from breathing large quantities of the dust. (Handbook of Chem. & Phys., 3)

silicon alloys Silicon bronze is a noncorroding alloy with copper and tin. Silicon copper (70% to 80% copper and 20% to 30% silicon) is an alloy added to molten copper or brass to remove oxygen. Silicon iron is a grain-improved iron, corrosion resistant. See also: *ferrosilicon.* (Pryor, 1963)

silicon borides Two compounds have been reported: SiB_4, oxidation resistant to 1,370 °C; SiB_6, melting point 1,950 °C. A special refractory has been made by reacting silicon and boron in air, the product containing SiB_4 and Si in a borosilicate matrix; it is stable in air to at least 1,550 °C and has good thermal shock resistance. Syn: *boron silicides.* (Dodd, 1964)

silicon copper A rich copper alloy added to molten copper in order to secure clean, solid castings free from blowholes, swellings, etc. (Crispin, 1964)

silicon dioxide *silica.*

siliconize To unite or cause to unite with silicon, as in the combination of iron with silicon in certain metallurgical processes. (Standard, 1964)

silicon-oxygen tetrahedron A fundamental structural unit of silicate minerals formed by four oxygen ions surrounding one silicon ion such that lines connecting the four oxygen nuclei show the outline of a geometric tetrahedron. It is commonly written SiO_4 with the electronic charge of minus 4 for the unit assumed. See also: *silicate.*

silicon spiegel (a) A spiegeleisen containing 15% to 20% manganese and 8% to 15% silicon used in making certain special steels. (Webster 3rd, 1966) (b) A form of pig iron. (Henderson, 1953)

silicon steel A variety of steel containing up to 5% silicon. It is very hard, but is brittle and difficult to work.

silicosis (a) Lung disease caused chiefly by inhaling rock dust from air drills. (Nichols, 1976) (b) A condition of massive fibrosis of the lungs marked by shortness of breath and resulting from prolonged inhalation of silica dusts by those—such as stonecutters, asbestos workers, miners—regularly exposed to such dusts. See also: *pneumoconiosis; simple silicosis; mining disease.* (Webster 3rd, 1966)

silicotuberculosis Complication of tuberculosis by silica. (Hartman, 1961)

silification *silicification.*

silk Microscopically small, needlelike inclusions of rutile crystals in a natural gem, such as ruby, sapphire, or garnet, from which subsurface reflections produce a whitish sheen resembling that of silk fabric.

silklay A finely ground-plastic clay of high refractoriness used as a bond for molding sands. (Osborne, 1956)

silky luster (a) The luster of silk, peculiar to minerals having a fibrous structure. The fibrous form of gypsum and satin spar are good examples of silky luster. (Nelson, 1965) (b) A type of mineral luster characteristic of some fibrous minerals, such as chrysotile and gypsum.

sill (a) Applied in mining to flat-bedded strata of sandstone or similar hard rocks. (AGI, 1987) (b) A concordant sheet of igneous rock lying nearly horizontal. A sill may become a dike or vice versa. (Nelson, 1965) (c) The floor of a gallery or passage in a mine. (Standard, 1964) (d) Fireclay, used for making slate or sill pencils, Coal Measures. Cf: *stone sill.* (Arkell, 1953) (e) See also: *floor sill.* (f) A submarine ridge or rise at a relatively shallow depth, separating a basin from another basin or from an adjacent sea and causing the basin to be partly closed, e.g., in the Straits of Gibraltar. (AGI, 1987) (g) A ridge of bedrock or earth material at a shallow depth near the mouth of a fjord, separating the deep water of the fjord from the deep ocean water outside. Syn: *threshold.* (AGI, 1987) (h) The upper limit of any variogram model that has such a limit, i.e., that tends to "level off" at large distances. The spherical, gaussian, exponential, and nugget models have sills. For the linear model, "sill/range" is used merely to define the slope.

sill depth Greatest depth at which there is free, horizontal communication between two ocean basins. (Hy, 1965)

sillenite An isometric mineral, $Bi_{12}SiO_{20}$; forms greenish, earthy or waxy masses; at Durango, Mexico.

sillimanite (a) An orthorhombic mineral, Al_2SiO_5; trimorphous with kyanite and andalusite; forms long, slender, needlelike crystals commonly in wisplike or fibrous aggregates (fibrolite) in gneisses and schists at granulite grade; also occurs in alluvial deposits. It forms at the highest temperatures and pressures of a regionally metamorphosed sequence and is characteristic of the innermost zone of contact-metamorphosed sediments. Cf: *mullite; fibrolite.* (b) Loosely used for the aluminum silicate minerals sillimanite, kyanite, andalusite, dumortierite, topaz, and mullite. (c) A high heat-resisting ceramic material containing a maximum amount of mullite, developed from the alteration of andalusite during firing to 1,550 °C and used for special porcelain shapes, furnace patches, and refractories.

sillimanite schist A schist containing an appreciable amount of sillimanite (fibrolite). (Sinkankas, 1959)

sillite Gumbel's name for a rock from the Bavarian Alps, variously referred to by others as gabbro, diabase, mica syenite, and mica diorite.

sills Strong timbers laid horizontally to support posts or other tunnel timbers. (Stauffer, 1906)

silo A tall tower, usually cylindrical and of reinforced concrete construction, in which grain, cement, coal, or similar bulk material is stored. (Hammond, 1965)

silt (a) In anthracite terminology, the accumulation of waste fine coal, bone, and slate settled out of breaker water. It is made up of particles ranging in size from 3/32-in (2.4-mm) round-opening to the finest slime. The material is also called sludge, culm, fines, slush, and mud. It is the partly dewatered solids content of what has been defined as slurry. See also: *rock flour.* (Mitchell, 1950) (b) In bituminous coal terminology: Syn: *coal sludge.* (c) Material passing the No. 200 U.S. standard sieve that is nonplastic or very slightly plastic and that exhibits little or no strength when air-dried. (ASCE, 1958) (d) Breaker waste composed of water, coal, slate, pyrite, and clay. (Korson, 1938)

siltation *silting.*

silt box A loose iron box fitted in the bottom of a gulley for collecting deposited silt. It can be removed periodically for emptying and flushing. (Hammond, 1965)

silt displacement A system of using a shield for driving a tunnel in silts that are nearly fluid. (Hammond, 1965)

silting (a) The deposition or accumulation of silt that is suspended throughout a body of standing water or in some considerable portion of it; esp. the choking, filling, or covering with stream-deposited silt behind a dam or other place of retarded flow, or in a reservoir. The term often includes sedimentary particles ranging in size from colloidal clay to sand. Syn: *siltation.* (AGI, 1987) (b) Sedimentation in water that results in the deposition of somewhat fine material, which is suspended in the entire body of water or in some considerable portion of it. (AGI, 1987) (c) Filling with soil or mud deposited by water. See also: *hydraulic mine filling.* (Nichols, 1976)

silting up The filling, or partial filling, with silt, as of a reservoir that receives fine-grained sediment brought in by streams and surface runoff. The term has been used synonymously with sedimentation without regard to any specific grain size. (AGI, 1987)

siltite *siltstone.*

siltstone An indurated silt having the texture and composition of shale but lacking its fine lamination or fissility; a massive mudstone in which the silt predominates over clay; a nonfissile silt shale. It tends to be flaggy, containing hard, durable, generally thin layers, and often showing various primary current structures. Syn: *siltite.* See also: *mudstone; claystone.* (AGI, 1987)

silt trap A settling hole or basin that prevents water-borne soil from entering a pond or drainage system. (Nichols, 1976)

Silurian A period of the Paleozoic, thought to have covered the span of time between 440 and 400 million years ago; also, the corresponding system of rocks. The Silurian follows the Ordovician and precedes the Devonian; in the older literature, it was sometimes considered to include the Ordovician. It is named after the Silures, a Celtic tribe. (AGI, 1987)

silver (a) A white metallic element that is very ductile and malleable. Symbol, Ag. Occurs native and in ores such as argentite and horn silver; lead, lead-zinc, copper, gold, and copper-nickel ores are its principal sources. Used for jewelry, photography, dental alloys, and coinage. (Handbook of Chem. & Phys., 3) (b) An isometric or hexagonal mineral, Ag, native silver; commonly alloyed with Hg or Au; soft; metallic; sp gr, 10.5; in oxidized zones of hydrothermal deposits.

silver amalgam (a) A solid solution of mercury and silver crystallizing in the cubic system. The percentage of silver is usually about 26%, but in the variety arquerite it reaches 86%. It is of rare occurrence, and is found scattered either in mercury or silver deposits. (CMD, 1948) (b) A naturally occurring "amalgam," an isometric mineral, (Ag,Hg); rarely found scattered either in mercury or silver deposits. See also: *amalgam.*

silver bonanza A rich silver mine. (Mathews, 1951)

silver-copper glance *stromeyerite.*

silver freighter A wagoner who hauls silver ore. (Mathews, 1951)

silver glance The native silver sulfide, argentite. (Fay, 1920)

silver halides Silver bromide, AgBr; silver iodide, AgI; silver chloride, AgCl; and silver fluoride, AgF. The bromide and chloride are sensitive to light and are of basic importance in photography. (CMD, 1948)

silvering (a) A plating or covering of silver or an imitation of it, as applied to any surface; as, the silvering on the back of a mirror. (Standard, 1964) (b) The art or process of coating surfaces with silver. (Standard, 1964)

silver lead Lead containing silver. (Standard, 1964)

silver lead ore The name given to galena containing silver. When 1% or more of silver is present, it becomes a valuable ore of silver. Syn: *argentiferous galena.* (CMD, 1948)

silver minerals Occurs native, alloyed with gold as electrum, as sulfide argentite, Ag_2S, proustite, pyrargyrite, and horn silver, AgCl, or cerargyrite. Main source is argentiferous ores of lead, zinc, and copper where it is extracted as a byproduct. Bulk of production is used for coinage, electrical alloys, photographic chemicals, and the arts. (Pryor, 1963)

silver sand (a) A sharp, fine sand of a silvery appearance used for grinding lithographic stones, etc. (b) Specially pure silica. (Pryor, 1963)

sima (a) The basic outer shell of the Earth; under the continents it underlies the sial, but under the oceans it directly underlies the water. Originally, the sima was considered basaltic in composition with a specific gravity of about 3.0. In recent years, it has been suggested that the sima is peridotitic in composition with a specific gravity of about 3.3. First used in its present form and spelling by Suess. (AGI, 1987) (b) A petrologic name for the lower layer of the Earth's crust, composed of rocks that are rich in silica and magnesia. It is equivalent to the oceanic crust and to the lower portion of the continental crust, underlying the sial. Etymol: an acronym for silica + magnesia. Adj: simatic. Cf: *sial; sialma.* Syn: *intermediate layer; basaltic layer.* (AGI, 1987)

Simbal breathing apparatus An improved liquid oxygen breathing apparatus, weighing 33 lb (15 kg) and approved for use in British mines. Air is fed to the wearer at a temperature of 50 °F (10 °C) rising to 70 °F (21.1 °C) in just over 1/2 h and is still only 80 °F (26.7 °C) after 2-1/2 h. (Nelson, 1965)

simetite A deep-red to light orange-yellow variety of amber having a high content of sulfur and oxygen and a low content of succinic acid; occurs in the waters off Sicily.

similar fold A fold in which the orthogonal thickness of the folded strata is greater in the hinge than in the limbs, but the distance between any two folded surfaces is constant when measured parallel to the axial surface. Thus, if the shape of one bed is that of a sine curve, all the beds show the same shape. Similar folds show thinning on the limbs and thickening at the axes. Cf: *parallel fold; supratenuous fold.* Syn: *concentric fold.* (AGI, 1987)

similor A golden-colored variety of brass. Also called Mannheim gold; Prince Rupert's metal.

simonyite *blödite.*

simple beam A simply supported beam. (Hammond, 1965)

simple bending The bending of a beam that is freely supported, having no fixed end. (Hammond, 1965)

simple explosives These explosives consist of one simple chemical compound. The explosive heat is liberated with the breaking down of the molecules and the atoms recombining to form water, carbon dioxide, nitrogen and other gases, and possibly solid substances such as carbon. To this group belong explosives in the proper sense of the word, such as nitroglycerin, nitroglycol, nitrocellulose, trotyl, and cyclonite (RDX). Also includes "molecular explosives.". (Fraenkel, 1953)

simple kriging A variety of kriging that assumes that local means are relatively constant and equal to the population mean, which is well known. The population mean is used as a factor in each local estimate along with the samples in the local neighborhood. This is not usually the most appropriate method for environmental situations.

simple mineral A mineral found in nature, as distinguished from rocks, which, in the scientific sense, are mixtures of minerals. Calcite and hematite are simple minerals, while granite is a mixture of three simple minerals—quartz, feldspar, and mica. (Standard, 1964; Fay, 1920)

simple ore Ore that yields a single metal. Cf: *complex ore.*

simple pneumoconiosis Pneumoconiosis of the lungs that can be related to the amount (and possibly the nature) of the dust breathed by miners over the years. See also: *complicated pneumoconiosis.* (Nelson, 1965)

simple silicosis Silicosis that is not complicated by tuberculosis; a condition that may remain almost stationary for many years. See also: *mining disease.* (Nelson, 1965)

simple split seam A coal seam that has separated into two layers of coal some distance apart vertically. See also: *multiple splitting.* (Nelson, 1965)

Simplex pump A reciprocating single- or double-action piston pump having one water cylinder. (Long, 1960)

simplotite A monoclinic mineral, $CaV_4O_9{\cdot}5H_2O$; dark green.

simpsonite A trigonal mineral, $Al_4(Ta,Nb)_3(O,OH,F)_{14}$; yellow-brown.

Simpson's rule A rule for estimating the area of an irregular figure after dividing it into an even number of parallel strips of equal width. See also: *trapezoidal rule.* (Hammond, 1965)

simulated insert bit A core bit in the face of which are deeply cut, closely spaced waterways to produce the superficial appearance of an insert-type bit. Also called Thedford crown bit. (Long, 1960)

simulated workplace protection factor A surrogate measure of the workplace protection provided by a respirator. (NIOSH, 1987)

simultaneous filling Filling in which the mined-out area or room is filled immediately after mining out only a small part of the deposit. (Stoces, 1954)

simultaneous shot firing The concurrent firing of a round of shots using instantaneous detonators. (BS, 1964)

sincosite A tetragonal mineral, $Ca(VO)_2(PO_4)_2{\cdot}5H_2O$; a vanadium analog of the meta-autunite group; green; a source of vanadium.

sing A hissing noise often made by gas and water when a seam of coal is cut into. (Fay, 1920)

singing (a) Resonance phenomenon that is frequently observed on marine seismograms. (Schieferdecker, 1959) (b) A seismic resonance phenomenon that is produced by short-path multiples in a water layer. Syn: *reverberation; ringing.* (AGI, 1987)

single-acting ram *one-way ram.*

single-action pump A pump valved so as to discharge liquid at only one end of the water cylinder. Cf: *double-action pump.* (Long, 1960)

single-approach pit bottom A pit-bottom layout at a mine where development branches off on one side only. The empty cars are returned from the opposite side by a loop, shunt back, turntable, or traverse. See also: *loop-type pit bottom.* (Nelson, 1965)

single-bench quarrying Quarrying a rock ledge as a single bench the full height of the quarry face.

single block A block with one pulley or sheave. (Long, 1960)

single consignment A quantity of coal that is to be sampled to a specified accuracy. It is used in contradistinction to the sampling of a coal received regularly at a given point. See also: *isolated consignment.* (Nelson, 1965)

single-core shot-firing cable *shot-firing cable.*

single crystal A crystal, usually grown artificially, in which all parts have the same crystallographic orientation. (McGraw-Hill, 1994)

single-cut sprocket For double-pitch roller chains, a sprocket having one set of effective teeth. (Jackson, 1955)

single-deck screen A screen having one screening surface, not necessarily limited to one size or shape of aperture. (BS, 1962)

single entry A system of opening a mine by driving a single entry only, in place of a pair of entries. The air current returns along the face of the rooms, which must be kept open.

single-entry room-and-pillar mining *room-and-pillar.*

single-entry zone test A test in which coal dust is placed only in a single entry. (Rice, 1913-18)

single-grained structure An arrangement composed of individual soil particles; characteristic structure of coarse-grained soils. See also: *soil structure; flocculent structure; honeycomb structure.* (ASCE, 1958)

single-hand drilling Rock drilling by hand; e.g., in narrow reefs. A drill steel held in the left hand is struck blows with a 4-lb (1.8-kg) hammer, the drill being turned between the blows. The drilling is very slow and laborious. (Nelson, 1965)

single-inlet fan A centrifugal fan in which air enters the impeller at one side only. (BS, 1963)

single-intake fan A ventilating fan that takes or receives its air from one side only.

single jack (a) A lightweight hammer, usually 4 lb (1.8 kg) or less. When used in hand drilling holes in rock, the hammer is held in one hand and the drill is held in the other. (Long, 1960) (b) Sometimes incorrectly used to designate a sinker drill. (Long, 1960) (c) A drill column having a single jackscrew in the bottom end. See also: *drill column.* (Long, 1960)

single-layer bit *surface-set bit.*

single opening Any underground opening separated from a free surface by a distance greater than three times the size of the opening in the direction of the free face.

single outlet For safety reasons, mines with only a single outlet (shaft) are subjected to restrictions in the numbers of mine workers employed at a time. (Beerman, 1963)

single packing The conventional method of strip packing on a longwall face, in which the widest pack is along the roadside. A single packing system varies but may have a 10-yd (9.1-m) roadside pack, then 7 yd (6.4 m) of waste, followed by 4-yd (3.7-m) packs and 5 yd (4.6 m) of waste repeated across the entire face. See also: *double packing; strip packing.* (Nelson, 1965)
single-phase circuit A two-wire circuit using alternating current. (Kentucky, 1952)
single-pulley-drive conveyor A conveyor in which power is transmitted to the belt by one pulley only. (NEMA, 1961)
single refraction Light refraction in an isotropic crystal or amorphous substance according to Snell's law, as opposed to the birefringence of an anisotropic crystal.
single-road stall S. Wales. A system of working coal by narrow stalls. (Fay, 1920)
single-roll breaker A coal-crushing machine in which the roll teeth crack downward on the lump and the roll itself compresses the coal against the breaker plates. Teeth of two or more designs are used generally on the same roll, some for the slugging, cracking, or blow action and others for a pulling and splitting force. The breakers are not easy to stall by choking, since they will pass a heavy overload, partly because of the action of a relief mechanism, with which they are all equipped. (Mitchell, 1950)
single-roll crusher A crushing machine consisting of a rotating cylinder with a corrugated or toothed outer surface that crushes material by pinching it between the teeth and stationary breaking bars. Cf: *double-roll crusher.* (ACSG, 1963)
single-rope friction pulley *Koepe sheave.*
single-rope haulage A system of underground haulage in which a single rope is used, the empty trip running in by gravity. Engine-plane haulage.
single-round nose The cross-sectional view of the cutting-face portion of a core bit when the profile is an arc having a radius equal to or greater than the wall thickness. See also: *profile.* Cf: *double-round nose.* (Long, 1960)
single-round-nose bit *single-round nose.*
single-row blasting The drilling, charging, and firing of a single row of vertical holes along a quarry or opencast face. The holes may be fired simultaneously or by delay detonators to give a peeling action starting at one end of the face. See also: *multiple-row blasting.* (Nelson, 1965)
single shot A charge in one drill hole only fired at one time, as contrasted with a multiple shot where charges in a number of holes are fired at one time.
single-shot blasting unit A single-shot blasting unit is a unit designed for firing only one explosive charge at a time. Syn: *blasting unit.*
single-shot exploders Exploders of the magneto type that are operated by the twist action given by a half-turn of the firing key. A magneto exploder consists essentially of a small armature that can be rotated between the poles of a set of permanent magnets. The armature is rotated by means of toothed gear wheels actuated by the movement of the firing key. The electric circuit between the exploder and the detonator is completed by means of an automatic internal switch operating at the end of the stroke, or contact may be made by means of a pushbutton. (McAdam, 1958)
single-shot instrument A borehole surveying instrument that records only one measurement of the bearing and of the inclination of a hole on a single trip into the borehole. (Long, 1960)
single-shot survey A borehole survey made with a single-shot instrument. (Long, 1960)
single sling A sling that has a single hook at one end and an iron or steel ring at the other. See also: *two-leg sling.* (Hammond, 1965)
single-speed floating control system In flotation, floating control in which the manipulated variable changes at a fixed rate, increasing or decreasing depending on the sign of the actuating signal. (Fuerstenau, 1962)
single-spot method One of three recognized methods of determining the average velocity of airflow along a mine roadway by anemometer. A velocity reading is taken at the center of the airway and the result is multiplied by the center constant, whose value ranges between 0.8 and 0.9, to give the average velocity of flow. Alternatively, the reading is taken along a midway line at a position some one-seventh to one-third of the width of the airway, measured from the side. At this position, the mean velocity is obtained, so that no adjustment is required. Cf: *division method; traversing method.* (Roberts, 1960)
single-stage pump A centrifugal pump with a lift of 100 ft (30.5 m) per stage. (Hammond, 1965)
single-stall working *room-and-pillar.*
single stamp mill (a) A mill possessing batteries of one stamp each, like the Nissen, instead of the usual five. (b) A mill possessing only one stamp, after the Lake Superior fashion, where one big stamp does the work of 150 ordinary gravity stamps. (Fay, 1920)
single-toggle jaw crusher A jaw crusher with one jaw fixed, the other jaw oscillating through an eccentric mounted near its top. This type of jaw crusher has a relatively high output, and the product is of fairly uniform size. (Dodd, 1964)
single-unit panel A longwall conveyor face from about 80 to 200 yd (73 to 183 m) long developed between two gate roads, one serving as an intake airway and usually for coal haulage, while the other acts as a return airway and for bringing in supplies to the face. See also: *double-unit conveyor.* (Nelson, 1965)
single vein A single ore deposit of identical origin, age and character throughout. A single small vein is weighed and measured by the same law and entitled to the same consideration as the mother lode, and very often is far more valuable in the eyes of the miner. (Ricketts, 1943)
singulosilicate A slag with a silicate degree of 1. (Newton, 1959; Newton, 1938)
sinhalite An orthorhombic mineral, $MgAlBO_4$; structurally analogous to olivines; yellow.
sinistral fault *left-lateral fault.*
sink (a) To excavate or drive a shaft or slope. (b) The depression in a shaft made by a center blast. (Standard, 1964) (c) A water lodgment. See also: *sump.* (Nelson, 1965) (d) To put standpipe or casing down through overburden by rotation or by driving, chopping, or washing—these methods being employed singly or in combination. (Long, 1960) (e) To drill or put down a borehole. (Long, 1960) (f) A depression in the land surface, esp. one having a central playa or saline lake with no outlet; a hollow in a limestone region communicating with a cavern or subterranean passage so that waters running into it disappear. Also called sinkhole; swallow hole. (Webster 3rd, 1966) (g) Lanc. Natural cavity found in iron mines.
sinker (a) A rock drill for drilling blasting holes in a sinking shaft. (Nelson, 1965) (b) A special movable pump used in shaft sinking. (c) *sinker bar.* (d) A person who sinks mine shafts and puts in supporting timber or concrete. (Webster 3rd, 1966)
sinker bar (a) A heavy rod used to increase the snatching effect of the sliding jars in rope drilling. (BS, 1963) (b) A short bar or stem placed above the drill jars to give force to the upward jar in well drilling with cable tools. (Webster 3rd, 1966)
sinker drill (a) A one-person drill that can be held in the hand but is frequently mounted. This drill has found wide application in sinking shafts and is made in several sizes, each suited for a particular kind of work. Also called plugger drill. (Lewis, 1964) (b) A rock drill of the jackhammer type commonly used in shaft sinkings. Also called sinker. (Webster 3rd, 1966) (c) A hand-held compressed-air rock drill used in boring down holes such as in shaft sinking. (Pryor, 1963)
sink-float processes Processes that separate particles of different sizes or composition on the basis of specific gravity. When ore or coal particles are introduced into a liquid (or into a medium: a solid suspension), those having a specific gravity higher than that of the liquid will sink, while those that are lighter than the liquid will float. (Chem. Eng., 1949)
sink-float separation *dense-media separation.*
sinkhole A circular depression in a karst area. Its drainage is subterranean, its size is measured in meters or tens of meters, and it is commonly funnel shaped. Syn: *doline; sink; leach hole.* Partial syn: collapse sink. (AGI, 1987)
sinking (a) The process by which a shaft is driven. (BCI, 1947) (b) Extending excavations downward at or near the vertical plane. See also: *raising; shaft sinking.* (Nelson, 1965)
sinking fire A forge in which wrought-iron scrap or refined pig iron is partly melted or welded together by means of a charcoal fire and a blast.
sinking head *deadhead.*
sinking in rock Shaft sinking in rock usually comprises the following cycle: drilling a round of holes, blasting, removing the broken rock, trimming the shaft to form, placing the sets or concrete in position, and then preparing to drill the next round. (Lewis, 1964)
sinking kibble A large bucket for raising the stones, etc., from a shaft being sunk. Sometimes called bowk; hoppett. (Peel, 1921)
sinking lift A lift (pump) of small size with esp. heavy castings to resist the force of blasting; used in shaft sinking. A sinking pump, which is also sometimes called a sinker. (Standard, 1964; Fay, 1920)
sinking plant In a shaft, a sinking plant consists of the headframe, hoisting equipment, air-compressor for drills, concrete-mixing equipment, and suitable pumps. It may be temporary or permanent. (Lewis, 1964)
sinking platform A scaffold or staging designed for use during shaft sinking, particularly during lining operations. (Nelson, 1965)
sinking pump A long, narrow pump designed for keeping a shaft dry during sinking operations. It is usually large enough to deal with 1,000 gal/min (3,780 L/min) from the greatest depth at which water will be encountered. A sinking pump must be slung from the surface and be fairly easy to raise and lower when shot firing takes place at the shaft bottom. Most are of the electrically driven centrifugal type and allow for additional stages to be fitted as the shaft depth increases. It may be

suspended by a single-drum, worm-driven, capstan engine with a very slow speed. See also: *borehole pump; water barrel.* (Nelson, 1965)

sinks Fractions with a defined upper limit of specific gravity and so described, e.g., sinks 1.60 specific gravity. (BS, 1962)

sinople A red or brownish-red variety of quartz containing inclusions of hematite. Also spelled sinopal; sinopel.

sinter (a) A chemical sedimentary rock deposited as a hard incrustation on rocks or on the ground by precipitation from hot or cold mineral waters of springs, lakes, or streams; specif. siliceous sinter and calcareous sinter (travertine). The term is indefinite and should be modified by the proper compositional adj., although when used alone it usually signifies siliceous sinter. Etymol: German sinter, cinder. Cf: *tufa.* (AGI, 1987) (b) A ceramic material or mixture fired to less than complete fusion, resulting in a coherent mass when cooled. (c) A process for agglomerating ore concentrate in which partial reduction of minerals may take place and some impurities be expelled prior to subsequent smelting and refining. (d) To heat a mass of fine particles for a prolonged time below the melting point, usually to cause agglomeration. (ASM, 1961) (e) A process commonly used in making diamond bits, whereby powdered metal is compacted in a diamond-set mold or die, and the temperature is raised to a point just below melting, thus fusing the entire mass together. Also called sintered. (Long, 1960)

sinter bit A bit, the crown of which is formed by applying heat and pressure to a mixture of powdered metals covering diamonds set inside a mold or die-shaped to the form of a bit crown. The bit crown thus formed may be a surface-set, multilayer, or impregnated type. (Long, 1960)

sinter cap A zone of sinter typically positioned at the top of epithermal systems.

sintered carbide Sintering as used in powder metallurgy consists of mixing metal carbide powders having different melting points, and then heating the mixture to a temperature approximating the lowest melting point of any metal included. In sintered carbides, powdered cobalt, having the lowest melting point, acts as the binder, holding together the unmelted particles of the hard carbides. See also: *cemented carbide.* (CTD, 1958)

sintered carbide-tipped pick The pick generally used in coal cutters and cutter loaders, in which the sintered tip is brazed in various ways to the shank of the pick. In the external-tip type, which is widely used, the sintered tip is brazed externally to the shank, which is usually a forging. It is self-gaging and as the tip wears down, the cutting edge maintains its shape and clearance. In the slotted type, the tip is brazed into a slot cut in the shank of the pick. In the inserted-rod type, the sintered carbide takes the form of a rod inserted into a hollow in the shank of the pick, which is a forging. These picks are widely used in the soft coal mines in Germany. See also: *tungsten carbide bit; coal-cutter pick.* Syn: *inserted rod-type pick.* (Nelson, 1965)

sintered matrix A bit-crown diamond-embedment metal or alloy produced by a sinter powder-metal process. See also: *sinter; sinter bit.* (Long, 1960)

sintered-metal bit *sinter bit.*

sintering A heat treatment for agglomerating small particles to form larger particles, cakes, or masses; in case of ores and concentrates, it is accomplished by fusion of certain constituents. (CTD, 1958)

sinter plant A plant in which sintering is carried out. (Nelson, 1965)

sinter set *sinter bit.*

sinter-set bit *sinter bit.*

sinuous flow *turbulent flow.*

Sioux Falls jasper A decorative brown jasperlike, fine-grained quartz, from Sioux Falls, SD. Used for tables and interior architectural trim.

siphon An arrangement of closed pipes and valves to conduct water from one level to a lower level over an intervening ridge. The difference of level between the inlet and outlet ends of the pipe column must be sufficient to provide a head great enough to overcome the frictional resistance of the pipe column. The siphon was often used in the earlier days of mining when pumps were too costly or power was not available. See also: *suction head; inverted siphon.* (Nelson, 1965)

siphonage The action or operation of a siphon. (Fay, 1920)

siphon separator An apparatus for the sizing of pulverized ores in an upward current of water. (Fay, 1920)

siphon tap *Arents tap.*

sipylite *typrite.*

Sirocco fan A centrifugal fan, invented by Samuel Davidson in 1898, with 64 narrow blades curved forward, mounted at the periphery of a braced, open drum. It is a high-speed, small-diameter fan, usually direct driven. It was a popular fan in Great Britain for many years. See also: *Waddle fan.* (Nelson, 1965)

siserskite Former name for iridosmine.

sismondine Former name for Mg-rich chloritoid.

site exploration (a) The investigation and testing of the surface, subsoil, and any obstruction at a site to obtain the full information necessary for designing a complete structure with its foundations. (Hammond, 1965) (b) *site investigation.*

site investigation The collection of basic facts about, and the testing of, surface and subsurface materials (physical properties, distribution, and geologic structure) at a site, for the purpose of preparing suitable designs for a mine, an engineering structure, or other use. (AGI, 1987)

site rivet A rivet driven on a construction site. See also: *shop rivet.* (Hammond, 1965)

size To separate minerals according to various screen meshes.

size analysis *particle-size analysis.*

size consist Screen analysis of particle size (coal). (Bennett, 1962)

size distribution (a) *particle-size distribution.* (b) Analysis of crushed or ground materials on the basis of particle size. (Bennett, 1962) (c) In sizing analysis of sands, the percentage of the sample retained on each laboratory sieve in the range examined. (Pryor, 1963)

size-distribution curve A graphical representation of the size analysis of a mixture of particles of various sizes, using an ordinary, logarithmic, or other scale. (BS, 1962)

sized variation The variation in dimensions of any ceramic product from the intended dimensions.

size fraction Portion of a sample of sand lying between two size limits, the upper being the limiting and the lower the retaining mesh. (Pryor, 1963)

size-frequency distribution *particle-size distribution.*

size range That between upper (limiting) and lower (retaining) mesh sizes with reference to screened or classified material. (Pryor, 1965)

size reduction The breaking of large coal, ore, or stone by primary breaker, or the breaking of smaller sizes by grinding. See also: *reduction ratio; primary breaker.* (Nelson, 1965)

size selector A device attached to the intake of a dust-sampling instrument to remove the bulk of the particles above 5 to 10 µm in size; thus the resulting sample is more representative of the health-hazard size range of dust— mainly 5 µm and smaller. See also: *dust sampling.* (Nelson, 1965)

sizing (a) The arrangement, grading, or classification of particles according to size; e.g., the separation of mineral grains of a sediment into groups each of which has a certain range of size or maximum diameter, such as by sieving or screening. (AGI, 1987) (b) *screening.*

sizing punch A punch used for pressing of the sintered compact during the sizing operation. (Osborne, 1956)

sizing screen A screen or set of screens normally used for dividing a product (for example, washed coal) into a range of sizes. Also called grading screen and classifying screen (undesirable usage). (BS, 1962)

sjögrenite (a) A hexagonal mineral, $Mg_6Fe_2(CO_3)(OH)_{16}{\cdot}4H_2O$; manasseite group; dimorphous with pyroaurite of the hydrotalcite group. (b) The name was formerly used for the phosphates dufrenite and chalcosiderite.

S-joint *longitudinal joint.*

S.J. table A pneumatic table of American design, for the drycleaning of coal. A sizing ratio of 2:1 is desirably the maximum range of sizes that the table can separate in one operation. See also: *Kirkup table.* (Nelson, 1965)

skares *scares.*

skarn An old Swedish mining term for silicate gangue (amphibole, pyroxene, garnet, etc.) of certain iron ore and sulfide deposits of Archean age, particularly those that have replaced limestone and dolomite. Its meaning has been generally expanded to include lime-bearing silicates, of any geologic age, derived from nearly pure limestone and dolomite with the introduction of large amounts of Si, Al, Fe, and Mg. In American usage, the term is more or less synonymous with tactite. See also: *calc-silicate hornfels.* (AGI, 1987)

skate-wheel conveyor A type of wheel conveyor making use of a series of skate wheels mounted on common shafts or axles or mounted on parallel spaced bars on individual axles.

skeletal crystal Crystals that develop under conditions of rapid growth and high degrees of supersaturation so that atoms or ions are added more rapidly to edges and corners of growing crystals, resulting in branched "dendrites" or hollow stepped depressions, "hoppers." Ice on windowpanes (frost) and pyrolusite on agate (moss agate) are dendrites; halite and gold may form hopper crystals. Cf: *dendrite; hopper crystal.*

skeleton crystals Hollow or imperfectly developed crystals formed by rapid crystallization. (Fay, 1920)

skeleton sheathing Consists of a continuous wood frame with sheathing planks placed vertically at intervals, usually of about 4 ft (1.2 m), behind it. Used where the banks consist of compacted, stable soils, primarily to prevent initial yield at the top. Cf: *close sheathing; tight sheathing.* (Carson, 1961)

skeleton tubbing A temporary method of supporting a circular shaft sinking. It consists essentially of iron curbs or rings. Each ring consists of segments of wrought iron, 3 to 5 in (7.62 to 12.7 cm) deep, 5/8 to 7/8 in (1.59 to 2.22 cm) thick, and from 6 to 8 ft (1.82 to 2.44 m) in length. The segments are bent to the curvature of the shaft and bolted together, each ring being suspended from the ring above. Laggings or backing deals are wedged behind the rings. Every fourth ring or so is supported on steel strata bolts driven into holes drilled in the rock sides. See also: *temporary shaft support.* (Nelson, 1965)

skelp Mild steel strip, often of Bessemer steel, from which tubes are made by drawing it through a welding bell, at welding temperature, to produce butt-welded or lap-welded tubes or pipes. See also: *butt weld.* (Hammond, 1965)

skelp iron Wrought iron rolled into flat bars suitable for making pipe. (Mersereau, 1947)

skerry (a) An old English term meaning, variously, thin layers of micaceous sandstone in the Coal Measures, a thin ferruginous sandstone crowded with fossils, or a hard gritty clay used in making firebricks. (b) Eng. Thin layers of sandstone in the Keuper Marls, Leicestershire, and Derbyshire coalfield. (Arkell, 1953) (c) Micaceous sandstones in the Coal Measures of Wales and the Midlands. (Arkell, 1953) (d) Eng. A thin band of ferruginous, micaceous rock, crowded with fossils, Middle Lias, Sutton Basset, Rutland. (Arkell, 1953)

skerrystone Mid. Hard, thin-bedded sandstone.

skew (a) An irregular discontinuous vein striking out from the principal vein in an uncertain direction, lying in a slanting and irregular position. (b) The angled support from which an arch is sprung. (AISI, 1949)

skewed On a horizontal angle, or in an oblique course or direction. (Nichols, 1976)

skewed roller conveyor A roller conveyor having a series of rolls skewed to direct objects laterally while being conveyed.

skewness In coal sampling, a lack of symmetry in the particle-size distribution. It is a tendency of the observed data to extend farther to one side of the average than to the other. (Mitchell, 1950)

skew plate *bloomery.*

skews (a) Stones cut to form the coping of a gable. (Arkell, 1953) (b) Cracks on irregular joints in a mine, sometimes indicating danger from falls. (Arkell, 1953)

skiagite A hypothetical garnet "molecule," $Fe_3Fe_2(SiO_4)_3$.

skialith A vague remnant of country rock in granite, obscured by the process of granitization. Cf: *schlieren; xenolith.* (AGI, 1987)

skid (a) An iron shoe or clog attached to a chain and placed under a wheel to prevent its turning when descending a steep hill; a drag. (Webster 3rd, 1966) (b) An arrangement upon which certain coal-cutting machines travel along a working face. (Fay, 1920) (c) A metal slide placed under a mine car wheel temporarily to restrict the speed of a trip on a descending grade. (BCI, 1947) (d) A metal plate placed under a shortwall cutting machine to control it while cutting. (BCI, 1947) (e) The sledlike platform forming the base on which a machine or structure is set and slid or skidded into position; also, the sled runner, bottom-most part of the base of a drill or other machine. (Long, 1960)

skid-mounted A term applied when a drill or other machine is attached permanently to a skid. (Long, 1960)

ski-lift conveyor A method of transporting miners to the coal face in special chairs that move continuously on an endless conveyor. (Nelson, 1965)

skim gate A gating arrangement designed to prevent the passage of slag and other undesirable materials into a casting. (ASM, 1961)

skimmer (a) A single-bucket excavator in which the bucket travels along the boom, which is kept almost horizontal during operation. (Dodd, 1964) (b) An excavator base machine equipped with jib and bucket for digging and loading from a shallow face above track level. (Nelson, 1965)

skimmer equipment A digging bucket mounted to slide along the boom of an excavator so that the bucket can be used to trim various angles of slope from the horizontal to about 60° elevation. (Hammond, 1965)

skimming The removal of the top layer of soil or of irregularities in the ground surface at new mine or opencast sites. See also: *skimping.* (Nelson, 1965)

skimming gate A channel in a sand mold having over it a bridge that removes the dross from the molten metal as it passes through. See also: *skimmer.* (Standard, 1964)

skimming ladle Any ladle used in skimming; specif., a ladle used for pouring molten metal, having its lip covered with a guard to retain the dross. (Standard, 1964)

skimping (a) The skimmings of the dross from the ore in the vat or tank. (Nelson, 1965) (b) *jigging.*

skin effect (a) Tendency of an alternating current to concentrate at the surface of a conductor. (Schieferdecker, 1959) (b) The frequency-dependent reduction of resistivity log measurements in conductive formations due to inductive interaction between the current paths; the induction logs now operating at about 20 kHz are most affected. (AGI, 1987) (c) The reduction of formation permeability in the vicinity of a well bore caused by drilling and completion operations. (AGI, 1987) (d) The concentration of alternating current in a conductor towards its exterior boundary. (AGI, 1987)

skin flotation (a) A concentration process in which adhesion is effected between a free water surface and particles, usually larger than those involved in froth flotation. (Gaudin, 1939) (b) In skin flotation processes, the separation between minerals and gangue is accomplished at the surface of a body of water, or, in other words, at the air-water interface. Use is made of the surface tension of the water and of the fact that certain minerals, such as sulfides and hydrocarbons, resist being wetted by water. See also: *flotation.* (Mitchell, 1950)

skin friction (a) Friction between a fluid and the surface of a solid moving through it, or between a moving fluid and its enclosing surface. (Webster 3rd, 1966) (b) The frictional resistance developed between soil and a structure. (ASCE, 1958) (c) Resistance of ground to the movement of a pile or caisson, generally proportional to the area in contact. (Hammond, 1965) (d) *wall friction.*

skin rock The thin band of rock immediately surrounding an excavation. (Spalding, 1949)

skip (a) A guided steel hoppit, usually rectangular, with a capacity up to 50 st (45.4 t), which is used in vertical or inclined shafts for hoisting coal or minerals. It can also be adapted for personnel riding. The skip is mounted within a carrying framework, having an aperture at the upper end to permit loading, and a hinged or sliding door at the lower end to permit discharge of the load. The cars at the pit bottom deliver their load either directly into two measuring chutes located at the side of the shaft or into a storage bunker from which the material is fed to the measuring chutes. (Nelson, 1965) (b) A large hoisting bucket, constructed of boiler plate that slides between guides in a shaft, the bail usually connecting at or near the bottom of the bucket so that it may be automatically dumped at the surface. (c) An open iron vehicle or car on four wheels, running on rails and used esp. on inclines or in inclined shafts. Sometimes spelled skep. (d) A thin slice taken off a breast, pillar, or rib along its entire length or part of its length. Also called slab. (e) A truck used in a mine. (Gordon, 1906) (f) A small car that conveys the charge to the top of a blast furnace. (Mersereau, 1947)

skip bucket The tub or bucket used for containing the material conveyed by a skip hoist.

skip haulage A method of underground haulage sometimes adopted in steep workings, where the gradient is 1:2 to 1:1.5. There are two kinds: (1) a skip carriage on which the tub is placed in a horizontal position; and (2) a self-dumping system in which the skip, which is permanently attached to the rope, is discharged automatically at the top of the incline and then returned for reloading. (Nelson, 1965)

skip hoist A bucket or car operating up and down a defined path, receiving, elevating, and discharging bulk materials.

skip loader I In metal mining, one who loads ore into a skip (large can-shaped container) from skip pockets (underground storage bins) at different shaft stations in a mine, operating a mechanical device to open and close the gates of the loading chutes. Also called skipman; skipper. (DOT, 1949)

skip miner In bituminous coal mining, one who drills holes into pillars of coal supporting the roof, charges holes with explosives, and blasts out slabs (skips) of coal to widen haulageway or working place. (DOT, 1949)

skip operator In the iron and steel industry, one who controls the skip hoist by which a skip car containing coke, limestone, or ore is hauled up an inclined runway to the furnace top and dumped into the charging bell of the furnace. (DOT, 1949)

skipping The working of 2 to 10 yd (1.8 to 9.1 m) of coal along the side or sides of a narrow stall or heading to gain coal and make room for ripping stone. See also: *rib-side pack.* (Nelson, 1965)

skipping the pillar (a) To take a slice off a pillar before abandoning the workings; to rob. (Fay, 1920) (b) Widening a gangway or entry. (Fay, 1920)

skip shaft A mine shaft esp. prepared for hauling a skip. (Standard, 1964)

skip system A system used for moving material from a quarry floor to a plant located at a considerable elevation. This system utilizes two parallel inclined tracks with a skip car operating on each track. The cars are operated by cables controlled by a winding gear at the head of the incline. The quarry trucks or cars deliver their loads to the skips through a chute at the base of the incline. A hopper at the top of the incline receives the loads from the skips and feeds the rock to a crusher. (Pit and Quarry, 1960)

skirt A vertical strip placed at the side of a conveyor belt to prevent spillage or to increase capacity. (Nichols, 1976)

skirting In pillar extraction, it refers to a stall or roadway working a slice or lift of coal along the side of a pillar. (Nelson, 1965)

skirt plates Steel sideplates that overlap a conveyor belt slightly and assist in settling the coal on the belt at the tail end or at a transfer point. (Nelson, 1965)

skirts That which bounds and limits a vein's breadth, Derbyshire, U.K.

skirt-type core spring A core lifter, usually a split-ring type, having a split, thin tubular extension attached above the beveled portion of the core spring, which slides upward and inside the lower end of the inner tube of a core barrel. (Long, 1960)

skleropelite A rock produced by low-grade metamorphism of an argillaceous sediment without the development of cleavage. See also: *hornfels.*
sklodowskite A monoclinic mineral, $(H_3O)_2Mg(UO_2)_2(SiO_4)_2{\cdot}2H_2O$; structurally similar to uranophane and cuprosklodowskite; strongly radioactive; citron-yellow. Also spelled sklodovskite.
skrin Derb. Cross fissures in limestone, sometimes containing small quantities of ore. Also called scrin.
skull cracker A heavy iron ball allowed to drop from a height to break up, or crack, hard substances, as aloxite, rock, etc. (Mersereau, 1947)
skull drop A place where heavy ladle skulls are broken. (Fay, 1920)
skulls *sculls.*
skutterudite An isometric mineral, $CoAs_{2\text{-}3}$, having Co replaced by Ni toward nickel-skutterudite; metallic; tin white to silver gray showing iridescent tarnish; a minor source of cobalt. Syn: *cobalt skutterudite.*
slab (a) A piece of metal, intermediate between ingot and plate, with the width at least twice the thickness. (ASM, 1961) (b) Cleaved or finely parallel jointed rocks, which split into tabular plates from 1 to 4 in (2.54 to 10.16 cm) thick. Slabs are seldom as strong as flags. Syn: *slabstone.* (Fay, 1920)
slabbing (a) Close timbering between sets of timber. (b) Lagging placed over bars. Also called slabs. (c) Cutting a slice or slab from the side of a pillar.
slabbing cut A drill hole pattern suitable for a wide rectangular tunnel; e.g, 8 ft by 15 ft (2.4 m by 4.6 m) wide. The entire face is fired in three separate rounds of shots, the first or cut holes providing a free face for the remaining shots. The face is broken in successive lifts or slabs from one side to the other. Also called swing cut. (Nelson, 1965)
slabbing machine (a) A power-driven, mobile cutting machine, which is a single-purpose cutter in that it cuts only a horizontal kerf at variable heights. Syn: *arcwall machine.* (b) A coal-cutting machine designed to make cuts in the side of a room or entry pillar preparatory to skipping or slabbing the pillar. (Jones, 1949)
slabbing method A method of mining pillars in which successive slabs are cut from one side or rib of a pillar after a room is finished, until as much of the pillar is removed as can safely be recovered. This system has the disadvantage that the open area is always increasing and the loaders are working away from the solid pillars toward the goaf. (Lewis, 1964)
slab entry An entry widened or slabbed to provide a working place for a second miner.
slabstone A rock that readily splits into slabs; flagstone. See also: *slab.* (AGI, 1987)
slack (a) Fine-grained coaly material resulting from weathering, screening, or washing of coal. (b) To disintegrate rapidly when exposed to weathering. (AGI, 1987) (c) Commonly used to describe the smaller sizes of coal passing through screen openings, approx. 1 in (25.4 mm) or less in diameter. (Mitchell, 1950) (d) The process by which soft coal disintegrates when exposed to the air and weather; also to slake, as lime. (e) Small coal, usually less than 1/8 in (3.2 mm). It has a high ash content and is difficult to clean in the washery. High-ash slack is being used increasingly in special boilers and power stations. See also: *culm; duff.* (Nelson, 1965)
slack adjuster In air brakes, the connection between the brake chamber and the brake cam. (Nichols, 1976)
slack box Aust. A bin in which fine coal (slack) is stored.
slack-brake switch *hoist slack brake switch.*
slacken In metal smelting, the scoria of previous operations, mixed with the ores to retard or prevent fusion of the nonmetallic portions. Also spelled slakin. (Standard, 1964)
slack hauler In bituminous coal mining, one who hauls small cars of slack (fine coal) from tipple to boiler room of power plant at mine to maintain fuel supply. (DOT, 1949)
slacking (a) Degradation in size (coal). (Bennett, 1962) (b) Coals having a pronounced tendency to disintegrate or slack on exposure to weather, particularly when alternately wetted and dried or subjected to hot sunshine. Coals that slack readily contain relatively large amounts of moisture. When exposed to the weather, such coals lose moisture rapidly. As the coal loses moisture at the surface, the moisture from the interior of the piece gradually drifts outward to the surface. If the loss of moisture at the surface proceeds at a faster rate than that at which it is replaced by moisture from the interior of the piece, then the shrinkage of the coal at the surface is greater than that in the interior; consequently, stresses are generated in the surface coal. These stresses cause the coal to crack and disintegrate. Also called weathering. See also: *weathering; weathering index.* (Mitchell, 1950)
slacking index *weathering index.*
slackline cableway (a) A cableway having one low and one high tower and a track cable with adjustable tension suspended between them. One end of the track cable is attached to a hoist drum by means of which the tension on the cable can be rapidly changed so as to position, lower, or raise the digging skip. (b) A cable excavator having a track cable that is loosened to lower the bucket and tightened to raise it. (Nichols, 1954)
slack quenching The process of hardening steel by quenching from the austenitizing temperature at a rate slower than the critical cooling rate for the particular steel—resulting in incomplete hardening and the formation of one or more transformation products in addition to or instead of martensite. (ASM, 1961)
slack water The state of a tidal current when its velocity is near zero, the moment when a current reverses direction. Sometimes considered the intermediate period between ebb and flood currents during which the velocity of the currents is less than 0.1 knot (0.16 km/h). (Hy, 1965)
slag (a) Material from the iron blast furnace, resulting from the fusion of fluxstone with coke ash and the siliceous and aluminous impurities remaining after separation of iron from the ore. Slag is also produced in steelmaking. Formerly a solid waste, slag is now utilized for various purposes, chiefly in construction. (AGI, 1987) (b) A scoriaceous or cindery pyroclastic rock. (AGI, 1987) (c) A British term for a friable shale with many fossils. (AGI, 1987) (d) A substance formed in any one of several ways by chemical action and fusion at furnace operating temperatures: (1) in smelting operations, through the combination of a flux, such as limestone, with the gangue or waste portion of the ore; (2) in the refining of metals, by substances such as lime added for the purpose of effecting or aiding the refining; or (3) by chemical reaction between refractories and fluxing agents, such as coal ash, or between different types of refractories. (Harbison-Walker, 1972) (e) Partially fused mixture of spilled batch, overflowed glass, breeze coal, and clay from the siege. (ASTM, 1994) (f) The top layer of the multilayer melt formed during some smelting and refining operations. In smelting, it contains the gangue minerals and the flux; in most refining operations, the oxidized impurities. (CTD, 1958) (g) Oxide liquids (exclusive of the commercial glasses) with a high melting temperature. (Van Vlack, 1964) (h) *blast-furnace slag; granulated slag.*
slag blanket The coating of slag, or scum, that forms on the top of the bath in the open-hearth furnace. (Mersereau, 1947)
slag buggy A very large pot for holding slag obtained in the smelting or ores. It is mounted on a railway truck or the like, so as to permit easy dumping. See also: *slag pot.* (Standard, 1964)
slag car Iron vessel on wheels used to transport molten slag from furnace to dump. Also called slag buggy. (Pryor, 1963)
slag dump A dumping place for the shell or cone that forms in a slag pot. (Standard, 1964)
slaggable Capable of becoming or forming into a slag. (Fay, 1920)
slagging Destructive chemical reaction between refractories and external agents at high temperatures, resulting in the formation of a molten liquid. (Henderson, 1953)
slagging of refractories Destructive chemical reaction between refractories and external agencies at high temperatures, resulting in the formation of a liquid.
slag hearth A hearth, on the principle of the Scotch hearth, for the treatment of slags, etc., produced by lead smelting in the reverberatory furnace. The English slag hearth has one tuyere; the Castillian or Spanish, three. (Fay, 1920)
slag inclusion Slag (dross) entrapped in a metal. (ASM, 1961)
slag lead Lead obtained by a resmelting of gray slag. (Fay, 1920)
slag pot A vessel for the disposal of slag at furnaces. Small pots are mounted on wheels and moved by hand, while the larger ones are mounted on trucks for mechanical transportation. See also: *slag buggy; slag car.* (Fay, 1920)
slag runoff Tapping off excess slag after the ore boil in the basic open-hearth process of steelmaking. (Bennett, 1962)
slake (a) To crumble or disintegrate, such as coal or lime. (b) Small coal. (c) To mix with water, so that a chemical combination takes place, as in the slaking of lime. (d) To crumble in water. (Van Vlack, 1964)
slaking (a) The crumbling and disintegration of earth materials upon exposure to air or moisture; specif. the breaking up of dried clay or soil when saturated with or immersed in water, or of coal or clay-rich sedimentary rocks when exposed to air. (AGI, 1987) (b) The disintegration of the walls of tunnels in swelling clay, owing to inward movement and circumferential compression. (AGI, 1987) (c) The treating of lime with water to give hydrated (slaked) lime. (AGI, 1987)
slam Eng. Thin slurry and mud, Yorkshire lead mines. (Arkell, 1953)
slant (a) Any short, inclined crosscut connecting the entry with its air course to facilitate the hauling of coal. Commonly called a dip switch when the coal is not level. See also: *shoo-fly.* (Fay, 1920) (b) A heading driven diagonally between the dip and the strike of a coal seam. Also called a run. See also: *counter; stone drift; surface drift.* (Fay, 1920)
slant chute Chute driven diagonally across to connect a breast manway with a manway chute. See also: *slant.*
slant drilling *directional drilling.*
slants Eng. A set of joints in slate parallel to the main cleavage, Denbighshire. (Arkell, 1953)

slant vein York. One vein crossing another at an acute angle. (Arkell, 1953)
slape back Eng. *back.*
slash (a) There are many successive ridges of shingle running in varying directions, and often with narrow strips of marsh enclosed between successive ridges. Such bands of marsh have been given the very appropriate name of slashes in New Jersey. (AGI, 1987) (b) Swampy land, overgrown with dense underbrush. Local in the Northeast. (AGI, 1987) (c) An open or cutover tract in a forest strewn with debris, as from logging; also such debris. (AGI, 1987)
slat bucket A digging bucket of basket construction, used in handling sticky, chunky mud. (Nichols, 1976)
slate (a) A compact, fine-grained metamorphic rock that possesses slaty cleavage and hence can be split into slabs and thin plates. Most slate was formed from shale. (AGI, 1987) (b) A coal miner's term for any shale accompanying coal; also, sometimes the equivalent of bone coal. (AGI, 1987) (c) Dark shale lying next to the coalbeds. It contains impressions of the plant life of distant ages, proving the vegetable origin of coal. (Korson, 1938) (d) A fine-grained metamorphic rock that breaks into thin slabs or sheets. Usually gray to black, sometimes green, yellow, brown, or red. Slates are composed of micas, chlorite, quartz, hematite, clays, and other minerals. Found in Pennsylvania, Vermont, Maine, Virginia, California, Colorado; Europe. Used for roofing; decorative stone; various building applications; in crushed form on shingles; abrasive; pigment. See also: *clay slate; phyllite; spotted slate.* (CCD, 1961)
slate chute (a) A chute for the passage of slate and bony coal to a pocket from which it is loaded into dump cars. (Fay, 1920) (b) A chute driven through slate. (Fay, 1920)
slate cutter In the stonework industry, one who operates an upright drilling machine to drill holes into slate so that the slabs may be fastened in place with wires or rods when installed in a building. (DOT, 1949)
slate ground A term used in southern Wales for a dark fissile shale, resembling slate. (AGI, 1987)
slate handler In bituminous coal mining, a laborer who shovels up falls of slate or rock along haulageways in a mine and loads it into cars. Also called rock handler. (DOT, 1949)
slate larryman *slate motorman.*
slateman In anthracite and bituminous coal mining, a general term for a worker handling slate or rock as distinguished from coal. Usually designated according to type of activity, as rock driller; rock loader; slate motorman; slate picker; slate shooter. Also called rockman; slate handler. See also: *rockman; slate handler.* (DOT, 1949)
slateman helper *slate-shooter helper.*
slate motorman In anthracite and bituminous coal mining, a person who operates a mine locomotive to haul trains of cars loaded with slate or shale, underground and at the surface of a mine. Also called larryman; slate larryman. (DOT, 1949)
slate saw *marble saw.*
slate shooter In bituminous coal mining, one who drills holes into the slate roof of haulageways, and charges and sets off explosives to blast down slate to increase height or improve the safety of the roof. (DOT, 1949)
slate-shooter helper In bituminous coal mining, a laborer who assists the slate shooter in removing the slate and rock from the roof, ribs, and face of haulageways. Also called rockman helper; slateman helper. (DOT, 1949)
slate spar A variety of crystallized calcite. Also called shiver spar. (Standard, 1964; Fay, 1920)
slat gate A gate, for controlling water, composed of two upright grooved posts with boards between the boards of slats being removed or added to regulate the height of water.
slaty band Scot. Ironstone and flaky blaes. (Nelson, 1965)
slaty cleavage A pervasive, parallel foliation of fine-grained, platy minerals (mainly chlorite and sericite) in a direction perpendicular to the direction of maximum finite shortening, developed in slate or other homogeneous sedimentary rocck by deformation and low-grade metamorphism. Most slaty cleavage is also axial-plane cleavage. Syn: *flow cleavage.* (AGI, 1987)
slave cylinder A small cylinder whose piston is moved by a piston rod controlled by a larger cylinder. (Nichols, 1976)
slave piston A small piston having a fixed connection with a larger one. (Nichols, 1976)
slave unit A machine controlled by or through another unit of the same type. (Nichols, 1976)
sleck (a) Eng. A kind of reddish sandstone. (b) Eng. Small pit coal. (Arkell, 1953) (c) Newc. Mud deposited by water in a mine. (d) *slack.*
sled A drag used to convey coal along the road to where it is loaded into cars, or to the chute. Also called sledge; slype. (Fay, 1920)
sledger (a) In bituminous coal mining, one who digs out dirt, rock, or coal with a long-handled pick (sledge) in a strip mine so that it may be loaded into cars by hand or with a power shovel. (DOT, 1949) (b) In the quarry industry, one who breaks up large stone into small pieces suitable for use in building work. Also called laborer, stone; rock breaker; spawl beater; stone sledger. (DOT, 1949)
sledging roll Crushing roll with projection that breaks the rock, instead of fracturing it by squeezing. Also called slugging roll. (Pryor, 1963)
sleeping rent A fixed rent stated in leases of coal mines, as distinguished from royalty or share of profits. (Standard, 1964)
sleeve catcher A skirt-type core lifter. (Long, 1960)
slender beam A beam, which if overloaded, will fail by buckling of the compression flange. (Hammond, 1965)
slew (a) To turn around (slue). (Mason, 1951) (b) *slough.*
slewing (a) The rotation of a crane jib so that the load moves through the arc of a circle on a horizontal line. (Hammond, 1965) (b) To turn or twist.
slice (a) To remove ore by successive slices from the top of an orebody of considerable lateral extent and thickness. The slices may be 6 ft, 12 ft, 20 ft, or 40 ft (1.8 m, 3.6 m, 6.1 m, or 12.2 m) thick. (b) A thin broad piece cut off, such as a portion of ore cut from a pillar or face.
slice bar A thin, wide iron tool for cleaning clinkers from the grate bars of a furnace. (Standard, 1964)
slice drill In sublevel caving, the crosscuts driven between every other slice from 18 to 36 ft (5.47 to 10.94 m) apart. (Lewis, 1964)
slicing In continuous mining, slicing consists of driving up some four to six places the set or desired distance, which may be 1,000 ft (304.8 m) or more, and then pulling the pillars on retreat. After completion of one slice, the unit moves over and mines another along the gob. See also: *top slicing.* (Coal Age, 1966)
slicing-and-filling system *overhand stoping.*
slicing method Removal of a horizontal layer from a massive orebody. In top slicing extraction retreats along the top of the orebody, leaving a horizontal floor that becomes the top of the next slice. A timber mat separates this from the overburden, which caves downward as the slices are made. Other methods attack from the bottom (sublevel caving) or side. (Pryor, 1963)
slick Ore in a state of fine subdivision. Syn: *slime.* Also called slickens.
slickens Extremely fine-grained material, such as finely pulverized tailings discharged from hydraulic mines or a thin layer of extremely fine silt deposited by a stream during a flood. (AGI, 1987)
slickensided clay *stiff-fissured clay.*
slickensides The striations, grooves, and polish on joints and fault surfaces. Cf: *striation.* See also: *fault striae; fault gouge.*
slicker A small implement used in a foundry for smoothing the surface of a mold. (Standard, 1964)
slick hole A hole column loaded with explosive, without springing. (Nichols, 1976)
slicking A narrow vein of ore. (Standard, 1964)
slick top In coal mining, a term used to describe the roof of the coal vein when it is very smooth. (Kentucky, 1952)
slide (a) An upright rail fixed in a shaft with corresponding grooves for steadying the cages. (Fay, 1920) (b) A trough used to guide and to support rods in a tripod when drilling an angled hole. Also called rod slide. (Long, 1960) (c) The bottom of a gold-washing cradle. (Standard, 1964) (d) As used by churn drillers, a fault plane or opening encountered in a hole that deflects the bit. (Long, 1960) (e) A mass movement of descent resulting from failure of earth, snow, or rock under shear stress along one or several surfaces that are either visible or may reasonably be inferred; e.g., landslide, snowslide, and rockslide. The moving mass may or may not be greatly deformed, and movement may be rotational or planar. A slide can result from lateral erosion, lateral pressure, weight of overlying material, accumulation of moisture, earthquakes, expansion owing to freeze-thaw of water in cracks, regional tilting, undermining, and human agencies. (AGI, 1987) (f) The mass of material moved in or deposited by a slide. (AGI, 1987)
slide coupling *slip joint.*
slide rail A mounting of steel or cast iron for a belt driven machine enabling it to be moved along as the belt stretches in order to take up the slack. (Hammond, 1965)
slide rock Rock making up the mass of material in a landslide or talus.
sliding angle Angle at or above which rock in movement will continue to slide, but less than the angle needed to initiate movement from rest. (Pryor, 1963)
sliding friction Sliding friction is the resistance offered when one body slides over another body. The amount of friction or resistance is dependent on the laws of friction. (Morris, 1958)
sliding gate A crest gate that has a high frictional resistance to opening and is therefore suitable only for small gates. See also: *roller gate.* (Hammond, 1965)
sliding scale A method of paying for the coal in proportion to the amount of lump coal it contains. (Fay, 1920)
sliding-scale system A system which regulated colliers' wages by the ascertained selling price of coal. (Nelson, 1965)

sliding shoe A metal plate that serves as partial or total support for devices used with shaker conveyors where the device must move or slide on the bottom. The shovel trough of a duckbill and certain types of swivels or angle troughs use this device. (Jones, 1949)

slime (a) Extremely fine sediment (0 mesh), produced in the processing of ore or rock, esp. phosphate rock, which remains suspended in water indefinitely. Consists chiefly of clay. (b) A material of extremely fine particle size encountered in ore treatment. (ASM, 1961) (c) Anode slimes are the metals or metal compounds left at, or falling from, the anode during electrolytic refining of metals. See also: *anode slime.* (Pryor, 1965) (d) A mixture of metals and some insoluble compounds that forms on the anode in electrolysis. (ASM, 1961) (e) A product of wet grinding containing valuable ore in particles so fine as to be carried in suspension by water; chiefly used in the plural. (Webster 3rd, 1966) (f) In metallurgy, ore reduced to a very fine powder and held in suspension in water so as to form a kind of thin ore mud; generally used in the plural. (g) Primary slimes are extremely fine particles derived from ore, associated rock, clay, or altered rock. They are usually found in old dumps and in ore deposits that have been exposed to climatic action; they include clay, alumina, hydrated iron, near-colloidal common earths, and weathered feldspars. Secondary slimes are very finely ground minerals from the true ore. (Pryor, 1960)

slime coating In mineral processing, adherence of an impalpably fine layer of particles of another (for example, calcite on galena), therefore hindering or preventing true surface reaction in leaching or flotation. (Pryor, 1963)

slime deliveryman In beneficiation, smelting, and refining, a laborer who washes slime from cloth strainers, electrolysis tank debris, and collection barrels into a settling tank, using a water spray preparatory to recovery of precious metals from slime. Becoming obsolete. (DOT, 1949)

slime leaching A leaching method in which the slime and the leach solution are agitated in one or more agitators until the ore minerals have been dissolved. Some agitators have mechanically driven paddles or elevators inside an agitation tank, which serve to keep the pulp in circulation until dissolution is complete. This method may be either continuous or intermittent. (Newton, 1959)

slime pit A tank or large reservoir of any kind into which the slimes are conducted in order that they may have time to settle, or in which they may be reserved for subsequent treatment. See also: *slime.* (Fay, 1920)

slimer A machine that makes slime; e.g., a tube mill. (Fay, 1920)

slime sludge (a) The pulp or fine mud from a drill hole. (b) *slime.*

slime table (a) A table for the treatment of slime; a buddle. (b) A shaking table used in gravity concentration of finely ground coal or ore, characterized by special riffles and shallow pools in which stratification is gently produced. (Pryor, 1963)

slime tin Cassiterite too finely ground to be readily concentrated by the use of gravity treatment. Usually associated with hydrated iron. (Pryor, 1963)

slime water Water defiled in washing ore. (Standard, 1964)

slim hole (a) A rotary borehole having a diameter of 12.7 cm or less. (AGI, 1987) (b) A drill hole of the smallest practicable size, often drilled with a truck-mounted rig; used primarily for mineral exploration or as a stratigraphic or structure test. See also: *structure test hole.* (AGI, 1987)

slim-hole drilling and casing Use of the smallest feasible drill hole and casing size. (Williams, 1964)

sliming Overgrinding in a ball mill. (Newton, 1959)

sline (a) Mid. Potholes in a mine roof. (b) The principal cleat in coal. (c) A natural transverse cleavage of rock; a joint.

sling (a) A rope or chain put around stones or heavy weights for raising them. (Zern, 1928) (b) A lifting hold consisting of two or more strands of chain or cable. (Nichols, 1954) (c) A ropelike device used to give additional support to lengths of drill rod too long to stand in the drill derrick without sagging unduly. (Long, 1960) (d) A short loop or length of cable with small loops at either end. (Long, 1960)

sling block A frame in which two sheaves are mounted so as to receive lines from opposite directions. (Nichols, 1976)

sling hygrometer *Storrow whirling hygrometer.*

sling psychrometer A hygrometer held on a short length of cord and whirled around, the observer standing sideways to the air current. The wet bulb is thereby rapidly reduced to its final reading. (Hammond, 1965)

slink Scot. A wide clayey joint; a stage.

slip (a) Landslip, or subsiding mass of rock or clay in a quarry or pit; a minor landslide. (b) A small fault. (c) The relative displacement of formerly adjacent points on opposite sides of a fault, measured in the fault surface. See also: *dip slip; strike slip.* Partial syn: shift. Syn: *total displacement.* (AGI, 1987) (d) A joint or cleat in a coal seam. (e) *kettle bottom.* Syn: *horseback; kettleback.* (f) A joint in coal upon which there may have been no perceptible movement. (g) *back.* (h) A joint or pronounced cleavage plane. (Mason, 1951) (i) A sudden descent of a hanging or sticking charge in a blast furnace. (j) One of a set of serrated-face wedges that fits inside the spider of a drill-rod clamping device. See also: *spider; slips.* (Long, 1960) (k) The percentage of water leaking through valves, expressed as a percentage of the volume swept out by the bucket or ram, a measure of the volumetric efficiency of a pump. Generally for normal pumping speeds, the slip is between 5% and 10% but it may rise to 20% with higher pumping speeds. (Sinclair, 1958) (l) Under stress, minerals deform plastically along specific crystallographically determined slip planes in slip directions, analogous either to the sliding in a deck of playing cards, such as in quartz, or to a bundle of pencils, such as in olivine. Deformation lamellae may be preserved or destroyed by annealing. Although such slip is referred to as glide, it is not to be confused with a crystallographic glide—the combined symmetry element of translation and reflection. Cf: *glide; glide direction; glide plane; gliding.*

slip bowl A spider. See also: *spider and slips.* (Long, 1960)

slip-casting process One in which clay, or other slip, is poured into plaster molds that absorb the water, leaving a body the shape of the mold.

slip clay An easily fusible clay containing a high percentage of fluxing impurities, used to produce a natural glaze on the surface of clayware. See also: *slip.* (AGI, 1987)

slip cleavage (a) That variety of foliation along which there has been visible displacement, usually shown by bedding that is cut by the cleavage. Such displacements are commonly shown along many adjacent cleavage planes. (Billings, 1954) (b) Microscopic folding and fracture accompanied by slippage; quarrymen's false cleavage. (c) S. Wales. The cleat of coal in planes parallel with slips or faults. (d) A type of cleavage that is superposed on slaty cleavage or schistosity, and is characterized by finite spacing of cleavage planes between which occur thin, tabular bodies of rock displaying a crenulated cross-lamination. Syn: *shear cleavage; strain cleavage; strain-slip cleavage; close-joints cleavage.* (AGI, 1987)

slip clutch A friction clutch that protects a mechanism by slipping under excessive load. (Nichols, 1976)

slip dike A dike that has been intruded along a fault plane. (AGI, 1987)

slip direction The crystallographic direction in which the translation of slip takes place. See also: *glide direction.* (ASM, 1961)

slipes S. Staff. Sledge runners, upon which a skip is dragged from the working breast to the tramway.

slip fiber Veins of fibrous minerals, esp. asbestos, in which the fibers are more or less parallel to slickensided vein walls. Cf: *cross fiber.* (AGI, 1987)

slip-fiber amphibole *anthophyllite.*

slip fold *shear fold.*

slip grip A hold or grip on a drill rod, casing, or pipe by means of serrated-face steel wedges or slips. (Long, 1960)

slip joint (a) A contraction joint between two adjoining sections of wall, or at the horizontal bearing of beams, concrete slabs, and precast units, to allow slight movement in relation to one of the other. (Hammond, 1965) (b) A splined connection loose enough to allow its two parts to slide on each other to change shaft length. (Nichols, 1976)

slip line Line that appears on the polished surface of a crystal or crystalline body that has been stressed beyond the elastic limit. In quantity, they represent the intersection of the surface by planes on which shear stress has produced plastic slip or gliding. Syn: *Lüders line.* (Roark, 1954)

slip maker *clay maker.*

slip mixer *clay maker.*

slippage *slip.*

slippery parting Eng. *back.*

slipping cut In blasting underground, a cut used in a wide tunnel face, in which each successive vertical line of shots (round) breaks to the face made by the previous round, so that the relieving cut moves across the end being blasted. Also known as slabbing cut; swing cut. (Pryor, 1963)

slip plane (a) Closely spaced surfaces along which differential movement takes place in rock. Analogous to surfaces between playing cards. Syn: *glide plane; gliding plane.* (Billings, 1954; AGI, 1987) (b) The crystallographic plane in which slip occurs in a crystal. (ASM, 1961)

slip process *wet process.*

slips (a) Small faults. (BS, 1964) (b) A tool used at the mouth of a borehole to grip the drill rods or the casing, as these are being inserted or withdrawn. (BS, 1963) (c) *backs; slickensides.*

slip spear A tool for extracting tubing from a borehole.

slip surface The surface along which an earth bank is liable to fail under load. (Hammond, 1965)

slip surface of failure In a bank of homogeneous earth or clay, the slip surface of failure closely follows the arc of a circle that usually intersects the toe of the bank. Stability depends upon fixing the position of the center of rotation of the slip surface along which the greatest shearing resistance would be required for equilibrium. See also: *circular slip; landslide.* (Nelson, 1965)

slip switch A sensor installed on a conveyor drive pulley snub roller, on a return roller, on a bend roller, or on a head pulley that will detect a slowing of the conveyor, resulting in conveyor shut-down to avoid conveyor belt fires or overloading of the conveyor because of the slipping.

slip-type core lifter A device used like a core spring, consisting of a series of tapered wedges contained in slotted recesses in a circular ring or sleeve; as the core enters the inner tube, it lifts the wedges along the taper, and when the barrel is lifted, the wedges are pulled tight against the core. (Long, 1960)

slip vein A mineral vein accompanied by faulting or dislocation.

slip velocity The rate, expressed in feet (meters) per minute, at which a given size and shape of rock particle will descend or settle in water; e.g., the slip velocity in water of a round, flat particle of rock, 1/2 in (1.27 cm) in diameter, is about 54 ft/min (16.5 m/min). (Long, 1960)

slit-side solid sampler A solid-tube sampler with a slight twist on the bottom and an offset slit in the side. When rotated, the lip of the slit scrapes a sample from the side of a borehole. (Long, 1960)

slitter Eng. A pick. (Fay, 1920)

slitting disk Circular saw used in preparing rock specimens. The cutting edge incorporates diamond dust, and the thin steel disk revolves at high speed. (Pryor, 1963)

slitting shot A shot put into a large mass of coal detached by a previous blast.

slocking stone Eng. A piece of rich ore used to tempt persons into a mining enterprise. See also: *salting a mine.* (Webster 2nd, 1960)

slope (a) The entry, passage, or main working gallery of a coal seam that dips at an angle. See also: *incline.* (b) Inside slope. (c) See gradient. (d) An inclined passage driven from the dip of a coal vein. *CF: slant.* When not open at one end to the surface, it is known as an inside slope. *See also: incline; plane.* (e) The inclination of a mine roadway or coal seam. (Nelson, 1965) (f) The main working gallery or entry of a coal seam that dips at an angle and along which mine cars are hauled. (Nelson, 1965) (g) An entrance to a mine driven down through an inclined coal seam; also, a mine having such an entrance. An inside slope is a passage in the mine driven through the seam by which coal is brought up from a lower level. (Korson, 1938) (h) In a mining statute or in mining parlance, an inclined way, passage, or opening used for the same purpose as a shaft. Sometimes used to embrace the main haulage passageway, whether inclined or level. (Ricketts, 1943) (i) The degree of inclination to the horizontal. Usually expressed as a ratio, such as 1:*25, indicating one unit rise in 25 units of horizontal distance; or in a decimal fraction (0.04); degrees (2° 18'); or percent (4%). It is sometimes expressed as steep, moderate, gentle, mild, flat, etc. Also called gradient. (Hunt, 1965) (j) In surface mining, the steepest possible slope of an excavation that is consistent with safety of working. (Mining) (k) An inclined passageway (tunnel) from the surface, through the strata, that intersects the coal bed to be developed.*

slope air course A passageway parallel to the haulage slope used for the passage of the air current.

slope cage A truck on which the cars are raised at slopes or steep dips. Also called slope carriage. See also: *carriage.*

slope conveyor (a) Usually a troughed belt conveyor used for transporting coal or ore through an inclined passage to the surface from an underground mine. See also: *apron conveyor; belt conveyor; flight conveyor.* (b) Generally less than 1,000 ft (304.8 m) in length, the conveyor is designed to raise or lower material on steep grades and is commonly used to transport material from discharge bins or a main haulage conveyor to the outside. It is often used as a transfer conveyor from a lower to a higher entry or to a gangway in a pitching seam. (NEMA, 1961)

slope correction A calculation of deduction from a length as measured on a slope to bring it to its true horizontal length. See also: *tape corrections.*

slope engineer In anthracite and bituminous coal mining, one who operates a hoisting engine to haul loaded and empty mine cars along a level or inclined haulage road (slope or plane) in a mine on a level, or from a lower to an upper level, or to the mine surface. Also called drag engineer; dragline engineer; drumman; plane engineer; plane tender; slope tender. (DOT, 1949)

slope failure (a) The downward and outward movement of rock or unconsolidated material as a unit or as a series of units. Also called slump. (Leet, 1958) (b) Failure of the mass of soil beneath a natural slope or a slope of an embankment by the formation of a slide. (Huntington, 1957) (c) Slope failure may take place by one or more of three processes: (1) raveling, in which the material will assume an angle of repose approx. equal to the angle of friction of the material, and within limits the stable slope is independent of the weight of the mass, the height of the slope, and the size of the fragments. It is, however, characteristic of each rock material and is dependent on angularity, grading, and mineral content; (2) transitional failure, in which failure occurs mostly along existing fault planes or other planes of weakness. Stability is a function of rock cohesion, the angle of internal friction, the angle of dip of the slip plane, the length of the slip surface, and the total weight of the block; and (3) rotational or base failure, which is uncommon in open-pit mines or rock cuts because of geologic structure. However, in the case of homogeneous cohesive material—very deep excavation or low rock strength—failure may occur along a cylindrical surface. Four types of slope failure are rockfall, rock flow, plane shear, and rotational shear. See also: *rockfall; rock flow; plane shear; rotational shear.* Syn: *base failure.* (Lewis, 1964; Woodruff, 1966) (d) Gradual or rapid downslope movement of soil or rock under gravitational stress, often as a result of man-caused factors; e.g., removal of material from the base of a slope. (AGI, 1987)

slope gage A staff gage placed on an incline and graduated to indicate vertical heights. (Seelye, 1951)

slope hoist *direct-rope haulage.*

slope mine (a) A mine opened by a slope or incline. (b) A mine with an inclined opening used for the same purpose as a shaft or a drift mine. It resembles a tunnel, a drift, or a shaft, depending on its inclination. (Kentucky, 1952)

slope stability (a) The resistance of any inclined surface, as the wall of an open pit or cut, to failure by sliding or collapsing. (b) The resistance of a natural or artificial slope or other inclined surface to failure by landsliding. (AGI, 1987)

slope stake (a) Stake set at the point where the finished side slope of an excavation or embankment cuts the surface of the ground. It is usually placed on a line at right angles to the center line and passing through the station point. (Seelye, 1951) (b) A stake marking the line where a cut or fill meets the original grade. (Nichols, 1976)

slope staking Marking the ground surface by pegs at points where proposed new slopes in cut or fill coincide with the orginal surface. (Hammond, 1965)

slope tender *slope engineer.*

slope test A test to determine whether, and to what extent, the course of a well deviates from vertical. Syn: *angularity test.* (Williams, 1964)

sloshing loss A loss occurring when there is a fluid in the pores of the rock. This loss arises from the relative movement of the fluid and solid as the elastic waves pass through the rock. (Wyllie, 1963)

slot A narrow, vertical opening generally too small to permit traverse by a person. (AGI, 1987)

slot-and-wedge bolt A special rod designed for use in roof bolting. It consists of a mild steel rod, threaded at one end, the other end being split into halves for a length of about 5 in (12.7 cm). When the bolt is driven into the hole, a wedge opens the slotted end, thus forming the anchorage. See also: *bolt; wedge-and-sleeve bolt.* (Nelson, 1965)

slot dozing A method of moving large quantities of material with a bulldozer. Each trip is made in the same path; thus the spillage from the sides of the blade builds up along each side. All material pushed into the slot is retained in front of the blade; bigger loads are handled. (Nelson, 1965)

slotted duct sampler An instrument for sampling airborne dust consisting of a wide horizontal duct through which mine air enters in a streamline flow, so that dust particles will be deposited on the duct floor according to their falling speeds as derived from Stokes' law. The instrument combines the duties of monitoring and measuring airborne dust concentrations in mine roadways. See also: *thermal precipitator.* (Nelson, 1965)

slough Fragmentary rock material that has crumbled and fallen away from the sides of a borehole or mine working. It may obstruct a borehole or be washed out during circulation of the drilling mud. Pron: sluff. (AGI, 1987)

sloughing Minor face and rib falls. (Coal Age, 1966)

sloughing-off cone Large cone (e.g., Callow) in pulp flow line designed to remove fine slimes as overflow while delivering a thickened spigot product containing the coarser particles. (Pryor, 1963)

slovan (a) Corn. The outcrop or back of a lode. This generally applies to the appearance of a lode in a marshy place. (b) A gallery in a mine; day level; esp. applied to damp places. (Standard, 1964)

slow-banking device An appliance for use in conjunction with the Lilly controller for controlling the landing speed of less than 5 ft/s (1.52 m/s) when workers are being transported by winding. On each dial of the Lilly controller, an auxiliary dial is bolted to carry a slow-banking cam engaging near the end of the wind with a roller arm. The action of the appliance depends on the relative rate of movement of this roller and that of the piston in an oil dashpot cylinder. (Sinclair, 1959)

slow gear (a) When applied to speed at which the drill motor rotates the drill stem, the transmission gear position giving the lowest number of bit revolutions per minute; thus, slow gear corresponds to low gear in an automobile. (Long, 1960) (b) When applied to a screwfeed-type drill, the pair of feed gears in the feed mechanism that advances the bit the least amount for each revolution of drill drive rod and/or the coupled drill stem; e.g., a 400-feed gear is slower than a 100-feed gear. Cf: *feed ratio.* (Long, 1960)

slow igniter cord It consists of a plastic incendiary composition extruded around a central copper wire. An iron wire is added to give greater strength, and the whole is then enclosed in a thin extruded plastic coating. The diameter of slow cord is 0.07 in (1.8 mm). (McAdam, 1958)

slow powder Blackpowder, often called gunpowder. Also, some of the slow-acting dynamites. (Nichols, 1976)

sludge (a) Refuse from a coal-washing plant. (Standard, 1964) (b) In diamond drilling, the portion of core ground finely by accident or defect in drilling, and therefore reducing the reliability of the portion of the sample in which it happened. Mineral mud, slurry too thick to flow. (Pryor, 1963) (c) Rock cuttings produced by a drill bit. (BS, 1963) (d) *cuttings.* (e) A semifluid, slushy, murky mass of sediment resulting from treatment of water, sewage, or industrial and mining wastes; often appearing as local bottom deposits in polluted bodies of water. (AGI, 1987) (f) *slime.* Sometimes called slurry.

sludge abatement The control of the discharge into watercourses (or on adjacent land), of mineralized or impure water, or sludge, or mining debris. (Nelson, 1965)

sludge assay The chemical assaying of drill cuttings for a specific metal or group of metals. (Long, 1960)

sludge barrel (a) *sediment tube.* (b) *calyx.*

sludgebound Any part of the drill-string equipment clogged by impacted cuttings. (Long, 1960)

sludge box (a) A wooden box in which the sludge is allowed to settle from the mud flush and sometimes retained for examination. (Nelson, 1965) (b) *settling box.*

sludge bucket *calyx.*

sludge channel A tailrace for conveying the tailings away after the gold has been extracted from alluvial beds.

sludge mill A machine in which the sludge (slime) from another mill is washed; as, e.g., a slime table.

sludge paddocks Collecting or settling areas for the slurry which results from hydraulicking overburden. (Austin, 1964)

sludge pit *sump.*

sludge pump Short iron pipe or tube fitted with a valve at the lower end, with which the sludge is extracted from a borehole. Also called bailer, mud pump. See also: *mud hog; sand pump.*

sludger (a) A long cylindrical tube, fitted with a valve at the bottom and open at the top, used for raising the mud that accumulates in the bottom of a boring during the sinking process. Also called sand pump; shell pump; sludge pump. (CTD, 1958) (b) A scraper for clearing mud out of a shothole. (CTD, 1958) (c) A centrifugal pump designed for dealing with sand and slime. See also: *bailer.* (Nelson, 1965)

sludge sample (a) Samples of mud from a rotary drill, or sand from a churn drill, or fine materials from diamond drilling used to obtain information about the formation being drilled. (Nichols, 1976) (b) All or part of the drill cuttings collected, dried, and saved for assaying or chemical analysis. (Long, 1960) (c) The mud and chippings made during boring with a diamond or churn drill and sometimes used for sampling purposes. Little reliance can be placed on the assay of the sludge, and it is not regularly saved for assay, except occasionally when drilling in weathered or friable ore zones. Sludge tanks are often used to collect sludge samples. (Nelson, 1965)

sludge sampler (a) An individual responsible for collecting and preparing drill cuttings for the purpose of examining, assaying, chemical analysis, or storage. Also called sampler. (Long, 1960) (b) A device used to collect and to split drill cuttings. See also: *riffle.* (Long, 1960)

sludge sampling Process of collecting and preparing drill cuttings as samples. (Long, 1960)

sludge-saver A device for collecting all the drill cuttings from a given interval of borehole. (Long, 1960)

sludge splitter *rifle.*

sludge water *return water.*

sludging Filling or choking the waterways of a bit with drill cuttings; mudding up.

sludging formation A formation from which it is nearly impossible to recover core, so that sampling is done by collecting the drill cuttings or sludge. (Long, 1960)

sludging up *sludging.*

slue To turn, twist, or swing about. To slide and turn or slip out of course. In cutting the coal, the machine moves from right to left, the back part moving faster than the front. It is necessary at intervals to stop the machine and straighten it, or "slue" it, as called by miners.

sluff (a) Mud cake detached from the wall of a borehole. (Long, 1960) (b) A variant, incorrect spelling of slough. (Long, 1960) (c) The falling of decomposed, soft rocks from the roof or walls of mine openings.

slug (a) A piece of alluvial gold up to about 1 lb (0.45 kg) weight. (Gordon, 1906) (b) A lump of metal or valuable mineral. (c) To inject a borehole with cement slurry or various liquids containing shredded materials in an attempt to restore lost circulation by sealing off the openings in the borehole-wall rocks. (Long, 1960) (d) Small, shaped pieces of hard metal that can be brazed or handpeened in slots or holes cut in the face of a blank bit. Slugs may or may not contain diamonds. Cf: *insert.* (Long, 1960) (e) A mass of half-roasted ore. (Webster 3rd, 1966)

slug bit *insert bit.*

slugga An Irish term for a hole in the ground surface, caused by the falling-in of limestone over a subterranean stream. Cf: *sinkhole.*

slugger A tooth on a roll-type rock crusher. (Nichols, 1976)

sluice (a) To mine an alluvial deposit by hydraulicking. (Nelson, 1965) (b) A conduit or passage for carrying off surplus water, often at high velocity. It may be fitted with a valve or gate for stopping or regulating the flow. (AGI, 1987) (c) A gate, such as a floodgate. (AGI, 1987) (d) A body of water flowing through or stored behind a floodgate. (AGI, 1987) (e) A long troughlike box set on a slope of about 1:20, through which placer gravel is carried by a stream of water. The sand and gravel are carried away, while most of the gold and other heavy minerals are caught in riffles or a blanket on the floor. See also: *box sluice; ground sluice; placer mining.* (Nelson, 1965) (f) *flume.* (g) An opening in a structure for passing debris. (Seelye, 1951) (h) A channel, drain, or small stream for carrying off surplus or overflow water. (Craigie, 1938)

sluicebox Long, inclined trough or launder containing riffles in the bottom that provide a lodging place for heavy minerals in ore concentration. The material to be concentrated is carried down through the sluices on a current of water. Sluiceboxes are widely used in placer operations for concentrating elements such as gold and platinum, and minerals such as cassiterite, from stream gravels. (Newton, 1959)

sluice fork A form of fork having many tines, used to remove obstructions from a sluiceway.

sluice gate The sliding gate of a sluice. (Webster 3rd, 1966)

sluice head Aust. A supply of 1 ft^3/s (0.028 m^3/s) of water, regardless of the head, pressure, or size of orifice. Cf: *miner's inch.*

sluice tender In metal mining, a laborer who tends sluiceboxes (troughs) used in placer mining to separate gold from the sand or gravel in which it occurs; and removes wood and other obstructions to see that the gravel and water run freely through the sluices and that the riffles (cleats) are clear, so that the gold will be caught and held when settling to the bottom. (DOT, 1949)

sluiceway An artificial channel into which water is let by a sluice. (Webster 3rd, 1966)

sluicing Concentrating heavy minerals by washing unconsolidated material through a box (sluice) equipped with riffles that trap the heavier minerals on the floor of the box. (AGI, 1987)

slum (a) The very finely divided clayey portion of the residue overflowing from a sluice box, particularly applied to deep lead mining. (Nelson, 1965) (b) A short roadway to the dip in coal mines used solely to stock spare cars or the spake until required at the end of the shift. (Nelson, 1965) (c) A soft clayey or shaley bed of coal. Also spelled slumb. (d) Used in the plural for the discharge or waste from hydraulic mines. See also: *tailing; slime.*

slumgullion A usually red, muddy deposit in mining sluices. (Webster 3rd, 1966)

slump (a) A landslide characterized by a shearing and rotary movement of a generally independent mass of rock or earth along a curved slip surface (concave upward) and about an axis parallel to the slope from which it descends, and by backward tilting of the mass with respect to that slope so that the slump surface often exhibits a reversed slope facing uphill. Syn: *slumping.* See also: *slide.* (AGI, 1987) (b) The mass of material slipped down during, or produced by, a slump. (AGI, 1987)

slump bedding A term applied loosely to any disturbed bedding; specif. deformed bedding produced by subaqueous slumping or lateral movement of newly deposited sediment. (AGI, 1987)

slumping The downward movement, such as sliding or settling, of a slump. Syn: *slump.* (AGI, 1987)

slung cartridges Cartridges of explosive lowered into position in drill hole blasting at the end of a length of strong twine (not wire). As detonating cord is normally used for ignition in drill holes, the primed cartridge is lowered first in each charge by using a detaching hook or a length of twine, and it is followed by the remainder of the charge. (Nelson, 1965)

slurry (a) The fine carbonaceous discharge from a mine washery. All washeries produce some slurry, which must be treated to separate the solids from the water in order to have a clear effluent for reuse or discharge. See also: *sludge.* Syn: *water gel.* (Nelson, 1965) (b) Fine particles concentrated in a portion of the circulating water (usually by settling) and waterborne to treatment plant of any kind. (BS, 1962) (c) A thin watery suspension; e.g., the feed to a filter press or other filtration equipment. (CCD, 1961)

slurry blasting agents Dense, insensitive, high-velocity explosives of great power and very high water resistance. They are usually mixtures of an explosive such as TNT, which is a reducing agent, or an oxidizing agent, such as ammonium nitrate and/or sodium nitrate, and water. A thickening or jelling substance such as guar gum is usually added.

Slurries may also be made with nonexplosive reducing agents, including finely divided metals such as magnesium or aluminum, and with organic compounds such as sugar, molasses, or emulsified oil. These slurries are used chiefly in open pit mines where rock is hard and/or holes are wet. Also called DBA (dense blasting agent). (Nichols, 1976)

slurry helper A laborer who makes slurry for use by furnace sprayer, mixing specified amounts of silica, clay, and water in large drums with air-pressure hose. (DOT, 1949)

slurrying The process of filling in joints with slurry. (Osborne, 1956)

slurry man *furnace sprayer.*

slurry pond Any natural or artificial pond or lagoon for settling and draining the solids from washery slurry. (BS, 1962)

slurry screen A screen to recover a granular product from the circulating water in a washer, usually after a preliminary concentration of the solids and with or without the use of water sprays. (BS, 1962)

slush (a) To fill mine workings with sand, culm, or other material, by hydraulic methods. See also: *hydraulic mine filling.* (b) Silt. (c) To move ore or waste filling with a scraper (slusher) hoist.

slusher (a) *scraper loader.* (b) A mechanical dragshovel loader. (Mason, 1951) (c) A mobile drag scraper with a metal slide to elevate the bucket to dump point. (Nichols, 1976)

slusher drift Drift in stope block above haulage level, down which scraper loader conveys broken ore to loading chutes, which are usually without gates. (Pryor, 1963)

slusher operator In metal mining, one who operates the hoisting engine of a scraper loader, known as a slusher, to load ore into cars or to scrape it into chutes, or to move sand or rock fill in the stopes. (DOT, 1949)

slushing Term sometimes applied to hydraulic stowing and also to scraper loader operations. (Nelson, 1965)

slushing drift A drift in an orebody, equipped wth a scraper loader, for loading ore directly into cars in the haulage level. The drift is formed at right angles to the haulage level and over it so that the ore drops into the cars. (Nelson, 1965)

slushing oil Used to coat metals, machine parts, etc., to prevent corrosion. It usually is nondrying oil or grease, which coats the metal very well but is easily removed when desired. (Crispin, 1964)

slush pit An excavation dug near a drill to form a reservoir in which the returns from a borehole are collected and stored. Also called drill sump, mud pit, sludge pit, slush pond, sump. (Long, 1960)

slush pond *slush pit; sump.*

sly bed Eng. Soft, black, bituminous shale with a white efflorescence on exposure. A hard calcareous band near the base, Middle Purbeck Beds, Swanage. (Arkell, 1953)

small coal (a) Coal broken into small pieces, usually smaller than stove size; slack. (Standard, 1964) (b) Coal with a top size less than 3 in (7.6 cm). (BS, 1960)

small colliery Gr. Brit. Generally a colliery producing less than 1,000 st/d (907 t/d). *large colliery.* (Nelson, 1965)

small-diameter blastholes Multiple row blasting with holes 1 to 3 in (2.5 to 7.6 cm) in diameter in low-face quarries. Short-delay blasting is usually adopted using explosive factors similar to those of large diameter holes. With smaller diameter blastholes, the explosive charges can be brought up higher in the holes and so provide better breakage in blocky ground. (Nelson, 1965)

Smalley process A method for desulfurizing iron and steel with metal hydrides in which a molten slag is floated on a mass of molten ferrous metal and at least one metal hydride is introduced into the mass. The molten ferrous metal is separated from the slag, the metallic hydrides breaking down into metal and nascent hydrogen. Hydrides of the alkali metals have been found very satisfactory and are readily available. (Osborne, 1956)

small mine A coal mine employing not more than 50 persons below ground.

small ore Eng. Copper, lead, and zinc ore dressed to a small size. See also: *smalls.*

smalls (a) Small coal; slack. (Standard, 1964) (b) One of the three main size groups by which coal is sold by the National Coal Board of Great Britain. It embraces all coals with no lower size limit and ranges from a top size of 2 to 1/8 in (50 to 3.3 mm), and can be either untreated, i.e., have received no preparation other than dry screening; or treated, having been washed or dry cleaned. See also: *graded coal; large coal.* (Nelson, 1965) (c) Small particles of mixed ore and gangue. (Standard, 1964) (d) *small ore.*

small-stone bit In mineral exploration drilling, a diamond bit set with 100% size or smaller diamonds. (Long, 1960)

small tin Eng. Tin recovered from slimes.

smaltine An arsenide of cobalt, often containing nickel and iron. Also called gray cobalt; tin-white cobalt. See also: *smaltite.* (Fay, 1920)

smaltite (a) Arsenic-deficient variety of skutterudite. Syn: *smaltine; tin-white cobalt; gray cobalt; white cobalt; speisscobalt.* (b) A general term for undetermined, apparently isometric, arsenides of cobalt or for a mixture of cobalt minerals.

smaragd (a) A precious stone of light green color; a variety of beryl. (Fay, 1920) (b) *emerald.*

smaragdite A thin-foliated variety of amphibole, near actinolite in composition but carrying some alumina. It has a light green color resembling much common green diallage. (Fay, 1920)

smart aleck A limit switch that cuts off power if a machine part is moved beyond its safe range. (Nichols, 1976)

smashup A wreck, usually of haulage equipment.

smectic state Liquid-crystalline state of some fatty acids and soaps in which bundles of long molecules are oriented into parallel layers. Smectic liquid crystals are composed of a series of planes. They glide rather than flow. When less completely oriented (nematic), they are distributed at random and flow. (Pryor, 1963)

smectite (a) Any clay mineral with swelling properties and high cation-exchange capacities; an expansive clay. (b) A term originally applied to fuller's earth and later to montmorillonite; also to certain clay deposits that are apparently bentonite, and to a greenish variety of halloysite. (c) The mineral group beidellite, hectorite, montmorillonite, nontronite, pimelite, saponite, sauconite, sobotkite, stevensite, and swinefordite.

smeddum (a) Fine particles of coal or ore. (b) Fine coal slack. Also spelled smiddam; smiddum; smitham; smithem; smitten; smytham. (Standard, 1964)

smelt *smelting.*

smelter (a) Person engaged in smelting or works in an establishment where ores are smelted. (Fay, 1920) (b) An establishment where ores are smelted to produce metal. (Fay, 1920) (c) A furnace in which the raw materials of the frit batch are melted. See also: *batch smelter; continuous smelter; rotary smelter.* (ASTM, 1994)

smelter returns In a contract, returns from the ore, less the smelting charges, without deducting transportation charges. (Ricketts, 1943)

smelting (a) The chemical reduction of a metal from its ore by a process usually involving fusion, so that earthy and other impurities separate as lighter and more fusible slags and can readily be removed from the reduced metal. An example is the reduction of iron ore (iron oxide) by coke in a blast furnace to produce pig iron. Smelting may also involve preliminary treatment of ore, such as by calcination and further refining processes, before the metal is fit for a particular industrial use. (Rolfe, 1955) (b) A process distinct from roasting, sintering, fire refining, and other pyrometallurgical operations. The two most important types are reduction smelting, which produces molten metal and molten slag, and matte smelting, which produces molten matte and molten slag. Smelting may be conducted in a blast furnace, a reverberatory furnace, or an electric furnace. Reduction smelting is usually performed in blast furnaces and matte smelting in reverberatories, but there are exceptions in both cases. Syn: *smelt.* (Newton, 1959)

smelting furnace A blast furnace, reverberatory furnace, or electric furnace in which ore is smelted for the separation of a metal. (Standard, 1964)

smelting works An establishment in which metals are extracted from ores by furnaces. (Standard, 1964)

smiddam Derb. Lead-ore dust. A variation of smeddum.

smiddum Eng. A variation of smeddum.

smiddum tails Eng.; Scot. Ore sludge; ore slime. A variation of smeddum.

Smidth agglomerating kiln A rotary kiln providing an alternative method to sintering for the treatment of fine ores and flue dust. (Osborne, 1956)

smith forge A small, open-hearth furnace utilizing an air blast with coal or coke for fuel. It is commonly used for heating small metal parts previous to manual, hot-working operations. (Henderson, 1953)

smithite A monoclinic mineral, $AgAsS_2$; dimorphous with trechmannite; red.

Smith process (a) A variation of the series system of copper refining in which the plates are placed horizontally, the top surface of each one acting as cathode, the lower as anode. Linen diaphragms must be placed between the plates to catch the slime. When these diaphragms break and allow the slime to drop on the cathode, it is difficult to remedy any short circuits without dismantling the tank. (Liddell, 1918) (b) A process for sponge iron production that is carried out in vertical ovens or retorts, similar to coke ovens in design. Crushed ore or iron oxide material is mixed with carbonaceous material and charged into the oven, where it is heated and cooled by means of horizontal flues. It is preheated in the upper part of the oven by the waste gases; then the charge enters the reduction zone, and is subsequently cooled by the incoming air for combustion in the heating flues. (Osborne, 1956)

smithsonite (a) A white to yellow, or to brown and rarely green or blue variety of zinc carbonate. See also: *zinc carbonate.* (Bennett, 1962) (b) A trigonal mineral, $ZnCO_3$, with Zn replaced by Fe; of the calcite group; rhombohedral cleavage; varicolored; commonly reniform, botryoidal, stalactitic, or granular; commonly altered from sphalerite in oxidized zones of limestone replacement; a source of zinc; distinguished from hemimorphite by its effervescence in acids. Syn: *dry-bone* *ore; calamine; zinc spar; szaskaite.* (c) A term sometimes used as a syn. of hemimorphite.

smitten Fine gravellike ore, occurring free in mud openings, or derived from the breaking of the ore in blasting. A variation of smeddum.

smoke The exhalation, visible vapor, or material that escapes or is expelled from a burning substance during combustion; applied esp. to the volatile matter expelled from wood, coal, peat, etc. together with the solid matter that is carried off in suspension with it. That which is expelled from metallic substances is generally called fume or fumes. See also: *fume; metallurgical smoke.*

smokeless powder Nitrocellulose containing 13.1% nitrogen. Produced by blending material of somewhat lower (12.6%) and slightly higher (13.2%) nitrogen content; converting to a dough with an alcohol-ether mixture; extruding; cutting; and drying to a hard, horny product. Small amounts of stabilizers (amines) and plasticizers are usually present, as well as various modifying agents (nitrotoluene and nitroglycerin salts). (CCD, 1961)

smoke stick A means of making a smoke cloud to measure the velocity of air. Fuming sulfuric acid or anhydrous tin tetrachloride are favorite smoke producers. (Zern, 1928)

smokestone *cairngorm; smoky quartz.*

smoke technique A technique used to measure only very low-speed air velocity. The release of smoke enables the fluid motion to be observed with the eye. If the smoke is timed over a measured distance along an airway of constant cross section then the velocity of flow can be determined. Usually a spot reading, that of maximum velocity, is obtained. (Roberts, 1960)

smoke tester One who tests the efficiency of the Cottrell plant and flue recovery method by determining the rate of discharge of gases and solids from the smelter smokestack. (DOT, 1949)

smoke tube To determine the presence of moving air, the direction of flow, and the approximate velocity of flow, the smoke tube method is commonly used, particularly in metal mines. The device consists of an aspirator bulb, which discharges air through a glass tube containing a smoke-generating reagent. Usually pumice stone saturated with anhydrous tin or titanium tetrachloride is employed. The dense white cloud of smoke released when the bulb is squeezed travels with the air current; the approximate air velocity in a mine airway is determined by timing how long the cloud takes to travel between two points. (Hartman, 1982)

smoke washer A device in which smoke is forced upward against a downward spray of water in order to remove the solid particles in the smoke. (Webster 3rd, 1966)

smoke zone The area surrounding a smelting plant in which the smoke or fumes damage vegetation, or in which it may be classed as a public menace or nuisance. (Fay, 1920)

smoky quartz A light to very dark brown variety of quartz sometimes used as a semiprecious gemstone. Syn: *cairngorm; smokestone.*

smoky topaz A trade name for smoky quartz used for jewelry.

smooth blasting A technique used in surface and underground blasting in which a row or closely spaced drill holes are loaded with decoupled charges (charges with a smaller diameter than the drill hole) and fired simultaneously to produce an excavation contour without fracturing or damaging the rock behind or adjacent to the blasted face. See also: *controlled blasting.*

smooth drilling A rock formation in which a high recovery of core can be attained at a high rate of penetration. (Long, 1960)

smoother bar A drag that breaks up lumps behind a leveling machine. (Nichols, 1976)

smooth-faced drum Plain-faced drum without grooves. (Hammond, 1965)

smooth head York. A smooth plane of cleavage. See also: *bright head.*

smoothing trowel Trowel used by plasterers and cement workers for finishing surfaces. (Crispin, 1964)

smooth roll A crusher in which the material passes between a moving set of rolls with smooth surfaces. See also: *roll.* (ACSG, 1963)

S.M.R.E. combustible gases recorder This methane recorder is a combination of the principles used in the Ringrose and McLuckie detectors. Two combustion chambers are used, each of which operates every 6 min, the operation being staggered so that 20 determinations are made per hour. Samples are drawn into the instrument by means of an electrically driven pump, the operation of the pump and the combustion filaments being controlled by cams on a shaft driven by the motor. Pressures are measured by an aneroid cell and recorded on a clockwork driven drum. (Roberts, 1960)

smut (a) Soft, inferior coal. (b) S. Staff. Bad, soft coal, containing much earthy matter. See also: *muck.* (c) Coal smuts. (CTD, 1958) (d) Worthless outcrop material of a coal seam. (CTD, 1958) (e) Poor, dull, sooty portions of a coal seam. (Gordon, 1906) (f) A reaction product sometimes left on the surface of a metal after a pickling or etching operation. (ASM, 1961)

smytham Lead ore that has been stamped or pounded down to a sand or powder to remove rock and earth from the ore. (Hess)

smythite A trigonal mineral, $(Fe,Ni)_9S_{11}$ or $(Fe,Ni)_{13}S_{16}$; physical properties are almost identical with monoclinic pyrrhotine.

snag boat A boat equipped with a hoist and grapple for clearing obstacles from the path of a dredge. (Nichols, 1976)

snake fashion A method of boxing core. Beginning in the upper-right-hand corner of the core box the core is run from right to left in the first row, from left to right in the second row, left to right in the third, etc., until the box is filled. Cf: *reverse book fashion.* (Long, 1960)

snakehole (a) A borehole driven horizontally or nearly so and approx. on a level with the quarry floor. (b) A borehole driven under a boulder for containing a charge of explosives. In quarry work, it is called a lifter. (c) Nearly horizontal holes drilled at the bottom of the face of a bench. The holes are not quite horizontal but are inclined slightly downward so the bottoms will be a few feet below grade. (Lewis, 1964) (d) A hole driven into a toe for blasting, with or without vertical holes. (Nichols, 1976)

snakeholing (a) A method of blasting boulders to break them up, by boring a hole under a boulder and firing a charge in it; this is more efficient but slower than using plaster shots. See also: *plaster shooting.* (Hammond, 1965) (b) A horizontal bore on the quarry floor. (Pryor, 1963) (c) Drilling under a rock or face in order to blast it. (Nichols, 1976)

snake line A line used to skid a drill rig from place to place using a block and tackle or cable, one end of which is attached to a deadman and the other wrapped around the hoisting drum. (Long, 1960)

snake statement A monthly statement by a coal company on which a crooked line in red ink was drawn to show a miner's indebtedness. The company checked off rent, supplies, and groceries, which often added up to more than a miner's monthly earnings. (Korson, 1938)

snaking (a) The progressive sliding forward of an armored flexible conveyor, by means of hydraulic rams, as the coal is removed by a cutter loader. See also: *self-advancing supports.* (Nelson, 1965) (b) Moving a drill rig by the use of its own cathead or hoist unit. See also: *bulldog.* (Long, 1960) (c) Towing a load with a long cable. (Nichols, 1976) (d) Inserting a tow or hoist line under an object without moving the object. (Nichols, 1976)

snaking conveyor *armored flexible conveyor.*

snaphead rivet A rivet having a hemispherical head. (Hammond, 1965)

snapper (a) A car coupler, trip rider, or brakeman. (b) *grab sampler.*

snatch block (a) A pulley in a case that can be easily fastened to lines or objects by means of a hook, ring, or shackle. (Nichols, 1976) (b) A single-rope sheave set in a housing provided with a latch link, which can be opened for admission of a rope or line without the necessity of threading the end of the rope through the block. (Long, 1960) (c) A block or sheave with an eye through which lashing can be placed for fastening to a scaffold or pole. (Hammond, 1965) (d) A sheave in a case having a pull hook or ring. (Nichols, 1976)

sneck (a) The latch or catch of a door. (Webster 3rd, 1966) (b) To lay (rubblework) with spalls and fragments to fill the interstices. (Webster 3rd, 1966)

Snell's law (a) The concept that the ray path of sound or light undergoes certain specific changes as it passes through different layers of water; the ratio of the sine of the angle of incidence to the sine of the angle of refraction is the same for all angles of incidence and is equal to the index of refraction. (Hy, 1965) (b) Ordinary refraction where the index of refraction equals the sine of the angle of incidence divided by the sine of the angle of refraction. Syn: *law of refraction.* Cf: *ordinary ray.*

snorehole The hole in the lower part of the windbore of a mining pump to admit the water.

snort valve A butterfly valve opening from the cold-blast main of a blast furnace to the atmosphere. It allows casting at the furnace without shutting down the blowing engines; operated by large wheel or lever in the cast house. (Fay, 1920)

snow Vapor-deposited skeletal ice crystals without a substrate; their accumulation; used for recreation and as a coolant. *ice.*

snow gage *rain gage.*

snub (a) To increase the height of an undercut by means of explosives or otherwise. (Fay, 1920) (b) To check descent of a car, by a turn of a rope around a post. Cf: *jig chain.* (Fay, 1920) (c) To check the descent of any object being lowered by hand.

snubber In bituminous coal mining, a laborer who follows in the wake of a coal-cutting machine as it undercuts the face of coal, and breaks down the front of the working face above the channel with a pick so that the coal will drop freely when it is blasted. (DOT, 1949)

snubbing (a) A term applied by bluestone quarrymen to the process of forcing a cross break in the absence of an open seam. (b) Increasing the height of an undercut by picking or blasting down the coal, just above the undercut.

snub-drive conveyor A conveyor in which a snub pulley is used. (NEMA, 1956)

snub pulley An idler pulley so mounted as to increase the arc of contact between a belt and a drive pulley. When used in a wrap drive, it has the added function of changing the direction of the return belt travel. (NEMA, 1961)

Snyder sampler A mechanical sampler consisting of a cast-iron plate revolving in a vertical plane on a horizontal axis, and having an inclined sample spout passing through the flange. The ore to be sampled comes to the sampler by way of an inclined chute and impinges upon the flange of the sampling disk. Whenever the sample spout comes in line with the ore stream, the ore passes through the plate and into the sample; at all other times the ore is deflected from the plate and drops into the reject. Generally, the sampler makes from 10 to 30 rpm, and removes about 1/5 of the ore stream. (Newton, 1959; Newton, 1938)

soaking (a) A phase of a heating operation during which metal is maintained at the requisite temperature until the temperature is uniform throughout the mass. (CTD, 1958) (b) Prolonged holding at a selected temperature. (ASM, 1961)

soap earth *steatite.*

soaprock *soapstone.*

soapstone (a) Massive talc. Syn: *steatite; soaprock.* (b) A metamorphic rock of massive, schistose, or interlaced fibrous or flaky texture and soft, unctuous (greasy) feel; composed essentially of talc with variable amounts of mica, chlorite, amphibole, and pyroxene; alteration product of ultramafic rock; may be carved into art objects or sawn into dimension stone for use where chemical resistance or high heat capacity is needed. (c) A miners' and drillers' term for any soft, unctuous rock. Syn: *agalmatolite; Manchurian jade; talcum.* (d) *saponite; talc.*

soapy feel Unctuous; said of talc and other magnesium minerals. (Nelson, 1965)

soapy heads Eng. In stones, joints filled with saponaceous or talclike mineral.

sobotkite An aluminum-rich variety of saponite(?).

socket (a) A device fastened to the end of a rope by means of which the rope may be attached to its load; the socket may be opened or closed. (Zern, 1928) (b) In blasting, the hole left after firing. (Pryor, 1963) (c) A hollow tool for grasping and lifting tools that have been dropped into a well boring. (Standard, 1964) (d) The point in a borehole, usually in bedrock, at which the bottom end of a string of casing or drivepipe is set. (Long, 1960) (e) To lower casing or drivepipe into, and seat it in, a borehole. (Long, 1960) (f) An overshot. (Long, 1960) (g) A fishing tool designed to encircle and grasp a cylindrical object. (Long, 1960) (h) To spring a borehole. See also: *camouflet.* (Long, 1960) (i) *bootleg.*

soda (a) The normal carbonate of sodium, Na_2CO_3, soda ash; the latter being the common name of the commercial product used in chemical industries. (b) Sodium carbonate, Na_2CO_3; esp. the decahydrate, $Na_2CO_3 \cdot 10H_2O$. Loosely used for sodium oxide, sodium hydroxide, sodium bicarbonate, and even for sodium in informal expressions such as soda spar. A prefix added to the names of igneous rocks to indicate that they contain soda pyroxenes and/or soda amphiboles; e.g., soda rhyolite, soda trachyte, soda granite, etc. (AGI, 1987) (c) A former name for natron.

soda alum (a) An alum of aluminum and sodium. Occurs in nature as the mineral mendozite. See also: *mendozite.* (Standard, 1964) (b) *sodium alum.*

soda-and-cleanup man A laborer who sifts soda ash for use in refining copper. (DOT, 1949)

soda ash Commercial term for sodium carbonate, Na_2CO_3. (AGI, 1987)

soda feldspar A sodium-aluminum silicate occasionally used as a refractory raw material in the manufacture of porcelain enamels, giving a softer enamel when used to replace potash feldspar in equal weights. See also: *albite.* (Enam. Dict., 1947; CTD, 1958)

soda hornblende *arfvedsonite.*

soda lakes (a) Salt lakes, the water of which contains a high content of sodium salts (chiefly chloride, sulfate, and acid carbonate). These salts also occur as an efflorescence around the lakes. (CMD, 1948) (b) For soda lakes where the water has evaporated leaving behind evaporite salts, the term "dry soda lakes" may be used.

soda lime (a) A granular mixture of calcium hydroxide with sodium hydroxide, or potassium hydroxide, or both, and sometimes with other substances (such as kieselguhr). Used to absorb moisture and acid gases, esp. carbon dioxide, as in gas masks. (b) Can refer to a type of glass container in which the principal ingredients are soda ash and lime.

soda-lime sinter process Process for recovering alumina from red mud by mixing it with soda ash (Na_2CO_3) and ground limestone, and sintering in a rotary kiln at temperatures of 1,800 to 2,000 °F (980 to 1,090 °C). This breaks up the sodium aluminum silicate and forms an insoluble calcium silicate and sodium aluminate. The sinter is leached with water to recover the sodium silicate, which is then treated in the same way as in the standard Bayer process. (Newton, 1959)

sodalite (a) An isometric mineral, $Na_8Al_6Si_6O_{24}Cl_2$; sodalite group; typically blue or blue-violet; in various sodium-rich igneous rocks. (b) The mineral group hauyne, lazurite, nosean, and sodalite.

sodalitite An urtite composed chiefly of sodalite, with smaller amounts of acmite, eudialyte, and alkali feldspar. (AGI, 1987)

soda mica *paragonite.*

soda microcline A microcline in which sodium replaces potassium. (Dana, 1959)

soda niter *nitratine; sodium nitrate.*

soda nitrate $NaNO_3$, widely used as a fluxing raw material in enamels, usually in conjunction with soda ash. A small percentage is beneficial in oxidizing any organic impurities. Also called Chile nitre; saltpeter. Syn: *sodium nitrate.* (Enam. Dict., 1947)

soda orthoclase Apparently monosymmetric feldspars with a notable amount of soda may be called soda orthoclase. When the soda equals or exceeds the potash, the crystals exhibit triclinic symmetry and are soda microcline. (Spencer, 1955)

soda prairie An extensive level barren tract of land covered with a whitish efflorescence of sodium carbonate (natron), as in parts of southwestern and western United States and Mexico. Syn: *salt prairie.* (AGI, 1987)

soda richterite To replace the name astochite. See also: *astochite.* (English, 1938)

soda spar An informal term for sodic feldspar, i.e. albite, or for a feldspar mixture assaying at least 7% Na_2O. Syn: *Na-spar.* Cf: *potash spar.* (AGI, 1987)

soddyite An orthorhombic mineral, $(UO_2)_2SiO_4 \cdot 2H_2O$; pale yellow; in pegmatites with malachite, in fissure fillings with curite and sklodowskite. Also spelled soddite.

Soderberg anode A continuously formed anode for aluminum production in which the mixture of petroleum coke and coal-tar pitch is continuously added to a steel casing and is baked as it passes through the heated casing, such that the baked anode emerging into the cell continuously replaces the anode being consumed.

Soderberg electrode A continuously formed electrode used in a metallurgical electrical furnace, in which a mixture of petroleum coke and coal-tar pitch is continuously added to a steel casing and is baked as it passes through the heated casing, such that the baked electrode emerging into the furnace continuously replaces the electrode being consumed; e.g., used in aluminum and ferroalloy production; may be oriented either vertically or horizontally.

sodium (a) A soft, bright, silvery metallic element; one of the alkali metals. Symbol, Na. It is a very reactive element and is never found free in nature. The most common compound is sodium chloride. Sodium compounds are important to the paper, glass, soap, textile, petroleum, chemical, and metal industries. Metallic sodium should be handled with great care; it should be kept in an inert atmosphere and contact with water avoided. (Handbook of Chem. & Phys., 3) (b) For some minerals with the "sodium" prefix, search under the root mineral name.

sodium alum An isometric mineral, $NaAl(SO_4)_2 \cdot 12H_2O$. Syn: *soda alum.*

sodium arc light *sodium light.*

sodium autunite A tetragonal mineral, $Na_2(UO_2)(PO_4)_2 \cdot 8H_2O$; meta-autunite group; radioactive; yellow.

sodium bentonite *Wyoming bentonite.*

sodium-calcium feldspar *plagioclase.*

sodium ethylxanthate Pale green; NaC_2H_5OCSS; molecular weight, 144.19; and soluble in water and in ethyl alcohol. An ore-flotation agent. (Bennett, 1962; Handbook of Chem. & Phys., 2)

sodium feldspar *albite.*

sodium hydroxide *caustic soda.*

sodium illite A sodium-rich illite. *brammallite.*

sodium light Yellow light emitted by the glowing vapor of sodium, consisting of the D lines of sodium at wavelengths 5,890 Å and 5,896 Å; produced by special lamps and filters. Syn: *sodium vapor light; sodium arc light.*

sodium niobate $NaNbO_3$; a compound believed to be ferro-electric and having potential use as a special electroceramic. The Curie temperature is 360 °C. (Dodd, 1964)

sodium nitrate Colorless; transparent; odorless; hexagonal; $NaNO_3$; molecular weight, 84.99; slightly bitter taste; sp gr, 2.267; melting point, 308 °C; no boiling point because it decomposes at 380 °C; soluble in water, in ethyl alcohol, and in methyl alcohol; very soluble in ammonia; only slightly soluble in acetone; and slightly soluble in glycerol. Used as an oxidizer in solid rocket propellants, a flux, in glass manufacture, in military explosives, and in enamel for pottery. Also used in enamel frit batches to prevent reduction of any easily reducible ingredients, such as lead oxide. The function of sodium nitrate, when used in glass batches, is to oxidize organic matter and to prevent reduction of some of the batch constituents. Syn: *soda niter.* See also: *soda nitrate.* (CCD, 1961; Handbook of Chem. & Phys., 2; Lee, 1961)

sodium nitrate gelignites These explosives are sometimes referred to as straight gelatins. They are really modifications of blasting gelatin in which varying percentages of nitroglycerin are replaced by sodium nitrate and combustible material to give a range of gelatinous explosives of varying strengths. The nitroglycerin content may be from about 30% to 80%. Straight gelatins are characterized by their plastic consistency; high densities of 1.5 to 1.6 g/cm^3; medium velocity of detonation of 2,500 m/s; good resistance to the effects of water, which also gives them good storage properties, and they possess fume characteristics suitable for underground workings. All the various requirements for metal mining, tunneling, and quarrying operations are covered by this wide range of gelatinous explosives. Their high resistance to water makes them particularly useful in wet conditions, and their relatively high density is advantageous where a powerful concentration of explosive energy is required. (McAdam, 1958)

sodium regulations A permit, not including more than 2,560 acres (1,037 ha) in one State, is granted to prospect for chlorides, sulfates, carbonates, borates, silicates, or nitrates of sodium, and the royalty and rentals are similar to those for potash. Where necessary to secure the most economical mining, a person, association, or corporation may be permitted to hold up to 15,360 acres (6,221 ha) in one State. (Lewis, 1964)

sodium sesquicarbonate Colorless, gray, or yellowish-white; monoclinic; $Na_3(CO_3)(HCO_3){\cdot}2H_2O$; mol wt, 226.03; sp gr, 2.112, but ranges from 2.11 to 2.147 in the mineral; Mohs hardness, 2.5 to 3.0; soluble in water. Extensive deposits of sodium sesquicarbonate, variously known as the mineral natrona, trona, or urao, occur in California and Wyoming; in Hungary and Egypt; and in the deserts of Africa, Asia, and South America. Syn: *natrona; trona; urao.* (Handbook of Chem. & Phys., 2; CCD, 1961)

sodium tannate Sometimes used as a deflocculant for clay slips; the effect is marked, only a small proportion being required. The material used for this purpose is generally prepared from NaOH and tannic acid; the former should be in excess and the pH should be about 8 to 9. (Dodd, 1964)

sodium tantalate $NaTaO_3$; a ferro-electric compound having the ilmenite structure at room temperature; the Curie temperature is approx. 475 °C. Of potential interest as a special electroceramic. (Dodd, 1964)

sodium uranate $Na_2(UO_2)_2(ASO_4)_2{\cdot}8H_2O$; has been used as a source of uranium for uranium red. (Dodd, 1964)

sodium uranospinite A tetragonal mineral, $(Na_2,Ca)(UO_2)_2(AsO_4)_2{\cdot}5H_2O$; meta-autunite group; radioactive; yellow-green to yellow; forms fine, tabular to elongated crystals or radial fibrous aggregates; in the oxidation zone of primary hydrothermal deposits. It is the most abundant secondary uranium mineral.

sodium vanadate $NaVO_3$; a ferro-electric compound having potential use as a special electroceramic. The Curie temperature is approx. 330 °C. (Dodd, 1964)

sodium vapor light *sodium light.*

sodium xanthate (a) Yellowish; amorphous; $NaS{\cdot}CS{\cdot}OC_2H_5$; molecular weight, 144.19; and soluble in water and in ethyl alcohol. Used as a flotation agent. (Bennett, 1962; Handbook of Chem. & Phys., 2) (b) *sodium ethylxanthate.*

soffioni An emanation, from the Earth, of vapors that are principally boric acid; also, the opening from which the vapors issue. See also: *solfatara; fumarole; mofette.*

soft (a) Bituminous as opposed to anthracite coal. (b) In mineralogy, usually refers to minerals readily scratched by a needle or knife blade. (Hess) (c) Tender, friable, or full of slips and joints. (d) As applied to a glass or glaze, refers to a low softening temperature; such a glass or glaze, when cold, is also likely to be relatively soft; i.e., of lower than average hardness, in the normal sense. (Dodd, 1964)

soft coal Bituminous coal as opposed to anthracite. See also: *bituminous coal.* (Fay, 1920)

soften To heat ore so that the minerals are cracked and fissured, permitting easier crushing. (Fay, 1920)

softening (a) Treatment in which metal is heated below its critical point and then slowly cooled. (Pryor, 1963) (b) Of lead, the removal of antimony and other impurities. (Fay, 1920)

softening point (a) Certain materials do not have a definite melting point but soften over a range of temperatures. In certain refractory substances, the softening point is measured as the pyrometric cone equivalent (PCE), which is the number of that standard pyrometric cone whose tip would touch the supporting plaque simultaneously with a cone of the refractory material being investigated. (b) When referring to glass, the temperature at which the viscosity is $10^{7.6}$ P ($10^{0.6}$ MPa·s); this viscosity corresponds to the temperature at which tubes, for example, can be conveniently bent. Also known as the 7.6 temperature; Littleton softening point. (Dodd, 1964)

soft formation *soft ground.*

soft ground (a) That part of a mineral deposit that can be mined without drilling and shooting. It is commonly the upper, weathered portion of the deposit. (AGI, 1987) (b) Heavy ground. Rock about underground openings that does not stand well and requires timbering.

soft-ground boring tool Drilling tool used in soft ground, such as overburden, clay, soft shale, etc. (Long, 1960)

soft inclusion Applied in the grading of quartz crystals to feathery or fernlike types of foreign inclusions, which look soft (no implication of physical hardness). (Am. Mineral., 1947)

soft iron Iron which can be worked with ordinary cutting tools or which can be readily abraded with files. It is darker gray in color than the harder cast iron. (Crispin, 1964)

soft mica Mica which, when slightly flexed or distorted with thumb pressure, generally shows a tendency toward delamination. Such mica, in thick pieces, generally gives a dull sound when tapped against a hard surface. (Skow, 1962)

softness Tendency to deform easily. It is indicated in a tensile test by low ultimate tensile stress and large reduction in area. Usually the elongation is also high. In a notched bar test, specimens bend instead of fracturing, and energy absorbed is relatively small. See also: *toughness; brittleness.* (CTD, 1958)

soft ore A term used in the Lake Superior region for an earthy, incoherent iron ore mainly composed of hematite or limonite (goethite) and containing 45% to 60% iron. (AGI, 1987)

soft phosphate A term which is applied arbitrarily to anything phosphatic that is not distinctly hard rock.

soft radiation Ionizing radiation of long wavelength and low penetration. (NCB, 1964)

soft rock (a) A term used loosely for sedimentary rock, as distinguished from igneous or metamorphic rock. (AGI, 1987) (b) Rock that can be removed by air-driven hammers, but cannot be handled economically by pick. Cf: *hard rock.* (AGI, 1987)

soft-rock geology A colloquial term for geology of sedimentary rocks, as opposed to hardrock geology. (AGI, 1987)

soft skin A soft outer skin developed on burned diamonds. (Long, 1960)

soft steel A general term applied to steels of low carbon content that do not temper. See also: *mild steel.* (Crispin, 1964)

soft vector A plane or direction in a diamond or other mineral having less resistance to abrasion than that of the hard-vector planes. Cf: *hard vector.* (Long, 1960)

soft water Water free from calcium carbonate and calcium sulfate. Cf: *hard water; hardness.* (Crispin, 1964)

soil (a) All unconsolidated materials above bedrock. This is the meaning of the term as used by early geologists and in some recent geologic reports, and has been vigorously advocated by Legget (1967, 1973). It is the common usage among engineering geologists (see, e.g., compaction; soil mechanics). In recent years the approx. syn. regolith has come into wide geological use. (AGI, 1987) (b) The natural medium for growth of land plants.—Etymol: Latin solum, ground. See also: *sounding.* (AGI, 1987)

soil analysis *soil survey.*

soil catena A related sequence of soil profile types created by changes from one drainage condition to another. These changes are usually transitional. (Hawkes, 1962)

soil cement The addition of cement to a soil, as a binding agent, and converting it into a weak form of concrete. See also: *cement stabilization.* (Nelson, 1965)

soil classification tests The tests are of two main types, namely: (1) mechanical analysis, performed by sieving or sedimentation, to determine the size-distribution of the constituent particles; and (2) index property tests, for soils passing a 36-mesh British Standard sieve, by means of which the type is deduced from the moisture content at standard consistencies. See also: *index properties.* (Nelson, 1965)

soil core A cylindrical sample of soil for tests and examination. Undisturbed samples may be obtained by the use of special appliances, which allow extraction of soil cores of diameter usually 1-1/2 in (3.8 cm) or 4 in (10.2 cm). The core barrel, forming part of the coring tool, is detachable and is capped and the core hermetically sealed for delivery to the laboratory. The natural moisture content and other properties are preserved for examination. Individual soil cores, up to 3 ft (0.9 m) in length, may be obtained by continuous coring if necessary. See also: *borehole samples; undisturbed sample.* (Nelson, 1965)

soil creep The gradual, steady downhill movement of soil and loose rock material on a slope that may be very gentle but is usually steep. Syn: *surficial creep.* (AGI, 1987)

soil flow *solifluction.*

soil fluction *solifluction.*

soil formation The processes whereby fragmental material resulting from rock weathering is transformed into a medium that can support plant growth.

soil-forming factors Factors, such as parent material, climate, vegetation, topography, organisms, and time, involved in the transformation of an original geologic deposit into a soil profile. (ASCE, 1958)

soil horizon A layer of a soil that is distinguishable from adjacent layers by characteristic physical properties, such as structure, color, or texture, or by chemical composition, including content of organic matter or degree of acidity or alkalinity. Soil horizons are generally designated by a capital letter, with or without a numerical annotation, e.g.

A horizon, A2 horizon. Syn: *horizon; soil zone; pedologic horizon.* (AGI, 1987)

soil mechanics The application of the principles of mechanics and hydraulics to engineering problems dealing with the behavior and nature of soils, sediments, and other unconsolidated accumulations of solid particles; the detailed and systematic study of the physical properties and utilization of soils, esp. in relation to highway and foundation engineering and to the study of other problems relating to soil stability. (AGI, 1987)

soil penetrometer A sounding instrument, which may be used to supplement the vane test. It consists essentially of a rod inside a tube. When the appliance is mechanically jacked into the ground, the point or cone of the rod records all differences in resistance to penetration. It thus determines quickly the soil strength profile in depth and detects any soft beds in advance of the vane tests. See also: *penetration log.* (Nelson, 1965)

soil physics The organized body of knowledge concerned with the physical characteristics of soil and with the methods employed in their determinations. (ASCE, 1958)

soil profile A vertical section of a soil that displays all its horizons. Syn: *profile.* (AGI, 1987)

soil sampler (a) A tube driven into the ground so as to obtain an undisturbed soil sample. In sands, such tubes would be fitted with a core catcher. (Hammond, 1965) (b) One of a number of different mechanical devices used for taking samples of an unconsolidated material. See also: *solid-barrel sampler; split-tube barrel; split-tube sampler.* (Long, 1960)

soil science The study of the properties, occurrence, and management of soil as a natural resource. Generally it includes the chemistry, microbiology, physics, morphology, and mineralogy of soils, as well as their genesis and classification. Syn: *pedology.* (AGI, 1987)

soil shredder A machine employed in soil stabilization comprising two nearly touching half drums, which rotate in opposite directions and break up the soil. (Hammond, 1965)

soil stabilization Chemical or mechanical treatment designed to increase or maintain the stability of a mass of soil or otherwise to improve its engineering properties. See also: *cement-modified soil; cement stabilization; processing; pulverization.* (ASCE, 1958)

soil structure The arrangement and state of aggregation of soil particles in a soil mass. See also: *flocculent structure; honeycomb structure; single-grained structure.* (ASCE, 1958)

soil survey (a) A detailed investigation of the soils at a site, including boreholes and tests to determine their nature, thickness, strength, and depth to bedrock. See also: *site investigation; soil mechanics.* (b) Geochemical prospecting term for the chemical analysis of systematically collected samples of soil and weathered rock.

soil test The laboratory procedure followed in examining and determining the physical characteristics of a soil sample. (Long, 1960)

soil zone *soil horizon.*

Soisson Rodange process A process for the manufacture of high-quality killed basic Bessemer steel in which the steel, after dephosphorization, is poured into another ladle containing the solid components of a basic oxidizing and fluid slag. Blowing for 30 to 40 s generates sufficient heat to promote mixing and to avoid skull. Phosphorus contents are readily lowered and high-quality killed steel is produced with low additional cost. (Osborne, 1956)

sol (a) A homogeneous suspension or dispersion of colloidal matter in a fluid (liquid or gas). (AGI, 1987) (b) A completely mobile mud. A sol is in a more fluid form than a gel. (AGI, 1987)

solar (a) A platform in a Cornish Mine shaft and esp. between a series of ladders; a longitudinal partition forming an air passage between itself and the roof in a mine. Usually spelled sollar or soller. See also: *sollar.* (Webster 3rd, 1966) (b) A colloquialism used by surveyors to mean an observation on the sun.

solar salt Salt obtained by solar evaporation of seawater or other brine in shallow lagoons or ponds.

solder (a) To unite the surfaces of metals. (Nelson, 1965) (b) An alloy of uniting metals. Brazing solders are alloys of zinc and copper, while soft solders are alloys of tin and lead. (Nelson, 1965)

soldier frame Frame set into the inside of a shaft prior to breaking through for a heading. (Stauffer, 1906)

sole (a) The under surface of a rock body or vein, esp. the bottom of a sedimentary stratum. (AGI, 1987) (b) The fault plane underlying a thrust sheet. (AGI, 1987) (c) The middle and lower parts of the shear surface of a landslide. (AGI, 1987) (d) The major fault plane over which other beds ride forward as a group during distributive faulting. (e) The lowest thrust plane in an area of overthrusting. Commonly rocks above are imbricated. (AGI, 1987) (f) The bottom of a level.

Solenhofen stone A lithographic limestone of Late Jurassic age found at Solenhofen (Solnhofen), a village in Bavaria, West Germany. It is evenly and thinly stratified and contains little clays. See also: *lithographic limestone.* (AGI, 1987)

solfatara A type of fumarole, the gases of which are characteristically sulfurous. Etymol: the Solfatara volcano, Italy. (AGI, 1987)

solfataric Applied to a dormant or decadent stage of volcanic activity characterized by the emission at the surface of gases and vapors of volatile substances.

solid (a) Coal that has not been undermined, shear cut, or otherwise prepared for blasting. Used in the expression, "shooting off the solid.". (Fay, 1920) (b) That part of the coal that cannot be thrown out by a single shot, or the coal beyond the loose end. Used in expressions describing holes drilled for blasting, such as "three feet into the solid," or "on the solid.". (Fay, 1920) (c) Unmined; ungot. (Mason, 1951) (d) A rock having few open cracks, crevices, or joints and relatively unaffected by the weakening effects of weathering. (Long, 1960) (e) A diamond free of cracks discernible by eye. (Long, 1960) (f) The rock near underground openings that stands well without artificial support. (Long, 1960)

solid-barrel sampler A straight-walled cylinder with or without a valve on the bottom. Used for taking soil samples. Cf: *soil sampler; split-tube sampler.* (Long, 1960)

solid bearing A one-piece bushing. (Nichols, 1976)

solid bit *noncoring bit.*

solid bituminous material Material having a penetration at 77 °F (25 °C), under a load of 100 g applied for 5 to 10 s. (Urquhart, 1959)

solid car A mine car equipped with a swivel coupling and generally used with a rotary dump. One or more cars are pulled into the rotary dump, which turns through 180° and the coal is emptied out. See also: *mine car.* (Kentucky, 1952)

solid couplings Generally of either the flanged-face or the compression type. They are used to connect two shafts to make a permanent joint and usually are designed to be capable of transmitting the full load capacity of the shaft. This coupling has no flexibility, either torsional, angular, or axial, hence it is limited to those installations where rigid connections are suitable, particularly in line shafts and extension shafts. (Pit and Quarry, 1960)

solid crib timbering Shaft timbering with cribs laid solidly upon one another.

solid-crown bit *noncoring bit.*

solid-drawn Drawn from hollow ingots, or otherwise, on mandrels of successively decreasing diameters; said of certain seamless metal tubes. (Standard, 1964)

solid drilling In diamond drilling, using a bit that grinds the whole face, without preserving a core for sampling. (Nichols, 1976)

solid explosives These explosives are employed to a certain extent in the form of a powder in cartridges, or as a light-running granulated mass, or as solid sticks. They have the disadvantage that the density of charging will be small, which means that the cost of drilling will be comparatively high. (Fraenkel, 1953)

solid fuels Any fuel that is a solid; such as wood, peat, lignite, bituminous, and anthracite coals of the natural variety and the prepared varieties such as, pulverized coal, briquettes, charcoal, and coke. Divided into two broad classes: naturally occurring and manufactured. (Newton, 1959)

solidification range Temperature range over which mixtures (alloys, fluxes) melt. In a constitutional diagram, the area between liquidus and solidus. (Pryor, 1963)

solidification shrinkage The decrease in volume of a metal during solidification. (ASM, 1961)

solid loading Filling a drill hole with all the explosive that can be crammed into it, except for stemming space at top. (Nichols, 1976)

solid map Gr. Brit. A geological map showing the extent of solid rock, on the assumption that all surficial deposits, other than alluvium, are absent or removed. (AGI, 1987)

solid packing *solid stowing.*

solid road Any roadway driven through the solid coal seam with rib sides. See also: *narrow stall; narrow work.* (Nelson, 1965)

solid rock Gr. Brit. Bedrock. (AGI, 1987)

solids handling pump Usually a centrifugal pump designed to resist abrasion and used for pumping sand, gravel, fine coal, and ore tailings. Rubber linings are generally used, which last longer than steel or iron. See also: *pulsometer.* (Nelson, 1965)

solid smokeless fuel A solid fuel, such as coke, which produces comparatively no smoke when burned in an open grate. See also: *anthracite; briquette; coke.* (Nelson, 1965)

solid solubility The extent to which one metal is capable of forming solid solutions with another. This varies widely between different pairs of metals, some of which are mutually soluble in all proportions, while others are practically insoluble in each other. (CTD, 1958)

solid solution (a) A single crystalline phase that may be varied in composition within finite limits without the appearance of an additional phase. Syn: *mix-crystal; mixed crystal.* (AGI, 1987) (b) Partial or total miscibility between two or more crystal structures; e.g., ionic substitution. Syn: *crystal solution.* Cf: *isomorphous mixture; isomorphous series.*

solid stowing The complete filling of the waste area behind a longwall face with stone and dirt. The packing operation may be by hand or mechanical methods, e.g., pneumatic stowing. See also: *double packing; stowing method; strip packing*. (Nelson, 1965)
solidus On a temperature-composition diagram, the locus of points in a system at temperatures above which solid and liquid are in equilibrium and below which the system is completely solid. In binary systems without solid solutions, it is a straight line; in binary systems with solid solutions, it is a curved line or a combination of straight and curved lines; in ternary systems, it is a flat plane or a curved surface. (AGI, 1987)
solid web The web of a steel beam consisting of a rolled section or a plate as distinct from a lattice. (Hammond, 1965)
solid woven conveyor belt A construction of conveyor belt consisting of multiple plies of fabrics woven into one piece, which is done on looms designed for this purpose. Stripes are woven into the belt to show the number of plies, which range from 2 to 10. Impregnating and coating treatments are frequently employed.
solifluction The slow, viscous downslope flow of waterlogged soil and other surficial material, normally at 0.5 to 5.0 cm/yr; esp. the flow occurring at high elevations in regions underlain by frozen ground (not necessarily permafrost) that acts as a downward barrier to water percolation, initiated by frost action and augmented by meltwater resulting from alternate freezing and thawing of snow and ground ice. The term has been extended to include similar movement in temperate and tropical regions; also, it has been used as a syn. of soil creep, although solifluction is generally more rapid. It is preferable to restrict the term to slow soil movement in periglacial areas. Also spelled solifluxion. Syn: *soil flow; soil fluction; sludging*. (AGI, 1987)
sollar (a) Landing stage in a mine shaft. Also spelled soller, solar. (Pryor, 1963) (b) A timber staging, alongside a haulage level, for piling ore ready for loading into mine cars. (Nelson, 1965) (c) A staging between ladderway sections in a shaft. (Nelson, 1965) (d) The plank flooring of a gallery covering a gutterway beneath. (e) A longitudinal partition forming an air passage between itself and the roof in a mine working. (Webster 3rd, 1966) (f) A platform from which trammers shovel or throw the ore or rock into a car. (g) A wooden platform fixed in a shaft for ladders to rest on. See also: *air sollar*. (Gordon, 1906)
soller *sollar.*
solonetz soil A soil occurring most commonly under arid conditions, but may also be found in semiarid and subhumid regions. Usually found in depressions where it has originated by evaporation under shallow ground-water conditions. Characterized by sodium carbonate as the predominant salt and a dark-colored B horizon, which is strongly alkaline in reaction. (Hawkes, 1962)
solubility (a) The extent that one material will dissolve in another, generally expressed as mass percent, or as volume percent or parts per 100 parts of solvent by mass or volume. The temperature should be specified. (ASTM, 1994) (b) The weight of a dissolved substance that will saturate 100 g of a solvent. (CTD, 1958) (c) Concentration of a substance in a saturated solution; i.e., in equilibrium between dissolved and undissolved phases at given temperature. (Pryor, 1963)
solubility product concentration In a saturated solution of an electrovalent compound having limited solubility, the product of the ionic concentrations, at the exponential value shown in the stoichiometric equation for its dissociation, is constant at a given temperature. (Pryor, 1963)
soluble Capable of being dissolved in a fluid. (Crispin, 1964)
soluble anode An anode that goes into solution during an electrolytic process. (Osborne, 1956)
soluble glass Solid sodium silicate or potassium silicate. Syn: *water glass*. (CTD, 1958)
solute (a) The substance dissolved in a solution, as distinguished from the solvent. (Standard, 1964) (b) A substance dissolved in a liquid.
solution A substance that is a homogeneous mixture and has a continuous variation of composition up to a solubility limit.
solution breccia A collapse breccia formed where soluble material has been partly or wholly removed by solution, thereby allowing the overlying rock to settle and become fragmented; e.g. a breccia consisting of chert fragments gathered from a limestone whose carbonate material has been dissolved away. See also: *evaporite-solution breccia*. Syn: *ablation breccia*. (AGI, 1987)
solution cavity A cavity formed in certain rocks in which percolating sollutions have filled with valuable minerals; cavities formed in certain rocks, such as limestones, where portions have been dissolved by percolating waters. See also: *mineralization*. (Nelson, 1965)
solution injection Artificial cementing of loose soils or strata to increase their load-bearing capacity. (Hammond, 1965)
solution mining (a) The in-place dissolution of water-soluble mineral components of an ore deposit by permitting a leaching solution, usually aqueous, to trickle downward through the fractured ore to collection galleries at depth. It is a type of chemical mining. Cf: *in situ leaching*. (AGI, 1987) (b) The mining of soluble rock material, esp. salt, from underground deposits by pumping water down wells into contact with the deposit and removing the artificial brine thus created. (AGI, 1987) (c) Hydrometallurgical treatment of ore for recovery of the mineral values at the mine site, in conjunction with conventional open pit or underground mining procedures. See also: *surface leaching*. (SME, 1992)
solution pipe A vertical cylindrical hole, formed by solution and often without surface expression, that is filled with detrital matter. (AGI, 1987)
solution plane A crystallographic direction of chemical solubility in a crystal, possibly due to exsolution of a second crystalline phase on falling temperature and/or pressure or to unannealed slip during plastic deformation. Solution planes may lead to parting in some mineral specimens.
solution ripple *scallop.*
solution subsidence Gradual subsidence of nonsoluble strata due to the solution of underlying rock. (AGI, 1987)
solvate A chemical compound consisting of a dissolved substance and its solvent, e.g. hydrated calcium sulfate. (AGI, 1987)
Solvay process Manufacture of sodium carbonate or soda ash, Na_2CO_3 from salt (sodium chloride), ammonia, carbon dioxide, and limestone by a sequence of reactions involving recovery and reuse of practically all the ammonia and part of the carbon dioxide. Limestone is calcined to quicklime and carbon dioxide. The carbon dioxide is dissolved in water containing the ammonia and salt, with resulting precipitation of sodium bicarbonate. This is separated by filtration, dried, and heated to form sodium carbonate. The liquor from the bicarbonate filtration is heated and treated with lime to regenerate the ammonia. Calcium chloride is a major byproduct. (CCD, 1961)
solvent (a) A substance used to dissolve another substance. (Crispin, 1964) (b) That component of a solution that is present in excess; or the physical state of which is the same as that of the solution. (CTD, 1958)
solvent extraction (a) A method of separating one or more substances from a mixture, by treating a solution of the mixture with a solvent that will dissolve the required substances, leaving the others. (Hammond, 1965) (b) A process in which one or more components are removed from a liquid mixture by intimate contact with a second liquid, which is itself nearly insoluble in the first liquid and dissolves the impurities and not the substance that is to be purified. Syn: *liquid-liquid extraction*. (CCD, 1961)
sonar The method or equipment for determining, by underwater sound, the presence, location, or nature of objects in the sea. The word sonar is an acronym derived from the expression SOund NAvigation and Ranging. (Hunt, 1965)
sonar boomer seismic system A complete continuous seismic profiling system consists of the boomer unit, sonar recorder, transducer fish, receiving hydrophone, preamplifier, if necessary, and variable filter. Sonar boomer units are available from 1000 W-second models up to 13,000 W-second (experimental models). The standard boomer consists of a power supply, capacitor bank and transducer. Boomers are used for marine geological studies and dredging surveys. The power supply output is fed to the capacitor bank, which is discharged into the transducer producing a precisely repeatable pressure pulse in the water. (Hunt, 1965)
sondalite A metamorphic rock composed of cordierite, quartz, garnet, tourmaline, and kyanite. (Holmes, 1928)
sonde The elongate cylindrical tool assembly used in a borehole to acquire a well log. It contains various energy-input devices and/or response sensors. The sonde is lowered into the borehole by a multiconductor cable, or wire line. (AGI, 1987)
sonic drilling Core drilling through high-powered vibrations transmitted down the drill casing to a cutting shoe. The casing and the shoe vibrate into the ground and through the rock, resulting in an undisturbed core of unconsolidated material. (SME, 1992)
sonic gage *acoustic-strain gage.*
sonic log An "acoustic log" showing the interval-transit time of compressional seismic waves in rocks near the well bore of a liquid-filled borehole. Used chiefly for estimating porosity and lithology.
sonic method A method of measuring underground rock pressure by determining the velocity of sound through the rock. Sonic velocity is a function of the elastic modulus of the rock traversed by the wave, and this, in turn, is a function of the pressure. A hammer blow on the rock face is used to initiate the sound waves, which are picked up by a microphone placed at the site of the blow and by a second microphone at the other end of the path through the rock under test. The difference between the times of the signals received from the two microphones will equal the time taken by the sonic pulse to pass through the rock. The signals are converted into visible waveforms on the screen of an oscillograph, and these are photographed or otherwise recorded to form a permanent record. (Issacson, 1962)
sonic pile driver A driver based on the principle of delivering vertical vibrations to the head of a pile in alternating up and down cycles at a

rate of 100 Hz. These vibrations set up high-amplitude waves of tension and compression in the pile, producing alternate expansion and contraction in minute amounts. The elongation of the pile in expansion displaces the soil at the pile tip, and the weight of the pile, hammer, and added loads shoves the pile into the miniscule void. Since this action is occurring at the rate of 100 Hz, the individual movements need not be of great magnitude to produce rapid penetration of the pile. (Carson, 1965)

sonigage An ultrasonic testing instrument used primarily for the measurement of the thickness of materials. (Osborne, 1956)

sonims Solid, nonmetallic inclusions in metal. (Henderson, 1953)

soniscope An inspection instrument, which sends, by electronic means, pulses of high frequency through the material to be tested and measures the time of travel from the transmitter on one face to the receiver on the distant face of the material. This method of inspection is known as pulse testing. (Osborne, 1956)

sonograph Seismograph developed by Frank Rieber for the application of reflection methods to areas of complex geology and steeply dipping beds. The ordinary oscillograph traces are replaced by sound tracks of variable transparency on a moving picture film. The analyzer adds up impulses which are in phase while the random effects tend to cancel one another. (AGI, 1987)

sonolite A monoclinic mineral, $Mn_9(SiO_4)(OH,F)_2$; humite group; dimorphous with jerrygibbsite; red-orange.

sonometer An instrument for measuring rock stress. Piano wire is tuned between two bolts cemented into drill holes in the rock, and a change of pitch after destressing is observed and used to indicate stress. (Pryor, 1963)

sonoprobe A type of echo sounder that generates sound waves and records their reflections. It is used in subbottom profiling. (AGI, 1987)

Sonstadt solution A solution of mercuric iodide in potassium iodide with a density of 3.2 g/cm^3 and used as a "heavy liquid." Cf: *bromoform; Klein solution; methylene iodide; Clerici solution.* Syn: *Thoulet solution.*

soot A black substance, consisting essentially of carbon from the smoke of wood or coal, esp. that which adheres to the inside of the chimney, containing also volatile products condensed from the combustion of the wood or coal, including certain ammonia salts. (Standard, 1964)

sooty chalcocite A black, pulverent variety of chalcocite of supergene origin.

sooty streamers Fine and often unburned dust settling out in steamy, warm conditions in stagnant situations. (Sinclair, 1958)

sop (a) Cumb. A hematite iron-orebody of circular or oval plan and conical section, formed in a swallow hole. (Arkell, 1953) (b) Cumb. A nest or pocket of black lead. (Arkell, 1953)

Sophia-Jacoba process *Barvoys process.*

sopwith staff A telescopic self-reading staff dividing into three sections, set one above the other when the staff is at its full extent of 14 ft (4.3 m). Graduations are marked in feet, tenths and hundredths of a foot, and the thickness of the horizontal lines is 0.01 ft (3.05 mm), alternately black and white. (Hammond, 1965)

sordavalite An old name for the glassy salbands of small, diabase dikes formerly regarded as a mineral. It is derived from Sordavala, Finland.

Sorel cement Calcined magnesite or magnesia mixed with a solution of magnesium chloride. It sets to a hard mass within a few hours. The basis of artificial flooring cements. (CTD, 1958)

sorelslag A titanium slag containing about 80% TiO_2. It is made by electric-furnace smelting of iron-titanium ores. (Newton, 1959)

Soret's principle If differences of temperature are induced in a solution of sodium chloride or some other substance in water, the dissolved material will become relatively more concentrated in those portions in which the temperature is lowest. (Fay, 1920)

soroche A mountain sickness that attacks miners who are newcomers in high altitudes. Symptoms are headaches, nausea, vomiting, and nosebleed. If the symptoms do not soon pass, there is nothing to be done except return to a lower altitude. (Hoover, 1948)

sorption Any type of retention of a material at a surface, esp. when the mechanism is not specified. Adsorption is then restricted to the physical process that leads to the formation of a unimolecular surface layer; chemisorption refers to the corresponding chemical process; and absorption to the entrance of the sorbed material within the solid. (Miall, 1940)

sorted *graded.*

sorting (a) In a genetic sense, it may be applied to the dynamic process by which granular or fragmental material having some particular characteristic, such as similar size, shape, specific gravity, or hydraulic value, is selected from a larger heterogeneous mass. (b) The degree of similarity, in respect to some particular characteristic, of the component parts in a mass of material. (c) A measure of the spread of a distribution on either side of an average. (d) The separation of coal or ore as mined into valuable material and waste. (e) *handpicking.*

sorting coefficient (a) A coefficient used in describing the distribution of grain sizes in a sample of unconsolidated material. It is defined as $S_0 = \sqrt{Q_1/Q_3}$, where Q_1 is the diameter that has 75% of the cumulative size-frequency (by weight) distribution smaller than itself and 25% larger than itself, and Q_3 is that diameter having 25% of the distribution smaller and 75% larger than itself. (Hunt, 1965) (b) Dimensionless measure for degree of sorting. (Schieferdecker, 1959)

soude emerald *emerald triplet.*

sound (a) The act of striking a mine roof with a metal testing bar to ascertain whether or not it is strong and safe. (Hudson, 1932) (b) A dam or barrier in a mine in which the frictional resistance to the passage of water is high. Such a dam permits little water to pass through it and is said to be "sound.". (Sinclair, 1958) (c) Elastic waves in which the direction of particle motion is longitudinal; i.e., parallel with the direction of propagation. The term is sometimes restricted to such waves in gases, particularly air, and in liquids, particularly water, but it is also applied to wave motion in solids. It is the type of wave motion most often used in reflection-seismic exploration. (AGI, 1987)

sound channel Sound waves in the surface layers of the ocean tend to be refracted downward due to decrease of temperature with depth. In deep waters the sound waves are refracted upward by high and increasing pressure, which has a greater influence on the resulting refractive index of these layers than the decreasing temperature. The result is the formation at mid-depth of a wave guide that permits compressional waves of acoustic frequencies to travel great distances. (Hy, 1965)

sounding (a) *roof testing.* (b) Rapping on a pillar to signal a person on the other side or to enable the person to estimate its width. (c) A rough method of judging by sound the direction and distance apart of two roadways driven in coal to meet each other. The sounding is made by giving two slow and three sharp knocks on the solid coal, which is answered in similar manner from the opposite roadway. The method is sometimes called chap. (Nelson, 1965) (d) Subsurface investigation by observing the penetration resistance of the subsurface material without drilling holes. This can be done by driving a rod into the ground or by using a penetrometer. See also: *penetrometer; soil.* (Long, 1960)

sounding lead In measuring water depth, the hand lead used in sounding. See also: *hand lead.* (Hammond, 1965)

sounding rod A closed pipe, 1 in (2.5 cm) in diameter, with a flush point and a driving tip, used in sounding. See also: *sounding.* (Long, 1960)

sounding the top Tapping the roof with a pick or bar to test its soundness. (Korson, 1938)

sound intensity In a specified direction at any point, the average rate of sound energy transmitted in the specified direction through a unit area normal to this direction at the point considered. (Hy, 1965)

sound velocity The rate at which a sound wave travels through a medium. Velocity is equal to the square root of the bulk modulus divided by density. (Hy, 1965)

sound wave (a) Sometimes used interchangeably with shock wave; technically, a wave motion in the air which affects the human ear as sound. Owing to the reflection of waves from the various surfaces, the sound is more or less prolonged, and as it reaches a given point is really a series of vibrations. (Rice, 1913-18) (b) The wave of compression emanating from any sound source. Since sound travels as a wave of compression it heats the water as it passes through a water mass. Density and compressibility thus influence the velocity of sound; increasing density due to temperature, salinity, and pressure changes increases the velocity. High frequency waves above 10,000 Hz have very short range because absorption is high. Low frequencies have greater range. *longitudinal wave.* (Hy, 1965)

sour (a) Having an acid or tart taste; applied to minerals having the taste of sulfuric acid. (b) Said of crude oil or natural gas containing significant fractions of sulfur compounds. Cf: *sweet.* (AGI, 1987)

source (a) In seismic prospecting either: (1) the point of origin or shot from which elastic waves are propagated, or (2) the formation, horizon, interface, or boundary at which a seismic wave is refracted and/or reflected and returned to the surface. In earthquake seismology, the point of origin of an earthquake. In neutron logging, the source of neutrons at one end of the logging tool. (AGI, 1987) (b) A radioactive material packaged to produce radiation for experimental or industrial use. (Lyman, 1964) (c) The point of origin or procurement. (Webster 3rd, 1966)

source area The area from which the sedimentary material is derived.

source-receiver product In seismic prospecting, the product of the number of detectors per trace and the number of sources used simultaneously. (AGI, 1987)

source rock The geological formation in which oil, gas, and/or other minerals originate. (AGI, 1987)

sourdough (a) Old-fashioned and seasoned prospector. (Pryor, 1963) (b) A miner who has lived in Alaska more than one season. (von Bernewitz, 1931)

sour gas Slang for either natural gas or a gasoline contaminated with odor-causing sulfur compounds. In natural gas the contaminant is usually hydrogen sulfide, which can be removed by passing the gas mixture through carbonate solutions containing special metal or organic activators. In gasolines, the sour contaminents are usually mercaptans, which are removed in the doctor treatment or by ethylene oxide with a phenolic catalyst. The improved gas or gasoline is known as sweet gas. (CCD, 1961)
souring (a) An alternative term for aging. See also: *aging*. (Dodd, 1964) (b) The storage for a short time of the moistened batch for making basic refractories; some magnesium hydroxide is formed and this acts as a temporary bond after the bricks have been shaped and dried. If souring is allowed to proceed too far, cracking of the bricks is likely during drying and the initial stages of firing. The high pressure exerted by modern brick presses generally gives sufficient dry strength without the bricks being soured, and souring is therefore now generally omitted. (Dodd, 1964)
South African diamond Whole-stone diamonds having outside faces that are smooth as contrasted to the pebbly, encrusted surface of a Congo diamond; also applied to diamonds produced in South African mines as contrasted to those found in the Sierra Leone, Congo, Brazil, etc. (Long, 1960)
southing A distance measured southwards from an east-west reference line. (Hammond, 1965)
south-seeking pole Cf: *north-seeking pole.*
South Staffordshire method *room-and-pillar.*
southwestern cell Pneumatic flotation machine consisting of a long tank with V-shaped base into which pipes connected with a low-pressure source deliver compressed air. Internal baffles provide an air lift and the cell discharges a mineralized froth along one or both sides, the tailing leaving at the far end. The Britannia cell is a modified form. See also: *Britannia cell*. (Pryor, 1963)
souzalite A monoclinic mineral, $(Mg,Fe)_3(Al,Fe)_4(PO_4)_4(OH)_6{\cdot}2H_2O$; forms a series with gormanite; green.
sovereign gold Standard 22 carat gold, containing 91.7% gold and 8.3% copper. (Osborne, 1956)
sövite A carbonatite that contains calcite as a dominant phase. The name, given by Brögger in 1921, is for Söve, Fen complex, Norway. (AGI, 1987)
sow (a) Mold of larger size than a pig. (Webster 3rd, 1966) (b) A channel or runner that conducts molten metal to the rows of molds in a pig bed. (Webster 3rd, 1966) (c) A mass of metal solidified in such a channel or mold. (Webster 3rd, 1966) (d) An accretion that frequently forms in the hearth or crucible of a furnace; it consists mainly of iron. Also called salamander, bear, or shadrach. (Fay, 1920)
Soxhlet thimble A dust-sampling instrument. The apparatus gives a gravimetric mass sample, with no information on the size distribution of the dust collected. Is useful where a sample is needed for determining mass concentration and for collecting dust for chemical, petrological, and X-ray analysis. See also: *Hexhlet sampler; tyndallometer*. (Roberts, 1960)
spaad A fibrous talc. From the German spath.
spaced loading (a) Loading so that cartridges or groups of cartridges are separated by open spacers that do not prevent the concussion from one charge from reaching the next. (Nichols, 1976) (b) *deck loading.*
space frame A three-dimensional frame that is stable against wind pressure without being braced against any other structure. (Hammond, 1965)
space group In crystallography, any one of 230 independent combinations of the 14 essentially different kinds of three-dimensional periodicity (Bravais lattices) with other symmetry operations (point groups, screws, glides). Cf: *plane group; Bravais lattice; point group; screw; glide.*
space lattice (a) A three-dimensional regularly repeating set of points so arranged as to determine sets of equally spaced parallel planes in various directions forming polyhedral cells (as in a honeycomb). Specif., a set of such points occupied by the atoms of a crystal. (b) The pattern formed by the spatial distribution of atoms or radicals in a crystal. See also: *crystal pattern; lattice*. (Hackh, 1944) (c) Any one of 14 infinite three-dimensional arrays of points such that each point is in an identical point environment. Syn: *Bravais lattice.* Cf: *net.*
spacer (a) A piece of metal wire twisted at each end so as to form at one end a guard to keep the explosive in a shothole in place and at the other end another guard to hold the tamping in its place, thus providing an open space between explosive and tamping. When this is provided, the charge constitutes a cushion shot. (Zern, 1928) (b) Piece of wood doweling which is interposed between charges to extend the column of explosive. (Nelson, 1965) (c) A marker block. Also called spacer block. (Long, 1960) (d) The tapered section of a pug joining the barrel to the die; in this section beyond the shaft carrying the screw or blades, the clay is compounded before it issues through the die. (Dodd, 1964)
spacing The distance between adjacent shotholes in a direction parallel to the quarry or other face. (Nelson, 1965)
Spackman System *coal constituent classification.*

spad (a) A means of marking an underground survey station that consists of a flat spike in which is drilled a hole for the threading of a plumbline. (BS, 1963) (b) *spud.*
spade drill *flat drill.*
spade-end wedge A type of deflecting wedge in boreholes. See also: *deflecting wedge*. (Long, 1960)
spadiard Corn. A worker in the tin mines. Also called spalliard. (Standard, 1964)
spake A term used in South Wales for a train of personnel carriages, for use on the main slants at the beginning and end of each shift. The seats are so arranged that they are horizontal when the carriage is on the inclined slant. See also: *man-riding car; man-riding conductor*. (Nelson, 1965)
spall (a) A relatively thin, commonly curved and sharp-edged piece of rock produced by exfoliation. (b) To break off in layers parallel to a surface. (c) To break ore. Pieces of ore thus broken are called spalls. Also spelled: spawl.
spalliard Eng. A pickman; a working miner. A laborer in tinworks. Also spelled spallier.
spalling The chipping, fracturing, or fragmentation, and the upward and outward heaving, of rock caused by the action of a shock wave at a free surface or by release of pressure. Syn: *exfoliation*. (AGI, 1987)
spalling floor A place for breaking ore with a 4- to 5-lb (1.8- to 2.3-kg) sledge hammer.
span (a) The horizontal distance between the side supports or solid abutments along sides of a roadway. See also: *abutment; pressure arch.* (Nelson, 1965) (b) The horizontal distance between the supports of a bridge, arch, beam, or similar structural member. See also: *clear span; effective span*. (Hammond, 1965)
spangle gold Aust. Smooth, flat scales of gold.
spangolite A trigonal mineral, $Cu_6Al(SO_4)(OH)_{12}Cl{\cdot}3H_2O$; soft; vitreous; dark green; in Cochise County, AZ.
Spanish chalk A variety of steatite in Aragon, Spain.
Spanish emerald Emerald of the finest quality (presumably from South America). (Schaller, 1917)
Spanish lazulite Iolite, a variety of cordierite.
Spanish topaz (a) Any orange, orange-brown, or orange-red variety of quartz resembling the color of topaz; e.g., heat-treated amethyst. Cf: *topaz; Madeira topaz.* (b) A wine-colored or brownish red citrine from Spain.
spantone *gypsum.*
spar (a) A term loosely applied to any transparent or translucent light-colored mineral that is readily cleavable having a vitreous luster; e.g., Iceland spar (calcite), fluorspar (flourite), feldspar, heavy spar (barite). (b) Applied locally by miners to small clay veins found in coal seams. (AGI, 1987) (c) Corn. Quartz. (d) *Iceland spar.*
sparable ore A nonmetallic tin ore occurring in small granules. (Osborne, 1956)
sparagmite A collective term for the late Percambrian fragmental rocks of Scandinavia, esp. the feldspathic sandstones of the Swedish Jotnian, consisting mainly of coarse arkoses and subarkoses, together with polygenetic conglomerates and graywackes. Etymol: Greek sparagma, fragment, thing torn, piece. (AGI, 1987)
spare N. of Eng. A wedge from 6 to 8 in (15 to 20 cm) long for driving behind plates when adjusting them to the circle of the shaft. Also called spear wedge.
spare face *standby face.*
sparite (a) A descriptive term for the crystalline transparent or translucent interstitial component of limestone, consisting of clean, relatively coarse-grained calcite or aragonite that either accumulated during deposition or was introduced later as a cement. It is more coarsely crystalline than micrite, the grains having diameters that exceed 10 μm. (AGI, 1987) (b) A limestone in which the sparite cement is more abundant than the micrite matrix. (AGI, 1987)
spark absorber *absorber.*
spark chamber An instrument for detecting and measuring nuclear radiation; analogous to the cloud chamber. It consists of numerous electrically charged metal plates mounted in a parallel array, the spaces between the plates being occupied by inert gas. Ionizing radiation causes sparks to jump between the plates along its path through the chamber. See also: *bubble chamber; cloud chamber*. (Lyman, 1964)
sparker A marine seismic source that employs an electrical spark discharge to form the outgoing signal and produces a trace in the recorder showing the subbottom strata. (Hy, 1965)
sparkle metal A matte containing about 74% copper. (Webster 3rd, 1966)
spark test Identification of type of iron alloy by appearance of sparks emitted when rubbed on grindstone. (Pryor, 1963)
sparry (a) Pertaining to, resembling, or consisting of "spar"; e.g., sparry vein or sparry luster. Like spathic; as in feldspathic. See also: *spar.* (b) Pertaining to "sparite" esp. in allusion to the relative clarity, both in thin section and hand specimen, of the calcite cement; abounding with sparite, such as a sparry rock.

sparry iron *siderite.*
sparry lode A lode filled with spar, e.g., fluorspar, calcspar, or heavy spar.
sparse vitrain A field term to denote, in accordance with an arbitrary scale established for use in describing banded coal, a frequency of occurrence of vitrain bands comprising less than 15% of the total coal layer. Cf: *abundant vitrain; dominant vitrain; moderate vitrain.* (AGI, 1987)
spartaite A variety of calcite containing some manganese. (Fay, 1920)
spartalite A former name for zincite.
spate (a) Corn. In mining, to fine for disobedience of orders. (b) A variation of spall.
spathic Resembling spar, esp. in regard to having good cleavage. Syn: *spathose.*
spathic iron A native ferrous carbonite, also called siderite, containing 48% iron and usually traces of manganese. It is the best native ore for making steel tools by the direct method formerly used. See also: *siderite.* (Sandstrom, 1963)
spathic iron ore *siderite.*
spathization Widely distributed crystallization of sparry carbonates, such as calcite and dolomite; development of relatively large sparry crystals that have good cleavage. (AGI, 1987)
spathose *spathic.*
spathose iron Carbonate of iron, $FeCO_3$. See also: *siderite.* (Nelson, 1965)
spawl *spall.*
spawl beater *sledger.*
SP curve *spontaneous potential curve.*
spear (a) One of several types of fishing tools designed to be driven and wedged inside of bits, rods, etc., lost in a borehole. Cf: *fishing tap.* (Long, 1960) (b) A rodlike fishing tool having a barbed-hook end, used to recover rope, wire line, and other materials from a borehole. (Long, 1960) (c) Eng. A wooden pump rod cut into lengths of about 40 ft (12 m) , and, for heavy work, often measuring 16 in (40.6 cm) square. Wrought iron spears are also used. (Fay, 1920)
spearhead (a) The point of convergence of two cross faces set off in the form of the letter V. (TIME, 1929-30) (b) A conical head on a wire-line core barrel, engaged by the dogs on the overshot assembly for the purpose of removing the inner tube of the core barrel from a borehole. (Long, 1960)
spear pyrite A marcasite in twin crystals resembling the head of a spear. *marcasite.* (Webster 3rd, 1966)
special *special rounds.*
special flexible rope A wire rope composed of 6 strands of 37 wires each. (Lewis, 1964)
special rounds Sometimes used to designate a very high quality or grade of drill diamonds. (Long, 1960)
specialty steel A steel containing alloys that provide special properties, such as resistance to corrosion or to heavy load. Also called alloy steel. (Hammond, 1965)
species A mineral distinguished from others by its unique chemical and physical properties; it may have varieties. (AGI, 1987)
specific Word used with a special meaning in mineral dressing, where minerals of the same species often exhibit differences in their reactions. "Specific to" warns the observer that the process in hand is empirical in some ways, designed to apply to one specific orebody. (Pryor, 1963)
specific adhesion The chemical bond between glued or cemented surfaces as distinct from any form of mechanical bond. (Hammond, 1965)
specific adsorption Selective adsorbing action. (Pryor, 1963)
specific damping capacity A measure of the vibrational energy absorbed by the rock and may be considered to be a measure of the internal friction. It is determined by the sharpness of resonance that is evident when a specimen is vibrated through a range of frequencies centered on the fundamental longitudinal resonant frequency. Damping of the dry type (coulomb damping) is commonly assumed to be independent of the velocity, and thus independent of the frequency, and is somewhat sensitive to moisture content. Syn: *coulomb damping.* (Lewis, 1964)
specific extraction of rock broken Quantity of broken rock (ore) in volume or weight per foot drilled or fired per quantity of explosive. (Fraenkel, 1953)
specific gravity (a) The ratio between the weight of a unit volume of a substance and that of some other standard substance, under standard conditions of temperature and pressure. For solids and liquids, the specific gravity is based upon water as the standard. The true specific gravity of a body is based on the volume of solid material, excluding all pores. The bulk or volume specific gravity is based upon the volume as a whole—i.e., the solid material with all included pores. The apparent specific gravity is based upon the volume of the solid material plus the volume of the sealed pores. See also: *apparent specific gravity.* (Harbison-Walker, 1972) (b) Ratio of densities of a gas and air, based on dry air = 1. (Hartman, 1982) (c) The weight of a substance compared with the weight of an equal volume of pure water at 4 °C. Specific gravity is numerically equal to density given in grams per cubic centimeter or milliliter. Cf: *density.*
specific-gravity hydrometer A hydrometer indicating the specific gravity or relation of the weight of a given liquid per unit volume to the weight of a given unit volume of water. See also: *hydrometer; Marsh funnel.* (Long, 1960)
specific gravity of soil grains This is measured in a calibrated glass bottle with special precautions against the inclusion of air. Such testing is applied in many soil problem computations. See also: *pycnometer.* (Hammond, 1965)
specific heat (a) Heat required to raise the temperature of a unit weight of air 1 °F (0.56 °C). Usually, the specific heat at constant pressure is used in air conditioning. For ordinary concrete and steel it is 0.22 Btu/lb/°F and 0.12 Btu/lb/°F (0.92 kJ/kg/°C and 0.50 kJ/kg/°C), respectively. (Hartman, 1982) (b) The ratio of the amount of heat required to raise a unit weight of a material 1 degree to the amount of heat required to raise the same unit weight of water 1 degree. (Brantly, 1961) (c) The heat in calories required to raise the temperature of 1 g of a substance 1 °C. (Webster 3rd, 1966)
specific humidity (a) The mass of moisture per unit mass of dry air. (Roberts, 1960) (b) Absolute humidity, or weight of water vapor contained per unit weight of dry air. Cf: *relative humidity.* See also: *humidity.* (Hartman, 1961)
specific mineral *essential mineral.*
specific population Number of particles in unit volume of pulp. (Pryor, 1965)
specific resistance *resistivity.*
specific retention The ratio of the volume of water that a given body of rock or soil will hold against the pull of gravity to the volume of the body itself. It is usually expressed as a percentage. Cf: *field capacity.* (AGI, 1987)
specific speed (a) A factor by which the performance of any particular design of impeller for a centrifugal pump or water turbine can be computed. It is the speed in revolutions per minute at which a geometrically similar impeller of suitable diameter will rotate to deliver 1 gal/min (3.785 L/min) at 1 ft (30.5 cm) head in the case of a pump. In a water turbine, the specific speed is that at which a geometrically similar runner of suitable diameter will turn to develop 1 hp (746 W) under a head of 1 ft (30.5 cm). (Hammond, 1965) (b) The particular speed at which a fan achieves its maximum efficiency. (Roberts, 1960)
specific surface (a) The surface area per unit of volume of soil particles. (ASCE, 1958) (b) The ratio of the total surface of a substance (as an adsorbent) to its volume; surface area (as of a finely divided powder) per unit mass. (Webster 3rd, 1966)
specific volume (a) Volume of one gram at specified temperature. (Pryor, 1963) (b) Volume per unit weight of dry air. Not equal to the reciprocal of density, which is based on unit volume of mixture. Measured in cubic feet per pound (cubic meters per kilogram). (Hartman, 1982)
specific weight of sediment The dry weight per unit volume of the sediment in place. See also: *dry density.* (Nelson, 1965)
specific yield The ratio of the volume of water that a given mass of saturated rock or soil will yield by gravity to the volume of that mass. This ratio is stated as a percentage. Cf: *effective porosity.* (AGI, 1987)
specimen (a) A sample, as of a fossil, rock, or ore. Among miners, it is often restricted to selected or handsome samples, such as fine pieces of ore, crystals, or fragments of quartz showing visible gold. Cf: *hand specimen.* (AGI, 1987) (b) A small mass of coal, rock, or, mineral, or soil, which gives, roughly, an idea of the kind and quality of the deposit from which it was derived. In the case of ore in particular, the specimen should admit of the identification of the various minerals present. A specimen cannot be viewed as a sample. (Nelson, 1965)
specimen boss An employee whose duty is to watch carefully, in all parts of the mine, for the appearance of high-grade mineral, and when it is likely that such spots will be opened up, should be the first person at the face after the blast to prevent high-grading (the theft of valuable samples). (Hoover, 1948)
specimen hunting Another name for high-grading. (Hoover, 1948)
specimens In mineral dressing, unusually rich pieces of ore or characteristic constituents thereof in coarsely crystalline form—not representative samples. (Pryor, 1965)
speck A small piece of alluvial gold weighing up to 1 oz or 2 oz (28 g or 56 g). (Gordon, 1906)
speckstone An early name for talc or steatite. Etymol: German "Speckstein", "bacon stone," alluding to its greasy feel.
spectacle A two-handled frame for carrying well-boring tools. (Standard, 1964)
spectacle stone An early popular name for selenite, alluding to its transparency. (Fay, 1920)
spectral gamma-ray log Record of the radiation spectrum and relative intensities of gamma rays emitted by strata penetrated in drilling. Because of their different energies the relative amounts of radioactivity

contributed by different elements can be determined. Cf: *gamma-ray well log.* See also: *radioactivity log.* (AGI, 1987)

spectrometer An optical instrument similar to, but more versatile than, the simple spectroscope. Scales are provided for reading angles. A wavelength spectrometer is one designed or equipped in a manner to measure the wavelengths at which absorption bands occur in an absorption spectrum.

spectrophotometer An instrument to detect very slight differences in color of solutions of different chemicals and thus measure the quantity of the chemical present. It consists of a light source, an optical prism for providing monochromatic light; i.e., light of a single wavelength only, and a device for measuring the intensity of the light beam after it has passed through the solution. Traces of aluminum in steel may be determined in this way. Also called spekker. (Nelson, 1965)

spectrum (a) A band of light showing in orderly succession the rainbow colors or isolated bands or colors corresponding to different wavelengths, as seen through a spectroscope or photographed in a spectrograph. The visible spectrum is only a small region in the vast spectrum of electromagnetic waves, which extend from the longest radio waves to the minutely short waves (gamma rays) emitted by radioactive elements. See also: *emission spectrum; continuous spectrum; absorption spectrum.* (Anderson, 1964) (b) An array of visible light ordered according to its constituent wavelengths (colors) by being sent through a prism or diffraction grating. (AGI, 1987) (c) An array of intensity values ordered according to any physical parameter, e.g. energy spectrum, mass spectrum, velocity spectrum. (AGI, 1987) (d) Amplitude and phase response as a function of frequency for the components of a wavetrain, such as given by Fourier analysis, or as used to specify filter-response characteristics. Pl: "spectra." Adj: "spectral."

spectrum colors The hues or wavelengths into which white light is separated upon passing through a transparent prism, six of which are readily distinguished by the normal human eye: red, orange, yellow, green, blue, and violet. See also: *visible spectrum.* By extension, any small range of wavelengths outside the visible range.

specular Mirrorlike, as specular iron ore, which is a hard variety of hematite. See also: *specularite.* (Fay, 1920)

specular hematite *specularite.*

specular iron Variety of hematite, Fe_2O_3, black, lustrous, metallic gleam. Mohs hardness, 5.5 to 6.5. May be micaceous in form. Contains 70% iron; sp gr, 4.9 to 5.3. See also: *specularite; hematite.* (Pryor, 1963)

specular iron ore A variety of hematite with brilliant black color and metallic luster. *hematite.* (CCD, 1961)

specularite A black or gray variety of hematite with splendant metallic luster, often showing iridescence; occurs in micaceous or foliated masses or in tabular or disklike crystals. Syn: *specular hematite; gray hematite.* See also: *hematite; specular iron; iron glance.*

specular schist Metamorphosed oxide-facies iron formation characterized by a high percentage of strongly aligned flakes of specular hematite. Cf: *itabirite.* (AGI, 1987)

specular stone Mica. (Standard, 1964)

speculative resources Undiscovered resources that may occur either in known types of deposits in favorable geologic settings where mineral discoveries have not been made, or in types of deposits as yet unrecognized for their economic potential. If exploration confirms their existence and reveals enough information about their quantity, grade, and quality, they will be reclassified as identified resources. (USGS, 1980)

speed The length of belt, chain, cable, or other linkage which passes a fixed point within a given time. It is usually expressed in terms of feet per minute. In the case of the rolling chain conveyor, the load is moved at a rate double the chain speed. In screw conveyors, the speed is expressed in terms of revolutions per minute and the speed at which the material is conveyed is dependent upon speed, pitch of the screw, type of flight, angle of inclination, nature of material, etc.

speedy moisture tester A calcium carbide method for the quick determination of moisture. A pressure gage is calibrated to give direct values of moisture content percent of soil samples. (Nelson, 1965)

speiss Metallic arsenides and antimonides smelted from cobalt and lead ores. (Pryor, 1963)

speisscobalt *smaltite.*

speleologist One who scientifically investigates caverns.

speleology The scientific study or exploration of caverns and related features.

speleothem A secondary mineral deposit formed in caves.

spell A rest period for crews at furnace, stock house, etc., or a period of work in drilling the taphole; a change or turn. (Fay, 1920)

spellerizing Subjecting the heated bloom to the action of rolls having regularly shaped projections on their working surface, then subjecting the bloom while still hot to the action of smooth-faced rolls. The surface working is said to give a dense texture to pipe made from the bloom, adapting it to resist corrosion. (Liddell, 1918)

spelter Zinc of under 99.6% purity. (Hammond, 1965)

Spence automatic desulfurizer An improved maletra furnace provided with automatic rakes. (Fay, 1920)

Spence furnace A furnace of the muffle or reverberatory type, the ore being supported on shelves and stirred mechanically. (Fay, 1920)

spencerite (a) A monoclinic mineral, $Zn_4(PO_4)_2(OH)_2{\cdot}3H_2O$; pearly white. (b) The synthetic material $(Fe,Mn)_3(C,Si)$.

spencite *tritomite.*

spend (a) To break ground; to continue working. (b) To exhaust by mining; to dig out; used in the phrase, to spend ground. (Standard, 1964)

spent fuel Nuclear-reactor fuel that has been irradiated to the extent that it can no longer effectively sustain a chain reaction. Fuel becomes spent when its fissionable isotopes have been partially consumed and fission-product poisons have accumulated in it. Syn: *depleted fuel.* (Lyman, 1964)

spent shot A blasthole that has been fired, but has not done its work.

spergenite A calcarenite that contains ooliths and fossil debris (such as bryozoan and foraminiferal fragments) and that has a quartz content not exceeding 10%. Type locality: Spergen Hill, situated a few miles southeast of Salem, Indiana, where the Salem Limestone (formerly the Spergen Limestone) is found. Syn: *Bedford Limestone.* (AGI, 1987)

sperrylite An isometric mineral, $PtAs_2$; pyrite group; tin white; sp gr, 10.6; occurs with heavy-metal ores, also in placers.

Sperry process A process for manufacturing white lead in which softened and desilverized lead anodes, preferably containing some bismuth, are placed in the Sperry cells. Direct current dissolves the lead from the anodes, and carbon dioxide is used to precipitate white lead (basic lead carbonate) from the solution. The Sperry process slime, which contains the impurities from the anodes, is washed, dried, and melted to an impure bismuth bullion, which goes to the bismuth refinery. (USBM, 1956)

spessartine An isometric mineral, $Mn_3Al_2(SiO_4)_3$; garnet group with Mn replaced by Fe and Mg; crystallizes as dodecahedra and trapezohedra; in skarns and granite pegmatites; may be of gem quality. Also spelled spessartite. Cf: *emildine.* Syn: *manganese-aluminum garnet.*

spessartite A "lamprophyre" composed of phenocrysts of green hornblende or clinopyroxene in a ground mass of sodic plagioclase, with accessory olivine, biotite, apatite, and opaque oxides. Named for Spessart, Germany. Syn: *spessartine.*

spew The cauliflowerlike blowout or outcrop of a lode that extends beyond the limits of the defined vein deeper down.

sphaerocobaltite A trigonal mineral, $CoCO_3$; calcite group; forms peach-blossom red spherical masses. Also spelled spherocobaltite. Syn: *cobaltocalcite.*

sphaerolitic *spherulitic.*

sphaerosiderite Alternate spelling of spherosiderite.

sphagnum peat Peat composed mainly of bog moss. It is characterized by an open texture, is lightweight, has a high absorbing power, good isolating properties, a clean appearance, and freedom from black dust. (Tomkeieff, 1954)

sphalerite (a) An isometric mineral, ZnS, with Zn replaced by Fe with minor Mn, As, and Cd; trimorphous with wurtzite and matraite; perfect dodecahedral cleavage; resinous to adamantine; occurs with galena in veins and irregular replacement in limestone; a source of zinc. Syn: *blende; zinc blende; jack; blackjack; steel jack; false galena; pseudogalena; mock ore; mock lead.* See also: *beta zinc sulfide.* (b) The mineral group coloradoite, hawleyite, metacinnabar, sphalerite, stilleite, and tiemannite.

sphene *titanite.*

sphenoid An open crystal form of two nonparallel faces which intersect two or three crystallographic axes. Cf: *dome; disphenoid.*

sphenolith A wedgelike igneous intrusion, partly concordant and partly discordant. (AGI, 1987)

sphenomanganite *manganite.*

Sphenopterias A fernlike tree of the coal forest characterized by round-lobed pinnules that are contracted at the base. See also: *Neuropteris.* (Nelson, 1965)

sphere ore *cockade ore.*

spheric A general term applied to rocks made up of spherules. It includes such textures as oolitic, pisolitic, spherulitic, variolitic, orbicular, etc. (Johannsen, 1931-38)

spherical dam A brick or concrete seal or stopping built into a roadway to close off an area against water. The convex surface is from the water side of the roadway. The construction is built well into the ground and all crevices cemented. (Nelson, 1965)

spherical model A function frequently used when fitting mathematical models to experimental variograms, often in combination with a nugget model.

spherical wave A seismic wave propagated from a point source whose front surfaces are concentric spheres. (AGI, 1987)

spherical wave front Spherical surface that a given portion of a seismic impulse (in an isotropic medium) occupies at any particular time. (AGI, 1987)

spherical weathering *spheroidal weathering.*

spherite (a) The preferred spelling for sphaerite. (English, 1938) (b) Spherical grains, including ovulite with concentric structure and spherulite with radial structure. (Spencer, 1943)

spherocobaltite *sphaerocobaltite.*

spherocrystal A homogeneous spherulite formed of minute crystals branching outward from the center. See also: *spherulite.* (Standard, 1964; Fay, 1920)

spheroid In general, any figure differing but little from a sphere. In geodesy, a mathematical figure closely approaching the geoid in form and size, and used as a surface of reference for geodetic surveys.

spheroidal (a) Having the shape of a spheroid. (AGI, 1987) (b) Composed of spherulites. (AGI, 1987) (c) Said of the texture of a rock composed of numerous spherulites. (AGI, 1987)

spheroidal jointing *spheroidal parting.*

spheroidal parting A series of concentric and spheroidal or ellipsoidal cracks produced about compact nuclei in fine-grained, homogeneous rocks. Cf: *exfoliation; spheroidal weathering.* Syn: *spheroidal jointing.* (AGI, 1987)

spheroidal structure *orbicular structure.*

spheroidal weathering A form of chemical weathering in which concentric or spherical shells of decayed rock (ranging in diameter from 2 cm to 2 m) are successively loosened and separated from a block of rock by water penetrating the bounding joints or other fractures and attacking the block from all sides. It is similar to the larger-scale exfoliation produced usually by mechanical weathering. See also: *spheroidal parting.* Syn: *onion-skin weathering; concentric weathering; spherical weathering.* Cf: *exfoliation.* (AGI, 1987)

spherosiderite A variety of siderite occurring in globular concretionary aggregates of bladelike crystals radiating from a center, generally in a clayey matrix (such as those in or below underclays associated with coal measures). It appears to be the result of weathering of waterlogged sediments in which iron, leached out of surface soil, is redeposited in a lower zone characterized by reducing conditions. Also spelled sphaerosiderite.

spherulite (a) A rounded or spherical mass of acicular crystals, commonly of feldspar, radiating from a central point. Spherulites may range in size from microscopic to several centimeters in diameter. Also spelled: sphaerolite. (AGI, 1987) (b) Any more or less spherical body or coarsely crystalline aggregate with a radial internal structure arranged around one or more centers, varying in size from microscopic grains to objects many centimeters in diameter, formed in a sedimentary rock in the place where it is now found; e.g., a minute particle of chalcedony in certain limestones, or a large carbonate concretion or nodule in shale. Cf: *spherite; variole.* (AGI, 1987) (c) A small (0.5 to 5 mm in diameter), spherical or spheroidal particle composed of a thin, dense calcareous outer layer with a sparry calcite core. It can originate by recrystallization or by biologic processes. (AGI, 1987)

spherulitic Said of the texture of a rock composed of numerous spherulites; also, said of a rock containing spherulites. Cf: *variolitic; radiated.* Syn: *globular; sphaerolitic.* (AGI, 1987)

spider (a) A ring inserted at the joints of the suspension column of a borehole pump. Radial vanes from the ring support a central sleeve, which acts as a steady bearing from the pump shaft. (BS, 1963) (b) The bowl part of a spider and slips. Also called bowl. See also: *spider and slips.* (Long, 1960) (c) *drum horn.* (d) Assembly of radiating tie rods on the top of a furnace. (ASTM, 1994)

spider and slips A gripping device used to grip and hold rods or casing while coupling or uncoupling them as they are being run into or pulled from a borehole. Also called bowl and slips. See also: *spider.* (Long, 1960)

spider gear A differential gear that rotates on its shaft in a rotating case. (Nichols, 1976)

spiderweb rock A local term in Ohio for sandstone beds that show crossbedding on a small scale, which is complicated by intricate interlacing of fine-bedding planes. Frequently seen in sawed stones, esp. where the lamination is slightly oblique or irregular. It is very like the grain of wood that shows in a planed board.

spiegeleisen An alloy containing 10% to 25% manganese. Is used in steelmaking as a deoxidizing agent and to raise the manganese content of the steel. Also called spiegel and psiegel iron. (CTD, 1958; Henderson, 1953; Camp, 1985)

spigot product In ore dressing, the material discharged at the bottom of the hydraulic classifier. (Newton, 1959)

spike amygdule A cylindrical amygdule whose longer axis is at right angles to the bedding.

spike driver In bituminous coal mining, one who drives a team of two or more draft animals in tandem for hauling wagons or cars of coal. Also called spike-team driver. (DOT, 1949)

spike team (a) A team consisting of three draft animals, two of which are at the pole while the third pads. (Fay, 1920) (b) Three mules, two abreast and one in the lead, used in a mine to haul coal cars. (Fay, 1920)

spike-team driver *spike driver.*

spiking A term used in the United States for the operation of adding ferromanganese, silicomanganese, or other deoxidizing agent, to an open hearth bath for the immediate stoppage of all oxidizing reactions. (Osborne, 1956)

spile (a) A large timber driven into the ground, used as a foundation; a pile. (Crispin, 1964) (b) A plank driven ahead of a tunnel face for roof support. Also called forepole. (Nichols, 1976) (c) A temporary lagging driven ahead on levels in loose ground. See also: *spill.* (d) A short piece of plank sharpened flatways and used for driving into watery strata as sheet piling to assist in checking the flow.

spiler An ironstone miner who excavates and sets timber supports in roadways through wastes and disturbed ground. (Nelson, 1965)

Spilhaus-Miller sea sampler An instrument resembling the bathythermograph and operating in a similar fashion, with the additional ability of obtaining water samples at discrete depths within the limit of operation. Basically, a bathythermograph to which 12 small seawater sampling bottles are attached, it performs the same functions as a cast of Nansen bottles and reversing thermometers to limited depths, but with less accuracy. It is useful for studies of shallow water areas, bays, and estuaries, where rapidity of sampling is of greater importance than the degree of accuracy of temperatures. (Hunt, 1965)

spiling (a) Forepoling over timber and steel supports in weak, loose beds. See also: *spilling.* (Nelson, 1965) (b) Driving timbers ahead of an advancing tunnel through treacherous, loose, watery ground. (Pryor, 1963)

spilite An altered basalt, characteristically amygdaloidal or vesicular, in which the feldspar has been albitized and is typically accompanied by chlorite, calcite, epidote, chalcedony, prehnite, or other low-temperature hydrous crystallization products characteristic of a greenstone. Spilite often occurs as submarine lava flows and exhibits pillow structure. Adj: spilitic. The name was given by Brongniart in 1827. (AGI, 1987)

spilitic suite A group of altered extrusive and minor intrusive basaltic rocks that characteristically have a high albite content. The group is named for its type member, spilite. (AGI, 1987)

spilitization Albitization of a basalt to form a spilite. (AGI, 1987)

spill Any of the thick laths or poles driven ahead of the main timbering to support the roof or sides in advancing a level in loose ground, or to support the sides of a shaft when sinking through a stratum of loose ground. Syn: *spile.* See also: *forepole.*

spilla Ore, pulp, circulating liquor inadvertently discharged from flow line and requiring appropriate means of recovery or removal. Syn: *spillage.* (Pryor, 1965)

spillage *spilla.*

spillage conveyor A small, short conveyor to lift coal out of a spillage pit and deliver it into a mine car or onto the main conveyor. It is usually a chain conveyor run either continuously or intermittently. (Nelson, 1965)

spillage pit An opening below the loading point of a trunk or gate conveyor, to receive spillage. If the output is small, the pit may be emptied by hand once or twice in a shift, but if big tonnages are loaded, a spillage conveyor is installed. (Nelson, 1965)

spilling A method of tunneling through loose, running ground by driving spills (sharp-edged thick boards or steel rods) ahead and around timber or steel frames. (Nelson, 1965)

spill pit *runoff pit.*

spill trough A trough to retrieve melted brass that may be spilled in pouring from a crucible into a flask. (Standard, 1964)

spilosite A rock representing an early stage in the formation of adinol or spotted slate. Cf: *adinole.* (AGI, 1987)

spindle (a) Shaft of a machine tool on which a cutter or grinding wheel may be mounted. (ASM, 1961) (b) Metal shaft to which a mounted wheel is cemented. (ASM, 1961) (c) In founding, a rod or pipe used in forming a core. (Standard, 1964)

spindle conveyor A chain-on-end conveyor in which the chain pins are extended in a vertical plane, usually of enlarged diameter in that portion above the chain, on which special revolvable fixtures can be rotated, for the purpose of spraying or drying. Outboard rollers or sliding shoes support the chain and product.

spindle speed (a) Same as bit rotational speed. (Long, 1960) (b) The number of times the drive rod of a gear-feed-drill swivel head must turn to advance the attached drill string 1 in (2.54 cm). (Long, 1960)

spindle stage A graduated horizontal rotation axis attached to a microscope stage for observation and measurement of optical properties of small crystals. Cf: *goniometer; universal stage.*

spinel (a) An isometric mineral, $MgAl_2O_4$; crystallizes as octahedra; colorless to pale tints; Mohs hardness, 7.5 to 8; in high-temperature metamorphic rocks, contact metamorphosed limestones, serpentinites, and ultramafic rocks; may be of gem quality. (b) The spinel series gahnite, galaxite, hercynite, and spinel. (c) The mineral group brunogeierite, chromite, cochromite, coulsonite, cuprospinel, franklinite, gahnite, galaxite, hercynite, jacobsite, magnesiochromite, magnesioferrite, magnetite, manganochromite, nichromite, ringwoodite, spinel, trevorite, ulvöspinel, and vuorelainenite. (d) Minerals with the spinel structure, such as the linnaeite group. (e) A synthetic crystal with spinel structure that is used as a gemstone, a refractory, or for instrument bearings; e.g., ferrospinel. Also spelled spinelle; spinell; spinelite. See also: *magnesium aluminate.*

spinel emery A mixture of spinel, corundum, and magnetite, the corundum being present in variable proportions. It is a heavy, black, fine-grained aggregate. Dark gray crystals of corundum appear in the best varieties.

spinellid *spinel.*

spinellite A medium- to coarse-grained hypidiomorphic-granular titaniferous magnetite-rich igneous rock with spinel from a few percent to 20%.

spinifex texture Interpenetrating lacy elongate olivine crystals in komatiite, commonly considered to have been formed by quenching. Their disposition resembles the intermesh of an Australian grass for which the texture is named. (AGI, 1987)

spinning cable A flexible wire or plant-fiber cable or rope used as a spinning chain. See also: *spinning chain.* (Long, 1960)

spinning chain Link chain wrapped several times around drill rod, casing, or pipe and used on the drum to spin up or spin out such equipment when it is being pulled or run into a borehole. A rope or flexible wire cable may be used in lieu of a chain. (Long, 1960)

spinning fiber Asbestos suitable for the spinning of asbestos fabrics.

spinning rope A plant-fiber rope used for the same purpose as a spinning chain. See also: *spinning chain.* (Long, 1960)

spin out To unscrew lengths of drill rod casing, or pipe by mechanical means, using a spinning chain, rope, or cable in conjunction with power derived from the cathead or other rotating device. (Long, 1960)

S-P interval In earthquake seismology, the time interval between the first arrivals of transverse (S) and longitudinal (P) waves, which is proportional to the distance from the earthquake source. (AGI, 1987)

spinthariscope Screen coated with zinc sulfide or other fluorescing substance, on which scintillations are observable when bombarded by radioactive rays. (Pryor, 1963)

spin up To screw lengths of drill rod, casing, or pipe together by mechanical means by using a spinning chain, cable, or rope in conjunction with power derived from the cathead or other rotating device. (Long, 1960)

spiral (a) A spiral coal chute that mechanically separates the slate from the coal. The lighter, irregularly-shaped coal falls over the edge of the spiral while the flatter and heavier slate adheres somewhat to the chute surface and is carried down to a special pocket. (b) The Humphrey's spiral is successfully used in recovering chromite from chrome; sands, rutile, ilmenite, and zircon from beach sands (Florida): and tantalum minerals and lepidolite from crushed ores. Also used in concentrating some iron ores and coal. See also: *Humphrey's spiral.*

spiralarm methanometer An instrument that depends on the heat output of burning methane for its action. There are two varieties: those having a controlled flame burning within them and those in which combustion of the methane occurs on electrically heated filaments. A combustible gas alarm making use of the flame principle is the Naylor Spiralarm. (Roberts, 1960)

spiral classifier *Akins' classifier.*

spiral cleaner A device for removing dirt from a conveyor belt. (Nichols, 1976)

spiral coal cleaner A spiral chute in which the plate is inclined towards the center of the spiral. The stone tends to flow down centrally, the coal tends to slide off around the outside of the spiral. They are seldom used as coal cleaners. (Nelson, 1965)

spiral concentrator A sluice formed in five or six tight spirals, in which centrifugal force aids the separating effect of sluice action. See also: *Humphrey's spiral.* (Pryor, 1963)

spiral conveyor *screw conveyor.*

spiral core A piece of core the outside surface of which is rifled. See also: *rifle.* (Long, 1960)

spiral curve (a) A curve of gradually increasing radius that allows an easy transition between a circular arc and a straight on a road or railway. (Hammond, 1965) (b) In railroad or highway surveying, a curve of progessively decreasing (or increasing) radius used in joining a tangent with a simple circular curve or in joining two circular curves of different radii. Syn: *transition curve.* (Seelye, 1951)

spiral grooving *rifling.*

spiral gummer A screw device, attached to a longwall coal cutter, for removing the holings and depositing them either in the track of the machine or at the side clear of the face. It is designed in two types: end discharge and side discharge. (Nelson, 1965)

spiral hole A borehole that follows a corkscrewlike course. Cf: *rifle.* (Long, 1960)

spiraling (a) Rifling. (Nichols, 1954) (b) A drill hole twisting into a spiral around its intended centerline. (Nichols, 1954)

spiral level A section of convex glass tube containing fluid and an air bubble. When level, the air bubble centers itself on an etched line on the tube.

spiral system In open pit mining, a haul road arranged spirally along the perimeter walls of the pit so that gradient of road is more or less uniform from the bottom to the top of the pit.

spiral track A track layout for rail or road transport from large opencast pits. The track is arranged spirally along the steep rise from the coal or ore benches so that the gradient is moderate throughout. (Nelson, 1965)

Spiral Vane Disk Cutter Trade name for a cutter loader incorporating a special variety of shearer head. The Mark II model consists of section plates welded together to form a composite whole spiral. The disk is made in different sizes to give cutting diameters of from 31 to 72 in (79 to 183 cm) with a maximum web depth of 30 in (76 cm). A single spiral is used for softer coals and a double spiral for hard coals. A plow attached to the machine throws the coal onto the conveyor. (Nelson, 1965)

spiral worm A device to withdraw broken rods from a borehole. It is lowered down the hole and the screw is turned around until it grips the broken rod below the joint. See also: *wad hook; wad coil.* (Nelson, 1965)

spire A kind of fuse. The tube carrying a train to the charge in a blasthole. Also called reed or rush, because these, as well as spires of grass, are used for the purpose. See also: *reed.* (Fay, 1920)

Spirelmo smoke helmet A helmet in which the crown and frontpiece are blocked out of rawhide, and the front shield is fitted with two mica windows in hinged aluminum frames. It has a twin-tubed air feed on each side of the helmet and a valve for the escape of excess and vitiated air. Air is supplied through an armored hose from double-acting bellows or a blower worked by a second person at the fresh air base. Airtightness is obtained by means of a soft leather apron secured in position about the neck and shoulders. Illumination is provided by portable electric lamps, and communication with the wearer by an approved type of mine telephone. (Sinclair, 1958; Mason, 1951)

spirit of copper Acetic acid obtained by distilling copper acetate.

spirits of niter (a) Nitric acid. (Osborne, 1956) (b) A solution of ethyl nitrite in alcohol. (Osborne, 1956)

spirits of sulfur Sulfurous acid. (Osborne, 1956)

spirits of verdigris Acetic acid. (Osborne, 1956)

spirits of vinegar Dilute acetic acid. (Osborne, 1956)

spirits of wine Ethyl alcohol. (Osborne, 1956)

spirits of wood Methyl alcohol. (Osborne, 1956)

spiroffite A monoclinic mineral, $(Mn,Zn)_2Te_3O_8$; red to purple.

spitted fuse Slow-burning fuse that has been cut open at the lighting end for ease of ignition. A small quantity of the plastic explosive used in the hole is sometimes inserted in the cut. (Pryor, 1963)

spitting (a) Lighting the fuse for a blast. (b) An action of or appearance on the surface of slowly cooled, large masses of melted silver or platinum, in which the crust is forcibly perforated by jets of oxygen, often carrying with them drops of molten metal. Also called sprouting. (Standard, 1964)

spitting rock A rock mass under stress that ejects small fragments with considerable velocity on breaking.

spitzkasten A series of hopper-shaped or pointed boxes for separating mineral-bearing slimes according to fineness, in which the width of each box is double that of its predecessor, while the lengths increase by arithmetical progression. As used in flotation, it is the froth-separating compartment of mechanical-agitation-type flotation machines. Also called spitz. See also: *funnel box.* (Hess)

spitzlutten Hydraulic classifiers shaped like the spitzkasten, but having provision for pressure water to flow upward from near the apex, thus improving efficiency of separation. (Pryor, 1963)

splasher (a) A plate lined with firebrick and placed over the iron trough next to the taphole to keep down flame that blows from the taphole during a cast. (Fay, 1920) (b) A water spray system for the protection of the metal structure immediately above the tapping hole during the tapping of a blast furnace. (Henderson, 1953)

splash man A laborer who shovels charcoal over the surface of molten copper being poured from a reverberatory furnace into a tilting ladle to prevent excess oxidation of metal. (DOT, 1949)

splay One of a series of divergent small faults at the extremities of a major fault. Splays are typically associated with rifts.

splay faulting Minor faulting that diverges from a longer dislocation at an acute angle.

splendent Applied to the degree of luster of a mineral, reflecting with brilliancy and giving well-defined images, such as hematite or cassiterite.

splice (a) A joint made in a broken haulage rope. Splicing is a skilled job and the rope ends are unlaid for a length on each side of the break and reformed to a definite pattern. (Nelson, 1965) (b) Generally used to designate an insulated reconnection of wires of an electric cable after it has been cut. (BCI, 1947) (c) To unite two ropes by interweaving the strands.

splice box An enclosed connector permitting short sections of cable to be connected together to obtain a portable cable of the required length.

spliced Said of a vein that pinches out and is overlapped by another parallel vein.

splint coal A miner's term long used in Eastern United States and Scotland for certain hard dull coals with a distinctive type of fracture. Splint coals are irregular and blocky, with an uneven rough fracture, grayish black in color and of granular texture. Splint coals are banded coals. Coals containing more than 5% of anthraxylon and more than 30% of opaque attritus determined by microscopic examination are classed as splint coal. The content of anthraxylon and opaque matter is determined perpendicular to the bedding across the entire thin section (2 to 3 cm in width). The opaque attrital portion of the splint coal may be intercalated with fine, hairlike streaks of anthraxylon. It occurs mainly as bands and benches in otherwise bright-banded coal and is wide-spread in bituminous coal seams. Corresponds either to duroclarite or, more frequently, to clarodurite according to the ratio of vitrinite and inertinite. May also correspond to vitrinertite. (IHCP, 1963)

splintery fracture The property shown by certain minerals or rocks of breaking or fracturing into elongated fragments like splinters of wood.

split (a) To divide the air current into separate circuits to ventilate more than one section of the mine. Cf: *air split*. (BS, 1963) (b) The workings ventilated by that branch. (c) A bench separated by a considerable interval from the other benches of a coal bed. (d) The upper or lower portion of a divided coal seam. (CTD, 1958) (e) To divide a pillar or post by driving one or more roads through it. (f) A layer of coal which has separated from its parent seam. See also: *split seam; ventilation; splitting*. (Nelson, 1965) (g) The process of dividing a core lengthwise, dividing a granular material into several representative parts for sending samples to several interested parties or reducing either core storage space or the quantity of material retained as a sample. (Long, 1960) (h) The division of a bed of coal into two or more horizontal sections by intervening rock strata. (Hudson, 1932)

split-barrel sampler A drive-type soil sampler with a split barrel; also a swivel-type double-tube core barrel, the inner tube of which is split. Syn: *split-tube sampler*. (Long, 1960)

split brilliant A brilliant split apart at the base of its pyramidal forms, so as to make two gems. (Standard, 1964)

split bushing A bushing made in two pieces, for ease of insertion and removal. (Nichols, 1976)

split check A system of leasing practiced at Cripple Creek, Colorado, whereby the miners and company divide the profits. (von Bernewitz, 1931)

split coal Coalbed separated by clay, shale, or sandstone parting that thickens so that both benches cannot be mined together. (AGI, 1987)

split core A core that has been split lengthwise into halves or quarters. (Long, 1960)

split core barrel A type of core barrel which can be opened longitudinally to remove the core. (BS, 1963)

split inner-tube core barrel A double-tube core barrel with the inner tube split lengthwise. (Long, 1960)

split lagging Drum lagging made in two pieces to allow changing it without dismantling the drum. (Nichols, 1976)

split-ring core lifter A hardened steel having an open slit, an outside taper, an inside or outside serrated surface. In its expanded state, it allows the core to pass through it freely, but when the drill string is lifted, the outside taper surface slides downward into the bevel of the bit or reaming shell, causing the ring to contract and grip tightly the core which it surrounds. Also called core catcher; core gripper; core lifter; ring lifter; split-ring lifter; spring lifter. (Long, 1960)

split-ring lifter *split-ring core lifter.*

split rock A rock possessing tabular structure, or which cleaves easily in the lines of lamination, and is consequently suitable for flagging and curbstones.

splits In mine ventilation, airways connected in parallel. (Hartman, 1982)

split seam A coal seam that has separated into two or more layers which may, or may not, rejoin some distance away. Syn: *coal split*. See also: *multiple splitting*. (Nelson, 1965)

split shovel A device for sampling fine ore, consisting of a fork in which the prongs are separate scoops, each scoop being the same width as the open spaces between.

split spread A type of seismic spread in which the shot point is at the center of the arrangement of geophones. It is commonly used for continuous profiling and for dip shooting. Syn: *straddle spread; symmetric spread*. (AGI, 1987)

split sprocket A two-piece sprocket that can be assembled on a shaft without removing the shaft bearings. (Nichols, 1976)

split system (a) A system of ventilation in which air is split along the airways or at the face. See also: *natural splitting; controlled splitting*. (b) Historically, a combination of warm air heating and radiator heating. Also used for other combinations, such as hot water steam, steam warm air, etc. (Strock, 1948)

split the air *split.*

splitting (a) Parting of a coalbed into two or more benches separated by other rocks. (AGI, 1987) (b) In mine ventilation, the practice of connecting airways in parallel by dividing the total air flow among them. (Hartman, 1982) (c) Abrasion of a rock fragment resulting in the production of two or three subequal parts or grains. (AGI, 1987) (d) The property or tendency of a stratified rock of separating along a plane or surface of parting. (AGI, 1987) (e) The sampling of a large mass of loose material (e.g., a sediment) by dividing it into two or more parts; e.g., quartering. (AGI, 1987)

splitting knife A knife used for splitting leather or for diamond cleaving. (Standard, 1964)

splitting method A method of mining pillars seldom followed. A room is first driven through the pillars, splitting them into smaller blocks. The pockets are turned at right angles and are driven into the blocks. This method is really gouging the pillars and is wasteful. (Lewis, 1964)

splitting of air *ventilation; splitting.*

splitting shot Arkansas. A shot put into a large mass of coal detached by a previous blast. See also: *block hole*.

split-tube barrel *soil sampler; split-barrel sampler.*

split-tube sampler *soil sampler; split-barrel sampler*. Cf: *solid-barrel sampler*.

spodumene A monoclinic mineral, $LiAlSi_2O_6$; pyroxene group; prismatic cleavage; in granite pegmatites in crystals up to scores of meters long (called logs); a source of lithium; may be of gem quality (lavender kunzite, green hiddenite). Formerly called triphane. Syn: *lithia amethyst; lithia emerald*.

spoil Overburden, nonore, or other waste material removed in mining, quarrying, dredging, or excavating. See also: *waste*. (AGI, 1987)

spoil bank (a) A term common in surface mining to designate the accumulation of overburden. (BCI, 1947) (b) Underground refuse piled outside. (BCI, 1947) (c) That part of a mine from which coal has been removed and the space more or less filled up with waste. (BCI, 1947) (d) To leave coal and other minerals that are not marketable in a mine. See also: *spoil heap*. (BCI, 1947)

spoil dam An earthen dike forming a depression in which returns from a borehole can be collected and retained. (Long, 1960)

spoil heap (a) The pile of dirt produced by mining operations and stacked at the surface of a mine either in conical heaps or in layered deposits. Syn: *dump; tip*. (Nelson, 1965) (b) A pile of refuse material from an excavation or mining operation; e.g., a pile of dirt removed from, and stacked at the surface of, a mine in a conical heap or in layered deposits, such as a tip heap from a coal mine. See also: *spoil bank*. (AGI, 1987)

spoil-heap fire The spontaneous heating and burning of small coal, carbonaceous shale, and perhaps iron pyrites in spoil heaps. (Nelson, 1965)

spoil pile *waste dump.*

spoil pool The reservoir formed by a spoil dam in which the returns from a borehole collect and are retained. Cf: *sludge pit; sump*. (Long, 1960)

spoils (a) *cuttings*. (b) The debris or waste material from a mine. (Long, 1960)

sponge (a) A form of metal characterized by a porous condition, which is the result of the decomposition or reduction of a compound without fusion. The term is applied to forms of iron, the platinum-group metals, titanium, and zirconium. Metal has appearance of a sponge due to high porosity. (ASM, 1961) (b) Hafnium produced by the Kroll process. (Thomas, 1960)

sponge iron powder Ground and sized sponge iron, which may have been purified or annealed or both. (ASTM, 1994)

sponge metal A form of metal characterized by a porous condition, which is the result of the decomposition or reduction of a compound without fusion. Metal has the appearance of a sponge due to high porosity. (ASM, 1961)

spongy Said of a vesicular rock structure with thin partitions between the vesicles, thus resembling a sponge.

spongy iron *reduced iron.*

spontaneous Used to describe the driving potential that causes electric currents to circulate in boreholes. These currents are not in any way deliberately induced by the well-logging equipment. Also called self-potential; SP curve. See also: *spontaneous potential curve*. (Wyllie, 1963)

spontaneous combustion (a) The heating and slow combustion of coal and coaly material initiated by the absorption of oxygen. The two main factors involved are (1) a coal of a suitable chemical and physical nature; and (2) sufficient broken coal and air leaking through it to supply the oxygen needed. The heat generated is retained with consequent rise in temperature. See also: *gob fire; hydrogen sulfide; open fire; weathering of coal*. (Nelson, 1965) (b) The outbreak of fire in combustible material that occurs without the direct application of a flame or a spark. It is usually caused by slow oxidation processes (such as atmospheric oxidation or bacterial fermentation) under conditions that do not permit the dissipation of heat. (c) Ignition that can occur when certain materials such as tung oil are stored in bulk, resulting from the generation of heat, which cannot be readily dissipated; often heat is generated by microbial action. Also known as spontaneous ignition. (McGraw-Hill, 1994)

spontaneous polarization (a) Electrochemical reactions of certain orebodies causing spontaneous electrical potentials. (Schieferdecker, 1959) (b) *self-potential method; spontaneous potential method.*

spontaneous potential curve The electric log curve that records changes in natural potential along an uncased borehole. Small voltages are developed between mud filtrate and formation water of an invaded bed, and also across the shale-to-mud interface. These electrochemical components are augmented by an electrokinetic potential (streaming potential) developed when mud filtrate moves toward a formation region of lower fluid pressure through the mud cake. Where formation waters are less resistive (more saline) than drilling-mud filtrate, the spontaneous-potential curve deflects to the left from the shale baseline. First used about 1932, the curve was added to the resistivity log to make up the basic electric log of well-logging practive. Syn: *SP curve; self-potential curve.* (AGI, 1987)

spontaneous-potential method *self-potential method.*

spontaneous potential method An electrical method in which a potential field caused by spontaneous electrochemical phenomena is measured. Syn: *self-potential method; spontaneous polarization.* (AGI, 1987)

spool (a) Cast iron distance piece placed between timbers. (Hammond, 1965) (b) To wind rope or cable on a hoist drum. (Long, 1960) (c) The drum of a hoist. (Long, 1960) (d) The movable part of a slide-type hydraulic valve. (Nichols, 1976) (e) To wind in a winch cable. (Nichols, 1976)

spool-type roller conveyor A type of roller conveyor in which the rolls are of conical or tapered shape with a diameter at the ends of roll larger than at the center.

spoon A tool for cleaning dust or sludge from quarry blasting holes. Syn: *scraper.* See also: *pneumatic blowpipe.* (Nelson, 1965)

spooner In bituminous coal mining, a laborer who scoops drillings out of boreholes in which explosives are to be charged for blasting down coal, using a slender iron rod with a cup-shaped projection bent at right angles to the handle. (DOT, 1949)

spooning Many mineral raw materials, such as petroleum, cementation water (water containing dissolved copper or iron sulfates or other metal compounds), or brine are extracted by pumping through boreholes. In spooning, a long spoon (a hollow cylinder with a bottom equipped with a clap valve, or ball valve, and open above) is attached to a cable that is let down into the boreholes where the cylinder fills with the liquid; this is emptied out after the cylinder is raised. (Stoces, 1954)

spoon proof Test-ladle specimen taken during various stages of melting and fining. (ASTM, 1994)

spoon sampler A rotating soil sampler, fitted with an auger-type cutting shoe. (Long, 1960)

spore Part of the reproduction organs of many coal measures' plants. There are two kinds, namely, megaspores (female) and microspores (male). They are found in most coal seams, particularly the dull layers. Megaspores vary from 1 to 5 mm in size, and microspores (or pollen grains) from about 0.01 to 0.1 mm. (Nelson, 1965)

spore coal (a) Coal in which the attritus contains a large amount of spore matter along with transparent attritus. See also: *cannel coal.* (b) Coal formed out of the spores of lycopods.

sporinite A maceral of the exinite group consisting of spore exines generally much flattened parallel to stratification. Cf: *cutinite.* See also: *resinite.* (AGI, 1987)

sporinite coal This type of coal consists of more than 50% of spores (microspores and megaspores). Other structured components uniformly distributed through the coal are cuticles, resin bodies, and gelified and fusinized tissues. Hand specimens of low rank sporinite coal are brownish, with matt or granular surfaces. The coal may have high or low ash and occurs in seams of different geological, age but is particularly common as bands of limited thickness in seams of the Lower Carboniferous. Sporinite coal naturally admixed with medium rank gelitocollinite coals is used for coking. (IHCP, 1963)

sporogelite A colloidal form of aluminum hydroxide, $Al_2O_3{\cdot}H_2O$, occurring as one of the constituents of bauxite. Also called cliachite; alumogel. Syn: *diasporogelite.* (English, 1938)

spot (a) To mark the site at which a borehole is to be drilled, a piece of equipment placed, or a structure built. (Long, 1960) (b) To set a drill or piece of machinery at a preselected site. (Long, 1960) (c) An inclusion in a diamond. (Long, 1960) (d) To direct to the exact loading or dumping place. (Nichols, 1976)

spot-bolting The use of one or just a few roof bolts at spot locations. (USBM, 1966)

spot cooler Low capacity, semiportable refrigeration unit of 150,000 to 500,000 Btu/h (44 to 146.5 kW) cooling capacity that is used in cooling sites of limited extent, such as an underground enginehouse or the face of a development end. The refrigerant used is nontoxic, and an electric or compressed-air drive is applied to a reciprocating compressor. (Roberts, 1960)

spot level The reduced level of any survey point. (Hammond, 1965)

spot log A log or marker placed to show a truck driver the spot to stop to be loaded. (Nichols, 1976)

spotted schist *spotted slate.*

spotted slate A slaty or schistose argillaceous rock whose spotted appearance is the result of incipient growth of porphyroblasts in response to contact metamorphism of low to medium intensity. Cf: *desmosite; spilosite; adinole.* See also: *fleckschiefer; fruchtschiefer; garbenschiefer; maculose.* Syn: *spotted schist; knotted schist; knotted slate.* (AGI, 1987)

spotter (a) In truck usage, the person who directs the driver into loading or dumping position. (Nichols, 1976) (b) In a pile driver, the horizontal connection between the machinery deck and the lead (pile guide). (Nichols, 1976) (c) *car pincher.*

spot tests Simple and speedy qualitative tests used to identify minerals species when prospecting, valuing a deposit, or testing mill products. (Pryor, 1963)

spotting hoist A small haulage engine used for bringing mine cars into the correct position under a loading chute, feeder or other point. See also: *pickrose hoist.* (Nelson, 1965)

spotty ore Ore in which the valuable material is concentrated irregularly as small particles; e.g., coarse gold in low-grade rocks. (Nelson, 1965)

spout delivery pump A pump, similar to a diaphragm pump, that is not capable of delivering water above its own height. See also: *force pump.* (Hammond, 1965)

spoutman Person who directs the pouring of slag from a ladle through a spout into a reverberatory furnace used for smelting. (DOT, 1949)

sprag (a) A short wooden prop set in a slanting position for supporting the coal during the operation of holding. (b) To chock or stop, as a vehicle or wheel, by a sprag; prop. See also: *spur.* (Standard, 1964) (c) The horizontal member of a square act of timbers running parallel to the axis of a heading. (Stauffer, 1906) (d) *rod spear; stell.*

spragger In anthracite and bituminous coal mining, a laborer who rides trains of cars and controls their free movement down gently sloping inclines by throwing switches and by poking sprags (short, stout, metal or wooden rods) between the wheel spokes to stop them. (DOT, 1949)

spragging The act of checking a mine car with a sprag. (Korson, 1938)

sprag road A mine road having such a sharp grade that sprags are needed to control the descent of a car; hence, two-, three-, or four-sprag road. See also: *sprag.*

spray In a hydrocyclone, the discharge from the apex in spray form, showing that the cyclone is not overloaded. (Pryor, 1963)

spraying machine A machine that applies a spray under pressure on mine timber supports to preserve and fireproof them. It may also be used for limewashing and water spraying of dust. The machine is mounted on wheels and operated by compressed air. (Nelson, 1965)

spraying screen A screen used for the removal, by spraying, of fine solids present among or adhering to larger particles. (BS, 1962)

sprays Appliances to damp deposits of dust in tunnels and workings before and after shotfiring and loading operations. Water sprays are also used along dusty roadways. Various types of mist projectors and atomizers are used and effect considerable improvement, but the dust trapped consists chiefly of the coarser particles. In many dusty mines, a water pipe system extends throughout the workings and sprays are employed at all loading and other dusty points. Sprays are also used to suppress dust at coal and ore processing plants. See also: *whale type jib.* (Nelson, 1965)

spray water *rinsing water.*

spread (a) The area covered at a given thickness by a given quantity of such materials as chippings or road binder. (Hammond, 1965) (b) The surface in proportion to the depth of a stone. (Hess) (c) The surface or width at the girdle in proportion to the depth of a cut stone, such as a diamond. (d) The layout of geophone groups from which data from a single shot are recorded simultaneously. Syn: *seismometer spread; seismic spread.*

spreader (a) A horizontal timber below the cap of a set, to stiffen the legs and to support the brattice when there are two air courses in the same gangway. (b) A piece of timber stretched across a shaft as a temporary support of the walls. (c) A tool used in sharpening machine drill bits. (Fay, 1920) (d) A strut in a tunnel or heading timber sets. (Nelson, 1965) (e) A machine which spreads dumped material with its blades. (Nelson, 1965)

spreader chains Chains joining the end of the tail chain to ends of the spreader. (Zern, 1928)

spreader operator *tripper man.*

spread recorder An instrument used in bridge testing to measure any outward spread of an abutment under load. See also: *rotation recorder*. (Hammond, 1965)
spring (a) To enlarge the bottom of a drill hole by small charges of a high explosive in order to make room for the full charge; to chamber a drill hole. See also: *camouflet*. (b) To chamber. See also: *chamber*. (Long, 1960) (c) A general name for any natural discharge of hot or cold pure or mineralized water.
spring auxillary cylinder A heavy tension spring, enclosed in a cylinder, which is connected to the panline of certain types of shaker conveyors to keep the conveyor in tension. It is attached to the conveyor by a driving chain and to a prop by a fixing chain. Keeping the conveyor in tension, it is claimed, will save the conveyor connections and increase the output. (Jones, 1949)
spring constant Force that produces a unit elongation of the spring used in geophysical instruments. (AGI, 1987)
spring core lifter *core lifter*.
spring dart (a) A tool used to retrieve lost boring gear. (Pryor, 1963) (b) A device to withdraw the steel casing from a borehole when finished. The casing is cut into convenient lengths and then the spring dart is lowered to bring up each length separately. The dart springs open immediately when it meets a cut or recess in the casing, which length it then grips and lifts to the surface. (Nelson, 1965)
springing (a) A quarry blasting method in which a succession of charges is fired in a borehole to open up a chamber. (BS, 1964) (b) Enlarging the bottom of a drill hole by exploding a small charge in it. (Nichols, 1976) (c) In certain types of rock, large quantities of stone can be blasted down by the method known as springing the shothole. The technique requires that the rock contains well-defined bedding or jointing planes, such as are found in most sedimentary and some igneous rocks, particularly granite. The principle of springing is to drill a borehole with a heavy burden and then explode a succession of gradually increasing charges of black powder so that the bedding planes or joints are opened up to permit the placing of a large final charge. Syn: *bullying; overspringing*. (McAdam, 1958)
springing a hole *springing*.
spring lifter *core lifter*.
spring line The meeting of the roof arch and the sides of a tunnel. (Nichols, 1976)
spring-loaded Held in contact or engagement by springs. (Nichols, 1976)
spring-roll crusher A crushing machine similar to the double-roll crusher with the difference that springs are fixed to the bearings of one roll. (Nelson, 1965)
spring rolls Crushing rolls used in ore breaking. Two parallel cylinders, mounted horizontally, are held apart by shims, and pressed together by powerful springs. Crushable rocks falling between them are drawn down as the cylinders revolve, but unbreakable material causes the springs to yield and let it pass without damage. (Pryor, 1963)
spring washer A washer consisting of a steel ring cut through and bent into helical form, which prevents a nut from unscrewing. (Hammond, 1965)
sprinkling An act of spraying water into the atmosphere and on coal surfaces to allay coal dust. (BCI, 1947)
sprocket A gear that meshes with a chain or crawler track. (Nichols, 1976)
sprocket gear A gear that meshes with a roller or silent chain. (Nichols, 1976)
sprue hole A pouring hole in a mold; a gate. (Standard, 1964)
spud (a) To break ground with a drilling rig at the start of well-drilling operations. Syn: *spud-in*. See also: *spad*. (AGI, 1987) (b) To bore, as the first 50 ft (15 m) of an oil well, by the use of a bull wheel. (Standard, 1964) (c) To commence drilling operations by making a hole. (Wheeler, 1958) (d) To begin the drilling of a borehole with a spud or diamond-point bit. (Long, 1960) (e) An offset type of fishing tool used to clear a space around tools stuck in a borehole. (Long, 1960) (f) A cabletool drill bit. (Long, 1960) (g) An anchorage during dredging provided by a steel post underneath a dredger that can be lowered by a toothed rack or by ropes until it is secured in the seabed, riverbed, or dredge pond. (Hammond, 1965)
spud bit (a) A mud or diamond-point bit used to drill through overburden or soil down to bedrock. (Long, 1960) (b) A broad, dull, chisel-face drilling tool for working in earth down to rock with a churn or cabletool drill. (Long, 1960)
spudded-in A term applied to a borehole that has been started and the hole has reached bedrock and/or the standpipe has been set. (Long, 1960)
spudder (a) A churn drill, churn-drill operator, or the special bit used to begin a borehole by rotary, diamond, or churn drills. (Long, 1960) (b) A colloquialism for a small drilling-rig. (Williams, 1964)
spudder drill *churn drill*.
spudding (a) The operation, in rope drilling, of boring through the subsoil at the start of a hole. (BS, 1963) (b) In diamond and/or rotary drilling, a general term applied to drilling through overburden with a fishtail bit, drag bit, or diamond-point bit. (Long, 1960) (c) Sinking a conductor, standpipe, or casing with a churn- or cable-type drill rig. (Long, 1960)
spudding bit (a) A broad dull drilling tool for working in earth down to the rock. (Standard, 1964) (b) A heavy chisel bit used in percussive drilling to drill through subsoil. (BS, 1963) (c) *spud bit*. (d) The bit used to start the hole. When the hole is deep enough, regular drilling tools are substituted. (Williams, 1964)
spudding boreholes The working of a cable drill up and down on a short length of rope, when passing through the superficial deposits down to bedrock. This section of hole is cased. (Nelson, 1965)
spudding drill A drill that makes a hole by lifting and dropping a chisel bit. Syn: *churn drill*. (Nichols, 1976)
spudding driller In petroleum production industry, one who uses a lightweight, portable drilling rig (spudder) for the drilling of shallow wells, or a regular cable drilling rig to drill the first few feet of a well. Also called spudder or spud driller. Syn: *spud driller*. (DOT, 1949)
spudding drum In a churn drill, the winch that controls the drilling line. (Nichols, 1976)
spudding tool Tool used to begin a borehole in earthy materials with a diamond or rotary drill; also, a drilling tool used by a cable tool or churn drill. (Long, 1960)
spud drill *churn drill*.
spud driller *spudding driller*.
spud-in *spud*.
spuds On a dredge, steel tubes pointed at the bottom and provided with lifting tackle at the top that are used to hold and to move the dredge. (Nichols, 1954)
spud setter A mine surveyor. See also: *spud*. (Fay, 1920)
spud well On a dredge, a pair of guide collars for a spud. (Nichols, 1976)
spur (a) A brace or prop. See also: *sprag*. (b) A small vein branching from a main one. (AGI, 1987) (c) A rock ridge projecting from a sidewall after inadequate blasting. (Nichols, 1976) (d) A relatively short and small vein of quartz that cuts across the bedding, in contrast to a saddle reef that more or less follows the bedding. (Nelson, 1965)
spur-end facet *triangular facet*.
spurrite A monoclinic mineral, $Ca_5(SiO_4)_2(CO_3)$; dimorphous with paraspurrite; forms light gray granular masses resembling limestone; at Velardena, Durango, Mexico; Scawt Hill, Ireland; Luna County, NM; and Crestmore, CA.
spurt Forest of Dean. A disintegrated stone.
spur track In railroading, a short sidetrack connecting with the main track at one end only. (Standard, 1964)
spur valley A short branch valley. (Nichols, 1976)
square A term used in the slate industry with reference to roofing slate. A square is a sufficient number of any size to lay 100 ft^2 (9.29 m^2) of roof, allowing the standard 3-in (7.62-cm) lap. The estimated weight of a square of 1/4-in (6.4-mm) slate is 1,000 lb (454 kg).
square drill collar A long stabilizer of rectangular shape that when used properly gives a super-packed-hole effect. Square drill collars are made primarily from 30-ft (9.14-m) steel bar stock with a diagonal measurement greater than the hole diameter in which the collar will eventually be used. This collar has proved successful in controlling rapid directional and deviational changes in wells drilled in a disturbed-belt-type area. (API, 1963)
square-mile-foot A unit of measure representing the volume of water 1 ft (0.3 m) deep over an area of 1 $mile^2$ (2.6 km^2). See also: *acre-foot*. (Hammond, 1965)
square set A set of timbers used to provide support in a stope or an underground mine. Each timber set consists of a vertical post and two horizontal members known as a cap and girt. The timber ends are sawed to allow adjoining timbers to interlock. They are framed at mutual right angles, and when joined with other sets form a continuous timber framework that conforms to the irregular shape of the stope. The posts are 6 to 7 ft (1.83 to 2.13 m) high, while the caps and girts are 4 to 6 ft (1.22 to 1.83 m) long. Caps and girts are placed on top of the posts, a line of caps being at right angles to a line of girts. Square sets vary in dimensions at different mines, but in general should give a clear opening of at least 5 ft (1.52 m) each way between posts to afford sufficient working space in the stope, and a clear height of 6-1/2 ft (1.98 m) is about the minimum height desirable. This system of timber support can be adapted to large and irregular orebodies resulting in an elaborate network extending the full height and width of a stope.
square-set and fill *square-set stoping*.
square-set block caving A method of block caving in which the caved ore is extracted through drifts supported by square sets. A retreating system is adopted. (Nelson, 1965)
square-set slicing *top slicing and cover caving*.

square-set stoping A method of stoping in which the walls and back of the excavation are supported by a system of interlocking framed timbers (square set). A square set of timber consists of a vertical post and two horizontal members set at mutually right angles. The mining process is slow and only enough ore is excavated to provide room for installation of each successive set of timber. The stopes are usually mined out in floors or horizontal panels, and the sets of each successive floor are framed into the top of the preceding floor. Syn: *alternate pillar and stope; overhand vertical slice; underhand vertical slice.* Cf: *back-filling system.*

square-set system A method of mine timbering in which heavy timbers are framed together in rectangular sets, 6 to 7 ft (1.83 to 2.13 m) high, and 4 to 6 ft (1.22 to 1.83 m) square, so as to fill in as the orebody is removed by overhand stoping. (Webster 3rd, 1966)

square-set underhand *square-set stoping.*

square thread (a) A screw thread the cross section of which is square. (Long, 1960) (b) A robust type of screw thread that can transmit thrust in both directions. (Hammond, 1965)

square work *sublevel stoping.*

square work and caving *sublevel stoping.*

squat lads! Fall flat down on the floor. In the early days of coal mining, igniting the gas was a very common thing; so, whenever an explosion took place, the colliers shouted to one another, "Squat, lads!". (Fay, 1920)

squealer A shot that breaks the coal only enough to allow the gases of detonation to escape with a whistling or squealing sound; also called a whistler. (Zern, 1928)

squeal-out Arkansas. *seam-out; squealer.*

squealy coal Arkansas. Seamy coal from which the powder gases escape with a squealing sound. (Fay, 1920)

squeeze (a) A crushing of coal or other materials with the roof moving nearer to the floor, due to the weight of the overlying strata. (Lewis, 1964) (b) The settling, without breaking, of a mine roof over a considerable area of workings. Also called creep; crush; pinch; nip. (c) The effect of the closure of stope walls on supports placed between them. (Spalding, 1949) (d) A passageway in a cave that is very narrow and can be passed by a person only with great difficulty. (AGI, 1987) (e) Applied to sections in coal seams where they have become constricted by the squeezing in of the overlying or underlying rock as a result of pressure during folding or other movements. (AGI, 1987) (f) A pinch of a vein in passing through hard bands of rock. (Gordon, 1906) (g) To inject a grout into a borehole under high pressure. (Long, 1960) (h) The plastic movement of a soft rock in the walls of a borehole or mine working that reduces the diameter of the opening. (Long, 1960) (i) Pumping cement back of casing under high pressure to block off or re-cement channeled areas. (Wheeler, 1958) (j) The rapid or gradual closing of a mine working by the displacement of weak floor strata from beneath supporting pillars into adjacent mine rooms. See also: *want.* (AGI, 1987)

squeezer A mine tub controller that acts by squeezing the tub or the wheels. (Mason, 1951)

squeeze riveter A single-stroke, compressed-air cylinder for closing rivets through the medium of a toggle mechanism. (Hammond, 1965)

squeezing The slow increase in weight on pillars or solid coal eventually resulting in such things as crushing of the coal, heaving of the bottom, and the driving of pillars into soft floor or top. The cause normally is leaving pillars or other supports which, after considerable area is opened up, prove to be inadequate, permitting the top to settle gradually with transfer of the weight to active places and solid coal. (Coal Age, 1966)

squib (a) *electric squib.* (b) A thin tube filled with black powder, forming a slow-burning fuse to explode a stemmed charge of black powder. (BS, 1964) (c) A small charge of powder exploded in the bottom of a drill hole, to spring the rock, after which a heavy shot is fired. A springing shot. (d) In well boring, a vessel, containing the explosive and fitted with a time fuse, that is lowered into a well to detonate the nitroglycerin charge. (e) A firing device that will burn with a flash which will ignite black powder. (Nichols, 1976)

squib shot A blast with a small quantity of high explosives fired at some point in a borehole for the purpose of dislodging some foreign material that has fallen into it. (Fay, 1920)

squinted vein Derb. A mineral vein cut by a dike and thereby thrown out of alinement on the two sides of the dike. (Arkell, 1953)

squirrel cage fan A centrifugal blower with forward-curved blades. (Strock, 1948)

squirrel-cage motor An alternating current electric motor with many applications. The rotor is made of strong parallel copper or aluminum bars on the perimeter, joined to end rings of the same metal. (Hammond, 1965)

squotting A stage in the heating of clay when so much of the material has fused that the mass begins to lose its shape and becomes viscous. See also: *fusion of clay; vitrifying.* (Nelson, 1965)

stab (a) To guide a pipe, casing, or drill rod so that the threads will engage properly. (Long, 1960) (b) To recover a drill tool lost in a borehole by using a spear-shaped or pointed fishing tool. (Long, 1960) (c) In adding to a drill string, the action of lining up and catching the threads of the loose piece. (Nichols, 1976)

stability (a) The resistance of a structure, slope, or embankment to failure by sliding or collapsing under normal conditions for which it was designed; e.g., bank stability and slope stability. See also: *bank slope stability.* (AGI, 1987) (b) In thermodynamics, an equilibrium state to which a system will tend to move from any other state under the same external conditions. (AGI, 1987)

stability series A grouping of minerals arranged according to their persistence in nature; i.e., to their resistance to alteration or destruction by weathering, abrasion during transportation, and postdepositional solution; e.g., olivine (least stable), augite, hornblende, biotite (most stable). The most stable minerals are those that tend to be at equilibrium at the Earth's surface. Syn: *order of persistence.* (AGI, 1987)

stabilized coupling A rod coupling built up to reaming-shell size by welding on an abrasion-resistant metal, applied in ridges parallel to the long axis of the drill rod. (Long, 1960)

stabilized tray conveyor *over-and-under conveyor.*

stabilizer (a) A hardened, splined bushing, sometimes freely rotating, slightly larger than the outer diameter of a core barrel. Also called ferrule; fluted coupling. (Long, 1960) (b) A misnomer for guide rod. (Long, 1960) (c) Any powdered or liquid additive used as an agent in soil stabilization. See also: *processing.* (Hammond, 1965)

stable (a) Not readily decomposed or deformed. Cf: *unstable.* (Nelson, 1965) (b) A short drivage, room, or space excavated at the end of a longwall face to accommodate a coal cutter or cutter loader. The stable provides room for turning the machine where this is necessary, and also exposes a buttock for the machine to start its cut across the face. (Nelson, 1965)

stable gravimeter An instrument that uses a high order of optical and/or mechanical magnification so that an extremely small change in the position of a weight or associated property can be accurately measured. (AGI, 1987)

stable hole conveyor A short belt or other conveyor for use in stables in advance of the longwall face. The conveyor is usually about 18 in (46 cm) wide and driven at the tail end by a combined electric motor and drive pulley. The unit can be transported by sliding on steel skids, and is useful where coal or stone has to be moved short distances in confined spaces. See also: *shortwall.* (Nelson, 1965)

stable isotope A nuclide that does not undergo radioactive decay. (Lyman, 1964)

stable lead Any of the nonradioactive isotopes of lead. (Handbook of Chem. & Phys., 2)

stable relict A relict mineral that was not only stable under the conditions prevailing while it was formed but also under newly imposed conditions of metamorphism. Cf: *unstable relict.*

stack (a) To stand and rack drill rods in a drill tripod or derrick. (Long, 1960) (b) Chock; a chock built of old timber. (Mason, 1951) (c) A shaft furnace. (d) Any structrue or part thereof that contains a flue or flues for the discharge of gases. See also: *inwall.* (ACSG, 1963)

stack effect The impulse of a heated gas to rise in a vertical passage, as in a chimney, a small enclosure, or building. Syn: *chimney effect.* (Strock, 1948)

stacker (a) A conveyor, mounted on a long steel boom, for carrying tailings beyond the stern of a gold or tin dredge to avoid silting it up. (Nelson, 1965) (b) A machine for blending ore before processing. (c) A conveyor adapted to piling or stacking bulk materials, packages, or objects. (d) With a blending system, the stacker operates over the stocking conveyor in a manner similar to a wing belt tripper to build layered piles or beds of material parallel to the stocking conveyor. See also: *boom conveyor; portable conveyor; wing belt tripper; apron conveyor; belt conveyor; flight conveyor; portable conveyor.* (e) One who controls conveyor belt moving molds containing molten lead through water spray to stamping and discharge tables. (DOT, 1949) (f) One who stacks coal, etc. (Fay, 1920) (g) Leic. A miner who looked after the unloading of the coal on the bank, on behalf of the miners, in the earlier days of mining. (Fay, 1920) (h) A machine for blending ore that beds the ore before reclaiming for processing.

stack height The height of a convector enclosure measured from the bottom of the enclosure to the top of the outlet. (Strock, 1948)

stacking fault A type of "plane defect" in a crystal structure, caused by one or more closest-packed layers added to or removed from a normal cubic closest-, hexagonal closest-, or other regular closest-packed sequence.

stadia (a) A surveying technique or method using a stadia rod in which distances from an instrument to the rod are measured by observing through a telescope the intercept on the rod subtending a small known angle at the point of observation, the distance to the rod being proportional to the rod intercept. The angle is usually defined by two fixed lines in the reticle of the telescope. (AGI, 1987) (b) *stadia rod.* (c)

An instrument used in a stadia survey; esp. an instrument with stadia hairs.—Pl: stadias. The term is also used as an adj. in such expressions as stadia surveying, stadia distance, and stadia station. (AGI, 1987)

stadia hairs Horizontal cross hairs equidistant from the central horizontal cross hair; esp. two horizontal parallel lines or marks in the reticle of a transit telescope, arranged symmetrically above and below the line of sight, and used in the stadia method of surveying. Syn: *stadia wires.* (AGI, 1987)

stadia rod A graduated rod used with an instrument having stadia hairs to measure the distance from the observation point to the place where the rod is positioned. Syn: *stadia.* See also: *telemeter rod.* (AGI, 1987)

stadia surveying The process of measuring distances and elevations by observing through a telescope the distance intercepted on a rod between two horizontal cross-hairs. These hairs are carried on the same ring as the regular horizontal crosshair, and are equidistant from it. (Zern, 1928)

stadia tables Mathematical tables from which may be found, without computation, the horizontal and vertical components of a reading made with a transit and stadia rod.

stadia wires *stadia hairs.*

stadia work Tacheometric survey, in which points sighted from a survey station are oriented as with a theodolite, and their distance is read by means of a vertically held leveling staff on stadia wires. (Pryor, 1963)

staff (a) A surveyor's leveling rod. (Standard, 1964) (b) An iron puddler's rabble or rabbler. (Fay, 1920)

staff gage Graduated scale marked on a rod or a metal plate, or on the masonry of a bridge pier or similar structure, from which the depth of water in a canal, dock, or river can be read. (Hammond, 1965)

staff hole A small hole in a puddling furnace through which the puddler heats the staff. See also: *staff.* (Fay, 1920)

staff man The person who carries a leveling staff for a surveyor. See also: *target rod; telemeter rod.* (Hammond, 1965)

Staffordian Series The so-called transition group of the British Coal Measures, between the Middle and Upper Coal Measures in the Carboniferous System. They include the Newcastle-under-Lyme Group and the Etruria Marl, and the Blackband Group in north Staffordshire, England. (CTD, 1958)

staflux A material made by sintering together lumps of limestone and certain kinds of iron oxide, such as iron ore or mill scale, at a temperature of 1,450 °C in a rotary furnace. Though fusion does not occur, the iron oxide rapidly penetrates the limestone completely and forms dicalcium ferrite. (Osborne, 1956)

stage (a) A landing, such as in a shaft mine. (b) A platform on which mine cars stand. (c) A step in a process. (d) A time-stratigraphic unit next in rank below a series and corresponding to an age; it generally consists of several biostratigraphic zones. It is the most important unit for long-range correlation.

stage addition In flotation, this refers to deliberate use of insufficient reagent in the early part of the treatment to increase selectivity of conditioning, followed by further addition at a later point in the process. (Pryor, 1965)

stage compression *compound compression.*

stage crushing A method of crushing in which there is a series of crushers, each one crushing finer than the one preceding.

stage grinding Comminution in successive stages. (Pryor, 1965)

stage loader *feeder conveyor.*

stage plumbing A precise method of orienting underground workings in which plumblines are transferred down a deep shaft in stages of 400 to 600 ft (120 to 185 m). While shaft sinking is in progress, the lines can also be employed to orient the shaft itself and to keep it plumb. (BS, 1963)

stage pumping Draining a mine by means of two or more pumps placed at different levels, each of which raises the water to the next pump above or to the surface.

stage treatment In mineral processing, development of the desired condition of the particles by defined states, such as comminution to successively fine sizes (possibly coupled with staged concentration or gangue elimination) between such stages of communition. (Pryor, 1963)

stage winding Winding, usually in compound shafts, where the wind is divided into two or more stages, and underground winding engines are installed to deal with the lower stages. (Sinclair, 1959)

stage working A system of working minerals by removing the strata above the beds, after which the various beds are removed in steps or stages.

staggered blastholes When shot firing in thick coal seams, two rows of holes may be necessary. These are usually staggered to a triangular pattern to distribute the burden. A similar pattern is often adopted in quarry wellhole blasting. (Nelson, 1965)

staggered holes To arrange boreholes in a row, in such a manner that those in one row are placed opposite the spaces between the boreholes in the next row. (Long, 1960)

stag hole Usually a short hole drilled, charged, and fired to shatter the rock near the collars of the cut holes. (Nelson, 1965)

staging (a) A temporary flooring or scaffold, or platform. (Zern, 1928) (b) One or more working platforms, fixed at defined levels in deep trenches or similar excavations, on to which excavated earth is thrown by shovel. (Hammond, 1965)

stained stone A gemstone with color altered by a coloring agent, such as a dye, or by impregnation with a substance, such as sugar, followed by chemical or heat treatment, which usually produces a permanent color; e.g., green chalcedony. Cf: *burnt stone.* See also: *altered stone; heated stone.*

stainierite *heterogenite.*

stainless steel Iron-base alloy containing enough chromium to confer a superior corrosion resistance. (ASM, 1961; Newton, 1959; CTD, 1958; Camp, 1985)

stains Inclusions and intergrowths in mica arising from foreign materials, resulting in a partial or total loss of transparency. (Skow, 1962)

stake (a) Grubstake. (b) A pointed piece of wood driven into the ground to mark a boundary, survey station, or elevation. (c) *sprag.* (d) An iron peg used as power electrode to transfer current into the ground in electrical prospecting. This term is also used to include all power and search electrodes, such as iron pegs, copper coils, and copper screens; also, a station marker used by field parties. (AGI, 1987) (e) A permanent interest, as in an enterprise or a mine.

staking out The physical act of locating a lode or placer mining claim.

stalactite A conical or cylindrical mineral deposit that hangs from the ceiling of a cave. See also: *stalagmite.* Cf: *helictite.* (AGI, 1987)

stalagmite A conical or cylindrical mineral deposit that is developed upward from the floor of a cave by the action of dripping water. See also: *stalactite.* Syn: *dropping stones.* Cf: *helictite.* (AGI, 1987)

stalagmometer An apparatus for determining surface tension. The mass of a drop of a liquid is measured by weighing a known number of drops or by counting the number of drops obtained from a given volume of the liquid. (Lowenheim, 1962)

stalch Eng. A mass of ore left in a mine.

stall (a) A narrow coal drivage in pillar-and-stall. See also: *narrow stall; stallman.* (Nelson, 1965) (b) A working place at the coal face; a term associated with narrow workings. (Mason, 1951)

stall-and-breast *room-and-pillar.*

stall-and-room working A pillar method of working a relatively thick coal seam by a system of compartments; a modification of pillar-and-stall. (Nelson, 1965)

stalling angle The blades of axial-flow fans are of aerofoil section, which when inclined at a small angle (known as the angle of attack) to the air stream produce a large lift or raising force for a small drag or retarding force. The lift force is the useful one which gives the thrust to the air in an axial-flow fan. The lift increases with increase in the angle of attack until a point is reached when the lift begins to fall. This angle is the "critical" or "stalling angle." Syn: *critical angle.* (Sinclair, 1958)

stallman A collier who works at the face of a narrow stall or a longwall stall. The collier is paid according to a pricelist of so much per ton of coal loaded out and for other work, such as timbering. A stallman usually has another miner alongside. (Nelson, 1965)

stall roasting The roasting of ore in small enclosures of earth or masonry walls. The enclosures are called stalls and may be open or closed. (Fay, 1920)

stamler *feeder breaker.*

stamp (a) To break up ore and gangue by machinery, for washing out heavier metallic particles. (b) A heavy pestle raised by water or steam power for crushing ore. A stamp in which the blow of the pestle is caused by its mere weight is called a gravity stamp. (Webster 3rd, 1966)

stamp battery *stamp mill.*

stamp duty The amount of ore (tons) that one stamp will crush in 24 h.

stamper box A stamp-mill mortar box. (Fay, 1920)

stamp hammer A power hammer that moves vertically. (Webster 3rd, 1966)

stamp head A heavy and nearly cylindrical cast-iron head fixed on the lower end of the stamp rod, shank, or lifter to give weight in stamping the ore. The lower surface of the stamp head is generally protected by a cheese-shaped shoe of harder iron or steel that may be removed when worn-out. These shoes work upon dies of the same form laid in the bottom of the mortar or stamper box. See also: *stamp.* (Fay, 1920)

stamping (a) Reducing to the desired fineness in a stamp mill. The grain is usually not so fine as that produced by grinding in pans. (Fay, 1920) (b) A general term covering almost all press operations. It includes blanking, shearing, hot or cold forming, drawing, bending, and coining. (ASM, 1961) (c) A process for application, by hand or by machine, of decoration to pottery ware; a rubber stamp with a sponge backing is used. Stamping is particularly suitable for the application of backstamps and for some forms of gold decoration. See also: *backstamp.* (Dodd, 1964)

stamping maundrill Leic. A heavy pick.

stamp mill An apparatus, and the building containing it, in which rock is crushed by descending pestles (stamps), operated by water power or steam power. Amalgamation is usually combined with crushing when gold or silver is the metal sought, but copper, tin, and other ores are stamped to prepare them for dressing. The technique is obsolete. Syn: *stamp battery.*

stamp shoe The heavy, chilled-iron casting, attached to the lower end of a stamp piston, which does the actual crushing of rock in a stamp mill. It drops on a round steel block called a die. (Weed, 1922)

stampsman Person who attends or operates a stamp or stamp battery. (Fay, 1920)

stampwork A term used in the Lake Superior region for rock containing disseminated native copper.

stanchion (a) A vertical prop or strut. (Zern, 1928) (b) A support or post of iron or wood. (Crispin, 1964)

stand (a) Two or more lengths of drill rod or casing coupled together and handled as a unit length as they are taken from a borehole and set upright in a drill tripod or derrick. See also: *double; forble; treble.* (Long, 1960) (b) A drill floor. (Long, 1960) (c) To allow a cement slurry to remain undisturbed in a borehole until it hardens or sets. (Long, 1960) (d) To set a string of casing in a borehole. (Long, 1960)

standage (a) Reservoir or storage capacity, said of water and of mine cars. (Mason, 1951) (b) *sump.* (c) The capacity of a sump or lodge. (BS, 1963) (d) *lodge.*

standage room A length of roadway provided near a shaft bottom to stock loaded mine cars and/or empty cars: (1) during peak hours when the coal reaches the pit bottom at a faster rate than the shaft can wind; and (2) during emergency periods, such as plant breakdown at the surface, thus permitting coal production to continue. In general, the standage room accommodates 45-min to 1-h winding capacity. See also: *bunker conveyor; bunkering capacity.* (Nelson, 1965)

standard air density In mine ventilation, the standard density of air for mine ventilation work is considered to be 0.075 lb/ft^3 (1.2014 kg/m^3). This is based upon the weight of 1 ft^3 (0.028 m^3) of dry air at 70 °F (21.1 °C) at a sea-level pressure of 29.9 in (759 mm) of mercury.

standard bit A bit the size and design of which are as specified in standards accepted by the drilling industry. (Long, 1960)

standard conditions In refrigeration, an evaporation temperature of 5 °F (-15 °C), condensing temperature of 86 °F (30 °C), liquid temperature before the expansion valve of 77 °F (25 °C), and suction temperature 14 °F (-10 °C). (Strock, 1948)

standard copper Practically any brand of 96%, or higher, fineness.

standard core bit *standard bit.*

standard electrode reference Electrode used as a standard in measurements of electrode potential, because its potential is constant and reproducible; used for pH measurements, polarographic analysis, etc.

standard ignition test A method developed for testing coal dust to obtain the limits of explosibility. (Rice, 1913-18)

standard impinger For many years, the Greenburg-Smith impinger was the routine dust sampling instrument in this country. It is still relied upon as a standard, but because of its size and weight, is little used underground today. (Hartman, 1961)

standard lay *regular lay.*

standard mineral *normative mineral.*

standard mix Concrete mixed in the proportions of 1 part cement, 2 parts sand, and 4 parts coarse material. See also: *aggregate; cement.* (Nelson, 1965)

standard of ventilation An adequate amount of ventilation to dilute and render harmless all noxious and flammable gases to such an extent that all roads and workings in a mine shall be kept in a fit state for working or passing therein. (Mason, 1951)

standard parallel (a) A parallel of latitude that is selected as a standard axis on which to base a grid system; specif. one of a set of parallels of latitude (other than the base line) of the U.S. Public Land Survey system, passing through a selected township corner on a principal meridian, and on which standard township, section, and quarter-section corners are established. Standard parallels are usually at intervals of 24 miles north or south of the base line, and they are used to limit the convergence of range lines that intersect them from the south so that nominally square sections and townships can be laid out. Syn: *correction line.* (AGI, 1987) (b) A parallel of latitude that is used as a control line in the computation of a map projection; e.g., the parallel of a normal-aspect conical projection along which the principal scale is preserved. (AGI, 1987) (c) A parallel of latitude on a map or chart along which the scale is as stated for that map or chart. (AGI, 1987)

standard penetration test (a) A soil-sampling procedure to determine the number of blows by a drive hammer, freely falling a distance of 30 in (0.76 m) per blow, needed to drive a standard sampling spoon 1 ft (0.3 m). The first 6 to 7 in (15.24 to 17.78 cm) of penetration is disregarded, but the blows required to drive the sample the ensuing foot are counted. (Long, 1960) (b) *penetration test.*

standard plow The original coal plow; a heavy double-ended machine with fixed blades. Its length is 6-1/2 ft (1.98 m); its height ranges from 14 to 31 in (35.6 to 78.7 cm). The depth of cut can be varied from 2 to 6 in (5.1 to 15.2 cm). The rapid plow has evolved from this relatively slow-moving machine. (Nelson, 1965)

standard pressure (a) A term applied to valves and fittings suitable for a working steam pressure of 125 psi (862 kPa). (Strock, 1948) (b) *normal pressure.*

standard rig (a) An archaic term for a cable-tool drilling rig. (AGI, 1987) (b) A common misnomer for cable-tool rig, churn-drill rig. (Long, 1960)

standard section A geologic section showing as completely as possible a sequence of all the strata in a certain area, in their correct order, thus affording a standard for correlation. It supplements (and sometimes supplants) the type section, esp. for time-stratigraphic units. (AGI, 1987)

Standard Temperature and Pressure Atmospheric pressure of 760 mm of mercury, at 0 °C. (Pryor, 1963)

standard tin Tin of 99.75% or greater purity. (Bennett, 1962)

standard wire gage Gage number defining the diameter of wire. Abbrev.: SWG. (Pryor, 1963)

standby face A spare conveyor face, of normal length, that could be worked should another face cease production due to faults, washouts, roof collapse, water, gas, or any other unforeseen impediment. Syn: *spare face.* (Nelson, 1965)

standdown Gr. Brit. The sending of miners home because they cannot be usefully employed due to any reason outside the control of the management. In some cases, coal mining awards confine this right to certain occurrences, e.g., breakdown of plant or machines. (Nelson, 1965)

standing (a) Used by drillers to denote that work has been stopped for a considerable time. (Long, 1960) (b) Drill rods or casing stacked vertically in the drill tripod or derrick. (Long, 1960)

standing column The column of drilling liquid left in the hole when the drill tools have been removed. (Long, 1960)

standing fire A fire in a mine continuing to smoulder for a long time, often many years. (Fay, 1920)

standing gas A body of combustible gases known to exist in a mine, but not in circulation; sometimes fenced off. (Fay, 1920)

standing shot The result of a small or undercharged shot wherein the coal is loosened so that it is easily mined by pick. The term is a misnomer, as it applies to the result and not the "shot" or "charge.". (Fay, 1920)

standing time Gr. Brit. The period when face workers are idle due to the lack of empty cars, etc. Payments are made to miners on piecework for time lost. See also: *lying money.* (Nelson, 1965)

stand of drill rods *stand.*

standoff (a) A short length of core attached to and left standing upright in the bottom of the borehole when the core barrel is pulled. (Long, 1960) (b) On taper-tool or drillpipe joints, the space between the pin- and box-thread shoulders before wrenching up. (Long, 1960)

standpipe (a) A relatively short length of pipe driven into the upper soillike portion of the overburden as the first step of collaring or spudding-in a borehole. Also called conductor; conductor pipe. (Long, 1960) (b) A short piece of pipe wedged or cemented into a borehole after completion to act as a marker and keep collar free of cave. Cf: *surface string.* (Long, 1960)

standpiping Driving pipe deep enough through overburden to keep soil, sand, etc., out of a borehole. See also: *standpipe.* (Long, 1960)

stands Connected joints of drill pipe racked in the derrick while changing the bit. (Wheeler, 1958)

stank (a) To make watertight; to seal off; an airtight and watertight wall against old mine workings. See also: *seal; sealed-off area.* (Nelson, 1965) (b) A small cofferdam constructed of timber and made watertight with clay. (Hammond, 1965) (c) *stanking.*

stanking (a) A watertight stopping or bulkhead. (BS, 1963) (b) The application of a waterproofing material to a stopping or bulkhead. (BS, 1963)

Stanley compensating diaphragm A specially designed theodolite used as a direct reading tacheometer. See also: *Beaman stadia arc.* (Hammond, 1965)

stanley header *header.*

stannary (a) A tin mine or tin works. (Webster 2nd, 1960) (b) One of the regions in England containing tinworks and formerly placed under jurisdiction of special courts. Usually used in plural. (Webster 3rd, 1966)

stannary courts Eng. Courts in Cornwall and Devonshire for the purpose of regulating the affairs of tin mines and tin miners.

stannatores An early name applied to Cornish tin miners.

stanniferous Relating to or containing tin; as, stanniferous ore.

stannine *stannite.*

stannite (a) A tetragonal mineral, Cu_2FeSnS_4; zinc may replace iron; tannite group; metallic; in granular masses in veins associated with cassiterite. Syn: *tin pyrites; bell-metal ore.* (b) The mineral group briartite, cernyite, famatinite, hocartite, kuramite, luzonite, permingeatite, pirquitasite, sakuraiite, and stannite. (c) Impure cassiterite. Syn: *tin pyrites; stannine.*

stannopalladinite A hexagonal mineral $(Pd,Cu)_3Sn_2$(?).

stannous chloride *salt of tin.*

stannum *tin.*

stantienite A black variety of retinite having a very high oxygen content (23%). Syn: *black amber.*

Stanton diagram Historically, a plot of the airflow friction coefficient against the Reynolds number is referred to as a Stanton diagram.

staple (a) A shaft that is smaller and shorter than the principal one and joins different levels. (Webster 3rd, 1966) (b) An internal shaft connecting two coal seams. Also called staple pit. Cf: *winze.* (CTD, 1958)

staple shaft (a) An underground shaft, which does not penetrate to the surface. (Fraenkel, 1953) (b) A relatively small vertical pit connecting a lower seam to an upper seam. It corresponds to a rise or winze in metal mining. A staple shaft is an important drivage in horizon mining and may be used for dropping coal or stowing dirt to a lower level. It is often equipped with a spiral chute or an auxiliary winder system with a single cage and counterweight. See also: *subincline.* (Nelson, 1965)

star In minerals, the presence of needlelike oriented inclusions aligned along crystallographic axes, generally in the plane normal to the *c* axis in the hexagonal and trigonal crystal systems. Syn: *asteriated.*

star antimony Refined metallic antimony characterized by crystalline patterns resembling stars or fern leaves on its surface. Also called star metal. (Webster 3rd, 1966)

starch Used as depressant in flotation process. Alkaline starch (starch dissolved in dilute sodium hydroxide) is a flocculating agent used in purifying the water in coal-cleaning plants. Also known as amylum. (Pryor, 1963)

star drill A tool with a star-shaped point used for drilling in stone or masonry. (Crispin, 1964)

star facet Small triangular facet situated between the bezel facets and the table on the crown of an American (Tolkowski theoretical) brilliant-cut diamond.

star feeder A rotating feeder consisting of a horizontal shaft fitted with radial blades running within a close-fitting cylindrical chamber provided with an inlet and an outlet. Also called star gate; star valve (undesirable usage). (BS, 1962)

star garnet Variant of almandine. (Hey, 1955)

starkeyite A monoclinic mineral, $MgSO_4{\cdot}4H_2O$; rozenite group. Named for the Starkey Mine, Madison County, MO. Syn: *leonhardtite.*

Starlite Trade name given by Kunz to artificially-colored blue zircon from Thailand. (English, 1938)

star metal *star antimony.*

star quartz A variety of quartz containing within the crystal whitish or colored starlike radiations along the diametral planes. The asterism is due to the inclusion of submicroscopic needles of another mineral arranged in parallel fashion. In star blue quartz in Virginia and Texas, the other mineral is rutile. See also: *asteriated quartz.*

star reamer A star-shaped tool for regulating the diameter of, or straightening, a borehole.

star ruby A semiopaque to semitransparent asteriated variety of ruby normally having six chatoyant rays. Cf: *corundum cat's eye.*

star sapphire A semiopaque to semitransparent asteriated variety of sapphire normally with six rays resulting from the presence of microscopic crystals (e.g., rutile needles) in various orientations with the gemstone. See also: *asteria.* Cf: *corundum cat's eye.*

star stone (a) An asteria; esp. a star sapphire. (AGI, 1987) (b) Less correctly, any asteriated stone, including even petrified wood containing small starlike figures in its more transparent parts. (AGI, 1987)

starter (a) A slightly larger drill used for making the beginning of a hole, the remainder of the hole being made with a drill of smaller gage known as a follower. (Fay, 1920) (b) Pennsylvania. The miner who ascends to the battery to start the coal to run. See also: *battery starter.* (Fay, 1920) (c) Protective equipment to ensure that an electric motor does not receive too high a current when starting. (Hammond, 1965)

starter bar A steel reinforcing bar embedded in the concrete and projecting through a construction joint to bind adjoining masses of concrete together. (Hammond, 1965)

starting barrel A short core barrel used to begin coring operations when the distance between the drill chuck and the bottom of the hole, or to the rock surface in which a borehole is to be collared, is too short to permit use of a full 1.5-m-long or 3-m-long core barrel. (Long, 1960)

starting casing barrel A short piece of casing to which a casing bit and shell are attached and used under the same conditions as a starting barrel. See also: *starting barrel.* (Long, 1960)

starting sheet A thin sheet of metal used as the cathode in electrolytic refining. (ASM, 1961)

starting submergence In an air lift, the distance below the static head at which the air picks up water. (Lewis, 1964)

starvation (a) In comminution, avoidance of crowding in the machine by restricting rate of feed. (b) In conditioning for flotation, use of threshold quantity of collector agent to aid in selective adsorption by the desired species of mineral.

Stassfurt deposits A series of saline minerals, found in the Triassic rocks at Stassfurt, Saxony, Germany, which include halite, anhydrite, kieserite, gypsum, and boracite.

stassfurtite A massive variety of boracite found in Germany. It resembles a fine-grained white marble; sometimes has a subcolumnar structure. (Fay, 1920; CTD, 1958)

statement of performance A statement describing the scope and duty of a plant in terms, e.g., of the tonnage of coal treated per hour, the processes used, the separations effected and sizes produced; sometimes also used to express the results of plant operation. (BS, 1962)

state mine inspector *inspector; mine inspector.*

state point The psychrometric properties of air at given conditions, e.g., dry bulb temperature, wet bulb temperature, and barometric pressure. (Hartman, 1982)

stathmograph An apparatus that records automatically, in the form of a graph, the loss of weight during the whole reduction of iron ores. (Osborne, 1956)

static air mover *air mover.*

statically determinate frame A structural frame in which the bending moments and reactions can be determined by the laws of statics alone. (Hammond, 1965)

statically indeterminate frame A redundant frame in which the bending moments and reactions cannot be calculated from statical equations alone. See also: *perfect frame.* (Hammond, 1965)

static balance A condition of rest created by inertia (dead weight) sufficient to oppose outside forces. (Nichols, 1976)

static efficiency Is calculated in the same way as fan efficiency, but using a reading of static pressure at some point instead of total pressure. Was formerly widely quoted, and is still used to some extent, in relation to mine fans. See also: *fan efficiency.* (Roberts, 1960)

static E.P. The electrode potential measured when no current is flowing between the electrode and the electrolyte. (Lowenheim, 1962)

static grizzly A grizzly in the form of a stationary bar screen, often improvised from bars or rails set longitudinally, without cross bars. If used as a chute it has a slope of 35° to 45°. It may allow suitable pieces of coal or ore to pass over, and the unwanted small sizes drop through, or it rejects oversize pieces while allowing suitable material to drop through. See also: *power grizzly.* (Nelson, 1965)

static head (a) The height of a standing column of water as measured from the bottom of a borehole upward. Sometimes expressed in units of weight as measured at the bottom of the borehole. (Long, 1960) (b) In an air lift, the distance from the surface or top of the well casing to the normal surface of the water when not pumping. (Lewis, 1964) (c) The sum of the suction and discharge heads. (Carson, 1961) (d) *hydrostatic head; static level.*

static level The water level of a well that is not being affected by withdrawal of ground water. (AGI, 1987)

static load (a) The basal pressure exerted by the weight of a mass at rest, such as the load imposed on a drill bit by the weight of the drill-stem equipment or the pressure exerted on the rocks around an underground opening by the weight of the superimposed rocks. Syn: *dead load.* (Long, 1960) (b) A load that is at rest and exerts downward pressure only, such as a hydrostatic load. (Nichols, 1976)

static metamorphism A variety of regional metamorphism brought about by the action of heat and solvents at high lithostatic pressures, not at pressures induced by orogenic deformation. See also: *load metamorphism.* Cf: *thermal metamorphism.* (AGI, 1987)

static moment The static moment of a section about an axis, Y, is also termed the first moment of the area about the axis. It is the sum of the products obtained by multiplying each component of an area, A, by its distance, X, from Y. See also: *moment of inertia.* (Hammond, 1965)

static penetration test A penetration test in which the testing device is pushed into soil with a measurable force, as distinct from a dynamic penetration test in which the testing device is driven into the ground by blows from a standard hammer. See also: *penetrometer; soil.* (Hammond, 1965)

statics That branch of mechanics dealing with the relations of forces that produce equilibrium among material bodies. (Webster 3rd, 1966)

static switch A device giving contactless control of a circuit; e.g., a transistor, thyratron, saturable reactor, etc. (NCB, 1964)

static tube A static tube has a shaped, solid nose, on the downstream side of which a number of small holes are positioned around the circumference. The holes are so placed that the pressure in the tube is that of the undisturbed airstream. Unlike the Pitot tube, the measured pressure is affected considerably both by the position of the stem of the tube in relation to the pressure holes, and by the distance between the holes and the nose tip. The static tube is considerably more sensitive to yaw than is the Pitot tube. See also: *Pitot tube; Pitot-static tube.* (Roberts, 1960)

static water level The level of water in a well or borehole when pumping is not in progress. (BS, 1963)

station (a) *underground station; tank station.* (b) A reference point in surveying, marked at the surface by a metal plate set in concrete, or by a plug drilled into the roof of an underground working. (Pryor, 1963) (c) A length of 100 ft (30.5 m), measured along a given line, which may be straight, broken, or curved. (Seelye, 1951) (d) Any point on a straight, broken, or curved line whose position is indicated by its total distance from a starting point, or zero point. For example, station 4+47.2 identifies a point 447.2 ft (135.3 m) from the starting point, the distance being measured along a given line. (e) A location on a conveyor system where bulk material is received or discharged. (f) Any one of a series of stakes or points indicating distance from a point of beginning or reference. (g) A setup point; i.e., a marked point on the ground, over which an instrument is to be placed.

stationary bar screen A large-capacity screening or sorting appliance for coal or ore. It consists of a series of heavy metal bars arranged side by side and spaced at a definite distance apart. The bars are set at an angle so that material delivered at the upper end will just slide, and chutes are arranged to receive oversize at the lower end and undersize passing between the bars. The stationary bar screen is still used at many small mines. See also: *Bradford breaker; resonance screen; screen.* (Nelson, 1965)

stationary block The relatively undeformed rocks beneath the plane of an overthrust fault. See also: *autochthon.*

stationary dredge (a) A dredge that is not self-propelled, the dredged material from which is discharged into either a hopper barge or a pipeline. (Hammond, 1965) (b) A fixed vessel with equipment for digging, washing, and concentrating alluvial deposits. See also: *dredge.* (Nelson, 1965)

stationary engine An engine located on a fixed foundation, as distinguished from a portable engine. (Crispin, 1964)

stationary equipment Stationary equipment is installed in a given location and is not moved from that location in performing its function. This includes equipment such as substations, pumps, and storage-battery charging stations.

stationary grizzly The simplest of all separating devices and the cheapest to install and maintain. It consists of a series of fixed bars or rails spaced the required distance apart in order that the "undersize" may drop through. The use of a stationary grizzly is limited to coarse screening of dry material (aperture 2 in or 5.1 cm and larger), although it is sometimes used with openings as small as 3/4 in (1.9 cm), the efficiency dropping off in proportion. It is not satisfactory for moist or sticky material. (Pit and Quarry, 1960)

stationary inner-tube core barrel *rigid-type double-tube core barrel.*

stationary jaw The fixed jaw of a safety clamp or wrench. Syn: *stationary slip.* Cf: *anvil.* (Long, 1960)

stationary mass In some seismometers, a heavy weight, either suspended or supported, that, because of inertia, tends to remain quiescent during an earthquake. Syn: *steady mass.* (AGI, 1987)

stationary-piston drive sampler A piston-type sampler in which the position of the piston relative to the sample remains constant during the sampling operation. (Long, 1960)

stationary slip *stationary jaw.*

station foreman In metal mining, a laborer who supervises the haulage and handling of ore, timber, and mining supplies at a shaft station. (DOT, 1949)

stations Permanently marked points on the centerline of a tunnel. These stations may be outside of the tunnel and used for projecting the centerline into the tunnel, or they may mark the centerline inside the tunnel. (Stauffer, 1906)

station yards haul Equals the number of cubic yards multiplied by the number of 100-ft (30.5-m) stations through which it is moved. (Nichols, 1976)

statistical uniformity A term describing that variation in quality of materials of manufactured goods that is stable and determinate, so that statistical analysis and prediction can be applied to it. See also: *representative sample; level of control.* (Hammond, 1965)

statistics The collection, tabulation, and study of numerical facts and data. In industry, statistics indicate trends that would be almost impossible to establish by other means. The statistical method is useful in: (1) estimating the real value of work done, goods, or machines in terms of useful service and maintenance costs; and (2) estimating and forecasting profits and markets. See also: *parameter.* (Nelson, 1965)

stator In a torque converter, a set of fixed vanes that change the direction of flow of fluid entering the pump or the next stage turbine. (Nichols, 1976)

statuary marble A fine-grained saccharoidal marble used by sculptors. The best qualities are pure white and free from markings.

staurolite A monoclinic mineral, $Fe_2Al_9Si_4O_{22}(OH)_2$; pseudo-orthorhombic; Mohs hardness, 7.5; a common accessory in medium-grade regional metamorphic rocks; may be of gem quality; cruciform twins called fairy crosses. Syn: *staurotide; cross-stone; grenatite; fairy stone; lucky stone.*

stauroscope A type of polariscope used to determine the direction of light polarization in a crystal for accurate measurement of angles of extinction. See also: *polariscope.*

staurotide *staurolite.*

staurotypous In mineralogy, having crosslike markings. (Standard, 1964)

stave (a) A ladder step. (Zern, 1928) (b) A wedge-shaped section placed around the die of a stamp to take up the side wear.

stavrite An obsolete term for a type of biotite amphibolite.

stay A diagonal brace or tie bar to stiffen or prevent movement of a structural component. (Hammond, 1965)

stay-bolt tap A type of combination reamer and tap used extensively in locomotive-boiler work. (Crispin, 1964)

stead S. Wales. Very thin bands of ironstone in coal measures. (AGI, 1987)

Stead's reagent An etching reagent, used in metallographic examination of steels, containing 100 mL methyl alcohol, 18 mL water, 2 mL concentrated hydrochloric acid, 1 g copper chloride, and 4 g magnesium chloride. (Osborne, 1956)

steady-flow process A flow process in which none of the variables of flow changes with time. (Hartman, 1982)

steady-head tank In connection with use of moderate pressure hydraulic water (e.g., in classification of ore pulps), a reservoir set above the draw-off points of the system, which maintains a full supply of water at a set height and therefore constant pressure. (Pryor, 1963)

steady mass *stationary mass.*

steady point A pointed steel bar that can be locked in a clamp, and is used to brace a drill frame against the ground. (Nichols, 1976)

steady-state creep *secondary creep.*

steady-state velocity The constant maximum detonation velocity achieved by an explosive charge of a given diameter, mixture and density; it is the velocity at which a detonation will sustain itself through a column of explosive.

stealite Chiastolite, a variety of andalusite.

steamboat rolls Those rolls in an anthracite breaker that are set farthest apart to break the coal into steamboat coal. (Standard, 1964)

steam gas Highly superheated steam. (Webster 2nd, 1960)

steam hammer A heavy hammer, moving between vertical guides, actuated by steam pressure. (Crispin, 1964)

steam-hoist man *hoistman.*

steam infusion The injection of steam into the coal seam by infusion tubes, connected to a small boiler through high-pressure hose pipes, to suppress the dust in situ. The technique and equipment are somewhat similar to water infusion. Owing to technical, safety, and other problems, water infusion is preferred. (Nelson, 1965)

steam jet (a) A system of ventilating a mine by means of a number of jets of steam at high pressure kept constantly blowing off from a series of pipes in the bottom of the upcast shaft. (b) A jet of steam to moisten the intake air current and thus keep the coal dust in the mine wet. (Zern, 1928)

steam jet refrigeration A method of cooling involving the use of steam nozzles to reduce the pressure in a water chamber so that the water boils at a low temperature; since heat is drawn from the water, it is thus cooled. (Strock, 1948)

steam main A horizontal pipe for carrying live steam from a boiler to radiators, a steam engine, or other steam consuming device. (Crispin, 1964)

steam point *point.*

steam shovel An excavating machine in which a large dipper is operated by steam power. Used for stripping purposes and in open-pit mining, esp. for iron and coal. A similar shovel is now operated by electricity, gasoline, and diesel engines. (Standard, 1964)

steam shovel mine An opencut mine in which steam shovels or other power shovels are used for loading cars. (Hess)

steam stamp A crushing machine consisting of a vertical stamp shaft that is forced down to strike its blow, and lifted up preparatory to striking the next, by a steam piston. (Fay, 1920)

steam thawing A method of dredging permanently frozen ground in Alaska and the Yukon Territory in which steam is forced through pipes that are fitted with steel points on one end and a driving head on the other end so that the pipes can be hammered into the frozen gravel. Thawing by steam is a slow and costly process. See also: *thawing.* (Lewis, 1964)

steam winder The most common type of steam winder is the two-cylinder double-acting horizontal engine driving direct on the drum shaft. These engines, which are made with cylinders up to 42 in (1.1 m) in diameter and with a 84-in (2.1-m) stroke, possess the merit of simplicity and ease of control. The two cylinders act on cranks set at 90° to each other and are large enough for either to start the engine from rest against a full load, since one may happen to stop at dead center (i.e., with the piston at the end of its stroke, in which position it can exert no turning moment on the crank). (Mason, 1951)

Steart fan A propeller or axial-flow fan developed by Steart in Australia. See also: *fan.* (Nelson, 1965)

steatite (a) A compact, massive, fine-grained, fairly homogeneous talc-rich rock. (b) Gray-green or brown massive impure talc that is carved easily into ornamental objects. Syn: *lardite; lard stone; soapstone; soap earth.* See also: *talc.*

steatite talc A relatively pure or high-grade variety of talc suitable for use in electronic insulators, the purest commercial form of talc. Syn: *French chalk.*

steatitization Introduction of, alteration to, or replacement by, talc (steatite); esp. the act or process of hydrothermal alteration of ultrabasic rocks that results in the formation of a talcose rock (such as steatite, soapstone, or relatively pure concentrations of talc). (AGI, 1987)

Stebinger drum A delicate vertical-angle adjustment for the vernier on the alidade, graduated in hundredths of a revolution. See also: *gale alidade.* Cf: *tangent screw.* (AGI, 1987)

steel (a) An iron-base alloy, malleable in some temperature range as initially cast, containing manganese, usually carbon, and often other alloying elements. In carbon steel and low-alloy steel, the maximum carbon is about 2.0%; in high-alloy steel, about 2.5%. The dividing line between low-alloy and high-alloy steels is generally regarded as being at about 5% metallic alloying elements. Steel is to be differentiated from two general classes of irons: the cast irons, on the high-carbon side and the relatively pure irons, such as ingot iron, carbonyl iron, and electrolytic iron, on the low-carbon side. In some steels containing extremely low carbon, the manganese content is the principal differentiating factor, steel usually containing at least 0.25%; ingot iron contains considerably less. (ASM, 1961) (b) The borer, consisting of shank, shaft, and bit or cutting edge; used for rock-drilling with drifters or jackhammers. (CTD, 1958) (c) In air hammers, the hollow or solid steel bar that connects the hammer with the cutting tool. (Nichols, 1954)

steel arch Curved length of steel, usually of H-section, used for supporting mine roadways. Two-, three-, or four-segment arches are available, with straight leg, splayed leg, horseshoe, or circular design; in double radius or with welded baseplate. See also: *Usspurwies arch; Toussaint-Heintzmann arch; steel support; steel ring.* (Nelson, 1965)

steel band belt A belt of relatively thin carbon or stainless strip steel alloyed and heat treated to withstand continued flexing over pulleys.

steel belt Thin, flat, steel belts ranging from 0.008 to 0.035 in (0.02 to 0.09 cm) in thickness and from 7/8 to 8 in (2.2 to 20.3 cm) in width have been successfully used. The pulleys should be faced with a thin layer of cork. Steel belts can be run at speeds as high as 10,000 ft/min (3.0 km/min). It has been claimed that a 4-in (10.2-cm) steel belt will transmit as much power as a 19-in (48.3-cm) leather belt. (Crispin, 1964)

steel bit The cutting tool at the end of the drill steel. Various bit shapes are used, the three commonest being the single chisel bit (used only for hand drills); the double chisel bit (used for fairly soft rock), and the cross bit (used for hard rock and for general purposes). See also: *tungsten carbide bit; chisel bit; cross-chopping bit.* (Nelson, 1965)

steel boy A youngster who carries drills to the miners, and collects dull drills and sees that they are returned to the blacksmith shop. (Fay, 1920)

steel cable A flexible rope, the strands of which are steel wires. See also: *cable.* (Long, 1960)

steel-cable conveyor belt A rubber conveyor belt in which the carcass is composed of a single plane of steel cables that acts as a longitudinal tension-carrying member and includes two or more plies of fabric to provide transverse strength and hold the cables together.

steel casing A pipe to support the walls of a borehole in loose ground. The casing is secured in position by a concrete block or by the cross beams of the platform. It is driven down from the surface and follows the drilling operation closely or sometimes even precedes the borehole in sand or very loose ground. See also: *borehole casing.* (Nelson, 1965)

steel centralizer On a wagon drill, a guide to hold the starting steel in proper alignment. (Nichols, 1954)

steel erector A skilled member of a team specially trained to erect steel framed buildings, bridges, and other steel structures. (Hammond, 1965)

steel guides Steel rails, rods, or bars fixed in a vertical shaft to guide the cage and prevent it from swinging. See also: *fixed guides.* (Nelson, 1965)

steel jack (a) A screw jack esp. suitable in mechanical mining. Under headers at or near the face, steel jacks or posts are used for upright timbers to be replaced as equipment advances. Also called steel post. (Kentucky, 1952) (b) *sphalerite.*

steelmaking The process of making steel from solid or molten pig iron, with or without admixture with steel scrap. The processes used are the Bessemer, open-hearth, crucible, electric arc, high-frequency induction, and duplex. (CTD, 1958)

steel mill A mill where steel is made, processed, and shaped. (Webster 3rd, 1966)

steel needle An instrument used in preparing blasting holes; used before the safety fuse was invented. (Fay, 1920)

steel ore A name given to various iron ores and esp. to siderite, because it was supposed to be esp. adapted for making steel by the earlier and direct process. (Fay, 1920)

steel plate conveyor *plate conveyor.*

steel press A machine for compressing molten steel in casting to improve the quality of the product. (Standard, 1964)

steel prop A steel upright or post used to support the nether roof at a longwall or other face. It usually incorporates a yielding device. See also: *hydraulic prop; mechanical yielding prop; self-advancing supports.* (Nelson, 1965)

steel puller A hinged clamp on the bottom of a hand drill. (Nichols, 1976)

steel rectangular shaft supports A shaft support consisting of H-beams, I-beams, angles, and sheeting. The design is somewhat similar to that used for timber. Bolts, rivets, and fastening angles are used to connect and secure the steel members. The fastening angles are riveted to the beams. The addition of galvanized-iron corrugated wall sheets (or laggings) form a secure and fireproof shaft. See also: *permanent shaft support; barring; bunton; wallplate; lagging.* (Nelson, 1965)

steel ring Ring- or horseshoe-shaped support for underground traveling way. Also called arch ring. See also: *steel arch.* (Pryor, 1963)

steel scrap Miscellaneous pieces of steel, old and new, used in the bath for steel making, esp. in the open-hearth furnace. (Mersereau, 1947)

steel separation door A steel door specially erected for the purpose of being closed only in an emergency, such as a fire or an explosion. Steel is necessary for strength and to avoid destruction by fire. Steel doors may also be used as separation doors in the vicinity of the pit bottom or fan drift. Also called safety door; emergency door. See also: *separation door.* (Nelson, 1965)

steel sets Used in main entries of coal mines and in shafts of metal mines in the United States. The sections have I-beams for caps and H-beams for posts or wall plates, the H-section giving equal stiffness in two directions at right angles to each other. Steel sets of various shapes are coming into wide use in deep European coal mines where pressures are so great that timber would not be satisfactory. See also: *steel tunnel support.* (Lewis, 1964)

steel sheet piling Piling composed of interlocking rolled steel sections driven vertically into the ground with guide walings in place before excavation starts. (Hammond, 1965)

steel shot Chilled cast iron drops. Syn: *chilled shot.* (Bennett, 1962)

steel support A straight or curved length of steel, usually of H or channel section, used for support purposes in mine roadways, faces, or shafts. A steel support (1) possesses a high degree of permanency or long service; (2) ensures a minimum area of excavation for given dimensions in the clear; and (3) is fireproof. In return airways and shafts, a chrome-nickel-copper steel is sometimes used to counteract the corrosive air. For high-strength roof bars best results are obtained by the use of heat-treated low-alloy steels of the carbon manganese type. See also: *arch girder; steel arch; straight girder support; support; Usspurwies arch; Toussaint-Heintzmann arch.* (Nelson, 1965)

steel tunnel support Tunnel-support systems made of steel are roughly of five types: continuous rib; rib and post; rib and wallplate; rib, wallplate, and post; and full circle rib. (1) A continuous rib system is usually made in two pieces for maximum speed of erection, lowest first cost, and lowest erection cost. Sometimes used in three or four pieces to meet special conditions and the following methods of attack: full face, side drift, and multiple drift. (2) A rib and post system is employed with the following methods of attack: full face in tunnels whose roof arch makes an angle with the sidewall; multiple drift and side drift in tunnels of such large size that two-piece continuous ribs cannot be shipped and/or handled; and heading and bench and top heading for support in the drift (with truss panels) for early support to roof. (3) In a rib and wallplate system, the rib is also usually made in two pieces for maximum speed of erection, lowest first cost, and lowest erection cost. It is sometimes used in three or more pieces to meet special conditions and with the following methods of attack: heading and bench, top heading, and full face. This type is esp. applicable to circular and high-sided tunnel sections where only a light roof support is needed. (4) In a rib, wallplate, and post system, these elements of support are used with the following methods of attack: heading and bench and top heading—for quick support to a roof; side drift—in large tunnels with bad rock conditions requiring quick support; and full face—for favorable rock where support is not needed tight to the face, for a tunnel whose roof makes an angle with the sidewall, and where post and rib spacing differ; and (5) A full circle rib—this method is used with the following attack: full face—in tunnels in squeezing, swelling and crushed rock, or any rock that imposes considerable side pressure, also where bottom conditions make it impossible to carry roof loads on foot blocks, and in earth tunnel conditions sometimes encountered in rock tunnels; and heading and bench—under earth tunnel conditions with joints at spring line. The inverted strut is used where mild side pressures are encountered and also to prevent the bottom from heaving. A full circle with ribs closely spaced is heavily lagged for heavy loads associated with squeezing conditions. See also: *steel sets.* Syn: *tunnel support.* (Lewis, 1964)

steel wire rope *wire rope.*

steenstrupine A trigonal mineral, $Na_{14}Ce_6Mn^{+2}Mn^{+3}Fe_2(Zr,Th)(Si_6O_{18})_2(PO_4)_7{\cdot}3H_2O$.

steep *brasque.*

steep gradient In general, in coal mining, an inclination (of a roadway, working, or coal seam) steeper than 1:4. (Nelson, 1965)

steeply inclined Deposits and coal seams having a dip of from 40° to 60°. (Stoces, 1954)

steering brake A brake that slows or stops on one side of a tractor. (Nichols, 1976)

steering clutch A clutch that can disconnect power from one side of a tractor. (Nichols, 1976)

Stefan-Boltzmann Law (a) The energy radiated in unit time by a black body is given as $E=K(T^4-T_0^4)$, where T is the absolute temperature of the body, T_0 the absolute temperature of the surroundings, and K is a constant. (Osborne, 1956) (b) The statement that the radiant flux of a black body is equal to the absolute temperature to the fourth power times the Stefan-Boltzmann constant of $(5.6696\pm0.001)\times10^{-8}$ W $(m)^{-2}(K)^{-4}$. (AGI, 1987)

steigerite A monoclinic mineral, $AlVO_4{\cdot}3H_2O$; weakly radioactive; forms coatings on highly weathered sandstones of the Colorado-Utah carnotite region.

stein Stonework used to secure the sides of a shaft. (Gordon, 1906)

steinmannite A variety of galena with part of the lead replaced with antimony and arsenic. (Standard, 1964)

stele In coal, primarily the vascular tissues of the axis of a vascular plant. It consists of two parts: the xylem that carries water from the roots, and the phloem that carries the food. (Hess)

stell *sprag.*

stellate Said of an aggregate of crystals in a starlike arrangement; e.g., wavellite.

stell prop A steel or timber prop fixed firmly between the roof and the floor at the end of a longwall face and from which a coal cutter is hauled by rope when cutting. A stell prop may also be used as part of a belt-tensioning arrangement or a return sheave. Syn: *anchor prop.* See also: *conveyor creep.* (Nelson, 1965)

stem (a) To insert and pack stemming in a shothole. See also: *tamp.* (BS, 1964) (b) The assemblage of drill rods in a borehole connecting a drill bit and core barrel to the drill machine. (Long, 1960) (c) The heavy iron rod acting as the connecting link between the bit and the balance of the string of tools on a churn rod in a borehole connecting a drill bit drill. (Long, 1960) (d) Frequently used as a syn. for ram or tamp. *stemming.*

stem bag Fire-resisting paper bag, about 8 in (20.3 cm) long, filled with dry sand for stemming shotholes in coal or hard headings. *water-ampul stemming.* (Nelson, 1965)

stemmer A wooden rod used by shot firers for inserting the explosive cartridges and stemming material in shotholes. The stemmer must be long enough to reach the back of the shothole, and has a diameter 1/8 in (3.2 mm) larger than the cartridges. Metal is not permitted in any part of a stemmer used in British coal mines. Also called tamping rod or stick; beater. See also: *tamping; scraper; break detector.* (Nelson, 1965)

stemming (a) The material (limestone chippings or sand and clay) used to fill a shothole, after the explosive charge has been inserted, to prevent the explosion from blowing out along the hole. In tunnels and hard headings, the stemming may be blown in by a hurricane air stemmer. (Nelson, 1965) (b) The act of pushing and tamping the material in the hole. See also: *water-ampul stemming; tamp.* (Nelson, 1965) (c) Inert material packed between the explosive charge and the outer end of the shothole, or between adjacent charges in deck charging. (BS, 1964) (d) *tamping; stem.*

stemming rod A nonmetallic rod used to push explosive cartridges into position in a shothole and to ram tight the stemming. Syn: *stemming stick.* See also: *tamping stick; tamping rod.* (BS, 1964)

stemming stick *stemming rod.*

stench A substance with a distinctive, disagreeable odor put in the air current to warn underground workers of fire or other emergency; ethyl mercaptan is commonly used. (Hess)

stench capsule A fire-warning device designed to be bolted to a flat surface that may rise to a dangerous temperature. It consists of a cavity filled with 20 cm^3 of ethyl or butyl mercaptan alone or with other stench agents and is sealed with a fusible plug in a brass container with a hexagonal head, arranged to liberate the stench agent at any temperature chosen. Tests in pits have shown that a strong smell could be detected 1.7 miles (2.7 km) from the discharge point 25 to 30 min after the device operated. (Sinclair, 1958)

stenonite A monoclinic mineral, $(Sr,Ba,Na)_2Al(CO_3)F_5$.

stent (a) The amount of work expected from a coal miner in a day or week. See also: *stint.* (Nelson, 1965) (b) *pitch.* (c) Corn. Tourmaline and quarz veins in kaolinized granite. (d) U.K. Rubble; waste. (e) U.K. Extent or limit, as of a pitch or bargain.

stenting N. of Eng. *stenton.*

stenton A connecting roadway between two adjacent roadways that may be used for ventilation purposes. Also called air slit; crosscut; cross hole; thirling; througher; spout. Syn: *crosscut.* See also: *air slit; breakthrough; pillar-and-stall.* (BS, 1963)

step (a) Fault; a small fault; a small fault in a stepped series of faults. (Mason, 1951) (b) A small offset on a piece of core or in a drill hole resulting from a sudden sidewise deviation of the bit as it enters a hard, tilted stratum or rock underlying a softer rock. Cf: *kick.* (Long, 1960) (c) One of several terracelike or stairstep concentric configurations on the crown of a diamond bit. See also: *step-face bit.* (Long, 1960) (d) A treatment of one part of a sample in a sample divider (thus a pass consists of one or more steps). (BS, 1960) (e) The action of setting a lock gate into a vertical position. (Hammond, 1965)

step cut (a) A mode of cutting gems in steplike facets. (Standard, 1964) (b) A form of cutting employed for stones not deeply colored when they are not cut as brilliants; a simple typical form is that of a stepped pyramid with the apex sliced off. Also called trap cut. (Hess) (c) A style of cutting, widely used on colored gemstones, in which long, narrow, four-sided facets form in a series or row parallel to the girdle and decrease in length as they recede above and below the girdle, giving the appearance of steps. The number of rows, or steps, may vary, although it is usually three on both the crown and pavilion. Different shapes of step cuts are described by their outline; e.g., rectangular or square step cut. Syn: *trap cut.* Cf: *emerald cut.*

step-face bit A thin-nosed bit with diamonds set in several concentric terracelike rows that form the outside wall. (Long, 1960)

step fault (a) One of a set of parallel, closely spaced faults over which the total displacement is distributed. Cf: *fault zone.* Syn: *multiple fault; distributive fault.* (AGI, 1987) (b) One of a series of low-angle thrust faults in which the fault planes step both down and laterally in the stratigraphic section to lower glide planes. Step faulting is due to variation in the competence of the beds in the stratigraphic section. (AGI, 1987) (c) A series of parallel faults that, all inclined in the same direction, gives rise to a gigantic staircase; hence these are called step faults. Each step is a fault block and its top may be horizontal or tilted. (AGI, 1987)

stephanite An orthorhombic mineral, Ag_5SbS_4; soft; metallic; sp gr, 6.2 to 6.3; in veins; a source of silver. Syn: *brittle silver ore; black silver; goldschmidtine.*

Stephenson lamp An early type of coal miners' lamp. It had a glass chimney surrounded by a wire gauze about 2 in (5.1 cm) in diameter. The glass chimney was covered by a perforated copper cap, and the air was fed to the flame from below through small holes and wire gauze in a lateral extension of the oil vessel. The lamp was unsafe; it passed flame when the velocity of the air current exceeded about 8 ft/s (2.4 m/s). See also: *safety lamp.* (Nelson, 1965)

steppe An extensive, treeless grassland area in the semiarid mid-latitudes of southeastern Europe and Asia. It is generally considered drier than the prairie which develops in the subhumid midlatitudes of the United States. (AGI, 1987)

stepped foundation *benched foundation.*

stepped longwall A system of longwall stalls in which the faces are carried forward in a steplike formation, one stall about 5 yd (4.6 m) in advance of the next stall. It is claimed to have advantages when the roof is friable. See also: *top holes.* Syn: *hitch-and-step.* (Nelson, 1965)

stepped stope The term implies that mining at one face is stepped aside from that below so as not to hinder the work at it. Syn: *advance stope.* (Stoces, 1954)

stepping ahead Term used in dredging operations when the digging spud is dropped, the other spud is raised, and the dredge is ready to begin a new cut. (Lewis, 1964)

step reef *step vein.*

step socket A special form of socket for use on locked-wire rope. (Zern, 1928)

step up To increase the voltage of (a current) by means of a transformer. (Webster 3rd, 1966)

step vein A vein alternately cutting through the strata of country rock and conforming with them. Syn: *step reef.*

stercorite A triclinic mineral, $H(NH_4)Na(PO_4){\cdot}4H_2O$. Syn: *microcosmic salt.*

stereocomparator A stereoscope for accurately measuring the three space coordinates of the image of a point on an aerial photography; it is used in making topographic measurements by the comparison of stereoscopic photographs. (AGI, 1987)

stereogram (a) A graphic diagram on a plane surface, giving a three-dimensional representation, such as projecting a set of angular relations; e.g., a block diagram of geologic structure, or a stereographic projection of a crystal. (AGI, 1987) (b) A stereoscopic pair of photographs correctly oriented and mounted for viewing with a stereoscope. Syn: *stereographic projection; stereograph*. (AGI, 1987)
stereograph *stereogram.*
stereographic projection (a) A map projection in which meridians and parallels are projected onto a tangent plane, with the point of projection on the surface of the sphere diametrically opposite to the point of tangency of the projecting plane. Any point of tangency may be selected (at a pole, on the equator, or a point in between). (AGI, 1987) (b) A similar projection used in optical mineralogy and structural geology, made on an equatorial plane (passing through the center of the sphere) with the point of projection at the south pole. Syn: *stereogram.* (AGI, 1987)
stereometric map A relief map made by the application of the stereoscopic principle to aerial or terrestrial photographs. Syn: *stereotopographic map.*
stereoscopic principle The formation of a single, three-dimensional image by simultaneous vision with both eyes of two photographic images of the same terrain taken from different camera stations. (AGI, 1987)
stereoscopic vision Simultaneous vision with both eyes in which the mental impression of depth and distance is obtained, usually by means of two different perspectives of an object (such as two photographs of the same area taken from different camera stations); the viewing of an object in three dimensions. Syn: *stereoscopy; stereovision*. (AGI, 1987)
stereoscopy *stereoscopic vision.*
stereosphere (a) That part of the Earth's crust that lies above the level of compensation, or the top of the asthenosphere. See also: *asthenosphere.* (AGI, 1987) (b) The relatively strong outer shell of the Earth. (AGI, 1987) (c) A term that was originally proposed for the innermost shell of the Earth's mantle, but is also used as equivalent to the lithosphere. (AGI, 1987)
stereotopographic map *stereometric map.*
stereovision *stereoscopic vision.*
sterilized coal That part of a coal seam that, for various reasons, is not mined. (BS, 1963)
sterling silver A silver alloy containing at least 92.5% silver, the remainder being unspecified but usually copper. (ASM, 1961)
sternbergite An orthorhombic mineral, $AgFe_2S_3$; cubanite group; forms tabular crystals or soft flexible laminae. Syn: *flexible silver ore.*
sterny Scot. Rough; coarse-grained or crystalline, for example, sterny limestone.
sterrettite *kolbeckite.*
stetefeldite A somewhat uncertain compound containing silver, copper, iron, antimony, sulfur, and water [$Ag_2Sb_2(O,OH)_7$] (?).
Stetefeldt furnace A furnace for the chloridizing and roasting of silver ores, and also for roasting fine copper ores low in sulfur. Provision is made for an auxiliary fireplace. (Fay, 1920)
stevensite A monoclinic mineral, $(Ca/2)_{0.3}Mg_3Si_4O_{10}(OH)_2$; smectite group; with no tetrahedral substitution of Al for Si, its layer charge arises from octahedral vacancies. Syn: *aphrodite.*
stewartite (a) A triclinic mineral, $MnFe_2(PO_4)_2(OH)_2{\cdot}8H_2O$; dimorphous with laueite; forms minute crystals and tufts of fibers in pegmatites. (b) A steel-gray, ash-rich, fibrous variety of bort containing iron, having magnetic properties, in the diamond mines of Kimberley, South Africa.
stey Scot. Steep; highly inclined.
stibianite *stibiconite.*
stibiconite (a) An isometric mineral, $Sb_3O_6(OH)$; pale yellow or chalky white; in antimony ore deposits as an alteration of stibnite. Syn: *stibianite; stiblite.* (b) The mineral group blindheimite, lewisite, partzite, romeite, stetefeldtite, and stibiconite.
stibiocolumbite An orthorhombic mineral, $SbNbO_4$; forms a series with stibiotantalite.
stibiopalladinite A hexagonal mineral, Pd_5Sb_2; sp gr, 9.5; occurs with platinum ores; a source of platinum and palladium. Syn: *allopalladium.*
stibiotantalite An orthorhombic mineral, $SbTaO_4$; forms a series with stibiocolumbite.
stibium The ancient name for antimony and stibnite and now used in pharmacy for the metal. *stibnite.* (Dana, 1914)
stiblite *stibiconite.*
stibnite An orthorhombic mineral, Sb_2S_3; dimorphous with metastibnite; soft; metallic; may contain gold and silver; occurs in massive forms and in vertically striated prisms having perfect cleavage, in low-temperature veins and around hot springs; the chief source of antimony. Syn: *antimonite; antimony glance; gray antimony; stibium.* See also: *stibnium.*
stibnium An ancient name for "stibnite" used (as in Egypt) as a cosmetic for painting the eyes. See also: *stibnite.*
stichtite A trigonal mineral, $Mg_6Cr_2(CO_3)(OH)_{16}{\cdot}4H_2O$; hydrotalcite group; dimorphous with barbertonite; lilac colored; in Dundas, Tasmania, Australia; Transvaal, South Africa; Cunningsburgh, Shetland Islands; and Quebec, Canada.
stick A cartridge of explosive. (BCI, 1947)
sticking (a) A small vein (a scrin) not wide enough for shoulder room, Derbyshire, U.K. (Arkell, 1953) (b) The selvage of mineralized country rock at the side of a vein. (Arkell, 1953) (c) A rib or ore in a vein, or a small rake vein crossing the main vein, Derbyshire, U.K. (Arkell, 1953) (d) U.K. A thin vein of ore or thin seam of clay in an ore vein.
sticking scrins Eng. Small veins that do not afford shoulder room.
stick loading A technique used in trench blasting, etc., in which a lower concentration of charge is obtained by placing wooden pegs between every cartridge in the hole thus halving the concentration. (Langefors, 1963)
stickup *standoff.*
sticky A term applied when drilling rock or a formation so soft that the drill bit tends to penetrate too rapidly and the circulation fluid is unable to clear the cuttings away fast enough to prevent their adhering to and compacting on the surfaces of the bit and other downhole drilling equipment and/or the borehole sidewalls. Cf: *balling formation; gummy.* (Long, 1960)
sticky limit The lowest water content at which a soil will stick to a metal blade drawn across the surface of the soil mass. (ASCE, 1958)
stiff clay Clay of low plasticity.
stiffener A steel angle or bar riveted or welded across the web of a built-up girder to stiffen it. (CTD, 1958)
stiff-fissured clay A clay that is firm when dry at depth but is intersected by cracks through which water will seep easily. Clay deposits of this type are liable to slips on hillside slopes. Syn: *slickensided clay.* (Nelson, 1965)
stiff mud A plastic mix of clay of very stiff consistency, as extruded from an auger machine.
stiffness The ability of a metal or shape to resist elastic deflection. For identical shapes, the stiffness is proportional to the modulus of elasticity. (ASM, 1961)
stifle Scot. Noxious gas resulting from an underground fire.
stilbite (a) A monoclinic and triclinic mineral, $Na_2Ca_4[Al_{10}Si_{26}O_{72}]{\cdot}34H_2O$; zeolite group, with K replacing Na; forms sheaflike crystal aggregates and radiated masses in cavities in igneous rocks, as an alteration of plagioclase, or in hydrothermal veins. (b) Ger. *heulandite.*
still (a) An apparatus in which a substance is changed by heat into vapor, with or without chemical decomposition. The vapor is then liquefied in a condenser and collected in another part of the apparatus. (Standard, 1964) (b) *amalgam retort.*
still coke The residue left in the still on distilling crude shale oil to dryness. (Fay, 1920)
stilling well A chamber connected to a main body of water by a small inlet; such an arrangement is suitable for a recording gage. (Hammond, 1965)
Stillson wrench The pipe wrench of common use, named for its inventor. (Crispin, 1964)
stillwellite A trigonal mineral, $(Ce,La,Ca)BSiO_5$.
stilpnomelane A monoclinic and triclinic mineral, $K(Mg,Fe)_8(Si,Al)_{12}(O,OH)_{27}$(?); black to green-black; in micalike plates, fibrous forms, and velvety bronze-colored incrustations. Syn: *chalcodite.*
stilt A device to allow roadway steel arches a measure of yield under roof pressure to prevent buckling. The stilt may take the form of wooden extensions strapped to the legs of the arch, wire bags filled with dirt, a mechanical frictional appliance, or a hydraulic stilt. See also: *yielding support.* (Nelson, 1965)
stinger (a) A steel cylinder projecting beyond the face of a cutting bit that serves as a pilot or guide. See also: *pilot.* (Long, 1960) (b) The pneumatically actuated piston attached to a pointed rod that acts as a feed mechanism on a stoper drill. (Long, 1960)
stinger ream To ream a borehole using a reaming bit equipped with a pilot or stinger. (Long, 1960)
sting-out Hot air and flame exhausted through openings in furnaces or tanks due to positive internal pressure. (ASTM, 1994)
stinkdamp A mining term for hydrogen sulfide. The gas has an unpleasant smell, resembling that of rotten eggs, hence the name. The presence of this gas may indicate a gob fire in its early stages. Produced by the distintegration of iron pyrites. See also: *damp.* (Nelson, 1965; Hudson, 1932)
stinker (a) A British miner's term for inferior coal that stinks when burned. (Tomkeieff, 1954) (b) A Welsh term for dolomite. (Arkell, 1953)
stinkquartz A variety of quartz that emits a fetid odor when struck. (Fay, 1920)
stinkstein *stinkstone.*
stinkstone (a) A stone that emits an odor on being struck or rubbed; specif. a bituminous limestone (or brown dolomite) that gives off a

fetid smell (owing to decomposition of organic matter) when rubbed or broken. It may emit a sweet-and-sour smell if the carbonate rock is rich in organic-phosphatic material. See also: *anthraconite; bituminous limestone.* Syn: *stinkstein.* (AGI, 1987) (b) Boulders of phosphate rock from Tennessee.

stint A fixed work target on the coalface that every collier is expected to do in one shift or a single week. In longwall conveyor work, it is the length, area, or volume of face that the miner regularly clears or loads out. Also called stent; cut. Syn: *cut.* See also: *darg.* (Nelson, 1965)

stip *hitch.*

stirian An early name for nickel-bearing marcasite.

stitch To fasten a timber by toenailing.

stitched canvas conveyor belt A construction of conveyor belt made up of plies of cotton fabric stitched together. Stitched canvas belts may be untreated, impregnated, or coated. See also: *belt.*

stochastic Containing a random variable; word used to describe a system (e.g., sampling method) that has in it an element of randomness. (Pryor, 1963)

stock (a) A rarely used term for a chimneylike orebody. Syn: *pipe.* (AGI, 1987) (b) An irregular, metalliferous mass in a rock formation, such as a stock of lead ore in limestone. Cf: *boss.*

stock craneman Person who lifts and moves stock, such as limestone, scrap iron, or pig iron, for the open-hearth furnace. (DOT, 1949)

stockhouse larryman Person who drives an electric car to haul ore and limestone from the stockpiles to the blast furnace. (DOT, 1949)

stocking conveyor A belt conveyor in a blending system that receives bulk materials for delivery to the stacker conveyor.

stockpile (a) An accumulation of ore or mineral built up when demand slackens or when the treatment plant or beneficiation equipment is incomplete or temporarily unequal to handling the mine output; any heap of material formed to create a reserve for loading or other purposes. (Nelson, 1965) (b) The ore accumulated at the surface when shipping is suspended. (Standard, 1964) (c) Material dug and piled for future use. (Nichols, 1954)

stockwork A mineral deposit consisting of a three-dimensional network of planar to irregular veinlets closely enough spaced that the whole mass can be mined. Cf: *reticulate.* Syn: *network deposit; stringer lode.* (AGI, 1987)

stockyard A space reserved on the surface near the materials shaft for the temporary storage of steel, timber, and other bulky items of supplies for mine use. The yard is surfaced and a mine car is used throughout. (Nelson, 1965)

stoichiometric With reference to a compound or a phase, pertaining to the exact proportions of its constituents specified by its chemical formula. It is generally implied that a stoichiometric phase does not deviate measurably from its ideal composition. (AGI, 1987)

stoke Unit of kinematic viscosity. The cgs unit of kinematic viscosity being that of a fluid that has a viscosity of 1 P (100 mPa·s) and a density of 1 g/cm^3. (Webster 3rd, 1966)

stokehole A hole, as in a reverberatory furnace, for introducing a rabble or other tool for stirring. (Standard, 1964)

stoker A mechanical appliance for feeding coal, coke, or other fuel into a boiler or furnace. In hand stoking, the person who shovels the fuel into the furnace is known as the stoker. See also: *underfeed stoker; vibrating grate.* (Nelson, 1965)

stoker coal (a) A screen size of coal specif. for use in automatic firing equipment. (BCI, 1947) (b) This coal can be of any rank and the stoker is usually designed to fit the coal available. Factors of importance in the selection of coal for stoker use are size limits; size consist; uniformity of shipments; coking properties; ash-fusion characteristics; ash, sulfur, and volatile-matter percentages. (Mitchell, 1950)

stokesite An orthorhombic mineral, $CaSnSi_3O_9{\cdot}2H_2O$; forms acute pyramids; at Roscommon Cliff, St. Just, Cornwall, U.K.

Stokes' law (a) A formula expressing the rates of settling of spherical particles in a fluid. (AGI, 1987) (b) Gives the rate of fall of a small sphere in a viscous fluid. When a small sphere falls under the action of gravity through a viscous medium it ultimately acquires a constant velocity, V: $V = 2ga^2(d_1-d_2)/9\eta$ where g is gravitational acceleration, a is the radius of the sphere, d_1 and d_2 are the densities of the sphere and of the medium, respectively, and η is the coefficient of viscosity. V will be in centimeters per second if g is in centimeters per second per second; a will be in centimeters; d_1 and d_2 will be in grams per cubic centimeter; and η will be in dynes second per square centimeter, or poises. (Handbook of Chem. & Phys., 2) (c) The wavelength of light emitted by a fluorescent material is longer than that of the radiation used to excite the fluorescence. In modern language, the emitted photons carry off less energy than is brought in by the exciting photons; the details accord with the energy conservation principle. See also: *elutriator; sedimentation test; terminal velocity.* (Handbook of Chem. & Phys., 2) (d) At low velocities, the frictional force on a spherical body moving through a fluid at constant velocity is equal to 6π times the product of the velocity, the fluid viscosity, and the radius of the sphere. The wavelength of luminescence excited by radiation is always greater than that of the exciting radiation. (McGraw-Hill, 1994)

Stokes stretcher The simplest type of stretcher used for underground first aid. This basket-type stretcher acts as a splint for the whole body, and is constructed of tubular steel and strong wire mesh. Used for lifting or lowering injured persons in difficult places. This type of stretcher is used in metal mines or in coal mines where the coalbed has a steep pitch. See also: *stretcher.* (McAdam, 1955; Kentucky, 1952)

stoking *continuous sintering.*

stoltzite A tetragonal mineral, $4[PbWO_4]$; dimorphous with raspite; crystallizes in bipyramids; sp gr, 7.9 to 8.3; in oxidized zones of tungsten deposits.

stomp A short wooden plug fixed in the roof, to which lines are hung, or to serve as a bench mark for surveys. (Fay, 1920)

stone (a) A mineral or group of consolidated minerals either in mass or in a fragment of pebble or larger size. (AGI, 1987) (b) A stony meteorite. (c) A cut and polished gem or other precious mineral (but not a synthetic compound used in ornamentation). (d) Crushed or naturally angular particles of rock that will pass a 3-in (7.6-cm) sieve and be retained on a No. 4 U.S. Standard sieve. (ASCE, 1958)

stone band *dirt band.*

stone bind Eng. Interbedded layers of sandstone and shale, or for a rock (such as siltstone) intermediate between a sandstone and a mudstone. (AGI, 1987)

stone butter (a) A variety of halotrichite. Syn: *rock butter.* (b) A variety of alum. (c) A variety of clay.

stone clunch Eng. Very hard underclay (clunch) with interbeds of sand. (AGI, 1987)

stone coal (a) Wales. Anthracite, in lumps; also certain other very hard varieties of coal. (Fay, 1920) (b) Mineral coal, as distinguished from charcoal; esp., in England, hard or anthracite coal. (Standard, 1964) (c) Sometimes applied to anthracite on account of its hardness. See also: *anthracite.* (Nelson, 1965) (d) Early term for anthracite. (Korson, 1938)

stone concentration *diamond concentration.*

stone content *diamond content.*

stone count *diamond count.*

stone crane operator (a) In the quarry industry, a person who lifts blocks of stone and boxes of broken stone with a guy derrick in a quarry. Also called stone hoist operator. (DOT, 1949) (b) In the stonework industry, one who lifts and moves blocks of stone with an electric bridge crane in a stone mill. (DOT, 1949)

stonecutter (a) Person whose occupation is cutting stone; a stone mason. (Standard, 1964) (b) A gem cutter. (Standard, 1964) (c) A machine for facing stone. (Standard, 1964)

stone drift A drift excavated in rock, such as from the surface down to a coal seam. See also: *hard heading; rock drivage; slant.* (Nelson, 1965)

stone dust (a) In coal mines any inert dust spread on roadways as a defense against the danger of coal dust explosions. The stone dust used is of a type that does not cake in mine air. Ground limestone is satisfactory in most conditions. About 5% less limestone is required than shale, and gypsum has about two to three times the efficiency of shale. Gypsum appears to owe its high efficiency to its hydrate water. Inert dusts are effective because they absorb heat that the coal dust would otherwise receive from the flame. See also: *dust consolidation.* (Nelson, 1965) (b) Shale dust. (Mason, 1951) (c) Loosely applied to any incombustile dust used to render coal dust incombustible. (Mason, 1951)

stone-dust barrier A device erected at strategic points in mine roadways for the purpose of arresting explosions. Consists essentially of trays loaded with stone dust, which are upset or overturned by the pressure wave in front of an explosion ahead of the flame, producing a dense curtain or cloud of inert dust to blanket the flame and stop further propagation of the explosion. See also: *colliery explosion; flame inhibitor.* (Sinclair, 1958)

stone duster The person in charge of stone dusting in coal mines. (Nelson, 1965)

stone dusting The systematic distribution of stone dust along mine roadways to cover the coal dust and reduce its flammability. Although stone dusting may not always stop an explosion, it is less violent with stone dusting than without. Safety regulations in many countries require that stone dust be applied either mixed with the coal dust along mine roadways or as stone dust barriers, to reduce the liability of coal-dust explosions. (Nelson, 1965; Roberts, 1960)

stone exposure *diamond exposure.*

stone fields *block field.*

stone flax An early name for asbestos.

stone gobber In bituminous coal mining, person who removes stone and other refuse from coal mine floors and dumps refuse into mine cars for disposal. (DOT, 1949)

stone hammer A hammer for breaking or dressing stone. (Standard, 1964)

stonehead (a) U.K. A heading driven in stone or bind. (b) The solid rock first met in sinking a shaft. (c) *bedrock.*
stone intrusion Irregular masses of sandstone occurring within a coal seam or penetrating the seam from top to bottom, sometimes much distorted, but always connected with a similar sandstone in the roof or higher strata. Also called stone eye. See also: *drop.* (Raistrick, 1939)
stone land An area that is economically valuable for some variety of stone, such as granite or sandstone, that can be quarried. Cf: *mineral land.* (AGI, 1987)
stoneman (a) A miner employed on stonework. See also: *ripper.* (Nelson, 1965) (b) A worker who drills shotholes in rock in readiness for firing by the shotfirer. (Hammond, 1965)
stone mill (a) A stone crusher. (Standard, 1964) (b) A machine for dressing and finishing marble, slate, etc.; a stone dresser. (Standard, 1964)
stone mine (a) Scot. An ironstone mine or working. (b) Scot. A mine driven in barren strata.
stone ocher Ocher found in hard globular masses. (Webster 3rd, 1966)
stone of ore A piece of ore.
stone per carat The number of near-equal-size diamonds the weight of which is 1 carat, hence a relative measure of the size of diamonds. Syn: *diamonds per carat.* (Long, 1960)
stone pit A quarry where stones are dug. (Webster 2nd, 1960)
stone planer In the concrete products and stonework industry, person who shapes, smooths, squares, and removes excess material from the surfaces of blocks or slabs of building or ornamental stone (limestone, slate, sandstone, marble, and such concrete work as imitation marble) on a planer. Also called planer hand; planer man; planer operator; planing machine operator. (DOT, 1949)
stone polisher In the stonework industry, person who polishes to a high luster by hand methods those curbed, irregular, and straight surfaces of blocks and slabs of marble and granite that cannot be polished by machine. May be designated according to kind of stone, such as granite polisher or marble polisher. Also called stone finisher. (DOT, 1949)
stone pressure *diamond pressure.*
stones (a) Detached particles of rock usually smaller than 10 in (25 cm) in diameter. Stones are classed as gravel on bottom sediment charts. (Hunt, 1965) (b) In mica, small embedded crystals or holes resulting from stones. Also called stone holes. (Skow, 1962)
stone saw A stone-cutting apparatus having no teeth, being a simple iron band fed with sand and water, cutting by attrition. (Standard, 1964)
stone sawyer Person who cuts stone.
stone sill Cumb. Sandy fireclay. Cf: *sill.* (Arkell, 1953)
stone spavin York. Usually a stone bind or sandstone with rootlets. (Arkell, 1953)
stone tubbing Watertight stone walling of a shaft cemented at the back.
stone wall *hogback.*
stoneware clay A clay suitable for manufacture of stoneware (ceramic ware fired to a hard, dense condition and with an absorption of less than 5%); used for items such as crocks, jugs, and jars. It possesses good plasticity, fusible minerals, and a long firing range. (AGI, 1987)
stone weight *diamond content.*
stonework (a) All underground work involving the excavation, loading, or handling of rock or dirt, such as ripping, dinting, tunneling, packing, etc. In thin coal seams, stonework is a major cost item. See also: *dead work.* (Nelson, 1965) (b) The process of working in stone; the shaping, preparation, or setting of stone. (Webster 3rd, 1966)
stone yellow *yellow ocher.*
stony clunch Eng. Compressed clays with sandstone layers interspersed, Midlands. (Arkell, 1953)
stook The last stump or corner block of coal left when extracting pillars by means of lifts, in pillar methods of working. (Nelson, 1965)
stool (a) The point where a miner stops digging downward to work outward. (Hess) (b) The assembly carrying the return rollers and brackets for connecting standard sections together in belt conveyors. (Nelson, 1965)
stooled Eng. Applied to a vein cut vertically for some distance.
stool end A supporting pillar mine. (Webster 3rd, 1966)
stoop-and-room *bord-and-pillar.*
stooper A miner in pillar methods of working employed in pillar robbing; a practice now obsolete. (Nelson, 1965)
stoopway A passageway, the height of which requires a person to stoop or crouch in traversing it. (AGI, 1987)
stop (a) Any cleat or beam to check the descent of a cage, car, pump, pump rods, etc. (b) In mining, a variation of stope.
stopblocks A simple arrangement of two stout timbers, sliding on pivots, one of which can be placed across the track rail and held in position by the other block. They are often used on haulage landings to prevent trams running down the incline uncontrolled. Syn: *nubber.* (Nelson, 1965)
stope (a) An excavation from which ore has been removed in a series of steps. A variation of step. Usually applied to highly inclined or vertical veins. Frequently used incorrectly as a syn. for room, which is a wide-working place in a flat mine. (Standard, 1964) (b) To excavate ore in a vein by driving horizontally upon it a series of workings, one immediately over the other, or vice versa. Each horizontal working is called a stope because when a number of them are in progress, each working face under attack assumes the shape of a flight of stairs. When the first stope is begun at a lower corner of the body of ore to be removed, and, after it has advanced a convenient distance, the next is commenced above it. This is called overhand stoping. When the first stope begins at an upper corner, and the succeeding ones are below it, it is called underhand stoping. The term stoping is loosely applied to any subterranean extraction of ore except that which is incidentally performed in sinking shafts, driving levels, etc., for the purpose of opening the mine. (c) Commonly applied to the extraction of ore, but does not include the ore removed in sinking shafts and in driving levels, drifts, and other development openings. (Lewis, 1964) (d) The working above and below a level where the mass of the orebody is broken. A stope is the very antithesis of a shaft, tunnel, drift, winze, or other similar excavation in a mine. (Ricketts, 1943) (e) Any excavation in a mine, other than development workings, made for the purpose of extracting ore. The outlines of the orebody determine the outlines of the stope. The term is also applied to breaking ground by drilling and blasting or other methods. See also: *caving.* (Nelson, 1965) (f) A body of mineral left by running drifts about it. (Standard, 1964)
stope board A timber staging on the floor of a stope for setting a rock drill. The stage is tilted to enable the bottom holes being drilled in the same inclined direction. (Nelson, 1965)
stope development The driving of subsidiary openings designed to prepare blocks of ore for actual extraction by stoping.
stope driller In metal mining, one who operates compressed air, percussion-type rock drill in a stope (an underground opening from which ore is extracted in a series of steps). Also called stoper. (DOT, 1949)
stope fillings Broken mullock or rock or the broken low-grade portion of a lode or vein used to fill stopes on abandonment. (Nelson, 1965)
stope hoist A small portable compressed-air hoist for operating a scraper-loader or for pulling heavy timbers into position, often used in narrow stopes. (Nelson, 1965)
stope miner *miner.*
stop end The shuttering erected at the end of a length of concrete construction and shaped to provide a suitable joint for the next length. (Hammond, 1965)
stope pillar A column of ore left to support the stope. (Pryor, 1963)
stoper (a) A stoping drill. (b) A light percussive drill incorporating a pneumatic cylinder to provide support and thrust while drilling steeply upward. (BS, 1964) (c) *stope driller.*
stoperperson *roof bolter.*
stope sampling The sampling of exposure in the stopes, or of material coming from the stopes. This type of sampling permits a closer control of the grade of ore being won and is conducted largely at the working place or stope. (Truscott, 1962)
stope scraper In metal mining, one who shovels or rakes ore into mine cars in a stope (an underground opening in a vein from which ore is extracted in a series of steps). Syn: *hand scraper.* (DOT, 1949)
stope washings In gold mines, auriferous slimes washed down from the floor (footwall) of the stope and sent to the mill. (Pryor, 1963)
stoping (a) The act of excavating rock, either above or below a level, in a series of steps. In its broadest sense rock stoping means the act of excavating rock by means of a series of horizontal, vertical, or inclined workings in veins or large, irregular bodies of ore, or by rooms in flat deposits. It covers the breaking and removal of the rock from underground openings, except those driven for exploration and development. The removal of ore from drifts, crosscuts, shafts, winzes, and raises, which are excavated to explore and develop an ore deposit, is incidental to the main purpose for which stopes are driven and is not a stoping operation. Exploratory and development openings are driven to prepare a mine for extraction of the ore by stoping. See also: *stope.* (b) In civil engineering, an enlargement. (Fraenkel, 1953) (c) The loosening and removal of ore in a mine either by working upward (overhead or overhand) or downward (underhand). (AGI, 1987)
stoping-and-filling *overhand stoping.*
stoping drill A small air or electric drill, usually mounted on an extensible column, for working stopes, raises, and narrow workings.
stoping ground Part of an orebody opened by drifts and raises, and ready for breaking down.
stoping in horizontal layers *overhand stoping.*
stoping methods The classification of stoping methods adopted by the U.S. Bureau of Mines, devised largely on the basis of rock stability, is as follows: (1) stopes naturally supported—this includes open stoping with open stopes in small orebodies, and sublevel stoping; and open

stopes with pillar supports that includes casual pillars and room (or stope) and pillar (regular arrangement); (2) stopes artificially supported—this includes shrinkage stoping, with pillars, without pillars, and with subsequent waste filling; cut-and-fill stoping; stulled stopes in narrow veins; and square-set stoping; (3) cave stopes—this includes caving (ore broken by induced caving), block caving, including caving to main levels and caving to chutes or branched raises; sublevel caving and top slicing (mining under a mat that, together with caved capping, follows the mining downward in successive stages); and (4) combinations of supported and caved stopes (as shrinkage stoping with pillar caving, cut-and-fill stoping with top slicing of pillars, etc.)

stoping underhand Mining a stope downward in such a series that presents the appearance of a flight of steps.

stoping width (a) Width of lode broken during mining, including any barren rock. (Pryor, 1963) (b) Used in underground sampling and is estimated from direct measurement behind the stope face and reduced to allow for any waste stowed. With wide tabular deposits, there is little difference between the stoping width and the clean width. (Nelson, 1965)

stopoff To close off part of a mine by means of a brattice, wall, stopping, etc.

stopper hole In a puddling furnace, the hole through which the rabble is introduced. (Webster 2nd, 1960)

stopper maker One who forms and finishes fire clay stoppers for open hearth ladles, using a stopper press and a finishing machine. (DOT, 1949)

stopping (a) A brattice, or more commonly, a masonry or brick wall built across old headings, chutes, airways, etc., to confine the ventilating current to certain passages, and also to lock up the gas in old workings, and in some cases to smother a mine fire. (b) A permanent wall built to close off unused crosscuts to prevent the air from short circuiting. (Lewis, 1964) (c) A dam or seal to isolate old workings containing water or injurious gases. See also: *dam; inrush of water.* Syn: *ventilation stopping.* (Nelson, 1965)

stopping builder In bituminous coal mining, one who builds walls of concrete, stone, or brick and mortar, to close off old passageways or haulageways underground, to maintain ventilation in new workings. (DOT, 1949)

storage battery locomotive An underground locomotive powered by storage batteries. (Hammond, 1965)

Storrow whirling hygrometer A hygrometer in which the two thermometers are mounted side by side on a brass frame and fitted with a loose handle so that it can be whirled in the atmosphere to be tested. The instrument is whirled at some 200 rpm for about 1 min and the readings on the wet- and dry-bulb thermometers recorded; used in conjunction with Glaisher's or Marvin's hygrometrical tables. It gives more consistent and accurate results than the ordinary instrument. See also: *whirling hygrometer.* (Nelson, 1965)

stottite (a) A tetragonal mineral, $FeGe(OH)_6$; greasy; at Tsumeb, Namibia. (b) The mineral group jeanbandyite, mopungite, stottite, and tetrawickmanite.

stove (a) A large steel furnace or oven connected with the blast furnace to preheat the blast before it is introduced into the furnace proper. (Mersereau, 1947) (b) A kiln, as for firing pottery or drying minerals. (Webster 2nd, 1960)

stove coal (a) In anthracite only; two sizes of stove coal are made, large and small. Large stove, known as No. 3, passes through a 2-1/4- to 2-in (5.7- to 5-cm) mesh and over a 1-7/8- to 1-1/2-in (4.8- to 3.8-cm) mesh; small stove, known as No. 4, passes through a 1-7/8- to 1-3/8-in (4.8- to 3.5-cm) mesh and over a 1-1/8- to 1-in (2.9- to 2.5-cm) mesh. Only one size of stove coal is now usually made. It passes through a 2-in (5-cm) square mesh and over a 1-3/8-in (3.5-cm) square mesh. (Zern, 1928) (b) *anthracite coal sizes.*

stoved salt Stoved open-pan salt. (Kaufmann, 1960)

stovepipe Riveted seam or spiral welded seam, thin-wall pipe used as a conductor, standpipe, or casing in a borehole. (Long, 1960)

stovepipe casing *stovepipe.*

stow (a) To pack away rubbish into goaves or old workings. (b) To gob; to fill the waste; to put debris into the waste. (Mason, 1951)

stowage Scot. In longwall mining the space from which the mineral has been extracted and which has been filled with waste.

stowce Derb. A wooden landmark, placed to indicate possession of mining ground. Also spelled stowse.

stowing (a) A method of mining in which all the material of the vein is removed and the waste is packed into the space left by the working. (b) The debris of a vein thrown back of a miner and which supports the roof or hanging wall of the excavation. (Zern, 1928) (c) *solid packing.*

stowing method Any of several methods of working coal or ore deposits in which systematic stowing of the worked out areas is part of the system. See also: *coal mining methods; solid stowing; strip packing.* (Nelson, 1965)

straddle A vertical mine timber, esp. one supporting a set in a shaft.

straddle spread *split spread.*

strahlite From the German "strahlstein." (radiating stone), the original name for actinolite (Greek "aktis" radiating). See also: *actinolite.*

straight Common term for straight brick, as a 9-in (23-cm) straight.

straightaway In stripping, a pit that follows a straight line when projected on a horizontal plane. (Woodruff, 1966)

straight chopping bit *chopping bit.*

straight-cut gang frame In quarrying, a saw gang that slides back and forth on a bed, as contrasted with the ordinary saw gang that swings back and forth when suspended from above.

straight dynamites Dynamites composed of nitroglycerin, a combustible such as wood meal, sodium nitrate, and an antacid, such as calcium or magnesium carbonate, and are made in 15% to 60% strength, the percentage representing the proportion of nitroglycerin in the dynamites. They are powerful, quick acting, and fairly water resistant, but on detonation produce poisonous gases, esp. in the higher grades. Their relatively high cost, sensitivity to shock and friction, and high flammability, together with the dangerous fumes developed, make them less suitable for general use than more recently developed modifications. (Lewis, 1964; Carson, 1961)

straightedge leveling A system of leveling using a straightedge and a spirit level. (BS, 1963)

straight girder support An H-section girder used as a roadway or face beam support. The girder for spanning roadways is commonly 6 in (15.2 cm) deep and 5 in (12.7 cm) in width of flange. It is supported by wood or steel props or by brick or concrete sidewalls, the roof being made secure by timber or sheet lagging. Channel section girders are also used in special cases. See also: *steel support.* (Nelson, 1965)

straight point Aust. That straight portion of the inner main rail between the rails of a turnout.

straight-side core bit *straight-wall bit.*

straight stall A lateral excavation into a thick seam of coal. Also called straight coal. (Standard, 1964; Fay, 1920)

straight-type wedge A plain deflecting wedge, not equipped with a rose or stabilizing ring. (Long, 1960)

straight-wall bit An annular-shaped (core) bit the inner walls of which are parallel with the outer walls and not tapered to receive a core lifter. Syn: *straight-side core bit.* (Long, 1960)

straight-wall core shell (a) A reaming shell the outside walls of which are straight and not set with diamonds or hard-metal reaming points. (Long, 1960) (b) Sometimes used as a syn. for blank reaming shell. (Long, 1960)

straight work (a) Narrow headings in coal. (CTD, 1958) (b) A method of working coal by driving parallel headings and then removing the coal between them. (CTD, 1958)

strain (a) Change in the shape or volume of a body as a result of stress; a change in relative configuration of the particles of a substance. Syn: *deformation.* (AGI, 1987) (b) Deformation resulting from applied force; within elastic limits strain is proportional to stress. Cf: *stress.* (AGI, 1987) (c) A measure of the change in the size or shape of a body, referred to its original size or shape. Linear strain is the change per unit length of a linear dimension. True strain (natural strain) is the natural logarithm of the ratio of the length at the moment of observation to the original gage length. Conventional strain is the linear strain referred to the original gage length. Shearing strain (shear strain) is the change in angle (expressed in radians) between two lines originally at right angles. When the term strain is used alone, it usually refers to the linear strain in the direction of the applied stress. See also: *true strain.* (ASM, 1961) (d) There are, generally speaking, two kinds of strains: normal and shear. Normal strains are those that may result in the relative displacement of two particles along the line joining those particles, whereas in shear strains, the particles are displaced at a right angle to the line joining them. All possible deformations may be represented as a combination of these two types of strain. (Issacson, 1962)

strain bar An instrument well suited for measuring the strain on a rock face. It consists essentially of an invar steel bar with a fixed point at one end and a movable point attached to a rider at the other end. The rider may be moved along the bar between two stops, and the extent of movement is indicated by a dial gage attached to the bar. (Issacson, 1962)

strain break Fractures occurring in rock quarries where the rock is under compressive stress. This stress is relieved locally in the process of quarrying, resulting in the rending or fracturing of the rock mass.

strain burst Rock burst in which there is spitting, flaking, and sudden fracturing at the face, indicating increased pressure there. (Higham, 1951)

strain cleavage *slip cleavage.*

strain ellipsoid In elastic theory, a sphere under homogeneous strain is transformed into an ellipsoid with this property; the ratio of the length of a line, which has a given direction in the strained state, to the length of the corresponding line in the unstrained state, is proportional to the central radius vector of the surface drawn in the given direction.

The ellipsoid whose half axes are the principal strains. Cf: *reciprocal strain ellipsoid*. (AGI, 1987)

strain gage (a) A general term for a device with which mechanical strain can be measured, commonly by an electrical signal, e.g., a wire strain gage. (AGI, 1987) (b) An electrical, mechanical, or optical device for measuring movement of rock, cumulative loading of support props, opening cracks, etc. (Pryor, 1963) (c) An electromechanical device that transforms small displacements to changes in resistance that are proportional to the displacement. Strain gages are used in ocean bottom pressure measuring equipment. (Hunt, 1965)

strain relief method A technique for the determination of absolute (total) strain and stress within rock in situ. In this method, a smooth hole is bored in the rock and a gage is inserted to measure diametral deformation. The hole is then overcored with a large coring bit so that the cylinder of rock containing the deformation measuring gage is free to expand. The change in the diameter of the hole when the rock cylinder is free to expand is a function of the original stress in the rock and its elastic modulus. (Woodruff, 1966)

strain restoration method A technique for the determination of absolute (total) strain and stress within rock in situ. This method involves (1) installation of strain gages on the rock surface; (2) cutting of a slot in the rock between the strain gages so that the surface rock is free to expand; (3) installation of a "flat-jack" (hydraulic pressure cell) in the slot; and (4) application of hydraulic pressure to the flat jack until the rock is restored to its original state of strain. The original stress in the rock is presumed to be equal to the final pressure in the flat jack. (Woodruff, 1966)

strain rosette At any point on the surface of a stressed body, strains measured on each of three properly chosen intersecting gage lines make possible the calculation of the principal stresses at that point. Such gage lines, and the corresponding strains, are called a strain rosette. (Roark, 1954)

strain sheet (a) A skeleton drawing of a structure, as a roof of truss or a bridge, showing the stress to which each member will be subjected. (b) A quarryman's term for granite sheets produced by compressive strain.

strain-slip cleavage *false cleavage; slip cleavage.*

Straits tin One of the purest commercial forms of tin (99.89% purity) produced from alluvial ores in Malaysia.

strait work *straight work.*

strake (a) A relatively wide launder or sluice set at a slope and covered with a blanket or corduroy for catching comparatively coarse gold and valuable mineral. See also: *blanket strake*. (Nelson, 1965) (b) A trough in which ore, gravel, etc., are washed; a launder. (Standard, 1964) (c) The place where ore is assorted on the floor of a mine; a dressing floor. (Standard, 1964)

stranded Term for wire rope with one or more broken strands. (Pryor, 1963)

stranded rope *preformed rope; multistrand rope; flattened strand rope.*

stranskiite A triclinic mineral, $(Zn,Cu)_3(AsO_4)_2$; blue; at Tsumeb, Namibia.

strap (a) A bar; a beam; a coal face bar. (Mason, 1951) (b) A thin bar or metal plate, similar to a fishplate, used to secure together butt-jointed timber or steel members. (Hammond, 1965)

strap brake A brake generally used on small winding engines. (Sinclair, 1959)

strap fishplates Flattened bars of iron with holes punched through them for bolts. The holes are made somewhat larger than the bolt to permit rail expansion and contraction. (Kiser, 1929)

strap-rope haulage A system of haulage (usually endless rope) in which the engine is installed on the surface, and the power is transmitted to the haulage drums at the pit bottom by means of a rope, which is known as a strap rope, or a driving rope. This rope merely transmits power and is distinct from the haulage rope. (Nelson, 1965)

straps Thin metal support members from 5 to 20 ft (1.5 to 6 m) long are bolted to the mine roof to prevent roof deterioration between the bolts. Also known as roof mats or bacon skins.

strata Plural of stratum.

strata bolt A bolt or rod, from 2 to 5 ft (0.6 to 1.5 m) or more in length, set in drill holes in the strata for the support of curbs, skeleton tubbing, helical steel supports, etc., in shafts and staple shafts. In general, the weaker the ground, the longer the bolts. Cf: *roof bolting*. (Nelson, 1965)

strata-bound Said of a mineral deposit confined to a single stratigraphic unit. The term can refer to a stratiform deposit, to variously oriented orebodies contained within the unit, or to a deposit containing veinlets and alteration zones that may or may not be strictly conformable with bedding. (AGI, 1987)

strata control *roof control.*

strata gases These occur in the mineral deposit itself or in adjacent or nearby formations. Their origin may be in a particular formation in which they were laid down or formed subsequently by chemical action, or they may occasionally migrate into other formations, frequently because of release of pressure with mining. Water flow and rock porosity and fissures also allow gas migration. The principal strata gases are methane, carbon dioxide, nitrogen, sulfur dioxide, hydrogen sulfide, and radon. (Hartman, 1961)

stratascope An apparatus inserted in the drill hole that permits engineers to make a visual inspection of the strata. (Coal Age, 1966)

strata temperature The strata temperature is determined by the surface temperature, the diffusivity of the strata, and the emissivity of the surface. With rocks of high thermal conductivity, and thus, high diffusivity, as in metal mines, the increase in temperature with depth is small, i.e., the geothermic gradient is low. Where rocks have low thermal conductivity, as in coal measure strata, the geothermic gradient is steep. Syn: *geothermic gradient*. (Roberts, 1960)

strategic and critical material Materials that (a) would be needed to supply the military, industrial, and essential civilian needs of the United States during a national emergency, and (b) are not found or produced in the United States in sufficient quantities to meet such need. (Strategic, 1989)

strategic mica Ruby and nonruby block mica. Good-stained or better qualities, grade No. 6 or larger; ruby-stained A/B quality, grade No. 6 or larger; ruby and nonruby film, first and second qualities, grade No. 4 and smaller; muscovite and phlogopite splittings; and phlogopite block of high heat quality. (Skow, 1962)

strategic minerals Minerals essential to the national defense, for the supply of which, during war, we are wholly or in part dependent on foreign sources, and for which strict measures controlling conservation and distribution are necessary. For example, chromium- and tin-bearing minerals, quartz crystal, and sheet mica were some of the "strategic minerals" during World War II.

straticulate Characterized by numerous very thin parallel layers, whether separable or not, either of sedimentary deposition (as a bed of clay) or of deposition from solution (as in a stalagmite or banded agate). (AGI, 1987)

stratification (a) The formation, accumulation, or deposition of material in layers; specif. the arrangement or disposition of sedimentary rocks in strata. See also: *bedding*. (AGI, 1987) (b) A structure produced by deposition of sediments in strata; a stratified formation, or stratum. It may be due to differences of texture, hardness, cohesion or cementation, color, internal structure, and mineralogic or lithologic composition. (AGI, 1987) (c) The state of being stratified; a term describing a layered or bedded sequence, or signifying the existence of strata. (AGI, 1987) (d) A structure produced by deposition of sediments in beds or layers (strata), laminae, lenses, wedges, and other essentially tabular units. (AGI, 1987)

stratification of methane In relatively unventilated cavities in coal mines, such as wastes, it is frequently found that the methane percentage or concentration is higher at roof level, lower at midheight, and least at floor level. This is termed stratification of methane. See also: *combustible gases layer*. (Nelson, 1965)

stratification plane *bedding plane; stratification of methane.*

stratified Formed, arranged, or laid down in layers or strata; esp. said of any layered sedimentary rock or deposit. See also: *bedded*. (AGI, 1987)

stratified rock *sedimentary rock.*

stratified rocks (a) Derivative or stratified rocks may be fragmental or crystalline; those that have been mechanically formed are all fragmental; those that have been chemically precipitated are generally crystalline; and those composed of organic remains are sometimes partially crystalline. Syn: *sedimentary rocks*. (AGI, 1987) (b) Rocks arranged in layers. (Shell)

stratiform (a) Said of a special type of strata-bound deposit in which the desired rock or ore constitutes, or is strictly coextensive with, one or more sedimentary, metamorphic, or igneous layers; e.g., beds of salt or iron oxide, or layers rich in chromite or platinum in a layered igneous complex. (AGI, 1987) (b) Having the form of a layer, bed, or stratum; consisting of roughly parallel bands or sheets, such as a stratiform intrusion. Incorrect spellings: strataform, stratoform. (AGI, 1987) (c) Bedded or layered. (Bateman, 1951)

stratigrapher A geologist who specializes in stratigraphy.

stratigraphic Pertaining to the composition, sequence, and correlation of stratified rocks.

stratigraphic classification The arbitrary, but systematic arrangement, zonation, or partitioning of the sequence of rock strata of the Earth's crust into units with reference to the many different characters, properties, or attributes that the strata may possess (Hedberg, 1958). (AGI, 1987)

stratigraphic control (a) The influence of stratigraphic features on ore deposition, e.g., ore minerals selectively replacing calcareous beds. Cf: *structural control*. (AGI, 1987) (b) The degree of understanding of the stratigraphy of an area; the body of knowledge that can be used to interpret its stratigraphy or geologic history. (AGI, 1987)

stratigraphic geology *stratigraphy.*

stratigraphic heave (a) For normal faults, the width of the gap between two parts of a disrupted bed, measured in the direction of the faulted bedding plane. (Schieferdecker, 1959) (b) For reverse faults, the width of the overlap between two parts of a disrupted bed, measured in the direction of the faulted bedding plane. (Schieferdecker, 1959)

stratigraphic hole A borehole drilled specif. to obtain a detailed record of the character and composition of the rock formation penetrated and not for the purpose of locating a mineral deposit. See also: *record hole.* (Long, 1960)

stratigraphic section *geologic section.*

stratigraphic separation The thickness of the strata that originally separated two beds brought into contact at a fault. Syn: *stratigraphic throw.* (AGI, 1987)

stratigraphic sequence A chronologic succession of sedimentary rocks from older below to younger above, essentially without interruption; e.g., a sequence of bedded rocks of interregional scope, bounded by unconformities. (AGI, 1987)

stratigraphic throw *stratigraphic separation; throw.*

stratigraphy (a) The science of rock strata. It is concerned not only with the original succession and age relations of rock strata but also with their form, distribution, lithologic composition, fossil content, geophysical and geochemical properties; indeed, with all characters and attributes of rocks as strata; and their interpretation in terms of environment or mode of origin, and geologic history. All classes of rocks, consolidated or unconsolidated, fall within the general scope of stratigraphy. Some nonstratiform rock bodies are considered because of their association with or close relation to rock strata. Syn: *stratigraphic geology.* (AGI, 1987) (b) The arrangement of strata, esp. as to geographic position and chronologic order of sequence. (AGI, 1987) (c) The sum of the characteristics studied in stratigraphy; the part of the geology of an area or district pertaining to the character of its stratified rocks. (AGI, 1987) (d) A term sometimes used to signify the study of historical geology. (AGI, 1987) (e) That branch of geology that treats of the formation, composition, sequence, and correlation of the stratified rocks as parts of the Earth's crust. (f) That part of the descriptive geology of an area or district that pertains to the discrimination, character, thickness, sequence, age, and correlation of the rocks of the district.

stratometric survey A system whereby the in situ orientation of a core sample can be reproduced on the surface. A line is inscribed on the smoothed bottom of a borehole, and its azimuth relationship with a compass direction photographically recorded. When cored and removed from the borehole, the inscribed line can be used as a guide in orienting the core on the surface. (Long, 1960)

stratum (a) A bed or layer of rock; strata, more than one layer. (Fay, 1920) (b) A layer greater than 1 cm in thickness. (Pettijohn, 1964) (c) A tabular or sheetlike body or layer of sedimentary rock, visually separable from other layers above and below; a bed. The term is more frequently used in its plural form, strata. Cf: *lamina.* (AGI, 1987)

stratum plain *stripped plain.*

straw boss A generic term that sometimes includes all supervisory officials in a mine; a supervisor of a small group of miners usually working under a certified foreman. (BCI, 1947)

stray current (a) The use of direct current electric power in most coal mines presents a problem of corrosion caused by an electrolytic action on pipes. Where the track is employed as a conductor of electricity in this type of power system, the voltage drop produces a difference in potential between the track and earth or other structures that may serve as a conductor, such as pipelines. This condition may cause an electric current to flow in pipelines that form a parallel path with the track system. (b) Electric current that is introduced in the earth by leakage of industrial currents. (Schieferdecker, 1959)

streak (a) The color of a mineral when scraped on a white ceramic plate. (b) The color of a mineral powder usually obtained by scratching with steel or by rubbing on unglazed porcelain; it is more diagnostic for identification than bulk mineral color. (c) A long, narrow, irregular stretch of land or water. (d) A comparatively small and flattish or elongate sedimentary body, visibly differing from adjacent rock, but without the sharp boundaries typical of a lens or layer. (e) A long, narrow body of sand, perhaps representing an old shoreline; a shoestring. (f) The outcropping edge of a coal bed.

streaking *mineral streaking.*

streak plate In mineral identification, a piece of unglazed porcelain, hardness about 7 on the Mohs scale, used for rubbing a sample to obtain its powder color, or "streak."

stream-down sluice A sluice box placed to receive the material rejected from the tables of a dredge.

stream gold Gold in placer deposits of alluvial origin.

streaming (a) Separating ore from gravel by the aid of running water. (b) The working of alluvial deposits for the tin found in them; the washing of tin ore from the detrital materials; also, the reduction of stream tin. (Standard, 1964) (c) A property that combustible gas possesses as a result of its low density. If, in a sloping roadway having a smooth roof, combustible gas is released from a break at the lower end, the gas will cling to the roof and stream upwards forming a pool at the upper end. Pure streaming and diffusion can occur where the air currents are sluggish or nonexistent, such as in wastes that have not collapsed, and roadways of large cross-sectional area in which the ventilation air speed is low. (Roberts, 1960)

streaming potential Potential difference between a permeable diaphragm and the liquid passing through it. (Pryor, 1963)

streamline A hypothetical line that shows the velocity direction of the fluid stream at each point along the line. Streamlines do not therefore cross each other. A set of streamlines charts the flow pattern. If the flow is steady then the streamline pattern does not change with time. If, however, the streamlines are continually changing shape the flow is unsteady. (Roberts, 1960)

streamline flow *laminar flow.*

streamlining This involves the placing of fairing around or over obstructions to airflow, in such a way as to change that flow from turbulent to laminar. (Spalding, 1949)

stream tin Cassiterite occurring as waterworn pebbles in alluvial or placer deposits. Cf: *lode tin.* Syn: *alluvial tin.*

stream tube A tube of fluid the enveloping surface of which consists of streamlines. (Roberts, 1960)

streamworks (a) Corn. A name given by miners to alluvial tin deposits usually worked in the open air. (b) A place where ore, generally tin ore, is washed from alluvial deposits. (Standard, 1964)

strebbau Ger. The longwall system of coal mining. (Fay, 1920)

street ell A pipe elbow with male threads on one end, female on the other. (Nichols, 1976)

strek Corn. A trough for washing tin ore. A variation of strake.

strengite (a) An orthorhombic mineral, $FePO_4 \cdot 2H_2O$; variscite group; forms a series with variscite; may have Mn replacing Fe; dimorphous with phosphosiderite; pale red. (b) The mineral group mansfieldite, scorodite, strengite, and variscite.

strength The stress at which a rock ruptures or fails.

strength of current The number of amperes flowing through a circuit. Analogous to the flow of gallons per minute in a water pipe. (Crispin, 1964)

strength of materials The science that deals with the effects of forces in causing changes in the size and shape of bodies. (Crispin, 1964)

stress In a solid, the force per unit area, acting on any surface within it, and variously expressed as pounds or tons per square inch, or dynes or kilograms per square centimeter; also, by extension, the external pressure that creates the internal force. The stress at any point is mathematically defined by nine values: three to specify the normal components and six to specify the shear components, relative to three mutually perpendicular reference axes. Cf: *strain.* See also: *normal stress; true stress; shear stress.* (AGI, 1987)

stress analysis Determination of the stresses in the component parts of a structure when subjected to load. See also: *photoelasticity.* (Hammond, 1965)

stress circle A Mohr circle that reveals the distribution of stress. (Hammond, 1965)

stress-corrosion cracking Failure by cracking under combined action of corrosion and stress, either external (applied) or internal (residual). Cracking may be either intergranular or transgranular, depending on metal and corrosive medium. (ASM, 1961)

stress diagram *stress-strain diagram.*

stress difference The difference between the greatest and least of the three principal stresses. (AGI, 1987)

stress envelope The zone of extra stress around a cylindrical hole being actually cylindrical in form. Syn: *ring stress.* (Spalding, 1949)

stress field The state of stress, either homogeneous or varying from point to point, in a given domain. (AGI, 1987)

stress meter An instrument designed to measure pressure changes within rock as a result of mining operations. The instrument consists of a tongue of steel containing a groove filled with glycerin. When pressure is exerted upon the external surface of the tongue, the glycerin is partially squeezed from the groove and exerts pressure on a diaphram, which accordingly bulges outwards. A strain gage is cemented on the outer wall of the diaphragm, and as the curvature increases and the gage is strained, so does its resistance vary. (Issacson, 1962)

stress mineral A term suggested for minerals such as chlorite, chloritoid, talc, albite, epidote, amphiboles, and kyanite, whose formation in metamorphosed rocks is favored by shearing stress. Cf: *antistress mineral.*

stress-number (S-N) curve A curve obtained in fatigue tests by subjecting a series of specimens of a given material to different ranges of stress and plotting the range of stress against the number of cycles required to produce failure. In steel and many other metals, there is a limiting range of stress below which failure will not be produced even by an indefinite number of cycles. (CTD, 1958)

stress ring Stress rings are force lines drawn on the cross section of an excavation to indicate the distribution of the additional stress in the rock caused by that excavation. (Spalding, 1949)

stress solid The solid figure formed by surfaces bounding vectors drawn at all points of the cross section of a member and representing the unit normal stress at each such point. The stress solid gives a picture of the stress distribution on a section. (Roark, 1954)

stress-strain curve A curve similar to a load extension curve, except that the load is divided by the original cross-sectional area of the test piece and expressed as tons or pounds per square inch, while the extension is divided by the length over which it is measured and expressed in inches per inch. See also: *load-extension curve.* (CTD, 1958)

stress-strain diagram (a) A graph on which is plotted stress vs. strain. Such a graph may be constructed in any test during which frequent or continuous measurements of both stress and strain are made. It is commonly constructed for the compression, tension, and torsion tests. It is usually necessary for the determination of deformation energy, elastic limit, modulus of elasticity, modulus of rigidity, proportional limit, and yield strength. It is often useful in determination of elongation, modulus of rupture, ultimate strength and related properties. (Hunt, 1965) (b) The curve obtained by plotting unit stresses as ordinates against corresponding unit strains as abscissas. Syn: *stress diagram.* (Roark, 1954)

stress trajectory A line (in a stressed body) tangent to the direction of one of the principal stresses at every point through which it passes. Syn: *isostatic.* (Roark, 1954)

stress zone This is the zone of additional stress in the rock surrounding and caused by a stoped area. (Spalding, 1949)

stret The system of mining coal by headings or narrow work. See also: *bord-and-pillar.* Also spelled strett.

stretch A particular direction or course; as, the stretch of a coal seam. (Standard, 1964)

stretcher (a) A bar used for roof support on a roadway, which is either wedged against or pocketed into the sides of the roadway and not supported by legs or struts. (TIME, 1929-30) (b) A bar fixed across a narrow working place or tunnel to support a rock drill. (CTD, 1958) (c) A main backing deal or longitudinal bar in contact with three or more support bars or girders. (Mason, 1951)

stretchers In shaft-sinking, the crosspieces holding the walling apart. (Stauffer, 1906)

stretch fault *stretch thrust.*

stretch thrust A little-used term for a reverse fault formed by shear in the inverted limb of an overturned fold. Syn: *stretch fault.* (AGI, 1987)

stria (a) One of a series of parallel straight lines on the surface of a crystal, as in pyrite, indicative of an oscillation between two crystal forms; also, one of a series of such lines on the cleavage planes of a mineral, as of plagioclase, calcite, or corundum, indicative of polysynthetic twinning. Syn: *striation.* Pl: striae. (AGI, 1987) (b) Striation. (AGI, 1987) (c) A minute groove or channel; a threadlike line or narrow band (as of color) esp. when one of a series of parallel grooves or lines, such as glacial stria. (Webster 3rd, 1966)

striae (a) A line or furrow generally seen on the walls of a lode or fault. (Gordon, 1906) (b) Plural of stria. (AGI, 1987)

striated Adj. of striation.

striated cleavage A cleavage surface grooved with striae, e.g., plagioclase.

striated crystal A crystal face grooved with striae; e.g., pyrite or quartz.

striation (a) *stria.* (b) One of multiple scratches or minute lines, generally parallel, inscribed on a rock surface by a geologic agent, i.e., glaciers, streams, or faulting. Syn: *scratch.* Cf: *slickensides.* (AGI, 1987) (c) The condition of being striated; the disposition of striations.—Adj: striated; striate. (AGI, 1987)

striding level (a) A spirit level so mounted that it can be placed astride a surveying instrument and so supported that it can be used for precise leveling of the horizontal axis of the instrument or for measuring any remaining inclination of the horizontal axis. (AGI, 1987) (b) A demountable spirit level that can be attached to the telescope tube to level the line of sight. (AGI, 1987)

strike (a) The course or bearing of the outcrop of an inclined bed, vein, or fault plane on a level surface; the direction of a horizontal line perpendicular to the direction of the dip. Cf: *direction of strata; trace; trend.* (b) v. To find a vein of ore. n. A valuable discovery. (Webster 3rd, 1966) (c) *course; level course; fault strike.* (d) To withdraw supports. See also: *line of bearing.* Syn: *strike a lead.* (Mason, 1951)

strike a lead *strike.*

strike cut In separating blocks of stone in a quarry, the cut that is parallel to the strike of the rock strata.

strike fault A fault whose strike is parallel to the strike of the strata. Cf: *dip fault; oblique fault.*

strike shift In a fault, the shift or relative displacement of the rock units parallel to the strike of the fault, but outside the fault zone itself; a partial syn. of strike slip. See also: *shift.* (AGI, 1987)

strike-shift fault *strike-slip fault.*

strike slip (a) In a fault, the component of the movement or slip that is parallel to the strike of the fault. Cf: *dip slip; oblique slip.* Syn: *horizontal displacement; horizontal separation.* Partial syn: strike shift. (AGI, 1987) (b) A horizontal component of the slip parallel with the fault strike. (Schieferdecker, 1959)

strike-slip fault (a) A fault on which the movement is parallel to the fault's strike. Cf: *dip-slip fault.* See also: *transcurrent fault.* Syn: *strike-shift fault.* (AGI, 1987) (b) A fault in which the net slip is practically in the direction of the fault strike. Syn: *transcurrent fault.* (AGI, 1987)

strike valley A valley eroded in and parallel to the strike of the underlying rocks of a region. Syn: *longitudinal valley.*

strike working Where the dip of the coal seam is about 1 in 10 or less, the opencast method of working usually employed is to excavate along lines parallel to the outcrop. This is termed the strike or opencast method. See also: *box-cut method.* (Nelson, 1965)

striking Electrodepositing, under special conditions, a very thin film of metal that will facilitate further plating with another metal or with the same metal under different conditions. (ASM, 1961)

striking hammer A quarryman's hammer for striking a rock drill. (Standard, 1964)

striking plates Two horizontal timbers separated by striking wedges and supporting an arch center. The latter is lowered by slacking the wedges. (Stauffer, 1906)

striking solution A dilute solution of silver cyanide, containing potassium cyanide, in which articles to be silver-plated are dipped before being immersed in the silver bath proper. (Standard, 1964)

string (a) Drilling bit, jars, drill stem, rope socket, and other tools connected to the lower end of a drilling cable in standard or percussion drilling. Also used for the rig and complete drilling equipment. Syn: *drill string; string of tools.* (AGI, 1987) (b) A measurement of depth of a drill hole obtained by stringing over the length of cable from the drilling floor to the crown pulley on top of the derrick or mast. (AGI, 1987) (c) A very small vein, either independent or occurring as a branch of a larger vein; a stringer.

stringer (a) A mineral veinlet or filament, usually one of a number, occurring in a discontinuous subparallel pattern in host rock. See also: *stringer lode.* (AGI, 1987) (b) A thin layer of coal at the top of a bed, separating in places from the main coal by material similar to that comprising the roof. (c) A heavy timber or plank, usually horizontal but sometimes inclined, supporting other members of a structure; also, the horizontal crosspiece in square set timbering. (d) *roof stringer.*

stringer lead A small orebody—generally, a vein leading to a more valuable one. (AGI, 1987)

stringer lode A shattered zone containing a network of small nonpersistent veins. Syn: *stringer zone.* See also: *stringer.*

stringer sets Mine timbering in which the caps reach across two or more sets in a drift or stope. (Hess)

stringer zone *stringer lode.*

string loading Filling a drill hole with cartridges smaller in diameter than the hole, without slitting or tamping them. (Nichols, 1976)

string of tools (a) In a churn drill, the tools suspended on the drilling cable. (Nichols, 1976) (b) The entire downhole drilling assembly. See also: *string.* (Long, 1960)

string survey (a) A rough method of transferring points from upper to lower levels in very narrow steep workings by suspending strings from point to point and measuring offsets. The excavation dimensions may be similarly obtained. See also: *plumbing.* (Nelson, 1965) (b) The use of stretched strings in awkward underground workings to provide survey lines and a basis for offsets. In the lost-thread method of survey, a string or thread is paid out over a measuring device as a traverse line is walked. (Pryor, 1963)

strip (a) In mining, to remove the earth, rock, and other material from the mineral to be mined, usually by power shovels. Generally practiced only where the mineral lies close to the Earth's surface. (BCI, 1947) (b) To remove from a quarry, or other open working, the overlying earth and disintegrated or barren surface rock. See also: *baring.* (c) To mine coal, alongside a fault, or barrier. (d) To fill prepared coal from a coal face. *stripping.* Syn: *stripping a mine.* (Mason, 1951)

Stripa process (a) A heavy medium process developed in Sweden for concentrating iron, copper, and chromite ores. The rate of supply of water over a shaking bed of wet sand effects the separation of heavy and light fractions. (b) A method of gravity treatment of coarse sands, in which feed is shaken along horizontal launder and at the same time kept in teeter by hydraulic water. Constituent minerals stratify into separable layers. (Pryor, 1963)

strip-borer drill A skid- or crawler-mounted drill operated by electric motor or diesel engine. It is used at quarry or opencast sites for drilling horizontal blast holes 3 to 6 in (7.6 to 15.2 cm) in diameter, and up to 100 ft (30 m) in length, without the use of flush water. It cannot penetrate strong strata. (Nelson, 1965)

stripe The series of bands of variation in color or texture in a rock mass, or the course of the planes of such bands, as indicative of the course of the bedding plane when that is otherwise obscure. See also: *ribbon.* (Standard, 1964)

Stripkolex Trademark for a dynamite for coal stripping operations. (CCD, 1961)

strip mine A stripping; an opencut mine in which the overburden is removed from a coalbed before the coal is taken out. See also: *opencast; opencut; openpit mine.* (Hess)

strip mining The mining of coal by surface mining methods as distinguished from the mining of metalliferous ores by surface mining methods; the latter is commonly designated as openpit mining. See also: *opencast method; openpit mining; surface mining.* (Woodruff, 1966)

strip packing An arrangement of alternate packs and wastes built in a direction parallel to the gate roads in longwall conveyor mining. A common practice is to allow 5-yd (4.5-m) wastes between 4-yd (3.7-m) packs, or both are made 5 yd wide. The dimensions vary with local conditions. See also: *stowing method; double packing; single packing.* (Nelson, 1965)

stripped illite *degraded illite.*

stripped plain A plain composed of flat-lying or gently tilted sedimentary rocks from which sediments have been removed down to some resistant bed that has controlled the depth of erosion. Cf: *dip slope.*

stripper (a) A nearly depleted well whose income barely exceeds operating cost of production. (Wheeler, 1958) (b) In the quarry industry, a laborer who cleans up dirt left by the power shovel in stripping overlying ground from rock, using a shovel and wheelbarrow. (DOT, 1949)

stripping (a) The removal of earth or nonore rock materials as required to gain access to the desired coal, ore, or mineral materials; the process of removing overburden or waste material in a surface mining operation. (b) The earth, rock, or soil so removed. (c) The loading or clearing away of coal from a longwall face after shotfiring. (Nelson, 1965) (d) Opencast mining. (Nelson, 1965) (e) *stripping the quarry; strip.* (f) In chemical extraction of minerals, treatment of pregnant solution to remove dissolved values. (Pryor, 1963)

stripping a gutter Removing the headings from off the wash dirt, which is left undisturbed.

stripping a jig Aust. The forming of a jig, by enlarging a cut-through on an incline. See also: *jig.*

stripping a mine (a) *strip.* (b) Robbing a mine of its best ore.

stripping area In stripping operations, an area encompassing the pay material, its bottom depth, the thickness of the layer of waste, the slope of the natural ground surface, and the steepness of the safe slope of cuts. (Nichols, 1976)

stripping a shaft (a) Taking out the timber from an abandoned shaft. (b) Trimming or squaring the sides of a shaft.

stripping-pit limits The strip area that includes area of pay material plus enough area beyond the limits of the ore pit to provide for a bench. The total volume of stripping will be that vertically above the limits of the ore pit plus that outside of the ore pit necessary to maintain safe strip-pit slopes and benches and provide working approaches to the pit. (Jackson, 3)

stripping ratio The unit amount of spoil or overburden that must be removed to gain access to a unit amount of ore or mineral material, generally expressed in cubic yards of overburden to raw tons of mineral material.

stripping salt *abraumsalze.*

stripping shovel A shovel with an esp. long boom and stick that enables it to reach further and pile higher. (Nichols, 1954)

stripping-shovel operator In bituminous coal mining, one who operates a power shovel in a strip mine to strip back overlying ground and to load coal into cars. Also called coal-loading-shovel engineer; loading-shovel engineer. Syn: *boom cat.* (DOT, 1949)

stripping solution In solvent extraction, the aqueous solution used to re-extract the metal from the pregnant solution.

stripping system The removal of the overburden and mining of the ore in one or more benches, the ore face being broken by blasting and the broken ore loaded by hand, shoveling machine, or steam shovel. The name "terrace or bench open-pit working" has been suggested.

stripping the quarry The removal of all dirt and unwanted disintegrated material from the quarry face. *stripping.* (Nelson, 1965)

strip pit A coal or other mine worked by stripping. An open-pit mine.

strip sample A sample, making a notch or groove, cut from roof to floor of a coal seam, or from hanging wall to footwall of a vein. See also: *channel sample.*

strip thrust *décollement.*

stroke (a) The distance traveled by a piston in a pump or a piston in a hydraulic-feed mechanism on a drill. (Long, 1960) (b) The maximum distance a piston moves within a cylinder before the direction of its travel is reversed. (Long, 1960) (c) The distance a churn-drill stem and bit are raised for dropping while drilling. (Long, 1960)

stroke of crusher The difference between the open and closed positions measured at the throat of the crusher. For small crushers it is about 3/8 to 3/4 in (0.95 to 1.9 cm), and for large crushers from 1 to 2 in (2.5 to 5 cm). Also called throw of crusher. (Newton, 1938)

stromatite A type of mixed rock or chorismite in which the units of the fabric (granitic material and metamorphic or sedimentary rock) form a system of layers, strata, or bands so intricately united that the whole rather than the individual layers constitutes the geological field unit.

stromatolite A structure produced by sediment trapping and/or precipitation as a result of the growth of cyanophytes (blue-green algae). It has a variety of gross forms, from nearly horizontal to markedly columnar, domal, or subspherical. Syn: *algal stromatolite; stromatolith.* (AGI, 1987)

stromatolith (a) *stromatolite.* (b) A complex, sill-like igneous intrusion that is interfingered with sedimentary strata. (AGI, 1987)

stromeyerite An orthorhombic mineral, AgCuS; metallic; soft; steel gray with blue tarnish; sp gr, 6.2 to 6.3; in copper-silver veins; a source of copper and silver. Syn: *silver-copper glance.*

strong (a) Hard and thick; said of dikes. (Standard, 1964) (b) Important or rich; said of veins. (Standard, 1964) (c) Referring to the character of bind, meaning that the argillaceous material is largely mixed with the arenaceous or siliceous material. (d) Scot. Hard, not easily broken, e.g., strong coal, strong blaes. (e) Said of large or important mineral veins or faults. (AGI, 1987)

strongback A heavy timber or metal beam or bar for taking a strain. (Webster 3rd, 1966)

strong lode A large, persistent lode. At Alston moor, England, applied to lodes lying in a fault plane in which the difference of level between similar strata is considerable.

strontianite An orthorhombic mineral, $4[SrCO_3]$; aragonite group; in hydrothermal veins associated with limestones, less commonly with eruptive rocks in California, New York, Washington, Germany, and Mexico; a source of strontium. Syn: *carbonate of strontium.*

strontioginorite A monoclinic mineral, $(Sr,Ca)_2B_{14}O_{23}{\cdot}8H_2O$; forms a series with ginorite. Cf: *ginorite.*

strontium A silvery-white, alkaline-earth metal. Symbol, Sr. It does not occur naturally; found chiefly as celestite ($SrSO_4$) and strontianite ($SrCO_3$). Major use at present is for color television picture tubes. (Handbook of Chem. & Phys., 3)

strontium-apatite A hexagonal mineral, $Sr_5(PO_4)_3(OH)$; apatite group; vitreous; light green; in sugary albite filling interstices between crystals of aegirine and eckermannite in veins in alkalic pegmatites from Inagil massif, southern Yakutia, Russia.

strontium minerals Used primarily in television face-plate glass, in ceramic ferrites, and in pyrotechnics. The United States imports most of its celestite, the chief source of strontium, from Mexico and Germany.

strontium stannate $SrSnO_3$; sometimes used as an additive to titanate bodies, one result being a decrease in the Curie temperature. (Dodd, 1964)

strontium titanate $SrTiO_3$; isometric; and melting point, 1,670 °C. Used in ceramic dielectric bodies, either alone or in combination with barium titanate or other titanates. (Lee, 1961; Handbook of Chem. & Phys., 2)

strontium zirconate $SrZrO_3$; melting point, 2,700 °C; sp gr, 5.48. Sometimes used in small amounts (3% to 5%) in ceramic dielectric bodies, one effect being to lower the Curie temperature. (Dodd, 1964)

struck capacity (a) The capacity of a mine car, tram, hoppit, or wagon to the flat surface at the edges; i.e., the volume of water it would hold if of watertight construction. (Nelson, 1965) (b) In scraper loading, the maximum volume of liquid that the bowl can hold. Cf: *heaped capacity.* (Carson, 1961)

struck out Corn. The termination of a vein or lode by a fault.

structural Of or pertaining to rock deformation or to features that result from it. (AGI, 1987)

structural analysis The initial stage of structural design, in which all the forces carried by the various parts of a structure are determined. (Hammond, 1965)

structural basin A low area in the Earth's crust, of tectonic origin, in which sediments have accumulated, e.g., a circular centrocline such as the Michigan Basin, a fault-bordered intermontane feature such as the Bighorn Basin of Wyoming, or a linear crustal downwarp such as the Appalachian Basin. Such features were drainage basins at the time of sedimentation, but are not necessarily so today. (AGI, 1987)

structural bottoming *bottoming.*

structural color *pseudochromatism.*

structural control The influence of structural features on ore deposition, e.g., ore minerals filling fractures. Cf: *stratigraphic control.* (AGI, 1987)

structural dome *dome.*

structural drilling Drilling done specif. to obtain detailed information delineating the location of folds, domes, faults, and other subsurface structural features undiscernible by studying strata exposed at the surface. Cf: *structure drilling.* (Long, 1960)

structural fabric *fabric.*

structural geology The branch of geology that deals with the form, arrangement, and internal structure of the rocks, and esp. with the description, representation, and analysis of structures, chiefly on a moderate to small scale. The subject is similar to tectonics, but the latter is generally used for the broader regional or historical phases. (AGI, 1987)

structural high High point of a structure. See also: *high.*

structural load The load due to the structure itself as distinguished from the imposed load. (Crispin, 1964)

structural log A record of the breaks, fractures, faults, and physical properties of rocks within a formation.

structural low Low point of a structure. See also: *low.*

structural petrology The analysis of fabric on the thin-section or micro scale. It includes the study of grain shapes and relationships (microstructure) and the study of crystallographic preferred orientations. The transmission electron microscope is also employed to examine the substructures of deformed crystals. (AGI, 1987)

structural plain *stripped plain.*

structural relief (a) The vertical distance between stratigraphically equivalent points at the crest of an anticline and in the trough of an adjacent syncline. (AGI, 1987) (b) More generally, the difference in elevation between the highest and lowest points of a bed or stratigraphic horizon in a given region. (AGI, 1987)

structural steelwork Rolled steel sections or other fabricated members assembled to form structural frames by riveting, welding, bolting, or a combination of all three. (Hammond, 1965)

structural terrace (a) A local shelf or steplike flattening in otherwise uniformly dipping strata, composed of a synclinal bend above and an anticlinal bend at a lower level. (AGI, 1987) (b) A terracelike landform controlled by the structure of the underlying rocks; esp. a terrace produced by the more rapid erosion of weaker strata lying on more resistant rocks in a formation with horizontal bedding. (AGI, 1987)

structural valley A valley that owes its origin or form to the underlying geologic structure. (AGI, 1987)

structural vitrain *phyllovitrinite.*

structure (a) The parts or members of any building that carry the loads and transmit them to the foundations. Structures in mining areas may suffer some subsidence and are designed accordingly. (Nelson, 1965) (b) Geologically, the disposition of the rock formations; i.e., the broad dips, folds, faults, and unconformities at depth. (Nelson, 1965) (c) In petrology, one of the larger features of a rock mass, like bedding, flow banding, jointing, cleavage, and brecciation; also, the sum total of such features. Cf: *texture.* (d) *soil structure.*

structure contour A contour that portrays a structural surface such as a formation boundary or a fault. Syn: *subsurface contour.* Cf: *contour.* (AGI, 1987)

structure-controlled shoot An ore shoot that is localized by geologic structure. Changes in strike and dip of fissures are favorable sites for ore shoots.

structure drilling (a) Exploratory drilling to determine the geological structure with reference to the coal or minerals sought. Rotary or diamond drilling is usually employed to yield cores at key horizons. (Nelson, 1965) (b) A form of drilling practiced in the Lake Superior iron district to sample soft iron formations by countercirculation-wash boring methods. Cf: *structural drilling.* (Long, 1960)

structure section A diagram to show the observed geological structure on a vertical face or, more commonly, to show the inferred geological structure as it would appear on the side of a vertical trench cut into the earth. The vertical scale is often exaggerated.

structure test hole A hole drilled for geologic structure alone, although other types of information may be acquired during the drilling. This type of hole is drilled to a structural datum, which is normally short of known or expected producing zones. See also: *seismic shothole; slim hole.* (Williams, 1964)

strunzite A triclinic mineral, $MnFe_2(PO_4)_2(OH)_2{\cdot}6H_2O$; pseudomonoclinic; forms straw-yellow radiating fibers from the alteration of triphylite.

strut (a) A piece of wood or steel inserted between each pair of steel or timber supports on roadways to resist buckling and to maintain the proper spacing between the sets. (Nelson, 1965) (b) A mine prop to sustain compression, whether vertical or inclined. (Zern, 1928) (c) An inside brace. (Nichols, 1976) (d) A diagonal brace between two legs of a drill tripod or derrick; also, a vertical-compression member in a structure or in an underground timber set. (Long, 1960)

strut tenon A tenon, such as is used on a diagonal piece or strut, usually on heavy timbers. (Crispin, 1964)

Struve ventilator A pneumatic ventilating apparatus consisting of two vessellike gas holders, which are moved up and down in a tank of water. By this means, air is sucked out of a mine. (Zern, 1928)

struvite An orthorhombic mineral, $(NH_4)MgPO_4{\cdot}6H_2O$; soft; has one perfect and one good cleavage.

stub entry A short, narrow entry turned from another entry and driven into the solid coal, but not connected with other mine workings; a dead end. Syn: *dead end.* (Fay, 1920)

Stub's gage A gage for measuring the size of wire. Also known as Birmingham gage. (Crispin, 1964)

stub switches Switches used to some extent on narrow-gage industrial tramways; consisting of a pair of short switch rails, held only at or near one end and free to move at the other end to meet rails of straight or diverging track. (Urquhart, 1959)

stud (a) A bolt having one end firmly anchored. (Nichols, 1976) (b) A threaded rod or a bolt without a head. (Hammond, 1965) (c) An upright beam or scantling as in the framework of a dwelling. (Crispin, 1964)

studdle (a) A strong crossbeam in a shaft collar set. (Nelson, 1965) (b) Corn. A prop to support the middle of a stull. (c) A distance piece between successive frames of timbering. (d) A vertical member of a shaft-timber set. The sets are placed at each corner and at the intersection of the dividers and the wallplates. (e) An upright prop supporting a platform in a mine, usually one of a set of four. (Standard, 1964)

stud-type chain A roller chain in which the inner (block) links are connected solidly by nonrotating bushings. (Nichols, 1976)

stuffed (a) A mineral having foreign ions in large interstices in its structure; e.g., garnet with an extra cation. (b) A mineral structure derived from another by coupled replacement of a cation with two cations of lower valence, one of which occupies a structural cavity; e.g., feldspars derived from coesite by aluminum ions replacing silicon in the framework coupled with alkali or alkali earth ions in structural cavities.

stuffed mineral A mineral having large interstices in its structure may accommodate various foreign ions in these holes; such a mineral is then said to be stuffed. The stuffing may have considerable consequences on the stability of the mineral. (AGI, 1987)

stuffing box A chamber designed to contain packing and to maintain a fluid-tight joint about a piston rod where it enters a cylinder or around a drill rod where it enters the casing at the collar of a borehole. (Long, 1960)

stull (a) A timber prop set between the walls of a stope, or supporting the mine roof. (b) A timber platform on which valueless rock or mineral is deposited. (c) Corn. A platform (stull-covering) laid on timbers (stullpieces), braced across a working from side to side, to support workers or to carry ore or waste. (d) A round timber used to support the sides or back working of a mine. (Mersereau, 1947)

stull covering A platform resting on stulls in a stope as a stage for miners or for holding rock or mineral. (Nelson, 1965)

stulled stopes Stopes in which the roof is supported by stulls, square-set timbering, or concrete columns. (Stoces, 1954)

stull piece A piece of timber placed slanting over the back of a level to prevent rock falling into the level from the stopes above. (Standard, 1964)

stull stoping The walls of narrow veins frequently are supported by stull timbers placed between the foot and hanging walls, which constitute the only artificial support provided during the excavation of the stopes. Stulls may be placed at irregular intervals to support local patches of insecure ground, in which case the stopes are virtually open stopes. Sometimes the stulls are placed at regular intervals both along the stope and vertically, in which case stull stoping should be considered a distinctive method.

stull timbering The support of walls in shrinkage stoping by setting stulls. (Nelson, 1965)

stulm An approx. horizontal passageway into a mine; an adit. Taken from the German term stollen. (Webster 3rd, 1966)

stump (a) Entry pillars; small portion of room pillars left for pick mining. (BCI, 1947) (b) A small pillar of coal left between the gangway or airway and the breasts to protect these passages; any small pillar. (Fay, 1920) (c) A narrow pillar of coal. (Lewis, 1964)

stumper A narrow heavy dozer attachment used in pushing out stumps. (Nichols, 1976)

stump prop Short posts set under the crown bars of a tunnel. (Stauffer, 1906)

stump pulling *pillar robbing.*

stunning A quarry worker's term for the formation of fractures caused by the cutting bars of a channeling machine striking the rock with excessively heavy blows.

Sturtevant balanced rolls Rolls in which all four boxes are moveable in position by springs to divide the thrust whenever the springs yield and thus reduce internal stresses. (Liddell, 1918)

Sturtevant grinder A disk grinder in which one disk is stationary and the other rotates. The stationary disk is moved out of center from time to time, so that any groove that forms can be ground out. (Liddell, 1918)

Sturtevant ring roll crusher A crusher similar to the Kent roller mill. (Liddell, 1918)

Sturtevant roll jaw crusher A crusher in which the motion of the upper part of the jaws is like that of the Dodge crusher, while the lower parts of the jaws, of cylindrical surfaces of varying radii, grind the ore between them. (Liddell, 1918)

Sturzelberger iron reduction process A process evolved for dealing with pyrite roasting residues rich in zinc (8% to 10%). The direct reduction takes place in a short rotary drum that has a rammed tar-dolomite lining. Lime is added to produce a highly basic slag; the pyrite cinders, precalcined, are mixed with coke breeze and fired with pulverized coal burners. The zinc is recovered from the waste gases. The drum works discontinuously in 7-h heats, and the capacity of such a plant is limited in comparison with a blast furnace. The product is a liquid pig iron. (Osborne, 1956)

Stygian deposits A general term for ore deposits formed underground by waters of atmospheric origin.

stylolite A surface or contact, usually in carbonate rocks, that is marked by an irregular and interlocking penetration of the two sides: the columns, pits, and teethlike projections on one side fit into their counterparts on the other. As usually seen in cross section, it resembles a suture or the tracing of a stylus. The seam is characterized by a concentration of insoluble constituents of the rock; e.g., clay, carbon or iron oxides, and is commonly parallel to the bedding. Etymol: Greek stylos, pillar, + lithos, stone. Syn: *crowfoot; suture joint.* (AGI, 1987)

stylotypite Syn: *tetrahedrite.* Formerly applied to tetrahedrite containing appreciable silver.

Styrian jade Pseudophite, a compact massive variety of clinochlore.

stythe (a) Gr. Brit. Carbonic acid gas, often found in old workings and given off in most shallow mines. Also spelled stithe. (Tomkeieff, 1954) (b) *blackdamp.*

sub- A prefix denoting under, below, or less than. Containing only a relatively small proportion or less than the normal amount of (such) an element or radical; not used systematically. (Webster 3rd, 1966)

subaeration cells *sub-A flotation cells.*

subaeration method In flotation, a method employing an impeller, of which the principal function is to keep the pulp in suspension, and a port for admission of air below the surface of the pulp, this port of entry being in the vicinity of the impeller. (Gaudin, 1957)

subaerial Said of conditions and processes, such as erosion, that exist or operate in the open air on or immediately adjacent to the land surface; or of features and materials, such as eolian deposits, that are formed or situated on the land surface. The term is sometimes considered to include fluvial. Cf: *subaqueous; subterranean.* See also: *surficial.* (AGI, 1987)

sub-A flotation cells Those in which air is supplied direct to a rotary agitator, mechanically driven, situated at depth in the flotation cell so that air is churned with ore pulp. Two types are (1) that in which air is drawn down by impeller; and (2) that in which low-pressure air is blown in. Syn: *subaeration cells.* (Pryor, 1963)

subalkalic (a) A group term applied to rocks of the tholeiitic and calc-alkaline series. (AGI, 1987) (b) Said of an igneous rock that contains no alkali minerals other than feldspars. (AGI, 1987) (c) Used to describe an igneous rock of the Pacific suite. (AGI, 1987)

subangular Somewhat angular, free from sharp angles but not smoothly rounded; specif. said of a sedimentary particle showing definite effects of slight abrasion, retaining its original general form, and having faces that are virtually untouched and edges and corners that are rounded off to some extent. (AGI, 1987)

subaqueous Said of conditions and processes, or of features and deposits, that exist or are situated in or under water, esp. freshwater, as in a lake or stream. Cf: *subaerial.* (AGI, 1987)

subaqueous disposal A mine waste management practice whereby mill tailings or mine waste rock are emplaced under a body of water to minimize sulfide oxidation and acid formation, and eliminate airborne particulate pollution. If a natural sedimentation basin is utilized, erosion hazards inherent to subaerial impoundments are eliminated.

subaqueous mining Surface mining in which the material mined is removed from the bed of a natural body of water. (AIME, 2)

subarid *semiarid.*

subarkose A sandstone that does not have enough feldspar to be classed as an arkose, or a sandstone that is intermediate in composition between arkose and pure quartz sandstone. Cf: *arkose.* (AGI, 1987)

subaudible noise Noise the intensity of which is so low that it can only be detected by means of a microphone and suitable amplifying equipment. How these subaudible noises, or microseisms, originate has not definitely been established, but it is believed that they are produced by incipient cracking or intermovement between fragments of crystalline aggregates in the rock. It has been found that almost invariably a period of increased microseismic activity precedes any large-scale ground movement or failure. (Issacson, 1962)

subbase (a) A layer of material laid on the natural ground under a road base for purposes of strengthening. Cf: *subgrade.* (Hammond, 1965) (b) The lowest part of a base. (Crispin, 1964)

subbituminous A coal The rank of coal, within the subbituminous class of Classification D 388, such that, on the moist, mineral-matter-free basis, the gross calorific value of the coal in British thermal units per pound is equal to greater than 10,500 (24.42 MJ/kg) but less than 11,500 (26.75 MJ/kg) and the coal is nonagglomerating. (ASTM, 1994)

subbituminous B coal The rank of coal, within the subbituminous class of Classification D 388, such that, on the moist, mineral-matter-free basis, the gross calorific value of the coal in British thermal units per pound is equal to greater than 9,500 (22.10 MJ/kg) but less than 10,500 (24.42 MJ/kg) and the coal is nonagglomerating. (ASTM, 1994)

subbituminous C coal The rank of coal, within the subbituminous class of Classification D 388, such that, on the moist mineral-matter-free basis, the gross calorific value of the coal in British thermal units per pound is greater than 8,300 (19.31 MJ/kg) but less than 9,500 (22.10 MJ/kg), and the coal is nonagglomerating. (ASTM, 1994)

subbituminous coal (a) Black lignite or lignitic coal. (Fay, 1920) (b) Coal of rank intermediate between lignite and bituminous. In the specifications adopted jointly by the American Society for Testing and Materials (D388-38) and the American Standards Association (M20.1-1938), subbituminous coals are those with calorific values in the range 8,300 to 13,000 Btu (19.3 to 30.2 MJ/kg), calculated on a moist, mineral-matter-free basis, which are both weathering and nonagglomerating according to criteria in the classification. (Stokes, 1955)

subconchoidal Partially or indistinctly conchoidal. (Webster 3rd, 1966)

subcooling Cooling of a liquid refrigerant below the condensing temperature at constant pressure. (Strock, 1948)

subcritical area of extraction An area of goaf too small to cause full subsidence at the surface. See also: *critical area of extraction.* (Nelson, 1965)

subdeposit level *group level.*

subdrift caving *sublevel caving.*

subdrifting and caving *top slicing combined with ore caving.*

subdrilling *underdrilling.*

subduction The process of one lithospheric plate descending beneath another. See also: *obduction.* (AGI, 1987)

subduction zone A long, narrow belt in which subduction takes place; e.g., along the Peru-Chile trench or in the volcanic arc belts of the western Pacific Ocean. (AGI, 1987)

subeconomic resources The part of identified resources that does not meet the economic criteria of reserves and marginal reserves. (USGS, 1980)

suberain A subvariety of provitrain in which the corky origin of the cellular structure is microscopically visible. Cf: *periblain; xylain.* (AGI, 1987)

suberinite (a) A variety of provitrinite characteristic of suberain and consisting of corky tissue. (AGI, 1987) (b) A maceral of brown coal and lignite derived from the suberin layer in corkified cell walls of some Mesozoic and younger plants. Cf: *periblinite; xylinite; telinite.* (AGI, 1987)

subfeldspathic (a) Said of a mature lithic wacke (or lithic graywacke) in which quartz grains and fragments of siliceous and argillaceous rocks predominate, and feldspars make up less than 10% of the rock and may be altogether lacking. Such rocks have also been called subgraywackes. (AGI, 1987) (b) Said of a mature lithic arenite containing abundant quartz grains and fragments of the more stable rocks (such as cherts), and less than 10% feldspar grains. (AGI, 1987)

subgrade A layer, stratum, or material immediately beneath some principal surface; specif. a layer of earth or rock that is graded to receive the foundation of an engineering structure. Often it is the soil or natural ground that is prepared and compacted to support, and that lies directly below, a road, pavement, building, airfield, or railway. Cf: *subbase.* (AGI, 1987)

subgrade surface The surface of the earth or rock prepared to support a structure or a pavement system. (ASCE, 1958)

subgraywacke The most common type of sandstone, intermediate in composition between orthoquartzite and graywacke (Pettijohn, 1957). Composition is typically 30% to 65% quartz and chert, less than 15% clay matrix, more than 25% unstable materials such as feldspar grains and rock fragments, and voids or mineral cement exceeding the amount of clay matrix. The rock is lighter colored and better sorted, and has less matrix, than graywacke. Cf: *graywacke.*

subhedral (a) Said of a mineral grain that is bounded partly by its own rational faces and partly by surfaces formed against preexisting grains as a result of either crystallization or recrystallization. (AGI, 1987) (b) Said of the shape of such a crystal, intermediate between euhedral and anhedral. (AGI, 1987)

subhydrous macerals Macerals having a low hydrogen content, such as fusinite. (Tomkeieff, 1954)

subincline An inclined shaft along the footwall of a reef on the Rand. It develops and extracts ore from areas below the main haulage level. See also: *staple shaft.* (Nelson, 1965)

subindividual One of the small crystals that often unite in parallel growths to build up larger crystals of the same general habit. (Standard, 1964)

subjacent Said of a stratum situated immediately under a particular higher stratum or below an unconformity. Ant: superjacent. Syn: *underlying*. (AGI, 1987)

subjective brightness The subjective brightness of a surface is determined by two factors, the light flux radiated from the surface and the sensitivity of the eye under the conditions in which the surface is seen. The sensitivity of the eye is partly controlled by the contrasts presented over the visual field, but is mainly dependent on its adaptation brightness level. (Roberts, 1958)

sublevel (a) A secondary level for working ore in top slicing and sublevel caving; a companion heading. (Nelson, 1965) (b) An intermediate level opened a short distance below the main level; or, in the caving system of mining, 15 to 20 ft (4.6 to 6.1 m) below the top of the orebody, preliminary to caving the ore between it and the level above. See also: *sublevel stoping; caving system.*

sublevel backstoping *sublevel stoping.*

sublevel caving (a) A stoping method in which relatively thin blocks of ore are caused to cave by successively undermining small panels. The ore deposit is developed by a series of sublevels spaced at vertical intervals of 18 to 25 ft or 30 ft (5.5 to 7.6 m or 9.1 m) and occasionally more. Usually only one or two sublevels are developed at a time, beginning at the top of the orebody. The sublevels are developed by connecting the raises with a longitudinal subdrift from which timbered slice drifts are driven right and left opposite the raises to the ore boundaries or to the limits of the block. Usually alternate drifts are driven first, and caving back from them is begun and continued while the intermediate slices are being driven. The caving is begun at the ends of the slices by blasting out cuts and retreating in the same manner toward the raises. The broken and caved ore formerly was shoveled into cars and trammed to the raises, but in recent years it is dragged to the raises by power scrapers. Successively lower sublevels are developed and caved back until the entire block has been mined. This method is intermediate between block caving and top slicing, since part of the ore is mined as in top slicing and part is caved. See also: *top slicing combined with ore caving.* (b) Similar to top slicing from which it is thought to have been developed. The general plan of operations is to mine every other slice by driving crosscuts (slice drifts) from 18 to 36 ft (5.5 to 11.0 m) apart. The ore between the crosscuts as well as that in the slice above is then mined, thus causing the overlying material to cave. The method is applicable to irregular and steeply dipping orebodies that cannot be worked by top slicing. The present tendency is to sink vertical shafts in the footwall rather than inclined shafts as formerly done. Also called subdrift caving. (Lewis, 1964)

sublevel drive A drive often made in a section, esp. in gently inclined deposits, that divides the deposit into narrower panels and zones. They are narrower, and the support and equipment for them is more simple than that required in level drives. (Stoces, 1954)

sublevel method *sublevel stoping.*

sublevel slicing *top slicing combined with ore caving; sublevel stoping.*

sublevel stoping (a) In this mining method, the ore is excavated in open stopes, retreating from one end of the stope toward the other. The orebody is developed first by a series of sublevel drifts above the main haulage level. The sublevels are connected by a starting raise at one end of the stope and by a passageway raise for entrance to them and the stope face at the other end. Chute raises connect the haulage level to the lowest sublevel, at which the tops of the chute raises are belled out to form mill holes. Beginning at the starting raise the ore is benched down from the sublevels; the broken ore falls into the mill holes, where it is drawn off through the chutes. The stope face is kept nearly vertical as it is benched backward toward the passageway raise. (b) A mining method involving overhand, underhand, and shrinkage stoping. Its characteristic feature is the use of sublevels. The sublevels are worked simultaneously, the lowest on a given block being farthest advanced and the subs above following one another at short intervals. The uppermost sublevel underneath the cover is partly caved. The caved cover follows down upon the caved ore. The broken ore is in part drawn from the level, and a part remains in the stope to give lateral support to the walls and to prevent admixture of cover and ore. The breaking faces are developed by crosscuts, which are extended from wall to wall from the end of the sublevel. The method can also be looked upon as a retreating method, the orebody being worked from the top down and the individual blocks upon a given level being worked from their ends to the center. Modifications of this method are chamber-and-pillar system; chambers without filling; combination of subslicing and stoping; drift stoping; filling system; Mitchell slicing system; pillar robbing; pillar robbing and hand filling; room-and-pillar system; square work and caving; square work, pillar robbing, and hand filling; sublevel back stoping; sublevel method; sublevel slicing system; substoping. (c) A method of mining best adapted to steeply inclined deposits that have strong ore and strong walls. The ore is usually blocked out by two horizontal drifts separated vertically by 100 to 200 ft (30 to 61 m) and raises between the two horizontal drifts, the latter separated by comparable distances. Vertical pillars may be left between stopes on the same level, and horizontal ones to support the main haulage. After the main blocks of ore have been completely mined, it is common practice to rob the pillars, and the walls of the stope may collapse after the pillars have been robbed. (Lewis, 1964) (d) Of lodes, open-stope mining in which ore is blasted and drawn through footwall openings to a gathering level in the country rock below. Used with strong containing walls and wide lodes. (Pryor, 1963) (e) Of massive deposits, working simultaneously of a series of sublevels echeloned vertically, the lowest leading and the uppermost being partly caved as the covered rock descends. (Pryor, 1963) (f) *sublevel backstoping; substoping.*

sublimate (a) A coating or deposit formed in a glass tube or on charcoal as a result of heating certain minerals. (b) The product of sublimation. (CTD, 1958) (c) A solid deposit by a gas or vapor; commonly used in reference to material deposited by volcanic gases. (AGI, 1987)

sublimation (a) The process by which a solid substance vaporizes without passing through a liquid stage. Cf: *evaporation.* (AGI, 1987) (b) The process of ore deposition, as of sulfur or mercury, by vapors; the volatilization and transportation of minerals followed by their deposition at reduced temperatures and pressures. Sublimation deposits are generally associated with fumarolic activity. (AGI, 1987)

sublimation theory The theory that a vein was filled first with metallic vapors.

sublime To cause to pass from the solid state to the vapor state by the action of heat and again to condense to solid form.

submarginal land Generally means land not very good for farming. Tens of millions of tons of coal are removed each year by stripping the surface of this type land. (Kentucky, 1952)

submarine blast A charge of high explosives fired in boreholes drilled in the rock underwater for dislodging dangerous projections and deepening channels. (Fay, 1920)

submarine canyon An elongated, steep-walled cleft running across or partially across the Continental Shelf, the continental borderland and/or slope, the bottom of which grades continually downwards. (AGI, 1987)

submarine drilling Drilling from the surface of a body of water with a drill mounted on an anchored tower, platform, or barge. (Long, 1960)

submarine mines Workings that follow the mineral under the sea. Syn: *undersea workings.* (Zern, 1928)

submarine packing Special heavy paper shells in which dynamite is packed in underwater blasting. (Carson, 1965)

submarine throat A throat with the fluid level below the bottom of a melter. See also: *throat.*

submerged unit weight The weight of the solids in air minus the weight of water displaced by the solids per unit of volume of soil mass; the saturated unit weight minus the unit weight of water. See also: *unit weight.* (ASCE, 1958)

submergence (a) In an air lift, the distance below the water level, during pumping, at which the air picks up water. (Lewis, 1964) (b) A term that implies that part of a land area has become inundated by the sea but does not imply whether the sea rose over the land or the land sank beneath the sea. (AGI, 1987)

submetallic luster A luster between metallic luster and nonmetallic luster. (Hurlbut, 1964)

subophitic Said of the texture of an igneous rock in which the feldspar crystals are approx. the same size as the pyroxene and are only partially included by them. The term ophitic generally includes such textures. (AGI, 1987)

suboutcrop (a) S. Afr. Rock that would have been the outcrop on the surface, but is covered by other formations. (Beerman, 1963) (b) Area of intersection of a geologic feature with the surface of bedrock beneath the regolith. (Hawkes, 1957)

subsample In coal and coke sampling, part of the sample consisting of a number of increments spaced evenly over the unit. (BS, 1960)

subsequent ore deposit *epigenetic.*

subsidence The sudden sinking or gradual downward settling of the Earth's surface with little or no horizontal motion. The movement is not restricted in rate, magnitude, or area involved. Subsidence may be caused by natural geologic processes, such as solution, thawing, compaction, slow crustal warping, or withdrawal of fluid lava from beneath a solid crust; or by human activity, such as subsurface mining or the pumping of oil or groundwater. See also: *shift; cauldron subsidence; settlement.* Syn: *land subsidence; bottom subsidence.* (AGI, 1987)

subsidence area The area affected by subsidence over areas where minerals or other substances have been removed. The area is larger than the mined-out area below. (AGI, 1987)

subsidence basin A shallow troughlike depression at the surface resulting from subsidence. (Nelson, 1965)

subsidence break A fracture in the rocks overlying a coal seam or mineral deposit as a result of its removal by mining operations. The subsidence break usually extends from the face upward and backward over the unworked area. (Nelson, 1965)

subsidence factor Full subsidence expressed as a fraction of the thickness of coal seam extracted. See also: *maximum subsidence.* (Nelson, 1965)

subsidiary fracture Minor breaks sometimes developed in the rocks along a fault plane. They often indicate the general direction of movement and were caused by differential tension in the rocks contiguous to the main fault plane. *tension fracture.* (Nelson, 1965)

subsidiary survey An underground survey made to determine the position of a faceline or goafline or some other specific feature. (BS, 1963)

subsidiary transport The conveying or haulage of coal or mineral along the working faces and outward to a junction or loading point. See also: *main transport; underground haulage.* (Nelson, 1965)

subsieve analysis In powder metallurgy, size distribution of particles all of which will pass through a 44-µm (No. 325) standard sieve, as determined by specified methods. (ASM, 1961)

subsieve material In mineral dressing, material finer than 400 mesh that must be sized by elutriation in rising currents of water or air, by microscopic counts, or other methods. (Newton, 1959)

subsieve sizes Particle sizes too small for efficient grading on screens; usually minus 200-mesh material. They are examined by elutriation (beaker decantation, sedimentation, infrasizing in air, turbidimetry, permeability). (Pryor, 1963)

subsilicate A basic silicate.

subsilicic A term proposed by Clarke (1908) to replace basic. Cf: *persilicic; mediosilicic.* (AGI, 1987)

subslicing *top slicing combined with ore caving.*

subsoil drainage The removal of subsoil water by open intercepting ditches and drain pipes. The distance between subsoil drains is a maximum in sandy soils and a minimum in clay. See also: *catchwater drain.* (Nelson, 1965)

subsoiling The firing of small charges of dynamite 2 ft or 3 ft (0.6 m or 0.9 m) below the surface for breaking up impervious strata of soil, clay, etc., for aerating, draining, and moistening the soil.

subsoil plow A one-tooth ripper designed for agricultural work. Also called pan breaker. (Nichols, 1976)

substage (a) A subdivision of a stage; the rocks formed during a subage of geologic time. (AGI, 1987) (b) In a microscope, a mechanism for holding polarizers or other attachments below the stage. (AGI, 1987)

substation An electrical installation containing generating or power-conversion equipment and associated electric equipment and parts, such as switchboards, switches, wiring, fuses, circuit breakers, compensators, and transformers.

substitute Any substance represented to be, or used to imitate, a gemstone; e.g., plastic, glass, doublet, synthetic ruby, or natural spinel; all could be substitutes for natural ruby. (AGI, 1987)

substitution A chemical defect wherein one ion replaces another in a crystal structure. Substitution may be partial; e.g., iron for zinc in sphalerite up to 30%, or complete; e.g., manganese and iron in the series rhodochrosite-siderite.

substoping (a) An open stope method of mining employed in wide orebodies with strong walls. (Nelson, 1965) (b) *sublevel stoping.*

substrate (a) A layer of metal underlying a coating, regardless of whether the layer is basis metal. (ASM, 1961) (b) The true lattice of a crystal, as distinct from its discontinuity lattice, or surface. (Pryor, 1960)

substructure (a) That part of any structure that is below ground, more particularly the foundations. The latter may take many forms, according to the nature and bearing strength of the ground. (Hammond, 1965) (b) The lower portion of a structure upon which something else is built up. (Crispin, 1964)

subsurface (a) The zone below the surface, in which geologic features, principally stratigraphic and structural, are interpreted on the basis of drill records and various kinds of geophysical evidence. (AGI, 1987) (b) Rock and soil materials lying beneath the Earth's surface.—adj. Formed or occurring beneath a surface, esp. beneath the Earth's surface. Cf: *surficial.* See also: *subterranean.* (AGI, 1987) (c) An underground workplace. (Long, 1960)

subsurface contour *structure contour.*

subsurface correlation Correlation of rock units and structures that do not appear at the surface, by means of well logs, mine maps, and geophysical data. (Stokes, 1955)

subsurface corrosion Formation of isolated particles of corrosion products beneath the metal surface. This results from the preferential reaction of certain alloy constituents by inward diffusion of oxygen, nitrogen, and sulfur. (ASM, 1961)

subsurface geology Geology and correlation of rock formations, structures, and other features beneath the land or sea-floor surface as revealed or inferred by exploratory drilling, underground workings, and geophysical methods. Cf: *surface geology.* (AGI, 1987)

subsurface injection *subsurface waste disposal.*

subsurface map (a) A map depicting geologic data or features below the Earth's surface; esp. a plan of mine workings, or a structure-contour map of a petroleum reservoir or an underground ore deposit, coal seam, or key bed. (AGI, 1987) (b) A plane surface representation, generally in horizontal projection, of geologic data or features beneath the Earth's surface. There are many types of subsurface maps, such as structure contour maps, isopachous maps, and maps showing variations in lithology, or proportions of different types of lithology in rocks not exposed at the surface. (Stokes, 1955)

subsurface waste disposal Waste disposal in which manufacturing wastes are deposited in porous underground rock formations. Disposal wells should be at least 200 ft (61 m) deeper than the deepest water-bearing formation, and they must be sealed with cement from top to bottom. Also called subsurface injection.

subsurface water Water in the lithosphere in solid, liquid, or gaseous form. It includes all water beneath the land surface and beneath bodies of surface water. Syn: *subterranean water; underground water; ground water.* (AGI, 1987)

subtense bar A horizontal bar used in the subtense system of surveying by tacheometry. It is held at a distant point and its distance is calculated from its known length and the angle that it subtends at the observer's eye. See also: *tachymeter.* (Hammond, 1965)

subterrain *subterrane.*

subterrane The bedrock beneath a surficial deposit or below a given geologic formation. Syn: *subterrain.* Adj. subterranean. (AGI, 1987)

subterranean Formed or occurring beneath the Earth's surface, or situated within the Earth. Cf: *subaerial.* See also: *subsurface.* Syn: *subterrestrial; subterrane.* (AGI, 1987)

subterranean stream A body of subsurface water flowing through a cave or a group of communicating caves, as in a karstic region. (AGI, 1987)

subterranean water *ground water; subsurface water.*

subterrestrial *subterranean.*

subtranslucent *semitranslucent.*

subtransparent Imperfectly or partially transparent; semitransparent. (Webster 3rd, 1966)

subvitreous Not quite vitreous. (Webster 3rd, 1966)

subvolcanic *hypabyssal.*

subweathering Below weathering. Pertaining to the consolidated material-bedrock or high-velocity weathered layer or zone. This velocity is distinctly greater than that in the weathered zone. (AGI, 1987)

succinite (a) An old name for amber, esp. amber mined in former East Prussia (Poland) or recovered from the Baltic Sea. (b) A light yellow, amber-colored variety of grossular garnet.

suck The shape of the bottom of a cutting edge or tooth that tends to pull it into the ground as it is moved. (Nichols, 1976)

sucked stone Corn. A honeycombed or porous stone.

sucking pump A suction pump. (Standard, 1964)

suction (a) Atmospheric pressure pushing against a partial vacuum. (Nichols, 1976) (b) The pull of a pump. (Nichols, 1976) (c) Adhesion of a mass of mud to the underside of an object being lifted out of it. (Nichols, 1976)

suction anemometer An anemometer that measures wind velocity by the degree of exhaustion caused by wind blowing through or across a tube. (Standard, 1964)

suction bailer *sand pump.*

suction basket The strainer at the foot of the suction pipe of a pump or of a suction hose. (Standard, 1964)

suction blast *backlash.*

suction chamber A suction chamber is designed to provide a trough of low pressure between the sealed area and the intake, so that air that would otherwise be drawn into the sealed area, through fissures and pores surrounding the seal, is drawn instead into the chamber. If the pressure within the sealed-off mine area is less than that outside, it is necessary to reduce the pressure within the chamber. This can be done by using an ejector or fan to draw air through a pipe in the outer wall of the chamber; a second pipe, fitted with a control valve, serves as an intake. Adjustment of the valve gives regulation of the chamber pressure within fine limits. (Roberts, 1960)

suction cutter (a) In dredging, use of pump fed by pipe with power-rotated cutting blades to lift spoil. (Pryor, 1958) (b) In alluvial dredging, use of power-rotated cutting shoe to detach minerals from deposit, followed by their delivery by suction and elevation through a centrifugal pump. (Pryor, 1963)

suction-cutter dredge A dredge in which rotary blades dislodge the material to be excavated, which is then removed by suction, as in a sand-pump dredger. (CTD, 1958)

suction dredge (a) A dredge that digs by means of powerful suction pumps, the semiliquid spoil thus raised being frequently conveyed away in a floating pipeline. (Hammond, 1965) (b) *sand-pump dredger.*

(CTD, 1958) (c) A dredge in which the material is lifted by pumping through a suction pipe.
suction fan A fan that sucks or draws the air toward it through airways or air pipes. The term generally used is exhaust fan. (Nelson, 1965)
suction head The head or height to which water can be raised on the suction side of the pump by atmospheric pressure. See also: *lift pump.*
suction lift In pump nomenclature, it exists when the liquid level is below the pump centerline and/or when a gage on the suction would show a vacuum.
sudburite An augite-bearing hypersthene basalt characterized by pillow structure and containing bytownite and magnetite. It differs from normal basalts in containing neither glass nor olivine and in having an equigranular texture (Johannsen, 1937). Its name, given by Coleman in 1914, is derived from the Sudbury District, ON, Canada. Not recommended usage. (AGI, 1987)
sugar sand A variety of sandstone that breaks up into granules resembling sugar.
sugar spar Corn. Friable granular quartz. See also: *sugary quartz.* (Fay, 1920)
sugar stone (a) Eng. An ironstone in Norfolk, so-named from its rich brown color. (Arkell, 1953) (b) Compact white to pink datolite from the Michigan copper district.
sugar-tube method A dust-sampling technique that measures airborne dustiness on a mass basis. In the glass tube is placed a layer, 1-1/2 in (3.8 cm) deep, of sized granulated sugar weighing 100 g. By means of a suitable pump, air is drawn through the sugar tube at a rate of approx. 1 ft^3/min (0.0283 m^3/min). The dust is retained in the sugar tube, which is then stoppered and sent to the laboratory for analysis. The sugar is dissolved and the dust is caught on a filter paper that is incinerated to give the weight of dust. (Greenburg, 1922)
sugary quartz A granular and somewhat friable and massive variety of quartz. Syn: *sugar spar.* (Fay, 1920)
suite (a) A set of apparently comagmatic igneous rocks. (AGI, 1987) (b) A collection of rock specimens from a single area, generally representing related igneous rocks. (AGI, 1987) (c) A collection of rock specimens of a single kind; e.g., granites from all over the world. (AGI, 1987) (d) In stratigraphy, the lithodemic unit next higher in rank to lithodeme. It comprises two or more associated lithodemes of the same class (e.g., plutonic, metamorphic). For cartographic and hierarchical purposes, suite is comparable to group. The name of a suite combines a geographic term, the term suite, and an adj. denoting the fundamental character of the suite; e.g., Idaho Springs Metamorphic Suite. (AGI, 1987) (e) A major group of Tertiary and younger igneous rocks thought by Harker (1909) to characterize regions around each of the world's great oceans. Because there is such a wide variety of tectonic environments and associated rock types in these regions, the terms are now seldom used. See also: *Atlantic suite; Pacific suite; Mediterranean suite.* Cf: *lithodeme.* (AGI, 1987)
suites (of igneous rocks) *consanguinity.*
sulfate-bearing soils On silts where the groundwater contains more than 0.1% of SO_3, or where a clay soil has more than 0.5% of this substance, the use of lime-free cement in all concrete work is essential to prevent the resulting chemical attack. High-alumina cement gives complete protection, but a sulfate-resisting cement to British Standard 12 may be used where conditions are less severe. (Hammond, 1965)
sulfate sulfur The inorganic sulfur in coal other than the pyritic sulfur. (BS, 1961)
sulfate test Sulfate in soil can be precipitated as barium sulfate and measured so as to indicate whether water or soil will have a harmful effect on concrete. (Hammond, 1965)
sulfating roast A roast in which conditions in a furnace allow sulfur in feed to recombine with calcined products to form sulfates. Syn: *sulfatizing roast.* (Pryor, 1963)
sulfatizing The chemical reaction that takes place in many roasting operations in which metallic sulfates form instead of oxides. (Newton, 1959)
sulfatizing roast *sulfating roast.*
sulfhydryl-collector method In flotation, a method for treatment of various oxygen ores where sulfhydryl collectors are used to float the base-metal minerals from associated minerals. (Gaudin, 1957)
sulfide (a) A mineral compound characterized by the linkage of sulfur with a metal or semimetal; e.g., galena, PbS, or pyrite, FeS_2. (AGI, 1987) (b) A mineral with S^{2-} or S_2^{2-} as its anion.
sulfide enrichment Enrichment of a deposit by replacement of one sulfide by another of high value, as pyrite by chalcocite. (AGI, 1987)
sulfide of iron An iron ore with sulfur as its main impurity. (Mersereau, 1947)
sulfide ore Ore in which the sulfide minerals predominate. See also: *mixed ore.*
sulfide zone (a) That part of a sulfide deposit that has not been oxidized by near-surface waters. See also: *oxidized zone; protore.* (AGI, 1987) (b) A generally manto-shaped deposit in which secondary sulfide enrichment has occurred as a part of ore-deposit oxidation. Syn: *secondary sulfide zone.* (AGI, 1987)
sulfidization In conditioning a flotation pulp, addition of soluble alkaline sulfides in aqueous solution to produce a sulfide-metal layer on an oxidized ore surface. (Pryor, 1963)
sulfidizing method In flotation, a method for treatment of various oxide ores in which the desired base-metal minerals are sulfidized, then the ore is floated as if it were a sulfide ore. It is useful in treating lead carbonate ores, less useful if other lead minerals are present, and of limited utility in connection with copper and zinc ores. (Gaudin, 1957)
sulfite A salt or ester of sulfurous acid; a compound containing the radical, SO_3^-. (AGI, 1987)
sulfo-antimonite A mineral in which sulfur and antimony are united chemically with a metal. (Weed, 1922)
sulfoarsenite An ore mineral of any metal or metals with which sulfur and arsenic are united chemically. (Weed, 1922)
sulfonic acid A hydrophilic group. In the sulfate ester, an oxygen link is provided, forming the acid group of many commercial surface-active agents in which a little sulfonic acid and much long-chain sulfate ester are linked to fatty acids and oils. Examples of reagents used in mineral processing include Lissapol LS (sodium salt of oleic acid chloride and p-anisidine sulfonic acid); and Aerosol (sulfosuccinic acid diester). (Pryor, 1963)
sulfonite Light green to yellow vaselinous variety of sulfur-containing bitumen. (Tomkeieff, 1954)
sulfophile Elements that occur preferentially in minerals free of oxygen (or fluorine or chlorine); i.e., mostly as sulfides, selenides, tellurides, arsenides, antimonides, intermetallic compounds, native elements, etc. This group includes some of the chalcophile and some of the siderophile elements as classified by Goldschmidt. (AGI, 1987)
sulfur (a) Sulfureted hydrogen, H_2S; stinkdamp. (Fay, 1920) (b) One of the elements present in varying quantities in most bituminous coal as part of the ash and deleterious to coke for steelmaking. (BCI, 1947) (c) An orthorhombic mineral, 128[S]; native sulfur; dimorphous with rosickyite; soft; yellow; around volcanic fumaroles, in salt deposits associated with limestone, gypsum, and anhydrite; a source of elemental sulfur. Also called brimstone. (d) A mining term for iron sulfide (pyrite) in coal seams and with zinc ores in Wisconsin and Missouri. Formerly spelled sulphur. (e) *crude sulfur.*
sulfur ball (a) An accumulation of sulfur in the form of iron pyrites sometimes found in coal seams, often hard enough to break the bits on cutting machines. (BCI, 1947) (b) A concretionary form of the sulfide of iron occurring as both pyrite and marcasite. This material seems to crystallize or grow within the coal as a result of the action of waters bearing sulfuric acid acting upon compounds of iron. This iron is then taken into solution as iron sulfates and subsequently converted to sulfides that form into the sulfur ball. (Kentucky, 1952)
sulfur dome An inverted container, holding a high concentration of sulfur dioxide gas, used in die casting to cover a pot of molten magnesium to prevent burning. (ASM, 1961)
sulfuret Pac. The undecomposed metallic ores, usually sulfides. Chiefly applied to auriferous pyrites. Concentrate and sulfide are preferable. An old syn. for sulfide. Obsolete.
sulfur group The group VI elements sulfur, selenium, tellurium, and oxygen.
sulfur mining Thick sulfur-bearing deposits may be worked by a network of tunnels and general caving. The bed is extracted in a series of thick slices, horizontal if the dip is great, or parallel to the dip if it is moderate. See also: *Frasch process.* (Nelson, 1965)
sulfur ore Pyrite, often roasted for its sulfur. (Webster 2nd, 1960)
sulfydril Mercapto thiol. -SH, the monovalent radical. (Pryor, 1963)
sullage (a) Mud and silt deposited from flowing water. (Hammond, 1965) (b) Scoria on molten metal in the ladle. (Webster 3rd, 1966)
Sullivan angle compressor A two-stage compressor in which the low-pressure cylinder is horizontal and the high-pressure cylinder is vertical. It is a compact compressor and is driven by a belt, or it can be directly connected to an electric motor or diesel engine. (Lewis, 1964)
sulphur *sulfur.*
sulvanite An isometric mineral, Cu_3VS_4; forms a series with arsenosulvanite; bronze yellow; at Burra Burra, South Australia; and Mercur, UT.
summary of reinforcement A cutting list with details of reinforcing bars. (Hammond, 1965)
summer black oil A black lubricating oil of 540 °F (282 °C) fire test, used as a heavy tempering oil and for waterproofing cement.
sump (a) An excavation made underground to collect water, from which it is pumped to the surface or to another sump nearer the surface. Sumps are placed at the bottom of a shaft, near the shaft on a level, or at some interior point. (Lewis, 1964) (b) An excavation smaller than and ahead of the regular work in driving a mine tunnel or sinking a mine shaft. (Webster 3rd, 1966) (c) A hole sunk in a drift to a depth of 2 to 3 yd (1.8 to 2.7 m). (Hess) (d) To undercut coal preliminary to placing a shortwall machine in position for cutting along the working

face. Sometimes called a sumping cut. (e) To test the load in depth. (Gordon, 1906) (f) *cut.* (g) To drill diagonally. (Mason, 1951) (h) *jib in.* (i) A pit or basin in which the returns from a borehole are collected and stored and in which the cuttings settle before recirculating the cuttings-free drilling fluid. (Long, 1960) (j) A cellar under a drill floor. (Long, 1960)

sump cleaner In bituminous coal mining, one who shovels up accumulations of coal, rock, dirt, and refuse at the bottom of a shaft and loads it into buckets that are hoisted to the surface or an upper level for dumping. (DOT, 1949)

sumper (a) A shothole drilled diagonally. (Mason, 1951) (b) In bituminous coal mining, a person who oils and greases coal-cutting machines. Also called machine sumper. (DOT, 1949)

sump fuse A waterproof fuse for use in a sump. (Standard, 1964)

sumping (a) Forcing the cutter bar of a coal cutter into or under the coal. Also called sumping cut. (b) A small square shaft, generally made in the airheadings when crossing faults, etc., or made to prove the thickness of coal, etc.

sumping bar An angle iron about 8 ft (2.4 m) long with flanges about 4 in (10 cm) high, weighing about 75 lb (34 kg). Its function is to guide the cutter bar on an electric coal-cutting machine. (Fay, 1920)

sumping cut *sump.*

sumping-in *jibbing-in.*

sumpman In metal mining, a person who installs sets of timbers to support the walls of a shaft, working with the shaft sinking crew and installing the timbers as the work advances. (DOT, 1949)

sump shaft The shaft in a mine at the bottom of which is the sump. (Standard, 1964)

sump shot A blast made in a shaft that is being sunk, to make a collecting place for water. (Standard, 1964)

sump winze A winze sunk in the bottom of the lowest level, to explore the lode below and ascertain whether the sinking of the main shaft is advisable. (Standard, 1964)

sun bed (a) A sublithographic limestone or mudstone at the top of the White Lias, the upper surface of which bears polygonal cracks attributed to sun drying. (b) Alternatively a corruption of "sound" bed because, when dry, the rock rings when broken with a hammer. Cf: *guinea bed; ring stone.*

sun cheek The south side of a vein. (Arkell, 1953)

sun crack *mud crack.*

Sunday stone A calcareous deposit formed inside pipes carrying waste water from collieries. It is composed of alternating dark and light bands corresponding to the day and night shifts and a broader light band corresponding to Sunday. (Arkell, 1953)

Sundberg method In electrical prospecting, an inductive method in which the current flows through an insulated copper cable connected to a source of alternating current and run along the surface in a rectangle 1 mile by 1/2 mile (1.6 km by 0.8 km) in dimensions. A series of transverse profiles are laid out perpendicular to and crossing the cable, and the magnetic part of the electromagnetic field is measured at discrete points along the profiles by special search coils consisting of several hundred turns of wire. The magnitude and direction of the induced field observed by the coils can be related to the inductive effect of the subsurface material directly below. (Dobrin, 1960)

sundiusite (a) A monoclinic mineral, $Pb_{10}(SO_4)Cl_2O_8$; forms plumose aggregates of adamantine crystals having one perfect cleavage; at Långban, Sweden. (b) A name proposed for a hypothetical amphibole end-member composition, but never accepted by the IMA nor widely used by mineralogists.

sundtite *andorite.*

sun gear The central gear in a planetary set. (Nichols, 1976)

sunk Drilled or excavated downward. (Long, 1960)

sunken pit *dig-down pit.*

sunk shaft A shaft that is driven from the top downward (vertical or inclined) (Fraenkel, 1953)

sun opal *fire opal.*

sunshine A name of a soft grade of paraffin wax with a low melting point. It can be burned in an ordinary miners' lamp with a nail (usually copper) in the wick and gives little smoke. Also called miners' sunshine.

sunstone (a) An aventurine feldspar, generally a brilliant, translucent variety of oligoclase that emits a reddish or golden billowy reflection from minute scales or flakes of hematite spangled throughout and arrayed parallel to planes of repeated twinning. Cf: *goldstone; moonstone.* Syn: *heliolite.* (b) A red-to-yellow variety of bytownite caused by minute inclusions of copper and found in basalt from Oregon. See also: *aventurine.*

sun vein N. of Eng. Ore vein discovered on the south side of a hill. Sun is synonymous with south, so sun veins are south veins.

super airflow cleaner In this cleaner, the coal is fed from a raw coal hopper by an oscillating plunger feeder ensuring uniform feed throughout the entire width of the machine deck. A rotating shutter in the inlet duct provides a pulsating air current, which is much more effective as a separating current than a uniform airstream. The coal and refuse stratify as they move toward the discharge end of the deck with the refuse falling to the bottom against the deck. The perforated deck has four refuse draws; normally the products of the first two draws are discharged as refuse and the third and fourth draws as middlings. The product of the third draw may be refuse or middlings, depending on the ash content of the feed. Generally, this cleaner is an effective machine for cleaning coal from 3/4 in (1.91 cm) to 1/4 in (0.64 cm) by 28 mesh if it contains less than 5% of surface moisture. It will not clean effectively below 20 mesh. (Kentucky, 1952)

superalloy An alloy for very high-temperature service in which relatively high stresses (tensile, vibratory, and shock) are encountered and oxidation resistance is frequently required. (ASM, 1961)

supercapacity bucket elevator A type of continuous bucket elevator employing supercapacity elevator buckets. See also: *supercapacity elevator bucket.*

supercapacity elevator bucket A type of continuous elevator bucket used with a pair of chains in which the back of the bucket at the bottom extends backward into space between the up and down runs to provide additional capacity without increase in length or projection. See also: *supercapacity bucket elevator.*

supercharger A blower that increases the intake pressure of an engine. (Nichols, 1976)

superconductivity The abrupt and large increase in electrical conductivity exhibited by some materials at very low temperatures. (ASM, 1961)

supercritical area of extraction An area of goaf of sufficient extent to cause full subsidence at more than one point on the surface. (Nelson, 1965)

superdip An instrument of limited sensitivity for measuring changes in the total intensity of the magnetic field. (Schieferdecker, 1959)

superdusting The presence of a very large excess of coal dust in the air over that required for consumption of the oxygen. (Rice, 1913-18)

superexchange The antiparallel alignment of unpaired *d* electrons of metal cations through coupling with *p* electrons of intervening anions to produce magnetic order. Cf: *ferrimagnetism; antiferromagnetism.*

superficial Pertaining to, or lying on or in, a surface or surface layer; e.g., superficial weathering of a rock, or a superficial structure formed in a sediment by surface creep. The term is used esp. in Great Britain; the syn. surficial is more generally applied in the United States. (AGI, 1987)

superficial compaction Compaction of soil in layers generally not greater than 6 in (15 cm) deep, by various methods, including the use of the frog rammer, vibration, sheepsfoot roller, pneumatic tired rollers, or similar machines. See also: *compaction equipment.* (Hammond, 1965)

superficial deposit The most recent of geological formations; unconsolidated detrital material lying on or near the surface, generally unstratified. *surface deposit.* (Fay, 1920)

supergene Said of a mineral deposit or enrichment formed near the surface, commonly by descending solutions; also, said of those solutions and of that environment. Cf: *hypogene; mesogene.* (AGI, 1987)

supergene enrichment A mineral deposition process in which near-surface oxidation produces acidic solutions that leach metals, carry them downward, and reprecipitate them, thus enriching sulfide minerals already present. Syn: *enrichment; secondary enrichment.* See also: *oxidized zone.* (AGI, 1987)

superheated steam Steam above a temperature of 100 °C; produced by heating water under a pressure greater than standard atmosphere. (Nelson, 1965)

superheater (a) A device that superheats, esp. steam or other gases; esp. a coil or other device through which steam from a boiler passes to be superheated. (Webster 3rd, 1966) (b) A refractory lined chamber in a water-gas plant that ensures completion of the decomposition of the oil vapors begun in the carburetor. (Dodd, 1964)

superimpose In geology, to establish (a structural system) over, independently of, and eventually upon underlying structures; said of terranes, rivers, drainage systems, valleys, etc.; as, a superimposed valley. (Standard, 1964)

superimposed halo A dispersion pattern formed in the regolith by the movement of material in subsurface waters. (Hawkes, 1957)

superintendent-of-tanks One who supervises the feeding and maintenance of glass-melting furnaces (tanks), and the operation of reheating ovens (lehrs) for fire-glazing glass articles. Directs unloading and storage of raw materials and crushing and washing of waste glass (cullet) used as ingredients in the manufacture of new glass. (DOT, 1949)

superjacent pattern A dispersion pattern developed more or less directly over the bedrock source. (Hawkes, 1962)

superjacent roadway system *methane drainage.*

superlattice X-ray diffractions, in addition to fundamental diffractions, which appear as an alloy inverts from disorder to perfect order of its constituent atoms at lattice sites; called superlattice diffractions and attributed to a superstructure of ordered atoms. Superlattice diffractions also appear where long-range order obtains in the stacking of silicate structural units, such as in phyllosilicates. *superstructure; long-range order.*

superpanner A mechanism that simulates rocking, bumping, and sluicing action used in panning and gives precise information as to the possibility of gravity treatment of sands. It is used in rapid assays and as a research aid. (Pryor, 1960)

superphosphate Any of various commercial phosphate fertilizers obtained as white to gray granules or powders by acidulating ground insoluble phosphate rock, such as: (1) a product made by acidulating with sulfuric acid, consisting essentially of primary calcium phosphate, calcium sulfate, and smaller quantities of secondary calcium phosphate, and containing usually about 20% of available phosphoric acid; or (2) a product made by acidulating with phosphoric acid, consisting essentially of primary calcium phosphate and containing usually 40% to 50% of available phosphoric acid. (Webster 3rd, 1966)

superposition (a) A principle stating that if a body is subjected to several stresses acting simultaneously, then each stress produces its own strain or strains, and these strains may be superimposed to give the complete state of strain of the solid. Similarly, two separate stress distributions in a body, due to the application of two separate stresses, may be superimposed to give the stress distribution due to the simultaneous application of these two stresses. (Issacson, 1962) (b) The order in which rocks are placed or accumulated in beds one above the other, the highest bed being the youngest. (AGI, 1987)

superprint *overprint.*

superpushing Using extra large pushers, or two or even three standard units in tandem, to increase the speed and size of loading. (Nichols, 1976)

supersaturated solution A solution that contains more of the solute than is normally present when equilibrium is established between the saturated solution and undissolved solute. (AGI, 1987)

super section Two sets of mining equipment operating simultaneously and sharing a common dumping point on the same section, with each set being ventilated by a separate split on intake air. (FR 95, 1992)

superstructure A crystal structure resulting from large unit cells when an alloy inverts from disordered occupancy of lattice sites by "averaged" constituent atoms in small unit cells to individual atomic species occupying specific lattice sites and having long-range order; e.g., disordered Cu_3Au has a primitive cubic unit cell with averaged "Cu-Au atoms" at each corner, while ordered Cu_3Au has a face-centered cell with Au atoms its corners and Cu its face centers, the large face-centered unit cell containing eight of the small primitive unit cells. *superlattice; long-range order.*

supplementary twinning Twinning by which a crystal simulates the symmetry of a crystal class with higher grade in the same system.

supplied-air respirator An atmosphere-supplying device that provides the wearer with respirable air from a source that is outside of the contaminated area. (Best, 1966)

supply cager In anthracite and bituminous coal mining, one who directs and assists the loading of mine supplies on a cage (elevator) at a mine having a separate shaft or shaft compartment for handling supplies. (DOT, 1949)

supply-hoist engineer *supply-hoist operator.*

supply-hoist operator In anthracite and bituminous coal mining, one who operates the hoisting machinery that serves the shaft or shaft compartment of a mine in which mining supplies are lowered into the mine. Also called supply-hoist engineer. (DOT, 1949)

supply motorman In bituminous coal mining, a person who operates a mine locomotive to haul trips (trains) of cars, loaded with timbers, rails, explosives, and other supplies, into a mine. (DOT, 1949)

supply pump *bank pump.*

support A general term for any timber, steel, concrete, brick, or stone structure erected to counteract the subsidence of the roof strata when undermined. See also: *self-advancing supports; steel arch; steel support.* (Nelson, 1965)

support roller In a crawler machine, a roller that supports the slack upper part of the track. (Nichols, 1976)

supports Materials placed in stopes for the purpose of ground control, that is to arrest or regulate the closure of the walls. (Spalding, 1949)

suppressed weir A measuring weir having its sides flush with the sides of the channel, so that there are no end contractions. (Hammond, 1965)

supratenuous fold A pattern of fold in which there is thickening at the synclinal troughs and thinning at the anticlinal crests. It is formed by differential compaction on an uneven basement surface. Cf: *similar fold; parallel fold.* (AGI, 1987)

surcharge (a) Any load including earth that is supported above the level of the top of a retaining wall. See also: *relieving platform; active earth pressure.* (Hammond, 1965) (b) The algebraic sum of the losses and gains of a cornet of gold during cupellation and solution. (Fay, 1920)

surface (a) The top of the ground. As used in the conveyance of coal in place, or in a conveyance of land reserving the minerals, includes not merely the surface within the boundary lines, without thickness, but includes whatever earth, soil, or land lies above the superincumbent upon the coal or mineral reserved. (b) *cover; drift; mantle; overburden.* (c) In geology, usually refers to (1) the boundary surface between one bed or mass of rock and another immediately adjacent, such as a bedding surface, a fault surface, a surface of unconformity, a surface of igneous compact, or (2) an imaginary surface, such as the axial surface of a fold. (Challinor, 1964)

surface action Any kind of action that affects a surface; e.g., action of smoke fumes, moisture, etc., on a painted surface. (Crispin, 1964)

surface-active agent Chemical compound that modifies physical, electrical, or chemical characteristics of surface of solid, also surface tensions of solids or liquids. Used in froth flotation and in detergency. Characteristically, its heteropolar molecules are attracted to a specific type of surface in a mixture where one group forms polar monolayer attachments while the rest of the molecule points outward and changes the relations between the surface and the ambient phase. These relations may change lyophilic and aerophilic attraction, surface tension, intermiscellar grouping, emulsification, and froth foaming. Surface-active agents include cleaners (e.g., soaps); water repellants (e.g., greases); dispersants and emulsifiers (e.g., glue); and additives adsorbed at interfaces between liquids (usually aqueous) and external gas, liquid or solid phases, with resulting change in interfacial tension. Three electrochemical types are unionized molecule, anion, and cation. Important characteristics of surface-active agents are solubility in the medium and effects of specific adsorption at interfaces. Such agents either provide anchorage between phases or form a barrier, according to their flocculating or dispersing effect. (Pryor, 1963)

surface activity The property possessed by certain solid substances to influence the surface tension of liquids. See also: *depressant; flotation agent; surface tension.* (Hammond, 1965)

surface air leakage The amount of surface air entering a fan through the casing at the top of the upcast shaft, the airlock doors, and the fan-drift walls. The extent of leakage will depend on the fan-drift water-gage method of construction, the number and type of entrances, and whether the upcast shaft is used for winding. The surface leakage at airlocks may vary from about 25,000 ft^3/min (708 m^3/min) at 5-in (12.7-cm) water gage to about 55,000 ft^3/min (1,557 m^3/min) at 25-in (63.5-cm) fan-drift water gage. See also: *volumetric efficiency.* (Nelson, 1965)

surface area Of a particle, area calculated from data obtained by a specified method, such as: (1) adsorption measurement, (2) calculation, (3) permeability measurement, (4) microscopic observation, or (5) close screening and averaging from study of a number of particles. (Pryor, 1963)

surface break *surface damage.*

surface bunker A large capacity hopper or standage room to store coal or mineral coming from the winding shaft. The provision tends to equalize the run of mine going to the preparation plant and smooth out any minor breakdowns in the plant. (Nelson, 1965)

surface charges All expenses incurred on the surface of a mine that have to be charged against the mineral.

surface circuit The mine car track layout from the shaft to the preparation plant and back to the shaft. The term includes all the equipment necessary to move and control the movement of cars, such as creepers, retarders, back shunts, traversers, and turntables. (Nelson, 1965)

surface clay An unconsolidated, unstratified clay occurring on the surface. (ASTM, 1994)

surface conductance The heat transmitted from (or to) a surface to (or from) the fluid in contact with the surface in a unit of time per unit of surface area per degree temperature difference between the surface and fluid. Measured in units of watts per square meter Kelvin. (Strock, 1948)

surface damage (a) Any damage done to land surface during mine exploration or development operations. Syn: *surface break.* (b) Scot. Ground occupied and damaged by colliery operations; the compensation for such.

surface deposit (a) Orebody that is exposed and can be mined from the surface. (Zern, 1928) (b) *superficial deposit.*

surface dressing The covering of an existing surface with a coating of bituminous binder covered by a layer of chippings or fine aggregate. (Nelson, 1965)

surface drift A drift (usually inclined) from the surface to the coal seam or orebody to be developed. See also: *drift mining; slant.* (Nelson, 1965)

surface drilling Boreholes collared at the surface of the Earth, as opposed to holes collared in mine workings or underwater. See also: *surface rig.* (Long, 1960)

surface energy Product of surface tension (dynes per centimeter) and surface area (centimeters), expressed in ergs. Work required to increase surface area by unit area. (Pryor, 1963)

surface factor *fineness factor.*

surface geology (a) Geology and correlation of rock formations, structures, and other features as seen at the Earth's surface. Cf: *subsurface geology.* (AGI, 1987) (b) *surficial geology.*
surface leaching A mining practice that represents an alternative to conventional ore beneficiation in that the mineral values are hydrometallurgically extracted rather than concentrated in solid form for further processing. See also: *solution mining.* (SME, 1992)
surface lift A term used in the freezing method of shaft sinking. The surface around the shaft tends to heave owing to the formation of ice and the variation of temperatures. This uplift is sufficient to throw surface structures, such as winding towers, out of alignment. To enable corrections to be made, the tower bases may be mounted on grillages with facilities for jacking to keep the towers level. (Nelson, 1965)
surface lines The boundary lines of a mining claim as indicated by the locator.
surface loss Losses of air occurring in the restricted area around the tub and caused by the increased velocity along the walls of the airway, and from the friction and turbulent effects due to the surfaces of the tub. For an empty tub, a portion of the turbulent effect is due to eddying and shock within the tub. (Roberts, 1960)
surface mining (a) The mining in surface excavations. It includes placer mining, mining in open glory-hole or milling pits, mining and removing ore from opencuts by hand or with mechanical excavating and transportation equipment, and the removal of capping or overburden to uncover the ores. (b) Mining at or near the surface. This type of mining is generally done where the overburden can be removed without too much expense. (Kentucky, 1952) (c) The obtaining of coal from the outcroppings or by the removal of overburden from a seam of coal, as opposed to underground mining; or any mining at or near the surface. Also called strip mining; placer mining, opencast; opencut mining; open-pit mining. (BCI, 1947; Standard, 1964)
surface plan A map of the surface layout of a mine. (BS, 1963)
surface reinstatement The restoration of the surface after opencast mining operations have been completed. The work may involve leveling the hill-and-dale formation, drainage, and relaying of the original topsoil, also known as resoiling. (Nelson, 1965)
surface rig A drill rig designed specif. and used only for surface drilling operations. See also: *surface drilling.* (Long, 1960)
surface rights (a) The ownership of the surface of land only where mineral rights are reserved. (Weed, 1922) (b) The right of a mineral owner or an oil and gas lessee to use so much of the surface of land as may be reasonably necessary for the conduct of operations under the lease. (Williams, 1964) (c) Those reserved to the owner of the land beneath which ore is being mined. (Pryor, 1963)
surface-set bit A bit containing a single layer of diamonds set so that the diamonds protrude on the surface of the crown. Also called single-layer bit. Cf: *multilayer bit.* (Long, 1960)
surface slope The inclination of the water surface expressed as change of elevation per unit of slope length; the sine of the angle that the water surface makes with the horizontal. The tangent of that angle is ordinarily used; no appreciable error resulting except for the steeper slopes. (Seelye, 1951)
surface speed *peripheral speed.*
surface string A large diameter drivepipe sunk through the uppermost part of the overburden. Cf: *conductor; standpipe.* (Long, 1960)
surface tension (a) In the flotation process, the contractile force at the surface of a liquid whereby resistance is offered to rupture. (b) Interfacial tension between two phases, one of which is a gas. (ASM, 1961) (c) A condition that exists at the free surface of a body (such as a liquid) by reason of intermolecular forces unsymmetrically disposed about the individual surface molecules and is manifested by properties resembling those of an elastic skin under tension. Specif., the force per unit length of any straight line on the surface that the surface layers on opposite sides of the line exert upon each other. See also: *surface activity.* (Webster 3rd, 1966)
surface texture The aggregate of the surface features of sedimentary particles, independent of size, shape, or roundness; e.g., polish, frosting, or striations. (AGI, 1987)
surface thrust *erosion thrust.*
surface water Water that rests on the surface of the lithosphere. (AGI, 1987)
surface-water drain Any pipe laid in the ground for carrying away surface water. See also: *separate system.* (Hammond, 1965)
surface wave An elastic wave in which the energy is confined to the surface or a narrow region just below the free surface of an extended solid. These waves readily follow the curvature of the part being inspected and are reflected only from sharp changes of direction at the surface. (ASM, 1961)
surface working *surface mining.*
surfactant Surface active agent, a substance that affects the properties of the surface of a liquid or solid by concentrating in the surface layer. (Brantly, 1961)
surfeit Gr. Brit. Afterdamp or chokedamp; or pressure exercised by a pent-up gas resulting in its escape with or without rupture of strata. (Tomkeieff, 1954)
surficial Pertaining to, or occurring on, a surface, esp. the surface of the Earth. Cf: *subsurface.* See also: *subaerial.* Syn: *superficial.* (AGI, 1987)
surficial creep *soil creep.*
surficial geology Geology of surficial deposits, including soils; the term is sometimes applied to the study of bedrock at or near the Earth's surface. See also: *surface geology.* (AGI, 1987)
surf zone The area between the outermost breaker and the limit of wave uprush. Syn: *breaker zone.* (Schieferdecker, 1959)
surge (a) To move sideways; to fleet. (Mason, 1951) (b) The uneven rate of flow and regular variations in pressure caused by time lags between pressure strokes on a piston-type pump. (Long, 1960) (c) In fluid flow, long-interval variations in velocity and pressure that are not necessarily periodic and may even be transient. (AGI, 1987)
surge bin (a) In salt mining, a generally large bin above the crusher into which the mine-run salt is dumped prior to being discharged into the primary crusher. A feeder at the bottom of the surge bin facilitates transfer of the mine-run material to the crusher. (Kaufmann, 1960) (b) A compartment for temporary storage, which will allow converting a variable rate of supply into a steady flow of the same average amount. (Nichols, 1976)
surge bunker A large-capacity storage hopper, installed near the pit bottom or at the input end of a processing plant to provide uniform feeding of material from bulk deliveries. Surge bunkers are generally required either on the surface or underground to act as a buffer between the shaft and the coal preparation plant and the working faces. See also: *bunker; surge hopper.* (Nelson, 1965; Sinclair, 1959)
surge hopper A hopper (bunker) designed to receive a feed at fluctuating rate and to deliver it at some predetermined rate. Syn: *surge bunker.* (BS, 1962)
surge pipe Open-topped standpipe to release pressure from surge. (Hammond, 1965)
surge pulley The pulley used on a tension carriage in endless-rope haulage. See also: *balanced direct-rope haulage.* (Nelson, 1965)
surge tank (a) A standpipe or storage reservoir at the downstream end of a closed aqueduct or feeder pipe, as for a water wheel, to absorb sudden rises of pressure and to furnish water quickly during a drop in pressure. (Webster 3rd, 1966) (b) An open tank to which the top of a surge pipe is connected so as to avoid loss of water during a pressure surge. (Hammond, 1965) (c) In pumping of ore pulps, a relatively small tank that maintains a steady loading of the pump. When new pulp is in short supply, a float valve causes recirculation of part of the load, therefore avoiding settlement. Alternatively, a float may vary the speed of the pump, vary the rate of delivery to the tank, or divert the flow to a parallel system. (Pryor, 1963)
surging The flapping of a moving rope. Syn: *whipping.* (Zern, 1928)
surturbrand An Icelandic term for a peatlike variety of brown coal or lignite occurring in the Pliocene deposits and sometimes under the volcanic overflows of Iceland.
survey (a) The orderly and exacting process of examining and delineating the physical or chemical characteristics of the Earth's surface, subsurface, or internal constitution by topographic, geologic, geophysical, or geochemical measurements; esp. the act or operation of making detailed measurements for determining the relative positions of points on or beneath the Earth's surface. (AGI, 1987) (b) The associated data or results obtained in a survey; a map or description of an area obtained by surveying. (AGI, 1987) (c) An organization engaged in making surveys; e.g., a government agency such as the U.S. Geological Survey. (d) *borehole survey.* (e) To determine and delineate the form, extent, and position of (as a tract of land, a coast, or a harbor) by taking linear and angular measurements and by applying the principles of geometry and trigonometry. (Webster 3rd, 1966) (f) A measured plan and description of a portion of an area or of a road or line through an area obtained by surveying. (Webster 3rd, 1966) (g) The information plotted from a borehole survey. (Long, 1960)
surveying (a) The act or process of making a borehole survey. See also: *borehole survey.* (Long, 1960) (b) Specif., in civil engineering, the science or art of making such measurements as are necessary to determine the relative position of points on or beneath the surface of the Earth, or to establish such points. (Seelye, 1951) (c) The art of making a survey; specif. the applied science that teaches the art of making such measurements as are necessary to determine the area of any part of the Earth's surface, the lengths and directions of the boundary lines, and the contour of the surface, and of accurately delineating the whole on paper. (AGI, 1987) (d) The act of making a survey; the occupation of a person that surveys. (AGI, 1987)
survey meter An instrument sensitive to ionizing radiations used in prospecting for radioactive deposits. (Webster 3rd, 1966)
surveyor One who applies special knowledge and techniques gained through experience or training to make surface and underground surveys at a mine, locating himself/herself on the Earth's surface by

taking instrument shots on the sun or stars and making necessary calculations, surveying and calculating the volume of material in dumps, carrying survey lines underground by shaft plumbing (cord or wire with attached bob is suspended from the shaft surface) and instrument shots taken on the bob at a shaft station, controlling by underground surveys and calculations the driving and connection of underground passages on and between various levels, computing the volume of coal in portions of the mine from survey notes, and drafting maps of the mine workings. Also called spud setter; underground surveyor. Syn: *mine surveyor.* (DOT, 1949)

surveyor's compass An instrument used in surveying for measuring horizontal angles. Syn: *surveyor's dial.* Cf: *circumferentor; semicircumferentor.* (Webster 3rd, 1966)

surveyor's cross A simple instrument made of two bars forming a right-angled cross with sights at each end and used in setting out right angles in surveying. (Webster 3rd, 1966)

surveyor's dial *surveyor's compass.*

surveyor's level A leveling instrument consisting of a telescope (with cross hairs) and a spirit level mounted on a tripod, revolving on a vertical axis, and having leveling screws that are used to adjust the instrument to the horizontal. (AGI, 1987)

surveyor's measure A system of measurement used in land surveying, having the surveyor's chain (one chain = 4 rods = 66 ft = 20.1 m = 100 links = 1/80 mile) as a unit. Cf: *Gunter's chain.* (AGI, 1987)

Surwell clinograph Well-surveying device for large-diameter holes that determines the departure of the borehole from vertical. Uses a gyroscope and spherical level with a photographic record made on 16-mm moving-picture film, which includes a photograph of a small watch by which the depth is determined from correlation with a synchronized depth-time record made at surface. (AGI, 1987)

susannite A trigonal mineral, $Pb_4(SO_4)(CO_3)_2(OH)_2$; trimorphous with leadhillite and macphersonite; in the Susanna Mine, Leadhills, Scotland.

susceptibility (a) Property of a material that defines the extent to which it will be magnetized in a given external field. (Schieferdecker, 1959) (b) The ratio of the electric polarization to the electric intensity in a polarized dielectric. The ratio of induced magnetization to the strength H of the magnetic field causing the magnetization. Syn: *magnetic susceptibility; volume susceptibility.* (AGI, 1987)

suspect terrane A terrane whose spatial and genetic relations with respect to adjacent terranes during their time of formation is unknown or uncertain. Inasmuch as most terranes fall into these categories, the term may be considered redundant. See also terrane. (AGI, 1987):

suspended matter Particles from the feed, of specific gravity equal or close to that of a separating medium, and which are therefore relatively difficult to remove from the bath. (BS, 1962)

suspended solids Solids that can be separated from a liquid by filtration. (Bennett, 1962)

suspended tray conveyor A vertical conveyor having one or more endless chains with suitable pendant trays, cars, or carriers that receive objects at one elevation and deliver them to another elevation. Syn: *corner-fastened tray conveyor; corner-hung tray conveyor; suspended tray elevator; suspended tray lift.*

suspended tray elevator *suspended tray conveyor.*

suspended tray lift *suspended tray conveyor.*

suspended tubbing A permanent method of lining a circular shaft in which the tubbing (German type) is temporarily suspended from the next wedging curb above. Slurry is run in behind the tubbing by means of a funnel passing through the holes provided in the segments. No temporary supports are required. See also: *tubbing.* (Nelson, 1965)

suspended water *vadose water.*

suspension (a) The condition of a mixture of solid particles and water or air in which the solid particles are individually supported, normally by means of an upwardly moving current and sometimes with the assistance of mechanical agitation. (BS, 1962) (b) A mode of sediment transport in which the upward currents in eddies of turbulent flow are capable of supporting the weight of sediment particles and keeping them indefinitely held in the surrounding fluid (such as silt in water or dust in air). (AGI, 1987) (c) The state of a substance in such a mode of transport; also, the substance itself. (AGI, 1987) (d) A method of rock bolting employed to secure loose fragments or sections of rock that may fall from a mine roof. (Lewis, 1964; Fraenkel, 1953)

suspension current *turbidity current.*

suspension dryer *flash coal dryer.*

suspensoid particles Small solid particles suspended in a liquid; they exhibit the Brownian movement and do not settle by themselves, but can be readily coagulated. (Bennett, 1962)

suspent Material in suspension that is to be filtered out. (Bennett, 1962)

sussexite A white mineral: $(Mn,Mg)BO_2(OH)$. It is isomorphous with szaibelyite. (AGI, 1987)

Sutton, Steele, and Steele dry table A concentrator of the Wilfley type in motion, but instead of using water, stratification is by means of rising currents of air. Heavy grains are pushed forward by the head motion, while lighter grains roll or flow down the slope toward the tailing side. (Liddell, 1918)

suture joint *stylolite.*

svabite A hexagonal mineral, $Ca_5(AsO_4)_3F$; apatite group.

svanbergite A trigonal mineral, $SrAl_3(PO_4)(SO_4)(OH)_6$; beudantite group; at Wermland and Skane, Sweden.

svitalskite *celadonite.*

swab (a) A pistonlike device provided with a rubber cap ring that is used to clean out debris inside a borehole or casing. (Long, 1960) (b) *bailer; sand pump.* (c) In well drilling, to pull the drill string so rapidly that the drill mud is sucked up and overflows the collar of the borehole, thus leaving an undesirably empty borehole. (Long, 1960) (d) Procedure for applying suction within the casing or tubing to draw fluid from a reservoir rock. (Wheeler, 1958) (e) A rod with flexible rubber suction cups working inside the pipe on a wire line. (Wheeler, 1958) (f) A hemp brush used in founding, esp. for holding water, moistening mold joints, spraying on edges, or spreading blacking on dry-sand molds. (Webster 3rd, 1966)

swad (a) Newc. A thin layer of stone or refuse coal at the bottom of the seam. (b) N. of Eng. Impure shaly coal or black shale. (Arkell, 1953)

swag (a) A shallow, water-filled hollow produced by subsidence, resulting from underground mining. (AGI, 1987) (b) A digger's roll of blankets, containing spare clothes, food, etc. (Gordon, 1906)

swage block A large rectangular block of cast iron used by a blacksmith. It is pierced through with numerous holes, both round and square in section, for the reception of work that requires shouldering. (Crispin, 1964)

swallet hole *sinkhole.*

swallow Derb. A loose, broken, or porous place in a vein. It derives its name from the ease with which water sinks through the loose material.

swamp As applied to a mining claim, to clear a narrow strip along the boundary line, where the location is on timberland.

swamp buggy Any vehicle with very large, low-pressure tires enabling it to be used in swamps.

swamp ore *bog iron; bog iron ore.*

swarf (a) In diamond grinding operations, a relatively dry dust derived either from grinding operations where no coolant or lubricant is applied to the grinding operation or where the coolant (kerosene, an aqueous solution, or an emulsion of oil and water) is sprayed on the wheel as a fine mist. (Chandler, 1964) (b) Fine metallic particles removed by a cutting or grinding tool; chippings and shavings from soft iron castings used as a reducing agent in various chemical syntheses. (Webster 3rd, 1966) (c) Scot. A tool for widening boreholes. (Fay, 1920) (d) An intimate mixture of grinding chips and fine particles of abrasive and bond resulting from a grinding operation. (ASM, 1961)

swarm *dike swarm.*

swartzite A monoclinic mineral, $CaMg(UO_2)(CO_3)_3 \cdot 12H_2O$; radioactive; green; forms an efflorescence with gypsum, schröckingerite, bayleyite, and andersonite.

swaugh Derb. A soft clay in the vein.

S wave A body wave that travels through the interior of an elastic medium with particle motion perpendicular to the direction of propagation. Originally applied to earthquake seismology, where it was the second (hence: S) type of wave to arrive at a recording station. Syn: *distortional wave; equivolumnar wave; rotational wave; secondary wave; shear wave.* See also: *transverse wave.* (AGI, 1987)

sway A sideways movement, such as sidesway, in a structural frame. (Hammond, 1965)

swaying of a bank York. Undergoing disturbance due to weight of the roof. A settling of the mine roof.

sway rod A diagonal brace designed to resist wind or other horizontal force acting on a structural framework. (Hammond, 1965)

sweal (a) Eng. To burn slowly. (b) To melt and run down; to waste away without feeding the flame. A candle is said to sweal when the grease runs down owing to its burning in a strong current of air or being improperly carried or fixed.

sweat (a) To gather surface moisture in beads as a result of condensation. The roof of a mine is said to sweat when drops of water are formed upon it, by condensation of steam formed by the heating of the waste or goaf. (Webster 3rd, 1966) (b) To exude nitroglycerin; said of dynamite in which nitroglycerin separates from its adsorbent. (Webster 3rd, 1966)

sweaters *thawing.*

sweating (a) The condensation of moisture and distillation products on the surface of a roast heap, forming a damp and sticky crust. (Fay, 1920) (b) *exudation.*

sweating out Bringing small globules of one of the low-melting constituents of an alloy to the surface during heat treatment, such as lead out of bronze. (ASM, 1961)

swedenborgite A hexagonal mineral, $NaBe_4SbO_7$; at Långban, Sweden.

Swedish bit *chisel bit.*

Swedish iron An iron of highest quality owing to the freedom from phosporus and sulfur of the Swedish ore. (Crispin, 1964)

Swedish mining compass A compass in which a magnetic needle is suspended on a jewel and a stirrup so that it can rotate about both a horizontal and a vertical axis. (AGI, 1987)

sweep (a) Aust. That part of a branch that reunites with the main vein farther on. (b) In founding, a profile pattern, used esp. in forming molds for cylindrical or other symmetrical articles. (Standard, 1964) (c) A form or template used for shaping sand molds or cores by hand. (ASM, 1961) (d) A curved metal blade projecting from the central shaft of a pug mill to force clay through holes at the bottom. (Standard, 1964)

sweeping *sweeps.*

sweeping table Stationary circular buddle provided with rotating brushes that prevent formation of channels as pulp flows radially across. (Pryor, 1963)

sweep plates Eng. Curved plates for barrowways at a turn. A turnsheet.

sweeps (a) The dust of the workshops of jewelers, goldsmiths, silversmiths, and assayers and refiners of gold and silver. Also called sweeping. (b) Brushwood arms on round buddles that rotate slowly and break down channels as ore slime runs across the surface. (Pryor, 1963)

sweepwashings Valuable metal washed from sweeps. (Standard, 1964)

sweet (a) Eng. Free from combustible gases or other gases, or from fire stink. (b) Applied to potable water and to oil and gas free of hydrogen sulfide. (c) Said of crude oil or natural gas that contains few or no sulfur compounds. Cf: *sour.* (AGI, 1987)

sweetish astringent Applied to those minerals that have the taste of alum.

sweet roast *dead roast.*

sweet roasting Complete roasting or until arsenic and sulfur fumes cease to form. See also: *roasting.* (Fay, 1920)

sweet water *fresh water.*

swell (a) A space in a seam from which the coal has been eroded and its place filled with clay or sand. Syn: *horseback.* (b) A local enlargement or thickening in a vein or ore deposit, as opposed to a pinch. (c) The tendency of soils, on being removed from their natural, compacted beds, to increase in volume owing to an increase in void ratio; i.e., the space between soil particle increases. (Carson, 1961) (d) A low dome or quaquaversal anticline of considerable areal extent. (e) Waves caused by the wind but no longer being activated. (Schieferdecker, 1959) (f) Long and generally symmetrical waves, period approx. 10 s, produced by storm and wind remote from the point of observation. These are gravity waves and contribute to the mixing processes in the surface layer and thus to its sound transmission properties. (Hy, 1965) (g) In geology, a large-scale submarine topographic feature rising above the surrounding surface and having nearly equal length and width. (Hy, 1965)

swelled coupling A rod coupling having a considerably larger outside diameter than the drill rods to which it is threaded, such as BW rod outside diameter with AW rod threads. Syn: *oversize coupling.* (Long, 1960)

swelled ground (a) A soil or rock that expands when wetted. (Long, 1960) (b) Soil or rock that flows into mine workings as a result of pressure.

swelling number A numerical expression to indicate the relative swelling properties of a sample when heated under standardized conditions. (Nelson, 1965)

swelling of shale When a shaft is sunk through a thick, dry deposit of shale, the absorption of water may cause the shale to swell and damage the shaft lining. Again, when shale is exposed to weathering, the lamina tends to separate and the material swells. When wet, the disintegrated mass still further swells and eventually becomes a plastic clayey deposit. (Nelson, 1965)

swelling pressure (a) The pressure that heated and softened coal exerts when it is obstructed from free swelling. (b) The pressure exerted by a contained clay when absorbing water in a confined space.

swilley A depression in a mine road from which the road rises both ways. (Mason, 1951)

swimming stone *floatstone.*

swinestone (a) A variety of marble that gives off a fetid odor when broken or rubbed. Also called stinkstone. See also: *bituminous limestone.* (b) *fetid calcite.* (c) *anthraconite.*

swing (a) In power-shovel nomenclature, the rotation of the superstructure on the vertical shaft in the mounting. (Carson, 1961) (b) In revolving shovels, to rotate the shovel on its base. (Nichols, 1976) (c) *swing radius.* (d) In churn drills, to operate a string of tools. (Nichols, 1976)

swing angle The distance in degrees that a shovel must swing between digging and dumping points. (Nichols, 1954)

swing cut *slabbing cut.*

swing-hammer crusher (a) A rock breaker in which crushing force is generated by hammers loosely mounted on a rapidly revolving shaft. Rock entering the crushing chamber is hit and rebounds against liner plates of walls or against other rock, until small enough to escape through a grid. (Pryor, 1963) (b) A machine in which size reduction is effected by elements loosely pivoted to disks fitted on a rotating horizontal shaft mounted in a surrounding casing. Also called pulverizer; swing-hammer mill; swing hammer. See also: *Jeffrey crusher.* (BS, 1962)

swing-hammer regulator A simple method of regulating the flow of lump ore in a chute. It consists of several heavy pivoted hammers that allow fine ore to pass through, but check the passage of lumps. (Nelson, 1965)

swinging a claim The adjustment of the boundaries of a mining claim to more nearly conform to the strike of the vein. A reasonable time is allowed the discoverer to explore the vein or lode to find out its strike and make the adjustment.

swinging-electrode controller This controller is made up of three fixed electrodes consisting of groups of parallel plates of noncorroding alloy fixed at the bottom of curved troughs of insulating material of uniform width and varying depth. The trough is deep at the end, corresponding to full speed and minimum resistance; it is shallow at the maximum-resistance starting position. The moving electrodes, of similar construction, are joined to form the star point of the rotor and are moved toward or away from the fixed electrodes, giving a wide range of resistance. (Sinclair, 1959)

swinging-gate anemometer An instrument of the steady deflection type where speed is read off directly from the scale of the instrument. This is most useful for measuring low speeds, since it permits a spot reading. This instrument does not integrate and is used extensively in work connected with the ventilation of building interiors and to a fair extent underground. (Roberts, 1960)

swinging plate An amalgamated copper plate hung in a sluice to catch float gold. (Fay, 1920)

swinging-vane anemometer This instrument consists essentially of a damped, pivoted vane that is deflected when placed in an airstream. As the weight of the vane is constant, the angle of inclination will be dependent upon the rate of change of momentum of the impinging airstream. The instrument gives a direct reading and can be calibrated for use over a wide range of velocities, from 20 to 2,000 ft/min (6.1 to 609.6 m/min). In underground airways, it can be used without attachments. Its main use is the measurement of air velocity in ducts and the rate of air discharge from ventilating grills. (Roberts, 1960)

swing-jib crane A crane with one horizontal boom on which there is a counterweight. It can swing through a full circle. See also: *tower crane.* (Hammond, 1965)

swing loader A tractor loader that digs in front and can swing the bucket to dump to the side of the tractor. (Nichols, 1976)

swing loose Arkansas. To gradually loosen over a considerable area and sag; said of the rock over a mine working.

swing of a lathe The largest diameter of work that can be carried between the centers of a lathe. In England, the swing refers to radius. (Crispin, 1964)

swing parting Arkansas. A parting some distance from the mouth of an entry. The loaded cars are left by the gathering driver to be taken out by a swing driver.

swing radius *swing.*

swing roller In a revolving shovel, one of several tapered wheels that roll on a circular turntable and support the upper works. (Nichols, 1976)

swing shift (a) Workday from 4 p.m. to midnight. (Nichols, 1976) (b) Occasionally refers to the 12 midnight to 8 a.m. shift. (Nichols, 1976) (c) Working arrangement in a three-shift continuously run plant that changes working hours at regular intervals. During swing the old morning shift becomes the new afternoon shift. The afternoon shift of the first period must work the morning shift of the next with only an 8-h break on the first day of change. (Pryor, 1963)

Swiss lapis A fraudulent imitation of lapis lazuli (lazurite) obtained by staining pale-colored jasper or ironstone with ferrocyanide. Also known as German lapis. See also: *false lapis.*

switch (a) A mine switch is a device for enabling a car or a trip of cars to pass from one track to another. The term switch is also frequently used in a loose sense to apply to the whole side track or turnout, and a car standing on a side track is frequently said to be standing on the switch. See also latches. (Kiser, 1929) (b) Eng. A mechanical device for opening and closing an electric circuit; a mechanism for shifting a moving body in another direction. (CTD, 1958):

switchback (a) A zigzag arrangement of a roadway (or rail tracks) for surmounting the grade of a steep hill or the slope wall of a surface or open-pit mine. Common in mountainous mining districts. (b) A hairpin curve. (Nichols, 1976) (c) *shunt back.*

switchgear This is a general term applied to switching, interrupting, controlling, metering, protective, and regulating devices, as well as assemblies of these devices with associated interconnections, accessories, and supporting structures. The term is used primarily in connection with generation, transmission, distribution, and conversion of electric power.

switch plate An iron plate on tramroads in mines used to change the direction of movement. Syn: *turnsheet*. (Standard, 1964)

switch point A movable tongue or rail for diverting a train from one track to another.

switch throw (a) The arrangement of levers by means of which a switch is thrown for the straight track or the turnout. (Jones, 1949) (b) The handle or lever by which a switch is operated.

swither A colloquial term used in the Wisconsin lead-mining region for an offshoot or branch of a main lode. (AGI, 1987)

swivel coupling (a) A coupling where one link is made so that it can be rotated independently of other links. When such a coupling is used, one or more cars can be rotated on a revolving dump without uncoupling from the rest of the trip. (Zern, 1928) (b) A coupling that gives complete rotary freedom to a deflecting wedge-setting assembly in boreholes. (Long, 1960)

swivel head (a) The assembly of a spindle, chuck, feed nut, and feed gears on a diamond-drill machine that surrounds, rotates, and advances the drill rods and drilling stem. On a hydraulic-feed drill, the feed gears are replaced by a hydraulically actuated piston assembly. Also called boring head; drill head; drilling head; gate. (Long, 1960) (b) In a diamond drill, the mechanism that rotates the kelly and drill string. (Nichols, 1976)

swivel-head bevel gear The bevel gear mounted on the outside of the drive quill in the swivel head of hydraulic-feed and/or some types of gear-feed diamond drills. The gear meshes with, and is driven by, a matching gear on the drill-motor shaft. (Long, 1960)

swivel hoisting plug *hoisting plug.*

swivel neck Syn. for a water or a mud swivel in borehole drilling. Also called gooseneck. (Long, 1960)

swivel plug *hoisting plug.*

swivel trough A short, adjustable-angle trough that permits turning the conveyor panline any amount up to 30°, either to the right or to the left. The position of the swivel is controlled by a roof jack and a pendulum. (Jones, 1949)

swivel vise A bench vise that may be rotated on its base to bring the work that it holds into better position. (Crispin, 1964)

sycee silver Pure, uncoined, lump silver of various sizes, usually stamped with a banker's or assayer's seal; used by the Chinese as a medium of exchange and reckoned by weight. The larger lumps, sometimes called shoes, are boat shaped and weigh about 1 lb troy (370 g). (Standard, 1964)

syenite A group of plutonic rocks containing alkali feldspar (usually orthoclase, microcline, or perthite), a small amount of plagioclase (less than in "monzonite"), one or more mafic minerals (esp. hornblende), and quartz, if present, only as an accessory; also, any rock in that group; the intrusive equivalent of "trachyte." With an increase in the quartz content, syenite grades into "granite." Its name is derived from Syene, Egypt. A.G. Werner in 1788 applied the name in its present meaning; the Egyptian rock is a granite containing much quartz.

syenodiorite A group of plutonic rocks intermediate in composition between syenite and diorite, containing both alkali feldspar (usually orthoclase) and plagioclase feldspar, commonly more of the former; also, any rock in that group. Generally considered a syn. of monzonite, but may also include both monzonite and rocks intermediate between monzonite and diorite (Streckeisen, 1967). See also: *monzonite.* (AGI, 1987)

syenogabbro A plutonic rock differing in composition from gabbro by the presence of alkali feldspar. (AGI, 1987)

sylvane Former name for native tellurium and for sylvanite.

sylvanite (a) A monoclinic mineral, $2[AuAgTe_4]$; soft; metallic; commonly in implanted crystals resembling written characters; sp gr, 8.1; in quartz veins; a source of gold and silver. Also spelled silvanite. Syn: *graphic ore; graphic tellurium; white tellurium; yellow tellurium; goldschmidtite.* (b) An old name for native tellurium. (Dana, 1914)

sylvester A hand-operated device for withdrawing supports from the waste or old workings. The appliance enables a leverage of about 30 to 1 to be applied. A long chain allows it to be positioned a safe distance from the support to be extracted. It may also be used for applying tension or for moving machines short distances. See also: *tension end; monkey winch.* (Nelson, 1965)

Sylvester process A three-step method for the recovery of manganese and iron from open-hearth slag and low-grade ores. (Osborne, 1956)

sylvine The name for potassium chloride found native in the salt deposits at Stassfurt, Germany. See also: *sylvite.* (Cooper, 1963)

sylvinite A mining term for the mixtures of sylvite and halite occurring in the Prussian salt deposits; mined as potassium ore. (English, 1938; AGI, 1987)

sylvite An isometric mineral, $4[KCl]$; cubic cleavage; bitter salty taste; soft; white; in evaporite deposits and around fumaroles; the chief source of potassium. Syn: *sylvine; leopoldite.*

symbiosis Two or more organisms living together to the mutual benefit of both. (Rogoff, 1962)

symbol A diagram, design, letter, color hue, abbrev., or other graphic device placed on maps, charts, and diagrams, that by convention, usage, or reference to a legend is understood to represent a specific characteristic, feature, or object, such as structural data, rock outcrops, or mine openings. (AGI, 1987)

symbols of crystal faces (a) The mathematical expressions for designating the position of crystal faces on coordinate axes. See also: *Miller indices; plane.* (b) Any sign or letter used in crystallography to designate a group of smaller faces. (AGI, 1987)

symmetrical dispersion In optical mineralogy, the dispersion that produces an interference figure with color distribution symmetrical to the trace of the axial plane and also to a line normal to it.

symmetrical fold A fold whose limbs have the same angle of dip relative to the axial surface. Cf: *asymmetric fold*. Syn: *normal fold*. (AGI, 1987)

symmetric spread *split spread.*

symmetry (a) Symmetry in crystallography results from periodic repetition. A symmetry element is the geometrical locus about which a group of repeating operations acts. It may be a center, a mirror plane, or a rotation axis. (b) A symmetry group is a collection of symmetry elements that may intersect at a point (point group) or that may be distributed in three-dimensional space (space group). See also: *point group.* (c) The symmetry of a fabric is the combined symmetry of all the elements making up the fabric. There are five possible symmetries: (1) spherical, for fabrics having the symmetry of a sphere; (2) axial, for fabrics having the symmetry of a spheroid; (3) orthorhombic, for fabrics having the symmetry of a triaxial ellipsoid; (4) monoclinic, for fabrics having one unique plane of symmetry; and (5) triclinic, for fabrics having no planes of symmetry.

symmetry axis *axis of symmetry; crystal axis.*

symon Red shale, Shropshire, United Kingdom. See also: *calamanco.*

symond strings Thin veins of calcium carbonate running through the coal. Also spelled simon strings.

symon fault A syn. of horseback, named after such a structure in the Coalbrookdale coalfield of England that was originally thought to be a large fault. See also: *horseback*. (AGI, 1987)

Symon's cone crusher A modified gyratory crusher used in secondary ore crushing that consists of a downward-flaring bowl within which is gyrated a conical crushing head. The main shaft is gyrated by means of a long eccentric that is driven by bevel gears. See also: *gyrasphere crusher.* (Newton, 1959)

Symon's disk crusher A mill in which the crushing is done between two cup-shaped plates that revolve on shafts set at a small angle to each other. These disks revolve with the same speed in the same direction and are so set as to be widest apart at the bottom. Feed is from the center, and the material is gradually crushed as it nears the edge and is then thrown out by centrifugal force. (Liddell, 1918)

sympathetic detonation (a) Detonation of an explosive material by means of an impulse from another detonation through air, earth, or water. (Dick, 1983) (b) The initiation of an explosive charge without a priming device by the detonation of another charge close by. Syn: *flash over*. (Meyer, 1981)

symplectic Said of a rock texture produced by the intimate intergrowth of two different minerals; sometimes the term is restricted to such textures of secondary origin. One of the minerals may assume a vermicular habit. Also, said of a rock exhibiting such texture, or of the intergrowth itself; i.e., symplectite. Also spelled: symplektic; symplectitic; symplektitic. Cf: *dactylitic.* (AGI, 1987)

symplectite An intimate intergrowth of two different minerals, sometimes restricted to those of secondary origin; also, a rock (igneous or thermally metamorphosed) characterized by symplectic texture. Also spelled: symplektite. Cf: *pegmatite.* (AGI, 1987)

symplesite A triclinic mineral, $Fe_3(AsO_4)_2{\cdot}8H_2O$; vivianite group; dimorphous with parasymplesite; soft.

symptomatic mineral *diagnostic mineral.*

synadelphite A triclinic mineral, $(Mn,Mg,Ca,Pb)_9(AsO_3)(AsO_4)_2(OH)_9{\cdot}2H_2O(?)$; pseudo-orthorhombic.

synantetic Proposed by Sederholm and applied to those primary minerals in igneous rocks that are formed by the reaction of two other minerals, as in kelyphite rims, reaction rims, etc. (Johannsen, 1931-38)

synchisite A rare, weakly radioactive, orthorhombic or monoclinic mineral, $Ca(Ce,Nd,Y,La)(CO_3)_2F$, usually found in pegmatites associated with aegirite, microcline, astrophyllite, fluorite, gadolinite, xenotime, cordylite, and catapleiite. May be related to parisite. Also spelled synchysite. (Crosby, 1955)

synchromesh A silent-shift transmission construction in which hub speeds are synchronized before engagement by contact of leather cones. (Nichols, 1976)
synchronal *synchronous.*
synchroneity The state of being synchronous or simultaneous; coincident existence, formation, or occurrence of geologic events or features in time, such as glacial synchroneity. Syn: *synchronism.* (AGI, 1987)
synchronic *synchronous.*
synchronism The state when the phase difference between two or more periodic quantities is zero; they are then said to be in phase. *synchroneity.* (NCB, 1964)
synchronous Occurring, existing, or formed at the same time; contemporary or simultaneous. The term is applied to rock surfaces on which every point has the same geologic age, such as the boundary between two ideal time-stratigraphic units in continuous and unbroken succession. It is also applied to growth (or depositional) faults and to plutons emplaced contemporaneously with orogenies. Cf: *isochronous; diachronous.* Syn: *synchronal; synchronic.* (AGI, 1987)
synchronous motor This type of motor has a stator similar to a squirrel cage motor, but the rotor has a direct-current field winding with salient poles equal in number to the stator poles. The direct current is supplied to the field winding through slip rings. In addition to the direct-current field windings, the rotor normally has a squirrel cage (amortisseur) winding that is used for starting. (Pit and Quarry, 1960)
synchysite Orthorhombic minerals, synchysite-(Ce) $CaCe(CO_3)_2F$, synchysite-(Nd) $CaNd(CO_3)_2F$, and synchysite-(Y) $CaY(CO_3)_2F$ (doverite) of the bastnäsite group; pseudohexagonal. *doverite.*
synclinal An obsolete form of syncline.—adj. Pertaining to a syncline. Cf: *anticlinal.* (AGI, 1987)
synclinal axis In geology, the central line of a syncline, toward which the beds dip from both sides. See also: *axis.* (Fay, 1920)
synclinal mountain A mountain whose geologic structure is that of a syncline. Cf: *anticlinal mountain.*
synclinal valley A valley whose geologic structure is a syncline. Cf: *anticlinal valley.*
syncline A fold in which the core contains the stratigraphically younger rocks; it is generally concave upward. Cf: *anticline.* See also: *synform; synclinal.* (AGI, 1987)
synclinorium (a) A compound syncline; a closely folded belt, the broad general structure of which is synclinal. Also called synclinore. (b) A major syncline composed of many smaller folds. See also: *geosyncline.* (Ballard, 1955)
syndicate man In bituminous coal mining, a person who works with a party of miners who operate machines for undercutting, drilling, and loading coal into cars at the working face and are paid on a basis of tonnage of coal mined. (DOT, 1949)
synergism (a) Action of two agents, usually two chemicals, to produce an end effect greater than or different from the sum of the effects of the two agents acting separately. (b) Used in metallurgy with reference to reagent combinations to obtain the maximum possible recovery of ore or metal. (Nelson, 1965)
synform A fold whose limbs close downward in strata for which the stratigraphic sequence is unknown. Cf: *syncline; antiform.* (AGI, 1987)
syngenetic (a) Said of a mineral deposit formed contemporaneously with, and by essentially the same processes as, the enclosing rocks. Cf: *epigenetic.* (AGI, 1987) (b) Said of a primary sedimentary structure, such as a ripple mark, formed contemporaneously with the deposition of the sediment. (AGI, 1987)
syngenetic deposit A deposit formed contemporaneously with the parent rock and enclosed by it. There are two types of syngenetic deposits, igneous and sedimentary. Some examples are nickeliferous sulfides, nontitaniferous magnetite, diamond, chromite, and corundum. (Lewis, 1964)
synkinematic *syntectonic.*
synneusis A mechanism by which small plagioclase crystals float into growing phenocrysts of potassium feldspar. Also, said of the texture of a rock showing such crystals. Etymol: Greek, to swim together. (AGI, 1987)
synorogenic Said of a geologic process or event occurring during a period of orogenic activity; or said of a rock or feature so formed. Cf: *syntectonic.* (AGI, 1987)
synorogenic pluton An igneous intrusion emplaced during a period of orogenic activity. (AGI, 1987)
syntaxy An intergrowth between two mineral species in which single or multiple unit cells coincide in size and shape, e.g., bastnaesite-synchisite. Cf: *distaxy; epitaxy; topotaxy; polycrystal.*
syntectic The adj. of syntexis.
syntectite A rock formed by syntexis.
syntectonic Said of a geologic process or event occurring during any kind of tectonic activity, or of a rock or feature so formed. Cf: *synorogenic.* Syn: *synkinematic.* (AGI, 1987)
syntexis (a) The formation of magma by melting of two or more rock types and assimilation of country rock; anatexis of two or more rock types. (AGI, 1987) (b) Modification of the composition of a magma by assimilation. (AGI, 1987) (c) Any kind of reaction between a rising body of magma and the crustal rocks with which it comes into contact.—Adj: syntectic. See also: *anatexis.* (AGI, 1987)
synthesis The production of a chemical compound by the union of elements or of simpler compounds or by the degradation of a complex compound, esp. by laboratory or industrial methods.
synthetic *synthetic stone.*
synthetic diamond (a) A diamond produced artificially by subjecting a carbonaceous material to extremely high temperature and pressure; currently and commonly called MM and/or manmade diamond. See also: *manmade diamond.* (Long, 1960) (b) A misnomer for sintered tungsten carbide. (Long, 1960)
synthetic gem One artificially made from chemicals. (Schaller, 1917)
synthetic mineral An artificial substance having all the properties of a mineral. (Hurlbut, 1964)
synthetic ore Material that is the equivalent of, or better than, natural ore, can be put to the same uses, and is produced by means other than ordinary concentration, calcining, sintering, or nodulizing.
synthetic ruby In chemical composition and in all their physical characters, including optical properties, synthetic ruby and synthetic sapphire are true crystalline ruby or sapphire, but they are produced in quantity in the laboratory by fusing pure precipitated alumina with a predetermined amount of pigmentary material. They can be distinguished from natural stones only by the most careful examination. Syn: *synthetic sapphire.*
synthetic sapphire *synthetic ruby.*
synthetic stone A man-made stone that has the same physical, optical, and chemical properties, and the same chemical composition, as the genuine or natural stone that it reproduces. Many gem materials have been made synthetically as a scientific experiment, but only corundum, spinel, emerald, rutile, garnet, quartz, chrysoberyl (alexandrite), opal, and turquoise have been made commercially and cut as gemstones for the jewelry trade. Syn: *imitation.* See also: *synthetic.*
syntron feeder A feeder placed under a bin, hopper, or ore pass opening (or raise) that vibrates by the use of magnetic force to distribute ore evenly onto a moving conveyor belt. It can by adjusted to regulate the flow through various degrees of vibration.
syphon brick A brick for tapping metal from the cupola, the primary object of which is to eliminate the tapping and botting up of the cupola tap hole each time metal is drawn off. With the syphon brick, the orifice from which the metal is drawn is continually open to the atmosphere, and the flow of metal is controlled by shutting the blast on and off. The case of control permits the use of quite small ladles at the cupola, so that there is no need for redistribution from large to small ladles. (Osborne, 1956)
Syrian garnet Trade name for almandine garnet, of gem stone quality. (CTD, 1958)
sysertskite Former name for iridosmine.
system (a) A standard, worldwide division; contains rocks formed during a fundamental chronologic unit, a period. An example is the Devonian system. (AGI, 1987) (b) The fundamental time-rock unit is the system. (AGI, 1987) (c) In crystallography, the division of first rank, in the classification of crystals according to form. The six systems ordinarily recognized are the isometric, tetragonal, hexagonal, orthorhombic (or rhombic), monoclinic, and triclinic; some divide the hexagonal system into hexagonal and trigonal. (d) Applied to the sum of the phases that can be formed from one, two (binary system), three (ternary system), or more components under different conditions of temperature, pressure, and composition. (Holmes, 1928) (e) The term system or general system of work means simply that the work, as it is commenced on the ground is such that, if continued, will lead to a discovery and development of the veins or orebodies that are supposed to be in the claim, or, if these are known, that the work will facilitate the extraction of the ores and mineral. (Ricketts, 1943)
systematic error Any error that persists and cannot be considered as due entirely to chance, or an error that follows some definite mathematical or physical law or pattern and that can be compensated, at least partly, by the determination and application of a correction; e.g., an error whose magnitude changes in proportion to known changes in observational conditions, such as an error caused by the effects of temperature or pressure on a measuring instrument or on the object to be measured. Cf: *random error.* (AGI, 1987)
systematic sampling error (a) An error that arises from some basic defect in the sampling or preparation process such that the result obtained is always either higher or lower than the true figure. Systematic errors are additive; i.e., if there are two sources of error, the total error is obtained by adding the individual errors. (Nelson, 1965) (b) An error due to some known physical law by which it might be predicted; those errors produced by the same cause affect the mean in the same sense and do not tend to balance each other but rather give a definite bias to the mean. An error that results from some bias in the

measurement process and is not due to chance, in contrast to random error. (McGraw-Hill, 1994)

systematic support The setting of timber or steel supports regularly at fixed intervals irrespective of the condition of the roof and sides; a support in accordance with a system specified in rules made by the manager of the mine. (Nelson, 1965)

systematic timbering Placing mine timbers according to a predetermined plan, regardless of roof conditions. (Zern, 1928)

systems of crystals The seven large divisions into which all crystallizing substances can be placed, namely isometric (or cubic), tetragonal, hexagonal, trigonal, orthorhombic (or rhombic), monoclinic, and triclinic. This classification is based on the degree of symmetry displayed by the crystals. (CMD, 1948)

szaibelyite A white to yellow acicular mineral, $MgBO_2(OH)$. Probably orthorhombic. Occurs in nodules; related to camsellite. Syn: *ascharite.* (Larsen, 1934; Dana, 1944)

szaskaite *smithsonite.* From Szaska, Hungary. (English, 1938)

szomolnokite A monoclinic mineral, $FeSO_4{\cdot}H_2O$; kieserite group.

T

taaffeite A hexagonal mineral, $BeMg_3Al_8O_{16}$; högbomite group; a dimorph of musgravite; violet-red.
tab Token, check, tally. Syn: *teller.* See also: *tally.* (Mason, 1951)
tabasheer Translucent to opaque and white to bluish white opaline silica of organic origin (deposited with the joints of a bamboo shoot), valued in the East Indies as a medicine and used in native jewelry. Also spelled tabaschir or tabashir.
tabby A mixture of lime with shells, gravel, or stones in equal proportions, with an equal proportion of water, forming a mass that when dry becomes as hard as rock; a substitute for bricks or stone in building. (Fay, 1920)
Tabbyite Trade name for a variety of solid asphalt found in veins in Tabby Canyon, UT.
table (a) *rotary table.* (b) A concentrating machine with a flat surface for separating finely crushed particles of ore or coal from gangue. (Weed, 1922) (c) In placer mining, a wide, shallow sluice box designed to recover gold or other valuable mineral from screened gravel. See also: *undercurrent.* (Hess)
table cut (a) An early style of fashioning diamonds in which opposite points of an octahedron were ground down to squares to form a large cutlet and a larger table, the remaining parts of the eight octahedral faces being polished. (b) A term used somewhat loosely to describe any one of the variations of the "bevel cut," provided it has the usual large table of that cut. Syn: *bevel cut.*
table diamond A relatively flat diamond of table cut. (Webster 3rd, 1966)
table flotation Flotation process practiced on a shaking table. Ore or coal is ground, deslimed, conditioned, and fed to table as thick slurry. Flotable particles become glomerules, held together by minute air bubbles and edge adhesion. These roll across and are discharged nearly opposite the feed end. The process is helped by jets of low-pressure air from piping set across table. Tailings work along deck to discharge end. (Pryor, 1963)
tableland (a) A general term for a broad, elevated region with a nearly level or undulating surface of considerable extent; e.g., South Africa. (AGI, 1987) (b) A plateau bordered by abrupt clifflike edges rising sharply from the surrounding lowland; a mesa. (AGI, 1987)
table spar *tabular spar; wollastonite.*
tabling Separation of two materials of different densities by passing a dilute suspension over a slightly inclined table having a reciprocal horizontal motion or shake with a slow forward motion and a fast return. (Bennett, 1962)
tabular (a) Said of a feature having two dimensions that are much larger or longer than the third, such as a dike, or of a geomorphic feature having a flat surface, such as a plateau. (AGI, 1987) (b) Said of the shape of a sedimentary body whose width to thickness ratio is greater than 50:1, but less than 1,000:1 (Krynine, 1948); e.g., a graywacke formation in a geosynclinal deposit. Cf: *blanket; prism.* (AGI, 1987) (c) Said of a sedimentary particle whose length is 1.5 to 3 times its thickness (Krynine, 1948). Cf: *prismatic.* (AGI, 1987) (d) Said of a metamorphic texture in which a large proportion of grains are tabular and have approximately parallel orientation (Hart, 1977). (AGI, 1987) (e) Said of a crystal form that shows one dimension markedly smaller than the other two. (f) Tabular crystals, such as wollastonite, may occur in tables, plates, disks, foliae, and scales. (Schieferdecker, 1959; AGI, 1987) (g) A mineral showing broad, flat surface, e.g., wollastonite. (Nelson, 1965) (h) *tabular deposit.* (i) Refers to crystals that have three distinctly different dimensions. Cf: *acicular; anisodesmic; equant.*
tabular crystal A crystal flattened parallel to any face. (Standard, 1964)
tabular deposit A flat tablelike or stratified bed; e.g., a coal seam. *tabular.* (Nelson, 1965)
tabular spar Wollastonite in tabular form. Syn: *table spar.*
tabular structure The structure of a mineral or rock that makes it tend to separate into plates or laminae. (AGI, 1987)
Taby cut A modified double-spiral cut, the benefits of which are that the holes are located vertically below one another on one and the same line to facilitate drilling. (Langefors, 1963)
tacharanite A monoclinic mineral, $Ca_{12}Al_2Si_{18}O_{51}{\cdot}18H_2O$.
tacheometer *tachymeter; subtense bar.*
tachometer An instrument for measuring speed. In mining it is used on hoists lifting cages, cars, or skim in shafts or slope. (Zern, 1928)
tachyhydrite A trigonal mineral, $CaMg_2Cl_6{\cdot}12H_2O$. Also spelled tachydrite or tachhydrite.
tachymeter A surveying instrument designed for use in the rapid determination from a single observation of the distance, direction, and elevation difference of a distant object; esp. a transit or theodolite with stadia hairs, or an instrument in which the base line for distance measurements is an integral part of the instrument. Syn: *multiplying constant; tacheometer.* (AGI, 1987)
tachymetry A method of rapid surveying using the tachymeter; e.g., the stadia method of surveying used in United States. (AGI, 1987)
tack (a) A small pillar of coal. (Fay, 1920) (b) Veinstone; gangue, etc. See also: *taking.* (Fay, 1920)
tackey Having a rough, catchy surface. (Gordon, 1906)
tackle An assemblage of ropes or wire cables and pulleys arranged for hoisting or pulling. (Long, 1960)
Tacoma process An electrolytic method for the production of iron powder. (Osborne, 1956)
taconite (a) A local term used in the Lake Superior iron-bearing district of Minnesota for any bedded ferruginous chert or variously tinted jaspery rock, esp. one that enclosed the Mesabi iron ores (granular hematite); an unleached iron formation containing magnetite, hematite, siderite, and hydrous iron silicates (greenalite, minnesotaite, and stilpnomelane). The term is specif. applied to this rock when the iron content, either banded or disseminated, is at least 25%. Also spelled taconyte. (AGI, 1987) (b) Since World War II, a low-grade iron formation suitable for concentration of magnetite and hematite by fine grinding and magnetic treatment, from which pellets containing 62% to 65% iron can be produced. (AGI, 1987)
taconite ore A type of highly abrasive iron ore now extensively mined in the United States. (Sandstrom, 1963)
taconyte *taconite.*
tactite A rock of complex mineralogical composition, formed by contact metamorphism and metasomatism of carbonate rocks. It is typically coarse-grained and rich in garnet, iron-rich pyroxene, epidote, wollastonite, and scapolite. Approximate syn: skarn. See also: *calc-silicate hornfels.* Cf: *garnetite.* (AGI, 1987)
taeniolite A monoclinic mineral, $KLiMg_2Si_4O_{10}F_2$; mica group; at Narsarsuk, Greenland, and Magnet Cove, AR. Also spelled tainiolite.
taenite An isometric mineral, (Fe,Ni); contains 27% to 65% nickel; occurs in iron meteorites as lamellae or strips flanking bands of kamacite. Also called gamma-nickel-iron (face-centered cubic). See also: *nickel iron.*
tag A numbered piece that a miner attaches to or places on each car a miner loads. The tag is removed at the tipple where the car is credited to the miner. See also: *tally; ticket.* Cf: *wedge rock.* (Fay, 1920)
tagged atom *tracer.*
Tagg's method (a) A graphical method of determining the resistivity of the ground. (AGI, 1987) (b) A method of interpretation of resistivity sounding data obtained with a Wenner array over a two-layered Earth. (AGI, 1987)
tagilite *pseudomalachite.*
tagline A line from a crane boom to a clamshell bucket that holds the bucket from spinning out of position. (Nichols, 1976)
tail (a) (also plural). The inferior, less valuable, or refuse part of anything; foots, bottoms, dregs; sediment. See also: *tailings.* (Fay, 1920) (b) The poor grade of ore slime at the lower end of the slime box as it flows from the stamps. (Standard, 1964) (c) The unexposed end of a brick or stone in a wall; a tailing. (Standard, 1964) (d) The rear of a shovel deck. (Nichols, 1976) (e) The anchor end of a cable excavator. (Nichols, 1976) (f) A bar or barrier formed behind a small isle or a skerry. Also called trailing spit; banner bank. (Schieferdecker, 1959)
tail anchor The anchor for a track cable, or the turn point for a backhaul line in a cable excavator. (Nichols, 1954)
tail beam A joist or beam that abuts against the header joist. Syn: *tail joist.* (Crispin, 1964)
tailblock (a) The boom foot and idler sprocket assembly on a ladder ditcher. (Nichols, 1976) (b) The block used to pull a slusher to the face.

tailboard *tailgate.*
tail chain A chain used in mine haulage; also, tail rope. (Korson, 1938)
tail crab In mining, a crab or winch for operating a tail rope. (Standard, 1964)
tailend (a) That part of a mining belt conveyor that consists of the tail section and, when required, a belt takeup, a telescopic section, and a loading station. (NEMA, 1961) (b) The end of a conveyor remote from the delivery point. See also: *tension end.* (Nelson, 1965)
tailend loading station *loading station.*
tailgate A subsidiary gate road to a conveyor face as opposed to a main gate. The tailgate commonly acts as the return airway and supplies road to the face. Syn: *barrier gate; tailboard.* See also: *bottom gate.* (Nelson, 1965)
tail house The buildings in which tailings are treated. (Fay, 1920)
tailing (a) Giving the proper angle, or elevation, in driving the poling boards in a heading. (Stauffer, 1906) (b) The part of a projecting brick or stone inserted in a wall. (Crispin, 1964)
tailing pit *catch pit.*
tailing pond Area closed at lower end by constraining wall or dam to which mill effluents are run. Clear water may be returned after settlement in dam, via penstock(s) and piping. (Pryor, 1963)
tailings (a) The gangue and other refuse material resulting from the washing, concentration, or treatment of ground ore. See also: *tail.* (Webster 3rd, 1966) (b) Those portions of washed ore or coal that are regarded as too poor to be treated further. (Standard, 1964) (c) Applied to sectional residue, e.g., table tailings, which is the residue from shaking screens and tables. (Nelson, 1965) (d) The reject from froth flotation cells. (BS, 1962)
tailings dam One to which slurry is transported, the solids settling while the liquid may be withdrawn. (Pryor, 1963)
tailings machine A machine for sifting the tailings and collecting the gold from the detritus after it has passed through the washer. (Nelson, 1965)
tailings settling tank A vessel to remove solids from the tailings effluent as in a coal washery. The tank is about 60 ft (18 m) in diameter and 10 ft (3 m) deep. The tailings are fed in at the center with a flocculant. As the suspension travels from the center to overflow at the perimeter of the tank, the solids settle out and the clear water overflows, is collected, and is returned to the washer for reuse. (Nelson, 1965)
tailings wheel A wheel carrying buckets or compartments on the periphery and used in conveying liquid, pulp, or sand from a lower to a higher level. (Fay, 1920)
tail joist A joist that has one end terminating against a header joist. *tail beam.* (Crispin, 1964)
taillight A light carried at the back end of a car, train, trip, or movable machinery.
tail of water The edge of water standing in mine workings. (BS, 1963)
tail pipe The suction pipe of a pump. (Fay, 1920)
tail pulley (a) The terminal pulley at the end of the conveyor opposite the normal discharge end. It is usually an idler pulley but may be a drive pulley. (NEMA, 1961) (b) The pulley or roller in the tail or foot section of a belt conveyor around which the belt runs. Also known as foot-section pulley. (Jones, 1949)
tailrace A trough or channel used for conveying the tailings; a channel for conducting water away from any plant or works. (Nelson, 1965)
tail rope (a) The rope that passes around the return sheave in main-and-tail haulage or a scraper loader layout. See also: *main rope.* (Nelson, 1965) (b) The rope used to draw the empties back into a mine in a tail-rope haulage system. (Zern, 1928) (c) A counterbalance rope attached beneath a cage when the cages are hoisted in balance. (Zern, 1928) (d) A hemp rope used for moving pumps in shafts. (Fay, 1920)
tail-rope boy *tail-rope coupler.*
tail-rope coupler In bituminous coal mining, one who works on a tail-rope haulage system, removing the haulage-cable hook from the rear of a train of empty cars that has been lowered down an inclined haulageway, and attaching the hook to the front of a train of loaded cars to be hauled to the surface. Also called tail-rope boy. (DOT, 1949)
tail-rope engineer In bituminous coal mining, one who operates a hoisting engine that draws the cable of a tail-rope haulage system used to raise and lower mine cars on tracks between the surface and a level in a mine. (DOT, 1949)
tail rope fireman In bituminous coal mining, a person who fires the boiler supplying steam for the engine that powers the tail-rope haulage system, a type of cable haulage used to raise and lower cars along an incline between the surface and a level in a mine. (DOT, 1949)
tail-rope haulage A single track system of rope haulage in which a double-drum haulage engine at the unloading terminal winds the main rope on one drum. The train or trip of cars is connected to the other end of the main rope and also to a tail rope that extends to the inner terminal in the mine, around a tail sheave, and then back over idler sheaves at one side of the haulageway to the other drum at the haulage engine. See also: *rope haulage.* (Lewis, 1964)
tail-rope man *tail-rope rider.*
tail-rope rider In bituminous coal mining, one who works on trains of cars hauled by tail-rope haulage system at mine, coupling and uncoupling cars, and hooking and unhooking cable to and from trains. Also called tail-rope man. (DOT, 1949)
tail-rope system A method of haulage in which one rope—the main rope—is attached to the front end of a trip of cars and another rope—the tail rope—is attached to the rear end of the trip. It is operated by a hoisting engine and two separate drums. (Hudson, 1932)
tails Can. Portion of tailings containing some mineral that cannot be economically removed. This is constantly assayed as it leaves the treatment plant so that recovery can be known and controlled at all times. Cf: *heads.* (Hoffman, 1958)
tail section The part of a mining belt conveyor that consists of the tail pulley, the framing, belt idlers if included, and means for attaching a belt takeup. (NEMA, 1961)
tail shaft The shaft in the tail or foot section of a belt or chain conveyor that supports either the tail pulley or the tail sprocket. (Jones, 1949)
tail sheave (a) An arrangement whereby a sheave is placed at the bottom of a shaft, and a rope is fastened to the bottom of one cage and then passed down around the sheave and up to the bottom of the cage in the other compartments; thus practically complete balancing is effected. (Lewis, 1964) (b) The pulley around which the tail rope of a scraper loader or main-and-tail haulage passes. See also: *turn pulley.* (Nelson, 1965)
tail swing The clearance required by the rear of a revolving shovel. (Nichols, 1976)
tail-track system The simplest form of track layout for car or trip loading. In this system, the track can merely be extended down the heading, or it can be turned right or left, and then turned back, U-fashion, in an adjacent heading. The major disadvantage is that trips must come out the same way they go in, meaning increased loss of time unless the changing track is very close. (Coal Age, 1966)
tail water The water downstream from a structure, as below a dam. (AGI, 1987)
tainiolite *taeniolite.*
take (a) A mineral-bearing area that a mine is permitted to work. Also called holding; parcel; taking. (BS, 1963) (b) The area or extent of coal that a coal mine owner has the right, under a lease, to mine and extract. See also: *concession system; royalty.* (Nelson, 1965) (c) Eng. The extent or area of a lease of mineral property, often several thousand acres. (Fay, 1920) (d) Lanc. To show or reveal gas. (Fay, 1920)
takeout Cumb. An outcrop. As a verb, to crop out. (Fay, 1920)
take the air (a) To measure the ventilating current. (Fay, 1920) (b) Applied to a ventilating fan as working well, or working poorly. (Fay, 1920)
takeup (a) In a belt-conveyor system, a tensioning device such as a carriage-mounted weight free to run downslope or a takeup pulley with weights hanging vertically below the belt near the feed end. (Pryor, 1963) (b) Any device for taking up slack or removing the looseness of parts due to wear or other cause. (Crispin, 1964) (c) *chain takeup.*
takeup pulley An idler pulley so mounted that its position is adjustable to accommodate changes in the length of the belt as may be necessary to maintain proper belt tension. (NEMA, 1961)
taking Eng. A mineral-land lease. See also: *take; tack.* (Fay, 1920)
takyr A surface depression containing clay and evaporites in south-central Asia. Also spelled takir.
Talbot continuous process A pig iron and ore process that depends upon the rapid oxidation of the impurities contained in pig iron by a liquid, highly ferruginous slag and that is carried out in the basic open-hearth furnace, generally of the tilting type. The essential feature of the process is to retain a certain amount of metal in the furnace (1) to dilute the impurities contained in the additions of pig iron, and (2) to supply the heat necessary to keep the slag very fluid. (Osborne, 1956)
Talbot process A process for protecting the inside of cast-iron pipes with a coating of sand and bitumen. (Hammond, 1965)
talc (a) A monoclinic and triclinic mineral, $2[Mg_6(OH)_4(Si_8O_{20})]$; basal cleavage; soft; has a greasy or soapy feel; easily cut with a knife; occurs as hydrothermal alteration of ultramafic rocks, low-grade metamorphism of siliceous dolomites in foliated, granular, or fibrous masses; an insulator, ceramic raw material, and lubricant. Originally spelled talck. See also: *steatite; soapstone.* (b) In commercial usage, a talcose rock; a rock consisting of talc, tremolite, chlorite, anthophyllite, and related minerals. Syn: *talcum.*
talcite (a) A massive variety of talc. (Fay, 1920) (b) Damourite, a soapy-feeling hydromuscovite.
talcoid Resembling talc, as talcoid schist. (Fay, 1920)
talcose (a) Pertaining to or containing talc; e.g., talcose schist. (b) Resembling talc; e.g., a talcose rock that is soft and soapy to the touch.
talc schist A schist in which talc is the dominant schistose mineral. Common associates are mica and quartz. (AGI, 1987)
talcum *talc; soapstone.*

talking Applied to a series of small bumps or cracking noises within the walls. Bumping, talking, and spitting are signs that the rock is beginning to yield to the stresses and indicate a change in conditions within the rock. (Spalding, 1949)
tallow drop A style of cutting precious stones in which the stone is domed on one or both sides. (Fay, 1920)
tallow peat Ir. A variety of highly flammable peat. (Tomkeieff, 1954)
tallow top A precious stone with a very rounded front and a flat back. (Standard, 1964)
tally (a) A mark or number placed on every car of coal or ore a miner sends out, usually a metal ticket. By counting these, a tally is made of all the cars the miner sends out. See also: *tab; tag; ticket*. (Fay, 1920) (b) A brass tag attached to a chain at every tenth link, and so marked or shaped as to enable the position of the tally along the chain to be immediately read. (CTD, 1958)
tally boy *tally shouter.*
tallyman *chute checker.*
tally shouter A laborer who calls out the number chalked on each loaded mine car, as it is run on scales for weighing, so that the weighmaster can identify for pay purposes the miner who loaded the car. Syn: *tally boy*. (DOT, 1949)
talus Rock fragments of any size or shape (usually coarse and angular) derived from and lying at the base of a cliff or very steep, rocky slope. Also, the outward sloping and accumulated heap or mass of such loose broken rock, considered as a unit, and formed chiefly by gravitational falling, rolling, or sliding. See also: *scree*. Syn: *rubble*. (AGI, 1987)
talus creep The slow downslope movement of talus, either individual rock fragments or the mass as a whole. (AGI, 1987)
talus fan *alluvial fan.*
tam-o-shanter A very fine-grained, soft, gritty, natural stone found in Scotland. (Fay, 1920)
tamp (a) To tightly pack a drilled hole with moist, loose material after the charge has been placed. (Hudson, 1932) (b) To fill a charged shothole with clay or other stemming material to confine the force of the explosion. See also: *stemming*. (CTD, 1958) (c) To ram or pound down ballast on a railway track, or road metal. (CTD, 1958) (d) *stem.*
tamper (a) In bituminous coal mining, one who fills drill holes in which explosives have been charged, by machine driller or miner, with clay or some other tamping material, using a tamping bar. Also called shot tamper. (DOT, 1949) (b) An implement for tamping or compacting material; a tamping iron or tamping bar. Sometimes made of wood, copper, or iron with a copper tip. See also: *tamping bar*. (Standard, 1964; Fay, 1920) (c) One who tamps. (Standard, 1964) (d) A tool for compacting soil in spots not accessible to rollers. (Nichols, 1976)
tamping (a) The act of inserting and packing explosives and stemming in a shothole. See also: *stemmer*. (Nelson, 1965) (b) The act of packing a drilled hole around a cartridge with fine dirt from the floor of a mine before blasting, to prevent a misdirection of the force of the blast. (Korson, 1938) (c) The material placed over a charge in a borehole, to better confine the force of the explosion to the lower part of the hole. (Stauffer, 1906) (d) Ramming down, as of ballast. (Crispin, 1964) (e) The operation of compacting freshly placed concrete by repeated blows. (Taylor, 1965) (f) The shaping of a semidry powder, e.g., of refractory material, in a mold by repeated blows delivered mechanically on the top mold plate. (Dodd, 1964)
tamping bag A paper bag that is filled with good stemming material such as sand for use in horizontal and upward-sloping holes. Plastic bags are also available for this purpose. See also: *tamp*. (Hammond, 1965)
tamping bar A piece of wood the size of a broom handle for pushing explosive cartridges and stemming into shotholes. See also: *tamper*. (Hammond, 1965; von Bernewitz, 1931)
tamping plug A plug of iron or wood used instead of tamping material to close up a loaded blasthole. (Standard, 1964)
tamping pole A pole of nonsparking material used to tamp in blast holes. Syn: *loading pole.*
tamping rod *stemmer; stemming rod.*
tamping roller One or more steel drums, fitted with projecting feet, and towed by means of a box frame. (Nichols, 1976)
tamping stick *stemming rod.*
tandem (a) A double-axle drive unit for a truck or grader. A bogie. (Nichols, 1976) (b) A pair in which one part follows the other. (Nichols, 1976)
tandem drive A three-axle vehicle having two driving axles. (Nichols, 1976)
tandem-drive conveyor A conveyor having a belt drive mechanism in which the conveyor belt is in contact with two drive pulleys, both of which are driven with the same motor. (NEMA, 1961)
tandem hoisting Hoisting in a deep shaft with two skips running in one shaft. The lower skip is suspended from the tail rope of the upper skip. Both are loaded and discharged simultaneously. The upper one discharges at the surface and is loaded at a pocket halfway down the shaft. The lower skip is loaded at the shaft bottom and discharges at the half-way pocket. Thus, the rope on the winding drum is only equal to half the full depth. See also: *two-stage hoisting.*
tandem hydroseparator A two-celled hydroseparator with troughs. The raw coal feed is conveyed through a trough by water under pressure where the refuse stratifies to the bottom. The action in the first cell is that of a forceful upward current, which results in the removal of the heavy refuse. In the second cell a lighter current permits the settling of lighter and smaller refuse. The refuse settles to a perforated cell deck where it joins the slowly moving slate bed to the discharge. Refuse discharge is controlled by a refuse gate or hinged plate at the end of the cell bed. (Kentucky, 1952)
Tandem support system A trade name for a longwall steel support system. It consists of two 50-st (45-t) chocks in line at right angles to the face and linked together with a double-acting ram. In operation, the front chock is lowered and advanced with the conveyor and reset to the roof; the rear chock is then lowered and brought forward. (Nelson, 1965)
tandem unit panel A longwall conveyor face with two face conveyors of different capacities, one delivering on to the other—tandem fashion. The layout has the disadvantage that the whole tonnage of coal must be transported along the second conveyor, and any breakdown on the second conveyor will affect the output of the entire face. (Nelson, 1965)
tanette A small hill covering a residual surface of laterite, frequently ore bearing. (Hess)
tangawaite *bowenite.*
tangeite Former name for calciovolborthite.
tangent (a) A straight line that touches, but does not transect, a given curve or surface at one and only one point; a line that touches a circle and is perpendicular to its radius at the point of contact. (AGI, 1987) (b) The part of a traverse included between the point of tangency (the point in a line survey where a circular curve ends and a tangent begins) of one curve and the point of a curvature (the point in a line survey where a tangent ends and a circular curve begins) of the next curve. (AGI, 1987) (c) A great-circle line that is tangent to a parallel of latitude at a township corner in the U.S. Public Land Surveys system. (AGI, 1987) (d) A term sometimes applied to a long straight line of a traverse whether or not the termini of the line are points of curve. (AGI, 1987) (e) The ratio of the length of the leg opposite an acute angle in a right-angled triangle to the length of the leg adjacent to the angle. Adj. said of a line or surface that meets a curve or surface at only one point. (AGI, 1987)
tangent distance The distance from the point of curvature to the point of intersection (vertex), or from the point of intersection to the point of tangency. See also: *intersection point*. (Seelye, 1951)
tangential stress (a) Stress parallel to the tangent to the boundary of any opening. (Obert, 1960) (b) *shear stress.*
tangent point The point at which a curve meets a straight line or another curve. (Hammond, 1965)
tangent screw A very fine, slow-motion screw giving a tangential movement for making the final setting to a precision surveying instrument (such as for completing the alignment of sight on a theodolite or transit by gentle rotation of the reading circle about its axis). Cf: *Stebinger drum*. (AGI, 1987)
tangiwai A variety of serpentine used by the Maoris for ornaments. Similar to bowenite. See also: *bowenite*. (Dana, 1914)
tangiwaite A term used by the Maoris of New Zealand for bowenite, a massive variety of antigorite. Also spelled tangiwai or tangawaite.
tangle sheet Mica with intergrowths of crystals or laminae resulting in books that split well in some places but tear to produce a large proportion of partial films. (Skow, 1962)
tank (a) A large vessel or receptacle, made either of wood or of metal, intended to contain a fluid or gas, as water tank, gasoline tank. Syn: *vat*. (Fay, 1920) (b) A melting unit, in which the container for the molten glass is constructed from refractory blocks. (ASTM, 1994)
tankage (a) The act or process of storing oil, etc., in a tank. (Fay, 1920) (b) The price charged or paid for storage in a tank. (Fay, 1920) (c) The capacity of a tank or tanks. (Fay, 1920) (d) The waste residue deposited in lixiviating vats or tanks. (Fay, 1920)
tank furnace Essentially a large box of refractory material holding from 6 to 200 st (5.4 to 181 t) of glass, through the sides of which are cut ports fed with a combustile mixture (producer gas and air, coke oven gas and air, or oil spray and air), so that flame sweeps over the glass surface. With the furnace is associated a regenerative or recuperative system for the purpose of recovering part of the heat from the waste gas. (CTD, 1958)
tank station *station.*
tantalite A mineral series ferrotantalite-manganotantalite; unless specified it refers to ferrotantalite, an orthorhombic mineral, $FeTa_2O_6$; black; in pegmatites; the main source of tantalum.
tantalum A rather brittle, lustrous, hard, heavy, gray metallic element. Symbol, Ta. Occurs principally in the mineral columbite-tantalite, $(Fe,Mn)(Nb,Ta)_2O_6$. Widely used to fabricate chemical process equipment, nuclear reactors, and aircraft and missile parts. Used to make

electrolytic capacitors, vacuum furnace parts, and surgical appliances. (Handbook of Chem. & Phys., 3)

tantalum borides Several borides are known, including the following: TaB_2, melting point, 3,200 °C; sp gr, 12.5; thermal expansion, 5.5×10^{-6}; TaB, melting point, 2,400 °C; sp gr, 14.3; Ta_3B_4, melts incongruently at 2,650 °C; sp gr, 13.6. (Dodd, 1964)

tantalum nitrides Two nitrides are known: TaN, melting point, 3,090±50 °C; Ta_2N, which loses nitrogen at 1,900 °C. (Dodd, 1964)

tanteuxenite An orthorhombic mineral, $(Y,Ce,Ca)(Ta,Nb,Ti)_2(O,OH)_6$; black. Syn: *delorenzite.*

tantite A possibly triclinic mineral, Ti_2O_5.

Tanzanite A blue to violet gem variety of zoisite.

tap (a) To cut or bore into old workings or water-bearing strata for the purpose of proving or extracting gas or water. (BS, 1963) (b) To intersect with a borehole and withdraw or drain the contained liquid as water from a water-bearing formation or from underground workings. (Long, 1960) (c) To drive one passageway into another. (Hudson, 1932) (d) To win coal in a new district. (Fay, 1920) (e) A threaded cone-shaped fishing tool. It may be either an inside or an outside tap, depending on whether the tap fits into or over the outside of a piece being fished. Syn: *tapered tap.* (Long, 1960) (f) A quantity of a liquid, as molten metal from a furnace, run out at one time. (Webster 3rd, 1966) (g) To drain a furnace. (ASTM, 1994) (h) To remove excess slag from the floor of a pot furnace. (ASTM, 1994)

tap bar A pointed bar by which a blast-furnace tap-hole is opened or the metal in a melting pot, etc., is tested. Syn: *tapping bar.* (Standard, 1964)

tape A continuous ribbon or strip of steel, invar, dimensionally stable alloys, specially made cloth, or other suitable material, having a constant cross section and marked with linear graduations, used by surveyors in place of a chain for the measurement of lengths or distances. (AGI, 1987)

tape corrections These are applied as a routine matter to slope, temperature, sag, standardization, gravity and sea level effect when measuring a length accurately with a tape. See also: *precision; tension correction.* (Hammond, 1965)

taper A gradual and uniform decrease in size, as a tapered socket, a tapered shaft, a tapered shank. (Crispin, 1964)

taper bit A long cone-shaped noncoring bit used in drilling blastholes and in wedging and reaming operations. When the nose of the bit is rounded and the overall shape resembles the silk end of a corncob, the bit often is called a corncob bit. Cf: *bullnose bit.* (Long, 1960)

tapered core bit A core bit having a conical diamond-inset crown surface tapering from a borehole size at the bit face to the next larger borehole size at its upper, shank, or reaming-shell end, as from EX to AX, or BX to NX. (Long, 1960)

tapered end An end of rope having a reduced diameter to facilitate threading the rope through fittings and over pulleys. (Hammond, 1965)

tapered-flange beam The common form of rolled steel joist in which the inner surfaces of the flanges are tapered, normally at an angle of 98° to the web. See also: *web.* (Hammond, 1965)

tapered reamer A reamer having a conical diamond-inset surface tapering from any borehole size at its lower (bit) end to the next larger borehole size at its upper (core barrel) end, such as EX to AX, AX to BX, BX to NX. (Long, 1960)

tapered step-core bit *tapered step-face bit.*

tapered step-face bit A tapered core bit having the cutting face set in the same manner as a step-face bit. (Long, 1960)

tapered tap *tap.*

taper-lock sprocket A sprocket with a split tapered bushing for rigid mounting on a shaft. (Jackson, 1955)

taper of thread Measurement in inches of taper or slope for a 1-ft (0.3-m) length of the threaded section of rod or pipe. (Brantly, 1961)

taper pin A straight-sided pin that is smaller at one end than at the other. (Nichols, 1976)

taper rope A rope that has a gradually diminishing diameter from the upper to the lower end. The diameter of the rope is decreased by dropping one wire at a time at regular intervals. Both round and flat ropes may be made tapered, and such ropes are intended for deep shaft hoisting with a view to proportioning the diameter of the rope to the load to be sustained at different depths. (Zern, 1928)

taper-type dropper A device by which straight-type wedge can be attached to a diamond-drill rod, lowered, and set in a borehole. (Long, 1960)

tape-triangulation method A method of measuring mine roadway area in which a tape is stretched diagonally across the roadway. Offsets to the roof, floor, and sides are taken at right angles to the tape and on both sides of it. Alternatively, the floor of the cross section is divided into equal increments and vertical offsets to the roof are made at each division. Horizontal offsets to the sidewalls are made from the nearest adjacent vertical offsets. The measurements so obtained are plotted to scale, and the area of the resulting diagram is determined from the plot. Syn: *mine roadway area measurement.* (Roberts, 1960)

taphole A hole at or near the bottom of a furnace or ladle through which molten metal, matte, or slag can be tapped or drawn from a furnace. Also called tapping hole. Syn: *metal notch.* (Webster 3rd, 1966)

tapiolite The mineral series ferrotapiolite-manganotapiolite.

tappet (a) A sliding member working in a guide, interposed between a cam and the push rod or valve system that it operates, to eliminate side thrust on the latter. (CTD, 1958) (b) The collar under which a cam is inserted so as to lift a stamp. Also called disk. (Fay, 1920)

tapping (a) Opening the outlet of a melting furnace to remove molten metal. (ASM, 1961) (b) Removing molten metal from a furnace. (ASM, 1961) (c) The act of boring a hole into old workings to release gradually any accumulation of water and gas. This may be followed by driving an advance heading into the area. As the heading is extended, boreholes are kept in advance of the face to prevent the sudden breakthrough of water. See also: *inrush of water.* (Nelson, 1965)

tapping assembly A mechanical device consisting of a short piece of casing cemented in the collar of a borehole at the upper end of which is affixed a gate or large plug valve followed by a rod stuffing box. Utilizing this assembly, underground drilling can be accomplished safely in areas of high hydrostatic pressure. (Long, 1960)

tapping bar *tap bar.*

tapping clay A plastic clay used in plugging the tap-hole of a smelting furnace. (Standard, 1964)

tapping old workings Boring a hole into old workings to release gradually any accumulation of water and gas. The borehole tapping may be followed by driving an advance heading into the area. As the heading is extended, boreholes are kept in advance of the face to prevent the sudden breakthrough of water. See also: *inundation; inrush of water.* (Nelson, 1965)

tar (a) A thick brown to black viscous organic liquid, free of water, which is obtained by condensing the volatile products of destructive distillation of coal, wood, oil, etc. It has a variable composition, depending on the temperature and material used to obtain it. (AGI, 1987) (b) Any of various dark brown or black, bituminous, usually odorous, viscous liquids or semiliquids that are obtained by the destructive distillation of wood, coal, peat, shale, and other organic material, and yield pitch on distillation. (Webster 3rd, 1966) (c) Soft pitch or thickened petroleum, found in cavities of some limestones. (Fay, 1920)

taramellite An orthorhombic mineral, $Ba_4Fe_4(B_2Si_8O_{27})O_2Cl$; forms a series with titantaramellite; forms reddish-brown radiating fibrous aggregates; at Candoglia, Piemont, Italy.

taramite A monoclinic mineral, $Na_2CaFe_3Al_2(Si_6Al_2)O_{22}(OH)_2$; amphibole group with magnesium replacing iron toward magnesiotaramite; black; at Wali-tarama, Mariupol, Ukraine.

tarapacaite An orthorhombic mineral, K_2CrO_4; has the olivine structure; yellow.

tarasovite An interstratified mica and clay.

tarbuttite A triclinic mineral, $Zn_2(PO_4)(OH)$; forms a series with paradamite; at Broken Hill, Kabwe, Zambia. Syn: *salmoite.*

tare (a) To weigh mine cars when empty in order to determine the weight of coal in a car when the loaded car is weighed, done at specific intervals so that miners paid on a tonnage basis may receive proper credit for coal that they have loaded. (BCI, 1947) (b) Allowance for weight of packing or container in which goods are moved. The difference between gross and net weight. (Pryor, 1963)

target (a) Sliding weight on a leveling rod used in surveying, to enable the staffman to read the line of collimation. In underground leveling, a bead on a hanging plumbline used for the same purpose; distance from this to the roof or working is then measured. (Pryor, 1963) (b) The point a borehole or exploration work is intended to reach. (c) The distinctive marking or instrumentation of a ground point to aid in its identification on an aerial photograph. It is a material marking so arranged and placed on the ground as to form a distinctive pattern over a geodetic or other control-point marker, on a property corner or line, or at the position of an identifying point above an underground facility or feature. (AGI, 1987) (d) The image pattern on an aerial photograph of the actual mark or target placed on the ground prior to photography. (AGI, 1987) (e) The vane or sliding sight on a surveyor's level rod; a device, object, or point upon which sights are made. (AGI, 1987)

target rod A type of leveling staff provided with a sliding target, which can be moved by the staffman, under direction from the leveler, to a position in which it is in line with the line of sight of the level, the staff reading being recorded by the staffman. (CTD, 1958)

tarmacadam Asphalt that is made artificially from grit, crushed stone, or gravel and is bonded or coated with tar or a tarbitumen mixture. See also: *premix.* (Nelson, 1965)

tarnish A thin alteration film that forms on mineral surfaces, esp. on copper minerals, with color and luster different from that on fresh fractures.

Tarnowitz process A metallurgical process in which large charges of lead ore are roasted at low temperatures in furnaces and treated substantially, such as in the Carinthian process. The residual containing considerable lead is remelted in special furnaces. (Fay, 1920)
tarring (a) The coating of piles used for permanent work with prepared acid-free tar before driving. The tar is obtained from the high-temperature carbonization of coal in horizontal retorts. (Nelson, 1965) (b) The act of coating, (as of a pipe) with tar.
task The number of tons or the amount of ore or material that can or should be loaded either by mechanical loaders or by hand loaders. Also called score. (Jackson, 1936)
task training Specific training given to a miner prior to performing a task where the worker has had no previous work experience. Cf: *new miner training; refresher training.* (Federal Mine Safety, 1977)
Tasmanian alexandrite Alexandrite (chrysoberyl) of good gem quality from Tasmania, Australia.
tasmanite An impure coal, transitional between cannel coal and oil shale. Syn: *combustible shale; yellow coal; Mersey yellow coal; white coal.* (AGI, 1987)
tasmanite shale *tasmanite.*
tatarskite An orthorhombic mineral, $Ca_6Mg_2(SO_4)_2(CO_3)_2Cl_4(OH)_4{\cdot}7H_2O$; with anhydrite in saline strata of the Caspian depression.
tator butt Shrop. Fragile sandstone.
tautline cableway The tautline cableway differs from the aerial tramway in that its operation is limited to the distance between two towers (not more than 3,000 ft or 915 m apart), it has only one carrier, and the traction cable is reeved at the carrier so that loads can be raised and lowered. Also, the tautline cableway is not restricted to a fixed position; the towers can be mounted on trucks or crawlers, and the machine then can be shifted across a wide area. The machine will hoist loads from any point under the span, convey these loads in either direction, and lower these loads at any point under the span. By using movable towers, an area of any length can be traversed. Equipped with slings, this machine will pick up and carry unwieldy loads of every kind; then by exchanging the slings for a skip, it will handle large chunks of ore, stone, etc., or it can be equipped with a dump bucket to handle any bulk material, including semifluid mixtures. (Pit and Quarry, 1960)
tautomeric Descriptive term for amphoteric substance, able to react in accordance with two oppositely directed structural arrangements of its atoms. (Pryor, 1963)
tavistockite An old name for an apatite mineral.
tavorite A triclinic mineral, $LiFe(PO_4)(OH)$; amblygonite group; forms green or yellow fine-grained aggregates; in Brazil.
tawmawite A yellow to green to dark-green variety of epidote containing chromium; occurs in Tawmaw, upper Burma (Myanmar).
taxoite Serpentine from Chester County, PA. (Schaller, 1917)
taylorite (a) An obsolete name for bentonite; named after William Taylor who made the first commercial shipments of the clay from the Rock Creek district, Wyoming. (b) A variety of arcanite containing ammonium found in guano beds on islands off Peru.
Taylor producer A furnace used for the manufacture of producer gas. (Fay, 1920)
Taylor-White process A process for heat-treating high-speed steels. (Webster 3rd, 1966)
T-bolt A bolt with a T-shaped head, made to fit into a T-shaped slot in a drill swivel head; by means of it the swivel head can be turned to any angle of inclination to drill a borehole. Also, a similar bolt made to fit into a T-slot in the bed of a machine, for the purpose of holding a piece of metal to be machined or to fasten a machine to its base. (Long, 1960)
tchesa stick (a) An igniting stick used to light powder fuses when firing a round of shots. Also called a fire stick. (Pryor, 1963) (b) A paper shell about 1/4 in (0.64 cm) in diameter and 8 in (20.3 cm) long, filled with a balanced combustible that gives a strong spitting flame of 1 min duration. This device requires the individual lighting of each fuse. (Lewis, 1964) (c) *fuse lighter.*
T-chisel A boring tool with its cut-edge made in the form of the letter T. (Fay, 1920)
teallite An orthorhombic mineral, (Pb,Sn)S; perfect basal cleavage; soft; blackish gray; sp gr, 6.4; in veins; a source of lead and tin.
teardrop set A surface-set damond-bit crown molded in a die, prepared so that each inset diamond is backed by a raised teardrop-shaped mound of matrix metal. (Long, 1960)
tear fault A steep to vertical fault associated with a low-angle overthrust fault and occurring in the hanging wall. It strikes perpendicular to the strike of the overthrust; displacement may be horizontal, and there may be a scissor effect. It is considered by some to be a type of strike-slip fault. (AGI, 1987)
teary ground (a) Ground easily broken and worked. (Gordon, 1906) (b) Corn. A lode or stratum that breaks easily by reason of many joint planes. (Fay, 1920)
teasing rods Light iron rods, about 2 ft (0.6 m) long, hinged together to form one continuous length of 40 to 60 ft (12 to 18 m). They are pushed up inside a drainage borehole casing to clear stoppages of pebbles and gravel, thus allowing the drainage water to flow freely. (Eng. Min. J., 1938)
teaze hole The opening of a glass furnace through which fuel is introduced. (Standard, 1964)
tectogene (a) A long, relatively narrow unit of downfolding of sialic crust considered to be related to mountain-building processes. (AGI, 1987) (b) The downfolded portion of an orogen. Syn: *downbuckle.* (AGI, 1987)
tectogenesis *orogeny.*
tectonic Said of or pertaining to the forces involved in, or the resulting structures or features of, tectonics. Syn: *geotectonic.*
tectonic breccia A breccia formed as a result of crustal movements, usually developed from brittle rocks. Cf: *fault breccia; fold breccia; crush breccia.* (AGI, 1987)
tectonic conglomerate *crush conglomerate.*
tectonic gap *lag fault.*
tectonic map A map that portrays the architecture of the outer part of the Earth. It is similar to a structure-contour map, which primarily shows dipping strata, folds, faults, and the like, but the tectonic map also presents some indication of the ages and kinds of rocks from which the structures were made, as well as their historical development. (AGI, 1987)
tectonics A branch of geology dealing with the broad architecture of the outer part of the Earth; i.e., the regional assembling of structural or deformational features, a study of their mutual relations, origin, and historical evolution. It is closely related to structural geology, with which the distinctions are blurred, but tectonics generally deals with larger features. Adj: tectonic. Syn: *geotectonics.* (AGI, 1987)
tectonism *diastrophism.*
tectonite Any rock whose fabric reflects the history of its deformation; a rock whose fabric clearly displays coordinated geometric features that indicate continuous solid flow during formation (Turner and Weiss, 1963). Also spelled tektonite. (AGI, 1987)
tectonometer A radar apparatus that detects changes in rock structure, particularly faults. (Nelson, 1965)
tectonophysicist One who studies clastic deformation, flow, and rupture of constituent materials of the Earth's crust and mantle to make deductions concerning the forces causing these deformations (changes). (DOT, 1949)
tectonosphere The outer part of the Earth above the level of isostatic equilibrium, in which the dynamic processes are thought to occur that cause orogenesis near and at the surface. Cf: *crust.* (Schieferdecker, 1959)
tedge In founding, an ingate in a mold. (Standard, 1964)
tee (a) Eng. A crossvein meeting a main vein without intersecting it. (b) A sleeve with a third opening in the side, usually at right angles, to allow a branch line to be connected to the main pipeline. (Kentucky, 1952) (c) A fitting, either cast or wrought, that has one side outlet at right angles to the run. A single outlet branch pipe. (Strock, 1948)
tee-beam A rolled steel section in the shape of the letter T, the flat top being the table. (Hammond, 1965)
tee-bolt *T-bolt.*
teem To pour molten metal from a ladle into ingot molds. The term applies particularly to the specific operation of pouring either iron or steel into ingot molds. Also called teeming. (ASM, 1961)
teemer (a) A pourer of metal. (Standard, 1964) (b) One who controls the rate of pouring (teeming) stainless steel into molds. (DOT, 1949) (c) The person who teems or casts the pot of glass. See also: *casting.* (ASTM, 1994)
teeming (a) Shaping glass by pouring it into or on molds, tables, or rolls. (b) *casting.*
teeming trough Lanc. A cistern (or trough) into which the water is pumped from a mine. (Fay, 1920)
teepleite (a) A tetragonal mineral, $Na_2B(OH)_4Cl$. (b) *burkeite; gauslinite.* From Borax Lake, CA. (English, 1938)
teeter (a) Dancing or boiling movement of small particles in a rising fluid column, when the velocity is too high to let them fall and too low to sweep them clear. Characteristic zone in hydraulic classifiers. (Pryor, 1963) (b) The condition of a suspension of solids in an upward-moving current of water or air, whereby the support given to the particles reduces the internal friction between them to such an extent that the suspension acquires fluid or partially fluid properties. (BS, 1962)
teineite An orthorhombic mineral, $CuTeO_3{\cdot}2H_2O$; blue; at the Teine Mine, Japan.
tekoretin Fossil hydrocarbon similar to fichtelite. (Tomkeieff, 1954)
tektite Glass spheroid, often with aerodynamic shape, found in strewn fields and associated with impact craters; each cluster of tektites is named for its locality, such as moldavites and australites. A tektite has been shaped by flight through the atmosphere while chilling and ablating and melted by meteorite impact. See also: *water chrysolite.*
tektonite *tectonite.*

telain (a) Anglicized from the German telit. Greater fragments of plant tissues, which are completely soaked with vitrain; i.e., the cell walls as well as the cell cavities. (AGI, 1987) (b) Used in the names of transitional coal lithotypes, e.g., clarotelain. (AGI, 1987)

telegraph A vertical rectangular timber or steel chute for the transfer of coal to a lower level. Strips of wood placed crosswise in the chute retard the downward flow, and the chute is kept full for the same purpose. (Nelson, 1965)

telemagmatic Said of a hydrothermal mineral deposit located far from its magmatic source. Cf: *apomagmatic; cryptomagmatic.* See also: *telethermal.* (AGI, 1987)

telemeter rod A leveling staff used in connection with stadia work. See also: *stadia rod.* (Hammond, 1965)

teleoperation The remote manual operation of equipment that is usually not within the direct eyesight of the operator, yet the operator requires and is provided with sensory information (sight, sound, accelerations, etc.) for effective manual control.

telephoto lens A combination of positive and negative lenses designed to obtain larger magnification of distant objects than is possible with ordinary lenses. (Seelye, 1951)

telescopic derrick A drill derrick divided into two or more sections, made so that the uppermost sections nest successively into the lower sections. In use, the sections are extended and locked into place to form a tall derrick and when moved are nested to form a unit length transportable on a single truck. (Long, 1960)

telescopic drill rig A mobile electric, hydraulic, four-drill rig for boring blasting holes in quarries and opencast pits. All drills, percussive and rotary, can be simultaneously or independently raised, lowered, or slewed, enabling the rig to serve a working face 32 ft (9.8 m) high and 24 ft (7.3 m) wide. (Nelson, 1965)

telescopic loading trough A shaker conveyor trough of two sections, one nested in the other, used near the face for advancing the trough line without the necessity of adding either a standard or a short length of pan after each cut. C-clamps hold the two sections together in any desired length. (Jones, 1949)

telescopic section That section of a rigid side-framed conveyor that is (1) adjustable in length, (2) immediately adjacent to the tail section, and (3) so designed that it forms a continuous framing and cover for the return belt when the tail section is pulled back to tension the belt. (NEMA, 1961)

telescoping conveyor A type of conveyor, the length of which may be varied by telescoping frame members. See also: *extendable conveyor.*

telethermal Said of a hydrothermal mineral deposit formed at shallow depth and relatively low temperatures, with little or no wall-rock alteration, presumably far from the source of hydrothermal solutions. Also, said of that environment. See also: *telemagmatic.* Cf: *hypothermal deposit; mesothermal; epithermal; xenothermal; leptothermal.* (AGI, 1987)

telinite (a) This term was proposed by W.J. Jongmans (1935) to designate a vitrinite showing cellular structure. The Nomenclature Subcommittee of the International Committee for Coal Petrology decided in 1957 to use the term telinite only for the cell walls seen in vitrinite. Only in this manner can telinite be rightly included among the macerals. Telinite shows more or less clearly defined cell structure (wood, periderm, etc.) sometimes deformed. The cells are generally filled with collinite, but the structure is better shown when the cells are either empty or filled by material such as resinite, fine micrinite, clay minerals, etc. (IHCP, 1963) (b) A maceral of coal within the vitrinite group, characteristic of vitrain and consisting of cell-wall material. Cf: *suberinite; xylinite.* (AGI, 1987)

teller *tab.*

telltale (a) A simple device for indicating selected conditions of loading, flow, direction, etc. (Pryor, 1963) (b) A device for keeping a check on employees (as factory hands, drivers, check takers), esp. a time clock. (Webster 3rd, 1966)

telluric bismuth A former name for tetradymite. See also: *tetradymite.*

telluric current Natural electric current that flows on or near the Earth's surface in large sheets. Syn: *earth current.* (AGI, 1987)

telluric-current prospecting A geophysical prospecting technique utilizing natural Earth currents as a source instead of artificially generated currents injected into the ground. (Dobrin, 1960)

telluric ocher The mineral tellurite, TeO_2. (Fay, 1920)

telluric silver *hessite.*

telluride A mineral that is a compound of a metal and tellurium, such as hessite, Ag_2Te.

tellurides Ores of the precious metals (chiefly gold) containing tellurium. (Gordon, 1906)

telluriferous Yielding or containing tellurium. (Standard, 1964)

tellurite An orthorhombic mineral, TeO_2; dimorphous with paratellurite; colorless to yellow; occurs in hydrothermal veins.

tellurium A trigonal mineral, Te, native tellurium; soft; sp gr, 6.2; semimetallic; in pyrite, sulfur, or in the fine dust of gold-telluride mines.

tellurium glance *nagyagite.*

tellurium mineral Primarily native tellurium, Te; tellurite, TeO_2; tetradymite, Bi_2Te_2S; or hessite, Ag_2Te.

tellurnickel *melonite.*

tellurobismuthite A trigonal mineral, Bi_2Te_3; tetradymite group; forms a series with tellurantimony.

tellurometer Trade name of a rugged portable electronic device that measures ground distances precisely by determining the velocity of a phase-modulated, continuous, microwave radio signal transmitted between two instruments operating alternately as master station and remote station. It has a range up to 35 to 40 miles (56 to 64 km). Cf: *geodimeter.* (AGI, 1987)

telpher An electric hoist that hangs from a power-driven wheeled cab rolling on an overhead rail; it is often referred to as a monorail. See also: *aerial ropeway.* (Hammond, 1965)

telpherage Automatic aerial transportation, as by the aid of electricity, esp. that system in which the carriages having independent motors are run on a stout wire conducting an electric current. (Standard, 1964)

Telsmith breaker A type of gyratory crusher often used for primary crushing. It has a fixed spindle; i.e., the spindle is not suspended from above, but is mounted in a long eccentric sleeve. Rotation of the sleeve imparts a gyratory motion to the crushing head, but gives a parallel stroke; i.e., the axis of the spindle describes a cylinder rather than a cone as in the suspended spindle gyratory. Adjustment for set in the Telsmith breaker is accomplished by placing shims between the bottom of the breaking head and an adjusting plate—the addition of shims at this point raises the crushing head and increases the throat opening. (Newton, 1959; Newton, 1938)

Telsmith gyrasphere A type of secondary crusher that utilizes the gyratory principle; it has a hemispherical crushing head. (Newton, 1959)

temblor *earthquake.*

temiskamite *maucherite.*

temperature (a) The heat content of a body as measured on a definite scale based on some observable phenomenon; e.g., the expansion of mercury on heating. See also: *absolute temperature; Celsius; centigrade; critical temperature; Fahrenheit; Kelvin temperature scale; Rankine scale.* (Webster 3rd, 1966) (b) A degree of hotness or of coldness measured on one of several arbitrary scales based on some observable phenomenon; e.g., the expansion of mercury on heating. The degree of a material substance that is a linear function of the kinetic energy of the random motion of its molecules. The degree of a vacuum that depends upon the density of the radiant energy within it. Abbreviations and symbols, temp; T; t; *T*; *t*. Cf: *absolute zero.* (Webster 3rd, 1966)

temperature coefficient A numerical value indicating the relation between a change in temperature and a simultaneous change in some other property (e.g., solubility). Specif., the factor *a* in the equation $R_t = R_0 (1 at)$, in which R_t equals the resistance of a conductor at t °C, and R_0 equals its resistance at 0 °C. Syn: *positive temperature coefficient.* (Webster 3rd, 1966)

temperature colors Colors shown to the eye by incandescent bodies at different temperatures.

temperature dew point Temperature at which condensation of water occurs; a saturation temperature. (Hartman, 1982)

temperature gradient (a) The rate of change of temperature with distance in a specified direction. Also called lapse rate. (AGI, 1987) (b) A curve showing the temperature at different distances from the hot face, in a refractory wall.

temperature logging The measurement of temperature in boreholes by use of a delicate thermometer that will record temperature anomalies of as much as 7 °F (3.9 °C) for thin coal seams in coal measures according to the thermal conductivity of the rocks concerned. (Sinclair, 1963)

temperature profile recorder A portable unit consisting of a thermistor sensing element, 6-V power supply, amplifier, and recorder. The recorder is geared to a drum containing an electrical cable to which the bead is fastened. When the bead is lowered into the water, the paper on the recorder is moved accordingly. Depth is measured by the amount of wire paid out. This device is used in shallow water, particularly in lakes. (Hunt, 1965)

temperature-regulating equipment Any equipment used for heating and cooling the rectifier together with the devices for controlling and indicating its temperature. (Coal Age, 1960)

temperature standards For normal measurement, 0 °C (regarding gas properties). For thermodynamics and physical properties, either 18 °C or 25 °C, as defined in each stated case. (Pryor, 1963)

temperature steel Reinforcement introduced into a concrete slab or other member to minimize any cracks arising from shrinkage or from temperature stresses. (Hammond, 1965)

temperature stress Stress in a structural member due to a rise or fall of temperature. See also: *temperature steel.* (Hammond, 1965)

temperature survey (a) A geophysical prospecting method that measures either (1) temperature anomalies in boreholes or (2) temperature trends and concentrations along the ground surface. For example, a temperature survey across a salt dome may give peak values in the central area, due to the high thermal conductivity of the buried salt mass. See also: *temperature logging*. (Nelson, 1965) (b) Measurement of temperature in drill holes. An absolute accuracy of about 0.05 °C and a precision of about 0.005 °C can be obtained. Maps of isotherm surfaces can be constructed that help to detect anomalies in geologic structure or subsurface ground-water conditions. (AGI, 1987)

tempered In brickmaking, (1) moistened and worked to the proper consistency, as clay for bricks or molding, and (2) capable of being cut with ease, as bricks made of such clay. (Standard, 1964)

tempered steel Steel that has been hardened and subsequently tempered by a second lower heating. (Fay, 1920)

temperer One who or that which tempers; specif., a machine for mingling and thoroughly working potter's clay, brick clay, mortar, plaster, or other materials. (Standard, 1964)

temper hardening A term applied to alloys that increase in hardness when heated after rapid cooling; also to the operation of producing this. Also called artificial aging; distinguished from aging or age hardening, which occurs at atmospheric temperature. Both processes are covered by the term precipitation hardening. (CTD, 1958)

tempering bar *furgen.*

tempering furnace A furnace for heating articles in the process of tempering. (Standard, 1964)

tempering machine A machine for giving large steel plates a uniform and thorough tempering without permitting them to bend or buckle; usually by pressing them between hot masses of iron, or by firmly clamping them between jaws or plates while immersing them in a tempering bath. (Standard, 1964)

tempering oven An oven for heating glass in the process of annealing; a leer. (Standard, 1964)

template (a) A form for building tunnel inverts. (Stauffer, 1906) (b) A pattern device used as a guide to mark points at which boreholes are to be collared in ring drilling. (Long, 1960)

temporary adjustment An adjustment, such as leveling or focusing, made to a surveying instrument at each setup. See also: *permanent adjustment.* (Hammond, 1965)

temporary hardness *carbonate hardness.*

temporary hardness of water The carbonate component of water hardness, which can be destroyed by boiling. See also: *hard water.* (Nelson, 1965)

temporary roof support In coal mining; during roof bolting process, vertical posts are installed tight with wedges near the area where the next roof bolt will be installed. These are installed by reaching from a bolted area of the roof. Installed to support potentially loose roof to prevent fall onto persons.

temporary shaft support A timber or steel lining inserted for a limited period until a permanent shaft support is installed. See also: *skeleton tubbing.* (Nelson, 1965)

temporary splice According to the Federal Coal Mine Safety Act, a temporary splice is one that does not have a rubber or neoprene jacket vulcanized over the splice and bonded to the cable jacket. (USBM, 1963)

tenacity (a) The property of the particles or molecules of a substance to resist separation; tensile strength. (AGI, 1987) (b) The force of strength with which the particles (or molecules) of a mineral or rock hold together or resist separation. The terms commonly used to describe the tenacity of a mineral are friable, brittle, sectile, malleable, flexible, elastic, and tough. (Stokes, 1955)

tender (a) Said of roof shale that tends to break up or crush under pressure into small fragments and that will not hold in any span over a few inches. Also called short. (Raistrick, 1939) (b) The formal offer by the tenderer to carry out the work described in the drawings and/or specification for a certain sum of money. See also: *agreement*. (Nelson, 1965)

tennantite An isometric mineral, $(Cu,Fe)_{12}As_4S_{13}$; tetrahedrite group; forms a series with tetrahedrite; may contain zinc, silver, or cobalt replacing copper; in veins; an important source of copper. Syn: *gray copper ore.*

tenon A projecting tongue fitting into a corresponding cavity called a mortise. (Fay, 1920)

tenor *grade.*

tenorite A monoclinic mineral, CuO; occurs in gray scales, black powder, or earthy masses; sp gr, 6.4; in oxidized zones of copper deposits; a source of copper. Syn: *melaconite; black copper.* See also: *black copper ore.*

tensile force A force such as the force applied when a haulage rope pulls a set of tubs. (Morris, 1958)

tensile strength The maximum applied tensile stress that a body can withstand before failure occurs. Syn: *tenacity.* See also: *ultimate tensile stress.* (AGI, 1987)

tensile stress A normal stress that tends to cause separation across the plane on which it acts. Cf: *compressive stress.* (AGI, 1987)

tensile test A test in which material is subjected to an increasing tensile pull until it fractures. A stress-strain curve may be plotted, and the limit of proportionality, proof stress, yield point, ultimate tensile stress, elongation, and reduction in area can be determined. (CTD, 1958)

tension (a) In subsidence, the amount of lengthening per unit of measurement. (Nelson, 1965) (b) In engineering, a pulling force or stress; metals in tension are strong, while concrete and masonry are weak. (Nelson, 1965) (c) A system of forces tending to draw apart the parts of a body, esp. of a belt, a line, a cord, or a sheet, combined with an equal and opposite system of resisting forces of cohesion holding the parts of the body together. The stress caused by pulling; opposite of compression and distinguished from torsion. (Standard, 1964) (d) Sometimes used in place of voltage or electromotive force. See also: *tension zone.* (CTD, 1958)

tension carriage A bogie or frame carrying a pulley around which the rope of an endless rope haulage passes to be tensioned or tightened. The bogie moves on rails and may be kept taut by balance weights or placed on an inclined roadway (with sufficient weights) to move up or down according to the tension in the endless rope. A tension device is necessary to take up any slack rope created by varying loads on the haulage system. (Nelson, 1965)

tension-control cylinder A hydraulic piston and cylinder mechanism that can be attached to a rotary-drill feedoff line and adjusted to allow the drill stem to feed downward while maintaining a constant preset tension on the drill string. See also: *tension drilling.* (Long, 1960)

tension correction The correction that must be applied to a tape if it is being used at a tension different from that at which it was standardized. See also: *tape corrections.* (Hammond, 1965)

tension drilling Drilling with part of the weight of the drill string supported by the drill swivel head or suspended on a drilling line, as opposed to drilling with the entire weight of the string imposed on the bit. See also: *tension-control cylinder; weight indicator.* (Long, 1960)

tension end The tail end or receiving end of a belt conveyor. It consists of a return drum carried in a boxlike structure. A scraper, plow, or brush is attached to remove as much as possible of the spillage on the bottom belt before it passes on to the return drum. The tension end is drawn back by two sylvesters attached to staking anchor props; this enables adequate, but not excessive, tension to be imparted to the belt. See also: *sylvester; tailend.* (Sinclair, 1959)

tension fault (a) A generic term for any fault caused by tension. (AGI, 1987) (b) Geological fault due to tension, which separates rock strata; unlike gravity or normal fault, since strata may reappear on other side of gap caused by fall of intervening section to lower level when fissure opened. (Pryor, 1963)

tension flange The side of a beam in tension, being the lower side in the general case of a simple beam supported at both ends. (Hammond, 1965)

tension fracture (a) A fracture that is the result of tensional stress in a rock. Cf: *shear fracture.* See also: *extension fracture; tension joint.* (AGI, 1987) (b) A fracture that is the result of stresses that tend to pull material apart. (Billings, 1954) (c) *subsidiary fracture.*

tension jack A type of jack equipped with a jackscrew for wedging against the roof, which also has a ratchet device for applying tension on a chain to be attached to the tail or foot section of a belt conveyor. The jacks and tension chains pull the tail section back until the belt is at the proper tension. (Jones, 1949)

tension joint A joint that is a tension fracture. Syn: *cross joint; tension fracture.*

tension linkage A chain application in which linear motion is not continuous in direction. (Jackson, 1955)

tension zone The surface area affected by tensile strain. Cf: *compression zone; neutral zone; tension.* (Nelson, 1965)

tepee butte A conical hill or knoll resembling a Native American tepee; esp. an isolated, residual hill formed by a capping of resistant rock that protects the underlying softer material from erosion. Also spelled: teepee butte. (AGI, 1987)

tepetate (a) An evaporite consisting of a calcareous crust coating solid rocks on or just beneath the surface of an arid or semiarid region; a deposit of caliche. (AGI, 1987) (b) Mex. A volcanic tuff, or a secondary volcanic or chemical nonmarine deposit, very commonly calcareous. Etymol. Mexican Sp., from Nahuatl (Aztec) tepetatl, stone matting. (AGI, 1987)

tephra A general term for all pyroclastics of a volcano. (AGI, 1987)

tephrite A group of extrusive rocks, of basaltic character, primarily composed of calcic plagioclase, augite, and nepheline or leucite as the main feldspathoids, with accessory alkali feldspar; also, any member of that group; the extrusive equivalent of theralite. With the addition of olivine, the rock would be called a basanite. (AGI, 1987)

tephritoid Said of a rock resembling tephrite.

tepla-masse A coal-tar pitch, for protecting the outside of steel tubes against corrosion and bacteria. (Osborne, 1956)

terlinguaite A monoclinic mineral, Hg_2ClO; yellow; at Terlingua, TX.

terminal curvature A sharp, local change in the dip of strata or cleavage near a fault. Not commonly used in the United States. Cf: *drag.* (AGI, 1987)

terminal velocity The constant velocity acquired by a particle falling in water or air when the frictional resistance is equal to the gravitational pull. See also: *equal-falling particles; Stokes' law.* (Nelson, 1965)

termination In mineralogy, the end of a crystal, esp. crystal faces that intercept the crystallographic axis, as distinguished from a broken or polished end. Crystals are singly terminated if faces appear on one end as in attached crystals or ones lacking the symmetry to require faces on both ends to complete a crystal form; they are doubly terminated if faces appear on both ends.

terms of reference Schedule that defines responsibilities and area of activity delegated to and/or accepted by subsection, department, or subordinate official in organization working on line-and-staff system of large company where harmonious cooperation might otherwise be endangered. (Pryor, 1963)

ternary (a) A phase system that may be defined in terms of three components. (b) Any phase in a ternary system consisting of all three components. (c) A eutectic, peritectic, or other singular point in a ternary system.

ternary steel An alloy steel that contains one alloying element; the term is synonymous with a simple alloy steel. It contains the one element plus the iron and carbon, hence ternary. (Fay, 1920)

terne Sheet iron or steel coated with an alloy of about four parts lead to one part tin. (Webster 3rd, 1966)

terneplate Sheet steel covered with a tin-lead alloy.

Terni furnace A modification of the open-hearth furnace in which the essential feature is the port design. The air ports gradually increase in cross section until they are as large as the hearth itself, thus practically eliminating turbulent flow in the furnace. (Osborne, 1956)

terosin *fichtelite.*

terpene Hydrocarbon present (30% to 60%) in pine oil, which is widely used as frother in the flotation process. (Pryor, 1963)

terpineol The principal frothing agent in pine oil. (Pryor, 1963)

terra alba (a) Finely pulverized powder, $CaSO_4 \cdot 2H_2O$, made from gypsum and used in the manufacture of paper, paints, artificial marble, and composition plastics. (CCD, 1961) (b) Any of several white mineral substances, such as (1) gypsum ground for a pigment; (2) kaolin used esp. as an adulterant of paints; (3) burnt alum; (4) magnesia; and (5) blanc fixe. (Webster 3rd, 1966)

terrace (a) A level or nearly level plain, generally narrow in comparison with its length, from which the surface slopes upward on one side and downward on the other side. Terraces and their bounding slopes are formed in a variety of ways, some being aggradational and others degradational. (Fay, 1920) (b) A flaw in marble, commonly cored out and filled up. (Webster 2nd, 1960) (c) A raised portion of an ancient riverbed or a bank on which alluvial deposits may be found. (Nelson, 1965) (d) A bench in quarry or opencast mining. (Nelson, 1965) (e) A ridge, a ridge and hollow, or a flat bench built along a ground contour. (Nichols, 1976) (f) A narrow, gently sloping constructional coastal strip extending seaward or lakeward, and veneered by a sedimentary deposit; esp. a wave-built terrace. (AGI, 1987) (g) Loosely, a stripped wave-cut platform that has been exposed by uplift or by lowering of the water level; an elevated wave-cut bench. (AGI, 1987) (h) Any long, narrow, relatively level or gently inclined surface, generally less broad than a plain, bounded along one edge by a steeper descending slope and along the other by a steeper ascending slope; a large bench or steplike ledge breaking the continuity of a slope. The term is usually applied to both the lower or front slope (the riser) and the flattish surface (the tread), and it commonly denotes a valley-contained, aggradational form composed of unconsolidated material as contrasted with a bench eroded in solid rock. A terrace commonly occurs along the margin and above the level of a body of water, marking a former water level; e.g., a stream terrace. (AGI, 1987) (i) A term commonly but incorrectly applied to the deposit underlying the tread and riser of a terrace, esp. the alluvium of a stream terrace; this deposit should more properly be referred to as a fill, alluvial fill, or alluvial deposit, in order to differentiate it from the topographic form. (AGI, 1987) (j) *structural terrace.*

terrace placer *bench placer.*

terra cotta clay A term applied loosely to any fine-textured, fairly plastic clay that acquires a natural vitreous skin in burning and that is used in the manufacture of terra cotta. It is characterized by low shrinkage, freedom from warping, strong bonding, and absence of soluble salts. (AGI, 1987)

terra-cotta model maker One who makes metal profiles and wooden forms for use in casting plaster terra-cotta block mold. (DOT, 1949)

terrain A tract or region of the Earth's surface considered as a physical feature, an ecologic environment, or a site of some planned human activity, e.g., an engineering location; or in terms of military science, as in terrain analysis. Not to be confused with terrane. (AGI, 1987)

terrain coefficient A terrain coefficient is a number expressing the ratio of actual ground displacement by elastic waves to that which the same waves would produce in rock. The terrain coefficient for rock is thus 1; for unconsolidated materials it ranges upward to as high as 30, depending on the thickness of the material. (Leet, 1960)

terrain correction A correction applied to observed values obtained in geophysical surveys in order to remove the effect of variations in the observations due to the topography near observation sites. Syn: *topographic correction.* (AGI, 1987)

terrain slope Used to describe quarries when located in low slopes. Cf: *hillside.* (Streefkerk, 1952)

terrane (a) A group of strata, a zone, or a series of rocks; used in the description of rocks in a general, provisional, or noncommittal sense. (Fay, 1920) (b) A region considered in relation to its fitness for some purpose; an extent of ground or territory. (Standard, 1964) (c) A fault-bounded body of rock of regional extent, characterized by a geologic history different from that of contiguous terranes. A terrane is generally considered to be a discrete allochthonous fragment of oceanic or continental material added to a craton at an active margin by accretion. See also: *suspect terrane.* (AGI, 1987) (d) Informally, a region where a particular rock or group of rocks predominates. Not to be confused with terrain. (AGI, 1987)

terra ponderosa Literally, heavy earth; another name for heavy spar or barite. (Fay, 1920)

terra rossa A reddish-brown residual soil found as a mantle over limestone bedrock, typically in the karst areas around the Adriatic Sea, under conditions of Mediterranean-type climate. Also spelled: terra rosa. Etymol: Italian, red earth. (AGI, 1987)

terras In marble working, a defective or disfigured place in a marble block, which is cut out and filled with a composition. Also spelled terrace. (Standard, 1964; Fay, 1920)

terra sienna *ocher.*

terra silicea *infusorial earth.*

terrazzo Small chips or pieces of stone, usually marble or limestone, about 1/2 to 3/4 in (1.3 to 1.9 cm) in diameter, made by crushing and screening. Terrazzo chips are used with portland cement in making floors, which are smoothed down and polished after the cement has hardened. (USBM, 1960)

terrestrial (a) Pertaining to the Earth. (AGI, 1987) (b) Pertaining to the Earth's dry land. (AGI, 1987)

terrestrial deposit (a) A sedimentary deposit laid down on land above tidal reach, as opposed to a marine deposit, and including sediments resulting from the activity of glaciers, wind, rainwash, or streams; e.g., a lake deposit, or a continental deposit. (AGI, 1987) (b) Strictly, a sedimentary deposit laid down on land, as opposed to one resulting from the action of water; e.g., a glacial or eolian deposit. (AGI, 1987) (c) A sedimentary deposit formed by springs or by underground water in rock cavities. Cf: *terrigenous deposit.* (AGI, 1987)

terrestrial magnetism The natural magnetic field within and surrounding the Earth and the factors affecting it. (Hy, 1965)

terre verte (a) Fr. "green earth". Glauconite or other phyllosilicate used as artist's pigment. (b) Collective name for various pale bluish-green earths formed by the disintegration of minerals, principally those of the hornblende type. Also called green earth.

terrigenous Derived from the land or continent. (AGI, 1987)

terrigenous deposit Shallow marine sediment consisting of material eroded from the land surface. Cf: *terrestrial deposit.* (AGI, 1987)

terrigenous sediments Sediments derived from the destruction of preexisting rocks on the Earth's surface, as distinguished, e.g., from sediments of organic or volcanic origin. (Stokes, 1955)

territe A plastic explosive that consists of the constituents blasting gelatin+BNT+sodium nitrate+ammonium perchlorate. The explosive has a relatively low rate of detonation and is very insensitive, on which account care must be taken to ensure that its initiation is extremely powerful. (Fraenkel, 1953)

territorial sea A belt of sea, not exceeding 12 nmi (22.2 km) in breadth, lying beyond its land territory and internal waters and, in the case of an archipelagic State, its archipelagic waters, in which a coastal State has sovereignty. (United Nations, 1983)

terroite An extra-strong high explosive of the nitroglycerin type. (Standard, 1964)

Tertiary The first period of the Cenozoic Era (after the Cretaceous of the Mesozoic Era and before the Quaternary), thought to have covered the span of time between 65 million years and 3 to 2 million years ago. It is divided into five epochs: the Paleocene, Eocene, Oligocene, Miocene, and Pliocene. It was originally designated an era rather than a period; in this sense, it may be considered to have either five periods (Paleocene, Eocene, Oligocene, Miocene, Pliocene) or two (Paleogene and Neogene), with the Pleistocene and Holocene included in the Neogene. (AGI, 1987)

tertiary crushing The preliminary breaking down of run-of-mine ore and sometimes coal. In metal mines, the tertiary crushing may be performed at a central point underground. See also: *primary breaker.* (Nelson, 1965)

tertiary grinding When a particularly fine grinding of ore is needed, two and even three ball mills may be used in a series to attain the degree of fineness. The successive stages are referred to as primary, secondary, and tertiary grinding. (Newton, 1959)

tertiary shaft The shaft that extends a mine downward from the bottom of the secondary shaft. (Spalding, 1949)

tervalent Having three different valences. Syn: *trivalent.* (Pryor, 1963)

tessellated (a) A surface divided into squares, or figures approaching squares, by joints or natural divisions. (Fay, 1920) (b) Composed of tesserae—small cubes of stone, marble, glass, or terra cotta variously colored and arranged in artistic design: inlaid; mosaic; as tessellated pavement. (Standard, 1964)

tesseral In crystallography, the same as isometric. (Standard, 1964)

test (a) To search for mineral deposits in an unproved area by means of boreholes. (b) To obtain samples of soil or rock from which the physical characteristics of the soil or rock can be determined, such as in foundation testing. (c) An exploratory borehole.

test bore Drilling to test subsoil and rocks when considering foundations of buildings, dams, and heavy plant. (Pryor, 1963)

test boring As used by foundation engineers, the act or process of sinking holes into the overburden (sometimes to considerable depth into bedrock) with rotary or drive sampling equipment for the purpose of recovering samples from which information on the physical characteristics of the materials penetrated can be obtained; also applied to the sample or samples so obtained. Syn: *borehole; drill hole; drilling.* (Long, 1960)

test core Core removed from a concrete structure by diamond core drilling and tested in a laboratory to determine the strength and other physical properties of the concrete. Also, core removed from a borehole drilled in search of oil and used to determine the porosity of the core and whether oil is present. (Long, 1960)

test detonator An instantaneous detonator that has a strength equivalent to that of a detonator with a base charge of 0.40 to 0.45 g PETN. (CFR, 4)

tester (a) A sampling instrument. (Nelson, 1965) (b) A person responsible for carrying out ventilation, dust, or other tests. (Nelson, 1965) (c) Service company representative who supervises borehole testing operations. (Wheeler, 1958)

test hole (a) Generally, any borehole drilled to obtain samples whereby the structural and physical characteristics of the rocks penetrated can be determined; more specif., a hole produced by rotary or driving soil-testing tools in the course of obtaining samples used in soil- and foundation-testing work. (Long, 1960) (b) Usually a small hole drilled ahead and flanking in a working place to ascertain proximity of old workings and to determine water or air content of same. (BCI, 1947) (c) A drill hole or shallow excavation for testing an orebody; a test pit. (Fay, 1920) (d) A taphole, as in a cementation furnace. (Standard, 1964)

testing bedrock *bedrock test.*

testing flame The lowered flame of a miner's flame safety lamp, which is used to detect the presence of small percentages of combustible gases in mine air. (Nelson, 1965)

testing machine A machine used for applying test loads to standard test pieces or to structural members. Machines are available for carrying out tensile, compressive, impact, and fatigue tests. (Hammond, 1965)

test lead Lead free from any silver, and often finely granulated; used in testing or cupelling, assaying, etc. (Webster 3rd, 1966)

test paper Paper (as litmus paper) cut usually in strips and saturated with an indicator or other reagent that changes color in testing for various substances. (Webster 3rd, 1966)

test piece A piece of material prepared in a suitable shape so that it can be tested in a testing machine. (Hammond, 1965)

test pit (a) A shallow shaft or excavation made to determine the existence, extent, or grade of a mineral deposit, or to determine the fitness of an area for engineering works, such as buildings or bridges. (Stokes, 1955) (b) *test hole; trial pit.*

test ring An oval iron frame for holding a test or movable cupelling hearth. (Fay, 1920)

test stone Basanite. Used for testing streak of precious metals.

tetartohedral Said of a point group or of specific crystal forms in the isometric and tetragonal crystal systems that have but one-fourth the crystal faces generated for the equivalent crystal form in the holohedral (most symmetric) crystal class. These complimentary merohedral forms are designated plus or minus and left or right, four forms being required to show all the faces of the holohedral form because each form lacks both a center and mirror planes of symmetry. See also: *merohedral.*

tetraboron carbide *boron carbide.*

tetrad A crystallographic axis of rotation of 90°, four-fold. Cf: *axis of symmetry.*

tetradymite (a) A trigonal mineral, Bi_2Te_2S; forms foliated masses in auriferous veins, commonly with tellurobismuthite; a source of bismuth. Syn: *telluric bismuth; bismuth telluride.* (b) The mineral group kawazulite, paraguanajuatite, skippenite, tellurantimony, tellurobismuthite, and tetradymite.

tetraethyllead An antiknock constituent of gasoline; $Pb(C_2H_5)_4$.

tetragonal (a) Designating or belonging to a system of crystallization having all three axes at right angles and the two lateral axes equal. This system is called tetragonal system. (Fay, 1920) (b) The crystal system in which crystals have one four-fold symmetry axis. (Hurlbut, 1964) (c) The crystal system characterized by three orthogonal crystallographic axes, the principal or *c* axis being a tetrad (4-fold axis), longer or shorter than the two lateral *a* axes. See also: *crystal systems.*

tetragonal trisoctahedron *trapezohedron.*

tetrahedral Having the symmetry or shape of a tetrahedron.

tetrahedrite (a) An isometric mineral, $(Cu,Fe)_{12}Sb_4S_{13}$, having copper replaced by zinc, lead, mercury, cobalt, nickel, or silver; forms a series with tennantite and freibergite; metallic; crystallizes in tetrahedra; occurs in hydrothermal veins and contact metamorphic deposits; a source of copper and other metals. Syn: *gray copper ore; gray copper; panabase; panabasite; stylotypite.* (b) The mineral group freibergite, giraudite, goldfieldite, hakite, tennantite, and tetrahedrite.

tetrahedron An isometric crystal form of four faces, each an equilateral triangle; the alternate faces of an octahedron. Adj: tetrahedral.

tetrahexahedron An isometric crystal form {hk0} of 24 faces, each an isosceles triangle, so arranged that four faces appear to replace each face of a cube (hexahedron).

tetravalence An atom, or group, having four valence bonds. (Pryor, 1963)

tetravalent (a) Having a valence of 4. (Webster 3rd, 1966) (b) Having four valences; e.g., chlorine, which has valences of 1, 3, 5, and 7. (Webster 2nd, 1960; Handbook of Chem. & Phys., 2)

tewel (a) A hole; bore; a chimney, as for smoke. (Webster 3rd, 1966) (b) The tuyère of a furnace. (Webster 2nd, 1960)

texture The general physical appearance or character of a rock, including the geometric aspects of, and the mutual relations among, its component particles or crystals; e.g., the size, shape, and arrangement of the constituent elements of a sedimentary rock, or the crystallinity, granularity, and fabric of the constituent elements of an igneous rock. The term is applied to the smaller (megascopic or microscopic) features as seen on a smooth surface of a homogeneous rock or mineral aggregate. The term structure is generally used for the larger features of a rock. The two terms should not be used synonymously, although certain textural features may parallel major structural features. Confusion may arise because in some languages, e.g., French, the usage of texture and structure are the reverse of the English usage. Cf: *structure.* (AGI, 1987)

thalenite A monoclinic mineral, $Y_3Si_3O_{10}(OH)$(?); red to pink; in pegmatites.

thallium A metallic element resembling lead in physical properties; the metal is silvery-white, but turns bluish-gray in air. Symbol, Tl. Occurs in crooksite, lorandite, and hutchinsonite. It is also present in pyrites and is recovered from the roasting of this ore in the manufacture of sulfuric acid and from the smelting of lead and zinc ores. Used in low-melting glasses, photocells, and infrared detectors. (Handbook of Chem. & Phys., 3)

thalweg (a) The line of continuous maximum descent from any point on a land surface; e.g., the line of greatest slope along a valley floor, or the line crossing all contour lines at right angles, or the line connecting the lowest points along the bed of a stream. Etymol: German Talweg, valley way. Also spelled: talweg. (AGI, 1987) (b) In physical geography, a term adopted into English usage signifying the line of greatest slope along the bottom of a valley; i.e., a line drawn through the lowest points of a valley in its downward slope. It thus marks the natural direction of a watercourse. (AGI, 1987) (c) In hydraulics, the line joining the deepest points of a stream channel. (AGI, 1987) (d) By many geomorphologists, the term is used as a syn. for valley profile. (AGI, 1987) (e) The center line of the principal navigational channel of a waterway constituting a boundary between political subdivisions. (Hunt, 1965)

thanite (a) A mixture of kainite and halite. (English, 1938) (b) Carbon oxysulfide, COS, as a natural gas.

thaumasite A hexagonal mineral, $Ca_6Si_2(CO_3)_2(SO_4)_2(OH)_{12}\cdot 24H_2O$; white; fibrous.

thaw house A small building, designed for thawing dynamite, of such size as to provide enough thawed dynamite for a day's work. (Fay, 1920)

thawing (a) A method of working permanently frozen ground in which water at a temperature of from 50 to 60 °F (10 to 15.6 °C) is pumped through pipes down into the frozen gravel. The pipes through which the water is pumped are called sweaters. See also: *steam thawing; thaw pipe.* (Lewis, 1964) (b) In dynamiting, warming to reduce risk of premature explosion that might originate from rupture of frozen crystal. Performed in thaw house or thawing kettle using steam or hot water. With modern methods of explosive manufacture, the need has practically disappeared. (Pryor, 1963)

thawing kettle A double kettle, built somewhat like a farina boiler, having two compartments, an outer compartment, which is filled with hot water and which entirely surrounds the inner compartment that contains the dynamite to be thawed. It is provided with a lid for retaining the heat. (Fay, 1920)

thaw pipe A string of pipe lowered into a string of drill rods that is frozen in a borehole drilled into permafrost, through which water is circulated to thaw the ice and free the drill rods. See also: *thawing.* (Long, 1960)

thaw shed operator In the coke products industry, one who thaws frozen materials in railroad cars by heating sections of shed where cars are spotted. (DOT, 1949)

THDM *translucent humic degradation matter.*

thelotite A carbonaceous constituent of torbanite, occurring in the form of a solidified clear, jellylike substance, something like solidified dopplerite, but probably of unlike chemical composition. (Tomkeieff, 1954)

thematic mapper A cross-track scanner deployed on Landsat that records seven bands of data from the visible through the thermal IR regions. *multispectral scanner.* (SME, 1992)

thenardite An orthorhombic mineral, Na_2SO_4; soft; forms masses and crusts in evaporite deposits and around fumaroles; a source of sodium sulfate. Also called verde salt.

theodolite A precision surveying instrument that is used for measuring angular distances in both vertical and horizontal planes.

theoretical depression The water gage produced by an imaginary fan that is perfect and is connected to an evasé chimney of infinite height to eliminate kinetic losses at discharge. Its calculated value depends only on the speed of the blade tips and on the shape of the blades. See also: *manometric efficiency; initial depression.* (Nelson, 1965)

theoretical fan depression The depression that can be produced by a perfect fan. (BS, 1963)

theoretical tonnage *probable reserves.*

theoretical yield The maximum yield (as shown by the washability curve) of a product with a specified percentage of ash. (BS, 1962)

theory of lateral secretion The theory that the contents of a vein or lode are derived from the adjacent country rock by a leaching process, in which either superficial water or thermal water is involved.

theory of machines Comprises the study of the relative motion between the parts of a machine and the study of the forces that act on those parts. See also: *machine design.* (Nelson, 1965)

theralite A group of mafic plutonic rocks composed of calcic plagioclase, feldspathoids, and augite, with lesser amounts of sodic sanidine and sodic amphiboles and accessory olivine; also, any rock in that group; the intrusive equivalent of tephrite. Theralite grades into nepheline monzonite with an increase in the alkali feldspar content, into gabbro as the feldspathoid content diminishes, and into diorite with both fewer feldspathoids and increasingly sodic plagioclase. The term, defined by Rosenbusch in 1887, is derived from the Greek word for eagerly looked for, not from the island of Thera (Santorini). (AGI, 1987)

therm Equals 100,000 Btu (105,500 kJ). (Newton, 1959)

thermal Hot or warm; applied to springs that discharge water heated by natural agencies. (Fay, 1920)

thermal-acceptance ratio A physiological method of assessing the effect of a given climate upon workers, that is based on the ratio between the heat actually lost by the body via the skin, lungs, etc., and the maximum that can be lost in the prevailing conditions. (Roberts, 1960)

thermal analysis (a) A method for determining transformations in a metal by noting the temperatures at which thermal arrests occur. These arrests are manifested by changes in slope of the plotted or mechanically traced heating and cooling curves. (ASM, 1961) (b) The study of chemical and/or physical changes in materials as a function of temperature, i.e., the heat evolved or absorbed during such changes. (AGI, 1987)

thermal boring Use of high-temperature flame to fuse rock in drilling. Heat comes from ignition of kerosene with oxygen or other fuel system, at bottom of drill hole, and water with compressed air may be used to flush out the products. (Pryor, 1963)

thermal capacity (a) Heat required to raise the temperature of a body 1 °C. Syn: *heat capacity.* (Bennett, 1962) (b) The amount of heat that a clay product will absorb, usually expressed in British thermal units per degree Fahrenheit (kilojoules per degree Celsius). (ACSG, 1963)

thermal conductivity (a) The time rate of transfer of heat by conduction, through unit thickness, across unit area for unit difference of temperature. (AGI, 1987) (b) A measure of the ability of a material to conduct heat. Typical values of thermal conductivity for rocks range from 3 to 15 mcal/cm/s/°C (12.6 to 62.8 kJ/cm/s/°C). Syn: *heat conductivity; thermal diffusivity of strata.* (AGI, 1987)

thermal cutout A device fitted in hydraulic power systems underground so that temperatures cannot rise above about 85 °C. It is a safeguard against fire risk due to a rapid rise in temperature if the fluid circuit is interrupted by wrong manipulation of valves, etc. (Nelson, 1965)

thermal diffusivity of strata *thermal conductivity.*

thermal drying (a) The evaporation of water by thermal means from a mixture of coal and water. See also: *McNally-Vissac dryer; multilouvre dryer; Raymond flash dryer; cascade coal dryer; flash coal dryer; fluidized bed dryer.* (Mitchell, 1950) (b) The application of heat (generally hot-air currents) to wet coals and other materials and the evaporation of the free moisture and also part of the inherent moisture. (Nelson, 1965)

thermal efficiency The ratio of the electric power produced by a powerplant to the heat value of the fuel consumed; thus, a measure of the efficiency with which the plant converts thermal energy to electric energy. Symbol, η. (Lyman, 1964; Handbook of Chem. & Phys., 2)

thermal electromotive force The electromotive force generated in a circuit containing two dissimilar metals when one junction is at a different temperature from the other. (ASM, 1961)

thermal emissivity The thermal emissivity, or heat transfer coefficient, of a rock surface is the rate at which heat will flow from rock to air, per unit area for 1 degree temperature difference. This varies with color and other surface characteristics. Measured in $W/m^2 \cdot K$. (Roberts, 1960)

thermal expansion The increase in linear dimensions and volume that occurs when materials are heated and that is counterbalanced by a contraction of equal amount when the materials are cooled. (Harbison-Walker, 1972)

thermal gradient In the Earth, the rate at which temperature increases with depth below the surface. In a mine, this is usually estimated at 5.3 °F/1,000 ft (1 °C/100 m) of shaft depth. Some variability with time of day is typically observed. (SME, 1992)

thermal metamorphism A type of metamorphism resulting in chemical reconstitution controlled by a temperature increase, and influenced to a lesser extent by confining pressure; there is no requirement of simultaneous deformation. See also: *pyrometamorphism.* Cf: *contact metamorphism; load metamorphism; static metamorphism.* Syn: *thermometamorphism.* (AGI, 1987)

thermal precipitator An instrument for obtaining information regarding the number of particles present in unit volume of a dust cloud, together with their size distribution. This is probably the most efficient instrument used for dust counts since its efficiency is practically 100% for all particles from 0.2 to 10 µm in diameter. See also: *Hexhlet sampler; dust sampling.*

thermal probe A device used for measuring the heat flow out of ocean bottom sediment. (Hunt, 1965)

thermal prospecting A system of geophysical prospecting based on measuring underground temperatures or temperature gradients and relating their irregularities to geological deformation. (AGI, 1987)

thermal reactor A nuclear reactor in which the fission chain reaction is sustained primarily by thermal neutrons. Most reactors are thermal reactors. (Lyman, 1964)

thermal-release water spray A device that brings water sprays into action when heating occurs on roadway belt conveyors. The local heat fuses an element holding taut wires. On release by fusion, the wires allow spray valves to open and the water cools the affected area. (Nelson, 1965)

thermal resistance *resistance.*

thermal resistivity *resistivity.*

thermal separation The separation of minerals and metals by heat. The method is used, e.g., to remove impurities from rock salt. The crude salt is first exposed to radiant heat. The impurities absorb the heat and become warm, while the rock salt transmits the radiant heat and remains cool. The warm, impure particles adhere slightly to a belt covered with heat-sensitive resin, while the salt remains free. Separation takes place at the end of the belt. The cool salt is thrown into one container, while the adhering impurities drop directly into another. (Nelson, 1965)

thermal shock Failure of a material, esp. a brittle material, due to the thermal stress of rapidly rising or falling temperature. (AGI, 1987)

thermal shock resistance The ability to withstand sudden heating, cooling, or both without cracking or spalling.

thermal spalling The chipping or spalling of ceramic ware by repeated heating and cooling.

thermal spring A spring whose water temperature is appreciably higher than the local mean annual atmospheric temperature. (AGI, 1987)

thermal value (volatile matter) The calorific value of volatile matter in therms per ton of dry coke, of the gas given off when dry coke, ground to pass a 36-mesh B.S. test sieve, is heated under standard conditions. (BS, 1961)

thermal water Water, generally of a spring or geyser, whose temperature is appreciably above the local mean annual air temperature. (AGI, 1987)

thermic boring (a) A method of boring holes in concrete under the high temperature generated by a burning steel tube, known as a lance. This is packed with steel wool, through which a jet of suitable gas flows to ignite the end of the lance and keep it burning. (Hammond, 1965) (b) *jet piercing.*

thermic drilling *jet piercing.*

thermistor An electrical resistor made of a material whose resistance varies sharply in a known manner with the temperature. Thermistors are commonly used for shipboard oceanographic temperature measurements because of their percentage response to unit temperature change and their great sensitivity. (Hunt, 1965)

thermite (a) Any fossil combustible substance. (Tomkeieff, 1954) (b) An intimate mixture of aluminum powder and powdered iron oxide that when caused to react by strong heating emits a great deal of heat and yields alumina and a white-hot molten mass of metallic iron. (Webster 3rd, 1966)

thermit process The energetic action of finely divided aluminum on a metallic oxide, when heated together, is utilized for the production of metallic iron manganese, chromium, tungsten, molybdenum, uranium, etc. The aluminum combines directly with the oxygen of the oxide, and the heat emitted by the reaction is sufficient to promote the fusion of the reduced metal. (Nelson, 1965)

thermoanemometer *thermometer anemometer.*

thermocline Frequently used in geophysics to describe the decrease in temperature that always occurs at great depths. (Hunt, 1965)

thermocouple Two conductors of different metals joined together at both ends, producing a loop in which an electric current will flow when there is a difference in temperature between the two junctions. Abbrev., tc. (Lyman, 1964; Zimmerman, 1949)

thermodynamics The mathematical treatment of the relation of heat to mechanical and other forms of energy. (AGI, 1987)

thermoelectric conversion The conversion of heat into electricity by the use of thermocouples. (Lyman, 1964)

thermoelectricity Electricity involved in thermoelectric phenomena. Specif., electricity accumulated or put in motion by thermoelectric action. (Webster 3rd, 1966)

thermoelectric metals Metals or alloys used in thermocouples for measuring high temperatures. Platinum, nickel, copper, rhodium, etc., are much used. (Crispin, 1964)

thermograph A self-recording thermometer that gives a continuous trace of air temperature on a rotating drum worked by clockwork. It is mainly used for recording variations in temperature rather than actual temperatures. (Hammond, 1965)

thermoluminescence The property of minerals to emit light when heated. It results from the release of energy stored by displaced electrons trapped in a crystal structure. See also: *calorescence.* Cf: *luminescence.*

thermometamorphism *thermal metamorphism.*

thermometer An instrument for determining temperature usually by means of a scale graduated directly in temperature units and consisting typically of (1) a device having a bimetallic element, the expansion or contraction of which indicates a change in temperature, or (2) a glass bulb attached to a fine tube of glass with a numbered scale etched on it or fastened to it and containing a liquid (as mercury or cooled alcohol) that is sealed in and rises and falls with changes of temperature and that indicates the temperature by the number at the top of the column of liquid. (Webster 3rd, 1966)

thermometer anemometer An anemometer consisting of two thermometers, one with an electric heating element (battery powered) connected to the bulb. The heated bulb cools in an airstream, and the difference in temperature as registered by the heated and unheated thermometers can be translated into air velocity by a conversion chart. It is nondirectional and can be made safe (6 V) for use in explosive atmospheres. Syn: *thermoanemometer.* (Hartman, 1982)

thermometer float This instrument is used for studying the temperature structure in the upper 10 m of water. The instrument is in two sections; a float, which contains a spooling winch from which the sensing unit is lowered, and an indicator case, which contains the remote indicating equipment and remote control system. The two sections are connected by an electric cable and flexible shaft, supported by net floats. The indicator case is clear lucite. It contains the indicating meter and electric circuitry for temperature determination. It also contains a Veeder-Root counter, which indicates directly in centimeters the depth at which the sensing unit is located. This instrument makes it possible to read temperature to within 0.1 °C and to know the depth of the sensing unit to within ±0.5 cm. It is intended to be used at sea from a skiff or tender rather than from the research vessel itself. (Hunt, 1965)

thermometer scales Two thermometer scales are in general use, the Fahrenheit, which is generally used in engineering, and the Celsius, which is almost universally used in scientific work. The Fahrenheit scale has the freezing point at 32 °F and the boiling point at 212 °F, whereas the Celsius scale has the freezing point at 0 °C and the boiling point at 100 °C. The Celsius scale is commonly called the centigrade scale. (Nelson, 1965)

thermometric fan test A method of assessing the efficiency of a mine fan by comparing the temperature rise in an ideal isentropic fan for a given fan pressure with the measured temperature rise actually occurring in the fan under consideration when producing the same fan pressure. The ratio of isentropic temperature rise to the actual temperature rise across the fan gives the fan efficiency. The method gives an accuracy of ±5%. See also: *overall ventilation efficiency.* (Nelson, 1965)

thermonatrite An orthorhombic mineral, $Na_2CO_3{\cdot}H_2O$; forms flat, white, water-soluble crystals in some lakes and alkali soils, also a saline residue.

thermo-osmosis Natural migration of moisture from a relatively warm part of a mass of soil toward a cooler part. See also: *electro-osmosis.* (Hammond, 1965)

thermopile An apparatus consisting of a number of thermoelectric couples (as of antimony and bismuth or of copper sulfide and German silver) combined so as to multiply the effect used; (1) to generate electric currents for various purposes, and (2) in a very sensitive form for determining intensities of radiation due esp. to its heating effect. (Webster 3rd, 1966)

thermoplastic In plastics, rigid material that temporarily becomes soft when heated and can then be molded into a shape that it will retain on cooling. Ant. for thermosetting, a material that reacts chemically on heating (curing) and is then resistant to deformation when reheated. (Pryor, 1963)

thermoscopic bar Small ceramic bar of specified composition that softens at certain temperatures. See also: *Holdcroft thermoscope bar.* (Osborne, 1956)

thermostat An automatic device for regulating temperature (as by opening or closing the damper of a heating furnace or by regulating the supply of gas) and commonly utilizing either the differential expansion of solids or the vapor pressure of liquids. (Webster 3rd, 1966)

Thetis hairstone Coarsely crystalline quartz containing inclusions of asbestiform amphibole, esp. hornblende or actinolite, that may be tangled or wound into a ball. See also: *Venus hairstone; sagenitic quartz.*

thick-bedded A relative term applied to sedimentary beds variously defined as more than 2.5 in (6.4 cm) to more than 40 in (100 cm) in thickness; specif. said of a bed whose thickness is in the range of 2 to 4 ft (60 to 120 cm), a bed greater than 120 cm being very thick-bedded. Cf: *tight-bedded; thin-bedded.* (AGI, 1987)

thickener A vessel or apparatus for reducing the proportion of water in a pulp by means of sedimentation. (Nelson, 1965)

thickening (a) The process of concentrating a relatively dilute slime pulp into a thick pulp, i.e., one containing a smaller percentage of moisture, by rejecting liquid that is substantially solid free. Settling is another name for the same operation. (Taggart, 1927) (b) The concentration of the solids in a suspension with a view to recovering one fraction with a higher concentration of solids than in the original suspension. (BS, 1962)

thickness (a) The distance at right angles between the hanging wall and the footwall of a lode or lens. (Standard, 1964) (b) As used in mine subsidence, the thickness of a bed or seam of mineral is the distance from its roof to its floor, measured at right angles to the plane of stratification. (Briggs, 1929) (c) That dimension designed to lie at right angles to the face of the wall, floor, or other assembly. (ASTM, 1994)

thickness contour *isopach.*

thickness line *isopach.*

thick seam In general, a coal seam over 4 ft (1.22 m) in thickness. See also: *medium-thickness seam.* (Nelson, 1965)

thick-wall sampler A soil sampler made from a steel tube having a wall thickness greater than 16 gage. See also: *drive sampler.* Cf: *Shelby tube; Shelby-tube sampler.* (Long, 1960)

Thiess process A chlorination process for recovering gold from its ore. For each ton (0.9 t) of ore in a revolving drum, 130 gal (492 L) of water, 30 lb (13.6 kg) of lime chloride, and 36 lb (16.3 kg) of concentrated sulfuric acid are added, and the drum is revolved for some time. A solution of gold chloride is thus obtained. (Fay, 1920)

thill (a) The floor of a coal seam. See also: *underclay.* (Nelson, 1965) (b) Eng. Seat earth or pavement of underclay directly underlying a coalbed, Newcastle. (Raistrick, 1939) (c) A thin stratum of fireclay. (Webster 3rd, 1966)

thimble An oval iron ring around which a rope end is bent and fastened to form an eye. (Zern, 1928)

thimble joint A sleeve joint packed to allow longitudinal expansion. A slip expansion joint. (Fay, 1920)

thin-bedded A relative term applied to sedimentary beds variously defined as less than 1 ft (30 cm) to less than 0.4 in (1 cm) in thickness; specif. said of a bed whose thickness is in the range of 2 to 24 in (5 to

60 cm), a bed less than 5 cm but more than 1 cm thick being very thin-bedded. Cf: *tight-bedded; thick-bedded.* (AGI, 1987)

thinolite A tufa deposit of fibrous calcite pseudomorphous after an unknown precursor occurring on an enormous scale in northwestern Nevada; also occurs about Mono Lake, CA. It forms layers of interlaced pale yellow or light brown crystals, commonly skeletal with pyramidal terminations. Syn: *thinolitic tufa.*

thinolitic tufa *thinolite.*

thin out To grow progressively thinner in one direction until extinction. The term is applied to a stratum, vein, or other body of rock that decreases gradually in thickness so that its upper and lower surfaces eventually meet and the layer of rock disappears. The thinning may be original or due to truncation beneath an unconformity. Syn: *pinch out; wedge out.* (AGI, 1987)

thin seam In general, a coal seam 2 ft (0.6 m) and under in thickness. See also: *economic coal reserves.* (Nelson, 1965)

thin-seam miner An adaptation of the auger surface miner, the thin-seam miner cuts an entry 8 ft (2.4 m) wide and up to 5 ft (1.5 m) high into coal located under the highwall in surface mines. (SME, 1992)

thin section (a) A fragment of rock or mineral mechanically ground to a thickness of approx. 0.03 mm, and mounted between glasses as a microscope slide. Rocks and most minerals except the oxides and sulfides of the metals are translucent to transparent in thin section, and the optical properties of each mineral can be studied with the microscope. See also: *polished section.* (AGI, 1987) (b) A rock or mineral slice cut for study by transmitted light with a polarized-light microscope. It may also be polished for study with a reflected-light microscope. (c) A coal seam less than its normal thickness. (Nelson, 1965)

thin stock Slabs of stone employed for wainscoting, flooring, etc. (Fay, 1920)

thin-wall bit A coring bit the kerf or wall thickness of which is about one-half or less that of the wall thickness of the same outside-diameter-size standard coring bit. (Long, 1960)

thin-wall drive sampler *Shelby tube; thin-wall sampler.*

thin-wall sampler A soil-sampling barrel made from steel tubing having approx. a 16-gage wall thickness. Syn: *thin-wall drive sampler; thin-wall tube sampler.* Cf: *Shelby tube.* (Long, 1960)

thin-wall tube sampler *thin-wall sampler.*

Thiobacillus ferrooxidans A bacterial strain that catalyzes the oxidation of ferrous iron (Fe^{+2}) to ferric iron (Fe^{+3}). (SME, 1992)

Thiobacillus thiooxidans A bacterial strain that catalyzes the oxidation of sulfur to sulfate (SO_4^{-2}). (SME, 1992)

thio-carbanilid A derivative from thio-urea. In the flotation process, it is used as a collector agent of low solubility in water; it is sometimes used in copper or galena flotation. (Pryor, 1963)

thiocarbonates Powerful collector agents in the flotation process where xanthates fail. (Pryor, 1963)

thiofuran *thiophene.*

thiokerite Kerite containing 9% sulfur. (Tomkeieff, 1954)

thiophene A liquid; C_4H_4S; analogous to furan and pyrrole in its heterocyclic structure and resembles benzene both physically and chemically except for its greater reactivity. Found in small amounts (as up to 0.5% by weight) in benzene from coal tar unless it has been removed by treatment with sulfuric acid. Used chiefly in organic synthesis. Syn: *thiofuran.*

thiophile elements Equivalent to chalcophile. Literally, sulfur-loving. See also: *chalcophile.* (AGI, 1987)

thiophosphates Sulfydric flotation agents, produced by reacting phosphorus pentasulfide with phenols, alcohols, etc., and marketed as Aerofloats. (Pryor, 1963)

third-class lever A lever to which force is applied between the fulcrum and the work point. (Nichols, 1976)

Third Theory of Comminution The Third Theory states that the specific work input required for size reduction is inversely proportional to the square root of the product size, less the work required to form the feed. (Pit and Quarry, 1960)

thirl (a) A cross hole or ventilation passage between two headings. See also: *thurl; thurling; thirling.* (Standard, 1964) (b) To cut out the lost coal between two workings or headings. (CTD, 1958) (c) To cut through from one working into another. (CTD, 1958)

thirling (a) The driving of a proposed roadway from two points some distance apart to meet each other; the connecting of underground roadways or shafts. (Nelson, 1965) (b) *thirl; crosscut; through cut. (c) holing; stenton.*

this vein A notice claiming a location upon this vein has only one meaning. It raises an inference that the notice was posted upon or in close proximity to a vein or lode, although, as a fact, no vein or lode then was exposed. (Ricketts, 1943)

thixotropic fluid Clays termed thixotropic are those that reveal this property by weakening when they are remolded and by increasing in strength when allowed to stand undisturbed. See also: *remolding.* An important characteristic of oil-well drilling fluids. (Hammond, 1965)

thixotropy (a) The property of certain colloidal substances, such as a thixotropic clay, to weaken or change from a gel to a sol when shaken, but to increase in strength upon standing. (AGI, 1987) (b) The property of a material that enables it to stiffen in a relatively short time on standing, but upon agitation or manipulation to change to a very soft consistency or to a fluid of high velocity, the process being completely reversible. Used in muds for drilling deep oil wells since, if the drill stops, rock chips on the way to the surface are held in suspension instead of settling to the bottom where they might jam the drilling bit. See also: *gel; plastic deformation.* (ASCE, 1958)

tholeiite A silica-oversaturated (quartz-normative) basalt, characterized by the presence of low-calcium pyroxenes (orthopyroxene and/or pigeonite) in addition to clinopyroxene and calcic plagioclase. Olivine may be present in the mode, but neither olivine nor nepheline appear in the norm. Cf: *basalt.* (AGI, 1987)

thomaite A variety of siderite that is found massive and in pyramidal crystals. (Standard, 1964)

Thomas converter A bottom-blown basic pneumatic converter having a basic bottom and lining, usually dolomite, and employing a basic slag. (ASM, 1961)

Thomas-Gilchrist process Conversion of iron to steel in basic-lined Bessemer converter. Phosphorus combines with dolomite in this lining to produce basic slag. (Pryor, 1963)

Thomas slag The finely powdered basic slag obtained in the Thomas-Gilchrist process. It consists of phosphates and is used as a fertilizer. (Hackh, 1944)

Thomas steel Steel made in a Bessemer converter using a basic refractory lining. The process was developed by Thomas and Gilchrist. In Europe such steel is known as Thomas steel; in Great Britain, Thomas-Gilchrist steel; and in the United States, basic Bessemer steel. (Henderson, 1953)

Thompson arc cutter *arc cutter.*

thompsonite An orthorhombic mineral, $Na_4Ca_8[Al_{20}Si_{20}O_{80}]\cdot 24H_2O$; zeolite group; pseudotetragonal; occurs in amydules and crevices in basalts and tuffs; also an alteration of anorthite. Also called ozarkite.

Thompson pilot shoulder reamer A reaming or coring bit with an articulated steel pilot protruding about 36 in (91 cm) beyond the face of the bit. The diameter of the pilot is slightly smaller than the set inside diameter of the bit; its upper end is a piston fitted tightly inside a single-tube barrel with its attached coring bit. When lowered into a borehole in which a deflection wedge has been set, the pilot section forces the coring bit to ream out the first part of the deflected hole at a point about 20 in (51 cm) above the tip of the wedge. Reaming is continued to about 6 in (15 cm) below the wedge tip, at which point the pilot shoulder reamer is withdrawn and replaced by a bullnose or deflection bit. (Long, 1960)

Thompson wedge A retrievable type of deflecting wedge. See also: *deflecting wedge.* (Long, 1960)

thomsenolite A monoclinic mineral, $NaCaAlF_6\cdot H_2O$; dimorphous with pachnolite; forms small, white prismatic crystals on cryolite.

thoreaulite A monoclinic mineral, $SnTa_2O_6$; forms a series with foordite; brown; in pegmatite at Katanga, Congo.

thoria A rare refractory oxide, ThO_2.

thorian gummite *thorogummite.*

thorianite An isometric mineral, ThO_2; commonly contains lanthanides; has fluorite structure; forms a series with uraninite; highly radioactive and metamict; sp gr, 9.7; in pegmatites; in placers associated with zircon, ilmenite, geikielite, thorite, and other heavy minerals; an important source of uranium and thorium. Also called isometric thorium.

thorian uraninite A variety of uraninite containing thorium in partial substitution for uranium. (Crosby, 1955)

thorium borides Two borides are known: ThB_4 (gray) and ThB_6 (deep red). More attention has been paid to the tetraboride, the properties of which are melting point, >2,200 °C (but oxidizes slowly above 1,000 °C); thermal expansion, 5.9×10^{-6} (20 to 1,000 °C); sp gr, 8.45 g/mL; modulus of rupture (20 °C), 20,000 psi (138 MPa) . Some properties of ThB_6 are melting point, 2,200 °C; sp gr, 7.1. (Dodd, 1964)

thorium carbides Two carbides are known: ThC, melting point, 2,625 °C; ThC_2, melting point, 2,655 °C. These special carbides are of potential interest in nuclear engineering. (Dodd, 1964)

thorium disintegration series The series of radioactive elements produced as successive intermediate products when the element thorium (thorium 232) undergoes its spontaneous natural radioactive disintegration into stable lead (lead 208). (CCD, 1961)

thorium mineral One of several minerals including monazite, in which Th replaces rare earths; thorite, $ThSiO_4$; and thorianite, ThO_2.

thorium nitride Three thorium nitrides have been reported: ThN, Th_2N_3, and Th_3N_4. (Dodd, 1964)

thorium sulfides Three thorium sulfides have been reported: Th_4S_7, Th_2S_3, and ThS. Crucibles made of these sulfides have been used as containers for molten cerium. (Dodd, 1964)

thorogummite A tetragonal mineral, $Th(SiO_4)_{1-x}(OH)_{4x}$; may contain up to 31% uranium; has zircon structure. Differs from thorite (1) in being secondary, formed by the alteration of primary thorium minerals including thorite itself; (2) in not being metamict but forming crystalline aggregates; and (3) in containing essential (OH). Syn: *thorian gummite; mackintoshite; maitlandite.*

thoron A name for radon 220, a member of the thorium disintegration series; symbol, Tn; emits alpha particles; and half-life, 5.5 s. Also called emanation; thorium emanation. (Handbook of Chem. & Phys., 2)

thorosteenstrupine A mineral, $Na_2(Th,RE)(Mn,Ta,Fe)H_2[(Si,P)O_4]_3$; metamict; dark brown to black.

thorotungstite *yttrotungstite.*

thorough joints Vertical joints affecting all the strata, as opposed to cricks or looses, oolite quarries, Northamptonshire, United Kingdom. Syn: *upright joints.*

thortveitite A monoclinic mineral, $(Sc,Y)_2Si_2O_7$; may contain up to 42% yttrium replacing scandium; weakly radioactive; grayish green where fresh, altering to white to reddish gray; in pegmatites associated with monazite, euxenite, beryl, struvite, and possibly fergusonite; the only known mineral rich in scandium.

Thoulet's law Heavy mineral suites increase in complexity with decreasing geologic age.

Thoulet solution A yellowish-green, transparent, aqueous solution of potassium mercuric iodide, having a maximum specific gravity of 3.19. Also known as Sonstadt solution. Used in the sink-float process of mineral separation. (Holmes, 1928)

thread (a) An extremely small vein, even thinner than a stringer. (Fay, 1920) (b) A more or less straight line of stall faces, having no cuttings, loose ends, fast ends, or steps. (Fay, 1920) (c) To reeve rope or cable through a sheave or block and tackle. (Long, 1960)

thread cutter A name sometimes applied to a diamond crystal having the shape of an octahedron. (Long, 1960)

thread-lace scoria A scoria in which the vesicle walls have burst and are represented only by an extremely delicate three-dimensional network of glass threads. See also: *reticulite.* (AGI, 1987)

three-circle goniometer *goniometer.*

three-cone bit *roller rock bit.* Also called tricone bit.

three-dimension dip In seismic prospecting, the true dip of a reflection or refraction horizon found by exploration and calculation. *true dip.* (AGI, 1987)

threefold An axial symmetry operation requiring three repetitions to complete 360° or return to identity. Syn: *triad.*

three-high mill Consists of three horizontal rolls, one above the other, each rotating continuously in one direction only, the piece being rolled between the bottom and middle, and middle and top rolls alternately. (Osborne, 1956)

three-hinged arch An arch hinged at its abutments and at its crown with the advantage that each half can sink in relation to the other without damaging the arch. (Hammond, 1965)

three-jaw chuck A drill chuck having three serrated-face movable jaws that can be made to grip and hold fast an inserted drill rod. See also: *chuck.* (Long, 1960)

threeling In crystallography, a crystal of three parts united by the same twinning law. Syn: *trilling.*

three-part line A single strand of rope or cable doubled back around two sheaves so that three parts of it pull a load together. (Nichols, 1976)

three-phase circuit Usually a three-wire circuit using alternating current with three equal voltages. This should not be confused with the three-wire service supplying 110 V and 220 V. This latter is merely a two-voltage single-phase circuit arrangement and is used almost universally to provide power and lighting to homes and small business establishments. (Kentucky, 1952)

three-phase current Alternating current in which three separate pulses are present, identical in frequency and voltage, but separated by 120°. (Pryor, 1963)

three-phase inclusion An inclusion consisting of a liquid with a gas bubble and a crystal within a crystal. See also: *negative crystal; inclusion.*

three-piece set A set of timber used in ground that requires greater support than a two-piece set or stull will provide. A cap is supported by two posts often spread apart at the bottom to give greater stability. See also: *timber set; four-piece set.* (Lewis, 1964)

three-point problem (a) In surveying, a method used to orient underground workings via three plumblines suspended in a vertical shaft. (Pryor, 1963) (b) The problem in plane table surveying of locating precisely the point at which the table is set up, using three fixed points that are visible from the plane table. (Hammond, 1965) (c) The problem of determining dip and strike of a plane from elevations determined at three known points not in a straight line.

three-product washing A method in which the cleanest fraction of the coal with an ash content of 1% to 2% (for hydrogenation, etc.) is separated; the remainder giving coal with an ash content of 10% to 15% (for boiler firing, etc.) and finally incombustible shale. (Nelson, 1965)

three-shift cyclic mining A system of cyclic mining on a longwall conveyor face, with coal cutting on one shift, hand filling and conveying on the next, and ripping, packing, and advancement of the face conveyor on the third shift. The system restricts coal production to one shift. (Nelson, 1965)

three-term process controller Instrument that automatically regulates a process in proportional, integral, and derivative terms, thus neutralizing and removing the errors that arise during operation. (Pryor, 1963)

three-throw ram pump This type of pump consists essentially of three single-acting ram pumps side by side, either vertical or horizontal, and driven from a triple crankshaft with cranks set at angles of 120°. The three-throw pump can deal with heads up to 1,000 yd (900 m) in a single lift. (Sinclair, 1958)

three-wire system A voltage improvement system that consists of the series operation of two generators. One circuit of a mine is fed from one generator, and a second circuit is fed from the other generator. The mine track or return is common to both generators and is connected between the generators. This method provides high voltage for transmission of power, yet the individual circuits provide normal low-voltage power. In effect, this is the Edison three-wire system, wherein the rails form the neutral third wire. The third wire (mine track) carries only the unbalance between the loads on the two separate circuits. (Kentucky, 1952)

threshold (a) In geochemical prospecting, the limiting anomalous value below which variations represent only normal background effects and above which they have significance in terms of possible mineral deposits. (Hawkes, 1957) (b) In analytical chemistry, the limiting sensitivity of an analytical method, the detection limit. (Hawkes, 1957)

Threshold Limit Value A time-weighted average concentration under which most people can work consistently for 8 hours a day, day after day, with no harmful effects. A table of these values and accompanying precautions is published annually by the American Conference of Governmental Hygienists. (NSC, 1996)

throat (a) The part of a blast furnace at the top of the stack. (Dodd, 1964) (b) The zone of decreased cross section found between the port area and the furnace chamber in some designs of open-hearth steel furnaces. (Dodd, 1964) (c) The submerged passage connecting the melting end to the working end of a glass tank furnace; the refractory blocks forming the sides of the throat are known as throat cheeks, sleeper blocks, or dice blocks; the refractories for the top are the throat cover. See also: *submarine throat.* (Dodd, 1964) (d) The least thickness of a weld, the calculation of its strength being based on the thickness at the throat. (Hammond, 1965)

throat of crusher Point at which the rock is discharged. Its short dimension varies, depending upon whether the swing jaw is in the open or closed position. (Newton, 1938)

throttle To obstruct the flow of, as steam to an engine esp. by a throttle valve. (Webster 3rd, 1966)

throttle valve A valve designed to regulate the supply of a fluid (as steam or gas and air) to an engine. (Webster 3rd, 1966)

through A passage cut through a pillar to connect two rooms. (Fay, 1920)

through cut An excavation between parallel banks that begins and ends at original grade. See also: *thirling.* (Nichols, 1976)

througher *stenton.*

throughput Quantity of material passed through the mill or a section thereof in a given time or at a given rate. (Pryor, 1963)

through ventilation The normal ventilation produced in a mine as the air flows from the intake to the return, as opposed to ventilation produced locally by auxiliary fans. (Nelson, 1965)

throw (a) The amplitude of shake of a vibratory screen, concentrating table, jigger conveyor, etc. (Nelson, 1965) (b) Lateral displacement of a screen, shaking table, or crushing surface in motion. (Pryor, 1965) (c) The projection of broken rock during blasting. See also: *flyrock.* (Standard, 1964; Fay, 1920) (d) The distance from an air supply opening measured in the direction of air flow, from the opening to the point where the air velocity is 50 ft/min (15.2 m/min). (Strock, 1948) (e) The amount of vertical displacement up (upthrow) or down (downthrow) produced by a fault; sometimes, loosely, a dislocation not vertical, the direction being specified. See also: *heave; perpendicular throw; stratigraphic throw.* (Fay, 1920) (f) The vertical component of the net slip on a fault. (AGI, 1987)

thrower belt *boxcar loader.*

throwing clay Clay plastic enough to be shaped on a potter's wheel.

thrown (a) Faulted or broken up by a fault. (Fay, 1920) (b) Turned, as a piece of ceramic ware on a potter's wheel. (Standard, 1964)

throw of crusher *stroke of crusher.*

throwoff switch Aust. A switch by means of which an obstruction is thrown across the rails of a track, causing the derailment of the trucks. A derailing switch.

throwout bearing (a) A bearing, sliding on a clutch jackshaft, that carries the engage and disengage mechanism. (Nichols, 1976) (b) A bearing that permits a clutch throwout collar to slide along the clutch shaft without rotating with it. (Nichols, 1976)

thrust (a) An overriding movement of one crustal unit over another, such as in thrust faulting. (AGI, 1987) (b) *thrust fault; fault.* (c) A crushing of coal pillars caused by excess weight of the superincumbent rocks, the floor being harder than the roof. Cf: *creep.* (Fay, 1920) (d) The ruins of a fallen roof, after pillars and stalls have been removed. (Fay, 1920) (e) The weight or pressure applied to a bit to make it cut. (Long, 1960)

thrust arm A cable-controlled bar that can slide by power in two directions. (Nichols, 1976)

thrust bearing (a) A bearing that resists attempts of a shaft to move along its axis. (Pryor, 1963) (b) A bearing designed to carry axial loads on a shaft. (Shell)

thrust block The antifriction part of the thrust yoke attached to the drive rod in the swivel head of a diamond-drill machine. Also called cage; friction head; thrust collar. See also: *thrust sheet.* (Long, 1960)

thrust borer Mechanism for forcing a hole through an embankment for the insertion of pipes or cables. (Hammond, 1965)

thrust fault (a) A fault with a dip of 45° or less over much of its extent, on which the hanging wall appears to have moved upward relative to the footwall. Horizontal compression rather than vertical displacement is its characteristic feature. Cf: *normal fault.* Partial syn: reverse fault. Syn: *thrust; overthrust.* (AGI, 1987) (b) A reverse fault that is characterized by a low angle of inclination with reference to a horizontal plane. (AGI, 1987) (c) A reverse fault heading at a high angle. (BS, 1964)

thrust nappe *thrust sheet.*

thrust plane (a) The surface of a thrust fault, when the surface is planar. (AGI, 1987) (b) The plane of a thrust or reversed fault. Syn: *overthrust plane.*

thrust plate The upper and/or lower race parts of the thrust bearing in the thrust block or cage on the drive rod in a diamond-drill swivel head. Syn: *thrust race.* See also: *thrust sheet.* (Long, 1960)

thrust race *thrust plate.*

thrust sheet The body of rock above a large-scale thrust fault whose surface is horizontal or very gently dipping. Syn: *thrust block; thrust nappe; thrust plate.* (AGI, 1987)

thrust washer A washer that holds a rotating part from sideward movement in its bearings. (Nichols, 1976)

thrust yoke The part connecting the piston rods of the feed mechanism on a hydraulic-feed diamond-drill swivel head to the thrust block, which forms the connecting link between the yoke and the drive rod, by means of which link the longitudinal movements of the feed mechanism are transmitted to the swivel-head drive rod. Also called back end; cage. (Long, 1960)

thulite An intense pink variety of zoisite containing manganese; an ornamental stone.

Thum-Balbach process A silver refining process using carbon cathodes, doré anodes, and a silver nitrate-nitric acid electrolyte. The silver is scraped off the bottom as crystals. (Liddell, 1918)

thumb-marked fracture The minute ripples or thumbmarks characteristic of the fractured surface of amethyst. (CMD, 1948)

Thum furnace A gas-fired furnace esp. for the treatment of zinc ore that is high in lead. (Fay, 1920)

thunderbolt (a) A stone or stony concretion, esp. if elongated and tapering, found in the ground and ignorantly supposed to have fallen from the sky. (Standard, 1964) (b) A nodule or mass of iron pyrites found in English chalk formations. (Standard, 1964)

thunder egg A popular term for a small, geodelike body of chalcedony, opal, or agate that has weathered out of the welded tuffs or lava, particularly from central Oregon. (AGI, 1987)

thuringite A monoclinic mineral, $[Fe(OH)_2(AlSi_3O_{10})]{\cdot}3[(Mg,Fe)(OH)_2]$; pseudohexagonal; forms monoclinic plates having micaceous cleavage; an iron-rich chamosite.

thurl S. Staff. To cut through from one working into another. Also spelled thirl. (Fay, 1920)

thurling A passage cut from room to room, in post-and-stall working. See also: *thirl; thirling.* (Fay, 1920)

thwarting A short tunnel driven between two or more veins where they are nearly vertical. (Nelson, 1965)

thyratron A gas-filled valve or tube in which the initiation of current in an ionized gas or vapor is controlled by the voltage applied to a control electrode. (NCB, 1964)

Thyssen gravimeter An early gravity meter of the unstable equilibrium type. (AGI, 1987)

Tibet stone A mixture of aventurine quartz and quartz porphyry that may be of various colors and has been cut as ornamental or curio stones, in Russia.

ticket (a) A sealed bid for ore to be sold. (Webster 3rd, 1966) (b) The numbered check that the miner puts on his loaded car to inform the weighmaster to whom the coal belongs. See also: *tag; tally.* (Fay, 1920)

tick hole A small cavity in a rock; a vug. (Fay, 1920)

tidal flat An extensive marshy or barren tract of land that is alternately covered and uncovered by the tide, and consisting of unconsolidated sediment (mostly mud and sand). It may form the top surface of a deltaic deposit. See also: *sabkha.* Syn: *tidal marsh; tide flat.* (AGI, 1987)

tidal marsh *tidal flat.*

tide flat *tidal flat.*

tidelands Technically lands overflowed during floodtide, but the term, by reason of the so-called Tidelands cases, has been used to describe that portion of the continental shelf between the shore and the claimed boundaries of the States—3 miles or 9 miles (4.8 km or 14.4 km) at sea. (Williams, 1964)

tie (a) A beam, post, rod, or angle to hold two pieces together; a tension member in a construction. (Webster 3rd, 1966) (b) One of the transverse supports to which railroad rails are fastened to keep them to line, gauge, and grade. (Webster 3rd, 1966) (c) Linear or angular measurements or a combination of the two made for the purpose of locating other points from points of known position. Ties may be made to connect physical objects with the survey line, or to locate the instrument point with reference to physical objects so that it can be reestablished if lost. To tie in is to close a survey on itself or on another survey, or to locate a point by means of ties. (Seelye, 1951)

tieback (a) A beam serving the purpose similar to a fend-off beam, but fixed at the opposite side of the shaft or inclined road. (Fay, 1920) (b) The wire ropes or stayrods that are sometimes used on the side of the tower opposite the hoisting engine, in place of or to reinforce the engine braces. (Fay, 1920)

tie bar (a) A bar used as a tie rod. (Webster 3rd, 1966) (b) A rod between two railway switch rails to hold them to gauge. (Webster 3rd, 1966)

tied retaining wall A retaining wall tied into the adjoining ground by means of a deadman (wood block) or other suitable anchorage. (Hammond, 1965)

tie line (a) A line at constant temperature that connects any two phases that are in equilibrium at the temperature of the tie line. Syn: *conode.* (AGI, 1987) (b) A line measured on the ground to connect some object to a survey; e.g., a line joining opposite corners of a four-sided figure, thereby enabling its area to be checked by triangulation. (AGI, 1987)

tiemannite An isometric mineral, HgSe; has sphalerite structure; dark gray to black.

tie plate A metal plate used under rails where they rest on ties. The rail is spiked to the tie through holes in the plate. (Jones, 1949)

tie point (a) A point to which a tie is made; esp. a point of closure of a survey either on itself or on another survey. (AGI, 1987) (b) An image point identified on oblique aerial photographs in the overlap area between two or more adjacent strips of photography. They tie individual sets of photographs into a single flight unit and adjacent flights into a common network. (AGI, 1987)

tie pumping When track is not adequately drained and water enters the ballast and roadbed, tie pumping occurs. Under the action of the rolling stock, pressure on the tie discharges water to the surface, washing the ballast from beneath and around the tie.

tie rod A round or square iron rod passing through or over a furnace and connected with a buckstay to assist in binding the furnace together. (Fay, 1920)

tierra Sp. Any rock or mineral; tierra blanca (Mex.), a calcareous tufa; tierra de batan, fuller's earth; tierra de fluor (Venez.), a bed of reddish clayey earth; tierra de porcelana, china clay; tierra pesada, heavy spar. (Fay, 1920)

tierra blanca A Spanish term for white ground or white earth, and applied to white calcareous deposits such as tufa, caliche, and chalky limestone. (AGI, 1987)

tiff (a) Sparry calcite in Wisconsin and southwestern Missouri zinc fields. (b) Sparry barite in southeastern Missouri.

tiffanyite A hydrocarbon present in certain diamonds causing phosphorescence.

tiger A device, as a fork, for supporting a continuous series of well-boring rods or tubes while raising or lowering them in the hole. (Standard, 1964; Fay, 1920)

tigereye Alternative spelling of tiger's-eye.

tiger's-eye (a) A usually yellow-brown chatoyant stone that is much used for ornament and is a silicified crocidolite in which the fibers penetrating the quartz are changed to oxide of iron. Also spelled tigereye. Cf: *hawk's-eye.* See also: *occidental cat's-eye; oriental cat's-eye.* (Webster 3rd, 1966) (b) Crocidolite asbestos replaced by quartz to yield a yellow-brown chatoyant stone used for ornament. (c) Ceramic glaze resembling tiger's-eye.

tight (a) Soil or rock formations lacking veins of weakness. Syn: *tight formation.* (Nichols, 1976) (b) Blasts or blastholes around which rock cannot break away freely. (Nichols, 1976) (c) Inadequate clearance or the barest minimum of clearance between working parts. (Long, 1960) (d) Unbroken, crack-free, and solid rock in which a naked hole will stand without caving. (Long, 1960) (e) A borehole made impermeable to water by cementation or casing. (Long, 1960) (f) An impermeable rock formation. (Long, 1960) (g) An underground opening having limited space in which to work. (Long, 1960) (h) Lacking in porosity; impervious. (Wheeler, 1958)

tight-bedded In sandstone quarrying, a term used to describe the rock if it is massive, showing no open-bed seams. Cf: *thin-bedded; thick-bedded.* (AIME, 1960)

tight-burning clay A clay that is dense or approaches vitrification after firing.

tight formation *tight; tight rock.*

tight hole (a) A borehole the diameter of which is too small for adequate clearance between the drill-stem equipment and/or inserted casing. (Long, 1960) (b) A borehole the wall rocks of which are impermeable to water or have been made tight by cementation or insertion of casing. (Long, 1960) (c) A borehole-drilling operation, access to which and information about which are not released except to authorized persons. (Long, 1960)

tight lagging Lagging placed touching each other. See also: *tight sheathing.*

tight rock (a) Rock formation in which the joints, cracks, or crevices are sealed and impermeable to water. (Long, 1960) (b) Rock composed of tightly cemented grains of very fine, even-sized crystals. (Long, 1960) (c) Rock that does not chip easily under the impact of cable tools. (Long, 1960) (d) A tough, resilient rock. (Long, 1960) (e) Can. Without evidence of shearing or mineralization. Syn: *tight formation.* (Hoffman, 1958)

tightset A quarrymen's term, equivalent to blind seam, or incipient joint. (Fay, 1920)

tight sheathing The most complete sheathing using wood timbering. Used where water or fine wet soils must be retained. The frame is designed for this use and is generally stronger than that required for other types of sheathing. A specially edged plank, generally tongue-and-grooved, eliminates the crevices existing in close sheathing. Cf: *close sheathing; skeleton sheathing.* See also: *tight lagging.* (Carson, 1961)

tight shot An explosive shot that has been set off to loosen coal in a seam that has not been previously cut or sheared. (BCI, 1947)

tilasite A monoclinic mineral, $CaMg(AsO_4)F$; isomorphous with isokite; violet-gray; at Långban, Sweden, and Kajlidongri, Central India.

tile copper Copper obtained by roasting and refining the metal bottoms that collect under the regulus in smelting certain impure ores; usually cast in flat, rectangular plates, hence its name. See also: *bottoms.* (Standard, 1964)

tile machine A machine for making tubular or arch-shaped tiles from clay, operating by forcing the raw material through a die, in a continuous stream, which is cut into suitable lengths by wires. (Standard, 1964)

tile ore An earthy variety of cuprite, brick red because of admixed iron oxides.

tiler (a) A kiln or oven for baking tiles. (Standard, 1964) (b) A maker or layer of tiles. (Standard, 1964)

tile shoe A device that permits laying tile directly behind a ditcher. Also called tile box. (Nichols, 1976)

tile works A tilery or tile field. (Standard, 1964)

tilgate stone Beds of calcareous sandstone or ironstone, near Hastings, England. (Fay, 1920)

till Dominantly unsorted and unstratified drift, generally unconsolidated, deposited directly by and underneath a glacier without subsequent reworking by meltwater, and consisting of a heterogeneous mixture of clay, silt, sand, gravel, and boulders ranging widely in size and shape. See also: *boulder clay; moraine.* Syn: *glacial till; ice-laid drift.* (AGI, 1987)

tiller rope A flexible wire rope composed of six small ropes, usually of seven wire strands each laid about a hemp core. (Zern, 1928)

tilleyite A monoclinic mineral, $Ca_5(Si_2O_7)(CO_3)_2$; forms white grains in rock at Crestmore, CA.

tillite A consolidated or indurated sedimentary rock formed by lithification of glacial till, esp. pre-Pleistocene till (such as the Late Carboniferous tillites in South Africa and India). (AGI, 1987)

tilt (a) The angle at the perspective center of an aerial photograph between the plumb line and the perpendicular from the interior perspective center to the plane of the photograph. (AGI, 1987) (b) The lack of parallelism (or the angle) between the plane of the photograph from a downward-pointing aerial camera and the horizontal plane (normal to the plumb line) of the ground. (AGI, 1987) (c) In aerial photography, the angle between the lens axis and a vertical through the exposure station (rear nodal point of lens). It is seldom more than 3° and can generally be kept to 1°. This is regarded as satisfactory for vertical photographs. (Seelye, 1951; Hammond, 1965) (d) To hammer or forge with a tilt hammer; as, to tilt steel to render it more ductile. (Standard, 1964)

tilt hammer A hammer for shingling or forging iron, arranged as a lever of the first or third order, and tilted or tripped by means of a cam or cog gearing and allowed to fall upon the billet, bloom, or bar. (Fay, 1920)

tilting dozer A bulldozer whose blade can be pivoted on a horizontal center pin to cut low on either side. (Nichols, 1976)

tilting furnace Open-hearth furnace swung about its major axis when pouring out the melted product. (Pryor, 1963)

tilting gate A crest gate for dam spillways designed so that water pressure acting upon it will do so only at a definite level. It closes automatically when the water level falls to normal. (Hammond, 1965)

tilting idlers An arrangement of idler rollers on a conveyor in which the top set is mounted on vertical arms which pivot on spindles set low down on the frame of the roller stool. This permits the entire carrier frame to lean forward slightly in the direction of belt travel. In the event of the belt not running true, the tilting idlers guide it back to its correct course again. (Nelson, 1965)

tilting level A surveying instrument with sighting telescope so mounted that it can be raised or lowered through a limited arc without impairing accuracy of reading, though axis of rotation is not precisely horizontal. The bubble tube is usually mounted alongside the telescope and is viewed from the eyepiece and through an optical sighting arrangement, which either brings opposite halves of the bubble image into coincidence or the end of the bubble to a reference line. (Pryor, 1963; Mason, 1951)

tilting mixer A concrete mixer with a rotating drum, which is tilted to discharge its contents. (Hammond, 1965)

timber (a) Any of the wooden props, posts, bars, collars, lagging, steel joists or beams, etc., used to support mine workings. (Fay, 1920) (b) To set or place timbers in a mine. (Fay, 1920) (c) Applied to rough blocks of natural rock as it comes from a quarry before being shaped into sharpening stones. (Mersereau, 1947)

timber boss *timberman.*

timber drawer (a) An appliance for withdrawing timber supports from wastes, e.g., a sylvester. (Nelson, 1965) (b) A miner engaged in timber drawing. (Nelson, 1965) (c) *timber puller.*

timber drawing *drawing timber.*

timbered stope Stope in which square-set timbering and its variations are employed. As a rule, the ground is broken by overhand methods, the face being advanced by successive small excavations, each one timbered before the next is begun. (Higham, 1951)

timberer One who cuts, frames, and/or puts in place any of the timbers used in a shaft, slope, mine, or tunnel. Also one who draws props, posts, etc. Syn: *timberman.* (Fay, 1920)

timber foreman *timberman.*

timbering (a) The operation of setting timber supports in mine workings or shafts to support the roof or the face of a tunnel during excavation and lining. The term "support" would cover the setting of timber, steel, concrete, or masonry supports. See also: *setting; timber set; face timbering; pin timbering.* (Nelson, 1965) (b) Timber work taken collectively in a mine. (Standard, 1964) (c) Protecting against falls of roof formation of a mine, by means of horizontal timbers or caps extending across the passageway just under the roof, the ends of such timbers resting upon the vertical timbers or posts. (Ricketts, 1943) (d) Timber to support the roof or the face of a tunnel during excavation and lining. (Stauffer, 1906)

timbering machine An electrically driven machine to raise and hold timbers in place while supporting posts are set after being cut to length by the machine's power-driven saw.

timbering set A tunnel support consisting of a roof beam, or arch, and two posts. (Nichols, 1976)

timber jack A jack to raise and hold crossbars against the roof while props are being set. (Hess)

timberman (a) In bituminous coal mining, a head timberman is a foreman who supervises workers installing timbers in a mine to support the roof and walls of haulageways, passageways, and the shaft. Also called timber boss; timber foreman. (DOT, 1949) (b) A miner skilled in notching, erecting, and securing timbers set in mine workings. The craft of the timberman is gradually becoming extinct with the advent of power tools and steel as a support. (Nelson, 1965) (c) *roof bolter; timberer.*

timberman helper In mining, a laborer who assists a timberman in erecting supports for the roof of a mine, using posts, headers, cap pieces, and wedges. (DOT, 1949)

timber mat Broken timber forming roof of ore deposit being extracted by caving methods such as top slicing. It separates the downward gravitating overburden and rock strata from the ore. (Pryor, 1963)

timber packer (a) A laborer who delivers timber to the working place in a pitching or inclined coal seam. (Fay, 1920) (b) *pack builder.*

timber pickling A method to assist timber preservation; e.g., creosoting. See also: *timber preservation.* (Nelson, 1965)

timber preservation Any treatment of mine timber for the purpose of extending the useful life of the timber. Various preservatives are used, such as creosote, zinc chloride, sodium fluoride, and other chemicals. See also: *brush treatment; guniting; open-tank method; pressure process; Bethell's process; timber pickling; seasoning timber.* (Lewis, 1964)

timber puller (a) A piece of equipment used in removing the supports or timbers in a mine. A timber puller should be constructed so that the operator will be under safe roof while drawing the timber. A sylvester is an example of this type of equipment. Also called timber drawer. (Kentucky, 1952) (b) *timber robber.*

timber rights The right to cut timber on the public domain for commercial use.

timber robber In anthracite and bituminous coal mining, a laborer who pulls out and recovers timbers and props in working places from which all coal has been mined. Also called prop drawer; timber puller. (DOT, 1949)

timber set A timber frame to support the roof, sides, and sometimes the floor of mine roadways or shafts. For a mine roadway, the simplest timber set consists of a crossbar, cap, or collar supported on two upright posts or arms with round or board lagging. Such a set will resist roof pressure and moderate side pressure and is erected at intervals of from 2 to 6 ft (0.6 to 1.8 m). The timbers are about 5 to 10 in (12.7 to 25.4 cm) in diameter. In South Wales, such a timber set is known as double timber. See also: *timbering; bar timbering; two-piece set; three-piece set; four-piece set.* (Nelson, 1965)

timber trolley A strong carriage of low height for transporting timber from the surface stockyard to underground workings. It consists of a timber or steel base, mounted on wheels, with U-shaped arms in which the timber is lashed with chains. See also: *bogie.* (Nelson, 1965)

timber truck Any truck or car used for hauling timbers inside a mine. In conveyor work, it is applied to the small truck mounted on wheels that is designed to run in the panline of a shaker conveyor for the purpose of carrying timber and other materials to the face. (Jones, 1949)

time-and-motion study The coordination and analysis of the data provided by time study and motion study. (Nelson, 1965)

time at shot point In seismic exploration, the time required for the seismic impulse to travel from the charge in the shothole to the surface of the Earth. Syn: *uphole time.* (AGI, 1987)

time break An indication on a seismic record showing the instant of detonation of a shot or charge. Cf: *time signal.* Syn: *shot moment; shot instant; time mark.* (AGI, 1987)

time correlation Correlation of rocks in one area with those of another area on basis of time equivalence or contemporaneity of origin.

time-delay relay Relay that does not operate until a predetermined time has elapsed. The time ratings are usually adjustable, but some time-delay relays have the time rating built into them. (Coal Age, 1966)

time-depth chart A graphical expression of the functional relation between the velocity function and the times observed in the seismic method of geophysical exploration. It permits time increments to be converted to corresponding depths. Syn: *time-depth curve.* (AGI, 1987)

time-depth curve *time-depth chart.*

time-distance curve In refraction seismic computations, a graph, usually with arrival times of distinctive seismic signals plotted as ordinates and with distances along the surface of the Earth plotted as abscissas. In earthquake studies, the times of arrival of seismic waves at recording stations may be known, but the time of initiation of the waves may be unknown. As data are accumulated from different recording stations, a time-distance graph may be constructed. If it is possible to extrapolate this graph to the origin on the time and distance coordinates, it becomes a travel-time curve. (AGI, 1987)

time-distance graph In refraction seismic computations, a plot of the arrival times of refracted events against the shot-point-to-detector distance. The reciprocal slopes of the plotted segments are the refraction velocities for the refracting bed. (AGI, 1987)

time gradient In the reflection seismic methods applied to dipping reflectors, the travel time curves may not be straight lines; i.e., the apparent velocity observed varies with the spread from shot point to detectors. The time gradient is the reciprocal of the apparent velocity. In seismic prospecting, also the rate of change of travel time with depth. (AGI, 1987)

time lag (a) In refraction seismic interpretation, where arrival times are plotted against shot-detector distances, if some of the paths from shot point to detector include a low-speed bed, the corresponding arrival times will be abnormally long, and the departure from normal travel time is called a time lag. Also, in seismic prospecting, time delays in arrivals due to phase shifts in filtering, to shot-hole fatigue, etc. (AGI, 1987) (b) A delay in the arrival time of seismic energy from the time expected. Time lags may be produced by an abnormal low-velocity layer, phase shifts in filtering, or other factors. (AGI, 1987)

time leads In a method of interpretation of refraction seismic records where the arrival times are plotted against shot-detector distances, if some of the paths from shot point to detector include a high-speed segment, the corresponding travel times will not fall on a smooth curve. The departure in this case from the curve is called a time lead, and it is proportional to the horizontal extent of the high-speed segment. Used in salt-dome exploration. (AGI, 1987)

time mark (a) *time break.* (b) Mark corresponding to a particular time (e.g., hour, minute, and second time marks) on a seismic recording.

timer lines Lines on a seismogram that mark increments of time. (Schieferdecker, 1959)

time-rock unit *time-stratigraphic unit.*

time scale *geologic time scale.*

time signal In geophysics, a signal used to indicate the time of explosion in a shothole and successive intervals of time on the recording. Cf: *time break.*

time-stratigraphic unit A body of rock established to serve as the material reference for all rocks formed during the same span of time. Each of its boundaries is synchronous. Chronostratigraphic units in order of decreasing rank are eonothem, erathem, system, series, stage. Syn: *chronostratic unit; chronolithologic unit; time-rock unit; chronolith.* (AGI, 1987)

time study A detailed investigation in which the average time taken to do each operation of a complete cycle is recorded. See also: *method study; motion study.* (Nelson, 1965)

time tie In seismograph continuous profiling, a coincident travel path for seismic energy initiated at opposite ends of the path. The use of such coincident travel paths on adjacent reflection layouts facilitates correlation from one layout to the next as the shot point or recording position is changed. (AGI, 1987)

time-transgressive *diachronous.*

time value The interval of geologic time represented by or involved in producing a stratigraphic unit, an unconformity, the range of a fossil, or any geologic feature or event. (AGI, 1987)

timing The time elapsing between two successive exposures of an aircraft camera when taking vertical photographs. See also: *photogrammetry.* (Hammond, 1965)

timing line One of a series of marks or lines placed on seismic records at precisely determined intervals of time (usually at intervals of 0.01 or 0.005 s) for the purpose of measuring the arrival time of recorded events. (AGI, 1987)

tin (a) A tetragonal mineral, Sn: rare; soft; malleable: bluish white. (b) The metal extracted from cassiterite; used as a coating to protect iron and copper, such as a foil, and in solder, bronze, and other alloys. Commercially, tin is available in three grades: Grade A must assay 99.75%; grade B must assay 99.7%; and grade C, or common tin, must assay 99% tin. (c) To coat with tin, such as to tin iron; tinplate. (Standard, 1964) (d) Metallic element that has a highly crystalline structure. Symbol, Sn. Found chiefly in cassiterite, SnO_2. Used in alloys such as soft solder, type metal, fusible metal, pewter, bronze, and bell metal and as a crystalline tin-niobium alloy. (Handbook of Chem. & Phys., 3) (e) *zinn.*

tin bound (a) Corn. To mark a limit, as on a tract of waste land, within which one claims or reserves the right to mine unworked tin ore. (Standard, 1964) (b) Land so reserved. (Standard, 1964)

tincal The name given, since early itmes, to crude borax obtained from salt lakes in Kashmir, India, and Xizang (formerly Tibet), China. Also spelled tinkal. Syn: *borax.* (CTD, 1958; Fay, 1920)

tincalconite A trigonal mineral, $Na_2B_4O_5(OH)_4{\cdot}3H_2O$. Syn: *mojavite; octahedral borax.*

tin-can safety lamp A Davy lamp placed inside a tin can or cylinder having a glass in front, airholes near the bottom, and an open top. (Fay, 1920)

tinder ore An impure variety of jamesonite. (Standard, 1964)

tin dish A pan used by prospectors for washing gold-bearing materials and extracting the gold. See also: *pan.* (Fay, 1920)

tin dredging The extraction of tin-bearing ore from placers by means of dredges. (Nelson, 1965)

tine The actual excavating tooth or point of a grab bucket, scraper loader, dragline, or excavator bucket. (Hammond, 1965)

tin floor (a) Corn. A thin flat mass of tinstone between beds of rock. (Fay, 1920) (b) A flat mass of tin ore. (Standard, 1964)

tinge A color designation. A faint trace of a hue that modifies another hue, as a blue with a tinge of green, i.e., blue tinged with green or, stated differently, very slightly greenish-blue.

tin ground Corn. Tin-bearing alluvium, stream works. (Arkell, 1953)

tin hat A head covering made of reinforced sheet aluminum or plastic-impregnated fabric and shaped somewhat like a sun helmet; worn for protection and/or to reduce the severity of head injuries from falling objects. Also called hard hat. Syn: *safety hat.* (Long, 1960)

tinkal *tincal.*

tinker Derb. Laminated carbonaceous shale. (Fay, 1920)

tin minerals Virtually all the industrial supply of tin comes from cassiterite, SnO_2, though a little has been won from the sulfides stannite, cylindrite, and franckeite. Bulk of cassiterite comes from alluvial workings. Main market is in tin plating, tin foil, solders, bearing metals,

bronze and other alloys, and such salts as opacifiers and dye mordants. (Pryor, 1963)
tinned sheet iron *tin plate.*
tinner A tinsmith. (Webster 3rd, 1966)
tinning metal An alloy of equal parts of tin and lead; used by electrotypers for coating copper shells before backing. (Fay, 1920)
tin ore *cassiterite.*
tin plate Sheet iron or steel, cleaned by pickling in acid and then passed through bath of molten tin to produce coating. Three grades are charcoal plate, coke plate, and crystallized. Tin is also deposited by electroplating. Syn: *tinned sheet iron.* (Pryor, 1963)
tin pyrites *stannite.*
tin salt *salt of tin.*
tin spar Syn: *cassiterite.* Also called tinstone.
tinstone *cassiterite.*
tin stone *cassiterite.*
tin-white cobalt *smaltite.*
tinworks A place or an establishment where tin is manufactured or mined. (Standard, 1964)
tip (a) The point at which loaded mine cars are dumped on the surface. Also called tipple. (Hudson, 1932) (b) A piece of tool material secured to a cutter tooth or blade. (ASM, 1961)
tipple (a) Originally the place where the mine cars were tipped and emptied of their coal, and still used in that sense, but more generally applied to the surface structures of a mine, including the preparation plant and loading tracks. (BCI, 1947) (b) The dump; a cradle dump. (c) Aust. The tracks, trestles, screens, etc., at the entrance to a colliery where coal is screened and loaded. Cf: *dump.*
T-iron (a) An angle iron having T-shaped cross section. (Fay, 1920) (b) T-rails used in a mine, as distinguished from wooden rails. (Fay, 1920)
titan *titanium; titanite.*
titanaugite *titanian augite.* Found in basaltic rocks.
titanhornblende A titaniferous variety of hornblende. (English, 1938)
titania (a) *titanium dioxide.* (b) TiO_2. Also called titanium oxide; titanic dioxide; titanic oxide. A common constituent of iron ores. Used as a pigment, and it replaces zinc oxide in manufacturing white rubber and as a filler for paper; can be used alone as a refractory and as an electrical insulator. Its crystals show marked piezoelectric effects and have a greater brilliance and a higher refractive index than diamond. (c) The minerals tetragonal rutile, anatase, octahedrite, and orthorhombic brookite. (d) Commonly refers to synthetic white titanium dioxide that is produced mainly from ilmenite, $FeTiO_3$, that contains 50% to 54% TiO_2.
titanian augite *titanaugite.*
titanic anhydrite A white pulverulent titanium oxide, TiO_2, found native as brookite, octahedrite, and rutile, and a common constituent of iron ores. Also called titanic oxide. (Fay, 1920)
titanic dioxide *titanium dioxide.*
titanic iron ore *ilmenite.* Also called titaniferous iron ore.
titanic oxide *titanium dioxide.*
titanic schorl *rutile.*
titaniferous magnetite Magnetite containing titanium. (Bateman, 1951)
titanite A monoclinic mineral, $CaTiOSiO_4$; Ca is replaced by Sr, Ba, Na, Mn, Th, or rare earths; Ti is replaced by Al, Fe, Mg, Nb, Ta, V, or Cr; up to 1/5 of O_{2-} may be replaced by $(OH,F)^-$; weakly radioactive: forms wedge-shaped crystals; a common accessory in felsic plutonic rocks, in gneisses, schists, and marbles; a source of titanium. Formerly called sphene.
titanium A silvery-gray or iron-gray, metallic element. Symbol, Ti. Found in nature only in combined form; occurs chiefly in ilmenite ($FeTiO_3$), and in rutile and titanite. Used as an alloying agent with aluminum, molybdenum, manganese, iron, and other metals. Used in aircraft and missiles and has potential for use in desalination plants. (Handbook of Chem. & Phys., 3)
titanium carbide A compound produced by fusing titanium dioxide with carbon or calcium carbide. Has a melting point in the range of 3,140 to 3,160 °C. This very hard, refractory material is used for wear-resistant applications and where good thermal shock resistance is needed, as in bearings, nozzles, and special refractories under either neutral or reducing conditions. (Lee, 1961)
titanium dioxide (a) Also called titanium oxide, titanic dioxide, titanic oxide, titania, rutile, anatase, or brookite. Colorless, white, pale yellow or yellowish-red, reddish-brown, brown, blue or bluish, violet, and black; tetragonal and orthorhombic; TiO_2; molecular weight, 79.90; sp gr, 3.82 to 5.13 depending on crystal system and crystal form; Mohs hardness, 5.5 to 6.5; melting point, 1,825 to 1,850 °C; boiling point, 2,500 to 3,000 °C; insoluble in water and in most acids; and soluble in hot concentrated sulfuric acid and in alkalies. Titanium dioxide occurs as the minerals rutile (tetragonal); anatase or octahedrite (tetragonal); and brookite (orthorhombic). Titanium dioxide is a common constituent of iron ores. (Handbook of Chem. & Phys., 2; CCD, 1961) (b) Titanium dioxide as rutile: colorless, pale yellow, reddish-brown, red, bluish, violet, and black; tetragonal; adamantine to submetallic luster; refractive indexes, 2.616 and 2.903; sp gr, 4.26 and ranges from 4.18 to 5.13; Mohs hardness, 6.0 to 6.5; and the same melting points, boiling points and solubility characteristics as above. Titanium dioxide possesses the gredatest hiding power of all the white pigments. Used in glassware and in ceramics, in enamel frits, in welding rods, and single crystals are used as high-temperature transducers. Syn: *titania; octahedrite.*
titanium dioxide pigments Any of three grades of titanium-dioxide-based pigments used in the production of paints, paper, and many other products requiring a white pigment with a high hiding power and chemical stability. Rutile and anatase grades are more or less pure titanium dioxide, but owing to a difference in crystal structure, they differ slightly in hiding power and chalking quality. Titanium dioxide of pigment quality is manufactured principally by treating finely ground ilmenite or titanium slag with concentrated sulfuric acid. Also used in ceramics and fiberglass, and in making titanium gems. (USBM, 1960)
titanium nitride TiN; a special refractory material (melting point, 2,930 °C). It can readily be produced from $TiC1_4$ and NH_3. (Dodd, 1964)
titanium oxide (a) TiO_2; used as an opacifier, particularly in vitreous enamels, and as a constituent of some ceramic colors. Titania and titanate electroceramics, for use in the radio frequency field, are based on this oxide and its compounds. Titania occurs in three crystalline forms: anatase, brookite, and rutile. (Dodd, 1964) (b) *titanium dioxide.*
titanium silicide Ti_5Si_3; sp gr, 4.2. This special ceramic has good resistance to high temperature oxidation, but not to thermal shock. (Dodd, 1964)
titanium sponge The metal product from reducing titanium tetrachloride with magnesium in the Kroll process. It is called sponge because of its spongelike appearance. Sodium-reduced metal also is referred to as sponge. (USBM, 1965)
titanmagnetite *titanomagnetite.*
titanmelanite A titaniferous andradite approaching schorlomite in composition; garnet group.
titanomagnetite (a) A titaniferous variety of magnetite with titanium in crystal solution. Syn: *titanmagnetite.* (b) A term for mixtures of magnetite, ilmenite, and ulvöspinel. Cf: *mogensenite.*
titan process A process of concentrating iron ore that comprises the steps of (1) effecting a dry, thermal, partial reduction of the iron in the ore to the metallic state to a degree of reduction of between 50% and 80%, (2) subjecting the reduced product to a magnetic separation, and (3) recovering the magnetic concentrate. (Osborne, 1956)
tithe ore Eng. A portion of ore set aside for the payment of rental or royalty on mineral lands. (Fay, 1920)
title The right to enter, develop, and work a coal or mineral deposit. See also: *claim; lease.* (Nelson, 1965)
TLV *Threshold Limit Value.*
toadrock *toadstone.*
toad's-eye Eng. Shelly pink limestone in the Corbula Beds of the Purbeck Beds at Durlston Bay. So called because it is full of the small gastropods Pachychilus manselli that resemble eyes when seen in transverse section. Cf: *rabbit-eye.* (Arkell, 1953)
toad's-eye tin A reddish or brownish variety of cassiterite in botryoidal or reniform shapes with internal concentric or fibrous structure.
toadstone (a) Applied earlier to various stones or stonelike objects likened in color or shape to toads (batrachites, bufonites, crapodius). Syn: *toadrock; fiery dragon.* (Arkell, 1953) (b) Eng. A kind of traprock. (c) An old Derbyshire name for amygdaloidal basalt lava in the Carboniferous Limestone. (d) A fossilized object, such as a fish tooth or palatal bone, once thought to have formed within a toad and frequently worn as a charm or an antidote for poison. Syn: *toadrock.*
tobacco rock A term used in the Southwest United States for a favorable host rock for uranium, characterized by light yellow or gray color and by brown limonite stains. (AGI, 1987)
tobermorite An orthorhombic mineral, $Ca_9Si_{12}O_{30}(OH)_6{\cdot}4H_2O$; the principal cementing compound in Portland cement.
toddite A mixture of columbite and samarskite.
todorokite A monoclinic mineral, $(Na,Ca)_{0.5}(Mn,Mg)_6O_{12}{\cdot}4H_2O$; forms black spongy banded and reniform aggregates composed of minute lathlike crystals; in Hokkaido, Japan.
toe (a) The base of the coal, ore, or overburden face in a quarry or opencast mine. (Nelson, 1965) (b) The front end of a frog, opposite the heel, in a car track. (c) The lowest part of a slope or cliff; the downslope end of an alluvial fan. (AGI, 1987)
toeboard A raised edging around the perimeter of a work platform in drilling to prevent hand-tools from being accidentally kicked or knocked off the platform. (Long, 1960)
toe cut In underground blasting, the cut obtained by the use of single cut holes inclined downward. (Lewis, 1964)

toehole A blasting hole, usually drilled horizontally or at a slight inclination into the base of a bank, bench, or slope of a quarry or open pit mine.

toeing-in A quarry term for the wedging-in of the end of a granite sheet under an overhanging joint, probably in consequence of the faulting of the sheets along the joint. It is also applied to the overlapping of lenticular sheets. (Fay, 1920)

toe of a shot The distance from the inner end of the hole to the adjacent free face measured at right angles to the direction of the hole; or that portion of the hole that is filled with powder; or that part of the seam to be broken lying between the powder and a free face. (Zern, 1928)

toernebohmite *törnebohmite.*

toe-to-toe drilling The drilling of large-diameter blasting holes in quarries and opencast pits. They are put down vertically from top to bottom of the quarry face. Deck loading is often adopted, with half to two-thirds of the total charge at the bottom and the remainder in one or more deck charges as required. (Nelson, 1965)

to gauge (a) Made to gauge, or a size as specified, esp. as applied to the outside set diameter of bits and reaming shells and the inside diameter of a borehole. (Long, 1960) (b) To determine, by measurement or other test, the capacity, quantity, or dimension. (Long, 1960)

toggle action Application of crushing force so that the distance moved diminishes without change of input strength, between gape and set. Thus greatest speed of movement of the approaching faces is applied with weakest thrust and vice versa. (Pryor, 1965)

toggle joint A joint having a central hinge like an elbow, and operated by applying the power at the junction of motion, as from horizontal to vertical, and giving enormous mechanical advantage; a mechanism common in many forms of presses and in stone crushers. (Standard, 1964)

toggle mechanism A mechanism utilized to apply heavy pressure from a small applied force, such as in a jaw breaker, and other machinery. (Hammond, 1965)

toise An old French unit of length used in early geodetic surveys and equal to 6 French ft, 6.396 U.S. ft, or 1.949 m. (AGI, 1987)

tolerance A specified allowance (either plus or minus) of the given dimensions of a finished product to take care of inaccuracies in workmanship of parts to be fitted together. The amount allowed as tolerance is generally small as compared with the standard dimension of the part; e.g., the tolerance allowed in the set diameters of a diamond bit is ±0.02 mm. (Long, 1960)

tolerance limits In control of a measured value in a process, limited drift from optimum or norm (e.g., pH 7.6 ± 0.2). (Pryor, 1963)

toll Ches. Royalty on rock salt, or other mineral. (Fay, 1920)

toll enrichment A proposed arrangement by which uranium owned privately could be enriched in uranium-235 content in U.S. Government facilities upon the payment of a service charge. (Lyman, 1964)

toll refining Situation in which the owner of ore or concentrate contracts the refining of the metal to another party for a fee, but the refined metal remains under the original ownership for final sale or disposition.

tom (a) An inclined trough in which gold-bearing earth or gravel is crudely washed; usually called long tom because it is longer than the rocker. (Webster 2nd, 1960) (b) Cumb. A parting of black shale in a coal seam. (Arkell, 1953)

Tomassi process An electrolytic process for refining lead in which the electrolyte is a solution of a double acetate of lead and potassium or sodium. The anodes are cast from crude argentiferous lead, and the cathodes are in the form of large disks of copper or aluminum bronze and are about half immersed in the electrolyte. (Fay, 1920)

tommy bar A short rod used as a lever or handle for turning a jackscrew or a spanner by being inserted loosely in the hole provided for that purpose. (Long, 1960)

tomography A division of radiography dealing with the photography of a particular plane in an object while leaving out undesired detail in other planes. Although this technique was developed for medical radiography, it is recommended for certain purposes in work with metals where it is essential that the location of faults be exactly known. (Osborne, 1956)

ton-cap screen Commercial brand of wire screen cloth with long rectangular meshes. (Pryor, 1963)

tone (a) That attribute of a color that determines its position in a scale from light to dark. Thus white, and also light gray, are light tones, and dark gray is the dark tone of the same color sensation; pink is a light tone of red, and maroon a dark tone. A light tone is usually known as a tint, a dark tone as a shade. See also: *intensity.* (b) A monochromatic frequency of vibration, such as in a violin string or vibrations of bodies of finite size and shape.

tong die A hard, replaceable, serrated metal insert in pipe tongs, which comes in contact with and grips the outside of a pipe, casing, or drill rod. Also called tong key. (Long, 1960)

tongs One of the various tools or wrench devices that can be made to fit and grasp drill rods, casing, or drivepipe. (Long, 1960)

tongue (a) A branch or offshoot of a larger intrusive body. See also: *epiphysis.* Syn: *apophysis.* (AGI, 1987) (b) A minor lithostratigraphic unit of limited geographic extent, being a subdivision of a formation or member, and disappearing laterally (usually by facies change) in one direction; a member that extends outward beyond the main body of a formation. Cf: *lentil.* (AGI, 1987) (c) A lava flow that is an offshoot from a larger flow; it may be as much as several kilometers in length. (AGI, 1987)

tongue joint In welding, a split joint formed by inserting a wedge-shaped piece into a corresponding split piece and welding the two together. (Fay, 1920)

tongue plate An adjustable plate that controls the quantity of feed entrapped by the rolls of a double-roll press. (BS, 1962)

tongue test A test by which crystals and crystalline gemstones can be distinguished from glass that feels warmer in comparison when held to the tongue because of its lower thermal conductivity.

tonite A blasting explosive consisting of a mixture of guncotton with a nitrate and sometimes a nitro compound. (Webster 3rd, 1966)

ton-kilometer A unit of measurement often used for the work done in transport. The number of ton-kilometers is the weight in tons of material transported multiplied by the number of kilometers driven. (Stoces, 1954)

Ton MPH A system that permits calculating how hard an earth-moving tire should work and how much work it is doing. The Ton MPH for any tire is determined by multiplying the mean load and average speed. The resulting figure provides an index to the work a tire is doing. The system enables the operator to determine which type of tires to use to get top performance without overheating. (Coal Age, 1966)

tonnage factor Cubic feet of ore per ton in deposit. (Pryor, 1963)

tonnage man In anthracite coal mining, a person who is paid at a certain rate per ton of coal mined. (DOT, 1949)

ton of refrigeration The extraction of 200 Btu/min (211 kJ/min), 12,000 Btu/h (12.7 MJ/h), or 288,000 Btu/d (3.1 GJ/d). The last is also called a ton-day of refrigeration. A ton of refrigeration is equal to 3.5168 kW of heat removal. (Strock, 1948)

Tonpilz machine An apparatus for measuring damping values. (Osborne, 1956)

tonstein A compact argillaceous rock containing the clay mineral kaolinite in a variety of forms together with occasional detrital and carbonaceous material; commonly occurring as a thin band in a Carboniferous coal seam (or locally in the roof of a seam); often used as an aid in correlating European strata of Westphalian age. (AGI, 1987)

tool box *powder box.*

tool box miner A lazy miner, specif. one who rests on a tool box while another miner does the work. (Korson, 1938)

tooldresser The driller's helper on a cable tool rig. Syn: *cable-tool dresser; toolie.* (AGI, 1987)

tool extractor An implement for grasping and withdrawing drilling tools when broken, detached, or lost in a borehole. A fishing tool. Also called tool grab. (Long, 1960)

tool grinder In stonework industry, one who grinds the cutting tools for stoneworking planers and lathes to a keen edge of the desired shape. (DOT, 1949)

tool heat treater One who hardens and tempers tools, dies, and fixtures.

toolie A worker who sharpens churn-drill bits; a dresser. See also: *tooldresser.* (Long, 1960)

toolmaker A worker skilled in the making of jigs, fixtures, gauges, etc. (Crispin, 1964)

tool nipper A person who carries powder, drills, and tools to the various levels of a mine and brings dulled tools and drills to the surface. (Fay, 1920)

toolpusher (a) The head driller or drill foreman. (Long, 1960) (b) The general supervisor of operations on a drilling rig. More commonly used in petroleum drilling. (AGI, 1987)

toolstone Industrial diamond used for wire-drawing dies, indentor points, shaped-diamond tools, glaziers, and dressers. Toolstones approach gem diamonds in perfection, although not in color. The finer grades may be identical with diamonds sold as low-grade gems. The lower grade toolstones are also sometimes used as drill diamonds. Cf: *drill diamond.* (Long, 1960)

tooth (a) Steel projections on a tool, such as a saw or excavation bucket, designed to provide a cutting or increased digging action. (b) A projection on the circumference of a wheel (gear), designed to engage corresponding projections on another wheel (cog), and thereby transmit force.

tooth base (a) The inner part of a two-piece tooth on a digging bucket. (Nichols, 1976) (b) Occasionally, the socket in which a tooth fits. (Nichols, 1976)

tooth brake A brake used to hold a shaft by means of a tooth or teeth engaging with fixed sockets. Not used for slowing or stopping. (Nichols, 1976)

toothed roller bit *roller bit.*
toothed-shoe cutter A drivepipe or casing shoe with a serrated or toothed cutting edge. (Long, 1960)
tooth turquoise Odontolite; fossil tooth material.
too wet A mine-safety expression used to describe those mines or areas of mines that are too wet to propagate explosions even though they are not rock dusted. Too wet is when water exudes from a ball of dust when it is squeezed in the hands.
top (a) The surface around a mine shaft; the outside. (Jones, 1949) (b) A mine roof; the upper part of a coalbed separated from the rest by a seam or parting. (Fay, 1920) (c) The apex of a vein. (Hess) (d) *cap; blue cap; overburden.*
top and apex The words top and apex as applied to mineral veins were not a part of the miner's terminology prior to the adoption of the U.S. Mining Law of 1872, but were words used by legislators to convey the intent of the formulators of that law. Cf: *apex.* (Ricketts, 1943)
topaz A triclinic mineral, $Al_2(F,OH)_2SiO_4$; pseudo-orthorhombic; colorless to pale blue, pale yellow to pinkish-beige ("sherry" topaz); defines hardness 8 on the Mohs scale; in cavities in granites, granite pegmatites, and rhyolites, and in surrounding metamorphic rocks; may be of gem quality. Cf: *false topaz; Scotch topaz; Spanish topaz.*
topazfels *topazite.*
topazite A hypabyssal rock composed almost entirely of quartz and topaz (Johannsen, 1920). Syn: *topazfels; topazoseme; topaz rock.* (AGI, 1987)
topazolite A yellow variety of andradite resembling topaz.
topazoseme *topazite.*
topaz-quartz *gold topaz.*
topaz rock *topazite.*
top bed Eng. Often applied to the highest bed in a quarry. In Dorset and Somerset, the upper division of the Inferior Oolite. (Arkell, 1953)
top benching The method by which the bench is removed from above, as with a dragline. See also: *benching.* (AIME, 2)
top brick Fireclay brick for use in lining the top section of a blast furnace. (ARI, 1949)
top cager A worker at the top of a shaft to superintend the operation of lowering and raising of a cage. At most mines, duties include removing loaded cars from the cage and placing empty cars on the cage. See also: *cager.* (Fay, 1920)
top canch That part of a mine roof that has to be taken down to give headroom on roadways. (Fay, 1920)
top crystals Standard grade of diamonds. (Hess)
top cut (a) A machine cut made in the upper limit of the workable section of a coal seam. See also: *overcut.* (Nelson, 1965) (b) A horizontal cut or groove made in coal at or near the top of the working face. See also: *middle cut; bottom cut.*
topcutter A cutting machine designed esp. for cutting through the seam at a high level above the footwall. (Stoces, 1954)
top frame A frame set at or just below ground level as a preliminary to the main timbering in an excavation. See also: *setting.* (Hammond, 1965)
top gate A gate road at the upper end of an inclined longwall conveyor face; usually a tailgate. See also: *bottom gate.* (Nelson, 1965)
top heading A method of driving used for adits, tunnels, and drifts. The upper part or top heading is driven to the full length, before the enlargement of the rest of the section is carried out. (Fraenkel, 1953)
top holes An earlier system of working coal between two levels in an inclined coal seam. The top holes are driven to the full rise, and the face-line is usually stepped. Only the coal is worked, and no ripping is done in the top holes. The coal gravitates into trams in the lower level. See also: *stepped longwall.* (Nelson, 1965)
top hooker *lander.*
tophus *tufa.* Etymol: Latin. Pl: tophi. (AGI, 1987)
top kick *top shot.*
top lander *lander.*
top lease A lease granted by a landowner during existence of a recorded mineral lease that is to become effective if and when the existing lease expires or is terminated. (Williams, 1964)
topman In mining, a worker who is employed at surface jobs around the mine plant.
topographic contour An imaginary line on the ground, all points of which are at the same elevation above (or below) a specified datum surface. (AGI, 1987)
topographic correction *terrain correction.*
topographic map A map showing the topographic features of a land surface, commonly by means of contour lines. It is generally on a sufficiently large scale to show in detail selected man-made and natural features, including relief, and such physical and cultural features as vegetation, roads, and drainage. Cf: *planimetric map.* (AGI, 1987)
topographic profile *profile.*
topographic quadrangle Map upon which is shown a portion of land having elevations indicated by a series of separate lines, each of which passes through a specified elevation; the sinuosity and spacing or crowding together of the contour lines, as they are called, indicate slope and relief of the terrain.
topographic unconformity (a) The relationship between two parts of a landscape or two kinds of topography that are out of adjustment with one another, due to an interruption in the ordinary course of the erosion cycle of a region; e.g., a lack of harmony between the topographic forms of the upper and lower parts of a valley, due to rejuvenation. (AGI, 1987) (b) A land surface exhibiting topographic unconformity. (AGI, 1987)
topography (a) The general configuration of a land surface or any part of the Earth's surface, including its relief and the position of its natural and manmade features. Cf: *relief.* Syn: *lay of the land.* (AGI, 1987) (b) The natural or physical surface features of a region, considered collectively as to form; the features revealed by the contour lines of a map. In nongeologic usage, the term includes manmade features (such as are shown on a topographic map). (AGI, 1987) (c) The art or practice of accurately and graphically delineating in detail, as on a map or chart or by a model, selected natural and manmade surface features of a region. Also, the description, study, or representation of such features. Etymol: Greek topos, place, + graphein, to write. (AGI, 1987)
topotaxy A recrystallization in which the crystallographic orientation of the parent crystal determines that of the product crystal, e.g., goethite to hematite with $a \rightarrow c/3$, $b \rightarrow 2\,a$, and $c \rightarrow a$ T3/3. Cf: *epitaxy; syntaxy.*
topping (a) The contents of a loaded mine car above water level. (Hudson, 1932) (b) Fine material forming a surface layer or dressing for a road or grade. (Nichols, 1976) (c) A finishing layer of fine concrete, usually 2 in (5.1 cm) thick, laid over the base concrete of a ground floor or over the structural components of a solid or hollow suspended floor or roof. (Hammond, 1965)
toppings Eng. The first regular layers of flints in the Brandon flint mines. (Arkell, 1953)
top ripping Roof ripping. (Nelson, 1965)
top rod Scot. The rod connecting the uppermost pump rod to the bellcrank (lever) on a bellcrank drive. (Fay, 1920)
top shot An explosion or puff of gas at a furnace top. Syn: *top kick.* (Fay, 1920)
topside (a) On the surface as opposed to underground. (Long, 1960) (b) Above the drill rig in the derrick or tripod. (Long, 1960) (c) The inlet end of the hydraulic cylinder of a hyraulic-feed mechanism on a diamond drill. (Long, 1960)
top slice A horizontal block of ore extracted by top slicing. The dimensions vary in different mines. (Nelson, 1965)
top slicing A method of stoping in which the ore is extracted by excavating a series of horizontal (sometimes inclined) timbered slices alongside each other, beginning at the top of the orebody and working progressively downward; the slices are caved by blasting out the timbers, bringing the capping or overburden down upon the bottom of the slices that have been previously covered with a floor or mat of timber to separate the caved material from the solid ore beneath. Succeedingly lower slices are mined in a similar manner up to the overlying mat or gob, which consists of an accumulation of broken timbers and lagging from the upper slices and of caved capping. See also: *block caving.* (Jackson, 1936)
top slicing and caving *top slicing and cover caving.*
top slicing and cover caving A mining method that entails working the orebody from the top down in successive horizontal slices that may follow one another sequentially or simultaneously. The whole thickness of the slice is worked. The ore may be broken by overhand or underhand stoping in each unit. The overburden or cover is caved after mining a unit. Syn: *mining ore from top down; top slicing and caving; transverse slicing with caving.* (Fay, 1920)
top slicing combined with ore caving A method of working an orebody from the top down in successive slices. Instead of taking the full height of the slices, only the lower part is taken and the upper part is caved. After removing this portion of the ore, the cover is caved. A timber mat is used in most cases to separate the broken cover from the ore and for safety. Also known as caving system subdrifting and caving; subslicing; slicing under ore with back cave; sublevel caving; sublevel slicing. (Fay, 1920)
top testing *roof testing.*
top wall *hanging wall.*
top water Water introduced with the raw coal feed to assist the transport of material through the washbox. Also called transport water. Cf: *flush water.* (BS, 1962)
tor A high, isolated crag, pinnacle, or rocky peak; or a pile of rocks, much jointed and usually granitic, exposed to intense weathering, and often assuming peculiar or fantastic shapes, e.g., the granite rocks standing as prominent masses on the moors of Devon and Cornwall, England. (AGI, 1987)
Torbane Hill mineral *torbanite.*
torbanite (a) A variety of algal or boghead coal from Torbane Hill, Scotland. It is layered, compact, brownish-black to black in color, very tough, and difficult to break. On distillation, torbanite gives a high

yield of oil. Also called bathvillite. See also: *bitumenite; boghead cannel; boghead coal; boghedite; kerosine shale.* Syn: *Torbane Hill mineral.* (Tomkeieff, 1954) (b) A dark-brown variety of cannel coal. (c) An oil shale mined in South Africa. (Beerman, 1963)

torbernite A tetragonal mineral, $Cu(UO_2)_2(PO_4)_2{\cdot}12H_2O$; autunite group; strongly radioactive; green; in tabular or foliated crystals, associated with other uranium minerals; commonly in parallel growth with autunite; a secondary mineral resulting from alteration of uraninite. Also called chalcolite, copper uranite, cuprouranite. Syn: *uranmica; uranphyllite.* Cf: *metatorbernite.*

torch An oil-burning, wick-fed, miners' lamp of tin or copper, with a long spout. (Fay, 1920)

torendrikite A monoclinic mineral, $(Na,Mg,Fe)_7Si_4O_{11}(OH,F)$ approx.; an amphibole intermediate between richterite and glaucophane; dark blue; at Itorendrika, Madagascar, and near Tine, Wadai, Africa.

torf dolomite *coal ball.*

torf-dopplerit *dopplerite.*

tormentor (a) A wooden axle, studded with iron spikes, for puddling auriferous clay as it spins or turns in a trough. See also: *puddler.* (Nelson, 1965) (b) A device somewhat similar to a log washer. (Fay, 1920)

Tornado The Tornado crusher is based on the principle of central impeller shoes spinning to hurl particles of gravel against breaker plates at tremendous speed. The impact literally "explodes" the rock, causing it to cleave across the grain as well as with the grain, producing the most desirable cubical product. (Pit and Quarry, 1960)

törnebohmite A monoclinic mineral, $(Ce,La)_2Al(SiO_4)_2(OH)$; weakly radioactive; green to olive; in contact zones at Bastnäs, Sweden. Also spelled toernebohmite.

torpedo An encased explosive charge that is slid, lowered, or dropped into a borehole and exploded to clear the hole of obstructions or to open connections with passage ways to an oil or water supply. Also called a bullet. (Long, 1960)

torque The effectiveness of a force that tends to rotate a body; the product of the force and the perpendicular distance from its line of action to its axis. (AGI, 1987)

torque bar Square or vertically fluted bar run on one type of auger drill to rotate, raise, and lower the auger. See also: *torque rod.* (Long, 1960)

torque converter A hydraulic coupling that utilizes slippage to multiply torque. (Nichols, 1976)

torquemeter A device for measuring the actual torque transmitted to the drilling head and/or to the drill-rod string. (Long, 1960)

torque rod A bar having the function of resisting or absorbing twisting strains. Syn: *torque bar.* (Nichols, 1976)

torque thickener Tank thickener in which bottom rakes rise when overloaded with settled material. (Pryor, 1963)

torr The pressure exerted per square centimeter by a column of mercury 1 mm high at a temperature of 0 °C where the acceleration of gravity is 980.665 cm/s^2. (Hy, 1965)

torrents Beds of quicksand encountered below the chalk marl in the Anzin Coalfield, in France.

torreyite A monoclinic mineral, $(Mg,Mn)_9Zn_4(SO_4)_2(OH)_{22}{\cdot}8H_2O$; related to mooreite and lawsonbauerite; at Sterling Hill, NJ.

torsional center If a twisting couple is applied at a given section of a straight member, that section rotates about some point in its plane. This point, which does not move when the member twists, is the torsional center of that section. It is sometimes defined as though identical with the flexural center, but the two points do not always coincide. Syn: *center of twist; center of torsion; center of shear.* Cf: *flexural center; elastic center; elastic axis.* (Roark, 1954)

torsional shear test A shear test in which a relatively thin test specimen of solid circular or annular cross section, usually confined between rings, is subjected to an axial load and to shear in torsion. In-place torsion shear tests may be performed by pressing a dentated solid circular or annular plate against the soil and measuring its resistance to rotation under a given axial load. (ASCE, 1958)

torsion anemometer *Ree's torsion anemometer.*

torsion balance A geophysical prospecting instrument that is used to determine distortions in the gravitational field. It consists of a pair of masses suspended by a sensitive torsion fiber and so supported that they are displaced both horizontally and vertically from each other. A measurement is made of the rotation of the suspended system about the fiber; the rotation is caused by slight differences in the gravitational attraction on the two masses. Syn: *Eötvös torsion balance.* (AGI, 1987)

torsion break A break in the drill core caused by an accumulation of chips at the bit face. When drilling is stopped to rechuck, these chips grip the core, and the core is twisted and broken. Cf: *torsion fracture.* (Long, 1960)

torsion fracture A spiraled crack in a drill core caused by torque in a blocked bit or core barrel. Cf: *torsion break.* (Long, 1960)

torsion seismometer A seismograph with which the horizontal component of the earthquake can be defined making use of the torsion of a vertical suspension thread on which a stationary mass is fastened offcenter. (Schieferdecker, 1959)

tortoise Sometimes applied by miners to structures such as pots, bells, kettles, and other rock masses that tend to fall easily from the roof of a coal mine. See also: *pot bottom; camel back.* (AGI, 1987)

tortuous flow *turbulent flow.*

tossing The operation of raising the grade or purity of a concentrate by violent stirring, followed by packing, in a kieve or open dolly tub. Chimming is a similar process on a smaller scale. *tozing.* (CTD, 1958)

total acidity Acidity to phenolphthalein. Total acidity of mine water indicates the complete capacity of water to produce chemical change by acid reaction. It is the total amount of acid held in solution or the sum of the quantities of both the ionized and the un-ionized portions of actual acid and the potential quantity of acid that can be formed from mineral salts held in solution. Total acidity is customarily reported in equivalent parts per million (ppm) by weight of calcium carbonate. The indicated total acidity of mine water found by currently accepted methods of analysis generally is greater than the actual total acidity. (Felegy, 1948)

total ash Residue of the mineral matter obtained by incinerating coal under standard conditions. (BS, 1962)

total bit load A drilling term describing the total amount of any load or pressure, expressed in kilograms, pounds, or tons, that is applied to a bit when it is in use. (Long, 1960)

total cap lag A blasting term describing the total time between application of current and the detonation. (Streefkerk, 1952)

total carbon The sum of the free and combined carbon (including carbon in solution) in a ferrous alloy. (ASM, 1961)

total cooling effect The difference between the total heat content of the air-steam mixture entering a conditioner per hour and the total heat of the mixture leaving per hour. (Strock, 1948)

total cooling load The sum of the sensible and latent heat components that must be removed from a space to maintain desired conditions. (Hartman, 1982)

total critical load (a) The total load or pressure that must be applied to a bit for its optimum rate of penetration in a specific rock. (Long, 1960) (b) The maximum load that can be applied to a bit without causing damage to the bit. Cf: *critical pressure.* (Long, 1960)

total displacement *slip.* Cf: *dip slip.*

total dynamic head The total of the static head (the suction discharge heads), the friction head, together with any discharge head that must be overcome by a pump is termed the total dynamic head. (Carson, 1961)

total energy The total energy at any section in a moving fluid consists of the sum of the internal static, velocity, and potential energies at that section. (Hartman, 1982)

total hardness *hardness.*

total hardness of water All waters contain two forms of hardness, i.e., temporary (or carbonate) hardness, and permanent (noncarbonate) hardness. The combination of the two is referred to as total hardness. See also: *hard water.* (Nelson, 1965)

total head The sum of the elevation head, pressure head, and velocity head of a liquid. For ground water, the velocity-head component is generally negligible. (AGI, 1987)

total heat (a) The total heat of atmospheric air is the heat contained in the same amount of dry air (known as sensible heat) plus the latent heat of the contained water vapor plus the sensible heat of the water vapor above the wet-bulb temperature. This is called the sigma function or sigma heat. True total heat or enthalpy is the sigma heat plus the heat of the water below the wet-bulb temperature. The latter is a very small quantity, and in mining work sigma heat is always used. Sigma heat is usually measured above 0 °F (-18 °C), so that it is the heat that would be given up if all moisture were condensed out and removed and the air cooled to 0 °F. Some engineers use 32 °F (0 °C) as the basic temperature so care should be taken to verify the base used. If absolute total heat is specified, it is measured above absolute zero (-459 °F or -273 °C). (Spalding, 1949) (b) The sum of sensible heat and latent heat in a substance or fluid above a base point, usually 32 °F or 0 °F. (Strock, 1948)

total lift In an air lift, the distance water is elevated during pumping. Total lift equals drop plus static head plus elevation. Also called lift. (Lewis, 1964)

total magnetic intensity The vector resultant of the intensity of the horizontal and vertical components of the Earth's magnetic field at a specified point. (Hy, 1965)

total moisture (a) Free moisture plus moisture in air-dried coal, both being expressed as percentages of the sample as received. (BS, 1962) (b) The moisture in the coal or coke as samples. (BS, 1961)

total of correctly placed material The sum of the weights of material correctly included in the products of a sizing or density separation, expressed as a percentage of the weight of the feed to the separator (and equal to 100 minus the "total of misplaced material"). (BS, 1962)

total of misplaced material The sum of the weights of the misplaced material in the products of a sizing or density separation, expressed as a percentage of the weight of the feed. When three products are made in a single separator, the total of misplaced material will be the sum of the weight of material wrongly placed in each of the three products, expressed as a percentage of the feed to the separator. (BS, 1962)
total porosity *porosity.*
total pressure (a) The total ventilating pressure in a mine, usually measured in the fan drift. (Nelson, 1965) (b) The algebraic sum of static pressure and velocity pressure at any particular point. (BS, 1963) (c) The pressure in a soil mass due to overlying material and any superimposed loads. (Nelson, 1965) (d) The pressure on any horizontal plane in a mass of soil as calculated from the weight of the material above the plane, or the soil together with any applied loads. (Hammond, 1965)
total reflection In gemology and optical mineralogy, total reflection occurs in a transparent solid where a light ray strikes the surface of a medium of lower refractive index at any angle greater than its critical angle as defined sin r = 1/n, where r is the critical angle and n is the refractive index of the solid (or n_2/n_1, where n_1 represents the lower refractive index if other than 1 for air). Cf: *critical angle; reflection; law of refraction.*
total reflectometer An instrument for measuring the critical angle in a transparent solid.
total resistance The total resistance (R) or friction of a ventilation system is calculated from the total ventilating pressure (P) and the total quantity of air at the fan (Q). Thus: $R= P/Q^2$. (Nelson, 1965)
total stress The total force per unit area acting within a mass of soil. It is the sum of the neutral and effective stresses. (ASCE, 1958)
total tonnage Tonnage of ore or product shipped plus tonnage of ore, waste, and tailings dumped. (Hoover, 1948)
total value Value of ore or product shipped plus value of the ore, waste, and tailings dumped. (Hoover, 1948)
total ventilating power The sum of the natural ventilating power plus the effective (or air) horsepower of all fans in series. When the circuit is divided and fans are in parallel, the total ventilating power of each split is worked out separately, the natural and fan powers being added; finally the power of each circuit is summed to give the total ventilating horsepower. (Spalding, 1949)
total ventilating pressure The pressure required to overcome the static and potential energy head losses and to provide the velocity head to move a quantity of air through a network. See also: *ventilating pressure; mine total head; fan total head; water gage.*
Totco test A test to determine the deviation of a well from the vertical, employing an instrument known as a Totco. (Williams, 1964)
touchstone A black, flinty stone, such as a silicified shale or slate, or a variety of quartz allied to chert or jasper, whose smoothed surface was formerly used to test the purity or fineness of alloys of gold and silver by comparing the streak left on the stone when rubbed by the metal with that made by an alloy of predetermined composition. Syn: *Lydian stone; basanite.* (AGI, 1987)
tough The exact state or quality of texture and consistency of refined copper. See also: *tough cake.* (Fay, 1920)
tough cake Refined or commercial copper. See also: *cake copper; tough.* (Fay, 1920)
toughness (a) A property of a material that denotes, nominally, an intermediate value between softness and brittleness. Tensile tests show a tough material to have a fairly high tensile strength accompanied by moderate values of elongation and reduction of area. (Henderson, 1953) (b) The amount of work required to deform a body to its rupture point. (AGI, 1987)
toughness index The ratio between the index of plasticity and the flow index of a soil. (Hammond, 1965)
toughness of refractories Resistance to crumbling, to abrasion, or to coarse particles being dislodged from the brick structure.
tough pitch (a) A term used in electrolytic copper refining to designate copper that has set, from the molten condition, with a level surface. See also: *underpoled copper; overpoled copper.* (Fay, 1920) (b) A term applied to copper in which the oxygen content has been correctly adjusted at 0.03% to 0.06% by poling. Distinguished from overpoled and underpoled copper. (CTD, 1958)
tough pitch copper Copper containing from 0.02% to 0.05% oxygen, obtained by refining copper in a reverberatory furnace. (ASM, 1961)
tough tom Soft tenacious clay floor of coal seams. (Arkell, 1953)
tough way A quarryman's term for the third easiest direction of rock fracture after the rift and the grain. Syn: *hard way.* Also called head.
tour A work-shift. Sometimes incorrectly spelled tower. (Long, 1960)
tourmaline (a) Any member of the trigonal mineral group, $XY_3Z_6(BO_3)_3Si_6O_{18}(OH,F)_4$ where X is Na partially replaced by Ca, K, Mg, or a vacancy, Y is Mg, Fe^{2+}, Li, or Al, and Z is Al and Fe^{3+}; forms prisms of three, six, or nine sides; commonly vertically striated; varicolored; an accessory in granite pegmatites, felsic igneous rocks, and metamorphic rocks. Transparent and flawless crystals may be cut for gems. (b) The mineral group buergerite, dravite, elbaite, ferridravite, liddicoatite, schorl, and uvite.
tourmalinization Introduction of, or replacement by, tourmaline. (AGI, 1987)
Toussaint-Heintzmann arch A channel-type, steel-arch support consisting of three elements or sections set close to the face of a tunnel. These elements overlap and yield by sliding one upon the other under the constraint of bolted clamps. The center or crown element is usually foreset to give temporary protection until the complete arch is erected. See also: *Usspurwies arch; steel arch; steel support.* (Nelson, 1965)
tow conveyor An endless chain supported by trolleys from an overhead track or running in a track at (above, flush with, or under) the floor with means for towing trucks, dollies, or cars.
towed grader *grader.*
tower (a) *tour.* (b) A misnomer for derrick and an incorrect spelling of tour. (Long, 1960)
tower crane A swing-jib or other type of crane mounted on top of a tower, the base of which may sometimes move on rails. These cranes are esp. effective on congested sites. See also: *monotower crane; swing-jib crane.* (Hammond, 1965)
tower engineer In anthracite coal mining, one who operates a hoist to raise loaded mine cars from the surface of a mine to the top of the breaker, where the coal is dumped, crushed, and prepared for market. (DOT, 1949)
tower excavator A cableway excavator designed specif. for levee work, but which is used extensively in the stripping of overburden, spoil, or waste in surface mining. The unit is basically an excavator with towers either fixed or movable. With the headtower located on the spoil pile and the tail tower on the unexcavated wall, it is possible to dig pits of almost unlimited width.
tower loader A front-end loader whose bucket is lifted along tracks on a more or less vertical tower. (Nichols, 1976)
Towers magnetic stirrer A device utilizing a rotating field of magnetic force to induce a vigorous rotary movement in a small magnetized bar totally enclosed in a polythene or glass tube, and placed in the liquid to be stirred. (Osborne, 1956)
township The unit of survey of the U.S. Public Land Survey system, representing a piece of land that is bounded on the east and west by meridians approx. 6 miles (9.6 km) apart (exactly 6 miles at its south border) and on the north and south by parallels 6 miles apart, and that is normally subdivided into 36 sections. Townships are located with reference to the initial point of a principal meridian and base line, and are normally numbered consecutively north and south from a base line (e.g., township 14 north indicates a township in the 14th tier north of a base line). The term township is used in conjunction with the appropriate range to indicate the coordinates of a particular township in reference to the initial point (e.g., township 3 south, range 4 west indicates the particular township that is the 3rd township south of the base line and the 4th township west of the principal meridian controlling the surveys in that area). Abbrev. (when citing specific location): T. (AGI, 1987)
township line One of the imaginary boundary lines running east and west at 6-mile (9.6-km) intervals and marking the relative north and south locations of townships in a U.S. public land survey. Cf: *range line.* (AGI, 1987)
toxic dusts Dusts poisonous to body organs, tissue, etc. They include ores of beryllium, arsenic, lead, uranium, radium, thorium, chromium, vanadium, mercury, cadmium, antimony, selenium, manganese, tungsten, nickel, and silver (principally the oxides and carbonates). (Hartman, 1982)
toxicity symptoms In geochemical exploration, a collective term for the abnormal colors and morphological features of a plant caused by a poisonous element in the nutrient solution. (Hawkes, 1957)
toxic mine drainage Water that is discharged from active or abandoned mines or other areas affected by mineral exploration or surface mining and reclamation operations that contains a substance that through chemical action or physical effects is likely to kill, injure, or impair biota commonly present in the area(s) to which it might be exposed.
tozing *tossing.*
trace (a) A concentration of a substance that is detectable, but too minute for accurate quantitative determination. (AGI, 1987) (b) A quantity of precipitation that is insufficient to be measured by a gauge. (AGI, 1987) (c) A sign, evidence, or indication of a former presence; specif. a mark left behind by an extinct animal, such as a trace fossil. (AGI, 1987) (d) The record of the output of one geophone group with time after the shot, displayed on paper, film, or magnetic tape. (AGI, 1987) (e) The intersection of a geological surface with another surface, e.g., the trace of bedding on a fault surface, or the trace of a fault or outcrop on the ground. Cf: *trend; strike.* (AGI, 1987) (f) A very small quantity of a chemical constituent or component, esp. when not quantitatively determined because of extremely low concentrations. (g) To follow the lode on the surface, and to lay it open by long pits. (Fay,

1920) (h) Recording on the seismogram of a single seismometer station. (Schieferdecker, 1959)

trace-by-trace plotting A procedure used in seismic reflection where reflection times from all traces, or sometimes alternate traces, are plotted at the reflecting point positions (midway between shot and detector). When no correction has been made for "normal moveout," the plotted times appear to lie along arcs that are convex upward and straddle the shot positions symmetrically. Cf: *center-trace time.* (Dobrin, 1960)

trace element (a) An element that is not essential in a mineral, but that is found in small quantities in its structure or adsorbed on its surfaces. Although not quantitatively defined, it is conventionally assumed to constitute significantly less than 1.0% of the mineral. Syn: *accessory element; guest element.* (AGI, 1987) (b) An element that occurs in minute quantities in plant or animal tissue and that is essential physiologically. Syn: *minor element; microelement.* (AGI, 1987)

tracer An element or compound that has been made radioactive so that it can be followed (traced) easily in industrial and biological processes. Radiation emitted by the tracer (radioisotope) pinpoints its location. Abbrev., tcr. (Lyman, 1964; Zimmerman, 1949)

tracer gas A gas introduced in small quantities into the main body of air to determine either the air current or the leakage paths in a ventilation system. (BS, 1963)

tracer-gas technique This method, as applied for the measurement of airflow in headings, can be used for determining velocities below the working range of the vane anemometer. The tracer used is normally nitrous oxide. The technique consists of releasing a quantity of tracer gas, either instantaneously or over a timed interval. The tracer then diffuses throughout the airstream until a position is reached where it is uniformly dispersed over the cross section of the airway. At such a position, samples are taken and these are put through the analyzer to determine the gas concentration. (Roberts, 1960)

tracer test In ground water hydrology, a field technique for estimating transport parameters by injecting a tracer (e.g., dye, radioactive substance, or chemical not naturally present in the flow system) and monitoring the time and concentrations at a downgradient location. (Domenico, 1990)

trace slip In a fault, that component of the net slip that is parallel to the trace of an index plane, such as bedding, on the fault plane. See also: *trace-slip fault.* (AGI, 1987)

trace-slip fault A fault on which the net slip is trace slip, or slip parallel to the trace of the bedding or other index plane. (AGI, 1987)

trachyandesite An extrusive rock, intermediate in composition between trachyte and andesite, with sodic plagioclase, alkali feldspar, and one or more mafic minerals (biotite, amphibole, or pyroxene). See also: *mugearite.* (AGI, 1987)

trachybasalt An extrusive rock intermediate in composition between trachyte and basalt, characterized by the presence of both calcic plagioclase and alkali feldspar, along with clinopyroxene, olivine, and possibly minor analcime or leucite. Approx. synonymous with hawaiite. (AGI, 1987)

trachyte A group of fine-grained, generally porphyritic, extrusive rocks having alkali feldspar and minor mafic minerals (biotite, hornblende, or pyroxene) as the main components, and possibly a small amount of sodic plagioclase; also, any member of that group; the extrusive equivalent of syenite. Trachyte grades into latite as the alkali feldspar content decreases, and into rhyolite with an increase in quartz. Etymol: Greek trachys, rough, in reference to the fact that rocks of this group are commonly rough to the touch. (AGI, 1987)

trachytic (a) A textural term applied to volcanic rocks in which feldspar microlites of the groundmass have a subparallel arrangement corresponding to the flow lines of the lava from which they were formed. Cf: *trachytoid; pilotaxitic.* (AGI, 1987) (b) Pertaining to or composed of trachyte. (AGI, 1987)

trachytoid A textural term originally applied to phaneritic igneous rocks by analogy with the trachytic texture of some lava flows. In such rocks (e.g., many nepheline syenites), the feldspars have a parallel or subparallel disposition; trachytoid is now used for all similar textures, regardless of the composition of the rock in which they occur. Cf: *trachytic.* (AGI, 1987)

tracing float A process in which float is followed back to its origin. A prospector moves up slope looking for pieces of float until no more are to be seen. If, at this point, no outcrop is visible, the probabilities are that soil or loose surface material covers a hidden outcrop, which can be sought by digging pits or trenches. (Lewis, 1964)

track (a) The groove cut in a rock by a diamond inset in the crown of a bit. (Long, 1960) (b) A pattern applied to setting diamonds in a bit crown, in which the diamonds are arranged in concentric circular rows so that the diamonds in a specific row follow the track cut by a preceding diamond. (Long, 1960) (c) The slide or rack on which a diamond-drill swivel head can be moved to positions above and/or clear of the collar of a borehole. (Long, 1960) (d) *conveyor track.* (e) A crawler track. (Nichols, 1976)

track bolt A chair bolt or coach screw used in fastening rails. (Hammond, 1965)

track braking Track brakes, similar to those used on surface tramcars, may be installed on heavy downgrades underground to supplement other braking systems; they apply blocks to the rails by mechanical, pneumatic, or electromagnetic power. The normal shoe brake must be designed to work in conjunction with the track brakes so that the wheels are not skidded when the track brakes utilize part of the weight of the locomotive. Electromagnetic track brakes may utilize the braking currents produced in rheostatic braking to excite the electromagnets, which are then pulled down onto the rails and produce a strong retarding pull. (Sinclair, 1959)

track cable Steel wire rope, usually a locked-coil rope that supports the wheels of the carriers of a cableway. (Hammond, 1965)

track cable scraper This type of excavator operates in general the same as a slackline cableway, except that it uses a bottomless scraper bucket that must convey its load over the ground instead of through the air. Like the slackline cableway, this machine is operated by a two-drum hoist that controls a track cable that spans the working area and a haulage cable that leads to the front of the bucket. Both cables are reeved through sheave blocks attached to a high guyed mast or tower at the head end of the installation. When the bucket comes in with a load and reaches the desired dumping point, a few rotations of the rear drum of the hoist serve to tighten the track cable and lift the bucket off its load; then the brake is released on the front drum that controls the haulage cable, permitting the empty bucket and carrier to glide back down the inclined track cable. (Pit and Quarry, 1960)

track channeler In quarrying, a rock channeler designed to operate from a track on which it is mounted; frequently a combined locomotive and channeling machine. (Standard, 1964)

track cleaner In mining, a laborer who cleans mine track and switches by shoveling coal, ore, rock, mud, and refuse, and throwing it to one side or loading it into a mine car. Also called road cleaner. (DOT, 1949)

track cleaners A machine to remove gob from railroad tracks, between rails, and to a distance of 48 in (1.2 m) from the track centers. Digger plates remove gob to the top of ties, while wings on either side of the machine gather it from the sides of the track into the track, to be moved by conveyors to a car at the machine's rear. One person operates the towing motor and observes the digging, while another controls the digging plate height, adjusts wing plows, and observes loading. (Best, 1966)

track diamonds Diamonds set in the face or lead portion of the drill-bit crown. Syn: *track stones.* (Long, 1960)

track frame In a crawler mounting, a side frame to which the track roller and idler are attached. Also called truck frame. (Nichols, 1976)

track gage (a) The distance between the inside edges of installed railway rails. (BCI, 1947) (b) The minimum track gage that should be used on a modern haulage system is 3 ft 6 in (1.07 m). This gage has been established as a standard by the American Mining Congress as a result of the consensus of opinion of mining engineers and manufacturers in the United States. See also: *rail gage.* (Wheeler, 1946)

track haulage Movement or transportation of excavated or mined materials in cars or trucks running on rails.

tracklaying tractor A tractor moving on crawler tracks. (Hammond, 1965)

trackless mine A mine with no rails. In such mines, rubber-tired vehicles operate independently of tracks and are used for haulage and transport. (Stoces, 1954)

trackless tunneling A method of tunneling using loaders mounted on crawler tracks, and a diesel- or battery-powered dump-truck haulage system. See also: *rubber-tired haulage.* (Nelson, 1965)

track-mounted Referring to the operation of equipment on tracks, such as track-mounted cutting machines, track-mounted loaders, etc. (Jones, 1949)

track-moving machine operator In metal mining, one who operates a machine that moves and lays track mechanically in open pit mines, picking up a section of track and moving and laying it in the desired position without having to detach rails from ties. Also called track-laying machine operator. Syn: *hydraulic jack operator.* (DOT, 1949)

track pin A hinge pin connecting two sections or shoes of a crawler track. (Nichols, 1976)

track resistance The total rolling friction of a train on straight level track. It is generally taken as 30 lb per short ton weight (15 kg/t) of the train for cars having plain bearings and 20 lb/st (10 kg/t) for cars with roller bearings. These figures may be increased by 10 to 15 lb (5.0 to 7.5 kg) if the track is in poor condition and may be less for track in excellent condition. (Lewis, 1964)

track roller In a crawler machine, the small wheels that are under the track frame and that rest on the track. Also called truck rollers. (Nichols, 1976)

track shifter A machine or appliance used in shifting a railway track laterally. Also the operator of such a device or machine. See also: *shifter.*

track spike A heavy steel nail of square section that is driven into a wooden sleeper to hold a flanged rail. (Hammond, 1965)
track stones *track diamonds.*
track wheel One of a set of small flanged steel wheels resting on a crawler track and supporting a track frame. (Nichols, 1976)
traction (a) The act of drawing a vehicle over a surface and the force exerted in so doing. Traction is the friction developed between tracks or tires and the surface of the ground on which they are moving. (Carson, 1961) (b) The total amount of driving push of a vehicle on a given surface. (Nichols, 1976)
tractive efficiency A measure of the proportion of the weight resting on tracks or drive wheels that can be converted into vehicle movement. (Nichols, 1976)
tractive effort (a) The effort exerted by a locomotive at the rim of its driving wheels; it is a function of its weight, the nature of its tires, and the condition of the track. It is equal to the weight of the locomotive times the adhesion of the locomotive to the track. For steel wheels on clean dry track the adhesion is 25%, and the tractive effort is therefore 500 lb/st of weight (250 kg/t). For cast-iron wheels the adhesion is 20%. (Lewis, 1964) (b) The necessary drawbar pull plus the resistance of the locomotive itself. (Kentucky, 1952) (c) *tractive force.*
tractive force (a) The pull exerted by a haulage rope on the drawbar of a car to overcome the frictional resistance of the car and the force of gravity acting on it. Also, the force available at the wheels of a locomotive to move the machine and its attached load. Syn: *tractive effort.* (Nelson, 1965) (b) The pull that a locomotive is capable of producing at its drawbar. (Hammond, 1965) (c) In hydraulics, drag or shear developed on the wetted area of a streambed, acting in the direction of flow. As measured per unit of wetted area, unit tractive force equals the specific weight of water times hydraulic radius times slope of channel bed (Chow, 1957). (AGI, 1987)
tractive power The weight of the vehicle multipled by the coefficient of traction; it is the total pounds of pull that can be exerted before slippage occurs. (Carson, 1961)
tractive resistance The resistance to motion due to friction per unit weight hauled. (Hammond, 1965)
tractor A self-propelled vehicle—which may be mounted on crawler tracks, on wheels with large pneumatic tires, or on a mixture of both—intended for moving itself and other vehicles. See also: *wheeled tractor; mine tractor.* (Hammond, 1965)
tractor drills These drills have a crawler mounting that supports the feed guide bar on an extendable arm. Small air motors control the movements of the arm used in tilting and turning the guide bar and provide power for crawler movement. The compressor is a separate unit, which can be towed by the tractor drill. Also called a Jumbo.
tractor loader A tractor equipped with a bucket that can be used to dig, and to elevate to dump at truck height. Also called tractor shovel; shovel dozer. (Nichols, 1976)
tractor pan operator In beneficiation, smelting, and refining, one who scrapes up bauxite ore from stockpiles, hauls the ore to the crusher, and dumps it into the crusher hopper, using a combination bulldozer carryall. (DOT, 1949)
tractor shovel One of the names applied to a class of excavating equipment that has a bucket supported from the front end of a tractor. See also: *loader.* (Carson, 1961)
traffic marks Abrasions that result from metal to metal contact and vibration during transit. These abrasions are usually dark in appearance because of the presence of a dark powder, which consists of aluminum and aluminum oxide fines produced by the abrasive action of surfaces rubbing together. (Light Metal Age, 1958)
trailer (a) A towed carrier that rests on its own front and rear wheels. Also called full trailer. (Nichols, 1976) (b) *drag.*
trailing cable (a) A cable for carrying electricity from a permanent line or trolley wire to a movable machine, such as used in mining or quarrying. The cable is heavily insulated and protected with either galvanized steel wire armoring, extra stout braiding hosepipe, or other material. It is usually paid out from a reel as the machine advances. See also: *individually screened trailing cable.* (Nelson, 1965) (b) A flexible insulated cable used for transmitting power from the main power source, such as a trolley wire, nipping station, or junction box, to a mobile machine. It includes cables between the nipping station and distribution center.
trailing cable coupler An assembly of two restrained-type sockets for coupling together two trailing cables fitted with restrained plugs. (BS, 1965)
trail of a fault Crushed material of a bed or vein that indicates the direction of fault movement; it may be valuable as a guide to the main vein. See also: *drag ore.*
train A number of empty or loaded mine cars, coupled together, for transport by rope haulage or locomotive. See also: *journey.* (Nelson, 1965)
train dispatcher *motor boss.*
training face Usually a longwall coal face where new colliery entrants or trainees can gain experience, skills, and confidence in the winning and working of coal. (Nelson, 1965)
training gallery A short tunnel or chamber attached to a rescue station in which rescue workers may receive training in an atmosphere of dense smoke. It is also used to give rescue workers experience in stretcher drill, the use of stretchers, reviving apparatus, tube-breathing apparatus, artificial respiration, etc. (Nelson, 1965)
training the belt Adjusting the troughing idlers, return idlers, or tail or head pulleys forward or back so that conveyor belting will run in the desired position on the conveyor frame to prevent spilling of ore.
Trainite The trade name for a mixture of vashegyite with a colloidal zeolitic mineral, used as a gemstone; originates near Manhattan, NV. Originally described as banded variscite. Syn: *Sabalite.* (English, 1938)
trainman In metal mining, a laborer who loads ore into railroad cars. (DOT, 1949)
train-mile One mile (1.6 km) traversed by one train; used as a unit in railroad accounting. (Webster 3rd, 1966)
train of rolls A series of mills, one after the other, each successive mill approaching more nearly the size of the finished piece. (Mersereau, 1947)
train resistance The grade resistance plus the track resistance. (Kentucky, 1952)
trajectory Line that intersects wave fronts at right angles. (Schieferdecker, 1959)
tram (a) A trip of coal cars. (Jones, 1949) (b) Generally, to move a self-propelled piece of equipment other than a locomotive; tramcar. (BCI, 1947) (c) A boxlike wagon of steel, running on a tramway or railway in a mine, for conveying coal or ore. Also called tramcar, tub, cocoa pan, corve, corf, or hutch. (CTD, 1958; Webster 3rd, 1966) (d) To haul or push trams or cars about in a mine. (Fay, 1920) (e) A four-wheeled truck to carry a tub, corve, or hutch. (Fay, 1920)
tramcar Eng. A car used in coal mines. Syn: *tram.*
tramlines (a) An overfill appearing as two parallel lines on rolled bars. (Osborne, 1956) (b) Long, straight marks due to drawn-out inclusions on rolled sheet. (Osborne, 1956)
trammel (a) A board with two grooves intersecting at right angles, in which the two ends of a beam compass can slide and describe an ellipse. (Hammond, 1965) (b) *beam compass.* (c) A pivoted rod, used to keep brick in alignment in lining circular kilns.
trammer (a) A person who loads broken rock on tramcars and delivers it at the shaft. (Fay, 1920) (b) Trammers work as assistant miners in all the work a miner does. They load the broken mineral onto shaker or belt conveyors, fill and haul the mine cars, bring in the mine timber and other materials to support and equip the mine workings, serve the mining and transport machines, and work also as auxiliary mine timbermen. (Stoces, 1954) (c) One who transports coal, ore concentrate, or flux to roasting furnaces or bins. (DOT, 1949)
tramming The practice of pushing tubs, mine cars, or trams, by hand. Tramming was an earlier practice in longwall stall mining, but is now largely obsolete. On the surface, tramming means moving material in skips or wagons running on light railway track. (Nelson, 1965)
tramming motor (a) An electric locomotive used for hauling loaded trips. (Jones, 1949) (b) The motor in a cutting machine that supplies the power for moving or tramming the machine. (Jones, 1949)
tramp iron (a) Stray metal objects, such as coal-cutter picks or bolts, that have become mixed with the run-of-mine coal or ore. (Nelson, 1965) (b) Any loose piece of metal in a borehole. (Long, 1960)
tramp oversize Ore that is too large to be handled efficiently by the machine into which it is fed. (Pryor, 1963)
tramrail Eng. A rail for a tram. A light railroad rail distinguished from tram plate by being rolled while the latter is cast. (Webster 3rd, 1966)
tramroad A mine haulage road. (Jones, 1949)
tram rope A hauling rope, to which cars are attached by a clip or chain, either singly or in trips. (Zern, 1928)
tram vibrator *vibrating platform.*
tramway (a) A roadway having plates or rails on which wheeled vehicles may run. Syn: *tramroad.* (Standard, 1964) (b) A suspended cable system along which material, such as ore or rock, is transported in suspended buckets. See also: *aerial tramway.* (c) A system in which carriers are supported by cable and in which the movement is continuous over one or more spans.
transcurrent fault A large-scale strike-slip fault in which the fault surface is steeply inclined. Syn: *transverse fault; transverse thrust.* See also: *strike-slip fault.* (AGI, 1987)
transducer (a) A device actuated by one transmission system and supplying related waves to another transmission system; the input and output energies may be of different forms. Ultrasonic transducers, e.g., accept electrical waves and deliver ultrasonic waves, the reverse also being true. (ASM, 1961) (b) A device that measures physical quantities in a system—such as ground displacement, velocity, or accelera-

tion—and converts them into related or proportional units of electronic outputs (e.g., voltage, current). Cf: *geophone*. (NCB, 1964)
transfer A vertical or inclined connection between two or more levels; used as an ore pass. (Nelson, 1965)
transfer car (a) A quarry car provided with transverse tracks on which the gang car may be conveyed to or from the saw gang. (Fay, 1920) (b) A car equipped with rails, used to transfer a drier or kiln car from one set of tracks to another. (ACSG, 1963)
transfer case A transmission or gearset that provides drive to secondary drives, winches, etc.
transfer chute A chute used at a transfer point in a conveyor system. The chute is designed with a curved base or some other feature so that the load can be discharged in a centralized stream and in the same direction as the receiving conveyor. (Nelson, 1965)
transfer conveyor Conveyor generally 50 to 300 ft (15 to 90 m) in length. It is used to transport material only from one conveyor to another. (NEMA, 1961)
transfer-gang-car system A system used in quarries to save time in handling stone blocks and slabs. In this system, a transfer car that runs on a depressed track in front of the gangs is provided with a short section of track across the top. A gang car loaded with marble, sandstone, etc., is placed on this track and when moved to proper position is shifted beneath the gangsaw. Similarly, a gang car loaded with sawed slabs may be quickly moved from beneath the saws to the transfer car for transportation to the shops. (AIME, 1960)
transfer gear Self-actuating mechanism at shaft head by which a skip is emptied and its contents moved away to the next stage of handling. (Pryor, 1963)
transfer impedance (a) The transfer impedance of a network made up of a source and a load connected by a transducer is the ratio of the phasor representing the source voltage to the phasor representing the load current of the load. (Hunt, 1965) (b) The complex-valued ratio of voltage at one pair of terminals to the current at another pair in a four-terminal network. (AGI, 1987)
transfer point The point where coal or mineral is transferred from one conveyor to another. See also: *loading point*. (Nelson, 1965)
transfer table A table connected with rolling mills for laterally transferring work from one mill to the other. (Mersereau, 1947)
transformation A constitutional change in a solid metal; e.g., the change from gamma to alpha iron, or the formation of pearlite from austenite. (CTD, 1958)
transformation temperature The temperature at which a change in phase occurs. The term is sometimes used to denote the limiting temperature of a transformation range.
transformed flow net A flow net whose boundaries have been properly modified (transformed) so that a net consisting of curvilinear squares can be constructed to represent flow conditions in an anisotropic porous medium. (ASCE, 1958)
transformist A proponent of the theory that all granites had a metasomatic or palingenic origin. Cf: *magmatist*. Syn: *granitizer; antimagmatist*. (AGI, 1987)
transfusion The entry and exit of any gaseous or hydrothermal fluid in solid rock to produce such rocks as granite. Cf: *granitization*. (AGI, 1987)
transgranular *intracrystalline*.
transgression (a) The spread or extension of the sea over land areas, and the consequent evidence of such advance (such as strata deposited unconformably on older rocks, esp. where the new marine deposits are spread far and wide over the former land surface). Also, any change (such as rise of sea level or subsidence of land) that brings offshore, typically deep-water environments to areas formerly occupied by nearshore, typically shallow-water conditions, or that shifts the boundary between marine and nonmarine deposition (or between deposition and erosion) outward from the center of a marine basin. Ant: regression. Cf: *onlap*. Syn: *invasion; marine transgression*. (AGI, 1987) (b) A term used mostly in Europe for discrepancy in the boundary lines of continuous strata; i.e., unconformity. See also: *unconformity*. (AGI, 1987)
transgressive overlap *onlap*.
transient velocity A velocity, different from the steady state velocity, which a primer imparts to a column of powder to start detonation.
transistor A device for controlling or amplifying electric currents by means of potential probes through a crystal of a semiconductor, commonly silicon or germanium.
transit (a) A theodolite in which the telescope can be reversed (turned end for end) in its supports without being lifted from them, by rotating it 180° or more about its horizontal transverse axis. (AGI, 1987) (b) The act of reversing the direction of a telescope (of a transit) by rotation about its horizontal axis.—v. To reverse the direction of a telescope (of a transit) by rotating it 180° about its horizontal axis. Syn: *plunge*. (AGI, 1987)
Transite A trade name for a material of asbestos fiber and Portland cement molded under high pressure. Used for fireproof walls, roofing, and in lining ovens, etc. (Crispin, 1964)
transition belt A short belt carrying material from a loading point to a main conveyor belt. (Nichols, 1976)
transition curve (a) A curve designed to effect a gradual change between a straight and a circular curve. (Hammond, 1965) (b) *spiral curve*.
transition elements (a) In the periodic system, those elements characterized by the increment of inner *d* shells of electrons which may become involved in secondary or hybrid bond formation. (b) Elements having atomic numbers 21 (Sc) to 30 (Zn), 39 (Y) to 48 (Cd), and 57 (La) to 80 (Hg).
transition metals Elements in the middle of the long periods of the periodic table. Usage varies, but most commonly the transition elements are taken to include those from scandium to zinc in the first long period, from yttrium to cadmium in the second, and from lanthanum to mercury (excluding the 14 rare-earth metals from cerium to lutecium) in the third. (AGI, 1987)
transition point A single point at which different phases of matter are capable of existing together in equilibrium. Syn: *inversion point*. (Webster 3rd, 1966)
transition temperature The temperature at which the change from tough to brittle fracture occurs in a notched bar impact test, or sometimes in other forms of test, e.g., notched tensile test. Syn: *inversion point*. (Osborne, 1956)
transitman One who operates a surveyor's transit. The person need not necessarily be a graduate engineer. (Crispin, 1964)
translational fault A fault in which there has been translational movement and no rotational component of movement; dip in the two walls remains the same. It can be strictly applied only to segments of faults. Syn: *translatory fault*. (AGI, 1987)
translational movement Apparent fault-block displacement in which the blocks have not rotated relative to one another, so that features that were parallel before movement remain so afterwards. Cf: *rotational movement*. See also: *translational fault*. Less-preferred syn: translatory movement. (AGI, 1987)
translation lattice *crystal lattice*.
translatory fault *translational fault*.
translucency A term used to describe mineral crystals sufficiently transparent to transmit light. Also called subtransparency. Cf: *transparency*. (Pryor, 1963)
translucent The optical property of a mineral to transmit light without objects being visible. Cf: *opaque*.
translucent attritus This term was first used by R. Thiessen in 1930 referring to the attritus of ordinary humic coal, which is ordinarily composed largely of transparent humic matter, with spores, cuticles, resins, and opaque matter in minor proportions. Translucent attritus consists of the complex residual organic matter, exclusive of anthraxylon, in bituminous lower rank coal that transmits light in thin section. The following macerals of the Stopes-Heerlen nomenclature are included in translucent attritus: vitrinite less than 14 μm thick; sporinite; cutinite; alginite; resinite; and those parts of semifusinite, micrinite, and sclerotinite that are weakly reflecting, that is semitranslucent. Translucent attritus is a collective term and is not comparable with any of the microlithotypes of the European system of nomenclature. (IHCP, 1963)
translucent glass Glass that will admit rays of light to pass but through which objects cannot be seen. (Mersereau, 1947)
translucent humic degradation matter Transparent humic matter consisting of irregular particles varying greatly in shape and size. Although rounded or ovoid particles are not rare, humic matter is mostly flattened in form and usually of frayed or tattered appearance. The particles are of the same deep red color as the anthraxylon strips, becoming lighter in thinner sections. Syn: *THDM*. (IHCP, 1963)
transmissibility Term used to describe the ability of a system either to amplify or to suppress an input vibration. It is the ratio of the response amplitude of the system in steady-state forced vibration to the excitation amplitude. The ratio may be between forces, displacements, velocities, or accelerations. (Hy, 1965)
transmission A mechanism that provides a variety of gear ratios for different load conditions, speed, or direction of rotation.
transmission reversing A transmission that has only a forward and reverse shift. (Nichols, 1976)
transmissometer An instrument that measures the capability of a fluid to transmit light; esp. one that measures the turbidity of water by determining the percent transmission of a light beam. See also: *turbidimeter*. (AGI, 1987)
transmittance Rate of heat flow per unit area per unit temperature difference. (Strock, 1948)
transmitted light Light that has passed through an object, as distinguished from light reflected from a surface. The nonopaque minerals in rocks are commonly studied in transmitted plane-polarized light for

characterization and identification by use of a polarized-light microscope. Gems are usually examined for imperfections by transmitted light. Cf: *birefringence; index of refraction.*

transmutation The transformation of one element into another. Radioactive decay is an example. Transmutation can also be accomplished by bombardment of atoms with high-speed particles. (AGI, 1987)

transmute In alchemy, the transformation of baser metals into more precious metals. (Gordon, 1906)

transopaque The property of a mineral to transmit light in part of the visible spectrum while being opaque in another. Cf: *opaque.*

transparency (a) The degree to which visible light is transmitted through a solid. A mineral is termed transparent if objects can be clearly seen through it, as through glass; e.g., rock crystal (quartz), selenite (gypsum), Iceland spar (calcite), and gem diamond. Syn: *diaphaneity.* Cf: *translucency; opaque.* (b) The capacity of seawater to transmit light; the depth to which water is transparent may be measured by use of a Secchi disc. (c) A positive image, either black and white or in color, on a clear base (glass or film), intended to be viewed by transmitted light; a diapositive.

transparent (a) Permitting the passage of electromagnetic radiation. (b) Things that may be seen through; e.g., rock crystal (quartz), Iceland spar (calcite), selenite (gypsum). Cf: *opaque.*

transparent glass Glass through which objects can be seen. (Mersereau, 1947)

transponder An automated receiver and/or transmitter for transmitting signals when triggered by an interrogating signal. (Hy, 1965)

transport (a) A mining term used to cover vehicular transport, hydraulic transport, and conveyors. See also: *conveyor; haulage; hydraulic pipe transport.* (Nelson, 1965) (b) Syn: *transportation.* The term is favored in British usage, and often occurs in combined terms such as sediment transport and mass transport. (AGI, 1987)

transportable equipment Machines or equipment that can be moved from one part of a mine to another by mechanical means, such as but not by self-propulsion, on a track, or on attached wheels. See also: *mobile equipment.* (Nelson, 1965)

transportable substation A transformer equipped with switchgear and mounted upon wheels or skids. Syn: *portable substation.* (BS, 1965)

transportation (a) In geology, the shifting of material from one place to another on the Earth's surface by moving water, ice, or air. The carriage of mud and dissolved salts by rivers, the passage of a dust-laden whirlwind across a desert, the inland march of sand dunes from a seashore, and the creeping movement of rocks on a glacier are all examples of transportation. See also: *transport.* (Fay, 1920) (b) The hauling or moving from one place to another of material, such as ore, coal, rock, etc. (Fay, 1920)

transport controller A person stationed in a central position, at a large opencast pit or quarry, to observe all the excavations. By means of signals, the individual is informed when and where vehicles or cars are required and can instruct the lorry or locomotive drivers accordingly. (Nelson, 1965)

transported gossan Some transported gossans are simply colluvial accumulations of fragments of normal gossan that have moved down the slope from the site of weathering. Another variety of an entirely different origin is effectively a fossil spring or seepage deposit, where at one time iron-rich ground water has precipitated massive limonite at or near the daylight surface. (Hawkes, 1962)

transport number In electrolysis, a proportional fall in the number of ions concentrated at an electrode n=loss of concentration at cathode, and/or loss at cathode and anode. This phenomenon is caused by differences in the rate of travel of various species of ion. (Pryor, 1963)

transuranic (a) Of, or pertaining to, radioactive substances produced by bombarding uranium with neutrons. (Bennett, 1962) (b) Having an atomic number higher than that of uranium; having an atomic number higher than 92. (Webster 3rd, 1966)

Transvaal emerald *African emerald.*

Transvaal garnet A green garnet, possibly grossular or a variety of andradite.

transverse fault A fault that strikes obliquely or perpendicular to the general structural trend of the region. See also: *transcurrent fault.* (AGI, 1987)

transverse gallery An auxiliary crosscut made in thick deposits across an orebody for the purpose of dividing it into sections along strike. It is used only a short time and runs across the thickness of the deposit only. (Stoces, 1954)

transverse joint A joint that is transverse to the strike of the strata or schistosity.

transverse loading The loading on a beam. (Hammond, 1965)

transverse pitch The lateral distance between the center lines of each strand of a multiple-strand chain, or between the tooth profiles on a sprocket for a multiple-strand roller chain. (Jackson, 1955)

transverse riffles *Hungarian riffles.*

transverse slicing with caving *top slicing and cover caving.*

transverse strength (a) A measure of the capability of a stone bar (or beam) supported at its ends to bear a weight or load at its center. (Fay, 1920) (b) The strength of a specimen tested in transverse bending; normally synonymous with modulus of rupture but also used to refer to breaking load. (Taylor, 1965)

transverse thrust *transcurrent fault.*

transverse trace A trace on the ground motion record representing the component of motion in a horizontal plane at a right angle to the seismic wave travel direction.

transverse ventilation *peripheral ventilation.*

transverse wave (a) In seismology, a wave motion in which the motion of the particles, or the entity that vibrates, is perpendicular to the direction of progression of the wave train. (AGI, 1987) (b) In geophysics, a body seismic wave advancing by shearing displacements. (AGI, 1987) (c) A wave in which the direction of propagation of the wave is normal to the displacements of the medium; e.g., a vibrating string. The gravity wave in which fluid parcels move in circular orbits is an example of a mixed transverse-longitudinal wave. The Rossby wave is also mixed, except in the case of zero current speed, when it is a transverse wave. Syn: *distortional wave; secondary wave.* See also: *S wave.* (Hunt, 1965)

transverse with filling *overhand stoping.*

trap (a) Any dark-colored fine-grained nongranitic rock, such as a basalt, peridotite, diabase, or fine-grained gabbro; also applied to any such rock used as crushed stone. Syn: *trapp; traprock; trappide.* Cf: *whinstone.* (AGI, 1987) (b) A device for separating suspended sediment from flowing water; e.g., a sand trap. (AGI, 1987) (c) A door used for cutting off a ventilating current and that is occasionally opened for haulage or passage. See also: *trapdoor.* (Fay, 1920) (d) A device to separate denser material from less-dense material, such as entrained water in a stream or a compressed-air line. (Long, 1960) (e) That portion of any mass of porous, permeable rock that is sealed on top and down the sides by relatively nonporous and impermeable rock and that lies above the intersection of a horizontal plane passing through the lowest point of complete sealing.

trap cut A gem with a row or rows of steplike facets around the table and culet (or small lower terminus of the gem, parallel to the table), or around the culet alone. See also: *step cut.* (Fay, 1920)

trapdoor (a) A door in a mine passage to regulate or direct the ventilating current. Also called weather door. See also: *trap.* Syn: *mine door.* (Fay, 1920) (b) *air door.*

trapezohedron (a) An isometric crystal form of 24 faces, each face of which is ideally a four-sided figure having no two sides parallel, or a trapezium. Syn: *leucitohedron.* (AGI, 1987) (b) A crystal form consisting of six, eight, or twelve faces, half of which above are offset from the other half below. Each face is, ideally, a trapezium. The tetragonal and hexagonal forms may be right- or left-handed. (AGI, 1987)

trapezoid A quadrilateral that has only two sides parallel. (Jones, 1949)

trapezoidal rule A rule for estimating the area of an irregular figure by dividing it into parallel strips of equal width, each strip being a trapezium. See also: *Simpson's rule.* (Hammond, 1965)

trapp *trap.*

trapper (a) An employee, normally an apprentice, used to open and close mine doors. Also called trapper boy; nipper; door tender; doorman. Syn: *door boy.* (BCI, 1947; Fay, 1920) (b) An employee who assists the dispatcher by throwing switches and attending telephone at an inside station. (BCI, 1947)

trapper boy *trapper.*

trappide *trap.*

trap points Points placed on a railway line to derail a train that has been incorrectly signaled. (Hammond, 1965)

traprock *trap.*

trash screen Protective screen for removing detritus from the pulp stream ahead of a processing unit. (Pryor, 1965)

trass A common name in the older literature for unwelded massive ash and pumice-flow deposits. Trass has been used in the production of pozzolan cement. Cf: *ignimbrite.* Syn: *amause.* (AGI, 1987)

traveling angle of draw The angle of draw advancing with a moving face. (Nelson, 1965)

traveling apron *apron.*

traveling block (a) The movable unit, consisting of sheaves, frame, clevis, and/or hook, connected to, and hoisted or lowered with, the load in a block-and-tackle system. Also called floating block; running block. (Long, 1960) (b) The pulley block that hangs below the crown block and is used for lifting the drilling column. (BS, 1963) (c) A frame for a sheave or a set of sheaves that slides in a track. Syn: *traveling sheave.* (Nichols, 1976)

traveling compartment The section of a mine shaft used for raising and lowering miners. (Stoces, 1954)

traveling gantry A movable gantry built on wheels for traveling on rails and supporting a hoisting device. (Hammond, 1965)

traveling road A roadway used by miners for walking to and from the face; i.e., from the shaft bottom or main entry to the workings and back. Syn: *traveling way.* (Nelson, 1965)

traveling sheave A sheave block that slides in a track. See also: *traveling block.* (Nichols, 1976)

traveling way *traveling road.*

traveling weight The portion of the overlying strata at the coal face that is supported and controlled by face supports. This weight "advances" as the face line moves forward. See also: *underweight.* (Nelson, 1965)

traverse (a) A sequence or system of measured lengths and directions of straight lines connecting a series of surveyed points (or stations) on the Earth's surface, obtained by or from field measurements, and used in determining the relative positions of the points (or stations). (AGI, 1987) (b) A line surveyed across a plot of ground. (AGI, 1987) (c) To make a traverse; to carry out a traverse survey. See also: *traverse survey.* (AGI, 1987) (d) A vein or fissure in a rock, running obliquely and in a transverse direction. (AGI, 1987) (e) A line across a thin section or other sample along which grains of various minerals are counted or measured. (AGI, 1987)

traverser A platform superimposed upon or forming part of the rail track that is free to roll or slide sideways so that a car can be moved bodily from one track to another parallel to it. See also: *inclined traverser.* (Nelson, 1965)

traverser system The basic idea of this system is to confine the mine-car circuit to the smallest possible compass near the mine shaft. This avoids locking up cars on the surface that are better employed underground and reduces labor requirements. In this system, instead of the use of shunt backs on car circuits, the direction of car travel is changed by running onto a portable platform that then moves the car bodily in a transverse direction. The system can be applied, in conjunction with the necessary lifts, to multideck cages. This system may be operated electrically, hydraulically, or pneumatically. (Sinclair, 1959)

traverse survey A survey in which a series of lines joined end to end are completely determined as to length and direction, these lines being often used as a basis for triangulation. It is used esp. for long narrow strips of land (such as for railroads) and for underground surveys. Syn: *traverse.* (AGI, 1987)

traverse tables Published tables giving the differences of latitude and departure for different angles. (Hammond, 1965)

traversing method One of three recognized methods for determining the average velocity of airflow along a mine roadway by an anemometer. This is the general routine procedure applied when measuring air velocities in mine roadways. While the instrument is running, it is slowly and steadily moved up and down a series of imagined vertical lines, so as to cover equal areas in equal time. The total period is usually 1 min for a medium-sized roadway. The integrated reading is then the mean velocity for that section. Cf: *division method; single-spot method.* (Roberts, 1960)

travertine (a) A dense, finely crystalline, massive or concretionary limestone; generally white, tan, or cream; commonly having a fibrous or concentric structure and splintery fracture; formed by rapid chemical precipitation of calcium carbonate from solution in surface and ground waters, such as by agitation of stream water or by evaporation around the mouth or in the conduit of a spring, esp. a hot spring. It also occurs in limestone caves, where it forms stalactites, stalagmites, and other deposits; as a vein filling; along faults; and in soil crusts. The spongy or less compact variety is tufa. See also: *cave onyx; onyx; onyx marble.* Syn: *calcareous sinter; calc-sinter.* (b) A term that has been applied to any cave deposit or calcium carbonate. (c) A term used inappropriately as a syn. of kankar. (d) Etymol. Italian tivertino from the old Roman name of Tivoli, a town near Rome, where travertine forms an extensive deposit. Also spelled travertin. Syn: *travertine marble.*

travertine marble *travertine.*

trawley A small truck or car conveying material about a furnace or iron mill; sometimes applied to trucks, in mines, etc. See also: *trolley.* (Fay, 1920)

tray (a) A car, carrier, or pallet, usually suspended from the moving element of a conveyor. (b) A section of gravity conveyor, chain conveyor, or shaker conveyor. See also: *pan; trough.* Syn: *tray carrier.* (Nelson, 1965)

tray carrier *tray.*

tray thickener A drying unit that differs from the ordinary round tank in that it houses several horizontal trays that divide it into compartments. Each has its own set of rakes and its own underflow for settled material and peripheral overflow. Used where space is limited or in subarctic conditions that call for antifreeze housing. (Pryor, 1963)

tread (a) The ground contact surface on a tire or a track shoe. (Nichols, 1976) (b) Occasionally, a high-friction lagging on a belt pulley. (Nichols, 1976) (c) The pit in which brickmakers soak their clay before putting it into the pug mill. (Standard, 1964)

tread tractor A form of locomotive that is serviceable over rough roads. A trailer, with capacity up to 16 yd^3 (12.2 m^3), with bottom discharge or two-way side discharge is used with the tractor. The maximum speed is about 6 m/h. (Nelson, 1965)

treasure box A pocket of very rich ore.

treated stone A gemstone that has been heated, stained, oiled, or coated; treated by various types of irradiation, in order to improve or otherwise alter its color; or laser-drilled to make flaws inconspicuous. Also, a stone that has been preserved from dehydration, such as an opal whose cracks have been filled with oil or other liquid; or one in which special effects have been produced; e.g., amber with "spangles" (tension cracks). See also: *altered stone.*

treatment The reduction of ores by any process whereby the valuable constituent is recovered. (Fay, 1920)

treble Three standard lengths of drill rod or drill pipe connected together and handled and stacked in a drill tripod or derrick as a unit length of rod on borehole round trips. Also incorrectly spelled thribble; thrible; tribble; trible. See also: *stand; rod stand.* (Long, 1960)

treble coursing In mining, the system of dividing a ventilating current into three coursings. (Standard, 1964)

trechmannite A trigonal mineral, $AgAsS_2$; forms minute scarlet-vermillion rhombohedral crystals; at Binenthal, Switzerland.

tree (a) Visible projection of electrodeposited metal formed at a site of high current density. (ASM, 1961) (b) A thick log used as a prop in heavy ground. A prop, leg, or puncheon. See also: *treed.* (Fay, 1920) (c) The fulcrum for the lever used in boring. (Fay, 1920) (d) A tree-like aggregate of crystals, which forms from solution on a suspended substrate that induces crystallization. (Webster 2nd, 1960)

tree agate A moss agate with dendritic markings resembling trees. Syn: *tree stone.*

treed Supported by props, such as a mine roof. See also: *tree.* (Fay, 1920)

treeling *trilling.*

treenail A hardwood plug drilled so as to allow a track spike to be driven through it into a timber sleeper. (Hammond, 1965)

tree ore A high-grade uranium ore consisting of buried carbon trash that has been replaced or enriched with uranium-bearing solutions. (AGI, 1987)

tree stone *tree agate.*

tremolite A monoclinic mineral, $2[Ca_2Mg_5Si_8O_{22}(OH)_2]$; amphibole group with magnesium replaced by iron, and silicon by aluminum toward actinolite; white to green; long-bladed or stout prismatic crystals; may show columnar, fibrous, or granular masses or compact aggregates; in low-grade metamorphic rocks such as dolomitic limestones and talc schists; the nephrite variety is the gemstone jade; the asbestiform variety is byssolite. Cf: *actinolite.*

tremolitic Pertaining to or characterized by the presence of tremolite, as tremolitic marble.

tremor tract (a) An area of intensely jumbled coal and associated beds. The contortions contain sharp folds, thrusts, and glides. The mode or origin is controversial. A theory that has gained some favor is that the disturbance was initiated by a seismic shock, causing the coal seam and beds to crack and heave. Later, lateral forces appear to have produced the final complicated structures. (Nelson, 1965) (b) In coal mining, an area of complex folding, faulting, and gliding of coal seams and associated rocks. It may be formed by seismic shocks during the deposit's semicompacted state. (AGI, 1987)

trench (a) A long, straight, commonly U-shaped valley or depression between two mountain ranges. (AGI, 1987) (b) A narrow, steep-sided canyon, gully, or other depression eroded by a stream. (AGI, 1987) (c) Any long, narrow cut or excavation produced naturally in the Earth's surface by erosion or tectonic movements. Also, a similar feature produced artificially, such as a ditch dug in prospecting for minerals. (AGI, 1987) (d) An elongated but proportionally narrow depression, with steeply sloping longitudinal borders, one of which (the continental) rises higher than the other (the oceanic). Trenches are the ends of unsymmetrical basins and lie beside the continental border or island chains. Syn: *marginal trench.* (AGI, 1987) (e) A long but narrow depression of the deep-sea floor having relatively steep sides. (AGI, 1987) (f) A long, narrow, intermontane depression occupied by two or more streams (whether expanded into lakes or not), alternately draining the depression in opposite directions. (AGI, 1987) (g) A narrow ditch. (AGI, 1987) (h) In geological exploration, a narrow, shallow ditch cut across a mineral deposit to obtain samples or to observe character. (i) A long, narrow excavation in the ground, as a trench dug for the laying of pipes. (Crispin, 1964) (j) A temporary scar in which a conduit is placed and then covered over. Cf: *ditch.* (Carson, 1961)

trench excavation Excavation in which the width of operations and, generally, the depth are limited. Trenching may be performed in any soil and will sometimes fall into the category of limited-area, vertical excavation. (Carson, 1961)

trench excavator A self-propelled machine generally mounted on crawler tracks designed for digging trenches or ditches. It is equipped with either a bucket ladder or buckets mounted around the periphery of a circular wheel. (Hammond, 1965)

trenchman *ripper.*

trench sampling A slight refinement of grab sampling in which the material to be sampled is spread out flat and channeled in one direction with a shovel, and the material for the sample is taken at regular intervals along the channel. The procedure is repeated with several other channels in different directions until a sample of the proper size has been secured. Also called channel sampling. (Newton, 1959)

trend (a) A general term for the direction or bearing of the outcrop of a geological feature of any dimension, such as a layer, vein, orebody, fold, or orogenic belt. Cf: *strike.* Syn: *direction.* (AGI, 1987) (b) The direction or rate of increase or decrease in magnitude of the individual members of a time series of data when random fluctuations of individual members are disregarded; the general movement through a sufficiently long period of time of some statistical progressive change. (AGI, 1987) (c) The direction or bearing of a bed, dike, sill, etc., or of the intersection of the plane of a bed, dike, joint, fault, or other structural feature with the surface of the ground. (Fay, 1920) (d) The direction or bearing of a fold or series of folds in rocks, of the axes of the folds, of subsurface structures, of oriented or elongated structures indicated by geological surveys, or of topographic features that are consequent on the geologic structure. (As used in either sense, the trend may or may not coincide with the strike, depending on the structural relations at the place of observation.) Cf: *trace.* (Fay, 1920)

trent agitator An agitator with paddle-wheel-type arms; they are hollow, and the pulp solution or air is discharged from nozzles on these arms, thus causing the stirrer to rotate. (Liddell, 1918)

Trenton (a) A subdivision of the American Ordovician sometimes considered as the equivalent of the whole Middle Ordovician and sometimes restricted to a portion of this series. (Webster 3rd, 1966) (b) Formerly, a division of the lower Silurian.

Trent process Agglomeration process sometimes used in coal cleaning and briquetting. Raw coal crushed to minus 65 mesh is agitated with water and oil. Coal agglomerates and ash-forming fraction are removed in aqueous solution. (Pryor, 1963)

trepan (a) A boring tool once used in the Kind-Chaudron shaft-sinking method. (Nelson, 1965) (b) A boring machine used for shaft sinking through water-bearing strata.

trepanner A cutter loader for continuous mining in longwall faces. Its main cutting unit is the trepanner wheel with cutting arms, one at each end of the machine to enable it to cut in both directions on the face. Also fitted are a vertical, back-shearing jib; a floor-cutting jib, duplicated to enable cutting in either direction; and, if necessary, a roof-cutting disk. The machine is used in conjunction with an armored flexible conveyor on a prop-free front face; suitable for seams between 3 ft and 4 ft (0.9 m and 1.2 m) thick, although it can work in thicker seams if the top coal falls freely. (Nelson, 1965)

trepanning A type of boring in which an annular cut is made into a solid material, with the coincidental formation of a plug or solid cylinder. (ASM, 1961)

trepan shearer A cutter loader in which the trepanner head is incorporated into the shearer-loader while the cutting drum is retained to dress the floor and back of the cut. (Nelson, 1965)

trespass Working coal from the property or take of another coal mine owner. See also: *encroachment.* Syn: *bootleg.*

trestle (a) A bridge, usually of timber or steel, that has a number of closely spaced supports between abutments. (Nichols, 1976) (b) A bent of timber, reinforced concrete, or steel, supporting a temporary or permanent structure. (Hammond, 1965)

trestleman Person who unloads coke, limestone, and ore, and keeps bins poked down. (Fay, 1920)

tret Allowance to purchaser for waste. (Pryor, 1963)

trevorite An isometric mineral, $NiFe_2O_4$; magnetite series of the spinel group; black with greenish tint, at Barberton, Transvaal, South Africa.

triad In crystallography, an element of symmetry characterized by a rotational axis requiring three operations to return to identity. Isometric symmetry contains four diagonal triads; trigonal symmetry requires a unique triad. Cf: *threefold.* See also: *trigonal.*

trial face *experimental face.*

trial pit A shallow hole, 2 to 3 ft (0.6 to 0.9 m) in diameter, put down to test shallow minerals or to establish the nature and thickness of superficial deposits and depth to bedrock. See also: *test pit; pit sampling.* (Nelson, 1965)

trial shots The experimental shots and rounds fired in a sinking pit, tunnel, opencast, or quarry to determine the best drill-hole pattern to use. This is carried out when hard rocks are exposed. (Nelson, 1965)

triamorph A chemical compound crystallizing in three different crystal structures.

triangle cut The characteristic feature of this cut lies in the fact that the drill holes are arranged in zigzag. In this way a larger opening is obtained because the drill holes can break out between the preceding rows. Each vertical row of holes breaks out a layer. If the front holes do not break out to the full depth, the burnt-out holes indicate the direction of break for the following row of holes since the holes are arranged in zigzag. The name, triangle cut, is due to the distribution of the holes at the working face and the form of the initial opening. (Fraenkel, 1953)

triangle shooting A refraction type of seismic shooting used to facilitate the separation of intercept times into constituent delay times. Three profiles can be laid out as sides of a triangle. If intercept times are obtained at each of the vertices of the triangle from shots at the other two vertices, one can solve for the delay times at the three corners. Delay times along the sides of the triangle can be determined by taking differentials in the intercept times with respect to the delay times established at the vertices. (Dobrin, 1960)

triangular core The strand core of a flattened strand rope. (Hammond, 1965)

triangular facet A physiographic feature having a broad base and an apex pointing upward; specif. the face on the end of a faceted spur, usually a remnant of a fault plane at the base of a block mountain. A triangular facet may also form by wave erosion of a mountain front or by glacial truncation of a spur. Syn: *spur-end facet.* (AGI, 1987)

triangular method A method of ore reserve estimation based on the assumption that a linear relationship exists between grade difference and the distance between all drill holes. (Krumlauf)

triangular texture In mineral deposits, texture produced when exsolved or replacement mineral crystals are arranged in a triangular pattern, following the crystallographic directions of the host mineral. (AGI, 1987)

triangulate To divide into triangles; esp. to use, survey, map, or determine by triangulation. Etymol: back-formation from triangulation. (AGI, 1987)

triangulation (a) A trigonometric operation for finding the directions and distances to and the coordinates of a point by means of bearings from two fixed points a known distance apart; specif. a method of surveying in which the stations are points on the ground at the vertices of a chain or network of triangles, whose angles are measured instrumentally, and whose sides are derived by computation from selected sides or base lines, the lengths of which are obtained by direct measurement on the ground or by computation from other triangles. Triangulation is generally used where the area surveyed is large and requires the use of geodetic methods. Cf: *trilateration.* Syn: *trigonometrical survey.* (AGI, 1987) (b) The network or system of triangles into which any part of the Earth's surface is divided in a trigonometric survey. (AGI, 1987)

triaxial compression test A test in which a cylindrical specimen of rock encased in an impervious membrane is subjected to a confining pressure and then loaded axially to failure. See also: *unconfined compression test.* Syn: *triaxial shear test.* (AGI, 1987)

triaxial shear test *triaxial compression test.*

tribe A subdivision of the rock association or kindred. A tribe is made up of clans. See also: *rock association.* (AGI, 1987)

tribocouple Two chemically dissimilar metals in mutual electrical contact. The friction produced by the mechanical agitation of the two members of the couple results in the flow of an electric current. The power of a tribocouple is the magnitude of the current that it will generate under specified conditions of friction. (Osborne, 1956)

Triboelectrostatic separation process An electrostatic process under development, in which fine-size dry coal is blown rapidly past a copper baffling device that imparts positive triboelectric charges on the coal and negative charges on the associated mineral matter. The material is introduced into an electrostatic separator, where it is separated.

triboluminescence (a) The property of some specimens of zinc sulfide of emitting sparks when scratched. (Fay, 1920) (b) Luminescence in which electrons in a trapped state are released by abrasion or crushing. Cf: *luminescence.*

tribrach (a) Three leveling screws and footplate used to attach a theodolite or surveyor's level to its tripod, level the instrument, and center it precisely over its mark. (Pryor, 1963) (b) The frame below a theodolite on which three foot screws are mounted. See also: *limb.* (Hammond, 1965)

tribromomethane *bromoform.*

tribute work In mining, work on shares. (Standard, 1964)

tributing (a) A system under which a syndicate of miners delivers coal at the pithead at an agreed price. This system may be used where ore deposits are too small and scattered to conduct normal mining activities. "Tributors" work and deliver their ore to the owner and receive payment based on the ore's ascertained value. (Nelson, 1965) (b) Working on a sharing basis. (Pryor, 1963)

tricalcium pentaluminate A compound, $3CaO{\cdot}5Al_2O_3$, formerly believed to be present in high-alumina hydraulic cement. It is now known that a melt of this composition consists of a mixture of $CaO{\cdot}2Al_2O_3$ and $CaO{\cdot}Al_2O_3$, the latter compound being responsible for the hydraulic properties. (Dodd, 1964)

tricalcium silicate A compound, $3CAO{\cdot}SiO_2$; dissociates at approx. 1,900 °C to form CaO and $2CaO{\cdot}SiO_2$. This compound is the principal

cementing constitutent of Portland cement, small quantities of MgO and Al_2O_3 usually being present in solid solution. Tricalcium silicate is also present in stabilized dolomite refractories. (Dodd, 1964)
trichalcite A former name for tyrolite. See also: *tyrolite.*
trichite A straight or curved hairlike crystallite, usually black. Trichites occur singly or radially arranged in clusters and are found in glassy igneous rocks. (AGI, 1987)
trichloroethylene Colorless; stable; low-boiling; heavy; mobile; toxic; liquid; $CHCl:CCl_2$. (CCD, 1961)
trichroism Characteristic of a crystal showing three different colors in transmitted light; limited to crystallization in the orthorhombic, monoclinic, and triclinic systems. Cf: *pleochroism; dichroism.*
trickle drain A pond overflow pipe set vertically with its open top level with the water surface. (Nichols, 1976)
trickle scale Scale that has become detached from a pack of sheets in pack rolling, trickling in between the pack and becoming embedded in the surface of the sheets during further rolling. (Osborne, 1956)
triclinic (a) In crystallography, periodicity requiring three crystal axes having no further constraint on the interaxial angles designated alpha, beta, and gamma. (b) The least symmetric of the seven (or six) crystal systems, requiring all three axial vectors *a*, *b*, and *c* and three interaxial angles α, β, and γ as lattice parameters. Of its two crystal classes, one has a center of symmetry and the other does not. Feldspars and axinites crystallize in the triclinic system. Syn: *anorthic (obsolete); triclinic system.*
triclinic block In quarrying, a term applied to a block of stone bounded by three pairs of parallel faces, none of which intersect at right angles. (Fay, 1920)
triclinic crystal Crystal having no symmetry elements, or only an inverse center. The typical crystal has three unequal axes, no two of which are perpendicular. (Henderson, 1953)
triclinic system *triclinic.*
tricone bit A roller bit having three cone-shaped cutters in the head of the bit. See also: *roller bit.* Syn: *tricone roller bit.* (Long, 1960)
tricone roller bit *tricone bit.*
tridimite *tridymite.*
tridymite A monoclinic and triclinic mineral, $4[SiO_2]$; pseudohexagonal; polymorphous with coesite, cristobalite, quartz, and stishovite; colorless to white; in felsic volcanics and refractories. Also spelled tridimite.
Triger process A method of sinking through water-bearing ground in which a shaft is lined with tubbing and provided with an air lock so that work proceeds under air pressure. Cf: *Kind-Chaudron process.* (Webster 3rd, 1966)
trigger circuit A circuit having a number of states of electrical condition (which are either stable or quasi-stable) or that is unstable with at least one stable state and so designed that a desired transition can be initiated by the application of a suitable trigger excitation. By quasi-stable state is meant a state that persists during the time of interest. (NCB, 1964)
trigger effect When rock is subjected to increasing stresses there comes a time when it is on the point of failure. In some circumstances it may remain at that point for a considerable time. Any small external influence, such as a seismic wave, may then be sufficient to precipitate the failure. This is known as the trigger effect. (Spalding, 1949)
triggers Term applied to any number of things that may initiate or trigger rock bursts. Such triggers include blasting, changes of temperature, sudden influxes of water, and even rock bursts themselves, which sometimes act as a trigger impulse to initiate a second burst. (Issacson, 1962)
trigonal (a) Describes a crystal form or structure with a unique triad of symmetry. (b) A symmetry operation requiring three repetitions to return to identity. Syn: *triad.* (c) Characteristic of, pertaining to, or belonging in the trigonal system or in the rhombohedral division of the hexagonal system. (Not all trigonal point groups may be characterized by rhombohedral axes, hence the incongruence between the trigonal system and the rhombohedral division of the hexagonal system.) Cf: *rhombohedral division; hexagonal system.*
trigonal coordination In a crystal structure, a cation with three anions around it at the points of a triangle, e.g., C in a CO_3^{2-} group.
trigonal system In assigning point groups to seven crystal systems, the trigonal system is characterized by a unique triad, that element of rotational symmetry for which three operations of 120° return a lattice to identity. Cf: *rhombohedral division; hexagonal system.*
trigonite A monoclinic mineral, $Pb_3Mn(AsO_3)_2(AsO_2OH)$; forms sulfur-yellow triangular wedge-shaped crystals; at Långban, Sweden.
trigonometrical leveling Basically this method consists of determining the vertical heights by measurement of distances and angles of inclination. Angles of inclination are measured either by hand instruments or more accurately by theodolite. (Mason, 1951)
trigonometrical survey *triangulation.*
trigonometry Measurement of three-angled figures, or measurement by use of three-angled figures. (Jones, 1949)
trihydrocalcite A hydrous calcium carbonate, $CaCO_3{\cdot}3H_2O$. A moldlike incrustation on chalk marl. From Nova-Alexandria, Poland. (English, 1938)
trikalsilite A hexagonal mineral, $(K,Na)AlSiO_4$; polymorphous with kaliophilite, kalsilite, and panunzite.
trilateration A method of surveying in which the lengths of the three sides of a series of touching or overlapping triangles are measured (usually by electronic methods) and the angles are computed from the measured lengths. Cf: *triangulation.* (AGI, 1987)
trill *trilling.*
trilling A cyclic crystal twin consisting of three individuals. Cf: *fourling; fiveling; eightling.* Syn: *treeling; trill.* See also: *threeling.*
trilobite A primitive, extinct crustacean, occurring throughout the Paleozoic and abundant in the earlier Paleozoic periods, characterized by a segmented body divided by longitudinal grooves into three lobes.
trimerite A monoclinic mineral, $CaMn_2Be_3(SiO_4)_3$.
trimetric A solid figure in which the three axes are all unequal, but intersect one another at right angles. Syn: *orthorhombic.* (Gordon, 1906)
trim holes Unloaded drill holes closely spaced along a line to limit the breakage of a blast. Syn: *relief holes.* (Nichols, 1976)
trimmer (a) A shothole bored slightly outward to trim the drivage to the shape required. (Mason, 1951) (b) An apparatus for trimming a pile of coal into a regular form (such as a cone or prism). (Webster 3rd, 1966) (c) One who uses a shovel to distribute loose material—such as coal, rubbish, sand, or other substances—in railroad cars or holds of ships and barges during or after loading. May be designated according to material trimmed, such as a coal trimmer. (DOT, 1949)
trimmer conveyor A self-contained, light-weight portable conveyor, usually of the belt type, for use in unloading and delivering bulk materials from trucks to domestic storage, and for trimming bulk materials in bins or piles. See also: *portable conveyor.*
trimmer holes These complete the breaking out of the ground. The positioning and number of trimmer shots are governed by the size of the drift, the hardness of the ground, and the fragmentation required for the loading-out method to be adopted. (McAdam, 1958)
trimmers (a) The shotholes drilled around the periphery of a shaft or tunnel that break or trim the sides of the excavation to the shape and size required. See also: *cut holes.* (Nelson, 1965) (b) The top row of holes in a tunnel face. (Stauffer, 1906)
trimming shed *mica house.*
trimorphism The property of a chemical compound to crystallize in one of three different crystal structures. Cf: *dimorphism; polymorphism.*
trimorphous A chemical compound that may crystallize in one of three different crystal structures.
trinascol Dense asphaltic petroleum containing 9% sulfur. (Tomkeieff, 1954)
trinitrate glycerol *nitroglycerin.*
trinitrin *nitroglycerin.*
trinitrotoluene-ammonium nitrate explosive An explosive containing ammonium nitrate sensitized with trinitrotoluene. A proportion of aluminum powder or calcium silicide may be added to increase power and sensitiveness. (Nelson, 1965)
trinkerite A resinous substance occurring in large amorphous masses of a hyacinth-red to chestnut-brown color in brown coal near Albona, Istria (former Yugoslavia). Resembles tasmanite in composition. (AGI, 1987)
trioctahedral Pertaining to a layered-mineral structure of the kaolinite-serpentine, talc-pyrophyllite, mica, or chlorite groups in which all three of the positions with octahedral coordination are occupied, mainly by divalent cations, such as Mg, or by a mix of monovalent and trivalent cations, such as Li and Al. Cf: *dioctahedral.*
trip (a) A small train of mine cars. (Korson, 1938) (b) The number of cars moved at one time by a transportation unit. (Hudson, 1932) (c) The operation in rotary drilling of pulling out (trip out) and running in (trip in) the drill string, as required to replace a worn bit, extract a core, or recover a fish. Syn: *round trip.* (AGI, 1987) (d) An automatic arrangement for dumping cars; a tipper, a kickup. (Fay, 1920) (e) A release catch. (Nichols, 1976)
trip change A term used in mine transportation for the period during which the loads (loaded mine cars) are taken away and a fresh trip of empties is brought back. This period is known as trip change in contrast to car change. In this interval a great deal of potential loading time can be lost. (Kentucky, 1952)
trip coil A device for opening protective equipment or a circuit breaker, operated by a solenoid. (Hammond, 1965)
tripestone (a) A mineral form with the outward appearance of tripe or intestines, e.g., stalactitic calcite, crumpled gypsum laminae, contorted concretionary anhydrite, barite. Also spelled tripe stone. (b) Stalactite resembling intestines. (Arkell, 1953) (c) A variety of gypsum formed of crumpled, alternating laminae of pure white gypsum and gray argillaceous gypsum. (Arkell, 1953) (d) A contorted concretionary variety of anhydrite. (Fay, 1920)

trip hammer A power hammer operated by a tripping mechanism that causes the hammer to drop. (Crispin, 1964)
triphane A former name for spodumene.
triphylite An orthorhombic mineral, $4[LiFe^{2+}(PO_4)]$, with manganese replacing iron toward lithiophilite; bluish to greenish gray; in granite pegmatites.
trip lamp A removable self-contained mine lamp, designed for marking the rear end of a train (trip) of mine cars.
triple entry A system of opening a mine by driving three parallel entries for the main entries. See also: *main entry.* (Fay, 1920)
triple-entry room-and-pillar mining *room-and-pillar.*
triple point An invariant point at which three phases coexist in a unary system. When not otherwise specified, it usually refers to the coexistence of solid, liquid, and vapor of a pure substance. (AGI, 1987)
triplet An assembled stone of two main parts of gem materials bonded by a layer of cement or other thin substance (the third part of the triplet), which gives color to the assemblage. Cf: *doublet; emerald triplet.*
tripletine A name for emerald-colored beryl triplet. See also: *emerald triplet.*
triple-tube core barrel A special core barrel used to take soil samples, as in foundation testing. The inner tube is swivel mounted and nonrotating and extends through, and a short distance beyond, the bit. Hence, the bit only cuts clearance for the outer tube or core-barrel assembly, and the core taken by the inner tube is cut by a spudding action. The triple or core tube is mounted inside the inner tube to receive the core and is split longitudinally to facilitate removal of the core. Also called clay barrel. (Long, 1960)
Triplex glass A patented form of laminated glass. See also: *safety glass.* (CTD, 1958)
triplexing A method of steelmaking that involves the use of three processes, e.g., a sequence of melting in a cupola, blowing in a Bessemer converter, and finishing in a basic electric furnace, or a combination of the acid Bessemer converter, the basic open-hearth furnace, and the basic electric furnace. (Osborne, 1956)
triplex pump A positive-displacement piston pump having three water cylinders mounted side by side. It may be either a single- or double-action type. Cf: *duplex pump.* (Long, 1960)
triplite A monoclinic mineral, $8[(Mn,Fe,Mg,Ca)_2(PO_4)(F,OH)]$; brown; forms fibrous masses; in granite pegmatites. Syn: *pitchy iron ore.*
triploidite A monoclinic mineral, $(Mn,Fe)_2(PO_4)(OH)$; perfect cleavage.
trip maker A device to elevate cars on an inclined track as received from a kickback. (Zern, 1928)
tripod A three-legged support for a rock drill, hoisting drum, magnetometer, or any other piece of equipment.
tripoli *diatomite.*
tripolite A term that has been applied as a syn. of diatomite, in reference to the material from the north African location of Tripoli. It has also been used, less correctly, as a syn. of tripoli. See also: *diatomite.* (AGI, 1987)
tripper (a) A device in the run of a conveyor comprising two free drums around which the belt passes S-fashion. (Nelson, 1965) (b) A device for discharging material from a belt conveyor. (Nelson, 1965) (c) A double pulley that turns a short section of a conveyor belt upside down in order to dump its load into a side chute. (Nichols, 1976) (d) A device or mechanism that trips, as a device for causing the load on a conveyor to be discharged into a hopper or other receptacle. (Webster 3rd, 1966) (e) An automatic car dump. (f) A device for tipping and dumping the skip at the top of the blast furnace. (Mersereau, 1947)
tripper man Person who unloads grain or ore from conveyor belt into bins or processing equipment by operating a tripper. Syn: *conveyor-operator tripper; conveyor-tripper operator; spreader operator.* (DOT, 1949)
tripping (a) The process of pulling and/or lowering drill-string equipment in a borehole. (Long, 1960) (b) To open a latch or locking device, thereby allowing a door or gate to open to empty the contents of a skip, bailer, etc. (Long, 1960)
trippkeite A tetragonal mineral, $CuAs^{+3}{}_2O_4$; excellent prismatic cleavage permitting crystals to be broken into flexible fibers; blue-green; at Atacama, Chile.
trip recorder *hoist trip recorder.*
trip sender In bituminous coal mining, a laborer who switches cars to various tracks, couples and uncouples trains, and attaches and detaches cars to and from the haulage cable at a mine where there are several sidetracks on the haulageway. (DOT, 1949)
tripuhyite A tetragonal mineral, $FeSb_2O_6$; forms dull greenish-yellow microcrystalline aggregates; in the cinnabar-bearing gravels of Tripuhy, Ouro Preto, Minas Gerais, Brazil.
trisilicate In metallurgy, a slag with a silicate degree of 3. (Newton, 1959; Newton, 1938)
trislope screen A screen in which each section of the deck is flatter than the preceding one. The rate of feed is reduced on succeeding sections to maintain proper bed depth for rapid stratification. It is designed for fine dry screening of 3/4-in (19-mm) by 0-in, 1/2-in (13-mm) by 0-in, and 1/4-in (6 mm) by 0-in moist coal or other material. See also: *varislope screen.* (Nelson, 1965)
trisoctahedron A crystal form of 24 faces in the isometric system with the gross appearance of each face of an octahedron being replaced by three faces arrayed around a diagonal triad. Each face of a trigonal trisoctahedron is an isosceles triangle, while each face of a tetragonal trisoctahedron is a trapezoid. Syn: *trapezohedron.*
trisodium phosphate Na_3PO_4, a chemical compound used in some enamel frit compositions. (Enam. Dict., 1947)
tristetrahedron (a) An isometric hemihedron included under 12 trapeziform faces; a tetragonal tristetrahedron. (Standard, 1964) (b) An isometric hemihedron included under 12 isoceles triangular faces; a trigonal tristetrahedron. (Standard, 1964)
tritium The radioactive isotope of hydrogen having two neutrons and one proton in the nucleus. Being hydrogen-3, it is heavier than deuterium (heavy hydrogen or hydrogen-2). (Lyman, 1964; Handbook of Chem. & Phys., 2)
tritomite A trigonal mineral, $(Ce,La,Y,Th)_5(Si,B,Al)_3(O,OH,F)_{13}$(?); moderately radioactive; metamict; dark brown; in syenite with leucophanite, analcime, mosandrite, aegirine, and catapleiite. Formerly called spencite.
triton value The number of grams of TNT required to produce the same angle of recoil of the ballistic mortar as 10 g of the explosive under test. (McAdam, 1958)
triturate (a) A powder produced from a solid by grinding, usually with the addition of some liquid. (Gaynor, 1959) (b) To grind to a powder, usually with the addition of some liquid. (Gaynor, 1959)
trituration Reduction to a fine powder by grinding. Syn: *comminution.* Cf: *levigation.*
trivalent (a) Having a valence of 3. (Webster 3rd, 1966; Standard, 1964) (b) Having three valences; e.g., chromium, which has valences of 2, 3, and 6. Syn: *tervalent.* (Webster 2nd, 1960; Handbook of Chem. & Phys., 2)
troctolite A gabbro that is composed chiefly of calcic plagioclase (e.g., labradorite) and olivine with little or no pyroxene. Such rocks commonly are speckled like trout. Syn: *forellenstein; troutstone.* (AGI, 1987)
troegerite *trögerite*
trof-dopplerit *dopplerite.*
trögerite A very rare tetragonal mineral, $H(UO_2)(AsO_4)·4H_2O$; autunite group; strongly radioactive; lemon-yellow; in veins with walpurgite, zeunerite, uranospinite, pitchblende, and other uranium minerals. Also spelled troegerite.
Tröger's classification A quantitative mineralogic classification of igneous rocks proposed by E. Tröger in 1935. (AGI, 1987)
trogtalite An isometric mineral, $CoSe_2$; pyrite group; dimorphous with hastite; violet.
troilite A hexagonal mineral, FeS; a meteorite mineral related to pyrrhotite, $Fe_{1-x}S$, in terrestrial rocks.
Trojan coal powder High explosive used in mines. (Bennett, 1962)
trolley (a) The grooved wheel, fixed in bearings at the end of a trolley pole, pressed upward in rolling contact with the overhead trolley wire to take off the electric current for operating the locomotive or other piece of motorized equipment. A trolley glider is frequently used in place of the wheel, making a sliding contact with the wire. Also called trolley wheel. (Jones, 1949) (b) A low carriage, mounted on wheels, for carrying timber, supplies, and machines underground. See also: *trawley; bogie.* (Nelson, 1965)
trolley conveyor A series of trolleys supported from or within an overhead track and connected by an endless propelling medium, such as a chain, a cable, or other linkage, with loads usually suspended from the trolley. Trolley conveyors may be designed for single or multiple plane operation. Syn: *overhead conveyor; overhead trolley conveyor.*
trolley locomotive (a) A mine locomotive operated by electricity drawn from overhead conductors. Small grooved wheels or gliders are held in contact with the conductors, and the current passes down a trolley arm to the motor. It is very efficient where heavy loads are hauled up relatively steep gradients. Generally restricted to intake airways not nearer than 300 yd (274 m) to a working face. (Nelson, 1965) (b) A mine locomotive operated by electricity drawn from overhead trolley wires. (Nelson, 1965)
trolley voltage Although not actually set by law, the generally accepted maximum direct current trolley voltage is considered to be 300 V. The use of alternating current voltages above 220 V in mines is usually permitted, provided the conductors are properly insulated and the cables end in suitable terminal boxes. (Kentucky, 1952)
trolley wire The means by which power is conveyed to an electric trolley locomotive. It is hung from the roof and conducts power to the locomotive by the trolley pole. Power from it is sometimes also used to run other equipment. (BCI, 1947)
trolley wire guards Coverings for exposed trolley wires in mines and other locations where transportation power wires are within reaching height and are a constant source of danger to all personnel. Coverings, made of rubber or some other insulating material, guard workers from

severe burns or electrocution by direct contact with the wire. (Best, 1966)

trombe An apparatus for producing a blast of air by means of a falling stream of water, which mechanically carries air down with it, to be subsequently separated and compressed in a reservoir or drum below. See also: *water blast*. Syn: *trompe*. (Fay, 1920)

trommel A revolving cylindrical screen used in size classification of coarsely crushed ore, coal, gravel, and crushed stone. The material to be screened is delivered inside the trommel at one end. The fine material drops through the holes; the coarse material is delivered at the other end. Also called, according to its various uses, sizing trommel, washing drum, and washing trommel. Also spelled tromel. See also: *revolving screen; rotary breaker; rotary screen; shaking screen*. (ASM, 1961; Liddell, 1918; Standard, 1964)

trommel screen A screen in which the screening surface is formed into a cylinder or frustum of a cone, mounted upon a rotating shaft or on revolving rollers. See also: *revolving screen*. (BS, 1962)

tromp curve *ash curve*.

tromp cut point *partition density*.

Tromp distribution curve A curve showing the float-sink percentage of each density fraction of the feed coal. The quantity of clean coal recovered is plotted against the mean densities of the density fraction. From this curve the specific gravity of separation and the sharpness of the separation can be determined. Such a curve is independent of the coal being washed and is characteristic of a specific coal-washing device.

trompe *trombe*.

tromp error curve *error curve*.

Tromp process (a) The Tromp process was the first to introduce (about 1938) the use of magnetite suspension in dense-medium washing. The magnetite is ground to about minus 1/250 in (0.1 mm) and added to water. The process makes use of an unstable suspension with horizontal currents of differing densities at intermediate levels. The process operates within the size range 6 to 200 mm and in practice is used for raw coal down to 1/4 in (6.4 mm). It gives a reasonably accurate three-product separation. (Nelson, 1965) (b) A dense-medium process that utilizes a rapidly settling suspension of finely powdered magnetite or sintered roasted pyrite. This process may be used on any size of coal from 10 to 1/4 in (254 to 6.4 mm) and for any specific gravity from 1.3 to 1.9. The grain size of the magnetite or pyrite is minus 0.1 mm. The quick settling of the magnetite particles gives a higher specific gravity in the lower layers of the wash box, which makes it possible to obtain three products: clean coal, middlings, and refuse. (Mitchell, 1950)

trona A monoclinic mineral, $Na_3(Co_3)((HCO_3)\cdot 2H_2O)$; soft; vitreous; colorless to white; alkaline tasting; in saline lake deposits and desert soils; a major source of sodium compounds from extensive deposits at Searles and Owens Lakes, CA, and in Wyoming, Hungary, Egypt, Africa, and Venezuela. See also: *sodium sesquicarbonate*.

Trona process The method used for the separation and the purification of soda ash (anhydrous sodium carbonate), anhydrous sodium sulfate, boric acid, borax, potassium sulfate, bromine, and potassium chloride from brine at Searles Lake, San Bernardino County, CA. (CCD, 1961)

trondhjemite A light-colored plutonic rock composed primarily of sodic plagioclase (esp. oligoclase), quartz, sparse biotite, and little or no alkali feldspar. Its name, given by Goldschmidt in 1916, is derived from Trondhjem, Norway. Also spelled: trondjemite; trondheimite. (AGI, 1987)

trondjemite *trondhjemite*.

troostite (a) A manganoan variety of willemite occurring in large reddish crystals. (b) A previously unresolvable, rapidly etching, fine aggregate of carbide and ferrite produced either by tempering martensite at a low temperature or by quenching a steel at a rate slower than the critical cooling rate. Preferred terminology for the first product is tempered martensite; for the latter, fine pearlite. (ASM, 1961)

troostitic structure Fine aggregates of ferrite and cementite in steel; emulsified ferrite. (Pryor, 1963)

Tro-Pari survey instrument Trade name of a single-shot borehole surveying instrument combining a compass and inclinometer, which is locked in place by the action of a preset time clock. (Long, 1960)

tropic pack A special type of packing to protect explosives from deteriorating when subjected to hot, humid atmospheric conditions, such as in tropical areas. The explosives, after being sealed with paraffin, are packed in cartons, which are then wrapped in waxed paper and also sealed with paraffin. The filled cartons are then placed inside a satchel-type case liner of bitumen-laminated paper reinforced with sisal fiber and completely sealed with a waterproof adhesive. (Nelson, 1965)

trough (a) A channel, open or covered, that contains coal or ore being conveyed on a chain or shaker conveyor. The shape of the cross section depends on the type of conveyor involved. See also: *tray*. (b) A hollow or undulation in a mineral field, or in a mineral working. In geology, synonymous with basin; synclinal. See also: *graben*. (Fay, 1920) (c) The lowest point of a given stratum in any profile through a fold. Cf: *trough line*. (AGI, 1987) (d) A line occupying the lowest part of a fold; the line connecting the lowest parts on the same bed in an infinite number of cross sections. See also: *trough plane*. (McKinstry, 1948)

trough banding Rhythmic layering or alignment of minerals in an igneous rock, confined to troughlike depressions and considered to have been produced by currents set up in the magma during cooling. (AGI, 1987)

trough conveyor A pan conveyor or gravity conveyor. (Nelson, 1965)

troughed belt A belt conveyor in which the carrying side is made to form a shallow trough by means of troughing idlers. (Nelson, 1965)

troughed belt conveyor A belt conveyor with the belt edges elevated on the carrying run to form a trough by conforming to the shape of the troughed carrying idlers or other supporting surface. See also: *belt conveyor*.

troughed roller conveyor A roller conveyor having two rows of rolls set at an angle to form a trough over which objects are conveyed. See also: *el conveyor*.

trough fault A fault, generally a normal fault, that bounds a graben or other structural depression. (AGI, 1987)

troughing (a) A structural section shaped like a wide U; riveted or welded to form a bridge deck with the U-shaped sections turned alternately upwards and downwards. (Hammond, 1965) (b) Making repeated dozer pushes in one track, so that ridges of spilled material hold dirt in front of the blade. (Nichols, 1976) (c) Eng. In Derbyshire, toadstones filling fissures.

troughing idler A belt idler having two or more rolls arranged to turn up the edges of the belt so as to form the belt into a trough. (NEMA, 1956)

troughing rolls The rolls of a troughing idler that are so mounted on an incline as to elevate each edge of the belt to form a trough. (NEMA, 1956)

trough line (a) The line occupying the lowest part of the fold, or, more precisely, the line connecting the lowest parts of the same bed in an infinite number of cross sections. Cf: *trough*. (Stokes, 1955) (b) The line joining the trough points of a given stratum. (AGI, 1987)

trough plane The plane that joins the troughs of a series of beds in a syncline; generally, but not necessarily, the same as the axial plane. See also: *trough*.

trough vein A trough-shaped ore deposit formed between sedimentary beds in the troughs of synclinal structures.

trough washer (a) A washer applying the principle of alluviation in troughs. (BS, 1962) (b) In its simplest form, a trough washer is a sloping wooden trough, 1-1/2 to 2 ft (0.46 to 0.6 m) wide, 8 to 12 ft (2.4 to 3.7 m) long, and 1 ft (0.9 m) deep, open at the tail end, but closed at the head end. It is used to float adhering clay or fine material from the coarser portions of ore or coal. A log washer.

troutstone *troctolite*.

trow A wooden channel for air or water. (Fay, 1920)

troy ounce One-twelfth of a pound of 5,760 grains (troy pound), or 480 grains. A troy ounce equals 20 pennyweights, 1.09714 avoirdupois oz, or 31.1035 g. It is used in all assay returns for gold, silver, and platinum-group metals. (Fay, 1920; Zimmerman, 1949)

troy pound A unit of weight that equals 5,760 grains, 12 tr oz, 240 pennyweights, 13.1657 avoirdupois oz, 0.82286 avoirdupois lb, or 373.2509 g. (Fay, 1920; Zimmerman, 1949)

troy weight These are the weights used for precious metals. The equivalents are 24 grains = 1 pennyweight; 20 pennyweights = 1 oz; 12 oz = 1 lb. The troy grain is the same as the avoirdupois grain, but the ounce is larger on the troy scale; 1 tr oz = 31.103 g; 1 avoirdupois oz = 28.35 g. (Anderson, 1964)

truck Any wheeled vehicle, usually self-propelled, used to transport heavy articles or materials. In mining, usually applied to dump and/or bottom-dump semitrailers used to transport mined waste and ore materials. The number of types of these haulage units varies widely from the small 2-st (1.8-t) standard dump truck to the unit with capacity 200 st (181 t) or greater. For larger stripping operations, where the haulage conditions are not too rugged, a diesel tractor pulling a bottom-dump semitrailer of capacity 40 to 60 st (36 to 54 t) is most common. The newer trucks are equipped with power steering, power brakes, torque converters, and automatic transmissions.

truck mixer A concrete mixer, generally mounted on a lorry, or crawler-type tracks, which mixes concrete during the journey from the batching plant to the construction site. (Hammond, 1965)

truck roller *track roller*.

trudellite A mixture of chloraluminite and natroalunite.

true azimuth The azimuth measured clockwise from true north through 360°. (AGI, 1987)

true bearing The bearing expressed as a horizontal angle between a geographic meridian and a line on the Earth; esp. a horizontal angle measured clockwise from true north. Cf: *magnetic bearing*. (AGI, 1987)

true depth The actual depth of a specific point in a borehole measured vertically from the surface in which the borehole was collared. Syn: *true vertical depth.* (Long, 1960)

true dip (a) A syn. of dip, used in comparison with apparent dip. Syn: *full dip.* (AGI, 1987) (b) The angle at which veins, strata, etc., dip, as measured vertically downward from the horizon along a line at right angles to the strike of the veins, strata, etc.; also, the dip of a vein, strata, etc., as determined on oriented core. See also: *core orientation; oriented core; apparent dip.* (Long, 1960) (c) The maximum angle which an inclined bed makes with a horizontal plane. It is the direction in which water would flow if poured on the smooth upper surface of the bed at the outcrop. Also called dip. See also: *level course.* (Nelson, 1965) (d) *three-dimension dip.*

true lode *fissure vein.*

true middlings (a) *bone coal.* (b) Comparatively high-ash material so nearly homogeneous that its quality cannot readily be improved by crushing and cleaning. (BS, 1962)

true strain The integral, over the whole of a finite extension, of each infinitesimal elongation divided by the corresponding momentary length. It is equal to $\log_e(1 + \varepsilon)$, where ε is the strain as ordinarily defined. See also: *strain.* (Roark, 1954)

true stress For an axially loaded bar, the load divided by the corresponding actual cross-sectional area. It differs from the stress as ordinarily defined because of the change in area due to loading. See also: *stress.* (Roark, 1954)

true vein An occurrence of ore, usually disseminated through a gangue of veinstone, and having more or less regular development in length, width, and depth. See also: *vein; fissure vein.* (Fay, 1920)

true vertical depth *true depth.*

true whiting A finely divided calcium carbonate prepared by wet grinding and levigating natural chalk; a variety of limestone. (USBM, 1965)

true width (a) The width or thickness of a vein, stratum, etc., as measured perpendicular or normal to dip and strike. The true width is always the width of the vein, etc., at its narrowest point. Cf: *apparent width.* (Long, 1960) (b) The true width of a vein in sampling may be found by $w = h \sin a$, where h = horizontal width, w = true width, and a = angle of dip. In this formula, angle a is known from previous observations, and the horizontal width can be measured with a level. It is important that horizontal width is measured at right angles to strike. (Hoover, 1948)

truffite Fibrous nodular lignite which when struck emits an odor like that of truffles. It occurs in large nodular masses inside a normal lignite of Cretaceous age in France. (Tomkeieff, 1954)

truing-machine operator One who grinds the surfaces of refractory blocks to reduce them to standard dimensions, using a truing machine. (DOT, 1949)

trumpeting Eng. A channel or passage partitioned off from a shaft or left behind the lining, usually running along one corner of the latter. Used for ventilation. (Webster 2nd, 1960)

trumpet log *microlaterolog.*

truncated spur A spur that projected into a preglacial valley and was partially worn away or beveled by a moving glacier as it widened and straightened the valley. See also: *faceted spur.* (AGI, 1987)

trunk (a) A long, narrow, inclined box in which fine ore is separated from impurities. (Fay, 1920) (b) A launder for conveying slimes, etc. (Webster 2nd, 1960) (c) To separate slimes by means of a trunk for further treatment. (Webster 2nd, 1960)

trunk conveyor A high-capacity main road conveyor, usually a belt conveyor. It may extend from the main inby loading point to the shaft bottom or along levels or drifts to the surface. It varies from 42 to 60 in (1.07 to 1.52 m) wide and is powered by a motor of about 200 hp (149 kW). See also: *conveyor; gathering conveyor.* (Nelson, 1965)

trunkline (a) A detonating cord line used to connect the downlines or other detonating cord lines in a blast pattern. Usually runs along each row of blastholes. (Dick, 2) (b) The line of detonating cord that is used to connect and initiate other lines of detonating cord, used on the ground surface to initiate downlines.

trunk pumping engine A pump that commands the drainage of underground waters over a considerable area of mine workings, being a substitute for a number of smaller, independent pumps. (Fay, 1920)

trunk roadway The main development heading from the pit bottom, usually driven along the strike of the coal seam. Because it will carry heavy traffic and large volumes of air, a trunk roadway is at least 14 ft (4.3 m) wide and wide enough for two rail tracks. At intervals, cross-headings are excavated for opening out conveyor panels in the coal seam. Trunk roadways are usually driven in pairs for ventilation, storage space, and access. (Nelson, 1965)

trunnion (a) Either of two opposite pivots, journals, or gudgeons, usually cylindrical and horizontal, projecting one from each side of a piece of ordnance, the cylinder of an oscillating engine, a molding flask, a converter, etc., and supported by bearings, to provide a means of swiveling or turning. (Webster 2nd, 1960) (b) An oscillating bar that allows changes in angle between a unit fastened to its center, and another attached to both ends. (Nichols, 1976) (c) A heavy horizontal hinge. Also called walking beam; walking bar. (Nichols, 1976)

trunnion axis The horizontal axis about which the telescope of a theodolite can be rotated. (Hammond, 1965)

trunnion plate A metal plate lining the bearings or recesses in which the trunnions rest. (Webster 2nd, 1960)

truscottite A hexagonal mineral, $(Ca,Mn)_{14}Si_{24}O_{58}(OH)_8{\cdot}2H_2O$; in spherical aggregates of white scales; at Benkulen, Sumatra.

truss (a) An assemblage of members, such as beams, bars, and rods, typically arranged in a triangle or combination of triangles to form a rigid framework, such as for supporting a load over a wide area that cannot be deformed by the application of exterior force without deformation of one or more of its members. (Webster 3rd, 1966) (b) A framed structure built up entirely from tension and compression members, arranged in panels so as to be stable under load; used for supporting loads over long spans. (CTD, 1958)

trussed beam A beam of timber or other material that is stiffened so as to reduce deflection. (Hammond, 1965)

tsavolite A green gem variety of garnet. Also spelled tsavorite.

tscheremchite *cheremchite.*

tschermakite A monoclinic mineral, $Ca_2(Mg,Fe)_3Al_2(Si_6Al_2)O_{22}(OH)_2$; amphibole group, having Mg/(Mg + Fe) = 0.5 to 1.0; forms a series with ferrotschermakite.

tsilaisite A synthetic manganese tourmaline or manganoan elbaite.

tsumebite A monoclinic mineral, $Pb_2Cu(PO_4)(SO_4)(OH)$; brackenbuschite group; in small, tabular, emerald-green crystals; at Tsumeb, Namibia.

tub (a) A tram, wagon, corf, or corve. (CTD, 1958) (b) A small rail-track vehicle for carrying coal or minerals, with a capacity ranging from 10 to 25 hundredweights (453.6 to 1,134 kg). Tub is the term used in most English mines; tram is used in South Wales; and hutch is used in Scotland. (Nelson, 1965) (c) A box or bucket in which coal or ore is sent up a shaft. A keeve. (Webster 3rd, 1966) (d) To line, such as in a mine shaft, with tubbing; to keep back water by tubbing. See also: *tubbing.* (Webster 3rd, 1966) (e) A large circular base that provides the maximum practical bearing area, and on which is mounted the revolving frame or subbase of a walking dragline. (Austin, 1964)

tub-and-stall *bord-and-pillar working.*

tubber A double-pointed pickax; a beele. (Standard, 1964)

tubber man A person who uses a tubber. (Standard, 1964)

tubbing (a) The watertight cast-iron lining of a circular shaft built up of segments that are fixed together with flanges. The flanges are internal and bolted in German tubbing; they are external in English tubbing, which gives a smooth inner face because the segments are wedged and not bolted. The space outside the tubbing is grouted to add strength and improve watertightness. See also: *tub; suspended tubbing.* (Nelson, 1965) (b) Eng. A lining of timber or metal for a shaft, as in a mine, esp. a watertight shaft lining consisting of a series of cast-iron cylinders bolted together and used to sink through water-bearing strata. (Webster 3rd, 1966) (c) A shaft lining of casks or cylindrical caissons of iron or wood. See also: *plank tubbing; wooden tubbing.* (Fay, 1920)

tubbing wedge A small wooden wedge hammered between the joints of tubbing plates. (Zern, 1928)

tube-axial fan An airfoil (propeller) or disk fan within a cylinder and including driving-mechanism supports either for belt drive or direct connection. (Strock, 1948)

tube blower A person who cleans boiler tubes. (Fay, 1920)

tube clamp (a) A clamp or clip for gripping a tube or pipe; esp., a jawed tool used in hoisting and lowering well tubes in drilling. (Standard, 1964) (b) A misnomer for casing clamp. (Long, 1960)

tube mill A revolving cylinder, usually lined with silex, nearly half filled with glacial or water-worn flints, used for fine grinding of certain ores, preliminary to further treatment. The material to be ground, mixed with water, is fed through a trunnion at one end and passes out the opposite trunnion as a slime. This is an exceptionally long mill with a relatively small diameter. Syn: *cylindrical mill.* (Fay, 1920; Newton, 1959)

tubercle texture In mineral deposits, a texture in which gangue is replaced by automorphic minerals. Cf: *atoll texture.* (AGI, 1987)

tuberose A mineral exhibiting very irregular rounded surfaces, often giving rise to gnarled, rootlike shapes. (Nelson, 1965)

tub hooker The person who hooks or unhooks the hoisting rope to or from the buckets. (Hess)

tubing (a) The tube lining of boreholes; casing. (Fay, 1920) (b) Hollow cast-iron segments placed in a shaft to dam water or sink through quicksand. Also spelled tubbing. (Fay, 1920) (c) *ventilation tubing.* (d) A small-diameter removable pipe, suspended and immobilized in a well inside a large-diameter casing and opening at a producing zone, through which fluids are produced (brought to the surface). (AGI, 1987) (e) The act or process of placing tubing in a well. (AGI, 1987) (f) The act of lining a deep borehole by driving down iron tubes. See also: *casing.* (Fay, 1920) (g) A misnomer for casing. (Long, 1960) (h)

Small-diameter removable pipe through which oil and gas are produced from the well. (Wheeler, 1958)
tubular screw conveyor *screw conveyor.*
tucanite *scarbroite.*
tucking frame A frame in timbering in which the poling boards are supported by walings at their upper and lower ends. (Hammond, 1965)
tucking space The space between the blocks separating the cap in a heading set from the poling driven. This space provides for driving a second set of poling boards. (Stauffer, 1906)
tue irons Blacksmith's tongs. (Standard, 1964)
tufa A chemical sedimentary rock composed of calcium carbonate, formed by evaporation as a surficial, spongy, porous, semifriable incrustation around the mouth of a hot or cold spring or seep, or along a stream carrying calcium carbonate in solution, and exceptionally as a thick, bulbous, concretionary or compact deposit in a lake or along its shore. It may also be precipitated by algae or bacteria. The hard, dense variety is travertine. The term is rarely applied to a similar deposit consisting of silica. It is not to be confused with tuff. Etymol: Italian tufo. Cf: *sinter.* Syn: *calcareous tufa; calc-tufa; tophus; tuft; petrified moss.* (AGI, 1987)
tufaceous Pertaining to or like tufa. Not to be confused with tuffaceous. (AGI, 1987)
tuff A general term for all consolidated pyroclastic rocks. Not to be confused with tufa. Adj: tuffaceous. Cf: *crystal tuff.* (AGI, 1987)
tuffaceous Said of sediments containing up to 50% tuff. Cf: *tufaceous.* (AGI, 1987)
tuff breccia A pyroclastic rock consisting of more or less equal amounts of ash, lapilli, and larger fragments. (AGI, 1987)
tuffite A tuff containing both pyroclastic and detrital material, but predominantly pyroclasts. (AGI, 1987)
tuff lava Applied to consolidated, lavalike tuff consisting primarily of lenses of black and gray obsidian lying in a tuffaceous matrix that displays a streaky, varicolored banding or eutaxitic structure. Rocks of this sort are generally considered to be the product of ash flows or nuëes ardentes. Syn: *welded tuff.* (AGI, 1987)
tuft Eng. Any porous or soft stone, such as the sandstone in the Alston district of Cumberland; tufa. (AGI, 1987)
tuft stone (a) Eng. Tufa near Newport, Monmouthshire, and Dursley, Gloucestershire. (b) Eng. Toadstone, Derbyshire. See also: *toadstone.* (Arkell, 1953)
tugger *air hoist.*
tugger hoist An air hoist for mines. (von Bernewitz, 1931)
tugger man *tugger operator.*
tugger operator In mining, a person who operates a small portable or semiportable hoist (tugger), powered by compressed air or electricity, to raise coal, ore, rock, or supplies in a shaft or stope or along an incline inside a mine. Also called tugger man. (DOT, 1949)
tugtupite A tetragonal mineral, $Na_4AlBeSi_4O_{12}Cl$; in the Ilimaussaq massif, southwest Greenland. Formerly called beryllosodalite.
tuiles The working openings at the discharging end of a glass furnace. (Mersereau, 1947)
Tukon hardness test A method of determining the hardness of microconstituents by using the Knoop or Vicker's type of diamond indenter. See also: *microhardness; Vickers hardness test.* (Henderson, 1953)
tumble To smooth, clean, or polish, as castings, by friction with each other or with a polishing material in a rotating box or barrel; to rattle. (Standard, 1964)
tumbled Semiprecious and precious stones, cleaved carbon, or other diamonds, the sharp edges and corners of which have been rounded and blunted by tumbling action in a barrel-shaped vessel. (Long, 1960)
tumbler (a) A projecting piece on a revolving shaft or rockshaft for actuating another piece. In dredges, both an upper and a lower tumbler support the bucket line. (Fay, 1920) (b) Any piece of equipment that polishes gemstones by a tumbling action.
tumbler test Test for determining relative friability of a particular size of sized coal. (Bennett, 1962)
tumbling An operation in which the work, usually castings or forgings, is rotated in a barrel with metal slugs or abrasives to remove sand, scale, or fins. It may be done dry or with an aqueous solution. Sometimes called rumbling or rattling. (ASM, 1961)
tumbling barrel A revolving barrel, cask, or box in which objects or materials (such as small metal parts, castings, plastics, leather, or clothing) undergo a process (such as finishing, polishing, coating, softening, or drying) by being whirled about and so brought into vigorous frictional contact. Also called rattler; rumble; scouring barrel. (Webster 3rd, 1966)
tumbling box A tumbling barrel for small objects. (Webster 3rd, 1966)
tumbling mill Any horizontally mounted cylindrical mill in which contents are tumbled when rotating. Name often used in connection with cleaning of objects. (Pryor, 1963)
tumbling shaft The camshaft used in stamp mills. (Fay, 1920)
tumbling stone N. of Eng. Boulders or detached masses of rock.
tumescence The swelling of a volcanic edifice due to accumulation of magma in the reservoir. It may or may not be followed by an eruption. Syn: *inflation.* (AGI, 1987)
tundra A treeless, level or gently undulating plain characteristic of arctic and subarctic regions. It usually has a marshy surface, which supports a growth of mosses, lichens, and numerous low shrubs and is underlain by a dark, mucky soil and permafrost. (AGI, 1987)
tundra placer *gravel plain placer.*
tunellite A monoclinic mineral, $SrB_6O_9(OH)_2{\cdot}3H_2O$; subvitreous to pearly; colorless; forms compact fine-grained secondary nodules; also prismatic and tabular crystals; at Kramer and in the Furnace Creek area of Death Valley, CA.
tune work Labor paid for by the day or the hour, in contrast to piecework. (Fay, 1920)
tungstate A mineral containing the radical $(WO_4)^{2-}$, in which the hexavalent tungsten ion and its four oxygens form a flattened square rather than a tetrahedron, e.g., the wolframite series, $(Fe,Mn)WO_4$. Tungsten and molybdenite may substitute for each other.
tungsten A hard, brittle, white or gray metallic element. Symbol, W. Also known as wolfram. Found combined in certain minerals such as wolframite, $(Fe,Mn)WO_4$; scheelite, $CaWO_4$; huebnerite, $MnWO_4$; and ferberite, $FeWO_4$. Tungsten and its alloys are used extensively for filaments for electric lamps, electron and television tubes, X-ray targets, and numerous space missile and high-temperature applications. See also: *wolframite.* Syn: *wolfram.* (Handbook of Chem. & Phys., 3)
tungsten alloy An alloy used in drill-bit-crown matrices and in making bit and reaming-shell inserts by powder methods in which the principal constituent is tungsten, generally in the form of carbide. Tungsten carbide powder usually is mixed with a powdered cobalt or other metal to bind it together in a cohesive mass. (Long, 1960)
tungsten carbide A mixture consisting of 85% to 95% tungsten carbide and 5% to 15% cobalt; sp gr, 12 to 16; Mohs hardness, about 9.0; it is not affected by severe high industrial temperatures. Used for machine tools and for abrasives for machining and grinding metals, rocks, molded products, porcelain, and glass. (CCD, 1961)
tungsten carbide bit A drilling bit tipped with tungsten carbide. A 9% cobalt carbide generally gives the best results, and comparisons are usually referred to bits of this standard. Tests with tungsten carbide bits indicate that efficient drilling is possible only up to a hardness of about 55 Shore; beyond this, wear increases rapidly until, at 62 Shore, the cost becomes prohibitive. Several factors affect the cutting life of the bits, including the grade of carbide used, the rake angle of the cutters, the length of cutting edges, and support of cutters. See also: *steel bit; coal-cutter pick; Shore hardness test; sintered carbide-tipped pick.* (Nelson, 1965)
tungsten carbide insert (a) A small plate or slug of tungsten carbide alloy mounted in the crown or shank of a bit or in grooves on the outside surface of a reaming shell to provide wear-resistant or rock-cutting surfaces or edges. The term is sometimes incorrectly applied to diamond-set plates of tungsten carbide alloy inset as reaming surfaces in reaming shells. (Long, 1960) (b) In mining, a slug composed of tungsten carbide alloy shaped and mounted in the bit face so that the slug acts as the cutting edge of the bit. (Long, 1960) (c) Hemispherical-ended cylinders of sintered carbide are inserted in place of the usual teeth to give 10 to 15 times the total footage and 2 to 3 times the cutting rate. However, hard rocks are drilled more economically by diamond boring. (Nelson, 1965)
tungsten direct-from-ore process An electrowinning method developed by the U.S. Bureau of Mines for producing high-quality tungsten powder directly from ore. A strong electric current separates the metal from the ore, which has been placed in solution, and deposits it as a pure powder on an electrode. Electrowon tungsten compares favorably with hydrogen-reduced tungsten.
tungstenite A trigonal or hexagonal mineral, WS_2; forms dark, lead-gray, minute, foliated or earthy scales; at the Emma Mine, Utah.
tungstic ocher *ferritungstite; tungstite.* Also called wolfram ocher.
tungstite An orthorhombic mineral, $WO_2(OH)_2$; yellow to green; in oxidized zone of tungsten deposits. Syn: *tungstic ocher; wolframine.*
tunna A Welsh term for a hoisting bucket; a bowk; a kibble. (Fay, 1920)
tunnel (a) A horizontal or inclined stone drivage for development or to connect mine workings, seams, or shafts. It may be open to the surface at one end and used for drainage, ventilation, or haulage or as a personnel egress (walking or riding) from the mine workings. See also: *tunneling.* (Nelson, 1965) (b) *crut.* (c) A leaden tube used in making sulfuric acid to connect adjoining chambers in a series. (Standard, 1964) (d) A long, narrow subterranean passageway. (e) A horizontal or nearly horizontal underground passage that is open at both ends. The term is loosely applied in many cases to an adit. An adit, if continued through a hill, would be a tunnel. Any level or drift in a mine open at one end, or which may serve for an adit. Often used as a syn. for adit; drift; gallery. See also: *adit.* (Lewis, 1964) (f) To penetrate

with or as if with a tunnel; to make a passage through or under; to make or use a tunnel; to undermine. (Webster 3rd, 1966)

tunnel blast (a) A blast effected by the detonation of great quantities of explosive, loaded in small tunnels driven into the face at the level of the quarry floor or at the level of the terrain at the foot of the slope of the deposit. This blasting method is called tunneling. (Streefkerk, 1952) (b) *heading blast.*

tunnel blasting A method of heavy blasting in which a heading is driven into the rock and afterwards filled with explosives in large quantities. This is similar to a borehole on a large scale, except that the heading is usually divided into two parts on the same level at right angles to the first heading. This forms a T, the ends of which are filled with explosives and the intermediate parts of which are filled with inert material like an ordinary borehole. Similar to gopher hole blasting. See also: *gopher hole blasting.* (Fay, 1920)

tunnel borer Any boring machine for making a tunnel; often a ram armed with cutting faces operated by compressed air. (Standard, 1964)

tunnel carriage A rapid tunneling procedure, consisting of a combined drill carriage and manifold for water and air so that immediately when the carriage is at the face, drilling may commence with no lost time for connecting up, waiting for drill steels, etc. The air is supplied at pressures of 95 to 100 psi (655 to 690 kPa). (Nelson, 1965)

tunnel claim When a lode or vein is discovered in a tunnel, the tunnel owner is called upon to locate the area containing the vein or lode on the surface and thus create a mining claim. (Ricketts, 1943)

tunnel column A heavy bar used for mounting machine drills in large drifts or tunnels, and usually holding two machines. (Fay, 1920)

tunnel excavation Excavation carried out completely underground and limited in width and height. (Carson, 1961)

tunnel face The working face in an excavation or tunnel or other working place from which driving is carried out. (Fraenkel, 1953)

tunneling The operation of excavating, driving, and lining tunnels. (Nelson, 1965)

tunnel kiln A long tunnel-shaped furnace through which the charge is generally moved on cars, passing progressively through zones in which the temperature is maintained for preheating, firing, and cooling. (ARI, 1949)

tunnel-kiln operator One who controls the operation of a tunnel kiln in which bricks are fired, and a preheating chamber in which bricks are heated prior to firing and after drying. (DOT, 1949)

tunnel lining (a) The timber, brick, concrete, or steel supports erected in a tunnel to maintain dimensions and safe working conditions. See also: *steel tunnel support; lining.* (Nelson, 1965) (b) *ring; tunnel support.*

tunnelman In anthracite coal mining, one who drives a tunnel in rock from one coal seam to another or through a fault (the movement of the earth having separated a once continuous seam into two sections). (DOT, 1949)

tunnel miner A miner experienced in the use and handling of rock drills and shovel loaders, and in tunnel-blasting methods. Such a miner is wholly employed on tunneling and is usually paid a fixed rate per shift with perhaps a bonus payment for high rates of tunnel advance. (Nelson, 1965)

tunnel right A right to enter upon and occupy a specific piece of ground for the purpose of carrying out work in a tunnel and extracting waste rock or earth necessary to complete the tunnel, and making such use after completion as may be necessary to work the mining ground or lode owned by the party running the tunnel. By implication, the grant of such a right carries with it every incident and appurtenant thereto, including the right to dump the waste rock at the mouth of the tunnel on the land owned by the grantor at the time of the conveyance of the tunnel right, such right or easement being necessary for the full and free enjoyment of the tunnel right. (Ricketts, 1943)

tunnel set Timbers of sufficient strength to support the roof of the tunnel. They are sometimes set upon sills and usually capped with short crosspieces. (Fay, 1920)

tunnel shaft A shaft sunk, such as in a hill, to meet a horizontal tunnel. Also called tunnel pit. (Standard, 1964)

tunnel site (a) An area for a tunnel. The locator of a tunnel site is given the right to all veins cut by the tunnel within 3,000 ft (915 m) of its portal, and 1,500 ft (457 m) on the strike of each blind vein cut; this length may be all on one side of the tunnel or divided as desired. The veins must be blind lodes not previously known to exist. (Lewis, 1964) (b) There is no distinction between a tunnel claim under which a tunnel is run for the development of veins or lodes already located, and one where a tunnel is projected for blind veins or lodes. (Ricketts, 1943)

tunnel support *steel tunnel support; tunnel lining.*

tunnel system A method of mining in which tunnels or drifts are extended at regular intervals from the floor of the pit into the orebody. The extension of the drift beyond the working face is made great enough to facilitate the handling of several cars at a time. The ore is mined above the drift level, and the cars are loaded by lifting short boards that span an opening, through the lagging on and above the centerline of the drift. The method avoids the construction of raises and chutes and facilitates the filling of the cars. (Fay, 1920)

tup The ram or monkey, or falling weight, of a piledriver, drophammer, etc.; specif., the heavy head of a steam hammer in which the upper pallet is secured. (Webster 2nd, 1960)

turanite An orthorhombic(?) mineral, $Cu_5(VO_4)_2(OH)_4$(?); weakly radioactive; green; forms reniform crusts and spherical concretions having a radial fibrous structure with other vanadium and uranium minerals; in cavities in limestone; at Tyuya Muyun, Fergana, Turkistan.

turbid Stirred up or disturbed, such as by sediment; not clear or translucent, being opaque with suspended matter, such as of a sediment-laden stream flowing into a lake; cloudy or muddy in physical appearance, such as of a feldspar containing minute inclusions. (AGI, 1987)

turbidimeter An instrument for measuring or comparing the turbidity of liquids in terms of the reduction in intensity of a light beam passing through the medium. See also: *transmissometer.* (AGI, 1987)

turbidimetry Measurement of the amount of suspended or slow-settling matter in a liquid; the measurement of the decrease in intensity of a light beam passed through a medium. Cf: *nephelometry.* (AGI, 1987)

turbidite A sediment or rock deposited from, or inferred to have been deposited from, a turbidity current. It is characterized by graded bedding, moderate sorting, and well-developed primary structures. (AGI, 1987)

turbidity (a) The state, condition, or quality of opaqueness or reduced clarity of a fluid, due to the presence of suspended matter. (AGI, 1987) (b) A measure of the ability of suspended material to disturb or diminish the penetration of light through a fluid. (AGI, 1987)

turbidity current A density current in water, air, or other fluid, caused by different amounts of matter in suspension, such as a dry-snow avalanche or a descending cloud of volcanic dust; specif. a bottom-flowing current laden with suspended sediment, moving swiftly (under the influence of gravity) down a subaqueous slope and spreading horizontally on the floor of the body of water, having been set and/or maintained in motion by locally churned- or stirred-up sediment that gives the water a density greater than that of the surrounding or overlying clear water. Such currents are known to occur in lakes, and are believed to have produced the submarine canyons notching the continental slope. They appear to originate in various ways, such as by storm waves, tsunamis, earthquake-induced sliding, tectonic movement, oversupply of sediment, and heavily charged rivers in spate with densities exceeding that of sea-water. The term is applied to a current due to turbidity, not to one showing that property. Syn: *suspension current.* (AGI, 1987)

turbidity size analysis A kind of particle-size analysis based upon the amount of material in turbid suspension, the turbidity decreasing as the particles settle. (AGI, 1987)

turbine pump A pump with a shrouded impeller and receiving the water at its center. A diffusion ring containing vanes surrounds the impeller and directs the impeller discharge into a circular casing, which delivers into the eye of the next impeller in series. The diffusion ring converts the high-velocity discharge of the impeller into pressure head. The turbine pump is widely used in mines. See also: *diffuser chamber.* (Nelson, 1965)

turboaxial fan An axial flow fan with a turbine rotor-type impeller. (BS, 1963)

turbocompressor The type of machine commonly installed at a colliery today where a large volume of compressed air is required. A single unit can deliver 10,000 ft^3/min (283 m^3/min) or more of free air, and the floor space occupied is a minimum for these capacities. It is also ideally suited for direct drive by a steam turbine, and this combination is commonly found at collieries. The compressor consists essentially of a number of impellers keyed to a shaft and running in a fixed casing with specially shaped passages. Each impeller is in the form of a hollow wheel, the two sides being united by curved vanes. See also. *air-conditioning process.* (Mason, 1951)

turbodrill In rotary drilling, a drill bit that is directly rotated by a turbine attached to the drill pipe at the bottom of the hole and driven by drilling mud pumped under high pressure. It was developed in the former U.S.S.R. for drilling deep oil wells. (AGI, 1987)

turbodrilling A system of drilling in which the bit is directly driven by a turbine at the bottom of the hole. (BS, 1963)

turbulence *turbulent flow.*

turbulent flow (a) Water flow in which the flow lines are confused and heterogeneously mixed. It is typical of flow in surface-water bodies. Cf: *laminar flow.* Syn: *tortuous flow; turbulence.* (AGI, 1987) (b) Fluid motion in which random motions of parts of the fluid are superimposed upon a simple pattern of flow. All or nearly all fluid flow displays some degree of turbulence. Opposite of streamline flow. (Hunt, 1965) (c) A fluid flow in which there is an unsteady motion of the particles, the motion at a fixed point being inconstant. Turbulent flow occurs at a speed above the critical velocity of Reynolds. Also called tortuous flow; sinuous flow; eddy flow. (Nelson, 1965) (d)

When air flows over roughnesses on the sides of the airway or passes obstructions at over a certain velocity, eddies are set up in the air and its flow becomes turbulent. Opposite of laminar flow. (Spalding, 1949) (e) When the fluid particles are moving in directions other than in a straight line parallel to the axis of the pipe or duct. (Strock, 1948)

turbulent resistance Resistance that causes vortices and eddies to form behind a moving particle because of the rapid displacement of the liquid when the body moves through it. Cf: *viscous resistance.* (Newton, 1959)

turf (a) Same as peat. There are several varieties, as white, brown, black, stone, gas, or candle turf. (Fay, 1920) (b) Sod, the upper strata of topsoil filled with the roots of grass and other small plants.

turgite An iron ore and sandstone cement consisting of hematite with adsorbed water. It is fibrous and red in mass with an orange tint where powdered. Also spelled turjite. Syn: *hydrohematite.*

Turkey slate A whetstone or honestone. See also: *Turkey stone.* (Fay, 1920)

Turkey stone A very fine-grained siliceous rock, containing up to 25% calcite, quarried in central Turkey and used as a whetstone; novaculite. Syn: *Turkey slate. novaculite; turquoise.* (AGI, 1987)

turkis A turquoise. (Standard, 1964)

turmeric paper Paper impregnated with an extract of turmeric. Used as a test for alkaline substances, which turn it from yellow to reddish-brown, and for boric acid, which turns it red-brown. (Webster 3rd, 1966)

turn (a) A curve into a pillar. (BCI, 1947) (b) The time or period during which coal, etc., is raised from a mine. Also called run; shift. (c) To open rooms, headings, or chutes off from an entry or gangway. (d) The number of cars allowed each miner. Good turn means many cars for each miner. (Fay, 1920) (e) To draw or wind coal up a shaft or up an inclined plane to the surface. (Fay, 1920) (f) Curved tramrails, often made of cast iron, laid round a corner or turn. (Fay, 1920) (g) To set undried bricks on edge to facilitate drying. (Standard, 1964)

turn angles To measure the angle between directions with a surveying instrument. (Nichols, 1976)

turn bat A wooden stick used in turning the tongs that hold a bloom under the hammer. (Fay, 1920)

turn bolt A bolt turned in a lathe to a close tolerance and used in steel-to-steel connections. (Hammond, 1965)

turned vertical shaft A shaft sunk vertically in the hanging wall block until it intersects a reef, after which it is sunk down at an angle in the footwall parallel to the reef. This unusual practice is sometimes adopted on the Rand because it enables the mine to become productive at an earlier stage. See also: *incline shaft.* (Nelson, 1965)

turnerite A yellowish-brown variety of monazite. (Standard, 1964)

turnhouse (a) A point where workings turn from a crosscut to a level along the lode. (Gordon, 1906) (b) The first cutting on a lode after it is cut in a crosscut. Syn: *house.* (Fay, 1920)

turning effect *moment of force.*

turning over and packing shift On mechanized longwall faces, the shift during which face conveyors are moved over, and the operations of ripping, packing, and drawing supports from the wastes are performed. (Mason, 1951)

turning point (a) A surveying point on which a level rod is held, after a foresight has been made on it, and before the differential-leveling instrument is moved to another station so that a backsight may be made on it to determine the height of instrument after the resetting; a point of intersection between survey lines, such as the intervening point between two bench marks upon which rod readings are taken. It is established for the purpose of allowing the leveling instrument to be moved forward (alternately leapfrogging with the rod) along the line of survey without a break in the series of measured differences of elevation. Abbrev: TP. (AGI, 1987) (b) A physical object representing a turning point, such as a steel pin or stake driven into the ground. (AGI, 1987)

turning vane Curved strips placed in a sharp bend or elbow in rectangular duct to direct the air around the bend in a streamlined flow. (Strock, 1948)

turn keeper *motor boss.*

turnout (a) The branching off of one rail track from another. (Nelson, 1965) (b) A contrivance for passing from one track to another. (Zern, 1928) (c) A siding or bypass in an underground haulageway. (d) A switch on a mine railroad. (Korson, 1938)

turnover (a) The distance the conveyor is advanced during each cycle of operations; i.e., approx. the depth of machine cut. See also: *conventional machine mining.* (Nelson, 1965) (b) A device used to rotate an object through approx. 180° so that its carrying surface is changed to the opposite side. (c) *move-up.*

turn pulley A sheave fixed at the inside end of an endless or tail-rope hauling plane, around which the rope returns. See also: *tail sheave.* (Zern, 1928)

turns A term used with any device used to change the direction of a shaker conveyor trough line; e.g., curved trough turn, adjustable angle turn, right angle turn, etc. The angle turn corresponds to the bell crank drive in principle of operation. (Jones, 1949)

turnsheet *flat sheet.*

turquoise (a) A triclinic mineral, $1[CuAl_6(OH)_8(PO_4)_4{\cdot}4H_2O]$ with Fe replacing Al toward chalcosiderite; forms waxy blue-green reniform masses having a botryoidal surface, rarely with minute crystals; occurs in arid regions where surface water acted on aluminous rock; may be a gemstone. Also spelled turquois. Syn: *Turkey stone; calaite.* (b) The mineral group aheylite, chalcosiderite, coeruleolactite, faustite, planerite, and turquoise.

turret coal cutter A coal cutter in which the horizontal jib can be adjusted vertically to cut at different levels in the seam, for example, an overcut. The center of gravity of such a machine makes it top heavy and less stable than the ordinary undercutter. See also: *overcutting machine; universal coal cutter.* (Nelson, 1965)

turret jib A vertical rotating jib fitted with cutter picks and driven from the end sprocket of the bottom jib of a coal cutter. The turret jib is satisfactory in seams where the coal parts readily from the roof and is not too hard. See also: *mushroom jib; curved jib.* (Nelson, 1965)

turtle back A name for chlorastrolite (pumpellyite), esp. the green variety with patches of color; also, turquoise matrix or variscite matrix.

turtle stone *septarium.*

turtlestones Large, nodular concretions found in certain clays and marls. In form, they have a rough resemblance to turtles, and this appearance is increased by their being divided into angular compartments by cracks filled with spar, reminding one of the plates on the shell of a turtle. (Fay, 1920)

Tuscarora quartzite An important source of raw material for silica refractories. A typical analysis is 97.8% SiO_2, 0.9% Al_2O_3, 0.7% Fe_2O_3, and 0.4% alkalies. (Dodd, 1964)

tusiite A former name for calciocopiapite.

tutenag (a) A white alloy, resembling German silver, used in making tableware, etc., with varying proportions of copper, zinc, nickel, and sometimes a little lead or iron. (Standard, 1964) (b) Zinc or spelter, esp. that from China and the East Indies. (Standard, 1964)

Tutogen A foam-producing agent used in fire extinguishers. (Sinclair, 1958)

tutwork Sometimes used for piecework or contract work. (Nelson, 1965)

tuxtlite A pyroxene mineral, $(Na,Al,Ca,Mg)Si_2O_6$; midway in composition between diopside and jadeite; pea-green; massive; at Tuxtla, Mexico. Formerly called diopside-jadeite. Also spelled tuxlite.

tuyère A tube or opening in a metallurgical furnace through which air is blown as part of the extraction or refining process. In a blast furnace, the tuyeres are water-cooled metal tubes which pass through the refractory lining of the bosh (tube). (Dodd, 1964)

tuyère arch An arch in a blast furnace to admit a tuyère. See also: *tuyère.* (Standard, 1964; Fay, 1920)

tuyère brick A refractory shape containing one or more holes through which air and other gases are introduced into a furnace. (ARI, 1949)

tweel A counterweighted furnace door, opening vertically. Also spelled tuille. (ASTM, 1994)

twig (a) *divining rod.* (b) Thin strip of plastic fire clay used in ceramic modeling, esp. in imitation basketwork. (Standard, 1964)

twill cloth Weave used in screens and filters, in which two or more warp threads interweave one wood thread. (Pryor, 1963)

twin axis The crystal axis about which one individual of a twin crystal may be rotated (usually 180°) to bring it into coincidence with the other individual. It cannot be coincident with the axes of twofold, fourfold, or sixfold symmetry. Syn: *twinning axis.*

twin colors *dichroic colors.*

twin-core shot-firing cable *shot-firing cable.*

twin crystal A composite of two or more crystal individuals having a definite crystallographic relationship to each other. The orientation of one individual may be the mirror image of the other across a twin plane, or an orientation that can be derived by rotating the twin portions about a twin axis, or some other rational twin law. Twinned individuals in a twin crystal commonly show reentrant angles between crystal faces or on cleavage planes. A twin crystal may exhibit symmetry higher than that of its crystal individuals.

twin crystals Crystals in which one or more parts, regularly arranged, are in reverse position with reference to the other part or parts. They often appear externally to consist of two or more crystals symmetrically united, and sometimes have the form of a cross or star. They also exhibit the composition in the reversed arrangement of part of the faces, in the striae of the surface, and in re-entering angles; in certain cases, the compound structure can only be surely detected by an examination in polarized light. (Fay, 1920)

twin entry A pair of parallel entries, one of which is an intake air course and the other a return air course. Rooms can be worked from both entries. Often called double entry. (Fay, 1920)

twin laminae The laminae or thin plates in repeated or polysynthetic twins. Cf: *polysynthetic twinning; repeated twinning.*

twin law A statement or statements of the symmetrical relationships between the members of a twin crystal, e.g., the twin plane or the twin axis that resolves one crystal individual into congruity with another in a twin crystal. The twin law cannot be an element of symmetry of the point group of the twin parts, although it may be an element in a point group with higher symmetry in the same crystal system.

twinning axis *twin axis.*

twinning law The special and characteristic method according to which twin crystals of any mineral are formed. (Fay, 1920)

twinning plane In a twin crystal, a plane normal to the twinning axis. (Fay, 1920)

twinoriascope A type of instrument used to detect and mark twinning and determine the sense of orientation in etched sections. (Am. Mineral., 1947)

twinoscope An instrument employing a directed beam of light used to examine etched wafers for twinning. (Am. Mineral., 1947)

twin packer A packer designed so that a borehole can be sealed simultaneously at two separated points. (Long, 1960)

twin plane In a twin crystal, a plane through which one twin individual forms a mirror image of the other, or the plane at right angles to the axis about which one individual is rotated with respect to the other. The twin plane is commonly the composition plane across which the individuals are joined. With rare exceptions, a twin plane is a possible crystal face for the crystal individuals.

twist conveyor An L-shaped conveyor in which the carrying surface and guard gradually exchange their functional duties.

twist drill A drill made by twisting a length of steel of rectangular or oval section into a spiral form, hence the term twist drill. Many hand-operated coal drills are of this type, and the rotation of the drill spiral removes the cuttings from the hole. See also: *auger; coal auger.* (Nelson, 1965)

twisted-loop splice Splice made by holding the bared wires side by side. Half of their length is bent back to form a loop at the end. The loop is then twisted around the main shank of wire. (Carson, 1961)

twister operator In the asbestos products industry, one who twists together two or more strands of wire and asbestos yarn for use in weaving asbestos products, such as brake linings. (DOT, 1949)

twisting force A force, such as the force on the shaft of a rotating motor. (Morris, 1958)

twistoff The breaking off of a member of the drill string, caused by excessive torsional stress. (Long, 1960)

two-circle goniometer A device permitting rotation of a small crystal about two orthogonal axes for optical observation. Cf: *goniometer.*

two-component explosives Consist of two or more unmixed, commercially manufactured, prepackaged chemicals, including oxidizing chemicals, flammable liquids, or solids that are not independently classified as explosives. When combined, however, the mixture is classified as an explosive and is stored, transported, and handled as an explosive. (Cote, 1991)

two-cone bit *roller rock bit.*

two-fan auxiliary ventilation An arrangement, using two auxiliary fans, for ventilating a mine tunnel or hard heading. It consists of an exhausting fan with rigid ducting to within about 100 ft (30 m) of the face, and a forcing fan using a flexible duct discharging air about 20 ft (6 m) from the face. The ducts of the two units overlap by at least 30 ft (9 m) to minimize the recirculation of air. The air delivered by the forcing fan does not exceed about one-third of that removed by the exhaust fan. See also: *overlap auxiliary ventilation; auxiliary ventilation; recirculation of air; reversible auxiliary ventilation.* (Nelson, 1965)

twofold A symmetry axis requiring two repetitions to complete 360° and return to identity. Syn: *diad.*

two-high mill Contains two horizontal rolls, one above the other. In some two-high mills the direction of rolling can be reversed, and these are known as reversing mills; i.e., when a piece has passed through the rolls, the rolls are stopped and then rotated in the opposite direction, thus imposing another pass on the steel, the operation being repeated until the desired reduction is attained. Between passes, adjustment is made to the height of the top roll, and/or the piece is moved sideways by means of manipulators, to be in line with other grooves in the rolls. (Osborne, 1956)

two-hinged arch A rigid frame hinged at both supports. It may have an arched or rectangular form. (Hammond, 1965)

two intakes The provision of two intake airways, generally side by side, to a ventilating area of a mine. (Nelson, 1965)

two-jaw chuck A chuck equipped with two movable clamping or holding devices by means of which the motion of the chuck is imparted to the drill rods. (Long, 1960)

two-leg sling A sling having two chains or ropes which hang from a thimble. See also: *single sling.* (Hammond, 1965)

two-liquid differential manometer Consists of two concentric glass tubes, each expanded into a large bulb at the upper end. The lower end of the outer tube is sealed, and the inner tube reaches nearly to the bottom of the outer tube. Two liquids are used, and the movement of the interface between the two liquids down the central tube is used as an index against which the change in pressure is measured. See also: *manometer.* (Roberts, 1960)

two-mica granite A granite containing both dark mica (biotite) and light mica (muscovite). This rock was called true granite by Rosenbusch and binary granite by Keyes. Cf: *aplogranite.* (AGI, 1987)

two-part line A single strand of rope or cable doubled back around a sheave so that two parts of it pull a load together. (Nichols, 1954)

two-piece set A set of timbers consisting of a cap and a single post. If the ground is loose and must be supported over the side or back, lagging, commonly of 2-in (5.1-cm) boards, is used. These boards extend from the center line of the post or cap to the middle of the next post or cap. If they are placed touching each other, such an arrangement is called tight lagging; if a few inches apart (which depends on the nature of the ground to be held back), it is called open lagging. See also: *timber set.* (Lewis, 1964)

two-plane idler A troughing idler in which the troughing roll shafts are in a vertical plane separate from but parallel to a vertical plane through the shaft of the center roll or rolls. (NEMA, 1956)

two-process washer A method of cleaning raw coal in which the material from 8 to 1/16 in (203 to 1.6 mm) in size is treated in heavy-medium washers, and the fines below 1/16 in are treated by froth flotation. (Nelson, 1965)

two-speed differential A differential having a high-flow gearshift between the drive shaft and the ring gear. (Nichols, 1976)

two-stage compression Air compression carried out in two stages as is usual for pressures exceeding 60 psi (414 kPa) or for outputs greater than 100 hp (74.6 kW). See also: *intercooler.* (Hammond, 1965)

two-stage hoisting Deep-shaft hoisting with two winders, one at the surface and the other at middepth in the shaft. The surface engine winds minerals from the middepth pocket, and the other winds from the pit bottom to the middepth point. The arrangement is often adopted in turned vertical shafts. See also: *tandem hoisting.* (Nelson, 1965)

two-step control system In flotation, a system in which the manipulated variable alternates between two predetermined values. (Fuerstenau, 1962)

two-stroke cycle A working cycle of a piston in an internal combustion engine consisting of two strokes, in which the piston during the first stroke compresses the fuel mixture on one side while receiving the expansive thrust of previously compressed gases on the other side, and during the second draws in a fresh charge on one side while expelling burnt gases on the other. (Webster 3rd, 1966)

two-way ram A hydraulic cylinder in which fluid can be supplied to either end, so the piston can be moved by power in two directions. Syn: *double-acting ram.* (Nichols, 1954)

tychite An isometric mineral, $Na_6Mg_2(CO_3)_4(SO_4)$, having iron replacing magnesium toward ferrotychite; isomorphous with northupite, $Na_6Mg_2(CO_3)_4Cl_2$, with which it occurs; forms small white octahedra; at Borax Lake, San Bernadino County, CA.

tying across and behind Systematic exploration by a mine rescue team of all intersecting and adjacent passageways so that the team is never forward (toward the working face) of an accessible, unexplored area. (MSHA, 1992)

Tyler sieve *sieve.*

Tyler Standard series The series of carefully woven, square-mesh wire screens most commonly used in the United States in screening ores. (Newton, 1959)

tymp Eng. A horizontal roof timber in a coal mine; a cap or lid. (Standard, 1964)

tymp stone A large clay plug filling an open space in the front jackets of a smelting furnace, through which the taphole passes. (Standard, 1964)

tyndallascope *tyndallometer.*

Tyndall effect The scattering or reflection of a strong beam of light by suspended colloids; no such scattering or reflection comes from true solutions.

tyndallometer An instrument that measures the intensity of the light scattered at an angle from the incident beam by a dust cloud. It correlates well with the concentration determined by a thermal precipitator and surface area calculated from such a count. It needs to be calibrated for each type of dust against the thermal precipitator. Syn: *tyndalloscope.* See also: *dust sampling.* (Nelson, 1965)

tyndalloscope *tyndallometer.*

type (a) *rock type.* (b) A coal classification based on the constituent plant materials. Cf: *rank; grade.* (AGI, 1987) (c) Those differences in coals that are due to variations in the kind of plant material of which the coal is composed, whereby such varieties as common banded coal, cannel coal, algal coal, and splint coal are produced. (AGI, 1987) (d) A kind, particularly in petrology (rock type); either general (for example, basalt is a rock type) or particular (for example, a particular basalt from a particular locality is a unique type specified by a description). (Challinor, 1964)

type-D drift indicator A single-shot borehole-surveying instrument utilizing photographic paper on which is recorded the compass bearing and inclination of the course of a borehole. The type-D instrument, when mounted in a special thin-walled protective container, is small enough to be used in an AX-size hole. (Long, 1960)

type-M drift indicator A single-shot borehole-surveying instrument that records the compass bearing and inclination of the course of a borehole through the action of a strong beam of light directed through the plumb bob onto a light-sensitive paper disk. It is similar to, but larger than, a type-D drift indicator. (Long, 1960)

type of coal (a) The concept type provides a means for classifying standard varieties of coal microscopically on the basis of simple proportions of anthraxylon or anthraxylon and opaque attritus, including their subdivision into banded and non-banded coals. (IHCP, 1963) (b) A type of coal is a variety initially determined by the nature of the ingredient matter, the conditions of deposition, and the extent of operation of the first or biochemical process of coal making. (IHCP, 1963)

type section The originally described sequence of strata that constitute a stratigraphic unit. It serves as an objective standard with which spatially separated parts of the unit may be compared, and it is preferably in an area where the unit shows maximum thickness and is completely exposed (or at least shows top and bottom). (AGI, 1987)

type specimen A specimen or individual designated as type of a species or lesser group and serving as the final criterion of the characteristics of that group. (Webster 3rd, 1966)

type-W drift indicator A mechanical single-shot borehole-surveying instrument for use where exceptionally high temperatures are encountered in a hole. It records the compass bearing and inclination of a borehole by making a dot on a special paper by means of a plumb bob, incorporating a depressible stylus. (Long, 1960)

typhonic rocks Rocks that have come from the depths of the Earth; i.e., plutonic and eruptive rocks. (Fay, 1920)

typomorphic mineral A mineral that is typically developed in only a narrow range of temperature and pressure. Cf: *index mineral.*

typrite (a) A variety of fergusonite found near Arendal, Norway. (Fay, 1920) (b) *sipylite.*

tyre valve Adjustable annular ring, made of plastic, used to control the aperture area at the apex of a hydrocyclone. (Pryor, 1963)

tyrolite An orthorhombic mineral, $Ca_2Cu_9(AsO_4)_4(OH)_{10}{\cdot}10H_2O$; forms green crystal aggregates having foliated micaceous structure. Formerly called trichalcite.

tyuyamunite An orthorhombic mineral, $Ca(UO_2)_2(VO_4)_2{\cdot}8H_2O$; soft; waxy; yellow; fluoresces yellow-green; in secondary encrustations in limestones, sandstones, or concentrated by organic matter; associated with malachite, ferghanite, turanite, barite, calcite, carnotite, and vanadium minerals; widely distributed in the Colorado Plateau area; source of uranium and vanadium. Formerly called calciocarnotite.

U

U-bit A popular type of rotary bit used in British mining practice. It has two cutting legs, although some American and German bits employ three legs. A core is formed between the legs, which is broken off as cutting proceeds; some bits have a core-cutting device consisting of a tungsten carbide tip in the center. (Fraenkel, 1953)

Udden grade scale A logarithmic grade scale devised by Johan A. Udden (1859-1932), U.S. geologist; it uses 1 mm as the reference point and progresses by the fixed ratio of 1:2 in the direction of decreasing size and of 2:1 in the direction of increasing size, such as 0.25, 0.5, 1, 2, 4. See also: *Wentworth grade scale*. Syn: *grade scale*. (AGI, 1987)

ugrandite The group of calcium garnets **u**varovite, **gr**ossular, **andr**adite, goldmanite, hibschite, kimzeyite, and schorlomite

uhligite An isometric mineral, $(Ca,Ti,Al,Zr)_2O_3$; forms black octahedra in a nepheline syenite near Lake Magad, Tanzania.

uigite A discredited term equal to thomsonite. (Am. Mineral., 1945)

uintaite A variety of natural asphalt occurring in the Uinta Valley, Utah, as rounded masses of brilliant black solid hydrocarbon. Syn: *gilsonite; mineral rubber*. (CMD, 1948)

ulexite A triclinic mineral, $NaCaB_5O_6(OH)_6{\cdot}5H_2O$; soft; forms silky white, saline crusts and masses of extremely fine acicular crystals; in saline lake deposits as in Nevada and Chile. Also called cotton ball; boronatrocalcite; natronborocalcite; natroborocalcite.

ullmannite A triclinic mineral, NiSbS; cobaltite group; pseudocubic; metallic; steel-gray to silver-white; in veins; a source of nickel. Also called nickel-antimony glance.

Ullrich magnetic separators These machines have powerful electromagnets in a wedge section. The material is treated on rolls on which magnetism is induced; they consist of alternate disks of soft iron and some nonmagnetic material. The ore is fed over the first roll, which removes the most magnetic material, and the tailings go on to the second, which is weaker, where a second separation is made. (Liddell, 1918)

ulmain A kind of euvitrain that consists completely of ulmin but that is not precipitated from solution. Cf: *collain*. (AGI, 1987)

ulmification The process of peat formation. (Tomkeieff, 1954)

ulmin brown *Vandyke brown*.

ulminite (a) A maceral of brown coal within the huminite group, consisting of gelified plant-cell walls (ICCP, 1971). (AGI, 1987) (b) A variety of euvitrinite characteristic of ulmain and consisting of gelified but not precipitated plant material. Cf: *collinite*. (AGI, 1987)

ulrichite A hypabyssal rock composed essentially of large phenocrysts of alkalic feldspar, sodic pyroxene, amphibole, and nepheline with smaller phenocrysts of accessory olivine. Feldspar, pyroxene, and amphibole recur in the groundmass. A porphyritic variety of olivine-bearing phonolite. Syn: *uraninite*. (AGI, 1987)

ultimate analysis (a) The determination of the elements contained in a compound, as distinguished from proximate analysis, which is the determination of the compounds contained in a mixture. (Standard, 1964) (b) In the case of coal and coke, the determination of carbon and hydrogen in the material, as found in the gaseous products of its complete combustion, the determinations of sulfur, nitrogen, and ash in the material as a whole, and the calculation of oxygen by difference. (ASTM, 1994) (c) The principal reason for the ultimate analysis of coal is for the classification of coals according to rank, although it is often used for commercial and industrial purposes when it is most desirable to know the sulfur content of coal. Also known as total analysis of coal. (Cooper, 1963)

ultimate bearing capacity The average load per unit of area required to produce failure by rupture of a supporting soil mass. See also: *bearing capacity*. (ASCE, 1958)

ultimate bearing pressure The pressure under which a foundation will settle with no increase of load. See also: *plate bearing test*. (Hammond, 1965)

ultimate CO_2 The percent of carbon dioxide that would appear in the flue gases if combustion were perfect. Varies with the fuel. (Strock, 1948)

ultimate compressive strength That point at which failure by crushing occurs. (Pryor, 1963)

ultimate elongation The percentage of permanent deformation remaining after tensile rupture, measured over an arbitrary length including the section of rupture. (Roark, 1954)

ultimately controlled variable In mineral processing, the variable whose control is the end purpose of the automatic control system. (Fuerstenau, 1962)

ultimate strength The ultimate strength of a material in tension, compression, or shear, respectively, is the maximum tensile, compressive, or shear stress that the material can sustain, calculated on the basis of the ultimate load and the original or unstrained dimensions. It is implied that the condition of stress represents uniaxial tension, uniaxial compression, or pure shear, as the case may be. (Roark, 1954)

ultimate tensile stress The load at which a test piece breaks, divided by its original area. See also: *tensile strength*. (Hammond, 1965)

ultrabasic Said of an igneous rock having a silica content lower than that of a basic rock. Percentage limitations are arbitrary; the upper limit was originally set at 44%. The term is frequently used interchangeably with ultramafic. Although most ultrabasic rocks are also ultramafic, there are some exceptions; e.g., monomineralic rocks composed of pyroxenes are ultramafic but are not ultrabasic because of their high silica content. A monomineralic rock composed of anorthite would be considered ultrabasic (SiO_2 = 43.2%) but not ultramafic. Ultrabasic is one subdivision of a widely used system for classifying igneous rocks on the basis of silica content; the other subdivisions are acidic, basic, and intermediate. Cf: *ultramafic; silicic*. (AGI, 1987)

ultrabasite A germanium-bearing variety of diaphorite from Freiberg, Saxony, Germany.

ultraflotation A recently developed process for use in fine-particle flotation. The underlying principle is the use of a finely ground (minus 325-mesh) auxiliary mineral as a carrier for the fine particles to be floated. The fine particles form a slime coating on the carrier mineral; the carrier mineral is then floated, and the fines are piggybacked into the froth. (Fuerstenau, 1962)

ultramafic Said of an igneous rock composed chiefly of mafic minerals, e.g., monomineralic rocks composed of hypersthene, augite, or olivine. Cf: *hypermelanic; ultrabasic*. (AGI, 1987)

ultramafites Collective name for igneous rocks containing 90% or more mafic minerals; includes picrites, peridotites, and pyroxenites.

ultramarine A name for synthetic lazurite; extended to related compounds. Also applied to the durable brilliant blue pigment made from its powder. Syn: *lapis lazuli*.

ultramarine yellow A lemon-yellow pigment consisting of barium chromate. (Fay, 1920)

ultrametamorphism Metamorphic processes at the extreme upper range of temperatures and pressures, at which partial to complete fusion of the affected rocks takes place and magma is produced. The term was originated by Holmquist in 1909. (AGI, 1987)

ultramicroscope A microscope in which a strong beam of light (Tyndall beam) is viewed at right angles. Individual soluble particles too small to be seen under a normal microscope then appear as bright spots against a dark background. Ultramicroscopy operates below 0.25 µm. (Pryor, 1963)

ultramylonite An ultracrushed variety of mylonite, in which primary structures and porphyroclasts have been obliterated so that the rock becomes homogeneous and dense, with little if any parallel structure. Cf: *protomylonite; pseudotachylyte*. Syn: *flinty crush rock*. (AGI, 1987)

ultrasima The supposedly ultrabasic layer of the Earth below the sima, immediately below the Mohorovicic discontinuity. (AGI, 1987)

ultrasonic drilling A vibration drilling technique that can be used in drilling, cutting, and shaping of hard materials. In this method, ultrasonic vibrations are generated by the compression and extension of a core of electrostrictive or magnetostrictive material in a rapidly alternating electric or magnetic field. The most easily assembled is a magnetostrictive transducer, and the most common magnetostrictive materials, which change in dimension when magnetized, are nickel and vanadium permandur. (Min. Miner. Eng., 1965)

ultrasonic inspection A nondestructive method of testing, based upon the fact that ultrasonic waves are reflected and refracted at the boundaries of a solid medium, from which it is possible to obtain the echoes of a wave transmitted from the surface of a test piece. In addition to being reflected from the boundary of the specimen at which they are directed, the waves are also reflected back by any flaws that lie in the path of the wave. Syn: *ultrasonic testing*. (Osborne, 1956)

ultrasonic testing *ultrasonic inspection.*

ultrasonic tests Tests in which high-frequency vibrations (inaudible to a human ear) are used to assist in determining wall thicknesses, pulp densities, etc. (Pryor, 1963)

ultrasonography A modification of the use of ultrasonic waves for the detection of internal flaws in metals. By using a persistent screen cathode ray tube and causing the echoes to brighten the trace instead of deflecting it, an ultrasonic image is produced that can be examined and interpreted like a radiograph. (Osborne, 1956)

ultraviolet (a) Of radiation, beyond the visible spectrum at its violet end; having a wavelength shorter than those of visible light and longer than those of X-rays. (Webster 3rd, 1966) (b) Relating to, producing, or employing ultraviolet radiation. Cf: *visible light*. (Webster 3rd, 1966)

ultraviolet rays Electromagnetic waves in the wavelength between visible light rays and X-rays. Ultraviolet light furnishes a quick method of finding and identifying certain metals. (Nelson, 1965)

ultrawet Alkylated monosodium benzene sulfonate, a wetting agent. (Pryor, 1963)

ulvite *ulvöspinel.*

ulvöspinel An isometric mineral, $8[Fe_2TiO_4]$; spinel group; in mafic igneous rocks as fine exsolution lamellae in magnetite. Named for the Ulvö Islands, Sweden. Syn: *ulvite.*

umangite A tetragonal mineral, Cu_3Se_2; dark red, tarnishing to violet.

umbauhobel A plow developed from the Anbauhobel machine to allow the conversion of a Lobbe Hobel to give the plow an independent drive. (Nelson, 1965)

umber A brown earth that is darker than ocher and sienna, consisting of iron oxide and oxyhydroxide with manganese oxides, clay, and lime. Highly valued as a permanent pigment, it may be used in its greenish brown natural state (raw umber) or in the dark or reddish brown calcined state (burnt umber). Cf: *ocher; sienna.*

umbrella Protective hood over a hoisting cage in a mine shaft. (Pryor, 1963)

umohoite A monoclinic and orthorhombic mineral, $UO_2MoO_4{\cdot}4H_2O$; black to blue-black; secondary; the only known uranium mineral to contain molybdenum.

umpire An assay made by a third party to settle a difference found in the results of assays made by the purchaser and the seller of ore. See also: *control assay*. (Fay, 1920)

U.M. plate Universal mill plate, or plate that is rolled to width by vertical rolls as well as being rolled to thickness by horizontal rolls. (Osborne, 1956)

unakite An epidote-rich granite, which also contains pink orthoclase, quartz, also minor opaque oxides, apatite, and zircon. The name is derived from the type locality, the Unaka Range, Great Smoky Mountains, in eastern Tennessee. (AGI, 1987)

unbalanced cutter chain A cutter chain that carries more picks along the bottom line than the topline. Most chains for cutting at floor level are unbalanced to assist in keeping the jib down. See also: *balanced cutter chain.* (Nelson, 1965)

unbalanced hoisting The method of hoisting in small one-compartment shafts where only one cage is in operation, as opposed to balanced winding. (Nelson, 1965)

unbalanced shothole A shothole in which the explosive charge breaks down the coal at the back of the machine cut while leaving the front portion standing or in large blocks. This may happen with a deep bottom cut in a thin seam where the vertical distance from the explosive to the inner end of the cut is shorter than the horizontal distance to the exposed face of the seam. (Nelson, 1965)

unchuck To disengage the drill chuck from the drill stem. (Long, 1960)

unclassified excavation Excavation paid for at a fixed price per yard, regardless of whether it is earth or rock. (Nichols, 1976)

unconfined compression appliance A portable appliance for carrying out uniaxial compression tests at a site. Syn: *unrestrained compression apparatus.* (Nelson, 1965)

unconfined compression test A special condition of a triaxial compression test in which no confining pressure is applied. See also: *triaxial compression test; crushing test.* (AGI, 1987)

unconfined compressive strength *compressive strength.*

unconformability The quality, state, or condition of being unconformable, such as the relationship of unconformable strata; unconformity. See also: *unconformity.* (AGI, 1987)

unconformable Said of strata or stratification exhibiting the relation of unconformity to the older underlying rocks; not succeeding the underlying rocks in immediate order of age or not fitting together with them as parts of a continuous whole. In the strict sense, the term is applied to younger strata that do not conform in position or that do not have the same dip and strike as those of the immediately underlying rocks. Also, said of the contact between unconformable rocks. Cf: *conformable.* Syn: *discordant.* (AGI, 1987)

unconformity (a) A substantial break or gap in the geologic record where a rock unit is overlain by another that is not next in stratigraphic succession, such as an interruption in the continuity of a depositional sequence of sedimentary rocks or a break between eroded igneous rocks and younger sedimentary strata. It results from a change that caused deposition to cease for a considerable span of time, and it normally implies uplift and erosion with loss of the previously formed record. (AGI, 1987) (b) The structural relationship between rock strata in contact, characterized by a lack of continuity in deposition, and corresponding to a period of nondeposition, weathering, or esp. erosion prior to the deposition of the younger beds, and often marked by absence of parallelism between the strata; strictly, the relationship where the younger overlying stratum does not conform to the dip and strike of the older underlying rocks, as shown specif. by an angular unconformity. Syn: *unconformability; transgression.* (AGI, 1987)

unconsolidated strata Rocks consisting of loosely coherent or uncemented particles, whether occurring at the surface or at depth.

unconsolidated surface deposits Surface deposits such as moss, peat, sand, gravel, silt, or mud. (BS, 1963)

unconventional mineral deposit A mineral deposit of such unusual grade, mineralogy, or geologic setting that experienced mining personnel would not consider it to be similar to any known deposit type. (Barton, 1995)

unctuous Greasy or soapy to the touch, as certain magnesian minerals. (Standard, 1964)

uncut (a) A diamond the original shape of which has not been altered artificially. (Long, 1960) (b) Unadulterated.

undation theory A theory proposed by Van Bemmelen (1933) that explains the structural and tectonic features of the Earth's crust by vertical upward and downward movements caused by waves that are generated by deep-seated magma. (AGI, 1987)

under *undermanager.*

underboom sprays Sprays located on the rear corners of the shovel at the sides of the continuous miner and aimed towards the front of the gathering arms to suppress dust by wetting the coal. (SME, 1992)

underbreak Rock that remains unbroken inside the neat lines in a tunnel or shaft after firing a round of explosive shots. Cf: *overbreak.* (Nelson, 1965)

underbreaking *underhand stoping.*

underburden Insufficient burden of rock in relation to the explosive charge, resulting in a blown-out shot or a premature shot through shock of a neighboring charge of a blast pattern, often yielding less work than expected.

undercast (a) An air crossing in which one airway is deflected to pass under the other. (BS, 1963) (b) The lower airway of an air crossing. See also: *air crossing.* (BS, 1963) (c) An undercast is nothing more than an inverted overcast. Undercasts are not considered to be as efficient as overcasts, owing to the tendency of water to collect in them. Cf: *overcast.* (Kentucky, 1952)

underchain haulage Haulage in which the chains are placed beneath a mine car at certain intervals with suitable hooks that thrust against the car axle. (Stoces, 1954)

underchaining A drive is underchained when it incorporates a chain of substantially lower rating than that indicated by normal selection procedures. (Jackson, 1955)

underclay (a) A layer of fine-grained detrital material, usually clay, lying immediately beneath a coalbed or forming the floor of a coal seam. It represents the old soil in which the plants (from which the coal was formed) were rooted, and it commonly contains fossil roots (esp. of the genus *Stigmaria*). It is often a fireclay, and some underclays are commercial sources of fireclay. Syn: *underearth; seat earth; seat clay; root clay; thill; warrant; coal clay; warrant clay.* (AGI, 1987) (b) A bed of clay, in some cases highly siliceous, in many others highly aluminous, occurring immediately beneath a coal seam, and representing the soil in which the trees of the Carboniferous swamp forests were rooted. Stigmarian roots commonly occur as fossils in underclays, many of which are used as fireclays. See also: *fireclay.* (CTD, 1958)

underclay limestone A thin, dense, nodular, relatively unfossiliferous fresh-water limestone underlying coal deposits, so named because it is closely related to underclay. (AGI, 1987)

undercliff S. Wales. An argillaceous shale forming the floor of many coal seams.

underconsolidated soil deposit A deposit that is not fully consolidated under the existing overburden pressure. Not yet in equilibrium with existing physical environment. Still being compacted. (ASCE, 1958)

undercurrent A short sluice much wider than the main sluice and set on a steeper grade, generally at right angles to the main sluice. It is designed to save fine gold that does not readily settle. See also: *table.* (Lewis, 1964)

undercut (a) To remove a horizontal section of kerf in the bottom of a block of coal to facilitate its fall. See also: *underhole; undercutting.* (BCI, 1947) (b) To undermine, to hole, or to mine. To cut below or in the lower part of a coalbed by chipping away the coal with a pick or mining machine. Undercutting is usually done on the level of the floor of the mine. See also: *cut; undermine.* (Fay, 1920) (c) A machine cut along floor level in a coal seam to ease its removal by hand, machine, or shotfiring. See also: *holding.* (Nelson, 1965) (d) Excavation of ore from beneath a larger block of ore to induce settling under its own weight. (Nelson, 1965) (e) In stoping, removal either of footwall or of the lower part of a a flattish lode, bed, or seam of ore or coal, thus facilitating detachment of the portion left hanging. Method used in block caving to induce caving in. (Pryor, 1963) (f) To enlarge a drillhole at a depth that has been previously drilled.

undercut atomizer A device for passing a fine spray of compressed air and water to dilute the combustible gases and allay the coal dust in the track of a coal-cutter jib. The spray is passed through a modified whale-type jib. (Nelson, 1965)

undercut ignition The ignition of an explosive mixture of combustible gases and air in the undercut of a coal cutter due to frictional sparking. Combustible gases in dangerous quantities often exist in a machine undercut. See also: *whale-type jib.* (Nelson, 1965)

undercut quarry A quarry in which the walls slant outward (overhang the working face) so as to make the floorspace wider with increasing depth. (Fay, 1920)

undercutter (a) In salt mining, an electrically driven machine somewhat like a gigantic chain saw. It has a long, thin horizontal bar, about which revolves an endless chain with cutting bits. The most common type is an adaptation of the shortwall coal cutter, a drag-type machine with continuous pick-filled chains to cut at the floor or bottom of the seam. It can make a rapid, continuous cut across the entire width of the face. (Kaufmann, 1960) (b) *machineman.*

undercutting (a) The process of cutting under the face of a coal seam with a machine so the coal can be shot down readily. (b) A quarrying method intermediate between open pit and adit. Channel cuts, or separations made by wire saws or other means along the quarry walls, are slanted outward; thus, the floorspace is enlarged gradually. Wings or buttresses of stone may be left at intervals for wall support. (c) The making of a cut, by hand or coal cutter, along the floor level in a coal seam to ease its working by hand or breaking by explosive. See also: *undercut; floor cut; holing.* (Nelson, 1965)

undercutting machine An electrically driven machine used to make a cut about 10 ft (3.0 m) deep near the bottom of a coalbed. (Hudson, 1932)

undercutting of old workings A method of mining a vein that has been worked out above and in which the shaft is further sunk, with a crosscut being made to the vein at a depth below the previous workings. (Stoces, 1954)

underdrilling Drilling below the theoretical blasting bottom. Syn: *subdrilling.*

underearth A hard fireclay forming the floor of a coal seam. *underclay.* (AGI, 1987)

underedge stone The material that forms the floor of an ironstone mine. (Nelson, 1965)

underfeed To advance a diamond or other type of rock-drilling bit into rock at a lesser rate than that warranted by the condition of the rock and/or the condition of the bit. (Long, 1960)

underfeed stoker A mechanical stoker suitable for small boilers, such as the vertical, water-tube, and locomotive types. Coal is conveyed direct from a bunker or hopper by a feed worm, which forces the fuel up through the bottom of the retort in which it is burned. Volatiles driven off must pass through the ignited fuel, thus eliminating smoke. An underfeed stoker operates most successfully on graded coals with an upper size limit of 1 to 2 in (2.5 to 5.1 cm). See also: *stoker; vibrating grate.* (Nelson, 1965)

underfire (a) In ceramics, to fire (as brick) insufficiently. (Webster 3rd, 1966) (b) To fire from beneath. (Webster 3rd, 1966)

underflow The oversize material leaving a classifier. (Nelson, 1965)

underground bunker (a) Arrangements, such as high-capacity supplementary conveyors, staple pits, hoppers, or standage room for cars, positioned at key points between the faces and pit bottom. The object is to enable costly power-loading machines to operate continuously when there are surface or shaft delays. See also: *bunker conveyor; gate road bunker.* (Nelson, 1965) (b) A large-capacity hopper to absorb peak deliveries and provide an even rate of feed to main transport systems, or winding shaft. See also: *bunker.* (Nelson, 1965)

underground cable A single or multiple conductor cable sheathed in lead or other waterproof materials, carried in a duct beneath the surface of the ground. (Crispin, 1964)

underground coal gasification *underground gasification.*

underground connections Mines or areas that are connected underground shall be considered as a single mine if the underground connections between previously separate mines or areas subject the workers in the respective mines or areas to a reasonable likelihood of danger from mine fires or the products of fires, explosions or the forces and products of explosions, mine inundations, or personnel accidents.

underground dam Seal against water or spread of fire. (Pryor, 1963)

underground exploration (a) The driving of advance exploring headings and up-and-down boring to establish the continuity and thickness of coal seams or other mineral deposits. See also: *exploratory drilling.* (Nelson, 1965) (b) Extensions of a known ore deposit may be probed along its strike or dip in which shafts, drifts, or crosscuts may be driven. A study of the habits of known ore shoots, by mapping, surveying, and sampling, is a desirable preliminary to underground exploration. (Nelson, 1965)

underground fires There are two types of underground fires: (1) those that involve exposed surfaces and are known as open, freely burning fires and (2) those that may be wholly or partly concealed and are invariably caused by spontaneous heating of the coal itself, known as gob fires. (Mason, 1951)

underground garage *locomotive garage.*

underground gasification A method of burning the coal in place to produce a combustible gas that can be burned to generate power or processed into chemicals and fuels. Air and/or steam is blown underground to support the controlled combustion in the coal seam. The resultant gaseous mixture is a low-heating-value fuel gas. See also: *blind borehole process.* Syn: *underground coal gasification.* (Kentucky, 1952; USBM RI)

underground geology Usually implies direct evidence derived from shafts, wells, and borings, or obtained by geophysical methods. Also called subsurface geology. (Challinor, 1964)

underground glory-hole method A method used in large deposits with a very strong roof. In this method, a deposit is divided by levels and on every level chutes are raised to the next one. Mining starts from the mouth of the chutes in such a way as to develop a funnel-shaped excavation (mill, glory) with slopes so steep that the broken ore falls into the chutes and thus to the cars on the lower level. A sufficiently strong pillar is left for protection at the higher level. Syn: *underground milling.* (Stoces, 1954)

underground haulage The transportation of coal or minerals from the working face to the shaft bottom. Haulage usually implies trams, tubs, or mine cars drawn by horses, locomotives, electric or compressed-air haulage engines. Conveyors are not generally regarded as a haulage method. See also: *gravity haulage; haulage; locomotive haulage; main transport; subsidiary transport.* (Nelson, 1965)

underground milling *underhand stoping; underground glory-hole method.*

underground mine conveyor Sectional conveyor, usually of the troughed belt type, capable of being lengthened or shortened as mining operations advance or retreat, as contrasted to an above-ground conveyor having a fixed length for reasonably permanent installation. According to location in the mine or usage, underground conveyors may be known as face, room, gathering, main haulage, or intermediate haulage conveyors. See also: *belt conveyor; conveyor; haulage conveyor; flight conveyor; mother conveyor.* Syn: *main conveyor; entry conveyor.*

underground opening Natural or manmade excavation under the surface of the Earth. (Obert, 1960)

underground ore bin *measuring chute.*

underground shaft A shaft sunk from an adit, tunnel, or working level, through which mining operations are conducted. The upper end terminates underground. A winze or raise becomes an underground shaft when equipped and used for hoisting and the conduct of other mining operations. Cf: *winze.* (Fay, 1920)

underground station (a) An enlargement of an entry, drift, or level at a shaft at which cages stop to receive and discharge cars, workers, and material. (Fay, 1920) (b) An underground station is any location where stationary electric equipment is installed for the utilization of electricity. This includes pump rooms, compressor rooms, hoist rooms, battery-charging rooms, etc. (c) Excavation housing special equipment. (Pryor, 1963)

underground surveying Distinctive features of underground surveying are that stations are usually in the roof instead of the floor; the object to be sighted and the crosshairs of the telescope must be illuminated; distances are usually measured on the slope; either the transit tripod has adjustable legs or a trivet is used; and often an auxiliary telescope is attached to the transit, either at one end of the horizontal axis or above the main telescope, with the line of sight of the auxiliary telescope parallel to that of the main telescope. Horizontal and vertical distances are computed from slope distances and vertical angles. The transit is set up at one station, being centered by plumb, and the vertical distance from the station to the horizontal axis of the transit is measured. A plumb bob is hung at the next station, with a point on the plumbline marked by some form of clamping target. The vertical angle to the point so marked is measured, and the distance from horizontal axis to the target is taped. (Urquhart, 1959)

underground transformer A flameproof, air-filled transformer of a size up to 300 kV·A, which can be used inby near the face in safety-lamp mines. A nitrogen-filled transformer for mining use is the latest trend. (Nelson, 1965)
underground transportation The transporting of ore, rock, people, materials, and supplies through shafts and haulageways, including the loading of ore or rock into cars and carrying it to the surface. (Jackson, 3)
underground water *ground water; subsurface water.*
underground workshop An underground room prepared at an accessible spot in an underground mine in which repairs can be made on the mining equipment used underground.
underhand longwall Underground mining technique based on underhand stoping and cut-and-fill stoping with a single advancing face. See also: *underhand stoping; cut-and-fill stoping.*
underhand stope A stope made by working downward from a level.
underhand stoping The working of a block of ore from an upper to a lower level; mining downward. The method is particularly suitable for narrow, highly inclined deposits. Syn: *horizontal-cut underhand; underbreaking; underground milling.* Cf: *overhand stoping.* See also: *underhand longwall.* (AGI, 1987)
underhand vertical slice *square-set stoping.*
underhand work Picking or drilling downward. (Fay, 1920)
underhole (a) To cut away or mine out the lower portion of a coal seam or a part of the underclay so as to win or get the overlying coal. (Craigie, 1938) (b) To mine out a portion of the bottom of a seam, by pick or powder, thus leaving the top unsupported and ready to be blown down by shots, broken down by wedges, or mined with a pick or bar. In England, the terms jad, hole, undercut, kirve, and bench are synonymous. See also: *undermine; undercut.* (Fay, 1920)
underlay The extension of a vein or ore deposit beneath the surface; also, the inclination of a vein or ore deposit from the vertical; i.e., hade. Syn: *underlie.* See also: *hade.* (AGI, 1987)
underlay shaft (a) A shaft sunk in the footwall and following the dip of a vein. Also called underlier. (Fay, 1920) (b) Shaft that slopes at the dip angle of a lode, but is carried below the ore. Also called a footwall shaft. See also: *incline shaft.* (Pryor, 1963)
underlevel A development level or drift, driven from the surface, in ironstone mining. (Nelson, 1965)
underlie (a) The angle at which stulls or posts are set between walls. The setting angle is slightly steeper than the perpendicular to the vein, and the post thus tightens with the downward settlement of the hanging wall. (Nelson, 1965) (b) To lie or be situated under, to occupy a lower position than, or to pass beneath. The term is usually applied to certain rocks over which younger rocks (usually sedimentary or volcanic) are spread out. Ant: overlie. Syn: *underlay.* (AGI, 1987)
underloading Insufficient charging of a ball mill for proper grinding of enamel slip. (ACSB, 1948)
underlying Lying under or beneath; fundamental. See also: *seat earth.* (Webster 3rd, 1966)
underlying beds The rocks situated under a deposit or other strata. (Stoces, 1954)
undermanager In Great Britain, the underground mining engineer and senior executive official. Everything that has to do with the underground must be subject to his or her control, esp. regarding safety and health of personnel. Also called underlooker. Syn: *under.* (Mason, 1951)
undermine To excavate the earth beneath, esp. for the purpose of causing to fall; form a mine under. See also: *undercut; underhole.* (Webster 3rd, 1966)
underpinning (a) Building up the wall of a mine shaft to join that above it. (Fay, 1920) (b) The act of supporting a superior part of a wall, etc., by introducing a support beneath it. (Fay, 1920)
underpoled copper To reduce to metal the cuprous oxide from blister copper produced in a converter, green poles are pushed into the molten material to bring the percentage of dissolved oxygen below 0.5. If this reaction (which results in flat-topped ingots having a smooth fracture) is incomplete, the ingot fractures too readily and has a darker color. This shows it to have been insufficiently reduced, or underpoled. See also: *tough pitch.* (Pryor, 1963)
underream To enlarge or ream a borehole below the casing. See also: *underreaming.* (Long, 1960)
underreamer A tool or device having cutters that can be expanded or contracted by mechanical or hydraulic means and used to enlarge or ream a borehole below the casing or drivepipe. Also called expansion bit; expansion reamer. See also: *hydraulic underreamer; underreamer bit.* (Long, 1960)
underreamer bit The assembled device consisting of the lugs or jaws attached to an expanding mechanism used to enlarge or ream a borehole below a string of casing. See also: *underreamer.* (Long, 1960)
underreamer cutter *underreamer lug.*
underreamer lug A diamond set or other type of expansible or contractable jaw on an underreamer bit. Syn: *underreamer cutter.* See also: *cutter.* (Long, 1960)
underreaming The widening out of the foot of a bored hole or of certain types of foundation piers to increase the load-bearing area. See also: *underream.* (Hammond, 1965)
underreaming bit An expanding bit used to enlarge the diameter of the hole below the casing to allow the casing to be lowered farther down the borehole. (BS, 1963)
underrope haulage An endless rope haulage in which the ropes run under the cars. The cars are attached singly or in short sets by clips. See also: *overrope haulage.* (Nelson, 1965)
undersaturated (a) Said of an igneous rock consisting of unsaturated minerals; e.g., feldspathoids and olivine. (AGI, 1987) (b) Said of a rock whose norm contains feldspathoids and olivine, or olivine and hypersthene. Cf: *unsaturated; saturated; saturated rock.* (AGI, 1987)
underscreen water Water that is fed into the cells of a washbox below the level of the screen plate. (BS, 1962)
undersea prospecting The driving of exploring headings seaward from landside mine workings combined with up-and-down boring to establish higher or lower seams. Also, a technique developed by the National Coal Board, Great Britain, for exploration drilling from a tower that can be floated out to sea and grounded on the seabed. (Nelson, 1965)
undersea workings *submarine mines.*
undershot wheel (a) A vertical waterwheel into the circumference of which are set blades that are pushed by water passing underneath. (Webster 3rd, 1966) (b) A waterwheel used for low heads, in which the power is obtained almost entirely from the impulse of the water on the vanes. (CTD, 1958)
undersize (a) Particles in a screen overflow that are smaller than the normal dimensions of the screen apertures. (BS, 1962) (b) The smaller of two classified products. In the case of ore pulp or fine coal, the undersize is the overflow and the oversize is the underflow. See also: *classifier.* (Nelson, 1965) (c) A drill hole that is not to size because of gage loss on the bit and/or the reaming shell with which it was drilled. (Long, 1960) (d) A bit or reaming shell, the diametric dimensions of which are less than specified as standard. (Long, 1960) (e) That part of a crushed material that passes through a screen. (f) Material in a product of size smaller than the reference size; may be expressed as a percentage of the product. (BS, 1962)
undersize control screen A screen used for the removal of undersize from a product. (BS, 1962)
undersize core Core the outside diameter of which is less than standard. (Long, 1960)
underthrust A low-angle reverse fault resulting from the sliding of the footwall beneath a relatively passive hanging wall. Cf: *overthrust.*
undertow The seaward return flow, near the bottom of a sloping beach, of water that was carried onto the shore by waves. Cf: *rip current.* (AGI, 1987)
undertub system The endless-rope system generally used on moderate and constant gradients where the floor is good. In this system, the rope runs underneath the tubs or cars in the center of the rails. Curves are negotiated by a series of small vertical pulleys between the rails and are best of large radius. Clips are generally preferred to lashing chains, and the system suits automatic clipping and unclipping. Cf: *overtub system.* (Sinclair, 1959)
undervoltage relays Like undercurrent relays, undervoltage relays indicate when voltage is not up to the level it should be. Undervoltage values result in the breaker tripping and staying out until the undesirable condition is corrected. (Coal Age, 1966)
underweight (a) The weight of that portion of the strata overlying a coal seam at the face, which is supported by the timber or steel props. See also: *nether roof; overarching weight; traveling weight.* (Nelson, 1965) (b) A diamond bit, the crown of which is inset with diamonds so widely spaced that part of the crown is without cutting points and the bit cannot be made to cut. (Fay, 1920)
underwinding A rope or cable wound and attached so that it stretches from the bottom of a drum to the load. (Nichols, 1954)
undeveloped land Land remaining in its natural state, not disturbed by mineral exploration or extraction activities. (SME, 1992)
undiscovered resources Resources, the existence of which are only postulated, comprising deposits that are separate from identified resources. Undiscovered resources may be postulated in deposits of such grade and physical location as to render them economic, marginally economic, or subeconomic. To reflect varying degrees of geologic certainty, undiscovered resources may be divided into hypothetical resources and speculative resources. (USGS, 1980)
undisturbed sample A sample that is as undisturbed as humanly possible, as distinct from a sample disturbed by boring tools. Special appliances are used to obtain such samples from boreholes, and the material is preserved in its natural state in airtight containers. Undisturbed samples are required so that the in-place (in situ) properties of

the soil may be determined. It is difficult to obtain undisturbed samples of sandy soils without considerable preparation. See also: *soil core.* (Nelson, 1965)

undulatory extinction In polarized-light microscopy, said of a mineral that fails to go extinct as a unit under crossed polars, but exhibits waves of extinction sweeping across the grain upon rotation of the microscope stage. Syn: *wavy extinction.* Cf: *extinction; zoning.*

unequal angle A metal angle section with two legs of unequal length. (Hammond, 1965)

uneven fracture A general type of mineral breakage that produces rough and irregular surfaces. Cf: *fracture.*

unfaced quartz A name given to defaced masses of raw quartz used in the oscillator industry.

ungemachite A trigonal mineral, $K_3Na_8Fe(SO_4)_6(NO_3)_2{\cdot}6H_2O$; in thick, tabular rhombohedral crystals; at Chuquicamata, Chile.

ungotten Unworked rock of any kind. (Arkell, 1953)

uniaxial In optical crystallography, those anisotropic crystals having one direction of apparent isotropy, i.e., one optic axis, corresponding to the unique direction of axial symmetry in the hexagonal, trigonal, and tetragonal crystal systems. Uniaxial crystals are positive if their extreme refractive index (extraordinary ray) is less than their axial refractive index (ordinary ray), negative if greater. Cf: *isotropic; isotropy; anisotropy; biaxial.*

uniaxial stone One that has crystallized in the tetragonal, trigonal, or hexagonal crystal system and, therefore, has only one direction or axis of single refraction. Cf: *biaxial stone.*

unicline An obsolete syn. of monocline. Cf: *anticlinal bend.* (AGI, 1987)

unidimensional consolidation A test in which the volume change in a soil sample is observed when subjected to increasing increments of load. A soil sample, which may be 3 in (7.6 cm) in diameter and 0.8 in (2 cm) thick, is compressed between two porous stones, and the movement under load is noted. (Hammond, 1965)

unidirectional ventilation (a) A form of air travel; the air enters at one point, passes through the workings, and goes out at a distant point. Recirculation is impossible where unidirectional ventilation is used; thus the mine is safer than with recirculated air. (Lewis, 1964) (b) Ventilation in which air in adjacent openings flows in the same direction and is entirely fresh or exhaust air. (Hartman, 1982)

uniformitarianism The fundamental principle or doctrine that geologic processes and natural laws now operating to modify the Earth's crust have acted in the same regular manner and with essentially the same intensity throughout geologic time, and that past geologic events can be explained by phenomena and forces observable today; the classical concept that the present is the key to the past. The doctrine does not imply that all change is at a uniform rate and does not exclude minor local catastrophes. Syn: *principle of uniformity.* (AGI, 1987)

uniformity coefficient An expression of variety in sizes of grains that constitute a granular material. (AGI, 1987)

Unifrax A low-density nitroglycerin powder type of equivalent sheathed explosive. (McAdam, 1958)

unilateral transducer A transducer that cannot be actuated at its outputs by waves in such a manner as to supply related waves at its inputs. (Hy, 1965)

un-ionized (a) State of a substance that has dissolved without dissociating into ions; solute retaining its compounded state. (Pryor, 1963) (b) State of a substance (in solid, gaseous, or liquid state) with atoms that have neither an excess nor a deficiency of electrons, so that they have no net charges.

union shop A shop or mine run according to the requirements of a trade union. Cf: *open shop.*

unique diameter A direction in a crystal parallel to the highest symmetry axis; i.e., the *c* axis in trigonal, tetragonal, and hexagonal systems; the three reference axes in the orthorhombic system; and the *b* (or *c* in the first setting) axis in the monoclinic system.

unit Smelter contracts make frequent use of the work unit. A unit means 1%. Since a short ton contains 2,000 lb (907.2 kg), a unit is equivalent to 20 lb/t (10.0 kg/t) of ore. (Lewis, 1964)

unit cell A parallelepiped unit of atomic dimensions within a crystal structure, containing all the atoms in the chemical unit in definite, fixed positions; and all the symmetry elements of the space lattice from which the whole crystal or crystal structure is built by regular repetitions of this unit in three dimensions. Cf: *asymmetric unit.*

unit coal (a) Applied to prepared coal as for analysis, and being the pure coal substance considered altogether apart from extraneous or adventitious material (moisture and mineral impurities), which may by accident or through natural causes have become associated with the combustible organic substance of the coal. (AGI, 1987) (b) The pure or actual coal substance as derived from taking into consideration the corrected ash. The differentiation between the noncoal substance of a sample being analyzed and the coal itself. It is expressed by the formula: Unit coal = 1.00 - (W + 1.08 A + 0.55 S), where W = water, A = ash, and S = sulfur. (AGI, 1987)

united veins Where two or more veins unite, the oldest or prior location takes the vein below the point of union, including all the space of intersection. (Ricketts, 1943)

unit operation Recognition, study, application, and control of the principles and factors utilized in a distinct and self-contained process (for example, filtration). This avoids the duplication of effort that attends study of filtration of oil, sugar, ore pulp, etc., as though each involved a unique set of principles. (Pryor, 1963)

unit pressures The total pressure divided by the number of area units on which the load is imposed, such as the diamonds in a diamond-bit crown, usually expressed as pounds per square inch, tons per square foot, pounds per diamond, etc. (Long, 1960)

unit process Distinct and self-contained operations that can be studied individually. In mineral processing, unit processes include crushing, grinding, classification, gravity treatment, pulp conditioning, flotation, thickening, leaching, and filtration. (Pryor, 1963)

unit retardation plate *selenite plate.*

unit strain Unit tensile strain is the elongation per unit length; unit compressive strain is the shortening per unit length; unit shear strain is the change in angle (radians) between two lines originally at right angles to each other. (Roark, 1954)

unit stress The stress or load per unit of area, usually taken per square inch of section. For instance, if a bar is 1 in by 2 in (2.5 cm by 5.1 cm) in section, the unit stress of the bar will be 1,000 divided by 2 (sectional area) or 500 psi (3.4475 MPa). (Zern, 1928)

unit train A system delivering coal more efficiently in which a string of cars, with distinctive markings, and loaded to "full visible capacity," is operated without service frills or stops along the way for cars to be cut in and out, from the loading place to the point of delivery.

unit value The monetary value of a mineral or rock product per ton or other unit of measurement. See also: *value.* (AGI, 1987)

unit ventilation A system of ventilation in which each working face is ventilated by a separate air current. (BS, 1963)

unit weight (a) Weight per unit of volume. See also: *dry unit weight; effective unit weight; maximum unit weight; saturated unit weight; submerged unit weight; unit weight of water; wet unit weight; zero air voids unit weight.* (b) The density of a material. See also: *density.* (Hammond, 1965)

unit weight of water The weight per unit volume of water; normally equal to 62.4 lb/ft^3 or 1 g/cm^3. See also: *unit weight.* (ASCE, 1958)

univalent (a) Having a valence of one. (Webster 3rd, 1966) (b) Having one valence; e.g., calcium, which has only a valence of two. (Webster 2nd, 1960)

universal arc-shearing machine A machine with a rotating jib head so that vertical or shearing cuts can be made in addition to the arc wall cut. An arcwall machine will cut a 12-ft (3.7-m) heading to a depth of 6 ft (1.8 m) in 10 min. (Mason, 1951)

universal clamp A clamping device used on a drill column by means of which a horizontal arm can be affixed at any point on the vertical section of a drill column. (Long, 1960)

universal coal cutter A coal cutter with a jib capable of cutting at any height or angle. It may be mounted on crawler tracks. See also: *universal machine; turret coal cutter.* (Nelson, 1965)

universal coupling Coupling that joins two driving shafts that rotate about differently slanted axes. (Pryor, 1963)

universal gas mask Designed as an all-purpose mask for protection against a great variety of toxic gases, vapors, and smokes, including carbon monoxide. It is equipped with an indicator that shows at a glance the remaining service time of the canister for carbon monoxide. This mask is particularly effective for mines, fire fighting, and general industrial uses where the contaminants are of relatively low content. (Best, 1966)

universal-joint couplings These couplings are used where shafts intersect at any angle or where pivoted members must be driven. If there is an offset in the two shafts, or if one or both shafts must change location during operation, two of these couplings are used, and one is fitted with a splined joint mating with the connecting intermediate shaft to allow axial movement. (Pit and Quarry, 1960)

universal lay rope *lang lay rope.*

universal machine A power-driven coal cutter that will not only cut horizontal kerfs, but will also cut vertical kerfs at any angle, and is designed for operation either on track, crawler treads, or rubber tires. Syn: *universal coal cutter; arc shear machine.*

universal motor An electric motor rated at less than 1 hp (746 W) output that operates on either direct or alternating current. (Hammond, 1965)

universal pH indicator Mixture of several indicator dyes, each of which changes color through a specific pH range, so that by suitably combining a wide color change and pH, a reading can be obtained. (Pryor, 1963)

universal plant indicator Indicator plant that is restricted exclusively to rocks or soils of a definite mineral content and not found under any other conditions. Cf: *local indicator plant.* (Hawkes, 1962)

universal stage A stage of three, four, or five axes of rotation, attached to the rotating stage of a polarized-light microscope, that enables a thin section or mineral grain under study to be turned in a precisely controlled fashion about three mutually orthogonal axes. It is used to determine precisely the optical properties and parameters of minerals or their orientation in a set of external coordinates. Syn: *U-stage; Federov stage*. Cf: *goniometer; spindle stage.*

universal testing machine An instrument so designed that it is capable of exerting a tensile, compressive, or transverse stress on a specimen under test. Further, it can be adapted for the determination of Brinell hardness, ductility, cold bend, and other properties. The machine consists essentially of three systems: loading, weighing, and indicating, the loading being applied either mechanically or hydraulically. (Osborne, 1956)

universal train A roll train having adjustable horizontal and vertical rolls, so as to produce sections of various sizes. (Fay, 1920)

unkeying In attacking a rock face, the first effort of the miner is directed toward making a cut that will permit the succeeding shots to exert the greatest force with the minimum charge of explosive. In doing this unkeying, the miner takes advantage of any persistent seam in the rock face. (Stauffer, 1906)

unlimited pump A deep-well pump operated from the level of the ground above. (Standard, 1964)

unloader A machine that unloads iron ore from boats and cars, by power, generally electric. (Mersereau, 1947)

unloading conveyor Any of several types of portable conveyors adapted for unloading bulk materials, packages, or objects from conveyances. See also: *portable conveyor.*

unloading trough A short section of trough, designed for insertion in a standard shaker trough, which will allow the coal to be unloaded at that point by being diverted to either side by the unloading trough. C-clamps are used to hold the unloading section in place. (Jones, 1949)

unlocking *liberation.*

unmix Proprietary flotation collector agent based on emulsified tall oil, fuel oil, and water-soluble aryl-alkyl sulfonate used to treat hematite ores. (Pryor, 1963)

unmixing *exsolution.*

unoriented (a) Said of a rock specimen whose original position in space, when collected, is unknown. (b) Said of a rock fabric that shows no ordered spatial arrangement. Cf: *preferred orientation.* (Stokes, 1955) (c) Said of a map or surveying instrument whose internal coordinates are not coincident with corresponding directions in space. (Stokes, 1955)

unpatented claim (a) Mining claim to which a deed from the U.S. Government has not been received. A claim is subject to annual assessment work, to maintain ownership. (Weed, 1922) (b) A claim that requires $100 of work to be done each year. A claim cannot be patented until $500 has been spent on it. (von Bernewitz, 1931) (c) A mining claim for which the holder has no patent. Under the Multiple Surface Use Act of 1955, discoveries of common varieties of sand, stone, gravel, pumice, cinders, and clay cannot be located as mining claims; however, it does not affect the validity of a discovery in these materials based on the presence of other valuable minerals. (Lewis, 1964)

unproductive development The drifts, tunnels, and crosscuts driven in stone, preparatory to opening out production faces in a coal seam or orebody. Horizon mining is characterized by a heavy outlay on the initial unproductive development. See also: *dead work; in-the-seam mining; productive development.* (Nelson, 1965)

unproven area An area in which it has not been established by drilling operations whether oil and/or gas may be found in commercial quantities. (Williams, 1964)

unreserved mineral A mineral that belongs to the owner of the land on which or in which it is located. The owner of the land is its exclusive owner and can deal with it freely. Examples include limestone, dolomite, barite, fluorite, fireclay, plastic clay, glass sand, marble, and gypsum. (Stoces, 1954)

unrestrained compression apparatus *unconfined compression appliance.*

unripe diamond *rock crystal.*

unsaturated (a) Applied to minerals that are incapable of crystallizing from rock magmas in the presence of an excess of silica. Such minerals are said to be unsaturated with regard to silica and include feldspathoids, analcime, magnesian olivine, melanite, pyrope, perovskite corundum, calcite, and perhaps spinel. Cf: *undersaturated; saturated rock.* (AGI, 1987) (b) Applied to air that contains less water vapor than the maximum or saturation content for the conditions pertaining. (Spalding, 1949)

unscreened coal (a) Coal for which no size limits are specified. (BS, 1960) (b) Aust. Run-of-mine coal.

unsealing The recovery of a sealed-off mine area that had been sealed to extinguish a fire. Two general systems may be employed: (1) recovering the fire area in successive blocks by means of air locks, and (2) reventilation of the fire area after there is conclusive evidence that the fire has been extinguished. See also: *sealed area.* (Kentucky, 1952)

unsoiling The act or process of removing soil, as in opening a quarry. (Standard, 1964)

unsoundness (a) A quarry term that refers to all cracks or lines of weakness, other than bedding planes, that may cause rock to break before or during the process of manufacture. Various types of unsoundness are known locally as joints, headers, cutters, hairlines, slicks, seams, slick seams, dry sewns, dries, and cracks. (Fay, 1920) (b) The condition of a solid metal that contains blowholes or pinholes due to gases, or cavities resulting from the liquid-to-solid contraction (that is, contraction cavities). See also: *gas evolution.* (CTD, 1958)

unstable (a) Said of a constituent of a sedimentary rock that does not effectively resist further mineralogic change and that represents a product of rapid erosion and deposition (as in a region of tectonic activity and high relief); e.g., feldspar, pyroxene, hornblende, and various fine-grained rock fragments. (AGI, 1987) (b) Said of an immature sedimentary rock (such as graywacke) consisting of unstable particles that are angular to subrounded, poorly to moderately sorted, and composed of feldspar grains or rock fragments. Cf: *labile.* (AGI, 1987) (c) Said of a radioactive substance. Cf: *stable.* (AGI, 1987)

unstable isotope *radioisotope.*

unstable relict A relict that is unstable under newly imposed conditions of metamorphism, but persists in a perhaps altered but still recognizable form owing to the low velocity of transformation. A preferable term would be metastable relict. Cf: *stable relict.* See also: *armored relict.* (AGI, 1987)

unstratified Not formed or deposited in strata; specif. said of massive rocks or sediments with an absence of layering, such as granite or glacial till. (AGI, 1987)

unwatering Pumping or draining the water from mines. Cf: *dewatering.* (Hess)

unweathered *fresh.*

upcast (a) The opening in a mine through which the return air ascends and is removed. Ant: downcast; intake. See also: *air shaft.* (b) An upward current of air passing through a shaft, or the like. (c) Material that has been thrown up, such as by digging. (Webster 3rd, 1966) (d) Same as upthrow, such as the upcast side of a fault; opposite of downthrow or downcast. (Standard, 1964) (e) The lifting of a seam or bed by a dike. Syn: *uptake.* (CTD, 1958)

upcast shaft (a) A shaft through which air leaves the mine. (BS, 1963) (b) The shaft up which the ventilating current of air returns to the surface or to the fan. The term corresponds to main return or return drift in drift mining. Also called fan shaft. Syn: *uptake.* See also: *downcast shaft.* (Nelson, 1965)

upconing *coning.*

upgrade (a) To increase the commercial value of a coal by appropriate treatment. (BS, 1962) (b) To increase the quality rating of diamonds beyond or above the rating implied by their particular classification. (Long, 1960) (c) To increase the quality of grades.

upgrading *aggradation.*

uphill shaker conveyor Any shaker conveyor that is so designed as to have the proper stroke for shaking the maximum amount of coal up a grade. On certain types of shaker conveyors, this requires the replacement of certain parts of the drive to secure the desired stroke, rather than replacing the entire drive unit. (Jones, 1949)

up hole (a) A borehole collared in an underground working place and drilled in a direction pointed above the horizontal plane of the drill-machine swivel head. (Long, 1960) (b) A shothole drilled in rock at an upward angle. (Pryor, 1963)

uphole shooting In seismic exploration, the setting off of successive shots in a shothole at varying depths to determine velocities and velocity variation of the materials forming the walls of the hole. (AGI, 1987)

uphole time Used to denote the observed travel time of a seismic wave from the point of generation at a given depth in a shothole to a detector at the surface; the observed time equivalent of the corresponding shot depth. Syn: *time at shot point.* (AGI, 1987)

uplift (a) Any force that tends to raise an engineering structure and its foundation relative to its surroundings. It may be caused by pressure of subjacent ground, surface water, expansive soil under the base of the structure, or lateral forces such as wind. (AGI, 1987) (b) A structurally high area in the crust, produced by positive movements that raise or upthrust the rocks, as in a dome or arch. Cf: *depression.* (AGI, 1987)

up-over Designating a method of shaft excavation by drifting to a point below and then raising. (Webster 2nd, 1960)

upper (a) Pertaining to rocks or strata that are normally above those of earlier formations of the same subdivision of rocks. The adjective is applied to the name of a chronostratigraphic unit (system, series, stage) to indicate position in the geologic column and corresponds to late as applied to the name of the equivalent geologic-time unit; e.g., rocks of the Upper Jurassic System were formed during the Late Jurassic Period. The initial letter of the term is capitalized to indicate a formal subdivision (e.g., "Upper Devonian") and is lowercased to indicate an informal subdivision (e.g., "upper Miocene"). The informal term may be used where there is no formal subdivision of a system or series (counterpart of lower). Cf: *lower; middle.* (AGI, 1987) (b) *up hole.*

Upper Barren Coal Measures The part of the Carboniferous strata of the Appalachian field that is now assigned to the Dunkard group of the Permian series. Obsolete. (Fay, 1920)

upper break The upper bend of a terrace or monocline. Also called head.

upper explosive limit of flammability The highest quantity of combustible gases that, when mixed with a given quantity of air (or oxygen), will just support a self-propagating flame. (Francis, 1965)

Upper Productive Coal Measures The part of the Carboniferous strata of the Appalachian field that is now assigned to the Monongahela group of the Pennsylvanian series. Obsolete. (Fay, 1920)

upraise An auxiliary shaft, a mill hole, carried from one level up toward another. See also: *rise; raise.* (These are better terms.). (Fay, 1920)

upright *post.*

upright fold A fold having an essentially vertical axial surface. Syn: *vertical fold.* (AGI, 1987)

upright joints Eng. Vertical joints. See also: *thorough joints.* (Arkell, 1953)

upset (a) A narrow passage driven on a slope, leaving a wider pillar which is to be mined by slabbing or otherwise. (Hess) (b) A narrow working place driven from one pair of entries to another for the development of a long face in semilongwall or longwall mining. (Hess) (c) A tubular part such as a drill rod, the wall thickness of which has been increased by hot forging for a short distance on one or both ends, thereby reinforcing the area in which screw threads are cut. See also: *inside upset; outside upset.* (Long, 1960) (d) To increase the diameter of a rock drill by blunting the end. (Fay, 1920)

upsetting A means of increasing the diameter of a red-hot steel bar during forging by striking it on the end, a state that also occurs in riveting. (Hammond, 1965)

uptake *upcast; upcast shaft.*

upthrow (a) The upthrown side of a fault. (AGI, 1987) (b) The amount of upward vertical displacement of a fault. Cf: *downthrow; heave.* (AGI, 1987)

upward-current washer A washer in which separation takes place under the influence of an upward current of water or dense medium. (BS, 1962)

uraconite A name used for amorphous, yellow, hydrous uranium sulfates of unknown composition. Syn: *uranic ocher.*

uralborite A monoclinic mineral, $CaB_2O_2(OH)_4$; dimorphous with vimsite; forms radiating fibrous aggregates; at a skarn deposit in the Turinsk area of the Urals, Russia.

Uralian emerald (a) Emerald from near Sverdlovsk in the Ural Mountains, Russia. (b) A green variety of andradite garnet (demantoid), occurring as nodules in ultramafic rocks in the Nizhniy-Tagilsk district of the Ural Mountains; may be of semiprecious gem quality, although rather soft. Also known as Bobrovska garnet.

uralite A fibrous amphibole pseudomorphous after pyroxene. A trade name for a fireproof material, chiefly of asbestos.

uralitization The development of amphibole from pyroxene; specif. a late-magmatic or metamorphic process of replacement whereby uralitic amphibole results from alteration of primary pyroxene. Also, the alteration of an igneous rock in which pyroxene is changed to amphibole; e.g., the alteration of gabbro to greenstone by pressure metamorphism. (AGI, 1987)

uramphite An orthorhombic(?) mineral, $(NH_4)(UO_2)(PO_4)\cdot 3H_2O$; meta-autunite group; forms bottle-green flakes in the oxidation zone of a uranium-coal deposit.

uran Combining form meaning containing uranium; e.g., uranothorite. Syn: *urano-.* (Webster 3rd, 1966)

urania ceramics Ceramic products containing appreciable amounts of UO_2 (or thorium) that are used in atomic reactors. They are stable against corrosion.

uranian opal A variety of opal having an apple-green fluorescence reputedly caused by the presence of minute amounts of uranium. (Crosby, 1955)

uranic ocher *uraconite.*

uraninite An isometric mineral, UO_2, commonly impure with actinide and lanthanide rare earths, radium, helium, and zirconium; strongly radioactive; metamict; generally black; sp gr, 10.9; in pegmatites and veins with lead, tin, and copper minerals; a source of uranium called pitchblende where massive and metamict. See also: *pitchblende.* Syn: *ulrichite; coracite.*

uranite A general term for any mineral consisting of uranyl phosphate and arsenate of the autunite, meta-autunite, and torbernite groups.

uranium A radioactive, silvery-white, metallic element. Symbol, U. Occurs in numerous minerals such as pitchblende, uraninite, carnotite, autunite, uranophane, davidite, and tobernite. It is also found in phosphate rock, lignite, and monazite sands. Uranium and its compounds are highly toxic, both chemically and radiologically. Uranium is of great importance as a nuclear fuel; it is used as ballast for missile reentry vehicles, as a shielding material, and for production of high-energy X-rays. (Handbook of Chem. & Phys., 3)

uranium borides Three borides are known: UBr_2, UBr_3, and UBr_4. The most attention has been paid to the tetraboride, the properties of which are: melting point, >2,100 °C (but oxidizes rapidly above 600 °C); sp gr=9.38 g/mL; thermal expansion, 7.1×10^{-6} (20 to 1,000 °C); modulus of rupture (20 °C), 60,000 psi (414 MPa); electrical resistivity, 3×10^{-5} ohm·cm. (Dodd, 1964)

uranium disintegration series The series of nuclides resulting from the decay of uranium-238. The mass numbers of all members of the series are given by 4n+2, where n is an integer; therefore, the sequence is also known as the 4n+2 series. It is also known as the uranium-radium series. (Glasstone, 1958)

uranium galena Galena containing Pb 206, the lead isotope produced by radioactive decay of U 238.

uranium-lead *radium G.*

uranium minerals More than 150 uranium-bearing minerals are known to exist, but only a few are common. The five primary uranium-ore minerals are pitchblende, uraninite, davidite, coffinite, and brannerite. These were formed by deep-seated hot solutions and are most commonly found in veins or pegmatites. The secondary uranium ore minerals, altered from the primary minerals by weathering or other natural processes, are carnotite, tyuyamunite and metatyuyamunite (both very similar to carnotite), torbemite and metatorbernite, autunite and metaautunite, and uranophane. (Pearl, 1961)

uranium oxide The important oxides of uranium are UO_2, UO_3, and U_3O_8. The dioxide (melting point 2,880 °C) is used as a nuclear-fuel element. Uranium oxide has been used to produce red and yellow glazes and ceramic colors. (Dodd, 1964)

uranium trioxide Red, orange, and yellow; UO_3; an intermediate product in the refining of uranium.

uranmica *torbernite; uranite.*

urano- *uran.*

uranocircite A tetragonal mineral, $Ba(UO_2)_2(PO_4)_2\cdot 10H_2O$; autunite group; yellow-green.

uranolite A meteorite. (Fay, 1920)

uranophane A monoclinic mineral, $Ca(UO_2)_2[(SiO_3)(OH)]_2\cdot 5H_2O$; dimorphous with uranophane-beta; radioactive; soft; yellow; as secondary coatings, commonly associated with autunite and torbernite around uranium deposits; a source of uranium.

uranopilite A monoclinic mineral, $(UO_2)_6(SO_4)(OH)_{10}\cdot 13H_2O$; radioactive; yellow; secondary on uraninite; associated with gypsum and metauranopilite.

uranorthorite A variety of thorium silicate; thorite containing a small percentage of oxide of uranium. (Fay, 1920)

uranosphaerite A monoclinic mineral, $BiUO_3(OH)_3$; radioactive; orange to red; an oxidation product of pitchblende. Also spelled uranospherite.

uranotantalite *samarskite.*

uranothallite *liebigite.*

uranothorianite Thorianite with uranium in partial substitution for thorium. (Crosby, 1955)

uranothorite A uranian variety of thorite.

uranotile *uranophane.*

uranphyllite *torbernite.*

uranpyrochlore An isometric mineral, $(U,Ca,Ce)_2(Nb,Ta)_2O_6(OH,F)$; pyrochlore group; radioactive; yellow-brown; with samarskite in pegmatites in Mitchell County, NC, and at Hybla, ON, Canada. Formerly called hatchettolite. Cf: *betafite.*

urao A mixture of trona and thermonatrite. See also: *sodium sesquicarbonate.*

urea A tetragonal mineral, $CO(NH_2)_2$. Carbonyl diamide; also called carbamide.

Ure's process The treatment of quicksilver ores by heating in iron retorts with admixture of lime. (Fay, 1920)

Urgonian A division of the European Lower Cretaceous characteristically developed in certain parts of France and Belgium. (Standard, 1964)

urnel Eng. Kentish term for ragstone. Also spelled urnell or ournal.

urolith A pathogenic precipitate occurring in humans and other animals. They are very complex and include numerous biominerals, chiefly phosphates and oxalates. Syn: *kidney stone.*

urtite A light-colored member of the ijolite series that is composed chiefly of nepheline and 0% to 30% mafic minerals, esp. acmite and apatite. Cf: *melteigite.* The name, given by Ramsay in 1896, is for Lujavr-Urt (now Lovozero), Kola Peninsula, Russia. (AGI, 1987)

Uruguay amethyst (a) A deep violet, very transparent amethyst. (b) Any amethyst originating along the border between Uruguay and Brazil.

usable diamond A resettable salvage diamond. See also: *usable stone; usables.* (Long, 1960)

usable iron ore The product of a mine, or of a beneficiating or agglomerating plant, which is shipped without further processing to the consumer. (USBM, 1965)
usables Salvaged diamonds considered as being fit for resetting and reuse in another bit or tool. See also: *usable diamond.* (Long, 1960)
usable stone *usable diamond.*
used bit A diamond bit so dulled by use that it is no longer of any value as a cutting tool. (Long, 1960)
useful area Working area of a screen. The nominal area, less any area occupied by fixings or supports that obstruct the passage of material over or through the screen deck. (BS, 1962)
useful pressure For a mine fan, the natural ventilation pressure deducted from the total ventilation pressure required to circulate air through the mine. (Roberts, 1960)
Usspurwies arch An articulated yielding arch, provided by a single bolted joint at the crown. The joint is so designed that the bolt is not subjected to shear stress. The yield element is a rectangular box in which the foot of the arch rests. The resistance to yield is by means of a piece of crushing timber placed in the box from the bottom before setting. See also: *steel arch; steel support; Toussaint-Heintzmann arch.* (Nelson, 1965)
U-stage *universal stage.*
ustarasite A mineral, $Pb(Bi,Sb)_6S_{10}$; forms gray prismatic crystals in bismuth ore at the Ustarasaisk deposit in western Tyan-shan, Siberia.
usual mining privileges By this term in a deed, the grantee has and may enjoy the right to go upon the land and explore for, open, and operate mines, take out and sell the products, and do all things incident to that work. (Ricketts, 1943)
utahite A discredited term for jarosite or natrojarosite. See also: *natrojarosite.* (Am. Mineral., 1945)
utahlite A compact, nodular variscite from Lewiston, Cedar Valley, UT. (English, 1938)
U-tube manometer The vertical U-tube is the simplest type of pressure gage and consists either of a single U-shaped glass tube having a uniform bore with vertical arms or two separate glass tubes connected to a cistern. The level of the liquid in the vertical U-gage can be read easily to 0.1 in (2.5 mm) and in well-made instruments to 0.05 in (1.3 mm) water gage. See also: *manometer.* (Roberts, 1960)
uvanite An orthorhombic(?) mineral, $(UO_3)_2V_6O_{15}\cdot 15H_2O$(?); radioactive; brownish-yellow; in asphaltic sandstone with carnotite, rauvite, hewettite, metatorbernite, hyalite, and gypsum; from Utah; resembles carnotite.
uvarovite An isometric mineral, $Ca_3Cr_2Si_3O_{12}$; garnet group; crystallizes in emerald green dodecahedra and trapezohedra; in serpentinite and in skarns. Also spelled ouvarovite or uwarowit. Syn: *chrome garnet; chromium garnet.*
uvite The trigonal mineral, $(Ca,Na)(Mg,Fe)_3Al_5Mg(BO_3)_3Si_6O_{18}(OH,F)_4$, of the tourmaline group.
uwarowit *uvarovite.*
uzbekite *volborthite.*

V

vacancy A point defect in a crystal structure where an atom or ion is missing from its expected position. Syn: *Schottky defect.*
vacant land As defined by its opposite, land is not vacant when occupied as a mining claim without discovery by one who is diligently prospecting it for minerals that it may contain. (Ricketts, 1943)
Vacquier-Steenland method A numerical method used in gravity interpretation for calculating the depth to the source of many typical total field anomalies. The method involves the computation of the curvature of the observed total intensity by superposition of a special grid over the intensity contours. The curvature is proportional to the second vertical derivative of the magnetic intensity. (Dobrin, 1960)
vacuole *vesicle.*
vacuum A method of producing ventilation by exhausting air from a mine. See also: *vacuum fan.* (Fay, 1920)
vacuum casting (a) The casting of metals in vacuum. Also called suction casting. See also: *vacuum metallurgy.* (Henderson, 1953) (b) Slip casting in which the slip is de-aired before casting.
vacuum common fine salt VCF; evaporated salt made in vacuum pans; of ordinary purity and ordinary screen analysis. (Kaufmann, 1960)
vacuum concrete Concrete poured into a framework that is fitted with a linen filter in the form of a vacuum mat. As a result of the process ensuing, the concrete attains its 28-day strength in 10 days and has a 25% higher crushing strength. (Hammond, 1965)
vacuum deposition Condensation of thin metal coatings on the cool surface of work in a vacuum. (ASM, 1961)
vacuum fan A fan for creating suction or partial vacuum. An exhaust fan. See also: *vacuum.* (Fay, 1920)
vacuum filter (a) A form of filter in which the air beneath the filtering material is exhausted to hasten the process. (b) One in which the pulp is drawn into contact with a porous septum by means of a moderately high vacuum. Solids are arrested and filtrate drawn through. In the drum and disc types, filtration is continuous. Vacuum is produced by means of a pump. See also: *pulp.* (Pryor, 1963)
vacuum filtration (a) The separation of solids from liquids by passing the mixture through a filter and where, on one side, a partial vacuum is created to increase the rate of filtration. It may be used to extract fine coal from the suspension or cyanide solution for reuse. (Nelson, 1965) (b) See also: *filter.*
vacuum lifting Lifting by a crane fitted with a suction pad, employed for such items as precast concrete components, large panes of glass, and sheet steel. (Hammond, 1965)
vacuum metallurgy The processing of metals at elevated temperatures, usually by induction heating in high vacuum. See also: *vacuum casting.* (Henderson, 1953)
vacuum method In flotation, a method in which the pulp, saturated with air at atmospheric pressure, is allowed to rise to a height above the normal hydrostatic level of the pulp. In the course of this ascent, the dissolved gases precipitate from solution and form a vast number of very tiny bubbles that attach themselves selectively to the hydrophobic solids. (Gaudin, 1957)
vacuum method of testing sand A method of carrying out a triaxial test on a sand sample by maintaining a partial vacuum in the rubber bag containing the sample. (Hammond, 1965)
vacuum pump (a) A centrifugal or reciprocating pump that extracts steam or air from a chamber or pipe to create a partial vacuum. A vacuum pump, hand or power operated, is part of a pump station equipment where gravity flow is absent. (Nelson, 1965) (b) *pulsometer.* (c) A pump for exhausting air or other gas from an enclosed space to a desired degree of vacuum. (Webster 3rd, 1966) (d) A pump in which water is forced up a pipe by the difference of pressure between the atmosphere and a partial vacuum. Cf: *air pump.* (CTD, 1958)
vacuum system A two-pipe, steam-heating system equipped with vacuum pumps to permit maintenance of pressure below atmospheric within the radiators. (Strock, 1948)
vadose water Water of the zone of aeration. Syn: *suspended water.* (AGI, 1987)
vaesite An isometric mineral, NiS_2; pyrite group.
vake Soft, compact, mixed claylike material with a flat, even fracture, found most often in volcanic terrains. Not recommended. *wacke*
valaite A pitch-black resin of unknown composition. Found in thin crusts on dolomite and calcite in the Coal Measures of Moravia, Czechoslovakia. (Tomkeieff, 1954)
Valantin conveyor cutter A cutter chain on an armored flexible conveyor that cuts its own stable holes; pushed by pulsating rams; height, 18 in (45.7 cm); minimum workable seam, 20 in (50.8 cm); on gradients 0° to 20°; maximum length of face, 45 yd (40 m). (Nelson, 1965)
valence The degree of combining power of an element or a radical.
valence bond Linkage of pairs of electrons so as to unite their atoms as a molecule. When an element has more than one valence, its commonest combination is called the principal valence. (Pryor, 1963)
valence crystals Crystals whose atoms are held in position by covalent bonds; e.g., diamond and silicon. (Newton, 1959)
valencianite A variety of adularia in a silver mine at Valencia, Guanajuato, Mexico.
Valentine scale Pocket-sized beam scale of Chinese origin used in valuation of alluvial tin gravels. The beam is so calibrated as to read in catties per cubic yard when concentrates from washing of 1/4 ft^3 (0.007 m^3) are weighed. (Pryor, 1963)
valentinite An orthorhombic mineral, Sb_2O_3; soft; dimorphous with senarmontite; an oxidation product of antimony ores. Syn: *white antimony.*
valentite Antimony trioxide, Sb_2O_2, in orthorhombic crystals. Syn: *white antimony; antimony trioxide.*
valleriite A hexagonal mineral, $1.53[(Mg,Al)(OH)_2]\cdot[(Fe,Cu)S]$; an irrational but discrete interlayer complex of hydroxide and sulfide layers; massive; soft; resembles pyrrhotite in color.
vallevarite A light-colored monzonitic igneous rock composed chiefly of andesine, microcline, and antiperthite, with small quantities of clinopyroxene, biotite, and apatite. The name, given by Gavelin in 1915, is for Vallevara, Sweden. Obsolete. (AGI, 1987)
valley brown ore A local name for comparatively pure high-grade limonite or brown iron ore in Cambro-Ordovician limestones in the Shanendoah Valley of Virginia. See also: *mountain brown ore.*
valley fill A fill structure consisting of any material other than coal waste and organic material that is placed in a valley where side slopes of the existing valley measured at the deepest point are greater than 20°, or the average slope of the profile of the valley from the toe of the fill to the top of the fill is greater than 10°.
Vallum diamond Rock crystal (quartz) from the Tanjore District, India.
Val separator A launder used for cleaning buckwheat, rice, and barley sizes of anthracite. It has three distinguishing features: (1) a mixing tank at the head end of the machine, (2) a baffle in the bottom of the machine next to the mixing tank to facilitate the stratification of the solids in specific gravity layers, and (3) the use of a screen and bed of slate in the free discharge boxes. (Mitchell, 1950)
valuation (a) The act or process of valuing, or of estimating the value or worth; appraisal. (Webster 3rd, 1966) (b) The value or estimated price set upon a thing. (Webster 3rd, 1966)
value The valuable constituents of an ore; their percentage in an orebody, or assay grade; their quantity in an orebody, or assay value. See also: *assay grade; assay value; unit value.* (AGI, 1987)
valve tower A tower built up within a reservoir to house the control valves of supply pipes drawing off water at different levels. (Hammond, 1965)
vamping The debris of a stope, which forms a hard mass under the feet of a miner. (Fay, 1920)
van (a) To separate, such as ore from veinstone, by washing it on the point of a shovel. (Fay, 1920) (b) A shovel used in ore dressing. (Fay, 1920)
vanadate A salt or ester of vanadic acid; a compound containing the radical $(VO_4)^{3-}$ (ortho) or $(VO_3)^-$ (meta). Cf: *arsenate; phosphate.*
vanadic ocher A native yellow vanadium oxide found near Lake Superior. (Standard, 1964)
vanadinite A hexagonal mineral, $Pb_5(VO_4)_3Cl$; apatite group; soft; varicolored; sp gr, 6.7 to 7.1; in oxidized zones of lead ore deposits; in New Mexico, Arizona, Africa, Scotland, and Russia. Syn: *vanadite.*
vanadite *descloizite; vanadinite.*

vanadium A gray or white, malleable, ductile, metallic element. Symbol, V. Found in about 65 different minerals, among which are carnotite, roscoelite, vanadinite, and patronite; also found in phosphate rock, certain iron ores, and some crude oils. About 80% of the vanadium now produced is used as a ferrovanadium or as a steel additive; also used in ceramics, as a catalyst, and in the production of a superconductive magnet. (Handbook of Chem. & Phys., 3)

vanadium minerals Those most exploited for industrial use are patronite (MoS_2 with vandium sulfide), roscoelite (vanadium mica), vanadinite, carnotite, and chlorovanadinite. Metal is silvery and whitish; melting point is 1,720 °C; used in high-speed steels and shock-resistant alloys, chemicals, ceramics, and textiles. (Pryor, 1963)

vanadium ore Most vanadium is obtained from vanadium-bearing magnetite (1% to 2.2% V) in South Africa. Other sources include patronite, carnotite, roscoelite, vanadinite, descloizite, and volborthite.

vanadium-zirconium turquoise *zirconium-vanadium blue.*

vanalite A monoclinic mineral, $NaAl_8V_{10}O_{38}{\cdot}30H_2O$; forms bright-yellow incrustations on weathered shales; in northwest Kara-Tau, Kazakhstan.

Van Allen radiation zone Powerful doughnut-shaped zone of radiation 1,000 to 3,000 miles above the Earth's surface and parallel with the Equator. (AGI, 1987)

van Arkel and de Boer process *iodide process.*

vandenbrandeite A triclinic mineral, $Cu(UO_2)(OH)_4$; radioactive; dark green to black; secondary; associated with kasolite, sklodowskite, malachite, goethite, chalcocite, chalcopyrite, uraninite, curite, uranophane, and cobalt wad; occurs at Karungwe, Katanga, Republic of the Congo. Also spelled vandenbrandite.

vandenbrandite A very rare, strongly radioactive, triclinic, dark green to almost black mineral, $Cu^{2+}(UO_2)(OH)_4$; a secondary mineral found associated with kasolite, sklodowskite, malachite, geothite, chalcocite, chalcopyrite, and uraninite; also found associated with curite, uranophane and cobalt wad; from Karungwe, Katanga, Zaire. Also spelled vandenbrandeite. (Crosby, 1955; Hey, 1955)

vandendriesscheite An orthorhombic mineral, $Pb(UO_2)_4(OH)_9{\cdot}2H_2O$; radioactive; forms small, amber-orange pseudohexagonal crystals, commonly barrel shaped; at Katanga, Congo.

van der Kolk method A test used in microscopy to determine the index of refraction of a mineral grain relative to that of an immersion liquid. When transmitted light is blocked, the grain acts as a lens and the ocular inverts the image of the obstacle, causing a shadow to appear on the same side as the obstacle when the grain has the higher refractive index, but on the opposite side when the grain has the lower index. Syn: *oblique illumination method.* Cf: *Becke test.*

van der Waals bond Weak forces in crystal structures caused by induced dipoles resulting from juxtaposition of molecules; e.g., the bonding between electrostatically neutral layers in the talc structure.

van Dorn sampler This sediment sampler consists of a Plexiglas cylinder closed at each end by an ordinary rubber force cup. The two cups are connected by a length of surgical rubber tubing inside the cylinder, prestressed enough to permit the force cups to retain the sample in the cylinder. In the armed position, the two cups are pulled outside the cylinder, where they are restrained by a releasing mechanism attached to the outside wall. Two short loops of wire connect the cups to the releasing mechanism. The cups are released underwater by sending a messenger down the hydrographic wire. This sampler does not invert, which prevents use of reversing thermometers in conjunction with sampling. (Hunt, 1965)

Vandyke brown (a) A naturally occurring pigment derived from indefinite mixtures of iron oxide and organic matter. Obtained from bog earth and peat deposits, or from ochers containing bituminous matter. (CCD, 1961) (b) Etymol: its use by the 17th-Century Flemish painter Van Dyck. Syn: *ulmin brown.* (AGI, 1987)

vane The target of a leveling staff; one of the sights of a compass or quadrant. (Webster 3rd, 1966)

vane anemometer (a) A small windmill-type instrument used to measure air velocity and to infer air volume movement. The vane anemometer has been the primary instrument for airflow measurement since the early 1900's. (Hartman, 1982) (b) See also: *anemometer.* (c) Consists of several light, flat vanes, usually eight in number, mounted on radial arms that are attached to a horizontal spindle. This rotor drives, through a suitable gear train, a counting mechanism that indicates the revolutions of the rotor. The indicating dial, usually graduated in feet of air, may be located either concentrically with the rotor, or in a plane at right angles to the plane of rotation. By observing the number of revolutions over a timed interval, the velocity of flow is found. The instrument is available in a number of forms to cover velocities ranging from 30 to 6,000 ft/min (9.1 to 1,830 m/min). (Roberts, 1960)

vane-axial fan An airfoil (propeller) or disk fan within a cylinder and equipped with air-guide vanes either before or after the wheel; includes driving-mechanism supports for belt drive or direct connection. (Strock, 1948)

vane shear test An in-place shear test in which a rod with thin radial vanes at the end is forced into the soil and the resistance to rotation of the rod is determined. (ASCE, 1958)

vane shear tester A device used in soil testing, consisting of flat blades affixed to the end of a rod. It is forced into the soil, and the torque required to shear the soil, in situ, is determined as a measure of the shear strength of the zone tested by rotating the device. Syn: *vane tester.* (Long, 1960)

vane test An in-place test to measure the shear strength of fine-grained cohesive soils and other soft deposits. A rod with four flat radial blades, or vanes, projecting at 90° intervals is forced into the soil and rotated; the torque required to rotate the rod is a measure of the material's shear strength. (AGI, 1987)

vane tester *vane shear tester.*

vanner grease belt In ore dressing, smelting, and refining, used for separation of valuable mineral from the gangue (waste minerals) in an ore. (DOT, 1949)

vanoxite A mineral, $V_6O_{13}{\cdot}8H_2O$(?); a mixed-valence oxide; weakly radioactive; black; in the Colorado Plateau area as a sandstone cement; also a massive wood replacement associated with carnotite, gypsum, hewettite, pintadoite, tyuyamunite, and pyrite. See also: *kentsmithite.*

vanthoffite A monoclinic mineral, $Na_6Mg(SO_4)_4$; colorless; at Wilhelmshall, Stassfurt, Germany.

vanuralite A monoclinic mineral, $Al(UO_2)_2(V_2O_8)(OH){\cdot}8H_2O$; yellow; at Mounana, Gabon.

vapart mill A centrifugal grinder for pulverizing ore, coal, and coke. (Fay, 1920)

vapor (a) A substance in the gaseous state as distinguished from the liquid or solid state. (Webster 3rd, 1966; Zimmerman, 1949) (b) Foul air in a mine. Cf: *gas.* (Fay, 1920)

vapor barrier A material intended to prevent the passage of water vapor through a building wall to prevent condensation within the wall. (Strock, 1948)

vapor density The relative density of a gas or vapor as compared with some specific standard (as hydrogen). Abbrev., v d. (Webster 3rd, 1966; Zimmerman, 1949)

vaporization *evaporation.*

vapor pressure The pressure at which a liquid and its vapors are in equilibrium at a definite temperature. If the vapor pressure reaches the prevailing atmospheric pressure, the liquid boils. Symbol, *p*. (Hackh, 1944; Handbook of Chem. & Phys., 2)

vapor system A steam-heating system operating at pressure very near that of the atmosphere. (Strock, 1948)

vara Any of various old Spanish units of length used in Latin America and the Southwestern United States, equal in different localities to between 31 in and 34 in (78.7 cm and 86.4 cm); e.g., a unit equal to 33.3333 in (84.666 cm) in Texas, to 33.372 in (84.764 cm) in California, to 33.00 in (83.82 cm) in Arizona and New Mexico, and to 32.9931 in and 32.9682 in (83.802 cm and 83.739 cm) (among others) in Mexico. For other values, see ASCE (1954). (AGI, 1987)

V-arching Rock failure above a tunnel due to ring stresses. These cause rock to crack across a weakness plane and fall. The final shape is a reentrant V rather than a rounded arching. Syn: *arching to a weakness.* See also: *arching.* (Pryor, 1963)

Varian nuclear magnetometer This magnetometer is available in two models, one for airborne surveys and the other for use on the ground. Both measure the total magnetic field of the Earth rather than its components. (Dobrin, 1960)

variation The angle by which the compass needle deviates from the true north; subject to annual, diurnal, and secular changes. Called more properly declination of the needle. See also: *declination.* (Standard, 1964; Fay, 1920)

variation compass A compass of delicate construction for observing the variation of the magnetic needle. (Webster 2nd, 1960)

variegated Said of a sediment or sedimentary rock, such as red beds or sandstone, showing variations of color in irregular spots, streaks, blotches, stripes, or reticulate patterns. (AGI, 1987)

variegated copper ore Bornite, erubescite. (Pryor, 1963)

variegated ore *bornite.*

varietal mineral A mineral that is either present in considerable amounts in a rock or characteristic of the rock; a mineral that distinguishes one variety of rock from another. Syn: *distinctive mineral.* (AGI, 1987)

variety In mineralogy, a mineral showing differences in color, other physical properties, or minor variations in composition from the material considered typical of the species. An example is emerald, the green-colored gem beryl. (Hess)

variogram A plot of the variance (one-half the mean squared difference) of paired sample measurements as a function of the distance (and optionally of the direction) between samples. Typically, all possible sample pairs are examined, and grouped into classes (lags) of approx. equal distance and direction. Variograms provide a means of quantifying the commonly observed relationship that samples close together will tend to have more similar values than samples far apart.

variole A pea-size spherule, usually composed of radiating crystals of plagioclase or pyroxene. This term is generally applied only to such spherical bodies in basic igneous rock, e.g., variolite. Cf: *spherulite.* (AGI, 1987)

variolite A fine-grained igneous rock of basic composition containing varioles. Not recommended usage. (AGI, 1987)

variolitic Said of the texture of a rock, esp. a basic igneous rock, composed of pea-size spherical bodies (varioles) in a finer-grained groundmass. Cf: *spherulitic.* (AGI, 1987)

variometer A geophysical device for measuring or recording variations in terrestrial magnetism; a variable inductance provided with a scale. (AGI, 1987)

variscite (a) An orthorhombic mineral, $AlPO_4 \cdot 2H_2O$; dimorphous with metavariscite; waxy; forms nodular masses in cavities where phosphate water acts on aluminous rock. Variscite is a popular gem material for cabochons and various carved objects, commonly substituting for turquoise. Syn: *lucinite; utahlite.* (b) The mineral group mansfieldite, scorodite, strengite, and variscite. Cf: *amatrice.*

varislope screen A suspended multiple deck screen with increased slopes in the second and third deck. It is used principally for coal and other large feed materials and combines scalping and sizing operations. The oversize lump material is removed on the top deck, egg or range size on the second, and nut size on the bottom deck. See also: *trislope screen; vibratory screen.* (Nelson, 1965)

varlamoffite A mineral that is perhaps a variety of cassiterite.

varulite A monoclinic mineral, $(Na,Ca)_2(Mn,Fe)_3(PO_4)_3$; alluaudite group; forms dull olive-green granular masses; at Varutrask, Sweden.

varve (a) A sedimentary bed or lamina or sequence of laminae deposited in a body of still water within 1 yr's time; specif. a thin pair of graded glaciolacustrine layers seasonally deposited, usually by meltwater streams, in a glacial lake or other body of stillwater in front of a glacier. A glacial varve normally includes a lower summer layer consisting of relatively coarse-grained, light-colored sediment (usually sand or silt) produced by rapid melting of ice in the warmer months, which grades upward into a thinner winter layer, consisting of very fine-grained (clayey), often organic, dark sediment slowly deposited from suspension in quiet water while the streams were ice bound. Counting and correlation of varves have been used to measure the ages of Pleistocene glacial deposits. (AGI, 1987) (b) Any cyclic sedimentary couplet, as in certain shales and evaporites.—Etymol: Swedish. (AGI, 1987)

varve clay *varved clay.*

varved clay A distinctly laminated lacustrine sediment consisting of clay-rich varves; also the upper, fine-grained, winter layer of a glacial varve. Syn: *varve clay.* (AGI, 1987)

varves Clayey soil containing thin alternate layers of different particle sizes; often combine the undesirable properties of clay and silt; formed from seasonal deposits from glacial streams. (Nelson, 1965)

vashegyite An orthorhombic mineral, $Al_4(PO_4)_3(OH)_3 \cdot 13H_2O$(?); at Vashegy, Hungary, and near Manhattan, NV.

vat (a) A vessel or tub in which ore is washed or subjected to chemical treatment, as cyanide vat and chlorination vat. Syn: *tank.* (b) Salt pit. (AGI, 1987) (c) A term used in the Southwestern United States for a dried and encrusted margin around a waterhole. (AGI, 1987)

vaterite A hexagonal mineral, $CaCO_3$; trimorphous with calcite and aragonite; at Ballycraigy, Larne, Northern Ireland.

vaughanite A pure, dense, homogeneous, dove-colored, fine-textured limestone that breaks with a smooth and more or less pronounced conchoidal fracture that contains relatively few fossils, and that typically has a white, chalky appearance on weathered surfaces. Named after T. Wayland Vaughan (1870-1952), U.S. paleontologist. (AGI, 1987)

vauquelinite A monoclinic mineral, $Pb_2Cu(CrO_4)(PO_4)(OH)$; fibrous.

vauxite A triclinic mineral, $FeAl_2(PO_4)_2(OH)_2 \cdot 6H_2O$; less hydrous than metavauxite or paravauxite; blue.

V-bob A strong frame shaped like an isosceles triangle, turning on a pivot at its apex and used as a bell crank to change the direction of a main pump rod. It is used with Cornish pumping engines. (Webster 3rd, 1966; Fay, 1920)

V-box Sloughing box, used to separate slime (as overflow) from faster-settling portion of solids in pulp. (Pryor, 1963)

V-bucket conveyor elevator *gravity-discharge conveyor elevator.*

VCR *vertical crater retreat.*

V-cut (a) In mining and tunneling, a cut where the material blasted out in plan is like the letter V; usually consists of six or eight holes drilled into the face, half of which form an acute angle with the other half. (Fay, 1920) (b) In underground blasting, a type of cut employed in which the cut holes meet in a V to pull the cut to the bottom of the holes properly. A single pair of holes may do in one kind of rock, but in another, two or three sets of V-holes entirely across the face may be needed. Also See also: *wedge cut.* Syn: *plow cut; vee cut.* (Lewis, 1964)

veatchite A triclinic and monoclinic mineral, $Sr_2B_{11}O_{16}(OH)_5 \cdot H_2O$; trimorphous with p-veatchite and veatchite-A.

vector (a) An entity represented as a directed magnitude, such as velocity, which is defined as consisting of a speed and a direction. See also: *scalar.* (Hy, 1965) (b) *hard vector.*

vee cut *wedge cut; V-cut.*

vees (a) A layer of soft clay or earth on the sides of a fault or dike. Syn: *veez.* (Nelson, 1965) (b) The acute angle between the fault plane and a coal seam; e.g., working the coal to the vees of the fault. (Nelson, 1965)

vee table A pneumatic table, of U.S. design, for the drycleaning of coal, an improved form of the S.J. table. (Nelson, 1965)

veez *vees.*

vegasite *plumbojarosite.*

vegetable jelly Same as fundamental jelly, carbohumin, etc. Syn: *jelly.* (Tomkeieff, 1954)

veil (a) A removable plate to cover a screen, the action of which is not desired. (Zern, 1928) (b) An aggregate of minute bubbles creating a whitish or cloudlike appearance in quartz. (c) A variously formed weblike or netlike film in a radiolarian; e.g., patagium.

vein (a) An epigenetic mineral filling of a fault or other fracture in a host rock, in tabular or sheetlike form, often with associated replacement of the host rock; a mineral deposit of this form and origin. Cf: *lode.* See also: *true vein.* (AGI, 1987) (b) A narrow waterway or channel in rock or earth. Also, a stream of water flowing in such a channel. (AGI, 1987) (c) A thin, sheetlike igneous intrusion into a fissure. Not recommended usage. (AGI, 1987) (d) A coal seam or a bed of slate or other rock. Not recommended usage. (BS, 1964) (e) A zone or belt of mineralized rock lying within boundaries clearly separating it from neighboring rock. It includes all deposits of mineral matter found through a mineralized zone or belt coming from the same source, impressed with the same forms and appearing to have been created by the same processes. (Ricketts, 1943) (f) A mineral deposit, usually steeply inclined. Used to describe a body that is usually smaller and has better defined walls than a lode. (Nelson, 1965) (g) A rock fissure filled by intruded mineral matter. Many valuable minerals are codeposited with gangue stuff in veins. Usually the formation is steep to vertical, unlike a bedded deposit in which values are sandwiched horizontally. Vein is typically long, deep, and relatively narrow. (Pryor, 1963) (h) The term lode is commonly used synonymously for vein. (i) The filling of a fissure or fault in a rock, particularly if deposited by aqueous solutions. When metalliferous, it is called by miners a lode; when filled with eruption material, a dike. A bed or shoot of ore parallel with the bedding. Also called blanket deposit. (Standard, 1964) (j) A comparatively thin sheet of igneous rock injected into a crevice in rock. When this intrusion is large, it is called a dike. (k) An irregular, sinuous, igneous injection, or a tabular body of rock formed by deposition from solutions rich in water or other volatile substances. (Holmes, 1928) (l) A mineral body, thin in relation to its other dimensions, which cuts across the bedding and in which the minerals are later than the country rock. (BS, 1964) (m) Sometimes used for a bed; e.g., a coal seam or a bed of slate. (BS, 1964) (n) A layer, seam, or narrow irregular body of material different from surrounding formations. See also: *vein or lode claim.* (Nichols, 1976)

veined gneiss A composite gneiss with irregular layering. The term is generally used in the field and has no genetic implications (Dietrich, 1960). Cf: *venite; composite gneiss.* (AGI, 1987)

vein intersection (a) The depth in the borehole at which the hanging and/or footwall of a vein is encountered. (Long, 1960) (b) The place where two or more veins cross or meet. (Long, 1960)

vein material *veinstuff.*

vein miner A miner experienced in the winning and working of mineral veins. See also: *metal mining.* (Nelson, 1965)

vein or lode claim The terms "vein or lode" and "vein or lode claim" are used indiscriminately and interchangeably, and it follows that the term "vein or lode" is intended to be synonymous with the term "vein or lode claims." See also: *vein; lode; lode claim.* (Fay, 1920)

vein quartz A rock composed chiefly of sutured quartz crystals of pegmatitic or hydrothermal origin and commonly of variable size. (AGI, 1987)

veinstone The valueless stone that occurs with the valuable minerals in lodes and veins. Also called lodestuff; matrix; vein mineral; veinstuff. Sometimes mistakenly called gangue. (Nelson, 1965; Fay, 1920)

veinstuff (a) All the minerals occurring in a vein. See also: *lodestuff.* (b) Gangue. Syn: *vein material.*

vein system An assemblage of veins of a particular area, age, or fracture system, usually inclusive of more than one lode. (AGI, 1987)

velardenite *gehlenite.*

velikhovite A variety of pyrobitumen having a shining conchoidal fracture and occurring in the form of veins. It is partly soluble in organic solvent; sp gr, 1.2. In many ways it is similar to grahamite. It is assumed that it represents a weathering product of albertite. From the South Urals, Russia. (Tomkeieff, 1954)

velocities in pipes Experience has proved that the following are allowable velocities in pipes: air, 30 to 50 ft/s (9.1 to 15.2 m/s); compressed air, 25 to 40 ft/s (7.6 to 12.2 m/s); steam, 160 to 250 ft/s (46.8 to 76.2 m/s); water, 5 to 10 ft/s (1.5 to 3.0 m/s). (Hammond, 1965)

velocity (a) In explosives, the speed (in meters or feet per second) at which the detonating wave passes through a column of explosives. A high-velocity explosive renders a shattering effect, whereas a low-velocity explosive has a pushing or heaving effect. Syn: *high velocity*. (Kentucky, 1952) (b) Linear flow rate of air per unit time. Measured in meters per second. (Hartman, 1961) (c) A vector quantity that indicates a time rate of motion. (AGI, 1987)

velocity determination The determination of velocities and average velocities within the earth by seismic measurements. (AGI, 1987)

velocity discontinuity An abrupt change in the velocity of propagation of seismic waves within the earth, as at an interface. (AGI, 1987)

velocity distribution Relationship between seismic wave velocity and depth. (Schieferdecker, 1959)

velocity head (a) The constant difference of height of a liquid between a level surface in a tank and a uniformly flowing jet through an orifice. (Standard, 1964) (b) The distance a body must fall under the force of gravity to acquire the velocity it possesses. See also: *kinetic energy*. (Seelye, 1951) (c) The energy possessed per unit weight of a fluid owing to its velocity. If at a given point the velocity is v feet per second, the velocity head at this point, $v^2/2g$, g being the acceleration due to gravity in feet per second squared. Also called kinetic head. Syn: *velocity pressure*. (CTD, 1958)

velocity lag *distance lag*.

velocity meter A seismometer used to record the velocity of ground motions. (AGI, 1987)

velocity of air current The higher the velocity of the air current, the greater will be the resistance to airflow. The resistance is nearly proportional to the velocity squared. (Mason, 1951)

velocity of approach (a) The average velocity of water in a channel at the point where the depth over a flow measuring weir is recorded. (Hammond, 1965) (b) The mean velocity in the conduit immediately upstream from a weir, dam, Venturi throat, orifice, or other structure. (Seelye, 1951)

velocity of detonation (a) The velocity with which the shock wave traverses an explosive charge on detonation. (BS, 1964) (b) The velocity of detonation of an explosive was previously determined by what is known as the Dautriche test. The basis of this test is that a length of Cordtex detonating fuse detonates at a uniform speed, and if the two ends of a length of Cordtex are detonated simultaneously, the detonation waves will meet at the middle of the length of fuse. Similarly, if the two ends are detonated at different times, the distance from the middle of the fuse to the point where the two detonation waves meet is directly proportional to the interval of time between the detonations of the two ends of the Cordtex fuse. Furthermore, if the distance can be measured, the interval of time between the detonation can be calculated, since the velocity of detonation of Cordtex is known. Syn: *Dautriche test*. (McAdam, 1958)

velocity of retreat An average velocity of flow of a liquid just downstream of a measuring weir. (Hammond, 1965)

velocity pressure (a) The pressure equivalent of the air velocity at any particular point. This is always positive. (BS, 1963) (b) The pressure exerted by a moving fluid in the direction of its motion. It is the difference between the total pressure and the static pressure. (Strock, 1948) (c) In Mine ventilation, the pressure exerted by the kinetic energy of air movement. Syn: *velocity head*. (d) The algebraic difference between the total head and the static pressure. (Roberts, 1960)

velocity profile A linear arrangement of sensors used to record reflections over a large range of shot-to-geophone distances, which is used to determine seismic velocity from the time-distance relationship. (AGI, 1987)

velocity ratio The ratio of the distance through which the force applied to a machine moves, and the distance through which the load moves. See also: *mechanical advantage*. (Hammond, 1965)

velocity reducing collector This type of collector is designed to remove very large dust particles. It is often used ahead of other collectors to reduce the dust load, and to remove the particles most likely to cause abrasion. The velocity reducing collector has no moving parts and, in most instances, can be installed in front of the induced draft fan, reducing the abrasion of the fan blades. This type of collector also can be used under many high-temperature conditions. (Pit and Quarry, 1960)

velometer A small portable direct-reading instrument used to measure the velocity of air at a point. (Hartman, 1982)

velvet Profit; easily earned money. (Fay, 1920)

velvet copper ore Cyanotrichite in bright blue velvetlike druses and spherical forms. Syn: *lettsomite; cyanotrichite*.

vend Products sold by coal mine annually. (Pryor, 1963)

vendeennite A variety of fossil resin from Vendee, France. (Tomkeieff, 1954)

Venetian red A high-grade ferric-oxide pigment of a purer red hue than either light red or Indian red. Obtained either native as a variety of hematite red or more often artificially, by calcining copperas in the presence of lime. The composition ranges from 15% to 40% ferric oxide and from 60% to 80% calcium sulfate. The 40% ferric oxide is the pure grade, and sp gr, 3.45. (CCD, 1961)

venite Migmatite of which the mobile portion(s) were formed by exudation (secretion) from the rock itself (Dietrich & Mehnert, 1961). Cf: *veined gneiss; composite gneiss*. Not widely used. (AGI, 1987)

vent (a) In explosives, a small passage made with a needle through stemming, for admitting a squib to enable the charge to be lighted. (b) A hole, extending up through the bearing at the top of the core-barrel inner tube, that allows water and air in the upper part of the inner tube to escape into the borehole or into the annular space between inner and outer barrels. (Long, 1960)

vent bag An enclosed airway to direct airflow to a given area or location. (MSHA, 1992)

ventilating column *motive column*.

ventilating current A current of air traveling in mines. (Peel, 1921)

ventilating fan *mine ventilating fan*.

ventilating pressure (a) The total head in pascals or kilopascals required to overcome the friction of the air in mines and to provide some final or exit velocity to discharge the air to the atmosphere. (b) The total pressure exerted on the atmosphere by the mine fan to overcome the resistance of the mine to the passage of a required volume of air throughout the mine necessary for its ventilation. See also: *mine total head; fan total head; total ventilating pressure*. (Kentucky, 1952)

ventilation Mine workings are usually subdivided to form a number of separate ventilating districts. Each district is given a specified supply of fresh air and is free from contamination by the air of other districts. Accordingly, the main intake air is split into the different districts of the mine. Later, the return air from the districts reunite to restore the single main return air current at or near the upcast shaft. See also: *compound ventilation; fan drift; regulator*. (Nelson, 1965)

ventilation department A department for the purpose of planning adequate and economic ventilation for all future projects and to provide frequent information on existing ventilation systems. (Nelson, 1965)

ventilation doors A door constructed to restrict the flow of ventilation air while permitting the passage of personnel and equipment. See also: *door; separation door*.

ventilation ducts Two kinds are available, flexible and rigid ducts. Flexible ducts generally consist of flexible tubes made from fabrics coated with rubber or polyvinyl chloride, a nonflammable substance. They are available in varying lengths. Flexible ducting is suited for face ventilation in a variety of mining methods. It is suited also to crooked workings of limited extent. It has a higher resistance and a greater tendency to leak than rigid ducting. Rigid ducts are made of steel or fiberglass in lengths suitable for underground transport. This type of duct does not have to be accurately aligned and is therefore used, in the smaller sizes, in subsidiary work, particularly in crooked headings. For main tunnels where leakage must be minimized, flanged joints are used with suitable gaskets. (Roberts, 1960)

ventilation efficiency One measure of the efficiency of a mine ventilation system is the ratio of the total amount of air actually reaching the working faces to the total amount (volume in cubic feet per minute) of air handled by the fan. See also: *overall ventilation efficiency; thermometric fan test; ventilation standards; volumetric efficiency*. (Kentucky, 1952)

ventilation mason A worker who erects by rough masonry or cement work, partitions of stone, brick, or concrete blocks to control proper circulation of air through passageways outby working places. (DOT, 1949)

ventilation plan A plan or drawing, required by law, that shows the ventilation air currents in a mine and the means of controlling them. (BS, 1963)

ventilation planning When a new mine is projected or a new seam is to be worked from an existing mine, plans are prepared to show the proposed ventilating system, including the quantities of air and pressures and the principal appliances to control and distribute the air. Investigations and calculations are made to select a fan of the necessary type and size for the ventilation required. All of this important work comes within the general term ventilation planning. See also: *air requirements; pressure survey; ventilation survey*. (Nelson, 1965)

ventilation pressure (a) The pressure or head producing ventilation in a mine and measured by the height of a column of water it will support. The instrument used for this purpose is a water gage. (b) Pressure producing the flow of air, measured by a water gage, or the difference in level between the two ends of the water column in a vertical U-shaped tube, one end of which is connected to the air under pressure—for example, in the passageway leading to the fan—the other end being to the atmosphere. In some cases, the ventilating pressure is reported as meters of water in the U-tube; in other cases, the head is given in pascals or kilopascals.

ventilation regulator *regulator.*

ventilation standards The standards prescribed by regulations to provide air underground of a certain degree of purity. See also: *ventilation efficiency.* (Nelson, 1965)

ventilation stopping *stopping.*

ventilation survey (a) Systematic observation of air pressure, quantity and quality, throughout a mine or part of a mine, to allow a detailed analysis of the ventilation system. Syn: *pressure-quantity survey.* (Sinclair, 1958; BS, 1963) (b) To distribute the air in a mine efficiently and economically, ventilation surveys are conducted. They may be classified as qualitative, quantitative, and pressure surveys. Qualitative surveys determine the proportion of flammable or poisonous gas, or dust, in the air that is being circulated through the mine. In hot and humid mines, they determine the conditions of air temperature and humidity. Quantitative surveys determine the quantity of air being circulated through the mine workings for a variety of reasons. This is done by measuring the volume of air passing at different points in the circuit by means of an anemometer, to investigate the existing air distribution, particularly to the individual faces; the location of leakage; and the possibility of its reduction or elimination. Pressure surveys measure the pressure absorbed and the resistance of the roadways and faces included in the survey. This enables determination of the power required to circulate the air in the different sections of the circuit and that is expended in ventilating individual districts. The total power expended in ventilating the mine may then be summed and the cost estimated. See also: *ventilation planning.* (Sinclair, 1958; BS, 1963)

ventilation symbols A set of standard letters, signs, or marks used on mine ventilation plans to represent certain appliances or constructions to direct and control the flow of air underground. (Nelson, 1965)

ventilation tubing Sheet steel or canvas piping 12 to 24 in (0.3 to 0.6 m) in diameter for conducting air to or from a tunnel, hard heading face, or sinking pit. The tubing extends from an auxiliary fan to within a few yards of the face to be ventilated. See also: *tubing; auxiliary ventilation.* (Nelson, 1965)

ventilator (a) A mechanical apparatus for producing a current of air underground, as a blowing or exhaust fan. (b) A furnace for ventilating a mine by heating the upcast air. (c) A device for providing fresh air to a room or other space by introducing outside air or by exhausting foul air. (Crispin, 1964)

vent pipe *vent tube.*

vent tube (a) Hose or piping conducting air-ejected drill cuttings from the borehole collar to a point some distance from the drill. (Long, 1960) (b) An exhaust pipe or tube. (Long, 1960) (c) Tubing suspended from a wire in a mine opening to supply fresh air to a working place. Syn: *vent pipe.* (Long, 1960)

ventubes Tubes of steel, fiberglass, or coated fabric with thin walls that can be easily connected. They are used in mine ventilation to lead air wherever it is needed. Also called ventilation tubing. (Stoces, 1954)

Venturi A contraction in a tube or duct to accelerate the flow and lower the static pressure. It is used for metering and other purposes.

Venturi blower (a) A device resembling a Venturi meter that directs a jet of compressed air for ventilating short headings. The device is commonly made at a mine, and one well-proved type is called the Modder Deep. These blowers are mainly used in conjunction with ventilation ducting for the ventilation of headings several hundred feet in length. (Roberts, 1960) (b) An apparatus to induce a flow of air or gas in a duct by means of a jet of compressed air or water from a small nozzle in the duct. (BS, 1963)

Venturi flume (a) A type of open flume with a contracted throat that causes a drop in the hydraulic gradeline; used for measuring flow. (Seelye, 1951) (b) A control flume that comprises a short constricted section followed by one expanding to normal width. See also: *control.* (Hammond, 1965)

Venturi meter A trademark for a form of the Venturi tube arranged to measure the flow of a liquid in pipes. Small tubes are attached to the Venturi tube at the throat and at the point where the liquid enters the converging entrance. The difference in pressure heads is shown on some form of manometer and from this difference and a knowledge of the diameters of the tubes, the quantity of flow is determined. (Webster 2nd, 1960)

Venturi tube A closed conduit that is gradually contracted to a throat causing a reduction of pressure head by which the velocity through the throat may be determined. The contraction is generally followed, but not necessarily so, by gradual enlargement to original size. Piezometers connected to the pipe above the contracting section and at the throat indicate the drop in the pressure head, which is an index of flow. (Seelye, 1951)

vent wire A wire used by founders to make a hole in a sand mold for the escape of air or gases. (Standard, 1964)

venus hairstone *rutilated quartz.*

Venus's hairstone Quartz containing needle-shaped crystals of rutile. See also: *Thetis hairstone; sagenitic quartz; sagenite.* (Standard, 1964)

verde antique A dark-green rock composed essentially of serpentine (hydrous magnesium silicate) usually crisscrossed with white veinlets of marble. Found in California, Georgia, Maryland, Massachusetts, New York, and Virginia. Used as an ornamental stone. In commerce, it is often classed as a marble. (CCD, 1961; Sanford, 1914)

verdelite A green variety of elbaite (tourmaline).

verde salt *thenardite.*

Verdet's constant The rotation of the plane of polarization per centimeter per unit magnetic field in the Faraday effect. The value of the constant varies with temperature and is approx. proportional to the square of the wavelength of the light. (CTD, 1958)

verdite A deep green, relatively soft metamorphic rock of green fuschite (chromian muscovite) and clay with scattered grains of rutile; occurs in Transvaal and Zimbabwe; is carved for ornamental use.

verdolite Talcose-dolomitic breccia rock from New Jersey. (Schaller, 1917)

verifier (a) A tool used in deep boring for detaching and bringing to the surface portions of the wall of the borehole at any desired depth. (Fay, 1920) (b) In gas testing, an apparatus by which the amount of gas required to produce a flame of a given size is measured; a gas verifier. (Standard, 1964)

vermeil (a) An orange-red garnet. Syn: *vermilion.* (b) A reddish brown to orange-red gem variety of corundum. (c) An orange-red spinel. Syn: *vermeille.*

vermeille *vermeil.*

vermicular quartz Quartz in wormlike intergrowths with feldspar. See also: *myrmekite.*

vermiculite A monoclinic mineral, $(Mg,Fe,Al)_6(Si,Al)_8O_{20}(OH)_4 \cdot 8H_2O$; mica group; basal cleavage; soft; pearly; a hydrothermal or weathering alteration of biotite; expands 6 to 20 times by thermal exfoliation; occurs in clay sizes in soils and as crystals and megacrysts in ultramafic rocks; in Montana, North Carolina, South Carolina, Wyoming, Virginia, Colorado, and South Africa. Syn: *lernilite.*

vermilion (a) A red pigment used in enormous quantities. Usually made from mercuric sulfide, HgS, tinted with puranitraniline. Also spelled vermillion. (Crispin, 1964) (b) A bright-red pigment consisting of mercuric sulfide. Prepared synthetically (as by the reaction of mercury, sulfur, and sodium hydroxide), but formerly obtained from the mineral cinnabar. Color ranges from crimson when coarse grained to nearly orange when finely divided. Both spellings are correct. (Webster 3rd, 1966) (c) *alpha mercuric sulfide; cinnabar; vermeil.*

vermillion In the Lake Superior region, the lowest of the stratified schists; the crystalline schists. (Fay, 1920)

vernadskite Antlerite in aggregates of minute crystals or as pseudomorphs after dolerphanite. See also: *antlerite.*

Verneuil process A technique developed by Auguste V.L. Verneuil (1856-1913), French mineralogist and chemist, for the manufacture of large crystals of corundum and spinel in which powdered alumina with appropriate oxide dopants is melted in an oxyhydrogen flame to produce boules of synthetic gems. See also: *boule.*

vernier closure meter An instrument used to measure strain.

vernier compass A surveyor's compass with a vernier, used for measuring angles without the use of the magnetic needle by means of a compensating adjustment made for magnetic variation. (AGI, 1987)

vernier-reading manometer This series of manometers covers a range of pressures from 0.001-in (25.4 μm) water gage to 40-in (1.02-m) water gage. Syn: *micrometer-reading manometer.* (Roberts, 1960)

verrankohle Rolled fragments of brown coal found on the coast of Norway. (Tomkeieff, 1954)

vertical (a) A term used to define a direction that is perpendicular to a horizontal, or level, plane. (AGI, 1987) (b) Local usage for vertical fractures, esp. in the Black Hills, South Dakota. (AGI, 1987) (c) Said of deposits and coal seams with a dip of from 60° to 90°. (Stoces, 1954) (d) In aerial photographic mapping, a vertical line through the exposure station or rear nodal point. (Seelye, 1951) (e) Orientation of a plumb-bob on a string under the force of gravity. (Seelye, 1951)

vertical aerial photograph A photograph taken from an aircraft for purposes of aerial mapping or aerial geophysical prospecting; special cameras and techniques are employed. See also: *profile flying.* (Nelson, 1965)

vertical angle Angle of elevation or depression, measured from the true horizontal plane. (Seelye, 1951)

vertical auger drill A mobile-type rotary drill used on opencast sites with no hard rock for drilling vertical blasting holes. It can drill a hole of 5 in or 6 in (12.7 cm or 15.2 cm) in diameter to depths of about 30 ft (9.1 m). Drilling is by means of a rotary cutting head with interchangeable cutting bits, the auger removing the cuttings from the hole. An overall speed of 30 ft/h (9.1 m/h) can be obtained. See also: *horizontal auger.* (Nelson, 1965)

vertical balance An instrument for measuring variations in the vertical component of the terrestrial magnetic field, usually by balancing the torque on a magnet system by means of a counter gravitational torque acting on counterweights. (AGI, 1987)

vertical chain conveyor Opposed-shelf type that has two or more vertical elevating conveying units opposed to each other. Each unit consists of one or more endless chains whose adjacent facing runs operate in parallel paths. Thus, each pair of opposing shelves or brackets receive objects (usually dish trays) and deliver them to any number of elevations.

vertical circle (a) Graduated circle on theodolite or tacheometer, by use of which the slope of the collimation line through sighting telescope is measured in survey work. (Pryor, 1963) (b) Any great circle of the celestial sphere passing through the zenith. (AGI, 1987)

vertical collimator An instrument in which the telescope sights vertically (upward or downward); used chiefly for centering a theodolite on a tower exactly over a station mark on the ground. It may be used for any vertical sight. (Seelye, 1951)

vertical component That part, or component, of a vector that is perpendicular to a horizontal or level plane. (AGI, 1987)

vertical crater retreat (a) A blasting method employed in underground sublevel mining. Initially a vertical slot extending across the width of the stope is mined. The remaining portion of the stope is then blasted by section into the vertical slot following sublevel procedures. (SME, 1992) (b) A variation of the sublevel stoping method that uses basic crater blasting models. Blasting is carried out at the base of vertical boreholes, making horizontal cuts and advancing upwards. A spherical charge is placed at an optimal distance from the stope back so that a maximum volume of rock is broken in the shape of an inverted cone. Borehole spacing is determined so that overlapping fragmentation cones do not disturb adjacent explosive charges. Abbrev. VCR.

vertical crater retreat mining A patented mining method in which large, parallel, vertical drillholes permit placement of nearly spherical explosive charges, such that horizontal slices of ore are then broken into an undercut; applicable to ore of only moderate strength. (SME, 1992)

vertical curve (a) The curve between two lengths of a straight roadway that possess different gradients. The curve provides a gradual change for haulages from one inclination to the other. The curve leading to the top or brow of an inclined plane would be convex and at the bottom would be concave. (Nelson, 1965) (b) The graduated curve connecting two lengths of a railway or road, which are at different slopes. (Hammond, 1965) (c) The meeting of different gradients in a road or pipe. (Nichols, 1976)

vertical cut *shear cut.*

vertical dip slip *vertical slip.*

vertical drains Usually column of sand used to vent water squeezed out of humus by weight of fill. (Nichols, 1976)

vertical exaggeration (a) A deliberate increase in the vertical scale of a relief model, plastic relief map, block diagram, or cross section, while retaining the horizontal scale, to make the model, map, diagram, or section more clearly perceptible. (AGI, 1987) (b) The ratio expressing vertical exaggeration; e.g., if the horizontal scale is 1 in to 1 mi and the vertical scale is 1 in to 2,000 ft, the vertical exaggeration is 2.64. Abbrev: VE. (AGI, 1987) (c) The apparent increase in the relief as seen in a stereoscopic image. (AGI, 1987)

vertical excavation limited area This method of excavation is used in loose or wet soils—unconsolidated formations—where the banks must be supported by shoring or sheathing. The material must, out of necessity, be lifted out vertically. (Carson, 1961)

vertical fold *upright fold.*

vertical gradient The rate of change of a quantity in the direction of the vertical.

vertical gradiometer An instrument for measuring the vertical gradient of gravity. (AGI, 1987)

vertical guide idler An idler roller of about 3 in (7.6 cm) in diameter so placed as to make contact with the edge of the belt conveyor should the latter run too much to one side. Although vertical guide rollers are effective, they cause edge wear on the belting and their use is not favored. (Nelson, 1965)

vertical intensity The vertical component of the magnetic field; usually considered positive if downward, negative if upward. Cf: *horizontal intensity.* (Hy, 1965)

vertical line One that is exactly upright, or it points straight up and down. (Jones, 1949)

vertical load-bearing test *load-bearing test.*

vertical mill A rolling mill in which the rolls are oriented vertically. (Osborne, 1956)

vertical photograph An aerial photograph made with the camera axis vertical (camera pointing straight down) or as nearly vertical as possible in an aircraft. (AGI, 1987)

vertical pump This pump is often of the single-acting bucket or ram type with single or double cylinders and either with or without a flywheel. Vertical pumps may be used where headroom is adequate but area restricted, although horizontal reciprocating pumps are more generally used. (Sinclair, 1958)

vertical reciprocating conveyor A power or gravity-actuated unit that receives objects on a carrier or car bed usually constructed of a power or roller conveyor. The object is then elevated or lowered to other elevations.

vertical sand drain (a) A boring through clay or silty soil that is filled with sand or gravel to facilitate drainage of liquid from the soil. (Hammond, 1965) (b) *perched water table.* (Nelson, 1965)

vertical screw conveyor A screw conveyor that conveys in a substantially vertical path. See also: *screw conveyor.*

vertical seismograph An instrument that registers the vertical component of ground motion. (Schieferdecker, 1959)

vertical separation In a fault, the distance measured vertically between two parts of a displaced marker such as a bed. Cf: *horizontal separation.* (AGI, 1987)

vertical shaft A shaft sunk at an angle of 90° with the horizon or directly downward toward the center of the Earth. (Weed, 1922)

vertical shear Reference is to a beam, assumed for convenience to be horizontal and to be loaded and supported by forces, all of which lie in a vertical plane. The vertical shear at any section of the beam is the vertical component of all forces that act on the beam to the left of the section. The vertical shear is positive when upward and negative when downward. (Roark, 1954)

vertical shift In a fault, the vertical component of the shift. See also: *shift.* (AGI, 1987)

vertical slip In a fault, the vertical component of the net slip; it equals the vertical component of the dip slip. Cf: *horizontal slip.* Syn: *vertical dip slip.* (AGI, 1987)

vertical takeup A mechanism in which the takeup or the movable pulley travels in a vertical plane. (NEMA, 1961)

vertical theory The earliest view of subsidence in which it was supposed that the lines of break (limiting lines) were more or less vertical. Pillars left for support were accordingly formed immediately under the object to be protected, the question of dip being disregarded. (Briggs, 1929)

vertical trace A trace on the ground motion record representing the component of motion in a vertical plane and in the direction of the seismic wave travel direction.

vesicle A cavity of variable shape in a lava, formed by the entrapment of a gas bubble during solidification of the lava. Syn: *vacuole.* (AGI, 1987)

vesicular Said of the texture of a rock, esp. a lava, characterized by abundant vesicles formed as a result of the expansion of gases during the fluid stage of the lava. Cf: *cellular; scoriaceous.* (AGI, 1987)

vesuvian (a) Original spelling of vesuvianite. See also: *leucite.* (Hey, 1955) (b) A mixture of calcite and hydromagnesite. (Hey, 1955)

vesuvianite A mineral, $Ca_{10}Mg_2Al_4(SiO_4)_5(Si_2O_7)_2(OH)_4$. Tetragonal. Common in contact-metamorphosed limestones. A massive light green variety is known as californite. Syn: *idocrase.* (AGI, 1987; Dana, 1959; Fay, 1920)

vesuvian jade A jadelike variety of vesuvianite (idocrase). Also called californite. (English, 1938)

veszelyite A monoclinic mineral, $(Cu,Zn)_3(PO_4)(OH)_3{\cdot}2H_2O$; greenish-blue. Formerly called arakawaite.

Vezin's sampler A mechanical sampling device that automatically selects one twenty-fifth or one sixty-fourth of the ore passing through. (Fay, 1920)

V-flume A V-shaped flume, supported by trestlework and used by miners for bringing down timber and wood from the mountains, at the same time using the water for mining purposes. (Fay, 1920)

vibracone A vibrating ore screen in which the feed is from a saucer-shaped distributer onto a conical surface kept in vibration by a ratchet motion. (Liddell, 1918)

vibrate To have a swinging or oscillating motion; to move or swing back and forth, such as a pendulum does; to have a period of vibration; to fluctuate; to vacillate; to sound, such as a voice vibrates in an ear; to throb. (AGI, 1987)

vibrating conveyor (a) A trough or tube flexibly supported and vibrated at relatively high frequency and small amplitude to convey bulk material or objects. See also: *oscillating conveyor.* (b) A metal trough mounted on flexible supports and free to move in a vertical plane. It is vibrated at an angle of about 30° to the horizontal. The material being conveyed moves in a series of gentle pitches and catches that blend to produce continuous, uniform flow. There is no tumbling or sliding of the material to cause wear of the trough. There are two basic types of vibrating conveyors: (1) the natural frequency types (those supported by heavy-duty stiff coil or leaf springs), and (2) forced vibration types (those supported by rocker arms or rods pivoted at the trough and at the base connections). Materials can be moved downward, horizontally, or up to 10° slopes. It can convey coal, limestone, sand, coke, granite, gravel, etc. See also: *shaker conveyor.* (Nelson, 1965)

vibrating coring tube A sediment coring tube designed to vibrate in such a way as to overcome the resistance of compacted ocean floor sediments, sands, and gravel. (Hunt, 1965)

vibrating grate A stoker developed in Germany and used increasingly in that country and in the United States. The hearth consists of a rigid water-cooled matrix. Coal is fed on to this at one end and is moved across it by the vibrating motion to discharge as ash at the other end. The vibrations, with an amplitude of about 1/8 in (3 mm) and in progress for about 5 s every 2 min give a satisfactory feed rate. The rate of feed is controlled by altering the duration of the vibrations. See also: *stoker; underfeed stoker.* (Nelson, 1965)

vibrating grease table This type table is used at the Kimberly Mines in South Africa for concentrating the -3.33- to +0.59-mm fraction of pan concentrate and other material of +0.59-mm size. Efficiency is 99%, and the ratio of concentration 50,000:1.

vibrating grizzlies Bar grizzlies mounted on eccentrics so that the entire assembly is given a forward and backward movement at a speed of some 100 strokes a minute. This is the type of grizzly now generally used ahead of a primary crusher. (Pit and Quarry, 1960)

vibrating platform A loading stage or structure with a double vibrating action that causes the coal or minerals to settle down in a mine car while being loaded. This settlement increases the car-carrying capacity and reduces spillage during transit. Syn: *tram vibrator.* (Nelson, 1965)

vibrating screen (a) A commercial screen in which the cloth, wire, or bar deck is vibrated by solenoid or by magnetostriction, or mechanically by eccentrics or unbalanced spinning weights. (Pryor, 1963) (b) A screen oscillated either by mechanical or electrical means. The amplitude of movement of the vibrating screen is smaller than that of the jigging screen and its speed of oscillation is higher. (BS, 1962) (c) A screen that is vibrated to separate and move pieces resting on it. (Nichols, 1976) (d) Machines of this type consist of one or more slightly inclined screening surfaces mounted in a robust frame. To increase the capacity and prevent blinding of holes, the screening surfaces are caused to vibrate. This may be done by mounting the screen on powerful springs and causing it to bear down on the underside of the frame. An alternative method used in the Hummer screen is to stretch the wire screen to a high tension and mount an electromagnet actuated by an alternating current at some convenient point on the frame. The magnet works against the springs on which the screen is mounted, and in this way very rapid vibration can be secured and blinding greatly reduced. (Miall, 1940)

vibrating screens (heated) Wire-mesh screens that are vibrated and heated electrically to increase efficiency. See also: *screens.* (ACSG, 1963)

vibrating wire strain gage This consists of a thin steel wire stretched between knife edges, one being free to move longitudinally. The wire is maintained vibrating at its natural frequency by an electrical method. The knife edges are held firmly against the girder under test, a change of strain in the girder varying tension in the wire and hence its natural frequency. This gage is used in conjunction with a reference instrument of fixed frequency; electrical impulses from both instruments are superimposed to produce beats having a frequency equal to the difference between the frequencies of the two instruments. Changes in the frequency of the test gage caused by variations in strain result in identical changes in the beat frequency. The joint output from these two instruments is applied to the plates of a cathode-ray tube, leading to an oscillation of the electron beam with a frequency equal to that of the higher of the two applied frequencies, with an amplitude that increases and decreases with the same frequency as that of the beats. (Hammond, 1965)

vibration (a) The act of vibrating; oscillation. Vibrations may be free or forced; longitudinal, transverse, torsional, or dilatational; also classified according to kind, such as acoustical, electrical, flexural, etc. (AGI, 1987) (b) The undesirable oscillatory movements of a drill string. (Long, 1960)

vibration drilling Drilling in which a frequency of vibration in the range of 100 to 20,000 Hz is used to fracture rock. Ultrasonic drilling is one of the better known methods of vibration drilling. (Min. Miner. Eng., 1965)

vibration gravimeter A device that measures gravity by observation of the period of transverse vibration of a thin wire tensioned by the weight of a known mass; useful for observation at sea. (AGI, 1987)

vibration meter A seismometer that is used for measuring vibrations of structures from other than seismic causes. (AGI, 1987)

vibration method of roof testing A person's fingertips are placed against the roof, and then the roof is struck a sharp heavy blow. Such a blow usually sets up easily felt vibrations in an unsound roof.

vibration of foundations The foundations of machinery installed in a building should be so designed that the frequency of the machine is two times the natural frequency of the combined system of machines and foundations. (Hammond, 1965)

vibration test An approximate grading test for coarse-grained soils. A flat paper-covered board is inclined at a slope of 1:24. The dry and powdered sample of soil is spread in a thin layer across the top of the board. The board is tapped sharply and repeatedly. The soil will travel down the board, the largest particles traveling faster and further than the smaller ones. Dependent on the degree in which the soil spreads out, a grading can be allotted to the soil. (Nelson, 1965)

vibrator (a) A mechanism imparting vibration to screens, concrete consolidators, and shaking tables. (b) A tool that vibrates at 3,000 to 10,000 cycles per minute. It can be inserted into wet concrete or attached to formwork to compact the concrete. (c) A device for attachment to bins or chutes to produce vibration and thus assist in gravity flow of contained material.

vibratory screen A sizing screen similar to the shaking screen, but the reciprocating movement imparted to it is of greater frequency and much smaller amplitude—1,000 rpm and 1/4 in (6.4 mm) being typical. High-frequency vibration is more effective than the slow movement of the shaker in preventing blinding of holes, and the screening is more effective. It may contain one, two, or three screen decks with water sprays for washing products when screening. Five products ranging from plus 3/4 in (19.1 mm) to minus 1/8 in (3.2 mm) are possible from a double-deck screen. In general, the screen is inclined at from 12° to 14° for the coarser sizes and 17° to 21° for the finer sizes with counterflow operation. See also: *pool washing screen; varislope screen.* (Nelson, 1965)

Vibrex grease table A device to concentrate and separate diamonds from gangue material. It is based on the principle that short, sharp vibrations in rapid succession transmitted to a greased surface cause diamonds to become imbedded in the grease, while water washes away other materials.

Vibroflotation The trade name for a geotechnical process that uses vibration to compact clean sands and gravels. The vibration is combined with a water jet to give a high degree of compaction. (Hammond, 1965)

vibrograph An instrument for recording the ground vibrations caused by heavy quarry blasts. The relationship between the amount of vibration, the distance from the blast, and the weight of explosive fired may be expressed thus: $A = (K\sqrt{E})/D$, where A = maximum amplitude in thousandths of an inch; K = constant depending on the quarry site; E = weight of explosive in pounds; and D = distance in feet. The constant K can be determined by firing a specimen blast of a given size at a given distance and measuring the amplitude of the record obtained. Amplitudes in excess of 0.04 in (1 mm) may give rise to damage. A movement of 0.008 in (0.2 mm) can be felt, and if used an excessive number of times may give rise to complaints of nuisance and damage. Short-delay blasting methods with small diameter holes reduce vibration hazards. See also: *falling-pin seismometer; seismograph.* (Nelson, 1965)

vicinal face (a) One of the facets modifying normal crystal faces; they usually lie nearly in the plane of the face they modify. (CMD, 1948) (b) One of the crystal faces with complex Miller indices in apparent violation of the Bravais law requiring high densities of lattice points parallel to prominent faces. Vicinal faces are small and diverge from major faces by very small angles. Syn: *vicinal form.*

vicinal form *vicinal face.*

Vickers' diamond hardness tester A small impression machine, capable of testing very hard metals, finished components, and very thin sheets. The diamond is similar to that used in the diamond pyramid hardness test. The duration of application of the load is controlled automatically, being always applied and removed in exactly the same manner. This machine may also be used with a ball indenter for the Brinell hardness test. (Hammond, 1965)

Vickers hardness test A test of resistance to deformation of metals or minerals in which a pyramid-shaped diamond is forced into a polished surface of the specimen to be tested under various static loads. The result is a function of the average length of the diagonals of the resulting indentation. Cf: *Tukon hardness test; Brinell hardness test.* (AGI, 1987)

victualic coupling A development in which a groove is cut around each end of a pipe instead of the usual threads. Two ends of pipe are then lined up and a rubber ring is fitted around the joint. A pair of semicircular bands, forming a sleeve, are placed around the ring and are drawn together with two bolts. These have a ridge on both edges that fits into the groove of the pipe. As they are tightened, the rubber ring is compressed, making a watertight joint, while the ridges fitting in the grooves make it strong mechanically. Victualic pipe is faster to lay because in large sizes it does not have to be aligned perfectly and screwed in. (Kentucky, 1952)

victualic joint A proprietary pipe joint that allows the pipes to move through several degrees after fixing but yet to remain watertight. This joint is designed to allow about 12° of movement without causing leakage. The pipes have specially shouldered ends that are contained by a circumferential rubber washer held by a special circumferential-type flange. The water has access to the inner part of the washer, on which it exerts pressure and thereby seals the joint. (Hammond, 1965; Mason, 1951)

Vielle-Montagne furnace A mechanical roasting furnace similar to the Ross and Welter type. (Fay, 1920)

Vienna turquoise An amorphous turquoise imitation once manufactured in Austria, Czechoslovakia, France, and England. Having approx. the same chemical composition, hardness, density, and fracture, it is more difficult to detect than the various blue-stained minerals since used as turquoise substitutes.

vierendeel girder An open-frame N-truss without diagonal members, with rigid joints between the top and bottom chords and the verticals. Known also as open-frame girder. (Hammond, 1965)

viese Scot. The line of fracture of a fault or the soft earth in a fissure or on the sides of a fault. Also spelled vise.

vignite A magnetic iron ore. (Fay, 1920)

vigorite An explosive resembling dynamite No. 2 and consisting of nitroglycerin with a more or less explosive dope. (Fay, 1920)

Vigorite No. 5 Permissible explosive; used in mines. Also called L.F. Vigorite No. 5. (Bennett, 1962)

villamaninite An isometric mineral, $(Cu,Ni,Co,Fe)(S,Se)_2$; pyrite group; forms small iron-black cubo-octahedra and radiating nodular masses; at Villamin, Spain.

Villela's reagent An etching reagent consisting of 95 ml of ethyl alcohol, 5 ml of hydrochloric acid, and 1 g of picric acid. (Osborne, 1956)

villiaumite An isometric mineral, NaF; soft; deep carmine; forms small crystals and grains in nepheline syenite; in Islands of Los, Guinea.

vinney Copper ore, with a green efflorescence like verdigris. (Fay, 1920)

vinogradovite A monoclinic mineral, $(Na,Ca,K)_4Ti_4AlSi_6O_{23}(OH)\cdot 2H_2O$; white to colorless; forms crystals and spherical aggregates in nepheline syenite in the Kola Peninsula, Russia.

vinyl acetal resins Prepared from polyvinyl acetate. Properties are toughness, adhesiveness, imperviousness to moisture, and stability toward light and heat. Used as an interlayer in safety glass and as a bonding resin. (Crispin, 1964)

violaite A highly pleochroic variety of clinopyroxene found in the Caucasus Mountains, Russia.

violan A violet variety of diopside found at St. Marcel, Piemont, Italy.

violite *copiapite.*

virgate To branch in diverging lines. (Webster 3rd, 1966)

virgation (a) A divergent, branchlike pattern of fault distribution. The term is used in Russian literature. (AGI, 1987) (b) A fold pattern in which the axial surfaces diverge or fan out from a central bundle. (AGI, 1987) (c) A sheaflike pattern, as shown on a map, of mountain ranges diverging from a common center. Ant: syntaxis. (AGI, 1987)

virgin (a) Unworked or untouched; said of areas where there has been no mining. (Mason, 1951) (b) An unexploited area or rock formation in which boreholes have not been drilled. (Long, 1960) (c) *primary metal.*

virgin clay Fresh clay, as distinguished from that which has been fired. (AGI, 1987)

virgin coal An area of coal that is in place (in situ) and unimpaired by mining activities. (Nelson, 1965)

virgin metal Pure metal obtained directly from ore. See also: *primary metal.* (ASM, 1961; Newton, 1959)

viridine A green manganese-rich variety of andalusite.

viridite (a) A ferruginous chlorite in chloritic iron ore. (b) A general term formerly applied to indeterminable and obscure green alteration products occurring in scales and threads in the groundmass of porphyritic rocks.

virtual value The calibration of alternating current instruments is based upon what is called the virtual value, and this corresponds to the direct-current value, which would produce the same heating effect in a given resistance. The peak value of alternating voltage or current is 1.4 times greater than the virtual value. (Mason, 1951)

viscometer An instrument used to measure the viscosity. Syn: *viscosimeter.* (AGI, 1987)

viscosimeter *viscometer.*

viscosity The property of a fluid to offer internal resistance to flow; its internal friction. Specif., the ratio of the shear stress to the rate of shear strain. (AGI, 1987)

viscosity coefficient A numerical factor that measures the internal resistance of a fluid to flow; it equals the shearing force in dynes per square centimeter transmitted from one fluid to another that is 1 cm away, and generated by the difference in fluid velocities of 1 cm/s in the two planes. The greater the resistance to flow, the larger the coefficient. Syn: *absolute viscosity; dynamic viscosity.* (AGI, 1987)

viscountess Building slate 18 in by 10 in (45.7 cm by 25.4 cm). (Pryor, 1963)

viscous (a) Adhesive or sticky, having a ropy or glutinous consistency. (Webster 3rd, 1966) (b) Imperfectly fluid; designating a substance that, like tar or wax, will change its form under the influence of a deforming force, but not instantly, as more perfect fluids do. (Standard, 1964)

viscous damping Viscous damping is the dissipation of energy that occurs when a particle in a vibrating system is resisted by a force the magnitude of which is a constant, independent of displacement and velocity, and the direction of which is opposite to the direction of the velocity of the particle. (Hy, 1965)

viscous flow A type of fluid flow in which there is a continuous steady motion of the particles; the motion at a fixed point always remains constant. Also called streamline flow; laminar flow; steady flow. (CTD, 1958)

viscous resistance The effect of surface friction between a particle and a liquid when the particle moves through the liquid. Cf: *turbulent resistance.* (Newton, 1959)

Visean Upper Lower Carboniferous. (AGI, 1987)

viseite An isometric mineral, $Ca_{10}Al_{24}(SiO_4)_6(PO_4)_7O_{22}F_3\cdot 72H_2O(?)$; white; forms wartlike masses; at Vise, Belgium.

visibility meters The general principle of such meters is to observe a portion of the visual field against its background and then to bring about a condition such that the observed difference in brightness reaches a threshold value so that it is only just discernible. The instruments differ in their means by which this end is achieved. The threshold may be produced quite simply by interposing a light-absorbing medium, such as an optical wedge, in the field of view. Other methods include reducing the contrast between the object and its background by superimposing a veiling brightness over the observed field. (Roberts, 1958)

visible light (a) The light of the visible spectrum. (b) Electromagnetic radiation, with wavelength range approx. 4,000 to 7,000 Å, which a normal human eye can detect. Cf: *invisible light; ultraviolet; infrared.*

visible spectrum That portion of the electromagnetic spectrum in which the waves normally produce, upon the human eye, color sensations of red, orange, yellow, green, blue, violet, or their intermediate hues, or of white light if the rays are combined. Distinguished from radio, infrared, ultraviolet, gamma, and X-rays.

Vissac jig An air-operated pulsator jig in which air is alternately compressed and allowed to expand to produce pulsation. This jig has been used principally on sized bituminous coal. (Mitchell, 1950)

visual indicator A device by which the winding or haulage engineman can see on a dial or panel the position of the cages in a shaft or the journey on the haulage plane. See also: *depth indicator.* (Nelson, 1965)

Vitasul A trade name for a chemical additive that eliminates or reduces considerably the danger of diesel locomotive fumes underground. Tests have established that the chemical, added to diesel fuel, reduces the carbon-monoxide danger from diesel locomotive exhausts to negligible proportions. (Nelson, 1965)

vitiated air Air that has been rendered impure by the breath of workers and horses, or by being mixed with the various gases given off in mines. It is frequently called return air. (Peel, 1921)

vitrain (a) Designation of macroscopically recognizable, very bright bands of coals. Very bright bands or lenses, usually a few millimeters (3 to 5) in width; thick bands are rare. Clean to the touch. In many coals, the vitrain is permeated with numerous fine cracks at right angles to stratification and consequently breaks cubically, with conchoidal surfaces. In other coals, the vitrain is crossed by only occasional perpendicular cracks. In the macroscopic description of seams, only the bands of vitrain having a thickness of several millimeters are usually noted. Examination with the microscope shows vitrain to consist of microlithotypes very rich in vitrinite. After clarain, vitrain is the most widely distributed and common macroscopic constituent of humic coals. Occurs in lenticular bands, each derived from a single piece of original vegetable growth. When it constitutes 30% to 60% of total seam, it is termed abundant; more than 60%, dominant; between 15% and 30%, moderate; below 15%, sparse. (IHCP, 1963; Pryor, 1963) (b) A coal lithotype characterized macroscopically by brilliant, vitreous luster, black color, and cubic cleavage with conchoidal fracture. Vitrain bands or lenticles are amorphous, usually 3 to 5 mm thick, and their characteristic microlithotype is vitrite. Cf: *clarain; durain; fusain; vitrite.* Syn: *pure coal.* (AGI, 1987)

vitreous (a) In minerals, a luster typical of that of quartz or calcite. Cf: *adamantine; pearly; resinous.* (b) That degree of vitrification evidenced by low water absorption. See also: *impermeable; nonvitreous; semivitreous.* (ASTM, 1994) (c) Amorphous. (d) Noncrystalline, such as volcanic glass. (e) Consisting of or resembling glass. (Kinney, 1962)

vitreous copper *chalcocite.*

vitreous copper ore *chalcocite.* (Pryor, 1963)

vitreous fusion Gradual fusion; having no sharp melting point. (Webster 3rd, 1966)

vitreous silica Silica glass.

vitreous silver *argentite.*

vitric Said of pyroclastic material that is characteristically glassy; i.e., contains more than 75% glass. (AGI, 1987)

vitrics (a) Fused siliceous compounds, such as glasses and enamels, as distinguished from ceramics, or fused aluminous compounds. (Standard, 1964) (b) The art or history of glass production. (Standard, 1964)

vitric tuff A tuff that consists predominantly of volcanic glass fragments. Cf: *crystal tuff; crystal-vitric tuff.* (AGI, 1987)

vitrifacture The manufacture of vitreous or vitrified wares, as glass. (Standard, 1964)
vitrifiable Of or pertaining to a substance that can be vitrified. (Webster 3rd, 1966)
vitrification An act, or instance, or the process of vitrifying or making glassy; the condition of being vitrified; a vitrified body. (Webster 3rd, 1966)
vitrification spalling That resulting directly or indirectly from the permanent physical changes caused by vitrification.
vitrified (a) That characteristic of a clay product resulting when the temperature in a kiln is sufficient to fuse all the grains and close all the pores of the clay, making the mass impervious. (ACSG, 1961) (b) Converted into glass. (Kinney, 1962)
vitrified brick A very hard paving brick burned to the point of vitrification and toughened by annealing. (Crispin, 1964)
vitriform Having the form or the appearance of glass; glassy. (Webster 3rd, 1966)
vitrify To change into glass or into a glassy substance by heat and fusion. To make vitreous; esp. to produce (as in a ceramic ware) enough glassy phase or close crystallization by high-temperature firing to make nonporous. To undergo vitrification or vitrifaction; to become vitreous. (Webster 3rd, 1966)
vitrifying A stage in the heating of a clay when some of the ingredients have melted and have partially or completely closed the pores, as in stoneware and porcelain. The completion of this stage occurs at the point of maximum shrinkage without loss of shape. See also: *baking; squotting.* (Nelson, 1965)
vitrinertite A coal microlithotype that contains a combination of vitrinite and inertinite totalling at least 95%, and containing more of each than of exinite. It generally occurs in high-ranking bituminous coals. (AGI, 1987)
vitrinite A group name comprising collinite and telinite. Differentiation between collinite and telinite depends in part on the method of observation. The distinction is more easily made in thin section or after etching a polished surface. Often there is uncertainty of distinction by reflected light, and in such cases, it is proper to use the general term vitrinite. See also: *collinite; telinite.* (IHCP, 1963)
vitrinization The process in coalification that results in the formation of vitrain. See also: *coalification.* Cf: *incorporation; fusinization.* (AGI, 1987)
vitrinoid Vitrain and similar material in coal. (AGI, 1987)
vitriol A sulfate of any of various metals (such as copper, iron, or zinc,); esp. a hydrate (as the heptahydrate) of such a sulfate having a glassy appearance or luster. (Webster 3rd, 1966)
vitriol ocher *glockerite.*
vitrite A coal microlithotype group that contains vitrinite macerals totalling at least 95%. Cf: *liptite; vitrain.* (AGI, 1987)
vitro- Prefix meaning glassy.
vitroclarain A rock-type coal consisting of vitrinite (collinite or telinite) and other macerals, mainly exinite, and in which the other macerals exceed vitrinite in quantity. Cf: *clarovitrain.* (AGI, 1987)
vitroclastic Pertaining to a pyroclastic rock structure characterized by fragmented bits of glass; also, said of a rock having such a structure. (AGI, 1987)
vitrodurain Durain in which much vitrain is present. Judged obsolete by the Heerlen Congress of 1935. Cf: *durovitrain.* (AGI, 1987)
vitrofusain A coal constituent transitional between vitrain and fusain, and showing plant cell structure. The cell walls are soaked with vitrain, where the cell cavities are empty. It is not a mixture but a transition. Accepted by the Heerlen Congress of 1935 to designate material transitional between vitrain and fusain with fusain being predominant. Cf: *fusovitrain.* (AGI, 1987)
vitrophyre Any porphyritic igneous rock having a glassy groundmass. Adj: vitrophyric. Cf: *felsophyre; granophyre.* Syn: *glass porphyry.* (AGI, 1987)
vitrophyric Said of a porphyritic igneous rock having large phenocrysts in a glassy groundmass. (AGI, 1987)
vivianite (a) A monoclinic mineral, $2[Fe_3(PO_4)_2{\cdot}8H_2O]$; colorless where fresh; turning blue on oxidation; soft; a secondary mineral found in ore deposits and pegmatites, in clays associated with bone and other organic remains, and in anaerobic lake sediments. Syn: *blue iron earth; blue ocher; Prussian blue.* (b) The mineral group annabergite, baricite, hoernesite, koettigite, parasymplesite, and vivianite.
vlasovite A monoclinic and triclinic mineral, $Na_2ZrSi_4O_{11}$; forms colorless crystals in the contact zone of the Lovozero massif, Kola Peninsula, Russia. It is related to narsarsukite.
V-method of roasting The introduction of a supplementary roast heap between each two regular heaps, so that, if left untouched, there would be a continuous and unbroken roast heap the entire length of the roast yard. (Fay, 1920)
vogesite A lamprophyre composed of hornblende phenocrysts in a groundmass of alkali feldspar and hornblende. Clinopyroxene, olivine, and plagioclase feldspar also may be present. Vogesite contains less biotite than minette. The name, given by Rosenbusch in 1887, is for the Vosges Mountains, France. (AGI, 1987)
vogle *vug.*
voglianite (a) A soft, green, basic uranium sulfate, found in nodules or as earthy coatings. (Standard, 1964) (b) Validity of species is doubtful. All existing specimens, upon examination, have proved to be cuprosklodowskite. (Crosby, 1955; Dana, 1944) (c) A variety of zippeite(?).
voglite A monoclinic mineral, $Ca_2Cu(UO_2)(CO_3)_4{\cdot}6H_2O$(?); strongly radioactive; emerald to grass green; an alteration product of uraninite associated with liebigite.
void (a) A general term for pore space or other openings in rock. In addition to pore space, the term includes vesicles, solution cavities, or any primary or secondary openings. Syn: *pore; interstice.* (AGI, 1987) (b) That portion of a borehole from which the core could not be recovered. (Long, 1960)
void ratio The ratio of the volume of void space to the volume of solid substance in any material consisting of voids and solid material, such as a soil sample, sediment, or sedimentary rock. Symbol e. Syn: *voids ratio.* (AGI, 1987)
voids ratio *void ratio.*
vol *vole.*
volatile Readily vaporizable. (AGI, 1987)
volatile combustibles *volatile matter.*
volatile fluxes The volatile constituents of a magma.
volatile matter In coal, those substances, other than moisture, that are given off as gas and vapor during combustion. Standardized laboratory methods are used in analysis. Syn: *volatiles; volatile combustibles.* (AGI, 1987)
volatile ratio In coal, the ratio of the volatile matter to the sum of the volatile matter and the fixed carbon. (Federal Mine Safety, 1977)
volatiles The volatile constituents (or rest magma) remaining after the less volatile ores have crystallized as igneous rocks. Syn: *volatile matter.* (AGI, 1987)
volborthite A monoclinic mineral, $Cu_3V_2O_7(OH)_2{\cdot}2H_2O$; radioactive; has one perfect cleavage; dark olive to yellow-green; a secondary mineral with carnotite in sandstone. Syn: *uzbekite.*
volcanic Characteristic of, pertaining to, situated in or upon, formed in, or derived from volcanoes. See also: *extrusive.* (Fay, 1920)
volcanic ash *ash.*
volcanic breccia (a) A pyroclastic rock that consists of angular volcanic fragments that are larger than 64 mm in diameter and that may or may not have a matrix. (AGI, 1987) (b) A rock that is composed of accidental or nonvolcanic fragments in a volcanic matrix. Syn: *alloclastic breccia; lava breccia.* (AGI, 1987)
volcanic clay *bentonite.*
volcanic conglomerate A water-deposited conglomerate containing more than 50% volcanic material, esp. coarse pyroclastics. (AGI, 1987)
volcanic dust *ash.*
volcanic earthquake A seismic disturbance that is due to the direct action of volcanic force, or one whose origin lies under or near a volcano, whether active, dormant, or extinct. (AGI, 1987)
volcanic focus The subterranean seat or center of volcanism of a region or of a volcano. (AGI, 1987)
volcanic glass A natural glass produced by the cooling of molten lava, or a liquid fraction of it, too rapidly to permit crystallization. Examples are obsidian, pitchstone, tachylyte, and the glassy mesostasis of many extrusive rocks. Cf: *glass.* (AGI, 1987)
volcanicity *volcanism.*
volcanic ore deposits The major group of ore deposits of magmatic origin, designated as young by European mineralogists, which have been formed under near-surface conditions and very often in Tertiary or younger volcanic rocks. In a strict sense, deposits formed in relation to surface eruptions. (Schieferdecker, 1959)
volcanic plain Surface formed by extensive lava or ash flows that cover topographic irregularities. (AGI, 1987)
volcanic rift zone *rift zone.*
volcanic rock (a) A generally finely crystalline or glassy igneous rock resulting from volcanic action at or near the Earth's surface, either ejected explosively or extruded as lava; e.g., basalt. The term includes near-surface intrusions that form a part of the volcanic structure. Syn: *volcanite.* (AGI, 1987) (b) A general term to include the effusive rocks and associated high-level intrusive rocks; they are dominantly basic. (AGI, 1987)
volcanics A general collective term for extrusive igneous and pyroclastic material and rocks.
volcanic water Water in, or derived from, magma at the Earth's surface or at a relatively shallow level; juvenile water of volcanic origin. (AGI, 1987)
volcanism The processes by which magma and its associated gases rise in the crust and are extruded onto the Earth's surface and into the atmosphere. Also spelled vulcanism. Syn: *volcanicity.* (AGI, 1987)

volcanite An obsolete term variously used to denote a volcanic rock and selenian sulfur.

volcano (a) A vent in the surface of the Earth through which magma and associated gases and ash erupt; also, the form or structure, usually conical, that is produced by the ejected material. (AGI, 1987) (b) Any eruption of material; e.g., mud, that resembles a magmatic volcano. Obsolete var; vulcano. Pl: volcanoes. Etymol: the Roman deity of fire, Vulcan. (AGI, 1987)

volchonskoite *volkonskoite.*

vole The place where tin ore is stored to be dried before being put into a smelting furnace. Syn: *vol.* (Nelson, 1965)

voler reductol A lubricant for enclosed gear units; composed of high-quality mineral oil with a suspension of superfine colloidal graphite and silicone foam inhibitor. (Nelson, 1965)

volgerite A discredited term equal to stibiconite. (Am. Mineral., 1945)

volkonskoite A monoclinic mineral, $Ca_{0.3}(Cr,Mg,Fe)_2(Si,Al)_4O_{10}(OH)_2 \cdot 4H_2O$; smectite group. Also spelled volchonskoite.

volley The act of exploding blasts in sections. A round of holes fired at any one time. (Standard, 1964; Fay, 1920)

volt The practical meter-kilogram-second (mks) unit of electrical potential difference and electromotive force (emf) that equals the difference of potential between two points in a conducting wire carrying a constant current of 1 A when the power dissipated between these two points equals 1 W. It equals the potential difference across a resistance of 1 Ω when 1 A of current is flowing through it; the standard in the United States. (Webster 3rd, 1966)

voltage Electromotive force. (Nichols, 1976)

voltaite An isometric mineral, $K_2Fe_9(SO_4)_{12} \cdot 18H_2O$; has both ferric and ferrous iron.

Volta's list A list or series of metals, such that any one will be at a higher electrical potential when put in contact with any of those that follow, and at a lower potential if in contact with any metal before it in the series. (Fay, 1920)

voltmeter An instrument for determining voltage. (Crispin, 1964)

voltzite Wurtzite mixed with an organometallic zinc compound.

volume defect A crystal structure deviating from ideality by having two or more chemical species in one or more crystal sites. Cf: *crystal defect; disorder.*

volume susceptibility *susceptibility.*

volumetric Chemical analysis based upon the reaction of a volume of standard solution with the material being analyzed. (ASTM, 1994)

volumetric analysis Quantitative chemical analysis in which known weight of sample is dissolved and reacted with a standard chemical solution of strength proportional to its normality or hydrogen equivalent. Completion of reaction (end point) is judged by change of color, incipient precipitation, or effect on an indicator. See also: *dry assay; wet assay.* (Pryor, 1963)

volumetric efficiency (a) The ratio of the total quantity of air passing along the faces to the quantity flowing in the fan drift. See also: *overall ventilation efficiency; surface air leakage.* (Roberts, 1960) (b) The ratio of the volume of air discharged to the displacement for a fan or compressor. (Hartman, 1982) (c) The volume of water that enters a pump cylinder for each piston stroke divided by the volume swept by the piston (piston area times stroke). (Nelson, 1965)

volumetric shrinkage The decrease in volume, expressed as a percentage of the soil mass when dried, of a soil mass when the water content is reduced from a given percentage to the shrinkage limit. Also called volumetric change. (ASCE, 1958)

volute (a) A spiral casing to a mine fan to provide an area of passage, which gradually increases in proportion to the increasing area of discharge from the fan. See also: *evase.* (Nelson, 1965) (b) A spiral casing for a centrifugal pump or a fan designed so that speed will be converted to pressure without shock. (Hammond, 1965)

volute pumps This type of centrifugal pump is the most commonly used. The impellers may be open, closed or semienclosed, single suction, double suction, or nonclogging. They discharge into casings that are progressively expanding spiral designs of one or more stages (multistage). The casings housing the rotating elements may be vertically or horizontally split, and a few designs have casings divided on an angle from the horizontal. Pumps in this class usually have a specific speed below 4,000 rpm with single-suction impellers and a specific speed of 5,000 rpm with double-suction impellers. (Pit and Quarry, 1960)

von Neumann spike The pressure peak leading the detonation wave prior to the establishment of the C-J state. (Van Dolah, 1963)

vonsenite An orthorhombic mineral, Fe_3BO_5; ludwigite group; contains both ferrous and ferric iron with ferrous iron replaced by magnesium toward ludwigite; black; forms coarse granular masses; at Riverside, CA.

von Sterneck-Askania pendulum A device for measuring the vertical component of gravity, characterized by the use of four pendulums in a single case. (AGI, 1987)

von Wolff's classification A quantitative chemical-mineralogic classification of igneous rocks proposed in 1922 by F. von Wolff. (AGI, 1987)

vooga hole *vug.*

Vooys process A coal-cleaning process using a heavy suspension, consisting of clay and finely ground barite (-150 or 200 mesh) in water. A coal containing as little as 3.3% to 3.4% ash is steadily produced, with a yield practically equal to the theoretical float-and-sink yield. (Gaudin, 1939)

vorobievite A white or rose-colored variety of beryl from the Ural Mountains, Russia, and Madagascar. Syn: *morganite.* Also spelled vorobyevite.

vortex finder Tube projecting into central vortex of hydrocyclone or dense medium cyclone through which the classified fines or lighter specific gravity fraction of pulp leaves the system.

vraibite An orthorhombic mineral, $Tl_4Hg_3Sb_2As_8S_{20}$; forms gray-black crystals or thin red splinters; commonly intergrown with realgar and orpiment; occurs at Salonika, Macedonia, Greece.

vug A small cavity in a rock, usually lined with crystals of a different mineral composition than the enclosing rock. Adj: vuggy. Cf: *druse; miarolitic cavity; geode.* Syn: *bug hole; vogle; vooga hole.* See also: *cavity; vuggy porosity; loch.* Etymol: Cornish vooga, cavern or cavity.

vugg A misspelling of vug. (Long, 1960)

vuggy Pertaining to a vug or having numerous vugs. Syn: *vugular.* (AGI, 1987)

vuggy lode A lode or vein in which vugs or drusy cavities are of frequent occurrence. (Fay, 1920)

vuggy porosity Porosity due to vugs in calcareous rock. The term vugular is used by some writers but condemned by others. See also: *vug.* (AGI, 1987)

vugh-arching When a pocket of rock in the periphery of an excavation is weaker than the remainder, it may fail under the ring stress. Fragments split away or fall out until all the weak rock is removed, forming an artificial vugh. This is called vugh-arching. (Spalding, 1949)

vug hole *vug.*

vugular *vuggy.*

vulcan coal powder Explosive; used in mines. (Bennett, 1962)

vulcanism Volcanism.

vulcanite (a) An orthorhombic mineral, CuTe; with rickardite and native tellurium, it forms coatings on rocks; occurs at the Good Hope Mine, Vulcan, CO. (b) A dark-colored, hard variety of vulcanized India rubber that differs from the softer rubber in having been vulcanized at a high temperature; ebonite. It takes a high polish, and is used for making combs, ornaments, etc., and in electrical work because of its fine insulating properties. (Standard, 1964)

vulcanites A general name for igneous rocks of fine grain size, normally occurring as lava flows, and thus in direct contrast with plutonites. (CTD, 1958)

vulcanized rubber A rubber that has been heated with sulfur to change its properties. (Nelson, 1965)

vulcanizing machine Consists essentially of two heavy metal plattens that are placed one on each side of the previously prepared joint and clamped firmly together. Each platten is heated, and this combined application of heat and pressure over a period completes the joint. These machines are used to vulcanize the belt joints of conveyors.

vulcan powder High explosive composed of 30% nitroglycerin, 52.5% sodium nitrate, 10.5% charcoal, and 7% sulfur. (Pryor, 1963)

vulpinite A scaly, granular variety of anhydrite; may be admixed with silica; cut and polished for ornamental purposes.

V-vat (a) A funnel box; also, having a groove or grooves of a triangular section. (Webster 2nd, 1960) (b) *spitzkasten.*

vysotskite A tetragonal mineral, (Pd,Ni)S; isomorphous with braggite; in minute grains or prisms; at Norilsk, Russia.

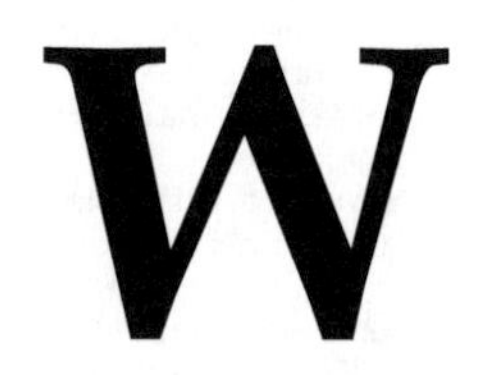

W.8 methanometer A dual-scale, direct-reading instrument for measuring the combustible gases percentage in mine air. It gives combustible gases readings over the range from 0.2% to 5% and is graduated 0.1% per division on the scale. (Nelson, 1965)

wabanite Banded cream to black and gray to purple chocolate-colored slate from Massachusetts. (Schaller, 1917)

wacke (a) A dirty sandstone that consists of a mixed variety of angular and unsorted or poorly sorted mineral and rock fragments, and of an abundant matrix of clay and fine silt; specif. an impure sandstone containing more than 10% argillaceous matrix. The term is used for a major category of sandstone, as distinguished from arenite. (AGI, 1987) (b) A clastic sedimentary rock in which the grains are almost evenly distributed among the several size grades; e.g., a sandstone consisting of sediment poured in to a basin of deposition at a comparatively rapid rate without appreciable selection or reworking by currents after deposition, or a mixed sediment of sand, silt, and clay in which no component forms more than 50% of the whole aggregate. (AGI, 1987) (c) A term commonly used as a shortened form of graywacke. This usage is not recommended. (AGI, 1987) (d) Originally, a term applied to a soft earthy variety of basalt, or to the grayish-green to brownish-black claylike residue resulting from the partial chemical decomposition of basalts, basaltic tuffs, and related igneous rocks. Syn: *vake*.—Etymol: German Wacke, an old provincial mining term signifying a large stone or stoniness in general. (AGI, 1987)

wacken Rocklike clay, formed by the decomposition of basalts in situ. Cf: *graywacke*. (Arkell, 1953)

wad (a) An earthy, dark brown to black mineral material consisting chiefly of an impure mixture of manganese oxides and oxyhydroxides with variable amounts of copper, cobalt, and iron oxides and oxyhydroxides and silica plus 10% to 20% adsorbed water. It is commonly soft (soiling hands), but may be hard and compact, and has a low density. Wad results from the decomposition of other manganese minerals and accumulates in marshy areas or other zones of groundwater emission where it is an ore of manganese. See also: *asbolan*. Cf: *psilomelane*. Syn: *bog manganese; black ocher; earthy manganese; manganese hydrate*. (b) A general term for massive, fine-grained manganese oxides and oxyhydroxides of low density, but not further identified. (c) In drilling, a term applied to rock cuttings that tend to ball and adhere to drill-string equipment and borehole walls in lumps. (Long, 1960)

wad coil Eng. A tool for extracting a pebble or broken tool from the bottom of a borehole. It consists of two spiral steel blades arranged something like a corkscrew. See also: *spiral worm*. Also called wad hook.

wadding Paper or cloth placed over explosives in a hole. (Nichols, 1976)

Waddle fan An earlier type of centrifugal fan. It had no external casing, but delivered directly to the atmosphere all around its periphery. The veins were curved backwards in the direction of rotation and the air was led into the fan by a curved inlet passage or throat. It was usually driven by steam at about 70 rpm; efficiency about 40%; external diameter of about 30 ft (9.1 m). See also: *Sirocco fan*. (Nelson, 1965)

wadeite A hexagonal mineral, $K_2CaZ{\cdot}(SiO_3)_4$; forms hexagonal plates; in Western Australia.

wad hook *wad coil; spiral worm.*

wadi (a) A term used in the desert regions of Southwestern Asia and Northern Africa for a stream bed or channel, or a steep-sided and bouldery ravine, gully, or valley, or a dry wash, that is usually dry except during the rainy season, and that often forms an oasis. (AGI, 1987) (b) The intermittent and torrential stream that flows through a wadi and ends in a closed basin. (AGI, 1987) (c) A shallow, usually sharply defined, closed basin in which a wadi terminates.—Etymol: Arabic. Variant plurals: wadis; wadies; wadian; widan. See also: *arroyo; nullah*. Also spelled: wady; waddy. Syn: *oued; widiyan*. (AGI, 1987)

Waelz process A process by which low-grade ores, slags, or residues from retorts may be treated either for the recovery of zinc alone or for the recovery of zinc, lead, and tin. It employs a rotary kiln, and the zinc-bearing material mixed with fine coal is fed into the kiln and heated, so that the zinc is vaporized and converted to oxide fume. (Newton, 1959)

wafer (a) A name given to the rough slice obtained by sawing directly from a mother crystal or section. The process of manufacturing wafers is variously known as wafering, wafering from the crystal or slab, wafering from the mother crystal, and baloney slicing. (Am. Mineral., 1947) (b) Small sheet of electroceramic material 0.001 to 0.01 in (0.025 to 0.25 mm) thick for use in electronic equipment, particularly in miniature capacitors, transistors, resistors, and other circuit components.

waggon *wagon.*

wagnerite A monoclinic mineral, $(Mg,Fe)_2(PO_4)F$; magnesium is replaced by ferrous iron or calcium; imperfect cleavage.

wagon (a) An underground coal car. (Korson, 1938) (b) A mine car. (c) Any vehicle for carrying coal or debris. (Mason, 1951) (d) A trailer with a dump body. (Nichols, 1976)

wagon arrester An appliance that can bring a wagon completely to rest and is usually used near the departure end of mine sidings. It can be rendered inoperative by remote control if required. (Nelson, 1965)

wagon booster-retarder An appliance that reduces the speed of wagons traveling above the design value, but for wagons traveling at speeds less than this, it releases energy by thrusting against the wheel flanges, therefore speeding up the vehicle. See also: *wagon retarder*. (Nelson, 1965)

wagon breast (a) Rooms or wide coal roadways into which mine cars or wagons are taken. (Nelson, 1965) (b) A pillar method of working a relatively thick, flattish coal seam. (Nelson, 1965)

wagon drill A drilling machine mounted on a light, wheeled carriage. (BS, 1964)

wagon pinch bar A device for moving railway wagons and locomotives short distances by hand. It consists of a cast-steel wedge-shaped tip with a wood handle. The tip is placed over the rail and under the wagon wheel and the up-and-down movement of the handle exerts sufficient pressure on the wheel to move the wagon. (Nelson, 1965)

wagon rerailer A device for bringing a derailed wagon back onto the track. It usually consists of ramp elements, which can be fixed at intervals along the track or temporarily fitted to the track just beyond the end of the wagon. The wagon is then pulled to cause the wheels to ride up the ramp and back on to the rails. (Nelson, 1965)

wagon retarder An appliance that reduces the speed of a wagon traveling in excess of a designed value (e.g., 3-1/2 mph), while having no effect on wagons moving at speeds less than this figure. The appliance is a self-contained hydraulic unit. (Nelson, 1965)

wagon rooms Rooms driven in inclined seams in such a way that an adequate gradient is secured for cars, which are often hauled to the heads of the rooms. (Stoces, 1954)

wagon spotter A wagon spotting appliance. It may be a "bogey" that is hauled backwards and forwards on a separate track installed between the main track rails by a winch. A forward pull on the bogey raises a pair of arms to engage in the wagon axle, and a reverse pull lowers the arms to enable the bogey to be drawn back under the next wagon ready for the next pull. (Nelson, 1965)

wagon tippler A power-operated structure for discharging coal or other material from a railway wagon. (Nelson, 1965)

wairakite A monoclinic mineral, $CaAl_2Si_4O_{12}{\cdot}2H_2O$; zeolite group; pseudocubic; colorless to white; in tuffaceous rocks in geothermal areas.

wairauite An isometric mineral, CoFe; forms minute grains with awaruite in the Red Hills serpentines; at Wairau, South Island, New Zealand.

waiver Involves the notion of an intention entertained by the holder of some right, to abandon or relinquish instead of insisting on the right. It is a question of fact. Proof of waiver must include proof of knowledge of the facts upon which the waiver is based. (Ricketts, 1943)

Wakefield sheet pile Consists of three boards bolted or spiked together with the center board offset. This arrangement produces a tongue and groove that makes Wakefield sheet piling fairly watertight if the piles are properly driven and tightly fitted together. (Urquhart, 1959)

walaite A variant spelling of valaite. (Tomkeieff, 1954)

walchowite A honey-yellow variety of retinite containing little nitrogen, occurs in brown coal at Walchow, Moravia, Czech Republic.

waling Eng. Cleaning coals by picking out refuse.
walk To deviate from the intended course, such as a borehole that is following a course deviating from its intended direction. Also called deviating; war; wandering. Syn: *walking.* Cf: *deviate; drift; wander.* (Long, 1960)
Walker balance A type of counterpoised beam balance. Cf: *Westphal balance.*
walker's earth *fuller's earth.* Etymol: German Walkererde. (AGI, 1987)
walking (a) The movement forward or backward of a dredge by first winding up on one side and then the other, swinging the boat from side to side and thereby advancing with a slight offsetting to the side. (Fay, 1920) (b) *walk.*
walking bar A trunnion or walking beam. (Nichols, 1976)
walking beam (a) The beam used to impart a reciprocating movement to the drilling column in percussive drilling. Syn: *oscillating beam; rocking beam.* (BS, 1963) (b) On cable tool and churn drill rigs, the beam that carries the string of drilling tools at one end and is connected to a cranked drive wheel at the other. The rotation of the wheel causes the tool string to lift and drop; thus the hole is drilled by concussion. (Long, 1960)
walking crane A light crane traveling on an overhead channel iron and a single rail vertically beneath this in the floor. (Webster 3rd, 1966)
walking dragline (a) A dragline that is equipped with apparatus that permits it to "walk" by the alternate power movement of vertical booms fastened to large outrigger platforms so arranged as to push the equipment forward as work progresses. (BCI, 1947) (b) An excavator of very large capacity, equipped with walking beams operated by eccentrics in place of crawler tracks. Such machines can excavate 1,650 st/h (1,500 t/h) of overburden to a depth of 100 ft (30 m). (Hammond, 1965)
walking miner *joy walking miner.*
walking props *self-advancing supports.*
walking support *self-advancing supports.*
walkout Act of walking out or leaving; specif., a labor strike. (Webster 2nd, 1960)
wall (a) The side of a level or drift. (Fay, 1920) (b) The country rock bounding a vein laterally. The side of a lode; the overhanging side is known as the hanging wall and the lower lying side as the footwall. See also: *hanging wall; footwall.* Syn: *walls of a vein.* (Fay, 1920) (c) The face of a longwall working or stall, commonly called a coal wall. (Fay, 1920) (d) A rib of solid coal between two rooms; also, the sides of an entry. (BCI, 1947)
wall accretions Material adhering to the inner walls of a blast furnace between the water jackets and the feed door. (Fay, 1920)
Wallace agitator Mixing device, driven by an impeller, used in pulp mixing and aeration in cyanidation where strong agitation is needed. (Pryor, 1963)
wall boss (a) A person who supervises a crew of workers operating a face conveyor. (Hess) (b) *room boss.*
wall cake *cake.*
wall cavitation The development of enlarged sections in a borehole as the result of caving, erosive action of the circulated liquid, or erosion caused by drill rods rubbing against the borehole walls. (Long, 1960)
wall clearance The distance between the wall of the borehole and the outside of a piece of drill-string equipment when the string is centered in the borehole. (Long, 1960)
wall closure *closure.*
wall-controlled shoots Ore shoots that occur adjacent to certain favorable wall rocks that presumably influenced deposition from the mineralizing fluids. (Stokes, 1955)
wall drag The amount of friction resulting from the drill rods rubbing against the walls of a borehole or the inside surface of the casing lining a borehole. (Long, 1960)
waller Laborer who builds walls to support backfilling. *pack builder.* (Fay, 1920)
wall face Scot. The face of the coal wall; the working face. (Fay, 1920)
wall friction (a) The drag created in the flow of a liquid or gas because of contact with the wall surfaces of its conductor, such as the inside surfaces of a pipe or drill rod or the annular space between a drill string and the walls of a borehole. (Long, 1960) (b) The drag resulting from compaction of loose materials around the outside surfaces of drive pipe, casing, etc. Also called skin friction. (Long, 1960)
walling (a) The brick or stone lining of shafts. (Fay, 1920) (b) Derb. Stacking or setting up ironstone, etc., in heaps, preparatory to being measured or weighed. (Fay, 1920)
walling curb *curb; foundation curb; water ring.*
walling scaffold *bricking scaffold.*
walling stage A movable wooden scaffold suspended from a crab on the surface, upon which the workers stand when walling or lining a shaft. (Fay, 1920)
walling up The building up of a layer of mud cake or compacted cuttings on the borehole sidewalls; the filling of cracks or caved portions of the borehole walls with cement. (Long, 1960)
wall off To seal cracks, crevices, etc., in the wall of a borehole with cement, mud cake, compacted cuttings, or casing. (Long, 1960)
wall packing The compaction of sticky cuttings that collect and adhere to the walls of a borehole. (Long, 1960)
wallplate (a) A horizontal timber supported by posts resting on sills and extending lengthwise on each side of a tunnel. Roof supports rest on wallplates. Syn: *pad.* (Stauffer, 1906) (b) A horizontal member, usually of wood, bolted to a masonry wall to which the frame construction is attached. Also called headplate. (ACSG, 1961)
wallplate anchor A machine-bolt anchor with a head at one end and threaded at the other, and fitted with a plate or punched washer so that when embedded in the masonry it will be securely anchored and will hold a wallplate in place. (ACSG, 1961)
wall rock (a) *country rock.* (b) The rock forming the walls of a borehole. (Long, 1960) (c) The rock adjacent to, enclosing, or including a vein, layer, or dissemination of ore minerals. It is commonly altered. The term implies more specific adjacency than host rock or country rock. Syn: *walls of a vein.* (AGI, 1987) (d) The rock mass comprising the wall of a fault. (AGI, 1987)
wall-rock halo A dispersion pattern formed in the rock adjoining mineral deposits where the chemical composition has been modified by the ore-forming fluids. *halo.* (Hawkes, 1957)
wall-rock pattern A channel dispersion pattern in which the minor elements of the walls of the channels have been modified. Wall-rock dispersion patterns of importance usually are those formed at the time the orebodies were being deposited. (Lewis, 1964)
walls (a) Coal roadways in pillar-and-stall mining. (Nelson, 1965) (b) The side of an orebody defining where the ore ceases and the country rock begins. Walls may be definite or indefinite. See also: *footwall; hanging wall.* (Nelson, 1965)
wall saltpeter *nitrocalcite.*
wallscraper bit A rotary bit used to enlarge the diameter of a borehole. (BS, 1963)
Wallsend Eng. A grade of coal for household purposes: originally from Wallsend, on the Tyne, but now from any part of a large district in and near Newcastle. (Standard, 1964)
walls of a vein *wall; wall rock.*
walpurgite A triclinic mineral, $Bi_4O_4(UO_2)(AsO_4)\cdot 2H_2O$; radioactive; yellow-orange; associated with trögerite, zeunerite, uranspharite, torbernite, and uranospinite. Also spelled walpurgin.
waltherite A discredited term equal to walpurgite. (Am. Mineral., 1945)
Walton filter An emerald glass or beryloscope mounted to resemble a hand loupe. Observed through it the filament of an incandescent lamp appears reddish yellow, and this color is not changed when also passing through most genuine emeralds; but a Brazilian emerald from Minas Geraes appears green, an epidote red, and a dioptase green. Syn: *emerald loupe.* See also: *emerald glass.*
wander (a) An unintentional change in the course of a borehole. Cf: *deviate; walk; warp.* (Long, 1960) (b) *band wander.*
wane A defect in a timber or plank. (Crispin, 1964)
want A zone in which the coal of a coal seam is missing, owing to a low-angle normal fault or a washout, squeeze, or roll. Cf: *nip; pinch.* Syn: *cutout.* See also: *washout; squeeze; roll.* (AGI, 1987)
Ward drill A hand drill that can be used in a river on a barge or on a platform built on two large canoes. Basically, it consists of four straight poles, 5 to 7 in (12.7 to 17.8 cm) in diameter at the large end, which are set into notches in planks to prevent their sinking into the ground. The poles are joined at the top by a shaft that holds the pulley for the drill wire or rope. The walking beam is activated by 8 to 10 persons lining up on the crossarm. They pull down to raise the tools and vary their manner of movement, depending on whether they are driving casing, drilling, or pulling casing. The Ward drill is most efficient in shallow ground, but can be used in depths up to 90 ft (27.4 m). (Mining)
warden (a) A term used in south Wales for a strong massive sandstone associated with coal. (AGI, 1987) (b) In Australia an officer under the Mining Act with magisterial and executive authority over a goldfield.
wardite A tetragonal mineral, $NaAl_3(PO_4)_2(OH)_4\cdot 2H_2O$; bluish-green; forms oolitic or crystalline encrusting layers; in Cedar Valley and near Fairfield, UT.
Ward-Leonard control (a) A method of controlling the speed of electric winding or other large direct-current motors, employing a variable voltage generator to supply the motor armature, and driven by a shunt motor. See also: *automatic cyclic winding; Ilgner system.* (Nelson, 1965) (b) In a modification, the Ward-Leonard-Ilgner system incorporates a heavy flywheel on the shaft of the generator, which smooths out surges in the system. (Pryor, 1963)
waringtonite A hydrated sulfate of copper that shows an emerald green color. Syn: *brochantite.* (Hey, 1955)
wark Eng. Black slaty stone overlying coal seams, Somerset Coalfields. Also spelled werk.

warning lines The lines drawn on working plans to indicate the limit beyond which workings should not extend; e.g., because of the proximity of disused or abandoned workings. (BS, 1963)

warp (a) The amount a borehole has wandered off course. Cf: *wander*. (Long, 1960) (b) A general term for a bed or layer of sediment deposited by water; e.g., an estuarine clay, or the alluvium laid down by a tidal river. See also: *warping*. (AGI, 1987)

warped fault A fault, usually a thrust fault, that has been slightly folded. (AGI, 1987)

warping The slight flexing or bending of the Earth's crust on a broad or regional scale, either upward (upwarping) or downward (downwarping); the formation of a warp. (AGI, 1987)

warrant (a) A general term for the clay floors of coal seams, particularly when hard and tough. See also: *underclay*. (Nelson, 1965) (b) The document of title to metal stored in an LME registered warehouse. The warrant is a bearer instrument and states the brand of metal, its weight, the number of pieces, and the rent payable. (Wolff, 1987)

warrant clay *underclay*.

warren earth *fireclay*. Corruption of warrant. (Arkell, 1953)

Warren girder A triangulated truss made up only of sloping members between the horizontal top and bottom members with no verticals. See also: *N-truss*. (Hammond, 1965)

warrenite (a) A general term for gaseous and liquid bitumens consisting mainly of a mixture of paraffins and isoparaffins: a variety of petroleum rich in paraffins. (b) A pink cobaltoan variety of smithsonite. (c) A discredited name for owyheeite or jamesonite.

Warrington Strand construction in which one layer of wires is composed of pairs of large and small wires, thus 6 x 19 (6 and 6/6/1) equal laid. See also: *equal lay*. (Hammond, 1965)

Warrington rope A wire rope comprising 7 wires of the same size covered by 12 wires alternately large and small. (Lewis, 1964)

warwickite An orthorhombic mineral, $(Mg,Ti,Fe,Al)_2(BO_3)O$; forms dull, brownish-black prismatic crystals having perfect cleavage.

Warwick safety device A safety appliance placed near the upper end of an inclined haulage road to stop a tram running wild down the incline. It consists of a heavy beam longer than the height of the roadway and is normally held up entirely at roof level, but is hinged at the lower end. In the event of a tram running away from above, a haulage hand can pull a rope that releases the upper end of the beam that drops and stops the tram. (Nelson, 1965)

Warwickshire method A method of mining contiguous seams. See also: *bord-and-pillar*. (Fay, 1920)

wash (a) Loose or eroded surface material (such as gravel, sand, silt) collected, transported, and deposited by running water, such as on the lower slopes of a mountain range; esp. coarse alluvium. Syn: *wash stuff*. (AGI, 1987) (b) An alluvial placer. (AGI, 1987) (c) In coal mining, a washout. (Nelson, 1965) (d) The wet cleaning of coal or ores. (Nelson, 1965) (e) Auriferous gravel. (Fay, 1920) (f) To clean cuttings or other fragmental rock materials out of a borehole by the jetting and buoyant action of a copious flow of water or a mud-laden liquid. The similar ejection of core or drill spring equipment from a borehole. See also: *alluvial cone*. (Long, 1960)

washability Coal properties determining the amenability of a coal to improvement in quality by mechanical cleaning. (BS, 1962)

washability curve A curve or graph showing the results of a series of float-and-sink tests. A number of these curves are drawn to illustrate different conditions or variables, usually on the same axes, thus presenting the information on one sheet of paper. Washability curves are essential when designing a new coal or mineral washery. There are four main types of washability curves: characteristic ash curve, cumulative float curve, cumulative sink curve, and densimetric or specific gravity curve. (Nelson, 1965; BS, 1962)

wash boring (a) Drilling by use of jet water applied inside a casing pipe, in unconsolidated ground. (Pryor, 1963) (b) A test hole from which samples are brought up mixed with water. (Nichols, 1976)

wash-boring drill A drill rig utilizing the jet action of a high-pressure stream of water to produce a borehole in soft or unconsolidated material. (Long, 1960)

wash bottle (a) A bottle or flask fitted with two glass tubes passing through the stopper, so that on blowing into one tube a stream of water issues from the other tube. The stream may be directed upon anything to be washed or rinsed. Newer wash bottles are made of flexible plastic with a single tube. (Hess) (b) A bottle for washing gases by passing them through liquid contained in it. (Hess)

washbox In coal preparation, the jig box in which feed is stratified and separated into fractions (heavier below and lighter above). A feldspar washbox has a bedding of that mineral. (Pryor, 1963)

washbox air cycle The valve-timing cycle determining the periods of air admission and exhaust. (BS, 1962)

washbox cells The individual portions into which the part of a washbox below the screen plate is divided by transverse division plates, each being capable of separate control. (BS, 1962)

washbox center sill A sill fitted over a center extraction chamber. (BS, 1962)

washbox center weir An adjustable plate situated between the feed end and the discharge end of a washbox and serving to regulate the forward movement of material through the box. (BS, 1962)

washbox compartments The sections into which a washbox is divided by transverse division plates that extend above the screen plate to form a weir; each compartment usually comprises two or more cells. (BS, 1962)

washbox discharge sill That part of the washbox over which the washed coal passes out of the box. Usually the discharge sill is a part of the discharge-end refuse extraction chamber. (BS, 1962)

washbox feed sill That part of the washbox over which the feed passes when it enters the box. Usually the feed sill is a part of the feed-end refuse extraction chamber. (BS, 1962)

washbox screen plate (a) The perforated plate or grid that supports the bed of material being treated. (BS, 1962) (b) Also called grid plate; sieve plate; bedplate.

washbox slide valve A washbox air valve operated by means of a reciprocating motion. Also called washbox piston valve. (BS, 1962)

wash dirt (a) The tailings or material discarded in the operation of washing an alluvial deposit for gold. (Nelson, 1965) (b) Gold-bearing earth worth washing. Also called wash stuff; wash gravel. See also: *washing stuff*. (Fay, 1920)

washed coal (a) Coal from which impurities have been removed by any form of mechanical treatment. (BS, 1960) (b) Coal produced by a wet-cleaning process. (BS, 1962)

washed out Said of a coal seam when the bed thins out. (Mason, 1951)

washery (a) A place at which ore, coal, or crushed stone is freed from impurities or dust by washing. Also called wet separation plant; washing plant. See also: *washing apparatus; wash place; coal-preparation plant; dense-medium washer; efficiency of separation*. (Webster 3rd, 1966) (b) A coal preparation plant in which a cleaning process is carried out by wet methods. (BS, 1962) (c) A building resembling a breaker used in reclaiming culm and fine coal from old banks. (Korson, 1938)

washery effluent Surplus water discharged from a washery, usually to waste (after settlement of solids in suspension). (BS, 1962)

washery products The final products from a washery. (BS, 1962)

washery pump A pump generally of simple construction and heavy design since slurry presents a difficult pumping problem owing to its erosive action. This type pump is generally of the single-stage type as heads are small, with a solid casing of steel or cast iron about twice the normal thickness to provide against erosive action. (Sinclair, 1958)

washery refuse The refuse removed at preparation plants from newly mined coal.

washery water The water used in the wet separation of coal from shale by differences in density. See also: *recirculation of water*. (Nelson, 1965)

wash gold *placer gold*.

wash gravel Gravel washed to extract gold. Cf: *wash dirt*. (Webster 3rd, 1966; Fay, 1920)

washhouse A building on the surface at a mine where the miners can wash before going to their homes. A changehouse. A dryhouse.

washing (a) That which is retained after being washed; as, a washing of ore. (b) The selective sorting, and removal, of fine-grained sediment by water currents. (AGI, 1987) (c) Erosion or wearing-away by the action of waves or running water. (AGI, 1987) (d) The act or process of cleaning, carrying away, or eroding by the buoyant action of flowing water. (Long, 1960) (e) Ore mineral, such as gold dust, that is retained after being washed.

washing apparatus (a) Machinery and appliances erected on the surface at a coal mine, for extracting, by washing with water, the impurities mixed with the coal dust or small slack. (Zern, 1928) (b) Machinery for removing impurities from coals and ores. See also: *washery*. (Zern, 1928)

washing drum *trommel*.

washing hutch *hutch*.

washing machine Scot. A machine for separating impurities from coal by means of water. (Fay, 1920)

washing plant *washery*.

washing screen Flat screen or trommel on which passing ore is exposed to sprays or jets of water to remove as undersize any adherent mud or other fine material. (Pryor, 1963)

washing screws Devices in which continuous helical blades arranged about shafts force the material up an inclined trough against a stream of water introduced at the higher end. This action carries away the soluble material occurring with the material and dumps the washed product over the higher end of the trough. (Pit and Quarry, 1960)

washing stuff An earthy deposit containing gold that may be extracted by washing. See also: *wash dirt*. Syn: *wash stuff*. (Webster 3rd, 1966; Fay, 1920)

washing trommel *trommel*.

wash metal Molten metal used to wash out a furnace, ladle, or other container. (ASM, 1961)

Washoe canary A miner's slang term for a donkey; burro. (Standard, 1964)

Washoe process The process of treating silver ores by grinding in pans or tubs with the addition of mercury, and sometimes of chemicals, such as blue vitriol and salt. Named from the Washoe District, NV, where it was first used. (Webster 3rd, 1966)

wash ore Crude iron ore containing readily liberated particles of pure iron ore, loosely agglomerated with sands from which they can be separated by scrubbing treatment. See also: *natural ore.* (Pryor, 1963)

washout (a) A channel or channellike feature produced in a sedimentary deposit by the scouring action of flowing water and later filled with the sediment of a younger deposit. Syn: *scour and fill.* (AGI, 1987) (b) A channel cut into or through a coal seam at some time during or after the formation of the seam, generally filled with sandstone—or more rarely with shale—similar to that of the roof. See also: *cutout; horseback; want; low.* (Raistrick, 1939; AGI, 1987) (c) Barren, thin, or jumbled areas in coal seams in which there is no actual disruption and no vertical displacement of the coal and strata. These disturbances may be divided into three main types; namely (1) classical washouts, (2) pressure belts, and (3) tremor tracts. Authentic washouts should be restricted to the first group. Also called rock fault; nip. See also: *roll.* (Nelson, 1965) (d) Local thinning or disappearance of a coal seam due to erosion during or shortly after its formation. (BS, 1964) (e) Channellike features that cut or transgress the stratification of the underlying beds; may be small scour-and-fill structures or large erosional channels. Also called cutout. (Pettijohn, 1964)

washout valve Valve in a pipeline or a dam that can be opened occasionally to clear out sediment. (Hammond, 1965)

washover (a) To wash away or remove material from around the outside of casing pipe, drill stem, junk, or tramp materials in a borehole. See also: *washover shoe.* (Long, 1960) (b) Material deposited by the action of overwash; specif. a small delta built on the landward side of a bar or barrier, separating a lagoon from the open sea, produced by storm waves breaking over low parts of the bar or barrier and depositing sediment in the lagoon. Cf: *blowover.* (AGI, 1987)

washover shoe A casing shoelike bit used to drill downward around a piece of drilling equipment stuck in a borehole. See also: *washover.* (Long, 1960)

wash pan A pan for washing pay dirt in placer mining. (Standard, 1964)

wash pipe The pipe that ejects the jet of water through the bit, used in wash boring. (Long, 1960)

wash place A place where ores are washed and separated from the waste; usually applied to places where the hand jigs are used. See also: *washery.* (Fay, 1920)

wash plain *alluvial plain.*

washpot In tin-plate manufacturing, a pot containing melted tin into which the plates are dipped to be coated. (Webster 3rd, 1966)

wash rod A heavy wall pipe used in lieu of drill rods to conduct water downward inside and to the bottom of a drivepipe being sunk through overburden by a wash-and-drive method. Syn: *wash tube.* (Long, 1960)

wash sale A practice in which promoters, through the connivance of brokers who pretend to carry through transactions and thus obtain false quotations, create a fictitious flurry of activity in the stock market. (Hoover, 1948)

wash stuff *wash; washing stuff.*

wash table An inclined table used for cleaning coal or ore in which the lighter material or gangue is washed away by water. The coal or ore is fed onto the table and water is allowed to flow down the table carrying away the impurities.

wash trommel Rotating horizontal drum that receives ore at one end and water at the other. Ore is tumbled countercurrent to the water so that coarse solids are discharged continuously while water now charged with mud and fine material overflows at the feed end. (Pryor, 1963)

wash tube *wash rod.*

wash water Water circulated through the drill string, past the bit, and then out of the borehole between the rods and the walls of the hole while drilling or during washing operations. See also: *water wash.* (Long, 1960)

wasite An altered variety of allanite. (Standard, 1964)

waste (a) The part of an ore deposit that is too low in grade to be of economic value at the time of mining, but which may be stored separately for possible treatment later. (Lewis, 1964) (b) Refuse and impurities removed in mining and treating coal; also, the coal left in a mine as pillars. (c) Gangue. (d) Tailings. (e) Overburden. (f) The refuse from ore dressing and smelting plants. Gob; goaf; old workings; also, the fine coal made in mining and preparing coal for market; culm; coal dirt; also used to signify both the mine waste (such as coal left in pillars) and the breaker waste. (g) A working or shaft which has been abandoned and filled with refuse (goaf or gob), or with material from the fall of the hanging wall. Syn: *condie.* (Standard, 1964) (h) *spoil.*

waste blasting On some coal faces, the stone overlying the seam does not always fall in the wastes after withdrawing the supports. To avoid excessive weight on the face, which would cause dangerous roof conditions, it is desirable to blast down the stone in the wastes. Also, in thick seams, the overlying strata requires breaking down to provide sufficient stone for building packs. The holes for waste shots must be drilled from the face side so that the driller is working under a supported roof. Care must be taken to ensure that the holes are not drilled up into the solid strata and that the burden on the shot is not excessive. (McAdam, 1958)

waste drainage The controlled leakage of air through a waste to ensure that large concentrations of mine gases do not accumulate in it. (BS, 1963)

waste dump The area where mine waste or spoil materials are disposed of or piled.

waste edge support A row of rigid timber or steel props or chocks set along the edge of the waste and parallel to the longwall face to induce the roof beds to break and to secure caving of the waste area. See also: *breaker props.* (Nelson, 1965)

waste-filled stopes In these methods, support for walls and for workers and machines is furnished by waste rock, tailing sand, etc., called filling or gob. In true waste filling, the orebody is excavated in sections alternating with filling, and it is sometimes referred to as cut-and-fill stoping. (Higham, 1951)

waste filling Material used for support in heavy ground and in large stopes to prevent failure of rock walls and to minimize or control subsidence and to make it possible to extract pillars of ore left in the earlier stages of mining. Material used for filling includes waste rock sorted in the stopes or mined from rock walls, mill tailing, sand and gravel, smelter slag, and rock from surface open cuts or quarries. (Lewis, 1964)

waste-heat boiler A boiler that uses the heat of exit gases from furnaces to produce steam or to heat water. (Pryor, 1963)

wasteman In anthracite and bituminous coal mining, a laborer who looks after and keeps clean the airways, haulageways, or working places of a mine. Also called cleanup man; dirt shoveler; sweeper. See also: *jerry man.* (DOT, 1949)

waster (a) Tinplate below the standard weight and quality. (Standard, 1964) (b) Spoiled or imperfect casting or machined part that must be discarded although partly processed. (Pryor, 1963) (c) A brick, structural or refractory, that is defective as drawn from the kiln; wasters in the refractories industry are crushed and reused as grog. (Dodd, 1964)

waste raise An excavation in a mine in which barren rock and other material is broken up for use as filling at the stope. (Stoces, 1954)

waste rock Barren or submarginal rock or ore that has been mined, but is not of sufficient value to warrant treatment and is therefore removed ahead of the milling processes. (Pryor, 1965)

wastes The unfilled or unpacked portions of workings in a mine. (TIME, 1929-30)

waste water Excess water allowed to run to waste from the water circuit. (BS, 1962)

wasteway (a) A channel for carrying off superfluous water. (Webster 3rd, 1966) (b) The channel required to convey water discharged into it from a spillway, escape, or sluice; a spillway. (Seelye, 1951)

wasting asset Property (as mines or lumber tracts) subject to depletion. (Webster 3rd, 1966)

wastrel Eng. A tract of wasteland, or any waste material. (Fay, 1920)

watchmen Weak coal pillars left in workings to give warning of an impending collapse. (Briggs, 1929)

water (a) Clear, colorless liquid. (b) A rarely used term referring to the color and clarity of a precious stone or pearl, and esp. of a diamond. (AGI, 1987)

water ampule A fire-resistant plastic container of water that is used as a safety precaution in shotholes. (BS, 1964)

water-ampul stemming A water cartridge for stemming shotholes in coal or rock. The ampul consists of a plastic (polyvinylchloride) bag, 1-1/4 in (3.2 cm) in diameter and 18 in (45.7 cm) in length. When filled with water and the neck of the bag tied off, the filled ampul is about 15 in (38.1 cm) in length and holds slightly over 1/2 pint (0.24 L) of water. Compared with dry clay or sand, the use of water ampuls for stemming effects substantial reductions in both the airborne dust and the nitrous fumes produced by shot firing. This applies to both coal and rock blasting. Also called a water dummy. Syn: *cushion firing. stem bag.* (Nelson, 1965)

water-avid surface A term used to describe a surface that seems to prefer contact with water to contact with air. In flotation, minerals with a water-avid surface will not float, while those with an air-avid surface will. The object of reagent additions in flotation is to form a water-repellent surface on the minerals to be floated and a water-avid surface on the minerals that are not to float (hydrophilic). Cf: *air-avid surface.* (Newton, 1959)

water bailer *bailer.*

water balance An obsolete water-raising apparatus consisting of a swinging frame carrying a double series of troughs ascending in zigzag lines, and so adjusted to each other that as the frame rocks in either direction water may be passed to a higher level. (Standard, 1964)

water barrel A barrel-shaped hoppit designed to collect and hoist water from the bottom of a sinking shaft. Water barrels are now obsolete. See also: *pneumatic water barrel; sinking pump; water kibble.* Also called barrel; bailer. (Nelson, 1965)

water barrier (a) An area of solid material left unworked to protect a mine, or part of a mine, against entry of secondary water. (BS, 1963) (b) *barrier pillar.*

water-base mud A drill mud in which the solids are suspended in water. (Long, 1960)

water blast (a) The expulsion of water under pressure, in mine workings, caused by trapped air expanding as the water level is lowered. (BS, 1963) (b) Explosion caused by a sudden inrush of water. (Mason, 1951) (c) The discharge of water down a shaft to produce or quicken ventilation. See also: *trombe.* (d) A water-actuated ventilating device. (CTD, 1958)

water blasting Pulsed infusion shot firing. (Nelson, 1965)

water block (a) A sudden stoppage of water-flow past the face of a bit while drilling is in progress. (Long, 1960) (b) A hollow box or block of iron through which water is circulated to protect part of a furnace wall.

water boss Aust. The owner or holder of water or water rights who sells the same for mining purposes.

waterbound A general term indicating that water is the medium used to assist in filling the voids between mineral fragments and to improve compaction. (Nelson, 1965)

water box (a) A rectangular wooden pipe used in shafts for conveying water between garlands. (BS, 1963) (b) A square, open, wooden tank car used for removing small amounts of water from low places in a mine. Also, a tank car used for sprinkling the roadways to settle the dust.

water break A break in the continuity of the water film upon a metal when it is withdrawn from a bath. Cf: *wetting.* (ASM, 1961)

water cage A special cage running in guides in a special compartment of a shaft with a separate winding engine.

water cartridge A waterproof cartridge surrounded by an outer case, the space between being filled with water, which is employed to destroy the flame produced when the shot is fired, thereby lessening the chance of an explosion should gas be present in the place.

water chamber A water reservoir in a mine, usually located at the lowest place, commonly near the shaft station. Also called sump. (Stoces, 1954)

water chrysolite Moldavite. See also: *tektite.* (CMD, 1948)

water color The apparent color of the surface layers of the sea caused by the reflection of certain components of the visible light spectrum coupled with the effects of dissolved material, concentration of plankton, detritus, or other matter. Color of oceanic water varies from deep blue to yellow and is expressed by number values that are a variation of the Forel scale. Plankton concentrations may cause a temporary appearance of red, green, white, or other colors. See also: *Forel scale.* (Hy, 1965)

water content (a) Of a bottom sediment, a ratio obtained by multiplying the weight of the water in the sample by 100 and dividing the results by the weight of the dried sample; expressed as a percentage. (Hunt, 1965) (b) *moisture content.*

water core A hollow core through which water circulates in a mold used for cooling the interior of a casting more rapidly than the outside while the metal is solidifying, such as in casting a cannon. (Webster 3rd, 1966)

water coupling *water swivel.*

watercourse (a) A natural or artificial channel for the passage of water, as a river, canal, flume, or drainage tunnel. (b) A subsurface opening or passage in rocks through which groundwater flows.

water creep The movement of water under or around a structure, such as a dam, built on a semipermeable foundation. See also: *piping.* (AGI, 1987)

water curb *garland.*

water cushion A water load pumped into drill pipe during a drill-stem test to retard fillup and prevent collapse of pipe under sudden pressure changes. (Wheeler, 1958)

water-cutoff core barrel A core barrel having a device in its head part that closes and stops the flow of drill-circulation liquid when a core block occurs in the inner tube of the core barrel. (Long, 1960)

water cycle *hydrologic cycle.*

water dam A permanent stopping to seal off a large body or feeder of water. It consists usually of a block of concrete between two brick end walls and these are extended well into the surrounding ground. The contact points and all breaks in the strata are sealed by cement injection. The various pipes, pressure gages, etc., may be left through the stopping. (Nelson, 1965)

watered Eng. Containing much water—full of springs or feeders; e.g., heavily watered mines, heavily watered measures, etc.

waterfinder *divining rod.*

water flush A system of well boring in which percussive drills are used in connection with water forced down to the bottom of the hole through the drill rods. This water jet makes the tools cut better and washes the detritus up out of the hole.

water gage (a) An instrument for measuring the difference in pressure produced by a ventilating fan or air current. See also: *manometer.* (CTD, 1958) (b) An instrument for measuring the ventilation pressure. One-inch (2.54-cm) water gage is equivalent to a pressure of 62.5/12 = 5.2 psi (35.85 kPa). See also: *inclined water gage; total ventilating pressure.* (Nelson, 1965) (c) A measure of ventilating pressure, expressed in terms of the height of a column of water. (BS, 1963) (d) A device that measures the pressure at which water is discharged by a pump or the volume of water flowing through a pipe or other conductor. (Long, 1960) (e) An instrument used to measure the depth or quantity of water, such as in a steam boiler or water storage tank. (Long, 1960) (f) A manometer used with a Pitot tube to indicate air pressure. (Pryor, 1963)

water garland *garland.*

water gel (a) An explosive material containing substantial portions of water, oxidizers, and fuel, plus a cross-linking agent. (FR 249, 1993) (b) An explosive material containing substantial portions of water, oxidizers and fuel, plus a cross-linking agent. Syn: *slurry.*

water gin Scot. A gin actuated by a water wheel.

water glass A concentrated and viscous solution of sodium silicate or potassium silicate in water. Used as an adhesive, a binder, a protective coating, in waterproofing cement, and in bleaching. Colorless; amorphous; $Na_2O{\cdot}XSiO_2$, in which X = 3 to 5; deliquescent. Syn: *waterglass; soluble glass; liquid glass.* (Handbook of Chem. & Phys., 2)

waterglass *water glass.*

water grade (a) An entry inclination that is just sufficient to drain off water. (b) A grade determined by keeping the working place nearly parallel to the edge of a pool of water standing upon its floor. Water grade is sometimes incorrectly called water level. Cf: *water level.*

water groove *waterway.*

water hauler (a) One who collects in a water box (car) water that accumulates in low places—at the mine entrance, along haulageways, or at the working face—bailing it into a car with a bucket or using a small hand pump. Also called waterman; water monkey; water tender. (DOT, 1949) (b) A laborer who hauls water cars into a mine to supply water for sprinkling haulage roads and working places. (DOT, 1949)

water hoist A simple method of disposing of mine water using tanks with an engine or a motor on the surface. The machinery can be easily repaired and the plant is in no danger of being flooded. The high cost of this system and the fact that the shaft cannot be used for other purposes while water is being hoisted are important disadvantages. Water is delivered intermittently and at a decreasing rate as the depth of hoisting increases. This method is less economical than pumping but is useful as an emergency measure in reclaiming a flooded mine. See also: *drainage.* (Lewis, 1964)

water-holding capacity The smallest value to which the water content of a soil can be reduced by gravity drainage. (ASCE, 1958)

water inch (a) The discharge from a circular sharp-edged orifice 1 in (2.54 cm) in diameter with a head of one line above the top edge that is commonly estimated at 14 pints/min (6.6 L/min), and that constitutes an old unit of hydraulic measure. (Webster 3rd, 1966) (b) *miner's inch.*

water infusion A technique being used abroad to suppress or prevent the formation of dust, in advance of mining a coal seam. Water (or sometimes foam or steam, which is costlier but more effective) is injected into the coal ahead of the face through long drill holes, as many as four to six per face and 6 to 20 m in length. The liquid infuses into the seam along fractures and cracks and, under pressure, penetrates a considerable distance from the hole radially, wetting the coal well. It has proved very effective in reducing dust concentrations during subsequent mining—in some instances, as much as 80%. Water infusion originated in Great Britain (it is used in 25% of the dusty mines) and has been tried experimentally with some success in the United States. See also: *pulsed infusion shot firing.* (Hartman, 1961)

water infusion gun A special tube that acts as a borehole seal in the water infusion process. The tube has two separate passages: one for the infusion water and one for admitting hydraulic fluid, which actuates a piston, expanding the seal in the borehole. The infusion water is supplied by a power pump and the hydraulic fluid is supplied by a hand pump. Syn: *infusion gun.* (Nelson, 1965)

water infusion method A method of removing methane from mines. It consists of injecting water under pressure into a coal seam to push out the gas. Holes are drilled horizontally into the coal face and water is pumped into some of the holes at pressures varying from 200 to 650 psi (1.4 to 4.5 MPa). This forces the methane out through the other holes and also from the exposed part of the coal seam. Syn: *methane removal.*

water infusion pump A power pump, mounted on wheels, used to supply high-pressure water for coal seam infusion. It consists of an oil hydraulic circuit that drives two reciprocating rams, which in turn are directly coupled to the two rams of the water pump. (Nelson, 1965)

water inrush A heavy and sudden inflow of water into mine workings or shafts. See also: *inrush of water*. (Nelson, 1965)

water jacket Cast- or wrought-iron sections of a furnace so constructed as to allow free circulation of water for keeping the furnace cool. Also called water block and water box.

water-jet drilling The drilling of boreholes in unconsolidated or earthy formations using the erosive power of a small-diameter stream of water forcefully ejected as the cutting tool. See also: *jet*. (Long, 1960)

water kibble A large iron bucket with a valve in the bottom for self-filling; sometimes used in hoisting the water from a mine. See also: *water barrel*. (Standard, 1964)

water level (a) A level roadway, constructed with an impervious seal or barrier on the dip side, to divert the flow of water along the level and prevent its seepage to workings on the dip side. The level dips slightly outward to allow gravity flow. See also: *drain tunnel*. (Nelson, 1965) (b) The level at which, by natural or artificial drainage, water is removed from a mine or mineral deposit. (c) A drift at the water level. See also: *water grade*. (d) The level of underground waters in a mine, or the elevation to which water will rise in a mine, when the mine is not being drained. (Statistical Research Bureau, 1935)

water leyner A type of rock drill in which water is fed into the drill hole through the hollow drill steel to remove the drill cuttings and, at the same time, allay the dust. Also known as Leyner-Ingersoll drill. (Fay, 1920)

waterlime *hydraulic limestone*.

water load S. Wales. The head, or pressure per square inch, of a column of water in pumps, etc.

water lodge An underground reservoir. (CTD, 1958)

waterlogged (a) Said of workings or mines that have become filled with water because of abandonment or stoppage of operations. See also: *inrush of water*. (Nelson, 1965) (b) Referring to land where the water table is permanently located at or near ground level. See also: *water table*.

water loss The amount of drill water that escapes into porous or fractured borehole wall rocks and hence does not return and overflow at the collar of the borehole. (Long, 1960)

water machine Scot. A pump or other appliance actuated by a water wheel for raising water. (Fay, 1920)

water mains In coal mining, pipes made of cast iron or steel for the conveyance of water. (Mason, 1951)

waterman (a) A laborer who quenches coke with water so that it may be drawn from the oven, using a sprinkling system of perforated pipes. (DOT, 1949) (b) *water hauler*.

watermelon tourmaline A variety of elbaite in pink, green, and colorless prismatic crystals.

water monkey *water hauler*.

water of capillarity The water held in the soil above the standing-water level by reason of capillary attraction. Also called held water. See also: *capillary water*. (Hammond, 1965)

water of compaction Water furnished by destruction of pore space owing to compaction of sediments.

water of constitution Water as an essential component in a mineral or other compound, either as water of crystallization; e.g., $CaSO_4{\cdot}2H_2O$ gypsum, or water more tightly bound as (OH) groups; e.g., FeO(OH) goethite. Also called H_2O+ in wet chemical analyses.

water of crystallization Water that combines with salts when they crystallize and remains as H_2O molecules in the crystal structure; given off upon heating to 100 to 200 °C. Also called H_2O in wet chemical analyses.

water of hydration Water that is chemically combined in a crystalline substance to form a hydrate, but that may be driven off by heat. (AGI, 1987)

water opal (a) *hyalite*. (b) Any transparent precious opal.

water packer An expandable device that is placed in a borehole to bar entry of water into the lower part of a hole or to separate two distinct flows of water from different strata. See also: *packer*. (Long, 1960)

water pocket A small, bowl-shaped depression on a bedrock surface, where water may gather; esp. a water hole in the bed of an intermittent stream, formed at the foot of a cliff by the action of falling water when the stream is in the flood stage. (AGI, 1987)

waterpower (a) The power of water derived from its gravity or its momentum as applied or applicable to the driving of machinery. (Standard, 1964) (b) A descent or fall in a stream from which motive power may be obtained; specif., in law, the fail in a stream in its natural state, as it passes through a person's land or along the boundaries of it. (Standard, 1964)

water privilege (a) The right to the use of the water of a certain stream. (Standard, 1964) (b) The right to the possession and use of a fall of water for mechanical purposes. See also: *water right*. (Standard, 1964)

waterproofed stone dust The proofing of stone dust to prevent the particles from caking or becoming sticky in humid atmospheres. Waterproofing is considered essential if stone-dust barriers are to operate effectively at humidities above 85%. (Nelson, 1965)

waterproof electric blasting cap A cap specially insulated to ensure reliability of firing when used in wet work. Syn: *blasting cap*.

water-quenched Cooled with water, as in hardening steel. (Standard, 1964)

water rate The weight of dry steam consumed by a steam engine for each horsepower per hour. The result is stated in either indicated horsepower or brake horsepower. (Brantly, 1961)

water-repellent surface *air-avid surface*.

water resistance A qualitative measure of the ability of an explosive or blasting agent to withstand exposure to water without deteriorating or becoming desensitized.

water right (a) The right to use water for mining, agricultural, or other purposes. See also: *water privilege*. (b) The right to appropriate water granted to miners by Federal laws; however, this right applies only to water on public domain. Rights of way are granted over public lands for ditches, canals, flumes, and for the construction of a reservoir to one who has a right to water. (Lewis, 1964) (c) When one has legally acquired a water right, the person has a property right therein that cannot be taken away for public or private use, except by due process of law and upon just compensation being paid therefor. One who has acquired a legal water right can only be deprived of it by the voluntary act of conveying it to another, by abandonment, forfeiture under some stature, or by operation of law. A water right is an independent right and is not a servitude upon some other thing, and is an incorporeal hereditament, being neither tangible nor visible. (Ricketts, 1943)

water ring A special form of cast-iron bricking curb whereby space is provided for building up the walling and also a channel or groove for collecting water running down the shaft sides. Rings are built into the shaft lining at intervals and pipes are arranged to conduct the water to the next lower ring or a sump. See also: *curb; garland*. Syn: *walling curb*. (Nelson, 1965)

water-rolled Said of round, smooth sedimentary particles that have been rolled about by water. (AGI, 1987)

water sapphire (a) A light-colored blue sapphire. (b) An intense blue variety of cordierite occurring in water-worn masses in river gravels (such as in Sri Lanka) that may be used as a gemstone. Syn: *saphir d'eau*. (c) A term applied to water-worn pebbles of topaz, quartz, and other minerals in Sri Lanka.

water seal A water accumulation in a depression in an underground roadway or in a pipe, sufficient to form a seal. (BS, 1963)

water separation *elutriation*.

water shutoff The sealing off of salt-water-bearing formations to prevent harmful underground water pollution. This is ordinarily done by cementing. (Williams, 1964)

water slot A groove incised in the face and outside wall of a noncoring bit that serves as a waterway. Syn: *waterway*. (Long, 1960)

water smoke To heat a kiln slowly to dry out the moisture from the bricks, before firing. (Standard, 1964)

water softening Removal of excess calcium and magnesium, from water through precipitation of their carbonates (ion exchange) to remove ionized calcium, magnesium, etc. (Pryor, 1963)

water-soluble oils Oils having the property of forming permanent emulsions or almost clear solutions with water. (Fay, 1920)

water stemming bags Water-filled plastic bags with a self-sealing valve classified as a permissible stemming device by the U.S. Mine Safety and Health Administration. (Atlas, 1987)

waterstone (a) A stone whose cutting crystals break away rapidly from its bond. The use of water forms a gritty paste which acts in much the same way as oil when used on an oilstone. The Queer Creek and Hindostan stone are good examples. (b) Forest of Dean. A shale, so called in consequence of the wet soil that is found wherever it appears at the surface. (c) The formation name for certain flaggy micaceous sandstone and marls in the Keuper of the Midlands. (d) Eng. Quarrymen's name for the lowest bed in a Portland stone quarry at Long Crendon, Buckinghamshire. (e) A whetstone requiring water instead of oil.

water string Casing used to shut off water-bearing formations encountered in the drilling of a well. (Williams, 1964)

water swivel A device connecting the water hose to the drill-rod string and designed to permit the drill string to be rotated in the borehole while water is pumped into it to create the circulation needed to cool the bit and remove the cuttings produced. Also called gooseneck; swivel neck; water coupling. Cf: *air swivel*. (Long, 1960)

water table The surface between the zone of saturation and the zone of aeration; that surface of a body of unconfined ground water at which the pressure is equal to that of the atmosphere. Syn: *ground-water surface; ground-water table; plane of saturation; saturated surface; level of saturation; phreatic surface; ground-water level; free-water elevation; free-water surface*. See also: *waterlogged*. (AGI, 1987)

water-table contour A line drawn on a map to represent an imaginary line in the water table of a definite level. These contours are constructed from the data provided by the water-table levels, corrected for differences in surface level at the respective boreholes. A site investigation or opencast plan sometimes show water-table contours. (Nelson, 1965)
water-table level Level showing the depth of the water table below the surface; the depth at which water is encountered in trial pits or boreholes. (Nelson, 1965)
water-table map A contour map of the upper surface of the saturated zone. (Stokes, 1955)
water-table stream Concentrated ground water flow at the water table in a formation or structure of high permeability.
watertight (a) A borehole in which the conditions are such that no loss of the circulated drill fluid occurs. (Long, 1960) (b) A connection, container, or rock strata so tight as to be impermeable to water. (Long, 1960)
water-to-cement ratio The ratio between the weight of water and the weight of cement in mortar or concrete. The lower the water-to-cement ratio, the higher will be the strength of the concrete. (Hammond, 1965)
water tower (a) A standpipe or its equivalent, often of considerable height, giving a head to a system of water distribution. (Standard, 1964) (b) A tower in which a falling spray of water is used to wash gas, etc. (Standard, 1964) (c) A tower containing tanks in which water is stored, built at or near the summit of an area of high ground in cases where the ordinary water pressure would be inadequate for distribution to consumers in the area. (CTD, 1958)
water transport Water is used for transport in some mines, esp. in placers and in claypits, and generally in mines in an elevated position and with a loose mineral. Also filling material is often transported into the mine by water. The mixture of water and solid material can also be conveyed by pumps horizontally or raised to a small height. (Stoces, 1954)
water turbine A prime mover coupled to an alternator, using a purely rotary motion to generate an alternating current. The main types of water turbines are (1) the Pelton wheel for high heads, (2) the Francis turbine for low to medium heads, and (3) the Kaplan turbine for a wide range of heads. (Hammond, 1965)
water tuyère A water-jacketed tuyère. (Webster 3rd, 1966)
water vein (a) Groundwater in a crevice or fissure in dense rock. (AGI, 1987) (b) A term popularly applied to any body of groundwater, in part because dowsers commonly describe water as occurring in veins. The term is little used among hydrologists. (AGI, 1987)
water wash The use of water to remove the soluble constituents of a mill product before further treatment. See also: *wash water*. (Fay, 1920)
waterway (a) A groove or slot incised in the surface of a bit or other piece of drill-string equipment to provide a channel through which the circulated drilling fluid can flow. Also called watercourse; water groove; water passage; water slot. (Long, 1960) (b) A way or channel, either natural (as a river) or artificial (as a canal), for conducting the flow of water. (AGI, 1987) (c) A navigable body or stretch of water available for passage; a watercourse. (AGI, 1987)
water wheel A wheel so arranged with floats, buckets, etc., that it may be turned by flowing water; used to drive machinery, raise water, etc. The overshot and undershot water wheel, the breast wheel, and the tub wheel are now largely discarded in favor of the turbine. (Standard, 1964)
water witch A device for determining the presence of water, usually electrically. Cf: *divining rod*. (AGI, 1987)
water witching *dowsing*.
waterworn stone Gem minerals, esp. crystals, rounded by action of water rolling them against rocks or gravels in beds of rivers, lakes, or the ocean.
water yardage Extra payment to miners who work in a wet place, either by the yard of progress or the ton of coal mined. (Fay, 1920)
watt The absolute meter-kilogram-second (mks) unit of power that equals 1 absolute joule per second; the standard in the United States; equals 1/746 hp. Abbrev., w and W. (Webster 3rd, 1966; Zimmerman, 1949)
wattevillite An orthorhombic or monoclinic mineral, $Na_2Ca(SO_4)_2{\cdot}4H_2O$(?); forms hairlike crystals. Also spelled watteville-ite.
watt-hour A unit of measurement of electrical work that equals 1 W expended for 1 h. Abbrev., whr and wh. (Crispin, 1964; Zimmerman, 1949)
wattless current An alternative name for the reactive component of an alternating current. (CTD, 1958)
wattmeter An instrument for measuring electric power in watts, the unit of electrical energy, volt times amperes; therefore, combining the functions of a volmeter and an ammeter. Abbrev., wm. (Crispin, 1964; Zimmerman, 1949)
watt-second A unit of measurement of electrical work that equals the rate of 1 W expended for 1 s. Abbrev., wsec. (Crispin, 1964)
Waugh drill *rock drill*.
wave diffraction *diffraction*.
wave front In seismology, the surface of equal time elapse from the source point to the position of the resulting outgoing signal at any given time after the source charge has been activated. In a more restricted sense, the surface along which phase is constant at a given instant. (AGI, 1987)
wave-front chart In seismology, a diagram of a series of lines showing equal times from the point of detonation. In its construction, velocity information must be known or assumed. Charts are usually constructed so that the horizontal and vertical scales are in length, but they can be constructed so that the horizontal scale is in length and the vertical scale is in time. (AGI, 1987)
wave interference The phenomenon that results when waves of the same or nearly the same frequency are superposed, characterized by a spatial or temporal distribution of amplitude of some specified characteristic differing from that of the individual superposed waves. (Hy, 1965)
wavelength (a) The linear distance between successive wave crests or other equivalent points in a waveform or harmonic series. It is equal to the velocity divided by the frequency—measured in cycles per second—and may be represented by the wave number in reciprocal units, e.g., cm^{-1}. (b) In symmetrical, periodic tectonic fold systems, the distance between adjacent antiformal or synformal axial planes. For asymmetrical and nonperiodic systems, various definitions have been proposed.
wavellite An orthorhombic mineral, $Al_3(PO_4)_2(OH,F)_3{\cdot}5H_2O$; pearly; forms hemispherical radiating aggregates in low-grade aluminous phosphatic rocks; also occurs in veins; a source of phosphorus.
wave meter An instrument to measure and record the wave spectra. (Hy, 1965)
wave period The time interval between the appearance of two consecutive wave repetitions at a given point, usually expressed in seconds. The wave segments considered must be the same; i.e., the crests, troughs, etc. (Hy, 1965)
wave propagation Radiation, such as from an antenna of r-f energy into space or of sound energy into a conducting medium. (Hy, 1965)
wave refraction (a) The process by which a water wave, moving in shallow water as it approaches the shore at an angle, tends to be turned from its original direction. The part of the wave advancing in shallower water moves more slowly than the part still advancing in deeper water, causing the wave crests to bend toward parallel alignment with the shoreline. (AGI, 1987) (b) The bending of wave crests by currents. (AGI, 1987)
wave spectrum A concept used to describe the distribution of energy among waves of different period. Wave speed increases with wave length, so distant storms may be detected by the increase of energy in long period waves. Sea is fully developed when all possible wave frequencies possess energies appropriate to the spectrum for the prevailing wind speed. (Hy, 1965)
wave velocity A quantity that specifies the speed with which a wave travels through a medium. (Hy, 1965)
wavy extinction *undulatory extinction*.
wavy vein A vein that alternately enlarges or pinches at short intervals. (Fay, 1920)
wax (a) A solid, noncrystalline hydrocarbon of mineral origin, such as ozocerite and paraffin; composed of the fatty acid esters of the higher hydrocarbons. (b) Soft or puddled clay used for dams or stoppings in a mine.
wax opal An early name for yellow opal with a waxy luster. (Fay, 1920)
wax stone Crude ozokerite associated with earthy matter. (Tomkeieff, 1954)
wax wall A wall of clay built around the gob or goaf to prevent the entry of air or egress of gas. (CTD, 1958)
wax walling The building of clay lumps as a lining to the pack to reduce leakage. If about 15% to 20% of calcium chloride is added to the clay, it will remain plastic. (Sinclair, 1958)
way The rails, sleepers, etc., upon which cars, tubs, or corves run. (Fay, 1920)
wayboard (a) Eng. A thin layer or band that separates or defines the boundaries of thicker strata; e.g., thick beds of limestone separated by wayboards of slaty shale, sandstone separated by wayboards of clay. Also spelled weighboard. (Fay, 1920) (b) Leic. Beds of green marl among sandy shales in the Trias. (Arkell, 1953)
waygate The tailrace of a mill. (Fay, 1920)
waylandite A trigonal mineral, $(B,Ca)Al_3(PO_4,SiO_4)_2(OH)_6$; crandallite group; white; in Uganda.
way shaft *winze*.
weak ground Roof and walls of underground excavations that are in danger of collapse unless suitably supported. (Pryor, 1963)
wearing strip A strip of metal applied to any particular device to take the wear of moving parts or objects. In conveyor work, it refers to the

strip that is sometimes applied to the chain conveyor troughs and on which the drag chain rides. (Jones, 1949)

wear plate Sections of hardened steel plates of various thicknesses (as desired) which are welded (or bolted) in areas of severe wear, such as the insides of feeders, crushers, transfer chutes, or any other area where mined material passes, creating abrasion, and threatens wearing of these areas.

wear rib Hard metal ridge applied to the outside surface of bottom-hole equipment and built up as close as practicable to the set-outside-diameter size of a reaming shell, which serves as a wear pad. (Long, 1960)

weather (a) To undergo changes, such as discoloration, softening, crumbling, or pitting of rock surfaces; brought about by exposure to the atmosphere and its agents. (AGI, 1987) (b) To undergo or endure the action of the elements; to wear away, disintegrate, discolor, or deteriorate under atmospheric influences. (Webster 3rd, 1966)

weather door A door in a mine level whose purpose it is to regulate ventilation currents. A trapdoor. (Fay, 1920)

weathered layer (a) In seismic work, a zone extending from the surface to a limited depth, usually characterized by a low velocity of transmission, which abruptly changes to a higher velocity in the underlying bedrock. The name is erroneous, and the zone is more properly called the low-velocity layer. (AGI, 1987) (b) In seismology, the zone of the Earth that is immediately below the surface, characterized by low seismic-wave velocities. (AGI, 1987)

weathered rock Rock, the character of which has been changed by exposure to decaying conditions found in the zone of weathering. (Long, 1960)

weathering (a) The destructive processes by which earthy and rocky materials on exposure to atmospheric agents at or near the Earth's surface are changed in color, texture, composition, firmness, or form, with little or no transport of the loosened or altered material; specif. the physical disintegration and chemical decomposition of rock that produce an in-situ mantle of waste and prepare sediments for transportation. Most weathering occurs at the surface, but it may take place at considerable depths, as in well-jointed rocks that permit easy penetration of atmospheric oxygen and circulating surface waters. (AGI, 1987) (b) Exposing ore to the atmosphere for long periods in order that a part, at least of the sulfide content, may become oxidized and washed away by the rain. (Osborne, 1956) (c) *seasoning.*

weathering correction In seismic exploration, a correction applied to reflection and refraction data for variations in travel time produced by irregularities in a low-velocity or weathered layer near the surface. Syn: *low-velocity-layer correction.* (AGI, 1987)

weathering index A measure of the weathering characteristics of coal, according to a standard laboratory procedure. Syn: *slacking index.*

weathering map In seismic work, a map on which the low-velocity layer, or weathered layer, is plotted and contoured to show areal variations. (AGI, 1987)

weathering of coal The slow disintegration of coal into fires in surface stockpiles under the action of the weather, particularly frost after a wet period. See also: *spontaneous combustion.* (Nelson, 1965)

weathering of roadways The disintegration or scaling of exposed surfaces of mine roadways, particularly in the case of clay or shale rocks. Gunite has been used for roadway protection against weathering. See also: *guniting.* (Nelson, 1965)

weathering shot In seismic exploration, the detonation of a small explosive charge in the weathering or low-velocity layer to determine its velocity characteristics and thickness. (AGI, 1987)

web The slice or thickness of coal taken by a cutter loader when cutting along the face. The thickness of a web varies from a few inches with plow-type machines to up to about 6 ft (1.8 m) with the A.B. Meco-Moore. The term web tends to be restricted to thin or medium slices of coal. See also: *buttock.* (Nelson, 1965)

weber Magnetic flux equivalent in the meter-kilogram-second (mks) system of the maxwell in the centimeter-gram-second (cgs) system. One weber equals 10^8 maxwells. Abbrev., wb. (Pryor, 1963)

weberite An orthorhombic mineral, Na_2MgAlF_7; pale gray; forms grains in cryolite; in Greenland.

Weber process A method of manufacturing pig iron in which the ore is mixed with a proportionate amount of coal sufficient to smelt it; after adding a binder the mixture is briquetted by means of a roller press into ovoids, which are subjected to low temperature carbonization between 550 °C and 600 °C, followed by smelting in a low shaft furnace. (Osborne, 1956)

weddellite A tetragonal mineral, $Ca(C_2O_4){\cdot}2H_2O$; in mud at the bottom of the Weddell Sea, Antarctica; also as urinary calculi. Cf: *whewellite.*

wedge (a) A wedge-shaped piece of wood used to tighten timber sets against the roof and sides. See also: *lag; lid.* (Nelson, 1965) (b) A tapered piece of material used to initiate the deflection of a borehole. See also: *deflecting wedge; Hall-Rowe wedge.* (Long, 1960) (c) Tapered piece of core that tends to bind and block a core barrel. (Long, 1960) (d) A piece of mica that, on splitting, yields pieces thicker at one end than at the other. (Skow, 1962) (e) The shape of a stratum, vein, or intrusive body that thins out; specif. a wedge-shaped sedimentary body, or prism. See also: *prism.* (AGI, 1987)

wedge-and-sleeve bolt A bolt designed for use in roof bolting. It consists of a rod 1-3/4 in (4.45 cm) in diameter with one end threaded and the other end shaped to form a solid wedge. A loose split sleeve with an outside diameter of 1-1/2 in (3.81 cm) is fitted over the wedge. Anchorage is provided when the bolt is pulled downward in a hole and the sleeve is held by a thrust tube. Split by the wedge head of the bolt, the sleeve expands until it grips the sides of the hole. See also: *bolt; slot-and-wedge bolt.* (Nelson, 1965)

wedge bit A tapered-nose noncoring bit, used to ream out a borehole alongside a steel deflecting wedge in hole-deflection operations. Also called: bullnose bit; wedge reaming bit. Syn: *wedging bit.* (Long, 1960)

wedge capping A winding rope capping consisting of two tapered iron wedges that encircle the rope, the end of which is prevented from unravelling by casting onto it a small block of white metal. The wedges are contained by a steel bow, over which four or five wrought-iron hoops are driven. The greater the pull on the rope, the more the wedges grip it as they are drawn into the encircling hoops. (Mason, 1951)

wedge clinometer An end clinometer the bottom end of which is shaped to match the wedge-guide pin on the drive wedge; hence the two can be fastened together with copper shear rivets. When a drive wedge is driven into a wooden plug in a borehole the copper rivets break; after the clinometer has been removed from the borehole, the relation of the bearing and inclination readings to the flat face of the projection on the bottom of the clinometer case can be used to orient and place the deflection wedge in a manner so as to direct the deflected hole to follow the desired course. Cf: *plain clinometer; line clinometer.* (Long, 1960)

wedge core lifter A core-gripping device consisting of a series of three or more serrated-face, tapered wedges contained in slotted and tapered recesses cut into the inner surface of a lifter case or sleeve. The case is threaded to the inner tube of a core barrel. As the core enters the inner tube, it lifts the wedges up along the case taper. When the barrel is raised, the wedges are pulled tight, gripping the core. (Long, 1960)

wedge cut (a) A cut in which the central holes are positioned to cause the breakout of a wedge-shaped section of strata when fired. (BS, 1964) (b) A drill-hole pattern with the cut holes converging to form a V or wedge. The other holes are drilled to break to the opening made by the cut holes. See also: *V-cut.* (Nelson, 1965) (c) A type of geometry for blasting pattern often used in drifting work. It can be adopted for all strata conditions. A wedge cut consists of pairs of holes, usually drilled horizontally, that meet or finish close together at the back of the cut so that a wedge-shaped section of the rock face will be removed on blasting. The holes should be drilled at an angle of approx. 60° to the face line. Accordingly, the depth of pull that can be obtained with this type of cut is governed by the width of the drift, as this determines the maximum length of drill steel that can be used. Typically, the pull ranges from 5 ft (1.5 m) in a 12-ft-wide (3.7-m-wide) drift up to 6-1/2 ft (1.8 m) in a 15-ft-wide (4.6-m-wide) drift. (McAdam, 1958)

wedge guide One of several implements used collectively to arrest a cage or skip in the event of an overwind using a multirope friction winder. Frictional forces, which gradually increase by virtue of the wedging action of the guides, are relied upon to bring the cage to rest and hold it in a stationary position with the aid of jack catches. See also: *detaching hook.* (Nelson, 1965)

wedge off To deviate or change the course of a borehole by using a deflecting wedge. See also: *bypass.* (Long, 1960)

wedge-out The edge or line of pinch-out of a lensing or truncated rock formation. (AGI, 1987)

wedge out *thin out.*

wedge pilot That part of the bottom end of a Hall-Rowe deflecting wedge that matches the guide pin on the upper end of the Hall-Rowe drive wedge and by means of which the deflecting wedge can be oriented to direct the deflected borehole in the intended direction. (Long, 1960)

wedge reaming bit A tapered or bullnose rotary bit used to restart drilling after a deflection wedge has been fitted into a borehole. See also: *wedging reamer; wedge bit. (BS, 1963)*

wedge roaster Multiple-hearth vertical furnace. Rabbles rotating on each circular horizontal hearth work the continuously fed material across alternately to the periphery and then, on the next hearth below, toward the center, so that it gravitates through either a central or a peripheral opening and is at the same time exposed to rising heat or air blown through rabble arms. (Pryor, 1963)

wedge rock An expression used on the Comstock lode to designate rock better than waste but too poor to be classed as "pay ore" or even "second-class ore." It became a custom to throw a wooden wedge onto a car containing very low grade ore, hence the term wedge rock. Syn: *pin.* Cf: *tag.*

wedge rose bit A serrated-face, hardened-metal, noncoring, cone-shaped bit used primarily to mill off part of the stabilizing or rose ring on the top end of a Hall-Rowe deflecting wedge. (Long, 1960)
wedge-set A diamond bit with wedge-shaped configurations in the crown that are inset with diamonds.
wedge shot *wedge cut.*
wedge socket fitting A wire rope attachment in which the rope lies in a too-small groove between a wedge and housing, so that pull on the rope tightens the wedge. (Nichols, 1976)
wedge theory The analysis formulated by Coulomb in 1776 of the force tending to overturn a retaining wall. Its basis is the weight of the wedge of earth that will slide forward if the wall fails. (Hammond, 1965)
wedge-wire deck A screen deck comprising wires of wedge-shaped cross section spaced from each other at a fixed dimension; the underflow thus passes through an aperture of increasing cross section. Also called wedge-wire sieve (undesirable usage). (BS, 1962)
wedge-wire screen A screen designed to reduce or eliminate clogging of material. It consists of wedge-shaped parallel wires with their wide edge upward. This type of screen is used for dewatering coal on vibrating, shaking, or stationary screens. (Nelson, 1965)
wedging (a) A method used in quarrying to obtain large, regular blocks of building stone, such as syenite, granite, marble, and sandstone. In this method, a row of holes is drilled, either by hand or by pneumatic drill, close to each other to create a longitudinal crevice. A gently sloping steel wedge is driven into this crevice. Usually several wedges are driven, and the block of stone can be detached without shattering. (Stoces, 1954) (b) The act of changing the course of a borehole by using a deflecting wedge. Syn: *whipstocking.* (Long, 1960) (c) The lodging of two or more wedge-shaped pieces of core inside a core barrel and therefore blocking it. (Long, 1960) (d) The material, moss, or wood used to render a shaft lining tight. (Fay, 1920) (e) The splitting, breaking, or forcing apart of a rock as if by a wedge, such as by the growth of salt or mineral crystal in interstices; specif. frost wedging. (AGI, 1987)
wedging and blocking *blocking and wedging.*
wedging bit *wedge bit.*
wedging crib In a circular mine shaft, a steel ring made of segments wedged securely to rock walls for use as a foundation for masonry lining. Also called: wedging ring; wedging curb. (Pryor, 1963)
wedging curb *curb.*
wedging down Breaking down coal at the face with hammers and wedges instead of by blasting. (Fay, 1920)
wedging reamer A reaming bit used to ream down alongside and pass the deflection wedge when deviating a borehole. Also called wedge bit; wedge reaming bit. (Long, 1960)
wedging shot An opening shot. A center cut. (Fay, 1920)
weeksite An orthorhombic mineral, $K_2(UO_2)_2Si_6O_{15}{\cdot}4H_2O$; resembles uranophane; in the Thomas range, Juab County, UT. Syn: *gastunite.*
weeks manometer An inclined water gage. (Nelson, 1965)
weeper A small feeder of water. (BS, 1963)
weeping rock A porous rock from which water oozes.
Weg rescue apparatus Oxygen supply fed automatically to a rescue worker through a valve controlled by the worker's breathing action, forming part of a portable outfit. (Pryor, 1963)
wegscheiderite A triclinic mineral, $Na_5(CO_3)(HCO_3)_3$; forms fibrous aggregates of tiny acicular to bladed crystals; Green River formation, WY.
wehrlite (a) A mixture of pilsenite and hessite. (b) A peridotite composed chiefly of olivine and clinopyroxene with accessory opaque oxides common.
weibullite An orthorhombic mineral, $Pb_6Bi_8(S,Se)_{18}$(?); differs from galenobismutite in possessing two distinct cleavages; steel-gray.
Weichbraunkohlen-dopplerit *dopplerite.*
weigh batcher Batching plant in which all ingredients for a concrete mix are measured by weight. (Hammond, 1965)
weigher-and-crusher man Person who weighs zinc ore and other materials to be sintered, and crushes sintered ore preparatory to further reduction. (DOT, 1949)
weighing feeder A device that handles the feed continuously over a belt that is balanced to weigh a stream of ore. (Pit and Quarry, 1960)
weighing-in-motion system An electronic system which weighs individually loaded coal wagons as they roll over a rail scale, feeding the information into a totalizing printer, which automatically prints out gross and net weights. It employs hermetically sealed load cells which accurately convert physical weight into electrical impulses. See also: *weightometer.* Syn: *electronic weighing.* (Nelson, 1965)
weigh larry A traveling hopper for receiving, weighing or measuring, and distributing bulk materials. Usually fitted with a scale, either manually operated or of the automatic recording type. Weigh larries may be suspended between overhead tracks or carried on rails mounted below them. They may be hand-pushed or power-propelled, and some designs provide a riding platform or cab for the operator. A remote-control device for operating the bunker or bin gates is usually mounted on the larry chassis.
weight (a) Roof movement, esp. when it can be seen or heard. (Mason, 1951) (b) Fracturing and lowering of the roof strata at the face as a result of mining operations. See also: *crush.* (Nelson, 1965)
weight batching Weighing and combining the correct proportion of cement and aggregate in a concrete mix. (Nelson, 1965)
weight break Cracks developing from the upper or tension side of a stratum caused by bending moment over a coal seam. Such breaks are induced in the nether roof in consequence of the moment exercised by the weight of the roof-stratum overhanging the face, cantilever fashion. The weight break differs from a shear plane in the more ragged and uneven character of the fracture and in hanging over the waste instead of over the unmined coal seam. Also called first break. (Briggs, 1929)
weight dropping A seismic technique by which energy can be sent downward into the ground without the necessity of drilling shotholes. This technique involves lifting a weight, then permitting it to fall and strike the ground. The waves from the impact are then recorded. In areas where drilling is difficult, otherwise undesirable, or unduly expensive, this technique may be highly advantageous. (Dobrin, 1960)
weighted average Value calculated from a number of samples, each of which has been assigned an importance in accord with its position and general trustworthiness. In this connection a sample which was cross-checked by others would be more reliable than one which was isolated, particularly if the latter showed abnormal values or was for any other reason suspect. The cross-checked sample is sometimes called a weighted sample in mineral valuation of a deposit. (Pryor, 1963)
weighted average depth A method of comparing the average depth of mine workings. It is based on the average depth from which the output is obtained, weighted for the tonnage produced. (Nelson, 1965)
weighted flowsheet A materials flowsheet including a statement of the capacity in tons per hour at principal points in the plant. (BS, 1962)
weight indicator An apparatus for recording and indicating the tension on a drilling line of a diamond or rotary drill. See also: *tension drilling.* (Long, 1960)
weighting The occurrence of fracturing of the upper roof, with consequent rapid increase of the weight carried on the timber and packs supporting the roof; distinct from the fracturing of the nether roof. (TIME, 1929-30)
weightometer An appliance for the continuous weighing of coal or other material in transit on a belt conveyor. See also: *weighing-in-motion system.* (Nelson, 1965)
weight pit The pit below the shaft station where heavy weights are attached to the guide ropes to keep them taut. Roughly, the weight needed is 1 st for each 1,000 ft (1.97 t per 1,000 m) of rope. (Higham, 1951)
weight strength The strength of an explosive per unit weight, expressed as a percentage of the value for blasting gelatin as a standard. See also: *absolute weight strength.* (BS, 1964)
Weinig flotation cell Square type of machine in which air is blown down to join pulp entering the cell below a mechanically driven impeller. (Pryor, 1963)
weir (a) A small dam in a stream, designed to raise the water level or to divert its flow through a desired channel. (AGI, 1987) (b) A notch in a levee, dam, embankment, or other barrier across or bordering a stream, through which the flow of water is regulated. (AGI, 1987)
weir head The depth of water in a measuring weir as measured from the bottom of the notch to the surface of the water upstream of the weir. The velocity of approach is not included. (Hammond, 1965)
weir table A device to estimate the quantity of water that flows during a given time over a weir of a given width at different heights of water. (Fay, 1920)
Weisbach triangle (a) A configuration used in the surveying of a mine shaft. (Pryor, 1963) (b) The highly attenuated triangle formed by the plan position of two shaft plumblines and one observation station. (BS, 1963)
weissite An isometric(?) mineral, Cu_5Te_3; bluish-black; at Vulcan, CO.
weiss quadrilateral The quadrilateral formed by the plane position of two shaft plumblines and two observation stations. (BS, 1963)
welded dike A term applied to pegmatitic and aplitic dikes, the boundaries of which have been obliterated by continued growth of the minerals of the granite into which the dikes have been injected. (Holmes, 1928)
welded tuff A glass-rich pyroclastic rock that has been indurated by the welding together of its glass shards under the combined action of the heat retained by particles, the weight of overlying material, and hot gases. It is generally composed of silicic pyroclasts and appears banded or streaky. Syn: *tuff lava.* Cf: *ignimbrite.* (AGI, 1987)
weldment A base or frame made of pieces welded together, as contrasted with a one-piece casting or a bolted or riveted assembly. (Nichols, 1976)
Weldon mud *Weldon process.*

Weldon process A process used formerly for the recovery of manganese dioxide in making chlorine from hydrochloric acid in a stoneware still, by adding lime to the still liquor and oxidizing with air to precipitate a mud containing calcium manganite and yielding chlorine when recirculated and treated with hydrochloric acid. (Webster 3rd, 1966)

well (a) A borehole or shaft sunk into the ground for the following purposes: obtaining water, oil, gas, or mineral solutions from an underground source; introducing water, gas, or chemical reagent solutions under pressure into an underground formation; or removing the leachate from such an operation. See also: *borehole mining.* (b) A slot in the front of a hydraulic dredge hull in which the digging ladder pivots. (Nichols, 1976) (c) A hollow cylinder of reinforced concrete, steel, timber, or masonry built in a hole as a support for a bridge or building. (Webster 3rd, 1966) (d) Commonly used as a syn. for borehole or drill hole, esp. by individuals associated with the petroleum-drilling industry. (Long, 1960) (e) A wall around a tree trunk that protects it from fill. (Nichols, 1976) (f) An artificial excavation (pit, hole, tunnel), generally cylindrical in form and commonly walled in, sunk (drilled, dug, driven, bored, or jetted) into the ground to such a depth as to penetrate water-yielding rock or soil and to allow the water to flow or to be pumped to the surface; a water well, originally applied to natural springs or pools, esp. mineral spas. See also: *artesian well; deep well.* (g) The crucible of a furnace or a cavity in the lower part of some furnaces to receive falling metal. (h) The small dark nonreflecting area in the center of a fashioned stone, esp. in a colorless diamond cut too thick. (i) A vertical opening through the hull of a ship in which drill pipe or mining machinery is lowered to the seafloor, rather than being lowered over the side of the ship. Also called: moonpool.

well-conditioned triangle A triangle that is equilateral or nearly so. In such a triangle any error in the measurement of an angle will be reduced to a minimum. (Hammond, 1965)

well core A sample of rock penetrated in a well or other borehole obtained by use of a hollow bit that cuts a circular channel around a central column or core. (AGI, 1987)

well cuttings Rock chips cut by a bit in the process of well drilling and removed from the hole by pumping or bailing. Well cuttings collected at closely spaced intervals provide a record of the strata penetrated. (AGI, 1987)

well-drill hole Hole drilled by means of an apparatus known as a well drill, or similar to that, and used in groups for blasting on a comparatively large scale. Such holes are usually 5 in or 6 in (12.7 cm or 15.2 cm) in diameter and from 30 to 150 ft (9.1 to 45.7 m) deep.

wellglass fitting A transparent lighting device used as a main lighting device in mines. Its distribution characteristics are simple, since there is no control of light other than that provided by the interior of the upper part of the fitting, which absorbs a large proportion of the upward flux component from the lamp and reflects the remainder. It is usually to be found spaced at intervals of 20 to 50 ft (6.1 to 15.2 m), 6 to 7 ft (1.83 to 2.13 m) high, along the centerline of roads 12 to 14 ft (3.66 to 4.27 m) wide. In general, the most noticeable effect of such a layout is glare. (Roberts, 1958)

wellhole (a) A large-diameter (about 6 in or 15 cm) vertical hole used in quarries and opencast pits for heavy explosive charges. (Nelson, 1965) (b) Change room. (Hess)

wellhole blast A method of quarry blasting in which the explosive charges are placed in rows of vertical holes. The loading ratio varies from about 3 st of rock per pound (6 t/kg) of explosive up to about 7 st (14 t/kg) under favorable conditions. Deck loading is usually employed, and a powerful gelatinous explosive is loaded at the bottom of the holes. (Nelson, 1965)

wellhole blasting This type of blasting is virtually benching on a large scale. The depth and burden of the holes are much greater, and in consequence, the hole diameter must also be increased to ensure sufficient concentration of the explosive charge. Wellhole blasting is used in limestones esp. if the beds are horizontal and well-defined. It is not often used in highly abrasive igneous rocks because of the cost of drilling. (Fraenkel, 1953)

Wellington formula The Engineering News formula for calculating the load-bearing capacity of driven piles. (Hammond, 1965)

well log A graphic record of the measured or computed physical characteristics of the rock section encountered in a well, plotted as a continuous function of depth. Measurements are made by a sonde as it is withdrawn from the borehole by a wire line. Several measurements are usually made simultaneously, and the resulting curves are displayed side by side on a common depth scale. Both the full display and the individual curves are called logs. Well logs are commonly referred to by generic type, such as resistivity log and radioactivity log, or by specific curve type, such as sonic log and gamma-ray log. See also: *borehole log; sample log; driller's log.* Syn: *borehole survey; geophysical log; wire-line log.* (AGI, 1987)

well logging (a) A widely used geophysical technique that involves probing of the Earth with instruments lowered into boreholes, their readings being recorded at the surface. Among rock properties currently being logged are electrical resistivity, self-potential, gamma-ray generation (both natural and in response to neutron bombardment), density, magnetic susceptibility, and acoustic velocity. (Dobrin, 1960) (b) In deep bores, measurement of resistivity of the formations drilled through. Electrodes are plunged into the drilling mud at controlled spacings. See also: *electric log.* (Pryor, 1963) (c) The lowering of sensors into a borehole and recording of physical information as a function of depth. Numerous types of geophysical logs can be recorded, the more common being rate of penetration, temperature, gamma-ray, electric, and caliper (for hole diameter). (Nelson, 1965)

well point A hollow vertical tube, rod, or pipe terminating in a perforated pointed shoe and fitted with a fine-mesh wire screen; connected with others in parallel to a drainage pump; driven into or beside an excavation to remove underground water, to lower the water level and thereby minimize flooding during construction, or to improve stability. (AGI, 1987)

well-point pump A centrifugal pump that can handle considerable quantities of air; used for removing underground water to dry up an excavation. (Nichols, 1976)

well seismometer Special type of seismometer that is used when recording in a borehole. (Schieferdecker, 1959)

well shooting In seismic work, a method or methods of logging a well so that average velocities, continuous velocities, or interval velocities are obtained by lowering geophones into the hole. Shots are usually fired from surface shot holes, but may be fired in the well itself, or perforating-gun detonations may be used. In continuous logging, a sound source is lowered in the hole together with recording geophones. (AGI, 1987)

well-sorted As used by geologists, it applies to material composed of grains of approx. uniform size. As used by engineers, it applies to material containing approx. equal amounts of several grain sizes. (Stokes, 1955)

well tube A tube or tubing used to line wells. (Standard, 1964)

well-tube filter A strainer on a driven well tube to keep out grit. (Standard, 1964)

well-tube point A point at the end of a perforated tube used for sinking wells. (Standard, 1964)

well-velocity survey Method of determining the velocity distribution by recording in a borehole. (Schieferdecker, 1959)

Welshman A heavy steel ring, about 3 in or 4 in (7.6 cm or 10.2 cm) inside diameter, used in withdrawing a bar stuck or frozen in a skull or iron. The ring is placed on the bar, a wedge inserted, and the bar backed out by sledging on the wedge. (Fay, 1920)

Welsh notch A form of joint between the tubing used to line a collar (or a crossbar) and the arms (or uprights) of a timber set; developed in the Welsh coal mines. A Welsh notch is designed to be equally effective in resisting side and roof pressure. (Nelson, 1965)

wem *whim.*

wenkite A hexagonal mineral, $Ba_4Ca_6(Si, Al)_{20}O_{39}(OH)_2(SO_4)_3 \cdot nH_2O$ (?); cancrinite group.

Wentworth grade scale An extended version of the Udden grade scale, adopted by Chester K. Wentworth (1891-1969), U.S. geologist, who modified the size limits for the common grade terms but retained the geometric interval or constant ratio of 1/2. The scale ranges from clay particles (diameter less than 1/256 mm) to boulders (diameter greater than 256 mm). It is the grade scale generally used by North American sedimentologists. See also: *Udden grade scale; phi grade scale; grade scale.* (AGI, 1987)

Wentworth scale A logarithmic grade scale for size classification of sediment particles, starting at 1 mm and using the ratio 1/2 in one direction (and 2 in the other), providing diameter limits to the size classes of 1,-1/2,-1/4, etc., and 1, 2, 4, etc. Syn: *grade scale.* See also: *Udden grade scale.* (AGI, 1987)

Wenzel's law Applies to the dissolution of a solid in a liquid. The rate of dissolution is proportional to the surface area of the solid exposed to the action of the solvent. (Newton, 1959)

wernerian Of or relating to Abraham G. Werner (1749-1817), German mineralogist and geologist, who classified minerals according to their external characteristics, advocated the theory of neptunism, and postulated a worldwide age sequence of rocks based on their lithology. Also, said of one who is a great, but dogmatic, teacher of geology. n. An adherent of wernerian beliefs; a neptunist.

wernerite A common scapolite. A mineral of the scapolite group, intermediate in composition between meionite and marialite. Syn: *scapolite.* (Fay, 1920)

Wesco coal powder Nongelatinous permissible explosive; used in coal mines. (Bennett, 1962)

weslienite *romeite.*

Westfalia pillar plow A plow designed for the extraction of coal pillars and for short, rapidly advancing development faces. The plow is guided along a panzer conveyor, but the pulling forces are not transmitted to the conveyor structure. The plow is automatically advanced at each end of its short run, giving a rapid face advance. This requires the use of self-advancing supports in conjunction with hydraulic anchorages. (Nelson, 1965)

Westfalia plow *hard-coal plow.*

Westfalia tandem plow A plow designed for use in a seam with sticky coal. It consists essentially of two shortened plow bases connected by a heavy tension spring and carrying two adjustable booms fitted with cutting bits and connected at the apex to a cutterhead, which plows at roof level or at any lower level that will bring down the top coal. (Nelson, 1965)

westfalite A blasting explosive composed of ammonium nitrate and resin. (Webster 2nd, 1960)

Weston photronic cell This consists of a small box containing an iron disk thinly coated with the rare element selenium. When electromagnetic radiation in the form of light waves falls on this surface it sets up a potential difference between the iron and the selenium, which in turn causes a minute electric current to flow through the sensitive microammeter connected between them. The magnitude of this current is proportional to the intensity of the light, and it can be used as a measure. For use with miners' hand lamps the cell is mounted in a box, the microammeter being housed in the top or side. The lamp to be tested is placed in the box on a turntable, and the candlepower is read off directly since the instrument is already calibrated. (Standard, 1964)

Westphal balance In mineral analysis, a balance used to determine the specific gravity of heavy liquids which are in turn used to determine the specific gravity of mineral grains by a sink-float method. It is a modification of a Walker steelyard (beam) balance. Also called: Mohr balance. Cf: *Jolly balance; Walker balance.*

West's solution A liquid consisting of eight parts of white phosphorous and eight parts of sulfur to one part of methylene iodide. Useful in obtaining the refractive index by the Becke method. Refractive index, 2.05.

weta material A refractory suitable instead of porcelain and quartz glass. Powdered silicon carbide mixed with silicates and certain metals; resists acids, alkalis, and temperature shock; is not easily broken.

wet analysis A method of estimating the effective diameters of particles smaller than 0.06 mm by mixing a sample in a measured volume of water and checking its density at intervals with a sensitive hydrometer. A test may take several days. (Nelson, 1965)

wet- and dry-bulb thermometer *hygrometer; wet-bulb thermometer.*

wet-and-dry screening In sizing analysis of fine material, preliminary screening by washing a weighed sample on a 200-mesh screen, perhaps with use of a dispersing agent, such as sodium silicate. Removal of the minus fraction (which is settled and later brought into account) is followed by drying of the on-size and by standard sizing analysis. Screen action is thus rendered more efficient. (Pryor, 1963)

wet assay Any type of assay procedure that involves liquid, generally aqueous, as a means of separation. Cf: *dry assay.* See also: *volumetric analysis.* (AGI, 1987)

wet blasting Shot firing in wet holes. Special explosives are available for wet conditions, and the detonator wires must be well insulated to prevent short-circuiting and misfires. (Nelson, 1965)

wet-bulb temperature (a) The temperature of the air as measured by a wet-bulb thermometer; it is lower than the dry-bulb temperature (for all cases except when the air is saturated)—ininversely proportion to the humidity. (Strock, 1948) (b) The lowest temperature which can be produced in a given air by the evaporation of moisture into that air. (Spalding, 1949) (c) Temperature at which water evaporating into air can bring the air to saturation adiabatically at that temperature—a measure of the evaporating capacity of air. Indicated by a thermometer with a wetted wick. Measured in °C. (Hartman, 1982)

wet-bulb thermometer An instrument that measures the evaporating capacity of air. (Lewis, 1964)

wet cleaning A coal-cleaning method that involves the use of washers plus the equipment necessary to dewater and heat-dry the coal. This method is generally used when cleaning the coarser sizes of coal. It is a more expensive method than air cleaning and creates the additional problem of water pollution. Coal can, however, be cleaned more accurately by this method than by air cleaning. See also: *froth flotation; washery.* (Kentucky, 1952)

wet clutch A clutch that operates in an oil bath. (Nichols, 1976)

wet criticality Reactor criticality achieved with a coolant. (Lyman, 1964)

wet cutting A method of dust prevention in which water is delivered onto a moving cutter chain, through water pipes, and is carried into the cut where it is intimately mixed with the cuttings. This method is successful in seams up to 4 ft (1.22 m) thick. (Mason, 1951)

wet cyclone *cyclone.*

wet drilling In rock drilling for blasting purposes, injection of water through a hollow drill shank to the bottom of a hole, to allay dust and danger of pneumoconiosis.

wet gold-silver ore Lead ore with high silver content. During smelting, the lead trickles through the mass and collects gold and/or silver, which are later recovered. (Nelson, 1965)

wet grinding (a) Any milling operation carried on in water or a liquid. (Enam. Dict., 1947) (b) The practice of applying a coolant to the work and the wheel to facilitate the grinding process. (ACSG, 1963) (c) Comminution of ore in aqueous suspension; typically practiced in the ball milling of finely crushed rock. See also: *grinding.* (Pryor, 1963)

Wetherill's furnace A furnace with a perforated iron bottom, under which a blast is introduced, and upon which zinc ore (red oxide) is reduced. A muffle furnace for roasting zinc ores. (Fay, 1920)

Wetherill's magnetic separator An apparatus for separating magnetic minerals from nonmagnetic minerals. It consists of two flat belts, the upper of which is the wider, run parallel to each other and over long magnets set obliquely to the belts. Consequently, magnetic particles are drawn up against the upper belt, and as they pass beyond the influence of the magnets, fall from the edge past the other belt into a bin. Another form operates by belts moving across the line of travel of the main belt. (Liddell, 1918)

Wetherill vacuum casting process In this process, a mold, arranged for bottom feeding, is placed inside a vacuum bell; the bottom of the mold is connected by a tube to a ladle containing the molten metal, which is sucked into the mold cavity when a vacuum is formed. (Osborne, 1956)

Wethey furnace A multiple-deck, horizontal furnace for calcining sulfide ores. It resembles the Keller furnace. (Fay, 1920)

wet metallurgy *hydrometallurgy.*

wet method Any hydrometallurgical process, such as the cyanide process, flotation process, etc. See also: *wet process.* (Fay, 1920)

wet milling Comminution of ores and materials in the presence of a liquid in a suitable mill, either by rods, balls, or pebbles, or autogenously, by the material itself. See also: *dry milling.*

wet-milling plant A mill in which a wet process is employed. (Fay, 1920)

wet-mill man *wet-pan operator.*

wet mining (a) A system of mining in which water is sprayed into the air at all points where dust is liable to be formed, and no attempt is made to prevent the air from picking up moisture. It therefore soon becomes saturated and remains so throughout the ventilation circuit. (Spalding, 1949) (b) Mining for salt and other water soluble minerals as brine rather than in the dry state. (Kaufmann, 1960)

wet mixer *clay maker.*

wet-pan charger Person who adds water to mixture, in addition to clay, shale, or brick, in grinding pans in order to make it plastic. May be designated according to clay ground in pans, such as silica-wet-pan charger. Syn: *wet-pan feeder.* (DOT, 1949)

wet-pan feeder *wet-pan charger.*

wet-pan operator Person who tends and supervises loading of wet pans used for grinding and tempering clay, performing essentially the same duties as described under dry-pan operator. Syn: *clay temperer; wet-mill man.* (DOT, 1949)

wet-plant operator Person who works as a member of a crew performing any one or a combination of duties concerned with extracting cadmium, lead sulfate, and zinc oxide from dust recovered in Cottrell precipitators. (DOT, 1949)

wet process A metallurgical process in which the valuable contents of an ore are dissolved by acid or other solvents; a leaching or lixiviation process. Opposite of dry process. See also: *wet method.* (Fay, 1920)

wet puddling The ordinary process of puddling in which a furnace is lined with material rich in oxide of iron. (Fay, 1920)

wet rot (a) Timber decay set up when mine props have not been treated with zinc sulfate, etc., and are exposed to alternations of moisture and drying out. (Pryor, 1963) (b) Timber decay caused by alternating wet and dry periods.

wet screening The addition of water to a screen to increase its capacity and improve its sizing efficiency. Water may be introduced either by adding it to the feed stock or by spraying it over the material on the screen deck. The latter method is also used in rinsing or washing ores, etc., to recover minerals. (Nelson, 1965)

wet separation A term used in connection with coal washing or other processes using fluid. See also: *coal-preparation plant; washery.* (Nelson, 1965)

wet sieve analysis An American Society for Testing and Materials standard method of test, recommended for determining the grain sizing of materials in which slaking would occur.

wet sphere device An instrument for assessing climatic conditions in mines, such as wet- and dry-bulb temperatures, air velocity, barometric pressure, and radiation. The field of usefulness of the instrument is limited to mines where the workers are normally sweating freely and wear few clothes. (Roberts, 1960)

wettability The ability of a liquid to form a coherent film on a surface, owing to the dominance of molecular attraction between the liquid and the surface over the cohesive force of the liquid itself. (AGI, 1987)

wettable sulfur Sulfur treated so that it is easily dispersed in water. (USBM, 1965)

wetted perimeter The total length of surface in a channel or pipe that is in actual contact with water. See also: *hydraulic mean depth.* (Hammond, 1965)

wetterdynamite Originally, only guhr dynamites to which were added salts containing water of crystallization, such as Glauber's salts, ammonium oxalate, etc., with the view of making them available in mines containing combustible gases. (Fay, 1920)

wetting A phenomenon involving a solid and a liquid in such intimate contact that the adhesive force between the two phases is greater than the cohesive force within the liquid. Thus, a solid that is wetted, on being removed from the liquid bath, will have a thin continuous layer of liquid adhering to it. Foreign substances, such as grease, may prevent wetting. Addition agents, such as detergents, may induce wetting by lowering the surface tension of the liquid. Cf: *water break.* (ASM, 1961)

wetting agent (a) A reagent to reduce the interfacial tension between a solid and a liquid, so facilitating the spreading of the liquid over the solid surface. (BS, 1962) (b) A chemical promoting adhesion of a liquid (usually water) to a solid surface. (Pryor, 1960) (c) *plasticizer.* (d) Substance that lowers the surface tension of water and thus enables the water to mix more readily with other substances and so facilitates the spread of the liquid over a solid surface. (BS, 1962)

wetting coal dust The spraying of mine roadways with water or treatment with a wetting agent in order to (1) increase the difficulty of raising the dust deposit into the air to take part in an explosion, and (2) reduce the flammability of the dust raised in an explosion. (Nelson, 1965)

wetting effect (a) When surface-active agents cause water to displace air at the surface of a solid-water-air system, the surface tension of the solid-water phase has been lowered in comparison with that of the solid-air phase, by adsorption of the wetting agent (adhesional wetting). For a small particle, wetting can be separated into three stages (adhesion, immersion, spread). Wetting is a preliminary step in deflocculation. (Pryor, 1963) (b) A condition in which the surface tension of the solid-water phase is lowered in comparison with surface tension in the solid-air phase.

wet unit weight The weight (solids plus water) per unit of total volume of soil mass, irrespective of the degree of saturation. Syn: *mass unit weight.* See also: *unit weight.* (ASCE, 1958)

whaler A horizontal beam in a bracing structure. (Nichols, 1976)

whale-type jib A coal-cutter jib that enables water to be taken, during cutting, to the back of the cut for dust suppression and prevention of gas ignition from frictional sparking. With an undercutting jib, it consists of a feed water pipe and four or five distributor pipes terminating in jets all arranged in the top plate of the jib. See also: *undercut ignition; dust-suppression jib.* (Nelson, 1965)

wheal Corn. A mine. (Nelson, 1965)

Wheatstone-bridge-type instrument An instrument that makes use of electrically heated filaments that burn methane and measure the heat output by resistance pyrometry. One or more filaments form the arms of a Wheatstone bridge circuit, the out-of-balance current being a function of the methane percentage. (Roberts, 1960)

wheelbase The distance between the contact points on the front and back wheels of any vehicle and the surface upon which the vehicle travels. (CTD, 1958)

wheel brae A flat or landing on the top of an incline. (Fay, 1920)

wheel conveyor A series of wheels supported in a frame over which objects are moved manually or flow by gravity. See also: *el conveyor; gravity conveyor; gravity wheel conveyor; portable conveyor.*

wheel ditcher (a) A wheel equipped with digging buckets, carried and controlled by a tractor unit. (Nichols, 1976) (b) A machine that digs trenches by rotation of a wheel fitted with toothed buckets. (Nichols, 1976)

wheel dresser A tool for cleaning, resharpening, and truing the cutting faces of grinding wheels. (Crispin, 1964)

wheeled tractor A tractor, fitted with large rubber-tired wheels, which can travel comparatively fast over rough ground. See also: *tractor.* (Hammond, 1965)

wheeler *pusher.*

wheelerite (a) A yellowish resin, found in the Cretaceous beds of northern New Mexico, filling the fissures of the lignite, or interstratified in thin layers. It is soluble in ether. (Fay, 1920) (b) A yellowish variety of retinite that is soluble in ether and fills fissures in, or is thinly interbedded with, lignite beds in northern New Mexico.

wheel excavator A large-capacity machine for excavating loose deposits, particularly at opencast coalpits. It consists of a large digging wheel that rotates on a horizontal axle and carries large buckets on its rim. (Nelson, 1965)

wheelman *rollerman.*

wheel ore *bournonite.*

wheel pit A pit in which the lower part of a flywheel runs. (Webster 2nd, 1960)

wheelrace The place in which a water wheel is set. (Webster 3rd, 1966)

wheel runner *incline man.*

wheel scraper A scraper mounted on an axle supported by a pair of wheels. It affords an easy means of conveying a loaded scraper to a dumping ground. See also: *bowl scraper.* (Fay, 1920)

wherk Eng. A small unexpected turning in the stone, side, or ore, often accompanied by a small joint; Derbyshire. (Arkell, 1953)

wherryite A monoclinic mineral, $Pb_4Cu(CO_3)(SO_4)_2(Cl,OH,O)_3$; pale green; finely granular; related to caledonite; at the Mammoth Mine, AZ.

whetstone Any hard fine grained rock, commonly siliceous, that is suitable for sharpening implements such as razors, knives, and other blades; e.g., novaculite.

Whetwell stove A firebrick hot-blast stove on a regenerative system. (Fay, 1920)

whewellite A monoclinic mineral, $Ca(C_2O_4){\cdot}H_2O$; soft; in coal seams, incrustations on marble; also in human urinary tracts. Syn: *oxacalcite.* Cf: *weddellite.*

which-earth Eng. A whitish earth found at Thame, Waterperry, and Adwell; mixed with straw, and used for sidewalls and ceilings; a natural mixture of lime and sand, flakes in water (like gypsum) without any heat. (Arkell, 1953)

whim (a) A large capstan or vertical drum turned by horsepower or steam power. Used to raise coal, water, etc., from a mine. Also called: whimsey; whim gin; horse gin. (Fay, 1920) (b) Drum on which a hoisting rope is coiled. Also spelled wem. (Pryor, 1963)

whim driver A laborer who drives a draft animal at the surface of a mine to supply power to a whim (large drum on which a cable is wound) used to hoist ore, coal, or rock in a shallow shaft. Becoming obsolete. (DOT, 1949)

whim gin *whim; horse gear.*

whim shaft A shaft through which coal, ore, water, etc., are raised from a mine by means of a whim. (Zern, 1928)

whin (a) Igneous rock. When parallel to the bedding planes, it is called a whinsill; when cutting across the strata, a whin dike. (Nelson, 1965) (b) A hard, compact rock. (Gordon, 1906) (c) Whinstone or whinrock. In Nova Scotia, miners apply this term to a thick-bedded rock composed of grains of quartz with argillaceous or feldspathic matter that resembles graywacke. (Fay, 1920) (d) A whim or winch. (Webster 2nd, 1960) (e) *dolerite.*

whin float (a) Scot. A horizontal sill or lava flow of igneous rock. (Arkell, 1953) (b) Scot. A kind of greenstone, basalt, or trap, occurring in coal measures. (Fay, 1920)

whinny Resembling or abounding in whinstone. (Fay, 1920)

whinstone A colloquial British term for dolerite, basalt, and other dark fine-grained igneous rocks. The term is derived from the Whin Sill in northern England. Cf: *trap.* (AGI, 1987)

whintin Cumb. Spotted schist. Cf: *calamanco; linsey.* (Arkell, 1953)

whip (a) A rope that passes over a pulley and is pulled by a horse to hoist ore. Modern form occasionally seen has a car or lorry working backward and forward for shallow pitting. Called whip-and-derry when used with a derrick or gin. (Pryor, 1963; Fay, 1920) (b) One who operates such a hoisting apparatus. (Standard, 1964) (c) *rod slap.*

whip-and-derry *whip.*

whip gin A gin block for use as a whip, as in hoisting. (Standard, 1964)

whip out The enlargement of a portion of a borehole caused by the eccentric rotation and slap of the drill rods. (Long, 1960)

whipper Person who raises coal, merchandise, etc., with a whip, as from a ship's hold. (Standard, 1964; Fay, 1920)

whipping (a) The thrashing about of a moving rope, such as a hoisting cable in a mine shaft. Syn: *surging.* (Fay, 1920) (b) Hoisting ore, coal, or other material by means of a whip. (Fay, 1920)

whipping hoist A hoist worked with a whip, esp. if by steam power. (Standard, 1964)

whipstock Wedge-shaped device used to deflect and guide a drill bit away from vertical; procedure of deflecting a hole. Cf: *arc cutter.* See also: *deflection wedge.* (Wheeler, 1958)

whipstocking A term commonly used by petroleum-field drillers to designate the act or process referred to by diamond drillers as deflecting a borehole. See also: *wedging.* (Long, 1960)

whipstock point The point within a borehole at which a deflection or change in course is desired. (Long, 1960)

whirling hygrometer In mining, a hygrometer used to obtain wet-bulb temperatures. The hygrometer is spun round and round at a speed of about 200 rpm, for at least 1 min, and then read as quickly as possible. See also: *Storrow whirling hygrometer.* (McAdam, 1955)

whirlstone Eng. Usually applied to sandstone (not a freestone) in the Carboniferous Limestone Series of Cumberland. Sometimes applied to a dolomite, limestone, or shale. (Arkell, 1953)

whistler *squealer.*

white The color of the mixture of electromagnetic radiation visible to the normal human eye. White is commonly, but incorrectly, applied to colorless minerals and gemstones.

white agate *chalcedony.*

white alkali An older term for accumulation of salts with high levels of sodium that may develop as a crust. Cf: *black alkali.* (AGI, 1987)
white antimony *valentinite; valentite.*
white arsenic Commercially called arsenic. Arsenic trioxide, As_2O_3; the most important compound of arsenic. Obtained from the roasting of arsenical ores. See also: *arsenic trioxide.* (CTD, 1958)
white Bengal fire A very brilliant light produced by burning pure metallic arsenic. (Fay, 1920)
white cast iron Cast iron that gives a white fracture because the carbon is in combined form. See also: *iron.* (ASM, 1961)
white clay *kaolin.*
white coal (a) *tasmanite.* (b) Water power: a French designation (houille blanche). (Standard, 1964)
white cobalt A name frequently applied to smaltite and cobaltite. See also: *cobaltite; smaltite.* (Fay, 1920)
white copper A white alloy of copper; paktong. Usually German silver. Syn: *domeykite.* (Webster 2nd, 1960; Fay, 1920; Hey, 1955)
white copperas *coquimbite; goslarite.*
white copper ore *kyrosite.*
white-countered gutta percha A safety fuse in which the powder is enclosed in a thin tube of gutta percha, which in turn is enveloped in a waterproof textile covering. Abbreviated W.C.G.P. (Higham, 1951)
white damp Carbon monoxide, CO. A gas that may be present in the afterdamp of a gas- or coal-dust explosion, or in the gases given off by a mine fire; also one of the gases produced by blasting. It is an important constituent of illuminating gas, supports combustion, and is very poisonous because it is absorbed by the hemoglobin of the blood to the exclusion of oxygen. See also: *damp.* (Fay, 1920; Hudson, 1932)
white feldspar *albite.*
white flat Small, pale, rough nodule in hard shale in Shropshire. (Arkell, 1953)
white garnet (a) *leucite.* (b) A translucent variety of grossular resembling white jade in appearance.
white gold Gold alloyed with nickel or palladium to give it a white color. (CTD, 1958)
white gunpowder A mixture of 2 parts potassium chlorate, 1 part potassium ferrocyanide, and 1 part sugar. Syn: *white powder.* (CCD, 1961)
white heat A common division of the color scale, generally given as about 2,800 °F (1,540 °C).
white horse A term used by quarrymen to denote a light-colored gneiss, aplite, or pegmatite. (Fay, 1920)
white-hot A state in which a material is heated to full incandescence so as to emit all the rays in the visible spectrum in such proportion as to appear dazzling white. (Fay, 1920)
White-Howell furnace A revolving, cylindrical furnace for calcining calamine. (Fay, 1920)
white iron An extremely hard cast iron, resulting when a casting is chilled in a metallic mold. See also: *gray iron.* (Crispin, 1964)
white iron ore An early name for siderite. See also: *siderite.* (Fay, 1920)
white iron pyrites *marcasite.*
white latten An alloy of copper, zinc, and tin in thin sheets. (Standard, 1964)
white lead Basic lead carbonate or lead hydroxycarbonate. (CTD, 1958)
white lead ore (a) *cerussite.* (b) A decomposition product of sphalerite. See also: *white vitriol.* Syn: *lead carbonate.* (CTD, 1958)
White Lias The Upper Rhaetic, as opposed to the overlying Blue Lias. (Arkell, 1953)
white limestone The principal limestone division of the Great Oolite Series. (Arkell, 1953)
white metal (a) A general term covering a group of white-colored metals with relatively low melting points (lead, antimony, bismuth, tin, cadmium, and zinc). Includes the alloys based on these metals. (ASM, 1961) (b) A copper matte of about 77% copper obtained from smelting sulfide copper ores. (ASM, 1961) (c) Usually a tin-base alloy (more that 50% tin); used for lining bearings and winding rope cappels. (Nelson, 1965)
white-metal cappel A cappel in which the end of the winding rope (separated into a brush) is embedded in a plug of white metal inside a socket. See also: *capping.* (Nelson, 1965)
white-metal capping A winding rope capping formed by opening out the wires of the rope for a length equal to that of the socket, cleaning them thoroughly of all grease, cutting out the hemp core, and finally drawing them into the warmed socket and running in the white metal, which is an alloy of lead, antimony, and tin. (Mason, 1951)
white mica *muscovite.*
white mineral press A machine for briquetting flue dust. (Fay, 1920)
white mundic *arsenopyrite.*
white nickel *rammelsbergite; nickel-skutterudite.*
white nickel ore *chloanthite.*
white olivine The mineral forsterite, Mg_2SiO_4. (Fay, 1920)
white opal Precious opal of any light color, as distinguished from black opal; e.g., a pale blue-white gem variety of opal.
white ore *cerussite.*
white powder *white gunpowder.*
white pyrite *marcasite; arsenopyrite.*
white salt (a) Salt dried and calcined; decrepitated salt. (Fay, 1920) (b) Salt refined and prepared mainly for household use. Also, vacuum pan salt used for salting skins; distinguished from rock salt. (Kaufmann, 1960)
white sand Sand that is usually quartzitic and pure enough to resist heat and slags; used for the final layer in Siemen's steel furnaces. (Nelson, 1965)
white schorl *albite.*
white smoker A plume of hydrothermal fluid issuing from the crest of an oceanic ridge; e.g., the East Pacific Rise at the mouth of the Gulf of California. The fluid is clouded by white precipitates, mostly barite and silica. It issues at rates of tens of cm/second and at temperatures of 100 to 350 °C (Macdonald et al., 1980). Cf: *black smoker.* (AGI, 1987)
white tellurium *sylvanite; krennerite.*
white tin Metallic tin after smelting, in contradistinction to black tin or cassiterite. (Fay, 1920)
white tombac A variety of brass made white by the addition of arsenic. (Standard, 1964)
white vitriol The mineral goslarite, $ZnSO_4{\cdot}7H_2O$. Also called: salt of vitrol; zinc vitrol. (CTD, 1958; Fay, 1920)
whiting substitute A finely ground calcium carbonate, about 98% pure, contaminated by magnesia, silica, iron, or alumina. This material should not be confused with chalk whiting or precipitated chalk. (USBM, 1965)
Whiting system A system in which two parallel, grooved sheaves are coupled. These sheaves are driven directly from a hoisting engine. One sheave is set at a slight angle to the vertical so that the rope will pass freely from one sheave to the other. The hoisting rope passes around the drive sheaves and then to a takeup sheave, which is mounted on a horizontally movable carriage to take up slack or to change the hoist from one level to another. A tail rope is used in this system, although it may not be absolutely necessary. (Lewis, 1964)
whitlockite A trigonal mineral, $Ca_9(Mg,Fe)H(PO_4)_7$; colorless; forms rhombohedra; at North Groton, NH. Syn: *merrillite.*
Whitney stress diagram Diagram showing the stress distribution in a reinforced concrete beam in accordance with the theory of ultimate load. (Hammond, 1965)
Whitwell stove A firebrick, hot-blast stove on a regenerative system. (Fay, 1920)
whizzer mill *Jeffrey crusher.*
whole body counter A device used to identify and to measure the radiation in the body (body burden) of humans and animals. Uses heavy shielding to keep out background radiation, ultrasensitive scintillation detectors, and electronic equipment. (Lyman, 1964)
whole-circle bearing A bearing that defines the direction of a survey line by its horizontal angle measured clockwise from true north. (Hammond, 1965)
whole diamond A diamond (as mined), the shape of which has not been modified artificially. Syn: *whole stone.* (Long, 1960)
whole stone *whole diamond.*
whole-stone bit A bit, the crown of which is either surface set or impregnated with whole diamonds, as opposed to an impregnated or surface-set bit in which the inset diamonds are fragmented diamonds. (Long, 1960)
wich Celtic for salt spring; often used in England as the termination of names of places where salt is or has been found; e.g., Droitwich, Nantwich. Syn: *wych.*
wichert A subsoillike chalk, Haddenham, near Thame, U.K.
wick To place a soft twisted-cotton string between rod joints as they are made up or coupled. (Long, 1960)
wicket (a) N. Wales. A kind of pillar-and-stall, or bord-and-pillar, system of working a seam of coal, with pillars up to 15 yd (13.7 m) apart and stalls up to 24 yd (22.0 m) wide. Also called wicket work. (Fay, 1920) (b) A wall built of refractories to close an opening into a kiln or furnace; it is of a temporary nature, serving as a door; e.g., in intermittent or annular kilns. (Dodd, 1964)
wicket conveyor A conveyor comprising two or more endless chains connected by crossbars and to which vertical rods are attached at spaced intervals. The crossbars are also provided with spaced projections at the same level to form in effect a continuous carrying surface through which product cannot fall.
wicking The soft twisted-cotton string used to wick drill-rod joints; the act of placing the cotton string on the rod joints. See also: *wick.* (Long, 1960)
wide-mouthed socket A fishing tool similar to a bell-mouth socket, but lacking a latch. (Long, 1960)
wide opening An underground excavation whose width is greater than two or three times its height. (Woodruff, 1966)

widiyan *wadi.*

widowmaker A rock drill operated without water and hence, produces a lot of dust, leading to silicosis; often fatal to a miner.

width of lode The thickness of ore measured at right angles to the dip. The term "true width" is often used to describe this value, in which case width of lode is used to denote the thickness passed through, irrespective of the angle of dip. (Nelson, 1965)

Wiedgerite Trade name for a soft bitumen resembling elaterite, but containing much sulfur and water. (Tomkeieff, 1954)

wiggle stick *divining rod.*

wiggle tail (a) A rock-cutting tool or bit, used to deflect a borehole, that has an articulated pilot part, which also can be attached to a knuckle-jointed device and coupled to the bottom end of a drill string. Also called whipstock. (Long, 1960) (b) Nickname for hand-rotated stoper drill.

wightmanite A monoclinic mineral, $Mg_5(BO_3)O(OH)_5{\cdot}2H_2O$; colorless; forms pseudohexagonal prisms; at Commercial quarry, Crestmore, CA.

Wilcoxian Lower or upper lower Eocene. (AGI, 1987)

wildcat (a) A borehole and/or the act of drilling a borehole in an unproved territory where the prospect of finding anything of value is questionable. It is analogous to a prospect in mining. (Long, 1960) (b) A mining company organized to develop unproven ground far from the actual point of discovery. Any risky venture in mining.

wildcat drilling The drilling of boreholes in an unproved territory. Also called: cold nosing; wildcatting. (Long, 1960)

wildcatter (a) An individual or corporation devoted to exploration in areas far removed from points where actual minerals or other substances of value are known to occur. Also called cold noser. See also: *wildcat.* (Long, 1960) (b) One who locates a mining claim far from where a deposit has been discovered or developed.

wilderness (a) An area or tract of land that is uncultivated and uninhabited by people. (b) North American stage: Middle Ordovician (above Porterfield, below Barnveld; it includes uppermost Black River and Rockland rocks). (c) Mottled red and gray grit at Forest of Dean, U.K.

wild heat A heat of molten steel that is boiling violently, and so, if poured, honeycombs an ingot with contained gases. (Webster 2nd, 1960)

wild lead *sphalerite.*

wild steel (a) Steel in, or made from, a wild heat. (Webster 2nd, 1960) (b) Said of liquid steel, esp. rimmed steel, that is producing considerable effervescence. (Henderson, 1953)

wild work A type of bord-and-pillar system of coal mining in which the very narrow pillars left to support the roof are not recovered. (Fay, 1920)

Wiles' process A method of reducing iron ores in which an electric furnace is fitted with two or more hollow electrodes, through which the finely divided ore, intimately mixed with reducing materials, is introduced. (Osborne, 1956)

Wilfley slimer A form of shaking canvas table that is given a vanner motion. (Liddell, 1918)

Wilfley table Long-established and widely used form of shaking table; rectangular; mounted horizontally and can be sloped about its long axis. It is covered with linoleum (occasionally rubber) and has longitudinal riffles tapering at the discharge end to a smooth cleaning area, triangular in the upper corner. A compound eccentric is used to create a gentle and rapid throwing motion on the table, longitudinally. Sands, usually classified for size range, are fed continuously and worked along the table with (1) the aid of feedwater, and (2) across riffles downslope by gravity tilt adjustment and added wash water. At the discharge end, the sands have separated into bands: the heaviest and smallest uppermost; the largest and lightest lowest. The Dodd, Cammet, Hallett, and Woodbury are similar types of tables. See also: *Woodbury table.* (Pryor, 1963; Liddell, 1918)

Wilkinite Trade name for a colloidal bentonite (jelly rock) used in papermaking. Also spelled Wilconite.

willemite A trigonal mineral, Zn_2SiO_4; white to pale tints; fluoresces bright yellow; in zinc deposits in New Jersey, New Mexico, Africa, and Greenland; a source of zinc. Syn: *belgite.*

willen stone (a) Eng. An oolitic freestone used for cornices and chimney pieces; also, a good paving stone, Halston, Northamptonshire. (Arkell, 1953) (b) A massive yellow to green, impure variety of antigorite resembling jade; used for decorative purposes. It commonly contains specks of chromite. (c) An old "misspelling" of willemite.

Williams' hinged-hammer crusher A crusher with a rotating central shaft, carrying a number of hinged hammers that fly out from centrifugal force, crushing the feed against the casing. Syn: *hinged-hammer crusher.* (Liddell, 1918)

williamsite (a) An apple-green impure variety of serpentine. See also: *jade.* (Sanford, 1914) (b) A translucent bright green serpentine, usually containing specks of chromite, used as a semiprecious stone. (Dana, 1914) (c) An old misspelling of willemite. (Dana, 1914)

Wilmot jig A basket-type jig, the basket being suspended in a tank of water. Pulsations are effected by moving this basket upward and downward by means of eccentrics. Has been used extensively in the preparation of anthracite of all sizes and, to a lesser extent, of bituminous coal. (Mitchell, 1950)

Wilputte oven A byproduct coke oven having two outer zones in the heating system and one double inner zone. In this oven, the gas is alternately burned upwards in the two outer zones with the products of combustion being carried down through the double inner zone and, on reversal, burned upwards in the double inner zone with the products of combustion being carried down through the two outer zones. Known as a double-divided oven. (Camp, 1985)

wilsonite (a) A purplish-red material consisting of an aluminosilicate of magnesium and potassium; represents an altered scapolite. (AGI, 1987) (b) A tuff composed of fragments of pumice and andesite in a matrix of vitric and granular material (Holmes, 1928). (AGI, 1987)

Wilton stopper Automatic arrangement that arrests a mine car when it runs away downslope. The car displaces a pendulum beyond its nonactivating limit of swing and a pivoted rail then falls between the rail tracks below. (Pryor, 1963)

wiluite (a) A green variety of grossular garnet. (b) A greenish variety of vesuvianite.

win (a) To extract ore or coal. To mine, to develop, to prepare for mining. (b) To recover (as metal) from ore. (Webster 3rd, 1966)

winch (a) A small hand- or power-operated drum haulage used for light-duty work in surface and underground mines. A heavy-duty power winch fitted to the rear of a tractor. See also: *pickrose hoist.* (Nelson, 1965) (b) Syn: *hoist.* Formerly, a manually powered hoisting machine, consisting of a horizontal drum with rank handles. (Long, 1960) (c) A drum that can be rotated so as to exert a strong pull while it is winding the attached line. (Nichols, 1976) (d) A small drum haulage or hoist. A gear train is interposed between the handle and the drum. The handle is turned several revolutions to one revolution of the drum. Two handles and a pawl and ratchet may be fitted to prevent the load from running back should the pressure on the handle be reduced. See also: *monkey winch.* (Nelson, 1965)

winchellite *mesolite.*

winchman In metal mining, a person who operates a power-driven winch on a gold dredge to move it from one working position to another during dredging operations, winding up the cables anchored at points in advance of the dredge. (DOT, 1949)

winch operator *hoist operator.*

wind To hoist or raise coal or ore; to spool rope or cable on the drum of a hoist.

wind beam A beam incorporated into a structure for the sole purpose of resisting wind pressure. (Hammond, 1965)

wind blast A blown-out or "windy" shot. (Fay, 1920)

wind box In blast furnace operation, the compartment in the bottom of the converter that receives the blast and delivers it to the tuyeres. (Mersereau, 1947)

winder (a) An electrically driven winding engine for hoisting a cage or cages up a vertical mine shaft. (CTD, 1958) (b) *card tender.*

winder brake An appliance or piece of equipment capable of retarding or stopping cages in a shaft in an emergency. See also: *brake; post brake; semiautomatic control.* (Nelson, 1965)

wind furnace Any form of furnace using the natural draft of a chimney without the aid of a bellows or blower. (Fay, 1920)

wind gage An anemometer for testing the velocity of air in mines.

wind hatch In mining, an excavation or opening for removing ore. (Standard, 1964)

winding The operation of hoisting coal, ore, miners, or materials in a shaft. The conventional method is to employ two cages actuated by a drum type of winding engine. Steel ropes are attached at either end of the drum, one over and the other under it, so that as one cage ascends, the other descends. Thus the cages arrive at pit top and bottom simultaneously. See also: *automatic cyclic winding; balanced winding; balance rope; deep winding; Koepe winder; winding cycle; winding rope.* (Nelson, 1965)

winding apparatus The machinery and equipment used to lower and raise loads through a shaft. (Nelson, 1965)

winding cycle In general, a cycle refers to any series of changes or operations in which any part of the system return to its original state or position. In winding, the term usually refers to a complete wind, which is comprised of three phases: (1) acceleration to full speed; (2) full-speed running; and (3) retardation to rest. The period of the winding cycle is the sum of winding time and decking time in seconds. See also: *winding speed.* (Nelson, 1965)

winding drum For haulage to the surface through a mine shaft, the surface gear includes a winding drum of cylindrical or cylindroconical form on which the winding rope (hoisting rope) is coiled as the cage, or skip or kibble, is raised, and from which it is paid off as the return journey is made. Two such receptacles are usually worked simultaneously in balanced hoisting, one rising as the other descends, from a

compound drum. The drum is driven by the winding engine. See also: *bicylindroconical drum; conical drum; Koepe sheave; parallel drum; winding engine.* (Pryor, 1963)
winding engine (a) A steam or electric engine at the top of a shaft that powers the winding drum, thus hoisting and lowering a cage or skip by means of a winding rope. In metal mining, the winding engine is usually called a hoist. Also called: winder. See also: *hoist.* (Nelson, 1965) (b) *winding drum.*
winding engineman A skilled person in charge of the steam or electric winding engine at a mine. Also called hoistman. (Nelson, 1965)
winding guide The purpose of a winding guide is to permit winding to proceed safely at relatively high speeds by preventing collisions between the cages and between cages and the side of, or fittings in, a shaft. It must be (1) rigid enough to prevent material deviation of the cages or skips from the vertical; (2) strong, since a broken guide causes danger from damage; (3) smooth, so as to offer as little resistance to the movement of a cage as possible; and (4) firmly supported and maintained vertical. Guides may be of two types, rigid or flexible. The former may be of timber, and in new shafts have generally been replaced by steel channels, steel rails or angles; the latter are steel ropes of round or semilocked section steel rods. Rigid guides are adopted in shafts of rectangular cross section; these and the shaft sides and fittings are small. Rope guides are used in circular and elliptical shafts where adequate clearances can be provided. Rope guides maintain the vertical automatically, and expand and contract with temperature variation without complication. (Sinclair, 1959)
winding pulley *winding sheave.*
winding rope The rope that carries a cage, skip, or hoppit in a shaft. The wires are twisted together symmetrically according to a definite geometrical pattern. See also: *wire rope; winding drum.* Syn: *lay of rope.* (Nelson, 1965)
winding sheave A grooved pulley wheel, mounted on plummer blocks, at the top of the headgear. The winding rope passes from a cage or skip around the sheave and onto the winding drum. For normal loads, the sheave rim and boss are made of castiron, and the spokes are made of round, mild steel. Winding sheaves range up to about 24 ft (7.32 m) in diameter. Sheaves up to 8 ft (2.44 m) in diameter are usually made in one piece, but above this size, they are built in halves and bolted together. To give efficient service, the sheave diameter should be at least 96 times the winding rope diameter. See also: *head sheave; Koepe sheave.* Syn: *hoisting sheave.* (Nelson, 1965)
winding speed The velocity at which a winding engine lifts a cage or skip in a shaft. Winding speeds reach up to 6,000 ft/min (1.8 km/min) for deep mines. The normal maximum speed for deep shafts is 3,500 to 4,000 ft/min (1.1 to 1.2 km/min) and for geared winders, 1,500 to 3,000 ft/min (0.46 to 0.91 km/min). (Nelson, 1965)
windlass (a) A device used for hoisting; limited to small-scale development work and prospecting because of its small capacity. (Lewis, 1964) (b) A drum or a section of tree trunk set horizontally on rough bearings above a shallow pit or shaft; used to raise or lower buckets of spoil in exploratory work. Handles at each end of the drum allow for manual rotation. (Pryor, 1963)
wind method An air-blowing system of separating coal into various sizes, and extracting waste from it, which in principle depends on the specific gravity or size of the coal and the strength of the current of air.
windmill anemometer An anemometer in which a windmill is driven by the air stream, and its rotation is transmitted through gearing to dials or other recording mechanism. In some instruments, the rotating vanes and dials are in the same plane; i.e., both vertical, while in others the dial is horizontal. In the windmill type, the operation of air measurement involves readings of dials at the beginning and end of a measured period. Windmill instruments may be fitted with an extension handle, providing a form of remote control; used to measure air speed in an otherwise inaccessible spot. (Mason, 1951)
window An eroded area of a thrust sheet that displays the rocks beneath it. Syn: *fenster.* (AGI, 1987)
window pipe A dredge discharge pipe with one or more openings in the bottom. (Nichols, 1976)
window-type sample *door-type sampler.*
wind pressure The pressure on a structure due to wind, which increases with wind velocity approx. in accordance with the formula $p = 0.003\ v^2$, where p is the pressure in pounds per square feet of area affected, and v is the wind velocity in miles per hour. (Hammond, 1965)
wind road Underground ventilation road. See also: *airway.* (Pryor, 1963)
wind rose A diagram which shows the proportion of winds blowing from each of the main points of a compass at a given locality, recorded over a long period. The prevailing wind with its average strength is thereby revealed at a glance. (Hammond, 1965)
windrow (a) A row of peats or sod set up to dry, or cut in paring and burning. (Standard, 1964) (b) A ridge of soil pushed up by a grader or bulldozer. (Hammond, 1965)
winds *winze.*
windup The amount of twist occurring in a string of drill rods when the string is rotated during drilling. There can be as many as several complete revolutions of the rod at the collar before the bottom member of the string begins to rotate. Also called wrap-up. (Long, 1960)
windy shot A blast in a coal mine which—due to improperly placed charges, the wrong kind or quantity of explosives, or insufficient stemming—expends most of its force on the mine air; it sometimes ignites a gas mixture, coal dust, or both, thus causing a secondary explosion, which may or may not spread throughout the mine; a shot that blows out without disturbing the coal; a shot that is not properly directed or loaded; a blown-out shot. (Fay, 1920)
wing (a) The side or limb of an anticline. (Fay, 1920) (b) *catch; chair; dog; rests; wing.*
wing belt tripper A belt conveyor tripper having auxiliary conveyors extending laterally to one or both sides to provide wider distribution of bulk material being discharged.
wing dam *pier dam.*
winged pillar Scot. Pillar of coal that has been reduced in size. (Fay, 1920)
wingwall A wall that guides a stream into a bridge opening or a culvert barrel. (Nichols, 1976)
winklestone Warty, elongate pyrite nodule, Essex, U.K.
winning (a) The excavation, loading, and removal of coal or ore from the ground; winning follows development. (Nelson, 1965) (b) The operation of mining an ore and opening up a new portion of a coal seam. (CTD, 1958) (c) The portion of a coalfield laid out for working. (Fay, 1920) (d) The combined process of excavating and transporting a raw material such as clay to a brickworks or stockpile. (Dodd, 1964)
winning heading A development heading off which oblique headings and conveyor panels are formed and worked (longwall); any of the development drivages in the solid coal, about 15 yd (13.7 m) apart, and of which bords and pillars are formed (pillar method of working). (Nelson, 1965)
winnowing gold Air blowing. Tossing up dry powdered auriferous material in air, and catching the heavier particles not blown away. (Fay, 1920)
winter dumps A term used in Alaska to describe gold-bearing gravel mined during the winter and stored on the surface for sluicing in the spring and summer. (Fay, 1920)
winze (a) A vertical opening driven downward connecting two levels in a mine. When one is standing at the top of a completed connection the opening is referred to as a winze, while when standing at the bottom, the opening is a raise, or rise. Syn: *winds.* Cf: *underground shaft.* (b) A subsidiary shaft that starts underground. It is usually a connection between two levels. Syn: *way shaft.* (Higham, 1951) (c) Can. Interior mine shaft. (Hoffman, 1958)
wire (a) A continuous length of metal drawn from a rod. (Hammond, 1965) (b) War. A haulage rope. (Fay, 1920) (c) *capillary.*
wirebar A cast shape, particularly of tough pitch copper, which has a cross section approx. square with tapered ends; designed for hot rolling into a rod for subsequent drawing into wire. (ASM, 1961)
wire cloth Screen composed of wire or rod woven or crimped into a square or rectangular pattern. (Nichols, 1976)
wired glass A form of sheet glass produced by rolling wire mesh into a ribbon of glass so that it acts as a reinforcement and holds the fragments together in the event of the sheet being fractured. (CTD, 1958)
wire drag A buoyed wire towed by a ship at a given depth to determine whether any isolated rocks, small shoals, etc., extend above that depth, or for determining the least depth of an area. (Hunt, 1965)
wire gage (a) A gage for measuring the diameter of wire. (Fay, 1920) (b) A standard series of sizes used in the manufacture of wire (diameter) or sheet metal (thickness) and indicated by arbitrary numbers. (Fay, 1920) (c) A notched plate having a series of gaged slots, numbered according to the sizes of the wire and sheet metal manufactured; used for measuring the diameter of wire. The gage most widely used in the United States is the U.S. Standard Steel Wire, which name has official sanction, without legal effect. The Birmingham gage is recognized in acts of Congress for tariff purposes. Two gages (American Gage; Browne and Sharpe's) are used for copper wires and all nonferric metal wires. (Crispin, 1964)
wire gauze A gauzelike texture of fine wire, such as that used for the chimneys of flame safety lamps. (Fay, 1920)
wire hanger The hanger from which wire or cable is suspended.
wire line (a) As used in a general sense, any cable or rope made of steel wires twisted together to form the strands. Specif., a steel wire rope 5/16 in (7.9 mm) or less in diameter. See also: *cable.* (Long, 1960) (b) A general term for any flexible steel line or cable drill connecting a surface winch to a tool assembly lowered in a well bore. Also spelled wireline. (AGI, 1987)
wire-line barrel *wire-line core barrel.*

wire-line cable A wire rope 3/16 in or 1/4 in (4.8 mm or 6.4 mm) in diameter; used to handle the inner tube of a wire-line core barrel. See also: *cable.* (Long, 1960)

wire-line core barrel Double-tube, swivel-type core barrel; available in various outside diameters corresponding to sizes of diamond- and rotary-drill boreholes; designed so that the inner-tube assembly is retractable. At the end of a core run, the drill string is broken at the top joint so that an overshot latching device can be lowered on a cable through the drill-rod string. When it reaches the core barrel, the overshot latches onto the retractable inner-tube assembly, which is locked in the core barrel during the core run. The upward pull of the overshot releases the inner tube and permits it to be hoisted to the surface through the drill rods; it is then emptied and serviced and dropped or pumped back into the hole, where it relocks itself in the core barrel at the bottom. Syn: *wire-line barrel.* (Long, 1960)

wire-line coring The act or process of core drilling with a wire-line core barrel. See also: *wire-line core barrel.* (Long, 1960)

wire-line dredging In this method, digging tools or buckets are suspended on a steel cable and lowered to the sediment surface, where they are loaded and retrieved. Includes the use of drag-bucket and clamshell dredges, and generally to a depth of not more than 500 ft (152 m) below sea level. (Mero, 1965)

wire-line drilling The drilling of boreholes with wire-line core-barrel drill-string equipment. (Long, 1960)

wire-line drill rod Drill rod having couplings that are nearly flush on the inside and designed so that the inner tube of a wire-line core barrel and overshot assembly can be run inside the rod. (Long, 1960)

wireline drill-rod coupling A rod coupling designed for use on wire-line drill rods. See also: *wire-line drill rod; wire-line core barrel.* (Long, 1960)

wire-line drum A winding drum or hoist on which the wire line is wound when handling the inner tube and overshot assemblies of a wire-line core barrel. Syn: *wire-line hoist.* (Long, 1960)

wire-line hoist *wire-line drum.*

wire-line log *well log.*

wire-line socket The socket connecting the wire line to a wire-line core barrel overshot assembly. (Long, 1960)

wireman In mining, a person who installs and repairs underground power, light, and trolley lines, making extensions into new working places as openings advance. Also called lineman; mine wireman; wire hanger. (DOT, 1949)

wireman helper In mining, one who assists a wireman in installing and repairing underground power, light, telephone, and trolley lines. Also called wire-hanger helper. (DOT, 1949)

wire-mesh conveyor belt A woven-wire conveyor belt composed of various combinations of flattened-helical coils of wire, which may or may not be joined by straight or crimped members.

wire-mesh reinforcement Expanded metal, wire, or welded fabric used as reinforcement for concrete or mortar. See also: *mesh.* (Hammond, 1965)

wire pack A circular pack consisting of waste stone built within woven fencing fixed to light props. The pack is still effective as support after the props have failed. Wire packs, e.g., are used for small openings in the Rand mines in South Africa. (Nelson, 1965)

wire rod Hot-rolled coiled stock that is to be cold-drawn into wire. (ASM, 1961)

wire rope (a) A rope made of twisted strands of wire for winding in shafts and underground haulages. Wire ropes are made from medium carbon steels. See also: *cable; flattened strand rope; locked coil rope; multiple-strand rope; winding rope.* (Long, 1960) (b) Various constructions of wire rope are designated by the number of strands in the rope and the number of wires in each strand. The following are some common terms encountered: airplane strand; cable-laid rope; crane rope; elevator rope; extra-flexible hoisting rope; flat rope; flattened-strand rope; guy rope; guy strand; hand rope; haulage rope; hawser; hoisting rope; Lang lay rope; lay; left lay rope; left twist; nonspinning rope; regular lay; reverse-laid rope; rheostat rope; right lay; right twist; running rope; special flexible hoisting rope; standing rope; towing hawser; transmission rope. (Hunt, 1965)

wire-roper anchor A device for tieing off and securing the tension in the wire ropes of a wire-rope, side-framed, intermediate section. (NEMA, 1961)

wire-rope side-framed intermediate section An intermediate section consisting of interchangeable increments or parts in which the carrying idlers are supported by one or more steel wire ropes. Return idlers may or may not be supported from the ropes. (NEMA, 1961)

wire-rope spreader That part of a wire-rope, side-framed intermediate section that maintains a horizontal plane a fixed distance between wire ropes but does not support the wire ropes. (NEMA, 1961)

wire-rope support That part of a wire-rope, side-framed intermediate section that positions the wire rope or ropes with respect to the roof or floor. It may or may not have provisions for mounting a return idler. (NEMA, 1961)

wire ropeway A ropeway using a wire cable or cables. Used for conveying ore and supplies in rough mountainous districts; a wire tramway. See also: *aerial tramway.* (Fay, 1920)

wire saw A saw consisting of one- and three-strand wire cables up to 16,000 ft (4.9 km) long running over pulleys. When fed by a slurry of sand and water and held against rock by tension, the saw cuts narrow, uniform channels by abrasion. This saw is used for cutting granite, slate, marble, limestone, or sandstone blocks. (USBM, 1965)

wire-saw operator *wire sawyer.*

wire sawyer In a stonework industry, a person who operates a wire saw to cut very large blocks of granite, limestone, marble, slate, or sandstone into smaller blocks that can be handled on gang or circular saws. Also called wire saw operator. (DOT, 1949)

wire setter Person who tends electrically powered unwinding machine that supplies wire netting to be embedded in sheet glass. (DOT, 1949)

wire silver Native silver in the form of wires or threads. (AGI, 1987)

wire strand Several steel wires twisted together to form one strand of a wire rope or cable. (Long, 1960)

wire-strand core A core in which the number of wires shall not be less than the number of wires in a main strand of the wire rope, and the individual wires shall be of an appropriate grade of steel in accordance with the best practice and design, either bright (uncoated), galvanized, or drawn galvanized wire. See also: *independent wire rope core.*

wiry Occurring as thin wires, often twisted like the strands of a rope; e.g., native copper.

witching stick *divining rod.*

withamite A red to yellow variety of epidote with a little manganese; in andesites at Glencoe, Scotland. Cf: *piemontite.*

withdraw To draw off; to take out supports. (Mason, 1951)

withdrawal Segregation of particular lands from the operation of specified public land laws, making those laws inapplicable to those lands. Lands may be withdrawn from all or any part of the public land laws, including the mineral location and mineral leasing laws. (SME, 1992)

witherite An orthorhombic mineral, $BaCO_3$; aragonite group; colorless to milky; in low-temperature hydrothermal veins. Syn: *carbonate of barium.*

witness corner A marker set on a property line leading to a corner; used where it would be impracticable to maintain a monument at the corner itself. (Seelye, 1951)

witness mark A mark or stake set to indicate the position (approximate or exact) of a property corner, instrument station, or other survey point. A witness may be a rock, tree, or other object; e.g., a blazed tree on the bank of a river to indicate a corner at the intersection of some survey line with the center line of the stream, which, therefore, cannot be marked directly; a stake driven so as to stand out; and a stake marked with a station number, driven flush with or below the surface of the ground. (Seelye, 1951)

witness post Satellite beacon used to mark a claim when the correct boundary post is inaccessible. (Pryor, 1963)

wittichenite An orthorhombic mineral, Cu_3BiS_3; gray to tin-white; at Wittichen, Baden, Germany. Also called wittichite.

wittite A monoclinic mineral, $Pb_9Bi_{12}(S,Se)_{27}$; lead-gray; at Falun, Sweden.

Witton-Kramer magnet A circular magnetic separator suspended over a conveyor head pulley to extract small pieces of tramp iron. (Nelson, 1965)

Witwatersrand The gold-mining district, now usually called the Rand, in South Africa. (Nelson, 1965)

wobble wheel roller A skip body mounted on nine or more oscillating, smooth, rubber-tired wheels for compaction and fine rolling of soil. (Nelson, 1965)

wodginite A monoclinic mineral, $(Ta,Nb,Sn,Mn,Fe)_{16}O_{32}$; dimorphous with ixiolite; occurs in granite pegmatites; named for Wodgina, Australia.

wöhlerite (a) A monoclinic mineral, $Ca_2NaZr(Si_2O_7)(O,OH,F)_2$. (b) The organic matter in carbonaceous chondrites. Also spelled woehlerite.

Wöhler test A fatigue test in which one end of a specimen is held in a chuck and rotated in a ball bearing placed on the other end. The ball bearing carries a weight and, as the specimen rotates, the stress at each point on its surface passes through a cycle from a maximum in tension to a maximum in compression. (CTD, 1958)

wold A range of hills produced by differential erosion from inclined sedimentary rocks; a cuesta. Cf: *cuesta.* (AGI, 1987)

Wolf (a) The name of a naphtha-burning flame safety lamp. (Jones, 1949) (b) The name of carbide and electric lamps.

wolfachite A mineral (not established as a species), Ni(As,Sb)S, intermediate between gersdorfite and ullmannite; occurs at Wolfach, Baden, Germany.

wolf cut point *equal-errors cut point.*

wolfeite A monoclinic mineral, $(Fe,Mn)_2(PO_4)(OH)$; forms a series with triploidite; dimorphous with satterlyite.

Wolf nickel-cadmium battery While other nickel-cadmium batteries generally adopt either tubular or pocketed positive plate construction, the Wolf battery has individual features of interest. The supporting medium for the active materials consists of strips of compressed corrugated nickel foil. The method of construction is to perforate strips of the foil, which are pasted with active material. The strips are folded into corrugations and compressed into a cake. Two or more cakes are mounted in a pure nickel frame to form the finished plate. This method of construction results in a plate of satisfactory electrical conductivity, and no admixture of graphite or flake nickel in the active material is necessary. (Roberts, 1958)

Wolf process A flotation process invented by Jacob D. Wolf in 1903. He used sulfochlorinated or other oils and aimed to secure a high extraction with a low grade of concentrate in the first step, and by washing with hot water to concentrate the concentrate in a second step. Apparently no commercial use was made of it. (Liddell, 1918)

wolfram *wolframite; tungsten.*

wolframine *tungstite; tungstic ocher; wolframite.*

wolframite A monoclinic mineral, $(Fe,Mn)WO_4$; within the hübnerite-ferberite series; pseudo-orthorhombic due to twinning. See also: *tungsten.* Syn: *wolfram.*

wolfram lamp A tungsten lamp. (Webster 3rd, 1966)

wolfram ocher *tungstic ocher.*

wolfsbergite *chalcostibite; jamesonite.*

wollastonite A triclinic mineral of the pyroxenoid group: $CaSiO_3$. It is dimorphous with parawollastonite. Wollastonite is found in contact-metamorphosed limestones, and occurs usually in cleavable masses or sometimes in tabular twinned crystals; it may be white, gray, brown, red, or yellow. It is not a pyroxene. Symbol, Wo. Syn: *tabular spar.* (AGI, 1987)

wollongite *wollongongite.*

wollongongite A coallike shale similar to torbanite. It is named from its type locality, Wollongong, New South Wales, Australia. Also spelled wollongite; wallongite. (AGI, 1987)

wölsendorfite An orthorhombic mineral, $(Pb,Ca)U_2O_7 \cdot 2H_2O$; forms bright red crusts on fluorite; at Wölsendorf, Bavaria, Germany.

Wommer safety clamp A type of foot operated drill-rod safety clamp, the operation of which is similar to a bulldog safety clamp. Also called automatic spider. (Long, 1960)

wonder metal Applied to metals, such as beryllium, magnesium, titanium, and zirconium, that were put into expanded use following World War II. (Pearl, 1961)

wood agate A term used for agatized wood, esp. agate formed by siliceous permineralization of wood. (AGI, 1987)

Woodbury jig A jig with a plunger compartment at the head end, so that the material is given a classification in the jig. (Liddell, 1918)

Woodbury table A table of the general Wilfley-Overstrom-Card type, with riffles parallel to the tailings side, and a hinged portion without riffles (unlike the Card). The table top is a rhomboid, and the riffles gradually shorten as they near the tailings side. See also: *Wilfley table.* (Liddell, 1918)

wood chain S. Staff. A hoisting chain, the iron links of which are filled with small blocks of wood. (Fay, 1920)

Wooddell scale A scale of resistance to abrasion based on the following method: if specimens of different materials are mounted so that they present surfaces substantially in the same plane, and if the surfaces are subjected to a lapping operation with a properly selective abrasive, the harder materials will stand out in relief, whereas the softer ones will be cut or worn to a depth, depending upon their hardness. By averaging several readings, a scale of hardness was established by which the quantitative values of the hardness of various materials could be determined. On the Woodell scale, diamond has approx. 2 times the hardness of boron carbide, 3.5 times that of tungsten carbide, and nearly 5 times that of corundum. (Chandler, 1964)

wooden tubbing Consists of wooden staves driven down in soft ground during sinking to keep back water. The lining is stated to be capable of withstanding pressures up to a maximum of 130 psi (896 kPa). The lining resembles the sides of a wooden tub and the word tubbing is doubtless derived from this similarity. See also: *tubbing.* (Nelson, 1965)

wood hematite A finely radiated variety of hematite exhibiting alternate bands of brown or yellow of varied tints. (Fay, 1920)

woodhouseite A trigonal mineral, $CaAl_3(SO_4)(PO_4)(OH)_6$; beudantite group; forms small colorless rhombohedra.

wood iron A fibrous variety of chalybite (siderite), $FeCO_3$. (Fay, 1920)

wood iron ore Corn. Fibrous limonite; Land's End district. (Arkell, 1953)

wood opal A variety of opal with woody texture by replacement. Syn: *opalized wood; xylopal; lithoxyl.*

wood piling A method of sinking a shaft through loose surface deposits by driving a ring of wood piles down vertically. As the piles are rammed downward, the loose material is removed. Frames are set to prevent the piles being forced inward. Each new set of piles reduces the dimensions of the excavation. To avoid this reduction in size, the piles are driven at an angle away from the shaft space. (Nelson, 1965)

Wood process A flotation process utilizing the surface tension of water, either fresh, acid, or salt.

woodrock A variety of asbestos resembling wood. (Standard, 1964)

woodruffite A monoclinic mineral, $(Zn,Mn)Mn_3O_7 \cdot H_2O$; related to todorokite; at Sterling Hill, NJ.

woodstave piping Piping formed from wood boards fitted and strapped together by encircling steel bands. (Pryor, 1963)

wood stilt A piece of wood attached to the leg of steel girders to provide a measure of yield and prevent premature distortion and damage to the ring. See also: *arch girder; stilt.* (Nelson, 1965)

woodstone *silicified wood.*

wood tin A nodular variety of cassiterite, or tinstone, of a brownish color and fibrous structure, and somewhat resembling dry wood in appearance. Syn: *dneprovskite.* (Fay, 1920)

woodwardite A mineral, $Cu_4^{2+}Al_2(SO_4)(OH)_{12} \cdot 2\text{-}4H_2O(?)$; greenish-blue; closely akin to cyanotrichite.

Worden gravimeter A compact temperature-compensated gravity meter, in which a system is held in unstable equilibrium about an axis, so that an increase in the gravitational pull on a mass at the end of a weight arm causes a rotation opposed by a sensitive spring. The meter weighs 5 lb (2.25 kg) and has a sensitivity of less than 0.1 mgal. (AGI, 1987)

work (a) The process of mining coal. (BCI, 1947) (b) To crumble and yield under the action of a squeeze. Applied to pillars or roof of a coal mine. (c) To be slowly closing under the action of a squeeze. Applied to portions of mine workings. (d) Denoting that creep or squeeze is taking place. (Hudson, 1932) (e) The product of a force in terms of weight and the lineal distance through which it acts. (Hammond, 1965) (f) To undergo gradual movement, such as heaving, sliding, or sinking; said of rock materials. (AGI, 1987) (g) Ore before it is beneficiated. (Fay, 1920) (h) A place where industrial labor of any kind is carried on. Usually in the plural as saltworks, ironworks, etc. (Webster 3rd, 1966) (i) Objects that are to be, are being, or have been treated, such as in cleaning or finishing. (ASM, 1961)

workable A coal seam or orebody of such thickness, grade, and depth as to make it a good prospect for development. In remote and isolated locations, other factors influence minability, such as access, water supply, transport facilities, etc. See also: *economic coal reserves.* (Nelson, 1965)

workable bed Any bed or vein that is capable of being mined, but usually applied to a coal seam or ore deposit that can be mined profitably.

workable tonnage *probable reserves.*

work arm The part of a lever between the fulcrum and the working end. (Nichols, 1976)

work capacity The limit of energy expended or absorbed, within which a body is not unduly fatigued. (Brantly, 1961)

worked out (a) A mine, or large section of a mine, from which all minable coal or ore has been taken. (BCI, 1947) (b) Exhausted; said of a coal seam or ore deposit.

worker cage A special cage for raising and lowering workers in a mine shaft. See also: *worker car.* (Fay, 1920)

worker car A type of car for transporting miners up and down mine shafts; also cars used to transport miners from the shaft to working areas. See also: *worker cage.* (Fay, 1920)

worker door (a) A small door in a stopping to allow the passage of workers. (Nelson, 1965) (b) Scot. A small trapdoor on a traveling road. (Fay, 1920)

worker-hoist engineer In mining, a person who operates the hoisting engine that serves the shaft in which only workers are raised from and lowered into a mine. (DOT, 1949)

worker trip (a) A trip made by mine cars and locomotives to take workers, rather than mined material or supplies, to and from the working places. (BCI, 1947) (b) A similar trip made by a worker cage in a shaft.

work index *Bond's third theory.*

working (a) When a coal seam is being squeezed by pressure from the roof and floor it emits creaking noises and is said to be "working." This noise often serves as a warning to miners that additional support is needed. Sagging roof emitting noises and requiring additional timbering. (BCI, 1947) (b) A working may be a shaft quarry, level, opencut, or stope, etc. Usually in the plural. See also: *labor; workings.* (Fay, 1920)

working a claim Activities such as extracting ore, building structures, and otherwise developing a mining claim after ore is discovered. (Ricketts, 1943)

working capital The amount of money available to finance the operations of a company beyond the amount required for the purchase of fixed assets, such as property. (Truscott, 1962)

working cycle A complete set of operations. In excavation, it usually includes breaking, loading, moving, dumping, and returning to the loading point. (Nichols, 1976)

working face The place at which mining is being done in a breast, gangway, airway, chute, heading, drift, adit, crosscut, etc. See also: *face.*

working gullet The immediate excavation needed for opencast working of ore. (Nelson, 1965)

working-hole In glassmaking, a small opening over pots enabling workers to introduce or withdraw material required.

working home Mining toward the main shaft while extracting ore or coal, such as in longwall retreating. See also: *longwall.* (Fay, 1920)

working interest The operator's mineral ownership involving the costs of drilling, completion, equipment, and producing in contrast to the (free) royalty interest. (Wheeler, 1958)

working level (WL) Any combination of radon daughters in one liter of air that result in the ultimate emission of 1.3×10^5 million electrons volts (MeV) of alpha energy. (NSC, 1996)

working load The maximum weight a hoist line or other rope or cable can carry under working conditions without danger of straining. (Long, 1960)

working on air A pump is said to be working on air when air is sucked up with the water. (Fay, 1920)

working-on-the-walls The eroding or corroding of blast furnace linings. (Fay, 1920)

working out Mining away from the main shaft while extracting ore or coal, such as in longwall advancing. See also: *advancing.* Cf: *working home.* (Fay, 1920)

working pit A mine shaft through which ore and miners are carried, as distinguished from one used only in pumping. (Standard, 1964)

working place (a) A place in a mine at which coal or ore is actually being mined. See also: *working face.* (Fay, 1920) (b) A miner's room or chamber. (Hudson, 1932)

working room Generally the space between the working face and the area being backfilled. (Stoces, 1954)

workings (a) Any area of development; usually restricted in meaning to apply to breasts, etc., in contradistinction to gangways and airways. Often used in a broader sense to mean all underground developments. See also: *working.* (Fay, 1920) (b) The entire system of openings in a mine. Typical usage restricts the term to the area where coal, ore, or mineral is actually being mined. (Nelson, 1965) (c) Colloquial term for an anthracite mining operation. (Korson, 1938)

working section All areas of a coal mine from the loading point of the section to and including the working faces. (FR 95, 1992)

working stress (a) The stress considered to be a safe maximum for a particular material under ordinary conditions. See also: *yield stress; load factor.* (Hammond, 1965) (b) The maximum unit stress to which the parts of a structure are to be subjected. (Zern, 1928) (c) *allowable stress.*

working the broken The extraction of coal pillars in a pillar method of mining. See also: *pillar extraction; broken working; second working; robbing pillars.* (Nelson, 1965)

working the whole The driving of the narrow coal headings to form pillars in a pillar method of working. (Nelson, 1965)

work lead (a) Impure pig lead that is to be desilverized or refined. See also: *base bullion.* (Standard, 1964; Fay, 1920) (b) The electrical conductor connecting the source of arc welding current to the work. Also called welding ground; ground lead. (ASM, 1961)

workplace protection factor A measure of the protection provided in the workplace by a properly functioning respirator when correctly worn and used. (NIOSH, 1987)

work platform A board or small platform placed at a suitable height in a drill tripod or derrick so that a worker standing on it can handle the drill rod stands. (Long, 1960)

work shaft A shaft that is in daily use for hoisting coal, ore, or miners. See also: *air shaft.* (Nelson, 1965)

work stone A plate in the bottom of a blast hearth or ore hearth having a groove down its center for conducting away the molten lead. (Standard, 1964)

work study Techniques for analyzing methods used in performing an operation and measuring the work involved. Work study fosters better use of materials, plant, and labor, thus ensuring higher productivity. See also: *method study.* (Nelson, 1965)

worm A spiral tool, shaped like a carpenter's wood-boring auger, with the bottom end shaped like the cutting end of a diamond point or mud bit. The tool is rotated inside a casing to loosen and clean out debris or to loosen and drill through tough clay at the bottom of a borehole. Also called worm auger; worm-type auger. (Long, 1960)

worm auger *worm.*

worm conveyor (a) A conveyor consisting of a spiral plate encircling and fastened to a shaft lying longitudinally within a trough; rotation of the spiral pushes the material forward. Also called screw conveyor. (BS, 1962) (b) *helical conveyor.*

worm wheel A modified spur gear with curved teeth that meshes with a worm. (Nichols, 1976)

wough The side of a mineral vein.

wound rotor motor A wound rotor induction motor differs from a squirrel cage induction motor only in the construction of its rotor. The rotor, instead of having short-circuited copper bars, has a definite winding connected for the same number of poles as the stator with the leads brought out to slip rings. The stator and rotor are commonly called the primary and secondary, respectively, because under locked rotor conditions, the motor becomes a transformer with a given ratio. This ratio depends upon motor design and is not standardized. Frequently called a slip ring motor. (Pit and Quarry, 1960)

woven-wire vibrating screen Ore screening machine whose screen is woven of steel wire and stretched tightly on a metal frame. Near the center of the screen is fastened the vibrating element of a high-speed vibrator, which produces a vibratory motion at right angles to the plane of the screen. (Newton, 1959)

WP-cut A tunnel blasting method in which holes are arranged in a geometrical figure as an incomplete pyramid and not parallel in the planes of the sides. For tunnel widths less than 25 ft (7.5 m), the WP-cut provides a greater advance than V-cuts and fan cuts.

wracking force A horizontal force tending to distort a rectangular shape into a parallelogram. (Hammond, 1965)

wrap-drive conveyor A conveyor in which the return strand of the belt is driven by a wrap drive which combines a drive pulley with a snub pulley. (NEMA, 1961)

wrap-up Same as windup, as applied to the twist in a drill-rod string. See also: *windup.* (Long, 1960)

wreath In glassmaking, a wavy appearance in glass, esp. flint glass, due to defective manufacture. (Standard, 1964)

wrecking bar A steel bar usually from 1 to 2 ft (0.3 to 0.6 m) in length, with one end drawn to a thin edge, the other curved to a claw. (Crispin, 1964)

wrench fault A transverse strike-slip fault that is more or less vertical.

wrist action In a bucket, the ability to change its digging or dumping angle by power. (Nichols, 1976)

wrought alloy Type of alloy suitable for forming by mechanical means at temperatures below the melting point. (Light Metal Age, 1958)

wrought iron (a) A low-carbon iron containing a relatively high proportion of residual slag that gives it ductility and toughness. (Strock, 1948) (b) A commercial form of iron containing less than 0.3%, and usually less than 0.1%, carbon; also carrying 1.0% or 2.0% of slag mechanically mixed with it and originally made directly from ore (as in the Catalan forge) but subsequently by puddling. See also: *iron.* Cf: *ingot iron.* (Webster 3rd, 1966)

wrought metal A metal that has been worked by cold rolling, forging, pressing, drawing, or extension. (Hammond, 1965)

Wuensch process In metallurgy, a heavy suspension method for the concentration of ores in which the waste has a specific gravity of 2.7 or more. Minerals having a specific gravity in excess of 5.25 must be used, since a suspension containing over 40% solids by volume is too plastic for use. Galena (sp gr, 7.4 to 7.6) and ferrosilicon (sp gr, 6.7 to 7.0) have been used. (Hess)

wulfenite A tetragonal mineral, $PbMoO_4$; prismatic cleavage; soft; resinous; yellow to brown; sp gr, 6.5 to 7.0; in oxidized zones of lead-molybdenum veins; a source of molybdenum. Syn: *yellow lead ore.*

wurtzilite A black, massive, asphaltic pyrobitumen; sectile and infusible; closely related to uintahite; insoluble in turpentine; derived from metamorphosed petroleum; occurs in veins in Uinta County, UT. See also: *elaterite.*

wurtzite A trigonal and hexagonal mineral, (Zn,Fe)S; dimorphous with sphalerite; resinous; brownish black; forms hemimorphic pyramidal crystals or radiating needles and bundles within lamellar sphalerite. See also: *alpha zinc sulfide; zinc sulfide.*

wurtzite-8H; wurtzite-10H Two polytypes, 8ZnS and 10ZnS, respectively, of wurtzite found at Joplin, MO; hexagonal. The wurtzite polytypes evidently form a homologous series (2H, 4H, 6H, etc.) resulting from growth phenomena based on screw dislocations. (Am. Mineral., 1945)

wüstite A mineral, FeO. Artificially prepared specimens are characteristically deficient in iron. Also spelled wustite. Syn: *iozite.* (AGI, 1987)

wyartite An orthorhombic mineral, $Ca_3U(UO_2)_6(CO_3)_2(OH)_{18}\cdot 3\text{-}5H_2O$; violet-black; occurs with ianthinite altered from uraninite; at Shinkolobwe, Katanga, Republic of the Congo.

wych *wich.*

wye (a) Cumb. The beam-end connection above the pump rods of a winding and pumping engine. (Fay, 1920) (b) A cast or wrought fitting that has one side outlet at any angle other than 90°. The angle is usually 45°, unless another angle is specified. The fitting is usually indicated by the letter Y. (Strock, 1948)

Wyoming bentonite A swelling type of bentonite that absorbs about eight times its dry volume of water to form a gel. See also: *bentonite.* Syn: *sodium bentonite.*

xalostocite A rose-pink variety of grossularite garnet. Also called rosolite; landerite. From Xalostoc, Morelos, Mex. (English, 1938)
xanthate Common specific promoter used in flotation of sulfide ores. A salt or ester of xanthic acid made of an alcohol, carbon disulfide, and an alkali.
xanthiosite A monoclinic mineral, $Ni_3(AsO_4)_2$; sulfur-yellow.
xanthitane (a) An alteration product of titanite (sphene). (b) *anatase.*
xanthochroite Amorphous cadmium sulfide. See also: *greenockite.*
xanthoconite A monoclinic mineral, Ag_3AsS_3; dimorphous with proustite; brilliant red, orange-yellow to brown. Syn: *rittingerite.*
xanthophyllite *clintonite.*
xanthorthite A yellow altered variety of allanite that contains considerable water. (Standard, 1964)
xanthoxenite A triclinic mineral, $Ca_4Fe_2(PO_4)_4(OH)_2{\cdot}3H_2O$; wax-yellow; forms thin plates; at Rabenstein, Bavaria, Germany.
xanthus An early name for heliotrope. (Hey, 1964)
***x*-axis** (a) The axis of abscissas in a plane Cartesian coordinate system. Commonly written x-axis. (Webster 3rd, 1966) (b) One of the three optic axes (x, y, and z) in a biaxial crystal. The x-axis is the axis of a greatest ease of vibration. Light vibrating parallel to the x-axis travels with maximum velocity and is called the fast ray, the x-ray (not to be confused with the penetrating X-rays of extremely short wavelength), and the α-ray. The lowest index of refraction n_α. in biaxial minerals is the index of the fast ray vibrating parallel to the x-axis. (c) One of three axes in a three-dimensional coordinate system. Crystallographers customarily use a right-handed system with the z-axis oriented positive upward, the y-axis positive to the right, and the x-axis positive toward the viewer. For mineral parameters the x-axis is labeled a with periodic translations t_1.
x-coordinate (a) An abscissa in a plane Cartesian coordinate system. (Webster 3rd, 1966) (b) One of the three coordinates in a three-dimensional rectangular coordinate system. (Webster 3rd, 1966)
***x*-direction** One of three orthogonal optic directions in biaxial crystals. Light with its electric vector (vibration direction in early terminology) parallel to the x-direction has the lowest refractive index (n_α) for a given crystal and is called the fast ray. In orthorhombic crystals, the x-axis is constrained by symmetry to correspond to one of the crystallographic directions, the correspondence determined empirically. In monoclinic crystals, one optic direction, commonly the y-direction, is constrained by symmetry to correspond to the unique diad. In triclinic crystals, there is no symmetrical constraint relating optic directions to crystallographic axes.
xeno- A prefix meaning strange or foreign.
xenoblast A crystal that has grown during metamorphism without the development of its characteristic faces. Cf: *idioblast.* See also: *crystalloblast.*
xenoblastic Applied to a texture of metamorphic rocks in which the constituent mineral grains lack proper crystal faces.
xenocryst A crystal in an igneous rock that is foreign to the body of rock in which it occurs. Cf: *xenolith.* Syn: *accidental inclusion; chadacryst.*
xenogenous A little-used syn. of epigenetic. (AGI, 1987)
xenolite A silicate of aluminum, related to fibrolite. (Fay, 1920)
xenolith A foreign inclusion in an igneous rock. Cf: *autolith; xenocryst.* Syn: *inclusion; exogenous inclusion; accidental inclusion.* (AGI, 1987)
xenomorphic (a) Said of the holocrystalline texture of an igneous or metamorphic rock, characterized by crystals not bounded by their own faces but with their forms impressed upon them by adjacent mineral grains. Also said of a rock with such a texture. (AGI, 1987) (b) (Syn: *anhedral.* in European usage.) Syn: *allotriomorphic; anidiomorphic; xenomorphic-granular.* Cf: *automorphic.* (AGI, 1987)
xenomorphic-granular *xenomorphic.*
xenon A zero-valent, very heavy, inert gaseous element; one of the so-called noble gases. Symbol, Xe. Obtained by the fractional distillation of liquid air. The gas is used in making electron tubes, stroboscopic lamps, and bactericidal lamps; also used in the atomic energy field in bubble chambers, probes, and other applications where its high molecular weight is of value. (Handbook of Chem. & Phys., 3)
xenothermal Said of a hydrothermal mineral deposit formed at high temperature but shallow depth; also, said of that environment. Cf: *telethermal; epithermal; mesothermal; hypothermal deposit; leptothermal.* (AGI, 1987)
xenotime A tetragonal mineral, YPO_4; commonly includes small quantities of thorium, uranium, and rare-earth elements; resembles zircon in form, structure, and occurrence; shows pale tints; an accessory in granites and pegmatites; also occurs in placers.
X-frame brace A brace of a square set in which two diagonal pieces of timber cross to form an X. (Fay, 1920)
xonotlite A monoclinic and triclinic mineral, $Ca_6Si_6O_{17}(OH)_2$; light-gray to pink; fibrous; in serpentinites in Santa Barbara, CA, and on Isle Royale, MI.
X-ray diffraction Reflection at definite and characteristic angles from space lattices of crystals of X-rays that have been caused to bombard them, thus giving data for identification of characteristic lattice structure of a given species of mineral. See also: *reciprocal lattice; diffraction.* (Pryor, 1963)
xylain (a) A subvariety of provitrain in which the woody origin of the cellular structure is microscopically visible. Cf: *periblain; suberain.* (AGI, 1987) (b) Those constituents of coal derived from lignified tissues in which structures were retained. (AGI, 1987)
xylanthite A variety of fossil resin. (Tomkeieff, 1954)
xylenite A maceral composed of xylain. (Tomkeieff, 1954)
xylinite (a) A variety of provitrinite. The micropetrological constituent, or maceral, of xylain. It consists of wood (xylem or lignified tissues) almost jellified in bulk but still showing faint traces of cell walls and resin contents under the microcope. (AGI, 1987) (b) A distinction of telinite, based on botanical origin (xylem or lignified tissues). To be used if desired but considered unnecessary by the Heerlen Congress of 1935. Cf: *suberinite; telinite.* (AGI, 1987)
xylocryptite *scheererite.*
xyloidin An explosive compound produced by the action of nitric acid on starch or woody fiber. Resembles guncotton. (Fay, 1920)
xylopal *wood opal.*
xyloretinite A white hydrocarbon similar to hartite found in fossil wood. (Tomkeieff, 1954)

YAG Acronym and abbrev. for yttrium-aluminum garnet. (USBM, 1965)

yard The British standard of length, equal to 36 in, 3 ft, or 0.9144 m. (Hammond, 1965)

yardage (a) The extra compensation a miner receives in addition to the mining price for working in a narrow place or in deficient coal. Usually at a certain price per yard (or meter) advanced. (b) A system of payment to workers in accordance with the number of yards (or meters) driven, repaired, or packed; the length in yards (or meters) of a drivage or face which a miner or contractor has excavated in a week or from one measuring day to the next. Also called yard work. See also: *piecework*. (Nelson, 1965) (c) Relates to cubic yards (or meters) of earth excavated. (Crispin, 1964) (d) Price paid per yard (or meter) for mining or cutting coal, usually by contract agreement, not on a tonnage basis. (BCI, 1947)

yardage man In bituminous coal mining, a laborer who pries down loose roof rock with a bar after coal has been blasted from the working face; the worker picks out seam partings (layers of rock) in the coal working face prior to blasting, using a long handled pick. (DOT, 1949)

yard price The price paid per yard driven (in addition to tonnage prices) for roads of certain widths and driven in certain directions. See also: *yardage*. (Fay, 1920)

yard work *yardage.*

yavapaiite A monoclinic mineral, $KFe(SO_4)_2$; at the United Verde copper mine, Jerome, AZ. Named for the local Yavapai Native Americans.

***y*-axis** (a) One of the three optic axes (x,y, and z) in a biaxial crystal. The y-axis is the intermediate optic axis, at right angles to the plane containing optic axes x and z. Light vibrating parallel to the y-axis is called the intermediate ray, the y-ray, and the β-ray. The middle-value index of refraction n_β in biaxial minerals is the index of the intermediate ray vibrating parallel to the y-axis. (b) One of three axes in a three-dimensional coordinate system. Crystallographers customarily use a right-handed system with the z-axis oriented positive upward, the y-axis positive to the right, and the x-axis positive toward the viewer. For mineral parameters, the y-axis is labelled b with periodic translations t_2. In monoclinic crystal systems, mineralogists conventionally adopt the second setting and designate the unique diad y.

y-coordinate (a) An ordinate in a plane Cartesian coordinate system. (Webster 3rd, 1966) (b) One of the three coordinates in a three-dimensional rectangular coordinate system. (Webster 3rd, 1966)

***y*-direction** (a) One of three orthogonal optic directions in biaxial crystals. Light with its electric vector (vibration direction in early terminology) parallel to the y-direction has a unique intermediate refractive index (n_β) for a given crystal and is called the fast ray, relative to light with its electric vector parallel to the z-direction, and the slow ray, relative to light with its electric vector parallel to the x-direction. In orthorhombic crystals, the y-direction is constrained by symmetry to correspond to one of the crystallographic directions, the correspondence determined being empirically. In monoclinic crystals, one optic direction, commonly the y-direction, is constrained by symmetry to correspond to the unique diad. In triclinic crystals, there is no symmetrical constraint relating optic directions to crystallographic axes. (b) One of three orthogonal optic directions in biaxial crystals. Light with its electric vector (vibration direction) parallel to the y-direction has a unique intermediate index of refraction (n_{beta}) for a given crystal and is fast or slow depending on crystal orientation.

yeast Fungi belonging to the ascomycetes, in which the usual and dominant growth form is unicellular. (Rogoff, 1962)

yeatmanite A triclinic mineral, $Mn_9Zn_6Sb_2Si_4O_{28}$; forms brown crystals; at Franklin Furnace, NJ.

yellow antimony (a) Yellow allotropic form of antimony. Obtained by oxidizing antimony hydride at a low temperature. (Bennett, 1962) (b) An unstable form of antimony. It can be obtained during the electrolysis of antimony trichloride, SbCl. As yellow antimony (alpha antimony) is deposited on an electrode, it forms a solid solution in the antimony chloride. When this solution is scratched or heated, metallic antimony (beta antimony) and clouds of antimony chloride form instantaneously, giving rise to the term explosive antimony. There are four allotropic forms of antimony: yellow antimony; black antimony; explosive antimony; and metallic antimony or ordinary antimony. These allotropic forms are also designated alpha antimony (yellow antimony); beta antimony (metallic antimony); and gamma antimony. (CCD, 1961; Handbook of Chem. & Phys., 2)

yellow arsenic *orpiment.*

yellow boy Deposit from the acid waters of a mine or partial neutralization. Ferrous anhydride and other impurities including fine clay carried down with it. (Zern, 1928)

yellow cake (a) Applied to certain uranium concentrates produced by mills. It is the final precipitate formed in the milling process. Usually considered to be ammonium diuranate or sodium diuranate, but the composition is variable and depends on the precipitating conditions. (USBM, 1965) (b) A common form of triuranium octoxide, the powder obtained by evaporating an amnonia solution of the oxide. (CCD, 1961)

yellow coal *tasmanite.*

yellow copper *chalcopyrite.*

yellow copperas *copiapite.*

yellow copper ore *chalcopyrite.*

yellow dog Field name for a drill tripod or derrick lamp, consisting of a metal container with two spouts holding cotton wicks, on which burning oil gives a very yellow light. (Long, 1960)

yellow earth (a) Impure yellow ocher. (Webster 3rd, 1966) (b) Loess of northern China.

yellow gravel Eng. The lower subdivision of the Aptian sponge gravel; Faringdon, Berkshire. (Arkell, 1953)

yellow ground Oxidized kimberlite of yellowish color found at the surface of diamond pipes (e.g., South Africa), above the zone of blue ground. (AGI, 1987)

yellow heat A division of the color scale. generally given as about 2,000 °F (1,090 °C).

yellow lead ore *wulfenite.*

yellow ocher (a) A mixture of limonite usually with clay and silica. Used as a pigment. (Webster 3rd, 1966) (b) A moderate orange yellow that is yellower and darker than deep chrome yellow. (Webster 3rd, 1966) (c) *goethite; limonite.*

yellow ore *carnotite; chalcopyrite.*

yellow ozokerine A product resembling vaseline, but less homogeneous; produced from crude ozokerite. (Fay, 1920)

yellow pyrite *chalcopyrite.*

yellow ratsbane *orpiment.*

yellow sands Eng. The basal part of the Permian, Durham. (Arkell, 1953)

yellow tellurium *sylvanite.* Also spelled yellow tellurim.

yenite A rejected syn. for ilvaite. (CTD, 1958)

yield (a) The proportion of coal or ore obtained in mining; the product of a metallurgical process; extraction; recovery. (Fay, 1920) (b) The percentage of "run-of-mine" material that is marketable. (Hudson, 1932) (c) The amount of a product obtained from any operation expressed as a percentage of the feed material. (BS, 1962)

yielding arch Steel arch installed in an underground mine. Arches are employed to support loads caused by changing ground movement or faulted and fractured rock. They are designed so that when the ground load exceeds the design load of the arch as installed, yielding takes place in the joint of the arch, permitting the overburden to settle into a natural arch of its own, thus tending to bring all forces into equilibrium. (Lewis, 1964)

yielding floor A soft floor that heaves and flows into open spaces when subjected to heavy pressure from packs or pillars. See also: *creep*. (Nelson, 1965)

yielding prop An adjustable steel prop that incorporates a sliding or flexible joint that comes into operation when roof pressure exceeds a set load or value. See also: *hydraulic chock; hydraulic prop*. (Nelson, 1965)

yielding support A support that incorporates a sliding or flexible joint or stilt to accommodate early pressure and thus delays damage and distortion of the support. Friction or hydraulic devices may be used so that a support, when subjected to a load above its set load, yields mechanically rather than by distorting. See also: *stilt*. (Nelson, 1965)

yield loss The difference between the actual yield of a product and the yield theoretically possible (based on the reconstituted feed) of a product with the same properties (usually percentage of ash). Also called washing error. (BS, 1962)

yield-pillar system A method of roof control whereby the natural strength of the roof strata is maintained by relieving pressure in working areas and controlling transference of load to abutments that are clear of workings and roadways. The method consists of causing certain coal pillars to yield in small amounts. See also: *double packing.* (Nelson, 1965)

yield point The differential stress at which permanent deformation first occurs in a material. Syn: *yield stress; yield strength.* (Roark, 1954)

yield strength The stress at which a material exhibits a specified deviation from proportionality of stress and strain. An offset of 0.2% is used for many metals. See also: *yield point.* (ASM, 1961)

yield stress (a) The lowest stress at which extension of a tensile test piece increases without increase in load. It is determined by observing the fall of the testing lever and checked by a pair of dividers on the original gage length. Many materials do not indicate a defined yield stress and in such cases the proof stress is used. See also: *working stress; high-tensile steel; mild steel; yield point.* (Hammond, 1965) (b) Minimum stress required to shear (that is, exceed elastic recovery) a suspension, such as the dense media used in mineral concentration. (Pryor, 1963)

YIG Acronym and abbrev. for yttrium-iron garnet. (USBM, 1965)

yoderite A monoclinic mineral, $(Mg,Al)_8Si_4(O,OH)_{20}$; purple; highly pleochroic; in a quartz-yoderite-kyanite-talc schist at Mautia Hill, Kongwa, Tanzania.

Yogo sapphire Dark-blue corundum from Yogo Gulch, MT. (Schaller, 1917)

yoke (a) An interconnecting link between the twin cylinders of a hydraulic-feed diamond drill through which the action of the hydraulic-feed cylinders is transmitted to the drill rods and bit. (Long, 1960) (b) A clamp fitted to the casing at the collar of a drill hole, which when anchored by means of wedge bolts prevents grout pressure from forcing the casing out of the hole. (Long, 1960)

yolk Nodule; occurs at Forest of Dean, U.K.

Yorkian A term which has been proposed (instead of Westphalian) for the Coal Measures strata between the Lanarkian and the Staffordian. (Nelson, 1965)

yoshimuraite A triclinic mineral, $(Ba,Sr)_2TiMn_2(SiO_4)_2$ $(PO_4,SO_4)(OH,Cl)$; forms orange-brown tabular crystals or stellate groups; occurs in an alkali pegmatite at the Noda-Tamagawa Mine, Iwate Arefecture, and the Taguchi Mine, Aichi Prefecture, Japan.

yowah nut Walnut-to almond-size pebble with an opal center; ironstone covered with a thin band of opal, or hollow; near Yowah station, western Queensland, Australia. A subvariety of boulder opal.

ytterbite *gadolinite.*

ytterbium A rare-earth element that has a bright silvery luster, is soft, malleable and quite ductile. Symbol, Yb. Occurs with other rare-earths in a number of rare minerals. Commercially recovered from monazite sand. Has a possible use in improving the mechanical properties of stainless steel; few other uses have been found. (Handbook of Chem. & Phys., 3)

yttergranat A calcium-iron garnet containing a small amount of yttria. A variety of andradite. (Fay, 1920)

yttrialite A possibly hexagonal mineral, $(Y,Th)_2Si_2O_7$; olive green tarnishing to orange-yellow; occurs in Texas.

yttrium A rare-earth element that has a silvery-metallic luster. Symbol, Y. Occurs in nearly all of the rare-earth minerals. Recovered commercially from monazite sand and from bastnasite. Widely used: in color television tubes; to reduce the grain size in chromium, molybdenum, zirconium, and titanium; to increase the strength of aluminum and magnesium alloys; as a deoxidizer for vanadium and other nonferrous metals; in nuclear technology for its high neutron transparency. (Handbook of Chem. & Phys., 3)

yttrium-aluminum garnet Synthetic; $Y_3Al_5O_{12}$; has useful magnetic properties. Actually not a true garnet and should not be confused with any of the silicate minerals called garnets in the garnet group of minerals. Manufacture of yttrium-aluminum garnets is a commercial use of yttrium. Used in lasers and in microwaves and other electronic applications. Acronym and abbrev., YAG. (Handbook of Chem. & Phys., 2)

yttrium garnet A variety of garnet containing a small amount of yttrium earths. Syn: *yttrogarnet.* See also: *yttergranat.* (Fay, 1920)

yttrium-iron garnet Synthetic; $Y_3Fe_5O_{12}$; has useful magnetic properties. Actually not a true garnet and should not be confused with any of the silicate minerals called garnets in the garnet group of minerals. Manufacture of yttrium-iron garnets is a leading commercial use of yttrium. Used as electronic transmitters, as filters for selecting or tuning microwaves, and as transmitters and transducers of acoustic energy. Acronym and abbrev., YIG. (Bennett, 1962; CCD, 1961; Handbook of Chem. & Phys., 2)

yttrocrasite An orthorhombic mineral, $(Y,Th,Ca,U)(Ti,Fe)_2(O,OH)_6$; also contains Th, U, Fe; radioactive; black altering to a dull brown coating; in granite pegmatites.

yttrogarnet *yttrium garnet.*

yttromicrolite A discredited mineral name for an amorphous mixture of calcium sulfate, tantalite, and heterogeneous microlite.

yttro-orthite *allanite.*

yttrotantalite An orthorhombic mineral, $(Y,U,Fe)(Ta,Nb)O_4$; radioactive; black to brown: in pegmatites.

yttrotungstite A monoclinic mineral, $YW_2O_6(OH)_3$. Syn: *thorotungstite.*

yugawaralite A monoclinic mineral, $CaAl_2Si_6O_{16}{\cdot}4H_2O$; zeolite group; colorless to white; in low-grade metamorphosed tuffs near the Yugawara hot spring, Kanagawa, Japan.

yukonite (a) A noncrystalline mineral: $Ca_2Fe_3(AsO_4)_4OH{\cdot}12H_2O$. (AGI, 1987) (b) An obsolete term originally assigned to an igneous rock intermediate in composition between a tonalite and an aplite. It is named after the Yukon River, Alaska. (AGI, 1987)

Z

zaffer Mixed arsenates and oxides of cobalt produced by roasting sulfide ores. (Pryor, 1963)
zanthochroite Amorphous cadmium sulfide; greenockite is the crystalline form. (Am. Mineral., 1945)
zaratite An isometric mineral, $Ni_3(CO_3)(OH)_4{\cdot}4H_2O$; emerald green; forms incrustations and compact masses in mafic and ultramafic rocks.
zarnec *zarnich.*
zarnich Native sulfide of arsenic, including sandarac and orpiment. Syn: *zarnec.* (Webster 2nd, 1960)
zavaritskite A tetragonal mineral, BiOF; at Sherlova Gory, East Transbaikal, Russia.
z-axis (a) One of the three optic axis (x, y, and z) in a biaxial crystal. The z-axis is the axis of least ease of vibration. Light vibrating parallel to the z-axis travels with minimum velocity and is called the slow ray, the z-ray, and the γ-ray. The highest index of refraction n_γ, in biaxial minerals is the index of the slow ray vibrating parallel to the z-axis. (b) One of three axes in a three-dimensional coordinate system. Crystallographers customarily use a right-handed system with the z-axis oriented positive upward, the y-axis positive to the right, and the x-axis positive toward the viewer. For mineral parameters, the z-axis is labelled c with periodic translations t_3. If one crystallographic axis is symmetrically unique, it is labelled z, except in the monoclinic system, where mineralogic convention uses the second setting labelling the unique diad y.
z-coordinate One of the three coordinates in a three-dimensional rectangular coordinate system. (Webster 3rd, 1966)
z-direction One of three orthogonal optic directions in biaxial crystals. Light with its electric vector (vibration direction in early terminology) parallel to the z-direction has the highest refractive index (n_γ) for a given crystal and is called the slow ray. In orthorhombic crystals, the z-direction is constrained by symmetry to correspond to one of the crystallographic directions, the correspondence determined empirically. In monoclinic crystals, one optic direction, commonly the y-direction, is constrained by symmetry to correspond to the unique diad. In triclinic crystals, there is no symmetrical constraint relating optic directions to crystallographic axes.
zeasite Wood opal, formerly a name for fire opal.
zebra dolomite A term used in the Leadville district of Colorado for an altered dolomite rock that shows conspicuous banding (generally parallel to bedding) consisting of light-gray coarsely textured layers alternating with darker finely textured layers. See also: *zebra rock.* (AGI, 1987)
zebra rock (a) A term used in the Colville district of NE Washington State for a dolomite that shows narrow banding consisting of black layers (indicative of organic matter) alternating with white, slightly coarse-grained, and somewhat vuggy layers. See also: *zebra dolomite.* (AGI, 1987) (b) A term used in Western Australia for a banded quartzose rock of Cambrian age. (AGI, 1987)
Zeiss konimeter A portable dust-sampling instrument. See also: *konimeter.* (Nelson, 1965)
Zellweger furnace A long-hearth reverberatory furnace used at Iola, KS. (Fay, 1920)
Zemorrian Lower Lower Miocene. (AGI, 1987)
zeolite A generic term for class of hydrated silicates of aluminum and either sodium or calcium or both, of the type $Na_2O{\cdot}Al_2O_3{\cdot}nSiO_2{\cdot}xH_2O$. The term originally described a group of naturally occurring minerals. The natural zeolites are analcite, chabazite, heulandite, natrolite, stilbite, and thomsonite. Artificial zeolites are made in a variety of forms, ranging from gelatinous to porous and sandlike, and are used as gas adsorbents and drying agents as well as water softeners. Both natural and artificial zeolites are used extensively for water softening. The term zeolite now includes such diverse groups of compounds as sulfonated organics or basic resins, which act in a similar manner to effect either cation or anion exchange. (CCD, 1961)
zeolite mimetics The dachiardite group of zeolites.
zeolite process (a) A base exchange method of treating hard water, in which zeolites, contained in a tank, remove salts. The zeolite layer is regenerated by backflushing with brine. (Nelson, 1965) (b) *base exchange.*
zeolitic deposit A deposit, particularly of native copper, that occurs in basalt accompanied by minerals of the zeolite group.
zeolitization Introduction of, alteration to, or replacement by, a mineral or minerals of the zeolite group. This process occurs chiefly in rocks containing calcic feldspars or feldspathoids, and is sometimes associated with copper mineralization. (AGI, 1987)
zeophyllite A triclinic mineral, $Ca_4Si_3O_8(OH,F)_4{\cdot}2H_2O$; pseudohexagonal; forms white spherical, radiated folia; at Cross-Preisen, Bohemia, Czech Republic.
zero air voids curve The curve showing the zero air voids unit weight as a function of water content. Syn: *saturation curve.* (ASCE, 1958)
zero air voids unit weight The weight of solids per unit volume of a saturated soil mass. See also: *unit weight.* (ASCE, 1958)
zero-length spring Special type of gravimeter spring for which the length is proportional to the applied force. (Schieferdecker, 1959)
zero-point energy The kinetic energy remaining in a substance at a temperature of absolute zero. (Webster 3rd, 1966)
zero potential The actual potential at the surface of the Earth taken as a point of reference. (Webster 3rd, 1966)
zero time When conducting a mine ventilation pressure survey, zero time is the time of the commencement of the survey from the base station, and the reading of the control barometer there is taken as the pressure datum to which subsequent pressures are referred. (Sinclair, 1958)
zero-zero gel A condition wherein a drilling fluid fails to form measurable gels during a quiescent 10-min time interval. (Brantly, 1961)
zeta potential The potential difference across an electric double layer, usually between a solid surface and a liquid. Syn: *electrokinetic potential* See also: *surface.* (Webster 3rd, 1966)
zeta-potential layer The zone of shear surrounding a particle immersed in an electrolyte. (Pryor, 1965)
zeunerite A tetragonal mineral, $Cu(UO_2)_2(AsO_4)_2{\cdot}10H_2O$; autunite group.
zeylanite *ceylonite.*
zhemchuzhnikovite A trigonal mineral, $NaMg(Al,Fe)(C_2O_4){\cdot}8H_2O$; an oxalate occuring in green crystals that appear violet in artificial light; in veinlets in coal in the Chaitumusuk deposits, Siberia, Russia.
Ziervogel process The extraction of silver from sulfide ores or matte by roasting in such a way as to form sulfate of silver, leaching this out with hot water, and precipitating the silver by means of metallic copper. (Fay, 1920)
zietrisikite Incorrect spelling of pietricikite, a waxlike hydrocarbon similar to ozokerite. (Tomkeieff, 1954)
zigzag car loader A form of vertical chute in which the chute is divided into independent sections that can be raised or lowered on a track arrangement. It is flexible and can be lowered to the bottom of the car, giving a solid stream of coal from loading pocket to car. (Mitchell, 1950)
zigzag fold *chevron fold.*
zigzag rule A wooden rule (generally 2 m or 6 ft long, folded zigzag fashion in 15-cm or 6-in lengths), used by drillers, craftsmen, etc., to measure short distances. The rule usually is graduated in centimeters or in feet, inches, and fractions of an inch (sometimes in feet, tenths of a foot, and hundredths of a foot). (Long, 1960)
zigzag transformer A zigzag transformer is a transformer intended primarily to provide a neutral point for grounding purposes. Syn: *grounding transformer.* (USBM, 1960)
ziment water Water impregnated with copper; found in copper mines. (Standard, 1964)
Zimmermann's rule A graphical method for finding the lost part of a vein on the other side of a fault. (Nelson, 1965)
zinc (a) The native metallic element, Zn. (b) A bluish-white, lustrous metal. Employed to form numerous alloys with other metals including brass, nickel silver, commercial bronze, spring brass, soft solder, and aluminum solder. Used extensively by the automotive, electrical, and hardware industries. (Handbook of Chem. & Phys., 3)
zincaluminite A light-blue mineral, $Zn_6Al_6(SO_4)_2(OH)_{26}{\cdot}5H_2O$.
zinc blende *sphalerite.*
zinc bloom *hydrozincite; zinc oxide.*

zinc box Wooden or enamel-ware rectangular box, with a bottom grid that supports zinc shavings. Used in the cyanide process to precipitate dissolved gold from a pregnant solution. Its place was taken by use of zinc dust. See also: *Merrill-Crowe Process.* (Pryor, 1963)

zinc carbonate White; $ZnCO_3$; soluble in acids, in alkalies, and in ammonium salt solutions; insoluble in water; and dissociates losing carbon dioxide at 300 °C. Used in ceramics and as a pigment. Also used less accurately to refer to any of several basic carbonates of zinc, which include the zinc-ore mineral hydrozincite, $Zn_5(CO_3)_2(OH)_6$ or $Zn_5(OH)_6(CO_3)_2$ and synthetically prepared pigments of the same or similar composition. Syn: *smithsonite; calamine; zinc spar.* (Handbook of Chem. & Phys., 2)

zinc chloride White; hexagonal; $ZnCl_2$; poisonous. Obtained by the solution of zinc, or zinc oxide, in hydrochloric acid, or by burning zinc in chlorine. Used in galvanizing iron, as a catalyst, as a dehydrating agent, as a condensing agent, as a wood preservative, as an ingredient in soldering fluxes, in burnishing and polishing compounds for steel, in electroplating, in glass-etching compositions, in petroleum refining, and in pigments. (Handbook of Chem. & Phys., 2)

zinc chromate $ZnCrO_4$; a golden-yellow pigment. Of variable composition; the chemically pure zinc chromate, a yellow crystalline powder is said to be zinc chromate heptahydrate, $ZnCrO_4{\cdot}7H_2O$. Called zinc yellow, but this term is also applied to hydrated zinc chromate and to hydrated zinc-potassium chromate. Another zinc chromate is dark green to black and has a different composition; isometric. $ZnCr_2O_4$. Syn: *zinc chrome; zinc yellow.* (Handbook of Chem. & Phys., 2)

zinc chrome *zinc chromate; zinc yellow.*

zinc dust Finely divided zinc that usually contains small amounts of zinc oxide and impurities. Also called powdered zinc; zinc gray. (Fay, 1920)

zinc gray *zinc dust.*

zinc hydrosulfite White; amorphous; ZnS_2O_4; and soluble in water. Used as a depressant in flotation. (CCD, 1961)

zincite A red to yellow brittle mineral; (Zn,Mn)O. It is an ore of zinc, as in New Jersey where it is associated with franklinite and willemite. Syn: *red zinc ore; red zinc oxide; ruby zinc; spartalite.* (AGI, 1987)

zinckenite *zinkenite.*

zinckiferous Carrying zinc. (Weed, 1922)

zinc-magnesia chalcanthite A variant of chalcanthite with the formula, $(Cu,Zn,Mg)SO_4{\cdot}5H_2O$. (Hey, 1955)

zinc melanterite A member of the monoclinic melanterite group, in which iron is partially replaced by zinc. (English, 1938)

zinc-melanterite A monoclinic mineral, $(Zn,Cu,Fe)SO_4{\cdot}7H_2O$.

zinc minerals The principal ore is sphalerite, ZnS; other important ore minerals are smithsonite, $ZnCO_3$; hemimorphite, $Zn_4Si_2O_7(OH)_2{\cdot}H_2O$; franklinite, $(Zn,Mn,Fe)(Mn,Fe)_2O_4$; willemite, Zn_2SiO_4; and zincite, ZnO.

zinc oxide White or yellowish-white; white turns yellow on heating; ZnO; odorless; absorbs carbon dioxide from the air; used in pigments, ceramic glazes, and opaque glass and in the manufacture of magnetic ferrites and specialized ceramics. Zinc oxide is a common constituent in high-grade fluoride opal glass, in tank-window glass, and in some optical glass. Commonly used in dry-process, cast-iron enamels. Syn: *zincite; zinc bloom; zinc white.* (Handbook of Chem. & Phys., 2)

zincrosasite A monoclinic mineral, $(Zn,Cu)(CO_3)(OH)_2$; rosasite group; at Tsumeb, Namibia.

zinc scum The zinc-silver alloy skimmed from the surface of the bath in the process of desilverization of lead by zinc. (Fay, 1920)

zincsilite A monoclinic mineral, $Zn_3Si_4O_{10}(OH)_2{\cdot}4H_2O$(?); smectite group; in Batystau, Kazakhstan.

zinc spar *smithsonite; zinc carbonate.*

zinc spinel *gahnite.*

zinc sulfate Orthorhombic $ZnSO_4$; used in flotation.

zinc sulfate heptahydrate Colorless; needles; $ZnSO_4{\cdot}7H_2O$; odorless; astringent, metallic taste; effloresces in air; sp gr, 1.9661; melting point, 50 °C if heated rapidly; soluble in water and in glycerol; insoluble in alcohol; and its solutions are acid to litmus. Used in preparing zinc chemicals. Orthorhombic; molecular weight, 287.54; sp gr, 1.957 (at 25 °C, referred to water at 4 °C) and ranges from 1.9 to 2.1; Mohs hardness, 2.0 to 2.5; melting point, 100 °C; loses $7H_2O$ on heating to 280 °C; and slightly soluble in alcohol and in glycerol. Occurs as the mineral goslarite, which is white or yellowish; formed by the oxidation of sphalerite (ZnS) in damp locations, esp. in the presence of iron sulfides. (CCD, 1961; Handbook of Chem. & Phys., 2)

zinc sulfide (a) Alpha zinc sulfide is the hexagonal mineral wurtzite. See also: *alpha zinc sulfide; wurtzite.* (b) Beta zinc sulfide is the isometric mineral sphalerite. See also: *beta zinc sulfide.*

zinc sulfide monhydrate Colorless, white, or yellowish; crystalline; $ZnSO_4{\cdot}H_2O$; molecular weight, 115.45; sp gr, 3.98; melting point, 1,049 °C; insoluble in water; and soluble in acids. Used as a pigment and in white glass and in opaque glass. (Handbook of Chem. & Phys., 2; CCD, 1961)

zinc vitriol *goslarite.*

zinc white Used as a pigment. It is the whitest of all pigments; permanent, not poisonous, but lacks the opacity and covering power of white lead or titanium dioxide. See also: *zinc oxide.*

zinc yellow A greenish-yellow pigment. See also: *zinc chromate.* Syn: *zinc chrome.* (Webster 3rd, 1966)

zinkazurite A mineral found in small, blue crystals; probably a mixture of sulfate of zinc and carbonate of copper. (Fay, 1920)

zinkenite A hexagonal mineral, $Pb_9Sb_{22}S_{42}$; in steel-gray crystals or exceptionally thin folia in fibrous masses. Also spelled zinckenite. Syn: *keeleyite.*

zinkite *zincite.*

zinkosite Orthorhombic $ZnSO_4$.

zinn Native tin. Syn: *gediegen.* See also: *tin.* (Hess)

zinnwaldite A monoclinic mineral, $KLiFeAl(AlSi_3)O_{10}(F,OH)_2$; mica group; basal cleavage; pale violet, yellowish or grayish brown; in granites, pegmatites, and greisens.

zippeite An orthorhombic mineral, $K_4(UO_2)_6(SO_4)_3(OH)_{10}{\cdot}4H_2O$; radioactive; orange-yellow to bright yellow; an alteration product of uraninite occuring with gypsum, uranopilite, and limonite. Formerly called dauberite.

Zipper Trade term for a conveyor belt lacing or fastening appliance, which can be applied to any thickness and width of belt. A lever mechanism is used to supply the pressure. The appliance is totally enclosed and grease-filled to protect against dust and damage. (Nelson, 1965)

zircon A tetragonal mineral, $ZrSiO_4$; occurs widely in granite, granite pegmatite, other felsic igneous rocks, and placers; the chief source of zirconium; a refractory; if cut and polished, the colorless varieties provide exceptionally brilliant gemstones. Syn: *azorite; zirconite; hyacinth; jacinth.*

zircon flour Finely milled zircon sand. (CCD, 1961)

zircon group Zircon, along with rutile, ilmenite, and monazite, constitutes a group of heavy minerals that are usually considered together because of their occurrence as black sands in natural beach and dune concentrations. For marketing, complete separation of the rutile and zircon as high-grade products is necessary. Separation is effected by combinations of electromagnetic and electrostatic processes, together with gravity concentration.

zirconia brick Brick containing zirconium oxide; used in metallurgical furnaces. (Bennett, 1962)

zirconiferous (a) Containing zircon or yielding zircon. (Webster 3rd, 1966) (b) Containing zirconium or yielding zirconium. (Webster 3rd, 1966)

zirconite A gray or brownish variety of zircon.

zirconium A grayish-white lustrous metallic element. Symbol, Zr. Occurs widely, but only in combined form, esp. in the minerals zircon, ($ZrSiO_4$), and baddeleyite, (ZrO_2). Uses include resisting corrosion, as a structural material in nuclear reactors, as an alloying agent, deoxidizer, bonding agent, refractory material, and in low-temperature superconductive magnets. (Handbook of Chem. & Phys., 3)

zirconium minerals Principal ore minerals are zircon and baddeleyite. Main uses are as refractories, ceramics, opacifiers, abrasives, enamels, insulators, and alloys. They are also a source of hafnium. Transparent zircon is a gemstone.

zirconium phosphate Normal zirconium phosphate, ZrP_2O_7, has a reversible inversion at 300 °C and at 1,550 °C dissociates into zirconyl phosphate, $(ZrO)_2P_2O_7$, with loss of P_2O_5 as vapor. Zirconyl phosphate is stable up to about 1,600 °C and has a very low thermal expansion—1 x 10^{-6} (20 to 1,000 °C). (Dodd, 1964)

zirconium silicate $ZrSiO_4$, natural silicate of zirconium found in Brazil and elsewhere. One such deposit known as brazilite is said to contain about 80% zirconium oxide in a semimanufactured form. This product is employed as a refractory in the making of "zirkite" bricks and cement. In rare cases, used as a cobalt groundcoat constituent. Zirconium silicate is used in formulation of zircon enamels that depend upon crystallization of zircon compounds for opacity development. See also: *zircon sand.* (Enam. Dict., 1947)

zirconium-vanadium blue A pigment for use in ceramic glazes. The composition is (parts by weight): ArO_2, 60% to 70%; SiO_2, 26% to 36%; V_2O_5, 3% to 5%. Alkali must also be present; e.g., 0.5% to 5% Na_2O. In the absence of alkali, a green color is produced. Syn: *vanadium-zirconium turquoise.* (Dodd, 1964)

zirconolite A discredited syn. of zirkelite.

zircon sand (a) A very refractory mineral, composed chiefly of zirconium silicate, having low thermal expansion and high thermal conductivity. (ASM, 1961) (b) A natural zircon-bearing material found in Australia, India, and Florida. See also: *zirconium silicate.* (Enam. Dict., 1947)

zirkelite (a) A monoclinic mineral, $(Ca,Th,Ce)Zr(Ti,Nb)_2O_7$; pseudocubic; dimorphous with calciobetafite of the pyrochlore group; radioactive; occurs in pyroxenite and alluvial deposits. (b) An obsolete name for altered basaltic glass.

Zirkite A trade name for a mixture of zircon and baddeleyite. (USBM, 1965)

zirklerite A trigonal mineral, $(Fe,Mg)_9Al_4Cl_{18}(OH)_{12}·14H_2O$; massive fine granular; at Hanover, Germany.

zoisite An orthorhombic mineral, $CaAl_3(SiO_4)_3(OH)]$; epidote group; dimorphous with clinozoisite; vitreous or pearly; varicolored; a common rock-forming mineral in medium-grade metamorphic rocks. The rose-red variety has been called thulite.

zonal axis *zone axis.*

zonal structure A term used esp. in miscroscopic work to describe those minerals whose cross sections show their successive concentric layers of growth. (Fay, 1920)

zonal theory A theory of hypogene mineral-deposit formation, and the spatial distribution patterns of mineral sequences to be expected from change in a mineral-bearing fluid as it migrates away from a magmatic source. It also deals with thermal-chemical gradients associated with the genesis of ore deposits, whether of direct magmatic origin or not, and with metallogenic zoning on a regional scale. (AGI, 1987)

zone (a) A belt, band, or strip of earth materials, however disposed; characterized as distinct from surrounding parts by some particular property or content; e.g., zone of saturation, fault zone, or a zone of secondary enrichment. Cf: *belt.* (AGI, 1987) (b) *aureole.* (c) A minor interval in any category of stratigraphic classification. There are many kinds of zones, depending on the characteristics under consideration—biozones, lithozones, chronozones, mineralized zones, metamorphic zones, zones of reversed magnetic polarity, etc. The term should always be preceded by a modifier indicating the kind of zone to which reference is made. (AGI, 1987) (d) A metal zone is equivalent to a mineral zone, yet the terms mineral and metal are not synonymous. (Ricketts, 1943) (e) The ground or mass bounded by horizontal or inclined planes or curved surfaces in which given chemical or physical conditions exist, such as zone of saturation or zone of weathering. (Nelson, 1965) (f) A group of beds characterized by the presence of one or more specific fossils; e.g., zonal fossil or fossils. (BS, 1964) (g) Geologically, a distinctively mineralized area, region, or level. In a specific lode or other deposit, the progressive change from upper to lower horizons. At the top is outcrop or gossan, oxidized or weathered. Next is the leached zone, impoverished by dissolution of its values (or part of them), which may be redeposited below in the zone of secondary enrichment. Below this is the primary, or unaltered zone, which consists of the original sulfide formation. (Pryor, 1963) (h) An area or region more or less clearly set off or characterized as distinct from surrounding or adjoining parts; e.g., the mineral zone in a metalliferous region. (Webster 2nd, 1960) (i) A series of faces of a crystal whose intersection lines with each other are all parallel. (Webster 3rd, 1966) (j) *zones.*

zone axis That crystallographic direction through the center of a crystal which is parallel to the intersection edges of the crystal faces defining the "crystal zone." Syn: *crystal axis; zonal axis.*

zone melting Highly localized melting, usually by induction heating, of a small volume of an otherwise solid piece. By moving the induction coil along the rod, the melted zone can be transferred from one end to the other. In a binary mixture where there is a large difference in composition on the liquidus and solidus lines, high purity can be attained by concentrating one of the constituents in the liquid as it moves along a rod. (ASM, 1961)

zone of accumulation *B-horizon.*

zone of aeration A subsurface zone containing water under pressure less than that of the atmosphere, including water held by capillarity, and containing air or gases generally under atmospheric pressure. This zone is limited above by the land surface and below by the water table. It contains vadose water. Cf: *zone of capillarity.* (AGI, 1987)

zone of capillarity A subsurface zone that overlies the zone of saturation in which capillary voids hold water above the zone of saturation by molecular attraction acting against gravity. Cf: *zone of aeration.*

zone of cementation The layer of the Earth's crust below the zone of weathering in which percolating waters cement unconsolidated deposits by the deposition of dissolved minerals from above. (AGI, 1987)

zone of discharge A term suggested for that part of the zone of saturation having a means of horizontal escape. (AGI, 1987)

zone of enrichment *zone of secondary enrichment.*

zone of faces All faces, belonging to one or more forms, the normals to which lie in one plane (the zone plane) and whose edges of intersection are parallel to the zone axis. See also: *zone axis.*

zone of flow *zone of plastic flow.*

zone of fracture (a) The outer, rigid part of a glacier, in which the ice is much fractured. (AGI, 1987) (b) The upper, brittle part of the Earth's crust in which deformation is by fracture rather than by plastic flow; that region of the crust in which fissures can exist. Cf: *zone of plastic flow.* Syn: *zone of rock fracture.* (AGI, 1987)

zone of illuviation *B-horizon.*

zone of influence The zone of rock surrounding an excavation, in which the additional stresses caused by the excavation are above a certain arbitrary value, is termed the zone of influence. (Spalding, 1949)

zone of mobility *asthenosphere.*

zone of oxidation *oxidized zone.*

zone of plastic flow That part of the Earth's crust that is under sufficient pressure to prevent fracturing, i.e., is ductile, so that deformation is by flow. Cf: *zone of fracture.* Syn: *zone of flow; zone of rock flowage.* (AGI, 1987)

zone of rock flowage *zone of plastic flow.*

zone of rock fracture The upper part of the lithosphere in which rocks are under stresses less than the stresses required to close their interstices by deformation of the walls of the interstices. (AGI, 1987)

zone of saturation A subsurface zone in which all the interstices are filled with water under pressure greater than that of the atmosphere. This zone is separated from the zone of aeration (above) by the water table. Syn: *saturated zone; phreatic zone.* (AGI, 1987)

zone of secondary enrichment The zone in which supergene enrichment has taken place.

zone of substantial deformation *destressed area.*

zone of weathering The superficial layer of the Earth's crust above the water table that is subjected to the destructive agents of the atmosphere and in which soils develop. Cf: *zone of cementation.* (AGI, 1987)

zone refining A purification technique in which a molten or high-temperature zone is moved along a length of material to be purified to bring about impurity segregation; the impurities become concentrated in the slowly moving hot zone, leaving behind the cooler solidified material that has a higher purity than the original material. The technique depends on differences in composition of liquids and solids in equilibrium and may be repeated to attain high degrees of purity. It usually is applied to crystalline materials, such as germanium or silicon.

zones In a shaft furnace, the different portions (horizontal sections) are called zones, and are characterized according to the reactions that take place in them, as the zone of fusion or smelting zone, the reduction zone, etc. (Fay, 1920)

zones of lode A lode may be divided into three main zones: (1) unaltered ore at depth; (2) gossan or altered surface portion of the lode, which contain native metals, oxides, and oxysalts that result from weathering of the ore; and (3) secondary enrichment which lies between the first two zones, where interaction between waters from the gossan and the unaltered ore have produced new materials, often of considerable economic value. (Nelson, 1965)

zone time Standard time applied at sea in which the surface of the globe is divided into 24 zones of 15°, or of 1 h, each. The "0 zone" extends 7.5° east and west of the meridian of Greenwich (England), (the Prime Meridian); the zones are designated by the number of hours that must be applied to the local time to obtain Greenwich time. Abbrev., ZT. (Webster 3rd, 1966)

zoning (a) In ore deposits, the spatial distribution patterns of elements, minerals, or mineral assemblages. (AGI, 1987) (b) A variation in the composition of a crystal from core to margin, due to a separation of the crystal phases during its growth, by loss of equilibrium in a continuous reaction series. The higher-temperature phases of the isomorphic series form the core, with the lower-temperature phases toward the margin. Syn: *zonal structure.* (AGI, 1987) (c) Concentric layering parallel to the periphery of a crystal, shown by color banding, such as in tourmaline, and by differences in optical reactions to polarized light, such as in plagioclase feldspar. Cf: *undulatory extinction.* (d) In a mineral deposit, the occurrence of successive minerals or elements outward from a common center. (e) The development of areas of metamorphosed rocks that may exhibit zones in which a particular mineral or suite of minerals is predominant or characteristic, reflecting the original rock composition, the pressure and temperature of formation, the duration of metamorphism, and whether or not material was added or removed.

zoning of ore deposits Spatial distribution patterns of elements, minerals, or mineral assembalages; paragenetic sequences, either syngenetic or epigenetic. Zoning is esp. well developed in the mineralization-alteration assemblages about subvolcanic occurrences such as porphyry base-metal deposits. See also: *zonal theory.* Syn: *mineral zoning.* (AGI, 1987)

zonite A name that has been used in Arizona for locally occurring jasper or chert of various colors.

zonochlorite A light and dark green variety of pumpellyite in green pebbles of banded structure; occurs in the Lake Superior region, MI.

Zonolite A trade name for a light, flaky material obtained by roasting vermiculite, which swells to 15 times its original volume, forming golden yellow scales; from Libby, MT. A titanium-bearing jefferisite from Westcliffe, CO., is similar. (English, 1938)

zooplankton The animal forms of plankton, e.g., jellyfish. They consume phytoplankton. (AGI, 1987)

zorgite A brass-yellow metallic mineral with dark yellow streak; possibly clausthalite with umangite.

Z reagent Any of the Dow series of xanthate flotation reagents. (Pryor, 1963)

zunyite An isometric mineral, $Al_{13}Si_5O_{20}(OH,F)_{18}Cl$; occurs in transparent tetrahedral crystals, also minute cubes; at the Zuni Mine, Silverton, CO. Also spelled zungite.

zurron A rawhide sack that holds about 150 lb (68.1 kg); used by miners for carrying ore. (Nelson, 1965)

zwieselite A monoclinic mineral, $(Fe,Mn)_2(PO_4)F$; forms a series with triplite; clove-brown.

References Cited

A

ACGIH, 1 American Conference of Governmental Industrial Hygienists. 1993-1994; *Threshold Limit Values For Chemical Substances & Physical Agents & Biological Exposure Indices 1993-94*; 1993.

ACGIH, 2 American Conference of Governmental Industrial Hygienists. *Threshold Limit Values and Biological Exposure Indices for 1987-88*; 1987.

ACSB, 1 American Ceramic Society Bulletin. *Clay Definitions.* V. 18, No. 6, June 1939, pp. 213-215. Reprinted by permission of the American Ceramic Society.

ACSB, 2 American Ceramic Society Bulletin. *Glass Glossary.* V. 27, No. 9, Sept. 15, 1948, pp. 353-362. Reprinted by permission of the American Ceramic Society.

ACSG, 1 Van Schoick, Emily C. *Tentative Ceramic Glossary,* Part I. American Ceramic Society, Alfred, N.Y., Jan. 10, 1961, 64 pp. Reprinted by permission of the American Ceramic Society.

ACSG, 2 Van Schoick, Emily C. *Ceramic Glossary.* American Ceramic Society, Columbus, Ohio, 1st ed., 1963, 31 pp. Reprinted by permission of the American Ceramic Society.

Acuff Acuff, A.D., et al.; *Compilation of Laws Related to Mineral Resource Activities on the Outer Continental Shelf,* V. II; Bureau of Land Management, 361 pp.; 1981.

AGI American Geological Institute. *Glossary of Geology,* Alexandria, VA, 3rd ed., 1987, 788 pp.; Glossary of Geology and Related Sciences, 1957, 325 pp; supplement, 1960, 72 pp.

AIME, 1 American Institute of Mining, Metallurgical, and Petroleum Engineers. *Industrial Minerals and Rocks (Nonmetallics Other Than Fuels).* New York, 3rd ed., 1960, 934 pp.

AIME, 2 American Institute of Mining, Metallurgical, and Petroleum Engineers. *Technical Publication No. 604.*

AISI American Iron and Steel Institute. *Definitions of Refractory Terms.* Contributions to the Metallurgy of Steel, No. 24, January 1949, 22 pp.

Am. Mineral., 1 American Mineralogist. *Glossary of Terms Used in the Quartz Oscillator Plate Industry.* V. 30, No. 1-2, January-February 1945, pp. 461-468.

Am. Mineral., 2 *American Mineralogist.* V. 32, No. 11-12, Nov.-Dec. 1947, p. 702

Anderson Anderson, Basil W. *Gem Testing.* Temple Press Books, Ltd., London, 7th ed., 1964, 377 pp. Includes a glossary, pp. 349-354.

ANSI ANSI definitions are reproduced with permission from American National Standard ANSI Z88.2 copyright 1992 by the American National Standards Institute. Copies of this standard may be purchased from the American National Standards Institute, 11 West 42nd Street, New York, NY 10036.

API, 1 American Petroleum Institute. *Glossary of Terms Used in Petroleum Refining.* New York, 1953, 188 pp. Reprinted courtesy of the American Petroleum Institute.

API, 2 American Petroleum Institute. *Drilling and Production Practice,* 1963. Reprinted courtesy of the American Petroleum Institute.

Aplan Aplan, F.F., et al. *Solution Mining Symposium*; American Institute of Mining, Metallurgical and Petroleum Engineers; February 25-27, 1974.

Arbiter Arbiter, Nathaniel (ed.). *Milling Methods in the Americas.* Gordon & Breach Science Publishers, New York, 1964, 486 pp.

Aref Aref, K., A. Moss and K. Durston. *Design Issues for Low Moisture Content Backfill,* RMSCC Proceedings--CIM 94th AGM. Montreal, Quebec, April 26-30, 1992.

ARI American Refractories Institute. *Proposed Glossary of Terms Relating to Refractories, Their Manufacture and Use.* Bull. 87, October 1949, n.p. (Taken from ASTM's "Manual of ASTM Standards on Refractory Materials," February 1952, pp. 161-170.)

Arkell Arkell, W.J., and S.I. Tomkeieff. *English Rock Terms Chiefly as Used by Miners and Quarrymen.* Oxford University Press, London, 1953, 139 pp. Reprinted with the permission of the Oxford University Press.

Army Corps of Eng. U.S. Army Corp. of Engineers. *Beneficial Uses of Dredged Material,* Engineering Manual 1110-2-5026. U.S. Army Corp. of Engineers. 1987.

ASCE American Society of Civil Engineers. *Glossary of Terms and Definitions in Soil Mechanics.* Proceedings, V. 84, No. SM4, Paper 1826, October 1958. pp. 1-43.

Ash Ash, S.H., W.L. Eaton, Karl Hughes, W.M. Romischer, and J. Westfield. *Water Pools in Pennsylvania Anthracite Mines.* U.S. Bureau of Mines Technical Paper 727, 1949. 78 pp., 50 figs.

ASM, 1 ASM International (formerly American Society for Metals). *Metals Handbook.* V. 1. Properties and Selection of Metals. Metals Park, Ohio, 8th ed., 1961, 1300 pp. Includes a glossary of Definitions Relating to Metals and Metalworking, pp. 1-41.

ASM, 2 ASM International; *ASM Handbook,* V. 4; ASM International; 1991.

ASPRS American Society for Photogrammetry and Remote Sensing. *Multilingual Dictionary for Remote Sensing and Photogrammetry,* American Society for Photogrammetry and Remote Sensing, 1984.

ASTM American Society for Testing and Materials. *Compilation of ASTM Standard Definitions.* American Society for Testing and Materials, Philadelphia, PA. 8th ed., 1994, 595 pp. Reprinted, with permission, from the Annual Book of ASTM Standards, copyright American Society for Testing and Materials, 100 Barr Harbor Drive, West Conshohocken, PA 19428-2959.

Atlas Atlas Powder Company; *Explosives and Rock Blasting*; ICI Explosives USA Inc. (formerly Atlas Powder Co.), 1987.

Austin Austin, J. *Dragline Excavators for Strip Mining.* Mining Magazine, V.3, No. 5, November 1964, pp. 296-311. Includes a glossary, p. 311.

B

Ballard Ballard, Thomas J., and Quentin E. Conklin. *The Uranium Prospector's Guide.* Harper & Brothers, New York, 1955, 251 pp. Includes a glossary, pp. 178-205.

Barger Barger, Harold, and Sam H. Schurr. *The Mining Industries, 1899-1939; A Study of Output, Employment and Productivity.* National Bureau of Economic Research, Inc., Pub. 43, 1944, 452 pp. Includes a glossary of Minerals and Mining Terms, pp. 409-422.

Barton Barton, P.B.; D.A. Brew; R.A. Ayuso; B.M. Gamble; D.A. John; S.D. Ludington; D.A. Lindsey; E.R. Force; R.J. Goldfarb; and K.M. Johnson. *Recommendations for Assessments of Undiscovered Mineral Resources,* Open File Report 95-0082. U.S. Geological Survey; 1995.

Bateman, 1 Bateman, Alan M. *Economic Mineral Deposits.* John Wiley & Sons, Inc., New York, 2nd ed., 1950, 916 pp.

Bateman, 2 Bateman, Alan M. *The Formation of Mineral Deposits.* John Wiley & Sons, Inc., New York, 1951, 371 pp. Includes a glossary, pp. 351-359.

BCI National Mining Association (formerly Bituminous Coal Institute and American Mining Congress). *Glossary of Current and Common Bituminous Coal Mining Terms.* Washington, D.C., January 1947, 26 pp.

Becker Becker, Pierre. *Phosphates and Phosphoric Acid.* Marcel Dekker, Inc. 1983.

Beerman Beerman Publishers, Ltd. *Beerman's All Mining Yearbook,* 1963. Capetown, Republic of South Africa, 1963, 473 pp. Includes a glossary of Technical and Mining Terms, pp. 451-461.

Bennett Bennett, H. (ed.). *Concise Chemical and Technical Dictionary.* Chemical Publishing Co., Inc., New York, 2nd ed., 1962, 1039 pp; addenda, 119 pp.

Best, 1 Best Co., Inc. (Alfred M.). *Best's Safety Maintenance Directory,* Morristown, N.J., 11th ed., 1966, 772 pp.

Best, 2 A.M. Best Company. *Best's Safety Directory*, V. 1. A.M. Best Company, Oldwick, NJ, 1993.

Billings Billings, Marland P. *Structural Geology*. Prentice-Hall, Inc., New York, 2nd ed., 1954, 514 pp.

Bonham-Carter Bonham-Carter, Graeme F. *Geographic Information Systems for Geoscientists: Modelling with GIS*. Pergamon Press. 1987. Reprinted with kind permission from Elsevier Science, Ltd., The Boulevard, Langford Lane, Kidlington 0X5 1GB, UK.

Bookhout Bookhout, Theodore A. *Research and Management Techniques for Wildlife and Habitats*. The Wildlife Society, 1994.

Boynton Boynton, Robert S. *Chemistry and Technology of Lime and Limestone*. Interscience Publishers, New York, 1966, 520 pp. Includes a glossary of Definitions and Properties of Limes, pp. 165-167.

Brady, 1 Brady, George Stuart. *Materials Handbook; an Encyclopedia for Purchasing Agents, Engineers, Executives, and Foremen*. McGraw-Hill Book Co., New York, 9th ed., 1963, 968 pp.

Brady, 2 Brady, George Stuart. *Materials Handbook; an Encyclopedia for Purchasing Agents, Engineers, Executives, and Foremen*. McGraw-Hill Book Co., New York, 4th ed., 1940, 591 pp.

Brantly, 1 Brantly, J.E. *Rotary Drilling Handbook*. Palmer Publications, New York, 6th ed., 1961, 825 pp. Includes a glossary of Mud Drilling Terms, pp. 312-317.

Brantly, 2 Brantly, J.E. *Rotary Drilling Handbook*. Palmer Publications, New York, 6th ed., 1961, 825 pp. Includes a general glossary, pp. 599-603.

Brauer Brauer, Roger L. *Safety and Health for Engineers*. Van Nostrand Reinhold, New York, 1994.

Briggs Briggs, Henry. *Mining Subsidence*. Edward Arnold & Co., London, 1929, 215 pp.

Bryant Bryant, E.E. *Porcelain Enameling Operations*. Enamelist Publishing Co., Cleveland, OH, 1953, 106 pp. Includes a glossary of Porcelain Enameling Defects, pp. 95-97; Glossary of General Terminology, pp. 99-101.

BS, 1 British Standards Institution (London). *Methods for the Analysis and Testing of Coal and Coke*, Part 16 M-Reporting of Results. British Standard 1016, 1961, 32 pp. Includes definitions, pp. 6-8. Note that this Standard has been superseded by a later edition. The current version must be consulted before these definitions are used. British Standards Institution, 389 Chiswick High Road, London W4 4AL, UK.

BS, 2 British Standards Institution (London). *The Sampling of Coal and Coke*. Part 1--Sampling of Coal. British Standard 1017, 1960, 124 pp. Includes a glossary, pp. 12-14. Note that this Standard has been superseded by a later edition. The current version must be consulted before these definitions are used. British Standards Institution, 389 Chiswick High Road, London W4 4AL, UK.

BS, 3 British Standards Institution (London). *The Sampling of Coal and Coke*. Part 2--Sampling of Coke. British Standard 1017, 1960, 106 pp. Includes a glossary, pp. 11-13. Note that this Standard has been superseded by a later edition. The current version must be consulted before these definitions are used. British Standards Institution, 389 Chiswick High Road, London W4 4AL, UK.

BS, 4 British Standards Institution (London). *Glossary of Coal Terms*. British Standard 3323, 1960, 19 pp. Note that this Standard has been superseded by a later edition. The current version must be consulted before these definitions are used. British Standards Institution, 389 Chiswick High Road, London W4 4AL, UK.

BS, 5 British Standards Institution (London). *Glossary of Terms Used in Coal Preparation*, British Standard 3552, 1962, 44 pp. Note that this Standard has been superseded by a later edition. The current version must be consulted before these definitions are used. British Standards Institution, 389 Chiswick High Road, London W4 4AL, UK.

BS, 7 British Standards Institution (London). *Glossary of Mining Terms*. Section 1, Planning and Surveying. British Standard 3618, sec. 1, 1963, 11 pp. Note that this Standard has been superseded by a later edition. The current version must be consulted before these definitions are used. British Standards Institution, 389 Chiswick High Road, London W4 4AL, UK.

BS, 8 British Standards Institution (London). *Glossary of Mining Terms*. Section 2, Ventilation. British Standard 3618, sec. 2, 1963, 16 pp. Note that this Standard has been superseded by a later edition. The current version must be consulted before these definitions are used. British Standards Institution, 389 Chiswick High Road, London W4 4AL, UK.

BS, 9 British Standards Institution (London). *Glossary of Mining Terms*. Section 3, Boring and Exploration. British Standard 3618, sec. 3, 1963, 14 pp. Note that this Standard has been superseded by a later edition. The current version must be consulted before these definitions are used. British Standards Institution, 389 Chiswick High Road, London W4 4AL, UK.

BS, 10 British Standards Institution (London). *Glossary of Mining Terms*. Section 4, Drainage. British Standard 3618, sec. 4, 1963, 11 pp. Note that this Standard has been superseded by a later edition. The current version must be consulted before these definitions are used. British Standards Institution, 389 Chiswick High Road, London W4 4AL, UK.

BS, 11 British Standards Institution (London). *Glossary of Mining Terms*. Section 5, Geology. British Standard 3618, sec. 5, 1964, 18 pp. Note that this Standard has been superseded by a later edition. The current version must be consulted before these definitions are used. British Standards Institution, 389 Chiswick High Road, London W4 4AL, UK.

BS, 12 British Standards Institution (London). *Glossary of Mining Terms*. Section 6, Drilling and Blasting. British Standard 3618, sec. 6, 1964, 20 pp. Note that this Standard has been superseded by a later edition. The current version must be consulted before these definitions are used. British Standards Institution, 389 Chiswick High Road, London W4 4AL, UK.

BS, 13 British Standards Institution (London). *Glossary of Mining Terms*. Section 7, Electrical Engineering and Lighting. British Standard 3618, sec. 7, 1965, 10 pp. Note that this Standard has been superseded by a later edition. The current version must be consulted before these definitions are used. British Standards Institution, 389 Chiswick High Road, London W4 4AL, UK.

C

Camm Camm, Frederick James (ed.). *A Dictionary of Metals and Their Alloys*. George Newnes, Ltd., London, 1940, 244 pp.

Camp Lankford, William T., Jr., et al. *The Making, Shaping, and Treating of Steel*. Association of Iron & Steel Engineers. 10th ed, 1985, 1572 pp.

Campbell Campbell, W.J. et al. *Selected Silicate Minerals and their Asbestiform Varieties*. USBM Information Circular 8751.

Carson, 1 Carson, A Brinton. *General Excavation Methods*. F.W. Dodge Corp. (a McGraw-Hill company), New York, 1961, 392 pp.

Carson, 2 Carson, A. Brinton. *Foundation Construction*. McGraw-Hill Book Co., New York, 1965, 424 pp.

CCD, 1 Reinhold Publishing Corp. *Condensed Chemical Dictionary*. New York, 3rd ed., 1942, 756 pp.

CCD, 2 Reinhold Publishing Corp. *Condensed Chemical Dictionary*. New York, 6th ed., 1961, 1256 pp.

CFR, 1 Code of Federal Regulations. *30 CFR, part 50, July 1992; 30 CFR, pt. 1-199, July 1, 1993; 30 CFR, part 75.201; MSHA, 30 CFR part 90.2, July 1, 1993*. Office of the Federal Registry, Washington, DC.

CFR, 2 Code of Federal Regulations. *30 CFR, part 32.2; Mineral Resources*. Office of the Federal Registry.

CFR, 3 Code of Federal Regulations. *30 CFR 56.2 and 57.6133; Mineral Resources*. Office of the Federal Registry.

CFR, 4 Code of Federal Regulations. *30 CFR, Ch. 1*. Office of the Federal Registry.

CFR, 5 Code of Federal Regulations. *30 CFR, V. 56, No. 13; MSHA, 30 CFR V. 567, No. 13, January 18, 1991*. Office of the Federal Registry.

CFR, 6 Code of Federal Regulations. *30 CFR, std. 75.1714-1a*. Office of the Federal Registry.

CFR, 7 Code of Federal Regulations. *30 CFR, Title 40, Part 440*. Office of the Federal Registry.

Challinor Challinor, John. *A Dictionary of Geology*. University of Wales Press, Cardiff, 2nd ed., 1964, 289 pp.

Chandler Chandler, Henry P. Industrial Diamond. *A Materials Survey*. U.S. Bureau of Mines Information Circular 8200, 1964. 149 pp., 20 figs. Includes appendixes on (1) Government wartime orders, (2) U.S. patents on industrial diamond and (3) glossary.

Chem. Eng. *Chemical Engineering*, V. 56, No. 1. McGraw-Hill, Inc. Jan., 1949.

Chem. Indust. *Chemistry & Industry*, V. 58. Chemistry & Industry. March 18, 1939.

CIPW Cross, Whitman; Joseph Iddings, Louis Pirsson; and Henry Washington. *Quantitative Classification of Igneous Rocks*. University of

Chicago Press, Chicago, IL, 1903, 286 pp. Includes a glossary, pp. 261-284.

CMD Chemical Publishing Co. *Chamber's Mineralogical Dictionary.* New York, 1948, 47 pp.

Coal Age, 1 Coal Age. *Glossary of Electrical Terms.* Intertec Publishing, Chicago, IL. V. 65, No. 6, June 1960, pp. 115-117.

Coal Age, 2 *Coal Age.* Intertec Publishing, Chicago, IL. V. 66, No. 3.

Coal Age, 3 *Coal Age.* Intertec Publishing, Chicago, IL. V. 71, No. 8, August 1966.

Connolly Connolly, Randall E. *Overview of Research on the Management of Mining Wastes, from the Proceedings of the 2nd International Symposium on Mine Planning and Equipment Selection, 1990.*

Cooper Cooper, Terence. *An Introduction to Mining Chemistry.* Leonard Hill Books, Ltd., London, 1963, 439 pp. Chapter 1, Chemical Definitions and Terminology, pp. 1-27.

Cote Cote, Arthur E. and John L. Linville. *Fire Protection Handbook,* National Fire Protection Association, Quincy, MA, 17th ed., 1991.

Craigie Craigie, William A., and James R. Hulbert (eds.). *A Dictionary of American English on Historical Principles.* University of Chicago Press. Chicago, IL, 1938, 1940, 1942, 1944, 4 volumes.

Crispin Crispin, Frederic Swing. *Dictionary of Technical Terms.* The Bruce Publishing Co., Milwaukee, WI, 10th ed., 1964, 455 pp.

Crosby Crosby, James W. III. *A Descriptive Glossary of Radioactive Minerals.* Washington State Institute of Technology, Bull. 230, series B, 1955, 148 pp.

Cruickshank Cruickshank, M.J., et al. *Marine Mining on the Outer Continental Shelf.* Minerals Management Service, OCS Report 87-0035, 1987.

CTD Tweney, C.F., and L.E.C. Hughes (eds.). *Chamber's Technical Dictionary.* MacMillan Co., New York. 3rd ed. 1958. 1028 pp.; supplement, pp. 952-1028.

Cumming, 1 Cumming, James D. *Diamond Drill Handbook.* J.K. Smit & Sons of Canada, Ltd., Toronto, Canada, 1951, 501 pp. Includes a glossary of Diamond Drilling and Mining Terms. pp. 472-492.

Cumming, 2 Cumming, James D. and A.P. Wicklund. *Diamond Drill Handbook,* J.K. Smit and Sons, Toronto, Canada. 3rd ed., 1981.

D

Dana, 1 Dana, James Dwight. *The System of Mineralogy of James Dwight Dana,* 1837-1868. John Wiley & Sons, New York, 6th ed., 1914, (c1892), 1134 pp. Appendix I, completing work to 1899, 75 pp. Appendix II, completing work to 1909, 114 pp. Appendix III, completing work to 1915, 87 pp. Appendix III published separately in 1915.

Dana, 2 Dana, James Dwight, and Edward Salisbury Dana. *The System of Mineralogy of James Dwight Dana and Edward Salisbury Dana,* rewritten and enlarged by Charles Palache, Harry Berman, and Clifford Frondel. John Wiley & Sons, Inc, New York. 7th ed., 1944, 1951, 1962, 3 volumes.

Dana, 3 Hurlbut, Cornelius S., Jr. *Dana's Manual of Mineralogy.* John Wiley & Sons, Inc., New York, 1955, p. 520. Copyright © John Wiley & Sons, Inc. This material is used by permission of John Wiley & Sons, Inc.

Dana, 4 Hurlbut, Cornelius S., Jr. *Dana's Manual of Mineralogy.* John Wiley & Sons, Inc., New York, 17th ed., 1959. 609 pp. Includes a Mineral Index, pp. 597-609. Copyright © John Wiley & Sons, Inc. This material is used by permission of John Wiley & Sons, Inc.

Davis Davis, C.W. and H.C. Vacher. Revised by John E. Conley. *Bentonite: Its Properties, Mining, Preparation, and Utilization.* U.S. Bureau of Mines Technical Paper 609, 1940. 83 pp., 2 figs.

Desai Desai, C.S. and H.J. Siriwardene. *Constitutive Laws for Engineering Materials with Emphasis on Geologic Materials.* Prentice Hall Inc., Englewood Cliffs, NJ. 1984.

DeVerle DeVerle, Harris. *An Investigation of the Estimation Process of Predictive Metallogeny.* Geological Association of Canada. Geoscience Canada. V. 10, No. 2, 1983.

Dick, 1 Dick, Richard A., Larry A. Fletcher and Dennis V. D'Andrea. *Explosive Blasting Procedure Manual.* U.S. Bureau of Mines. 1983.

Dick, 2 Dick, Richard A., Larry R. Fletcher and Michael A. Peters. *Blasting Training Manual for Metal/Non-Metal Mines,* Appendix B. U.S. Bureau of Mines.

Dobrin Dobrin, Milton B. *Introduction to Geophysical Prospecting.* McGraw-Hill Book Co., New York, 2nd ed. 1960, 446 pp.

Dodd Dodd, A. E. *Dictionary of Ceramics: Pottery, Glass, Vitreous Enamels, Refractories, Clay Building Materials.* Cement and Concrete, Electroceramics, Special Ceramics. Philosophical Library Inc., New York. 1964, 327 pp.

Domenico Domenico, P.A. and F.W. Schwartz. *Physical and Chemical Hydrogeology.* John Wiley & Sons. 1990. Copyright © John Wiley & Sons, Inc. This material is used by permission of John Wiley & Sons, Inc.

DOT U.S. Department of Labor. Bureau of Employment Security. *Dictionary of Occupational Titles.* 2nd ed., Washington, D.C., Government Printing Office, 1949. Prepared by the Division of Occupational Analysis, United States Employment Service. 2 volumes; supplement 1, March 1955. 341 pp.

Driscoll Driscoll, F.G. *Groundwater and Wells.* Johnson Division, St. Paul, MN. 1986.

Du Pont, 1 Du Pont de Nemours & Company (E.I.). *Blasters' Handbook; a Manual Describing Explosives and Practical Methods of Use.* Wilmington, DE, 15th ed., 1966, 524 pp.

Du Pont, 2 DuPont. *Blaster's Handbook,* 175th Anniversary ed. E.I. Dupont de Nemours and Co., 1977 (International Society of Explosives Engineers).

Duvall Duvall, Wilbur I. and Thomas C. Atchison. *Rock Breakage by Explosives.* U.S. Bureau of Mines Reports of Investigations 5356, 1957. 52 pp., 33 figs.

E

Enam. Dict. Ferro Enamel Corp. *The Enameler's Dictionary.* Cleveland, Ohio, 1947.

Encyclopaedia Britannica Encyclopaedia Britannica. *Britannica Book of the Year, 1964.*

Eng. Min. J., 1 *Engineering and Mining Journal.* V. 139, No. 4. Intertec Publishing. Apr. 1938.

Eng. Min. J., 2 *Engineering and Mining Journal.* V. 165. Intertec Publishing. Nov. 1964.

Eng. Min. J., 3 *Engineering and Mining Journal.* V. 167, No. 66. Intertec Publishing. June 1966.

English English, George Letchworth. *Descriptive List of the New Minerals,* 1892-1938; Containing All New Mineral Names not Mentioned in Dana's System of Mineralogy, 6th ed., 1892. McGraw-Hill Book Co., New York, 1st ed., 1939, 258 pp.

F

FR 166 MSHA, *Federal Register,* V. 54, No. 166. U.S. Department of Labor. Aug. 29, 1989.

FR 95 *Federal Register,* V. 57, No. 95. Office of the Federal Registry. May 15, 1992.

FR 249 MSHA, *Federal Register,* V. 58, No. 249, December 30, 1993.

Fay Fay, Albert H. *A Glossary of the Mining and Mineral Industry.* U.S. Bureau of Mines Bull. 95, 1920, 754 pp.

Federal Mine Safety *Federal Mine Safety Act of 1977.*

Felegy Felegy, E.W., L.H. Johnson, and J. Westfield. *Acid Mine Water in the Anthracite Region of Pennsylvania.* U.S. Bureau of Mines Technical Papers 710, 1948. 49 pp., 13 figs.

Fleischer Fleischer, Michael and Joseph A. Mandarino. *Glossary of Mineral Species.* 7th ed. The Mineralogical Record, Inc., Tucson, 1995, 280 pp.

Forrester Forrester, James Donald. *Principles of Field and Mining Geology.* John Wiley & Sons, Inc., New York, 1946, 647 pp.

Fowler Fowler, C.M.R. *The Solid Earth, An Introduction to Global Geophysics.* Cambridge University Press, New York. 1990.

Fraenkel Fraenkel, K.H. (ed. in chief). *Manual on Rock Blasting.* Atlas Copco Aktiebolag, Stockholm, Sweden, and Sandvikens Jernverks Aktiebolag, Sandviken, Sweden, 2nd ed., 1953 with supplements through 1961, 3 volumes. V. 1: Glossary of Terminology, Article 4, pp. 4:01-1 to 4:01-27.

Francis, 1 Francis, Wilfrid. *Coal: Its Formation and Composition.* Edward Arnold Publishers, Ltd., London, 1954, 567 pp.

Francis, 2 Francis, Wilfrid. *Fuels and Fuel Technology*. Pergamon Press, New York, 1965, 2 volumes.

Freeman Freeman, Correll H. *Natural Bonded Moulding Sands of Canada*. Canada, Department of Mines No. 767, 1936, 144 pp. Appendix II - Glossary of Foundry Terms, pp. 133-134.

Freeze Freeze, R.A. and J.A. Cherry. *Groundwater*. Prentice Hall Inc., 1979.

Fuerstenau Fuerstenau, D.W. (ed.). *Froth Flotation, 50th Anniversary Volume*. American Institute of Mining, Metallurgical, and Petroleum Engineers, Inc., New York, 1962, 677 pp.

G

Gaudin, 1 Gaudin, A. M. *Principles of Mineral Dressing*. McGraw-Hill Book Co., New York, 1st ed., 1939, 544 pp.

Gaudin, 2 Gaudin, A.M. *Flotation*. McGraw-Hill Book Co., New York, 2nd ed., 1957, 573 pp.

Gaynor Gaynor, Frank. *Concise Dictionary of Science, Physics, Mathematics, Nucleonics, Astronomy, Chemistry*. Philosophical Library Inc., New York, 1959, 546 pp.

Glasstone Glasstone, Samuel. *Sourcebook on Atomic Energy*. D. Van Nostrand Co., Princeton, N.J., 2nd ed., 1958, 641 pp.

Goldman Goldman, Marcus I. Memoir 50. *Deformation, Metamorphism, and Mineralization in Gypsum-anhydrite Cap Rock, Sulpher Salt Dome, Louisiana*. Geological Society of America, 1952, 169 pp.

Gordon Gordon, Henry A. *Mining and Engineering and Miners' Guide*. John Mackay, Government Printer, Wellington, New Zealand, 3rd ed., 1906, 615 pp. Includes a glossary of Scientific and Mining Terms, pp. 574-594.

Greenburg Greenburg, Leonard and George W. Smith. *A New Instrument for Sampling Aerial Dust*. U.S. Bureau of Mines Reports of Investigations 2392, 1922. 3 pp.

Griffith Griffith, S.V. *Alluvial Prospecting and Mining*. Mining Publications, Ltd., London, 1938, 142 pp.

H

Hackh Hackh, Ingo W.D. *Grant & Hackh's Chemical Dictionary*, rev. and ed. by Julius Grant. The Blakiston Co., Philadelphia, PA, 3rd ed., 1944 (reprinted with changes and additions in 1946), 925 pp.

Hammond Hammond, Rolt. *Dictionary of Civil Engineering*. George Newnes, Ltd., London, 1965, 253 pp.

Handbook of Chemistry and Physics, 1 *Handbook of Chemistry and Physics*. CRC Press, Inc. (formerly Chemical Rubber Co.), Boca Raton, FL, 42nd ed., 1960-61 (1960), 3481 pp.

Handbook of Chemistry and Physics, 2 *Handbook of Chemistry and Physics*. CRC Press, Inc. (formerly Chemical Rubber Co.), Boca Raton, FL, 45th ed., 1964-65 (1964), v.p.

Handbook of Chemistry and Physics, 3 *Handbook of Chemistry and Physics*. Robert C. Weast (ed.). 69th ed. CRC Press Inc., Boca Raton, FL. 1988- 89.

Hansen Hansen, J.E. *A Manual of Porcelain Enameling*. The Enamelist Publishing Co., Cleveland, OH, 1937, 513 pp. Includes a glossary, "The Enameler's Dictionary," pp. 468- 495.

Harbison-Walker
Harbison-Walker Refractories Co. *Modern Refractory Practice*. Pittsburgh, PA, 5th ed., 1972.

Harrington Harrington, D. and Sara J. Davenport. *Review of Literature on Effects of Breathing Dusts, with Special Reference to Silicas*. U.S. Bureau of Mines Bulletin 400, 1937. 305 pp.

Harrison Harrison, H.L.H. *Alluvial Mining for Tin and Gold*. Mining Publications, Ltd., London, 1962, 313 pp.

Hartman, 1 Hartman, Howard L. *Mine Ventilation and Air Conditioning*. The Ronald Press Co., New York, 1961, 398 pp. Copyright © John Wiley & Sons, Inc. This material is used by permission of John Wiley & Sons, Inc.

Hartman, 2 Hartman, Howard L., *Mine Ventilation and Air Conditioning*, Second Edition, 1991, Krieger Publishing Co., Malabar, Florida. Copyright © John Wiley & Sons, Inc. This material is used by permission of John Wiley & Sons, Inc.

Hawkes, 1 Hawkes, H.E. *Principles of Geochemical Prospecting*. U.S. Geol. Survey, Bull. 1000-F., 1957, pp. 225-355. Includes a glossary, pp. 336-338.

Hawkes, 2 Hawkes, H.E., and J.S. Webb. *Geochemistry in Mineral Exploration*. Harper & Row Publishers, New York, 1962, 415 pp.

Henderson Henderson, J.G., and J.M. Bates. *Metallurgical Dictionary*. Reinhold Publishing Corp., New York, 1953, 396 pp.

Hess Hess, Frank L. Definitions furnished by Frank L. Hess, Mining Engineer, U.S. Bureau of Mines. Deceased.

Hey, 1 Hey, Max H. *An Index of Mineral Species and Varieties Arranged Chemically With an Alphabetical Index of Accepted Mineral Names and Synonyms*. The British Museum, London, 2nd ed., 1955, 728 pp. Alphabetical Index of Mineral Names and Synonyms. pp. 317-658.

Hey, 2 Hey, Max H. *Twenty-Second List of New Mineral Names*. Mineralogical Magazine, V. 32, Dec. 1961, pp. 941-991.

Hey, 3 Hey, Max H. *Twenty-Third List of New Mineral Names*. Mineralogical Magazine, V. 33, Dec. 1964, pp. 1125-1158.

Higham Higham, S. *An Introduction to Metalliferous Mining*. Charles Griffin & Co., Ltd., London, 1951, 337 pp.

Hoffman Hoffman, Arnold. *Free Gold: The Story of Canadian Mining*. Associated Book Service, New York, 1958 (copyright 1947), 420 pp. Includes a glossary of Canadian Mining Terms, pp. 393-397.

Holmes, 1 Holmes, Arthur. *The Nomenclature of Petrology; With References to Selected Literature*. Thomas Murby & Co., London, 1920, 284 pp.

Holmes, 2 Holmes, Arthur. *The Nomenclature of Petrology; With References to Selected Literature*. Thomas Murby & Co., London, 2nd ed., 1928, 284 pp.

Holser Holser, A.F. *Offshore Lands of the U.S.A.: Marine Policy*. January 1988.

Hoover Hoover, Theodore Jesse. *The Economics of Mining*. Stanford University Press, Stanford, CA, 3rd ed., 1948, 551 pp.

Houlsby Houlsby, A.C. *Construction and Design of Cement Grouting*. John Wiley & Sons. 1990. Copyright © John Wiley & Sons, Inc. This material is used by permission of John Wiley & Sons, Inc.

Hudson Hudson Coal Co. *The Story of Anthracite*. New York, 1932, 425 pp. Includes a glossary of Mining Terms, pp. 401-409.

Hunt Hunt, Lee M., and Donald G. Groves (eds.). *A Glossary of Ocean Science and Undersea Technology Terms*. Compass Publications, Inc., Arlington, VA, 1965, 172 pp.

Huntington Huntington, Whitney Clark. *Earth Pressures and Retaining Walls*. John Wiley & Sons, Inc., New York, 1957, 534 pp.

Hurlbut Hurlbut, Cornelius S., Jr., and Henry W. Wenden. *The Changing Science of Mineralogy*. D.C. Heath & Co., Boston, MA, 1964, 117 pp. Includes a glossary, pp. 109-113.

Hy Data Publications. *Hydrospace Buyers' Guide*, 1965. Washington, D.C., V. 3, 1965 (1962), pp. 85-116. Glossary of Hydrospace Projects, Terms, Abbreviations, and Weaponry.

I

IHCP International Committee for Coal Petrology. *International Handbook of Coal Petrography*. Centre National de la Recherche Scientifique, Paris, France, 2nd ed., 1963, n.p. Parts I and II in one volume. Part I - Alphabetical Lineup of Coal Nomenclature.

Indust. Miner. Lefond, Stanley J. (editor-in- chief). *Industrial Minerals and Rocks*, V. 1. Society for Mining, Metallurgy, and Exploration Inc. 5th ed., 1983.

Inst. Petrol. Institute of Petroleum (London). *A Glossary of Petroleum Terms*. London, 3rd ed., 1961, 39 pp.

Issacson Issacson, E. de St. Q. *Rock Pressure in Mines*. Mining Publications, Ltd., London, 2nd ed., 1962, 260 pp.

J

Jackson, 1 Jackson & Moreland, Boston. *Design Manual for Roller and Silent Chain Drives*. Association of Roller and Silent Chain Manufacturers, Boston, MA, 1955, 95 pp. Includes a glossary, pp. 92-93.

Jackson, 2 Jackson, Chas. F. and E.D. Gardner. *Stoping Methods and Costs*. U.S. Bureau of Mines Bulletin 390, 1936. 296 pp., 78 figs.

Jackson, 3 Jackson, Chas. F. and J.H. Hedges. *Metal-Mining Practice*. U.S. Bureau of Mines Bulletin 419, 1939. 512 pp., 156 figs.

Jessop Jessop, James A., et al. *Evaluation of Stope Leaching Site using Geotomography.* (Conference on In situ Minerals Recovery II. Santa Barbara, CA, Oct. 25-30, 1992) U.S. Bureau of Mines.

Johannsen Johannsen, Albert. *A Descriptive Petrography of the Igneous Rocks.* University of Chicago Press, Chicago, IL, 1st ed., 1931-38, volumes 2-4; 2nd ed., 1939, 318 pp.

Jones, 1 Jones, Donald C., and Joseph W. Hunt. *Coal Mining.* Pennsylvania State College, 3rd ed., 1949, 2 volumes. Volume 1 includes a glossary of Mining Terms, pp. 322-354.

Jones, 2 Jones, Donald C. *Mining Mathematics.* Pennsylvania State College, Miner. Ind. Extension Services, 2nd ed., 1949, 186 pp.

K

Kaufmann Kaufmann, Dale W. (ed.). *Sodium Chloride; the Production and Properties of Salt and Brine* (American Chemical Society Monograph 145). Reinhold Publishing Corp., New York, 1960, 743 pp. Includes a glossary of Words Pertaining to Salt, pp. 687-693.

Kelly Kelly, E.G. and D.J. Spottiswood. *Introduction to Mineral Processing.* Wiley-Interscience, 1982.

Kentucky Kentucky, Mayo State Vocational School (Paintsville), and the Kentucky Mining Institute. *Elements of Practical Coal Mining.* The Dunne Press. 1952, 436 pp.

Kinney Kinney, Kay. *Glass Craft.* Chilton Co., Philadelphia, PA, 1962, 178 pp. Includes a glossary, pp. 167-169.

Kirk Kirk, Raymond E., and Donald F. Othmer. *Encyclopedia of Chemical Technology.* Interscience Encyclopedia, Inc., New York, 1947, 15 volumes. Copyright © John Wiley & Sons, Inc. This material is used by permission of John Wiley & Sons, Inc.

Kiser Kiser, A.B. *Coal Cutters, Loaders, and Conveyors; Trackwork.* International Textbook Co., Scranton, PA, 1929. 97 pp. in two parts: Pt. 1, Coal Cutters, Loaders, and Conveyors, 39 pp. Pt. 2, Trackwork, 58 pp.

Knopf Knopf, Eleanora Bliss, and Earl Ingerson. *Structural Petrology.* Geological Society of America, Memoir 6. 1938. 270 pp.

Kordosky Kordosky, Gary A. *Copper Solvent Extraction: The State of the Art.* JOM (formerly called "Journal of Metals"), May 1992.

Korson Korson, George. *Minstrels of the Mine Patch; Songs and Stories of the Anthracite Industry.* University of Pennsylvania Press, Philadelphia, PA, 1938, 332 pp. Includes a glossary of anthracite technical and colloquial words and phrases, pp. 311-320.

Krumlauf Krumlauf, Harry E. (ed.). *Proceedings of Symposium on Surface Mining Practices.* College of Mines, University of Arizona, Tucson, AZ, Oct. 17-19, 1960, 131 pp.

L

Langefors Langefors, U., and B. Kihlstrom. *The Modern Technique of Rock Blasting.* John Wiley & Sons, Inc., New York, 1963, 405 pp.

Larsen Larsen, Esper S., and Harry Berman. *The Microscopic Determination of the Nonopaque Minerals.* U.S. Geol. Survey, Bull. 848, 2nd ed., 1934, 266 pp.

Lee Lee, P. William. *Ceramics.* Reinhold Publishing Corp., New York, 1961, 210 pp. Includes a glossary of Raw Materials, pp. 165-198.

Leet, 1 Leet, L. Don, and Sheldon Judson. *Physical Geology.* Prentice-Hall, Inc., Englewood Cliffs, N.J., 2nd ed., 1958, 502 pp. Includes a glossary, pp. 437-464.

Leet, 2 Leet, L. Don. *Vibrations From Blasting Rock.* Harvard University Press, Cambridge, Mass., 1960, 134 pp.

Legrand Legrand, Harry E. *Geology and Ground-Water Resources of Pittsylvania and Halifax Counties.* Virginia Division of Mineral Resources, Bull. 775, 1960, 86 pp. Includes a glossary, pp. 80-81.

Leick Leick, Alfred. *GPS Satellite Surveying.* John Wiley and Sons. 1989. Copyright © John Wiley & Sons, Inc. This material is used by permission of John Wiley & Sons, Inc.

Lewis Lewis, Robert S., and George B. Clark. *Elements of Mining.* John Wiley & Sons, Inc., New York, 3rd ed., 1964, 768 pp. Copyright © John Wiley & Sons, Inc. This material is used by permission of John Wiley & Sons, Inc.

Liddell Liddell, Donald M. *The Metallurgists and Chemists' Handbook: a Reference Book of Tables and Data for the Student and Metallurgist.* McGraw-Hill Book Co., New York, 2nd ed., 1918, 656 pp.

Light Metal Age *Light Metal Age,* V. 16, No. 9. Fellom Publishing Co., San Francisco, CA. Oct., 1958.

Litton Litton, Charles D., Ronald S. Conti, and John G. Tabacchi. *Evaluation of a Nitric-Oxide-Compensated Carbon Monoxide Free Sensor.* U.S. Bureau of Mines Information Circular 9339, 1993. 10 pp., 7 figs.

Lombardi Lombardi, J.A. *"In Situ Liners". Proceedings, Institute of Shaft Drilling Technology Conference (Env. Technology Session).* Apr. 18-21, 1994.

Long Long, Albert E. *A Glossary of the Diamond-Drilling Industry.* U.S. Bureau of Mines Bull. 583, 1960, 98 pp.

Lowenheim Graham, A. Kenneth (ed.). *Electroplating Engineering Handbook.* Reinhold Publishing Corp., New York, 2nd ed., 1962, 774 pp. Includes a glossary by Frederick A. Lowenheim, pp. xiii-xxii.

Lyman Lyman, James D., and Benjamin S. Loeb. *Nuclear Terms: A Brief Glossary.* U.S. Atomic Energy Commission, Div. of Tech. Inf., Oak Ridge, TN, Apr. 1964, 35 pp.

M

MacCracken MacCracken, Helen Dolman. *Basic Earth Science.* L.W. Singer Co., Syracuse, NY, 1964, 442 pp. Includes Science definitions, pp. 425-435; Oceanographic terms (between pp. 154-155-Insert).

Marsden Marsden, R.W. *Geology of the Iron Ores of the Lake Superior Region in the United States in J.D. Ridge (ed.), Ore Deposits in the United States 1933/1967,* AIME Graton-Sales Volume. 1968. pp. 489-506.

Mason Mason, E. (ed.). *Practical Coal Mining for Miners.* Virtue & Co., Ltd., London, 2nd ed., 1951, 2 volumes. Glossary of terms, V. 2, pp. 769-775.

Mather Mather, Kirtley F. *The Earth Beneath Us.* Chanticleer Press, Inc., New York, 1964, 320 pp. Includes a glossary, pp. 313-315.

Mathews Mathews, Mitford M. (ed.). *A Dictionary of Americanisms on Historical Principles.* University of Chicago Press, Chicago, IL, 1951, 2 volumes, 1946 pp.

McAdam, 1 McAdam, R., and D. Davidson. *Mine Rescue Work.* Oliver and Boyd, London, 1955, 183 pp.

McAdam, 2 McAdam, R., and R. Westwater. *Mining Explosives.* Oliver and Boyd, London, 1958, 187 pp.

McGraw-Hill, 1 McGraw-Hill. *Dictionary of Scientific and Technical Terms,* 5th ed., 1994.

McGraw-Hill, 2 McGraw-Hill. *Encyclopedia of Science and Technology.* 1960

McKinstry McKinstry, Hugh Exton. *Mining Geology.* Prentice-Hall, Inc., New York, 1948, 680 pp. Includes a glossary of Mining and Geological Terms, pp. 631-659.

Mero Mero, John L. *The Mineral Resources of the Sea.* Elsevier Publishing Co., Amsterdam, Netherlands, 1965 (Copyright 1964), 312 pp.

Mersereau, 1 Mersereau, Samuel Foster. *Materials of Industry.* McGraw-Hill Book Co., New York, 1941, 578 pp.

Mersereau, 2 Mersereau, Samuel Foster. *Materials of Industry.* McGraw-Hi11 Book Co., New York, 4th ed., 1947, 623 pp.

Meyer Meyer, Rudolf. *Explosives.* Verlag Chemie GmbH, Weinheim, Germany. 1981.

Miall Miall, Stephen (ed.). *A New Dictionary of Chemistry.* Longmans, New York, 1940, 575 pp.

Min. Miner. Eng., 1 *Mining and Minerals Engineering.* V. 1, No. 5, Jan. 1965

Min. Miner. Eng., 2 *Mining and Minerals Engineering.* V. 2, No. 2, Feb. 1966, p. 65.

Mining Institution of Mining and Metallurgy. *Symposium on Opencast Mining, Quarrying, and Alluvial Mining,* London. Nov. 16-19, 1964. Papers 6-9, 13, 14 and 17.

Mitchell Mitchell, David R. (ed.). *Coal Preparation.* American Institute of Mining and Metallurgical Engineers, New York, 2nd ed., 1950, 830 pp.

Morris Morris, W.J., and T. Cooper. *An Introduction to Mining Science.* George G. Harrap & Co., Ltd., London, 1958, 288 pp.

Mowrey Mowrey, Gary L. *Promising Coal Interface Detection Methods.* Mining Engineering Magazine, Jan. 1991.

MSHA, 1 Mine Safety and Health Administration. *Mine Ventilation - Safety Manual No. 20,* 1986.

MSHA, 2 Mine Safety and Health Administration. *MSHA Health Handbook.*

MSHA, 3 Mine Safety and Health Administration. *The Radiation Hazard in Mining.* Safety Manual No. 7, 1986.

MSHA, 4 Mine Safety and Health Administration. *Metal and Nonmetal National Mine Rescue Content Rules.* 1992.

MSHA, 5 Mine Safety and Health Administration. *MSHA Handbook Series PH90-IV-4.* U.S. Department of Labor. 1990.

N

NCB Great Britain, National Coal Board. *A Glossary of Automation and Remote Control As Applied to the Coal-Mining Industry.* London, 1964, 34 pp.

Nelson Nelson, A. *Dictionary of Mining.* Philosophical Library Inc., New York, 1965, 523 pp.

NEMA, 1 National Electrical Manufacturers Association. *NEMA Standards Publication for Mining Belt Conveyors.* NEMA Pub. MB1-1956, 1956, 34 pp. Includes definitions, pp. 7-17.

NEMA, 2 National Electrical Manufacturers Association. *NEMA Standards Publication for Mining Belt Conveyors.* NEMA Pub. MB1-1961, 1961, v.p.

New South Wales New South Wales, Department of Mines. *Prospectors Guide.* Sydney, Australia, 7th ed., 1958, 222 pp. Includes a glossary of Mining Terms, pp. 191-196.

Newton, 1 Newton, Joseph. *Extractive Metallurgy.* John Wiley & Sons, Inc., New York, 1959, 532 pp. Copyright © John Wiley & Sons, Inc. This material is used by permission of John Wiley & Sons, Inc.

Newton, 2 Newton, Joseph. *Introduction to Metallurgy.* John Wiley & Sons, Inc. 1938.

Nichols, 1 Nichols, Herbert L., Jr. *Moving the Earth; the Workbook of Excavation.* North Case Books, Greenwich, CT, 3rd ed., 1976, v.p. Includes a glossary, pp. G-1 to G-27.

Nichols, 2 Nichols, Herbert L., Jr. *How to Operate Excavation Equipment.* McGraw-Hill, Inc., New York, NY, 1954, 150 pp. Includes a glossary, pp. 146-150.

Nichols, 3 Nichols, Herbert L., Jr. *Modern Techniques of Excavation.* McGraw- Hill, Inc., New York, NY, 1956, v.p.

NIOSH National Institute for Occupational Safety and Health. *DHHS Publication No. 88-101.* October 1987.

Noke Noke, Charles J., and Harold J. Plant. *Pottery.* Sir Isaac Pitman & Sons, Ltd., London, 1927, 148 pp. Includes a glossary, p. 135.

NSC, 1 National Safety Council. *Accident Prevention Manual for Business & Industry.* NSC. 10th ed., 1992.

NSC, 2 National Safety Council. *Fundamentals of Industrial Hygiene.* NSC. 4th ed., 1996.

O

Obert Obert, Leonard, Wilbur I. Duvall, and Robert H. Merrill. *Design of Underground Openings in Competent Rock.* U.S. Bureau of Mines Bulletin 587, 1960. 36 pp., 27 figs.

Osborne Osborne, A. K. *An Encyclopaedia of the Iron & Steel Industry.* Philosophical Library Inc., New York, 1956, 558 pp.

OSHA OSHA. *Federal Register,* V. 58, No. 9, Appendix C to 1910.146. Jan. 14, 1993.

P

Padan Padan, John. *Marine Heavy Metal Project Offshore Nome, Alaska. U.S.B.M. Tech. Progress Report #4.* U.S. Bureau of Mines, August 1968.

Pearl Pearl, Richard M. *Successful Mineral Collecting and Prospecting.* McGraw-Hill Book Co., Inc., New York, 1961, 164 pp.

Peel Peel, Robert. *An Elementary Textbook of Coal Mining.* Blackie and Son, Ltd., London, 20th ed., 1921, 420 pp. Includes a glossary, pp. 408-415.

Peele Peele, Robert (ed.). *Mining Engineers' Handbook.* John Wiley & Sons, Inc., New York, 3rd ed., 1941, 2 volumes, v.p. Copyright © John Wiley & Sons, Inc. This material is used by permission of John Wiley & Sons, Inc.

Peters Peters, William C. *Exploration and Mining Geology.* John Wiley & Sons, Inc. 2nd ed., 1987. Copyright © John Wiley & Sons, Inc. This material is used by permission of John Wiley & Sons, Inc.

Petroleum Age *Petroleum Age.* V. 11, Jan. 15, 1923.

Pettijohn, 1 Pettijohn, Francis John, and Paul Edwin Potter. *Atlas and Glossary of Primary Sedimentary Structures.* Springer-Verlag, Inc., New York, 1964, 370 pp. Includes a glossary of Primary Sedimentary Structures. pp. 283-353.

Pettijohn, 2 Pettijohn, Francis John. *Sedimentary Rocks.* Harper & Brothers, New York, 2nd ed., 1957, 718 pp.

Pit and Quarry Pit and Quarry Publications, Inc., *Pit and Quarry Handbook and Purchasing Guide for the Nonmetallic Minerals Industries.* Chicago, IL, 53d ed., 1960, v.p.

Porter Porter, Hollis Paine. *Petroleum Dictionary for Office, Field, and Factory.* Gulf Publishing Co., Houston, Tex., 1930, 234 pp.

Pryor, 1 Pryor, Edmund J. *Economics for the Mineral Engineer.* Pergamon Press, New York, 1958, 254 pp. Includes a glossary, pp. 217-245.

Pryor, 2 Pryor, Edmund J. *Mineral Processing.* Mining Publications, Ltd., London, 2nd ed., 1960, 814 pp. Includes a glossary, Appendix A, pp. 789-798.

Pryor, 3 Pryor, Edmund J. *Dictionary of Mineral Technology.* Mining Publications, Ltd., London, 1963, 437 pp.

Pryor, 4 Pryor, Edmund J. *Mineral Processing.* Elsevier Publishing Co., London, 3rd ed., 1965, 844 pp. Includes a glossary, Appendix A, pp. 809-818.

R

Raistrick Raistrick, A., and C.E. Marshall. *The Nature and Origin of Coal and Coal Seams.* The English Universities Press, Ltd., London, 1939, 282 pp.

Rampacek Rampacek, Carl, W.A. McKinney, and P.T. Waddleton. *Treating Oxidized and Mixed Oxide-Sulfide Copper Ores by the Segregation Process.* U.S. Bureau of Mines Reports of Investigations 5501, 1959. 28 pp., 13 figs.

Rice, 1 Rice, C.M. *Dictionary of Geological Terms (Exclusive of Stratigraphic Formations and Paleontologic Genera and Species).* Edwards Brothers, Inc., Ann Arbor, MI, 1960, 465 pp.

Rice, 2 Rice, George S., L.M. Jones, W.L. Egy, and H.P. Greenwald. *Coal-Dust Explosion Tests in the Experimental Mine 1913-18,* Inclusive. U.S. Bureau of Mines Bull. 167, 1922, 639 pp. Includes a glossary of terms developed to describe conditions in the experimental mine, pp. 544-551.

Ricketts Ricketts, A.H. *American Mining Law With Forms and Precedents.* California, Division of Mines, Bull. 123, 4th ed., enl. and rev., February 1943, 1018 pp. Includes a glossary of Mining Terms and Phrases, pp. 1-48. Also includes a glossary of Oil Mining Terms and Phrases, pp. 49-58.

Roark Roark, Raymond J. *Formulas for Stress and Strain.* McGraw-Hill Book Co., New York, 3rd ed., 1954, 381 pp. Includes definitions, chapter 1, pp. 3-15.

Roberts, 1 Roberts, A. (ed.). *Mine Ventilation.* Cleaver-Hume Press, Ltd., London, 1960, 363 pp.

Roberts, 2 Roberts, A. *Underground Lighting in Mines, Shafts and Tunnels.* Technical Press, London, 1958, 292 pp.

Rogoff Rogoff, Martin H., Irving Wender, and Robert B. Anderson. *Microbiology of Coal.* U.S. Bureau of Mines Information Circular 8075, 1962. 85 pp., 7 figs.

Rolfe Rolfe, R.T. *A Dictionary of Metallography.* Chemical Publishing Co., New York, 1955, 287 pp.

Rosenthal Rosenthal, Ernest. *Pottery and Ceramics.* Penguin Books, Harmondsworth, Middlesex, England, 1949, 304 pp. Includes a glossary, pp. 292-301.

S

Sandstrom Sandstrom, Gosta E. *Tunnels.* Holt, Rinehart and Winston, New York, 1963, 427 pp. Includes a glossary, pp. 411-414.

Sanford Sanford, Samuel, and Ralph W. Stone. *Useful Minerals of the United States.* U.S. Geol. Survey, Bull. 585, 1914, 250 pp. Includes a glossary, pp. 218-250.

Schaller Schaller, Waldemar T. *Gems and Precious Stones*. Chapter in Mineral Resources of the United States, 1917, Part II - nonmetals. U.S. Geol. Survey, 1920, 1293 pp. Includes an alphabetical lineup, pp. 145-168.

Scheffaur Scheffaur, F.C. *The Hopper Dredge*. U.S. Government Printing Office. 1954.

Schieferdecker Schieferdecker, A.A.G. (ed.). *Geological Nomenclature*. J. Noorduijn en Zoon, 1959, 521 pp. (Royal Geological and Mining Society of the Netherlands).

Schlain Schlain, David. *Corrosion Properties of Titanium and Its Alloys*. U.S. Bureau of Mines Bulletin 619, 1964. 228 pp., 102 figs.

Schmuck Schmuck, Carl. *Cable Bolting at Home Stake Gold Mine*. Mining Engineering Magazine. December 1979.

Seelye, 1 Seelye, Elwyn E. *Data Book for Civil Engineers. V. 2. Specifications and Costs*. John Wiley & Sons, Inc., New York, 2nd ed., 1951, 506 pp. Includes a glossary of Hydraulics, pp. 408-418.

Seelye, 2 Seelye, Elwyn E. *Data Book for Civil Engineers. V. 2. Specifications and Costs*. John Wiley & Sons, Inc., New York, 2nd ed., 1951, 506 pp. Includes a glossary of Surveying, pp. 484-491.

Shanz Schanz Jr., John J. and John G. Ellis. *Assessing the Mineral Potential of the Federal Public Lands*. Congressional Research Service. May 1983.

Shell Shell Oil Co. *Simple Vocabulary of Petroleum Geology, Production, Manufacture, and Application*. 43 pp.

Sinclair, 1 Sinclair, John. *Environmental Conditions in Coal Mines (Including Fires, Explosions, Rescue, and Recovery Work)*. Sir Isaac Pitman & Sons, Ltd., London, 1958, 341 pp.

Sinclair, 2 Sinclair, John. *Geological Aspects of Mining*. Sir Isaac Pitman & Sons, Ltd., London, 1958, 343 pp.

Sinclair, 3 Sinclair, John. *Planning and Mechanized Drifting at Collieries*. Sir Isaac Pitman & Sons, Ltd., London, 1963, 358 pp.

Sinclair, 4 Sinclair, John. *Water in Mines and Mine Pumps*. Sir Isaac Pitman & Sons, Ltd., London, 1958, 130 pp.

Sinclair, 5 Sinclair, John. *Winding and Transport in Mines*. Sir Isaac Pitman & Sons, Ltd., London, 1959, 370 pp.

Sinclair, 6 Sinclair, John. *Winning Coal*. Sir Isaac Pitman & Sons, Ltd., London, 1960, 406 pp.

Sinclair, 7 Sinclair, W.E. *Asbestos; Its Origin, Production and Utilization*. Mining Publications, Ltd., London, 2nd ed., 1959, 512 pp.

Sinkankas Sinkankas, John. *Gemstones of North America*. D. Van Nostrand Co., Inc., Princeton, NJ, 1959, 650 pp. Includes a glossary, Appendix 2, pp. 620-637.

Skow Skow, Milford L. *Mica. A Materials Survey*. U.S. Bureau of Mines Information Circular 8125, 1962, 241 pp. Includes a glossary of Mica Terminology, pp. 169-171.

SME, 1 Hartman, Howard L. *SME Mining Engineering Handbook*, Two Volumes. Society of Mining, Metallurgy, and Exploration, Inc. 2nd ed., 1992.

SME, 2 Weiss, N.L. (ed.). *SME Mineral Processing Handbook*, V. 2. Society of Mining, Metallurgy and Exploration Inc. 1985.

Smith Smith, Allan F., Henry P. Barthe, and Samuel P. Polack. *Effects of Hydraulic Fluids in Spontaneous Heating of Coal*. U.S. Bureau of Mines Reports of Investigations 6221, 1963. 16 pp., 10 figs.

SMRB Great Britain, Safety in Mines Research Board. *A Glossary of Mining Terms*. SMRB Paper 61, 1930, pp. 110-115.

South Australia South Australia, Department of Mines. *Handbook on Quarrying*. W.L. Hawkes, Government Printer, Adelaide, South Australia, 1961, 185 pp.

Spalding Spalding, Jack. *Deep Mining; an Advanced Textbook for Graduates in Mining and for Practicing Mining Engineers*. Mining Publications, Ltd., London, 1949, 405 pp. Includes a glossary, pp. 389-392.

Spencer, 1 Spencer, L.J. *Fifteenth List of New Mineral Names*. Mineralogical Mag., V. 25, Sept. 1940, pp. 621-648.

Spencer, 2 Spencer, L.J. *Sixteenth List of New Mineral Names*. Mineralogical Mag., V. 26, Sept. 1943, pp. 334-343.

Spencer, 3 Spencer, L.J. *Seventeenth List of New Mineral Names*. Mineralogical Mag., V. 27, Sept. 1946, pp. 266-276.

Spencer, 4 Spencer, L.J. *Nineteenth List of New Mineral Names*. Mineralogical Mag., V. 29, Sept. 1952, pp. 974-997.

Spencer, 5 Spencer, L.J. *Twentieth List of New Mineral Names*. Mineralogical Mag., V. 30, Sept. 1955, pp. 727-749.

Spencer, 6 Spencer, L.J. *Twenty-First List of New Mineral Names*. Mineralogical Mag., V. 31, Sept. 1958, pp. 951-975.

Standard, 1 Funk & Wagnalls Co. *Funk & Wagnalls New Standard Dictionary of the English Language*. New York, 1959, 2815 pp.

Standard, 2 Funk & Wagnalls Co. *Funk & Wagnalls New Standard Dictionary of the English Language*. New York, 1964 (Copyright 1963), 2816 pp.

Statistical Research Bureau Statistical Research Bureau. *Mining Bureau Mining Manual for 1935*. Los Angeles, CA, 1935, 320 pp. Includes a glossary of Mining Terms, pp. 28-38, by Julian Boyd.

Stauffer Stauffer, David McNeely. *Modern Tunnel Practice*. Engineering News Publishing Co. New York, 1906, 314 pp. Includes a glossary of some of the more unusual terms used in tunneling, pp. 301-307.

Stoces Stoces, Bohuslav. *Introduction to Mining*. Lange, Maxwell & Springer, Ltd., London, 1954, 2 volumes.

Stokes Stokes, William Lee, and David J. Varnes. *Glossary of Selected Geologic Terms With Special Reference to Their Use in Engineering*. The Colorado Scientific Society, Denver, CO, 1955, 165 pp.

Strategic *Strategic and Critical Materials Stockpiling Act* (50 U.S.C. 98 et seq.) as amended by the National Defense Authorization Act for Fiscal Year 1989 (P.L. 100-456), sec. 12.

Streefkerk Streefkerk, H. *Quarrying Stone for Construction Projects*. Uitgeverij Waltman, Delft, The Netherlands, 1952, 159 pp.

Strock, 1 Strock, Clifford. *Heating and Ventilating Engineering Databook*. The Industrial Press, New York, 1st ed., 1948, v.p. Includes a glossary of Piping Terms, Sec. 3, pp. 6-9.

Strock, 2 Strock, Clifford. *Heating and Ventilating Engineering Databook*. The Industrial Press, New York, 1st ed., 1948, v.p. Includes a glossary of Terminology, Sec. 10, pp. 1-8.

Stutzer Stutzer, Otto, and Adolph C. Noe. *Geology of Coal*. The University of Chicago Press, Chicago, IL, 1940, 461 pp.

T

Taggart, 1 Taggart, Arthur F. *Handbook of Ore Dressing*. John Wiley & Sons, Inc., New York, 1927, 1679 pp.

Taggart, 2 Taggart, Arthur F. *Handbook of Mineral Dressing. Ores and Industrial Minerals*. John Wiley & Sons, Inc., New York, 1945, v.p.

Taylor Taylor, W.H. *Concrete Technology and Practice*. American Elsevier Publishing Co. New York, 1965, 639 pp. Includes a glossary of terms, pp. 611-623.

Thomas Thomas, D.E., and E.T. Hayes (eds.). *The Metallurgy of Hafnium*. U.S. Atomic Energy Commission, 1960, 384 pp. Includes a glossary, Appendix D, p. 369.

TIME Institution of Mining Engineers (London). *Control of Room and Support of Mine Workings*. Trans., V. 79, 1929-30, pp. 363-366, Appendix II, "Glossary".

Toenges Toenges, Albert L. and Frank A. Jones. *Truck vs. Rail Haulage in Bituminous-Coal Strip Mines*. U.S. Bureau of Mines Reports of Investigations 3416, 1938. 54 pp., 14 figs.

Tomkeieff Tomkeieff, S.I. *Coals and Bitumens and Related Fossil Carbonaceous Substances; Nomenclature and Classification*. Pergamon Press, Ltd., London, 1954, 122 pp. Includes an alphabetical glossary, pp. 21-97.

Trist Trist, E.L. *Organizational Choice; Capabilities of Groups at the Coal Face Under Changing Technologies; the Loss, Rediscovery and Transformation of Work Tradition*. Tavistock Publications, London, 1963, 332 pp. Includes a glossary of Mining Terms (Northwest Durham), pp. 299-305.

Truscott Truscott, S.J. *Mine Economics*. Mining Publications, Ltd., London, 3rd ed., 1962, 471 pp.

U

United Nations United Nations. *The Law of the Sea: Official Text of The United Nations Convention on the Law of the Sea* with Annexes and Index - Articles 2-16, 55-75, 76-85. 1983.

Urquhart Urquhart, Leonard Church (ed. in chief). *Civil Engineering Handbook*. McGraw-Hill Book Co., New York, 4th ed., 1959, v.p.

USBM, 1 U.S. Bureau of Mines. *American Standard Practice for Rock-Dusting Underground Bituminous Coal and Lignite Mines to Prevent Coal-Dust Explosions* (ASA Standard M13.1- 1960, UDC 622.81). U.S. Bureau of Mines Information Circular 8001, 1960. 5 pp.

USBM, 2 U.S. Bureau of Mines Health and Safety staff. *Recommended Standards for Alternating Current in Coal Mines.* U.S. Bureau of Mines Information Circular 7962, 1960. 25 pp., 11 figs.

USBM, 3 U.S. Bureau of Mines Health and Safety Activity staff. *The Federal Coal Mine Safety Act and Federal Mine Safety Codes: Interpretations and Applications.* U.S. Bureau of Mines Information Circular 8149, 1963. 26 pp.

USBM, 4 U.S. Bureau of Mines. *Instructions for Disaster, Fatal-Accident, and Miscellaneous Health and Safety Reports,* April 1966, Chapter 5.1 and Chapter 11.5, p. 63.

USBM, 5 U.S. Bureau of Mines. *Mineral Facts and Problems.* U.S. Bureau of Mines Bulletin 556, 1956. 86 chapters. 1042 pp., 47 figs.

USBM, 6 U.S. Bureau of Mines. *Mineral Facts and Problems.* U.S. Bureau of Mines Bulletin 585, 1960. 1016 pp., 69 figs.

USBM, 7 U.S. Bureau of Mines. *Mineral Facts and Problems,* 1965 ed. U.S. Bureau of Mines Bulletin 630, 1965. 1118 pp., 55 figs.

USGS, 1 Cox, Dennis P. and Donald A. Singer. *Mineral Deposit Models,* Bulletin 1693. U.S. Geological Survey. 1986.

USGS, 2 U.S. Bureau of Mines and U.S. Geological Survey. *Principles of a Resource/Reserve Classification for Minerals,* Circular 831. U.S. Geological Survey. 1980.

V

Van Dolah Van Dolah, Robert W., Michael G. Zabetakis, David S. Burgess, and George S. Scott. *Review of Fire and Explosion Hazard of Flight Vehicle Combustibles.* U.S. Bureau of Mines Information Circular 8137, 1963. 80 pp., 52 figs.

Van Vlack Van Vlack, Lawrence H. *Physical Ceramics for Engineers.* Addison-Wesley Pub. Co., Reading, MA, 1964, 342 pp. Includes a glossary of terms as applied to Ceramic Materials, pp. 285-295.

Van Zyl Van Zyl, D.J.A., I.P.G. Hutcheson, and J.E. Kiel. *Introduction to Evaluation Design and Operation of Precious Metal Heap Leaching Projects.* Society of Mining, Metallurgy, and Exploration Inc. 1988.

Virginia Polytechnic Virginia Polytechnic Institute, *Mineral Industries Journal,* V. 7, No. 1, Mar. 1960.

von Bernewitz von Bernewitz, M.W. *Handbook for Prospectors.* McGraw-Hill Book Co., New York, 2nd ed., 1931, 359 pp. Includes a glossary of terms used in Mining, pp. 304-341.

W

Webster 2nd Merriam-Webster, Inc. *Webster's New International Dictionary of the English Language, Second Edition, Unabridged.* G. & C. Merriam Co., Springfield, MA, 1960 (1959), 3194 pp.

Webster 3rd Merriam-Webster, Inc. *Webster's Third New International Dictionary of the English Language, Unabridged.* G. & C. Merriam Co., Springfield, MA, 1966, 2662 pp.

Weed, 1 Weed, Walter Harvey. *The Mines Handbook; an Enlargement of the Copper Handbook, volume XIII.* W.H. Weed, New York, 1918, 1896 pp. Includes a glossary and description of all copper-bearing minerals, pp. 29-47.

Weed, 2 Weed, Walter Harvey. *The Mines Handbook; Succeeding the Copper Handbook, volume XV.* The Mines Handbook Co., Tuckahoe, NY, 1922, 2248 pp. Includes a glossary of Mining Terms, pp. 1-22.

Wheeler, H.R. Wheeler, H.R. *A Manual of Modern Underground Haulage Methods for Mining Engineers.* Charles Griffin & Co., Ltd., London, 1946, 67 pp.

Wheeler, R.R.
Wheeler, Robert R., and Maurine Whited. *Oil: From Prospect to Pipeline.* Gulf Publishing Co., Houston, TX, 1958, 115 pp. Includes an Oil Dictionary, pp. 75-86.

Williams Williams, Howard R., and Charles J. Meyers. *Oil and Gas Terms.* Matthew Bender & Co., New York, 2nd ed., 1964, 449 pp. Copyright 1995 by Matthew Bender & Co., Inc. Reprinted with permission from Oil and Gas Law. All rights reserved.

Wolff Rudolf Wolff & Co Ltd. *Wolff's Guide to the London Metal Exchange.* Metal Bulletin Books Ltd. 3rd ed., 1987.

Wood Wood, F.W., and R.A. Beall. *Studies of High-Current Metallic Arcs.* U.S. Bureau of Mines Bulletin 625, 1965. 84 pp., 46 figs.

Woodruff Woodruff, Seth D. *Methods of Working Coal and Metal Mines.* Pergamon Press, Oxford, England, 1966, 3 volumes.

Wyllie Wyllie, M.R.J. *Fundamentals of Well Log Interpretation.* Academic Press, New York, 3rd ed., 1963, 238 pp.

Z

Zern Zern, E.N. *Coal Miners' Pocketbook.* McGraw-Hill Book Co., New York, 12th ed., 1928, 1273 pp. Includes a glossary of mining terms, pp. 1201-2251; glossary of rope terms, pp. 755-757.

Zimmerman Zimmerman, O.T., and Irvin Lavine. *Scientific and Technical Abbreviations, Signs and Symbols.* Industrial Research Service, Dover, N.H., 2nd ed., 1949, 541 pp.